ENVIRONMENTAL SCIENCE

ENVIRONMENTAL SCIENCE

Dr S C Santra MSc, PRS, PhD
Professor, Department of Environmental Science
University of Kalyani, Nadia, West Bengal

New Central Book Agency (P) Ltd

Co-Published by

Bijay Publications

E-mail : bijaypublications@gmail.com

Exclusively Distributed by

Bijay Pustak Bhandar (P) Ltd

90/6A M. G. Road, Bhawani Dutta Lane, Kolkata-700007

✆ : (033) 4605-7856, 8777898576, 8274916104

E-mail : bijaypustakbhandar@gmail.com

NCBA

REGD OFFICE
8/1 Chintamoni Das Lane, Kolkata 700 009, India
email: ncbakolkata@gmail.com

OVERSEAS
NCBA (UK) Ltd, 149 Park Avenue North, Northampton, NN3 2HY, UK
email: ncbauk@yahoo.com

CORPORATE OFFICE
212 Shahpur Jat, New Delhi 110 049
email: delhincba@gmail.com

BIJAY PUBLICATIONS
90/6A M. G. Road, Bhawani Dutta Lane, Kolkata-7
E-mail : bijaypublications@gmail.com

BIJAY PUSTAK BHANDAR (P) LTD
90/6A M. G. Road, Bhawani Dutta Lane, Kolkata-7
E-mail : bijaypustakbhandar@gmail.com

ENVIRONMENTAL SCIENCE • S C Santra

First Published: April 2001
Second Edition: July 2005
Reprinted: 2006, 2008, 2010, 2011, 2012
Thoroughly Revised and Updated Third Edition: April 2013
Reprinted: 2014, 2016
Thoroughly Revised and Updated Fourth Edition: 2020
Reprinted: 2022, 2024

PUBLISHER • TYPESETTER • PRINTER
New Central Book Agency (P) Ltd
8/1 Chintamoni Das Lane, Kolkata 700 009

CO-PUBLISHED BY
Bijay Publications
90/6A M G Road, Bhawani Dutta Lane, Kolkata 700007

PROJECT TEAM : Prabir Ghosh, Pradip Baidya, Pradip Biswas and Prabhat Jas

ISBN: 978 81 7381 404 4

Price ₹ 1450.00

CONTENTS

SECTION I : PRINCIPLES AND CONCEPTS

SECTION II : RESOURCES AND ECOSYSTEM

SECTION IV : CURRENT ENVIRONMENTAL ISSUES

SECTION V : ENVIRONMENTAL HEALTH

SECTION VI : ENVIRONMENTAL QUALITY AND ENVIRONMENTAL IMPACT ASSESSMENT

SECTION VIII : ENVIRONMENTAL LAWS, RULES AND REGULATIONS

SECTION IX : ENVIRONMENTAL MODELLING, INDUSTRIAL POLLUTION MANAGEMENT AND ENVIRONMENTAL PLANNING

SECTION X : ANALYTICAL TECHNIQUES, STATISTICAL APPLICATIONS IN ENVIRONMENTAL SCIENCE, INSTRUMENTATION TECHNIQUES

Preface to the Fourth Edition

With the appreciation of readers of wide range of subjects (Life Sciences, Environmental Science and Earth Sciences), this edition is thoroughly revised.

The entire publication team of M/s New Central Book Agency (P) Ltd., Kolkata extend their full support for this edition.

Kolkata
12 January 2020

S C Santra

PREFACE TO THE FIRST EDITION

Knowledge about the environment, especially its holistic nature and the laws that govern the ongoing processes in the environment—particularly in an ecosystem and the interdependence of different elements in the environment is a priority need for a better understanding, protection and improvement of the environment. This knowledge needs be provided through formal and nonformal environmental education to decision-makers, politicians, students, teachers and the general public.

It is further interesting to note the fact that every human being should be aware, conscious and knowledgeable about his or her role in the environment. There is almost a balanced relationship between the human being and the environment and it has to be maintained. For a long time human beings cherished the concept that they were the masters of the environment. This concept and its associated actions gave rise to an idea that humanity and environment are against each other. This has been the most erroneous and damaging concept developed in respect to humanity-environment relationship. Environmental education is a necessary tool for taking appropriate decisions concerning the solution and prevention of environmental problems which are considered a disequilibrium caused by certain factors in the established relationship between human beings, animals, plants, and others. In the last few decades the impact of human activities on the environment has become more accelerated and pronounced. Pressures from rapid population growth, uncontrolled and lavish consumption, urbanization, industrial expansion and advances in science and technology and their applications, coupled with huge energy utilization, have caused these accelerated changes—culminating in serious environmental problems. Man is the cause and maker of the current environmental problems. So human beings need be educated for the understanding, solution and prevention of these problems.

`Environmental Science' education at graduate level has now been made compulsory in all Indian universities. So the present attempt to bring out a comprehensive textbook on the concerned subject has been made.

In the first edition, a wide variety of issues have been described in different chapters for the students, teachers and the common readers. I shall sincerely appreciate the feedback from all kinds of readers in near future for further improvement of the book.

During the preparation of the manuscript, I have received a great deal of help from various teachers, researchers and professionals. I am thankful to all of them. Nevertheless, I am thankful to the publisher New Central Book Agency (P) Ltd, its artists, and the editorial staff associated with this publication.

I sincerely appreciate the cooperation received from all concerned.

Kolkata
15 February 2001

S C SANTRA

SECTION I

PRINCIPLES AND CONCEPTS

1 ENVIRONMENTAL SCIENCE: GENERAL PERSPECTIVES

The 'environment'—defined as that whole outer physical and biological system in which man and other organisms live—is a whole, albeit a complicated one with many interacting components. The wise management of that environment depends upon an understanding of those components: of its rock, minerals and waters, of its present and potential vegetation, of its animal life and potential for livestock husbandry and of its climate. It demands positive and realistic planning that balances human needs against the potential environment has for meeting them ...

Source: UNEP, 1976

1.0 ENVIRONMENTAL SCIENCE — A DISCIPLINE BY ITSELF

It has long been held that environmental science deals with the study of the atmosphere, the land, the oceans and the great chemical cycles that flow through the physical and biological systems. Currently it was noticed that modern environmental science is increasingly becoming interdisciplinary, preparing people for global citizenship and training them to be flexible, yet competent, analytic and decision-takers. Three branches of sciences interact, along with their own developmental arena; the attractive research areas are by no means found only on the margins where an element of joint disciplinarity takes place (Fig. 1.1). But it is possible for the climate system to be studied this way, with much success, and for the emerging field of environmental economics and ecology. A more formal multidisciplinarity is often found in the ever-widening area of Environmental Impact Assessment (EIA).

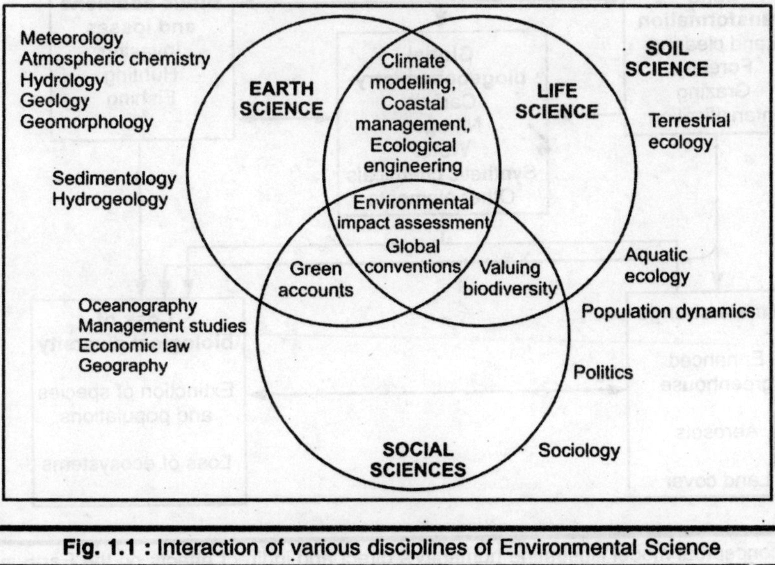

Fig. 1.1 : Interaction of various disciplines of Environmental Science

1.1 MAN AND ENVIRONMENT INTERACTION

All organisms modify their environment, and humans are no exception. As the human population has grown and the power of technology has expanded, the scope and nature of this modification has changed drastically. Until recently, the term "human-dominated ecosystems" would have elicited images of agricultural fields, pastures or urban landscapes; now it applies with greater or lesser force to all of Earth. Many ecosystems are dominated directly by humanity, and no ecosystem on Earth's surface is free of pervasive human influence.

The growth of the human population and growth in the resource base used by humanity is maintained by a suite of human enterprises such as agriculture industry, fishing and international commerce. These enterprises transform the land surface (through cropping, forestry and urbanisation), alter the major biogeochemical cycles, and add or remove species and genetically distinct populations in most of Earth's ecosystems. Many of these changes are substantial and reasonably well quantified. All are ongoing. These relatively well-documented changes in turn entrain further alterations to the functioning of the Earth system, most notably by driving global climatic change and causing irreversible losses of biological diversity. The human alteration in the Earth system as operating through the interacting processes are summarised in Fig. 1.2.

Recognition of the global consequences of the human enterprise suggests three complementary directions. First, we can work to reduce the rate at which we alter the Earth system. Humans and human-dominated systems may be able to adapt to slower change, and ecosystems and the species they support

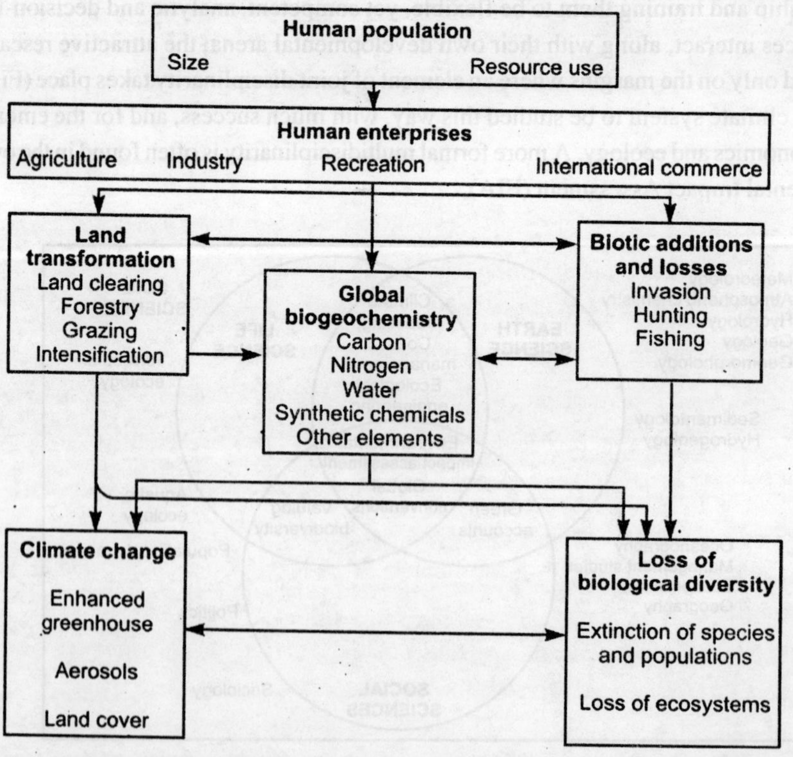

Fig. 1.2 : A conceptual model illustrating humanity's direct and indirect effects on the Earth system

may cope more effectively with the changes we impose—if those changes are slow. Our footprint on the planet (54) might then be stabilized at a point where enough space and resources remain to sustain most of the other species on Earth, for their sake and our own. Reducing the rate of growth in human effects on Earth involves slowing human population growth and using resources as efficiently as is practical. Often it is the waste products and by-products of human activity that drive global environmental change.

Second, we can accelerate our efforts to understand Earth's ecosystems and how they interact with the numerous components of human-caused global change. Ecological research is inherently complex and demanding: It requires measurement and monitoring of populations and ecosystems; experimental studies to elucidate the regulation of ecological processes; the development, testing, and validation of regional and global models: and integration with a broad range of biological, earth, atmospheric, and marine sciences. The challenge of understanding a human-dominated planet further requires that the human dimensions of global change—the social, economic, cultural, and other drivers of human actions—be included within our analyses.

Finally, humanity's dominance of Earth means that we cannot escape responsibility for managing the planet. Our activities are causing rapid, novel, and substantial changes to Earth's ecosystems. Maintaining populations, species, and ecosystems in the face of those changes, and maintaining the flow of goods and services they provide humanity (55), will require active management for the foreseeable future. There is no clearer illustration of the extent of human dominance of Earth than the fact that maintaining the diversity of "wild" species and the functioning of "wild" ecosystems will require increasing human involvement.

1.2 GLOBAL ENVIRONMENTAL PROBLEMS

Environmental problems are always interrelated. Often solution to one problem actually creates another problem! For example, when people are sick and dying from disease, it is natural to want to improve human health. When health is improved and mortality is reduced, a population explosion may result. To feed this growing population, natural habitats are often converted into farmland. In turn, the destruction of natural habitats eliminates the native wild species of plants, predatory animals and parasites. The outbreak of insect pests become more common. Subsequently, farmers use pesticides to control the pests and protect the crops, but they also pollute the environment. The development of this entire cycle in itself consumes irreplaceable fossil fuel supplies and, when fuels are burned, air pollutants are generated in turn!!

On the whole, the major environmental problems can be divided into following categories:

1. **Over-population**: It causes human suffering directly and amplifies all other environmental problems;
2. **Pollution**: It makes vital resources less useful and reduces the quality of life;
3. **Depletion of resources**: It makes things that are vital to human existence more expensive or impossible to obtain;
4. **Global changes**: This results from human activities and may permanently alter the Earth in unpredictable ways;
5. **War**: This may be caused by or may cause all other environmental problems—modern (chemical/biological) warfare threatens the survival of the human species.

Thus, individuals act in ways that promote their own short term welfare, which often conflicts directly with the long-term environmental interests of present and future generations. This "tragedy of commons" is the main factor that limits the effectiveness of voluntary action as a solution to environmental problems.

At present the world is very crowded, more polluted, less stable, ecologically fragile and vulnerable to disruption. Serious stresses involving population, resources and environment are clearly visible ahead. Despite greater material output, the world's people will be poorer in many ways than they are today. Regional water shortage will become more severe. Significant losses of world forests will continue over the next 20 years as demand for forest products and fuel wood increases. The world's forests are currently disappearing at the rate over 20 million hectares per year, with most of the loss occurring in the humid tropical forests of Africa, Asia and South America.

Serious deterioration of agricultural soils will occur worldwide, due to erosion, loss of organic matter, desertification, salinization, alkalinization and water-logging. Atmospheric concentrations of carbon dioxide and ozone depleting chemicals are expected to increase at rates that could alter the world's climate and upper atmosphere significantly by 2050. Acid rain from increased combustion of fossil fuels threatens and causes damage to lakes, soils and crops. Radioactive and other hazardous materials present health and safety problems in increasing numbers of countries. Extinction of plant and animal species will increase dramatically. More than 20 per cent of all species on earth will be irretrievably lost from their habitats and thus vanish, especially in tropical forests.

1.3 ENVIRONMENT PRIORITIES IN INDIA

During past three decades, enormous efforts were made for environmental protection and also some attempts has already been taken up for sustainable development. A good number of areas have been identified where priority action is needed:

1. **Population Stabilization**: On 1 March 2011, India had a population of 1,210 million. Can we afford this? We account for 17.51 per cent of the world population, with only 2.4 per cent of the world area available to us. The crux of the matter is that future population growth has to be related to the resource base. Resources shrink as people multiply and demographic pressures lead to economic pressures. Thus, our immediate task is to know the carrying capacity of our country. In any case, a zero population growth holds the key to our prosperity, and to achieve this the government must be prepared for a measure of unpopularity. This has to be accomplished in the next decade or so. In turn, it takes to the birth control strategies, education and vocation—for both men and women—which involve considerable scientific and technological inputs including immunology and biotechnology. A policy on population has to be enunciated in a comprehensive manner. Population forecast of India is shown in Box 1.1:

Box 1.1: India population forecast

Year	Population	Yearly % Change	Yearly Change	Migrants (net)	Median Age	Fertility Rate	Density (P/km^2)	Urban Pop %	Urban Population	Country's Share of World Pop	World Population	India Global Rank
2020	1,383,197,753	1.11%	14,828,755	– 490,000	28.2	2.30	465	34.0%	470,726,086	17.74%	7,795,482,309	2
2025	1,451,829,004	0.97%	13,726,250	– 400.000	29.8	2.19	488	36.2%	525,459,048	17.74%	8,185,613,757	1
2030	1,512,985,207	0.83%	12,231,241	– 400.000	31.4	2.10	509	38.5%	583,038,483	17.69%	8,551,198,644	1
2035	1,564,570,223	0.67%	10,317,003	– 400.000	33.0	2.02	526	41.1%	642,296,787	17.59%	8,892,701,940	1
2040	1,605,355,574	0.52%	8,157,070	– 400.000	34.5	1.95	540	43.7%	701,357,654	17.43%	9,210,377,004	1
2045	1,636,496,308	0.38%	6,228,147	– 400.000	36.1	1.90	550	46.4%	758,961,289	17.22%	9,504,209,572	1
2050	1,658,978,162	0.27%	4,496,371	– 400.000	37.5	1.86	558	49.1%	814,398,506	16.98%	9,771,822,753	1

Source: Worldometers (WWW.Worldometers.info(1))

2. **Natural Resource Assessment and Management**: Land is one of the important components of the life support system and has been overused and abused. There are a number of competing demands on land, like agriculture, forestry, grasslands, urban and industrial development. We need a time-bound

micro-level land use survey, starting with the village indicating our long-term requirements for competing land uses and land capability. Obviously, land use will have to be related to soil quality use and not land economics. The total scenario of land use will need to be built up meticulously and backed by appropriate legislations, because the existing legislations are indeed very weak.

Water resource is another important natural resource which needs to be assessed and conserved in appropriate manner for sustaining development of the society. Connected with land use planning is watershed management, a problem of utmost urgency on account of the fact that our water regimes in the mountain ranges are threatened resulting in the depletion of water resources along with the other associated problems. We have to evolve environmentally sound water conservation and management strategies for future decades ahead, otherwise this will be a major problem of the entire country in near future. The river valley projects envisaged in future and those that are ongoing need to be looked into with their environmental impacts and due provisions made for environmental safeguards. A basin-wise approach needs to be followed and the engineering component should be considered in conjunction with catchment and command areas development. In addition, unplanned irrigation has caused more problems and losses than solutions and gains.

The country has performed exceedingly well in agriculture and production has increased from 50.8 million tonnes in 1950-51 to 155.5 million tonnes in 1984-85 and finally to over 275 million tonnes in the year 2000 to feed a billion people. Thus, there is an urgent need to boost productivity per unit area per unit time through overcoming the gap between actual and potential yields. In spite of the adverse environmental effects of high-input high-output agricultural technology, for the foreseeable future, the country will have to depend on this model, which cannot be suddenly changed in favour of environmentally sustainable but low input models like organic farming, no-till or minimum-till agriculture.

The loss of top soil, quantitywise, is the maximum in India, being 18.5 per cent of the total soil loss at the global level. Strategies to reduce the top soil need to be developed on an urgent basis, otherwise land degradation will be faster. Degraded mines and other types of wasteland should be reclaimed and put to some productive use.

Regarding forest wealth, the country is facing a serious problem. The effective forest cover today is very low i.e., about 23.41 per cent as revealed by recent remote sensing survey, but the national requirement is about 33 per cent in plains and 60 per cent in hills. Thus there is great need for existing forest protection, as well as promotion of all sorts of efforts relating to creation of new forest area. To achieve these objectives, three broad types of forestry—conservation, production and social forestry—need to be practiced. There is need for considerable inputs in forestry with regard to science and technology, education and training, and demonstration and extension, if it has to meet the comprehensive objectives.

The biological wealth of the country is fairly considerable with nearly 45,000 plant and 65,000 animal species (except insects). However, at present our efforts of conservation have been limited to a few vertebrates and tree species. It is now felt as extremely important that the conservation effort on an ecosystem basis will be much appropriate than on a species basis. On paper we have over 550 sanctuaries and 96 national parks, but in practice, we have no idea of the holdings in these areas. There has been considerable adhocism in declaring sanctuaries and national parks. Though there is a reasonable understanding of the threatened animals and plants, yet appropriate strategies for their conservation is not framed so far.

Regarding fossil fuel and mineral resource exploration and their subsequent use pattern, there is no short term or long term policy of resources exploitation. The productivity of resource exploitation is only linked with market force economy.

3. **Strategies for Environmental Pollution Monitoring and Control**: All kinds of environmental pollution are man-made problems, which arise from anthropogenic, industrial and agricultural sources. During last couple of decades enormous amount of pollutants contaminated the land, water and air. The nature of pollutants and their quantification were made by regular monitoring programmes. Minimum National Standards (MINAS) with respect to major industries have been established. Implementation of MINAS would control pollution at source.

For strict control of pollutants discharge/emission to environment, regular monitoring and surveillance is highly essential. A number of environmental laws and the implementing agencies like—the State Pollution Control Board, Central Pollution Control Board and National Green Tribunal—need to perform more active role on strict enforcement of the pollution regulations. But to achieve the goal, there is need for participatory approach for effective action plan implementation.

Conventional energy generation is principally a polluting process and energy is a very important input for development. Thus, there is a great need for harvesting non-polluting renewable energy system for major developmental activities. Thus, a perspective in the energy science is yet to be developed. In the immediate future, the country will have to conserve—to the maximum extent—the commercial sources on the one hand, and replace and supplement the same with non-conventional sources on the other.

Solid waste and effluent waste water must be recycled for appropriate use and residue must be disposed in eco-friendly manner. Considerable attempts has already been taken up over the years on these lines.

4. **Human Settlement and Public Health**: Due to continued rise of population, the demand for housing, drinking water, sewerage disposal and various other public health support systems was increased enormously. At present, there is shortage of at least 25 million houses, 18 million (75 per cent) in rural areas and 7 million (25 per cent) in urban areas. There is scarcity of drinking water supply and the public sanitary system. As a consequence, there is rise of infectious diseases, continuous recurrence of mosquito-borne diseases, and gastroenteritis.

There is a great need for future improvement of drinking water supply and substantial improvement of sanitary conditions, particularly in urban habitats. Improvement of slums through various programmes is undertaken in urban centres, but its faulty action plan with short-term approach does not improve the quality of life to a greater extent.

5. **Environmental Laws and Regulations**: There is a plethora of Central and State laws/enactments which have direct or indirect relevance to environment. In spite of these laws, the improvement of environmental quality is not substantial. There are a lot of fallacies in the laws that have to be amended —otherwise effective implementation could not be possible.

In general, the citizens of India, by and large, are not too law abiding. As such, for effective implementation, both judiciary and executive bodies should work together in future days ahead.

6. **Environmental Education and Awareness**: Since 1970, many conferences—national and international—have been held, but the fact remains that while most people recognize the urgent need for environmental education, only some have clear ideas about what needs to be done, and very few have either the actual experience or the knowledge of the courses that need to be taught. Thus, there needs an objective orientation of environmental education—either in formal system or in non-formal sector. India is one of the major countries having large number of NGOs working on environmental awareness campaign.

The role of mass media and various other public institutions cannot be ignored thereby. Time has come when we have to be very realistic to make economic planning and environmental protection must be compatible and deeply integrated in the planning process.

Box 1.2: SUMMARY

The principles of environmental management will have to be applied to all developmental activities. Inherently, environmental management is interdisciplinary and also interdepartmental. Furthermore, in our country, land, water and forest resource exploitation and management are State subjects which means a very high degree of interface between the Central and the State governments. There is, therefore, an urgent need for a national body, outside the Department of Environment, to advise the Government on pressing national environmental issues. Such type of bodies, like Scientific Advisory Committee, already existed in the past, but their achievement is far from the requirement as yet. Thus the national planning for environmental co-ordination and action implementation, like Environmental Protection Agency (EPA) of USA, has to be reviewed seriously.

The debate on environmental protection within India shows striking parallel to the larger debate between North and South. An ecologically based causal paradigm, stressing over-population and non-sustainable recources use, is in conflict with a moral paradigm, that attributes environmental degradation to the consumption pattern of the rich. Adherents of both causal models agree that sustainable development is the answer to India's environmental problems, but they interpret the concept in different ways. Thus elements of the Indian case suggest that within a single country even such deep divisions of world view and ideology can be negotiated with the help of appropriately constructed institutions.

1.4 ENVIRONMENTAL ETHICS

The Earth is remarkable and valuable for both the nature and the culture that occur on it. Evolutionary history has been going on for billions of years, while cultural history is only about a hundred thousand years old. But certainly, from here onward, culture increasingly determines what natural history shall continue. The debate about ethics as applied to nature asks whether the primary values about which we should be concerned are cultural, that is, anthropocentric, or whether there is also intrinsic natural value, independent of humans, which humans ought to consider. Although all deliberate human behaviours differ from the processes of spontaneous nature, some are healthy for humans because they agree with the natural systems with which their cultural decisions interact. In a relative sense, what humans do can be nature conservation values, but are not the only values; there are numerous values autonomous to cultures. Some of these can be gained by the sacrifice of natural values. So the environmental ethics of this century will increasingly have to ask whether and why cultures should preserve any natural values at all and what kind of balance ought to be reached?

It is true that the Earth is now in a post-evolutionary phase. Culture is the principal determinant of Earth's future, more now than nature—we have passed into a century when this will be increasingly obvious. Indeed, some say that will be the principal novelty of this new millennium—Earth will be a managed planet. Meanwhile, the technosphere remains in the biosphere; we are not in a post-ecological phase. The management of the planet must conserve environmental values. Hopefully such policy can, in places, let nature take its course.

At this point, on the eve of the 21st century, it has become clear that humankind is facing one of the greatest challenges of recorded history; how to reconcile continued economic growth and all that this implies with the constraints imposed by a shrinking resource base and an increasingly degraded environment. If we do not want merely to cast ourselves on the seas of changes we should use our knowledge of the historical processes that helped shape human technological powers to orient ourselves.

The human race and nature are products of evolutionary processes, subject to the final test of survival. And the evidence of human resourcefulness in the art of environmental husbandry can be charted throughout our evolution from Neolithic food-gathering bands to agro-industrial society. We shall, therefore, hereunder consider the relationship between the inner dynamics of beliefs, values and goals and the outer dynamics of agricultural, industrial and ecological transformation in order to discover how we have survived through the creative values of our inner evolution.

The World Commission for Environment and Development was called upon by the General Assembly of the UN to draw up a global agenda for change in our patterns of environmental use. The clear message of the report, named after the chair of the commission, former Prime Minister Brundtland of Norway, reminds all that we have—only one Earth for our survival. The report touched the sensibilities of many people, on development, but more importantly, it paved the way for the heightened global concern that preceded the Earth Summit of 1992.

Scientific innovation is an indispensable step towards the search for sustainable development. UNESCO's Man and Biosphere Programme (MAB), launched in 1968, is one such endeavour which seeks to match science and ideas in the search for human survival on Earth. On the whole, people are rapidly approaching a crossroads of crisis. But drawing from past and the new found activities, driven by environmental ethics and scientific innovation within individual nations and the international community, there is yet some hope for man and planet Earth.

1.5 MAN AND ENVIRONMENT

Man cannot be considered in isolation from his environment. Over the world, the needs of people still differ enormously. At one time the environmental problems debated in international organizations would predominantly have been those recognized by the developed countries—the need to control pollution and the desirability of conserving samples of the ecological and genetic richness and the natural beauty of the earth. Since the United Nations Conference of the Human Environment; however, it has been increasingly realized that environmental issues are also of vital concern to developing countries and that over much of the world the environmental problems are still those associated with poverty—poor housing, bad public health, malnutrition and inadequate employment.

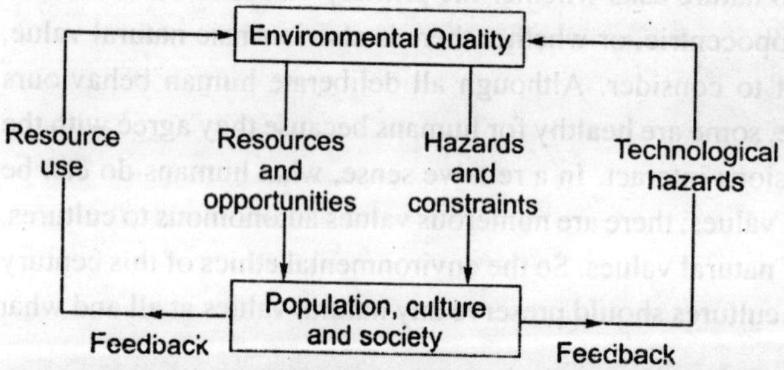

Fig. 1.3 : Two-way relationship between people and environment. The relationship between people and their environment is symbiotic, involving both resources (opportunities) and hazards (constraints)

Both the creation and the recognition of environmental problems depend closely on the way society is organised and on its values and objectives. Changes on the relationship between man and his physical environment depend to a large degree on changes in the organisation and aims of society. If man is to escape from a situation in which much energy and resources are devoted to correcting part mistakes, his aim must be to build a society which is intrinsically compatible with its environment (Fig. 1.3).

The net effect of these changes, particularly during the second half of the twentieth century, was an increase in exposure to many hazards and increased potential for catastrophic losses. More than 1.4 million people have died as a result of natural disasters over the past 50 years, with earthquakes by far the biggest killers. Hazards can cause major financial problems, as well as killing many people and damaging such property. In 1998, for example, financial losses from natural disasters worldwide were greater than US$ 93 billion.

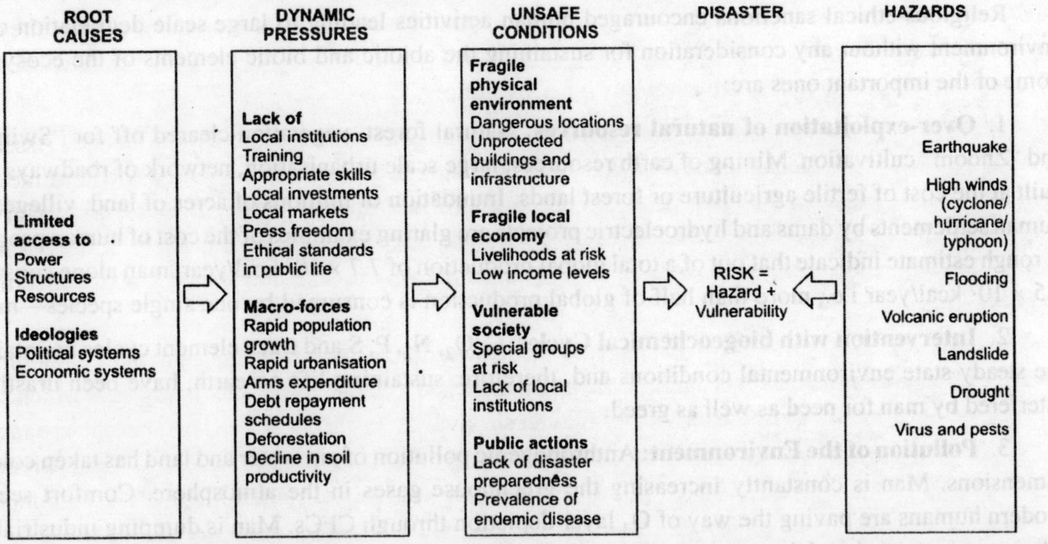

Fig. 1.4 : Progression of human vulnerability to disasters. Disasters are caused by hazard events, but the impacts on people are strongly influenced by degree of vulnerability. This, in turn, is a product of unsafe conditions, which result from some important root causes and are compounded by dynamic pressures

Analysis of recent trends shows significant regional disparities in losses, particularly between developed and developing regions. Financial losses associated with natural hazards are highest among the developed countries, such as the USA, where natural hazard losses exceed those of many other national social problems, including fire and crime. In the developing world, in contrast, the costs are largely measured in terms of human suffering and hardship. Many low-income populations are forced to occupy illegal settlements on low-lying lands, steep hillsides, floodplains or other hazard-prone areas. They are very vulnerable to significant health risks from flooding, landslides, mud slides and other natural hazards, and their dwellings and infrastructure are subject to accidents, massive damage and collapse (Fig. 1.4).

Three technological advancements of man were highly detrimental to environment—especially air, land and water. These are (i) lighting of fire, (ii) agricultural production of food, and (iii) industrial revolution. These opened the floodgate of anthropogenic pollution of air, water and land and threatening the very existence of life on mother earth.

With the development of religious concepts several religions—particularly Judaism, Christianity and Islam—viewed the created order as existing for human exploitation (Chapman and Reiss, 2000). In Genesis (1:27) it is said that man was created in the image of God and was set over nature and had authority to do much as he liked. In Genesis (1:28), Adam and Eve were told 'Be fruitful and increase, fill the earth and subdue it, rule over the fish in the sea, the birds of heaven, and every living thing that moves upon the earth'. J. L. White (1967), in analyzing the ecological crisis of earth, wrote "Christianity is the most anthropocentric religion ... it insisted that it is God's will that man exploit nature for his proper ends". In the Vedic literature, mother earth is personified as the Goddess Bhumi or Prithvi. The abundant mother showers her mercy on her children. The resources are, therefore, endless and can be exploited for human benefit. Exploitation of nature or other types are done under the pretext of God's will (You do it Oh mother, but people say that I do it). In Buddhism, there is very strong emphasis on how we should relate to the natural world—for example, there is a prohibition on animal slaughter.

Religious-ethical sanctions encouraged human activities leading to large scale degradation of the environment without any consideration for sustaining the abiotic and biotic elements of the ecosystem. Some of the important ones are:

1. **Over-exploitation of natural resources**: Natural forest, vegetation cleared off for "Swinden" and "Zhoom" cultivation. Mining of earth resources, large scale urbanisation, network of roadways were built at the cost of fertile agriculture or forest lands. Inundation of millions of acres of land, villages and human settlements by dams and hydroelectric projects are glaring examples of the cost of human progress. A rough estimate indicate that out of a total global production of 7.7×10^7 kcal/year, man alone consumes 4.5×10^7 kcal/year i.e., more than half of global production is consumed by one single species—man.

2. **Intervention with biogeochemical Cycles**: C, O_2, N_2, P, S and trace element cycles maintaining the steady state environmental conditions and, therefore, sustaining life on earth, have been drastically interfered by man for need as well as greed.

3. **Pollution of the Environment**: Anthropogenic pollution of air, water and land has taken colossal dimensions. Man is constantly increasing the greenhouse gases in the atmosphere. Comfort seeking modern humans are paving the way of O_3 layer depletion through CFCs. Man is dumping industrial and city sewage wastes into lakes, rivers and seas. Toxic chemicals used in modern agriculture for combating pests, diseases and weeds plus synthetic fertilizers are silently killing useful microbes maintaining the biogeochemical cycles, useful insects, birds, butterflies of the forests and fishes in the streams and lakes (Rachel Carlson: Silent Spring).

At least five activities of man may lead to global cataclysm killing all life on the earth:

1. **Continuous greenhouse gas emission**: Emission of CO_2, CH_4, N_2O and CFCs by the civilized world, if not reduced to 75% of current emission rate by 2030, there will be global warming and large scale O_3 depletion.

2. **Hazardous chemicals of agriculture and industry**: More than 100,000 Xenobiotic chemicals are now being used. Many of these are recalcitrant compounds. They are accumulating in the environment causing threat of cancer epidemics and total extinction of birds, fishes, butterflies, bees and trillions of soil microbes essential for geo-chemical cycles.

3. **Man-made nuclear arsenals**: Several countries possess nuclear bombs with efficient delivery system for target sites. As compared to 'Hiroshima type' these bombs are 5, 25, 100 or 500 times more powerful. If 10 of 100 megaton bombs are exploded in different parts of the world, whole fabrics of life-forms including man will be extinct.

4. **Man-made radioactive wastes**: Nuclear wastes generated in reactors are real threat to life on earth. $Plutonium^{239}$ produced in reactors are used in bombs. It has half life of 24,400 years and after that half of it decays to U^{235} with a half-life of 750,000 years. We have accumulated 750,000 kg of plutonium. Even if bombs are not made and those already existing are dismantled (another 100,000 kg of Plutonium) — what to do with these wastes?

5. **Biotechnological misuse**: The deliberate production of extremely powerful disease-producing bacteria, viruses or fungi for biological warfare is another great environment threat to mankind. These super-powered pathogens, when released, will disseminate through air or water and can cause catastrophic epidemics to man and his domestic animals.

Man—supposedly the most intelligent life-form—by misuse and over-use of science and technology has become the potential terminator of mankind and other life-forms. Gaia theory suggests coopeation between men of different nations for not allowing further pollution and cooperative approach for ameliorating the pollutants already accumulated. By cooperation between the nations—not by competition—we can avoid and prevent total annihilation of mother earth and its offsprings including the human beings.

On the whole, the environmental problems can only be solved through development. But that development, producing more food and drawing on the still great resources of the planet, needs to be environmentally wise, and to be based upon thorough evaluation of the potential uses of the different regions of this highly variable earth. Short-term solutions may all too easily—as they have in many countries—lead to long-term losses which a growing world population cannot afford.

UNESCO's programmes on environment and natural resource management (MAB) aim at providing the scientific basis and trained personnel needed for solving the environmental problems of our times. Traditionally, MAB has involved research by interdisciplinary teams on ecological and social systems, field training and the application of a system approach to understanding the relationship between natural and human components of development. Today, the overall goal of MAB is to help establish the scientific basis for sustainable development and to assist countries in the development of their human resources.

Although there were 14 defined areas of research, the financial and human resources of MAB were concentrated—more or less—on the following main areas for the initial phase of the programme: Coastal areas and islands, humid and sub-humid tropics, arid and semiarid zones, temperate and cold zones, urban systems, and biosphere reserves.

Some of the more notable and recent achievements of MAB includes:

- MAB established an international network of biosphere reserves, which are protected areas representative of the world's major ecosystem types.
- MAB evolved, through the work in biosphere reserves and specialized projects, interdisciplinary, cross-sectoral approaches to sustainable development, indifferent ecological and socio-cultural settings, including arid and semi-arid zones, Mediterranean systems, small islands, mountain areas etc.
- MAB set up networks for comparative studies on topics which scientists around the world consider of vital importance.
- MAB promoted regional cooperation on natural resource issues through regional field projects and technical seminars.
- MAB published technical syntheses useful for scientists and resource managers.
- MAB contributed professional trainees through various courses.
- MAB mobilized resources for building up the capacities of developing countries for integrated approaches to resource management.
- MAB prepared a series of 36 posters — "Ecology in Action" — as an experiment in communicating scientific information to a broad public audience.

1.6 EARTH'S CARRYING CAPACITY

It takes no stretch of the imagination to see that the human species is now an agent of change of geologic proportions. We literally move mountains to mine the earth's minerals, redirect rivers to build cities in the desert, torch forests to make way for crops and cattle, and alter the chemistry of the atmosphere in disposing of our wastes. At humanity's hand, the earth is undergoing a profound transformation—one with consequences we cannot fully grasp.

It may be the ultimate irony that, in our efforts to make the earth yield more for ourselves, we are diminishing its ability to sustain life of all kinds—humans included. Signs of environmental constraints are now pervasive. Cropland is scarcely expanding any more, and a good portion of existing agricultural land is losing fertility. Grasslands have been overgrazed and fisheries overharvested, limiting the amount of additional food from these sources. Water bodies have suffered extensive depletion and pollution, severely restricting future food production and urban expansion. And natural forests—which help stabilize the climate, moderate water supplies, and harbor a majority of the planet's terrestrial biodiversity—continue to recede.

Biologists often apply the concept of "carrying capacity" to questions of population pressures on an environment. Carrying capacity is the largest number of any given species that a habitat can support indefinitely. When that maximum sustainable population level is surpassed, the resource base begins to decline some time thereafter, so does the population.

The earth's capacity to support humans is determined not just by our most basic food requirements but also by our levels of consumption of a whole range of resources, by the amount of waste we generate, by the technologies we choose for our varied activities, and by our success at mobilizing to deal with major threats. In recent years, the global problems of ozone depletion and greenhouse warming have underscored the danger of overstepping the earth's ability to absorb our waste products. Less well recognized, however, are the consequences of exceeding the sustainable supply of essential resources, and how far along that course we may already be.

As a result of our population size, consumption patterns, and technology choices, we have surpassed the planet's carrying capacity. This is plainly evident by the extent to which we are damaging and depleting natural capital. The earth's environmental assets are now insufficient to sustain both our present patterns of economic activity and the life-support systems we depend on. If current trends in resource use continue, and if world population grows as projected, by 2010 per capita availability of rangeland will drop by 22 per cent and the fish catch by 10 per cent. Together, these provide much of the world's animal protein. The per capita area of irrigated land, which now yields about one-third of the global food harvest, will drop by 12 per cent. And cropland area and forestland per person will shrink by 21 and 30 per cent, respectively.

Since 1950, three trends have contributed most directly to the excessive pressures now being placed on the earth's natural systems—the doubling of world population, the quadradoubling of global economic output, and the widening gap in the distribution of income. The environmental impact of the world's population (2010 numbering 6.9 billion) has been vastly multiplied by economic and social systems that strongly favour growth and ever-rising consumption over equity and poverty alleviation; that fail to give women equal rights, education, and economic opportunity, and thereby perpetuate the conditions under which poverty and rapid population growth persist; and that do not discriminate between means of production that are environmentally sound and those that are not.

Table 1.1: Global Income Distribution, 1960-2010

Year	Share of Global Income Going to		Ratio of Richest to Poorest
	Richest 20%	Poorest 20%	
	(per cent)		
1960	70.2	2.3	30 to 1
1970	73.9	2.3	32 to 1
1980	76.3	1.7	45 to 1
1990	83.5	1.4	60 to 1
2000	85.2	1.3	70 to 1
2010	86.1	1.29	75 to 1

Source: United Nations Development Programme, *Human Development Report*, 2010

- *Growing Inequality in Income*: Of the three principal driving forces, the growing inequality in income between rich and poor stands out in sharpest relief. In 1960, the richest 20 per cent of the world's people absorbed 70 per cent of global income; by 2010 (the latest year for which comparable figures are available), the wealthy people's share had climbed to nearly 86 per cent. The poorest 20 per cent, meanwhile, saw their share of global income drop from an already meager 2.3 per cent to just 1.29 per cent. The ratio of the richest fifth's share to the poorest's thus grew from 30 to 1 in 1960 to 75 to 1 in 2010 (Table 1.1). In 2015, it was reported that 13% of people globally are middle income group.

Table 1.2: Grain Consumption in Selected Countries, 2010

Country	Grain Consumption Per Annum (Thousand Metric Tons)
Canada	23,235
United States	22,172
Russia	31,950
Mexico	36,100
Japan	10,485
China	13,770
Brazil	38,904
India	32,100
South Africa	9,273
Indonesia	7,700
Saudi Arabia	7,614
Malayasia	2,550
World Average	605,390

Source: World Grain Database, 2005

Disparities in food consumption are revealing as well (Table 1.2). As many as 700 million people do not eat enough to live and work at their full potential. The average African, for instance, consumes only 87 per cent of the calories needed for a healthy and productive life. Meanwhile, diets in many rich countries are so laden with animal fat as to cause increased rates of heart disease and cancer. Moreover, the meat-intensive diets of the wealthy usurp a disproportionately large share of the earth's agricultural carrying capacity, since producing one kilogram of meat takes several kilograms of grain. If all people in the world required as much grain for their diet as the average American does, the global harvest would need to be 2.6 times greater than it is today—a highly improbable scenario.

- *Economic Growth*: The second driving force—economic growth—has been fuelled in part by the introduction of oil onto the energy scene. Since mid-twentieth century, the global economy has expanded fivefold. As much was produced in two-and-a-half months of 1990 as in the entire year of 1950. World trade, moreover, grew even faster. Exports of primary commodities and manufactured products rose elevenfold.

  Unfortunately, economic growth has most often been of the damaging variety—powered by the extraction and consumption of fossil fuels, water, timber, minerals, and other resources. Between 1950 and 1990, the industrial roundwood harvest doubled, water use tripled, and oil production rose nearly sixfold. Environmental damage increased proportionately.

- *Population Growth:* Compounding the rises in both poverty and resource consumption in relation to the worsening of inequality and rapid economic expansion, population growth has added greatly to pressures on the earth's carrying capacity. The doubling of world population since 1950 has meant more or less steady increases in the number of people added to the planet each year. Whereas births exceeded deaths by 37 million in 1950, the net population gain in 1993 was 87 million—roughly equal to the population of Mexico.

The current world population is around 7 billion, the pattern of world population growth to date and the projections based on the growth trends allows one to speculate about the implications associated with it. Fig. 1.5 shows the growth of human population from the origin to the present times. We can see, for example, the dramatic reduction in the time required for the population to increase in increments of 1 billion if present trends continue. It also depicts that the population growth rates have increased tremendously from a doubling time of 150 years (1750 - 1900) to a doubling time of 35 years (1930 - 1965), and considering the present rate of growth (with downward trends) the next doubling in 2010, that is within a time span of 45 years (1965-2010).

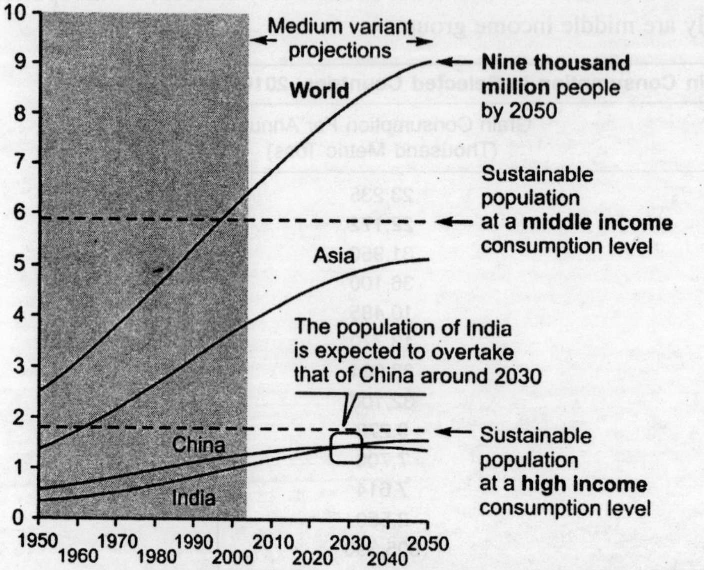

Fig. 1.5 : Population in thousand million

Source: Population Division of the Department of Economic and Social Affairs of the United Nations Secretariat, World Population Prospects: The 2004 Revision, Global Footprint Network, 2005

The annual growth rate was 1.7 per cent in 1982, peaked in 1965 at 2 per cent and the present rate is about 1.5 percent. It is expected that with this rate the world population will cross the 8 billion mark (box 1.3) by the year 2025 (Fig. 1.5).

Table 1.3: Global Population Trends

Year	Time span (years)	Population (billions)	Doubling time — years	Growth rate billion yr^{-1}	Growth rate percent yr^{-1}
1600 - 1750	—	0.75		—	—
1750 - 1800	50	1.00		0.005	0.66
1800 - 1900	100	1.60	150	0.006	0.60
1900 - 1930	30	2.00		0.013	0.83
1930 - 1960	30	3.00		0.033	1.66
1960 - 1965	5	3.30	65	0.60	2.00
1965 - 1975	10	4.00		0.70	2.12
1975 - 1987	12	5.00		0.83	2.08
1987 - 1998	11	6.00		0.90	1.82
1998 - 2000	2	6.20	35	0.100	1.70
2000 - 2025	25	8.25		0.82	1.50

Data based on World Resources, 1989, Merrick et al, 1986, UN 1988, 1989, World Projection 1974 and FAO Production Yearbook 1990

Some most populous countries and densely populated countries are shown in Table 1.4 and Table 1.5.

Table 1.4: 10 most populous countries

Rank	Country / Territory	Population	Date	Approx. % of world population
1	China	1,387,120,000	October 25, 2017	18.3%
2	India	1,323,020,000	October 25, 2017	17.5%
3	United States	325,992,000	October 25, 2017	4.3%
4	Indonesia	261,600,000	October 31, 2016	3.45%
5	Pakistan	209,354,000	October 25, 2017	2.76%
6	Brazil	208,172,000	October 25, 2017	2.75%
7	Nigeria	188,500,000	October 31, 2016	2.49%
8	Bangladesh	163,365,000	October 25, 2017	2.16%
9	Russia	146,773,226	June 1, 2017	1.94%
10	Japan	126,750,000	July 1, 2017	1.67%

Approximately 4.3 billion people live in these ten countries, representing around 58% of the world's population as of March 2016. Approximately 3.7 billion people, or half of the world population, live in the six most populous countries.

Table 1.5: 10 most densely populated countries (with population above 5 million)

Rank	Country	Population	Area (km^2)	Density (Pop. per km^2)
1	Singapore	5,607,300	710	7,898
2	Bangladesh	163,360,000	143,998	1,134
3	Taiwan	23,554,803	36,190	651
4	South Korea	51,446,201	99,538	517
5	Lebanon	6,082,000	10,452	582
6	Rwanda	11,553,188	26,338	439
7	Burundi	11,552,561	27,816	415
8	Netherlands	17,170,000	41,526	413
9	Haiti	10,911,819	27,065	403
10	India	1,323,020,000	3,287,240.	402

Box 1.3: Calculating the future population and doubling time

Now suppose we were to find out the global population that would be in the year 2020 taking the following assumptions:

(a) population in 1990 is 5 billion = W

(b) population growth rate is 1.7 per cent = P

To answer this question we take the equation = $q = 1 + \frac{P}{100}$.

Then $q = 1 + \frac{1.7}{100} = 1 + 0.017 = 1.017$

The growth of the population in 30 years can be calculated as:

$\log q^{30} = 30 \log q = 30 \log 1.017 = 30(.0073) = 0.219$ antilog = 1.655

Hence the population will increase by 65.5% of that in 1990 or by 3.275 billion. So the population in the year 2020 will be about (5 + 3.275) 8.275 billion.

If we wish to find the doubling time after 1990 then we take:

$$q = 1.017 \text{ and } q^n = 2$$

$$q^n = n \log q = \log 2$$

$$n = \frac{\log 2}{\log q} = \frac{0.3010}{0.0073} = 41.23 \text{ years}$$

that is, the doubling will occur in the year 2032 (1990 + 41.23 years)

The U.N. median population projection now shows world population reaching 8.9 billion by 2030, and levelling off at 11.5 billion around 2150.

The outer limit of the planet's carrying capacity is determined by the total amount of solar energy converted into biochemical energy through plant photosynthesis minus the energy those plants use for their own life processes. This is called the earth's net primary productivity (NPP), and it is the basic food source for all life.

Prior to human impacts, the earth's forests, grasslands, and other terrestrial ecosystems had the potential to produce a net total of some 150 billion tons of organic matter per year. Stanford University biologist Peter Vitousek and his colleagues estimate, however, that humans have destroyed outright about 12 per cent of the terrestrial NPP and now directly use or co-opt an additional 27 per cent. Thus, one species—*Homo sapiens*—has appropriated nearly 40 per cent of the terrestrial food supply, leaving only 60 per cent for the millions of other land-based plants and animals.

It may be tempting to infer that, at 40 per cent of NPP, we are still comfortably below the ultimate limit. But this is not the case. We have appropriated to 40 per cent that was easiest to acquire. It may be impossible to double our share, yet theoretically that would happen in just 60 years (i.e., between 1900 to 1960) if our share rose in tandem with population growth. And if average resource consumption per person continues to increase, that doubling would occur much sooner.

- *Cropland*: Cropland area worldwide expanded by just 2 per cent between 1980 and 1990. That means that gains in the global food harvest came almost entirely from raising yields on existing cropland. Most of the remaining area that could be used to grow crops is in Africa and Latin America; very little is in Asia. The most sizable near-term additions to the cropland base are likely to be a portion of the 76 million hectares of savanna grasslands in South America that are already accessible and potentially cultivable, as well as some portions of African rangeland and forest. These conversions, of course, may come at a high environmental price, and will push our 40 per cent share of NPP even higher.

  Moreover, a portion of any cropland gains that do occur will be offset by losses. As economics of developing countries diversify and as cities expand to accommodate population growth and migration, land is rapidly being lost to industrial development, housing, road construction, and the like. Canadian geographer Vaclav Smil estimates, for instance, that between 1957 and 1990, China's arable land diminished by at least 35 million hectares—an area equal to all the cropland in France, Germany, Denmark, and the Netherlands combined. At China's 1990 average grain yield and consumption levels, that amount of cropland could have supported some 450 million people, about 40 per cent of its population.

Box 1.4: Carrying Capacity of India

In August 1999, the population of India reached 1 billion. This is an almost threefold increase since the country achieved independence in 1947. The population of India is currently growing at a rate of about 1.0 per cent a year, and within a few decades India is expected to overtake China to become the world's most populous country. By 2016, India is expected to house more people than all of the industrialised countries in the world. Population density in India is more than double that of China and more than ten times that of the United States.

This rapid population explosion has had significant impacts on the environment, and continues to do so through deforestation, soil erosion, water shortages, pollution of air and water, and declining biodiversity.

India has struggled for many years to reduce population growth, but it has fallen short of the extreme approach adopted by China. Many experts insist that the future stability and prosperity of India will depend largely on the country's ability to stabilise its population.

Box 1.4: Thomas Malthus on Human Population Problems

In 1798, Thomas Malthus published 'An Essay on the Principle of Population'. In that book, he argued that populations tend to grow geometrically while the resources available to support them tend to grow arithmetically. As a result, population grows much faster than food supply, so a population will almost inevitably outgrow the supply of food available to feed it. Indeed, Malthus stressed that population growth was already outpacing the production of food supplies in eighteenth century England. He predicted that this imbalance of population and resources would lead to degradation of the land, and this, in turn, would lead to massive famine, disease and war. The disaster that Malthus believed imminent did not occur, mainly because of improvements in agriculture and the technological innovations introduced by the Industrial Revolution (1760 on). Nonetheless, many experts argue that his ideas still have great value today.

Table 1.6: Human-induced Soil Degradation for the World, expressed in million hectares, 2010

	Type	Light	Moderate	Strong	Extreme	Total
Wt	Loss of Topsoil	301.2	454.5	161.2	3.8	920.3
Wd	Terrain Deformation	42.0	72.2	56.0	2.8	173.3
W	WATER	343.2	526.7	217.2	6.6	1093.7 (55.6%)
Et	Loss of Topsoil	230.5	213.5	9.4	0.9	454.2
Ed	Terrain Deformation	38.1	30.0	14.4	—	82.5
Eo	Overblowing	—	10.1	0.5	1.0	11.6
E	WIND	268.6	253.6	24.3	1.9	548.3 (27.9%)
Cn	Loss of nutrients	52.4	63.1	19.8	—	135.3
Cs	Salinization	34.8	20.4	20.3	0.8	76.3
Cp	Pollution	4.1	17.1	0.5	—	21.8
Ca	Acidification	1.7	2.7	1.3	—	5.7
C	CHEMICAL	93.0	103.3	41.9	0.8	239.1 (12.2%)
Pc	Compaction	34.8	22.1	11.3	—	68.2
Pw	Waterlogging	6.0	3.7	0.8	—	10.5
Ps	Subsidence organic soils	3.4	1.0	0.2	—	4.6
P	PHYSICAL	44.2	26.8	12.3	—	83.3 (4.2%)
TOTAL		749.0 (38.1%)	910.5 (46.4%)	295.7 (15.1%)	9.3 (0.5%)	1964.4 (100%)

Sources: Worldwatch Institute, based on "The Extent of Human-Induced Soil Degradation", Annex 5 in L.R. Oldeman et al, *World Map of the Status of Human-Induced Soil Degradation* (Wageningen, Netherlands: United Nations Environment Programme and International Soil Reference and Information Centre, 2010).

In addition, much of the land we continue to farm is losing its inherent productivity because of unsound agricultural practices and overuse. The "Global Assessment of Soil Degradation,", a three-year study involving some 250 scientists, found that more than 550 million hectares are losing topsoil or undergoing other forms of degradation as a direct result of poor agricultural methods (Table 1.6).

On balance, unless crop prices rise, it appears unlikely that the net cropland area will expand more quickly over the next two decades than it did between 1980 and 1990. Assuming a net expansion of 5 per cent (which may be optimistic), total cropland area would climb to just over 1.5 billion hectares. Given the projected 33-percent increase in world population by 2010, the amount of cropland per person would decline by 21 per cent (Table 1.7).

- *Pasture and Rangeland*: They cover some 3.4 billion hectares of land, more than twice the area in crops. The cattle, sheep, goats, buffalo, and camels that graze them convert grass (which humans cannot digest) into meat and milk (which they can). The global ruminant livestock herd, which numbers about 4.4 billion, thus adds a source of food for people that does not subtract from the grain supply, in contrast to the production of pigs, chickens, and cattle raised in feedlots.

Much of the world's rangeland is already heavily overgrazed and cannot continue to support the livestock herds and management practices that exist today. According to the "Global Assessment of Soil Degradation," overgrazing has degraded some 680 million hectares since mid-twentieth century. This suggests that 20 percent of the world's pasture and range is losing productivity and will continue to do so unless herd sizes are reduced or more sustainable livestock practices are put in place.

During the 1980s, the total range area increased slightly, in past because land deforested or taken out of crops often reverted to some form of grass. If similar trends persist over the next two decades, by 2010 the total area of rangeland and pasture will have increased 4 percent, but it will have dropped 22 percent in per capita terms. In Africa and Asia, which together contain nearly half the world's rangelands and where many traditional cultures depend heavily on livestock, even larger per capita declines could significantly weaken food economies.

Table 1.7: Population Size and Availability of Renewable Resources, since (1990-2010)

	1990	2010	Total Change	Per Capita Change
	(million)		(percent)	
Population	5,290	7,030	+ 33	
Fish Catch (tons)[1]	85	102	+ 20	– 10
Irrigated Land (hectares)	237	277	+ 17	12
Cropland (hectares)	1,444	1,516	+ 5	– 21
Rangeland and Pasture (hectares)	3,402	3,540	+ 4	– 22
Forests (hectares)[2]	3,413	3,165	– 7	– 30

1. Wild catch from fresh and marine waters, excludes aquaculture.
2. Includes plantations; excludes woodlands and shrublands.

Sources: FAO Report 2010

- *Fisheries*: Another natural biological system that humans depend on to add calories, protein, and diversity to human diets is our fisheries. The annual catch from all sources (including aquaculture) is about 150.0 million metric ton in 2000, — about 7 per cent of the protein humans consume. Fish account for a good portion of the calories consumed overall in many coastal regions and island nations.

  The world fish catch has climbed rapidly in recent decades, expanding nearly fivefold since 1950. But it peaked at just above 100 million tons in 1989. Although catches from both inland fisheries and aquaculture (fish farming) have been rising steadily, they have not offset the decline in the much larger wild marine catch, which fell from a historic peak of 82 million tons in 1989 to 77 million in 1991, a drop of 6 per cent.

  With the advent of mechanized hauling gear, bigger nets, electronic fish detection aids, and other technologies, almost all marine fisheries have suffered from extensive overexploitation. Under current practices, considerable additional growth in the global fish catch overall looks highly unlikely. Indeed, the U.N. Food and Agriculture Organization (FAO) now estimates that all seventeen of the world's major fishing areas have either reached or exceeded their natural limits, and that nine are in serious decline.

  FAO scientists believe that better fisheries management might allow the wild marine catch to increase by some 20 per cent. If this could be achieved, and if the freshwater catch increased proportionately, the total wild catch would rise to 102 million tons; by 2010, this would nonetheless represent a 10 per cent drop in per capita terms.

- *Fresh Water*: It may be even more essential than cropland, rangeland, and fisheries: without water, after all, nothing can live. Signs of water scarcity are now pervasive. Today, twentysix countries have insufficient renewable water supplies within their own territories to meet the needs of a moderately developed society at their current population size. And populations are growing fastest in some of the most water-short countries, including many in Africa and the Middle East. Rivers, lakes, and underground aquifers show widespread signs of degradation and depletion, even as human demands rise inexorably.

  Water constraint already appears to be slowing food production, and those restrictions will only become more severe. Agricultural lands that receive irrigation water play a disproportionate role in meeting the world's food needs. The 237 million hectares of irrigated land account for only 16 per cent of total cropland but more than one-third of the global harvest (2010). For most of human history, irrigated area expanded faster than population did, which helped food production per person to increase steadily. In 1978, however, percapita irrigated land peaked, and it has fallen nearly 6 per cent since 2010.

Box 1.6: Human Impacts on Environmental Systems: The Situation in 2000

- Up to half of the land surface has already been transformed or degraded by land-use changes, including agriculture and urban development.
- More than half of the accessible surface fresh water is already being exploited.
- Atmospheric concentrations of CO_2 have increased by 30 per cent as a result of air pollution.
- More than half of terrestrial nitrogen fixation is caused by human activity including the use of nitrogen fertilisers, the planting of nitrogen-fixing crops and the release of reactive nitrogen from fossil fuels into the atmosphere.
- About 20 per cent of the plant species found on continents have been introduced to their present locations by human; on many islands the figure is as high as 50 per cent.
- About 20 per cent of bird species have become extinct over the last 200 years,, mostly because of human activity.
- About 8 per cent of the primary productivity of the oceans is used for food.
- Over 20 per cent of marine fisheries are over-exploited or depleted,, and an additional 40 per cent are at the limit of exploitation.

- *Forests and Woodlands*: They are the last key components of the biological resource base. They contribute a host of important commodities to the global economy—logs and lumber for constructing homes and furniture, fibre for making paper, fruits and nuts for direct consumption, and, in poor countries, fuelwood for heating and cooking. More important even than these benefits, however, are the ecological services forests perform—from conserving soils and moderating water cycles to storing carbon, protecting air quality, and harboring millions of plant and animal species.

  As on 2000, forests cover 24 per cent less area than in 1700—3.4 billion hectares compared with an estimated 4.5 billion about 300 years ago. Most of that area was cleared for crop cultivation, but cattle ranching, timber and fuelwood harvesting, and the growth of cities, suburbs, and highways all claimed a share as well. In 1990s, assessments suggest that the world's forests declined by about 130 million hectares between 1980 and 1990, an area larger than Peru.

A successful global effort to lighten humanity's load on the earth would directly address the three major driving forces of environmental decline—the grossly inequitable distribution of income, resource-consumptive economic growth, and rapid population growth—and would redirect technology and trade to buy time for this great movement. Although there is far too much to say about each of these challenges to be comprehensive here, some key points bear noting.

Wealth inequality may be the most intractable problem, since it has existed for millennia. The difference today, however, is that the future of rich and poor alike hinges on reducing poverty and thereby eliminating this driving force of global environmental decline. In this way, self-interest joins ethics as a motive for redistributing wealth, and raises the chances that it might be done.

Important actions to narrow the income gap include greatly reducing Third World debt—much talked about in the 1980s but still not accomplished—and focusing foreign aid, trade, and international lending policies more directly on improving the living standards of the poor. If decision-makers consistently asked themselves whether a choice they were about to make would help the poorest of the poor—that 20 per cent of the world's people who share only 1.4 per cent of the world's income—and acted only if the answer were yes, more people might break out of the poverty trap and have the opportunity to live sustainably.

A key prescription for reducing the kinds of economic growth that harm the environment is the same as that for making technology and trade more sustainable—internalizing environmental costs. If this is done through the adoption of environmental taxes, governments can avoid imposing heavier taxes overall by lowering income taxes accordingly. In addition, establishing better measures of economic accounting is critical. Since the calculations used to produce the gross national product do not account for the destruction or depletion of natural resources, this popular economic measure is extremely misleading. It tells us we are making progress even as our ecological foundations are crumbling. A better beacon to guide us toward a sustainable path is essential. The United Nations and several individual governments have been working to develop better accounting methods, but progress with implementation has been slow.

In September 1994, government officials gathered in Cairo for the "International Conference on Population and Development," the third such gathering on population. This is a timely opportunity to draw attention to the connections between poverty, population growth, and environmental decline; and to devise strategies that simultaneously address the root causes. Much greater efforts are needed, for instance, to raise women's social and economic status and to give women equal rights and access to resources. Only if gender biases are rooted out will women be able to escape the poverty trap and choose to have fewer children.

The challenge of living sustainably on the earth will never be met, however, if population and environment conferences are the only forums in which it is addressed. Success hinges on the creativity and energy of a wide range of people in many walks of life. The scientists' "Warning to Humanity" ends with a call to the world's scientists, business and industry leaders, the religious community, and people everywhere to join in the urgent mission of halting the earth's environmental decline.

1.7 ENVIRONMENTAL CRISIS

Inevitably, there has been heated debate about what has caused the environmental crisis, and many different (and at times conflicting) suggestions have been made about what the most important root causes might be. There are certainly a number of important factors (Box 1.7) that, added together, help to account for observed changes in the way in which the environment has been used.

While in recent years there have been great advances in sicentific and public understanding of the relevance and seriousness of the environmental crisis, the debate about root causes remains largely unresolved. This is perhaps inevitable, given the many different factors involved and the wide variety of disciplines (including history, social science, economics, psychology and religious studies) that have taken part in the debate.

Box 1.7: Root Causes of the Environmental Crisis

A number of factors are involved, including:

- developments in technology throughout history, which have given people a better ability to use the environment and its resources for their own ends (particularly since the Industrial Revolution);
- the rapid increase in human population in recent centuries, which has significantly increased population densities in many countries;
- a significant rise in human use of natural resources, particularly over the last century;
- the emergence of free-market economies, in which economic factors play a central role in decision making about production, consumption, use of resources and treatment of wastes;
- attitudes towards the environment, particularly among Western cultures, which regard it as freely available for people to do whatever they like with;
- the short-term time horizon over which many people, companies and countries make decisions, which means that short-term profit maximisation has generally been taken more seriously than long-term sustainable use of the environment.

COMMON RESOURCES

An important prespective on the environment crisis, which sets it into context even if it does not fully explain it, is the idea of the environment as a commons (common property resource) which is freely available—at least in theory—to anybody who wants to use it.

The way in which herdsmen treated their medieval commons has many similarities with the way we treat the environment today:

- we think of it as freely available without restriction;
- we regard it as our right to be able to use it;
- each of us wants to get as much as we can out of it;
- we forget that it can tolerate only a certain level of use before it is damaged;
- we compete for a larger share of it than other people.

The 'tragedy of the commons' analogy applies just as well to common property resources, including atmosphere outside 'national air space', the ocean outside 'territorial limits' and wilderness (unspoiled land that is not privately owned). Growth, competition and private ownership are hallmarks of modern Western society, yet these are the very forces that threaten and damage the commons. UNEPs Global Environment Outlook 2000 stresses the need for a much better understanding of the interrelationships between economic, social and environmental change as a basis for a more sustainable future.

Little wonder that many scientists and politicians have recently argued that these forces must be replaced by cooperation and sustainable use, and that the commons debate was rekindled at the 1992 Rio Earth Summit (Box 1.8). Critics of modern development practices often argued that many environmental and socio-economic problems (including proverty) can be traced to enclosure of the commons and the domination and dispossession of local communities by enclosure. The principle of permanent sovereignty over natural resources (PSNR)—which emerged in the 1950s as a legal concept to formally recognise the sovereignty of developing countries over their resources—is also part of this wider debate.

Box 1.8: The Rio Earth Summit

The United Nations Conference on Environment and Development, based at the Rio Centro, was accompanied by a series of events held in the city between 3 and 14 June 1992, all designed to raise awareness of environmental issues. There were many unofficial events, displays, concerts and fringe meetings, which added to the colourful carnival atmosphere.

The conference, described by some as 'the biggest show on Earth', was a record breaker in many ways:

- 172 countries were represented and took part;
- the summit at the end was attended by 120 heads of state,, making it the largest ever gathering of world leaders;
- there were more than 350 scheduled meetings;
- 1,600 non-government organisations (NGOs) were represented;
- the conference and associated events attracted an estimated 450,000 visitors;
- it was by far the largest conference ever held about the environment;
- there were almost 100 press conferences, and nearly 9,000 journalists covered the events.

Although most of the world's media attention was focused on the conference itself, the debates that took place were effectively open forums where the views of different countries were reported and the draft documents were finally approved. Most of the hard work of drafting documents and position papers, preparing agreements for signing, ironing out many political and bureaucratic differences between the countries represented, and agreeing on the main agenda items had been completed during the preparatory meetings.

Environmental Objectives: The key outcomes of the Earth Summit are incorporated in five formal documents—two conventions, two statements of principle and an action programme:

- a convention on climate change
- a convention on biodiversity
- a statement on forest principles
- the Rio Declaration
- Agenda 21 (the action programme).

The Convention on Climate Change: One of the most important product of UNCED, this convention calls on countries to commit themselves to stabilise emissions of greenhouse gases by the year 2000. Oil producing countries were concerned that limitations on the burning of fossil fuels (a principal source of carbon dioxide) would be a direct threat to their economies, so there was heated debate, and a great deal of behind-the-scenes negotiating was required to get the convention accepted. Five years after Rio, in 1997, the debate was still raging (Box 1.8).

The Convention on Biological Diversity: This convention is designed to preserve plant and animal life. It was also hotly debated, partly because it defined wild (as well as improved) species as having economic value, and it proposed that genetic resources be recognised as national assets that can only be exploited on the basis of agreements between signatory states. US President George Bush did not sign the convention, allegedly out of consideration for the country's biotechnology industry.

The Forest Principles document: The proper title of this document is the 'Non-legally binding authoritative statement of principles for a global consensus on the management, conservation and sustainable development of all types of forest'. It includes important principles on replanting, establishing new forests and protecting the rights of indigenous people.

The Rio Declaration: This declaration is a general statement of intent that centres on the well-being of people, on the right of states to control their own natural resources and on their obligation not to damage the environment of other countries. Among other things, it recommends the precautionary principle, internalisation of environmental costs, use of environmental impact statements and the pollutery-pays principle.

Agenda 21: This is the action plan, which·179 states agreed to that is designed to turn the theory into practice. Agenda 21 is a national and international 'blueprint for action' on environment and development, an agenda for the twentyfirst century. It is over 500 pages long and contains forty chapters, each of which focuses on a specific theme and is divided into programme areas setting out goals, activities and provisions on funding and implementation.

2 ENVIRONMENTAL COMPONENTS

2.0 THE ENVIRONMENT AS A SYSTEM: AN OVERVIEW

The entire environment of planet Earth can be divided into four spheres: the biosphere (living organisms of Earth), atmosphere (the gaseous envelope surrounding the planet), hydrosphere (the liquid water on the surface of Earth), and lithosphere (the stony or rocky matter composing the bulk of the surface of Earth). Matter cycles and energy flows through these spheres (Fig. 2.1). A systems approach provides a convenient overview and manner of understanding how our environment works.

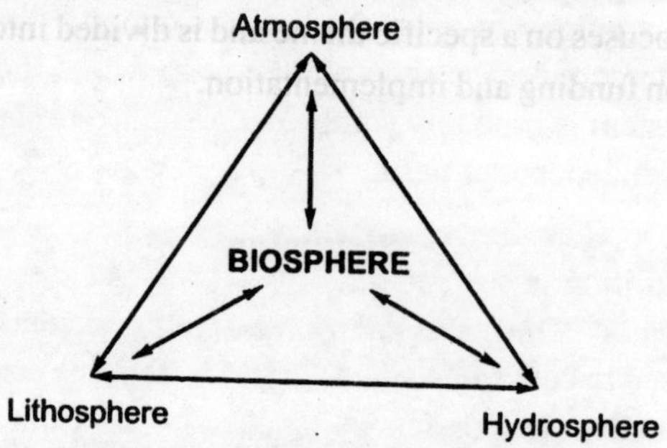

Fig. 2.1 : Interaction of environmental components

What is a System?

A **system** can be technically defined as a "set of components functioning together as a whole". A systems view allows us to isolate a part of the world and focus on those aspects that interact more closely than others. For example, a cell in the system we call a human body generally interacts much more closely with other cells in the body than with the outside world. By focusing only on those cells that function in digestion, we confine our view further to the digestive system. The key point here is that most systems are *hierarchical*: they are composed of smaller sets of systems made of smaller interacting parts.

Three Key Traits of the Environmental System

We can analyze the global environment in terms of three system traits: openness, integration, and complexity.

Openness refers to whether a system is isolated from other systems. An **open system** is not isolated in that it exchanges matter and/or energy with other systems. A **closed system** is isolated and exchanges nothing.

The law of entropy means that energy cannot be fully recycled; rather, "high-quality" energy is degraded to "lower quality" energy (such as waste heat); therefore, any system that does not have a renewing supply of energy from outside will eventually cease to exist. Not surprisingly then, the Earth is an open system in terms of energy. Fig. 2.1 shows how energy flows from the Sun and is often radiated back into space. In contrast, the Earth, for all practical purposes, is a closed system in terms of matter. If we discount the relatively small amount of matter added from meteorites and other space debris, the Earth contains all of the matter it will even have. Driven by energy from the Sun, this matter cycles over and over among the four spheres, often moving back and forth among the gaseous, liquid, and solid states and participating in the metabolism of living things.

Integration refers to the strength of the interactions among the parts of the system. For instance, the human body is a highly integrated system whose cells are interdependent and in close communication. The loss of certain cells, such as those composing the heart or brain, can result in death of all of the other cells in the system (the whole organism) because the cells are so interdependent. At the other extreme are systems with very weak integration, such as the cells in a colony of single-celled organisms (such as the green algae, chlamydomonas). Removal of many cells (to a point) will have little effect on the remaining cells because they are less dependent on each other.

The degree of integration of the global environmental system is under debate. At one extreme are scientists who argue that the global system is a superorganism: many complex pathways intimately interconnect the lithosphere, hydrosphere, atmosphere, and biosphere. According to this **Gaia hypothesis**, the Earth is similar to an organism, and its component parts are so integrated that they are like cells in a living body. Many scientists, however, believe that the global environment is less integrated than the Gaia hypothesis argues. This does not mean that the environment is "unconnected" or even as weakly connected as a colony of cells. As we describe in future chapters, many kinds of matter cycles and energy flows interconnect the spheres and cycle within the spheres as well. The true level of integration in the global system is probably somewhat less than a "superorganism" but considerably more than a loose collection of independent parts.

Complexity is often defined as how many kinds of parts a system has. This definition conforms to our intuition: a tiny insect seems more complex to us than a large rock because it has many more types of "parts". The insect has more complex molecules, as well as more different types of molecules, and these are used to construct cells and organs. This example also illustrates that complexity is often hierarchical, with smaller components being used to construct larger ones.

As you would expect, the environment is enormously complex. The four spheres, with their matter cycles and energy flows, have trillions and trillions of different components operating at many spatial and temporal scales. Organisms, soils, rainwater, air, and many other components interact in complicated ways. The individual spheres themselves are equally complex. Even with advanced computers, no one has been able to predict accurately and precisely the weather, or even climate, very far into the future because the atmosphere is so complex. Indeed, the many interactions make unpredictability a basic characteristic of complex system.

2.1 ATMOSPHERE

The earth's atmosphere, a complex fluid system of gases and suspended particles, did not have its origin in the beginning of the planet. The atmosphere as of today has been derived from the Earth itself by chemical and biochemical reactions. Although the fluid system forms a gaseous envelope around the Earth, its boundaries are not easily defined. They can be arbitrarily defined as the Earth's atmosphere interface and space interface.

The gases like Nitrogen, Oxygen, Argon, Carbon dioxide and Water vapour etc. together make up the total volume of atmosphere. Together with suspended particulates, viz. dust and soot, they constitute the gaseous turbidity particularly in troposphere. However, the composition of atmosphere and so also its structure is variable in time and space.

The vertical structure of atmosphere is very much related to radiant energy absorption and thus can be described in terms of variable of temperature. Below 60 km. there are two main zones of absorption at the Earth's surface and in the Ozone layer. The absorbed energy is redistributed by radiation, conduction and convection. There are, therefore, two temperature maxima: at the Earth's surface and at an elevation of 50 km. Above each of these maxima there is mainly convectional mixing. Temperatures in these mixing layers decrease with height above the heat source. The lower of these two zones is referred to as *troposphere* and the upper the *mesosphere*. These are separated by a layer of little mixing in which the atmosphere tends towards a layered structure referred to as the *stratosphere*. Between the troposphere and the stratosphere is the tropopause which marks the approximate upper limit of mixing in the lower atmosphere. The average height of this is usually given as 11 km., but this varies over the earth. In tropical latitude its average height is 16 km. and in polar latitude it is only 10 km. There is one further zone of heating, above the mesosphere and more than 90 km. from the Earth's surface, where shortwave ultraviolet radiation is absorbed by any oxygen molecules present at this height. This is referred to as *thermosphere*. Within this layer, ionization occurs which produces charged ions and free electrons. Beyond the thermosphere, at a height of approximately 700 km., lies the exosphere where the atmosphere has an extremely low density. At this level there are increasing numbers of ionization particles which are concentrated into bands referred to as the *Van Allen radiation belts*.

However, this simple model of vertical structure can be simplified to provide a model of the atmosphere as two concentric shells, the boundaries of which are defined by the *stratopause*, at approximately 50 km.

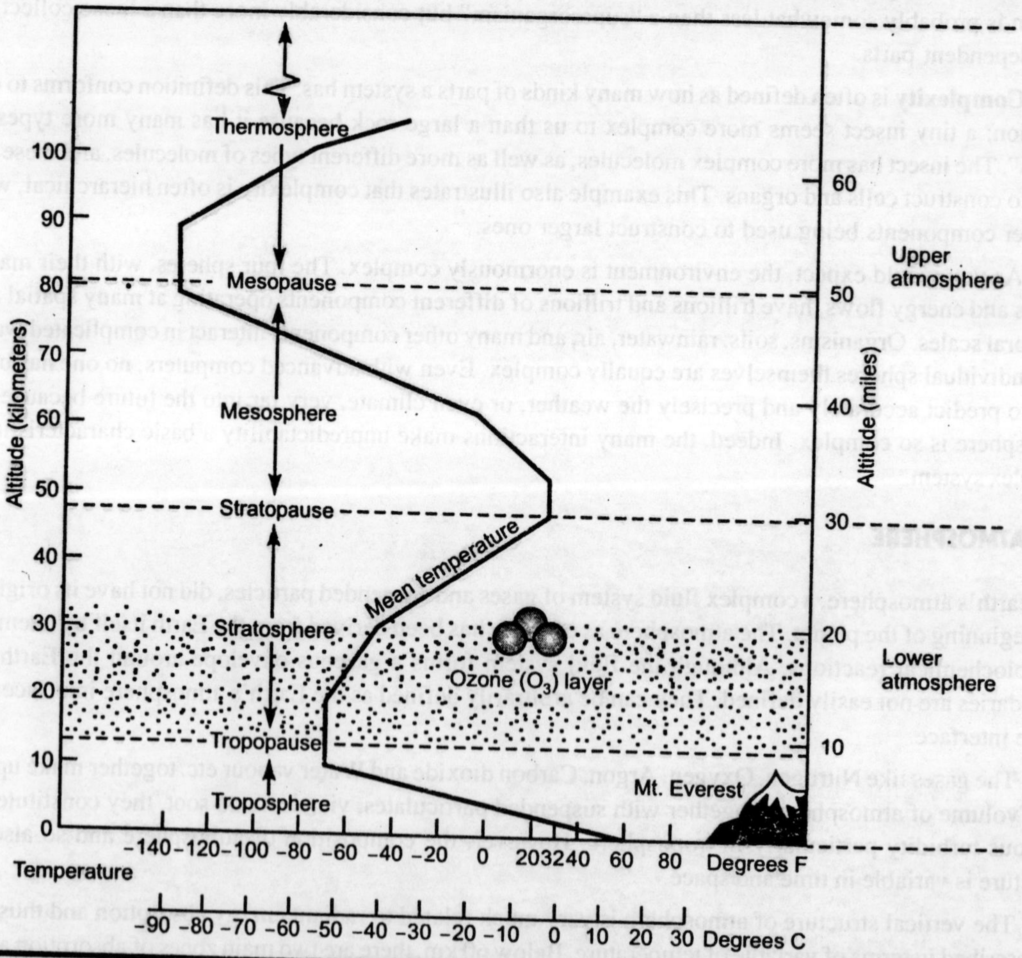

Fig. 2.2 : The structure of the Earth's atmosphere. The tropopause, stratopause and mesopause are the boundaries between the troposphere and stratosphere, stratosphere and mesosphere, and mesosphere and thermosphere respectively

above the Earth's surface and a hypothetical outer limit of the atmosphere, at approximately 80,000 km. Below the stratopause, in the stratosphere and troposphere, there is 99% of the total mass of the atmosphere and it is at this level that atmospheric circulatory systems operate. Beyond the stratopause a layer of nearly 80,000 km. thick contains only 1% of total atmospheric mass and experiences ionization by high-energy, short wavelength solar radiation. The temperature profile of atmospheric layer are given in Fig. 2.2.

2.2 HYDROSPHERE

It includes the surface water and its surrounding interface. It is vital for life molecule to survive. Water possesses a number of physical and chemical properties that help the molecule to act as best suited medium for life activities. The movement of water from earth surface to atmosphere through hydrological cycle appears to be a close system. Water is the most abundant substance on the Earth's surface. The oceans cover approximately 71% of the planet, glaciers and ice caps cover additional areas; and water is also found in lakes and streams, in soils and underground reservoirs, in the atmosphere, and in the bodies of all living organisms. Thus, water in all its forms—ice, liquid, water and water vapour—is very familiar to us.

Human being use water in the home, in industry, in agriculture, and for recreation. These applications differ widely in the quantities and quality of the water that they require. In one way or another, we use all available sources—inland waters, ground water, and even ocean water. We pollute it, repurify it, and reuse it, over and over again.

The demand for global water resources increased day-by-day though pure fresh water availability decreased severely. Thus, there is need for precious use of available pure fresh water and their fruitful storage and conservation. A simplified outline of hydrological cycle is given in Fig. 2.3.

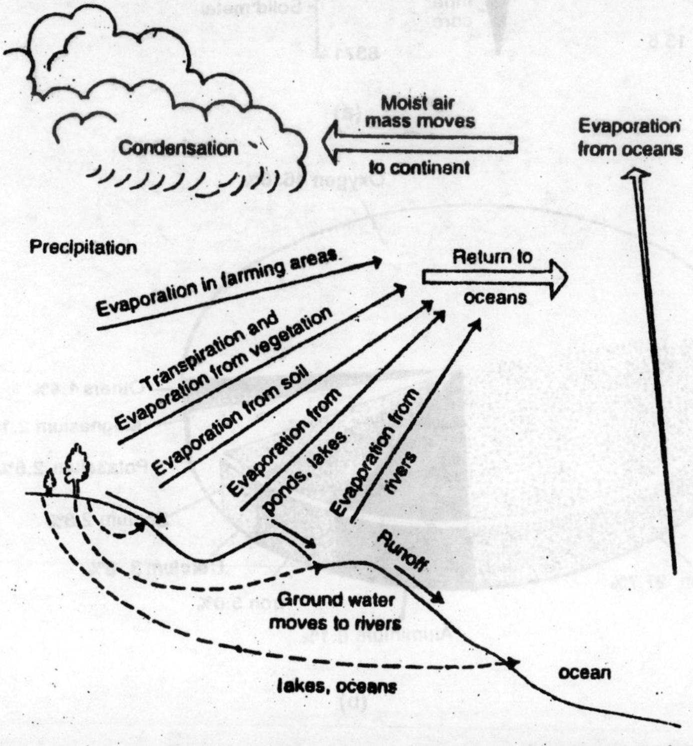

Fig. 2.3 : A simplified outline of hydrological cycle

2.3 LITHOSPHERE

It is the outer boundary layer of solid earth and the discontinuity within the mantle. The outer boundary forms a complex interface with the atmosphere and hydrosphere and is also the environment in which life has evolved. The inner boundary is adjacent to rock, which is near its melting point and is capable of motion

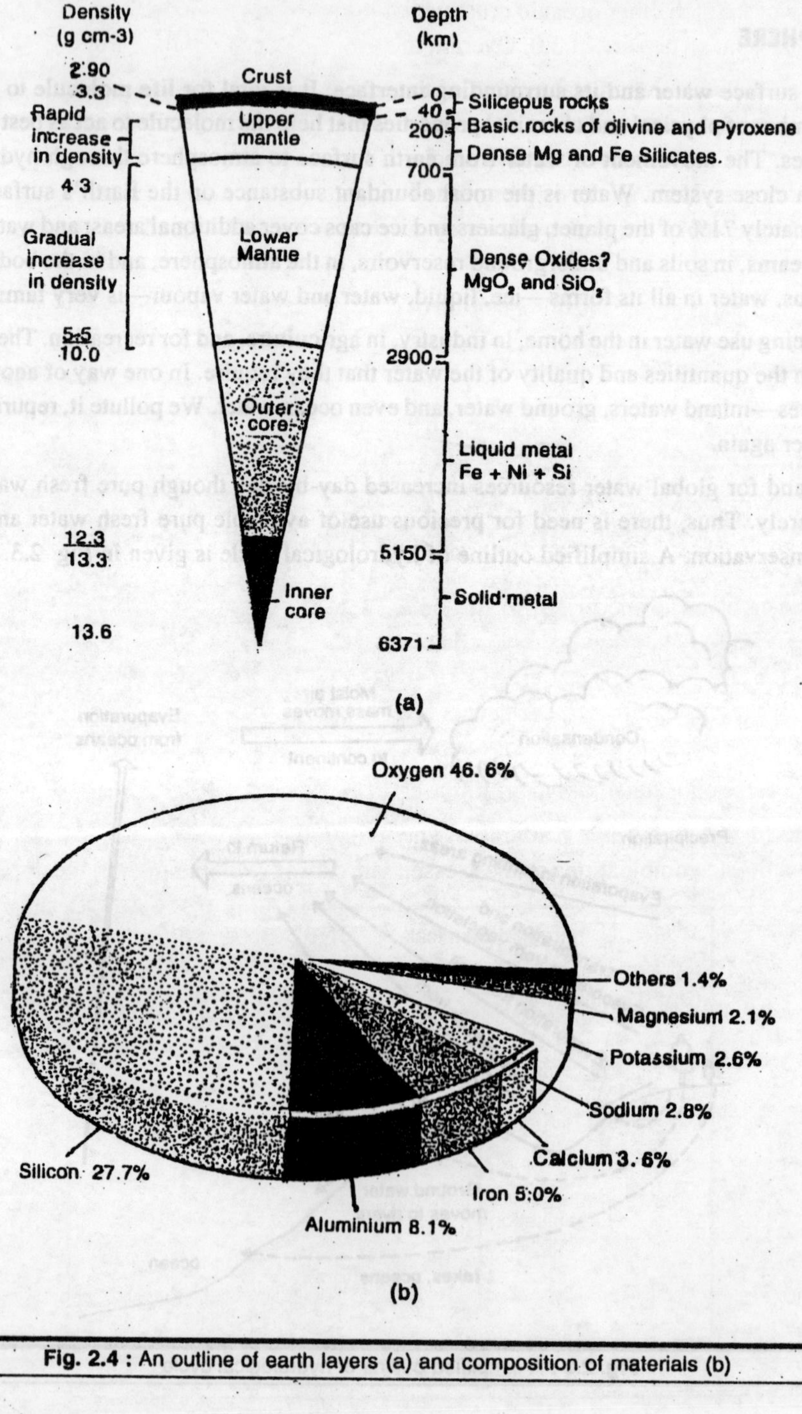

Fig. 2.4 : An outline of earth layers (a) and composition of materials (b)

relative to the lithosphere above. Basically lithosphere is nothing but a crustal system composed of various layers: core, mantle and outer crust. Various elements constitute such crustal layer in mixture of different proportion. In general the earth crust is composed of three major classes of rocks as classified on the basis of their mode of origin: Igneous rocks, sedimentary rocks and metamorphosed rocks. There are two types of crusts—continental crust which is composed of granitic rocks in silicon-aluminium and with a mean density of 2.8; the other oceanic crust which is basaltic in composition, consisting of more basic minerals and has a mean density of 3.0. Overall, the average density of the earth is 5.5 gm/c.c.

Interaction between the crustal system of the lithosphere and the atmosphere and biosphere takes place where continental crust is exposed above sea-level. At the land/air interface crustal material becomes exposed to inputs of solar radiant energy, precipitation and atmospheric gases. These inputs are often modified by or operate through the effects of the living systems of the biosphere. Under the influence of these inputs, crustal rocks are broken down by weathering processes and are transferred to fine porus crustal layer called *soil*.

An outline of earth layers and composition of crustal materials is given in Fig. 2.4.

2.4 BIOSPHERE

The biosphere encompasses all the zones on the Earth in which life is present, i.e., entire bioresources of the earth. It develops on earth since 4.5 billions year through evolutionary processes. At the top of the lithosphere, throughout the hydrosphere and into the lower atmosphere life of diverse type exists. These bioresources and their surrounding constitute the "Biosphere", where mankind acting as the most evolved creature.

The steps involved in the origin of life on earth is very complex and requires several centuries. Considerable uncertainty surrounds the details of atmospheric composition, the processes involved and even the sequence of some events leading to formation of living cells (Fig. 2.5). The conventional view has been that the earliest organism on the planet were heterotrophic prokaryotic bacteria. Subsequently autotrophic prokaryotes & eukaryotes start appearing as stepwise evolutionary changes.

The major steps of origin of life in primitive earth are depicted in Fig. 2.5.

Life on Earth requires water, a source of energy (sunlight) and various nutrients found in the soil, water and air. Suitable combinations of these essentials cannot be found high in the upper atmosphere or

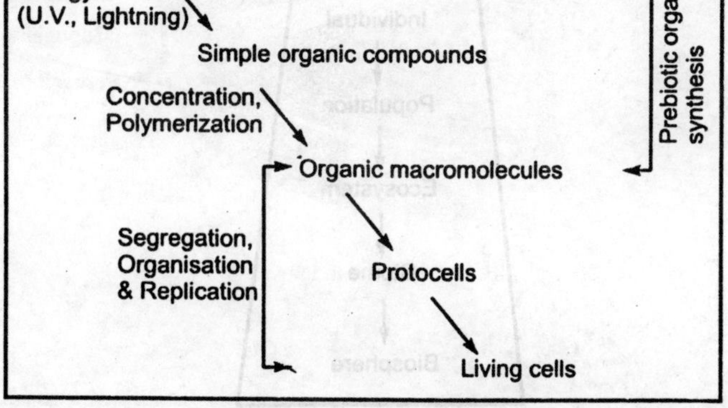

Fig. 2.5 : Steps to origin of life

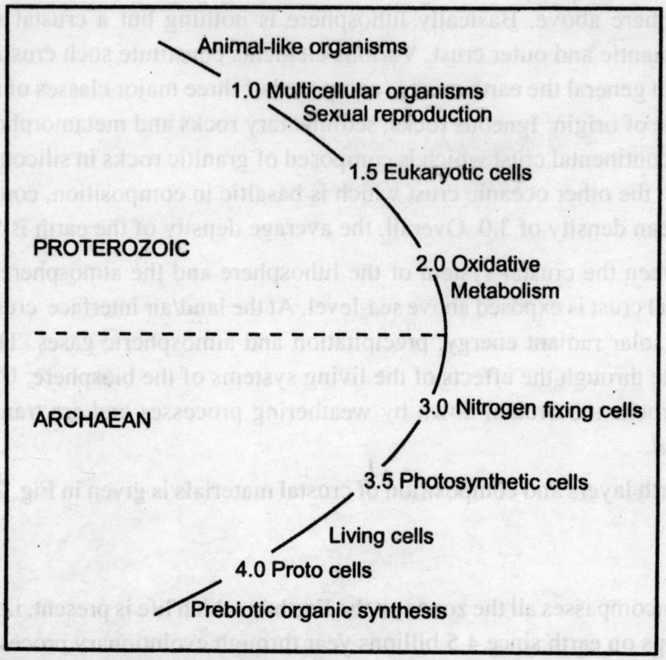

Fig. 2.6 : Suggested approximate timing of first appearance in the biosphere during the Archaean & Proterozoic aeons (Nos. in billion years)

deep underground. They exist only in a narrow layer near the surface of the Earth. This biosphere layer extends over most of the surface of the Earth. It includes the upper layers of the Earth's crust and the thin layer of soil that supports plant life. This zone of life also extends about 8 km. up into the atmosphere (air borne biota) and as much as 8 km. down into the depths of the sea. Living organisms are not distributed uniformly on globe : few organisms live on polar ice caps and glaciers, whereas many live in tropical rainforests (Fig. 2.6).

Within the biosphere, there are several major regions containing specific types of ecosystems. These major regions are called *biomes*. Biomes are then recognized by the types of dominant ecosystems—tropical rainforests, temperate forests, prairies, deserts, and Arctic Tundra. The ecosystems again composed of populations and population composed of individuals. The detailed subdivisions of biospheres are depicted in Fig. 2.7.

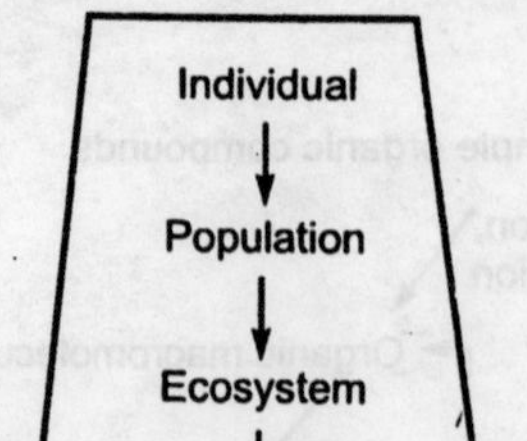

Fig. 2.7 : Subdivisions of a typical biosphere

The global estimate of species of both prokaryotes and eukaroytic life forms are given in Tables 2.1 and 2.2. These figures imply the fact that how diverse is our biological world on earth:

Table 2.1: Estimated total microorganism species in the world

Group	Known species	Total (expected) species	Percentage Known (%)	Percentage in culture collection (%)
Algae	40,000	60,000	67	2.5
Bacteria	3,000	30,000	10	7.0
Fungi	69,000	1,500,000	5	0.8
Protozoa	40,000	100,000	40	—
Virus	5,000	130,000	4	2.0
Total	157,000	1,820,000	9	

Sources: Cerliss (1991), Hawksworth, (1991).

Table 2.2: Number of Species by class (from World Resources Inst., 1986)

Class	Identified species	Estimated species
Mammals	4,170	4,300
Birds	8,715	9,000
Reptiles	5,115	6,000
Amphibians	3,125	3,500
Fishes	21,000	23,000
Invertebrates	1,300,000	4,400,000
Vascular plants	250,000	280,000
Nonvascular plants	150,000	200,000
Rounded total	1,742,000	4,926,000

* minimum figure

3 ORIGIN AND EVOLUTION OF BIOSPHERE—MATTER, ENERGY AND LIFE

3.0 INTRODUCTION

Biological organisms come in a wonderful variety of sizes, shapes and complexity. Besides providing us with the food, shelter and oxygen—essentials for life, the living creatures with which we share the planet give us pleasure in their bright colours, lively songs and graceful movements. All living organisms of the earth, as noticed today, evolved through ages in this planet using solar energy and matter from the environment to build organized structures and to carry out the dynamic processes that define life sensing and responding to external stimuli, maintaining internal identity, integrity and reproductivity.

Every organism is a chemical factory that captures matter and energy from its environment and transforms them into structures and processes that make life possible. Matter consists of unique chemical forms—elements, molecules and compounds. Organisms use some elements in abundance, some in trace amounts and others not at all. Certain substances are hoarded within cells, while others are actively excluded. Carbon is particularly important because chains and rings of carbon atoms form the skeletons of *organic compounds*, the material of which biomolecules, and, therefore, living organisms are made. The four major categories of bioorganic compounds—carbohydrates, proteins, fats and nucleic acids—constitute the basic material for life. These compounds organized in a proper fashion to constitute the structural and functional units of living organisms i.e., *cells*. Microscopic organisms—such as bacteria, some algae and protozoa—are composed of single cells. But higher organisms are multicellular, usually with many different cell varieties. Every cell is surrounded by a thin but dynamic membrane of lipid and protein that receives information about the exterior world and regulates the flow of materials between the cell and its environment. Inside, cells are subdivided into tiny organelles and subcellular particles that provide the machinery for life. All of the chemical reactions required to create these various structures and provide them with energy and materials to carry out their functions, dispose of wastes, and perform other functions of life at the cellular level are carried out by a special class of proteins called *enzymes*.

A steady influx of solar radiation provides the heat and light energy needed to support life in the biosphere. Water, which covers approximately three-fourth of the earth's surface, is readily available to lifeforms. Because of its unique characteristics, it stabilizes the biospheric temperature and provides the medium in which life processes occur. The earth's atmosphere provides gases necessary for life, helps maintain surface temperatures, and filters out dangerous radiation.

The population of different species that live and interact within a particular area make up biological communities. The interactions of a community with the physical factors of its environment comprise that ecosystem level of organization and study. Matter and energy are processed through the tropic levels of an ecosystem via food chains and food webs. Thus biosphere is a source of large quantities of essential elements, and in a given ecosystem these elements are constantly used and reused by living organisms (Fig. 3.1).

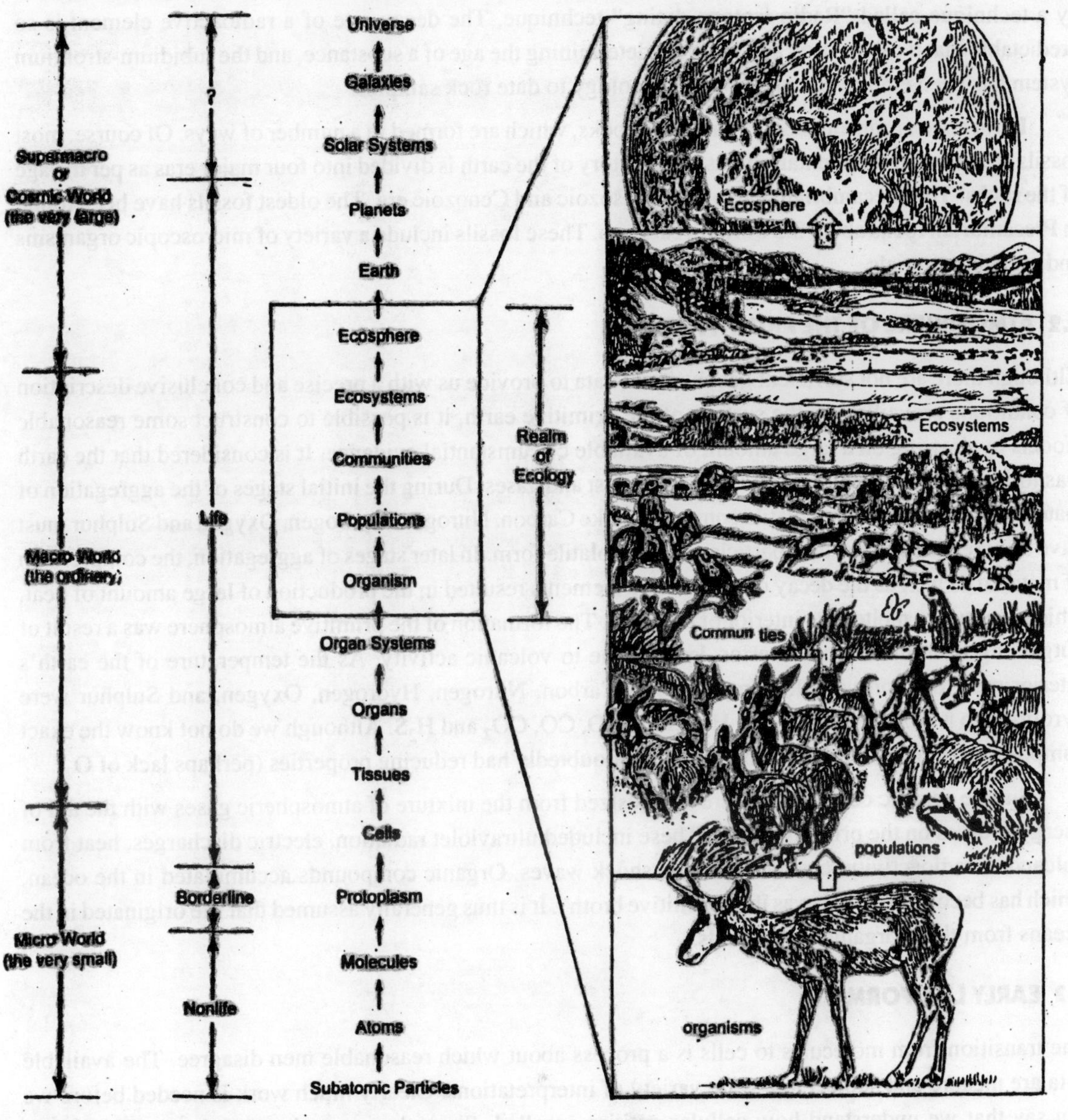

Fig. 3.1 : Level of organization of matter according to size and function

3.1 ANCESTRAL LIFEFORMS : FOSSIL

In order to trace and to understand the evolutionary process by which we believe living systems arose on earth, it is quite essential to know the early lifeforms of earth surface, i.e., the fossil remnants of the past lifeforms. The fossil record, consisting of remains or impressions of organisms which have been preserved in the earth's crust, tells us something about the types of organism in present at specific times during the past history of the earth.

Fossils are date by determining the age of the rock strata in which they are found. This is often done by a technique called "Radio isotope dating" technique. The decay rate of a radioactive element is so predictable that it can be used as a basis for determining the age of a substance, and the rubidium-strontium system has been employed extensively in geology to date rock samples.

Earth crust is constituted of a variety of rocks, which are formed in a number of ways. Of course, most fossils are found in sedimentary rocks. The history of the earth is divided into four major eras as per the age of the rocks: The Precambrian, Paleozoic, Mesozoic and Cenozoic era. The oldest fossils have been found in Precambrian rocks about 3.2 billion years ago. These fossils include a variety of microscopic organisms and organic materials.

3.2 ATMOSPHERE OF THE PRIMITIVE EARTH

Although there are not sufficient geochemical data to provide us with a precise and conclusive description of conditions prevailing on the surface of the primitive earth, it is possible to construct some reasonable models by drawing on a large amount of available circumstantial evidence. It is considered that the earth was formed from a turbulent cloud of cosmic dust and gases. During the initial stages of the aggregation of material forming the earth, volatile compounds like Carbon, Nitrogen, Hydrogen, Oxygen and Sulphur must have been retained by the dust particles in non-volatile form. In later stages of aggregation, the compression of material as well as the decay of radioactive elements resulted in the production of large amount of heat, which eventually melted the interior of the earth. The formation of the primitive atmosphere was a result of outgassing from the earth's interior, largely due to volcanic activity. As the temperature of the earth's interior rose, the non-volatile compounds of Carbon, Nitrogen, Hydrogen, Oxygen, and Sulphur were pyrolyzed to a mixture of NH_3, N_2, H_2, CH_4, H_2O, CO, CO_2 and H_2S. Although we do not know the exact composition of the primitive atmosphere, it undoubtedly had reducing properties (perhaps lack of O_2).

Simple organic compounds were synthesized from the mixture of atmospheric gases with the aid of energy sources on the primitive earth. These included ultraviolet radiation, electric discharges, heat from volcanoes, radioactivity, cosmic rays, and shock waves. Organic compounds accumulated in the ocean, which has become referred to as the "primitive broth". It is thus generally assumed that life originated in the oceans from these organic compounds.

3.3 EARLY LIFE-FORMS

The transition from molecules to cells is a process about which reasonable men disagree. The available data are meagre and subject to a wide variety of interpretations. Clearly much work is needed before we can say that we understand how cellular entities enrolled. Since the atmosphere of the earth probably lacked free oxygen, it seems reasonable to assume that the first cells were *anerobic heterotroph*—they existed in the absence of oxygen and depended for survival on organic compounds synthesized by abiotic means. Gradually, they evolved an increased ability to synthesize organic molecules from precursors available in the environment. As long metabolic pathways on the cells evolved, cells became less dependent on the environment for supplies of nutrient compounds. The evolution of these pathways was favoured by natural selection, since a cell that evolved the capacity to synthesize any necessary compound could survive without obtaining that compound as a nutrient from the environment and hence, possessed a competitive advantage over metabolically less versatile cells. One of these metabolic innovations that eventually evolved was the ability to degrade glucose and to use the energy released in the process to form ATP.

A particularly significant metabolic breakthrough was the appearance of anaerobic autotrophs, some species of chemosynthetic bacteria and a wide range of photosynthetic prokaryotes. Photosynthesis endowed organisms with the ability to use the energy of solar radiation (light) to produce "energy-rich" ATP. Some of these photosynthetic organisms produce molecular oxygen as a byproduct of the photosynthetic process. Oxygen productivity by photosynthetic prokaryotes—some of which resemble modern forms of blue-green algae—were probably firmly established as early as 2.7 billion years ago, shortly before the beginning of the Middle Precambrian. The establishment of oxygen-producing photosynthesis had far reaching effects. Eventually, the atmosphere of the earth became drastically changed as a result of the release of free oxygen into the air.

As oxygen accumulated in the atmosphere is a byproduct of photosynthesis, anaerobic prokaryotes that were not capable of surviving in the presence of O_2 were eliminated or were restricted to local environments that lacked oxygen. Facultative anaerobes, anaerobic organisms capable of living in either the absence or presence of free O_2, probably became widely distributed. When the oxygen concentration in the air reached high levels, aerobic heterotrophs and aerobic autotrophs became the dominant types: these organisms required oxygen for survival.

This scheme is supported by the wide range of metabolic types found among present day prokaryotes. These include heterotrophs; autotrophs, some of which produce O_2 as a byproduct of photosynthesis and some of which do not; strict anaerobes; facultative anaerobes, & strict aerobes.

About 1.5 billion years ago, eukaryotic cells arose in some unknown way, probably from prokaryotic ones. By the time eukaryotic cells appeared, it is probable that all the basic metabolic patterns now found in present-day cells had already evolved. These eukaryotic cells, aerobic and hence adapted to survival in an oxygen-containing environment, became the structural unit of all autotrophic and heterotrophic multicellular forms.

3.4 ORIGIN OF LIFE: CHEMICAL BASIS

The question of the origin of life reinvented from the scientific to the philosophical realms, a position it still holds in the minds of many. In the twentieth century, Alexander Oparin (1924) and J.B.S. Haldane (1929) proposed that life originated as an interaction between organic compounds formed under the anoxic conditions that prevailed on the early earth. After these proposals were published, some scientists began to consider the question of the origin of life as an experimental problem. Miller's (1953) pioneering discovery of the early and rather rapid formation of amino acids from inorganic precursors ushered in a new field of laboratory science referred to as "Primitive-Earth-Model Experiments", "Prebiotic Chemistry" or "Experimental Chemical Evolution".

From a source of Carbon, Hydrogen, Oxygen, Nitrogen & Energy—Hydrocarbons, Fatty acids, Benzene & Phenolic compounds, Amino acids, and many other familiar organics may be easily produced in the laboratory, enhance confidence in the hypothesis that life arose from cosmic constituents under anoxic conditions. In fact, the organic compounds found in meteorites are quite similar to those produced in simulation experiments.

Although there is no unanimity about the source of energy for organic compounds formation, it is generally conceded that the Archean ultraviolet radiation flux at the Earth's surface prior to the accumulation of atmospheric O_2 was probably the major source of energy, both for prebiotic organic synthesis and for early biological processes such as the induction of lysis and transduction by bacteriophage.

The problem of the production of organic components of small molecular weight seems more or less solved, but how was organic matter aggregated? How were monomers polymerized? How was the information transmitted to new aggregations of organic matter and reliable heritability established? The gap between the most complex mixture of organic chemicals and the simplest cell is still to be crossed in theory and in the laboratory.

The linking of amino acid residues into polymers including both peptide and nonpeptide linkages has been reported to occur under conditions presumed of the early earth (Fox & Harada, 1960; Fox, 1972). The mechanism of the spontaneous as opposed to enzymes mediated, formation of polypeptide and nucleotide polymers from monomers is a mistery.

All life, using the same basic membrane-wrapped machinery of nucleic-acid-determined protein synthesis, reproduces with incredible fidelity. Although the question of how the first bacteria formed from aggregates of organic compounds cannot be answered yet in details, but it is assumed that prokaryotes evolved extensively by direct filiation in the primitive earth.

The earliest microbes were heterotrophs reproducing in environments supplied with abiotically produced nutrients and minerals. The autotrophs were details from these heterotrophs after the evalution of O_2 on the earth surface. The conspicuous forms of life today require O_2 to live. The eukaryotic cells are derived from the prokaryotes in course of time.

Although life is patchingly distributed, a veritable scum that reaches mainly to within a meter of the Earth's watery or solid surface, it extends over an area of about 500 million square kilometers. The next advance in the expansion of life may be above the troposphere to the stark cold reaches of the rest of the solar system and perhaps, beyond. But we should be warned; for its own survival, life must moisten and modify its environment. Without maintaining species diversity, the capacity for rapid reproduction, interactions, flexibility and redundancy, life will not maintain itself. That life can survive in space has been amply demonstrated, but survival is not reproduction. Whether the biosphere will ever expand into space is surely an open question.

3.5 ACCUMULATION OF ORGANIC MATERIAL AND FORMATION OF THE FIRST CELL

A precondition to any form of biological evolution requires the accumulation of organic material and formation of the cell. There are at least two contrasting theories as to how organic matter accumulated in the prebiotic environment: terrestrial accretion and cometary infall.

The oldest and most widely cited theory is that of terrestrial accretion. This theory, first proposed by Miller in 1953, is based on the principle that, under laboratory conditions, when gases such as hydrogen, carbon dioxide, methane, and ammonia are heated with water, and energized by electrical discharge for at least 24 hours, approximately half the carbon originally present in the methane gas will be converted into amino acids, sugars, and other organic molecules, including purines and pyrimidines (required to make nucleotides). Amino acids and nucleotides can associate to form polymers including polypeptides (proteins) and polynucleotides (RNA and DNA) (Miller, 1992) thus forming the first building blocks of the cell.

An alternative theory is that the first organic material on Earth came from other planets and entered the Earth's atmosphere via meteoritic input. This idea was originally viewed as better suited to the realms of science fiction. However, there is increasing evidence to support the theory that meteoritic and cometary infall provided the first organic material on Earth, analysis of the present day meteorites on the ice beds of the Antarctic (carbonaceous chrondrites), for example, indicate that they typically contain between 1 and 4% carbon, mainly as graphite but also as much as 1% organic molecules. Organic compounds contained within the carbonaceous chrondrites include hydrocarbons, amino acids, carbon, hydrogen cyanide and amphiphilic molecules.

Another crucial event in the formation of the first cells would have been development of the outer cell membrane, since without this there would be no means of simple containment of the macromolecules within the cell. There is also the question of how organic compounds become sufficiently concentrated to form more complex molecules. An extremely important environment for both processes to occur would have been shallow tidal pools, where cycles of solar heating and cooling, combined with tidal wetting and drying, would have provided the necessary conditions for concentration and molecular self-assembly.

3.6 THE FIRST PROKARYOTES: THE GEOLOGICAL EVIDENCE

Of all the organisms presently on the Earth, prokaryotes are the simplest in structure, smallest physically, and most abundant in terms of number of individuals (Raven *et al*, 1992). They range between approximately 1 and 10 μm in size, are single celled (although many types have cells joined together within a mucilaginous sheath), lack an organized nucleus surrounded by a nuclear envelope, and reproduce by binary fission (i.e. each cell increases in size and divides into two). Prokaryotes also tend to have one of three basic cell shapes: a straight rod shape (bacilli); a spherical shape (cocci); or a long, coiled shape (spirilli).

Fossil microorganisms have also been extracted from other early sedimentary deposits, including cherts and shales. Some of the earliest examples are found in the Warrwoona Group in Western Australia, where filamentous and colonial fossil structures have been reported from carbonaceous cherts dated between 3500 and 3300 Ma. These fossils, which are predominantly small spheres and filaments, indicate close morphological comparison with extant prokaryotes, based on both cell size and also structure (i.e., whether single cells or in colonies, whether the colonies are sheath enclosed, etc.). The *microorganisms* contained within the Warrwoona Group sediments have therefore been 'classified' as photoautotrophic cyanobacteria (oxygen-producing blue-green algae) and taken to indicate the presence of organisms capable of oxygen-producing photosynthesis as far back as 3500 million years ago (early Archaean). Cell size and structure have not been the only factors used to identify these early *microorganisms* as prokaryotes; evidence of their reproductive mode has also been discovered. There are now many examples in the early fossil record of cells that have been preserved while in a state of cell enlargement and division, suggesting binary fission.

3.7. EVOLUTION OF THE EUKARYOTES

Eukaryotes differ from prokaryotes in that they have a membrane-bound nucleus in which the DNA is contained, organelles including mitochondria, integrated multicellularity, and sexual reproduction (sometimes). They constitute the three major groups of multicellular organisms (plants, animals, and fungi), along with many groups of the Protista, including species of red, green, and brown algae. Because of the diversity and importance of eukaryotic organisms to life on Earth, it is often stated that the evolution of eukaryotes was one of the most important events in the history of life.

Evidence for eukaryotes (from fossil organisms and ancient biomolecules) suggests that they were probably present on Earth from as early as 2700 Ma. Three records of early eukaryotes of particular interest to the evolution of plants include the oldest recorded alga *Grypania*, dated to 2100 Ma, fossil bangiacean red algae, dated to approximately 1200 Ma (Butterfield *et al*, 1990), and cladophoralean green algae, dated to between 800 and 700 Ma.

3.8 POSSIBLE TRIGGERING MECHANISMS OF EUKARYOTIC EVOLUTION

A number of researchers have suggested that the combined geological/molecular record demonstrates a general trend of episodic increases in biological diversity (Knoll, 1992) through the Archaean and Proterozoic (~ 3500 – 540 Ma). Six major biological events, which are specific to the plant fossil record are recognized, including:

(i) the origin of life on Earth and diversification of anaerobic archaebacteria and eubacteria at or around 3500 Ma;
(ii) origin of photosynthetic organisms around 3300 Ma;
(iii) the appearance of organisms capable of aerobic metabolism and cellular acquisition of the organelles mitochondria and, later, chloroplasts, between 2800 and 2400 Ma;
(iv) the appearance of complex eukaryotic organisms in the geological records from approximately 2100 Ma;
(v) the large increase in diversity of organisms and radiation of acritarchs—a group of organic-walled microfossils, the majority of which are though to represent reproductive cysts of green algae or alage cysts, from approximately 1000 Ma;
(vi) the large increase in diversity of planktonic algae from approximately 540 Ma.

3.9 EVOLUTION OF LIFEFORMS THROUGH AGES

Over the years in the geological part, the Earth's geomorphology, climate and biotic community changed gradually. In early Palaeozoic, some 540 million years ago, three separate land masses existed: Asia, North America and Europe, and Gondwanaland (which includes present day Africa, South America, Australia, New Zealand and Antarctica, etc.). During the Palaeozic, 420 million years ago, North America and Africa lay close together around the south pole and the rest of Gondwanaland lay on the far side of the south pole, pointing towards the equator. Slowly the land mass moved northward, so that by the Carboniferous age, 340 million years ago, the whole of Africa had moved across the south pole and Antarctica lay in the region of the south pole. During the Permian, these three blocks joined, forming a single land mass, Pangaea, which further moved northward, and began to break apart slowly by mid-Mesozoic age.

By the mid-Cretaceous, about 100 million years ago, Africa and South America had split apart. Thus, by the end of the Cretaceous, Gondwanaland had broken up, the only intact land mass was North America —thus, collectively known as Laurasia. Until the lower Eocene, Laurasia remained intact. North America was connected to Europe by Greenland and Scandinavia. But in the mid-Eocene, the North Atlantic joined the Arctic Ocean, separating Laurasia into North America and Europe.

The formation, breakup and northward drift of continents resulted in broad climatic changes and the formation of geological barriers that affected evolving plant and animal life. Between 2,700-2,000 million years ago, prokaryotes developed, but eukaryotes evolved sometime later. The Precambrian era came to a close with the sudden rise of Metazoa. The Palaeozoic saw the evolution of the earliest known fish, the first amphibians, the first reptiles, the first insects and the first land plants as well as the rise of the great coal forests. The end of the Palaeozoic witnessed the expansion of primitive reptiles, and the end of Cretaceous saw breaking of land due to land mass drifting and climatic warming. With continuous climatic changes and land mass shifting results, and with the disappearance of many plants and animal species, simultaneously some new species start appearing during environmental selection.

A detailed scheme for 'Geological time scale' is given in Table 3.1.

Table 3.1: Lifeforms through Geological Time Scale

Era	Period	Epoch	Age (Millions of years)	Dominant Life: Plants	Dominant Life: Animals
	Quaternary	Recent	0.01	Agricultural plants	Domesticated animals
		Pleistocene	2		Ice Age — modern humans; mixture and then thinning out of mammalian faunas
Cenozoic The age of mammals	Tertiary	Pliocene	10	Herbaceous plants rise; forests spread	Culmination of mammals; radiation of apes
		Miocene	25	First extensive grasslands	
		Oligocene	35		Modernization of mammals; mammals become dominant
		Eocene	55		Mammals become conspicuous
		Paleocene	70		Expansion of mammals; extinction of dinosaurs
Mesozoic The age of reptiles	Cretaceous		135	Angiosperms or flowering plants rise: gymnosperms decline	Dinosaurs reach peak; first snakes appear
	Jurassic		180	Cycads prevalent	First birds and mammals appear
	Triassic		230	Gymnosperms rise; seed ferns die out	First dinosaurs; reptiles prominent
Palaeozoic	Permian		280	Conifers become forest trees: cycads important	Great expansion of primitive reptiles
	Carboniferous Pennsylvanian		310	Lepidodendron: Sigillaria and Calamites dominant: the swamp forest	Age of cockroaches; first reptiles
	Mississippian		345	Lycopods and seed ferns abundant	Peak of crinoids and bryozoans
	Devonian		405	First spread of forests	First amphibians; insects and spiders
	Silurian		425	First known land plants	First land animals (scorpions)
	Ordovician		500	Algae, fungi, bacteria	Earliest known fishes; peak of trilobites
	Cambrian		600	Algae, fungi, bacteria; lichens on land	Trilobites and brachipods; marine invertebrates
Precambrian	Late			Algae, fungi, bacteria	First known fossils
	Early		4,500	Bacteria	No fossils found

The preceding cursory review of past changes on the Earth thus serves to emphasize the fact that plants and animals have evolved through the times. As environmental conditions slowly changed, flora and fauna responded—either adjusting to a new set of conditions or perishing. To survive as a species, organisms must adapt to the changing environment. This is the case of natural selection.

3.10 CHANGING ENVIRONMENTAL CONDITIONS ASSOCIATED WITH GLACIATIONS

Geological evidence suggests that as many as four major glacial periods occurred in the late Proterozoic (~ 1000 – 540 Ma), the final one, the Varanger ice age, taking place between approximately 610 and 590 Ma (Harland *et al,* 1990; Briggs, 1995). During the Varanger ice age, more acritarch extinctions occurred than originations, resulting in a decline in overall species number (75% reduction). Following the ice age, there was a large increase in species number and many new and highly ornamented acritarchs appeared. A tentative link has therefore been suggested between the acritarchs and environmental/oceanographic changes (e.g. changing patterns of marine sedimentation) that accompanied the end of the Varanger glaciation (Knoll and Lipps, 1993; Knoll, 1994a, b).

Although just two examples are provided above, there are many other suggestions in the literature of bursts of evolutionary activity occurring at or around the same time as significance geological events in this early period of Earth history (e.g. Knoll, 1992; Knoll and Lipps, 1993; Schopf, 1994). However, it is also highly probable that in these early environments there would have been spiralling ecological relationships (Knoll, 1991, 1994). Without a doubt, these would also have had a major impact on species diversity and number, resulting in the development of a complex marine ecosystem of plants and animals by the beginning of the Cambrian period, 590 Ma.

Natural selection results when environmental forces—both abiotic and biotic—favour certain genotypes, as expressed by phenotypes, over others. It functions through non-random reproduction within a population. Not every individual in a population is able to contribute its share of genetic characteristics to the next generation or leave surviving offspring. Some fail to survive to reproduce, others fail to mate or produce offspring. Their contribution to the gene pool or fitness increases at the expense of others.

On the whole, natural selection tends to eliminate the less fit alleles and favour the most fit. If it favours those close to the population mean, selection is stabilizing and produces little change in the population. It is characteristic of more or less stable environments. If it favours the extremes, it is disruptive or diversifying. Such selection may occur when segments of a population are subjected to different selection pressures environmentally induced. Or selection may favour one extreme, that is, the population mean in that direction. Characteristic of a changing environment, directional selection produces the most rapid phenotypic and genotypic changes.

Although natural selection may seem to imply a steady improvement or perfection of a species through the times, but that is not the case. Selection pressure may only enable individuals to maintain their fitness over time. Fitness is a response to current environmental conditions and not to some future condition. Individuals with the greatest fitness in some future situation happen to have the genetic variability that enables them to survive in a changed environment. Nor does natural selection work for the long-term good of the species. While it may improve fitness over the short-term, natural selection may actually work against the long-term survival of a species, especially in situations where natural selection favours speciation.

The great diversity of plants and animals in the world causes one to wonder how all these species arose. Each is adapted to an ecological niche in the ecosystem to which it belongs, and each is genetically independent. The process by which this has come about, by which one form becomes genetically isolated from the others, is 'speciation'—the multiplication of species. Speciation is accomplished in most plants and animals by an interaction of heritable variation, natural selection, and barriers to gene flow.

Of major importance to natural selection and adaptation are genetic or inherited variations within a population—the variations that arise from mutations and especially from the shuffling of genes and chromosomes in sexual reproduction. Variations found in a population are transmitted to the next generation through sexual reproduction when a sample of the gene pool contained in gamete joins to form zygotes.

The most widely accepted mechanism of speciation involves geographic isolation of one part of a population from another. Each part experiences different selection pressures. The two populations diverge and may ultimately become two distinct species. If geographic barriers break down, the two populations may hybridize, or they may evolve strong, usually premating, isolating mechanisms that reinforce their apartness. That is known as 'allopatric species'. Another type of speciation mechanism among populations of organisms that have low vagility and inhabit contiguous, that is, different and spatially separated habitats, is 'parapatric speciation'. It is probably most common among plants. In parapatric speciation both reproductive isolation and speciation take place within differentiation and the process takes place within a population. The outcome is 'sympatric speciations'.

How rapidly evolution and speciation take place is, of course, a matter of debate. Sometimes heritable variation takes place within a shortest possible time while in other situation, some species apparently have never changed for millions of years. On the whole evolution among some populations is always taking place somewhere on Earth, even though evolution may not be a continuous process among all populations.

3.11 DIVERSIFICATION OF THE BIOSPHERE

Where does individual variation come from? This question is crucial because without the initial variation in the population, natural selection would have nothing to act on. This question troubled Darwin himself

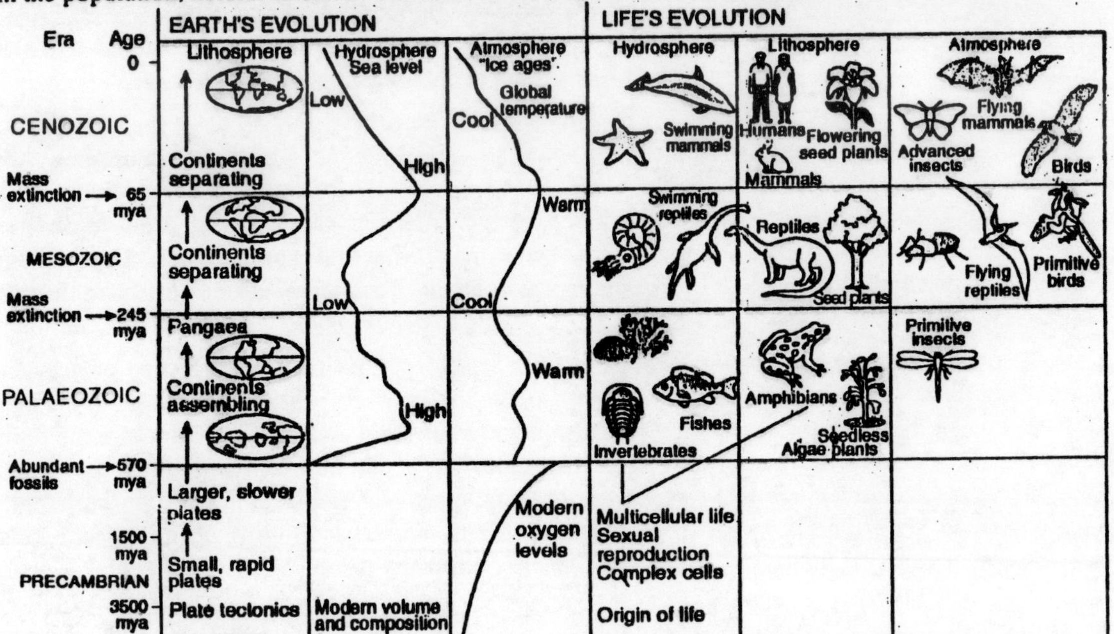

Fig. 3.2 : The left side of this diagram summarizes the changes in the Earth's continents, oceans, and atmosphere, whereas the right side summarizes the evolution of life on Earth

because he was not aware of the work of Gregor Mendel who is credited with discovering the laws of inherited variation in 1865. Indeed, initially Mendel's work was overlooked, and scientists only rediscovered it in the early 1900s. Mendel discovered that genes, which are the basic units of heredity, pass on traits. According to the Human Genome Project, humans have an estimated 30,000 to 40,000 genes that determine our traits. Variation in a population, or gene pool, occurs because individuals possess different sets of genes that produce different traits. How do different sets of genes arise? Reproduction is one way. Genes are shuffled when sperm and egg cells are fused, causing offspring from the same parents to have different genes. A second cause of genetic variation is **mutation**, which is a spontaneous change in a gene. Genes are composed of DNA molecules and mutations occur when DNA molecules are altered, such as during DNA replication. Mutation is the ultimate source of all genetic variation.

Evolution through natural selection has produced an increasingly diverse biosphere, with the total number of species becoming greater through time. Initially, evolution was relatively slow. The right side of Figure 3.2 summarizes the evolution of the life on Earth.

SECTION II

RESOURCES AND ECOSYSTEMS

4 NATURAL RESOURCES

4.0 CONCEPT OF RESOURCES

All means of satisfying human needs, at a given time and place, are *resources*. Thus "Resources" are means for attaining individual and social welfare. Natural components like land, water, minerals, forests, wildlife, energy—or even man himself—are considered as resources as well as resource-creating factors.

A major global concern has been whether the present rate of global population growth will be sufficient to meet up the resource needs for mankind's survival and comfort? Thus there lies the quest for sustainable use of resources to meet up the ever-increasing needs of human populations. Recent technological innovations helped a lot in solving the problems of resource depletion at a faster rate.

4.1 TYPES OF RESOURCES

There are four major categories of resources:

(a) **Continuous resources**: These continue to be available, and with the possible exception of solar energy, the receipt of which could be affected by atmospheric pollution, cannot be degraded, even with gross mismanagement. Examples: solar energy, wind, gravity, tidal energy, geothermal energy.

(b) **Renewable resources (flow resources)**: These resources are those capable of natural regeneration into useful products within a timespan relevant to man. These resources are potentially renewable and could be indefinitely available, provided their capacity to regenerate is not damaged by natural catastrophe or human activities. Once degraded beyond a certain critical point, a renewable resource may never recover. Examples: clean water, flora, fauna, soil and clean air.

(c) **Non-renewable resources (stock resources)**: These are available only in finite quantities, or else the rate of renewal is so slow that they must be regarded as available only in fixed quantities. Examples: minerals and ground waters etc.

(d) **Extrinsic resources**: These are fickle resources and prone to breakdown or degradation, yet can be continuous resources if well-managed. Examples: human skills, institutions, management abilities etc.

4.2 CURRENT STATUS OF MAJOR RESOURCES

Among the global resources the major components are water, land, forest, energy and minerals. Global survivality depends on the availability of such resources. It is well-known that the requirement for early man was very limited, but with the advancement of industrialisation, man's requirement changed significantly with times.

The strength of nations—social, economic or political—is chiefly determined by the resources they conquered and their capacity to utilize and conserve these resources. The developing countries produce a large fraction of the world's raw materials and control an insignificant fraction of international trade in finished good. Their resources have been subjected to constant exploitation through the market mechanism of the developed countries. What is more striking is the fact that while over-utilisation of natural resources by the developed nations got reflected in the adverse shift of the ecological balance of the world, the less developed countries are still living in conditions of severe poverty because of under-utilisation of their resources and compelled to live in an environment polluted by the waste generating economics of the industrialised nations.

(A) Water Resources

Water is the most vital resource for life. In our planet earth, approximately 97.2% water lies in oceans as salt water, while 2.15% in frozen ice form and the remaining 0.65% remains as fresh water either on surface or as ground water (Fig. 4.1). Available fresh water resources are very limited. The details of water distribution on earth and their subsequent removal process and also renewal periodicity are depicted in Tables 4.1 and 4.2.

The demand for fresh water has increased day by day and will increase with the rapid growth of population, agriculture and industry. As a result the fresh water reserve depletes day by day too. The requirement of clean water per person is about 2.7 liter per day, thus the global requirement is about 16.5 billion lit per day only for drinking purpose. Agriculture is also one of the major consumer of fresh water resources.

Table 4.1: Distribution of Water on the Earth

	Volume, km³	Mass, g
Sea water	1.37×10^9	1.41×10^{24}
Fresh water lakes	1.3×10^5	1.3×10^{20}
Saline lakes and Inland seas	1.0×10^5	1.0×10^{20}
Rivers	1.3×10^3	1.3×10^{18}
Ground water	8.4×10^6	8.4×10^{21}
Icecaps	2.9×10^7	2.9×10^{22}
Water vapour	1.3×10^4	1.3×10^{19}

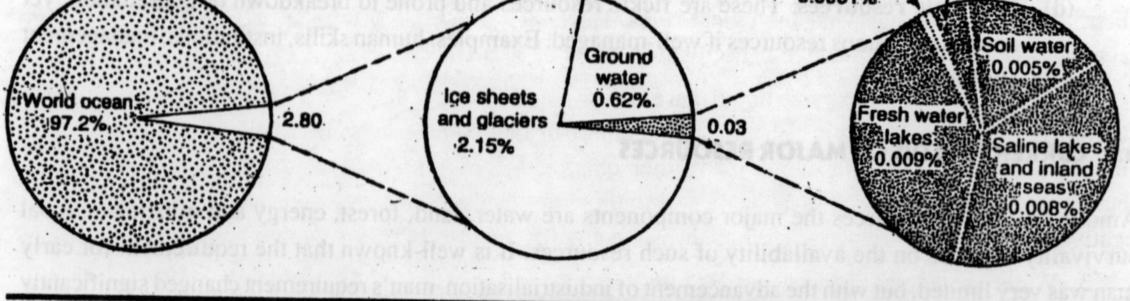

Fig. 4.1 : Global water resouces status (percentage-wise)

Rivers constitute an insignificant amount (0.001 or 0.1%) of the land surface. Only 0.0001% of water of the planet earth occurs in river channels. In spite of these low quantities, running waters are of enormous significance to humans. The mean water flow and drainage area of the major River Basins are shown in the Table 4.2.

Table 4.2: Major River Basins (with catchment size)

Rivers by continent	Drainage area (10^3 sq km)	Mean annual flow (m^3s^{-1})
North America		
1. Mississippi	3,222	17,300
South America		
2. Amazon	5,578	212,000
3. Panama	2,305	14,900
Europe		
4. Danube	817	6,200
Africa		
5. Congo	4,015	40,000
6. Nile	2,980	2,800
Asia		
7. Ganges	1,060	19,000
8. Ob-Irrysh	2,403	12,000
9. Yangtze	1,943	22,000
10. Indus	927	5,600

Source: Szestay (1982)

Water Planet

If travellers from other solar systems were to visit our lovely, cool, blue planet, they might call it Aqua rather than Terra because of its outstanding feature: the abundance of streams, rivers, lakes, and oceans of liquid water. This is the only place that we know of anywhere in the universe where water exists in liquid form in any appreciable quantity. Liquid water covers nearly three-fourths of the earth's surface, and during the winter, snow and ice cover a good deal of the rest. Not only is water essential for cell structure and metabolism, but water's unique physical and chemical properties have a direct impact on the earth's surface temperatures, its atmosphere, and the interactions of life-forms with their environments. Water has many unique, almost magical qualities. Without the wonderful properties of water, life would not be possible here. Let's look in more detail at some of these properties and how they affect life:

1. Water is the primary component of cells and makes up 60 to 70 per cent (on average) of the weight of living organisms. It plumps out cells, thereby giving form and support to many tissues. It is not just "filler", however, but has vital biological roles. Water is the medium in which all of life's chemical reactions occur, and it is an active participant in many of these reactions.

2. Water is the only inorganic liquid that exists in nature, and it is the solvent in which most substances must be dissolved before cells can absorb, use, or eliminate them. These substances include food molecules, mineral nutrients, gases, hormones and other chemical communicators, and waste by-products of metabolism. When molecules dissolve in water, they often have a tendency to break into positively and negatively charged particles, called ions. Ions are significant participants in cellular reactions.

3. Water molecules themselves can ionize, breaking into H^+ (hydrogen ions) and OH^- (hydroxyl ions). These ions help maintain the acid/base balance in cells, helping to offset (buffer) fluctuations caused by the release of other ions during metabolism. The term pH refers to the relative abundance of H^+

ions in a solution. On a pH scale from zero to fourteen, seven is neutral, values lower than seven are acidic, and those higher than seven are basic (Fig. 4.2). A proper pH balance is critical to healthy cellular functioning.

4. Water molecules are cohesive, tending to stick together. You have experienced this property if you have ever done a bellyflop off a diving board. This cohesion causes capillary action, which is the tendency of water to be drawn into small channels. Without capillary action, movement of water and nutrients into groundwater reservoirs and through living organisms might not be possible. Because of its cohesiveness, water has the highest surface tension of any common, natural liquid. The surface of water, where it meets the air, acts like an elastic skin strong enough to support small insects.

5. Because water has a high heat capacity, it helps protect us from temperature fluctuations. It can absorb or release large amounts of heat energy before its own temperature changes. For this reason, large bodies of water such as the oceans and the Great Lakes have a moderating effect on their local climates. Without the presence of liquid oceans, the surface temperature of the earth would undergo wide temperature fluctuations between day and night as do the moon and Mars.

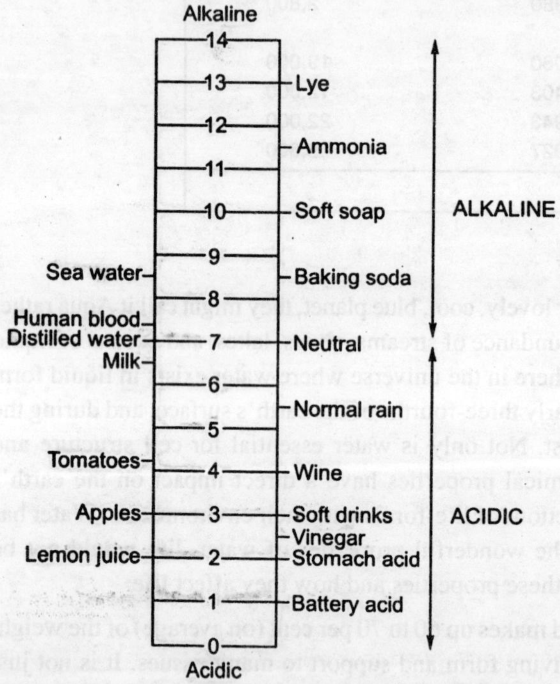

Fig. 4.2 : The pH scale

6. Water exists as a liquid over a wide temperature range that for most of the world (at least during summer months) corresponds to the ambient temperature range. For most substances, the freezing point is only a few degrees lower than the boiling point. This means that they exist as either a solid or a gas, so that inorganic liquids are relatively rare. Organisms synthesize organic compounds such as oils and alcohols that remain liquid at ambient temperatures because they are so valuable to life, but the original and predominant liquid in nature is water.

7. Water is unique in that it expands when it crystallizes. Most substances shrink as they change from liquid to solid. Ice floats because it is less dense than liquid water. When temperatures fall below freezing, the surface layers of lakes, rivers, and oceans cool faster and freeze before deeper water. Floating ice then insulates underlying layers, keeping most water bodies liquid throughout the winter in most places. Without this feature, lakes, rivers, and even oceans in high latitudes would freeze solid and never melt.

Water has a high heat of vaporization. Because of the amount of heat it absorbs in changing from a liquid to a vapour state, evaporating water is an effective way for organisms to shed excess heat. Many animals pant or sweat to moisten evaporative cooling surfaces. Why do you feel less comfortable on a hot, humid day than on a hot, dry day? Because the water vapor-laden air inhibits the rate of evaporation from your skin, thereby impairing your ability to shed heat. The heat absorbed when water vaporizes is released when condensation occurs. This accounts for a large transfer of heat from the oceans over the continents as water vapour turns into rain.

Altogether, these unique properties of water not only shape life at the molecular and cellular level, they also determine many features of both the biotic and abiotic components of our world.

A number of unwise use of fresh water resources that are depleting availability are :

(a) Extraction of ground water at rates higher than its recharging allowing the incursion of saline water in aquifers and subsidence of ground;

(b) Pollution with persistent toxic chemicals;

(c) Inefficient technology for water use; and

(d) Change of water runoff and infiltration patterns by deforestation and soil erosion.

Table 4.3: World's Water Resources, 2000

Resources	Volume (W), Thousands of km³	Annual Rate Removal (Q) (in Thousand of km³) and process of removal		Renewal period (T) (T = W/Q)
Total water on Earth	1,460,000	520	Evaporation	2,800 years
Total water in the Ocean	1.370,000	449	Evaporation	3,100 years
		37	Difference between precipitation & evaporation	37,000 years
Free gravitational waters in the Earth's crust (to a depth of 5 km)	60,000	13	Underground runoff	4,600 years
Amount of water present in the zone of active water exchange	4,000	13	D0	300 years
Lakes	750	—		16,000 years
Glaciers & permanent snow	29,000	1.8	Runoff	
Soil & subsoil moisture	65	8.5	Evaporation and underground runoff	
Atmospheric moisture	14	520	Precipitation	9 days
River waters	1.2	36.3	Runoff	12(20) days

Source: UNESCO, 2000

Therefore, in coming decades, stress will be laid on management of water use pattern through developing water-efficient technologies in agriculture and industry and also imposition of water pricing system.

Water Footprint

The water footprint measures the amount of water used to produce each of the goods and services we use. It can be measured for a single process, such as growing rice, for a product, such as a pair of jeans, for the fuel we put in our car, or for an entire multinational company. The water footprint can also tell us how much water is being consumed by a particular country- or globally- in a specific river basin or from an aquifer.

• *"The water footprint is a measure of humanity's appropriation of fresh water in volumes of water consumed and/or polluted".*

The water footprint allows us to answer a broad range of questions for companies, governments and individuals. For example:

- where is the water dependence in my company's operations or supply chain?
- how well are regulations protecting our water resources?

- how secure are our food or energy supplies?
- can I do something to reduce by own water footprint and help us manage water for both people and nature?

Depending on the question you are asking, the water footprint can be measured in cubic metres per tonne of production, per hectare of cropland, per unit of currency and in other functional units. The water footprint helps us understand for what purposes our limited freshwater resources are being consumed and polluted. The impact it has depends on where the water is taken from and when. If it comes from a place where water is already scarce, the consequences can be significant and require action.

The water footprint has three components: green, blue and grey. Together, these components provide a comprehensive picture of water use by delineating the source of water consumed, either as rainfall/soil moisture or surface/groundwater, and the volume of fresh water required for assimilation of pollutants.

Green water footprint is water from precipitation that is stored in the root zone of the soil and evaporated, transpired or incorporated by plants. It is particularly relevant for agricultural, horticultural and forestry products.

Blue water footprint is water that has been sourced from surface or groundwater resources and is either evaporated, incorporated into a product or taken from one body of water and returned to another, or returned at a different time. Irrigated agriculture, industry and domestic water use can each have a blue water footprint.

Grey water footprint is the amount of fresh water required to assimilate pollutants to meet specific water quality standards. The grey water footprint considers point-source pollution discharged to a freshwater resource directly through a pipe or indirectly through runoff or leaching from the soil, impervious surfaces, or other diffuse sources.

WATER RESOURCE OF INDIA

In India, over 4000 BCM of water is received from rainfall, which is perhaps among the highest in the world. Fourteen major river systems share 83% of the drainage basin, account for 85% of the surface flow and serve 80% of the total population of the country. There are also 44 medium and 55 minor rivers which mostly originate in the coastal mountains and 80% of their discharge occurs during the monsoon months. The estimated utilisable surface water is around 690 BCM.

With respect to ground water reserve in India, the estimated available ground water to be over 433 billion cu. m. and the annual exploitable potential is estimated 3.99 BCM (GOI, 2006) of which only one-fourth is utilised currently. In spite of fairly high amount of available freshwater, India has a very low per capita availability of drinking water, about 1720.29 cu m per year.

There were a large number of swamps and wetlands in this country, which are essential for a balanced ecosystem, but these areas are currently disappearing for human transformation. The water resource crises is no longer a futuristic prediction for India. It is an increasing prevalent feature, resulting from an overall disruption of the hydrological cycle. Thus the vital need for a National water policy to ensure sustainable availability and use of water for diverse objectives.

The average annual availability and also utilisable surface water in different water basin of India in shown in Table 4.4.

The other categories of Inland water resources of India is shown in Table 4.5 (statewise database).

In India, there are four major sectors where water requirement is the prime concern. As per water use pattern, water demand in India by 2010 AD might follow the following pattern viz., for 'agriculture & livestock' 80%, for 'power generation' and other industrial uses 12.8%, 'domestic' 7.5%.

Table 4.4: Utilisable Flow in Different River Basins

Sl. No.	River Basin	Water Resources (kilometre cube/per year)			
		Catchment Area (square kilometre)	Available fresh water		Utilisable Surface Water
			1993	2005	
1.	Indus	321289	73.31	73.31	46.00
2.	Ganga Brahmaputra-Barak (Meghna)	862769	525.02	525.02	250.00
	(a) Ganga Sub-basin	197316	537.24	629.05	24.00
	(b) Brahmaputra and Barak sub-basins	41157	48.36	48.36	24.00
3.	Subernarekha	29196	12.37	12.37	6.81
4.	Brahmai-Battrarni	51822	28.48	28.48	18.30
5.	Mahanadi	141589	66.88	66.88	49.99
6.	Godavari	312812	110.54	110.54	76.30
7.	Krishna	258948	78.12	69.41	58.00
8.	Pennar	55213	6.32	6.32	6.86
9.	Cauvery	87900	21.36	21.36	19.00
10.	Tapi	65145	14.88	14.88	14.50
11.	Narmada	98796	45.64	45.64	34.50
12.	Mahi	34842	11.02	11.02	3.10
13.	Sabarmati	21674	3.81	3.81	1.93
14.	West flowing river of Kutch, Saurashtra and Luni	334390	15.1	15.1	14.98
15.	West flowing river South of Tapi	113057	200.94	200.94	36.21
16.	East flowing rivers between Mahanadi and Krishna	49570	17.08	17.08	13.11
17.	East flowing rivers between Godavari and Krishna	12289	1.81	1.81	13.11
18.	East flowing rivers between Krishna and Pennar	24649	3.63	3.63	13.11
19.	East flowing rivers between Pennar and Cauvery	64751	9.98	9.98	16.73
20.	East flowing Rivers South of Cauvery	35026	6.48	6.48	16.73
21.	Area of North Ladakh not draining into India	28478	0	0	NA
22.	River draining into Bangladesh	10031	8.57	8.57	NA
23.	River draining into Myanmar	26271	22.43	22.43	NA
24.	Drainage areas of Andaman, Nicobar and Lakshadweep Islands	8280	0	0	NA
	Total	3287260	1869.37	1952.87	690.32

Source: Central Water Commission, New Delhi, 2005

Table 4.5: Inland Water Resources of India

States/UTs	Rivers and Canals (in kms.)	Reservoirs (in lakh ha.)	Tanks and Ponds (in lakh ha)	Floodplain Lakes and Derelict Water (in lakh ha)	Brackish Water (in lakh ha)	Total replenishable groundwater resource (mha m/year)
Andhra Pradesh	13891	2.34	4.63	—	1.50	—
Arunachal Pradesh	2000	—	2.50	2.0	—	0.14385
Assam	4820	0.02	0.26	1.27	—	2.24786
Bihar	2000	0.06	0.64	0.40	—	2.69796
Chhattisgarh	3573	0.83	0.63	—	1.47	1.60705
Goa	250	0.03	0.03	—	—	0.02182
Gujarat	3865	2.43	0.71	0.12	3.76	2.0367
Haryana	5000	0.09	0.10	—	0.30	1.11794
Himachal Pradesh	3000	0.42	0.01	—	—	0.02926
Jammu and Kashmir	27781	0.07	0.24	0.06	—	0.44257
Jharkhand	1200	0.94	0.29	—	—	0.66045
Karnataka	9000	2.11	2.90	—	0.08	1.61750
Kerala	3092	0.30	0.30	2.43	2.43*	0.79003
Madhya Pradesh	17088	2.27	0.60	—	—	3.48186
Maharashtra	16000	2.29	0.59	—	0.10	3.78677
Manipur	3360	0.01	0.05	0.04	—	0.31540
Meghalaya	3194	0.08	0.02	Neg	—	0.05397
Mizoram	1750	—	0.02	—	—	
Nagaland	1600	0.03	0.50	Neg	—	0.07240
Orissa	7219	1.96	1.16	1.80	4.18	2.01287
Punjab	15270	Neg	0.07	—	—	1.81923
Rajasthan	6802	1.20	1.80	—	—	1.26021
Sikkim	900	1.20	—	0.03		—
Tamil Nadu	7420	0.52	2.56	0.07	0.56	2.64069
Tripura	1200	0.05	0.13	—	—	0.06634
Uttar Pradesh	28500	1.38	1.61	1.33	—	8.25459
Uttaranchal	2686	0.20	Neg	Neg	—	0.28411
West Bengal	2526	0.17	2.76	0.42	2.10	2.30914
Andaman and Nicobar Islands	115	0.01	0.03		0.32	—
Chandigarh	2	—	Neg	Neg	—	0.00297
Dadra and Nagar Haveli	54	0.05	—	—	—	0.00422
Daman and Diu	12	—	Neg	—	Neg	0.00071
Delhi	150	0.04	—	—		0.02916
Lakshadweep		—	—	—		0.03042
Puducherry	247	—	Neg	0.01	0.01	0.01746
India	195567	21.64	25.15	9.98	16.85	43.38593

Note. UT – Union Teritories, Neg – Negligible, Kms – kilometres, ha – hectares, mha m – million hectare metre

Source: Annual Report 2003-04. Department of Animal Husbandry and Dairying, Ministry of Agriculture, Government of India

The river basinwise ground water potentials are given in Table 4.6.

Table 4.6: Groundwater potential in River Basins of India (pro-rata basis) (in km³/year)

Sl. No.	Basin	Total replenishable ground water resource	Provision for domestic industrial & other uses	Available ground water for irrigation	Net Draft	Balance ground-water potential	Level of of ground-water develop-ment (%)
1.	Brahmani with Baitarni	4.05	0.61	3.44	0.29	3.16	8.45
2.	Brahmaputra	26.55	3.98	22.56	0.76	21.80	3.37
3.	Chambal composite	7.19	1.08	6.11	2.45	3.66	40.09
4.	Cauvery	12.30	1.84	10.45	5.78	4.67	55.33
5.	Ganga	170.99	26.03	144.96	48.59	96.37	33.52
6.	Godavari	40.65	9.66	30.99	6.05	24.94	19.53
7.	Indus	26.49	3.05	23.43	18.21	5.22	77.71
8.	Krisnhna	26.41	5.58	20.83	6.33	14.50	30.39
9.	Kutch & Saurashtra composite	11.23	1.74	9.49	4.85	4.64	51.14
10.	Madras & South Tamil Nadu	18.22	2.73	15.48	8.93	6.55	57.68
11.	Mahanadi	16.46	2.47	13.99	0.97	13.02	6.95
12.	Meghna	8.52	1.28	7.24	0.29	6.95	3.94
13.	Narmada	10.83	1.65	9.17	1.99	7.18	21.74
14.	Northeast composite	18.84	2.83	16.02	2.76	13.26	17.20
15.	Pennar	4.93	0.74	4.19	1.53	2.66	36.60
16.	Subranarekha	1.82	0.27	1.55	0.15	1.40	9.57
17.	Tapi	8.27	2.34	5.93	1.96	3.97	33.05
18.	Western Ghat	17.69	3.19	14.50	3.32	11.18	22.88
	TOTAL	431.43	71.08	360.35	115.21	245.13	31.97

Source: Central Ground Water Board, 2005

Dependence on groundwater has increased considerably over the last two decades. Groundwater sources contribute more than 85 per cent of the drinking water requirements in rural areas, about 76 per cent of irrigation requirements and more than 50 per cent of the urban and industrial water supplies. Excessive withdrawal of groundwater has resulted in the drying up of aquifers in many parts of the country. In the search for water, deeper tubewells are sunk to access water from fossil aquifers which cannot be replenished. In the Indus Basin as a whole, groundwater pumping is estimated to exceed recharge by 50 per cent.

Statewise dynamic fresh groundwater resources are given in Table 4.7:

WATER CONSERVATION

Water is utilised in enormous quantity over the year—particularly in agriculture, industries and domestic purposes. In tropics, water demand increased manyfold during past few decades. At the same time, due to improper collection of surface water, water bodies get polluted or lost through run off or evaporation. Waste water generated in industries of municipalities/cities are also thrown off into rivers. More and more ground water is withdrawn for domestic and agricultural uses. These situations aggravates the entire water demand and supply problem. Thus there is a great need for comprehensive water conservation planning in all countries.

Table 4.7: State-wise ground water resources availability, utilization and stage of development, India

No.	States/Union Territories	Annual replenishable ground water resource					Natural discharge during non-monsoon season	Net annual ground water availability	Annual ground water draft			Projected demand for domestic and industrial uses up to 2025	Ground water availability for future irrigation	Stage of ground water development (%)
		Monsoon season		Non-monsoon season		Total			Irrigation	Domestic and industrial uses	Total			
		Recharge from rainfall	Recharge from other sources	Recharge from rainfall	Recharge from other sources									
1	2	3	4	5	6	7	8	9	10	11	12	13	14	15
	States													
1	Andhra Pradesh	17.25	6.29	5.38	6.97	35.89	3.32	32.57	13.18	1.33	14.51	2.81	16.97	45
2	Arunachal Pradesh	3.36	0.00	1.15	0.00	4.51	0.45	4.06	0.002	0.001	0.003	0.01	4.05	0.08
3	Assam	17.90	1.64	8.64	0.34	28.52	2.73	25.79	2.86	0.64	3.49	0.78	22.14	14
4	Bihar	19.54	3.95	3.40	2.44	29.34	2.47	26.86	10.25	1.70	11.95	2.51	14.10	44
5	Chhattisgarh	9.90	0.70	0.87	0.94	12.42	0.79	11.63	3.43	0.62	4.05	0.76	7.44	35
6	Delhi	0.11	0.10	0.02	0.08	0.31	0.02	0.29	0.14	0.25	0.39	0.26	0.01	137
7	Goa	0.16	0.008	0.01	0.07	0.24	0.10	0.145	0.01	0.03	0.04	0.04	0.10	28
8	Gujarat	12.79	2.55	0.00	3.23	18.57	0.98	17.59	10.75	1.11	11.86	1.48	5.87	67
9	Haryana	3.65	2.77	1.01	3.35	10.78	0.99	9.79	12.35	0.71	13.05	0.76	– 3.31	133
10	Himachal Pradesh	0.39	0.02	0.10	0.05	0.56	0.03	0.53	0.25	0.13	0.38	0.13	0.15	71
11	Jammu & Kashmir	1.45	2.06	0.36	0.37	4.25	0.43	3.83	0.20	0.61	0.81	0.76	2.87	21
12	Jharkhand	4.75	0.13	1.06	0.36	6.31	0.55	5.76	1.31	0.55	1.86	0.76	3.69	32
13	Karnataka	6.81	4.17	2.67	3.38	17.03	2.22	14.81	8.59	0.82	9.41	1.06	6.53	64
14	Kerala	4.85	0.06	0.63	1.15	6.69	0.61	6.07	1.30	1.53	2.84	1.71	3.07	47
15	Madhya Pradesh	28.22	1.17	0.79	4.87	35.04	1.75	33.29	17.48	1.35	18.83	1.91	13.90	57
16	Maharashtra	22.36	1.68	1.84	8.07	33.95	1.80	32.15	16.15	1.03	17.18	1.97	14.28	53
17	Manipur	0.23	0.01	0.19	0.01	0.44	0.04	0.40	0.0033	0.0007	0.004	0.05	0.35	1.02
18	Meghalaya	1.68	0.03	0.07	0.005	1.78	0.18	1.60	0.0015	0.0002	0.0017	0.232	1.37	0.08
19	Mizoram	0.0257	Negligible	0.005	Negligible	0.030	0.003	0.027	0.00	0.001	0.001	0.002	0.025	3.52
20	Nagaland	0.40	Negligible	0.21	Negligible	0.62	0.062	0.55	0.00	0.03	0.03	0.04	0.51	6.13
21	Odisha	11.29	2.53	1.33	2.63	17.78	1.09	16.69	3.81	0.92	4.73	1.24	11.64	28
22	Punjab	5.82	10.64	1.33	4.74	22.53	2.21	20.32	34.17	0.71	34.88	0.98	– 14.83	172
23	Rajasthan	8.75	0.68	0.28	2.20	11.94	1.11	10.83	13.13	1.71	14.84	1.89	0.91	137
24	Sikkim		–		–	–	–	0.44	0.003	0.009	0.011	0.01	0.031	26
25	Tamil Nadu	7.38	10.28	1.69	2.18	21.53	2.15	19.38	13.17	1.76	14.93	1.82	4.39	77
26	Tripura	1.248	0.000	0.740	0.598	2.587	0.229	2.358	0.093	0.069	0.163	0.200	2.065	7
27	Uttar Pradesh	42.13	11.57	5.15	18.34	77.19	5.53	71.66	48.74	4.04	52.78	6.56	19.64	74
28	Uttarakhand	1.09	0.26	0.20	0.49	2.04	0.04	2.00	1.10	0.03	1.13	0.09	0.80	57
29	West Bengal	18.53	5.72	1.42	3.58	29.25	2.67	26.58	9.72	0.97	10.69	1.48	15.38	40
	Total States	**252.11**	**68.99**	**40.56**	**70.44**	**432.11**	**34.55**	**397.60**	**222.21**	**22.66**	**244.85**	**32.28**	**154.34**	**62**
	Union Territories													
1	Andaman & Nicobar	0.262	Nil	0.046	Nil	0.308	0.022	0.286	0.001	0.012	0.013	0.014	0.272	4.44
2	Chandigarh	0.015	0.001	0.005	0.001	0.022	0.002	0.019	0.000	0.000	0.000	0.000	0.000	0
3	Dadra & Nagar Haveli	0.043	0.003	0.009	0.007	0.062	0.003	0.059	0.007	0.006	0.013	0.010	0.042	22
4	Daman & Diu	0.014	0.002	0.000	0.002	0.018	0.001	0.017	0.014	0.002	0.016	0.003	0.000	97
5	Lakshdweep	0.000	0.000	0.000	0.000	0.011	0.007	0.0035	0.000	0.0023	0.0023	0.000	0.000	67
6	Puducherry	0.089	0.060	0.008	0.032	0.189	0.019	0.170	0.124	0.029	0.153	0.032	0.057	90
	Total Uts	**0.42**	**0.07**	**0.07**	**0.04**	**0.61**	**0.05**	**0.56**	**0.15**	**0.05**	**0.20**	**0.06**	**0.37**	**36**
	Grand Total	**252.53**	**69.06**	**40.63**	**70.48**	**432.72**	**34.60**	**398.16**	**222.36**	**22.71**	**245.05**	**32.34**	**154.71**	**62**

Source: CWC – 2014

April 2002, Government of India declared "National Water Policy" where greater emphasis was given on surface water collection, storage and rational utilisation, as the ground water level depleted very fast in different parts of the country.

In addition, water scarcity leads to excessive ground water withdrawal which, in many places, is contaminated with flouride or arsenic.

With the rising water demand in agricultures the irrigation requirement enhanced very much during past four decades. This situation accelerates more water crisis. The details of irrigation potential are shown in Table 4.8.

In tropics, monsoon is fairly widespread, so the scope for collection and storage of rain water is enormous. Unfortunately, over the years, such attempts were grossly ignored and for that reason ground water consumption was enhanced very much. Thus there is a great need for water storage facilities and efficient utilisation for multifarious purpose.

Table 4.8: State-wise Irrigation Potential from Major and Medium Irrigation ('000 hectares)

State/UTs	Major & Medium Irrigation	Surface Water	Ground Water	Total
Andhra Pradesh	5000	2300	3960	11260
Arunachal Pradesh	0	150	18	168
Assam	970	1000	900	2870
Bihar*	6500	1900	4947	13347
Goa	62	25	29	116
Gujarat	3000	347	2756	6103
Haryana	3000	50	1462	4512
Himachal Pradesh	50	235	68	353
Jammu & Kashmir	250	400	708	1358
Karnataka	2500	900	2574	5974
Kerala	1000	800	879	2679
Madhya Pradesh*	6000	2200	9732	17932
Maharashtra	4100	1200	3652	8952
Manipur	135	100	369	604
Meghalaya	20	85	63	168
Mizoram	0	70	0	70
Nagaland	10	75	0	85
Odissa	3600	1000	4203	8803
Punjab	3000	50	2917	5967
Rajasthan	2750	600	1778	5128
Sikkim	20	50	0	70
Tamil Nadu	1500	1200	2832	5532
Tripura	100	100	81	281
Uttar Pradesh*	12500	1200	16799	30499
West Bengal	2300	1300	3318	6918
UTs	98	41	5	144
India	58465	17378	64050	139893
Percentage	56	71	71	65

Note: *Figures include Jharkhand, Chhattisgarh, Uttaranchal

Source: Central Water Comission and Lok Sabha Unstarred Question No 144, dated 19.11.2001. Annual Report 2001-02, Ministry of Water Resource

Watershed management and appropriate adoption of irrigation technologies are the two prime actions needed for water conservation. Conservation of lakes, oxbow lakes or other categories of wetlands are the prime requisite of rainwater storage. At household level, rain water also harvested by a number of techniques. These are:

(a) Construction of traditional ring well (kua),

(b) Roof-top rainwater harvesting and storage.

Waste water treatment and reuse for irrigation, aquaculture or even industrial purposes is also a part of water conservation strategies. In earlier days river water lifting was made for public water supply, industrial use and also for agricultural purposes. But due to scarcity of river water, we are more and more dependent on the groundwater.

In recent decades some parts of India and about half of Bangladesh are facing crisis of ground water which is contaminated with arsenic. Thus there is great need for surface water collection, storage, treatment and supply at household level for domestic uses. It was suggested that roof top rainwater harvesting and storage into underground tank, followed by chlorination may lead to better domestic water supply.

In different parts of our country, rainwater harvesting was attempted in different ways. Some illustrations are given here:

In Madhya Pradesh, more than 1,000 check dams, 1,050 tanks and 1,100 community lift irrigation schemes have been implemented in Jhabua district. The result—drought proofing achieved and food production increased by 38% in between 1990 to 1995. Moreover, micro-watershed project with people's participation in Ghelhar Choti village, Jhabua district has been implemented. The result is that cultivated areas increased and yield per hectare doubled. Low cost small farm reservoirs along with improved crop and soil management systems tried in erstwhile Chhattisgarh state and it augmented ground water storage, saved paddy from water stress during extended dry spells in 1990-91 and 1991-92. District administration in Dewas made roof top rain water harvesting mandatory for all houses having tubewells and banned tubewell drilling. This resulted in improved soil moisture and recharged first aquifer.

In Andhra Pradesh, percolation tanks and check dams constructed in chronically drought-affected Rayalseema region and it helped in drought proofing.

In Maharashtra, more than 7,000 percolation tanks built after the severe drought of 1971-72. All areas under the influence of the percolation tanks converted into green lands.

The town plannars in Chennai realised the utility of rain water harvesting in 1993, when the building by-laws were amended to ensure all new structures to have arrangements for rain water harvesting. No water or sewage connection would be given if a new building did not have provisions for rain water harvesting as per the new rules. The decision has led to a rise in ground water level.

In Chandigarh, roof top rain water harvesting recharges ground water through injection well in one of the CSIO buildings. This caused a rise of 2 metres in water level.

In Jodhpur district of Rajasthan, Gramin Vigyan Vikash Samity motivated the residents of 25 villages and built 2000 storage tanks (tankas). Each house has a tanka (a water collection structure) lined with lime and alum to keep the water fresh for 4 to 5 months.

Roof top rain water harvesting and recharging of wells as a movement initiated by the Saurashtra Lok Manch Trust in Mandlikpur village of Rajkot district in Gujarat has prevented drying up of wells. Aga Khan Rural Support Programme in Junagadh and Surendranagar districts of Gujarat, Saurasthra, harvested rain water by check dams and percolation ponds involving beneficiary farmers. As a result, optimum utilisation of harvested rain water was achieved. Vivekananda Research and Training Institute in Kutch, Bhavnagar (Gujarat) and Amreli districts constructed rain water harvesting structures. This helped in improving ground water quality and controlling the declining water level.

In Amravati district of Maharashtra three percolation tanks and ten cement plugs benefiting an area of 280 ha and 100 ha, respectively, have been constructed.

The following are the proposed policy measures adopted for rainwater harvesting:

(a) Provide at least one roof-top rain water harvesting structure for every 200 sq. m. plot in urban areas.
(b) Revive or rehabilitate all village ponds.
(c) Subject to technical feasibility, provide at least one check dam/KT weir/sub-surface dyke in each streamlet with a catchment of 1 to 3 sq. km.

(d) Ban construction of irrigation wells/tubewells withing a radius of 200 m or less (depending on scientific criteria) of the drinking water supply well.

The Action Plan during 2000-2003 included:

(a) Prepare national and state level water harvesting perspective plans.
(b) Develop plans and implement roof-top rain water harvesting measures using 100,000 wells (existing, defunct and/or operative wells) to be used in the first instance.
(c) Provide rural drinking water wells with recharge facilities, covering 1 lakh wells.
(d) Harvest and recharge city storm water in 100 towns.
(e) Revive and rehabilitate 1,000 dying village ponds.
(f) Design and construct 200 percolation tanks, 5,000 check dams/bhandaras and 1,000 sub-surface dykes.
(g) Recycle secondary treated urban waste waters through aquifers at 5 centers.
(h) Identify potential aquifers in drought-prone areas and declare these as Ground Water Sanctuaries.

A number of donor agencies are currently involved in India's Water Sector Development. The details given in the Table 4.9 below:

Table 4.9: Donor Agency Involvement in India's Water Sector

Agency	Important Projects
DANIDA (Danish International Development Assistance)	• Karnataka Watershed Development Project • Watershed Development Project—Tirunelveli and Ramanathapuram, Tamil Nadu • Watershed Development Project—Malwa region, Madhya Pradesh • Koraput Watershed Project—Orissa • DANIDA fluoride removal projects, Karnataka
KFW and GTZ	• Watershed Organisation Trust, Maharashtra • Indo-German Watershed Programme
SIDA (Swedish International Development Agency)	• With UNICEF and World Bank, supporting National Drinking Water Project that is mainly being implemented in the rural areas. • With OXFAM INDIA, supporting river basin projects in southern states
Royal Netherlands Embassy Development Cooperation Cell	• North Bengal Terai Development Project (NBTDP), West Bengal • Andhra Pradesh Groundwater Bore Well Irrigation Schemes (APWELL) • Community managed Rural Water Supply and Sanitation Project – Andhra Pradesh • Project to reclaim saline lands in Haryana • Project on controlling pollution in Ganga river, Uttar Pradesh • Project to save the lakes of Hyderabad, Andhra Pradesh • Project for creating a database on water resources
Department for International Development (DFID)	• KRIBHCO, Rajasthan, Gujarat, Madhya Pradesh • Andhra Pradesh Rural Livelihoods Programme
India Canada Environment Facility (ICEF)	• Arsenic Mitigation and Technology Park, West Bengal • Loktak Lake Development, Manipur • Tank rehabilitation with PALMYRA, Puducherry • Integrated Watershed Management and Water Storage in Pushkar Lake Area, Rajasthan • Coastal Wetlands Mangrove Conservation and Management (East Coast), Tamil Nadu

(B) Land Resources

Land is another most vital resource as it will be used for production of crops and other biological materials needed for food, fodder, medicine, fibre and associated materials. The volume of production depends on the extent and yield capacity of the cultivable land, which, in turn, is flexible according to human attitude. The physical limits of cultivable land are temperature, moisture, topography and soil. Its availability and cultivability also vary according to the human social pattern, particularly the use pattern of energy.

India has a total land area of 2.4% of the world total, but supports a population over 17% of the world. Thus the per capita land availability was only 0.48 ha. as against 8.43 ha in former USSR (undivided) and 0.98 ha in China. The major land use classification of India is given in the Table 4.10.

Table 4.10: Land use classification in India, (2005-2006)

Sl. No.	Classification	2000-01	2001-02 (P)	2002-03 (P)	2003-04 (P)	2004-05 (P)	2005-6 (P)
	1	2	3	4	5	6	7
I.	Geographical area	328.73	328.73	328.73	328.73	328.73	328.73
II.	Reporting area for land utilisation statistics (1 to 5)	305.08	305.01	305.24	305.32	305.23	305.27
	1. Forests	69.62	69.51	69.64	69.67	69.67	69.79
	2. Not available for cultivation (a+b)	41.55	41.78	42.08	42.23	42.30	42.51
	(a) Non-agricultural uses	23.81	24.07	24.28	24.66	24.72	25.03
	(b) Barren and unculturable land	17.74	17.71	17.80	17.57	17.58	17.48
	3. Other uncultivated land excluding fallow land (a+b+c)	27.71	27.37	27.41	26.98	27.00	26.92
	(a) Permanent pastures and other grazing land	10.83	10.59	10.51	10.45	10.43	10.42
	(b) Land under miscellaneous tree crops and groves not included in net area sown	3.32	3.37	3.36	3.39	3.38	3.38
	(c) Culturable wasteland	13.56	13.41	13.54	13.14	13.19	13.12
	4. Fallow land (a+b)	25.03	24.94	33.46	25.48	24.94	24.17
	(a) Fallow land other than current fallows	10.19	10.30	11.76	11.20	10.72	10.50
	(b) Current fallows	14.89	14.64	21.70	14.28	14.22	13.67
	5. Net area sown (6-7)	141.16	141.42	132.66	140.95	141.32	141.89
	6. Gross cropped area	185.70	189.75	175.66	190.37	190.91	192.80
	7. Area sown more than once	44.54	48.33	43.00	49.42	49.59	50.90
	8. Cropping intensity*	131.60	134.20	132.40	135.10	135.10	135.90
III.	Net irrigated area	54.84	56.30	53.88	56.00	58.54	60.20
IV.	Gross irrigated area	75.82	78.07	72.89	77.11	79.51	82.63

P : Provisional

* : Cropping intensity is obtained by dividing the gross cropped area by the net area sown.

Note : The decline in net area sown in 2002-03 reflects the impact of the severe drought of 2002-03 on agriculture operations.

Source: Agricultural Statistics at a glance 2008. Ministry of Agriculture.

(C) Biological Resources

Although our understanding of the earth's organisms — its *biological resources* — is still imperfect, there is no doubt that the abundance add and diversity of living organisms provide many benefits and make our world a beautiful and interesting place to live. Unfortunately, rapidly expanding human populations and resource consumption now threaten to deplete three types of biodiversity: the number of species, the genetic richness within species, and the variety of biological communities in the world.

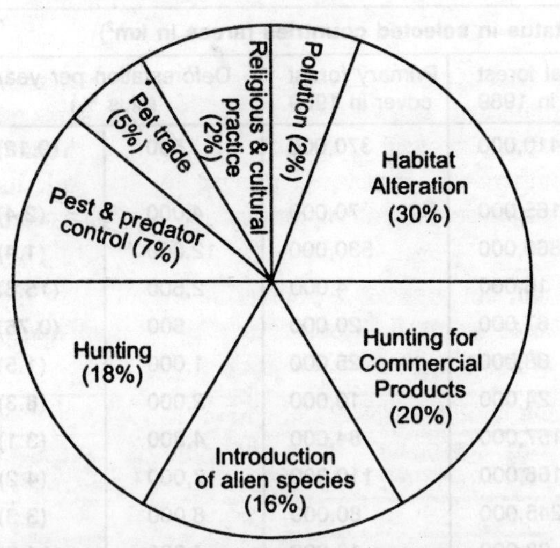

Fig. 4.3 : Estimated contribution of various human activities in biological resources extinction

At the end of the great exploration era of the nineteenth century, we now believe that more than 1.7 million species presently known, this is a just small fraction of the total number as expected to exist on earth. Major fraction of the organisms are yet to be discovered and clarified.

Biological resources are diminished or destroyed in a number of ways. Natural changes in the environment eliminate once successful species or reduce their numbers to mere remnant populations. Humans disrupt ecosystems and extirpate species, both deliberately and accidentally. There are three kinds of losses of species as noticed today, viz. depletion of *abundance* of a once plentiful species, *species extinction* and ecosystem disruption. There is an increasing gradient of seriousness in these losses. Humans have a long history of biological resource depletion. We have reduced biological abundance or driven species into extinction in a number of ways (Fig. 4.3).

Forests play a vital role in regulating climate, controlling water runoff, providing shelter and food for wildlife, and purifying the air. They produce valuable materials, such as wood and paper pulp, on which we all depend. Furthermore, forests also have scenic, cultural and historic values that deserve to be protected. Before large scale human disturbances during past couple of decades, forests probably covered 6 billion hectares (15 billion acres). About two-thirds of the forest is classified as closed canopy forest (over 20 per cent or more cover) and the rest is open canopy forest or woodland (canopy cover less than 20% of the ground). Broadly speaking, there are six principal types of forest, viz. cool coniferous forest, temperate mixed forest, warm-temperate moist forest, tropical moist evergreen forest, tropical moist deciduous forest, and dry forest. The distribution of different forest types primarily depends on the temperature and rainfall regime of the concerned region.

The richest and most diverse terrestrial ecosystems on the earth are the tropical forests. Although they now occupy less than 10 per cent of the earth's land surface, these forests are thought to contain more than two-thirds of all plant biomass and at least one-half of all plant, animal and microbial species of the world. Brazil has by far both the largest total area and the highest rate of deforestation in the world, estimated to have been 6.8 million hectares of forest cut annually between 1986 and 1991. The rate of destruction of tropical forests are most prevalent than any other region.

The existing forest cover area of the major countries particularly in tropics and their rate of deforestation (as on 1989) is given in the Tables 4.11 and 4.12.

Table 4.11: Tropical moist forest : present status in selected countries (areas in km²), 2000

Country (plus area)	Original forest cover	Total forest cover in 1989	Primary forest cover in 1989	Deforestation per year (plus %)	
Bolivia (1,098,581)	90,000	70,000	45,000	1,500	(2.1)
Brazil (8,511,960)	2,860,000	2,200,000	1,800,000	50,000	(2.3)
Cameroon (475,442)	220,000	164,000	60,000	2,000	(1.2)
Central America (522,915)	500,000	90,000	55,000	3,300	(3.7)
Colombia (1,138,891)	700,000	278,500	180,000	6,500	(2.3)
Congo (342,000)	100,000	90,000	80,000	700	(0.8)
Ecuador (270,670)	132,000	76,000	44,000	3,000	(4.0)
Gabon (267,670)	240,000	200,000	100,000	600	(0.3)

Contd.

Table 4.11: Tropical moist forest : present status in selected countries (areas in km²)

Country (plus area)	Original forest cover	Total forest cover in 1989	Primary forest cover in 1989	Deforestation per year (plus %)	
Guyanas (French Guiana, Guyana and Suriname) (469,790)	500,000	410,000	370,000	500	(0.12)
India (3,287,000)	1,600,000	165,000	70,000	4,000	(2.4)
Indonesia (1,919,300)	1,220,000	860,000	530,000	12,000	(1.4)
Ivory Coast (322,463)	160,000	16,000	4,000	2,500	(15.6)
Kampuchea (181,035)	120,000	67,000	20,000	500	(0.75)
Laos (236,800)	110,000	68,000	25,000	1,000	(1.5)
Madagascar (590,992)	62,000	24,000	10,000	2,000	(8.3)
Malaysia (329,079)	305,000	157,000	84,000	4,800	(3.1)
Mexico (1,967,180)	400,000	166,000	110,000	7,000	(4.2)
Myanmar (Burma) (696,500)	500,000	245,000	80,000	8,000	(3.3)
Nigeria (924,000)	72,000	28,000	10,000	4,000	(14.3)
Papua New Guinea (461,700)	425,000	360,000	180,000	3,500	(1.0)
Peru (1,285,220)	700,000	515,000	420,000	3,500	(0.7)
Philippines (299,400)	250,000	50,000	8,000	2,700	(5.4)
Thailand (513,517)	435,000	74,000	22,000	6,000	(8.4)
Venezuela (912,050)	420,000	350,000	300,000	1,500	(0.4)
Vietnam (334,331)	260,000	60,000	14,000	3,500	(5.8)
Zaire (2,344,886)	1,245,000	1,000,000	700,000	4,000	(0.4)
TOTAL	13,626,000*	7,783,500**	5,321,000***	138,600	(1.8)

Notes

* = 97 per cent of estimated total original extent of tropical forests, around 14 million km²

** = 97.5 per cent of present total extent of tropical forests, 8 million km²

*** = 67 per cent of total remaining tropical forests, 8 million km²

Table 4.12: Country-by-country estimates of the deforestation release of carbon, 1989

Country	Deforestation (km²)	Carbon Release (million tonnes)	(% of total)
Brazil	50,000	454	32.1
Indonesia	12,000	124	8.9
Myanmar (Burma)	8,000	83	5.9
Mexico	7,000	64	4.6
Thailand	6,000	62	4.4
Colombia	6,500	59	4.2
Nigeria	4,000	57	4.1
Zaire	4,000	57	4.1
Malaysia	4,800	50	3.6
India	4,000	41	2.9
Vietnam	3,500	36	2.6
Papua New Guinea	3,500	36	2.6
Ivory Coast	2,500	36	2.6

Contd.

Country	Deforestation (km²)	Carbon Release (million tonnes)	(% of total)
Peru	3,500	32	2.3
Central America	3,300	30	2.1
Philippines	2,700	28	2.0
Cameroon	2,000	28	2.0
Madagascar	2,000	28	2.0
Ecuador	3,000	27	1.9
Bolivia	1,500	14	1.0
Venezuela	1,500	14	1.0
Laos	1,000	10	0.7
Congo	700	10	0.7
Gabon	600	9	0.6
Kampuchea	500	5	0.4
Guyanas	500	4	0.3
TOTAL	1,386,000	1,398	100.0

Table 4.12: Country-by-country estimates of the deforestation release of carbon, 1989

IMPACT OF DEFORESTATION IN TROPICAL REGION

Biodiversity Impacts

Although tropical forests cover only about 7 percent of the Earth's dry land, they probably harbor about half of all species on earth. Many species are so specialized to microhabitats within the forest that they can only be found in small areas. Their specialization makes them vulnerable to extinction. In addition to the species lost when an area is totally deforested, the plants and animals in the fragments of forest that remain also become increasingly vulnerable, sometimes even committed, to extinction. The edges of the fragments dry out and are buffeted by hot winds; mature rainforest trees often die standing at the margins. Cascading changes in the types of trees, plant, and insects that can survive in the fragments rapidly reduces biodiversity in the forest that remains. People may disagree about whether the extinction of other species through human action is an ethical issue, but there is a little doubt about the practical problems that extinction poses.

First, global markets consume rainforest products that depend on sustainable harvesting: latex, cork, fruit, nuts, timber, fibers, spices, natural oils and resins, and medicines. In addition, the genetic diversity of tropical forests is basically the deepest end of the planetary gene pool. Hidden in the genes of plants, animals, fungi, and bacteria that have not even been discovered yet may be cures for cancer and other diseases or the key to improving the yield and nutritional quality of foods—which the U.N. Food and Agriculture Organization says will be crucial for feeding the nearly ten billion people the Earth will likely need to support in coming decades. Finally, genetic diversity in the planetary gene pool is crucial for the resilience of all life on Earth to rare but catastrophic environmental events, such as meteor impacts or massive, sustained volcanism.

Soil impacts

With all the lushness and productivity that exist in tropical forests, it can be surprising to learn that tropical soils are actually very thin and poor in nutrients. The underlying "parent" rock weathers rapidly in the tropics high temperatures and heavy rains, and over time, most of the minerals have washed from the soil. Nearly all the nutrient content of a tropical forest is in the living plants and the decomposing litter on the forest floor.

When an area is completely deforested for farming, the farmer typically burns the trees and vegetation to create a fertilizing layer of ash. After this slash-and-burn deforestation, the nutrient reservoir is lost, flooding and erosion rates are high, and soils often become unable to support crops in just a few years. If the area is then turned into cattle pasture, the ground may become compacted as well, slowing down or preventing forest recovery.

Social Impacts

Tropical forests are home to millions of native (indigenous) people who make their livings through subsistence agriculture, hunting and gathering, or through low-impact harvesting of forest products like rubber or nuts. Deforestation in indigenous territories by loggers, colonizers, and refugees has sometimes triggered violent conflict. Forest preservation can be socially divisive, as well. National and international governments and aid agencies struggle with questions about what level of human presence, if any, is compatible with conservation goals in tropical forests, how to balance the needs of indigenous peoples with expanding rural populations and national economic development, and whether establishing large, pristine, uninhabited protected areas—even if that means removing current residents—should be the highest priority of conservation efforts in tropical forests.

In India, the forest cover area ranging from about 6% to 25% as varies from state to state. In 1989, an assessment of forest cover was attempted by using NRSA remote serving data (Table 4.10). This indicate that the average forest cover area is about 19.47% as against non-forest area of 78.40%. Again among the forest area only 11.51% area having dense forest (i.e., > 40% crown cover).

Table 4.13: Status of forest cover of India in 2007

Class	Area (km^2)	% of geographical area
Forest Cover		
Very Dense Forest	83,510	2.54
Moderately Dense Forest	319,012	9.71
Open Forest	288,377	8.77
Total Forest Cover*	**690,899**	**21.02**
Non-forest		
Scrub	41,525	1.26
Non-forest**	2,554,839	77.72
Total Geographical Area	**3,287,263**	**100.00**

* Includes 4,639 km^2 under mangroves

** Excludes scrubs and includes water bodies

The forest cover in India is about 19% of the total geographical area. In most cases the good, dense and virgin forests are lacking very much due to rapid extraction of forest resources from different parts of the country. But the diversity of forest type is very much unique in India. Over 16 major types of forests are distributed in different biozones of the country (Table 4.14).

Table 4.14: Types of forest in India

I. Moist Tropical forests:

1. Tropical wet evergreen Western Ghats forest (Kerala, Karnataka, Maharshtra);
2. Tropical Semi Evergreen-Lower hills of Western Ghats forest (Karnataka, Kerala);
3. Tropical moist deciduous forest (Dehradun,, Mahabaleswar, Eastern Ghats);
4. Littoral and Swamp forests (Sunderban, Andaman, Other deltas).

II. Dry Tropical forests:

5. Tropical dry deciduous forest (MP, UP);
6. Tropical Thorn forest (Delhi, Punjab, Gujrat);
7. Tropical Dry Evergreen forest (Eastern Ghats).

III. Montane Sub-Tropical forests:

8. Sub tropical broad leaved forest (Shillong, Nilgiris);
9. Sub tropical prime forests (Khasia, Kashmir, Uttarakhand);
10. Sub tropical dry evergreen forest (foothills of Himalayas).

IV. Montane Temperate Forest:

11. Mantane wet temperate forest (Nilgiris, Palni hills);
12. Himalayan wet temperate forest (Simla, Assam, Himachal Pradesh);
13. Himalayan dry temperate forest (NW Kashmir).

V. Sub-Alpine forests:

14. Sub-Alpine forest (Laddakh, Sikkim).

VI. Alpine Scrubs:

15. Moist Alpine scrub forest (High Himalayan forest);
16. Dry Alpine scrub forest (Sikkim).

Source: Champion & Seth, 1968.

In 2007, survey data shows the different distributional pattern of forest types (state-wise) in India (Table 4.15(a)). Among the different states, form viz. Andaman & Nicober, Arunachal Pradesh, Mizoram and Nagaland have over 80% forest cover zones. However the tree cover area in forest zone is much less. The details of tree cover area of each state is given in Table 4.15(b).

Table 4.15: (a) Extent of dense forest, open forest and mangrove in States/UTs (2015 Assessment)

(sq. km)

State/UT	Geographical area	Very dense forest	Mod. dense forest	Open forest	Total forest	Percent of geographical area	Change in forest cover wrt ISFR 2013	Change percent	Scrub
Andhra Pradesh	160,204	375	13,093	10,956	24,424	15.25	67	0.27	8,562
Arunachal Pradesh	83,743	20,804	31,301	15,143	67,248	80.30	– 73	– 0.11	264
Assam	78,438	1,441	11,268	14,914	27,623	35.22	– 48	– 0.17	384
Bihar	94,163	248	3,376	3,664	7,288	7.74	– 3	– 0.04	120
Chhattisgarh	135,191	4,152	34,846	16,588	55,586	41.12	– 35	– 0.06	117
Delhi	1,483	6.94	57.15	124.68	188,77	12.73	8.96	4.75	2
Goa	3,702	542	580	1,102	2,224	60.08	5	0.22	0
Gujarat	196,022	376	5,220	9,064	14,660	7.48	7	0.05	1,490
Haryana	44,212	27	452	1,105	1,584	3.58	– 2	– 0.13	151
Himachal Pradesh	55,673	3,224	6,381	5,091	14,696	26.40	13	0.09	301
Jammu & Kashmir*	222,236	4,061	8,815	10,112	22,988	10.34	450	1.96	625
Jharkhand	79,714	2,588	9,663	11,227	23,478	29.45	5	0.02	685
Karnataka	191,791	1,781	20,063	14,577	36,421	18.99	289	0.79	3,191
Kerala	38,863	1,523	9,301	8,415	19,239	49.50	1317	6.85	36
Madhya Pradesh	308,245	6,629	34,902	35,931	77,462	25.13	– 60	– 0.08	6,383
Maharashtra	307,713	8,712	20,747	21,169	50,628	16.45	– 4	– 0.01	4,157
Manipur	22,327	727	5,925	10,342	16,994	76.11	4	0.02	1,182
Meghalaya	22,429	449	9,584	7,184	17,217	76.76	– 71	– 0.41	348
Mizoram	21,081	138	5,858	12,752	18,748	88.93	– 306	– 1.63	0
Nagaland	16,579	1,296	4,695	6,975	12,966	78.21	– 78	– 0.60	622
Odisha	155,707	7,023	21,470	21,861	50,354	32.34	7	0.01	4,499
Punjab	50,362	0	735	1,036	1,771	3.52	– 1	– 0.06	37
Rajasthan	342,239	76	4,426	11,669	16,171	4.73	85	0.53	4,302
Sikkim	7,096	500	2,160	697	3,357	47.31	– 1	– 0.03	311
Tamil Nadu	130,058	2,993	10,469	12,883	26,345	20.26	2501	9.49	414
Telangana	114,865	513	12,712	8,366	21,591	18.80	– 168	– 0.78	1,874
Tripura	10,486	113	4,609	3,089	7,811	74.49	– 55	– 0.70	55
Uttar Pradesh	240,928	2,195	4,060	8,206	14,461	6.00	112	0.77	803
Uttarakhand	53,483	4,754	13,602	5,884	24,240	45.32	– 268	– 1.11	307
West Bengal	88,752	2,948	4,172	9,708	16,828	18.96	23	0.14	133
Andaman & Nicobar	8,249	5,686	685	380	6,751	81.84	40	0.59	1
Chandigarh	114	1.36	14.09	6.58	22.03	19.32	4.77	21.65	0.03
Dadar & Nagar Haveli	491	0	80	126	206	41.96	– 7	– 3.40	5
Damman & Diu	112	1.4	5.82	12.39	19.61	17.51	10.34	52.73	0.55
Lakshadweep	32	0	17.22	9.84	27.06	84.56	0	0.00	0
Puducherry	480	0	29.68	25.70	55.38	11.54	5.32	9.61	0
Grand Total	3,287,263	83,904	315,374	300,395	701,673	21.34	3,775	0.54	41,362

* Includes Jammu and Kashmir areas outside LoC that is under illegal occupation of Pakistan and China.

Source: State of Forest Report, 2015

Table 4.15: (b) State/UT-wise Tree Cover Estimates

Sl. No.	State/UT	Geographic Area	Tree Cover	
			Area (km²)	% of Geo. Area
1.	Andhra Pradesh	275,069	7,191	2.61
2.	Arunachal Pradesh	83,783	592	0.71
3.	Assam	78,438	1,590	2.03
4.	Bihar	94,163	2,495	2.65
5.	Chhattisgarh	135,191	4,027	2.98
6.	Delhi	1,483	123	8.29
7.	Goa	3,702	286	7.73
8.	Gujarat	196,022	8,390	4.28
9.	Haryana	44,212	1,409	3.19
10.	Himachal Pradesh	55,673	638	1.15
11.	Jammu & Kashmir	222,236	6,764	3.04
12.	Jharkhand	79,714	3,032	3.80
13.	Karnataka	191,791	5,683	2.96
14.	Kerala	38,863	2,801	7.21
15.	Madhya Pradesh	308,245	6,871	2.23
16.	Maharashtra	307,713	9,466	3.08
17.	Manipur	22,327	197	0.88
18.	Meghalaya	22,429	542	2.42
19.	Mizoram	21,081	172	0.82
20.	Nagaland	16,579	300	1.81
21.	Odissa	155,707	4,435	2.85
22.	Punjab	50,362	1,699	3.37
23.	Rajasthan	342,239	8,274	2.42
24.	Sikkim	7,096	20	0.28
25.	Tamil Nadu	130,058	4,968	3.82
26.	Tripura	10,486	171	1.63
27.	Uttar Pradesh	240,928	7,381	3.06
28.	Uttarakhand	53,483	665	1.24
29.	West Bengal	88,752	2,458	2.77
30.	Andaman & Nicobar Islands	8,249	44	0.53
31.	Chandigarh	114	11	9.65
32.	Dadar & Nagar Haveli	491	27	5.50
33.	Damman & Diu	112	9	8.04
34.	Lakshadweep	32	4	12.50
35.	Puducherry	480	34	7.08
	Total	3,287,263	92,769	2.82

Source: Annual Report 2009-2010, MOEF (Government of India)

In the assessment of 2015, it shows that 15 States/UTs have above 33 percent of the geographical area under forest cover. Out of these states and UTs, seven states have more than 75 percent forest cover while eight states have forest cover between 33 percent and 75 percent. Table 4.15(c) gives the details of forest cover of these states in the descending order of the percentage of the forest cover.

Table 4.15: (c) States/UTs with forest cover more than 33 percent (2015 Assessment)

(sq. km)

State/UT	Geographical area	Very dense forest	Mod. dense forest	Open forest	Total forest	Scrub	Forest cover percent
Mizoram	21,081	138	5,858	12,752	18,748	88.93	– 306
Lakshadweep	32	0	17.22	9.84	27.06	84.56	0
Andaman & Nicobar	8,249	5,686	685	380	6,751	81.84	40
Arunachal Pradesh	83,743	20,804	31,301	15,143	67,248	80.30	– 73
Nagaland	16,579	1,296	4,695	6,975	12,966	78.21	– 78
Meghalaya	22,429	449	9,584	7,184	17,217	76.76	– 71
Manipur	22,327	727	5,925	10,342	16,994	76.11	4
Tripura	10,486	113	4,609	3,089	7,811	74.49	– 55
Goa	3,702	542	580	1,102	2,224	60.08	5
Kerala	38,863	1,523	9,301	8,415	19,239	49.50	1317
Sikkim	7,096	500	2,160	697	3,357	47.31	– 1
Uttarakhand	53,483	4,754	13,602	5,884	24,240	45.32	– 268
Dadar & Nagar Haveli	491	0	80	126	206	41.96	– 7
Chhattisgarh	135,191	4,152	34,846	16,588	55,586	41.12	– 35
Assam	78,438	1,441	11,268	14,914	27,623	35.22	– 48

Other than forest, there is vast biological resources found in aquatic systems, viz, lake, reservoir, river, estueary, sea and ocean. Till date the global community is primarily dependent on biological resources of land, except fishes and some other seaweed from aquatic system. In the present century more attention perhaps will be focussed on an aquaculture as the land productivity is almost at maximum level.

The distribution of diverse biological forms in the globe are both horizontally and vertically over the year. The existence of life is noticed from arctic to autistic zones, in varied forms. Identically, there is also a great extent of vertical distribution of life in both terrestrial and aquatic environment (Fig. 4.4). Up to 10,000 meter microbial life exists on earth (i.e., Parabiopheric zone). Similarly, in water, up to a depth of 5,000 meters, (i.e., abyssal zone) life exists, though Euphotic zone i.e., surface water showed maximum productivity.

(D) Minerals

A variety of minerals both metals & non-metals were exploited by mankind over centuries. Most metals used for industrial purposes occur in the earth's crust (i.e., 10-70 km thickness). These metals can be mined economically where considerable explorable deposit exists. There are three categories of metals in the earth crust: (a) structural metals like iron, aluminium, (b) scarce metals like Cu, Pb, Zn, Mg, Ni, Hg etc. and (c) precious metals like Au, Pt and Ag etc.

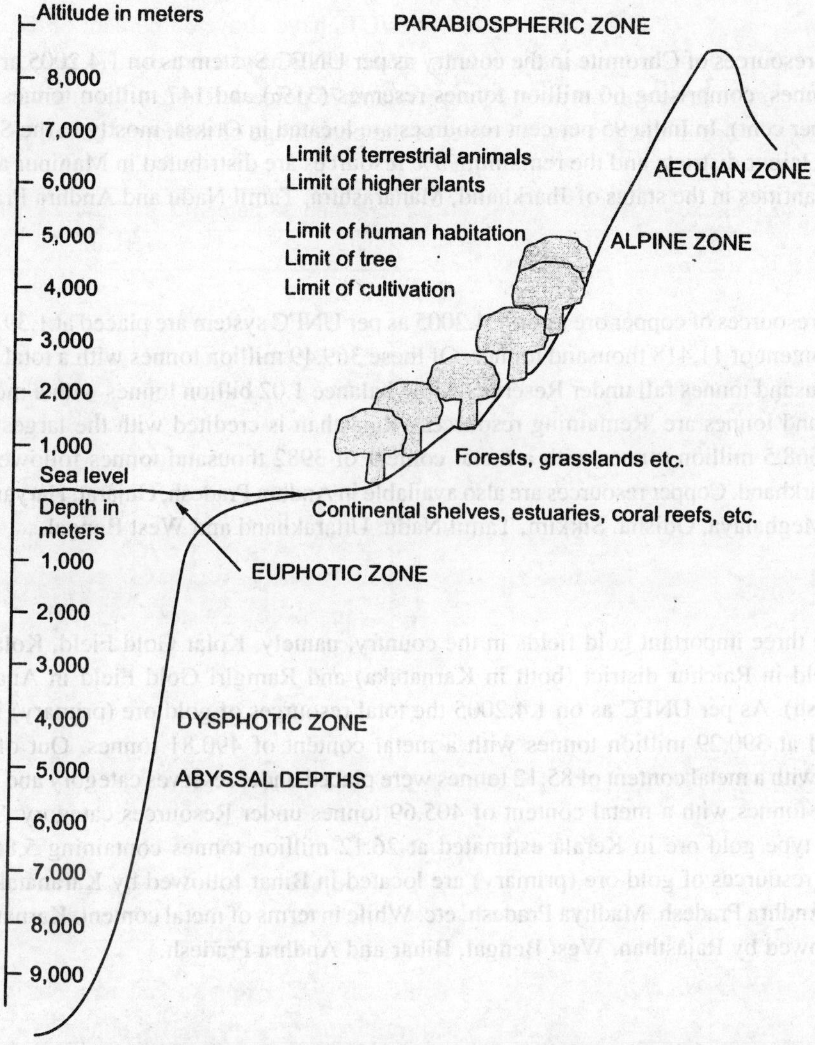

Fig. 4.4 : Vertical extent of the biosphere. The parabiospheric zone represents altitudes where dormant forms of life, such as bacterial and fungal spores, exist. The euphotic zone in aquatic environment is the zone of active photosynthesis. (*Redrawn from Hutchinson*, 1970)

MINERAL RESOURCES IN INDIA

These principal minerals found in the country along with their estimated reserves/resources are given below:

BAUXITE

The total resources of Bauxite as per United Nations Framework Classification (UNFC) in the country are placed at about 3,290 million tonnes as on 1.4.2005. These resources include 899 million tonnes of Reserves and 2,391 million tonnes of Remaining resources. Odisha, Andhra Pradesh, Gujarat, Chhattisgarh, Madhya Pradesh, Jharkhand and Maharashtra are the principal states where bauxite deposits are located. Major deposits are concentrated in the East Coast Bauxite deposits of Odisha and Andhra Pradesh.

CHROMITE

The total resources of Chromite in the country as per UNFC System as on 1.4.2005 are estimated at 213 million tonnes, comprising 66 million tonnes reserves (31%) and 147 million tonnes of remaining resources (69 per cent). In India 95 per cent resources are located in Orissa, mostly in the Sukinda valley in Cuttack and Jaipur districts and the remaining 5% resources are distributed in Manipur and Karnataka and meagre quantities in the states of Jharkhand, Maharashtra, Tamil Nadu and Andhra Pradesh.

COPPER

The total resources of copper ore as on 1.4.2005 as per UNFC system are placed at 1.39 billion tonnes with a metal content of 11,418 thousand tonnes. Of these 369.49 million tonnes with a total metal content of 4383.97 thousand tonnes fall under Reserves while balance 1.02 billion tonnes with a metal content of 7033.75 thousand tonnes are 'Remaining resources'. Rajasthan is credited with the largest resources of copper ore at 668.5 million tonnes with a metal content of 3982 thousand tonnes followed by Madhya Pradesh and Jharkhand. Copper resources are also available in Andhra Pradesh, Gujarat, Haryana, Karnataka, Maharashtra, Meghalaya, Odisha, Sikkim, Tamil Nadu, Uttarakhand and West Bengal.

GOLD

There are three important gold fields in the country, namely, Kolar Gold Field, Kolar district and Hatti Gold Field in Raichur district (both in Karnataka) and Ramgiri Gold Field in Anantpur district (Andhra Pradesh). As per UNFC as on 1.4.2005 the total resources of gold ore (primary) in the country were estimated at 390.29 million tonnes with a metal content of 490.81 tonnes. Out of these, 19.25 million tonnes with a metal content of 85.12 tonnes were placed under reserves category and the remaining 371.03 million tonnes with a metal content of 405.69 tonnes under Resources category. The resources include placer-type gold ore in Kerala estimated at 26.12 million tonnes containing 5.86 tonnes gold metal. Largest resources of gold ore (primary) are located in Bihar followed by Karanataka, Rajasthan, West Bangal, Andhra Pradesh, Madhya Pradesh, etc. While in terms of metal content. Karnataka remained on the top followed by Rajasthan, West Bengal, Bihar and Andhra Pradesh.

IRON ORE

Iron & Steel is the crux for industrial development in a country. The vitality of the iron & steel industry largely influences the economic status of a country Iron ore being the essential raw-material for Iron & Steel Industry, its mining arguably is the cynosure of all mining activities undertaken by any country. With the total resources of over 28.5 billion tonnes of hematite (Fe203) and magnetite (Fe304), India is one of the leading producers as well as exporters of iron ore in the world.

As per UNFC system, the total provisional resources of hematite as on 1.4.2010 are estimated at 17,882 million tonnes of which 8,093 million tonnes (45%) are under reserve' category and the balance 9,789 million tones (55%) are under 'remaining resources' category. By grades, lumps constitute about 56% followed by fines (21%), lumps with fines (13%) and the remaining 10% are black iron ore, others and not known grades. Major resources of hematite are located in Odisha-5,930 million tonnes (33%), Jharkhand-4,597 million tones (26%), Chhattisgarh-3,292 million tones (18%), Karnataka-2,159 million tonnes (12%) and Goa-927 million tonnes (5%). The balance resources of hematite are spread in Andhra Pradesh, Assam, Bihar, Maharashtra, Madhya Pradesh, Meghalaya, Rajasthan and Uttar Pradesh.

Magnetite is another principal iron ore that also occurs in the form of oxide of iron either in igneous or metamorphosed banded magnetite-silica formation, possibly of sedimentary origin. As per UNFC system, the total resources of magnetite as on 1.4.2010 provisional are estimated at 10,644 million tonnes of which 'reserves' constitute a mere 22 million tonnes while 10,622 million tonnes are placed under 'remaining resources'. Classification on the basis of grades show 21% resources of metallurgical grade while 77% resources belong to unclassified, not-known and other grades. The resources of coal washery and foundry grades constitute meagre proportions. India's 97% magnetite resources are located in its four states, namely, Karnataka-7812 million tonnes (73%) followed by Andhra Pradesh-1,464 million tonnes (14%), Rajasthan-527 million tonnes and Tamil Nadu-507 million tonnes (5% each). Assam, Bihar, Goa, Jharkhand, Kerala, Maharashtra, Meghalaya and Nagaland together account for the remaining 3% resources.

LEAD-ZINC

Lead-Zinc resources are located in Rajasthan, Bihar, Maharashtra, Madhya Pradesh, Andhra Pradesh, Gujarat, Uttarakhand, West Bengal, Odisha, Sikkim, Tamil Nadu and Meghalaya. The total resources of lead and zinc ores as on 1.4.2005 as per UNFC are estimated at 522.58 million tonnes with a metal content of 7207 thousand tonnes of lead metal and 24260 thousand tonnes of zinc metal. Of these, 125.75 million tonnes with a metal content of 2591 thousand tonnes of lead metal and 11093 thousand tonnes of zinc metal fall under 'Reserves' while balance 396.83 million tonnes are with a metal content of 4617 thousand tonnes lead metal and 13167 thousand tonnes of zinc metal classified as 'Remaining resources'.

MANGANESE

The total resources of manganese ore in the country as on 01.04.2010 are placed at 430 million tonnes as per UNFC system. Out of these, 142 million tonne are categorised as reserves and the balance 288 million tonnes are in the remaining resources category. Gradewise, ferro-manganese grade accounts for only 8%, medium grade 11%, BF grade 34% and the remaining 47% are of mixed, low, others, unclassified, and not-known grades including 0.35 million tonnes of battery/chemical grade. Statewise, Odisha tops the total resources with 44% share followed by Karnataka 22%, Madhya Pradesh 13%, Maharashtra 8%, Andhra Pradesh 4% and Jharkhand & Goa 3% each. Rajasthan, Gujarat and West Bengal together shared about 3% of the total resources.

NICKEL

Nickel, when added in small quantity to iron, increases its properties manifold and makes the product hard and stainless. The reason behind the demand of 66% primary nickel in the entire world is for the production of stainless steel. When it is used in plating, it makes the surface tarnish-resistant and provides polished appearance.

As per UNFC, as on 1.4.2010, the total provisional resources of nickel ore have been estimated at 189 million tonnes. About 92% resources;i.e., 175 million tonnes are in Odisha. The remaining 8% resources are distributed in Jharkhand (9 million tonnes), Nagaland (5 million tonnes) and Karnataka (0.23 million tonnes).

TUNGSTEN

The total resources of tungsten ore in the country, as per UNFC system, as on 1.4.2010 have been estimated at 87.4 million tonnes containing 142,094 tonnes W03 content. All these resources are placed under remaining resources' category. Resources are mainly distributed in Karnataka (42%), Rajasthan (27%), Andhra Pradesh (17%) and Maharashtra (9%). Remaining 5% resources are in Haryana, Tamil

Nadu, Uttarakhand and West Bengal. At Degana, Rajasthan, W03 value in vein deposits varies from 0.25 to 0.54% while in gravel deposit, it is, on an average 0.04%. In Sirohi deposit, Rajasthan, W03 content ranges from 0.02 to 2.2%. In West Bengal, Bankura deposit, Rajasthan, W03 content ranges from 0.02 to 2.2%. In West Bengal, Bankura Deposit contains, on an average, 0.1% W03. In Kuhi-Khobana-Agargaon belt, GSI has identified seven mineralised zones in Sakoli basin in Bhandara and Nagpur districts, Maharashtra. The analysis showed 0.01 to 0.19% W03 in Kuhi block, 0.13 to 0.38% W03 in Khobana block and 0.48% W03 in Pardi Dahegaon-Pipalgaon block. The deposit contains 0.17% W03 on an average. Gold area at Mysore mine of BGML in Karnataka has been reckoned as a potential source of scheelite. The tailing dumps at Kolar Gold Fields contain about 0.035 to 0.18% W03.

BARYTES

The total resources of barytes in India as on 1.4.2005 as per UNFC system are placed at 74 million tonnes of which about 46% (34 million tonnes) are in 'Reserves' category and 54% (40 million tonnes) are in 'Remaining Resources' category. The Mangampet deposit in Cuddapah district (Andhra Pradesh) is the single largest barytes deposit in the world. Andhra Pradesh alone accounted for more than 94 per cent country's resources. Minor occurrences of barytes are located in Rajasthan, West Bengal, Madhya Pradesh, Tamil Nadu, Himachal Pradesh, Maharashtra, Jharkhand, Uttarakhand, Karnataka and Haryana.

DIAMOND

Diamond deposits occur in three types of geological settings such as kimberlite pipes, conglomerate beds and alluvial gravels. The main diamond bearing areas in India are Panna belt in Madhya Pradesh, Munimadugu-Banganapalie conglomerate in Kurnool district, Wajrakarur kimberlite pipe in Anantapur district, the gravels of Krishna river basin in Andhra Pradesh and damondiferous kimberlite in Raipur, Bastar and Raigarh districts in Chhattisgarh. Reserves have been estimated in Panna belt, Madhya Pradesh;Krishna Gravels in Andhra Pradesh;and in Raipur district, Chhattisgarh. As per the UNFC system as on 1.4.2005 diamonds are placed at around 4582 thousand carats, out of which about 1206 thousand carats are under Reserve category and remaining 3376 thousand carats are under remaining Resources category.

DOLOMITE

Total resources of dolomite as per UNFC system as on 1.4.2005 are placed at 7533 million tonnes, out of which Reserves are 985 million tonnes and the balance i.e. 6548 million tonnes are in the 'Remaining Resources'. Dolomite occurrences are widespread in almost all parts of the country The major share of about 90 per cent resources is distributed in the states of Madhya Pradesh, Andhra Pradesh, Chhattisgarh, Odisha, Karnataka, Gujarat, Rajasthan and Maharashtra.

FIRECLAY

Fireclay occurs as a bedded deposit, mostly associated with coal measures of Gondwana and Tertiary periods. Important deposits are associated with Jharia and Raniganj coalfields in Jharkhand and West Bengal, Korba coalfield in Chhattisgarh and Neyveli Lignite field in Tamil Nadu. Notable occurrences of fireclay not associated with coal measures are known in the state of Gujarat, Jabalpur region of Madhya Pradesh and Belpahar-Sundergarh areas of Orissa. The total resources of fireclay as per UNFC system as on 1 April 2005, are about 705 million tonnes in India out of which 59 million tonnes and under Reserve category and about 646 million tonnes are under remaining Resources category. It is necessary to assess the fireclay reserves on priority basis, especially those associated with coal measures in the leasehold areas. The reserves of fireclay are substantial but resources of high grade (non-plastic) fireclay containing more than 37 per cent alumina are limited.

FLUORSPAR

The total resources of fluorite as per UNFC system as on 1.4.2005 were estimated at 20.16 million tonnes. Out of these, 9.21 million tonnes were placed under 'Reserves' category and the remaining 10.95 million tonnes under Remaining Resources category. Major deposits of Fluorspar are located in Gujarat, Rajasthan, Chhattisgarh and Maharashtra.

GYPSUM

The total resources of mineral Gypsum as per UNFC system as on 1.4.2005 were estimated at 1,237 million tonnes. Of these 69 million tonnes have been placed under Reserve and 1,168 million tonnes under 'Remaining Resources' The main occurrences of gypsum are located in Rajasthan, Jammu and Kashmir, Gujarat and Tamil Nadu. Rajasthan alone accounts for more than 80 per cent country resource. Minor occurrences of gypsum are in Andhra Pradesh, Himachal Pradesh, Karnataka, Madhya Pradesh and Uttarakhand.

GRAPHITE

As per the UNFC the total resources of graphite in the country as on 1.4.2005 are placed at about 168.77 million tonnes comprising 10.75 million tonnes in the Reserve category and remaining 158.02 million tonnes under Resources category. Out of total resources, Arunachal Pradesh accounts 43% followed by Jammu and Kashmir (37%), Jharkhand (6%), Tamil Nadu (5%) and Odisha (3%). However, in term of reserves, Tamil Nadu has major share of about 37%.

ILMENITE

The resources of Ilmenite are 461.37 million tonnes as per Department of Atomic Energy. Ilmenite occurs mainly in beach sand deposits right from Ratnagiri (Maharashtra) to coast in Kerala, Tamil Nadu and Orissa. The mineral is also found in Andhra Pradesh, Bihar and West Bengal.

KAOLIN

India possesses fairly large resources of china clay. The total resources as per UNFC system as on 1.4.2005 are about 2596 million tonnes out of which, 222 million tonnes are placed in Reserves category. The occurrences of china clay are distributed in Kerala, West Bengal, Rajasthan, Odisha, Karnataka, Jharkhand, Gujarat Meghalaya, Andhra Pradesh and Tamil Nadu.

LIMESTONE

The total resources of limestone of all categories and grades as per UNFC system as on 1.4.2005 are estimated at 175345 million tonnes. Of which 12715 million tonnes are under 'Reserves' category and 162630 million tonnes are under 'Remaining Resources' category. Karnataka is the leading state followed by Andhra Pradesh, Gujarat, Rajasthan, Meghalaya, Chhattisgarh, Madhya Pradesh, Odisha, Maharashtra and Uttarakhand.

MICA

Important mica bearing pegmatite occurs in Andhra Pradesh, Jharkhand, Maharashtra, Bihar and Rajasthan. The total resources of Mica in the country as per UNFC system as on 1.4.2005 are estimated at 393855 tonnes, out of which only 68570 tonnes are placed under 'Reserves' category. 'Remaining resources' are placed at 325285 tonnes. Rajasthan accounts for about 51 per cent resources, followed by Andhra Pradesh Maharashtra and Bihar.

MAGNESITE

The total resources of magnesite as per UNFC system as on 1.4.2005 are about 338 million tonnes, of which Reserves and Remaining resources are 76 million tonnes and 262 million tonnes, respectively. Substantial quantities of resources are established in Uttarkhand (68%) followed by Rajasthan (16%) and Tamil Nadu (13%). Andhra Pradesh, Himachal Pradesh, Jammu and Kashmir, Karnataka and Kerala contribute for the balance.

KYANITE AND SILLIMANITE

The total resources of kyanite and sillimanite as per UNFC system as on 1.4.2005 are 103 million tonnes and 74 million tonnes, respectively. Out of these the Reserves categories are 1.4 million tonnes for kyanite and 11 million tonnes for sillimanite. Kyanite deposits are located in Maharashtra, Karnataka, Jharkhand, Rajasthan, Andhra Pradesh, Kerala, Tamil Nadu and West Bengal. Sillimanite resources are located mainly in Odisha, Tamil Nadu, Uttar Pradesh, Kerala, Andhra Pradesh, Assam and West Bengal with minor occurrences in Jharkhand, Karnataka, Madhya Pradesh, Maharashtra, Meghalaya and Rajasthan.

PHOSPHATE MINERALS

Deposits of phosphorites are located in Madhya Pradesh, Rajasthan, Uttarakhand, Jharkhand, Uttar Pradesh and Gujarat. Besides, apatite deposits of commercial importance are reported from Jharkhand, West Bengal, Andhra Pradesh, Tamil Nadu and Rajasthan. The total resources of apatite as per UNFC system as on 1.4.2005 are placed at 26.86 million tonnes, out of which 6 million tonnes are under Reserves category and about 21 million tonnes are under Remaining resources category. Out of the total resources, the bulk 61% are located in West Bengal. The total resources of rock phosphate as per UNFC system as on 1.4.2005 are placed at 305 million tonnes, out of which 53 million tonnes are placed under reserves and 252 million tonnes under remaining resources category. Bulk of reserves are located in Rajasthan and Madhya Pradesh. (*Source:* India Year Book)

In addition to crustal layers, metals are also deposited as polymetallic nodules at the oceanic beds. Many countries in Asia and Europe have considerable mineral deposits. A major quantity was already mined out for human utility. Among the mineral rich countries, India is one. The states like Rajasthan, Bihar, Orissa, Sikkim, West Bengal, Assam, AP, MP, Maharastra, Karnataka have a number of mineral deposits. As per estimates made by Geological Survey of India, India has more than 22 different types of minerals in considerably high quantity. The details are given in Table 4.16. In most cases, they are already mined out partially. If the current rate of exploitation continues and if there is no further new exploration of deposit then our mining activity might be completed by 2020. At present most of the minerals are mined out on land deposit, not mere coast line or sea bed deposits.

(E) Energy

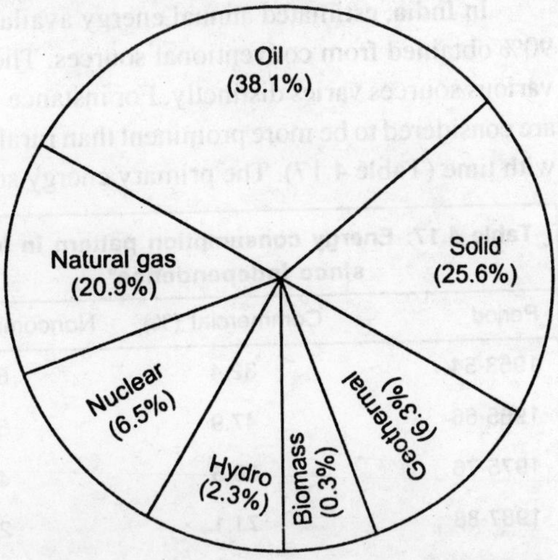

Fig. 4.5 : World share of primary fuel, 1991

In general, the world shares of primary fuels in 2000, was dominated by oil (38.1%) followed by solid like coal and lignite (25.6%) and natural gas (20.9%). The total renewable gross consumption makes up 8.9 per cent of the total primary energy. Once again these composite indicators hide wide variations in fuel mix. It is anticipated that biomass and solid fuel energy will continue to decline as oil—gas, nuclear and renewable energy penetrate the economics of NICS and developing countries.

Most of the world's energy comes from fossil fuels, viz., oil, coal, natural gas or nuclear materials. Crude oil is a mixture of liquid hydrocarbon compounds sometimes found permeating sedimentary rocks (Fig. 4.5). There are some 600 sedimentary basins in the world. About 200 of them are unexplored for petroleum, because they are in the polar regions, deep water, the remote interior of continents or are restricted for political reasons, another 240 have been explored to some extent without making commercial discoveries and the remaining 160 are commercial producers.

Coal is another important fossil fuel found in carboniferrous deposits. About 10 major countries have 92% of the 2000s estimated reserves. It is estimated that 113×10^{12} m^3 of recoverable reserves of natural gas has an energy value equivalent to about 103×10^{10} tonnes of oil. Identically the stipulated world uranium deposit varies widely for country to country. Crude oil, coal, natural gas and uranium are all primary raw materials which need processing before they can be used as fuel.

Table 4.16: Energy Status of the World, 2008

Sl. No.	Categories	Estimated reserve (Joules)
(A)	**Non-renewable forms:**	
	1. Fossil fuels	
	(a) Coal	20×10^{22}
	(b) Oil	1.5×10^{22}
	(c) Gas	1.1×10^{22}
	2. Nuclear fuels	
	(a) Uranium	10^{22}
	(b) Thorium	10^{24}
	(c) Deuterium	3×10^{31}
	(d) Lithium	10^{24}
(B)	**Renewable forms:**	
	1. Solar	3.8×10^{24} J/yr
	2. Wind	7.5×10^{20} J/yr
	3. Biomass	1.5×10^{22} J/yr
	4. Hydroelectricity	9×10^{22} J/yr
	5. Tidal	10^{20} J/yr
	6. Wave	10^{18} J/yr
	7. Geothermal	9.5×10^{20} J/yr

An overall estimate of global energy resource availability is shown in Table 4.16.

In India, estimated annual energy availability lies somewhat between 2,50,000 MW, of which over 90% obtained from conventional sources. The relative energy requirement in urban and rural areas from various sources varies distinctly. For instance in urban area transport, industry and household requirement are considered to be more prominent than rural areas. In addition energy consumption pattern also changed with time (Table 4.17). The primary energy sources of India is also shown in piechart.

Table 4.17: Energy consumption pattern in India, since independence*

Period	Commercial (%)	Noncommercial (%)
1953-54	32.4	67.6
1965-66	47.9	52.1
1975-76	56.5	43.5
1987-88	71.1	28.9
2007-2008 (Prorata basis)	90.5	9.5

*Planning Commission of India, 2010

Primary Energy Sources of India

Primary Energy Sources (%)

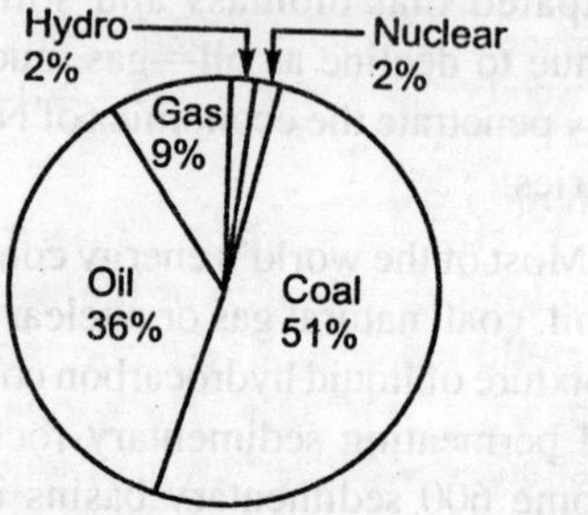

Source: Planning Commission of India, 2006

WORLD PRIMARY ENERGY CONSUMPTION

Estimates of world primary energy consumption in 2002, and the shares of the principal energy sources, are shown in chart below:

In accordance with the conventions currently used until 1994 in compiling UK energy statistics,, the primary energy equivalent of hydroelectricity production is expressed here in terms of the thermal energy content of the fossil fuel that would need to be burned in a conventional thermal power plant to generate the same amount of electricity. The primary energy equivalent of nuclear electricity is expressed in the same way. This convention is also used by the World Energy Council (WEC, 2003).

Other sources of energy data may quote slightly different figures because:

(a) they assume slightly different values for the thermal energy content of fuels;

(b) they adopt different conventions for estimating the primary energy equivalent of hydro and nuclear electricity;

(c) they may adopt different estimates of the contribution made by traditional biomass fuels, or even ignore them completely.

In 1994, the UK Department of Trade and Industry (DTI) changed the basis on which the primary energy equivalent of electricity from some renewables is calculated. From 1994 onwards, the primary energy equivalents of energy from those renewables that produce 'primary electricity', such as hydroelectricity or wind power, are

Chart : Estimated annual world primary energy consumption, 2002, by source

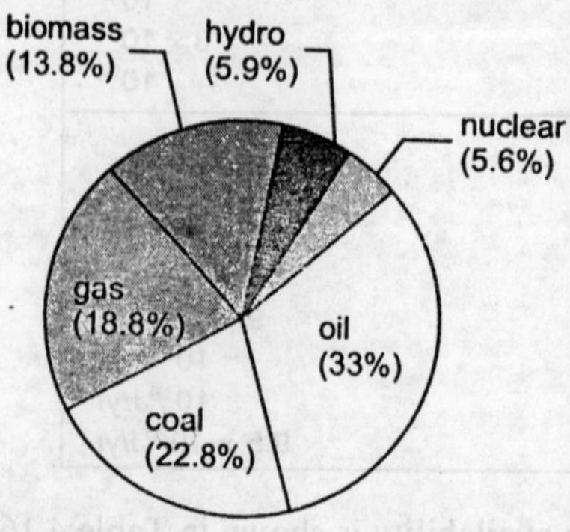

oil	: 131 EJ (3128 mtoe)
coal	: 91 EJ (2164 mtoe)
gas	: 75 EJ (1781 mtoe)
biomass	: 55 EJ (1310 mtoe)
hydro	: 24 EJ (561 mtoe)
nuclear	: 22 EJ (532 mtoe)
Total	**: 398 EJ (9476 mtoe)**

Contd.

no longer calculated on the 'substitution' basis described above but on an 'energy supplied' basis, where production is expressed in terms of the energy content of the electricity produced. (Department of Trade and Industry, 1994). The new approach is consistent with that adopted by the International Energy Agency and the European Union. This change has the overall effect of reducing by 73% the primary energy contributions of renewables sources such as hydroelectricity, wind power and photovoltaics. (The primary energy contribution of nuclear electricity generation is relatively unaffected by these chages.)

However, the DTI continues to publish some primary energy figures calculated on the former 'substitution' basis, alongside the new 'energy supplied' figures, in order to facilitate estimates of the quantities of fossil fuel consumption that have been replaced by 'primary electricity' sources.

If reflected in the Figure above, the change would have the effect of reducing the estimated hydro contribution to world primary energy needs by more than two thirds, from nearly 6% to less than 2%.

(F) Wildlife

The biological resources of the world is fast depleting due to natural resources exploitation and man-made pollution of earth's environment. As such life in the wild (wildlife), both plants and animals are disappearing very fast. The global estimate of species extinction rate is fairly high. This situation necessitates the conservation of wildlife either through in-situ or ex-situ programmes. In-situ programme involves the deceleration of different categories of protected areas, viz., biosphere, sanctuaries, national park, reserve forest, International heritage site etc. Such a programme has a good legal protection as stated in Wildlife Protection Act, 1974.

In the world, there are over 230 biospheres in about 65 countries covering an area of over 125 million hectares. India alone has thirteen biosphere reserves already demarcated (Table 4.18). In addition there were considerably a good number of sanctuaries, national park, reserve forest etc. for conservation of wildlife (Table 4.18). However as per latest information a total 89 national parks and 500 wildlife sanctuaries located in India (MOEF Annual Report 2002-2003).

Table 4.18: Biosphere Reserves

Sl. No.	Name of the site & area in sq. km.	Date of Notification	Location (State) and Bio-geographic zones
1.	Nilgiri (5,520)	01.08.86	Part of Waynad, Nagarhole, Bandipur and Mudumalai, Nilambur, Silent Valley and Siruvani Hills (Tamil Nadu, Kerala and Karnataka)—Western Ghats.
2.	Nanda Devi (5,860.69)	18.01.88	Part of Chamoli, Pithoragarh & Almora Districts (Uttranchal)—West Himalayas
3.	Nokrek (820)	01.09.88	Part of Garo Hills (Meghalaya)—East Himalayas.
4.	Manas (2,837)	14.03.89	Part of Kokrajhar, Bongaigaon, Barpeta, Nalbari, Kamrup and Darang Districts (Assam)—East Himalayas
5.	Sunderbans (9,630)	29.03.89	Part of delta of Ganges and Brahmaputra river system (West Bengal)—Gagantic Delta.
6.	Gulf of Mannar (10,500)	18.02.89	Indian part of Gulf of Mannar between India and Sri Lanka (Tamil Nadu)—Coasts
7.	Great Nicobar (885)	06.01.89	Southern most islands of Andaman and Nicobar (A & N Islands)—Islands.
8.	Similipal (4,374)	21.06.94	Part of Mayurbhanj district (Orissa)—Deccan Peninsula.
9.	Dibru-Saikhowa (765)	28.07.97	Part of Dibrugarh and Tinsukia districts (Assam)—East Himalayas.
10.	Dehang Debang (5,122)	02.09.98	Part of Siang and Debang valley in Arunachal Pradesh—East Himalayas.
11.	Pachmarhi (4,926.28)	03.03.99	Part of Betul, Hoshangabad and Chindwara districts of Madhya Pradesh—Semi-Arid-Gujarat Rajputana.
12.	Kanchanjunga (2,619.92)	07.02.2000	Parts of Kanchanjunga Hills and Sikkim—East Himalayas.
13.	Agasthyamalai (1701)	12.11.2001	Neyyar, Peppara and Shenduruny Wildlife Sanctuaries and their adjoining areas in Kerala.

Contd.

14.	Achanakmar Amarkantak (3,835.51)	30.03.2005	Parts of Anuppur and Dindori districts of MP and parts of Bilaspur district of Chattisgarh State (MP & Chattisgarh)
15.	Kochchh (12,454)	29.01.2008	Parts of Kochchh, Rajkot, Surendranagar and Patan civil districts of Gujarat State.
16.	Cold Desert (7,770)	28.08.2009	Pin Valley National Park and surroundings; Chandratal and Sarchu & Kibber wildlife sanctuary in Himachal Pradesh
17.	Seshachalam Hills (4755.997)	20.09.2010	Sechachalam Hill ranges covering parts of Chittoor and Kadpa district of Andhra Pradesh
18.	Panna (2998.98)	25.08.2011	Part of Panna and Chattarpur districts in Madhya Pradesh

In spite of all efforts our wildlife resource gets depleted very fast for commercial interest. This can only be prevented by proper peoples' participation in wildlife protection activities.

Table 4.19: National Parks and Wildlife Sanctuaries in India

States	National Parks		Wildlife Sanctuaries		Total area (sq. km.)
	National Parks	Area (sq. km.)	National Parks	Area (sq. km.)	
Andhra Pradesh	4	3,314.50	13	12,084.59	15,844.59
Arunachal Pradesh	2	2,468.23	11	7,114.45	9,582.86
Assam	5	1,173.71	18	939.88	2,113.59
Bihar	1	335.65	12	2,068.25	2,403.90
Chattrisgarh	3	1,458.00	11	3,419.41	4,877.41
Delhi	0	0	6	27.60	27.60
Goa	1	107.00	6	647.96	754.96
Gujarat	4	479.67	23	16,422.71	16,902.38
Haryana	2	1.43	8	278.32	279.75
Himachal Pradesh	2	1.429.40	28	5,736.85	7,166.25
Jammu & Kashmir	4	4,650.07	15	10,172.15	14,822.22
Jharkhand	1	231.67	11	1,822.00	2,053.57
Karnataka	5	2,472.18	30	3,930.61	6,402.79
Kerala	6	536.52	17	2,143.36	2,679.88
Madhya Pradesh	9	5,016.69	25	7,284.64	12,301.33
Maharashtra	6	955.93	42	14,387.78	15,343.71
Manipur	2	81.80	2	184.85	266.65
Meghalaya	2	267.48	4	34.20	301.68
Mizoram	2	250.00	8	634.00	884.00
Nagaland	1	202.02	3	24.41	226.43
Odissa	2	990.70	19	6,971.15	7,961.85
Punjab	0	0	13	317.79	317.79
Rajasthan	5	3,856.53	25	5,712.83	9,569.36
Sikkim	1	1,784.00	6	265.10	2,049.10
Tamil Nadu	5	307.85	20	2,602.07	2,909.92
Tripura	2	36.7	4	603.62	603.62
Telengana	3	19.62	9		
Uttaranchal/Uttarakhand	6	4,920.82	7	2,396.39	7,317.21
Uttar Pradesh	1	490.00	23	5,198.15	5,688.15
West Bengal	6	1,692.65	15	1,103.48	2,796.13
Andaman & Nicobar Island	9	1,157.14	96	372.17	1,529.27
Chandigarh	0	0	2	26.01	26.01
Dadra & Nagar Haveli	0	0	1	0	0
Daman & Diu	0	0	1	2.18	2.18
Lakshadweep	0	0	1	0	0
Puducherry	0	0	0	0	0
TOTAL	102	34,684.53	535	124,164.58	158,849.11

Source: MOEF Annual report, 2008

(G) Ocean

Ocean is the treasure of wealth in the globe. It has vast array of biological resources along with minerals and energy sources in the form of tides and waves. The diverse biological form of ocean includes phytoplank- tons (Fig. 4.6), dinoflagellates (Fig. 4.7), zooplanktons (Fig. 4.8), nieroplanktons (Fig. 4.9), fishes and several benthic flora and fauna.

Ocean space has been used by mankind from the beginning of civilization as a source of food and tools and also as an area for commerce, colonisation and welfare. Its vast living resources, however, have been utilised only fractionally. There is tremendous scope of improvement in exploiting and enhancing these resources, on a sustainable basis.

The resource potential of conventional marine fisheries has been estimated to be 100 MT per year. Appropriate aquaculture in coastal areas might enhance the productivity in much higher extent in future. There is little exploitation of the coastal resources like mangroves, sea weed and sea grass and also coral reefs.

In addition to over-exploitation, pollution, and habitat destruction, many marine resources gets depleted very much with time. Such a situation necessitates the appropriate steps of protection of coastal and marine environment.

In India, oceanic resources particularly along the coast line is fairly high, i.e., 46% of the total exploited living resources from Indian Ocean lies in Indian share. Though our exploitation capacity as yet is very low, but it has future prospect. Indian hydrocarbon reserve is also fairly high i.e., over 12.70 billion tons, of which 8.70 billion tons are estimated from the sedimentary offshore basin.

Over 6.3% of the coastline of Peninsular India is covered by mangrove. Coral reefs form 7% of the coastline. More and more biological forms were identified as novel drug sources from coastal and marine sources. So we have to curb our activities on coastal areas so that valuable biological wealth could be conserved for human use.

(H) Human Resources

Human being are at the centre of two different equations. First it is for them that other resources are being utilised and second, they themselves are resources and if adequately developed are capable of unleashing an unprecedented scale of productive forces. Yet, this probably is the most under-utilised resource in our country. A sizable part of our labour force does not engage in any form of work and most of those who do work, work only for a fraction of a day, week, month and year, and that too not efficiently. Such a state of affairs has drained our agriculture—except in plantation and a few market-oriented sub-sectors — of much of its vitality. These attitudes towards work could survive over a long period of history only because of the traditional institutional framework of our society. The situation has become more and more serious as these attitudes and institutions are handed down to an ever-increasing population from their predecessors. Though more determined steps have been taken in the field of family planning in the last couple of years, the labour force will continue to increase somewhere between 2-3 per cent, till the end of the twentieth century. As a consequence of high population, human resources are bulkly under-utilised

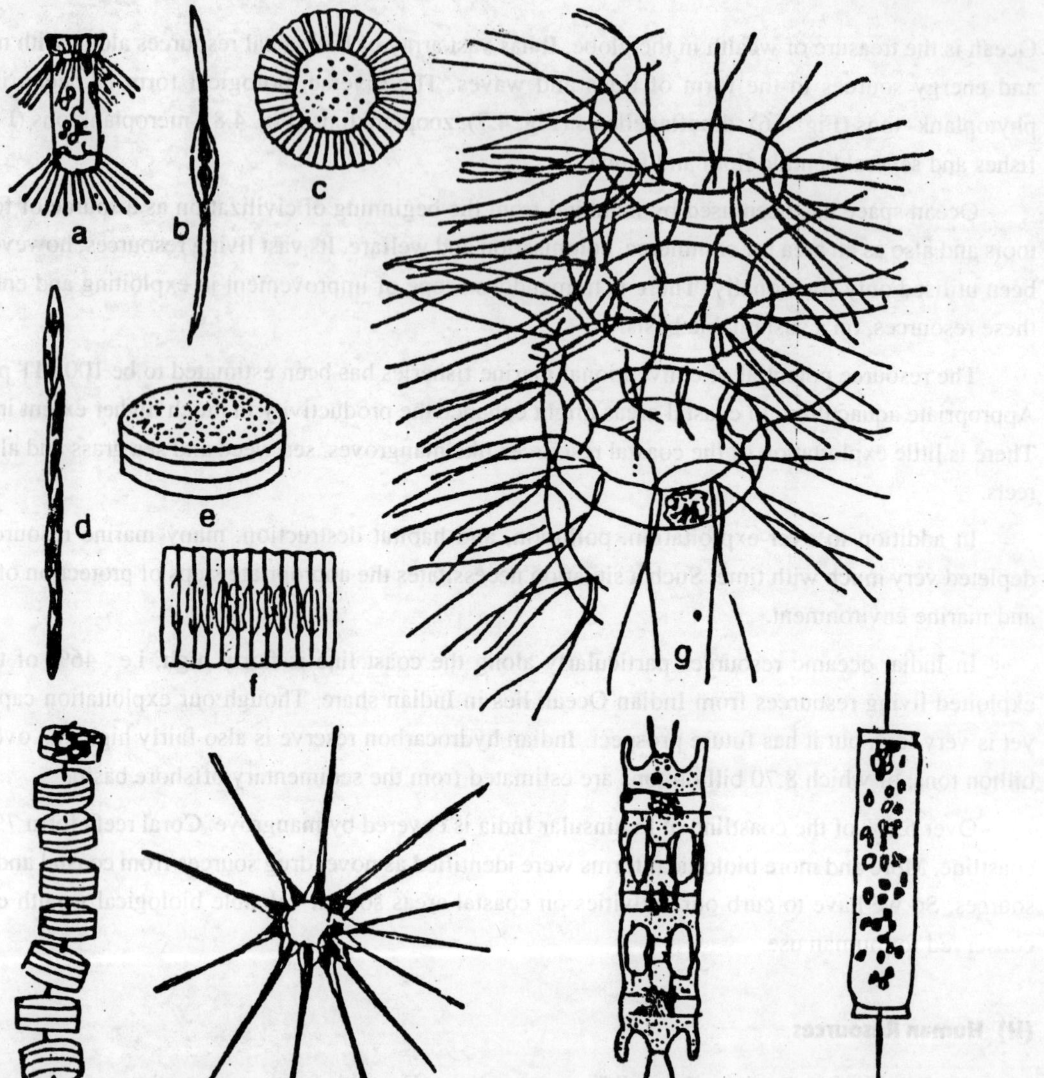

Fig. 4.6 : Phytoplankton producers of the sea : **Diatoms** (a) Corethron; (b) Nitzschia closterium; (c) Planktoniella; (d) Nitzschia seriata; (e) Coscinodiscus; (f) Fragilaria; (g) Chaetoceros; (h) Thalassiosira; (i) Asterionella; (j) Biddulphia; (k) Ditylum; (l) Thalassiothrix; (m) Navicula; (n) & (o) Rhizosolenia semispina, summer and winter forms

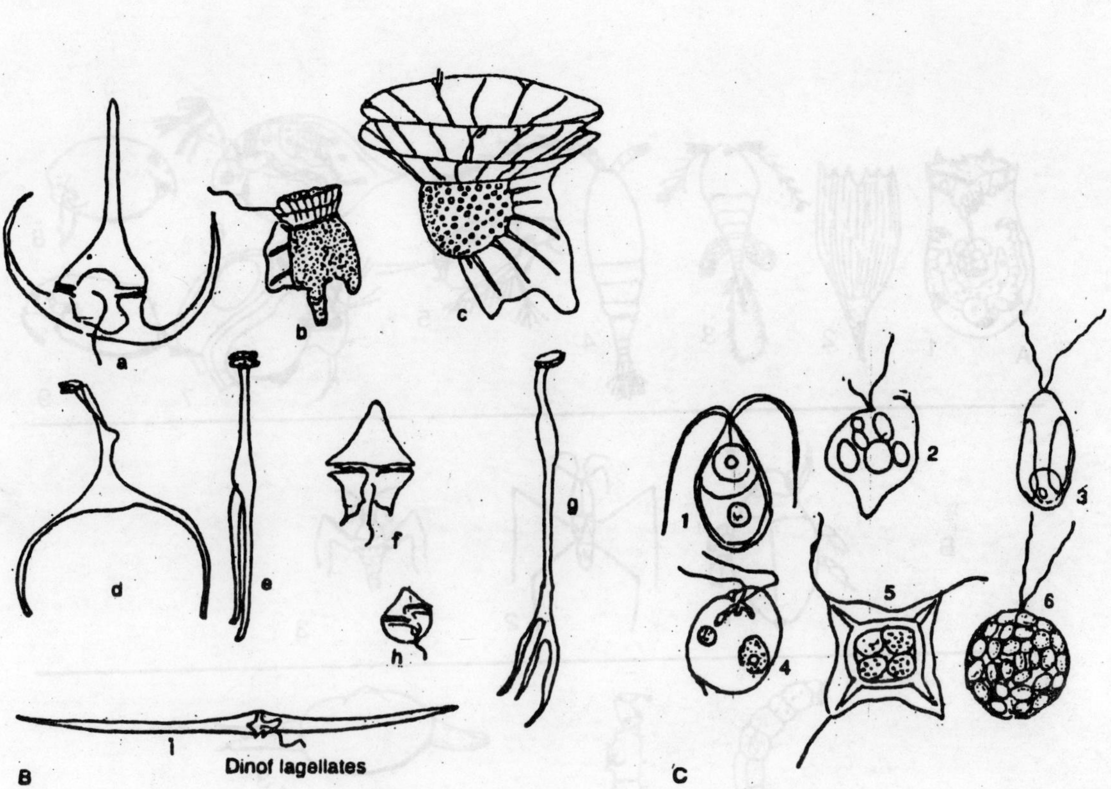

Fig. 4.7 : Dinoflagellates: (a) Ceratium; (b) Dinophysis; (c) Orithocercus; (d) (e) Triposolenia, front and side views; (f) Peridinium; (g) Amphisolenia; (h) Goniaulax; (i) Ceratium. **Microflagellates**: 1. Dunaliella (phytomonad); 2. Chloramoeba (xanthomonad); 3. Isochrysis (chrysomonad); 4. Proto-chrysis (cryptomonad); 5. Dictyocha (silicoflagellate); 6. Pontospharea (coccolithophore)

over the years. Thus on the subject of human resource development, the appropriate emphasis should be given on concept of 'Investment in Man' which is of recent origin and has not been given due recognition both in theory and practice, particularly in developing countries. There is hardly any doubt that the interdependent and self-supporting system of attitudes and social institution plays a dominant role in the utilisation of human resources and, in fact, all resources. Over past couple of decades attempts have been made for improvement of human resources in all tier through primary mass education followed by job-oriented training as well as high education, better skill development for national and international competition. As yet the success is very meagre, there is a need to define the problem more explicitly and to set up appropriate institutions for monitoring of investment on human development.

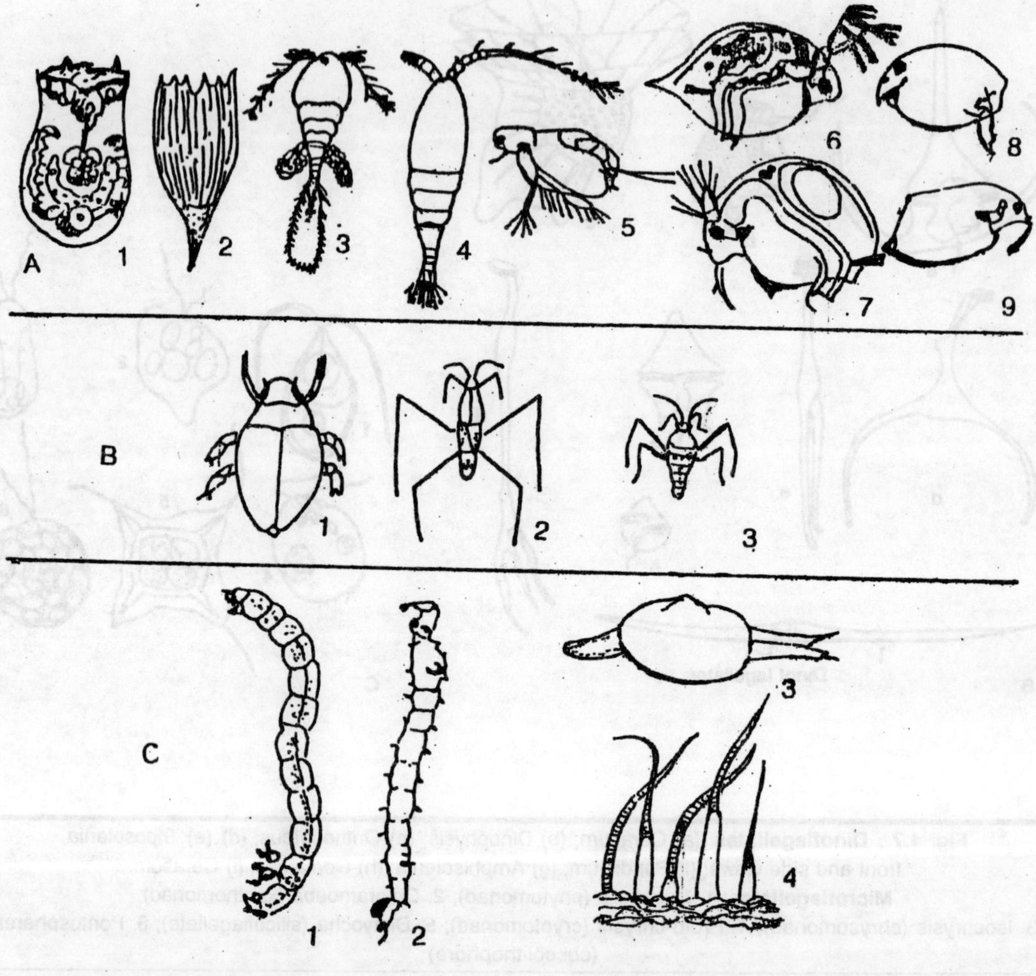

Fig. 4.8 : A. Representative zooplankton. Rotifers : 1. Asplanchna; 2. Notholca (lorica only). Copepods : 3. a cyclopoid copepod, Macrocyclops; 4. a calanoid copepod, Senecella. Cladocera (representative of each of five families); 5. Diaphanosoma (Sididae); 6. Daphnia (Daphnidae); 7. Bosmina (Bosminidae); 8. Pleuroxus (Chydoridae); 9. Acantholeberis (Macrothricidae).

B. Zooneuston. 1. a whirligig beetle, Dineutes (Gyrinidae); 2. a water strider, Gerris (Gerridae); 3. a broadshouldered water strider, Rhagovelia (Veliidae).

C. Some characteristic profundal zone types. 1. a chironomid larva, or blood worm, Tendipes (note prolegs and abdominal gills); 2. a "phantom larva", Chaoborus (note two air sacs which apparent aid the animal in performing vertical migrations; 3. a "pea-shell" clam, Musculium (Sphaeriidae), with "foot" and two branchial siphons extended; 4. Tubifex, a red annelid which builds tubes on the bottom and vigorously waves its posterior end around in the water. (Redrawn from Robert W. Pennak, "Fresh-water Invertebrates of the United States", 1953, The Ronald)

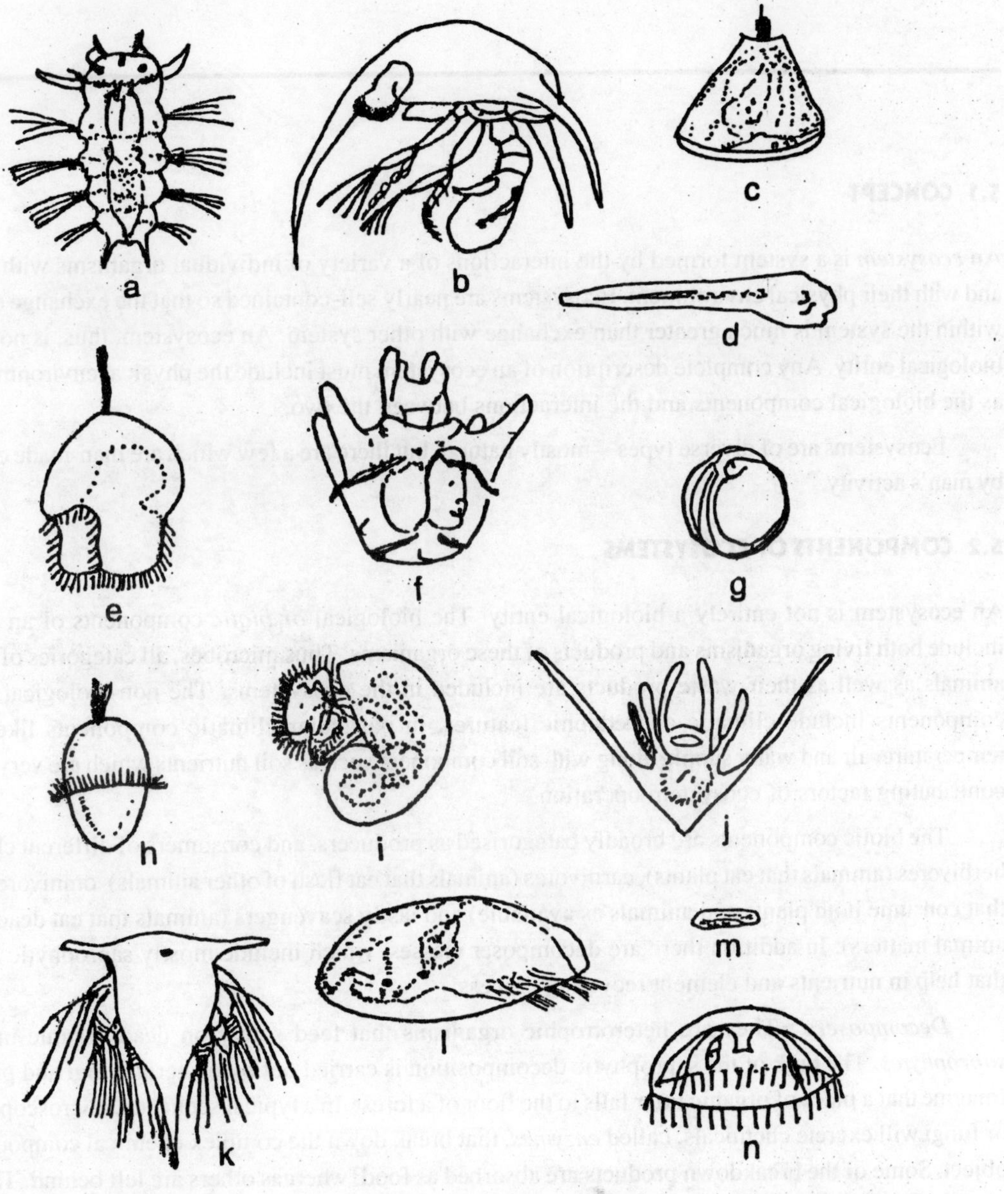

Fig. 4.9 : Larvae which make up the temporary plankton (**meroplankton**) of the sea. (a) chaetate larva of the annelid Platynereis; (b) zoea of the sand crab Emerita; (c) cyphonautes larva of a bryozoan; (d) tadopole larva of a sessile tunicate; (e) pilidium larva of a nemertean worm; (f) advanced pluteus larva of a sea urchin; (g) fish egg with embryo; (h) trochophore larva of a scaleworm; (i) veliger larva of a snail; (j) pluteus larva of a brittle starfish; (k) nauplius larva of a barnacle; (l) cypris larva of a branacle; (m) planula larva of a coelenterate; (n) medusa stages of a sessile hydroid

5 ECOSYSTEM

5.1 CONCEPT

An *ecosystem* is a system formed by the interactions of a variety of individual organisms with each other and with their physical environment. Ecosystems are nearly self-contained so that the exchange of nutrients within the system is much greater than exchange with other system. An ecosystem, thus, is not entirely a biological entity. Any complete description of an ecosystem must include the physical environment as well as the biological components and the interactions between the two.

Ecosystems are of diverse types—mostly natural, but there are a few which are man-made or modified by man's activity.

5.2 COMPONENTS OF ECOSYSTEMS

An ecosystem is not entirely a biological entity. The biological *or biotic* components of an ecosystem include both living organisms and products of these organisms. Thus microbes, all categories of plants and animals as well as their waste products are included in the ecosystems. The non-biological or *abiotic* components include climatic and edaphic features, in particular climatic components like sunlight, temperature, air and water supply along with soil component such as soil nutrients which are very important contributing factors of ecosystem operation.

The biotic components are broadly categorised as producers, and consumers of different classes viz., herbivores (animals that eat plants), carnivores (animals that eat flesh of other animals), omnivores (animals that consume both plants and animals as available) and lastly scavengers (animals that eat dead plant and animal matters). In addition there are decomposer classes, which include mostly saprophytic organisms that help in nutrients and element recycling process.

Decomposers : They are heterotrophic organisms that feed solely on dead organic matter, i.e., *saprophytes*. The bulk of the saprophytic decomposition is carried out by bacteria, fungi and protozoans. Imagine that a piece of organic litter falls to the floor of a forest. In a typical sequence, microscopic bacteria or fungi will excrete chemicals, called *enzymes*, that break down the complex chemical compounds in the object. Some of the breakdown products are absorbed as food, whereas others are left behind. These serve as a food supply for other organisms that carry the decomposition one step further. Eventually the waste products of the final line of decomposers are energy-poor mineral nutrients that are reabsorbed, and thus recycled, by plants — Fig. 5.1 (see next page).

5.3 THE LAWS OF THERMODYNAMICS AND ENERGY FLOW IN ECOSYSTEM

The application of the two fundamental laws of thermodynamics to energy and matter transformations at the cellular level was discussed over the years through various ecosystem modes.

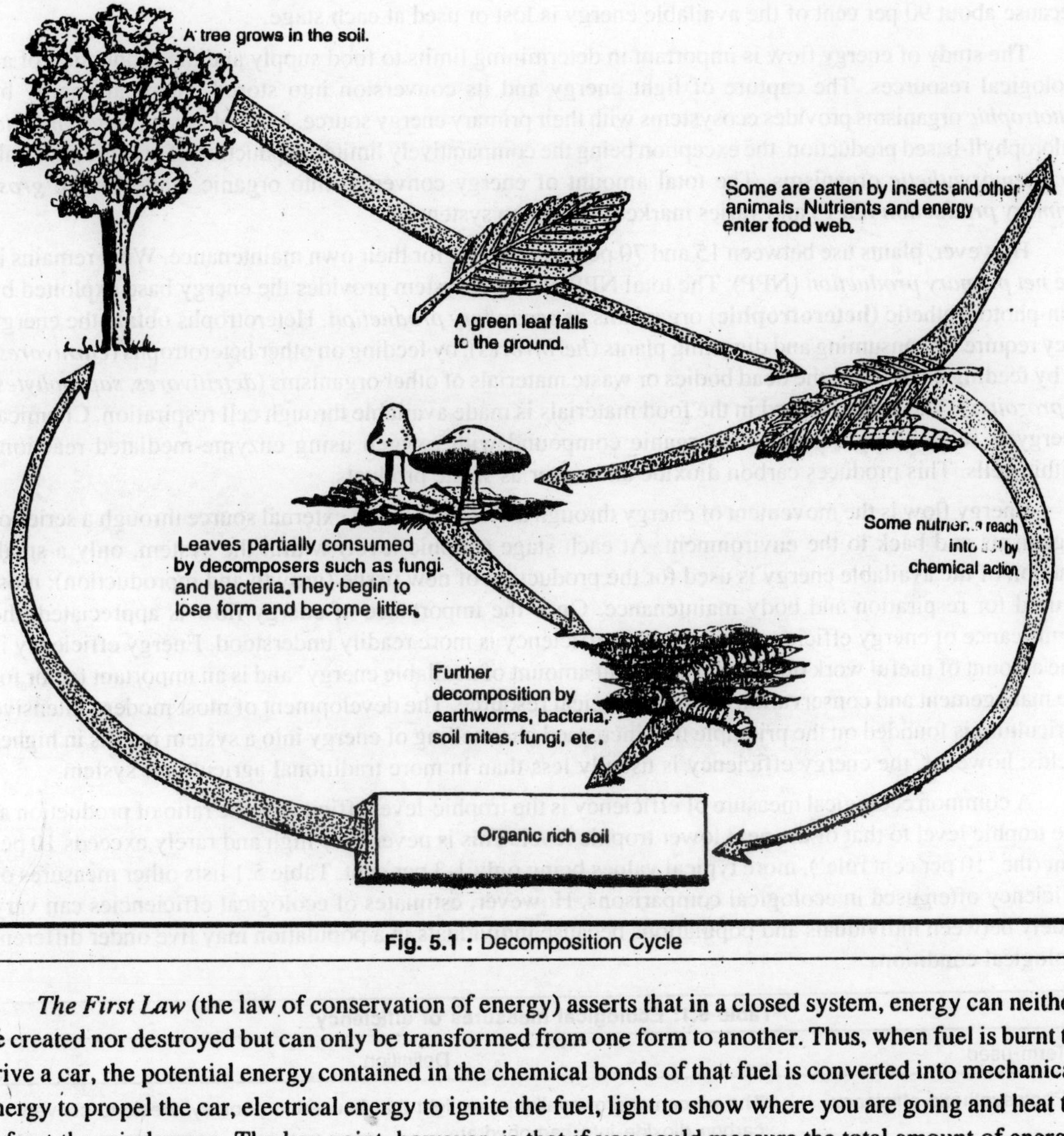

Fig. 5.1 : Decomposition Cycle

The First Law (the law of conservation of energy) asserts that in a closed system, energy can neither be created nor destroyed but can only be transformed from one form to another. Thus, when fuel is burnt to drive a car, the potential energy contained in the chemical bonds of that fuel is converted into mechanical energy to propel the car, electrical energy to ignite the fuel, light to show where you are going and heat to defrost the windscreen. The key point, however, is that if you could measure the total amount of energy consumed and compare it with the total amounts being produced in these various other forms, the two would be equal. Energy conversions such as these also take place in biological systems. Photosynthetic organisms such as plants capture and transform light energy from the sun and transfer this energy throughout the system subject only to the consequences of the Second Law.

The Second Law asserts that disorder (entropy) in the universe is constantly increasing and that during energy conversions, energy inevitably changes to less organised and useful forms, i.e., it is degraded. Think of this as energy always going from concentrated to less concentrated forms, the least useful (i.e., least concentrated) being heat energy. The consequences of this are very significant biologically. During each conversion stage, some energy is lost as heat. Therefore, the more conversions taking place between the capture of light energy by plants and the trophic (feeding) level being considered, the less the energy available to that level. The efficiency of the transfer along food chains is generally less than 10 per cent because about 90 per cent of the available energy is lost or used at each stage.

The study of energy flow is important in determining limits to food supply and the production of all biological resources. The capture of light energy and its conversion into stored chemical energy by *autotrophic* organisms provides ecosystems with their primary energy source. Most of this is *photosynthetic*, chlorophyll-based production, the exception being the comparatively limited production of organic materials by *chemosynthetic* organisms. The total amount of energy converted into organic matter is the *gross primary production* (GPP) and varies markedly between systems.

However, plants use between 15 and 70 per cent of GPP for their own maintenance. What remains is the *net primary production* (NPP). The total NPP of an ecosystem provides the energy base exploited by non-photosynthetic (**heterotrophic**) organisms as *secondary production*. Heterotrophs obtain the energy they require by consuming and digesting plants (*herbivores*), by feeding on other heterotrophs (*carnivores*) or by feeding on detritus, the dead bodies or waste materials of other organisms (*detritivores, saprophytes, saprozoites*). The energy stored in the food materials is made available through cell respiration. Chemical energy is released by burning the organic compound with oxygen using enzyme-mediated reactions within cells. This produces carbon dioxide and water as waste products.

Energy flow is the movement of energy through a system from an external source through a series of organisms and back to the environment. At each stage (trophic level) within the system, only a small fraction of the available energy is used for the production of new tissue (growth and reproduction); most is used for respiration and body maintenance. Once the importance of energy flow is appreciated, the significance of energy efficiency and transfer efficiency is more readily understood. Energy efficiency is 'the amount of useful work obtained from a unit amount of available energy' and is an important factor for the management and conservation of any biological resource. The development of most modern intensive agriculture is founded on the principle that increased channelling of energy into a system results in higher yields; however, the energy efficiency is usually less than in more traditional agricultural system.

A common ecological measure of efficiency is the trophic-level efficiency, the ratio of production at one trophic level to that of the next lower trophic level. This is never very high and rarely exceeds 10 per cent (the '10 per cent rule'), more typical values being only 1-3 per cent. Table 5.1 lists other measures of efficiency often used in ecological comparisons. However, estimates of ecological efficiencies can vary widely between individuals and populations because individuals in a population may live under different ecological conditions.

Table 5.1: Ecological measures of efficiency

Term used	Definition
Photosynthetic efficiency	The percentage of available sunlight that is used in photosynthesis to convert carbon dioxide into carbohydrate
Exploitation efficiency	The percentage of production at one trophic level that is ingested by the trophic level above it
Assimilation efficiency	The percentage of energy ingested that is actually absorbed by the body rather than egested
Growth efficiency	The percentage of assimilated energy that is used for growth rather than being respired or used for reproduction
Reproductive efficiency	The percentage of assimilated energy that is used for reproduction rather than being used for respiration or growth
Production efficiency	The percentage of assimilated energy that is used for production (growth and reproduction) rather than respiration
Trophic efficiency	The efficiency of energy transfer from a trophic level to the next one above

5.4 ECOSYSTEM PRODUCTIVITY

During the process of production and consumption, energy is passed along, or flows, from one organism to another. For example, solar energy is converted to chemical energy within the leaves of green plants. The leaves can then be eaten by some herbivore, and the herbivore may, in turn, be eaten by a carnivore. Consider a hypothetical ecosystem that receives 1,000 Kilocalories of light energy in a given day (Fig. 5.2). Most of this energy is not absorbed at all. Some is simply reflected back into space. Of the energy that is absorbed, most is stored as heat or used for evaporation of water. A small amount is assimilated by plants.

The productivity in an ecosystem are of two kinds—primary productivity and secondary productivity. The primary productivity of an ecosystem is the rate at which organic matter is produced during photosynthesis. Some amount of photosynthetic material is subsequently utilized for respiratory purpose. The initial photosynthetic product which is formed at first is called gross primary productivity which is expressed as Kcal/m^2/yr.

Only about half of the gross productivity accumulates as new plant matter, because the rest of the chemical energy is metabolized by the plant's own respiration and released to the environment as heat. The net gain in plant matter is called the net primary productivity.

Thus the **Net primary productivity = Gross primary productivity – Plant respiration**

On the whole, productivity depends on a variety of factors such as sunlight, temperature, rainfall and the availability of nutrients. The total quantity of organic matter present at any one time in an ecosystem is called the *biomass*. The biomass equals the total organic matter gained through net primary productivity over a period of time minus the quantity of material that is consumed and lost during respiration by animals. Let us return to the hypothetical ecosystem that receives 1,000 Kcal of sunlight (Fig. 5.2).

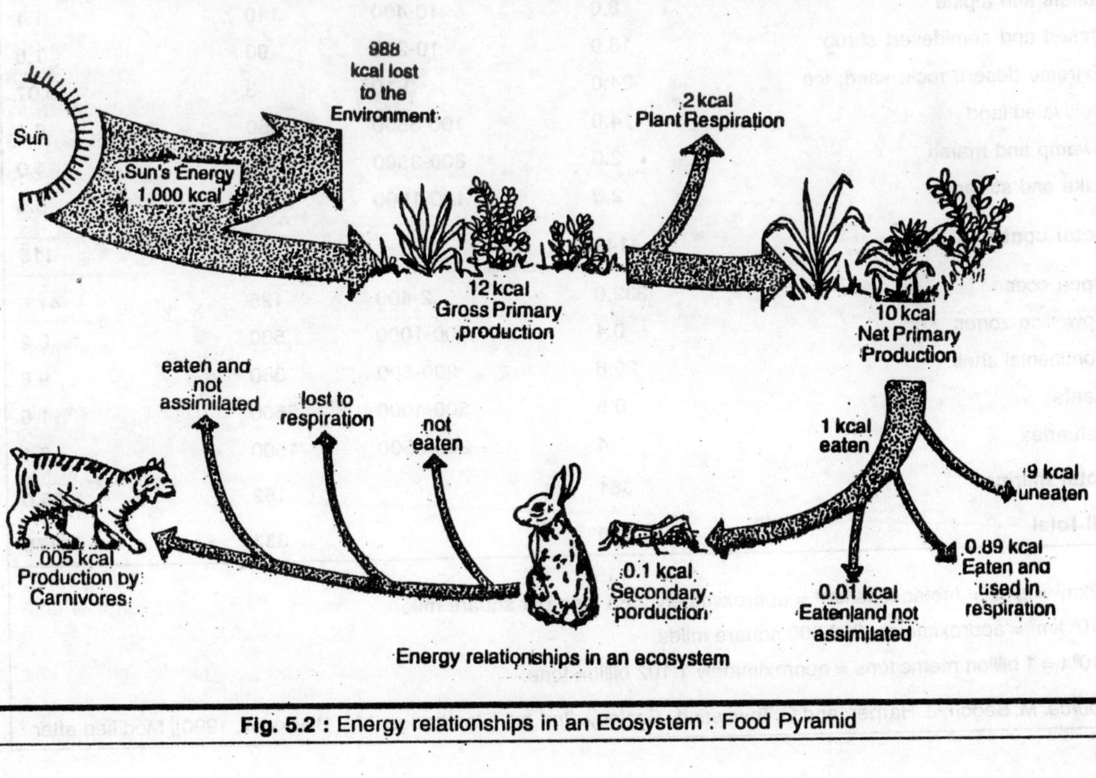

Fig. 5.2 : Energy relationships in an Ecosystem : Food Pyramid

Although the efficiency of energy transfer varies from ecosystem to ecosystem, as an average value, of the 1,000 Kcal absorbed, only about 12 Kcal are utilized during photosynthesis. Of these 12 Kcal, 2 Kcal are used for plant respiration and about 10 Kcal are stored in the plant tissue as energy-rich material, which animals can use for food.

1,000 Kcal of sunlight → 988 Kcal lost to the environment + 12 Kcal stored as plant tissue

The net primary productivity of various ecosystems varies distinctly. Among them tropical rain forests are most productive and rock, ice and sandy areas are least productive ecosystems (Table 5.2).

Table 5.2: Ecosystems and Productivity

Ecosystem Type	Appoximate Area (10^6 km²)[b]	Net Primary Productivity, Per unit area per year (g/m² or t/km² per year)[a] Normal Range	Mean	World Net Primary production (10^9t per year)[c]
Tropical rain forest	17.0	1000-3500	2200	37.4
Tropical seasonal forest	7.5	1000-2500	1600	12.0
Temperate evergreen forest	5.0	600-2500	1300	6.5
Temperate deciduous forest	7.0	600-2500	1200	8.4
Boreal northern forest	12.0	400-2000	800	9.6
Woodland and shrubland	8.5	250-1200	700	6.0
Savanna	15.0	200-2000	900	13.5
Temperate grassland	9.0	200-1500	600	5.4
Tundra and alpine	8.0	10-400	140	1.1
Desert and semidesert shrub	18.0	10-250	90	1.6
Extreme desert, rock, sand, ice	24.0	0-10	3	0.07
Cultivated land	14.0	100-3500	650	9.1
Swamp and marsh	2.0	800-3500	2000	4.0
Lake and stream	2.0	100-1500	250	0.5
Total continental	**149**		**773**	**115**
Open ocean	332.0	2-400	125	41.5
Upwelling zones	0.4	400-1000	500	0.2
Continental shelf	26.6	200-600	360	9.6
Reefs	0.6	500-4000	2500	1.6
Estuaries	1.4	200-3500	1500	2.1
Total marine	**361**		**152**	**55.0**
All total	**510**		**333**	**170**

[a] t/km² = g/m² = metric tons/km² = approximately 2.85 tons per square mile.

[b] 10^6 km² = approximately 386,000 square miles.

[c] 10^9 t = 1 billion metric tons = approximately 1.102 billion tons.

Source: M. Begon, J. Harper, and C. Townsend, *Ecology*, 2nd ed. [Cambridge, MA: Blackwell, 1990]. Modified after.

Table 5.3: Productive regions of the oceans	
Region	Mean productivity (mg C. m^{-2} . day^{-1})
Blue waters of subtropical gyres	70
Transitions	140
Equatorial divergence and subpolar	200
Inshore waters	340
Shallow shelves	1,000

The variation in global patterns of primary production illustrates the range and variety of ecosystem character and functioning. The main factors affecting global patterns of primary production are light, heat, water, carbon dioxide and oxygen, and nutrient elements.

Occupying most of the Earth's surface, the oceans provide an excellent example of how productivity varies with regional conditions. In the oceans there are very well-mapped and predictable patterns of primary productivity (Table 5.3). Russian oceanographic research work during the 1960s defined five main productive regions of the oceans. In general, productivity is highest where there is a strong circulation of water, with upwelling currents brining deep water to the surface. Nutrient concentrations tend to the greater in deep waters due to the continuous rain of material from shallow waters, down past the euphotic zone (where autotrophs can recycle the nutrients into the food chain). High productivity is commonest in shallow waters, where plants can grow to larger sizes.

Human impacts upon primary production are very important in many ecosystems. The broad pattern of natural environmental factors which control primary production at a global scale is shown in Table 5.4. This, however, provides only a generalised overview of ecosystem productivity (at biome level). The pattern of influence of the major limiting factors shown is frequently different at regional or local scales.

Table 5.4: Environmental controls on primary production		
Type of ecosystem	Primary production levels	Major limiting factors
Oceanic (open ocean)	Very low	Macro-nutrient supply
Coastal and oceanic upwellings	Very high	Light availability below immediate surface layer
Freshwater lakes	Low-moderate	Macro-nutrient supply
Tropical rainforest	Very high	Macro-nutrient supply
Savannah grassland	Intermediate	Seasonal water supply
Desert	Very low	Perennial water supply
Tropical wetland	Very high	Macro-nutrient supply
Temperate grassland	Intermediate	Seasonal heat conditions and water supply
Boreal forest	Intermediate	Seasonal heat conditions
Tundra	Very low	Perennial heat conditions

Secondary productivity is defined as the rate of formation of new organic matter by heterotrophs of the net primary productivity available in a forest, herbivores (e.g. insects, deer) eat only about 1-3%. In grassland, as much as 15% of the vegetation may be eaten by herbivores. In the ocean, 80% of the net primary productivity is consumed by herbivores. Very little of the plant matter that is consumed is actually

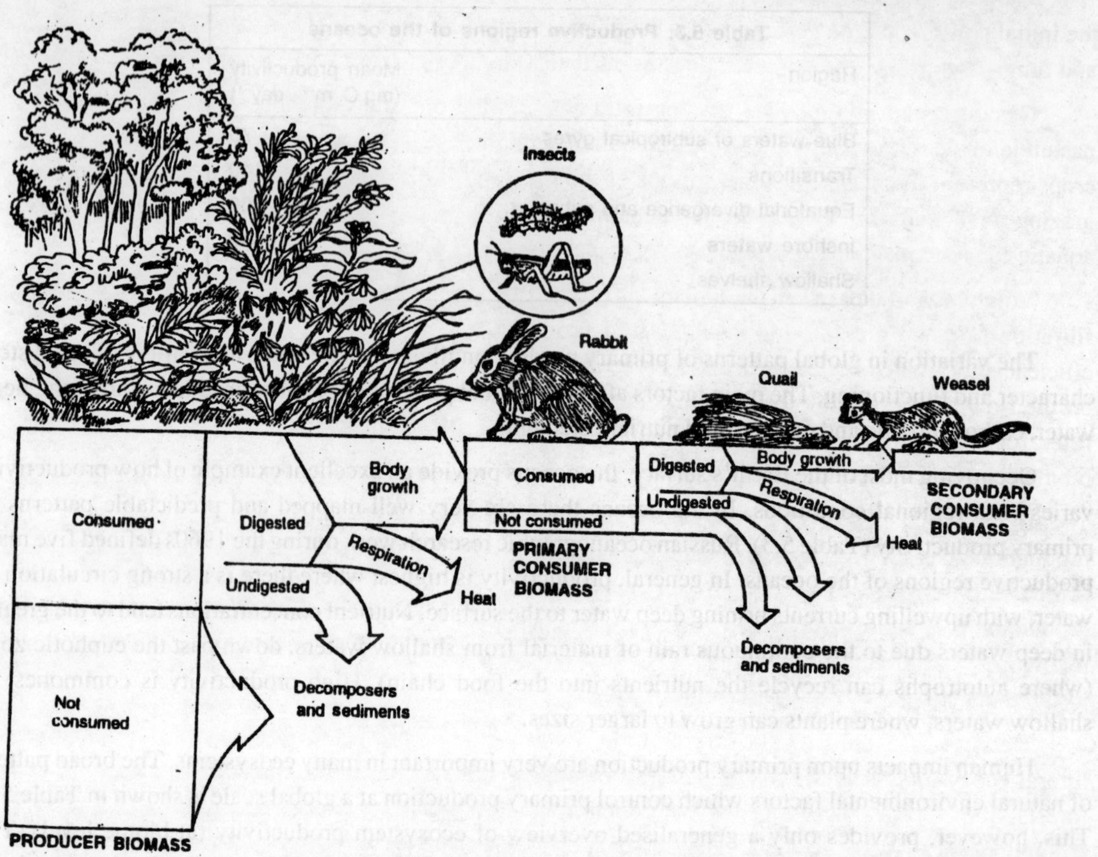

Fig. 5.3 : The Energy Pyramid

converted to animal tissue. In terms of energy content, the conversion is only about 10 per cent. To summarise, although there are large variations from ecosystem to ecosystem, as a generalisation, for every 10 Kcal of plant tissue available to herbivores, about 1 Kcal will be eaten, and only about 0.1 Kcal will be stored in the form of body weight. Carnivores that eat herbivores are likewise inefficient in converting food to body weight, so the energy available to the carnivore is even less. It is obvious, then, that the amount of usable energy decreases, as it is transferred from sunlight to plants to animals.

Thus there is a decreasing quantity of energy available at each trophic level. As a consequence, there will be a decreasing mass of organisms at each level. This relationship can be shown as Fig. 5.3 and called "energy pyramid".

5.5 FOOD CHAIN AND FOOD WEBS

The flow of food energy in an ecosystem progress through a *food chain* in which one step follows another—primary consumers eat producers, secondary consumers eat primary consumers, and so on.

Within any ecosystem there are two major food chains, the grazing food chain and the detrital food chain (Fig. 5.4). The two-food chains are distinguished by their source of energy or food for the initial consumers. In grazing ecosystems autotrophs, or living plant tissues, are the primary source of energy for the initial consumers, the herbivores. In the detrital food chain, the initial consumers, primarily bacteria and fungi, use dead organic matter, detritus, as their source of energy.

Grazing food chains: The grazing food chain is the one more obvious to us. Cattle grazing on pastureland, deer browsing in the forest, rabbits feeding in old fields, and insect pests feeding on garden crops represent the basic consumer groups of the grazing food chain. Although highly conspicuous, the grazing food chain is not the major food chain in terrestrial and many aquatic ecosystems. Only in some aquatic ecosystems do the grazing herbivores play a dominant role in energy flow.

Voluminous data exist on phytoplankton productivity, filtration rates by grazing zooplankton, and production efficiencies of zooplankton. Few data, however, are available on the flow of energy, rate of grazing, biomass turnover rates for phytoplankton, and turnover of zooplankton biomass within the same aquatic system.

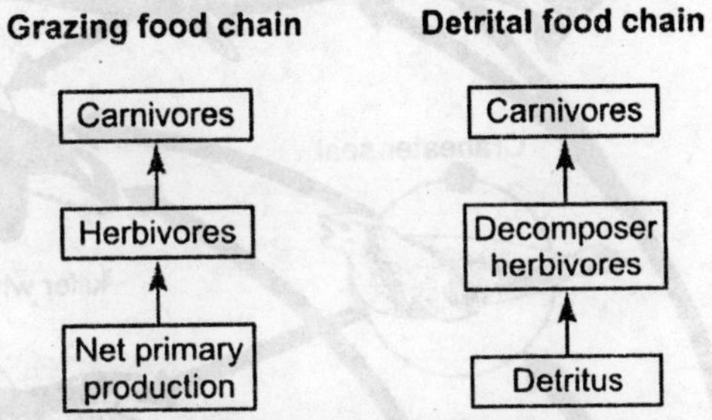

Fig. 5.4 : Two parts of any ecosystem: (a) a grazing food, and (b) a detrital food chain

In terrestrial systems, a small proportion of primary production goes by way of the grazing food chain. Over a three year period, only 2.6 per cent of net primary production of a yellow-poplar forest was used by grazing herbivores, although the holes made in the growing levels resulted in a loss of 7.2 per cent of photosynthetic surface.

Detrital food chains: The detrital food chain is common to all ecosystems, but in terrestrial and littoral ecosystems it is the major pathway of energy flow. In yellow-poplar (*Liriodendron tulipifera*) forests, 50 per cent of gross primary production goes into maintenance and respiration, 13 per cent is accumulated as new tissue, 2 per cent is consumed by herbivores, and 35 per cent goes to the detrital food chain (Edwards, unpublished, cited by O'Neill 1976). Two-thirds to three-fourths of the energy stored in a grassland ecosystem that is ungrazed by cattle is returned to the soil as dead plant material, and less than one-fourth is consumed by herbivores (Hyder 1969). Of the quantity consumed by hervivores, about one-half is returned to the soil as feces. In the salt march ecosystem, the dominant grazing herbivore, the grasshopper, consumes just 2 per cent of the net production available to it.

Supplementary food chains: Other feeding groups, such as the parasites and scanvengers, form supplementary food chains in the community. Parasitic food chains are highly complicated because of the life cycle of the parasites. Some parasites are passed from one host to another by predators in the food chain. External parasites (ectoparasites) may move from one host to another. Other parasites are transmitted by insects from one host to another through the blood-stream or plant fluids.

However, natural systems are rarely so orderly and linear. Many organisms occupy several trophic levels simultaneously. For example, a raven can be a primary consumer when it eats corn, a secondary consumer when it eats grasshoppers, and a tertiary consumer if it manages to catch a shrew or small snake. In addition, ravens will eat dead animals and are, therefore, also scavengers. In nearly all natural ecosystems, the patterns of consumption are so complicated that the term *food web* is more descriptive because there are many cross-links connecting the various organisms (Fig. 5.5).

Once food is eaten, its energy follows a variety of pattern through the organisms. Not all food can be fully digested and assimilated. Hair, feathers, insect exoskeletons, cartilage and bone in animal foods and cellulose and ligin in plant foods cannot be digested by most animals. These materials are either egested by defecation or regurgitated in pellets of indigested remains.

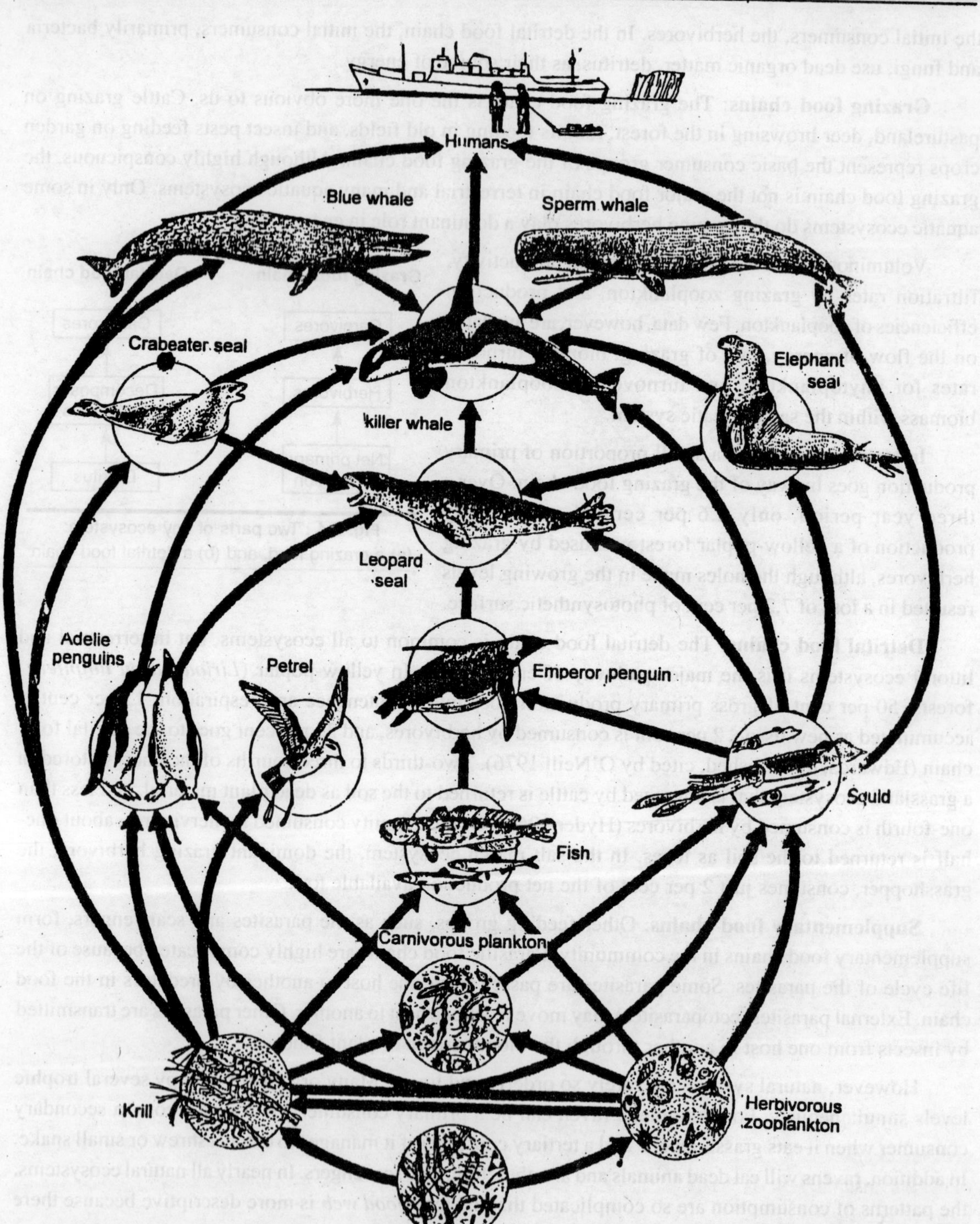

Fig. 5.5 : Simplified food web in the Antarctic

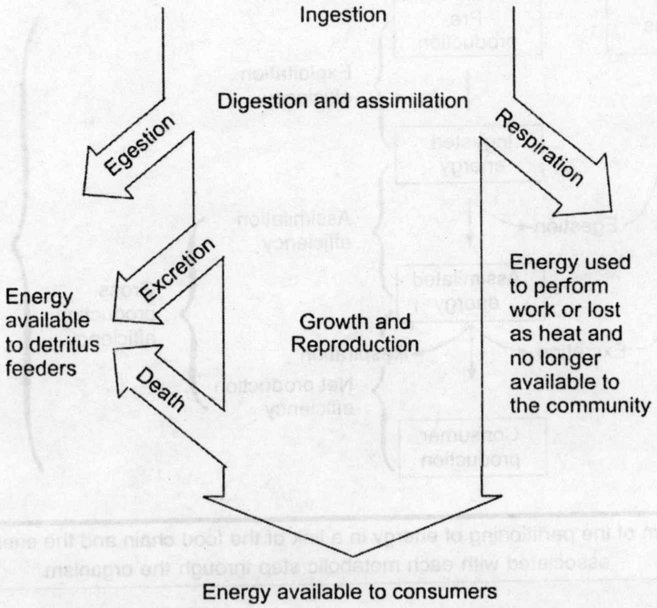

Fig. 5.6 : Partitioning of Energy within a link of a food chain

Organisms use most of the food energy that they assimilate into their bodies to fulfil their metabolic requirements: performance of work, growth and reproduction. Because biological energy transformations are inefficient, a substantial proportion of metabolized food energy is lost, unused, as heat. Organisms are no different from man-made machines in this respect. Most of the energy in gasoline is lost as heat in a car's engine rather than being transformed into the energy of motion. In natural communities, energy used to perform work or dissipated as heat cannot be consumed by other organisms and is forever lost to the ecosystem.

Assimilated energy that is not lost through respiration or excretion is available for the synthesis of new biomass through growth and reproduction. Populations lose some biomass by death, disease or annual leaf drop, where they enters the detritus pathways of the food chain. The remaining biomass is eventually consumed by herbivores or predators and its energy thereby enters the next higher trophic level in the community (Fig. 5.6).

5.6 ENERGETIC EFFICIENCIES

The movement of energy through the community depends on the efficiency with which organisms consume their food resources and convert them into biomass. This efficiency is referred to as the *food chain* or *ecological efficiency*. Ecological efficiencies are determined by both internal, physiological characteristics of organisms and their external, ecological relationships to the environment. To understand the biological basis of ecological efficiency, one must dissect the individual link of the food chain into its component parts (Fig. 5.7). Ecological efficiency depends on the efficiencies of three major steps in energy flow exploitation, assimilation and net production (Table 5.5). The product of the assimilation and net production efficiencies is the gross production from efficiency—the percentage of the exploitation and gross production efficiencies is the food chain or ecological efficiency—the percentage of available food energy in prey converted into consumer biomass.

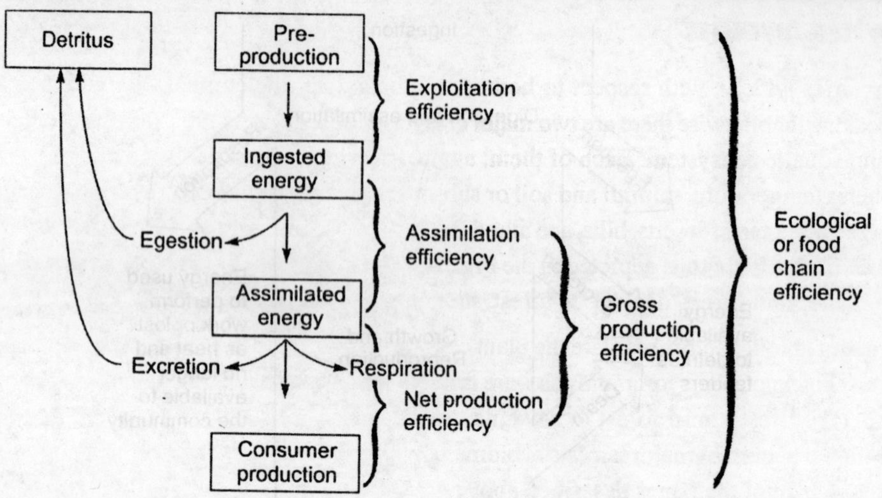

Fig. 5.7 : Diagram of the partitioning of energy in a link of the food chain and the energetic efficiencies associated with each metabolic step through the organism

Table 5.5: Depictions of various energetic efficiencies

1. Exploitation efficiency = $\frac{\text{Ingestion of food}}{\text{Prey production}}$
2. Assimilation efficiency = $\frac{\text{Assimilation}}{\text{Ingestion}}$
3. Net production efficiency = $\frac{\text{Production (growth of reproduction)}}{\text{Assimilation}}$
4. Gross production efficiency = 2 × 3 = $\frac{\text{Production}}{\text{Ingestion}}$
5. Ecological efficiency = 1 × 2 × 3 = $\frac{\text{Consumer production}}{\text{Prey production}}$

The measurement of energy flow through an entire community is a complex and virtually impossible task. Yet the total flux of energy and the efficiency of its movement determine the basic trophic structure of the community, number of trophic levels, relative importance of detritus and predatory feeding, steady state values for biomass and accumulated detritus and turnover rates of organic matter in the community. Each of these properties, in turn, influences the inherent stability of the community.

The study of energy transformation by plants and animals does, however, provide a useful insight into the basis of ecological efficiency and the trophic structure of the community. Plants are terrible predators as they harness light energy, animals, however, have different problems—they have feeding ability and thus consume organic mass and energy assimilation takes place thereafter. Energy flow between trophic levels represents the sum of the feeding activities of many species. Individual populations usually consume only a small fraction of available food resources. Carnivores, seed-eaters and aquatic herbivores are most efficient; the commonest species consume 10 to 100% of the food available to them. Terrestrial herbivores usually consume only 1 to 10% of the leafy vegetation.

5.7 ECOSYSTEM DIVERSITY

Ecosystem varies widely, with respect to both its biotic and abiotic composition and their interactions. Broadly speaking, habitatwise there are two major categories of ecosystem, viz., terrestrial or land ecosystem or biome and aquatic ecosystem. Each of them, again, is divided into several ecosystem types depending on atmospheric temperature, rainfall and soil or substrate nature. The habitat condition varies widely from coastal areas, to mountains, deserts, hills, and alluvial plains in both tropical and subtropical zones. Habitatwise, major land ecosystems are thus depicted in the Fig. 5.8. The major ecosystem types in terrestrial habitat are these—grassland, rain forest, deciduous forest, mountain, desert, coastal, alpine and glacier zones etc.

Two major abiotic factors regulate the plant community structure in terrestrial habitat viz., rainfall and temperature. The major plant communities were tropical rain forest, temperate rain forest, tropical seasonal forest, temperate forest, thorn forest or savanna, woodland or grassland, taiga, desert and tundra zones. The distributional pattern of major terrestrial biomes are given in Fig. 5.9. The overall topography and biotic composition of each of the zones described above are quite unique and distinguishable one from another (Fig. 5.10). In many cases human interference on such ecosystems leads to alteration of climate, landscape and so also the composition of species.

Though plants were widely adapted to climatic changes, yet major climatic factors govern the successional pattern of biological species in different habitat zones. Three distinctive successional profiles of plant communities is depicted in Fig. 5.11, to show the influence of climatic gradients on community formation.

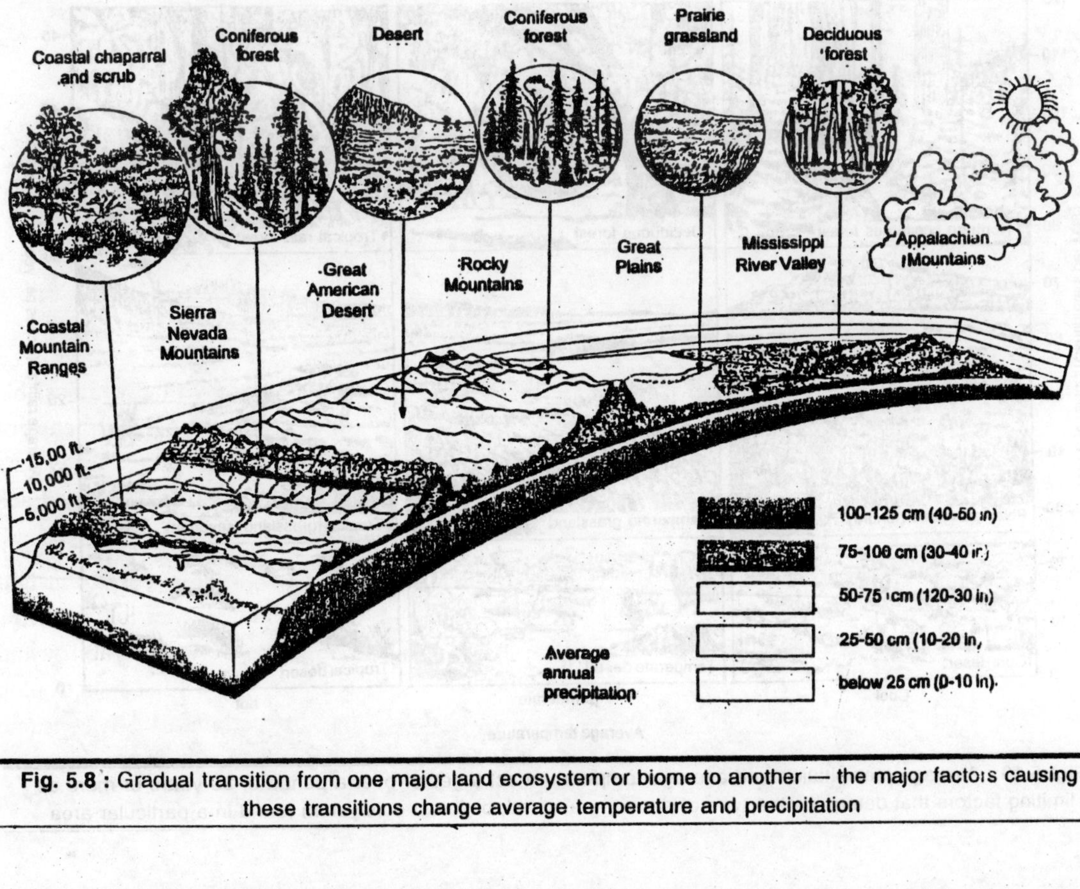

Fig. 5.8 : Gradual transition from one major land ecosystem or biome to another — the major factors causing these transitions change average temperature and precipitation

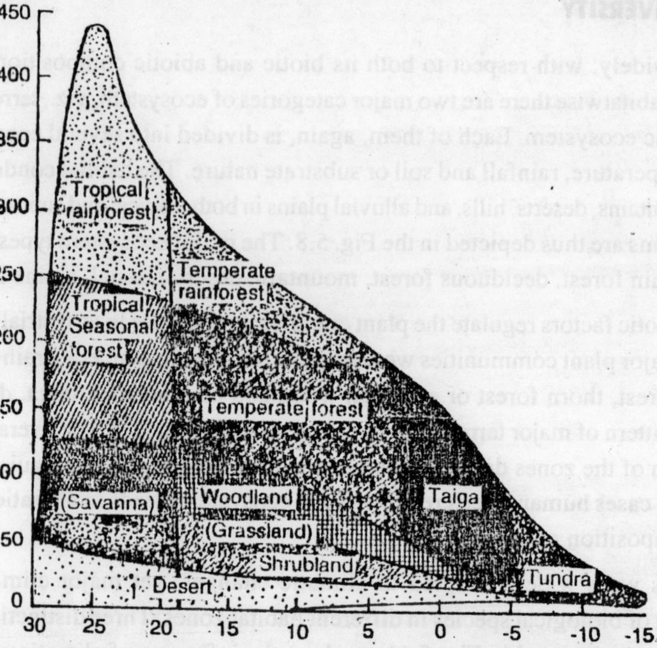

Fig. 5.9 : Simplified scheme of the major terrestrial biomes

Fig. 5.10 : Average precipitation and average temperature act together over a period of 30 years or more as limiting factors that determine the type of desert, grassland, or forest ecosystem found in a particular area

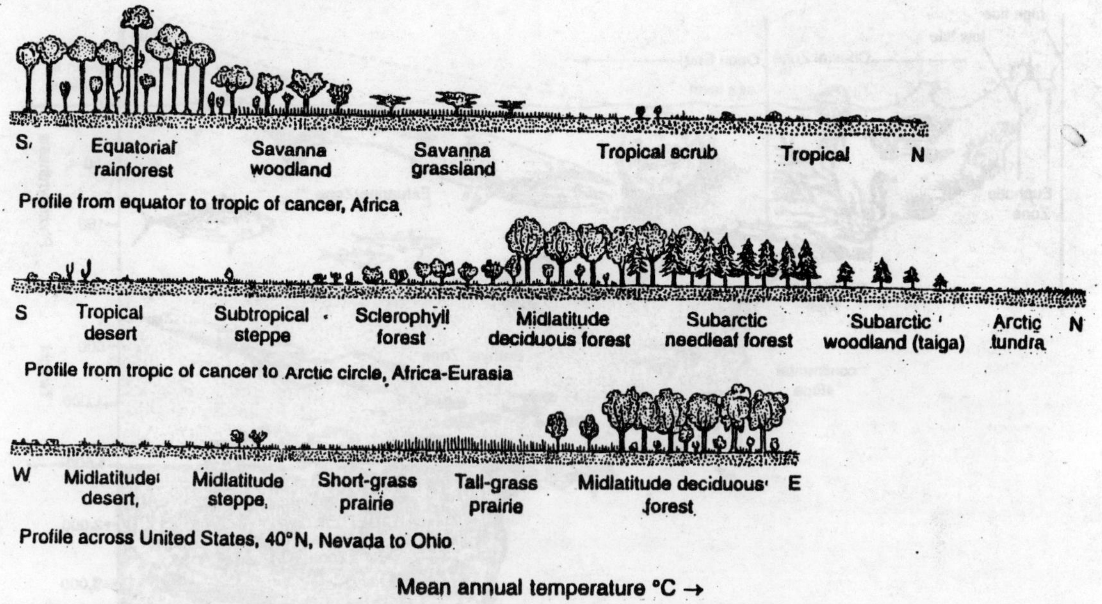

Fig. 5.11 : Three schematic profiles showing susing of plants

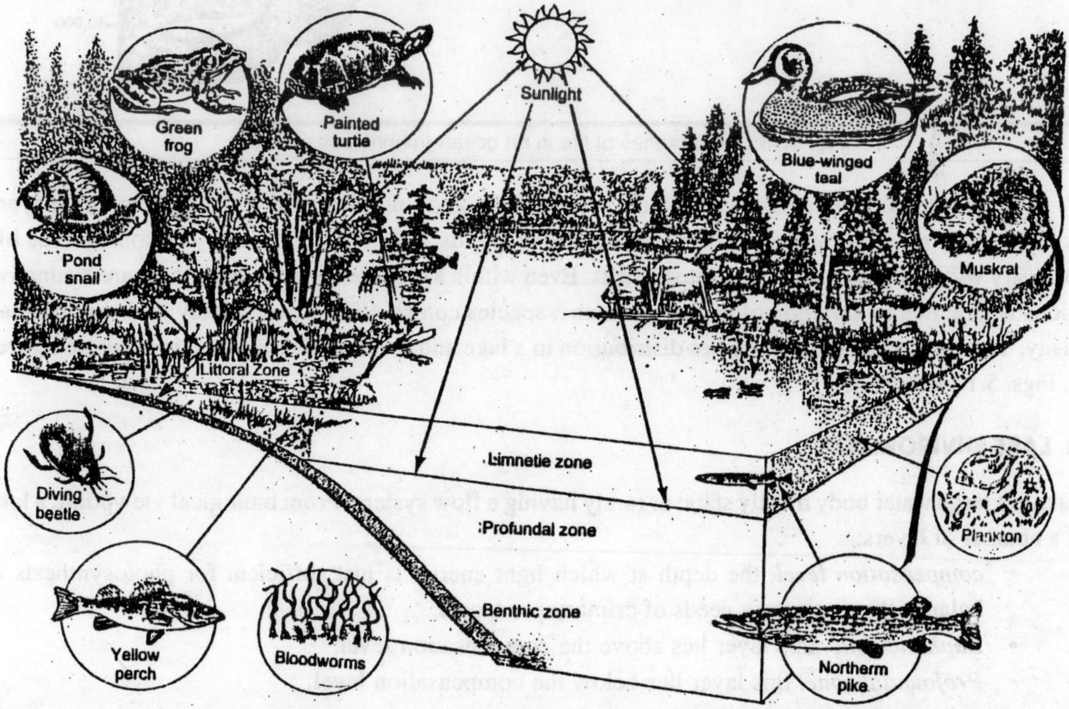

Fig. 5.12 : Major zones of life in a Lake

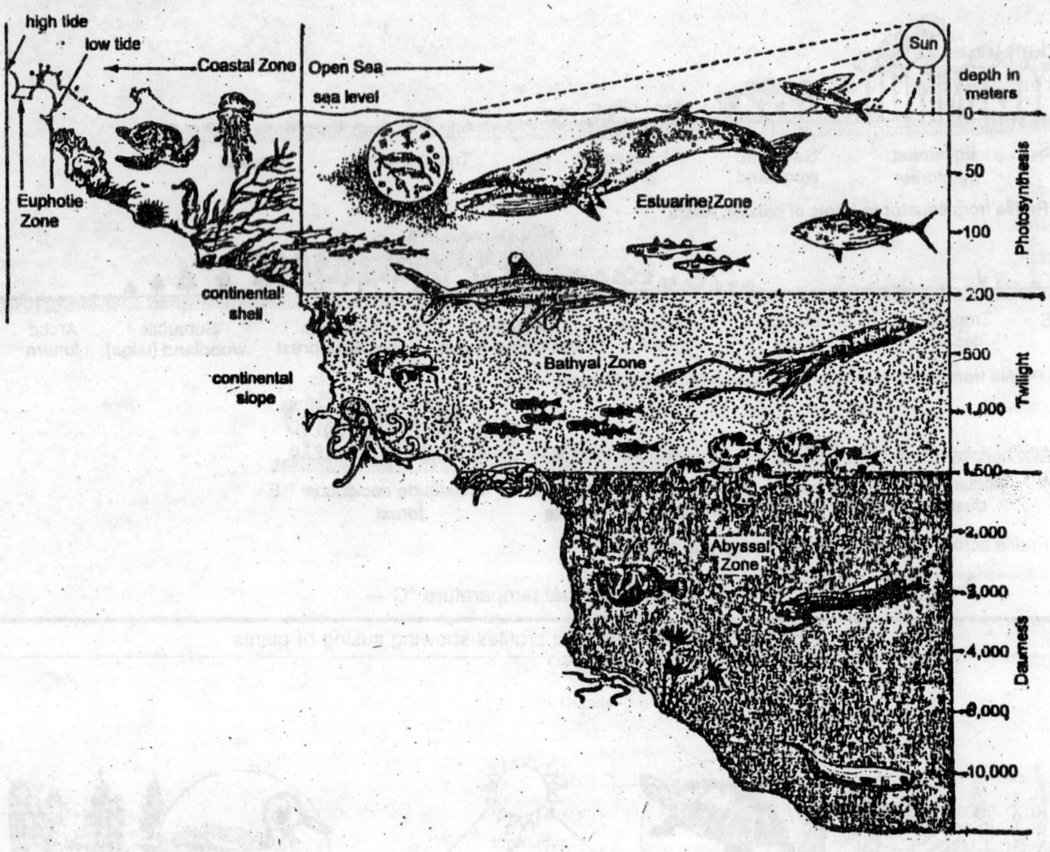

Fig. 5.13 : Major zones of life in an ocean (marine ecosystem)

With respect to aquatic ecosystem there were quite a good number of diversities. The major types were lake, open sea, coastal water, coral habitats, rivers, ponds, reservoir and swampy region etc. The life forms vary widely in each of the ecosystem zones. Even within a single ecosystem zone there are distinctive regions where, due to variation of ecological niche, species composition varies widely in diversities and density. Two clear examples of species distribution in a lake and in ocean are depicted in separate figures i.e., Figs. 5.12 and 5.13.

5.8 LAKE ENVIRONMENT

A lake is a large water body mostly static or rarely having a flow system. From biological viewpoint, a lake has a number of layers:

- *compensation level*, the depth at which light energy is just sufficient for photosynthesis to balance the respiratory needs of primary producers.
- *Euphotic zone*, this layer lies above the compensation level.
- *Profoundal zone*, this layer lies below the compensation level.

Within the *Euphotic zone*, again, there are two subdivisions—the shallow *littoral zone*, deeper *limnetic zone*. The phytoplanktons are mostly dominant in limnetic zone and abundant in littoral zone. There are occasional blooms also. Zooplanktons and swimming nektons are also abundant in limnetic zone.

The profoundal zone, by definition, lacks sufficient light per photosynthesis (i.e., <1% of total incident light), it is, therefore, dominated by consumers. This zone has greater amount of debris, i.e., more of organic load. The most productive lakes are those with a large littoral zone in relation to their volume. Here, sunlight reaches most of the lakes waters, providing an energy source for the primary producers in addition to significant warming during the growing season. Shallowness of the water allows the development of a large biomass of lightly productive rooted aquatic plants. Phytoplankton blooms are characteristic because of high inorganic nutrient concentrations produced as bottom micro-organisms degrade large volumes of organic matter. Since profoundal waters have low oxygen concentrations, stagnation of bottom water is relatively frequent. Such shallow, highly productive lakes are termed *Eutrophic lake*.

In contrast, *Oligotrophic lakes* are much less productive. An oligotrophic lake is usually deep and steep-sided, with a narrow littoral zone. Concentrations of inorganic nutrients are low, and so is phytoplankton density. Blooms are rare, for the intense competition for nutrients keep population levels low. Dissolve oxygen is high in the hypolimnion; its large volume, low temperatures and low input of organic matter serve to keep oxygen depleting microbes in check. Cold water bottom pushes such a lake trout or landlocked salmon find suitable habitats in the profoundal zone.

5.9 MATERIAL CYCLE IN ECOSYSTEMS

Matter in organisms and ecosystems serve two functions:

- *First,* matter can serve to store chemical energy as carbohydrate, protein and fats.
- *Second,* matter can serve to make up physical structures that support the biochemical activities of life.

Life is only possible with molecules that intercept and transform energy from one form to another. Life also requires molecules that contain and provide the physical and chemical environment necessary for those energy transforming processes. As molecules are formed and reformed by chemical and biochemical reactions within an ecosystem, the atoms that compose them are not changed or lost. Matter can thus be conserved within an ecosystem, and atoms and molecules can be used and reused or cycled within ecosystems.

Atoms and molecules move through ecosystems under the influence of both physical and biological processes. The pathways of a particular type of matter through the earth's ecosystem comprise a material cycle (may also be referred to as biogeochemical cycle or nutrient cycle).

The living world depends on the flow of energy and the circulation of matter through ecosystems. Both influence the abundance of organisms, the rate of their metabolism, and the complexity and structure of the ecosystem. Energy and matter flow through the ecosystem together as organic matter; one cannot be separated from the other (Fig. 5.14). The link between energy and matter begins in the process of photosynthesis. Solar energy is utilized in the fixation of CO_2 into organic carbon compounds. Organic matter, the tissues of plants and animals, is composed not only of carbon, but a variety of essential nutrients.

There are two types of material cycle—the gaseous cycle and the sedimentary cycle. In the gaseous cycle, the element or compound can be converted to a gaseous form, diffuse through the atmosphere, and they arrive over land or sea, to be reused by the biosphere, in a much shorter time. The primary constituents of living matter—carbon, hydrogen, oxygen and nitrogen—all move through gaseous cycle. In the sedimentary cycle, the compound or element is released from rock by weathering, then follows the movement of running water either in solution or as sediment to the sea. Eventually, by precipitation and sedimentation, these materials are converted into rock. When the rock is uplifted and exposed to weathering, the cycle is completed (Fig. 5.15).

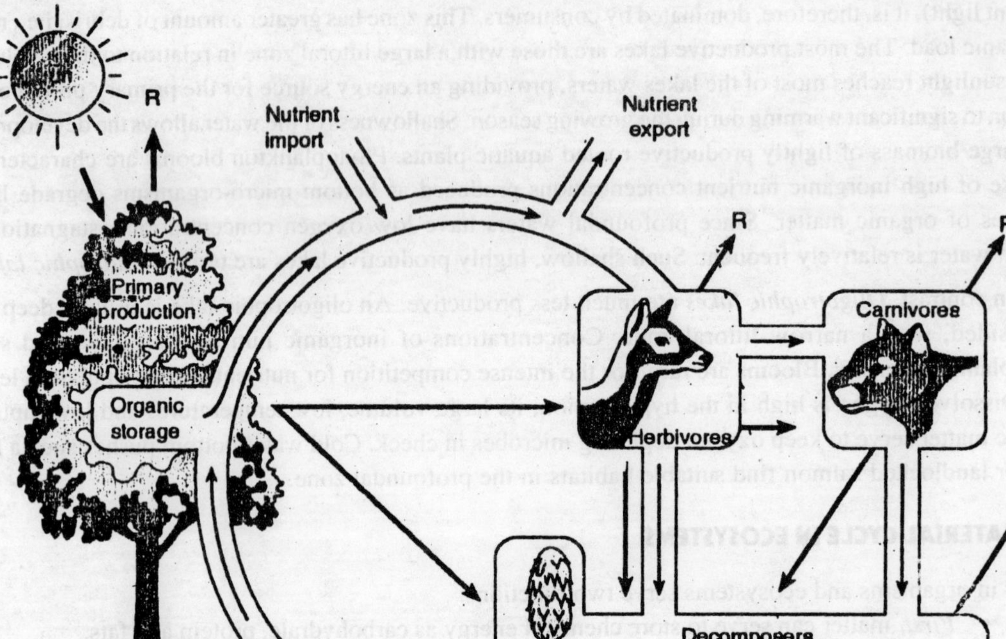

Fig. 5.14 : The interrelationship between nutrient cycling and energy flow in the ecosystem

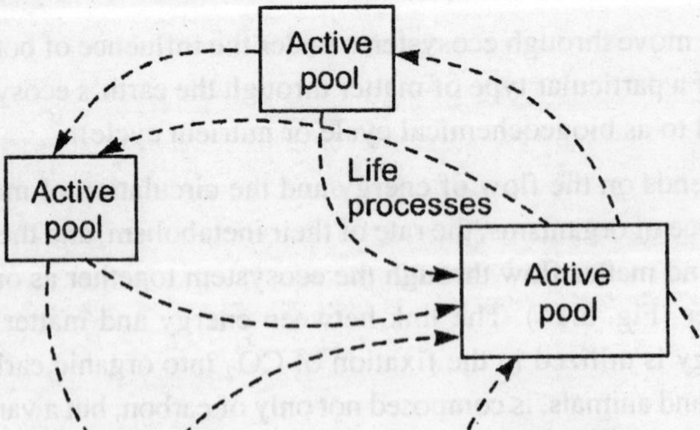

Fig. 5.15 : Material cycle

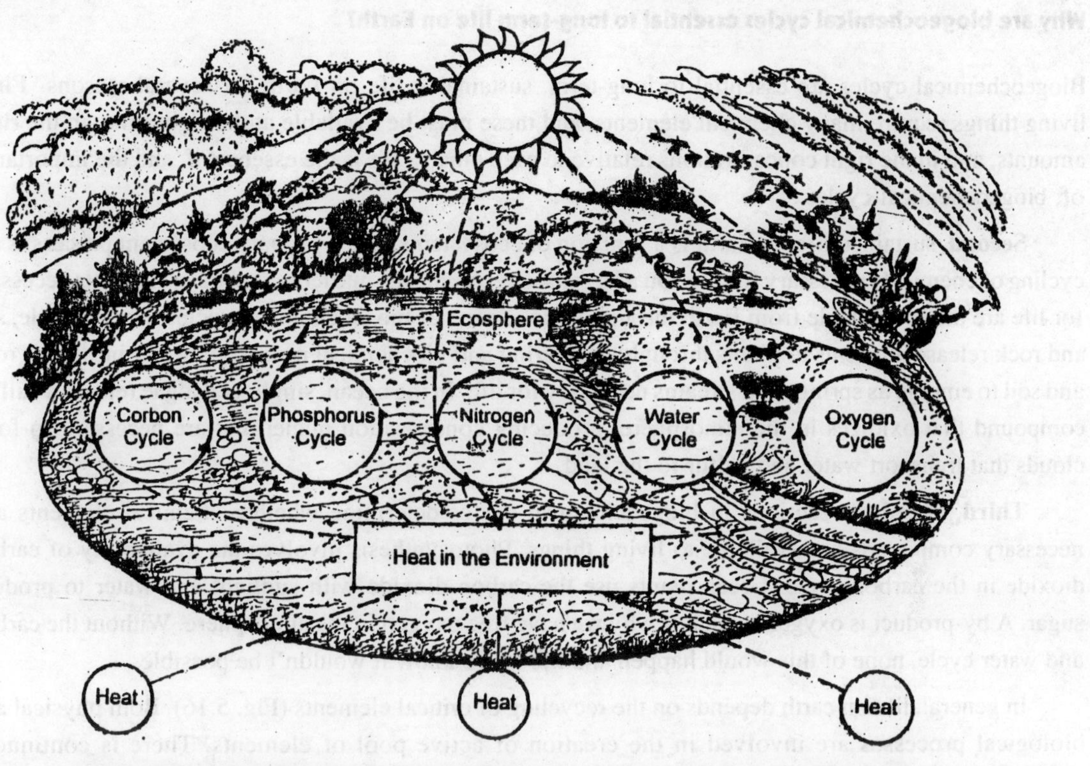

Fig. 5.16 : Life on earth depends on the recycling of critical chemicals (solid lines) and the one-way flow of energy through the ecosphere (Dashed lines)

The major features of a material cycle are given in Fig. 5.16, where there are two types of pools: *active pools*, i.e., materials are in forms and places easily accessible to life processes; and *storage pools*, i.e., materials are more or less inaccessible to life. A system of pathways of material flows connects the various active and storage pools within the cycle. Pathways between active pools are usually controlled by life processes, whereas pathways between storage pools are usually controlled by physical processes.

Elemental Cycle: Living organisms require several essential elements—either in large quantities or in trace quantities. The major essential elements are C, H, O, N, P, K, Ca, Mg and S. These elements are available to organisms in the form of a variety of different chemical compounds. For example, nitrogen as N_2, Carbon as CO_2 and Oxygen as O_2 are present in the atmosphere. Water (H_2O) is present throughout the biosphere in its liquid, solid or vapour form. Compounds of phosphorus, sulfur and other elements are incorporated into the chemical structure of various rocks. Under proper weathering process these elements are released as available elements in soil and water. On the whole, the movements of nutrients within an ecosystem are called *biogeochemical cycles.*

Nutrients are sometimes recycled rapidly through ecosystems, as in grasslands, where the above-ground vegetation dies back each year and its nutrients are made available again the following season. In other cases, nutrients may spend many years or millennia away from the activities of the biological world. For example, coal or oil deposits were formed from partially decayed organic matter that has been isolated from the biosphere for millions of years. Every nutrient element has a somewhat different fate, depending on its physical and chemical properties and its role in living organisms.

Why are biogeochemical cycles essential to long-term life on Earth?

Biogeochemical cycles are essential to long-term, sustainable life on Earth for several reasons. **First**, living things require many chemical elements, and these must be available at the right time, in the right amounts, and in the right concentrations relative to each other. This is the essence of, and the importance of, biogeochemical cycles.

Second, sustained life on Earth is a function of ecosystems, and a primary ecosystem process is the cycling of chemicals necessary for life. On a larger scale, chemicals, nutrients, and trace elements necessary for life are made available from Earth through various parts of biogeochemical cycles. For example, soil and rock release nutrients to plants through weathering and biochemical processes; water infiltrates rock and soil to emerge as springs and streams necessary for life. In the ocean, single-cell algae release a sulfide compound that oxidizes in the atmosphere, producing condensation nuclei that are necessary to form clouds that transport water and sulfur to the land.

Third, chemical reactions in biogeochemical cycles determine whether chemical elements and necessary compounds are available to living things. Photosynthesis involves the availability of carbon dioxide in the carbon cycle. Green plants use the carbon dioxide with sunlight and water to produce sugar. A by-product is oxygen, which is why we have free oxygen in the atmosphere. Without the carbon and water cycle, none of this would happen and life as we know it wouldn't be possible.

In general, life on earth depends on the recycling of critical elements (Fig. 5.16). Both physical and biological processes are involved in the creation of active pool of elements. There is continuous transformation of elements from storage pool to active pool, or *viceversa.*

The major elemental cycles are described separately:

(a) ***Carbon cycle*:** The carbon cycle is very complex due to the fact that carbon can exist in a wide variety of different types of compounds, occupying in plants, animals, rocks, liquid, sediments and air. The most important carbonaceous materials are CO_2, CO, carbonates, and organic carbonaceous materials. In the atmosphere 0.03 per cent CO_2 gas is present. Major amount of carbon materials lie as dissolved CO_2 in fresh water and saline water. They form carbonate or bicarbonate and thereby make deposition as sediment or crustal deposits like limestone and marble stone. However a major amount of CO_2 is fixed as organic carbon by the green plants through photosynthesis in both terrestrial and aquatic ecosystem. A portion of them are deposited as fossil fuel like coal and oil. On the other hand, CO_2 is liberated back to atmosphere from burning of fuel, respiration of biota, volcanoes, and decomposition of organic waste. An overall scheme of carbon cycle is given in Fig. 5.17.

In some situations, CO_2 cycle is very rapid as in plants which fix CO_2 within an hour by photosynthesis, and are also capable of releasing the same by respiration in same rapidly. But non-biological processes or complex biological and abiotic processes leading to carbonate rock formation, or fossil organic fuel deposit creation requires millions of years. In modern times, human activities are making a significant impact on the global carbon cycle. When fossil fuels are burned, geological reserves of carbon are converted into CO_2 and released into the atmosphere. Today the relase of CO_2 or discharge of CO_2 into the atmosphere is steadily increased, owing both to the burning of fossil fuels and to the destruction of forests. A simplified model of carbon flow in the environment is given in Fig. 5.18.

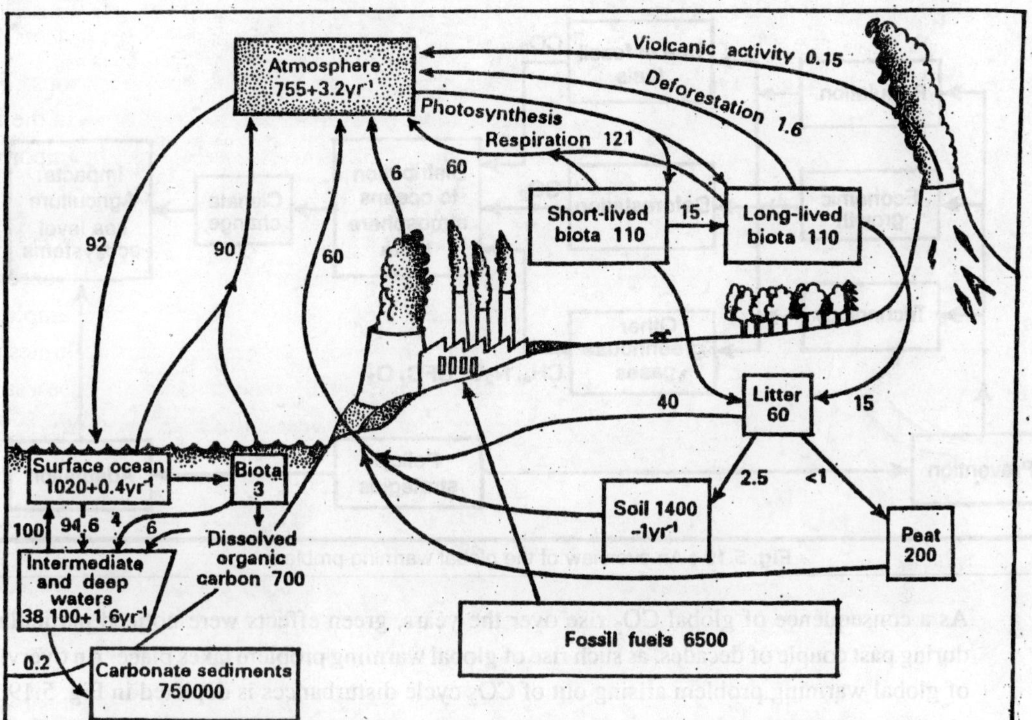

Fig. 5.17 : Global carbon cycle (units in boxes Gt. Yr^{-1}) is the reservoirs of carbon. Figures beside arrows denote flux rates in Gt. Yr^{-1}

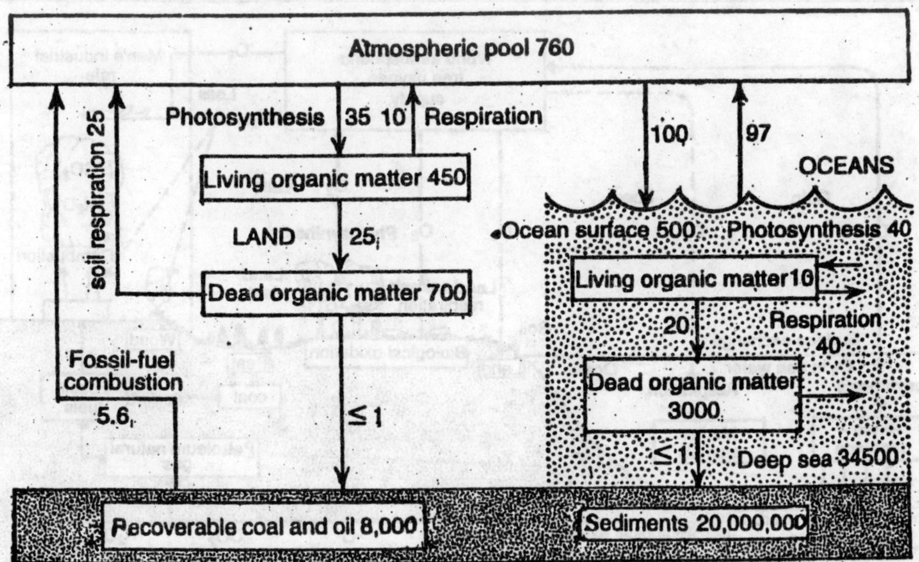

Fig. 5.18 : Simple model of carbon flow (quantities are either in billions of tons of carbon (G + C) for standing crops or billions of tons per year

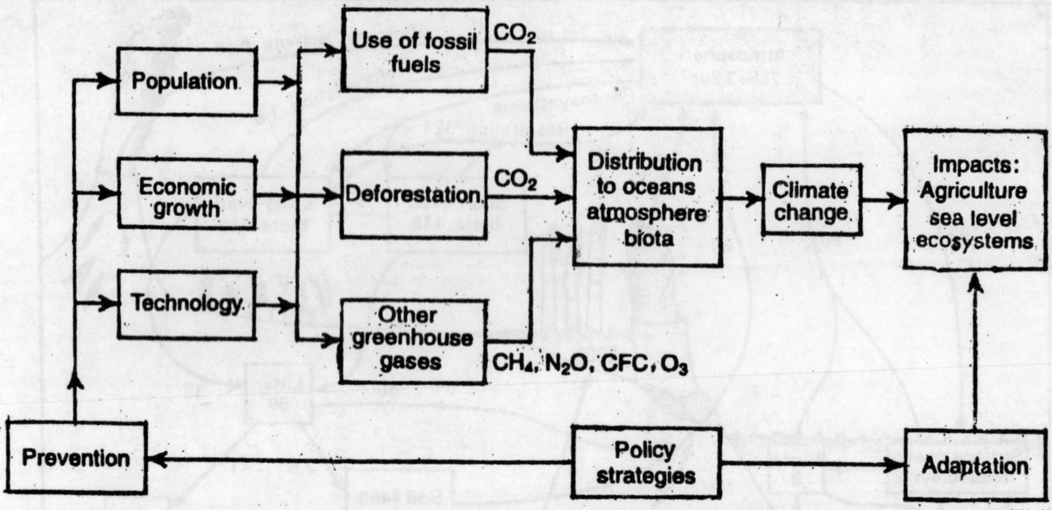

Fig. 5.19 : An overview of the global warming problem

As a consequence of global CO_2 rise over the years, green effects were noticed particularly during past couple of decades, as such rise of global warming problem takes place. An overview of global warming problem arising out of CO_2 cycle disturbances is depicted in Fig. 5.19.

(b) ***Oxygen Cycle*:** This is one of the major gaseous components of atmosphere. The oxygen cycle involves movements and storages in different forms and its oxidative reactions with various elements. The world supply of O_2 is shown in Fig. 5.20.

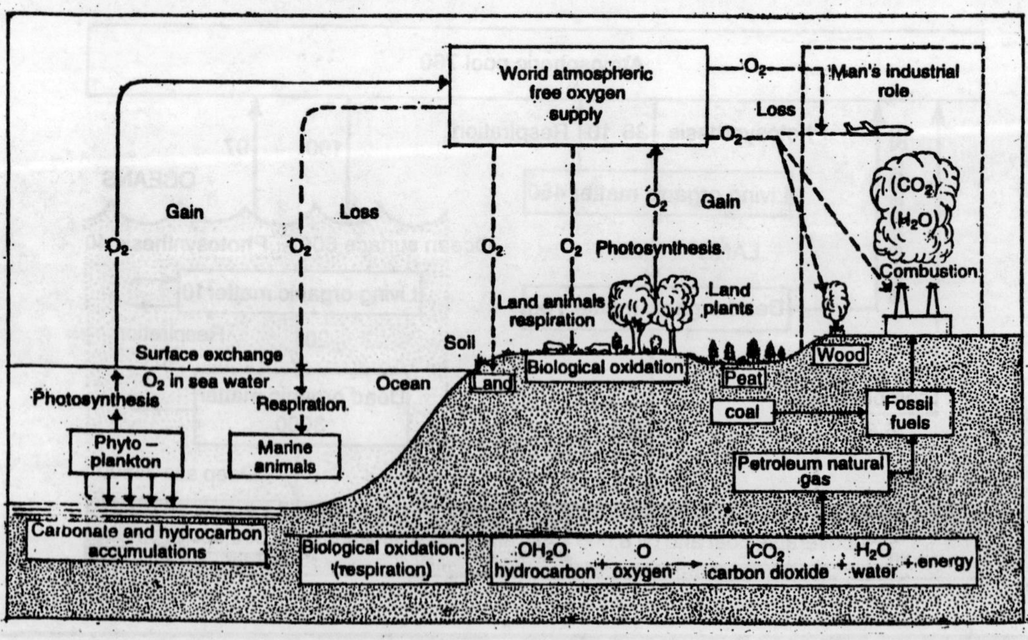

Fig. 5.20 : The oxygen cycle

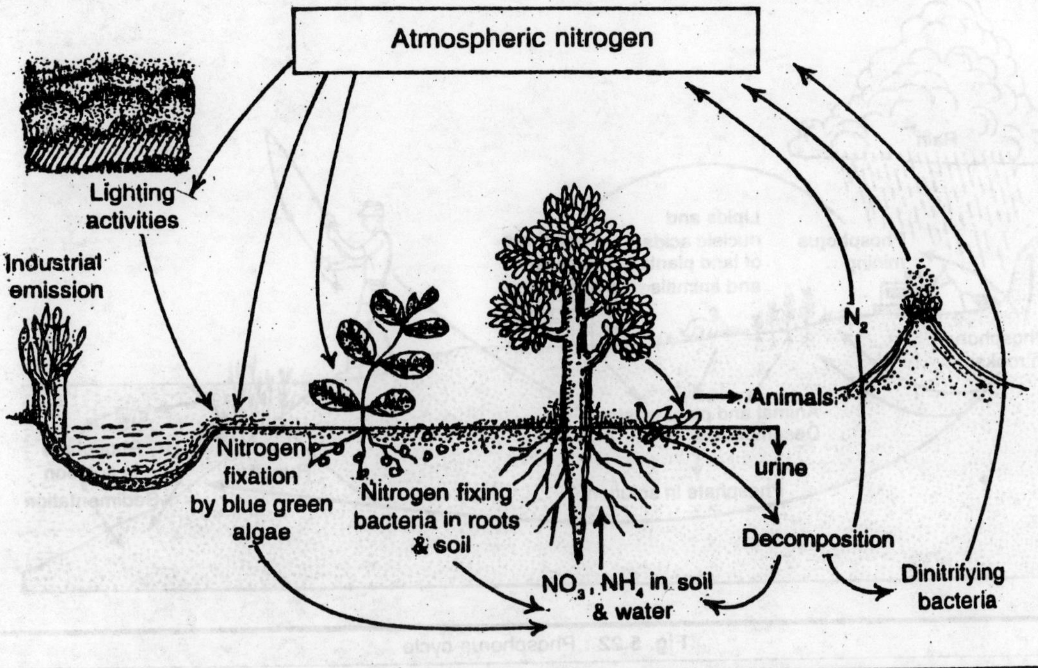

Fig. 5.21 : The Nitrogen cycle

Oxygen enters this storage pool through release in photosynthesis—both in the oceans and on the lands. Balancing the input to the atmospheric storage pool is loss through organic respiration and mineral oxidation. The oceans serve as a major storage pool for dissolved gaseous oxygen. Some oxygen is continuously placed in storage in mineral carbonate form in ocean floor sediments.

(c) ***Nitrogen Cycle***: Nitrogen is one of the very important elements in biosphere. It remains in various forms in different environmental components, viz., as gases N_2 and NO_x: as ions in soil and water NO_3, NO_2^-, NH_4^+, or as organic material or insoluble nitrogenous materials. Among these, atmosphere possesses 78% of nitrogen (by volume). Even though atmosphere has such a high amount of nitrogen, the growth of many organisms is limited by a shortage of nitrogen. This shortage occurs because very few organisms can utilize the atmospheric nitrogen (N_2) directly. Most plants can absorb this vital nutrient only when it is in the form of nitrites (NO_2^-), nitrates (NO_3^-) or ammonium ions (NH_4^+). Only some selected bacteria—either in symbiotic or free living state—fix atmospheric nitrogen to NH_3 and then utilize the same for amino acid biosynthesis. The animals obtained the nitrogen from plants as food materials. The organic nitrogen thus available as extra from plants and animals or dead remains of organisms is decomposed, which may release N_2 gas, NO_X gas, NO_3 or NO_2^- or NH_4^+ ions. Various processes are thus involved in such cycling of nitrogen from environmental components to biological communities (Fig. 5.21).

(d) ***Phosphorus Cycle***: Phosphorus is a major constituent of biological membranes, energy rich compounds and nucleic acids. In addition, many animals also need large quantities of this element to make shells, bones and teeth. Since phosphorus does not occur naturally as gas, its cycle, unlike those of carbon and nitrogen, is a ***sedimentary cycle***. Many rocks contain phosphorus, usually in the form of phosphates (PO_4^-) that are bound into the mineral structure.

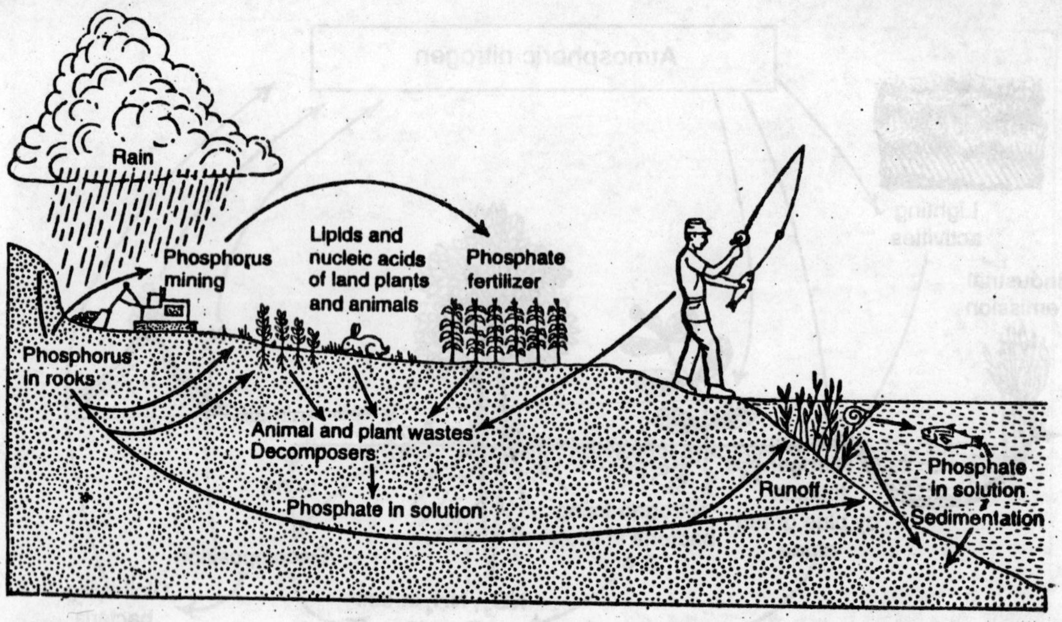

Fig. 5.22 : Phosphorus cycle

When rocks are weathered, minute amounts of these phosphates dissolve and become available to plants. Animals then absorb this element when they eat plants or other animals. Much of the phosphorus excreted by animals is also in the form of phosphate, which plants can reuse immediately. Thus, on land, phosphate cycles circulate from plants to animals and back again. Land ecosystems preserve phosphorus efficiently, since both organic and inorganic soil particles absorb phosphates, providing a local reservoir of this element.

In an undisturbed ecosystem, the intake and loss of phosphorus are small compared with the amounts of phosphorus that are internally recycled in the day-to-day exchange among plants and animals. Some phosphorus is inevitably lost by leaching and erosion of the soil into stream and rivers. When an ecosystem is disturbed, as by mining or farming, erosion can become so significant that large quantities of phosphorus and other nutrients are washed away. When phosphate reaches the ocean, it reacts with other minerals and sediment at ocean floor. This is why in some open seas, available open sea have low phosphate level. An overall scheme for phosphorus cycle is given in Fig. 5.22.

(e) ***Sulfur Cycle***: Sulfur is one of the essential elements of life system and its major reservoir lies in the earth's crust together with oceanic sediment. It acts as a component of several compounds like amino acids, protein, enzymes etc. Natural fuel like coal and oils also have some sulfurs (1 to 3%). Burning these materials leads to formation of the gas SO_2. The reactions of this gas with other compounds in the atmosphere are extremely complex, but one of the important products of its solution and further oxidation is sulfuric acid, together with sulfurous acid (both of which occur in combination with rain) forms *acid rain*. An overall scheme of global sulphur cycle showing major reservoirs and the flux rates between these are presented in Fig. 5.23. (All values expressed as Tg (g × 10^{12} or tonnes × 10^6) sulphur.)

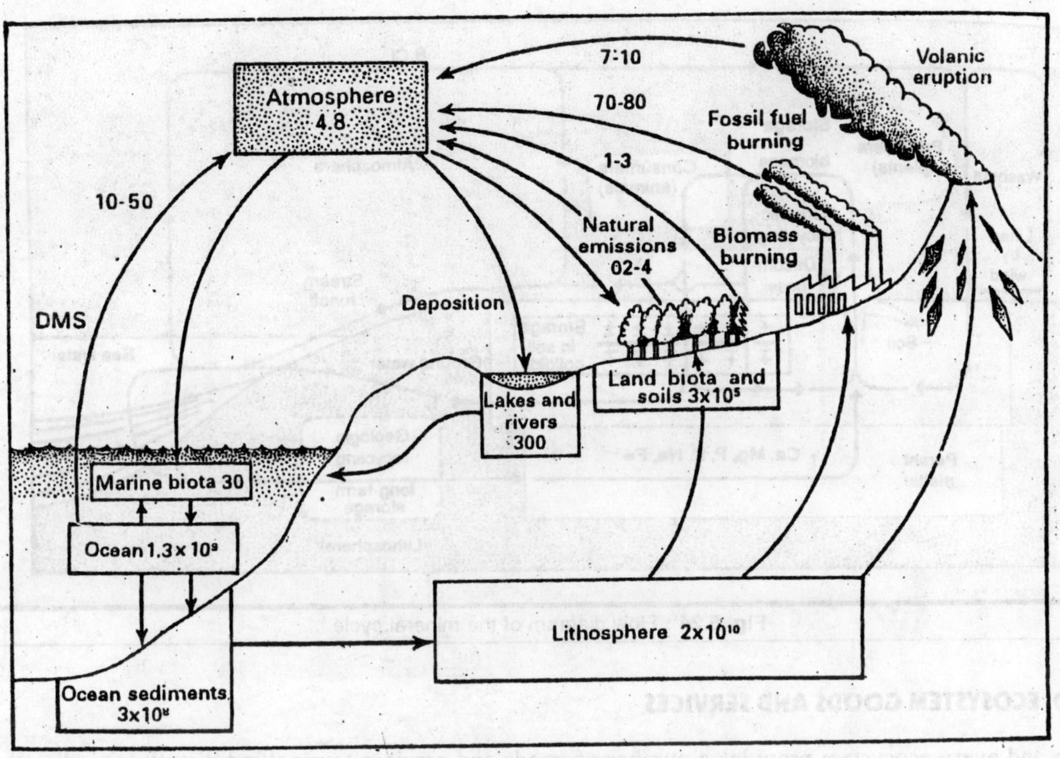

Fig. 5.23 : Sulphur cycle

There is a strong analogy between sulfur and nitrogen in the way that micro-organisms influence their biogeochemical cycling. When organic sulfur compounds are decomposed by bacteria, the initial excreted sulfur product is often hydrogen sulfide, H_2S. Many marine phytoplanktons produce compounds that break down to produce dimethyl sulfide $(CH_3)_2S$. Dimethyle sulfide is a major biogenically produced sulfur compound, releasing about 20–40 Tg Sa^{-1} from the oceans.

$$\underset{\text{Organic hydrosulfide}}{R\text{–}SH} \xrightarrow{\text{Bacteria}} H_2S + \underset{\text{Organic compound}}{RH}$$

(f) ***Mineral Cycle***: Quite a good number of minerals are essential for biological life processes. They are either used as major or minor essential elements. They are mostly stored in rocks and soils (i.e., lithosphere). They are absorbed by the plants and used in metabolism, while animals obtain them from plants as food sources. They may be available in water bodies as soluble ions. In organic combination they may be stored in biomass. In particulate matters, minerals are also present in substantial amounts and are thereby absorbed by plants. A schematic flow diagram of mineral cycle is given in Fig. 5.24.

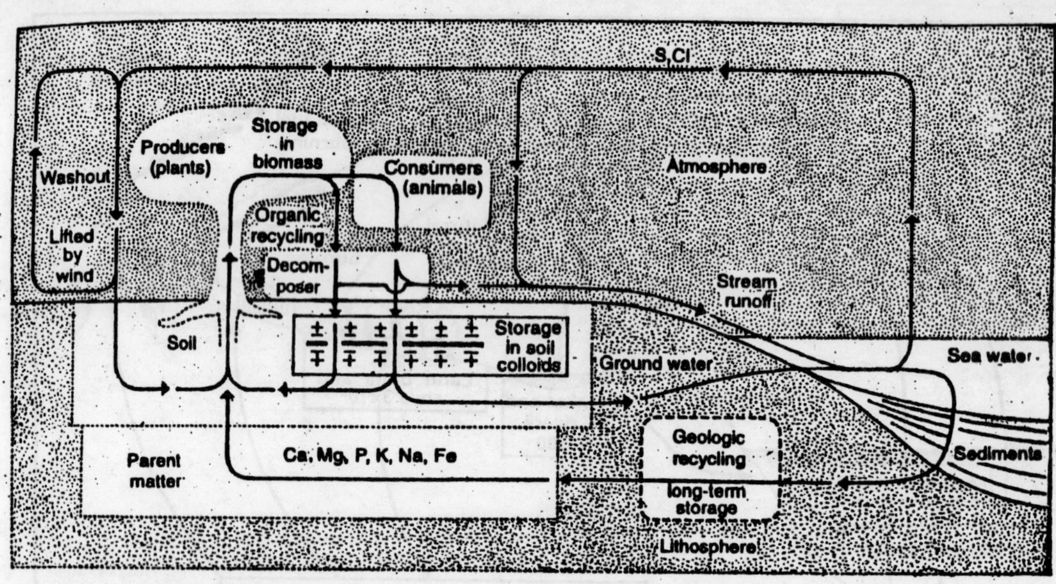

Fig. 5.24 : Flow diagram of the mineral cycle

5.10 ECOSYSTEM GOODS AND SERVICES

Each and every ecosystem provides a number of goods and services to mankind directly or indirectly. These includes material supply like food, fiber, fuel, medicines and biochemicals; or it helps in control of flood, erosion, habitat protection, waste recycling, recreation and so on. An example of provisioning of goods and services of estuarine ecosystem is shown in Table 5.6:

Table 5.6: Goods and services of estuarine ecosystem

	Rank	Biotic contributors	Abiotic regulators
Provisioning Services			
Plants as food	–1 to 1	grazers, pathogens	oxygen, circulation, substrate, nutrients
Animals as food	–3 to 3	fish, invertebrates, all zoo- and phytobenthos, pathogens	oxygen, circulation, substrate
Other biological products	–3 to 3	bait worms	sediment type
Biochemical/medicine/ models for human research	2	microbes, natural products, invertebrate models	temperature, chemical availability, sediment type
Fuels/energy	1	microbes, peat, mangroves	temperature, time
Fiber	1	sponges, sea grasses, mangroves	water flow, sediment type
C sequestration	1	microbes, peat	sediment type, redox
Nonliving materials (geological effects)	2	bioturbators, microbes, infauna, maerl, shellgravels	hydrodynamic processes, sediment type
Clean seawater	3	seagrasses, saltmarsh plants, biofiltration, bioturbators	hydrodynamic processes, redox, sediment type

Contd.

Table 5.6: Goods and services of estuarine ecosystem

	Rank	Biotic contributors	Abiotic regulators
Regulation Services			
Sediment formation: biodeposition	–1 to 3	microbes, lithothamnia, biogenic sediments, vegetation, filter feeders, infauna	hydrodynamic processes, fresh water & land runoff, sediment type
Nutrient cycling	3	microbes, bioturbators, macro-fauna, fishes, phytobenthos	hydrodynamic processes, temperature
Biological control: disease, invasive species resistance	?	?	oxygen, eutrophication, sedimentation, salinity
Detoxification, waste disposal	– 3 to 3	microbes, zoo- & phytobenthos, biofilters & bioturbators	circulation, resuspension, sedimentation
Climate regulation (C sequestration)	1	bioturbators, microbes, infauna, mobile fauna	hydrodynamic processes, upwelling, resuspension, sedimentation
Food web support processes	3	entire benthos	hydrodynamic processes, upwelling, resuspension, sedimentation, oxygen
Atmosphere composition	2	microbes, kelps, wetlands	oxygen, substrate, turbulent mixing wind
Flood and erosion control	3	vascular plants, biostabilizers	
Redox processes	3	bioturbators, microbes	oxygen
Habitat maintenance Services			
Landscape linkages & structure/habitat/refugia	3	vegetation, biogenic reefs, migrating fauna	oxygen, temperature, depth, substrate
Aesthetic Services			
Spiritual / cultural	3		
Aesthetic	3		
Recreation	3		
Scientific understanding	3	ecological paradigms & education	

5.11 HOMEOSTASIS ECOSYSTEM

Ecosystems are capable of self-maintenance and self-regulation as are their component populations and organisms. Thus *cybernetics*—the science of controls, has important application in ecology, especially since man increasingly tends to disrupt natural controls or attempts to substitute artificial mechanisms for natural ones. 'Homeostasis' is the term generally applied to the tendency for biological systems to resist change and to remain in a state of equilibrium. However, the control depends on 'feedback', either in positive or negative ways. The positive feedback is deviation accelerating, while negative feedback is deviation counteracting. Both natural and man-made activities induced the changes in the ecosystem, but regulated feedback mechanism tends to resist the same; thus, somewhat, homeostasis is maintained.

In every ecosystem, there are three basic components: inputs, internal cycling and output, those are influenced by soil & climate.

The **input** of nutrients to the ecosystem depends on the type of biogeochemical cycle. Nutrients with a gaseous cycle, such as carbon and nitrogen, enter the ecosystem via the atmosphere. In contrast, nutrients such as calcium and phosphorus have sedimentary cycles, with inputs dependent on the weathering of rocks and minerals.

Primary productivity in ecosystems depends on the uptake of essential mineral (inorganic) nutrients by plants and their incorporation into living tissues. Nutrients in organic form, stored in living tissues, represent a significant proportion of the total nutrient pool in most ecosystems. As these living tissues senescence, the nutrients are returned to the soil or sediments in the form of dead organic matter. Various microbial decomposers transform the organic nutrients into a mineral form, a process called mineralization, and the nutrients are once again available to the plants for uptake and incorporation into new tissues. This process is called **internal cycling** and is an essential feature of all ecosystems. It represents a recycling of nutrients within the ecosystem.

The export of nutrients i.e., **output** from the ecosystem represents a loss that must be offset by inputs if a net decline is not to occur. Export can occur in a variety of ways, depending on the nature of the specific biogeochemical cycle. Carbon is exported to the atmosphere in the form of CO_2 via the process of respiration by all living organisms. Likewise, a variety of microbial and plant processes result in the transformation of organic and inorganic nutrients to a gaseous phase that can subsequently be transported from the ecosystem in the atmosphere. Examples of these processes will be provided in the following sections, which examine specific biogeochemical cycles.

A generalised model is shown in Fig. 5.25. The feedback systems as available in internal cycling balances the homeostasis of ecosystem.

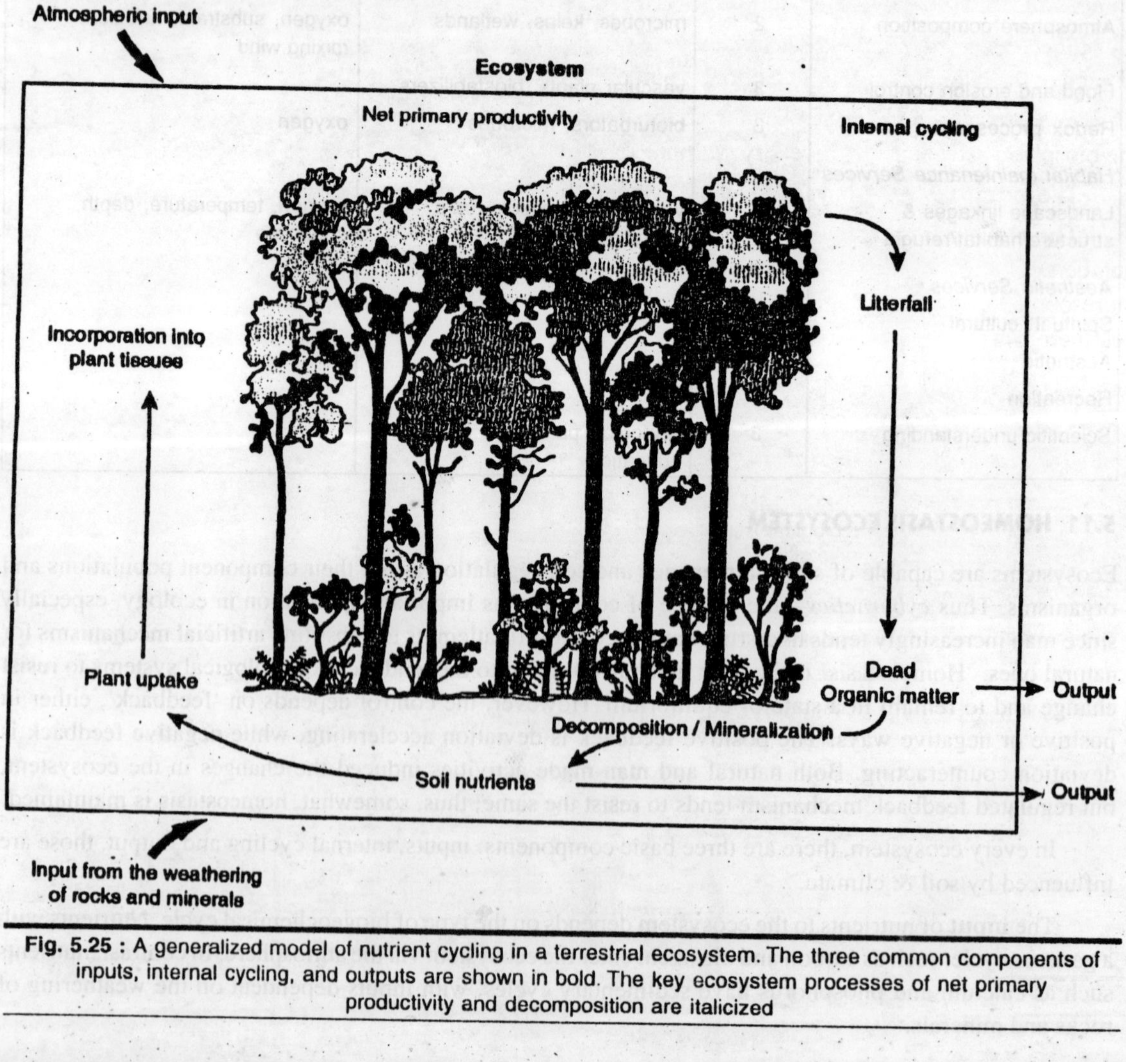

Fig. 5.25 : A generalized model of nutrient cycling in a terrestrial ecosystem. The three common components of inputs, internal cycling, and outputs are shown in bold. The key ecosystem processes of net primary productivity and decomposition are italicized

6 BIOLOGICAL RESOURCES (BIOLOGICAL DIVERSITY): ASSESSMENT AND MANAGEMENT

6.0 BIOLOGICAL DIVERSITY

6.1 CONCEPT

The term *biological diversity*, commonly shortened to 'biodiversity', is used to describe the number, variety and variability of living organisms. Thus, it may be described in term of genes, species and ecosystems, corresponding to three fundamental levels of biological organisation:

Genetic diversity is the total genetic information contained in the genes of individuals of plants, animals and micro-organisms. Each species is the repository of an immense amount of genetic information.

Species diversity implies the variability of population with respect to the reproductive distinctiveness of each of the individuals. It is a taxonomic unit of the biological world.

Ecosystem diversity relates to the variety of habitats, biotic communities and ecological processes in the biosphere as well as the diversity within ecosystems.

The overall composition and level of biological diversity is given in Fig. 6.1.

The global biodiversity wealth is vast and only a small fraction is known as yet. The approximate numbers of known species of plants and animals are listed in Table 6.1. With respect to global wealth, Indian bioresources are also considered to be fairly rich, many of which appears to be endemic (Table 6.2). But, as yet, there is lack of proper assessment of microbes and invertebrates in diverse ecosystems.

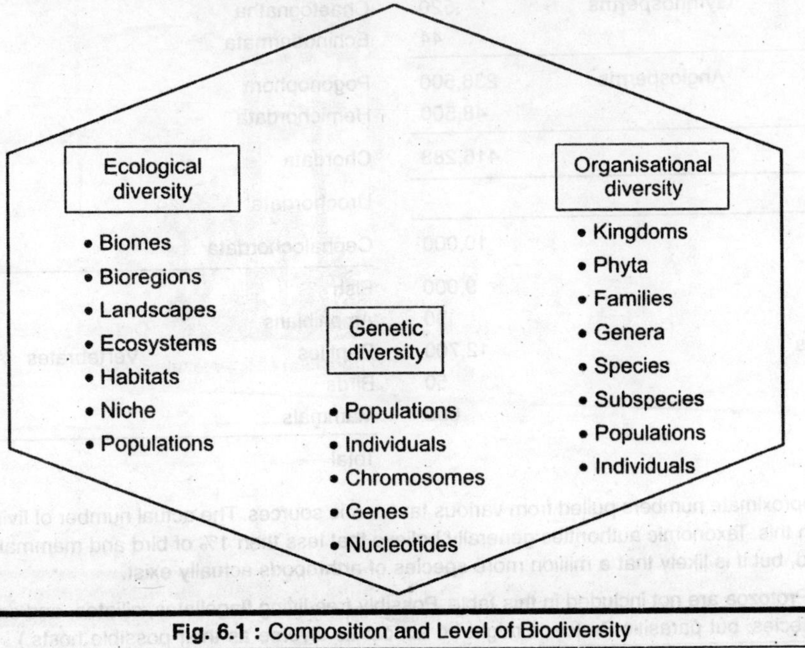

Fig. 6.1 : Composition and Level of Biodiversity

In general, the Indian subcontinent is floristically richer than its fauna. Many cultivated species also originated from this region. Among the various ecosystems, Western Ghats and Eastern and North-eastern Himalayas appear to be very important habitats for high floristic diversity, i.e., "hot spot" region.

Table 6.1: Living Species of Plants and Animals, 2000

Plants		
Bacteria		1,600
Slime molds		500
Fungi		100,000
Blue-green algae	Algae	1,500
Euglenids		450
Golden algae, diatoms		425
Yellow-green algae		5,800
Dinoflagellates		1,000
Brown algae		1,500
Red algae		4,000
Green algae		7,000
Liverworts	Bryophytes	
Mosses		23,000
Hornworts		
Psilopods	Ferns and Allies	3
Lycopods		1,200
Horsetails		40
Ferns		10,000
Cycads	Gymnosperms	100
Ginkgos		1
Conifers		520
Gnetum		44
Dicots	Angiosperms	236,500
Monocots		48,500
Total		416,283
Animals		
Porifera		10,000
Coelenterata		9,000
Ctenophora		90
Platyhelminthes		12,700
Mesozoa		50
Rhynchocoela		650
Animals (continued)		
Gnathostomulida	Invertebrates	43
Rotifera		1,500
Gastrotricha		175
Kinorhyncha		64
Nematoda		10,000
Nematomorpha		230
Acanthocephala		500
Entoprocta		60
Priapulida		8
Sipunculida		250
Mollusca		128,000
Echiurida		60
Annelida		8,700
Terdigrada		350
Onychophora		65
Arthropoda		923,000
Pentastomida		70
Phoronida		70
Bryozoa		4,000
Brachiopoda		260
Chaetognatha		50
Echinodermata		5,300
Pogonophora		80
Hemichordata		80
Chordata		39,000
Urochordata		1,600
Cephalochordata		1
Fish	Vertebrates	30,700
Amphibians		2,600
Reptiles		6,500
Birds		8,600
Mammals		4,060
Total		1,208,366

(*These are approximate numbers pulled from various taxonomic sources. The actual number of living species must be greater than this. Taxonomic authorities generally believe that less than 1% of bird and mammal species remain to be described, but it is likely that a million more species of arthropods actually exist.

Heterotrophic Protozoa are not included in this table. Possibly free-living flagellates, ciliates, and sarcodina number about 1,000 species, but parasitic Protozoa might be almost as diverse as their possible hosts.)

Table 6.2: Estimated species richness & endemism : Plants & vertebrates in India, 1998

Categories	Estimated no. of species	No. of Endemism
(A) Plants:		
(i) Flowering plants	17,548	5,783
(ii) Ferns	1,200	193
(B) Vertebrates:		
(i) Mammals	390	42
(ii) Birds	1166	50
(iii) Reptiles	456	214
(iv) Amphibia	209	128

India is exceptionally rich in biodiversity and is one of the 17 mega-diversity centres of the world. With 10 bio-geographic zones and 25 biotic provinces, all major ecosystems are represented here. The bio-diversity hotspots include mainly the Eastern Ghats, the Western Ghats and the North-Eastern hills.
The estimates figure out to over 45,000 plant species and 81,000 animal species, representing 7% of world's flora and 6.5% of world's fauna, of which 33% flora and 62% fauna are endemic.

6.2 ORIGIN OF BIODIVERSITY

There are two conditions that cause population diversity. *First*, new genotypes are constantly cropping up in a population through mutation, recombination, and related genetic phenomena; and through immigration of individuals, their gametes or their propagules. *Secondly*, diversity is the population eliminated by natural selection and lost through emigration of individuals. Every genetic variation, from gene mutations to entire species, will disappear eventually. This loss can be a very fast process, or the variants can survive for a long period. Species that have survived for extended periods include horseshoe crabs, which have been around for 200 million years and cockroach, which originates even earlier (350 million years) in the carboniferous period.

The speed at which new variants originate in relation to the rate at which they are eliminated determines the actual diversity of the system. Thus diversity is the result of two opposite actions: the processes that produce new genotypes, new varieties and new species and the processes that eliminate mutations, variants and species from the system.

Natural selection is primarily responsible for the reduction of biodiversity, it acts through differential reproduction and differential mortality.

6.3 VALUES OF BIODIVERSITY

Biological diversity is so widely accepted as being valuable that the reasons why it is valued are rarely carefully analysed. Though not all levels of variation are equally valuable for all reasons, yet the values can be accounted for various reasons as stated below:

	Value categories	Level of operation
1.	Commodity value: The utilisation value of an organism to humans as food, medicine etc. Populations that are not used may possess genes that improve the value of exploited population.	Population & species diversity
2.	Amenity value: The value of humans of visiting, viewing or learning about natural communities.	Species & Ecosystem diversity
3.	Ecological integrity value: The importance of maintaining all functionally critical aspects of an ecosystem.	Functional diversity
4.	Ethical value: The moral obligation on humans not to drive other species to extinction.	Species & Ecosystem diversity
5.	Option value: The future evolutionary potential of a species of the potential for future exploitation of a species, commodity value.	Genetic and population value

It is estimated that it would cost around 30 billion per year to expand this to 10% of the terrestrial surface, properly protected and with compensation to indigenous people. It is further noted the fact that 1% of global GWP is needed to ensure the continuing delivery of ecosystem services. Now time has come to assess the valuation of benefit of biodiversity from utilitarian stand point. Currently, 25% of the drugs on the shelves in the pharmacy derive from a mere 120 species of plants. But throughout the world, the traditional medicinal plants used by the native peoples is around 25,000 species (about 10% of the total number of plant species)—we have much to learn. In addition, there are ecosystem values or ethical arguments too.

It is a general belief that species diversity is essential for the proper functioning of communities and for the emergence of community level properties. Just as many different DNA encoded enzymes are needed for a complex organism to function properly, so, scientists believe, are many kinds of species necessary to maintain community structure. But is any diversity sufficient, or are specific mixes of species necessary for the communities and ecosystems to function?

There is a very old question in ecology, and two opposing views exist. One view is that a community is formed by the species that happened to arrive first—that the mix of species in a community is a matter of chance. The vegetation of an area is merely the resultant of two factors—the fluctuating and fortuitous immigration of plants—and an equally fluctuating and variable environment. According to the opposite view: "In any fairly limited area, only a fraction of the forms that could theoretically do so actually form a community at any one time. Thus the community really is an organised association of species in that it has a limited membership."

Communities with high species diversity may cope with long term environmental fluctuations better than do communities with few species. But this is not always true! Terrestrial communities in the climatically variable midlatitudes are less diverse than tropical communities in more uniform environments.

We often derecognised the values of biodiversity in many tiny ecosystems, but as the time passes we realised its multifaceted values. An example is given below with respect to estimated economic benefits of soil organisms (Table 6.3).

Table 6.3: Total estimated economic benefits of biodiversity with special attention to the services that soil organism activities provide worldwide (modified from Pimentel et al, 1997)

Activity	Soil biodiversity involved in the activity	World economic benefits of biodiversity (× US$ 10^9/year)
Waste recycling	Various saprophytic and litter feeding invertebrates (detritivores), fungi, bacteria, actinobacteria, and other micro-organisms	760
Soil formation	Diverse soil organisms, e.g., earthworms, termites, fungi, eaubacteria, etc.	25
Nitrogen transformations	Biological nitrogen fixation by diazotroph bacteria, conversion of NH_4 to NO_3 by nitrifying bacteria, conversion of NO_3 to N_2 by denitrifying bacteria	90
Bioremediation of chemicals	Maintaining biodiversity in soils and water is imperative to the continued and improved effectiveness of bioremediation and biotreatment	121
Biotechnology	Nearly half of the current economic benefit of biotechnology related to agriculture involves nitrogen-fixing bacteria, pharmaceutical industry, etc.	6
Biocontrol of pests	Soil provides microhabitats for natural enemies of pests, soil organisms (e.g., mycorrhizae) that contribute to host plant resistance and plant pathogen control	160
Pollination	Many pollinators may have edaphic phase in their life cycle	200
Wild food	For example mushrooms, earthworms, small arthropods, etc.	180
Total		**1,542**

6.4 LOSS OF BIODIVERSITY

Speciation and extinction are natural processes. Traditionally, from the Darwinian perspective, extinction is the fate of species which lose in the struggle for survival. But habitat loss due to human intervention for various reasons are the prime concern for species' extinction. In fact, there is no precise extinction of the numbers of species which are being lost over the decades. The reasons are very simple—in fact there was no estimate of actual species member in each category earlier and vast majority of the species are not monitored. Only the larger plants and animals were mostly monitored for over past couple of decades.

Various estimates of the rate of loss are shown in Table 6.4.

Table 6.4 : Estimated Rates of Extinction (adapted from Mawdsley and Stork, 1995)

Estimate	% of global loss per decade	Method of estimation	Reference
One million species between 1975 and 2000	4	Extrapolation of past exponentially increasing trend	Myers (1979)
15-20% of species between 1980 and 2000	8 – 11	Estimated species-area curve; forest loss based on Global 2000 projections	Lovejoy (1980)
50% of species by 2000 or soon after 100% by 2010-2025	20 – 30	Various assumptions	Ehrlich and Ehrlich (1981)
9% extinction by 2000	7 – 8	Estimates based on Lovejoy's calculations using Lanly's (1982) estimates of forest loss	Lugo (1988)
12% of plant species in neotropics	–	Species area curve (z = 0.25)	Simberloff (1986)
15% of bird species in Amazon basin			
2000 plant species per year in tropics and subtropics	8	Loss of half the species in area likely to be deforested by 2015	Raven (1987)
25% of species between 1985 and 2015	9	As above	Raven (1988a, b)
At least 7% of plant species	7	Half of species lost over next decade in 10 'hot-spots' covering 3.5% of forest area	Myers (1988)
0.2-0.3% per year	2 – 6	Half of rain forest species assumed lost in tropical rain forests due to local endemics and becoming extinct with forest loss	Wilson (1988, 1989, 1992)
2-13% loss between 1990 and 2015	1 – 5	Species-area curve (0.15 < z < 0.35); range includes current rate of forest loss and 50% increase	Reid (1992)
Red Data Books: selected species: 50% extinct in 50-100 years (palms), 300-400 years (birds & mammals)	1 – 10	Extrapolating current recorded extinction rates and by the dynamics of threatened status	Smith et al (1993)
Red Data Books: selected vertebrate species	0.6 – 5	Fitting of exponential extinction functions based on IUCN categories of threat	Mace (1995)

Most of these estimates are based on estimate of loss of habitat. The procedure estimates potential losses of species based on extrapolation of rates of habitat destruction and calculation of associated extinctions using species area curves.

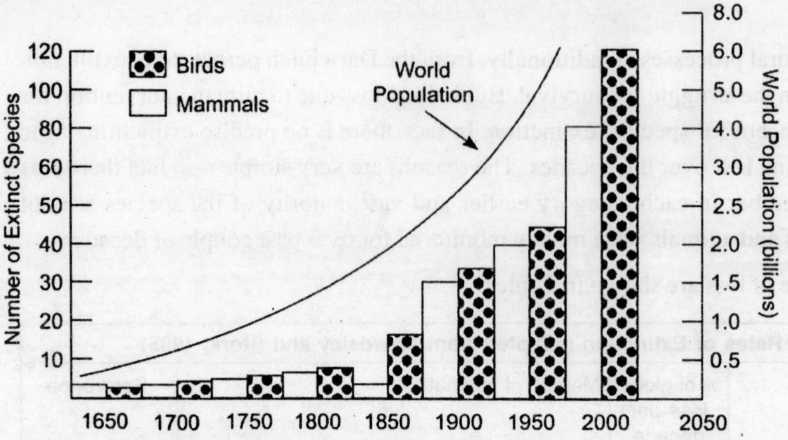

Fig. 6.2 : The growth of human population is paralleled by the increase in the extinction rate of birds and animals, 2000

It seems likely that major episodes of species extinction have occurred throughout the past 250 million years at average intervals of approximately 26 million years. However, according to Wilson (1988), the current reduction of diversity seems likely to approach that of the great natural catastrophes at the end of the Paleozoic and Mesozoic eras, the most extreme in the last 65 million years.

The major causes of extinction of species are:

- Habitat destruction;
- Hunting of wildlife;
- Introduction of new species;
- International trade of biotic resources;
- Pollution of various habitats;
- Catastrophic processes; and
- Several other accidental causative.

It is very interesting to note the fact that the growth of human population is paralleled by the increase in the extinction rate of birds and animals (Fig. 6.2). There were some assessments made to assess the relative importance of various extinction factors on the disappearance of mammals and birds (Fig. 6.3). On the whole, habitat destruction followed by hunting appears to be the major threat of extinction.

Considering the relative importance of the threat of species extinction, IUCN categorised six major classes of threated species (Table 6.4) as incorporated in their "Red Data Book". This categorisation involves both flora and fauna as exists in their wild habitats. The priority of conservation of various species thus made on the basis of their relative threat potentialities.

The approximate number of species extinct so far globally is given in Table 6.5. This number is changed day by day.

Table 6.4: Categories of threatened species

Extinct	A species not definitely located in the wild during the past 50 years
Endangered	A taxon in danger of extinction and whose survival is unlikely if the causal factors continue to operate
Vulnerable	A taxon believed likely to move into the Endangered category in the near future if the causal factors continue operating
Rare	A taxon with a small world population that is not at present Endangered or Vulnerable but is at risk
Indeterminate	A taxon *Known* to be Endangered,, Vulnerable or Rare but where there is not enough information to say which of the three categories is appropriate
Insufficiently known	A taxon that is *suspected* but not definitely known to belong to any of the above categories because of lack of information

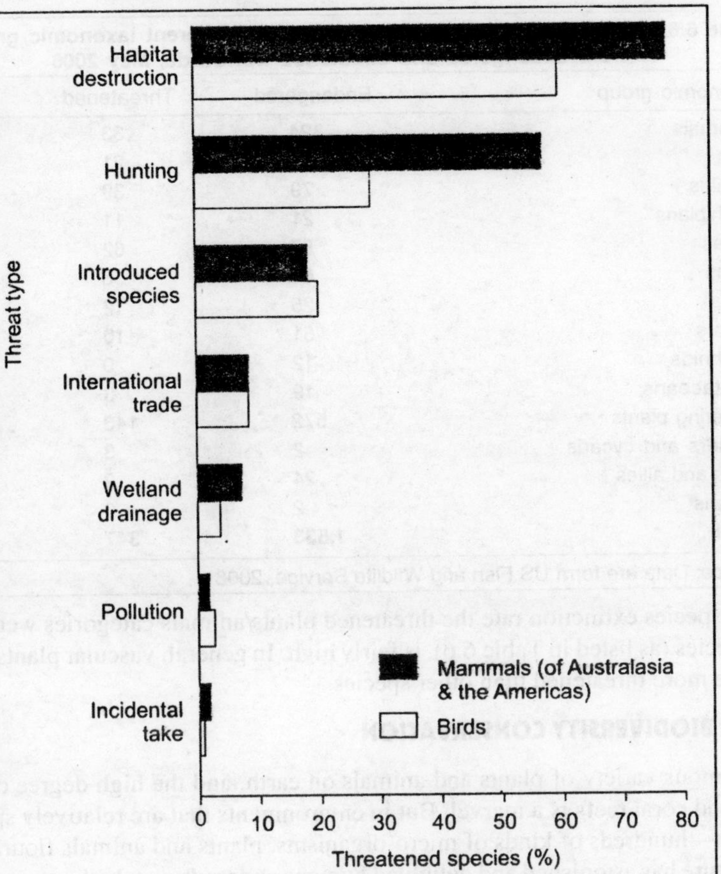

Fig. 6.3 : Threats to mammals and birds (in WCMC, 1992, Global Biodiversity: Status of the Earth's Living Resources. Champman & Hall)

Table 6.5: Record extinctions of vertebrates and plants, 1600 to 2000

Many additional species have presumably gone extinct without scientists having recorded them. Many more species are now threatened with extinction in the next few years. Except for mammals and birds, most threatened percentages are probably greatly underestimated.

	Species Extinct	Threatened	Total species	Percentage extinct	Percentage threatened
Mammals	87	1,130	4,600	1.9	24.6
Birds	131	1,183	9,500	1.4	12.5
Reptiles	22	296	6,300	0.3	4.7
Amphibians	5	146	4,200	0.1	3.5
Fishes	91	750	19,100	0.5	3.9
Plants	90	5,611	250,000	0.04	2.2

Source: Data from International Union for Conservation of Nature and Natural Resources. The World Conservation Union Red List 2000

Table 6.6: Number of Species and populations in different taxonomic groups that are endangered and threatened worldwide, May 2006

Taxonomic group	Endangered	Threatened	Total
Mammals	324	33	357
Birds	251	21	272
Reptiles	79	39	118
Amphibians	21	11	32
Fishes	87	62	149
Clams	64	8	72
Snails	25	12	37
Insects	51	10	61
Arachnids	12	0	12
Crustaceans	19	3	22
Flowering plants	572	143	715
Confiers and cycads	2	3	5
Ferns and allies	24	2	26
Lichens	2	0	2
Total	**1,533**	**347**	**1,880**

Source: Data are form US Fish and Wildlife Service, 2006

To restrict species extinction rate the threatened plants/animals categories were made. The number of threatened species (as listed in Table 6.6), is fairly high. In general, vascular plants, large animals, birds and reptiles were more threatened than other species.

6.5 NEEDS FOR BIODIVERSITY CONSERVATION

There is an enormous variety of plants and animals on earth, and the high degree of species richness in tropical forests and coral reefs is a marvel. But in environments that are relatively species poor—such as the ocean bottom—hundreds of kinds of micro-organisms, plants and animals flourish. This tremendous prodigality of nature has astonished and delighted humans and is ultimately the source of their sustenance. Yet, in using plants and animals for food and clothing, for building houses and for medicine, humans are endangering this immense richness of species. The current rate of loss of species diversity and the reduction in the genetic variety of crops and wild species could seriously affect human welfare.

Box 6.1: MEDICINAL PLANTS IN DECLINE IN WESTERN HIMALAYA

There is a long tradition of use and trade of medicinal plants from the western Himalayas. However, commercial extraction began to adversely affect the distribution and abundance of various medicinal plants only in the Twentieth century.

Heavy extraction and local processing by traders of Berberis roots from Mandi district led to local scarcity in some areas.

Similarly, unregulated harvest of *Taxus baccata* leaves and bark in the Sainj valley of Kulu in the sixties led to the disappearance of the tree in some pockets. Both Berberis and Taxus extractions were a result of increased market demand. While a number of species were known and used locally in Himachal Pradesh, escalating consumer demand in the 1970s for medicinal and cosmetic products in urban India and outside, has led to commoditising of many species in the last 25 years.

Shinglmingli (*Dioscorea deltoidea*) has been the worst affected by this onslaught. While parts of Chamba report local extinction of this herb following heavy extraction, it has also become rare throughout the rest of the state. In the last one decades, Berberis and *Taxus baccata* have faced heavy extraction once again.

The collection of the plants is carried out by local villagers on the basis of rights over forest produce, whereas the ownership of lands over which these occur rests with the Government. Traders are also involved in illicit collection through migrant labour resulting in conflicts with local collectors.

Lack of security of tenure for the local collectors is a major reason for breaking down of customary regulations for sustainable harvests. While earlier, gram panchayats had the powers to levy collection fees for harvesting medicinal plants, the power was discontinued with the constitution of Himachal Pradesh as a full state in 1971. As of now, neither the local collectors and local bodies, nor the traders have any say in ensuring a sustainable harvest. The government machinery, on the other hand, has failed to rise to the occasion.

Box 6.2: Biodiversity Heritage Sites

"Biodiversity Heritage Sites" (BHS) are well defined areas that are unique, ecologically fragile ecosystems - terrestrial, coastal and inland waters and, marine having rich biodiversity comprising of any one or more of the following components: richness of wild as well as domesticated species or intra-specific categories, high endemism, presence of rare and threatened species, keystone species, species of evolutionary significance, wild ancestors of domestic/ cultivated species or their varieties, past pre-eminence of biological components represented by fossil beds and having significant cultural, ethical or aesthetic values and are important for the maintenance of cultural diversity, with or without a long history of human association with them. (*Source*: NBA, India).

List of Biodiversity Heritage Sites (As on July, 2017)

Sl. No.	Name of the Site	Name of the District	Taluk	Area	Importance of the area	Government Notification
1.	Nallur Tamarind Grove	Bengaluru	Deva-nahalli	54 acres	It is popularly believed to be a relic of the Chola Dynasty that ruled nearly 800 yrs ago, is spectacle of awesome wonder and a freakish site. This BHS spread over 54 acres comprising a population of nearly 300 trees, is a picture of dynamic pattern of plant diversity. The significant component of this popular structure is a group of old plants standing like ageless sentinels, firmly rooted to the ground with their gigantic trunks, along with large picturesque crowns spread very high and aloft like open wings.	Aapaji 154
2.	Hogrekan	Chikmagalur	Kadur	2508.15 acres	The area has unique Shola vegetation and grass land with number of floral species which are unique and having lot of medicinal value. Hogrekan is moderately wooded land and its vegetation is of dry deciduous type and has a link with Bababudanagiri and Kemmangundi, adjoining Bhadra Wildlife Sanctuary and Yemmedode Tiger Reserve and serving as "Wildlife Corridor" between Kudremukha and Bhadra Wildlife Sanctuary.	No.FEE.35 ENV 200 dt. 4.9.2010
3.	University of Agricultural Sciences, GKVK Campus, Bengaluru	Bengaluru		167 hectares	The GKVK campus is considered one of the greenest areas in Bengaluru. Biological diversity of this campus constitutes a critical repository of various forms of flora and fauna (including 13 sp of mammals, 10 sp of reptiles, 165 sp of birds and 530 sp of plants) which needs to be protected nurtured to posterity.	No.FEE.132 ENV 200 dt. 2.9.2010
4.	Ambaraguda	Shimoga		3857.12 hectares	It is a revenue land located between Sharavathi Wild Life Sanctuary and Someshwara Wildlife Sanctuary. It has Shola vegetation which is primitive vegetation in the Western Ghat and also has grasslands.	2011
5.	Glory of Allapalli	Gadchiroli		6 hectares	It is a reserved forest being preserved as natural forest having biological, ethinical and historical values.	No.WLP.0914/ C.R.317/M-1 dt. 15.07. 2014
6.	Tonglu BHS under the Darjeeling Forest Division	Darjeeling		230 hectares	It is a Medicinal Plant Conservation Areas	No.716-ENT/ T-11-7/003-ii/ 2003 dt. 20.3.2015
7.	Dhotrey BHS under the Darjeeling Forest Division	Darjeeling		180 hectares	It is a Medicinal Plant Conservation Areas	No.716-ENT/ T-11-7/003-ii/ 2003 dt. 20.3.2015
8.	Dialong Village	Tamenglong		11.35 Sq.km	—	No.24/3/2017- For & ENV dt. 23.5.2017

6.6 BIODIVERSITY CONSERVATION STRATEGIES: GLOBAL SCENARIO

Considering the immense values of biodiversity and subsequent rapid loss of the same, during past couple of decades, enormous conservation efforts were made for restoring the biodiversity of various habitats. The detailed objective and management strategies are given in Table 6.7.

Table 6.7: Protecting biological diversity

On-site protection		*Off-site protection*	
Ecosystem maintenance	*Species management*	*Living collections*	*Germplasm storage*
Management Systems			
National Parks Natural research areas	Agroecosystems Wildlife refuges	Zoological Parks Botanic gardens	Seed and pollen banks Semen, ova, and embryo banks
Marine sanctuaries Resource development	In situ gene banks Games parks and reserves	Field collections Captive breeding programs	Microbial culture collections Tissue-culture collections
Conservation Objectives			
To establish a reservoir of genetic resources	To promote genetic interaction between semi-domesticated species	To maintain breeding stock	To provide a convenient germplasm source for breeding programs
To preserve evolutionary potential	To maintain viable wild populations for sustainable exploitation	To facilitate field research and the development of new varieties and breeds	To preserve germplasm of uncertain or threatened species
To preserve various ecological processes	To safeguard viable populations of threatened species	To facilitate off-site cultivation and propagation	To maintain reference type collections as standard for research and patenting
To preserve speices	To preserve species that provide important indirect benefits (as for pollination or pest control)	To maintain captive breeding stock of populations threatened in the wild	To provide access to germplasm from wide geographic areas
To preserve representative ecosystems	To provide "keystone" species with ecosystem support	To make wild species readily available for research, education, and display	To preserve genetic materials from endangered species

Various kinds of on site (in-situ) and off-site (ex-situ) protection strategies were already taken up as a part of global biodiversity conservation programme. Most of the developed and developing countries have already taken up such conservation programme as a national priority task. In many cases, IUCN supports these programme and periodically monitor them.

6.7 BIODIVERSITY PROBLEM IN AGRICULTURE: CONVERGENCE ON SPECIALISED VARIETIES OF SPECIES

It is well-known that out of vast number of vascular plants only 35-45 species were domesticated and cultivated for human consumption directly. Such domesticated species were cultivated through modern agriculture for higher yield over the years. Due to monoculture of several species, biodiversities of both plants and animals are gradually shrinking with time. Even the cultivated old traditional varieties are now replaced by only few yielding forms (Table 6.8). This is a very important problem of modern agriculture vis-a-vis progressive civilization.

Table 6.8: Examples of genetic uniformity in selected crops

Crop	Country	Number of Varieties
Rice	Sri Lanka	From 2,000 varieties in 1959 to 5 major varieties in 2000. 75% of varieties descended.
Rice	India	From 30,000 varieties to 75% of production from less than 10 varieties.
Rice	Bangladesh	62% of varieties descended from one maternal parent.
Rice	Indonesia	74% of varieties descended from one maternal parent.
Wheat	USA	50% of crop in 9 varieties.
Potato	USA	75% of crop in 4 varieties.
Cotton	USA	50% of crop in 3 varieties.
Soyabean	USA	50% of crop in 6 varieties.

6.8 BIODIVERSITY AND FOOD SECURITY

In a recent estimate it was speculated that over 25 per cent of the world's plant species might be lost by the year 2025, if the current rate of plant genetic erosion continues. Preserving these germ pool is an integral part of food security. It is evident that preservation of wide range of germ pool is an integral part of breeding programme. If we are unable to combat the problems of genetic erosion, it may lead to losing sources of resistance to pests, diseases and climatic stress and, finally, leading to crop failure in future. It is well-known that out of over 20,000 edible species only a few dozen of plants are domesticated and now feed most of the people. These pillar crops include cereals like rice, wheat, barley, maize, millet, sorghum; root crops like cassava, potato; handful of legumes like bean, peanuts, soyabean; fruits like banana, mango, jackfruits and other species like coconut, sugarcane etc. Identically, in forestry, out of 50,000 tree species only about a dozen of species were cultivated for commercial purpose. The noteworthy species among them were sal, teak, eucalyptus, subabul, akasmoni, and poplars etc.

With respect to animals, the world has over 3,500 species of mammals, 9,000 species of birds and considerable numbers of insects and other animals. But almost all meat, milk, eggs and other animal products today come from just five categories of animals viz., cattles, pigs, goats, sheep and poultry birds.

Some of the major crop failures viz., wheat failure in Eastern India, in 1943, cassava mealy bug attack in Zaire in 1973; black sigotaka attack on banana plantation of Fiji in 1920; leaf blight fungus attack on rubber plantation of South America in 1950s, have been experienced as long ranging consequences. Thus agricultural scientists and professional forester require a special attention on this subject for appropriate mitigatory plan formulation. Many wild relatives of present day cultivated crops have unique gene pool that could be used for saving the crop failures from insect pest attack or even from natural calamities like drought and flood hazards. The noteworthy examples are:

Ireland's potato blights resistance source was found in a wild potato growing in Mexico. Wild rice varieties of Silent Valley (Kerala) is shown to carry disease resistance genes. Central America's banana industry was saved from devastating Panama disease from a banana plant collected in a botanic garden in Saigon that carried the resistance gene. Chinese large-leafed spinach provided the resistance genes for blight and with genes saving the US spinach crop.

Considering their importance in 1983, Food and Agricultural Organisation (FAO) made a convention on International Undertaking on Plant Genetic Resources which stressed that genetic resources were a common heritage. This convention felt that local farmers were grossly ignored for their credits of preserving traditional crop varieties, many of which proved to be useful in breeding programmes in recent years. This led to promulgation of Article 15(1) of the Convention on Biological Diversity (CBD): "Recognising the sovereign rights of states over the natural resources, the authority to determine access to genetic resources rests with the national governments and is subject to national legislation." A conceptual connection can be made between intellectual property rights and conservation. A common application is the charging of royalties for use, meaning that genetic resources have perceived value, providing an incentive to conserve them.

In coming decades, increased agricultural output is necessary to feed the teeming millions but, at the same time, conversing biodiversity is a must. Hence both have to be made partners for sustainable development. Though no single path can be laid, various policies and practices can be devised to promote biodiversity conservation in managed landscapes and to enhance greater use of biodiversity in agriculture.

Box 6.3: Major Conventions on Various Issues Relating to Biodiversity

(i) Conventions on Wetlands of International Importance as Waterfowl Habitat (1971).
(ii) Convention for the Protection of the World Cultural and Natural Heritage (1972).
(iii) Convention on International Trade in Endangered Species of Wild Fauna and Flora (1973).
(iv) Convention on the Conservation of Migratory Species of Wild Animals (1979).
(v) United Nations Convention on the Law of the Sea (1982).
(vi) Vienna Convention for the Protection of the Ozone Layer (1985).
(vii) Montreal Protocol on Substances that Deplete the Ozone Layer (1987).
(viii) Basel Convention on the Control of Transboundary Movements of Hazardous Wastes and their Disposal (1989).
(ix) Convention on Biological Diversity (1992).
(x) United Nations Framework Convention on Climate Change (1992).
(xi) International Tropical Timber Agreement (1994).
(xii) United Nations Conventions to Combat Desertification in those Countries Experiencing Serious Drought and/ or Desertification, Particularly in Africa (1994).

6.9 BIODIVERSITY CONSERVATION FUND

India as a nation state is already investing substantial amounts of state revenue in conservation efforts. Its population is further contributing by accepting costs of conservation, whether it be through foregoing some economically more attractive development options, or tolerating crop damage by elephants. What we propose is that these financial inputs be organized as a national biodiversity conservation fund, perhaps with contributions from international sources such as Global Environment Facility. These funds could then be rationally allocated to the various local communities spread across the nation in relation to the levels of conservation value of biological communities in their respective territories.

Over and above this, all countries could agree to levying a biodiversity conservation cess on (a) all products of biodiversity based enterprises, and (b) polluting industries responsible for depletion of biodiversity. A portion of this cess could be used to promote conservation measures within the country. Another portion could be deposited in a Global Biodiversity Conservation Fund which may come to replace the currently functional Global Environment Facility.

The fraction of the national biodiversity conservation cess deposited in the Global Biodiversity Conservation Fund should: (a) increase with the per capita GNP of a country, and also (b) increase with the

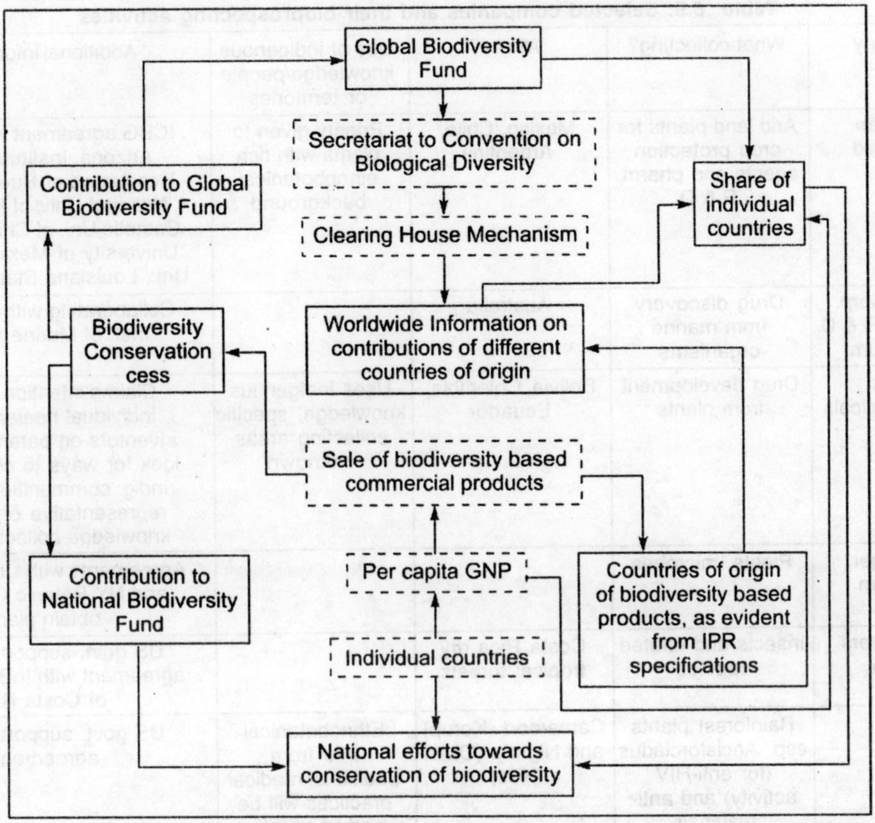

Fig. 6.4 : Financial mechanisms to Promote Worldwide Biodiversity Conservation Effort

extent of a country's dependence on use of biodiversity resources of origin outside the country. The Global Biodiversity Conservation Fund should be used to support conservation programmes in gene-rich developing countries in particular to reward local communities for their continuing participation in conservation efforts (Fig. 6.4).

Any country's share in grants from the Global Biodiversity Conservation Fund should be proportional to the extent of use of biodiversity resources for which it is a country of origin in the patents being worked by biodiversity enterprises throughout the world. This would encourage all countries to promote the use of their genetic resources by biodiversity based enterprises. Industrialised countries may voluntarily forego claiming their shares of this fund in the interest of helping the developing world undertake conservation measures more vigorously.

6.10.1 BIOPROSPECTING

This is a kind of commercial venture for utilising plants or plant products in large scale community as an usable commodity. Over past couple of decades enormous studies were made in such kind of bioprospecting activities. Various companies in USA/ Europe already marketed a number of such products, and many companies now actively engage their manpower on R & D related to bioprospecting potentiality assessment of both terrestrial and aquatic resources. A list of some selected companies of the world which already started bioprospecting activities is given in Table 6.9.

Table 6.9: Selected companies and their bioprospecting activities

Company	What collecting?	Where?	Use of indigenous knowledge/people or territories	Additional information
American Çyanamid	Arid land plants for crop protection agents and pharm. R & D	Mexico, Chile, Argentina	Priority given to plants with rich ethnobotanical background	ICBG agreement with: Uni. of Arizona, Institute of Biol. Resources of Buenos Aires, National Univ. of Patagonia, Catholic Uni. of Chile, National University of Mexico, Purdue Uni., Louisiana State University.
AMEAD Corp. (American R & D Consortium)	Drug discovery from marine organisms	Australia		Collaborating with Australian Inst. of Marine Science
Andes Pharmaceuticals (USA)	Drug development from plants	Bolivia, Colombia, Ecuador	Uses indigenous knowledge, specific collecting areas unknown	Claims intention to name individual healers as co-inventors on patents and will look for ways to compensate indig. communities through representative orgs. when knowledge collectively held
Boehringer Ingelheim	Plants, microbes			Agreements with Uni. of Illinois and NY Botanic Garden to obtain plants
Bristol Myers Squibb	Insects and related species	Costa Rica dry tropical forests		US govt. supported ICBG agreement with InBio and Uni. of Costa Rica
	Rainforest plants esp. Ancistorciadus (for anti-HIV activity) and anti-malarials	Cameroon (Korup) and Nigeria (Oban Hills)	Ethnobotanical info. from traditional medical practices will be used to prioritize plant collection	US govt. supported ICBG agreements
	Fungi, microbes, plants, marine organisms			Ranked 2nd largest Pharm. co. in the US. Contracts with third parties to collect specimens, incl. Scripps Institute and Oncogen
	Rainforest plants for drug devt.; non-medicinal plants for sust. harvest,	Surinam	Uses of plants by indigenous people to be documented. Specific terms of benefit sharing agreement not made public. Clause will set up 'Shamans Apprentice' programme and indigenous people fund	US. govt. sponsored ICBG agreement with Virginia Polytechnic and State Uni. of Blacksburg, Missouri Botanical Garden, National Herbarium of Surinam, Bedrijf Geneesmiddelen & Conservation International (CI)
Glaxo Group (UK)	Plants, fungi, microbes, marine organisms	Asia, Latin America, poss. elsewhere		Materials obtained from Kew Royal Botanic Gdns, Biotics Ltd, Uni. of Illinois, National Cancer Institute, contracts with Carnivore Preservation Trust to collect plants in Laos
Johnson & Johnson (USA)	Novel chemical compounds			Funds chemical prospecting at Cornell Uni. and training of Southern scientists in prospecting

Table 6.9: Selected companies and their bioprospecting activities

Company	What collecting?	Where?	Use of indigenous knowledge/people or territories	Additional information
Magainain Pharmaceuticals (USA)	African reptiles marine fish and organisms			Developing human drugs from African clawed frog and antibiotic steroid from dogfish shark
Marine Biotechnology Institute (Japan)	Marine organisms	Micronesia		Consortium of Japanese govt. and 21 Japanese corporations
Maxus Ecuador, Inc. (subsid. of Maxus Petroleum, USA)	1,200 plant species gathered, 18 new to world	Ecuador primary trop. rain forest	Plant collection and inventory traverses Yasuni Natl. Park and Waorani ethnic reserve	Contracts w/Missouri Botanic Garden for plant collection & inventory during construction of 120 km road in tropical moist forest
Merck and Co.	Fungi, microbes, marine orgs., plants	Latin America	Indig. knowledge from Urueu-wau-wau of Brazil; holds patent on anticoagulant derived from their plant material	Major pharmaceutical corp. Contracts with NY Botanic Gardens, MYCOSearch; high-profile contract with InBio (Costa Rica)
Parcelsian & Pacific Liaisons (USA)	Plants, food	China	Focusing on traditional medicinal plants	Pacific Liaisons has provided > 1,000 samples of trad. Chinese medicinal compounds to major US p'ceutical co. Will launch in-house screening
Pfizer Inc. (USA)	Plants	USA	Plant collection based partly on existing ethnobotanical leads	3-yr, $2 million research in collab with NYBG
Pharmaco-genetics (USA)	Natural products for drug development	Latin America	Hopes to rely entirely on leads for indig. people. Interest in developing line of cosmetics based on indig. people's products	Founded 1993; partly owned by non-profit Pan American Development Foundation that works with rural and indigenous groups in Latin America. Will use these connections to organize plant collection and indentification activities
PharmaMar (Spain)	Bioactive marine materials for AIDS and cancer	Worldwide		PharmaMar researchers travel aboard the ships of Pescanova, one of the largest fishing fleets in the world

Adapted from: Pirating Indigeneous Plants, RAFI and Indigenous Peoples' Biodiversity Network, RAFI Occasional Paper Series Vol. 1. No. 4, November 1994

6.10.2 COMMERCIAL VALUES OF BIODIVERSITY

Estimating the 'market value' or 'global sales' of biotechnology products is extremely difficult. To determine exactly which products have a strong biotechnology component would entail a company-by-company and product-by-product assessment. Not only would these figures be too fragmented and detailed to gather and analyse, but national statistics, figures from trade associations and reports by market analysts do not, as a rule, even estimate them, and may use different definitions when they do.

A much truer value of biotechnologies would be gathered by calculating the value of the biotechnology products to their customers—the 'problem owners'. There is a multiplier effect between the price of the product sold by a biotechnology company and its economic value in industrial processes. Thus an enzyme sold to a food company may make an important contribution to that company's production, effectively earning it many hundreds of thousands of dollars. This 'multiplier' is hard to gauge, and varies for each application of biotechnology. Based on data from the European Association for Bioindustries, the OECD has estimated a multiplier of 6 to 9 (OECD, 1998).

Finally, available statistics for products or market sectors do not reveal what proportion is attributable to biotechnology. For example, global trade in forest products is worth US$ 140 billion a year, and in cotton US$ 30 billion a year (FAO Yearbook, 1995). There are no data on the proportion of these products produced using biotechnology, which are negligible today, but likely to grow significantly in the coming decades.

Despite all these difficulties in assessing the size of markets, rough estimates of the global market for biotechnology products (other than in healthcare and agriculture) from the information we have been able to find, are set out in Box 6.4.

Box 6.4: Rough Estimates of Markets for Various Biotechnology Products

Product	Annual market (US$ bn)	Source
Environmental biotechnology	40 (early 1990s)	OECD, 1994
	75 (by 2000) or :	
	56-120	OECD, 1998
... of which soil remediation	10-25	In EC, Japan, USA, OECD, 1998
Industrial enzymes		
Detergent enzymes	0.7	
Starch enzymes	0.16	
Textiles enzymes	0.13	
Baking enzymes	0.09	
Beverages enzymes	0.09	
Dairy enzymes	0.06	
Other (leather, tanning, metals, oilfields, etc.)	0.24,	
Biocatalysts	0.02-0.03	Freedonia Group, 1997
	0.05-0.1	Global catalyst market US$ 9 bn, CMR
... of which PCR enzymes	0.05-0.1	UNEP/CBD/SBSTTA 2/15
Total industrial enzymes and biocatalysts:	1.5-1.6	Freedonia Group, 1997
	> 3 (in 2001)	Freedonia Group, 1997
Diagnostics: DNA polymerase alkaline phosphatase glucose oxidase		
Total (mostly DNA polymerase)	0.15-0.20	Freedonia Group, 1997
... of which 5% in areas other than healthcare	< 0.01	Personal communication Dr. Ken Murray, Biozyme, 5 March 1999
Biomaterials: forest products and fibres such as cotton produced using biotechnology	Negligible today Great potential	
Bioenergy	Negligible today Great.potential	
Rough estimates of total:	Low : 60 High : 120	

Box 6.5: TRIBALS AND TRADERS

Nestled amongst the Western Ghats in Pune district is the village of Shilimb. It lies in a tract of heavy rains near the origin of Pavana, whose waters eventually drain into the river Krishna.

People of Shilimb depend on cultivation of paddy in the valley, millets on the hill slope and husbanding livestock collection and sale of a variety of forest produce. Its Inhabitants belong to 13 different communities, Including Katkaris, who are a tribe traditionally dependent on hunting-gathering; Dhangars who hired buffaloes, and now some cattle and sheep as well, and Brahmins, traditionally priests, but also landowners and traders. Put together, members of these 13 communities have narrated 226 distinct uses of 173 plant species. Listed in the accompanying Table are uses in 10 major categories known to all or most of the 13 communities, and more specifically to the Katkaris, Dhangars and Brahmins. Katkaris clearly know much more than the other two. They are aware of very many medicinal uses not known to most other communities. Brahmins—on the other hand—have no special knowledge, they merely share in what members of most other communities know.

But it is the Brahmins who share formal learning and have market savvy. So it is a Brahmin from this village who has organised the trade in medicinal plants. He has set up a shop in the city of Pune after obtaining information on the identity of medicinal plants in demand in the urban market by practitioners of Ayurveda. He also manufactures in the village one of the commonest Ayurvedic medicines, *triphala churna*. Katkaris collect these plants in lieu of very low wages; all the value added goes to the Brahmin entrepreneurs.

Usage of 173 Species of Plants in Shilimb

Use	All 13 communities	Most communities	Katkaris	Dhangars	Brahmins
Human medicine	2	9	57	16	9
Veterinary medicine	1	1	1	3	1
Food	45	46	56	49	45
Fodder (cattle)	0	0	0	5	0
Fodder (sheep)	0	0	0	4	0
Agricultural implements	2	13	8	13	13
Ropes	3	4	4	4	4
Fuel wood	4	4	4	4	4
Fish poison	0	0	4	0	0
Other commercial uses	0	1	7	1	1

6.11 BIOPIRACY

This is a kind of procurement of biological wealth from other countries and subsequent patentisation of their products for commercial purpose. This is an unethical transaction of live materials and their subsequent patentisation of the products as made by transactional companies (TNCs) began in the wake of Trade Related Intellectual Property Rights (TRIPS) and WTO regime.

Most of the recent Western patents on biodiversity, whose raw materials are obtained from Third World countries and thus their claim of novelty in the properties and uses which they patentized is fraudulent. Such knowledge mostly lies within the indigenous communities over centuries. Thus, for example, Ayahuasca (*Banisteriopsis coapi*) of the Amazon and mustard (*Brassica campestis*) of India have been patented by TNCs for their novel properties, claimed to have been invented by TNCs. Over 110 patents on 29 plants from India, raised a serious concern of marketing various bioresources of India. These include spices like kalojeera (*Cyminum cuminum*), gol marich (*Piper nigrum*), ada (*Zingiber officinale*); edible fruits like jackfruit (*Artocarpus integrifolia*) and ber (*Zizyphus jujuba*) and medicinal plants like neem (*Azadirachta indica), ghritakumari* (*Aloe barbadensis*) and dudhi (*Euphorbra hirta*), and also the globally renowned fragrant rice "Basmati".

These are clear cases of 'biopiracy' as these patents clearly take the rights of marketing and exclusive use for commercial purpose though the centre of origin and their distribution lies in other countries.

The first campaign against biopiracy was the "Neem campaign". It was challenged in US Patent & Trademark office and European Patent office to disapprove the WR. Grace's neem patent (No. 5124349) and European patent (No. 436257BI), respectively. On the whole, these patents were found to violate the criteria of "novelty" and "prior knowledge" criteria that must be fulfilled before granting patents.

The Govt. of India challenged the US patent on turmeric and basmati and is preparing to challenge two more patents on ginger and karala. In the course of lawsuit on turmeric, a patent on turmeric was revoked. But there are numerous such patents on various elements of Indian biodiversity (36 on neem, 13 on mustard and so on) and there seems to be much difficulty in dismantling the basmati patent. Thus, it appears that by an introduction of a simple anti-biopiracy law in India, one could derecognize any patents on life forms of this country will appear to be more effective in upholding rights of indigenous people to access to, use of and trade and in all indigenous materials. A list of traditional Indian drugs targeted for biopiracy is given Table 6.10.

Box 6.6: BIOSAFETY

The biosafety protocol under the CBD Article 19 (3) was to be an important product emerging from international negotiations over a two-year period but, faced with US and UK opposition, it could not receive global approval.

At the Second COP, in 1995, the Parties to the CBD urged establishment of an open-ended ad hoc working group of experts who can help develop a biosafety protocol for the Convention. The group met several times after its establishment in July 1996 to develop the protocol which was to be adopted by the CBD at an extra-ordinary COP in February 1999. The provisions in the protocol include those relating to safety in biotechnology for research and development of Living Modified Organisms (LMOs), their use, distribution, handling, exchange, movement, risk assessment and risk management procedures, trade, socio-economic and environmental impacts of LMOs and the related. One of the contentious issues was the liability and compensation clause in case of any adverse effect.

India's position on Biosafety matters has always been positive with a precautionary approach as envisaged by the Preamble of the CBD. India already has developed a regulatory framework for recombinant DNA research within India which is being implemented through the Department of Biotechnology (DBT) and the Department of Science and Technology. There are provisions for establishing committees to monitor and advise safety in Biotechnology at the institutional level as well at the national level through the Recombinant DNA Advisory Committee which has scientists as members. Though implementation of the provisions is carried out scientifically, one of the major criticisms has been that the process is not transparent. Also there is no due representation of the civil society in the decision-making process.

The outcry on Monsanto's Bollgard cotton trials in India is an example of the need for careful attention if there is to be public acceptance of the technology. The regulators must take into confidence the public in the implementation process by ensuring representation of civil society and the media in the committee. The DBT need to publish, in the vernacular press, any future permits for field testing of LMOs so that the public is informed about the process. To prevent any entry of unwanted technologies like the 'Terminator technology' into the country there is a need for enhanced awareness and training of our quarantine officials.

Efforts by the DBT in creating such awareness programmes for scientists by conducting workshops and seminars is a way forward indeed for the implementation of the biosafety protocol, once it is in place.

Table 6.10: List of Traditional Indian Drugs Targeted by Bio-piracy

Common Name	Indigenous use	Patent Number	Patentee	Purpose
1. *Turmeric* [*Curcuma longa*]	Heal wounds and rashes	US Patent 5401504	University of Missisipi Medical Center, Mississipi	Wound-healing agent
2. *Neem* [*Azadirachta indica*]	Antiseptic tooth-brush, Insect repellent, Diabetes	EP436527	US Corporation W.R. Grace Company and US Department of Agriculture	Neem Extracts
		US Patent 5047242	Native Plant Institute (NPI), Salt Lake City	Azadirachtin derivative insecticides
		US Patent 4943434	Rohm and Haas Company, Philadelphia	Insecticidal hydrogenated *neem* extracts
		US Patent 5110591	PPG Industries, Inc., Pittsburgh	*Neem*-oil-emulsifier
		US Patent 5405612	W. R. Grace & Co. –Conn, New York	Hydrophobic extracted *neem* oil–a novel insecticide
		US Patent 4556562	Vikwood, Ltd., Sheboygan	Stable anti-pest *neem* seed extract
3. *Anar* [*Punica granatum*]	Anti diarrhoea, womicidal	US Patent 5411733	Toyoharu Hozumi, Takao Matsumoto, HaruoOoyama, Tsuneo Namba, Kimiyasu Shiraki, Hattori Masao, Masahiko Kurokawa, ShigetoshiKadota, Japan	Anti-viral agent
4. *Karela, Jamun,* Brinjal	Control of Diabetes	US Patent 5900240	Cromak Research Inc.	Anti-diabetic properties
5. *Basmati* Rice	Unique aroma and Flavour	US Patent 5663484	RiceTec	New variety of basmati lines
6. *Amla* [*Phyllanthus emblica*]	Greyness of hair	JP9110661	Unilever in the Japanese patent office	Hair blackening composition extracted from the *amla* fruit
7. *Guggul* [*Commifora mukul*]	Lowering fat	US Patent 643699	SABINSA Corporation, US	Nutritional / Nutraceutical improvement of cardiovascular disease
		US Patent 6113949	Prolab Nutrition, Inc.	Weight control product and method of treating hyperlipidemia
8. *Kala jeera* [*Nigella sativa*]	Jaundice, Halitosis, Skin diseases	US Patent 6042834	Baraka, Mohamed Wasif	Herbal composition for diabetes and method of treatment
		US Patent 5653981	Medenica, Rajko D	To increase immune function
9. *Aswagandha* [*Withania somnifera*]	Treatment of depression, insomnia, gastritis, gastric ulcer and convulsions	EP 1906980	Natreon Inc., US Multinational	Method of treatment or Management of stress
10. *Bel* [*Aegle marmelos*]	Leaves useful in diabetes	US Patent 5886029	Dhaliwal, Kirpal S	Method and composition for treatment of diabetes
11. *Latjira* [*Achyranthes aspera*]	Skin diseases and anti-inflammatory	US Patent 6080401	Reddy; Malireddy S Englewood, CO	Enhancement of the curative action of the drugs
12. *Amaltas* [*Cassia fistula*]	Prevention of dysentery	US Patent 5411733	ToyoharuHozumi, Japan	Antiviral agent
13. *Tulsi* [*Ocimum sanctum*]	Skin diseases	US Patent 6264995	Newmark; Thomas Schulick; Paul	Herbal composition for reducing inflammation

Contd.

Table 6.10: List of Traditional Drugs Targeted by Bio-piracy

Common Name	Indigenous use	Patent Number	Patentee	Purpose
14. *Kalimirch* [*Piper nigrum*]	Dermatopathy, Used in the preparation of pipla	US Patent 6346539	Raman; Amala Lin; Zhixiu Robert; Charles Hider, London	Treatment of skin conditions
		US Patent 5536506	Majeed, Muhammed Badmaev, Vladimir Sabinsa Corporation	Use of piperine to increse the bioavailability of nutritional compounds
15. *BhoomiAmla* [*Phyllanthus niruri*]	Cures liver disorders used for treating everything from jaundice to sluggish livers	US Patent 4673575	Baruch Blumberg, Venkateswaran; Pinayur S, Fox Chase Cancer Center	Composition, Pharmaceutical preparation and method for treating viral hepatitus
16. *Sharifa* [*Annona squamosal*]	Inflammation, constipation, urinary infection, infertility	US Patent 4762716	Moeschler; Heinrich F, Pfluger; Wolfgang Wendisch; Detlef	Insectide Annonin obtained from pulverized *annona squamosa* extract
		US Patent 5955497	Mclaughlin, Jerry L. Hopp, David C	Novel acetogenins isolated from Annona squamosal exhibit cytotoxicity to human tumor cell lines
17. *Harad* [*Terminalia chebula*]	For dysentery and diarrhoea, stomach complaints, ulcers, vomiting and worms Flatulence	US Patent 6187313	Alvin Burton Segelman, Nature's Sunshine Products, USA	Composition and method for treating and preventing helicobactor-pylori-associated stomach gastritis, ulcers and cancer
18. *Imli* [*Tamarindus indica*]	Fruit Drink	US Patent 5474791	Linda zablock; Suzanne Pecore; The Nutra-Sweet Company, (Deerfield)	Beverages using tamarind extract and method of making such beverages
19. *Manjistha* [*Rubia cordifolia*]	Skin diseases	US Patent 6258344	The Procter & Gamble Company, Cincinnati	Skin lightening compositions
20. *Chandrabhaga* [*Rauwolfia serpentina*]	Useful in epilepsy	US Patent 6323236	Susan McElroy; University of Cincinnati, Cincinnati	Use of sulfamate derivatives for treating impulse control disorders
		US Patent 2870140	Marvin R. Thompson, Robert E. Thompson	Therapeutic agents derived from *rauwolfia serpentine*
21. *Banaba* [*Lagerstroemia speciosa*]	Used as a remedy for the symptoms associated with elevated blood glucose levels	US Patent Applications 2006019890	Alex Moffett, Parag Shah	Pharmaceutical, therapeutic, and dietery compositions derived from *Lagerstroemia speciosa L,* plant
		US Patent 6589573	Ito En, Ltd., Tokyo, Japan	Xanthine oxidase inhibitor and method for producing the same
22. *Arjuna* [*Terminalia arjuna*]	Useful in bilious affections	US Patent 5411733	HozumiToyoharu Matsumoto Takao et. al.	Antiviral agent containing crude drug
23. *Bahera* [*Terminalia bellirica*]	Germicidal	US Patent 5693327	Eladevi Shah	Herbal compositions
24. *Adrak* [*Zingiber officinale*]	In rheumatism and inflammation of the liver	US Patent 6274177	Tian-Shung Wu; National Science Council, Taipel	Method of preparing an extract potent in anti-inflammation and anti-platelet aggregation from *Zingiber officinale*
	promotes digestive power	Patent application number 20090104293	Medical and Pharmaceutical Industry Technology and Development Center, Taipei	Use of an extract from rhizomes of *Zingiber officinale* in treating a disease associated with Helicobacter pylori

6.12 INDIA AS A MEGA DIVERSITY NATION

India is known for its rich heritage of biological diversity, having already documented over 91,000 species of animals and 45,500 species of plants in its 10 biogeographic regions. Nearly 6,500 native plants are still used prominently in indigenous healthcare systems. Thousands of locally adapted crop varieties grown traditionally since ancient times, and nearly 140 native breeds of farm livestock, continue to thrive in its diversified farming systems. The country is recognized as one of the eight Vavilovian centres of origin and diversity of crop plants, having more than 300 wild ancestors and close relatives of cultivated plants still growing and evolving under natural conditions.

Biodiversity is not distributed evenly across the globe. Certain countries, lying mostly in the tropics, are characterised by high species richness and more number of endemic species—called Megadiversity nations. About 17 countries/Nations belongs to this category as on today. India is one of them. India, a megadiversity country with only 2.4% of the land area of the world, but accounts for 7.8% of the recorded species of the world. The detailed biodiversity system of India is given in Tables 6.11 to 6.14.

Table 6.11: Recorded plant species

Taxonomic group	Number of species		% of the world flora
	World	India	
Angiosperms	250,000	17,500	7.0
Gymnosperms	650	48	7.4
Pteridophytes	10,000	1,200	12.0
Bryophytes	14,500	2,850	19.7
Lichens	13,500	2,075	15.0
Fungi	70,000	14,500	20.7
Algae	40,000	6,500	16.3
Virus/Bacteria	8,050	850	10.6
Total	**406,700**	**45,523**	**11.8**

Source: India's Third National Report to CBD, 2006.

Table 6.12: Recorded animal species

Taxonomic group	Number of species		% in India
	World	India	
PROTISTA (Protozoa)	31,250	2,577	8.24
ANIMALIA			
Mesozoa	71	10	14.08
Porifera	4,562	500	10.70
Cnidaria	9,916	842	8.49
Ctenophora	100	12	12.00
Platyhelminthes	17,500	1,622	9.22
Nemertinea	600		
Rotifera	2,500	330	13.20
Gastrotricha	3,000	100	3.33
Kinorhyncha	100	10	10.00

Contd.

Table 6.12: Recorded animal species

Taxonomic group	Number of species		% in India
	World	India	
Nematoda	30,000	2,850	9.50
Nematomorpha	250		
Acanthocephala	800	229	28.62
Sipuncula	145	35	24.14
Mollusca	66,535	5,072	7.62
Echiura	127	43	33.86
Annelida	12,700	840	6.61
Onychophora	100	1	1.00
Arthropoda	**970,670**	**69,903**	**7.20**
Crustacea	35,534	2,934	8.26
Insecra	861,696	61,151	7.10
Arachnida	73,440	5,818	7.96
Pycnogonida	600	16	2.67
Pauropoda	360		
Chilopoda	3,000	100	3.33
Diplopoda	7,500	162	2.16
Symphyla	120	4	3.33
Merostomata	4	2	50.00
Phoronida	11	3	27.27
Bryozoa (Ecroprocta)	4,000	200	5.00
Entoprocta	60	10	16.66
Brachiopoda	300		1.00
Pogonophora	80		
Priapulida	8		
Pentastomida	70		
Chaetognatha	111	30	27.02
Tardigrada	514	30	5.83
Echinodermata	6,223	765	12.29
Hemichordata	120	12	10.00
Chordata	**48,451**	**4,994**	**10.40**
Protochordata	2,106	119	5.65
Pisces	21,723	2,546	11.72
Amphibia	5,150	240	4.66
Reptilia	5,817	460	7.91
Aves	9,026	1,232	13.66
Mammalia	4,629	397	8.58
Total (Animalia)	1,191,208	88,730	7.45
Grand Total (Protista + Animalia)	1,222,458	91,307*	7.46

*Updated up to January 2007

Source: Annual Resources of India, 2007, Zoological Survey of India

Table 6.13: Wild relatives of crop plants in India

Crop	Number of wild relatives
Cereals & Millets	46
Pulses	81
Fruits	91
Spices and Condiments	28
Vegetables	76
Fibre crops	15
Oilseeds	14
Miscellaneous plants	28
Total	379

Table 6.14: Indian native breeds of domesticated animals

Group	Number
Cattle	30
Buffalo	10
Sheep	42
Goat	20
Camel	9
Horse	6
Donkey	2
Poultry	18
Total	137

6.13 BIOLOGICAL SAMPLE ANALYSIS

A. Sampling pattern options for plant species and communities

Random sampling (unrestricted): Observations are made at random points within the site or study area. It is theoretically the most statistically acceptable method, and many ecologists argue that it should be used whenever possible. However, it assumes that species have random distributions, which is rarely, if ever, true, and it has two serious disadvantages: (a) unless a large sample is taken, the chance observations are unlikely to provide representative data (e.g., they could be clustered in one small area), and (b) the results do not normally permit the detection of community gradients within the sampling area.

Systematic (regular) sampling: Provided that care is taken to avoid regular features that can be the most cost-effective pattern for obtaining representative samples of a site or study area, and of sampling community gradients. Options include:

- Line transects which can be useful for reconnaissance or application of the line-intercept method;
- Interrupted belt transects which can consist of (a) lines with sampling points (stations) at regular intervals (and intermediate stations if required), or (b) selected (e.g., alternate) squares within which sampling is conducted. They can be particularly useful for sampling community gradients;

- A grid system which can cover the site or selected study areas, with observations taken at intersections (stations) or within grid squares. All or a proportion of (e.g., alternate) stations or grid squares may be sampled. A grid can provide the most representative sample of a study area, though precision is affected by the size of the squares which should be as small as practicable.

• *Restricted random sampling*: This is sampling at random locations within areas defined by a systematic pattern, e.g., the squares of a belt transect or grid system. It is considered to be a reasonable compromise between unrestricted random sampling and systematic sampling (Greig-Smith 1983). It does not detect community patterns within the defined areas, e.g., grid squares, which should, therefore, be small.

Stratified sampling: This involves the selection of study areas, and sampling each of these according to its size. It is particularly useful for large sites where adequate systematic sampling of the whole area is unnecessary or impracticable. The study areas can be defined by features such as fairly discrete vegetation boundaries, land-use boundaries, or tracks, and can be sampled systematically or randomly. Stratified random sampling should not be confused with restricted random sampling.

Selective sampling: This involves the subjective (not random or systematic) selection of sampling points thought to be representative of a community type or to contain special feature such as a particular species under study. It may be the only practicable method, e.g., when access to sampling points is difficult or when a species only occurs in scattered locations (Goldsmith et al 1986). However, it is also employed in the relieve method in which samples are deliberately obtained within apparently homogeneous vegetation patches that are thought to be representative of a community type. In this case, a requirement is that the area sampled should be large enough to include all the community's species. A popular procedure to ensure this is estimation of minimal area, though it is only reliable in truly homogeneous vegetation (see Kent & Coker 1992, Kershaw & Looney 1985). An advantage of the relieve method is that the samples should represent communities amenable to formal classification, e.g., in the NVC; disadvantages are the subjective selection of sampling areas and the deliberate disregard of community gradients.

B. Measures of abundance for plant species and communities

Semi-quantitative abundance ratings =	Visually estimated measures made using rating system such as DAFOR or ACFOR.
DAFOR	D = dominant; A = abundant; F = frequent; O = occasional; R = rare
ACFOR	A = abundant; C = common; F = frequent; O = occasional; R = rare

With DAFOR the most abundant species is assumed to be dominant; ACFOR may be preferred because it does not require this assumption (although the estimated dominant can be noted). Conversely, it may be difficult to distinguish between C and F in ACFOR. Either system can be supplemented by ratings such as co-dominant and subdominant, and by distributional information such as widespread or local (e.g., LA = locally abundant). The measures permit rapid survey but are subjective, approximate, and have limited potential for analysis and presentation (although the ratings can be converted to a 5-point numerical scale). Consequently, there is little point in using them if equivalent quantitative data can be obtained.

Number of individuals: Numbers are usually counted in quadrats, and expressed as density (number per unit area) or population size (number in site or overall study area). These are appropriate for estimating abundance of selected species, provided that individuals can be readily counted, but generally not applicable in community studies because number of individuals has little meaning when comparing species of widely differing size.

Cover (%): This is the percentage of ground occupied by a perpendicular projection onto it of the aerial parts of a species. Cover can be measured by the following methods:

- Visual estimation *in quadrats* is the most flexible and popular method, but is subjective and prone to observer error, especially in dense or tall vegetation, and tends to under-value species having small scattered individuals. For these reasons, values are usually recorded using incremental cover-abundance scale.
- *Line-intercept method*, which can be accurate and quite rapid in simple vegetation, but is more difficult and time-consuming in complex vegetation.
- *Point-intercept (point quadrat) method*, which can be the most objective and precise method, but is time-consuming and only readily applicable in short vegetation.

Individual observations are quantitative and should be representative; but the time needed with any of the methods can discourage the collection of an adequate overall sample of a study area.

Cover-abundance: This aims to avoid underestimating the importance of small species with scattered individuals by using cover for species with cover 4% or 5%, but abundance in the strict sense (of numbers) for species with cover 4% or 5%.

Frequency (%): This is the percentage of observation in a sample that contain the species, i.e., is derived from presence/absence observations (usually in quadrats). Disadvantages are: it is strictly a measure of distribution rather than abundance, and tends to over-represent small species; it fails to discriminate between high density (with many individual present in quadrats) and density that is just sufficient for at least one individual to be present in a large proportion of quadrats; and frequency values increase with increasing quadrat size, so results from surveys using different sized quadrats, and should not be calculated from sets of less than 20 observations. In spite of its limitations, frequency can be the most cost-effective method for obtaining large representative samples in community studies because it is relatively rapid and free from observer error.

C. Methods for sampling invertebrates

(i) Observer dependent methods

- *Direct observation and identification*: General searching and recording species found, usually in areas defined as having "interesting" vegetation. Such methods are not normally quantitative, can easily lead to misidentification, only record species that are active when the observation limited to species that are either common in the study area or extremely conspicuous, e.g., dragonflies (Moore & Corbet 1990).
- *Transect walking*: It involves the observation, identification and enumeration of species on a set route, undertaken within prescribed time and weather conditions; usually restricted to butterflies and day-flying moths (Pollard 1977, Pollard et al 1975, Brooks 1993b).
- *Sweep netting*: A handheld net is swept through vegetation up to 1 m in height. Most invertebrates are swept off the vegetation and can be collected and transferred to a preservative for later identification. The method can be quantitative if a standard number of sweeps are taken, but sweeps in different vegetation types are not directly comparable, because different vegetation type differentially resist the net. Invertebrates occupying the basal parts of vegetation, are not normally sampled, and active flying individuals often escape from a net before they are captured. The method is not suitable for thorny or wet vegetation.

- *Swish sampling*: It is like sweep netting but is restricted to the air boundary immediately above vegetation. It is especially good at collecting Diptera (flies) and Hymenoptera (bees and wasps).
- *Suction sampling*: A portable vacuum is used to collect invertebrates from the ground layer and/or basal parts of vegetation. It can be an efficient collecting method in dry conditions and where there is little vegetation litter, and can provide quantitative data if a set number of samples are obtained.
- *Soil samples*: It can be taken for identification and enumeration of soil invertebrates. A variety of physical or chemical extraction methods are used to extract the organisms from the soil samples.
- *Beating*: A stout stick is used to knock invertebrates off vegetation onto a sheet from which they are collected. It is usually used to sample the fauna of individual species of tree. With care, the method can be used to obtain quantitative data, but it is not practical in wet conditions.
- *Subsidiary methods*: Most experts have favoured methods for detecting those invertebrate groups with which they are most familiar. Methods include observing flower visitors hand-searching vegetation for plant grazers (especially molluscs), stone turning (especially for beetles, molluscs and millipedes), and investigating litter and dead wood for decomposers.

(ii) **Observer-independent methods**

- *Pitfall traps*: A good quantitative method used chiefly for ground-dwelling beetles, which fall into the traps. Pitfalls are placed on a regular grid within selected areas. They usually contain a fluid that kills and preserves the invertebrates caught.
- *Malaise traps*: Flying insects are intercepted by a net and funnelled into a collection vessel by their flight mechanism. The method can collect large numbers of insects, especially Diptera and Hymenoptera (Disney et al, 1982), and is efficient for obtaining quantitative data. It does not discriminate between insects resident in, or simply flying through, the area.
- *Sticky traps*: Usually consists of a mesh screen on which a viscous oil is applied. They can be used like malaise traps, or placed within vegetation. Fragile invertebrates may become damaged in trying to escape from the trap, and sampled invertebrates have to be removed by a solvent.
- *Water traps* : Rely on the fact that a variety of flying insects (especially flower visitors) are attracted to coloured surfaces. They are simple to use but selective (see Disney et al, 1982, Usher 1990).
- *Light traps*: Night-flying insects are attracted to light sources, especially if these emit ultra-violet wavelengths. They are useful, but (a) they require a power source and are not easily transported, and (b) they may sample species that are not associated with the site but which are flying over it (see Waring 1994 for information on light traps and other types of moth trap).
- *Emergence traps*: Usually consist of a closed mesh canopy placed over vegetation, and a collecting vessel. They are designed to collect most adult flying insects which were in a developmental stage on the vegetation or in the soil when the trap was erected. They can be used quantitatively, because each will have a known basal area, but must be in place for long periods.

D. Bird census method

(i) **General aspects** : All methods are time-consuming, involve extensive site walking, and require expertise, e.g., in the recognition of bird calls/singing. They are affected by (a) seasonal variations, e.g., breeding seasons, in which different species sing, and seasons of migrant visitors, and (b) time of day, e.g., shortly after dawn being generally the best time in relation to bird song. They require repeat sampling, e.g., early morning visits at weekly intervals, variation in routes and directions taken so that particular parts of the site are not always visited at the same time of day, and records of all contacts (sightings or call/song registrations) and the times these are made. Factors affecting census accuracy include:

- birds are easier to find in some habitats than in others, e.g., they are less easily detected in dense scrub and woodland than in open habitats.
- some species are less conspicuous and/or noisy than others and therefore may be overlooked.
- recognizing different species, counting individuals and determining territories may be difficult where birds occur at high density.
- in wet and windy weather birds may be less active and skulk out of sight.

(ii) **Territory mapping**: Can be used to determine densities, locations and territories. It is best conducted during the breeding season when most species are territorial, and territories are often marked by singing, displays and disputes with neighbours. It usually involves walking field boundaries (on agricultural land) or recording within defined plots (e.g., quadrats). It is widely accepted in Britain but requires specialist expertise and is time-consuming, which can lead to error, e.g., if small plot sizes are used to save time. Consequently it may have to be restricted to those species which are rare or uncommon on a local, regional or national scale. The territory map should be presented in association with a habitat map, especially if impacts via habitat loss or change are to be predicted.

(iii) **Line transect method**: Involves walking transects of fixed length (e.g., 1 km, though this can be sub-divided) and location (arranged randomly or systematically, e.g., as a grid), at a standardised speed, e.g., c. 2 km per hour. The method can be used to estimate densities, e.g., by counting in relation to a prescribed band each side of the transect, is more rapid than mapping, and can be carried out throughout the year; but its value may be limited in small and/or heterogeneous sites.

(iv) **Point count method**: Involves randomly located points at which observations are made. It can be useful in small sites and heterogeneous vegetation, and permits good analysis of association of bird species (including densities) with given habitat/vegetation types.

(v) **Plotless method**: Involve the selection of observation points in relation to specific aims, e.g., at central points in the range of selected species. This can be useful in the study of scarce species that tend to be neglected by the other methods.

6.14 WILDLIFE MANAGEMENT

6.14.1 Concept of Wildlife Management

Wildlife includes all free-ranging vertebrates in their naturally associated environments. But, truly speaking, wildlife includes all plants, animals in wild ecosystems.

Most wildlife management is directed toward birds and mammals. Fish management has developed quite separately, and amphibians, reptiles and plants have received little attention in wildlife management

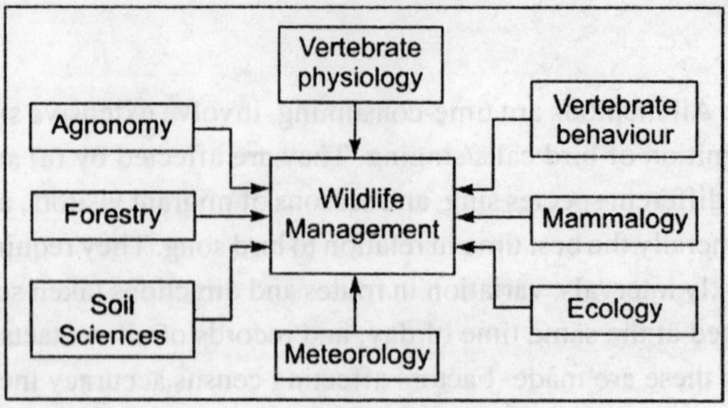

Fig. 6.5 : The Principles of Wildlife Management

until the recently increased concern for endangered species. Wildlife management is the art of making land produce valuable populations of wildlife. It is implied that this definition includes the control of pest populations to limit negative value of wildlife. Wildlife management involves direct population management (control of harvest, transplanting etc.) and indirect management of populations through habitat manipulation to favour or inhibit target species.

The principles of wildlife management include some that are specific of the profession and many that are shared with other professions and sciences. Therefore, the education of a wildlife manager should include study not only of wildlife biology and management, but also of basic sciences, such as chemistry and meteorology, and applied sciences related to land use, such as forestry, agriculture and economics (Fig. 6.5).

Wildlife management is a part of wildlife conservation. Most wildlife managers participate in several activities in the conservation process because some of their efforts involve administration, education, law enforcement and research.

6.14.2 Concept of Wildlife Conservation

Wildlife conservation is a social process encompassing both lay and professional activities that define and seek to attain wise use of wildlife resources and maintain the productivities of wildlife habitats. Wildlife managers, research biologists, administrators, extension agents, and also mining company lobbyists, non-governmental organisations etc. participate in this process. The professional activities in wildlife conservation are research, education, administration, law enforcement and management (see Fig. 6.6).

6.14.3 Wildlife Values

The values and benefits of wildlife can be classified in various ways (Dixon and Sherman 1990; Ledec and Goodland 1988). A distinction is often made between consumptive and non-consumptive uses of wildlife. Consumptive use is one form of direct use, as described in Table 6.15, and may be of either a traditional (e.g., bushmeat and harvesting of medical plants) or non-traditional (e.g. safari hunting) nature. Non-consumptive use generally refers to non-use, or non-consumptive values of wildlife which may also be of either a traditional or non-traditional nature. Traditional non-consumptive values embrace the cultural use of wildlife, including the reverence of certain species and the use of wildlife imagery in art, folklore and religion. Non-traditional, non-use (non-consumptive) values include the existence values of wildlife as perceived by Western conservationists.

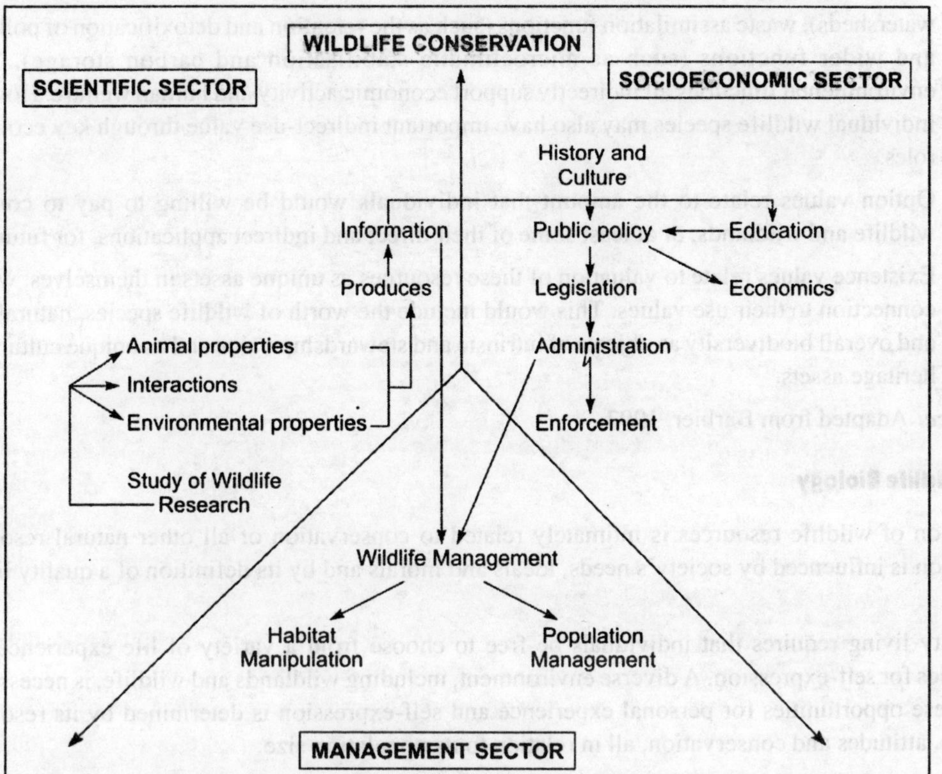

Fig. 6.6 : Professional activities in the wildlife conservation process

Table 6.15: Economic values of wildlife and wildlands

Wildlife values			
Use values			Non-use values (d)
Direct-use values (a)	Indirect-use values (b)	Option values (c)	
Harvested products for traditional and non-traditional consumptive use	Ecological functions	Future uses as for direct- and indirect-use values	Biodiversity
	Protection functions		Cultural heritage
Recreation and tourism	Waste assimilation functions		
Genetic material	Microclimatic functions		Spiritual
Education	Carbon store functions		
Human habitat			

Notes :

(a) Direct-use values are the resources and services provided by directly harvesting and exploiting wildlife and by natural areas.

(b) Indirect-use values comprise mainly the environmental functions of natural areas—ecological functions (such as nutrient cycling), protection functions (such as ground cover for key watersheds), waste assimilation functions (such as the retention and detoxification of pollution) and wider functions (such as microclimatic stabilisation and carbon storage). These environmental functions all indirectly support economic activity and human welfare. However, individual wildlife species may also have important indirect-use value through key ecological roles.

(c) Option values relate to the amount that individuals would be willing to pay to conserve wildlife and wildlands, or at least some of their direct and indirect applications, for future use.

(d) Existence values relate to valuation of these resources as unique assets in themselves, with no connection to their use values. This would include the worth of wildlife species, natural areas and overall biodiversity as objects of intrinsic and stewardship value and as unique cultural and heritage assets.

Source: Adapted from Barbier, 1992.

6.14.4 Wildlife Biology

Conservation of wildlife resources is intimately related to conservation of all other natural resources. Conservation is influenced by society's needs, ideals and morals and by its definition of a quality life for humans.

Quality living requires that individuals be free to choose from a variety of life experiences and opportunities for self-expression. A diverse environment, including wildlands and wildlife, is necessary to provide these opportunities for personal experience and self-expression is determined by its resources, knowledge, attitudes and conservation, all in relation to its population size.

Wildlife biologists manage land ecosystems supporting a great variety and abundance of plants and animals that are interrelated in many complex ways, including—but not limited to—foodwebs and competition. The complexity of an ecosystem managed by a wildlife biologist is at least as great as that of any system managed by any profession. This complexity defies complete understanding, limits confidence in predictions and requires caution in management prescriptions.

Wildlife have several values—commercial, recreational, biological, scientific, social, esthetic and negative. Commercial, recreational, and negative values can be measured in monetary terms. The other values are difficult or impossible to measure. Since wildlife values are not commensurable, the total value of a wildlife population can be estimated by its cost of replacement.

Wild animals have numerous anatómical, physiological, and learned and genetical behavioural adaptions that enhance their abilities to live in certain ways in certain kind of environments.

Good nutrition enhances wildlife reproduction and reduces wildlife susceptability to many forms of mortality. Quality wildlife foods are digestible and provide water, energy, and several required nutrients. The quantities and qualities of foods required by wildlife are highly variable among species of wildlife, among the sex-age classes of animals, among seasons and related physiological functions of animals, and among weather conditions and geographical regions. Food shortages are not uncommon among carnivores. Herbivores have strong forage preference related to forage quality. Herbivores tend to experience nutritional problems, not for lack of foliage, but, whenever available, forage is of poor quality.

The quantities and qualities of food available to wildlife are highly variable among areas, seasons and years. Food quantities are influenced by weather and many site factors. For herbivores, forage qualities are influenced by weather, site factors, genetics of plants and the parts of plants eaten.

Full understanding of the nutritional dynamics of wildlife requires periodic assessment of food abundance, analysis of forage qualities, study of food habits and preferences, and study of the nutritional physiology of wildlife species. Biologists seldom have full understanding of the nutritional dynamics of wildlife population they manage.

Awareness of the zoogeographic history of a wildlife species can enhance understanding of its ecological adaptations and limitations, its requirements for habitat resources, and its abilities to withstand competition in various environments. Although natural barriers—especially climate and salt water—have delayed dispersal of terrestrial vertebrates, the dominant pattern of the part has been for successive newly evolved taxa of disperse and replace—rather than supplement—existing fauna. Thus, transplants of competing animals from other continents present a threat of ultimate extinction to elements of a native fauna. This threat is greatest when animals are transplanted from continents onto oceanic islands where faunal diversity and competition have been low.

6.14.5 Wildlife Management Process

For wildlife management purposes, it is useful to classify the numerous and complexly interrelated factors operating in ecosystems as **welfare factors, deciminating factors** and **environmental influences**. Welfare factors, the habitat requirements of wildlife species, can be defined at several levels of resolution. At the lowest level of resolution, (1) all species require oxygen, foods, cover types, species needs, interspersion and space. With increasingly greater resolution, the habitat requirements of a species can be defined, (2) empirically as a list of needed habitat types, (3) as a list of specific welfare factors and (4) as a list of welfare factor components. Suitable habitat must supply welfare factors, regardless of sex, age or class of animals, during all seasons and weather conditions. Detecting limiting welfare factors is basic to efficient programmes of habitat management. Two methods for detecting limiting factors are (1) observing correlations between trends of welfare factors and trends in population performance, and (2) observing symptoms of welfare factors deficiencies in the animals and symptoms of welfare-factors overcome in the habitat.

Wildlife habitats are not stable. Much change occurring in habitats is due to biotic succession, retrogression, or to rather sudden natural or man-caused disturbances such as by fire, logging or flooding. These changes alter food, cover and other habitat resources for all wildlife species and are fairly predictable. Much wildlife habitat management is, therefore, the management of succession, retrogression and disturbance.

Weather affects wildlife directly as a cause of mortality and indirectly by restricting animal movements and influencing the abundance and availabilities of habitat resources and the abundance of competitions, predators and disease organisms.

Wildlife management consists of controlling the number, distribution, and quality of wild animals, either directly — as by manipulating wildlife habitat. Selecting a data basis for a management programme is one of the management biologist's most important decisions. Selection will be influenced by the availability and precision of methods for measuring population and habitat characteristics and by budget constraints; but the selected data base should have a meaningful relation to management objectives.

The types of wildlife measurements are: (1) population indices, (2) population censuses, (3) habitat measurements, and (4) indices of ecological density including population-condition and habitat-condition indices.

Wildlife management is the art of making land produce populations of wildlife, for harvest or other values. In simplest terms, wildlife management consists of a series of decisions — whether to have a long hunting season or a short one, whether to plant food patches and pine trees or to manipulate food and cover with fire or herbicides, whether to spend money on a grain-check station or on a forage survey, whether to compromise biologically optimum goals with prevailing public opinion, whether to improve

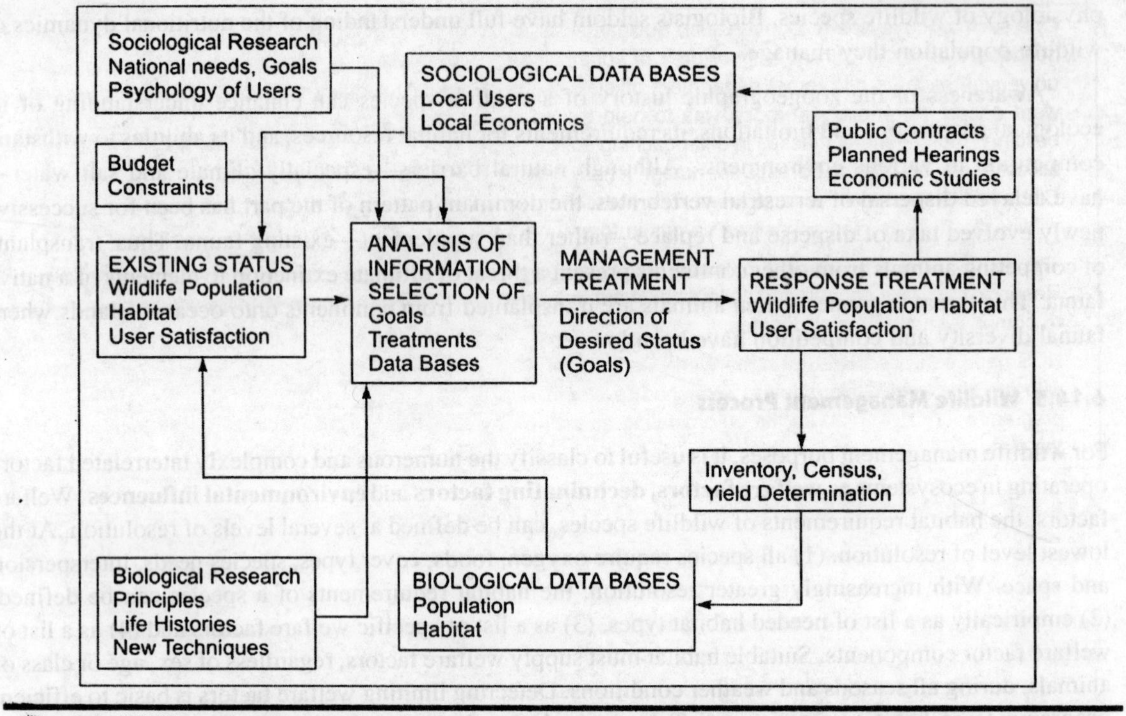

Fig. 6.7 : Wildlife management and cyclic, incremental process. (At each cycle, the information bases increase, permitting more intensive management. Management activities and responsibilities are indicated within ellipses. The process is based on concepts in King 1941 and Lindblom 1959)

existing habitat or to purchase more land, and whether to feed deer during severe winters or to reduce the deer herd or to improve the natural supply of winter foods. These decisions must be made every year.

An overall scheme for wildlife management processes is given in Fig. 6.7.

Box 6.7: ECOSYSTEM PEOPLE

In West Champaran district of Bihar lies Sariyaman, a horseshoe-shaped lake with a spread of 9 sq. km. formed by a shift in the course of the river Narayani. Manjharia is on the bank of the lake, with 250 refugee families from Bangladesh, primarily of the Das and Namasudra communities who were resettled in what was scrubland with scattered paddy fields. They remain totally dependent on locally produced biomass for all of their requirements; the one school shed in the village being the only brick, cement and asbestos structure. The village is surrounded on three sides by the lake, and on the fourth by the forest. This permits them to raise three crops of paddy, that is supplemented by fish from the lake for their food. The naturally growing bamboo is excellent construction material, and the elephant grass serves to thatch the roofs. They stall feed the cattle on grass collected from the lake shore and use the dung and straw as fuel. So they have no need of fuelwood. They neither poach on the wild animals, nor encroach on the forests for cultivation.

As a result, large flocks of waterfowl take refuge in the lake adjoining Manjharia. Their only problem is the damage that wild pigs and cheetal inflict on their crops. They are also concerned with the overfishing and poaching of wildlife by other villagers on the bank of Sariyaman.

Box 6.8: VOLUNTARY EFFORTS

All over India there exist highly motivated groups dedicated to nature conservation working at the grassroots. These could come to play a valuable role in organising decentralised, people-oriented conservation efforts throughout the countryside.

Contd.

One such group is Nature Conservation Society (NCS) in Bihar. NCS is based in Daltonganj, the commission headquarters of the Chhota Nagpur area of south Bihar. It was founded in 1976 by a group of college and university teachers and forest department officials. Later on, more members from other sections of society were added. Its primary purpose was to help the newly-created Palamau Tiger Reserve by awareness-building among general masses in anti-poaching activities and in interaction with local communities. It works in full collaboration with the Forest Department. It has organised college and school students in the form of Nature Clubs in urban as well as rural areas. The Nature Clubs have surveyed tribal villages, run competitions on wildlife and forests, and sponsored camps in forest areas.

The documentation by these clubs since 1978 is now the basis of monitoring the biodiversity of Palamau Tiger Reserve. NCS has encouraged college teachers to undertake wildlife research and has helped eight of them to complete their doctoral degrees.

NCS has been associated with the World Bank in eco-development activities in the Palamau Tiger Reserve, with the Bihar State Government in watershed development programmes, and with WWF in the Tiger Conservation programme. NCS served as the co-ordinating agency for the People's Biodiversity Project (PBR) project for the state of Bihar.

6.14.6 Protected Area Concept

There are two basic mode of wildlife management process. These are named as in-situ conservation i.e., on site preservation; and ex-situ conservation i.e., experimental situation in monitored condition. For in-situ conservation of species of threatened plants and animals, various types of protected area demarcation is suggested. The details of protected area categories and their objectives are given in Table 6.15.

In India, such a protected area network comprises of about 104 National Parks, 550 Sanctuaries, 18 Biosphere reserves, several other protected areas, constituting around 5.02% of the total geographical area of the country. In addition there are large number of preservation plots, reserve forests, reserve plantation plots, sacred groves. By and large all types of protected area constitute over 15% of the total forest area of the country. This network of protected areas covers most of the representative habitat types in the country and affords protection both to the wild flora and fauna (WII, 2019).

(a) **Biosphere Reserves**: Area rich in biodiversity and encompassing unique and representative ecosystems are identified and designated as "Biosphere Reserves". The goal is to facilitate conservation of representative landscapes and India's immense biological diversity which is estimated to be over 47,000 plant species and 81,000 animal species representing about 7% of the world's flora and 6.5% of world's fauna, respectively. Biosphere reserve network programme was launched by UNESCO in 1971 under its Man and the Biosphere Programme. The prime objectives of this programme are as follows:

- Conserve representative samples of ecosystems;
- Provide long term in-situ conservation of genetic diversity;
- Promote and facilitate basic and applied research and monitoring;
- Provide opportunities for education and training;
- Promote appropriate sustainable management of the living reserve;
- Disseminate the experience so as to promote sustainable development elsewhere; and
- Promote International Co-operation.

In each of the Biosphere reserve there are two zones—"core areas", which is strictly protected in order to maintain their ecological integrity and "Buffer areas" where some activities of resource exploitation and also towards developments can be accounted for.

Eighteen Biosphere Reserves set up so far not only aim at protecting representative ecosystems, but also serve as laboratories for evolving alternative models of development. A list of Biosphere reserves set up so far is also given in Table 4.16. Eleven out of eighteen biosphere reserves are a part of the world network of biosphere reserve based on UNESCO MAB Programme (Table 4.17).

Table 6.16: Biosphere reserves of India (18 sites)

Year	Name	Location	State	Type	Key fauna	Area (km²)
1. 1986	Nilgiri Biosphere Reserve	Part of waynad, Nagarhole, Bandipur and Mudumalai, Nilambur, Silent Valley	Tamil Nadu, Kerala and Karnataka	Western Ghats	Nilgiri tahr, lion-tailed macaque	5520
2. 1988	Nanda Devi Biosphere Reserve	Parts of Chamoli District, Pithoragarh District & Bageshwar District	Uttarakhand	Western Himalayas	Snow Leopard, Himalayan Black Bear	5860
3. 1989	Gulf of Mannar	Indian part of Gulf of Mannar extending from Rameswaram island in the North to Kanyakumari in the South of Tamil Nadu and Sri Lanka	Tamil Nadu	Coasts	Dugong	10500
4. 1988	Nokrek	In west Garo Hills	Meghalaya	East Himalayas	Red panda	47.48
5. 1989	Sundarbans	Part of delta of Ganges and Brahmaputra river system	West Bengal	Gangetic Delta	Royal Bengal tiger	9630
6. 1989	Manas	Part of Kokrajhar, Bongaigaon, Barpeta, Nalbari, Kamrup and Darrang Districts	Assam	East Himalayas	Asiatic Elephant, Tiger, Assam roofed turtle, hispid hare, Golden Langur, pygmy hog	2837
7. 1994	Simlipal	Parts of Mayurbhanj district	Odisha	Deccan Peninsula	Gaur, royal Bengal tiger, Asian elephant	4374
8. 1998	Dihang-Dibang	Parts of Siang and Dibang Valley	Arunachal Pradesh	Eestern Himalaya	Mishmi Takin, Musk Deer	5112
9. 1999	Pachmarhi Biosphere Reserve	Parts of Betul District, Hoshangabad District and Chhindwara District	Madhya Pradesh	Semi-Arid	Giant squirrel, flying squirrel	4981.72
10. 2005	Achanakmar-Amarkantak Biosphere Reserve	Part of Annupur, Dindori and Bilaspur districts	Madhya Pradesh, Chhattis-garh	Maikala Hills	Four-horned antelope, Indian wild dog, Sarus crane, White-rumped vulture, Philautus (Sacred grove bush frog)	3835
11. 2008	Great Rann of Kutch	Part of Kutch, Morbi, Surendranagar and Patan Districts. It is the largest Biosphere Reserved in India	Gujarat	Desert	Indian wild ass	12454

Table 6.16: Biosphere reserves of India (18 sites)

	Year	Name	Location	State	Type	Key fauna	Area (km²).
12.	2009	Cold Desert	Pin Valley National Park and surroundings; Chandratal and Sarchu & Kibber Wildlife Sanctuary	Himachal Pradesh	Western Himalayas	Snow Leopard,	7770
13.	2000	Khangchendzonga	Parts of Kangchenjunga	Sikkim	East Himalayas	Snow leopard, red panda	2620
14.	2001	Agasthyamalai Biosphere Reserve	Neyyar, Peppara and Shenduruny Wildlife Sanctuary and their adjoining areas	Kerala, Tamil Nadu	Western Ghats	Nilgiri tahr, elephants	3500.08
15.	1989	Great Nicobar	Southern most islands of Andaman and Nicobar islands	Andaman and Nicobar islands	Islands	Saltwater crocodile	885
16.	1997	Dibru-Saikhowa	Part of Dibrugarh and Tinsukia districts	Assam	East Himalayas	White-winged wood duck, water buffalo, black-breasted parrotbill, tiger, capped langur	765
17.	2010	Seshachalam Hills	Seshachalam Hill Ranges covering parts of Chittor and Kadapa districts	Andhra Pradesh	Eastern Ghats	Slender Loris elephant	4755
18.	2011	Panna	Parts of Panna District and Chhatarpur District	Madhya Pradesh	Catchment. Area of the Ken River	Tiger, chital, chinkara, sambhar and sloth bear	2998.98

Table 6.17: List of Biosphere reserves of India in UNESCO

	Name	States	Year
1.	Nilgiri Biosphere Reserve	Tamil Nadu, Kerala, Karnataka	2000
2.	Gulf of Mannar Biosphere Reserve	Tamil Nadu	2001
3.	Sundarbans Biosphere Reserve	West Bengal	2001
4.	Nanda Devi Biosphere Reserve	Uttarakhand	2004
5.	Nokrek Biosphere Reserve	Meghalaya	2009
6.	Pachmarhi Biosphere Reserve	Madhya Pradesh	2009
7.	Simlipal Biosphere Reserve	Odisha	2009
8.	Great Nicobar Biosphere Reserve	Great Nicobar	2013
9.	Achanakmar-Amarkantak Biosphere Reserve	Chhattisgarh, Madhya Pradesh	2012
10.	Agasthyamalai Biosphere Reserve	Kerala and Tamil Nadu	2016
11.	Khangchedzonga National Park	Sikkim	2018

(b) **Sanctuaries**: These are protected areas generally species oriented. For example, Citrus sanctuary, Napenttus sanctuary, Orchid sanctuary, Great Indian Bustard sanctuary, Bird sanctuary etc. The size of sanctuary area ranges from 0.61 to 7818 sq. km. This is a notified area as per Wildlife Protection Act, 1972. The boundaries in sanctuaries are not sacrosant. The biotic interference in sanctuaries is limited, but tourism is permissible.

(c) **National Parks**: These are hitched to the habitat for particular wild animals. The size of National Parks varies from 0.04 to 3162 sq. km. The boundaries of National Parks and also Biosphere reserves are circumscribed by an act of State Legislature, and are, therefore, alterable only by the Legislature. There should not none whatsoever activities in National Park. Limited tourism is permitted in National Park also.

(d) **Tiger Reserves**: As per the recommendations of a special task force of the Indian Board of Wildlife, Project Tiger was launched with the following objectives:

- To ensure maintenance of available population of tiger in India for scientific, economic, aesthetic, cultural and ecological values;
- To preserve, for all times, the areas of such biological importance as a national heritage for the benefit, education and enjoyment of the people.

In the year 2009, there are 29 Tiger Reserves spreading over in 14 states and covering an area of about 37,761 sq. km area. Locational details of various Tiger reserves is given in Table 6.18(a).

Table 6.18(a): Name of the Tiger Reserves in Tiger Range State with year of Creation and Area

Sl. No.	Year of Creation	Name of Tiger Reserve	State	Total Area in Sq. Kms.
1.	1973-74	Bandipur	Karnataka	866
	1999-2000	Nagarhole (extension)		643
2.	1973-74	Corbett	Uttar Pradesh/Uttarakhand	1,316
3.	1973-74	Kanha	Madhya Pradesh	1,945
4.	1973-74	Manas	Assam	2,840
5.	1973-74	Melghat	Maharashtra	1,677
6.	1973-74	Palamau	Bihar	1,026
7.	1973-74	Ranthambhore	Rajasthan	1,334
8.	1973-74	Similipal	Odissa	2,750
9.	1973-74	Sundarbans	West Bengal	2,585
10.	1978-79	Periyar	Kerala	777
11.	1978-79	Sariska	Rajasthan	866
12.	1982-83	Buxa	West Bengal	759
13.	1982-83	Indravati	Madhya Pradesh	2,799
14.	1982-83	Nagarjunsagar	Andhra Pradesh	3,568
15.	1982-83	Namdapha	Arunachal Pradesh	1,985
16.	1987-88	Dudhwa	Uttar Pradesh	811
	1999-2000	Katermaghat (extension)		551
17.	1988-89	Kalakad-Mundanthurai	Tamil Nadu	800

Contd.......

Table 6.18(a): Name of the Tiger Reserves in Tiger Range State with year of Creation and Area

Sl. No.	Year of Creation	Name of Tiger Reserve	State	Total Area in Sq. Kms.
18.	1989-90	Valmiki	Bihar	840
19.	1992-93	Pench	Madhya Pradesh	758
20.	1993-94	Tadoba-Andheri	Maharashtra	620
21.	1993-94	Bandhavgarh	Madhya Pradesh	1,162
22.	1994-95	Panna	Madhya Pradesh	542
23.	1994-95	Dampha	Mizoram	500
24.	1998-99	Bhadra	Karnataka	492
25.	1998-99	Pench	Maharashtra	257
26.	1999-2000	Pakhui-Nameri	Arunchal Pradesh-Assam	1,206
27.	1999-2000	Bori, Satpura, Panchmari	Madhya Pradesh	1,486
28.	1999-2000	Namari	Assam	344
29.	2006-2007	Kaziranga	Assam	859
Total				**38,620**

As per notification of MOEFCC (2017), the tiger reserve number goes up to 50. A new list is appended here [Table 6.18(b)]

Table 6.18(b): Recent Tiger Reserves Area

Sl. No.	Name of Tiger Reserve	State	Total Area in sq. kms.
1.	Anamalai	Tamil Nadu	1479.87
2.	Udanti Sitanadi	Chattisgarh	1842.54.
3.	Satkosia	Odisha	963.87
4.	Achanakmar	Chattisgarh	914.017
5.	Dandeli Anshi	Karnataka	1097.514
6.	Sanjay Dubri	Madhya Pradesh	1674.502
7.	Mudumalai	Tamil Nadu	688.59
8.	Nagarahole	Karnataka	1205.76
9.	Parambikulam	Kerala	643.662
10.	Sahyadri	Maharashtra	1165.57
11.	Biligiri Ranganatha Temple	Karnataka	574.82
12.	Kawal	Telengana	2019.12
13.	Sathyamangalam	Tamil Nadu	1408.40
14.	Mukandra hills	Rajasthan	759.99
15.	Nawegaon-Nagzira	Maharashtra	653.674
16.	Amrabad	Telengana	2611.39
17.	Pilibhit	Uttar Pradesh	730.2498
18.	Bor	Maharashtra	138.12
19.	Rajaji Tiger Reserve	Uttarkhand	1075.17
20.	Orang Tiger Reserve	Assam	492.46
21.	Kamlang Tiger Reserve	Andhra Pradesh	783.10

(e) **Elephant conservation areas**: A good number areas are now declared under elephant conservation sites. The detailed in Table 6.19.

Table 6.19: List of Elephant Reserves

Sl. No.	Elephant Range	Elephant Reserve date of notification	State	Total Area (sq. km)
I.	Eastern India (South West Bengal-Jharkhand-Odisha)	1. Mayurjharna ER (24.10.02)	West Bengal	414
		2. Singhbhum ER (26.09.01)	Jharkhand	4,530
		3. Mayurbhanj ER (29.09.01)	Odisha	3,214
		4. Mahanadi ER (20.07.02)	Odisha	1,038
		5. Sambalpur ER (27.03.02)	Odisha	427
		6. Baitami ER #	Odisha	1,755
		7. South Odisha ER #	Odisha	4,216
		8. Lamru #	Chattisgarh	450
		9. Badalkhol-lamorphingla #	Chattisgarh	1048.30
			Total	**17,092.30**
II.	North Brahamputra (Arunachal-Assam)	10. Kameng ER (19.06.02)	Arunachal	1,892
		11. Sonitpur ER (06.03.03)	Assam	1,420
			Total	**3,312**
III.	South Brahamputra (Assam-Arunachal)	12. Dihing-Patkai ER (17.04.03)	Assam	937
		13. South Arunachal ER (29.02.08)	Arunachal	1,957.50
			Total	**2,894.50**
IV.	Kaziranga (Assam-Nagaland)	14. Kaziranga-Karbi Anglong ER (17.04.03)	Assam	3,270
		15. Dhansiri-Lungding ER (19.04.03)	Assam	2,740
		16. Intanki ER (28.02.05)	Nagaland	202
			Total	**6,212**
V.	Eastern Dooars (Assam-West Bengal)	17. Chirang Ripu ER (07.03.03)	Assam	2,600
		18. Eastern Dooars ER (28.08.02)	West Bengal	978
			Total	**3,578**
VI.	E. Himalayas (Meghalaya)	19. Garo Hills ER (31.10.01)	Meghalaya	3,500
		20. Khasi Hills ER #	Meghalaya	1,331
			Total	**4,831**
VII.	Nilgiri Eastern Ghat (Karnataka-Kerala-Tamil nadu-Andhra Pradesh)	21. Mysore ER (25.11.02)	Karnataka	6,724
		22. Wayanad ER (02.04.02)	Kerala	1,200
		23. Nilgiri ER (19.09.03)	Tamil Nadu	4,663
		24. Rayala ER (09.12.03)	Andhra Pradesh	766
			Total	**13,353**
VIII.	South Nilgiri (Kerala-Tamil Nadu)	25. Nilambur ER (02.04.02)	Kerala	1,419
		26. Coimbatore ER (19.09.03)	Tamil Nadu	566
			Total	**1,985**
IX.	Western Ghat (Tamil Nadu-Kerala)	27. Anamalai ER (19.09.03)	Tamil Nadu	1,457
		28. Anamudi ER (02.04.02)	Kerala	3,728
			Total	**5,185**
X.	Periyar (Kerala-Tamil Nadu)	29. Periyar ER (02.04.02)	Keralal	3,742
		30. Srivilliputtur ER (19.09.03)	Tamil Nadu	1,249
			Total	**4,991**
XI.	NorthernIndia (Uttaranchal-UP)	31. Shivalik ER (28.10.02)	Uttaranchal	5,405
		32. Uttar Pradesh ER (09.09.09)	Uttar Pradesh	744
			Total	**6,149**
			TOTAL	**69,582.80**

Marine protected area

In addition to the above discussed protected areas, in coastal zone and islands large number of marine protected area are currently declared and conserved for biodiversity protected. This include 25 sites in peninsular part and 106 sites in islands zone specially in A & N and Lakshadweep.

An additional, 106 important coastal and marine areas which are identified as probable biodiversity conservation area by Wildlife Institute of India (WII), Dehradun.

In these protected areas, there are a number of traditional and non-traditional management practices which could be undertaken for maintaining adequate wildlife population. In cases, threatened wildlife population not only increased but also help to get back the consumptive uses also. The details of management regimes are given in Tables 6.20 & 6.21.

Table 6.20: Protected-area categories and management objectives

Category	Type	Objective
(I)	Scientific reserve/strict nature reserve	Protect nature and maintain natural processes in an undisturbed state. Emphasise scientific study, environmental monitoring and education, and maintenance of genetic resources in a dynamic and evolutionary state.
(II)	National Park	Protect relatively large natural and scenic areas of national or international significance for scientific, educational and recreational use.
(III)	National monument/ natural landmark	Preserve nationally significant natural features and maintain their unique characteristics.
(IV)	Managed nature reserve/ wildlife sanctuary	Protect nationally significant species, groups of species, biotic communities, or physical features of the environment when these require specific human manipulation for their perpetuation.
(V)	Protected landscapes	Maintain nationally significant natural landscapes characteristic of the harmonious interaction of people and land while providing opportunities for public recreation and tourism within the normal lifestyle and economic activity of these areas.
(VI)	Resource reserve	Protect natural resources for future use and prevent or contain development that could affect resources pending the establishment of managed objectives based on appropriate knowledge and planning.
(VII)	Natural biotic area/ anthropological reserve	Allow societies to live in harmony with the environment, undisturbed by modern technology.
(VIII)	Multiple use managment area/managed resource	Sustain production of water, timber, wildlife, pasture and outdoor recreation. Conservation of nature oriented to supporting economic activities (although specific zones can also be designed within these areas to achieve specific conservation objectives).

Source: IUCN, 1985

Table 6.21: Wildlife management regimes traditional and non-traditional consumptive use

Management type	Traditional	Non-traditional
Wildlife cropping (a)	Substance hunting of wild populations, regulated by rights of access and cultural taboos	Sustained yield cropping of wild populations, regulated by ecologists
Game ranching		Cropping of free-ranging but captive populations for sustainable yield
Wildlife farming (b)		Husbandry of semi-domesticated animals an intensified type of game ranching
Wildlife culling		Cropping of populations to regulate their size at or below a theoretical carrying capacity
Sport/ceremonial hunting (c)	Hunting of animals to symbolise passage to manhood (e.g. Maasai, Samburu), chiefs' royal hunting rights, annual village hunts. Regulated by custom. Trophy and sport hunting regulated by official quotas and licences	Trophy and sport hunting regulated by official quotas and licences
Problem animal control	Killing predators that prey on domestic livestock	Killing of animals to protect crops, control tsetse fly and other diseases
Non-timber forest product collection	Collection of foods, medicines, fuelwood, materials for household goods. Regulated by rights of access and cultural taboos	Harvest for large-scale commercial trade in charcoal, rattans, chewsticks, etc.
Timber harvesting	Subsistence felling for building materials, cultural symbols, canoes, etc.	Commercial logging for local and export markets.

Notes:

(a) *Wildlife cropping* is the management of large savannah mammals by periodic culling, whilst producing a sustainable yield of products such as meat, hides and skins. Cropping (or culling) is also done on an ad hoc basis to solve over-population of wildlife. Cropping can be a high-cost exercise because of the logistics of harvesting game, retrieving the carcasses and organising carcass inspection if the meat is intended for human consumption. The animals should be easily accessible. In some cases, it is necessary to invest in abattoir and refrigeration facilities. Cropping for meat is economically viable only if there is a large internal market for game meat products or if high-value products, such as zebra skins, are involved. Parker (1984) and Bell (1984) have pointed out that the most economically efficient croppers in Africa are poachers. They have an in-depth knowledge of the area and species, have low maintenance costs, use appropriate technology and do not have to satisfy stringent hygiene requirements.

(b) *Wildlife farming* is the rearing and harvesting of wild animals in artificial conditions. The animals are usually kept at high population densities and receive special diets and veterinary care. Species farmed include ostrich, crocodile and eland. It involves high initial and recurrent costs and requires good knowledge of the biology of species involved. Often the products are destined for luxury markets and have high value.

(c) *Sport/Ceremonial hunting* is a lightly consumptive utilisation option. Animals are tacked and hunted in a 'wilderness environment' and the prestige lies in taking home the trophy (e.g., head or skin). If well regulated and managed, safari hunting can deliver quick and substantial profits. This type of hunting may be less environmentally damaging than wildlife tourism, since it involves smaller numbers of tourists. A hunting operation requires a professional hunter and staff, basic infrastructure and equipment. Profits depend partly on the availability of a suitable range of 'plains game' (e.g., zebra, various antelope species) and 'big game' (e.g., lion, leopard, elephant). Returns from hunting are high and land owners/communities can charge a fee for each animal hunted. Communities can demand that the outfitter employs local labour wherever possible, and trains local people in skinning and trophy preparation.

Box 6.9: SACRED GROVES

Mundas, Oraons and Santhals are tribal groups spread over many forested districts of Madhya Pradesh, Bihar, Jharkhand, Orissa and West Bengal. Several of our study sites in Bihar, Jharkhand and Orissa fall in their tract. Here the tribals still display many traditional conservation practices. These include protection of sacred groves from which no plant material is removed. These sacred groves are called Saranas, and harbour their deities in natural forms such as stones.

These Saranas fulfil a variety of functions welcoming entrants to the village, favouring a good harvest of crops, protection against wild animals and epidemics, guarding against evil spirits, etc.

Many tribal festivals are linked to plants; sal (*Shorea robusta*) with sarhul, karam (*Adina cordifolia*) with karama, simal (*Bombax malabarica*) and seedha (*Lagerstroemia parviflora*) with weddings.

Traditionally, tribals practice an annual ritual hunt after worshipping the forest god. In this hunt they spare pregnant females and all immature animals. The herbal medicine men belonging to these communities promote protection of medicinal plants from overharvesting, fire and grazing.

These traditions are now being eroded due to two forces: organised religions like Hinduism and Christianity and commercial interests who are often helped by the Forest Department. Our investigations suggest that about 90 per cent of the total harvests of plant and animal materials from the area are for export and that those engaged in such export have little concern for sustainability.

Contd.

All the above stated conservation programmes are in-situ or on site conservation. But there are a number of cases where the conservation is made outside their habitats i.e., ex-situ conservation. These activities imply conservation of species by perpetuating sample population in genetic resources centres, zoos, botanical gardens, culture collections centre, gene bank, and seed bank etc.

Botanical Gardens play an important role in the conservation of plant species so much so that there are several instances when plants believed to be extinct were found living only in botanical gardens. Many endemic species, rare and endangered species could be maintained in botanical garden as a germplasm stock. The Indian Green Book prepared by Botanical Survey of India lists 100 such plant species which are rare, endangered or endemic but all growing in living state in some botanic gardens of the country.

Zoological Park/Garden also acts as a centre for conservation of animals also. It is also the favourable site for the captive breeding of rarest animals. There are a number of zoo gardens set up at different cities of the country for public recreation as well as research and development activities on selected animals.

Seed Bank is a mode of germplasm conservation by storing seeds at cold storage (at 4 to 15°C). A good number of seed banks are already set up in the country for storing cultivated crop species. By this method recalcitrant seeds (i.e. seeds which cannot be stored at low moisture level) cannot be stored for long time. Dry storage or cryogenic storage may be the alternative methods. Storage of tissues through callous embryo, pollen and sperm etc are also the various modes of germplasm conservation.

A **national Gene Bank** is set up as national plant/animal genetic resource centre, maintaining many different germplasm samples of current and potential interest to people working in research nationally. In recent FAO report (1996), there are over 1,308 gene banks found in the world of which 496 banks are located in Europe, 101 in USA, 293 in Asia, and 227 in Latin American countries. National Bureau of Plant Genetic Resources (NBPGR), India, having more of 175,000 accessions of plant types stored in gene bank. This number is increased by 1,500 to 2,000 per year.

Box 6.10: BISHNOIS—An Eco-religion

The conservation of biodiversity has now become a global concern. It was internationalised at the Earth Summit held in 1992. Different countries are presently evolving strategies for economic development and the sustainable use of their biodiversity by trying to involve indigenous communities. But can one imagine that more than 450 years ago, one simple villager from a remote desert area, without the benefit of even basic formal education, clearly understood the importance of preserving biodiversity? They not only understood it themself but had the wisdom of influencing generations of people to preserve it by weaving it with their religion.

Guru Jambeshwarji or Jamboji, as he is now affectionately referred to by his followers, was a great saint and a philosopher of the medieval period of Indian history. The rulers and religious heads of that time exploited the poor in the name of rituals. Jamboji raised his voice against this exploitation. He launched a new sect and prescribed 29 tenets. The followers of these 29 tenets are called Bishnois (literally meaning 'twenty-niners' in Hindi).

Of the 29 tenets laid down by him, 10 are directed towards personal hygiene and maintaining good basic health. These provide simple instructions like drinking filtered water, taking a daily bath and improving sanitary conditions.

Eight tenets have been prescribed to preserve biodiversity and encourage good animal husbandry. These include a ban on killing of all animals and felling green trees and providing protection to all life forms. The followers are even directed to see that the fire-wood is devoid of small insects before burning it in their hearths. Even the wearing of blue clothes is prohibited, because the dye for colouring them is obtained by cutting a large quantity of shrubs.

Seven tenets provide directions for healthy social behaviour. These direct the followers to live a simple, truthful life and avoid making false arguments. Criticising others is prohibited and in the event of being criticised by others, tolerance is to be shown. The followers are advised to donate for a good cause but the donations should be given only to the right type of people.

Five tenets provide guidelines for worshiping God daily and to always remember that God is omnipresent. On every *amavasya* (new moon night), a fast should be observed and at regular intervals *havan* (lighting of holy fire) is to be performed for the salvation of the soul. In a Bishnoi temple, no statue of any deity is kept. Only a lamp filled with "ghee" (clarified butter) is kept lit in front of a painting of Guru Jamboji.

Contd.

While praying, no elaborate ceremonies or rituals are performed by the priests. With folded hands, they repeat the teachings of their guru Jamboji or the *shabd* (words) spoken by him. Bishnois are basically worshipers of Lord Vishnu.

Because of its simplicity, a large number of people, mostly farmers of the region, adopted the Bishnoi sect willingly. Over the years it spread to other parts of India and the number of followers increased. Presently, Bishnois are spread over the western parts of Rajasthan and parts of Haryana and Punjab.

Perhaps because of their eco-friendly life, the Bishnois are more prosperous than other communities living in the Thar desert. With plenty of trees and other vegetation and herbs of antelopes roaming freely near their homes, their villages are easily distinguishable. They generally raise only one crop of *bajra* (*Pennisetum typhoides*) during the monsoon season. The crop is entirely dependent on the uncertain rainfall. The fields are ploughed with simple ploughs using bullocks or camels. This ploughing causes least damage to the fragile desert eco-system. It does not interfere with the small bushy growth of xerophytic vegetation like the ber and phog which grow in the fields. These bushes protect the loose sand from wind erosion and provide the much needed fodder for animals during a famine.

The Bishnois not only live an eco-friendly life, they are eco-friendly even in their death rites. Despite being staunch Hindus, they do not burn their dead. The dead are buried to save precious wood and trees. They are strict vegetarians, teeto tallers and non-smokers.

Bishnoi men are well-built, hard-working and friendly. They normally wear white shirts, dhotis and a loosely tied white turban on their heads. This dress is ideal for the hot dry desert climate.

The Bishnoi women are beautiful and sharp-featured. They are fond of donning gorgeous gold ornaments like the borla (on the forehead), nose rings, earrings and necklaces. They also put on heavy silver anklets. Like their men, the women too are hard-working. They collect fuel, fetch water, grind flour, cook meals and churn curd for butter. During the crop season, they also help their men in hoeing and weeding.

The traditional Bishnoi houses have a conical thatched roof and mud-plastered walls. These houses provide remarkable protection against the hot summer. There are separate chambers for living and cooking. Water is stored in underground tanks called "tankas" by collecting rain water. Cattle are kept in a separate compound adjoing the house. A shed made of thatch is provided in the compound.

Bishnoi history, although not well-chronicled, was recorded by *bhaats* (nomads) who kept family records of all marriages, births and deaths, etc. The *bhaats* were paid by the family in kind. According to them, in 1737, an official of Jodhpur State started felling a few Khejri trees in Khejerli village for burning of lime needed for the State. This was opposed by the Bishnois of the village. The initiative in this case was taken by a woman. Amritadevi, a mother of three minor daughters, was the first to sacrifice her life by hugging the tree that was being axed. She defied the orders of the State and tried to prevent the felling of a Khejri tree. She was killed. Her example was followed first by her three daughters and later on by a long chain of Bishnoi men, women and children.

For every tree felled in Khejerli village, a man, woman or a child made the extreme sacrifice. In all, 363 Bishnois from Khejerli and adjoining villages sacrificed their lives.

When this news reached the Maharaja of Jodhpur, he apologised for the mistake committed by his official and issued a royal decree, engraved on a copper plate, prohibiting the cutting of trees and hunting of animals in all Bishnoi villages in the State. It was also ordered that if by mistake any individual violate this order, he would be prosecuted by the State and a severe penalty imposed. Even the members of the ruling family did not shoot animals in or even near the Bishnoi villages. A monument and a temple has been raised at this site in Khejerli village in memory of these 363 martyrs.

Every year, on the day of this sacrifice, Bishnois in their thousands assemble here to observe the extreme sacrifice made by their people for the sake of their faith and religion. Khejri trees and the antelopes, particularly the blackbuck and chinkara, are aggressively protected by Bishnois even now. There is a saying that goes "*Sir santhe rookh rahe to bhi sasto jaan*" (If a tree is saved from felling at the cost of one's head, it should be considered as a good deed). The antelopes are not only protected by them but they also share their food and water with them. In a number of villages the animals are hand-fed by the Bishnois. There will hardly be any Bishnoi village in Rajasthan where stories of trigger-happy hunters are not narrated with pride.

Contd.

The sound of a gunfire in a Bishnoi village immediately attracts men, women and children. They surround the hunter and make his escape impossible. In case the poachers escape after leaving the dead antelope in an agricultural field, the owner of the field will mourn its death like that of a relative. He will not eat food or drink water till the last rites for the dead animal are performed. In their bid to escape, the hunters sometimes fire at these people also. At times, the Bishnois are injured and even killed by the hunters. The stories of such sacrifices are narrated with a sense of pride by the villagers. Recently, in Lohawat village, a school teacher named Birbal Bishnoi lost his life in an attempt to catch a poacher who had killed a chinkara. With the increased use of four-wheel drive jeeps by poachers, the Bishnois now come out of the houses at the sound of a jeep in their fields.

It is because of this rigid protection provided by the entire Bishnoi community that the blackbuck and chinkara roam fearlessly in Bishnoi settlements. It is for this environmental awareness and commitment that the Bishnois stand apart from countless other sects and communities in India.

6.15 CBD IMPLEMENTATION PROCESS IN INDIA

Two substantive developments in India have been the development of a draft Bill on the Biodiversity Act and the preparation of national status report which, unfortunately, India failed to submit at the Bratislava meeting of the COP.

The draft Bill is entitled "Biological Diversity Act, 1998". One feature of the Act is that it proposes the establishment of a National Biodiversity Authority which will comprise a chairperson, representatives of the Ministry of Environment and Forests, Department of Agricultural Research and Education, Department of Biotechnology, Department of Ocean Development, five non-official members who are experts and scientists familiar with biodiversity issues and the provisions of the CBD and representatives of Indian industry.

Functionally, the Authority shall look at issues of conservation, IPRs, traditional knowledge and will frame guidelines for access and benefit-sharing. For implementation of the Act, State/Union Territory Biodiversity Boards and Biodiversity Management Committees in Local Bodies have to be set-up. These units will advise the respective people on issues relating to biodiversity as well as help implement the Act.

Other important clauses of the Act include:

- Prohibition of transfer of biological material and/or associated knowledge by non-Indians without the prior approval of the National Authority.
- Stipulation that anyone seeking a patent or any kind of IPR needs to seek prior permission of the Authority.
- Provision for a levy of fees and royalties on such applications.
- Provision for respecting and protecting the knowledge of local people through measures such as registration of such knowledge at local, state and national levels and development of and adjustments to the *sui-generis* system for intellectual property protection of such knowledge.
- Provision to ensure that those who obtain prior informed consent will honour terms and conditions that will ensure a fair and equitable benefit-sharing process. Such processes will be notified to the public.
- Provisions for establishing Biological Diversity Heritage Sites recommended by the State Boards in consultation with local units of self-government.
- Proposes the establishment of National, State and Local level Biodiversity Funds to be created from fees, royalties and monetary benefits resulting from approvals granted by the National Authority.
- Envisages strict punishment with a fine up to Rs. 10 lakhs and/or imprisonment up to 5 years for those who infringe the provisions in the Act.

The implementation of the provisions in the CBD has begun in India but what is required is to 'catch-up' with the process and ensure that the interrelationship of the CBD does not run counter to the TRIPS provisions and other agreements. There is an increasing role which the NGO community can play in the process. Unfortunately, there is still no dialogue between the Government, NGOs and other stake-holders on the issues and positions that India needs to take at a COP. This is very much required so that the view can be balanced. It will also ensure that 'lobbying' at the COP is more effective.

6.16 LEGAL COVERAGE OF BIODIVERSITY CONSERVATION AND USE IN INDIA

Each country needs to assess the legal coverage of biodiversity conservation and use within its boundaries. Such an assessment has been broadly carried out for India (Kothari and Singh 1992). This assessment first provides a conceptual framework within which to examine each law, and deals not only with conventional conservation laws but also other laws relating to development, use and distribution of resources.

For the sake of the analysis, Kothari and Singh (1992) categorised biodiversity components under five headings: wild fauna, wild flora, domesticated fauna (livestock), domesticated flora (crops and other cultivated plants), and genetic materials (seeds, eggs, germplasm, semen). They also distinguished 11 distinct aspects of biodiversity:

1. *Identification*: The process of locating, and thereafter establishing the identity, distribution, occurrence, status, and value of biodiversity components. Relevant to both wild and domesticated fauna and flora.

2. *Protection in situ*: Efforts to protect and conserve biodiversity, including the habitats within which biodiversity components survive and evolve. Applies especially to wild fauna and flora, but also to domesticated fauna and flora at the sites of their use, in so far as these may need to be protected from disease/infestation, displacement by new varieties and destruction from human activities.

3. *Protection ex-situ*: Attempts at preserving living species or genetic material in gene banks, zoological parks, botanical gardens and sites other than their natural habitats. Relevant to all bioidiversity components.

4. *Access/Extraction*: Controls on the kind and amount of access to fauna and flora in the wild, to *ex-situ* collections, and to genetic material; also the modes of extraction of these components, relevant expecially to *in-situ* stock. Overlaps with intellectual property rights (see aspect number 11 next page).

5. *Use*: Attempts at utilizing biodiversity for subsistence, commercial, scientific or other purposes. Relevant to all biological components.

6. *Breeding/Cultivation/Multiplication*: Artificial or induced measures for regenerating flora-fauna populations in captive or closed conditions. Relevant to both wild and domestic fauna and flora.

7. *Trade*: Activities relating to the barter, sale, import, export and other forms of exchange of biodiversity components.

8. *Introduction/Augmentation/Re-introduction*: Measures to deliberately stock an ecosystem or area with species that did not previously exist there (introduction) or where they have declined or need to be increased in number (augmentation) or where they have died out (reintroduction). Relevant to both wild and domesticated fauna and flora, as also to genetic material (e.g., seeds) used for the purpose.

9. *Release*: Disposal of biodiversity components, without the specific purpose of introducing, augmenting, on reintroducing them, into the environment. Relevant to all biodiversity components.

10. *Movement*: Measures to transport, by human means, biodiversity components from one location to another, especially relevant in cases of transportation across ecosystems or bioregions, involving the possibility of accidental release of components. Does not include natural or voluntary movement of animals, or movement of seeds or other floral parts by natural means. Relevant to all biodiversity components.

11. *Intellectual property rights*: Measures pertaining to the rights of the state, organisations, or individuals over biological and biotechnological knowledge, including patents, royalties, intellectual property rights, farmers, rights and breeders' rights.

Over 40 National Acts were then assessed with respect to their coverage (or lack of it) of each of the 11 aspects listed above for each of the five biodiversity categories (i.e., wild flora, wild fauna, domesticated flora, etc.). The results of these analysis are presented in Box 6.11.

Box 6.11

Type of activity	Wild flora	Wild fauna	Domestic flora	Domestic fauna	Genetic materials
1. Identification	N	N	N	N	N
2. Protection *in-situ*	W	W	N	N	N
3. Protection *ex-situ*	W	P	N	N	N
4. Access/Extraction	P	P	N	N	N
5. Use	W	P	W	N	N
6. Trade	W	W	P	W	P
7. Breeding/Cultivation/Multiplication	W	P	P	P	P
8. Introduction/Augmentation/Re-introduction	P	P	P	P	P
9. Release	N	N	N	N	P
10. Movement	W	P	P	P	P
11. Intellectual property rights	N	N	N	N	N

N = Not covered; P = Partially covered; W = Well-covered
Source: Kothari and Singh (1992).

Selected rare, endangered plants/animals of different states are presented in Table 6.22 & Table 6.23.

Table 6.22: Selected rare, endangered and endemic plant species of India (Statewise)

Name of the states	Rare, endangered and threatened plants	Endemic plants
Andaman & Nicobar Islands	*Crinum pusillum* *Dioscorea vexans* *Garcinia calycina* *Henslowia erythrocarpa* *Hypoestes andamanica* *Uvaria nicobarica*	*Antidesma andamanica* *Artabotrys nicobarianus* *Bentinckia nicobarica* *Eulophia nicobarica*
Andhra Pradesh	*Andrographis beddomei* *Argyrea arakuensis* *Nilgiranthus circarensis* *Pimpinella tirupatiensis* *Gnetum ula*	*Bupleurum andhricum* *Kalanchoe cherukondensis* *Neuracanthus neesianus*
Arunachal Pradesh	*Abies delavayi* *Buddleja yunnanensis* *Coptis teeta* *Eria discolor* *Galeola falconri* *Rheum nobile* *Rhododendron santapaui*	*Albizia arunachalensis* *Eurya arunachalensis* *Hopea shinkgkeng* *Leycesteria dibangvalliensis* *Primula subansirica* *Rhododendron santapaui*

Contd.

Assam	*Melocalamus mastersii* *Miscanthus wardii* *Robynslochloa goalparensis*	*Acacia diadenia* *Chonemorpha assamensis* *Dendrobium assamicum* *Diospyros cacharensis* *Glochidion assamicum*
Bihar	*Berberis asiatica* *Chloris montana* *Clematis gouriana* *Leea alata* *Murdania blumei*	*Carum villosum* *Chrysopogon hamiltonii* *Sophora bakeri*
Chandigarh		
Dadra & Nagar Haveli		
Daman and Diu		*Artiplex stocksii* *Crotalaria filipes* *Hibiscus talbotii* *Ixora brachiata* *Neanotis rheedi*
Delhi	*Crotalaria hirta* *Cuscuta chinensis* *Decaschistia crotonifolia* *Lotus corniculatus* *Melhania futteyporensis* *Petunia parviflora* *Vahlia digyna,*	
Goa	*Canarium strictum* *Ceropegia fantastica*	*Glyphochloa goaensis* *Glyphochloa talbotii* *Hyphaene dichotoma*
Gujarat	*Alysicarpus ovalifolius* *Blumea bovei* *Canscora concanensis* *Carum villosum* *Cyperus dwarkensis* *Dignathia hirtella* *Exacum lawii* *Nepata hindostana*	*Aglaia haselettiana* *Canscora khandalensis* *Cyperus drawkensis* *Helichryum cutchicum* *Pycnocyclea glauca* *Rhynchosia hainesiana* *Tamarix kutchensis* *Tephrosia jamnagarensis*
Haryana	*Erigeron sublyratus* *Erythrina suberosa* *Exacum pedunculatum* *Hemigraphis latebrosa* *Vicoa vestita*	*Crotalaria mysorensis* *Pueraria tuberosa* *Trema politoria*
Himachal Pradesh	*Aconitum heterophyllum* *Astragalus munori* *Ceropegia longifolia* *Limosella aquatica*	*Falconeria himalaica* *Orobanche hansii* *Viola biflora*
Jammu & Kashmir	*Berberis huegeliana* *Crepis nanifoma* *Puccinellia thomsonii*	*Aconitum kashmiricum* *Allium gilgiticum* *Festuca lavingei*
Karnataka	*Brachystelma rangacharii* *Heliotropium rottleri* *Jasminum wightii* *Parahyparrhenia bellariensis* *Streptocaulon kleinii*	*Acanthopale jogensis* *Christisonia flammea* *Cryptolepis elegans* *Hoppea canarensis* *Leuca angustissima*

Contd.

Kerala	*Acacia wightii* *Butea purpurea* *Crotalaria barbata* *Eria albiflora* *Hoya wightii* *Ixora johnsonii*	*Adhatoda beddomei* *Aglaia canarensis* *Aporosa bourdillonii* *Gluta travancorica*
Lakshadweep	*Acalypha wilkesiana* *Acorus calamus* *Euphorbia articulata* *Lepturus repens* *Pancratium zeylanicum* *Thuarea involuta* *Entada pursaetha*	
Madhya Pradesh	*Acacia donaldi* *Alstonia venenata* *Andropogon pumilus* *Calpurnia aurea* *Cleome felina* *Flemingia grahaminana* *Hypericum gaittii* *Nogra dalzelli*	*Alysicarpus vasavadae* *Berberis hainesii* *Curcuma sulcata* *Ficus cupulate* *Grewia hainesiana* *Hibiscus pachmarhiana*
Maharashtra	*Achyranthes coynei* *Arisaema caudatum* *Canscora concanensis* *Dimeria blatteri* *Seghagiria sahyadrica* *Vepris bilocularis*	*Bidaria khandalensis* *Derris thothathrii* *Euphorbia katrajensis* *Habenaria panchganiensis* *Vigna khandalensis*
Manipur	*Acer laevigatum* *Aconitum elwesii* *Begonia obversa* *Canarium strictum* *Dalhousiea bracteata* *Hedychium greenii*	*Aldrovanda vesiculosa* *Clematis apiculata* *Ilex embelioides* *Sapria himalayana*
Meghalaya	*Anthoxanthum clarkei* *Gastrodia exilis* *Panax pseudo-ginseng* *Pleione maculata* *Sapindus rarak* *Unona dumosus* *Vanda coerulea*	*Cyathocalyx martabanicus* *Habenaria khasiana* *Ilex khasiana* *Nepenthes khasiana*
Mizoram	*Arundina chinensis* *Arundina phar* *Crotalaria kurzii* *Coelogyne nitida* *Dendrobium falconeri*	*Acampe ochracea* *Begonia lushaiensis* *Cymbidium macrorhizon* *Cyanotis lushiensis* *Derris lushiensis* *Cynotis lushiensis* *Vernonia parryae*
Nagaland	*Cymbidium tigrinum* *Kalanchoe roseus* *Livistona jenkinsiana* *Michelia punduana* *Renanthera imschootiana* *Senecio rhabdos*	*Chaerophyllum orientalis* *Begonia watti* *Berberis micropetala* *Corydalis boweri* *Coelogyne hitendrae* *Calamagrostis nagensis* *Themeda huttonensis*

Contd.

Odissa	*Cerbera odollam* *Ceropegia hirsuta* *Mollugo disticha* *Nothopegia heyneana* *Polyalthia korintii* *Tinospora sinensis* *Tripogon roxburghianus*	*Bulbophyllum panigrahiana* *Dimeria orissae* *Eria meghasaniensis* *Eriocaulon echinulatum* *Heritiera kanikensis*
Punjab	*Sesamum mulayanum* *Withania coagulens*	*Argyrolobium album* *Hibiscus hoshiarpurensis* *Rumex punjabensis*
Rajasthan	*Ammania desertorum* *Citrullus colocynthis* *Commiphora wightii* *Tecomella undulata*	*Cenchrus rajasthanesis* *Dicliptera abuensis* *Euphorbia jodhpurensis* *Pulicaria rajputanae* *Tribulus rajasthanensis*
Sikkim	*Acronema nervosa* *Acronema wolffiana* *Aphyllorchis parviflora* *Begonia rubella* *Bulleyia yunnanensis* *Calanthe alpina* *Carex kingiana* *Ceropegia lucida* *Cheirostylis whiteana* *Lloydia himalensis* *Phoenix rupicola* *Pimpinella tongloensis* *Pternopetalum radiatum*	*Arneria thangoensis* *Astragalus zemuensis* *Aster sikkimensis* *Blumea sikkimensis* *Carex kingiana* *Elaeocarpus sikkimensis* *Gentiana prainii* *Gentiana recurvata* *Juncus sikkimensis* *Ligularia yakla* *Primula hemispherica*
Tamil Nadu	*Acacia hohenackeri* *Bentickia condapanna* *Ceropegia barnesii* *Clematis theobromina* *Crotalaria kodaiensis* *Dalbergia tinnevelliensis*	*Berberis nilghiriensis* *Dalbergia coromandeliana* *Eugenia singampattiana* *Indigofera tirunelvelica* *Sageretia coimbatorensis* *Syzygium tamilnadensis*
Tripura	*Begonia surculigera* *Jasminum listeri* *Mangifera sylvatica* *Ophiorrhiza villosa* *Wallichia caryotoides,*	
Uttar Pradesh	*Androsace garhwalicum* *Carex nadadeviensis* *Derris kanjilalii* *Eulophia ucbii* *Spiraea panigrahiana* *Themeda dacruzii*	*Carex nadadeviensis* *Festuca nandadevica* *Polygonum kumaonum* *Pygaeophyton garhwalensis*
West Bengal	*Acer osmastonii* *Aldrovanda vasiculosa* *Bulleyia yunnanensis* *Christella clarkii* *Ceriops tagal* *Heritiera fomes* *Nypa fruticans*	*Bulbophyllum roxburghii* *Lindenbergia titensis* *Cuscuta sharmanum* *Dalbergia duarensis* *Hydrocotyle himalaica*

Table 6.23: Rare and Endangered Animal Species in India

The following is the list of highly endangered or rare mammals, reptiles and amphibians and birds according to Schedule One of the Wildlife (Protection) Act, 1972 as updated in 2001

MAMMALS

1. Andaman Wild Pig
2. Bharal
3. Binturang
4. Black Buck
5. Blue Whale
6. Brown-antlered Deer
7. Capped Langur
8. Caracal
9. Cetecean species
10. Cheetah
11. Chinese Pangolin
12. Chinkara or Indian Gazelle
13. Clouded Leopard
14. Crab-eating Macaque
15. Desert Cat
16. Dugong
17. Fishing Cat
18. Four-Horned antelope
19. Dolphin (Gangetic)
20. Flying Squirrels
21. Golden Cat
22. Golden Langur
23. Himalayan Ibex
24. Himalayan Thar
25. Hispid Hare
26. Hog Badger
27. Hoolock Gibbon
28. Hump-backed Whale
29. Indian Elephant
30. Indian Lion
31. Indian Wild Ass
32. Indian Wolf
33. Kashmir Stag
34. Leaf Monkey
35. Leopard or Panther
36. Leopard Cat
37. Lesser or Red Panda
38. Lion-tailed Macaque
39. Loris
40. Lynx
41. Malabar Civet
42. Malay or Sun Bear
43. Marbled Cat
44. Markhor
45. Mouse Deer
46. Musk Deer
47. Nilgiri Tahr
48. Ovis Ammon or Nyan
49. Pallas's Cat
50. Pangolin
51. Pygmy Hog
52. Ratel
53. Rhinoceros
54. Rusty-Spotted Cat
55. Serow
56. Sloth Bear
57. Slow Loris
58. Small Travancore Flying Squirrel
59. Snow Leopard
60. Spotted Linsang
61. Swamp Deer
62. Takin or Mishmi Takin
63. Tibetan Antelope or Chiru
64. Tibetan Gazelle
65. Tibetan Wild Ass
66. Tiger
67. Urial or Shapu
68. Wild Buffalo
69. Wild Yak

AMPHIBIANS AND REPTILES

1. Agra Monitor Lizard
2. Atlantic Ridley Turtle
3. Barred, Oval or Yellow Monitor Lizard
4. Crocodiles
5. Gharial
6. Ganges soft-shelled Turtle
7. Green Sea Turtle
8. Hawksbill Turtle
9. Himalayan Newt or Salamander
10. Indian egg-eating Snake
11. Indian soft-shelled Turtle
12. Indian Tent Turtle
13. Large Bengal Monitor Lizard
14. Leathery Turtle
15. Logger-head Turtle
16. Olive-back logger-head Turtle
17. Peacock-marked soft-shelled Turtle
18. Pythons
19. Three-keeled Turtle
20. Tortoise
21. Viviparous Toad
22. Water Lizard

BIRDS

1. Andaman Teal
2. Assam bamboo Partridge
3. Bazas
4. Bengal Florican
5. Black-necked Crane
6. Blood Pheasants
7. Brown headed Gull
8. Cheer Pheasant
9. Eastern White Stork
10. Forest spotted Owlet
11. Great Indian Bustard
12. Great Indian Hornbill
13. Hawks
14. Hooded Crane
15. Hornbills
16. Houbara Bustard
17. Humes bar-backed Pheasant
18. Indian Pied Hornbill
19. Jerdon's Courser
20. Lammergeier
21. Large Falcons
22. Large Whistling Teal
23. Monal Pheasants
24. Mountain Quail
25. Narcondom Hornbill
26. Nicobar Megapode
27. Nicobar Pigeon
28. Osprey or Fish eating Eagle
29. Peacock Pheasant
30. Peafowl
31. Pinkheaded Duck
32. Scalater's Monal
33. Siberian While Crane
34. Spur Fowl
35. Tibetan Snow-cock
36. Tragopan Pheasants
37. White-bellied Sea Eagle
38. White-eared Pheasant
39. White Spoonbill
40. White-winged Wood Duck

Box 6.12: Bhimbetka Rock Shelters & Its Painting

The Rock Shelters of Bhimbetka are in the foothills of the Vindhyan Mountains on the southern edge of the central Indian plateau. Within massive sandstone outcrops, above comparatively dense forest, are five clusters of natural rock shelters, displaying paintings that appear to date from the Mesolithic Period right through to the historical period. The cultural traditions of the inhabitants of the twenty-one villages adjacent to the site bear a strong resemblance to those represented in the rock paintings.

There is widespread evidence from nearly all over India of the workmanship of the Homo erectus, the earliest human ancestor, in the form of Palaeolithic tools. Direct and indirect evidences so far indicate the earliest presence of this ancestral form in the country approximately 160,000 years ago (based on evidences from the Didwana region, Rajasthan).

About 40 kms from Bhimbetka in the alluvial cemented gravels of the river Narmada at the site Hathnora (district Sehore), a cranium of the the Narmada hominid was found along with a large number of Lower Palaeolithic tools. The evidences together suggest a high density of human occupation in and around the Narmada - at a conservative estimate for the zone - about 100,000 years ago. As all these discoveries have been made from alluvial gravels, they are secondary in nature and do not reveal where they were made and used. Primary sites are rare, and of the Old World those of the Pleistocene Age (28,00,000 BP to 10,000 BP) constitute less than 1 per cent of all discovered sites. Bhimbetka is one such rare primary site. The discovery of Bhimbetka, therefore, is not only of vital significance for the prehistorian but in many ways provides evidences which are unique in the entire world.

Set in the Vindhyan range in Madhya Pradesh in central India, the site complex is an integral part of the natural undulating formation with an undisturbed geomorphological character and eco-system - spread over the seven low sedimentary hills of Bineka, Bhonrawali, Bhimbetka, Lakha Juar (East), Lakha Juar (West), Muni Baba ki Pahari and Jaora. Massive naturally sculpted rock formations through wind erosions crest the Bhimbetka hill making it an easily identifiable visible landmark for kilometres around. Covered with protected forests and waterways of the Ratapani Wildlife Sanctuary, within the northern periphery of which the site falls, are found natural clusters of rock shelters integrated with the landscape spread over the upper parts of the hills which exhibit a history and evolution of man through several cultural layers.

While the contents of the shelters have revealed a continuity, elements of these are also observed in the continuing traditional expressions in the lifestyles of the surrounding adivasi settlements of the Gonds, Pradhans and the Korkus. The site complex is a magnificent repository of rock paintings within natural rock shelters with archaeological evidences of habitation and lithic industry, from the Palaeolithic and Mesolithic periods through the Chalcolithic to the Mediaeval period. Some of the world's oldest floors and walls have been found here. All of these have remained, providing an undisturbed continuous sequence of history of stone age cultures to the historical period. A consistent use of these shelters over this long span of time, apart from habitational deposits, is also indicated by paintings which have been superimposed on earlier existing ones, by inscriptions between the second century BC and seventh century AD, and more recently, by paintings of the Mediaeval period.

Criterion (iii): The significance of Bhimbetka is connected to the way people have interacted with the landscape and how aspects of that interaction have persisted over a very long time-span. The rock art images demonstrate hunting and gathering traditions that still persist in a modified form in the local villages. The tradition of painting symbols and pictures, seen in huge qualities and spanning many millennia in the caves, is still carried on the local villages on shrines and on houses. The Bhimbetka landscape thus bears testimony to a cultural tradition closely linked to the locality of the caves. The quantity and quality of the rock art make that testimony exceptional.

Criterion (v): The Bhimbetka landscape is closely associated with a hunting and gathering tradition, which has an extremely long connection with the area (as documented by the rock art), and is still part of the culture of the local adivasi villages surrounding the rock art site. Persistence of hunting and gathering traditions is now extremely rare anywhere in the world and yet once they were widespread. Bhimbetka is thus of value for the way it can still represent this way of life, although hunting and gathering is no longer a dominant part of the economy.

Source: CPR Environmental Education Centre, Chennai.

6.17 STATUS OF BIOPROSPECTING IN INDIA

India's ratification of the CBD in 1994 resulted in its first national law regulating bioprospecting. The Biodiversity Act of 2002 and the Biodiversity Rules of 2004 regulate bioprospecting of Indian biological resources and associated traditional knowledge. India is a significant source country for bioprospecting as evidenced by the report of the Traditional Knowledge Task Force established by the World Intellectual

Property Organization (WIPO) in 2001. The Task Force carried out a search on international patent databases on patents relating to traditional knowledge systems. The search uncovered in excess of 5000 patent references relating to 90 medicinal plants in the US Patent and Trademark Office alone. 80 percent of the patents relating to these 90 medicinal plants came from seven plants, all seven of which were on Indian origin.

To further underscore the extensive bioprospecting of Indian plants in drug discovery, 50 percent of the drugs reported in the British Pharmacopoeia have their origins in medicinal plants from the Western Himalayan region alone. This is not surprising considering that the genetic diversity of the Western Himalayas informs 80 per cent of *Ayurvedic*, 46 per cent of *Unani* and 33 per cent of allopathic system of medicines.

Despite being an early entrant among countries developing bioprospecting legislation, bioprospecting in India continue to proceed unregulated. Large amounts of medicinal plants are exported from unregulated local markets in Mumbai, Delhi, Chennai and Tuticorin that function through supply chains that are indifferent to regulatory requirements. Furthermore there is a lack of sufficient data and trained personnel to monitor this trade. The reasons for this include a lack of inventories on Indian medicinal plants, unavailability of reliable systems matching trade names to botanical names and the fact that the medicinal plant trade occurred in dried form making these plants difficult to identify. While the inventorying gaps are gradually being overcome the greater challenge of distinguishing between bioprospecting and commodity trade and developing practical ways of regulating the different stages of bioprospecting still remain.

Besides the pharmaceutical and the cosmetics sector, Indian plant genetic resources for food and agriculture is widely used by agribusinesses including for the development of genetically modified varieties and other proprietary lines. Many of these agribusinesses have Indian germplasm in their collections, which they use to develop new proprietary varieties. However, despite the existence of the Biodiversity Act since 2002, none of these companies have disclosed any information regarding their use of Indian germplasm nor have they made nay concrete offers to share any benefits arising from such use. The Biological diversity Act (2002) and its rules (2004) regulate the access to Indian biological resources and associated knowledge. The central regulatory agency the National Biodiversity Authority (NBA). The BD Act also devolves certain powers of oversight to the State Biodiversity Boards at the state level and the Biodiversity Management Committees at the local level.

The BD Act and Rules bring the following activities in relation to Indian biological resources and associated knowledge under the purview of Access benefit sharing (ABS):

(a) Commercial utilization of Indian biological resources develop products such as drugs, industrial enzymes, food flavours, fragrance, cosmetics, emulsifiers, oleoresins, colours, extracts and genes used for improving crops and livestock through genetic intervention. However, this commercial utilization does not include conventional or traditional practices that are used in agriculture, horticulture, poultry, dairy farming, animal husbandry or bee keeping.

(b) Research which means any study or investigation or technological application of biological resources or their derivatives to develop products or processes;

(c) Biosurvey or bioutilization which means survey or collection of biological resources for any purpose;

(d) Application for any intellectual property rights based on Indian biological resources and associated knowledge.

Any other uses of biological resources for e.g. direct consumption, trading, as commodities etc. do not attract the provisions of the BD Act and Rules.

The BD Act and Rules require non-Indians to get the prior approval of the NBA when engaging in the aforementioned activities. However, there are certain situations where Indians are also required to get the approval of the NBA. Non-Indians as per the BD Act are individuals who are not citizens of India, nonresident Indians (as per Section 2(30) of the Income Tax Act) and body corporates, associations or organizations not incorporated or registered in India or having non-Indian participation in its share capital or management. Nevertheless, both Indians and non-Indians are required to get the previous approval of the NBA when they apply for intellectual property rights or transfer for monetary consideration research relating to Indian biological resources and associated knowledge.

The reasoning behind the distinction between Indians and non-Indians in the BD Act has to do with incentivising domestic industry by giving them a free pass at least to the point where they secure proprietary rights over innovations based on Indian biological resources and associated knowledge. This distinction however has been increasingly harder to maintain since a large number of Indian companies have non-Indian participation in its share capital and management.

Moreover differences between the NBA and the State Biodiversity Boards (SBBs) regarding the interpretation of the BD Act and Rules regarding benefit-sharing obligations of Indian companies are becoming increasingly public and frequent. This is exemplified by actions of insufficiently funded SBBs interpreting the BD Act as providing them with the discretion to require Indian companies to engage in benefit sharing where they use Indian biological resources from the jurisdiction of the SBBs.

The processing of ABS applications involves various steps as laid down by the NBA in the decision making map below. There are 13 steps that need to be undertaken with each of these steps being time consuming for various reasons ranging from incomplete applications submitted by the bioprospector to time taken to consult the BMCs or to organize meetings of the Expert Committee or the NBA to approve or reject the applications.

In reality, the critical step where decisions regarding terms of access and benefit sharing are made is step 6 in the decision chart where the Expert Committee on ABS applies its mind regarding a bioprospecting application. Prior to submitting a bioprospecting application for the consideration of the Expert Committee, the legal and technical advisors (step 4) are expected to have completed all enquiries regarding the application, ensured that the application is complete in terms of the necessary documentation and facilitated the consultation with the relevant bodies such as the BMCs and the SBBs (step 5).

It is only after these processes have been completed by the legal and technical advisors within the NBA will a bioprospecting application be placed before the expert committee. Once the expert committee makes its decision, then decision is placed before the secretary and chairman of the NBA, who in turn place it for final approval before a full meeting of the NBA.

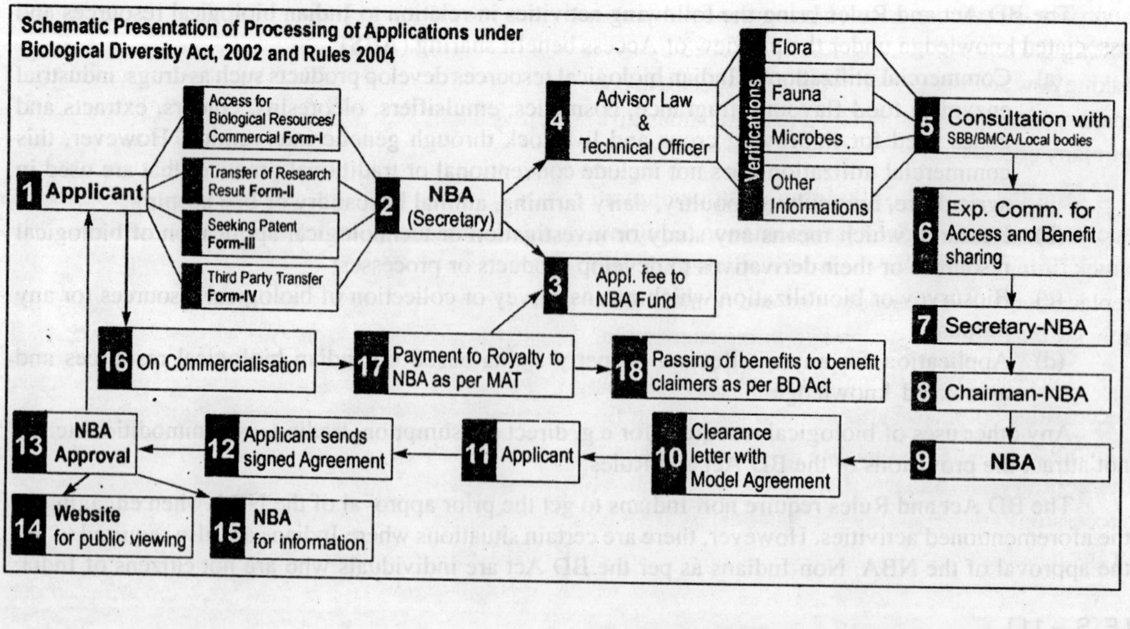

7 ECOLOGICAL SUCCESSION

7.1 CONCEPT

Ecological succession is the gradual change that occurs in an ecosystem of a given area of the Earth's surface on which populations succeed each other. The most familiar ecological successions are those on abandoned farmland, like the succession from weeds to forest. A series of plant communities occupies the fields in roughly predictable order while the habitat progressively changes. Identically, newly formed marshlands or sand-dunes are colonised by regular successions of communities even as the habitats change, most strikingly when moist soil develops on dry dunes.

The forcing functions of ecosystem changes often seem to be provided by vegetation, as the biomass of developing plant communities changes the physical state of the ecosystem and provides the habitat for animal or decomposers. But animal communities also change in succession. They may modify the habitat for plants and replace other animal communities in a colonising dynamics of their own. These facts of ecological succession have attracted ecologists principally because they suggest ordered development.

7.2 SUCCESSION PROCESS

Succession has the appearance of a cooperative enterprise. Plants and animals come and go in sequence, each apparently making its contributions to ecosystem change and then departing. The final result is a perfected complexity that persists indefinitely.

Ecologists use several approaches to study processes of community building. One is to model the changing rates of colonisation and extinction when communities are built on isolated raw habitats, the approach is known as island biogeography. A particular aim of this approach is to understand the conditions setting the equilibrium eventually reached by the number of species. Other approaches are evolutionary, asking how accommodations between species in complex communities come about. This approach addresses matters of mutualism, mimicry and the relationship between predators and their prey that allow the coexistence of many species.

But the process of building plant communities in relatively short span of time through ecological succession is yet a separate phenomenon requiring its own explanations. Communities are built not in one quick fitting together of coevolved species but from pools of local species through the more or less regular replacements of succession. The end point is the complex community dominated by a few characteristic species that typically arrive late in the succession.

Succession traditionally is divided into primary and secondary: autogenic and allogenic. Primary successions colonize bare sites and lead to the first occupation of the habitat by the climax community. Examples are successions on sand-dunes, volcanic mud-flows, glacial till, filled in lakes and marshes. On the contrary, secondary successions replace a climax community following a disturbance. Old field successions are secondary successions, as are the successions that replace forest in gaps after hurricanes or fires.

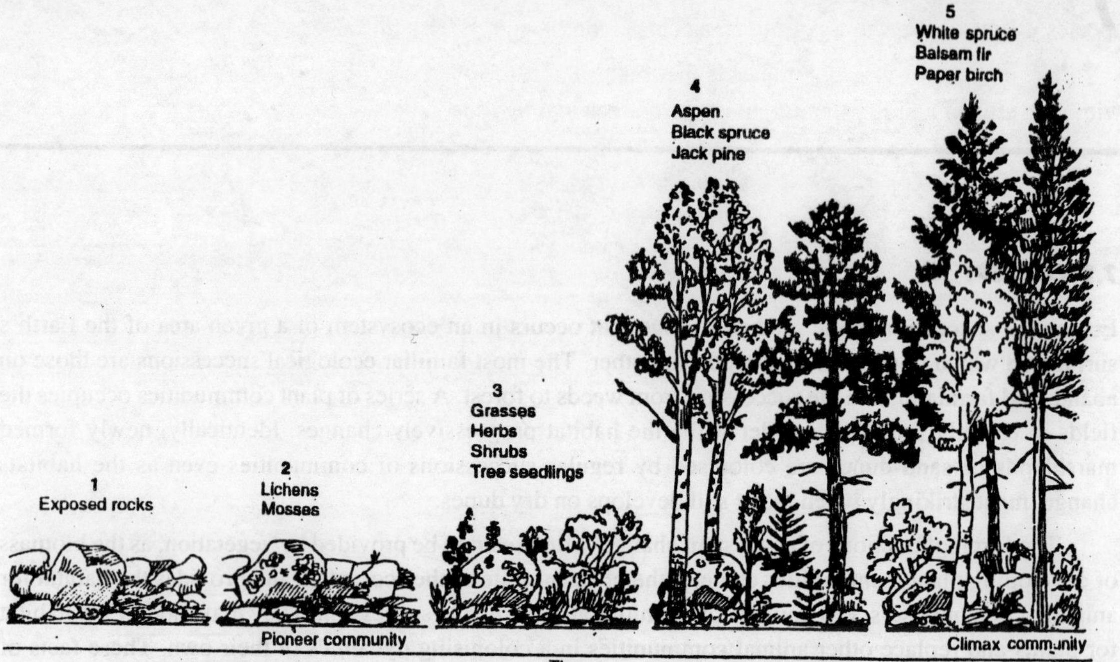

Fig. 7.1 : Primary succession on a terrestrial site (shown in five stages)

Autogenic succession is the succession directed from within the ecosystem itself. The term particularly refers to habitat changes brought about by the biota, as when soil is built and nutrients collected. If these enrichments of soil promote the next community replacement then the succession is autogenic. On the other hand, allogenic succession is succession driven by forces outside the ecosystem, as when a progressive fall in water-table due to stream down-cutting leads to a succession of plant communities suited to progressively drier habitats. The communities themselves may have had no influence on the critical habitat changes. The conceptual differences between autogenic and allogenic successions are significant.

A particular recognised succession is called a sere. A hydrosere is a successional sequence on wetland. Wetland successions in general are called hydrarch successions. In a like manner, successions on dry sites like sand-dunes are called xerarch successions.

The communities of a sere are called seral communities. The earliest seral community on a site is the pioneer community. The succession is said to end with the climax, which is the community with which a succession ends and in which species perpetuate themselves through reproduction. The composition of the climax community depends on the physical properties of the habitat and of the local climate. The climax for each local habitat type can be recognized by its characteristic dominant plants, like the maples of beech-maple forest of the American Midwest or the oaks of English woodlands or the rainforests of India.

A typical primary succession on a terrestrial site is shown in Fig. 7.1. Many of the population replacements of a typical sere can be understood as a necessary consequence of opportunist and equilibrium species living together in the same area. This is the 'r' and 'k' hypothesis of succession.

species living together in the same area. This is the 'r' and 'k' hypothesis of succession.

The essence of this hypothesis is that both opportunist and equilibrium species are able to coexist within an area of country that might be called reasonable dispersal distance. No evolution of fresh taxa is required for ecological succession because this is a comparatively short-term event. It happens in ecological time rather than in evolutionary time. We can discount from a hypothesis of causes both invasions from remote places and the production of fresh species by speciation. Both are possible, of course, but their appearance cannot affect our general explanation. Thus this hypothesis predicts that the plants of the pioneer seral community will be extreme r-strategists, adapted for dispersal and rapid growth but not for sustained existence at a species equilibrium. They should be highly fecund, short-lived and without elaborate storage organs and the more costly predator defenses. If these plants exist in a neighbourhood, it can be predicted that they will first occupy vacant habitat. This prediction is upheld by the plants of pioneer communities which are the typical annual weeds of an old field. But the hypothesis predicts that the pioneers will certainly be displaced as k-strategists invade the site from surrounding complex communities. This is what we observe, most notably in seres that lead to a forest climax. The progressive invasions of perennial herbs, shrubs and trees are by species whose life-history phenomena are of progressively longer life, with all the diversion of resources from reproduction to storage and defence that these life histories-demand. The final number of species at climax will depend on the rate of continued invasion and the rate of local extinction, which eventually should be very slow.

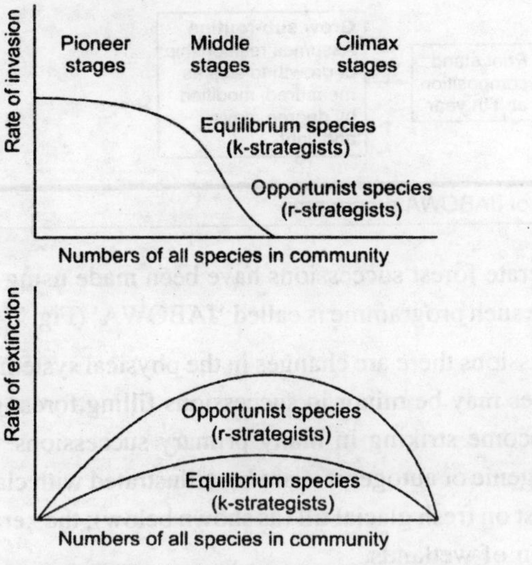

Fig. 7.2 : Model of succession as a function of rate of invasion. Opportunist species invade early and rapidly (top), but their extinction by competition is also early and rapid (bottom)

The hypothesis that succession is the predictable replacement of r-strategists by k-strategists is readily convincing for all seral stages up to forest. But when the whole habitat is covered by large trees, all following a k-strategy of sorts, the continued successional change might not be so self-evident. And yet in classic old field successions to forest a long succession of tree populations continues between the first establishment of forest and the final slow change of climax (Fig. 7.2).

Grime (1979) had suggested that a series of three base strategies should be recognised for plants instead of just the opportunist and equilibrium strategies of r and k. These three strategies would be that of coloniser, competitor and stress competitor—called ruderal, competitor and stress tolerant strategies or R, C and S-

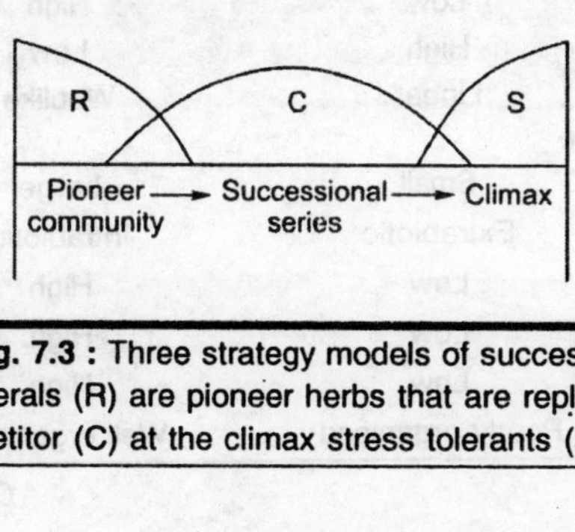

Fig. 7.3 : Three strategy models of succession : Ruderals (R) are pioneer herbs that are replaced by competitor (C) at the climax stress tolerants (S) invade

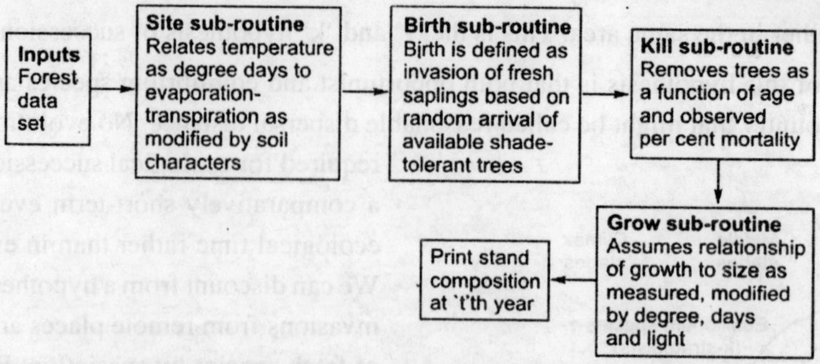

Fig. 7.4 : Partial flow chart of JABOWA programme

Successful computer simulations of some temperate forest successions have been made using data for invasion, growth, and death of individual trees. One such programme is called 'JABOWA' (Fig. 7.4).

Alongside the population changes of many successions there are changes in the physical system that can be called 'ecosystem development'. These changes may be minor in successions filling forest gaps but they can be large in old field succession and become striking in many primary successions. The possibilities of ecosystem development—whether allogenic or autogenic—are best illustrated with classic examples of primary successions; development of forest on fresh glacial till (as shown below); the xerarch succession on sand-dune; and the hydrarch succession of wetlands.

Age	Plant invasions	Soil pH
0	Moss polsters, firewood	8.4
50	Alder thicket	5.0
100	Spruce forest	-do-
150	Hemlock forest	-do-
200	Climax spruce-hemlock forest	-do-

On the whole, Odum (1969) summarised the overall consequences of succession, which are given in Table 7.1:

Table 7.1: Consequences of Succession

Sl. No.	Ecosystem Attributes	Developmental Stages	Mature Stages
	(A) Community energetics		
1.	Gross production/community respiration (P/R ratio)	> 1 or < 1	1 (approx.)
2.	Gross production/standing crop biomass (P/B ratio)	High	Low
3.	Biomass supported/unit energy flow (B/E ratio)	Low	High
4.	Net community production (yield)	High	Low
5.	Food-chain	Linear	Weblike
	(B) Community structure		
6.	Total organic matter	Small	Large
7.	Inorganic nutrients	Extrabiotic	Intrabiotic
8.	Species diversity-variety component	Low	High
9.	Species diversity-equitability component	Low	High
10.	Biochemical diversity	Low	High
11.	Stratification and spatial heterogeneity	Poorly organised	Well organised

Contd.

Table 7.1: Consequences of Succession

Sl. No.	Ecosystem Attributes	Developmental Stages	Mature Stages
	(C) Life history		
12.	Niche specialisation	Broad	Narrow
13.	Size of organism	Small	Large
14.	Life cycle	Short, simple	Long, complex
	(D) Nutrient cycling		
15.	Mineral cycles	Open	Closed
16.	Nutrient exchange rate between organisms and environment	Rapid	Slow
17.	Role of detritus in nutrient regeneration	Unimportant	Important
	(E) Selection Presser		
18.	Growth form	For rapid growth (r-selection)	For feedback control (k-selection)
19.	Production	Quantity	Quality
	(F) Overall homeostasis		
20.	Internal symbiosis	Undeveloped	Developed
21.	Nutrient conservation	Poor	Good
22.	Stability (resistance to external perturbations)	Poor	Good
23.	Entropy	High	Low
24.	Information	Low	High

7.3 STRATEGIES OF SPECIES SURVIVABILITY

Consciousness of the importance of different strategies to the study of populations grew out of the debate over the importance of density dependent and density independent control of populations. The debate made clear that there were two extreme types of life history, common in nature, the one suited to marginal habitats with fluctuating weather and the other for more equable places where a population equilibrium seemed more likely. The two kinds of species were at first known as opportunist and equilibrium species, but later named as r- and k-selection species.

In marginal habitats many life histories are clearly adapted to rapid population growth during short favourable seasons, with other adaptations letting individuals survive the hostile times that inevitably come. These are the life histories of opportunist species. Necessary to the opportunist lifestyle is the ability to disperse rapidly so that remote habitats can be colonised or recolonised quickly from the places of refuge.

Another way of looking at dwellers in marginal habitats is to say that they are fugitives from more desirable places, for one who journeys to the desert must have been from a more fertile place. Thus they can be treated as fugitive species.

On the contrary, an equilibrium species represents the alternative strategy to that of opportunists or fugitive species. This is the strong competitor that sends the opportunist away to the desert, or to ephemeral habitats, as a fugitive. Essential to life at a species equilibrium are adaptations that allow persistence. Each individual of an equilibrium species should compete strongly with its own kind and with others, and be programmed for life in a crowded habitat. Dispersal is less important than persistence. The result is a life history distinctly different from that of opportunist or fugitive species.

The adaptations that result in an equilibrium species are essentially opposite to those of an opportunist. This is most clearly seen by considering the calorie budget of the animals. If an animal, throughout its life history, is made to invest calories in mechanisms for defence or endurance, then these calories are not available for extra reproduction. An animal can use the strategy of putting calories into structures that lead to persistence or the strategy of putting all possible calories into reproduction so that genes persist even though individuals perish. Different strategies represent direct trade-offs.

r- and k- selection

There are a number of instances of the formation of young volcanic islands, raised from the sea. This island is devoid of terrestrial plants and animals initially, but then the island will be colonised rapidly by migrants arriving from across the sea. There will be plants coming as wind-blown seeds, birds and insects coming on their own wings, seeds and small animals clinging to the birds, and many things from small mammals and lizards to living plants drifting out on floating flotsam. But all these immigrants will have in common the ability to disperse. The overwhelming number of arrivals, perhaps all, will be opportunist or fugitive species well-provided with mechanisms for dispersal.

Once settled on the island, the opportunist immigrants, adapted as they are to high fecundity, should achieve rapid population growth. Soon the island will be crowded with their descendants and there ought to establish population equilibria. But these animals and plants are not adapted to crowded lives at equilibrium; they are usually fugitives from other species that do better in crowds.

On their island, weather may not strike them down and the better competing equilibrium species cannot reach them. Natural selection should then work to find the most competitive varieties from the island's opportunist. This is the process now known as 'k-selection'.

The term r-selection can be equivalent to opportunists and fugitives and k-selection to equilibrium species. Pianka (1970) gave a summary statement of how r- and k-selection could be used to organise so many ideas in contemporary ecology (Table 7.2).

Table 7.2: Characteristics of r- and k-selection species

Variables	r-selection	k-selection
Climate	Variable and/or unpredictable : uncertain	Fairly constant and/or predictable : more certain
Mortality	Often catastrophic, nondirected, density independent	Mostly directed, density dependent
Population size	Variable in time, non-equilibrium : usually well below carrying capacity of environment; unsaturated communities or portions thereof : ecological vacuums : recolonization each year	Fairly constant in time, equilibrium : at or near carrying capacity of the environment ; saturated communities : no recolonisation necessary
Infra and interspecific competition	Variable, often lax	Usually keen
Relative abundance	Often does not fit MacArthur's broken stick model	Frequently fits the MacArthur model
Selection favours	Rapid development, high reproductive maxima, early reproduction, small body size, semelparity : single reproduction	Slower development, greater competitive ability, lower resource thresholds, delayed reproduction, large body size, iteroparity : repeated reproduction
Length of life Lead to	Short—usually less than one year— Productivity	Longer—usually more than one year— Efficiency

The ecological niche: a multidimensional hypervolume called home

The term 'niche' (Elton 1927) is used in a variety of contexts such as:

- *Spatial*, which includes habitat, distribution and physical location.
- *Functional*, which includes trophic levels and other food web relationships
- *Behavioural*, which refers to mode of life such as predator, competitor, etc.
- *Abstract*, as a multidimensional space composed of interactions with diverse environmental factors, both abiotic and biotic.

In all natural communities, successful functioning depends upon the operation and integration of different 'professions', each of which occupies this multidimensional space called a niche.

A niche represents the sum of:

- An organism's adaptations.
- Its use of resources and
- The lifestyle to which it is fitted.

Each niche is occupied by a different species, although ecologically equivalent species may occupy comparable niches in different or similar ecosystems in other parts of the world.

Some niches occur in every major community because they perform an indispensable function such as primary production. Others are widespread but restricted by certain physical conditions. Thus *bioturbation* processes (the vital biological aeration and mixing of soils and sediments) are typically carried out by earthworms in moist soils but not in dry soils because earthworms cannot live in dry soils. Evolution can result in the development of highly specialised niches such as those occupied by organisms that provide a valet service for larger animals. Certain birds carry out this function in terrestrial regions and on various, host species while coral reefs have their own equivalents such as cleaner wrasse and shrimps.

Niches may be divided into:

- **Fundamental niches**, the theoretical multidimensional niche that would be occupied in the absence of other species and
- **Realised niches**, which are the actual niches occupied in a community where there are interactions and sharing of resources with other competitive species.

No two species can occupy exactly the same multidimensional niche in the same community for any length of time because *competitive exclusion* will occur, one species eliminating its competitor. However, fundamental niches can overlap, leading to species coexisting and competing. It is this competition between different species that defines a species realised niche.

However there are three distinct pathways for plant colonisation in any habitat. The details are given in Fig. 7.5 (next page).

7.4 POPULATION ECOLOGY

Population ecology is concerned with interrelationships of co-actions between individuals within and between species. Co-actions may either be beneficial to the participants, cooperation, or harmful to them. Interspecific cooperation includes mutualism, commensalism, and many of the interactions that establish the community as a dynamic unit. Co-actions that are harmful, to at least one of the participants, include parasitism, predation and competition.

An early manifestation of cooperation in the evolution of animals and plants is the grouping of free-living protozoans to form colonies, and the further development of such colonies into multi-cellular metazoans thereafter behave and respond as unit organism. It is impossible to say whether the first gathering of protozoan cells to form colonies developed for better protection from some predator or environmental condition, for improved utilisation of food supplies or more efficient reproduction.

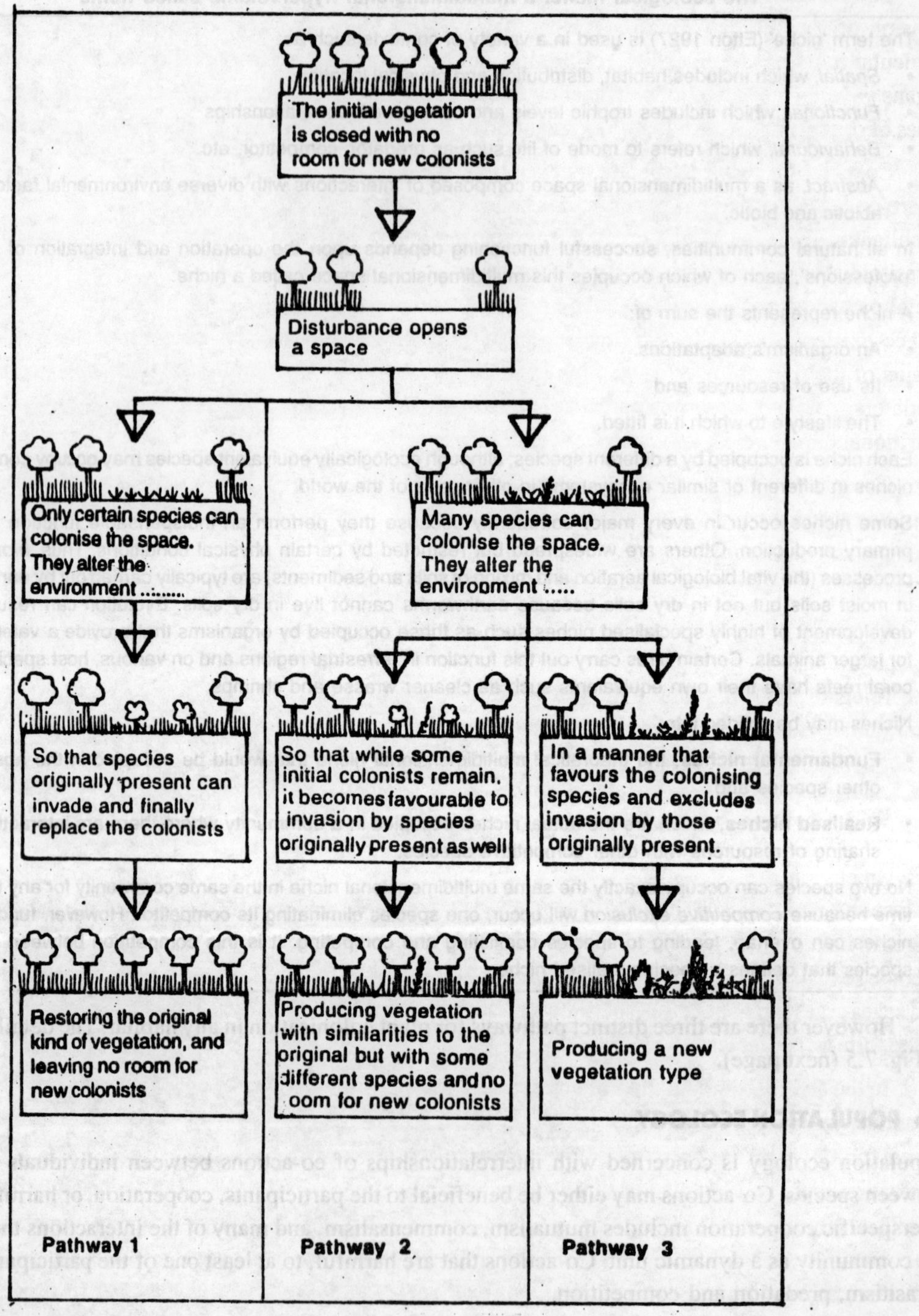

Fig. 7.5 : Pathways of plant colonization

Colonisation quickly led to division of labour between somatic and reproductive cells and, later, to division of somatic cells themselves, so that different cells or organs became specialised to serve the particular functions of digestion, respiration, circulation and so on. Cooperation among cells, tissues, and organs gave greater metabolic efficiency to the whole individual and resulted in evolution of the highest types of organisms. Similarly, the aggregation of individuals must have survival value, because it persists.

Cooperation between species that is intimate and beneficial to both participants is called mutualism; where only one participant benefits, it is called commensalism. These relations may be either facultative or obligatory. Where one or more of the participants is harmed there is no cooperation, of which parasitism, parasitoidism, competition and predation are the examples. Distinction is made between true parasites, social parasites, and parasitoids. True parasites and their hosts have evolved adaptive interrelations so that co-existence occurs for varying lengths of time. However, the host is generally weakened and virulent strains of the parasite may cause high mortalities. Causes of mortality or disease among organisms are predators, parasitoids, worm parasites, protozoan parasites, bacteria, viruses, fungi, external parasites, nutritional deficiencies, toxication, physiological stress and accidents.

Competition may be exerted directly through interference in the activities of one organism by another, or indirectly in the form of excessive exploitation of natural resources. It may be either intraspecific or interspecific. Allelochemic secretions from one organism that affect the growth, behaviour, or health of other organisms are a subtle means of competition, cooperation or intercommunication.

Competition may result in establishment of social hierarchies, establishment of territories, regulation of population size, segregation of species into different niches or speciation. The overall effect of competition is to relegate the individual and species to an orderly place in the structure and organisation of the community with the result that there is decrease in tension and disturbance.

7.4.1 Reproductivity and Population Structure

The rate at which a species reproduces, and the frequency of its population turnover, can affect the speed with which it occupies new areas, becomes adapted to new niches, or evolves into new races. In order to analyse the population dynamics of a species, it is necessary to know its life history. This involves the stages in its life cycle, mortality rates of each stage, longevity, sex and age ratios, age at which individuals become sexually mature, fecundity, factors causing mortality, and so forth. The proportion of different ages and sexes gives the population a definite structure. All these essential data may be conveniently summarised in the form of life tables.

In the stabilised population of any species, whatever the number of eggs or young produced per pair of adults, the average number of offspring reaching reproductive status can never be greater than two sexual forms—which is the number required to replace the parents on their death. Therefore, with each new generation there is a population turnover, with newly born individuals replacing the adults that die. In a stabilised population the rate of increase of a population through the course of several reproductive cycles must equal the death rate, so that the value of one factor is also a measure of the other. Either factor is indicative both of the rate of population turnover and of the intensity of environmental resistance.

Ratios of young animals/plants to adults often indicate whether a population is expanding, contracting or is stabilised. In stabilised populations the number of offspring reaching reproductive maturity can never be greater or less than the number of adults themselves. The number of young ones that must be produced to permit such a population turnover gives a measure of the rigour of the environment, and how well-adapted a species is to its niche.

7.4.2 Population Growth

The characteristic sigmoid shape of the cumulative growth curve of organisms was demonstrated by Verhulst in 1839 and the most widely used description of the curve is the logistic equation. This curve has been used to describe the population growth of such diverse organisms as Yeast, Paramoecium, Drosophila and man and even the growth of communities in particular habitats as represented by the increase in number of species. Under natural conditions, however, growth of a population is subject to many variable factors, including the change from one morphological stage to another in the life cycle of many species, and the change in the physical environment both daily and seasonally. The curve is often not fully expressed, even though its trend is present inherently (Slobodkin, 1961).

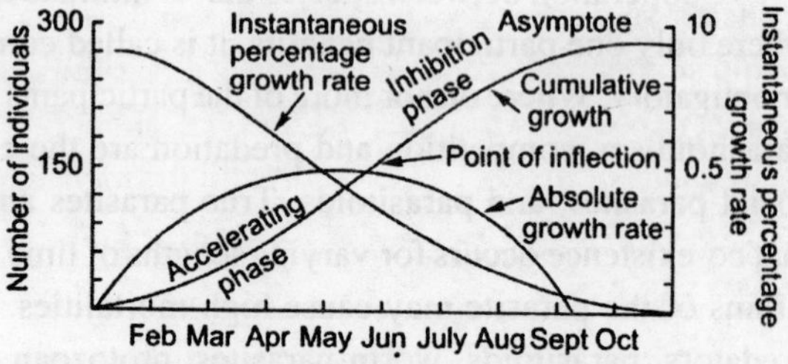

Fig. 7.6 : Growth phases vs growth rate

The sigmoid curve shows that a finite population grows slowly at first, then at an accelerating rate which is at maximum at the point of inflection, after which the population continues to increase but at a decelerating rate, finally becoming stabilised at the upper asymptote. Logistic growth curves are symmetrical, and the point of inflection is one-half the value of the asymptote. The lower concave part of the curve is called the accelerating phase of growth.

If the number of new individuals added during each unit of time—absolute growth rate—is plotted against time midway in each period, a bell-shaped curve is obtained, the peak of this curve coinciding with the point of inflection on the sigmoid curve. In contrast the instantaneous growth rate (r) is the rate of growth at a point on a time scale and is usually expressed in terms of increase per individual or unit biomass per unit of time (Fig. 7.6).

7.4.3 Population Regulation

No single mechanism is responsible for the control of natural population. Some populations sometimes can exist close to equilibrium, with birth rates balanced by death rates, but more often numbers fluctuate between wide limits of changing or repetitive environments, persistently alter the balance between births and deaths. But the possible causes of death vary from species to species and from time to time.

Failures to be born or to find a mate are as important as actual death, being equally effective at removing a unit of reproduction from the next generation. Birth rates vary in some species, actually being kept low by patterns of behaviour that work to regulate births according to supplies of resources. This birth regulation always gives advantage to individuals by increasing the eventual number of surviving offspring over what would result from more initial births. Natural birth regulation, therefore, always works to increase the rate of population growth. Selection for group birth control is highly unlikely. Although a plausible mechanism has been put forward for the selection of groups of predators that are below maximum efficiency as hunters, the consensus of ecologists is that all regulation of births works to achieve optimum reproductive success and that all control of populations is by some form of environmental resistance. For minority species that attain lasting population equilibria, competition for vacant niche spaces is the decisive mechanism; short-lived opportunistic species, or those suffering heavy predation, may persist with constantly fluctuating local populations. Some of the longest-lived animals and plants, particularly those of conservative breeding strategies, live in populations that are always changing very slowly and that may reach neither equilibrium nor extinction before the onset of climatic or environmental change.

A struggle for existence inevitably follows from the high rate at which all organic beings tend to increase. Every being, which during its natural lifetime produces several eggs or seeds, must suffer destruction during some period of its life, and during some season or occasional year; otherwise, on the principle of geometrical increase, its numbers would quickly become so inordinately great that no country could support the produce. Hence, as more individuals are produced that can possibly survive, there must in every case be a struggle for existence, either one individual with another of the same species, or with the individuals of distinct species, or with the physical conditions of life.

On the whole a population can thus be controlled by invoking any or all the seven ways of death:

(i) Starvation (outright energy deficiency: food, calories or light);

(ii) Malnutrition (lack of nutrients or water or essential components);

(iii) Predation;

(iv) Parasitic diseases;

(v) Accident (probably weather-induced);

(vi) Failure to be born; and

(vii) Failure to find a mate.

The importance of these various ways of death varies widely with different life-history strategies, producing quite different patterns of population control. The overall fitness stakes of each species are shown in Fig. 7.7:

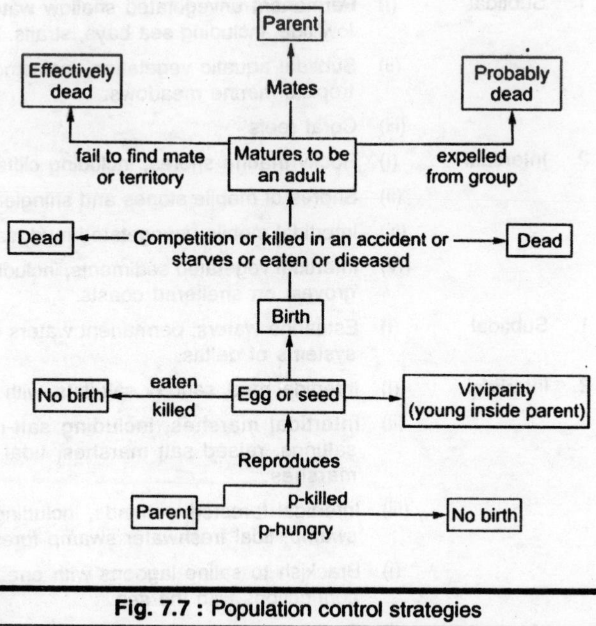

Fig. 7.7 : Population control strategies

8 WETLANDS

8.1 CONCEPT OF WETLANDS

'Wetlands' are those areas inundated or saturated by surface or ground water at a frequency and duration sufficient to support and that cover with water under normal circumstances. It also supports a prevalence of vegetation typically adapted for life in saturated soil conditions. Wetlands generally include swamps, marshes, bogs and similar areas. In general, wetlands are areas of land that remain waterlogged for a substantial period of the year.

Tropical wetlands cover 2.64 million km^2 worldwide whereas wetlands in temperate and boreal regions occupy about 5.72 million km^2. They support a wide variety of plant and animal species restricted to such environments. Wetland ecosystems are among the one of the most threatened of all environmental resources. Much of the physical loss of wetland area has been because of the conversion to industrial, agricultural and residential use.

Over the time, wetlands are classified in various ways. One of the classification made by Dugan (1990) is given in the Table 8.1:

Table 8.1: Classification of Wetlands (Source : Dugan 1990)

1. Salt Water			
1.1. Marine	1. Subtidal	(i)	Permanent unvegetated shallow water less than 6 m depth at low tide, including sea bays, straits.
		(ii)	Subtidal aquatic vegetation, including kelp beds, sea grasses, tropical marine meadows.
		(iii)	Coral reefs
	2. Intertidal	(i)	Rocky marine shores, including cliffs and rocky shores.
		(ii)	Shores of mobile stones and shingle.
		(iii)	Intertidal mobile unvegetated mud, sand or salt flats.
		(iv)	Intertidal vegetated sediments, including salt marshes and mangroves, on sheltered coasts.
1.2. Estuarine	1. Subtidal	(i)	Estuarine waters; permanent waters of estuaries and estuarine systems of deltas.
	2. Intertidal	(i)	Intertidal mud, sand or salt flats, with limited vegetation.
		(ii)	Intertidal marshes, including salt-marshes, salt meadows, saltings, raised salt marshes, tidal brackish and freshwater marshes.
		(iii)	Intertidal forested wetlands, including mangrove swamp, Nipa swamp, tidal freshwater swamp forest.
1.3. Lagoonar		(i)	Brackish to saline lagoons with one or more relatively narrow connections with the sea.
1.4. Salt lake		(i)	Permanent and seasonal, brackish, saline or alkaline lakes, flats and marshes.

Contd.

Table 8.1: Classification of Wetlands (Source : Dugan 1990)

2. Freshwater			
2.1. Riverine	**Perennial**	**(i)**	**Permanent rivers and streams, including waterfalls.**
		(ii)	**Inland deltas.**
	Temporary	**(i)**	**Seasonal and irregular rivers and streams.**
		(ii)	**Riverine floodplains, including river flats, flooded river basins, seasonally flooded grassland.**
2.2. Lacustrine	**Permanent**	**(i)**	**Permanent freshwater lakes (8 ha), including shores subject to seasonal or irregular inundation.**
		(ii)	**Permanent freshwater ponds (8 ha).**
	Seasonal	**(i)**	**Seasonal freshwater lakes (8 ha), including floodplain lakes.**
2.3. Palustrine	**Emergent**	**(i)**	**Permanent freshwater marshes and swamps on inorganic soils, with emergent vegetation whose bases lie below the water table for at least most of the growing season.**
		(ii)	**Permanent peat-forming freshwater swamps, including tropical upland valley swamps dominated by Papyrus or Typha.**
		(iii)	**Seasonal freshwater marshes on inorganic soil, including sloughs, potholes, seasonally flooded meadows, sedge marshes, and dambos.**
		(iv)	**Peatlands, including acdophilous, ombrogenous, or soligenous mires covered by moss, herbs or dwarf shrub vegetation, and fems of all types.**
		(v)	**Alpine and polar wetlands, including seasonally flooded meadows moistened by temporary waters from snowmelt.**
		(vi)	**Freshwater springs and oases with surrounding vegetation.**
		(vii)	**Volcanic fumaroles continually moistened by emerging and condensing water vapour.**
	Forested	**(i)**	**Shrub swamps, including shrub-dominated freshwater marsh, shrub carr and thickets, on inorganic soils.**
		(ii)	**Freshwater swamp forest, including seasonally flooded forest, wooded swamps on inorganic soils.**
		(iii)	**Forested peatlands, including peat swamp forest.**
3. Man-Made Wetlands			
3.1. Aquaculture/Mariculture		**(i)**	**Aquaculture ponds, including fish ponds and shrimp ponds.**
3.2. Agriculture		**(i)**	**Ponds, including farm ponds, stock ponds, small tanks.**
		(ii)	**Irrigated land and irrigation channels, including rice fields, canals and ditches.**
		(iii)	**Seasonally flooded arable land.**
3.3. Salt Exploitation		**(i)**	**Salt pans and salines.**
3.4. Urban/Industrial		**(i)**	**Excavations, including gravel pits, borrow pits and mining pools.**
		(ii)	**Wastewater treatment areas, including sewage farms, settling ponds and oxidation basins.**
3.5. Water-storage areas		**(i)**	**Reservoirs holding water for irrigation and/or human consumption with a pattern of gradual, seasonal, draw down of water level.**
		(ii)	**Hydro-dams with regular fluctuations in water level on a weekly or monthly basis.**

Dugan, P. J. (ed.) 1990. Wetland Conservation : A Review of Current Issues and Required Actions, IUCN, Switzerland, 96 pp.

8.2 THE RAMSAR CONVENTION

The convention on Wetlands of International Importance especially as Waterfowl habitat often known as the Ramsar Convention from its place of adoption in Iran in 1971—is an intergovernmental treaty which provides the framework for international cooperation for the conservation of wetlands.

By 2002, more than 133 countries had joined the Convention, thereby agreeing to accept a number of obligations, including the following:

- To designate wetlands of international importance for inclusion in a list of so called 'Ramsar Sites'.
- To maintain the ecological character of their listed Ramsar Sites.
- To recognise their planning so as to achieve the wise use of all the wetlands on their territory.
- To designate wetlands as nature reserves.

Already more than 1201 wetland sites have been added to the Ramsar list, covering more than 105.8 million hectares of wetland habitat. India joined the convention in 1982 and till November 2010 had designated 26 wetlands as Ramsar sites. A list of selected Ramsar wetlands of India with their importance and threat is given in the Table 8.2.

Table 8.2: Ramsar Wetlands: Importance and Threats

Ramsar Site	Importance	Problems/Threats
Keoladeo Ghana National Park, (2,873 ha) Bharatpur, Rajasthan	Only wintering site in India for Siberian cranes and a large numbers of aquatic birds from Central Asia. It supports five species of ungulates, four species of cats, and two species of primates, as well as diverse plants, fish and reptiles	Scarcity of water resulting in parched situation for wild life. Conflict over water between parks needs and local village irrigation requirements
Loktak Lake (26,600 ha), Manipur	Plays an important role in the economy of Manipur—source for hydropower, irrigation and drinking water. The Keibul Lamjao National Park is a unique floating wildlife reserve and the only home of the endangered Manipur brow-antlered deer or *sangai*.	Due to deforestation in the catchment area, the rate of soil erosion has increased. Development activities threatening lake. Loss of vegetal cover in the catchment.
Harike lake (4,100 ha), Punjab	Recharges ground water and provides irrigation water. This wetland sustains a large number of waterfowl.	Silt had started depositing near the barrage. For irrigation, more water has to be released leaving the wetland bare.
Wular Lake (2,400 ha), Jammu and Kashmir	This wetland is the source of drinking water for Srinagar. The Wular is an important habitat for the fish fauna of the region.	The dimensions of the lake in the past were estimated at 20,200 ha. Nearby rivers and streams regularly deposit large quantities of silt that has already claimed about 90 per cent of the lake.
Sambhar Lake (24,300 ha) Rajasthan	This saline lake is important for a variety of wintering water birds like flamingos and pelicans. Human activities consist of salt production and livestock grazing.	Grazing pressure from domestic animals. Illegal hunting by local people.

Contd.

Table 8.2: Ramsar Wetlands: Importance and Threats

Ramsar Site	Importance	Problems/Threats
Deepor Beel (4,000 ha), Assam	Is the only major storm water storage basin for Guwahati. Some globally threatened birds are supported here—Spotbilled Pelican, Lesser and Greater Adjutant Stork and Baer's Pochard.	It is now threatened with large-scale encroachment, brick making factory and soil cutting within the beel ecosystem. Poaching and hunting.
Point Calimere Wildlife and Bird Sanctuary (38,500 ha), Tamil Nadu	Some 257 species of birds have been recorded, 119 of them water birds, including the vulnerable species of Spoonbill Sandpiper, Grey Pelican and some 30,000 Greater and Lesser Flamingos.	Collection of firewood and forest produce gathered by lopping off tree branches. Increased brackish groundwater caused by expansion of the salt works, and decreasing inflow of freshwater.
Kolleru Lake (300 sq km), Andhra Pradesh	It provides habitat for the vulnerable Grey Pelican. Sustains fisheries, and agriculture.	In April 2003, the lake went nearly dry.
Chilika Lake (116,500 ha), Odisha	This brackish lake has a rich bird life with over 150 recorded migratory and resident species and supports staging for 33 species of waterbirds. There are 118 species of fish, including commercially important species.	It faced problems caused by siltation and sedimentation. Changes in salinity level.
Pong Dam Lake (15,662 ha), Himachal Pradesh	More than 220 bird species have been identified, with 54 species of waterfowl. Hydrological values include monsoon-season flood prevention, groundwater recharge, silt trapping and prevention of soil erosion.	Slow increase of human activities can convert into threat for the wetland.
Tso moriri Lake (12,000 ha), Laddakh, Jammu and Kashmir	The site is said to represent the only breeding ground outside China for the Black-necked crane and the only breeding ground for Bar-headed geese in India. The barley fields at Korzok have been described as the highest cultivated land in the world.	Increased human activities likely to affect the breeding waterfowl population. Tourists making noise, and disposing garbage.
Ropar (1,365 ha), Punjab	The site is an important breeding place for smooth Indian otter, hog deer, sambar, and several reptiles, and the endangered Indian pangolin is thought to be present.	Siltation. Agricultural activity along with fertiliser use is on the rise.
Sasthamkotta (373 ha), Kerala	It is a source of drinking water for half a million people in the Kollam district. Some 27 freshwater fish species are present.	Filling of lake for cultivation. Agricultural and domestic waste.
Vembanad-Kol (151,250 ha), Kerala	It supports the third largest waterfowl population in India in winter. Over 90 species of resident birds and 50 species of migratory birds are found here.	Land reclamation and industrial sewage.
Ashtamudi Wetland (61,400 ha), Kerala	The site supports a number of mangrove species as well as over 40 associated plant species, and 57 species of birds, including six that are migratory.	Population density and urban pressures pose threats to the site. Pollution from oil spills from thousands of fishing boats, and industries.

Contd.

Table 8.2: Ramsar Wetlands: Importance and Threats

Ramsar Site	Importance	Problems/Threats
Bhoj Wetlands (3,201 ha), Bhopal, Madhya Pradesh	Lake is an important source of drinking water for Bhopal. The lakes are very rich in macrophytes, phytoplankton, zooplankton, resident and migratory birds, insects, reptiles and amphibians.	Discharge of sewage and silt. Urban pressure.
Bhitarkanika (65,000 ha), Odisha	Home to the largest salt water crocodile, numerous birds. Has the largest species of mangroves in India.	Rising population increasing conflicts between human and crocodile. Pressure on mangroves.
Kanjli (183 ha), Punjab	Diversity of aquatic, mesophytic, and and terrestrial flora and fauna in the region. Acts as a key regulator of groundwater discharge and recharge.	Excessive weed growth. Deforestation and erosion.
East Kolkata Wetlands (12,500 ha), Kolkata, West Bengal	Natural wastewater treatment system for Kolkata. Pisciculture and vegetable production is a thriving activity	Under threat from urbanisation and real estate development. Discharge of tannery and other waste.
Compiled by WWI Research Team, ha : hectares, sq km : square kilometres		

Other Ramsar sites are—Chandertal wetland (Himachal), Renuka wetland (Himachal), Ropar lake (Punjab), Rudrasagar lake (Tripura), Sasthamkotta lake (Kerala), Surinsar–Manas lakes (J & K), and upper Ganga river (Brijghat to Narora stretch) (Uttar Pradesh) are considered as notified wetlands of India.

8.3 IMPORTANCE OF WETLANDS

Wetlands have played a crucial role in human history, major stages in the evolution of life itself probably took place in nutrient-rich coastal waters. Some of the first prehistoric cultures—such as those of the early Mesolithic settlements around the post-glacial lake margins and coastals of Europe and those of the coastal Indian communities in North America—were dependent on wetlands for food and materials for building, shelter and clothing.

- Among the many vital functions of wetlands, **flood control** is one of the most important. Wetlands act like sponges, storing and slowly releasing rainfall and runoff, thus reducing flood peak. This can reduce the need for expensive dams and other engineering structures.
- The **binding effect of their vegetation** is another important aspect of wetlands; it helps in the stabilisation of banks and shores. Not only that—in some cases it helps in the accretion of sediment, thus counteracting forces of erosion, subsidence and sea level rise. Both these functions are vital.
- Wetlands have a further key role to play in what is known as **groundwater recharge** and discharge. Recharge occurs when water moves from the wetland into the underground aquifer-rock, such as sandstone, which holds water. The wetland acts as a filter for certain kinds of waste and soluble contaminants. The process is important for controlling storm water run-off, for replenishing supplies of water for human consumption and also in maintaining the flow of groundwater which may support other wetlands at the point of discharge. Discharge of wetland stored groundwater may be important in sustaining the agricultural production of surrounding land.
- Regular deposition of nutrient rich silt contributed to the **success of agriculture** along large rivers. Sediment is also vital for maintaining aquatic fertility and physical stability of floodplains and deltas.
- Tampering with wetlands can also damage their ability to deal with various contaminants and nutrients. Biological, chemical and physical processes in wetlands are often able to **immobilize**

and transform a wide range of environmental contaminants and nutrients, which, in excess, would cause severe eutrophication and pollution. Heavy metals, pesticides and industrial wastes, for instance, can be bound to soil and sediment particles, and there be rendered more or less inert.

- Wetlands can **act as a 'sink', preventing nitrate build-up** which could lead to eutrophication. It prevents nitrates from reaching freshwater streams and lakes. Nitrate run-off from fertilised agricultural areas can be recycled to harmless nitrogen gas by this mechanism.
- Wetland flora in particular are able to **sink nutrients and contaminants**. Nutrients can be decomposed and needed in agriculture thereafter.

A generalised values of wetland is shown in Table 8.3:

Table 8.3: Wetland values

	Estuaries (without mangroves)	Mangroves	Open coasts	Floodplains	Freshwater marshes	Lakes	Peatlands	Swamp forest
Functions								
Groundwater recharge				•	•	•	O	O
Groundwater discharge	O	O	O	O	•	O	O	•
Flood control	O	•		•	•	•	O	•
Shoreline stabilisation/Erosion control	O	•	O	O	•			
Sediment/Toxicant retention	O	•	O	•	•	•	•	•
Nutrient retention	O	•	O	•	•		•	•
Biomass export	O	•	O	•	O	O		O
Storm protection/Windbreak	O	•	O	O				
Micro-climate stabilisation		O		O	O	O		O
Water transport	O	O		O		O		
Recreation/Tourism	O	O	•	O	O	O	O	O
Products								
Forest resources		•		O				•
Wildlife resources	•	O	O	•	•	O	O	O
Fisheries	•	•	O	•	•	•		O
Forage resources	O	O		•	•			
Agricultural resources				•	O	O	O	
Water supply				O	O	•	O	O
Attributes								
Biological diversity	*	O	O	•	O	•	O	O
Uniqueness to culture/heritage	O	O	O	O	O	O	O	O

Note: '•' Common and important value of that wetland type; 'O' less common/important.

Source: Dugan. P. J. (Ed.) 1990. Wetland Conservation: A Review of Current Issues and Required Action. IUCN, Gland, Switzerland.

8.4 WETLAND LOSS

Over the years, wetland loss is a global concern. The rate of wetland loss is mostly not quantified except a few countries, however, the causes of such losses were well-documented (Table 8.4). Over 90 wetlands of various countries were seriously threatened in Asia alone (Table 8.5). Thus there needs detailed conservation strategies for protection of such extinct habitats.

Table 8.4: The causes of wetland loss

	Estuaries	Open coasts	Floodplains	Freshwater marshes	Lakes	Peatlands	Swamp forest
Human actions							
Drainage for agriculture, forestry and mosquito control	•	•	•	•	O	•	•
Dredging and stream channelisation for navigation and flood protection	•			O			
Filling for solid waste disposal, roads, and commercial, residential	•	•	•	•	O		
Conversion for aquaculture/mariculture	•	O	O	O	O		
Construction of dykes, dams, levees, and seawalls for flood control, water	•	•		•			
Discharges of pesticides, herbicides, nutrients from domestic sewage	•	•	•	•	•		
Mining of wetlands for peat, coal, gravel, phosphate and other materials	O	O	O		•	•	•
Groundwater abstraction			O	•			
Sediment diversion by dams, deep channels and other structures	•	•	•	•			
Hydrological alterations by canals, roads and other structures	•	•	•	•			
Subsidence due to extraction of groundwater, oil, gas and other minerals	•	O	•	•			
Natural causes							
Subsidence	O	O			O	O	O
Sea-level rise	•	•					•
Drought	•	•	•	•	O	O	O
Hurricane and other storms	•	•				O	O
Erosion	•	•	O			O	
Biotic effects			•	•	•		

Note: '•' Common and important cause of wetland degradation and loss; 'O' present, but not a major cause of loss.

Source: Dugan, P.J. (Ed.) 1990. Wetland Conservation: A Review of Current Issues and Required Action. IUCN, Gland, Switzerland.

Table 8.5: Most seriously threatened wetlands in Asia

Bangladesh
1 Chalan Beel*
2 Haor Basin of Sylhet and Eastern Mymensingh
3 Dubriar Haor*
4 The Sundarbans
5 Wetlands in Pablakhali Wildlife Sanctuary
6 Chokoria Sunderbans*

Bhutan
7 Boomthang Valley

Burma (Myanmar)
8 Irrawaddy Delta

People's Republic of China
9 Yancheng Marshes
10 Shijiu Hu
11 Shengjin Hu and the lower Yangtze Lakes
12 Shengjin Hu
13 Xi Jiang (Pearl River) Delta*
14 Tuosu Hu (Kurlyk Nor) and Kuerhleiko Hu

India
15 Dal lake
16 Wular Lake
17 Harike Lake
18 Jheels in the vicinity of Haidergarh*
19 Dahar and Sauj (Soj) Jheels
20 Southern Gulf of Kutch
21 Gulf of Khambhat
22 Khabartal
23 Dipor (Deepar) Bheel
24 Logtak Lake
25 Salt Lakes Swamp*
26 The Sunderbans
27 Chilika Lake
28 Kolleru Lake
29 Estuaries of the Karnataka coast
30 Kaliveli Tank and Yedayanthittu Estuary
31 The Cochin Backwaters
32 Wetlands in the Andaman and Nicobar Islands

Indonesia
33 Banyuasin Musi River Delta
34 Muara Cimanuk*
35 Sukolilo
36 Cilacap and Segara Anakan
37 Danau Bankau and other swamps in the Barito Basin*
38 Banau Sentarum
39 Wetlands in Manusela Proposed National Park
40 Wasur and Rawa Biru

Japan
41 Shonai-Fujimae Tidal Flats and Inner Isle Bay
42 Lake Shinji and Lake Nakaumi

Republic of Korea
43 South Kanghwa and North Yongjong Mudflats
44 Mudflats of South Yongjong and adjacent islands
45 Namyang Bay
46 Asan Bay
47 Kum, Mankyung and Tangjin Estuaries

Malaysia
48 Sedili Kecil Swamp Forest
49 Klang Islands : Pulau Ketam*
50 Kapar Forest Reserve
51 North Selangor Swamp Forest
52 Marintaman Mengalong*
53 Tempasuk Plain
54 Lawas Mangroves
55 Trusan-Sundar Mangroves
56 Limbang Mangroves
57 Maludam Swamp Forest
58 Sarawak Mangrove Forest Reserve

Nepal
59 Begnas Tal*

Pakistan
60 Khabbaki Lake*
61 Siranda Lake*
62 Hawkes Bay/Sandspit Beaches and adjacent creeks
63 Clifton Beach
64 Korangi and Gharo Creeks
65 The Outer Indus Delta

Philippines
66 Pangasinan Wetlands*
67 Manila Bay*
68 Laguna de Bay
69 Tayabas Bay including Pagbilao Bay
70 Inabanga Coast
71 Mactan, Kalawisan and Cansaga Bays*
72 Agusan Marsh
73 Lake Leonard*
74 Davao Gulf
75 Liguasan Marsh

Singapore
76 Serangoon Esturary*

Sri Lanka
77 Thandamannar Lagoon*
78 Chundikkulam Lagoon
79 Chalai Lagoon*
80 Periyakarachchi and Sinnakarachchi Lagoons*
81 Mahaweli Ganga Floodplain System
82 Maha Lewaya and Karagan Lewaya
83 Lunama Kalapuwa and Kalametiya Kalapuwa*
84 Bellanwilla-Attidiya Marshes

Taiwan, R. O. C.
85 Tatu Estuary
86 Tungshih (Ton-Shou) Mangroves*

Thailand
87 Gulf of Thailand
88 Pak Phanang Estuary
89 Pa Phru

Socialist Republic of Vietnam
90 Red River Delta
91 Red River Estuary
92 Mekong Delta
93 Nam Can Mangrove Forest

Source: Scott, D. A. and Poole, C. M. 1989, A Status Overview of Asian Wetlands. No. 53. AWB, Kuala Lumpur, Malaysia.
Note : * Sites marked with an asterisk are considered to be already too degraded to merit any special conservation effort.

8.5 RESEARCHES ON INDIAN WETLANDS

In India, wetlands are distributed in various regions ranging from the cold arid zone of Ladakh through the wet Imphal in Manipur and the warm and arid zone of Rajasthan-Gujarat to the tropical monsoonic central India and the wet humid zone of the southern Peninsula.

On the basis of survey made in 1990, the Ministry of Environment and Forests made an inventory of wetlands in the country. According to this survey about 4.1 million hectares are covered by wetlands of different categories (Table 8.6). In addition, mangroves—the coastal wetlands—occupy an area of about 6740 km^2, about 80% of which is the Andaman and Nicobar Islands. Stretches of mangrove vegetation are also found in coastal Orissa, Andhra Pradesh, Tamil Nadu, Karnataka, Maharashtra, Gujarat and Goa.

Recently National Committee on Wetlands, Mangroves and Coastal Reefs identified tweenty four wetlands as major wetlands of national importance (Table 8.7). Among them management action plans for 10 sites has been prepared, based on problems identified so far.

Table 8.6: Distribution of wetlands in India, including wetlands of less than 100 ha, by State/Union Territory

State	Number	Natural area (ha)	Number	Man-made area (ha)
Andhra Pradesh	219	100,457	19,020	425,892
Arunachal Pradesh	2	20,200	NA	NA
Assam	1,394	86,355	NA	NA
Bihar	62	224,788	33	48,607
Goa	3	12,360	NA	NA
Gujarat	22	394,627	57	129,660
Haryana	14	2,691	4	1,079
Himachal Pradesh	5	702	3	19,165
Jammu & Kashmir	18	7,227	NA	21,880
Karnataka	10	3,320	22,758	539,195
Kerala	32	24,329	2,121	210,579
Madhya Pradesh	8	324	53	187,818
Maharashtra	49	21,675	1,004	279,025
Manipur	5	26,600	NA	NA
Meghalaya	2	NA	NA	NA
Nagaland	2	210	NA	NA
Odisha	20	137,022	36	148,454
Punjab	33	17,085	6	5,391
Rajasthan	9	14,027	85	100,217
Sikkim	42	1,101	2	35
Tamil Nadu	31	58,868	20,030	201,132
Tripura	3	575	1	4,833
Uttar Pradesh	125	12,832	28	212,470
West Bengal	54	291,963	9	52,564
Total	**2,164**	**1,449,338**	**65,250**	**2,587,965**
Union Territories				
Chandigarh	NA	NA	1	170
Puducherry	3	1,533	2	1,131
Grand Total	**2,167**	**1,450,861**	**65,253**	**2,589,266**

Table 8.7: Selected Wetlands under the National Wetland Programme

Wetlands	Area (Ha)	Types of Wetland	State/Union Territory
Kolleru	90,000	Fresh water	Andhra Pradesh
Wular	18,900	Fresh water	Jammu and Kashmir
Chilika	114,000	Brackish	Odisha
Loktak	26,600	Fresh water	Manipur
Bhoj	1,000	Fresh water	Madhya Pradesh
Sambhar	7,200	Saline	Rajasthan
Pichola	1,480	Fresh water	Rajasthan
Aswamudi	4,335	Brackish	Kerala
Sasthamkotta	325	Fresh water	Kerala
Harike	4,100	Fresh water	Punjab
Kanjli	14	Fresh water	Punjab
Ujni	35,700	Fresh water	Maharashtra
Sukhana	22,826	Fresh water	Chandigarh
Renuka	670	Fresh water	Himachal Pradesh
Kabar	6,737	Fresh water	Bihar
Nalsarovar	13,450	Brackish	Gujarat

Realising the importance of wetlands, conservation of wetland ecosystem has been taken up since Seventh Plan and a national committee thus formed with a view to make report on management of wetland areas in India.

In India, origin of wetlands is diverse. Kashmir has a wide range of natural lakes at altitudes varying from 1,585 to 1,600 m above sea level. These lakes occur mainly in the flood plain of river Jhelum, and are derived either from ox-bows or abundant flood channel. The lakes include Dal, Anchal, Mansbal and Wular. Many high altitude wetlands are formed by the glacial action, such as Kumaon lakes in Uttar Pradesh. Small lakes in limestone tracts of Khashi and Jaintia hills of Assam are believed to have been formed by dissolution of underlying rock. Pichola-Fates agar complex of Rajasthan are formed as a result of movement of fine loose clay and sand particles by wind. Similarly, salt lakes of Sambhar (Rajasthan) are depressions created by the shifting dumpes, forming swamps. Most wetlands of Bihar and West Bengal are of ox-bow type and are formed as a result of shifting of river courses in the geological past.

India has a vast coastline stretching over 3,600 km with sandy shores mangroves, backwaters, lagoons, mudflats and estuaries. Among the coastal wetland the most important one is that of Chilika (Odisha) is the largest brackish water lake. This is followed by Pulicat in Tamil Nadu and Ashtamudi in Kerala.

Among the man-made wetlands, the best one is the Keoladeo National Park of Bharatpur (Rajasthan) created about 250 years ago, by the then ruler of the State. It is a classic example of man-made conversion of natural rainfed depression into a permanent waterfowl reserve by containing and regulating water column by means of various canals and sluice gates. In addition the construction of several multipurpose projects in the states of Punjab, Haryana, U. P., H. P., M. P. and other states has resulted in the formation of large reservoirs, creating man-made wetlands.

In general, wetland communities are highly dynamic as they are located at the interface of terrestrial and aquatic ecosystem. The species richness and so also biodiversity in such an area is quite high, as it offers a specialised habitat for many macrophytes, planktons, invertebrates and vertebrates. The richness is mainly attributed to the "spatial heterogeneity, that is, the greater the number of niches, the greater is the opportunity for successful invasion by a species."

Nymphea naucheli

Nymphea lotus

Nymphea naucheli

Nymphea stellata

The basic components of a wetland ecosystem are: the water regime where the physical and chemical characteristic of water determine the nature of the wetland; the hydrological characters and water quality control the composition of the wetland communities; sediment and substrate which influence the ability of the vegetation to establish in such areas, and they serve to maintain a steady flow of nutrients and energy through the system; decomposers which release the nutrients required from the organic matter; and, finally, vegetation which support the major biota.

The species richness in macrophytes vegetation varies, depending on the period of flooding. Area flooded for long periods have lower species diversity than less frequently flooded areas. In presence of high concentration of salts and sulfides in salt marsh water such as Pulicat lake, the species diversity is low, but a definite zonation and distribution of vegetation is noticed. In contrast, flowing water is considered to act as a stimulus to diversity, probably due to its ability to renew minerals and reduce anaerobic conditions.

In the wetlands of Himalayas and Nilgiris, peat bogs are often formed. It has its own characteristic flora and fauna. Peat is often characterised by grasses, sedges and mosses. Apart from richness and diversity of flora, the presence of large number of 'wild' relatives of cultivated plants for desirable characters, make the wetland unique. The common examples are 'wild species' of *Oryza, Elettaria, Saccharum, Zinziber, Curcuma, Coffea, Cinnamomum* and *Dioscorea* (Nair and Daniel, 1986; Ramakrishna, 1990).

Due to presence of high amount of dissolved oxygen and nutrients, wetlands in general support a large invertebrate fauna, form a feeding ground for fishes and protect them from predators among submerged vegetation. Wetlands vegetation also harbours several air-breathing snails such as *Pila virens, Lymnaea* sp., *Indoplanorbis* sp. and *Bellamya* sp. Due to the presence of gastropods and fishes the wetland habitat forms an excellent site for visit of both residential and migratory bird population.

Some wetlands of West Bengal, Assam and Bihar in India were surveyed in the past for limnological investigation. Pathak (1989) made comparative studies of the limnological features of three beels (viz., Dhir beel of Assam, Kulia beel of West Bengal and Muktapur Mann of Bihar), in which he described the hydrochemical features, dynamics of chemical constituents and evaluation of productivity trends, nutrient cycles, organic bottom deposits, detritus energy, energy transformation and flow of energy in beels ecosystem. He found that the bottom deposits of decaying weeds contribute very rich organic carbon 2.8 - 9.0%, nitrogen (605 - 988 ppm) and phosphorus (40 - 185 ppm) in all the beels studies. This rich nutrient status of the beels shows immense production potential. Vaas (1989) discussed the productivity status, physico-chemical and biological factors of beels and their exploitation and productivity in West Bengal. Babu Lal (1989) also studied the energy flow in Kulia beel ecosystem in West Bengal and found the beel heavily choked with luxuriant growth of the macrophytes (species of *Eichhornia, Hydrilla, Ceratophyllum, Najas, Azolla* etc.) and very low fixation of solar energy by the primary producers are found (2500 cal m^2 day^{-1}) in the first year of the study. But in the second and third years—after the removal of the macrophytes— energy fixation increased many times (photosynthetic efficiency increased from 0.12% at the beginning to ten-fold immediately after the removal of the macrophytes). This was due to sudden increase in phytoplankton bloom, when the chlorophyll concentration in the beel also increased.

Various types of limnological studies were also conducted in adjoining countries like Bangladesh. These studies include the following aspects:

(a) Limnology, pollution and phytoplankton of ponds, lakes, beels and rivers; (Islam and Khatun 1966, Islam and Nahar 1967, Islam and Begum 1969, Islam and Saha 1975, Islam and Haroon 1975, Islam and Zaman 1975, Islam and Rahaman 1979, Islam et al 1980, 1991 and 1992, Khondker and Parveen 1992a, 1992b, Alam et al 1987a, Islam and Paul 1978, Ali et al 1978).

(b) Limnology, growth of benthic flora and fauna; (Islam and Paul 1977, Islam et al 1979; Islam and Mendes 1976, Islam and Paul 1978, Islam and Khondker 1991, Islam et al 1992, Miah et al 1983 and 1985, Ali 1980).

(c) Limnology, water quality and primary productivity of a hypertrophic lake; (Khondker et al 1988, Khondker and Parveen 1992, Khondker and Rahim 1991).

(d) Limnology of fish ponds, zooplankton and other animals; (Ameen et al 1988, Mumtazuddin et al 1982, Ali et al 1980, 1982 and 1989, Sufi and Farooque 1983, Oppenheimer et al 1978, Alam et al 1987b, Mahmood 1986, Begum et al 1989, Khaleque and Islam 1983).

(e) Limnology, macrophytes and algae of deep water rice-fields of Brahmaputra and Meghna floodplains (Whitton et al 1988a, 1988b, 1988c, Rother and Whitton 1988, Catling et al 1981).

(f) Limnology, pollution and domestic, agricultural and industrial pollutants in the river system. (Haider 1988, Khair 1988, Alam 1988, Malik 1988, Choudhury 1988, Sharif et al 1988, Safiullah and Mofizuddin 1988, Begum and Hossain 1993, Rahman 1988, Eunus 1988).

8.6 WETLAND MANAGEMENT FOR SUSTAINABLE DEVELOPMENT

Natural bodies of water are not absolutely pure. Various organic compounds and inorganic elements remain in dissolved form. Silt and clay remain suspended in the water for a considerable period. Many kinds of microscopic and a good number of macroscopic flora and fauna grow in different types of aquatic habitats. The physical and chemical quality of the waters vary according to nature of the basin shape and size, depth, light penetration, precipitation, location, temperature, chemical nature of the surrounding soil, dissolved minerals, and pH. Finally, on these the biological components of the habitat depend. If all the physical, chemical and biological parameters are in optimum condition and a balance between these is maintained, we may assume an ideal or 'normal' condition. If, for some reason, this balance is disturbed and anyone or several of these components are in excess, then we can say that the excess materials have polluted the environment. Under natural conditions if this happens, gradually the homeostatic condition will be naturally restored. But, if it is due to deliberate human activity, natural recovery will be difficult and ecological disasters may occur. In countries like India and Bangladesh, pollution problems mostly originate from 'abuse, misuse or cocktail' use of pesticides, and overdose and, untimely application of fertilizers and from domestic wastes.

Major pollution sources of the aquatic environments are:

(a) domestic wastes, sewage and garbages;

(b) agricultural wastes, pesticides, insecticides and fertilizers;

(c) industrial effluents from different heavy industries, mills, factories, cottage industries containing various pollutants; and

(d) biological origin, growth of excessive microbial organisms causing health hazards, oxygen depletion, etc.

Most of the industries and factories are situated on the banks of the rivers or very close to a river system and the effluents and wastes are mostly thrown directly into the river water without any treatment to make the effluent 'safe' from the biological standpoint. As a result, the depletion of the biotic components near the sources is observed. A preliminary survey near the tanneries in Dhaka on the river Buriganga will reveal this fact. As the rivers are connected with each other and different mills and factories are situated on their banks, the recovery time of the water from the effects of the effluent is very low, and, during the non-monsoon period, conditions become worse.

Besides chemical pollutants, pollution by the excessive growth of microbial organisms is posing a serious threat to the health conditions of humans and fishes and birds. One aspect of limnology is the study of water quality. Clean water is free from any excess organic, inorganic or microbial contamination. With regard to the latter component, bacteria, bluegreen and other algae play their important role, not only as decomposers and primary producers, but at times they create eutrophication, anoxic condition, fish death and even human illness, and are largely responsible for deterioration of the water quality for drinking, bathing and other domestic uses.

Raw sewage contamination in water systems is the single major factor for the recurrence of various diseases. Having innumerable water systems India and Bangladesh is perennially plagued with as many as 32 waterborne diseases including cholera, dysentery, diarrhea, typhoid, shigellosis etc. The diseases of fish, waterfowl, cattle and other animals are also known but not much studied. With respect to human diseases e.g., cholera, shigellosis, *E. coli* diarrhea and rotavirus diarrhea, dysentery etc. the relationship between hydrophytes and the pathogenic bacteria under eutrophication has been discussed by Islam and Islam (1993). It has been reported that several bacteria remain attached with some hydrophytes (Table 8.8) and these pathogens could spread by some free-floating hydrophytes and phytoplankton. Inter-village transmission of cholera pathogen takes place by the connecting waterways. Not only the diarrheal disease but diseases affecting skin and eyes are caused by different algae. Thus the role of hydrophytes and phytoplankton in the spread and cause of diseases in Bangladesh and India is an important aspect of limnology and should be studied with due emphasis.

Table 8.8: Relationship between Hydrophytes and Bacteria

Hydrophytes/Phytoplanktons	*Bacteria attached*
Eichhornia crassipes	*Vibiro spp., Plesiomonas shigelloides, Aeromonas hydrophyla*
Pistia stratiotes	*Aeromonas sobria*
Alternanthera philoxeroides	*Aeromonas caviae*
Anabaena variabilis	*Aeromonas spp. and Plesiomonas shigelloides*
	Vibiro cholerae 01

Source: Islam and Islam, 1993

No systematic study have been made on this aspect of limnology which includes sanitary engineering, and other aspects associated with the disposal of sewage and domestic wastes. The fruits of *Trapa bispinosa* (Paniphal), *Euryale ferox* (Makhna) and lotus and the petiole of water lily (Shapla) are eaten raw directly from the contaminated water by the people and they become the victims of diarrheal diseases. Many blue-green algae blooms or the blooms caused by *Euglena, Trachelomonas* or *Dinoflagellates* or diatoms may cause not only oxygen-depletion of the water and thereby creating problems for fish growth but several of these secrete toxigenic substances in the water which are responsible for various ailments in the human body such as skin irritation, eye trouble, or if they enter into the body, may affect the nervous system.

Pollutants from the industrial and agricultural sources (point sources) may be checked by adopting suitable treatments or diversion, but those of non-point source (i.e., entry to the beels through land runoff or atmospheric fallout) are difficult to control through conventional methods of treatment. These substances are highly persistent and thereby contaminate the entire biogeochemical cycle of the beels and tanks.

Biological factors also contribute to the ultimate effect of these pollutants, which may remain in low concentrations. However, through bioaccumulation and biomagnification by fish and other aquatic organisms their toxins ultimately enter the human body through the food chain with serious harmful effects. Metals and pesticides have the greatest potential for biomagnification in the aquatic food chain.

The toxicity of the pesticides has been tested on fish, prawn and fish-food organisms. It was found that the latter, i.e., plankton and benthos, are very sensitive to these chemicals as compared to fish, thus adversely affecting the latter also (Joshi 1989). Metals vary in toxicity, which depend on environmental conditions and their specification. In combination, metals show higher toxicity than individual metals. Zn, Cu and Cr in combination have shown several times more toxicity to Oreochromis mossambicus than the individual metals.

The wetland ecosystems have two complementary phases—aquatic and terrestrial (during dry season). During the latter phase it may be used for agriculture or for grazing and both benefit the fishery by enriching the aquatic phase during flood. The former may, however, lead to cultural eutrophication.

Freshwaters are perhaps the most vulnerable of habitats through their use by humans as:

(a) sinks for many products of human activity in their catchment;
(b) natural drains for the removal of waste to the sea; and
(c) a valuable and essential resource for various purposes.

Aquatic ecosystems have been changed by man for various reasons. For example:

(a) to control water-borne diseases;
(b) to establish farmland (but without the knowledge of the consequence of pollution and eutrophication); or
(c) to prevent flood damage to property built on natural floodplains through stupidity or greed (Moss 1980) i.e., against the law of nature and by violating ecological principles.

Conservation of natural ecosystems can be defined in various ways (Moss 1980). For example:

(a) it is a process of maintaining, through whatever means necessary or available, the maximum diversity of organisms and ecosystems;
(b) it is the total management of the rural areas for the fair and equal benefit of all groups which have a direct interest in their use;
(c) it amounts to making the best of what is left after waterways and their catchments have been used for the obvious needs of the society; and
(d) legislation is essential for reservation of particularly interesting areas.

In spite of these, more wetlands/waterways are being degraded and lost than created and restored. The agents of change are varied but two are of widespread significance:

(a) Drainage involving river dredging, building of flood banks and clearance of macrophytes; and
(b) Eutrophication causing threats of toxic pollution, overfishing, introduction of exotic species and recreation pressure.

The uses of water and associated biological problems have strong utilitarian and economic undercurrents. The conservation of natural aquatic communities, in contrast, seem something of a luxury. This is partly true to some extent, but the costs of seriously disrupting natural freshwaters may outweigh the benefits of doing so (Moss 1980). For conservation of wetland resources the following approach/system may be considered:

(a) An ecosystem approach is needed to ensure that the water use, land use pattern, pollution and modification of the river system do not impair the trophic structure and functions of the ecosystems.

In *beels* and *haors* ecosystems, which have a limited waterspread area, large scale abstraction of water for irrigation leads to drastic variations in the water quality (Ahmed et al 1990).

The wetland waters are used for irrigation, industry, thermal power generation, potable supplies and fish production under natural system. Of these uses, fish production will be given priority over other under the ecosystem approach and this should be followed by (i) urban and industrial waste management practices, (ii) judicious use of fertilisers in agriculture, and (iii) afforestation and social forestry programmes.

In addition, there is also a need for catchment modification for the control of soil incursion, transport of fertilisers and pesticides into the beels, haors, etc. through agricultural runoff. The best management practices (BMP) are needed to control and treat pollutants (Joshi 1989), and adopt suitable horticultural practices for making grassed waterways, retention ponds and terraces for checking soil erosion. These can reduce pollutant losses in runoff and these practices provide the best cost-effective and acceptable technology for the control of soil erosion and nonpoint source phosphorus losses.

Natural ecosystems are in a continuous process of change and adjustment through natural selection on the organisms comprising them and a system of maximum homeostasis is maintained under natural physical and chemical environment. There is nothing to 'improve' such an undisturbed ecosystem and under this condition fish growth is moderate or aquatic macrophyte beds are extensive. But attempts to increase the fish growth or clear the weed beds will always result in a chain of repercussions which is very expensive to counteract.

(b) Geomorphological systems should be taken into consideration. The river floodplain is essential for the efficient and natural disposal of occasional high discharges. But attempts to change its use from this purpose to farming (or otherwise) have necessitated the raising of the height of the river banks. In cost-benefit analysis, so far as the general public is concerned, the balance might be very unfavourable (Moss 1980).

For conservation measures each ecosystem or habitat with its geomorphological, physical and chemical background must be known by a limnologist to diagnose the problems and suggest solutions.

(c) It is relatively difficult to have these solutions implemented, because implementation is by and large a political and social matter. Those who are responsible for and are involved in taking decisions may be either unable or unwilling to grasp the technical issues and the long-term economic implications. This conservation and restoration of the freshwater wetlands in India and Bangladesh, as elsewhere, cannot be achieved without a study of politics and sociology and it should not be outside the scope of the scientist concerned with the fresh-water environment.

A major part of the wetlands in the eastern region of the country—particularly in the state of West Bengal—is waterlogged. In order to derive additional wealth from these unutilised and under-utilised wetlands it will be necessary to develop a separate kind of specialisation and planning as distinct from that of tree plantation schemes which have succeeded in the wastelands of arid regions.

Unenviably, wetlands, in most cases, are standing examples of environmental mismanagement and bear testimony of our ignorance about the environment that sustains us and the insults we have inflicted upon it. Therefore, any positive action towards wetland development will be a tangible step towards restoring the ecological balance rendered unstable by man's action on nature.

In India, significance of wetland development activities has gained a steep ascendancy since last few years. This can largely be attributed to a new generation of political pragmatism that is far more sensitive to the growing challenge of environmental issues than it was ever before. We, therefore, have one of the biggest wasteland development programmes in the world with twin objectives of poverty alleviation and environmental repair. The country's one hundred million hectares of under-productive marginal lands are devoted towards uplifting the condition of living of about two hundred million marginal population struggling under the poverty line to raise the prosperity of the rest of the countrymen.

For these poorer millions, wetlands are the terminal resources to fall back upon. This is also the reason for which the programme deserves the most careful and systematic planning attention.

An important feature of India's wetland development programme is that it combines the task of securing out a strategy for wetland development alternatives with the task of lending similar support to implement the projects. Such a unified programme ensures reduction of gap between expected benefit and actual outcome—a gap that has been a matter of serious consent in many of the lofty endeavours all over the world.

Basic Strategy

To ensure a sustainable development of available wetlands it will be imperative to adhere to the strategy outline by the Executive Director, UNEP, in this regard. In brief, the strategy pays emphasis on three key issues, viz., coordination, training and awareness campaigns. Coordination is necessary to pursue a cross-sectoral approach so that various implementing agencies and specialists—belonging to agriculture, forestry, agronomy, environment, hydrology, fisheries, ecology, geography, economics, and other relevant fields—can sit together and contribute to develop appropriate alternatives. Similarly, training facilities will aim at creating indigenous ability to carry out sustainable development work. And, finally, raising the level of awareness of environmental issues amongst the people is considered as a matter of vital importance in the whole range of activities in resources conservation.

The task of specialised coordination assumes a particular significance and is an intrinsic requirement in wetland management, the major components of coordination being:

1. Advanced scientific and technical information.
2. People's perception.
3. Development agencies active on wetlands.

It is important to remember that even if a project is economically viable and technically sound it has to be ecologically balanced and socially acceptable for the plan to succeed. A number of basin management plans did not work because people could not accept or conform to the anticipated land use. There have also been numerous failures from the point of view of maintenance of ecological balance. Advanced scientific and technological information are important for avoiding ecological imbalance and possible disaster, and also for implementing advanced technology in various wetland resource systems. Any change envisaged will have to be linked with the preference of the people and their participation will have to be ensured beforehand. Simultaneously, appropriate coordination with existing development agencies will have to be established. This will ensure the most appropriate and acceptable course of action for implementing desired modifications.

The following are some of the agencies active on wetlands:

(1) Dept. of Agriculture; (2) Dept. of Irrigation & Flood Control; (3) Dept. of Fisheries; (4) Dept. of Environment (5) River Sanitation (Ganga Actions); (6) Dept. of Urban Development; (7) Dept. of Rural Development; (8) Dept. of Land & Land Reforms; (9) Institutional Finance; (10) Wasteland Development; (11) Meteorological Observatories; (12) Zoological Survey of India; (13) Botanical Survey of India; (14) Geological Survey of India; (15) Research Institutes and University Programmes, and (16) Non-governmental Organisations (NGOs).

Task of Planning

Social and economic perceptions help to perfect the choice of technology options and design. Educated people's self-styled efforts in planning for rural development technologies meet with a success no more than what is due to these naive efforts. The scenario becomes so dismal that it falls contingent to introduce the concept of "Appropriate Technology" to distinguish the useless ones and also to give a direction to what is needed in the field of technology design and development. Today a major criterion of appropriate technology is to ensure sustainable change. Use of these advanced perception of planning need and planning process is particularly important while dealing with the problems of marginal area development. Yet, it will be proper to remember that fundamental understanding of the process and appropriate methods to bring change remain far from being achieved.

To assists the task of planning for natural resource systems a set of ground rules has been developed. These are:

1. Wetlands, in most cases, will have conflicting land use interests and any change contemplated cannot be universally good to all the contending interests. Therefore, designer will have to take side.

2. In land use studies no observation is independent of observer-object relationship and observer for that matter becomes a part of the object. This is why the participatory action research is the strongest methodological tool for understanding the use of land and design of wiser use.

3. The culture of conserving resources and using it as many ways as possible, has been found to thrive among the poor. There is prima facie an inverse relationship between affluence and affinity with recycling or reuse.

4. Information about natural resource systems, particularly about those of developing countries, are meagre. Taking decisions on wise use, therefore, will use cybernetic models in place of any exercise in pretentious optimisation.

5. All public decisions are political decisions.

6. Any misuse of natural biological systems will have its social, economic and environmental blacklash. However, such reactions are not instant and this time lag provides opportunities to the manipulators and speculators to inflict permanent damages to our ecosystems by forcing wrong land use changes on the society.

7. Environmental damages are identified not by observing the quality of the environment or measuring the external stresses alone but also by evaluating the internal strength and metabolic resistance of the particular group of people inflicted by that external stress.

It is easily agreed that the above postulates and the protocol developed in figure are of preliminary nature. Subjective ability to handle the challenge of sustainable development is still amateurish. This majority will only come through the ongoing experiments with change that many of us are involved in bringing in our respective areas of operation at various places of the world. These experiments with change are either in the form of a movement to preserve a biosphere or to design wise use of an ecosystem for the same goal of sustainable development.

Major concerns in developing rural technology in general and wetlands in particular may be summarised as follows :

- beneficiaries externalities in the process of planning,
- social feasibility of the proposed changes are ignored,
- search for alternative technologies are not rooted on the strength and weakness of the existing technologies.

It is, therefore, considered imperative to re-orient the search for development options through bench-mark studies of existing technology and use of land. The total chain of efforts has been described in Fig. 8.1. In addition state-wise list of wetlands of international importance is given in Table 8.9.

Table 8.9: State-wise list of wetlands of international importance in India under Ramsar Convention

Sl. No.	State/UT	Name of Ramsar Site	Area (ha.)	Date of Declaration
1.	Andhra Pradesh	Kolleru Lake	90,000	19.08.02
2.	Assam	Deepor Beel	4,000	19.08.02
3.	Himachal Pradesh	Pong Dam Lake	15,662	19.08.02
		Renuka Wetland	20	08.11.05
		Chandertal Wetland	49	08.11.02
4.	Jammu & Kashmir	Wular Lake	18,900	23.03.90
		Tsomoriri	12,000	19.08.02
		Hokera Wetland	1,375	08.11.05
		Surinsar-Mansar Lakes	350	08.11.05
5.	Kerala	Ashtamudi Wetland	61,400	19.08.02
		Sasthamkotta Lake	373	19.08.02
		Vembanad-Kol Wetland	151,250	19.08.02
6.	Madhya Pradesh	Bhoj Wetland	3,201	19.08.02
7.	Manipur	Laktak Lake	26,600	23.03.90
8.	Odisha	Chilika Lake	116,500	01.10.81
		Bhitarkanika Mangroves	65,000	19.08.02
9.	Punjab	Harike Lake	4,100	23.03.90
		Kanjili	183	22.01.02
		Ropar	1,365	22.01.02
10.	Rajasthan	Sambhar Lake	24,000	23.03.90
		Keoladeo National Park	2,873	01.10.83
11.	Tamil Nadu	Point Calimere Wildlife and Bird Sanctuary	38,500	19.08.02
12.	Tripura	Rudrasagar Lake	240	08.11.05
13.	Uttar Pradesh	Upper Ganga River (Brijighat to Narora Stretch)	26,590	08.11.05
14.	West Bengal	East Kolkata Wetlands	12,500	19.08.02
		Sundarban	426,000	01.01.19
15.	Gujarat	Nalsarobar Bird Sanctuary	123	29.09.12
		Total sites 27	**1092,354**	

Source: National Biodiversity Action Plan, 2008, MoEF

8.7 MANAGEMENT ACTION PLAN—WETLAND SITE

For successful action plan for wetland conservation the following aspects of wetland should be addressed in appropriate manner:

- Protection
- Siltation Control
- Biodiversity Conservation
- Fisheries Development
- Peoples' Participation
- Research
- Catchment Area Development
- Pollution Control
- Sustainable Resource Utilisation
- Generating Awareness
- Monitoring and Evaluation

Then only an integrated action plan can be drafted with active involvement of the surrounding community.

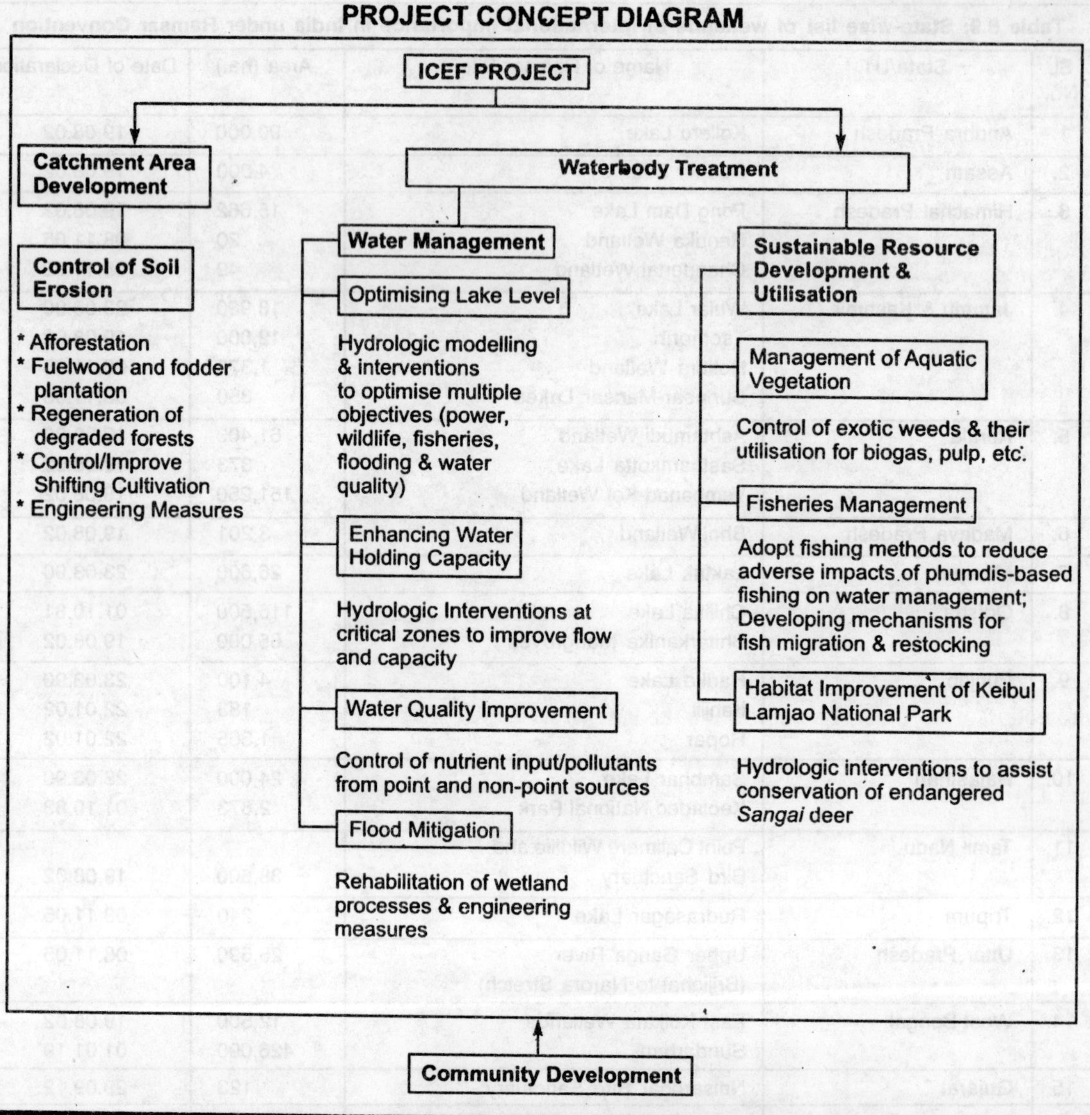

Fig. 8.1 : Linkage with rural development programmes (eg, Low-cost sanitation) & developing income generation schemes

8.8 NATIONAL WETLAND POLICY

With rapid realisation of wetland values, the national policy was formulated. The major goals of such policy are:

- Commitment by the Government to sustain and restore wetlands
- Build wetland conservation within the broader framework of cross-sectoral national planning
- Maintain biological diversity and ecological character while ensuring benefits to the local community
- Stop conversion of wetlands for any other land use purposes including practices which change or tend to change the ecological processes, functions and attributes of wetlands
- Rehabilitation of degraded wetlands and enhancement of their resource base

- Involvement of local communities, NGOs and concerned agencies for management of wetlands
- Application of EIA on all activities impacting the wetlands.

However, the existing Policies state the following:

- Directive Principles of State Policy — Articles 48A & 51A (g) state;

  '...to protect and improve the natural environment including the rivers, lakes and wildlife...'
- National Conservation Strategy and Policy Statement on Environment and Development (1992) highlights Conservation of Wetlands, Coastal Areas, Rivers and Island Ecosystems
- National Forest Policy amended in 1948 and National Wildlife Action Plan (1983) emphasises long-term conservation of wildlife

On assessment over situation of wetland conservation in this country one can easily feel the following lacunae in the policy:

- Lack of coordination and integration of policies
- Conflicts in land use planning/development
- Intersectoral or interdepartmental conflicts/claims including boundary disputes

Thus, there is a great need for adoption of a "National Wetland Policy" in this country.

8.9 LEGISLATION RELATING TO WETLANDS

Over 200 State and Central laws exist for regulating wetland resources of the world with respect to protection of — water, fish, birds and the land.

In addition there are International Laws and conventions for the said purpose. In India alone there are a number of Laws having some relevance to wetland habitat regulation. They are:

- The Environment (Protection) Act, 1986
- The CRZ Notification, 1991
- The Environment Impact Assessment Notification, 1994
- The Wildlife (Protection) Act, 1972
- The Tamil Nadu Aquaculture (Regulation) Act

Fig. 8.2 : Current action plan required for wetland protection

- The West Bengal Town and Country Planning Act
- The West Bengal Fishery Act, 1996
- The Wetlands (Conservation and Management) Rules, 2010.

But there is no specific legislation on wise use of wetlands. Thus, following actions are needed at present (Fig. 8.2):

- Comprehensive unified wetland conservation legislation based on cross-sectoral approach
- Include wetlands as a category under land use classification in its own right taking their structural and functional aspects into consideration
- Legislation relating to wetlands to include regulation of activities in the catchment area impacting waterbody
- Zone–based regulation of activities to ensure sustainable use of wetland resources.

Finally, the concept of wise use policy has to be reviewed on site specific criterion including people's need (Fig. 8.3):

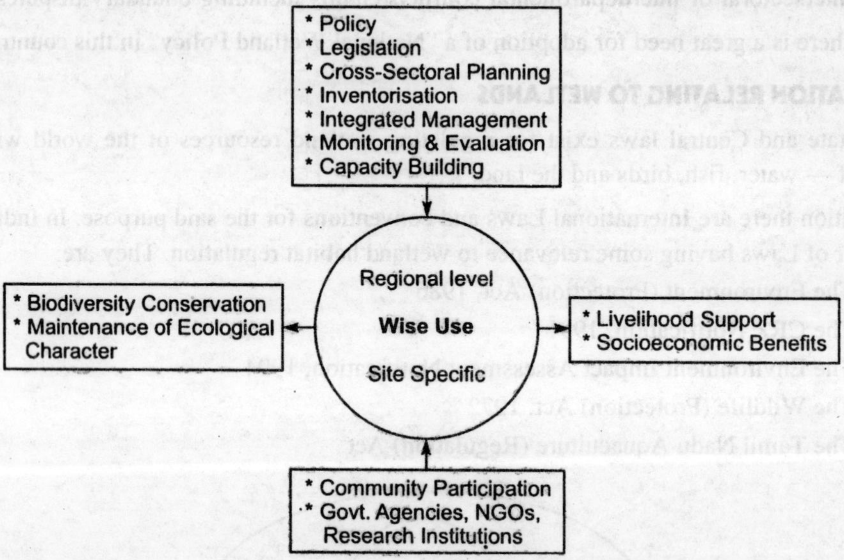

Fig. 8.3: 'Wise use' conceptual framework

8.10. NATIONAL LAKE CONSERVATION PLAN (NLCP)

A National Lake Conservation Plant (NLCP) is being implemented for conservation of polluted and degraded urban/semiurban lakes, leading to take rejuvenation in terms of improvement in water quality and biodiversity. Under this programme over 46 lakes have been taken up covering the entire country.

8.11. NATIONAL RIVER CONSERVATION PLAN (NRCP)

Similarly National River Conservation Plant (NRCP) is also under implementation in 160 towns along polluted stretches of 34 rivers spread over 20 states, the major rivers being Ganga, Yamuna, Gomti, Damodar, Satluj, Krishna, Cauveri and Godavari. The objective of NRCP is to check pollution in rivers through implementation of various pollution abatement schemes.

SECTION III

POLLUTION

9 ENVIRONMENTAL POLLUTION (GENERAL)

9.1 CONCEPT OF ENVIRONMENTAL POLLUTION

'Pollution' means the presence of undesirable substance in any segment of environment, primarily due to human activity—discharging by-products, waste products or harmful secondary products, which are harmful to man and other organisms. But it should be borne in mind that even in natural state no environmental component is pure in the scientific sense, there being several admixtures and impurities, but they are insignificant and do not cause appreciable harm. A clean air is a natural mixture of many gases, moistures and particulate matters but when foreign matters are added in significant quantities like oxides of nitrogen and sulfur or lead containing fumes or fluorides etc., air becomes harmful to plants, animals and man—directly or through food-chain—and thus the air becomes polluted.

9.2 SOURCES OF POLLUTION

Environmental pollution is now considered as a global phenomenon, which maximally attracts the attention of human beings for its severe long term consequences. Over past couple of decades, various sources of pollution were identified that altered the composition of water, air and soil of the earth environment. It is often stated that industrialisation is the principal cause of environmental pollution. But many other non-specified sources were often ignored earlier. There are altogether six principal sources of pollution identified so far. The details are given in Fig. 9.1. Depending on the site specific situation, the relative importance of all sources were considered. For instance, in urban environment, anthropogenic sources of pollutants are considered to have the major significance, while in rural agricultural areas, the agricultural sources are the principal agents as recognised by the scientific experts.

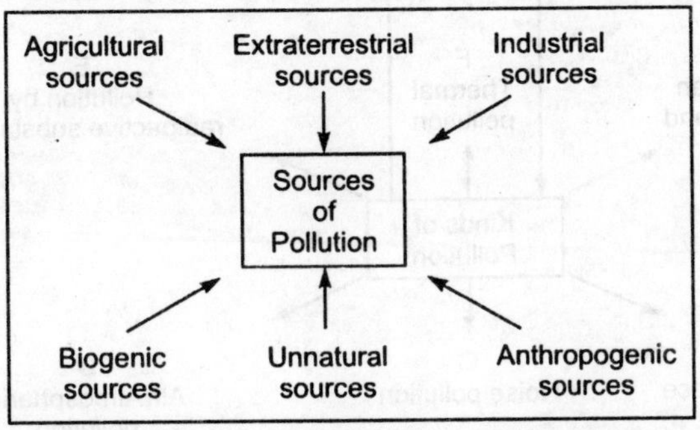

Fig. 9.1 : Sources of pollution

9.3 NATURE OF POLLUTANTS

A pollutant can be any chemical (toxic metal, radionuclide, organo-phosphorus compound, gases) or geochemical substance (dust, sediment), biological organism or product, or physical substance (heat, radiations, sound wave) that is released intentionally or inadvertently by man into the environment with actual or potential adverse, harmful or unpleasant or inconvenient effects. Such undesirable effects may be direct (i.e., affecting man) or indirect, being mediated via his resource organisms or climatic changes.

Depending on the diversities of pollutants' nature, there are four major categories viz., Physical, Chemical, Biological and pollutants of uncertain types. Again, in each category, there are various agents as described in details (Fig. 9.2). Depending on the nature of pollutants and also subsequent pollution of environmental components, the pollution may be categorised under six major heads:

(a) Water pollution
(b) Soil/Land pollution
(c) Noise pollution
(d) Air/Atmospheric pollution
(e) Pollution by radioactive substances, and
(f) Thermal pollution.

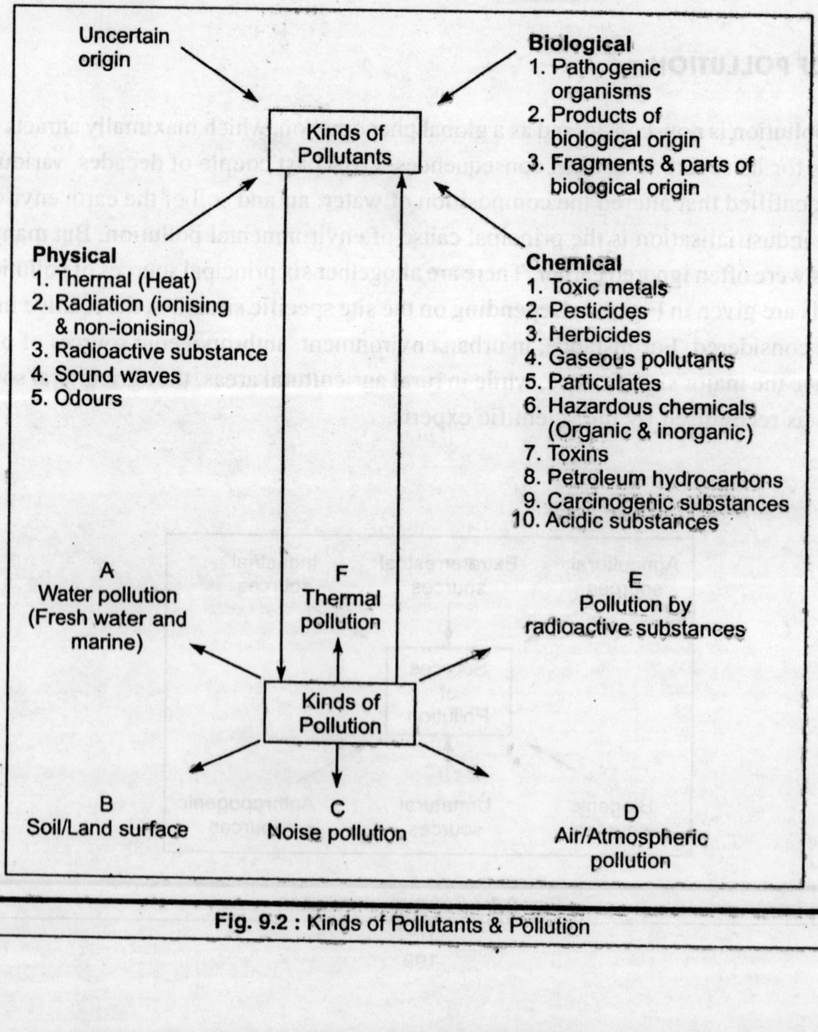

Fig. 9.2 : Kinds of Pollutants & Pollution

There is an increasing trend in environmental pollution. Water is polluted by four kinds of substances: traditional organic waste, waste generated from industrial processes, chemical agents for fertilizers and pesticides for crop protection and silt from degraded catchments. While it is estimated that three-fourths by volume of the waste water is generated from municipal sources, industrial waste—though small in volume—contributes over half of the total pollutants load, and the major portion of this is coming from large and medium industries. For Class I cities of the country, less than five per cent of the total waste generated is collected and less than one-fourth of this is treated!

Ambient air quality trends in the major cities indicate the levels of Suspended Particulate Matters (SPM) are higher than the prescribed standards or limits, especially in summer months. Levels of NO_x are increasing in urban centres with growing vehicle emissions. Environmental problems are becoming larger in scale. The chemical industry generates an increasing quantity of substance every year—adversely affecting essential aspects of the composition of the atmosphere, soil, and water. In the industrial high density areas, in addition to the effects on local health and impact on nature, we are confronted with damage to the social and economic functions of the environment.

In urban environment in particular, the noise pollution problem appears to be more acute day-by-day. The state of the environment continues to deteriorate. The growth of scientific and technical knowledge has made it possible to use an ever-increasing quantum of natural resources. The increase in population is further enhancing the pressure on the environment. The depletion of forests has been accompanied by increasing amount of pollution affecting the atmosphere, soil, and water.

9.4 POLLUTION MONITORING

The commitment of the Government on monitoring and abatement of pollution for preventing deterioration of the environment is started through a comprehensive approach.

Such an approach has the following objectives:

- Monitor and prevent pollution at sources;
- Encourage, develop and apply the best available practicable technical solutions;
- Ensure that the polluter pays for the pollution and control arrangements;
- Focus protection on heavily polluted areas and river stretches;
- Involve the public in decision-making.

In general, there are four major levels of pollution monitoring and exercising pollution abatement policies (Table 9.1). All our efforts has to be made at all levels depending on the degrees of the problem concerned.

Table 9.1: Pollution-monitoring agencies

Level	Types of organisations
Primary (Village level, Block level District level):	Science forums, Rural development agencies, various non-governmental bodies.
Secondary (State level):	State Pollution Control Board/Department of Environment/ Department of Science & Technology.
Tertiary (Country level):	Department of Environment/Environmental Protection Agency.
Quaternary (Inter-Governmental):	United Nation Environmental Program/World Health Organisation.

9.5 POLLUTANT STABILITY

Various pollutants of water, air, and soil have different degrees of stability. Though many pollutants degrade by chemical transformation or by microbial degradation or even being biotransformed after uptake by the biological system, yet there are a good number of pollutants which have pronounced stability. For instance, pesticides, PAHs, PCBs, various pollutant gases, and herbicides have different degrees of stability in nature. These are called non-degradable pollutants or persistant pollutants. The atmospheric residence time of major gaseous pollutants are shown in Table 9.2:

Table 9.2: Atmospheric residence time of major pollutant gases

	Air Pollutants	Residence Time
1.	CO_2	4 years
2.	CO	1 - 4 months
3.	He	3 - 16 years
4.	Halogenocarbon	Over 20 years
5.	SO_2	3 - 7 days
6.	H_2S	2 days
7.	NO_x	4 days
8.	NH_3	2 days
9.	Particulates	Variable

9.6 POLLUTION REDUCTION

Effective implementation of pollution prevention technologies can actually save companies' money by reducing process inefficiencies and eliminating costly end-of-pipe treatments. There are a number of basic approaches to the pollution reduction technologies as adopted currently.

These are:

- Identifying and evaluating alternative strategies;
- Specific waste treatment strategies for particulates, solid wastes and waste water;
- Source reduction;
- Separation and recycling of high value wastes;
- Adoption of waste management economics.

An overall flow sheet of pollution reduction techniques are shown in Fig. 9.3 (next page).

BIO-WARFARE AGENTS	
ANTHRAX	Bacteria in spore form which is resistant to sunlight, heat and disinfectant. Infection can cause death more than 90 per cent of the time. It's not contagious but is most likely spread by an aerosol mist. It can be spread through inhalation, ingestion or through skin abrasions. Initial flu-like symptoms lead to breathing problems. Treatment is through early administration of antibiotics and prevention by vaccines.

Contd.

BIO-WARFARE AGENTS	
BOTULISM	The most lethal toxin known. It can be inhaled or ingested through contaminated food or water. Symptoms can show up two weeks after exposure and starts at the top of the body—double vision, slurred speech, dry mouth—and works its way down causing muscle weakness. Death can be caused by paralysis of the breathing muscles within 24 hours. Treatment is through anti-toxins and prevention by vaccines.
PLAGUE	The causative agent is found in rodents. It can be deliberately distributed through aerosol cans, and it's contagious. The first signs are fever, headache and weakness, which can lead to death within two to four days. Treatment is through antibiotics.
SMALL POX	Though WHO has (1970) successfully eradicated it, there are limited stocks in the USA and Russia, reportedly for scientific purposes. The virus is deadly and highly contagious. It, too, can be released through aerosol cans, but if used by terrorists they have to risk the prospect of infecting themselves. Initial symptoms are fever, fatigue and body aches, followed by severe rash with abrasions and can lead to death in the first two weeks. There is no treatment, but can be prevented by vaccine. Vaccine immunity wears off in 10 years.
TULAREMIA	This is caused by the most infectious bacteria, *Francisella tularensis*, through bites from infected anthropods, inhalation or contaminated food or water. Initial symptoms appear like influenza that can appear up to two weeks after exposure. It progresses into systematic weight loss causing death. It's treated by antibiotics and there are no vaccines yet.
VIRAL HEMORRHAGIC FEVERS	VHF is a group of diseases caused by several families of viruses. They range from mild ones to the lethal Ebola virus. They can be spread through animals and insects and are contagious. Often flu-like symptoms are reported but they sometimes cause bleeding of internal organs and kidney failure. There is no treatment but they can be prevented by vaccines.

Contd.

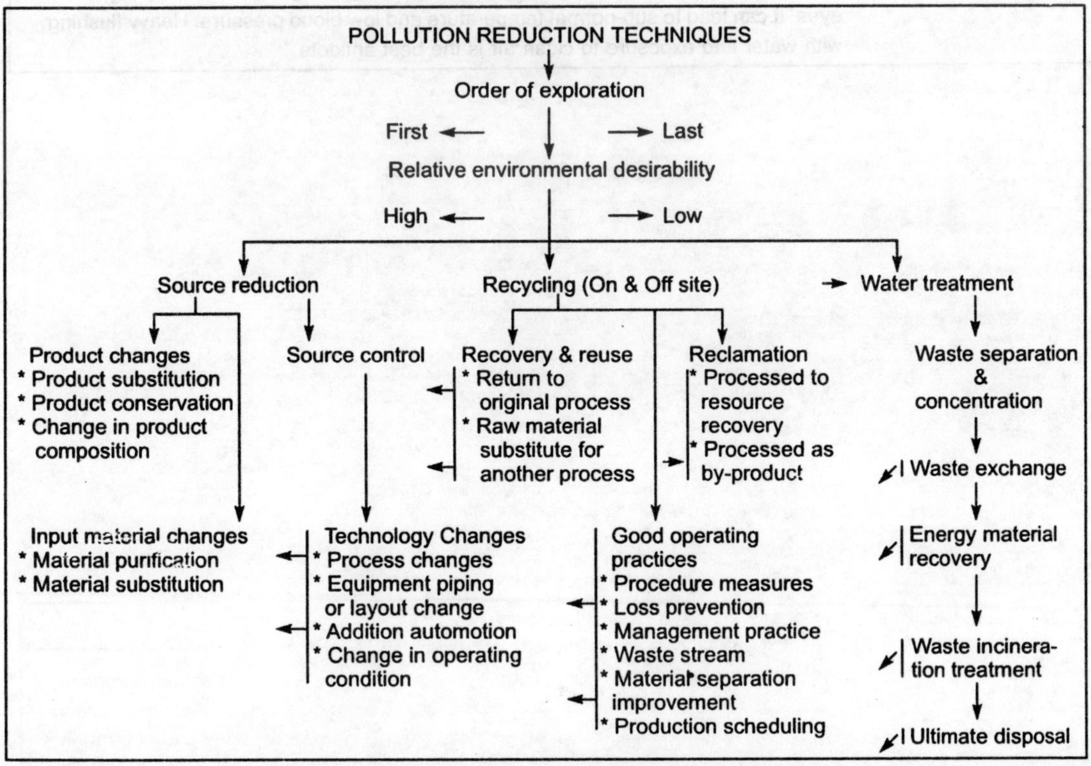

Fig. 9.3 : Pollution Reduction Techniques

CHEMICAL WARFARE AGENTS	
SARIN	This lethal nerve gas is a highly toxic liquid or vapour form of phosphorus ($C_4H_{10}FPO_2$) that immediately attacks the central nervous system. It can be inhaled or absorbed through skin and eyes. Watery eyes, drooling, sweating are the initial symptoms and death can be caused within minutes. Immediate treatment involves flushing the body and eyes with water. Damage control is swift decontamination. Sarin was developed during World War II (1939-1945).
VX	This is similar to Sarin but is 100 times more toxic. Both are easy to acquire and store but difficult to manufacture quantities for a largescale attack.
CHLORINE	This is a greenish-yellow pungent smelling gas. On interaction with some other compounds it can cause fires and explosions. When exposed, it causes blurred vision, burns and heavy breathing. High exposure may result in death. Immediate treatment involves flushing the body with water.
HYDROGEN CYANIDE	A highly inflammable gas or liquid, it causes irritation to eyes, skin and throat, causing burning of the skin, drowsiness and confusion leading to a collapse. It proceeds to attack the central nervous system. The best antidote is flushing with water and exposure to clean air.
SULPHUR MUSTARDS	Blistering agents having a yellow to brown colour, these can be absorbed through the eyes, skin and respiratory tracts causing injury in the process. The US Center for Disease Control says there is no antidote and the only effective damage control is swift decontamination of exposed areas.
LEWISITE	Another blistering agent, it causes violent sneezing, coughing and a burning of the eyes. It can lead to sub-normal temperature and low blood pressure. Heavy flushing with water and exposure to clean air is the best antidote.

10 AIR POLLUTION

10.1 COMPOSITION AND CHEMISTRY OF EARTH ATMOSPHERE

The Earth's atmosphere is 99.9% nitrogen, oxygen and argon, of the remaining 0.1%, roughly half (0.04%) is accounted for by the three transition gases CO_2, Ne and He. The remaining minute fraction of the atmosphere (0.66%) is composed of a complex mixture of hundreds of trace gases, the most abundant being methane at 1.7 ppm. Many gases at the parts per trillion level (10^{-12}v/v), are entirely man-made but may still have significant effect on the environment. Some trace gases control or affect the Earth's climate and stability. Long lived gases that are increasing at substantial rates because of human activities are of particular current interest since they lead eventually to stratospheric ozone depletion, global warming, and disturbances in atmosphere chemistry that many believe will be harmful to human life.

As per the present estimate the composition of the atmospheric gases are presented in Table 10.1. Substantial effects of human activities on the global cycles of trace gases have arisen only during the last century. Since 1800 the human population has increased from 1 to 7 billion, bringing with it fast changes in the surface of the earth and disturbing the global balance of the trace gases, changing in atmospheric composition that have occurred are unprecedented during the last several thousand years. The rate of change is much faster now than can occur from natural climatic cycles, although it is hard to prove that there were no times when trace gas concentrations did not rise rapidly and later fall. The changes that are occurring now have their origin in the Industrial Revolution (since 1760) and the expansion of technology. These changes are directly related to human events and activities. There is a prevalent belief that the changes man is causing to the atmospheric environment are unnatural and unhealthy.

From the earth surface atmosphere to the outer periphery of the planet, there were several layers of gaseous clouds which are variously designated as Troposphere, Stratosphere, Mesosphere and Thermosphere (Table 10.2). There were various types of natural processes like volcanic, earthquake, decomposition and climatic transformations that led to the changes in global atmospheric composition in addition to man-made activities like industrialisation, urbanisation and modernisation of agricultural activities. The common atmospheric pollutants (both primary and secondary) along with their major sources is presented in Table 10.3.

Table 10.1: Composition of the Atmosphere

Constituent	Abundance by volume
Nitrogen (N_2)	78.8%
Oxygen (O_2)	20.95%
Argon (Ar)	0.93%
Water vapour (H_2O)	trace (variable % ppmv)
Carbon dioxide (CO_2)	340 ppmv

Contd.

Table 10.1: Composition of the Atmosphere

Constituent	Abundance by volume
Neon (Ne)	18 ppmv
Helium (He)	5 ppmv
Krypton (Kr)	1 ppmv
Xenon (Xe)	0.08 ppmv
Methane (CH_4)	2 ppmv
Hydrogen (H_2)	0.5 ppmv
Nitrogen oxide (N_2O)	0.3 ppmv
Carbon monoxide (CO)	0.05 – 0.2 ppmv
Ozone (O_3)	variable
Ammonia (NH_3)	(0.02 – 10 ppmv)
Nitrogen dioxide (NO_2)	4 ppbv
Sulphur dioxide (SO_2)	1 ppbv
Hydrogen sulphide (H_2S)	0.05 ppbv

ppmv = parts per million by volume
ppbv = parts per billion by volume

Table 10.2: The structure and gross composition of atmospheric layers

Height (km)	Layer	Sub layer	Water vapour	Temperature	Gases
40,000	Thermosphere			Increase with height–slow	Rarefied
650					
500		Ionosphere	None	Increase with height–rapid	Atomic H, O, N, He
80	Mesosphere		None	Decreases with height	Ozone layer O_3
50	Stratosphere		Little	Increases with height	Diatomic O_2
12	Troposphere		Most with vapour	Decrease with height	(Composition % vol)
			Clouds		N_2 78.08 O_2 20.95 CO_2 0.03 O_3 0.00002 Kr 0.0001 He 0.0005 Ne 0.0018 Ar 0.93
			Earth surface		

Table 10.3: Common atmospheric pollutants

Physical state	Sources			
	Fuel combustion		Industrial process	Natural process (volcanic)
	Primary	Secondary		
Gaseous	CO_2 CO Hydrocarbon NO_X SO_2 Aldehyde NH_3 H_2S	PAN O_3	CFC Acid Aldehyde Fluorides Chlorides	CO_2 CO O_3 CH_3 H_2S SO_2 NO_X NH_3
* Particulates	Smoke dust Lead particles etc.		Grit dust metallic particles etc.	Dust
	1. Grit		500 m	
	2. Dust		Between 0.25 & 500 m	
	3. Smoke		2 m	
	4. Fumes		1 m	
	5. Mist (liquid droplets)		2 m	
	6. Aerosol (suspended)		1 m	

10.2 PARTICLES IN THE ATMOSPHERE

Particles in the atmosphere, which range in size from about 0.5 mm down to molecular dimensions, are made up an amazing variety of materials and discrete objects that may consist of either solids or liquid droplets. A number of terms are commonly used to describe various kinds of particles as:

	Terms	Characteristics
1.	Aerosol	Colloidal size atmospheric particles.
2.	Condensation Aerosol	Formed by condensation of vapours or reactions of gases.
3.	Dispersion Aerosol	Formed by grinding of solids or dispersion of dusts.
4.	Fog	High level of water droplets.
5.	Haze	Decreased visibility due to the presence of particles.
6.	Mist	Liquid particles.
7.	Smoke	Particles formed by incomplete combustion of fuel.

Particulate matter makes up the most visible or obvious forms of air pollution. Atmospheric **aerosols** are solid or liquid particles smaller than 100 μm in diameter. Pollutant particles in the 0.001 to 10 μm range are commonly suspended in the air near sources of pollution such as the urban atmosphere, industrial plants, highways and power plants. Carbon black, silver iodide, sea salt are forming very small particles (710 μm); while cement dust, soil dust, foundary dust form larger particulates. Liquid particulates form mist or fog. Particulate matter may be organic or inorganic.

There are sources of particulate matter viz., industrial, transport, domestic and also natural processes. Some particulates are of biological origin (bioparticulate) viz., bacteria, spores of fungi and other plants and pollens. Particles in Arctic, remote from sources of industrial pollution, is forming "Arctic Haze" from October to May each year. Typically, particles less than 2.5 μm is significant in forming respiratory diseases in man. Fly ash is the mineral particulate matter in a polluted atmosphere in the form of oxides and other compounds produced during the combustion of high-ash fossil fuel.

Some of the metals found predominantly as particulate matter in polluted atmosphere are known to be hazardous to human health. They are Pb, Hg, Cr, Cd, Ni, Cu, Zn, and arsenic etc. Some of the radioactivity detected in atmospheric particles is of natural origin. They are known as radionuclides viz., ^{14}Be, ^{14}C, ^{39}Cl, ^{3}H, ^{22}Na, ^{32}P, radon, and so on. Transport sources produces various levels of carbonaceous particles with high organic materials (hydrocarbon like Benzene, Toluene, Xylene, Polyaromatic hydrocarbons).

10.3 AIR POLLUTANTS

The pollutants which discharge from different sources may thereafter persist in atmosphere—transformed, deposited, or degraded. There were various kinds of estimates of pollutants liberated by different activities of both developed and developing countries. The total emission load of traditional primary pollutants load (i.e., particulates, SO_x, NO_x, CO, HCs etc.) are depicted in the Table 10.4. However, the countrywise data on the relative importance of various activities on pollutant generation is very much limited. The fuel computing combusting industries like power plants are the major SO_x and NO_x generation source followed by transportation and other industrial process activities. Table 10.5 shows the trends of annual emission of pollutants in USA. Among the various categories of industries, in general, nine major types of industries are considered to be the major air pollutant generating industries (Table 10.6), of which thermal power plants, steel industries, petroleum refineries, metal smelters are the major polluting industries.

Principally air pollutants are of two categories viz., particulates and gaseous pollutants (NO_X, SO_X, CO, NH_3, Cl, H_2S, HF, acid vapour, other VOC).

There are many air pollutants which are directly derived from sources and released into atmosphere. They are primary air pollutants, while, some air pollutants after release from their principal sources, may react and produces complex air pollutants (O_3, Pan, aldehyde radicals). They are secondary air pollutants. The principal reactions among NO, NO_2, and HNO_3 in the atmosphere is shown:

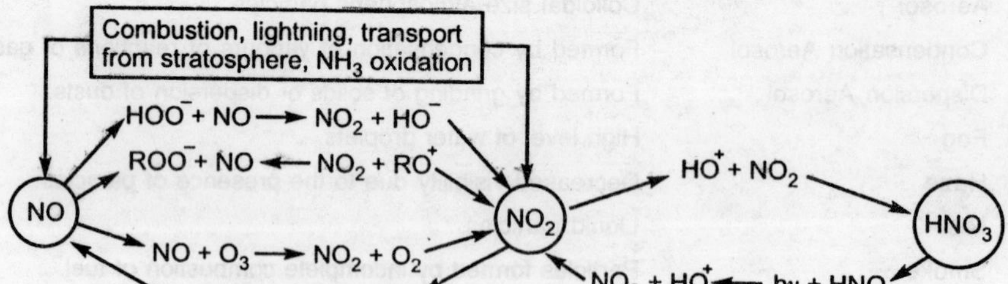

(ROO⁻ represent an organic peroxyl radical such as the methylperoxyl radical CH_3OO^-)

In addition, biogenic organic compounds like terpenes and ethylene are also emitted into the atmosphere by plants and other organisms. In addition, a number of complex organonitrogen compounds (amines, nitrites), organohalides (Dichloromethane, Chlorobenzene), Chlorofluorocarbons (Freous, Halens) organo-sulphur compounds (Dimethyl sulphide) are found in urban and industrial air. They are all persistent hazardous air pollutants.

Table 10.4: Total emission of traditional primary air pollutants, 1980 (in thousand tonnes/year)

	Country	Population (thousands)	Major primary pollutants				
			SO_x	Particles	NO_x	CO	HC
1.	USA	243,915	23,900	8,500	20,300	76,100	23,000
2.	Japan	122,090	1,259	133	1,340	—	—
3.	Germany	61,199	3,187	696	2,935	11,708	2,490
4.	UK	56,930	4,836	290	2,264	4,999	2,241
5.	Italy	56,664	3,211	386	1,585	5,487	1,566
6.	France	55,630	3,512	483	1,861	6,620	1,972
7.	Turkey	52,893	714	138	380	3,707	201
8.	Spain	38,832	2,543	1,521	973	3,780	843
9.	Canada	25,695	4,650	1,907	1,942	9,928	2,100
10.	Yugoslavia	23,410	1,175	—	—	—	
11.	Australia	16,264	1,479	271	915	3,704	423
12.	Netherlands	14,665	462	162	553	1,450	493
13.	Greece	9,994	546	40	217	695	130
14.	Belgium	9,834	856	267	317	839	339
15.	Portugal	9,744	266	119	166	533	159
16.	Sweden	8,398	502	170	318	1,250	410
17.	Austria	7,586	325	50	201	1,126	251
18.	Switzerland	6,619	126	28	196	711	311
19.	Denmark	5,127	452	47	245	577	197
20.	Finland	4,932	584	97	284	660	163
21.	Norway	4,187	150	28	203	608	159
22.	Ireland	3,543	217	94	71	497	62
23.	New Zealand	3,309	88	21	89	566	38
24.	Luxembourg	372	24	—	23	—	11
25.	OECD Europe	407,395	22,500	4,600	12,800	45,200	12,000
26.	EC	322,534	20,100	4,100	11,200	37,200	10,500
27.	North America	269,540	28,550	10,407	22,242	86,028	25,100
28.	Australasia	19,573	1,567	292	1,004	4,270	461
	OCED total	818,598	54,000	15,500	37,500	14,100	38,500
	World total	5,024,000	110,000	59,000	69,000	193,000	57,000

Table 10.5: US annual emissions of pollutants from various sources in 1975, 1984 and 1989 (million tonnes/year)

Pollutant & Source	1975	1984	1989
A. Total suspended particulates:			
Transportation	1.3	1.3	1.5
Fuel combustion	2.8	2.1	1.8
Industrial processes	5.2	2.8	2.7
Solid waste	0.6	0.3	1.0
Miscellaneous	0.7	0.9	1.0
Total	10.6	7.4	7.2
B. Sulphur oxides:			
Transportation	0.7	0.8	1.0
Fuel combustion	20.2	17.4	16.8
Industrial processes	5.0	3.3	3.3
Solid waste	0.0	0.0	0.0
Miscellaneous	0.0	0.0	0.0
Total	25.8	21.5	21.1
C. Carbon monoxide:			
Transportation	65.0	50.6	40.0
Fuel combustion	4.3	8.3	7.8
Industrial processes	6.9	4.7	4.6
Solid waste	3.1	1.9	1.7
Miscellaneous	4.8	6.3	6.8
Total	84.1	71.8	60.9
D. Oxides of Nitrogen:			
Transportation	9.3	8.8	7.9
Fuel combustion	9.3	10.2	11.1
Industrial processes	0.7	0.6	0.6
Solid waste	0.1	0.1	0.1
Miscellaneous	0.1	0.2	0.2
Total	19.5	19.9	19.9
E. Volatile Organic Compounds:			
Transportation	8.8	8.1	6.4
Fuel combustion	0.6	1.0	0.9
Industrial processes	8.3	8.8	8.1
Solid waste	0.9	0.6	0.6
Miscellaneous	2.5	2.7	2.5
Total	21.1	21.2	18.5
F. Lead (thousand tonnes/year):			
Transportation	122.6	34.7	2.2
Fuel combustion	9.3	0.5	0.4
Industrial processes	10.3	2.3	2.3
Solid waste	4.8	2.6	2.3
Total	147.0	40.1	7.2

Source: USA EPA, 1991.

Table 10.6: Sources of Industrial emission of air pollutants

	Industries	Air pollutants generated
1.	Steel mills	Particulates, smoke, CO. fluoride.
2.	Petroleum refineries	SO_2, hydrocarbon smoke, particulates, odour.
3.	Acid plants	SO_2, acid mist.
4.	Paper mills	SO_2, particulates, odours.
5.	Soap & detergent plants	Particulates, odour.
6.	Fertilizer plants	Particulates, ammonia, SO_2, NO_X, fluoride.
7.	Cement plants	Cement dust, SO_2, smoke.
8.	Thermal power plant	NO_X, SO_2, particulates.
9.	Metal smelters	SO_2, NO_X, particulates, smoke.

10.4 PHOTOCHEMICAL SMOG

Hydrocarbons are eliminated from the atmosphere by a number of chemical or photochemical reactions. These reactions are responsible for the formation of many noxious secondary pollutant products and intermediates from relatively innocuous hydrocarbon precursors. These pollutant products and intermediates make up **photochemical smog**.

Hydrocarbons and most other organic compounds in the atmosphere are thermodynamically unstable toward oxidation and tend to be oxidised through a series of steps. The oxidation process terminates with formation of CO_2, solid organic particulate matter that settles from the atmosphere, or water soluble products (viz., acids, aldehydes) that are removed by rain. Inorganic species such as O_3 or HNO_3 are by products of these reactions. A large number of specific reactions are involved in the overall scheme for the formation of **photochemical smog** (Fig. 10.1).

The harmful effects of smog occur mainly in the areas of (a) human health and comfort; (b) damage to materials, (c) effects on the atmosphere, and (d) toxicity to plants. The exact degree to which exposure to smog affects human health is not known, although substantial adverse effects are suspected. Pungent smelling smog produced ozone is known to be toxic. Materials are adversely affected by some smog composts specially rubber and other polymers. Smog produces harmful effects in plants and vegetation. Forest & crop production is highly affected.

The most serious air pollution episodes are the classical smog so closely associated with London. These episodes all occurred in cold winter months in areas where coal was a major fuel for both industry and heating. They are all characterised by high levels of sulphur dioxide and smoke particulate which built up under stagnant weather conditions lasting three days or more. Mortality and illness were significantly increased and, in the worst case in terms of the number of affected, in London in the week of 5-11 December 1952, there were 3,500 to 4,000 more deaths than the expected average for the time of the year. Those most susceptible were the elderly and the young, particularly individuals with a history of cardio-respiratory illness. No single primary pollutant could be singled out for blame in any of these incidents and it is probable that a synergistic association between SO_2 particulates and their reaction products sulphuric acid was operative. Nitrous oxides are similarly oxidised and hydrolysed to nitric acids also.

Still there is more complex smog termed 'photochemical smog'. This was first recognised as a problem in Los Angeles in 1943 and has since been detected in many cities of the world. Photochemical smog refers to the complex mixture of products formed from the interaction of sunlight with major components of automobile exhaust, nitric oxide (NO) and hydrocarbons. Other pollutant species present in the atmosphere —such as SO_2 and particulates—can also be participants, as can hydrocarbon and nitric oxide from other

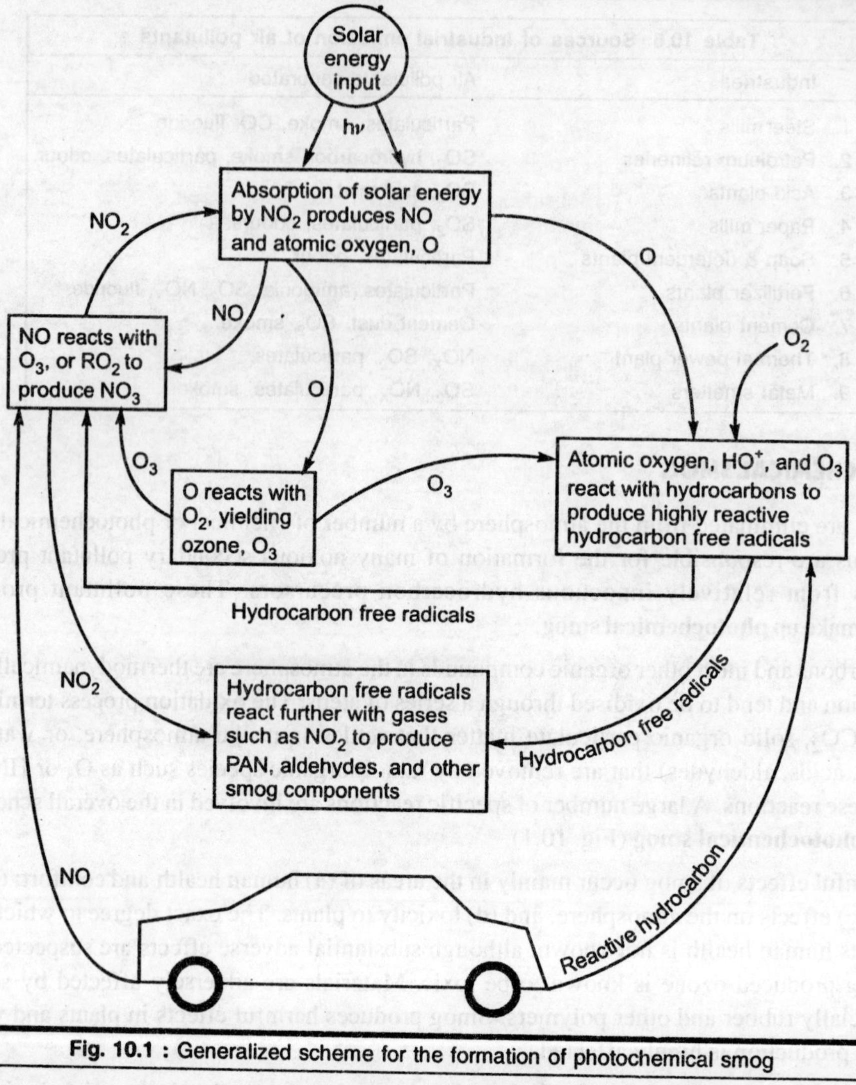

Fig. 10.1 : Generalized scheme for the formation of photochemical smog

sources, but they are not essential for the production of the characteristic high oxidant levels associated with photochemical smog formation. The details of smog formation reaction is given in Fig. 10.2 (next page). Smog is further favoured by stable meteorological conditions, when the urban emissions are held in the urban airshed by an inversion acting rather like a lid over a reaction vessel, maximising contact and reactions whilst preventing dispersion.

The several incidents occur when this inversion is steady for several days, allowing for further emissions and reactions adding to those of the previous days. The oxidant formed is predominantly ozone with varying amount of other oxidising components, including the notorious proxyacetyl nitrates (PAN). These minor components contribute to a variation in the observed effects for the same total oxidant concentration in different places. The variations in oxidant composition occur as a result of variations in the mix of hydrocarbons initially present and the length of time the particulate parcel of air has to react. These factors vary from airshed to airshed and from day to day within the same airshed.

The distinction between two kinds of smog formation is given in Table 10.7 (next page).

Table 10.7: Comparison of Los Angeles and London smog

Characteristics	Los Angeles	London
Air temperature	24 – 32°C	–1 to – 4°C
Relative humidity	< 70 per cent	85 per cent (+ fog)
Type of temperature inversion	Subsidence, at a few thousand meters	Radiation (near ground) at about 100 meters
Wind speed	< 3 m/s	Calm
Visibility	< 0.8 – 1.6 km	< 30 m
Months of most frequent occurrence	Aug. – Sept.	Dec. – Jan.
Major fuels	Petroleum	Largely coal
Principal constituents	O_3, NO, NO_2, CO organic matter	Particulate matter, CO, S compounds
Type of chemical reaction	Oxidative	Reductive
Time of maximum occurrence	Midday	Early morning
Principal health irritation effects (SO_2/smoke)	Temporary eye irritation (PAN)	Lung diseases
Materials damaged	Rubber cracked	Iron, concrete corroded

Source : R. W. Raiswell et al (1980)

THERMAL INVERSION

During the day, the sun warms the air near the earth's surface. Normally, this heated air expands and rises, carrying low-lying pollutants higher into the troposphere. Colder, denser air from surrounding high-pressure areas then sinks into the low-pressure areas created when the hot air rises [Fig. 10.2(a)]. This continual mixing of the air helps keep pollutants from reaching dangerous concentrations near the ground.

Sometimes, however, a layer of dense, cool air beneath can be trapped beneath a layer of less dense, warm air in an urban basin or valley, causing a phenomenon known as a **temperature inversion**, or a **thermal inversion** [Fig. 10.2(b)]. In effect, a lid of warm air covers the region and prevents ascending air currents (that would disperse and dilute pollutants) from developing. Although these inversions usually last for only a few hours, sometimes—when a high-pressure air mass stalls over an area—they last for several days, allowing air pollutants at ground level to build up to harmful and even lethal concentrations. Thermal inversions also enhance the harmful effects of urban heat islands and dust domes that build up over urban areas.

Thermal inversions trap pollutants in a pool of cool air that cannot rise to carry the pollutants away. Because of its topography, Los Angeles has frequent thermal inversions, many of them prolonged during the summer months.

Fig. 10.2(a) : Normal pollutants flow in atmosphere

Fig. 10.2(b) : Thermal inversion condition

10.5 EFFECTS OF AIR POLLUTION

The gaseous and particulate materials added to the atmosphere by the activities of man are considered to be pollutant when their concentrations are sufficient to produce harmful effects. The majority of man-made emissions to the atmosphere also have natural sources and in many cases these are larger than the pollutant ones. The world as we know it developed in the presence of these chemicals which cannot, therefore, be considered to be harmful gases but only if they produce unacceptable effects at concentrations above the natural background level.

After formation, pollutants are emitted to the atmosphere and dispersed. Once mixed with the air, some air pollutants—such as the inert fluorinated hydrocarbons used in sprays—persist unaltered and become mixed throughout the atmosphere where they potentially have a global influence. More reactive pollutants have a shorter lifetime in the atmosphere and are removed either by conversion to normal atmospheric constituents or by deposition on the surface of the earth. In the process they may react with other atmospheric constituents to form secondary pollutants, which are also removed by the same process. Both the primary pollutants and the secondary pollutants can cause alteration to the chemical composition of soils and waters, and direct damage to biological systems and property. In certain cases, a synergistic interaction occurs where the total effect is enhanced over and above the sum of the effects of the individual pollutants present.

Laboratory studies can provide information under controlled conditions on the effect of pollutants on materials, plants, animals and to a limited extent on humans. This approach also provides for controlled conditions and the removal of unwanted variable and allows for the unequivocal association of level of effect with concentration.

(A) *Effects on Vegetation*: Air pollution can affect plants to varying degrees. At the lowest levels, i.e., below the 'threshold', there is no effect, such as visible damage, cumulative chronic effects, genetic effects or even gradual changes in the composition of the plant community. However, even at this level, air pollutants can be stored in the plants and introduced into the food chain, affecting animals which eat the plants.

The entry of air pollutants to plants may take place directly by gaseous diffusion or from the contaminated soil, acidic air pollutant in particular. The direct entry of gaseous air pollutants like SO_X, NO_X, CO_2 etc. may take place directly by stomata of the foliages. Solid particulates are, however, adsorbed on the surface. In general, pollutants cause injury at lethal concentration on sensitive plants while tolerant plant species are capable of sinking the pollutants to a considerable extent without any injury. The air pollution induced injury on various plant species are given in Table 10.8.

Thus, various air pollutants have different types of injuries on exposed sensitive plants. Suspended particulates after deposition on foliages cause a number of damages to leaf functions, viz.,

(a) Changes in the sun's energy falling on the leaf surface affecting the energy exchange due to dust layer;
(b) Decreases chlorophyll content;
(c) Interruption in gaseous exchange due to clogging of stomata by dust particulates; and
(d) Dust deposition causes changes in the soil properties that support the plant growth.

Table 10.8: Effect of air pollutants on plants

Pollutants	Phytotoxic concentration (ppm/h)	Symptoms
(A) MAJOR POLLUTANTS:		
(a) Sulphur dioxide	0.1 – 0.5	Interveinal bleaching and marginal bleaching, chlorosis, discolouration (red/brown)
(b) Oxides of nitrogen	0.21 – 100	Chlorosis and marginal bleaching
(c) Ozone	0.04 – 0.70	Bronzing, Chlorosis, coloured fleck
(d) Fluorides	0.001 – 0.10	Chlorosis, particularly along the leaf margin, marginal necrosis from leaf tips to leaf bass

Contd.

Table 10.8: Effect of air pollutants on plants		
Pollutants	Phytotoxic concentration (ppm/h)	Symptoms
(B) MINOR POLLUTANTS:		
(a) Ammonia	0.001 – 0.10	Leaves turn dull green and the brown or black glating or silvering on the underside of leaves.
(b) Ethylene & propylene	0.0005 – 10	Leaf curling.
(c) Particulates (with toxic metals)	100 – 500	Chlorosis followed by reddening & yellowing, necrosis, bleaching of leaf margin.

The oxides of sulphur, nitrogen and carbon are absorbed by the plants through gaseous exchange and then either assimilate in such a manner so that there is no detrimental injury to the plant parts or it causes appreciable injury to various plant organs. In general, sensitive plant forms could be used for biomonitoring air pollutants. Several species of lichens and angiospermic plants are often used for biomonitoring of these pollutants.

(B) *Effects on Animals*: The indirect effect of air pollutants has been observed for a considerable time. It tends to occur near smelters treating non-ferrous ores, and near factories, such as phosphate fertilizer works, brick kilns and aluminium smelters, where fluorides are emitted and are concentrated in the grasses in surrounding fields. Identically, lead compounds from automobile exhausts are deposited near to roads, although the concentrations found in vegetation are much smaller than those near smelters and have so far been measured as below the accepted threshold for toxicity to animals. But, with time, such metals accumulate in animal bodies through food chain and finally pose danger to animal health. The signs of heavy metal poisoning are diarrhea, anemia and stiffness. Fluorosis is a more widespread problem, which affects ruminants—particularly dairy cows.

Of course there are couple of reports of direct effect of gaseous pollutants on animal but considerable systematic studies have yet to be made.

(C) *Effects on Man*: Air pollutants, along with breathing air, enters the nose, where fine hair filter out most particles greater than about 10 micrometers diameter. The air is then warmed and humidified, and is then passed through the windpipe into the bronchial tubes which subdivide the air stream and pass it into the lungs where there are a multiplicity of air sacs. From air sac through capillary diffusion gaseous pollutants enter into blood streams and particulates deposit in the alveolar sacs. The dangers of some small particles—such as silica and asbestos, which are common in mines, quarries and some industrial plants—are well-known over the years. They lead to specific occupational diseases such as silicosis or asbestosis.

The details of human respiratory system is given in Fig. 10.3. In general, air pollutants mostly enter through the respiratory passages and thereby cause mostly various types of health disorders. Limits are set as to the maximum concentrations to which healthy persons are permitted to be exposed for periods of up to 8 hours per day. These are called Threshold Limit Values (TLV). The WHO guidelines for such a threshold limit of exposure is given in Table 10.9:

Table 10.9: Health effects and air quality guidelines for major air pollutants			
Pollutant	WHO guidelines		Effects
	Annual mean ($\mu g/m^3$ of air)	98 percentile ($\mu g/m^3$ of air)[a]	
Sulphur dioxide	40 – 60	100 – 150	Exacerbations of respiratory illness from short-term exposures. Increased prevalence of respiratory symptoms, including chronic bronchitis from long-term exposures

Contd.

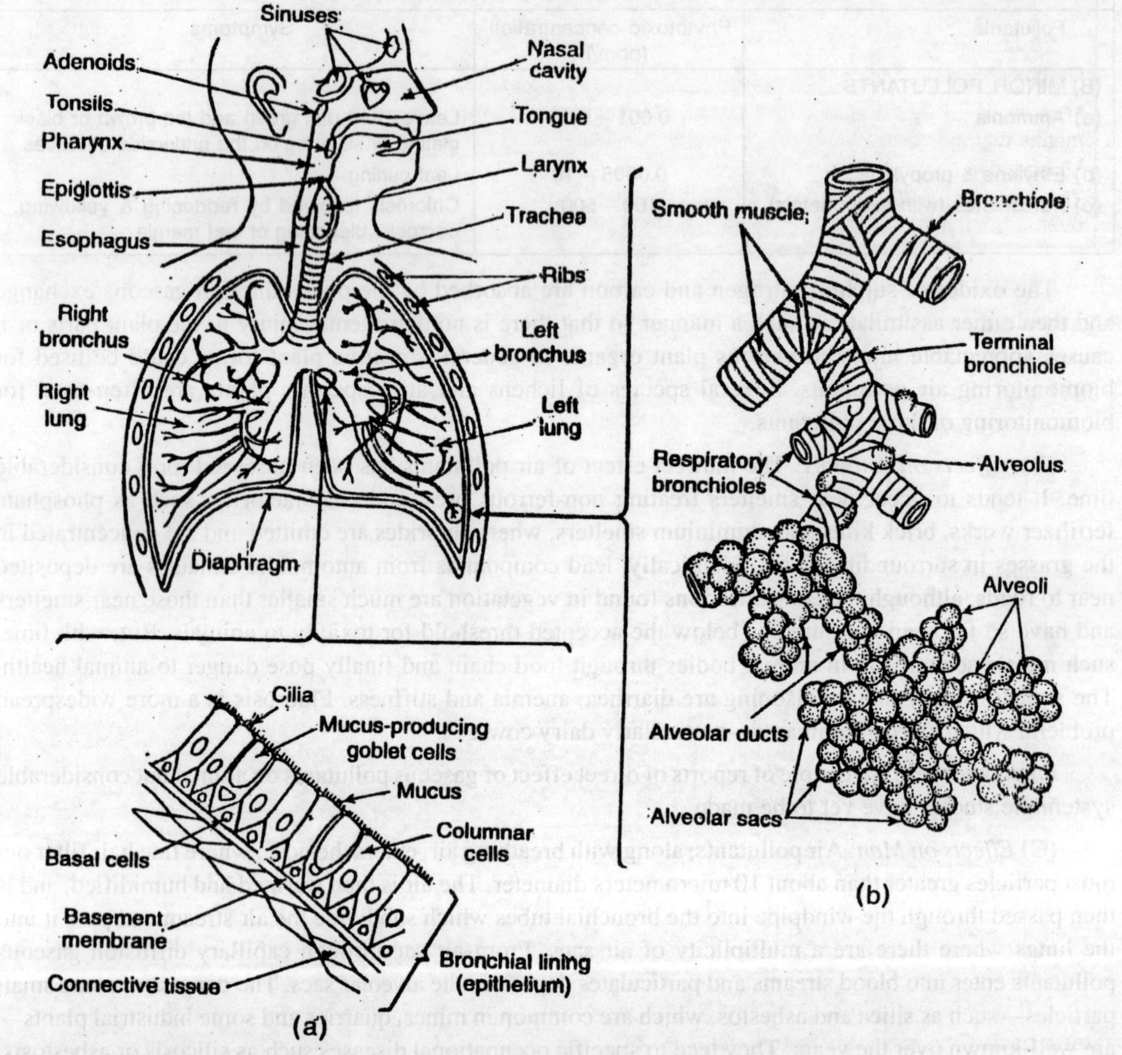

Fig. 10.3 : The Human Respiratory System : (a) Bronchial epithelium, and (b) Respiratory passages

Suspended particulate matter			
Black smoke	40 – 60	100 – 150	Same as for SO_2
Total suspended particulates	60 – 90	150 – 230	Combined exposure to SO_2 and SPM[b] may have pulmonary effects
Lead	0.5 – 1	—	Blood enzyme changes. Anemia. Hyperactivity and neurobehavioural effects
Nitrogen dioxide			
1 hour	400	—	Effects on lung function in asthmatics from short-term exposures
24 hours		150	
Carbon monoxide			
15 minutes	100,000	—	Reduced oxygen-carrying capacity of blood

Contd.

30 minutes	—	60,000
1 hour	30,000	—
8 hours	—	10,000
Carboxy-haemoglobin	—	2.5-3%

[a]The 98 percentile value stipulates that 98 per cent of the daily averages must fall below a given concentration. This means that less than 2 per cent, or less than 7 days per year, may exceed that concentration.
[b]Suspended particulate matter.
These are the guidelines used by the World Health Organisation and form the basis of air pollution control the world over.

In the atmosphere, such high concentrations are seldom met, but, on the other hand, the exposure of a population is over 24 hrs, rather than 8 hours, and all members of that population, including the weakest, are involved. In addition, even at long term low level exposure, pollutants, as noticed in city environment, may lead to a number of slowly grown detrimental health effects. An overall generalised health effect of various air pollutants are given in Table 10.10:

Table 10.10: Generalised Effect of air pollutants on human health

Air Pollutants	Concentration (ppm)	Health Disorders
1. CO	10	Impairment of judgement & vision
	100	Headache, dizziness
	250	Loss of consciousness
	> 700	Death (60% CoHb)
2. NO_2	50 – 100	Inflammation of lung tissue
	150 – 200	Bronchiolitis fibrosa
	> 500	Death
3. SO_2	> 5	Trachial irritation & cough
	5 – 10	Bronchial spasms
	> 200	Death
4. O_3 & PAN	> 5	Irritation of eyes & respiratory tract
	> 50	Death due to pulmonary edoma
5. Particulates	> 100	Respiratory hazard, Cardiovascular change

It was interesting to note the fact that low level exposure of air pollutants like NO_x and SO_2 on long term basis has significant health disorders. In general, an in-depth analysis with respect to SO_2 was made in the past, which is primarily based on various kinds of urban studies. The details are depicted in Table 10.11 (Page 205) and in Fig. 10.4. It is most interesting to note that even at low SO_2 concentration (0.01 ppm) exposure over a period of 1 year may lead to increased cardiovascular morbidity. There are also possibilities of air pollution induced carcinogenesis in urban environment as urban air pollutants contain a number of carcinogenic elements. Extracts of atmospheric particulates showed the existence of polyaromatic hydrocarbons which are known to be potent carcinogen.

(D) *Effects on Materials*: Air pollutants have a deleterious effect on materials: stone, paintwork, stained glass, fibre material and others. The soiling effect of particulate is obvious in industrial cities where building of light-coloured stones and bricks soon take on the characteristic black colour. The erosion of the stonework on buildings of great historic and architectural value is very serious indeed, other results of air pollution are the faster deterioration of clothing, curtains and wood, the corrosion of metals and the soiling and subsequent cracking of paintwork. The rapid deterioration of rubber acted on by ozone can be easily demonstrated in the laboratory. Silverware tarnishes rapidly due to hydrogen sulphide and sulphur dioxide in industrial cities. The details of different types of air pollution damage is given in Table 10.12 (Page 214).

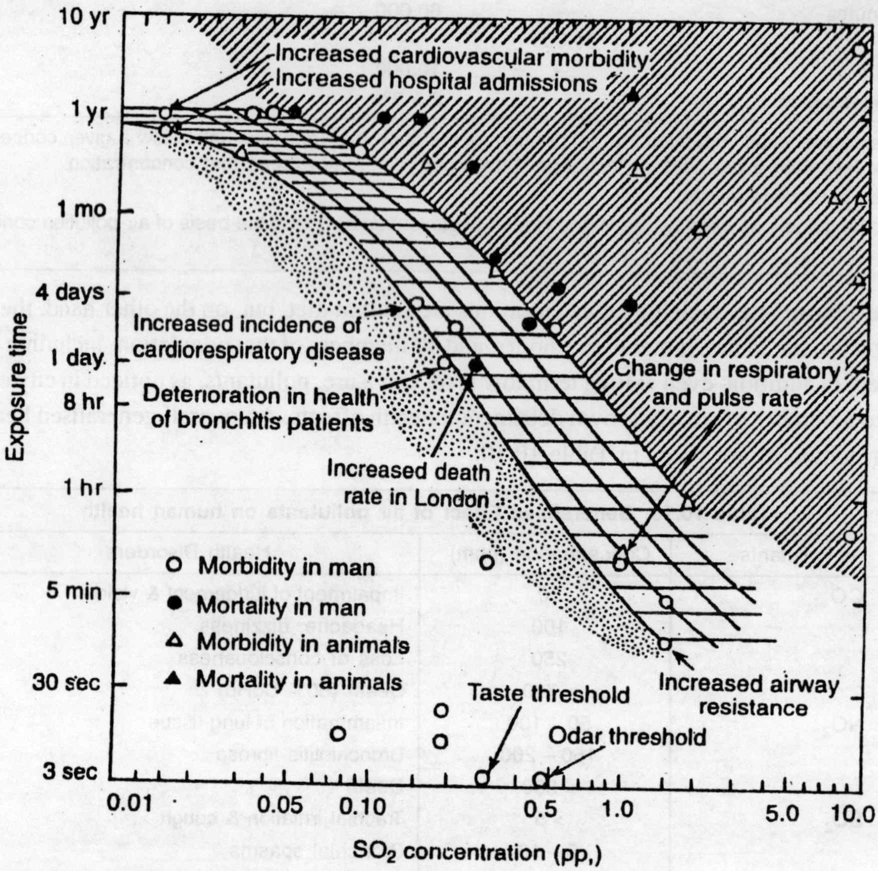

Fig. 10.4 : Effects of sulphur oxides on human health

In addition, among SPM, small particulates (PMO) induces a variety of human health effects after inhalation (Fig. 10.5):

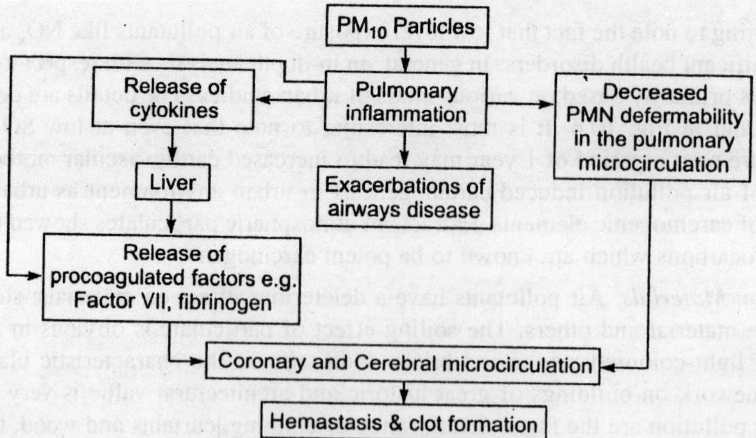

Fig. 10.5 : Effects of PMIO on human health

Table 10.11: Human Response to Different Levels of Sulphur Dioxide and Oxides of Nitrogen

Concentration (ppmv)	Period	Effect
(A) SO_2 Concentration(s):		
0.03 – 0.5	Continuous	Condition of bronchitic patients worsened
0.3 – 1	20 sec	Brain activity changed
0.5 – 1.4	1 min	Odour perceived
0.3 – 1.5	15 min	Increased eye sensitivity
1 – 5	30 min	Increased lung airway resistance, sense of smell lost
1.6 – 5	less than 6 hours	Constriction of nasal or lung passage
5 – 20	more than 6 hours	Lung damage reversible if exposure ceases
20 upwards	more than 6 hours	Water-logging of lung passageways and tissues, eventually leading to paralysis and/or death
(B) NO_X concentration(s):		
0 – 0.11		No effect
0.11 – 0.21		Slight odour detection
0.22 – 1.1		Some metabolic effects associated with either toxicity, adaptation or repair of lung tissues
1.1 – 2		Singnificant changes to respiratory rate and lung volume,, enhanced susceptibility to infection and evidence of tissue repair
2.1 – 5.3		Deterioration of lung tissues not balanced by repair mechanisms
above 5.4		Gross distortion of lung tissues and emphysema, possible death, if prolonged

The recent study (2017) on health impact by air pollutants on global basis showed the following facts:

- Exposure to $PM_{2.5}$, the leading environmental risk factor for death, accounting for about 4.2 million deaths, ranks 5th worldwide among all risks, including smoking, diet and high blood pressure.
- The highest concentration of $PM_{2.5}$ in 2015 related to combustion sources were in South and Southeast Asia, China, and Central and Western sub-Saharan Africa.
- Over 90% of the world's population lived in areas with unhealthy air in 2015.
- Among the 10 most populous countries and EU, Bangladesh and India now have the highest exposures to $PM_{2.5}$, having experienced the steepest increases since 2010.
- Coal-burning by industry, power plants and for home heating accounted for 40% of $PM_{2.5}$ exposures and an estimated 366,000 global death in 2013.
- In 2015, long-term exposure to $PM_{2.5}$ contributed to 4.2 million deaths and to loss of 103 million years of healthy life. China and India together accounted for 52% of the total global deaths attributed to $PM_{2.5}$.
- India and Bangladesh experienced some of the largest increases in $PM_{2.5}$ attributable mortality, on the order of 50% to 60%. India now approaches China in the number of deaths attributed to $PM_{2.5}$.
- Globally there was a 60% increase in ozone attributable deaths, with a striking 67% of this increase occurring in India.

(E) *Global Climate Changes*: Over past couple of decades, it was realised that air pollutants cause a considerable change in global climate and associated processes, viz., greenhouse effects, ozone depletion, acid precipitation and *El Nino* effects etc. These phenomena are described separately in later chapters.

An overall summarised effects of air pollution is given in Table 10.13 (next page).

Table 10.12: Air pollution Damage to Materials

Material	Damage	Responsible Air pollutants
Metals	Tarnishing, corrosion, structural weakening	Acidic gases*, moisture, salt
Masonry	Surface erosion, soiling	Acidic gases, moisture, particulate matter
Ceramic	Surface erosion, encrusting	Acidic gases, especially hydrogen fluoride
Paints	Surface erosion, discolouration	Acidic gases, H_2S, ozone, particulate matter, moisture
Paper	Embrittlement, discolouration	Acidic gases, moisture
Textiles	Reduced strength, soiling	Acidic gases, moisture
Textile dyes	Fading	Ozone and oxides of nitrogen in sunlight
Leather	Weakening	Oxides of sulphur
Rubber	Cracking	Ozone
Photographic film	Blemishes	Oxides of sulphur, moisture

*Unless otherwise noted, "acidic gases" refer to oxides of nitrogen and sulphur that react with moist air to form the strong acids H_2SO_4 (sulphuric acid) and HNO_3 (nitric acid).

10.6 MONUMENTS' DAMAGE BY AIR POLLUTANTS

The pollutants of the atmosphere, viz., CO_2, SO_2 and NO_2 often produces acid after dissolution with atmospheric moisture. As such, they produce bulk of carbonic acid and traces of sulphuric and nitric acid. Such acidic water may cause gradual corrosion of materials by which monuments were made of.

Primarily, limestone blocks (chalk to marble) of the monuments are affected. Rainfall containing carbonic acid slowly attacks limestone to give a solution of calcium bicarbonate, which then transforms to calcium carbonate. The action of oxides of sulphur in polluted air in the presence of moisture is to convert the calcium and magnesium carbonates of stone to calcium and magnesium sulphates, calcium sulfate is slightly soluble in water and magnesium sulfate is very soluble. On drying the solutions, the calcium and magnesium sulphates form hydrated crystals of greater volume than the original carbonates, with the result that encrustations and friable scale are formed to produce effects known as blistering, flaking and exfoliation.

Slates containing carbonates and the calcareons sandstones, often used as roofing materials, are attacked by pollutants, and atmospheres. Damage usually occurs mainly on the undersides, especially between the laps where water is held for some time as a thin film. Though brickwork can soon become badly discoloured by soot and grime, most types are fairly resistant to damage by sulphur oxides in the air, unless the bricks absorb sulphates produced by the action of the oxides of sulphur on adjacent mortar limestone.

Recent debate regarding the damage of internationally reputed monuments like Taj Mahal of Agra (India) and Victoria Memorial of Kolkata (India) are considered to be very significant in this regard.

Table 10.13: A summary of air pollution effects

Pollutants	Major Sources	Human Health	Vegetation	Materials	Aesthetics/ Nuisances	Comments
Carbon monoxide (CO)	Transportation, industrial processes	Reacts with hemoglobin reducing mental attentiveness, physical exertion, and exacerbating cardiovascular disease symptoms	None	None	None	Past knowledge was based on study of high exposure for short periods with healthy, young individuals. New data show possible health effects for susceptible persons at CO levels in the blood found in urban populations

Contd.

Table 10.13: A summary of air pollution effects

Pollutants	Major Sources	Human Health	Vegetation	Materials	Aesthetics/ Nuisances	Comments
Nitrogen oxides (NO_X)	Transportation, space heating/cooling, power generation	Interfere with respiratory functions producing long-term (chronic) disease symptoms	Reduction in growth of plants with broad leaves (eg. beans, tomatoes)	Accelerated deterioration of dyes and paints	Creation of a brownish colouring in urban air	Conclusions are based on limited exposure of healthy adults to low doses, extensive animal studies, and only limited data relevant to ambient conditions
Hydrocarbons (HC)	Transportation & industrial processes	See photo-oxidants	None	None	None	Indirectly polluting through the production of photochemical oxidants upon reaction with NO and NO_2 in the presence of sunlight
Photo-oxidants (O_3)	See nitrogen oxides and hydrocarbons	Interfere with respiratory functions & cause eye irritations	Severe reduction in growth and eventual death of leafy vegetables, field and forage crops, shrubs, fruit and forest trees caused by ozone and PAN	Ozone causes the cracking of rubber and the accelerated deterioration of nylon, rayon dyes and paints	Ozone has a distinct — although not terribly offensive — odour	Ozone (O_3) is the most common type and the key indicator for photo-oxidants. Health effects are based on limited and inadequate data. Ozone, peroxy-acetylnitrite (PAN), etc. are formed by atmospheric reactions
Sulphur oxides (SO_x)	Power generation, space heating/cooling, industrial processes	Little effect in the pure gas form; similar effects as particulates when combined with them	Reduction in growth of plants with broad leaves	Corrosion of iron, metals, accelerated deterioration of building stone, cotton, paper, leather, paints and other finishes	Scattering of sunlight to produce haze, production of unpleasant odours	Sulphur dioxide is readily converted to SO_3 and then to sulphuric acid (a particulate). Determining which effects are due solely to SO_2 is difficult in acid rain research
Particulates	Power generation, space heating/cooling, industrial processes, soil erosion	Interference with respiratory functions, possible contribution to lung cancer	Reduction in plant growth by physical blockage of light when deposited on leaf surface	Soiling of fabrics and buildings and corrosion of metals when combined with SO_2	Creation of smoke plumes, scattering of sunlight to produce haze and colourful sunsets, and formation of hydroscopic nuclei to produce fog	The effects of particulates are difficult to separate from those of sulphur dioxide
Heavy metals, radioactive agents, carbon dioxide, others	Power generation, industrial processes space heating	Specific to each pollutant	Fluoride causes long-term damage to selected field crops (and animals)	Tarnishing of metals by hydrogen sulphide	Hydrogen sulphide produces extremely unpleasant odours	Pollution from these agents can be intense at the source, but tends not to be widespread. However, carbon dioxide is the main cause of global 'greenhouse effect' and chlorofluorocarbons affect the ozone layer

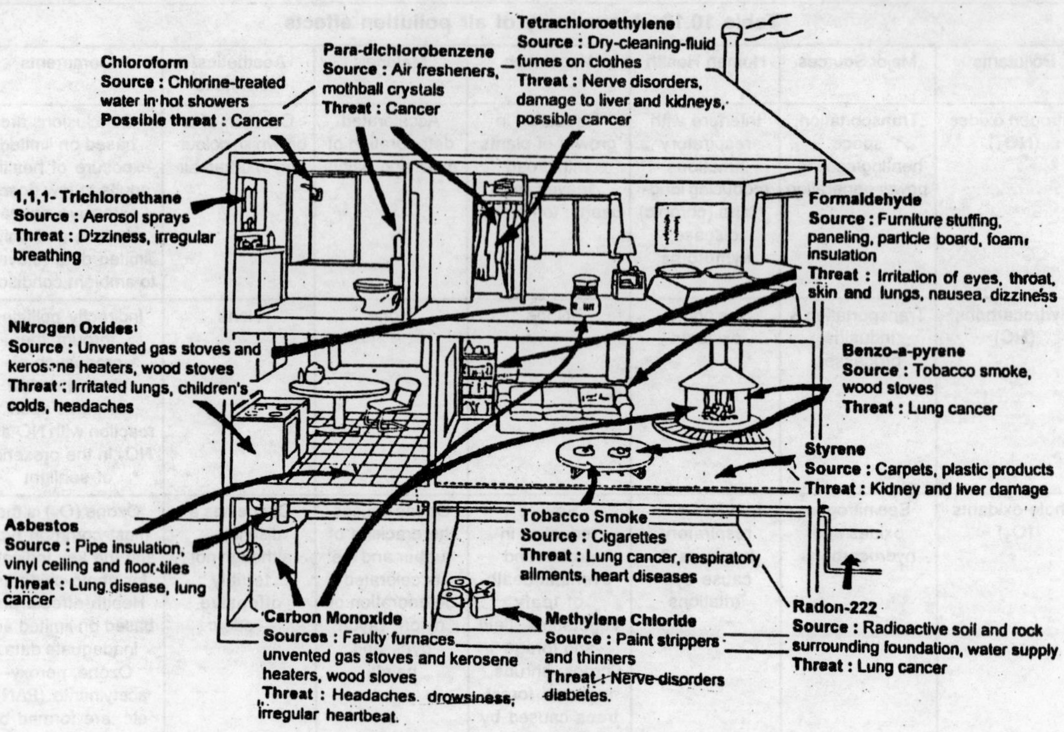

Fig. 10.6 : Some major indoor air pollutants

10.7 EFFECT OF INDOOR AIR POLLUTANTS

There are a number of chemical/gas exposures encountered in indoor environment as occupational exposure. These exposures on long term basis have serious health hazards. The details of such indoor exposure of air pollutants is shown in Fig. 10.6.

For a long time such exposures were ignored by mankind. But such indoor exposures are much more significant health hazards than outdoor air pollutant exposures. Thus there is an extreme necessity for protection of occupational exposure to diverse gases/chemicals in respective workplaces.

On the whole the overall harmful effects of air pollutants on plants, animals and ecosystems has manyfold short-term and long-term consequences. A generalised summary is presented in Fig. 10.7 (next page).

10.8 URBAN AIR POLLUTION

Air pollution problems in urban environment is a global problem over several decade. Since 1975, systematic monitoring of primary air pollutants (NO_x, SO_x, SPM, CO, HC & Pb) were attempted. Among the megacities of the world remarkable variation was noticed with respect to SO_2 & SPM level (Table 10.14). According to WHO guidelines of air pollutant level, many cities already cross the tolerance limit of human exposure of individual pollutants. The details ranking of pollution level is depicted in Fig. 10.8. The major causatives of such urban pollution is due to automobile exhaust and anthropogenic activities. Thus there is need for careful pollution control measures in major megacities of the world not only for local environmental protection but also for regulation of greenhouse gas emission.

Table 10.14: Trends in Emission of SO_2 and SPM in Selected Megacities*								
City	Mean Daily Concentrations (μg m^{-3})						Percentage change	
	1976-78		1979-81		1982-85		(1976-1985)	
	SO_2	SPM	SO_2	SPM	SO_2	SPM	SO_2	SPM
New York	69	68	59	60	49	54	− 29	− 21
Houston	9	106	16	89	23	77	+ 156	− 27
Chicago	—	117	—	95	30	—	—	—
St Louis	66	109	34	—	—	—	—	—
Toronto	26	72	20	85	10	65	− 62	− 10
Montreal	39	71	36	74	24	61	− 38	− 14
Frankfurt	62	32	55	25	50	36	− 19	+ 13
Brussels	83	29	68	25	50	21	− 40	− 28
London	86	30	69	30	40	23	− 53	− 23
Milan	243	—	222	—	137	—	− 44	—
Tokyo	62	61	48	58	30	52	− 52	− 15
Hong Kong	24	67	47	52	41	37	+ 71	− 45
Beijing	52	—	131	224	—	376	—	—
Kolkata	38	361	53	458	59	380	+ 55	+ 5
Rio de Janeiro	94	—	121	118	93	110	0	—
Santiago (Chile)	62	67	53	78	76	117	+ 23	43

*World Resource Institute (1988-89); '—' Data not available (WHO permissible standard for SO_2 = 40 – 64 μgm^{-3} & for SPM = 60 – 90 μgm^{-3})

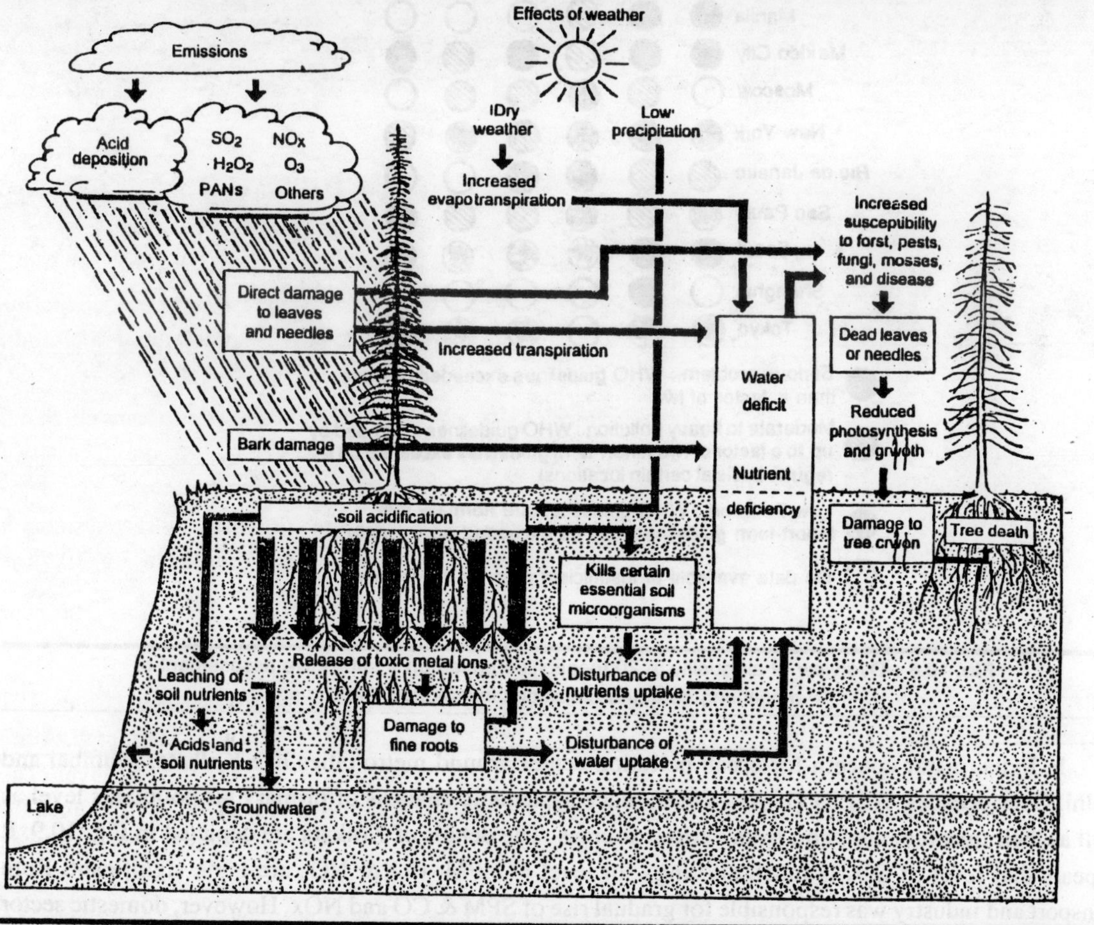

Fig. 10.7 : Harmful effects of air pollutants

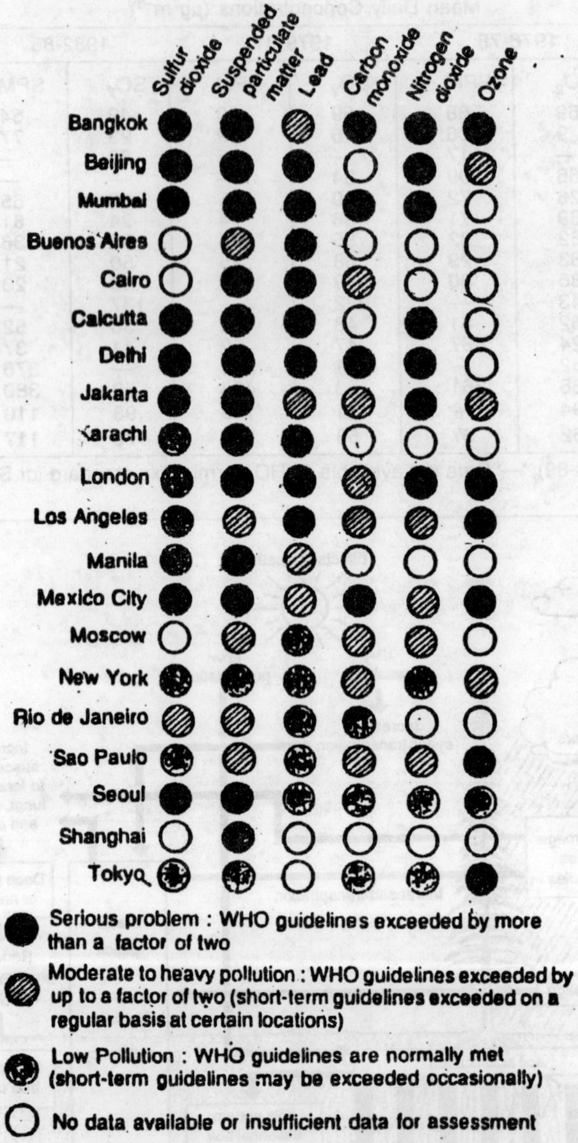

Fig. 10.8 : Overview of air quality in 20 megacities based on a subjective assessment of monitoring data and emissions inventories, 2000

In India, major concern for urban pollution center around metro cities like Kolkata, Mumbai and Delhi. Over decades, NEERI (1970 – 2000) made systematic analysis of primary air pollutants level as well as their quantum emission from various sources. The details of Kolkata are depicted in Fig. 10.9. It appears that level of SO_2 is much alluring, but level of SPM, CO & NO_x are quite significant. Both transport and industry was responsible for gradual rise of SPM & CO and NOx. However, domestic sector is also responsible for rise of CO level.

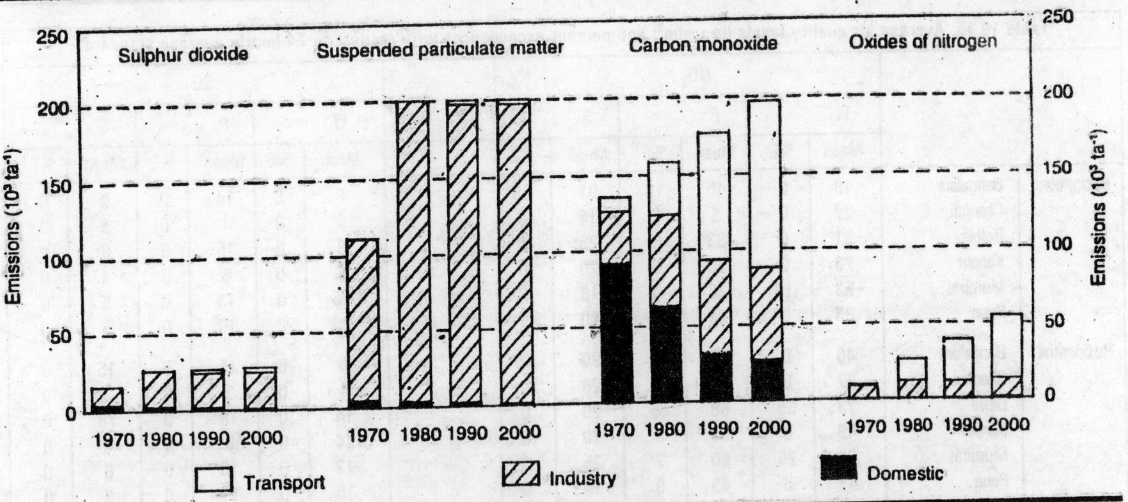

Fig. 10.9 : Trend in Kolkata's air quality and their quantum of emission source-wise, 2000

Recent air quality monitoring data (Table 10.15) in major Indian cities, show summary of SPM, PM_{10}, $PM_{2.5}$, NO_2, SO_2 level along with exceedance of 24-hr standard prevailed in 2007. Many cities showed high level of air pollution exceeding ambient air quality standard.

Table 10.15: Average air quality levels (in μg/m³) and percent exceedances with respect to 24-hourly average standard*

		SPM						PM_{10}						$PM_{2.5}$					
		W*		P**		S***		W		P		S		W		P		S	
		Mean	%E	Mean	%E	Mean	%E	Mean	%E	Mean	%E	Mean	%	Mean	%E	Mean	%E	Mean	%E
Background	Bangalore	110	0	82	0	83	0	47	0	105	32	66	10	27	0	23	0	27	0
	Chennai	117	17	76	0	178	22	55	0	88	50	71	31	35	14	39	0	34	14
	Delhi	549	100	546	100	517	100	355	100	300	100	232	100	–	–	–	–	131	100
	Kanpur	361	100	329	93	342	97	204	97	169	97	187	90	172	100	132	100	136	100
	Mumbai	246	63	204	57	159	17	184	97	139	86	91	39	92	67	60	33	29	0
	Pune	257	95	204	65	139	5	123	60	63	5	76	10	45	0	32	0	22	0
Residential	Bangalore	294	100	301	100	177	25	133	88	93	35	69	14	36	0	41	33	29	0
	Chennai	164	19	173	14	175	24	82	25	200	46	86	23	78	86	34	0	34	0
	Delhi	828	100	967	100	284	90	505	100	671	100	81	40	301	100	–	–	30	0
	Kanpur	429	100	373	97	422	100	226	100	195	100	217	100	208	100	161	100	190	100
	Mumbai	523	100	445	100	277	54	267	100	236	100	119	48	97	100	87	100	54	33
	Pune	499	100	362	95	206	50	165	95	128	72	103	58	58	0	35	0	28	0
Industrial	Bangalore	262	0	245	0	171	0	171	81	171	50	69	5	30	0	21	0	22	0
	Chennai	311	8	348	11	319	5	136	31	147	44	141	38	67	57	41	0	79	30
	Delhi	965	100	1239	100	611	70	546	100	781	100	229	8	197	100	314	100	52	100
	Kanpur	603	62	577	58	591	61	396	76	371	74	388	74	305	100	273	100	232	100
	Mumbai	395	3	388	0	238	3	271	100	218	96	99	7	127	100	87	100	17	0
	Pune	400	25	164	0	270	0	216	85	71	10	121	22	63	33	26	0	37	0
Kerbside	Bangalore	306	100	287	93	411	100	199	100	184	85	109	43	64	50	43	33	38	0
	Chennai	350	78	243	59	211	36	111	48	128	77	271	67	73	57	56	29	51	14
	Delhi	1082	100	2592	100	#####	100	451	100	941	100	337	100	306	100	361	100	107	100
	Kanpur	564	100	532	100	561	100	292	100	260	100	273	100	216	100	226	100	218	100
	Mumbai	383	100	383	100	314	8	256	100	234	100	124	65	119	100	126	100	41	18
	Pune	655	100	583	100	507	100	254	100	193	95	138	95	124	100	62	67	46	0

*Standard refer to 24 hourly average standards as prevailing in 2007; $PM_{2.5}$ standards refers to the proposed standards in 2007

Contd.

Table 10.15: Average air quality levels (in $\mu g/m^3$) and percent exceedances with respect to 24-hourly average standard*

		NO_2						SO_2					
		W		P		S		W		P		S	
		Mean	%E	Mean	%E	Mean	%E	Mean	%E	Mean	%E	Mean	%E
Background	Bangalore	18	0	45	18	91	56	6	6	14	0	9	0
	Chennai	27	0	8	0	14	0	3	0	1	0	5	0
	Delhi	31	0	33	0	25	0	8	0	15	0	8	0
	Kanpur	23	0	20	0	20	0	8	0	8	0	4	0
	Mumbai	53	10	38	0	18	3	15	0	13	0	5	0
	Pune	36	0	34	0	10	0	23	0	10	0	5	0
Residential	Bangalore	46	0	29	0	90	46	9	0	15	0	15	0
	Chennai	32	0	17	0	28	0	4	0	3	0	3	0
	Delhi	73	35	88	65	29	0	14	0	18	0	78	0
	Kanpur	49	0	32	3	19	0	14	0	8	0	4	0
	Mumbai	72	25	60	7	25	0	12	0	13	0	6	0
	Pune	41	6	43	0	14	0	18	0	11	0	6	0
Industrial	Bangalore	53	6	30	0	89	44	9	0	10	0	10	0
	Chennai	45	0	20	0	42	0	6	0	4	0	6	0
	Delhi	159	85	142	80	60	0	85	20	77	20	11	0
	Kanpur	35	0	24	0	23	0	26	0	19	0	15	0
	Mumbai	72	0	53	0	20	0	18	0	15	0	7	0
	Pune	55	0	17	0	22	0	40	0	16	0	22	0
Kerbside	Bangalore	94	62	105	65	66	26	10	0	19	0	13	0
	Chennai	45	0	33	0	43	0	6	0	1	0	4	0
	Delhi	109	85	121	95	47	0	20	0	20	0	12	0
	Kanpur	46	0	42	7	37	0	15	0	9	0	8	0
	Mumbai	82	43	64	20	33	2	14	0	15	0	6	0
	Pune	71	50	43	0	59	20	36	7	12	0	7	0

% Exceedance	0-25	25-50	50-75	75-100

*W: Winter; **P: Post Monsoon, Summer in case of Bangalore; ***S: Summer, Pre-Monsoon in case of Bangalore

On the whole, the health hazards of urban population is great concern for comming decades ahead (Table 10.16).

Table 10.16: Main Urban Air Pollutants and their Effects on Health

Pollutant	Action	Effect
1. Traditional (reducing) pollutants from coal/heavy oil combustion:		
Smoke/suspended particulates (some contribution from diesel traffic too)	Can penetrate to lungs; some retained: possible long-term effects. May also irritate bronchi	**London smog complex** *Short term effects*: sudden increases in deaths in hospital admissions and in illness among bronchitic patients; temporary reductions in lung function (patients and some normal people) *Long term effects*: increased frequency of respiratory infections (children); increased prevalence of respiratory symptoms (adults and children); higher death rates from bronchitis in polluted areas *Possible carcinogenic effects*: may take some part in the higher incidence of lung cancer in urban areas
Sulphur dioxide	Readily absorbed on inhalation: irritation of bronchi, with possibility of bronchospasm	
Sulphuric acid (mainly a secondary pollutant formed from sulphur dioxide in air)	Hygroscopic; highly irritant if impacted in upper respiratory tract. Acid absorbed on other fine particles may penetrate further to promote bronchospasm	
Polycyclic aromatic hydrocarbons (small contribution from traffic also)	Mainly absorbed on to smoke; can penetrate with it to lungs	

Contd.

2. Photochemical (oxidizing) pollutants from traffic or other hydrocarbon emissions:		
Hydrocarbons (volatile: petrol etc.)	Non-toxic at moderate concentrations	**Los Angeles smog complex** *Short term effect*: primarily eye irritation. Reduced athletic performance. Possibly small changes in deaths, hospital admissions
Nitric oxide	Capable of combining with hemoglobin in blood but no apparent effect in humans	
Nitrogen dioxide and ozone (mainly secondary pollutants formed in photochemical reactions)	Neither gas is very soluble; some irritation of bronchi but can penetrate to lungs to cause edema at high concentrations. Urban concentrations too low for such effects, but evidence of reduced resistance to infections in animals	*Long term effects*: increased onsets of respiratory illness (children); increased asthma attacks (adults). No clear indication of increased bronchitis
Aldehydes, other partial oxidation products, peroxyacetylnitrate	Eye irritation, odour	
3. Other from traffic		
Carbon monoxide (other sources contribute—smoking an important one)	Combines with haemoglobin in blood, reducing oxygen-carrying capacity	Possible effects on central nervous system (reversible unless concentrations are very high). Some evidence of effects on perception and performance of fine tasks at moderate concentrations
Lead (some industrial sources contribute to air lead; human intake often dominated by lead in food & drink)	Taken up in blood, distributed in soft tissues and some to bone	Possible effects on central nervous system (longer time scale than in case of CO and not necessarily reversible). Indications of neuro-psychological effects on children within overall environmental exposure

10.9 URBAN AIR QUALITY MANAGEMENT

The components of an urban air quality management system include an air quality monitoring network, emission inventories, numerical prediction models, air quality standards and public information bands, and a range of cost-effective pollution control policies and measures, together with the resources and powers to impose them. It should be borne in mind that they all form part of an integrated system (Fig. 10.10). If inadequate attention is given by the authorities to one component it is likely to result in limitations in the effectiveness of the entire system. Many cities in the world have developed some form of urban air quality management system, although many have not reached the sophistication and completeness ultimately needed by urban air quality managers.

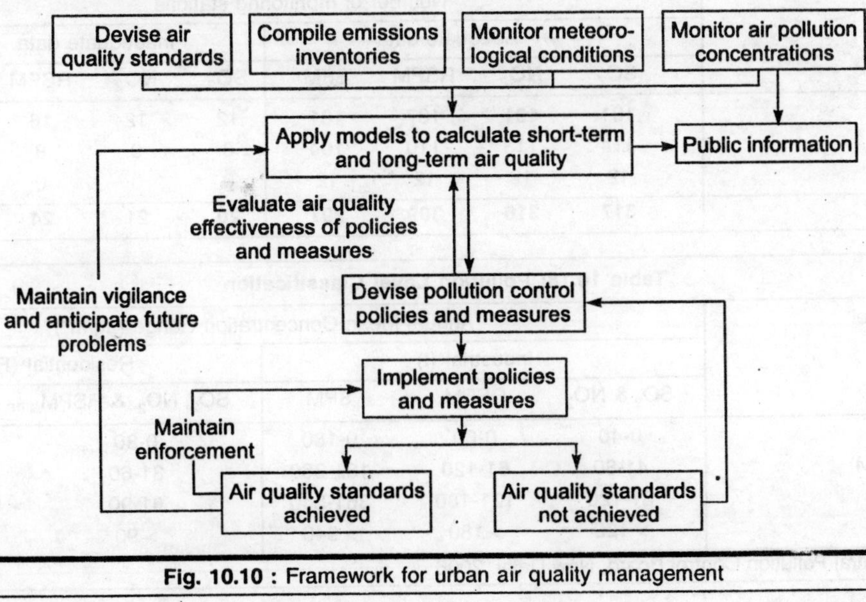

Fig. 10.10 : Framework for urban air quality management

10.10 NATIONAL AMBIENT AIR QUALITY STATUS

The air quality of different cities/towns has been compared with the respective NAAQS (2008). The air quality has been categorized into four broad categories based on an Exceedence Factor (the ratio of annual mean concentration of a pollutant with that of a respective standard).

The Exceedence Factor (EF) is calculated as:

$$\text{Exceedence Factor} = \frac{\text{Observed annual mean concentration of criteria pollutant}}{\text{Annual standard for the respective pollutant and area class}}$$

The four air quality categories are:

- Critical pollution (C): when EF is more than 1.5;
- High pollution (H): when the EF is between 1.0 - 1.5;
- Moderate pollution (M): when the EF between 05 - 1.0; and
- Low pollution (L): when the EF is less than 0.5.

It is obvious from the above categorization that the locations in either of the first two categories are actually violating the standards, although, with varying magnitude. Those falling in the third category are meeting the standards as of now but likely to violate the standards in future if pollution continues to increase and is not controlled. However, the locations in Low pollution category have a rather pristine air quality and such areas are to be maintained at low pollution level by way of adopting preventive and control measures of air pollution. Adequate data for annual average concentration (with 50 and more days of monitoring) for SO_2 was received for 317 stations and adequate data for NO_2 was received for 316 stations. Adequate data for RSPM was received for 309 stations and adequate data for SPM was received for 297 monitoring stations. The detail of number of stations for which data was adequate or inadequate is given in Table 10.17. Data of sixteen stations under State Ambient Air Quality Monitoring Programme (SAMP) were also included in analysis. The ambient air quality status with respect to annual mean concentration range for various parameters is given in Table 10.18.

Table 10.17: Details of Monitoring Stations where Ambient Air Quality Monitoring was carried out during 2008

Area type	Number of monitoring stations							
	Adequate data				Inadequate data			
	SO_2	NO_2	RSPM	SPM	SO_2	NO_2	RSPM	SPM
Residential	191	191	187	181	12	12	16	13
Industrial	114	113	110	104	8	9	8	8
Sensitive	12	12	12	12	0	0	0	0
Total	**317**	**316**	**309**	**297**	**20**	**21**	**24**	**21**

Table 10.18: Pollution Level Classification

Pollution level*	Annual Mean Concentration Range (μg/m³)				
	Industrial (I)			Residential (R)	
	SO_2 & NO_2	RSPM	SPM	SO_2, NO_2, & RSPM	SPM
Low (L)	0-40	0-60	0-180	0-30	0-70
Moderate (M)	41-80	61-120	181-360	31-60	71-140
High (H)	81-120	121-180	361-540	61-90	141-210
Critical (C)	> 120	> 180	> 540	> 90	> 210

Source: Central Pollution Control Board, New Delhi, 2006

The analysis of four criteria pollutants with respect to National Ambient Air Quality Standards (NAAQS) during 2008 revealed that number of locations falling with respect to sulphur dioxide in low pollution levels category is 114, moderate category is 2 and there is no locations fall under high and critical levels in industrial area calls if considered time weighted annual average concentrations. The 82 monitoring locations considering annual average concentrations of NO_2 for industrial area class is falling in low category, 32 locations in moderate category, and only 2 locations in high levels whereas no critical category obtained. The number of monitoring locations considering annual average concentration of RSPM for industrial area class fall in low category is 17, moderate category in 44 locations, high category in 31 locations and critical category in 24 locations of industrial area class of the country. The 35 locations considering annual average concentrations of SPM for industrial area class fall in low category, 49 numbers in moderate category, 23 numbers in high category whereas no critical category was found in industrial area class. The number of locations fall under four such category in industrial area class is depicted in Fig. 10.11. For residential area class 192 locations fall in low category with respect to Sulphur dioxide and 4 locations comes under moderate category whereas no high and critical levels were found. Locations with respect to NO_2 falls under low category are 136, while 52 locations indicated moderate

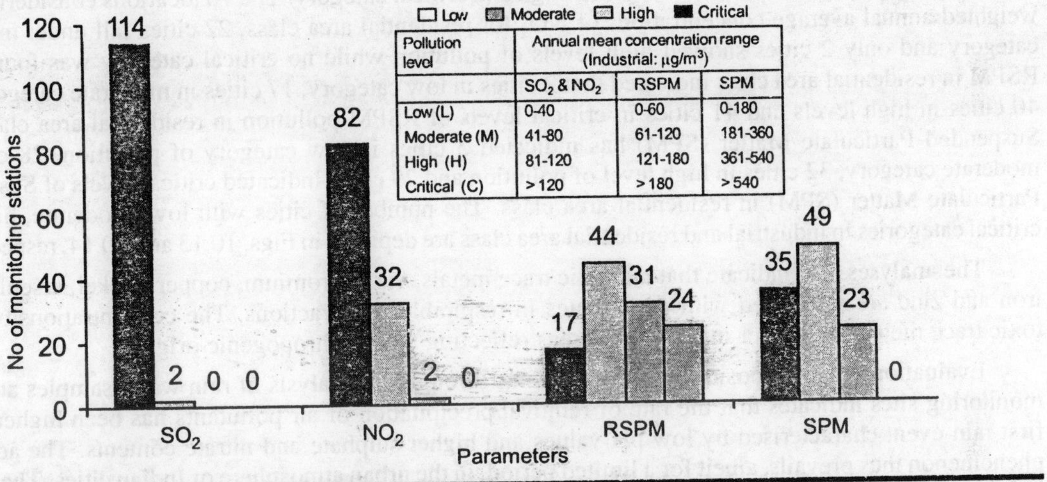

Fig. 10.11 : Number of locations with low, moderate, high and critical pollution levels in industrial areas

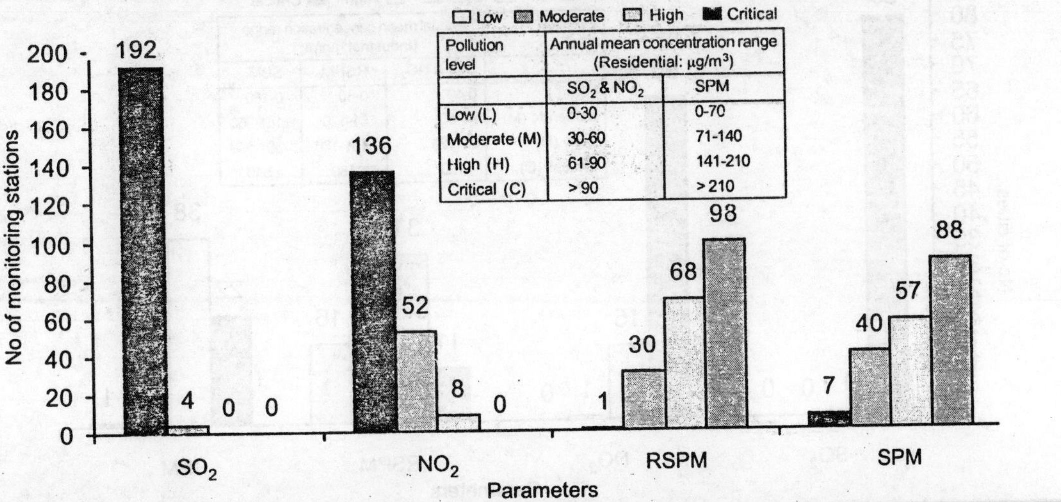

Fig. 10.12 : Number of locations with low, moderate, high and critical pollution levels in residential areas

category and 8 locations indicated high pollution levels in residential areas whereas no critical levels obtained in residential areas. RSPM indicated only one location in low category, 30 locations in moderate category, 68 locations in high category and 98 locations in critical category of residential area class of the country. For residential area class detailing is depicted in Fig. 10.12.

The analysis of four criteria pollutants with respect to National Ambient Air Quality Standards (NAAQS) during 2008 revealed that 80 cities falling with respect to Sulphur dioxide (SO_2) in low category and only one city fall under moderate category. Whereas no high and critical levels were found with respect of SO_2 in industrial area class. Nitrogen dioxide pollution levels, if considered in time weighted annual average concentrations indicated that low category in 65 cities and 15 cities fall under moderate category and only one city in high pollution levels of NO_2 in industrial areas. The RSPM in industrial area class indicated that 11 cities falls under low category, 34 cities in moderate category, 19 cities in high pollution levels category and 16 cities in critical category of industrial area class. The SPM showed 22 cities with low category, 38 cities with moderate category, 11 cities with high pollution level category and only one city indicated with critical category in industrial area class. The analysis for residential area class showed that 98 cities for Sulphur dioxide have low category and only 2 cities indicated only moderate category while no cities fall under high and critical category. The 76 locations considering time weighted annual average concentrations of NO_2 for residential area class, 22 cities fall under moderate category and only 2 cities showed high levels of pollution while no critical category was found. The RSPM in residential area class indicated that 2 cities in low category, 17 cities in moderate category, and 40 cities in high levels and 41 cities in critical levels of RSPM pollution in residential area class. The Suspended Particulate Matter (SPM) has indicated 3 cities in low category of pollution, 28 cities in moderate category, 32 cities in high level of pollution and 29 cities indicated critical levels of Suspended Particulate Matter (SPM) in residential area class. The number of cities with low, moderate, high and critical categories in industrial and residential area class are depicted in Figs. 10.13 and 10.14, respectively.

The analyses also indicate that the toxic trace metals, viz., chromium, copper, nickel, arsenic, lead, iron and zinc are associated with particulates in respirable size fractions. The concentrations of these toxic trace metals are higher in industrial areas reflecting their anthropogenic origin.

Evaluation of wet deposition of air pollutants through the analysis of rain water samples at all the monitoring sites indicates that the rate of removal/precipitation of air pollutants has been higher in the first rain event characterised by low pH values and higher sulphate and nitrate contents. The acid rain phenomenon thus prevails, albeit for a limited period, in the urban atmosphere of Indian cities. The lowest

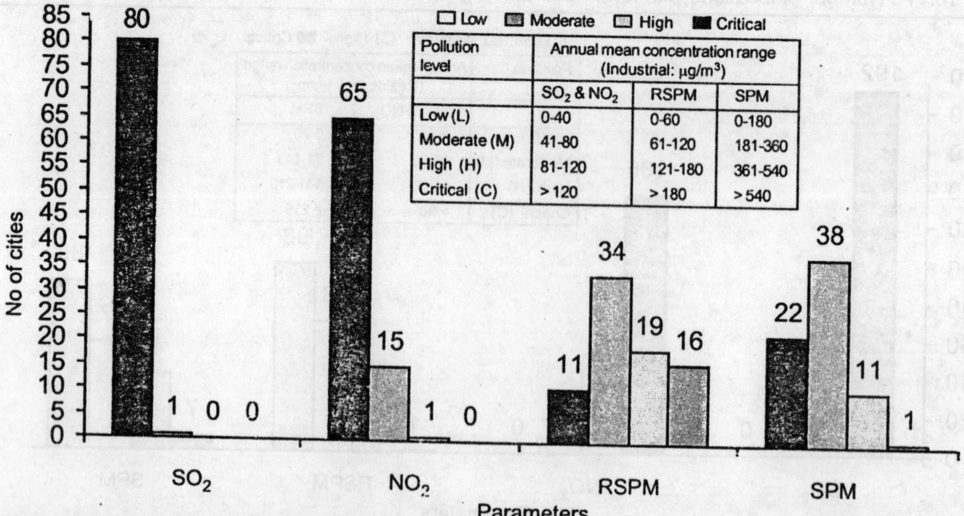

Fig. 10.13 : Number of cities with low, moderate, high and critical pollution levels in industrial areas

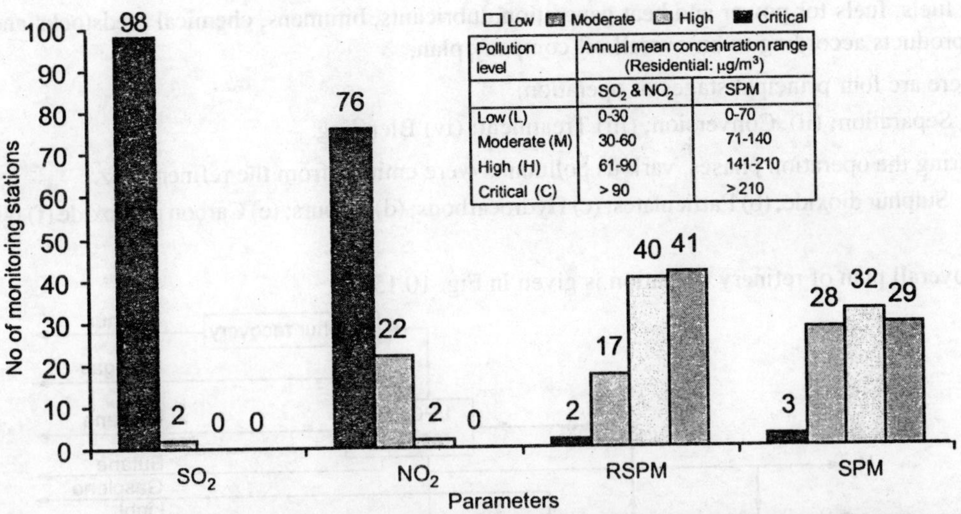

Fig. 10.14 : Number of cities with low, moderate, high and critical pollution levels in residential areas

pH of 4.8 was observed in the rain water samples at Delhi in industrial area. Conductivity of 178 μSiemens/cm was recorded in Kolkata at the commercial site in the first rain event of the season. Air quality exposure index (AQEI) has been used to assess the combined impacts of SMP/RSPM, SO_2 and NO_X concentration in air environment.

In addition to criteria pollutants, other pollutants such as Benzene, 1-3 Butadiene, Formaldehyde, Non-Methane and Total Hydrocarbon (NMHC & THC), expected in urban environment were also monitored. However, monitoring of these parameters was limited to once in each of the three seasons at each monitoring location. Summary of data of these pollutants is given in Table 10.19. The ambient concentration levels are primary governed by the contributing sources in the neighborhood, and vide variations were observed in different cities.

Benzene levels are higher in Bangalore, Pune and Kanpur. The values of formaldehyde are also matter of concern in Mumbai, Pune and Bangalore. However, more work is required to be done in future for proper understanding of these pollutants.

Table 10.19: Concentration of Organic Pollutants

	Benzene (μg/m³)			1,3-Butadiene (ppb)			Formaldehyde (μg/m³)			NMHC (ppm)			HC (ppm)		
City	Min	Max	Avg	Min	Max	Avg	Min	Max	Avg	Min	Max	Avg	Min	Max	Avg
Bangalore	7.4	237.8	119.1	0.41	3.7	2.18	8	35	20	8	12	11	9	15	13
Chennai	4	17	10.2	0.452	1.8	1.1	1.7	26	12.1	0.02	0.18	0.06	0.02	0.18	0.06
Delhi	1.92	11.19	4.96	0.2	1.6	0.78	1.94	19.0	11.27	0.2	1.7	0.9	2.6	5.3	3.7
Kanpur	4.88	68.11	26.86	Not monitored			4.39	18.43	10.08	0.06	0.255	0.14	0.07	0.26	0.15
Mumbai	Not monitored			Not monitored			8.8	93.0	32.6	32.6	0.1	24.6	1.5	25.5	4.6
Pune	28.14	96.53	57.30	0.4	2.5	1.2	3.77	41.12	17.12	1.32	3.82	2.55	1.74	4.22	2.96

10.11 AIR POLLUTION: CASE STUDIES IN OIL REFINERY

Crude oil is predominantly a complex mixture of hydrocarbons, but may contain minor proportions of compounds containing sulphur, oxygen and nitrogen as well as trace metals such as vanadium and nickel. No two crude oils are alike, but usually the percentages by weight of the main materials are: carbon 85-90, hydrogen 10-14, sulphur 0.1-7, oxygen 0.06-0.4 and nitrogen 0.01-0.9. The function of a refinery is to produce from this mixture a very wide range of products which meet market needs for automotive and aviation fuels, fuels for power and heat generation, lubricants, bitumens, chemical feedstocks and other special products according to an overall oil company plan.

There are four principal stages of operation:

(i) Separation; (ii) Conversion; (iii) Treatment; (iv) Blending.

During the operation phases, various pollutants were emitted from the refinery, viz.,

(a) Sulphur dioxide; (b) Particulates; (c) Hydrocarbons; (d) Odours; (e) Carbon monoxide (f) Nitrogen oxides.

A overall plan of refinery operation is given in Fig. 10.15.

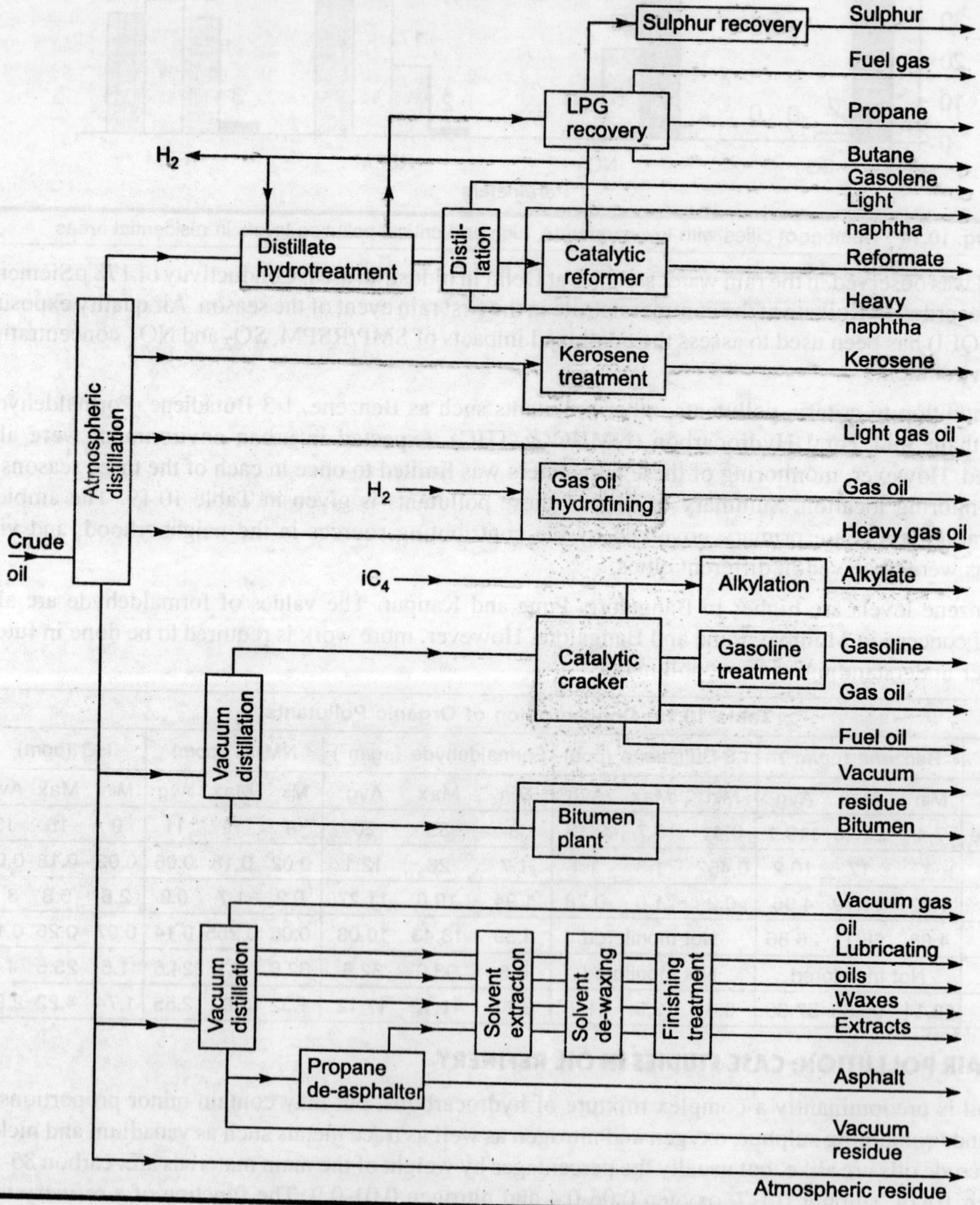

Fig. 10.15 : Plan of complex activities in refinery

Table 10.20, showed sulphur content of various crude oils and refinery products (per cent by weight):

Table 10.20

Products	Sulphur content (% by weight)				
	Kuwait	Iranian (heavy)	South Iraq	Abu Dhabi	Nigerian
Crude oil	2.5	1.58	1.95	1.38	0.21
Gasoline	—	0.14	0.013	—	—
Kerosene	—	0.26	0.057	—	—
Diesel oil	—	1.01	1.11	—	—
Heavy fuel oil	—	2.58	3.40	—	—

The pollution control of oil refinery was taken up in various steps :

1. Installation of desulphurisation plants.
2. Sulphur recovery process maintenance.
3. Control of particulate emission from catalyst regeneration.
4. Regulation of hydrocarbon spillage & evaporation.
5. Waste gas disposal by flaring up.
6. Treatment of oily waste water.

10.12 AIR POLLUTION MONITORING TECHNIQUES

Air quality monitoring protocol showed the following guideline:

Particulars	Pollutants						
	SPM	PM_{10}/RPM	SO_2	NO_2	$PM_{2.5}$	CO	VOC
Equipment	High Volume Sampler	Multi-speciation sampler/ Respirable dust sampler (RDS)	Impingers attached to HVS/RDS	Impingers attached to HVS/RDS	FRM sampler	Automatic Analyzer	VOC Sampler
Sampling period	8/24 hrly	8/24 hrly	8/24 hrly	8/24 hrly	8/24 hrly	4/24 hrly/ continuous	4/8/24 hrly
Sampling frequency	20/30 days continuous in each of the three seasons	20/30 days continuous in each of the three seasons	20/30 days continuous in each of the three seasons	20/30 days continuous in each of the three seasons	one week continuous in each of the three seasons	one week continuous in each of the three seasons	once in each of the three seasons

Primarily there are three kinds of air quality monitoring, viz.

(a) Stack emission monitoring;

(b) Fugitive emission monitoring; and

(c) Ambient air quality monitoring.

Each category of monitoring has different types of air sampling methodology, but the techniques of analysis remain uniform.

(a) *Stack emission monitoring equipment*: This equipment has two basic components:

(i) Particulate matter collection equipment assembly, and

(ii) Gaseous sampling train.

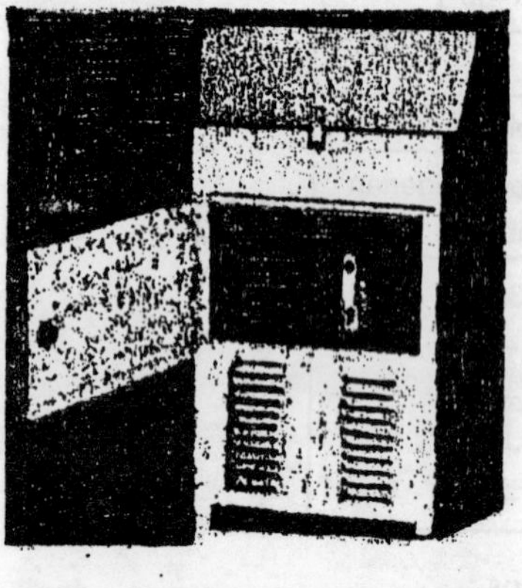

Fig. 10.16 : High volume sample

During gas sampling, samples were collected from isokinetic locations with the help of sampling ducts, particulates were measured by gravimetric method, while gaseous pollutants were measured by respective spectrophotometric methods.

(b) *Fugitive emission monitoring techniques*: When pollutants enter the ambient air without the aid of primary emission systems like stack, or any other air pollutant generating systems, then such emissions are called fugitive emission. Monitoring of this kind of emission is a major problem. Monitoring of fugitive emissions requires innovative methods, viz., quasi-stack sampling method, roof monitoring method and upwind-downwind sampling method. In general, upwind-downwind sampling method is most commonly and widely used method for estimation of fugitive emission. The sampling was done by using high volume sampler (both particulates and gaseous pollutants).

(c) *Ambient air quality monitoring*: Air quality samplings were made at pre-selected sites with the help of high volume samples. SPM was collected by filter paper placed against the suction pump. The clean gaseous samples are then collected in gas-impinger for absorption of pollutants using respective solutions. For SO_2 and NO_x measurement West and Gacke method and Jacobs and Hochheiser modified method were adopted, respectively.

A configuration of high volume sampler is given in Fig. 10.16.

Air quality index (AQI)

It i, computed by using the level of criteria air pollutants – NO_X, SO_X, PM_{10}, $PM_{2.5}$, CO and O_3. Generally government agencies use the index to characterise air quality of an area. United States Environmental Protection Agency (EPA) has developed an air quality index in the following manner.

Air quality index	Level of health concern	Colour indicator for public awareness
0 to 55	Good	Green
51 to 100	Moderate	Yellow
101 to 150	Unhealthy for sensitive groups	Orange
151 to 200	Unhealthy	Red
200 to 300	Very unhealthy	Purple
301 to 500	Hazardous	Maroon

The process of AQI computation varies widely in different countries and expressed as an hourly index, daily index and annual index. In recent study held by WHO (2016), it was reported that about 34 Indian cities figured in the list of the 100 most polluted ones, and 22 Indian cities found their names among the top 50 most polluted cities in the world.

10.13 BIOMONITORING OF AIR POLLUTION

The concept of monitoring of air quality by plants is a well-established fact. The plants used for this purpose are termed **indicator plants.** It is known that some plants are very sensitive to air pollutants, they are thus used as indicator species for biomonitoring of air quality. In recent years, increasing efforts are being made to use plants for detection of air quality, particularly the presence of SO_2, NO_x, H_2S, O_3 and HF. The use of lichens and moss for air pollution level mapping in urban and industrial area of Europe and also in North America are the finest examples of plant Biomonitoring of air quality (De Sloover & Le Blanc, 1968; Johnson and Sochting, 1973; Manning and Feder, 1980). The use of higher plants for monitoring of air pollution is, however, a recent development. A number of plant parameters either simply or in combination may be used for evaluating the pollution stress.

Table 10.21 shows the list of plants species used for biomonitoring of different air pollutants.

In India, considerable work has been done on the effects of air pollution on plants at BHU (Varanasi), JNU (New Delhi), NBRI (Lucknow), CU and KU (West Bengal) and AU (Vishakapattanam) in recent years. This method now appears to be a sensitive and low cost technique, only appropriate standardisation of techniques has to be made.

Table 10.21: Plants suitable for biomonitoring of air pollution

Plants	Air pollutants for which use is suggested
1. Tobacco (Ble - W_3 variety), Crape & Garden Bean	O_3
2. White pine, Moss & Lichens	SO_2
3. Lettuce & Bean	PAN
4. Tomato & Lettuce	MO_X
6. Orchids, Cucumber, & Marigold	C_2H_2
7. Apricot, Peach, & Gladiolas	HF
8. Moss & Lichens	SO_2, particulates & heavy metals

Identically, in Europe, similar type of plant species selection were made for air quality biomonitoring (Table 10.22).

Table 10.22: List of Plant species suitable for use as Air Pollution Bioindicators in Europe (After Steubing and Jager, 1982)

Pollutants	Plant Species	
	Scientific names	Common names
Sulphur dioxide	*Medicago sativa* L. cu. Du Puits.	Alfalfa
	Trifolium incarnatum L.	Clover
	Pisum sativum L.	Pea
	Fagopyrum esculentum Moench.	Buckwheat
Nitrogen dioxide	*Apium graveolens* L.	Wild celery
	Nicotiana bylutinosa L.	Ornamental tobacco
Ozone	*Nicotiana tobacum* L. cu Bel W_3	Tobacco
Peroxyacyl nitrates	*Urtica urens* L.	Small nettle
	Poa annua L.	Annual meadow grass
Hydrogenfluoride and fluorides	*Gladiolus gandavensis* L. cu. Snow Princess	
	Tulipa gesneriana L. cu. Blue Parrot.	Tulip
General accumulators	*Bassica oleracea* L.var, Acephala	Cabbage

The use of lichen indicators for air pollution monitoring was attempted in urban environment during recent decades by Santra and Mitra (1990). By examining the pattern of lichen patch occurrence, an index of air pollution was determined in different sites of Kolkata city (WB). Thus an air pollution zone map was prepared (Fig. 10.17 next page). Similar attempt was also made in Haldia industrial zone. The details are given in Box subsequently.

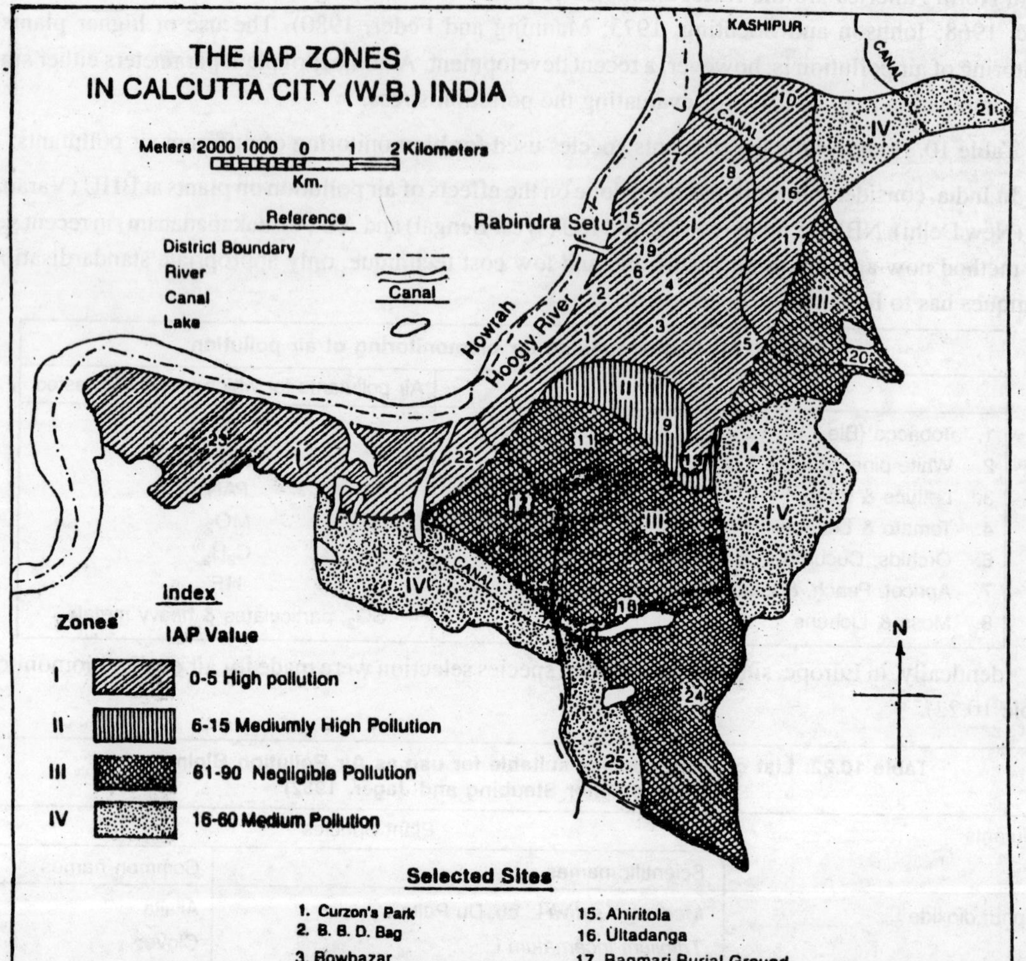

Fig. 10.17 : Zoning of Kolkata city by Lichen biomonitoring techniques

LICHENS ARE RELIABLE MONITORS OF AIR POLLUTION

Indian scientists have classified levels of atmospheric pollution using lichens.

Indian scientists find lichens can be used to monitor urban pollution levels. Lichens are disappearing within the core zone of the highly industrialised Haldia township in West Bengal due to increasing air pollution (*Pollution Research,* Vol 11, No 1).

Sensitivity of lichens—essentially symbiotic associations of fungi and algae, often found growing on tree trunks and rocks—to air pollution, particularly gases like sulphur dioxide and particulate matter like dust, is fairly well-known.

At Haldia, emission from companies like the Indian Oil, Eureka Chemicals, Hindustan Fertilizer, Shaw Wallace and Hindustan Lever and smoke from marine vessels pollute the atmosphere.

S. G. Mitra and S. C. Santra at the school of Environmental Sciences, University of Kalyani, and K. N. Roychowdhury of the Botanical Survey of India, Howrah, discovered more lichen species are present in areas where pollution levels are low. The scientists have drawn up an index of atmospheric purity (IAP) using lichens to classify levels of pollution in various areas of Haldia.

Both the lichen constituents are affected by gaseous and particulate pollutants. Matter accumulates in the tissues bringing about degeneration and a number of morphological changes. Mitra and Santra also discovered more pollution-tolerant species, like *Bacidia convexula* and *Parmelia caperata*, are found in areas with higher sulphur dioxide levels, while more sensitive species like *Trypethelium tropicum* and *Gyrostomium scyphiliferum* are found where pollution levels are lower.

LICHENS AS POLLUTION MONITORS

Pollution Zone I	Pollution Zone II	Pollution Zone III	Pollution Zone IV
Lecidea granifera	Lecidea granifera	Lecidea granifera	Lecidea granifera
Parmelia caperata	Parmelia caperata	Parmelia caperata	Parmelia caperata
Beacidio convexula	Pyrenula nitida	Bacidia convexula	Bacidia convexula
	Graphis scripta	Pyrenula nitida	Pyrenula nitida
		Graphis scripta	Graphis scripta
		Lecanactis premnea	Lecanactis premnea
		Arthonia antillarum	Arthonia antillarum
		Catilaria indica	Catilaria indica
			Trypethelium tropicum
			Gyrostomium scyphiliferum
			Phaeographis leprosulans
			Ramalina caricaris

Haldia town : Pollution level indications based on lichen population density

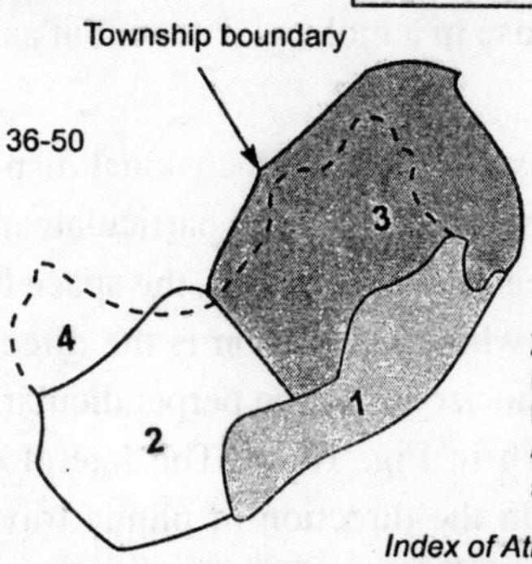

Zone 4
Clean air, IAP value : 36-50
Average SO_2 level : below 20 µg/cu.m.

Zone 2
Medium pollution
IAP value : 0.20
Average SO_2 level : 25-32 µg/cu.m.

Zone 1
High pollution
IAP value : 0.5
Average SO_2 level : above 32 µg/cu.m.

Zone 3
Negligible pollution
IAP value : 21-35
Average SO_2 level : 20-25 µg/cu.m.

Index of Atmospheric Purity (IAP) is computed on the basis of the density of occurence of lichens in the sample sites.

10.14 AIR POLLUTANTS DISPERSAL MODEL

Since the atmosphere is practically in continuous motion, it is now generally recognised that dispersal of the pollutants discharged from a single point source, for example, a stack, occurs into the atmospheric space under the influence of two components of atmospheric motion: (1) the mean wind field which transports the pollutants from one point to another, and (2) the turbulent motions, including eddy currents, which disperse the pollutants about some central position. Atmospheric turbulence produces pulsation velocities in all the three directions which ensure mixing of flue gases with atmospheric air through widening of the plume as it is carried away in the direction of the wind.

As the ground surface/atmosphere interface offers friction and as air in common with other fluids possesses viscosity, both the direction and speed of wind changes with height above ground level. Though the precise manner in which surface wind velocity could be used to compute the wind speed at stack height through quantifiable mathematical relationship is not known, power low types of relationships are described in many literatures. The dispersal of pollutants is affected by motions that occur in the atmosphere near the ground and are influenced by the friction of ground surface and surface heating and cooling.

The atmosphere above the ground surface can be divided into three zones:

(a) *Surface boundary layer* which extends up to 50-150 m, where wind motion is affected by surface roughness and vertical temperature gradients;

(b) *Transition layer* in space above 100 m up to a height of 500-1,000 m or more where wind motion is affected by surface friction, density or temperature gradient and earth's rotation.

(c) *Free atmosphere* where wind motions are unaffected by ground friction or by surface heating and cooling.

The surface boundary and the transition layers together are called the planetary boundary layer. Most of the motions and turbulence occur in the planetary boundary layer but once the pollutants reach the free atmosphere they are rapidly carried away to far off places. The entry of pollutants into the free atmosphere is a common occurrence and is confirmed by scientific monitoring experiments for the transport of pollutants discharged in the atmosphere.

Turbulent pulsations and eddy currents in surface boundary and transition layers encourage tendencies for particulate and gaseous pollutants to disperse in the atmosphere. Ideally, the gaseous constituents should also have a tendency to diffuse in a molecular state, but as yet the problem has defied rigorous mathematical development.

In view of the inherent difficulties of measuring three-dimensional dispersion coefficients, models for mathematical representation for the dispersion of gaseous and particulate matters in atmosphere based on statistical (Gaussian) distribution in space began to gain favour, the space being fully defined with the help of a three-dimensional coordinate system where x direction is the direction of plume travel in the direction of mean wind motion, y axis is in the horizontal plane perpendicular to the x axis and the z axis extends vertically. This is schematically shown in Fig. 10.18. The lateral dispersion coefficient (σ_y) represents the horizontal spread of the plume in the direction of plume travel : the vertical dispersion coefficient (σ_z) represents its spread in the vertical direction. The basic assumptions of this model are that the turbulence is everywhere the same, σ_y and σ_z are functions of distance from the source; the wind speed is constant throughout the layer where the plume is being transported, and the wind direction remains constant during travel.

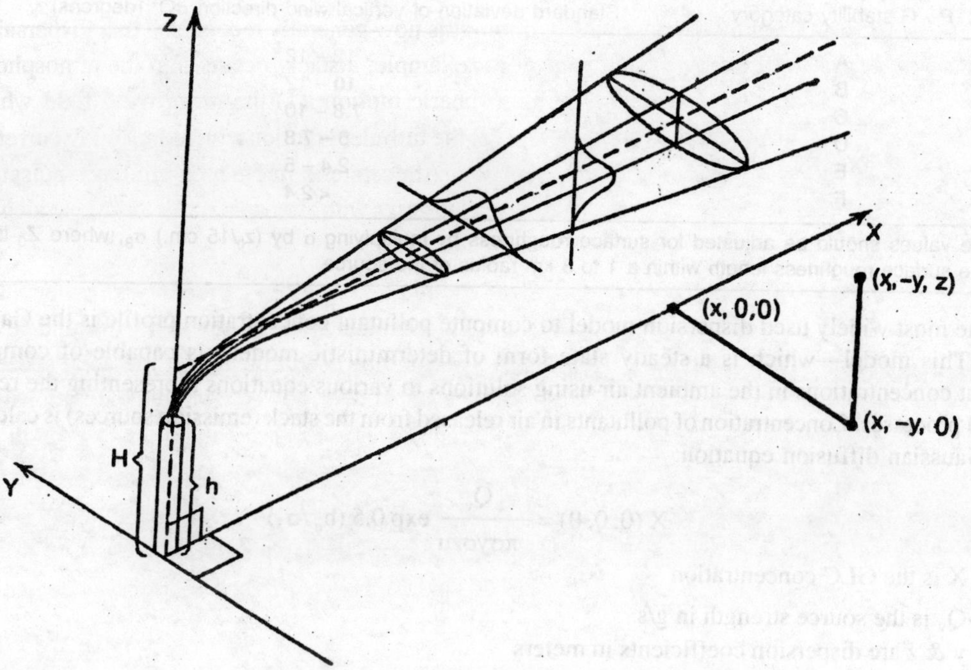

Fig. 10.18 : Plume dispersion coordination system

In terms of the Gaussian distribution theory, dispersion of pollutants is favoured by atmospheric turbulence : The magnitude of turbulence, however, varies a great deal during the course of a day and every day along the year. The widely used dispersion coefficients were obtained by Pasquill (1959). For the computation of dispersal model key to stability categories (Table 10.23) and vertical wind direction fluctuation (Table 10.24) is considered.

Table 10.23 : Key to stability categories

Surface wind speed (at 10 m) m/s	Day*			Night*	
	Incoming Solar radiation			Thinly overcast or : ≥ Low cloud ≤ cloud	
	Strong	Moderate	Slight		
< 2	A	A – B	B	—	—
2 – 3	A – B	B	C	E	E
3 – 5	B	B – C	C	D	F
5 – 6	C	C – D	D	D	D
> 6	C	D	D	D	D

* The neutral class, D, should be assumed for overcast conditions during day or night

Table 10.24 : Pasquill-Gifford stability categories versus vertical wind direction fluctuation

P - G stability category	Standard deviation of vertical wind direction σQ^+ (degrees)
A	> 12 – 12
B	10 – 12
C	7.8 – 10
D	5 – 7.8
E	2.4 – 5
F	< 2.4

* These values should be adjusted for surface roughness by multiplying σ by (z_0/15 cm.) σ_3, where Z_0 is the average surface roughness length within a 1 to 3 km radius of the source

The most widely used dispersion model to compute pollutant concentration profile is the Gaussian model. This model—which is a steady state form of deterministic model—is capable of computing pollutant concentrations in the ambient air using solutions to various equations representing the relevant physical processes. Concentration of pollutants in air released from the stack (emission sources) is calculated using Gaussian diffusion equation:

$$X(0, 0, 0) = \frac{Q_v}{\pi \sigma y \sigma z u} \exp 0.5 (h_e / \sigma_3)^2$$

where; X is the GLC concentration

Q_v is the source strength in g/s

y & z are dispersion coefficients in meters

u is the horizontal wind speed, m/s

Effective stack height 'h_e' is dependent on plume site 'h'.

$$h_e = \frac{0.84 (12.4 + 0.09 h_p)}{u} Q_h$$

where h_p = physical stack height

Δh = plume rise

u = mean wind speed at the stack top in m/s

Q_h = head release rate in Cal/sec.

10.15 AIR POLLUTION CONTROL DEVICES

In order to reduce the pollution load entering the environment from industries (stationary sources), several measures may be taken. These measures include replacement of burning fuel by electricity or solar energy or by improvement of fuel quality and fuel burning processes. Besides such innovations there are a number of mechanical devices that might be helpful for reduction of pollutant emission level from stationary sources.

There are two categories of devices that are often used by the polluting industries, viz., (a) those devices which help in reducing particulate matters; and (b) those devices which help in reducing gaseous pollutants.

(a) ***Control of particulate pollutants***: The important devices which are used to control particulate matters are :

(i) Gravity settling chamber,

(ii) Centrifugal collectors, viz., cyclone collectors and dynamic precipitators;

(iii) Wet scrubbers, viz., spray towers and venturi scrubbers;

(iv) Electrostatic precipitators, and

(v) Fabric filters.

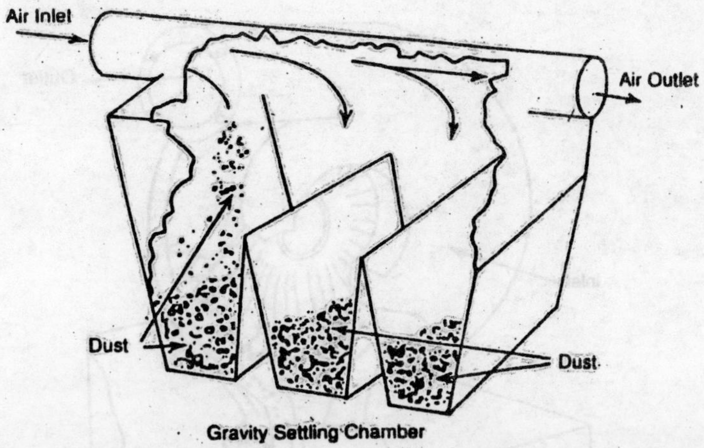

Fig. 10.19 : Gravity settling chamber

Gravity settling chambers: These are the oldest and simple type of particulate collector (Fig. 10.19). Particles of size between 40-100 μm in diameter are readily collected by this technique. Their efficiency is very poor on fine dust and decreases as the load increases. They have a very large physical size, but, on the other hand, they are inexpensive and reliable. The cost of equipment is very low and very easy to maintain.

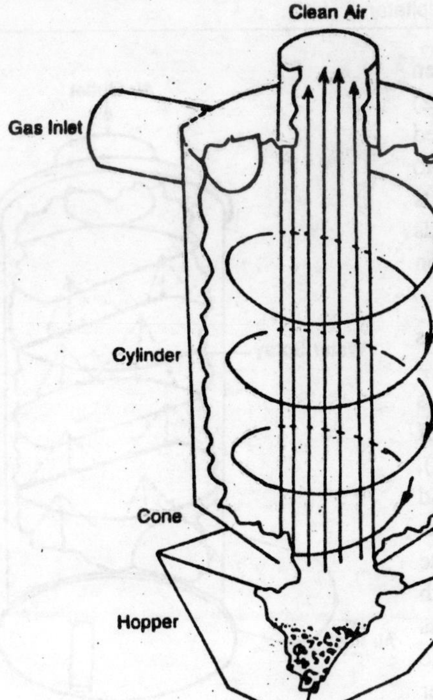

Fig. 10.20 : Cyclone collector

Cyclone collector: It operates on the principle of centrifugal force. As the carrier gas enters the cylindrically shaped collector, it takes a helical path and the inertia of the particles carries them to the walls from where they drop into a hopper at the bottom. The clean gas rises near the centre and is exhausted through the top. The cyclone collector is quite simple and reliable with low initial cost, easy maintenance and high temperature capabilities. Cyclone collectors tend not to be efficient on particles smaller than 10 μm and are best for 15-50 μm particles (Fig. 10.20).

Dynamic precipitators: Dynamic precipitators (Fig. 10.21) also operate on the principle of centrifugal force. They are really a combination of a specially designed centrifugal form and a dust collector. The centrifugal force generated by the rotating blades probes the particles in the air stream to the tips of the blades, from where they are drawn off in a concentrated stream. It is fairly efficient, even on particles 5-20 μm in diameter. However, this type of collector is unsuitable for sticky or fibrous materials, for the solids tend to build up on the blades.

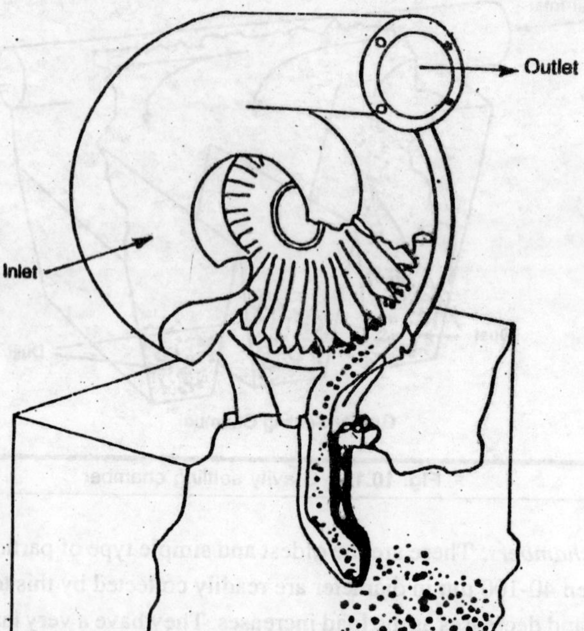

Fig. 10.21 : Dynamic precipitator

Spray Towers: There are two types of spray towers : open spray towers and packed towers. Open spray towers (Fig. 10.22) are best for coarse particulate (>10 μm in diameter) such as attained in the iron pyrite roasting step of steel production. They tend to be good for heavy particulate loads. While, a packed tower consists of a series of contact beds, through which the gases and liquids flow either counter current or crossflow. A typical application would be for the absorption of fluorine compounds.

Venturi scrubbers: In the venturi scrubbers, the dirty gas and scrubbing liquid are moving at very high velocities (100 - 450 mph), generating much turbulence and mixing, and thus exhibiting good collection efficiency. There are two types of venturi scrubbers: dry and wet type. The dry types (Fig. 10.23), are not really dry, where the dirty gas enters, concentrates, and speeds up due to the narrowing passageway, is dry, but as the gas passes through the constricted area, it is inundated by a transverse water spray. The wet types, on the other hand, require that both collector sections be wet. The water spray contacts the dust as soon as it enters the collector and the two mix immediately. One advantage of this immediate mixing is to allow collection of hotter gases. Venturi scrubbers are very efficient (better than 99%) on particles as small as 0.5 μm. As such, they are typically used in the steel industry for fine particulates, for acid mists and for lime and coal dust (whose size is often 1 μm or less).

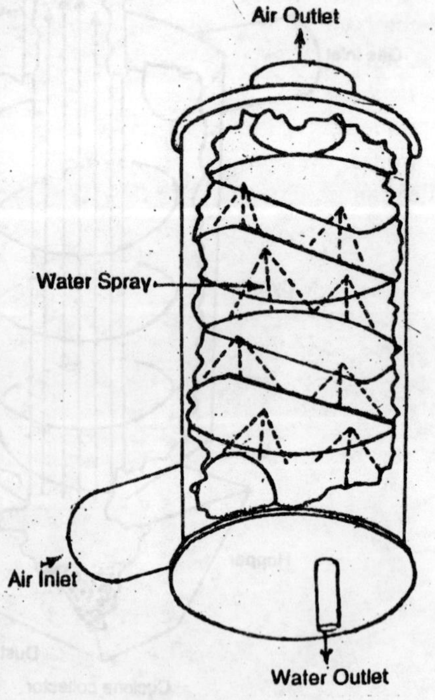

Fig. 10.22 : Open Spray Tower

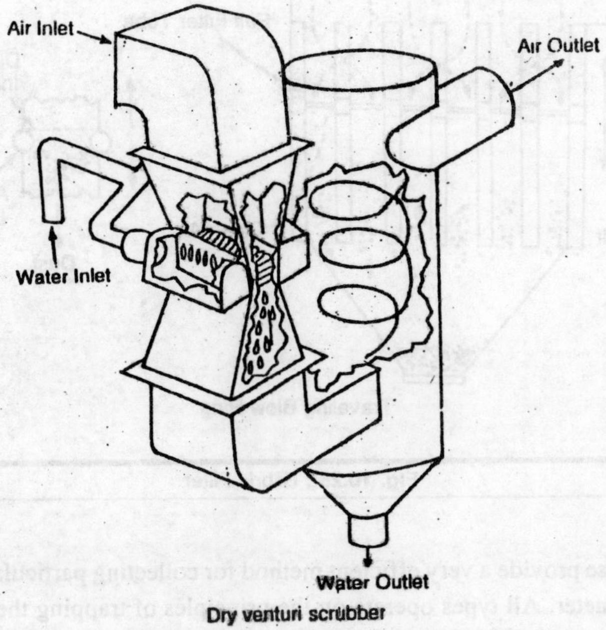

Fig. 10.23 : Dry Venturi Scrubber

Electrostatic precipitators: **The electrostatic precipitator is a very versatile and efficient type of collector, which can operate at over 1,000 °F (540 °C) and can be used for dry particulates or fumes, as well as mists. Electrostatic precipitators (Fig. 10.24) operate on the basis of electrostatic attraction. The dirty gas is channelled between two electrodes, a high voltage (between 40,000 v and 50,000 v) discharge electrode and a grounded collecting electrode. Usually a precipitator will have more than one set of electrodes, thus the dust can be collected in different "compartments". Precipitators allow for the dry collection of very fine particles. They are highly efficient, often approaching 99.9% efficiency.**

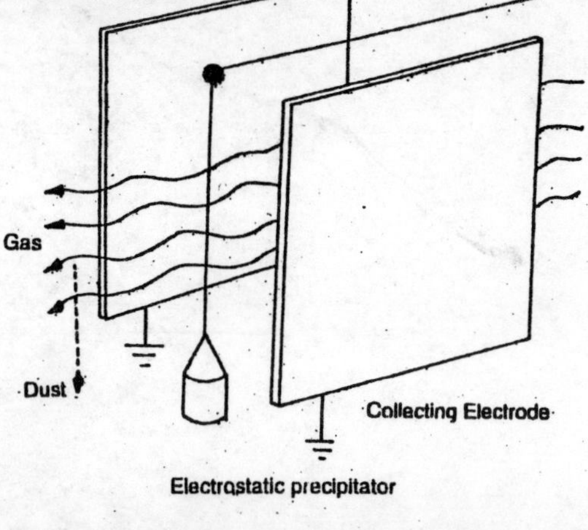

Fig. 10.24 : Electrostatic Precipitator

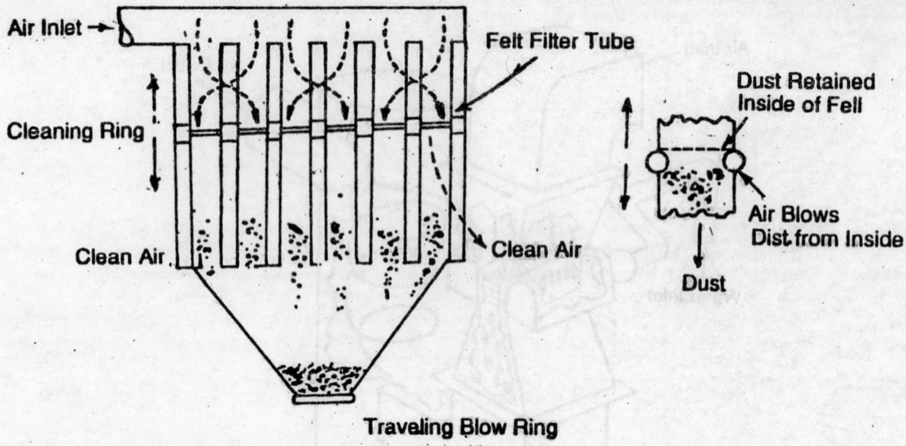

Fig. 10.25 : Fabric Filter

Fabric filters: These provide a very efficient method for collecting particulates, even those particles less than 0.5 μm in diameter. All types operate on the principles of trapping the dust on one side of the fabric, the dirty gas side, while allowing the air itself to pass through the interstices in the fabric. As the dust collects on the side of the fabric, the collector becomes more efficient, for the dust layer — the filter cake — is even more efficient as a collector than the cloth. This filter cake blocks the larger of the fabric interstices, which are typically about 100 μm in diameter, and thus helps to capture the finer dust particles. The fabric must be cleaned frequently, or eventually no gas would be able to pass through it. Various types of fabric filter collectors (Fig. 10.25) are often used in industries for pollution control.

The sizes of particles removed by these mechanical devices as described above, and their working principles with comparative merits and demerits of techniques are given in Table 10.25 (see next page).

Table 10.25 : Comparative Study of Various Mechanical Devices used in Particulate Emission Control

S.No.	Name of Mechanical Device	Minimum particle size that can be removed in μm	Efficiency %	Merits	Demerits	Principle of Working and Uses
(1)	(2)	(3)	(4)	(5)	(6)	(7)
1.	Gravitational Settling Chambers	> 50 μm	< 50%	Simple to design and maintain, and low pressure loss	Requires larger space, and collection efficiency is low. Only larger sized particles are separated out.	The emitted smokes, when made to pass through a settling chamber (Fig. 10.14), drop some of their larger sized particles in the chamber, under Stoke's law. The largest size particle (d) that can be removed with 100% efficiency in such a chamber of length L and height H is given by equation $d = C \cdot \sqrt{\frac{18a \cdot V_h \cdot H}{g \cdot L \cdot \rho_p}}$ where v_h = horizontal velocity of gas passing through the chamber, between 0.5 to 2.5 m/sec ρ_p = density of particles removed. μ = viscosity of air, at the given temperature in kg/m. sec. C = correction factor for existing non-quiescent conditions in the gas flow through the chamber, generally taken equal to 2.
2.	Centrifugal Collectors including cyclone collectors or Separators and Dynamic precipitators	5–25μm	50–90%	Relatively inexpensive, simple to design and maintain; requires less floor area; ensures dry continuous disposal of collected dusts; low to moderate pressure loss (2.5 to 20 cm); can handle large volumes of gases at temperatures up to 90°C. Cyclones are widely used in industries producing larger quantities of gas containing larger sized particles, like Cement and Fertilizer plants, Petroleum refineries, Asphalt, mixing plants, Grain mills, Cotton gins, etc.	Requires much head room; collection efficiency is low for smaller particles, quite sensitive to variable dust loadings and flow rates.	A cyclone collector (Fig. 10.15) is a specially designed closed chamber, in which the velocity of the inlet gas is transformed into spinning vortex, and the particles from the gas are thrown out under the centrifugal force. The particles thrown out on the walls of the chamber, slide down to the hopper, and are thus removed. It's efficiency depends on the generated centrifugal force, which, in turn, depends on mass of the particles (M_p), inlet gas velocity (v_i), and radius of cyclone (R), and is given by equation F_c = centrifugal force generated $M_p \cdot \frac{v_1^2}{R}$ A dynamic precipitator (Fig. 10.16) imparts centrifugal force to the entering gas with the help of rotating vanes, and is, thus, about 7 times more effective than an ordinary cyclone. Such a unit can work as an exhaust fan as well as a dust collector. They are widely used in Ceramics, Food and Pharmaceuticals, and Wood working industries. They cannot, however, handle wet fibrous material, which tend to accumulate on the moving vanes. A photoview of such a device is shown in Fig. 10.17.
3.	Wet Scrubbers or Collectors, including spray towers	> 10 μm	< 80%	(i) They can be made to remove gaseous pollutants also, along with remaining particulate matter.	(i) A lot of wastewater, needing disposal, may be produced.	In these devices, the flue gas is made to push up against a down-falling water (liquid) current. The particulate matter mixes up with water droplets and, thus, falls down and gets removed (Fig. 10.18).

Contd.

Table 10.25 : Comparative Study of Various Mechanical Devices used in Particulate Emission Control

S.No.	Name of Mechanical Device	Minimum particle size that can be removed in μm	Efficiency %	Merits	Demerits	Principle of Working and Uses
(1)	(2)	(3)	(4)	(5)	(6)	(7)
				(ii) Hot gases can be cooled down.	(ii) Maintenance cost is high, when corrosive materials were collected.	Water solutions, when replaced with other aqueous chemical solutions, like lime, potassium carbonate, slurry of MnO and MgO, etc. do help in removing gaseous pollutants also from the flue gases.
		> 0.5 μm	< 99%	(iii) Corrosive gases can be recovered and neutralised.	(iii) Wet outlet gases cannot rise high from the stack.	
				(iv) The separated gases through contact with aqueous chemicals may produce useful byproducts, as chemicals and fertilizers.	(iv) Poses freezing problems in cold countries.	
					(v) Plume may sometimes be visible in the sky due to presence of water vapours.	
4.	Electro-static Precipitators	> 1 μm	95-99%	(i) Particles may be collected wet or dry.	(i) Higher initial costs.	In electrostatic precipitators, the flue gas is made to pass through a highly ionised zone, where the particles get electrically charged and are separated out from the gas, with the help of electrostatic forces in the powerful electric field (Fig. 10.20).
				(ii) 99% and plus efficiency can be obtained.	(ii) Sensitive to variable dust loadings and flow rates.	
				(iii) Even small particles can be removed.	(iii) They use high voltage, and hence may pose risk to personal safety of the staff.	They are widely used in Thermal power plants, Pulp and Paper industries, Mining and Metallurgical industries, Iron and Steel plants, Chemical industries, etc.

Contd.

Table 10.25 : Comparative Study of Various Mechanical Devices used in Particulate Emission Control

S.No.	Name of Mechanical Device	Minimum particle size that can be removed in μm	Efficiency %	Merits	Demerits	Principle of Working and Uses
(1)	(2)	(3)	(4)	(5)	(6)	(7)
				(iv) Maintenance is nominal, unless corrosive and adhesive materials are present in flue gases	(iv) Collection efficiency reduces with time.	
				(v) They contain a few moving parts.		
				(vi) They can be operated at high temperatures up to 300-450 °C,		
5.	Fabric Filters	< 1 μm	> 99%	(i) Fabric filters can give high efficiency, and can even remove very small particles in dry state.	(i) High temp. gases need to be cooled to the range 100-450°C, within which the filters are stable.	In such a system, the flue gas is allowed to pass through a woven or felted fabric, which filters out the particulate matter and allows the gas to pass. Small particles are retained on the fabric, initially through interception and electrostatic attraction; and later on, when a dust mat is formed, the fabric starts collecting particles more efficiently (Fig. 10.21).
				(ii) Performance decreases, becomes visible, giving prewarning.	(ii) The flue gases must be dry, as, otherwise, there is a risk of condensation inside the filter, which can cause clogging.	
					(iii) The fabric is liable to chemical attack.	A baghouse filter unit, provided in an ordinary room of the factory, contains several vertically hanging fabric cylindrical bags (1.8 to 9 m long); the upper ends of the bags are closed, and lower ends are attached to a hopper, where also the inlet of flue gas is located. The upward moving gas drops out particulate matter in these bags, which settles down into the hopper, and cleaner gas goes out through the fabric filters. The framework, housing the hanging bags, is provided with an automatic shaking device for cleaning the bags of the collected dust.

(b) ***Control of Gaseous Pollutants:*** The important devices, which are used to control emission of gaseous pollutants in industries are various absorption units, viz., *spray towers, packed towers* and *venturi scrubbers*. In these absorption units different packed materials like activated alumina, silica gel, activated carbon etc. were used as absorbant. Air pollution control from automobiles i.e., diffused sources, may be suitably done by the installation of catalytic convertors, diesel particulate filter and also changing fuel quality.

10.16 BIOFILTRATION OF AIR POLLUTANTS

Biofiltration can be described as the removal and oxidation of harmful gases, mainly VOC's, from contaminated air. The contaminated air is passes through a support of compost or soil which is disposed into a horizontal or vertical reactor. Under the bed lies a distribution system, consisting of perforated pipes designed specially to provide a equal flow of the gas through the porous bed.

Gas treatment by biological processes can also be described as a diffusion of gaseous phase in to an aqueous phase, where microorganisms converts biodegradable pollutant components into harmless products. Biofilters are systems in which gas is blown through a bed of compost or soil, where the natural microorganisms presented into the support consume the gaseous organic pollutants or an inert support where a special microorganism or pool of microorganisms are cultivated. In these cases, the degradation of environmentally harmful compounds is due to an active biofilm deposited onto a package consisting of active and selected microorganisms. The aqueous phase is presented in the form of humidity, at the solid support.

Biofilters composed of a solid support, covered by an active biofilm, a bed through which the gas with harmful compounds pass to the environment a technology available and is being investigated nowadays. The mechanism of the process consists of removing the unwanted odorous organic compound from the gaseous phase, concurring its absorption or adsorption onto the porous solid phase of the biofilter or dissolution into the liquid phase, and then oxidation of this compounds by microorganisms.

A range of technologies is available to treat odorous air from wastewater treatment plants, sulfur effluents, industrial processes and industrial waste water such as organic compounds. Many kinds of apparatus can be designs and many kinds of media are available for this purpose. Biofiltration is advantageous because it does not require large amounts of energy during operation and besides produces a relatively low toxicity waste steam.

Advantages and disadvantages of biofiltration

The advantages and disadvantages are shown in the below table 10.26.

Table 10.26: Advantages and disadvantages of biofiltration

Advantages	Disadvantages
Lower capital cost Lower operation costs Lower chemical usage No combustion source	System is not fitted for compounds, which have low adsorption and degradation rates, mainly chlorinated voc's.
Biofiltration units can be designed to fit in space and size to the industrial unit setting, operation spaces.	Large biofilters units or large areas are required to treat treat contaminated sources with high chemical emissions.
System versatile to treat odors, toxic compounds and voc's. Efficiency > 90% for the low contaminant cocc. (< 1000ppm)	Source of emission that vary severely or produces spikes, can be detrimental to the biofilter performance and to the microbial population.
Possibility of different media, microorganism, and operational conditions, for many emission points.	Biofilters require long period of acclimation for microbial population, weeks or even months, mainly for voc treatment.

Because of its wide range of application, many studies have been done on biofiltration system in last few decades. However, theoretically it is still difficult to explain the behaviour of a biofilter. The growth of different types of microorganisms in different working conditions makes it impossible to generalize the microbial activities in a biofilter. The biofilters of operated at different filtration rates and influent characteristics can have diverse efficiency for different target pollutants. Besides, due to some of the operational drawbacks of the biofilter such as performance fluctuation, maintenance of biomass, and disinfection adequacy of the biofilter effluent, research on biofiltration process has become imperative. This paper mainly focuses on the theoretical and modeling aspects, and the performance of the biofilter in removing organics, nutrients, and some specific pollutants from water and wastewater. Biofiltration is as an economically viable treatment process, used capture pollutants from air & water. The contaminated air is passed through soil, peat, or compost, the naturally present microorganisms also begin to biodegrade the air contaminants, which led to the development of soil biofilters, in which soil with low clay and high organic carbon content was packed in a bed and contaminated air was passed through the soil bed to biodegrade the air contaminants.

However, as more research was conducted on this simple process, it became clear that the biodegradation rates were low and hence the size of the biofilter bed required to achieve high destruction efficiencies was very large. Since, compost had a higher concentration of microorganisms, compost became the media of choice for biofilters. Major problems encountered were setting of the compost, resulting in increase gas-phase pressure drop, availability of nutrients, such as nitrogen and phosphorus, pH maintenance, and drying of the compost material due to moisture transferring to the flowing gas phase. These problems were countered to some extent by adding wood chips, which provided mechanical support to minimize settling, humidifying the inlet air to maintain proper water content in the compost material, adding lime pellets for pH control, and fortifying the compost with fertilizers containing nitrogen and phosphorus compounds. Further, in compost beds, it was necessary to have shallow beds (height < 1.5m or 4.5 feet), to prevent compaction of the material and drying of the bed from the top surface. Currently, there are several companies that offer compost biofilters for treatment of odorous and volatile chemicals.

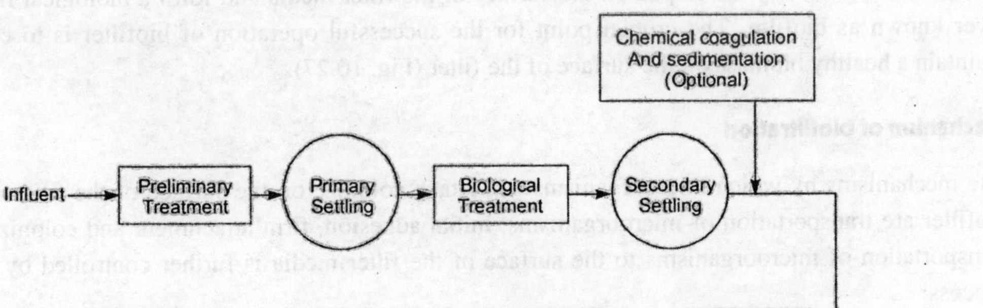

Fig. 10.26 : Schematics of biofiltration system for advanced waste water treatment

In advanced wastewater treatment, biofilter can be used along with conventional physico-chemical processes such as coagulation, flocculation, filtration and sedimentation. (Fig. 10.26). The conventional filter and the biofilter units can be combined together depending on the suspended solid concentration. Since the main purpose of the biofilter is to remove the dissolved organics, the suspended particles are removed in conventional filter before subjecting the waste water to the biofiltration system.

How biofiltration occurs?

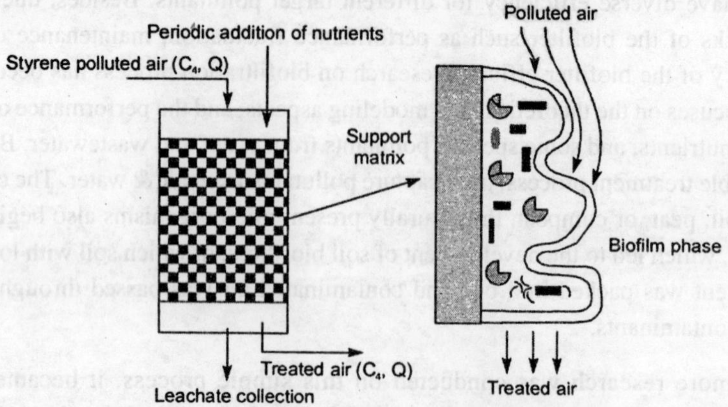

Fig. 10.27 : Process of biofiltration

Microorganisms fixed to a porous medium are used in the biofiltration process to break down pollutants present in the wastewater stream. The microorganisms growing a biofilm on the surface of the medium or are suspended in the water phase surrounding the medium particles. The filter bed medium consists of relatively inert substances which ensure large surface attachment areas and additional nutrient supply. The overall effectiveness of a biofilter is largely governed by the properties and characteristics of the support medium, which inclųde porosity, degree or compaction, water retention capabilities and the ability to host microbial populations. Critical biofilter operational and performance parameters include the microbial inoculation, medium pH, temperature, the medium moisture, and nutrient content. The progress of filtration process, microorganisms (aerobic, anaerobic, facultative, bacteria, fungi, algae, and protozoa) are gradually developed on the surface of the filter media and form a biological film or slim layer known as biofilm. The crucial point for the successful operation of biofilter is to control and maintain a healthy biomass on the surface of the filter (Fig. 10.27).

Mechanism of biofiltration

The mechanisms by which microorganisms can attach colonize on the surface of the filter media of a biofilter are transportation of microorganisms, initial adhesion, firm attachment and colonization. The transportation of microorganisms to the surface of the filter media is further controlled by four main process:

- Diffusion (Brownian motion), convection, sedimentation due to gravity, and active mobility of the microorganisms. The following parameters are taken into consideration to estimate the attachment of microorganisms on the surface of the filter media:
- The steric effect, hydrophobicity of the microorganisms, contact angle, and electrophoretic mobility values.

Regarding the microbiological potential of biofilters, different aspects have been recently studied: isolation and characterization, the use of pure cultures of bacteria, fungi, mixed microbial populations. Microorganisms presented in biofiltration systems are predominantly aerobic. Oxygen supply is abundant in the incoming air but it has to be dissolved into the water phase to be available to the micro flora in the biofilm.

11 WATER POLLUTION

11.1 INTRODUCTION

Water is one of the most important constituents of life support system. It is indeed a wonderful chemical medium which has unique properties of dissolving and carrying in suspension a huge varieties of chemicals. Thus it can get contaminated easily. Natural surface water bodies often have impurities from various sources. The impurities may be suspended particles, colloidal materials and may also be dissolved cationic and anionic substances. The details are given in Table 11.1. Various kinds of natural and man-made activities—industrial, domestic, agricultural and others—are day by day creating water pollution problem, particularly in fresh water system. Time has come to work together in war footing to combat water pollution challenges with new protections to give all our children the gift of clean and safe water in the 21st century.

Table 11.1: Impurities in Natural Water

Source	Particles Size Classification				
	Suspended	Colloidal	Molecules	Dissolved	
Atmosphere	— Dust —		Carbon dioxide, CO_2 Sulphur dioxide, SO_2 Oxygen, O_2 Nitrogen, N_2	Positive ions Hydrogen, H^+	Negative ions Bicarbonate HCO_3^- Sulphate, SO_4^{2-}
Mineral soil and rock	— Sand — — Clays — — Mineral soil particles —		Carbon dioxide, CO_2	Sodium, Na^+ Potassium, K^+ Calcium, Ca^{2+} Magnesium, Mg^{2+} Iron, Fe^{2+} Managanese, Mn^{2+}	Chloride, Cl^- Fluoride, F^- Carbonate, CO_3^{2-} Sulphate, SO_4^{2-} Bicarbonate, HCO_3^- Nitrate, NO_3^- Various phosphates Chloride, Cl^- Bicarbonate, HCO_3^- Nitrate, NO_3^-
Living organisms and their decomposition products	Algae Diatoms Bacteria		Carbon dioxide, CO_2 Oxygen, O_2 Nitrogen, N_2 Hydrogen sulfide, H_2S Methane, CH_4	Hydrogen, H^+ Sodium, Na^+ Ammonium, NH_4^+	
	— Organic soil (topsoil) —				
	Fish and other organisms	Viruses Organic colouring matter	Various organic wastes, some of which produce odour and colour		

Table 11.2: Potential source of ground water pollution arising from domestic, industrial and agricultural activities

Contaminant source	Contaminant characteristics
Domestic	
Septic tank	Suspended solids 100 - 300 mg/l BOD 50 - 400 mg/l Ammonia 20 - 40 mg/l Chloride 100 - 200 mg/l High fecal coliforms and streptococci. Trace organisms, greases
Storm water drains	Suspended solids 1,000 mg/l Hydrocarbons from roads, service areas. Chlorides or urea from de-icing. Compounds from accidental spillages. Bacterial contamination.
Industry	
Food and drink manufacturing	High BOD. High suspended solids. Colloidal and dissolved organic substances. Odours.
Textile and clothing	High suspended solids and BOD. Alkaline effluent.
Tanneries	High BOD, total solids, hardness, chlorides, sulphides, chromium.
Chemicals	
acids	Low pH
detergents	High BOD.
pesticides	High TOC, toxic benzene derivatives, low pH
synthetic resins and fibres	High BOD.
petrolium and petrochemical refining	High BOD, chloride, phenols, sulphur compound
Fertilizer process	High BOD, suspended solids, chloride, variable pH
Painting and metal finishing	Low pH. High content of toxic metals.
Engineering works	High suspended solids, cloride, variable pH.
	Pulverized fuel ash: sulphate, and may contain germanium and selenium.
Power generation	Fly ash and flue gas scrubber sludges: low pH, disseminated heavy metals.
Deep well injection	Concentrated liquid wastes, often toxic brines. Acid and alkaline wastes. Organic wastes.
Leakage from storage tanks and pipelines	Aqueous solutions, hydrocarbons, petrochemicals, sewage.
Agriculture	
Arable crops	Nitrate, ammonia, sulphate, chloride and phosphates from fertilizers. Bacterial contamination from organic fertilizers.
Livestock	Organo-chlorine compounds from pesticides. Suspended solids, BOD, nitrogen. High fecal coliforms and streptococci.
Silage	High suspended solids, BOD 1-6 × 10^4 mg/l. Carbohydrates, phenols.
Mining	
Coal mine drainage	High TDS (total dissolved solids), suspended solids. Iron. Low pH. Possibly high chloride.
Metals	High suspended solids. Possibly low pH. High sulphates. Dissolved and particulate metals.
Household wastes	High sulphate, chloride, ammonia, BOD, TOC and suspended solids from fresh wastes. Bacterial contamination. On decomoposition: initially TOC of mainly volatile fatty acids (acetic, butyric, propionic acids), subsequently changing to high molecular weight organics (humic substances, carbohydrates).

BOD is Biological Oxygen Demand, TOC is Total Organic Carbon, pH is log 10 (H^+)

Source: Adapted from Jackson (1980)

11.2 WATER CHEMISTRY

To understand water pollution, it is essential to know the chemical phenomena that occur in water. The chemical phenomena includes acid-base reaction, solubility, oxidation-reduction and complexation reactions. Acid-base phenomena in water involve loss and acceptance of H^+ ions. Many species act as acids in water by releasing H^+ ions, others act as bases by accepting H^+ ions, and the water molecule itself does both. An important species in the acid-base chemistry of water is bicarbonate ion, HCO_3^-, which may act as either an acid or a base:

$$HCO_3^- \rightleftarrows CO_3^{2-} + H^+ ; \; HCO_3^- + H^+ \quad CO_2\,(aq) + H_2O$$

Various gases viz., O_3, CO_2, NH_3, Cl, SO_X, NO_X etc. are soluble in water along with different salts and mostly maintain an equilibrium of ions. Oxidation-reduction phenomena are highly significant in the environmental chemistry of natural water and wastewater. In a lake, for example, the reduction of oxygen (O_2) by organic matter (represented by CH_2O). [$CH_2O + O_2 \rightarrow CO_2 + H_2O$], results in oxygen depletion that can kill fish. The rate at which sewage is oxidised is crucial to the operation of a waste treatment plant. Reduction of insoluble iron (III) to soluble iron (II), [$Fe(OH)_3(S) + 3H^+ + e^- \rightarrow Fe^{2+} + 3H_2O$] in a reservoir contaminates the water with dissolved iron, which is hard to remove in the water treatment plant. Oxidation of NH_4^+ to NO_3^- in water converts ammonium nitrogen to nitrate, a form more assistable by algae in the water.

$$NH_4 + 2O_2 \rightarrow NO_3^- + 2H^+ + H_2O$$

The chelating agents are common potential water pollutant and thus forms various metal complexes.

Table 11.3: Major routes of entry of pollutants to surface waters

Route	Major pollutants	Comments
Sewage outfalls	A very wide range of organic and inorganic pollutants from commercial and domestic sources; detergents generally present	Highly variable; dependent not on what sewage works receive but also on the treatments that sewage is given
Outfalls from commercial premises	Dependent on the commercial activity; wide range of pollutants from chemical industry; heavy metals from mining operations; pulp mills are important sources of pollutants in some areas	Concentration of pollutants in effluents must stay below statutory limit
Outfalls of nuclear power stations	Radionuclides	Subject to regular monitoring and close control in most countries
Run-off from land	A variety of pollutants dumped on land surface; pesticides	Generally uncontrolled and difficult to measure
From the air	(a) Precipitation with rain or snow	Sometimes pollutants transported over great distance
	(b) Direct application of biocides	Control of pests, parasites, vectors of disease and aquatic weeds
	(c) Accidental contamination by sprays or dusts	Aerial spraying a potential problem
Dumping at sea	Raw sewage; radiochemicals and toxic wastes in sealed containers dumped in deep ocean	Concern sometimes expressed about release from containers in the longer term when they degenerate
Release from oil rigs and terminals	Hydrocarbons	Sometimes accidental, sometimes as a result of war (e.g. Gulf War in Kuwait)
Shipwrecks	Hydrocarbons and some other organic pollutants	Wrecks of oil tankers particularly a problem

11.3 GROUND WATER POLLUTION

The ground water is most prime water which have multipurpose uses ranging from drinking to industrial and agricultural uses. The quality requirement varies distinctly with respect to the specific uses. For instance drinking water must have specified quality (Table 11.1), which is not at all essential for industrial purposes or other domestic uses. Though ground water appears to be less prone to pollutant mixing yet there are a number of potential sources of ground water pollution (Table 11.2). Ground water contamination with arsenic, fluoride, and nitrate recently possess serious health hazards to large sector of communities all over the world. Most illustrated accounts of ground water pollution is further dercribed in 11.2.

11.4 SURFACE WATER POLLUTION

Major lakes, rivers, reservoirs of the world are now getting polluted by various ways (Table 11.3), and thereby posing threat to the survivability of the life systems on these diverse water bodies. There are a number of routes of entry of pollutants to the surface water. Regular monitoring of these contaminating routes and their effective protective action plan has to be evolved for better conservation of surface water resources in future. Water quality status of some rivers of India is depicted in Table 11.4.

Table 11.4: Rivers of India: Water Quality

Name of river with length	Importance	Problems, protests and important actions initiated
Periyar 244 km	The river is surrounded by the Periyar Wildlife Sanctuary. Here are about 2,500 small-scale industries within its banks	There are more than 30 unauthorised effluent pipes going into the river. The river is dammed in 12 places making it one of the most dammed catchments in the world. Periyar Action Plan initiated.
Cauvery 770 km	It is the most important river in South India	The river receives fertilisers and pesticides used in the basin, mainly from coffee plantations, which contributes 4,700 tonnes of BOD each season. The sewage from Bangalore city is also dumped into the river.
Godavari 1,465 km	It is one of the largest drainage basins of the country, second only to Ganga. Its delta is one of the main rice growing areas in the country	Huge quantity of fertilizers used in the heavily cultivated basin, amounting to more than double of the national average fertilizer use and toxic pollutants and effluents from the industries and Jalna and Aurangabad towns find way to the river.
Krishna 1,400 km	Flows through three states of Maharashtra, Karnataka and Andhra Pradesh. It supports the sugar industries in Maharashtra	Industrial effluents, mainly from sugar belt of Satara to Sangli and Pimpri-Chinchwad area of Pune in Maharashtra, Dharwar and Bhadravati in Karnataka and Rangareddy and Vijayawada in Andhra Pradesh.
Tapti 724 km	The Tapti basin is predominantly agricultural. It supports several industries, especially near Surat in Gujarat, Jalgaon in Maharashtra and Khandwa in Madhya Pradesh	The river receives fertilizers, pesticides, industrial waste and untreated domestic waste from cultivations, urban centres and industries in the basin and along the river. At Surat, the drainage systems to pump in rain water is illegally used for discharging domestic waste into the river.
Brahmaputra 2,900 km	Originating in western Tibet, it is longer, and more voluminous than Ganga	It is less polluted compared to other rivers of the country. Petroleum refining units contribute most of the industrial pollution load.
Subarnarekha 450 km	It is a source of drinking water for many villages and towns.	The river receives tonnes of organic pollution every day. The biggest threat to the river is from uranium mining in Jadugada.

Contd.

Table 11.4: Rivers of India: Water Quality

Name of river	Importance	Problems, protests and important actions initiated
Damodar 563 km	Heavy industries like TISCO and TELCO are located here. Also a source of drinking water for 3 million people of Damodar Valley	Water is turbid due to discharge from coal washeries and thermal power plants. The river receives phenolic compounds, chromate, ammonia, naphthalene, and heavy metals from the industries. Damodar Action Plan has been initiated under NRCP.
Mahanadi 830 km	It is the life line of Odisha. The Hirakud Dam has been constructed in the middle reaches of the river	The river has been polluted by fertilizers. It has been alleged that DAP fertilizer plant of Oswal group has caused pollution and a committee in 2003 has recommended the closure of the plant.
Betwa 537 km	Drinking water source for Vidisha, MP	It is one of the most severely polluted rivers of the country due to sewage from industries and domestic waste from Bhopal. In 1996, people of Vidisha protested when they could not use river water, forcing the authorities to close down Som Distillery, because of sudden increase in pollution loads in the river.
Narmada 1,312 km	It is the largest West-flowing river in India and supports a number of urban centres.	The major source of pollution is agricultural run-off, domestic waste discharge, especially around Jabalpur-Hoshangabad-Khargone in Madhya Pradesh and Bharuch in Gujarat and industrial effluents from Ankleshwar-Bharuch industrial belt.
Sutlej 1,087 km	It supports many urban centres. The Bhakra dam is built on it.	River Budhinala, a tributary of Sutlej near the textile town of Ludhiana, has been receiving toxic effluents from Ludhiana for the past 20 years!
Indus 2,900 km	The lifeline of 5,000 years old Indus valley civilisation, it is the most important river of northwestern region, after which India derives its name.	The 100 km long Hudiara Drain, starting from Gurdaspur district in India and discharging into river Ravi in Pakistan which was meant to be a storm drain, now carries untreated waste containing organic material and toxic heavy metals.

Compiled by WWI Research Team, km: kilometre

11.5 NATURE OF WATER POLLUTANTS

There are diverse categories of water pollutants often encountered in nature. They are primarily high nutrients like PO_4, NO_3, CO_3 or SO_4; acidity or alkalinity of water, contamination of toxic metals, high particulates load, contamination with pathogenic microbes and pesticides, presence of oils, grease and other hydrocarbons etc. These contaminants causing detrimental effects on aquatic biota or its consumption in any form cause serious health hazards. In addition, higher water temperature may also cause detrimental effects on aquatic biotic life too. A generalised list of common water pollutants is given in Table 11.5.

Table 11.5: Water Pollutants

Pollutants	Origin	Damage	Preventive action needed
1. Suspended solids:	domestic and industrial waste	inhibit photosynthesis	treat waste
silt and colloids	soil erosion	silt up rivers and lakes	change flow rates
2. Dissolved mineral salts:	sewage	eutrophication—weed	
Phosphates	washing powders	growth foam	biodegradable washing powders
Nitrates	farmyard waste	toxic to young babies	
Calcium and Magnesium	fertilizers		divert waste flow
Bicarbonates	soil	temporary hardness	
Calcium and Magnesium sulphates	industrial waste-mining	permanent hardness	

Contd.

Table 11.5: Water Pollutants

Pollutants	Origin	Damage	Preventive action needed
3. Organic residues	sewage-farmyard waste industrial waste, e.g. paper pulp and food processing	increased BOD, damages fish, encourages anaerobic bacteria	proper sewage treatment
4. Toxic chemicals:			
DDT	pesticides	kill fish and birds	test first
Metal ions	industrial wastes	kill organisms and alter pH	remove from effluent
Phenols	disinfectants	kill organisms	
Oil	spillage from factories and ships	clog respiratory systems of animals, prevent O_2 uptake at water surface	remove with detergent
5. Heat	factory water coolers	less dissolved O_2 for organisms affected	increase flow
6. Pathogens:			
Entamoeba Histolytica		amoebic dysentery	proper disposal of feces, water purification
Shigella		bacillary dysentery	
Vibrio cholerae		cholera	
Salmonella sp.		typhoid, paratyphoid and food poisoning	hygienic food preparation immunisation
Leptospira	sewer rats	leptospirosis	destroy rats
Enterobias		nematode worms	hygienic food preparation
Schistosoma		bilharzia	destroy snails
Ancylostoma		hookworm	avoid contact with skin
7. Vectors-mosquitoes:	adults lay eggs in stagnant water	transmit filarial worms transmit malaria—*Plasmodium*	insecticides, spray oil on water surface, take drugs, e.g. Paludrine
Anopheles			
Aedes		transmit yellow fever virus	
Culex		transmit dengue virus	
8. Radiation:	power station wastes	mutation	use screens
Radionuclides	X-rays	mutation	nuclear testing ban
9. Trace elements:			
Arsenic	Mining by product & chemical waste	Toxic (carcinogenic)	Removal by precipitation
Chromium	Metal plating & Tannery	Toxic	DO
Copper	Metal plating industrial waste	Toxic above thresold limit	DO
Iron	Mining, industrial waste	DO	DO
Lead	DO	DO	DO
Manganese	DO	DO	DO
Mercury	DO	DO	DO
Selenium	Natural sources	DO	DO
Zinc	Industrial waste	DO	DO
10. Detergents : *Soap and other synthetic detergents*	Domestic and industrial waste	Toxic	Physical treatment
11. Specialised organic pollutents from pharma-ceutical industries:			
Biorefractory organic pollutants	Pharmaceutical industrial waste	Persitant toxics	Bio-treatment
Estrogenic subastance			

The eutrophication is a nutrient enrichment (mostly nitrate and phosphates) in water bodies. This is quite common in confined water bodies like lakes and reservoirs. The sign of eutrophication is manifested by a number of features viz., phytoplanktonic bloom, higher macrophyte growth and depletion of oxygen level of water.

The eutrophication is primarily caused by anthropogenic or cultural activities of mankind. Sewage contamination, waste dumping in water bodies or even agricultural run-off storage leads to eutrophication.

After the planktonic bloom or macrophytic growth nutrients are reduced due to plants uptake, then finally water bodies became much more less polluted. There is distinct loss of species diversity due to eutrophication of water bodies.

In lentic and lotic water bodies, there is an intimate relationship of nitrogen and phosphorus cycling and sunlight. For instance, if in a water body the ratio of phosphate to nitrate is 1:15, the aquatic plants will be able to use all phosphates but only about half of the nitrate i.e., the phosphate becomes limiting and nitrate in abundance. But keeping the nitrate at the same level, if phosphate enrichment takes place say to 4:15 phosphate; nitrate level, then all nitrates will be used by plants and it becomes limiting. The acceptable level of total inorganic phosphate in water is 0.03 to 0.40 mg/l. In most of the lakes and rivers of topics where eutrophication is encountered, the principal cause is excessive enrichment of water by phosphate and nitrates. In and around cities and industries, phosphate content has increased by 20 - 25 folds during the last 10 - 15 years. Formation of froths and foams in lentic and lotic waters are due to phosphate containing detergents now profusely used for washing clothes.

11.6 CAUSES OF WATER QUALITY DEGRADATION

On the whole, the quality of water degrades by a number of reasons. The principal reasons are stated in Table 11.6. It is very interesting to note the fact that due to the manner in which we have wasted our fresh water resources—either by the surface loss or by pollution—we have to face the crisis in couple of years. Portable water will be a precious commodity in the long run. Already we have seen that potable mineral water used for drinking is costlier than milk which is considered to be a 'balanced food'. This situation will further aggravate in coming decades. Multinational companies are already taking appropriate steps for marketing such potable mineral water in tropical countries.

Table 11.6: Major causes of water quality degradation

Major cause (pressure)	Agent causing impact
Agriculture/Aquaculture	
• Land clearing	Erosion → sediments
	Weeds
• Removal of riparian vegetation	Erosion → sediments
	Nutrients etc. (loss of riparian filter)
	Increase in light
	Loss of normal organic input (leaf litter, wood)
• Grazing	Bank erosion → sediments
	Mobilisation of sediments and nutrients
	Feces and urine — organics and nutrients
• Irrigation	Sediments
	Nutrients
	Pesticides
	Altered flows
• Horticulture	Sediments
	Nutrients
	Pesticides
• Intensive animal production	Sediments
	Nutrients

Contd.

Table 11.6: Major causes of water quality degradation

Major cause (pressure)	Agent causing impact
Mining	
• Land clearing; overburden removal	Sediments — inert or reactive
• Extraction	Sediments — inert or reactive Leaching of chemicals
• Infrastructure	Sediments
• On-site processing	Chemical disposal/leakage
• Dewatering	Altered flows and water quality
Industry	
• Waste disposal (effluent)	Toxic chemicals Organic materials Sediments
• Waste disposal (aerial)	Toxic gases Fine sediment
• Site activities	Oils, heavy metals Chemical spills
• Heat	Hot water
Urbanisation and infrasturcture	
• Land clearing	Sediments, weeds
• Residences	Fertilizers (nutrients) Pesticides Garden wastes: sediments
• Sewage disposal	Nutrients Organic materials
• Vehicles	Oils, heavy metals
• Garbage disposal sites	Chemical leachates
• Water resources development	Altered flow regimes
Recreation	
• Swimming	Nutrient and sediment input and mobilisation Sunscreens
• Boating	Oils Bank erosion leading to sedimentation

11.7 EFFECT OF SELECTED POLLUTANTS ON FRESHWATER FLORA AND FAUNA

1. **Organic matter (and associated deoxygenation of water)**: Decomposition of organic matter by microorganisms in water can lead to partial or total deoxygenation. Low oxygen levels are particularly damaging to river communities where fish and specialised river invertebrates require consistent high oxygen levels. Lakes may suffer if the bottom waters become highly deoxygenated, by (a) causing loss of bottom-dewelling biota, and (b) exacerbating the effects of eutrophication by promoting the release of phosphorus from the sediments. Small still-water bodies have highly variable oxygen levels, and support communities adapted to these conditions. These may also be harmed by organic pollution, but the effects are less easy to detect.

2. **Thermal pollution**: Temperatures above the normal range (e.g. near to power stations) can: (a) exacerbate the effects of organic matter pollution by increasing decay rates and hence deoxygenation; (b) disrupt the life cycle timing of native species; (c) cause stress to cold-blooded animals by causing above- normal rates of respiration, and (d) favour an abnormal biota (for the area) which may include exotic and sometimes nuisance species and allow these to become acclimatised to the region. The first three factors may make heated water, particularly that polluted with organic matter, uninhabitable by many forms of aquatic life.

3. **Acidification**: Low pH, and toxic materials (particularly aluminium) brought into solution at lower pHs, are directly injurious to many fresh-water animals, and have diverse biological effects including changes in the abundance, biomass and diversity of invertebrates, plants, fish and amphibians. Effects are greatest in the uplands and areas of lowland heaths and bogs—wherever there is high rainfall and/or a prevalence of acidic soils. The most significant commercial effect is the decline of fish populations.

4. **Eutrophication**: This results in enhanced plant growth, followed by the decay of this plant material and subsequent oxygen depletion of the water. In many cases enrichment by nitrogen and phosphorus is accompanied by addition of organic wastes which exacerbate the deoxygenation problem. Anoxic (oxygen-free) conditions may lead to the release of phosphate from bottom sediments, and hence allow the eutrophication process to become self-perpetuating, irrespective of future decreases in phosphorus inputs to the system. It is primarily a threat to communities of standing waters but may affect slow-flowing and highly regulated rivers. It can result in considerable loss of conservation value including loss of species diversity, and community changes which allow dominance by a few tolerant plants (particularly algae). Fish community composition may alter, with an initial increase in fish biomass, often followed by high mortality during periods when decay of high-standing crops of plants causes deoxygenation. Phosphorus is generally considered to be the principal eutrophicating agent in temperate fresh-waters. However, once a system is rich in phosphorus, nitrates may become the main factor controlling aquatic productivity, and these conditions tend to promote the growth of undesirable, nitrogen-fixing blue-green algae.

5. **Silt (fine organic and/or inorganic particles)**: Silt may contain organic matter and, hence, have a high biochemical oxygen demand (so causing deoxygenation). It often carries nutrients (particularly phosphates) and adsorbed pollutants (such as microorganics). Abrasive effects in rivers may kill fish through gill damage. Reduction of light by suspended sediment inhibits the growth of macrophytes, and may favour algal dominance. On settling, silt may (a) destroy salmon spawning beds, and the habitats of specialised plants and bottom-dwelling invertebrates, and (b) progressively seal water-bodies, isolating them from groundwater flows, and so potentially enhancing eutrophication.

6. **Metals, microorganics and other harmful materials**: Polluting effects are diverse. Toxicity data are available for very few chemicals, but many have detrimental effects on aquatic life, some at levels considerably below those which cause immediate death, e.g. sublethal levels may enhance the risk of disease, affect reproductive capacity, or alter community structure due to changes in competitive or foraging behaviour. Some toxins may accumulate up the food chain or have synergistic effects (where the effect of a pollutant cocktail is greater than the sum of the parts).

7. **Oils and grease**: They deoxygenate water as they are broken down. Oil can blanket the water surface, inhibiting oxygen diffusion, and may directly coat plants and animals, causing injury and death. Oils contain many carcinogens, such as polycyclic aromatics and phenols, which mix with water and poison aquatic life.

11.7 WATER QUALITY MONITORING

There are two basic kinds of water monitoring—chemical analysis and biological assessment. Normally, surface water quality was monitored by both ways. But the groundwater quality is monitored principally by chemical ways except a few bacteriological tests.

In India, river water quality was monitored regularly by joint efforts of Central Water Commission, State and Central Pollution Control Board. There are couple of stations of Global Environmental Monitoring Systems (GEMS) for monitoring of major river water quality. In contrary, groundwater quality was monitored by Central Ground Water Board and also State Water Investigation Departments. The primary objective of these monitoring was to assess the groundwater quality and also to assess the chemical or biological contaminations that are harmful for mankind.

Though over decades the routine analysis were made for water quality assessment, but, in recent years, the utility of biological approaches to water quality assessment were also realised. The principal biological approaches are described in Table 11.7.

Table 11.7: Principal biological approaches to water quality assessment : their uses, advantages & disadvantages

Criterion	Ecological methods		Microbio-logical methods	Physiological & biochemical methods	Bioassays & toxicity texts	Chemical analysis of biota	Histological & morpho-logical studies
	Indicator species	Community studies					
1. Principal organisms used	Invertebrates, plants & algae	Invertebrates	Bacteria	Invertebrates algae, fish	Invertebrates fish	Fish, shellfish, plant	Invertebrates Fish
2. Major assessment used	Basic surveys impact surveys trend monitoring	Impact surveys trend monitoring	Operational surveillance impact surveys	Early warning, monitoring, impact survey	Operational, surveillance early warning monitoring, impact surveys	Impact surveys, trend monitoring	Impact surveys, early warning, moni-toring basic surveys
3. Appropriate pollution sources or effects	Organic matter pollution, nutrient enrichment, acidification	Organic matter pollution, toxic wastes, nutrient enrichment	Human health risks, organic matter pollution	Organic matter pollution, toxic wastes, nutrient enrichment	Toxic wastes, pesticides, organic matter pollution	Toxic wastes, pesticides, Human health risks	Toxic wastes, organic matter pollution and pesticides
4. Advantages	Simple to perform, Relatively cheap, No special requirement	Simple to perform, Relatively cheap, No special requirement	Simple to perform, Relatively cheap, No special requirement	Simple to perform, Relatively cheap, No special requirement, usually very sensitive	Simple to perform, Relatively cheap, No special requirement, continuous monitoring possible		Simple to perform, Relatively cheap, No special requirement
5. Disadvantages	Localised use, knowledge of taxonomy required, susceptible to natural changes	Relevance of some meth-ods to aquatic systems not always tested, Susceptible to natural changes in aquatic environment	Organisms easily transported, therefore, may give false position	Specialised knowledge and techniques required for some methods	Laboratory based tests not always indicative of field condition	Analytical equipment and well-trained personnel necessary, Expensive	Specialised knowledge required, some special equipment needed for certain methods

11.9 BIOMONITORING OF WATER QUALITY

In addition to conventional water quality monitoring, there are various biological systems of water quality monitoring. These systems are also very sensitive and useful. The biotic resources of water bodies are diverse, these include a variety of macrophytes, phytoplanktons, zooplanktons, bacterial and other animal communities. Some well-known biological representatives of lentic communities is shown in Fig. 11.1. Identically, some representative animals of the littoral zones of ponds and lakes of tropics is shown in Fig.11.2. Among these biota, many forms are sensitive to pollutants of water or rather sensitive towards change in water quality. However, there are couple of forms which appears to be tolerant and acting as **'bioindicator'** species. According to Liebmann (1962) there are three distinct categories of tolerant aquatic species which are considered to be good indicator of pollution load of water, viz. oligosaprobic organisms (indicators of scarcely polluted waters), mesosaprobic organisms (indicators of moderate to highly polluted water) and polysaprobic organisms (indicator of extremely polluted water).

The detailed check list of indicator species of various saprobic categories are given in the Table 11.8:

Table 11.8: Indicators of various saprobic level (as modified after LIEBMANN, 1962)

OLIGOSAPROBIC ORGANISMS (Indicators of scarcely polluted waters)		MESOSAPROBIC ORGANISMS (Indicators of moderate to highly polluted waters)		POLYSAPROBIC ORGANISMS (Indicators of extremely polluted waters)	
Plants	Animals	Plants	Animals	Plants	Animals
Cyclotella sp.	*Vorticella sp.*	*Oscillatoria sp.*	*Stentor sp.*	*Socillatoria sp.*	*Chironomus sp.*
Synedra sp.	*Planaria sp.*	*Nitzchia sp.*	*Vorticella sp.*	*Spirulina sp.*	*Tubifex sp.*
Micrasterias sp.	*Larva of Oligoneuria sp.*	*Stephanodiscus sp.*	*Stratiomys sp.*	*Euglena sp.*	*Rotaria sp.*
Surirella sp.	*Larva of Perla sp.*	*Uronema sp.*	*Spirostomum sp.*		*Colpidium sp.*
Tabellaria sp.		*Closterium sp.*	*Sphaerium sp.*		
Bulbochaete sp.		*Asterionella sp.*	*Cloeon sp.*		
Ulothrix sp.		*Melosira sp.*	*Brachionus sp.*		
Cladophora sp.		*Scenedesmus sp.*	*Stylaria sp.*		
Euastrum sp.		*Pediastrum sp.*	*Cyclops sp.*		
Staurastrum sp.		*Tabelloria sp.*			
Batrachospermum sp.		*Spirogyra sp.*			
		Cladophora sp.			

There are a number of indicator species identified as important materials for water quality assessment. These species include phyto- and zooplanktons (Figs. 11.1, 11.2, 11.3, 11.4, 11.5, 11.6).

Usually, for quantitative way of water assessment, species diversity indices of diverse water bodies were computed in recent years. These exercise provide valuable clues for water quality monitoring. The diversity indices are calculated from the abundance data of the organisms and serve as a very good indicator of pollution. Some diversity indices are described:

(a) **The Shannon-Wiener Index:** Species richness may be compared between two communities or areas by simply counting the number of species either in the whole community—if that is practical—or in suitable samples. However, there is at least one more aspect of species diversity that should sometimes be considered—relative abundance or equitability. Imagine two communities each made up of two species. In one community there are 99 individuals of species A and one individual of species B, whereas, in the second community, there are 50 individuals of each species. The second community is more diverse in this sense; almost every individual sample is predictable in the first community but not in the second.

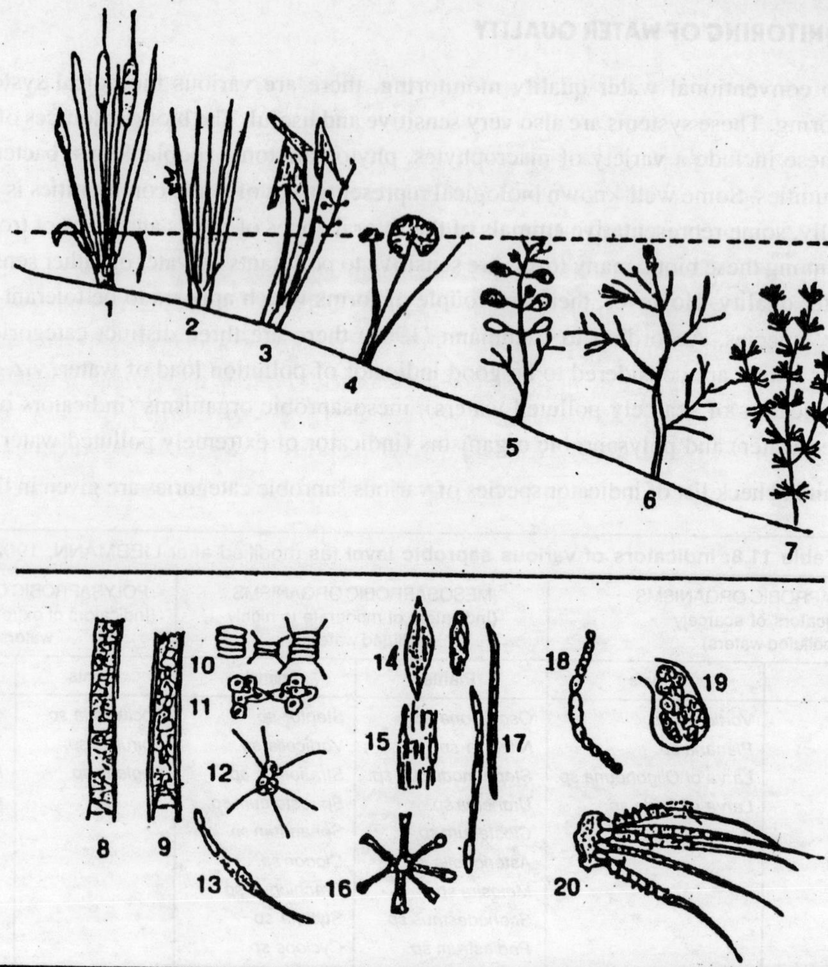

Fig. 11.1 : Some producers of lentic communities, including emergent, floating, and submergent rooted littoral plants (1-7), filamentous algae (8-9), and phytoplankton (10-20)

Organisms diagrammed are : (1) cattail (*Typha*); (2) bulrush (*Scirpus*); (3) arrow need (*Sagittaria*); (4) water lily (*Nymphaea*); (5) and (6), two species of pond weeds (*Potamogeton diversifolia, P. vectinatus*); (7) muskgrass- (*Chara*); (8) *Spirogyra*; (9) *Zygnema*; (10) *Scenedesmus*; (11) *Coelastrum*; (12) *Richteriella*; (13) *Closterium* (a desmid); (14) *Navicula*; (15) *Fragilaria*; (16) *Asterionella* (which floats in the water like a parachute); (17) *Nitzschia*; (18) *Anabaena*; (19) *Microcystis*; (20) *Gloeotrichia* (19 and 20 represent parts of colonies enclosed in a gelatinous matrix)

One widely used measure of diversity that combines species richness with equitability is the Shannon-Wiener index. It is supposed to measure the uncertainty involved in predicting the identity of the next individual. The formula for calculating the index H′ is

$$H' = -\sum_{i=1}^{s} P_i \log P_i$$

where S = number of species and P_i = proportion belonging to the i-th species. The logarithms used can be to any base. The base 2 is used by some because the answer H′ is then in bits; however, most people use base e because it is easy and not so far from 2.

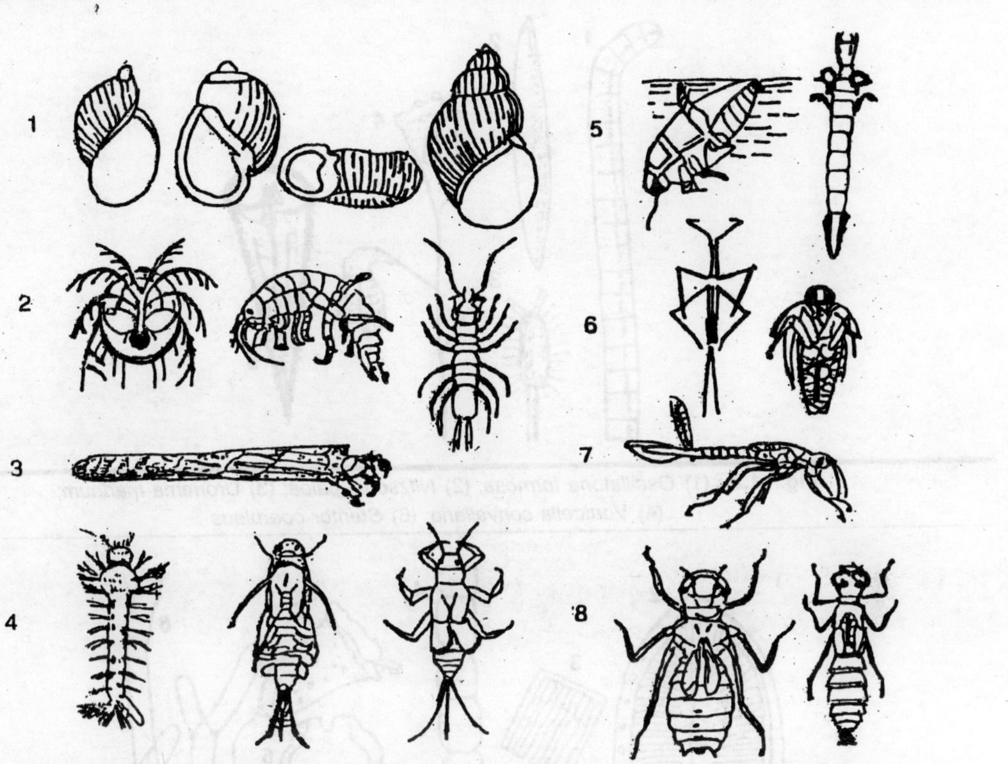

Fig. 11.2 : Some representative animals of the littoral zone of ponds and lakes. Series 1 to 4 are primarily herbivorous forms (primary consumers); series 5 to 8 are predators (secondary consumers). (1) Pond snails (left to right): *Lymnaea* (pseudosuccinea) *columella*; *Physa gyrina*; *Helisoma trivolvis*; *Campeloma decisum*. (2) Small arthropods living on or near the bottom or associated with plants or detritus (left to right): a water mite, or *Hydracarina* (Mideopsis); an amphipod (*Gammarus*); an isopod (*Asellus*). (3) A pond caddis fly larva (*Triaenodes*), with its thin, light portable case. (4) (left to right) A mosquito larva (*Culex pipiens*); a clinging or periphytic mayfly nymph (*Cloeon*); a benthic mayfly nymph (*Caenis*)—note gill covers which protect gills from silt. (5) A predatory diving beetle, *Dytiscus*, adult (right) larva. (6) Two predaceous Hemipterans, a water scorpion, Ranatra (*Nepidae*), and (right) a backswimmer, *Notonecta*, (7) a damsel fly nymph, Lestes (*Odonata-zygoptera*); note three caudal gills. (8) Two dragonfly nymphs (Odonata—*Anisoptera*); Helocordulia, a long-legged sprawling type (*benthos*), and (right) *Aeschna*, a slender climbing type (periphyton)

Calculation of H′ may be illustrated using some fabricated numbers from tropical African Savanna:

Species	No. of individuals	P_i	Log P_i	Pix (log P_i)
Zebra	50	0.556	– 0.5870	– 0.326
Hyena	30	0.333	– 1.0996	– 0.366
Lion	10	0.111	– 2.1982	– 0.244
Total	**90**	**1.00**	—	**– 0.936**

In the formula we take the negative sum of P_i (log P_i) for the practical reason of preferring to deal with positive numbers, H′ for this set of data is then 0.936, (If the index is needed in binary form, the value log P_i may be converted to log 2 P_i by dividing it by 0.6931. For example, log 2 0.556 equals – 0.5870/0, 6931 equals – 0.8469).

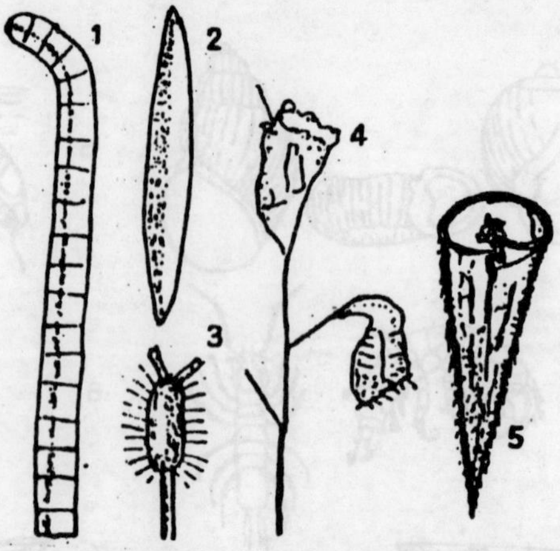

Fig. 11.3 : (1) *Oscillatoria formosa*; (2) *Nitzschia palea*; (3) *Uronema marinum*; (4) *Vorticella convallaria*; (5) *Stentor coeruleus*

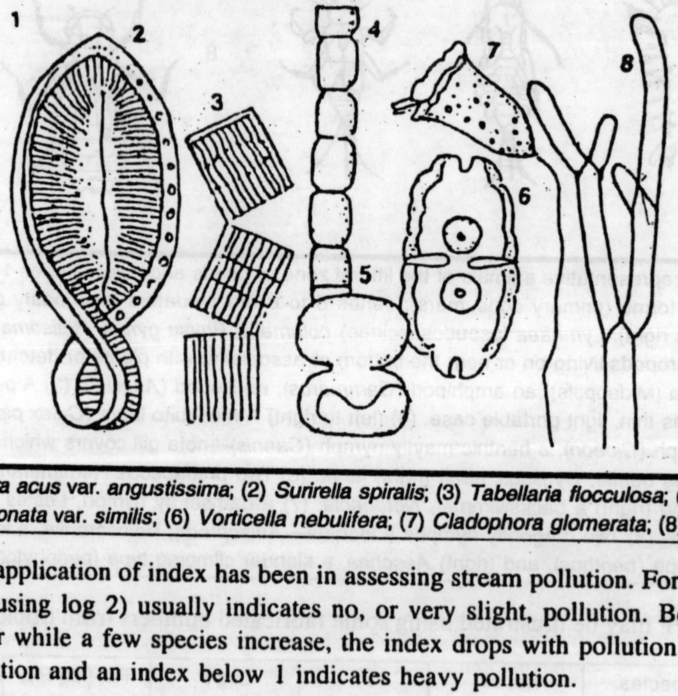

Fig. 11.4 : (1) *Synedra acus* var. *angustissima*; (2) *Surirella spiralis*; (3) *Tabellaria flocculosa*; (4) *Staurastrum punctulatum*; (5) *Ulothrix zonata* var. *similis*; (6) *Vorticella nebulifera*; (7) *Cladophora glomerata*; (8) *Euastrum oblongum*

One practical application of index has been in assessing stream pollution. For the bottom fauna an H′ greater than 3 (using log 2) usually indicates no, or very slight, pollution. Because some species decline or disappear while a few species increase, the index drops with pollution. An index of 2 to 3 indicates light pollution and an index below 1 indicates heavy pollution.

(b) **Kothe's species deficit index (1962)**: This index is based on the principle that in a flowing ecosystem the number of species decreases after they are exposed to some pollutant discharge. In this method the number of species of either a particular group (algae or macro invertebrates) or of all the groups are counted at the polluted and non-polluted points and then index is calculated as:

$$\text{Species deficit index} = \frac{A_1 - A_x}{A_1} \times 100$$

where, A_1 = Number of species at the unpolluted site

A_x = Number of species at the polluted site (downstream)

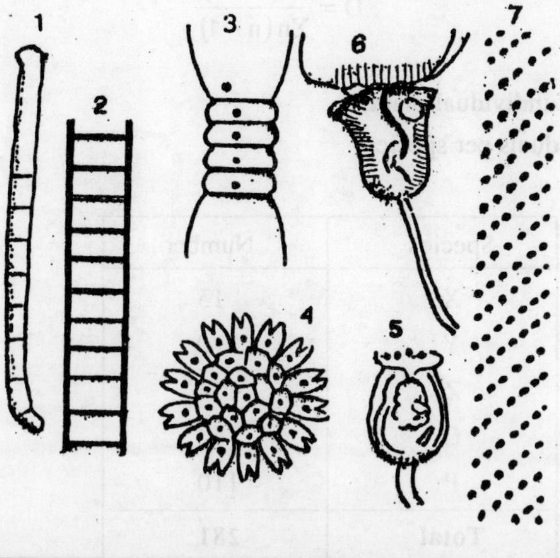

Fig. 11.5 : (1) *Oscillatoria rubescens*; (2) *Melosira varians*; (3) *Scenedesmus quadricauda*; (4) *Pediastrum boryanum*; (5) *Vorticella campanula*; (6) *Spirogyra crassa*; (7) *Brachionus urceus*

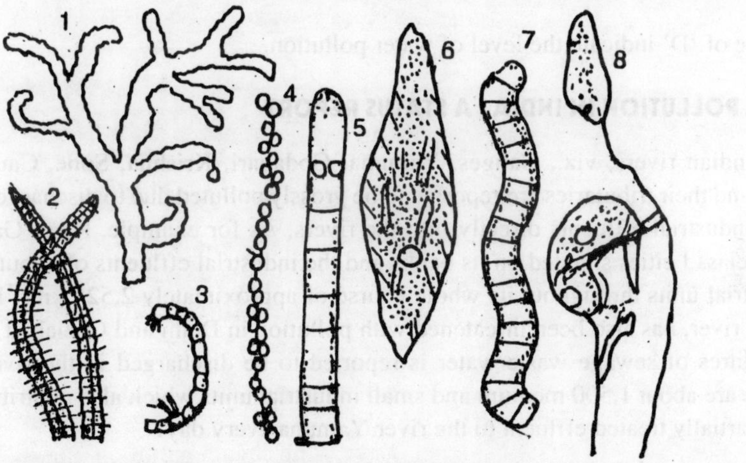

Fig. 11.6 : (1) *Zoogloea ramigera*; (2) *Beggiatoa alba*; (3) *Oscillaloria putrida*; (4) *Spirulina jenneri*; (5) *Euglena viridis*; (6) *Vorticella microstoma*; (7) *Tubifex rivulorum*; (8) *Chironomus thummi*

It gives the data in a percentage linear scale and is very useful in indicating the consequences of point sources of waste water discharges.

(c) Odum's Species index (1971): It is an excellent index to determine the level of pollution in both flowing and standing water bodies. It is calculated as:

$$\text{Index Value} = \frac{\text{Total number of species encountered in the sample}}{\text{Total number of individuals of all the species}} \times 1{,}000$$

(d) Simpson diversity index: It is also an important index, used widely for water quality monitoring. The formula used is:

$$D = \frac{N(N-1)}{\Sigma n(n-1)}$$

where D = Diversity index

N = Total number of individual plants

n = number of individuals per species

Σ = Sum

Species	Number
X	15
Y	10
Z	104
O	42
P	110
Total	**281**

$$D = \frac{281(281-1)}{15(15-1)+10(10-1)\cdots\cdots \text{etc}} = \frac{78,680}{16,552} = 4.8$$

The value of 'D' indicate the level of water pollution.

11.10 RIVER POLLUTION IN INDIA : A STATUS REPORT

Most of the Indian rivers, viz., Ganges, Yamuna, Godabari, Krishna, Sone, Cauvery, Damodar and Brahmaputra and their tributaries are reported to be grossly polluted due to discharge of untreated sewage disposal and industrial effluents directly into the rivers. As for example, River Ganges alone receives sewage of 29 class I cities situated on its banks and the industrial effluents of about 300 small, medium, and big industrial units throughout its whole course of approximately 2,525 kms. Identically, Yamuna, another major river, has also been threatened with pollution in Delhi and Gaziabad area. Approximately 5,15,000 kilolitres of sewage waste water is reported to be discharged in the river Yamuna daily. In addition, there are about 1,500 medium and small industrial units which also contribute huge amounts of untreated or partially treated effluent to the river Yamuna every day!

Similarly, many other rivers were surveyed during past three decades with respect to their pollutional status. In addition to domestic and industrial discharge into the rivers, there were continued surface run-off of agricultural areas, mines and even from cremation on the river banks.

Since 1980s, Govt. of India took up serious steps for checking water pollution of holy river Ganga and other major rivers of India. The major activity is Ganga Action Plan which was adopted over decades to minimise and check the sources of the pollution of the rivers. The creation of Farakka Barrage is one of such significant developments.

11.11 GANGA ACTION PLAN (GAP)

For ages the Ganga has remained the lifeline of India. Water of the river has been used for drinking, farming, washing, navigating etc. so much so that a sentimental bond has grown up with the river. People love and respect the river, and even worship it as divine entity. A large number of religious centres have grown up all along the length of the river viz., Gomukh at the point of origin and Kapil Dham (i.e., Ganga

Sagar) at the final destination. In between there are important places like Hardwar, Varanasi, Prayag in Allahabad, Patna and Kalighat in Kolkata city. The flow of the river extremely reduced over the years due to creation of dam and high rate of water abstraction in upstreams of river. As such, the pollution problem of the river enhanced manifold.

A comprehensive survey of the Ganga Basin was made by the Central Board for Prevention & Control of Water Pollution directorate that river, despite its extraordinary resilience, is heavily polluted at several places. Considering these facts, the Government of India constituted the Central Ganga Authority in February 1985 to evolve a long term programme for restoring the quality of the river.

Ganga flows through primarily three states—Uttar Pradesh, Bihar, and West Bengal. The main sources of pollution are urban and industrial wastes from 29 class I cities (i.e., those having population of over 1 lakh), 23 class II cities (i.e., those having population of over 50,000 & 1 lakh) and about 48 towns (i.e., those having population of less than 50,000). All these towns are located along the 2,525 kms stretch of the river from Gangotri to Gangasagar. It has been examined that bulk of the pollution of the river is caused by municipal wastes, particularly by untreated sewage.

Considering these facts National Ganga Action Plan was formulated with a view to take the following measures :

- Renovation of existing trunk sewers and outfalls to prevent the overflow of sewage into Ganga;
- Construction of interceptors to divert flow of sewers and other liquid wastes into Ganga;
- Renovation of existing sewage pumping stations and sewage treatment plants and installation of new sewage treatment plants to recover the maximum possible resources especially bio-energy to operate the pumping and treatment plants and derive maximum possible revenue;
- Biological conservation measures based on proven techniques; and
- Other sanitation schemes.

A monitoring committee was framed by Central Ganga Authority (CGA), the terms of reference of the said committee was:

- Technical analysis and review of the progress made in the implementation of the action programme approved by the Central Ganga Authority;
- Assessment of shortfalls and gaps in the implementation aspects;
- Additional points/suggestions to improve/expedite the implementation of the programme;
- Establishing a computer data based collection/analysis system;
- Providing technical advice/guidance to steering Committee on any of the issues which may arise as feedback during the implementation of the project; and
- Reporting to the Central Ganga Authority after every three months;

A multimedia programme is also being finalised by the Ministry of Information and Broadcasting to keep the public informed of the progress of various schemes under the Action Plan. Though the river cleaning programme of the Government of India was started in 1985 with the launching of Ganga Action Plan (GAP), yet since 1993 the said programme was extended to other major rivers like Yamuna, Gomti, Krishna, Godabari and so on.

Since 2000, National River Conservation Authority (NRCA) replaces the function of CGA and expanded its activities in 160 towns located alone 31 inter-state rivers in 20 states. About 2,257.06 MLD of sewage is targeted to be intercepted, diverted and treated. With the persistent efforts of NRCA, the water quality with respect to chemical constituents are improved substantially.

Namami Gange Programme

'Namami Gange Programme', is an Integrated Conservation Mission, approved as 'Flagship Programme' by the Union Government in June 2014 with budget outlay of Rs.20,000 Crore to accomplish the twin objectives of effective abatement of pollution, conservation and rejuvenation of National River Ganga.

Main pillars of the Namami Gange Programme are:

1. Sewerage Treatment Infrastructure	2. River-Front Development
3. River-Surface Cleaning	4. Bio-Diversity
5. Afforestation	6. Public Awareness
7. Industrial Effluent Monitoring	8. Ganga Gram

Its implementation has been divided into Entry-Level Activities (for immediate visible impact), Medium-Term Activities (to be implemented within 5 years of time frame) and Long-Term Activities (to be implemented within 10 years).

The key achievements under Namami Gange programme are:

1. **Creating Sewerage Treatment Capacity**: 63 sewerage management projects under implementation in the States of Uttarakhand, Uttar Pradesh, Bihar, Jharkhand and West Bengal. 12 new sewerage management Projects Launched in these states. Work is under construction for creating Sewerage capacity of 1187.33 (MLD). Hybrid Annuity PPP Model based two projects has been initiated for Jagjeetpur, Haridwar and Ramanna, Varanasi.
2. **Creating River-Front Development**: 28 River-Front Development projects and 33 Entry level Projects for construction, modernization and renovation of 182 Ghats and 118 crematoria has been initiated.
3. **River Surface Cleaning**: River Surface cleaning for collection of floating solid waste from the surface of the Ghats and River and its disposal are afoot and pushed into service at 11 locations.
4. **Bio-Diversity Conservation**: Several Bio-Diversity conservation projects are namely: Biodiversity Conservation and Ganga Rejuvenation, Fish and Fishery Conservation in Ganga River, Ganges River Dolphin Conservation Education Programme has been initiated. 5 Bio-Diversity center's at Dehradun, Narora, Allahabad, Varanasi and Barrackpore has been developed for restoration of identified priority species.
5. **Afforestation**: Forestry interventions for Ganga through Wildlife Institute of India; Central Inland Fisheries Research Institute and Centre for Environment Education has been initiated. Forestry interventions for Ganga has been executed as per the Detailed Project Report prepared by Forest Research Institute, Dehradun for a period of 5 years (2016-2021) at project cost of Rs. 2300 Crores. Work has been commenced in 7 districts of Uttarakahnd for medicinal plants.
6. **Public Awareness**: A series of activities such as events, workshops, seminars and conferences and numerous IEC activities were organized to make a strong pitch for public outreach and community participation in the programme. Various awareness activities through rallies, campaigns, exhibitions, shram daan, cleanliness drives, competitions, plantation drives and development and distribution of resource materials were organized and for wider publicity the mass mediums such as TV/Radio, print media advertisements, advertorials, featured articles and advertorials were published. Gange Theme song was released widely and played on digital media to enhance the visibility of the programme. NMCG ensured presence at Social Media platforms like Facebook, Twitter, You Tube etc.
7. **Industrial Effluent Monitoring**: Real Time Effluent Monitoring Stations (EMS) has been installed in 572 out of 760 Grossly Polluting Industries (GPIs). Closure notice have been issued to 135 GPIs so far and others have been given deadlines for compliance to stipulated norms and for installations of online EMS.

8. **Ganga Gram:** Ministry of Drinking Water and Sanitation (MoDWS) identified 1674 Gram Panchayats situated on the bank of River Ganga in 5 State (Uttarakhand, Uttar Pradesh, Bihar, Jharkhand, West Bengal). Rs. 578 Crores has been released to Ministry of Drinking Water and Sanitation (MoDWS) for construction of toilets in 1674 Gram Panchayats of 5 Ganga Basin States. Out of the targeted 15, 27,105 units, MoDWS has completed construction of 8, 53,397 toilets. Consortium of 7 IITs has been engaged in the preparation of Ganga River basin Plan and 65 villages has been adopted by 13 IITs to develop as model villages. UNDP has been engaged as the executing agency for rural sanitation programme and to develop Jharkhand as a model State at an estimated cost of Rs. 127 Crore.

INDIA, BANGLADESH SIGN 30-YEAR PACT (SINCE 1996) ON GANGA WATER : WATER SHARING TREATY

India to release water at Farakka during the 5-month lean period from Jan 1 to May 31 every year.

Bangladesh to get a minimum of 35,000 cusecs or 50% of water available of it's 70,000 cusecs or less.

India and Bangladesh to receive a guaranteed 35,000 cusecs in three alternate 10-day periods from March 1 to May 10 every year.

Bangladesh to get a maximum of 67,516 cusecs during Jan 1-10 and a minimum of 27,633 cusecs during April 11-20 every year.

India to get a maximum of 40,000 cusecs in seven 10 day periods during Jan, Feb and May 21-31 every year. It's minimum share will be 25,992 cusecs during April 21-30 every year.

If the flow at Farakka falls below 50,000 cusecs during any 10-day period, the two governments will discuss to make adjustments.

11.12 POLLUTION OF INDIAN LAKES

Among the surface water bodies, lakes and reservoirs are considered to be most valuable water resources. These surface water bodies are currently under serious pollution threat not only in India but also globally considered to be an important issue. Over past couple of decades national and international programmes on lake water quality assessment and their management in the perspective of conservation were attempted. Considerable studies were made in Himalayan lakes of J & K (Dal & Nagin Lake), Kumaun (Nainital and Bhimtal Lake), Madhya Pradesh (Bhopal Lake), Odisha (Chilika Lake), Manipore (Loktak Lake) and so on. In almost all cases lake water is highly polluted with silts, agrochemicals, fertilizer, organic wastes and industrial discharges etc. As a result, considerable biotic changes along with loss of productivity is noticed. In most of limnological studies of the lake system thus revealed the fact that most of the lakes slowly transformed into swampy marsh without any proper management for conservation. Since 1990s there is a national lake conservation policy adopted by the Ministry of Environment and Forest, Govt. of India.

The National Lake Conservation Plan (NLCP) was initiated by the Ministry of Environment and Forests in June 2001 with objectives to maintain the ecological health of lakes. Initially three lakes are taken as a model for this programme. They are: Powai lake (Maharashtra), Ooty and Kodaikanal lake (Tamil Nadu). Now about 49 lakes of 13 states have been considered for this programme.

The activities covered under NLCP include:

- Prevention of pollution from point sources by intercepting, diverting and treating the pollution loads entering the lake;
- In situ measures of lake cleaning such as desalting, deweeding, bioremediation, constructed wetland approach etc. depending upon the site conditions;
- Catchment area treatment and lake front eco-development which may include bunding, fencing, shoreline development, creation of facilities for public recreation and entertainment and public area;
- Public awareness and public participation;
- Other activities depend upon location-specific conditions including the interface with human population.

LAKE CHILIKA

Chilika is the largest brackish water lake in India. Spread over more the 1,000 sq.km, the lake is rapidly degrading. Every year, 113 million tonnes of silt are carried into the lake clogging all but two of the numerous channels connecting it to the Bay of Bengal. Of these two, there is water in the Magardha Muhana only during the monsoon. Decreased natural clearance of silt coupled with increasing silt inflow is filling up the lake.

The conservation problem in Chilika is fraught with conflicts. Chilika is the leading centre in Odisha for fish, prawn and crab hauls. For centuries, the lake was managed by the traditional fishing communities through their system of "caste panchayats." The panchayat imposed controls on various aspects of fishing. Numerous rules regarding the seasonal catch areas and community-related fishing equipment ensured that the lake was managed like a common property. There were festival and religion-linked "no fishing days" and also restricted fishing seasons. These provided safe breeding seasons when the lake could restock itself with fish and prawns.

Changing the rules

The entry of commercial interests into fishing changed all this. The new entrants in the fishing trade neither belonged to the traditional fishing communities nor obeyed their rules and regulations for managing the lake. For instance, the "no-fishing days" observed by fisherfolk during certain festivals were not respected and the highly-evolved system of dividing seasons into "catch" seasons and "growth" seasons was also ignored. Traditional methods of fishing and the various cultural factors connected with it were not respected by this community of non-traditional fishermen.

Until recently, fishing was considered a low-caste profession. So it was only communities belonging to the so-called lower castes which took up fishing as their major source of livelihood. Traditional fishing communities could fish throughout the year by paying a licence fee. People outside this community fished in the lake mostly for their household consumption.

Also, in the past, the commercial fishing trade was restricted by the fact that only the traditional fisherfolk knew the art of weaving nets. In the past few decades, however, readymade nylon nets have become available in finer mesh sizes than those woven by the fisherfolk. As a result, even fingerlings and immature prawn are trapped—thereby contributing to lower yields.

The commercialisation of the fish business has been the bane of fishermen and fish-eaters alike. A conspicuous impact of this is evident in the decline of the dried fish business and the high prices of dried fish. Dried fish is an important source of animal protein for the poor. Matsyajivi Mahasangha, a fishermen's cooperative, has been fighting to restore to the fishermen their rights over the lake. But the Government has not been sensitive to its demands.

In the past few decades, fishing has become modernised and dependent on heavy external inputs. Large areas of the lake have been encroached upon by local mafia, businessmen, top administrators and political leaders of the State. The preference for a large catch and bigger prawn farms has led to the sidelining of the interests and aspirations of the original fishermen. Despite the Supreme Court directive ordering the demolition of prawn farm embankments, it is yet to be implemented in most places. Elsewhere, only cosmetic changes have been made.

The mushrooming of prawn farms has led to various environmental problems. For the fishermen, a major problem is the fall in the yield from traditional fishing methods. These fishermen do not give in to the growing demand for prawn juveniles. However, illegal suppliers catch the juveniles at the mouth of the lake and sell them. This has led to a sharp fall in the quantum of juveniles growing inside the lake and consequently a drop in the availability of prawns. The prawn farms have thus increased their profits only at the cost of the traditional fishing community.

In the past, hereditary fishermen received preference over other people in the allocation of fishing areas (sairats). But now the government has taken away from these fishermen much of the area that was vested with them and leased it out to people who are not fisherpersons by tradition. The fishermen are losing their rights as a result of the systematic effort of the state to assert its sovereign rights over the lake.

One of the lake's major problems is the decrease in its depth. The rivers that empty into the lake snake through kilometers of hilly terrain that surround it. Heavy deforestation has resulted in greater erosion, thereby increasing the particulate matter in the river water. Due to the clogging of channels connecting the lake to the sea, silt is not carried in adequate quantities to the sea.

Contd.

This fall in depth has caused a decline in the availability of deeper shelter zones as breeding grounds. The overexploitation and loss of breeding grounds have together affected the fish population in the lake. Siltation and shallowing have also contributed to the growth of weeds.

The decreasing interaction with the sea has also reduced the salinity of the water. This has contributed to weed infestation and decreased suitability of the water for some varieties of fish. Consequently, the catch has gone down over time. The brackish water fish production in the State has declined by 55.8 per cent between 1988-89 and 1994-95 despite the increasing production from prawn aquaculture. This is evident from the fact that fish landings by traditional fishing have declined by more than 60 per cent.

Although the prices of prawns and fish have increased, this does not compensate for the loss in volume. The maximum available size and the average size of the fish have declined indicating a fall in the fish population. About 45 per cent of fish species have become extinct in Chilika. The report of the Zoological Survey of India revealed that at present, only 69 species of fish are found in the lake as against the 126 mentioned in the 1920 report (1995).

The fish and crab haul from the lagoon has consistently dropped from 8,872 tonnes in 1986-87 to a mere 780 tonnes in 1996-97. Among the threatened species are dolphin, khainga (Mugil Cephalus), sahala (Eleutheronema Tetradactylum) and crabs.

Pesticides used in agricultural fields situated in the coastal areas often get flushed into the Chilika lake. The chemicals react when they come into contact with salt water in the lake. This results in the loss of oxygen in the water, affecting the biotic life. Besides, a caustic alkali industry discharges its effluents directly into the lake causing problems of mercury accumulation in the lake ecosystem. Recommendations of a workshop on the Chilika lake, of which the Odisha Government was one of the organisers, mention the mercury accumulation problem in the lake. The 1996 report of the Zoological Survey of India also warns of the presence of toxic heavy metals like copper, lead, zinc and nickel in the lake.

The fishing community is the most affected by the deterioration in Chilika's environment dependent as it is on the lake for its survival. The fishermen have been thinking seriously about ways to stem the slide. Their suggestions regarding environmental conservation and sustainable utilisation of the lake are practical and pertinent.

The major areas of concern are how to protect this resource from siltation, weeds, pollution, unregulated prawn culture and increasing fishing intensity. Intensive fishing without an increase in the stock of fish will soon lead to the extinction of key species in the lake.

Action Needed

In terms of restrictions needed to conserve the resource, people feel that one day in the week should be observed as a non-fishing day, catching small fries and prawns below two inch size and fishing during the breeding season should be prohibited and the minimum mesh of nets should not be less than 1.5 inches. They stressed that motor boats should be used only for transportation and not for fishing. Unauthorised prawn embankments should be demolished and prawn culture restricted. The large number of bunds/enclosures which obstruct capture fisheries and the free circulation of water in the lake need to be demolished. Fishing in channels connecting the lake to the sea impedes the free movement of water and fish. Therefore, the channels need to be closed. People also demand the introduction of biological control agents like carps which are vigorous eaters of aquatic vegetation and can help reduce the growth of weeds.

Soil conservation measures in the catchment areas of the lake and tributary rivers were also recommended. The fishermen felt the need for breeding hatcheries to be set up for stocking prawns and fish in the lake through artificial release and also for supplying prawn culture farms. Strict measures to prevent unauthorised bird poaching were suggested. It was felt that gun licenses in the area should be kept to the minimum.

There was also a demand that tourism which suited the lake's resources be promoted and that the local youths should be trained as guides. The rights of traditional fishermen over the lake must be restored to bring back a sense of belonging and greater local control over the lake and to prevent overfishing and excessive prawn culture.

Proper demarcation of fishing sairats (zones) of fishermen and non-fishermen communities is a must to prevent the "might is right" regime that goes against the customary right of fishing communities. In this context

Contd.

it would be necessary to restore the rights and the role of the Matsyajivi Mahasangha and revitalise the cooperative societies to ameliorate the conditions of the fishermen.

Chilika is the sole life support system for its fishing communities. In thousands of homes the hearth burns only if the net catches something. This subsistence livelihood is under threat of being wiped out by the incursion of commercial interests into fishing in the area. The very survival of the lake is threatened. The need today is not only to save the lake but to save its fishing communities.

11.13 GROUND WATER POLLUTION: NITRATE, ARSENIC AND FLUORIDE

In general, ground water is mostly chemically and microbiologically nonpolluted and thus safe for drinking and cooking in addition to agricultural or industrial uses. But in recent decades there were enormous amount of reports of ground water contamination of either arsenic, fluoride or nitrate. These contamination causes serious health hazards in the long run.

It is well-known that ground water contamination with toxic metals principally takes place due to leaching from toxic waste dumps or from crustal layer of earth through biotransformation. Agricultural chemicals also leached into upper aquifers and thus contaminates the ground water.

Nitrate in Ground water: Nitrate in ground water is primarily derived from mineralisation of soil organic matter or from use of excessive nitrogen fertilizers. The nitrate content in supply water ranges from less than 1 mg NO_3^- per lit to around 50 mg NO_3 per lit. The limit of 50 mg NO_3 per lit was originally set to protect babies against *methanoglobinaemia*. This is a condition where more than about 10% of the haemoglobin in the blood is converted into the methanoglobin form. It is a reversible condition, but death can occur only if more than 40% of the haemoglobin in the blood is converted because of reduced oxygen carrying capability. The condition methaenoglobinaemia occurs in young infants with high nitrate level in well water used for making milk formulae.

It has also been suggested that a high nitrate can reduce the body assimilation of iodine, thus causing goitre. It has also been suggested that an increased nitrate intake through drinking water carries an increased risk for bearing a malformed child in pregnant women.

Arsenic in Ground water: Ground water arsenic contamination is known over the years in various parts of the world. However, the major incidence noted in Indian subcontinent is in West Bengal and Bangladesh region, where a vast tract is under the arsenic calamity. Several thousands of people are suffering from chronic arsenicosis through drinking of contaminated ground water.

The problem relating to arsenic contamination in ground water of West Bengal was first noticed in 1978 through a short study made by School of Tropical Medicine, Kolkata. Then, with the initiative of Public Health Engineering Dept., Govt. of West Bengal, an elaborate study was made during 1988-91 for making detailed report on the status of arsenic contamination of ground water. All Indian Institute of Health & Hygiene, Central Ground Water Board, Centre for Man and Environment, State Water Investigation Directorate, and School of Environmental Sciences (Jadavpur University) participated in the detailed investigation on the said aspect. The report indicates the fact that along a linear NNW-SSE tract extending from Kaliachak in Malda through Raninagar in Murshidabad, parts of Nadia and North 24 Parganas up to Baruipur in South 24 Parganas, the ground water between 10-80 m below ground level has been found to contain arsenic, at places above permissible level of 0.05 mg/liter. The area forms a part of the Ganga- Brahmaputra delta having a near surface succession of quaternary sediments of varying thickness. The arseniferrous belt lies entirely within the upper delta plain (UDP) characterised by a series of meander belts formed by rivers.

Arsenic content of ground water tapped at some depth within the same zone varies between wide limits. However, the maximum arsenic content occurs in ground water in the 20-60 m range, ranging from < 0.01 mg/lit up to a maximum of 2.0 mg/lit. There was no water samples (collected from ponds or dug-wells) containing arsenic above 0.01 mg/lit.

It has been established beyond doubt that water of the intermediate aquifer is polluted with arsenic. Neither the shallow first nor the deep (third) aquifer contain arsenic above the permissible limit. The ground water in arseniferous areas is characterised by high iron, arsenic, Ca, Mg and bicarbonate with low chloride, sulphate and fluoride. The pH is about 7 to 8, which is very ideal for leaching of arsenic.

Further, it is speculated from the lithological and minerological analysis that the sources of the arsenic in the groundwater is primarily in the clayey sediments immediately above and intercalated within the intermediate aquifer. These sediments were perhaps transported from the Chhotanagpur-Rajmahal highlands and deposited in sluggish meandering streams under reducing conditions. Under such conditions arsenic got deposited partly as arsenious minerals, partly absorbed on surfaces of minerals or clay particles and the rest as organic complexes.

Arsenic content in ground water as measured in some areas of West Bengal is given in Table 11.8:

Table 11.8: Ground water Arsenic content, 2000

Locations	Arsenic content in Ground Water (mg/lit)
1. Murshidabad District, Raninagar Block II, Mouza Katlamari	0.06 – 0.16
2. Nadia District, Nabadwip Block, Mouza Swarupganj	0.02 – 0.1
3. North 24-Parganas District, Habra Block II, Mouza Ashoknagar	0.38 – 0.67
Source: School Environmental Studies, Jadavpore University, Kolkata	

In many areas of West Bengal, the ground water—of shallow depth in particular—is contaminated with higher quantity of arsenic. It is now felt urgent to identify those sources and prevent the inhabitants not to consume such water. In contrary, deeper tubewell or dug well does not contain appreciable amount of arsenic. So it is wise to provide them either some deeper tubewell or dug well or alternative source of treated surface water for drinking.

As the dissolved inorganic arsenical substance can make complex with various alum, it can easily be precipitated and then filtered for removal of contaminated arsenic. Many trials has already reached a great deal of success. Yet much is needed to develop for promotion of efficacy at grass-root level. Chemical investigation with respect to routine urine and blood test for arsenic determination or its deposition in hairs or nails among the affected population is still quite inadequate.

Some arsenic removal technology currently adopted in arsenic affected area:

Table 11.9: Arsenic: Technology interventions

Technology	Advantages	Disadvantages
Co-precipitation process where aluminum and ferric salts are added to precipitate arsenic in raw water. These precipitates after setting down are removed.	It's a low cost venture, where the chemicals are easily available.	The disposal of sludge is a problem.
In adsorption process, chemicals such as activated alumina, iron coated sand; ion exchange resins are used to adsorb arsenic.	It is a highly efficient technology.	It needs a huge dosage of chemicals. The disposal of sludge is a problem.

Contd.

Table 11.9: Arsenic: Technology interventions

Technology	Advantages	Disadvantages
Dugwells as a source of drinking water is being tried in villages like Karnarpara. Field and lab studies indicated correlation between the iron content of dug well water and the lower levels of arsenic in the water. It was surmised that dug wells, with their higher exposure to air, allow iron and arsenic to co-precipitate out of the water.	It is a low cost technology	Risk of biological contamination of water; the well needs to be protected.
Four pitchers–The contaminated water in the first pitcher percolates down through second pitcher consisting of charcoal and brick chips, and the third pitcher filled with sand after which the last pitcher gets filled with arsenic-removed water. The system is based on the property that arsenic is adsorbed to iron by oxidisation.	The system does not require any chemical	Effective results cannot be expected unless groundwater is iron-rich.
In Two Bucket System tubewell water is poured into a red (top) bucket, potassium permanganate (oxidising agent) and aluminum sulfate (coagulant) are added to remove arsenic from the water. The tubewell water is then filtrated through the sand in a green (bottom) bucket.	It is an efficient process	Chemical process is complex.

Further, the status of arsenic in crops and vegetable grown under shallow irrigation was not at all assessed in depth. There were enormous scope for studying of arsenic in the ecosystem as a whole in near future.

Fluoride contamination in Ground Water: Like arsenic, fluoride contamination was also reported since mid 1970s, in some districts of West Bengal and Assam. Earlier there were reports of surface water (dug well) contamination of fluoride in Andhra Pradesh. Due to such contamination, exposed community suffers very much in both short-term and long term basis. There were reports of bony deformity and associated problems of locomotory organs. The children are more susceptible than the adults. There is a great need for identification of contaminated areas and also need for supply of clean non-contaminated water for cooking and drinking.

Government efforts to deal with the problem of fluoride contamination have focused primarily on providing safe drinking water through various schemes and the distribution of defluoridation plants at household and community levels. However, the main constraint to the success of these schemes has been lack of maintenance of these plants. The Rajib Gandhi National Drinking Water Mission set up 427 defluoridation plants across the country, many of which are not in operation today. DANIDA initiated a rural water supply and sanitation project scheme in Karnataka's four districts that include defluoridation.

Central Government also proposes to provide 50,000 household water filters to families and groups of people at subsidised rates of Rs. 1,500-2,500 each.

An interesting scheme is being implemented in Punjab, where Uruguay gifted a modular water-treatment plant, the first of its kind in India, to the village of Talwandi Sabo. It was successfully implemented and now the Punjab Government has signed a MoU with the Uruguay Government for installing 20 such plants at a cost of Rs. 20 lakh each. The compact plant requires little space, and purifies water with a mix of coagulation, sedimentation and filtration. According to the Ministry of Rural Development, fluoride mitigation centres are to be set up across the country. Currently in the first phase, centres are being proposed in three states of Andhra Pradesh, Rajasthan and Gujarat to increase the availability of potable water in villages.

Some technology intervention for removal of fluoride is given:

Table 11.10: Fluoride technology interventions

Technology	Advantages	Disadvantages
Nalgonda technology innovated in 1975 by Nagpur based NEERI in Nalgonda district of Andhra Pradesh. In this process, alum and lime is added to raw water and stirred for 10 minutes. Alum adsorbs fluoride and flocs are formed. These flocs are allowed to settle down for 90 minutes after which water is filtered.	The technology is a low cost method. It can be used at household and community levels.	Need of technical expertise, as alum dose needs calculations. It is a time consuming process. There is the problem of disposal of flocs as there is the risk of it leaching back to the ground water aqifer.
Activated alumina system developed by IIT, Kanpur, is also known as Matryi or Prasanti technology. It requires the user to pour water through a filter that consists of a container with a solution of caustic soda and hydrosulphuric acid. It has two chambers of stainless steel or plastic, each of 12 litres capacity. The upper chamber contains 4 litres of activated alumina, a baffle plate and a microfilter. Water poured in this chamber is removed of excess fluoride as it is adsorbed by the alumina and percolates down in the lower chamber. the activated alumina needs to be regenerated after every four months.	The units are easy to handle and maintain. The filters can be attached to handpumps or tubewells. Compared to the Nalgonda filter, sludge formation is less.	The cost is high. Again regular regeneration of activated alumina is cumbersome.
Reverse osmosis uses membrane technology allows only complete defluoridated water to pass.	No chemicals are added making it the most effective.	The problem with RO system is the cost that is high and regular maintenance is necessary.
Resin method, where a resin media absorbs the fluoride	The rate of fluoride removal is high.	Large amount of salt is required for resin regeneration. Resins need regular replacing.

Iron in ground water

In most places in India, iron is also a major contaminant of ground water, so also drinking water too. There is also need for iron removal.

Control measures include providing alternate sources of water free from iron or treating iron contaminated water to within permissible limit of 1 mg/l with the help of iron removal plants. For treating iron contaminated water, the Rajib Gandhi Drinking Water Mission has approved the installation of 16,415 plants, out of which 9,355 plants have already been commissioned.

Iron removal from water supply can involve complex choices. The selection of an appropriate iron removal equipment would depend on the type and quantity of iron in the water, the characteristics of the water supply, other water treatment equipment in use, and the user's requirements for cost, convenience of usage, and maintenance.

11.14 EFFECT OF DISCHARGE OF SEWAGE AND INDUSTRIAL EFFLUENTS ON ENVIRONMENT

Discharge of untreated sewage and industrial effluents leads to number of conspicuous effects on various environmental components (Table 11.11). The impact involves gross changes in water quality including a number of parameters viz., dissolved oxygen, ammonia, water turbidity, and blackening of substratum.

Such kinds of water quality change have both short and long term consequence on aquatic ecosystem. In general, aeration, interaction with sunlight and microbial degradation leads to faster transformation of pollutants.

Table 11.11: Environmental implications of the discharge of sewage and industrial effluents

Factor	Principal environmental effect	Potential ecological consequences	Probable severity	Remedial or ameliorative action	Comments
Degradable organic matter					
1. High biochemical oxygen demand (BOD) caused by bacterial breakdown of organic matter	Reduction in dissolved oxygen concentration	Elimination of sensitive species, increase in some tolerant species; change in community structure	Dependent upon degree of deoxygenation, often very severe	Pretreatment of effluent; ensure adequate dilution	BOD can be reduced by adequate treatment
2. Partial biodegradation of proteins and other nitrogenous material	Elevated ammonia concentrations; increased nitrite and nitrate levels	Elimination of intolerant species, reduction in sensitive species	Variable, locally severe	Improved treatment to ensure complete nitrification; nutrient stripping possible but expensive	Adequate treatment is best solution
3. Release of suspended solid matter	Increased turbidity and reduction of light penetration	Reduced photosynthesis of submerged plants; abrasion of gills or interference with normal feeding behaviour (see inert solids below)	Moderate, usually local	Provide improved settlement, ensure adequate dilution	
4. Deposition of organic sludges in slower water	Release of methane and hydrogen as sulphide matter decomposes anoxically; Modification of substratum by blanket of sludge	Elimination of normal benthic community; loss of interstitial species; increase in species able to exploit increased food source	Variable, may be severe	Discharge where velocity adequate to prevent deposition	Tends to be local
Other poisons					
1. Presence of poisonous substances	Change in water quality	Water directly and acutely toxic to some organisms, causing change in community composition; consequential effects on prey-predator relations; sub-lethal effects on some species (changes in behaviour, etc.)	Highly variable, depending upon substance and its concentration	Increase dilution	Difficult to generalise

Contd.

Table 11.11: Environmental implications of the discharge of sewage and industrial effluents

Factor	Principal environmental effect	Potential ecological consequences	Probable severity	Remedial or ameliorative action	Comments
Inert solids					
1. Particles in suspension	Increased turbidity. Possibly increased abrasion	Reduced photosynthesis of submerged plants. Impaired feeding ability through reduced vision or interference with collecting mechanisms of filter feeders (e.g. reduction in nutritive value of collected material). Possible abrasion	Variable, often moderate	Improve settlement	Inert solids change the character of the substrate and are unstable. They provide no additional nutrition
2. Deposition of material	Blanketing of substratum, filling of interstices and/or substrate instability	Change in benthic community, reduction in diversity (increased number of a few species)	Variable, often severe	Discharge where velocity adequate to ensure dispersion	

11.15 EFFECT OF WATER POLLUTION ON HUMAN HEALTH

On a worldwide scale, the pollution of water supplies is probably responsible for more human illness than any other environmental influence. The diseases so transmitted are chiefly due to microorganisms and parasites. Among the various water-borne diseases, the notable examples are 'cholera' and 'schistosomiases'. Cholera, an illness caused by ingestion of the bacterium *Vibrio cholerae*, is characterised by intense diarrhea, which results rapidly in massive fluid depletion and death in very large percentage of untreated patients. During the twentieth century it had been largely restricted in Asia, particularly the area of the Ganga basin in India. Thousands of Indians continue to die from it each year. Identically, in Africa, Latin America and some parts of Asia, worms infection disease 'schistosomiasis' is affecting more than 100 million people. Till today it is very difficult to control.

The major water borne microbial and parasitic diseases are given in Table 11.12.

(a) **Water Borne Pathogens**: A wide variety of human pathogens may be found in the excrement from humans as well as from other animals. Most human pathogens can be classified as either viruses, protozoans, helminths or bacteria. Both raw sanitary sewage and land runoff contain pathogenic organisms. Approximately 20% of the reported outbreaks and illness associated with waterborne diseases are attributed to bacterial pathogens belonging to the genera *Salmonella* and *Shigella*. Cholera is another serious and acute intestinal disease which is caused by the bacterium *Vibrio cholerae*. It occurs sporadically as epidemic diseases.

Numerous protozoan species normally inhabit the intestinal tract of warm blooded animals, including humans. *Entameba* and *Giardia* are the most common protozoan infective agents.

A good number of enteric viruses were found in water which happens to infect various organs of human being. These are namely poliovirus, Hepatitis A, Reovirus, Adenovirus and Rotavirus, etc. A variety of helminth diseases were found to occur as human beings—whose cysts were found to be present in the waste water—contaminated water bodies. Among the helminths, *Ascariasis* and *Taenia* infections were most prevalent. A summarised account of water related infections is given in Table 11.12.

Table 11.12: Classification of water-related infections

Category	Infection	Pathogenic agent
Fecal-oral (Waterborne or water-washed):	*Diarrheas and dysenteries*	
	Amebiases	P
	Campylobacter enteritis	B
	Cholera	B
	E. Coli diarrhea	B
	Giardiasis	P
	Rotavirus diarrhea	V
	Salmonellosis	B
	Shigellosis (bacillary dysentery)	B
Enteric:	*Enteric fevers*	
	Typhoid	B
	Paratyphoid	B
	Poliomyelitis	V
	Ascariasis (giant roundworm)	H
	Trichuriasis (whipworm)	H
	Tacnia solium taeniasis (pork tapeworm)	H
Water-washed:		
Skin and eye	Infectious skin diseases	M
	Infectious eye diseases	M
	Louse-borne typhus	R
Other Water-based:	Louse-borne relapsing fever	S
Penetrating skin	Schistosomiasis	H
Ingested	Dracunculiasis (guinea worm)	H
	Clonochiasis	H
	Others	H
Water-related insect vector:		
	Trypanosomiasis (sleeping sickness)	P
Biting near water	Filariasis	H
Breeding in water	Malaria	P
	Onchocerciasis (river blindness)	H
	Mosquito-borne viruses	
	Yellow fever	V
	Dengue	V
	Others	V

B, bacterium; P, protozoan; S, spirochaete; M, miscellaneous; H, helminth; R, rickettsia; V, virus.
(Adapted from Feachem 1984 and Nash 1993)

Descriptive accounts of some of water-borne diseases are given in Table 11.13:

Table 11.13: Water-borne diseases

Disease	Type of Disease	Transmission/Causes	Symptoms	Approximate Cases/Year	Prevention
Malaria	Viral Infection	Caused by the single-celled parasite *Plasmodium* and transmitted through the bite of female *Anopheles* mosquito. These mosquitoes breed in stagnant water bodies. In the human body, the malaria parasite undergoes various cycles and finally reaches a form when it transfers itself back to the mosquito when it bites the infected person. The parasite matures within 10 to 14 days and reaches sexual stage when it can infect human host again.	Attacks immune system, liver and blood cells. Malaria symptoms start appearing 9 to 14 days after an infectious mosquito bite. This effect happens in phases, each fever phase characterised by sudden and violent chills, high temperature and profuse sweating, lowering down the temperature. Initial infection leaves the patient exhausted, while over a period of time it can lead to anaemia. Some type of malaria cause severe headaches, convulsions and delirium. It can develop into cerebral malaria and is fatal. Some cause fever every 48 or 72 hours. Some parasites persist for three to five years and can cause frequent relapses.	More than 2 million	Maintaining cleanliness not allowing stagnant water bodies to form
Dengue	Viral Infection	Through the bites of infective female *Aedes* mosquitoes, which acquire the virus while feeding on the blood of an infected person. After 8-10 days, an infected mosquito is capable of transmitting the virus back to humans for the rest of its life. *Aedes aegypti* breeds primarily in man-made containers like jars, metal drums and concrete cisterns used for domestic water storage and items that collect rainwater.	Severe fever. Infants and young children may have a non-specific febrile illness with rash. Older children and adults may have sudden high fever, severe headache, pain behind the eyes, muscle and joint pains, and rash. Dengue haemorrhagic fever is a potentially deadly complication that is characterised by high fever, haemorrhagic phenomena often with enlargement of the liver and, in severe cases, circulatory failure.	More than 1,000. In 2002, the figure reached above 3,000	Combating mosquitoes by preventing accumulation of stagnant water maintaining cleanliness
Hepatitis A & E	Liver inflammation caused by virus	Hepatitis A and E viruses, though unrelated, are transmitted via water contaminated with excreta. Inadequate water supplies, poor sanitation and hygiene can also cause infection and inflammation of the liver. Hepatitis A could also be transmitted via food contaminated by infected food-handlers.	The illness starts with an abrupt onset of fever, body weakness, loss of appetite, nausea and abdominal discomfort, followed by jaundice within a few days. The disease may range from mild (lasting 1-2 weeks) to severe disabling disease (lasting several months). In areas highly endemic for hepatitis A, most infections occur during early childhood. The majority of cases may not show any symptoms; fatal cases due to acute hepatitis are rare.		Safe drinking water and better sanitation and hygiene. Vaccine is available for Hepatitis A
Typhoid & Paratyphoid fever	Bacterial infection of intestinal tract and blood-stream	Typhoid and parathyphoid fevers are caused by the bacteria *Salmonella typhi* and *Salmonella paratyphi* respectively. These germs are passed in the faeces and urine of infected people. People become infected after eating food or drinking water that have been handled by a person who is infected. Once the bacteria enter human body, they multiply and spread from the intestines, into the blood-stream.	Mild or sever and sustained fever as high as 39°-40°C, headache, constipation or diarrhoea, loss of appetitie, nausea, rose coloured spots on the chest area and enlarged spleen and liver. Most people show symptoms 1-3 weeks after exposure. Paratyphoid fever has similar symptoms to typhoid fever but is generally a milder disease.		Better personal hygiene, especially hand-washing after toilet use and before food preparation, safe drinking water and proper sanitation systems

Contd.

Table 11.13: Water-borne diseases

Disease	Type of Disease	Transmission/Causes	Symptoms	Approximate Cases/Year	Prevention
Diarrhoea	Gastointestinal infection	Diarrhoea is caused by a host of bacterial, viral and parasitic organisms most of which can spread by contaminated water. It is more common when there is a shortage of clean water for drinking or cooking. Water contaminated with human faeces can cause the illness. Animal faeces also contain micro-organisms that can cause diarrhoea. It can also spread from person to person, aggravated by poor personal hygiene. Food is another major cause of diarrhoea when it is prepared or stored in unhygienic conditions.	Frequent passing of liquid stool is the typical symptom. Diarrhoea may last a few days, or several weeks. Severe diarrhoea may be life threatening due to fluid loss, particularly infants and young children, and the malnourished. Diarrhoea is also associated with other infections such as malaria and measles.		Safe drinking water, improved sanitation and better personal as well as food hygiene
Dysentery	Gastrointestinal infection	Amoebic dysentery is caused by *Entamoeba histolytica* and most commonly spread by water or contaminated uncooked food or from carriers. Flies may carry the cysts to spread the amoeba from the faeces of infected people to food. Water contaminated by faeces of infected persons can cause infection. Bacillary dysentery is caused by certain non-motile bacteria of the genus *Shigella*. This dysentery is usually self-limiting and rarely manifests into a severe amoebic form. Bacillary dysentery is spread by contaminated water, milk, and food. Faeces from active cases as well as those from healthy carriers contain immense numbers of the disease-producing bacteria. Flies transmit the bacteria to food.	Frequent passage of small, watery stools, often containing blood and mucus, accompanied by severe abdominal cramps. Ulceration of the walls of the intestine may occur.	More than 8 million	Better hygiene and sanitation
Cholera	Intestinal infection	Cholera is an acute, diarrhoea illness caused by infection of the intestine by the bacteria *Vibrio cholerae*. Cholera is caused by drinking contaminated water or eating contaminated food. In an epidemic, the source of the contamination is usually the faeces of an infected person. The disease can spread rapidly in areas with inadequate treatment of sewage and drinking water. The cholera bacterium may also live in the environment in brackish rivers and coastal waters. Shellfish eaten raw have been a source of cholera. The disease is not likely to spread directly from one person to another.	The infection is often mild or without symptoms, but sometimes it can be severe. In 5-10 per cent cases, the patients develop watery diarrhoea, vomiting, and leg cramps within 6 hours to 5 days after the exposure to the bacteria. Rapid loss of body fluids leads to dehydration and shock. The person can die within hours if not treated.	More than 4,000 cases, maximum are during the monsoon months	Better personal hygiene and sanitation

Identically, chemical contamination also leads to serious health hazards. Toxic metals, pesticides and aromatics were often taken up by the human being through foods and drinks. The body burden of these toxicants increases day by day, so that a variety of complex physiological as well as pathological diseases manifest in course of time. Every country has stipulated a standard for water quality to be used for drinking purpose, where maximum allowable concentrations of various inorganic, organic and microbial substances were stipulated. Some of these may be acutely toxic, others produce chronic illness. The standard for drinking water as stipulated by World Health Organisation (WHO, 1971) and Bureau of Standard Institute (BSI, India, 1975) is given in Table 11.14.

Table 11.14: Drinking Water Standards

Characteristics	World Health Organisation 1971		Ministry of Works and Housing 1975	
	Highest Desirable	Maximum Permissible	Acceptable limit	Cause of Rejection limit
I. PHYSICO-CHEMICAL				
Turbidity, JTU	5.0	25.00	2.5	10.00
Taste and odour	Nothing	Disagreeable	Nothing	Disagreeable
Colour (Pt. scale)	5.00	50.00	5.00	25.00
pH	7.0 - 8.5	6.5 - 9.2	7.0 - 8.5	6.5 - 9.2
Total solids	500.0	1,500.0	500.0	1,500.0
Total hardness (as $CaCO_3$)	100.0	500.0	200.0	600.0
Magnasium	30.0	150.0	30.0	150.0
Iron (Fe)	0.1	1.0	0.1	1.0
Mangenese	0.05	0.5	0.05	0.5
Copper	0.05	1.0	0.05	1.5
Chloride	200.0	600.0	200.0	1,000.0
Sulphates (as SO_4)	200.0	400.0	200.0	400.0
Phenolic substances	0.001	0.002	0.001	0.002
Fluoride*	1.0	1.5	1.0	1.5
Nitrate	45.0	45.0	45.0	45.0
Zinc	5.0	15.0	5.0	15.0
Mineral Oil	0.01	0.30	0.01	0.30
Anionic detergents (as MBAS)	0.2	1.0	0.2	1.0
Arsenic	0.01	0.05	0.05	0.05
Hexavalent Chromium	—	0.01	0.05	0.05
Cyanide	—	0.05	0.05	0.05
Lead	—	0.10	0.10	0.10
Selenium	—	0.01	0.01	0.01
Cadmium	—	0.01	0.01	0.01
Mercury	—	0.001	0.001	0.001
PCB (μg/l)	—	0.2	0.2	0.2
Gross Alfa-activity (PCi/l)	—	3.0	3.0	3.0
Gross Beta-activity (PCi/l)	—	30.0	30.0	30.0

All the values are in mg/l, otherwise stated

*The acceptable Fluoride concentration varies as a function of ambient temperature.

Contd.

II. BACTERIOLOGICAL

W. H. O. (1971)

A. **Water entering distribution system:** If disinfected coliform count in any sample of 100 ml should be zero.

B. **Water in the distribution system:** Ideally, all samples taken from the distribution system including consumer's premises should be free from coliform organisms. Since, in practice, it is not always possible: hence the following standard.

(i) Throughout any year, 95% of the samples examined should not have any coliform organisms.

(ii) E. coli count in 100 ml of any sample should be zero.

(iii) Coliform organisms not more than 10 per 100 ml shall be present in any sample.

(iv) Coliform organisms should not be detectable in 100 ml of any two consecutive samples.

Min. of Works & Housing (1975)

A. Coliform count in any sample of 100 ml should be zero.

B. Water in the distribution system shall satisfy all the three criteria indicated below:

(i) E. coli. count in 100 ml of any sample should be zero.

(ii) Coliform organisms not more than 10 per 100 ml shall be present in any sample.

(iii) Coliform organisms should not be detectable in 100 ml of any two consecutive samples or more than 50% of the samples collected for the year.

11.16 WATER POLLUTION CONTROL

In most cases, water pollution takes place by the industrial or domestic waste water discharger or even by the surface run-off from agricultural field and mined area. Thus the principal methods of pollution control is centred around the waste water treatment processes.

The conventional waste-water treatment system has three major steps:

- Primary treatment;
- Secondary treatment;
- Tertiary treatment.

The domestic raw waste water (sewage) or industrial waste water (effluent) treatment processes have some resemblances. The primary objective of this treatment lies on the removal of the following pollutants so that the treated water could be safely discharged into river, lake or other surface water bodies:

> Total solids (dissolved and suspended),

> BOD,

> COD,

> Nitrate & phosphate,

> Oil & grease; and

> Toxic metals, etc.

11.17 WASTE WATER TREATMENT PLANT

When the waste water (sewage/effluent) reaches the treatment plant, it first passes through a series of screens that remove large objects. Then it passes through primary, secondary and tertiary treatment steps:

(a) **Primary treatment**: A series of settling chambers are designed to remove heavy grits including coarser organic materials. Then there is a sedimentation clarifier, where suspended particles are oxidised and sedimented at the bottom as sludge. In this process considerable amount of suspended solid and organic materials are removed; as such substantial lowering of total solids, BOD and COD are noticed.

(b) **Secondary treatment**: The organic matter and other impurities remaining after the primary treatment are in solution or in fine suspension that does not settle and cannot readily be filtered. There are two ways by which finer particles and dissolved materials are sedimented out viz., activated sludge (aerobic) treatment and trickling filter treatment. In the activated sludge process, sewage, after primary treatment, is pumped into an aeration tank, where it is mixed for several hours with air and with bacteria-laden sludge. The bacteria then decomposes organic waste along with oxidative sedimentation of suspended and dissolved materials, ultimately leads to sludge deposition at the bottom of the tanks. The effluent from the biological action is still laden with bacteria and is not fit for discharge into open waters.

An alternative technique is 'trickling filter' treatment. In this device, long pipes rotate slowly over a bed of stones, it offers its nutrients in the presence of air to an abundance of rather unappetizing forms of life. A fast-moving food chain is set in operation. Bacteria consume molecules of protein, fat and carbohydrates. Protists consume bacteria. Further up the chain are worms, snails, flies and spiders. Each form of life plays its part in converting high energy chemicals to low-energy chemicals. All the oxygen consumed at this stage represents oxygen that will not be needed later when the sewage is discharged to open water. There, this process constitutes a very significant purification.

Thus each step in the biological consumption of this waterborne waste, from sewage nutrients to bacteria to protozoa and continuing to consumers of higher orders (such as worms), represents a degradation of energy, a consumption of oxygen and a reduction in the mass of pollutant matter. Finally, when the microorganisms die, their bodies stick together to form aggregates large enough to settle out air in a reasonably short period of time. Some agglomeration also occur in the metabolic processes of the protozoa, so that their excreta are usually larger than the particles of food that they ingest. Thus the process of making big particles out of little ones is of prime importance in any system of waste water treatment. The mushy mixture of living and dead organisms and their waste products at the bottom of a treatment tank constitute the 'biologically active sludge'.

(c) **Tertiary or Advanced treatments**: Although considerable purification is accomplished by the time that waste-water have passed through the primary and secondary stages, these treatments are still inadequate to deal with some complex aspects of water pollution. First, many pollutants in domestic sewage are not removed. Inorganic ions, such as nitrates and phosphates, remain in the treated waters. These materials, as we have seen, serve as plant nutrients and are, therefore, agents of "eutrophication". If there is excess chlorine, it may react with industrial wastes such as organic solvents and produce chlorinated products that are more objectionable than the original pollutants.

Additionally, many pollutants originating from sources such as factories, mines, agricultural run-offs and even homes cannot be handled by municipal sewage treatment plants. Some synthetic organic chemicals from Industrial wastes are foreign to natural food-webs (non biodegradable). They not only resist the bacteria of the purification system but may also poison them and thereby nullify the biological oxidation. However, several troublesome pollutants which arise from different industrial units need special provision for treatment viz. oil separation, phenol recovery, lime treatment, and polyelectrolyte or alum treatment etc. A generalised outline of three stages of waste water treatment process is given in Fig. 11.7 (next page).

Waste-water treatment with aquatic macrophytes provided an option for selecting plants in lowcost treatment of municipal and domestic sewage waters. The capacity of ecosystems that are dominated by aquatic macrophytes to assimilate and decompose inputs of nutrients and organic matters have resulted in the extended use of such systems to different types of waste waters. In general, municipal and domestic sewage effluents are the rich source of organic and inorganic nutrients (organic carbon, nitrogen and phosphate etc.). Due to higher organic loading the dissolved oxygen level is very poor. Under such anoxic environment several aquatic macrophytes can grow and help in the treatment of waste water by removal of BOD, COD, phosphate and nitrate. Thus constructed wetland can be developed for such kind of waste water treatment using selected macrophytes. This system was practiced in many European countries.

Currently an assessment was made for domestic waste water generated and subsequent treatment facilities in various States of India. The data shows that over 3/4 of the total waste water disposed off without treatment (Table 11.15).

Table 11.15: Status of Sewage Generation and Treatment Capacity of Urban Population of India

Sl. No.	State	Sewage Generation of Urban population in 2015 (MLD)	Installed Treatment Capacity (As on March, 2015) (MLD)
1.	Andaman & Nicobar Islands	22	0
2.	Andhra Pradesh	2871	247.27
3.	Arunachal Pradesh	50	
4.	Assam	703	0.21
5.	Bihar	1879	124.55
6.	Chandigarh	164	314.5
7.	Chhattisgarh	951	0
8.	Dadra & Nagal Haveli	26	–
9.	Daman & Diu	29	–
10.	Goa	145	74.58
11.	Gujarat	4119	3062.92
12.	Haryana	1413	852.7
13.	Himachal Pradesh	110	114.72
14.	Jammu & Kashmir	547	264.74
15.	Jharkhand	1270	117.24
16.	Karnataka	3777	1304.16
17.	Kerala	2552	152.97
18.	Lakshadweep	8	–
19.	Madhya Pradesh	3214	482.23
20.	Maharashtra	8143	5160.36
21.	Manipur	132	–
22.	Meghalaya	95	1
23.	Mizoram	90	10
24.	Nagaland	92	–
25.	Delhi	4155	2693.7
26.	Odisha	1121	385.54
27.	Puducherry	136	68.5
28.	Punjab	1664	1245.45
29.	Rajasthan	2736	865.92
30.	Sikkim	24	31.88
31.	Tamil Nadu	5599	1799.72
32.	Telangana	1671	685.8
33.	Tripura	154	0.05
34.	Uttar Pradesh	7124	2646.84
35.	Uttaranchal	495	152.9
36.	West Bengal	4667	416.9
	Total	**61948**	**23277**

(*Source*: Sulabh, 2016)

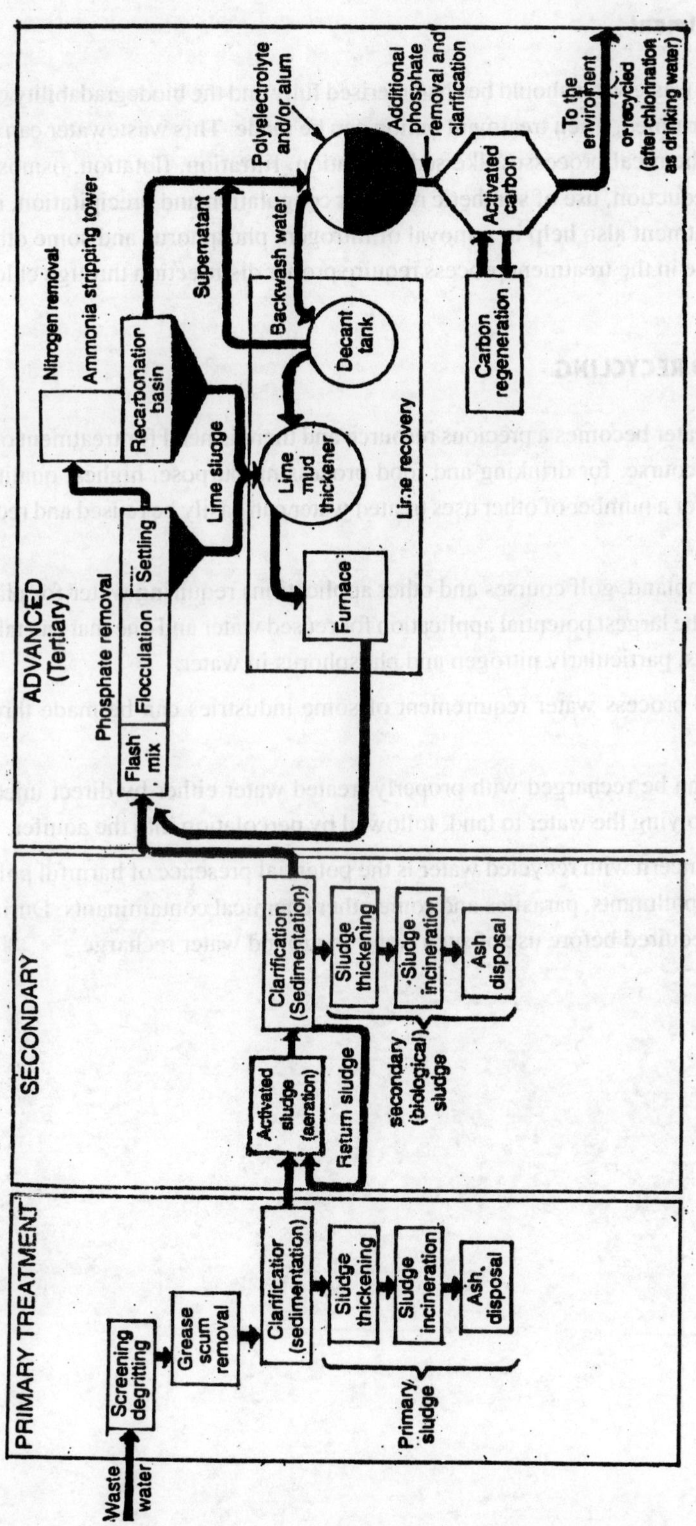

Fig. 11.7 Stages of Wastewater treatment process

Industrial Wastewater Treatment

Before treatment, industrial wastewater should be characterised fully and the biodegradability of wastewater constituents should be determined. Then treatment option can be made. This wastewater can be treated by a variety of physical and chemical processes like sedimentation, filtration, flotation, osmosis, acid-base neutralization, oxidation-reduction, use of synthetic resin for coagulation and precipitation, ion exchange process etc. Biological treatment also help in removal of nitrogen, phosphorus and some other dissolved salts. Sludge which generate in the treatment process require proper disinfection through chlorine dioxide or otherwise.

11.18 WATER REUSE AND RECYCLING

With the passage of time, water becomes a precious resource and there is need for treatment of wastewater and then reuse it back. Of course, for drinking and food processing purpose, highest quality of potable freshwater is required. But for a number of other uses treated water can easily be reused and recycled. Some examples are:

- Irrigation for cropland, golf courses and other applications requiring water for plant and grass growth. This is the largest potential application for reused water and one that can take advantage of plant nutrients, particularly nitrogen and phosphorus in water.
- For cooling and process water requirement of some industries can be made through treated wastewater.
- Ground water can be recharged with properly treated water either by direct injection into an aquifer or by applying the water to land, followed by percolation into the aquifer.

However the major concern with recycled water is the potential presence of harmful pollutants, viz., bacteria, persistent organic pollutants, parasites and some other chemical contaminants. During treatment appropriate monitoring is required before use of such water in ground water recharge.

12 MARINE POLLUTION

12.1 INTRODUCTION

The coastline is a complex region comprising bays, estuaries, and large semi-enclosed areas where human populations and industrial development are concentrated. It is the focus for contaminants from inland areas as well as from developments along its length. Most of the sources of contamination discussed in this chapter contribute directly or indirectly to problems in the immediate coastal zone, and arise from activities specifically located there.

Although a relationship between human population increase and environmental change has long been recognised, attempts have been made only recently to assess the cumulative impacts of land development in the coastal zone by recording their physical, chemical and biological consequences. This requires knowledge of trends in water quality and an understanding of the management of aquatic habitats. Equally important is the economic analysis of damage to natural resources and human health against which the cost of control measures will need to be justified. Many of the impacts recorded are common to most coastal developments, but it is useful to consider industrial and recreational activities separately.

The development and maintenance of ports and harbours is of prime concern to human populations. Water exchange in these areas is often limited and shipping activities introduce contaminants, including oily wastes, cargo escapement and human wastes released from shipboard. These are subject to national and international regulations, but contamination at ports is difficult to control since it enters the sea by many routes, including discharges from pipes, run-off from streets, roofs and parking areas and inputs from the atmosphere. Also, harbours are the first point of contact with the sea for many rivers, which add a wide variety and large quantity of land-derived material.

In addition to the activities within the harbour itself, all large ports are provided with a wide range of services and support facilities, including roads, railways, shipyards, power plants and local waste disposal sites, each of which add to the general contamination. Their collective effects may be well-recognized but are poorly studied.

After release or discharge to the sea, many contaminants become associated with sediments and may remain sequestered until resuspended by waves and currents or until the sediments are disturbed by dredging to maintain shipping channels. This dredging not only stirs up contaminated material and reintroduces it into the water column, it also brings the problem of dredge spoil disposal, and may give rise to changed patterns of water circulation. In general, harbour sediments carry particularly large chemical contaminant loads, significantly greater than in adjacent areas. Fish or shellfish harvested from these locations are usually tainted by petroleum. which affects their flavour adversely, and levels of contaminants in edible tissues may be high, and fail to meet health standards.

The recreational use of coastal waters for various leisure activities such as bathing, swimming, diving, boating, surfing, and fishing is increasing. In some areas this represents the major, or even the only, industry. With the world-wide pressure to open up new locations for tourist development, pristine wetlands and swamps are being developed as recreational beaches and harbours for small vessels, or as sites for commercial and domestic activity. As well as structural engineering alterations to beaches themselves, the immediate hinterland is built up with hotels and support infrastructure. The resulting restructuring along the coastline disrupts traditional fisheries, interferes with marine life and eliminates important habitats. Ironically, this environmental degradation and congestion may destroy the main natural assets on which the tourism development is based.

An intrinsic feature of such developments is the influx of large numbers of people, at least seasonally, resulting in an increase of sewage outflow such that locally available means of treatment and disposal may be overtaxed. There is also increased traffic and general disturbance from activity, noise and lights, which many users expect and accept even in areas where damage will be caused. Not surprisingly, inhabitants of coastal areas identified for development often support schemes that they see as enhancing their income and standard of living. Local authorities and national governments also tend to encourage the influx of foreign currency.

The widespread consequences of all these developments are increasingly recognised. As a result, areas of high sensitivity or of special interest are being designated for protection through zoning and planning procedures, and species particularly at risk, such as many birds which feed or nest in wetlands and sea turtles which lay eggs at the high water mark of sandy beaches, are being included in protocols for legal protection. Much more, however, remains to be done to reconcile conflicting demands when major alterations of the coastal strip are envisaged.

Indeed, planning the development of the coastline as a whole, as is already being done in a number of countries, should be undertaken more widely. International guidelines, including criteria and standards, would provide valuable assistance in such planning in different geographical areas, but awareness, resources and political will are needed if the health of the resident and the transient populations, the survival of marine wildlife and the functional integrity of the vital land-sea interface are to be maintained.

12.2 POLLUTION OF MARINE ENVIRONMENT

Contaminants from land reach the marine environment by a variety of pathways. Coastal outfalls discharge directly to estuaries, inshore waters, bays and open coastal areas. Storm-water flows may be too great for drainage and treatment facilities and when run-off is too fast, may exceed the assimilative capacity of the receiving waters. Rivers act as large-scale collectors and carriers of wastewaters from diverse sources within their drainage basins and off-load them to the sea. Thus, rivers can be regarded as major point sources of mixed contaminants, the inputs of which depend on the contaminant load of the rivers and on the physico-chemical and biological transformations taking place in the river itself and especially in the estuaries and the near-shore zone.

Non-point sources draining to coastal waters include surface run-off from agricultural areas, wash-out of agrochemicals and transport of sediment due to coastal erosion or to deforestation and desertification in the hinterland. Land management practices largely determine these various influxes.

River inputs and non-point sources are discussed elsewhere (Chapter 11). This chapter focuses on direct discharges from point sources along the coastline. Two are predominant : domestic sewage and industrial effluents. In urban areas both waste streams are often mixed in sewerage systems and reach the sea together as municipal wastewater. This then contains material of domestic origin as well as a variety of industrial and other effluents with higher concentrations of contaminants which have been discharged to the community sewerage systems.

Contaminants from these sources can be grouped as:

— microorganisms

— organic material affecting oxygen balance

— nutrients

— trace elements

— synthetic organic compounds

— petroleum-related compounds

— particulates/sediments

— heat.

It should be emphasised that effluents usually contain mixture of these, the composition varying in both the short and the long term.

A specific agent or compound within any of these classes can exert varying degrees of impact on the marine environment. The magnitude of this impact is a function of the chemical and biological characteristics of the compound, its form, total amount and concentration, persistence, bioaccumulation, recycling potential and of the characteristics of the receiving environment. The early identification of substances that may have adverse effects on human health, living resources and the marine ecosystem is particularly important. The possibility of synergistic action of substances in a mixed discharge must be taken into account, although evidence so far suggests that synergism is unusual and that an assumption of additive effects is usually realistic.

Wastewater discharges at the coast affect all compartments of the marine environment and may interfere with human uses. The most widespread effects of urban and industrial discharges are:

— health risks from the presence of sewage pathogens

— eutrophication and/or oxygen depletion due to nutrients and organic

— carbon

— toxic effects on marine organisms

— risk to human health caused by various chemicals in seafood.

In addition, both europhication and toxic effects on marine organisms can damage marine living resources.

Many human pathogens, once released to coastal waters, die rapidly owing to adverse conditions in the marine environment. They may, however, be taken up by filter-feeding shellfish and some are able to proliferate in an intermediate host organism. The health consequences of consuming uncooked contaminated shellfish and of bathing in waters containing fecal micro-organisms are further discussed in Chapter 11.

Noticeable algal blooms in coastal waters are often an early sign of excessive nutrient inputs and their decay adds to the problem of the oxygen demand due to organic materials from domestic sewage, livestock wastes and various agro-industries (e.g. pulp and paper mills, fish and food processing, sugar refineries). Effects are most severe where the effluents from densely populated industrial areas discharge to semi-enclosed or shallow receiving waters with reduced circulation, long residence times and, hence, limited scope for self-cleansing.

Major sources of industrial chemical discharges are pulp and paper mills, iron and steelworks, petroleum refineries, petrochemical industries, fertilizer factories, leather tanning and finishing and other chemistry-based production installations, including pharmaceutical plants. These discharges are complex and without appropriate treatment and properly sited outfalls, may be harmful to a variety of marine targets and, indirectly, to man. Localised fish kills (e.g. in fish farms), altered benthic communities and accumulated chemical residues are often the first warning of chemical pollution.

In the past, most pollution control strategies had been prompted, usually *post hoc*, by evident effects of, for example, excessive oxygen demand, or of discharges of nutrients, metals and pathogenic organisms, but recent and current policies try to forestall effects by controlling inputs, defining potentially hazardous substances and stringently limiting their disposal. However, toxicological information by itself cannot be the basis of control. Knowledge of discharge loading and dilution rates—as well as evidence of effects on specific organisms—is required.

Few municipal wastewater treatment plants handle domestic sewage alone; most have an industrially contributed load in addition, restricting the options for efficient waste treatment and sludge utilisation. The degree of contaminant removal and the nature of the effluent discharge depend upon the types of substances released from industry, the treatment process technology used, the design of the treatment plant and its operational efficiency. Most of the large urban/industrial centres in developing countries have no effective wastewater treatment system and the siting and design of their outfalls often do not provide adequate dilution and dispersion, thus potentially endangering human health and resources.

Even in developed countries, large population centres are still likely sources of contaminants through their untreated domestic effluents. Even when facilities are available, they may be inadequate when load is high owing to seasonal increases. In these cases additional treatment for peak load capacity can be costly to provide and overflow release may be preferred in the short term, but appropriately sited offshore outfalls are needed and these require surveillance to ensure that they are, and continue to be, effective.

In tropical and sub-tropical oceanic areas where long stretches of the coast may be rather sparsely populated, most of the sewage from small communities either does not reach the sea or is too diluted on discharge to cause detectable harm. Notable exceptions are population centres in coastal areas and estuaries in Latin/Central America, South-East Asia and Western Africa and some small but densely populated oceanic islands. During storms, domestic and industrial wastewaters from large metropolitan areas often reach the sea through storm-water overflow and open canals, receiving no treatment. Seafood harvested in these places constitutes one of the major public health hazards related to the marine environment.

12.3 SOURCES AND NATURE OF POLLUTANTS

(a) Dredged material and mine tailings

- About 80 to 90 per cent of all materials dumped at sea results from dredging. Between 1980 and 1990 the reports provided to the Secretariat of the London Dumping Convention (LDC) record an average of 230 million tonnes of dredged material dumped at sea annually, which represents some 20-22 per cent of all dredged material, the rest being disposed of elsewhere. Of the total material dredged, about two-thirds are associated with operations to keep harbours, rivers and other waterways from silting up; the remainder represents new works. Future dredging operations are expected to show the same proportions.
- Uncontaminated dredged material, if properly handled, causes few problems in the long term, and indeed can serve a variety of useful purposes, including land-fill, building of artificial reefs and reclamation of previously damaged coastal sites. If dumped at sea, its physical impact must be taken into account and careful selection and management of the dump sites is important.

- About 10 per cent of dredged materials is contaminated from a variety of sources, including shipping, industrial and municipal discharges and land run-off. Typical contaminants include oil, heavy metals, nutrients and organochlorine compounds. Dumped dredged material has liquid and suspended particulate phases, but the greatest potential for impact generally lies with the settleable or solid-phase material which may affect benthic organisms by smothering and physical disruption of habitats; bioaccumulation and toxicity from both soluble and suspended phases may also occur.
- Contaminated dredged material may slowly release its adsorbed burden and result in long-term exposure of local habitats to one or more contaminants. However, laboratory and field studies show that leaching into the water column of chlorinated hydrocarbons, petroleum and metals is slight. Nutrients are released at concentrations much greater than background but mixing processes tend to mitigate effects. The major impact at disposal sites with small current velocities and low wave-energy is the physical mounding of the material. Benthic recolonisation of these mounds is relatively rapid on fine-grained sediments and slower on coarse-grained material.
- No single method of sea disposal, or category of sea disposal site, is suitable for every type of dredged material or industrial waste. All alternatives need consideration at the planning stage to ensure that the waste has the smallest possible environmental impact. These include temporary containment or pre-treatment to reduce toxicity and oxygen demand, sea disposal beyond the edge of the continental shelf, and "capping" of dredged material or wastes by placing them in depressions or pits sealed with clean sediment. There are alternative strategies by which sea disposal is concentrated at a single site or repeated as a cyclic operation over a more extensive area. Cycling will be more damaging if the capacity or time scale for recovery or recolonisation of a slow-growing community is exceeded by too frequent discharges.
- Mine tailings are of particular concern. It has been a common practice at coastal mines to dispose off waste materials to the sea—either directly or by dumping into rivers from where the wastes are periodically swept to the sea during flooding. Some effluents are chemically inert, for example from china-clay mining, and the environmental effect is largely due to physical blanketing. For metal ore mining, on the other hand, operations extend from initial extraction-concentration activities to the final smelting-refining procedures. The resulting wastes are toxic, particularly those from the later stages, since the refined product often comes from intensive chemical treatment designed to isolate metals and their compounds. The main metal products from coastal mines and/or processing facilities are aluminium, copper, iron, lead, mercury, molybdenum, tin, and zinc.

(b) Industrial wastes and sewage sludge

- Industrial wastes dumped at sea can present much more varied and intractable problems. They may be highly acidic or alkaline; liquid or largely particulate and anything from relatively inert to extremely toxic. They include wastes from the chemical, petrochemical and pharmaceutical industries, from pulp and paper production, from smelters, from the food industry, from flue-gas washings and from military activities. Pre-treatment of wastes at industrial sites before disposal is judged to be essential in an increasing number of countries. LDC records show that in the period 1970-1985 the largest amount of industrial waste dumped at sea was 17 million tonnes in 1982, and the smallest, 6 million tonnes in 1984.
- The monitoring techniques used in assessing the impact of waste disposal to the sea involve physical or chemical tests and species-diversity studies, but some are procedurally complex and costly and all are difficult to interpret if not properly planned and carried out. Many of the data are not comparable between sites or studies because of the specificity of discharges,

disposal sites and targets and because standard procedures are not routinely used. Earlier studies are not always of direct practical use in environmental forecasting for new developments. Current monitoring strategies and practices in relation to the disposal of industrial wastes at sea should be reviewed to assess their effectiveness. Guidelines should be developed to ensure that all relevant variables and only the relevant ones are monitored.

- The sludge arising from the treatment of sewage can be used as fertilizer on agricultural land or for land reclamation—but only when it is not contaminated with high levels of metals, oils and organic chemicals. However, in some cases it may be more economic and environmentally preferable to dispose it at sea. Municipal sewage sludge normally does not contain contaminants in high concentrations, but excessive dumping may have harmful effects such as oxygen depletion and eutrophication, while health risks may arise from the presence of pathogens if dump sites are not carefully selected. Between 1980 to 1990, 30 million tonnes of sewage sludge were dumped at sea annually, although in recent years there have been strong moves in the U.S.A. and Western Europe to phase out marine dumping of sewage sludge. Accordingly, sewage disposal at sea is gradually decreasing, although this may continue to be more economic and environmentally preferable for those countries which currently do not have land-based sites for incineration or disposal.

(c) Incineration at sea

- For certain wastes, particularly organic materials in liquid form, the least damaging disposal option may be destruction by incineration. The technical requirements for complete destruction of the various wastes are known, including the temperature, residence time and level of oxygen supply in the furnace. When these requirements are met, the wastes are broken down to their basic components which are usually innocuous, and the output of the incineration process, apart from traces of any metals which may have been present, is largely NO_x, CO_2 and chlorine. Although this output can be minimised by scrubbing or neutralisation, incineration sites on land are often resisted by neighbouring communities.
- Incineration at sea is attractive in that remote sites can be designated and effects on the marine environment appear to be negligible. The optimal procedure for incineration at sea is to conduct it while the vessel is steaming clear of normal shipping routes in the open ocean. Liquid organohalogen compounds have been incinerated at sea since 1969 and in the period 1980-1988 LDC reports show that the average amount was 100,000 tonnes per year, mainly in the North Sea. Incineration at sea is carried out in accordance with "regulations for the control of incineration at sea of wastes and other matters" established under the LDC.
- The arguments against incineration at sea are chiefly that there may be some reformation of toxic materials in the flue gases, that emissions may damage sensitive species in the sea-surface microlayer and that the continued availability of this technology may act as a disincentive to reducing wastes at source. Fears have also been expressed that an accident may occur during transport to the burn site, although the likelihood of an accident is estimated (IMO) at one occurrence per 68,000 voyages from European ports to the North Sea incineration site.
- In 1988, all countries party to the Convention for the Prevention of Marine Pollution by Dumping from Ships and Aircraft, 1972 (Oslo Convention) agreed to phase out incineration at sea by 1994. The decision was not based on evidence of observed harmful effects but on the conviction that adequate means of disposal on land are or should soon be available to most of its Contracting Parties (mainly North Sea countries). Contracting Parties to the London Dumping Convention have also agreed in principle to discourage incineration at sea.

- On available evidence, the environmental implications of incineration at sea cannot be regarded as more significant than those of incineration on land. There are limitations and risks associated with either choice. In practice, the choice will depend on a careful analysis of technical, social and political as well as environmental factors.
- Global marine pollution problems are clearly onging. Although there is evidence of a decrease in the concentration of certain constituents of the four classes of critical contaminants in the global marine environment, petroleum hydrocarbons, halogenated hydrocarbons, heavy metals and radionuclide, other types of pollution (e.g. plastics and other forms of litter) are increasing.

(d) Disposal of plastic litter

- The sea is inevitably the recipient of solid matter, intentionally disposed of or accidentally lost. This can interfere with fishing, shipping and other marine activities, but can also support life in the sea by providing additional surfaces for encrusting organisms and shelter for mobile species. In the past, much of such solid matter disintegrated quickly, but resistant synthetic substances have been replacing many natural, more degradable materials and this trend is continuing.
- There is now growing concern among fishermen, scientists, seafarers, conservationists, and tourists about the increasing amounts of plastic material found at sea and on beaches. Such litter originates from both land-based and sea-borne sources. Most beaches near population centres are littered by a multitude of plastic residues washed up from the sea, contributed by rivers, ships and outfalls, dumped by illegal refuse operators, or left behind by beach-users.
- Since most synthetic materials are buoyant and persistent, they present a threat to living organisms and the natural environment. Plastic debris can be loosely classified in three groups: (1) fishing gear and equipment, such as nets and lines; (2) packing bands, straps and synthetic ropes; and (3) plastic litter, including bags, bottles, sheeting, packaging material and small pellets from which "user" plastics are manufactured. Such debris has been recorded, sometimes in large quantities, in the oceans, including the polar regions and from intertidal areas to abyssal depths.
- First, debris from fishing vessels causes a particular problem. Nets which are discarded or lost can continue to catch or entangle marine life and, being non-degradable, they may act in this way for years (ghost-fishing). They can trap marine organisms when floating on the surface, when snagged on the bottom, or when drifting at some intermediate level. Marine mammals, fish, sea birds and turtles are among the animals at risk. There is also a degree of human risk to underwater workers who may become entangled in abandoned netting. Debris can also seriously affect shipping by fouling propellers, damaging drive shafts, and clogging sea intakes and evaporators. The loss in productive time at sea and the costs of repair represent a clear economic detriment to maritime industries and, in the case of fishing vessels, can reduce earnings significantly.
- Second, plastic strapping and packing bands are another threat. They are used to hold palleted material, to secure cargo, to strap boxes and crates and to reinforce packing cases. When simply pulled off rather than cut, the bands float free in the sea and can encircle marine mammals or large fish in a girdle which becomes progressively tighter as the animal grows, inhibiting respiration and restricting the ability to eat or move. Plastic yokes for beverage packages and other containers represent a similar threat to birds and smaller fish. As with nets, there are many reports of animals caught in this fashion.
- The third type of plastic debris consists of bags, containers, sheeting, packing material, raw plastic pellets and other similar items. Turtles, marine mammals and birds are killed by ingesting such material, while the increasing quantities and wider distribution of small plastic particles

in the ocean are also causing concern, since they are ingested by marine organisms and although inert, reduce the nutritional value of their food intake and consequently the animals' growth. Recent estimate (2004) indicate that plastic waste kills up to 1 million sea birds; 100,000 sea mammals and countless fish each year.

- There is little argument that discarded plastics are harmful in the ocean environment, but agreement about the degree of harm and about effective steps to solve the problem is not easy to achieve because quantitative estimates of loss of marine life due to plastic pollution are difficult to obtain. Nor is it easy to compare, on present information, the amount of debris originating from land with that arising from fishing and shipping.
- More study is needed to define the magnitude of the problem and to explore solutions. Better estimates of the number of marine animals killed and of the impact on the fishing and shipping industries are required. Research is needed on procedures for storing and disposing of, synthetic materials, as well as on the design and manufacture of nets and other gear less harmful to the environment or more readily recoverable. Research is also required to develop alternative materials.
- Some practical steps can be taken immediately. Packing bands and beverage yokes could be cut after use. All hazardous plastic products could be labeled to indicate their potential for damage to the environment and marine life and the ways to minimise it. Fishing-net manufacturers might make parts of their nets, or the lines connecting them to floats or anchors, of degradable material to reduce the damage done by ghost fishing. Lastly, beach clean-up campaigns, which have recently been promoted in several countries, as well as removing litter, serve to raise public awareness, while providing evidence on the problem. The introduction of economic incentives and the development of guidelines and regulatory instruments should be explored as means of encouraging the production and use of alternative materials that are not harmful to the environment and to promote the collection and recycling of discarded plastics.
- The International Convention for the Prevention of Pollution from Ships (1973), as modified by the Protocol of 1978 relating thereto (MARPOL 73/78), in its Annexe V, contains regulations for the prevention of pollution by garbage from ships, which entered into force in December 1988. These include prohibition of discharge to the sea of all plastics, such as synthetic ropes, fishing nets and plastic bags. Guidelines for the implementation of Annex V to the Convention request governments to consider a series of measures, including reporting systems, record books on board ships, compliance incentive systems and educational programmes.
- The abandonment of trash such as worn vehicle tyres, discarded domestic equipment, cans and crates contributes to the unseemly aspect of many shores, especially those close to urban areas. In some instances, even medical and surgical equipment has been reported on beaches.
- Educational campaigns will be a powerful tool in bringing changes in public attitudes towards the environment and such campaigns have indeed been introduced since 1988 by many contracting parties to the conventions or protocols for the prevention of marine pollution by dumping of wastes. Nevertheless, more effort should be directed towards convincing the public and the authorities of the value and small personal cost of keeping the environment clean — at sea and on the beaches as much as along rivers, inside cities and on mountain slopes.

(e) Oil Pollution

(i) **Oil in the Biosphere**: Oil including crude and refined petroleum is the major energy fuel of the world today. Oils and petroleum gases are often mined from the prospecting areas of the world for human utilisation. During the extraction, transportation, storage and distribution of various petroleum products, a considerable amount has been lost in the environment—in soil and also in water bodies. Sometime this kind of loss of oil may be deliberate or accidental. Due to loss of oil in the soil and water bodies, the environment become polluted day be day and it has tremendous impact on living systems of soil and water bodies.

Petroleum contains five principal classes of hydrocarbons: normal paraffins, branched-chain paraffins, cycloparaffins, aromatics, and asphalties. The distribution of the various hydrocarbons may differ in a crude oil and depends on the origin. Paraffin-base crudes are predominantly paraffinic hydrocarbons, whereas asphalt base crudes are lower in parraffines and contain more high boiling point asphaltic hydrocarbons. Crude oil also contains oxygen (less than 3%), nitrogen (less than 1%) and sulpher (0.1 - 1%). There are varying amounts of heavy metals (i.e. vanadium, nickel, iron and copper) in crude oil. Other petroleum products like gasoline, distillate fuel oil, high boiling point residual tars etc. would not be considered a significant problem in the environment as because of the fact that most of them are insoluble and evaporate quickly on spilling.

The presence of petroleum or refined petroleum products in soil or aquatic systems is generally harmful for some physical reasons (coating, smothering etc.) and also for toxicity to the biota.

The accidental spillage of oil in large quantities has had a dramatic impact on people throughout the world. Some major spills past five decades are mentioned in Table 12.1 which provide tremendous concern about the short and long term impact of oil pollution particularly in the ocean.

Table 12.1: Some major oil spills of the world

Year	Ship	Site	Cargo	Quantity of oil spilled (+) in tons
1957	Tampic Maru	Baja, Mexico	Diesel oil	8,000
1967	Torrey Canyon	Milford Haven, England	Kuwait Crude oil	115,000
1969	Well A-21	Santa Barbara Channel, Glif	Crude oil	4,500
1970	Arrow	Chedabucto Bay, Nova Scotia	Bunker Crude oil	10,000
1974	Metula	Strait of Magellan, S.A.	Arabian Crude oil	52,000
1976	Argo Merchant	Nantucket Schoals, Mass	No. 6 fuel oil	20,000
1978	Amco Cadiz	Brittany Coast, France	Crude oil	220,000
1979	Kurdistan	Cape Breton, NB	Crude oil	7,900
1981	Deivos	Helgoland, Norway	Crude oil	1,000
1983	Swedish tanker	Kattegat, Denmark	Crude oil	500
1984	Puesto Rican	San Fransisco Bay	Crude oil	4,900
1987	Stuyvesant	250-300 km off nothern British Columbia	Crude oil	2,000
1988	Nestucea	Gray's harbour, WA	Crude oil	770
1989	Exxon Valdez	Alaska	Crude oil	36,400
1991	Tenyo Maru	British Columbia	Crude oil	300
1993	Braer	Shetland Islands	Crude oil	26,000,000
1996	Sea Empress	SW Wales	Crude oil	18,000,000
2001	Jessica	Galapagos Islands	Crude oil	240,000
2002	Yemen	Gulf of Aden	Crude oil	12,200
2002	Spain	Galicia	Crude oil	63,000
2003	Pakistan	Karachi	Crude oil	30,000
2004	United States	Unalaska Isand, Alaska	Crude oil	1,560
2005	United States	Pointe a la Hache, Louisiana	Crude oil	1,500
2005	United States	Meraux and Chatrolette, Louisiana	Crude oil	3,410
2005	United States	Cox Bay, Louisiana	Crude oil	12,000

Contd.

Table 12.1: Some major oil spills of the world

Year	Ship	Site	Cargo	Quantity of oil spilled (+) in tons
2006	Lebanon	—	Crude oil	30,000
2007	Mexico	Bay of Campeche	Crude oil	1,869
2007	South Korea	Yellow Sea	Crude oil	10,800
2008	United States	New Orleans, Louisiana	Crude oil	8,800
2010	United States	Gulf of Mexico	Crude oil	627,000
2011	Canada	Alberta	Crude oil	3,800
2011	Venezuela	Maturin	Crude oil	41,000
2012	United States	New Jersey	Crude oil	1,130
2013	Canada	Quebec	Crude oil	4,830
2014	Israel	Eilat (Pipeline)	Crude oil	4,300
2016	United States	Shelby County, Alabama	Crude oil	1,092
2017	Greece	Saronic Gulf, Salamis	Crude oil	2,500
2017	India	Chennai, Exrone Port	Crude oil	60

The accumulation of various kinds of oil in the ocean is increasing in a very fast rate from various oil exploration activities on land and sea, accidental spills on land and sea, deliberate and careless contamination, and natural production from marine biota.

In Table 12.2 an estimated annual gross input of petroleum hydrocarbons into the ocean is given:

Table 12.2: Average Gross Annual Input of Petroleum products into the Ocean

Origin	Inputs (MT/Yr.)
(A) Oil Extraction, Exploration and other activities	
(I) Terrestrial:	
River runoff	1.60
Atmospheric origin	0.60
Urban runoff	0.30
Municipal wastes (Coastal)	0.30
Industrial wastes (Coastal)	0.30
Coastal refineries	0.20
Total	3.30
(II) Marine:	
Tankers	1.08
Dry docking	0.25
Terminal operation	0.003
Bilges and bunkering	0.50
Tanker accidents	0.20
Non-tanker accidents	0.10
Offshore production	0.08
Offshore seeps	0.60
Total	2.813
(B) Produced by the living organisms	6.1
Net Total :	11.943

Source: National Academy of Sciences, Washington

However, it is apparent that man's activities have in recent years resulted in a great increase in the total input of oil to the marine environment, but what is more important, this tends to be concentrated in certain localities. Nevertheless, the absence of gross pollution on the sea surface and shores indicates that considerable destruction of oil takes place. Mechanisms considered to be responsible include solution and evaporation of light components, ingestion of oil by marine animals and photochemical and microbial oxidation. Only the last two processes have any significant capacity for oil destruction, although the other mechanisms may result in an increase in the susceptibility of the oil to degradative attack. Photochemical Oxidation is appreciable only in this oil films, whereas microbial attack is most significant when oil is dispersed into the bulk of the sea. Oil spilled into the sea rapidly accumulates as a surface slick. Microbial infection occur immediately, but at this stage any changes which may be caused by the organisms are far outweighed by oil losses due to photo-oxidation and to evaporation and solution of light fractions.

Modification of the physical form of the oil precedes and determines further chemical changes; under the influence of wave action and natural or synthetic surfactants oil emulsions are formed. These emulsions ensure that the oil becomes dispersed into the bulk of the sea to provide a vast surface area for microbial attack.

The impact of oil pollution on the marine environment in particular can be divided into two major categories:

(i) Acute effects observed when major spills occur, with immediate consequences to the biota and to the entire coastal recreation area; and

(ii) Chronic pollution of the entire marine ecosystem caused by constant spillage or drainage of petroleum into the environment. The chronic effects can occur when petroleum or its derivatives flow into the biosphere at a constant rate or at consistent intervals such that the flora and fauna have an insufficient time to recover between exposures to the petroleum. Modifications of marine ecosystem brought about by chronic pollution and other activities of the petroleum industry can alter populations through changes in the rate of birth, longevity and distribution of species causing an imbalance in the entire community. The fate of oil in the oceanic environment is given in Fig. 12.1.

12.4 OIL POLLUTION AND MARINE BIOTA

The ocean is a remarkably stable environment with relatively constant pH, temperature and chemical composition. The level of dissolved oxygen resulting from photosynthesis by planktonic organisms, at depths to which light can penetrate, and its solutions from the atmosphere is sufficient to support a vast biomass. All living species in the ocean are interdependent, some known effects of petroleum on the flora and fauna of the marine environment will now be considered in details.

1. *Phytoplankton*: The general impression emerging from various studies is that although natural phytoplankton communities are adversely affected by oil, these organisms are sufficiently prolific that individual spills have only short-term effects on their overall populations. However, since phytoplankton is largely responsible for the fixation of the energy utilised by marine ecosystems, it is vital to know how oil in the water column affects it both as large spills and as continuous additions in estuarine, coastal and oceanic environments. Unfortunately, it is very difficult to detect the effects of oil on species abundance and composition of phytoplankton in the field, as these organisms show drastic natural seasonal variations with one species of diatom or dinoflagellate taking over the predominant position held by another species as consequence of changes in temperature, light or the availability of nutrients.

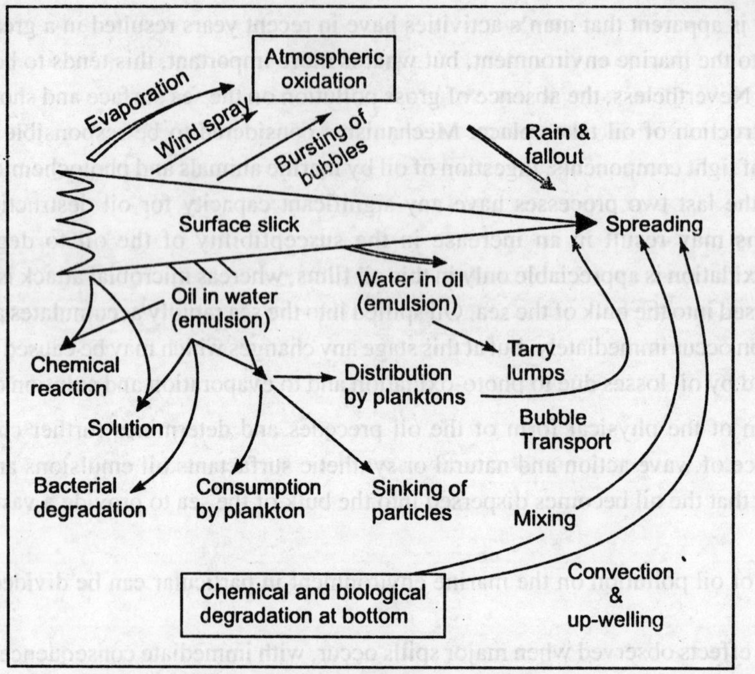

Fig. 12.1 : The fate of oil in the oceanic environment

However, there are several sensitive phytoplanktonic species. These species showed retarded cell division or death of cell even at very low oil concentration (0.1 to 0.00001ml/lit). The species of *Coscinodiscus, Skeletonema, Chlorella* and *Chlamydomonas* showed marked sensitivity to oil pollution in marine environment.

2. *Zooplanktons*: Copepods are most sensitive zooplanktons, young *Acartia clousi* and *Oithona sora* die after 3-4 days immersion in sea water containing 10 μg/lit. of oil, while the adults of these and three other copepod species suffer an accelerated death after longer exposures to 10 μg/lit. or after 5-60 minutes in 1,000 μg/lit.

3. *Macroscopic algae*: The growth and survibility of sea weeds like *Macrocystis, Fucus, Ascophyllum, Sargassum* are seriously affected by oil spill as reported after TORREY CANYON spill or TAMPICO MARU spill. The destruction of sporelings of *Polysiphonia* species, inhibition of gross photosynthetic activities of some green algae like *Enteromorpha* and *Chaetomorpha* are marked affected by oil pollution. However, some algal forms, particularly the blue-green algae, appear to be resistant to oil pollution and may even obtain nutrients from it (Spooner, 1970, Banker, 1971). For instance Oscillatoria—a typical blue-green algae — grows abundantly on trickling filters used to remove oil from refinery effluents (Crosby, Rudoufs and Henkelkian, 1954).

4. *Marine Grasses*: Much of the information regarding the effects of oil pollution on marine grasses and marsh plants has been gathered by workers at the Orielton Field station in Pembroke, Wales, where extensive studies have been made on the effects of oil on salt marsh plants. It was reported that *Spartina Sp., Sueda maritina, Salicornia Sp.* are very sensitive to oil pollution, while *Oenanthe lachesali,* a tolerant perennial marsh plants. Further it was also reported that the growth of some marsh grasses *(Pucinella maritinia* and *Festuca rubra)* may be stimulated by oil pollution.

5. *Benthic and intertidal organisms* : By far the greatest number of information concerning the effects of petroleum pollution on marine organisms pertains to benthic and intertidal organisms; that is, those organisms which spend a major portion of their lives at the sea bottom. These organisms include a large number of species of molluscs, crustaceans, echinoderms, polychaetes, coelenterates and hydroids. Many of these, notably lobsters, oysters, scallops and clams, not only constitute an important fisheries resources but are also amenable to mariculture. These creatures are very susceptible to oil pollution because many of them inhabit the intertidal zone where they become coated with oil and smothered in the event that heavy oil drifts onto the shore. Many of them are filter-feeders which indiscriminately extract fine particles of a certain size range from the water and thereby ingest oil present as droplets or adsorbed on the other particulate material.

In a large spill in the coastal area, molluscs frequently suffer heavy mortalities. Subsequent recolonisation needs several years together. It is, however, interesting to note that oysters, black sea mussel could withstand oil pollution up to a certain extent, as they could intake petroleum hydrocarbons from the environment into their lipid pool. Crude oils, oil derivatives and oil-dispersant emulsions are harmful to the larvae of some marine bivalve molluscs as shown by experiments. But common blue mussel *(Mytilus edulis)* was very much tolerant to oil pollution and these forms can readily take up the hydrocarbons.

The mobile crustaceans, lobsters, crabs etc., however, suffer heavy mortalities due to oil pollution. A wide variety of benthic invertebrates was killed by diesel oil spill into Puget Sound in 1971; with mortalities for 48 species of intertidal invertebrates ranging from 30-100%, the hardest bit belong to brittle stars, polychaetes, semertans, chitors, hermit crabs and limpets.

6. *Fish and Fisheries*: It is probable that fish swim away from heavy oil pollution and particularly from oil spills, where the oil spilled on the ocean causes a decrease in photosynthesis and, consequently, a lowering of oxygen levels in the water. The slimy mucus on the gills of fish make them somewhat resistant to oil. However, there is much evidence that fish are adversely affected by both spilled petroleum and chronic pollution by refined petroleum products. There were significant fish fall (at least 10%) following the oil spill from the Ocean Eagle in the harbour at Puerto Rico; spilled heavy fuel oil at Wave Island and West Falmonth, Massachusetts, also resulted in the death of many fishes. Fisheries are adversely affected by chronic petroleum pollution, and the catch of fish off the coast of Louisiana has decreased concomitantly with the development of the petroleum industries.

The greatest harm to fish life from petroleum hydrocarbons from both spillage and chronic pollution is the effect on eggs, larval stages, and fry. The destruction of young fish can have serious long-term consequences. After the Torrey Canyon accident (1967), 50 - 90% of the eggs of pilchards in Cornish waters were dead.

7. *Birds and Mammals*: The oiling of birds in the marine environment has received considerable publicity and is a serious problem associated with oil spills and oil slicks emanating from wasting of tankers. These oiling can be catastrophic for a population as migratory activity can bring a major proportion of a bird species to a confined area, oil fouling of birds can cause a disruption in the insulating capacity of feathers and result in death from pneumonia or it can cause a loss in buoyance and subsequent drowning. The estimated number of birds that died as a result of typical spills are as follows: Torrey Canyon 40,000-100,000; Santa Barbara blowout 3,600; San Francisco Bay spill 7,000; standing of the Gerd Maersk in the Elbe Estuary, 250,000-500,000. The birds, like walruses, sea lions and seals are potential victims of oil pollution. The mortality of sea birds resulting from oil spills is given in Table 12.3.

Table 12.3: Mortality of Seabirds Resulting from Oil Spills

Year	Vessel Name or Source of the Oil	Site	Spill Volume (t)	Oiled Birds No. Found	Oiled Birds Estimated Mortality
1991	Tenyo Maru	Off Vancouver Island, British Columbia	330	4,300	
1989[a]	Exxon Valdez	Prince William Sound, Alaska	36,400	31,000	390,000
1988[a]	Nestucca	Cray's Harbour, WA	770	12,535	56,000
1988	Barge MCN5	Anacortes, WA	240	None reported	—
1987	Stayvesant	150-300 km off northern British columbia	2,000	Not known	—
1986[a]	Apex Houston	S. California	87	4,198	10,577
1985	Arco Anchorage	Port Angeles, WA	800	1,917	4,000
1984[a]	Puerto Rican	San Francisco Bay	4,900	1,300	4,815
1984	Unknown	Whidbey Island, Puget Sound, WA	17	> 406	> 1500
1984	Mobil oil	Columbia River and WA coast	660	450	?
1983	Swedish tanker	Kattegat, Denmark	500		50,000
1981	Deivos	Helgoland, Norway	1,000	> 3,000	14,000
1979	Kurdistan	Cape Breton, NB	7,900	1,697	—
1979	Russian tanker	Ventpils, Sweden	5,500	3,053	
1978	Amoco Cadiz	Brittany, France	200,000	4,572	20,000
1978	Pantelis a Lemos	Cape Coast, South Africa	300	ca. 100	
1978	Outfall	Dounreay, UK	68	650	> 1,000
1976	Olympic Games	Delaware River, PA	450		
1976	Barge STC-101	Chesapeake Bay	833		20,000-50,000
1975	Olympic Alliance	Dover Spit, UK	2,000	> 199	
1974	Oriental Pioneer	Struisbaai, South Africa	200	"Thousands"	
1974	Metula	Magellan Strait	50,000	3,000	
1972	Dewdale	Cromarty Firth, UK	30		1,000
1972	Oswego Guardian, Texanita	Ystervark Point, South Africa	10,000	> 400	
1971	Barge U17	Padilla Bay, WA	767	> 374	Not known
1971	Collision	San Francisco	2,700	7,380	20,000
1971	Wafra	Cape Agulhas, South Africa	6,000-10,000	> 1,216	
1970	Kazu	Robben Island, South Africa	1,000	> 560	
1970	Arrow	Cape Breton, NB	10,000	567	7,000
1970	Irving Whale	SE Newfoundland	< 30	625	5,000
1969[a]	Hamilton Trader	Irish Sea, UK	700	4,400	5,900-10,600
1969	Palva	Uto. Finland	150	1,000	3,000
1968	Esso Essen	Cape Peninsula, South Africa	4,000	1,250	14,000-19,000
1968	Tank Duchess	Tay Estuary, UK	87	1,368	
1967	Torrey Canyon	English Channel	119,328	7,815	30,000
1966	Seestern	Medway, UK	1,700	2,772	5,000
1961	Collision	Poole, UK	300	487	
1956	Seagate	Olympic Peninsula, WA	Not known		> 3,000
1955	Gerda Maersk	Elbe, Germany	8,000		500,000
1952	Fort Mercer and Pendleton	Monomoy, Mass.	22,400		> 3,500
1937	Frand Back	San Francisco	11,800		10,000

Examples of studies in which experimental data and/or modeling were used to improve estimates of bird mortality.

Note: 50 barrels = 7 tonnes; 1 barrel = 0.14 tonnes; 1 tonne = approx 300 gallons (U. S.); 1 tonne = approx. 1 ton. 1 tonne = approx. 1,100 liters

Source: Burger A.E. Mar. Pollut. Bull. 26. 140, 1993

Coral Reefs: Coral reefs generally formed in tropical oceans are an environment of enormous biological diversity, complexity and productivity. The branched corals are particularly sensitive to oil and markedly so while exposed to air and many reefs in the Pacific Ocean and Indian Ocean protrude far above the water surface at low tide. It would be harmful to the reef if significant quantities of toxic petroleum were absorbed into this integral part of the biosystem.

Thus the overall effects of petroleum hydrocarbons in the marine environment may be summarised as follows:

(i) The introduction of high concentration of selected hydrocarbon factors can be lethal to seafood species. At sublethal levels, the seafood can contain hazardous concentrations of petroleum hydrocarbons or be tainted such that taste or aroma renders the food unfit for human consumption. This is of considerable consequence to the commercial seafood industry.

(ii) Oil slicks on water or oil deposited on beaches are unsightly and tar lumps carried by the tides onto beaches adjacent to major shipping lines (Florida Coast, Bermuda) are undesirable from an aesthetic standpoint.

(iii) A decrease in diversity and/or productivity of a species can bring about modifications that can have long-range deleterious effects on the entire marine ecosystem.

(iv) Extensive damage to wild life, generally caused by large spills or blowouts in offshore drilling, can significantly decrease population of seabirds and marine mammals.

A summarised view of oil spill effect on oceanic communities is given in Table 12.4:

Table 12.4: Summary of the Effects of Oil Spills on Oceanic Communities

Community	Effect	Period of Impact
Plankton		
Phytoplankton biomass and primary production	Increase due to diminished grazing; depression of chlorophylla	Days to weeks, during occurrence of slicks
Zooplankton	Population reduction; contamination	
Fish eggs	Decreased hatching and survival	
Benthos		
Amphipods, isopods, ostracods	Initial mortality; population decrease	Weeks to years, depending on oil-retentive characteristics of habitat
Molluscs, especially bivalves	Initial mortality; contamination; histopathology	
Opportunistic polychetes	Population increase	
Overall macrobenthic community	Decreased diversity	
Intertidal and littoral		
Meiofaunal crustaceans, crabs	Initial mortality; population decrease	Weeks to years, depending on oil-retentive characteristics of habitat
Molluscs	Initial mortality; contamination; histopathology	
Opportunistic polychetes	Population increase	
Overall community	Decreased diversity	
Algae	Decreased biomass; species replacement	
Phanerogams	Initial die-back	
Fish		
Eggs and larvae	Decreased hatching and survival	Weeks to months
Adults	Initial mortality; contamination; histopathology	
Birds		
Adults	Mortality; population decrease	Years

12.5 MICROBIAL DEGRADATION OF OIL AND PETROCHEMICALS IN THE SEA

A variety of microorganisms capable of growing on oil have been isolated from the marine environment. About 191 strains belonging to the genera Pseudomonas. Achromobacter, Acinetobacter, Arthrobacter, Flavobacteria, & Brevibacteria were isolated from sea water near coastal oil discharge sources. Several effective fungal strains were also isolated from sea water—these include *Cunninghamella elegans,*

Aspergillus versicolor, Cephalosporium Sp., Acremonium Sp. and *Penicillium Sp.* These fungi could grow on a wide range of hydrocarbon substracts.

As soon as oil enters the sea it becomes contaminated with a variety of microorganisms, a number of which can grow on the oil and its degradation products. Most workers consider that bacteria are by far the most important microorganisms involved in oil biodegradation in the sea. In the laboratory, several strains of Pseudomonas, Achromobacter, Flavobacterium were tested for their oil biodegradation efficiency. In the seawater and sediments of oil polluted areas, the frequency of the hydrocarbons oxidising bacteria were very much (10^8 - 10^9/ml of water or mud).

Microbially infected oils may undergo marked modifications in their chemical composition. The rate at which such changes occur in the sea is influenced by certain environmental factors (viz., water temperature, oxygen, nitrogen and phosphorus content of water) and also by the initial composition of the oil.

Crude oil contains 90 - 99% of hydrocarbons, the remainder comprising compounds containing sulphur, oxygen and nitrogen and also trace amounts of metals. Microorganisms can attack most hydrocarbons found in crude oils and elsewhere. The rate of biodegradation of oil at sea is highly variable, which depends on number of factors. The microbial flora in the oil film and oil deposits of the bottom are also rich in sulphate-reducing and denitrifying anerobic bacteria and sulphur oxidising aerolic bacteria like Thiobacillus.

Phenolic wastes are admittedly the most serious pollutants of oil refineries because even minute amounts of their chlorinated derivatives input objectionable taste and odour to drinking water. Several laboratories and companies established a number of microbial methods for disposal of such wastes.

The reaction leading to the decomposition of such wastes is:

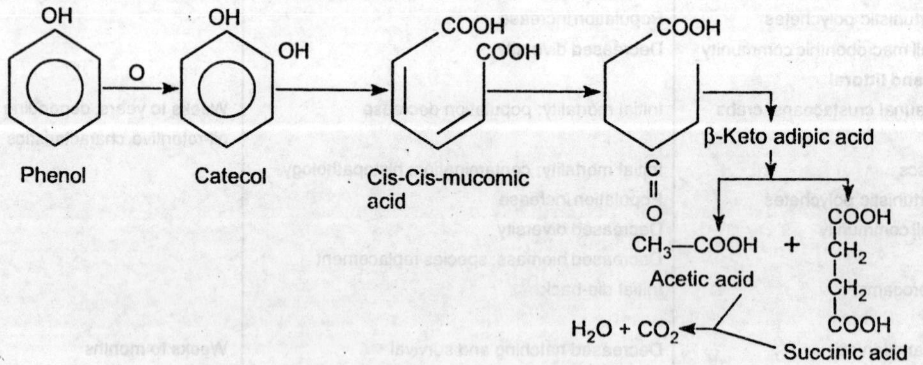

A number of bacteria viz. Bacillus subtilis, B. megaterium, E. coli, Chromobacter sp are quite efficient in phenol degradation. Thus disposal of refinery and oil field wastes into the river or sea needs proper microbial treatment for control of oil pollution into the sea.

In this context it is worthwhile to note the fact that the first engineered microorganisms with advanced degradative properties were generated by Dr Ananda Mohan Chakravorty and his colleagues in the 1970s. Plasmids were transferred into a bacterial strain that could degrade several compounds in petroleum. Subsequently Dr. Chakravorty obtained the first US patent for a genetically engineered microorganism.

12.6 METALLIC POLLUTANT AND AQUATIC BIOTA OF THE SEA

A large number of metallic pollutant are released into the sea through various kinds of effluent discharges. Among them chromium, cadmium, mercury, lead, copper, zinc, nickel, silver, etc. are very important. Most of these trace metals have tremendous effects on the growth and development of aquatic biota.

At sublethal concentration of metallic pollutant there are three kinds of reaction that takes place. Many organisms possess the ability to detect high levels of these pollutants in solution in the ambient aquatic environment. They maintain an avoidance reaction to overcome the pollution toxicity. Pollutant induced physiological changes may also be adjusted by some organisms. The predator-prey relationships may, however, be disturbed very much even at lower concentrations of trace metals.

Trace metal toxicity is very much pronounced in phytoplanktons, macroalgae, molluscs, echinoderms, crustaceans, and finfish.

12.7 FLORA AND FAUNA AS BIOINDICATORS TO OIL SPILL

Bioindicator of Marine Pollution: The selection of a suitable organisms is one of the first tasks in the preparation of a monitoring survey for pollutants, once the decision to use a biological indicator has been made. The basic prerequisites for this sort of study are:

(a) The organisms should accumulate the pollutant without being killed by the levels encountered in the environment.

(b) The organisms should be sedentary in order to be representative of the study area.

(c) The organisms should be abundant throughout the study area.

(d) The organism should be sufficiently long-lived to allow the sampling of more than one year class, if desired.

(e) The organisms should be of reasonable size, giving adequate tissue for analysis.

(f) The organism should be easy to sample and hardy enough to survive in the laboratory, allowing defecation before analysis and laboratory studies of pollutant uptake.

(g) The organism should tolerate brackish water.

(h) A simple correlation should exist between the pollutant content of the organism and the average pollutant concentration in the surrounding water.

(i) All organisms of a given species used in a survey should exhibit the same correlation between their pollutant content and the average pollutant concentration in the surrounding water, at all locations studies, under all conditions.

(j) It is also vital to examine the kinetics of pollutants in an organism which is being considered as a candidate for an indicator survey.

Whilst phytoplankton have attracted considerable interest in respect of their response to the toxic action of trace metals or organochlorines, they have been rarely used as true indicators for the comparison of pollutant abundance at more than one location. The principal reasons for this appear to be the difficulty in obtaining reasonable sample amounts free from other material or other organisms (zooplankton) and the acknowledged importance of species composition in the accumulation of pollutants by phytoplankton. Although the indicator ability of phytoplankton is suspected because of the effects of species composition the uptake of pollutants from sea-water by phytoplanktons is extremely important in the cycling of these chemicals through the ecosphere as it represents a bioconcentration and hence dictates the amounts of each pollutant passed on to higher organisms through the food chain.

As the macroalgae satisfy many of the basic requirements to be an indicator organism, it has been often used for detection of trace metals particularly.

Ascidians and Annelids can often be used as indicators for trace metals. Molluscs have certainly been the most commonly used indicators to date at least for trace metals; these are as monitors of organochlorines has not been as widespread as that of finfish.

Research into pollutant kinetics in finfish has been widespread and the use of finfish species to monitor trace elements and organochlorines has been no less common. The major reason for this emphasis has undoubtedly been the economic importance of finfish to man.

Bryan (1976), in his excellent review of trace metals in marine ecosystems, stated that the evidence to that point suggested that Cu, Mb, Zn, at least were regulated over a range of ambient concentrations in finfish.

Several marine mammals—viz., seals, dolphins and sea-lions—has potential significance to be a bioindicator with reference to trace metals or to organochlorines.

12.8 BIOLOGICAL IMPACT SYSTEM

The major biological impact processes are:

- The presence of petroleum in the marine environment decreases the surface tension at the organisms-water interface, accompanied by the impairment of osmoregulation mechanism;
- The dissolved fractions of petroleum products in sea water can block certain metabolic interactions of the organism across the marine environment and thus affect the biocommunication system in populations;
- There is loss of sensory connections and chemoreception;
- There is also accumulation of petroleum products in marine organisms like zooplanktons, crustacean, benthic organism, and fishes.

A summary of impact on biological system is given in Table 12.5:

Table 12.5: A summary of petroleum hydrocarbon effects on marine biota

Biotic communities	Effects of petroleum hydrocarbons & other products	Period of impact
Bacterial population (microbes)	Decrease in diversity * Prevalence of dominant degradable bacterial population (e.g. Pseudomonas, Bacillus, Vibrio)	Few weeks after oil spill
Phytoplanktons and Zooplanktons	* Decrease in diversity and population size * Decrease in primary productivity	From the days of oil spill to few weeks
Intertidal fauna	* Mortality and population decrease	About a month
Mangroves, sea grass & sea weeds	* Decrease in population size & productivity * Mortality enhanced	About a month
Benthos (Crustacean, Polychetes, Molluscs, Echinoderms etc.)	* Decrease in population size & diversity * Mortality enhanced except polychetes whose growth was enhanced	From a week to a year or so.
Fish and Mammals	* Mortality enhanced * Decrease in population	From a week to a year
Birds (Coastal & littoral)	* Mortality enhanced	From a week to year

Oil spill impact Assessment and management of Coastal Ecosystem with particular reference to living resources:

The pollution of marine environment due to oil spill is a well-established fact for last few decades. Its impact on living resources of the marine environment is also very much important.

Normally the impact of oil spill pollution was assessed:

(a) By measuring gross changes on the biological systems at a regular interval.

(b) By measuring the ecosystem structure before and after the pollution in any specified area.

(c) By estimating the loss of future economy from marine products.

(d) By estimating the loss of wet land which supports the ecosystem.

(e) By estimating the loss of other animal population as consequent changes of pollution of marine environment, (i.e., population of birds intertidal and benthic animals etc.).

(f) By assessing the aesthetic offenses.

All the above-mentioned aspects of oil spill pollution should be carefully assessed period to period to formulate any coastal resource management plan.

12.9 STATUS OF COASTAL AND ESTUARINE POLLUTION IN INDIA

India has a coastline of 6,000 k.m. and the coastal areas have provided excellent harbours which have been the nerve centres of great human settlements and industries. About 300 million people of India (about 25% of the total population) live in the coastal areas. Thus the marine environment as a whole is polluted from a variety of pollutants generated urban settlements, industrial discharges & oil spill etc. The industrial waste going into the coastal waters has been estimated to be about 50% of the total sewage outfall. Very large quantities of urban sewage is disposed off in coastal regions, much of it untreated. About 3,725 MLD industrial and sewage discharged to coastal water in 1990s.

In the past most of the Indian coastal waters were famous for the high biological productivity, but in recent years due to pollution of marine environment the productivity of marine resources is drastically reduced. The coastal polluted areas are shown in Table 12.6.

Table 12.6: Major polluted coastal areas in India

Place	Activity	Impacts
Alang, Gujarat	Ship breaking	Breaking 200 ships annually produces two million tonnes of steel and huge quantities of hazardous solid waste, which is either burnt on the same plot or disposed at the seashore. COD values are as high as 469 mg/l, and oil and grease rise up to 23 mg/l.
Area around Marine National Park, Jamnagar coast, Gujarat	Unloading imported crude oil and petroleum products, laying oil pipelines, tourism, world's fifth largest refinery is located here.	The Marine National Park houses 200 species of marine flora and fauna, which are becoming endangered due to the pollution.
Porbandar, Gujarat	Discharge of effluents from chloro-alkali industry.	Plankton species of Sargassum, Turbinaria and Gelidiella have become extinct.
Tuticorin, Tamil Nadu	Discharge of untreated domestic wastes and industrial effluents.	Significant levels of mercury and pathogenic bacteria.

Contd.

Table 12.6: Major polluted coastal areas in India

Place	Activity	Impacts
Ennore estuary off Cooum river, Tamil Nadu	Discharge of domestic wastes and industrial effluents.	Low level of dissolved oxygen.
Gautami-Godavari, Kakinada Bay and Kalingapatnam, Andhra Pradesh	Riverine runoff and sewage from neighbouring areas through Godavari and Vamsadhara rivers.	Low levels of oxygen and increased levels of ammonia, total phosphorus, nitrogen and suspended matter. In Kakinada Bay high BOD values ranging from 2 to 4 mg/l have been observed.
Visakhapatnam Harbour, Andhra Pradesh	Wastes from domestic and industrial units located around it.	Low values of DO and high values of phosphate, ammonia and metals.
Thana creek, Mumbai	Effluents of over 50 MLD.	High mercury levels are present in the water, sediments and organisms dwelling in the creek.
Ambarnath, Ulhasnagar and Kalyan, near Mumbai	Discharge from polluted river Kali	Mercury concentration exceeding 100 mg/l.
Haldia, West Bengal	Sewage and effluents	Up to 10 μg/g of pesticides, E. coli, shigella, salmonella and other human pathogens.
Mandovi-Zuari estuaries, Goa	30 MLD of partly treated domestic sewage and 15 MLD of industrial and agricultural effluents.	Agricultural productivity has reduced in the region.
Source: Complied by WWI Research Team		

The studies conducted at Mumbai, reveals the fact that the amount of industrial wastes mixing with the coastal waters has been considered to be of the order of 3.5 km^3 per year which has also resulted in the substantial pollution of nearshore water.

Not much data are available on the concentration of pollutants in the offshore areas of the Arabian Sea and in the Bay of Bengal. Toxicity of metals, pesticides and chlorinated hydrocarbons which are accumulated and transferred to higher trophic levels have been studied recently. The results have shown that the levels of toxic chemicals in water, fish and zooplanktons indicate expected natural background and there is no measurable deviation due to the wastes dumped in the coastal regions. However the distribution of metals in coastal water of India is shown in Table 12.7.

Transportation of about 60% of the world's crude oil in Indian Ocean has rendered Indian seas prone to oil pollution. In addition, the indiscriminate disposal of sewage from urban settlement and industrial wastes has created pockets of pollution along the Indian coast, thus affecting the ecology of our coastline.

12.10 OIL SPILL — CLEAN-UP OPTIONS

For protection of coastal vegetation—particularly in salt marsh and mangrove regions—certain clean-up measures could be adopted. They are illustrated in the Tables 12.8 & 12.9.

Table 12.7: Distribution of Heavey metals in Coastal Water of India

Water Samples	Metal (mg/lt)										References
	Cu	Zn	Pb	Cr	Mn	Ni	Co	Cd	Hg	As	
Arabian Sea	4.9	19.2	—	—	—	3.2	2.2	—	—	—	Sengupta, B. et al (1978) Ind. J. Mar. Sci, 7, 295
Bay of Bengal	2.7 to 6.8	2.6 to 15.3	—	—	—	0.8 to 30.3	—	0.3	0.05 to 0.3	—	Qusim, S.Z, and Sengupta S, 1983, Mar, Poll. Studies in India, South Asian Publishers Pvt. Ltd. New Delhi, P. 139
Bombay Harbour Bay	26.7 to 34.0	26.65 to 33.90	—		2.78	—	—	—	—	—	Matkar et al 1981 Indian. J. Mar. Sci Io, 35
Vellar estuary (Chennai)	8.0 to 26.5	25.5 to 95.5	—	—	—	—	—	—	—	—	Lyla 1991, Ph.D. thesis, Anamalai, TN Univ.
Rusikulya estuary (Odisha)	ND to 10.0	1.5 to 65.0	1.0 to 39.5	—	—	—	—	—	—	0.12 to 13.3	Mahapatra, B.C. 1994. Ph.D. thesis, Cochin Univ. of Science & Technology
Mandovi/Zuary estuary (Goa)	2.5 to 69.77	2.8 to 42.3	—	—	6.2 to 102.2	—		—	—	ND to 66.57	Zingade et al, 1976 Ind. J. Mar. Sci 5,212
Cochin backwater (Kerala)	2.0 to 40.0	25.0 to 165.5	ND to 60.5	—	—	—		—	—	—	Mahapatra B.C. 1994 Ph.D. thesis, Cochin Univ. of Sci. & Tech.
Sea Water	3.0	10.0	3.0	—	0.02	0.005	0.005	0.04	0.0003	—	Orr & Marshall, 1969. The Fertile Sea-fishing News Ltd. London, 13

'—' Not detected

Table 12.8: Recommended Clean-up Options for Salt Marshes

Clean-up	Marsh Fringe	Marsh Interior
Protection Measures	Offshore chemical dispersal Booms/Skimmers, Sorbent booms, Sorption	
Primary	Natural cleansing Low-Pressure flushing/recovery sorption vacuum pumping	Natural cleansing Low-pressure flushing/recovery
Secondary	Vegetation removal/flushing/recovery manual removal	Vegetation removal/flushing/recover
	Chemical dispersal, Mechanical clean-up, burning	
Future Use	Enhanced biodegradation	

Table 12.9: Recommended Clean-up Options for Mangrove Forests

	Mangrove Forest Types				
	Fringe Forests	Dwarf Forests	Riverine Forests	Basin Forests	Overwash Forests
Protective Measures (along shoreward edge of forests)	Booms, Sorbents, Sorbent booms				—
Clean-up Measures Forest Edge	Low-pressure flushing/recovery		Low-pressure flushing/recovery vaccum pumping, sorbents		—
Forest Interior	Natural cleansing				Natural wasteland
Avoid Use	Earthen barriers, Manual removal		Burning/Cutting, Steam cleaning		
Future Use	Chemical dispersal, Enhanced biodegradation				

13 LAND POLLUTION

13.1 CONCEPT

The global weathered earth crust form soil over the centuries that supports the variety of microscopic and macroscopic life-forms. The process of soil formation is very complex involving a number of physical, chemical and biological transformation (Fig. 13.1). The topmost layer of soil is comparatively more rich in nutrients and supports maximum bio-forms. It is composed of minerals of various sizes and organic matters along with pore space filled with air and water. The composition by volume of a typical top soil is given in Fig. 13.2. But in deeper layer soil, composition with respect to particles, nutrients and other associated properties varies distinctly. Thus, there are a number of zones one cannot notice when a soil profile is made

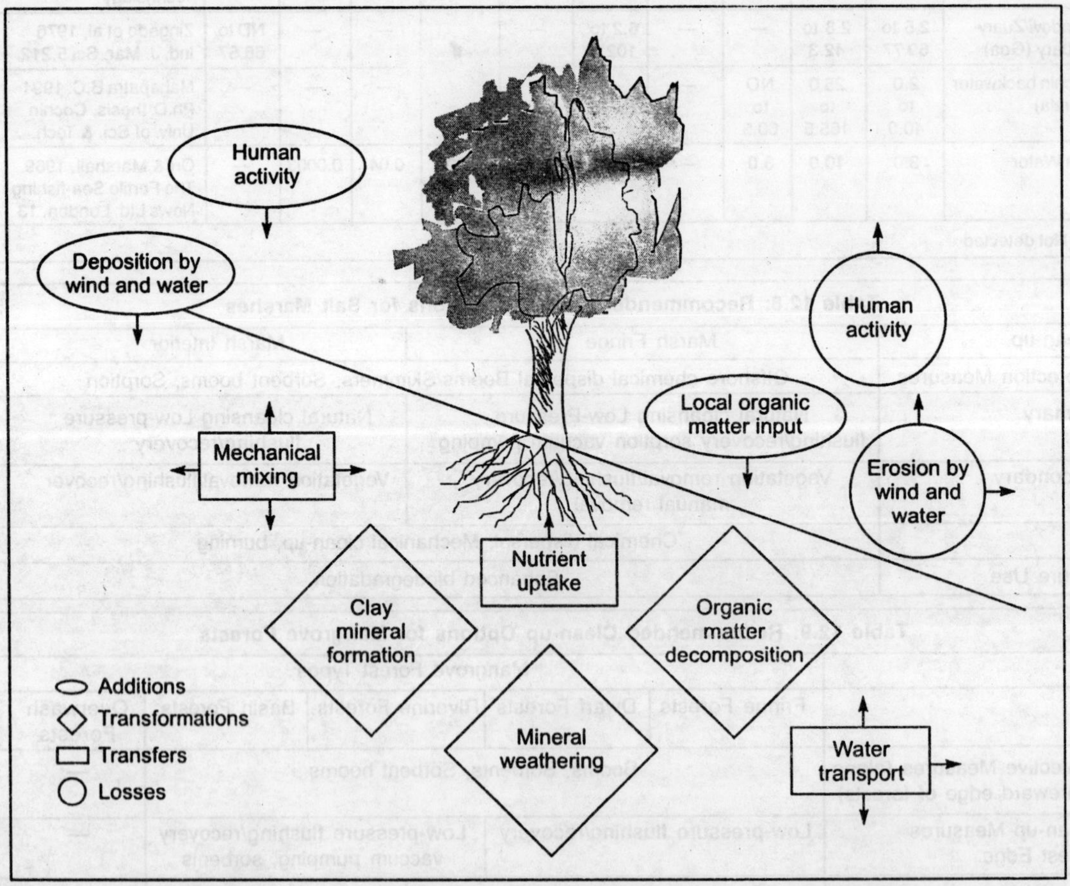

Fig. 13.1 : The major soil-forming processes (based on Simonson, 1978)

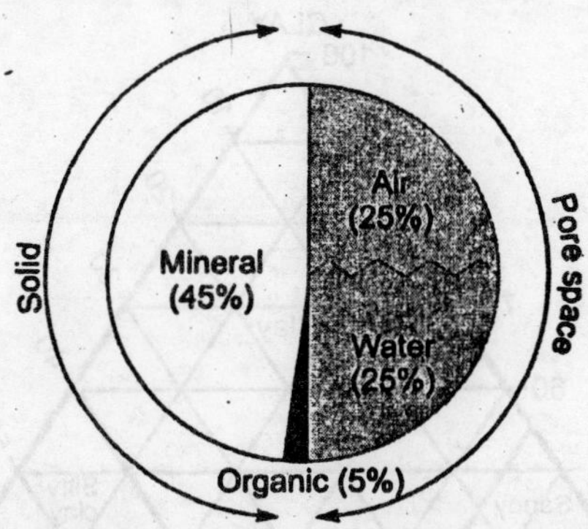

Fig. 13.2 : Composition by volume of a typical topsoil

for analysis of soil depth. The profile character varies distinctly from place to place, particularly with respect to their depth, colour and composition. A generalised profile of soil is shown in Fig. 13.3.

There are different classes of soil types primarily based on the particle size distribution pattern. The details are given in textural triangle of soil classes (Fig. 13.4).

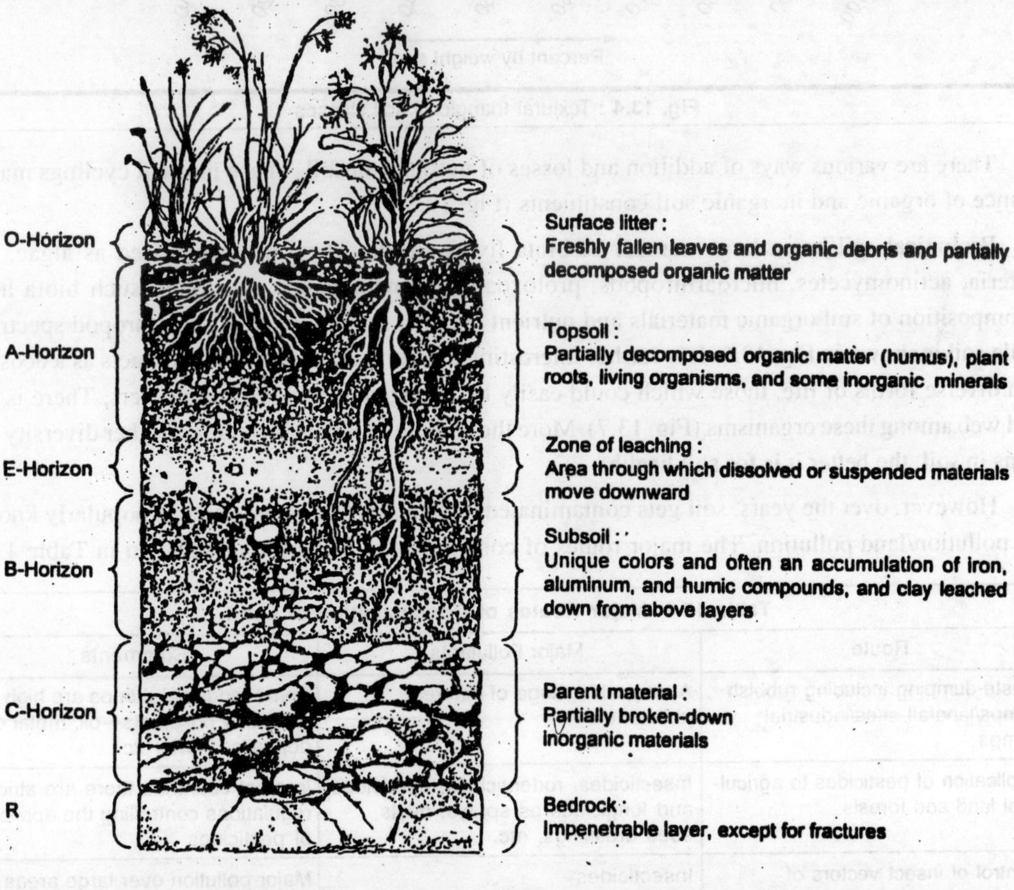

Fig. 13.3 : Generalized profile of soil : zones vary in number and composition

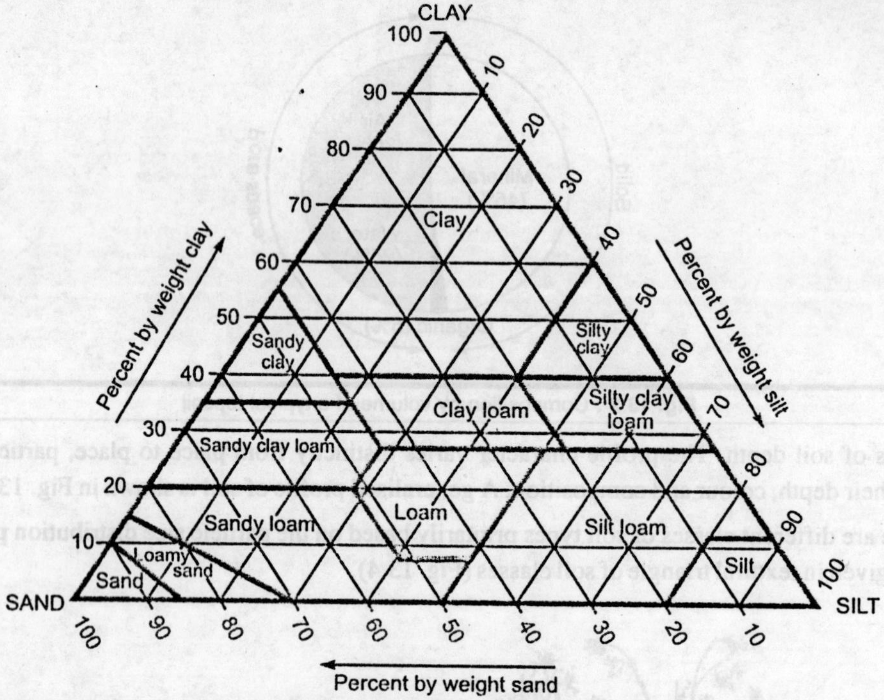

Fig. 13.4 : Textural triangle of soil classes

There are various ways of addition and losses of nutrients in soil. These nutrient cyclings make the balance of organic and inorganic soil constituents (Fig. 13.5).

Biological soil has a large number of biota living inside. They are categorised as algae, fungi, bacteria, actinomycetes, microarthropods, protozoans and nematodes, etc. Many such biota help in decomposition of soil organic materials and nutrient cycling. Some typical microarthropod spectrum of fertile soil is shown in Fig. 13.6. It is further interesting to note the fact that soil itself acts as a ecosystem with diverse forms of life, those which could easily be arranged in various trophic tiers. There is also a food web among these organisms (Fig. 13.7). More the complex food web vis-a-vis higher diversity of life forms in soil, the better it is for soil health.

However, over the years, soil gets contaminated by a number of ways which is popularly known as soil pollution/land pollution. The major routes of contamination of land/soil is shown in Table 13.1:

Table 13.1: Major Routes of Contamination of Land

Route	Major Pollutants	Comments
Waste-dumping including rubbish dumps/landfall sites/industrial dumps	A very wide range of different pollutants	Some industrial dumps are high in particular pollutants—oil, metal ore deposits, PCBs etc.
Application of pesticides to agricultural land and forests	Insecticides, rodenticides, herbicides and fungicides as sprays, dusts, seed dressings, etc.	In most countries there are strict regulations controlling the application of pesticides
Control of insect vectors of disease	Insecticides	Major pollution over large areas as a consequence of control measures against malarial mosquito and fly

Contd.

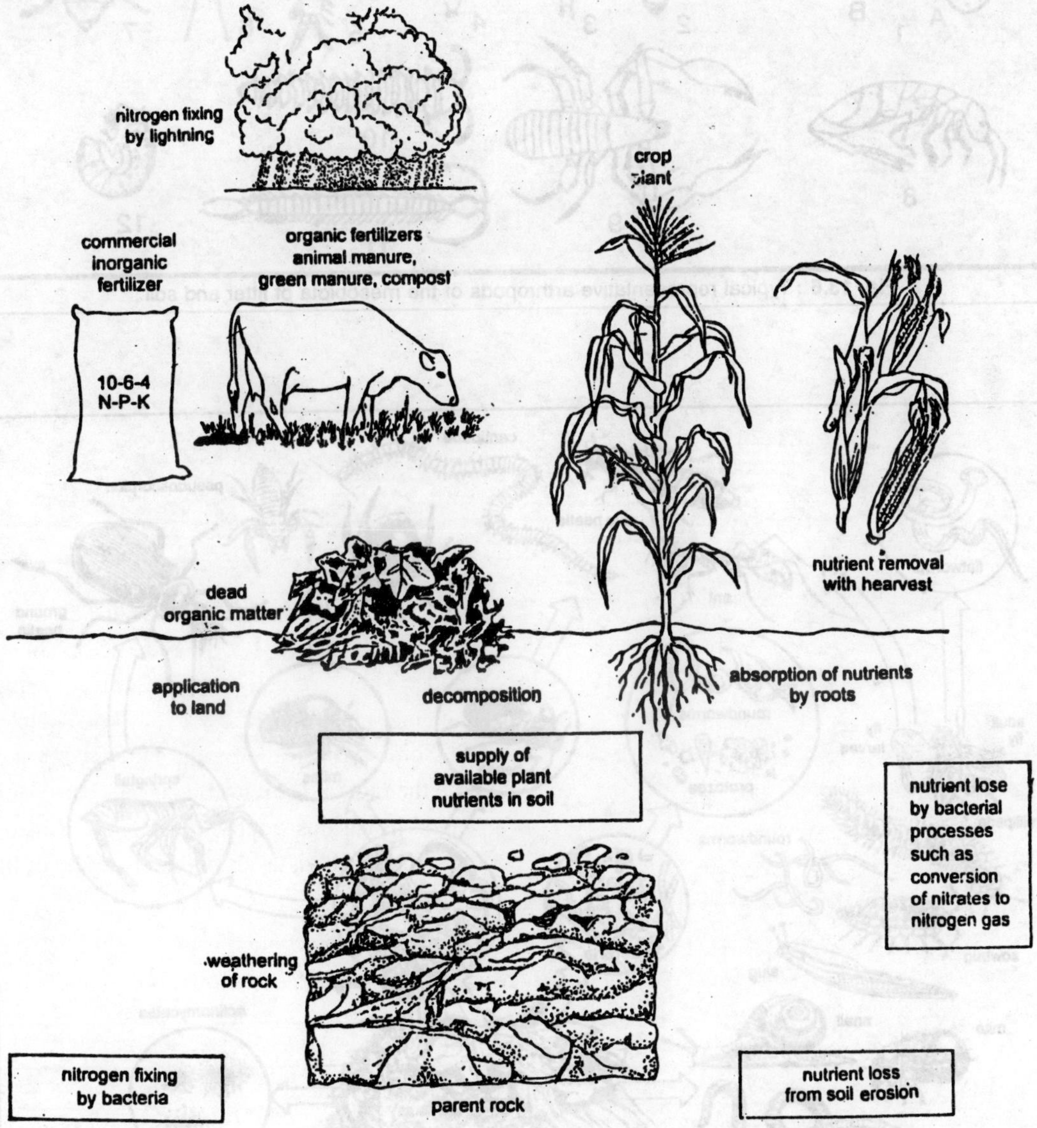

Fig. 13.5 : Addition and loss of plant nutrients in soils

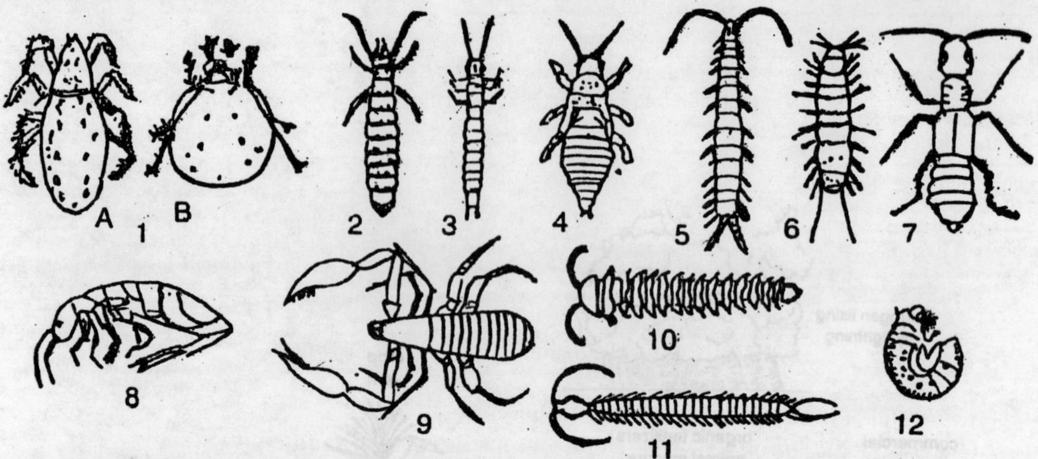

Fig. 13.6 : Typical representative arthropods of the mesobiota of litter and soil

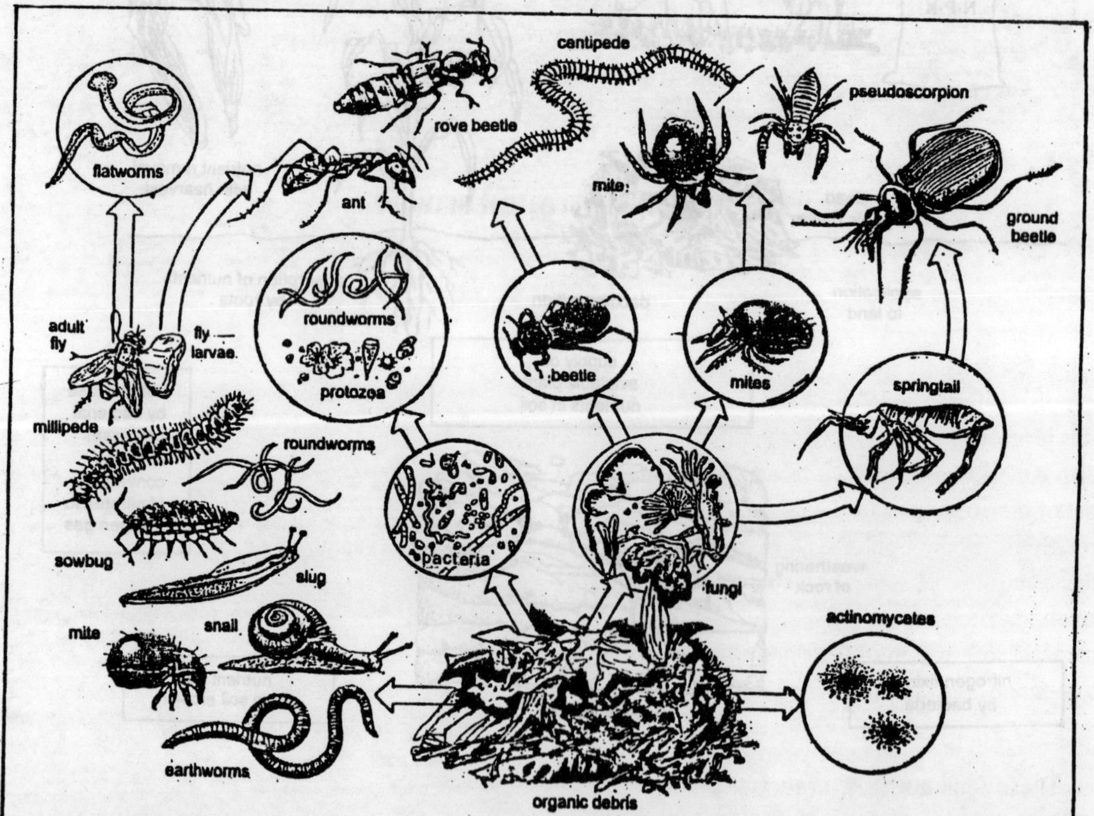

Fig. 13.7 : Greatly simplified food web of living organisms found in soil

Table 13.1: Major Routes of Contamination of Land

Route	Major Pollutants	Comments
Application of sewage to agricultural land	Heavy metals, nitrates, detergents	
Flooding by rivers or seas	A variety of pollutants including those associated with sewage	
Precipitation from air as dust or droplets or in rain or snow	Pollutants associated with soot and dust, acid rain, pesticides	Transport may be over short distances (spray drift, soot and dust from chimneys) or long distance (brought down especially by rain and snow)

There is a continuous elemental cycle operated in soil by microbes in addition to sustained addition and loss of nutrients from soil by various activities, resulting in the rapid transformation of soil nutrition in short and long-term basis (Fig. 13.7).

13.2 LAND/SOIL POLLUTION

Soil gets polluted by a number of ways. The major kinds of soil pollution are:

> Acidification,

> Salinisation and sodification,

> Agrochemical pollution,

> Contamination by metalliferous water.

(a) Acidification

Acidification has a number of natural and anthropogenic causes (Table 13.2). The main natural causes are long-term leaching and microbial respiration. The acids found in rainwater (carbonic acid) and in decomposing organic material (humic and fulvic acids) can stimulate leaching by dissociating into H^+ ions and their component anions which then displace or attract base cations from the soil exchange complex. Leaching of bases is most common where precipitation exceeds evapotranspiration and includes many eastern parts of North America and northwestern parts of Europe, where soils have been subjected to leaching throughout most of the 10,000-15,000 years of postglacial time. Microbial respiration also leads to soil acidification through the production of CO_2 (carbon dioxide), which is dissolved in soil water to form carbonic acid. Other natural processes associated with soil acidification are plant growth and nitrification. During plant growth, nutrient base cations are obtained through root systems in exchange for H^+ ions, thus leading to increased soil acidity. Nitrification is an oxidative process of organic decomposition whereby NH_4^+ (ammonium) ions are converted to NO_3^- (nitrate) ions by nitrifying bacteria, with H^+ ions as a byproduct:

$$NH_4^+ + 1.5O_2 \Rightarrow NO_3^- + 4H^+$$

These ions are then available for displacing and attracting base cations from the soil exchange complex, as mentioned above, thus leading to soil acidification.

The main anthropogenic causes of acidification include certain land-use practices, such as needleleaf afforestation, excessive use of inorganic nitrogen fertilizers, land drainage and acid deposition resulting from urban and industrial pollution. Needleleaf afforestation has been associated with the acidification of soils and surface waters for a number of reasons. First, needleleaf trees produce litter which is very acidic in comparison with most broadleaf species. Second, because of their high canopy surface area, needleleaf trees are able to 'scavenge' acid pollutants from the atmosphere, later releasing them into the soil via throughfall and stemflow. Third, due to modifications of the surface and soil hydrology by drainage channels and shallow root networks, water transfer is rapid and is concentrated either at the surface or in the uppermost layer of the soil. Under these conditions, the residence time of the water in the soil is limited, as is the depth to which it can percolate. Thus, the contributions of weathering and ion exchange reactions to the buffering process are limited (Bache 1983, Miller 1985).

Table 13.2: Causes of soil acidification

Source		H+ addition or equivalent (kg. $H^+ ha^{-1}a^{-1}$)
Natural	CO_2 in soil pH > 6.5 (calcareous soils)	7.2 – 12.8
	Organic acids in acid soils and from vegetation	0.1 – 0.7
Acid 'rain'	Wet deposition	0.3 – 1.0
	Dry deposition	0.3 – 2.4
	NH_3 and NH_4^+ oxidation	0.7
Land use	Cation excess in vegetation	0.2 – 2
	NH_4^+ oxidation (agricultural soils) and leaching	4 – 6
	Oxidation of N and S from organic matter and leaching	0 – 10

Source : Rowell and Wild 1985

Excessive use of inorganic nitrogen fertilizers in agricultural systems has also been associated with soil acidification, partly through the process of nitrification (see previous equation). If levels of NO_3^- ions in the soil are in excess of plant requirements, they will behave as mobile anions, thus encouraging the leaching process. The acidifying effect of nitrogen fertilizer is mobilised in the form of soluble organo-metallic complexes. If mineral acids predominate, however, then aluminium is mobilised in its ionic, labile-monomeric form (Al^{3+}). This form of aluminium is particularly toxic to many freshwater organisms including fish. In a similar way, the heavy metals, such as lead, zinc and cadmium, are more readily mobilised in acidic soils.

Acidification of soils and associated nutrient leaching has also been implicated in damage to trees in forested areas, particularly in Central Europe and Scandinavia. It is more likely, however, that nutrient leaching acts synergistically with other factors including ozone pollution, acid deposition, NH_4^+ (ammonium) uptake, drought and frost to product stress in the tree (Fig. 13.8). Signs of tree damage include needle discolouration, crown defoliation and deformed branch structures. Increased soil acidity and associated changes in aluminium mobility also have serious implications on the quality of surface waters which receive drainage from acidified soils.

Soils vary dramatically in their ability to buffer acidity, a characteristic known as the *buffering* or *acid-neutralising capacity*. The buffering capacity of a soil is defined as the amount of acid that needs to be added to cause a reduction in pH of one unit (Trudgill 1988). Soils which contain significant quantities of base-rich, weatherable minerals have a high buffering capacity whereas those which are dominated by quartz and similarly resistant minerals have a low buffering capacity. In Britain, for example, soils with the highest buffering capacities are found in eastern and southern areas where calcareous parent materials are

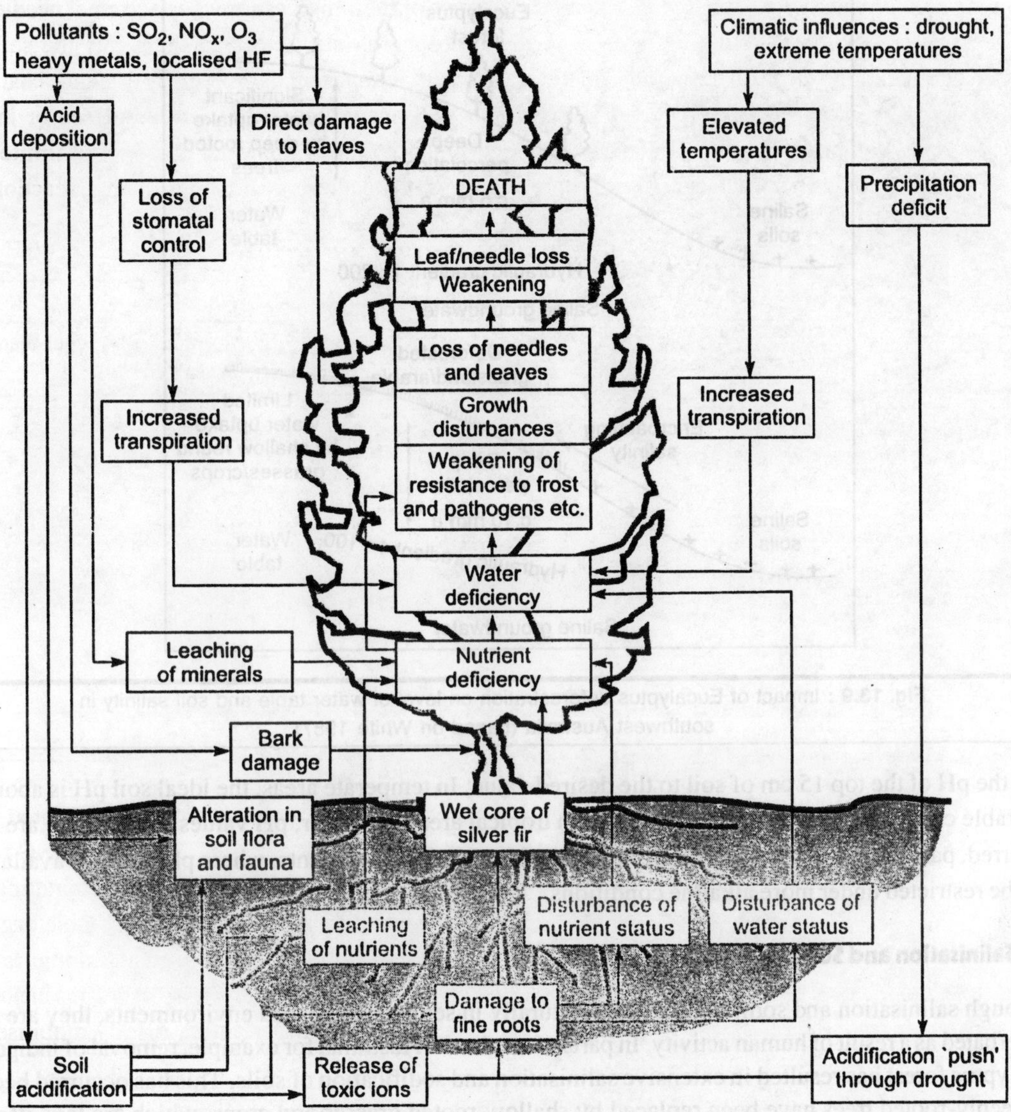

Fig. 13.8 : Tree damage and stress interactions (from Longhurst et al, 1987)

particularly common. Soils with the lowest buffering capacities are found in areas with base-poor, often siliceous, parent materials which predominate in upland areas in the west and north.

There are a number of approaches to the management and remediation of soil and surface water acidity, including liming, sensible forestry management and reduction of acid emissions into the atmosphere. Liming has long been practised on agricultural land to remedy the problems of acidity, slow organic matter turnover, poor modulation in some legumes, calcium and molybdenum deficiency and aluminium and manganese toxicity. Liming materials most commonly used include ground limestone, chalk, and basic slag, the main active constituent being calcium carbonate ($CaCO_3$); other minor components include quicklime (CaO), slaked lime ($Ca(OH)_2$) and magnesium carbonate ($MgCO_3$). The lime requirement of a soil varies depending on its buffering capacity and is usually expressed as the amount of $CaCO_3$ (t ha^{-1}) required to

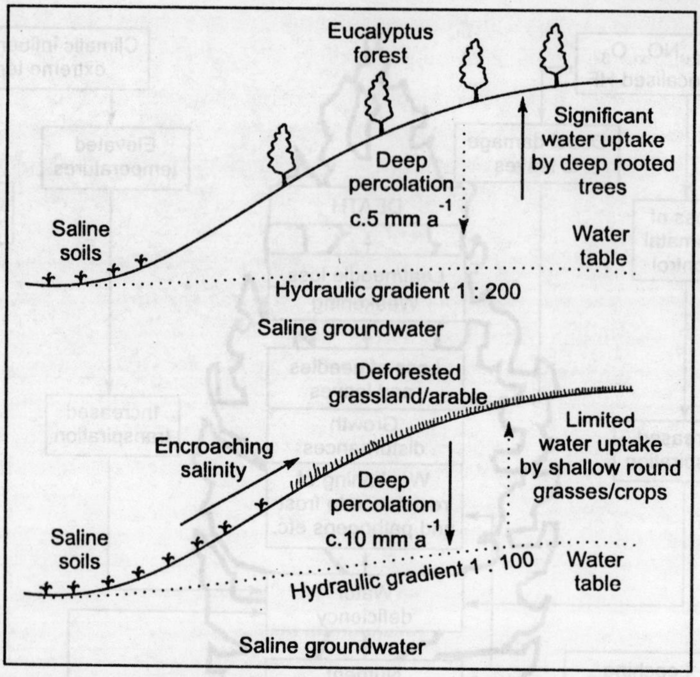

Fig. 13.9 : Impact of Eucalyptus deforestration on level of water table and soil salinity in southwest Australia (based on White 1987)

raise the pH of the top 15 cm of soil to the desired value. In temperate areas, the ideal soil pH is about 6.5 for arable crops and about 6.0 for grasslands. In tropical areas, however, pH values of about 5.5 are often preferred, particularly in soils with high exchangeable aluminium contents, where phosphorus availability may be restricted under more alkaline conditions.

(b) Salinisation and Sodification

Although salinisation and sodification occur naturally in semi-arid and arid environments, they are often exacerbated as a result of human activity. In parts of southwest Australia, for example, removal of indigenous eucalyptus forest has resulted in extensive salinisation and sodification of soils. This has occurred because the deeply-rooted trees have been replaced by shallow-rooted grasses and crops, which are less effective in lowering the ground-water level (Fig. 13.9). Capillary action is most intense, and salinity and sodicity are greatest, in soils where the water table is within about 2m of the surface. Another important cause of soil salinity and sodicity is poor irrigation practice. Overwatering leads to a rise in the water table which, in turn, causes enhanced capillary action. Similarly, poor maintenance of irrigation channels and canals results in leakage of water onto adjacent agricultural land. This has contributed to increased soil salinity and sodicity in parts of the Indus Valley in Pakistan, for example.

In the Nile Valley of Egypt, river impoundment resulting from construction of the Aswan Dam has been associated with a deterioration in soil quality in what were once very fertile floodplain areas. Prior to this, flooding would occur annually in response to heavy seasonal rains in the mountainous source regions of the river. The seasonal flood waters were particularly important to agriculture in the Nile Valley for three main reasons—they were an important source of irrigation water, they brought with them fertile silt deposits which were incorporated into the soils, and they helped to flush out salts from the soil. Since river impoundment, the benefits of seasonal flooding have been removed and the resulting decline in soil quality has been compounded by agricultural intensification, in order to meet both the needs of a rapidly growing population and the demands of overseas markets for cash crops.

Salinisation and sodification of soils are not only associated with arable land, but may also occur in pastoral landuse systems. Lavado and Taboada (1987), for example, examined the salt regimes of soils under grazed and enclosed (ungrazed) natural grassland in the Flooding Pampa of Argentina. Here, soils are affected to varying degrees by salts and Na^+ ions, the water table is high and flooding is common during the wet season. The salt concentration in the topsoil of the grazed land increased dramatically and episodically after flooding, whereas, on the ungrazed land, it did not (Fig. 13.9). This difference was attributed to compaction and impeded drainage and to reduced vegetation cover on the grazed land, which led to increased evaporation and enhanced capillary action in the soils.

Soil salinity and sodicity have a detrimental effect on both chemical and physical aspects of soil quality. In a chemical context, high salinity and sodicity are associated with elevated soil pH and under these conditions the availability of certain plant nutrients is reduced resulting in severe disturbance of the plant nutrient balance as whole. Elevated salt and Na^+ concentrations in soils are also highly toxic to many plants, although tolerance levels vary between different species. Crops such as barley, cotton and sugar beet, for example, have a relatively high tolerance level, whereas sugarcane, onion and lettuce have a very low tolerance level. As with arable crops, grasses vary in their tolerance to soil salinity and sodicity. Bermuda grass, for example, has a high tolerance level whereas Guinea grass is much less tolerant.

Soil salinity and sodicity management and remediation are complex issues and may involve a number of approaches. Appropriate irrigation and drainage techniques are critical in the effective management of saline, saline-sodic and sodic soils. In addition to the amount of water applied, the quality of the irrigation water is important. Use of water which contains high concentrations of salts can lead to increased salinity and sodicity in soils. As the water evaporates, ionic concentrations increase and ions in solution (particulary Na^+) are exchanged with those held on the soil exchange complex (notably Ca^{2+}) until a state of equilibrium is achieved. Saline-sodic soil, in particular, require careful management in this respect because, although structural stability is maintained if the concentration of salts in the soil solution is high, if soluble salts are depleted by leaching, structural deterioration will occur as Na^+ ions begin to dominate the soil exchange complex. Reclamation of such soils requires leaching with water of a low enough SAR to facilitate Ca^{2+} exchange for Na^+, but a sufficiently high total salt concentration to maintain soil structure and permeability. As the soil ESP is gradually reduced, the salinity of the leaching water may also be reduced until reclamation is complete. Intermittent leaching is more effective than continuous ponding as it allows time for salt to diffuse into the depleted outer regions of aggregates from where they are removed during the next leaching phase. During continuous ponding water flows preferentially through the larger inter-aggregate channels and does not remove salts trapped within the aggregates.

(c) Agrochemical Pollution

In recent decades, the use of inorganic fertilizers has increased dramatically at the expense of more traditional organic nutrient treatments. Between 1952 and 1985, the global use of fertilizers increased from 14 million tonnes to 125 million tonnes, an increase of almost 900 per cent. Inorganic fertilizers are used in preference to organic treatments because the nutrients are in a more readily available form and are released rapidly after applications. Organic material releases its nutrients slowly, through decomposition processes, and only when conditions are suitable (warm and moist), not necessarily when crops need them.

Fertilizers are applied in a variety of forms—solution, suspension, emulsion and solid. The solid forms vary in particle size from fine powder to coarse granules and are either spread evenly (broadcast) over the soil surface or mechanically placed, by drilling, into the rhizosphere; generally the rate of nutrient release decreases with increasing particle size. Fertilizers are based on compounds of plant macronutrients (e.g. nitrogen, phosphorus and potassium) and micronutrients (e.g., zinc, copper, boron and molybdenum) and a variety of nutrient combinations are available depending on the nature of the nutrient problem (Table 13.3).

There are five main fates of fertilizers applied to the soil—plant and animal uptake (immobilisation), adsorption and exchange in the soil (fixation) leaching and loss in soluble form through drainage, volatilisation and gaseous losses to the atmosphere (e.g. denitrification) and surface loss in solid form by runoff and erosion (Ross 1989).

In terms of plant uptake, the percentage recovery of nutrient varies markedly between the different types of fertilizers. During the first year of application, the recovery of nitrogen from inorganic nitrogen fertilizers is about 50-65 per cent, whereas from organic manures it is only 20-32 per cent (Fink 1982). Similarly, the recovery of phosphorus and potassium from inorganic fertilizers is about 5-15 per cent and 75 per cent respectively. In terms of fixation in the soil, many of the phosphorus and potassium fertilisers are of relatively low solubility and the nutrients released are often strongly adsorbed. In contrast, many of the nitrogen fertilizers are highly soluble, nutrient release is rapid and, in most soils, adsorption is limited. The differences in degree of immobilisation by plants and animals and in the extent of fixation in the soil, help to explain why leaching losses of nitrogen are usually far greater than those of phosphorus and potassium. Levels of nitrogen in drainage waters, for example, are often in the range 15-150 kg $ha^{-1}a^{-1}$, whereas levels of phosphorus are generally < 1 kg $ha^{-1}a^{-1}$; even in exceptional circumstances in eroded areas, phosphorus levels rarely exceed 10 kg $ha^{-1}a^{-1}$ (White 1987). Nitrogen losses of up to 30-35 per cent of that applied have been recorded for both leaching and denitrification. Leaching is most common in coarse-textured and well-drained soils, whereas denitrification losses are greatest in fine-textured, waterlogged and poorly aerated soils. Similarly, surface losses are greatest at sites which are most susceptible to surface runoff and erosion. In terms of the environmental problems associated with fertilizer use, perhaps the area which has received most attention in recent years, by way of research, is that of nitrate leaching.

Table 13.3: Composition of Some Commonly Used Fertilizers

Nutrient	Compound	Formulae	Composition
(a) Nitrogen			
solids			(%N)
	Ammonium sulphate	$(NH_4)_2SO_4$	21
	Potassium nitrate	KNO_3	13.8
	Ammonium chloride	NH_4Cl	26
	Ammonium nitrate (Nitro Chalk, if mixed with $CaCO_3$)	NH_4NO_3	35
	Urea (Carbamide)	$CO(NH_2)_2$	46
	Calcium cyanamide	$CaCN_2$	21-22
	Gold-N (sulphur-coated, slow-release fertilizer)		32
liquids	Ammonia gas liquors (dilute NH_3 solutions)		1-4
	Aqueous ammonia gas	NH_3	21-29
	Anhydrous ammonia gas	NH_3	82
(b) Phosphorus			(%P)
	Single superphosphate	$Ca(H_2pO_4)_2$; $CaSO_4$	8-9.5
	Triple superphosphate	$Ca(H_2PO_4)_2$	20
	Monoammonium phosphate	$(NH_4)H_2PO_4$	20-30
	Diammonium phosphate	$(NH_4)_2HPO_4$	21

Contd.

Table 13.3: Composition of Some Commonly Used Fertilisers

Nutrient	Compound	Formulae	Composition
	Basic slag		5-10
	Ground mineral phosphate (GMP)/ crushed rock phosphate (CRP)		12.5
(c) Potassium			(%K)
	Muriate of potash (potassium chloride)	KCl	50
	Sulphate of potash (potassium sulphate)	K_2SO_4	40-42
	Saltpetre (potassium nitrate)	KNO_3	36

Source: Ross 1989

In the last few decades, levels of nitrate in water supplies have increased dramatically, particularly in intensively cultivated areas where inputs of nitrogen fertilizers have been high. This association is not necessarily caused however, and may be indirect. Nitrogen fertilizers are not the only source of leached nitrate, as illustrated by the Broadbalk experiments carried out over a 150-year period at Rothamsted Experimental Station in southeast England. It was found that here much of the leached nitrate is derived from the vast reserves of organic nitrogen in the soil, rather than from fertilizers.

The main factors which influence the extent of nitrate leaching include land use, soil characteristics, and climate. In terms of land use, nitrate leaching tends to be greatest when there is no crop cover to utilise the nitrate released from fertilizers or organic reserves. With spring planted cereals, for example, the leaching risk is greatest during the wet autumn and winter period when the soil is left bare. In order to maximise uptake and minimise leaching, fertilizers should be applied just before and during the period of maximum crop growth and large applications at any one time should be avoided. Tillage practices also have an effect on nitrogen losses through leaching and denitrification. Tillage improves topsoil drainage and aeration and, as a result, rates of organic matter decomposition increase. Colbourn (1985) compared nitrogen losses by leaching and denitrification on conventionally cultivated and direct drilled land in southern England. Leaching losses were considerably greater from the conventionally cultivated land, whereas denitrification losses were smaller. For both tillage regimes, however, maximum nitrogen losses occurred during the autumn period after the harvest and during the spring following fertilizer application.

In addition to the environmental problems associated with fertilizer application, a number of problems arise from the use of pesticides. A wide range of pesticides has been developed (more than 450 compounds), the types most commonly used being insecticides, fungicides and herbicides (Table 13.4); other varieties include nematicides, miticides, rodenticides and molluscicides. Pesticides behave in a variety of ways following application. They may be degraded biologically or photochemically, adsorbed by organic matter, clay and oxides/hydroxides of iron and aluminium, washed into water courses by leaching (especially compounds with solubilities >10 mg such as simazine, bromacil and aldicarb) and surface runoff, or they may undergo volatilisation into the atmosphere (especially surface applied compounds and those of low solubility and vapour pressure such as organochlorines) (Fig. 13.10).

Ideally, pesticides should control only the target organism and persist for long enough to achieve this before degrading into harmless products. This is not always the case, however, and a number of environmental concerns arise from pesticide use. These include persistence in the environment, toxicity in soil, vegetation and water supplies and impact beyond the target organism including bioaccumulation and its implications for human health. In fact, the degradation products of some pesticides may be as toxic as the original source chemical but are not often measured by standard assays. The persistence and toxicity of many pesticide compounds and their degradation products are also dependent on a number of soil characteristics, notably clay and organic content and pH.

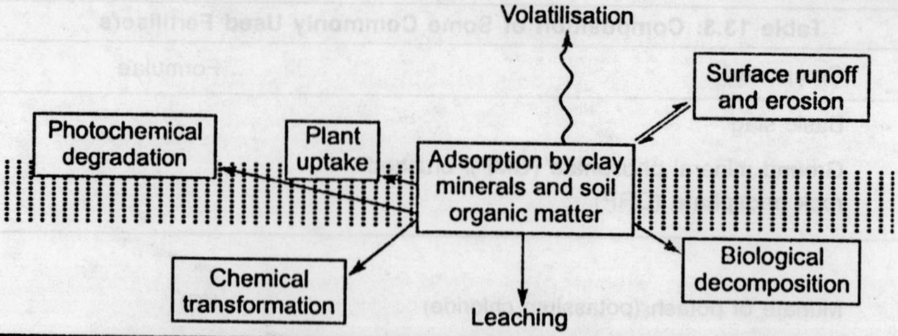

Fig. 13.10 : The fate of pesticides applied to soils (from Ross 1989)

Table 13.4: The Main Insecticide and Herbicide Groups

Pesticide groups	Specific pesticide types
Insecticides	
Organochlorines	DDT
	Aldrin
	Heptachlor
Organophosphates	Parathion
	Malathion
Carbamates	Carbaryl
	Carbofuran
Herbicides	
Phenoxyacetic acids	2, 4 – D
	2, 4, 5 – T
Toluidines	Trifluralin
Triazines	Artazine
	Simazine
Phenylureas	Fenuron
Bipyridyls	Diquat
	Paraquat
Glycines	Glyphosate

Source: Based on Ross 1989

The environmental impacts of pesticide use were felt most strongly with the early generation of pesticides, particularly the organochlorine compounds and those containing heavy metals. Generally, the organochlorine compounds are the most persistent of the pesticides and may survive for a number of years before they are degraded. Their residues have been found widely in soils, freshwater sediments, fish and cow's milk. Due to their high lipid solubility, they have also been found in the fatty tissues of animals. All of these findings have important implications for the ecology of food chains and, thus, for human health. An additional problem was that the effectiveness of organochlorine pesticides decreased as target organisms became resistant to them. The more recently developed organophosphate and carbonate pesticides are much less persistent than the organochlorines, with a half-life of about six months, but are often more toxic. The phenoxyacetic acids, 2,4-D and 2, 4, 5-T, are also rapidly degraded but have been associated with animal and human growth and reproductive abnormalities.

In terms of management and remediation of the environmental problems associated with pesticide use, continued research and monitoring—with the aim of minimising effective persistence and toxicity and maximising specificity—are essential. Attempts have been made to model the behaviour of pesticides in soils under different management regimes, with a view to reducing the risk of ground and surface water contamination. In addition, the use of granular slow-release pesticides is being explored, together with ultra-low volume application techniques. The environmental problems may also be alleviated by adopting alternative, non-chemical strategies of pest, disease and weed control. These include direct approaches, biological and cultural methods and habitat removal. Direct approaches are aimed at clearly identified animals and plants and specific practices include hunting and hand-weeding. Biological methods involve the use of predatory species, preferably with a wide environmental tolerance range, to control the specific target organism. Ideally, as the pest population grows, the predator population should follow, and vice versa; this is known as a *density-dependence relationship*. If such a relationship does not exist, however, the predator itself may become a pest. Cultural methods involve the use of tillage, crop rotation, fertilizer application, liming and drainage techniques. These help to disturb the cycle of pests, diseases and weeds and to improve the resistance of crops to them.

Habitat removal is perhaps the least acceptable approach to pest, disease and weed control. The basis of this approach is that semi-natural vegetation acts as a refuge for many pests, diseases and weeds and that control may, therefore, be achieved through the removal of such habitats. However, this approach fails to recognise both the conservation value of these habitats and the fact that they may be an important source of predators as well as pests. Woodland and hedgerows in lowland arable areas, for example, are an important habitat for ladybirds, essential in the control of greenfly populations and for bees, important in plant pollination. They also act as shelter belts, thus helping to control soil erosion.

(d) Urban and Industrial Pollution

Urban and industrial development has been associated with both physical degradation and chemical contamination of soils. Problems of physical degradation include erosion, compaction and structural damage resulting from construction activities and opencast mineral extraction. Similarly, chemical problems result form waste disposal activities, discharge and spillage of liquid effluents and atmospheric emissions, including acid deposition. Soils of urban and industrial environments are every bit as complex and variable in their characteristics as those in rural areas and merit classification in their own right. Hollis (1991), for example, suggests that the England and Wales soil classification system should be modified to include an 'anthropogenic' major soil group which could be subdivided on the basis of the types of contaminant materials present.

Physical disturbance and chemical contamination of soils in urban and industrial environments are not only recent phenomena. Archaeological investigations have revealed the extensive accumulation of building and domestic waste, although much of this was relatively harmless. Since the Industrial Revolution of the eighteenth and nineteenth centuries, however, the amount and variety of waste materials have increased dramatically. Furthermore, much of this waste is considerably greater than its early historical counterparts. It is particularly important that any survey of contaminated sites should include an assessment of this historical legacy, since the response of soils to stored pollutants may be delayed for long periods of time. This type of delayed response is potentially very serious and is often referred to as the 'chemical time-bomb' effect (Hekstra 1993).

Bridges (1991) identifies four major sources of soil contamination in urban and industrial environments—construction and demolition waste, metalliferous materials, power generation emissions and chemical and organic wastes. During construction and demolition a range of materials are disposed-off in the soil environment, including broken bricks, tiles, glass, timber, piping, wiring and cables, insulation materials, mortar, concrete and plaster. Once in the soil, these materials undergo a number of chemical changes. Plaster, for example, contains large amounts of gypsum and, at sites where the water table is high, the gypsum is dissolved and capillary actions may bring it into contact with new concrete structures, thus leading to serious corrosion problems. Similarly, although the use of asbestos and its removal from existing buildings is now carefully controlled, significant quantities can be present in soils and overburden at sites with a long history of construction and demolition.

Metalliferous wastes, particularly heavy metals (e.g. lead, zinc, cadmium, copper, and nickel), are commonly found in soils of areas where ore extraction and smelting have occurred. More locally, metal contamination occurs on land used for scrap metal dealing and munitions factories. In an attempt to provide systematic data or the extent of heavy metal contamination in urban soils in Britain, a national survey was commissioned by the Department of the Environment in 1981; 100 household gardens were sampled in each of 53 cities, towns and villages. In most of these localities, total lead levels in the soil exceeded 200 mg kg^{-1}, with values in excess of 500 mg kg^{-1} in industrial areas in the south and in former lead mining areas in the north. These values are considerably greater than the background levels of lead (usually < 100 mg kg^{-1}) normally found in uncontaminated soils.

Toxic metals may exist in the soil in a number of forms including adsorbed cations, attached to clay and humus colloids and organo-metallic chelates. Their availability to plants depends on a number of soil characteristics, particularly cation exchange capacity (CEC), pH and the interdependence effects of other metals. In soils with a low CEC, the metals are not retained effectively and are likely to be either leached from the soil or taken up by plants, while, in soils with a high CEC, they are likely to be fixed in the soil through adsorption processes. Similarly, the mobility and availability of heavy metals is considerably greater in acidic soils (pH < 5.5) than in near neutral or alkaline soils. Once mobilised, the metals may enter the food chain either through water supplies and aquatic organisms, or through arable produce and grazing animals. The effects of toxic metals in soils on human health are unclear and it is, therefore, difficult to establish threshold concentrations above which toxicity problems are likely to occur. However, a number of countries have devised such threshold values, although these vary markedly in response to differences in environmental policy and legislation. The Dutch government, for example, has a system with three levels — an acceptable reference or background value, an indicative value for further investigation, and an indicative value for clean-up.

A number of soil contaminants are derived from the power generation industry, including SO_2 (sulphur dioxide) from coal-fired power stations, and radionuclides from nuclear power stations and weapons testing. The anthropogenic radionuclides most commonly found in soils are those of caesium (^{137}Cs and ^{134}Cs). Much of the research in this field has occurred since the accident at Chernobyl in Ukraine in 1986, which resulted in large quantities of radionuclides being deposited across wide areas of Europe, including Britain. The behaviour of radionuclides in soils depends on a number of soil characteristics, particularly clay content and mineralogy, organic content, CEC, pH, NH_4^+ (ammonium) content and nutrient status. The radionuclidesa are immobilised most strongly in soils with a high ECE and near neutral pH values, where they are adsorbed on to clays, especially micaceous varieties, and humic materials. They are least well retained in acidic soils with a low CEC, where they may be available for plant uptake. Such conditions are widespread in upland areas of northern and western Britain where the milk and meat of grazing cattle and sheep were contaminated following the Chernobyl accident. In northern Scandinavia, reindeer herds grazing on slow growing lichen heath were similarly affected. The problem of radionuclide entry into the food chain from acid soils in Britain was not anticipated because much of the early research here had focused on clay soils with near neutral pH values where radionuclides are relatively immobile than the other contaminants. These include derivatives of detergents, fertilizers and pesticides, paints, dyestuffs, battery chemicals and leather tanning agents, to name but a few. Even silicon chip manufacture involves the use of chemical components and leakage of solvents used in this process led to the contamination of drinking water in Silicon Valley, California. In recent years, particular concern has arisen regarding the production of dioxins; these are a particularly toxic group of chemicals and are associated with the manufacture of several organic pesticides. In the Italian town of Seveso, for example, an area of 1,800 ha was affected by dioxin contamination in 1976 and 700 people were evacuated; subsequently, the contaminated soil had to be excavated and removed. A major event which publicised the issue of chemical contamination of soils was the Love Canal disaster near Niagara, USA. Here, extensive dumping of chemicals over a 30-year period led to severe soil contamination by migrating leachates and gases, with serious health implications for local people. This persuaded the US government to release funds to implement the Comprehensive Environmental Response Compensation and Liability Act of 1980 (CERCLA).

A number of organic wastes may lead to soil contamination. At Times Beach, Missouri, for example, the land was so badly contaminated by dioxins from waste oils, derived from the manufacture of the defoliant 'agent orange' that it was purchased by the Federal government (Bridges 1991a). Spillage of oils and related wastes is common on land used for storage and maintenance of motor vehicles and for fuel storage. Sewage sludge, often applied to agricultural land adjacent to urban and industrial areas, often contains high concentrations of heavy metals. The PCBs (polychlorinated biphenyls) are a group of organic solvents commonly implicated in soil contamination. These are used as dielectric fluids in transformers and are often released during the break-up of electrical equipment.

Management of contaminated soils is a difficult prospect as the guidelines on threshold toxicity levels of contaminant materials and on the vulnerability of soils to pollution, are still being developed and refined. Management options depend to a large extent on the geological and hydrological characteristics and size of the contaminated area. Such areas vary dramatically from large designated landfill sites with records of waste disposal activities, to former industrial sites with a long and unclear history of pollution. Management options also depend on the nature of the contamination problem and whether it can be contained at the affected site, or whether leachates and gases threaten areas beyond it. Furthermore, it may be possible to treat the contamination on site, or it may be necessary to remove and relocate the contaminated soil.

Bridges (1991) outlines a number of management approaches to the problem of contaminated soil. These include the use of physical stabilisation barrier, thermal and microbiological techniques. Stabilisation techniques involve treatment to reduce the solubility and mobility of the waste materials. This is achieved through the application of cement, lime, gypsum, silicate materials, epoxy-resins, polyesters or asphalt; these act as binding agents and help to stabilise the 'landform' within which the contaminated waste is stored. Barrier systems usually rely on physical containment using steel or concrete pilling to prevent downward and lateral migration of toxic materials. Similarly, layered cover systems are often used to prevent upward migration of contaminants. Thermal techniques involve heating contaminated soil in rotary kilns or furnaces, whereas microbiological techniques involve the inoculation of waste materials with microbiological communities. Both of these techniques aim to convert the toxic waste materials into less harmful forms. Other approaches to the management of contaminated soil include chemical treatments which may be used to hydrolyse or oxidise contaminants into less dangerous products; similarly, acidic or alkaline wastes may be neutralised. In addition, physical methods may be employed to separate out contaminants according to particle size or density.

USSAR SOIL

This is a special kind of saline and alkaline soil found in many parts of UP, MP, Haryana and Punjab. Normally this kind of soil is either unproductive or poorly productive (*Ushar*—Sanskrit = dry, barren yielder). To reclaim this kind of soil, more organic matters and other soil ameliorative substances can be added to improve the soil quality. It is known for quite some time that blue green algae like species of *Anabaena, Wollea, Scytonema*, and *Tolypothrix* etc. are able to improve the soil quality if Ussar soil is treated with above stated BAG culture for some time.

14 NOISE POLLUTION

14.1 CONCEPT OF SOUND, NOISE AND HEARING PROBLEMS

"Sound" is a form of energy, consisting of wave motion. It requires medium like gas, liquid or solid for propagation. Sound waves travel (at 330 m/s = 1,070 ft/sec = 1,220 kph or 760 mph) through the medium from the source where the sound is produced to the recipient or listener. Sound waves consist of variations in pressure or oscillations of the medium in which they travel. The rate of the oscillation is called the frequency of the sound and is measured in cycles per second or hertz (Hz). The frequency determines the pitch of the sound received by the listener. High pitched sounds have high frequencies and these are more disturbing to the individual than low frequencies, because of the varying sensitivity of the human ear.

The term "noise" is applied to the sound that cause irritation on hearing of healthy human being. In fact, there is very little demarcation of "sound" and "noise".

For a long time many individuals have accepted noise as a part of their environment. However, over the years, the general incidence of noise has been increasing. The development of various engines, technological machinery in industries, jet planes etc. — all contributed to an increasing noisy environment. Noise is now regarded by many people as a pollution component that contributes to a deterioration of the environment.

The human ear receives sound waves and these set up oscillations in the tympanic membrane or ear drum. These oscillations cause sympathetic movements of the three ossicles or small bones in the middle ear behind the ear drum. The oscillations then pass through the fluid in the inner ear to the auditory nerve and on to the brain. In the brain, the oscillations or sounds are identified and interpreted. The brain is able to select mixed sounds into different categories, for example, it can distinguish speech from background noises and, if desired, it can consciously suppress unwanted sounds depending upon their intensity. The ear is able to analyse sounds into frequency components and the range of an 18-year old person with perfect hearing is between 20 Hz and 20,000 Hz. Normally hearing is more acute in the frequency range 2,000 to 5,500 Hz. but it falls off rapidly below 200 Hz and above 10,000 Hz. However, the ear's sensitivity ranges vary considerably from person to person and also with ages of the person concerned. As such, aged people experience a progressive hearing loss of high frequency sounds which is called "presbycusis".

14.2 MEASUREMENT OF NOISE

There are two most important measurements of the noise, viz., sound pressure and sound intensity. The scientific acoustic unit in common use is the decibel (dB). It is not an absolute physical measurement unit comparable to the gram, volt or meter, but it is a ratio expressed as a logarithmic scale relative to a reference sound pressure level. The decibel is defined as:

$$\text{Sound intensity level} = 10 \log_{10} \frac{\text{intensity measured}}{\text{reference intensity}}$$

The reference intensity used is the threshold of hearing—which is a sound that can be first heard at a sound pressure of 2×10^{-5} N/m^2 or a sound intensity of 10^{-12} W/m^2.

The relationship between the three measurements is shown in Table 14.1:

Table 14.1: Intensities of sound pressures and decibels (dB) for sound in air at room temperature and sea level pressure

Intensity (w/m^2)	Pressure (N/m^2)	dB	Sources of Sound
100 M	200,000	200	Rocket take-off.
1.0	20	120*	Railway air horn, Boiler maker's shop.
10^{-2}	2.0	100	Siren; Popmusic concert.
10^{-4}	2×10^{-1}	80	Ringing alarm clock.
10^{-6}	2×10^{-2}	60	Normal conversation.
10^{-8}	2×10^{-3}	40	Public library, Rural house.
10^{-10}	2×10^{-4}	20	Church; quiet room.
10^{-12}	2×10^{-5}	0	Threshold of hearing.

*Threshold of pain

There are several instruments for the measurement of sound:

1. Sound level meter
2. Octave band analyser
3. Magnetic tape recorder
4. Statistical analyser
5. Noise analyser etc.

The sound level meter (Fig. 14.1) evaluates sound pressure linearly or on weighted scale while octave analyser determines average sound pressure of standard octave pressure bands. Magnetic tape recorder is used for storage of sound for latter analysis.

A single measurement of sound level is obviously not adequate for rating noise exposure for possible hearing damage. Instead, special integrating sound level meters are preferred instruments for this task. They are often called "noise dosimeters". Each dosimeter includes a small microphone that is connected to an electronic monitor by means of a connecting cable. The small microphone can be worn at the ear, on a collar, or on a shoulder. The electronic monitor is small enough to be carried in a shirt pocket. The monitor accumulates the dose according to the 5 dB exchange rate. After the full working day, the monitor is plugged into an indicator section. The noise dose can then be displayed on a 4-digit indicator.

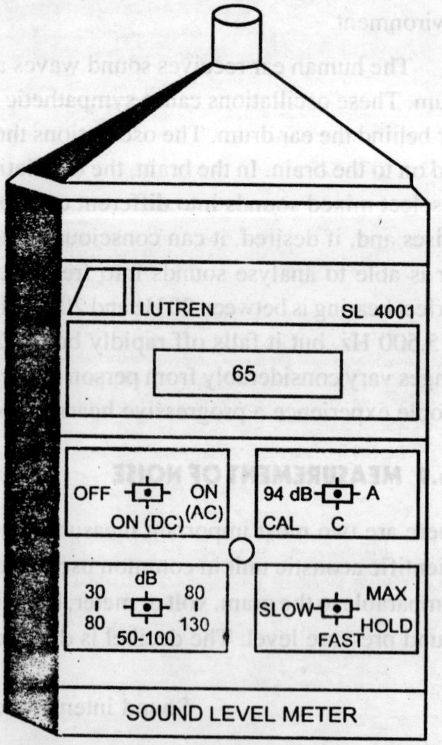

Fig. 14.1 : Sound Level Meter

A sound level meter, on the other hand, has a microphone, an attenuator, a preamplifier, an indicating meter and monitoring outputs, etc. In addition, octave band analysers possess passfilters, selected by means of a rotary switch.

The dBA scale is in general use, but it is not sufficiently refined to take account of peak noise levels, the duration of noise exposure and the quality of noise that are features of specific environmental noise situations.

Equivalent continuous sound pressure level (L_{eq}): The equivalent continuous sound pressure level is the level which, if generated continuously, would give the same energy content as some fluctuating signal over a given time period:

$$L_{eq} = 10 \log 10 \frac{1}{T} \int_0^T \frac{P_1^2}{P_0^2} dt \, (dB)$$

where, T = the time period under consideration

P_1 = the instantaneous sound pressure (n/m^2)

P_0 = Reference sound pressure level (2×10^{-5}n/m^2)

However, for calculation purpose of the A-weighted sound the simplest formulae thus used as:

$$L_{Aeq} = 10 \log 10 \frac{1}{T} \sum t\, 10^{p/10} \ldots\ldots dB(A)$$

[**Example**: In an 8 hour environmental noise study the following steady noise levels were observed for the period indicated:

1 hour at 66 dB(A)

2 hours at 58 dB(A)

2 hours at 50 dB(A)

1 hour at 45 dB(A)

2 hours at 63 dB(A)

The L_{eq} for the 8-hours-period would be given by:

$$L_{Aeq_{8h}} = 10 \log 10 \frac{1}{8}\left(1\times 10^{66/10} + 2\times 10^{58/10} + 2\times 10^{50/10} + 1\times 10^{45/10} + 2\times 10^{63/10}\right) dBA$$

$$L_{Aeq_{8h}} = 10 \log 10 \frac{1}{8}\left(1\times 10^{6.6} + 2\times 10^{5.8} + 2\times 10^{5} + 1\times 10^{4.5} + 2\times 10^{6.3}\right) dB(A)$$

$$L_{Aeq_{8h}} = 60.7\, dB(A)$$

Thus 60.7 dB(A) had been generated continuously over the 8-hour-period. The energy content of the noise would have been equivalent to that of the fluctuating noise pattern given.

Noise pollution level (L_{NP}): This is an expression of noise pollution in a given environment:

$$L_{NP} = L_{Aeq} + (L_{A_{10}} - L_{A_{90}})$$

where $L_{A_{10}}$ = This is the level exceeded for not more than 10% of the time.

$L_{A_{90}}$ = This is the level exceeded for not more than 90% of the time.

14.3 HEARING MECHANISM

The ear acts as a transducer converting sound energy into electrical energy which is sent to the receptive centres of the brain via electrical impulses (Fig. 14.2).

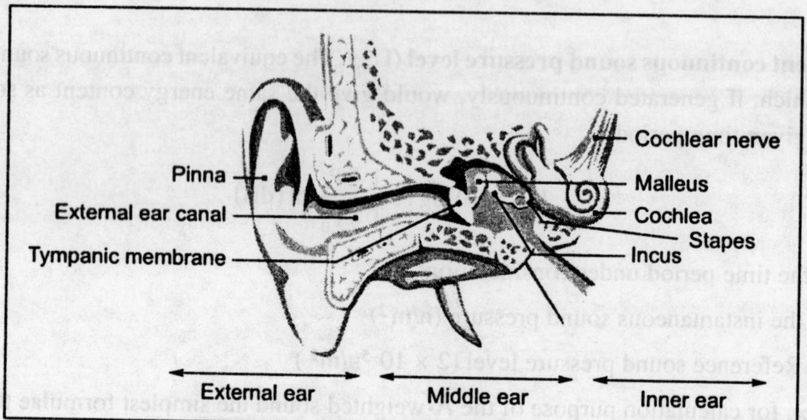

Fig. 14.2 : The Hearing Path

The ear comprises three main regions: (i) the outer ear (air-filled), (ii) the middle ear (air-filled), and (iii) the inner ear (fluid filled). Sound is directed into the ear via the pinna and the auditory canal and strikes the ear drum (tympanic membrane) which is set into vibration. The vibration is transferred mechanically across the middle ear by three small bones (ossicles) viz., malleus, incus and stapes. The ossicles acts as a system of levels transferring the sound pressure acting on the ear drum to the oval window at the footplate of the stapes. The amplified pressure acting on the oval window is sufficient to cause the fluid in the cochlea to move backwards and forwards. Then the hair cells of the basilar membrane of cochlea deflect the frequency content of the sound acting on the ear. Finally, the impulse is transferred to the respective centre of the brain via the auditory nerve.

There may be defect in hearing due to transmission of sound wave/impulse to the sensory centre. There are three kinds of deafness viz., conductive deafness (i.e. defects in pathways of sound transmission due to perforated drum, otoclenosis, blockage of air passage etc.), sensory neural deafness (defects in the cochlea or auditory nerve due to hair cell degeneration in high frequency sound), or central deafness (defects in hearing pathways in brain).

It is well-known that there is a particular sound pressure that is needed for hearing. But due to certain constrain, there is need for change of higher sound pressure level for hearing. This is threshold shift (TS). Threshold shifts can be classified as: (i) *Temporary Threshold Shifts* (TTS), resulting from exposure to high level sound for comparatively short period; (ii) *Permanent Threshold Shifts* (PTS), due to prolonged noise exposure, or age. Permanent threshold shifts which are due to years of exposure to high level of industrial noise are termed *Noise Included Permanent Threshold Shifts* (NIPTS).

14.4 SOURCES OF NOISE

The sound or noise could be generated from various sources like (a) Transport, (b) Industrial operation, (c) Domestic practices etc.

(a) **Transport noise** originates from road traffic (i.e. vehicular), aircraft and rail traffic etc. Noise from road vehicles produces disturbance to more people than any other noise source and this has been increasing over the last decades for a number of reasons. The total number of road vehicles and hence the density of road traffic is steadily increasing. One of the most important causes of noise on the roads is the traffic speed. The faster the traffic travels, the greater the noise volume and modern road development policy is encouraging higher speeds.

Road traffic noise fluctuates according to a number of operating factors. Noise is produced by all vehicles from the gear box and exhaust system. Heavy vehicles also produce rattles, squeaks and vibrations according to the degree of loading and age. All vehicle produce more engine noise at faster speeds and a doubling of engine speed can increase the noise level by 13 dBA. Also tyre noise increases with speed and a wet road can increase the noise by 10 dBA. Generally, a heavy goods vehicle produces twice as much noise as a private car, or, to a motorway observer, the noise from one heavy lorry equals that from 10 passing cars. The actual pattern of traffic noise on a main road is complex. There is a general noise level as long as any traffic is moving and this varies with the traffic density and the time of day.

The traffic noise can be accurately measured in dBA and noise levels are often expressed as on the L_{10} (18 hrs) index. The residential road having vehicle passing at a speed of 30 mph bears the noise level 60 dBA, while the busy main road having traffic speed of 48 kph generate noise at 70 dBA.

In addition, transport noise is also generated by the air-horn use. In fact, there is no standard of vehicle generated noise in many countries. In UK there is a noise generation standard by the vehicle (BS, 3539 : 1962).

Aircraft noise: Over the last decades there has been an increase in the noise nuisance from subsonic aircraft for several reasons. This noise is variable and intermittent and it is not continuous as in the case of road traffic noise. There are peak noise levels when aircraft are flying overhead or are taking-off and landing at airports and the peak frequency varies with the number and the type of aircraft and the operational height. The noise is mainly produced from aircraft engines. The noise generated by different types of aircraft is given in Table 14.2:

Table 14.2: Aircraft noise

	Aircraft type	Noise level as surveyed by BAA (dB)
1.	VC 10	110
2.	DC 8	112
3.	Boeing 707	115
4.	Boeing 737	110
5.	BAC 111	119
6.	Boeing 747	114
7.	DC 10	110
8.	HS Trident 3	119
9.	Lockheed Tristor 1093	105

Rail traffic noise: The noise from trains is not generally regarded as a serious nuisance by the majority of people. This may be because the noise is generally of a lower frequency than that of road vehicles and also most railway tracks run through rural areas. But where buildings are sited beside railway tracks, and specially where engine testing and shunting is carried out, there is noise disturbance.

(b) **Industrial noise** is complex and varies with the design, direction of movement of working parts and the method of mounting of machines. The noise is often produced in three stages. There is an initial disturbance at the point of origin of the sound; followed by amplification, often caused by the resonance of the machine parts, the workpiece or the floor and finally radiation of the sound to the surrounding environment.

Industrial noise can be classified into three types: (i) Impact and percussive noise is produced by presses; punch and stamp machines, pneumatic drills, milling machines, cutters and routers. (ii) The impact noise is caused when two surfaces meet each other, sometimes at high speed and vibration occurs at the point of contact followed by amplification and resonance. (iii) The third type of industrial noise is aerodynamic noise. Aerodynamic noise is produced by a blow lamp or torch, fans and dust extractors.

(c) **Domestic noise**: Noise is also generated from domestic appliances like washing machines, spin dryers, food mixers, sink waste grinders and vacuum cleaners. Both industrial and domestic noise sources are collectively referred as occupational noise. Some common occupational noise levels are given in Table 14.3:

Table 14.3: Occupational Noise Levels

	Domestic/Industrial sources	Noise level (dBA)
1.	Food Blender	90
2.	Washing machine	82
3.	Vacuum cleaner	70
4.	Waste grinder	83
5.	Dish washer	76
6.	Milling machine	90
7.	Newspaper press	101
8.	Circular saw	110
9.	Textile loom	112
10.	Pneumatic metal chipper	122
11.	Steel plate riveting	130

14.5 EFFECTS OF NOISE POLLUTION

The effect of noise on health are numerous. It can affect central nervous system, cause nausea, vomiting, deafness, loss of appetite, loss of sleep and even cardiac failure. All the hazardous effects of noise pollution are thus discussed separately.

(A) Hearing damage from noise exposure

There is no doubt that noise can damage the ear and cause temporary or permanent noise-induced hearing loss, depending upon the intensity and duration of the sound level. The most commonly occurring ear damage is caused by continuous periods of high intensity noise. If the ear receives a noise level of over 90 dB in the mid-high frequency range for more than a few minutes, then the auditory sensitivity is reduced. The effect is called the temporary threshold shift and can result in a sound of 4,000 Hz frequency requiring a noise level of 20 dB to be heard, instead of 5 dB under normal conditions. The typical hearing loss resulting from varied period of exposure in noisy environment is given in Fig. 14.3.

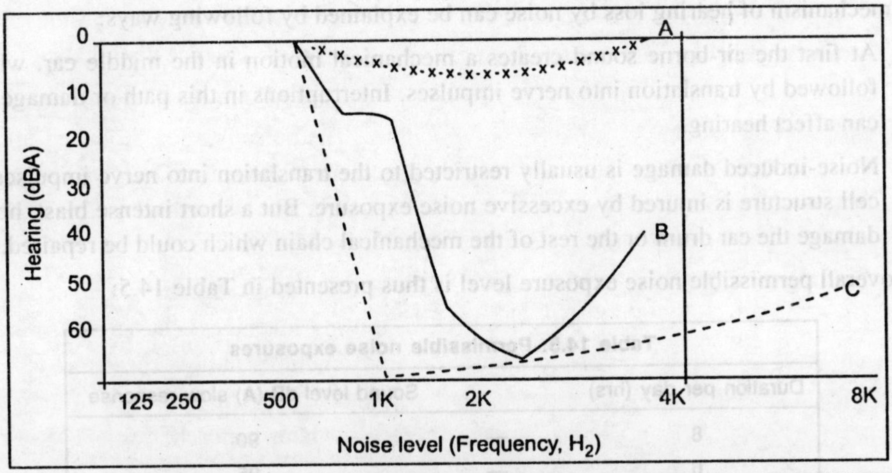

Fig. 14.3 : After a short period of exposure to noise; (B) After 20 years' exposure; and (C) After 35 years' exposure

Besides progressive hearing loss, there may be instantaneous damage or acoustic trauma. This is usually caused by very high intensity impulse type noise, that can result from an explosion, or sudden excessive noise of about 150 dB or more. Studies over many years have shown that, in general:

(i) Hearing damage from exposure to excessive noise is a cumulative process; both level and exposure time are important factors.

(ii) At a given level, low-frequency noise tends to be less damaging than noise in the mid-frequency range.

(iii) Individuals are not all equally susceptible to noise-induced hearing loss.

(iv) The hearing loss that results from noise tends to be most pronounced near 4,000 Hz but it spreads over the frequency range with increased exposure time and level.

Two types of hearing loss from noise exposure are recognized, temporary and permanent. Immediately after exposure to noise at a level of 100 dB, say, there is a marked increase in the minimum level that one can hear (threshold) compared to that there is a gradual recovery of hearing ability. But repeated exposures over extended periods will lead to incomplete recovery and some permanent hearing loss. This permanent hearing loss depends on noise level and the pattern of exposure and recovery time.

Noisy industrial processes and conditions produce hearing loss to the workers involved. Detailed occupational noise studies by a team from Manchester University in 1971 confirmed this (Table 14.4):

Table 14.4: Hearing loss by workers*

Occupation	Workers age (yrs)	Length of occupational service (yrs)	Hearing loss (dB)
Gravure Printers	55.3	21.8	35
Iron moulders	54.8	36.4	43
Bus drivers	53.7	21.2	44
Boiler makers	59.8	30.5	52
Weavers	55.0	31.5	55
Fettlers (carting, metal chipping)	53.5	21.3	65

*Average hearing loss 30 dB.

The mechanism of hearing loss by noise can be explained by following ways:

(i) At first the air-borne sound creates a mechanical motion in the middle ear, which is then followed by translation into nerve impulses. Interruptions in this path or damage to any part can affect hearing.

(ii) Noise-induced damage is usually restricted to the translation into nerve impulses. The hair-cell structure is injured by excessive noise exposure. But a short intense blast, however, can damage the ear drum or the rest of the mechanical chain which could be repaired.

The overall permissible noise exposure level is thus presented in Table 14.5:

Table 14.5: Permissible noise exposures

Duration per day (hrs)		Sound level dB (A) slow response
8	—	90
6	—	95
4	—	95
3	—	97
2	—	100
1	—	102
1/2	—	110
1/4 or less	—	115

The various degrees of hearing loss sometime poses a number of effects. For instance < 25 dB hearing loss does not pose any difficulty in hearing, but > 25 dB hearing loss causes hearing difficulty. Only hearing loss > 50 dB requires hearing aid. Complete deafness is noted above 90 dB. The risk factors for hearing loss is presented in Table 14.6.

(B) Pathological and Psychological Disorders

Pathological effects can result from particular noise frequencies causing vibration or resonance in materials or people's bodies. High frequencies, or ultrasonic sound above the normal audible range, can affect the semi-circular canals of the inner ear and cause nausea or dizziness. Alternatively, very low frequency noise can cause resonance in the body organs, producing the effects of decreased heart beat, variations in blood pressure and breathing difficulties. Mid-audible band frequencies are known to cause resonance in the skull and so affect the brain and nervous system with consequent effects upon thinking and coordination of the limbs.

Table 14.6: Risk Factors for Hearing Loss

A. Primary factors :
* Occupational noise exposure (85 dBA)
* Military history
* Noisy hobbies
— Musical instruments (percussion, brass etc.)
— Outdoor target shooting/hunting
— Gardening with gas mower etc.
— Work with cars/engines:
— Wood working/power tools.

Contd.

Table 14.6: Risk Factors for Hearing Loss

* Medication usage (may initiate as well as potentiate hearing loss)
  - Kanamycin, Cisplatin
  - Aspirin
  - Diuretics
* Age (may initiate potential hearing loss).
* Residential factors (highly urban environment, outdoor activities).

B. Secondary activities :

* Anatomic and genetic factors (History of hearing loss in families)
* Diabetes
* Elevated cholesterol (increased atherosclerosis)
* Cigarette smoking (increased atherosclerosis)
* Elevated blood pressure (increased atherosclerosis)

Non-pathological or psychological noise effects are also variable and very difficult to measure. The mildest effect is often physical or mental fatigue and lack of concentration. This effect is important in industrial situations, because it results in low efficiency, a reduced work rate, increased absenteeism, and a higher potential for accidents and injuries. In the non-work environment, psychological noise effects impinge upon sleep.

The sound level that cause damages of various body functions are:

Noise level (dBA)	Damages
100	Prolonged exposure causing permanent damage of hearing.
180	Ear drum rupture.
194	Lung damage.

Using headphones with high volume, or continuous mobile phones over the years may cause hearing loss too.

14.6 NOISE MAPPING

In order to assess the impact of noise on human health—no matter whether it is domestic noise or industrial noise—it is imperative to undertake sound survey of noise levels in different parts of city. It will be more appropriate to prepare sound maps—which are called *topographs*—of residential areas which are sound prone, like dwelling around factories, airport, railway station, cinema houses, and commercial plaza.

In India, such kinds of noise mapping is done in various townships, viz.,

Mumbai (SOCLEAN, 1979, 1983; Sahoo and Khopkor, 1987),

Kolkata (Roy, Santra, Mitra and Chanda, 1989; Chakravorty et al 1996),

Ahmedabad (Shrivastava, Shah and Desai, 1986),

Rourkela (Tiwari and Manzoor, 1988),

and Visakhapatnam (Rao and Rao, 1990), Vadodara, Delhi & Chennai.

From the studies, it appears that most of the Indian cities the traffic noise level already crossed the limits (> 75 dBA at Day and > 60 dBA at night) (Fig. 14.4).

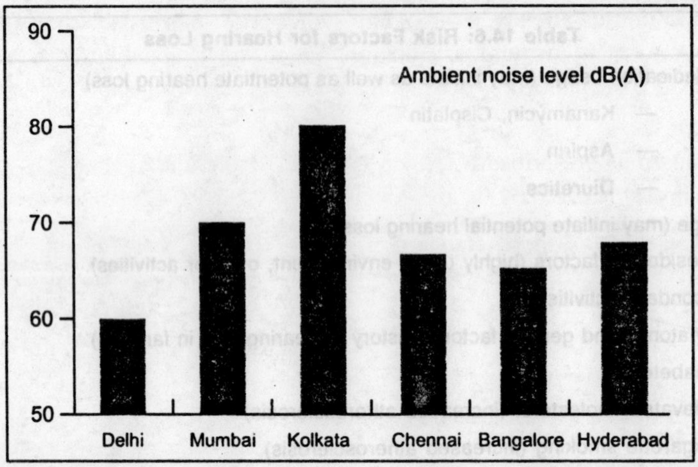

Fig. 14.4 : Ambient Noise Level in Residential Areas of Metropolitain Cities

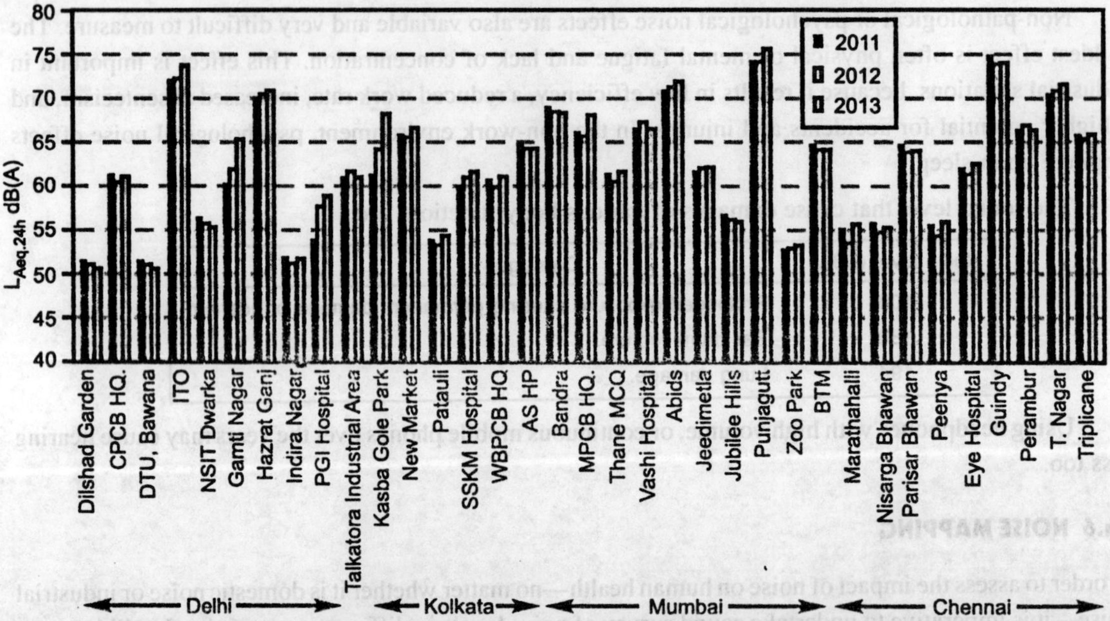

Fig. 14.5 : $L_{Aeq,24h}$ at 35 locations spread across seven major cities all over India during 2011-2013

In recent year, survey indicate that the city noise level in India gone up very much. Details are given in Fig. 14.5.

14.7 ABATEMENT OF NOISE POLLUTION

In general, ears have some protective reflexes which have only limited effect against loud noise. By modification of contraction of tympanic membrane, ear can accommodate high pitch sound for a moment. But prolong exposure of high noise level (> 60 dBA) leads to gradual hearing loss in almost all aged people.

On the whole, the variety of noise sources may cause disturbance and annoyance to the general public. It is an unjustifiable interference with ordinary human comfort and well-being. The total extent of the nuisance from various sources of noise in different countries are yet to be ascertained in details.

However, a few countries already set-up some standard for permissible noise level in the indoor environment and to some extent for outdoor situation. The number of statutes and regulations that have been brought into use since the late 1960s does indicate that Government is more aware of the problem of noise.

The general standard as prescribed by Environmental Protection Agency (US — EPA) as well as Department of Environment (DOE, UK) is given in Table 14.7:

Table 14.7: General standard of noise (dB): Indoor environment

Places/Locations	Noise level (dBA)
Bedrooms	15
Libraries	30
Living rooms	40
Classroom	55
Business office	60
Street traffic	80
School	35
Residential area at night	35
Restaurants	45
Shop	55
Workshops	70 - 75
Theaters	30

There are a number of measures that had been taken up to reduce noise generation at sources, like modification of car engines, modification of aircraft machines, banning of use of air horns, banning of use of loud speakers in residential areas or public places (Table 14.8).

Table 14.8: Noise limits for vehicles applicable at manufacturing stage applicable from 1st April 2005

	Types of vehicle	Noise limits dB(A)
1.	*Two wheeler*	
	• Displacement up to 80cc	75
	• Displacement more than 80cc but up to 175cc	77
	• Displacement more than 175cc	80
2.	*Three wheeler*	
	• Displacement up to 175cc	77
	• Displacement more than 175cc	80
3.	*Vehicles used for carriage of passengers and capable of having not more than nine seats, including the driver's seat*	74
4.	*Vehicles used for carriage of passengers having more than nine seats, including the driver's seat, and a maximum gross Vehicle Weight (GVW) of more than 3.5 tonnes*	
	• With an engine power less than 150KW	78
	• With an engine power 150KW or above	80
5.	*Vehicle used for carriage of passengers having more than nine seats, including the driver's seat: Vehicles used for carriage goods*	
	• With maximum GVW not exceeding 2 tonnes	76
	• With maximum GVW greater than 3 tonnes but not exceeding 3.5 tonnes	77
6.	*Vehicles used for transport of goods with a maximum GVW exceeding 3.5 tonnes*	
	• With an engine power less than 75KW	77
	• With an engine power of 75KW or above but less than 150KW	78
	• With an engine power of 150KW or above	80

In addition, there is stipulated noise standard for manufacture of domestic appliances (Table 14.9).

Table 14.9: Proposed noise limits for domestic appliances and construction equipments at manufacturing stage

Domestic appliance	Maximum permitted sound power level dB(A)
Refrigerator and freezer	55
Washing machine, spin dry	72
Microwave oven	70
Electrical fan (desk fan, stand fan, wall fan, ceiling fan)	70
Vacuum cleaner and dryer	70
Indoor heater	55
Air conditioners (split, indoor)	72
Air coolers	72
Hair dryer	75
Protable gensets (petrol or kerosene)	86
Food processor	90
Food blender	100
Lawn movers, welding generators, compressors	110
Handheld concrete breakers	120

In India, in fact, State and Central Pollution Control Broad entrusted for monitoring of noise problems particularly in the major cities. Over the year neighbourhood noise level increased enormously by several unsocial human activities. Even during festival loudness of microphones creates enormous panicy situation for the local residents.

Various non-governmental agencies now came forward for public awareness of noise reduction at sources and so also to aware people about the ill-effect of noise. The Society for Clean Environment, Mumbai, is noteworthy among them. However, no legal measures could be taken properly against the offender due to some lacunae of legal procedures.

However, Central Pollution Control Board Committee recommended the permissible level of noise at various places in the country (Table 14.10 vide Ambient Air quality standards in respect to noise—Notification under Schedule III of Part II — Sec. 3(i) of Extraordinary Gazette of India by the Ministry of Environment and Forests, Govt. of India. New Delhi, December 26, 1989). A revised ambient noise guideline is shown in the appendix.

Table 14.10: Noise Standard

Area	Noise level (dBA)	
	Day	Night
Industrial	75	70
Commercial	65	55
Residential	55	45
Silence Zone	50	40

Arrangement of proper traffic flow, control of traffic speeds by speed-breaker and banning silencerless vehicles etc. are considered to be adopted on an emergency basis in major Indian cities. It was further noticed that roadside planting is also able to absorb sound—at least 6-10 dBA. As such, the major traffic bound road should be planted with trees having dense foliage by which it can reduce the traffic noise by absorption.

A detailed chart on industrial noise control technique are given in Table 14.11:

Table 14.11 : Noise Control Techniques

Noise Level Monitoring and Control in Industry

Control at the source

(a) Siting
(b) Building layout
(c) Path Deflection
(d) Enclosures
(e) Absorption
(f) Impedance mismatch

Control of the Transmission path

Control at the Receiver

(a) Use of personal protective equipments
(b) Education & Publication
(c) Exposure control

Reduction of the amplitude of the exciting forces

(a) Reduction of impacts or implosive forces
(b) Balancing of moving mass
(c) Balancing of magnetic forces
(d) Use of dynamic compensation
(e) Reducing frictional forces
(f) Vibration isolation

Reduction of the response of various components to the exciting forces

(a) Alteration of the natural frequency of the resonant element
(b) Increasing the energy dissipation

Changes in operating procedure

(a) Siting
(b) Building layout
(c) Path deflection
(d) Providing enclosures
(e) Absorption
(f) Impending the sound by discontinuities between source & receiver

NOISE LEVELS AT MAJOR CITIES OF THE WORLD

The noise status of Kolkata is fairly high and considered to be one of the noisiest cities of the world (Table 14.12).

In general, average noise level (in L_{eq}) of silent zone, residential, commercial and Industrial areas of Kolkata are 65, 69, 78 & 80 dB(A), respectively in contrast to corresponding standard level of 50, 55, 65 and 75 dB(A).

Table 14.12: Noise level of selected cities of the world (1990)

Sl. No	Cities	Noise level (L_{eq}) in dB(A)
1.	Nanjing (China)	105
2.	Rome (Italy)	90
3.	Madrid (Spain)	89
4.	New York (USA)	88
5.	Kolkata (India)	85
6.	Mumbai (India)	82
7.	Hong Kong (Singapore)	80
8.	Delhi (India)	80
9.	Karachi (Pakistan)	80
10.	Kathmandu (Nepal)	75
11.	Colombo (Sri Lanka)	75

Many megacities of the world, facing noise induced hearing loss. According to WHO, 360 million people worldwide have disabling hearing loss, and 32 million of these are children. Exposure over 10 years at high noise area induced major hearing loss. Details shown in Table 14.13.

Table 14.13: Hearing loss in major cities of world

City	Country	Male hearing loss (years)	Female hearing loss (years)	Average hearing loss (years)	Hearing loss rank
Vienna	Austria	+ 10.53	+ 10.65	+ 10.59	0
Zurich	Switzerland	+ 10.43	+ 10.82	+ 10.63	0.02
Hamburg	Germany	+ 11.46	+ 11.46	+ 11.46	0.1
Portland	USA	+ 11.93	+ 11.70	+ 11.81	0.15
Stockholm	Sweden	+ 11.65	+ 12.08	+ 11.86	0.15
Oslo	Norway	+ 11.86	+ 12.02	+ 11.94	0.16

Contd.

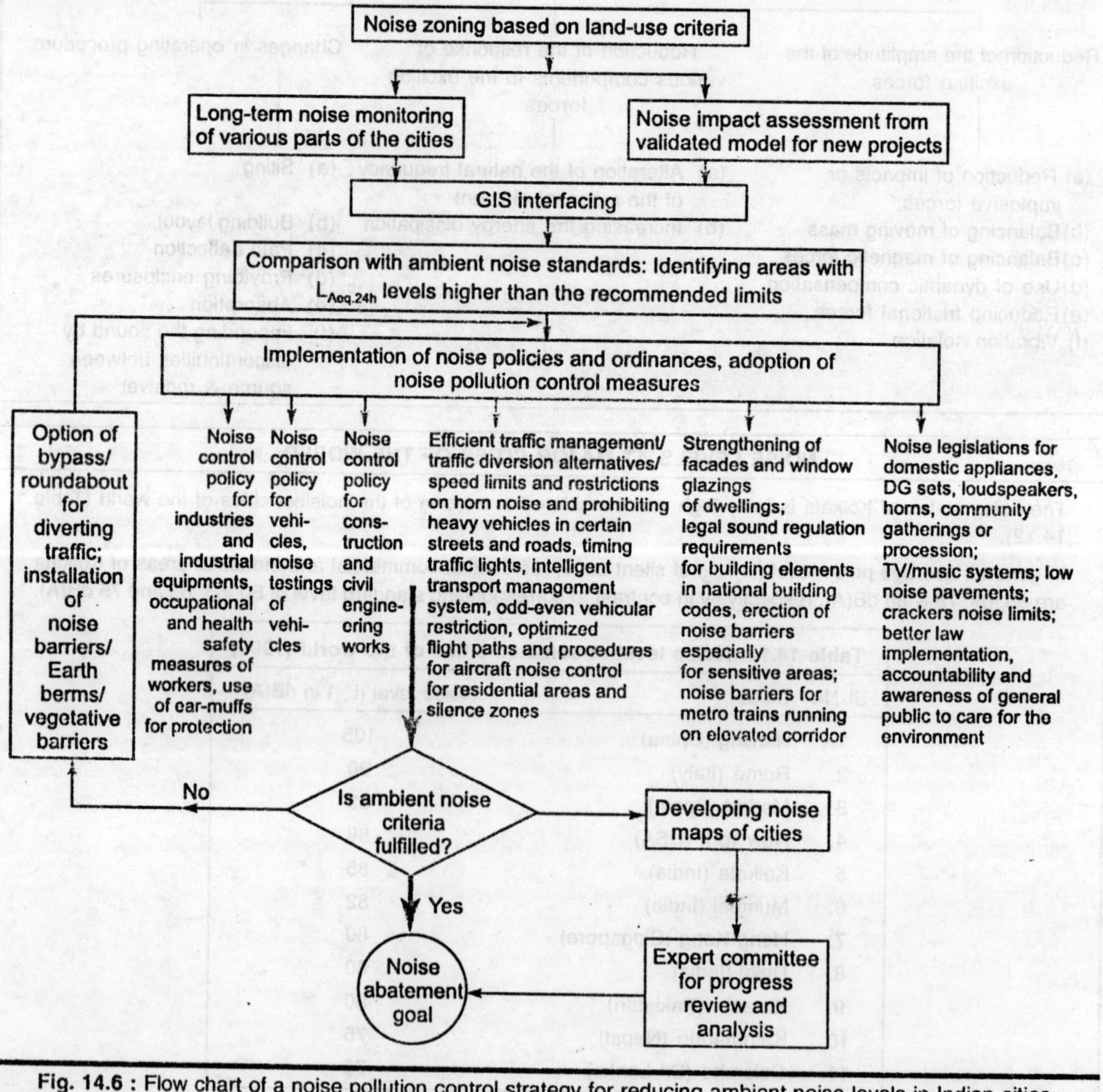

Fig. 14.6 : Flow chart of a noise pollution control strategy for reducing ambient noise levels in Indian cities

Table 14.13: Hearing loss in major cities of world

City	Country	Male hearing loss (years)	Female hearing loss (years)	Average hearing loss (years)	Hearing loss rank
Hanover	Germany	+ 12.08	+ 11.89	+ 11.98	0.17
Cologne	Germany	+ 10.96	+ 13.07	+ 12.01	0.17
Munich	Germany	+ 11.81	+ 12.24	+ 12.02	0.17
Delhi	India	+ 19.15	+ 19.53	+ 19.34	1
Mumbai	India	+ 18.40	+ 18.77	+ 18.58	0.98
Istanbul	Turkey	+ 17.94	+ 18.73	+ 18.33	0.93
Cairo	Egypt	+ 18.33	+ 17.73	+ 18.03	0.89
Guangzhou	China	+ 16.89	+ 17.98	+ 17.43	0.82
Saint Petersburg	Russia	+ 17.24	+ 17.06	+ 17.15	0.78
Taipei	Taiwan	+ 16.29	+ 17.30	+ 16.79	0.74
Seongnamsi	South Korea	+ 17.19	+ 15.89	+ 16.54	0.71
Buenos Aires	Argentina	+ 15.89	+ 17.19	+ 16.54	0.71
Madrid	Spain	+ 15.34	+ 17.18	+ 16.26	0.68

The overall noise control strategies to be adopted in Indian cities given schematic flow chart (Fig. 14.6 previous page).

APPENDIX

Ministry of Environment and Forests Notification

New Delhi, the 14 February, 2000

S.O. 123(E) — Whereas the increasing ambient noise levels in public places from various sources, inter-alia, industrial activity, construction activity, generator sets, loud speakers, public address systems, music systems, vehicular horns and other mechanical devices have deleterious effects on human health and the psychological well-being of the people, it is considered necessary to regulate and control noise producing and generating sources with the objective of maintaining the ambient air quality standards in respect of noise;

Whereas a draft of Noise Pollution (Control and Regulation) Rules, 1999 was published under the notification of the Government of India in the Ministry of Environment and Forests vide number S.O. 52(E) dated the 28th June, 1999 inviting objections and suggestions from all the persons likely to be affected thereby, before the expiry of the period of sixty days from the date on which the copies of the Gazette containing the said notification are made available to the public;

And whereas copies of the said Gazette were made available to the public on the 1 day of July, 1999;

And whereas the objections and suggestions received from the public in respect of the said draft rules have been duly considered by the Central Government;

Now, therefore, in exercise of the powers conferred by clause (ii) of sub-section (2) of section 3, sub-section (1) and clause (b) of sub-section (2) of section 6 and section 25 of the Environment (Protection) Act, 1986 (29.6.1986) read with Rule 5 of the Environment (Protection) Rules, 1986, the Central Government hereby makes the following rules for the regulation and control of noise producing and generating sources, namely:

14.8 THE NOISE POLLUTION (REGULATION AND CONTROL) RULES, 2000

1. **Short-title and commencement:**
   (a) These rules may be called the Noise Pollution (Regulation and Control) Rules, 2000.
   (b) They shall come into force on the date of their publication in the Official Gazette.
2. **Definition**: In these rules, unless the context otherwise requires—
   (a) "Act" means the Environment (Protection) Act, 1986 (20 of 1986);

(b) "Area/zone" means all areas which fall in either of the four categories given in the Schedule annexed to these rules;

(c) "Authority" means any authority or officer authorised by the Central Government, or as the case may be the State Government in accordance with the laws in force and includes a District Magistrate, Police Commissioner, or any other officer designated for the maintenance of the ambient air quality standards in respect of noise under any law of the time being in force;

(d) "Person" in relation to any factory or premises means a person or occupier or his agent, who has control over the affairs of the factory or premises;

(e) "State Government" in relation to a Union territory means the Administrator thereof appointed under Article 239 of the Constitution.

3. **Ambient air quality standards in respect of noise for different areas/zones:**

(a) The ambient air quality standards in respect of noise for different areas/zones shall be such as specified in the Schedule annexed to these rules.

(b) The State Government may categorize the areas into industrial, commercial, residential or silence areas/zones for the purpose of implementation of noise standards for different areas.

(c) The State Government shall take meansures for abatement of noise including noise emanating from vehicular movements and ensure that the existing noise levels do not exceed the ambient air quality standards specified under these rules.

(d) All development authorities, local bodies and other concerned authorities while planning development acitivity or carrying out functions relating to town and country planning shall take into consideration all aspects of noise pollution as a parameter of quality of life to avoid noise meanace and to achieve the objective of maintaining the ambient air quality standards in respect of noise.

(e) An area comprising not less than 100 metres around hospitals, educational institutions and courts may be declared as silence area/zone for the purpose of these rules.

4. **Responsibility as to enforcement of noise pollution control measures:**

(a) The noise levels in any area/zone shall not exceed the ambient air quality standards in respect of noise as specified in the Schedule.

(b) The authority shall be responsible for the enforcement of noise pollution control measures and the due compliance of the ambient air quality standards in respect of noise.

5. **Restrictions on the use of loud speakers/public address system:**

(a) A loud-speaker or a public address system shall not be used except after obtaining written permission from the authority.

(b) A loud-speaker or a public address system shall not be used at night (between 10.00 p.m. to 6.00 a.m.) except in closed premises for communication within, e.g. auditorium, conference rooms, community halls and banquet halls.

6. **Consequences of any violation in silence zone/area**: Whoever, in any place covered under silence zone/area commits any of the following offence, he shall be liable for penalty under the provisions of the Act:

(a) whoever, plays any music or uses any sound amplifiers,

(b) whoever, beats a drum or tom-tom or blow a horn either musical or pressure, or trumpet or beats or sounds any instrument, or

(c) whoever, exhibits any mimetic, musical or other performances of a nature to attract crowds.

7. **Complaints to be made to the authority**:

(a) A person may, if the noise level exceeds the ambient noise standards by 10dB(A) or more given in the corresponding columns against any area/zone, make a complaint to the authority.

(b) The authority shall act on the complaint and take action against the violator in accordance with the provisions of these rules and any other law in force.

8. **Power to prohibit etc. continuance of music, sound or noise:**

1. If the authroity is satisfied from the report of an officer incharge or a police station or other information received by him that it is necessary to do so in order to prevent annoyance, disturbance, discomfort or injury or risk of annoyance, disturbance, discomfort or injury to the public or to any person who dwell or occupy property on the vicinity, he may, by a written order issue such directions as he may consider necessary to any person for preventing, prohibiting, controlling or regulating:
   (a) The incidence or continuance in or upon any premises of:
      (i) any vocal or instrumental music;
      (ii) sounds caused by playing, beating, clashing, blowing or use in any manner whatsoever of any instrument including loudspeakers, public address systems, appliance or apparatus or contrivance which is capable of producing or re-producing sound, or
   (b) the carrying on in or upon any premises of any trade, avocation or operation or process resulting in or attended with noise.
2. The authority empowered under sub-rule (1) may, either on its own motion, or on the application of any person aggrieved by an order made under sub-rule (1), either rescind, modify or alter any such order:

Provided that before any such application is disposed of, the said authority shall afford to the applicant an opportunity of appearing before it either in person or by a person representing him and showing cause against the order and shall, if it rejects any such application either wholly or in part, record its reasons for such rejection.

SCHEDULE

(see Rule 3(1) and 4(1)

Ambient Air Quality Standards in respect of Noise

Area Code	Category of Area / Zone	Limits in dB(A) L_{eq}*	
		Day Time	Night Time
(A)	Industrial area	75	70
(B)	Commercial area	65	55
(C)	Residential area	55	45
(D)	Silence Zone	50	40

Notes:

Daytime shall mean from 6.00 a.m. to 10.00 p.m.

Night time shall mean from 10.00 p.m. to 6.00 a.m.

Silence zone is defined as an area comprising not less than 100 metres around hospitals, educational institutions and courts. The silence zones are zones which are declared as such by the competent authority.

Mixed categories of areas may be declared as one of the four abovementioned categories by the competent authority.

*dB(A) L_{eq} denotes the time weighted average of the level of sound in decibels on scale A which is relatable to human hearing.

A "decibel" is a unit in which noise is measured.

"A", in dB(A) L_{eq}, denotes the frequency weighting in the measurement of noise and corresponds to frequency response characteristics of the human ear.

L_{eq}: It is an energy mean of the noise level over a specified period.

14.9 NOISE POLLUTION CONTROL POLICY

The policy involved a number of stages and their appropriate execution. The legal provision for controlling noise pollution in India are: Indian Panel Code (Section 268, 290, 291); Criminal Procedure Code (Section 133), Factories Act 1948, Motor Vehicle Act 1988, Law of Torts, Air (Prevention and Control of Pollution) Act 1981, Environment (Protection) Act 1996. On the basis of the legal provisions a frameworks of community noise assessment and abatement is made (Fig. 14.7).

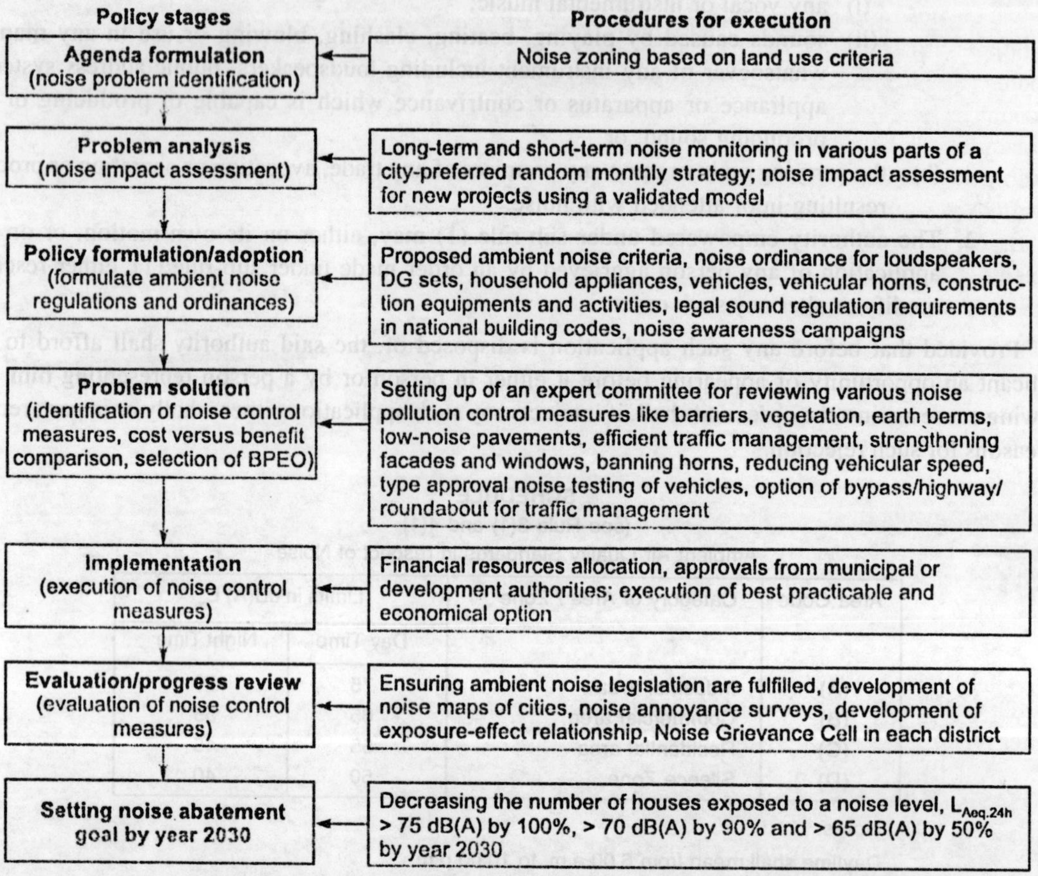

Fig. 14.7 : Proposed framework for community noise assessment and abatement in India

15 SOLID WASTES

15.1 INTRODUCTION

Population growth, increasing urbanisation and industrialisation and rising standards of living have all contributed to an increase in both the amount and variety of waste generated in most countries. Furthermore, many countries are now faced with dealing not only with greater volumes of—but also more dangerous—waste materials. This section of the 'UNEP Environmental Data Report' is concerned principally with the generation and disposal of solid waste arising and liquid effluents. Potential environmental and human health effects resulting from waste generation and disposal are briefly discussed. Gaseous wastes emitted to the atmosphere are dealt with in Part 1 : Environmental pollution, Atmosphere and international agreements relating to the transboundary movement of hazardous waste, in particular the Basel Convention, June 1989 are considered in Part 10 : International Cooperation.

The quality of statistics on solid waste arisings and disposal continues to be highly variable. The data presented in this report are taken selectively from statistical reports produced by government agencies, international organisations and trade associations, and from research papers published in the scientific literature.

Within UNEP, issue relating to the generation and disposal of wastes, in particular hazardous wastes, are addressed by the International Register of Potentially Toxic Chemicals (IRPTC), Geneva. The UNEP Industry and Environment Office (IEO), Paris, established in 1975, to facilitate the adoption of environmentally sound decisions regarding industrial development by governments, provides access to practical information and guidance relating to the safe disposal of industrial wastes and to waste minimisation technologies. Furthermore, UNEP in cooperation with partner organisations such as the World Health Organisation (WHO), the International Atomic Energy Agency (IAEA), the International Maritime Organisation (IMO) and the Organisation for Economic Cooperation and Development (OECD), support a number of programme activities relating to specific waste management problems.

15.2 WASTE GENERATION

Wastes, which arise from virtually all man's activities, can be classified conveniently with respect to their source. Major categories include household and consumer wastes (i.e., municipal wastes), industrial wastes, agricultural wastes, extraction wastes, energy production wastes and sewage sludges. Wastes can also be classified by hazard and by composition. Information on waste arisings, particularly on industrial and hazardous wastes, is often difficult to assemble. Inefficient data collection methods, infrequency of surveys, reluctance of industry to supply information, and confusion over definitions of hazardous wastes—are all contributory factors.

Table 15.1 (page 334) shows waste arising in five source categories: municipal, industrial, agricultural, mining (and quarrying) and sewage sludges. These data provide an incomplete picture of the total amount of wastes with which any of the listed countries may have to deal. Generally speaking, municipal wastes (in terms of mass) are a relatively small part of total waste arisings when compared with waste arisings from industry, agriculture and mining. However, the relative importance of the source categories for wastes largely depends on the economic base.

Table 15.1: Quantities of solid waste generated in selected countries by source category, 1985 ($10^3 ta^{-1}$)

Region/Country	Municipal	Industrial	Agricultural	Mining/ Quarrying	Sewage sludge
NORTH AMERICA					
Canada	16,000	61,000[a]	48,000	910,213	500
USA	178,000[b]	628,000[c]	1,400,000	1,300,000[d]	8,400[e]
ASIA					
Japan	41,530	312,000	90,544	26,017	2,003
Cyprus			3	190	
EUROPE					
Austria	2,052[e]	31,000[b,e]		466[e]	1,350[e]
Belgium	3,082[a]	8,000[a]	53,000[a]	7,069[a]	18[a]
Bulgaria	6,773		6,271	21,872	
Czech Republic & Slovakia				452,360[d]	
Denmark	2,161	1,317			82[f]
Finland	2,000	15,000	23,200	17,700	137[g]
France	15,000	50,000	399,400	60,160[h]	600
Germany, Fed. Rep.	19,387[g]	55,932[g]		3,454[g]	1,591[e]
Greece	2,500[a]	3,094		3,900	
Hungary	5,500	101,000	54,273	5,200	
Iceland	93				
Ireland	1,270[g]	1,580[g]	22,200[g]	1,930[g]	570[g]
Italy	15,000	35,000[b]	29,830[a]	57,000[a]	
Luxembourg	190	135			11
Netherlands	7,242[g]	3,942[ij]	100,501[g]	162[g]	250
Norway	1,970	2,186[a,k]	18,500[g]	9,000[g]	70[g]
Poland	7,900		107	101,400	
Portugal	2,246	11,200[a]		3,900[a]	
Spain	10,600	5,108[i]	47,600	180,000	10,000
Sweden	2,650	4,000[a]	17,000	28,000	372
Switzerland	2,500				255
UK[l]	16,700	50,000	2,50,000	2,30,000	30,000
USSR*			2,300		
OCEANIA					
Australia	10,000[a]	20,000[a,b]			
New Zealand	2,106[d]	300[d,j]			45[d]

a 1980
b Estimate
c Includes waste waters which meet the US definition of solid wastes.
d 1982
e 1983
f 1977
g 1984
h includes demolition waste.
i 1986
j Non-chemical wastes only.
k Chemical industry wastes on
l England and Wales only.

* Now splitted

15.3 MUNICIPAL WASTES

The term "municipal wastes" applies to those wastes generated by households and to wastes of similar character derived from shops, offices and other commercial units. Table 15.2 shows that levels of municipal waste production are related to levels of industrialisation and levels of income. Per capita waste generation varies between 2.75 and 4.0 kg per day in high-income countries, but is as low as 0.5 kg per day in those countries with lowest incomes.

Table 15.2 reflects the changes in the composition of municipal wastes in selected countries. Data are given on the percentage contribution of paper (and cardboard), glass, plastics, metals, organic matter and putricible matter to the total mass of municipal wastes in 1975, 1980 and 1985. It is generally very difficult to discern any trends in these figures, although in most developed countries it is generally accepted that the proportion of plastics in municipal wastes is growing (OECD, 1991). Packaging materials are becoming an increasingly important component of municipal wastes in developed countries. Recent estimates suggest that packaging materials account for about 30 per cent of municipal wastes in developing countries. They tend to have a higher organic and ash/grit content and also a higher moisture content.

Data reflecting the availability of municipal waste services in selected countries are presented in Table 15.2. It is evident that in OECD member countries more than 75 per cent of the total population—and certainly all urban populations—have access to municipal waste services. In contrast, in many developing countries, only a small proportion of the population in many urban areas has access to municipal waste services (WHO, 1990). Collection tends to be concentrated in commercial areas, markets and city centres. In other parts of the cities, waste collection and disposal is left to individuals or local communities or waste may be left to accumulate in the streets. For example, in small cities in Indonesia and Vientiane, the capital city of Laos, only about 30 per cent of solid waste generated is collected (WHO, 1990).

Many small business, commercial units and hospitals, especially those in developing countries, rely on municipal waste services. Consequently, the municipal stream may often contain some hazardous wastes, such as hospital wastes, paints, solvents and batteries. Careless handling and disposal of municipal wastes may thus pose a threat to human health in many developing countries. The practice of scavenging from municipal waste dump sites is also a matter of grave concern in some countries.

Table 15.2: Composition of municipal solid wastes in selected countries, 1975-1985

Region/Country	Paper and Cardboard (%)			Plastic (%)			Glass (%)		
	1975	1980	1985	1975	1980	1985	1975	1980	1985
NORTH AMERICA									
Canada	—	36.5	36.5	—	4.7	4.7	—	6.6	6.6
USA	32.5	29.7[a]	34.7[b]	3.0	5.3	6.7[b]	10.0	10.3	9.0[b]
ASIA									
Israel	20[c]	—	—	12[c]	—	—	3[c]	—	—
Japan[d]	30.9	27.4	38.3[b]	8.9	12.3	7.7[b]	6.1[e]	8.0[e]	1.3[b]
Pakistan[f]			10.1			9.9			1.2
EUROPE									
Austria	27.2[g]	30.7[a]	33.6[h]	6.0[g]	6.1[a]	7.0[h]	11.1[g]	10.4[a]	10.0[b]
Belgium	30[j]	35.0	—	5[j]	5.8	—	8[j]	8.2	—
Denmark	35[j]	34[k]	—	4[j]	7[k]	—	8[j]	6[k]	—
Finland	—	—	40	—	—	8	—	—	4
France	35[j]	28	27.5	5[j]	6	4.5	8[j]	11	7.5
German Federal Republic	25	19.9	17.9	8[l]	6.1	5.4	15	11.6	9.2

Contd.

Table 15.2: Composition of municipal solid wastes in selected countries, 1975-1985

Region/Country	Paper and Cardboard (%)			Plastic (%)			Glass (%)		
	1975	1980	1985	1975	1980	1985	1975	1980	1985
Greece	—	19.6	—	—	7.0	—	—	2.7	—
Ireland	33[j]	35[k]	24.5	4[j]	11[k]	14	8[j]	8[k]	7.5
Italy	20.4	22.5	22.3[h]	5.3	6.8	7.2[h]	6.4	6.7	6.2[m]
Luxembourg	25[j]		17.2	4.5[j]		6.4	5[j]		7.2
Netherlands	23.0	21.0	22.8	5.6	6.5	6.8	12.0	11.9	7.2
Norway[n]	31	31	31	4	4	4	3	3	3
Portugal	—	18.9	19	—	3.2	3	—	2.9	3
Spain	—	15	15	—	6	6	—	6	6
Sweden	—	43	35-45	—	10	8-10	—	5	6-8
Switzerland	—	30[o]	—	—	13[o]	—	—	9[o]	—
UK[p]	30	29		4	7	—	10	10	—
OCEANIA									
Australia[q]		26.0	—	—	6.1	—	—	15.1	—
New Zealand	—	3.6	—	—	3.0	—	—	2.5	—

a 1978
b 1984
c 1976
d 1975 and 1980 data are average values for four cities : 1985 data refer to Tokyo only.
e Glass and ceramics.
f Data refer to Karachi only.
g 1973
h 1986
i 1983
j 1977
k 1979
l Includes textiles
m OECD Secretariat estimate.
n 1975 and 1980 data refer to surveys conducted during 1973-1974 and 1979-1981, respectively.
o 1982
p England and Wales only.
q Data are based on sample figures for Sydney (1982) and Melbourne (1980-1981).

The composition of municipal wastes is described here in terms of the percentage contribution of "paper and cardboard", "plastic", and glass.

In India, it was estimated that municipal solid waste generated in small, medium and large cities and towns is about 0.1 kg. 0.3-0.4 kg and 0.5 kg per capita per day respectively. Subsequently, on careful analysis, it was found that average 60-65% waste is compostible and average 5-10% waste can be recycled. The typical composition is as follows:

		percent by weight
(A)	Compostable waste	
(i)	Vegetable, leaves	40.15
(ii)	Paper	0.81
(iii)	Miscellaneous	11.73
(B)	Recycleble waste	percent by weight
(i)	Glass	3.80
(ii)	Plastic	0.62
(iii)	Metals	0.64
(iv)	Stones & ashes	42.80
(v)	Ceramics	0.44

In India, rapid urbanization and uncontrolled growth rate of population are main reasons for MSW to became an acute problem. According to population size per capita waste generation rate and its growth during a decade are indicated Table 15.3 (Annepu, 2012). It is anticipated that population of India would be about 1,823 million by 2051 and about 300 million tons per annum of MSW will be generated that will require around 1,450 km^2 of land to dispose it in a systematic manner, if ULBs in India continue to rely on lanfill route for MSW management (Position paper on the solid waste management sector in India, 2009). However, these projections are on conservative side, keeping 1.33% annual growth in per capita generation of MSW (Bhide & Shekdar, 1998; CPCB, 2000a; Pappu, Saxena, & Asolekar, 2007; Shekdar, 1999). Therefore, with 5% annual growth in per capita generation landfill area required for disposal of waste could be many folds (CPCB, 2013).

Table 15.3: Statistics of MSW generated in different states in India

States	Municipal solid waste (TPD) 2000	Municipal solid waste (TPD) (2009-2011)	Collected (TPD) (2009-2011)	Treated (TPD) (2009-2011)	Growth (%)
1. Andhra Pradesh	4376	11500	10655	3656	163
2. Assam	285	1146	807	73	302
3. Delhi	4000	7384	6796	1927	85
4. Gujarat	NA	7379	6744	873	–
5. Karnataka	3278	6500	2100	2100	98
6. Kerala	1298	8338	1739	4	542
7. Madhya Pradesh	2684	4500	2700	975	68
8. Maharashtra	9099	19204	19204	2080	111
9. Manipur	40	113	93	3	182
10. Meghalaya	35	285	238	100	713
11. Odisha	655	2239	1837	33	242
12. Punjab	1266	2794	NA	Nil	121
13. Puducherry	69	380	NA	Nil	451
14. Rajasthan	1966	5037	NA	Nil	156
15. Tamil Nadu	5403	12504	11626	603	131
16. Tripura	33	360	246	40	991
17. Uttar Pradesh	5960	11585	10563	Nil	94
18. West Bengal	4621	12557	5054	607	172

Source: CPCB (2000b, 2013)

From household or communities solid waste (garbages) collected regularly and then disposed as land fill material in dumping sites adjoining cities.

But there are enormous scope for proper Municipal solid waste (MSW) management (details are given next page).

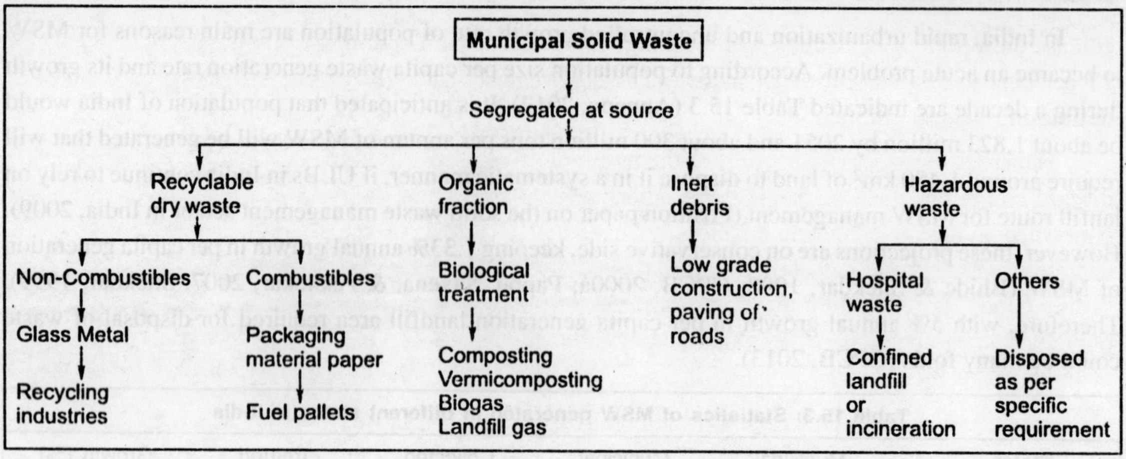

In September 99, Ministry of Environment & Forest, Govt. of India finalised the draft of the Municipal Solid Wastes (Management and Handling) Rules, where there is a need for supply of annual report by Municipal Authority about the collection, transportation and disposal methods of municipal solid wastes.

Now in major cities of India, there were concerted attempt for utilisation of compostable waste for preparation of biogas, compost or vermicompost. In some areas, garbage farming is also attempted. But segregation of recyclable waste till lies on the unorganised sector i.e. ragpickers only.

15.4 AGRICULTURAL WASTES AND SEWAGE SLUDGES

Waste produced by agricultural activities comprise animal slurries, silage effluents, end of spray residues and tank washing following pesticide use, and empty plastic packaging. Sewage sludges are produced as a result of the treatment of industrial and domestic wastes.

Animal manures and silage effluents are the main components of agricultural wastes. Like sewage sludges, these comprise a slurry of fine-grained, organic-rich particles. Sewage sludges may be contaminated with heavy metals, water-soluble organic chemicals, greases and oils depending on the source of effluent and efficiency of treatment.

15.5 INDUSTRIAL WASTES AND MINING WASTES

Industrial process wastes encompass a very wide range of materials and may include general factory rubbish, packaging materials, organic wastes, acids, alkalis and metalliferous sludges. Mining wastes arise as by-products of the extraction process and may include topsoil, rock and dirt, and may be contaminated with small quantities of such materials as metals and coal. The most important feature of industrial and mining wastes is that a significant proportion is regarded as hazardous and, as such, requires special treatment and disposal.

15.6 HAZARDOUS WASTES

It is generally accepted that the bulk of hazardous and toxic wastes are generated by process industries, the main sources being the chemical sector, mineral and metal processing industries and the engineering industries. Estimates of hazardous waste arisings are uncertain because of difficulties in providing a concise definition of hazardous waste arisings in the European Economic Community (EEC) vary from 17 per cent to 50 per cent of all industrial waste arisings depending on the interpretation of the EEC definition (Haines, 1989).

Hazardous wastes arise not only as the by-products of industrial processes, but also when consumers discard empty chemical packaging and other items at the end of their useful life. Many countries are concerned about the increasing quantities of hazardous materials in, for example, aerosol cans, "empty" chemical packages and batteries, that find their way into the municipal waste stream. In addition, many "white goods" (i.e., washing machines, refrigerators, etc.) contain small quantities of chemicals such as polychlorinatedbiphenyls (PCBs) and chlorofluorocarbons (CFCs).

Available estimates of the quantities of hazardous and special wastes generated in selected countries are summarized in Table 15.4. Caution should be exercised when making comparisons between countries as "hazardous" wastes are inconsistently defined. The data presented also illustrate the fact that some hazardous waste is moved between countries. Where export of wastes allows more environmentally-sound disposal, this is considered a valid practice. However, as environmental and safety laws in Europe and North America have become more stringent and the cost of hazardous waste disposal has mounted, an increase in exports to developing countries which often have neither the resources nor the knowledge for safe disposal has been reported.

Accurate information on the amounts and types of wastes exported, their origin and their destination is not widely available. However, it has been estimated that about 20 per cent of exported hazardous wastes are shipped from industrialized to developing nations. Table 15.4 below lists countries that have reported receiving hazardous wastes from developed countries between 1986-1988:

Table 15.4: Hazardous Waste Shipment Status (1987-88)

Region/country	Date of shipment(s)	Amount(t)
AFRICA		
Guinea	February 1988	15,240
Morocco	1988	4,87,680
Nigeria	1987-1988	4,064[a]
Sierra Leone	September 1987	635
South Africa	1986-1988	49.4
Zimbabwe	1986	0.3
NORTH AMERICA		
Haiti	February 1988	4,572[a]
Mexico	1987-1988	49,814
SOUTH AMERICA		
Brazil	1987-1988	29,758
Chile	1988	0.2
ASIA		
Lebanon	May 1988	2,450
Syria	December 1987	2,235
Turkey	1987	1,524
EUROPE		
German Dem. Rep.	1986-1988	3,048,000[a]
Hungary	1986	767[b]
Romania	1986	4,064[a]

[a] Estimate only
[b] Wagons.
Source: Vallette, 1988

Although proposals had been made to ship almost 18×10^6 t of waste per year from North America and Europe to less developed nations (more than 1×10^6 t per year were to be sent to Antigua, Barbuda, Benin, Guinea-Bissau, Panama, Peru, Surinam and Morocco alone), over 44 developing countries have since rejected these proposals. Furthermore, 33 countries have officially banned imports of foreign wastes (Vallette, 1988). The Basel Convention, which is scheduled to come into force 90 days after 20 countries have ratified it, uses the rule of prior written consent and notification to control the trade in hazardous wastes and its implementation should curb illegal dumping in developing countries in the future.

In India, the hazardous waste generated in the country is about 4.4 million tonnes, out of which 38.3 per cent is recyclable, 4.3 per cent is incinerable and the remaining 57.4 per cent is disposable in secure landfills. Twelve state of the country viz., Maharashtra, Gujarat, Tamil Nadu, West Bengal, Andhra Pradesh and Rajasthan account 87 percent for total waste generation.

15.7 RADIOACTIVE WASTES

Radioactive wastes arise primarily from nuclear power generation, smaller quantities are derived from military sources and a variety of uses in medical, industrial and university establishments. There are many types of radioactive wastes which can be classified either according to their radioactive properties or according to the sources from which they originated. Low-level radioactive wastes generally consist of contaminated laboratory debris, biological materials, building materials and uranium mine tailings. Spent fuels from nuclear power reactors, together with liquid and solid residues from reprocessing of spent fuels are classified as high-level radioactive waste.

15.8 BIOMEDICAL WASTE

"Biomedical" waste refers to any waste that includes anatomical waste, pathological waste, infectious waste, hazardous waste and other wastes generated in health care facilities and medical laboratories that require special handling. Previously, the terms "pathological" and "institutional" wastes were used to refer to what is now considered "biomedical" waste.

Classifications

The Ministry of Environment and Forest, Govt. of India, classify biomedical waste according to the categories presented in Table 15.5. The Ministry recommends that biomedical waste be segregated and packaged in leak-proof, colour-coded plastic bags to facilitate identification, handling, storage, decontamination and transportation.

Table 15.5: Ministry classification and colour coding for Biomedical waste

Classification	Description	Colour Code
Type A class 1	Human anatomical	Red
Type A class 2	Animal anatomical, infected	Orange
Type A class 3	(a) Non-anatomical infected	Yellow
	(b) Laboratory waste	Yellow
	(c) Wastes from DNA works	Yellow
Type B, class 1	Animal anatomical (non-infected)	Blue

Collection and Handling in Hospitals and Other Health Care Facilities

Waste collection and handling at most hospitals in Ontario, Canada, involve the transport of waste by employees from the point of generation (patients' beds, operating rooms, laboratories, etc.) to initial storage points (usually enclosed containers in utility rooms) in each hospital unit. At some hospitals, waste considered to be "contaminated," such as urine containers, histology laboratory cuttings, tubes and bags containing blood, spinal fluid containers and waste from isolated patients' rooms, are placed in colour-coded plastic bags for separate handling.

To minimise the potential risk of public exposure to hospital waste, some hospitals use rigid containers for transporting the colour-coded bags along routes within the hospital that are accessible to the general public. The use of rigid containers in this situation is considered good practice and is recommended for all hospitals.

Existing waste collection and handling procedures vary from hospital to hospital in Ontario and the practice of colour coding the disposal bags for separate handling is not consistent. It is, therefore, recommended that the colour-coding system be adopted so that all biomedical waste can be immediately identified. The use of the colour-code should be exclusive to biomedical wastes to preclude public exposure during general purpose disposal.

There are some guidelines for disposal of biomedical wastes (Schedule I)

SCHEDULE I
(See Rule 5)
CATEGORIES OF BIO-MEDICAL WASTE

Option	Waste Category	Treatment & Disposal
Category No. 1	Human Anatomical Waste (human tissues, organs, body parts)	incineration/deep burial
Category No. 2	Animal Waste (animal tissues, organs, body parts carcasses, bleeding parts, fluid, blood and experimental animals used in research, wastes generated by veterinary hospitals, colleges, discharge from hospitals, animal houses)	incineration/deep burial
Category No. 3	Microbiology & Biotechnology Waste (wastes from laboratory cultures, stocks of specific, of micro-organisms live or attenuated vaccines, human and animal cell culture used in research and infectious agents from research and industrial laboratories, wastes from production biologicals, toxins, dishes and devices used for transfer of cultures)	local autoclaving/microwaving/ incinerations
Category No. 4	Waste sharps (needles, syringes, scalpels, blades, glass, etc. that may cause puncture and cuts. This includes both used and unused sharps)	disinfection (chemical treatment/ autoclaving/microwaving and mutilation/shredding
Category No. 5	Discarded Medicines and Cytotoxic drugs (wastes comprising of outdated, contaminated and discarded medicines)	incineration/destruction and drugs disposal in secured landfills
Category No. 6	Solid Waste (Items contaminated with blood, and body fluids including cotton, dressings, soiled plaster casts, lines, beddings, other material contaminated with blood)	incineration autoclaving/ microwaving
Category No. 7	Solid Waste (wastes generated from disposable items other than the waste shaps such as tubings, catheters, intravenous sets etc.)	disinfection by chemical treatment/ autoclaving/microwaving and mutilation/shredding
Category No. 9	Incineration Ash (ash from incineration of any bio-medical waste)	disposal in municipal landfill
Category No. 10	Chemical Waste (chemicals used in production of biologicals, chemicals used in disinfection, as insecticides, etc.)	chemical treatment and discharge into drains for liquids and secured landfill for solids

15.9 SOLID WASTE GENERATION IN MAJOR CITIES OF INDIA

Due to rapid growth of industries and urban settlements, the quantity of waste from dwellings and commercial places is increasing at a faster rate. The urban centres of India produce 1,20,000 tonnes of solid water per day (Yadav Kushal et al., 2007, CPCB, 2000). The solid waste generated in major cities in as conducted by Central Pollution Control Board, is shown in Table 15.6.

Table 15.6: Municipal solid waste generation rates in major cities in India (2005)

Sl. No.	Name of City	Waste Quantity (TPD)	Compostables (%)
1.	Gangtok	13	46.52
2.	Itanagar	12	52.02
3.	Panjim	32	61.75
4.	Dhanbad	77	46.93
5.	Pondicherry	130	49.96
6.	Jammu	215	51.51
7.	Dehradun	131	51.37
8.	Kochi	400	57.34
9.	Raipur	184	51.40
10.	Bhubaneswar	234	49.81
11.	Tiruvanantapuram	171	72.96
12.	Chandigarh	326	57.18
13.	Guwahati	166	53.69
14.	Ranchi	208	51.49
15.	Vijaywada	374	59.43
16.	Srinagar	428	61.77
17.	Madurai	275	55.32
18.	Coimbatore	530	50.06
19.	Amritsar	438	65.02
20.	Allahabad	509	35.49
21.	Vishakhapatnam	584	45.96
22.	Faridabad	448	42.06
23.	Meerut	490	54.54
24.	Nashik	200	39.52
25.	Varanasi	425	45.18
26.	Jamshedpur	338	43.36
27.	Agra	654	46.38
28.	Patna	511	51.96
29.	Bhopal	574	52.44
30.	Nagpur	504	47.41
31.	Lucknow	475	47.41
32.	Jaipur	904	45.50
33.	Surat	1000	56.87
34.	Pune	1175	62.44
35.	Kanpur	1100	47.52
36.	Ahmedabad	1302	40.81
37.	Hyderabad	2187	54.20
38.	Banglaore	1669	51.84
39.	Chennai	3036	41.34
40.	Kolkata	2653	50.56
41.	Delhi	5922	54.42
42.	Greater Mumbai	5320	62.44

Source: Central Pollution Control Board Report, 2005

The quantity and quality of MSW generation is a major criterion to decide the richness of the city. As our richness increases the life-style changes, resulting in generation of more quantity of MSW.

In general, there are five distinct method of Municipal solid waste disposal done so far in many place. The current state of municipal solid waste dumping source countries of South East Asia is shown in Fig. 15.1.

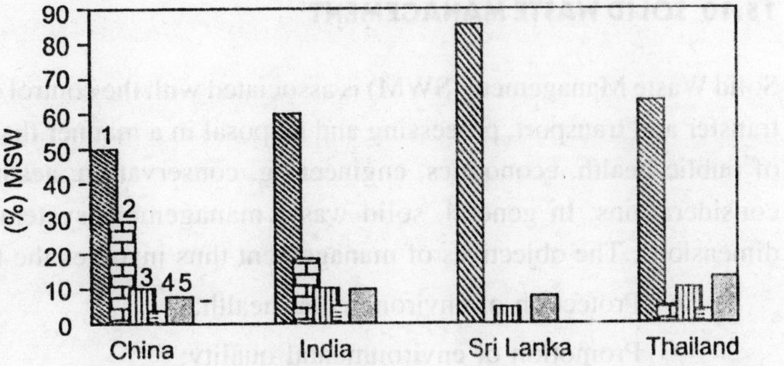

Fig. 15.1 : MSW disposal methods practiced in study countries

The method are as follows:

(a) Open dumping
(b) Landfilling
(c) Composting
(d) Incineration
(e) Others.

Among the major metrocities, Mumbai and Delhi generated the largest amount of municipal solid waste than Chennai and Kolkata (Table 15.7).

Table 15.7: Status of Municipal solid waste (MSW) generation in selected metrocities of India, 2004-05

Particulars	Kolkata	Chennai	Delhi	Mumbai
1. Area (sq km)	187.33	174	1484.48	437.71
2. Population (2001)	45,72,645	43,43,645	1,03,06,452	1,19,78,450
3. MSW generation (tonnes/day)	2,653	3,036	5,922	5,320

Source: Central Pollution Control Board, 2006

There is wide seasonal variation of composition of MSW as reported from various survey. The details of one such survey is given in Table 15.8.

Table 15.8: Seasonal Variation in Physico-Chemical characteristic of Municipal solid waste at Kolkata

Constituents	Seasons		
	Monsoon	Winter	Summer
Biodegradables (%)	41.17	55.03	59.20
Paper (%)	6.10	3.94	3.87
Plastics (%)	4.06	3.33	2.27
Stone, bricks, ash etc. (%)	35.00	26.90	18.57
Moisture (%)	63.54	60.87	60.31
Loss in ignite (%)	49.96	45.41	44.99
C/N ratio	35.05	27.72	25.62
Lower calorific value (KCal/kg.)	570.68	626.66	749.41

All values are in per cent. Physical parameters are on wet weight basis, while chemical parameters are on dry weight basis.

15.10 SOLID WASTE MANAGEMENT

Solid Waste Management (SWM) is associated with the control of waste generation, its storage, collection, transfer and transport, processing and disposal in a manner that is in accordance with the best principles of public health, economics, engineering, conservation, aesthetics, public attitude and environmental considerations. In general, solid waste management system has socioeconomic and environmental dimensions. The objectives of management thus involves the following aspects:

- Protection of environmental health;
- Promotion of environmental quality;
- Supporting the efficiency and productivity of the economy;
- Generation of employment and income.

The typical SWM system with its functional elements and linkages are shown in Fig. 15.2.

Waste generation

Wastes are generate from a variety of sources viz., households, commercial establishments, industrial processes, agricultural practices and also from so many other anthropogenic activities. It is estimated that solid waste generated in small, medium and large cities and towns are about 0.1 kg, 0.3 – 0.4 kg and 0.5 kg per capita per day respectively. It has also been estimated that recyclable waste content (glass, paper, plastics, metals and ceramics etc.) is around 13 to 15% and compostable material is around 60 – 65%. The remaining are stones and ashes. Composite wastes of residential and commercial areas are called garbage.

Waste Collection and Interim Storage

All these solid waste collected either by "Door-to-Door" collection processes or dumped by residents in a common community pits/vats/mechanised containers. Till now wastes at household levels are not segregated properly during / before collection or storage at community vat / containers. Ragpickers often collected recycled wastes (as stated above) from household or from community storage sites. In towns/ municipalities or in cities, there are organised manpower for the said collection jobs including the cleaning of wastes from streets on daily basis.

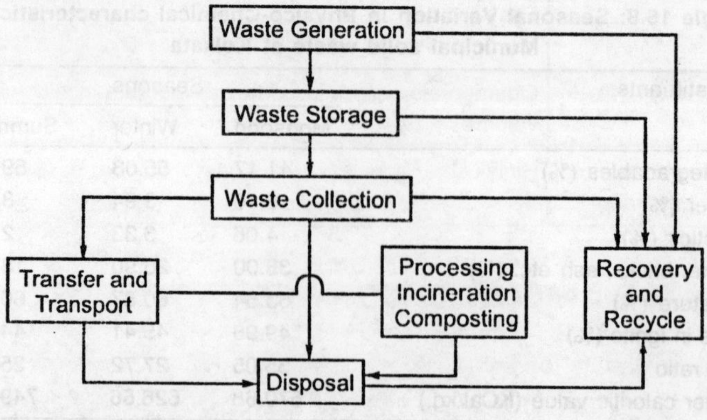

Fig. 15.2 : Typical SWM system : Functional elements

Transportation of Wastes

Transportation of waste from collection vats/pits to final disposal site is another important step in the management of garbage. It has been estimated that approximate 60% of waste is transported to disposal site, the rest is taken by ragpicker from community storage vats and some qualities always lost in transit. Usually once in a day garbage should be cleaned from storage vats / containers, which often non-complited due to shortage of transport carriers and absenteeism of crew etc.

Disposal of Garbage

Usually garbage is disposed in well defined disposal site as landfill. All transport fleets dump their waste, which often transport with cover from interim storage sites. All categories of wastes earlier dumped together in disposal sites.

Now there are some practices of processing introduced in many towns / municipalities. The processing is required to alter the physical and chemical characteristics of wastes for energy and resource recovery and recycling. The important activities in processing are screening and segregation of mixed waste, composting, incineration, biogas production and so on. These could be environment friendly sanitary landfill at disposal, but huge volume of solid waste does not permit such activities in general. In some areas, non processed garbages are spread over lowlying land and crops were grown up on such landfill areas. This is called garbage farming. In disposal areas of many cities, ragpickers often remove the recyclable materials from mixed contaminated infectious wastes.

During the recent years NGOs have taken up initiatives to work with local residents to improve the sanitation. NGOs can also play an important role in segregation of waste, its collection and handing over to local authorities. Various kinds of wasted disposal methods as adopted today are discussed in Table 15.9 with merits and demerits.

Table 15.9: Kinds of waste disposal methods

Method of disposal	Merits	Demerits
1. Landfilling	Easy operation; land gets labelled.	Required large area, chances of land and water contamination in adjoining areas.
2. Open land dumping	Easy for segregation by ragpicker, low cost of dumping	Environment pollution risk is high; large area required.
3. Burning / incineration	Operation is easy; Reduced waste volume	Air pollution problem in site.
4. Processing & bioconversion into compost / biogas / bioethanol	High useful ecofriendly process	More cost is involved.

15.11 WASTE DISPOSAL AND PROCESSING TECHNIQUES

Waste disposal is the final stage of the waste management cycle. About 90% of the municipal waste collected by the civic authorities in India is dumped in low-lying areas outside the city/town limits, which have no provision of leachate collection and treatment, and landfill gas collection and use. An over all composition of urban solid waste in Indian cities are given in Table 15.10 (next page).

Table 15.10: Composition of urban solid waste in Indian cities (% by weight), 2005

City	Paper	Metals	Glass	Textiles	Plastics	Ash & Dust	Organics	Others
	Components							
1. Chennai	5.90	0.70	0.30	7.07	1.56	16.35	56.24	13.74
2. Delhi	5.88	0.59	0.31	3.56	1.46	22.95	57.71	7.52
3. Kolkata	5.14	0.66	0.24	0.28	1.54	33.58	46.88	16.98
4. Bangalore	1.50	0.10	0.20	3.10	0.90	12.00	75.00	7.20
5. Ahmedabad	5.15	0.80	0.93	4.08	0.69	29.01	48.95	10.39
6. Mumbai	3.20	0.13	0.52	3.26	0.80	15.45	59.37	18.07

As a result, leachate containing heavy metals finds its way to the underground water, rendering it unfit for drinking. The landfill gas escapes into the atmosphere, adding to the greenhouse emissions, which otherwise could be used as thermal fuel. Solid waste management can be an income generating activity with cost benefit. However, since the Indian city refuse has a low calorific value with high moisture content and quantity of non-combustibles, it is not suitable for incineration, and currently none of the municipal corporations in India runs a full-scale incineration plant. The potential for power generation from urban municipal waste is tremendous, estimated to generate up to 1000 MW of electricity.

15.12 WASTE REDUCTION, PRODUCT RECOVERY AND RECYCLING

The fundamental principles of waste management is 3R, i.e. reduction, recovery and recycling. Source reduction of waste can serve several purposes viz., product reuse, toxicity reduction, increased product life time, decreased consumption and reduce materials volume use. Recycling is perhaps the most widely recognised form of source reduction involving the process of separating, collecting, processing, marketing and ultimately using a material that would have otherwise been discarded.

In a survey of Bangalore city, 66% of waste generated is collected for recovery of about 2172 tonnes per day. While 722 tonnes per day is reused, the rest (i.e. 1450 tonnes) goes for recycling. Paper and cardboard, glass, metals, plastic and batteries and tyres are often recycled.

15.13 BIOLOGICAL CONVERSION OF SOLID WASTE: COMPOST AND BIOGAS

Composting is one of the important components of solid waste management (SWM). It is a form of source reduction or waste prevention, as the materials are completely diverted from the disposal facilities and require no management or transportation. It is a biological process. A number of composting technologies are currently adopted in the country, viz., windrow, aerated static pile, in vessel composting and anaerobic processing. The first significant development in composing as a systemised process took place in India in 1925, when a process involving the anaerobic degradation of leave, refuse, animal manure and sewerage were placed in pits. This process took approximately six months to produce usable compost. However ICAR (Indian Council of Agricultural Research) improved this method substantially.

Biogasification is another biological transformation process of solid waste. This gasification is an anaerobic process. The gas composed of methane (40-70%), CO_2 (30-60%), other gases (1-5%). In India, the dissemination of large scale biogas plants began in the mid-seventies and the process has become consolidated with the advent of the National Project on Biogas Development (NPBD), initiated by the Ministry of Non-Conventional Energy sources in 1981. Over 3 million families and 3000 communities, institutional biogas plants are currently operated in India.

The compost produced and slurry from biogas plants is often used as a soil amendment in a variety of applications, after ascertaining the quality of the product.

15.14 INCINERATION AND ENERGY RECOVERY

Incineration is a process of energy recovery through thermal treatment. It is a chemical reaction which occurs when carbon, hydrogen and other elements in the waste mix with oxygen in the combustion zone and generate heat/steam, which is later converted to electricity also. Most of the municipal solid waste generated incineration currently practice energy recovery in the form of steam, which is used either to drive a turbine to generate electricity or directly for heating or cooling.

15.15 HAZARDOUS WASTE : MANAGEMENT AND TREATMENT

Hazardous wastes refer to wastes that may or tend to, cause adverse health effects on the ecosystem and human beings. These wastes pose potential risks to health or living organisms, due to the fact that they:

- are nondegradable or persistent in nature;
- can be biologically magnified;
- are highly toxic and even lethal at very low concentrations.

Hazardous wastes of diverse categories are generated in industries and also in households. The common hazardous wastes like radioactive substances, toxic chemicals, biological wastes, flammable wastes and explosives are collected separately and stored and transported in specified manner for processing at selected sites for final disposal. Physical, chemical and biological treatment processes are often adapted for hazardous waste treatment.

The total estimate of hazardous waste generated in India is 4,434,257 tonnes per annum. There are over 13,500 industrial units located in 350 districts, out of which about 11,500 units have been granted authorization for multiple disposal practices encompassing incineration, storage and land disposal options. However, small and medium sized enterprises (SMEs) are the major sources of hazardous wastes, but they have no ability to store or dispose their wastes of hazardous categories. They are now adviced to handover wastes to authorised company for treatment and final disposal on regular basis.

15.16 INTEGRATED WASTE MANAGEMENT (IWM)

Wastes are by products of human activity, physically, wastes contain the same material found in useful products, but differ only in their lack of value. To achieve both environmental and economic sustainability, an integrated waste management (IWM) system is necessary. IWM helps in achieving environmental objectives using economically sustainable systems tailored to the specific needs of a community or region.

IWM involves using a combination of techniques and programmes to manage the municipal waste stream. A hierarchy of waste management that can be used, to evaluate the components of IWMS vis-a-vis the needs of the community is shown in Fig. 15.3.

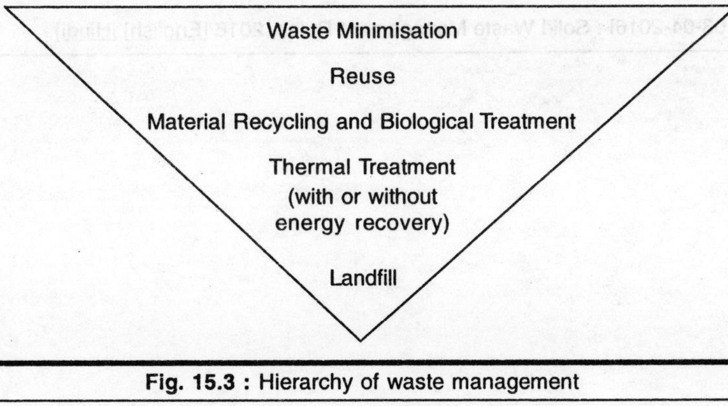

Fig. 15.3 : Hierarchy of waste management

Essentially, the overall objective of waste management should be to reduce the amount of waste generated and to manage the waste in a sustainable way by minimising the overall environmental burden associated with the waste management system.

15.17 WASTE MANAGEMENT POLICIES IN INDIA

National Waste Management Council (NWMC) was constituted in 1990 and one of its objective was municipal solid waste management.

The Ministry of Environment and Forest (MOEF) and the Central Pollution Control Board (CPCB) organised an interaction meet (March 1995) with municipal authorities and other concerned ministries to evolve a strategy for the management of municipal solid waste. Subsequently during disposal of a writ petition (C) No 286 of 1994, Hon'ble Supreme Court of India, directed to CPCB for drafting rules for Management of Municipal solid waste and Biomedical waste in 1996. Supreme court also constituted an expert committee to submit a report on Solid Waste Management in class I cities of India. Based on the report of the above said expert committee and various interactive meetings with CPCB and SPCBs (State Pollution Control Board), a draft rules on Municipal solid wastes management and handling was notified in 1999 under Environment (Protection) Act 1986 and finally the same was implemented in 2000 as Municipal solid waste (Handling and Management Rule). The rule profile the guideline of responsibility of Municipal authority to implement and terms and conditions laid down in schedules I, II, III & IV and also to submit annual report to SPCB as per format (Form I).

With respect to biomedical wastes and hazardous wastes management separate rules notified earlier by MOEF. The biomedical wastes (Management and handling) rules 1998 and the hazardous wastes (Management and handling) rules 1989 were notified as a part of Environment (Protection) Act 1986. Subsequently several other waste management rules are framed and old rules are amended.

The details of all major waste management rules are given in Table 15.11.

Table 15.11: Waste Management Rules under Environmental (Protection) Act 1986
• G.S.R. 320 (E) [18-03-2016] : Plastic Waste Management Rules, 2016 [English] [Hindi]
• G.S.R. 338 (E) [23-03-2016] : e-waste (Management) Rules, 2016 [English] [Hindi]
• G.S.R. 343 (E) [28-03-2016] : Bio-Medical Waste Management Rules, 2016 [English] [Hindi]
• G.S.R. 317 (E) [29-03-2016] : Construction and Demolition Waste Management Rules, 2016 [English] [Hindi]
• G.S.R. No. 395 (E) [04-04-2016] : Hazardous and Other Waste (Management and Transboundary Movement) Rules, 2016 [English] [Hindi]
• S.O. 1357 (E) [08-04-2016] : Solid Waste Management Rules, 2016 [English] [Hindi]

SECTION IV

CURRENT ENVIRONMENTAL ISSUES

16 GLOBAL WARMING AND GREENHOUSE EFFECT

16.1 INTRODUCTION

The Earth receives shortwave radiation from the sun (including the visible part of the spectrum), one- third of which is reflected while the rest is absorbed by the atmosphere, ocean, ice, land, and biota. The energy absorbed from solar radiation is balanced, in the long term, by outgoing radiation from the Earth and atmosphere. Terrestrial radiation is emitted in the form of long wave infrared energy. The balance between energy absorbed and emitted as longwave infrared radiation can change due to a number of factors: changes in the sun's energy output, slow variations in the Earth's orbit and the greenhouse effect. The greenhouse effect is one of the most important factors and is one which humankind has the capacity to change.

Shortwave radiation can pass easily through the atmosphere, whereas longwave terrestrial radiation emitted by the warm surface of the Earth is partially absorbed by a number of trace gases in the atmosphere. These trace gases are called greenhouse gases (GHG). The main natural GHGs are carbon dioxide (CO_2), methane (CH_4), nitrous oxide (N_2O), water vapour (H_2O) and ozone (O_3) in the troposphere and stratosphere. Further human activities in recent decades add new halocarbons (HFCs) as GHG.

The warming effect on the earth's surface by certain gases in the atmosphere was first recognized in 1827 by Jean-Baptiste Fourier, the famous French mathematician. Around 1860, the British scientist John Tyndall measured the absorption of infrared radiation by CO_2 and water vapour, and he suggested that the cause of the ice ages may be due to a decrease of atmospheric concentrations of CO_2. In 1896, the Swedish scientist Svante Arrhenius estimated that doubling the concentration of CO_2 in the atmosphere may lead to an increase of the earth's surface temperature by 5-6 °C.

Thus the greenhouse effect arises because the atmosphere is largely transparent to incoming solar radiation, while being quite heavily absorbing to outgoing thermal radiation from the planetary surface and the atmosphere. Since the beginning of the earth's origin, the effect exists as a natural process. But since the Twentieth century the effect is enhanced by man's activities that are liable to destabilise the natural balance. Different trace gases in the atmosphere contribute to the greenhouse effect, but among them the major gases are carbon dioxide (CO_2), chlorofluorocarbons (CFCs), methane (CH_4) oxides of nitrogen (NO_x), water vapour (H_2O) and perfluorocarbon (CF_4) and other trace gases.

The important greenhouse gases and their anthropogenic sources are given in Table 16.1, Fig. 16.1.

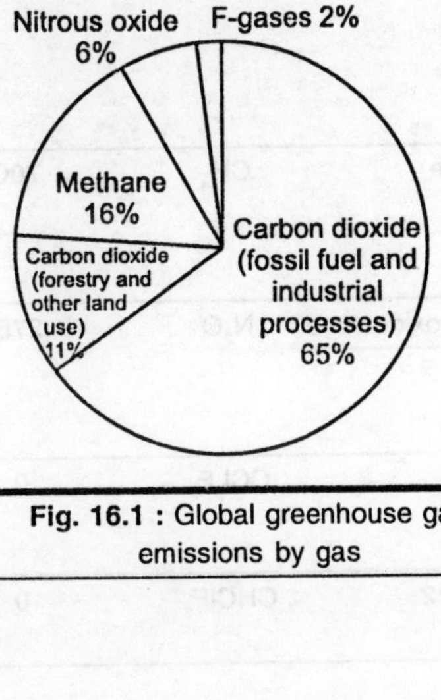

Fig. 16.1 : Global greenhouse gas emissions by gas

At the global scale, the key greenhouse gases emitted by human activities are:

Table 16.1: Greenhouse gases and their main anthropogenic sources

Gas	Sources
Carbon dioxide (CO_2)	Fossil-fuel burning Wood fuel Deforestation and land use change
Water vapour (H_2O)	Hydrological cycle induced by man
Methane (CH_4)	Release from gas, oil or coal production or transmission Enteric fermentation from ruminants (e.g., cattle, sheep, goats) Wetland rice cultivation Landfill waste sites Burning and decay of biomass
Chlorofluorocarbons (CFCs)	Used for solvents, refrigerants, aerosol spray propellants, foam packaging etc.
Perfluoromethane (CF_4)	Production of aluminium
Nitrous oxide (N_2O)	Fertilisers manufacture Fossil-fuel burning Tropical deforestation and wildfires Land conversion for agriculture
Gases involved in generation of low-level ozone :	
Nitrogen oxides	Fossil-fuel burning Biomass burning
Carbon monoxide	Fossil-fuel burning Biomass burning
Sulphur hexafluoride (SF_4)	Dielectric fluid
Non-methane hydrocarbons	Evaporation of solvents and liquid fuels

The relative effect of different gases on the greenhouse effects has been calculated by a number of authors. Since pre-industrial period (1750-1800) to present day (2011), IPCC scientific committee assessed the changes of major greenhouse gases affected by man's activities (Table 16.2) and opined that if current trend continues, an average global warming over the 21st century will be at 0.3°C per decade, with associated major changes in regional climate patterns and sea level rises averaging about 6 cm per decade.

Table 16.2: The Main Greenhouse Gases

Greenhouse gases	Chemical formula	Pre-industrial concentration (pphv)	Concentration in 1994 (pphv)	Atmospheric lifetime (years)*	Anthropogenic sources	Global warming potential (GWP)**
Carbon dioxide	CO_2	278,000	358,000	Variable	Fossil-fuel combustion Land use conversion Cement production	1
Methane	CH_4	700	1,721	12.2+/-3	Fossil fuels Rice paddies Waste dumps Livestock	21***
Nitrous oxide	N_2O	275	311	120	Fertilizer Industrial processes Combustion	310
CFC-12	CCl_2F_2	0	0.503	102	Liquid coolants Foams	62,00-7,100****
HCFC-22	$CHClF_2$	0	0.105	12.1	Liquid coolants	1,300-1,400****

Contd.

Table 16.2: The Main Greenhouse Gases

Greenhouse gases	Chemical formula	Pre-industrial concentration (pphv)	Concentration in 1994 (pphv)	Atmospheric lifetime (years)*	Anthropogenic sources	Global warming potential (GWP)**
Perfluoro-methane	CF_4	0	0.070	50,000	Production of aluminum	6,500
Sulfur hexa fluoride	SF_4	0	0.032	3,200	Dielectric fluid	23,900

Note : ppbv = parts per billion volume; 1 ppbv of CO_2 in the earth's atmosphere is equivalent to 2.13 million metric tons of carbon (www.cdiac.esd.ornl.gov, accessed on December 10, 2000).

* No single lifetime for CO_2 can be defined because of the different rates of uptake by different sink processes.

** GWP for 100 year time horizon.

*** Includes indirect effects of tropospheric ozone production and stratospheric water vapor production.

**** Net global warming potential (i.e., including the indirect effect due to ozone depletion).

Sources: United Nations Environment Programme's Introduction to Climate Change, accessed at www.grida.no/climate/vital/intro.html on April 17, 2001.

From time to time, it was attempted to assess the emission of greenhouse gases by different countries, some figures were given in Figs. 16.2 & 16.3. There are debate about the percentage responsibility of different countries on emission of GHGs. The summarised details are given in Fig. 16.5. Historically, it is true that developing countries have contributed relatively little to the risk of global warming, being responsible for only about 15% of world greenhouse gas emissions between 1870 to 1986, although some 75% of the world population lives in the Third World, the developing countries.

In a recent survey reports, GHG emission made by major countries are altered. The details shown in Fig. 16.4.

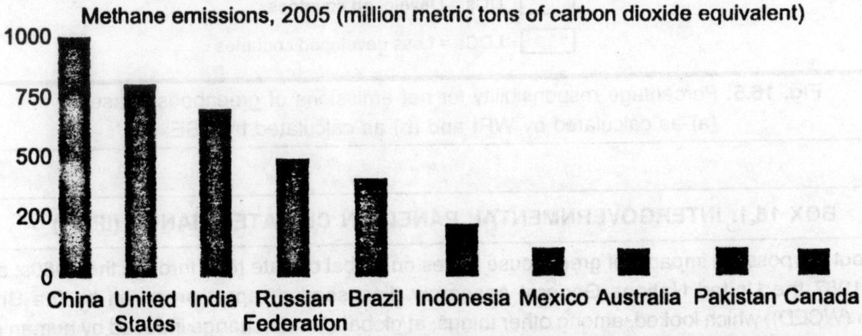

Fig. 16.2: The 10 largest contributors to methane emissions amount for about 62 percent of emissions

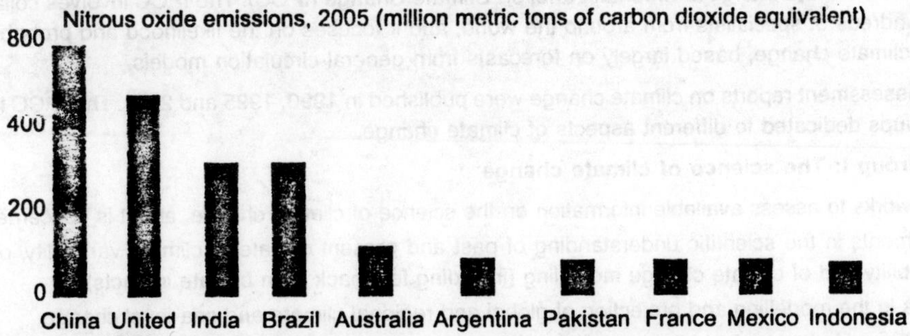

Fig. 16.3: The 10 largest contributors to nitrous oxide omissions amount for about 56 percent of emissions

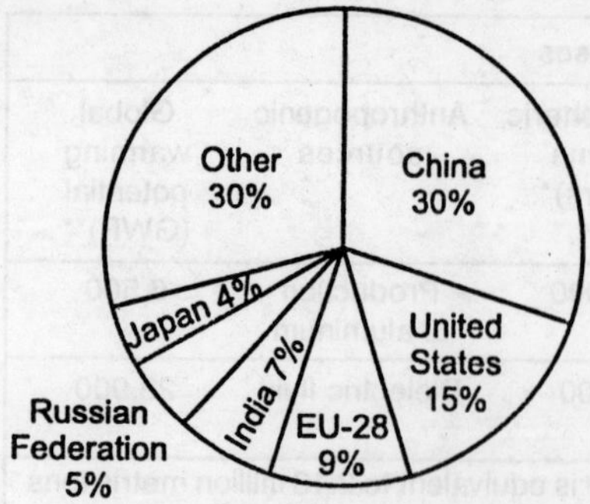

Fig. 16.4 : 2014 Global CO_2 emissions from fossil fuel combustion and some industrial processess

Source: Boden, T.A., Marland, G., and Andres, R.J. (2017). *National CO_2 emissions from fossil-fuel burning, cement manufacture and gas flaring*: 1751-2014, Carbon dioxide information analysis center, Oak Ridge National Laboratory, U.S. Department of Energy, doi 10.3334/CDIAC/00001_V2017

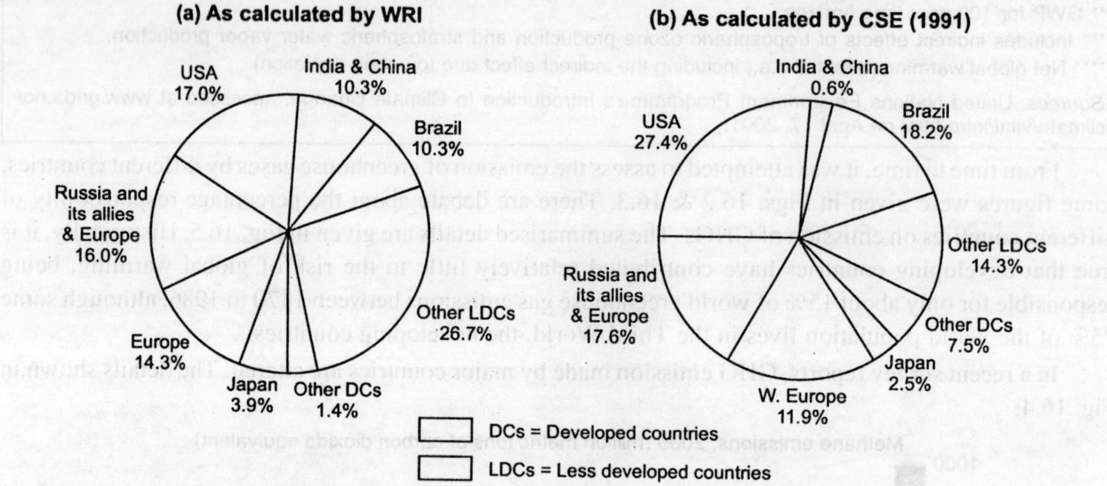

Fig. 16.5: Percentage responsibility for net emissions of greenhouse gases : (a) as calculated by WRI and (b) as calculated by CSE

BOX 16.1: INTERGOVERNMENTAL PANEL ON CLIMATE CHANGE (IPCC)

Concern about the possible impacts of greenhouse gases on global climate rose through the 1980s, and in the autumn of 1987 the United Nations General Assembly discussed a report prepared by the Brundtland Commission (WCED), which looked, among other things, at global climate change induced by human activities.

In 1988, the World Meteorological Organisation (WMO) and the United Nations Environmental Program (UNEP) established the Intergovernmental Panel on Climate Change (IPCC). The IPCC involves collaboration between hundreds of specialists from around the world, and it focuses on the likelihood and probable nature of induced climate change, based largely on forecasts from general circulation models.

The IPCC assessment reports on climate change were published in 1990, 1995 and 2001. The IPCC has three working groups dedicated to different aspects of climate change.

Working Group I: The science of climate change

This group works to assess available information on the science of climate change, and it is concerned with:

- developments in the scientific understanding of past and present climate of climate variability, of climate predictability and of climate change modelling (including feedback from climate impacts);
- progress in the modelling and projection of global and regional climate and sea-level change;
- observations of climate, including past climates, and assessment of trends and anomalies;
- gaps and uncertainties in current knowledge.

Contd.

Working Group II: Impacts, adaptation, vulnerability

This group's work focuses on assessing the scientific, technical, environmental, economic and social aspects of the vulnerability (sensitivity and adaptability) to climate change of, and the negative and positive consequences for, ecological systems, socio-economic sectors and human health, with an emphasis on regional sectoral and cross-sectoral issues.

Working Group III: Mitigation of climate change

This group is particularly concerned with the scientific, technical, environmental and economic and social aspects of mitigation of climate change.

In addition, the IPCC has a Task Force on Greenhouse Gas Inventories.

The chemical composition of troposphere and the stratosphere is an important factor determining the mean temperature of the planet's surface and, thus, its climate. The amount of heat trapped depends mostly on the concentrations of various heat trapping gases—known as greenhouse gases—in the troposphere. The principal greenhouse gases are carbon dioxide contributing the major share along with other trace gases as stated earlier.

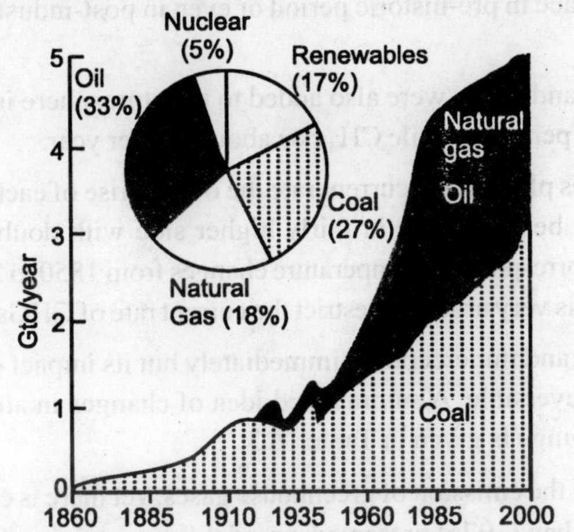

Fig. 16.6: Growth in fossil-fuel consumption since the Industrial Revolution

Until recently, most greenhouse gases were emitted and removed from the troposphere by the Earth's major biogeochemical cycles, without disruptive interferences from human activities. However, since the

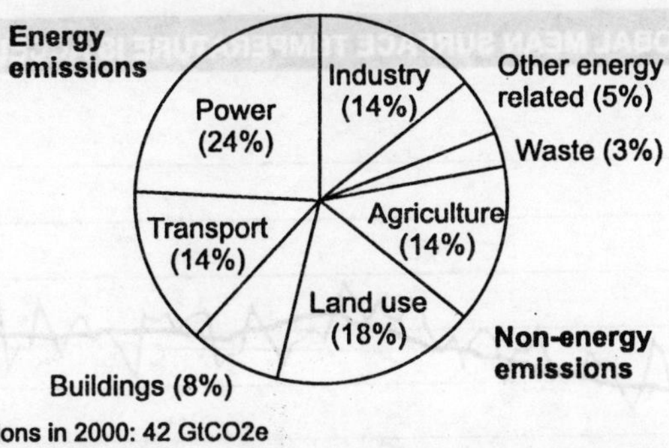

Total emissions in 2000: 42 GtCO2e

Energy emissions are mostly CO_2 (some non-CO_2 in industry and other energy related)

Non-energy emissions are CO_2 (land use) and non-CO_2 (agriculture and waste)

Source: Prepared by Stern Review, from data drawn from World Resources Institute Climate Analysis Indicators Tool (CAIT) on-line database version 3.0.

Fig. 16.7 : Global emissions of greenhouse gases

Industrial Revolution (1760) and especially since 1950, we have been putting enormous quantities of greenhouse gases into the atmosphere, primarily from the fuel burning (57%), the use of CFCs (17%), agriculture (15%) and deforestation (8%). There is growing concern that these gases can amplify the natural greenhouse effect and turn up the planets thermostat fairly rapidly.

Greenhouse gases emitted from various sectors are given in Fig. 16.7.

16.2 GREENHOUSE GASES AND GLOBAL CLIMATE CHANGES

The changes in CO_2 level—as measured at Mauna Loa observatory in Hawai—ranges from 315 ppm in 1958 to 380 ppm in 2006 with a narrow seasonal oscillation every year. There are two most important human activities that lead to gradual rise of CO_2 level in the atmosphere, viz., over increasing burning of fossil fuel and loss of CO_2 sink i.e., the gradual rise of destruction of forest cover, coal and oil consumption increased many fold over the years as chief energy sources (Fig. 16.6). The rise of CO_2 level have immediate effect on the rise of global temperature. There are clear examples that rise of CO_2 level in atmosphere have positive correlation with rise of temperature on earth surface in pre-historic period or even in post-industrialisation phase (Fig. 16.8).

Over the years other GHGs like NO_x, CH_4 and CFCs were also added to the atmosphere in alarming rate. As per NASA report, CFCs rise is about 5% per year, while CH_4 rise about 1% per year.

On the whole if the above rise of GHGs takes place in the current rate the overall rise of each pollutant gases that contribute the greenhouse effect will be somewhat in fairly higher side with doubling point sometime in 2030 (Figs. 16.9 & 16.10). Thus the corresponding temperature changes from 1850 to 2050 at the magnitude of 0.25°C to 3.5°C (Fig. 16.11). Thus it is very urgent to restrict the current rate of GHGs emission.

Though the GHGs effect on climate is slow and imperceptible immediately but its impact on climate change in the long run become alarming and irreversible. A summarised idea of changes in atmospheric trace gases and its relevance to global climate change is given in Table 16.3.

In spite of repeated global effects for abating the emission of greenhouse gases, yet there is cumulative rise of greenhouse gases over the years. Ramanathan (1976) presented a model for growth in some greenhouse gases, 1980-2030 (Table 16.3).

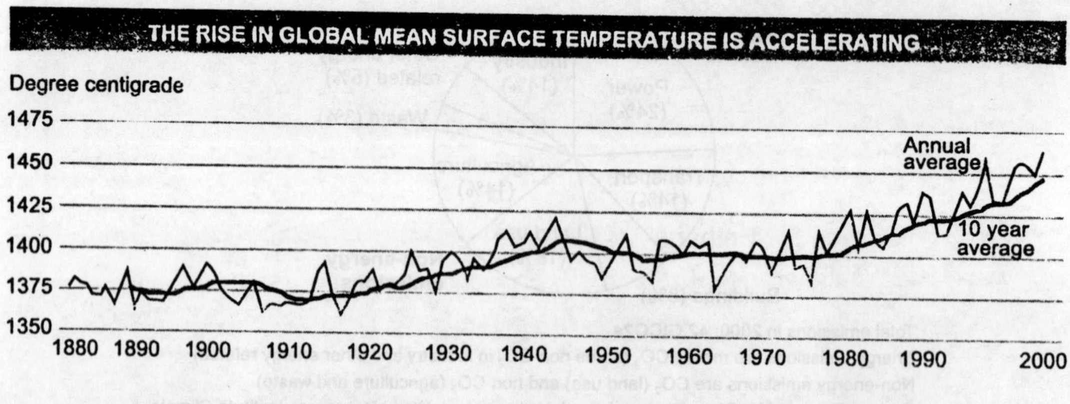

Fig. 16.8: Carbon dioxide and temperature are very closely correlated over the past 160,000 years (top) and to a lesser extent over the past 100 years (bottom)

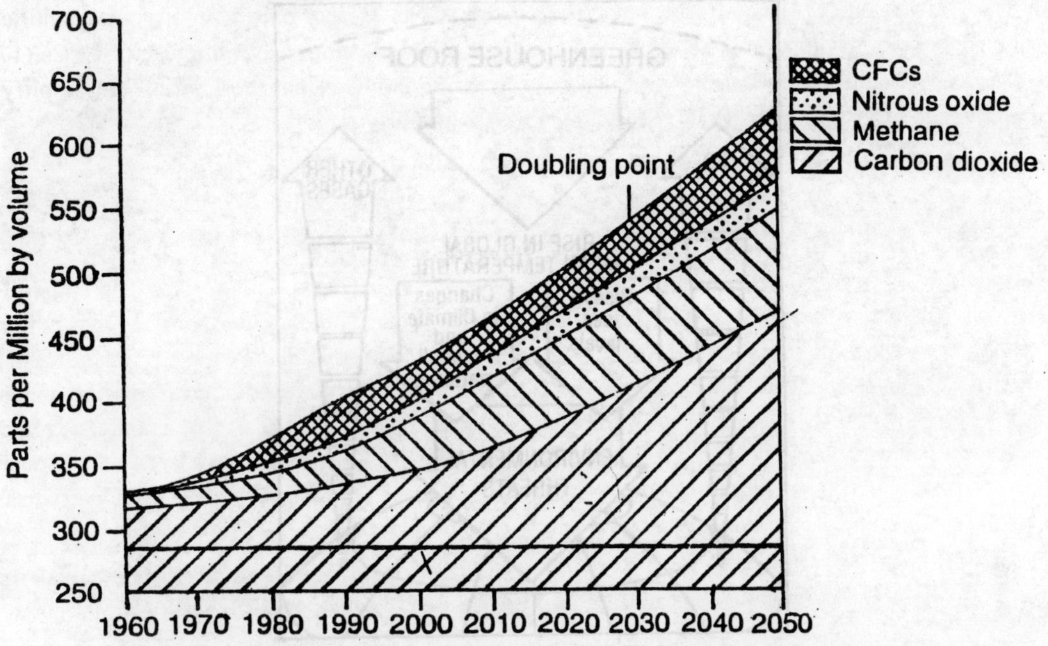

Fig. 16.9: Business-as-usual projection of the trend in effective GHGs

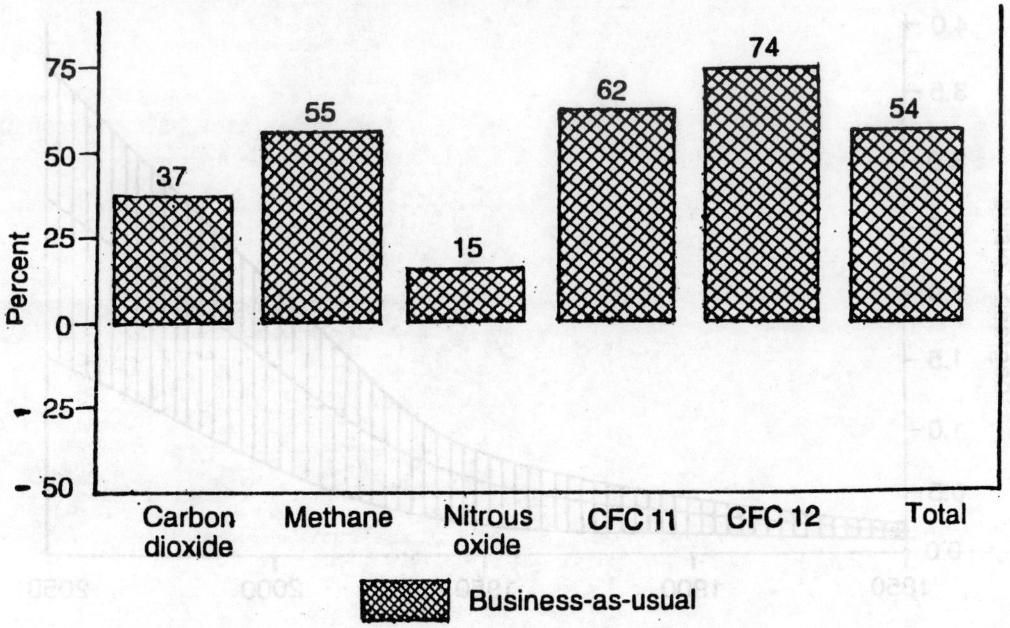

Fig. 16.10: Business-as-usual projection of the percentage change in greenhouse gas concentrations over the period 1990-2050

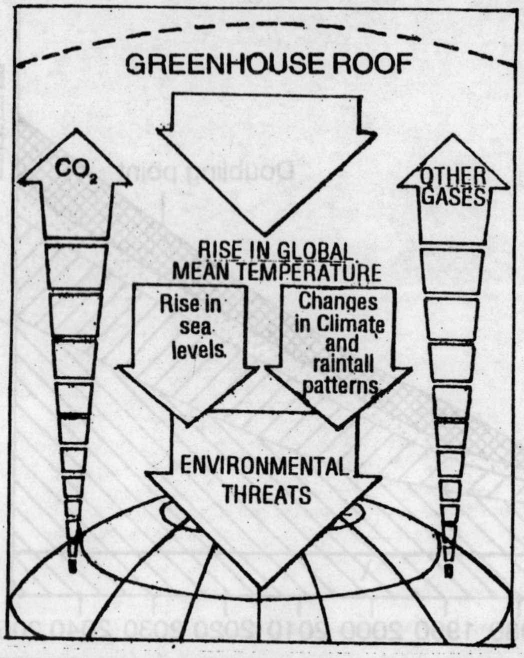

Fig. 16.11: A generalised scheme of environmental threats due to GHGs' accumulation

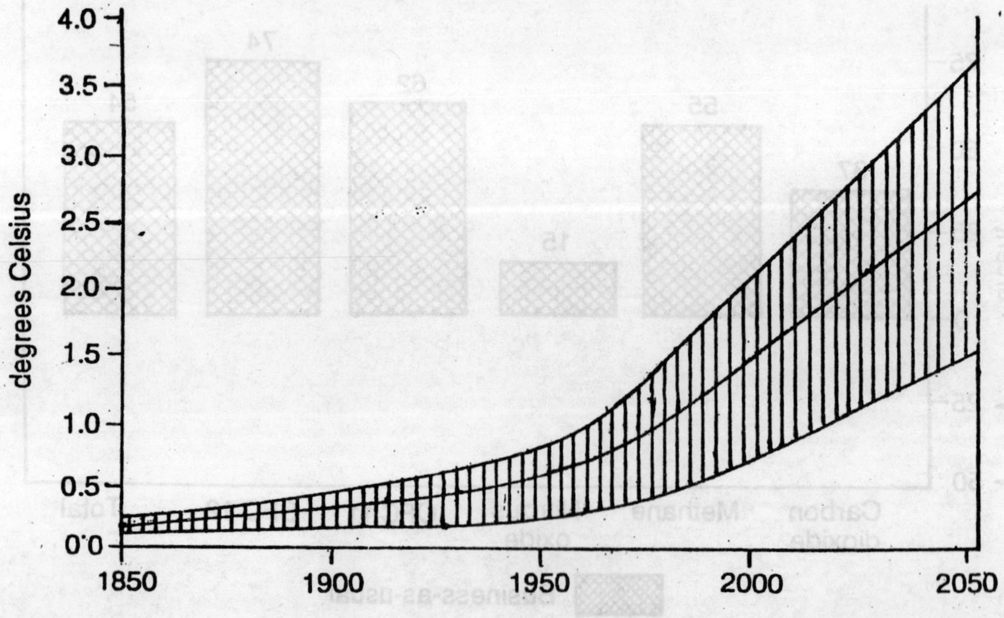

Fig. 16.12: Business-as-usual projection of the trend in global surface air temperature rise

BOX 16.2: FUTURE GREENHOUSE GAS EMISSION SCENARIOS

The most reliable forecasts of what the future might have in store are those based on a set of scenarios proposed by the IPCC :

Low-growth scenario

The lowest scenario assumes that world population will increase to 6.4 billion by 2010, the economy will grow by an average of 2 per cent a year by 2025 but slow down a great deal after that, and nuclear power costs will decline by 0.4 per cent a year. Under these assumptions, CO_2 emissions would increase from about 7.4 gigatonnes (Gt) in 2000 to 8.8 Gt in 2025 and then fall to 4.6 Gt by 2100. This would create an atmospheric concentration of CO_2 of 450 ppm by 2100, compared with 360 ppm in 2000.

High-growth scenario

The rapid-growth scenario assumes that world population grows to 11.3 billion people, the economy grows by 3 per cent a year over the next century, and nuclear power costs increase. Under these assumptions, CO_2 emissions would increase to 15.1 Gt by 2025 and to 35.8 Gt by 2100. This would create an atmospheric concentration of CO_2 of 900 ppm by 2100, compared with 360 ppm in 2000.

Middle-growth scenario

Under the middle scenario, annual emissions of CO_2 are expected to rise to 20.3 Gt by 2100, producing an atmospheric concentration of 700 ppm (double today's level) by 2100.

Table 16.3: Predicted changes in the growth of greenhouse gases (1980-2030)

Gases	Average residence time (Year)	1980 global average mixing ratio (ppb)	2030 estimated average mixing ratio (ppb)	Comments
CO_2	2	339,000	450,000	Assume 2.4% p.a. increase for next 5 years
NO_x	120	300	375	Combustion & fertiliser sources
CFCs	5-110	0.6-0.28	0.3-1.8	Depending on regulation
CH_4	5-10	1630	2340	—
O_3 (troposphere)	0.1-0.3	25-70	30-100	—

16.3 GLOBAL WARMING POTENTIAL

Global warming potential (GWP) is an index that attempts to quantify the climate impact of the same emission of a given greenhouse gas relative to that of a reference gas (usually CO_2), by taking into account the different heat trapping abilities on a molecule by molecule basis and the different rates of removal of the two gases. This is a calculation of possible warming effect on the lower atmosphere of each of the greenhouse gases relative to CO_2. The details of warming potential is:

Greenhouse gases	Direct global warming potential	Sign of indirect global warming potential	Long lifetime
CO_2	1	None	Yes
CH_4	11	Positive No	—
NO_x	270	Uncertain	Yes
CFC 11	3,400	Negative	Yes
CFC 12	7,100	Negative	Yes
HCFC 22	1,600	Negative	Mainly No
HFC 134a	1,200	None	Yes (?)

However, this computation is subject to change with time as our knowledge continuously improves.

16.4 POSSIBLE IMPACT OF GLOBAL WARMING

The rise of greenhouse gases and consequent global warming phenomenon have a number of effects on earth climate, ecosystems and biospheric processes. The major events are:

(a) **Sea Level Change:** One major consequence of global warming arising out of greenhouse effect is the sea level rise. Four major changes take place prior to sea level rise as consequence of global warming. They are: Thermal expansion, mountain glacier melting, Greenland ice sheet melting and Antarctic ice sheet melting. Tables 16.4, 16.5 & 16.6 provide an estimate of sea level changes in the past 100 years, between 1990-2030 and future trend if CO_2 emission rate continued as usual or not modified suitably.

Table 16.4: Estimated contributions to sea level rise over last 100 years (in cm)

	Low	Best estimate	High
Thermal expansion	2	4	6
Mountain glaciers	1.5	4	7
Greenland ice-sheet	1	2.5	4
Antarctic ice-sheet	– 5	0	5
Total	**– 0.5**	**10.5**	**22**

Table 16.5: Estimated contributions to sea level rise (in cm) from 1990-2030 according to business-as-usual scenario

	Low	Best estimate	High
Thermal expansion	6.8	10.1	14.9
Mountain glaciers	2.3	7.0	10.3
Greenland ice-sheet	0.5	1.8	3.7
Antarctic ice-sheet	– 0.8	– 0.6	0
Total	**8.8**	**18.3**	**28.9**

Source : IPCC (Inter governmental Panel on climate change)

Table 16.6: Future predictions of CO_2 concentrations rises in global temperature and sea level for various emission scenarios

(a) Business-as-usual, high emission scenario:

	1990	2030	2060	2100
CO_2 concentration	354	470	590	850
T(K)	0	1.1	2.0	3.25
Sea-level rise (cm)	—	18	38	65

(b) All emissions kept constant at 1990 rates:

	1990	2030	2060	2100
CO_2 concentration	354	420	465	520
T(K)	0	0.72	1.1	1.6
Sea-level rise	0	15	27	42

Contd.

Table 16.6: Future predictions of CO_2 concentrations rises in global temperature and sea level for various emission scenarios				
(c) 2% p.a. (compound) reduction in all emissions from 1990:				
	1990	2030	2060	2100
CO_2 concentration	354	388	395	390
T(K)	0	0.4	0.4	0.3
Sea-level rise	0	11	18	21
(d) 2% p.a. (compound) increase in all gas emissions 1990-2100, thereafter a 2% p.a. decrease				
	1990	2030	2060	2100
CO_2 concentration	354	436	458	464
T(K)	0	0.93	1.10	1.0
Sea-level rise	0	17	28	34

In spite of uncertainties about the degree of rise in sea level, any rise would pose a direct threat to low-lying coastal zones and islands and tropical coastal wetlands, where the mangrove ecosystems are under threat. Where rising sea level is combined with tectoxic subsidence and/ or human actions that may exacerbate the problem, the situation is potentially very serious. Table 16.7, indicates the synthesised results in sea level by the year 2100 based on the high estimate of global warming under the 1900 business-as-usual scenario. This may be regarded as extreme but down-scaling still implies substantial problems for countries such as Bangladesh and China, particularly where economic growth in terms of GNP remains at a low level.

Table 16.7: Estimates of impact of 1 meter rise in sea level (IPCC, WGII, 1996)

Country/	People affected		Land affected	
Source	People in million	% of total	Km²	% of total
China	72	7	35,000	—
Bangladesh	71	60	25,000	17.5
Japan	15	15	2,300	0.6
Netherlands	10	67	2,165	5.9
India	7.1	1	5,800	0.4
Egypt	4.7	9	5,800	1.0
Nigeria	3.2	4	18,600	2.0

There are also reports of

1. Rise in Arctic & Atlantic sea water levels due to meltwater lakes (as the one on the surface of the Petermann Glacier in Greenland in 2009).
2. Continuously lowering salinity level in the 'Brackish-water ecosystem' in the Baltic sea.

(b) **Crop Yield**: But with respect to crop yield it is expected to rise to 60-80% in some areas because of rise of CO_2 level. However, other factors could offset these effects. There is also a prediction of regional climate changes along the ecosystem.

Overall, the climate change is likely to be beneficial, due to the dominance of C_3 crops such as barley, wheat, rice and soybeans. The C_3 annual crops show yield increases of up to 30% at doubled (700 ppm) CO_2 concentrations under controlled experimental conditions (Table 16.8). This productivity could be further enhanced, since fourteen out of eighteen of the world's worst weeds are C_4 plants and would not directly benefit from the CO_2 fertilisation effect. Potentially limiting factors include changes in insect life cycles and an increase in the survival, growth and spread of pathogens.

Table 16.8: Trend in world crop productivity due to climate change

Crop	Change in crop yield (1967-97) FAO, 1998 report	Crop yield model : % change in yield with 2 × CO_2 climate change plus direct effects of CO_2
C3 : Barley	+ 35.8	– 30 (Uruguay)
Wheat	+ 96.1	+ 11
Rice	+ 74.5	– 2
Soyabeans	+ 56.2	+ 16
Potatoes	+ 20.3	+ 20 (W. Europe)
C4 : Maize	+ 70.3	– 15
Sorghum	+ 28.2	0 (USA)
Millet	+ 45.0	– 66 (Senegal)
Sugarcane	+ 19.6	+ 9 (Australia/Japan)

(c) **Water Balance**: Although changes in sea-level have received much publicity, problems of water availability are likely to be more serious and perhaps more expensive to solve. In future, warmer world will have water crisis in some parts while in other region it will be wetter than today.

There is uncertainty regarding regional forecasts of future precipitation as warming of the globe makes it difficult to predict. So also pattern of agricultural changes, or effects on ecosystems at large are fairly unpredictable.

(d) **Human Health**: In recent years, there were newer reports of major tropical disease spread with changing climate. More people will likely be affected on the globe by tropical diseases, in coming decades as globe will become warmer in future. The detailed prediction is given in Table 16.9.

On the whole, a generalised scheme for global environmental threat that arises out of green-house gas emission is depicted in Fig. 16.10.

Table 16.9: Major Tropical diseases likely to spread with global warming

Diseases	Vector	Population at risk	Prevalence of infection	Present distribution	Liklihood of altered distribution with warming
Malaria	Mosquito	2,100 million	270 million	(Sub) tropical	+++
Schistosomiasis	Water snail	600 million	200 million	(Sub) tropical	+
Filariasis	Mosquito	900 million	90 million	(Sub) tropical	+
Onchocerciasis (river blindness)	Black Fly	90 million	15 million	Africa/ Latin America	+
African trypanosomiasis (sleeping sickness)	Tsetse Fly	50 million	25,000 new cases/year	Tropical Africa	+
Dengue Fever	Mosquito	estimates unavailable	—	Tropics	+
Yellow Fever	Mosquito	estimates unavailable		Tropical South America & Africa	+

+++ Higher, + Lower

16.5 CLIMATE CHANGE

Climate can change naturally for a variety of reasons. Some of the driving factors of climate change operate at time scales of hundreds of million years, whereas others fluctuate over a time period of only a few years. The major driving factors of climate changes are:

- Changes in the composition of the earth's atmosphere (specially atmospheric CO_2 level);
- Changes in topography, land-sea geography and battymetry of ocean;
- Changes in solar luminosity;
- Changes in earth's orbit;
- Changes in earth's volcanic activity;
- Changes in internal variability of atmosphere—ocean system.

In addition human induced causes also accelerates the climate changes viz., increased concentrations of well-mixed GHGs; changes in the O_3 concentration, increasing concentration aerosol and changes in the land surface. However, the key component of climate changes is the heat trapping due to increase in CO_2 and other GHGs, which could easily reach 6-6 wm^2 by the end of the 21st century. This is much larger than the perturbation that can be reasonably expected due to natural fluctuation over a time span as short as 100 years. Furthermore the aerosol cooling effect is much lesser than GHGs heating effect. Thus major driving force of climate change of 21st century will be enhancement of GHGs. The climate change hazards are given in Table 16.10.

Table 16.10: Climate change hazards/risks

Primary threat		Secondary threat
• Extreme precipitation events within short spell	→	Flooding in river basin
	→	Increased erosion and landslides due to run off at hilly terrain
• Cyclones/Hurricanes/Storm events	→	Storm surges;
	→	High winds/storm damage
• Rising sea levels	→	Coastal and fluvial flooding
	→	Saline penetration in fresh water aquifers
• Increased temperature	→	Heat waves
	→	Rise of drought incidence
	→	Loss of glaciers
	→	Loss of permafrost sea ice
	→	Incidence of natural bush fire
• Ecosystem adaptation	→	Human adaptation/migration
	→	Distribution of diseases incidence high (Specially vector borne diseases)

16.6 GREENHOUSE EFFECT—POLICY RESPONSE

1. The prospect of future environmental security being compromised by global warming are now so real as to make adoption of a precautionary response imperative—in other words, policy should focus on buying insurance and policy-makers should be clearly aware that waiting in perpetuity for better scientific data entails the real risk of waiting until it is too late.

2. The main routes to surviving the greenhouse threat are energy efficiency, renewable forms of energy production of all CFCs and related gases, less greenhouse gas-intensive agriculture, stopping deforestation and reforestation.

3. The world community should strive to achieve a low energy future of the kind spelt out by the end-use global energy project, with a world energy conservation of not much more than around 12 terawatts (TW) by 2025, but spread through the world more equitably than the 10 TW consumed today, so the developing countries can enjoy standards of living akin to those enjoyed in the industrialised countries.

4. This pattern of consumption should be achieved by massive investment in energy efficiency in the industrialised countries and large-scale transfer or up to date energy efficiency and non-CFC dependent technologies from the industrialised countries to the developing countries, to enable them to leap-frog the kind of energy-intensive, polluting routes which the developed countries took during industrialisation.

5. Given suitable investment of funding and resources, the goal should be for the world's entire energy requirement to be produced by renewable forms of energy production as soon as possible in the 21st century.

6. Expansion of the nuclear power industry, or replacement of existing nuclear power stations when they come to the ends of their operating lifetimes, would involve diversity funds from cost-effective energy-solutions to the greenhouse crisis—especially when viewed in concert with safety waste disposal—decommissioning and nuclear weapons proliferation problems should have no rôle to play in the international policy response.

7. The detailed mix of policy prescriptions and their quantification should be subject to scientific research results during the 21st century, but essential first steps are :

(a) Implementation of immediate cuts in carbon dioxide emissions and the drawing-up of integrated strategies aimed at the phase-out of fossil fuels as early as possible in the 21st century;

(b) No effort be spared to arrest deforestation and halt the production or use of CFCs and related compounds, such as HFCs.

8. The large sums needed to fund these and other antigreenhouse strategies, plus the expanded scientific research effort which is essential in the 21st century to narrow down the remaining scientific uncertainties, can come from diverting the greater part of the $ 1,000 billion spent for armaments annually by the world community, as benefits the changing concepts of global security which the late 1980s have heralded.

9. GHG control strategic options needs to be implemented specially with respect to controlling CO_2 emissions. Emission reduction of CO_2 can be accomplished by a combination of several of the following approaches:

- End use efficiency improvements and conservation of energy.
- Energy supply side efficiency improvement;
- Capture and sequestration of CO_2 in subterranean reservoir or in the deep ocean;
- Utilisation of CO_2 for enhanced oil and natural gas recovery and for enhanced biomass production (photosynthesis);
- Shift to nonfossil energy sources.

The capture of CO_2 is only worthwhile in large power plants specially in coal fired power plants, where 1000 MG coal thermal power plant emits between 6 to 8 Mty^{-1} CO_2. The following CO_2 capture techniques can be adopted viz., Air separation CO_2 recycling, solvent absorption and membrane gas separation. After capture, the CO_2 needs to sequestered in a reservoir for an indefinite period, so it will not reemerge into the atmosphere. The following reservoirs are so far identified viz., deep aquifers, deep ocean and depleted oil and gas reservoirs. A number of such locations are identified in USA, Europe, Asia, Australia and Oceania. There are other ways of CO_2 utilisation viz., production of urea from CO_2, production of methanol from CO_2 and most important one is the biomass yield from CO_2 through enhanced photosyntheses.

Further, it goes without saying that the CO_2 caused climate change as induced by global warming can be ameliorated by shifting to nonfossil energy source. A combination of approaches needs to be implemented.

It is well-known that estimated higher share of GHGs producing countries are mostly developed world with about only 15% of the global population (Table 16.11, page 366). So the developed world should come forward to combat the problems in more meaningful manner. The UN-environment and development conference held in Brazil in 1992 was disappointing for those who hoped for early action to reduce our use of fossil fuels. As such, the GHGs gas emission increase over the years in unabated manner. The Kyoto Conference (1997) on climate change indicated the fact that countries which did not sign the earlier treaties emit enormous amount of GHGs, particularly CO_2. Thus there is a great need for GHGs emission sources by changing the energy generation pattern, as well as motivating people round the world for prevention of accelerated man-made climate changes (Table 16.12, page 366). Subsequently, sink area must be enhanced by increasing forested area, particularly in industrialised nations.

BOX 16.3: KYOTO PROTOCOL

Under the Climate Change Convention agreed at the Rio Earth Summit in 1992, industrialised countries agreed to stabilise their emissions of CO_2 at 1990 levels by the year 2000. But by 1996, most of these countries had accepted the scientific case for significant reductions in their emissions and promised to set reduction targets by the end of 1997. It then became very obvious that most countries would not be able to meet the 2000 target, and many countries abandoned the 1992 agreements. President Clinton postponed the stabilisation of US emissions to 2012 and ruled out real cuts before 2017. The European Union called for a 15 per cent cut by all industrialised countries by 2010.

In December 1997, scientists and diplomats from 160 countries gathered in Kyoto, Japan, for the UN Climate Convention. The aim was to find acceptable ways forward. Many coutries, including the United States, demanded 'flexibility measures', which would allow them to bank, borrow or trade spare emissions 'permissions'. It was also argued that flexibility would allow countries to emit more greenhouse gases if they planted more trees (which act as CO_2 sinks) or could demonstrate that they had reduced pollution in other countries (a procedure known as 'joint implementation'). The European Union negotiators were against flexibility measures and wanted to implement its targets collectively. The Kyoto Conference eventually settled for stability targets rather than cuts, and it adopted flexibility measures.

Under the Kyoto Protocol, rich industrial countries agreed to reduce their emissions of greenhouse gases by an average of 5.2 per cent by 2010. Developing countries were not set formal emission limits, partly because to do so would unfairly inhibit their pursuit of economic growth. But the political debate between countries continues, and the United States has insisted that some developing countries accept emission targets before it is willing to reduce its own greenhouse gas emissions. Most other industrialised countries (including the European Union and Japan) have promised to bring the Kyoto Protocol into law.

A new proposal, unveiled in June 1999, is designed to break this deadlock. The new plan requires developing countries to improve the carbon efficiency of their economies (by reducing the amount of carbon produced per dollar of their gross notional product) rather than requiring them to accept absolute limits on emissions. Some developing countries are already making great progress in this direction. China, for example, has increased the carbon efficiency of its economy by 47 per cent over the past 20 years, mainly as a result of measures designed to reduce urban air pollution. Supporters of the new plan point out that it reduces the scope for trading in rights to emit greenhouse gases.

Contd.

Trading of rights to emit greenhouse gases is a controversial theme, but one taken seriously under the Kyoto Protocol. Under current plans, developing countries would receive fixed targets above the their current emission levels to allow for economic growth. Rich countries could buy this excess capacity, and thus avoid the need to reduce their own output. Critics of the trading approach argue that it could well serve to increase overall emissions of greenhouse gases.

An alternative approach is the so-called clean development mechanism (CDM), an incentive shceme proposed at the 1998 Buenos Aires climate summit. The CDM approach seeks to give 'carbon credits' (credit to industrialised countries towards the emission targets agreed in the Kyoto Protocol) in developing countries by investing in renewable energy projects or by enlarging carbon sinks such as forests. But countries are divided over the question of which energy technologies should count (nuclear energy does not, for example) and the tendency to promote coal-fired power stations at the expense of renewable energy technologies.

A number of key issues surround the process and outcome of the Kyoto meeting:

- Global negotiations on climate control will never be easy, but they must be undertaken and within a unified frame work committed to seeking effective solutions.
- It must be understood that the main objective of the Kyoto Protocol—stabilising global climate at 'non-dangerous levels'—is a long-term objective that will inevitably take some time to achieve.
- Much more detailed assessment is required about what constitutes 'dangerous' climate change. The Kyoto Protocol proposals are designed without reference either to the level of climate change to which we can adapt or to the level of climate change that will cause significant damage.
- Given the history of global greenhouse gas emissions and the inertia of climate systems, we are already committed to further global warming.

Table 16.11: Estimated sharing of greenhouse gas emission

Region of the world	World population (%)	Responsibility for global warming (%)
The West	5	46
Central/Eastern Europe	7	19
Third World	78	35

Source : New International, April 1990. Adapted from Elliott (1994).

Table 16.12: Possible ways in which human activities may cause climatic change

* Gas emissions:	* Aerosol generation
— CO_2 — industrial and agricultural	* Thermal pollution
— Methane	* Albedo change:
— CFCs	— dust addition to ice caps
— NO_x	— deforestation
— Krypton-85	— overgrafting
— Water vapour	* Extension of irrigation
— Trace gases	* Alteration of ocean currents by constricting straits
	* Diversion of fresh water into oceans

16.7 CLIMATE CHANGE AND CLEAN DEVELOPMENT MECHANISMS

Climate change emerged on the political agenda in the mid 1980s with the increasing scientific evidence of human interference in the global climate system and with growing public concern about the environment. The UN general assembly formulate an international treaty on global climate protection which resulted in completion of the United Nations Frameworks Convention on Climate Change (UNFCCC) in May 1992. The legally binding Kyoto Protocol enforced since 1997 (Box 16.3). The Clean Development Mechanism

(CDM) is an arrangement under the Kyoto Protocol allowing industrialised countries with a greenhouse gas reduction commitment to invest in projects that reduce emissions in developing countries as an alternative to more expensive emission reductions in their own countries.

Clean Development Mechanism (CDM) was instituted in 2001 under the Kyoto Protocol (Article 12). It is designed as an element of the sustainable development strategy allowing industrialised countries invest in clean projects in developing countries also to gain emission credits. These credits are given in the form of **Certified Emission Reductions** (CERs), which, like all the other Kyoto accounting units, are expressed in tons of CO_2 equivalent. CERs are "certificates" just like a stock, given by the Executive Board in 15 different sectorial projects to be implemented in developing countries. This is often called **carbon trading**. As on 2008 India contribute maximum CDM project (i.e. over 400 projects) in energy, transport, waste management and afforestation sectors.

Thus approaches to dealing with climate change involve technological change coupled with political will and economic realities. A major step towards slowing global warming would be to increase the efficiency of energy utilization. More efficient use of fuels conserves the shrinking supplies of energy resources. It makes sense to increase energy efficiency, thus reducing CO_2 production, even if global warming is not a concern.

Another approach to the problem is to increase the amount of CO_2 removed from the atmosphere. If enough biomes is present the excess CO_2 can be used by vegetation during photosynthesis, thereby reducing the impact of CO_2 released by fossil fuel burning. Many countries have already undertaken the job of massive tree plantation of help for removal of CO_2 from the atmosphere under CDM projects.

BOX 16.4: UNITED NATIONS FRAMEWORK CONVETION ON CLIMATE CHANGE

Annex I countries

There are 41 Annex I countries and the European Union is also a member. These countries are classified as industrialized countries and countries in transition:

Australia, Austria, Belarus, Belgium, Bulgaria, Canada, Croatia, Czech Republic, Denmark, Estonia, Finland, France, Germany, Greece, Hungary, Iceland, Ireland, Italy, Japan, Latvia, Liechtenstein, Lithuania, Luxembourg, Malta, Monaco, Netherlands, New Zealand, Norway, Poland, Portugal, Romania, Russian Federation, Slovakia, Slovenia, Spain, Sweden, Switzerland, Turkey, Ukraine, United Kingdom, United States of America.

BOX 16.5: COPEHNHAGEN CLIMATE CHANGE SUMMIT 2009

This summit discusses the problems relating climate change and its consequences in 21st century and review the progress of Kyoto Protocol and CDM project. A total of 192 countries participated but comprehensive final accord could not be made. The voluntary commitment of some countries with respect to CO_2 emission reduction is:

- USA – 17% CO_2 emission reduction from 2005 level
- China – 40 to 45% reduction after 2005 level
- India – 20 to 25% reduction after 2005 level
- European Union – 20% reduction from 1990 level
- Japan – 25% reduction from 1990 level

However the negotiation will continue further.

BOX 16.6: DURBAN CLIMATE CHANGE SUMMIT 2011

Over 24,000 delegates including 10,590 government officials more than 13,000 representatives of UN bodies and agencies participated in this summit. Key issues that are considered is listed below:

- Kyoto and emission reduction targets are extended up to 2018;
- Creation of Green climate fund for taming the damaging effects of climate change in developing countries;
- Complete ban on unilateral trade measures on grounds related to climate change.

BOX 16.7: PARIS AGREEMENT ESSENTIAL ELEMENTS

The Paris agreement builds upon the convention and – for the first time – brings all nations into a common cause to undertake ambitious efforts to combat climate change and adapt to its effects with enhanced support to assist developing countries to do so. As such, it charts a new course in the global climate effort.

The Paris agreement's central aim is to strengthen the global response to the threat of climate change by keeping a global temperature rise this century well below 2 degrees Celsius above pre-industrial levels and to pursue efforts to limit the temperature increase even further to 1.5 degrees Celsius. Additionally, the agreement aims to strengthen the ability of countries to deal with the impacts of climate change. To reach these ambitious goals, appropriate financial flows, a new technology framework and an enhanced capacity building framework will be put in place, thus supporting action by developing countries and the most vulnerable countries, in line with their own national objectives. The agreement also provides for enhanced transparency of action and support through a more robust transparency framework. Further information on key aspects of the agreement can be found here.

Nationally determined contributions

The Paris agreement requires all Parties to put forward their best efforts through "nationally determined contributions" (NDCs) and to strengthen these efforts in the years ahead. This include requirements that all Parties report regularly on their emissions and on their implementation efforts.

Further information on NDCs can be found here.

In 2018, Parties will take stock of the collective efforts in relation to progress towards the goal set in the Paris agreement and to inform the preparation of NDCs.

There will also be a global stocktake every 5 years to assess the collective progress towards achieving the purpose of the agreement and to inform further individual actions by parties.

16.8 OCEAN-ATMOSPHERE INTERACTIONS

The ocean form the surface boundary conditions for 70 per cent of the atmosphere. This leads to a number of ocean and atmosphere interactions. One of the most well-known is the EL-Nino Southern Oscillation (ENSO). It is an inter annual change resulting from internal interaction within the Earth's ocean-atmosphere system. It is not the result of external forcing. The ENSO is an extremely important phenomenon which may have a global significance for the climate system. For many centuries Peruvian fishermen had noted that the ocean waters in which they fished occasionally experienced unusually high temperatures. The fish and birds disappeared and the usually dry desert land became fertile as rains fell. These years were called the 'years of abundance'. It did not occur regularly and consecutive events could be anything from 2 to 11 years apart. The warmer temperatures started around Christmas time and the Peruvian called the phenomenon El Nino, a Spanish word meaning Little Body or Christchild. The details of EL Nino climate events are given in box 16.8.

BOX 16.8: El NINO CLIMATIC CYCLES*

The El Nino event is caused by an irregular alteration of climatic conditions in the tropical Pacific region that can affect climatic patterns across the globe. The 1982-1983 El Nino was the strongest in the 20th century, causing droughts, floods, erosion and storms and having a severe effect on both bird and marine life. In Ecuador and Peru 100 inches of rain fell in six months and coastal deserts were transformed into grasslands. The insect, bird and amphibian populations soared as did the incidence of malaria. In other areas, coral reefs were exposed and destroyed, sea temperatures rose and typhoons changed paths. Monsoon rains fell in the central Pacific causing winter storms and flooding in California, while Indonesia and Australia sulfered droughts and fires. In total, the world economy lost an estimated $8 billion. Understanding these patterns and monitoring the necessary elements is, therefore, essential.

Contd.

The El Nino event generally occurs at irregular intervals, from two years to a decade. Under normal climatic conditions, easterly winds blow along the equator and south-easterly winds blow along the coasts of Ecuador and Peru raising the thermocline in the eastern Pacific and causing an upwelling of nutrient-rich, cool waters off the South American coast. This supports high levels of phytoplankton production, which, in turn, supports large fisheries. During El Nino years the easterly winds weaken and retreat eastward. The Pacific Ocean thermocline becomes flattened so that the cooler water lies deeper in the eastern Pacific. The coastal upwelling is unable to tap the nutrient-rich cooler waters, decreasing nutrient availability to the fisheries. The sea level flattens across the Pacific, dropping in west and rising in east and warm equatorial surface water flows eastward, further raising sea levels. The heavy rain, which in normal years falls over the western Pacific, moves east—causing droughts and fires in the west and heavy rain in the east.

In recent years, El Nino events have become both more frequent and stronger. For the first time on record there have been four events in four years (1990-1993). It is now imperative to understand what exactly precipitates El Nino and whether this apparent change in the events is part of a natural pattern or whether it is related to a real climatic change, possibly related to global warming.

There is another episode called La Niña. The little girl in Spanish La Niña is also sometimes called El Viejo, anti-El Nino, or simply "a cold event". La Niña episodes represent periods of below-average sea surface temperatures across the east-central Equatorial Pacific. Global climate La Niña impacts tend to be opposite those of El-Niño impacts. In the tropics, ocean temperature variations in La Niña also tend to be opposite those of El-Niño. During a La Niña year, winter temperatures are warmer than normal in the Southeast and cooler than normal in the Northwest.

(*Source :* Reports to the Nation on Our Changing Planet, 1994; Wuethrich, 1995)

17 GLOBAL OZONE PROBLEM

17.1 OZONE IN THE ATMOSPHERE

Ozone gas is perceptibly blue in colour and has a characteristic pungent smell. It occurs throughout the atmosphere, but only in small quantities. Indeed, if all the ozone contained in the first 60 km or so of the atmosphere could be brought down and assembled at the Earth's surface, it would form a layer only some 3 mm thick. Under average conditions, at ground level, each cm of air contains around 10^{19} molecules of all

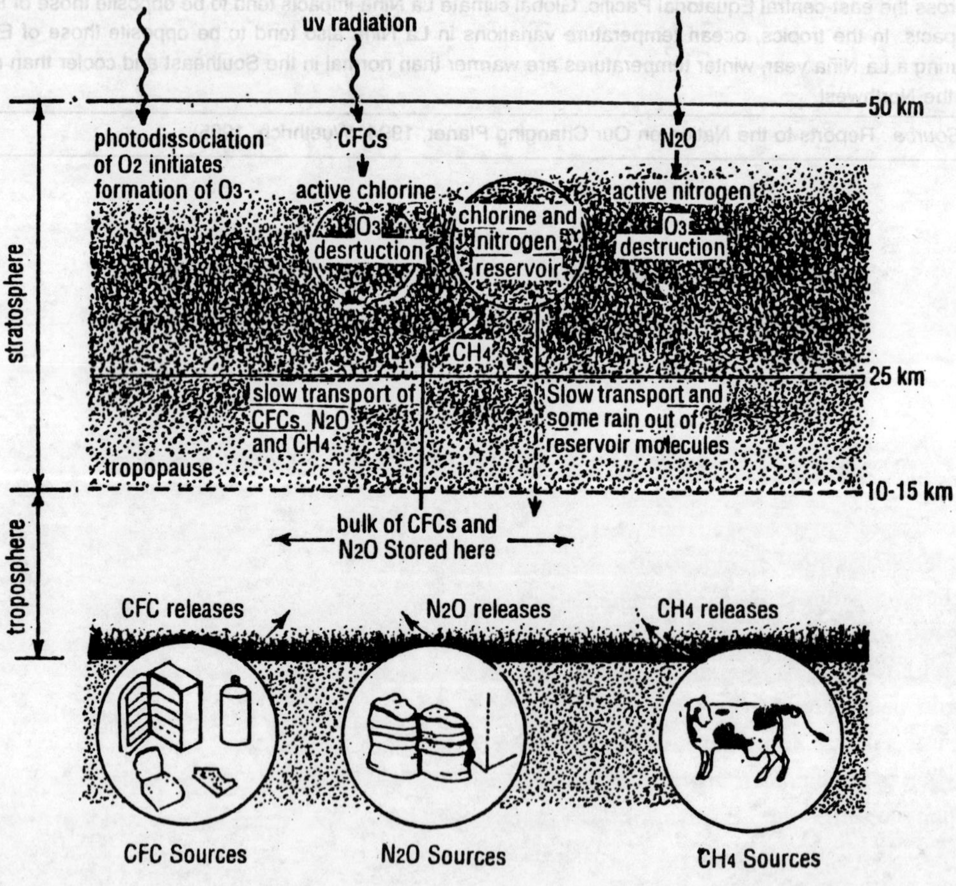

Fig. 17.1: Some of the key processes—both chemistry and transport, that determine the concentration of ozone in the atmosphere

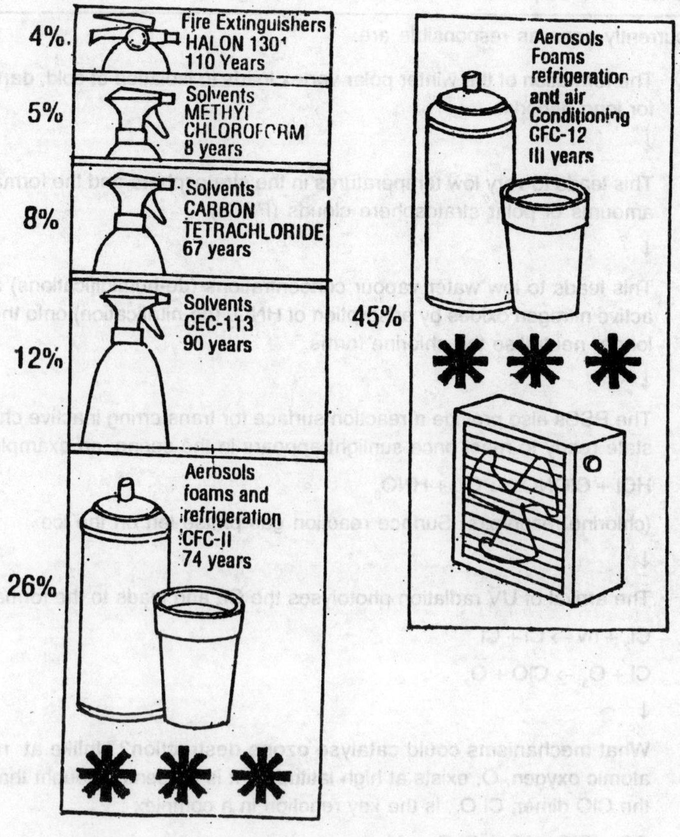

Fig. 17.2: Percentage contribution to ozone depletion of the most commonly used chemicals

the gases present of which ozone concentration is about 0.1 ppm. However, nearly 90% of the atmospheric ozone lies in the stratosphere, which is located in between 15-50 km above the earth surface.

Ozone is a form of oxygen, but whereas molecules of ordinary oxygen each contain two atoms, the ozone molecule has three. In industrial societies ozone is now being generated at ground level by the action of sunlight on gaseous pollutants, but the Earth's natural ozone factory is the stratosphere. Here the raw materials are ordinary oxygen seeping up from the troposphere, and sunlight, but the process involved depends on how both oxygen and ozone respond to that radiation. Ordinary oxygen absorbs UV with wavelengths below about 2.4×10^{-7}m; this provides the energy needed to split up or photodissociate the molecule into a pair of highly reactive oxygen atoms. Once released, an oxygen atom can combine with an intact oxygen molecule, forming ozone. This process not only produces ozone; it also thereby filters out most of the incoming solar UV rays with wavelengths less than 2.0×10^{-7}m and some of that in the 2.0×10^{-7} to 2.4×10^{-7}m region as well.

But up in the stratosphere, the odds are stacked against ozone. The crucial point is that ozone is constantly being created and destroyed in the stratosphere. There are a number of pollutant trace gases like NO, NO_2, Cl, ClO etc. known, which could easily react with O_3 and thus produce O_2 (Table 17.1). This is commonly known as "Ozone depletion"

Table 17.1: Sequence of processes forming the ozone hole

The series of processes currently seen as responsible are:	
Vortex formation	The formation of the winter polar vortex leads to isolation of cold, dark air over the pole for long periods. ↓
Very low temperature, PSCs	This leads to very low temperatures in the stratosphere and the formation of significant amounts of polar stratosphere clouds (PSCs). ↓
De-humidification de-nitrification	This leads to low water vapour concentrations (de-humidifications) and to removal of active nitrogen oxides by adsorption of HNO_3 (de-nitrification) onto the PSCs. These no longer neutralise the chlorine forms. ↓
Pre-conditioning	The PSCs also provide a reaction surface for transferring inactive chlorine to an active state ready to react once sunlight appears in the spring; an example reaction: $HCl + ClONO_2 \rightarrow Cl_2 + HNO_3$ (chlorine reservoirs) Surface reaction gas phase left on the ice ↓
Spring arrives : sunlight	The arrival of UV radiation photolyses the Cl_2 and leads to the formation of ClO : $Cl_2 + hv \rightarrow Cl + Cl$ $Cl + O_3 \rightarrow ClO + O_2$ ↓
Ozone destruction by catalysis	What mechanisms could catalyse ozone destruction? Unlike at mid-latitudes, little atomic oxygen, O, exists at high latitudes. It is presently thought that the photolysis of the ClO dimer, Cl_2O_2, is the key reaction in a complex : $CO + ClO + M \rightarrow Cl_2O_2 + M$ $Cl_2O_2 + hv \rightarrow Cl + ClOO$ $ClOO + M \rightarrow Cl + O_2 + M$ $2 \times (Cl + O_3 \rightarrow ClO + O_2)$ Net : $2O_3 + hv \rightarrow 3O_2$

Therefore, the overall effect of all these processes is to 'precondition' the Antarctic stratosphere so that when sunlight returns to the stratosphere in the spring, the net reaction :

$$2O_3 + hv \rightarrow 3O_2$$

occurs, i.e. photolysis of O_3. Once the spring is well-on, the high sun induces a dynamical breakdown of the stable vortex and the 'chemical containment vessel'—as it is sometimes called—is broken down, mixing of air with lower latitudes occurs, PSCs evaporate and the atmosphere returns to a more normal state.

Some of the key processes—both chemistry and transport—that determine the concentration of ozone in the stratosphere are given in Fig. 17.1. It is interesting to note the fact that some commonly used chemicals contribute substantially in the process of O_3 depletion also (Fig. 17.2).

In fact, the main ozone factory is at tropical latitudes (at altitudes of some 30-40 km), but it is transported from here downward and toward the poles in a circulation that is biased in favour of the summer hemisphere at the equinoxes, that is, towards the north pole in March/April and the south pole in September/October.

The longest data record of ozone measurements comes from a network of ground-based stations, the so-called Dobson Network, of which Halley Bay is a part. Although this network was extended in 1957, the distribution of stations remained quite uneven, with 42 out of 67 in northern temperate latitude. Since 1979 the problem of spatial coverage had been much alleviated by instruments borne on satellites (like the Total Ozone Mapping Spectrometer — TOMS — on borad NASA's Nimbus 7 satellite), but here the data record was too short to permit the identification of any statistically significant trend.

17.2 OZONE DEPLETION PROCESS

Ozone depletion in the stratosphere leads to an increase of UV-B on ground with its harmful effects on health, ecosystems, aquatic system, materials etc. It was estimated that about 3-5.5% O_3 depleted in northern hemisphere during 1969-1988.

In general, there are three principal ways of O_3 destruction:

(a) Hydrogen system (OH system)

(b) Nitrogen system (N_2O system)

(c) Chlorine system ($CFCl_3$ or CF_2Cl_2 system).

The OH system destroys only 10% of O_3. This reaction is seen above 40 km over the earth's crust. The details are:

$$H_2O + O(ID) \rightarrow 2OH$$

$$OH + O_3 \rightarrow H_2O + O_2$$

$$H_2O + O \rightarrow OH + O_2$$

$$\text{Net: } O + O_3 \rightarrow 2O_2$$

OH can also be formed from oxidation of CH_4 ($CH_4 + O(ID) - CH_3 + OH$). OH then reacts as above.

Sixty per cent of ozone destruction occurs through N_2O system. N_2O produced by bacterial action of microorganisms in ocean and soil (denitrification) diffuses upwards from troposphere to stratosphere where it reacts with 'O' in presence of light to produce NO, which then destroys O_3.

The detailed reaction may be expressed as :

$$N_2O + O(ID) \rightarrow 2NO$$

$$N_2O + h\nu \rightarrow NO + O$$

$$NO + O_3 \rightarrow NO_2 + O_2$$

$$NO_2 + O \rightarrow NO + O_2$$

$$O_3 + O \rightarrow 2O_2$$

Neutral chlorine contributes only very little to O_3 destruction. But $CFCl_3$, CF_2Cl_2 are the principal O_3 destroyers. They are inert in the troposphere but get dissociated into stratosphere:

$$CFCl_3, CF_2Cl_2 \rightarrow \frac{h\nu}{180-220\,\text{nm}} Cl_2$$

$$Cl + O_3 \rightarrow ClO + O_2$$

$$ClO + O \rightarrow Cl + O_2$$

$$\text{Net: } O + O_3 \rightarrow 2O_2$$

17.3 OZONE HOLE

Computer-based image processing from the TOMS data in 1987 indicates the fact that the O_3 depleted zone extends over the entire Antarctic Continent and beyond. Other measurements, from satellites and balloons have revealed that the depletion is mainly concentrated between 12 to 24 km in altitude—spanning much of the lower stratosphere at these latitudes. So the hole in the sky is actually more like a slice out of the heart of the ozone layer. The evidence came from a huge US-led experimental campaign during August and September—the period during which the ozone hole develops each year.

What causes the hole? This is a very controversial question. The overall consensus is that a sequence of stages is responsible for the peculiar efficiency with which chlorine destroys ozone over Antarctica. First there is the unique meteorology of the polar stratosphere. At the autumn equinox (the end of March), the sun sets for 6 months at the south pole, and an area of darkness spreads over the polar cap. It gets very cold. Up in the stratosphere, this rapid cooling sets up a pattern of very strong westernly wind that swirls around the pole—the so-called polar vortex, extending out of around 60°. This structure is extremely stable; it endures throughout the polar winter, weakens slightly with the return of sunlight in September, but does not finally break down until summer — usually sometime in November. An important effect of the vortex is to isolate the air within it (Fig. 17.3).

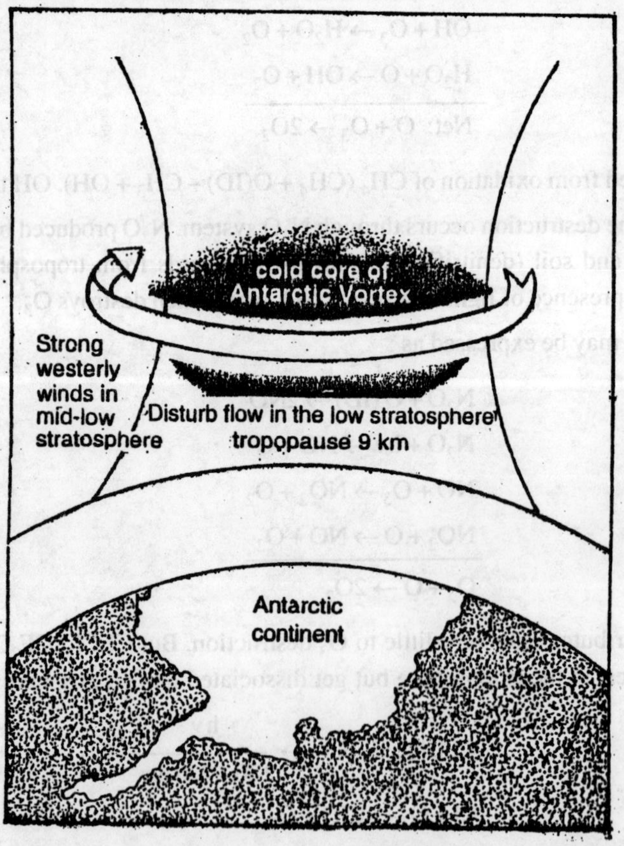

Fig. 17.3: A schematic diagram of our present understanding of Antarctic vortex in winter

The second key feature of the polar stratosphere is its odd chemistry. Once again, this stems from the isolation of the Antarctic vortex. Eventually, its frigid core becomes cold enough (temperature is lower than stratospheric clouds or PSCs).

They have an ethereal beauty but are believed to be deeply implicated in polar ozone loss. Full details of their role are only now beginning to emerge, but the central point is clear; they release chlorine from the holding tanks. This seems to involve special reactions that take place on the surfaces of the icy particles—surface compounds that are rapidly and efficiently broken up by sunlight. This so called pre-conditioning of the polar stratosphere apparently goes on as long as PSCs are around. It sets the scene for a burst of active chlorine radicals (Cl and ClO) as soon as sunlight returns. Once liberated, these appear to engage in special different ozone-destroying catalytic cycles, where the sharp onset of accelerated ozone loss in September is seen. This interplay between chemical and dynamic factors must also drive the year by year variability of the ozone hole. Thus, there is enough chlorine around already to wipe out virtually all the ozone in the Antarctic lower stratosphere, but what actually happens in a given year depends on the meteorological conditions that prevail; the stronger the vortex and the longer its frigid core is maintained after the return of sunlight, the deeper and more sustained will be the ozone loss, as in 1987 and again in 1989 and 1990. By contrast, 1988 saw poorly formed, much distorted vortex that began to dissipate early, and the ozone hole was correspondingly less severe. At present the detailed mechanisms whereby atmospheric dynamics set up conditions for the chemical depletion of ozone to occur are not well understood.

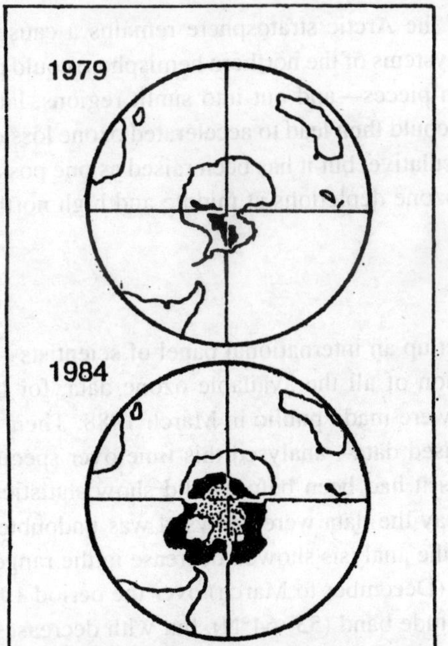

Fig. 17.4: Variation of Antarctic polar O_3 layer

In fact, the different distribution of land masses and oceans in the Arctic and Antarctica results in the north polar winter stratosphere being generally warmer than its southern counterpart. This has two important consequences. First PCSs do form over the Arctic, but they are usually less abundant. Second, the circumpolar vortex is weaker and may be disturbed, or even disrupted, during the winter by warm air pushing in from lower latitudes. As a result, the vortex and the PSCs within it rarely persist until sunlight returns.

Till late 1980s, no ozone hole as such had been detected over the north pole in spring. However, there is evidence that the chemical composition of the Arctic stratosphere in winter can be as highly perturbed in association with PSCs—as in Antarctica. So the region is primed for ozone depletion, even if no clear signal of major loss over the pole has so far been identified. The disturbing possibility is that this could be triggered by an unusually prolonged period of very cold winter temperatures—records suggest that this does occur in some years (Fig. 17.4).

The steps of O_3 destruction in polar regions is shown below:

$$O_3 + Cl^- \rightarrow O_2 + ClO^-$$

Atomic chlorine reconstitution:

Midstratosphere	Ozone hole / low stratosphere
$ClO^- + O \rightarrow Cl^- + O_2$	$2ClO^- \rightarrow ClOOCl$
	$ClOOCl + uv \rightarrow ClOO^- + Cl^-$
	$ClOO^- \rightarrow Cl^- + O_2$

Inactivation of chlorine:

$$Cl^- + CH_4 \rightarrow HCl + CH_3^+$$
$$ClO^- + NO_2^+ \rightarrow ClONO_2$$

Activation of chlorine on particle surfaces:

$$HCl\,(g) \xrightarrow{H_2O} H^+(aq) + Cl\,(aq)$$
$$H_2O\,(aq) + ClONO_2\,(g) \rightarrow HOCH\,(aq) + HNO_3\,(g)$$
$$Cl\,(aq) + HOCl\,(aq) \rightarrow Cl + OH\,(aq)$$
$$Cl_2\,(g) + \text{Sunlight} \rightarrow 2Cl^-\,(g)$$
$$H^+(aq) + OH^-(aq) = H_2O\,(aq)$$

Even without an ozone hole, the preconditioning of the Arctic stratosphere remains a cause for concern. For example, it is possible that the strong weather systems of the northern hemisphere could push the weak northern vortex off the pole—either bodily or in pieces—and out into sunlit regions, late in winter, but before the PSCs within it have evaporated. This could then lead to accelerated ozone loss over regions removed from the pole itself. The idea remains speculative, but it has been raised as one possible explanation for a disturbing report of unexpectedly large ozone depletions at middle and high northern latitudes during winter.

17.4 WORLDWIDE OZONE TRENDS

In 1986 NASA, in collaboration with WMO and UNEP, set up an international panel of scientists—the 'Ozone Trends Panel'—charged with a careful re-evaluation of all the available ozone data, for both ground-based and satellite instruments. Their conclusions were made public in March 1988. Their key finding was that a thorough reassessment of the ground-based data—analysed this time over specified latitude bands and by season, not globally and annually as it had been before—did show statistically significant long-term downward trends. This shift in the way the data were analysed was undoubtedly influenced by the results coming from Antarctica. In brief, the analysis showed decrease in the range of 2.3 to 6.2% between 30°N and 60°N for the winter months (December to March) over the period 1969-1986—the largest losses (some 7%) being in the higher latitude band (53- 64°N), but with decreases of around 5% as for south as 40°N. No significant trends suggested that the loss may be accelerating. Moreover the observed losses—although broadly consistent with the pattern of seasonal and latitudinal changes revealed by the best 2-D model calculations—are a factor of two or three times larger than the predicted decreases (Fig. 17.5).

Experimentally, the amount of uv-B reaching ground level increases by a factor of three to six in the Antarctica during the early part of the spring because of the appearance of the ozone hole. Even in the nonpolar region ozone depletion was detected seasonally. Such changes are highly correlation with increased level of uv at lower troposphere.

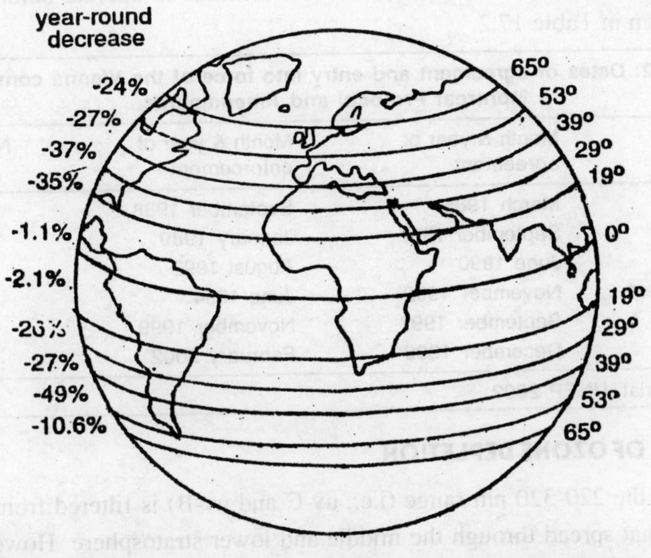

Fig. 17.5: Year-round decline in atmospheric O_3 for the period November 1978 through November 1987

The recent increase in levels of stratospheric chlorine and bromine is due primarily to the release into the atmosphere of organic compound containing chlorine and bromine which are anthropogenic. These compounds are causing depletion of O_3 in stratosphere. CFC derivaties, CH_4, NO_x pollutants are also consideration in as ozone depleting substance (ODS).

Surface ozone has been continuously recorded at the Mt. Cimone, a high mountain station (2,165 m above mean sea level) in the Mediterranean Basin and South to the Alpine region. Only in daytime there is O_3 level (0.28 ± 0.26 ppb/yr) from 2000 to 2009.

17.5 THE MONTREAL PROTOCOL

The diplomacy for ozone layer protection initiated since 1970, when first report of ozone depletion by NO_x came up. Since then global efforts are taken to protect ozone layer. However the accepted agreement came in Montreal (1987).

The protocol was agreed on 16 September 1987 and came into force on 1 January 1989, having been ratified by the required 11 countries representing two-thirds of global consumption. As originally agreed, it provided for a staged control of five CFCs and three halons. The overall effect would be to freeze consumption of the halons at 1986 levels in 1992, but to cut consumption of the CFCs to 50% of 1986 levels by the year 2000. The major highlights of the protocol:

- CFCs to be phased out by the year 2000, with cuts of 50% by 1995 and 85% by 1997;
- Halons to be phased out by 2000, with a 50% cut by 1995;
- Carbon tetrachloride to be phased out by 2000, with a cut of 85% by 1995;
- Methyl chloroform to be phased out by 2005, with a cut of 30% of 1990 levels by 2000.

Many factors contribute to the "incremental costs" associated with phasing out old CFC technologies. Such costs will vary widely from country to country — depending on differences in national development, on the scale of CFC and halogen producing and using industries and so on. Whatever the differences between individual nations, however, a broad distinction can be drawn between the developed world and the developing countries.

A number of agreement on ozone layer protection continue to operate before and after Montreal Protocol, these are given in Table 17.2.

Table 17.2: Dates of agreement and entry into force of the Vienna convention, Montreal Protocol and Amendments

Agreement	Month & year of agreement	Month & year of enforcement	Number of countries ratified up to 2001
Vienna Convention	March 1985	September 1988	184
Montreal Protocol	September 1987	January 1989	183
London Amendment	June 1990	August 1992	163
Copenhagen Amendment	November 1992	June 1994	141
Montreal Amendment	September 1997	November 1999	80
Beijing Amendment	December 1999	February 2002	33

Source: Ozone secretariat, UNEP 2002

17.6 CONSEQUENCES OF OZONE DEPLETION

The ultraviolet light in the 220-320 nm range (i.e., uv-C and uv-B) is filtered from sunlight mainly by ozone molecules, O_3, that spread through the middle and lower stratosphere. However, uv-A (320-400 nm) penetrates through O_3 layer mostly, when then screened through troposphere. Thus O_3 depletion in stratosphere leads to the loss of filtering ability of uv light, which are mostly harmful to biological systems in a variety of ways.

The most obvious cause for concern about ozone loss stems from its role as filter of the Sun's ultraviolet radiation. The band labeled uv-C ($2.0\text{-}2.9 \times 10^{-7}$ nm) is virtually eliminated by the atmosphere. This is just as well, because uv-C is lethal to microorganisms (whence its use in germicidal lamps) and can destroy both nucleic acids and proteins: in the range from $2.4 - 2.9 \times 10^{-7}$ nm. Protection from uv-C is due entirely to absorption by ozone.

However, more important as far as ozone loss is concerned is the band between 2.9×10^{-7} nm and 3.2×10^{-7} nm known as 'biologically active uv' or uv-B. Here the attenuation of the solar input is evident, due to ozone, but the effect is less complete; a fraction of uv-B penetrates all the way to the ground. Because of this, the proportions of uv-B reaching the surface should be highly sensitive to changes in the ozone column—an increase of around 2% for each 1% loss of ozone being generally accepted.

uv-B is known to have a magnitude of effects on humans, animals, plants and materials. Most of these effects are damaging—but few are sufficiently well understood at present for the impact of enhanced uv B to be quantified :

(a) **Human Health**: The link between uv-B and the incidence of skin cancer is particularly emotive: here, there are two main strands of evidence. First, skin cancer is predominantly a disease of white-skinned people and the dark pigment—melanin—is known to be an effective filter of uv-B. The second strand comes from epidemiology—a study of the factors that influence the occurrence of the disease in human populations.

Melanoma—the particular form of skin cancer as reported in many areas—is actually much rarer than other types, but it is the most serious with a substantial high mortality rate. Incidence of the more common non-melanoma skin cancers is losely correlated with long term uv-B exposure—occurring predominantly on light-exposed areas of the skin, in the elderly and in those who spend considerable time outdoors. These cancers are distressing, but can usually be

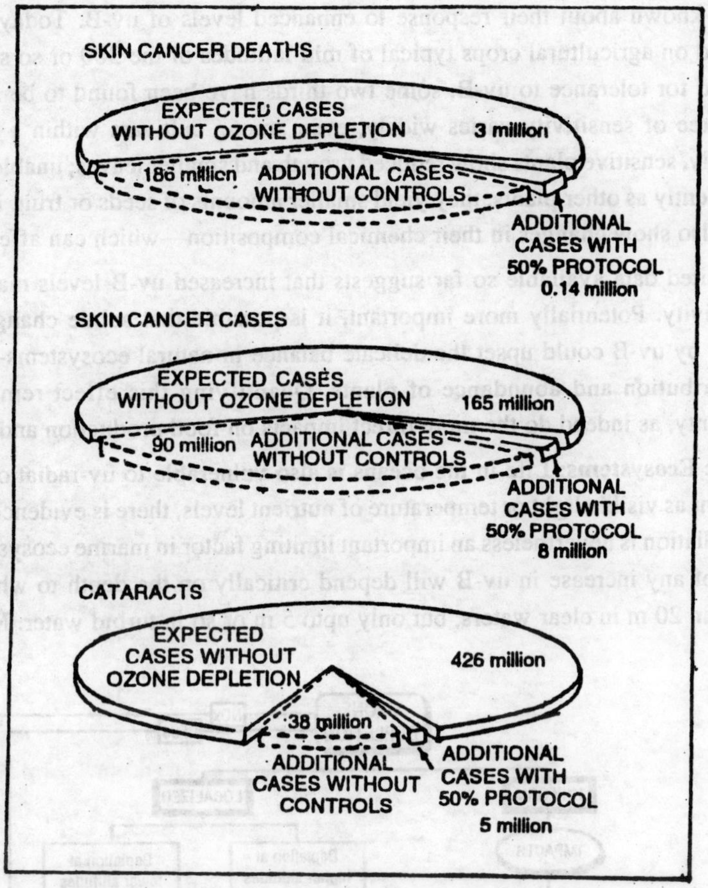

Fig. 17.6: Consequence of O_3 depletion on human health

treated successfully. By contrast, the epidemiology of melanoma is more complex: it affects relatively young people and, unlike other skin cancers, it has been increasing over the last few decades in all white-skinned populations studied. This and the observed correlation with indoor working and social class, have led to the suggestions that melanoma is associated with intermittent, but intense exposure to uv-B the sort of exposure associated with sunshine holidays and other recreational activities. Correlations between the attendant increase in uv-B and the estimated ozone loss may affect rates of skin cancer. Possibly the best current estimate comes from the EPA. This suggests that every 1% decrease of O_3 column will result in a 3% rise in the incidence of non-melanoma skin cancers, which translates into some 12-15,000 extra cases a year in the US, together with a possible 1% increase in mortality from melanoma (Fig. 17.6).

Exposure to enhanced levels of uv-B can also have other directly harmful effects on the human body, the two most serious being a tendency to suppress the body's immune responses and to cause damage to the eyes, especially in the development of cataracts. Although even more difficult to quantify it was noticed that these effects would touch all populations, with some consequences—possible increases in the incidence on severity of infectious diseases, for example—being particularly severe for people in tropical and subtropical areas.

(b) **Terrestrial Plants** : Plants are mostly adapted to present levels of visible radiation, but rather little is known about their response to enhanced levels of uv-B. Today, most studies have focussed on agricultural crops typical of mid-latitudes of the 300 or so species and cultivars screened for tolerance to uv-B, some two-thirds have been found to be sensitive—although the degree of sensitivity varies widely, even among cultivars within a given crop species. Typically, sensitive plants show reduced growth and smaller leaves; unable to photosynthesise as efficiently as other plants, they yield smaller amounts of seeds or fruit. In some cases, these plants also show changes in their chemical composition—which can affect food quality.

The limited data available so far suggests that increased uv-B levels may also affect forest productivity. Potentially more important, it is possible that subtle changes in plant growth induced by uv-B could upset the delicate balance in natural ecosystems—thereby changing the distribution and abundance of plants. Quantifying this effect remains a key area of uncertainty, as indeed do the more direct impacts on food production and forestry.

(c) **Aquatic Ecosystems:** Life in the oceans is also vulnerable to uv-radiation. Although not as important as visible light or temperature of nutrient levels, there is evidence that ambient solar uv-B radiation is nevertheless an important limiting factor in marine ecosystems. The potential impact of any increase in uv-B will depend critically on the depth to which it penetrates—more than 20 m in clear waters, but only upto 5 m or so in turbid water. Moreover, certainly

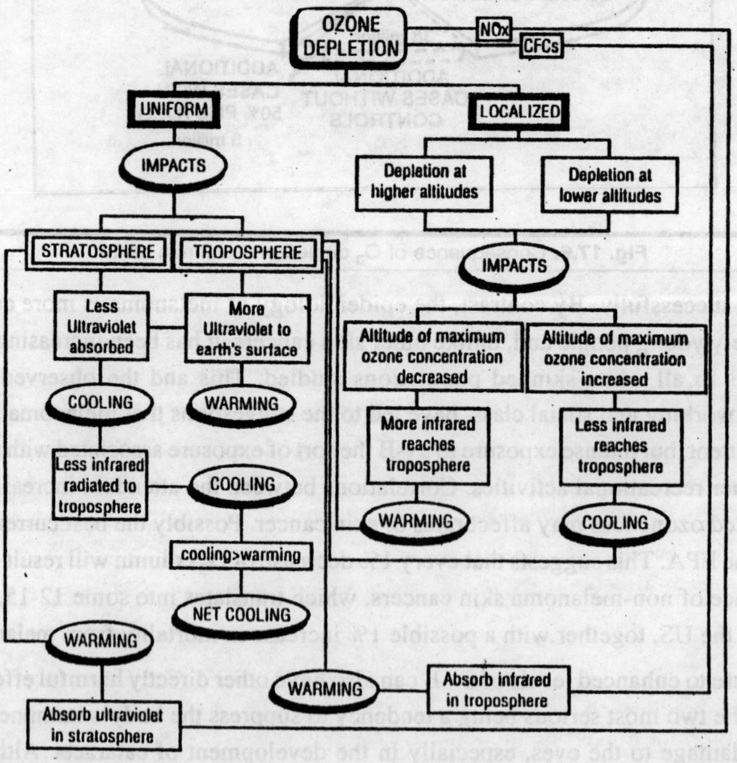

Fig. 17.7: Radiation and temperature changes accompanying the depletion of stratosphere ozone

enhanced uv-B has been shown to damage a range of small aquatic organisms—zooplanktons, larval crabs and shrimp and juvenile fish—as well as slowing photosynthesis in phytoplankton.

(d) **Climate**: A further area of concern derives from ozone's other major role in the atmosphere. As ozone cycles through its round of creation and destruction, there is an overall absorption of solar radiation, which is ultimately dumped as heat into the stratosphere. This warms the stratosphere and produces the temperature inversion at the tropopause: indeed, there would be no stratosphere without the ozone layer. Thus any depletion of stratospheric ozone is predicted to cool this region and, hence, change the temperature structure of the atmosphere to some extent. A schematic representation of climate change associated with O_3 depletion is presented in Fig. 17.7.

17.7 INDIA'S ROLE IN IMPLEMENTING MONTREAL PROTOCOL

The Government of India acceded to the Montreal Protocol along with its London Amendments on 19.06.92 and came into force from 17.09.92. The Government has not acceded to the Copenhagen Amendments. The Country Programme (CP) for phasing out of ODS submitted by Government of India was approved by the Executive Committee in its 11th meeting in November 1993. Country Programme (CP) was prepared with active involvement of industry, industry asociation, Government departments, research laboratories and other institutes. Action Plan has been drawn in CP for phase-out of ODS in country. The activities listed in the action programme are investment projects, information dissemination, study of small sector, incentives, ecomark, use restrictions, legislation, etc. The phase out of ODS will be through various non-ODS tecnology projects submitted by industries consuming of production enterprises.

The Ministry of Environment and Forests, Government of India has set up an Ozone Cell which is a focal point for coordinating with the industry and provide necessary support to the Steering Committee and its three Standing Committees. The Ozone Cell has organized workshops for projects preparation. So far, a number of projects have been submitted by different enterprises to the Ozone Cell for grants from Multilateral Fund. As on March 1994, 2 projects have been approved by the Executive Committee of Multilateral Fund. The implementation of the project will be initially monitored by the financial intermediary (IDBI) till their completion. To monitor the phase out of ozone depleting substance resulting from these projects, a Monitoring Standing Committee headed by the Chairman Central Pollution Control Board has been set up. The Monitoring Committee also helps the ozone Cell in collection and analysis of data relating to annual production, import and export of ODS.

The project proposals received by the Ozone Cell of the Ministry of Environment & Forests, which is located at Indian Habitat Centre, zone IV, East Court, 2nd floor, Lodi Road, New Delhi 110 003. The proposals are reviewed by the Technology and finance Committee and on approval of the Steering Committee, these projects are submitted to the Implementing agency for forwarding it to Multilateral Fund Executive Committee.

As per Montreal Protocol commitment, India also has to report data on ODS consumption of production import export of ODS every year. To streamline the process of data collection, as recommended by the Monitoring Committee, a notification under the Environment (Protection) Act, 1986 will be issued by the Ministry of Environment & Forests, Government of India whereby the concerned organizations/ enterprises in the country will be directed to furnish the relevant data on a regular basis is prescribed formats.

Subsequently, The Ozone depleting substances (Regulation and Control) Rules, 2000 was framed and implemented.

The rule covers the following aspects:

- **Regulation of production and consumption of ozone depleting substances;**
- **Prohibition on export to or import from countries specified or not specified;**
- **Regulation on purchase / sale of ozone depleting substances;**
- **Regulation on reclamation and destruction of ozone depleting substances, and so on.**

18 ACID RAIN

18.1 INTRODUCTION

With the rise of global industrialisation, increasing acidity in natural water and soils has become a problem. This acidity is associated with the transport and subsequent deposition of SO_2, NO_X and other acid oxidation products. The phenomenon is called "acid rain". Thus much concern was raised in the international platform about the future strategies for combating industrial emission of pollutants, regionally and locally.

The pH relationship of a number of substances are given in the following scale:

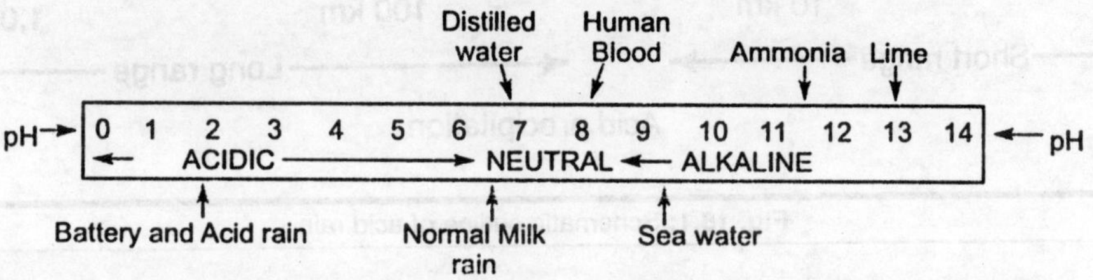

18.2 NATURE AND DEVELOPMENT OF ACID RAIN

Acid rain is now one of the most serious environmental problems in developed countries. Interest in the problem has increased significantly since the 1990s, especially since reports of extensive forest dieback in the then West Germany started to appear in 1975. Concern quickly spread to Scandinavia, where large scale fish deaths were believed to be caused by acid rain, and where more recently concern has focused on forest changes and heavy metal mobilisation in stream, lakes and groundwater.

Acid rain is widely believed to result from the washout from the atmosphere of oxides of sulphur and nitrogen. Although these oxides exist naturally in environmental cycles and ultimately have natural sources, the main sources today are coal-fired power stations and smelter and motor vehicle exhausts. These gaseous pollutants are light and imiscible and they can be carried hundreds—if not thousands—of kilometers by prevailing winds. The oxides may be mixed with other chemicals in the atmosphere to produce the poisonous and corrosive substances that either settle as dry fallout or are washed out by rain as acid deposition (Fig. 18.1). Emissions and fall out were previously extremely localised, but since the introduction of 'tall stacks' policies in both Britain (since 1958) and the USA (since 1971) paradoxically to disperse particulate pollutants and hence reduce local damage, emissions are now lifted into the upper air currents and carried long distances downwind. The tall stacks policies have thus turned a smoke problem into an acid rain problem. Neither winds nor acids respect political boundaries, so the pollutants are often carried across state and national frontiers. Acid rain thus represents a hidden export from one country to another — the exporter gains through not having to install costly pollution control activity to limit emissions and

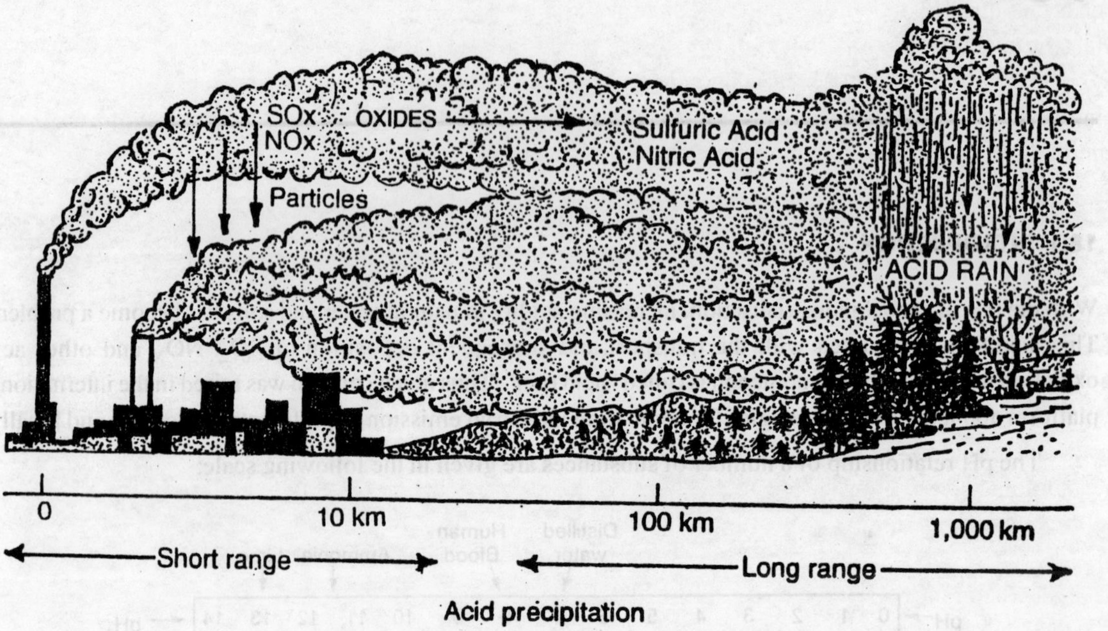

Fig. 18.1: Schematic outline of acid rain

the importer suffers from the adverse impacts on its environment. Thus acid rain is a critical problem for international relations and it rightly claims a high priority on the agenda of political debate—both within and between countries.

The acid rain debate now embraces many Western countries, including Canada, the United States, England, Scotland, Wales, Sweden, Norway, Denmark, Germany, the Netherlands, Austria, Switzerland and a growing number of eastern European countries including Russia, Poland and Czech Republic.

Those countries that are hardest hit by the problem (such as Canada and Scandinavian countries) are convinced that something must be done, without further delay, to reduce emissions and thus save their forests and lakes. With the belief that 'charity begins at home', over 21 governments resolved to cut their own emissions of SO_2 by at least 30% (over 1980 levels) by 1993 and hope to persuade the main producers of the culprit oxides to join them in a truly international attack on the problem.

The most suitable solutions to the problems of acid rain require prevention rather than cure and there is broad agreement in both the political and scientific communities on the need to reduce emissions of sulphur and nitrogen oxide to the atmosphere. The technology now exists to make this possible. However, clean-up programmes will be expensive and they might yield only qualified improvements. The problem of acid rain arises, strictly speaking, not so much from the rainfall itself as from its effects on the environment. Runoff affects surface water (such as rivers and lakes) and ground water, as well as soils and vegetation. Consequently, changes in rainfall acidity can trigger-off a range of impacts on the chemistry and ecology of lakes and rivers, soil chemistry and processes, the health and productivity of plants and building materials and metallic structures.

Human health might also be affected via the intake of food and water contaminated with toxic heavy metals mobilised by fallout acid rain. The significance of such impacts is more than purely scientific—they affect the quality of life for humans, they threaten environmental stability and the sustainability of food and timber reserves and they pose real economic problems. The most suitable solutions to the problems of acid rain require prevention rather than cure, and there is broad agreement in both the political and scientific communities on the need to reduce emissions of SO_2 and NO_x to the atmosphere.

18.3 ACID RAIN AND GEOLOGY

The impact of acid rain on the environment depends not only on the level of acidity in the rain, but also on the nature of the environment itself. Areas underlain by granitic or quartatic bedrock, for example, are particularly susceptible to damage, since soils and water are already acidic and lack the ability to buffer or neutralise additional acidity from the precipitation. Acid levels, therefore, rise, the environmental balance is disturbed and serious ecological damage is the inevitable result. In contrast, areas which are geologically basic—underlain by limestone or chalk for example—are much less sensitive and may even benefit from the additional acidity. The highly alkaline soils and water of these areas ensure that the acid added to the environment by the rain is very effectively neutralised. In areas covered by glacial drift or some other unconsolidated deposit the susceptibility of the environment to damage by acid rain will be determined by the nature of the superficial material rather than by the composition of the bedrock. In theory, it is important to establish background levels of acidity or alkalinity, so that the vulnerability of the environment to acidification can be estimated.

18.4 ACID RAIN AND AQUATIC ENVIRONMENT

The earliest concerns over the impact of acid rain on the environment were expressed by Robert Smith in England, as long ago as 1852, but modern interest in the problem dates only from the 1960s. Initial attention concentrated on the impact of acid rain on the aquatic environment, which can be particularly sensitive to even moderate increase in acidity and it was in the lakes and streams on both sides of the Atlantic that the effects were first apparent.

In general, there is a tendency for all lakes to become more acidic with time, as a result of natural ageing processes, but studies of acid sensitive lakes suggest that observed rates of change in pH value since the middle of the nineteenth century have exceeded the expected natural rates. The analyses of acid sensitive diatom species in lake sediments has allowed pH-age profile to be constructed for some 800 lakes in North America and Europe. Most of these profiles indicate that lake acidification is a relatively recent phenomenon, with little change in pH values indicated prior to 1850. In almost a century and a half since then pH values in the lakes studied have decreased by between 0.5 to 1.5 units. In contrast, the sensitive lakes—those with adequate buffering for example—showed little change. The consequence of lake acidification lead to a number of changes in lake biotic communities. The relative availability of inorganic carbon (as CO_3 or HCO_3) and phosphate in relation to water pH, altered appreciably in these lakes. The details are given in Figs. 18.2 and 18.3.

Most of the investigations into the impact of acid rain on aquatic communities have involved fish populations. Many fish species disappear with lake acidification. The sensitivity of water pH with respect to different species of fishes and their aquatic organisms are given in Fig. 18.4. In addition, fish in acid lakes also succumb to toxic concentrations of metals, such as aluminium, mercury, manganese, zinc and lead, leached from the surrounding rocks by the acids.

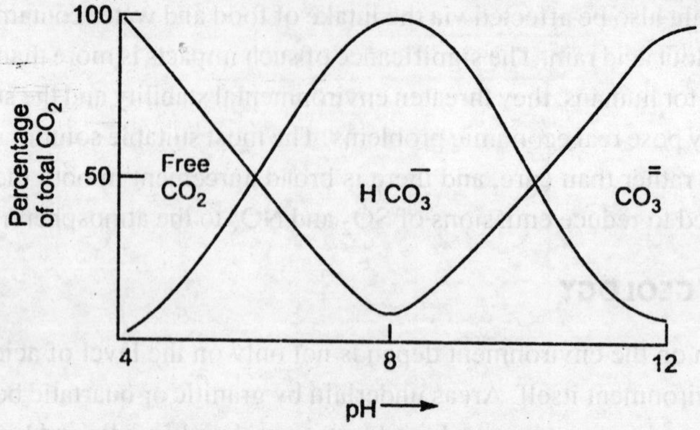

Fig. 18.2: The relative proportions of different forms of inorganic carbon in relation to the pH of water under normal conditions

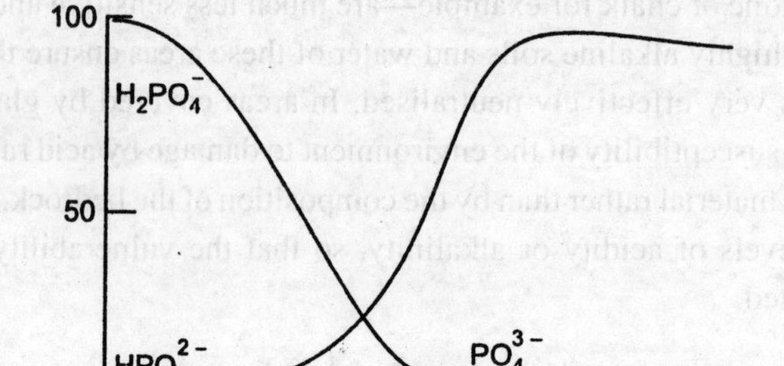

Fig. 18.3: The equilibrium of different forms of phosphate in relation to the pH of pure freshwater

18.5 ACID RAIN AND TERRESTRIAL ENVIRONMENT

Terrestrial ecosystem takes much longer period to show the effects of acid rain than aquatic ecosystem. As a result, the nature and magnitudes of the impact of acid precipitation on the terrestrial environment has been recognised only in recent decades. Acid precipitation damages the forests and other vegetation cover in a number of ways. The details are given in Fig. 18.5. The changes include crown dieback, necrosis of foliages, defoliation, soil acidification and various other related processes. As a result, vegetation cover or forest declines on long term basis. Mobilisation of toxic metals, such as aluminium, cadmium, zinc, mercury, lead and copper and iron is another feature which accompanies soil acidification. Thus detrimental effect of these toxic metals were seen with times in these affected areas.

18.6 ACID RAIN AND THE BUILD ENVIRONMENT

Acid rain also contributes to deterioration of the build environment—particularly in limestone and marble rock buildings. Several monuments thus get affected with time by continued acid precipitation in these areas. Crystals of calcium and magnesium sulphate begin to form on or beneath the surface of the stone, as corrosion effect induced by acid rain. By attacking the fabric of buildings, acid rain causes physical and economic damage, but it does more than that, it also threatens the world's cultural heritage (e.g. the Parthenon, the Sistine Chapel, the Taj Mahal).

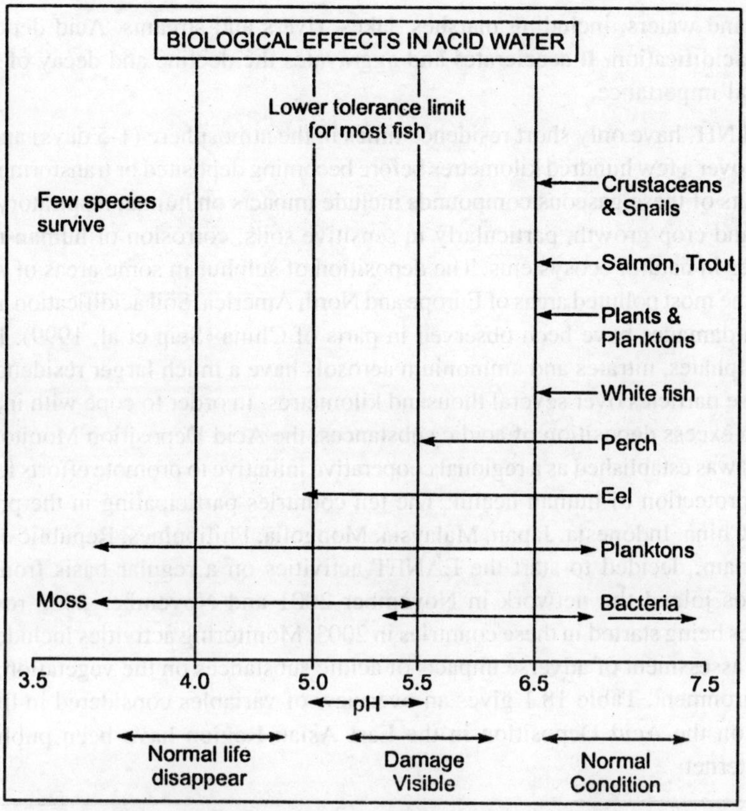

Fig. 18.4: Changes in tolerance of freshwater as water pH falls

18.7 ACID RAIN AND HUMAN HEALTH

The famous London Smog of 1952 developed as a result of meteorological conditions which allowed the build up of pollutants within the urban atmosphere. Smoke—produced by domestic fires, power stations and coal burning industries—was the most obvious pollutant, but the most dangerous was sulphuric acid, flowing free in aerosol form or attached to the smoke particles. Drawn deep into the lungs, the sulfuric acid caused or aggravated breathing problems and many of the 4,000 deaths attributed to the smog were brought about by the effect of sulfuric acid on human respiratory system. Moreover, elevated atmospheric acidity continues to cause chronic respiratory problems in these areas.

Heavy metals such as copper, cadmium, zinc, and mercury, etc. liberated from soil and bedrock by acid rain may eventually reach the human body via plants and animals in the food chain or through drinking water supplies. These changes pose threat to human health of the concerned region.

18.8 REGIONAL ACIDIFICATION

The rapid growth of cities in Asia has resulted in greatly increased emissions of SO_2 and NO_x. Acid deposition can take the form of wet precipitation, such as rain, fog, mist or snow, or dry deposition involving the deposition of acidic gases and aerosols. Acid deposition, also called 'acid rain' is caused mainly by SO_2 and NO_x, NH_3 emissions from fossil fuel combustion, industrial processes and agricultural practices. Acid wet deposition refers to rainwater with a pH of 5.6 or less. In the 1990s, acid rain as severe as that in Europe and North America was widely observed in Japan, with pH values as low as pH 4.3. Acid deposition has far-reaching impacts. It affects fish and other aquatic biota due to the progressive acidification of inland waters, including marshes, lakes, rivers and streams. Acid deposition threatens forests due to soil acidification. It accelerates and aggravates the decline and decay of monuments and buildings of cultural importance.

SO_2, NO_x and NH_3 have only short residence times in the atmosphere (1-5 days) and can, therefore, be transported only over a few hundred kilometres before becoming deposited or transformed into secondary fine particles. Effects of these gaseous compounds include impacts on human respiratory health; adverse impacts on forest and crop growth, particularly in sensitive soils; corrosion of human-made structures; and adverse changes in natural ecosystems. The deposition of sulphur in some areas of Asia is not equal to or higher than in the most polluted areas of Europe and North America. Soil acidification and acidification-induced vegetation damages have been observed in parts of China (Seip et al, 1999). In addition, acid aerosols such as sulphates, nitrates and ammonium aerosols have a much larger residence time allowing for transport of these particles over several thousand kilometres. In order to cope with increasing risks of problems related to excess deposition of acidic substances, the Acid Deposition Monitoring Network in East Asia (EANET) was established as a regional cooperative initiative to promote efforts for environmental sustainability and protection of human health. The ten countries participating in the preparatory phase activities, namely, China, Indonesia, Japan, Malaysia, Mongolia, Philippines, Republic of Korea, Russia, Thailand and Vietnam, decided to start the EANET activities on a regular basis from January 2001. Combodia and Laos joined the network in November 2001 and November 2002 respectively, with monitoring activities being started in these countries in 2003. Monitoring activities include wet deposition, dry deposition and assessment of adverse impacts of acidic substances on the vegetation, the soil and the inland aquatic environment. Table 18.1 gives an overview of variables considered in the project. Since 2001 data reports on the Acid Deposition in the East Asian Region have been published and made available on the internet.

Table 18.1: Monitored variables of EANET

Wet deposition	pH, EC, concentration of ions (SO_4^{2-}, NO_3^-, Cl^-, NH_4^+, Na^+, K^+, Ca^{2+}, Mg^{2+}, etc.)
Dry deposition	Gas concentrations (SO_2, NO/NO_2, O_3, HNO_3, NH_3, etc.)
Soil	pH, CEC, concentrations of exchangeable ions, etc.
Vegetation	Degree of decline of trees, abnormalities of leaves and branches, etc.
Inland aquatic environment	pH, EC, alkalinity, etc.

Notes: EC : electric conductivity; CEC: cation exchange capacity.
Source: EANET (2005)

18.9 MITIGATION OF ACID RAIN PROBLEMS

Although the cause-and-effect relationship between emissions of SO_2 and NO_x and acid rain damage is not universally accepted, most of the solutions proposed for the problem involve the disruption of that relationship. On the whole, the reduction of acid-forming gases is all that is required to slow down and eventually stop the damage being caused by the acidification of the environment. Among the acid-forming SO_2 emission, control is the most important aspect of the problem-solution. A number of measures were suggested for minimizing SO_2 emission into the atmosphere—these include desulphurification of fuel, flue gas emission control technology adoption and SO_2 recovery as H_2SO_4 etc. Several European governments paid considerable attention on this aspect for minimising acid precipitation. On the whole, progress towards the large scale abatement of acid emissions has been slow and methods for controlling NO_x lag behind those for dealing with SO_x. Of course, emissions of SO_x are beginning to decline in many areas. But lakes and forests already damaged require considerable time to recover back to the past conditions. A number of ameliorative measures were also undertaken for speedy recovery of these affected areas too.

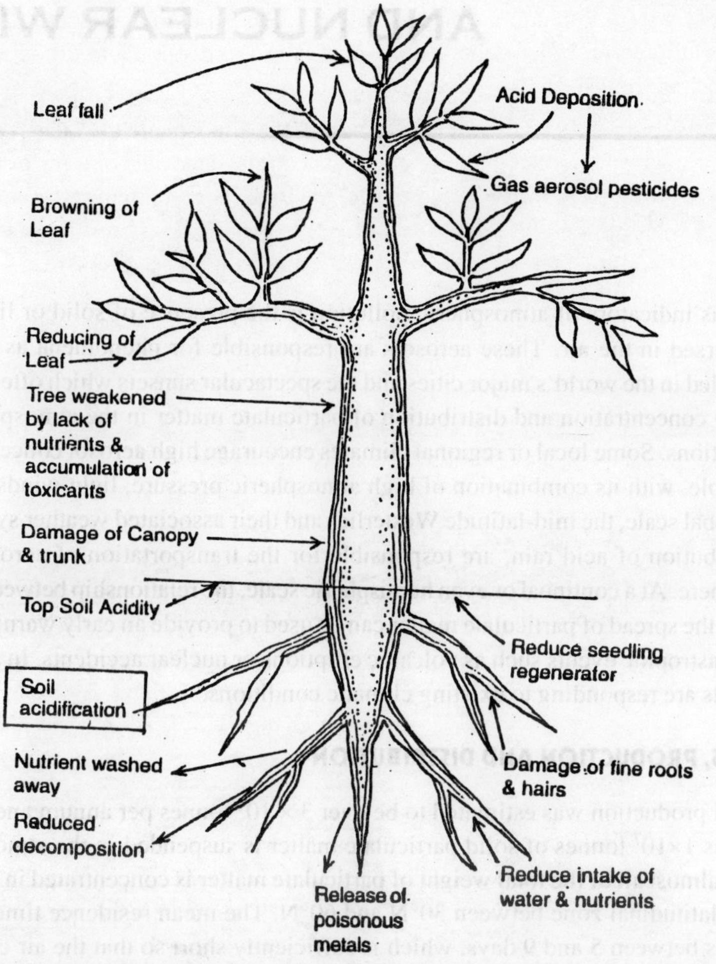

Fig. 18.5: Typical symptoms of damage in coniferous trees subject to acid rain

18.10 ACID RAIN IN INDIA

There have been many reports of acid rain in India in the past and that too have been only the episodic. Reported acid precipitations in India includes the acid rain in Chembur and Colaba industrial areas of Mumbai, in the vicinity of Singrauli Super thermal Power Plant (average pH value 5.3), at a rural site of Bhubaneswar (median pH value 5.0) and the Silent Valley (pH = 5.3). Latest reports on acid rain are at Kalyan (pH = 5.28), Chembur (pH = 4.8), Sinhagad (pH = 5.2), Delhi (pH < 5.6) and very recently at Panipat (pH < 5.6) of National Capital Region of Delhi. Although the pH value of rainwater at Pune has been reported to lie in the alkaline range, its value has shown decreasing trend from value of 7.5 in 1986 to 6.2 in 1998. The main reason is attributed to the decrease in the level of calcium ion and increase in sulfate and nitrate ions. Rain fall in Agra and Delhi regions have also shown decrease in pH value with the passage of time. Acid rain studies are being carried out by IITM since last three decades. The pH values are higher (pH > 7.0) in north & northwest parts of India. They are slightly lower ($6.0 \leq pH \leq 7.0$) in northeastern & southern parts of India. The higher pH is due to neutralization of acidic ions (SO_4 & NO_3) by soil originated cations (Ca, K & Mg) as well as by NH_4. However, the pH values are acidic ($pH \leq 5.56$) at some industrial, rural & remote locations which could be either due to anthropogenic emission (Kalyan & Singrauli), acidic soil (Goraur & Mohanbari) or due to thick cover of vegetation that prevents soil erosion (Trivandrum, Silent Valley & Tungnath). This work is being carried out at IITM, Pune by a team of scientist led by Dr. P.S.P. Rao.

19 ATMOSPHERIC TURBIDITY AND NUCLEAR WINTER

19.1 INTRODUCTION

One of the most obvious indicators of atmospheric pollution is the presence of solid or liquid particles, called *'aerosols'*, dispersed in the air. These aerosols are responsible for phenomena as diverse as the urban smogs that prevailed in the world's major cities and the spectacular sunsets which often follow major volcanic eruptions. The concentration and distribution of particulate matter in the atmosphere is closely linked to climatic conditions. Some local or regional climates encourage high aerosol concentrations, as in Los Angeles, for example, with its combination of high atmospheric pressure, light winds and abundant solar radiation. On a global scale, the mid-latitude Westerlies and their associated weather systems, already implicated in the distribution of acid rain, are responsible for the transportation of aerosols over long distances in the troposphere. At a continual or even hemispheric scale, the relationship between atmospheric circulation patterns and the spread of particulate matter can be used to provide an early warning of potential problems following catastrophic events such as volcanic eruptions or nuclear accidents. In such situations the atmospheric aerosols are responding to existing climatic conditions.

19.2 AEROSOL TYPES, PRODUCTION AND DISTRIBUTION

The total global aerosol production was estimated to be over 3×10^4 tonnes per annum and, on any given day, perhaps as many as 1×10^7 tonnes of solid particulate matter is suspended in the atmosphere. Under normal circumstances, almost all of the total weight of particulate matter is concentrated in the lower 2 km of the atmosphere in a latitudinal zone between 30°N and 60°N. The mean residence time for aerosol in the lower troposphere is between 5 and 9 days, which is sufficiently short so that the air can be cleansed in a few days once emissions have stopped. The equivalent time in the upper troposphere is about one month, and in the stratosphere the residence time increases to two to three years. As a result, anything added to the upper troposphere or stratosphere will remain in circulation for a longer time and its potential environmental impact will increase.

Aerosol can be classified in a number of ways, but most classifications include such elements as origin, size and development—sometimes individually, sometimes in combination. Most of the atmospheric aerosol content—perhaps as much as 90 per cent—is of natural origin, although anthropogenic sources may be dominant locally, as they are in urban areas. Dust particles created during volcanic activity or carried into the troposphere during dust storms are examples of natural aerosol.

Atmospheric aerosol comprise a very heterogeneous group of particles and the mix within the group changes with time and place. Following volcanic activity, for example, the proportion of dust particles in the atmosphere may be particularly high; in urban areas such as Los Angeles, photochemical action on vehicle emissions causes major increase in secondary particulate matter; over the oceans, 95 per cent of the aerosols may consist of coarse sea-salt particles. Such variability makes it difficult to establish the nature of the relationship between atmospheric aerosols and climate. It is clear, however, that the aerosols exert their influence on climate by disrupting the flow of radiation within the earth atmosphere system and there are certain elements which are central to the relationship. The attenuation of solar radiation caused by the presence of aerosols provides a measure of atmospheric turbidity, a property which, for most purposes, can be considered as an indication of the dustiness or dirtiness of the atmosphere.

Several things may happen when radiation strikes an aerosol in the atmosphere. If the particle is optically transparent the radiant energy passes through unaltered and no change takes place in the atmospheric energy balance. More commonly, the radiation is reflected, scattered or absorbed — in proportion which depend upon the size, colour and concentration of particles in the atmosphere and upon the nature of the radiation itself. Aerosol which scatter or reflect radiation increase the albedo of the atmosphere and reduce the amount of insolation arriving at the earth's surface. Absorbent aerosols will have the opposite effect. Each process, through its ability to change the path of the radiation through the atmosphere, has the potential to alter the earth's energy budget.

Among the atmospheric aerosols, desert dust and soot particles readily absorb the shorter solar wave lengths with soot—a particularly strong absorber across the entire solar spectrum. In addition to disrupting the flow of incoming solar radiation the presence of aerosols also has an effect on terrestrial radiation. Being at a lower energy level, the earth's surface radiates energy at the infrared end of the spectrum. Aerosols—such as soot, soil and dust particles—released into the boundary layer absorb infrared energy quite readily, particularly if they are larger than 1.0 μm in diameter and, as a result, these changes tend to raise the temperature of the troposphere.

19.3 ATMOSPHERIC TURBIDITY—NATURAL AND MAN-MADE SOURCES

The large volumes of particulate matter thrown into the atmosphere during periods of volcanic activity are gradually carried away from their sources to be redistributed by the wind and pressure patterns of the atmospheric airculation. Dust ejected during the explosive eruption of Krakatoa in 1883 encircled the earth in about two weeks following the original eruption and within 8 to 12 weeks had spread sufficiently to increase atmospheric turbidity between 35°N and 35°S. The diffusion of dust from the Mount Agurg eruption in 1963 followed a similar pattern and, in both cases, the debris eventually spread polewards until it formed a complete veil over the entire earth.

The volume of particulate matter produced by human activities cannot match the qualities emitted naturally. Estimates of the human contribution to total global particulate production vary from as low as 10% to more than 15% with values tending to vary according to the size-fraction included in the estimate. Human activities may provide as much as 22 per cent of the particulate matter finer than 5 μm. The development of the Arctic Haze in recent years is generally considered to be an indication of continuing arthropogenic aerosol loading of the atmosphere. The haze is not of local origin. It is created in mid-latitudes as atmospheric pollutions, which is subsequently carried polewards to settle over the Arctic.

Another anthropogenic activity is the war. For example during Gulf war in Feb '91, over 600 oil-wells were set alight by the retreating Iraqi army. These wells continued to burn for several months, kept alight by oil and gas brought to the surface under pressure from the underlying oil-fields. During the time they added massive amount of smoke, SO_x, CO_2, unburned HC and NO_x to the atmosphere. Most of these products were confined to the lower half of the troposphere, with the top of the plume never exceeding 5 km.

19.4 AVIATION AND CLIMATE CHANGE

The effect of aviation on the atmosphere is the well known contrail (condensation trail), which is line shaped ice crystal clouds formed behind aircraft from the initial emission of water vapour and particles from the engines. The potential effects on climate of contrails and their spreading into cirrus cloud coverage continues to exercise scientists in terms of characterizing their microphysical properties, optical effects, occurrence and ultimate effect on climate. In addition, emission if NO_x from aircraft were considered to be a potential problem in terms of stratospheric ozone (O_3) depletion. In the early 1990s, a number of publications dealt with potential impacts of aviation on climate, both NO_x effects on O_3, and contrails. It is further known that the recent supersonic aviation affects radiative forcing (RF) in a number of ways and induce climate change processes. It is also known that sulphate and black carbon particles (soot) are also involved in the formation of contrails along with water vapour. In fact CO_2 emission from aviation growth ranges from 2-2.5 per cent over past few year, but the major effects lies on change in overall RF induced by contrails and cirrus cloud formation. Although it is not possible to assess at this stage that real environmental consequences of future supersonic air transport present, knowledge indicates that there exists a real possibility of serious decrease in the atmospheric ozone shield due to the catalytic action of oxides of nitrogen, emitted in the exhaust of supersonic aircraft.

19.5 NUCLEAR WINTER

Despite recent agreements between the superpowers to limit the spread of nuclear weapons, the spectre of nuclear war continues to hover over the world, as it has done for the past 45 years or so. Terms such as thermonuclear device, first strike, fallout and ionizing radiation have become part of the modern lexicon. Added to these now is nuclear winter, perhaps the final blow for any survivors of a nuclear exchange (Fig. 19.1, page 391).

Environmental problems such as acid rain, ozone depletion and atmospheric turbidity would be intensified significantly by the multiple nuclear explosions which are a prerequisite for nuclear winter. Nuclear winter differs from these other issues, however, it is a potential prob'em rather than an existing one and, as a result, provides no elements capable of direct measurement. It has been necessary, therefore, to develop a theoretical approach using statistical and computerised modeling techniques.

Nuclear winter also differs from other issues in that its effects are catastrophic rather than gradual. The effects of acid rain or ozone depletion may become apparent after only years or even decades, with nuclear winter the effects are felt within days or, at most, weeks of the initial explosions. Thus there is little time to respond once the nuclear devices have been launched. If nuclear proliferation is to be avoided, it is important that nuclear conflict be avoided also. As a result of political developments, especially the break-up of USSR in 1991, however, the conflict seems less likely and the issue of nuclear winter is regarded as irrelevant by some. Nevertheless, the nuclear powers retain sufficient warheads to create nuclear winter many times over. A broad understanding of the potentially dire consequences of nuclear winter may help to ensure that they are never used.

The nuclear winter hypothesis was based on the assumption that smoke and dust thrown into the atmosphere during a nuclear war would increase atmospheric turbidity to such an extent that a high proportion of in-coming solar radiation would be prevented from reaching the lower atmosphere and the earth's surface. The net effect would be to drive land temperatures down.

The initial effect of a nuclear ground-burst would be the injection of large amounts of dust into the atmosphere, through the destruction of soil aggregates, the vaporisation of soil and rock and the incorporation of existing surface dust. Simulations using the TTAPS' (Turco, Toon, Ackerman, Pollack and Sagan — these investigators studied the process initally) baseline scenario resulted in an injection of more than 960 million tons of dust into the atmosphere of which 80% would reach the stratosphere. Such fine dust particles convected into the upper atmosphere would remain in place for more than a year, contributing to global cooling by scattering in-coming solar radiation.

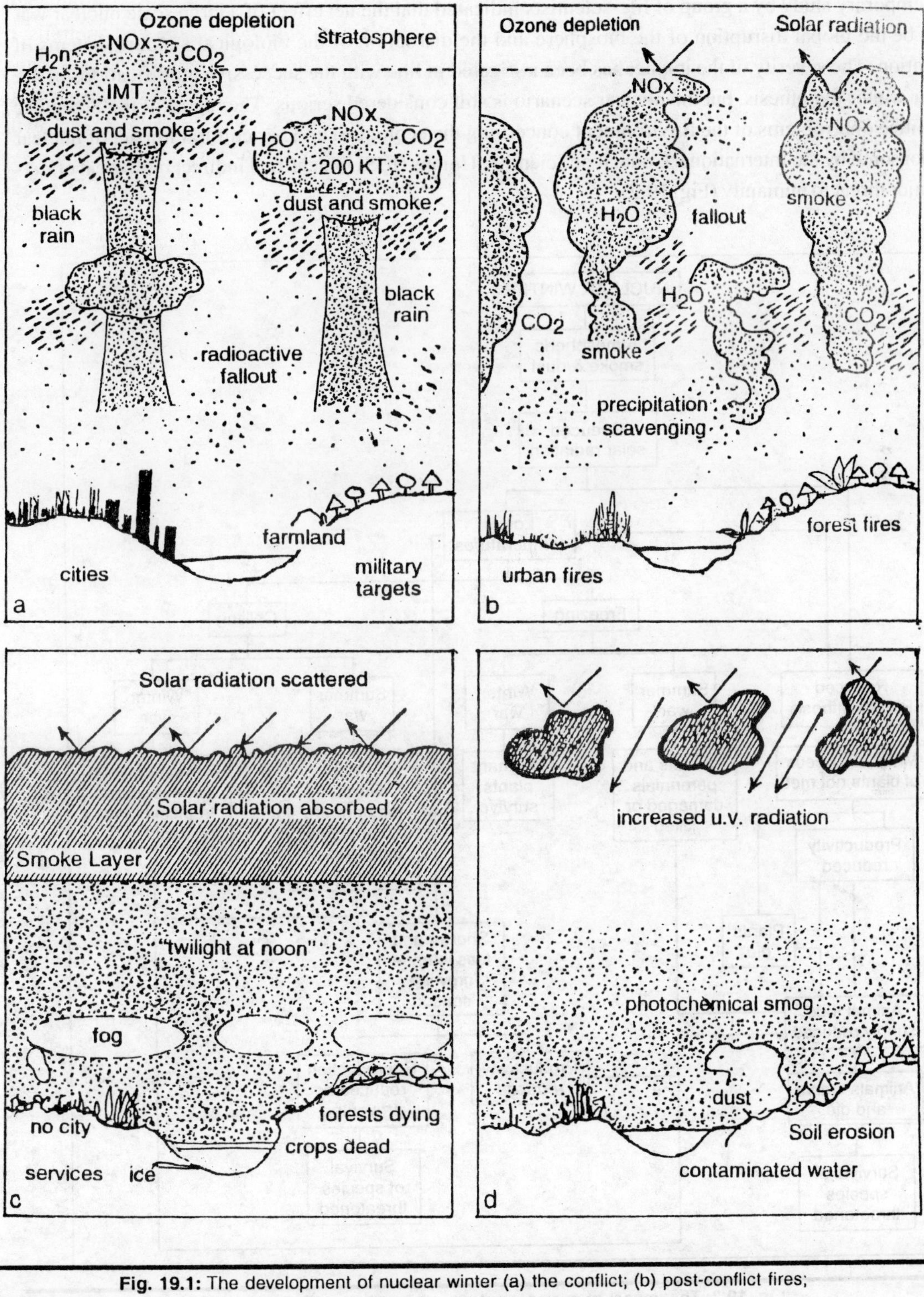

Fig. 19.1: The development of nuclear winter (a) the conflict; (b) post-conflict fires; (c) nuclear winter; (d) the after-effect

The environmental consequences of nuclear war received little attention in the TTAPS scenario, but a contemporary study by a group of life-scientists indicated that the net effect of a large scale nuclear war would be the global disruption of the biosphere and the disruption of the biological support systems of civilisation. The severity of the impact has been mitigated in line with the successive modifications of the nuclear winter hypothesis, but the nuclear scenario is still considered serious. The report of the Scientific Committee on Problems of the Environment concerning the ecological and agricultural impact of nuclear war sponsored by the International Council of Scientific Unions (ICSU) identified indirect biological effects as a major threat to humanity (Fig. 19.2).

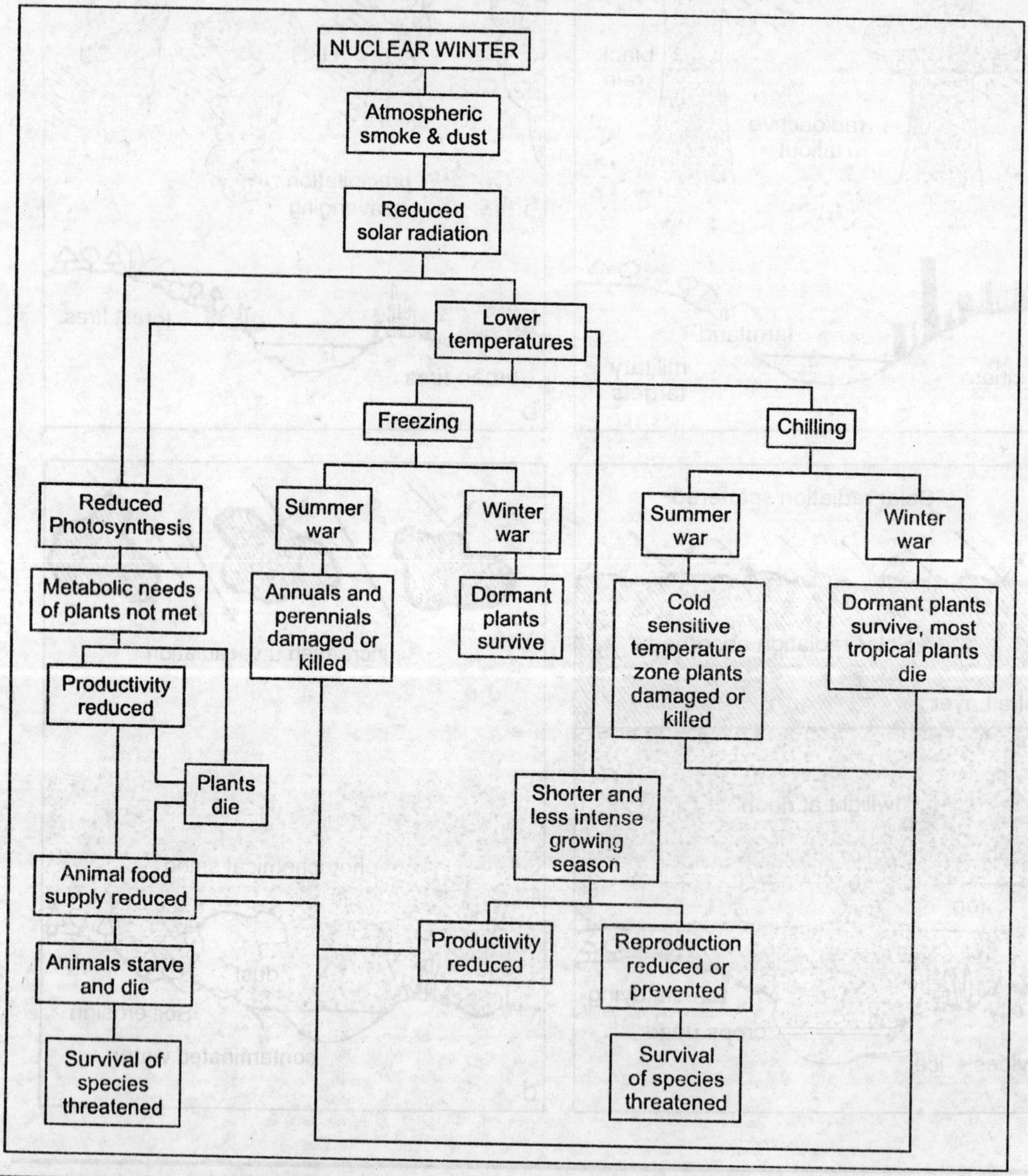

Fig. 19.2: The impact of nuclear winter on the natural environment

As a result of nuclear winter, it is likely that vegetation in tropical regions would suffer significant damage, whatever the season. Tropical plant systems flourish under conditions which include mild stable temperatures. They are susceptible to moderate declines in temperature and do not develop the resistance to cold which helps temperate plants to survive. Enrhich et al (1983) have suggested that forests in tropical regions might largely disappear if low light levels and low temperatures were to become widespread.

Marine ecosystems would be more capable of dealing with temperature extremes, because of the moderating influence of the oceans. Tropic webs would be seriously disrupted by the reduced productivity of phytoplankton during the low light period, however, and the increased storminess in coastal areas would cause damage to shallow-water ecosystems. Overall, it is considered the aquatic ecosystem have the potential for a more rapid recovery than terrestrial systems.

The damage to vegetation in natural ecosystems would be paralleled by damage to cultivated plant species. Tropical food crops viz., rice, banana, maize can be damaged by temperatures falling to 7-10°C for as little as a few days and even that moderate chilling would be sufficient to cause crop failure. Many tropical Third World countries already experience food shortages, which would be aggravated by the effects of chilling.

In addition to the low light levels, subfreezing temperatures and violent storms, the human survivors of the initial conflict would face continued radioactive fallout, high levels of toxic air pollution and enhanced levels of ultraviolet radiation. Combined with lack of food and drinking water, psychological stress and the disruption of support systems such as transportation, communications and medical care, this would probably ensure that fatality rates remained high even years after the conflict.

So it is very much debatable whether the earth is getting warmer day by day or cooler due to increasing atmospheric turbidity and nuclear winter processes. There is no realistic mode available so far to predict these future events.

BOX 19.1: ATMOSPHERIC BROWN CLOUD

The atmospheric brown cloud (ABC) is a regional phenomenon of aerosol and PM plumes which occurs across parts of Asia. In 2001 the Indian Ocean Experiment (INDOEX) documented the vast extent of a 3 kilometre thick brownish layer of pollutants hovering over most of tropical Indian Ocean, South, Southeast and East Asia. The haze consists of sulphates, nitrates, organics, black carbon and fly ash among several other pollutants, which can be transported far beyond their source region, particularly during the dry season. The potential direct effects include a significant reduction in the solar radiation reaching the surface; a 50-100 per cent increase in solar heating of the lower atmosphere; rainfall suppression; agricultural productivity decline; and adverse human health effects (UNEP, 2002).

The indirect effects include cooling of the land surface, increase in frequency and strength of thermal inversions which trap more pollution; evaporation reduction; soil drying; and disruption of monsoon rainfall. The regional and global impacts of the ABC are set to intensify over the next 30 years as the population of the Asian region and sources increase.

BOX 19.2: SOUTHEAST ASIA HAZE

Forestland conversion involving uncontrolled use of fire in land preparation has been identified as a major source of fires in the Southeast Asia region. Transboundary air pollution caused by these fires, used to clear land in Indonesia, remains a perennial problem. In the years when the El Nino phenomenon returns to the Southeast Asian region with its drier conditions, the worsening haze from forest and brush fires becomes an

Contd.

increasing challenge for adjacent countries. The impacts of the haze include major disruption of travel and transport, increasing risk of traffic accidents, public health risks through the transboundary transport of fine and ultrafine PM and environmental damage. PM pollution levels during the episode in 1997/1998 were extremely high in Indonesia (more than 1 mg/m^3 in Sumatera) and high (several hundreds μg/m^3) in Malaysia (Frankenberg et al, 2005; Sastry, 2002).

As a consequence of the 1997/1998 fire event, the Association of Southeast Asian Nations (ASEAN) and the Asian Development Bank (ADB) developed a Regional Haze Action Plan (RHAP). The primary objectives of the RHAP are to: (i) prevent forest fires through better management policies and enforcement; (ii) establish operations mechanisms to monitor land and forest fires; and (iii) strengthen regional land and forest firefighting capacity with other mitigation measures (ADB/ASEAN, 2001). At the same time WHO, in collaboration with UNEP and the World Meteorological Organization (WMO), developed the Health Guidelines for Vegetation Fire Events (WHO/UNEP/WMO, 1999). These guidelines emphasized the global occurrence of fires, analysed their causes, discussed the methods of observation and developed an early warning system for vegetation fire events with respect to the protection of public health.

In 2002, ASEAN adopted the ASEAN Agreement on Transboundary Haze Pollution to prevent and monitor transboundary haze pollution as a result of land and/or forest fires, and to control sources of fires (ASEAN, 2002).

20 GLOBAL CARBON DIOXIDE—RISE AND IMPACT ON BIOSPHERE

20.1 INTRODUCTION

During primeval times, fairly good amount of CO_2 was present in the earth's atmosphere, when water condensed to form oceans and CO_2 dissolved in the water forming carbonates. Subsequently, CO_2 level decreased with the evolution of photosynthetic organisms. In fact, the evolution of the earth's atmosphere has been intimately linked with the development of life on earth.

Since the Industrial Revolution (1760 onwards) there has been an increase in the concentration of CO_2 in the atmosphere. In 1991, the concentration of CO_2 in the atmosphere was estimated to be 355 ppm. The overall trend is ever-increasing with time, with periodic fluctuation due to seasonal utilisation of CO_2 and excess liberation from the biological world.

There are a couple of reasons for rise of global CO_2 level. The burning of fossil fuels and the changing of land use pattern are the two important events that are intimately associated with global CO_2 rise. However, natural disasters like volcanic eruption and decomposition of forest litters also contribute considerable amount of natural CO_2 emission processes.

Predicting how fast the concentration of CO_2 in the atmosphere will rise depends on both the scale of human activity and the response of the natural carbon cycle. Undoubtedly there will be an increase in CO_2 output as a consequence of growth in the world population and economic development. These factors will increase the demand for energy and cause further land use changes. The rate at which fossil fuels are burnt also depends on their availability and price, as well as the future development of alternative energy sources such as wind, solar and nuclear power. However, some of the greatest uncertainties lie on the reaction of the natural carbon fluxes—both oceanic and terrestrial: will photosynthesis increase as a consequence of the additional CO_2 in the atmosphere? How much will the oceans absorb and release? How might climate change feedback and increase or decrease these natural fluxes?

20.2 CONSEQUENCES OF GLOBAL CO_2 CHANGES

Atmospheric CO_2 constitute the major greenhouse gases which provide the insulation around earth surface as required for maintenance of life on earth. There is good evidence for a close relationship between the concentration of CO_2 in the atmosphere and the long-term climatic fluctuations. The impact of CO_2 as greenhouse gas is manifold. If the other climatic factors remain stable, the concentration of CO_2 rise will directly affect the function of plants. In addition, elevated CO_2 induced temperature rise of atmosphere will also affect plant growth. Quantitative and qualitative changes in primary production will influence the animal and microbial communities that depend on this productivity.

In plants, there are three categories—C_3, C_4 and CAM type—depending on the CO_2 fixation pathways. At elevated CO_2 rise, C_3 plants will be benefited more. The rise of photosynthesis in C_3 plants is directly proportional to atmospheric CO_2 rise and rate of photosynthesis will definitely enhance. The water use efficiency in C_3 plants will also be enhanced significantly under elevated CO_2 level.

BOX 20.1:

A carbon footprint is defined as:

The total amount of greenhouse gases produced to directly and indirectly support human activities, usually expressed in equivalent tons of carbon dioxide (CO_2).

In other words: When you drive a car, the engine burns fuel which creates a certain amount of CO_2, depending on its fuel consumption and the driving distance, (CO_2 is the chemical for carbon dioxide). When you heat your house with oil, gas or coal, then you also generate CO_2. Even if you heat your house with electricity, the generation of the electrical power may also have emitted a certain amount of CO_2. When you buy food and goods, the production of the food and goods also emitted some quantities of CO_2.

Your carbon footprint is the sum of all emissions of CO_2 (carbon dioxide), which were induced by your activities in a given time frame. Usually a carbon footprint is calculated for the time period of a year. The best way is to calculate the carbon dioxide emissions based on the fuel consumption. In the next step you can add the CO_2 emission to your carbon footprint. Below is a table for the most common used fuels:

Examples:

For each (UK-) gallon of petrol fuel consumed, 10.4 kg carbon dioxide (CO_2) is emitted.

For each (US-) gallon of gasoline fuel consumed, 8.7 kg carbon dioxide (CO_2) is emitted.

If your car consumes 7.5 liter diesel per 100 km, then a drive of 300 km distance consumes 3 × 7.5 = 22.5 liter diesel, which adds 22.5 × 2.7 kg = 60.75 kg CO_2 to your personal carbon footprint.

Fuel type	Unit	CO_2 emitted per unit
Petrol	1 gallon (UK)	10.4 kg
Petrol	1 liter	2.3 kg
Gasoline	1 gallon (USA)	8.7 kg
Gasoline	1 liter	2.3 kg
Diesel	1 gallon (UK)	12.2 kg
Diesel	1 gallon (USA)	9.95 kg
Diesel	1 liter	2.7 kg
Oil (heating)	1 gallon (UK)	13.6 kg
Oil (heating)	1 gallon (USA)	11.26 kg
Oil (heating)	1 liter	3 kg

Each of the following activities add 1 kg of CO_2 to your personal carbon footprint:

Travel by public transportation (train or bus) a distance of 10 to 12 km (6.5 to 7 miles)

Drive with your car a distance of 6 km or 3.75 miles (assuming 7.3 litres petrol per 100 km or 39 mpg)

Fly with a plane a distance of 2.2 km or 1.375 miles.

Operate your computer for 32 hours (60 Watt consumption assumed)

Production of 5 plastic bags

Production of 2 plastic bottles

Production of 1/3 of an American cheeseburger (yes, the production of each cheeseburger emits 3.1 kg of CO_2)

The pattern of carbon allocation is often altered when plants grow in elevated CO_2, but there seems to be great variation in the response of different species. In some species there is an increase in the allocation of resources to root growth at 700 ppm CO_2 than that of plant growth. Under conditions of non-limiting nutrient availability elevated CO_2 does appear to increase plant growth in a wide range of species, often by a considerable amount.

At elevated CO_2 level, the nitrogen use efficiency is reduced significantly that also limit the growth at certain time. In addition, CO_2-induced atmospheric temperature rise have a series of consequent changes on plants viz., alteration of photosynthetic rates, temperature induced plant growth process and productivity.

In contrast to plants, CO_2 is a waste product of metabolism in animals. CO_2 must be removed from the body parts for growth and survivality of species. There are a lot of biochemical changes in animals found due to elevated CO_2 level in atmosphere. The temperature changes in animal life induces the variation of species distribution.

The pathways by which temperature and changes in atmospheric CO_2 concentration could affect plant communities: (a) Effect of temperature, and (b) effect of high CO_2 level.

It is very difficult to predict the future response of whole communities to rising CO_2 concentration, mostly because there is no recent precedent for the changes in CO_2 concentration that are taking place. In contrast, there is a lot of evidence of the effect of temperature on communities.

The impact of CO_2 rise on agriculture can be considered at three levels. On the level of individual crop response, changes in temperature, precipitation, and CO_2 concentration can have direct effects on crop performance by reasonably well-characterised mechanisms. On regional scale, predicting changes in the yield potential of crops need to take into consideration the blend of environmental changes that occur in each region. Lastly, the extent to which agricultural practices can adapt to CO_2 and the implications for food security.

It is much more difficult to predict changes in aquatic life than in terrestrial environments, especially for large bodies of water. The main problem is that we have a very poor understanding of the relationship between global climate and the oceans. However, climatic changes like CO_2 level rise have impact on aquatic environment which not only affects the individual species population but also whole assemblage of species. A number of specialised aquatic environments will be affected by these changes viz., changes in coral reef life, changes in fishing ground, and overall productivity. The overall sea level changes have also direct impact on biotic life in aquatic system of ocean.

The higher atmospheric carbon dioxide levels will cause more acidity of sea water, and thus reduces the concentration of calcium availability in sea water. Many shelled animals and some types of phytoplankton that can fix calcium in their shells or skeletons and thus reduces their productivity. It may also shift the trophic relationship of the ecosystem.

With the rising CO_2 level induced global temperature change; there will be shift in plant communities, most probably the grasses, will encroach polewards, deciduous hardwoods and grasslands will follow, temperate mixed forests are expected to lose their conifers and become dominated by broadleaf trees. The extension of the monsoon rains towards the poles will increase the moist tropical forest cover in some areas, such as northern Australia. A higher frequency of natural forest fires in a drier Amazon basin will not only release significant amounts of carbon into the air, it will also reduce the pumping of water by transpiration. There will be shift in cropping pattern and crop productivity in northern latitudes.

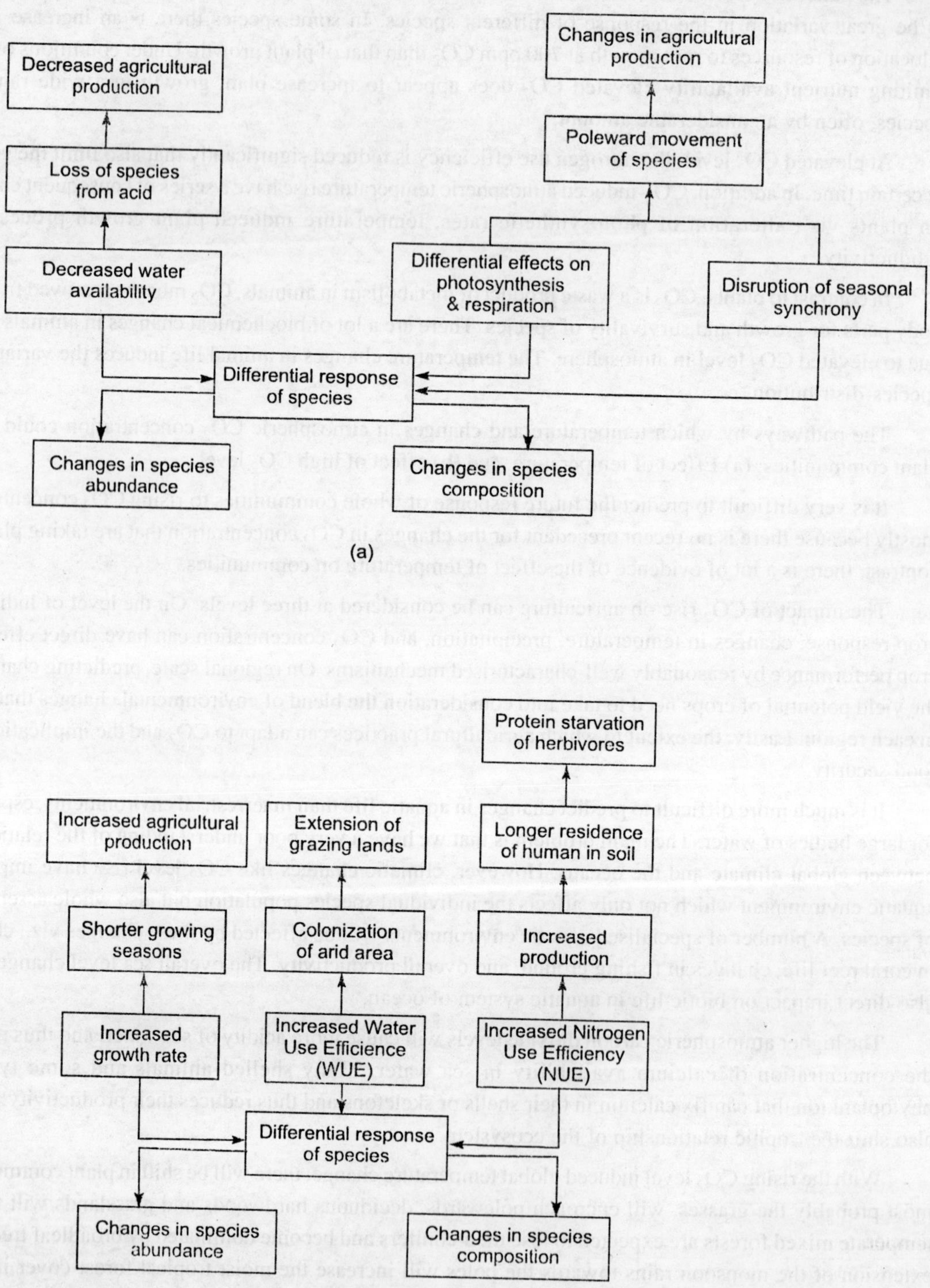

Fig. 20.1: (a) Effects of temperature on individual plants; (b) Effect of CO_2 rise on plants

20.3 STRATEGIES FOR CONSERVATION OF ENVIRONMENTAL CHANGES INDUCED BY CO_2 RISE

Over past couple of decades, it was felt that rate of extinction of species increased due to rapid environmental changes and associated human activities. On the whole, many different threats of extinction of many different species make the conservation of genetic species and ecosystem diversity an urgent priority. The threat of CO_2 is less immediate than others, but strategic decisions made now must take it into account. It is known that CO_2 can affect species in two ways: They may not be able to tolerate new conditions because of their physiology, and they may be affected by changing interactions with predators, parasites, competitors and mutualists. In fact, it is needed to understand how CO_2 and climate and these interactions among species determine range, size and abundance, the effects of small size on how populations persist and why populations become small in the first place.

The main mechanism for controlling the rise in atmospheric carbon dioxide is to limit our emissions. Next would be to limit out deforestation programmes. It is estimated twelve per cent CO_2 sink could be by 2100 through above stated processes. The proposed extension of forest cover up to 865 million hectare (globally) can only reduce 54 per cent higher level of CO_2 in a hundred year time. In addition, a variety of schemes are proposed to fix CO_2 by oceanic phytoplankton communities are proposed. But there is limitation of micronutrients like Fe^+, which could be supplementes through artificial application to increase the carbon sequestration level in oceans.

The combing of fossil fuel are to reduce CO_2 emission level is also equally important.

Can planned reforestation be combined with a conservation strategy that protects the rise CO_2 level and also the biodiversity of the tropics? This may only be realistic in those countries currently able to feed themselves. Tropical reforestation can be a realistic prospect only where the needs of local people are taken into account.

However, some general rules for the design of nature reserves are given in the Table 20.1.

Table 20.1: Basic design rules for nature reserves

Design rules	Variations during CO_2 rise
Large size.	Number of species increases with increasing land area, large patches have larger populations, reducing the risk of extinction.
Large number of habitats.	Allows species to adjust distribution within reserve in response to changes. Allows reserve to be effective for a long period.
Varied topography aspect and soil type.	Allows species to adjust distribution within reserve in response to change.
Wide corridors should connect reserves.	Allows species to adjust distribution by moving between large reserves. With increasing isolation the number of species decreases and there is a lower rate of colonization.
Corridors should contain large number of habitats.	Maximises diversity of species that move between reserves.
Network of corridors should be extensive and complete. Even small gaps or occasional fences are enough to prevent dispersal.	Maximises diversity of species that move between reserves.
Maintain reserves across the whole distribution of species.	Gives best changes of evolution in response to CO_2 rise.
Selective active management to promote some species and remove other according to stated goals.	Especially important during GEC because equilibrium is unlikely.
Studies on short term responses of individual species to GEC, showing what factors determine range and abundance and how large an area is required.	Required for effective land management inside and outside reserve.

21 VEHICULAR POLLUTION

21.1 INTRODUCTION

With the cultural and technological revolution came the automobile, totaling 53 million in the 1950s, world automobiles in 2010 exceed 550 million. About 19 million vehicles are added each year to the global total. Automobiles have come to mankind as a mixed blessing. While they enable mankind to conquer distances they also caused blatant destruction of the human environment, poisoned the atmosphere and made human beings vulnerable to several diseases, road accidents and fatalities. They are also a major consumer of oxygen. The most affected group of people are the urban habitants, the city-dwellers, drivers, hawkers, vendors, shopkeepers and residents in the busy city area. But perhaps the worst sufferers are the traffic policemen, who remain close to the fumes of vehicle exhaust and must inhale large quantities of their toxic pollutants every day, while on duty. Roughly, a man breaths 22,000 times and inhales 16 kg of air, daily.

Automobiles are a 'necessary evil', while they have made living easy and convenient, they have also made human life more complicated and vulnerable to both toxic emissions and an increased risk of accidents. Urban people are most affected and amongst the worst sufferers are traffic policemen who are particularly close to the fumes of automobile exhaust. Studies made in Jaipur, India, indicate that there is high rate of occurrence of respiratory, digestive, ocular and skin problems amongst the traffic policemen and a significant number of them become victims of lung disorders in the very first few months of their posting to a traffic department. Traffic policemen everywhere should wear 'pollution masks' for their own safety and to arouse public awareness of the risk of automobile pollution.

Urbanisation has a strong bearing on the travel demands in the country. Higher incomes, mobility, expanding cities and the proliferation of employment centres have increased the demand for motorised transport, resulting in a disproportionately high concentration of vehicles in urban centres. Irrational distances between homes and places of work, greater incentives for private transport and inadequate and poor quality public transport has further aggravated the problem. The Tata Energy Research Institute (TERI) estimates that of the total travel demand in 17 selected cities in India, the four metros generates nearly 56 per cent by 2010. This trend will negate any other efforts made to improve air quality and needs to be dealt with urgently. In 2005, for example, the vehicle population in India comprised about 25 million two wheelers, 5 million cars, jeeps and taxis and 4 million buses and goods vehicles. Of these 40% were concentrated in Delhi, 25% in Mumbai and 15% in Kolkata. However, the overall vehicular population distribution of major categories in metro-cities are depicted in Fig. 21.1.

The level of motorization in Asian countries has been growing at different rates and has been influenced by increasing affluence and population growth. The number of vehicles in Asia will continue to grow, with the possible exception of Singapore and Hong Kong, where active controls have been implemented to limit the number of vehicles and promote alternative public transport options.

The majority of additional vehicles will be new ones, while for some countries a substantial number of reconditioned vehicles, for which there is no adequate emission control in place, will be added to the fleet.

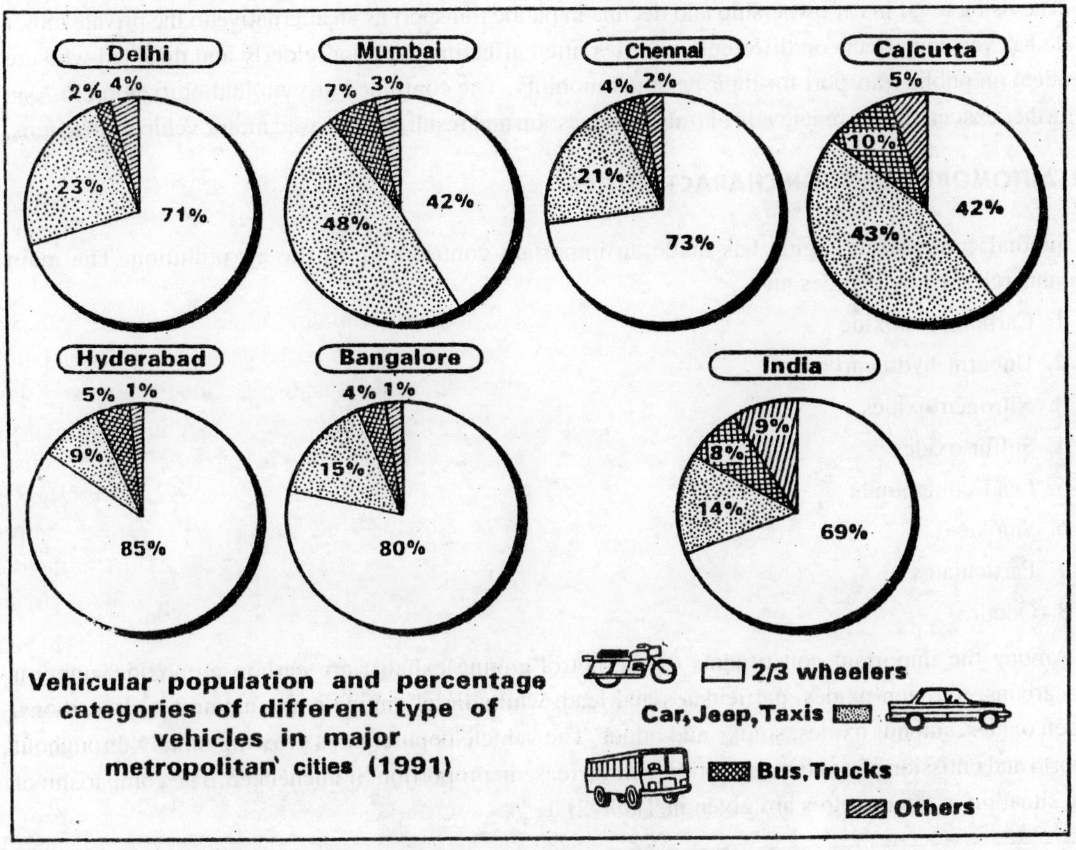

Fig. 21.1: Vehicular population

These are often buses or other public transport utility vehicles imported from other Asian countries where stringent in use emissions and safety requirements have made the engines obsolete. These are typically vehicles which are used intensively and not maintained very well. At present, vehicles in major industrialized Asian countries are being used for longer than was considered would be their normal useful life.

Two wheelers play an important role in Asia as they are accessible to a larger portion of the population who cannot afford to own a motor vehicle. This region has a very large population of two and three wheelers that dominate the emissions inventory for some compounds in most major cities. More than 75 per cent of the world's fleet of two wheelers and three-wheelers is in Asia. China alone accounts for approximately 50 per cent and India for 20 per cent of the world's fleet of two wheelers. In some cases, the inclusion of motorized two wheelers in total motorization rates brings these motorization rates up to the same level as cities with much higher average per capita (or per household) incomes. For example, when motorized two wheelers are included, Mexico City, with a GDP per capita 10 times higher than Chennai, has a lower motorization rate than Chennai. As the income of individuals rises, the two wheeler becomes accessible and therefore accelerates the motorization process in Asia.

A high proportion of the population in Asian cities currently depends on public transport and non-motorized forms of transport such as walking and cycling as the main mode of transport. As cities grow in area and decline in average density, the relative share of trips made by public transport and the absolute number of non-motorized trips are decreasing.

The rising trend in car ownership and decline in public transport as an alternative to the private motor vehicle has varying effects on different countries often affecting the poor, elderly and disabled who are dependent on public transport for their personal mobility. The continued growth in motorization in Asia will further exacerbate current levels of traffic congestion and result in increased motor vehicle emissions.

21.2 AUTOMOBILE EMISSION CHARACTERISTICS

The internal combustion engine has made an important contribution to the air pollution. The main pollutants released by engines are:

1. Carbon monoxide
2. Unburnt hydrocarbons
3. Nitrogen oxides
4. Sulfur oxides
5. Lead compounds
6. Smoke
7. Particulates
8. Odour.

Among the important constituents of the petrol engine exhaust are carbon monoxide, unburnt hydrocarbons, nitrogen oxides, particulates and lead, while diesel engine emits unburnt hydrocarbons, nitrogen oxides, sulphur oxides, smoke and odour. The vehicle population is growing rapidly throughout the world and emissions from transportation will increase in proportion, if unchecked. The comparison of traffic situation in three metros are given in Table 21.1.

Table 21.1: Comparison of traffic situation in three Indian metros (1993-94)

City	Contribution of vehicular	Vehicular pollution (tpd)	% of total vehicular population in India	Cars as % of total traffic	2-wheelers as % of total traffic	% of travel trips by buses	% of total trips by suburban rail	Average peak hour speed (kmph)
Delhi	64	1,300	29	20	69	62	1.0	10-15
Mumbai	52	659.57	11.5	42	38	72	30	15-20
Kolkata	30*	300.62	8.3	40	44	34	20	14

Notes: 2-whls: Two-Wheelers kmph: Kilometres per hour

tpd: tonnes per day: *: for 1988-89

** Vehicular population and percentage categories of different type of vehicles in major metropolitan cities of India is given in Fig. 21.1.

Among the list of automotive pollutions serial numbers 1 to 5 is decidedly dangerous to human health while serial nos. 6-8 are only nuisance. A comparison of exhaust constituents of petrol and diesel engines are given in Fig. 21.2.

The pollutants like carbon monoxide, unburnt hydrocarbons and nitrogen oxides have been the main targets of exhaust emission regulations throughout the world, because of their potential health hazard. To obtain the mass of exhaust emission from a vehicle, it is necessary to measure the concentration of the pollutants in the exhaust and the mass flow of air through the engine. On the whole, diesel engines are less of a health hazard than petrol engines, although the smoke and odour are certainly a nuisance.

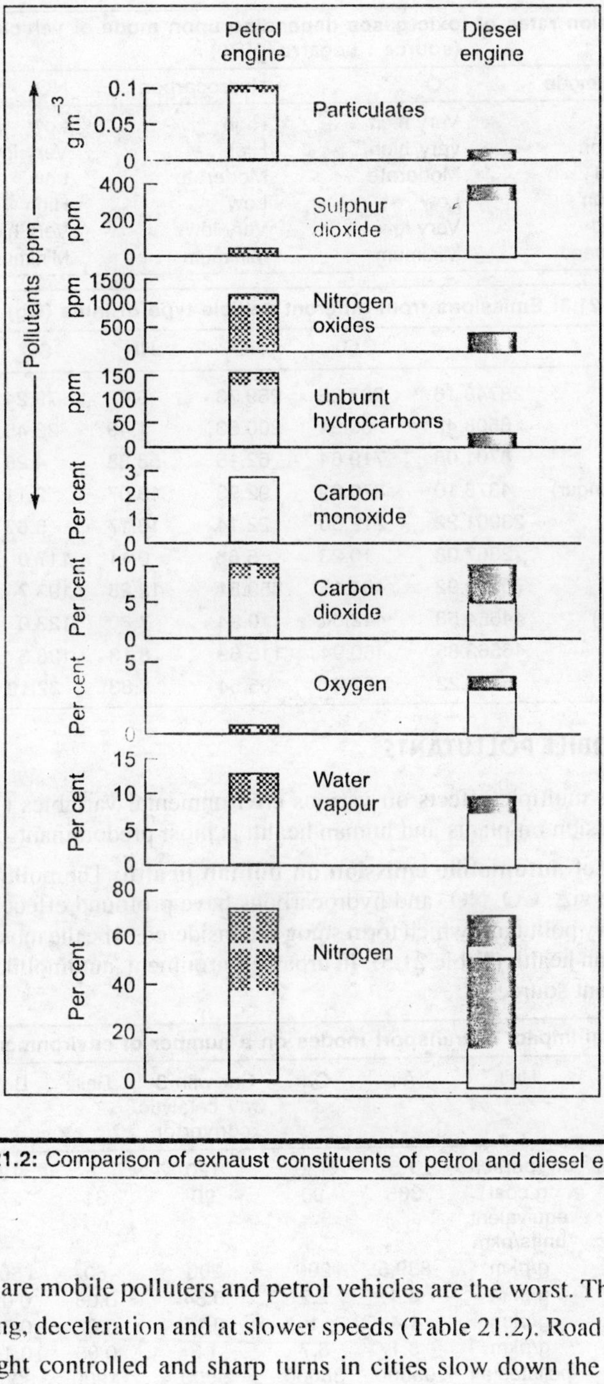

Fig. 21.2: Comparison of exhaust constituents of petrol and diesel engines

Thus automobiles are mobile polluters and petrol vehicles are the worst. The highest emission rates occur during motor idling, deceleration and at slower speeds (Table 21.2). Road intersections, especially those that are traffic light controlled and sharp turns in cities slow down the traffic and enhance the pollution emission rates. During idling and at slow speeds the greatest emission is of carbon monoxide and hydrocarbons. Ozone, the principal ingredient of urban smog, is formed when energy from sunlight causes hydrocarbons to react with Nitrogen oxides (NO_x). Sulfur dioxide (SO_2), Lead (Pb) and the Suspended Particulate Matter (SPM) are the other toxic pollutants from automobiles, all of which make the urban environment particularly unpleasant for the city dwellers (Table 21.3).

Table 21.2: Emission rates of toxic gases depending upon mode of vehicle operation (source : Legarra, 1970)

Operating mode	CO	Hydrocarbon	NO_x
Idling	Very high	High	Low
Deceleration	very high	High	Very low
Low speed	Moderate	Moderate	Low
Acceleration	Low	Low	High
High speed	Very low	Very low	Very high
Uniform speed	Minimum	Minimum	Minimum

Table 21.3: Emissions from different vehicle type of India (Gg)

Categories	CO_2	CO	NO_x	NH_4	SO_2	PM	HC
Bus	28748.16	207.26	269.73	5.02	79.24	31.36	51.72
Omni buses	8508.42	60.94	200.53	1.49	23.45	9.28	15.11
Two wheeler	8701.08	719.64	62.15	58.88	4.25	16.36	464.49
Light motor vehicles (Passenger)	4378.10	370.29	92.93	13.07	2.11	14.52	10.16
Cars and Jeeps	23901.22	212.30	22.14	18.17	5.67	3.22	28.01
Taxi	2367.08	10.23	5.68	0.11	117.0	0.80	1.48
Trucks and Lorries	70288.92	491.15	859.51	12.28	193.7	38.20	118.69
Light motor vehicles (Goods)	44654.58	442.04	110.94	7.80	123.0	17.33	12.13
Trailers and Tractors	46563.85	460.94	115.69	8.13	128.3	18.08	12.65
Others	5705.22	57.41	65.54	1.83	32.19	3.98	8.96

21.3 IMPACT OF AUTOMOBILE POLLUTANTS

Automobile pollutants have multiple effects on various environmental variables (Table 21.4). Among them the effect of auto-emission on plants and human health is most predominant.

(a) **Physical effects of automobile emission on human health**: The polluted gases as emitted from automobile viz., CO, NO_x and hydrocarbons have profound effect on human health. In addition secondary pollutants which form smog is considered to be the most severely affectively pollutant to human health (Table 21.5). In urban environment, automobile emission constitute the major pollutant source.

Table 21.4: Differential impact of transport modes on a number of environmental variables

Environmental variable	Unit	Air	Car	Car with 3-way catalytic converter	Rail	Bus	Bicycle	Pedestrian
Land use	m^2/person	1.5	120	120	7	12	9	2
Primary energy use	g coal equivalent units/pkm	365	90	90	31	27	0	0
Carbon dioxide emissions	g/pkm	839.5	200	200	60	59	0	0
Nitrogen dioxide emissions	g/pkm	6.4	2.2	0.34	0.08	0.02	0	0
Hydrocarbons	g/pkm	1.4	1	0.15	0.02	0.08	0	0
Carbon monoxide emissions	g/pkm	8.1	8.7	1.3	0.05	0.15	0	0
Air pollution	polluted air m^3/pkm	95000	38000	5900	1200	3300	0	0
Accident risks	Hours life lost/1000 pkm	1.4	11.5	11.5	0.4	1	0.2	0.01

Note : pkm = passenger kilometer. One passenger traveling 1 km is 1 pkm,, ten passengers traveling 50 km equals 500 pkm. The same logic used in freight transport where the unit of work done is the tonne kilometer (tkm)

Source : Teufel (1989)

(b) **Physical effects of automobile exhaust and NO_x on plants**: Plants that were kept on open field in close vicinity to busy highways reacted upon the pollutants with increased chlorophyll, sugar, free amino acid contents, an augmented ethylene production and peroxidase activity, a decreased β-carotene, ascorbate and auxin concentration and changes in fatty acid metabolism (Hoelloworth, 1981; Flueckinger et al, 1988). Automobile exhaust in concentration up to 1,000 μl/1 leads to a reduction in photosynthesis and impairment of stomata of Picea, Abies within only 15 min of treatment (Kammerbaner et al, 1986, 1987). These toxic effects of the automobile exhaust were traced back to NO_x, that are solubilised in the water saturated nitric acid (HNO_3). These compounds immediately dissociate into nitrate, nitrite and protons (Ronald et al, 1985) so that nitrate and nitrite accumulate in mesophyll cells of treated plants. This, in turn, induces the synthesis of a number of enzymes *viz.* nitrate (NR) and nitrite reductase (NiR) as well as glutamine synthetase (GS) and glutamate synthase (GOGAT). The effect of NO_x on photosynthesis and dark respiration is highly dependent on exposure time and doses. As a rule, a short-term treatment with NO_x at concentration beyond 0.3μl/1 increases chlorophyll content, dark respiration and net photosynthesis. Increasing concentrations of NO_x above 0.5μl/1 affect these parameters negatively.

Table 21.5: Effects of vehicular pollutants on humans

Substance	Health effects
CO (from gasoline cars, 2-wheelers, 3-wheelers)	Fatal in large doses, aggravates heart disorders; effects central nervous system; impairs oxygen carrying capacity of blood.
NO_x (from diesel vehicle)	Irritation of respiratory tract.
Ozone	Eye, nose and throat irritation; risk to asthmatics, children and those involved in heavy exercise.
Lead (from petrol vehicles)	Extremely toxic: affects nervous system and blood; can impair mental development of children; causes hypertension.
HC (mainly from 2-wheelers and 3-wheelers)	Drowsiness, eye irritation, coughing.
Benzene	Carcinogenic.
Aldehydes	Irritation of eyes, nose and throat, sneezing, coughing, nausea, breathing difficulties; carcinogenic in animals.
PAH (from diesel vehicle)	Carcinogenic.

Notes: CO: Carbon monoxide; HC: Hydrocarbons; NO_x: Nitrogen oxides; PAH: Polycyclic aromatic hydrocarbons

21.4 THE INDIAN SCENARIO

Automobile exhaust emissions contribute significantly to air pollution. In India production of vehicles increased from 210,000 pa in the 1970s to over 40 to 50 higher in 2008. This sudden increase in the number of vehicles has caused tremendous damage to the environment and has increased the intensity of air pollution in most of the big cities in the country. In the City of Mumbai, for example, there are 500,000 vehicles and these are responsible for about 90 per cent of the air pollution in this metropolis. The problem of vehicular pollution is increasing in almost all cities in India. Cities like Delhi, Kolkata, Ahmedabad, Chennai, Jaipur, Hyderabad, Kanpur, Nagpur, Lucknow, Agra, Ambala and Chandigarh are the worst affected by automobile pollution (Table 21.6).

Table 21.6: Estimated vehicular pollution load in the cities (2002)

City	Pollution Load in Metric tones per day			
	CO	NO_x	HC	PM
Delhi	421.84	110.45	184.37	12.77
Mumbai	189.55	46.37	89.93	10.58
Kolkata	137.50	54.09	47.63	10.80
Chennai	177.00	27.30	95.64	7.29
Bangalore	207.04	29.72	117.37	8.11
Hyderabad	163.95	36.89	90.09	8.00
Kanpur	28.73	7.25	11.70	1.91
Agra	17.93	3.30	10.28	0.91

Source : Auto Fuel Policy Report

Pollution levels in Delhi and Kolkata are at a greater level than that recommended by the World Health Organisation (WHO). The incidence of respiratory diseases in these cities is 12 times the national average and 30 per cent of their populations suffer from respiratory diseases due to vehicular air pollution. According to one estimate in 1995 (by the Department of Environment, Government of India), on average a million vehicles (diesel and petrol) spew from their exhaust daily, 250 t of carbon monoxide (CO), 400 t of hydrocarbons (HC), 6 t of sulphur dioxide (SO_2), 600 kg of lead (Pb) and large quantities of Suspended Particulate Matter (SPM) may comprise as much as 945 μgm^3 of the ambient air. Diesel vehicles give out black unpleasant smoke but their emissions are relatively less toxic.

Kolkata has 17 fixed active monitoring stations evenly distributed throughout the city which regularly monitor TSP, respirable particles (RSPM/PM_{10}), SO_2, NO_x and lead and are operated mostly by the WBPCB. Three stations are operated and maintained by the National Environment Engineering Research Institute (NEERI). In 2003 the WBPCB installed five automatic ambient air quality monitoring units in Kolkata and in the industrial areas of Howrah, Haldia and Durgapur. Since 1997 there has been a decreasing tendency in the annual mean concentrations of TSP and PM_{10} (RSPM); however, air quality standards for residential areas have been exceeded (140 $\mu g/m^3$ and 60 $\mu g/m^3$ respectively). In contrast, NO_2 concentrations have been increasing and exceeded annual standards for residential areas (60 $\mu g/m^3$) in 2001 and 2002 but complied with the standard in 2003 and 2004. Large and sudden variations in the NO_2 concentrations may indicate problems in data sampling. In the period 1998-2004 Kolkata achieved a significant reduction in annual SO_2 levels with average annual air quality standards for residential areas being met (60 $\mu g/m^3$).

Since the nationwide ban on leaded gasoline in India in 2000, lead concentrations have declined substantially and meet the stringent air quality standard (0.75 $\mu g/m^3$).

The Municipal Corporation of Greater Mumbai operates and maintains 22 monitoring stations in the city. These were manually operated on a noncontinuous basis to monitor the criteria air pollutants, namely SO_2, NO_2, TSP and ammonia. In addition, NEERI has operated three monitoring stations (World Bank, 1997).

Annual levels of TSP fluctuate around 250 μg/m^3 and do not comply with the Indian standard of 140 μg/m^3. PM_{10} (RSPM) levels in Mumbai have continued to exceed the national standard for annual averages in residential areas (60 μg/m^3). Between 1997 and 2002 PM_{10} levels had a decreasing tendency with the ambient concentrations for commercial, industrial and residential areas nearly approaching the annual ambient standard for residential areas. After 2002, concentrations of PM_{10} levelled off and show a slight increase. Between 1998 and 2004 the annual average concentrations of NO_2 in Mumbai declined and they are well below the India standard (80 μg/m^3). Between 1997 and 2004 SO_2 concentrations also declined. The annual mean levels are very low and comply with the standard (80 μg/m^3).

In Delhi, the CPCB and NEERI share the operation and maintenance of the existing 11 monitoring stations. All compounds, with the exception of NO_2, show a decreasing tendency. Roadside CO levels approach the national air quality standard (2000 μg/m^3) for residential 8-hour exposure to CO (GONCT, 2003). Ambient lead levels also decreased as a consequence of lead phase out in gasoline. The values shows the annual average concentrations of TSP, PM_{10}, NO_2 and SO_2 for the period 1994-2004. TSP levels fluctuate between 300 and 360 μg/m^3, well above the annual standard of 140 μg/m^3. Annual PM_{10} concentrations decreased from 270 μg/m^3 in 1994 to half this level in 2004, some 15 per cent above the Indian standard of 120 μg/m^3. PM_{10} levels, which have been recently monitored, also are double that of the national 60 μg/m^3 annual limit for residential areas. Since 2001 annual NO_2 concentrations have exhibited a slightly increasing tendency with values well below the national standard (60 μg/m^3) for residential areas (World Bank, 2004a). The annual SO_2 concentrations decreased steadily from 1994 to 2004 and are relatively low and comply with the air quality standard (60 μg/m^3) for residential areas.

Using the WHO Air Management Information System (AMIS) it is possible to calculate the average concentrations for selected pollutants over the period 1990-1999. In addition, annual air quality data provided by the 20 Asian cities were averaged to provide data for the period 2000-2004. Air pollutant concentrations for 2000-2004 compared with AMIS data (WHO, 2001) shows that there has been a general improvement in air quality. In 4 of the 13 cities, which are covered by AMIS and this study, air quality conditions have worsened for SO_2, NO_2 and PM_{10}. All the cities recorded an improvement in total suspended particulate (TSP) levels. For some cities, the data are insufficient to allow comparison (Table 21.7)

Table 21.7: Air quality data for 2000-2004 compared with 1990-1999 in selected Asian cities

Cities	SO_2	NO_2	SPM	PM_{10}
Bangkok	≥	>	<	<
Beijing	<	?	?	<
Busan	<	>	?	≥
Colombo	>	<	?	≤
Hong Kong	>	<	<	<
Kolkata	<	>	<	<
Manila	?	?	<	?
Mumbai	<	<	<	<
New Delhi	<	<	<	<
Seoul	<	>	<	<
Shanghai	<	<	<	?
Taipei, Rep. of China or Taiwan	<	<	?	>
Tokyo	<	≤	?	<

Notes: > (<) = Data for 2000-2004 larger (smaller) by more than 5 per cent than data for 1990-1999
> = approximately 5 per cent increase
< = approximately 5 per cent decrease
? = no data available

Source: WHO (2001), CAI-Asia (2006c)

21.5 AUTOMOBILE POLLUTION ABATEMENT

Over the years, the automobile population is increasing and thereby contributing substantialy to the total urban air pollution load. For abatement of such pollution, number of practices already adopted in various countries. These include: (i) Imposition of auto-emission standard i.e. declaration of permissible limit of pollutants emission by the different cars; (ii) Adoption to newer techniques for reduction of emission from engines.

The details of car emission standard as practiced in India and Europe is given in the separate Tables 21.8 & 21.9:

Table 21.8: Emission norms for Gasoline Powered Vehicles

Vehicle	Year	Emission Standards in g/km				
		CO	HC	NO_x	HC + NO_x	Deterioration n Factor DF
Two wheelers	1991	12-30	8-12	–	–	–
	1996	4.50	–	–	3.60	–
	2000	2.00	–	–	2.00	–
	2005 BS II	1.50	–	–	1.50	1.2
	2010 BS III	1.0	–	–	1.0	1.2
Three wheelers	1991	12-30	8-12	–	–	–
	1996	6.75	–	–	5.40	–
	2000	4.00	–	–	2.00	–
	2005 BS II	2.25	–	–	2.00	1.2
	2010 BS III	1.25	–	–	1.25	1.2
Car	1991	14.3-27.1	2.0-2.9	–	–	–
	1996	8.68-12.4	–	–	3-4.36	–
	1998	4.34-6.20	–	–	1.5-2.18	–
	2000	2.72	–	–	0.97	–
	BS II*	2.2	–	–	0.5	–
	BS II**	2.2-5.0	–	–	0.5-0.7	–
	BS III*	2.30	0.20	0.15	–	
	BS III**	2.3-5.22	0.20-0.29	0.15-0.21	–	1.2
	2010 BS IV*	1.00	0.10	0.08	–	
	2010 BS IV**	1.00-2.27	0.10-0.16	0.08-0.11	–	1.2
	2010 BS V n/a	1.00	0.10	0.06	–	–
	2010 BS VI	1.00	0.10	0.06	–	–

Note: For Catalytic Converter Fitter Vehicles

* Upto 6 seaters and Gross Vehicle Weight (GVW) up to 2500 kg

** More than 6 seaters and GVW up to 3500 kg

Table 21.9: Emission norms for Diesel Powered Vehicles

Vehicle	Year	Emission Standards					Type
		CO	HC	NO_x	HC + NO_x	PM	
	By Engine Dynamometer in g/kwh						
	1992	14.0	3.5	18	–	–	
	1996	11.20	2.40	14.4	–	–	
	2000	4.5	1.1	8.0	–	0.36/0.61 #	
	BS II	4.0	1.1	7.0	–	0.15	4 wheeler
	BS III	2.1	1.6	5.0	–	0.10	
	(or) By Chassis Dynamometer in g/km						
Diesel vehicles (GVM up to 3.5 tons)	1992	17.3-32.6	2.7-3.7	–	–		Light duty
	1996	5.0-9.0	–	–	2.0-4.0		
	2000	2.72-6.90	–	–	0.97-1.7	0.14-0.25	
	BS II	1.0-1.5	–	–	0.7-1.2	0.08-0.17	4 wheeler
	BS III	0.64-0.95	–	0.50-0.78	0.56-0.86	0.05-0.10	
	BS II 2005	1.00	–	–	0.85	0.10	
	BS III	0.50 DF 1.1	–	–	0.50 DF 1	0.05 DF 1.2	2 & 3 wheeler
	By Engine Dynamometer in g/kwh						
Diesel vehicles (GVM > 3.5 tons)	1992	17.3-32.6	2.7-3.7	–	–	–	
	1996	11.20	2.40	14.4			
	2000	4.5	1.1	8.0		0.36/0.36 #	
	BS II	4.0	1.1	7.0		0.15	
	2005 BS III	2.1	0.66	5.0		0.10/0.13	Smoke 0.8 m^{-1}
	2010 BS IV	1.5	0.46	3.5		0.02	Smoke 0.5 m^{-1}
	By Chassis Dynamometer in g/km						
	BS II	1.0	–	–	0.7	0.8	
	BS II**	1.0-1.5	–	–	0.7-1.2	0.08-0.17	
Cars	2005 BS III*	0.64	–	0.50	0.56	0.05	DF: CO 1.1,
	2005 BS III**	0.64-0.95	–	0.50-0.78	0.56-0.86	0.05-0.10	NO_x 1.0, PM 1.2
	2010 BS IV*	0.50	–	0.25	0.30	0.025	DF: CO 1.1,
	2010 BS IV**	0.50-0.74	–	0.25-0.39	0.30-0.46	0.025-0.06	NO_x 1.0, PM 1.2
	2010 BS V	0.50		0.17	0.23		
	2020 BS VI	0.50		0.08	0.17	–	–

For Engines with Power exceeding 85 kw/ For Engines with power up to 85 kw

* Up to 6 seaters and GVW up to 2500 kg

** More than 6 seaters and GVW up to 3500 kg

There is wide range of techniques which can be applied to the intake system, the combustion chamber and the exhaust systems of petrol and diesel engines to reduce exhaust emissions (Fig. 21.3).

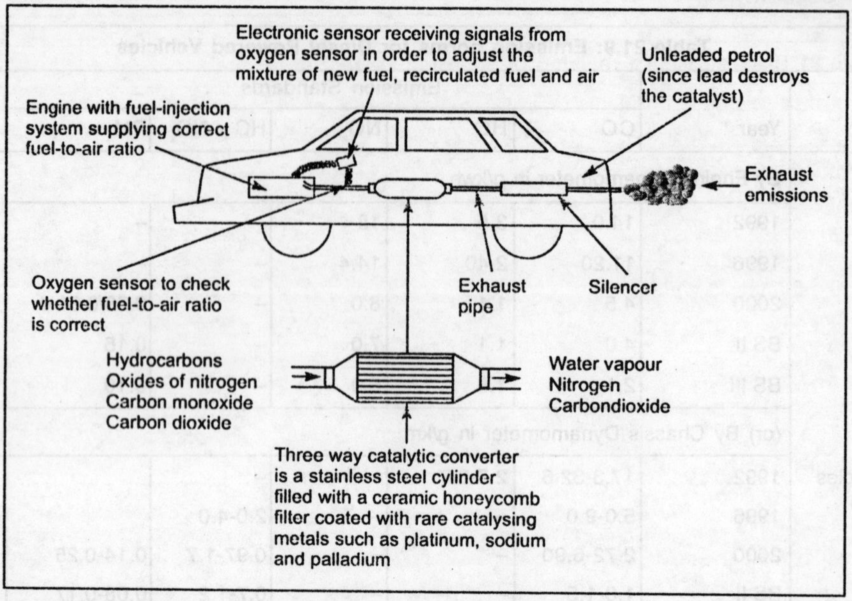

Fig. 21.3: Catalytic converter system to reduce exhaust emission in petrol & diesel engines

The outline of its activities is given as:

1. A positive crankcase ventilation valve is installed between crankcase and inlet system.
2. The fuel tank vent are connected to a coconister containing charcoal which absorb hydrocarbon molecules.
3. The control of combustion chamber hydrocarbons emission.
4. The oxidation of CO and unburnt hydrocarbons with air can be accelerated if a catalyst is present in the exhaust system. The most popular catalysts for vehicular exhaust emission control platinum and palladium. The catalyst can be introduced into the exhaust system by deposition on a ceramic honeycomb.
5. The emission of nitrogen oxides can also be reduced by attention to engine combustion.

MOEF proposed that the industry test emissions after fitting catalytic convertors, based on which government would finalise new standards for metros. Industry found that it could meet norms that were halfway between the 1996 and 2000 norms (Table 21.10) by using catalytic convertors. While the CO standard is 8.68 gm/km for 1996 and 2.72 gm/km for 2000, with catalytic convertors the industry has been able to bring it down to 5.64gm/km. Similarly, HC and NO_x has been brought down to 2.00 gm/km against the 1996 norm of 3.00 gm/km and 2000 norm of 0.97 gm/km. Encouraged by these results, new norms are now being contemplated for metro areas as an intermediate measure before 2000.

The automotive industry is unhappy with the fact that MOEF has allowed the Petroleum Ministry to drag its feet for so long, while it was being hammered with conditions. "To meet the new emission norms, the automotive industry will have to invest Rs. 50,000 crore in technology," says Doshi. "Unleaded petrol will only be a small fraction in the total petroleum bill. The government does not want to do its bit and spend some money on improving fuel quality". Eliminating lead from petrol and reducing the sulphur content of diesel, will address the specific problems of lead particulate matter pollution and facilitate the use of catalytic convertors. But there are several other shortcomings in fuel quality in India which interfere with efficient combustion and foul up the engine.

Table 21.10: Progressive revision of emission standards for Indian vehicles (1991-2000)

Types of vehicles and pollutants	April 1991 standard	April 1996 standard	April 2000 standard
Petrol vehicles: 2-wheelers			
CO (gm/km)	12-30	4.5	2.0
HC (gm/km)	8-12	—	—
HC + NO_x (gm/km)	—	3.6	1.5
Petrol vehicles: 3-wheelers			
CO (gm/km)	12-30	6.75	4.0
HC (gm/km)	8-12	—	—
HC + NO_x (gm/km)	—	5.40	1.5
Petrol vehicles: cars			
CO (gm/km)	14.3-27.1	8.68-12.40	2.72
HC (gm/km)	2.0-2.9	—	—
HC + NO_x (gm/km)	—	3.00-4.36	0.97
Diesel vehicles*			
CO (gm/kwh)	14.0	11.2	4.5
HC (gm/kwh)	3.5	2.4	1.1
NO_x (gm/kwh)	18.0	14.4	8.0
PM (gm/kwh)	—	—	0.36
Diesel vehicles **			
CO (gm/kwh)	14.3-27.1	—	5.0-9.0
HC + NO_x (gm/kwh)	2.7-6.9	—	2.0-4.0

Notes: * : gross vechicle weight > 3.5 tonnes
** : gross vechicle weight < 3.5 tonnes
kwh : Kilowatt hour
gm/km : grammes per kilometer

Bad quality fuel leads to gum deposits on the engine which can clog up the carburetor passage and disturb the air-fuel ratio, thus leading to improper combustion. Vehicle manufacturers suggest that multi-functional additives be used in the fuel, in order to clean up the engine passages. They claim that if fuel causes clogged passages, then engineering modifications like recirculation of blowby gas or exhaust gas for complete burning and cleaner crankcase emission will compound the problem, as the recirculated exhaust gas will leave behind more deposits.

The automotive industry complains that vehicles are tested with reference fuel, specifications for which have been separately notified. This reference fuel, a superior oil used for standardisation of refinery fuels across the country, is available only for testing purposes.

In 1996, National Environmental Research Institute, Nagpur (India) has developed a non-noble metal based, two-way catalytic converter technology for the control of exhaust emissions from petrol driven vehicles. The catalytic converter consists of a ceramic honeycomb support coated with a perovskite type catalytic housed in a suitable designed non-corrosive metallic housing. The device can be fitted to the exhaust system of a petrol driven vehicle and thus this device catalytically coverts the harmful carbon monoxide (CO) and hydrocarbon (HC) emissions to carbon dioxide (CO_2) and water (H_2O). The use of non-noble metal based catayst makes the device cost-effective and lead tolerant as compared to the noble metal based catalytic converters. The design of invented catalytic converter is shown in Fig. 21.4. For diesel vehicle the institute also developed a diesel particulate filter (DPF) for control of smoke from diesel engine exhaust. This filter consists of an aggregation chamber followed by different grades of filters, housed in suitable designed non-corrosive, metallic housing. The device can be fitted to the exhaust system in diesel driven vehicle for effective reaction of smoke (Fig. 21.5).

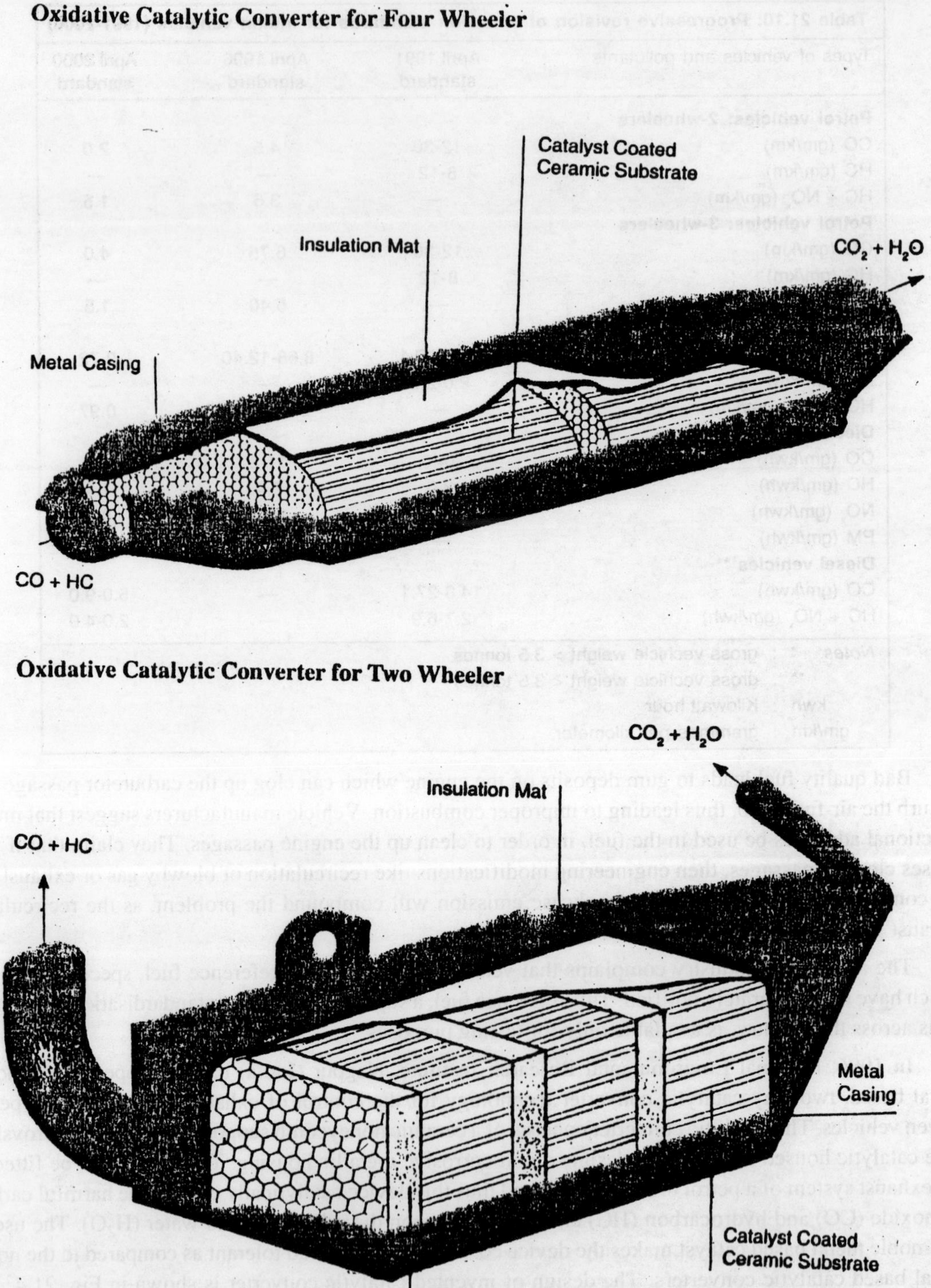

Fig. 21.4: NEERI invented Catalytic Converter

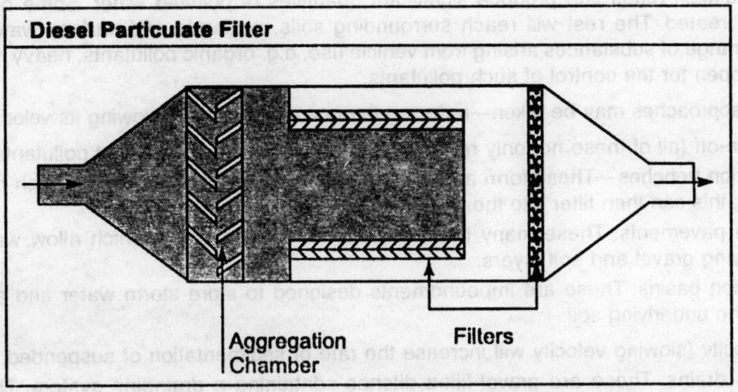

Fig. 21.5: Diesel particulate filter

Since 1990s, Ministry of Environment & Forest (MOEF), Govt. of India undertaken the following measures, for abatement of vehicular pollution:

- Low-lead petrol with a maximum lead content of 0.15 gms/litre has been introduced throughout the country. This measure is expected to reduce particulate lead emissions appreciably.
- Low-sulphur diesel (diesel with sulphur content of 0.5% or less) had been introduced in the four metros of Delhi, Mumbai, Kolkata and Chennai from 1.4.1996. Diesel with 0.25% sulphur is being supplied in the Taj Trapezium area from 1.9.1996.
- Tighter emission standards for vehicles at the manufacturing stage, notified by the Ministry of Surface Transport, have come into effect from 1.4.1996.
- Stricter emission norms for all categories of vehicles at the manufacturing stage are proposed to be notified by the Ministry of Surface Transport, which was proposed to be effective from 1.4.2000.
- Norms for 4-wheeler petrol vehicles fitted with catalytic converters have been finalised and recommended to the Ministry of Surface Transport for notification.
- A notification had been issued by the Ministry under the Environment (Protection) Act, 1986, on 2.4.96 specifying fuel quality for motor gasoline (petrol) and diesel. The Ministry of Petroleum and Natural Gas and the oil companies have been asked to meet the specifications as per the target dates.
- An organised standard was setup for vehicles manufactured on and after 1st April 2000.

POLLUTION FROM ROAD & ITS CONTROL

During heavy rainfall, roads can produce significant quantities of polluted water. Some of this enters storm sewers and is treated. The rest will reach surrounding soils, wetlands and surface waters. Such water is polluted with a range of substances arising from vehicle use, e.g. organic pollutants, heavy metals etc. A range of options are open for the control of such pollutants.

Three general approaches may be taken—reducing the quantity of run off, slowing its velocity, or treating it:

1. Reducing run-off (all of these not only reducing run-off, but also help filter out pollutants):
   (a) Infiltration trenches—These form a stone-filled reservoir along side a road, which retains much of the run-off; this can then filter into the subsoil;
   (b) Porous pavements: These many use particular forms of asphalt, which allow water penetration to underlying gravel and soil layers.
   (c) Infiltration basins: These are impoundments designed to store storm water and allow much of it to enter the underlying soil.
2. Slowing velocity (slowing velocity will increase the rate of sedimentation of suspended solids):
   (a) French drains: These are gravel-filled ditches containing a drainpipe system. The gravel reduces water velocity and helps to remove pollutants.
   (b) Swales: These are wide surface depressions alongside goods that are vegetated. The large surface area and surface roughner reduces water velocity and also helps to filter some pollutants.
3. Treatment:
   (a) Filter strips: These are areas of vegetated land designed to hold surface run-off and allow its seepage into the soil.
   (b) Detention ponds: These may either normally be dry or wet, but act to allow suspended solids to settle out.

There are two methods of treatment—Sedimentation tank / Lagoon system. The first system involves simple particulate settlement, while the second system involve complex lagoon system with layers of gravel and sand; partly overgrown by vegetation and supported by a thick plastic sheet.

Pollutants	% removal	
	Sedimentation tank	Lagoon
Suspended solids	52	92
Total lead	40	90
Total zinc	47	76
Oil	30	70

CABINET OKAYS NEW AUTO FUEL POLICY, EURO II NORMS BY 2005

The cabinet's decision to implement the recommendations of the expert committee on auto fuel policy would call for an investment of around Rs. 60,000 crore by the oil and the auto sector for upgrading facilities and technology as prescribed in the report.

In light of the huge investments to be made by the two sectors, the government is expected to provide them certain fiscal incentives. While the oil industry would be required to invest around Rs. 35,000 crore, the auto sector would have to make fresh investment of Rs. 25,000 crore.

As per the recommendations of the committee, set up by the government under the chairmanship of Dr R.A. Mashelkar, Director General, CSIR, Bharat Stage II emission norms are to be implemented in Bangaluru, Hyderabad and Ahmedabad by the end of year 2003 and the same norms have to be extended all across the country from April 1, 2005.

The committee has also recommended introduction of emission norms equivalen to Euro II in Delhi, Mumbai, Kolkata, Chennai, Hyderabad and Ahmedabad from April 1, 2005. Extension of these norms to the other parts of the country has been recommended from the year 2010.

The committee has left the choice of fuel and vehicular technology to the owners and operators of vehicles with the government prescribing only the emission standards. It has also called upon the government to encourage the use of alternative fuels like CNG, LPG, ethanol, electricity etc.

The interim report of the committee, whic comprised experts from the fields of environment, vehicular technology, petroleum policy, public finance and management, was submitted to the government on April 1, 2002.

21.6 LEGISLATIVE MEASURES

The following legislative coverage are being practiced to regulate vehicular pollution in India:

Legislation / Act	Authority	Responsibility / Notifications
The Environment (Protection) Act, 1986, amended 1991 • Environmental (Protection) Rules, 1986 (amended in 1992, 2001, 2002, 2003, 2004)	Ministry of Environment & Forests	Notification of standards for emission or discharge of environmental pollutants from the industries, operations or process. The notified standards related to vehicular are as follows: • Specification on Motors Gasoline for emission related parameters • Specification of Diesel for emission related parameters • Specifications of two- stroke engine oil • Standard for emission smoke, vapour etc. from motor vehicles • Noise limits for Automobiles at the manufacturing stage
The Central Motor Vehicles Act, 1988 • The Central Motor Vehicles Rules, 1989 (Second amendment 2009)	Ministry of Road Transport and Highways	Makes rules regulating construction, equipment and maintenance of motor vehicles and trailers as per section 110 of Motor Vehicle Act. The notified emission standards related to vehicular are as follows: • Relating to Emission of smoke and vapour from agricultural tractors driven by diesel engines • Relating to Diesel vehicles with original equipment fitment – Replacement of in use diesel engine by new LPG engine – Application emissions norms • Relating to Diesel driven Argicultural tractor for standards of gaseous pollutants • Relating to idling emissions standards for petrol/CNG/LPG driven vehicles • Mass emission standards Bharat Stage III for four wheeled vehicles in NCR & other cities • Mass emission standards Bharat Stage III for two and three wheelers manufactured on and from 1st April, 2010 • Mass emission standards Bharat Stage IV for M and N category vehicles

21.7 AUTO FUEL POLICY'S ROAD MAP for control of vehicular pollution from New Vehicles

Indian cities have different climatic conditions, human population, different vehicle population density and source of pollution. The Indian cities with a population of more than one million people with high vehicle population and the cities in which pescribed standards are violated in one or additional parameters in 72 non attainment cities, which require actions for vehicle pollution control ahead of the rest of the country. Since the year 2000, India started adopting European emission and fule regulations for four-wheeled light-duty and for heavy-duty vehicles. In respect of 2 & 3 wheeler norms, India is already ahead of most of the advanced world as far as the emission norms are concerned. The road map for vehicular emission norms Bhart Stage II, III and IV is given in Table 21.12:

Table 21.12: Road map for Vehicular Emission norms for new vehicles			
Emission Norms	**Date of effective**	**Entire country**	**Identified cities**
New Vehicles Except 2 & 3 wheelers			
Bharat Stage II	From Year 2002 & 2001	–	Delhi, Mumbai, Kolkata and Chennai
	From Year 2003	–	NCR, Bangalore, Hyderabad, Ahmedabad, Pune, Surat, Kanpur and Agra
	From April 01, 2005	Yes	–
Bharat Stage III	From April 01, 2005		For all private vehicles, city public service vehicles and city commecial vehicles in indentified cities*
	From April 01, 2010	Yes	–
Bharat Stage IV	From April 01, 2010		For all private vehicles, city public service vehicles and city commerical vehicles in indentified cities*
Bharat Stage IV	From April, 2017	Yes	Norms for all vehicles
Bharat Stage VI	From April, 2020	Yes	Norms for all vehicles (Skipping Bharat Stage V)
New 2 & 3 Wheelers			
Bharat Stage II	From April 01, 2005	Yes	
Bharat Stage III	Preferably from April 01, 2008 but not later than April, 01, 2010 in any case	Yes	

In addition, at manufacturing state noise limits for various kinds of vehicles are also imposed since 2005. The details are given in Table 21.13.

	Table 21.13: Noise limits for vehicles applicable at manufacturing stage applicable for 1st April, 2005 in India	
Sl No.	**Type of vehicle**	**Noise Limits dB(A)**
1.	Two wheeler • Displacement up to 80 cc • Displacement more than 80 cc but up to 175 cc • Displacement more than 175 cc	 75 77 80
2.	Three wheeler • Displacement up to 175 cc • Displacement more than 175 cc	 77 80
3.	Vehicles used for carriage of passengers and capable of having not more than nine seats, including the driver's seat	74
4.	Vehicles used for carriage of passengers having more than nine seats, incluing the driver's seat, and a maximum gross Vehicle weight (GVW) of more than 3.5 tonnes • With an engine power less than 150 KW • With an engine power of 150 KW or above	 78 80
5.	Vehicles used for carriage of passengers having more than nine seats, including the driver's seat: Vehicles used for carriage goods • With maximum GVW not exceeding 2 tonnes • With maximum GVW greater than 3 tonnes but not exceeding 3.5 tonnes	 76 77
6.	Vehicles used for transport of goods with a maximum GVW exceeding 3.5 tonnes With an engine power less than 75 KW With an engine power of 75 KW or above but less than 150 KW With an engine power of 150 KW or above	 77 78 80

22 RADIATION HAZARDS

22.1 INTRODUCTION

Man and all life forms on earth have evolved in a radiation environment. All non-living substances and generations of living beings are exposed to natural radiation. This is inevitable. The air we breathe, the water and beverages we drink, the food we eat are all parts of the same environment and contain radioactive materials, though in small amounts. Thus atomic radiation is an inseparable part of our environment. It has been there from time immemorial.

It is interesting to note that "Our bodies contain amounts of the radioactive isotopes potassium^{40} and carbon^{14}. About 38,000 atoms of potassium^{40} and 1,200 atoms of carbon^{14} explode in our body each second"!

22.2 ATOMIC RADIATION

Atoms of certain elements emit spontaneously very minute particles and electromagnetic radiations. These tiny particles may carry electric charge as in the case of alpha or beta rays; they may be neutral as in the case of neutrons. The electromagnetic radiations emitted are X-rays (from atom) and gamma rays (from nucleus) and they do not carry any electric charges.

Radioactivity is the result of spontaneous disintegration of an atom during which it emits alpha, beta or gamma particles. Simultaneously, the atom changes into another element which is itself radioactive, and this in turn disintegrates to become something else and the process continues. The sequence of changes is known as the transformation series. The disintegration decay activity of a radioactive element is entirely random. A radioactive element has a definite rate of decay known as the half-life period. *Half-life is defined as the time taken for half the atoms in any given sample of the substance to decay.*

An example of decay of uranimum^{238} is shown below:

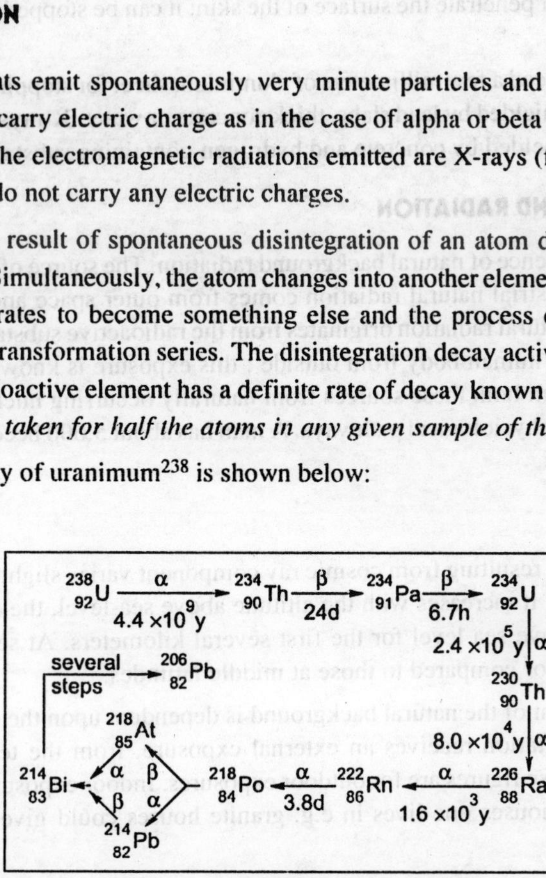

Alpha rays are helium nuclei, that is, helium atoms which have lost their two orbital electrons, and hence they have a net positive charge. They are all ejected with approximately the same velocity ranging from 1.4×10^9 to 1.7×10^9 cms per second. They have a range of several centimeters in air, but most are stopped by a very thin sheet of aluminum foil, about 0.1 mm thick.

Beta rays are streams of high energy electrons similar to cathode rays. They are emitted with variable velocities, approaching that of light (3×10^8 m/s). They are far more penetrating than the alpha rays and are stopped by a 1 cm thick aluminum sheet.

Gamma rays consist of electromagnetic radiations of very short wavelength and occupy a band among the X-rays which have the shortest known wavelengths. They have the same velocity as that of light. The highest energy gamma rays are very penetrating and approach absorption (or attenuation) only after traversing few centimeters of lead.

Alpha radiation consists of heavy, doubly positive charged particles emitted by atoms of elements such as uranium and radium.

Beta radiation consists of electrons. These electrons are much lighter than alpha particles and carry unit negative charge.

Gamma rays are a form of electromagnetic radiation similar to X-rays, light and radiowaves.

Neutrons are uncharged heavy particles contained in the nucleus of every atom heavier than hydrogen.

Alpha radiation may just penetrate the surface of the skin: it can be stopped completely by a sheet of paper.

Beta particles however need a few millimeters of aluminium sheet for stopping them. Gamma rays are more penetrating but can be shielded by lead slabs, thick concrete or water. X-rays are less penetrating than gamma rays. Neutrons are shielded by concrete and hydrogen-containing substances like wax, water etc.

22.3 NATURAL BACKGROUND RADIATION

Many are unaware of the existence of natural background radiation. The source of these are extraterrestrial and terrestrial. The extraterrestrial natural radiation comes from outer space and hence these are called cosmic rays. The terrestrial natural radiation originates from the radioactive substances in the earth's crust. These radiations irradiate the human body from outside : this exposure is known as External Exposure. 'Internal Exposure' arises from both these sources from naturally occurring nuclides which are taken up into the body through normal physiological pathways. A man has about 5,000 becquerel (Bq) radioactivity in his body mainly due to ^{40}K.

External Exposure

The external exposure of man resulting from cosmic ray component varies slightly with the geomagnetic latitude and, to a great extent, it increases with the altitude above sea-level, the dose rate approximately doubling for each 1.5 km above sea level for the first several kilometers. At sea level, the exposure is 10 per cent lower at the equator compared to those at middle latitudes.

The terrestrial component of the natural background is dependent upon the composition of the soils and rocks. On average, population receives an external exposure, from the terrestrial component, of 350 (± 100) μSv per year. These figures are for outdoor exposures. Indoor exposures are generally higher, depending upon the type of houses one lives in e.g. granite houses could give an order of magnitude higher exposure.

In India, external dose measurements in over 200 stations from different parts of the country indicate an annual average external exposure of 690 (± 200) μ Sv/yr.

Internal Exposure

Internal exposure results from radio-nuclides entering the body through ingestion or inhalation. These radio-nuclides are either cosmogenic (i.e. produced by the interaction of cosmic rays with atoms in the upper atmosphere) or primordial, in the sense that they existed in the earth's crust throughout its history. Internal exposure from cosmogenic nuclides (such as ^{4}H, ^{7}Be, ^{14}C, ^{22}Na) is very small. The short-lived ^{222}Rn and its decay products contribute a major fraction of internal dose. Other nuclides like ^{40}K the decay products of Thorium (^{220}Ra) and ^{210}Po contribute to a lesser extent. The internal exposure due to these primeordial nuclides is about 1,300 μSv per year. Total annual average exposure of man is 2,000 μSv/yr. These exposures are given in Table 22.1

Table 22.1: Estimated annual exposures of man to natural sources of radiation in areas of normal background

Source	Annual Exposure. μSv per year		
	External	Internal	Total
Cosmic Rays:			
Ionising component	280	—	280
Neutron component	20	—	20
Cosmogenic nuclides	—	15	15
Primordial nuclides in soils and rocks:			
Potassium-10	120	180	300
Rubidium-87	—	6	6
Uranium-238 series	90	950	1040
Thorium-232 series	140	190	330
Total (rounded)	650	1340	2000
Data source Adapted from UNSCEAR 1982			

Other Exposures

In addition to natural background radiation, man is exposed to sources of radiation that he himself has created; for example, X-rays and other types of radiation used for medical purposes: fallout from testing of nuclear weapons, occupational exposure from nuclear and other industries using ionising radiation and radioactive materials released in the course of nuclear power production. Table 22.1 shows the contribution of the various sources to the average annual dose to the population (a typical case). It is obvious that the contribution due to nuclear discharges is negligible (0.1%).

It may be interesting to compare the radiation exposures from various sources. An air travel in a jet plane gives an exposure of 8.5 μSv/hr while in a supersonic it is 16μSv/hr. One chest X-ray may give about 200μSv.

The radiation exposure near a nuclear power plant is only few tens micro sievert per year which is negligibly small and lies within the statistical variations in natural background radiation levels. But natural disaster like earthquake, tsunami, or several cyclone may lead to nuclear plant damage and explosion as seen March 2011 at Fukshima Daiichi, Japan.

Comparison of risks

Every human activity like smoking, rock climbing, driving and many such activities involve certain amount of risk. Table 22.3 (page 407) compares the probability of death per year in some of the human activities. It is very clear that work with radiation is one of the safest human activities involving least risk.

22.4 MEASUREMENT OF RADIOACTIVITY

Radioactivity can be measured by a Geiger-Müller counter. The type commonly used consists of an aluminium tube which acts as the negative electrode, while a wire down the centre forms the positive electrode (Fig. 22.1). The gas inside the tube consists of argon at low pressure with an added trace of bromine. A thin mica window at the end permits the entry of active particles. If one of these enter the tube, the gas is ionized. By the time the electrons from the ionization process reach the central electrode, they are moving so fast under the high potential gradient that they create an avalanche of extra ions by collision. This process of gas amplification as it is called increase the sensitivity of the tube, enabling it to record the entry of beta (β-) particles and gamma (γ) photons which, on their own, produce far fewer ion-pairs per centimetre of their path than do alpha (α) particles (Table 22.2).

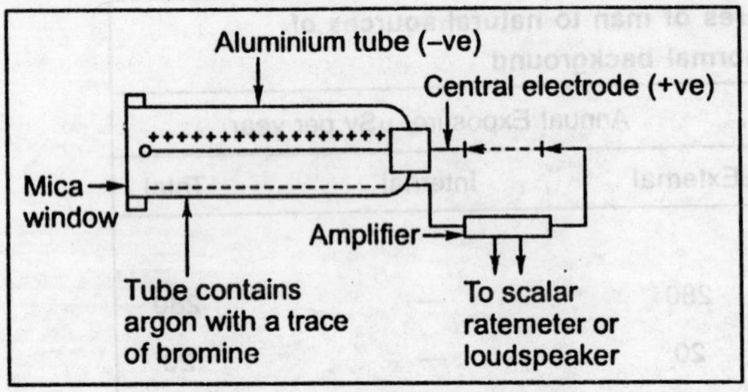

Fig. 22.1: The Geiger-Müller counter

Table 22.2: Representative emission energies (Wilson, 1966)

Nuclide		Energy (MeV)	Half-Life
beta emission			
^{234}Th	Thorium	0.19	24.1 days
^{234}Pa	Protactinum	2.31	1.18 min
^{210}Pb	Lead	0.063	22 yrs
^{35}S	Sulphur	0.167	87 days
^{36}Cl	Chlorine	0.71	3×10^5 years
^{40}K	Potassium	1.32	1.3×10^9 years
alpha emission			
^{238}U	Uranium	4.20	4.4×10^9 years
^{234}U	Uranium	4.77	2.4×10^5 years
^{236}Ra	Radium	4.78	1.6×10^3 years
^{222}Rn	Radon	5.48	3.8 days

The current pulses from the tube are amplified and used to operate either a dekatron counter or scalar or else a ratemeter. The former counts and records the pulses on a microammeter calibrated in counts per second.

The Geiger-Müller counter is connected to a combined power unit and scaler and the voltage adjusted to the recommended value specified for the tube. The scalar is switched on and allowed to warm up. A random count is recorded which is the background count and comes from radioactive material in the earth and nearby surrounding together with cosmic radiations which pernetrate the earth's atmosphere from outer space. The background count is timed over a period of at least 2 minutes and, after this, radiations from the source is measured. Nowadays portable battery operated Geiger tubes are available.

22.5 EFFECTS OF RADIATIONS

More is known today about effects of exposure to radiation than about any other physical and chemical agent in our environment. The health effects of radiation exposure are not unique. The main effects of radiation are cancer induction and genetic effects. In fact, out of the natural cancer incidence or fatality, only about 10 per cent can be accounted by the exposure to natural background radiation. Thus the effects that can be attributed to low level radiation exposure are also caused by large number of other environmental agents. Effects of radiation exposure to low level radiation should not be disregarded; but it should be recognised that they are very small compared to risks of similar effects from other known and unknown agents. At the radiation exposure standards set by nuclear industry for occupational workers as well as members of public, no discernible effects can arise in population groups.

Table 22.3: Comparison of risks

Cause	Chances of death per year
Smoking 20 cigarettes/day	1 in 200
Accidents in deep sea fishing	1 in 400
Natural causes — 40 years old	1 in 500
Accidents on the road	1 in 5,000
Accidents in the home	1 in 10,000
Accidents at work	1 in 20,000
Air crash	1 in 100,000
Radiation from nuclear facilities (1,000μ Sv per year)	1 in 100,000
From possible accidents due to every 100 nuclear power stations' operations	1 in 5,000,000,000

Data source: WHO Regional Publications, European Series No. 3. WASH-1400 and others.

Since the disaster caused by Nagasaki and Hiroshima atom bombs in August 1945, people all over the world became aware of the harmful effect of radioactive substances which emits radiations.

22.6 RADIOACTIVITY AND EFFECTS ON MAN

It is well-known that some chemical elements possess different kinds of atomic structure by modification of their nuclear composition (i.e. number of protons, neutrons, etc). Isotopes are capable of emitting radiant energy, which is called radioactivity. Some isotopes are, of course, non-emitting type. Uranium, Radium and Thorium are the most commonly known isotopes that are capable of emitting radioactive emissions. On emission of radiation from isotopic elements, there will be loss of weight of the element. The loss of weight and its conversion into energy is illustrated by Einstein's famous equation $E = mc^2$ (E = Energy, m = mass of the object, and c = speed of light emission). Thus mass can be converted into energy.

Environmental hazards from nuclear power plants is the release of radioactive substances which could happen at the slightest error or negligence. Radioactive isotopes are also known as radionuclides. These emitted radiations that are capable of damaging the protoplasmic system of living system. The

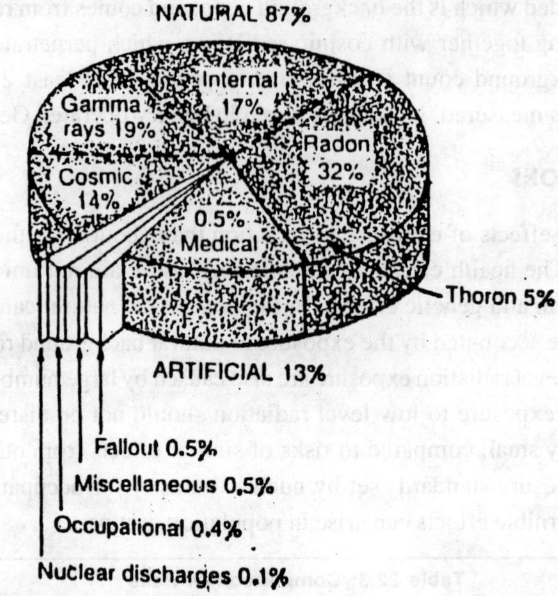

Fig.22.2: Different sources of radiation in Man's Environment

emitted radiations may be alpha, beta and gamma rays. These rays enter the environment thus from nuclear fallouts from nuclear power plant wastes and certain other sources.

The relative contribution of different sources of radiation in our environment is given in Fig. 22.2.

22.7 IMPACT OF RADIOACTIVE RADIATION

The radiation effect may be (i) instantaneous, (ii) prolonged, or (iii) delayed types. Even it could be carried to future generations or production of defective embryos. All kinds of impact depends on two major factors: doses of radiation exposure and duration of exposures.

The radiation effect is thus grouped at effects on different cellular component level.

The details are discussed separately under the following heads:

(a) Radiation effect on DNA

(b) Radiation effect on chromosome

(c) Radiation effect on cellular level, and

(d) Tissue response to radiation.

In addition there are a number of gross physiological effects on whole body including defects of circulatory, digestive and neurotic systems too.

(a) **Radiation Effects on DNA**: There are four categories of changes —

1. Base damage—change or loss of a base.
2. Single-strand breaks (SSB) — break in the backbone of one chain of the DNA molecule.
3. Double-strand break (DSB) — break in both chains of DNA molecule.

4. Cross-links — either within the DNA molecule (instrastrand) or from one molecule to another (DNA-inserted or DNA-protein)

   There may be an immediate or delayed effect depending on the doses of exposure.

(b) **Radiation Effects on Chromosomes**: These effects include a number of structural changes on chromosomes. The details are:

1. A single break in one chromosome or chromatid,
2. A single break in separate chromosomes or chromatids,
3. Two or more breaks in the same chromosome or chromatid.
4. "Stickiness" or dumping of the chromosomes.

The general consequence to the cell of these structural changes may be one of the following:

1. The broken ends may rejoin with no visible damage. This is restitution.
2. Loss of part of the chromosome or chromatid at the next mitosis giving rise to an aberration.
3. Rearrangement of the broken ends, which can produce a grossly distorted chromosome viz, ring chromosomes, dicentric chromosomes and anaphase bridges.
4. Rearrangement of the broken ends without visible chromosomal damage; the genetic material has been rearranged.

(c) **Radiation Effect on Cellular Level**: One of the three things can happen to a cell after irradiation:

1. It can be delayed from going through division; appropriately, the term used to define this response is division delay.
2. It can die before it divides, during interphase. This response is also appropriately named interphase death.
3. It can die when attempting mitosis; this response is termed reproductive failure.

   The details of extent of damage is shown below:

No Damage	Survival of normal reproductive capacity
Sublethal damage	Survival of normal reproductive capacity after time for recovery
Potentially lethal damage	Survival of normal reproductive capacity after time for recovery in stationary phase.
Non-lethal damage	Survival of reproductive capacity: but slow growth.
Lethal damage	Loss of reproductive capacity.

(d) **Tissue Response to Radiation**: The radiation effect at tissue level are:

1. Acute and chronic morphogenic effects on tissues.
2. Total body radiation syndrome viz., N-V-D syndrome (Nausea, Vomiting, Diarrhea).
3. Three general effects of radiation on the embryo and fotus: viz. lethality; congenital abnormalities present at birth and long-term effects on body after birth.
4. Late radiation effects, i.e., induction of carcinogenesis.

Radioactive fallouts

Radioactive fallouts are the input of radioactive dust from the atmosphere on ground surface. The source of such dusts are usually nuclear bombs. Atomic bombs based on fission of uranium and plutonium release fission products that cause enormous effect on living system. Radioactive dust thus generated accumulate on living system through food chain and thereby cause serious damage to the life processes.

22.8 RADIOACTIVE WASTE

From nuclear power plant radioactive waste generated—either as liquid or solid—should be treated properly, otherwise they will contaminate the environment. Radioactive waste water could be treated through flocculation, precipitation, sorption, filtration and ion exchange process. But the solid wastes could either be incinerated or disposed underground.

The solid waste products are stored in special shielded tanks. Long-term storage of nuclear waste relies on the use of multiple barriers to prevent radiation escape. One method of disposal consists of a series of chambers dug deep in the ground (Fig. 22.3) that has been checked to make sure it is free from geological faults. The chambers are at least 60 meters below the surface. Waste packed in steel drums and covered in concrete are loaded in outer cases and stacked in the underground chambers. The space around the cases are packed with concrete or earth and the filled chambers are then sealed-off.

Another method of storing high level radioactive waste is in the liquid form. This is reduced in volume by a process called vitrification which converts it into a glass like substance. These are then kept in repositories. An alternative way is to convert it into a rock-like substance termed as synrock.

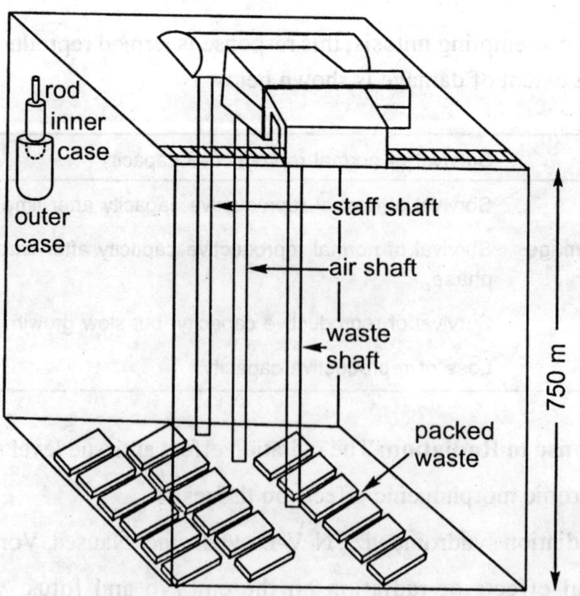

Fig. 22.3: Underground storage of radioactive wastes

22.9 IONISING RADIATION AND EFFECTS ON MAN

Like radioactive material exposure, ionising radiation also have deleterious effect on living being specially mankind. The degree of effect however depends on dose and duration of exposure (Table 22.4).

Table 22.4: Probable Health Effects Resulting from Exposure to Ionising Radiation

Dose in rems (whole body)	Health effects	
	Immediate	Delayed
1,000 or more	Immediate death, Frying of the brain	None
600-1,000	Weakness, nausea, vomiting and diarrhea followed by apparent improvement. After several days fever, diarrhea, blood discharge from the bowels, hemorrhage of the larynx, trachea, bronchi or lungs, vomiting of blood and blood in the urine.	Death in about 10 days. Autopsy shows destruction of hematopoietic tissues, including bone marrow, lymph nodes and spleen, swelling degenaration of epithelial cells of the intestines, genital organs and endocrine glands.
250-600	Nausea, vomiting, diarrhea, epilation (loss of hair), weakness, malaise, vomiting of blood, bloody discharge from the bowels or kidneys, nose bleeding, bleeding from gums and genitals, subcutaneous bleeding, fever, inflammation of the pharynx and stomach, and menstrual abnormalities. Marked destruction of bone marrow, lymph nodes and spleen causes decrease in blood cells—especially granulocytes and thrombocytes.	Radiation-induced atrophy of the endocrine glands including the pituitary, thyroid and adrenal glands. From the third to fifth week after exposure, death is closely correlated with degree of leukocytopenia. More than 50% die in this time period. Survivors experience keloids, ophthalmological disorders, blood dyscrasis, malignant tumours, and psychoneurological disturbances.
150-250	Nausea and vomiting on the first day. Diarrhea and probable skin burns. Apparent improvement for about two weeks thereafter. Fetal or embryonic death if pregnant.	Symptoms of malaise as indicated above. Persons in poor health prior to exposure, or those who develop a serious infection, may not survive. The healthy adult recovers to somewhat normal health in about three months. He or she may have permanent health damage, may develop cancer or benign tumours, and will probably have a shortened lifespan. Genetic and teratogenic effects.
50-150	Acute radiation sickness and burns are less severe than at the higher exposure dose. Spontaneous abortion or stillbirth.	Tissue damage effects are less severe. Reduction in lymphocytes and neutrophils leaves the individual temporarily very vulnerable to infection. There may be genetic damage to offsprings, benign or malignant tumours, premature ageing and shortened lifespan. Genetic and teratogenic effect.
10-50	Most persons experience little or no immediate reaction. Sensitive individuals may experience radiation sickness.	Transient effects in lymphocytes and neutrophils. Premature ageing, genetic effects and some risk of tumours.
0-10	None	Premature ageing, mild mutations in offspring, some risk of excess tumours. Genetic and teratogenic effects.

22.10 RADIATION PROTECTION

The protection against radiation hazards was not considered seriously as a major environmental threat for human health. However, the realisation of deleterious effects of radiation hazards in recent years necessitates the need for protection against radiation hazards.

Global banning of nuclear weapons, safe disposal of nuclear plant wastes and proper dumping of radioactive materials used in therapeutic purpose etc. are the major activities undertaken currently as a measure of radiation hazard protection. In addition, the workers handling radioactive materials require to undertake appropriate protection measures during working phase at the nuclear plants premises. International Atomic Energy Commission set up a standard guideline for the said purpose.

The International Commission in Radiological Protection (ICRP) makes recommendations to protect both workers in professions where radioactivity is a particular hazard, and the general public. They do not as yet set exact limits in all cases, but set forth the following general principles (Ottoway, 1980):

(a) no practices (which produce increased radiation) shall be adopted unless their introducing produces a positive benefit.

(b) all exposures shall be as low as reasonably achievable (ALARA), economic and social factors being taken into account.

(c) the dose to which an individual is exposed shall not exceed the limits recommended for the appropriate circumstances by the commission.

ICRP's general guidelines are that radiation workers should not be exposed to more than 50 m Gy (5 rads) a year, and the maximum exposure for the general public should be one tenth of this (5 m Gy or 0.5 rads).

Nuclear plants are located, designed, constructed and operated to conform to very stringent safety standards to ensure that the risk associated with atomic energy is extremely low. An exclusion zone of 1.6 km radius around the reactor is established which is free of habitation. An area in the annulus between the exclusion zone boundary and the 5 km radius is provided as sterilized zone. Off-site emergency plans are also worked out for an area of about 16 km radius to meet any contingency in the unlikely event of an accident.

23 ENVIRONMENTAL DEGRADATION

23.1 CONCEPT

Environmental degradation is a complex process involving transformation, alteration or material loss from any one of the environmental components. It may arise either by natural processes or by man-made activities. The major forms of environmental degradation and their consequences are shown in Table 23.1. Environmental pollution is also a component of degradation process.

Table 23.1: Major forms of environmental degradation and main consequences

Environmental degradation	Main consequences		
	Health	Productivity	Amenity
Water pollution and scarcity (including marine pollution)	Contribute to over 3 million deaths and billions of illnesses annually	Shortages constrain economic output; fisheries decline; high costs to provide safe water	Loss of recreation uses such as fishing, boating, swimming; offensive odours.
Air pollution	Causes many illnesses and deaths; many suffer from smoky indoor air.	High abatement costs; acid rain impact on forests and water bodies; corrosion of buildings.	Reduced visibility from smog; visual disamenity.
Global atmospheric changes	Risks from climatic natural disasters; possible shifts in vector-borne diseases; more skin cancer and eye damage from ozone depletion.	Damage to low-lying coastal assets; regional changes in agricultural output; disruption to marine food chain; threat to biodiversity; ecological impacts.	Species and ecosystems threatened by climate change.
Solid and hazardous wastes, including radiation	Toxic industrial and urban conditions; industrial accidents.	High collection costs; pollution of groundwater, high prevention costs.	Visual and odour disamenities.
Urban areas : congestion and noise	Infectious diseases; traffic accidents; hearing impairment.	Loss of productivity; costs of traffic management and noise abatement.	Overcrowding; landlord oppression.
Rural areas : soil quality		Loss of productivity, especially in tropical soils; off-site siltation of waterways and bodies.	
Rural areas : forests		Losses of forest products, soil productivity, genetic resources; watershed instability; carbon emissions.	Reduction in biodiversity; loss of recreation uses.
Ecosystems, other natural areas		Losses of tourism, genetic resources; ecological impacts.	Reduction in wilderness experiences, biodiversity loss.
Depletion of raw materials and energy resources		Reduction in economic growth.	

Several types of environmental degradation can change potentially renewable resource into nonrenewable, permanently extinct, or unusable resources :

- Covering of productive land with water, silt, concrete, asphalt, or buildings to such an extent that crop growth declines and places for wildlife to live (habitats) are lost.
- Cultivating land without proper soil management so that crop growth is reduced by soil erosion and depletion of plant nutrients. Each year the world's farmers must feed 95 million more people with 24 billion fewer metric tons (26 billion tons) of topsoil.
- Irrigating cropland without sufficient drainage so that excessive buildup of water (waterlogging) or salts (salinisation) in the soil decreases crop growth.
- Removing water from underground reservoirs (aquifers) and surface waters (rivers and lakes) faster than it is replaced by natural processes. Water scarcity is emerging on every continent.
- Removing trees from large areas (deforestation) without adequate replanting so that wildlife habitats are destroyed and long-term timber growth is decreased. Every minute a piece of tropical forest of the size of 20 city blocks disappears.
- Depletion of grass by livestock (overgrazing) so that soil is eroded to the point where productive grasslands are converted into unproductive land and deserts (desertification).
- Killing various forms (species) of wild plant and animal life through destruction of habitat, commercial hunting, pest control and pollution to the point where these species no longer exist (extinction).

The current status report of earth's vital resources are depicted in Table 23.2.

Table 23.2: Status Report for Some of the Earth's Vital Resources

Land	
Productive Land	About 8.1 million square kilometers (3.1 million square miles) of once-productive land (cropland, forests, grasslands) have become desert in the last 50 years. Each year almost 61,000 square kilometers (23,500 square miles) of new desert are formed.
Cropland Topsoil	Topsoil is eroding faster than it forms on about 35% of the world's cropland—a loss of about 24 billion metric tons (26 billion tons) of topsoil a year. Crop productivity on one-third of the earth's irrigated cropland has been reduced by salt buildup in topsoil. Topsoil waterlogging has reduced productivity on at least one-tenth of the world's cropland.
Forest Cover	Almost half of the world's original expanse of tropical forests has been cleared. Each year about 2,02,000 square kilometers (78,000 square miles) of tropical forests are destroyed and another 202,000 square kilometers (78,000 square miles) are degraded. Within 30 to 50 years there may be little of these forests left. One-third of the people on earth cannot get enough fuelwood to meet their basic needs and many are forced to meet their needs by cutting trees faster than they are being replenished. In MDCs 312,000 square kilometers (120,400 square miles) of forest have been damaged by air pollution. Also, many remaining areas of diverse, ancient forests are being cleared and replaced with more vulnerable tree farms that greatly reduce wildlife habitats and biodiversity.
Grasslands	Millions of hectares of grasslands have been overgrazed; some, especially in Africa and the Middle East, have been converted to desert. Almost two-thirds of US rangeland is in fair to poor condition.
Water	
Coastal and Inland Wetlands	Between 25-50% of the world's wetlands have been drained, built upon, or seriously polluted. Worldwide, millions of hectares of wetlands are lost each year. The United States has lost 56% of its wetlands in the lower 48 states and each year loses another 150,000 hectares (371,000 acres).
Oceans	Most of the wastes we dump into the air, water and land eventually end up in the oceans. Oil slicks, floating plastic debris, polluted estuaries and beaches and contaminated fish and shell-fish are visible signs that we are using the oceans as the world's largest trash dump.
Lakes	Thousands of lakes in eastern North America and in Scandinavia have become so acidic that they contain no fish; thousands of other lakes are dying; thousands are depleted of much of their oxygen because of inputs of various chemicals produced by human activities.

Data from Worldwatch Institute and World Resources Institute, 1997.

The overall consequence of environmental degradation is very significant in rural areas than urban location. Thus there is direct poverty connection with environmental degradation (Table 23.3).

Table 23.3: Causes and Consequences of Environmental degradation in Rural Areas

Consequence	Direct cause	Underlying cause
Deforestation	Shifting agriculture	Population growth
Loss of biodiversity	Permanent agriculture	Poverty
Soil erosion	Permanent pasture	Lack of land tenure
		Low-level agricultural inputs
		Mechanization of agriculture and/or the consolidation of agricultural lands
Flooding	Commercial logging	Destructive logging, lack of forest management and protection, poor reforestation
		Increased access to forests along loggging roads for farmers and ranchers
	Agriculture, ranching and logging	Production for export markets
		Fiscal policies and legislation, in part to promote exports of primary products due to need for foreign exchange to service debt.
		Inappropriate economic valuations of natural resources and biodiversity
	Use of biomass fuels	Inefficient use of fuelwood; overcutting of fuelwood resources
	Use of forest biomass for fodder	Shortages and lack of alternative sources of fodder
Desertification	Agricultural expansion onto fragile lands	Population growth
	Overgrazing	Poverty
	Burning of grasslands	Lack of land tenure
		Low-level agriculture and/or the consolidation of agricultural lands
	Use of biomass fuels	Inefficient use of fuelwood; overcutting of fuelwood sources
		Lack of access to higher quality fuels and stoves
	Climate change	Various; not well understood
Air pollution	Slash and burn agriculture	Population growth
	Burning of grasslands	Poverty
		Lack of land tenure
		Low-level agricultural inputs
Salinization and water-logging of irrigated lands	Poor planning and management	Lack of access to high-quality or alternative sources of fodder
		Cheap or free water contributing to inefficiency

Source: Office of Technology Assessment, UNEP, 1992.

23.2 LAND DEGRADATION AND SOIL EROSION

23.2.1 Introduction

All land use activities, particularly those which are poorly managed, involve destruction or disturbance, to a greater or lesser extent, of natural and seminatural ecosystems. Almost invariably, however, it is those ecosystems, in equilibrium with their environment, which offer most effective protection to the soil that supports them. A major consequence of ecosystem destruction and disturbance is that of soil degradation. This has been defined as the decline in soil quality caused through its misuse by human activity. More specifically, it refers to the decline in soil productivity through adverse changes in nutrient status, organic matter, structural stability and concentrations of electrolytes and toxic chemicals. Soil degradation incorporates a number of environmental problems, some of which are interrelated, including erosion, compaction, water excess and deficit, acidification, salinisation and sodification and toxic accumulation of agricultural chemicals and urban/industrial pollutants. In many instances, these have led to a serious decline in soil quality and productivity and it is only in recent decades that the finite nature of soil as a resource has become widely recognised.

Soil degradation is not a new phenomenon. Archeological evidence suggests that it has been on-going since the beginning of settled agriculture several thousand years ago. The decline of many ancient civilisations, including the Mesopotamians of the Tigris and Euphrates valleys in Iraq, the Harappans of the Indus valley in Pakistan and the Mayans of Central America, was due in part to soil degradation. More recently, an event of major significance was the dustbowl which occurred in the Great Plains of the American midwest during the 1930s. At this time, intensive agricultural practices, employed in the eastern states, were transferred to the drier midwest where the soils are lighter textured and more susceptible to erosion. A number of years of drought, combined with crop failure and destruction of the protective organic-rich topsoil, resulted in severe wind erosion.

According to the Global Assessment of Soil Degradation project, about 15 per cent of the global land area between 72°N and 57°S is degraded. Of this, an area slightly less than that of India (about 300 million hectares) is strongly degraded, largely as a result of deforestation (113 million hectares), inappropriate management of cropped land (83 million hectares) and overgrazing (75 million hectares). In recent decades, the global rate of soil degradation has increased dramatically and is likely to increase further as we approach the twentyfirst century; in 1983 it was estimated at 5-7 million ha a^{-1} and is set to rise beyond 10 million ha a^{-1} by the year 2000.

The effects of soil degradation are not restricted to the soil alone, but have a number of off-site implications. Soil erosion, for example, is often associated with increased incidence of flooding, siltation of rivers, lakes and reservoirs and deposition of material in low-lying areas (e.g., Morgan, 1986). These problems may be compounded in areas where infiltration capacity is reduced due to compaction, hardsetting or induration of soils. Salinisation and sodification of soils are often associated with poor quality irrigation water while soil acidification is commonly linked with acidification and aluminium contamination of surface waters. Leaching of fertilisers and pesticides from agricultural soils may also lead to contamination of surface and shallow ground waters. In addition, contamination of soils by urban and industrial pollutants, such as heavy metals and radionuclides, may lead to toxic accumulation in arable produce and in herbage for grazing animals, thus having important implications for human health (Thornton, 1991).

The extent of soil degradation is influenced by a number of factors, many of which are interrelated, namely soil characteristics, relief, climate, land use and socio-economic and political controls (Fig. 23.1). In many studies of soil degradation and its wider environmental implications, the socio-economic and

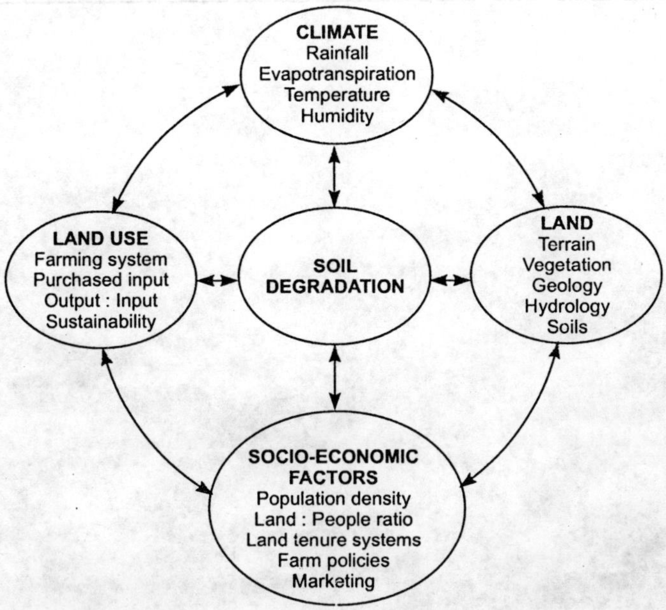

Fig. 23.1: The causal factors of soil degradation and their interactions (from Lal and Stewart, 1990b)

political controls are often overlooked, or at least not examined in any detail, perhaps because of the difficulties associated with the collection of reliable and comparable data. Increasingly, however, these controls on landuse systems are being viewed as central to the issue of soil degradation, particularly in the developing world.

Management of soil degradation, whether at a global, regional or local scale, is clearly a complex issue and represents one of our most challenging environmental problems. Emphasis should be placed on sustainable, rather than exploitative land use practices; this theme was highlighted by the World Soil Charter (FAO, 1981) which called for a commitment by governments, agencies and land-users to 'manage the land for long term advantage rather than short term expediency' (FAO, 1983). The problem requires a holistic, multidisciplinary approach involving the collaborative and coordinated efforts of ecologists, agronomists, soil scientists, hydrologists, engineers, sociologists and economists. Moreover, the involvement of government and non-government organisations, aid agencies and the farmers themselves is essential to the success of research and development in this area. Such involvement should facilitate the implementation of education, training and incentive programmes. Imposition from above of high-technology, high-cost solutions by technical experts from developed countries is certainly not the answer in the developing world. Inevitably, such solutions are not economically viable and low-technology, low-cost options—such as low external input agriculture, agroforestry and social forestry—are often the only answer. Hence, the approach to soil conservation has shifted in recent years from a rather technocentric standpoint to a more ecocentric position. Central to this approach are the concepts of land husbandry and sustainable development, which place emphasis on the land-users themselves rather than on the technical experts and advisors.

This chapter aims to examine a selection of the most pressing soil degradation problems and, in each case, the causal factors, on and off site effects and management strategies will be considered.

Land degradation due to deforestation

Soil erosion

Land degradation due to mining

Land degradation due to grazing

23.2.2 Soil Erosion

Soil erosion occurs when the rate of removal of soil by water and/or wind exceeds the rate of soil formation. Generally, rates of soil formation are very low, with profiles developing at a rate of about 1 cm every 100-400 years (FAO, 1983); assuming an average bulk density of 1.33g cm^{-3}, this equates to about 0.3 – 1.3 t ha^{-1} a^{-1}. It is important to differentiate between natural or background erosion and erosion which has been accelerated largely as a result of human activity. Background erosion rates are often similar to rates of soil formation at < 1.0 t ha^{-1} a^{-1}, although in mountainous areas they may be considerably higher (Morgan, 1986). In contrast, rates of accelerated erosion commonly exceed 10 t ha^{-1} a^{-1} and sometimes exceed 100 t ha^{-1} a^{-1}. Some of the highest soil erosion rates have been observed in the Loess Plateau area of China and in the Himalayan foothills of Nepal, where values in excess of 200 t ha^{-1} a^{-1} have been recorded. Similarly, in India, gully erosion results in a loss of about 8,000 ha of land per year (FAO, 1983).

The extent of soil erosion is governed by a number of factors. Those of particular importance include erosivity of the eroding agent, erodibility of the soil, slope steepness and length, landuse practices and conservation strategies. These factors are summarised in the Universal Soil Loss Equation which has been used widely in the modelling and prediction of soil erosion e.g., Colby-Saliba (1985):

$$E = R.K.L.S.C.P.$$

where, E = mean annual soil loss, R = rainfall erosivity index, K = soil erodibility index, L = slope length, S = slope steepness, C = cropping factor which represents the ratio of soil loss under a given crop to that from bare soil, and P = conservation practice factor which represents the ratio of soil loss where contouring and strip-cropping are practised to that where they unused (Morgan, 1986).

Although widely used, this model has been the subject of extensive criticism. For example, it assumes that a vegetation cover is always protective which is not necessarily the case; erosion on land with a good cover of crops planted in rows can be greater than on land which is sparsely vegetated. It is also water erosion based and cannot be used in areas affected extensively by wind erosion. More specifically, it focuses on rill and inter-rill erosion and is not easily applied to areas where gully and stream bank erosion are widespread. Its universal nature has also been questioned particularly in terms of its application to tropical soils. Furthermore, it should be emphasised that this model does not consider the wide range of socio-economic and political factors which play a crucial role in terms of their influence on the degree of soil erosion which will be examined later. Alternative models include SLEMSA (Soil Loss Estimator for Southern Africa) and CREAMS (Chemicals Run-off and Erosion Arising From Agricultural Management Systems).

Landuse is perhaps the most significant factor influencing soil erosion, for two main reasons. First, many landuse practices leave the soil devoid of a protective vegetation cover, or with only a partial cover, for significant periods of time and second, they involve mechanical disturbance of the soil. Specific aspects of landuse often associated with accelerated soil erosion include expansion and intensification of arable cultivation, overgrazing, deforestation, certain forestry practices, site clearance in preparation for urban and industrial construction and a number of recreational activities such as walking and skiing.

Arable cultivation has expanded and intensified dramatically in recent decades. Relatively steep slopes, formerly covered by grass or tress, have been converted to arable cropping, while an increased use of heavy agricultural machinery has resulted in compaction of the soil. This, in turn, has led to reduced infiltration capacity, particularly along wheel tracks, thus resulting in increased surface run-off and erosion.

Similarly, increased reliance on tillage activities, throughout the cropping cycle, has rendered soils more susceptible to erosion. This problem has been compounded by the decline in levels of soil organic matter and hence structural stability, largely in response to increased use of inorganic fertilisers. In addition, the tendency to increase field sizes on arable land has meant that there are fewer physical breaks and barriers in the landscape, such as tree lines, hedgerows and walls, to restrict erosion. Susceptibility to erosion is further increased if land is cultivated with the slope rather than parallel to the contours.

Overgrazing is particularly common in drought-affected parts of the developing world, such as the Sahel region of sub-Saharan Africa and the rangelands and communal lands of eastern and southern Africa. In a study of the impact of grazing on soils of the Savanna region of Nigeria, for example, Aweto and Adejumbobi (1991) attribute enhanced surface run-off and erosion to compaction of the soil and destruction of the protective vegetation cover by grazing animals and to the adoption of inappropriate burning strategies. Deforestation, largely for logging and woodfuel purposes, is also common in many parts of the developing world. Trees are well-known for their ability to protect soils from erosion, particularly on steeply sloping terrain. Their root systems and the organic material which they supply help to stabilise the soil, while water uptake and canopy interception serve to reduce the frequency and intensity of surface run-off.

In addition to deforestation, many forestry practices are associated with accelerated soil erosion, including the needleleaf forestry programmes which have become widespread in many areas of upland Britain. Here, erosion is most serious during the pre-planting stages of land preparation and drainage and after harvesting (Soutar, 1989). In relation to urban and industrial land use, construction and associated disturbance of land may lead to increased soil erosion. Even certain recreational activities have been implicated in this problem, including walking and skiing.

A number of socio-economic and political factors have been associated with accelerated soil erosion, particularly in the developing world. These include population pressure, skewed land resource distribution, poverty and marginalisation, increasing demand for woodfuel, inappropriate land tenure and farm policies, small size of land-holdings and poor infrastructure (McGregor and Barker, 1991; Thapa and Weber, 1991). In many developing countries, population growth is rapid and the demand for agricultural land and woodfuel is ever increasing (Table 23.4). Furthermore, agricultural systems are characterised by a skewed land resource distribution where a minority of affluent and powerful landowners control a majority of the land area. The poorest farmers are thus forced onto marginal land, which is particularly susceptible to erosion, and often end up in a vicious spiral of debt. Rural-urban migration, abandonment of land and increased soil erosion are often responsible to this poverty trap situation (Fig. 23.2).

Table 23.4: Woodfuel as a proportion of energy use in selected countries

Country	Woodfuel as a percentage of total energy consumption
Nepal	98
India	36
Thailand	63
Angola	74
Tanzania	94
Brazil	33
Chile	16

Source: Based on Thapa and Weber, 1991

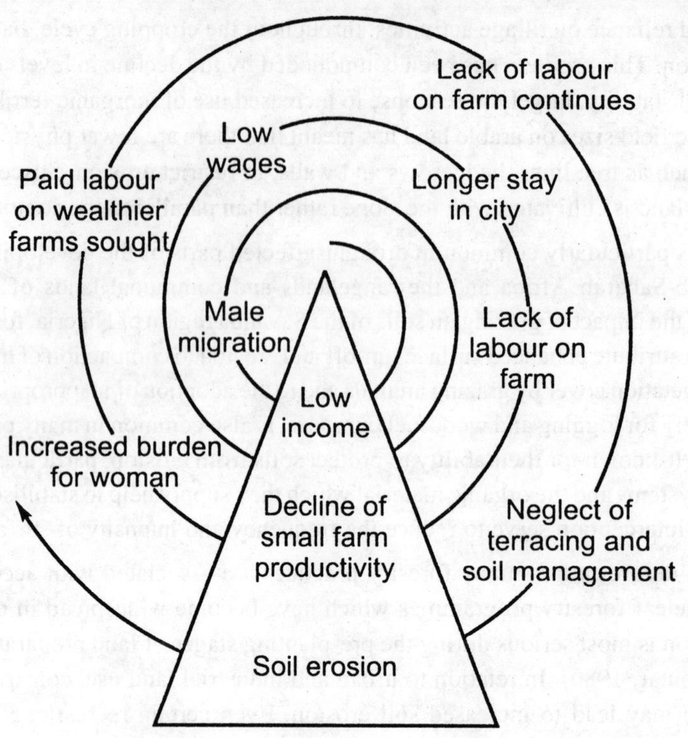

Fig. 23.2: The spiral of poverty, land abandonment and soil erosion in the developing world (from Millington *et al.* 1989)

In many parts of the developing world, large areas of land are utilised for mono-cultivation of cash crops, which are not necessarily best-suited to soil conditions, rather than for indigenous mixed food cropping. Such commercial pressure on agricultural systems, as well as contributing to the problem of marginalisation discussed above, has a detrimental effect on soil quality and is unlikely to be sustainable in the long term. There is also little political support in terms of education, training and incentive schemes to encourage farmers to adopt more sustainable landuse practices. The establishment of appropriate and comprehensive soil conservation and land husbandry programmes is further hindered by the small size of land-holdings and the large number of farmers involved (Table 23.5).

The on and off-site effects of soil erosion are considerable. At the global scale, it is estimated that unless soil conservation measures are introduced on all cultivated land, 544 million ha of potentially productive rain-fed crop land will be lost and agricultural production expected to decrease by almost 20 per cent, by the year 2000-2005. Undoubtedly, these effects will be felt most severely in those developing countries which are least able to cope with the problem. It should be noted that the deterioration in soil productivity is disproportionate to the amount of soil eroded, as it is the nutrientrich and structure-supporting constituents in the topsoil which are lost most readily.

Table 23.5: Distribution of agricultural land-holdings in selected developing countries by size of holdings and proportionate area

Country	Size of holdings (%)			Area (%)		
	< 5 ha	5–50 ha	> 50 ha	< 5 ha	5–50 ha	> 50 ha
Asia						
India	91	9	0	51	46	3
Philippines	86	14	0	51	37	12
Thailand	72	28	0	39	61	0
Africa						
Algeria	69	30	1	14	63	23
Zaire	99	1	0	60	5	35
Zambia	94	6	0	34	19	47
Latin America						
Argentina	15	39	47	0	2	98
Brazil	37	45	16	1	12	87
Columbia	60	32	8	4	19	78

Source: Thapa and Weber, 1991

23.3 IMPACT OF SOIL EROSION

23.3.1 Introduction

The rapid erosion of soil by wind and water has be… a problem ever since land was first cultivated. The consequences of soil erosion occur both on- and off-site. On-site effects are particularly important on agricultural land where the redistribution of soil within a field, the loss of soil from a field, the break-down of soil structure and the decline in organic matter and nutrient result in a reduction of cultivable soil depth and a decline in soil fertility. Erosion also reduces available soil moisture, resulting in more drought-prone conditions. The net effect is a loss of productivity which, at first, restricts what can be grown and results in increased expenditure on fertilizers to maintain yields, but later threatens food production and leads, ultimately, to land abandonment. It also leads to a decline in the value of the land as it changed from productive farmland to wasteland. Offsite problems result from sedimentation down stream or downwind which reduces the capacity of rivers and drainage ditches, enhances the risk of flooding, blocks irrigation canals and shortens the design life of reservoirs. Many hydroelectricity and irrigation projects have been ruined as a consequence of erosion. Sediment is also a pollutant in its own right and, through the chemicals adsorbed to it, can increase the levels of nitrogen and phosphorus in water bodies and result in eutrophication.

23.3.2 Process

Soil erosion is a two-phase process, consisting of the detachment of individual particles from the soil mass and their transport by erosive agents such as running water and wind. When sufficient energy is no longer available to transport the particles, a third phase—deposition—occurs.

Rainsplash is the most important detaching agent. The soil is also broken up by weathering processes, both mechanical, by alternate wetting and drying, freezing and thawing and frost action and biochemical. Soil is disturbed by tillage operations and by the trampling of people and livestock. Running water and wind are further contributors to the detachment of soil particles. All these processes loosen the soil so that it is easily removed by the agents of transport. The severity of erosion depends upon the quality of material supplied by detachment and the capacity of the eroding agents to transport it.

There are a number of factors that control erosion:

- Erosivity of the eroding agent;
- Erodibility of the soil;
- Slope of the land; and
- Nature of the plant cover.

In the field, soil erosion status may be surveyed and data are recorded as per proforma (Table 23.6) for further interpretation.

Table 23.6: Proforma for recording soil erosion in the field

Recorder : Date : Attitude :				Area : Air photo no. : Grid reference (if any) :
Present hand use				
Climate	Month	J F M A M J J A S O N D		Erosivity
	Rainfall (mm)			
	Mean Temp. (°C)			
	Maximum intensity			
Vegetation	Type	% Ground cover		% tree & shrub cover
Slope	Position	Degree	Distance from crest	Shape
Soil	Depth	Surface		Erodibility
		Permeability	Clay fraction	
Erosion				
Remarks				

EROSION	CODE	0	$\frac{1}{2}$	1	2	3	4	5	6	7	8	9	10

Code		Indicators
0	—	No exposure of tree roots, no surface crusting, no splash pedestals; over 70% plant cover (ground & canopy).
	—	Slight exposure of tree roots, no splash pedestals, soil level slightly higher on up slope, 30 - 70% plant cover.
1	—	Exposure of tree root, slight surface crusting, soil mound protected by vegetation, 30 - 70% plant cover.
2	—	Tree root exposure, crusting of the surface 30 - 70% plant cover.
3	—	Tree root exposure, less than 30% cover.
4	—	Tree root exposure, bare soil.
5	—	Gullies, bare soil, blowouts & dunes.

23.3.3 Measurement

Field measurements may be classified into two groups: those designed to determine soil loss from relatively small sample areas or *erosion plots* often as part of an experiment and those designed to assess erosions over a larger area such as a *drainage basin*.

In erosion plot, a standard size of 22 m long and 1.8 m wide are used. The plot edges are made of sheet metal, wood or any material which is stable, does not leak and is not liable to rust. At the downslope end is positioned a collecting trough or gutter, covered with a lid to prevent the direct entry of rainfall, from which sediment and runoff are chanelled into collecting tanks. For large plots or where run-off volumes are very high, the overflow from a first collecting tank is passed through a divisor which splits the flow into equal parts and passes one part, as a sample, into a second collecting tank. A flocculating agent is added to the mixture of water and sediment collected in each tank. The soil settles to the bottom of the tank and the clear water is then drawn-off and measured. The volume of soil remaining in the tank is determined and a sample of known volume is taken for drying and weighing. The sample weight multiplied by the total volume gives the total weight of soil in the tank. The total soil loss from the plot is the weight of the soil in the first tank plus, assuming one-fifth of the overflow from the first tank passes through the divisor into the second tank, five times the weight of soil in the second tank. Where automatic sediment sampling occurs, the sediment concentration is determined for each sample. Since the time that each sample was taken during the storm is known, the data can be integrated over time to give a sediment graph.

Investigation of sediment production in a catchment or drainage basin must be carried through an elaborate layout of erosion plot investigation in the stream slopes of various orders.

23.3.4 Soil Conservation

(a) Erosion Control

Soil conservation design most logically follows a sequence of events (Fig. 23.3) beginning with a thorough assessment of erosion risk, followed by designing a sound landuse plan based on what the land is best suited for under present or proposed economic and social conditions, land tenure arrangements and production technology and what is compatible with the maintenance of environmental stability.

However, the approach of soil conservation varies from place to place and also based on type of land use. For instance, erosion control in cultivated land is dependent upon good management which implies establishing sufficient crop cover and selecting appropriate tillage practices. Thus soil conservation relies strongly on agronomic methods combined with sound soil management whilst mechanical measures play only a supporting role. On the whole, the conservation strategies are aimed at establishing and maintaining good ground cover. The details are given in Table 23.7.

Further, it is recognised that strategies for soil improving traditional systems instead of imposing entirely new techniques from outside and on enhancing land husbandry (Fig. 23.4, Table 23.7).

In addition there are a number of mechanical field practices used to control the soil erosion. Three methods are nomally employed in conjunction with agronomic measures:

- *Contouring* i.e., carrying out ploughing, planting and cultivation on the contour can reduce soil loss from sloping land compared with cultivation up and down the slope.
- *Contour* bunds i.e., these are earth banks 1.5 - 2m wide thrown across the slope to act as a barrier to run-off, to form a water storage area on their upslope side.

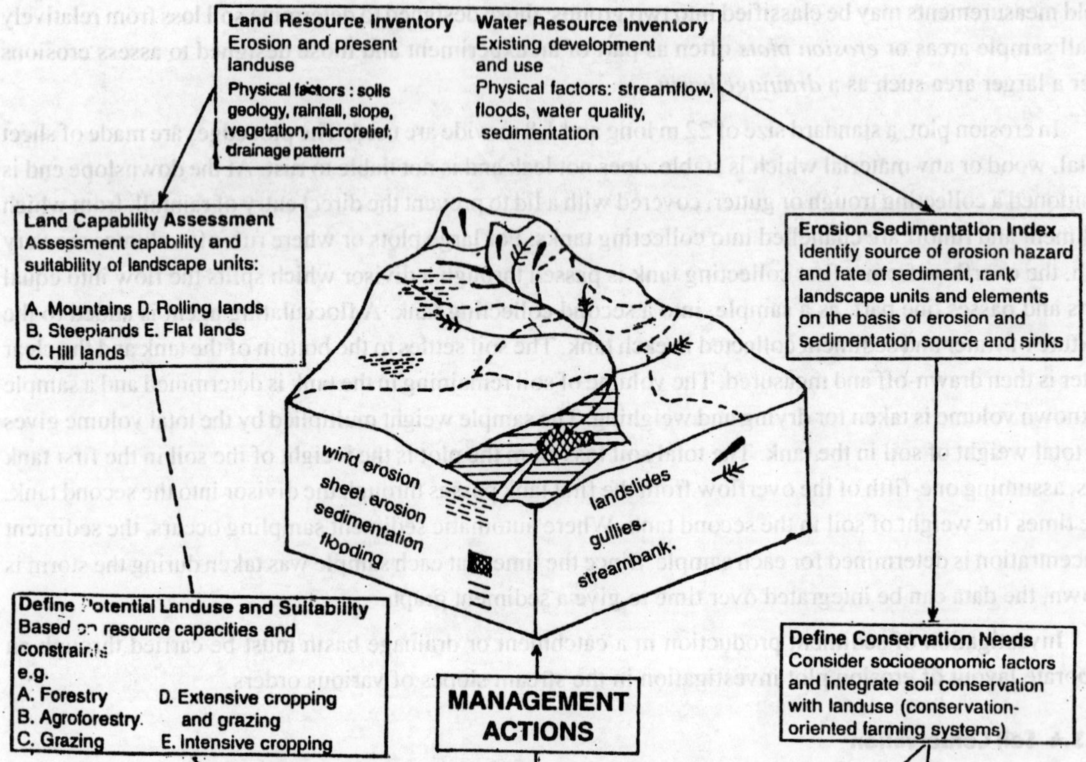

Fig. 23.3: Sequence of events in planning a soil conservation strategy (after Perrens and Trustrum, 1984)

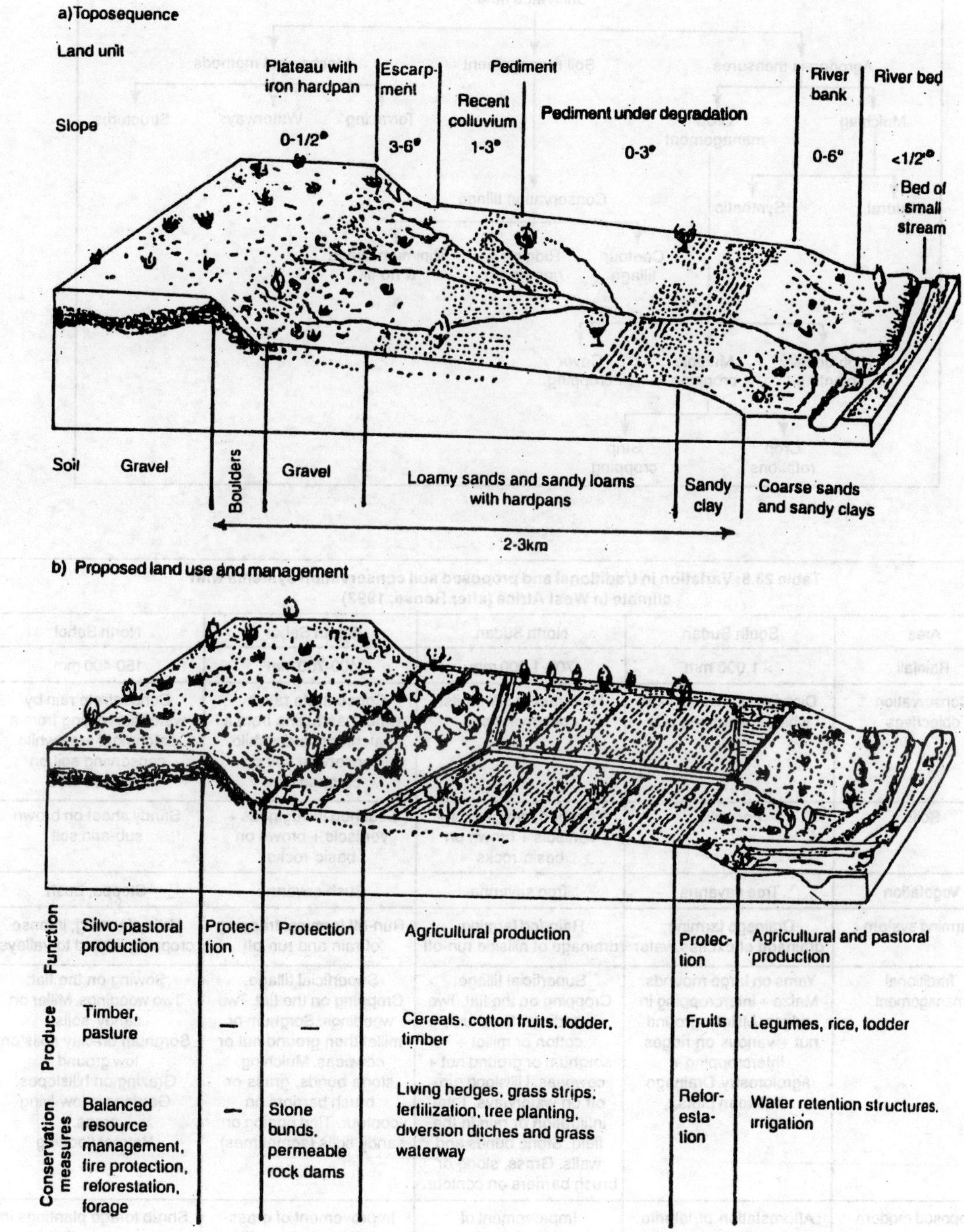

Fig. 23.4: Proposed soil conservation scheme on cultivated land in southern Mali (after Hijkoop, van der Poel and Kaya, 1994)

Table 23.7 : Soil conservation strategies for cultivated land

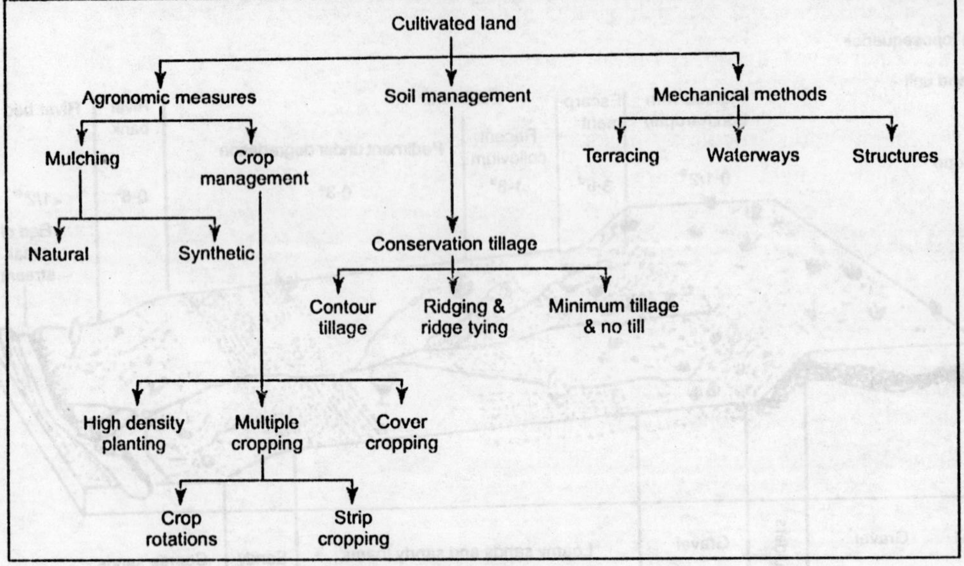

Table 23.8: Variation in traditional and proposed soil conservation systems with climate in West Africa (after Roose, 1992)

Area	South Sudan	North Sudan	South Sahel	North Sahel
Rainfall	> 1,000 mm	700-1,000 mm	400-700 mm	150-400 mm
Conservation objectives	Dispose of excess rain whilst conserving soil	Hold rain *in situ* whilst conserving soil	Concentrate rain by water harvesting from a catchment area while conserving soil on cropped land	Concentrate rain by water harvesting from a catchment area while conserving soil on cropped land
Soils	Ferralitic	Leached ferruginous + vertisols + brown on basic rocks	Leached ferruginous + vertisols + brown on basic rocks	Sandy sheet on brown sub-arid soil
Vegetation	Tree savanna	Tree savanna	Bush savanna	Steppe, bush
Farming system	Drainage farming; drainage of excess water	Rain-fed farming; drainage of hillside run-off	Run-off farming; trapping of rain and run-off	Valley farming; intense cropping limited to valleys
Traditional management	Yams on large mounds. Maize + intercropping in ridges. Millet + ground nut + various on ridges. Intercropping + agroforestry. Drainage between plots.	Superficial tillage. Cropping on the flat. Two weedings. Sorghum, cotton or millet + sorghum or ground nut + cowpeas. Hillslope run-off on waterways. Total infiltration or rain in the field. Stone bunds and walls. Grass, stone or brush barriers on contour.	Superficial tillage. Cropping on the flat. Two weedings. Sorghum or millet then ground nut or cowpeas. Mulching, stone bunds, grass or brush barriers on contour. Tied ridging on sandy soils (sometimes).	Sowing on the flat. Two weedings. Miller on sandy soils. Sorghum on clay soils on low ground. Grazing on hillslopes. Gardens in low-lying areas. Retreat flooding
Proposed modern management	Afforestation of laterite screes. Grassed buffer strips. Gully restoration. Protection of rice fields.	Improvement of grassland. Living hedges + stone bunds or grassed buffer strips. Grass waterways.	Improvement of grass-land. Pounds for cattle + supplementary irrigation of gardens. Stone bunds protected by grass. Gully management.	Shrub forage plantings in lunettes or ditches. Grassland management. Pounds for cattle. Trapping run-off on hillslopes. Stone bunds on pediment.

- Terraces—these are earth embankments constructed across the slope to intercept surface run-off and convey it to a stable outlet at a non-erosive velocity and to shorten slope length.
- Waterways—to convey run-off at a non-erosive velocity to suitable disposal part viz., diversion ditches, terrace channels, grass waterways etc.
- Stabilisation structures—this is a specialised structure build up to produce small dams (0.4 to 2 meter height) by locally available materials for gully erosion control.

23.4 WASTELAND

23.4.1 Introduction

Wastelands are lands which are economically unproductive, ecologically unsuitable and subject to environmental deterioration. The official estimate of wasteland in India in 2004 was nearly 63.85 million ha — more than 20% of the country's geographical area. The categorywise wasteland types of India is shown in Table 23.9.

Table 23.9: Category-wise Wastelands of India (2004)

(Area in sq. km.)

Sl. No.	Catetory	Total Wastelands	% of total Geographical Area covered
1.	Guilled and/or Ravinous land	20553.35	0.65
2.	Land with or without scrub	194014.29	6.13
3.	Waterlogged and Marshy land	16568.45	0.52
4.	Land affected by salinity/alkalinity-coastal/inland	20477.38	0.65
5.	Shifting Cultivation Area	35142.20	1.11
6.	Under utilized/degraded notified forest land	140652.31	4.44
7.	Degraded pastures/grazing land	25978.91	0.82
8.	Degraded land under plantation crop	5828.09	0.18
9.	Sands-Inland/Coastal	50021.65	1.58
10.	Mining/Industrial wastelands	1252.13	0.04
11.	Barren rocky/stony waste/sheet rock area	64584.77	2.04
12.	Steep sloping area	7656.29	0.24
13.	Snow covered and/or glacial area	55788.49	1.76
	Total Wasteland Area	638518.31	20.17

Note: 1,20,849.00 sq. km. in Jammu & Kashmir is not mapped and hence not considered for calculating the percentage.

Source: 1:50,000 scale wasteland maps prepared from Landsat Thematic Mapper/IRS LISS II/III Data.

According to the statewise break-up, the state with the greatest percentage of its area that is classified is Rajasthan with 45 per cent, followed by the hill states Jammu and Kashmir, Tripura, Himachal Pradesh, Manipur, Sikkim, Mizoram, Meghalaya and Nagaland. This predominance of the hill states at the top of the list is because of the extent of snow-covered and glacial land in these states (Table 23.10).

Table 23.10: Aerial Estimation of Wasteland, (Statewise) in 2000

		(Area in ha)
1. **Rajasthan**	Salt affected Land	497.640
	Gullied or Ravinous Land	8766.120
	Undulating Upland with or without scrub	7656
	Sandy Area	130193.938
	Barren Hill Ridge or Rock outcrop	7770.840
	Total	154884.538
	Percentage of State area	45.258
2. **Jammu & Kashmir**	Salt affected Land	38.280
	Undulating Upland with or without scrub	76.560
	Jhum or Forest Blank	2373.360
	Barren Hill Ridge or Rock outcrop	153.120
	Snow covered or Glacial Area	70054.141
	Total	72695.461
	Percentage of State area	32.710
3. **Tripura**	Undulating Upland with or without scrub	382.800
	Jhum or Forest Blank	2641.320
	Snow covered or Glacial Area	229.680
	Total	3253.800
	Percentage of State area	31.056
4. **Himachal Pradesh**	Undulating Upland with or without scrub	459.360
	Jhum or Forest Blank	995.280
	Snow covered or Glacial Area	13589.800
	Total	15044.440
	Percentage of State area	27.022
5. **Manipur**	Undulating Upland with or without scrub	306.240
	Jhum or Forest Blank	5014.680
	Barren Hill Ridge or Rock outcrop	344.520
	Total	5665.440
	Percentage of State area	25.340
6. **Sikkim**	Undulating Upland with or without scrub	765.600
	Snow covered or Glacial Area	1077.840
	Total	1843.440
	Percentage	25.256
7. **Mizoram**	Jhum or Forest Blank	3981.120
	Snow covered or Glacial Area	497.640
	Total	4478.760
	Percentage of State area	21.238

Contd.

Table 23.10: Aerial Estimation of Wasteland, (Statewise) in 2000

		(Area in ha)
8. **Meghalaya**	Jhum or Forest Blank	4555.320
	Barren Hill Ridge or Rock outcrop	191.400
	Total	4746.720
	Percentage of State area	21.106
9. **Nagaland**	Jhum or Forest Blank	3253.800
	Total	3253.800
	Percentage of State area	19.686
10. **Gujarat**	Salt affected Land	1110.120
	Gullied or Ravinous Land	191.400
	Jhum or Forest Blank	76.560
	Sandy Area	1492.920
	Barren Hill ridge or Rock outcrop	842.160
	Total	3713.160
	Percentage of State area	8.396
11. **Arunachal Pradesh**	Undulating Upland with or without scrub	191.400
	Jhum or Forest Blank	5512.320
	Snow covered or Glacial Area	7732.560
	Total	13436.280
	Percentage of State area	16.076
12. **Uttar Pradesh**	Salt affected Land	19407.961
	Gullied or Ravinous Land	8842.680
	Undulating Upland with or without scrub	880.440
	Jhum or Forest Blank	2794.440
	Sandy Area	918.720
	Barren Hill Ridge or Rock outcrop	880.440
	Snow covered or Glacial Area	7043.520
	Total	40768.201
	Percentage of State area	13.846
13. **Kerala**	Undulating Upland with or without scrub	803.880
	Jhum or Forest Blank	3483.480
	Barren Hill Ridge or Rock outcrop	918.720
	Snow covered or Glacial Area	114.840
	Total	5320.920
	Percentage of State area	13.690
14. **Tamil Nadu**	Salt affected Land	574.200
	Gullied or Ravinous Land	4210.800

Contd.

Table 23.10: Aerial Estimation of Wasteland, (Statewise) in 2000

		(Area in ha)
	Undulating Upland with or without scrub	6507.600
	Jhum or Forest Blank	3330.360
	Sandy Area	1914
	Barren Hill Ridge or Rock outcrop	1071.840
	Total	17610.800
	Percentage of State area	13.538
15. Odissa	Salt affected Land	153.120
	Undulating Upland with or without scrub	12326.160
	Jhum or Forest Blank	6622.440
	Sandy Area	382.800
	Total	19484.506
	Percentage of State area	12.506
16. Karnataka	Salt affected Land	267.960
	Gullied or Ravinous Land	2220.240
	Undulating Upland with or without scrub	12517.560
	Jhum or forest Blank	3100.680
	Barren Hill Ridge or Rock outcrop	3828
	Total	21934.440
	Percentage of State area	11.436
17. Andhra Pradesh	Salt affected Land	880.440
	Gullied or Ravinous Land	4517.040
	Undulating Upland with or without scrub	18642.359
	Jhum or Forest Blank	880.440
	Sandy Area	1033.560
	Barren Hill Ridge or Rock outcrop	4708.440
	Total	30662.279
	Percentage of State area	11.076
18. Madhya Pradesh	Gullied or Ravinous Land	21283.680
	Undulating Upland with or without scrub	13436.280
	Jhum or Forest Blank	7005.240
	Barren Hill Ridge or Rock outcrop	2411.840
	Total	44137.040
	Percentage of State area	9.966
19. Maharashtra	Salt affected Land	1186.680
	Gullied or Ravinous Land	7388.040
	Undulating Upland with or without scrub	19637.641
	Jhum or Forest Blank	1454.640

Contd.

Table 23.10: Aerial Estimation of Wasteland, (Statewise) in 2000

(Area in ha)

		Barren Hill Ridge or Rock outcrop	995.280
		Total	30662.281
		Percentage of State area	9.962
20.	Goa, Daman and Diu	Undulating Upland with or without scrub	306.240
		Jhum or Forest Blank	38.280
		Total	344.520
		Percentage of State area	9.034
21.	Haryana	Salt affected Land	1110.120
		Gullied or Ravinous Land	191.400
		Jhum or Forest Blank	76.560
		Sandy Area	1492.920
		Barren Hill Ridge or Rock outcrop	842.160
		Total	3713.160
		Percentage of State area	8.396
22.	Bihar	Gullied or Ravinous Land	1875.720
		Undulating Upland with or without scrub	8727.840
		Jhum or Forest Blank	2373.360
		Barren Hill Ridge or Rock outcrop	2335.080
		Total	15312.000
		Percentage of State area	8.806
23.	Punjab (including Chandigarh)	Salt affected Land	1301.520
		Sandy Area	2603.040
		Total	3904.560
		Percentage of State area	7.730
24.	West Bengal	Salt affected Land	1263.240
		Undulating Upland with or without scrub	4249.080
		Jhum or Forest Blank	535.920
		Barren Hill Ridge or Rock outcrop	306.240
		Total	6354.480
		Percentage of State area	7.232
25.	Assam	Undulating Upland with or without scrub	574.200
		Jhum or Forest Blank	2373.360
		Total	2947.560
		Percentage of State area	3.752

The subsequent survey of NRSA (2005) indicate the statewise distribution of wastelands in India (Table 23.11):

Table 23.11: State wise wastelands of India- NRSA (Information as on year 2003)

State	Wasteland (Area: In square km)
Andhra Pradesh	45267.15
Arunachal Pradesh	18175.95
Assam	14034.08
Bihar	5443.68
Chhattisgarh	7584.15
Goa	531.29
Gujarat	20377.74
Haryana	3266.45
Himachal Pradesh	28336.80
Jammu and Kashmir	70201.99
Jharkhand	11165.26
Karnataka	13536.58
Kerala	1788.80
Madhya Pradesh	57134.03
Maharashtra	49275.41
Manipur	13174.74
Meghalaya	3411.41
Mizoram	4469.88
Nagaland	3709.40
Odisha	18952.74
Punjab	1172.84
Rajasthan	101453.86
Sikkim	3808.21
Tripura	1322.97
Tamil Nadu	17303.29
Uttarakhand	16097.46
Uttar Pradesh	16984.16
West Bengal	4397.56
Union Territory	314.38
Total	**552692.26**

- About 120839 sq km of area of J & K is remain unsurveyed
- Total geographical area of India: 3287263 sq km

Source: Wasteland Atlas of India – 2005; NRSA

23.4.2 Causes of Wasteland Formation

Wasteland is formed either by natural ways or by man-made activities. The undulating upland, sandy area, snow covered area, coastal saline area etc., are natural formation while Gullied or Ravinous land, Jhum or Forest blank, Barren Hill-Ridges etc., are formed by man-made activities. There are four major anthropogenic activities that lead to the formation of wasteland areas viz., deforestation, overgrazing, overcultivation and unskilled irrigation. The first three activities strip the land of its protective vegetation cover, hastening the process of soil erosion and land degradation, while unskilled irrigation causes special problems of water logging and salinity. In addition, both cultivated and uncultivated lands suffer from soil erosion and degradation. Mining is another important activity that causes deforestation and land degradation too.

Desert is one of the major natural wasteland area and any activity relating to the process of desertification is considered also to be one of the major activities underlying the wasteland formation.

23.4.3 Reclamation of Wasteland

With the gradual rise of land resource demand, reclamation of wasteland appear to be one of the major tasks of the country. The reclamation and development of wasteland has four major ecological objectives:

1. to improve the physical structure and the quality of the soil,
2. to improve the availability and quality of water,
3. to prevent the shifting of soil, landslides and flooding, and
4. to conserve the biological resources of the land for sustainable use.

The capacity of the land to produce food, fodder, fuel and industrial raw materials is determined both by biological factors—the climate, soil, hydrological system etc.—and by social and economic factors such as land management practices and inputs like irrigation, fertilisers and seeds.

What is the potential productivity of wastelands and how can this be measured?—this depends on what is meant by productivity, which like many other ecological terms is often confused in common usage. One definition of the productivity of an ecosystem is the amount of biomass "fixed" in a given time. This is a continuous process, but is measured as the amount of food manufactured per unit of time (day, month, year).

Table 23.12: The productivity and energy yield of selected plant species in India

Species	Tonnes/ha./year	GJ/ha./year
Sweet sorghum	20	335
Exotic forage sorghum	68	1,250
Water hyacinth	154	2,680
Sugarcane	112	2,000
Hybrid-poplar (short rotation) three year old	20	342
Coppice crop — two year old	8.3	144
Eucalyptus species	39.1	680
Algae (fresh water pond culture)	88	1,460
Casuarina species (7-10 year rotation)	200	4,950
Lucerne (legume)	100 cu. m. per ha. per year	4,600

Source: Ghosh and Bisaria, 1981 (GJ = 10^9 Joules, measure of energy)

There are a number of ways and means by which wasteland can be reclaimed for productive purpose in more efficient manner. Among them watershed management, wasteland afforestation and social forestry programmes are noteworthy. In addition for improvement of land condition, control of soil erosion and enhancement of soil fertility a good number of measures has to be taken up.

Deforestation for timber and fuel wood extraction and subsequent conversion of forest land to agricultural or urban areas poses a major threat for land degradation and also quicken the process of wasteland formation. In wasteland reclamation process all these causative factors have to be addressed properly to arrive in a positive solution.

Some case studies of wasteland reclamation and development in India are described below.

(a) Microwatershed development in Sukhomajri (Punjab)

Sukhomajri is a small village which lies along the gorged river bed of the Sukhna Cho (hill torrent) in the Shivalik foothills 35 km from Chandigarh. For many years after the 1968 monsoon, when several hectares of fertile land plunged 12-15 metres into a deep gorge at one end of the village, Sukhomajri had been precariously perched, literally at the edge of a precipice.

Its 455 inhabitants, most of whom were very poor Gujjars, lived by grazing their cattle and goats on hill slopes sparsely covered with grass and shrubs. Many men went to work at tractor, machine tool and cement factories outside the village. Sukhomajri had no irrigation, water and electricity. Its only resource were its 10 ha. land (half owned by individual households) and over 400 goats and cattle. Although being aware that grazing the animals in certain areas was aggravating the erosion and affecting the production of their rainfed crops of maize, wheat and legumes, the villagers saw no alternatives.

In 1974-75 the Central Soil and Water Conservation Research and Training Institute at Chandigarh took on the task of desilting the city's famous Sukhna lake, 60 per cent of which had been filled with silt from the Shivaliks, brought down by the Sukhna Cho. P. R. Misra, Project Director of the Institute, surveyed the catchment of the lake and, while tracing the course of the Sukhna Cho, chanced upon Sukhomajri.

Misra devised a system of check dams which created three small reservoirs, with diameters varying from 60-120 meters. The heavy monsoon rain water is channeled into the reservoirs so that it can be used for drinking and irrigation. Misra convinced the villagers not to graze their cattle in the watershed areas by allowing them to cut grass from there instead. The plantation of trees and grasses in the catchment area helped to regenerate the watershed. Further erosion at the 15 meter gorge at one end of the village was also checked with vegetation.

Instead of one rainfed crop, two and even three crops are now being grown in a year and grass production has increased 12-fold. A unique method of sharing water has been evolved, by which each household irrespective of the amount of land it owns or the crops it grows is entitled to an equal-value water coupon. Those who do not want water, such as the landless, may loan or sell their coupons to others who want more water. This works in Sukhomajri because the village is homogeneous and the landholdings are fairly uniform.

Sukhomajri is a success story. However, irrigation has led to some inequalities. The larger farmers with lands in the command area have been using the water even in the kharif season as a result of which small farmers whose lands are further away from the reservoirs do not get enough water in the rabi season when water is scarce (CSE, 1982).

Although considerable amount of money was spent on it—till 1982, a sum of Rs. 80,4750.00 — perhaps more than will be available for many other villages, Sukhomajri shows the effectiveness of soil conservation in a catchment. The cost of restoring the catchment is also much cheaper than dealing with the consequences downstream. For instance, Rs. 27 million was spent on dredging Sukhna lake (CSE, 1982).

Those who worked on the project also point out that in the initial stages the villagers were not cooperative. It was only when the improved irrigation resulted in increasing the yield from their fields that they became interested in investing in conservation measures. The project took several years to tackle the technical problems (1974-79), to generate the trust of the villagers and to develop the system of water distribution (1980-83). All the key persons spent much time with the villagers, learning from them before introducing new methods. The entire process lasted more than a decade (SPWD, 1984).

(b) Afforestation programmes

Compared to the effort that went into improving the productivity of cultivable land and controlling soil erosion for improving agriculture, much less attention was paid to improving the productivity of forest lands. This was despite the many policy recommendations emphasising the need to improve the availability of firewood. For instance, the National Planning Committee (which was set up before independence with Jawaharlal Nehru as its Chairman but submitted its report in 1948) considered afforestation in relation to soil conservation problems. It pointed out that "the lack of forests within easy distance of the most populous districts" resulted in "the enormous waste of the only source of cheap manure by burning cowdung". A look at the expenditure of forestry shows how minimally policy statements, in fact, shaped programmes. Out of the total expenditure of Rs. 1177 million in the 1951-69 period, the farm forestry cum fuelwood-scheme (introduced in the Third Plan) accounted for only 2.2 per cent (NCA, 1976b).

The National Commission on Agriculture (NCA) in its interim report recognised the limitations of the farm forestry programmes and introduced the concept of social forestry.

This had five objectives (Ministry of Agriculture, 1973):

- firewood supply to rural areas and replacement of cowdung as fuel;
- small timber supply;
- fodder supply;
- protection of agricultural fields against wind;
- recreational needs.

The NCA's recommendations were adopted by the Fifth Plan (1974-1979) and the country's major social forestry programmes were launched soon after.

In the Sixth Plan (1980-1985) a number of new schemes were started in an attempt to come to grip with the ever-increasing problem of wastelands and the diminishing resources from these lands. A programme of Rural Fuelwood Plantations begun in 1980-81 to grow firewood as near as possible to the consumption points. Other schemes were the *Tree for Every Child Programme* and the formation of ecodevelopment task forces (mainly of ex-servicemen) to restore damaged hill ecosystems through soil conservation and afforestation.

In recognition of the enormity and the urgency of the problem of wastelands, the then Prime Minister Rajiv Gandhi in a broadcast to the nation (1985) called for a halt to deforestation by developing a people's movement for afforestation and announced the setting up of a National Wastelands Development Board with the objective of bringing 5 million ha of wasteland annually under firewood and fodder plantations. The National Land Board was reconstituted as the National Land Use and Wastelands

Development Council, under the chairmanship of the Prime Minister. This has two bodies under it. The National Wastelands Development Board, which is located in the Ministry of Environment, Forests and Wildlife and the National Land Use and Conservation Board, located in the Ministry of Agriculture and Rural Development, provide a national focal point for policy formulation, coordination, monitoring and evaluation of programmes.

(c) Social forestry : Trees for People

No government programme in India has been surrounded by as much confusion—or has received as much media attention—as social forestry. The programme, devised to increase the availability of firewood, fodder and small timber to the rural poor, is the subject of intense controversy. Both its approach and its achievements are matters of dispute. Tree planting for the rural poor is a component of the country's National Programme and the government claims spectacular progress, with targets being achieved or even exceeded (Ministry of Agriculture, 1984c). On the other hand, critics say that the programme has been distorted to benefit the upper classes and diverted away from its objectives (CSE, 1985).

Although the social forestry programmes vary from state to state, most include one or more of four components:

- strip plantations on road, rail and canal sides (government lands) by the Forest Department;
- village woodlands on government and community land by the Forest Department and by community self-help;
- rehabilitation of degraded forest land by forest departments;
- farm forestry on private land (including farm bunds) and leased government land (including wastelands) with assistance and saplings provided by the forest department.

The programme began in 1974 and during the Fifth Plan Rs. 373 million were allocated, roughly half of what had been recommended by the National Commission on Agriculture. Of the 1.2 million ha planted during the Fifth Plan period, 50 per cent was under the social forestry programme. In the Sixth Plan this was increased to Rs. 3,518 million, 51 per cent of the total forestry budget of Rs. 6,925 million (which itself was more than the total cumulative forestry budget in the previous 30 years — Rs. 4,820 million).

The Sixth Plan of social forestry programmes were carried out under three categories:

- Central Government programme for social forestry, which included a Rs. 500 million rural fuelwood plantations scheme in 157 fuel-deficit districts;
- Externally assisted projects in nine states (with a government allocation of Rs. 1,600 million);
- State government schemes (Rs. 1,418 million).

During the Seventh Plan a National Umbrella Project was envisaged, extending the foreign-assisted social forestry projects throughout the country and involving an investment of Rs. 800 million (Guhathakurta, 1984).

Though the planting activity under the various components of social forestry has on the whole been satisfactory, the programme has failed to achieve its broader objectives on three counts:

(a) It has subsidised the richer farmers;

(b) It has not produced wide-ranging social and environmental benefits;

(c) In some cases it has worsened the position of the poor.

Because tree-planting can be an attractive financial proposition where there is a market for the wood, the benefits from the programme are appropriated by bigger farmers. The initial success of social forestry in Gujarat, Uttar Pradesh, Haryana and Punjab was because of the response of medium farmers (with more than 2 ha.) and large farmers (more than 4 ha.), coupled with the rising price of wood. Surveys have shown that the percentage of farmers with holdings under 2 ha, who have received free or low-cost seedling, varies from 38.5 per cent in Haryana and 42.5 per cent in Gujarat to 64.7 per cent in Jammu and Kashmir and 80.8 per cent in West Bengal (Guhathakurta, 1984).

The failure of social forestry to fulfil the basic needs of the poor was highlighted by a study in Kolar district, Karnataka (Shiva et al, 1981). The eucalyptus plantations under the programme feed big rayon and paper industries; while some farmers benefit from the high prices the mills pay, agricultural labourers have been hard hit. Because the big farmers are growing eucalyptus instead of the traditional ragi, there is less food available locally, pushing food prices up, and because eucalyptus requires less care, agricultural workers are losing their jobs.

The social forestry programme in Tamil Nadu has been similarly criticised. In one instance at Acharapakkam village in Chengalpattu district in 1980, 5,000 eucalyptus hybrid saplings were uprooted by the villagers (CSE 1985). The state's ingenious programme to plant the exposed bed of the tanks (called "foreshores") also ran into trouble. Tamil Nadu has 38,000 tanks (rainwater collection reservoirs) irrigating a total area of 50,000 ha. The tanks are shallow with sloping banks and vary in size from small ponds to reservoirs of over 1,000 ha. Many were built in the Eleventh century in the late Chola period. The stored water is let out into the fields in summer, exposing an increasingly large area of foreshore. This land is already used for grazing and to grow vegetables, but the plan was to plant trees. The villagers opposed this. The poramboke or common lands belonging to the government are also not available for planting because they have been encroached upon for cultivation. Eventually, the government had to accept the position that there was no clear social justification in withdrawing land from the poor in order to provide them with fuel at a later date.

In contrast to large social forestry programmes which have problems in reaching the most needy, new schemes are being developed to reach the poorest directly. Under a social security scheme in Rajasthan, villagers are being leased degraded forest land to plant trees. The poorest individuals were identified and are to be given 2 ha. each, every year for 15 years, making 30 ha. in all. They were provided with 1,600 saplings for the initial planting on two hectares. Besides each gets a salary of Rs. 250 per month, Rs. 500 to build a house and a cash incentive for trees surviving into the second and third years. They are also entitled to 20 per cent of the profits from the trees they grow. In the Udaipur area, 435 individuals have joined the scheme. The scheme is now being revised and it is intended to give leases of one ha. each year, building up to a maximum of 10 ha. But 1,200 saplings will be planted per ha. Here too a problem has been that some of this land is used for grazing.

23.4.4 The West Bengal Experience : Joint Forest Management

One state which has succeeded in taking an ambitious social forestry programme to poorer sections of the village community is Marxist-ruled West Bengal.

The per capita forest area in West Bengal, 0.02 ha., is among the lowest in the country; the population density of 600 per square kilometer is second only to that of Kerala. Its livestock population is estimated at 19 million. The state's annual requirement of household energy is 10 million tonnes of fuelwood equivalent (f.w.e.). Of this one fifth comes from commercial sources and one tenth from Government forests. Of the rest of the 8 million tonnes of f.w.e., 55 per cent is firewood and charcoal, 20 per cent is agricultural waste and 25 per cent animal dung. To meet the cooking energy needs of the present population would require planting more than 1 million ha. (Guhathakurta, 1984).

In 1981 West Bengal launched a World Bank assisted social forestry project, to cover 93,000 ha. This includes plans to rehabilitate 15,000 ha. of degraded forest, strip plantations on 20,000 ha. and village woodlots on 6,000 ha. During the first three years of the programme 30 per cent of the 42,000 villages in the state had been covered. The area covered was 20,800 ha.

There are two reasons for the rapid success of social forestry in the state: the effective interaction between the village panchayats and the Forest Department and the allocation of government land to tribals and other landless poor. Like other successful community programmes, this one too has had a fairly long "gestation" period. In 1973 the West Bengal Panchayat Act introduced a three-tier system of elected village panchayats, block-level committees and district level panchayats. The panchayats are highly politicised, organised and effective. In 1975 the State Panchayat Department sponsored a social forestry scheme (under a local leadership scheme) to motivate both the panchayats and village youth to plant trees. By 1980, 1,200 ha. of community plantations had been established and 5 million seedlings had been distributed for individual planting.

For the World Bank-assisted project, 1,200 part-time village-level motivators (outside the regular forestry service) were appointed by the Forest Department. These motivators and the primary school teachers in the villages are the links between the Forest Department and the people. The sharp rise in the number of participants from 20,915 in 1981 to 85,000 in 1983 reflects the optimistic manner in which social forestry has spread among small land owners in the state (Guhathakurta, 1984).

Much of the success has been in the degraded forest areas in the south west of the state. Here the soils are 'poor in ingredients' or 'poor in quality' and laterised, with a hard layer of *kankar* 30 cm below the surface. In east Midnapore district 0.4 ha. plots of such wasteland have been allocated to landless people who may use it to grow trees or crops. Rain-fed agriculture, however, is very difficult in this area, but the return from the trees—mainly *Eucalyptus tereticornis* and *Acacia auriculiformis*—is expected to be extremely high. The landless do not get the title to the land but have full rights to the use of the land and the profit from it. They also continue their normal occupation as agricultural labourers, preparing the pits for plantations only in the dry season when they do not have other work. The Forest Department supplies the seedlings, a small amount of fertiliser and a cash incentive for surviving plants at the end of the second and third seasons.

As the tree-grower does not have to make any financial investment, the programme is attractive to the poor. Many small landowners have formed blocks of their lands for similar afforestation. Village panchayats are also converting their commonly owned wastelands into tree plantations. Assuming that in the seventh year each hectare of trees will be worth Rs. 1,000 the total value of the assets created would be over Rs. 200 million.

As on December 2001, there were 3,614 FPcs in the state comprising of total number of 415,200 members protecting the total forest areas over 529,945 ha.

23.5 DESERTIFICATION

'Desert' is an area of low rainfall i.e., less than 20 cm per year. It also experiences low temperature in winter and high temperature in summer. It has its own characteristic flora, fauna and microbial communities. The soil is often fertile enough to support agriculture if enough water can be supplied. Many desert areas throughout the world have been converted into productive farmland through irrigation. Thus the desert area is synonymous with 'arid region'. In India, such condition exists in 'Thar' covering an area of 1.3 million sq km. It is not an isolated desert, it has connection with the great Sahara desert of the world. Indian Thar desert cover a vast tract of Rajasthan, a portion of Gujarat, Punjab and Haryana.

On the contrary, desertification is systematic process of conversion of productive land to unproductive arid region. This involve the interplay of climate, edaphic or biotic factors. It is principally a man-made problem which leads to the environmental degradation of fertile lands. The major causes are :

> Inappropriate land use practices i.e., cultivation on the marginal lands affecting adjacent fertile land;
> Over-exploitation of water and land resources through intensive farming;
> Uncontrolled grazing and indiscriminate felling of trees.

There are large number of indicator systems for assessment of degrees of desertification. These are biological, agricultural or social indicator systems. The monitoring of extent of desertification is highly essential for developing combating technologies. There are a number of technologies adopted for combating the desertification problems under different edaphic and climatic conditions. These are shelterbelt creation, sand dune stabilisation, water resource conservation, use of saline water crop production and adoption of ecologically productive landuse.

23.6 DEFORESTATION

Deforestation or forest destruction is a global phenomenon. Every moment a vast majority of forest cover area cleared up for various reasons.

There are several reasons for deforestation in different parts of the world. There is no denial of the fact that deforestation is one of the major causes of land degradation. Thus strategies for regulating forest destruction is very important aspect of restoring land from degradation as stated earlier. The strategy for restoring forest from destruction is shown in the Table 23.13.

Table 23.12 : Forest Restoration Strategies from Destruction

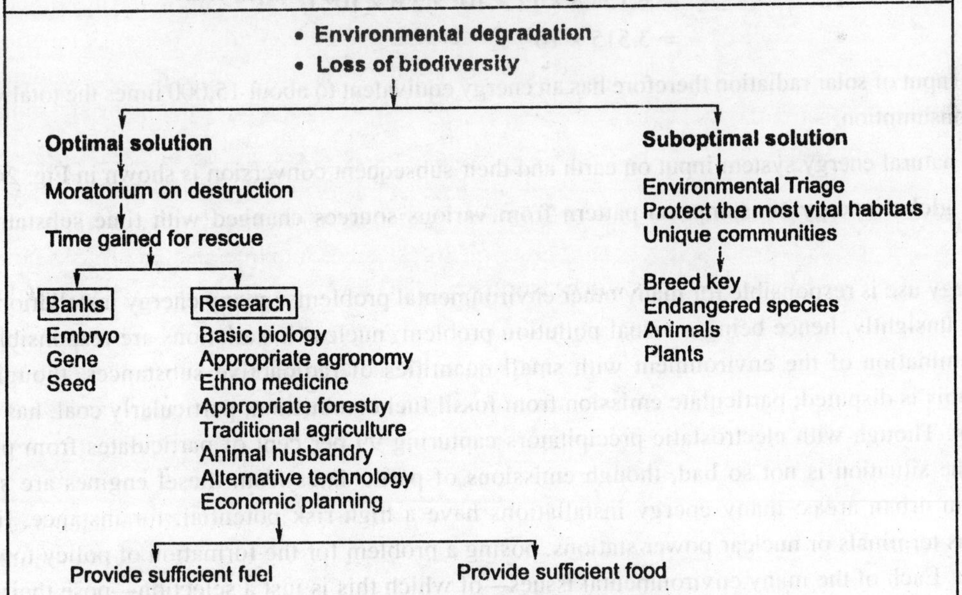

24 ENERGY MANAGEMENT

24.1. SOLAR ENERGY INPUT

Energy flows continuously through the surface of the earth and also through its atmosphere. The dominant component of the input energy is obtained from solar radiation. Each square metre of area at the atmospheric boundary intercepts normal (i.e. perpendicular) radiation at the rate of 1377 W/m^2. This so-called solar constant is not constant but subject to ±3.4% annual variation. The earth's diametric plane has an approximate area πr^2, where r is the mean earth radius in metres. The total solar radiation intercepted is therefore at the rate of 1,377 πr^2 and has a value of 1.73×10^{17} W.

In a year of 365.25 days the total input radiation energy is therefore

$$W_{\text{annual}} = 365.25 \times 24 \times 3{,}600 \times 1.73 \times 10^{17}$$

$$= 5.46 \times 10^{24} \text{ J}$$

This radiant energy is staggeringly large. The total commercial energy consumption in the world in 2005 was 325;835 PJ/year.

$$W_{\text{world consumption}} = 8{,}135.8 \text{ million tonnes of oil equivalent}$$

$$= 8{,}135.8 \times 12 \times 10^9 \text{ kWh}$$

$$= 8{,}135.8 \times 12 \times 10^9 \times 3.6 \times 10^6 \text{ J}$$

$$= 3.515 \times 10^{20} \text{ J}$$

The input of solar radiation therefore has an energy equivalent to about 15,000 times the total world energy consumption.

The natural energy system input on earth and their subsequent conversion is shown in Fig. 24.1.

The global energy consumption pattern from various sources changed with time substantially Fig. 24.2.

Energy use is responsible for many other environmental problems : many energy installations are large and unsightly, hence being a visual pollution problem; nuclear installations are responsible for the contamination of the environment with small quantities of radioactive substances, though the effect of this is disputed; particulate emission from fossil fuel combustion, particularly coal, has been a problem. Though with electrostatic precipitators capturing 99 per cent of particulates from power stations the situation is not so bad, though emissions of particulates from diesel engines are still a problem in urban areas; many energy installations have a high-risk potential, for instance, liquid natural gas terminals or nuclear power stations, posing a problem for the formation of policy towards such plant. Each of the many environmental issues—of which this is just a selection—pose their own individual problems, though CO_2 and acid emissions are the most intractable issues and certainly there are problems of control due to the scale of their effects.

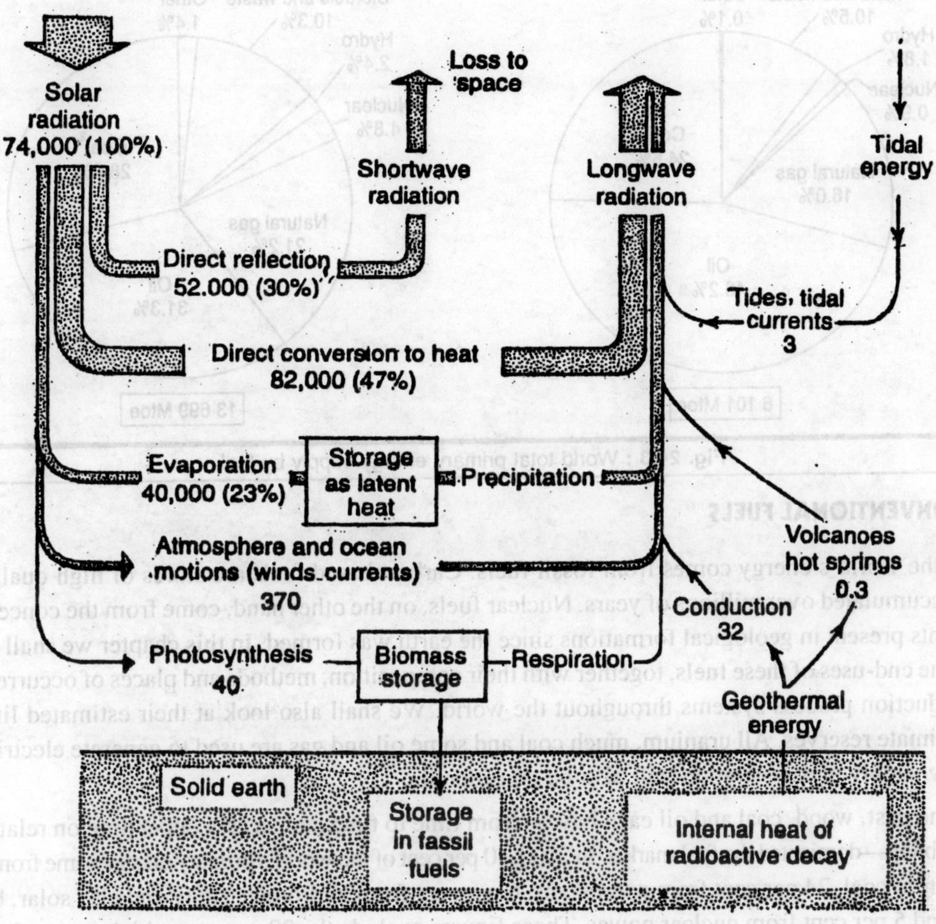

Fig. 24.1 : The natural energy system of the earth

Fig. 24.2 : Global final energy consumption shares – 2005

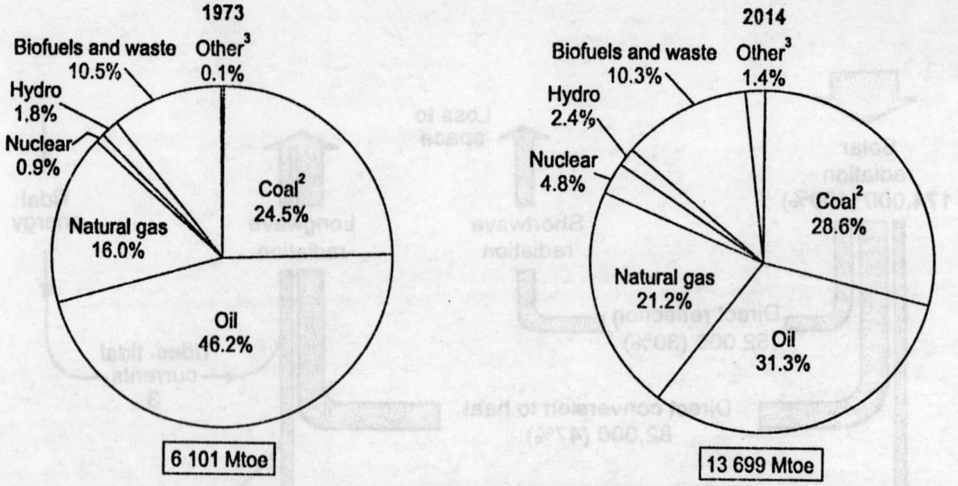

Fig. 24.3 : World total primary energy supply by fuel

24.2 CONVENTIONAL FUELS

Most of the world's energy comes from fossil fuels. Carbon-based fuels are stores of high quality solar energy accumulated over millions of years. Nuclear fuels, on the other hand, come from the concentration of elements present in geological formations since the earth was formed. In this chapter we shall discuss, briefly, the end-uses of these fuels, together with their composition, methods and places of occurrence and their production process systems throughout the world. We shall also look at their estimated life-times and at ultimate reserves. All uranium, much coal and some oil and gas are used to generate electricity. So electricity supply and generation will also be considered.

In the past, wood, coal and oil each have—from time to time and largely depending on relative cost and availability—dominated the fuel market. In 2000, 40 per cent of the world's primary energy came from oil, 25 per cent from coal, 24 per cent from natural gas, 6 per cent from hydro-electricity and also solar, biomass sources and 5 per cent from nuclear power. These figures exclude the 20 per cent which comes from non-traded fuels used in the developing countries. In the immediate future natural gas seems likely to become the most widely used and popular fuel. Large reserves have recently been discovered and its use reduces atmospheric pollution and releases less carbon dioxide than an equivalent amount of coal. (Fig. 24.3)

24.2.1 Oil

Occurrence

Crude oil is a mixture of liquid hydrocarbon compounds sometimes found permeating sedimentary rocks. Its elements are, by weight, carbon 82.2-87.1 per cent, hydrogen 11.7-14.7 per cent, oxygen 0.1-4.5 per cent, nitrogen 0.1-1.5 per cent and sulphur 0.1-5.5 per cent.

Different names, based on the number of carbon atoms in their compounds, are given to products derived from crude oil. They are called gasoline (C_4 to C_{10}), kerosene (C_{11} to C_{13}), diesel fuel (C_{14} to C_{18}), heavy gas oil (C_{19} to C_{25}), lubricating oil (C_{26} to C_{40}) and waxes (over C_{40}).

Oil is generated from organic matter in sedimentary rocks at depths of about 800-5,000 m at temperatures between 66°C and 150°C. Its predominant source material is probably marine organisms. Three steps are involved in the conversion of organic matter to petroleum. The regional share of crude oil production is shown in Fig. 24.4.

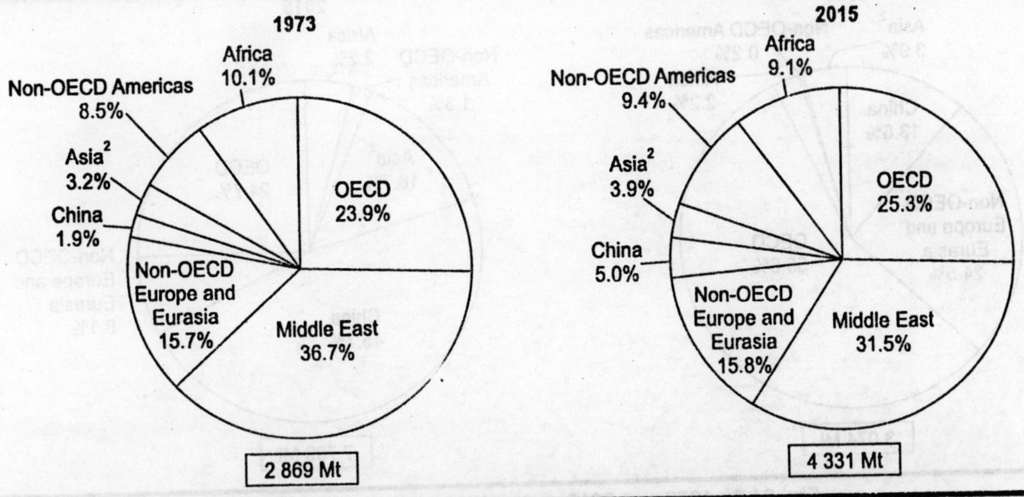

Fig. 24.4 : 1973 and 2015 regional shares of crude oil[1] production

24.2.2 Coal

Occurrence and Distribution

Coal is a complex organic material consisting of fused carbon rings held together by assorted hydrocarbon and other atomic (O, N, S) linkages. Its average composition is something like $C_{10}H_8O$ (this ratio of 10 carbon atoms to 8 of hydrogen can be contrasted with the ratio of 10 carbons to 17.5 hydrogens in crude oil).

It is formed from dead plant material which has accumulated in swamps—usually in estuarine deltaic deposits—and then has been consolidated and altered by increasing temperature and pressure. In a similar evolutionary pattern to oil, the first stage in the conversion process is an anerobic breakdown of the plant material which causes volatile products to be liberated and lost to give a compacted structureless mass of compounds enriched in carbon. The second stage is the process of coalification which proceeds through the ranks of peat, lignite, sub-bituminous coal, bituminous coal and anthracite to graphite. The proportion

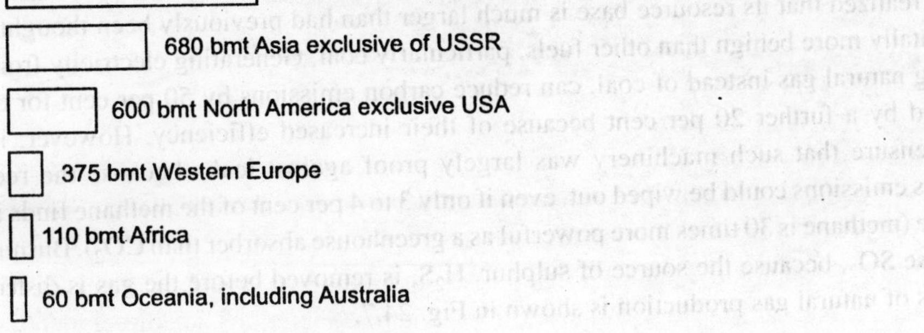

Fig. 24.5 : Estimated world coal reserves, 2000

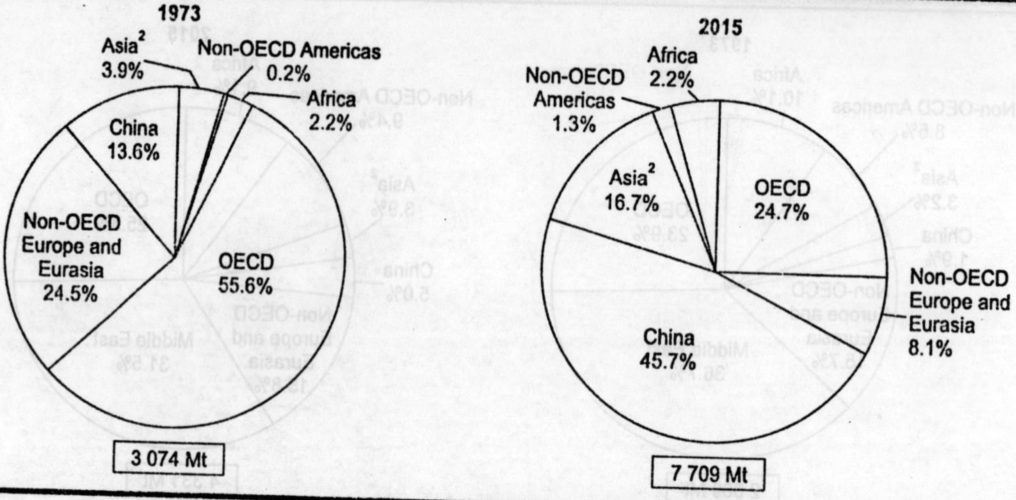

Fig. 24.6 : 1973 and 2015 regional shares of coal[1] production

of carbon is gradually increased in this progression. Calorific values of the various ranks range from 15-26 kJ g^{-1} for low rank lignites, through 31-35 kJ g^{-1} for bituminous coals to 30-33 kJ g^{-1} for anthracite.

Coal deposits are not found before the lower Carboniferous age (about 400 million years ago) and the most important and widespread date from the Carboniferous to the early Triassic age (345 to 200 million years ago) and from the Jurassic to the early Tertiary age (150 to 50 million years ago). In general, the older coals have the highest rank but, depending on the geological history of the deposits this is not necessarily so.

Although coal is widespread, the major deposits are unevenly distributed (Fig. 24.5). Ten countries account for 92 per cent of the currently estimated reserves and resources. Three countries alone, the US, the former Soviet Union and China, have 83 per cent of the resources and 60 per cent of the reserves. All three could become exporters of coal and in any case, compared with rates of use, the total reserves are so large that the uneven spread does not, at the moment, mean supply difficulties. The overall global coal production (regionwise) is shown in Fig. 24.6.

24.2.3 Natural Gas

Increasing interest is being shown in natural gas and there are two broad reasons for this. First, it has recently been realized that its resource base is much larger than had previously been thought. Second, it is environmentally more benign than other fuels, particularly coal. Generating electricity from advanced turbines, using natural gas instead of coal, can reduce carbon emissions by 50 per cent for each unit of production and by a further 20 per cent because of their increased efficiency. However, it would be important to ensure that such machinery was largely proof against leaks because the reductions in greenhouse gas emissions could be wiped out, even if only 3 to 4 per cent of the methane finds its way into the atmosphere (methane is 30 times more powerful as a greenhouse absorber than CO_2). Burning methane does not release SO_2, because the source of sulphur, H_2S, is removed before the gas is distributed. The regional shares of natural gas production is shown in Fig. 24.7.

Occurrence and global distribution of gas

Natural gas consists of hydrocarbons with one to five carbon atoms, together with small amounts of other gases as impurities. It was formed under essentially the same kind of conditions as oil, anerobic

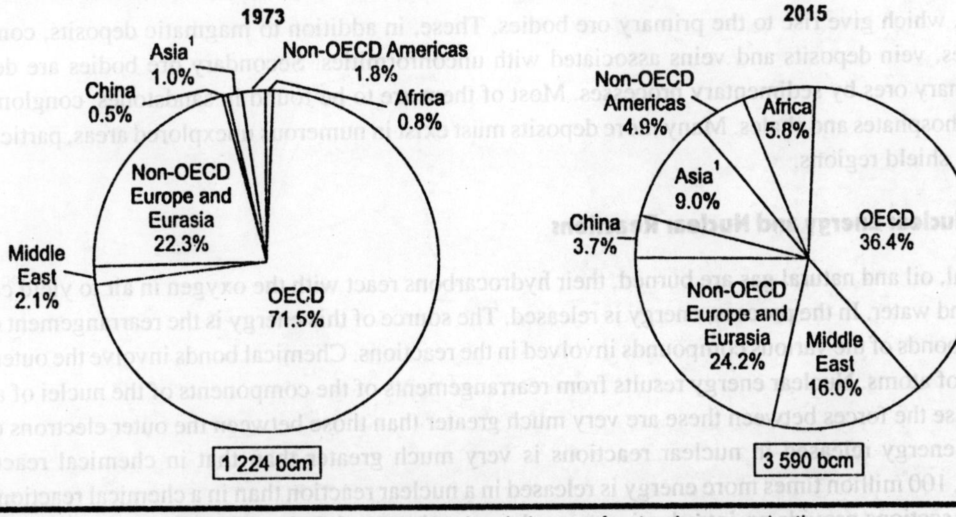

Fig. 24.7 : 1973 and 2015 regional shares of natural gas production

decomposition of organic matter under heat and pressure assisted by bacteria. Marine organisms are the primary source material for oil, but natural gas can be formed both from land plants and from marine organic material. Gas can be formed in very young deposits as, for example, marsh gas in swamps; it can also be formed in association with coal deposits, particularly those of the Permo-Carboniferous, with crude oil and as 'thermal' gas below the oil window. Therefore the depths and areas of sedimentary basins which may hold gas far exceed those of oil. Gas fields, like oil fields, are not distributed uniformly and differ in size and geographical concentration, but because of its more diverse origins gas is more widespread.

Gas, found on its own in 'dry' wells, is called 'non-associated gas'. It is also found dissolved under pressure in oil in a reservoir or as a 'gas cap' over an oil pool; in these cases it is called 'associated gas'. It has been estimated that about 72 per cent of world reserves are non- associated, 17 per cent are dissolved and 11 per cent are gas caps.

Natural gas fields are classified in terms of their size as supergiants with 10×10^{12} ft^3, (2.83×10^9 m^3), or more, of recoverable gas and giant (world class) gas fields with between 3×10^{12} ft^3 (8.5×10^{10} m^3) and 10×10^{12} ft^3 (2.8×10^{11} m^3) of recoverable gas.

24.2.4 Uranium

Uranium, in a variety of reservoirs with widely differing concentrations, occurs extensively in the earth's crust, with an average presence of about 2 ppm. Granites contain up to 20 ppm, currently exploited ores typically have over 350 ppm, coals commonly have around 20 ppm, (some are known with 500-2000 ppm). Sea water contains 0.0005 ppm. The estimate for the earth's crust as a whole is 2.5×10^{13} tonnes.

Uranium increases in quantity as the ore grade decreases, thus every tenfold decrease in ore grade leads to a three-hundredfold increase in the recoverable amount. The grade of ore mined at any particular time is determined by the price of the open market. Currently the price is not too high because, following the world-wide slow down in nuclear power plant construction, there is very little demand for it.

World distribution of uranium ore

Primary uranium ores are derived from Precambrian sources which have been buried and submitted to high temperatures and pressures. This results in the creation of uranium-rich magmas, or uranium-rich solutions, which give rise to the primary ore bodies. These, in addition to magmatic deposits, comprise pegmatites, vein deposits and veins associated with unconformities. Secondary ore bodies are derived from primary ores by sedimentary processes. Most of them are to be found in sandstones, conglomerate placers, phosphates and shales. Many more deposits must exist in numerous unexplored areas, particularly in remote shield regions.

24.2.5 Nuclear Energy and Nuclear Reactions

When coal, oil and natural gas are burned, their hydrocarbons react with the oxygen in air to yield carbon dioxide and water. In the process, energy is released. The source of this energy is the rearrangement of the chemical bonds of the various compounds involved in the reactions. Chemical bonds involve the outermost electrons of atoms. Nuclear energy results from rearrangements of the components of the nuclei of atoms and because the forces between these are very much greater than those between the outer electrons of the atom the energy released in nuclear reactions is very much greater than that in chemical reactions. Typically, 100 million times more energy is released in a nuclear reaction than in a chemical reaction. The following sections provide an introduction to nuclear reactions and their role in nuclear reactors.

There are two possible ways by which energy is released from radioactive materials:

1. *By fission* — whereby a nucleus with A greater than 120 is split into two nuclei which have A (A denotes atomic weight) around 60 and because the new nuclei have greater binding energies, energy must be released in the process.

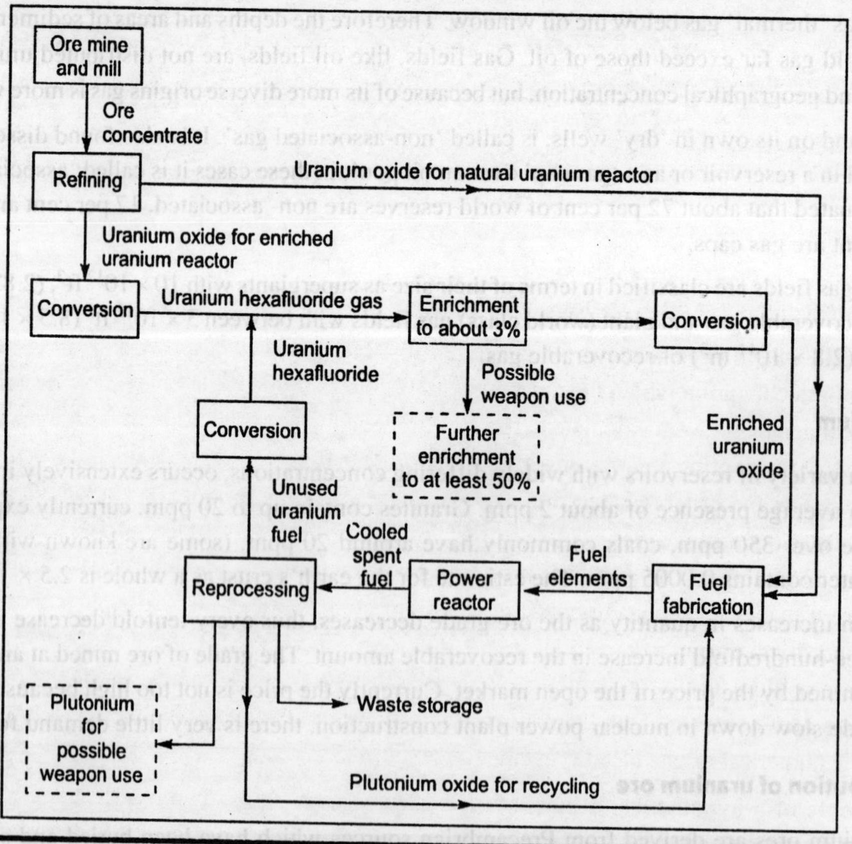

Fig. 24.8 : Steps involved in nuclear power production (*Source* : Coggle, 1983)

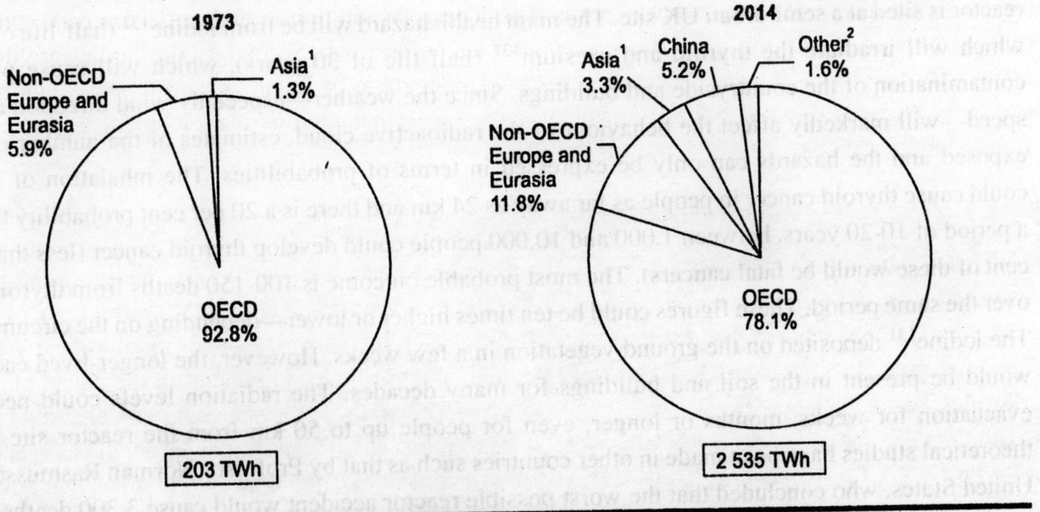

Fig. 24.9 : 1973 and 2014 regional shares of nuclear production

2. *By fusion* — whereby two light nuclei (A below about 20) fuse together to form a nucleus with A up to around 60. As the binding energy curve is much steeper for nuclei with low values of A than for those with values of A above 120, the energy released in the process of fusing together two light nuclei is much greater than that in the fissioning of a heavy nucleus.

The steps involved in the energy generation from enriched uranium shown in Fig. 24.8.

24.2.6 The Risk of Nuclear Accidents

Although accidents involving the release of radioactive material occur at any stage of the nuclear fuel cycle, most attention has been focused on reactor accidents. This is not because reactor accidents are more likely, but because the potential consequences for general public of some reactor accidents are much greater than from accidents at other stages in the fuel cycle.

Several studies have been made in which the probability reactor accidents of various degrees of severity were estimated, and in which the environmental impacts of the radiation release associated with these hypothetical accidents were calculated. The type of accident considered involves the reactor safety features that control the cooling and running of the reactor. If they fail, the temperature of the reactor core rises and the fuel melts and may breach the main containment above the reactor. Such a 'meltdown' results in the release of large quantities of radioactive material. It needs to be stressed that in a thermal reactor the fuel is near to the most active state and there is no way in which the nuclear assemble can go supercritical and explode like a nuclear weapon. In contrast, the plutonium fuel used in fast-breeder reactors could possibly go supercritical during a core collapse meltdown. In general, studies have suggested that a large accidental release could cause large numbers of early fatalities and large numbers of latent cancers, but the probability of such a release is low. However, it is now generally agreed that there is a large range of uncertainty in the numerical results quantifying the risks of an accident, as reactor accidents have highlighted.

One example of a hypothetical accident at a nuclear power station is provided by the UK Royal Commission on Environmental Pollution (1976). It assumes that 10 per cent of the gaseous and volatile fission products are released from a 1000 MW(e) nuclear reactor as a cloud of radioactive material. The reactor is sited at a semi-urban UK site. The main health hazard will be from iodine^{131} (half-life of 8 days), which will irradiate the thyroid and caesium^{137} (half-life of 30 years), which will cause prolonged contamination of the countryside and buildings. Since the weather—especially wind direction and wind speed—will markedly affect the behaviour of the radioactive cloud, estimates of the number of people exposed and the hazards can only be expressed in terms of probabilities. The inhalation of iodine^{131} could cause thyroid cancer in people as far away as 24 km and there is a 20 per cent probability that, over a period of 10-20 years, between 1,000 and 10,000 people could develop thyroid cancer (less than 10 per cent of these would be fatal cancers). The most probable outcome is 100-150 deaths from thyroid cancer over the same period. These figures could be ten times higher or lower—depending on the circumstances. The iodine^{131} deposited on the ground vegetation in a few weeks. However, the longer-lived caesium^{137} would be present in the soil and buildings for many decades. The radiation levels could necessitate evacuation for weeks, months or longer, even for people up to 50 km from the reactor site. Similar theoretical studies have been made in other countries such as that by Professor Norman Rasmussen in the United States, who concluded that the worst possible reactor accident would cause 3,300 deaths, 45,000 illnesses and 1,500 fatal cancers which would appear decades later (Hawkes et al, 1986).

The Chernobyl Accident, 1986

If the public's confidence in the safety record of the nuclear power industry faltered following the Three Mile Island (USA) accident, the accident at Chernobyl in the former Soviet Union caused nuclear reactor safety to be completely re-examined. After only 4,000 reactor years worldwide, Chernobyl approached the worst conceivable nuclear reactor accident, which was previously estimated to be of a probability somewhere near 1 in 10,000 or even 1 in a million per reactor year. The Chernobyl accident released 50-100 million curies of radioactivity, according to the official Soviet report, or as much as 287 million curies of radioisotopes with half-lives greater than one day according to Hohenemser et al, (1986). This makes it 100 times worse than the Windscale accident and 1,000 times worse than the Three Mile Island accident. The accident at the graphite moderated, light-water-cooled pressure tube reactor at Chernobyl occurred on 26 April 1986 when an explosion produced an uncontrollable fire which lasted several days and led to vast quantities of radionuclides being lifted high into the atmosphere. The plume of radionuclides was swept across Europe during the following 7-10 days, exposing up to 400 million people in 15 nations to high levels of ionising radiation (Fig. 24.10). In areas where heavy rainfall scavenged the radionuclides from the atmosphere, radiation levels peaked up to several hundred times higher than background levels.

The accident resulted in around 30 Soviet citizens immediately dying from radiation sickness (the majority of fatalities—most reactor workers or firemen—had received whole-body doses ranging from 600 rem. to 1,600 rem.). Another 237 people were treated for radiation sickness, some of whom needed bone-marrow transplants to restore their white blood-cell count. The number of casualties near the reactor was less than might be expected because the radiation plume was made especially buoyant by the intense fire and so, like the effect of a tall industrial stack, local deposition of radionuclides was reduced whereas the long-range transport of radionuclides was enhanced. Within a 30 km radius of Chernobyl, 135,000 people were evacuated. Evacuation was not prompt because the Soviet authorities were initially unwilling to publicise the seriousness of the accident and because few Soviet citizens had access to private transport. These people now face annual medical inspection for the rest of their lives to assess what the radiation may have bequeath to them in terms of a legacy of cancer or birth defects. Of the evacuated, 24,000 are estimated

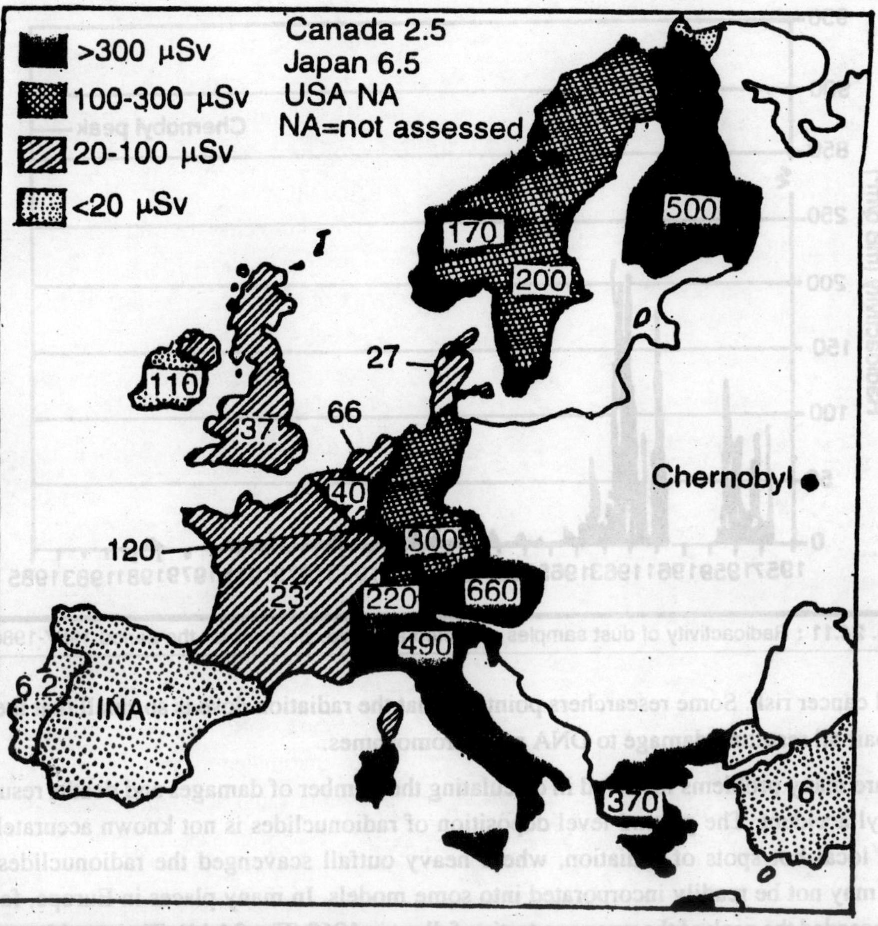

Fig. 24.10 : National radiation dose equivalents in various countries as a result of the Chernobyl accident in 1986 (*Source* : OECD 1987)

to have received an average 0.43 Sv (43 rem.). Some of the consequences for this group include over 120 spontaneous leukaemias during the next 50 years and increased risk of genetic disorders in future generations.

Estimates of the number of expected cancers resulting from exposure to the radionuclides vary quite markedly, according to the assumptions made in the model employed. Estimates range from several thousands to tens of thousands of thyroid cancers arising from the exposure to iodine^{131} and a similar number of cancers elsewhere in the body from caesium^{137}. Only a few per cent of the thyroid tumours would be fatal, whereas perhaps half of the cancers from caesium^{137} would be fatal. Hawkes et al, (1986) estimate the likely number of deaths in the Soviet Union at between 5,000 and 10,000. The official Soviet estimate of the projected cancer deaths in western Russia after Chernobyl is put at 4,000-5,000 due to exposure to iodine^{131} and as many as 40,000 from exposure to caesium^{137}. Hohenemser and Renn (1988) suggest 28,000 additional cancer deaths worldwide will be the result of Chernobyl contamination, including 10,000-15,000 in the Soviet Union, 1,500-3,000 in Poland, 4,000 in Western Europe and only 10-20 in the United States. Anspaugh et al, (1988) estimate 17,000 additional fatal cancers in Europe in the next 50 years but highlight that the Chernobyl legacy adds an increment of only 0.01 per cent to European

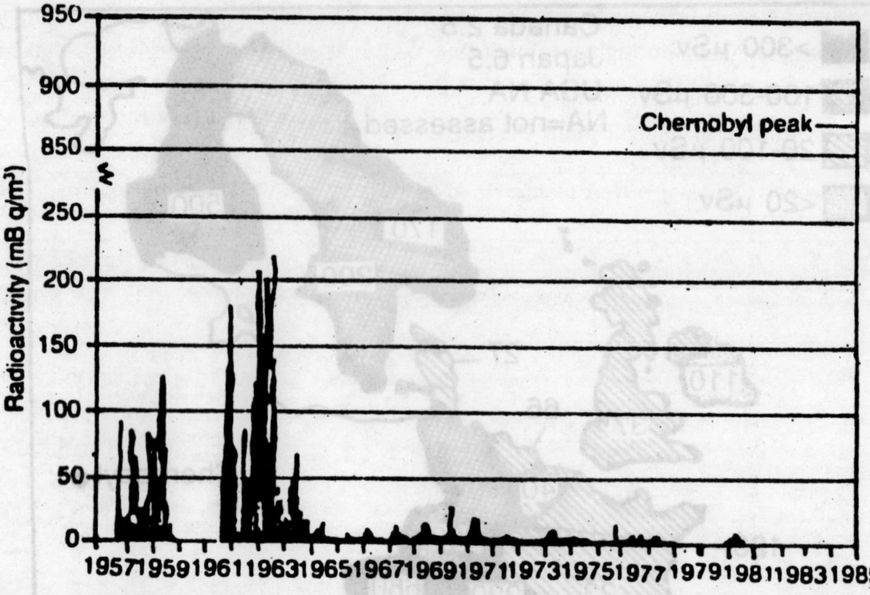

Fig. 24.11 : Radioactivity of dust samples collected at Uccle, Belgium for the period 1957-1986

lifetime fatal cancer risk. Some researchers point out that the radiation dose is so small that the body can perfectly repair all resulting damage to DNA and chromosomes.

There are many problems involved in calculating the number of damages and deaths resulting from the Chernobyl accident. The ground-level deposition of radionuclides is not known accurately into the existence of local hot spots of radiation, where heavy outfall scavenged the radionuclides from the atmosphere, may not be readily incorporated into some models. In many places in Europe, fallout from Chernobyl exceeded the peak of the weapons-testing fallout in 1963 (Fig. 24.11). Their problems associated with estimating the health effects of the Chernobyl radiation release include the difficulty of taking into account all of the radionuclides involved and all of the protective measures adopted by governments and individuals to reduce the radiation dose people received. Protective measures included staying indoors during the passage of the radiation cloud; the removal of the outer leaves and washing of vegetables and the washing and peeling of fresh fruit; the ban by Western European countries of the importation of fresh food from areas of Eastern Europe within 1,000 km of Chernobyl; the restriction in consumption of fresh milk and free-range eggs; the removal of cattle and sheep from open pasture or a ban on their sale (for example, 2.5 million sheep in parts of north Wales were prohibited temporarily from being moved or slaughtered; Kerr et al, 1989); the taking of prophylactic potassium iodide tablets to saturate the thyroid with iodine so as for minimise the amount of radioactive iodine being taken in (this measure was adopted most extensively in Poland); the advice not to drink fresh rainwater; and the warning to the public about possible contamination during outdoor activity.

As with most other air pollution disasters, the Chernobyl accident highlighted many inadequacies in the way in which the authorities reacted to the accident. The then Soviet government failed to provide a warning of the accident to European governments and when information was offered it was incomplete. Many European governments—such as France and the United Kingdom—failed either to provide adequate answers to the public's urgent questions or introduce precautionary measures quickly enough. Radiation monitoring networks were inadequate in many European countries for the purpose of rapidly pinpointing radiation hot spots, though, in hindsight, a rainfall map of Europe for the period provided useful approximation.

In spite of nuclear plant accidents dependence on nuclear reactors for energy harvesting increased day by day. An estimate of nuclear plant's establishment in various countries is depicted in Table 24.1.

Table 24.1: Nuclear power plants world-wide, in operation and under construction, as of November 28, 2016

Country	In operation		Under construction	
	Number	Electr. net output MW	Number	Electr. net output MW
Argentina	3	1,632	1	25
Armenia	1	375	–	–
Belarus	–	–	2	2,218
Belgium	7	5,913	–	–
Brazil	2	1,884	1	1,245
Bulgaria	2	1,926	–	–
Canada	19	13,526	–	–
China	36	31,402	20	20,500
Czech Republic	6	3,930	–	–
Finland	4	2,752	1	1,600
France	58	63,130	1	1,630
Germany	8	10,799	–	–
Hungary	4	1,889	–	–
India	22	6,225	5	2,990
Iran	1	915	–	–
Japan	43	40,290	2	2,650
Korea, Republic	25	23,133	3	4,020
Mexico	2	1,440	–	–
Netherlands	1	482	–	–
Pakistan	4	1,005	3	2,343
Romania	2	1,300	–	–
Russian Federation	36	26,557	7	5,468
Slovakian Republic	4	1,814	2	880
Slovenia	1	688	–	–
South Africa	2	1,860	–	–
Spain	7	7,121	–	–
Sweden	10	9,651	–	–
Switzerland	5	3,333	–	–
Taiwan, China	6	5,052	2	2,600
Ukraine	15	13,107	2	1,900
United Arab Emirates	–	–	4	5,380
United Kingdom	15	8,918	–	–
USA	99	98,868	4	4,468
Total	**450**	**391,915**	**60**	**59,917**

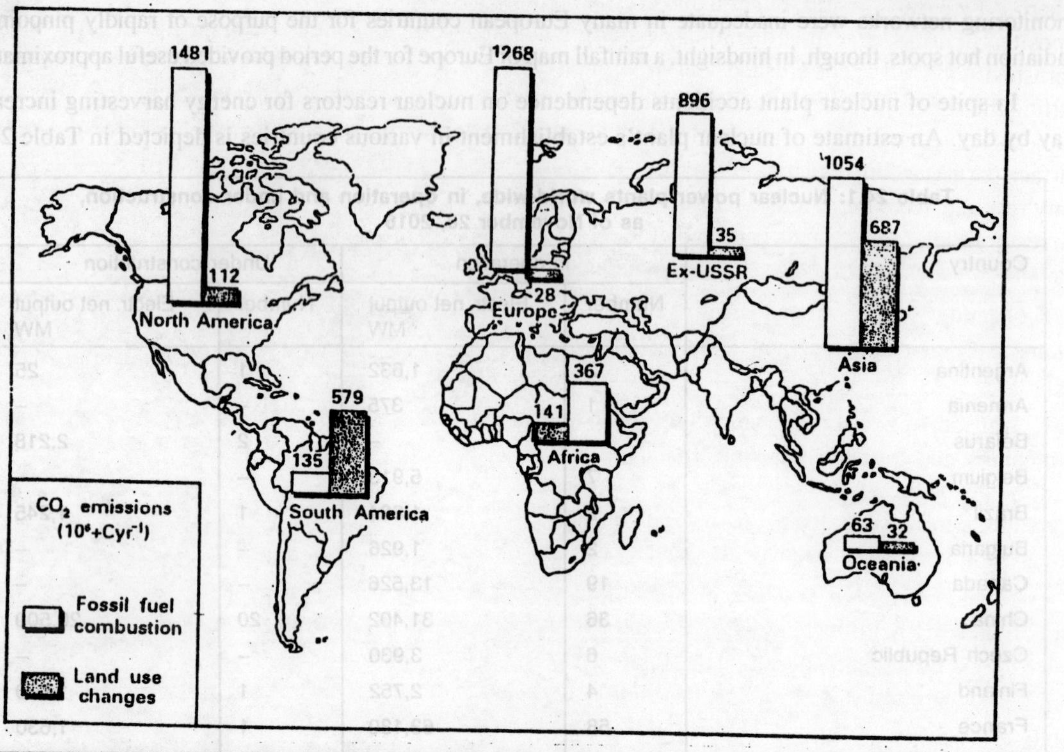

Fig. 24.12 : Emissions of CO_2 to the atmosphere as a result of human activity during 1980

On the whole, the CO_2 emission from various fossil fuel combustion and subsequent land use changes in different continents of the world is fairly significant (Fig. 24.12). In fact, two types of emissions are recognised here: those resulting from the burning of fossil fuels and those resulting from land use changes (e.g. forest clearance). However, the advantages and disadvantages to nuclear power plant is given:

Advantage & Disadvantages of Nuclear Power Plants

Advantages	Disadvantages
• Minimum air and water pollution except hot water discharge in higher volumes; • Efficient use of fuel resources; • Relatively low operating cost (but installation cost of a 1,000 Mw plant is over Rs. 3,000 crores.	• Radioactive waste disposal require special measures; • Possibility of accident produce serious health hazards; • Long term storage of waste & decommissioning of plants will involve high costs; • Requires an international security system to prevent diversion of nuclear materials to weapon use.

24.2.7 Environmental Effects of Fossil Fuel Use

The use of fossil fuels (coal, oil, or natural gas) almost always entails some environmental degradation and risk to human health. The negative impacts start at the mining phase, continue through transport and refining, and conclude with the fuel combustion and waste disposal process.

Underground (shaft) mining of coal has claimed thousands of lives throughout the centuries, because of explosions of methane gas in the mine shafts and also because of the inhalation of coal dust by the miners. At the mine mouth, mineral and crustal matter, called slag, is separated from the coal. The slag is deposited in heaps near the mine, thus despoiling the landscape. In recent times, most of the mined coal is crushed at the mine. The crushed coal is "washed" in a stream of water in order to separate by gravitational settling the adherent mineral matter, thereby beneficiating the coal prior to shipment. The "wash" usually contains heavy metals and acidic compounds, which if not treated, contaminates streams and groundwater. Surface (strip) mining, which is much more economical than shaft mining, causes scarring of the landscape. Only recently were there introduced some regulations in the United States and other countries to ensure the restoration of the wounds caused by the removal of overburden of the coal seams, and recovery of the pits and trenches after the coal has been exhausted.

On and off-shore oil and gas drilling produce piles of drilling mud, along with an unsightly vista of oil and gas derricks. Also, there is the risk of crude oil spills, and explosions or fire at oil and natural gas wells.

Transport of coal, oil, and gas by railroad, pipelines, barges, and tankers carry the risk of spills, explosions, and collision accidents. In the refining process, especially of crude oil, toxic gases are emitted into the air or flared. Usually, some liquid and solid byproducts are produced that may be toxic. Strict regulations must be enforced to prevent the toxic wastes from entering the environment and thereby threatening humans, animals, and vegetation.

The combustion of fossil fuels—whether coal, oil, or gas—inevitably produces a host of undesirable and often toxic byproducts: (a) gaseous and particulate emission into the atmosphere, (b) liquid effluents, and (c) solid waste. In many countries, strict regulations were enacted as to the maximum level of pollutants that can be emitted into the air or discharged into surface waters or the ground from large combustion sources, such as power plants, industrial boilers, kilns, and furnaces. However, the smaller and dispersed combustion devices, such as residential and commercial furnaces and boilers, are not regulated, and they do emit pollutants into the air. While great strides have been taken in many countries to control emissions from automobiles, trucks, and other vehicles, these mobile sources still contribute significantly to air pollution.

Perhaps the greatest longterm threat to the environment is the steadily increasing concentration of carbon dioxide in the atmosphere, for the most part a consequence of the combustion of fossil fuel. CO_2 ane some other so-called greenhouse gases may trap the outgoing thermal radiation from the earth, thereby causing global warming and other climate changes.

The consumption of vast quantities of fossil fuel by mankind causes many deleterious environmental and health effects. These effects start from the mining phase of the fossil fuels, through transportation, refining, combustion, and waste disposal. When coal is mined in deep shafts or in strip mines, mineral matter is separated at the mines by milling and washing. The residual slag may contain toxic, acidic, and sometimes radioactive material, which needs to be properly disposed of without endangering the environment and humans. Oil and gas well derricks, on and off-shore, are aesthetic eyesores and are the cause of oil spillage. Oil and gas pipes may rupture and leak their contents. Oil tankers and barges spill on the average 4 million tons per year of crude and refined petroleum on our waterways and oceans. Oil refineries are sources of: (a) toxic emissions through vents, leaks, and flaring; and (b) toxic liquid effluents and solid waste.

By far, the greatest environmental and health impact is due to fossil fuel combustion in furnaces, stoves, kilns, boilers, gas turbines, and the internal combustion engine, which powers our automobiles, trucks, tractors, locomotives, ships, and other mobile and stationary machinery. The combustion process emits pollutants through smoke stacks, chimneys, vents, and exhaust pipes, such as particulate matter, oxides of sulfur and nitrogen, carbon monoxide, products of incomplete combustion, and volatile toxic metals. Some of these pollutants are toxic to humans, animals, and vegetation per se, whereas others transform in the atmosphere to toxic pollutants, such as ozone, organic nitrates, and acids. The pollutants are advected by winds and dispersed by atmospheric turbulence over hundreds to thousands of kilometers, affecting sensitive population and biota far removed from the emission sources. Particles, besides containing toxic and potentially carcinogenic agents, often envelope whole subcontinental areas in a haze that reduces visibility and the enjoyment of the landscape and a starry sky.

Great strides have been taken, especially in the more affluent countries, for limiting the emissions of air pollutants from the combustion sources. For example, most particles can be filtered out of the flue gas by electrostatic precipitators or fabric filters. Sulfur oxide emissions can be reduced by wet or dry limestone scrubbers. Nitric oxide emissions can be reduced by catalytic and noncatalytic injections of chemicals into the flue gas. The catalytic converter, which is now applied to gasoline fueled automobiles in many countries, reduced the emissions of automobiles significantly compared to the uncontrolled predecessors. Diesel engines, with proper tuning, are also emitting less pollutants, although a magic box like the catalytic converter has yet to be found for diesel engines. Alas, little or no emission control devices are applied to the myriad of dispersed sources, such as residential furnaces, stoves, and fireplaces, and smaller industrial facilities.

24.3 NON-CONVENTIONAL ENERGY

The principal non-conventional energy sources are:

- Solar
- Wind
- Tidal
- Geothermal
- Bioenergy
  - — Biomass
  - — Bioalcohol
  - Biogas.

The estimated global availability of such energy sources are depicted in Table 24.2.

Table 24.2: Estimate of Global Non-conventional Renewable Energy Resources

Resources	Estimated as currently recoverable	Resource base
* Solar radiation	1000 TW	90,000 TW
* Wind	10 TW	1,200 TW
* Wave	0.5 TW	3 TW
* Tides	0.1 TW	30 TW
* Geothermal flow	—	30 TW
* Biomass standing crop	—	450 TW
* Geothermal heat stored	> 50 TW years	10^{13} TW years
* Kinetic energy stored in atmospheric and Oceanic Circulation	—	32 TW years

(1 TW year = 3.16×10^{19} Jules) Source: Sorensen, 1990

The estimates of potential capacities of renewable energy sources in India is given box below:

Source	Approx. Potential
Biomass energy	19,500
Solar energy	20,000
Wind energy	47,000
Small hydropower	15,000
Ocean energy	50,000

Source: India Ministry of Non-Conventional Energy Sources, 2005

24.3.1 Photovoltaics

Solar cells

A solar cell converts light to electricity. The cells produce both electric current and voltage by the 'photovoltaic effect' and the technology is often given the name 'photovoltaics'. Solar cells are electronic devices made from semiconductors such as silicon, usually in the form of thin slices (wafers) about 1/4 mm thick. The positive contact is a layer of metal on the back of the wafer, while the negative contact on top of the cell must collect the current but also allow as much light as possible to enter the device. The top contact is usually made in the form of a grid, as shown in Fig. 24.13.

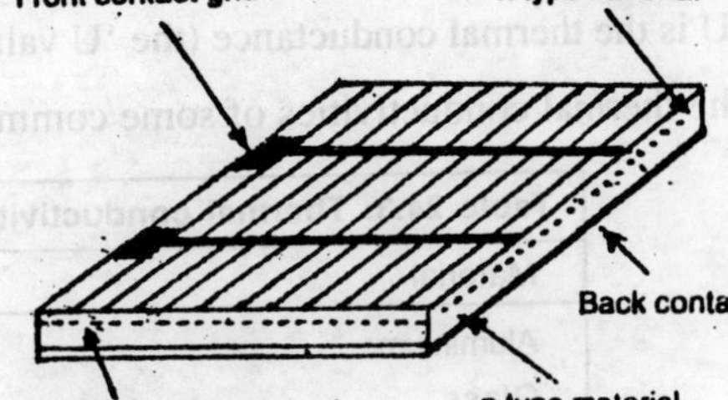

Fig. 24.13 : Schematic diagram of a solar cell

The process by which the absorption of light in a solar cell can produce DC electrical power is represented by Fig. 24.13. Note that a cell must produce both current and voltage to generate power, since power = current × voltage.

In bright sunlight, a 10 cm square cell will give an output of about $\frac{1}{2}$ volt and 3 amps, about $1\frac{1}{2}$ watts of power. Manufacturers quote the output of their cells for a sunlight intensity of 1 kilowatt per square metre (similar to that of the Sahara Desert at noon). This standard output is labeled 'peak watts' or 'W' and is measured at a standard temperature of 25°C.

The power output of a solar cell varies with the intensity of light. If the light is halved, then the current output will also halve but the voltage will drop by only a few per cent. The voltage also depends on the temperature of the cell and decreases by about 0.5 per cent for every one degree rise above 25°C.

24.3.2 Solar Heating

When an object absorbs sunlight it gets hot. This heat energy is usable in various ways, to provide space heating or cooling, to provide domestic or industrial hot water, to boil water or other fluids for industrial processes, or to drive engines. The solar heating or cooling of housing or working spaces can be accomplished without any machines or moving parts, but simply by the appropriate design of the buildings. Such buildings first appeared in Greece over 2000 years ago and were common throughout the last millennium in Islamic architecture. The advent of cheap and abundant fossil fuels led to the abandonment of these traditions, but they are now being re-established on a firm scientific basis and termed passive solar technologies.

Passive Solar Heating

All rooms with a window facing the sun are heated when the sunlight shines in. These unplanned solar gains contribute between 10-20 per cent of the annual space heating of a typical house in the UK. Some houses have conservatories or greenhouses which are designed to make use of solar heating and there is a growing number of houses and other buildings which are designed to minimize the total energy needed for space heating and to maximize the contribution which solar energy can make.

When the temperature inside a building is higher than that of outside, heat can be lost by conduction, convection and radiation. Conduction of heat depends on temperature gradients across, and the thermal conductivity of, a given material. For a given difference, ΔT, between inside and outside temperatures, the thicker the materials, for example the house walls, the lower the termperature gradient and, hence, the lower the heat losses. Also, the lower the thermal conductivity, k, of the material, the lower the conduction.

In general, the rate of energy loss through a wall of area, A, and thickness, L, is

$$P = kA\Delta T/L = UA\Delta T$$

where U is the thermal conductance (the 'U value') of the construction.

The thermal conductivities of some common building materials are shown in Table 24.3:

Table 24.3: Thermal conductivities of some common building materials

Material	k ($Wm^{-1}\ K^{-1}$)
Aluminium	211
Glass	1.05
Brick	0.6
Pine	0.14
Mineral wool	0.035
Polyurethane rigid form	0.025
Still air	0.026

A window pane of 3 mm glass has a U-value of 350. A 5 cm thick layer of still air between two sheets of glass reduces this to 0.52, showing the benefits of double glazing. A brick wall 10 cm thick would have a U-value of 6 and two such walls a U-value of 3. The cavity between the walls is wide enough to allow convection currents in the air, so heat is transported readily from the inner to the outer wall. If the cavity (say 10 cm wide) is filled with polyurethane foam, the U-value of the wall is reduced to 0.25. The U-value of a typical plastered plaster board ceiling is around 20. Laying 10 cm of mineral wool over the ceiling reduces the U-value to around 0.35. A well-designed and well-insulated house would have an overall U-value of around 0.2 to 0.3 $Wm^{-2}\ K^{-1}$.

Hot air rises while cold air sinks, thus producing convection currents. These are very efficient at transporting heat and are the main mechanism by which outside walls lose heat. Inside older houses, convection currents are called draughts and result in substantial heat loss.

As we remarked in the section on solar radiation, all objects radiate energy. The rate at which they do it and the wavelength of the radiation depends on the temperature of the object. The sun radiates mainly in the visible spectrum, while any surface at room temperature radiates in the infrared at a wave length of around 10 μm.

Rooms with large windows present a large area through which infrared radiation from the room can escape. On sunny days a south-facing window gains more energy from the sunlight than it loses through the infrared radiation from the surfaces in the room and larger the window the warmer the room. If the sun is not shining, however, larger the window the more infrared radiation escapes and the cooler the room. At night this effect can be mitigated by drawing curtains or blinds, but this is not sensible during the day. The solution is to use windows which transmit visible radiation but reflect infrared radiation back into the room. Glass itself reflects infrared more than visible light, but it is possible to enhance this effect considerably by special coatings on the glass. This then minimizes the loss of heat by infrared radiation while allowing the use of large south-facing windows to maximize the solar gain. In hot countries, other coatings can reflect much of the solar radiation while transmitting most of the infrared and so minimize the cooling needed to keep the room at a comfortable temperature.

From all this we can define the main requirements for a passive solar building. The south-facing glazed area should be large, to maximize solar gain, while the north-facing glazed area should be small to minimize radiation losses and the construction should have a low overall U-value. To avoid claustrophobia in the rooms on the north side of the building, the windows should be tall but narrow, so that the sky and ground are both visible, while keeping the window area small. The simplest type of passive solar building is shown in outline in Fig. 24.14 and such structures are said to provide 'direct solar gain'.

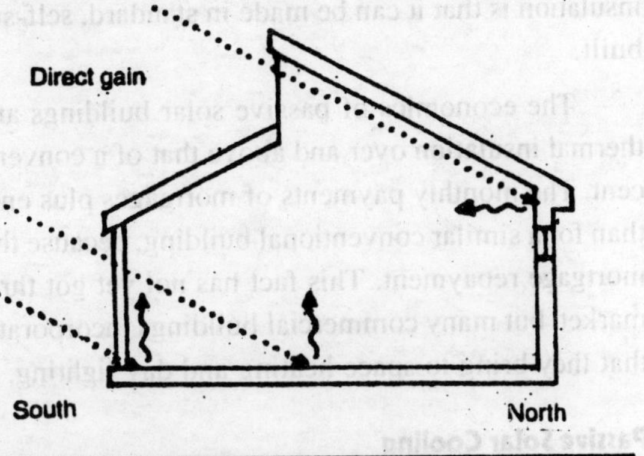

Fig. 24.14 : Passive solar gain

Additional solar gain can be provided in the morning by east-facing windows but west windows may need to be shaded to avoid unwanted solar gain on late summer afternoon. Excessive heating can be prevented by designing the roof structures which overhang the windows to provide shade against the high summer sun, while admitting the low winter sun. Alternatively, a plantation of deciduous trees in front of the south facade would provide considerable shading in the summer but little in winter when the leaves have been shed.

The solar collection area can be expanded to the entire south facade if 'indirect solar gain' techniques are used. The most common of these is a conservatory on the south wall with a means of circulating the warm air throughout the house (Fig. 24.15). A more effective, but expensive method is to glaze all or much of the south facing wall (Fig. 24.16). A black wall absorbs the sunlight and the air between the glazing and the wall rises as it is warmed and is distributed throughout the building. This type of structure was developed by Felix Trombe in the 1950s and is often called a 'Trombe Wall'. A recent development of Trombe Walls has been to use transparent insulation as the glazing, to reduce the heat otherwise lost by conduction and radiation.

Fig. 24.15 : Use of a conservatory for passive solar gain

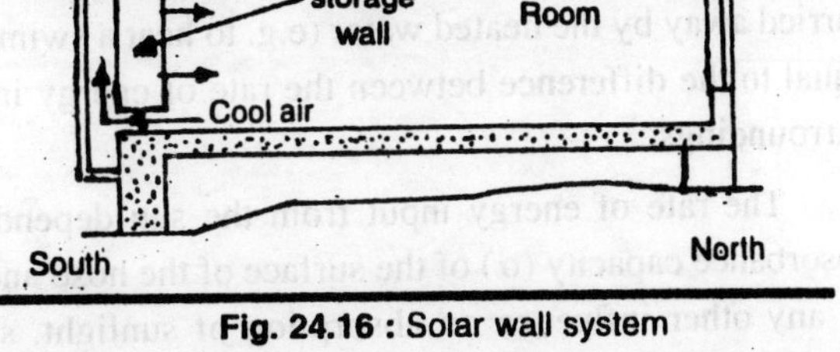

Fig. 24.16 : Solar wall system

Transparent insulation was first noticed in the coats of polar bears. Each strand of fur acts like an optical fibre, transmitting sunlight to their skin. Plastic fibres laid side by side can be glued together into sheets or other shapes. Light is transmitted down the fibres, so the sheet is transparent, but if the fibres are, say, 20 cm long, the U-value of the sheet would be about 0.5 or so. The great advantage of transparent insulation is that it can be made in standard, self-supporting shapes, like bricks, from which walls can be built.

The economics of passive solar buildings are attractive. The additional cost of passive solar and thermal insulation over and above that of a conventional new building is usually not more than 5-10 per cent. The monthly payments of mortgages plus energy costs are usually less for a passive solar building than for a similar conventional building, because the savings in energy costs are larger than the additional mortgage repayment. This fact has not yet got through to builders or mortgage lenders to the domestic market, but many commercial buildings incorporate atria and other passive solar features for the benefits that they bring to space heating and day lighting.

Passive Solar Cooling

Buildings can be designed so that the heat of the sun induces convection currents which draw cool air into the building and so reduce the inside temperature. Islamic architecture has used this principle for centuries and many of its buildings have a 'chimney' which draws up hot air and brings air into the buildings past north-facing surfaces which remain cool throughout the day. A modern variant of this uses a Trombe Wall to create the air movement. Instead of being circulated in the building, the hot air is vented to the atmosphere while the incoming air is cooled by underground, or north-facing, heavy masonry surfaces.

Active Solar Heating

Sunshine can be used to raise the temperature of some working fluid and the increases can vary from a few degrees to over 2,000°C, depending on the type of system used.

Low Temperature Water Heaters

If water is run through a black hose-pipe exposed to sunlight it will come out warmer than it went in. If it is filled with water and both ends are sealed and it is then coiled flat, the water will heat up, on a hot sunny day, to as much as 60°C. The water temperature will rise until the rate at which the hose loses heat to its surroundings is equal to the rate at which it gains heat from the sun. This equilibrium (also called the stagnation temperature) depends on the rate of heat loss as well as the rate of heat gain.

The rate of heat gain depends on the intensity of the sunlight and the efficiency with which it is absorbed by the surface of the hose-pipe. The rate of heat loss depends on the conduction, convection and radiation of energy from the surface of the hose. If water is flowing through the hose, the useful energy is carried away by the heated water (e.g. to heat a swimming pool). This useful energy is delivered at a rate equal to the difference between the rate of energy input from the sun and the rate of energy loss to the surroundings.

The rate of energy input from the sun depends on the intensity of solar radiation I(Wm^{-2}), the absorbance capacity (α) of the surface of the hose and a factor F (between 0 and 1), which takes account of any other influence on absorption of sunlight, such as reflections, surface roughness and surface geometry.

The rate of energy loss to the surroundings depends mainly on conduction at low water temperatures and so varies with the difference between water temperature T_w and ambient temperature T_a. It can be written as $U(T_w - T_a)$, where U is the 'U-value' of the solar collector system.

The rate of delivery of useful energy Q (Watts) can then be written as:

$$Q = \alpha FI - U(T_w - T_a)$$

This is known as the Hotel-Whillier-Bliss equation, after the three people who first derived the equation and used it to study the performance of solar heater cell.

The efficiency of a collector is the ratio of useful heat delivered to the incident solar radiation, i.e. Q/I. So the efficiency:

$$\eta_c = \alpha F - U/I\,(T_w - T_a)$$

To produce collectors with the highest efficiency, we should increase the absorption efficiency αF, reduce the heat loss factors and operate at the highest solar intensity. Fig. 24.17 shows a plot of collector efficiency against $(T_w - T_a)/I$, for different types of collectors. If the U-value of the collector was independent of temperature, the relationships would be straight lines. In fact, at higher temperatures, radiation begins to play a larger role in heat loss and the efficiency falls off more rapidly.

The standard solar water heater consists of a flat metal plate with a black upper surface and pipes, in good thermal contact, attached to the back. The black surface is exposed to the sunlight and so is heated. Water flowing through the pipes is warmed by heat conducted from the plate to the pipes. A great deal of heat would be lost if the plate and the pipes were exposed to winds or convection currents, so they are enclosed in a well-insulated box, with a glass or plastic front cover. For the highest efficiency, the front may be double glazed, although this adds to the cost.

The upper surface of the plate is coated with a black paint or other coating which absorbs sunlight efficiently. Most black surfaces are good absorbers of visible light and also efficient radiators of infrared. They therefore radiate heat through the glass front. Selective surfaces are good absorbers of sunlight but poor radiators of infrared and so reduce the radiation heat losses by factors of five or more. Such surfaces give rather higher efficiencies at low temperature, but, more importantly, they maintain efficiency up to much higher temperatures, as shown in curve C of Fig. 24.17.

The evacuated tube collector has the lowest heat loss of any of these devices. It consists of a glass tube, sealed at both ends and evacuated of all air, with a heat collector tube running down the middle. The bottom half of the outer tube is silvered on the inside, so sunlight is focused on to the inner collector tube. The collector tube may be hollow, with water flowing through it, or it may be a heat pipe (a device which transmits heat very efficiently along its length). Since there is no air between the collector tube and the outer tube, there can be no convection or conduction loss, except at the ends where the collector tube

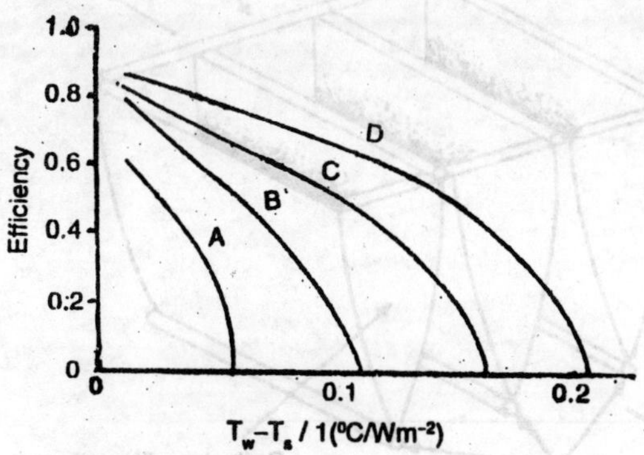

A Simple unshielded collector such as black hose-pipe.
B Standard flat-plate solar water heater.
C Highly insulated flat-plate collector with selective surface.
D Evacuated tube collector with selective surface.

Fig. 24.17 : Efficiency curves for solar water heaters

passes through the seals. Radiation losses are reduced by using selective black surfaces on the collector tube.

A solar collector will consist of 10 to 20 of those evacuated tubes side by side in a rectangular box, having a common water inlet and outlet. Although they are more expensive than the standard flat-plate collector they are much more efficient at higher temperatures and can easily produce low pressure steam in high solar intensities and produce very hot water even in winter sunshine.

In use the solar collector must be plumbed in to a building's water supply and heating system. The fluid in the solar collector is a mixture of water and anti-freeze, so it must be kept separate from the domestic water supply. When the fluid in the solar collector is hotter than the water in the pre-heat tank, the circulating pump is switched on. The hot fluid from the collector is sent through the heat exchanger to heat the water in the pre-heat tank, from where it is either used directly or stored in the hot water tank.

If very large areas of collector are required, it can be more cost-effective in some circumstances to use a solar pond. Any area of shallow water exposed to sunlight is heated by the solar energy absorbed by the water and the bed of the pond. Much of this heat is lost in normal ponds by convection currents which bring the hot water to the surface, where it loses heat to the atmosphere. The density of the water may be increased by adding salt. The concentration of salt is made to be higher at the bottom of the pond and decreased gradually toward the surface. Even if the water at the bottom of the pond is hotter than at the surface, it remains more dense because it has more salt dissolved in it. In this way, convection currents are suppressed. In these ponds the bottom layer of brine can become quite hot and the upper layers of water act as a good insulator, making an efficient solar collector. The concentrated brine can be pumped through a heat exchanger to heat water or other fluids which are then often used to drive a turbine and generating electricity.

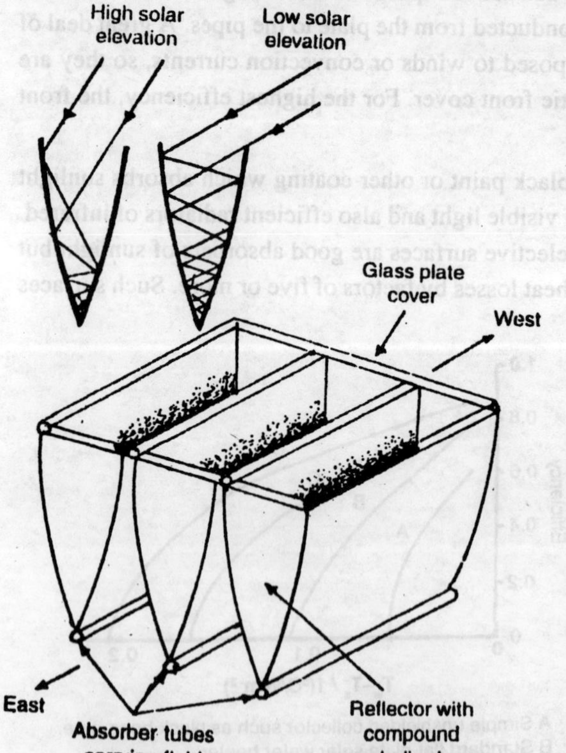

Fig. 24.18 : Compound parabolic reflector concentrating sunlight

Concentrating Solar Collectors

If higher temperatures are required that can be achieved by flat-plate collectors, then sunlight can be concentrated by mirrors or lenses. For low concentration ratios, mirrors in the shape of a long trough whose sides form compound parabolas can be used to collect sunlight wherever the sun is in the sky without having to move the mirrors. The troughs are oriented east-west (Fig. 24.18) so that the sunlight enters the mirrors throughout the day. The mirrors can collect sunlight from both high and low solar elevations, i.e. in both summer and winter (Fig. 24.19). The shape of these compound parabolic reflectors is quite complex and not so easy to manufacture. Furthermore, although the mirrors accept sunlight at low solar elevations, the energy density of the solar irradiance is reduced in proportion to the cosine of the angle of the sun.

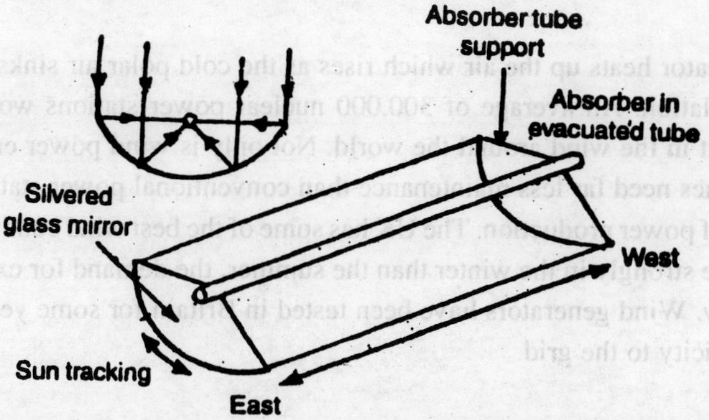

Fig. 24.19 : Tracking parabolic reflector

The large commercial systems use one-axis tracking of long parabolic trough-shaped mirrors. The absorber is usually a metal tube with a 'super black' surface coating enclosed in an evacuated glass tube. The working fluid is usually an oil with good chemical stability at high temperatures. The largest of such systems is in California where the collected energy is used to drive turbine generators to supply electricity. A total of 350 MW is installed, making this the largest solar electricity generating utility in the world.

Power Towers

Very high temperatures can be reached in solar furnaces, where square kilometres of mirrors are controlled to reflect sunlight on to a single absorber mounted high up on a tower—the 'power tower'. The mirrors, each around 10 m^2 in area, are mounted on a two-axis tracking system and each is controlled by computer to reflect and focus an image of the sun on to an absorber. There may be hundreds of such mirrors all focusing the sunlight on to one absorber, which can reach very high temperatures. The heat is carried away by a fluid, often a liquid metal is used to raise steam to generate electricity. These systems are extremely impressive—both technically and visually—but have not yet proved to be cost-effective.

The major solar application in India is given:

Major Solar Application Achievement in India	
Solar Lanterns	3,50,531 Nos.
Home Lighting Systems	1,52,180 Nos.
Street Lighting Systems	41,403 Nos.
PV Pumps	4,199 Nos.
Power Plants (stands alone) & other systems	1,132 kWp
Grid Connected Power Plants	1.6 MW
Solar Cookers	5,00,000 Nos.
Aditya Solar Shops	26 Nos.
Source: DNES, 2001	

24.3.3 Wind Energy

Solar radiation over the equator heats up the air which rises as the cold polar air sinks. This sets up the basic pattern of wind circulation. An average of 300,000 nuclear power stations would be needed to generate power equal to that in the wind around the world. Not only is wind power enormous, but it is pollution free and the turbines need far less maintenance than conventional power stations, all of which makes it an attractive form of power production. The UK has some of the best wind conditions in the world and, as the wind blows more strongly in the winter than the summer, the demand for extra energy can be matched by the extra supply. Wind generators have been tested in Britain for some years and supply an increasing amount of electricity to the grid.

Siting a wind generator

Windmills have existed for between one and two millennia and although modern turbines bear as little relation to them as a Concorde does to a coracle, what they do have in common is that the amount of energy either of them produces is heavily dependent on where they are sited. Placing a wind generator at the foot of a hill would be pointless, but if it is placed at the top where the wind has been forced to accelerate by the rise in the ground, it will work very successfully.

Ground drag also makes a large difference. Lesser the drag, faster the wind. A wind generator sited on a sheet of ice or concrete will produce its expected output, but put behind a forest it will not. We may notice that concrete council estates are windy while forests are calm. By the same token, wind generators are built quite tall—50 m above the ground is usually high enough to escape the drag factor.

Wind Energy Converters

Wind Energy Converters (WECs) convert the energy of the wind into practically useful energy, usually electricity, and they come in many shapes and kinds. For machines with a horizontal axis (Fig. 24.20) the dominant force is lift and the rotor blades may be in front of (upwind) or behind the tower (downwind). Upwind turbines need a tail or some other mechanism to point them into the wind. Downwind turbines may be quite seriously affected by the tower, which can produce wind shadow and turbulence in the path of the blades. Both kinds of machine with a capacity greater than about 50 kW are turned into the wind by an electric motor. Two or three bladed rotors are those most commonly used for generating electricity, multi-bladed rotors which have a high torque in light winds are used for pumping water and other tasks calling for low frequency mechanical power.

WECs with vertical axes have the advantage that they may, without adjustment, be driven by wind from any direction, but the torque from wind variation during each turn of the blades can produce unwanted vibrations. At present most working machines have horizontal axes, but this may change as machines with vertical axes, like the Musgrove rotor, are improved.

Cup anemometers are used to measure wind speed or to produce power for very small rotors. They depend on drag force and the shape of the cups produces an almost exact correspondence between wind and rotational speeds.

Savonius rotors can be homemade for simple water pumping as they can be made from two halves of an oil drum. As in the previous WEC, the motive force is drag produced by the complicated movement of wind—both through and around the curved sheet airfoils.

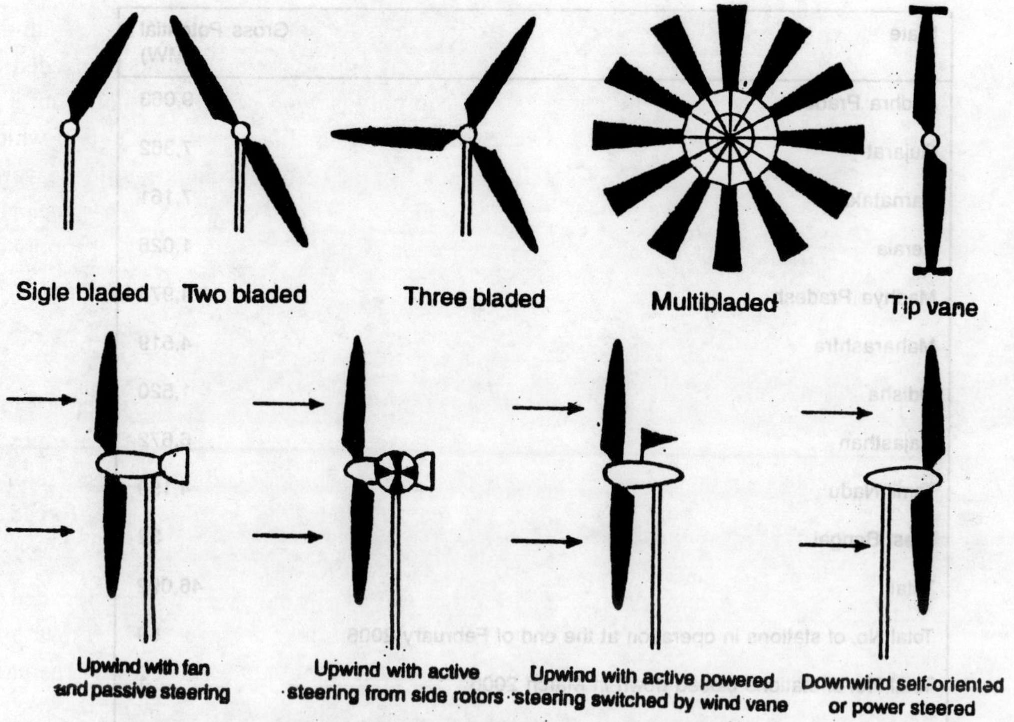

Fig. 24.20 : Horizontal axis wind machines of different types

The Darrieus rotor, otherwise known as the 'egg beater', has two or three thin blades with an airfoil section. The driving force is lift and maximum torque occurs when the blade is moving across the wind faster than the speed of the wind. But this rotor, which is used for generating electricity, is not, as a rule, self-starting and has to be started by the generator itself which calls either for complicated controls or constant supervision by an operator.

The Evans rotor is fitted with two vertical wing-shaped blades, fitted on a vertical axis, which create lift when the wind blows fast enough.

Advantages of Wind Power

- It is one of the most environment friendly, clean and safe energy resources.
- It has the lowest gestation period as compared to conventional energy.
- Equipment erection and commissioning involve only a few months.
- There is no fuel consumption, hence low operating costs.
- Maintenance costs are low.
- The capital cost is comparable with conventional power plants. For a wind farm, the capital cost ranges between 4.5 crores to 5.5 crores, depending on the site and the wind electric generator (WEG) selected for installation.

The estimated Wind Power Potential in India is given in box below:

State	Gross Potential (MW)
Andhra Pradesh	9,063
Gujarat	7,362
Karnataka	7,161
Kerala	1,026
Madhya Pradesh	4,978
Maharashtra	4,519
Odisha	1,520
Rajasthan	6,672
Tamil Nadu	4,159
West Bengal	32
Total	46,092
Total No. of stations in operation at the end of February 2006	59
Total No. of stations closed down in March 2006	1
Total No. of stations Commissioned in March 2006	1
Total No. of stations in operation as on 31st March 2006	59

The **wind power potential** on a national level, base data collected from 10 states considering only 1% of land availability, **is around 46,092 MW.**

24.3.4 Tidal Power To Energy

Hydropower Technologies

Hydropower relies on the conversion of the potential energy loss of water following from a higher to a lower level. This loss of potential energy each second is the power available and is given by P = MgH or P = ρVgH where P is the power (watts), M is the mass flow of water (kgs^{-1}), g is the acceleration due to gravity. H is the height in metres through which the water falls, ρ is the density of water (kgm^{-3}) and V is the volume flow of water (m^3s^{-1}). Taking g as 10 ms^{-2}, the power is then P = 10MH watts or P = 10 VHk Watts, since ρ = 11,000 kgm^{-3}.

There are two types of turbine which are used to convert this water power into practically useful power, usually electricity. These are:

1. Impulse turbines, in which the flow of water hits open turbine blades in a jet and the power derives from the kinetic energy of the water.

2. Reaction turbines in which the turbine blades are fully immersed in the water and the power comes from the pressure drop across the turbine.

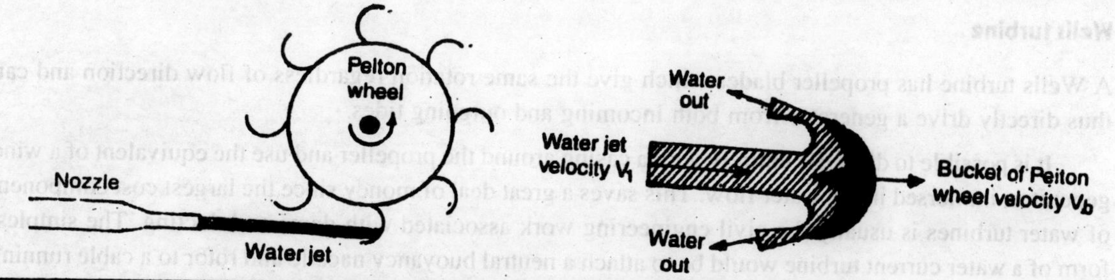

Fig. 24.21 : Pelton wheel showing shape of bucket

Pelton wheels

A Pelton wheel is an impulse turbine in which a jet of water hits a bucket attached to the wheel rim. The bucket is shaped as shown in Fig. 24.21 so that the jet is divided into two equal streams and deflected from the incoming jet and out of the bucket. The direction of the water flow is reversed, giving the water a change of momentum equal to twice the momentum of the jet relative to the bucket.

The force on the bucket is $F = 2M(V_j - V_b)$, M = Mass flow rate and the power developed is $P = FV_b$, i.e., Power $P = 2M(V_j - V_b)V_b$.

The power, in the ideal case, is equal to the total kinetic energy per second in the jet, i.e., the efficiency is 100 per cent. No real Pelton wheel can achieve this because of some friction of water flowing round the bucket and the momentum of the outlet water is not quite equal to the momentum of the jet. However, efficiencies around 90 per cent are achieved in practice.

Reaction turbines can be rather more efficient than a Pelton wheel, at the cost of greater mechanical and hydraulic complexity.

Francis turbine

The Francis turbine is shown in outline in Fig. 24.22. Water flows into a casing around the working parts. The water flow is guided by fixed vanes on to the rotating vanes and the water leaves via the central outlet.

To maximise the output of water, the machinery can be placed in a duct of the same diameter as the rotating vanes and the water then flows axially down the duct. The vanes are now similar to the propeller of a ship or aircraft and they rotate because of the stream of water passing over them.

Normal propeller design will give a change of rotation if the direction of water flow is altered. This would be very inconvenient for a turbine in a tidal flow, since the electrical generator must always be driven in the same direction or rotation.

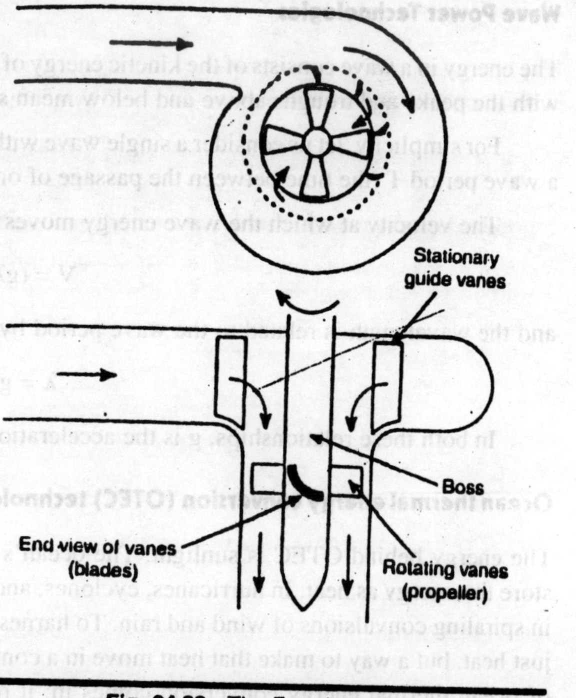

Fig. 24.22 : The Francis turbine

Wells turbine

A Wells turbine has propeller blades which give the same rotation regardless of flow direction and can thus directly drive a generator from both incoming and outgoing tides.

It is possible to dispense entirely with a casing around the propeller and use the equivalent of a wind generator, immersed in the water flow. This saves a great deal of money since the largest cost component of water turbines is usually the civil engineering work associated with dams and ducting. The simplest form of a water current turbine would be to attach a neutral buoyancy nacelle and rotor to a cable running from a sea bed anchor to a surface buoy.

In wide estuaries or rivers, many turbines could operate side by side. However, unlike a wind turbine, the structure would not experience extreme speeds in gales or gusts and so would not need to be engineered to cope with excessive loads. Since water is so much denser than air, a 2 ms^{-1} water current has the same power density as a 19 ms^{-1} air current and many tidal and river sites could generate useful power. Estimates of the power available in tidal races around the UK suggest a resource of about 15 GW, so around 5 GW of electrical power could be generated. Tidal currents are variable, being zero when the tide turns twice a day so the load factor on a tidal marine turbine may only be 20 per cent, i.e., its annual energy output is only 20 per cent of that which it could give if it is operated at full power all the time. A water turbine in a river whose flow varies little over the year might have a load factor of 80 per cent or so, giving four times the annual energy output of the same turbine in a tidal current. The economics of a river turbine would be correspondingly more attractive. The environmental impact of a submerged water turbine in a tidal estuary would be very small compared to that of a tidal barrage. In later studies 1993, a study of the technical and economic feasibility of tidal water turbines in the UK was undertaken and they may prove to be an attractive alternative to the Severn or Mersey barrages.

Wave Power Technologies

The energy in a wave consists of the kinetic energy of the moving water and the potential energy associated with the peaks and troughs above and below mean sea level.

For simplicity, let us consider a single wave with a wavelength λ (the distance from crest to crest) and a wave period T (the time between the passage of one crest and the following crest past a fixed point).

The velocity at which the wave energy moves is given by:

$$V = (g\lambda/8\pi)_{1/2}$$

and the wavelength is related to the wave period by:

$$\lambda = gT^2/2\pi$$

In both these relationships, g is the acceleration force.

Ocean thermal energy conversion (OTEC) technology

The energy behind OTEC is sunlight. The ocean's surface waters absorb it in prodigious amounts and store the energy as heat. In hurricanes, cyclones, and typhoons, this pent-up heat comes pouring back out in spiraling convulsions of wind and rain. To harness the energy for useful ends, OTEC systems need not just heat, but a way to make that heat move in a controlled way. That's where one of the great challenges of ocean thermal energy conversion comes in: it requires large amounts of cold water as well. OTEC systems tap into a store of cold water that's available around the world.

In theory, OTEC could meet all of today's electricity needs. But does the world need OTEC? Solar and wind energy have long track records and are already fast-growing industries. However, as OTEC supporters are quick to point out, those sources have a major limitation: so far, there's no easy or cheap way to store the energy. Solar energy only works during the day and when it's not too cloudy. Wind is intermittent and the turbines spin efficiently only when the weather cooperates.

Ocean thermal energy's big selling point is that it's always on, and it's always available. "Once you get the system going, it runs twenty-four/seven", says Harry Jackson, president of Ocean Engineering and Energy Systems (OCEES), a company based in Honolulu, Hawaii, that designs OTEC plants. It is consistent because the oceans act like a giant battery, storing up sunlight. "As an alternative energy source, it's one of the few to provide baseload power", says Reb Bellinger, vice president of Makai Ocean Engineering, also based in Honolulu, a firm that designs deep water pipes for OTEC.

The oceans' thermal energy can be effectively harnessed wherever the temperature difference between the warm surface water and the cold deep water is at least 20°C—which covers about one-third of oceans' area. OTEC's major downside, however, is that most of the resource is marooned far from land and far from people. But scores of specially outfitted ships could float in the open ocean, grazing on the energy and using it to synthesize fuels and chemicals that get shipped to shore—a futuristic dream that could be the key to unlocking the technology's potential.

But ocean thermal energy conversion (OTEC) still has to prove itself. So far it has been tested only in small pilot projects, and no one has built a commercial power plant. Now, with a few plants in various stages of planning, OTEC may finally get a shot at competing with other sources of clean energy (such as solar, wind, geothermal, and tidal) to show whether it can match up, or even beat them.

24.4 BIOLOGICAL ENERGY

24.4.1 Biomass and Biofuels

Any type of animal or plant material that can be converted into energy is called biomass. This includes trees and shrubs, crops and grasses, algae, aquatic plants, agricultural and forest residues plus all forms of human, animal and plant waste (1). When the material is used for energy production it becomes a biofuel. There are many forms of biofuel, existing in solid, liquid or gaseous categories.

An estimate of the world solid biomass standing in forests in 1979 was 1.8×10^{22} J. At that time this figure was comparable with the world's proven natural gas and oil reserves. The same biomass figure is about 50 times the value of world primary energy consumption in 1995. If the biomass values of grasslands and crops are included (although these are mainly used for food production—indirectly energy—rather than fuel supply) the solid biomass figure is about 200 times the global energy consumption.

Biomass may be used in a number of ways to produce energy. The most common methods are:

- Combustion
- Gasification
- Fermentation
- Anaerobic digestion

India is very rich in biomass. It has a potential of 19,500 MW (3,500 MW from bagasse-based congeneration and 16,000 MW from surplus biomass). Currently, India has 537 MW commissioned and 536 MW under construction. The facts reinforce the idea of a commitment by India to develop these resources of power production.

Following is a list of some States with most potential for biomass production:

- Andhra Pradesh (200 MW)
- Bihar (200 MW)
- Gujarat (200 MW)
- Karnataka (300 MW)
- Maharashtra (1,000 MW)
- Punjab (150 MW)
- Tamil Nadu (350 MW)
- Uttar Pradesh (1,000 MW)

The potential available and the installed capacities for Biomass and Bagasse

Source	Potential	Installed
Biomass	16,000 MW	222 MW
Bagasse (Co-generation) in existing sugar mills	3,500 MW	332 MW

Source: http://www.greenbusinesscentre.com/renenbio.asp

The main sources of raw material that constitute biofuels include:

(a) Natural vegetation
(b) Energy tree plantations
(c) Specific energy crops
(d) Wastes
(e) Water-based biomass.

Each of the above five categories is briefly discussed.

24.4.2 Natural Vegetation

In harvesting a natural vegetation site no energy costs are involved in clearing or replanting. Where an area may be unsuitable for agriculture the use of its vegetation for biomass would constitute a useful bonus biofuel source. There still remains the necessary costs of harvesting the vegetation and transporting it to a user site. A disadvantage is that even in fertile locations the yields are low—about half of the value that might be obtained from customised energy plantations.

24.4.3 Energy tree plantations

Trees and other types of lignocellulose materials may be grown specifically for burning as biofuels. By the choice of appropriate species, sites, planting densities and harvesting schedules biofuels can be grown at competitive costs. This process is sometimes referred to as "short rotation forestry".

24.4.4 Specific Energy Crops

Certain crops have high energy conversion efficiency. In appropriate locations, crops such as eucalyptus trees, rubber plants or sunflowers might be used because of their rapid growth and high energy content (3). Such crops do not have to be consumable by humans or animals and the entire crop can be used, including leaves, stalks and roots. The stored chemical energy can be converted directly to heat by combustion or processed into liquid or gaseous fuels.

Energy farming, like other agricultural operations, requires large quantities of water. In the developed countries of the western world there is competition for the use of land, water and nutrients with various forms of food production, including animal farming. An interesting economic issue is whether crops in growth command higher prices as fuel rather than food or fibre. Crop surpluses would provide low-cost biomass but is surplus food production the best overall use of the land and water?

There is a further potential disadvantage to the use of land-based energy farming. The hardy and fast-growing species required for energy use could become widespread nuisances if they escaped from the confines of the farm. They may then displace native plant species and impact on animal and insect life. Also, the methods necessarily used to control infestations and disease in the energy plants, especially if they are monocultures, could have adverse effects on neighbouring food production crops.

The same soil plantation may be used for two species of plants with out-of-phase growth patterns. If both species are in leaf together the leaf canopy cover profile is increased. Non-nitrogen-fixing crops can be grown side-by-side with legumes to reduce the need for nitrogenous fertilisers. Mixed cropping of this kind is less susceptible to damage by external parasites or predators than monoculture plantations.

There are some plants that produce high concentrations of "petroleum-like" products. These can be grown en masse and the "oils" extracted or squeezed out. Pilot schemes have been carried out in Mexico.

Widespread energy farming on marginal land implies the need for some level of overall control. A fast-growing plant that thrived in marginal conditions would create incentives to adapt the plant for food production. This is not necessarily bad but would create further pressure on land use for purely biomass reasons. In other words, the food versus biomass competition would increase.

It would seem sensible to think in terms of integrated growth mixed cropping, in which energy crops are developed alongside crops for food, fertilisers and chemicals.

The vertical distribution of primary productivity and biomass formation in forest is shown in Fig 24.23.

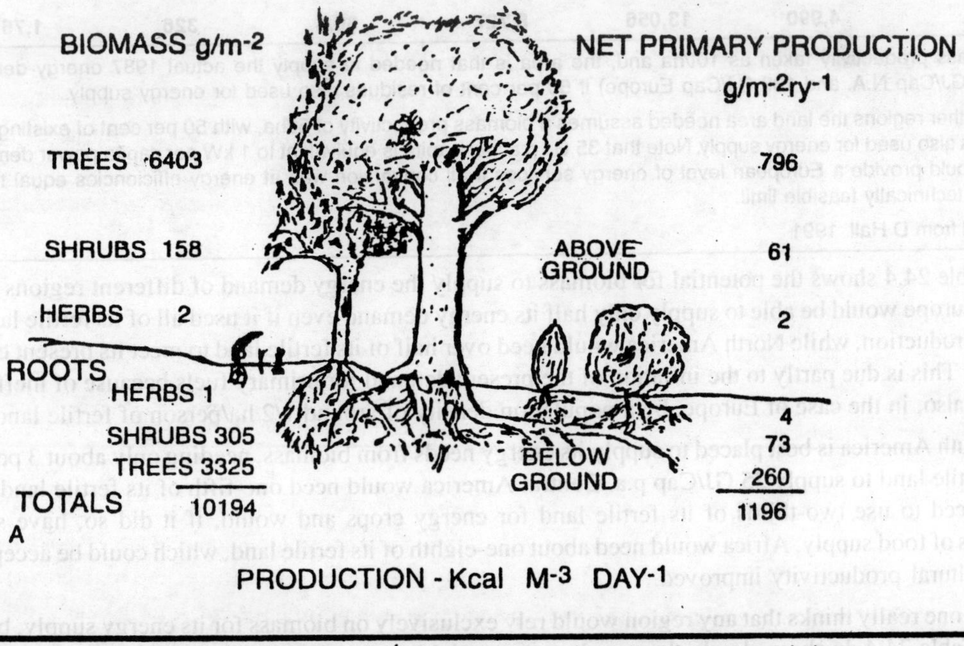

Fig. 24.23 : The biomass yield of various type of a typical forest association

24.4.5 Power from Biomass

Almost all lifeforms need sunlight for their energy (some creatures in volcanic ocean trenches use geothermal energy). Photosynthesis by green plants converts large amount of sunlight into biological material—grass, trees etc.—which are rich in energy and form the basis of food chains for other creatures. Human society derives products such as wood or alcohol from these basic photosynthetic materials and uses them to meet various human needs. The fossil fuels which we use at the moment are products of photosynthesis many million of years ago and form an 'energy bank' on which our present society is drawing heavily.

Biomass is used extensively throughout the world today. The use of fuelwood, crop residue or cow dung for cooking is widespread in developing countries as a free good and is usually the only option for their rural poor. The use of biomass is not, however, restricted to the poor of the Third World. For instances, the US obtains as much energy from biomass as it does from nuclear power, while Sweden derives 14 per cent of its primary energy from biomass.

The biomass resources are very large, the amount stored annually worldwide is equivalent to ten times the total world annual energy use. The total stored biomass in the world has an energy content equal to that of all the world's proven fossil fuel reserves.

Table 24.4: The Potential for the Use of Biomass (1987 data)

Region	Population (10^6)	Total land area 10^6 ha	Croplands forests pasture 10^6 ha	Energy content of residue 10^6 GJ	Total energy consumption 10^6GJ	Area to supply 35 GJ/Cap
N. America	276	1839	968	12	93	580[1]
Europe	497	473	283	9	66	470[1]
Asia	2,780	2,637	1,662	30.9	57.6	1,091
Africa	590	2,937	1,641	10.2	12.1	207
C. America	143	265	198	3.3	6.0	45
S. Amercia	279	1,751	1,541	11.6	10.8	53
World	**4,990**	**13,056**	**8,727**	**86.5**	**326**	**1,751**

1. Biomass productivity taken as 10t/ha and, the area is that needed to supply the actual 1987 energy demand (330 GJ/Cap N.A. and 133 GJ/Cap Europe) if 50 per cent of residues also used for energy supply.

For all other regions the land area needed assumes a biomass productivity of 5t/ha, with 50 per cent of existing crop residues also used for energy supply. Note that 35 GJ p.a. per capita is equivalent to 1 kW per capita power demand, which could provide a European level of energy services if all conversion was at energy efficiencies equal to the present technically feasible limit.

Adapted from D Hall. 1991

Table 24.4 shows the potential for biomass to supply the energy demand of different regions of the world. Europe would be able to supply only half its energy demand even if it used all of its fertile land for energy production, while North America would need over half of its fertile land to meet its present energy demand. This is due partly to the inflation of the present demand for primary fuels because of inefficient use, but also, in the case of Europe, by a population density of around 1/2 ha/person of fertile land.

South America is best placed to supply its energy needs from biomass, needing only about 3 per cent of its fertile land to supply 35 GJ/Cap p.a. Central America would need one-fifth of its fertile land. Asia would need to use two-thirds of its fertile land for energy crops and would, if it did so, have severe problems of food supply. Africa would need about one-eighth of its fertile land, which could be acceptable if agricultural productivity improved.

No one really thinks that any region would rely exclusively on biomass for its energy supply, but the data in Table 24.4 do show clearly the very large potential in many areas of the world.

There are four major attractions in the use of biomass; it is an indigenous resource in most parts of the world; it provides stored energy unlike most renewable sources; it is very flexible in use; and, during its growth, it captures CO_2. It also makes use of agricultural skills which are to be found throughout the world, and provides rural employment and thus, unlike most other energy technologies, does not increase the drift from rural to urban areas.

Use of Biomass

The simplest use of biomass is to burn it and the most common biomass used this way is wood, but corn stalks, cow dung and many other agricultural residues are also burnt on open fires for cooking, heating and other social purposes. The fireplace is usually formed by three stones, on which the cooking pot, or whatever, is balanced. Both the fuel and the stones are gathered locally, so the only cost is in time.

The next step in sophistication is to heat the wood in an enclosed space to make charcoal which is a very convenient fuel. It is much lighter than wood and so is much more easily transported. It burns without fumes and so is favoured in urban areas, but it is a processed fuel, sold commercially to those who can afford it.

In many parts of the world, animal—and sometimes human—wastes are used in biogas plants. The anerobic digestion of these wastes produces a methane-rich gas and leaves behind a benign residue which makes an excellent compost. There are millions of farms or village-sized biogas plants in operation around the world, particularly in China and India. But collecting waste from animals which roam freely can be so time-consuming that biogas production is no longer worthwhile and it is only really practicable where animals are kept in pens. None the less, enough gas to satisfy the family's cooking and lighting needs can be produced from the waste of the livestock owned by a typical peasant farmer. This is particularly true where fairly rapid cooking is the norm, as with the Chinese wok. In general, the richer the farmer, the more animals the family will have and hence the more likely it is that a biogas plant could supply the family with its cooking and lighting needs.

Table 24.5: Products and Services from Biomass Conversion

Process		End products	
	Electricity		
Bio-conversion	Methane	Ethanol	Fine chemicals
	Fuel gas	Methanol	Hydrocarbons
	Bio-oil		Hydrocarbons
Thermal	Charcoal		
	Slurry		
	Combustion		

In the industrialised countries and the urban conurbations of developing countries, the use of biomass would need to be on an industrial scale, far removed from farm or village sized charcoal or biogas plants. Just as farms can burn their own straw or dispose of animal waste in biogas plants, so municipalities could install waste burning and land-fill gas plants for energy production. The full potential of biomass as an energy source could, however, only be achieved by the use of energy crops and the associated processing plants.

The two major ways of processing woody biomass from energy crops are bio-conversion by a fermentation process and thermal conversion, usually by anerobic heating. Table 24.5 shows the products which can be derived from these processing steps.

Bio-conversion processing has been known for thousands of years and is used to produce those important solar products—beer, wine, and spirits. In the Brazilian bioalcohol programme a process similar

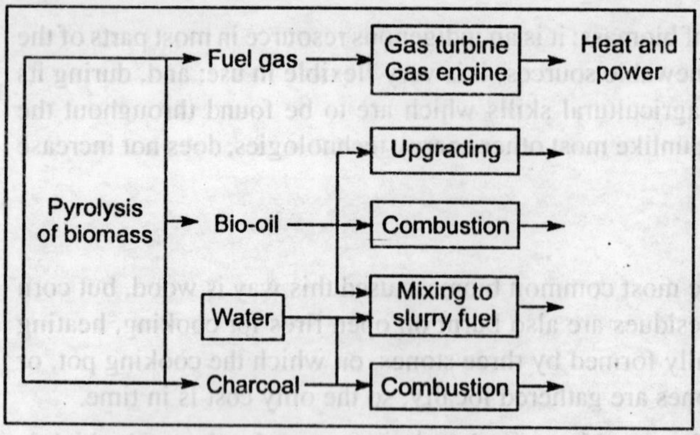

Fig. 24.24 : The products and processes of pyrolysis

to distilling spirits is used to ferment sugar cane and distill the ethanol; in the US, grain is used for the same purpose. To ferment woody biomass, produced by coppicing energy plantations, its cell structure must first be broken down by hydrolysis, acids or enzymes to allow fermentation to proceed efficiently.

There are four basic thermal conversion processes:

The oldest is direct combustion in air to produce heat. This can range from the domestic wood fire to large power stations fueled by peat, coconut fibres and other waste products.

Pyrolysis is the process of heating biomass in the absence of air and as it reduces wood to charcoal, it produces gaseous and liquid products which, in the traditional charcoal kilns are vented into the air. In fact, they are valuable products and in modern pyrolysis plant they are collected and used (Fig. 24.24).

The third thermal conversion process is the liquefaction of biomass to produce bio-oil which can be burnt to produce heat and power or can be upgraded to petrol or diesel fuel. Rapeseed oil, for instance, is becoming quite widely used as a transport fuel, either on its own or as an additive to diesel.

The fourth process is gasification, which produces either or both, medium heating value (MHV) or low heating value (LHV) gas. Each has a different chemical make-up, but they can both be burnt in air to give heat and power or they can be used as chemical feedstock. MHV gas can be synthesised into methanol and LHV gas into ammonia, both of which are important in the chemical industry.

24.4.6 Biomass Programmes—Experience and Promise

Biomass programmes can be divided into three broad classes. There are those which are cost-effective on purely financial accounting grounds, there are those in the demonstration stage which can or could become cost-effective and there are those which may never become cost-effective, but which bring such significant non-monetary benefits that they deserve continuing support.

This last category is the most difficult to appraise. It includes many social forestry and agroforestry projects. If the cost/benefit analysis includes only commercial transactions, then many projects seem quite unviable. However, their social and economic developmental benefits, though unquantifiable for accounting purposes are extremely important and often provide justification for continuing and even expanding or replicating them.

The second category includes many programmes for the development and dissemination of improved cooking stoves, small biogas plants in many parts of the world and the production of fuels by gasification or briquetting. The viability of these schemes depends crucially on local conditions, in agriculture as well as in social and economic structures. Biomass energy or products have no general cost, only local costs determined by local conditions and the cost/benefit analysis determined by outside agencies may be quite different from the costs and benefits perceived by local people. If these programmes are to be promoted by international agencies, an essential pre-condition for success is patience in working with and listening to local people so that the programme may be tailored to best meet their needs in their situation.

The first category, that of biomass projects judged on purely commercial grounds, is appropriate for Europe and other industrialised areas. Even so, it is not easy to produce definite cost-effectiveness judgments for them because there are so many 'spin-off' benefits. These lie in maintaining a vigorous rural economy and in improving the environment by replacing fossil fuels (whose price does not include the cost of environmental externalities) with biomass fuels.

The costs, in Sao Paulo in Brazil, of producing hydrous alcohol for mixing with petrol are shown in Table 24.6. The revenues are set at $0.18/litre and the residue (bagasse) from the process can be sold at the equivalent of $0.02/liter. The process is profitable on its operating costs, but makes a loss if the opportunity costs are included. The rate of return on investment and hence the opportunity cost, depends on the financial circumstances of the time and changes with the economic climate. The price of alcohol is set in relation to the cost of petrol, which depends on the international price of oil and the pricing policies of oil companies and the Brazilian government. The attractiveness of using sugar cane for alcohol rather than sugar production depends on the world price of sugar. However, if the large Brazilian producers were to switch from alcohol to sugar production there would be such a glut on the world market that prices would tumble.

It has been estimated that up to 10 GW could be produced from the bagasse residues from alcohol production and there are proposals in Sao Paulo to use it as a solid fuel for power stations. This would transform the economics of the alcohol programme. Electricity could be produced at under $0.045 k Wh and alcohol sold at $0.15/litre making it competitive even at $15/bbl for oil prices.

Table 24.6: Cost Breakdown of Production Costs of Hydrous Alcohol in Sao Paulo, Brazil

Item	Cost in $	
	Tonne	Litre
Total raw materials	5.832	
Cultivation	0.846	
Plant protection	1.400	
Collecting	1.771	
Transport	0.821	
Raw materials	4.837	
Distillation costs		0.125
Total direct costs	0.119	
Administrative costs	0.004	
Taxes and charges	0.003	
Opportunity costs		0.071
Break-even revenue requirement		0.196

The raw material costs feed into the distillation costs.
The opportunity costs relate to the return on investments.

Source: D Hall, 1992

The biomass programme in the US is also very effective and biomass could be a major component of the US energy mix. Table 24.7 shows the energy which could be supplied annually on a commercial cost-effective basis if the target cost can be met.

It is clear that many of these potential resources are at, or close to, commercial viability. Technical developments, particularly those permitting increases in scale are still needed but the major obstacles are structural and institutional. The exploitation of these resources requires the creation of new industries and this takes time, even when the technical and economic conditions are favourable.

There is, however, the problem of agricultural prices. Between 1980 and 1987, maize varied from \$1.41 per bushel to \$3.16 per bushel. By-product prices can vary by a factor of three. These variations are caused by market responses to worldwide supply and demand and are thus very difficult either to control or to predict. There is thus a considerable risk that biomass industries will go through periods of unprofitable operation as world agricultural prices fluctuate and they will need some assurance of stability through long-term financing and/or guaranteed pricing.

Table 24.7: Net raw biomass resource in the USA as estimated by Oak Ridge National Laboratory before conversion to fluid fuels or electricity

Feedstock	Biomass resource EJ/year	Cost (\$/GJ)	Current target
Residues	8.9		
Logging residues	0.8	> 3	< 2
Urban wood waste	1.2	2	2
Forest manufacturing residues	2.1	1	< 1
Agricultural residues	2.0	1-2	1
Municipal and food waste	2.4	2-3	< 1.5
Animal waste	0.5	4	3.5
Biomass from existing forests	9.5		
Commercial forest wood	4.5	< 2	< 2
Improved management	4.5		< 2
Shift 25% to energy forest	0.5	2	2
Biomass from energy crops	10.8		
Agricultural oil seeds	0.3	3	2
Wood energy crops	3.2	3	2
Herbaceous energy crops	5.9	4	2
Aquatic energy crops	1.4	3.5	2
Total	29.3		

1 EJ (10^8J) p.a. is equivalent to 32 GW continuous power.

Source: W Fulkerson et al, 1989

There is a very active programme of research, development and demonstration of biomass technologies in the European Union. Contemporary European biomass resources are estimated at 455 Mt p.a. and could rise to 820 Mt p.a. in the future. The breakdown of these resources is shown in Table 24.8.

The research programme of the European Commission has identified a wide range of crops which could be suitable for energy croping. These are shown in Table 24.9.

Table 24.8: The annual production of biomass in Europe

Resource	Production Mt p.a. (dry)	
	Now	Future
Wood	5	75
Other energy crops	—	250
Wood wastes	50	70
Agricultural wastes	250	250
Municipal wastes	60	75
Industrial/commercial wastes	90	100

Source: G Grassi, 1990

Of the species shown in Table 24.9 the front-runners for commercial energy cropping are sorghum for southern and central areas, eucalyptus for southern regions and poplar and willow for northern Europe. The costs of these resources are expected to be about 27-30 ECU/dry tonne in production, which is low enough to ensure that the end products of the bio or thermal conversion processes are cost-competitive.

Biomass production for non-agricultural purposes could be an important part of the European Common Agricultural Policy (CAP). Unless alternative profitable non-food uses for land can be found, the CAP schemes for limiting agricultural production will mean that it is abandoned, with all the consequent social and environmental costs. About 20 million hectares of agricultural land and 10-20 million hectares of marginal land are likely to become available for biomass production by 2010, so using it to produce biomass could provide the employment needed to maintain rural economies. Bio-energy, in the 21st century, could equal the present output of the North Sea oil fields or all the nuclear power stations so far installed and so could provide a significant fraction of the EU's demand for energy. It would also confer significant environmental benefits.

The distinctive characteristic of the European programme is its emphasis on an integrated approach to the use of biomass. The output is seen as a set of products, each of which contributes to the profitability of the enterprise, rather than focusing just on energy and treating others as by-products. The biomass industry can be targeted to meet needs in various sectors of industry, agriculture and energy in the different regions of Europe. The European programme has set target dates for demonstrating and commercialising the range of technologies needed for this integrated approach (Table 24.10).

Table 24.9: Present and future biomass productivity in different regions of Europe

	Productivity dry T/ha p.a.		
	Present	Future	European region
Agricultural crops			
Sweet sorghum	25	35	South/Central
Miscanthus etc.	20	30	South
Jerusalem artichoke	20	25	South/Central
Artichoke (tubes)	10	14	South/Central
Trees			
Eucalyptus	10/15	17	South
Poplar	12	16	North/West/Centre
Willow	10	15	North/West
Robinia	58		South
Conifers	5/6	8/10	North/West/Central

Source: G Grassi, 1990

24.4.7 Biomass and the Environment

Photosynthesis results in the absorption of carbon dioxide and the emission of oxygen, so the growth of plants counteracts the anthropogenic emissions of carbon dioxide from fossil fuels. The reduction in the area of high productivity plants in the tropical forests and grasslands is a cause for concern in its effect on the natural carbon cycle. The planting of trees is often regarded as a means of counteracting the increasing atmospheric concentrations of carbon dioxide. During their growth, the trees do absorb carbon dioxide, but this is re-emitted, possibly accompanied by methane, when they die and decay.

Table 24.10: Target dates for completion of demonstration and commercialisation of biomass technologies for Europe

Demonstration	Year	Commercialisation	Year
Short rotation forestry	1994	Advanced combustion for power	1996
Advanced combustion for power	1995	Compost	1998
Advanced pyrolysis	1995	Advanced pyrolysis	1998
Pulp for paper from sweet sorghum	1995	Short rotation forestry	2000
Compost	1996	Sweet sorghum production	2000
Sweet sorghum production	1996	Ethanol for fuel	2000
Char-slurry fired boilers, kilns etc.	1997	Advanced gasification	2003
Gas turbine fired by bio-oils or char-slurry	1997	Methanol synthesis	2003
Advanced gasification	1998	Hydrocarbon synthesis	2005
Hydrocarbon synthesis	2000	Hydrogen and ammonia production	2010

Source: G Grassi, 1990

David Hall has argued powerfully (Table 24.11) that the sustainable application of biomass within the commercial energy sector requires the substitution of biofuels for fossil fuels and has presented a scenario for the reduction in global CO_2 emissions which might be achieved by this substitution.

It is clear that biomass could offset much of the human-made emissions of CO_2 and play a very significant role in stabilising the global climate while allowing the increase in energy use which is essential if the people of the developing nations are to have good quality of life.

The substitution of fossil fuels by biomass energy sources has a number of environmental advantages other than reducing CO_2. Biomass contains little or no sulphur, so the acid emissions of sulphur compounds is avoided. The production of nitrogen oxides can also be reduced considerably. Burning biomass as, for instance, on a garden bonfire can produce large quantities of noxious chemicals, some of which are carcinogenic, but if burning is properly controlled, in a furnace with flue gas cleaning, the overall acid emissions can be very small. The more advanced processing of raw biomass can be designed to emit very little into air, water, or land.

A biomass industry would significantly benefit the rural environment. It produces compost and that, together with the improved agronomic practices, would help condition and stabilise the soil and avoid the desertification and forest fires beginning to occur in Southern Europe.

Social and political benefits would also accrue from the widespread use of biomass. The creation of profitable and sustainable rural employment would reduce the drift to urban areas and to maintain the vitality of rural communities. Biomass can be produced and used locally in most parts of the world, avoiding the concentration of supply which characterises fossil fuels. The promotion of local self-sufficiency and the geographical diversity of supply for trade will contribute to a more stable world-political environment to the benefit of all countries, but particularly to the Third World.

Table 24.11: Scenario for the reduction in anthropogenic carbon dioxide emissions by the substitution of fossil fuels by bio-energy sources

Year	Technology	Reduction Gtc pa a
2025	Electricity and alcohol from sugar cane	0.7
	Electricity from kraft pulp industry residues	0.2
	Energy from other sources	0.8
	Total reduction by 2025	1.7
2050	Electricity and alcohol from sugar cane	0.7
	Electricity from kraft pulp industry residues	0.2
	Energy from other residues	0.9
	Energy from biomass energy crops	3.6
	Total reduction by 2050	5.4

Source: D Hall, 1991

24.5 USE OF WASTES

The burning and biodegradation of wastes as a sources of energy is also significant.

24.5.1 Water-based Biomass

Aquatic plants do not require irrigation or appreciable rainfall, the temperatures fluctuate less than land temperatures and light absorption is high—giving rise to high photosynthetic efficiencies.

Algae, seaweed and other aquatic lower plants can be intensively grown in certain areas of the sea or in inland lakes or ponds. This would eliminate the competition for land with conventional agriculture. On average, the oceans are low in plant nutrients. Surface waters have low natural productivity but sometimes act as nutrient repositories due to runoff from the land. Deep ocean waters are rich in nutrients. If such deep water can be pumped to the surface a biomass crop such as giant kelp (i.e., large brown seaweed) could be grown.

A typical algae farm might contain a hectare of land, excavated to a depth of 18 inches (45 cm), flooded and algae plants introduced. The pond would be covered with plastic, injected with carbon dioxide and maintained at a temperature of 100°F (38°C).

24.5.2 Energy from Wastes

All human and industrial processes produce waste. Wastes that can form sources of biofuels include domestic refuse, industrial wastes, agricultural wastes, forestry residues, sewage and industrial effluents. It is highly desirable that the recovery of energy from waste should form part of an integrated approach to waste management, designed to maximise waste recycling and reclamation. Using wastes as a source of fuel can be highly cost effective, especially if the alternative disposal cost is discounted.

Wastes do have certain disadvantages when used as fuel sources. They may be difficult to handle and process and generally they have low energy density. Some wastes may be contaminated with non-fuel materials. It may be necessary to transport the waste from its source to a conversion site.

24.5.3 Solid Wastes

In a highly developed society like Great Britain or the USA, solid waste is produced at the rate of about one tonne per person per year (2000).

Most domestic waste is solid in form. The constituent proportions typical of UK municipal waste are shown in Table 24.12. Each year the UK produces 28 million tonnes of solid domestic wastes plus a similar amount of industrial wastes. Solid wastes are usually disposed of either by burning or by burial in landfill sites. In the UK a "landfill tax" of £7/ton has been introduced which affects both the costs and usage of landfill sites.

Table 24.12: Composition of typical UK municipal solid waste, 2000

Category	Per cent	Calorific value (MJ/Kg)
Paper and cardboard	30.5	14.6
Food waste	24	6.7
Metal	7.8	—
Glass	11.2	—
Dust and cinders	9.0	9.6
Textiles	4.9	16
Leather, rubber	8.3	17.6-37
Miscellaneous	4.3	17.6
Total	100%	

For combustion in a modern waste incinerator the refuse can be first sorted to separate out materials such as glass for recycling. The remainder is shredded to convert the burnable component into RDF (Refuse Derived Fuel) pellets. The RDF pellets burn like wood but produce less heat than coal and leave more ash. Modern municipal waste incinerators produce no smoke and reduce the bulk of the waste to about one-eighth. Some wastes, such as synthetic polymers (e.g. plastics produced from petrochemicals), produce toxic gases on combustion and these have to be safely contained within the incineration process.

General industrial waste consists mainly of paper, cardboard, wood, and plastics. It contains a lower moisture and lower ash content than municipal waste, making combustion easier to control and is less contaminated with non-fuel ingredients. Special forms of industrial waste such as batteries, motor tyres, poultry litter, and hospital wastes require special forms of treatment and are all useful biofuels for combustion systems.

Refuse incineration in municipal plants for energy recovery is a relatively new but growing technology in the UK, as illustrated in Table 24.13.

Most of the domestic waste produced in the UK is not burned but is buried in about 5,000 landfill sties. Decay of the waste produces a methane-rich gas called landfill gas. At the end of 1990 there were 33 landfill gas schemes in operation in the UK, including 18 MW_e of electricity generation. By 1995 there were 50 gas-producing sites in operation and an estimated further 400 sites that could be used. The largest current site in the UK produces 3,500 m^3 of methane per hour to make steam for use in paper production. A diagrammatic representation of a landfill gas system is given in Fig. 24.25.

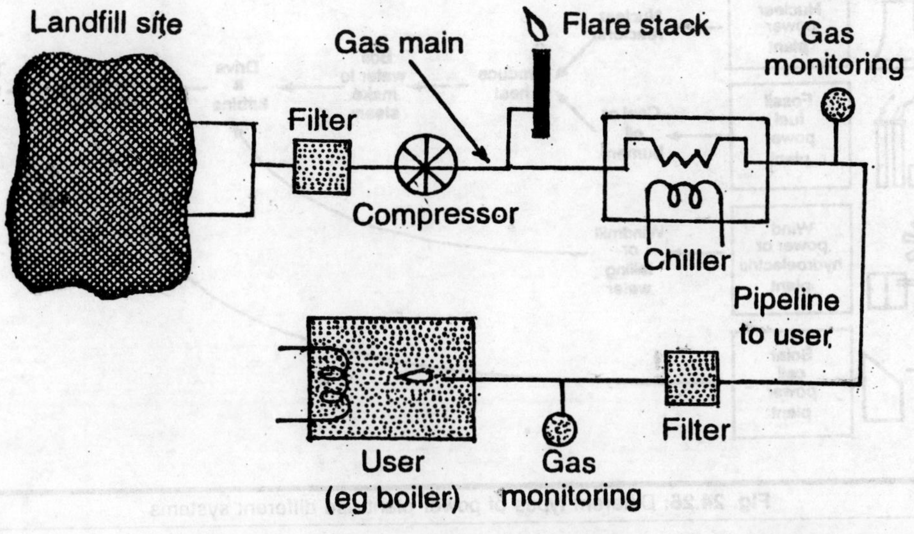

Fig. 24.25 : Production of landfill gas

Table 24.13: Refuse incineration in the European Community, 2000

Country	Number of plants	Total waste processed (Tonnes/day)	Energy recovered (MW_e) Hot water	Steam	Electricity
Belgium	6	3,650	3	1	2
Denmark	63	9,100	58	5	0
Eire	0	0	0	0	0
France	33	13,700	10	21	9
Germany	43	21,780	13	27	29
Greece	0	0	0	0	0
Italy	6	3,250	6	0	1
Luxemburg	2	800	0	0	2
The Netherlands	5	5,670	1	1	4
Portugal	0	0	0	0	0
Spain	4	1,350	0	4	2
UK	7	3,500	5	0	3
Total	169	62,800	96	59	52

The venting of landfill gases poses some environmental problems. Within the site vicinity there may be objectionable odours. Uncontrolled discharges from landfill using (say) power station cooling water, which is presently dumped into rivers. As the algae is harvested the nutrient-rich water used for growth is returned to the pond and re-seeded to recycle the operation.

The aquatic weed known as water hyacinth has been studied as a tropical water source of biogases, particularly by the US National Aeronautics and Space Administration (NASA). On a dry weight base, one kilogram of water hyacinth can produce 0.4 m^3 of biogas with a calorific value of 22 MJ/m^3. Aquatic weeds are a hazard in some waterways and have to be harvested, of necessity. The biofuel value is then a useful by-product.

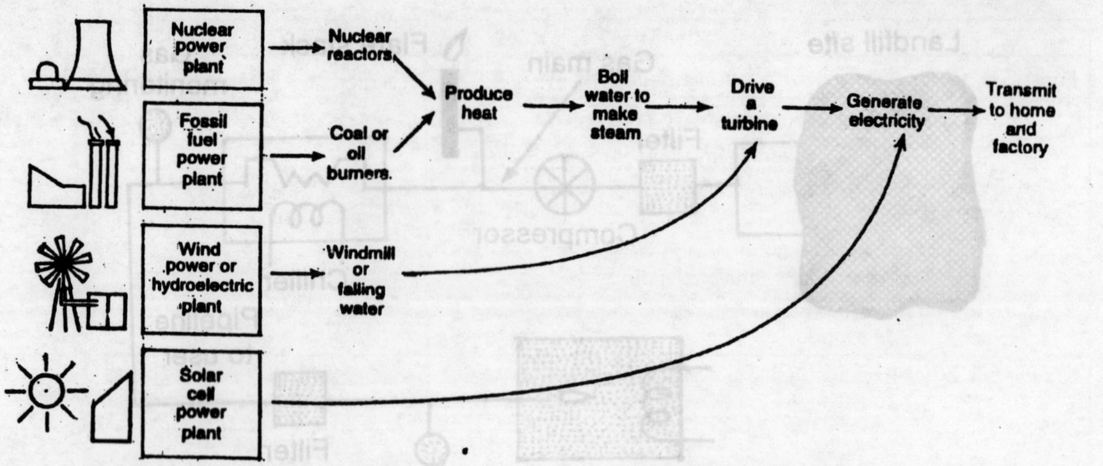

Fig. 24.26: Different types of power plant use different systems

However, there are a number of advantages and disadvantages of use of alternative fuels sources. The details are given in Table 24.14.

Finally different types of power plant operation and their energy generation systems are summarised in Fig. 24.26.

Table 24.14 : Pros and Cons of Alternative Fuels

Fuel	Advantages	Disadvantages
Methanol	Familiar liquid fuel.	Range as much as one-half less, or larger fuel tanks.
	Vehicle development relatively advanced.	Formaldehyde emissions a potential problem, especially at higher mileage, requires improved control.
	Organic emissions (ozone precursors) will have lower reactivity than gasoline emission.	More toxic than gasoline.
	Lower emissions of toxic pollutants, except formaldehyde.	M100 has nonvisible flame, explosive in enclosed tanks.
	Engine efficiency should be greater.	Costs likely somewhat higher than gasoline, especially during transition period.
	Abundant natural gas feedstock.	Greenhouse problem if made from coal.
	Less flammable than gasoline.	
	Can be made from coal or wood (as can gasoline), though at higher cost.	
	Flexfuel "transition" vehicle available.	
Ethanol	Familiar liquid fuel.	Higher cost than gasoline.
	Organic emissions will have lower reactivity than gasoline emissions (but higher than methanol).	Food/fuel competition possible at high production levels.
	Lower emissions of toxic pollutants.	Supply is limited, especially if made from corn or sugar.

Contd.

Table 24.14: Pros and Cons of Alternative Fuels

Fuel	Advantages	Disadvantages
	Engine efficiency should be greater.	
	Produced from domestic sources.	
	Flexfuel "transition" vehicle available.	Range as much as one-third less, or larger fuel tanks.
	Lower carbon monoxide with gasonol (10 per cent ethanol blend).	
	Enzyme-based production from wood being developed.	
Natural gas	Excellent emission characteristics except for potential of somewhat higher nitrogen oxide emissions.	Dedicated vehicles have remaining development needs.
	Gas is abundant worldwide.	Retail fuel distribution system must be built.
	Modest greenhouse advantage.	Range quite limited, need large fuel tanks with added costs, reduced space (liquefied natural gas (LNG) range not as limited, comparable to methanel; LNG disadvantages include fuel handling problems and related safety issues).
	Can be made from coal.	Dual fuel "transition" vehicle has moderate performance, space penalties.
		Slower recharging.
		Greenhouse problem if made from coal.
Electric	Fuel is domestically produced and widely available.	Range, power very limited.
	Minimal vehicular emissions.	Much battery development required.
	Fuel capacity available (for nighttime recharging).	Slow refueling.
	Big greenhouse advantage if powered by nuclear or solar.	Batteries are heavy, bulky, have high replacement costs.
	Wide variety of feedstocks in regular commercial use.	Vehicle space conditioning difficult.
		Potential battery disposal problem.
		Emissions for power generation can be significant.
Hydrogen	Excellent emission characteristics, minimal hydrocarbons.	Range very limited, need heavy, bulky fuel storage.
	Would be domestically produced.	Vechicle and total costs high.
	Big greenhouse advantage if derived from photovoltaic energy.	Extensive research and development effort required.
	Possible fuel cell use.	Needs new infrastructure.
Reformulated gasoline	No infrastructure change except refineries.	Emission benefits remain highly uncertain.
	Probable small to moderate emission reduction.	Costs uncertain, but will be significant.
	Engine modifications not required.	No energy security or greenhouse advantage.
	May be available for use by entire fleet, not just new vehicles.	

24.6 GEOTHERMAL ENERGY

It is a form of heat energy that confined within the earth interior through geological processes on a planetary scale. The presence of volcanoes, hot springs and other thermal phenomena must have led out ancestors to surmise that parts of the interior of the earth were hot. However, it was not until a period between the sixteenth and seventeenth century, when the first mechanical mines were excavated to a few hundred meters below ground level, that humankind deduced from simple physical sensations that the earth's temperature increased with depth. At present, it estimated that total flow of heat from earth could be available above 42×10^{12} w. This heat flow generate primarily from radioactive isotopes of earth interior (viz., uranium, thorium, and potasium).

The first attempt at generating electricity from geothermal steam was made at Larderello in 1904. By 1942 the installed geothermal electricity capacity had reached 127,650 kwe^{-}. During past two decades geothermal energy plays a significant role in electricity generation as well as non-electrical applications (Table 24.15). Electricity generation is the most important form of utilisation of high temperature geothermal resources (7150°C). The medium to low temperature resources (<150°C) are suited for many different non-electrical uses (viz., drying, heating, soil warning, distillation etc.).

Table 24.15: Installed geothermal electric capacity, 2016

	Country	Capacity (MW) 2007	Capacity (MW) 2010	Percentage of national electricity production	Percentage of global geothermal production
1.	United States	2687	3086	0.3	29
2.	Philippines	1969.7	1904	27	18
3.	Indonesia	992	1197	3.7	11
4.	Mexico	953	958	3	9
5.	Italy	810.5	843	1.5	8
6.	New Zealand	471.6	628	10	6
7.	Iceland	421.2	575	30	5
8.	Japan	535.2	536	0.1	5
9.	Iran	250	250		
10.	El Salvador	204.2	204	25	
11.	Kenya	128.8	167	11.2	
12.	Costa Rica	162.5	166	14	
13.	Nicaragua	87.4	88	10	
14.	Russia	79	82		
15.	Turkey	38	82		
16.	paapua-New Guinea	56	56		
17.	Guatemala	53	52		
18.	Portugal	23	29		
19.	China	27.8	24		
20.	France	14.7	16		
21.	Ethiopia	7.3	7.3		
22.	Germany	8.4	6.6		
23.	Austria	1.1	1.4		
24.	Australia	0.2	1.1		
25.	Thailand	0.3	0.3		
	TOTAL	9,981.9	10,959.7		

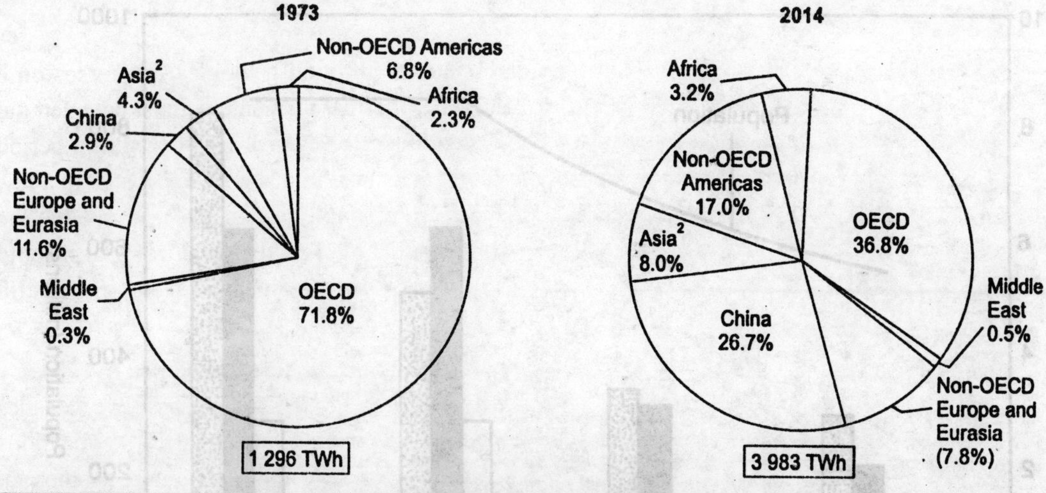

Fig. 24.27 : 1973 and 2014 regional shares of hydro production[1]

24.7 HYDROELECTRICAL POWER ENERGY

Various countries developed a large number of hydroelectrical power station through creation of dams over rivers. The largest one is the Three Gorge Dam of China alone produces 18,500 MW/day power. Of course mega dam creation for power generation is very much debatable. The regionwise shares of hydro power generation is shown in Fig. 24.27.

24.8 ENERGY USE PATTERN IN INDIA

The energy consumption in India was ranging between 356-425 Mtoe (or 14.9-17.8 EJ/year) in 1991, depending on the extent of biomass use considered for domestic sector. The share of biomass energy ranged from 36 to 46 per cent. Among the commercial sources, coal was the dominant source of energy followed by oil and among biomass fuels, the most primary energy use per capita was in the range 17.7-21.1 GJ/Yr, which is very low compared to the OECD countries (200 GJ/capita /Yr) (Table 24.16).

Table 24.16: Primary energy use in India (1991)

Fuels	Quantity	Using mean estimate for biomass		Using low estimate for biomass	
		Total energy (PJ/Tr.)	Per capita energy (GJ/Yr.)	Total energy (PJ/Yr.)	Per capita energy (GJ/Yr.)
Coal	217 Mt.	6325 (35.4)	7.49	6325 (42.2)	7.49
Oil	53.7 Mt.	2293 (12.8)	2.71	2293 (15.3)	2.71
Natural gas	17890 Mm³	665 (3.7)	0.79	665 (4.4)	0.79
Electricity	77789 GWh	280 (1.6)	0.33	280 (1.9)	0.33
Total Commercial		9563 (53.5)	11.32	9563 (63.8)	11.32
Fuel Wood	298 Mt	4470 (25.0)	5.29	3405 (22.7)	4.03
Crop residues	156 Mt	2267 (12.8)	2.68	1500 (10.0)	1.77
Cattle dung	114 Mt	1562 (8.7)	1.85	507 (3.5)	0.60
Total Biomass		8299 (46.5)	9.82	5412 (36.2)	6.40
Total energy		17862 (100) ≡ 425 Mtoe	21.14 ≡ 0.50 toe	14973 (100) ≡ 356 Mtoe	17.72 0.42 toe

Source: TERI (1992), CMIE (1994)

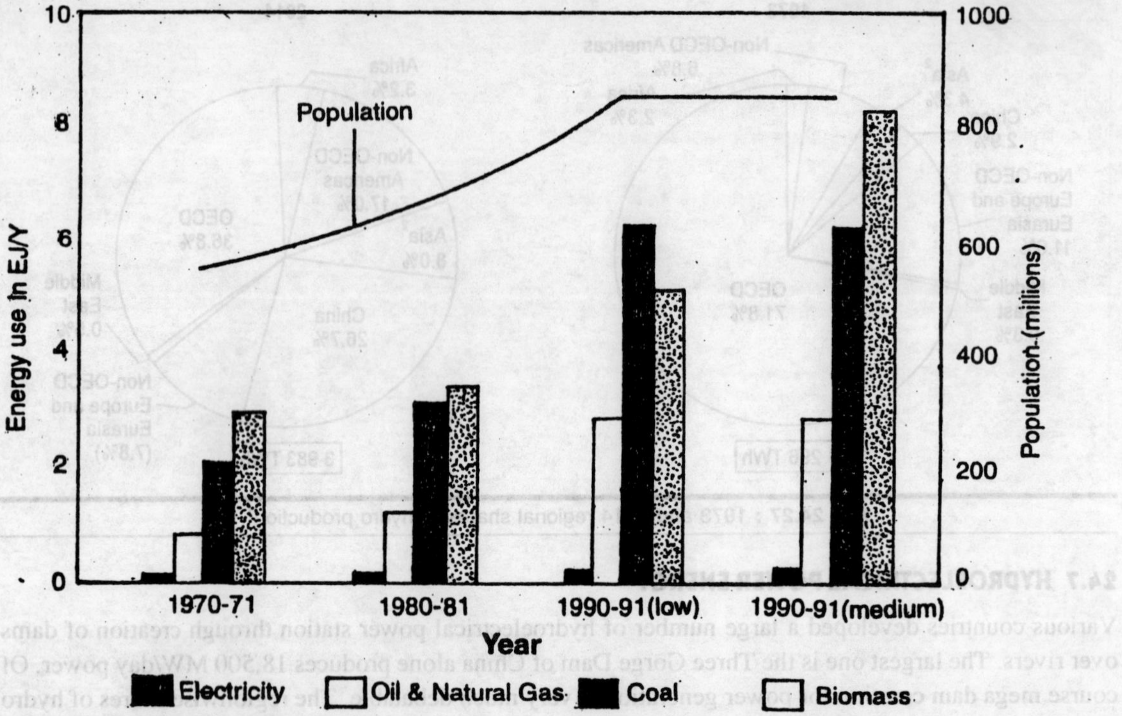

Fig. 24.28 : Population verses energy—the pattern in India (1970-90)

However, for all kinds of fuel use, demand is very much increasing with steady rise of population since 1970s. The details are given in Fig. 24.28. The current status of power grid connected installed capacity power (MW) is shown below:

Grid connected installed capacity from all sources as of 28 February 2017*		
Source	Installed Capacity (MW)	Share
Coal	189,047.88	59.93%
*RES MNRE (http://www.mnre.gov.in/)	50,018.00	15.86%
Large Hydro	44,413.43	14.08%
Gas	25,329.38	8.03%
Diesel	837.63	0.27%
Nuclear	5,780.00	1.83%
Total	**315,426.32**	**100.00%**
Source: *Renewable energy source		

The domestic sector dominates biomass energy use and fuel wood is the dominant source of energy in this sector in rural areas. Fuel wood is also used by urban households. Presently development efforts are mainly concentrated on fuel wood conservation and production, mainly for the domestic sector. Sectoral per capita biomass use in India is given in Table 24.17.

India is a large country with a geographical area of 328 M ha, with vast climatic variations and vegetation types ranging from tropical moist evergreen forest to dry deciduous forests and temperature maintain forests. We can thus expect variation in the types and extent of biomass use according to agro-climatic zones, since biomass uses likely to be determined by local climate (temperature and rainfall), vegetation type, fuel wood availability and access to fuel wood. The details are given in Table 24.18.

Coal and Lignite Reserve in India

- Coal deposits are mainly confined to eastern and south central parts of the country. The states of Jharkhand, Odisha, Chhattisgarh, West Bengal, Madhya Pradesh, Telangana and Maharashtra account for 98.58% of the total coal reserves in the country. The State of Jharkhand had the maximum share (26.29%) in the overall reserves of coal in the country as on 31st March 2016 followed by the State of Odisha (24.58%).
- As on 31.03.16, the estimated reserves of coal were 308.80 billion tonnes, an addition of 2.20 billion tonnes over the last year. There has been an increase of 0.7% in the estimated coal reserves during the year 2015-16 with Chattisgarh accounting for the maximum increase of 2.05%.

Estimated Reserves of Coal as on 31.03.2016

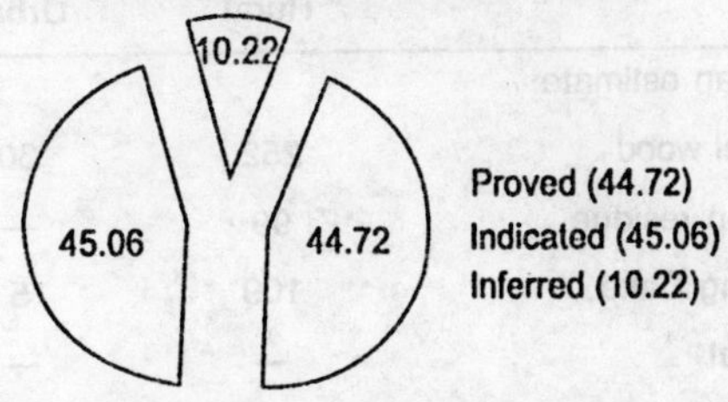

Total reserve = 308.80 billion tonnes

- The estimated total reserves of lignite as on 31.03.16 was 44.59 billion Tonnes against 44.12 billion tonnes on 31.03.15.

Estimated Reserves of Crude Oil in India as on 31.03.2016

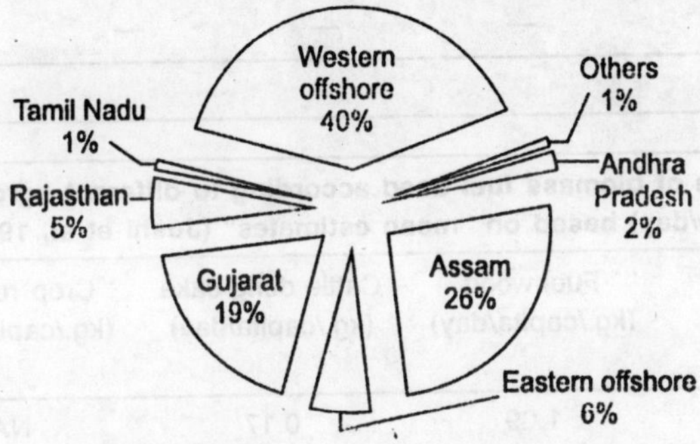

Total estimated reserves = 621.10 million tonnes

- There was decrease of 2.28% in the estimated reserve of crude oil for the country as a whole during 2015-16 as compared to the position a year ago. During the same period, estimated reserves of crude oil in Arunachal Pradesh, Rajasthan and Assam decreased by 44.75, 17.04 and 2.11% respectively, while the same in Tamil Nadu, Andhra Pradesh, Gujarat, Western Offshore and Eastern Offshore increased by 18.42%, 15.30%, 2.58%, 1.88% and 0.59% respectively.
- The estimated reserves of Natural Gas in India as on 31.03.2016 stood at 1227.23 Billion Cubic Meters (BCM) as against 1251.90 BCM as on 31.03.2015.

Estimated Reserves of Natural Gas in India as on 31.03.2016

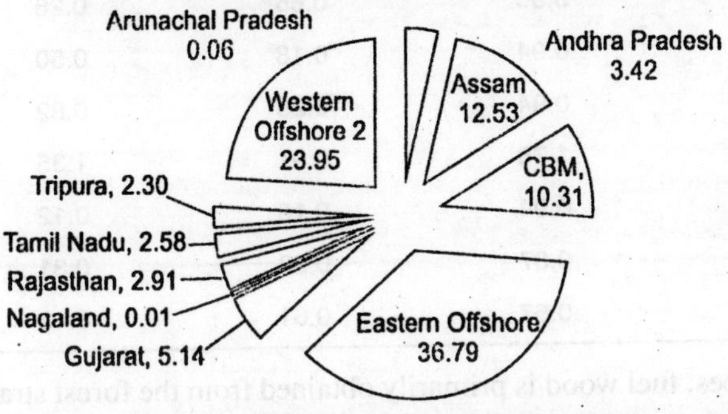

Total reserves = 1227.23 billion cubic meters

- The estimated reserves of Natural Gas decreased by 1.97% over the last year. The maximum contribution to this decrease has been from Western Offshore, followed by Eastern Offshore.

Table 24.17: Indian biomass use pattern

Types of biomass	Domestic sector (Mt/Yr.)		Industry and establishments (Mt/Yr.)		Total use (Mt/Yr.)	Per capita energy (GJ/Yr.)
	Rural	Urban	Rural	Urban		
Mean estimate:						
Fuel wood	252	30	6	10	298	5.29
Crop residue	99	—		57	156	2.68
Dung Cake	109	5			114	1.85
Total	—	—	—	—	—	9.82
Low estimate:						
Fuel wood	181	30	6	10	227	4.03
Crop residue	40	—	—	57	97	1.77
Dung	32	5			37	0.60
Total	—	—	—	—	—	6.40

Source: Joshi et al, (1992)

Table 24.18: Variation in type of biomass fuel used according to different agro-climatic zones in India (Kg./capita/day) based on "mean estimates" (Joshi et al, 1992 b)

Agro-Climatic Zone	Fuel wood (kg./capita/day)	Cattle dung cake (kg./capita/day)	Crop residue (kg./capita/day)	Energy value of biomass fuel use (GJ/capita/Yr.)
Western Himalayan Zone	1.09	0.17	NA	6.82
Eastern Himalayan Zone	1.90	0.20	0.37	13.16
Lower Gangetic Zone	1.93	0.56	0.62	10.83
Middle Gangetic Zone	0.45	0.41	0.30	5.94
Upper Gangetic Zone	0.29	1.29	0.04	8.23
Trans Gangetic Zone	0.38	0.57	0.34	6.54
Eastern Plateau and Hills	3.36	0.01	0.24	19.58
Central Plateau and Hills	0.95	0.65	0.26	9.69
Western Plateau and Hills	0.94	0.18	0.50	8.42
Southern Plateau and Hills	0.94	0.25	0.82	10.29
East Coast Plains and Hills	1.78	0.91	1.35	20.70
West Coast Plains and Hills	0.94	0.16	0.12	6.52
Gujarat Plains and Hills	0.87	0.23	0.31	7.38
Western dry Regions	0.57	0.61	0.13	6.79

Among the biomass types, fuel wood is primarily obtained from the forest strands and as such rapid forest destruction takes place during 1960s to 1980s. To meet up these challenges, afforestation programmes were adopted particularly since 1980s in massive forms. The sources of fuel woods as used in the domestic sector is given in Table 24.19.

Table 24.19: Summary table of sources of fuel wood used in the domestic sector mainly for cooking

Sources	Details	Total fuel wood contribution in Mt/Yr
1. Forests (Total forest strand 64Mha as on 1986)	1. Felling of trees	19.0
	2. Lopping of twigs & branches in existing forests	42.0
	3. Logging wastes	9.5
	Total	70.5
2. Tree plantation (17 Mha as on 1991)	Tops + twigs + branches + poles	40.0
3. Farm tree	Twigs + branches	46.0
4. Homestead garden	Twigs + branches	16.0
5. Degraded Land (52.8 Mha)	Bushes (@ 0.5 + wood/ha/Yr.)	26.4
6. Land for shrub (78.5 Mha)	Shrubs (@ 0.25 + /ha/Yr.)	19.6
Total wood supply		218.5

The importance of crop residues as a source of fuel for domestic use has also been well-recognised. Crop residues production and end use in India for fuel and fodder are given in Table 24.20.

Table 24.20: Crop residue production and use as a source of fuel in India (Mt dry/Yr. for 1985-1986)

Residue Types	Fodder for Livestock	Domestic fuel	Fuel in Industry and establishment	Other uses or unused
Cereals + millets	254	2	16	5.5
Pulses	19	10	—	—
Oil seeds	12	16	—	0.8
Fibre crops	—	12	—	—
Mulberry	—	2	—	—
Sugarcane	—	—	28	19.3
Coconut + arica nut		7	—	2.2
Coffee + Tea	—	3	—	—
Total	285	52	44	27.8

The population in India (121 crores in 2010) is growing at a rate of about 1.76 per cent annually and it is observed that biomass use is keeping pace with population growth. The use of biomass as a traditional fuel is associated with a lower quality of life and environmental degradation. It is thus necessary to understand what India is doing to alleviate such problems associated with the dependence on and use of biomass and to draw lessons therefrom before exploring the potential role for bioenergy in promoting sustainable development. During past two decades enormous efforts were made to harness bioenergy sources in India to meet up the growing demand of energy for domestic use in particular, as such Department of Non-conventional Energy Sources (DNES) under the Ministry of Power & Energy was set up on 1982.

The major bioenergy programmes of DNES are:

Conservation	Production	Conversion
Biogas	Fuel wood production	Biogas gasification
Improved stores	Energy plantation	Liquid fuels
		Cogeneration

India is a pioneer in biogass programmes, with the floating-drum biogas plant being developed in 1952. By 1993, about 1.85 million family biogas plants and 875 community biogas plants has been built.

The main goal of the biogas programme are as follows:

- Provide cooking energy in a clean & non-polluting form;
- Produce enriched manure to supplement the use of chemical fertiliser;
- Improve the quality of life of rural women; and
- Improve sanitation and hygiene.

Biogas (a mixture of about 60% methane and 40% CO_2) is a combustible gas which is the product of anaerobic fermentation of cellulosic materials, such as animal dung and plant leaves and also washes from food processing and the household. Biogas can he combusted directly as a source of heat for cooking (calorific value 23 MJ/m^3) or used with internal combustion engined for mechanical and electrical applications. The slurry produced after digestion can be used directly as valuable manure for crop production. Two types of biogas designs are promoted in India—floating drum and fixed dome. The characteristic features of these types are given in Table 24.21.

Table 24.21: Features of fixed and floating drum designs in India (Ravindranath et al., 1992)

Fixed drum	Floating drum
Digester and gas holder, masonry or concrete structure	Digester, masonry Gas holder, mild steel or fibre glass
Requires high masonry skills	Low masonry or fabricating skills
Low reliability due to high construction failure	High reliability, gas holder prefabricated
Variable gas pressure	Constant gas pressure
Digester could be inside the ground	Requires space above ground for three tanks: inlet, digester, outlet
Low cost (2m³ = Rs. 5,000)	High cost of plant due to gas holder (cost for 2m³ = Rs. 8,000)

In spite of enormous potentiality of biogas development in rural areas, yet it is not developed as expected due to a number of barriers. The crucial constrains are presented in Table 24.22.

Table 24.22: Barriers to utilising family biogas potential (Ravindranath et al, 1992)

	Demand-side barriers	Supply-side barriers
Economic barriers	High capital cost/first cost compared to zero cost wood stoves.	Inadequate funding for biogas at national level for organising dissemination programmes.
	Inadequate credit and high cost of credit, as own capital is scarce even for economic activities.	Inadequate subsidy arising out of low funding of the programme, especially for the plains.
	No income generation from investment in biogas or cash savings; competition from income-generating activities for investment.	Low market volumes for the entrepreneurs and government agencies, thus high cost of organising biogas programmes; cost per unit of construction and maintenance high.
	High repair and maintenance cost and failures demotivate users.	
	Easy availability of wood or no cash cost involved in procurement of wood.	

Contd.

Table 24.22: Barriers to utilising family biogas potential (Ravindranath et al., 1992)

	Demand-side barriers	Supply-side barriers
Technical barriers	Inadequate dung resource/animal ownership	Inadequate supply of raw materials and absence of quality control : cement, bricks, steel in rural areas.
	Problem of initial feeding of plant, which requires large amounts of dung; lack of information on options for initial feeding.	Inadequate workshop facilities for fabricating components.
		Inadequate or absence of trained manpower for construction and maintenance.
	Water scarcity for daily feeding of plant; especially in summer season and drought years in some regions.	Inadequate funding for research and development (R & D).
	High rates of failures and frequency of repairs in some regions.	
	No insurance against failures.	
Institutional barriers	Complex procedures for obtaining loan, subsidy, and repair services.	Absence of NGOs in most areas and limited scope for dissemination through NGOs.
	No instituions to provide back-up services for repair and maintenance problems.	Absence of private entrepreneurs, no incentives and rigid policies for their participation.
	No responsibility for failures leads to lack of faith by the users.	Target-orientied nature of programme leads to lack of quality control.
	Cooking energy; a problem for women but men take investment decisions.	Non-availability of women as builders and skilled persons for biogas dissemination programme.

Biomass, or more particularly wood, can be converted to a high-energy combustible gas for use in internal combustion engines for mechanical or electrical application. The process is generally known as gasification and the technology has been known for decades, but its application to power generation is of recent origin. A biomass gasifier consists of a reactor where, under controlled temperature and air supply, solid biomass is combusted to obtain a combusting gas (consists of H_2 and CH_4). This gas passes through a cooling and cleaning system before it is fed into a compression-ignition engine to run in dual fuel mode for generation of mechanical power or electricity (by coupling to a generator). An assessment of its potential concluded that India presents a unique opportunity for large scale commercial exploitation of biomass gasification technology to meet a variety of energy needs, particularly in the agricultural and rural sectors. The large potential of biomass gasification for water pumping and power generation for rural electrification was recognised and a research, development, and dissemination programme was launched during the period 1980-85.

Although initially a few programmes were launched based on imported designs, the Indian programme is now largely locally driven. A number of research and development (R & D) groups are working on the programme, and a number of designs of varying capacities have been developed. The biomass gasifier programme in India is along the following lines:

1. R & D in biomass gasification technology;
2. Pilot testing and evaluation, including prototype production;
3. Demonstration and evaluation of biomass gasifiers;
4. Training and manpower development;
5. Commercialisation and manufacturing.

The details of statewise list of commissioned biomass power/cogeneration projects shown in Table 24.23.

Table 24.23: State-wise/Year-wise list of commissioned biomass power/cogeneration project (as on 1.4.2016)

(IN MW)

State	Up to 31.03.2012	2012-13	2013-14	2014-15	2015-16	Total
1. Andhra Pradesh	363.25	17.5				380.75
2. Bihar	15.5	27.92				43.42
3. Chattisgarh	249.9		15	15		279.9
4. Gujarat	20.5	10	13.4	12.4		56.3
5. Haryana	35.8	9.5				45.3
6. Karnataka	441.18	50	112	111	158	872.18
7. Madhya Pradesh	8.5	7.5	10	9		35
8. Maharashtra	603.7	151.2	185.5	184	96.38	1220.78
9. Odisha	20					20
10. Punjab	90.5	34	16	15		155.5
11. Rajasthan	83.3	10	8	7		108.3
12. Tamil Nadu	532.7	6	32.6	31.6	39	626.9
13. Uttarakhand	10		20	20	13	50
14. Uttar Pradesh	644.5	132			93.5	842
15. West Bengal	16	10				26
Total	3135.33	465.6	412.5	405	400	4831.33

24.8.1 Research and Development

The ministry reports a number of achievements:

1. Wood gasifier designs of 5-100 kW capacity are developed and available for mechanical and electrical applications.

2. Gasifiers based on crop and plantation residues, such as rice husks, groundnut shells, cotton stalks and coconut shells are at various stages of development and field testing.

Demonstration programme

A demonstration programme was launched in 1986 and about 800 gasifier systems of various ratings were disseminated, with a total capacity of 6.5 MW, until 1992 (Table 24.24). The majority of the systems are small wood gasifiers for irrigations, water pumping and some are used for electricity generation.

Table 24.24: Dissemination of biomass gasifier systems in India (Subramanya et al., 1993)

	Number	Total capacity (kW)
Total number of systems installed (upto 1992)	800	6,500
Total number of systems with details	668	4,025
Capacity of systems		
3.7 kW	538	2,018
7.5 kW	89	667
20.0 kW	24	480
40.0 kW	14	560
100.0 kW	3	300

Though biomass use appears to be a potential material for energy producing raw materials in India for a long time, yet it is high time to evaluate the long term impact on environment arises due to biomass use.

It is necessary to understand the environmental implications of biomass use as an energy source via:

- Contribution to deforestation and forest degradation, leading to loss of biodiversity and soil degradation;
- Atmospheric pollution from emissions of greenhouse gases during combustion of wood, with its implications for climate change;
- Release of pollutants (indoor air pollution) leading to domestic health hazards (particularly for women during cooking); and
- Loss of nutrients due to combustion of cattle dung crop residues.

24.8.2 Renewable Energy

This is only sustainable form of energy in this country which has started receiving attention in recent decades. The overall renewable energy generation potential and the quality actually harnessed in recent years is given in Table 24.25.

India was the first country in the world to set up a ministry of non-conventional energy resources, in the early 1980s. India's overall installed capacity has reached 329.4 GW, with renewables accounting for 57.472 GW as of 14 June 2017. 61% of the renewable power came from wind, while solar contributed nearly 19%. Large hydro installed capacity was 44.41 GW as of 28 February 2017 and is administered separately by the Ministry of Power and not included in MNRE targets.

From 2015 onwards the MNRE began laying down actionable plans for the renewable energy sector under its ambit to make a quantum jump, building on strong foundations already established in the country. MNRE renewable electricity targets have been upscaled to grow from just under 43 GW in April 2016 to 175 GW by the year 2022, including 100 GW from solar power, 60 GW from wind power, 10 GW from bio power and 5 GW from small hydro power. The Ministry of Power has announced that no new coal-based capacity addition is required for the 10 years to 2027 beyond the 50 GW under different stages of construction and likely to come online between 2017 and 2022. The ambitious targets would see India quickly becoming one of the leading green energy producers in the world and surpassing numerous developed countries. The government intends to achieve 40% cumulative electric power capacity from non-fossil fuel sources by 2030.

Table 24.25: Grid connected renewable electricity
Installed grid interactive renewable power capacity (excluding large hydropower) in India as of 31 March 2017 (RES MNRE)

Source	Total Installed Capacity (MW)	2022 target (MW)
Wind power	32279.77	60,000.00
Solar power	12288.83	100,000.00
Biomass power (Biomass & Gasification and Bagasse Cogeneration)	8182.00	*10,000.00
Waste-to-Power	114.08	
Small hydropower	4379.85	5,000.00
Total	**57244.23**	**175,000.00**

**Source*: MNES Report, 2017.

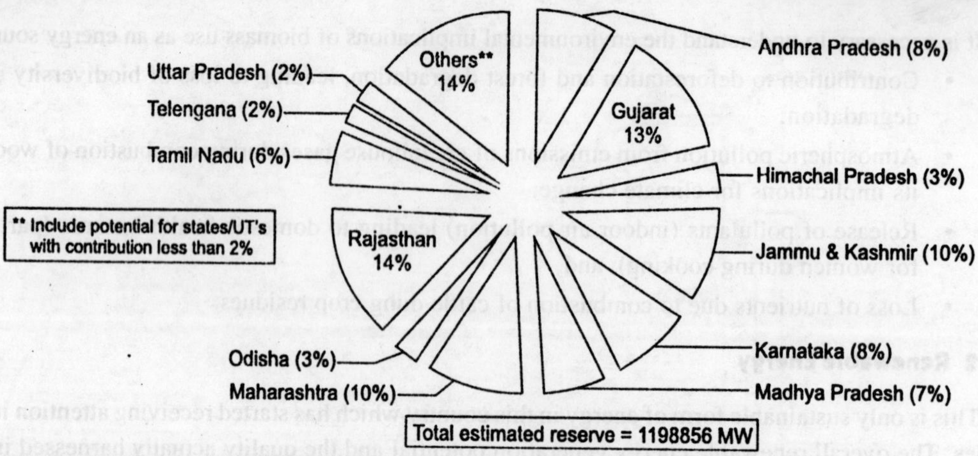

Fig. 24.29: Statewise estimated potential of renewable power in India as on 31.03.2016

In addition statewise estimated potential of renewable power is shown in Fig. 24.29.

24.8.3 Biogas — An Energy from Wastes

Fuel can also be produced from organic waste products viz., sewage, garbage, manure or crop residues. These wastes decompose in the absence of air—methane gas released. This methane gas acts as a biogas.

This is an alternative source of energy, which has immence utility in waste disposal and energy generation in environmentally sound manner. Domestic or community bases biogas plants are now effectively operated in various countries over past couple of decades.

24.8.4 Biogas Technology

A biogas plant supplies energy and fertiliser. It improves hygiene and protects the environment. 'Biogas' is produced by putrefactive bacteria which break down organic material under airless conditions. The process is called "anerobic digestion".

All feed materials consist of organic solids, inorganic solids and water. The waste of much feed materials can be used for biogas generation. A biogas plant is simple so that it can be operated by housewife also.

Biogass consists of about 60% methane and 40% carbon dioxide. It is somewhat lighter than air and has an ignition temperature of approximately 700°C (diesel oil 350°C; petrol and propane about 500°C). The temperature of the flame is 870°C.

Biogas digester can decompose a number of organic wastes. Usually, for better gas production the following percentage of organic wastes are mixed together. Cattle manure (65%), Poultry manure (55%), Straw (59%), Grass (70%), Leaves (58%), Kitchen waste (50%), Algae (63%), Water hyacinth (52%). The first gas from a newly filled biogas plant contains too little methane. The gas formed in the first 3 to 5 days must, therefore, be discharged unused.

During the digestion process, gaseous nitrogen (N_2) is converted to ammonia (NH_3). In this waste soluble form the nitrogen is available to the plant as a nutrient.

Three types of biogas plants are known: balloon plant, Fixed-drum plant and Floating drum plant. Fixed-dome plants are considered to be less expensive and easily operable, though there is scope for gas leakage and fluctuation of gas pressure.

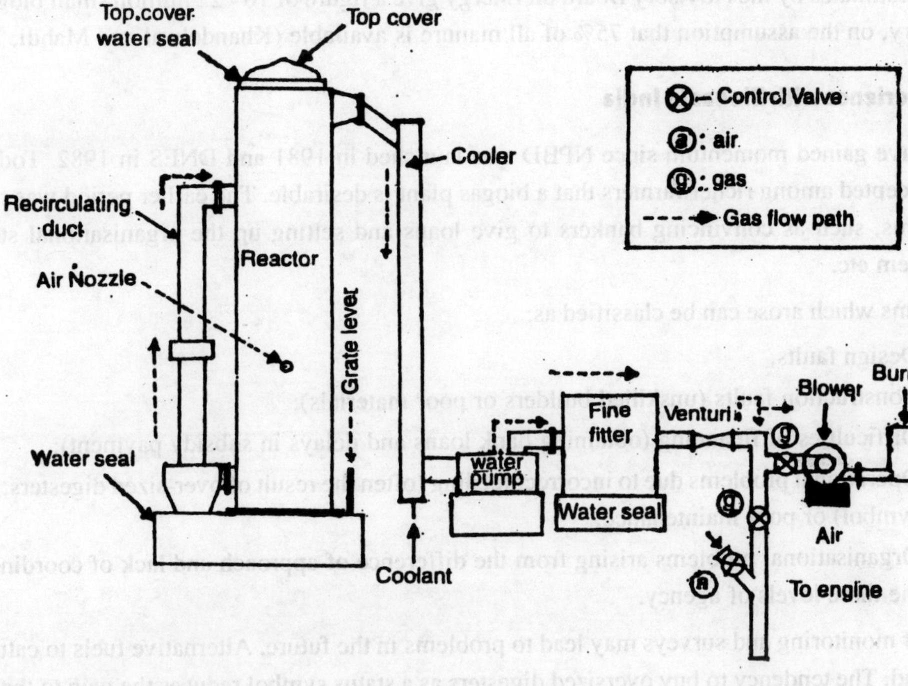

Fig. 24.30 : A 20 KW gasifier with cooling and cleaning system

24.8.5 Biogas Programmes in Developing Countries

Most countries became aware of biogas technology by the middle of the twentieth century. However, real interest in biogas was aroused from 1973 onwards, with the onset of the energy crisis, which drew general attention to the depletion of fossil fuel, energy resources and the need to develop renewable sources of energy, such as biogas. The importance of biogas as an efficient, non polluting energy source is now well-organised.

International organisation like ESCAP, FAO, UNBIDO, WHO, UNEP have done considerable work in disseminating and development of biogas technology in developing countries in particular Afghanisthan, Bangladesh, Bhutan, Myanmar, China, India, Indonesia, Iran, Israel, Korea, Malayasia, Nepal, Sri Lanka, Thiland, Philippines, Cuba, Chile, Brazil, Costa Rica, Mexico, Africa, Panama etc.

India has a total population of about 1.2 billion (1210.2 crores) people of whom 80% live in rural areas in some 576,000 villages. About 70% of them are landless. There are about 237 million head cattle, under the ownership of approximately 52 million households: of these 57% own 1-3 head of cattle, 27% own 4-6 head, 8.7% own 7-9 head and 6% own above this number.

According to rough estimates, 50% of India's total energy comes from non-commercial sources, upon which the majority of the rural population survives. These include: firewood (65%), Dungcakes (15%) and agricultural wastes (20%). On average, these items cover 84% of rural household energy requirements. Between 1/3 and 1/2 of all recoverable cattle dung is burned as fuel. The annual requirement for firewood has been estimated as 133 tonnes. Total annual production is 49 million tonnes, leaving an annual deficit of 84 million tonnes.

Latest estimates by the Advisory Board on Energy give a figure of 16 - 22 million small biogas units in the country, on the assumption that 75% of all manure is available (Khandelwal and Mahdi, 1986).

24.8.6 Experience with Biogas in India

Activities have gained momentum since NPBD was launched in 1981 and DNES in 1982. Today, it is generally accepted among richer farmers that a biogas plant is desirable. The earlier period was taken up with problems, such as convincing bankers to give loans and setting up the organisational structure, subsidy system etc.

Problems which arose can be classified as:

(a) Design faults;

(b) Construction faults (unskilled builders or poor materials);

(c) Difficulties of financing (obtaining bank loans and delays in subsidy payment);

(d) Operational problems due to incorrect feeding (often the result of over-sized digesters, a status symbol) or poor maintenance;

(e) Organisational problems arising from the difference of approach and lack of coordination at the three levels of agency.

Lack of monitoring and surveys may lead to problems in the future. Alternative fuels to cattle dung must be found. The tendency to buy oversized digesters as a status symbol reduces the gain to the user. It is clear that benefits derived from the effluent are 2-3 times higher than the direct benefits of biogas, though this and other points are assumptions that are not backed by proper research.

24.8.7 Biogas Plants

Up to 1986, a total of 642,900 digesters had been built: in 1985/6 alone, the total was 185,800. In view of the huge potential, targets have been gradually increased. Construction capacity almost doubles each year. At the current capacity, it would still take 50 years to saturate the market. However, this biogas programme, together with others (e.g., wood-saving stoves), has to compete with the pace of deforestation and other environmental hazards.

Community and Institutional Biogas Plants (IBP) are being constructed in India. Poor farmers and dalits are supposed to be involved and to participate in operation of community biogas plants (CBP).

24.8.8 Use of Biogas

The gas is commonly used for cooking and lighting. There are number of enterprises in each state that produce stoves and lamps. At some CPBs and IBPs, biogas operates engines or agricultural equipment. A number of enterprises in India manufacture or adapt diesel engines.

24.8.9 Utilisation of Effluent

The effluent is usually dried in the sun, either separately or in combination with agricultural wastes. Partial composing is performed, after which it is applied to the fields in a solid state. There is no information on how widespread the use of effluent is, how it is applied or in what quantities.

A study was made comparing its use with that of fresh dung, for various purposes as of October 1985. It was found that the sale value of digested spent slurry, scientifically composted, is 8 times higher than that of fresh manure sold to the owner of a digester.

24.8.10 Cost of Installation

Cost of installation varies according to the type and size of the plant, increasing by about 65% between 1981 and 1986 or about 13% per year. There are also variations from state to state and district to district. There is an effect of economy of scale in both digester types. A comparison of cost may be made on the basis of cost per m^3 of digester volume. For the same hydraulic retention time (HRT), the Janata plants are cheaper by 33% or more than KVIC digesters. The system of financial assistance is discussed below.

24.8.11 Annual Costs and Savings

Annual costs through depreciation, interest on loans, maintenance, repairs, overheads, and labour costs have been calculated comparatively for digesters of 2 m^3 gas/day for KVIC and Janata models (Myles 1985). In the first 5 years, annual costs amount to Rs. 2,480 for a KVIC model at 40 days HRT, and Rs. 1,770 for a Janata plant at the same HRT, the latter model costing Rs. 1,920 for the 55 days HRT type. On the other hand, annual savings through replacement of kerosene by biogas and fertiliser by composted effluent yield a marginal annual gain in the first 5 years. Thereafter income and profit are dubious: first, calculations are made on the assumption that kerosene is used as fuel, when wood is more common, and secondly, there is a tendency to buy digesters larger than necessary. Furthermore, figures on income and profit are based on the subsidy given to the farmers.

24.8.12 Financial Assistance from Government

In 2000, DNES provides financial assistance to:

(a) purchaser's subsidy;

(b) service charge to State Governments and the KVIC;

(c) turnkey construction fee;

(d) incentives to promoters;

(e) training programmes;

(f) repair of plants with structural problems.

24.8.13 Organisation of the biogas sector

India's "National Project on Biogas Development" (NPBD) for mass diffusion of digesters was launched at the end of 1981, using a "multi-agency, multi-model" approach. The programme is centrally administered by the Department of Non-Conventional Energy Sources (DNES), within the Ministry of Energy. DNES is responsible for the coordination of implementation and R & D of family-sized and community biogas digesters. It has to approve designs and to allocate budgets for training and subsidies. At State, District, Block and Village levels, staff is provided for the biogas scheme as shown in Table 24.26, though not in all cases is the scheme fully staffed.

Table 24.26: Organisation Structure (India)

No.	Level		Coordinating agencies Staff
19	State	14 States:	1 Joint Director 1 Sr. Agri Off 1 Accountant 1 UDC 1 Secretary
		5 States:	1 Sr. Agri Off 1 UDC
		All States:	National Bank for Agricultural Rural Development.
112	District		1 BDO 1 Supervisor 5 Technician
50 - 60% Dist.	Block		5 - 10 Rural Welfare Officer
30 -100% Block	Village		Village level workers

Khandelwal, K.C. and Mahdi, S.S. 1986. *Biogas technology: A Practical Handbook*. TMH, New Delhi.

24.8.14 Potential for biogas generation and digester construction

The only fuel available for family-sized digesters is cattle dung. This facilitates the assessment of potential, but, on the other hand, it restricts the use of biogas to cattle-owners. On the assumption that 4 head of cattle are required to generate 2m^3 gas (the cooking needs of a rural family of 8), there is a potential for 22 million digesters, though when cost to the family is taken into consideration, the maximum potential is 10 million plants, representing 19% of cattle-owning families. Given an average capacity of 4 m^3 gas per day, 40 million m^3 could be generated daily. This would involve about 71 million head of cattle (10 kg dung per day, 20% or 0.28 m^3 gas per kg total solids, about 30% of India's total cattle population).

Up to at state level, the organisation is called "Biogas Cell". 14 States, with a target of 10,000 digesters, are supposed to have a staff of 6, the remaining states a staff of 2.25 central departments in as many states have been set up, and others in a selected 100 Districts. In other Districts, State Governments have either set up such cells under the State Plan Sector or involved staff of allied schemes, e.g., minor irrigation programmes.

The broad terms of reference of the Biogas cell attached to the central department in each State Government includes:

(a) Overall planning of the execution of the programme in the state;
(b) State-level coordination of different departments/agencies;
(c) Institutional financing;
(d) Arrangements for raw materials;
(e) Monitoring of programme and the submission of progress reports to the Government of India;
(f) Maintenance of subsidy accounts and the submission of expenditure reports to the Central Government.

Agencies vary at State level: they may be State Department of Agriculture, Agro-Industries Corporation or State Department of Non-Conventional Energy, etc. They have varying levels of inolvement in extending technology.

At District level, the executive agencies are governmental: Khadi and Village Industries Commission (KVIC) and Action for Food Production (AFPRO). This multi-agency approach is necessary if targets are to be met. The construction of 20,000 digesters annually is channeled through KVIC (1985-1986 target). Since starting in 1974, they have constructed 161,000 floating drum digesters. KVIC has a technical staff of 300 (1 Director, 2 Assistant Directors, 40 Development Officers, 100 Assistant Development Officers and 160 Supervisors). In addition, many individual workshops have been recognised by KVIC.

AFPRO coordinates a network of NGOs at grassroot level, using the fixed dome digester (the Janata Model) exclusively (Janata = people). AFPRO concerns itself with institution building, placing emphasis on the organisation of many training courses in rural area, by competent NGOs. AFPRO has planned and initiated action to develop 80-100 Biogas Extension Centres (BEC), involving 60-100 NGOs. Up to 1985, 60 NGOs, with 90 such centres have been developed and are involved in construction activity. Their total construction capacity is about 9,000 digesters per year. Most of these NGOs, like AFPRO, promote several rural technologies, biogas among them. Each BEC is capable of constructing 100 biogas plants per year, as well as providing regular after-construction services to plant owners. Each BEC would have a staff of one supervisor and a master-builder. The former has the task of education and motivation of farmers, collection and processing of applications, levying cement and other materials, supervision and coordination. The latter acts as leader of the construction teams, professional rural builders hired by the farmers at daily rates. AFPRO has raised overseas funds (The Canadian Hunger Foundation) to support 80-100 BECs over a 3-5 year period. It is expected that, after this initial support, each centre will be self-sufficient.

Biogas Promotion — INDIA

- In India the First Biogas Plant was commissioned in the year 1962.
- This was known as KVIC Floating Dome Model (Popularly known as Indian Model).
- Till '79, the promotion was not in serious way; and in '79, the First Fixed Dome Model (Janata) was commissioned.
- In '82, Govt. of India launched 'National Biogas Programme' countrywide thro' Ministry of Non-conventional Energy Sources.
- Potential of Domestic Biogas Plants in India – 12 million.
- Achievement till end of 2005 – 3.8 million (31.66%).

Models that are Promoted

1. Floating Dome KVIC Model.
2. Fixed Dome – Deenbandhu Model.
3. Bag Type portable digester made of rubberized nylon fabric.
4. Low cost fixed dome models made of alternate construction materials, like the prefabricated Ferro Cement or Reinforced cement and in situ built Ferro cement materials.

All these are promoted by the states in the country by State Nodal Agencies of MNES (Ministry of Non-conventional Energy Sources, Govt. of India.

These are couple of off grid power station/wind power station developed in increasing during past one decade. The details are given below:

Off-grid power as of 31 July 2016 (MNRE) Capacities in $MW_{(EQ)}$	
Source	Total installed capacity (MW)
Biomass (non-bagasse) Cogeneration	651.91
SPV Systems	438.95
Biomass Gasifiers	186.88
Waste to Energy	164.45
Water mills / micro hydel	18.81
Aero-Generators / Hybrid systems	2.98
Total	**1463.98**

Wind farm	Producer	State	Current capacity (MW)
Muppandal wind farm	Muppandal Wind	Tamil Nadu	1,500
Jaisalmer Wind Park	Suzlon Energy	Rajasthan	1,275
Brahmanvel windfarm	Parakh Agro Industries	Maharashtra	528
Dhalgaon windfarm	Gadre Marine Exports	Maharashtra	278
Chakala windfarm	Suzlon Energy	Maharashtra	217
Vankusawade Wind Park	Suzlon Energy	Maharashtra	189
Vaspet Windfarm	ReNew Power	Maharashtra	144

24.9 INDIA'S NUCLEAR ENERGY PROGRAMME

The first two units of power, each with a capacity of 210 MW, were installed at Tarapur in 1969. They are enriched uranium reactors purchased on a turnkey basis from International General Electric of USA. Then in 1973, Rajasthan Atomic Power Station (RAPS) was set up under Canadian collaboration with two 220 MW units of PHWR designed by the Canadians. In 1983, the Madras Atomic Power Station (MAPS) was set up with two units of 220 MW each. Then series of power stations are set up at Narora, Kakrapar, Kaiga etc. By 2009 a total over 5000 MW nuclear power generation capacity was achieved by Nuclear Power Corporation of India Limited (NPCIL). However, India is planning to reach a target around 20,000 MW by the year 2020 with the support of various uranium donor countries like USA, Russia, Canada, France, Germany and so on. However the major debate of this power generation lies with disposal of nuclear spent fuel waste. There are a few options for disposal, but uncertainties and cost effectiveness is a big question.

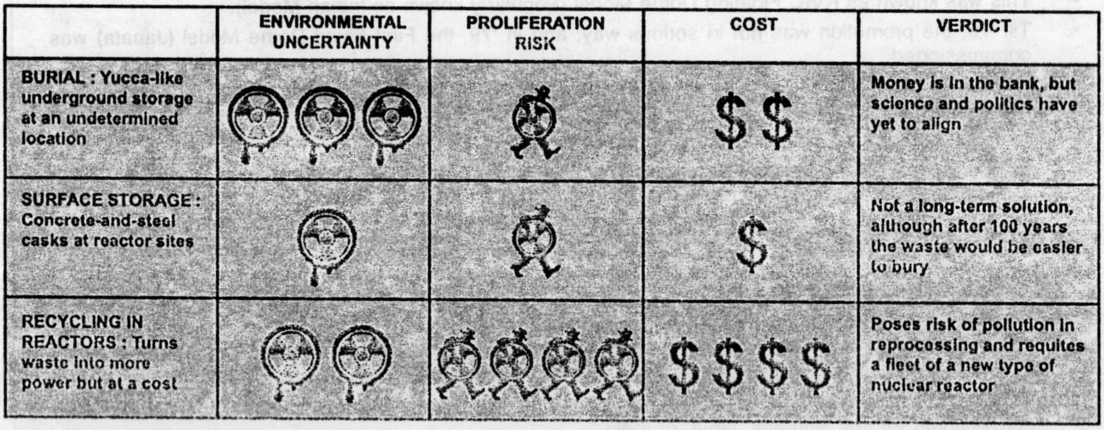

24.10 FUTURE ENERGY SCENARIO OF THE WORLD

Amongst the most optimistic of energy scenarios is that developed for the environmental pressure group Greenpeace International by the Stockholm Environment Institute, Boston Center, USA. The FFES adopts the UN prediction of a world population of 11 billion by 2100, with world GDP by this date showing a fourteen-fold increase from 1988. In the FFES, world primary energy demand increases to 1,000 EJ or about 24 Gtoe by 2100, and all fossil and nuclear fuels are assumed to have been phased out, to be replaced by a mixture of solar, hydroelectric, biomass, wind, and geothermal energy resources (Fig. 24.31). In the FFES, greenhouse gas emission specifically from fossil fuels are assumed to have been eliminated, resulting in a rise in global mean surface temperature of about 1.5°C, compared with the predicted 4°C if the use of fossil fuels continues at present rates (IPCC 2004).

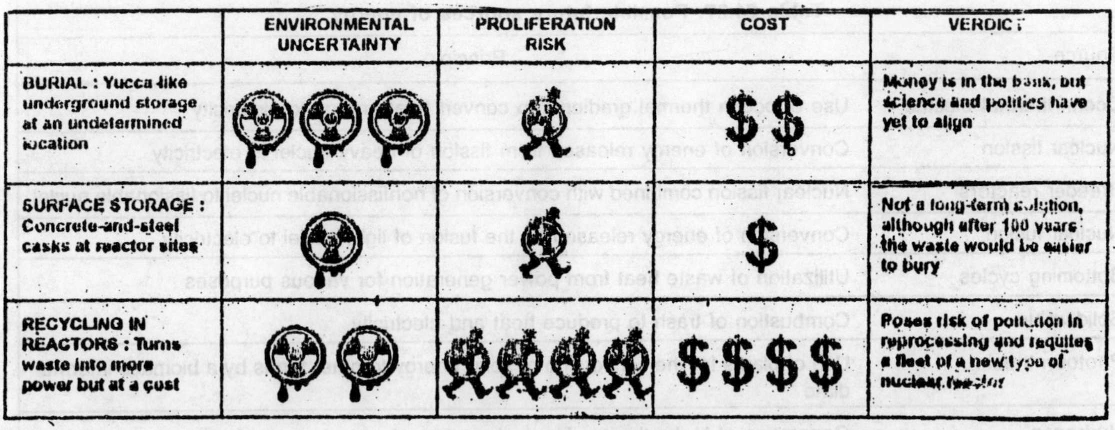

Fig. 24.31 : The Greenpeace fossil-free energy scenario (FFES), in which world energy demand increases by a factor of approximately 2.5 by 2100. By this date, all fossil and nuclear fuels are envisaged as phased out and replaced by a mix of solar, hydro, wind, biomass, and geothermal energy sources. (Modified after Lazarus et al, 1993)

As discussed in this chapter, a number of options are available for the supply of energy in the future. Some major possibilities are summarized in Table 24.26. There is a particular need for primary energy sources that do not emit carbon dioxide. Possible sources include carbon-sequestered fossil fuels, nuclear fission, nuclear fusion, nuclear fission–fusion hybrids, wind, ground-level solar, solar power satellites, and biomass. Other technologies that are not primary sources of energy but which have the potential to increase energy efficiency and facilitate energy transport or storage are increased efficiency of energy conversion devices, hydrogen (production, storage, and transport), and large-scale superconducting electrical grids.

Table 24.27: Possible future sources of energy

Source	Principle
Coal conversion	Manufacture of gas, hydrocarbon liquids, alcohol, or solvent-refined coal (SRC) from coal
Oil shale	Retorting petroleum-like fuel from oil shale
Geothermal	Utilization of underground heat
Gas turbine electric	Utilization of hot combustion gases in a turbine, followed by a topping cycle involving steam generation
Magnetohydrodynamics	Electricity generated by passing a hot gas plasma through a magnetic field
Thermionics	Electricity generated across a thermal gradient
Fuel cells	Conversion of chemical to electrical energy
Solar heating and cooling	Direct use of solar energy for heating and cooling through the application of solar collectors
Solar cells	Use of silicon semiconductor sheets for the direct generation of electricity from sunlight
Solar thermal electric	Conversion of solar energy to heat followed by conversion to electricity
Wind	Conversion of wind energy to electricity

Contd.

Table 24.27: Possible future sources of energy	
Source	Principle
Ocean thermal electric	Use of ocean thermal gradients to convert heat energy to electricity
Nuclear fission	Conversion of energy released from fission of heavy nuclei to electricity
Breeder reactors	Nuclear fission combined with conversion of nonfissionable nuclei to fissionable nuclei
Nuclear fusion	Conversion of energy released by the fusion of light nuclei to electricity
Bottoming cycles	Utilization of waste heat from power generation for various purposes
Solid waste	Combustion of trash to produce heat and electricity
Photosynthesis	Use of plants for the conversion of solar energy to other forms by a biomass intermediate
Hydrogen	Generation of H_2 by thermochemical or photochemical means for use as an energy-transporting medium

The overall cost estimates of renewable versus non-renewable energy is also attempted. The details are given in Table 24.28:

Table 24.28: Nonrenewable versus Renewable Energy : Costs and Tradeoffs*		
Technology	Cost / Kilowatt Hour (Electric Generation)	Qualitative Tradeoffs
Nonrenewables:		
Oil	\$.07 to \$.09/kWh	• high availability • affordable • traditional power plants available • lower emissions than coal
Natural Gas	\$.04 to \$.05/kWh	• higher availability outside OPEC • lower air emissions (e.g., SO_2, NO_x, Particulates) • affordable
Coal	\$.05 to \$.06/kWh	• not OPEC dependent • affordable, abundant • causes excess greenhouse gas emissions (i.e., CO_2)
Renewables:		
Hydroelectricity	N/A	• clean energy • availability limited by siting and relicensing conerns • high seasonal variance
Wind Power	\$.05 to \$.06/kWh	• clean energy • availability limited because wind turbines are dependent on location and wind conditions • high maintenance
Photovoltaic	\$17 to \$.18/kWh	• clean energy • requires sun-rich environment • high energy and purchase costs • suitable for smaller and remote users (e.g., substations, buildings, satellites)
Nuclear	\$.10 to \$.21/lWh	• clean energy • high public safety risk, yet very dependable • moderately high generating cost • difficult siting and licensing issues • toxic waste disposal
Solar Thermal	\$.08 to \$.10/kWh	• clean energy • excellent for developing countries

Contd.

Table 24.28: Nonrenewable versus Renewable Energy : Costs and Tradeoffs*

Technology	Cost / Kilowatt Hour (Electric Generation)	Qualitative Tradeoffs
Hydrogen Fuel Cells	\$.15 to \$.30/kWh	• clean energy • high energy cost yet excellent fuel-to-energy conversion efficiencies • technology unproven
Biomass Energy:		
Trees/wood	\$.04 to \$.08/kWh	• lower CO_2 emissions by offsetting CO_2 from coal • limited energy applications in urban areas • may be better applied to rural energy fuels (e.g., paper mill feedstock) • global deforestation may increase costs
Short rotation woody crops (e.g., sorghum alfalfa, and switchgrass)	\$.04 to \$.08/kWh	• can lessen soil erosion • less fertilizer soil amendment needed • higher crop value to localized farmers • lower CO_2 emissions by offsetting CO_2 from coal usage • high transportation costs for feedstocks

*Cost estimates in this table are based on data contained in Flavin, C., & Lenssen, N. (1994). Power surge : Guide to the coming energy revolution. Washington, DC : Worldwatch Institute; and Department of Energy. (1995). Programme plan for renewable energy generation of electricity. Prices of some renewable energy technologies are declining as their use becomes more prevalent.

24.11 RECENT COST COMPARISON OF ENERGY SOURCES 2017

While raw **forms of energy** are both free and practically infinite, the equipment and materials needed to collect, process, and transport the energy to the users are neither one. Currently, the RE costs are generally higher than that of fossil-based and nuclear energy. In addition to this, unlike well-established conventional designs, the advancement in different RE technologies still requires substantial investments. The economists often use so-called **levelized energy costs** (LEC) when comparing different technologies.

The LEC represents the total cost to build and operate a new power plant over its life divided to equal annual payments and amortized over expected annual electricity generation. It reflects all the costs including initial capital, return on investment, continuous operation, fuel, and maintenance, as well as the time required to build a plant and its expected lifetime. This table compares the US average levelized electricity cost in dollars per kilowatt-hour for both nonrenewable and alternative fuels in new power plants, based on US EIA statistics and analysis from Annual Energy Outlook 2017. Note, that the numbers for each source are given for a different capacity factor, which complicates direct comparison. Notwithstanding, I believe these figures are useful in comparing different power generation methods. Also note that the values shown in the table do not include any government or state incentives. In other words, they represent the actual cost to the society. We can see that at present natural gas, geothermal and coal are the most economic fuels. However, in future the price of coal-based electricity can nearly double due to government imposed cost on CO_2 emissions. Photovoltaic systems are still more expensive than fossil-based ones. The values in the chart represent just the cost of electricity production-the retail prices of courses are always higher.

Power Plant Type	Cost \$/kW-hr
Coal	\$ 0.11 – 0.12
Natural Gas	\$ 0.053 – 0.11
Nuclear	\$ 0.096
Wind	\$ 0.044 – 0.20
Solar PV	\$ 0.058
Solar Thermal	\$ 0.184
Geothermal	\$ 0.05
Biomass	\$ 0.098
Hydro	\$ 0.064

Adapted from US DOE, 2017

25 ENVIRONMENTAL TOXICOLOGY

25.1 CONCEPT

Our surrounding is filled up with a large number of toxicants in different forms. They contaminate our water, land and atmosphere where we live. Metals, pesticides and large number of complex organic chemicals, which are prepared through man-made activities, pose serious threat to the life of mankind. Most of the toxicants enter the human body through a number of routes—either through drinks, breath, food, medicines or even occupational or accidental exposures (Fig. 25.1).

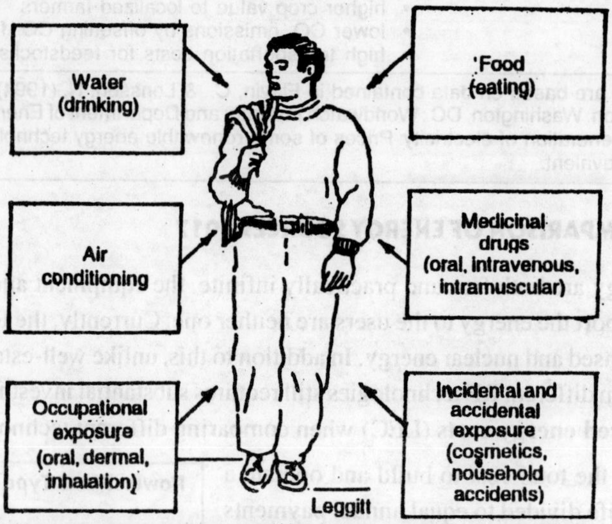

Fig. 25.1: Routes of exposure to toxic and hazardous environmental factors

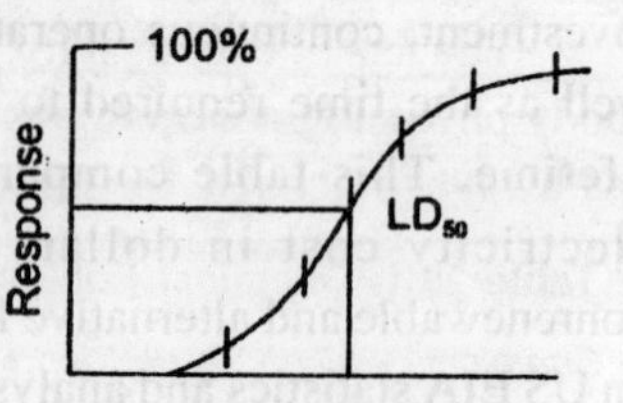

'Toxicology' is the study of chemical or physical agents that produce adverse responses in the biological systems with which they interact. The responses may be sudden and short span (i.e., acute toxicity) or long term constant or continuous (i.e., chronic toxicity). The chemical agent or toxicant that can produce an adverse effect in a biological system may do this either through an alteration of normal function or the destruction of life.

DOSE-RESPONSE RELATIONSHIP

A particular toxicity test is said to exhibit a dose-response relationship when a consistent mathematical relationship describes the proportion of test organisms responding for a given dosage interval for a given exposure period. Not only dose response to a chemical vary with different species of test organisms, response also varies within a group of test organisms of the same species. In general, responses of biological system was expressed against the cumulative long values of doses.

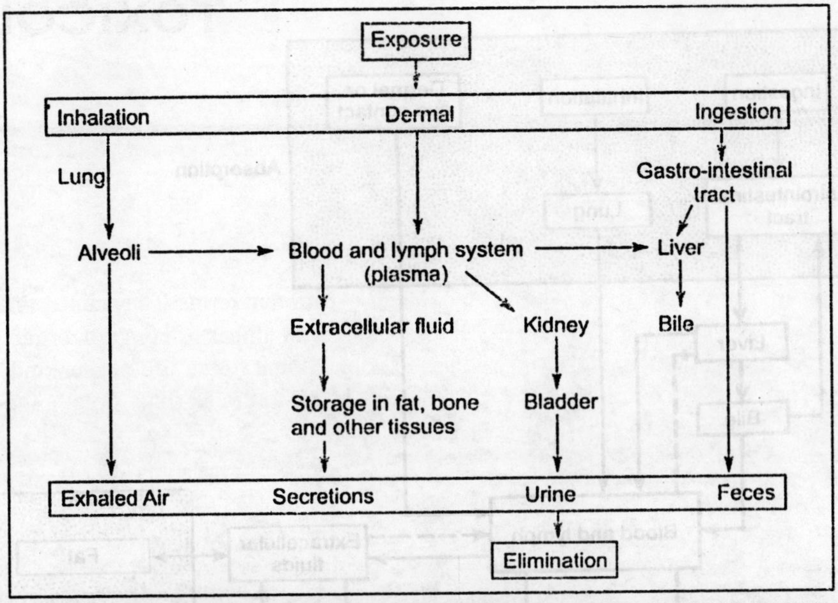

Fig. 25.2: Overview of absorption, distribution, storage, transformation and elimination of a toxic substance in the human body

Dosages are often described as lethal (LD), in test where the response is mortality; effective dose (ED) in tests where the response is a desirable effect, or toxic dose (TD) where the response is an undesirable toxicity other than death. LD_{50}, LC_{50} or LT_{50} can also be estimated form such dose response curve, where the dosage is lethal to 50% of the test organisms. The dose response curve has a number of utility values:

- Help in determining toxicity of substances;
- Identification of threshold limit of toxicity;
- Help in determination of LD_{50}, LC_{50}, LT_{50} & ED_{50} etc.

The dose-response relationship of any substance thus depends on the following aspects: (i) physical and chemical properties of substance, (ii) exposure conditions; (iii) genetic make up of test systems; (iv) environmental conditions.

The fate of chemical toxicants are diverse after entry into the body (Fig. 25.2, Table 25.1). In general, body physiological functions attempt to eliminate out the toxic substance through a variety of excretion pathways, otherwise the toxicants are deposited in storage sites.

Table 25.1: Chemical Pollutants within the Human Environment which are Hazardous to Human Health
Chemicals which can be found in food and water:
• lead (in food, in drinking water, especially where there is a combination of lead water pipes and acidic water);
• aflatoxins and other natural food toxicants;
• nitrates in drinking water (and their conversion into nitrites in the body);
• trace pollutants in water supply, many from agro-chemicals (for instance various halogenated organic chemicals);
• aluminium (food and drinking water);
• arsenic and mercury.
Chemicals commonly found in the indoor environment (home/workplace):
• carbon monoxide (incomplete combustion of fossil fuels);
• lead (paint—ingested by children);
• asbestos (usually from roofing insulation or air-conditioning conduits);

Contd.

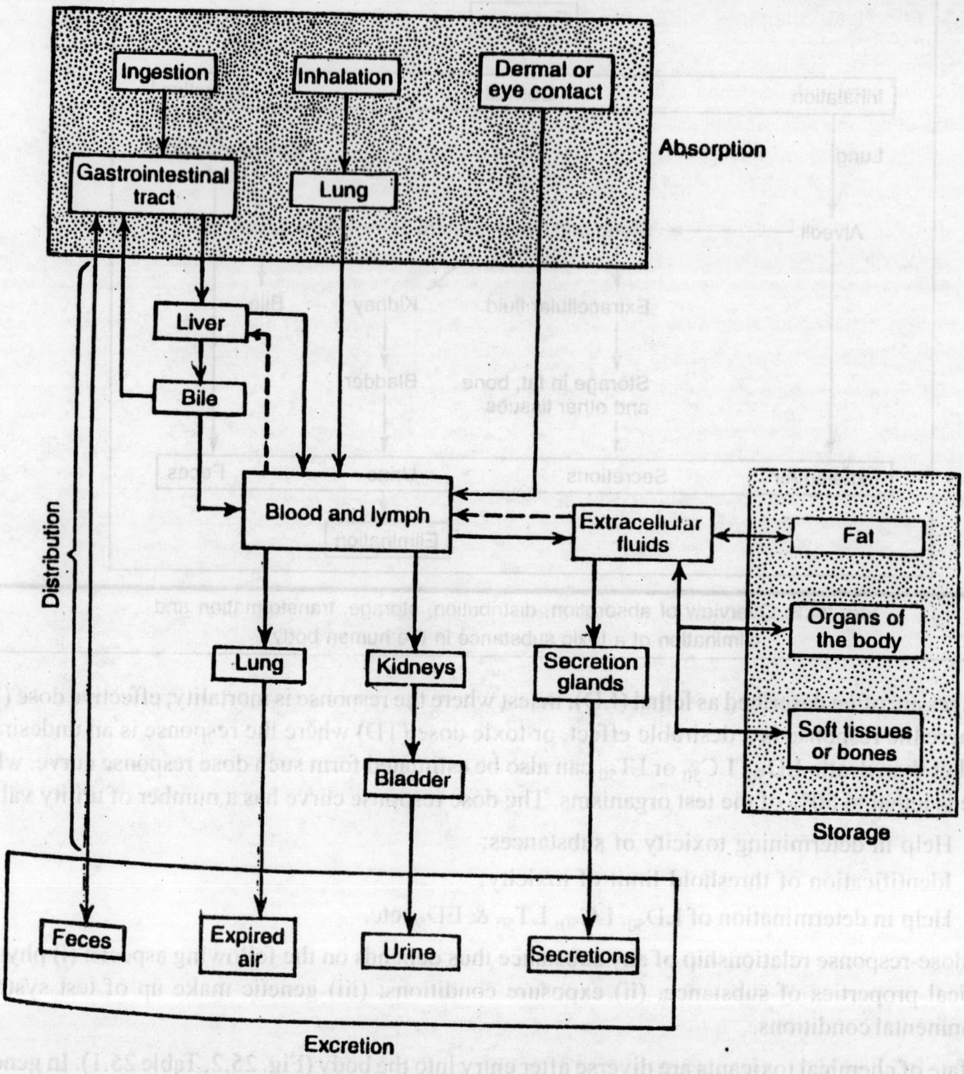

Fig. 25.3: Fate of chemical toxicant in the body

Table 25.1: Chemical Pollutants within the Human Environment which are Hazardous to Human Health
• smoke from combustion of coal and wood (or other biomass fuel); • tobacco smoke; • potentially dangerous chemicals used without health and safety safeguards (by home-workers and in occupational setting); • formaldehyde (mostly from insulation; also some wood preservatives and adhesives). *Chemicals found outdoors in urban areas in the air (ambient):* • lead (exhausts of motor vehicles using gasoline with lead additive, from external paint, some industrial emissions);

Contd.

Table 25.1: Chemical Pollutants within the Human Environment which are Hazardous to Human Health

- sulphur dioxide, sulphates and smoke/suspended particulates (mainly from coal or heavy oil combustion by industries, power stations and in some cities, households);
- oxides of nitrogen (in most cities, mostly from motor vehicle emissions; also some industries);
- hydrocarbons (motor vehicles, petrol stations, some industries);
- ozone (secondary pollutant formed by reaction of nitrogen dioxide and hydrocarbons in sunlight);
- Carbon monoxide (incomplete combustion of fossil fuels, mostly by motor vehicles);
- VOCs (Volatile Organic Compounds): there is a considerable range of such compounds, that are, or may be, hazardous.

Chemicals which may contaminate land sites:

- cadmium and mercury compounds and other heavy metal compounds (industrial wastes);
- dioxins, PCBs, arsenic, organochlorine pesticides (industrial wastes).

Also in both indoor and outdoor settings:

- micro-pollutants;
- mixtures each at trace level (with possible additive effects).

The toxicants not only affect human health, but also cause serious threat to the survivality of various biota. The details of route of uptake of different pollutants and their sources is shown in Table 25.2

Among the various pollutants two major groups were discussed at length—metals & pesticides.

Table 25.2: Major Routes of Uptake for Organic Pollutants

Type of organism	Route of uptake	Sources of pollutant
Terresterial vertebrates	Alimentary tract	Food and ingested water
	Skin	Contaminated surfaces droplets and particles in air
	Lungs	Vapour; droplets and particles in air
Terrestrial	Alimentary tract	Food and water
	Cuticle (Insects)	Contaminated surfaces
	Body wall (Slugs, worms)	Contaminated environment e.g.. soil
	Trachaea	Droplets and particles in air
Fish	Gills	Pollutants in ambient water, dissolved or suspended
	Alimentary tract	Principally food
Aquatic mammals and birds	Alimentary tract	Principally food
		Small amounts from ambient water or ingested water (birds)
Aquatic amphibia	Alimentary tract	Principally food
		Small amounts from ambient water
	Skin	Pollutants in ambient water, dissolved or suspended
Aquatic invertebrates	Alimentary tract	Principally food
		Some from ambient water
	Respiratory surfaces	Pollutants in ambient water, dissolved or suspended
Plants	Leaves	Pollutants in droplets or particles*
		Vapours
	Roots	Pollutants dissolved in soil water*

* Important routes of uptake for herbicides and for systemic insecticides and fungicides

25.2 TOXICOLOGICAL CHEMISTRY

A variety of chemical substances (gaseous, soluble or particulates) have distinct health disorders. Selected chemicals are discussed:

Ozone: It has toxic effects (above > 1ppm). Inhalation of ozone at 1 ppm level causes severe irritation and headache. Ozone irritates the eyes, upper respiratory system and lungs. Inhalation of ozone can sometimes cause fatal pulmonary oedema. Ozone generates free radicals in tissue. These reactive species can cause lipid peroxidation of sulfhydryl (–SH) groups and other destructive oxidation processes. The compound that protect organisms from the effects of ozone include radical scavengers, antioxidants, and compounds that protect organisms from the effects of ozone include radical scavengers, antioxidants and compounds containing sulfhydryl groups.

Carbon monoxide: It is a common cause of accidental poisonings. At CO levels in air of 100 ppm, impairment of judgement and visual perception occur; exposure to 100 ppm causes dizziness, headache, and weariness; loss of consciousness occurs at 250 ppm and inhalation of 1,000 ppm results in rapid death. Chronic long term exposures to low levels of carbon monoxide are suspected of causing disorders of the respiratory system and the heart.

After entering the bloodstream through the lungs, carbon monoxide reacts with haemoglobin (Hb) to convert oxyhaemoglobin (O_2Hb) to carboxyhaemoglobin (COHb).

Elemental halogens: Fluorine (F_2), Chlorine (Cl_2), Bromine (Br_2) and Iodine (I_2) are strong oxidant and that causes irritation on inhalation and subsequent respiratory distress and finally lead to death above > 1,000 ppm exposure.

Cyanide: Both hydrogen cyanide (HCN) and cyanide salt are rapidly acting poisons; a dose of only 60 to 90 mg is sufficient to kill a man.

Air pollutants viz., NO_x, SO_x, HF are also too toxic to human and plants, these are discussed in earlier chapters.

Phosphine (PH_3) : It is a colourless gas that undergoes a autoregulation at 100°C, is a potential hazard in industrial processes and in the laboratory.

Polyaromatic hydrocarbons (PAH_s): Various kinds of aromatic and polyaromatic hydrocarbons are carcinogenic. They also act by forming adducts with DNA.

Acrylamide: Toxic Potato chips?: In 2002, Swedish investigators reported finding potentially harmful amounts of acrylamide in fried and baked foods, but not in boiled food. Acrylamide in known to be a neurotoxin and is a carcinogen.

$$\text{Glucose} + \text{Asparagine} \xrightarrow[\;-H_2O \quad -CO_2\;]{\text{High temperature}} \text{Acrylamide}$$

Pyridine: It is a colourless & odorous chemical used in industrial processes. On inhalation, it causes anorexia, nausea, fatigue and chronic poisoning.

Acetonitrile (CH_3CN): It is a highly toxic chemical, which showns marked toxicity in man on inhalation for prolong period.

Nitrosamine: It is a carcinogenic chemicals found in beer, whisky and other similar liquors. Chronic exposure leads to carcinoma.

Toxic natural products: Bacterial toxins or fungal toxins often causes serious health disorder or ingestion. They may be neurotoxic or carcinogenic or teratogenic and thermostable.

US Department of Health and Human Services, Public Health Service Agency, published a document for toxic substances and disease registry (HTSDR's Toxicological Profile).

25.3 METALS IN ENVIRONMENT

Metals are introduced into aquatic systems as a result of the weathering of soils and rocks, from volcanic eruptions and from a variety of human activities involving mining, processing, or use of metals and/or substances that contain metal contaminants. Although some metals such as manganese, iron, copper and zinc are essential micronutrients, others such as mercury, cadmium and lead are not required even in small amounts by any organism. Virtually all metals, including the essential metal micronutrients, are toxic to aquatic organisms as well as humans if exposure levels are sufficiently high.

Table 25.3: Permissible Limits of Various Metals in Public Water Supplies as Recommended by the EPA (1972)

Metal	Maximum permissible concentration (μg/l)
Arsenic (As)	50
Cadmium (Cd)	10
Copper (Cu)	1,000
Chromium (Cr)	50
Lead (Pb)	50
Mercury (Hg)	2
Zinc (Zn)	5,000

The fact that metal concentrations in aquatic organisms are typically several orders of magnitude higher than concentrations of the same metals in water has led to some speculation that metals may become progressively concentrated at higher trophic levels in aquatic food chains due to biological magnification.

It is also true, in general, that the concentration of metals may differ greatly between one organism and another and between different parts of the same organism. In general, one ascribes these concentration differences to differences in the tendency of metals to bind to the various molecule groups formed within the cells of each organism, as well as to the degree of the organism's exposure to the metal as influenced by its metabolic characteristics and its position in the food chain (Table 25.3).

The term 'heavy metals' has been used extensively in the past to describe metals which are environmental pollutants. For a metal to be considered 'heavy' it must have a density relative to water greater than 5. However, this term is now replaced by toxic metals.

Metals are non-biodegradable. Unlike some organic pesticides, metals cannot be broken down to less harmful components. Detoxification by organisms consists of 'hiding' active metal ions within a protein such as metallothionein (binding covalently to sulphur), or depositing them to an insoluble form in intracellular granules for long-term storage or excretion in the feces.

Among the metals, some appear to be essential (at low concentration) for biota growth and reproduction, while there are a few non-essential but toxic, also a good number lies in the borderline:

Essential	Borderline	Non-Essential
Calcium	Zinc	Cadmium
Magnesium	Lead	Copper
Manganese	Iron	Mercury
Potassium	Chromium	Silver
Strontium	Nickel	
Sodium	Arsenic/Vanadium	

Table 25.4: Major industrial Sources and Concentrations of Metals contained in Waste Waters (Patterson, 1979)

Contaminant	Sources	Concentration range (mg/l)
Arsenic	Insecticide manufacturing; ore extraction; sulphuric acid and ammonia manufacturing; wood porducts preservation; arsenic trioxide manufacturing	10 – 1,000 (soluble > 100 mg/l)
Cadmium	Metallurgical alloying; ceramic and inorganic pigments; electroplating; textile printing; batteries; acid mine drainage	4 – 1,000
Chromium	Corrosion control additives; inks; pigments and dyes; tanning; aluminium anodizing; metal cleaning, treatment, and plating; preservatives	1 – 1,300 (wastes) (15 – 270,000 baths which may be batch dumped)
Copper	Metal cleaning, pickling and plating; rayon manufacturing; copper and brass manufacturing; copper sulphate production; primary and secondary copper manufacturing; electronics	10 – 900 (up to 7,000 in primary and secondary copper manufacturing)
Iron	Mine drainage; coal storage pile drainage; steel manufacturing and rolling, inorganic chemicals; metal processing and plating; titanium dioxide, paint and ink manufacturing	40 – 5,000 (pickle baths up to 100,000)
Lead	Battery manufacturing; electroplating; television tube manufacturing; glass and ammunition manufacturing; tetraethyl lead production; paint pigment; rubber hose manufacturing	< 1 – 400
Manganese	Mine drainage; refuse incinerator and industrial scrubber waters for ferromanganese and electrolytic manganese production; paint formulation	< 1– > 1,000 (1,000 – 6,700 in one case)
Mercury	Chlor-alkali plants; paper mills, smelting; water-based paint; acetaldehyde production; fertilizer mills; chemical manufacturing	< 0.3 – 20 (older references 2,000)
Nickel	Pickling and plating of nickel and nickel alloys; basic steel works. motor vehicles, and aircraft industries; business machine and tableware manufacturing	< 1 – 900
Selenium	Paint, pigment, and dye production; electronics and glass manufacturing; insecticide industry; pulp and paper production; non-ferrous metals smelting	< 1 – 10 (very limited data)
Silver	Jewellery and silverware manufacturing; metal alloying; food and beverage industry; photographics, ink formulation; electroplating	< 1 – 250
Zinc	Steel works (galvanizing); zinc and brass metal works; zinc and brass plating; groundwood pulp and newsprint paper production; inorganic pigments; electroplating and metal processing; zinc smelting; mine drainage	< 1 – 2,000 (plating, pickle and bonderizing baths 2,000-35,000)

Contamination of our environment by toxic metals may arises in a number of ways. Industrial waste disposal is one of the major routes for metal contamination in aquatic ecosystem (Table 25.4). In addition, there are other sources of metal contamination into water bodies. The global input of major metals annually is shown in Table 25.5.

Some metals were discussed in detail with respect to their sources and toxicity in preceding sections.

Table 25.5: Worldwide Inputs of Metals (Tonnes/year) in Aquatic Ecosystems (1990)													
Sources	Metals (thousand tonnes per year)												
	Ant.	As	Cd	Cr	Cu	Pb	Mn	Hg	Mo	Ni	Sel	Vd	Zn
1. Domestic wastewater	2.2	9.2	1.7	4.6	28	6.8	110	0.30	2.2	62	3.8	2.3	48
2. Sewage discharges	1.5	4.1	0.69	19	12	9.4	69	0.16	2.9	11	2.0	3.5	17
3. Manufacturing processes	9.3	7.0	2.4	51	34	14	21	2.1	4.2	7.4	4.3	0.55	85
4. Electric power station	0.18	8.2	0.12	5.7	13	0.72	11	1.8	0.65	11	18	0.30	18
5. Metallic ore mining and smelting	3.8	7.4	2.0	12	14	7.0	40	0.10	0.51	13	12	0.60	29
6. Atmospheric fall out	1.1	5.6	2.2	9.1	11	100	12	2.0	0.95	10	0.82	26	10
Total input	18.0	42.0	9.1	143.0	112.0	138.0	263.0	6.5	11.0	114.0	41.0	33.0	237.0

25.4 ARSENIC (As)

The element arsenic is widely distributed in nature, being highly concentrated in coals. It occurs mostly in the form of arsenides of true metals or as arsenopyrites. The average content of arsenic in freshwater is about 1.5 to 2 ppb and in sea water it is 0.15 to 6 ppb.

The inorganic forms of arsenic, as arsenate (5^+ oxidation state) and the more toxic arsenite (3^+ oxidation state), have been used extensively as herbicides and insecticides. Tonnes of arsenite, for example, have been added to freshwater lakes to kill aquatic plants. An analysis of environmental samples has shown the presence of dimethylarsenic acid and methylarsenic acid in natural water, seashells, bird eggshells, human urine and shellfish.

25.4.1 Biochemical Effects of Arsenic

Arsenic commonly occurs in insecticides, fungicides and herbicides. Among its compounds, those of As (III) are the most toxic.

As (III) exerts its toxic action by attacking –SH groups of an enzyme, thereby inhibiting enzyme action.

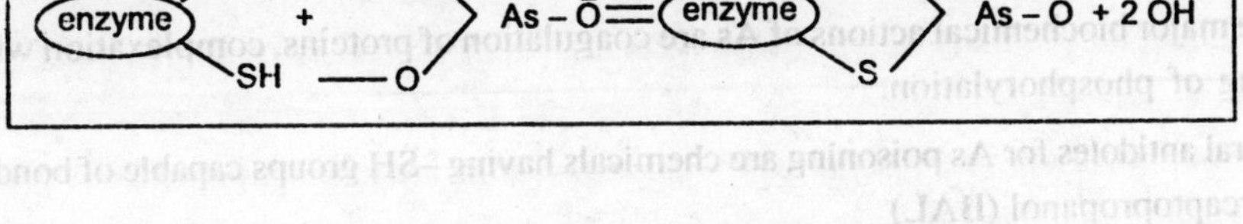

The enzymes which generate cellular energy in the citric acid cycle are adversely affected. The inhibitory action is based on inactivation of pyruvate dehydrogenase by complexation with As (III), whereby the generation of ATP is prevented.

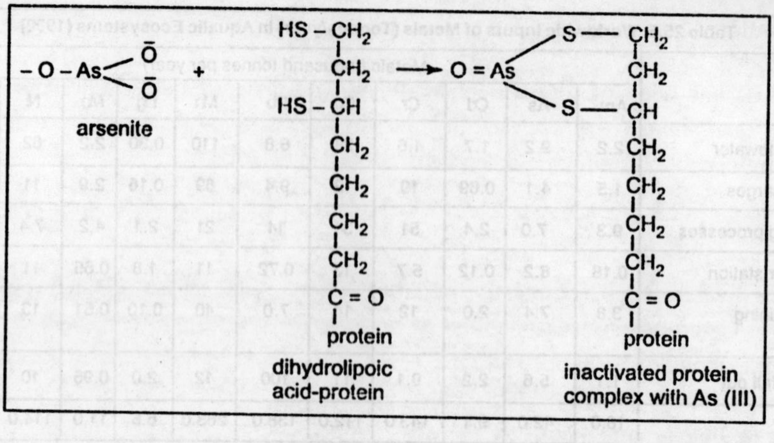

By virtue of its chemical similarity to P (phosphorus), As interferes with some biochemical processes involving P. This is observed in the biochemical generation of the key energy-yielding substance, ATP (adenosine triphosphate). An important step in ATP generation is the enzymatic synthesis of 1, 3-diphosphoglycerate from glyceraldehyde 3-phosphate. Arsenite interferes by producing 1-arseno-3-phosphoglycerate instead of 1, 3-diphosphoglycerate. Phosphorylation is replaced by arsenolysis which consists of spontaneous hydrolysis to 3-phosphoglycerate and arsenate.

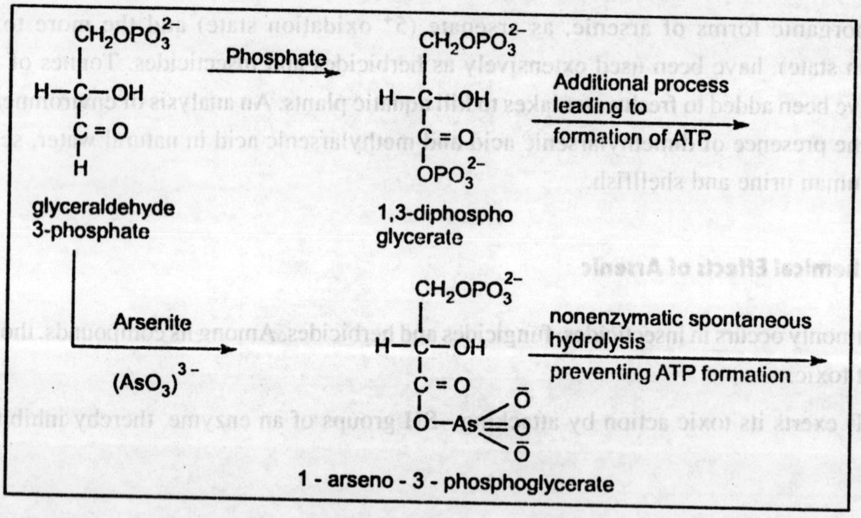

Arsenic (III) compounds at high concentrations coagulate proteins, possibly by attacking the sulphur bonds maintaining the secondary and tertiary structures of proteins.

The three major biochemical actions of As are coagulation of proteins, complexation with coenzymes, and uncoupling of phosphorylation.

The general antidotes for As poisoning are chemicals having –SH groups capable of bonding to As (III), e.g. 2, 3-dimercaptopropanol (BAL)

$$HSCH_2 - \underset{\displaystyle SH}{\underset{|}{CH}} - CH_2OH$$

25.4.2 Transformation

Toxic arsenite (III) has been oxidised to arsenate by a number of microorganisms like species of *Achromobacter, Pseudomonas arthrobacter*. In contrast, a yeast, *Candida*, an algae, *Chlorella*, and some marine bacteria are capable of reducing arsenate to the toxic arsenite.

25.5 CADMIUM

Cadmium (Cd) is a divalent metal, chemically similar to Zinc and Mercury. Although it is widely distributed in the lithosphere, Cadmium is usually found at quite low concentrations in crystal rocks, the mean concentration being about 200-300 ppb (OECD, 1975). Sometime Cd concentrations as high as 100 ppm are found in phosphatic rocks (due to presence of fossilised fish teeth). Cadmium may also be leached from sulfide ores of zinc, copper and lead. The world production of Cadmium is over 20,000 metric tons per year (2000), about 50% greater than the present rate of mercury production. The major producing countries are the US, Russia, Japan, Canada, Belgium, and France.

Cadmium is used in various industrial purposes viz.

(a) for electroplating of steel, iron, copper, brass and other alloys to prevent corrosion.

(b) for colouring pigment production which are used in fabrics, textiles and paints, etc.

(c) for uses as plastic stabilisers.

(d) for manufacture of alloys,

(e) for manufacture of Ni-Cd batteries,

(f) for production of certain pesticides, luminescent dials, X-ray screens, etc.

25.5.1 Emissions to the Environment

The estimates of Cd emissions to the environment in different countries vary widely. There are a number of sources of Cadmium emission. An estimate as measured in US during 1968 is presented in Table 25.6.

Table 25.6: Estimated Cd Emissions (in Metric Tons) to the Environment during 1968 in the United States

Sources	Air	Soil, water	Total
Extraction, refining and production	953	294	1,247
Industrial conversion	15	—	15
Consumption and disposal	186-927	909	1,095-1,836
Fossil fuel combustion	132-998	—	132-998
Application of fertiliser		23-227	23-227
Tires	5.2	—	5.2
Lubricating oils	0.8	?	0.8
Total	1,292-2,899	1,226-1,430	2,518-4,329

Source: OECD, 1975

25.5.2 Toxicity

The Cadmium ion (Cd^{2+}) is believed to exert no toxic effect. In human system the efficiency of cadmium absorption by the intestines is only about 5-6%, but once ingested, Cadmium is transported to all parts of the body by the blood stream. Although almost all organs probably absorb some Cd, the highest concentrations are invariably found in the liver and kidneys.

The toxic effects of cadmium are presumably associated with the metal's affinity for organic ligands containing sulphur, nitrogen or other electronegative functional groups. The metal appears to be particularly attracted to enzyme containing Zinc. The most serious effects of Cd poisoning usually involve damage to the kidney, particularly the renal tubes. These symptoms are associated with proteinurea, glucosuria and high alkaline phosphatase in the blood. Cd poisoning is also associated with bone softening, fatal lung damage, high rise of blood pressure and other heart diseases.

As such, as per the US drinking water standard, Cd levels should not exceed 10 ppb. Chronic toxicity of Cd is evident at 30-60 ppb in fresh water fish, 5 to 50 ppb at marine fishes. In general, Cd toxicity can be partially reduced by use of Zn.

In 1955, two Japanese physicians reported the occurrence of a mysterious disease in the Jintsu basin of Japan near the city of Toyama. The disease was characterised by severe pain in the back, joints and lower abdomen, development of a waddling or ducklike gait, kidney lesions, proteinurea, glycosuria and loss of calcium from the bones leading in some cases to multiple bone fractures. They designated the disease as 'Itai-itai' disease. Initial efforts to identify the cause of the disease were unsuccessful, but they tried to link the same disease with some nutritional toxicity. However, in 1970s, it was possible to identify that the said disease was due to chronic Cd toxicity (EPA, 1972, 1976).

25.5.3 Emission Control and Other Measures

The technology for removing Cd from industrial wastewater or from flue dust is well-established. In wastewater, dissolved Cd can be precipitated with sodium sulphide, cemented by the addition of Zinc or separated out by ion exchange. If the Cd is incorporated into particulates, the Cd can be dissolved by the addition of acid and then separated by one of the above techniques, or the solids can be settled out and the Cd removed with the sludge. Cadmium released to the atmosphere via smoke stacks is primarily in the form of fine particulate materials, which can be separated from the stack gases by wet scrubbers, fabric filters or electrostatic precipitators (OECD, 1975). Removal of Cd by any of the above techniques may lead to a solids disposal problem if Cd is not to be recovered or recycled. Dumping of Cd or Cd-containing compounds at sea or in fresh water is prohibited. Land disposal is the obvious solution but only in areas where the soil is neutral or basic so that the cadmium is not mobilised by percolating groundwater.

In many cases the use of Cd is substituted by other metallic compounds, viz.

(a) Zinc could be substituted for Cadmium in many electroplating applications;
(b) The Edison Cell (iron/KOH/Nickel oxide) is the chief alternative to the Ni-Cd battery;
(c) Yellow and red pigments made from iron oxides could be substituted for Cadmium for some processes.

25.5.4 Biochemical Effects of Cadmium

Cd occurs in nature in association with zinc minerals. Growing plants require Zn and they also take up and concentrate Cd with the same biochemical apparatus. The outbreak of Cd poisoning occurred in Japan in

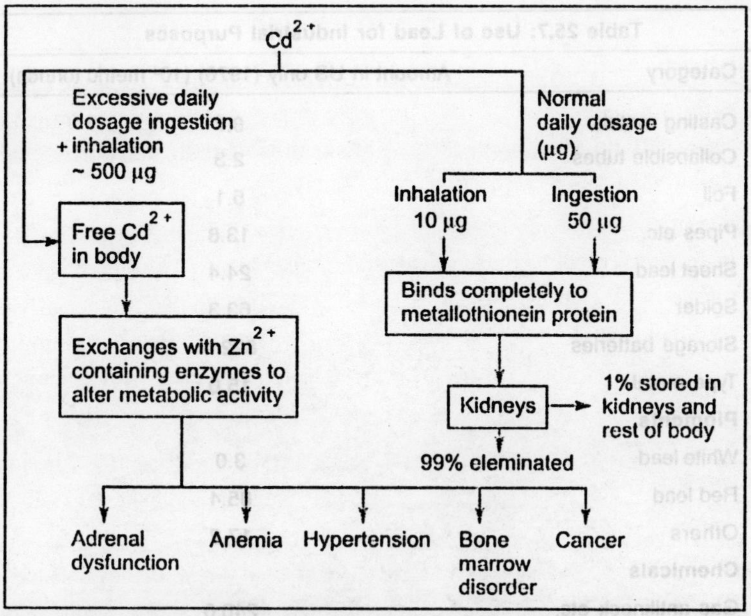

Fig. 25.4: Metabolism of cadmium

the form of *itai itai* "Ouch ouch" disease. Many people suffered from this disease in which their bones became fragile. At high levels, Cd causes kidney problems, anaemia and bone marrow disorders.

The major portion of Cd ingested into our body is trapped in the kidneys and eliminated. A small fraction is bound most effectively by the body proteins, *metallothionein*, present in the kidneys, while the rest is stored in the body and gradually accumulates with age. When excessive amounts of Cd^{2+} are ingested, it replaces Zn^{2+} at key enzymatic sites, causing metabolic disorders.

25.6 LEAD

Lead (Pb) is widely distributed in the rocks and soils of the earth's crust, although the mean concentration is only about 16 ppm. Lead is mined primarily from deposits of the mineral galena or lead sulphide (PbS). Since metallic lead can be separated from PbS by heating to temperatures easily achieved by burning wood or charcoal ($PbS + O_2 \xrightarrow{\text{heat}} SO_2 + Pb$), it was not difficult for early civilisations to extract lead.

Though there are a number of instances of lead use in domestic purposes since time immemorial, but the principal uses of lead today are of industrial nature (Table 25.7).

Table 25.7: Use of Lead for Industrial Purposes

	Category	Amount in US only (1976) (10^3 metric tonnes)
(A)	**Metal products**	
	Ammunition	73.5
	Bearing metals	13.1
	Brass & bronze	15.7
	Cable covering	15.9
	Calking lead	12.5

Contd.

Table 25.7: Use of Lead for Industrial Purposes

	Category	Amount in US only (1976) (10^3 metric tonnes)
	Casting metals	6.7
	Collapsible tubes	2.3
	Foil	5.1
	Pipes etc.	13.8
	Sheet lead	24.4
	Solder	63.3
	Storage batteries	822.4
	Type metal	15.0
(B)	**Pigments**	
	White lead	3.0
	Red lead	85.4
	Others	17.2
(C)	**Chemicals**	
	Gas antiknock etc.	240.0
(D)	**Miscellaneous**	
	Annealing	2.9
	Galvanising	1.3
	Lead plating	0.4
	Weights	22.4
	Others	32.5
	Total	1,490.1

Source: US Dept. of the Interior, 1978.

25.6.1 Emissions

There is good evidence that lead emissions to the environment have been greatly increased on a global scale as a result of human activities. Analysis of marine sediments has indicated that millions of years ago the flux of Pb to the oceans was about 13,000 mt per year, whereas river runoff alone contribute above 240,000 mt of Pb to the ocean each year. Lead is also deposited on the ice sheets through precipitation.

Lead released to the atmosphere from Pb smelting, refining of lead scrap and burning leaded gas subsequently falls out onto crops. It will contaminate or accumulate in foodgrains too. Some agricultural practices, including the ploughing and irrigation of fields and the use of lead pesticides, contribute to lead concentrations in water and soil. Lead pipes used to transport water pose an obvious water contamination problem.

The problem of air contamination by Pb in urban area is primarily derived from automobile exhaust fume. The typical Pb concentration in urban air is about 1.3 $\mu g/m^3$ in constrast to rural air which is 0.05 $\mu g/m^3$. Another major sources of global lead contamination has been the use of lead additives in gasoline.

25.6.2 Toxicity

Lead enters the human body primarily via inhalation and through the ingestion of contaminated food & water. Absorption of inhaled Pb is about 37% in lung and 25-30% and 8% via gastrointestinal tract in children and adults, respectively. Once absorbed into the blood stream, lead is transported to all parts of the body and incorporated into the tissues, particularly in bones, by replacing calcium. The highest concentrations of organic lead are often found in the brain and liver. Small amount of lead can be excreted out by feces, urine and sweat.

Lead is known to disrupt several enzymes and is also involved in blockage of haemoglobin biosynthesis. Thus, anaemia is developed. It damages nervous system—both nerve fibers and brain—and causes paralysis of the organs. Kidney damage characterised by atrophy of the renal tubules is a well-established effect of lead-poisoning. The damage is associated with elevated levels of amino acids, sugar and phosphate in the urine.

With respect to human health, in general, the greatest danger of lead poisoning is undoubtedly to young children, particularly those living in urban areas. Neurological damage caused by the poisoning of such children may be permanent and can result in impaired physical as well as mental development.

The toxicity of lead to aquatic organisms has not been examined thoroughly. However, chronic toxicity found in marine organisms is about 100 ppb of Pb. (EPA, 1972, 1975).

An average estimated of lead consumption from various sources is given in Table 25.8.

Table 25.8: Estimated Consumption of Pb from Various Sources for Americans

Source	Intake per day	Ingestion of Pb per day (μg)	Absorption of Pb per day (μg)
Food	2 kg	400	20
Water	1 kg	10	1
Urban air	20 m³	26	10
Rural air	20 m³	1	0.4
Tobacco smoke	30 cigarettes	24	10

25.6.3 Transformation

Recent evidence indicates that lead may also be methylated by bacteria like Pseudomonas, and Flavobacterium.

25.6.4 Biochemical Effects of Lead

Lead is a relatively abundant metal in nature, occuring in lead minerals. In the atmosphere it is relatively more abundant than other toxic metals. By far the major source of airborne Pb is the combustion of leaded petrol/gasoline. Pb is added in the form of tetraalkyl lead, primarily $Pb(CH_2)_4$ and $Pb(C_2H_5)_4$, together with the scavengers 1, 2-dichloroethane and 1, 2-dibromoethane. In common with other particulate pollutants, Pb is removed from the atmosphere by wet and dry deposition processes. As a result, street dusts and roadside soils become enriched with Pb, with concentrations typically of the order of 1,000-4,000 mg kg^{-1} on busy streets.

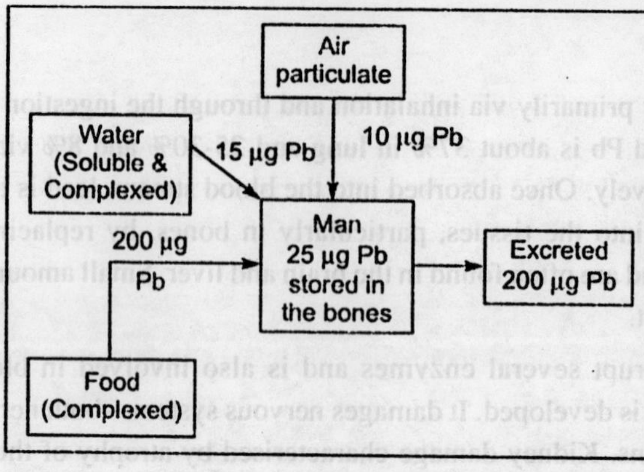

Fig. 25.5: Daily lead balance of a city resident

It may be noted that most of the Pb intake by a typical city dweller is from diet (about 200-300 µg per day), air and water adding a further 10-15µg per day each. Of this total intake, 200 µg of Pb is excreted while 25 µg is stored in the bones each day.

The major biochemical effect of Pb is its interference with heme-synthesis, which leads to haematological damage. Pb inhibits several of the key enzymes involved in the overall process of haeme-synthesis whereby the metabolic intermediates accumulate. One such intermediate is delta-amino levulinic acid. An important phase of haeme-synthesis is the conversion of delta-aminolevulinic acid to porphobilinogen.

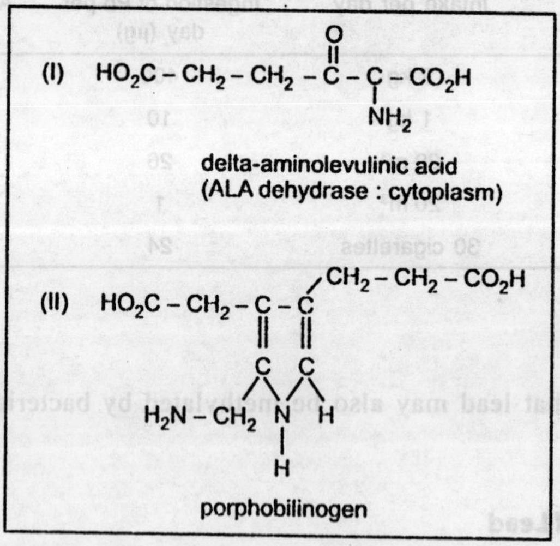

Pb inhibits the ALA-dehydrase enzyme (I) so that it cannot proceed further to form prophobilinogen (II). The overall effect is the disruption of the synthesis of hemoglobin as well as other respiratory pigments, such as cytochromes, which require heme. Finally, Pb does not permit utilisation of O_2 and glucose for life-sustaining energy production. This interference can be detected at a lead level in the blood of about 0.3 ppm. The detection of (I) provides a sensitive test for Pb in the body. At higher levels of Pb in the blood (at > 0.8 ppm) there will be symptoms of anaemia due to the deficiency of haemoglobin. Elevated Pb levels (0.5-0.8 ppm) in the blood cause kidney dysfunction and, finally, brain damage.

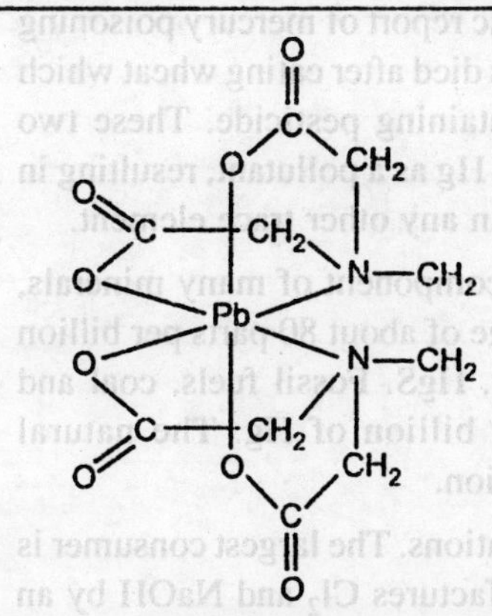

Pb-EDTA chelate

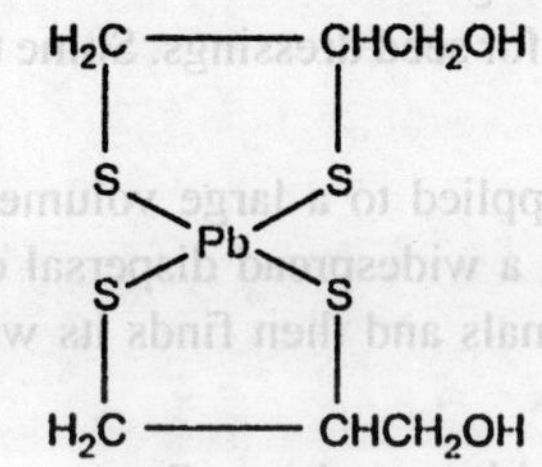

Pb-BAL chelate
(BAL = 2, 3-dimercaptopropanol)
BAL, abbrev. of British anit-lewisite

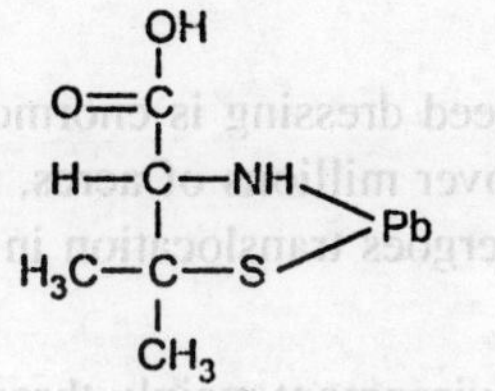

Pb-d-penicillamine chelate

Due to the chemical analogy of Pb^{2+} with Ca^{2+}, bones act as repositories for Pb accumulated by the body. Subsequently, this Pb may be remobilized along with phosphates from the bones which exert a toxic effect when transported to soft tissues.

Lead poisoning can be cured by treatment with chelating agents which strongly bind Pb^{2+}. Thus, calcium chelate in solution is fed to the victim of lead poisoning. Pb^{2+} displaces Ca^{2+} from the chelate and the resulting Pb^{2+} chelate is rapidly excreted in the urine. Three typical Pb chelates are shown above.

25.7 MERCURY

Mercury is one of the toxic metals in the ecosystem. It primarily arises out of industrial wastes or from metal mining activities. In nature, there are either inorganic forms of mercury or organic forms. All forms are not equally toxic to biota. Like all other toxic metals, mercury also accumulates through food-chain in ecosystem.

25.7.1 Biochemical Effects of Mercury

Mercury (Hg) is a well-known toxic metal which came to the limelight after the incidence of "Minamata disease" in 1953-60 in Japan. A total of 111 cases of mercury-poisoning were reported among people who had eaten mercury-contaminated fish from Minamata Bay. Among them, about 45 people died. Genetic defects were observed in 20 babies whose mothers had consumed seafood from the bay. The seafish in the Bay were found to have 27-102 ppm of mercury in the form of methyl mercury. This mercury source was the effluent from a vinyl chloride plant, Minamata Chemical Company, which used to discharge into the bay. This was followed by a more tragic report of mercury poisoning from Iraq, in 1972, where 450 villagers died after eating wheat which had been dusted with a mercury-containing pesticide. These two tragic events boosted the awareness of Hg as a pollutant, resulting in its being studied more extensively than any other trace element.

In nature, Hg occurs as a trace component of many minerals, continental rocks containing an average of about 80 parts per billion of Hg. The principal ore is Cinnabar, HgS. Fossil fuels, coal and lignite contain about 100 parts per billion of Hg. The natural abundance in soil is 0.1 parts per million.

Hg finds a wide variety of applications. The largest consumer is the chlor-alkali industry which manufactures Cl_2 and NaOH by an electrolytic process using Hg electrodes. The second largest consumption of Hg is in the production of electrical apparatus, e.g. Hg vapour lamp, electrical switches, Hg batteries etc. The third largest consumer is the agricultural industry using a large number of fungicides for seed dressings. Some typical compounds of this category are:

$CH_3-Hg-C\equiv N$
Methylmercurynitrile

$CH_3-Hg-N(H)-C(=NH)-NHCN$
Methylmercurydicyandiamide

$CH_3-Hg-C(=O)-OCH_3$
Methylmercuryacetate

$C_2H_5-Hg-Cl$
Ethylmercurychloride

The impact of seed dressing is enormous since it is applied to a large volume of seed, which is subsequently sowed over millions of acres, thereby causing a widespread dispersal of Hg compounds. Furthermore, Hg undergoes translocation in plants and animals and then finds its way into the human food chain.

Hg enters the environment mainly through human activities, as above. Sewage effluent sometimes contains up to 10 times the level of Hg in natural water (0.001-0.0001 ppm). Once Hg is absorbed on sediments of water bodies and streams, it is slowly released into the water and constitutes a reservoir which is likely to cause chronic pollution long after the original source of Hg is removed. Current rate of natural addition of Hg to the oceans is about 5,000 tons per annum and a further 5,000 tons is added via human activities.

25.7.2 Toxic Effects

The toxicity of Hg depends on its chemical species as shown in Table 25.9.

Table 25.9: Chemical Species of Mercury

Species	Chemical and biochemical properties
Hg	Elemental mercury: Relatively inert and nontoxic; vapour highly toxic when inhaled.
Hg_2^{2+}	Mercurous ion: Insoluble as chloride; low toxicity.
Hg^{2+}	Mercuric ion: Toxic but not easily transported across biological membranes.
RHg^+	Organomercurials: Highly toxic, particularly CH_3Hg^+ (methyl mercury); causes irreversible nerve and brain damage; easily transported across biological membranes; stored in fat tissue.
R_2Hg	Diorganomercurials: Low toxicity but can be converted to RHg^+ in acidic medium.
HgS	Mercuric sulphide: Highly insolube and non-toxic; trapped in soil in this form.

Elemental Hg is fairly inert and non-toxic. If swallowed, it is excreted without serious damage. It has a fairly high vapour pressure and so the vapour, if inhaled, is quite toxic. Hence, Hg should be handled only in well-ventilated areas and spills should be cleaned up as quickly as possible. Hg vapour, when inhaled, enters the brain through the blood stream, leading to severe damage of the central nervous system.

Hg_2^{2+} forms an insoluble chloride with chloride ions. As our stomachs contain a fairly high concentration of chloride, Hg_2^{2+} is not toxic. Hg^{2+} (mercuric ion), however, is fairly toxic. Because of its high affinity for sulphur atoms, it easily attaches itself to the sulphur-containing amino acids of proteins. It also forms bonds with haemoglobin and serum albumin, both of which contain sulphydryl groups. This ion, however, does not travel across biological membranes and, hence, does not get access into biological cells.

The most toxic species are the organomercurials, particularly CH_3Hg^+ (methyl mercury), which are soluble in fat, the lipid fraction of membranes and brain tissue. The covalent Hg–C bond is not easily disrupted and the alkyl mercury is retained in cells for prolonged periods of time. The most dangerous aspect is the ability of RHg^+ to move through the placental barrier and enter fetal tissues.

Attachment of Hg to cell membranes is likely to inhibit active transport of sugars across the membrane and allow the passage of K to the membrane. In case of brain cells, this will result in energy deficiency in the cell and disorders in the transmission of nerve impulses. This will explain why babies born to mothers subjected to methyl mercury poisoning suffer from irreversible damage to the central nervous system, including cerebral palsy, mental retardation and convulsions. Methyl mercury poisoning also leads to segregation of chromosomes, chromosome breakage in cells and inhibited cell division. All the symptoms of mercury poisoning set in at blood levels of 0.5 ppm of CH_2Hg^+.

25.7.3 Biological Methylation: Amplification in Food Chain

The Minamata Chemical Company discharged Hg into Minamata Bay, but the fish in the Bay were found to contain CH_3Hg^+. This missing link was filled up by subsequent research. Hg or its salts can be converted to methyl mercury by anaerobic methane-synthsizing bacteria in water. This conversion is facilitated by Co(III)-containing vitamin B_{12} conenzyme. A CH_3-group bonded to Co(III) on the coenzyme is transferred enzymatically by methyl cobalamin to Hg^{2+}, yielding CH_3Hg^+ or $(CH_3)_2Hg$:

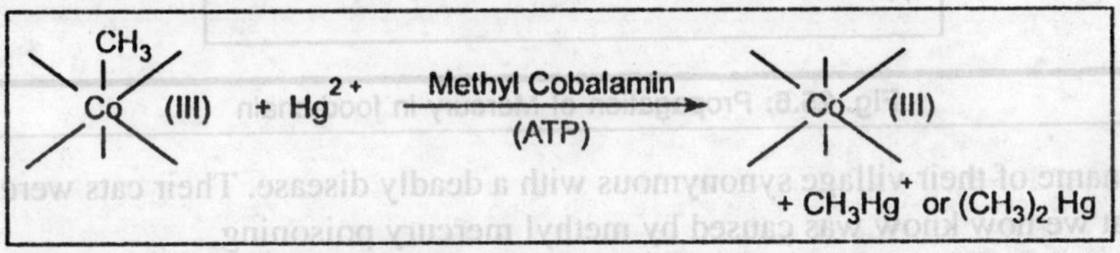

An acidic medium promotes the conversion of dimethyl mercury to methyl mercury which is soluble in water. It is methyl mercury which enters the food chain through plankton and is concentrated by fish by a factor of 10^3 or more as it passes up the food chain, as shown in Fig. 25.6 (page 536).

The Hg concentration builds up at each level of the food chain. This is valid even in uncontaminated waters. Hg has always been part of our environment and Hg cycles existed long before any industry developed. Large fish of ancient ages, preserved in some museums, have been found to contain significant level of Hg. However, Hg pollution considerably enhances the Hg concentration in each level of the food chain.

Soon after the Minamata disaster, it was reported that freshwater fish from Lake Erie and the St. Clair River showed high levels of Hg (0.1-3.5ppm), in the form of methyl mercury in their living tissues. As a result, important commercial fisheries in those areas were closed down.

Minamata Disease

In the early 1950s, people in the small coastal village of Minamata, Japan, noticed strange behavior that they called dancing cats. Inexplicably, cats would begin twitching, stumbling, and jerking about as if they were drunk. Many became "suicidal" and staggered off docks into the ocean. The residents did not realize it at the time, but they were witnessing an ominous warning of an environmental health crisis that

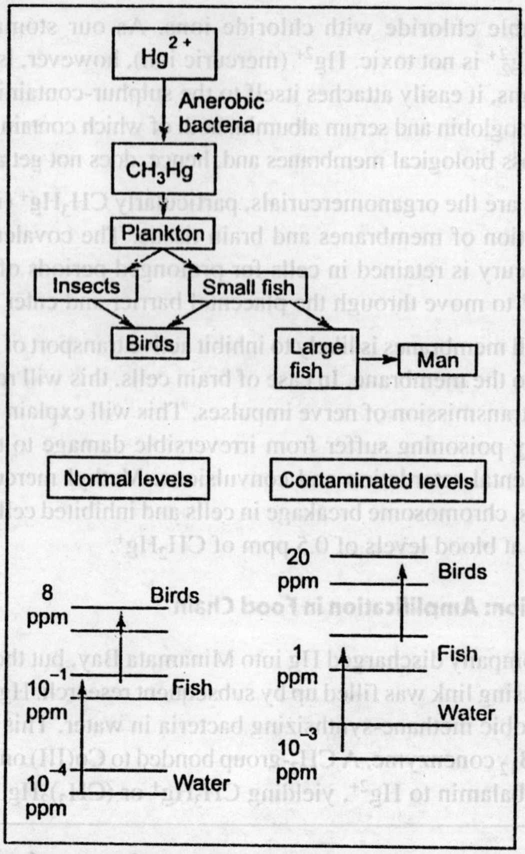

Fig. 25.6: Propagation of Mercury in food-chain

would make the name of their village synonymous with a deadly disease. Their cats were suffering from brain damage that we now know was caused by methyl mercury poisoning.

In 1956, the first human case of neurological damage was reported. A five-year-old girl who had suddenly lapsed into a convulsive delirium was brought into the local clinic. Within a few weeks there seemed to be an epidemic of nervous problems in the village, including numbness, tingling sensations, headaches, blurred vision, slurred speech, and loss of muscle control. For an unlucky few, these milder symptoms were followed by violent trembling, paralysis, and even death. An abnormally high rate of birth-defects also occurred. Children were born with tragic deformities, paralysis, and permanent mental retardation. Lengthy investigations showed that these symptoms were caused by mercury from fish and seafood that formed a major part of the diet of both humans and their cats.

For years the Chisso Chemical Plant had been releasing residues containing mercury into Minamata Bay. Since elemental mercury is not water soluble, it was assumed that it would sink into the bottom sediments and remain inert. Scientists discovered, however, that bacteria living in the sediments were able to convert metallic mercury into soluble methyl mercury, which was absorbed from the water and concentrated in the tissues of aquatic organisms. People who ate fish and shellfish from the bay were exposed to dangerously high levels of this toxic chemical. Altogether, more than 3,500 people were affected and about 50 died of what became known as Minamata Disease. After nearly twenty years of rancorous protests and litigation, the Chisso Company finally admitted that it was guilty of dumping the mercury and agreed to pay reparations to the victims.

Dumping of mercury into Minimata Bay was stopped twenty years ago. Mud containing mercury was dredged up and buried elsewhere so the bay is now considered safe for fishing. The minds and bodies of those people who ate the mercury-poisoned fish, however, can never be repaired. Have we learned from this tragedy how to anticipate and prevent future environmental disasters?

25.7.4 Remedial Measures

Further environmental pollution by Hg can be prevented by adopting the following measures, as recommended by the Environmental Protection Agencies of USA and Sweden :

1. All chlor-alkali plants must stop using Hg electrodes and switch to new technology.
2. All alkyl mercury pesticides must be banned.
3. All other mercurial pesticides must be restricted to some selected areas.

It should be noted that the already contaminated sediments of rivers and lakes will continue to yield highly toxic CH_2Hg^+ into the waters for many years to come. In Sweden, pilot experiments were undertaken for decontamination of sediments by covering the bottom sediments with fresh, finely divided materials having high adsorption capabilities and, alternatively, by burying the sediments under inorganic inert materials.

25.8 CHROMIUM

The major Chromium ore is Chromite, $FeOCr_2O_3$, from which Ferro-Chrom alloys and Chromium metal are obtained. Chromium is used in tanning industry and in corrosion protection in steel. Treatment of steel with CrO_4^{-2} may form an oxide film on the metal rendering it resistant to corrosion. There are two chromium ionic forms: hexavalent and trivalent forms.

Hexavalent Chromium (VI) has two main oxy-anion forms CrO_4^{-2} and $Cr_2O_7^{-2}$ which are involved in the equilibrium:

$$2CrO_4^{-2} + 2H_3O^- \leftrightharpoons Cr_2O_7^{-2} + 3H_2O$$

Chromium (VI) is toxic but its lifetime in the aqueous environment is limited as it is reduced to Chromium (III) by organic material. In fact, Chromium (VI) is used for the estimation of total organic Carbon in environmental studies of materials such as soil and water according to the reactions:

$$Cr_2O_7^{-2} + 14H^+ + 6e \rightarrow 2Cr^{+3} + 7H_2O$$

$$CH_2O + H_2O - 4e \rightarrow CO_2 + 4H^+$$

$$\text{i.e., } 2Cr_2O_6^{-2} + 3CH_2O + 16\,H^+ \rightarrow 4Cr^{+3} + 3\,CO_2 + 11H_2O$$

The Chromium (III) state, which is said to be non-toxic (though this may be because of its low concentrations in water), is mobilised in acid as the Cr^{+3} hydrated ion, while, in alkaline conditions, low soluble $Cr(OH)_3$ is produced. The hydroxy species of Cr(III) dominates in natural water systems. The recommended limit for Cr in potable water is 50 μg/l. A conceptual model depicting potential processes affecting inorganic and organic chromium in the aqeous environment is shown in Fig. 25.7 (page 538).

25.8.1 Sources

There are various sources of chromium entry into the human system:

1. **Drinking water**: Normally contains very low concentrations of Cr (i.e., 5 μg/l or less). Cr^{+3} rarely occurs in drinking water that is chlorinated. It is assumed that most waterborne Cr is in the hexavalent form.

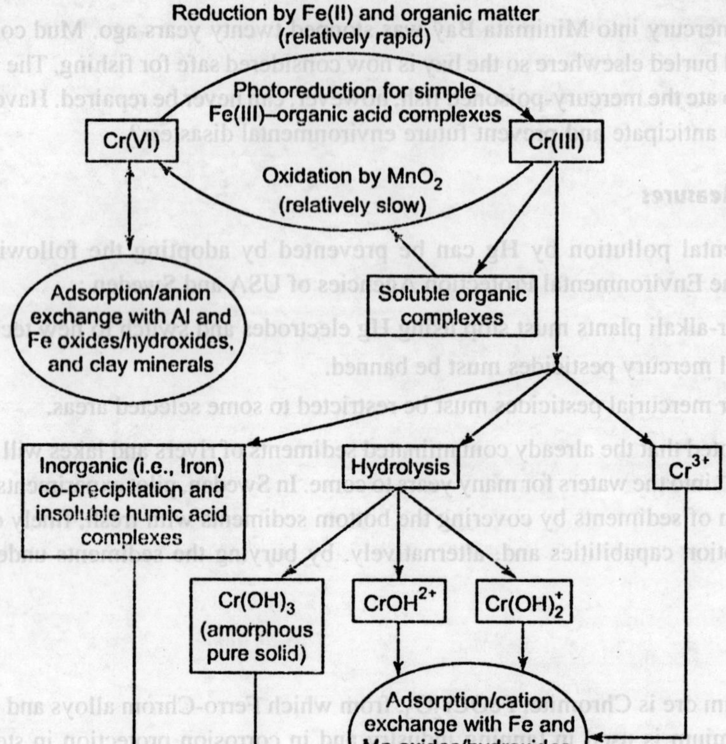

Fig. 25.7: Conceptual model depicting potential processes affecting inorganic and organic chromium in the aqeous environment (Adapted from USEPA, 2000)

2. **Food:** Foodstuffs vary considerably in chromium content which ranges from 20 to 590 μg/kg. Some seafoods appear to contain elevated levels of 0.02 - 0.21 μg/kg. Wines containing 'Cr' and concentrations up to 60 μg/litre have been reported.
3. **Air:** There is only a limited amount of information on the levels of chromium in the air. Cr concentrations in air in towns are about 0.02 μg/m³.
   Most of the Cr in the air will be in the form of fine particles of which, perhaps, one half of those inhaled could become deposited in the respiratory tract.
4. **Smoking:** Cigarettes contain traces of chromium—a value of 1.4 μg per cigarette have been reported and some of this will be inhaled and absorbed. The chromium retained in the lung by smoking 20 cigarettes per day would not exceed a few micrograms per day.
5. **Occupational Exposure:** Levels of airborne chromium in a number of industrial situations, particularly in plating plants and where welding occurs, can be very much higher than in the ambient environment. Concentrations as high as hundreds of micrograms per cubic meter of air have been reported.
6. **Ingestion of dirt, dust etc.:** Little is known about the levels of 'Cr' in dust but, in general, it is unlikely to be a very important source of exposure even where young children have the opportunity to ingest dirt and dust.

Chromium is absorbed through both the gastrointestinal and respiratory tracts. The amount absorbed differs in each system and depends on the form of chromium. Trivalent 'Cr' is an essential form of the element for human beings. Hexavalent is toxic. From 0.1% to 1.2% of trivalent 'Cr' salts are absorbed whereas 25% of Glucose Tolerance Factor (GTF), a chromium complex necessary for normal glucose tolerance is absorbed.

The largest stores of 'Cr' in man are in skin, muscle, and fat tissue levels and are a function of sex, age, and geographical location.

Chromium is excreted slowly—mainly in the urine but also in the feces.

The overall mobilisation of chromium is shown in Fig. 25.8.

25.8.2 Toxicity

At sublethal level, chromium has a number of metabolic effects. Some of these effects are:

The level of sugar in the blood is critical to the functioning of body tissues, especially the brain. This blood sugar level stays remarkably constant under normal conditions. Many factors help to control the amount of sugar in blood: one is the compound called Insulin. Insulin is secreted by the pancreas and works as a control in lowering the blood sugar level. When the pancreas fails to secrete enough insulin, the disease called diabetes mellitus results. Chromium plays a role in lowering the blood sugar level in the body by increasing effectiveness of insulin. This may explain why chromium deficiency produces symptoms similar to diabetes mellitus. A series of relevant experiments has shown that diets supplemented with inorganic chromium or high chromium brewers' yeast can significantly lower total serum cholesterol in human subjects. In addition to brewers' yeast, whole grain cereals and liver are good sources of chromium.

Hexavalent chromium at 10 mg/kg of body weight will result in liver necrosis, nephritis and death in man; lower doses will cause irritation of the gastrointestinal mucosa.

Hexavalent chromium in high doses has been implicated as the cause of digestive tract cancer in man and there is firm evidence that there is an increased risk of lung cancer for workers who are exposed to high level of Cr. Prostate cancer and maxillary sinus cancers have been reported in workers involved in Cr-planting. Other health effects related to industrial exposures have been reported, e.g. hexavalent chromium can produce cutaneous and nasal mucous membrane ulcers and mermatitis (from skin contact).

Higher concentration of chromium, however, cause much more deteriorous effects, viz., genetic damage and carcinogenesis.

25.9 COPPER

Copper is a low-toxicity, corrosion resistant metal widely used because of its workability, electrical conductivity and ability to conduct heat. Copper is extracted from its mined ore copper sulfide.

25.9.1 Biochemical effects of Copper

Copper is an essential trace element for both plants and animals. It acts as catalysts of enzyme reaction and also chelators of some organic compounds. In man, copper predominantly acts as a cofactor for a number of enzymes such as cerulo plasmin, cytochrome C-oxidase, dopamine β-hydroxylase, superoxide dismutase, and tyrosinase.

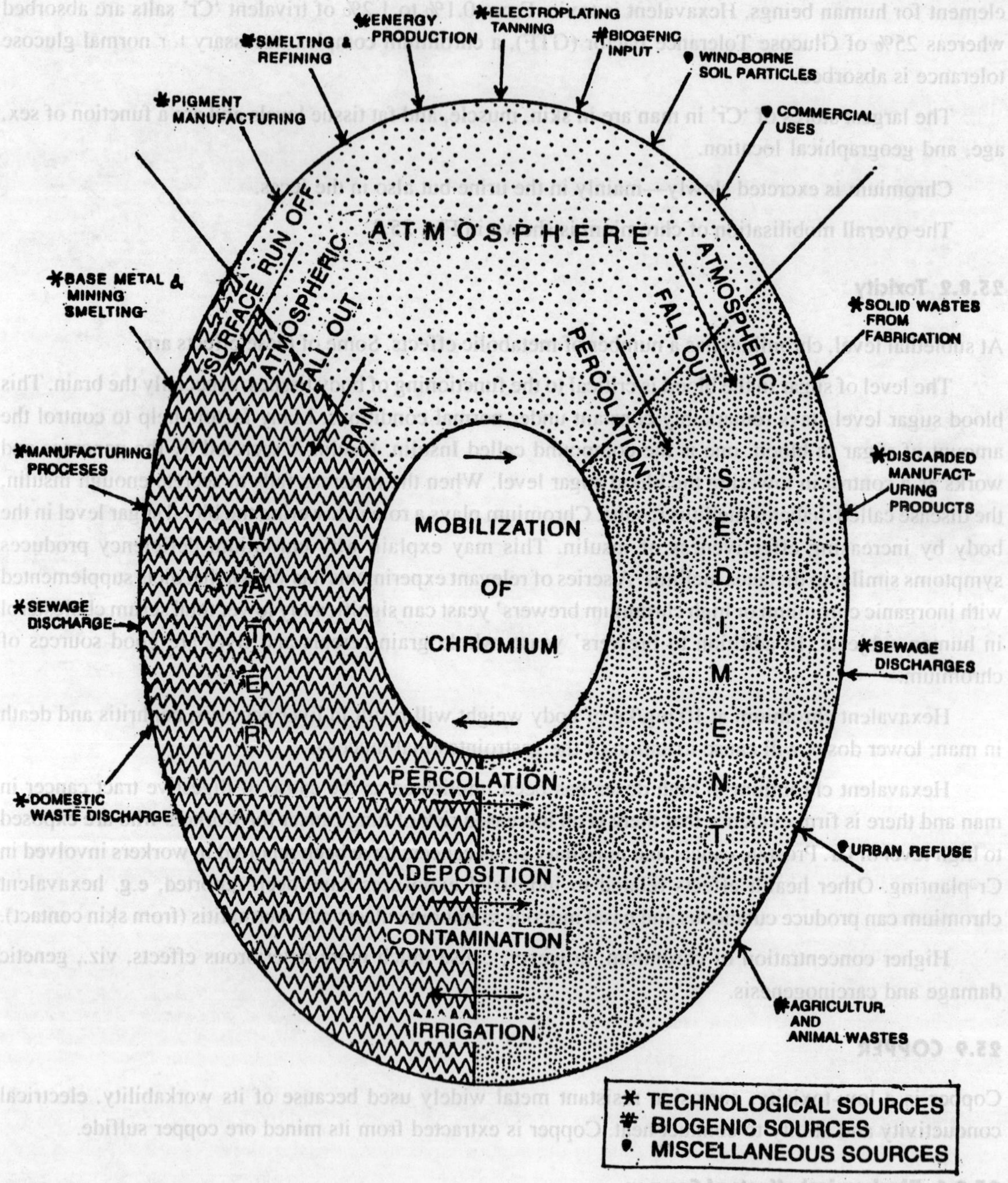

Fig. 25.8: Chromium status in environment

25.9.2 Toxic effects

Excess intake of copper induce genotoxicity. Thus autosomal mutation induced by copper is seen in man. The main sites of copper accumulation are the liver and the brain and consequently liver disease and neuropsychiatric sýmptoms are the main features that lead to diagnosis. The British physician Dr Samueal Alexander Kinnier Wilson (1878-1937), a neurologist, described the copper toxicity condition, including the pathological changes in the brain and liver in 1912. Since then this disease is called Wilson's disease.

Wilson's disease or hepatolenticular degeneration is associated with copper accumulation in tissues. This manifests as neurological or psychiatric symptoms and liver disease. It is treated with medication that reduces copper absorption or remove the excess copper form the body, but occasionally liver transplant is required.

In general, a diet low in copper-containing foods (mushrooms, nuts, chocolate, dried fruits, liver, and shellfish) is recommended. Penicillamine is the chelator first used in treatment of Wilson's disease. This binds copper (chelation) and leads to excretion of copper in the urine. Patients intolerant to penicillamine may instead commence on trientine hydrochloride, which has chelating properties. Sometimes zinc acetate may be used instead of chelators to maintain stable copper levels in the body. Zinc stimulates metallothiomin, a protein in gut cells that binds copper and prevents their absorption and transport to the liver.

25.10 ECOSYSTEM FLOW OF METALS: BIOACCUMULATION & BIOMAGNIFICATION

As the metals are non-biodegradable, they are absorbed by the biotic system and accumulates in higher order in different biota of the ecosystem. The most profound bioaccumulation is seen in aquatic system.

In recent year, accumulation of toxic metals like lead in sewage-fed ponds are noticed in a significant manner (Fig. 25.9, page 542). The accumulation through food chain takes place not only in a simple manner, but the level of toxic metals appears to be more higher in upper trophic level. This is a case of biomagnification of toxicants. The transport and biological magnification of some heavy metals in sewage-fed aquatic system (viz., jheels of Kolkata Port Trust), as studied in 2000, is shown in Fig. 25.10 (page 543).

25.11 PESTICIDES

25.11.1 Introduction

The term 'Pesticide' generally indicates any chemical, microbial agent or their mixture used as active ingredients of products for the control of crop pests and diseases, animal ectoparasites and pests in public health. The Pesticide Manual lists about 600 active ingredients although there may be two or three times as many chemicals labelled as pesticides throughout the world. The characteristics of major pesticides are given in Table 25.10 (page 542).

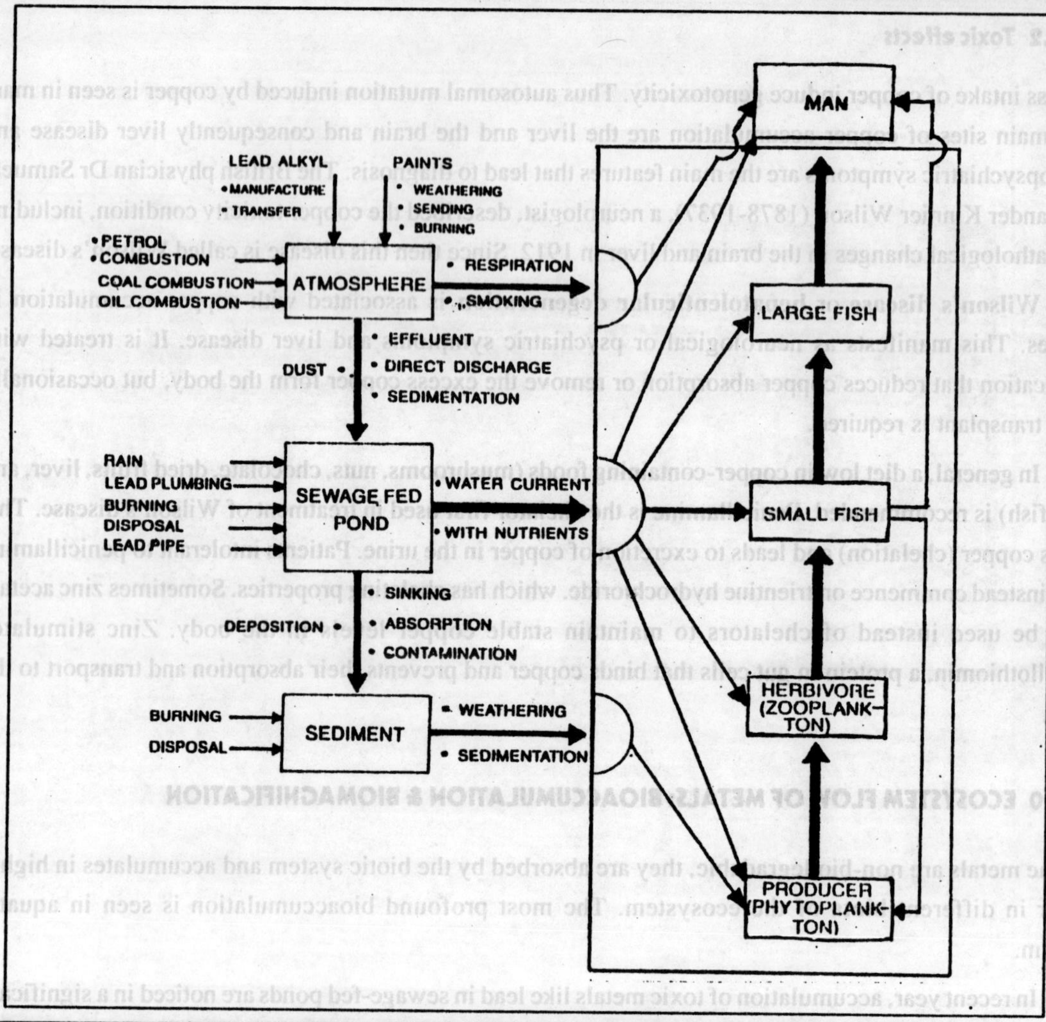

Fig. 25.9: Lead in sewage-fed pond ecosystem

Table 25.10: Pesticides

Type	Examples	Characteristics
Insecticides		
Inorganic chemicals	Mercury, lead, arsenic, copper, sulfate.	Highly toxic to many organisms, persistent, bioaccumulates.
Organochlorines	DDT, methoxychlor, heptachlor, HCH, pentachlorophenol, chlordane, toxaphene, aldrin, endrin, dieldrin, lindane.	Mostly neurotoxins, cheap, persistent, fast-acting, easy to apply, broad spectrum, bioaccumulates, biomagnific.
Organophosphates	Parathion, malathion, diazinon, dichlorvos, phosdrin, disulfoton, TEPP, DDVP.	More soluble, extremely toxic nerve poisons, fast-acting, quickly degraded, do not bioaccumulate, toxic to many organisms, very dangerous to farm workers.
Carbamates and urethanes	Carbaryl (Sevin), aldicarb, carbofuran, methomyl, Temik, mancozeb.	Quickly degraded, do not bioaccumulate, toxic to broad spectrum of organisms, fast-acting, very toxic to honey bees.

Contd.

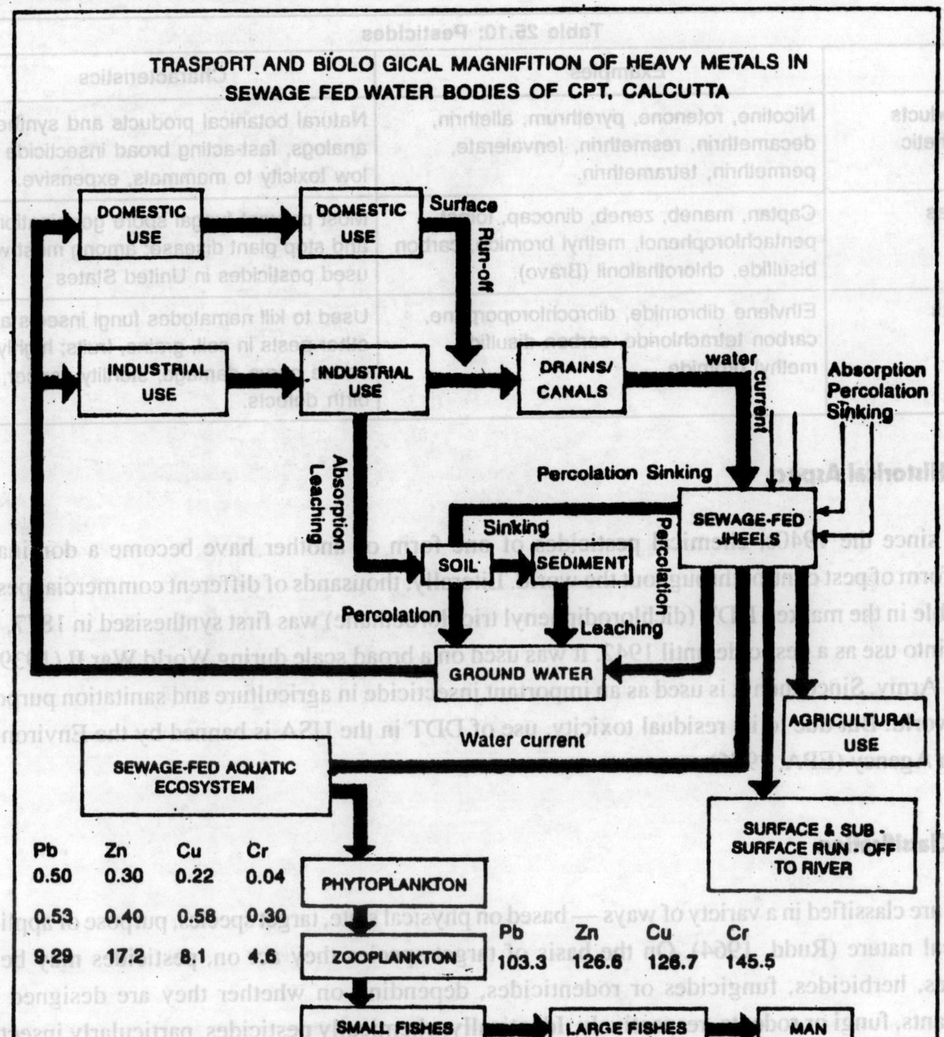

Fig. 25.10: Transport and biological magnification of heavy metals in sewage-fed water bodies of Kolkata

Table 25.10: Pesticides

Type	Examples	Characteristics
Formamidines	Amitraz, chlordimeform (Fundal and Galecron)	Neurotoxins specific for certain stages of insect development, act synergistically with other insecticides.
Microbes	Bacillus thuringensis Bacillus papillae Viral diseases	Kills caterpillars. Kills beetles. Attacks a variety of moths and caterpillars.

Contd.

Table 25.10: Pesticides

Type	Examples	Characteristics
Plant products and synthetic analogs	Nicotine, rotenone, pyrethrum, allethrin, decamethrin, resmethrin, fenvalerate, permethrin, tetramethrin.	Natural botanical products and synthetic analogs, fast-acting broad insecticide action, low toxicity to mammals, expensive.
Fungicides	Captan, maneb, zeneb, dinocap, folpet, pentachlorophenol, methyl bromide, carbon bisulfide, chlorothalonil (Bravo).	Most prevent fungal spore germination and stop plant disease; among most widely used pesticides in United States
Fumigants	Ethvlene dibromide, dibrochloroporpane, carbon tetrachloride, carbon disulfide, methyl bromide.	Used to kill nematodes fungi insects and other pests in soil, grains, fruits; highly toxic, cause nerve damage, sterility cancer, birth defects.

25.11.2 Historical Aspect

Roughly, since the 1940s, chemical pesticides of one form or another have become a dominant and essential form of pest control throughout the world. Literally, thousands of different commercial pesticides are available in the market. DDT (dichlorodiphenyl trichloroethane) was first synthesised in 1877, but did not come into use as a pesticide until 1942. It was used on a broad scale during World War II (1939-1945) by the US Army. Since then it is used as an important insecticide in agriculture and sanitation purposes all over the world. But due to its residual toxicity, use of DDT in the USA is banned by the Environmental Protection Agency (EPA, 1976).

25.11.3 Classification

Pesticides are classified in a variety of ways — based on physical state, target species, purpose of application, or chemical nature (Rudd, 1964). On the basis of target species they act on, pesticides may be either insecticides, herbicides, fungicides or rodenticides, depending on whether they are designed to kill insects, plants, fungi or rodents, respectively. Identically, chemically pesticides, particularly insecticides, may be of three kinds—(a) organic phosphate compounds, (b) chlorinated hydrocarbons, and (c) carbamates. In addition, there are pyrethroids and phenolics are acting as insecticides.

- **Organochlorine**: Organochlorine pesticides are hydrocarbon compounds containing multiple chlorine substituents. One of the most widely known organochlorine pesticides is DDT, an insecticide which was widely used during World War II. Because of its low toxicity to humans, DDT powder was widely applied by the US Army to the skin of soldiers to kill parasitic skin insects. Because of their low water solubilities and environmental persistence, DDT and other organochlorine pesticides also found widespread use as poisons for subterranean nests of termites. About 100,000 tons per year were produced in the late 1950s in the United States alone.
- **Organophosphate**: Organophosphate pesticides are used primarily as agricultural pesticides, and contain a chemically reactive phosphate ester side chain, consisting of a central phosphorous atom double bonded to either an oxygen or sulphur atom, and single bonded to two methoxy ($-OCH_3$) or ethoxy ($-OCH_2CH_3$) groups.

- **Carbamate**: Carbamate pesticides generally have the formula RHNCOOR, and are relatively polar, highly soluble in water, and chemically reactive. Some Carbamates of importance are Aldicarb, Carbaryl, Carbofuran, Ferbam, and Captan. Carbofuran, like many Carbamates, is primarily used in agriculture. Its usage pattern is consistent with general Carbamate pesticide application, since the Midwest and Southeast are the two regions with the greatest need for pesticides in agriculture.
- **Triazine**: Compounds of the s-Triazine family are among the most heavily used herbicides during the past 30 years. Although most have low acute toxicity in most animal species, chronic exposures to certain s-Triazines have been shown to be animal carcinogens. s-Triazine (for which this class is named) is a six member ring containing alternating nitrogen and carbon atoms. Common product derivatives include Atrazine, Simazine, Terbuthylazine, Ametryne, and Terbutryne.
- **Naturally occurring pesticides**: Natural pesticides are the products of plants or minerals. Nicotine (extracted from tobacco), Pyrethrum (extracted from **chrysanthemum** flowers), and Rotenone (extracted from the tuber of **Derris elliptica**) are plant derived, while pesticide like Boric acid, cryolite and Diatomamaceous Earth are mineral derived.

25.11.4 Application Potential

Most large-scale pesticide use is undertaken for reasons of public health, agricultural production or forestry management. It is well-known that more than a dozen important diseases are transmitted by mosquitoes, including malaria, yellow fever, encephalitis, etc. where DDT is used widely for control of mosquitoes for a considerable period, particularly in the tropics. Identically, in agriculture, a number of organophosphate, organochloride and carbamate compounds were used for control of disease pests all over the world. In forestry practices, a bulk amount of pesticides was used in the past where plantation forestry was widely practiced.

25.11.5 Limitation of Pesticide Uses

There is little doubt that pesticide use harmed nontarget species on a significant scale. A few examples are:

(a) The destination of speckled sea front in the Texas,
(b) Reduction of photosynthesis by the marine phytoplanktons,
(c) Residue accumulation in living system through food chain and subsequent destruction of population of birds, mammals, molluscs, etc.

25.11.6 Toxicology of Major Pesticides

In general, most of the pesticides are capable of inducing irreversible toxic effects in undesired biological systems. Unfortunately, since toxicity is often non-selective, adverse effects may develop in exposed humans or other non-target species.

There are three categories of chlorinated hydrocarbon insecticides, viz., DDT and its two other analogues. DDT is cheap and the most effective one. DDT has genotoxic and clastogenic functions. Normally DDT acts on central nervous system (CNS) by interfering with or blocking the transport of ions (Na^+, K^+ and Ca^{++}) across the nerve membrane. Identically, cyclodiene insecticides like Aldrin, Dieldrin, Heptachlor etc. are still more toxic than DDT compounds on CNS.

The organophosphorus insecticides is the single largest class of insecticides, accounting for over 40% of total registered pesticides in the world. Some of them are also used as herbicides.

The acute toxicity of some organophosphate compounds is very high. They act by phosphorylating and inhibiting acetyl cholinesterase in nervous tissues of man and insects. Carbamate insecticides have a similar—but not identical—mechanism of action to OP compounds (reversible inhibition of HehE) and, therefore, they induce quantitatively the same chollinergic form of toxicity.

Pyrethrum, the natural insecticide obtained from chrysanthemum flowers, contains at least six active ingredients generally known as pyrethrins. The mechanism of action of pyrethrins and pyrethroids on the nervous system is thought to be similar to that of DDT.

Herbicides differ widely as to their toxicity and selectivity. Some are toxic to plants and animals by different mechanisms, while some share a common mode of action against the two targets. There are three major types of herbicides—chlorophenoxy compounds (2, 4-D; 2-4-5-T), Bipyridylium compounds (paraquat), and Triazines (Atrazine, Simazine)—which are widely used in agriculture.

A summary of major pesticide effects and persistence is given in Table 25.11:

Table 25.11: Major Pesticides — Persistence & Effects

INSECTICIDES		
Type	Examples	Persistence
Chlorinated hydrocarbons	DDT, aldrin, dieldrin, endrin, heptachlor, toxaphene, lindane, chlordane, kepone, mirex.	High (12-15 years)
Organophosphates	Malathion, parathion, monocrotophos, methamidophos, methyl parathion, DDVP.	Low to moderate (normally 1-12 weeks, but some can last several years)
Carbamates	Carbaryl, maneb, priopoxor, mexicabate, aldicarb, aninocarb.	Usually low (days to weeks)
Pyrethroids	Pemethrin, decamethrin.	Usually low (days to weeks)
HERBICIDES		
Type	Examples	Effects
Contact	Triazines such as atrazone and paraquat.	Kills foliage by interfering with photosynthesis
Systemic	Phenoxy compounds such as 2, 4-D, 2, 5-T and silvex; substitute ureas such as diuron, norea, fenuron and other nitrogen-containing compounds such as diminozide (Alar), glyphsate.	Absorption creates excess growth hormones; plants die because they cannot obtain enough nutrients to sustain their greatly accelerated growth
Soil sterilants	Trifluralin, diphenamid, dalaphon, butylate.	Kills soil microorganisms essential to plant growth; most also act as systemic herbicides

Source: G. Tyler Miller Jr., *Living in the Environment* (Wadsworth, Belmont, California, 1990), p. 551

25.11.7 Pesticide usage in India

The use of synthetic pesticides started in 1948-49 with the use of DDT for malaria control and BHC for locust control. The Indian pesticides production industry started with the setting up of a BHC technical plant at Rishra near Kolkata in 1952. Shortly after, Hindustan Insecticides Ltd. set up two units to manufacture DDT. In 1969, Union Carbide set up a small plant (Union Carbide India Ltd (UCIL)) in Bhopal, the capital city of Madhya Pradesh, to formulate pesticides. The Bhopal facility was part of

India's Green Revolution aimed to increase the productivity of crops. The industry produced various pesticides—mainly sevin brand carbaryl insecticide and temikcbrand aldicarb pesticide. All the pesticides produced at UCIL were sold in the Indian market. The Union Carbide continued pesticide production till 1984 Bhopal disaster. Today, the Indian pesticides industry comprises of more than 125 basic producers of large and medium scale and more than 500 pesticide formulations. The pesticide formulation produced in the country are mainly of the conventional type currently, dusting powder has a major share (85%) in the market followed by water-soluble dispersible powder (12%) and emulsification concentrates (2%). The use of granules, which are advantageous in terms of lower drift, ease of application and safety to operate is still in infancy. Most of the units are in small scale industry (SSI) sector.

India is one of the few remaining countries still engaged in the large scale manufacture, use, and export of some of the toxic chlorinated pesticides, such as *p.p'*-dichlorodiphenyltrichloroethane (DDT), hexachlorocyclohexane (HCH) and pentachlorophenol (PCP). Even in the 1990s more than 70% of the gross tonnage of pesticides used in agricultural applications in India consisted of formulations which are banned or severely restricted in the east and west. According to a Green Peace Report, India is now producing 90,000 metric tons of pesticides as the largest industry in the whole of Asia and twelfth largest in the entire world (http://www.greenpeaceindia.org.nopesti.html). The cumulative consumption of the pesticide, hexachlorocyclohexane (HCHs), in India until 1985 was 575,000 tons and since then about 45,000 tons of HCHs has been used annually. The usage of DDT and HCH continued till recently. Apart from the US, India is the only country which has applied more than 100,000 tons of DDTs since its formulation, mainly in its agricultural and malarial control programmes until it was banned for agricultural use in 1989. Even though usage of technical HCH was finally banned completely in 1977, the Government of India is encouraging its replacement with Lindane (γ-HCH), an isomer which has all the hazardous properties of HCH. Even though DDT has been banned for agricultural use, India has sought exemption under Stockholm Convention for use of 10,000 tons of DDT for restricted use in the public health sector. The national malaria program (NAMP) used 3,750 tons of DDT in the year 2001, in rural and peri-urban areas for residual spraying.

The pattern of pesticide use differs significantly between the countries. The worldwide consumption of pesticide is about two million tons per year, of which 24% is consumed in the USA alone, 45% in Europe and 25% in the rest of the world. The usage of pesticides in India is only 0.5 kg ha^{-1}, while in Rep. Korea and Japan, it is 6.6 and 12.0 kg ha^{-1}, respectively. Among the various pesticides used in India, 40% of all the pesticides used belong to organochlorine class of chemical pesticides. The other major category is organophosphate pesticides. Monocrotophos, phorate, phosphamidon, methyl parathion, and dimethoate are some of highly hazardous pesticides that are continually and indiscriminately used in India. The three commonly used pesticides, lindane (γ-HCH), DDT, and malathion account for 70% of the total pesticides consumption. Regarding the usage of technical pesticides, insecticides account for 80% of total pesticides used in the country, followed by herbicides and fungicides.

The pesticides commonly used in India is given in Table 25.12 (page 548). Pesticides used in high quantity in irrigated areas and in few selected crops like cotton, rice, and potato.

Table 25.12: Pesticides commonly used in India

	Pesticide: common name	Chemical family
Ia.	Extremely hazardous[a]	
	1. Phorate	Organophosphate
Ib.	Highly hazardous[a]	
	2. Monocrotophos	Organophosphate
	3. Profenofos & Cypermethrin	Combination pesticide
	4. Carbofuran	Carbamate
II.	Moderately hazardous[a]	
	5. Dimethoate	Organophosphate
	6. Quinalphos	Organophosphate
	7. Endosulphan	Organochlorine
	8. Carbaryl	Carbamate
	9. Chlorpyrifos	Organophosphate
	10. Cyhalothrin	Pyrethroid
	11. Fenthion	Organophosphate
	12. DDT	Organochlorine
	13. Lindane	Organochlorine
III.	Slightly hazardous[a]	
	14. Malathion	
IV.	Unlikely to present acute hazard in normal use[a]	
	15. Carbendazim	Carbamate
	16. Atrazine	Triazine

[a]WHO classification [54]

25.11.8 Environmental Impact of Pesticides

A 1979 report of the United Nations Environment Programme (UNEP) summarised the environmental impact of the indiscriminate use of pesticides in the following words:

When carelessly applied, chemical pesticides can result in acute and long-term side-effects, including sickness and death of people, useful animals, fish, birds, and destruction of crops. Even when properly used, chemical pesticides have a number of unavoidable side-effects. Their persistence and ubiquitous nature, combined with a tendency of some compounds to concentrate in organisms as they move up the food chain, may increase their toxicity to fish, birds, and other forms of life including man and cause other harmful effects on man, his health and well-being.

The environmental fate of pesticides used in crop protection and vector control of communicable diseases is represented diagrammatically in Fig. 25.11 (page 528). Accumulation of pesticides in the environmental compartments is dependent on both the manner of their application and the chemical nature of the compounds used and their formulations. The most popular technique of application is spraying in the form of dusts, fumes, mists or vapours using manual or mechanical devices or even small aircraft which, while facilitating diffusion of the chemicals over a large area and making them more effective, leads inevitably to the persistence of potentially toxic chemicals in the environment, contamination of the environment relate to their non-degradable nature.

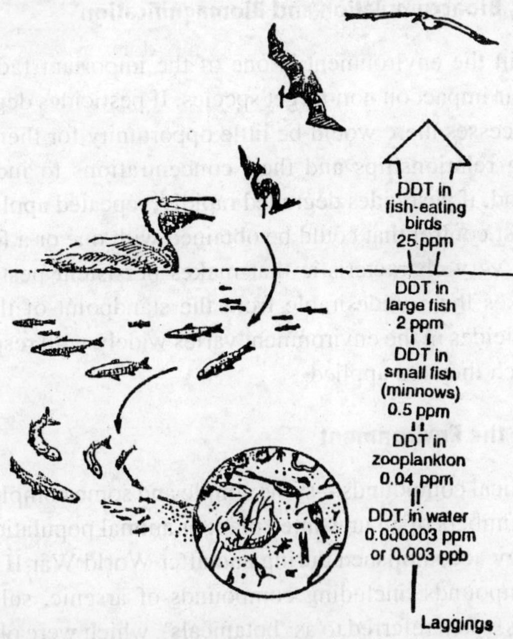

Fig. 25.11: Biomagnification of pesticides through food chain

It is apparent that by both precipitation and leaching pesticides may find their way eventually into water bodies. The transport and fate of pesticides in the aquatic environment is influenced by their concentration and degradation by biotic and biotic factors. Indeed, the two most important degradative forces acting on environmental chemicals including pesticides are sunlight and microbes. Most pesticides used in the agriculture and public health sectors are transferred to soil where their fate is determined not only by soil microflora and microfauna but also by the chemical environment of the soil. Very little is known about the interaction between chemical fertilizers and insecticides or the role of microbial flora in the accumulation of residues of persistent pesticides in soil.

The ecological risk inherent to pesticides arises from the differential vulnerability of different biotic communities to their toxic action and the differential response of organisms to their non-lethal but possible long-term effects. Besides the rapid development of resistance among arthopods, rodents, bacteria, fungi, and weeds, the two problems that have to be faced are the disruption of natural controls and the pesticide treadmill, entailing the use of more and more chemicals for more and more applications, escalating the cost of control.

The estimated annual impacts of pesticides in developing countries is given in Table 25.13.

Table 25.13: Estimated Annual Impact of Pesticides in Developing Countries, 2000

Population groups at risk	Character of exposure	Estimated overall annual impact
Pesticide formulators, mixers, applicators, pickers, and suicides and mass poisoning	Single and short-term very high-level exposure.	3,000,000 exposed 220,000 deaths.
Pesticide manufacturers, formulators, mixers, applicators and pickers.	Long-term high-level exposure.	735,000 suffer from specific chronic effects of long-term exposure (cancer).
All population groups.	Long-term, low-level exposure.	37,000 suffer from chronic effects of long-term exposure (cancer).

25.11.9 Pesticide Persistence, Bioaccumulation and Biomagnification

The persistence of pesticides in the environment is one of the important factors that determines their efficiency as pesticides and their impact on non-target species. If pesticides degraded rapidly to harmless substances through natural processes there would be little opportunity for them to be spread throughout the food web through feeding relationships and their concentrations to increase through biological magnification. On the other hand, if pesticides degraded rapidly, repeated applications may be necessary to achieve the same level of pest control that could be obtained with one or a few applications of a more persistent pesticide. Thus, the very characteristic that makes persistent pesticides desirable from the standpoint of pest control makes them undesirable from the standpoint of their impact on non-target species. The persistence of pesticides in the environment varies widely with respect to pesticide types and the nature of substrates on which they are applied.

25.11.10 Case Study—DDT in the Environment

DDT is one of many toxic chemical compounds—some simple and some complex—classed as pesticides, substances used to reduce the numbers of an unwanted plant or animal population. Before the widespread advances in chemical technology accomplished during and after World War II, two types of insecticides were in use: (a) inorganic compounds, including compounds of arsenic, sulphur, copper or cyanide, (b) complex organic compounds, often referred to as 'botanicals', which were obtained from plant tissues. Use of these compounds was not widespread and, thus, the impact was minimal. In post-War period, however, many new organic compounds were developed with toxic properties suitable for use as pesticides of one sort or another. One large class of pesticides thus developed is the chlorinated hydrocarbons of which DDT is a member.

Besides their apparent low toxicity to humans, the chlorinated hydrocarbons have two general characteristics that make them desirable as insecticides. First, they are toxic to large range of organisms and, therefore, can control many pests at once.

Second, they are relatively long-lasting and, therefore, need to be applied less frequently. Unfortunately, these characteristics also make the chlorinated hydrocarbons all the more dangerous to natural ecosystems. The danger lies in their biological magnification within the food chain, a phenomenon produced by the high solubility of these compounds in fats and oils and their storage within the fatty tissues of organisms.

A clear example of DDT accumulation on various organisms in salt marsh area of Great South Bay, New York is given in Table 25.14.

Table 25.14: DDT residues in biota of salt marsh of Great South Bay, NY

Sample	DDT residues (ppm)
Water	0.00005
Plankton	0.04
Shrimp	0.16
American eel	0.28
Atlantic needlefish	2.07
Cord grass-roots	2.80
Common tern	3.15
Herring gull-Brain	5.43
Double-crested cormorant	26.4
Ring-billed gull	75.5

DDT enters the marsh by aerial spraying for mosquito control. Because it is highly insoluble in water, most of the DDT remains in the marsh soil or settles to be incorporated in the bottom sediments of channels and pools. A very small amount, however, does dissolve in the water. The algae, phytoplankton and decomposer bacteria of the marsh are, thus, exposed to very low concentrations of DDT in their environment. Because DDT is much more readily soluble in fats and oils, it tends to accumulate in the bodies of these tiny organisms and reach concentrations may times those of the water itself. When these organisms are ingested in large numbers by zooplankton or detrital feeders such as clams and tiny crustaceans, the DDT they take in also tends to be retained because of its insolubility in water. In this way, the concentration of DDT in body tissues increase by many orders of magnitude up the food chain, from fingerling fish to larger fish to predatory birds, such as the herring gull and cormorant.

Physically, the DDT molecule is very stable and, unlike many other organic molecules, does not decompose under continued exposure to oxygen, sunlight or other reagents or energy sources available at the earth's surface. Similarly, from biological standpoint, DDT is also very stable. Only a few microbes seem to posses the ability to degrade DDT efficiently and the process appears to require anerobic conditions. In animals, DDT is metabolised very slowly by the liver into breakdown products DDE and DDD—both of which are still highly toxic. These factors combine to yield a half-life for DDT in the biosphere estimated at about 15 years.

At the present time, the total amount of DDT in the biosphere is estimated to be over 2 billion metric tons, of which 3/4 is distributed in terrestrial environments and the remaining 1/4 in marine environments. DDT enters the air largely through aerial spraying operations, but significant amounts enter the air by simply evaporating from water surfaces, plant surfaces and soils. In this fashion, about 1/4 of the total annual production of 100,000 metric tons/year reaches the ocean. Estimates of DDT reaching the ocean in run-off are very much lower, a result of the very low solubility of DDT in water (Fig. 25.12).

25.11.11 Pesticide Resistance

In recent years, pesticide resistance was noticed in different microbes and also in a number of insects. Thus, it appears that the resistant biota develop some mechanisms of resistance. The probable mechanisms of resistance are shown in Table 25.15 (page 552).

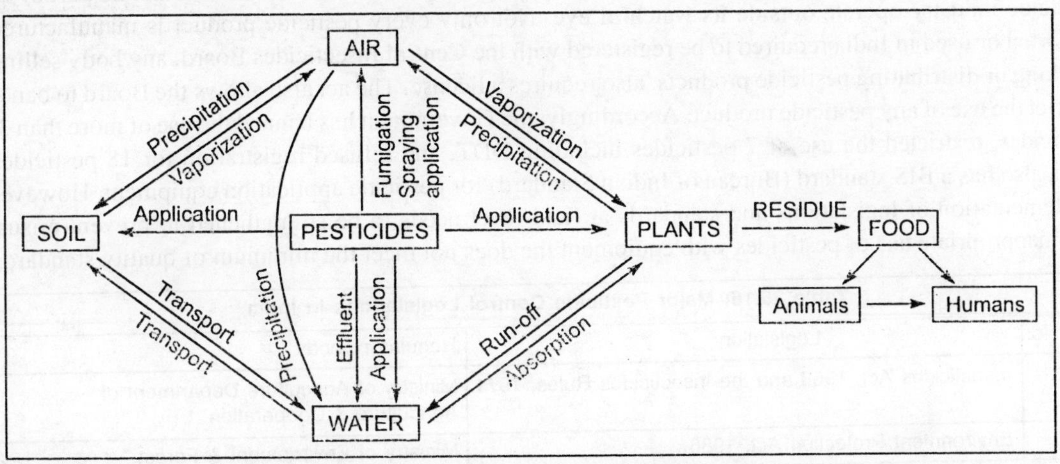

Fig. 25.12: Environmental impact of pesticide

Table 25.15: Primary Mechanisms of Pesticide Resistance*

Mechanisms	Insecticides affected
1. Behavioural	
increased sensitivity to insecticide	DDT
avoidance of treated microhabitats	Many
2. Increased detoxication	
DDT - dehydrochlorinase	DDT
microsomal monooxygenase	Carbamates
	Pyrethroids
	Organophosphorus compounds
glutathione transferase	Organophosphorus compounds
hydrolases, esterases	Organophosphorus compounds
3. Decreased sensitivity of target site,	Organophosphorus compounds
decreased sensitivity of acetylcholinesterase	Carbons
decreased nerve sensitivity (K_{dr})	DDT
	Pyrethroids
decreased nerve sensitivity	Dieldrin and other cyclodienes
(GABA receptor)	YHCH
4. Decreased cuticular penetration	Most insecticides

* Modified from Taylor (1986)

25.11.12 Regulations in Pesticide Application

In the regulation of pesticide application, government bodies have an important and major role because both producers and users are not likely to limit themselves in the sales and use of pesticides (Table 25.16). Quality control of pesticides is ensured through a rigorous registration procedure requiring testing in four different climatic conditions and making available toxicological data in Indian conditions. The import, manufacture, sale, transport, use, etc. of pesticides is being regulated under a comprehensive statute 'The Insecticides Act, 1968' and the rules framed there under, to ensure availability of quality, safe and efficacious pesticides to the farming community, comprehensive regulations to ensure that no part of the pesticide industry operate outside its watchful eye. Not only every pesticide product is manufactured, imported or used in India required to be registered with the Central Insecticides Board, any body selling, stocking or distributing pesticide products, also requires a license. The act also allows the Board to ban or restrict the use of any pesticide product. Accordingly, the Government has banned the use of more than 30 pesticides, restricted the use of 7 pesticides including DDT, and refused registration for 18 pesticides. India also has a BIS standard (Bureau of Indian Standard) for pesticide application equipment. However, implementation of legislations and standards at field level needs to be strengthened to prevent misuse and inappropriate use of pesticides with equipment the does not meet the minimum of quality standards.

Table 25.16: Major Pesticide Control Legislations in India

Sl. No.	Legislation	Regulatory body
1.	Insecticides Act, 1968 and the Insecticides Rules, 1971	Ministry of Agriculture Department of Agriculture & Cooperation
2.	Environment Protection Act, 1986	Ministry of Environment & Forest
3.	Prevention of Food Adulteration Act, 1954	Ministry of Health & Family Welfare

Under the Insecticides Act, compulsory registration of pesticides is provided. The use of chemical pesticides can be initiated only after the proper registration by the Registration Committee, after close scrutiny of the data about bio-efficacy and safety of human beings, wildlife, birds, domestic animals, beneficial parasites and predators. The Insecticide Rules takes care of the safety culture in pesticide handling and use. It covers periodical clothing, respiratory devices, antidotes and first aid medicines, training of workers and disposal of used packages, surplus materials and washing of insecticides. The Registration Committee reviews the pesticides from time to time and the recommendations are considered by the Ministry of Agriculture. The Committee as a policy has decided not to register WHO Class IA and IB pesticide unless there is sufficient justification. The Directorate of Plant Protection, Quarantine & Storage (DPPQS) has good schemes for training at the state level. Besides, training is imparted to the doctors of health centers of states by the medical experts of the Directorate of Plant Protection, Quarantine & Storage. The National Plant Protection Training Institute (NPPTI) at Hyderabad imparts training to the State Plant Protection functionaries. Farmers Field Schools (FFS) are regularly organized under the IPM programme in addition to season-long training for Masters Trainers under which State Extension Functionaries are trained for full cropping period of various crops. State Agriculture Universities (SAUs), Krishi Vigyan Kendras (KVKs), and State Department of Agriculture (SDA) also organize training to farmers on safe use of pesticides.

For the enforcement of the quality of pesticides, four important functionaries are notified under the provisions of the Insecticides Act/Rules viz., Licensing Officers, Appellate Authority, Insecticide Inspectors and Insecticide Analysts. A network of 46 pesticides testing laboratories, situated in 18 States and 1 Union Territory across the length and breadth of the country, with an annual analysis capacity of over 55,666 samples of pesticides is available in the country for continuously monitoring the quality of pesticides. One of the notified functionaries of the Government. Insecticide Inspectors can enter and search, at all reasonable times and with such assistance as he considers necessary in which he has the reason to believe that an offence under the Insecticides Act, 1968 and the Rules made thereunder has been or is being or is about to be committed, or for the purpose of satisfying himself that the provisions of this Act or the Rules made there under or the conditions of any certificate of registration or license issued thereunder are being complied with. Insecticide Inspectors also draw samples of pesticides, Insecticide analysts carry out their analysis. Besides, the Central Government has also established two regional pesticides testing laboratories to supplement the resources of the States/UTs, who do not either have a pesticide testing laboratory or adequate analysis capacity or adequate analysis facility for monitoring the quality of pesticides. Any disputes in the results of analysis are settled by a referral laboratory of the Central Government, called the Central Insecticides Laboratory (CIL). In order to strengthen the existing laboratories and to set up new pesticides testing laboratories, the Central Government also extends financial assistance to the States/UTs as grants-in-aid. Besides, State Governments also establish additional Pesticides Testing Laboratories with their own resources.

International Scenario

Pesticides have traditionally being the object of international legislation in several fields (Table 25.17). Laws and regulations on human health, environmental protection, agricultural practices, international trade and border control all address the proper use, manufacture, export, import and application technologies for pesticides. The international code of conduct and use of pesticides (Code of Conduct), adopted in 1985 by the 23rd session of the main governing body of the Food and Agriculture Organization of the United Nations (FAO), was formulated to provide universal standards of conduct for all parties, but especially national government and pesticide industry. Since then, the code of conduct has been amended once, in 1989, to include the prior informed consent procedure, and revised in 2002. Since 1985, several other international instruments, either dealing explicitly with pesticides or indirectly relating to their management, have come into force. The most relevant include the Rotterdam Convention on the prior informed consent procedure, for certain hazardous chemicals and pesticides in international trade (Rotterdam Convention), the Stockholm Convention on persistent organic pollutants (Stockholm Convention), the Basel Convention on the transboundary movements of hazardous wastes and their disposal (Basel Convention), the Montreal Protocol on substances that deplete the ozone layer (Montreal Protocol), the International Labour Organization Convention No. 184 on safety and health in agriculture (ILO Convention 184) and numerous standards on pesticide residues in food issues by the Codex Alimentarius Commission. Furthermore, a new globally harmonized system of classification and labeling of chemicals (GHS) has been designed to improve the protection of human health and the environment during the handling, transport and use of chemicals. However, it is very important to note that legislation alone is unlikely to balance effective pesticide management with an environmentally sustainable approach to pest control. Governments should consider adopting other policies and strategies to improve pesticide management, such as providing farmers with support and training in integrated pest management (IPM), allocating subsidies for the purchase of minimum-risk products and fostering scientific research, public education campaigns and training for both inspectors and professional users. A solid legislative and regulatory framework, however, underpins all of these.

Table 25.17: International instruments concerning pesticide use and application

Sl. No.	Convention	Adopted	Entry into force	Number of parties	Remarks
Legally binding					
1.	Rotterdam Convention on the Prior informed Consent Procedure for Certain Hazardous Pesticides and Industrial Chemicals in International Trade	1988	2004	117	Substances covered: specific types of pesticides (24 pesticides and 4 severely hazardous pesticide formulations) and 11 industrial chemicals
2.	Stockholm Convention on Persistent Organic Pollutants	2001	2004	149	Substances covered: persistent organic pesticides (9 of which are pesticides)
3.	Basel Convention on the Transboundary Movement of Hazardous Wastes and Their Disposal	1989	1992	170	Substances covered: hazardous wastes including pesticides, including pesticides at the end stages of their life cycle
4.	Montreal protocol on Substances that Deplete Ozone Layer	1987	1989	191	Substances covered: methyl bromide, a specific type of pesticides, which is also an ozone-depleting substances
5.	ILO Convention on Safety and Health in Agriculture, No. 184	2001	2003	8	Substances covered: not specified, addresses all aspects of health and safety for chemicals used in agriculture
6.	ILO Convention concerning Safety in the use of chemicals at Work, No. 170	1990	1993	15	Substances covered: (hazardous) chemicals used in all kinds of economic activities
Voluntary					
1.	FAO International Code of Conduct on the Distribution and Use of Pesticide	Adopted, 1985 Revised: 2002			Substances covered: pesticides (broadly defined)

PESTICIDE USE IN INDIA

There are 234 pesticides registered in India. Out of these, 4 are WHO Class Ia pesticides, 15 are WHO Class Ib pesticides and 76 are WHO Class II pesticides, together constituting 40% of the registered pesticides in India. In terms of consumption too, the greatest volumes consumed are of these poisons.

The following is a broad picture of the top pesticide-consuming states in India (total pesticides consumed, in metric tonnes of technical grade material, during 2005-06 to 2009-10, as per official data of the Directorate of Plant Protection, Quarantine and Storage, Govt of India.

	State	Total pesticides consumed
1	Uttar Pradesh	39948
2	Punjab	29235
3	Haryana	21908
4	Maharashtra	16480
5	Rajasthan	15239
6	Gujarat	13430
7	Tamil Nadu	12851
	All India	210,600

Most consumed pesticides in the country (during 2005-06 to 2009-10)

	Pesticide (Technical Grade)	Quantity consumed (metric tonnes)
1	Sulphur (fungicide)	16424
2	Endosulfan (insecticide)	15537
3	Mancozeb (fungicide)	11067
4	Phorate (insecticide)	10763
5	Methyl Parathion (insecticide)	08408
6	Monocrotophos (insecticide)	08209
7	Cypermethrin (insecticide)	07309
8	Isoproturon (herbicide)	07163
9	Chlorpyrifos (insecticide)	07163
10	Malathion (insecticide)	07103
11	Carbendazim (fungicide)	06767
12	Butachlor (herbicide)	06750
13	Quinalphos (insecticide)	06329
14	Copper oxychloride	06055
15	Dichlorvos (insecticide)	05833

CARCINOGENIC PESTICIDES: INDIAN SITUATION

The following is a list of 24 pesticides registered and used in India, classified as Potential Carcinogens by the US EPA: Acephate (C), Alachlor (B2), Atrazine (C), Benomyl (C), Bifenthrin (C), Captan (B2), Chlorothalonil (B2), Cypermethrin (C), Dichlorvos (C), Diclofop-Methyl (C), Dicofol (C), Mancozeb (B2), Methomyl (C), Metolachlor (C), Oxadiazon (C), Oxyflourfen (C), Permethrin (C), Phosphamidon (C), Propiconazole (C), Propoxur (B2), Thiodicarb (C), Thiophanate Methyl (C), Triadimefon (C), Trifluralin (C).

Poison Gas at Bhopal, India

Alarm sirens rang at 12:30 a.m. on the morning of December 3, 1984, in Bhopal, India, at a plant owned by Union Carbide. The internal siren was to warn workers at the plant that a toxic gas leak had occurred, and the public siren was to warn people living around the plant. Not realizing the extent of the emergency, the public siren was almost immediately shut off, and it was not until 45 minutes later that city authorities learned that people were fleeing the area and not until 3:00 a.m. that they learned that some people were already dead. Ultimately, it was to be recognized as the most deadly industrial accident in history, claiming thousands of lives, permanently disabling tens of thousands, and leaving perhaps as many as 100,000 people with some after-effects. Why do we not use more precise numbers? Well, perhaps some of the deaths that night or later were partly caused by other accidents or illness, possibly some of those disabled were really not all that badly injured and some of the disabled were not reported or found, and in a large industrial city in a nonindustrial country people are hard to count anyway.

What happened is slightly more clear than the effects. After operating in India for many years, in 1969 Union Carbide built a plant in Bhopal to manufacture pesticides and herbicides based on carbaryl. The plant was 50.9% owned by Union Carbide in the United States and 49.1% by Indian investors and was placed in Bhopal partly because the Indian Government and State and local authorities wanted to provide jobs and stimulate economic growth in the region. For the first 10 years of operation Union Carbide imported the major ingredients for pesticide manufacture and simply combined them at the Bhopal plant, but in 1979 the Indian Government overruled objections of authorities in Bhopal and gave Union Carbide permission to manufacture all of the necessary ingredients in Bhopal. By 1984 the Union Carbide plant was storing large amount of the two key ingredients, alpha naphthol and methyl isocyanate gas (MIC), and making carbaryl by the reaction:

$$\text{alpha naphthol} + \text{MIC} \rightarrow \text{Carbaryl}$$

which, using formulas, is:

$$C_{10}H_9OH + CH_3NCO \rightarrow C_{10}H_7OCH_2ONHCH_3$$

Unfortunately, for reasons that are still somewhat in dispute, in the early morning of December 3, 1984, water leaked into one of the tanks holding the MIC. This caused an immediate reaction that broke down some of the MIC to various other organic compounds and generated large amounts of CO_2 and heat (the reaction was exothermic). This combination of hot gases ruptured the control system and poured out into the streets of Bhopal. People who inhaled large quantities of the gas died immediately; their internal organs almost dissolved. People receiving lower doses had serious corrosion of their lungs (causing many later deaths), and many were blinded.

The ensuing financial and legal complications are as sickening in their own way as the gases themselves. They centered around two questions : Who was responsible for the accident, and who should be compensated for it? Union Carbide in America offered an immediate $200 million as a humanitarian gesture but disclaimed responsibility because the plant was operated by an Indian subsidiary and not under direct control of the parent company. The Indian government rejected the offer as insufficient and spent considerably less money to take care of the families of those who had died and those who were too ill to continue working. Hordes of US lawyers descended on Bhopal, attempting to get victims to let them bring suit in US courts, where any compensation awarded would presumably be larger than in an Indian court and, not incidentally, would pay the lawyers for the efforts. As we write 27 years later, neither of the two questions has been satisfactorily answered nor have most of the victims received adequate compensation, though Supreme Court asked the Indian Government to distribute the compensation fund to recorded victims of gas tragedy.

25.11.13 General Effects of Pollutants

Almost every part of the human body is affected by one pollutant. For example, lead and mercury (remember the Mad Hatter) affect the brain; arsenic, the skin; carbon monoxide, the heart; and fluoride, the bones. Wildlife is affected as well.

The lists of potential toxins and affected body sites for humans and other animals is shown in Fig. 25.13 (page 558).

Voluntary Exposure

People sometimes expose themselves to harmful pollutants on purpose. Voluntary exposure to toxins and potentially harmful chemicals is sometimes referred to as exposure to personal pollutants. The most common of these are tobacco, alcohol, and other so-called recreational drugs. Use and abuse of these substances have led to a variety of human ills, including death and chronic disease, criminal activity such as reckless driving and manslaughter, loss of careers, street crime, and the straining of human relations at all levels.

Risk potential of some of voluntary and accidental exposure is shown in Table 25.18.

Table 25.18: Risk of death from various activities

Activity	Result	Risk of Death	
Cigarette smoking (pack a day)	All causes: cancer, effect on heart, lungs, etc.	8 in 100	
Air in the home containing radon	Cancer	1 in 100	Naturally occurring
Automobile driving		1 in 100	
Death from a fall		4 in 1,000	
Drowning		3 in 1,000	
Fire		3 in 1,000	
Artificial chemicals in the home	Cancer	2 in 1,000	Paints, cleaning agents, pesticides
Sunlight exposure	Melanoma	2 in 1,000	Of those exposed to sunlight
Electrocution		4 in 10,000	
Air outdoors in an industrial area		1 in 10,000	
Artificial chemicals in water		1 in 100,000	
Artificial chemicals in foods		less than 1 in 100,00	
Airplane passenger (commercial airline)		less than 1 in 1,000,000	

Source: From *Guide to Environmental Risk*, 1991. U.S. Environmental Protection Agency. Region 5 publication no. 905/91/017

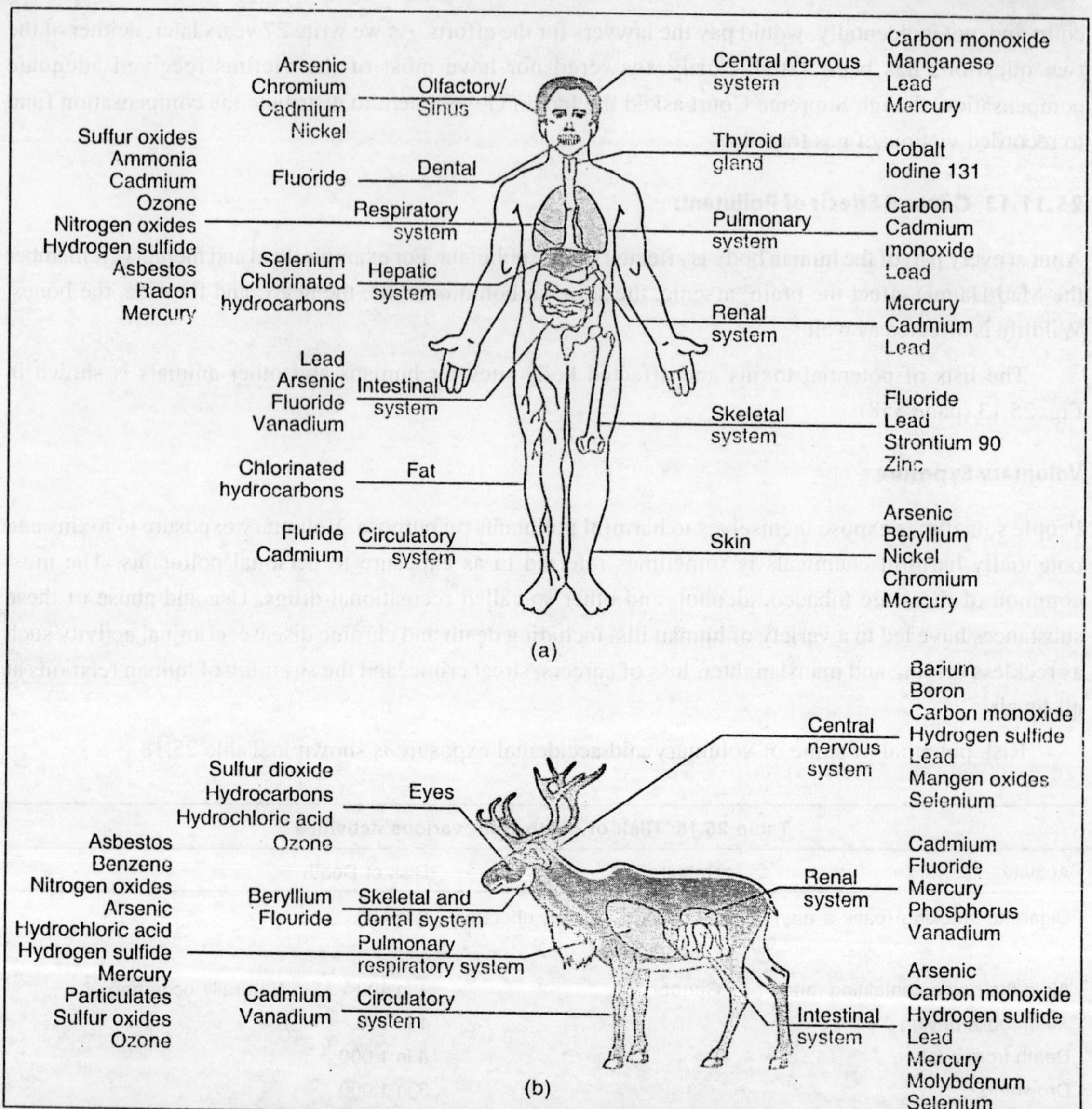

Fig. 25.13: Pollution effects on people and wildlife

(a) Effects of some major pollutants in human beings. (b) Known sites of effects of some major pollutants in wildlife.

Source: (a) G.L. Waldbott. *Health effects of environmental pollutants*, 2nd ed., St. Louis: Mosby, 1978. © 1978 by C.V. Mosby. (b) J.R. Newman, Effects of air emissions on wildlife resources. US Fish and Wildlife Services Programme, National Power Plant Team, FWS/OBS-80/40. Washington, DC: US Fish and Wildlife Service, 1980.]

26 REMOTE SENSING

26.1 PRINCIPLES OF REMOTE SENSING

Remote Sensing (RS) is generally defined as the science of collecting and interpreting information about a target without being in physical contact with the object under study. There are various categories of remote sensing starting from observation by naked eyes, photography by camera, photography from aircraft, and sensing by sensors from space satellite.

Depending on its physical features and chemical properties, different objects on the earth's surface reflect, reradiate or emit different amounts of electromagnetic energy in various wavelengths. The measurement of reflected or reradiated or emitted electromagnetic radiation forms the basis for understanding the characteristic of earth's surface features. Three typical responses are used to distinguish the objects from one another. Only selected portions of the electromagnetic spectrum which can pass through the earth's atmosphere with relatively little attenuation are used for remote sensing purpose. The selected region of the electromagnetic spectrum which are employed in remote sensing include: 0.4 to 0.7 μm, 0.7 to 3.0 μm (IR band), 3 to 5 μm, and 8 to 14 μm (TIR) and 0.1 to 30 cm (microwave).

26.2 TYPES OF REMOTE SENSING

Depending on the source of energy which illuminate the object under study, the remote sensing techniques are classified into two types, viz., (i) **Passive remote sensing**, and (ii) **Active remote sensing**. In passive remote sensing system, the naturally radiated or reflected energy from the earth's surface features is

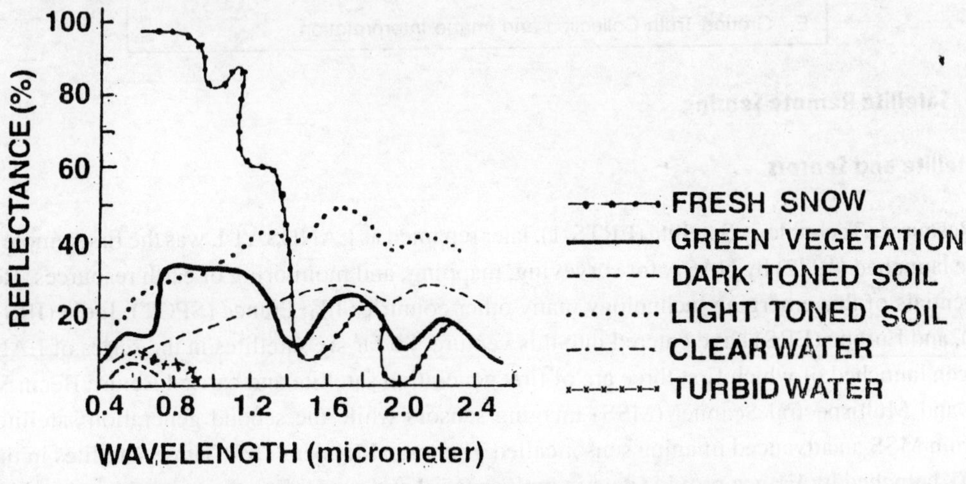

Fig. 26.1: Comparison of reflectance spectra of a few land cover objects

measured by the sensors operating in different selected spectral bands on board the air-borne/space-borne platforms (similar to photography in daytime without flash). An active remote sensing system supplies its own source of energy to illuminate the objects and measures the reflected energy returned to the system (similar to photography in night with flash).

26.3 SYSTEM OVERVIEW IN REMOTE SENSING

There are five distinct components in remote sensing process. The details are depicted in Table 26.1.

Two major steps are involved in this process. The first one is data acquisition and the second data processing and interpretation. Data acquisition is made by sensors from platforms. This information is used for the formation of products which are finally used for interpretation purpose by comparing with ground truth information.

Table 26.1: System Overview in Remote Sensing

A. Platforms	Ground stations
	Air-borne
	Space borne
B. Sensors	Image forming
	Active (Radar)
	Passive (Camera)
	Non-image forming
C. Products	Aerial photographs
	Computer compatable tape (CCT)
	Hard copy imagery (FCC/BW copy)
D. Processing	Backgroup correction
	Contrast stretching
	Digital processing
E. Ground Truth Collection and Image Interpretation	

26.3.1 Satellite Remote Sensing

(a) Satellite and Sensors

Earth Resources Technology Satellite (ERTS-1), later renamed as LANDSAT I, was the first remote sensing satellite launched (1972) by NASA for surveying, mapping, and monitoring of earth resources. Realising the potentials of this emerging technology many other countries like France (SPOT), India (IRS), Japan (JERS), and Europe (ERS) have entered into this venture. So far six satellites in the series of LANDSAT have been launched of which first three are of first generation satellite and carried Return Beam Vedicon (RBV) and Multispectral Scanner (MSS) imaging sensors while the second generation satellites carry apart from MSS an advanced imaging sensor called Thematic Mapper (TM). Three satellites in the series of SPOT launched by France provide data in multispectral and panchromatic spectral bands with normal and/or stereoscopic mode. Indian remote sensing satellites, IRS-IA, IRS-IB, and IRS-IC, the indigenously developed satellites, have been put into orbit in 1988, 1991, and 1995, respectively, with sensors like LISS-I, LISS-II, LISS-III, WIFS, and PAN (Figs. 26.2, 26.3, 26.4 & 26.5, page 561-562).

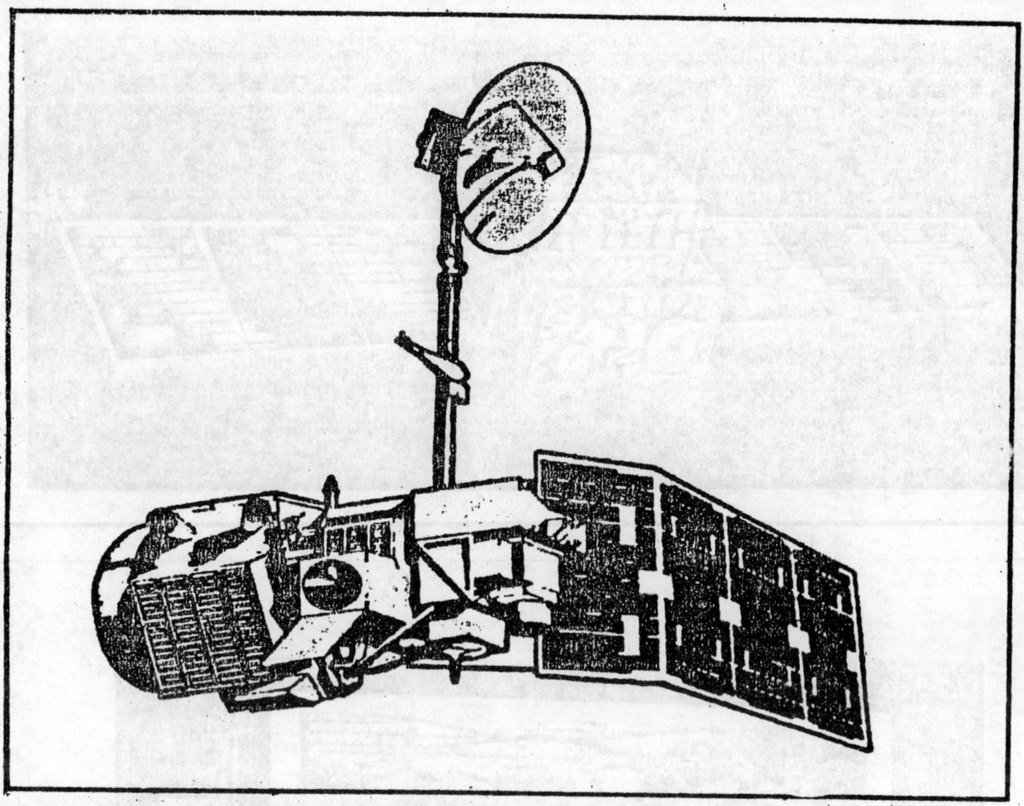

Fig. 26.2: LANDSAT

Fig. 26.3: Spot Satellite

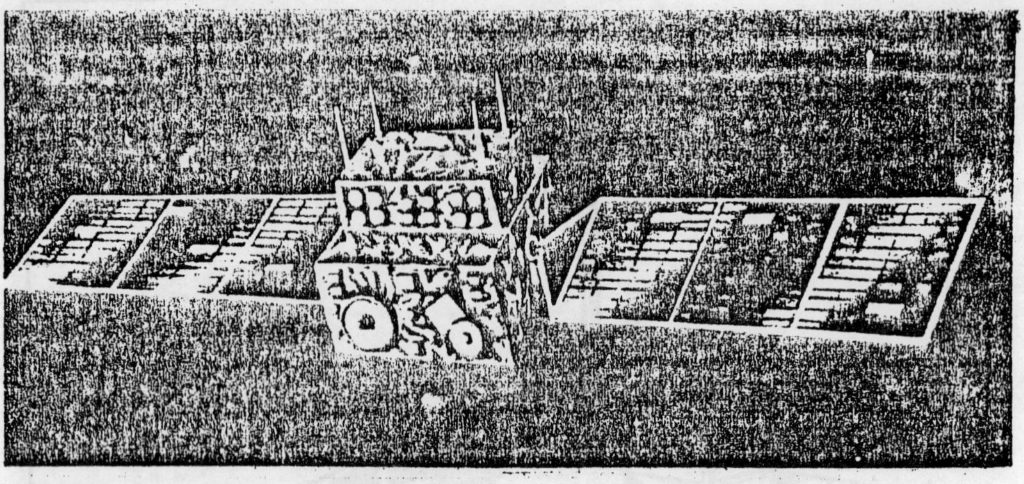

Fig. 26.4: An artist's concept of IRS-1B Spacecraft in orbit

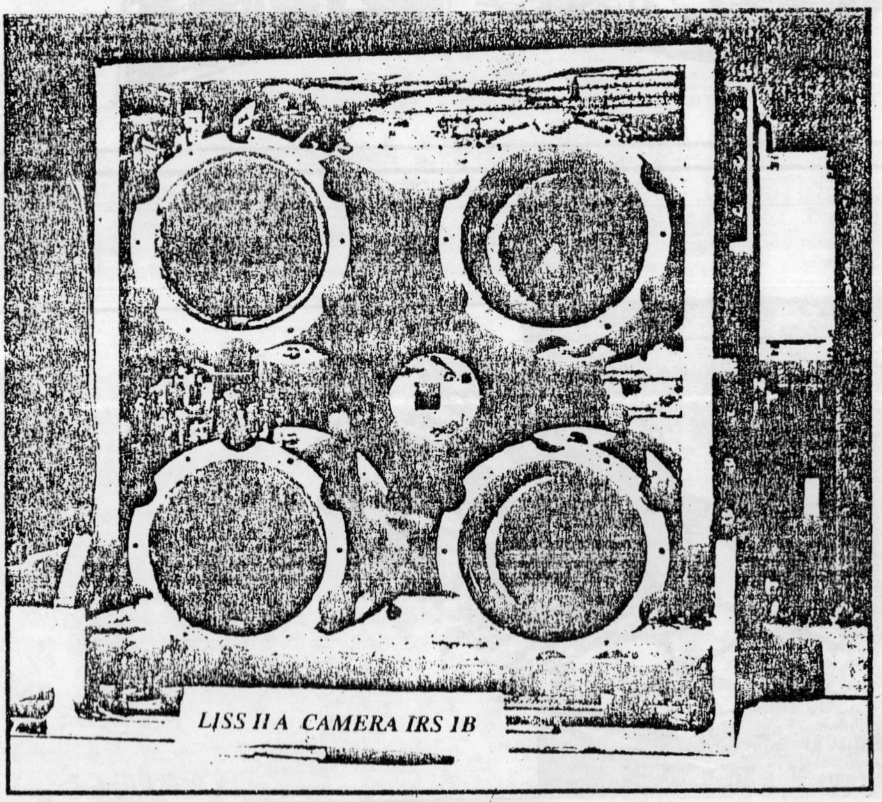

Fig. 26.5: Liss 11A Camera IRS-1B

Unlike geostationary satellites, these remote sensing satellites are sun-synchronous (crossing the equator/latitude at the same local time every day with descending node enabling the study of natural resources at various regions under the same illumination condition), and polar orbiting type with a repetitive cycle of 16 to 26 days enabling repeated collection of data at the same place at the same local time for continuous monitoring of the earth's resources. The imaging payloads of these satellites operate in different spectral bands, spatial resolutions (the smallest area in the ground being sensed by the sensors), and radiometric resolution (number of grey levels which are distinguishable).

IRS—An Overview

The image sensing characteristics of various sensors of IRS systems are:

Table 26.2: IRS-sensors

	LISS I	LISS II	LISS III	PAN	WIFS
1. No. of spectral bands	4	4	4	1	2
2. Spectral bands (microns)	$B_1$0.45 – 0.52 $B_2$0.52 – 0.59 $B_3$0.62 – 0.68 $B_4$0.77 – 0.86	Do	B_2 0.52 – 0.59 $B_3$0.62 – 0.68 $B_4$0.77 – 0.86 $B_5$1.55 – 1.70	0.50 – 0.75	B_1 – 0.62 – 0.68 B_2 – 0.77 – 0.86
3. Spatial resolution (m)	72.5	36.25	23.5 km for B_2, B_3, B_4 & 70.5 for B_5	> 10	188
4. Swarth	148	145	142 km for B_2, B_3, B_4 and 148 km for B_5	70 km	774 km
5. Rediometric levels	128	128	128	64	128

IRS satellite systems were launched into a polar sun-synchronous orbit, the orbit plane rotates at the same rate as the mean rate of the earth around the sun (i.e., 0.9856 deg/day). Thus, the satellite passes over a particular latitude approximately at the same local time. It enables the ground illumination conditions at sub-satellite regions to be constant throughout the mission. The equatorial crossing time of the descending node for IRS-IA is around 10.25 AM.

As the orbital period of IRS-IA is about 103 minutes, with the satellite completing 14 orbits/day, each successive orbit is shifted westward over the earth's surface by 25.798 degree of longitude, corresponding to 2,872 km at equator (Fig. 26.6, page 564). The satellite's path is shifted by 1.17 degree of longitude to the west every day corresponding to 130.54 km at the equator. The satellite completes one coverage cycle of the Indian subcontinent in 22 days (i.e., 307 orbits) (Fig. 26.6). In order to facilitate convenient and unique identification of any of the geographical regions of interest and cataloguing of data products, an image referencing scheme designated by path and row numbers has been evolved for the Indian subcontinent. The paths are sequentially numbered from east to west for all the 307 orbits in a coverage cycle. The row is the line joining the corresponding scene centres of different paths forming a contour parallel to the equator. The row number 1 falls around 81 degree North latitude. The Indian region is covered by row number 41 to 63 and path number 9 to 35 (Fig. 26.9, page 566).

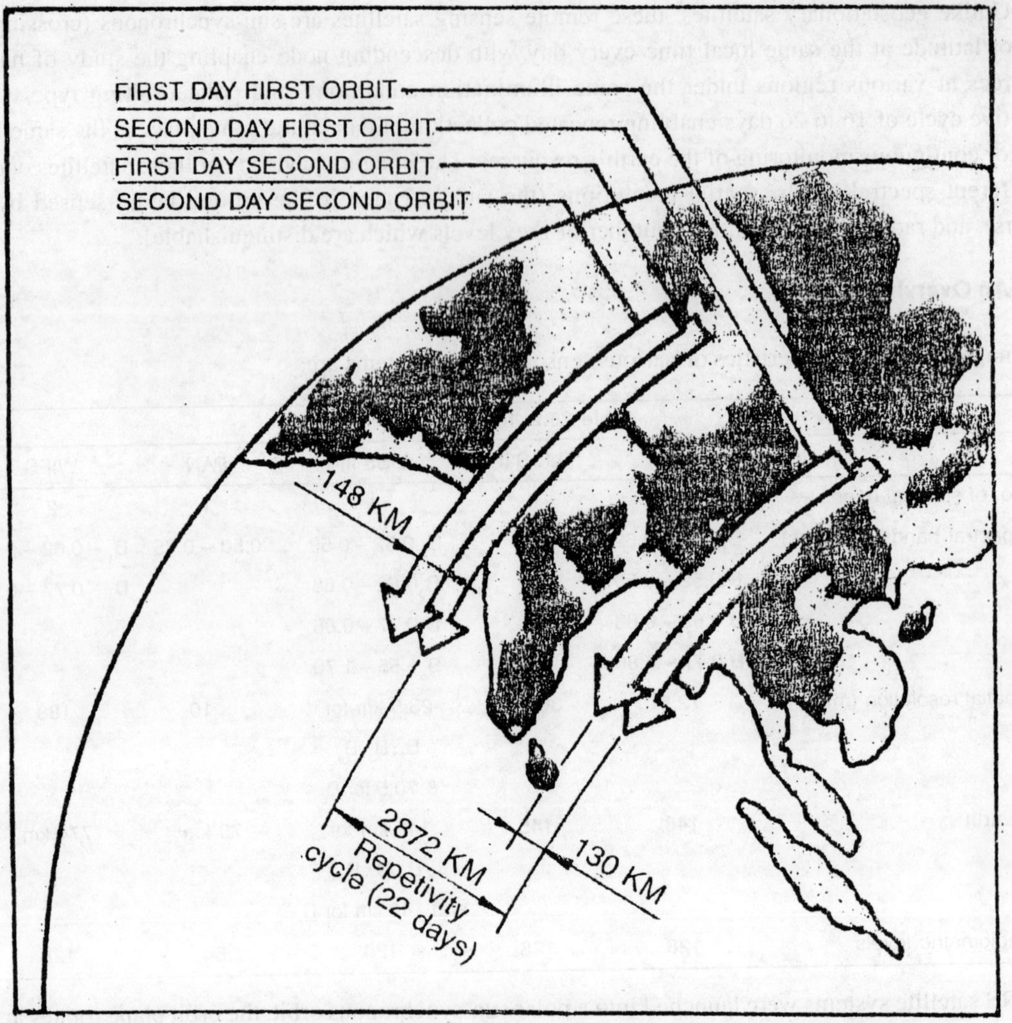

Fig. 26.6: Satellite orbit and coverage pattern

IRS-IA image sensors operate in pushbroom scanning mode using **Linear Image Self-scanning Sensors** (LISS). In this mode of operation, each line of the image is electronically scanned by a linear array of detectors, located in the focal plane of optical system and successive lines of image are produced by satellite movement (Fig. 26.7, page 565).

IRS-IA has two types of imaging sensors, one with a spatial resolution of 72.5 meters and designated as LISS-I and other with two separate imaging sensors designated as LISS-IIA and LISS-IIB with a spatial resolution of 36.25 meters each. LISS-I provides a swarth of 148 km, while a composite swarth of 145 km is attained by the two LISS-II sensors by suitable mounting of detectors in the focal plane of the system (Fig. 26.8, page 565).

In IRS system, the dual spectral resolution data available from LISS I & LISS II offer the capability of overview of larger areas, besides providing an in-depth look at finer areas. The selection of spatial and spectral characteristics of IRS image sensors enables the complementary and supplementary use of data from contemporary remote sensing satellite like LANDSAT-5, SPOT.

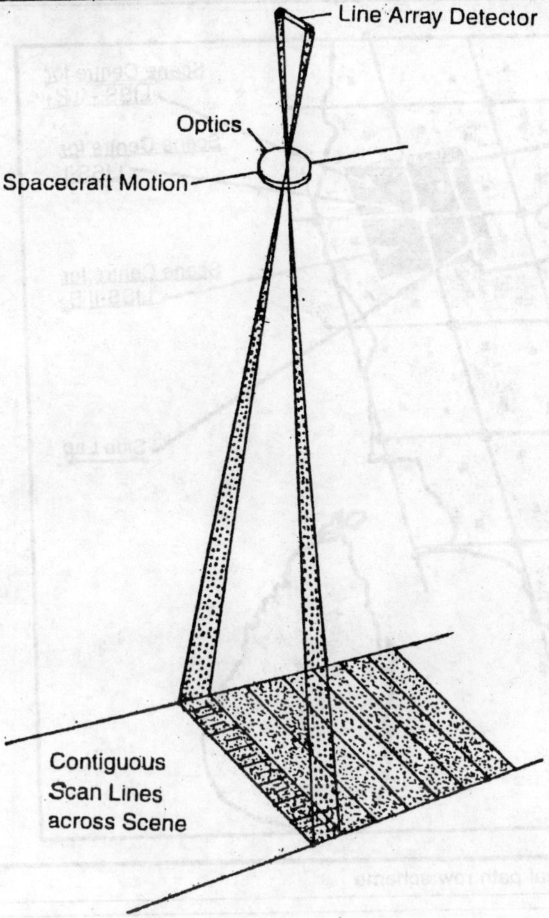

Fig. 26.7: Pushbroom Concept

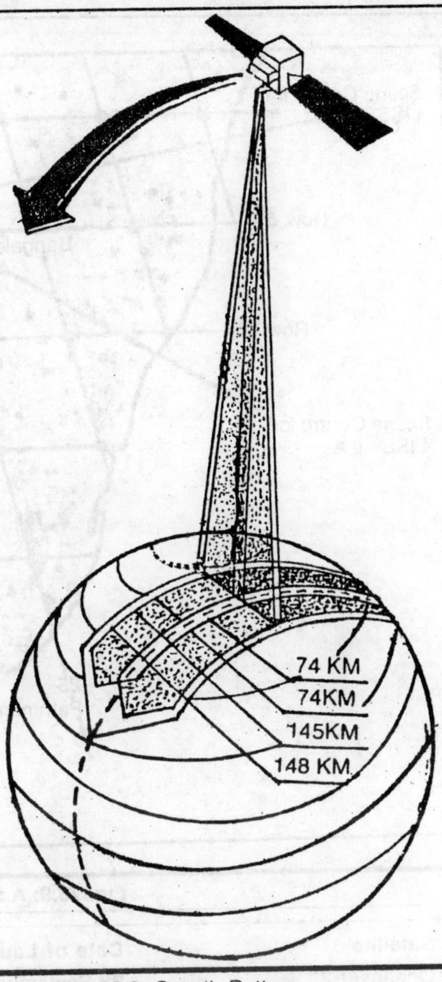

Fig. 26.8: Swarth Pattern

Box 26.1: IRS launch log

The initial versions are composed of the 1 (A,B,C,D). The later versions are named based on their area of application including OceanSat, CartoSat, ResourceSat. Some of the satellites have alternate designations based on the launch number and vehicle (P series for PSLV).

	Satellite	Date of Launch	Launch Vehicle	Status
1.	IRS-1A	17 March 1988	Vostok, USSR	Mission Completed
2.	IRS-1B	29 August 1991	Vostok, USSR	Mission Completed
3.	IRS-P1 (also IE)	20 September 1993	PSLV-D1	Crashed, due to launch failure of PSLV
4.	IRS-P2	15 October 1994	PSLV-D2	Mission Completed
5.	IRS-1C	28 December 1995	Molniya, Russia	Mission Completed
6.	IRS-P3	21 March 1996	PSLV-D3	Mission Completed
7.	IRS 1D	29 September 1997	PSLV-C1	Mission Completed
8.	IRS-P4 (Oceansat-1)	27 May 1999	PSLV-C2	Mission Completed
9.	Technology Experiment Satellite (TES)	22 October 2001	PSLV-C3	Mission Completed
10.	IRS P6 (Resourcesat-1)	17 October 2003	PSLV-C5	In Service
11.	IRS P5 (Cartosat 1)	5 May 2005	PSLV-C6	In Service
12.	IRS P7 (Cartosat 2)	10 January 2007	PSLV-C7	In Service
13.	Cartosat 2A	28 April 2008	PSLV-C9	In Service
14.	IMS 1	28 April 2008	PSLV-C9	In Service

Contd.

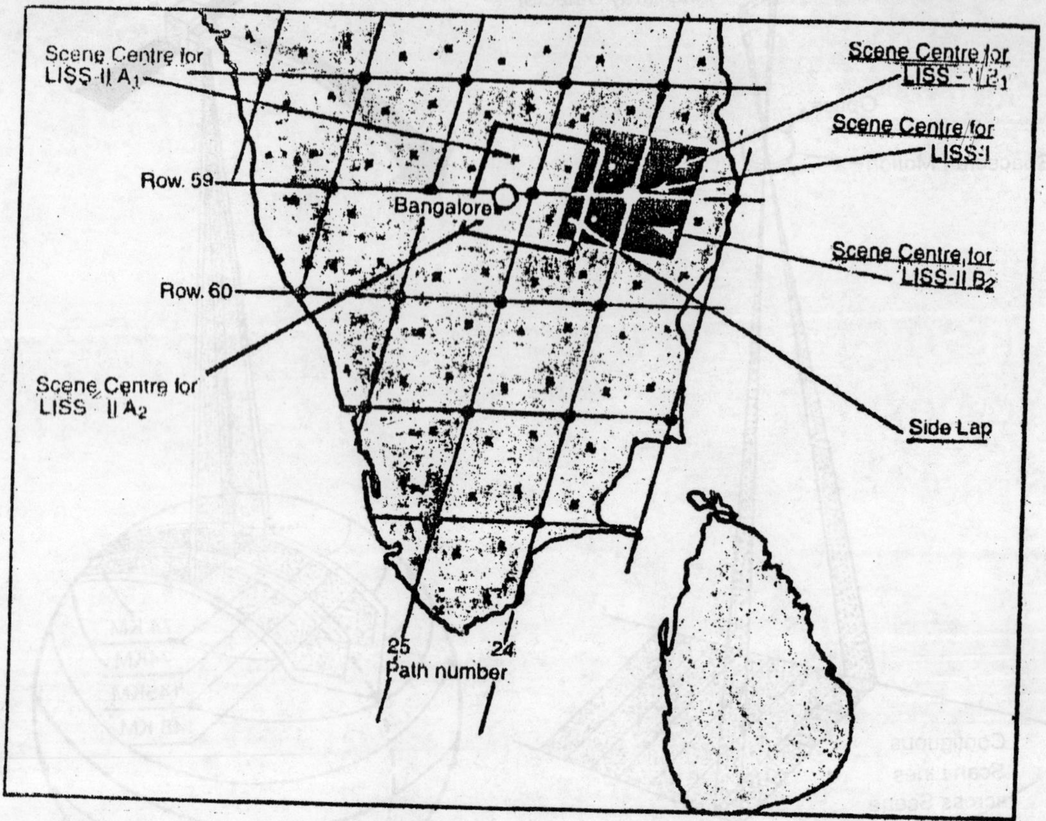

Fig. 26.9: A typical path row scheme

	Satellite	Date of Launch	Launch Vehicle	Status
15.	Oceansat-2	23 September 2009	PSLV-C14	In Service
16.	Cartosat-2B	12 July 2010	PSLV-C15	In Service
17.	Resourcesat-2	20 April 2011	PSLV-C16	In Service
18.	Megha-Tropiques	12 October 2011	PSLV-C18	In Service
19.	RISAT-1	26 April 2012	PSLV-C19	In Service
20.	SARAL	25 Feb 2013	PSLV-C20	In Service
21.	RESOURCESAT-2A	07 Dec 2016	PSLV-C36	In Service
22.	Cartosat-2D	15 Feb 2017	PSLV-C37	In Service
23.	Cartosat-2E	23 June 2017	PSLV-C38	In Service

Table 26.3

Band Nos.	Spectral range (microns)	Application areas
I	0.45-0.52	Coastal environment studies Soil/vegetation differentiation Coniferous/deciduous vegetation discrimination
II	0.52-0.59	Vegetation vigour Rock/soil discrimination Turbidity and bathymetry in shallow waters
III	0.62-0.68	Strong chlorophyll absorption leading to discrimination of plant species
IV	0.77-0.86	Delineation of water features Land form/geomorphic studies

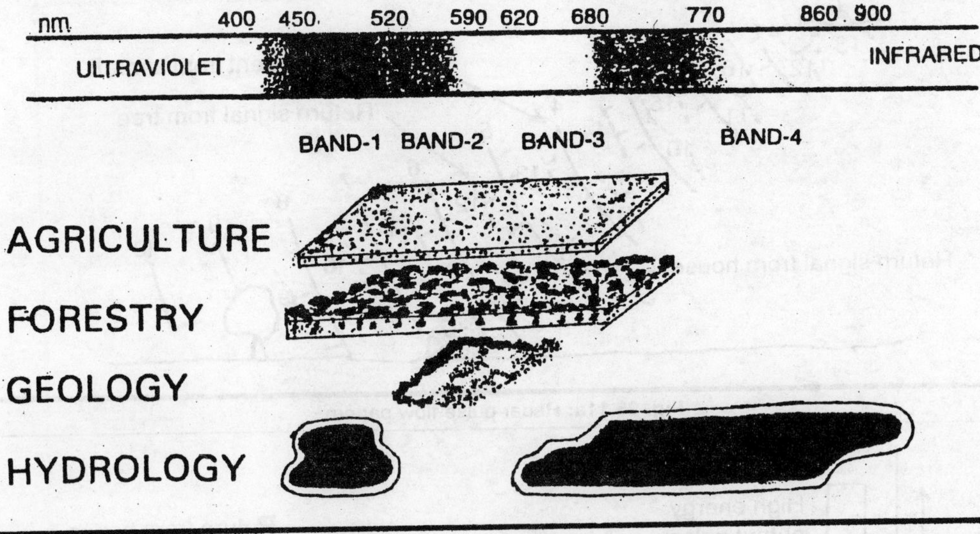

Fig. 26.10: Remote sensing utilization aspect

A synoptic accounts of the band utility for resource information assessment is also shown in the Fig. 26.10.

26.3.2 Microwave Remote Sensing: Radar

1. Introduction

Microwave portion of the spectrum includes wavelengths within the approximate range of 1mm to 1m. There are two distinctive features that characterise microwave energy from a remote sensing standpoint. Microwaves are capable of penetrating the atmosphere under virtually all conditions Depending on the wavelengths involved, microwave energy can 'see through' haze, light, rain and snow, clouds, smoke and so on. Microwave reflections or emissions from earth materials bear no direct relationship to their counterparts in the visible or thermal portions of the spectrum. For example, surfaces that appear 'rough' in the visible portion of the spectrum may be 'smooth' as seen by microwaves. In general, microwave responses afford us a markedly different 'view' of the environment—one face removed from the views experienced by sensing light or heat.

The word *radar* is an acronym for *Radio Detection and Ranging*. As its name implies, radar was developed (by Heinrich Hertz in the late 1880s) as a means of using radio waves to detect the presence of objects and determine their range (position). The process entails transmitting short bursts, or pulses, of a microwave energy in the direction of interest and recording the strength and origin of 'echoes' or 'reflections' received from objects within the system's field of view.

Radar systems may or may not produce images and they may be ground-based or mounted in aircraft or spacecraft. A common form of non-imaging radar is the type used to measure vehicle speeds. These systems are termed Doppler Radar systems because the Doppler Effect frequency shifts in the transmitted end return signals to determine an object's velocity. Doppler frequency shifts are a function of the relative velocities of a wave transmitter and a reflector. For example, we perceive Doppler shifts in sound waves as chosen in pitch, as in the case of a passing car horn or a train whistle.

The spatial resolution of a radar system is determined among other things by the size its antenna. For any given wavelength, the larger the antenna, the better the spatial resolution of an aircraft. It is very difficult to mount a rotating antenna that is very large. To circumvent the system, most airborne radar remote sensing techniques with systems use an antenna fixed on the aircraft and pointed to the sun's

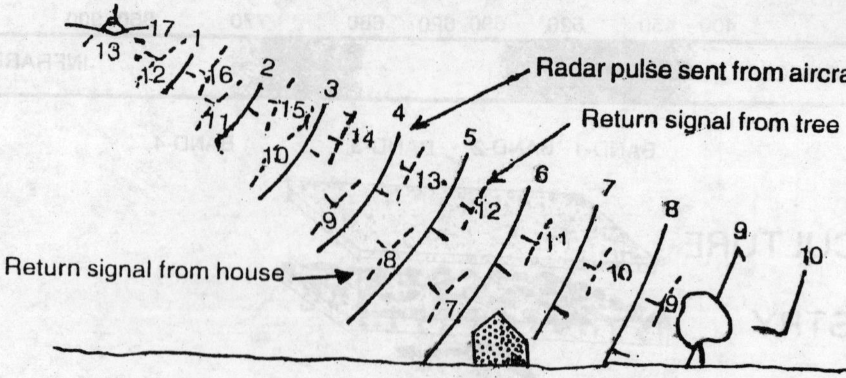

Fig. 26.11a: Radar pulse flow pattern

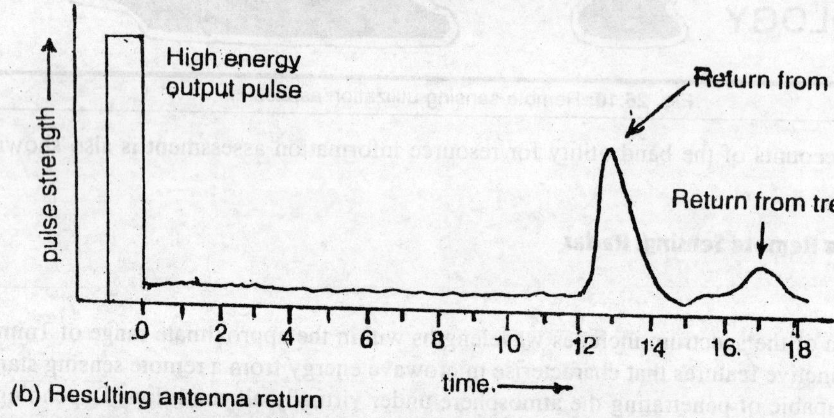

Fig. 26.11b: Opening principle of side looking airborne radar (SLAR)

pitch. The term Side-Looking Radar (SLR) or Side Looking Airborne Radar (SLAR) systems produce continuous strips of imagery which converge on ground areas near or adjacent to the aircraft's light line.

2. SLAR System Operation

The basic operation principle of SLAR system is shown in Fig. 26.11a, Fig. 26.11b. Microwave energy is transmitted from an antenna in very short bursts or pulses. These high energy pulses are emitted over a time period of the order of microseconds (10 μ sec.). In Fig. 26.11a propagation of one pulse is shown by indicating the wavefront locations. Because the tree is less reflective of radio waves than the house, a weaker response is recorded in Fig. 26.11b.

By electronically measuring the return time of signal echoes, the range, or distance, between the transmitter and reflecting objects may be determined. Since the energy propagates in air at approximately the velocity of light, c, the slant range, SR, to any given object is given by:

$$SR = ct/2 \quad (1)$$

where SR = Slant Range (Direct distance between transmitter and object)

c = speed of light (3×10^8 m/sec)

t = time between pulse transmission and echo reception.

(Note that the factor of 2 enters into the equation because the time is measured for the pulse to travel both the distance to and from the target, or twice the range.)

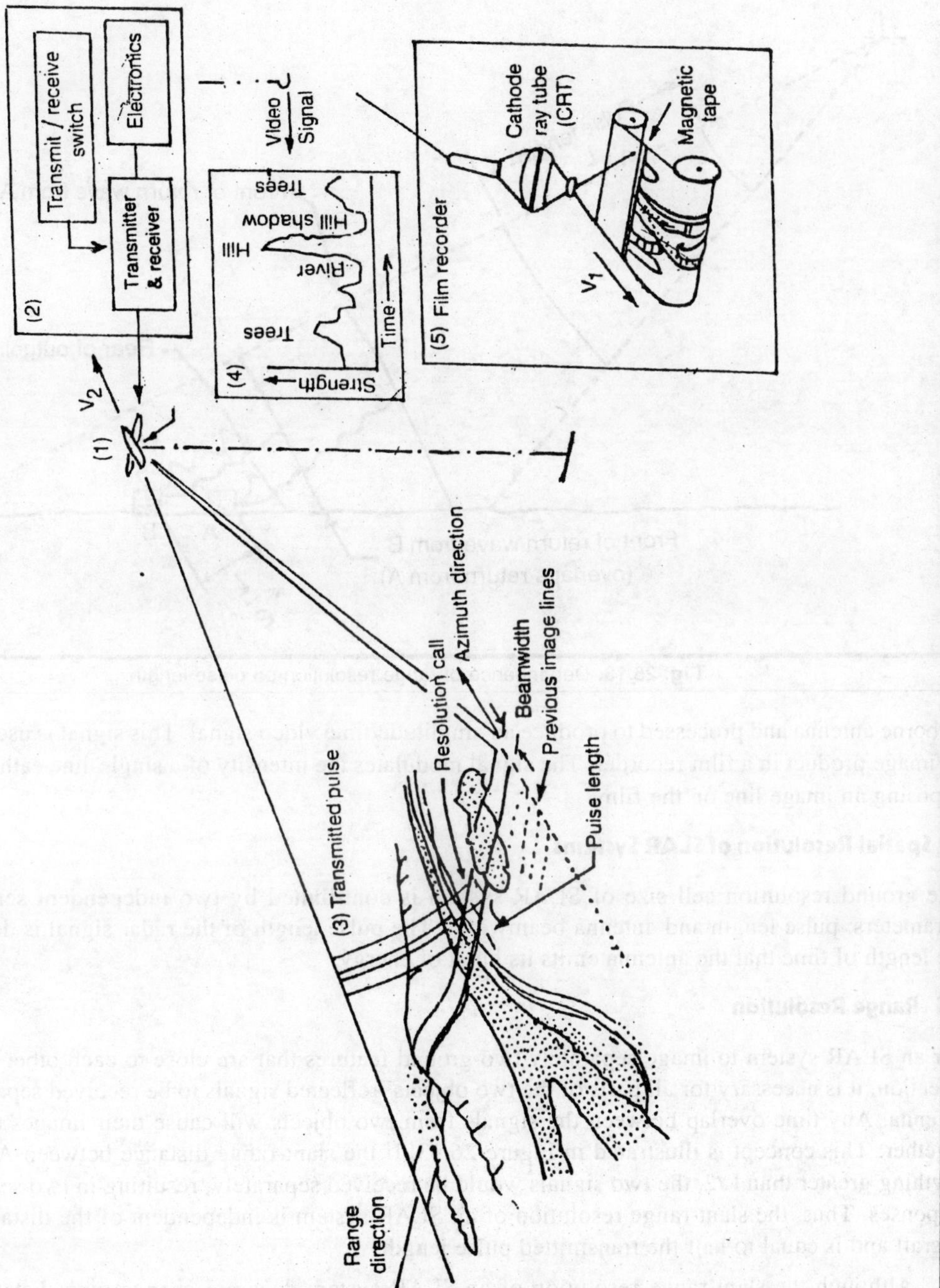

Fig. 26.12: Side looking airborne radar (SLAR) system operation

This principle of determining distance by electronically measuring the transmission-echo time is central to all imaging radar systems.

The manner in which SLAR images are created is illustrated in Fig. 26.12. As the aircraft advances, the antenna is continuously repositioned in the flight direction at the aircraft velocity V. The antenna is switched from a transmitter to a receiver mode by a synchronizer switch. Each transmitted pulse returns echoes from terrain features occurring along a single antenna beamwidth. The echoes are received by the

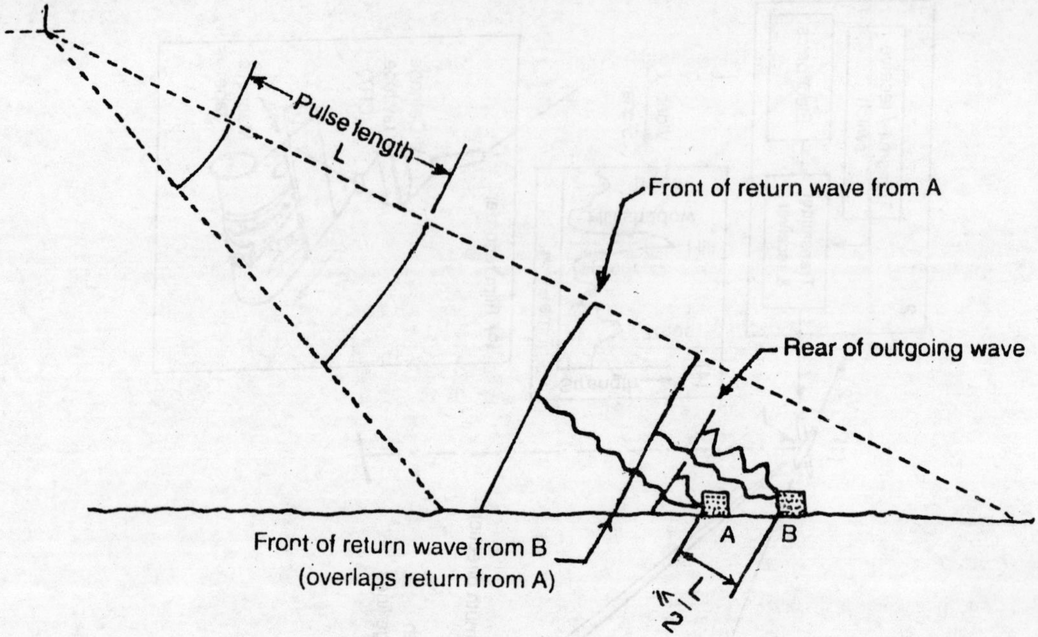

Fig. 26.13: Dependence of range resolution on pulse length

airborne antenna and processed to produce an amplitude/time video signal. This signal is used to generate an image product in a film recorder. The signal modulates the intensity of a single-line cathode ray tube, exposing an image line on the film.

3. Spatial Resolution of SLAR Systems

The ground resolution cell size of SLAR system is contributed by two independent sensing system parameters: pulse length and antenna beamwidth. The pulse length of the radar signal is determined by the length of time that the antenna emits its burst of energy.

3.1 Range Resolution

For an SLAR system to image separately two ground features that are close to each other on the range direction, it is necessary for all parts of the two objects' reflected signals to be received separately by the antenna. Any time overlap between the signals from two objects will cause their images to be blurred together. This concept is illustrated in Figure 26.13. If the slant-range distance between A and B were anything greater than L/2, the two signals would be received separately, resulting in two separate image responses. Thus, the slant-range resolution of an SLAR system is independent of the distance from the aircraft and is equal to half the transmitted pulse length.

Although the slant-range resolution of an SLAR system does not change with distance from the aircraft, the corresponding ground distance does. As shown in Fig. 26.14 (page 571) the ground resolution in the range direction varies inversely with the cosine of the angle between the horizontal ground plane and is called the depression angle, A.

Accounting for the depression angle effect, the ground resolution in the range direction, R_r, is found from:

$$R_r = c\acute{O} / 2 \, \text{Cos}A^0 \qquad (2)$$

where Ó is the pulse duration.

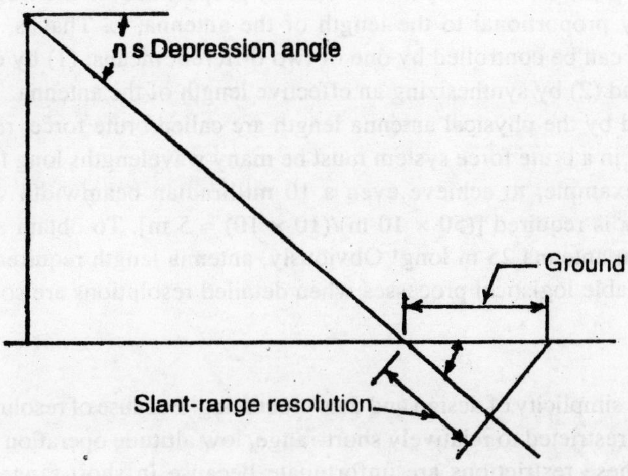

Fig. 26.14: Relationship between slant range resolution and ground range resolution

3.2 Azimuth Resolution

As shown in Fig. 26.15 the resolution of a SLAR system in the azimuth direction R is determined by the angular beamwidth, ß, of the antenna and the ground range (GR). As the antenna beam 'fans out' with increasing distance from the aircraft, the azimuth resolution deteriorates. Objects at points A and B would be resolved (imaged separately) at ground range (GR) but not at ground range (GR). That is at distance GR, A and B result in separate return signals. At distance GR, A and B would be in the beam simultaneously and would not be resolved.

Azimuth resolution R_a is given by:

$$R_a = GR.\ \beta \quad (3)$$

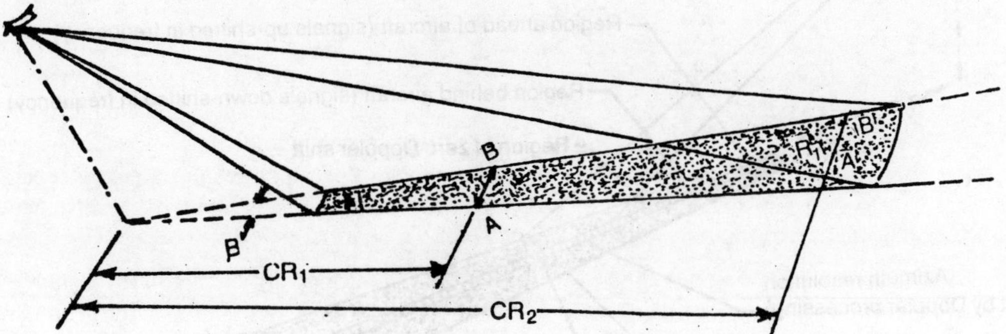

Fig. 26.15: Dependence of azimuth resolution (R_a) on antena beamwidth (β) and ground range (GR)

The beamwidth of the antenna of an SLAR system is directly proportional to the wavelength of the transmitted pulses and inversely proportional to the length of the antenna, L. That is, for any given wavelength, antenna beamwidth can be controlled by one of two different means: (1) by controlling the physical length of the antenna and (2) by synthesizing an effective length of the antenna. Those systems wherein beamwidth is controlled by the physical antenna length are called brute force, real aperture or noncoherent radars. The antenna in a brute force system must be many wavelengths long for the antenna beamwidth to be narrow. For example, to achieve even a 10 milliradian beamwidth with a 50 mm wavelength radar, a 5 m antenna is required [(50 × 10 m)/(10 × 10) = 5 m]. To obtain a resolution of 2 milliradians, we would need an antenna 25 m long! Obviously, antenna length requirements of brute force systems represent considerable logistical processes when detailed resolutions are sought.

4. SAR System

Brute force systems enjoy relative simplicity of design and data processing. Because of resolution problems, however, their operation is often restricted to relatively short-range, low altitude operation and the use of relatively short wavelengths. These restrictions are unfortunate because in short-range, low altitude operation time the area of coverage obtained by the system and short wavelengths experience these atmosphere attenuation and dispersion.

The deficiencies of brute force operation are overcome in synthetic aperture (or coherent) radar systems. These systems employ a short physical antenna, but through modified data recording and processing techniques they synthesise the effect of a very long antenna. The result of this mode of operation is a very narrow effective antenna beamwidth, even at far ranges, without requiring a physically long antenna or a short operating wavelength. For example, in a synthetic aperture system, a 2 meter antenna can be made, effectively, 600 m long!

The basic principle of synthetic aperture SLAR operation is illustrated in Fig. 26.16. In essence, return signals from the center portion of the beam-width are discriminated by detecting Doppler frequency

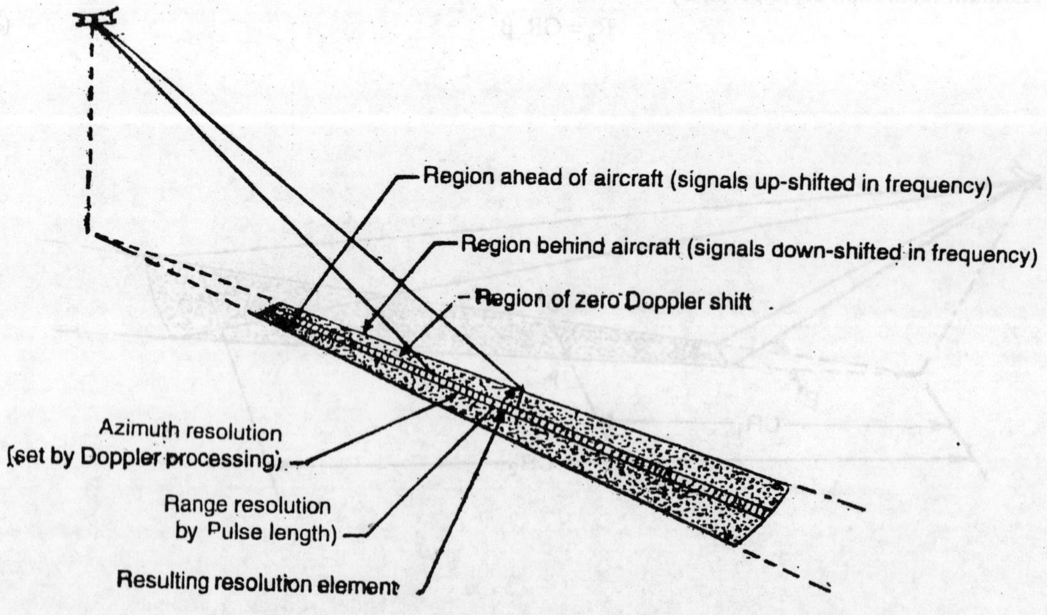

Fig. 26.16: Determinants of resolution in synthetic aperture SLAR

shifts. Recall that a Doppler shift is a change in wave frequency as a function of relative velocities of a transmitter and a reflector. Within the wide antenna beam, returns from features in the area ahead of the aircraft (Fig. 26.16, page 572) will have upshifted (higher) frequencies resulting from the Doppler Effect. Conversely, returns from the area behind the aircraft will have downshifted (lower) frequencies. Returns from features near the centerline of the beamwidth will experience little or no frequency shift. By processing the return signals according to their Doppler shifts, a very small effective beamwidth can be generated.

The details of synthetic aperture signal recording and processing are very complex and, therefore, beyond the scope of this discussion. The basic concept is to record both the amplitude and the frequency of signals returned from objects throughout the time period in which they are within the beam of the moving antenna. The frequency information is obtained by comparing the reflected return signals with a controlled frequency 'reference signal' generated internally by the system. The comparison is made by having the ground-reflected signal interfere with the reference signal. These signals interfere in various patterns, depending on their comparative frequencies. The patterns are generally recorded photographically, resulting in a 'signal film' (magnetic tape recording may also be used).

Because the signals received by a synthetic aperture system are recorded over a long time period, the aircraft translates the real antenna over a correspondingly long distance. This distance becomes the length of the 'synthetic' antenna. The azimuth resolution with the effective antenna length is greatly improved. Of interest is that this resolution is essentially independent of range because at long range an object is in the beam longer, hence returns from it are recorded over a longer distance.

5. Transmission Characteristics of Radar Signals

The two primary factors influencing the transmission characteristics of the signals from any given radar system are the wavelength and the polarization of the energy pulse used. Table 26.4 lists the common wavelength bands used in pulse transmission. The letter codes for the various bands (K, X, L etc.) were originally selected arbitrarily to ensure military security during the early stages of development of radar. They have continued in use as a matter of convenience and various authorities designate the various bands in slightly different wavelength ranges.

Table 26.4: Radar Band Designations

Band Designation	Wavelength ( ) in mm	Frequency ( ) = c in Megahertz (10^6 cycles sec)
K_a	7.5-11	40,000-26,500
K	11-16.7	26,500-18,000
K_u	16.7-24	18,000-12,500
X	24-37.5	12,500-8,000
C	37.5-75	8,000-4,000
S	75.0-150	4,000-2,000
L	150-300	2,000-1,000
P	300-1,000	1,000-300

Naturally, the wavelength of a radar signal determines the extent to which it is attenuated and/or dispersed by the atmosphere. Serious atmospheric effects on radar signals are confined to the shorter operating wavelengths (less than 3 cm). Even at these wavelengths, under most operating conditions the atmosphere only slightly attenuates the signals. As one would anticipate, attenuation generally increases as operating wavelength decreases and the influence of clouds and rain is variable. Whereas, radar signals are relatively unaffected by clouds, echoes from heavy precipitation can be considerable. Precipitation echoes are proportional, for a single drop, to the quantity D^6/λ^4, where D is the drop diameter. At the same time, the effect of rain is negligible with wavelengths of operation greater than 3 cm. So—short of the condition of a very heavy rainstorm—radar can be used through clouds, smoke, or fog. This feature, combined with day/night operation, makes radar a particularly valuable tool when time-dependent operations are undertaken.

Irrespective of wavelength, radar signals can be transmitted and/or received in different modes of polarisation, that is, the signal can be filtered in such a way that its electrical wave vibrations are restricted to a single plane perpendicular to the direction of wave propagation. (Unpolarized energy vibrates in all directions perpendicular to that of propagation). A SLAR signal can be transmitted in either a horizontal (H) or a vertical (V) plane. Likewise, it can be received in either a horizontal or vertical plane. Thus, we have the possibility of dealing with four different combinations of signal transmission and reception: H send, H receive; H send, V receive; V send, H receive; and V send, V receive. Like-polarized imagery is obtained from HV or VH combinations. Since various objects modify the polarization of the energy they reflect to varying degrees, the mode of signal polarization influences how the objects look on the resulting imagery.

6. Terrain Characteristics Influencing Radar Returns

There is a host of terrain characteristics that work hand-in-hand with the wavelength and polarisation of radar signals to determine the intensity of radar returns from various objects. These factors are many, varied and complex. Although many theoretical models have been developed to describe how various objects reflect radar energy, most of out practical knowledge in the subject has been derived from empirical observation. It has been found that the primary factors influencing objects' return signal intensity are their geometrical and electrical characteristics.

6.1 Geometrical Characteristics

One of the most readily apparent features of radar imagery is its side-lighted character when terrain of varying relief is imaged. This arises through variation in the relative sensor terrain geometry for various terrain orientations as illustrated in Fig. 26.17. Local variations in terrain slope result in varying angles of

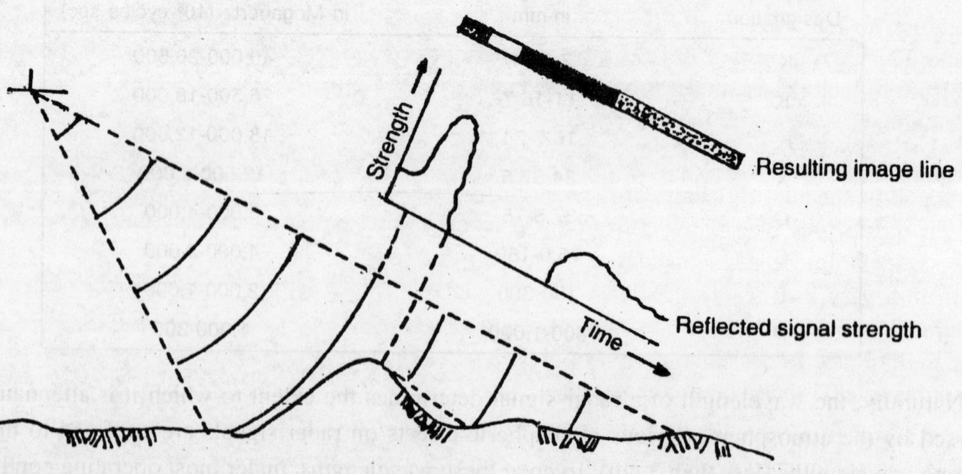

Fig. 26.17: Effect of sensor/terrain geometry on SLAR imagery

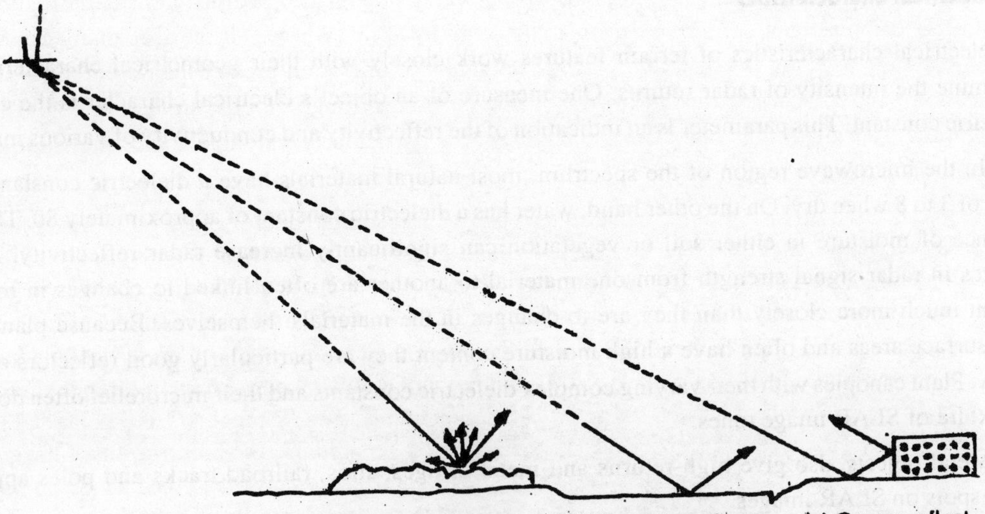

Fig. 26.18: Radar reflection from various surfaces

single incidence. This, in turn, results in relatively high returns from slopes facing the sensor and the level low returns or no returns from slopes facing away from the sensor. In Fig. 26.17 (page 574) the return strength versus time graph has been positioned over the terrain such that the signals can be correlated with the feature that produced them. Above the graph is the corresponding image line, in which the signal strength has been converted schematically to brightness values. The response from this radar pulse initially shows a high return from the slope facing the sensor. This is followed by a duration of no return signal from areas blocked from illumination by the radar wave. This radar shadow is completely black and sharply defined, unlike shadows in photography that are weakly illuminated by energy scattered by the atmosphere. Following the shadow, a relatively weak resource is recorded from the terrain that is not oriented towards the sensor.

The effect of the relative sensor/object geometry on the intensity of radar return signals is compounded by the effect of surface roughness. The roughness of an object's surface is a function of its relief variations in relation to the wavelength of the reflected energy. Surfaces with roughness essentially equal to or greater than the transmitted wavelength appear 'rough'. As shown in Fig. 26.18, rough surfaces tend to act as diffuse reflectors and scatter the incident energy in all directions, returning only a small portion of it to the antenna.

Targets with surface roughness much less than the wavelength of the radar energy (for example, /10) are 'smooth' specular reflectors of the energy. As shown in Fig. 26.18a & b, a smooth surface generally reflects most of the energy away from the sensor resulting in a low return signal. However, the sensor-object orientation must be considered as well, since a smooth surface oriented towards the sensor would result in a very intense return signal. A particularly bright response results from a corner reflector as illustrated in Fig. 26.18c. In this case, adjacent smooth surfaces cause a double reflection that yields a very high return. Because the corner reflectors generally cover only small areas of the scene, they often appear as bright 'sparkles' on the image.

It is worthy to note that some features, such as corn fields, might appear rough when seen both in the visible and the microwave portion of the spectrum. Other surfaces, such as roadways, may be diffuse reflectors in the visible region but specular reflectors of microwave energy. In general, SLAR images manifest many more specular surfaces than photographs do.

6.2 Electrical Characteristics

The electrical characteristics of terrain features work closely with their geometrical characteristics to determine the intensity of radar returns. One measure of an object's electrical character is the complex dielectric constant. This parameter is an indication of the reflectivity and conductivity of various materials.

In the microwave region of the spectrum, most natural materials have a dielectric constant in the range of 3 to 8 when dry. On the other hand, water has a dielectric constant of approximately 80. Thus, the presence of moisture in either soil or vegetation can significantly increase radar reflectivity. In fact, changes in radar signal strength from one material to another are often linked to changes in moisture content much more closely than they are to changes in the materials themselves. Because plants have large surface areas and often have a high moisture content they are particularly good reflectors of radar energy. Plant canopies with their varying complex dielectric constants and their microrelief often dominate the texture of SLAR image tones.

Metal objects also give high returns and metal bridges, silos, railroad tracks and poles appear as bright spots on SLAR images.

6.3 Relief Displacement

As in line scanner imagery, relief displacement in SLAR images is one-dimensional and perpendicular to the flight line. However, unlike scanner imagery and photography, the reflection of relief displacement is reversed. This is because radar images display ranges or distances, from terrain features to the antenna. When a vertical feature is encountered by a radar pulse, the top of the feature is reached before returns from the base of the future. This will cause a vertical feature to 'lay over' the closer features, making it appear to lean toward the nadir. This radar layover effect most severe at near range is compared to photographic relief displacement in Fig. 26.19.

Terrain slopes facing the antenna at near range are often displayed with a dramatic layover effect. This occurs whenever the terrain slope is steeper than a line perpendicular to the direction of the radar

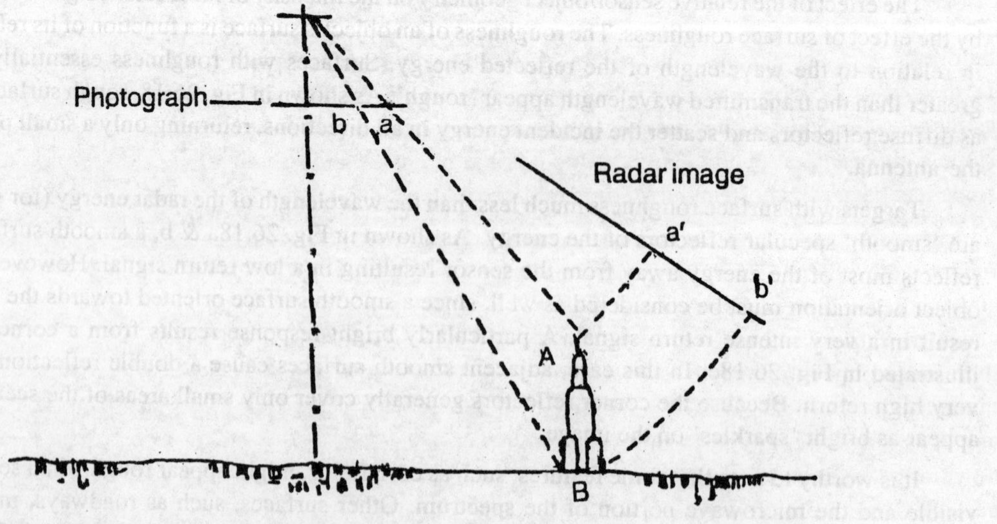

Fig. 26.19: Relief displacement of SLAR images versus photograph

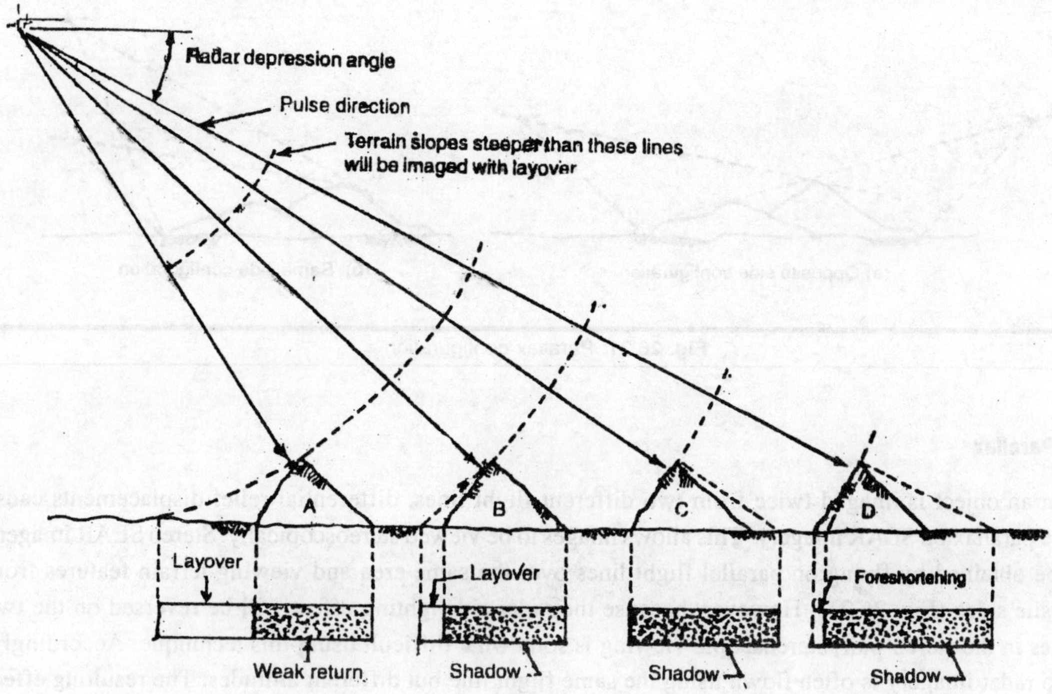

Fig. 26.20: Effect of terrain relief on SLAR images

pulse, expressed by its depression angle. This condition is met by the left sides of features A and B in Fig. 26.20. As such, the tops of these slopes will be imaged before their bases, causing layover. It can be seen in the image representations that the amount of layover displacement is greatest at short range, where the depression angle is greater.

When the slope facing the antenna is less steep than the line perpendicular to the depression direction, as in feature D in Fig. 26.20, no layover occurs. That is, the radar pulse reaches the base of the feature before the top. The slopes of the surfaces will not be presented in true size, however. As shown in feature D, the size of the sloped surface is compressed on the image. This foreshortening effect gets more severe as the slope's steepness approaches perpendicularity to the depression direction. In feature C, the front slope is precisely perpendicular to the depression direction and it can be seen that the image of the front slope has been foreshortened to zero length.

Foreshortening and layover are obviously interrelated to the previously described phenomenon of radar shadow. Slopes facing away from the radar side of feature A faces away from the aircraft, but it is less steep than the depression angle and will, therefore, be illuminated by the radar pulse. This will be weak, causing a fairly dark image area. In feature B, its right side is parallel to the depression angle and will, therefore, not be illuminated. As a result, the antenna will receive no return signals for a period of time and the image area will be black. When a slope faces away from the aircraft and is steeper than the depression angle, as in features C and D, the area of non-illumination will extend beyond the sloped area, masking downrange features in a radar shadow. As shown in Fig. 26.20, the shadow length increases with range because of the decrease in depression angle. Thus, a feature that casts an extensive shadow at far range (D) can be completely illuminated at close range (A).

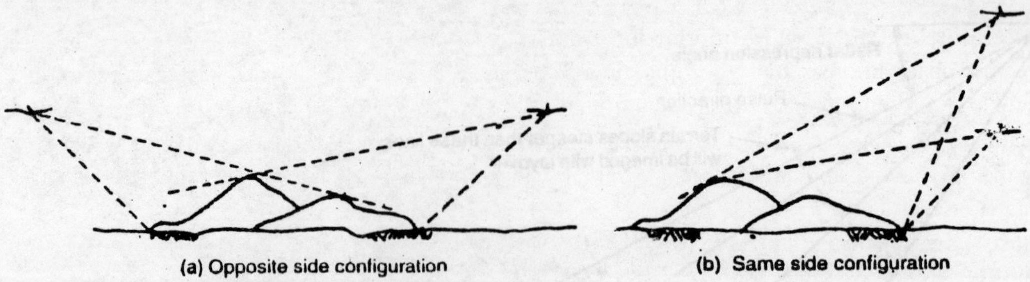

Fig. 26.21: Parallax configuration

6.4 Parallax

When an object is imaged twice from two different flight lines, differential relief displacements cause image parallax on SLAR imagery. This allows images to be viewed stereoscopically. Stereo SLAR imagery can be obtained by flying on parallel flight lines over the same area and viewing terrain features from opposite sides (Fig. 26.21). However, because the radar sidelighting effect will be reversed on the two images in the stereo-pair, stereoscopic viewing is somewhat difficult using this technique. Accordingly, stereo radar imagery is often flown using the same flight line but different altitudes. The resulting effect is called altitude parallax. In this case, the direction of illumination and the sidelighting effects will be similar on both images (Fig. 26.21).

26.4 APPLICATIONS OF REMOTE SENSING

The application of remote sensing can be divided into three phases: the first period, before 1950, constitutes the time of the initial applications of aerial photography to ecology; the second period, 1950 to 1972, represents the transition of applications from photography to unconventional imagery systems (for example, thermal infrared scenery and side-looking air-borne radars) and from low altitude aircraft to satellite platforms; from 1972 to the present, emphasis has been placed on the application of multi-spectral scanner and radars, meter data obtained on operational satellite platforms. Currently, remotely sensed data can be used for a great variety of practical applications, all of which relate, in general, to 'Earth Resources'. For overall discussion they are categorised into eight major categories:

1. Application to atmospheric studies.
2. Application to geospheric survey.
3. Application to biospheric survey.
4. Application to hydrospheric survey.
5. Application to cryospheric survey.
6. Overall application to environmental studies.
7. Application to geobotanical exploration.
8. Application to oceanography.

Application to Atmospheric Studies

Weather forecasters need access to information from large areas as quickly and as often as possible because weather observations rapidly become out of date. Satellites are best able to provide this kind of data that satisfy the requirements in terms of both coverage and immediacy. A good description of the current weather situation is essential to successful short-period weather forecasting, particularly for forecasting the movements and development of precipitation in the next six hours. Satellite pictorial data are particularly useful as they provide precision and detail for short-period weather forecasting. Thus, meteorologists have been making increasing use of weather satellite data as aids for the analysis of synoptic and small-scale weather systems.

Weather radars make it possible to track even small-scale weather patterns and individual thunderstorms. Many countries have now developed a good network of weather forecasting by the application of remotely sensed data, termed as METEOSAT imagery system.

In pictorial form, weather satellite data are also capable of revealing excellent resolution in the position, extent and intensity of cloud systems and troposphere temperature changes with adequate precision. Windspeed, cloud movement, tornado, etc. can also be forecasted by these datas. On the whole, remotely sensed data, a broad based data base including inventories of the following parameters, can be obtained:

1. The earth/atmospheric energy and radiation, particularly the net radiation balance at the top of the atmosphere, which is the primary driving force of the earth's atmospheric circulations.
2. Global moisture distribution in the atmosphere, which relates to the distribution of energy in the atmosphere.
3. Global temperature distribution over land and sea and, accordingly, the absorption and radiation of heat.
4. Distribution of cloud cover, which is a major influence on the albedo of the earth/atmosphere system and its component parts and indicates horizontal transport patterns of latent heat.
5. Sea surface temperature, which relates directly to the release of latent heat through evaporation.
6. Wind flow and air circulations, which relate to energy transfer within the earth/atmospheric system.
7. The climatology of synoptic weather systems including their frequencies and spatial distribution over extended periods.

The analysis of NIMBUS-7 satellite picture for determination of ozone hole in the atmospheric region during the month of October added a new dimension of application of remote sensed data in forecasting atmospheric changes.

The stratospheric aerosol measurement (SAM II) instrument on the NIMBUS-7 satellite provides a measurement of the optical depth of the atmosphere and accordingly is able to provide information on air quality.

Application in Geospheric Survey

During the past decades, LANDSAT imagery, in particular, laid a profound application in analysis of geological structures, identification of rock types and their mapping and also in determining the mineralised deposits on earth's surface. In addition to LANDSAT, TM data are also very useful for these kinds of studies.

Geothermal mapping using scanner data is mainly carried out in geothermal energy exploration studies as part of investigations of alternative energy resources or for the monitoring of active volcanoes—particularly in the case of eruption prediction studies.

Using infrared linescanner it is also possible to denote the extent of fire in coal mining industries. Radar imagery is also very useful for determination of lineament features. Since 1990s, SONAR, a major sensor, was used in sea floor mapping.

Application to Biospheric Survey and Analysis

Even in the most technologically advanced countries, up to date and accurate assessments of the total acreage of different crops in production, anticipated yields, stages of growth and condition (health and vigour) are often incomplete or non-timely in relation to the information needed by agricultural managers. These managers are continually faced with decision on planting, fertilizing, watering, control of pests and diseases, harvesting, storage, evaluation of crop quality, and planning for new cultivation areas. Remotely sensed information is used to predict marketing factors, to evaluate effects in case of crop failure, to assess damage from natural disasters or to aid farmers in determining when to plough, water, spray, or reap. The need for accurate and timely information is particularly acute in agricultural information systems because of the very rapid changes in the condition of agricultural crops and the influence of crop yield predictions on the world market. It is for these reasons that, as remote sensing technology has developed, the potential for the technologies to be used has received widespread attention. Previously aircraft surveys were used for crops to assist managers in gathering useful data; with the advent of multi-spectral repetitive satellite imagery, the possibility of an automonitored crop inventory is now approaching reality.

Colour-infrared film is sensitive to green, red and near infrared (500-900 μm) portions of the spectrum and is widely used in aerial and space photographic surveys for land use and vegetation analysis. Living vegetation reflects light in the green portion of the visible spectrum to which the human eye is sensitive. Additionally, it reflects 10 times as much in the near infrared (700-1,100 nm) portion of the spectrum which is just beyond the range of human vision. When there is a decrease in photosynthesis, whether caused by normal maturation or stress, there is a corresponding decrease in near infrared reflectance. Living or healthy vegetation appears as various hues of red colour in infrared film. If diseased or stressed, the colour response will shift to brown or yellow due to the decrease in near infrared reflectance. This film is also effective for haze penetration because blue light is eliminated by filtration.

Crops are best identified from computer-processed digital data that represents quantitative measures of radiance. In general, all leafy vegetations have a similar reflectance spectrum regardless of plant or crop species. The differences between crops by which they are separated and identified depend on the degree of maturity and percentage of canopy cover, although differences in soil types and soil moisture may serve to confuse the differentiation. However, if certain crops are not separable at one particular time of the year, they may be separable at a different stage of the season due to difference in planting, maturing, and harvesting dates. The degree of maturity and the yield for a giver crop also influence the reflectance at any stage of growth. This maturity and yield can be assessed by the history of any crop, traced in terms of its changing reflectance. The three year large crop inventory experiment, using LANDSAT MSS imagery demonstrated that the global monitoring of food and fibre production by satellite was possible. In forestry, the LANDSAT MSS have proved effective in recognising and locating the broadest classes of forest land finer and in separating deciduous, evergreen and mixed communities. More discrete classifications are possible with data from the Thermatric Mapper. Further possibilities include measurement of total hectareage given to forests and changes in these amounts. Timber volume, age of forests and presence of disease or pest infections can also be determined. Heat-sensitive channels on the LANDSAT Thermatric Mapper and the AVHRR are also used in forestry because of their ability to detect fires.

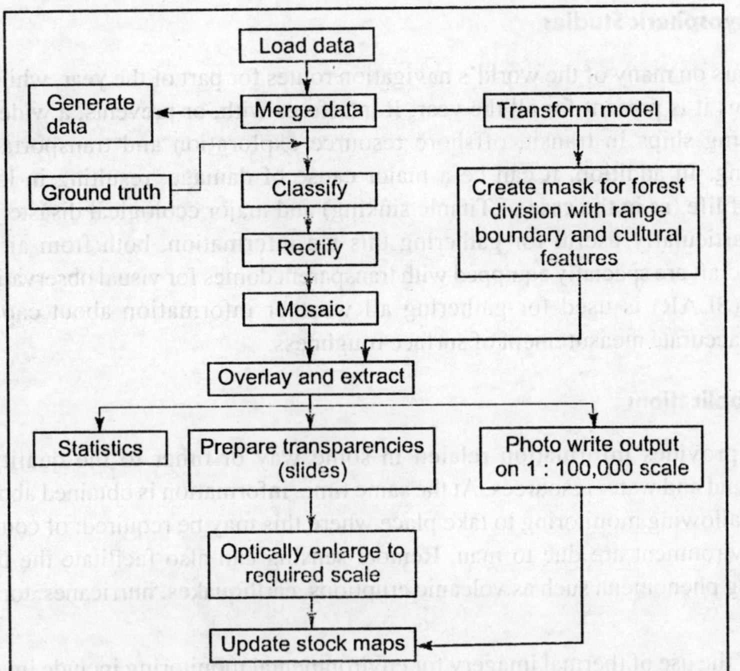

Fig. 26.22: Methodology for forest cover mapping

Recently IRS-IA LISS-II scenes were very often used for forest survey works in India.

A generalized scheme for forest cover mapping through image processing is given in Fig. 26.22.

Application to Hydrospheric Studies

Fortunately, many of the hydrological features of interest for improved water resource management are easily detected and measured with remote sensing systems. Although water sometimes reflects light in the visible wavelength in a similar manner to other surface features, it strongly absorbs light in the near infrared. As a consequence, standing water is very dark in the near infrared contrasting with soil and vegetation which appear bright in this part of the spectrum. Thus, in the absence of cloud, surface water can easily be distinguished and monitored in the optical wavebands.

Efficient management of marine resources and effective management of activities within the coastal zone is dependent, to a large extent, upon the ability to identify, measure and analyse a number of processes and parameters that operate or react together in the highly dynamic marine environment.

In this regard, measurements are required of the physical, chemical, geometrical, and optical features of coastal and open zones of the oceans. These measurements would include sea, ice, temperature, current suspended sediments, sea state bathymetry, and water and bottom colour. Different remote sensing capabilities exist for the provisions of the required information involving one of the combinations of measurement techniques.

The phenomenon of wind-driven coastal upwelling and the resulting high biological productivity off the west coast of North America were dramatically revealed in the images of sea-surface temperature and chlorophyll pigments from NOAA and NASA satellites.

Application to Cryospheric Studies

Floating ice appears on many of the world's navigation routes for part of the year, while in the case of the high Arctic regions it is present for all the year. It interferes with, or prevents, a wide variety of marine activities, including ships in transit, offshore resource exploration and transportation and offshore commercial fishing. In addition, it can be a major cause of damage, resulting in loss of vessels and equipment, loss of life (as in the case of Titanic sinking) and major ecological disasters. Remote sensing techniques are particularly useful for gathering this ice information, both from aircraft and satellite platforms. The aircraft are specially equipped with transparent domes for visual observations. Side Looking Airborne Radar (SLAR) is used for gathering all weather information about capability and lesser profilometers for accurate measurement of surface roughness.

Environmental Applications

Remote sensing provides information related in some way or other to the quality, protection and improvement of land and water resources. At the same time, information is obtained about man's effect on the environment, allowing monitoring to take place where this may be required: of course not all adverse effects on the environment are due to man. Remote sensing can also facilitate the timely response to naturally occurring phenomena such as volcanic eruptions, earthquakes, hurricanes, tornadoes, and forest fires.

Examples of the use of thermal imagery for environmental monitoring include investigations of heat loss from buildings, septic tank seepage into water-supply and sewage outfall. Thermal infrared image of the city of Dundee, Scotland, showed light grey tones depicting warm areas and dark tones depicting cool areas. The detention of contrasting thermal signals was useful for a cattle survey made in summer in west Texas.

Oil pollution through leakage in pipelines or accidental spillage is an all too familiar example of the harmful effects man can have on the environment. For oil pollution monitoring from satellites the spectral resolution is often a serious problem, although, under favourable conditions, it is possible to observe and monitor oil slicks using this data.

Spatial Information Systems, Land use and Land Cover Mapping

In recent years satellite data have been incorporated into sophisticated information systems of a geographical nature allowing the synthesis of remotely sensed data with existing information. The growing complexity of society has increased the demand for timely and accurate information on the spatial distribution of land and environmental resources, social and economic indicators, landownership and value and their various interactions. Land and Geographic Information System (GIS) attempt to model, in time and space, these diverse relationships so that, at any location, data on the physical, social and administrative environment can be accessed—interrogated and combined—to give valuable information to planners, administrators, resource scientists, and researchers.

Land use refers to the current use of the land surface by man for his activities, while land cover refers to the state or cover of the land only. Current remotely sensed data of the earth's surface generally provide information about the land cover, with interpretation or additional information being needed to ascertain the land use. Land use planning is concerned with achieving the optimum benefits for mankind in the development and management of the land for various purposes, such as food production, urbanisation, manufacture, supply of raw materials, farmer production, transport, and recreation. This planning aims to match land use with land capacity and specific uses with appropriate natural conditions so as to provide adequate food and material supply without significant damage to the environment. Land use planning has previously been severely hampered both by the lack of up to date maps, showing which categories are present and changing in an area or large region and the inadequacies of the means by which the huge quantities of data involved were handled. The cost involved in producing land cover, land use and land capabilities maps have prohibited their acquisition at useful working scales. Vast areas of Africa, Asia and South America remain poorly—and often incorrectly—mapped. Land use can be deduced or inferred indirectly from the identification and distribution patterns of vegetation; surface materials and cultural features are interpreted from imagery. With supplementary information which may be extracted from a appropriate data base or information system but is simply the accumulated knowledge of the interpreter, specific categories of surface features can be depicted in the map form.

Application of Remote Sensing in Detecting Geobotanical and Biogeochemical Anomalies

Remote sensing has immense role in the field of mineral exploration in vegetated terrain. The use of remote sensing in this context is a recent addition to the sciences of geobotany and biochemistry. Geobotany is concerned with the affinity of particular plants, plant associations or plant morphological aberrations to soil containing higher than normal contents of heavy metals, such as copper, lead, zinc, or tin. Biochemistry enables the analysis of plant tissue to assess its heavy metal content. Geobotanical indicator species, plant association changes and biogeochemical anomalies can all be distinguished from the synoptic view of remote sensing platform. So, during the last so many years, remote sensing, as an exploration tool, has become increasingly popular.

The introduction of high resolution spectro-radiometers to detect plant spectra associated with heavy metals has yielded results of great significance to mineral exploration in vegetated terrain. Healthy green leaves have a low visible and a high near infrared reflectance, while physiological disturbances—such as excess of heavy metals—cause changes in leaf pigmentation, structure and water content which lead to aberrant leaf reflectance. The most successful detection of biogeochemical stress has been achieved with high spectral resolution spectro-radiometers. Collins (1976) and Collins et al, (1979, 1980) have developed such an instrument which scans in 500 narrow bands between 0.4 μm and 1.1 μm. In addition to allowing studies of reflectance change in each narrow band this instrument also allows for an interpretation of reflectance changes that occur on the red edge.

Many case studies, which have utilized remote sensing as an aid to mineral exploration in vegetated terrain, have adapted an insular approach in that they have made studies of only single sets of imagery rather than taking multitemporal approach to the problem of anomaly detection.

Remote Sensing Application in Oceanography

Satellites with their potential to cover vast areas of land and oceanic areas on a repetitive basis are ideal for coastal and oceanic studies. Satellite data are also economical in comparison with other sources of data. For a nation like India with a long coastline and extensive exclusive economic zone, satellite based remote sensing can potentially provide substantial economic benefits; with the use of satellite data it is now possible to determine the location of potential fishing grounds, river meandering pattern and upwelling zone of coastal shelves. Since 1990s, with the use of NOAA, IRS, LISS-I data, many features of coastal water and land mass were detected. However, it is proposed to continue these efforts when the wide field sensor (WIFS) data becomes available in future from the IRS-IC/ID satellites. Once the technique is perfected and relationships between various levels of production are established, the remotely sensed data could be utilized for : (i) studying migratory patterns of the pelagic fish; (ii) preparation of spatial distribution of chlorophyll-a charts and fishery forecasting; and (iii) ecosystem modelling.

26.5 GEOGRAPHIC INFORMATION SYSTEM (GIS)

Introduction

In the late twentieth century, demands for data on the topography and specific themes of the earth's surface, such as natural resources, accelerated greatly. Stereo aerial photography and remotely sensed imagery allowed photogrammetrists to map large areas with great accuracy. The same technology also gave the earth resource scientists—the geologist, the soil scientist, the ecologist, the land use specialist—enormous advantages for reconnaissance and semi-detailed mapping. The resulting thematic maps were a source of useful information for resource exploitation and management. The study of land evaluation arose through the need to match the requirements for producing food and supporting populations to the available resources of climate, soil, water, and technology.

The need for spatial data and spatial analyses is not just the preserve of earth scientists. Urban planners and cadastral agencies need detailed information about the distribution of land and resources in towns and cities. Civil engineers need to plan the routes of roads and canals and to estimate construction costs, including those of cutting away hillsides and filling in valleys. Police departments need to know the spatial distribution of various kinds of crime, medical organisation and epidemiologists are interested in the distribution of sickness and disease, and commerce is interested in improving profitability through the optimization of the distribution of sales outlets and the identification of potential markets. The enormous infrastructure of what are collectively known as 'utilities'—that is water, gas, electricity, telephone lines, sewerage systems—all need to be recorded and manipulated as spatial data linked to maps. Today, there are many ways in which the political and economic processes responsible for our well-being need to be assisted such that democratic organization of the planet's resources can be used to the advantage of all. The needs are first to know what resources we have at our disposal—how much fertile land, how much energy, how many people? The second is to understand how these are distributed over the earth, and to know where they are, their extent, who owns them or has other rights to them, and how we can best use and manage them.

The history of using computers for mapping and spatial analysis shows that there have been parallel developments in automated data capture, data analysis, and presentation in several broadly related fields such as cadastral and topographical mapping, thematic cartography, civil engineering, geology, geography, hydrology, spatial statistics, soil science, surveying and photogrammetry, rural and urban planning, utility networks, and remote sensing and image analysis. Military applications have overlapped and even dominated several of these monodisciplinary fields. Consequently there has been much duplication of effort and a multiplication of discipline-specific jargon for different applications in different lands. This multiplicity of efforts in several initially separate—but closely related—fields has resulted in the emergence of the general purpose GIS.

Definitions of GIS

(a) **Toolbox-based definitions**: *'a powerful set of tools for collecting, storing, retrieving at will, transforming, and displaying spatial data from the real world'* (Burrough, 1986).

'a system for capturing, storing, checking, manipulating, analysing, and displaying data which are spatially referenced to the Earth' (Department of Environment, 1987).

'an information technology which stores, analyses, and displays both spatial and non-spatial data' (Parket, 1988).

(b) **Database definitions**: *'a database system in which most of the data are spatially indexed, and upon which a set of procedures operated in order to answer queries about spatial entities in the database'* (Smith et al, 1987).

'any manual or computer based set of procedures used to store and manipulate geographically referenced data' (Aronoff, 1989).

(c) **Organization-based definitions**: *'an automated set of functions that provides professionals with advanced capabilities for the storage, retrieval, manipulation, and display of geographically located data'* (Ozemoy, Smith, and Sicherman, 1981).

'an institutional entity, reflecting an organisational structure that integrates technology with a database, expertise and continuing financial support over time' (Carter, 1989).

'a decision support system involving the integration of spatially referenced data in a problem solving environment' (Cowen, 1988).

Components of a Geographical Information System

Geographical information systems have three important components—computer hardware, sets of application software modules, and a proper organisational context including skilled people—which need to be in balance if the system is to function satisfactorily.

Computer hardware: The computer had a hard disc drive for storing data and programmes, but extra storage can be provided via a network or by digital type cassettes, optical CD-ROMs, and other devices. A digitizer or scanner is used to convert maps and documents into digital form so that they can be used by the computer programme. A plotter or a printer or any other kind of display device is used to present the results of the data processing. The general hardware components of a geographical information system are presented in Fig. 26.23.

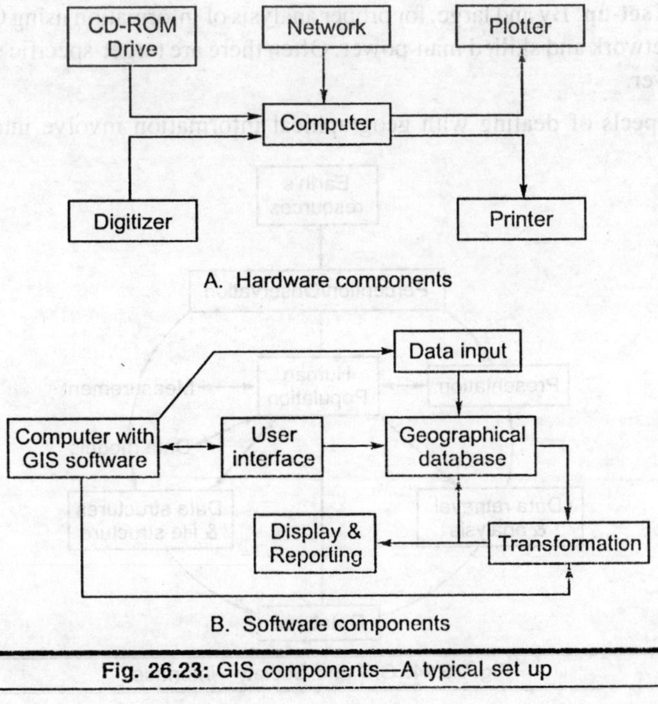

Fig. 26.23: GIS components—A typical set up

GIS software: These softwares may be split into five functional groups:

(a) Data input and verification.

(b) Data storage and database management,

(c) Data output and presentation,

(d) Data transformation,

(e) Interaction with the user.

Data input covers all aspects of capturing spatial data from existing maps, field observation, and sensors (including aerial photography, satellites and recording instruments) and converting them to a standard digital form. The main software components of a geographical information system is shown in Fig. 26.23.

Data storage and database management concerns the way in which data about the location, linkage (topology) and attributes of geographical elements (points, lines, areas, and more complex entities representing objects on the earth's surface) are structured and organised, both with respect to the way they must be handled in the computer and how they are perceived by the users of the system.

Data output and presentation concern the ways the data are displayed and how the results of analyses are reported to the users. Data may be presented as maps, tables, and figures in a variety of ways ranging from the ephemeral image on the computer screen, through hardcopy output drawn on printer or plotter to information recorded on magnetic media in digital form. Data transformation embraces two classes of operation, viz., (a) transformations needed to remove errors from the data or to bring them up to date or to match them to other data sets, and (b) the large array of analysis methods that may be applied to the data in order to achieve answers to the questions asked of the GIS.

GIS designers have realized that the requirements to users to retrieve and transform data are unlimited. Therefore most systems provide a range of interfaces by which the user can interact with the system. The simplest are menu-driven commands that can be selected by simply pointing and clicking with the mouse, and this is an efficient way of providing complex functionality for ordinary users.

Organisational set-up: By and large, for proper analysis of information using GIS require appropriate set-up of computer network and skilled man-power. Often there are target-specific training for developing appropriate man-power.

However all aspects of dealing with geographical information involve interactions with people. (Fig. 26.24).

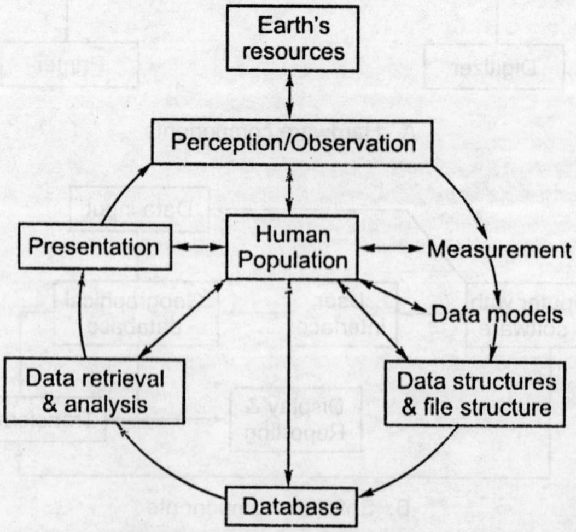

Fig. 26.24: GIS and Society—interface

Sources of Geographical Data: Creating a GIS database is a complex operation involving data capture, verification, and structuring process. Because raw geographical data are available in many different analogues or digital forms, such as maps, aerial photographs, satellite images, or tables, a spatial database can be built in several—not mutually exclusive—ways. These are: acquire data in digital forms from a data supplier.

- Digitize existing analogue data,
- Carry out one's own survey of geographic entities;
- Interplate from point observations to continuous surfaces.

In all cases the data must be geometrically registered to a generally accepted and properly defined coordinate system and coded so that they can be stored in the internal database structure of the GIS being used. For many people the most common source of geographical data is the paper or digital topograhic or thematic map, which is a graphical representation of the distribution of sptatial phenomena. Images derived from optical and digital remote sensing systems mounted in aircraft and satellite provide much spatial information over many levels of temporal and spatial resolution. A wide range of scanners mounted in satellites or aircraft also provide digital data directly. In addition, the traditional means of collecting geographical data is by ground or field surveys to record sample values at known locations, using instruments ranging from questionaries and soil augers to automated chemical probes. It is most important that all spatial data in a GIS are located with respect to a common frame of reference. It is further essential for GIS and spatial analysis that all data are referenced to the same coordinate system.

Georeferencing of data: In practice, ground and field surveys are georeferenced in many ways. Ground checks are needed to locate aerial photographs and satellite images correctly. The boundaries for cadastral systems are surveyed accurately on the ground using laser theodolites, as are the locations of infrastructure and utilities (roads, electricity and telephone cables, gas, water, and sewerage pipes). Sources of demographic and socio-economic data such as the census are linked to less accurate, cartographically defined areal units as the basic spatial references. A similar approach is adopted for data collected by administrative or municipal authorities. Subsequently the problem of defining and recording the location of a data point has been eased through the development of global position systems (GPS) which are being used increasingly in many types of data collection exercise, often in tandem with data-loggers.

Box 26.1: ACTIVE DOMAINS FOR GIS
Producers, kinds, and applications of Geographical Information
The main producers and sources
Topographical Mapping: National Mapping Agencies, Private Mapping Companies
Land Registration and Cadastre
Hydrographic Mapping
Military Organizations
Remote Sensing companies and satellite agencies
Natural resources surveys: Geologists; Hydrologists; Physical Geographers and Soil Scientists; Land Evaluators; Ecologists and Biogeographers; Meteorologists and Climatologists; Oceanographers
The main types of geographical data available
Topographic maps at a wide range of scales
Satellite and airborne scanner images and photographs
Administrative boundaries: Census tracts and census data; Postcode areas
Statistical data on people, land cover, land use at a wide range of levels.
Data from marketing surveys
Data on utilities (gas, water, electricity lines cables) and their locations.
Data on rocks, water, soil, atmosphere, biological activity, natural hazards, and disasters collected for a wide range of spatial and temporal levels of resolution.

Contd.

Some current applications	
Agriculture	Monitoring and management from farm to National levels
Archeology	Site description and scenario evaluation
Environment	Monitoring, modelling, and management for land degradation; land evaluation and rural planning; landslides; desertification; water quality and quantity; plagues; air quality; weather and climate modelling and prediction
Epidemiology and Health	Location of disease in relation to environmental factors
Forestry	Management, planning, and optimizing extraction and replanting
Emergency services	Optimizing fire, police, and ambulance routing; improved understanding of crime and its location
Navigation	Air, sea, and land
Marketing	Site location and target groups; optimizing goods delivery
Real Estate	Legal aspects of the cadastre,, property values in relation to location, insurance
Regional/Local planning	Development of plans, costing, maintenance, management
Road and rail	Planning and managment
Site evaluation and costing	Cut and fill, computing volumes of materials
Social studies	Analysis of demographic movements and developments
Tourism	Location and management of facilities and attractions
Utilities	Location, management, and planning of water, drains, gas, electricity, telephone, cable services

Application of GIS

The applications of GIS are:

- Conducting resource inventory as an application of standard classification techniques and simple integration of the information in a GIS to provide the required management information;
- Resource monitoring as a more advanced application of classification, with simple GIS integration to provide the required management information;

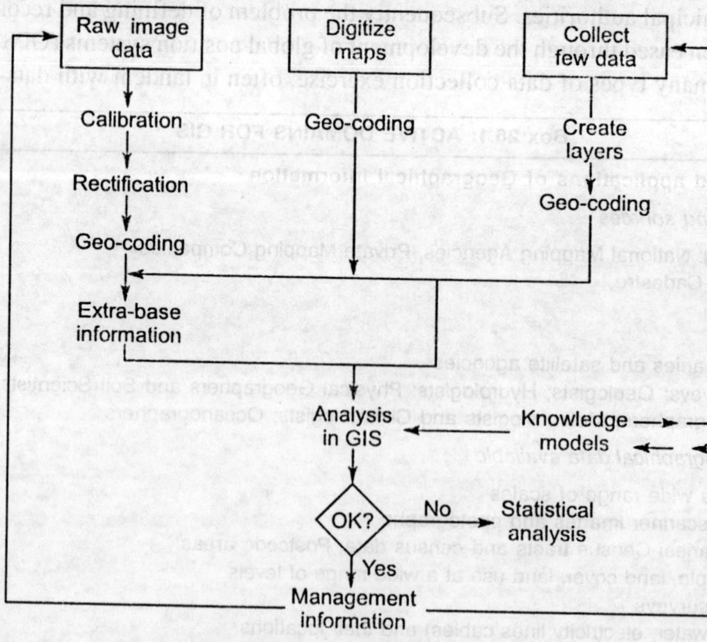

Fig. 26.25: GIS–process system

- Estimation of resource status using remote sensing;
- Resource modeling by using remote sensing and GIS to predict potential outcomes, to support the management of resources; and
- Estimation of crop production with the aid of GIS data.

Example: Accurate information on land and environmental resources is essential to effective land use planning. Key sources of information include topographic maps, aerial photos, satellite images, and data desired from geographic system (GIS). The process of plan compliance with land use planning involves the following steps:

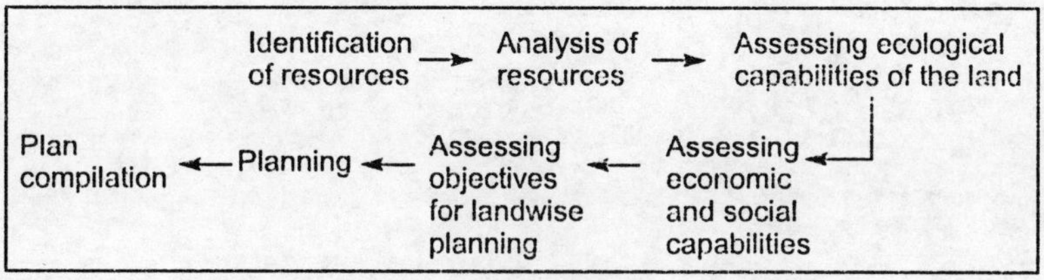

For comprehensive studies and for planning undertaken on a countrywide basis, aerial photography (RS method) are recommended as a way of collecting accurate and appropriate information, especially when the areas have not been adequately studied in the part. Aerial photography can save time and expenses, and the photographs taken can generally be merged with data gather from field operations.

26.6 Global Positioning System (GPS)

The satellite navigation system or global positioning system (GPS) is a new technique used by the geographers for locating spatial phenomena. This technique is based on the principles of trilateration, which have been used by surveyors for centuries. To understand how a GPS works, it is important to appreciate that the system has five components. These are:

- The satellite constellation (either American NAVSTAR or Russian GLONASS systems)
- The satellite control and monitoring stations;
- The national and regional GPS base station networks;
- The GPS receivers, and
- The users.

Usually, GPS has been used in field mapping, georeferencing, and also in digital elevation model verification.

Global Positioning Systems (GPS) Application: Traditionally, geographers have used numerous techniques to determine both the relative and actual location of spatial phenomena. These have ranged from the very accurate and precise techniques of the surveyors to the more general method of the social scientist. Recently, however, a new technique for locating spatial phenomena has found particular favour with geographers. This is the satellite navigation system or global positioning system (GPS). These portable locational devices can be mounted on a vehicle, carried in a backpack or held in the hand and used to record location at almost any point on the earth's surface.

GPS is a set of satellites and central systems that allow a specially designed receiver to determine its location anywhere on earth 24 hrs a day. Two main systems exists—the American (NAVSTAR) and the Russian (GLONASS) system.

The GPS framework has a satellite constellation, the satellite control and monitoring stations, the national and regional GPS base station network, receivers and the user ends.

By and large, GPS technology is feasible and effective in a variety of research areas where other surveying and mapping techniques would be impractical. For field mapping, or georeferencing of resource data, GPS technology is now used widely.

27 HYDROGEOLOGY

Precipitation—that is not intercepted or does not run off on the surface—infiltrates downward into the subsurface through pore spaces between mineral grains, cracks, solution cavities or vesicles. These water flows constitute the ground water. The physical properties of subsurface material that control the volume and movement of ground water are porosity (the percentage of the total volume of the rock consisting of voids) and permeability (the capacity of a rock to transmit fluids).

27.1 POROSITY

Porosity can be calculated by the formula:

$$\text{Porosity} = \frac{\text{volume of void space}}{\text{total volume}} \times 100$$

The nature of the void space varies considerably from one type of material to another. The most common types of pore spaces are:

- Void spaces between mineral grains
- Fractures
- Solution cavities
- Vesicles.

The porosity of unconsolidated sediments or different kinds of rocks varies widely, with respect to the constituents and packing of particles by which they are constituted.

27.2 PERMEABILITY

Permeability varies with the fluid's viscosity and hydrostatic pressure, the size of void space and the degree to which the open spaces are interconnected. Both porosity and permeability are very important features of ground water movement. In fact, porosity is a measure of how much water a rock can hold, whereas permeability is how easily water can be transmitted. Sediments or rocks with high porosity do not necessarily have high permeability.

The most copious producers of ground water are commonly coarse-grained, unconsolidated sediments that are noted for their high permeability. Conversely, clay has very low permeability. Mixtures of the two produce a wide-ranging continuum of permeability values for unconsolidated sediments. In fact, the permeability of a sediment depends largely on the size of pore spaces, which is directly related to grain size.

The permeability of rocks depends on the primary void space generated when the rock was formed and the secondary void space created after the rock was formed. Crystalline igneous, metamorphic and chemical sedimentary rocks typically have low porosity and low primary permeability because of tightly interlocking crystals which allow very few openings. However, high secondary permeability can develop in crystalline rocks as a result of fracturing. Some volcanic rocks have high primary porosity because of large numbers of vesicles. The permeability of these rocks are also high.

27.3 SPECIFIC YIELD AND SPECIFIC RETENTION

Specific yield (Sy) is the ratio of the volume of water that drains by gravity from saturated sediment or rock to the total volume of the material. The surface tension of water molecules causes them to cling to grain surfaces.

The *specific retention* (Sr) of a sediment or rock is the ratio of the volume of water a sample can retain against gravity drainage to the total volume of the rock. Specific yield (Sy) is the volume of water that a sediment or rock will yield by gravity drainage and specific retention (Sr) is the remaining volume. Thus, the sum of the two is equal to porosity:

$$\text{Porosity} = Sy + Sr$$

Because specific retention is greater for samples with small grain sizes, clay may have a porosity of 50 per cent and a specific retention of 48 per cent, yielding only a 2-per cent specific yield. Maximum specific yield occurs in sediments having medium to coarse sand sizes (0.5-1.0 mm).

Darcy (1856) demonstrated that the discharge of a saturated system is proportional to the hydraulic head between two points, to the cross-sectional area, and to the permeability of the material.

The relationship is now commonly expressed as *Darcy's law*:

$$Q = PIA$$

where Q = discharge

P = a permeability coefficient sometimes called the *hydraulic conductivity*

I = *hydraulic gradient*, represented by the change in hydraulic head between two points

A = cross-sectional area.

27.4 WATER TABLE

As water infiltrates soil at the ground surface, it is impelled downward by gravity through pore spaces that are only partially saturated. As long as the water continues to encounter empty pore spaces, movement of the water is dominantly vertical. Eventually, however, all of the pore spaces become completely filled with water, leaving no more space for continued vertical percolation and the material becomes totally saturated, producing a subsurface region known as the *zone of saturation*. The upper surface of the zone of saturation is the *water table*. Once water reaches the zone of saturation, it can no longer continue to move vertically downward because all of the pores beneath are already filled—so it moves laterally in the direction of the slope of the water table. The water table is not horizontal—it rises higher beneath hills and slopes in the direction of valleys where ground water is discharged into streams, approximately mimicking the surface topography. The slope of the water table (h/l) represents the hydraulic gradient whose inclination depends on the permeability of the material and the rate at which water is added to the zone of saturation. The distribution of ground water is shown in Fig. 27.1 (page 593).

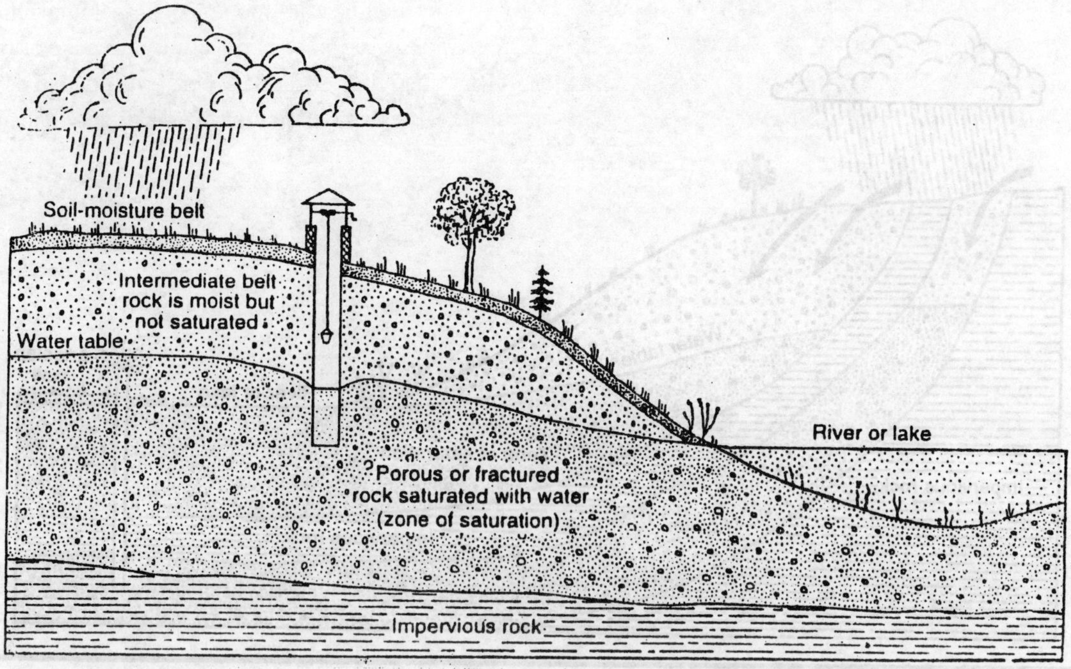

Fig. 27.1: Distribution of ground water

27.5 AQUIFERS

An *aquifer* is a geological unit that can store and transmit water. Unconsolidated sand and gravel, sandstone, limestone, lava flows and fractured pletonic and metamorphic crystalline rocks are examples of typical *aquifers*.

There are two kinds of aquifers:

(i) Unconfined aquifers, and

(ii) Confined aquifers.

Aquifers that extend continuously from the land surface downward through material of high permeability are known as *unconfined aquifers*. Recharge to the aquifers may be from percolation downward through the unsaturated zone, from later ground water flow, or from upward seepage through underlying material.

In case of *confined aquifers* subsurface material sometimes contains *confining layers*, having very low permeability, typically less than about 10^{-2} darcy. Ground water may move through confining layers, but the rate of movement is usually very slow. Confining layers can be subdivided into *aquitards*, *aquicludes* and *aquifuges*.

An *aquitard* is a leaky confining layer of low permeability that can store ground water and transmit it sluggishly from one aquifer to another.

An *aquiclude* also has low permeability but is situated in a position to form the upper or lower boundary of a ground water flow system.

An *aquifuge* is an almost totally impermeable body of rock or unconsolidated material.

Artesian water is confined ground water that is under high hydrostatic pressure. *Artesian aquifers* are overlain by an aquiclude. Recharge to them may occur where the aquifer is exposed at the surface or by slow, downward percolation through an aquitard, water from such an aquifer will rise above the top of

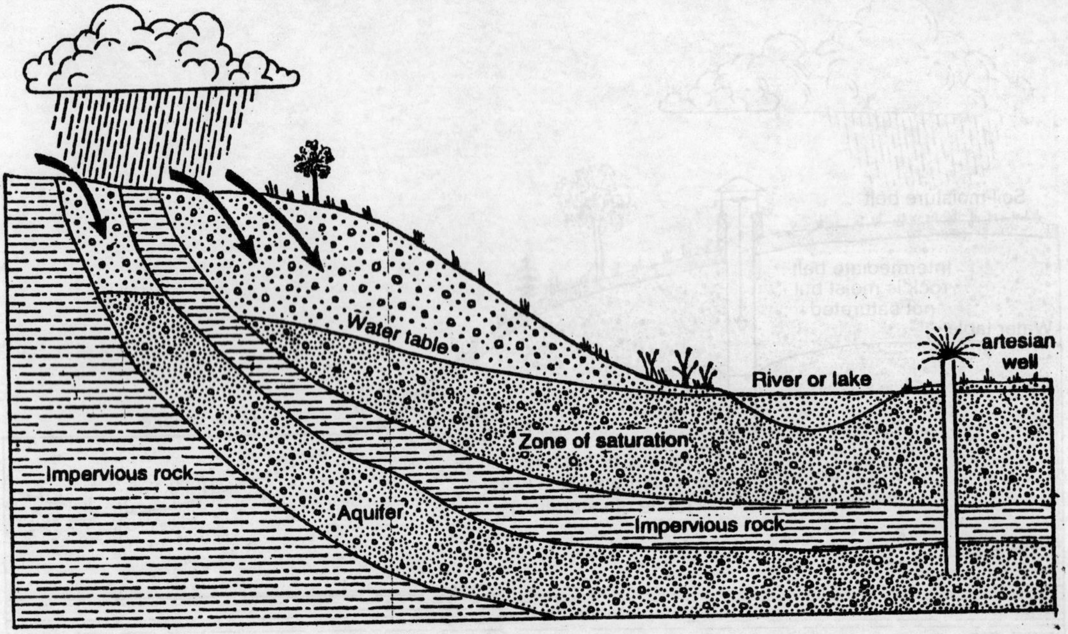

Fig. 27.2: Aquifer features

the confined aquifer where it is intersected by a well or otner conduit as a result of hydraulic head. The level to which water will rise in a well is the *potentiometric surface*. A potentiometric surface is included away from the recharge area because of frictional losses as water moves through the pore in an aquifer and because pressure is lost through leakage in fractures.

Where a layer of low-permeability material occurs as a lens within more permeable material, water moving downward through the unsaturated zone is intercepted and will collect on top of the impermeable lens. As saturation develops above the impermeable lens, a ***perched aquifer*** is formed above the main

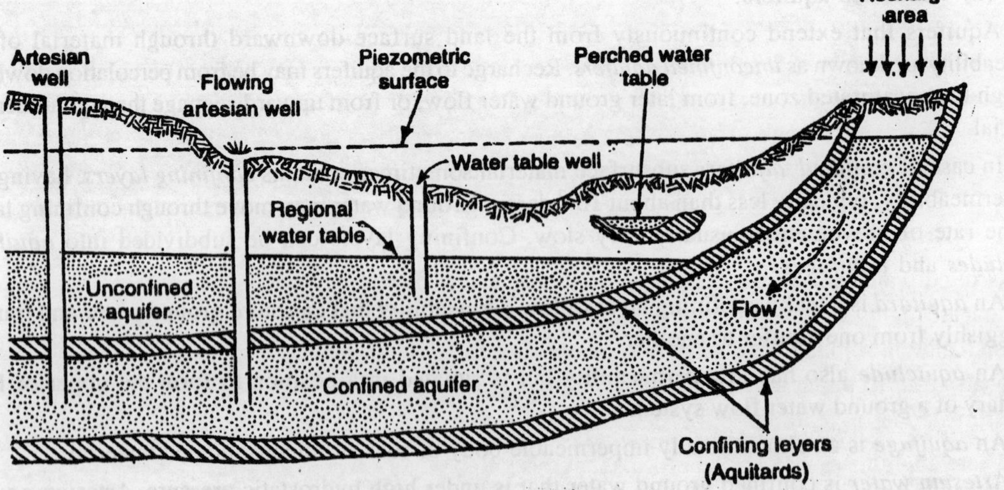

Fig. 27.3: Aquifer types

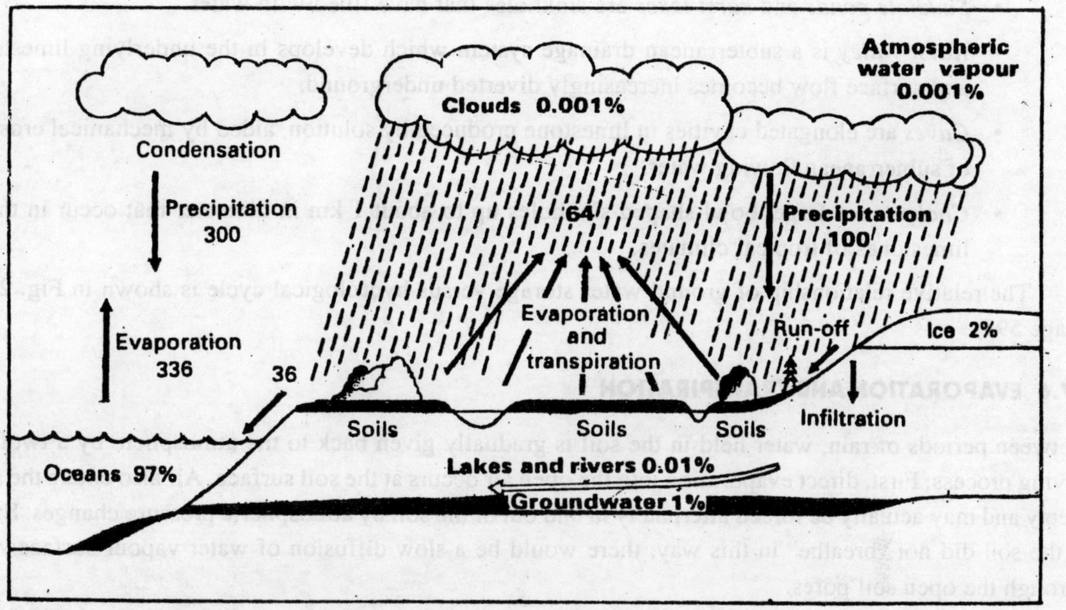

Fig. 27.4: Hydrological cycle

water table. However, under pumping, the removal of water causes the water table to be drawn down around the well, producing a *cone of depression* in the water table.

Different types of aquifers are shown in Fig. 27.3 (page 594).

- **Thermal Springs and Geysers**: In regions of unusually high geothermal gradients—especially those of recent volcanic activity—ground water becomes heated to high temperatures, creating *thermal springs* and *geysers* when the water discharges to the surface. If the water is hot enough, geysers may form when the water flashes into stream near the surface.
- **Ground Subsidence**: Withdrawal of large quantities of ground water or other fluids, such as petroleum, decreases the fluid pore pressure between grains. The decrease in pore pressure allows tighter grain-to-grain contact and leads to sediment compaction and subsidence of the ground surface.
- **Karst**: Although ground water is rarely seen, its effect on topography may be profound as a result of the dissolving of soluble rocks and the transportation of the dissolved material in solution. The effects of solution produce unique land forms that distinguish solution topography from other land forms.

Ground water is typically slightly acidic and reacts chemically with subsurface rocks, especially those that are highly soluble, such as carbonate rocks (limestone, marble), dolomite (magnesium carbonate), gypsum (calcium sulphate) and some evaporates. Extensive solution of such rocks produces extraordinary landforms grouped under general category of *karst*, characterised by numerous closed depressions, caves and various collapse features.

There are number of karst landforms features known today. These are:

- *Sinkholes* are small, shallow, circular to oval, closed depressions formed by downward solution of limestone from the surface or by collapse of the roof of a solution cavity.

- *Sinkhole ponds* and *karst lakes* are sinkholes that have filled with water.
- *Karst valley* is a subterranean drainage system which develops in the underlying limestone and surface flow becomes increasingly diverted underground.
- *Caves* are elongated cavities in limestone produced by solution, aided by mechanical erosion of subterranean flowing water.
- *Cockpits* are large, bowl-shaped sinkholes up to about 1 km in diameter that occur in thick limestones of tropical climates.

The relative relationship of ground water storage versus hydrological cycle is shown in Fig. 27.4 (page 595).

27.6 EVAPORATION AND TRANSPIRATION

Between periods of rain, water held in the soil is gradually given back to the atmosphere by a twofold drying process. First, direct evaporation into the open air occurs at the soil surface. Air also enters the soil freely and may actually be forced alternately in and out of the soil by atmospheric pressure changes. Even if the soil did not "breathe" in this way, there would be a slow diffusion of water vapour surfaceward through the open soil pores.

Second, plants draw the soil water into their systems through vast networks of tiny rootlets. This water, after being carried upward through the stem and branches into the leaves, is discharged through leaf pores into the atmosphere in the form of water vapour. The process is termed **transpiration.**

In studies of plant ecology and hydrology, the term **evapotranspiration** covers the combined moisture loss from direct evaporation and the transpiration of plants. The rate of evapotranspiration slows down as soil water supply becomes depleted during a dry summer period, because plants employ various devices to reduce transpiration. In general, the less water remains, the slower is the loss through evapotranspiration.

Fig. 27.5 illustrates the various terms explained up to this point and serves to give a more detailed picture of that part of the hydrologic cycle involving the soil. The soil layer from which plants can draw

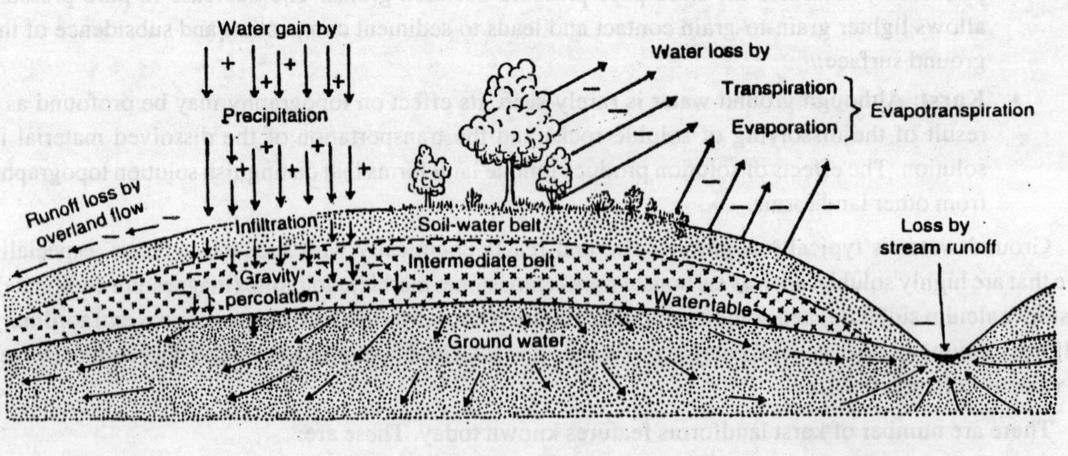

Fig. 27.5: Water flow pattern in soil

moisture is the **soil-water belt**. This belt gains water through precipitation and infiltration. As the minus signs show, the soil loses water through transpiration, evaporation and overland flow. Excess water also leaves the soil by downward **gravity percolation** to the ground water zone below. Between the soil-water belt and the ground water zone is an **intermediate belt**. Beneath upland areas, water held in the intermediate belt lies too deep to be returned to the surface by evapotranspiration, since it is below the level of plant roots.

27.7 ARTIFICIAL RECHARGE OF GROUND WATER

In most low rainfall areas of the country the availability of utilizable surface water is so low that people have to depend largely on ground water for agriculture and domestic use. Excessive ground water pumping in these areas, especially in some of the 91 drought prone districts in 13 states of India, has resulted in alarming lowering of the ground water levels. The problem has been further compounded due to large-scale urbanization and growth of mega cities, which has drastically reduced open lands for natural recharge. In hard rock areas there are large variations in ground water availability even from village to village.

In order to improve the ground water situation it is necessary to artificially recharge the depleted ground water aquifers. The available techniques are easy, cost-effective and sustainable in the long term. Many of these can be adopted by the individuals and village communities with locally available materials and manpower.

Following are the main advantages of artificially recharging the ground water aquifers:

- No large storage structures needed to store water. Structures required are small and cost-effective.
- Enhance the dependable yield of wells and hand pumps.
- Negligible losses as compared to losses in surface storages.
- Improved water quality due to dilution of harmful chemicals/salts.
- No adverse effects like inundation of large surface areas and loss of crops.
- No displacement of local population.
- Reduction in cost of energy for lifting water especially where rise in ground water level is substantial.
- Utilizes the surplus surface runoff which otherwise drains off.

The methods of artificial recharge can be broadly classified as:

- Spreading Method
  - Spreading within channel
  - Spreading stream water through a network of ditches and furrows
  - Ponding over large area
    - (a) Along stream channel viz., Check Dams/Nala Bunds
    - (b) Vast open terrain of a drainage basin viz., Percolation Tanks
    - (c) Modification of village tanks as recharge structures.
- Recharge Shafts
  - Vertical Shafts
  - Lateral Shafts
- Injection Wells

- Induced Recharge
- Improved Land and Watershed Management
  - Contour Bunding
  - Contour Trenching
  - Bench Terracing
  - Gully Plugging

In alluvial as well as hard rock areas, there are thousands of dug wells, which have either gone dry, or the water levels have declined considerably. These dug wells can be used as structures to recharge the ground water reservoir. Storm water, tank water, canal water etc. can be diverted into these structures to directly recharge the dried aquifer. By doing so the soil moisture losses during the normal process of artificial recharge, are reduced. The recharge water is guided through a pipe to the bottom of well, below the water level to avoid scouring of bottom and entrapment of air bubbles in the aquifer. The quality of source water including the silt content should be such that the quality of ground water reservoir is not deteriorated. Schematic diagrams of dug well recharge are given in Fig. 27.6.

In urban and rural areas, the roof top rainwater can be conserved and used for recharge of ground water. This approach requires connecting the outlet pipe from rooftop to divert the water to either existing wells/tubewells/borewells or specially designed wells. The urban housing complexes or institutional buildings having large roof areas can be utilised for harvesting roof top rainwater for recharge purposes.

Fig. 27.6: Recharge of dug wells through roof top rain water harvesting

27.8 GROUND WATER POLLUTION: GEOLOGICAL CONSIDERATION

Ground water is polluted by a number of reasons, viz., geological reasons, industrial contamination or even by man-made cultural activities.

The major geological conditions and the potential ground water contamination is shown in Table 27.1. The common ground water contaminants are iron, arsenic, fluoride, nitrate and several other trace metals. Urban and industrial waste dump sites also contaminate the ground water. Subsurface shallow ground water is also contaminated with pathogenic bacteria and parasites.

Table 27.1: Geological Conditions and Potential Ground Water Contamination

River and Stream Valleys	Karst Areas	Areas with a thin surficial Aquitard	Areas with a thick surficial Aquitard
Contamination Potential			
High: Alluvial Aquifer	High: Bedrock aquifer	High: Drift aquifers, Bedrock aquifers	High: Shallow drift aquifers
	Low: Deeprock aquifer	Low: Deeprock aquifer	Low: Deep drift aquifers, Bedrock aquifer, Deeprock aquifer
Zone of free contamination	Zone of contamination by surface run-off & infiltration	Zone of possible contamination by infiltration	

There are various possibilities of ground water sources by landfill sites through leaching processes as shown in Fig. 27.7.

In coastal areas, man-made activities like aquaculture induces the salt water intrusion to ground water aquifers. This is primarily due to change in water table by rapid withdrawal of ground water (Fig. 27.8).

27.9 WATER RESOURCE PLANNING FOR FUTURE

The drinking water crisis in Indian cities has reached explosive proportions. For years, urban dwellers have been silently suffering an intermittent supply of poor quality water. Now their patience has run out. Water riots have already occurred in several cities and could be repeated in many more.

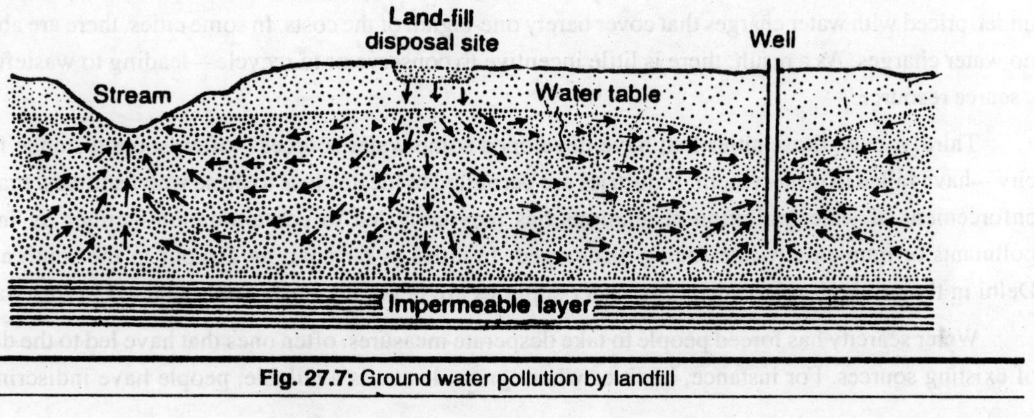

Fig. 27.7: Ground water pollution by landfill

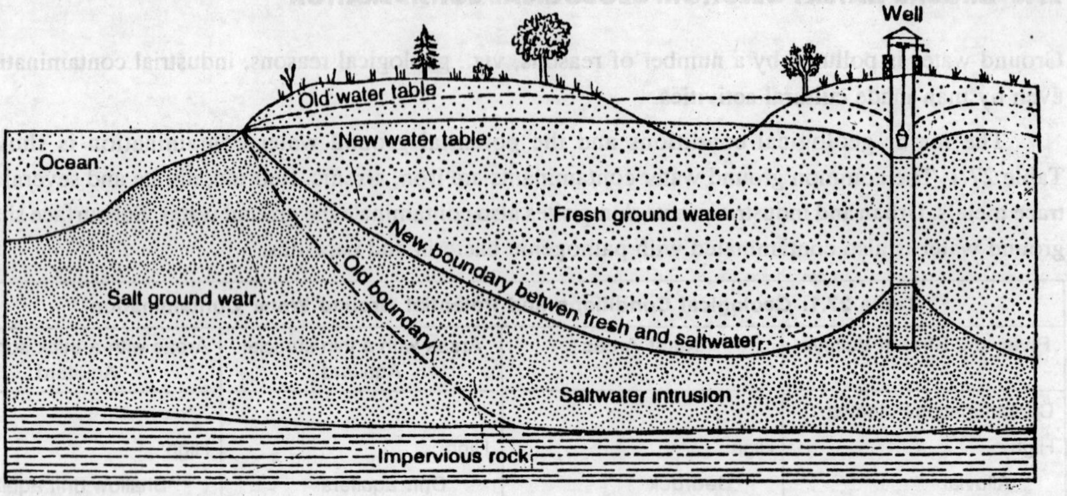

Fig. 27.8: Water table change due to rapid ground water abstraction

On paper, over 70 percent of the population of Class I towns—including the metropolitan cities—is supposed to be covered by the public water supply system. In fact, as the reports in this section illustrate, people living in even the most generously supplied city experience acute shortages largely due to an iniquitous water distribution network which supplies disproportionate quantities of water to richer areas while starving the poor.

The reasons for water scarcity in urban areas can be placed in three broad categories. First, a lack of foresight and bad planning. Most water supply systems were installed decades ago for a small urban population. Today, they are old, badly maintained and grossly inefficient. In some cities, up to 20 per cent of water is wasted due to leakages. Where sewage networks run alongside water pipes, there is cross-contamination leading to epidemics.

Secondly, inadequate financial resources have held up the modernisation and extension of urban water supply networks. But the fault lies not just with external funding sources but also with Central and state governments, who have failed to anticipate the rapid growth in urban populations and have also not devised strategies to recover their investment on water. In the majority of Indian cities, water is grossly under-priced with water charges that cover barely one-eighth of the costs. In some cities, there are absolutely no water charges. As a result, there is little incentive to conserve or to recycle—leading to wasteful use of a scarce resource.

Thirdly, even those cities that have developed some water sources—often some distance from the city—have failed to protect them. Industries are permitted to locate close to these water sources. Inadequate enforcement of water pollution regulations has inevitably led to contamination—either by industrial pollutants or municipal wastes being discharged. In cities as far apart as Tirupur in the south and New Delhi in the north, domestic drinking water supplies are being regularly contaminated in this way.

Water scarcity has forced people to take desperate measures, often ones that have led to the depletion of existing sources. For instance, in cities where groundwater is available, people have indiscriminately

Table 27.2 : Integrated Water Resources Environmental System & Planning or SYSTEM APPROACH

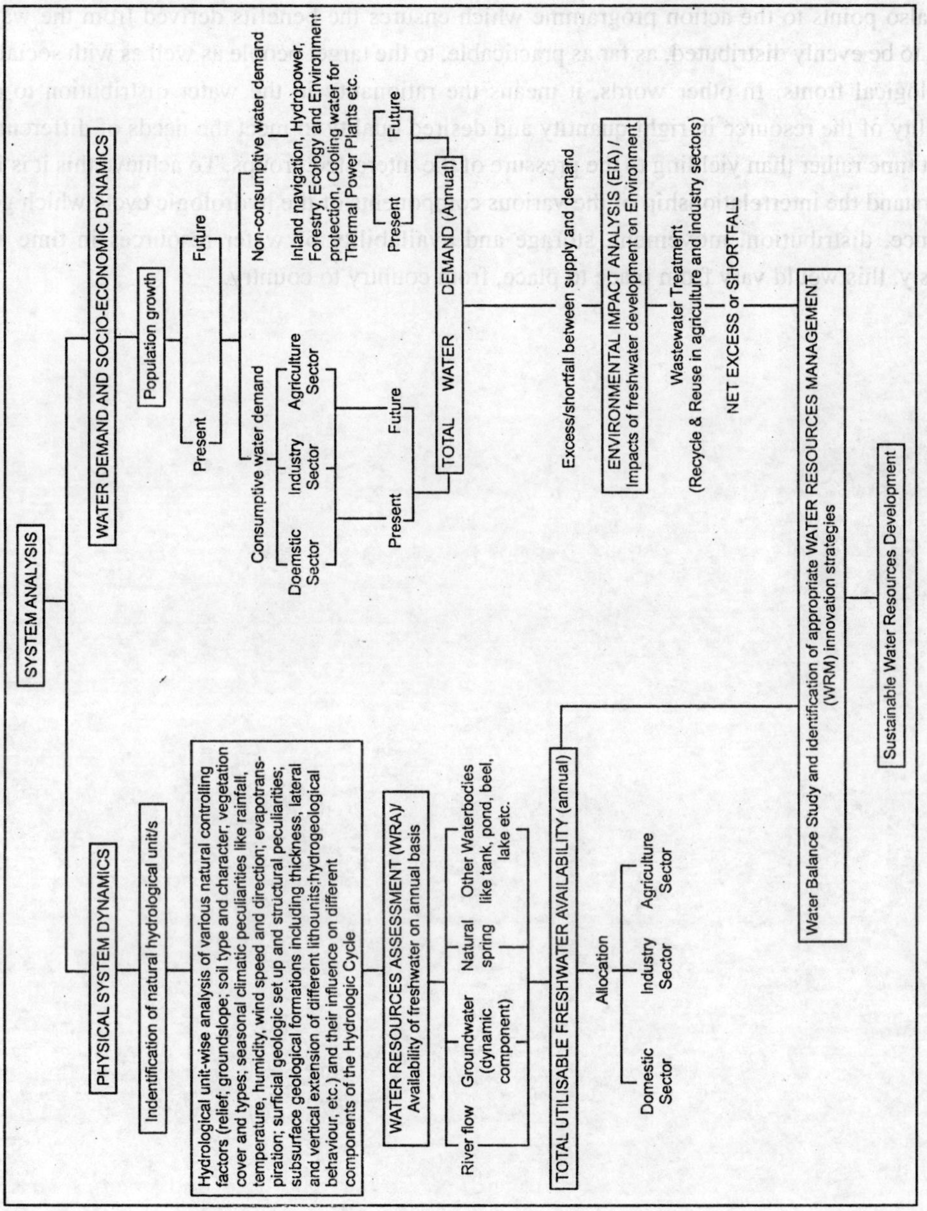

dug shallow and deep wells, leading eventually to groundwater supplies being exhausted. Elsewhere, selling water has become a lucrative business. On rickshaws, bicycles or in road tankers, private suppliers ferry water from outside city limits and sell it per bucket to the city's thirsty populace.

Water Resource Management means rational and effective management—at economically viable and ecologically sustainable development—of available resources in the contemporary setting without causing deliterious effects on the basic life support system and without endangering the resource base which is already under stress. An overall outline of Integrated water resource planning of an area is shown in Table 27.2.

It also points to the action programme which ensures the benefits derived from the water-related projects to be evenly distributed, as far as practicable, to the target people as well as with social, physical, and biological fronts. In other words, it means the rationality in the water distribution to ensure the availability of the resource in right quantity and desired quality to meet the needs of different sectors at the right time rather than yielding to the pressure of the interested groups. To achieve this it is imperative to understand the interrelationship of the various components of the hydrologic cycle which governs the occurrence, distribution, movement, storage and availability of water resources in time and space. Obviously, this would vary from place to place, from country to country.

28 ENVIRONMENT AND AGRICULTURE

28.1 INTRODUCTION

Over the course of the past 200 years, worldwide food production has generally kept pace with the exponential rise in the human population. The rapid increase in global food production has been brought about by very large increases—both in the productivity of existing farmland and in the amount of land that is being farmed. In addition, increased availability and efficiency of transportation and refrigeration have enabled people to ship food from one area or continent to another and, therefore, to alleviate problems arising from local drought or pestilence.

Though total food production of the earth has increased dramatically in recent years, yet, as the population rise is remarkably higher over past five decades (1950-2000), the per capita grain production will continue to decline. From the beginning of agriculture until the early 1900s, most increases in worldwide food production resulted from an expansion of the total area of farmland under cultivation. But, over the past couple of decades the mechanisation of farming processes lead to the sustained higher yield to meet up the growing food supply demand. High agricultural yields require specific varieties of seeds (high-yielding varieties), an active fertiliser programme, and pest control. In many regions, irrigation is also essential to high yields. The crop land area of world is given in Table 28.1.

Table 28.1: Cropland area – 2005

	Area (Mha)		Cropland as %
	Rainfed	Irrigated	of Total Land Area
North America	206	23	13%
Western Europe	76	13	22%
Pac OECD	54	6	7%
FSU	183	21	10%
Eastern Europe	55	10	37%
Africa	218	13	8%
Latin America	146	19	8%
Middle East	25	18	7%
China	102	59	14%
South Asia	124	81	50%
Southeast Asia	75	14	21%
World	1,266	277	12%

In the mid 1960s, an interdisciplinary team of scientists working in Mexico and the Philippines developed new varieties of wheat and rice that were adaptable to tropical climate and were capable of producing higher yields than any native grains. These varieties have short, upright leaves, so that plants can be grown closer together without shading each other. In addition, their stalks are short and thick so that they will not bend and break when the plant is heavily fertilized. The potential yields of these new varieties are so spectacular that many people heralded their introduction as the '*Green Revolution*'.

The processes of Green Revolution have three important steps:

(a) Developing and planting monocultures of selectively bred or genetically engineered high-yield varieties of key crops such as rice, wheat and corn;

(b) Lavishing fertiliser, pesticides and water on crops to produce high yields, and

(c) Often increasing the intensity and frequency of cropping.

Identically, during the past 50 years, the global livestock population has exploded as increased affluence has led to rising per capita consumption of meat and dairy products.

Environmentalists have now realised the cause and consequences of industrialisation in agricultural sectors (Table 28.2) and thus proposed sustainable agriculture. To many environmentalists, the key to reducing world hunger, poverty and the harmful environmental effects of both industrialized and traditional agriculture is to develop a variety of systems of sustainable agriculture.

The general guidelines for sustainable agricultural systems suggested by various analysis include:

Table 28.2: Environmental Impacts of Agriculture
Clearing vegetation
• Regional hydrology and probably local climate altered.
• Habitat damage leading to wildlife losses.
• Loss of biodiversity.
• Significant contribution to greenhouse effect.
• Air pollution.
Tillage
• Erosion (greater than stream silt loads).
• Altered groundwater (saline seeps).
• Dust.
• Soil structural changes.
• Loss of soil organic carbon.
• Altered soil flora and fauna.
• Altered albedo (weather changes?).
Irrigation
• Polluted return flows (stream pollution).
• Raised groundwater (waterlogging and salinity).
• Soil structural changes.
• Altered soil flora and fauna.
• Possibility of reduced soil erosion.
• Disease vector problems.
• Reduced streamflow and ground water due to extraction and evapotranspiration.
• Methane generation from paddy-fields, etc. (Greenhouse Effect).
Altered drainage
• Pollution from land drains.
• Polluted groundwater (saline seeps).
• Ground water raised (waterlogging and salinity).
• Ground water lowered (loss of soil organic carbon and subsidence).
• Genetic loses as wetlands destroyed.
Agrochemicals
• Pesticides: local, regional and global pollution possible.
• Pest resistance or loss of predators.
• Fertilisers: eutrophication of surface waters.

Contd.

Table 28.2: Environmental Impacts of Agriculture

Agrochemicals

- Nitrate/phosphate problems with ground waters.
- Alteration of soil structure.
- Alteration of soil flora/fauna.
- Greenhouse gases generated.
- Agricultural lime: alteration of soil flora/fauna and changes to aquatic flora/fauna (agrochemicals may affect marine life).
- Genetic losses as wildlife poisoned.

Agricultural 'wastes'

- Straw/sugarcane burning: air pollution and loss of soil nutrients, especially nitrogen—much depends on when and under what environmental conditions it is burnt.
- Processing crops: stream/ground water pollution.
- Livestock wastes: water pollution (genetic losses); smell; methane generation (greenhouse effect).
- Silage effluent: water/air pollution.
- Noise: aesthetic and wildlife damage.

Pest control/hunting

- Reduces wildlife (including malaria-vector control impacts).

Pasture damage

- Over/undergrazing.
- Burning to encourage sward.
- Receding/improving pasture.
- Genetic losses.

Marine and freshwater aquaculture

- There were 460 coastal salmon farms and 400 onshore trout farms in UK alone in 1992.
- Pesticides used to control fish-lice, also antibiotics, fungicides and additives to control growth (formaldehyde, nuran, dichlorovos pesticide and malachite green used).
- Waste nutrients may be a problem.
- Genetic erosion of wild stock when fish escape.

Biotechnology

- Risk of uncontrolled escape, terrorism or military use of organisms or genetic material.
- Risk of increased poverty for those who cannot adopt biotechnology leading to poor farming or use of marginal land, both leading to land degradation.
- Promotion of new crops may discourage growing of traditional varieties—loss of genetic diversity.
- Possibility that biotechnology might help reduce pressures on agricultural land.
- Combine traditional high-yield polyculture and modern monoculture methods for growing crops.
- Learn sustainable forms of crop-growing from successful traditional farmers and transmit knowledge of what works to other farmers.
- Grow more perennial crops.
- Shift to full-cost pricing that includes the harmful environmental effects of agriculture in food process.
- Minimise soil erosion, salinisation and waterlogging.
- Reduce destruction of natural forests, grasslands and wetlands for producing foods by emphasising increased yields per area of cropland using sustainable methods.
- Reduce water waste in irrigation.
- Reduce use and waste of fossil fuels.
- Increased use of organic fertilisers, solar, wind and biomass energy to grow and process crops.
- Increase use of biological pest control and integrated pest management.
- Subsidize sustainable farming and gradually withdraw subsidies for unsustainable farming.
- Reduce poverty.
- Slow population growth.

Box 28.1: Pesticides in India

Pesticides banned in some countries, but used in India[1]	Pesticides banned in India[2]
Aluminium Phosphide	Aldicarb
Benomyl	Aldrin
Captan	Benzene Hexachloride (BHC)
Carbaryl	Calcium cynide
Carbofuron	Captafol 80% powder*
Dicofol	Chlordane
DDT	Chlorobenzilate
Dimethoate	Cibromochloropropane
Diuron	Copper Acetoarsenite
Endosulfan	Dieldrine
Fenarimol	Endrin
Fenpropathrin	Ethylene Dibromide
Lindane	Ethyl Mercury Chloride
Linuron	Ethyl Parathion
Malathion	Heptachlor
Methomyl	Maleic Hydrazide
Methoxy Ethyl Mercury Chloride	Menazone
Methyl Parathion	Mehtomyl 12.5% L
Monocrotophos	Mehtomyl 24% L
Oxyfluorfen	Nicotine Sulphate*
Paraquat Dichloride	Nitrofen
Phorate	Paraquate Dimethyl Sulphate
Phosphamidon	Pentachloro nitrobenzene (PCNB)
Pretilachlor	Pentachlorophenol (PCP)
Sodium Cyanide	Phenyl Mercury Acelate (PMA)*
Triazophos	Phosphamidon 85% SL
Tridemorph	Sodium Methane Arsonate (MSMA)
Thiomethon	Tetradifon
Thiram	Toxafen
Zinc Phosphide	Trichloro acetic acid (TCA)
Ziram	

*These pesticides are manufactured in India for export only.

Source: 1 Lok Sabha Unstarred Question No. 2291, Dated 18.03.2002, LDPPQ & S, Faridabad

According to World Resource Institute, human exposure to pesticides in India is one of the highest in the world. According to worldwide estimates, about 3 million people suffer from acute pesticide poisoning every year, of which 200,000 die—mostly in countries like India. India accounts for one-third of the pesticide poisoning cases in the Third World.

Box 28.2: DEALING WITH AGRICULTURAL POLLUTANTS

Technologies to remove nitrates and pesticides from water sources are considered expensive even in the developed world. The most effective methods for pesticide removal are the activated carbon method, and reverse osmosis. In the carbon method, large scale treatment plants use granulated activated carbon or a combination of ozone followed by activated carbon. Ozone—being highly reactive—oxidises the pesticides in water and the products are removed by carbon. This is a high cost and high-energy consumption technology. In India, the conventional methods used for treatment of water cannot remove water-soluble pesticides. No estimates have been made as yet about the extent of the problem, the volume of water that would have to be treated and the costs that would be involved.

Preventive measures need to be undertaken urgently involving a gradual decrease in the use of chemical fertilisers and pesticides and promotion of bio-pesticides and organic fertilisers. Bio-pesticides are less expensive to produce as well as affordable for farmers. Several institutions, government and non-government are making efforts to increase awareness for better management of pests. The government has introduced integrated pest management at the central and state levels for the purposes of plant protection. Twenty-five (1975-2000) central, integrated pest-management centres have been established. Other strategies include elaborate testing for pesticides, periodic reviews of pesticides, transparent process of review and registration of pesticides, and monitoring and reporting by environment officials.

Similarly controlled and scientific application of fertilisers can prevent the contamination of water sources. Enabling policy changes related to fertiliser management and usage need to be implemented.

28.2 THE FAILURE OF THE GREEN REVOLUTION

Modern intensive agriculture has conspicuously failed to increase food production and to meet global food and nutrition needs. The claim that the Green Revolution has led to higher crop yields is highly exaggerated and does not reflect a fair and complex comparison with more ecologically sound systems:

- These claims are usually based on the measurement of yield as defined per acre or hectare of land. However, if one takes into account the hidden costs of input subsidies and non-renewable resources and the costs of ecological damage (leading to lower yields after some time) and, furthermore, measure yield against high fertiliser and water costs, then the Green Revolution techniques are highly inefficient. In contrast, the economic soundness of traditional and ecologically better varieties is striking.
- Even more seriously, the Green Revolution measurement of output is flawed because it only accounts for a single crop (e.g. rice) and even then only a single component of that crop (e.g. grain) whilst neglecting the uses of straw for fodder and fertiliser. Thus it neglects to take into account that there were many other biological resources (e.g. other crops, other non-grain uses of the measured crop and fish) within the same land in the traditional system that were reduced or wiped out with the Green Revolution.
- If output is measured in terms of total biomass, a more realistic picture of the performance of the Green Revolution will emerge.

Although yields of food crops in total have increased, less food is available to local populations. There are several reasons for this:

- There has been an increase in a few cereals (a large volume of which is fed to cattle in north India) at the expense of pulses and other crops.
- The increased dependency of Third World farmers and countries on intensive inputs has led to indebtedness and the breakdown of self-sufficiency.

- Much of the increased food production is exported, thus denying the food to local people.
- Many areas planted with "high-yielding varieties" (which are actually "high-response varieties" to the applied inputs, including chemical fertilisers and pesticides) are now experiencing diminishing returns.
- Ecological degradation (see below) is leading to reduced yields and to the abandonment of many areas of agricultural land.
- Losses during storage have increased markedly in many areas.
- The low prices paid for farm produce and the high prices charged for food in the shops, combined with increased levels of indebtedness, ensure that many farmers cannot afford to buy sufficient food for their families.

Box 28.3: Conventional farming

In the early Twentieth century advances in agricultural sciences led to the emergence of modern farming. At the end of World War II, modern farming had spread over the entire world. Because of its prevalence it is considered to be "conventional". Modern farming brought a departure from ancient practices that, once upon a time, were in alliance with nature.

Conventional farming aims at maximizing the output of a narrow range of species. This is achieved by giving plants exactly what they need and removing everything else. Plants are supplied with synthetic fertilizers that suit the needs of the species grown in respect of soil and climate. Herbicides remove unwanted weeds, pesticides eliminate pests and other less harmful insects that come in search of food and habitat. Fungi are kept in check with fungicides. This approach looks at the plant in isolation from nature in a mechanical if not simplistic way.

It is not surprising that the imbalance caused in the method backfires. On one hand we incur a substantial loss of soil fertility, increasing vulnerabilities to pest attacks and a drastic loss of biodiversity. On the other hand, the mass introduction of pesticides resulted in their progressive entry into the ground water and human food supplies where they compromise human health. This self-punishment does not appear entirely undeserved. The systemic imbalances of conventional farming are now imagined to be mended with even greater impedances: genetically modified crops, which today stand at the brink of commercialisation.

28.3 IMPACT OF MODERNISATION IN AGRICULTURE

The global ecological, social, economic, and nutritional consequences of modern intensive agriculture—and the development policies that underlie and encourage its promotion—have been devastating:

- It is noted that the countries like USA, Britain, and Germany are primarily responsible for these consequences, since most modern agriculture is located there; whilst institutions controlled by them were responsible for spreading the Green Revolution to the Third World.

The ecological impacts include:

- The breakdown of those ecological processes that maintain soil fertility;
- The degradation of agricultural lands through erosion, desertification, salinisation, waterlogging and compaction etc., and the loss of land to industrial development projects, infrastructure programmes, water development projects etc.
- The depletion of ground water and other water sources due to over-use (notably through the planting of water-intensive crops and crop varieties) and to the disruption of those hydrological regimes that ensure the recharge of water sources.

Box 28.4: Natural farming

Beginning to doubt the wisdom of modern agricultural science, microbiologist, farmer and philosopher **Masanobu Fukuoka**, born in 1913 in Japan, looked for alternatives. Around 1940 he worked out an answer, which aims at the reestablishment of the natural ecosystem. It is based on the insight that nature is intrinsically in harmony with itself as long as there is no interference of man. Known as natural farming, it is also referred to as Fukuoka method or, in India, as *Rishi Kheti* meaning agriculture adopted by the sages. The concept aims at reproducing natural conditions that improve soil productivity and growth of plants.

It surprises perhaps most because of its simplicity. It does not entail entirely new or innovative methods of growing. It disagrees with the general dictums of modern agricultural science. And it says that if you want to grow crops and vegetables in harmony with nature what you actually need is very, very little.

No pesticides: Natural farming is opposed to the use of pesticides. Apart from killing insects and other organisms, pesticides enter the soil and fruit and cause serious health problems when absorbed by humans.

No herbicides: In this method, weeds are used rather than killed. Wild growing grass such as rye and clover are grown for mulching. Natural farm orchards are green with grass growing between the trees. The grass prevents soil erosion, holds moisture, propagates microorganisms, produces organic fertilizers, improves soil ventilation and suppresses pests.

No fungicides: Natural farming recognizes that indigenous microorganisms and fungi are an integral part of the ecosystem. They have an important role to play and hence, instead of eliminating them, they need to be respected and protected.

No tillage: Land is not being ploughed. Instead, earthworms and small animals dig through the soil if they are not disturbed. And while mechanized ploughs drawn by tractors can plough up to 20 cm deep, fragile tiny earthworms dig a surprising 7 meters deep. The key to their services is to allow them to work by not interfering with them. The excretions of earthworms turn into the best soil. Because of the absence of tillage the grass seeds in the soil do not come up to the surface as they germinate and die. There are less weed problems than on tilled land.

No chemical fertilizers: Nitrogen, phosphorous, potassium, calcium and other common elements of chemical fertilizers are substituted with inputs from exclusively natural sources. For instance, fish amino acid provides nitrogen, eggshells supply calcium and animal bones provide phosphoric acid. These inputs are not only cheap but highly effective.

No animal waste: The liquid waste from animal stables is not discharged on the land or rivers outside. The treatment takes place right on the floor which is not made of concrete but of soil on which rice straw, sawdust, and fresh soil is strewn. Feces decompose quickly through inherent microorganisms. The combination of sunlight, air circulations and microorganisms keeps the floor dry and fluffy. There is no need of cleaning the area frequently as the waste does not pile up and it leaves very little smell. No installation of any expensive material or equipment is required and the problem of livestock wastewater entering water bodies does not arise.

Natural farming takes into account the nutritive cycle and identifies the growth stages of plants and animals. Fertilizers, soil improvers, pest controls, disease cures—all made by the farmers using natural materials—are supplied as and when needed. Farmers make what they need with ingredients available in nature. Nothing needs to be bought.

- The contamination of land and water supplies through chemical pollution.
- The eutrophication of waterways.
- The destruction of fisheries through pollution and erosion.
- The loss of habitat for wildlife.
- The erosion of genetic resources.
- An increase in the virulence of pest infestations and plant diseases, partly due to the increased resistance of pests to chemical pesticides.
- The raising of animals in disease and stress-ridden conditions through factory farming, requiring massive amounts of pesticides, antibiotics, hormones and energy.

- The dumping of dangerous agricultural chemicals forbidden in Europe.
- The generation—in the manufacturing process of pesticides and fertilisers—of highly toxic wastes which are dumped in the poorest among Third World countries.
- A growing increase in emissions of greenhouse gases. Such emissions from agriculture and industry are greatest in Europe, but are increasing in the Third World as agriculture becomes more dependent on fossil fuel-based inputs (chemical fertilisers, for example, are an important source of nitrous oxide) and as irrigated rice production is intensified (paddies are an important source of atmospheric methane).

Although they vary from region to region, the social impacts of modernising agriculture include:

- Undermining the confidence of farmers in their own abilities and in the value of their traditional knowledge, causing them to become increasingly dependent on outside expertise.
- The break-up of family farms and farming communities, impoverishing the social life of millions and fermenting growing social alienation.
- Increased rural unemployment.
- Increased rural-urban migration and rural depopulation.
- Increased landlessness.
- Increased indebtedness, due in part to the increasing dependence of farmers on purchased inputs.
- Increasing the burden on women. In many cultures, women have been displaced from their central role in food production. The access they previously enjoyed to land and other needed resources has been undermined and their participation in decision-making eroded.
- The transformation of independent farmers into often poorly paid and overworked plantation and industrial workers.
- The displacement of tribal and indigenous people from their traditional homelands, often due to the construction of large dams located on their lands, without adequate and fair alternative arrangements for their livelihood.

The economic impacts of modernising agriculture include:

- Increasing poverty in rural areas where farmers have been encouraged to adopt farming methods which have increased their dependence on external inputs and credit and that have now proven to be uneconomic and non-sustainable.
- Increasing national indebtedness in the Third World, leading to the imposition of structural adjustment programmes that further exacerbate the plight of the poor and dispossessed.
- Large spendings by Third World countries on subsidies for agricultural inputs, mainly imported from the industrialised countries, which hide the increasingly uneconomic returns of modern agriculture.
- Increasing the vulnerability of national economies, many of which are ever more dependent for their income on a small number of export crops.
- Increasing the domination of agriculture by a few international and national corporations and increasing their control over inputs, marketing, seeds, and land.
- Increasing the dependence of agriculture on imported non-renewable fossil fuels, chemicals, seeds, and machines.

Box 28.5: Organic farming

The fact is not widely known yet the roots of organic farming lie indeed in India. **Sir Albert Howard**, an English botanist born in 1873, worked in India as agricultural advisor between 1906 and 1931. Stationed in Indore he studied and gradually adopted cultivation practices from local peasants and began to promote composting to restore precious nutrients in the soil. Upon his return he developed the methods of organic farming further, documented them and began with their promotion through the Soil Association in England and the Rodale Institute in the USA.

In the 1940s organic farming emerged as a broad movement in response to the problems of conventional farming. It relies on "green" principles such as crop rotation, biological pest control, composting and manure that aim to maintain soil productivity. It promotes practices in line with nature, rather than against it. The use of synthetic pesticides, plant growth regulators, livestock feed additives and genetically modified organisms is banned, save a few and tightly controlled exceptions where "green" answers to controlling pests are yet to be found.

Organic agriculture contributes to sustainable development. The products, although more expensive, are generally tastier and sometimes more nutritious than their conventional counterparts. Local and international markets have tremendous growth prospects and offer creative producers opportunities to improve income and living conditions. Starting in the 1970s, a multitude of certification systems for organically grown produces were introduced in many countries to identify genuinely organic products in an increasingly complex market.

Now organic farming is being practiced in 100 countries of the world. The global organic farming area is 26 million hectare roughly alongwith 26 billion US$ market.

The nutritional and public health impacts of modern agriculture include:

- A reduction in the range of foods consumed, to the detriment of nutrition.
- An increasing incidence of pesticide poisonings—some of them fatal—on farms.
- Increasing cancer rates and other health problems, due to exposure to pesticide residues in food, water, and air.
- A reduction in the nutritional quality of food.
- An increase in the consumption of devitalised processed foods, a pattern that is encouraged by aggressive advertising and marketing.
- An increase in waterborne diseases in areas of intensive irrigation and associated water projects.
- The creation of new strains of bacteria which are resistant to antibiotics—thereby increasing the threat of human infectious diseases.
- An increase in the number of accidental and intentional poisonings due to the ready availability of agricultural poisons in the home.

The adverse impacts of modern agriculture in the Third World are likely to be exacerbated still further by current moves to:

- Further curtail proposals for equitable treatment of trade in agriculture under the General Agreement of Tariffs and Trade (GATT) and international patenting agreements.
- To implement a new regime on the use of patent and other intellectual property rights to tighten the control of corporations over resources and products.

This shift from common property use in traditional cultures to private ownership further reduces the access and control of ordinary farmers.

- Millennia of innovations and experience of farmers are turned into private property rights. This trend is most evident in the field of biotechnology.

Box 28.6: Biodynamic farming

Biodynamic agriculture was introduced by philosopher and educationalist **Dr Rudolf Steiner**, born 1861 in Austria. In a series of lectures in 1924 he offered suggestions to farmers concerned about the apparent loss of soil fertility in conventional farming. His approach, which predates the organic movement, focused squarely on the restoration and nourishment of soil fertility. While the spectrum of methods share a broad sameness with natural and organic farming, it also uses certain unique methods. These are based on the belief that a consonance of the relationships of soil, plants, animals and man as a self-nourishing system can be achieved better with the aid of medicinal properties and cosmic and terrestrial "forces".

Like natural and organic farming, biodynamic farming excludes synthetic chemicals and emphasizes the use of manure and compost. The use of specific compost preparations however is unique. These preparations comprise of selected plants (e.g. *Achillea millefolium, Matricaria recutita*) which for a period of time are buried underground alongside peat or animal intestines (minerals). Upon fermentation they are applied to compost in homeopathic quantities. Biodynamic farming follows an astrological calendar for planting, cultivation and harvesting. It considers soil and farm as living organisms.

Biodynamic farming has been adopted in most European countries, the USA, Australia and in some Asian countries. It was introduced to India in the 1990s and is practiced today by about 500 farmers across the country. Indian biodynamic farmers grow papaya, pepper, cocoa, vanilla, sugar cane, cotton, wheat and in Darjeeling and the Nilgiris, tea.

28.4 ALTERNATIVES EXIST

Numerous well-proven systems of ecological agriculture exist throughout the world, many of them combining the insights of modern holistic science with the wisdom of traditional practices. They include:

- Permaculture (*perma culture*) [i.e. permanent agriculture].
- No tillage systems of grain cultivation, as developed in Japan.
- Conventional organic farming systems.
- Bio-intensive agriculture (the "double digging" method).
- Bio-dynamic farming.
- Sustainable systems using perennial varieties.
- Systems employing companion planting, etc.

Such systems are economically viable and productive:

- In Nepal, the cumulative yield of well-developed permaculture systems are higher than conventional systems.
- In California, double-digging systems have been capable of feeding a family of six on a area of 100 square meter.
- In Japan, the yields of rice and wheat under 'no tillage systems' compare with high-tech modern systems.

In addition, because their success depends in large part on cooperation at the farm level and beyond, such systems have the potential to provide individual farmers and their families with a cohesive and supportive community. This will however depend critically on addressing the wider social, economic and political forces that underlie hunger. One should not forget that the slave plantations of the past used non-chemical systems of agriculture.

Box 28.7: Permaculture

Perhaps the youngest method of all, permaculture was developed during the 1960s. The Austrian farmer **Josef "Sepp" Holzer**, born 1942, was its first known practitioner. The theoretical development came only in the 70s and was led by two Australians: **Bill Mollison**, born 1928, and **David Holmgren**, born 1955. Their main interest was to develop a stable agricultural system in light of the problems intrinsic to conventional farming which had already acquired industrial proportions. Given its relatively young age and interest in a holistic, sustainable approach, it is not surprising that permaculture was inspired by the ideas and exponents of other postconventional agricultural methods. While Holzer appears to have worked on his own, Mollison and Holmgren benefitted from global networks in which experiences with non-conventional methods including natural farming were exchanged.

The result led to something larger then than a mere farming method: An approach to agriculture and human settlements modeled on the natural ecosystem and the interdependencies between its members. Permaculture strives for an integration of dwellings with plants, animals, soil and water into stable, productive and harmonious communities. The focus is not on any of these elements themselves, but on the quality of the relationships created between them.

28.5 SUSTAINABLE AGRICULTURE — AN AGROECOSYSTEM APPROACH

Over the few decades of modern agricultural activities, it was realised that sustainability of production and consequent environment damage was rampant. As such, there needs an alternative practice where a more ecological rationale will be adopted into sustaining agricultural production. This is a science of agroecosystem, where genetics, agronomy and edaphology will be discussed as deep understanding of ecological and social levels of coevolution. Agroecology encourages researchers to tap into farmers' knowledge and skills and to identify the unlimited potential of assembling biodiversity to create beneficial synergisms that provide agroecosystems with the ability to remain or return to an innate state of natural stability.

The contemporary use of the term agroecology dates from the 1970s, but the science and the practice of agroecology are as old as the origins of agriculture. As researches explore indigenous agricultures—which are modified relics of earlier agronomic forms—it is increasingly apparent that many locally developed agricultural systems routinely incorporate mechanisms to accommodate crops to the variability of the natural environment and to protect them from predation and competition. The mechanisms make use of regionally available renewable inputs and ecological and structural features of the agricultural field, fallows and surrounding vegetation.

Each region has a unique set of agroecosystem that results from local variations in climate, soil economic relations, social structure, history. Thus a survey of the agroecosystems of a region is bound to yield both commercial and subsistence agricultures, using high or low level of technology, depending on the availability of land, capital and labour.

A list of agroecosystem determinants is given in Table 28.3 (page 614). The objectives and processes in the design of a model of sustainable agroecosystem in shown in Fig. 28.1 (page 615). The importance of agroecological system in contrast to Green Revolution is shown in Table 28.4 (page 614-615).

Table 28.3: Agroecosystem Determinants that Influence the Type of Agriculture in Each Region

Types of Determinants	Factors
Physical	Radiation Temperature Rainfall, water supply (moisture stress) Soil conditions Slope Land availability
Biological	Insect pests and natural enemies Weed communities Plant and animal diseases Soil biota Background natural vegetation Photosynthetic efficiency Cropping patterns Crop rotation
Socio-economic	Population density Social organisation Economic (prices, markets, capital, and credit availability) Technical assistance Cultivation implements Degree of commercialisation Labour availability
Cultural	Traditional knowledge Beliefs Ideology Gender issues Historical events

Table 28.4: Comparison between Green Revolution and Agroecological Technologies

Characteristic	Green Revolution	Agroecology
Technical		
Crops affected	Wheat, maize, rice and few others	All crops.
Areas affected	Mostly flatlands and irrigated areas	All areas, especially marginal areas (rainfed, steep slopes)
Dominant cropping system	Monocultures, genetically uniform	Polycultures, genetically heterogeneous
Dominant inputs	Agrochemicals, machinery; high dependency on external inputs and fossil fuels	Nitrogen fixation, biological pest control, organic amendments, high reliance on local-renewable resources
Environmental		
Impacts and health hazards	Medium to high (chemical pollution erosion, salinization, pesticide resistance, etc.). Health risks in pesticide application and pesticide residues in food.	Low to medium (nutrient leaching from manure).
Crops displaced	Mostly traditional varieties and land races.	None.

Contd.

Table 28.4: Comparison between Green Revolution and Agroecological Technologies

Characteristic	Green Revolution	Agroecology
Economic		
Capital costs of research	Relatively high.	Relatively low.
Cash needs	High. All inputs must be purchased in the market.	Low. Most inputs are locally available.
Cash returns	High. Rapid results. High labour productivity.	Medium. Needs time to achieve highest yields. Low to medium labour productivity.
Institutional		
Technology development	Quasi-public sector, private companies.	Largely public; large NGO involvement.
Proprietary considerations	Varieties and products patentable and protectable by private interests.	Varieties and technologies under farmer's control.
Sociocultural		
Research skills needed	Conventional plant breeding and other disciplinary agricultural sciences.	Ecology and multidisciplinary expertise.

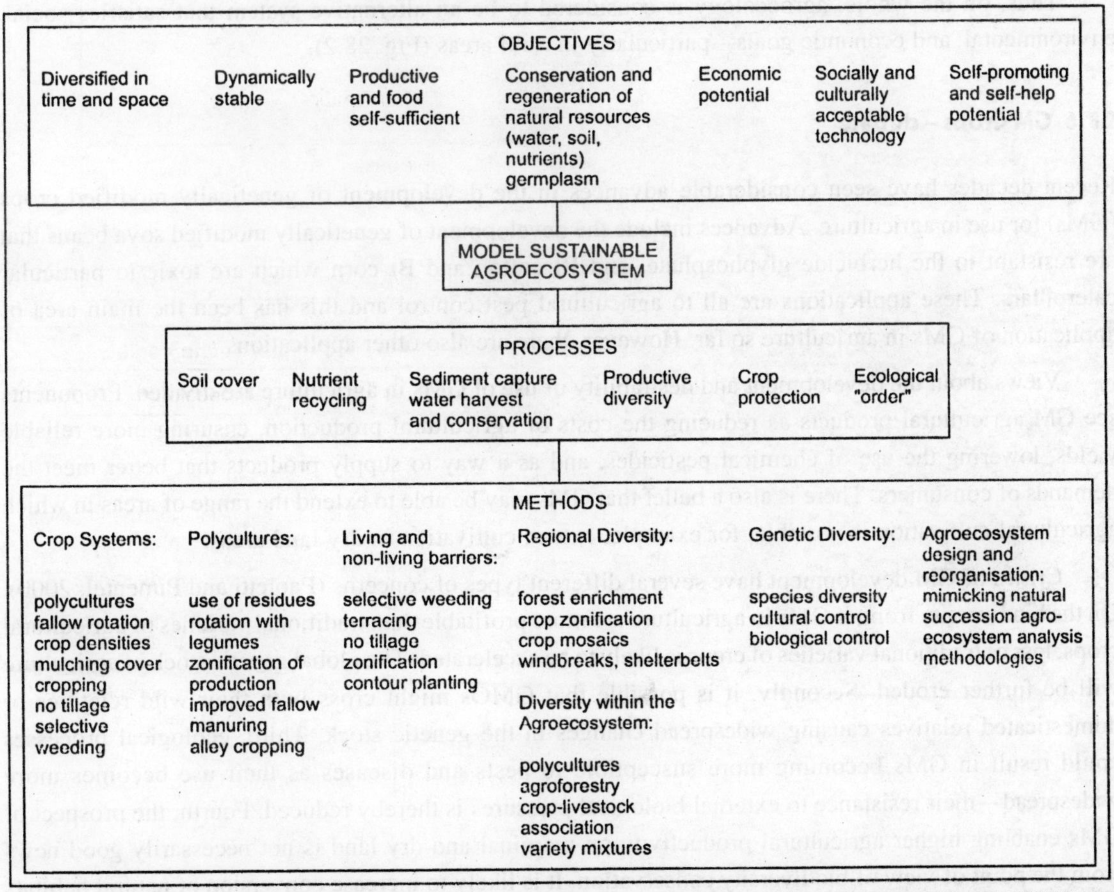

Fig. 28.1: Objectives and processes in the design of a model sustainable agroecosystem

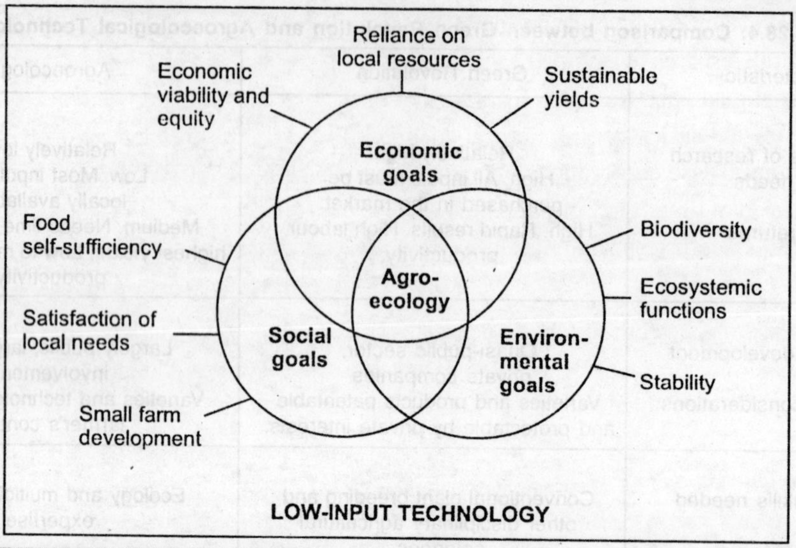

Fig. 28.2: Role of agroecology in satisfying social environmental and economic goals in rural areas

Thus, on the whole, agroecology is considered to be an alternative system that satisfies social, environmental, and economic goals—particularly in rural areas (Fig. 28.2).

28.6 GM Crops – debate

Recent decades have seen considerable advances in the development of genetically modified crops (GMs) for use in agriculture. Advances include the development of genetically modified soya beans that are resistant to the herbicide glyphosphate, and Bt cotton and Bt corn which are toxic to particular caterpillars. These applications are all to agricultural pest control and this has been the main area of application of GMs in agriculture so far. However, there are also other applications.

Views about the development and desirability of use of GMs in agriculture are divided. Proponents see GM agricultural products as reducing the costs of agricultural production, ensuring more reliable yields, lowering the use of chemical pesticides, and as a way to supply products that better meet the demands of consumers. There is also a belief that GMs may be able to extend the range of areas in which agricultural cultivation is possible, for example, extend cultivation in dry land areas.

Critics of GM development have several different types of concerns (Paoletti and Pimentel, 2000). On the biodiversity front, if GMs in agriculture are more profitable than traditional varieties of agricultural crops, loss of traditional varieties of crops is likely to be accelerated. The global genetic stock of agriculture will be further eroded. Secondly, it is possible that GMOs might cross with their wild relatives or domesticated relatives causing widespread changes in the genetic stock. Third, ecological processes could result in GMs becoming more susceptible to pests and diseases as their use becomes more widespread—their resistance to external biological pressures is thereby reduced. Fourth, the prospect of GMs enabling higher agricultural productivity on marginal and dry land is not necessarily good news from the point of view of biodiversity conservation. It is likely to increase conversion of natural habitats on such land to agriculture with increased loss of biodiversity in the wild.

Another major concern involves the intellectual property rights in GMs. In the Green Revolution period, agricultural research was mostly undertaken by governments and international public bodies such as the International Rice Research Institute and the results disseminated widely and free of charge. The results were treated basically as a public good. Now, however, developments are being undertaken by private corporations. These corporations obtain international private property rights in their innovations for a specified period of time. They provide their owners with an exclusive monopoly for a lengthy period of time. There are concerns especially in developing countries that the monopoly right could be used to exploit farmers.

Bt transgenic crop species (cotton, corn, potato) have already been commercialised with substantial benefit of farmers. Over a dozen of Bt crops under development and also for field use in India (Table 28.5). Only long time debate prevented to make from massive use of such GM in Indian agriculture except Bt cotton.

Table 28.5: Bt crops uner development and commercialisation in India

Sl. No.	GM Crop	Organisation (S)	Traits / Gene
1.	Brinjal	Mahyco, Mumbai (*Recommended for commercialization by GEAC in Oct* 2009 *meeting*) TNAU Coimbatore, IVRI Varanasi, UAS, Darwad, IARI, New Delhi, Sungro Seeds Ltd., New Delhi	Insect resistance / *cry* 1 Aa and *cry* 1 Aabc *cry* 1 Ac *cry* 1 Ac
2.	Cabbage	Nunhems India Pvt. Ltd.	Insect resistance / *cry* 1Ba and *cry* 1 CA
3.	Cauliflower	Sungro Seeds Ltd., New Delhi Nunhems India Pvt. Ltd.	Insect resistance / *cry* 1 Ac, *cry* 1 Ba and *cry* 1Ca
4.	Cotton	Mahyco, Monsanto, Rasi, Nuziveedu, Ankur, JK Seed, CICR, UAS-D	Insect Resistance, herbicide tolerance cry 1 Ac gene
5.	Groundnut	ICRISAT, Hyderabad	Virus resistance/Chitinase gene
6.	Maize	Monsanto, Mumbai	Shoot borer / *cry* 1 Ab gene
7.	Chickpea	ICRISAT	Insect Resistance/Pod borer, *cry* 1 Ac
8.	Mustard	UDSC, New Delhi	Hybrid seed, barnase / barstar gene
9.	Okra	MAHYCO, Mumbai, Beejo Sheetal, Jalna	Borer *cry* 1 Ac, *cry* 2 Ab
10.	Pigeon pea	ICRISAT, MAHYCO	Pod borer and Fungal pathogene, *cry* 1 Ac and chitinase
11.	Potato	CPRI, Shimla, NIPGR, New Delhi	*Amal* and *Rb* gene derived from *Solanum bulbocastanum*
12.	Rice	MAHYCO, Mumbai TNAU, Coimbatore	*cry* 1 B-*cry* 1 AC *fusion g* *cry* 1 Ac, *cry* 2 Ab *Rice chitinose gene* or *tobacco osmotin gene*
13.	Sorghum	NRCS, Hyderabad	Insect Resistance, Shoot borer
14.	Tomato	IARI, New Delhi MAHYCO, Mumbai NIPGR, New Delhi	*Antisense replicase* gene of tomato leaf curl virus *cry* 1 Ac

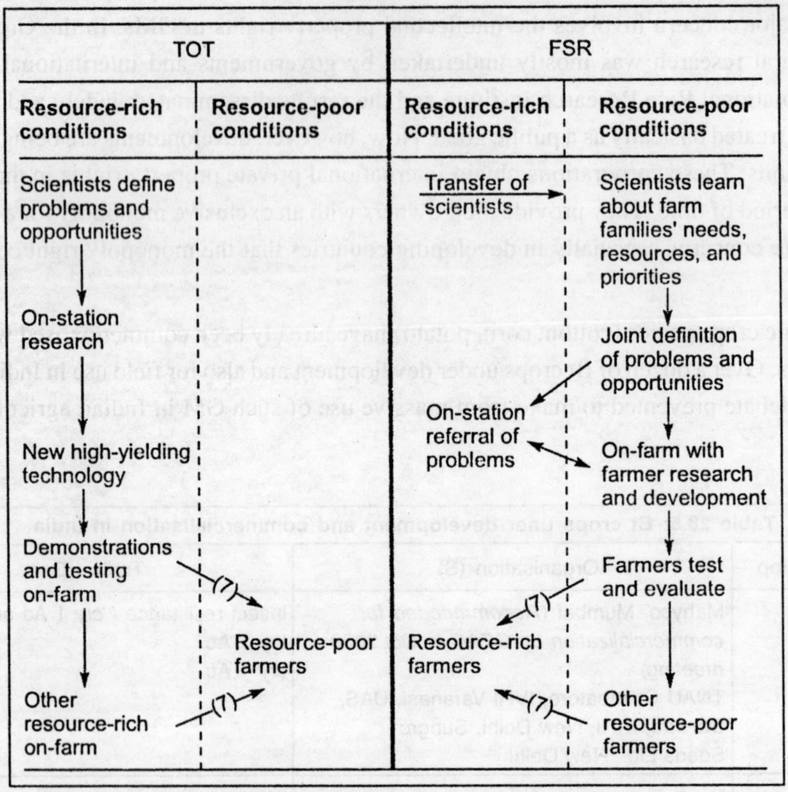

Fig. 28.3: A comparison between a typical top-down technology transfer approach (TOT) and a farming systems research approach (FSR) involving farmers (Chamber and Ghildyal, 1985)

Box 28.8: Biosafety: Cartegena Protocol

Cartagena Protocol on biosafety is an attempt to produce a globally harmonised regime for biosafety under the Convention of Biological Diversity (CBD). Named after the Colombian city where the final round of talks was launched, the Cartagena Protocol on Biosafety sets out a comprehensive regulatory system for ensuring the safe transfer, handling and use of Living Modified Organisms (LMOs) subject to transboundary movement. In every day usage LMOs are considered to be same as GMs although definitions and interpretations vary widely.

The Protocol deals primarily with LMOs that are to be intentionally introduced into the environment (such as seeds, trees or fish) and with genetically modified farm commodities (such as corn and grain used for food, animal feed or processing). It does not cover pharmaceuticals for humans addressed by other international agreements and organizations or products derived from LMOs, such as cooking oil from genetically modified corn.

ORIGIN

The objectives of the Convention on Biological Diversity (1992) are "the conservation of biological diversity, the sustainable use of tis components and the fair and equitable sharing of the benefits arising out of the utilisation of genetic resources". When developing the Convention, the negotiators recognised that biotechnology can make a contribution towards achieving the objectives of the Convention, if developed and used with adequate safety measures for the environment and human health. The Contracting Parties agreed to consider the need to develop appropriate procedures to address the safe transfer, handling and use of any Living Modified Organism (LMO) resulting from biotechnology that may have adverse effect on the conservation and sustainable use of biological diversity (Article 19.3 of the CBD). The Biosafety Protocol is the result of that process.

STATUS

The protocol entered into force from September 11, 2003. As of April, 2007, 140 countries have ratified the protocol.

28.7 PARTICIPATORY APPROACH OF MODERN AGRICULTURE

In addition to technological modification, several new concepts have emerged in modern agriculture. These are:

(A) Participatory approach of technology development.

(B) Top-down technology transfer approach involving farmers; and

(C) Emphasis on biological agriculture practices.

The participatory technological approach involves various levels of interaction among farmers, agricultural workers and concerned goverment departments to building up relationships for cooperation and identifying priority areas. The details are given in Table 28.6.

Table 28.6: Activities in Participatory Technology Development (after Reijntjes et al, 1992)

Activity	Description	Examples of operation	Examples of output indicators
Getting Started	Building relationships for cooperation. Preliminary situation analysis. Awareness mobilisation.	Organisational resources inventory. Community walk. Screening secondary data. Problem census. Community survey.	Inventories. Protocols for community participation. Core PTD network. Enhanced agroecological awareness.
Looking for things to try	Identifying priorities. Identifying local, community, and scientific knowledge/information. Screening options and choosing selection criteria.	Farmer expert workshop. Techniques to tap indigenous knowledge. Study tour. Options screening workshop.	Agreed research agenda. Improved local capacity to diagnose problems and identify options for improvement. Enhanced self-respect.
Designing experiments	Reviewing existing experimental practice. Planning and designing experiments. Designing evaluation protocols.	Improving natural experimentation. Farmer-to-farmer training. Design workshop. Testing alternative designs.	Manageable, valuable, reliable experimental designs. Protocols for monitoring and evaluation Improved local capacity to design experiments.

The typical technology transfer modalities for sustained yield has to be reevaluated in appropriate manner so that the objective achievement is obtained (Fig. 28.3, page 618).

Finally, the major emphasis on biological aspects of agriculture has to be considered in appropriate manner (Table 28.7, page 620).

On the whole, there is a need to bring birth rates well below death rates, increase food production while protecting the environment and distribute food to all who need it—**THIS IS THE GREATEST CHALLENGE OUR EARTH HAS EVER FACED.**

Table 28.7: A schematic outline of biological agriculture (after Coleman 1989)

- BIOLOGICAL AGRICULTURE
  - PEST MANAGEMENT (diseases, insects and weeds)
    - BIOLOGICAL
      - Crop Rotation / Soil Structure
        - allelopathy crop residue management
        - agrophytocoenosis
      - Nature Controls
        - Birds
        - insect and fungi
        - mammals
      - Cultivars / Soil Fertility
        - N-P-K
        - trace elements
        - organic matter
      - Soil pH
        - lime
        - sulphur
    - PHYSICAL
      - Tillage / Cultivation / Thermal
        - implements
        - planting systems
        - transplanting
        - undersowing
      - Environment / Inert Dusts
        - grazing management
        - mowing hedgerows
      - Temporal
        - planting harvest
        - cultural practices
      - Barriers / Traps / Manual Control
        - light adhesive
        - vacuum crop
    - CHEMICAL
      - Attractants / Botanicals / Repellents
        - rotenone pyrethrum
        - ryania sabadilla
        - quassia nicotine
        - Bt
        - allelopathic herbicide
  - SOIL FERTILITY MANAGEMENT
    - CULTURAL TECHNIQUES
      - Crop Rotation
        - vegetables grains
        - livestock fruits
      - Drainage
      - Green Manures
        - legumes
        - non-legumes
        - strip tillage
      - Shallow Tillage
      - Deep Tillage
        - rotary tiller chisel
        - plow subsoiler
      - Crop Residues
      - Undersowing
        - vegetables
        - grains, fruits
    - FERTILIZERS
      - Primary Minerals
        - rock phosphate
        - basalt dust
        - granite dust
        - greensand
        - lime gypsum
      - Organic Wastes
      - Organic Fertilizers
        - blood meal
        - bone meal
        - sea products
        - composted products
        - sul-po-mag
        - Chilean nitrate
      - Commercial Products
        - humates
        - soil conditioners
        - foliar sprays
        - soil testing services

29 ENVIRONMENTAL BIOTECHNOLOGY

29.1 CONCEPT

Virtually all biotechnological endeavours can be claimed, justifiably, to have environmental consequences. For example, the development of new crop plants and crop protectants will influence the use of land and our ability to feed a growing population. The replacement of the "dirty" manufacturing process by 'clean' biotechnological strategies can be regarded as environmentally important. On the whole, environmental biotechnology means the specific application of biotechnology to the management of environmental problems. It encompasses aspects of natural resources management, the treatment of waste and control of pollution (Fig. 29.1).

29.2 BIOTECHNOLOGICAL APPROACH OF ENVIRONMENTAL POLLUTION ABATEMENT

Over the decades, scientific understanding of the utility of bioresources for abatement of environmental pollutants from industrial effluents and toxic wastes were known in depth. This is popularly known as bioremediation process. Normally, organisms are capable of biodegrading the toxic materials, or sink the pollutants and, thereby, transfer the same into non-toxic organic forms.

A few examples of such biotreatment options are described:

(a) Biodegradation of Pollutants

Biological treatment of effluents is a long-established practice in many countries, but some constituents of these effluents are calcitrants and, thus, not amenable to conventional treatments. Biotechnology helps in overcoming this problem.

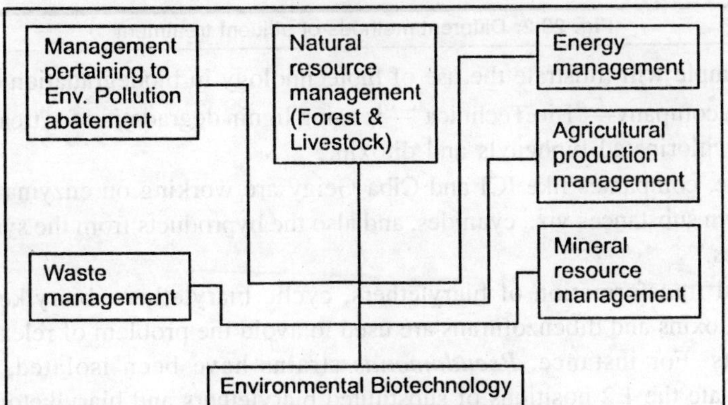

Fig. 29.1: Environmental Biotechnology

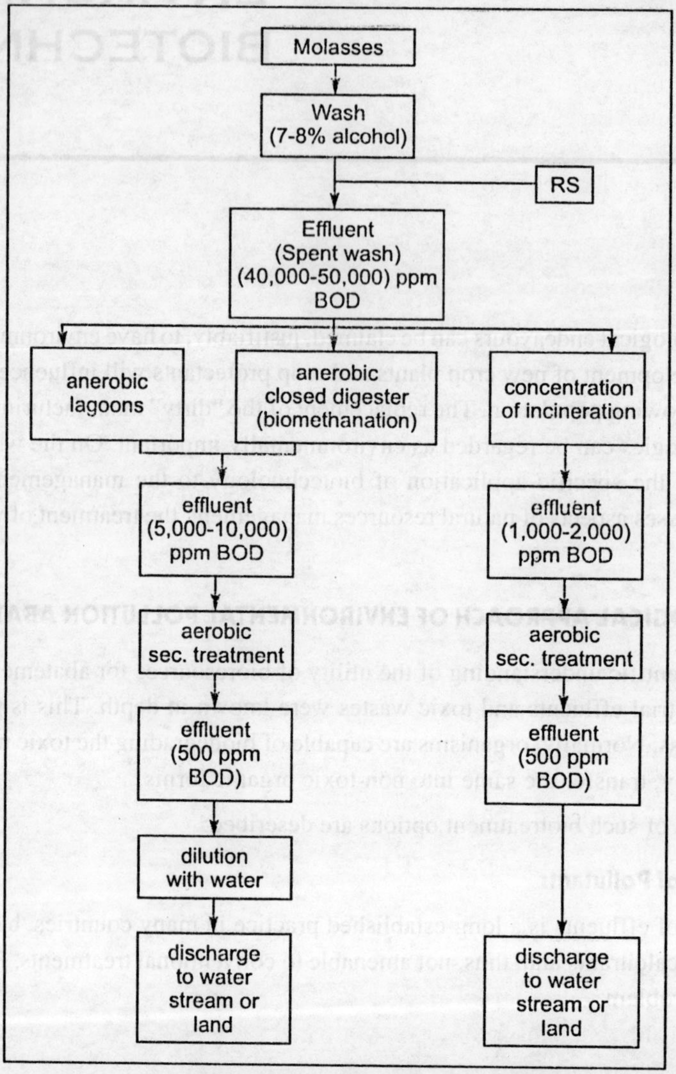

Fig. 29.2: Different methods of effluent treatment

Following example will illustrate the use of biotechnology in biodegradation of pollutants:

(i) In USA a company—"Bio Technica"—is using lignin degradation for treatment of substances like polychlorinated biphenyls and ditoxin.

(ii) In Europe, companies like ICI and Ciba-Geigy are working on enzymatic detoxification to breakdown substances viz., cyanides, and also the byproducts from the synthesis of S-triazine herbicides.

(iii) Microbial transformation of biarylethers, cyclic biarylethers, biarylketones, halogenated bibenzodioxins and dibenzofurans are used to avoid the problem of release of effluents with pollutants. For instance, *Pseudomonas* strains have been isolated, which selectively deoxygenate the 1,2 positions of substituted biarylethers and biarylketones.

(iv) Microbial degradation of chloro, dichloro and trichloromethanes and carbonate chloride is also used to deal with the problem.

The different methods of effluent treatment processes are shown in Fig. 29.2.

(b) Toxic Site Reclamation

Since the toxic constituents of industrial effluents are known in most cases, their treatment is easy. However, the waste disposal sites are more difficult to deal with, since their chemistry is often complex. Although incineration (drying and then burning to ashes in a furnace) or chemical treatments are suggested for treatment of these sites, biotechnological approaches involving biodegradation are considered more attractive. Companies like Bio Technica are working on treating polluted sites *in situ*. In this connection, it is interesting to note that the first patent ever awarded was given to Dr. Ananda Chakraborty for developing a microbe designed to biodegrade spilled oil. There has been a debate on whether or not to release genetically engineered microbes for treatment of toxic sites.

In view of the above, "Occidental Chemical Corporation" company of USA is conducting their work on effluents from toxic waste sites in closed reactors. One such experiment was conducted at Hyde Park landfill site at the north-west corner of Niagara, USA. This site occupies just over 6 hectares of land and contains an estimated 73,000 tonne of chemical waste. The site was used from 1953 to 1975 to dispose-off a wide range of waste materials including chlorinated organic materials. For treatment of this site, a clay cover was placed on the landfill in 1978 and a leachate collection system was installed around the perimeter in 1979.

From this system, 230 m^3 of leachate per week was shipped to a nearby treatment facility in Niagara where the treatment involved carbon absorption, thus making the effluent suitable for discharge. This solution was not found suitable on a long term basis. Therefore, the Hyde Park waste-water was subjected to biotreatment in sequencing batch reactors, where biodegradation rate of some more persistent chemicals was accelerated by adding bacterial strain isolated from the landfill site. Further genetic manipulation of these organisms produced more than 100 recombinant bacteria, which have been tested for their usefulness in reactors. It was demonstrated that significant cost savings will be possible if SBR treatment is used before carbon absorption stage.

(c) Removal of Spilled Oil and Grease Deposits

Oil spills from oil tankers on land surface and from distant oil spills have been recognised as a major environmental hazard. This spilled oil is believed to not only destroy the habitats of aquatic animals and fish, but also create health problems for local residents and causes long term damage to the environment. Therefore, remediation efforts in this connection have been made, using chemical dispersants. These chemical dispersants are believed to cause major pollution problems in shallow water due to their toxic nature. Now attempts are being made to use microbiological degradation of oil.

In order to achieve this objective, following methods have been used:

(i) **Use of a mixture of bacterial strains**: A mixture of bacterial strain has also been used to clean oil-contaminated water reserves and water supplies. This could be illustrated using the following example. When 800,000 gallons of oily water once accumulated in the bottom of the ship "Queen Mary", there was a danger that this oily water will be discharged at the harbour causing death to marine life and spoiling beaches. When a mixture of bacterial strains was introduced in the bottom of the ship, the oil was decomposed within 6 weeks. A mixture of bacterial strains is also being used for removal of grease deposited on the inner side of tanks in a ship. This technique may also be used for cleaning deposits of grease in pipes and vessels of a variety of industries.

(ii) **Use of Oleophilic fertilisers**: In early years, application of oleophilic (oil-loving) fertilisers as food for oil-utilising microbes was considered useful, since this would allow rapid growth and multiplication of indigenous microbes, thus speeding up the biodegradation process for removal of oil. When 6,000 gallons of petrol leaked from a company in Pennsylvania in USA, it posed a serious health problem due to contamination of underground water supplies. Addition of oleophilic fertilisers on the sites expedited the biodegradation of petrol. In recent years, oil-utilised microbes—*Pseudomonas aeruginosa*—have been developed through genetic engineering.

(d) Reducing Environmental impact of Herbicides, Pesticides, and Fertilisers

Majority of chemical herbicides, pesticides, and fertilisers that are used commercially cause environmental hazards. Following solutions to this problem of pollution are being tried under biotechnology programmes:

(i) **Weed Control & Herbicides**: Herbicides are being developed which will be environmentally safe. To allow the use of these herbicides for crop protection programmes, genetically engineered herbicide resistant plants have been produced in a number of crops. These plants are being field-tested. Similarly, genetically engineered insect-resistant plants have also been successfully produced in some crops, thus raising the *hope* for reduced use of insecticides in future.

(ii) **Pest Control and Biopesticides**: Bacterial pesticides are also being developed. *Heliothis* complex, which lives in close association with plant roots, consists of 2 major crop pests—budworm and ballworm. Biological insecticides against both these insects are being prepared by transfer of a gene from *Bacillus thuringiensis* (Bt) into either a naturally occurring soil bacterium, or in a strain of *Pseudomonas*. Bt is already in the market since 1970 and in future these will be modified through techniques of genetic engineering and will be utilised against a variety of insects. Several companies in USA—including Ecogen, Mycogen, Repligen and Zoe Con are involved in development and marketing of biological pesticides. Among these, Mycogen kills recombinant bacteria and applies them to the leaves of the crop plants.

(iii) **Viral Pesticides**: Have also been successfully developed and used for control of pest. These pesticides are environmentally safe and avoid risk of toxicity. These pesticides can also be used against the pest strains which have become resistant to chemical pesticides. A number of entomopathogenic viruses belonging to the family *Baculoviridae* (or nuclear polyhedrosis viruses = NPV) and family Reoviridae (cytoplasmic polyhedrosis virus = CPV) have been used as safe and effective pesticides. These viruses kill specific pest species and have no adverse effect on useful insect pollinators, warm-blooded mammals or even man.

Some viral pesticides, with the name of pests they control are:

Class of virus	Pest	Trade name	Crops
1. N.P.V.			
Gypsy moth baculovirus	*Lynantria dispar*	Gypcheck	Hard wood trees
2. CPV			
CPV of Pine	*Dendrolimus spectabilis*	—	Pine, spruce etc.
3. Granulosis virus (GVS)	*Cydia pomonellc*	—	Apple & vegetables

(iv) **Biofertiliser**: To reduce the impact of excess chemical fertilizers in the field of agriculture, the biofertiliser is a potential tool. The details are described in a separate section (29.4).

(e) Biosensors to detect Environmental Pollutants

Biosensors represent biophysical devices which will detect the presence and measure the quantities of specific substances in a variety of environments. These specific substances may include sugars, proteins or hormones in human body, pollutants in abiotic components of the environment including air, soil and water and a variety of toxins in the industrial effluents. In designing a biosensor an enzyme or an antibody or even microbial cells are associated with microchip devices which are used for quantitative estimation of a substance.

A biosensor equipment consists of:

(i) a biologically sensing agent.

(ii) a device for collecting the product obtained from the interaction between the substrate and the biosensor.

(iii) a device for measuring the quantity of this product, thus indirectly giving an estimate of the substrate.

The first reported application of biosensor dates back to 1860 when starch content was estimated with the help of malt extracts. Biosensor today is an analytical tool which consists of biological material in intimate contact with a transducer (Fig. 29.3). It combines high selectivity of the biological/chemical material with high accuracy of solid state devices. However, interfacing poses difficult problems. The biological material can be membrane, enzyme, antibody/antigen, receptor, protein, intact cells, tissue or whole organ. It is necessary that the analyte should reach the reaction site in the biological material.

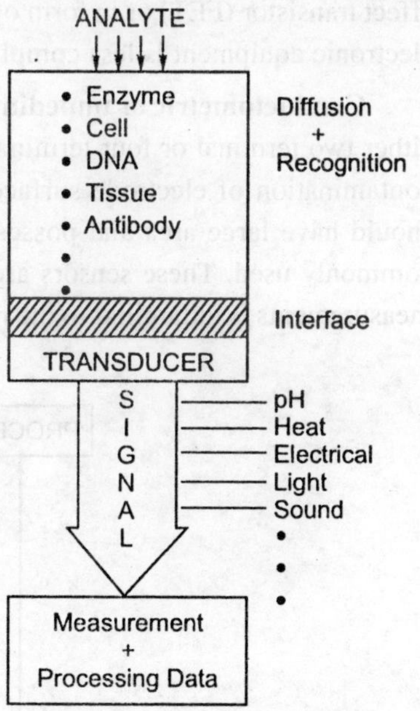

Fig. 29.3: Schematics of biosensor

The reaction center can recognize the analyte and generate signals which can be measured by an appropriate transducer. The choice of transducer depends upon the signal generated by the biological material. The signal can be heat generated, light emitted/absorbed, pH change, electrons released/absorbed or chemical species produced. The signal may or may not be amplified within a transducer. Weak signals require amplification prior to detection. Data processing can be done on a computer or recorded directly with a plotter. The first modern biosensor was assembled by Clark and Lyons (1962), who modified oxygen electrode to sense glucose concentration. Since then a variety of biosensors have been developed.

Biosensors can be classified according to transducers employed. The transducers are of different types:

- Electrochemical
- Optical
- Calorimetric
- Field effect transducers
- Piezoelectric
- Microbial biosensor
- Enzyme biosensor

Electrochemical transducers (potentiometric or amperometric) are most popular because of their familiarity to chemists and biologists as well as simplicity of operation. It has added advantage that the same device can be used to sense different types of redox molecules, provided their redox potentials are distinctly different from one another.

In optical biosensors, light is focussed on a membrane containing reagents which can combine with target molecules diffusing through membrane. As a result, reagents can absorb or fluorescence, which is, measured and thereby concentrations of target molecules are determined. There are two types of sensor materials used—bifurcated, and coated fiber divices.

Biosensors based on bioanalytical calorimetry rest upon the fact that most of the biological reactions are exothermic. Most of the reactions are associated with high molar enthalpy changes (20-100 KJ per mol) in a single enzymatic step. An enzyme immobilized electrode coupled to calorimetric device forms thermometric sensor. These are simple in operation, insensitive to optical properties of sample and possess high specificity. The most important aspect of this device is that it can be integrated with continuous flow systems and can easily detect 0.01 mM concentrations.

Semiconductor-based devices are ragged and can be mass-fabricated at low costs. The device consists of a chemically sensitive layer, integrated with solid state electronic circuits. It can be designed as a field effect transistor (FET) or in form of a capacitor. FET based device has built-in ampliation and associated electronic equipment is less complicated. It is immune to electromagnetic disturbances.

Conductometric or impedimetric measurements offer yet another possibility. The device can have either two terminal or four terminal measuring system. It is desirable that polarization of electrode and contamination of electrode surface due to oxide film formation should be minimum. The electrodes should have large area and possess reversible electrochemistry. Au, Pt, Pd, Ti, Al, Fe electrodes are commonly used. These sensors are not popular to biologists and industries as the theory behind these measurements is unfamiliar to the researchers.

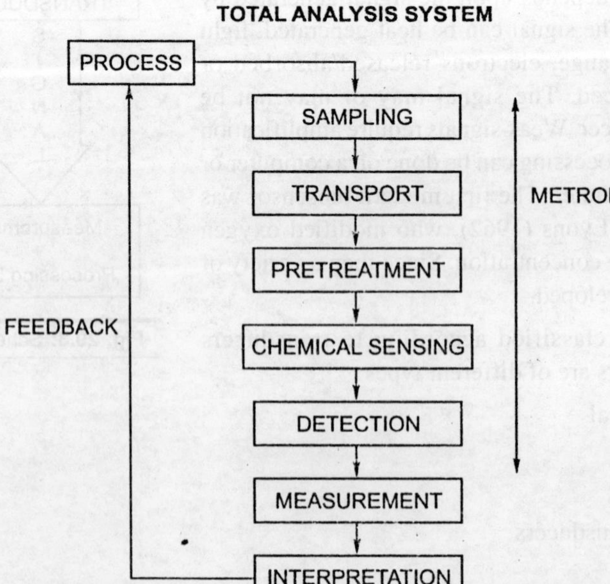

Fig. 29.4: Biosensor in flow injection analysis system

Microbial biosensors consist of a number of coupled or uncoupled enzymic steps. These are less sensitive to inhibitors and highly tolerant of variations in pH and temperature. These are cheaper and have long life times. But there are several disadvantages viz.,

(i) less selective and difficult to maintain;

(ii) response times are long and return to baseline at slower rate;

(iii) sensors are highly susceptible to contamination.

The immobilization of enzyme can be attained by physical entrapment, chemical attachment or by preparing thin solid films. Physical entrapment is a mild method and does not adversely affect catalytic activity of the enzyme. This can be achieved by immobilizing enzyme-substrate complex as a whole. Chemical attachment can be of ionic or covalent linkage type. The covalent coupling should not involve amino acid residues which are essential to catalytic activity. Lastly, the use of biofilm for immobilization of enzyme is also very effective. Such enzyme immobilized film is packed in a dialysis bag for effective biosensor functioning.

29.3 BIOTECHNOLOGICAL APPROACH OF ENERGY MANAGEMENT

It is well-known that use of conventional energy is linked with a variety of pollution problems. Thus, in recent decades, efforts were made for evolving non-polluting bioenergy sources or energy generation

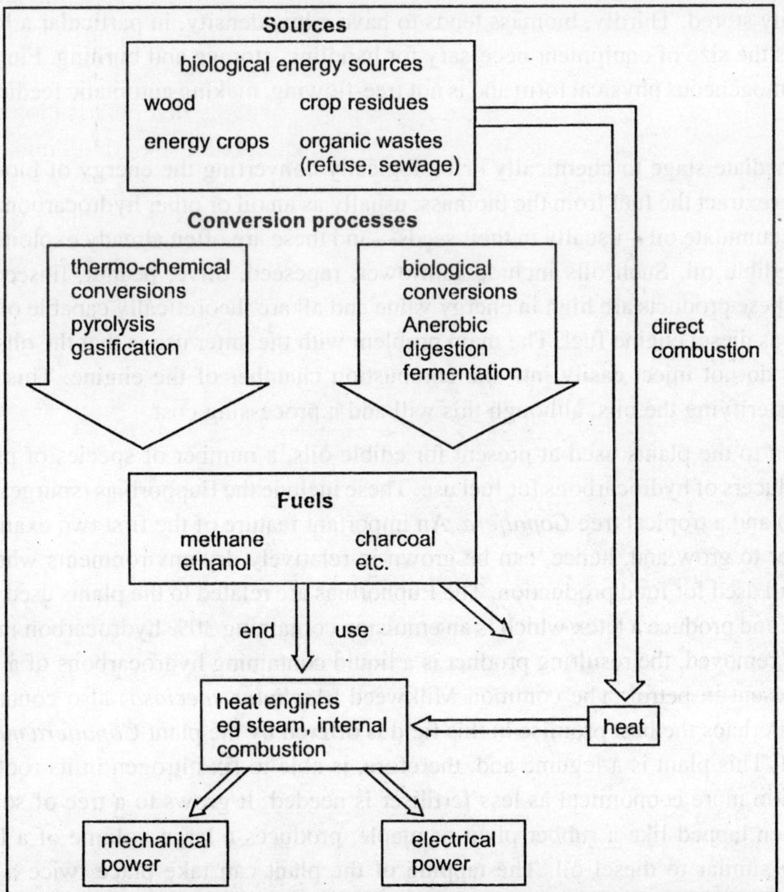

Fig. 29.5: Generalised schemes for the use of biological energy sources

from organic waste or biomass. These are all eco-friendly solution to the present day conventional energy-use pattern.

There are a variety of biological energy sources (Fig. 29.5). They can be combustible by direct burning or other conversion processes which are considered to be more efficient and non-polluting.

Biomass

Biomass is organic material of biological origin that has ultimately been derived from the fixation of carbon dioxide and the trapping of solar energy during photosynthesis. If it is to be used as a fuel, the biomass can either be burned, releasing the chemical energy directly, or it can be converted to a liquid or gaseous fuel that is higher in energy per kilogram than the original biomass. Although unconverted biomass can be used as a fuel and, indeed, plays an important role in countries such as India—where the collection and burning of firewood is important to the rural economy—the large scale use of such fuel suffers from a number of disadvantages over fossil fuels such as coal and oil.

There are several potential disadvantages of using biomass as a fuel; list any that you can think of.

Firstly, biomass usually has a lower thermal content than fossil fuels. Secondly, it often has a high moisture content, thus inhibiting easy combustion, resulting in a large energy loss on burning—mainly as latent heat of steam. The high moisture content also causes the material to be biodegradable so that it cannot be readily stored. Thirdly, biomass tends to have a low density, in particular a low bulk density, which increases the size of equipment necessary for handling, storage and burning. Finally, the material is rarely in a homogeneous physical form and is not free-flowing, making automatic feeding of combustion plants difficult.

An intermediate stage to chemically or biologically converting the energy of biomass to a higher energy fuel is to extract the fuel from the biomass, usually as an oil or other hydrocarbon. There are many plants which accumulate oil—usually in their seeds—and these are often already exploited commercially as sources of edible oil. Such oils include sunflower, rapeseed, olive, peanut, linseed, soyabean and safflower. All these products are high in energy value and all are theoretically capable of being burned in boilers or used as diesel engine fuel. The main problem with the latter use is that the oils are often rather viscous and so do not inject easily into the combustion chamber of the engine. This problem can be overcome by esterifying the oils, although this will add a processing cost.

In addition to the plants used at present for edible oils, a number of species of plant have shown promise as producers of hydrocarbons for fuel use. These include the Euphorbias (spurges), the milkweeds (*Asclepias* sps.) and a tropical tree *Copaifera*. An important feature of the first two examples is that they need little water to grow and, hence, can be grown in relatively dry environments where they will not compete for land used for food production. The Euphorbias are related to the plants used to commercially produce rubber and produce a latex which is an emulsion containing 30% hydrocarbon in water. After the water has been removed, the resulting product is a liquid containing hydrocarbons of a lower molecular weight than present in petrol. The common Milkweed (*Asclipias speciosa*) also contains around 30% hydrocarbons. Perhaps the best promise in this field is offered by the plant *Copaifera multijuga* which is found in Brazil. This plant is a legume and, therefore, is able to fix nitrogen in its root nodules, which makes cultivation more economical as less fertiliser is needed. It grows to a tree of some 30 meters in height and, when tapped like a rubber plant or maple, produces a large volume of a liquid which has properties very similar to diesel oil. The tapping of the plant can take place twice a year and so the potential for commercial exploitation is good. Another possible source of large scale production of hydrocarbons are certain freshwater and marine algae which are known to accumulate similar substances.

Although there is clearly potential for the production of biofuels by direct extraction from plants, at the present time (apart from small scale trials) these are only research projects or theoretical possibilities. There are, however, biofuels currently being produced by conversion of biomass to gaseous or liquid fuels and these will be considered in the following sections.

Biogas

Biogas is a mixture of gases—primarily methane and carbon dioxide which are produced when organic material is degraded under anerobic conditions.

It is important that you remember the significance of anerobic conditions. Briefly distinguish the nature of the products produced by microorganisms growing aerobically from those of microorganisms growing anerobically.

Under strictly aerobic conditions the principle products are cellular material, carbon dioxide and water. The cellular material will itself be recycled when the cell dies—generating more carbon dioxide and water. Under anerobic conditions the end-products are far more varied. They include carbon dioxide and water but also many reduced organic compounds such as acids, alcohols and so on.

Polymeric materials are hydrolysed by microorganisms—such as the cellulolytic and proteolytic bacteria—to smaller molecules which then act as substrates for a second group of microbes, the acetogens. These fermentative bacteria further degrade the higher fatty acids to low molecular weight fatty acids such as acetate and formate, alcohols such as methanol, methylamines, and gases such as hydrogen and carbon dioxide. Finally, these act as substrates for the methanogenic bacteria, which convert them to methane and carbon dioxide. The methanogenic bacteria are a group of strictly anerobic bacteria which requires highly reducing conditions (redoxpotential (Eh) < – 400 mV). The types of energy yielding reactions which methanogenic bacteria as a group are capable of are shown in Table 29.1.

Table 29.1: Energy Yielding reactions in Methanogenesis

From hydrogen and carbon dioxide:

$$4H_2 + CO_2 \rightarrow CH_4 + 2H_2O$$

From formate:

$$4HCOOH \rightarrow CH_4 + 3CO_2 + 2H_2O$$

From acetate:

$$CH_3COOH \rightarrow CH_4 + CP_2$$

From methanol:

$$4CH_3OH \rightarrow 3CH_4 + CO_2 + 2H_2O$$

From trimethylamine:

$$4(CH_3)_3N + 6H_2O \rightarrow 9CH_4 + 3CO_2 + 4NH_3$$

These microbes, unlike the fermentative acetogenic bacteria, are capable of anerobic respiration using carbon dioxide as the terminal electron acceptor and, in the process, produce methane. The resulting gas generated by the system—biogas—contains roughly 60% methane, 40% carbon dioxide and traces of hydrogen. Biogas burns readily, although it does not have as high energy value as natural gas which is almost pure methane. It is suitable as a fuel for boilers and kilns and for internal combustion engines, although impurities in the gas can cause corrosion problems in the latter case.

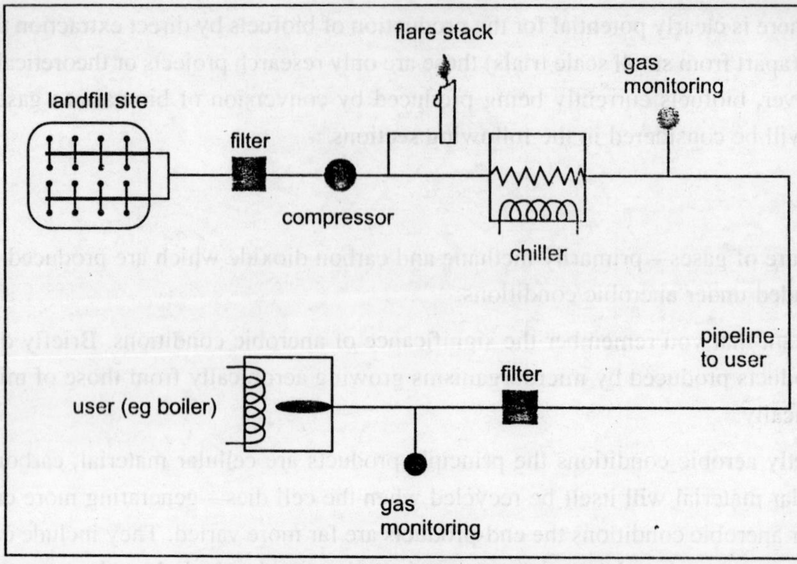

Fig. 29.6: Landfill gas abstraction plant

For anyone to seriously consider harvesting biogas from a landfill site the major criterion—after safety—is, obviously, economic viability. List and discuss some of the factors which have to be considered before embarking on commercial exploitation. Read our response carefully.

The process of gas abstraction at a landfill site is shown in Fig. 29.6. The gas is initially collected by sinking a number of 'gas wells' into the top of the landfill. These are constructed of perforated or slotted piping which is then interconnected via a network of pipes to a gas pump or compressor. If an existing landfill is being harvested for gas then there will be no internal structure within the landfill. With new sites which have been planned from the start for gas production, the gas wells are more complex and may have internal walls made of clay in order to trap the gas more effectively. The shape of these wells is the subject of research, but a number of different types have been tried including vertical wells, short horizontal, short vertical and bell-shaped. From the compressor, the gas is passed to a chiller, which cools the biogas in order to remove water from it, which would render it less combustible and cause corrosion problems. There is also a flare stack which is used to burn-off excess gas when production exceeds demand, or when there is a problem further down the line. Finally, the gas is filtered and its composition is determined by monitoring prior to end use.

There are three major uses for landfill gas and these are illustrated in Fig. 29.7. These end uses are:

- direct combustion in boilers, kilns or furnaces;
- generation of electricity;
- upgrading and export or compression or liquefaction.

Of these alternatives, the first is the most economically feasible. The most effective use of the landfill gas is burning at a site close to the landfill itself, since building gas pipelines is expensive. One of the most successful landfill gas schemes to date in the UK is that of the London Brick Company who are ideally positioned to exploit this technology since they have disused clay pits which can be used for landfill and they have a major use for the biogas in firing brick kilns produced near to the landfill. Other examples of direct combustion of landfill gas include the heating of greenhouses, the firing of cement kilns and the production of steam for various industrial uses.

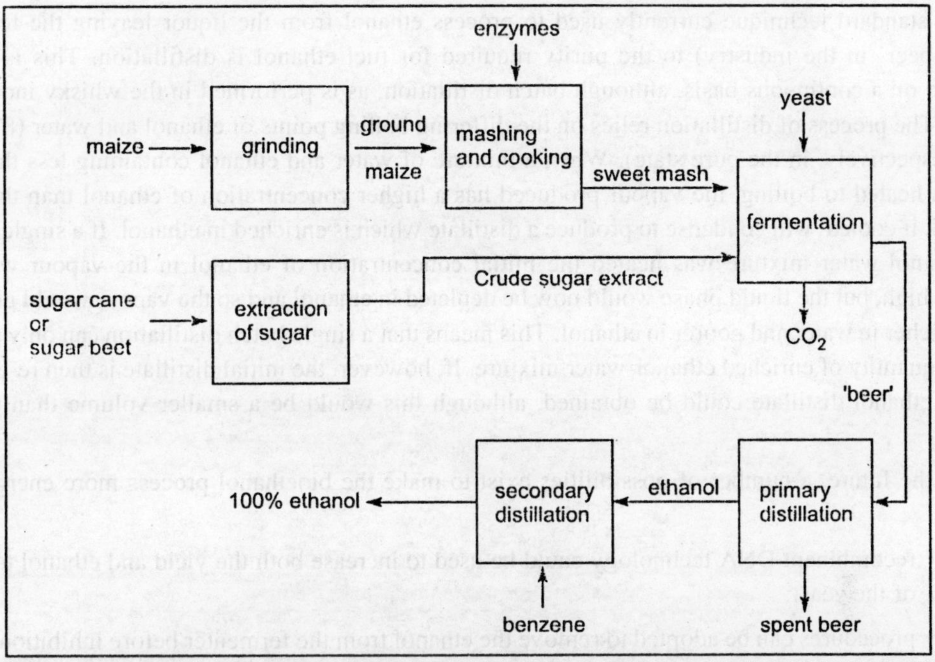

Fig. 29.7: Process flowsheet for the production of bioethanol. In the examples shown, maize, sugar cane or sugar beet serve as the primary feedstock

If a local end-user of the gas is not available, a second alternative is to generate electricity from the biogas. This can then be exported relatively cheaply by building electricity lines to connect in with the National Grid. At small scale, plants generating up to 1-2 MW, this usually involves the use of converted internal combustion gas engines, although larger schemes using gas turbines also exist. The main problem with this approach is that landfill gas is not purely a methane/carbon dioxide mixture, but contains traces of many other gases some of which are corrosive to metal parts of engines. This problem can be partially overcome by gas cleaning and by modifications to the generating engines and their lubricating oils. In the UK, in 1987, six out of the 20 landfill gas schemes in operation generated electricity, while, in the USA, the figure was 22 out of 54, with most new schemes opting for this approach.

The third alternative for landfill gas use is purification followed by export, either in a pipeline to a gas utility network or by bottling as compressed or liquefied gas. The process of gas cleanup to provide a suitably pure methane gas is complex and expensive and so this option is the least favoured of the three. There are, however, a number of purification plants operating, mainly in the USA, and one scheme in Santiago, Chile, where landfill gas is used to supplement the city's existing gas distribution system.

Bioethanol

The technology for the biological production of ethanol is as old as biotechnology itself since the production of alcoholic beverages by Man has been recorded for thousands of years. However, although the production of fuel ethanol has undoubtedly drawn upon this older technology, the processes involved are more complex. They are, however, less constrained since the product does not have to be potable. The basic process involves conversion of the feedstock to a fermentable form, followed by fermentation and then distillation. With some feedstocks, such as sugarcane, the initial stage can be omitted since the extracted substrate is readily fermentable. The unit operations involved in the production of ethanol from feedstocks such as maize and sugar cane are shown in Fig. 29.7.

The standard technique currently used to process ethanol from the liquor leaving the fermenter (termed 'beer' in the industry) to the purity required for fuel ethanol is distillation. This is usually performed on a continuous basis, although batch distillation, as is performed in the whisky industry, is possible. The process of distillation relies on the differing boiling points of ethanol and water (87°C and 100°C, respectively, in the pure state). When a mixture of water and ethanol containing less than 95% ethanol is heated to boiling, the vapour produced has a higher concentration of ethanol than the liquid phase and, if cooled, will condense to produce a distillate which is enriched in ethanol. If a single volume of an ethanol-water mixture was heated the initial concentration of ethanol in the vapour would be relatively high, but the liquid phase would now be depleted in ethanol and so the vapour would gradually become richer in water and poorer in ethanol. This means that a single batch distillation can only produce a limited quantity of enriched ethanol-water mixture. If, however, the initial distillate is then re-distilled, a further ethanol distillate could be obtained, although this would be a smaller volume than the first distillate.

For the future, a number of possibilities exist to make the bioethanol process more energetically feasible:

- recombinant DNA technology could be used to increase both the yield and ethanol tolerance of the yeast;
- procedures can be adopted to remove the ethanol from the fermenter before inhibition occurs. This could involve separating the yeast from the effluent by centrifugation and recycling the biomass to the fermenter. Alternatively, ethanol could be removed continuously from the fermenter by vacuum distillation in which a partial ,vacuum is maintained in the headspace above the fermentation broth so that the ethanol evaporates rapidly at the operating temperature of the fermenter;
- the separation of ethanol from water can be made more energy-efficient by using alternatives to distillation. These include reverse osmosis, where the ethanol is separated from the water by filtration through extremely small pores under high pressure; selective adsorption using solid adsorbants and the use of supercritical adsorbant.

Biohydrogen

Hydrogen is an important energy carrier in many biological systems, particularly in anerobic fermentation processes. Hydrogen can be produced by anerobic fermentative bacteria such as *Clostridia* as an 'electron sink', thereby providing a method of removing NADH from the system. When a substrate is oxidised during fermentative metabolism there must be a subsequent reduction and this usually generates NADH from NAD^+. The accumulation of NADH would upset the equilibrium and cause these oxidation reactions to stop and so the surplus NADH is channeled to hydrogen ions with the production of hydrogen, which is released from the cells:

$$NADH + H^+ \rightarrow NAD^+ + H_2$$

In anerobic environment, this hydrogen is an important source of energy. The methanogenic bacteria which can oxidise the hydrogen by using carbon dioxide as an electron acceptor to produce methane:

$$4H_2 + CO_2 \rightarrow CH_4 + 2H_2O$$

If the hydrogen-producing bacteria are grown in appropriate conditions in the absence of hydrogen-utilising bacteria then the hydrogen accumulates and can be collected. The feedstock for this fermentation could be cellulose, which can be hydrolysed to sugars and then fatty acids by certain anerobic bacteria

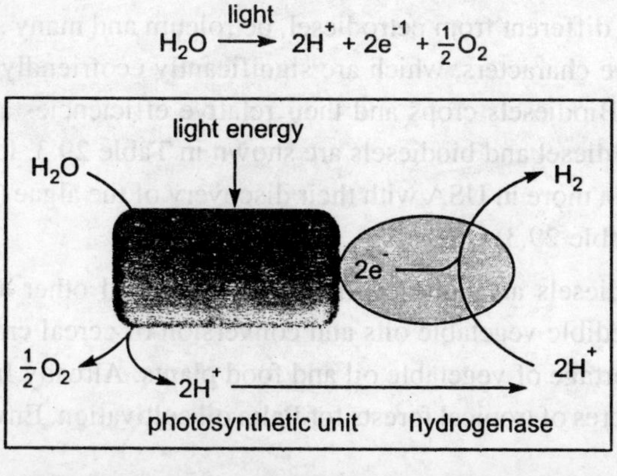

Fig. 29.8: Possible mechanism for production of hydrogen by linking photosynthesis with hydrogenase

providing a substrate for the hydrogen-generating clostridia. Some microscopic algae also produce hydrogen and there has been research into growing these organisms under sunlight in 'photobioreactors', with the additional advantage of biomass production which could be used as animal feed. A more direct approach has also been suggested in which the photosynthetic apparatus of plants is linked biochemically to the hydrogen-producing system. This possibility is shown in Fig. 29.8.

During photosynthesis the water molecule is split using light energy to yield hydrogen ions (protons), electrons and oxygen:

$$H_2O \xrightarrow{\text{light}} 2H^+ + 2e^{-1} + \frac{1}{2}O_2$$

PETROCROPS: (The fuel plants)—These are latex-bearing plants belonging to the families like Euphorbiaceae, Asclepiadaceae, Apocyanaceae, Cactaceae, which are often commercially cultivated for extraction of special group hydrocarbon from which petrol-diesel like transformed products can be made. Since middle of 1970s, commercial application of such plantation was attempted, though it is yet to be much more popularised. This is difinitely alternate non-polluting fuel except the production of CO_2 in substantial quantity.

Biodiesels are diesel like liquid fuels obtained from renewable biological substances. Any vegetable oil (edible or non-edible) and/or animal fats (lard or tallow) can be converted to biodiesel by methyl esterification and can be used in diesel engine vehicles. Biodiesels therefore are called vegetable oil methyl esters (VOME). Biodiesels are biodegradable and nontoxic. On combustion they produce 60% less net carbon dioxide emission than petro-diesel. As it is produced from atmospheric CO_2 via photosynthesis no fresh CO_2 is emitted, as it is the case with fossil fuels (Table 29.2).

Diesel engines are a type of combustion engine. Rudolf Diesel (1858-1913) originally designed the diesel engine to use vegetable oil as fuel to help agrarian society to compete with large petroleum industry in 1883 (patented). But American petroleum industry soon developed petro-diesel and developed modified diesel engines using petro-diesel, which was cheap and the first biodiesel was developed from groundnut oil. Profuse petrol supply from Middle East and cheap petrodiesel led to suppressions of biodiesel research and development. With gradual exhaustion of the petrol resources worldwide and because of environmental problems of CO_2 and N_xO gas emission from petrol or petro-diesel the possibility of alternative renewable biofuel searching started since 1970s.

Biodiesel is distinctly different from petrodiesel, petroleum and many synthetic diesels produced in recent time. Biodiesels have characters, which are significantly ecofriendly. Most vegetable oils can be converted into biodiesels. Biodiesels crops and their relative efficiencies are shown in Table 29.2 and relative efficiency of petroldiesel and biodiesels are shown in Table 29.3. It may be noted that biodiesel production potential is again more in USA with their discovery of the algae (trade secret) for commercial production of biodiesel (Table 29.3).

Environmentally biodiesels are quite ecofriendly but several other aspects are to be considered seriously. The scarcity of edible vegetable oils and conversion of cereal cropable lands to vegetable oil cultivation may lead to shortage of vegetable oil and food plants. Already Indonesia and the Philippines have cut down 10 million acres of tropical forests for Palm oil cultivation. Environmentalist and Statesmen must be careful about this.

Table 29.2: Biodiesel crops and efficiency

Crop	Kg oil/ha	L oil/ha
Corn	145	172
Rice (bran)	696	828
Soyabean	375	446
Sunflower	800	952
Pea-nut	890	1,059
Mustard	1,000	1,190
Castor	1,188	1,413
Jojoba	1,528	1,818
Jatropha carcus	1,590	1,892
Brazil nut	2,010	2,392
Coconut	2,260	2,689
Palm oil	5,000	5,950
Chinese tallow	5,500	6,545
Algae (green)*	79,832	95,000

*By biotechnology

Table 29.3: Comparison of Petrol, Petrodiesel & Biodiesel

	Petrol	P-Diesel	Biodiesel
Purification	Complex & costly	Simple & less costly	Very simple less costly
Density	15% less	High	Less than H_2O
Pre-ignition	High	Low	Very low
Greenhouse gas	Less	More than petrol	Low
Fuel economy	Low	High (20-40% more)*a	High
Boiling point	Very low	Low	High*b
Flash point	Low	High	Very high
Vehicles	Light	Heavy*c	All type*c
Quality	Low	High	Higher than diesel

*a Less consumption – overall less GH gas

*b Almost non-inflammable – less risky

*c Racing cars, airplanes, home heating

29.4 AGRICULTURAL PRODUCTION MANAGEMENT—BIOTECHNOLOGICAL APPROACH

It is well-known that modern agriculture is dependent on high-yield crop varieties, higher fertiliser and pesticide uses. Such an intensive agriculture has a number of disadvantages in the long run. Thus there needs a new approach in sustainable agriculture viz., application of biofertiliser and biopesticides. In addition, new seed varieties were developed through genetic engineering that have immense utility potential in agriculture.

(a) Biofertiliser

In recent years, use of microbial inoculants as a source of biofertilisers has become a hope for most countries, as far as economical and environmental viewpoints are concerned. Biologically fixed nitrogen is such a source which can supply an adequate amount of nitrogen to plants and other nutrients to some extent. Many free-living and symbiotic bacteria which fix atmospheric N_2 were used as biofertiliser materials as a substitute for nitrogenous fertiliser. In general, there are two kinds of biofertiliser used in modern agricultured viz., bacterial biofertiliser and algal biofertiliser. Nitrogen-fixing bacteria like *Azotobacters* sps. and *Rhizobium* sps. were mass cultured for the purpose of bacterisation. They may be used as seed-dressing materials or spray materials or even as granulated materials to be added during field preparation. Bacteria to be inoculated in soil as biofertiliser needs to be multiplied on artificial media to harvest on a large scale so that it can be supplied to farmers.

Identically, Japanese workers developed techniques for mass cultivation of BGA to be used as biofertiliser in paddy fields. Later it was established in Indian rice field also. *Azolla* water fern cultivation was also standardised as biofertiliser in rice field in various countries also.

Some of the important groups of biofertilizers are:

1. **Azolla-Anabaena Symbiosis**: *Azolla* is the globally distributed small aquatic fern comprising of 7 living species. In their leaves a heterocystous dinitrogenfixing blue green alga *Anabaena azollae* is always present as symbiont. *Azolla* is an alternative N source. The agronomic potential of Azolla as biofertilizer for rice has been recognized in many countries including India, Philippines, USA, Sri Lanka and Thailand. A trial under International Network on Soil Fertility and Fertilizer Evaluation for Rice (INSFER) conducted at 37 sites in 10 countries also confirmed the biofertilizer potential of Azolla.

2. **Nitrogen-fixing Bacteria *Rhizobium***: Symbiotic N_2-fixation by *Rhizobium* with legumes contribute substantially to total biological nitrogen fixation (BNF). *Rhizobium* inoculation is a well-known agronomic practice to ensure an adequate nitrogen nutrition of legumes in place of fertilizer nitrogen. The three rhizobial genera include *Rhizobium, Bradyrhizobium* and *Azorhizobium*. Root infection by rhizobia is a multistep process initiated by pre-infection events in the rhizosphere. Various events which lead to the formation of nodules include mutual recognition of host plant and *Rhizobium* species, rhizobial adherence to the root hairs, root hair curling, root hair infection, division of cortical cell to form root primordia and, finally, formation of nitrogen-fixing tissues in the nodules.

3. **Diazotrophs**: Diazotrophs which comprise a large group of aerobic chemolithotrophs (*Thiobacillus, Desulphovibrio*), anerbic photoautotrophs (members of Rhodospirillaceae, Chromatiaceae and Chlorobiaceae), aerobic to microaerobic heterotrophs (members of Azotobacteraceae and Bacillaceae). Corynebacteriaceae (*Azotobacter*) and Spirillaceae (*Azospirillum*) are the free-living as well as associative forms which fix nitrogen in the rhizosphere of a variety of crop species.

Diazotrophs like *Herbaspirillum* sps. grow endophytically in the stems and leaves of sugar cane and rice. *Azoarcus* inhibits roots of Kallar grass (*Leptochloa fusca*) and rice. Although many genera of bacteria are isolated from the rhizosphere of various cereals, many members of *Azotobacter* and *Azospirillum* genera have been widely tested to increase yields of cereals under field conditions.

4. **Frankia**: *Frankia* is a member of Actinomycetes or ray fungi. It forms nodules with higher plants like *Alnus* and *Elder*. There is a need to determine the defence response of a crop to *Frankia* and technologies to engineer the plant to nodule with *Frankia*. It produces true mycelium and the spores are not produced.

5. **Cyanobacteria**: Species of *Anabaena, Nostoc, Scytonema, Calothrix, Aulosira, Nodularia, Lyngbya* and *Mastigocladus*, etc., are able to fix nitrogen. Most of the crops which grow in plenty of water like rice, etc., have association of these blue green algae or cyanobacteria.

It is a common observation that if a soil is under rice cultivation there is native cyanobacterial flora and may not need fresh inoculation. If crops are changed—as is in most parts of the country—there is need to use fresh inoculum. Usually a mixture of these algae is used and not a monoculture. The cyanobacterial flakes are dried and mixed at the rate of 10^{15} kg ha^{-1} one week after rice is planted.

6. **Phosphate Solubilizing Bacteria**: A large number of heterotrophic and autotrophic soil microorganisms are now known to have the capacity to solubilize inorganic phosphates through their metabolic activities directly or indirectly. The solubilization of different types of insoluble phosphates varies with the type of microorganisms, the type of phosphates available and available carbon source.

Rhizosphere soil of wheat possesses a greater number of phosphate solubilizing bacteria than in the non-rhizosphere soil. *Bacillus* sps., *Pseudomonas* sp., *Brevibacterium* sp., *Acrobacter acrogenes, Serretia* sps., *Nitrobacter* and *Escherichia* freundii are important phosphate solubilizing and phosphate mobilizing microbes. Phosphatic biofertilizers were first prepared by (former) USSR scientists using *Bacillus megaterium* var. *phosphaticum* as phosphate solubilizing bacteria and the product was termed as 'Phosphobacterin'.

7. **Phosphate-Solubilizing Fungi**: Along with bacteria and actinomycetes certain fungi also solubilize phosphates. *Aspergillus awamori* has been found to be a better solubilizer than *Bacillus polymyxa* or *Pseudomonas striata*. There are several reports of found phosphate solubilizing bacteria (*Pseudomonas*) and fungi (*Aspergillus*) in the rhizosphere of coconut and cocoa.

8. **Plant Growth Promoting Rhizobacteria (PGPR)**: The role of PGPR in plant growth and disease management has been studied in 1980s. Several strains in Bacillus have been patented for commercial production in developed countries. The plant growth promoters belonging to pseudomonads and bacilli play a major role in plant productivity. These are usually noncolonizers. The spore-forming bacteria are promising organisms for microbial control. *Pseudomonas fluorescens* is registered as microbial pesticide against *Pythium* and *Rhizoctonia*.

9. **Mycorrhiza for Sustainable Agriculture**: The root-fungus symbiosis is caused by mycorrhizae. The association of mycorrhiza is wide-spread in bryophytes, a large number of pteridophytes, most or all species of gymnosperms and some 90 per cent or more of angiosperms.

Frank was the first to describe association of fungus and roots of higher plants and coined the term 'mycorrhiza'. Rees reported the connection between the mantle of the mycorrhizas and the basidiomata or *Elaphomyces granulatus*. Forest crops may have ectomycorrhizae while endomycorrhizae are present in most agricultural crops. These mycorrhizae differ in their structure and in the systematic position of the fungi involved. Trappe made a series of studies in *Pseudotsuga menziesii* (douglas fir) and Chilvers described 8 distinctive types of *Eucalyptus* ectomycorrhizas. Jayasinghe et al carried out observations on the mycorrhizas of *Pinus patula* in Sri Lanka and made an attempt to culture and identify the fungus. Typical ectomycorrhizas were synthesized on *Pinus ponderosa* by four fungi species viz. *Amanita muscaria, A. pantherina, Suillus granulatus* and *Lactarius deliciosus. Russula aeruginea* is another fungus forming ectomycorrhiza in *Picea*. Ectomycorrhizal fungi grow best at pH 5-6. Excess of inorganic fertilizers suppress ectomycorrhizal development. It has been found that optimum development of ectomycorrhiza in chir pine nurseries occurs when nutrients are present in the soil at half the normal level. Shading suppresses ectomycorrhizal development. The poor development of mycorrhizal root results in the stunted growth and chlorosis of leaves. The mycorrhizal plants are able to withstand better water stress condition that the nonmycorrhizal seedlings. The mycorrhizae are described as 'Biofertilizer' and 'Biocide'.

(b) Biopesticides

Pest control by biological antagonism appears to be very useful tool in recent years. A good number of organisms were already screened out (Table 29.4).

Table 29.4: Lists of Biopesticides

Organisms	Target arthopods
(a) **Bacteria**	
1. *Bacillus thuringenesis*	Lepidopteran
2. *B. cereus*	Do
3. *B. papillieae*	Scarab beetle
4. *B. sphaericus*	Mosquitoes
5. *Enterobacter aerogens*	Lepidopterans
6. *Proteus sereus*	Grasshoppers
7. *Pseudomonas septica*	Coleopterans
8. *P. aerugenes*	Grasshoppers
(b) **Fungi**	
1. *Beauveria bassiana*	Brown plant hopper
2. *Verticillium lecanil*	Aphids
3. *Nomerea rilevi*	Moths
4. *Hirsutelta thompsonii*	Citrus mite
(c) **Virus**	
1. *Cytoplasmic polyhydrosis virus*	Diptera, Lepidoptera & Coleoptera
2. *Baculovirus*	Lepidoptera, Diptera, Coleoptera
3. *Entomopox virus*	Lepidoptera, Diptera

(c) Vermicomposting Technology

It is a kind of composing of organic material by using earthworms. For past couple of decades, people are aware of the importance of different earthworm species in relative efficiency of composting of organic materials. Now the said technology was commercialised with suitable earthworm species use and moisture regulation during composting process.

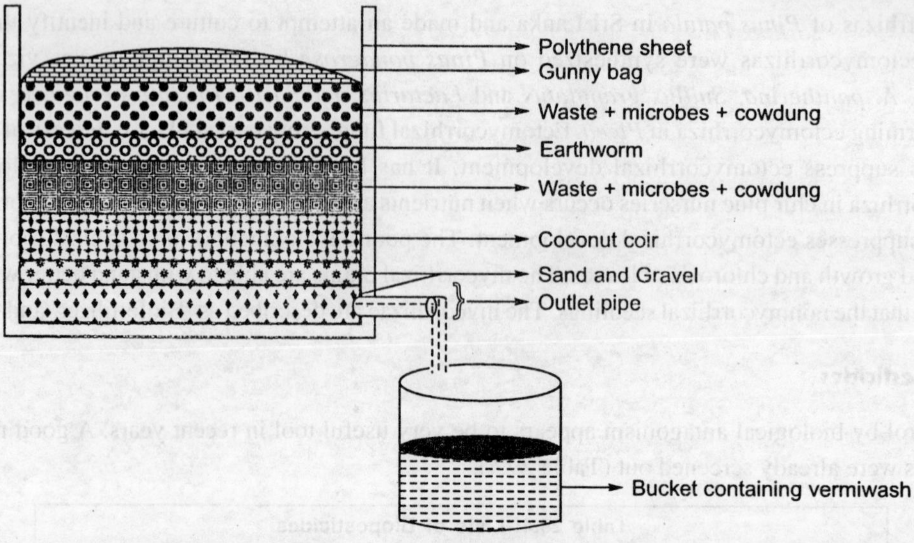

Fig. 29.9: Schematic structure of compost unit

The earthworm species *Eudrilus eugeniae* is widely used for this purpose. During the vermicomposting the substrate composition is also very important. The composted product is rich in nutrients and called vermicast and it also produces a shiny water called vermiwash which is also rich in nutrients:

(i) **Vermicomposting process**: A number of steps are involved for preparation of good quality vermicompost. Irrespective of type of compost unit, a diagrammatic representation of successive layers to be made for vermicomposting is shown in Fig. 29.9.

A black polythene sheet is spread on the pit surface uniformly. About 7.5cm layer of gravel and sand is spread on the polythene sheet. Above this, a layer of coconut coir is spread uniformly to make the vermibed. The waste materials are spread on the coconut coir to a depth of 15-20cm which is mixed with cowdung slurry and a broth culture of ligno-cellulolytic microorganisms are also mixed @ 300g/ton. Above the waste material a thin layer of well-rotted cowdung is spread. Above 200 earthworms of species *Eisenia foetida* are released on the surface of the cowdung. Each pit is covered with a gunny bag to avoid evaporation loss of water from the surface. The water is sprinkled time to time to maintain a moisture content of about 60% for the survival of earthworm and microorganisms. Earthworms feed on these wastes and consume 2-3 times their body weight and after using the feedstock for their growth, excrete the mucus-coated matter as vermicast. Vermicast consists of organic matter that has undergone physical and chemical breakdown through the activity of the muscular gizzard, which grinds the particle size to 1-2 micron. Following this process the compost gets matured in a period of 45-50 days. For large scale composting a series of compost pits can be constructed in sequence. The pits are covered with shed structures to avoid direct sunlight and rainwater.

(ii) **Vermiwash**: Besides solid vermicompost, another product is derived from composting process—liquid manure or vermiwash. At the end of the composting process pits are flushed out with water and the vermiwash is collected. In a similar way the vermiwash are collected in other compost units like metallic drums or fiberglass vats.

(iii) **Mobile compost unit for household purpose**: A metallic drum of 90 cm height and 56 cm diameter was chosen for development of a household compost unit. The mobile unit consisted of four rubber wheels and a handle. A polythene lining is provided in the drum for collection of leachate. At the base of the drum the vermibed is made by putting a layer of coconut coir and a layer of kitchen waste 30 cm in depth. Thereafter, a thin layer of well-decomposed cowdung is put and 100 earthworms were released. For collection of liquid manure, an outlet with plastic tube is made at the bottom of the drum. After maturity of the compost the same process is repeated. For maintaining aerobic condition for the survival of the earthworms, two hollow rods with perforations were inserted sideward inside the drum. Periodically air is pumped inside the drum for creating a congenial environment for survival of earthworms.

(iv) **Benefits of vermicompost and vermiwash**: The nutrients present in the vermicast or vermiwash are rich sources of macro and micro nutrients, vitamins, enzymes, antibiotics, growth hormones, etc. The nutrients present are readily available for uptake of plants. Controlled laboratory and field experiments showed that the decomposition process and maturity time of compost varies with type of substrate used. It is observed that the maturity of compost is attained faster in case of kitchen waste, followed by sesbania, paddy straw and water hyacinth. The application of mixed culture of microbes (*Paecilomyces fusisporus, Phenerocrete crysosporium, Trichoderma viride* and *Bacillus polymyxa*) has accelerated the decomposition process. The available nutrient status of the four substrates shows maximum in kitchen wastes followed by sesbania, paddy straw and water hyacinth. It is evident from the above finding that nutrient status of compost depends on the kind of waste material used and type of microbes introduced for composting. It is further observed that when earthworms were introduced in addition to microbial inoculant the maturity of compost becomes faster and the nutrient status of the product was also enhanced.

(v) **Method of application**: The vermicasts are applied at varying doses in various types of crops, i.e., agricultural, horticultural, forestry, etc. In case of field crops like cereals, legumes, oil seeds, sugar cane, cotton, vegetables, etc., the application rate varies 300 - 500 kg/acre. The material is either broadcasted or applied as side-dressing depending upon the crop and mixed with soil uniformly. In case of horticultural crops like coconut, rubber, arecanut, mango, cashew and other plantation crops, 1-3 solid vermicompost per tree in a year are applied. On area basis 1-2 metric tonnes/acre are applied for banana, flowers, grapes, tea, coffee and mulberry. In lawns or in nursery the vermicompost is applied @ 1 kg/m^2 on top soil and in potted soil 1 part of vermicompost, 1 part of sand and 1 part of garden soil are mixed well for plantation. For vermiwash the liquid extract is diluted 5-10 times depending upon the liquid concentration. It is generally applied by spraying to the standing crop, or it can be applied directly to soil.

29.5 BIOREMEDIATION

Concept

Bioremediation is a treatment technology that uses biodegradation of organic and metallic contaminants through stimulation of indigenous microbial populations by providing certain amendments, such as adding oxygen, limiting nutrients, or adding exotic microbial species. It uses naturally occurring or externally-applied microorganisms to degrade and transform hazardous organic constituents into compounds of reduced toxicity and/or availability. Specific technologies fall into two broad categories: (1) Ex situ technologies (e.g. slurry phase, land treatment, solid phase, composting), and (2) In situ technologies. Active remediation can include addition of amendments such as nutrients or oxygen while passive remediation utilizes natural attenuation to adequately characterize, model and monitor the site to evidence natural attenuation and protection of potential receptors.

Key Features of Bioremediation

- Most bioremediation treatment technologies destroy the contaminants in the soil matrix.
- These treatment technologies are generally designed to reduce toxicity either by destruction or by transforming toxic organic compounds into less toxic compounds.
- Indigenous microorganisms, including bacteria and fungi, are most commonly used. In some cases, wastes may be inoculated with specific bacteria or fungi known to biodegrade the contaminants in question. Plants may also be used to enhance biodegradation and stabilize the soil.
- The addition of nutrients or electron acceptors (such as hydrogen peroxide or ozone) to enhance growth and reproduction of indigenous organisms may be required.
- Field application of bioremediation may involve:
  - Excavation
  - Soil handling
  - Storage of contaminated soil piles
  - Mixing of contaminated soils
  - Aeration of contaminated soils
  - Injection of fluid
  - Extraction of fluid
  - Introduction of nutrients and substrates

Kinds of Bioremediation

There are several kinds of bioremediation processes, some of them are listed:

- Natural Attenuation
- Aerobic/Anaerobic biodegradation
- Biopiles
- Land Treatment
- Bioscrubbers
- Methanotrophic Process (in situ)
- Plant Root Uptake (Phytoremediation)
- Solid Phase Bioremediation
- Bio Wall for Plume Decontamination (in situ)
- Biodegradation
- Composting
- Bioreactors
- Dehalogenation
- Binding of Metals
- Fungi Inoculation Process
- Slurry Phase Bioremediation
- Bioventing (Chapter 7: BMPS for Vapor Extraction)
- Bioremediation of Metals (Changing the Valence)

Descriptive accounts of some bioremediation methods are given on next page.

Solid Phase Bioremediation

The solid phase bioremediation treatment can be conducted in lined land treatment units or in composting piles. A lined land treatment unit consists of a prepared bed reactor with a leachate collection system and irrigation and nutrient delivery systems. The unit may also contain air emission control equipment.

Bioventing

Bioventing uses relatively low-flow soil aeration techniques to enhance the biodegradation of soils contaminated with organic contaminants. Although bioventing is predominantly used to treat unsaturated soils, applications involving the remediation of saturated soils and groundwater (augmented by air sparging) are becoming more common. Generally, a vacuum extraction, an air injection, or a combination of both systems is employed. An air pump, one or more air injections or vacuum extraction probes, and emissions monitors at the ground surface level are commonly used (Fig. 29.10(b)).

Landfarming

Ex situ processes also include landfarming, which involves spreading contaminated soils over a large area. Bioremediation may also be conducted in a bioreactor, in which the contaminated soil or sludge is slurried with water in a mixing tank or a lagoon. Bioremediation systems require that the contaminated soil or sludge be sufficiently and homogeneously mixed to ensure optimum contact with the seed organisms (USEPA, 1980).

Bioreactors

Bioreactors function in a manner that is similar to sewage treatment plants. There are many ways in which a bioreactor can be designed; but most are a modification of one of two systems. In the first system, which is often referred to as a trickling filter or fixed media system, the aqueous waste stream is allowed to trickle over a solid support, such as rocks, that have been colonized extensively by microorganisms. As the liquid waste stream passes over the solids, the microorganisms break down the contaminants. Before the treated waste stream can be discharged, it must be clarified so that the number of microorganisms present in the discharge is in compliance with the regulations. This system can also be referred to as a biofilter, which is also used to treat contaminated gas streams. For this form of treatment to be effective, the contaminant must be volatile.

The process of bioremediation enhances the rate of the natural microbial degradation of contaminants by supplementing these microorganisms with nutrients, carbon sources or electron donors. This can be done by using indigenous microorganisms or by adding an enriched culture of microorganisms that have specifc characteristic that allow them to degrade the desired contaminant at a quciker rate. Ideally, bioremediation results in the complete mineralization of contaminants to H_2O and CO_2 without the build up of intermediates. Bioremediation processes can be broadly categorized into two groups: ex situ and in situ. Ex situ bioremediation technologies include bioreactors, biofilters, land farming and some composting methods. In situ bioremediation technologies include bioventing, biosparging, biostimulation, liquid delivery systems and some composting methods. The newly introduced phytoremediation treatment, involving plant uptake/degradation of contaminants, can be used either as an in situ or ex situ method. In situ treatments tend to be more attractive to vendors and responsible parties because they require less equipment, generally are more cost effective and cause less disturbance to the environment. However, the difficulties associated with implementing the in situ processes have limited their application in the field. Bioremediation using white-rot fungi to innoculate contaminated media is a promising technology that is currently being researched. This technology can be used for both ex situ or in situ treatment. Generally, this fungi is used to innoculate a composting process, but it does have other bioremediation applications.

In situ Biodegradation

The in situ biodegradation process is generally used in conjunction with the groundwater pumping and soil flushing systems to circulate nutrients and oxygen through a contaminated aquifer and the associated soil (Freeman and Harris, 1995). The process usually involves introducing aerated nutrient-enriched water into the contaminated zone through a series of infiltration galleries and injection wells and recovering the water downgradient. The recovered water can then be treated, if necessary, and reintroduced into the soil on site (where allowed by the applicable regulations). The in situ biodegradation system may also include aboveground treatment and conditioning of the infiltration water with nutrients and an oxygen source or another electron acceptor (Fig. 29.10a).

Ex situ Bioremediation

Ex situ processes also include landfarming, which involves spreading contaminated soils over a large area. Bioremediation may also be conducted in a bioreactor, in which contaminated soil or sludge is slurried with water in a mixing tank or a lagoon. Bioremediation systems require that the contaminated soil or sludge be sufficiently and homogeneously mixed to ensure optimum contact with the seed organisms (USEPA, 1990d). There are many variation of the basic bioreacter designs that have been applied for the treatment of pesticides and other hazardous wastes.

29.6 PHYTOREMEDIATION

Phytoremediation is an emerging technology which involves the use of different types of plants to degrade, extract, contain, or immobilize contaminants from the soil and water (USEPA, 2000). In recent years, phytoremediation has been used at a number of sites in the United States and other countries including India.

Phytoremediation can be applied to both organic and inorganic pollutants, present in solid substrates (e.g. soil), liquid substrates (e.g. water), and the air. Phytoremediation is currently divided into the following areas:

- **Phytoextraction**: the use of pollutant-accumulating plants to remove metals or organics from soil by concentrating them in the harvestable parts;

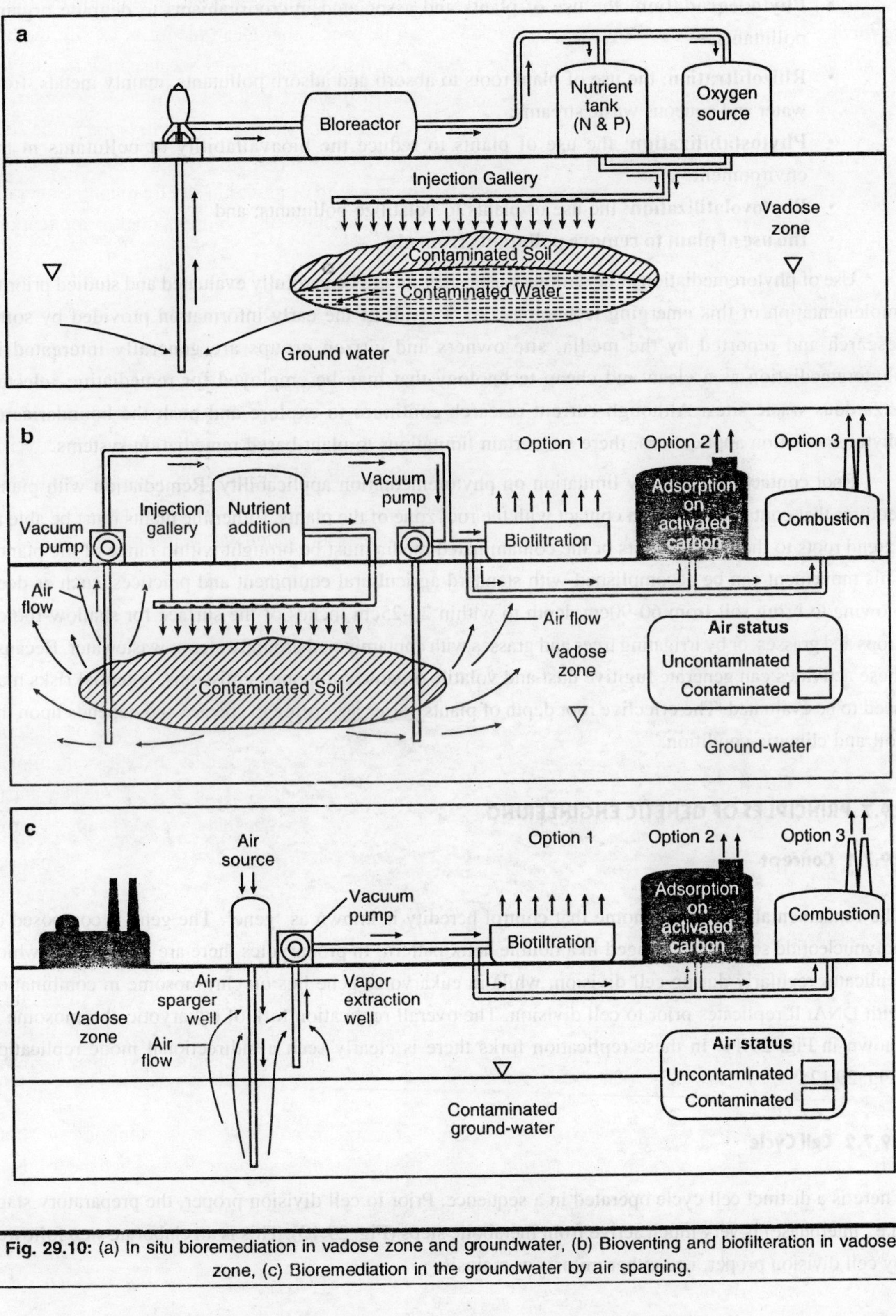

Fig. 29.10: (a) In situ bioremediation in vadose zone and groundwater, (b) Bioventing and biofilteration in vadose zone, (c) Bioremediation in the groundwater by air sparging.

- **Phytodegradation**: the use of plants and associated microorganisms to degrade organic pollutants;
- **Rhizofiltration**: the use of plant roots to absorb and adsorb pollutants, mainly metals, from water and aqueous waste streams;
- **Phytostabilization**: the use of plants to reduce the bioavailability of pollutants in the environment;
- **Phytovolatilization**: the use of plants to volatilize pollutants; and
- **the use of plant to remove pollutants from air**.

Use of phytoremediation at hazardous waste sites should be carefully evaluated and studied prior to implementation of this emerging technology. As a result of the early information provided by some research and reported by the media, site owners and citizen groups are generally interested in phytoremediation as a clean and cheap technology that may be employed for remediating selected hazardous waste sites. Although current research continues to explore and push the boundaries of phytoremediation applications, there are certain limitations to plant-based remediation systems.

Root contact is a primary limitation on phytoremediation applicability. Remediation with plants requires that contaminants be in contact with the root zone of the plants. Either the plants must be able to extend roots to the contaminants or the contaminated media must be brought within range of the plants. This movement can be accomplished with standard agricultural equipment and practices, such as deep plowing to bring soil from 60-90cm depth to within 20-25cm inches of the surface for shallow-rooted crops and grasses, or by irrigating trees and grasses with contaminated groundwater or wastewater. Because these activities can generate fugitive dust and volatile organic compound emissions, potential risks may need to be evaluated. The effective root depth of plants varies according to species and depends upon the soil and climatic condition.

29.7 PRINCIPLES OF GENETIC ENGINEERING

29.7.1 Concept

The fundamental unit of a genome that control heredity is known as 'gene'. The gene is composed of polynucleotide segments arranged in a double helix pattern. In prokaryotes there are naked DNA which replicates regularly during cell division, while in eukaryotes gene lies on chromosome in combination with DNA. It replicates prior to cell division. The overall replication fork of eukaryotic chromosome is shown in Fig. 29.11. In these replication forks there is clearly seen a bidirectional mode replication (Fig. 29.12).

29.7.2 Cell Cycle

There is a distinct cell cycle operated in a sequence. Prior to cell division proper, the preparatory stage (i.e., interphase) is very much active from metabolic steps (Fig. 29.13). This is an elaborate step followed by cell division proper, i.e., either mitosis or meiosis.

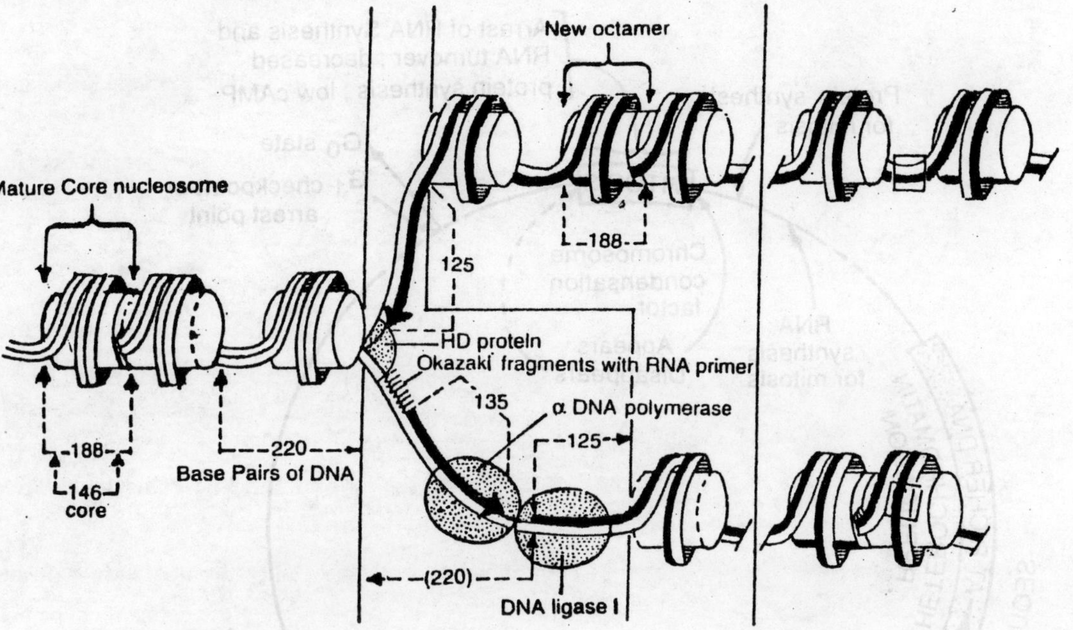

Fig. 29.11: Replication fork of an eukaryotic chromosome

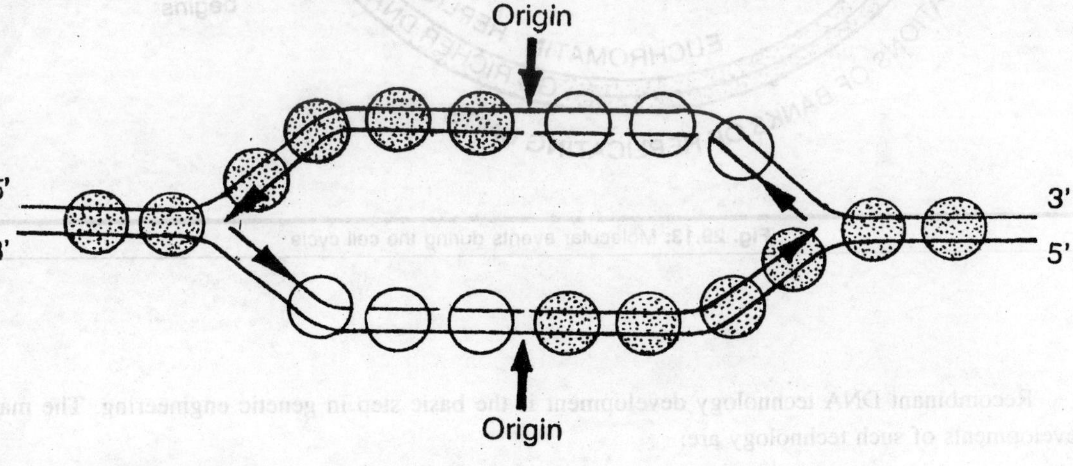

Fig. 29.12: Hypothetical segregation of new nucleosomes to the lagging strand at the replication fork

There are two distinct categories of cell division viz., vegetative division or mitosis with four steps (Fig. 29.14 A & B — Prophase; C & D — Metaphase; E — Anaphase; and F — Telephase), and the other meiosis or reductional divisions in germ cells with elaborate divisional phases. The details of two divisional stages with chromosomal behaviour is shown in Fig. 29.15.

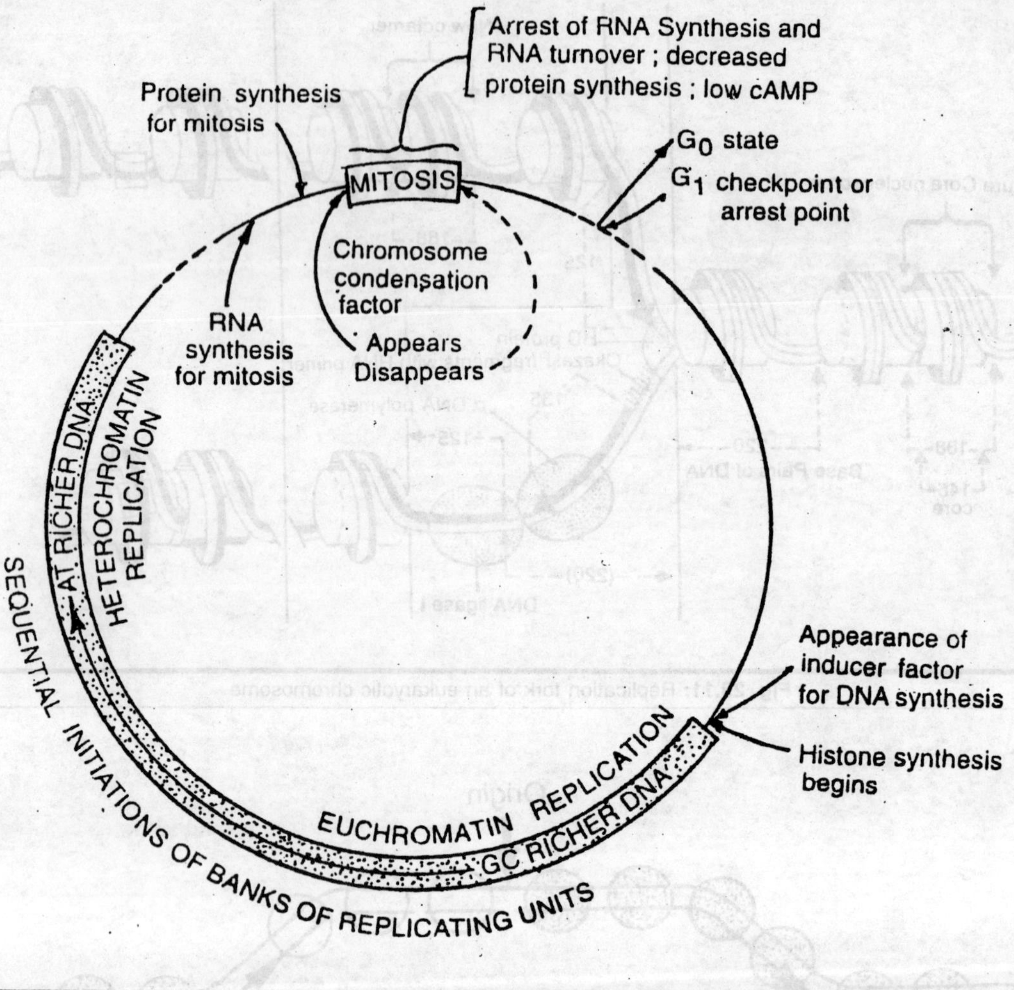

Fig. 29.13: Molecular events during the cell cycle

Recombinant DNA technology development is the basic step in genetic engineering. The major developments of such technology are:

Date (years)	Development
1970	Early discoveries of gene isolation and gene synthesis;
1978	Application of DNA technology;
1980	Direct DNA infection technique discovery;
1990	Trangenic plant/animal development.

The basic step in genetic engineering is, thus, gene cloning.

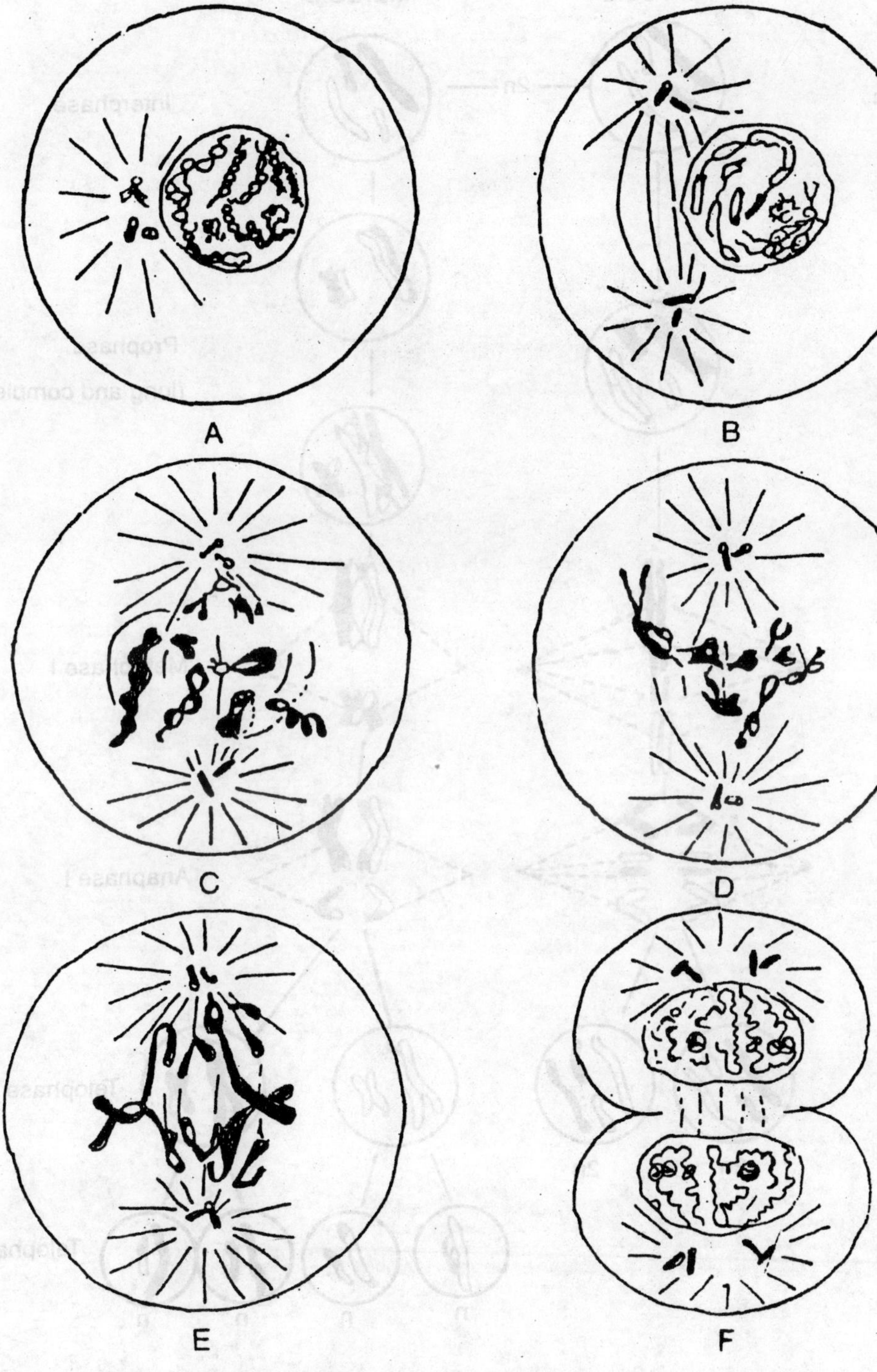

Fig. 29.14: General diagram of mitosis

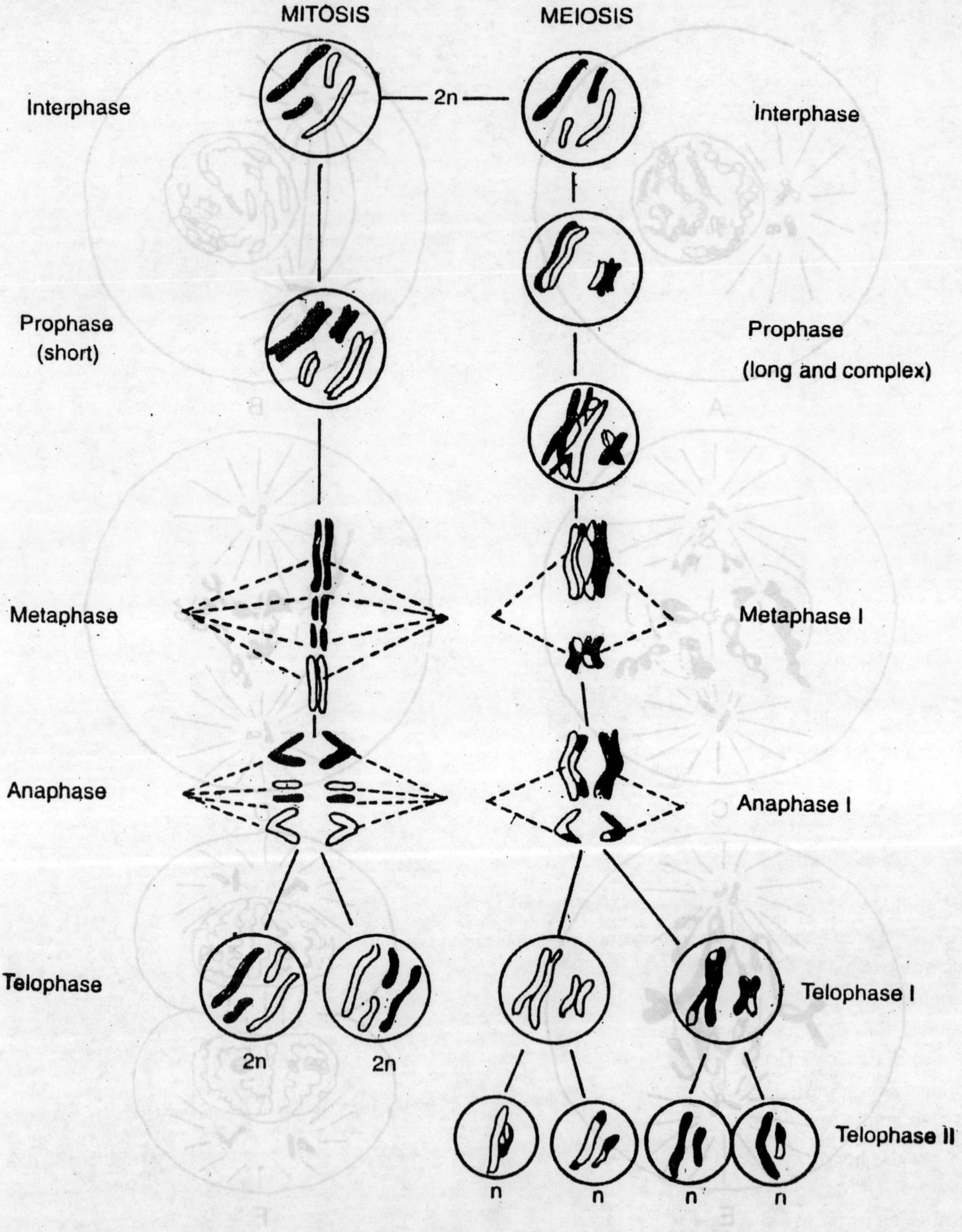

Fig. 29.15: Comparative diagram of mitosis and meiosis in idealized cells having four chromosomes (2n)

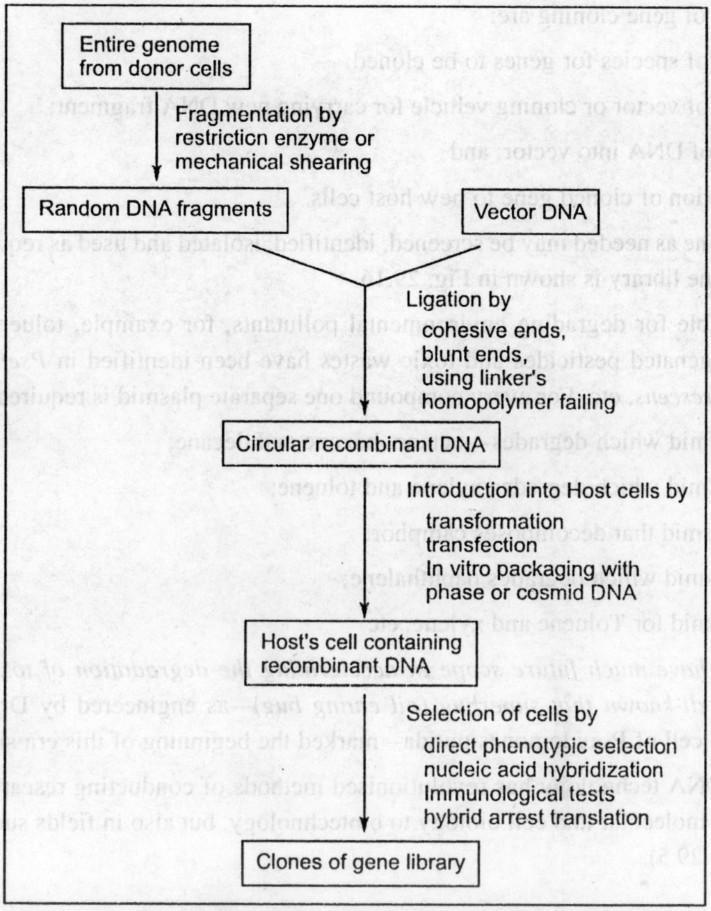

Fig. 29.16: Methods of construction of gene library

29.7.3 Gene Cloning

The cloning experiments of Herbert Boyer, Stanley Cohen, Paul Berg and their colleagues—in the early 1970s—ushered in the era of recombinant DNA technology. A gene can now be separated from the rest of the chromosomal DNA and transferred to a well-studied foreign host cell for further study. Genetic material introduced into foreign cells is replicated and passed on to progency cells. Today many methods are available for isolating and characterizing genes and proteins. Selected genes are being transferred to organisms such as plants, animals, bacteria and fungi for a variety of reasons. Commercially desirable products can be efficiently produced in host cells, genes and their proteins can be studied in ways not possible before and new medical biotechnologies are being explored.

At the heart of most molecular genetic technologies is the gene. A gene must be isolated and characterized before it can be used in genetic manipulations. One method of isolating and amplifying a DNA of interest is to clone the gene by inserting it into a DNA molecule that serves as a vehicle or vector. When these two DNAs of different origin are combined, the result is a recombinationant DNA molecule. The molecule is moved into a host–*Escherichia coli*, where it can be reproduced. Thus the DNA contained within the recombinant molecule is cloned. Gene cloning has many uses in agriculture, medical, pharmaceutical and environmental biotechnology.

The basic steps of gene cloning are:

- Selection of species for genes to be cloned;
- Selection of vector or cloning vehicle for carrying new DNA fragment;
- Insertion of DNA into vector; and
- Incorporation of cloned gene to new host cells.

In general, a clone as needed may be screened, identified, isolated and used as required. The method of construction of gene library is shown in Fig. 29.16.

Genes responsible for degrading environmental pollutants, for example, toluene, chlorobenzoic acids and other halogenated pesticides and toxic wastes have been identified in *Pseudomonas putida, P. aeruginosa, P. flurescens*, etc. For every compound one separate plasmid is required, for example:

- OCT plasmid which degrades—octane, hexane and decane;
- XYL plasmid which degrades xylene and toluene;
- CAM plasmid that decomposes camphor;
- NAH plasmid which degrades naphthalene;
- TOL plasmid for Toluene and xylene, etc.

Such plasmids have much future scope in accelerating the degradation of toxic wastes on our environment. It is well-known that superbug (oil eating bug)—as engineered by Dr. Ananda Mohan Chakravorty into the cell of Pseudomonas putida—marked the beginning of this era since 1970s.

Recombinant DNA technology has revolutionised methods of conducting research in a variety of fields, not only from molecular and cell biology to biotechnology, but also in fields such as ecology and anthropology (Table 29.5).

Table 29.5: Some application of recombinant DNA technology

Field		Application		Importance
Basic biology	→	DNA sequencing	→	Gene structure study
		Direct Mutagenesis	→	Gene function study
Medicine	→	Therapeutic proteins	→	Treatment of genetic diseases
		Gene therapy		
		Improved vaccines		
		Diagnosis		
		Veterinary medicine		
Industry	→	Altering microorganisms	→	Pharmaceutical industry
Agriculture	→	Altering plants, Altering farm animals	→	Improved breeds
Criminal investigation	→	DNA fingerprinting	→	Biological diagnosis
Environment	→	Biotreatment of toxicants	→	Bioremediation

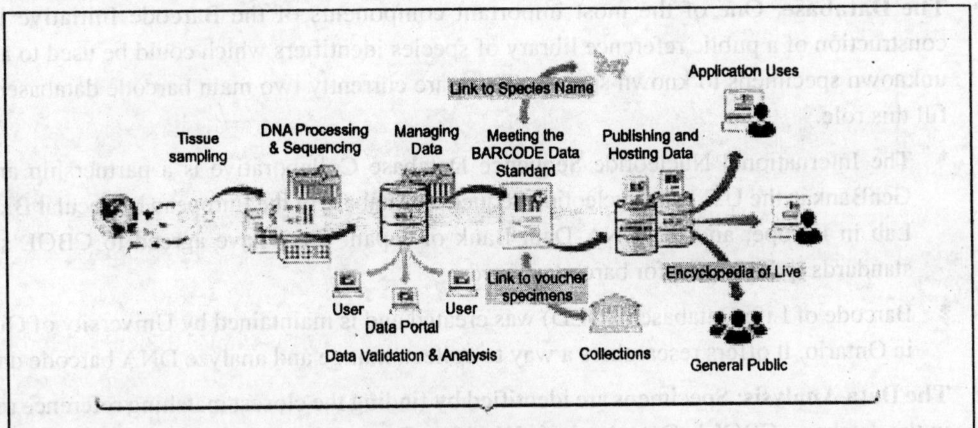

THE BARCODING PIPELINE

What is DNA Barcoding?

In 2003, Paul Hebert, researcher at the University of Guelph in Ontario, Canada, proposed "DNA barcoding" as a way to identify species. Barcoding uses a very short genetic sequence from a standard part of the genome the way a supermarket scanner distinguishes products using the black stripes of the Universal Product Code (UPC). Two items may look very similar to the untrained eye but in both cases the barcodes are distinct.

Until now, biological specimens were identified using morphological features like the shape, size and colour of body parts. In some cases a trained technician could make routine identifications using morphological "keys" (step-by-step instructions of what to look for) but in most cases an experienced professional taxonomists is needed. If a specimen is damaged or is in an immature stage of development, even specialists may be unable to make identifications. Barcoding solves these problems because even nonspecialists can obtain barcodes from tiny amount of tissue. This is not to say that traditional taxonomy has become less important. Rather, DNA barcoding can serve a dual purpose as a new tool in the taxonomists toolbox supplementing their knowledge as well as being an innovative device for non-experts who need to make a quick identification.

The gene region that is being used as the standard barcode for almost all animal groups is a 648 base-pair region in the mitochondrial cytochrome C oxidase 1 gene ("COI"). COI is proving highly effective in identifying birds, butterflies, fish, flies and many other animal groups. COI is not an effective barcode region in plants because it evolves too slowly, but two gene regions in the chloroplast, matK and rbcL, have been approved as the barcode regions for plants.

Barcoding projects have four components:

- **The Specimens**: Natural history museums, herbaria, zoos, aquaria, frozen tissue collections, seed banks, type culture collections and other repositories of biological materials are treasure troves of identified specimens.
- **The Laboratory Analysis**: Laboratory protocols (pdf; 400Kb) can be followed to obtain DNA barcode sequences from these specimens. The best equipped molecular biology labs can produce a DNA barcode sequence in a few hours. The data are then placed in a databased for subsequent analysis.

- **The Database**: One of the most important components of the Barcode Initiative is the construction of a public reference library of species identifiers which could be used to assign unknown specimens to known species. There are currently two main barcode databased that fill this role.
  - * The International Nucleotide Sequence Database Collaborative is a partnership among GenBank in the U.S., the Nucleotide Sequence Database of the European Molecular Biology Lab in Europe, and the DNA Data Bank of Japan. They have agreed to CBOL's data standards (pdf; 30 Kb) for barcode records.
  - * Barcode of Life Database (BOLD) was created and is maintained by University of Guelph in Ontario. It offers researchers a way to collect, mange and analyze DNA barcode data.
- **The Data Analysis**: Specimens are identified by finding the closest matching reference record in the database. CBOL's Data Analysis Working Group has created the Barcode of Life Data Portal which offers researchers new and more flexible ways to store, manage, analyze and display their barcode data.

30 NATURAL DISASTERS

30.1 INTRODUCTION

Natural disasters, with far-reaching and severe physical, ecological and socio-economic consequences, can occur suddenly like earthquakes, floods, tsunamis, volcanic eruptions, cyclones and landslides. They can also occur slowly, like drought and desertification which are caused by natural phenomena, or by natural phenomena coupled with overt effects of human actions. Natural disasters are becoming increasingly significant—both in terms of numbers of events and magnitude of impact. Evidence suggests that the frequency of natural disasters has increased in recent decades although there has also been a marked increase in awareness of phenomena such as earthquakes, volcanic eruptions, hurricanes, cyclones, floods and landslides due to direct and thorough reporting in the mass media. The description of natural events as disasters can be applied to those which have a major effect on populated areas or on those of economic importance to humankind.

The most widely accepted reason for the rise in number of natural disasters over the last few decades is that the "hit rate" has increased considerably due to the continuing growth of the world population (6 billion in 2010). This population growth has led to an increase in building density caused by the increasing concentration of people and economic assets in urban areas and by the large-scale migration of populations throughout the world into coastal regions that are generally more highly exposed to natural disasters. Already more than a third of the world's largest and fastest growing cities are located in regions of high seismic risk of natural disasters (Tyler, 1990). Many of these are in developing countries where poor construction methods exacerbate the risk. There is also some concern that global change throughout the twentieth century may be changing climatic conditions and increasing the severity and frequency of phenomena such as windstorms and floods (Berz, 1990).

Until relatively recently, there has been little collective response to the threat of natural disasters. One of the most effective means of reducing the adverse consequences of natural disasters is to ensure that communities are prepared for them. UNEP continues to promote the adoption of environmentally-based measures to develop community preparedness further. However, there is now a growing awareness of the need for documentation and data to supplement progress in science and technology and so aid political and economic decision-making leading to a significant reduction in disaster losses. Currently (after 1990s), there are only a few global databases for natural disasters. Of them, UNINET, the computer network for disaster-related information developed by the office of the United Nations Disaster Relief Coordinator (UNDRO), has been one of the most comprehensive and includes a facility for direct communication to aid disaster management. However, UNINET was suspended in 1988. UNDRO released, in 1990, a 'Preliminary Study of the Identification of Disaster-prone Countries Based on Economic Impact' (UNDRO, 1990). The United Nations Educational, Scientific and Cultural Organisation (UNESCO) and the World Meteorological Organisation (WMO) also support relevant data-gathering and research into an understanding of the basic phenomena causing disasters.

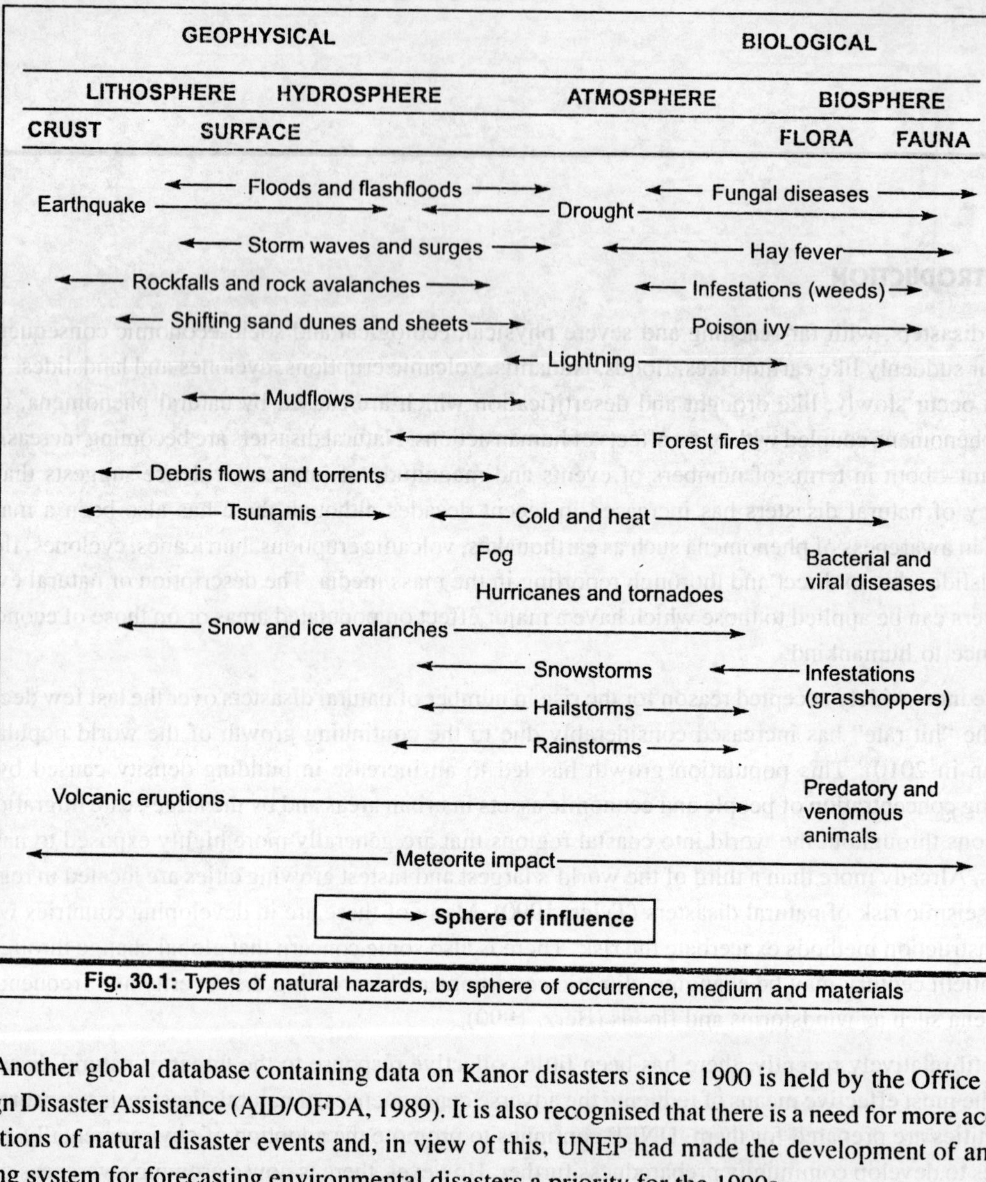

Fig. 30.1: Types of natural hazards, by sphere of occurrence, medium and materials

Another global database containing data on Kanor disasters since 1900 is held by the Office of US Foreign Disaster Assistance (AID/OFDA, 1989). It is also recognised that there is a need for more accurate predictions of natural disaster events and, in view of this, UNEP had made the development of an early warning system for forecasting environmental disasters a priority for the 1990s.

Quantifying the scale of natural disasters is not easy and is the cause of some debate. Almost all databases concentrate on three main criteria: the number of people killed, the number of people affected and the damage caused, estimated in financial terms (usually in US dollars). The last of these three has been subjected to the criticism that exchange and inflation rates render the figures meaningless and that they are particularly irrelevant when listed against numbers of dead, injured and homeless (Cruden, 1990). Published listings of disasters are often restricted to those in which at least 10 deaths were reported or property damage exceeded US$ 1 million. This approach necessarily results in the omission of numerous events which occur in remote regions but which can be recorded by global monitoring networks, such as those for seismic activity, and by satellite measurements.

The types of natural hazards and their sphere of influence is depicted in Fig. 30.1 and Table 30.1.

Table 30.1: Types of hazards/disasters

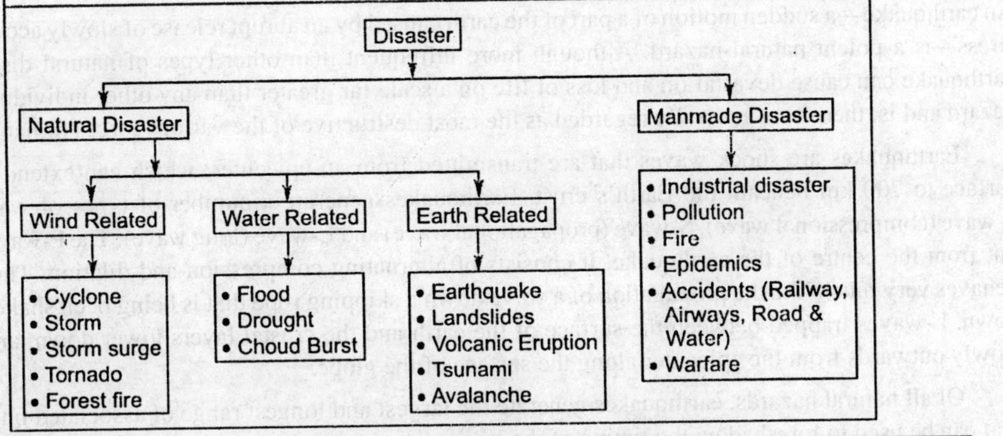

Box 30.1: Major natural disasters (1906 to 2011)

Year	Location	Disaster	Deaths
1906	San Francisco, USA	earthquake	3,000
1915	Avezzano, Italy	earthquake	32,610
1920	Gansu, China	earthquake	235,000
1923	Tokyo, Japan	earthquake	142,800
1931	Yangtze Kiang, China	flood	140,000
1935	Quetta	earthquake	35,000
1938	USA	hurricane	600
1939	Erzican, Turkey	earthquake	32,740
1942	India	cyclone	61,000
1954	Dongting area, China	flood	40,000
1960	Agadir, Morocco	earthquake	12,000
1965	Florida and Louisiana, USA	Hurricane Betsy	75
1970	East Pakistan	cyclone	300,000
1970	Chimbote, Peru	earthquake and landslide	67,000
1976	Guatemala City, Guatemala	earthquake	22,084
1976	Tangshan, China	earthquake	290,000
1985	Mexico City, Mexico	earthquake	10,000
1985	Mt. Nevado del Ruiz, Colombia	volcano	24,740
1988	Caribbean and Central America	Hurricane Gilbert	355
1988	Spitak, Armenia	earthquake	25,000
1989	San Francisco, USA	earthquake	68
1990	Western Europe	winter storms	230
1991	Bangladesh	Cyclone Gorky and storm surge	139,000
1991	Kyushu and Hokkaido, Japan	Typhoon Mirielle	62
1992	Florida and Louisiana, USA	Hurricane Andrew	62
1993	Maharashtra, India	earthquake	6,348
1995	Kobe, Japan	earthquake	100,000
1998	Nicaragua, Honduras	Hurricane Mitch	9,200
1999	Izmit, Turkey	earthquake	17,000
1999	Taichung, Taiwan	earthquake	2,400
2001	Gujarat, India	earthquake	20,000
2010	Port Au Prince, Haiti	earthquake	680,000
2011	Fukashima Daiichi, Japan	earthquake	10,000
2013	Uttarakhand Flashflood, India	Flood	5,000

30.2 EARTHQUAKES

An earthquake—a sudden motion of a part of the earth caused by an abrupt release of slowly accumulating stress—is a potent natural hazard. Although more infrequent than other types of natural disasters, an earthquake can cause devastation and loss of life on a scale far greater than any other individual natural hazard and is, therefore, generally regarded as the most destructive of the various forces of nature.

Earthquakes are shock waves that are transmitted from an epicenter, which can extend from the surface to 700 km beneath the Earth's crust. Earthquakes generate a number of types of waves, viz., P-wave (compressional wave), S-wave (propagational wave) and L-wave (long wave). The P-waves spread out from the centre of the earthquake. It consists of alternating compression and dilation. The S-wave behaves very much like the propagation of a wave down a skipping rope that is being been shaken up and down. L-waves trapped between the surface of the earth and the crustal layers lower down, and spread slowly outwards from the epicenter along the surface of the globe.

Of all natural hazards, earthquakes generate the largest and longest range of associated phenomena that can be used to foreshadow the timing of the impending events. Precursors of earthquakes group into the following five categories: land deformation, seismic activity, geomagnetic and geo-electric activity, ground water and natural phenomena (like catfish movement, aggressiveness of pigs, etc.). However, foolproof prediction of large earthquakes is still not possible.

Seismic activity is well-monitored and documented throughout the world by a number of national and international recording networks. Comprehensive listing of seismic events are held by several agencies such as the International Seismological Centre (ISC) in Newbury, UK, and the National Earthquake Information Centre at the World Data Centre A in Boulder, Colorado, USA. Much research has also been conducted on areas of the world particularly prone to earthquakes with a view to improving predictive capabilities.

At least 35 countries face a high probability of being stuck by earthquakes (NAS, 1987). Although there are thousands of relatively small earthquakes around the world each year, it is those of magnitudes 6-8 on the Richter Scale which generally cause the greatest damage.

Box 30.2: Major Earthquakes in India (occurs frequently due to the drifting of the Indian plate)

Year	Place	Death	Intensity (Richter scale)
1905 (Apr. 4)	Kangra (HP)	Major	8.0
1934 (Jan. 15)	Bihar-Nepal Border	Extensive	8.3
1941 (Jun. 26)	Andaman Islands	Major	8.1
1950 (Aug. 15)	Assam	Major	8.5
1975 (Jan. 19)	Lahul Spiti (J & K)	Extensive	6.2
1988 (Aug. 21)	Bihar-Nepal Border	900	6.5
1991(Oct. 20)	Uttarkashi (U.P.)	1600	6.6
1993 (Sept. 30)	Latur (Maharashtra)	10,000	6.3
1997 (May 22)	Jabalpur (MP)	40	6.0
1999 (Mar 29)	Chamoli (UP)	Extensive	6.8
2001 (Jan. 26)	Bhuj (Gujrat)	20,000	6.9

Risk Zone in India:

No Risk	Zone I — Bangalore, rest of Karnataka
Low Risk	Zone II — Bhopal, most of Central India
Moderate Risk	Zone III — Chennai, Lucknow, Kolkata
High Risk	Zone IV — Delhi, Patna, Mumbai, Ahmedabad
Very High Risk	Zone V — Chamoli, Hardwar, NE region, Sub-Himalayan region

Each year on average, earthquakes kill 10,000 people on average and cause $US 400 million property damage. The Calcutta (India) earthquake of 11 October 1737 wiped out half the city killing 300,000 people.

Tokyo (Japan) earthquake of 1923 totally destroyed the largest city of Japan, having a death toll of 143,000 people.

Tangshan (China) earthquake in 1976, is another worst case claiming the death of 250,000 people.

Seismic risk zone mapping was done through a variety of studies in many countries of Asia, Australia, and Alaska region. In India, risk zones (Zone I to V) were made depending on the geomorphic and other features of landforms. Appropriate land based activities and build structures were suggest as a mitigative measures.

30.3 TSUNAMI

Tsunami, or seismic sea waves, are large ocean waves generated by impulses from geophysical events occurring on the ocean floor or along the coastline, such as earthquakes, landslides and volcanic eruptions. Mostly occurring in the Pacific Ocean, tsunamis, although hardly noticeable at sea, can reach gigantic proportions as they reach shallow, coastal waters. In Hawaii and Japan, for example, tsunamis have been known to reach 30 m in height. At least 22 countries along the rim of the Pacific are estimated to be at risk from potential tsunami (NAS, 1978).

The fact that tsunamis can travel 10,000 km at velocities exceeding 900 km per hour with little loss of energy and are, therefore, capable of hitting areas not directly affected by the inducing event, has led to the establishment of a tsunami early warning service for the whole circum-Pacific area (Munich Reinsurance Company, 1988). However, only a few of the 22 countries most at risk are considered to have standard operating procedures for immediate evacuation or reliable, rapid communication systems capable of receiving real-time warnings from the Pacific Tsunami Warning Centre.

About 6,000 people have been killed by tsunami during 1977-1986 alone (NAS, 1987). Probably the best documented of these events is the occurrence at Noshiro, Japan, in 1983 which caused approximately 100 deaths and extensive property damage and flooding. The tsunami (Dec. 2004) in South East Asia lead to a death tool of over 2.5 lakhs peoples of Indonesia, Sri Lanka, Sumatra and India.

Tsunamis have multiple origin—16.5 per cent resulted from tectonic earthquakes associated with the eruption, 20 per cent from pyroclastic (ash) flows or surges hitting the ocean, 14 per cent from submarine eruptions, 7 per cent resulted from the collapse of the volcano and subsequent caldera formation, 5 per cent from landslides or avalanches, 3 per cent from atmospheric shock waves and 25 per cent had no discernible origin, but probably were produced by submerged volcanic eruptions.

A partial geographical distribution of tsunamis is given in Table 30.2:

Table 30.2: Percentage of tsunami in the world's Ocean and Seas

Location	Percent
Atlantic east coast	1.6
Mediterranean	10.1
Bay of Bengal	0.8
East Indies	20.3
Pacific Ocean	25.4
Japan–Russia	18.6
Pacific east coast	8.9
Caribbean	13.6
Atlantic west coast	0.4
Fukushima Daiichi, Northeast cost of Japan	5.0

Over past two thousand years there have been 10,00,000 deaths attributed to tsunami in the Pacific region alone. Earlier Pacific Tsunami warning system was established for forecasting the event. Now global network was established in all Oceans & Seas.

30.4 WINDSTORMS

Judged by the frequency with which they cause damage and by the surface area of the regions they strike, windstorms can be said to be the most significant of all natural hazards. Windstorms influence precipitation systems, floods and, most importantly, cause severe destruction to crops and properties. Severe tropical cyclones (called "hurricanes" in the Atlantic, Caribbean and northeastern Pacific; "typhoons" in the western Pacific; and "cyclones" in the Indian Ocean and in the sea around Australia), tornadoes, monsoons and thunderstorms between them affect every country in the world.

Today, increasing attention is being paid to windstorms, particularly tropical cyclones as some scientists see their incidence as being a possible indicator of global climatic change and predict an increase in their frequency. Have tropical cyclone frequencies or their intensities increased with global changes throughout the last century? At present, available evidence does not support this idea, perhaps because the warming is not yet large enough to make its impact felt (WMO/UNEP, 1990).

Global information on Kanor windstorms and their impact is collated by organisations such as UNDRO, UNEP and AID/OFDA. However, global listings of disasters rarely include those which occur in small states such as island states, which in areas such as the Caribbean, Indian Ocean and South Pacific are particularly prone to tropical cyclones (Lewis, 1990). This is because listings often set a criteria based on magnitude of impact with which small states cannot compete against larger countries. However, the proportional impact upon small states is often far greater in terms of population, housing and economics.

The impact of cyclones goes far beyond just deaths and building damage. In developing countries destruction of infrastructure and primary agriculture can lead to a decrease in exports and gross national product, while increasing the likelihood of forfeiture of international loan repayments. Contamination of water supplies and destruction of crops can also lead to disease and starvation.

Many mid-latitude cyclonic depressions can give rise to exceptionally heavy rain and widespread flooding and snow fall too. Dust storms are windstorms accompanied by suspended clay, silt materials, usually but not always without precipitation. Average 130-800 million tonnes of dust are entrained by winds each year.

Severe windstroms with high level of flush rain often called thunderstorms associated with lightning, hail and tornadoes cause massive destruction of properties and also human lives through out the world. Early warning and emergency relief operation are the major management activity.

30.5 FLOODS

Heavy rainfall within short period causes waterlogging and flood in many areas of the world. This is flash flooding.

The potential for major floods has increased in recent decades. Floods are not only caused by high amounts of precipitation, but also by man-made changes to the earth's surface. Poor management of agricultural land, increased deforestation and urbanization—all alter the retention properties of surface soil layers, resulting in greater run-off and increased erosion. Further human contributions to the causes of floods, particularly in developing countries, include reckless building, poor drainage and several other activities.

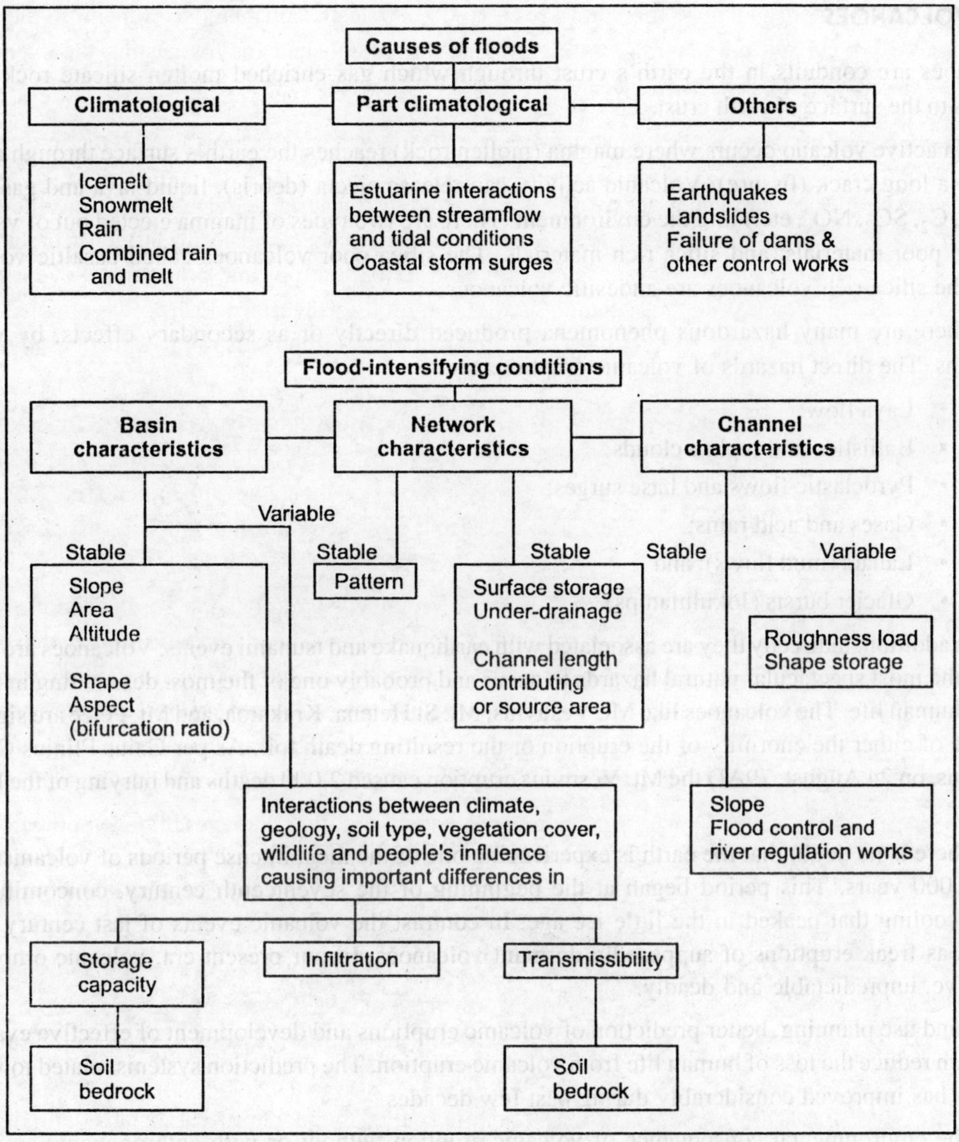

Fig. 30.2: The causes of floods and flood-intensifying conditions

Each year, flood kills thousands of people and causes tens of billions of dollars in property damage.

Various strategies have been developed to reduce the hazard of stream flooding, each with a mix of advantages and disadvantages. These measures are:

> Artificial levels or embankment development;

> Flood control, dam creation; and

> Water channelisation.

The overall cause and effects of flood is shown in Fig. 30.2. However, the large flood events have enormous potential to modify the landscape. Often flood causes devastating losses of lives and properties.

30.6 VOLCANOES

Volcanoes are conduits in the earth's crust through which gas enriched molten silicate rock magma reaches to the surface of earth crust.

An active volcano occurs where magma (molten rock) reaches the earth's surface through a central vent or a long crack (fissure). Volcanic activity can release ejecta (debris), liquid lava and gases (H_2O vapour, C_2, SO_2, NO_x, etc.) into the environment. There are two types of magma ejected out of volcanoes —silica poor materials, and silica rich materials. The silica poor volcanoes called basaltic volcanoes, while the silica rich volcanoes are andesitic volcanoes.

There are many hazardous phenomena produced directly or as secondary effects, by volcanic eruptions. The direct hazards of volcanic eruptions are:

- Lava flow;
- Ballistics and tephra clouds;
- Pyroclastic flows and base surges;
- Gases and acid rains;
- Lahars (mud flows); and
- Glacier bursts (Jökulhlamps).

In addition, indirectly they are associated with earthquake and tsunami events. Volcanoes are visually one of the most spectacular natural hazards to occur and probably one of the most devastating in terms of loss of human life. The volcanoes like Mt. Vesuvius, Mt. St Helena, Krakatoa, and Mt. Pelée are significant because of either the enormity of the eruption or the resulting death toll. As per Gaius Plinius Caecilius secundus, on 24 August, 79AD the Mt. Vesuvius eruption caused 2,000 deaths and burying of the Pompeii city.

There is no doubt that the earth is experiencing one of the most intense periods of volcanism in the last 10,000 years. This period began at the beginning of the seventeenth century, concomitant with global cooling that peaked in the little ice age. In contrast the volcanic events of last century may be viewed as freak eruptions of supposedly dormant volcanoes. In our present era, volcanic eruption are pervasive, unpredictable and deadly.

Land use planning, better prediction of volcanic eruptions and development of effective evacuation plans can reduce the loss of human life from volcanic eruption. The prediction systems related to volcanic activity has improved considerably during past few decades.

The environmental consequence of volcanic eruption without or with anthropogenic emission is shown in Fig. 30.3.

30.7 OTHER NATURAL HAZARDS

Drought, natural fire and landslide are three most natural hazards, which often causes severe distress in some local areas. Drought can be defined simply as an extended period of rainfall deficit during which agricultural biomass is severely curtailed. Drought onset is aperiodic, slow and insidious. Of all natural hazards, drought has the greatest impact beyond the period of its occurrence. In some countries spectacular forest fire causes serious damage to human properties, bioresources and even lives of wild animals. Most of cases the causes such fire is unknown.

Another important natural hazard is landslides or subsidence of land due to various anthropogenic activities. Both the phenomena is land instability hazards encompasses the most diverse range of processes and phenomena.

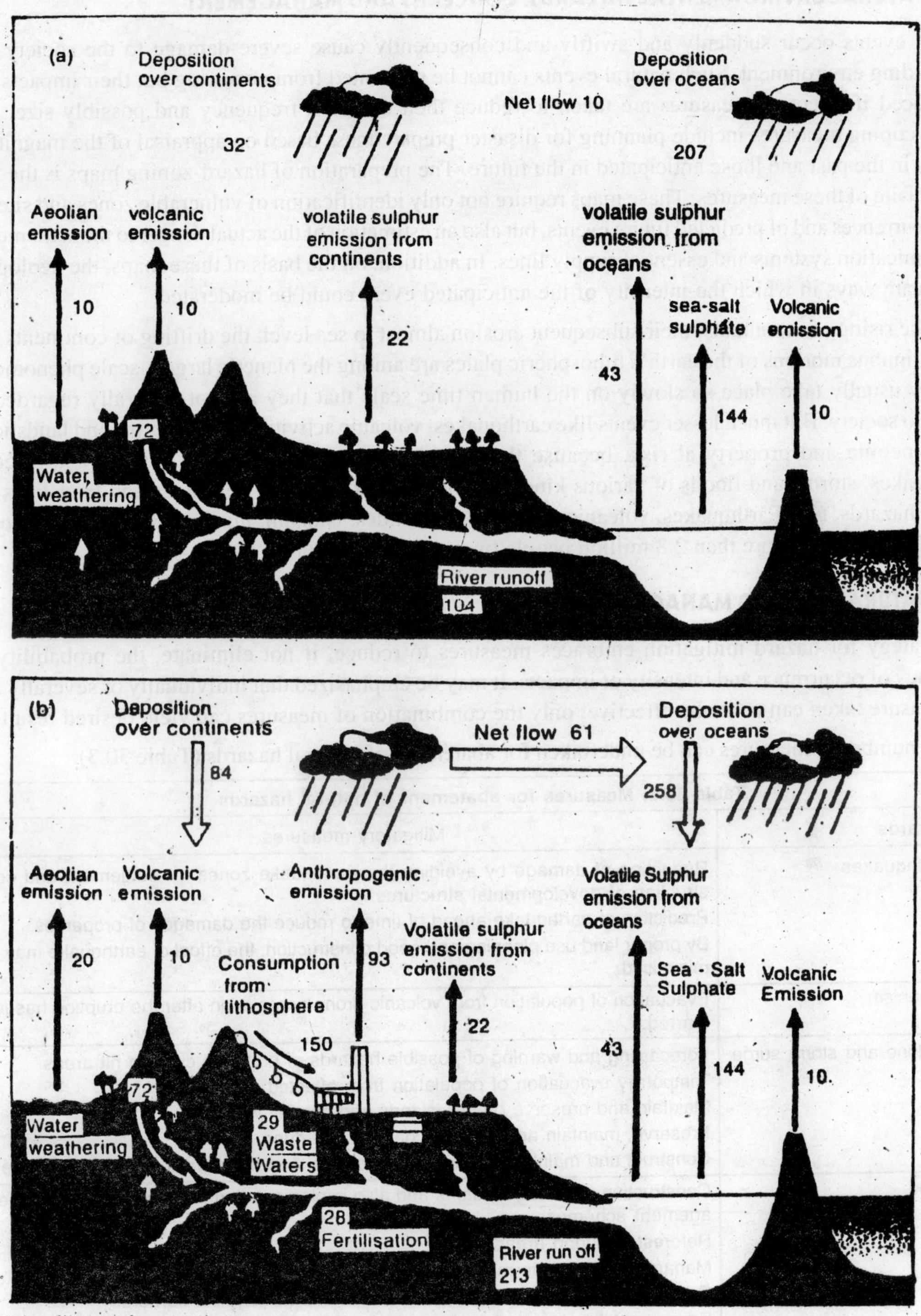

Fig. 30.3: Environmental consequence of volcanism (a) without anthropogenic activity; (b) with anthropogenic activity

30.8 NATURAL ENVIRONMENTAL HAZARDS: CONCERNS AND MANAGEMENT

Natural events occur suddenly and swiftly and consequently cause severe damage to the society and surrounding environment. Such natural events cannot be prevented from occurring but their impacts can be reduced if effective measures are taken to reduce their severity, frequency and possibly size. The hazard-coping measures include planning for disaster preparedness based on appraisal of the magnitude of risks in the past and those anticipated in the future. The preparation of hazard-zoning maps is the first prerequisite of these measures. These maps require not only identification of vulnerable zones and sites of past occurrences and of predicted future events, but also an estimation of the actual danger to the settlements, communication systems and essential supply lines. In addition, on the basis of these maps, the geologists investigate ways in which the intensity of the anticipated event could be moderated.

The rising of mountains, their subsequent erosion almost to sea level, the drifting of continents and the continuous motions of the earth's lithospheric plates are among the planet's largest-scale phenomena, but they usually take place so slowly on the human time scale that they are not generally regarded as threats to society. But much lesser events like earthquakes, volcanic activity, floods, storms and landslides, do put people and property at risk, because they usually occur suddenly and often unexpectedly. Earthquakes, storms and floods of various kinds are the major killers, but volcanoes and landslides are serious hazards, too. Earthquakes, volcanic eruptions, landslides, tsunami, hurricanes, tornadoes, floods and wildfires killed more than 2.8 million people over the period 1968 to 1988.

30.9 NATURAL HAZARD MANAGEMENT

The strategy for hazard mitigation embraces measures to reduce, if not eliminate, the probability or frequency of occurrence and intensity of impacts. It may be emphasized that individually or severally any one measure taken can never be effective; only the combination of measures can yield desired results.

A number of measures can be undertaken for abatement of Natural hazards (Table 30.3).

Table 30.3: Measures for abatement of natural hazards

Hazards	Mitigatory measures
1. Earthquakes	Reduction of damage by avoiding the earthquake zones for settlement and construction of developmental structures. Prediction of earthquake ahead of time to reduce the damages of properties. By proper land use planning and good construction, the effect of earthquake may be minimised.
2. Volcanism	Evacuation of population from volcanic prone areas even after the eruption has just started.
3. Cyclone and storm surge	Forecasting and warning of possible hazards of probable cyclone hit areas. Temporary evacuation of population from affected sites. Maintain and preserve barrier islands and coastal wetlands. Preserve, maintain and replace coastal sand dune. Construct and maintain break-waters, levees, and flood walls as physical barriers.
4. Flood	Construction of barrages, dams and also implementation of proper watershed management schemes. Reforest denuded areas. Manage watershed to minimise erosion. Construct and maintain levees and flood walls as physical barriers. Emergency flood control measures.
5. Landslides	Curtailment of human activities in landslide zones. Reforest the denuded areas. Reduce, prevent or control infiltration and erosion.

Priorities

The strategy for hazard mitigation embraces measures to reduce, if not eliminate, the probability or frequency of occurrence and intensity of impacts. It may be emphasized that individually or severally any one measure taken can never be effective, only a combination of measures can yield desired results. To ensure participation of the people of the affected/threatened area, which is extremely desirable, hazard-coping measures must be compatible with the local traditions and culture. What is effective in the USA or Philippines, for instance, may not be appropriate in the Kumaun Himalaya. It is also imperative that the cost of the measures should be within the economic capabilities of the local people or the government. To stabilize a slope, an Indian geologist would not necessarily recommend the same technology as a Swiss geologist would do.

As the vast majority (>80%) of hazard victims belong to the poorest section of the population, deaths are primarily due to the vulnerability of the locations of settlements and to the constructional weakness of their dwellings. Hazard-coping measures should therefore give priority to risks to the poorest sections of the community for whom the investment today is less than 5%, even though their fatality exceeds 80% (Davis, 1984). Recognizing the priorities of the local people and appreciating their sensitivity to tradition and culture, appropriate measures of rehabilitation should incorporate providing the affected people with key materials at (subsidized) prices which they can afford and offering them expertise for building houses and engineering structures for the prevention of hazard and protection of settlements.

Prediction, Warning and Public Information

By tapping the reserve of accumulated experience of generations, by reading various signs of instability in the natural systems, by analyzing various geophysical changes that take place on the ground, and by studying the erratic behaviours of insects, birds and animals who have an extraordinary perception of impending disasters, it may be possible to forewarn the people of a potentially hazardous event. Capability of predicting to some degree the occurrence of earthquakes has been developed in China, Japan and the USA. At Haicheng in China a major seismic event was predicted (February 4, 1975) which otherwise would have taken a toll of lakhs of people. However, after Haicheng, the Tangshan earthquake occurred on July 28, 1976 without warning, killing 240,000 people. In the USA, the success in prediction has been only partial—the place and magnitude have been successfully predicted, but not the time.

The confident prediction of likely event is to be followed by warning the people through various media and by enforcement of a hazard-preparedness programme—including evacuation, if necessary. The public must be educated simultaneously about the nature of the hazard, the risk involved, and the likely impacts on the natural environment. This education should spur the people to voluntary action.

Brochures on the hazard-preparedness plan should be distributed beforehand among various organizations and agencies identified for mobilization of relief work, including fire-fighting, medical care and sanitation, food supply, rehabilitation and shelter, as also law and order. The plan must ensure smooth and coordinated functioning of the various agencies.

Minimizing the Probability of Hazards

Measures should be taken to reduce the frequency of occurrence of hazards, such as developing effective drainage systems which prevent triggering of landslides, or constructing engineering structures to protect the people and their houses from the hazard. Table 30.4 offers various methods for the abatement of hazards of different kinds, as discussed in depth by Petak and Atkisson (1982).

Table 30.4: Measures for Abatement of Natural Hazards and Protection of Settlements

Hazard	Specific Action or Example	Purpose of Mitigation
Riverine Flood	1. Reforest or reseed denuded areas.	Decreases runoff coefficient for the treated area.
	2. Mechanical land treatment of slopes, such as contour farming, to minimize runoff.	
	3. Protect watersheds from forest and grass fires.	
	4. Control or eliminate large-scale clear-cutting of forest land, as in oak forests.	
	5. Manage watersheds to minimize erosion (e.g., by preventing overgrazing).	
	6. Design and construct new urban development so as to provide temporary on-site water retention.	
	7. Construct catchment basins for debris and sediment.	Decreases peak-flow downstreams.
	8. Clear debris and sediment from upper streams.	
	9. Create water-holding areas (i.e., small lakes).	
	10. Restore or preserve natural detention areas such as sloughs, swamps, small lakes.	
	11. Construct and maintain levees and flood walls as physical barriers.	Eliminates low return period floods.
	12. Construct and maintain channel improvements to increase flow capacity.	
	13. Emergency flood control.	Number of structures protected.
	14. Construct dams and reservoirs to store water.	
	15. Minimize failures of engineering works with inspection, rehabilitation, and removal of substandard structures.	
Headwater Flood	Local stream and storm drainage management with drains, street gutters, culverts, etc.	Decreases runoff coefficient, increases flow capacity (increases downstream flow volume).
Storm Surge	1. Seed hurricanes to reduce the pressure differential.	Decreases storm intensity.
	2. Maintain and preserve barrier islands and coastal wetlands.	Decreases water depth and/or reduces wave action.
	3. Preserve, maintain and replace coastal sanddunes.	
	4. Construct and maintain break-waters, levees and flood walls as physical barriers.	Eliminates low return period floods.
Earthquakes	Relieve earthstrain resulting from tectonic movement.	Reduces potential and/or severity of earthquake occurrence.
Expansive Soils	1. Cover sites with a blanket of nonswelling soil.	Counteracts swelling pressures.
	2. Insure proper drainage.	Reduces infiltration.
Landslides	1. Reforest and reseed denuded areas.	Reduces, prevents or controls infiltration and erosion.
	2. Protect hillside areas from forest and grass fires.	
	3. Manage hillside areas to minimize vegetation-removal and erosion.	

(Modified after W. J. Petak and A. A. Atkisson, 1982).

Public Policy for Hazard Management

Public policy for hazard management should include land use classification, building codes, capital investments by governments, public information, education and warning. Policy plans include *formulating and enforcing laws and regulations* for preventing or restricting development and use of the lands prone to hazards as indicated in the hazard-zoning maps. Since preventive and restrictive measures have proved ineffective in India, the most effective way of curbing the tendency to occupy hazardous tracts would be to impose a series of disincentives such as (i) denial of government assistance for development of roads, hospitals, schools, etc., in the identified hazard-prone areas, (ii) nonavailability of loans from the banks, (iii) denial of essential supplies such as water and electricity, and public services such as communication lines, sewer system, bus service, etc., and (iv) denial of insurance against hazards, natural or man-made. Alternatively, the government can acquire the areas for alternative landuse such as recreation parks, wildlife sanctuaries, afforestation or allowing the original owners to pursue agriculture but forbidding construction of buildings, etc. In this way productive use is made of land that is in short supply while the degree of risk is reduced.

The following policy actions involving legislative and/or administrative measures are suggested:

1. Formation of a national agency (which might be called the Natural Hazards Management Agency (NHMA)) to provide the community with all information relating to vulnerability of the areas to hazards, extent and magnitude of risks and likely impacts and the mitigation measures to be taken up in time. This can be conveyed through periodic bulletins on hazards or other media containing all the information, including the hazard-zoning maps. The Agency would plan and coordinate the efforts of different government agencies and voluntary organizations mobilized to cope with hazard relief and rehabilitation. It may be emphasized that each government agency—through its normal activities—is expected to provide leadership and take action to reduce the risk and minimize impacts. The chief of the NHMA will act as the coordinator (in the manner of the Chief Election Commissioner during elections) and help to mobilize governmental personnel and civil defence organization, and provide financial assistance for all measures, including relief, medicare, food, sanitation, shelter, unemployment allowances, loans for economic recovery, etc.

The NHMA would develop and promote an integrated programme of hazard-zone mapping and land use classification. This programme can be pursued through academic institutions, if need be, on a contractual basis. The various maps and comprehensive reports may be published and distributed among all agencies, organizations and the interested public.

The local governments should also be required to identify areas prone to any hazard and publish information regarding frequency of the hazard and magnitude of risks. Simultaneously, they should identify buildings and structures that are too dangerous for continued future use. The local governments would have their own hazard-management plan to be approved and coordinated by the NHMA. However, other kinds of problems may arise when such an agency starts functioning. The owner of the hazardous land, for instance, may claim that he is affected by loss of value of his property because of the agency's assertion that the land is dangerous and seek legal aid to counter. This is also true in the USA. Also of interest is the Swiss case where all property is inventoried and the existence and degree of risk are recorded in the legal offices of the authorities so that any possible buyer of land will see immediately if there is any natural hazard risk (Ives, 1985).

2. As already pointed out, the critical parts of the identified hazard zones should be compulsorily acquired by the governments and converted to parks, forests, fuelwood farms, fodder farms, horticultural gardens, etc. There is a strong need for legislation enforcing disincentives for use and occupation of the identified hazard-prone areas.

3. A distinction should be made between a critical structure and ordinary structure. A *critical structure*—such as a nuclear power plant or a high dam—is one whose destruction or severe damage by a natural disaster would cause such extensive damage that it should not be built even if the chance of hazard is relatively small. An *ordinary structure* (e.g., a building or a bridge) might fail or cause property damage or loss of life but the destruction would not be catastrophic.

4. Governments should lay down building codes in all populated areas identified as hazard-prone. Building standards should be enforced even by lending and mortgaging agencies, such as banks, which should not support or finance construction without first ascertaining that adequate safeguards against hazards have been incorporated into the construction plan. Implied in this regulation is the full declaration by the applicants of reports on the state-of-the-environment of the area, including its vulnerability to hazards.

There is also a need for legislation for mobilizing financial resources for monitoring, research and management of hazards.

Natural Disaster Management Policy in India

The National Policy framework has been prepared after due deliberation and keeping in view the National Vision to build a safe and disaster-resilient India by developing a holistic, proactive, multi-disaster and technology-driven strategy for DM. This will be achieved through a culture of prevention, mitigation and preparedness to generate a prompt and efficient response during disasters. The entire process will centre-stage the community and will be provided momentum and sustenance through the collective efforts of all government agencies and Non-Governmental Organizations'

In order to translate this vision into policy and plans, the NDMA has adopted a mission-mode approach involving a number of initiatives with the help of various institutions operating at the national, state and local levels. Central ministries, States and other stakeholders have been involved in the participatory and consultative process of evolving policies and guidelines.

This Policy framework is also in conformity with the International Strategy for Disaster Reduction, the Rio Declaration, the Millennium Development Goals and the Hyogo Framework 2005-2015. The themes that underpin this policy are:

- Community-based disaster management, including last mile integration of the policy, plans and execution.
- Capacity development in all related areas.
- Consolidation of past initiatives and best practices.
- Cooperation with agencies at the national, regional and international levels.
- Compliance and coordination to generate a multi-sectoral synergy.

From the national vision and the theme mentioned earlier, the objectives guiding the policy formulation have evolved to include:

- Promoting a culture of prevention and preparedness – by centre-staging DM as an overriding priority at all levels and at all times.
- Encouraging mitigation measures based on state-of-the-art technology and environmental sustainability.
- Mainstreaming DM concerns into the development planning process.

- Putting in place a streamlined institutional techno-legal framework in order to create and preserve the integrity of an enabling regulatory environment and a compliance regime.
- Developing contemporary forecasting and early warning systems backed by responsive and fail-safe communications and Information Technology (IT) support.
- Promoting a productive partnership with the Media, NGOs and the Corporate Sector in the areas of awareness generation and capacity development.
- Ensuring efficient response and relief with a caring humane approach towards the vulnerable sections of the society.
- Making reconstruction an opportunity to rebuild back better and construct disaster-resilient structures and habitats.

30.10 ANTHROPOGENIC ACTIVITY RELATED TO ENVIRONMENTAL CHANGES

Various manifestations of environmental stresses and changes are linked to one another—more or less directly—and these issues form the major concern for the present-day world. The polar-ozone hole, the greenhouse effect and the EI Nino all have links to each other. These current climatic issues are primarily man-made in many respects. Suddenly there is irrefutable evidence of major environmental changes, which man was not capable of predicting in advance, which seems likely to have roots in man-made pollution and which could lead to climatic effects and even economic effects. Now anthropogenic activity leading to environmental hazards are in the major political agenda and, thus, in global forum, inter-governmental agreements were made over the issues like reduction of greenhouse effect and ozone layer protection.

In addition, expansion of developmental and agricultural activity related to gradual rise of human population are also leading to major environmental degradation over past few decades.

30.11 MAN-MADE ENVIRONMENTAL DEGRADATION

Ever since mankind took to agriculture, the symptoms of deterioration of the environment have been apparent. The reclamation of forest and grasslands for cultivation initiated an unending cycle of erosion of soil. The farmlands were expanded at the cost of forests, leading inevitably to loss of protective vegetal cover, reduction in the number of plant and animal species in the cultivated tracts, thus rendering them vulnerable to disease and invasion of undesirable species. Nearly 1,000 animal species and 25,000 plant species all over the world were threatened with extinction. The destruction of forests in the developing countries has been so rapid and so alarming that only 0.2 hectare per capita forest is left in Asia and 0.5 hect./capita in Africa. This is very much comparable to the developed world, i.e., 2-2.7 ha/capita. India alone is losing forest cover at the rate of 0.161 million ha/year or 441 ha/day. The result of the degradation of forests is associated with the expansion of vulnerable dryland.

Climate and people interact in various ways. This is a classic example of a symbiotic relationship, because each affects the other. People affect climate, because human activities—particularly landuse changes, and air pollution by greenhouse gases and ozone depleters—alter the atmospheric system and change the pace and/or pattern of natural atmospheric processes.

Climate affects people both directly and indirectly. Climate affects people indirectly in numerous ways, because most environmental systems affect and are affected by climate (Box 30.3). Direct effects are experienced through the opportunities and constraints that climate creates, which can strongly influence what people do, and where, when, and how they do it. Many of the weather disasters affected people—both directly and indirectly.

Some places (including most of the world's great deserts) receive too little rainfall and are too dry to sustain abundant life. Other areas (such as many tropical rainforests) receive vast amounts of rainfall and are threatened by serious flooding, mass movement and soil erosion. Temperatures in some places (such as Antarctica) are too low for people to survive without the shelter of artificial environments, and other areas (around the equator, for example) are simply too hot for outdoor work under the scorching midday sun.

Many climatic hazards (including heatwaves, cold spells and flooding) affect people in quite specific ways; they start off quite slowly and tend to build up over a period of time; they often affect some groups of people (particularly the young, the old and the poor) more than others; and they damage health.

Box 30.3: Links between climate and other environmental systems

Climate exerts strong direct and indirect effects on such things as:

- the availability of water resources
- river flooding
- droughts
- desertification
- snow and ice cover
- the natural distribution of plant and animal species
- the development and distribution of soils
- patterns and processes of natural weathering.

Climate affects ocean currents and sea level and these can significantly affect people via their impacts on:

- weather systems
- storm surges
- coastal erosion
- ocean fisheries.

Some of the climatic effects are secondary impacts of other natural environmental processes, such as:

- variations in sunspot activity
- variations in atmospheric dust caused by volcanic activity
- variations in tropical ocean currents associated with El Nino events.

30.12 WILDLIFE STATUS

India harbours more than 17,000 species of higher plants and 94,000 species of animals, constituting, respectively, 5.2% and 4.3% of all the known plants and animals of the world (Gadgil, 1980). About 134 species of plants (99 species in Himalayan region only), 29 species of mammals, 29 species of birds and 3 species of reptiles are threatened. Deforestation has destroyed the habitats of birds and other vertebrates and also threatened the survival of many species of birds and mammals in particular. Significantly the extinction of birds and animals is directly related to the increase in human population in earth planet. Undoubtedly, in India, the extermination of rich wildlife is due to the demands of the fast multiplying hungry human population. While wildlife has been declining, the number of domesticated animals has increased phenomenally along with the human population. For instance, in India (with just 2.2% of geographical area of the world), supports 19% of the world cattle population.

30.13 THE OZONE HOLE

What is the ozone hole? In 1985 the ozone hole was discovered which is a depletion of the concentration of the global stratospheric ozone layer, which has been occurring locally over the Antarctic continent in the Austral spring, between September and November each year since about 1979. Stratospheric ozone is very important because it acts as a 'UV filter', cutting out shortwave ultraviolet radiation from the sun which, otherwise, would be damaging to plants and animals as they have evolved on Earth. In addition, as a result of absorbing this shortwave, highly energetic radiation, ozone causes a warming of the upper stratosphere, which influences global air motions. The growth in man-made pollution of the atmosphere by nitrogen oxides and, more seriously, by CFC and CH_4 has given rise to fears that the delicate ozone balance might be shifted, with possibly serious consequences. The processes leading to the ozone hole formation occur particularly in Antarctic region. This condition leads to serious biological impacts in Australia and New Zealand.

30.14 THE GREENHOUSE EFFECT

The greenhouse effect arises because the atmosphere is largely transparent to incoming solar radiation, while being quite heavily absorbing to outgoing thermal radiation from the planetary surface and the atmosphere (Fig. 30.4).

Of course, the greenhouse effect is a perfectly natural process which exists whether or not mankind is thought to be modifying the climate. It is the enhancement to the greenhouse effect resulting from man's activities that is liable to destabilise the natural balance. Many different gases in the atmosphere contribute to the greenhouse effect. Among the greenhouse gases, CO_2 has a major effect, in that a temperature rise of 0.53-0.79 k is predicted for just a 25% increase in concentration. Other gases, while not as effective as CO_2 in creating greenhouse conditions, nevertheless cause significant effects: these include the fluorocarbons CF_2Cl_2 and $CFCl_3$, N_2O and CH_4. It will also be noted from the table that while most of the gases have a warming effect if they increase in concentration, ozone, O_3, alone causes an increased cooling. This arises simply because ozone is the only one of the gases which exists predominantly in the

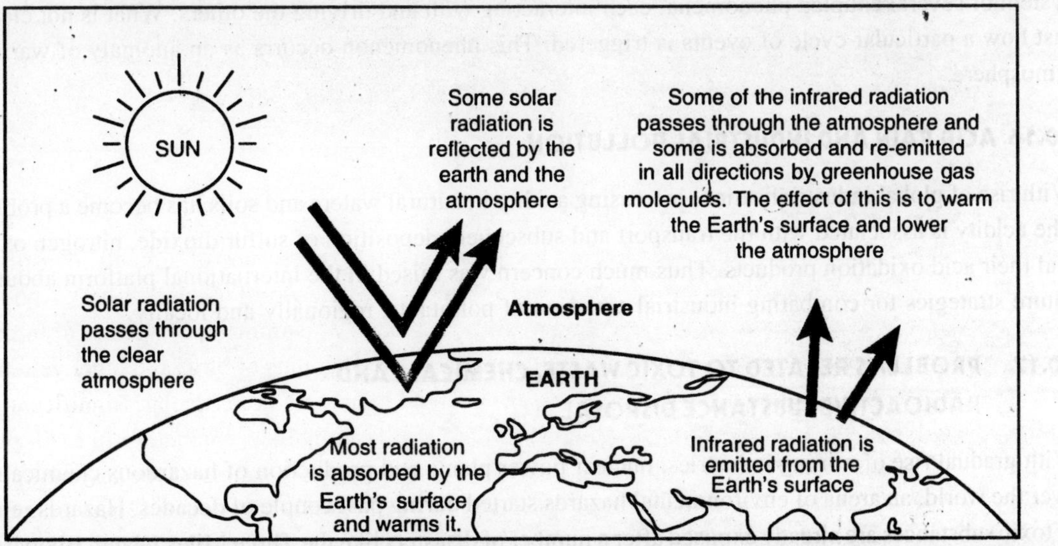

Fig. 30.4: The greenhouse effect

stratosphere. Because this is a region of cooling to space, an increase in ozone concentration would lead to an increased cooling to space, not a warming at lower levels as for the other gases.

Various ways of reducing greenhouse gas emissions is shown in Box 30.4.

Box 30.4: Ways of reducing greenhouse gas emissions

The energy sector

- introduce a carbon tax on electricity generation
- introduce a higher carbon dioxide tax and retain the existing energy tax on non-energy-intensive industry
- introduce an energy tax on combined heat and power
- introduce measures to save electricity (including standards for domestic appliances)

The transport sector

- new rules on company cars and tax (to reduce long-distance travel)
- expand public transport systems
- set carbon dioxide emission limits on light vehicles
- further develop environmental classification systems for vehicles and fuels
- reduce average speeds on roads
- subject all transport plans and infrastructure investments to environmental impact assessments
- experiment with the introduction of electric vehicles

Other greenhouse gases

- reduce the agricultural use of nitrogen fertilisers
- expand methane extraction from waste tips
- reduce process emissions of fluorocarbons from aluminium smelters and ban their use as chemicals; reduce use of hydrofluorocarbons

30.15 'EL NINO'

The El Nino phenomenon—'El Nino is a warm oceanic current affecting the Peruvian coast every 10 years or so'. (The words 'El Nino' mean 'the child'—from Child = Christ; as the current arrives at Christmas). It is not a process with a clear beginning and end, at least not on the basis of our present understanding. It is system of several coupled phenomena, each interacting with and driving the others. What is not clear is just how a particular cycle of events is triggered. This phenomenon occurrs as an anomaly of warming atmosphere.

30.16 ACID RAIN AND INDUSTRIAL POLLUTION

With rise of global industrialisation, increasing acidity in natural waters and soils has become a problem. The acidity is associated with the transport and subsequent deposition of sulfur dioxide, nitrogen oxides and their acid oxidation products. Thus much concern was raised in the international platform about the future strategies for combating industrial emission of pollutants, regionally and locally.

30.17 PROBLEMS RELATED TO TOXIC WASTE, CHEMICALS AND RADIOACTIVE SUBSTANCE DISPOSAL

With gradual rise of mining industries, nuclear power plants and production of hazardous chemicals all over the world, an arena of environmental hazards started during past couple of decades. Hazards related to toxic substances are already exposed after a number of disasters like the Three Miles Island, Bhopal Gas Tragedy, Chernobyl nuclear power plant accident, etc. Chemical and biological weapons used in wars also significantly deteriorates the global environment. Intensive agricultural practices since 1960s also added newer chemicals like insecticides and pesticides, etc. in the cropping areas. Many of these chemicals are not only hazardous to biological systems but also non-biodegradable—so that they accumulate in biological systems slowly and, finally, causes long-term damage to the systems.

30.18 FUTURE PLANNING FOR SUSTAINABLE DEVELOPMENT

There is no denial of the fact that developmental activities are always associated with environmental degradation. Time has come when sustainability in development has to enter in all our planning processes as one of the basic and permanent objectives. The very essence of sustainability in environmental restoration programme is based on a number of basic key issues. These are categorised in Table 30.5:

Table 30.5: Important steps towards sustainable development

1. Population stabilisation
2. Integrated landuse planning
3. Maintenance of sustainability in agriculture through restoration of healthy cropland and grassland
4. Conservation and rational exploitation of forest wealth and revegetation in wasteland and deforested areas
5. Conservation of biological diversity
6. Control of pollution of atmosphere (air) and hydrosphere (water)
7. Development of non-polluting renewable energy systems
8. Recycling of wastes and residues
9. Imposition of ecologically compatible human settlements and slum improvement
10. Introduction of environmental education for more awareness of basic environmental issues at both formal and non-formal level
11. Updating environmental laws and its strict imposition
12. Assessment of ecological security.

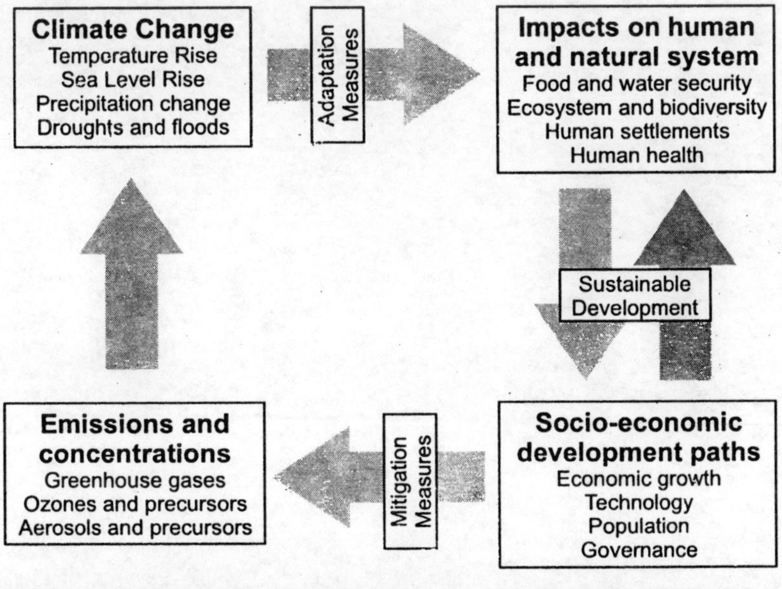

Fig. 30.5: Sustainable development pathways

The words "economics" and "ecology" have the same root, '*oikos*', which refers to house. In general, people through these two subjects are not compatible to each other. Economic goal will definitely damage the ecological goal as a common perception. But, in reality, if proper management can be made available, both could be compatible to each other. As, economics deals with "financial housekeeping" and ecology deals with "environmental house-keeping". This is the hour when one should realise that economic planning and environmental protection have identical goals of sustainable development which, therefore, must be deeply integrated in the planning process. On the whole, the principle of environmental management will have to be applied to all developmental activities in the present as well as in the future.

A generalised pathway of sustainbale development in climate change regime in given is Fig. 30.5 (previous page).

SECTION V

ENVIRONMENTAL HEALTH

31 HEALTH AND ENVIRONMENT

31.1 INTRODUCTION

There are two ways in which the environment affects human health. The first is directly—through pollutants discharged by industries into the air and the water and by automobiles. In urban areas, in particular, the current disease pattern is closely linked to the deterioration in the quality of air and water.

Human health is also affected by the quality of the environment in the place of work. Over the years, the authorities have recognised that the workers in certain occupations—such as textile mills or stone quarries—suffer from occupational diseases (tuberculosis). Although by law, the people running these factories and quarries are supposed to provide the workers with equipment to minimise their exposure to pollutants, in most cases such steps are not taken. As a result, an increasing number of workers— specially those in the informal sector—suffer for years from diseases, for which they receive little or no compensation. The World Health Organisation recognises a number of environmental factors that affect human health (Table 31.1).

Table 31.1: Environmental factors that affect human health, according to the World Health Organisation

Diseases	Polluted air	Poor sanitation and waste	Polluted water or poor water	Polluted food	Unhealthy housing	Global environmental change
Acute respiratory infections	✓				✓	
Diarrheal diseases		✓	✓	✓		✓
Other infections		✓	✓	✓	✓	
Malaria and other vector-borne diseases		✓	✓		✓	✓
Injuries and poisonings	✓		✓	✓	✓	✓
Mental health conditions						
Cardiovascular diseases	✓					✓
Cancer	✓			✓		✓
Chronic respiratory diseases	✓					

The deterioration of the environment—such as the existence of stagnant pools in villages and cities—provides fertile ground for the breeding of vectors. Many such disease-carrying vectors have developed immunity to pesticides. As a result malaria, which was once thought to have been eradicated, has reached endemic proportions in many parts of the country. What is worse, the malarial parasite has become immune to a growing number of pesticides and drugs. For instance, *P. falciparum*, which causes virulent cerebral malaria, is resistant to chloroquine, the drug most commonly used in the treatment of malaria. Similarly, diseases like kala-azar are difficult to eradicate because the organisms spreading it find increasing number of breeding places.

Overcrowding and unhygienic living conditions in cities, specially in slums, have also contributed to the spread of other communicable diseases like tuberculosis and cholera. Health interventions become almost meaningless so long as such conditions prevail. It has become abundantly clear that improving environmental conditions should form an important part of healthcare, because curative care alone will not suffice as more succumb to these diseases.

31.2 HUMAN DISEASES

A disease is a deleterious change in the body's condition in response to environmental factors that could be nutritional, chemical, biological and psychological. Diet and nutrition and infectious agents, toxic chemicals, physical factors and psychological stress all play roles in the onset or progress of human diseases. To understand how these factors affect us, let's look at some of the major categories of environmental health hazards (Table 31.2):

Table 31.2: Some major environmental health problems of the World (2000)

Diseases	New cases each year	Yearly death
Diarrhea	1 billion	10 million
Malaria	800 million	5-10 million
Parasitic worms including flukes	1 billion	Few people die
Anaemia	.375 million	"
Respiratory diseases	500 million	5-6 million
Trachoma	300 million	Few people die
Goiter and Cretinism	200 million	"
Tetanus	5 million	800,000
Polio	2 million	200,000

31.2.1 Infectious Diseases

For most people in the world, the greatest environmental health threat continues to be pathogenic (disease-causing) organisms. Although much of our attention is focused on toxic synthetic chemicals, we also should be aware of the biological hazards to which we are exposed. In the less developed countries, where nearly 80 per cent of the world population live, infectious agents, parasites and nutritional deficiencies still are the main cause of morbidity (illness) and mortality (death).

Gastrointestinal infections (diarrhea, dysentery and cholera) probably cause more deaths worldwide than any other group of diseases. Again, malnutrition and diarrhea create a vicious cycle. Poor nutrition makes people more susceptible to infection and infections, in turn, make it more difficult to obtain, absorb and retain food. Improved sanitation and better nutrition could prevent most, if not all. Similar other infectious diseases like malaria, parasitic worm infections, tuberculosis and allied respiratory hazards, tetanous, etc., took lives of millions, particularly in poor nations.

31.2.2 Chemicals

Toxic chemicals in the environment are becoming a source of increasing concern to people of industrialised countries, though human beings are often exposed to toxic substances in nature also. Chemical agents are divided into two broad categories—those that are *hazardous* and those that are *toxic*. Hazardous substances include inflammable chemicals, explosives, irritants, sensitizers, acids and alkalies, etc.; while toxic substances include poisonous materials that are cytotoxic in nature.

In addition allergens, mutagens, teratogens and carcinogens are also quite common in nature that causes severe health hazards in communities. Physical agents like radiation also causes serious health hazards to humans.

There are many sources of toxic and hazardous chemicals in the environment. The danger of each chemical can be determined by factors related to the chemical itself, its route or method of exposure and its persistence in the environment as well as characteristics of the target organism.

Cells have mechanism for *bioaccumulation*, selectively absorbing and storing a great variety of molecules. This allows them to accumulate nutrients and essential minerals, but they also may absorb and store harmful substances through this same mechanism. Subsequently, the effects of toxins are magnified in the environment through food chains. Biomagnification occurs when the toxic burden of a large number of organisms at a lower trophic level is accumulated and concentrated by a predator at a higher trophic level. Some such toxic chemicals/compounds are very unstable and degrade rapidly under most environmental conditions. For example, DDT and other chlorinated pesticides poses a serious threat to biotic communities due to its prolong persistence.

31.2.3 Diet

Diet also has an important effect on health. Malnutrition particularly poses a serious threat to the protection against infectious diseases as well as exposures to toxic substances. In addition to calories, we need specific nutrients such as proteins, vitamins and minerals in our diet.

The world's most prevalent diseases are shown in Table 31.3.

Some special categories of health problems are:

1. Water-borne pathogens;
2. Respiratory hazards due to air pollution in urban centers;
3. Climate changes and human health effects;
4. Vector-borne diseases;
5. Metal toxicity and diseases—e.g., Minamata disease; and
6. Radiation damage.

Table 31.3: The World's most prevalent diseases

Diseases	People Infected
Vaccine-preventable diseases	
Tuberculosis*	1,700 million
Measles*	49 million
Whooping cough (Pertussis)*	46 million
Tetanus*	1.5 million
Polio*	200,000
Diphtheria*	> 23,000
Tropical diseases	
Malaria*	270 million
Schistosomiasis*	200 million
Filariasis*	90 million
Chagas' Disease*	15-18 million
River Blindness*	18 million
Leishmaniasis*	12 million
Leprosy*	10-12 million
Guinea worm*	5-10 million
African sleeping sickness (new cases)	> 20,000

Contd.

Table 31.3: The World's most prevalent diseases

Diseases	People Infected
Diarrheal diseases	
Amebiasis*	480 million
Giardiasis*	200 million
Typhoid*	1 million
Cholera*	> 40,000
Sexually transmitted diseases (STDs) (new cases)	
Trichminiasis	120 million
Genital Chlamydia	50 million
Genital Papillomavirus	30 million
Gonorrhea	25 million
Genital herpes	20.5-10 million
Syphilis	3.5 million
Endemic Treponematoses	2.5 million
Chancroid	2 million
Other infectious and parasitic diseases	
Hepatitis B	2,000 million
Ascariasis*	700 million
Hookworm*	700-900 million
Trichuriasis*	500 million
Dengue*	30-60 million
Toxoplasmosis	420,000
Rabies*	35,000
Yellow Fever*	10,000-25,000
Endocrine and nutritional diseases	
Iodine deficiency*	200 million
Diabetes (Third World)	40 million
Vitamin A deficiency*	40 million
Others	
Cardiovascular diseases (deaths)	12 million
Pneumonia (deaths)*	4.8 million
Anaemia*	1,500 million
Mental disorders	200 million
Drug abuse	> 85 million
Trachoma*	6-9 million
Cancer (new cases)	6.4 million
Meningococcal Meningitis	> 65,000
Brucellosis	> 80,000

Source: WHO (2000)

31.2.4 Water-Borne Pathogens

A wide variety of human pathogens may be found in the excrement from humans as well as from other animals. Most human pathogens can be classified as either viruses, protozoans, helminths or bacterias. Both raw sanitary sewage and land runoff contain pathogenic organisms. Approximately 20% of the reported outbreaks and illness associated with water-borne diseases were attributed to bacterial pathogens belonging to the genera *Salmonella* and *Shigella*. Cholera is another serious and acute intestinal disease which is caused by the bacterium *Vibrio cholera*; occurs sporadically as epidemic disease.

Numerous protozoan species normally inhabit the intestinal tract of warm blood animals, including humans. *Entameba* and *Giardia* are the most common protozoan infective agents.

A good number of enteric viruses were found in water which happens to infect various organs of human being. These are namely Poliovirus, Hepatitis A, Reovirus, Adenovirus and Rotavirus, etc. A variety of helminth diseases were found to occur as human being whose cysts were found to be present in the waste water contaminated water bodies. Among the helminths, *Ascariasis* and *Taenia* infections were most prevalent. A summarised accounts of water related infections is given in Table 31.4.

Table 31.4: Classification of water-related infections

Category	Infection	Pathogenic agent
Fecal-oral (Water-borne or water-washed)	**Diarrheas and dysenteries**	
	Amebiases	P
	Campylobacter enteritis	B
	Cholera	B
	E. coli diarrhea	B
	Giardiasis	P
	Rotavirus diarrhea	V
	Salmonellosis	B
	Shigellosis (basillary dysentery)	B
Enteric fevers	**Enteric fevers**	
	Typhoid	B
	Paratyphoid	B
	Poliomyelitis	V
	Ascariasis (giant roundworm)	H
	Trichuriasis (whipworm)	H
	Tacnia solium taeniasis (pork tapeworm)	H
Water-washed : Skin and eye infections	Infectious skin diseases	M
	Infectious eye diseases	M
	Louse-borne typhus	R
Others	Louse-borne relapsing fever	S
Water-based : Penetrating skin	Schistosomiasis	H
Ingested	Dracunculiasis (Guineaworm)	H
	Clonochiasis	H
	Others	H
Water-related insect vector :		
	Trypanosomiasis (sleeping sickness)	P

Contd.

Table 31.4: Classification of water-related infections

Category	Infection	Pathogenic agent
Biting near water	Filariasis	H
Breeding in water	Malaria	P
	Onchocerciasis (river blindness)	H
	Mosquito-borne viruses	
	Yellow fever	V
	Dengue	V
	Others	V

B, bacterium; P, protozoan; S, spirochaete; M, miscellaneous; H, helminth; R, rickettsia; V, virus.
Adapted from Feachem, 1984 and Nash, 1993.

31.2.5 Respiratory Hazards

Due to rapid rise of air pollution problems, respiratory hazards increase day by day. Over past couple of decades, urban air pollution problems enhanced the variety of respiratory hazards ranging from bronchitis to pneumonia. The extent of hazards altered markedly in different seasons (Table 31.5).

31.3 CLIMATE CHANGE AND HUMAN HEALTH

Over the years, global climate has changed severely—particularly with respect to temperature. Such a change leads to serious health impact, particularly with respect to allergy, vector-borne diseases and also communicable diseases. A schematic presentation of such diseases is shown in Fig. 31.1.

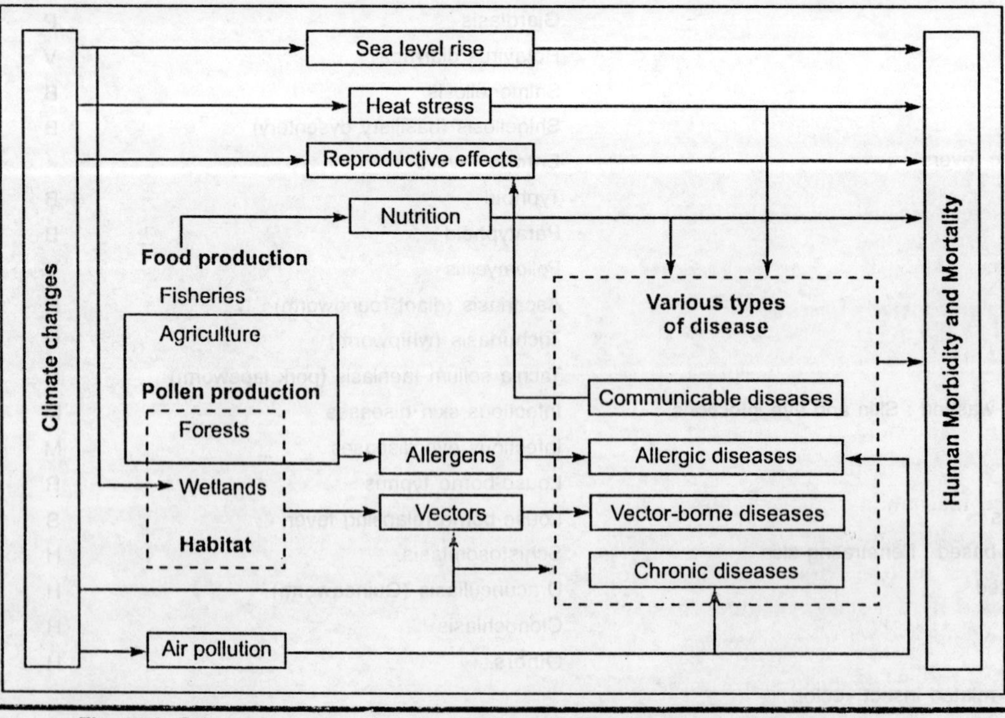

Fig. 31.1: Schematic presentation showing how climate change can affect human health

31.4 VECTOR-BORNE DISEASES

There are a large number of vector-borne diseases found in the tropics. These are Malaria, Kala-azar, Dengue, Encephalitis, and Yellow fever, etc.

Table 31.5: Recognised seasonal effects on respiratory illness

	Condition Affected	Possible Mechanism(s)	Source of Data
Winter	Acute bronchitis / bronchiolitis	Colder temperature	GP consultations
		— effect on bronchial responsiveness?	HAA
		— increased viral replication/ transmissibility?	Mortality tables (DHSS)
	Acute exacerbations of chronic airflow obstruction/ chronic bronchitis	Colder temperature	HAA
		— increased viral replication	Mortality tables
		— permissive effect on secondary bacterial infection	Use of the Emergency Bed Service
	Asthma requiring hospital admission	Viral induced asthma?	HAA Hospital attendance
	Pneumonia	Colder temperature	HAA
		Associated with relative humidity 1 week before?	Mortality tables
Summer	Asthma attacks	Pollen release	GP consultations
		Ozone?	
	Hay fever	Pollen release	GP consultations
Autumn	Acute bronchitis	Viral infections?	GP consultations
	Acute asthma	— return to school plus reduced temperature causing increased viral transmission	HAA GP consultations Hospital attendances

These diseases are spread by the various categories of mosquitoes. The incidence of mosquito-borne diseases increased significantly during past six (1950-2010) decades in various tropical countries. Due to rapid deterioration of sanitation and health-care systems influencing the gradual rise of mosquito-borne diseases (Fig. 31.2).

Malaria and Malaria Vaccine

Malaria is one of the major causes of mortality and morbidity worldwide, affecting nearly 40% of the world's population and accounting for about 3-5 million deaths and more than 500 million new cases annually. Wide spread resistance of the parasite to common and easily accessible drugs like chloroquine and insecticide resistance in the mosquito vector has made the malaria situation very alarming. It is believed that among other strategies, a successful vaccine to combat malaria will play a crucial role. However, although many vaccine constructs have been developed and taken to clinical trails, success has been at best limited. Most vaccine development strategies have targeted specific malaria antigens from the three different stages resulting in stage specific vaccines, i.e., pre-erythrocyte vaccines, blood stage vaccines, and sexual stage vaccines.

In spite of advancement of molecular biology, immunology and availability of complete genome sequences of the host, the parasites and the vector, the question remains unsolved at yet.

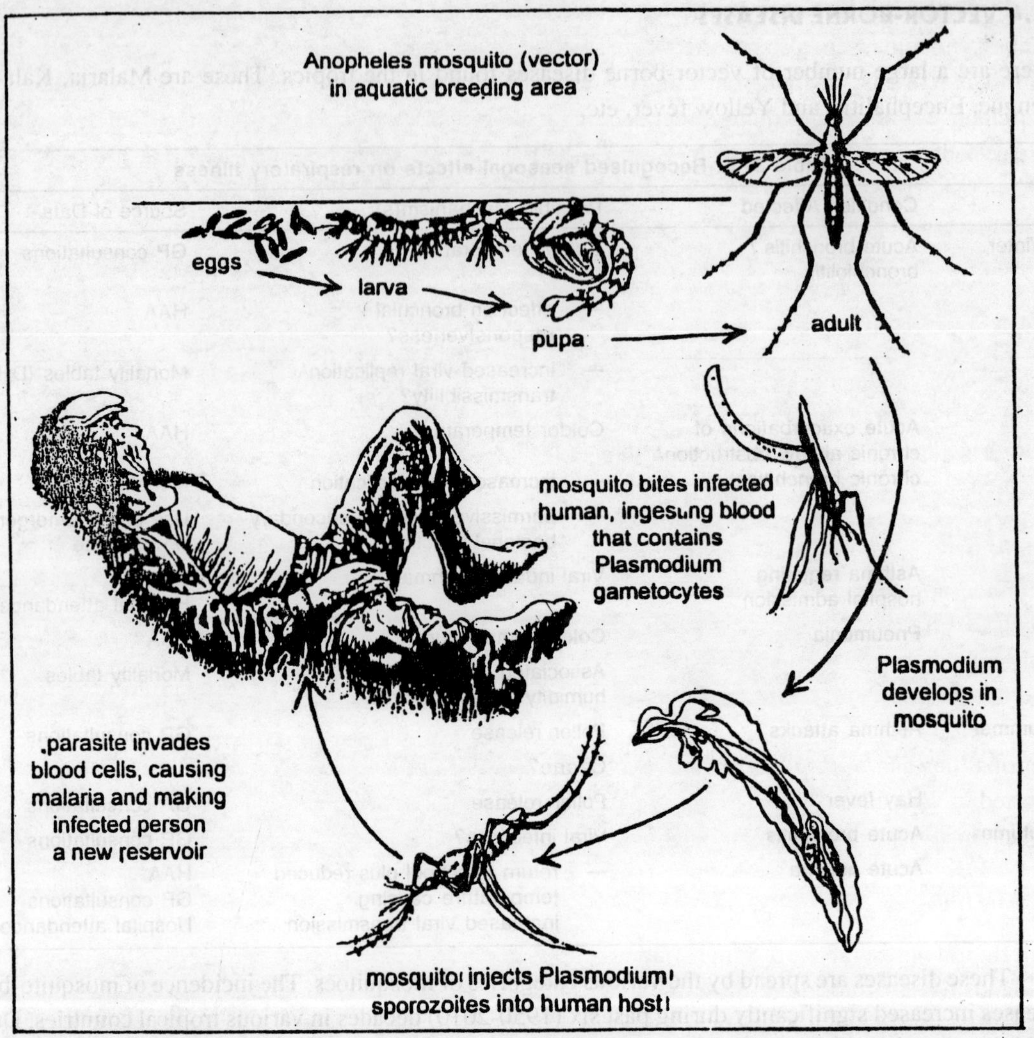

Fig. 31.2: The life cycle of malarial parasite

31.5 METAL TOXICITY

In general, a good number of heavy metals have toxicity symptoms. Among them toxicity caused by lead, mercury and arsenic are considered to be noteworthy. Minamata disease is one such metal toxicity disease.

31.5.1 Minamata Disease

It was in the early 1950s that fishermen and their families in the city of Minamata, Japan, first began to show the symptoms of what was to become known as the Minamata Disease. The first signs were loss of sensation at the extremities of the hands and feet and in areas around the mouth. These symptoms were followed by difficulty in walking, slurred speech, reduced vision and hearing loss. Unfortunately, many persons fell prey to eventual paralysis, followed by coma and death.

Public Health Department officials were at first unable to ascertain the cause of such new disease. Similar symptoms were also reported from sea birds and cats in Minamata. Later it was evidently proved that this is a disease of Industrial Toxicity, more precisely "mercury poisoning". This poisoning of the food chain was local and quite direct. The Chisso Corporation, a plastic manufacturer, was releasing mercury laden wastes into Minamata Bay. The mercury—in its toxic methyl form—was then concentrated in the predatory fish through food chain of the bay ecosystem. The fisherfolk were first to suffer the effect of the disease, for they subsisted largely on fish. By 1976, over 10,000 people were suffering from this disease.

31.5.2 Arsenicosis

Arsenic is one of the most important environmental global toxicants that cause acute and chronic adverse health effects, including cancer. Millions of people around the world have been exposed to arsenic through geologically contaminated drinking water. Epidemiological studies conducted since 1960s indicated that ingestion of inorganic arsenic is linked to internal cancers in lung, liver, skin, bladder and kidney of man. The evidence of health risk from arsenic contamination is so compelling that since 2000 WHO/EPA (USA) recommended lowering of the maximum contaminant level of arsenic from 50 to 10 µg/lit. In addition to cancer induction due to arsenic toxicity, cardiovascular disease, neurogenic effects, diabetes mellitus, hearing loss, fibrosis of lung and liver, anaemia and other hematological effects are seen as an effect of arsenic toxicity. The exposure pathways of arsenic could be inhalation of dusts containing arsenic, ingestion of food and drinks containing arsenic and also through dermal contacts. The potential role of carcinogenic effects of arsenic are considered as altered DNA/chromosomal abnormalities and/or through oxidative stresses. Various clinical cases of arsenic affected patients are reported in West Bengal viz., cancer of skin and kidney, keratosis, raindrop pigmentation, dry gangrene, hepatomagaly, splenomagaly, polyneuropathy, anaemia, dyspepsia etc.

31.5.3 Fluorosis

High fluoride level (above 1.5 mg/lit) in ground water led to fluorosis in several area of India, China and Thailand. Excess fluoride in drinking water over the years lead to dental and skeletal fluorosis (bony deformities). Around 90 districts of India have fluorosis problems. Several thousand population suffers from fluorosis, which needs to be treated with supply of fluoride free drinking water.

31.5.4 Goiter

Deficiency of iodine in water and diets leads to thyroid disfunction. As a result thyroxine hormone level is very low. During such hypothyroidic condition, goiter is seen. The sign of goiter is shown as swelling below the throat along with lack of proper assimilation. Thyroxine deficiency goiter often seen in foot hill region. Iodised salt is often prescribed as remedy of goiter protection.

31.6 RADIATION DAMAGE

It is well-known that radiation causes short and long-term damages to various body organs. Both cosmic rays and UV-radiation has detrimental effect on human health. One of the major health hazards as induced by radiation at low level persistent exposure is induction of cancer at various organs. The frequency of risk of different types of cancer varies widely with respect to dose of expose and also racial criterion. An overall example of cancer risk of human organs is shown in Fig. 31.3.

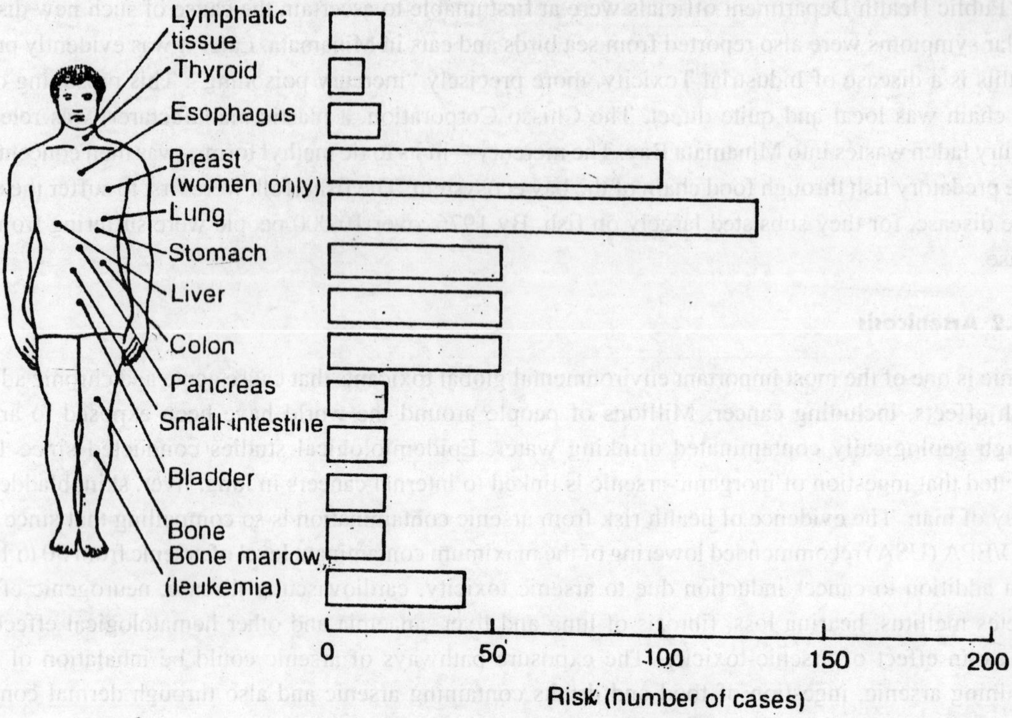

Fig. 31.3: Cancer risk of different human organs

31.7 HEALTH IMPACT ASSESSMENT

In general there are five major categories of health hazards viz., communicable diseases, non-communicable diseases, malnutrition injury and mental disorders (Table 31.6). The pattern of health impact assessment in any of these depends on types of health hazards, their causative and size of the population examined. However, a broad scheme for assessment is shown in Table 31.7.

Table 31.6: Types of health hazards

Health hazards	Examples & Causes
1. Communicable diseases	Malaria, diarrhea, respiratory infections
2. Non-communicable diseases	Poisoning, pollution, dust
3. Malnutrition	Micronutrient deficiency, lack of adequate food
4. Injury	Traffic, crashes & collisions, occupational injury, violence
5. Mental disorder	Abuse, stress

As of now, most environmental management programmes are restricted to sanitation and hygiene. The network of primary health centers are geared only to control traditional illness. The health morbidity data due to pollution-related problems are very scarce. The health laws, as they exist now, relate only to occupational safety and chemical safety.

Table 31.7: Assessment of Health Impact

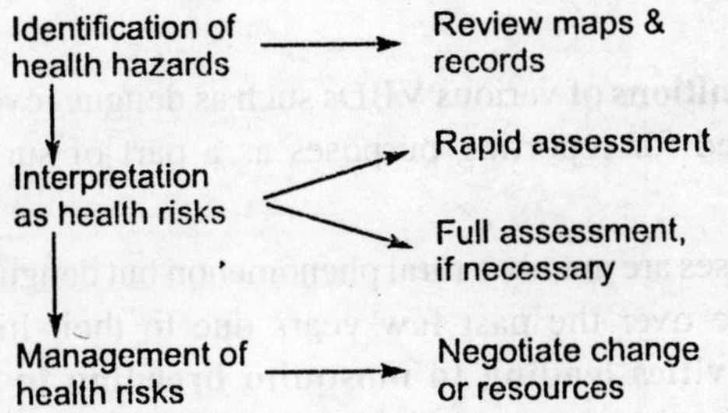

The existing environmental impact assessment procedures do not rigorously analyse the health and social impacts of developmental projects. Consequently, health continues to remain an extrinsic factor in all environmental analysis.

31.8 NATIONAL POLICY AND PROGRAMME OF VECTOR-BORNE DISEASES

1. While the public health system has to play a critical role through institution of preparedness and early warning mechanisms as well as a rapid containment capacity, comprehensive management of VBDs warrant a **broad multi-sectoral response** and social mobilization for full involvement of the community.

2. A medium-to long-term **strategic plan**, with allocation of adequate resources, should be developed including public health, environmental and legislative measures needed to prevent and control vector-borne diseases. Vector control or prevention of vector borne diseases should be among the **healthy public policies**, to ensure that before any developmental activity is planned or implemented, the public health concerns are taken into consideration.

3. **Partnership building** and enhancing the role of various sectors such as education, public works, tourism, industry, the private sector, NGOs, etc. is critical. Each sector should be sensitized about its role in the prevention and control of vector-borne diseases. Taking locally acceptable measures to prevent such diseases must receive a high priority. Active participation of inter-sectoral partners is needed, right from the programme planning stage to its implementation and evaluation.

4. **Operational research** on disease transmission and risk factors as well as on the health and socioeconomic impact of dengue are urgently needed. Data from **outbreak investigations** on disease distribution and determinants can also provide valuable information for formulating appropriate control measures. In addition, such data can help us in better understanding the epidemiology, as well as the pathogenesis of disease, clinical manifestations and response to management in Indian settings. The results of such studies can be gainfully employed in revising the policies and strategies.

Surveillance and outbreak response

5. The present surveillance system for VBDs need to be strengthened to make it more practical; therefore a **standardized passive and active laboratory-based surveillance** system of DF/DHF, with **emphasis on the early warning predictive capability** should be developed urgently

and implemented during the epidemic and inter-epidemic periods. The surveillance data should be effectively used to prevent vector-borne disease outbreaks or to take a pre-emptive epidemic response.

6. Standard **case definitions** of various VBDs such as dengue fever, chikungunya, etc., should be developed and used for reporting purposes as a part of surveillance and during outbreak investigation.

7. Vector-borne diseases are mainly a rural phenomenon but dengue and chikungunya has assumed greater importance over the past few years due to their high epidemic potential and to **development activities leading to mosquito breeding in urban areas.** In urban areas, surveillance requires that the multiplicity of health care providers must report regularly to a central agency using a standard case definition. In rural areas, dengue fever and vector surveillance should be a part of the Integrated Disease Surveillance Project (IDSP) and provide for an early warning system.

8. Vector surveillance including monitoring the presence and density of **vector species and their breeding sites** and the effectiveness and efficiency of vector control interventions should be carried out in each locality on an ongoing basis in order to ensure better planning and implementation of vector control strategies. Laboratory capacity building and networking within and outside the country are areas that require urgent attention.

Prevention through social mobilization

9. A prerequisite for prevention of vector-borne diseases is **scaling up the most effective interventions including integrated vector management** in the form of insecticide residual spray, bed nets and other personal protection measures. Sources reduction and measures to control the vector population are integral components of disease prevention and outbreak response.

10. Intensified efforts towards creating **public awareness and mobilizing the community** regarding the preventive measures they can take and regarding the early seeking of medical attention by those suffering from vector-borne disease are crucial. Communication messages using the mass media as well as interpersonal approaches can help bring about a change in the behaviour of the population. The services of a communications expert may be employed to formulate effective risk communication strategies.

    Similarly, school children can be involved in prevention activities through school health programmes.

11. Population groups such as **Residents Welfare Associations and Housing Societies** can play an important role in encouraging and enforcing preventive activities in their respective areas. Targeting these groups with risk communication can lead to a cascading preventive effect amongst many people.

12. The management of dengue and chikungunya can be enhanced through holding a **dry day every week** when all residents would dry up their coolers once a day in a week for only half day (preferably forenoon). This should be supported by a legal framework to ensure compliance. In residential areas, the dry days could be observed on Sundays and in offices or educational institutions on a weekday, i.e., Tuesdays. **A sanitation day**, as envisaged by the Govt. of India, in a year (i.e., 2nd October) should be observed throughout the SEA Region in order to create awareness regarding sanitation and hygiene for disease prevention and control.

Case management

13. Standard case management is **critical for reducing morbidity and preventing mortality** due to dengue haemorrhagic fever (DHF). For example, health care workers should ensure **appropriate and rational use of platelets** for case management. Development of standard guidelines, their dissemination and extensive training of physicians at all levels of the health care system are key elements to ensure good clinical practices. Since transfusion of platelets in DHF saves life, availability of safe blood products warrant strengthening the capacity of blood transfusion services to respond effectively during outbreaks.

14. **Laboratory support** for diagnosis and case management is not widely available. Access to quality reagents is limited. Forging a network of laboratories can assure quality and enhance access to diagnostic tools.

15. Emphasis should be placed on programmes to **educate physicians, nurses and others** in the medical community in dengue fever/dengue hemorrhagic fever, its diagnosis, management, prevention and control.

The road map

1. A national strategic plan for delivery of vector-borne disease prevention and control services with multi-sectoral involvement exists. However, it would require strengthening in view of recent outbreaks of chikungunya. A National Steering Committee should be set up to mobilize multi-sectoral involvement (March 2007). **Action**: Govt. of India

2. Adequate resources should be mobilized and made available to the National Vector Borne Disease Control Programme to plan and coordinate national preparedness and response to vector-borne diseases (ongoing). **Action**: Govt. of India

3. Vector-borne disease control adopted as one of the healthy public policies with an inbuilt mechanism for ensuring compliance (July 2007). **Action**: Govt. of India

32 WORLD HEALTH REPORT

32.1 INTRODUCTION

Water, food and soil are essential for life. But, when contaminated, they also transmit bacteria, viruses and parasites that cause some of the world's most menacing diseases, such as diarrhoea, cholera, typhoid, intestinal worms, hepatitis and tetanus. The organisms that are the main causes of these diseases are ingested with water or food or conveyed to the mouth by contaminated fingers. Where there is no proper sanitation, they find their way to other water sources from which they again infect other people. Soil can also be the breeding ground for a wide range of infections—to which children are particularly vulnerable—and for intestinal parasites.

Almost half the world's population suffers from diseases associated with insufficient or contaminated water and is at risk from water-borne and food-borne diseases. Their vital importance as a public health problem is often overlooked, because their true incidence is difficult to evaluate and the severity of their health and economic consequences is often not fully understood.

Diseases are classified as water-borne or food-borne based on epidemiological evidence that implicates consumption or use of water and ingestion of a common food as the source of infection. They are perhaps the most widespread health problems in the contemporary world and an important cause of low or lack of economic productivity. They range in severity from mild indispositions to life-threatening illnesses.

Apart from their immediate effects, food-borne diseases may cause serious and chronic health problems. Some of these diseases—for instance, listeriosis—are particularly dangerous during pregnancy, as they may either be fatal to the foetus or cause severe deformities. In the latter case, the child runs a risk of severe congenital, cerebral or ocular damage.

32.2 DISEASES CAUSED BY CONTAMINATED FOOD AND WATER

Diarrhoeal diseases caused more than 3 million deaths in 1995, of which more than 80 per cent were among children under age five. About 50 per cent of diarrhoeal deaths are due to acute watery diarrhoea, 35 per cent to persistent diarrhoea and 15 per cent to dysentery.

It was long thought that contaminated water supplies were the main source of pathogens causing diarrhoea but it has now been shown that food has been responsible for up to 70 per cent of diarrhoeal episodes. Infections due to pathogenic *E. coli* are the most common cause of diarrhea. Weaning food contaminated with pathogenic *E. coli* causes up to 25 per cent of all diarrheal episodes in infants and children, while campylobacteriosis and shigellosis account for another 5-15 per cent and 10-15 per cent, respectively.

Globally, large population uses traditional processes for drinking water and other domestic uses. The details are given in Table 32.1.

Table 32.1: Traditional systems for drinking water and domestic use

Region	Systems	Type	Use	Technique
Indo-Gangetic plains (mainly Punjab, Haryana, Uttar Pradesh, Bihar, and West Bengal)	Tapping groundwater	Dugwells	Domestic consumption	Very common for accessing groundwater for domestic use.
	Tapping rainwater and groundwater	*Baoli*	Domestic consumption	Step-well, either rock-hewn or with masonry work. Usually a square or oblong structure, often provided with stepped stone embankments.
		Dam	Domestic consumption	Damming a natural spring.
	Diverting river water to tank	*Dighi* (Delhi)	Drinking	Square or circular reservoir connected to a canal and with steps for entering. Usually a *kahar* or a *mashki* was hired to draw water.
Arid and semi-arid region (western, central and southern India)	Tapping groundwater from aquifers	*Baori* (Rajasthan, Gujarat)	Domestic consumption	Step-well, often elaborate structures with arches, sculptures and decoration. They are shallower than the wells. Most part of the structure is built underground.
		Jhalar (Rajasthan)	Bathing and religious rites	Rectangular structure with steps along three or all four sides.
	Tapping groundwater from seep springs	*Tanka* (western Rajasthan)	Drinking	Underground tanks. These were circular holes made in the ground and lined with fine polished lime. Such *tankas* were found in most houses. Used as a last resource.
		Beri/Kui (Rajasthan)	Drinking and animal consumption	Shallow and narrow wells about six to eight m deep, dug at par, a place where the flowing water accumulates and seeps into the earth.
		Toba (Rajasthan)	Drinking	Ground depression with a natural catchment area.
	Collecting rainwater from rooftop	*Tanka* (western Rajasthan)	Drinking	Rainwater falling on the sloping roofs was captured through a pipe into an underground tanka, sometimes as large a big room. The first spell rain was usually not collected as this cleaned the pipes and roof.
	Collecting rainwater within artificially created catchments and storage	*Kund* (western Rajasthan)	Drinking	Underground circular pits, and plastered with lime and ash. Usually covered with domical roof and provided with a pulley to lift water. Constructed at the centre of a saucer shaped catchment area. The catchment is created by using sealing material such as pond silt, charcoal, ash and gravel.
	Collecting rainwater for preventing mixture with saline ground water	*Virda* (Kutch)	Drinking	Shallow wells dug in low depressions called *jheels*. The structures reach down to the upper layer of fresh rainwater up to about 1 metre above ground water.
	Collecting rainwater	Pond/Lake/Tank	Domestic consumption	Circular/square/rectangular structure either with elaborate masonry work or with rudimentary design.
Islands (Andaman and Nicobar Islands)	Collecting water in pits through diversion channels	Bamboo/ Jackwells	Domestic consumption	Shompen and Jarawa tribes use longitudinally split bamboo, placed along the slopes to collect rainwater drop by drop into pits called jackwells. The bamboo channels are also placed under trees to harvest rain through leaves.

Contd.

Table 32.1: Traditional systems for drinking water and domestic use

Region	Systems	Type	Use	Technique
Hill and mountain region (Himalayan regions)	Collecting rainwater	*Chaal* (Uttaranchal)	Animal consumption	Natural formation or depression found along mountain ridge tops saddled between two adjacent crests.
		Khal (Uttaranchal)	Animal consumption	Large natural lake which can store several thousand cubic metres of water.
	Collecting water from subterranean seepages or spring	*Chuptyaula* (uttaranchal)	Animal consumption	Natural rudimentary structures found in the high-altitude areas.
	Collecting rain-water and water from subterranean seepages or springs	*Naula/Baori* (Uttaranchal)	Domestic water use	Shallow, four-sided stepped wells constructed in the form of an inverted trapezoid. Some are massive and ornate structures with rooms and platforms for bathing and washing clothes. Animals are prevented from entering the tank. The oldest functioning *naula,* the Badrinathji-ka-naula back to 7th century AD (Uttarakhand).
		Baori (Himachal Pradesh)	Domestic water use	Shallow step-well, usually covered structure. They are walled to keep animals out and to prevent unclean water from entering. Sometimes a trough is provided outside for animals.
		Nauns (Himachal Pradesh)	Bathing and washing clothes	Very large *baoris* are called *nauns.* They are usually uncovered structures.
	Diversion channels from hill streams or glaciers	*Dhara* (Uttaranchal)	Drinking (also used for irrigation)	These are basically water fountains. Water from springs or subterranean sources is channelled through outlets, which are in the shape of either a simple pipe or various figurines like animals, female forms etc. There are three types of *dhara* depending on the height above ground. These are seasonal.
		Nahun (Himachal Pradesh)	Domestic consumption	Water fountains, usually square or oblong and enclosed along the sides and back, but open in front with flooring of two massive stone beams overlaid with flat slabs laid diagonally.
		Chharedu (Himachal Pradesh)	Domestic consumption	Spring fountains wherein the spring water falls through a carved stone. It has a rectangular enclosure and a drain.
	Diversion channels from hill streams or glaciers leading to a storage tank	*Panihar* (Himachal Pradesh)	Domestic consumption	Cistern.
	Storage tanks/ponds collecting rainwater	Ponds (Jammu and Kashmir, Himachal Pradesh)	Domestic consumption	Rudimentary structures.
		Khola (Sikkim)	Drinking	Simple structure to collect the water. The water is transported through bamboo poles to individual houses.
	Rainwater collected on rooftops	Tanks (Mizoram)	Domestic consumption	Horizontal rain gutters are placed along the sides of the sloping roof. The pipes are connected to a storage tank.

Source: WWI Research Team

About 200 million people in Asia, Africa and Latin America have symptoms of the intestinal infection giardiasis; there are some 500,000 new cases a year, the vast majority of them among children. The infection causes acute and persistent diarrhoea, abdominal pain and rapid weight loss. Lack of sanitation and poor basic hygiene assist its spread. The main mode of transmission is in fecally-contaminated water, but food-borne and person-to-person spread also occur. The disease has become a serious problem in day-care centres in developed countries.

Epidemics of cholera and dysentery are frequent, striking adults as well as children. Cholera alone causes 120,000 deaths a year and is particularly deadly in Africa, where epidemics have become more widespread and more frequent since the 1970s, the death rates among sufferers are generally the highest. Seventynine million people are estimated to be currently at risk of cholera infection in Africa. A substantial proportion of cases and deaths have been among the displaced and refugee population in Somalia and Zaire.

A pattern of global cholera epidemics—called pandemics—that literally circle the world is being repeated. At least seven such pandemics have been recorded in the last 150 years (1850-2000). The seventh pandemic of the disease caused by *Vibrio cholerae* 01 biotype El Tor, which started in 1961 and then spread to other countries in South and Central America. In 1993, cholera was reported as endemic in some 80 countries, affecting several hundreds of thousands of people. In an important number of cases, food has been found to be the vehicle for transmission.

A new strain of cholera, code-named 0139, emerged in India in 1992. It spread west to Pakistan and east to China and in the early months of 1993 caused an estimated 100,000 cases and 1,000 deaths in southern Bangladesh. It has not spread rapidly since then, but remains a threat.

The economic impact of cholera epidemics, in losses of trade and tourism, can be enormous. For example, the 1991 epidemic in Peru cost that country a loss estimated at $770 million, almost a fifth of its total exports in a normal year.

Several countries, especially in Africa, have to cope with the emerging problem of recurrent epidemics of bacillary dysentery. The spread of epidemic dysentery started in 1979 in eastern Zaire and since then has invaded all African countries. For several months in 1994, dysentery was the main cause of death in refugee camps in Burundi, United Republic of Tanzania and Zaire. Epidemic dysentery also occurs in Asia and Latin America.

Typhoid fever causes about 16 million cases and more than 600,000 deaths a year. Almost 80 per cent of cases and deaths are in Asia and most of the others occur in Africa and Latin America. The socio-economic impact of the disease is huge, because typhoid survivors take several months to recover and resume work. Since 1950, the organism's resistance to antibiotics has been a growing problem; by 1989 resistance was reported in a number of countries, particularly in Asia and the Middle East.

The hepatitis E virus—discovered in the United States in 1990—exists in many parts of the world, causing outbreaks of hepatitis in Asia, North Africa and Central America, particularly in countries where environmental sanitation is inadequate. The virus is probably a major cause of acute hepatitis in these countries but does not lead to chronic disease. The infection is mainly transmitted by contaminated water, although person-to-person spread by the fecal-oral route is also likely. Preliminary evidence suggests that a hepatitis E vaccine could be developed in the not-too-distant future.

The problem of food-borne diseases is very serious in developing countries, but is not limited to them. The estimated annual incidence of food-borne diseases in the United States ranges from 6.5 million to 80 million cases. Surveys in several other industrialised countries suggest that annually, up to 10 per cent of the population may be suffering from a food-borne disease.

Salmonellosis especially has increased tremendously in industrialised countries on both sides of the Atlantic over the past few years. In many countries, poultry meat, eggs and foods containing eggs have been identified as the predominant sources of this pathogen. In certain countries, up to 60-100 per cent of poultry meat is contaminated with species of Salmonella and beef, frogs' legs, chocolate and milk have also been implicated. In 1985, some 170,000 to 200,000 persons were involved in an outbreak of salmonellosis in Chicago (United States) caused by contaminated pasteurised milk.

Hepatitis A is common all over the world: some 10 to 50 persons per 100,000 are affected annually. Shellfish grown in contaminated water have often been recognised as a source of this disease. An epidemic of shellfish-borne hepatitis A in China in 1988 affected some 292,000 persons (with 32 fatalities), and was related to consumption of contaminated clams. Food contaminated by food-handlers and not subsequently sufficiently heated may also transmit the disease. Many cases of hepatitis A are known to be linked to restaurants.

In India during past two decades, the Accelerated Urban Water Supply Programme (AUWSP) was initiated in 1993-94 to provide safe and adequate water supply facilities for towns having population less than 20,000 as per 1991 census. Under the scheme, the centre matches the funds provided by the state government which includes a 5 per cent town contribution. In special cases, 100 per cent finance is available as the central share. The AUWSP is being administered through the CPHEEO at the Centre. Until March 2002, 223 projects for 654 towns were approved at an estimated cost of Rs. 817.70 crore. An amount of Rs. 337.37 crore has been released by the Centre to the States for the implementation of these projects. According to the Planning Commission, 240 schemes have been completed. Among the main obstacles listed by the Planing Commission in the implementation of AUWSP are the frequent changes in priority given to different projects by the State Governments, delayed submission of reports and failure in conforming to guidelines. Another major problem is the absence of manpower for O&M activities. The availability of safe drinking water in urban area alone is shown in Table 32.2.

Table 32.2: Availability of safe drinking water per day in Metropolitan Cities of India

Towns and Cities	Per Capita Water Supply (lpcd)
Greater Mumbai	170
Kolkata	100
Delhi	155
Chennai	65*
Bengaluru	100
Hyderabad	90
Ahmedabad	116
Pune	220
Surat	180
Kanpur	118
Jaipur	97
Lucknow	264
Nagpur	130
Patna	107
Indore	80
Vadodara	135

Contd.

Table 32.2: Availability of safe drinking water per day in Metropolitan Cities of India

Towns and Cities	Per Capita Water Supply (lpcd)
Bhopal	130
Coimbatore	109
Ludhiana	140
Kochi	129
Visakhapatnam	58
Agra	134
Varanasi	103
Madurai	72
Meerut	185
Nashik	140
Jabalpur	95
Jamshedpur	90
Asansol	120
Dhanbad	70
Faridabad	120
Allahabad	111
Amritsar	135
Vijaywada	137
Rajkot	88

lpcd (liters per capita per day)

* Indicates current availability of water (due to consecutive failure of monsoon, the availability of water has gone down)

Source: Rajya Sabha Unstarred Question No. 226, dated 12.7.2004

32.3 SOIL-BORNE INFECTIONS

Diseases that come from the soil affect several million people annually. The most deadly of these diseases is neonatal tetanus, responsible for the deaths of at least 450,000 children every year (1958-1990), while intestinal worm infections are the most widespread.

Tetanus can affect all age groups but is a problem particularly for newborns and their mothers. It is a bacterial disease, but it does not spread from one individual to another. It occurs through the infection of a wound by the tetanus bacillus, the incubation period varying between 4 and 21 days.

Neonatal tetanus occurs as a result of contamination of the umbilical cord with tetanus spores at birth, caused by unsterile methods of handling or cutting the umbilical cord or dressing it with germladen substances such as ash, mud or animal dung (common practices in many areas of developing countries). Treatment of neonatal tetanus is difficult and generally unsuccessful, and almost all newborn babies affected die. The disease can be prevented by immunising women of reproductive age (especially pregnant women) with tetanus toxoid and by ensuring hygienic delivery and cord care practices in institutions and at home.

In the early 1980s, between 800,000 and 1 million newborns were estimated to die of neonatal tetanus every year (up to 25 per cent of infant deaths in many developing countries).

In 1989, WHO's member countries decided to eliminate neonatal tetanus as a public health problem by aiming to reduce its incidence to less than 1 case per 1,000 live births for each health district. If they succeed, 1 million newborn deaths and 50,000 maternal deaths due to tetanus will be avoided each year.

Of all soil-borne diseases, intestinal parasitic infections caused by worms account for by far the greatest proportion. Those caused by hookworm, roundworm and whipworm predominate. WHO estimates that as many as 3.5 billion people are infected by worms and that at any time some 450 million people, most of them children, are ill as a result. The number of those affected is increasing, and cases occur in all WHO regions.

While most of these conditions amount to no more than diarrhoea and abdominal pain, they can be fatal. It is estimated that in 1995 hookworm infection killed 65,000 people and roundworms killed 60,000. Chronic infections impair the physical and mental growth, nutrition and development of children in general and girls and young women in particular.

The infections are spreading rapidly in slums, shanty towns and squatter-settlements in and around many cities in developing countries. Poor sanitation and consequent contamination of the environment with human excreta are among the commonest hazards contributing to transmission. Soil is the breeding ground, when contaminated by infected individuals defecating on open spaces.

32.4 PREVENTING AND CONTROLLING FOOD-BORNE, WATER-BORNE AND SOIL-BORNE DISEASES

So much depends on how communities deal with human feces. At least 2.5 billion people in developing countries lack an adequate system for disposing of their feces. For the poor in many developing countries, feces deposited near their homes constantly threaten household hygiene, by direct contact with people or by being inadvertently carried into homes and kitchens by children, domestic animals or insects. Domestic, neighbourhood or district water supplies such as well, tanks and reservoirs may be contaminated by poorly designed or maintained sewage disposal systems.

The cleanliness of the water is not the only issue; the importance of an adequate water supply for household hygiene cannot be overemphasised. Washing hands after defecation and before preparing food is fundamental to controlling diseases, but even if this principle is understood fully, handwashing is impossible without sufficient water.

Improving the quality of drinking water, ensuring proper sewage disposal and providing more water for both personal and domestic hygiene are the keys to the prevention or control of major scourges such as diarrhoeal diseases (including cholera), typhoid, guinea-worm disease, schistosomiasis and giardiasis. The estimates, relating to the developing world for 1995, indicate that about 25 per cent of the population do not have access to safe water and over 66 per cent lack adequate sanitation.

Technological advances in water supply, sanitation, drainage and solid wastes management in recent years offer cost-effective solutions that can be adapted to local circumstances and can greatly improve health and environmental conditions. Personal hygiene is immensely important in the prevention of contamination of food and drinking water. The provision of a safe and adequate water supply requires partnership between communities and the organised public and private sectors, including NGOs. Government ministries concerned with education, development and industry have a major role to play. Communities can also take local initiatives for the provision of safe water, as demonstrated in Bolivia.

Deaths due to diarrhoeal diseases are particularly tragic because the great majority can be avoided by the prevention or treatment of dehydration, as the primary cause of death from acute diarrhoea is dehydration from the loss of fluids and electrolytes. Correct case management, both at home and in health facilities, can achieve immediate reduction in mortality from acute epidemic and non-epidemic diarrhoea in any age group.

This is achieved through the prevention of dehydration in the home by providing increased amounts of fluids and continuing to provide food or breast milk.

Dehydration can be treated through the use of oral rehydration salts, and the use of intravenous fluids for severe dehydration. Antibiotics should be used only for dysentery and severe cases of cholera.

Outbreaks of giardiasis can also be prevented by adequate filtration and treatment of water supplies. The illness itself can be treated with drugs, but control measures have been hampered by lack of knowledge about the biology and natural history of the giardia organism. There is increasing evidence that its natural host is to be found among animals.

Cholera, which is definitely linked to poverty, is likely to occur when there is overcrowding coupled with lack of adequate sanitation and safe drinking water. Therefore the best long-term strategy to prevent cholera is elimination of the factors that favour its transmission, especially by improving water supplies and sanitation.

The treatment of patients with rehydration therapy is the cornerstone of public health efforts to reduce cholera deaths. During acute emergencies, providing clean drinking water, basic sanitation and safe disposal of human waste are high-priority preventive interventions. Immunisation is unlikely to be useful in epidemics.

Safe drinking water systems and adequate sewage disposal facilities are essential for the prevention of typhoid. Scrupulous cleanliness is necessary in food preparation and handling, as is proper storage of salads and other foods served cold.

A vaccine exists, but it does not give complete protection. Because of the increasing emergence of multriresistant bacterial strains, there is a greater need for longer-lasting, more effective vaccines. Research in this field is continuing.

The prevention and control of food-borne infections depend on improving the hygienic quality of raw foodstuffs at the agricultural level: applying food processing technologies such as pasteurization, sterilisation, fermentation or irradiation; and, most critically, educating food-handlers in the principles of safe food preparation. Education of consumers and food-handlers is vital for the prevention of diarrhoeal diseases.

A vaccine for hepatitis A is available and is recommended for food-handlers. But because of its high cost, its uses has not been widely promoted.

Here too, intersectoral cooperation is essential. Without it, health promotion and disease prevention cannot make headway. The responsibility for providing safe food should be shared between governments and the food industry, from producers to retailers.

Tetanus can never be eradicated because the spores which cause it will always remain in the environment to pose the risk of infection. But it can be eliminated as a public health problem—which means reducing its incidence to less than 1 case per 1,000 live births in each health district where it occurs. This requires immunisation of pregnant women and all women of childbearing age in highrisk areas. Hygienic childbirth practices are also a key element of prevention and are being promoted through a WHO "Clean Delivery" campaign.

Babies born to immunised mothers have a natural immunity to retain for up to 12 weeks. Women need two doses of vaccine to gain up to three years protection and five doses to cover all their childbearing years of maximum immune response.

The probable number of women living in highrisk areas who need three doses—which give five years' protection—is 55 million. The estimated cost of immunising them all is 84 million—equal to 1.53 per woman in 2000.

If world efforts to immunise infants with three doses of DPT are maintained, reinforcing infant immunisation with a fourth dose of DPT at 15 to 24 months prolongs tetanus immunity for another five years. A fifth dose of tetanus toxoid (Given as Td of DT vaccine) at school entry will provide immunity for another 10 years. An additional doses when leaving school will ensure sufficient immunity for all childbearing age years.

In a perfect world there would be adequate sanitation for all and shoes for everybody. In the real world neither applies and controlling rather than eliminating soil-borne intestinal parasitic infections will remain the realistic goal until they do apply.

WHO promotes the regular use of single-dose, safe, cheap and effective drugs to combat the effects of intestinal worms infections. Research shows that in populations at risk, more of the worms are harboured by a small part of the population. This implies that it is most cost-effective to identify and treat the group that is most heavily infected.

In 2000, WHO studies in Zanzibar (United Republic of Tanzania) indicate that regular anthelminthic treatment with mebendazole (2.8 US cent a dose) significantly improves the iron status of school-children.

In areas where there is a high risk of hookworm infection and where many women have anaemia, WHO recommends that specific anit-hookworm chemotherapy should be included in strategies designed to improve the health, development and nutritional status of girls and women.

32.5 INSECT-BORNE DISEASES

Flourishing in conditions of heat and humidity, poverty and overcrowding, insects spread deadly and disabling diseases among almost half of the world's population. They prey particularly on small children, killing more than a million of them every year. They lay vast areas of land to waste, preventing productive settlement, destroying agricultural potential, and blighting whole communities and regions.

The Mosquito

But of all insects that transmit diseases, the mosquito represents by far the greatest menace. It remains mankind's most indomitable foe, resisting costly efforts to eradicate or even control it.

The mosquito is the vector for different diseases such as malaria, dengue fever and yellow fever, which together cause several million deaths and hundreds of millions of cases every year. Mosquitoes also spread lymphatic filarisis (elephantiasis), which infects about 120 million people up to 1990, and Japanese encephalitis, which takes its biggest toll among children.

Different species of mosquito are involved in the transmission of different diseases, each with its own particular habits: some feed by day and others by night; some prefer to feed indoors and others prefer to do so outside; some rest after feeding and others depart; some prefer to feed on humans and others on animals; and various species choose different places for laying their eggs. Mosquito control requires a knowledge of these habits.

Today, the mosquito is literally spreading its wings as a carrier of diseases and settling in new areas far from its original geographical boundaries. Some of the reasons—international travel, trade and migration—have already been outlined in the introduction to this chapter. In addition, expanding agriculture, the clearing of forests or the building of dams and irrigation schemes and unplanned urban development provide mosquitos with new breeding grounds, while at the same time bringing more people into contact with them.

Changes in climatic conditions are enabling mosquitoes and other disease carrying insects to survive and breed at more northern latitudes and higher altitudes.

One example illustrates the role of international travel and trade.

The Asian tiger mosquito, which can transmit deadly strains of encephalitis virus as well as dengue and yellow fever, has been introduced into the United States, Brazil and parts of Africa in cargoes of rubber vehicle tyres shipped from Asia. In the United States alone, this mosquito is established in at least 23 states and can survive winters as far north as Chicago (even at – 2°C).

Malaria is endemic in 91 countries, with about 40 per cent of the world's population at risk. Each year there are 300-500 million clinical cases of malaria, 90 per cent of them in Africa and between 1.5 million and 2.7 million deaths. Among all infectious diseases, malaria continues to be one of the biggest contributors to disease burdens in terms of deaths and suffering.

By undermining the health and capacity to work of hundreds of millions of people, it is closely linked to poverty and contributes significantly to stunting social and economic development.

In many parts of the world, malaria is becoming an even greater problem than before. Epidemics are recurring in areas where transmission had been interrupted and are generally associated with deteriorating social and economic conditions.

Malaria is a febrile disease caused by the four species of human malaria parasites belonging to the genus *Plasmodium* and naturally transmitted to people by the bite of an infected female mosquito of the genus *Anopheles*. Early symptoms include fever, shivering, aches and pains in the joints and headache. In Falciparum malaria, infected red cells can obstruct the blood vessels of the brain, causing cerebral malaria, which is often lethal. Other vital organs can also be damaged, with fatal consequences.

In the 1950s, there were hopes that malaria would be eradicated in a fairly short time, mainly through the intensive use of insecticides. However, malaria still remains a serious threat in endemic countries today.

According to National Vector Borne Disease Control Programme (NVBDCP) report estimates 25 million recorded cases of malaria with 30,000 deaths in 2009 against WHO's estimated 1.53 million cases and 15,000 deaths in global scale.

Those at greatest risk of dying from the disease are children under age 5 in malaria-endemic areas; pregnant women; people moving from non-malarious to malarious zones for reasons of work, migration, refuge, war or tourism; and travelers who visit endemic countries and return home with the disease.

WHO's global malaria strategy (which is intended to prevent mortality, reduce morbidity and lead to a decrease in social and economic loss) consists of providing early diagnosis and prompt treatment; planning and implementing selective and sustainable preventive measures, including vector control; detecting early, containing or preventing epidemics; and strengthening local capacities in basic and applied research to permit and promote the regular assessment of a country's malaria situation, in particular the ecological, social and economic determinants of the disease.

The main problems in malaria control are shortages of resources and resistance to pesticides and drugs.

Dengue fever, a severe influenza-like illness and dengue haemorrhagic fever are closely-related conditions caused by four distinct viruses transmitted by *Aedes aegypti* mosquitos. Dengue is the world's most important mosquito-borne virus disease. A total of 2,500 million people worldwide are at risk of infection. An estimated 20 million cases occur each year, of whom 500,000 need to be hospitalised up to 1990. There is no cross-immunity, which means that an individual who has recovered from one remains susceptible to the other three.

32.6 DIFFERENT VIRUSES

The haemorrhagic form of dengue (DHF) most often occurs in individuals who have previously had the infection but are infected again with a different virus type, although DHF may occasionally occur after one primary infection. The viruses move continuously between crowded urban centres, causing massive outbreaks of disease.

Dengue and especially dengue haemorrhagic fever, cause severe headache and pain behind the eyes which worsens with movement; patients have a high fever and muscle and joint pains. In the haemorrhagic form, there is bleeding from the nose, mouth and gums, excessive thirst and difficulty in breathing. There is no cure and, as yet, no vaccine.

Dengue affects more than 100 countries in all continents except Europe and is spreading rapidly in many areas.

It flourishes in conditions of poor housing, overcrowding and inadequate sanitation, especially in cities in tropical and subtropical regions of the Americas, Asia, Africa and the Western Pacific and in communities where there is accumulation of solid waste, especially discarded containers and tyres where mosquitoes breed.

In 1995, the worst dengue epidemic in Latin America and the Caribbean for 15 years struck at least 14 countries, causing more than 200,000 cases of dengue fever and almost 6,000 cases of dengue haemorrhagic fever. There was only limited dengue in the Western hemisphere until 1981, when the first outbreak occurred in Cuba.

Dengue epidemics are increasing due to growing urban population, expanding mosquito breeding sites and rapid transportation which involves both the movement of infected people and the spread of dengue-carrying mosquitos.

Japanese encephalitis is another mosquito-borne virus disease. Its transmission cycle involves wild birds and domestic pigs as amplifying hosts and the most common mosquito vector in *Culex tritaeniorhyncus*, which is associated with rice-growing regions. In recent years, the disease has appeared to spread into new parts of Asia, perhaps as a result of increased rice field agriculture.

Most infections are asymptomatic, with only one in a few hundred infections leading to clinical illness. Mild cases are often characterised by febrile headache or aseptic meningitis.

Severe infections are usually characterised by acute onset of headache, high fever, disorientation, coma, convulsions (especially in infants) and spastic paralysis. Among the severe cases, fatality rates may reach 60 per cent.

An estimated 43,000 cases of Japanese encephalitis occur globally each year, with 11,000 deaths and nearly 9,000 disabled. The vast majority of cases (about 85 per cent) occur among children less than 15 years of age.

Nearly 10 per cent of the cases are among those over 60 years, perhaps reflecting waning protective immunity. About three-quarters of the cases occur in the Western Pacific, primarily China and adjacent countries, with the reminder occurring in South East Asia, especially India. The disease is rare in other parts of the world and when seen, is generally associated with travelers returning from endemic areas.

Lymphatic filariasis or elephantiasis—the grotesquely disfiguring disease affecting the legs, arms and genitals—is the most notorious result of infection with filarial parasites transmitted by Culex mosquitos. About 120 million people are infected in tropical areas of Africa (40 million), India (45 million), South East Asia, the Pacific Islands and South and Central America up to 1990.

Lymphatic filariasis also causes acute fevers, inflammation of the lymphatic system and the bronchial-asthmatic condition known as "tropical pulmonary eosinophilia".

Plague

Plague is often regarded as a scourge of medieval times, from which the world is now largely free. But the latest evidence suggests that incidence of the disease is on the increase. During 1995, at least 1,400 cases of human plague (including at least 50 deaths) were notified to WHO. The disease occurs particularly in rodents. It spreads from rat to rat and from rats to humans mainly by rat fleas biting first a sick rat and then a person, thus transmitting *Yersinia perstis,* the bacterium of the disease.

Plague most commonly has two forms: bubonic and pneumonic. The more frequent form is bubonic, in which there is sudden onset of severe malaise, headache, shaking chills, fever and pain in the affected regional lymph nodes. Large and painful lumps appear under the skin, called buboes. The more dangerous form of the disease is pneumonic or pulmonary plague, which affects the lungs and can be transmitted from person to person by droplets in the air from sputum discharged by the infected individuals.

Some countries in Africa, the America and Asia report cases almost every year. Madagascar, United Republic of Tanzania and Zaire in Africa; Bolivia, Brazil, Peru and the United States in the Americas, and China, Kazakhstan, Mongolia, Myanmar and Vietnam in Asia. However, cases occur in areas that had apparently been free from the disease for many decades. For example, it has reappeared in Botswana, India and Malawi in recent years after "calm" periods of up to 30 years. Peru experienced a large outbreak of plague in 1984, followed by another in 1990 and again 1992. These outbreaks are linked to cyclical epidemics of plague in rodents.

Over the last 30 years, the mean annual global plague case fatality rate has been 9 per cent, ranging from just over 14 per cent in Africa to just under 6 per cent in the Americas. These high rates persist despite the availability of highly effective drugs against the disease.

Other factors are involved in the apparent increase in plague. In the United States, for example, rapid suburbanisation has resulted in increasing numbers of people living in or near areas where plague exists in nature. The number of states of that country reporting plague cases increased from three during 1944-1953 to 13 during 1984-1994. Surveillance of plague in rodents indicates that the disease has spread eastwards in the United States to areas believed to have been free of plague during the previous 50 years (up to 1990).

32.7 PREVENTION AND CONTROL OF INSECT-BORNE DISEASES

The five main diseases spread by mosquitoes—malaria, dengue, yellow fever, Japanese encephalitis and filariasis—do not lend themselves to a single, comprehensive approach. Each has to be tackled in its own right. At the same time, however, some methods of prevention or control are common to all of them. Vaccines against yellow fever and Japanese encephalitis already exist and others are being developed against malaria and dengue, although it will be several years at least before they become available. There is an effective drug against filariasis.

The two key measures to prevent the spread of these five diseases are the elimination of mosquito breeding places and the prevention of mosquito bites. The first involves community action, such as covering water storage containers and removing discarded containers from inside and around houses, house-to-house inspection; and control programmes using insecticide sprays.

The second involves people protecting themselves with insecticide-impregnated materials such as curtains and nets, repellents and screens.

Vector and disease surveillance are extremely important, as are rapid intervention and control when an outbreak occurs. Other important measures are health education campaigns and preparations to guard against outbreaks.

Following high-level political commitment for a Global Malaria Control Strategy made in 1992 at a Ministerial Conference in Amsterdam, a target was set for achieving a reduction in malaria mortality of at least 20 per cent by the year 2000 in at least 75 per cent affected countries. In 1994, the United Nations General Assembly called upon WHO, as the lead agency for health, to promote the international mobilisation of technical, medical and financial assistance to intensify the struggle against malaria. But malaria resurgence in tropical countries were reported in past three decades.

Drugs such as mefloquine and halofantrine have been developed and registered and artemisinin derivatives developed and brought close to registration.

In Africa, large-scale multicentre trials of the effectiveness of insecticide-treated bednets have demonstrated a dramatic reduction in mortality of children under the age of 4. These research-based results are now being translated into operational recommendations for national control programmes.

32.8 MODERN RESEARCH

Worldwide efforts in basic research in malaria vaccines are at last beginning to bear fruit. In the United Republic of Tanzania in 1994 a double-blind phase III trial of the Colombian SPf66 peptide vaccine demonstrated that it reduced the numbers of first malarial fevers in children by around a third. Continuing studies with this and other vaccines are yielding additional data that will form the basis for recommendations for their use. In the meantime, emphasis is being placed on strengthening national capacity to diagnose and treat malaria; epidemic preparedness and control; and mobilising sustainable international commitment for longterm malaria control.

In the countries of Africa, south of the Sahara, special attention is being given to strengthening general health services and enabling health care providers from other sectors and the communities themselves to provide early diagnosis and prompt treatment, i.e., disease management. In the rest of the world, the emphasis is on improving the provision of basic curative services at all levels of health care and the promotion of rational drug use; and implementing selective measures for disease prevention, including selective vector control.

A dengue vaccine is already developed in Thailand with WHO's help. The results of clinical trials in adult volunteers show that it is safe and its immunological response is encouraging. Trials of the vaccine in children have also begun.

A live, attenuated, highly efficacious vaccine for yellow fever is available and widely used, including in the Expanded Programme on Immunisation in many countries. Protective antibodies appear 7-10 days after immunisation and probably persist for life, although reimmunisation is currently recommended every 10 years. Other interventions centre around prevention of bites from infected mosquitos. Thus mosquito control activities, especially targeted to *Ae. aegypti,* are critical in the prevention of urban transmission.

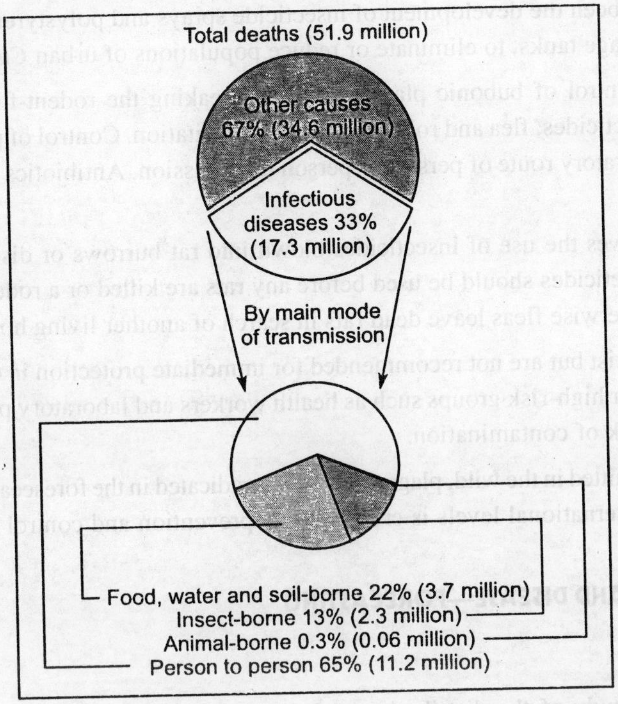

Fig. 32.1: Death due to selected infectious diseases, 1995 estimates

Urban transmission of yellow fever is essentially identical to that of dengue fever, a rapidly increasing public health problem in many countries of the tropics. Control of breeding of *Ae. aegypti* mosquitoes around houses should lessen the risk of urban yellow fever transmission as well as that of dengue fever.

Vaccines are available for Japanese encephalitis. They require an initial series of two injections, followed by a booster. Immunisation is the most effective means of disease prevention. Efforts are under way to develop a live, attenuated vaccine that will be an improvement on the existing inactivated vaccine.

In filariasis, four major breakthroughs have occurred, offering the prospect that—given sufficient political commitment and resources—the disease might eventually be well-controlled or even eliminated.

The first of these breakthroughs is the development of safe, single-dose, annual drug treatments. Trials have proved that a single dose of diethylcarbamazine (DEC) is very effective even two years after treatment. A single dose of invermectin has proved to be equally effective. A combination of single doses of both drugs reduced microfilaria more than 95 per cent two years after treatment.

Secondly, intensive local hygiene on an affected limb, with or without the use of antibiotic and antifungal creams, has been shown to have dramatic effects by halting the progression of—or even reversing—elephantiasis and lymphedema.

Thirdly, DEC-medicated table or cooking salt has been introduced in India. The carefully controlled addition of very low concentrations of DEC has long been recognised as an effective means of eliminating lymphatic filariasis infections in communities. However, the addition increases the price of the salt. The first commercially prepared DEC salt went on sale in India, at about twice the price of ordinary salt.

Finally, there has been the development of insecticide sprays and polystyrene beads to seal latrines and roof-top water-storage tanks, to eliminate or reduce populations of urban *Culex* mosquitos.

In general, the control of bubonic plague involves breaking the rodent-flea-human transmission chain by the use of insecticides, flea and rodent control and sanitation. Control of pneumonic plague aims at interrupting the respiratory route of person-to-person transmission. Antibiotics are effective in treating the disease.

Flea control involves the use of insecticides blown into rat burrows or distributed around houses frequented by rats. Insecticides should be used before any rats are killed or a rodent control campaign is carried out, because otherwise fleas leave dead rats in search of another living host such as humans.

Plague vaccines exist but are not recommended for immediate protection in outbreaks. Vaccination is only recommended for high-risk groups such as health workers and laboratory personnel who are more intensely exposed to risk of contamination.

Because it is transmitted in the wild, plague cannot be eradicated in the foreseeable future. Surveillance at local, national and international levels is crucial in the prevention and control of the disease.

32.9 EPIDEMIOLOGY AND DISEASE—FORECASTING

Concept

'Epidemiology' is the study of the distribution and determinants of health-related states or events in specified populations, and the application of this study to the control of health problems. Their basic understanding is essential for epidemiological studies viz., disease frequency, disease distribution and determinants of diseases.

Aims of Epidemiology

According to the International Epidemiological Association (IEA), epidemiology has three main aims:

(a) To describe the distribution and magnitude of health and disease problems in human populations;

(b) To identify the aetiological factors (risk factors) in the pathogenesis of diseases; and

(c) To provide the data essential to the planning, implementation and evaluation of services for the prevention, control and treatment of diseases and to the setting up of priorities among those services.

Epidemiological study approach

There are two ways of epidemiological survey as done regularly:

(a) Through questionnaire survey;

(b) Through field data collection on the basis of observation and then draw inferences on comparison basis.

The questionnaire may be related to health events or health actions, while field data collection actually based original observation.

Measurements in Epidemiology

Epidemiological measurement primarily involve the measurement of mortality and morbidity in human populations. However, the scope of measurements in epidemiology is very broad and unlimited and includes:

(a) Measurement of mortality;

(b) Measurement of morbidity;

(c) Measurement of disability;

(d) Measurement of natality;

(e) Measurement of the presence, absence or distribution of the characteristics or attributes of the disease;

(f) Measurement of medical needs, health care facilities, utilization of health services and other health-related events;

(g) Measurement of the presence, absence or distribution of the environmental and other factors suspected of causing the diseases;

(h) Measurement of demographic variables.

Traditionally and universally, most epidemiological studies begin with mortality data. Death certificate is the major record for obtaining such data. The death rate and the survival rate are the principal measurement factors. Morbidity has been termed as "any departure, subjective or objective, from a state of physiological well-being". Three aspects of morbidity are commonly measured viz., frequency, duration, and severity.

Epidemiological Methods

The primary concern of the epidemiologist is to study disease occurence in people, who, during the course of their lives, are exposed to numerous factors and circumstances, some of which may have a role in disease etiology.

Epidemiological studies can be classified as observational studies and experimental studies. The details are:

1. *Observational Studies*:
   - (a) Descriptive studies: Disease occurence (time, place, persons) and etiology.
   - (b) Analytical studies:
     - (i) Ecological or correlation with population;
     - (ii) Cross-sectional or prevalance with individuals;
     - (iii) Case control or case reference with individuals;
     - (iv) Coheret or follow up with individuals.
2. *Experimental Studies*:
   - (a) Randomized controlled trials or clinical trials;
   - (b) Field trials or community intervention studies.

Case Studies

Goiter: It is non-inflammatory, non-neoplastic enlargement of the thyroid gland. Hyperthyroid states can be recognised by this type of clinical symptoms, like goiter along with high basal metabolic rate, sleeplessness, heart rate and blood pressure increase, profuse sweating, and emotional instability.

Goiter is developed as a result of a lack of iodine availability to the thyroid gland. Lack of iodine causes decrease of thyroxine synthesis and output. This stimulates the anterior pituitary to secrete excessive thyrotrophic hormone (TSH). TSH activates the thyroid gland to produce hyperplasia of the thyroid vesicle to manufacture adequate thyroxine. However, this condition may be corrected by regulation of iodine supply to the diet.

Earlier days there were reports of goiter in the foothills, gradually it became prevalent in other parts also. As such, use of iodinized salt is now compulsory for all.

33 IMMUNOLOGY AND HUMAN DISEASES

33.1 INTRODUCTION

Immunology is the science that studies the structure and functioning of the immune system. The science began long before anyone knew about disease-causing microbes or even that individuals had an immune system that protected the body against diseases, e.g. in about c. 200 BC, Charak (the Indian 'Father of Medicine') gives a clear idea of Immunology just as Hippocrates (460-377 BC)—the European 'Father of Medicine' did.

There are two types of natural immune systems recognised in human bodies viz., humoral (antibodies) and cellular (immune cells) immune system (Fig. 33.1). These constitute the innate *immunity* of the body system. These innate immunity are non-specific and protects through two mechanisms viz., (a) non-specifically clears the infectious agents using preformed components, or (b) produces specific cells and molecules directed against the foreign invader. This is a natural resistance which operates non-specifically during the early phase of an immune response. The innate immunity serves as the first line of defense and includes both external and internal nonspecific responses.

The external innate immunity (Skin body secretions and mucous membranes) prevents the penetration of pathogens into host tissues. If a pathogen breeches the external innate defenses and invades the tissues, the internal defense mechanisms provide protection. Internal innate immunity includes three general mechanisms: (a) physiologic barriers, (b) phagocytosis and (c) inflammation. Human immune sites are shown in Fig. 33.2.

In contrast, acquired immunity develops during a host's lifeline and is based partly on the host's experiences. The exposure process is called *immunization*. Acquired immunity is the surveillance mechanism of vertebrates that *specifically* recognises foreign antigens and *selectively* eliminates them and, on reencountering the antigens, has an enhanced response. Once a host has been exposed to a specific disease, the host will probably not catch the disease again. The persistence of a foreign antigen in a host initiates, or induces, acquired immunity. The recognition of and response to the antigen are highly specific. In human system there are also two types of acquired immunity—Humoral and cell-mediated. Such may last few years as *active*, or disappear after a short period, i.e., passive only (Fig. 33.3).

There are six major characteristics of acquired immunity viz., (a) specificity, (b) inducibility, (c) diversity, (d) memory, (e) distinguishing self and non-self and (f) down regulation of itself (Fig. 33.4).

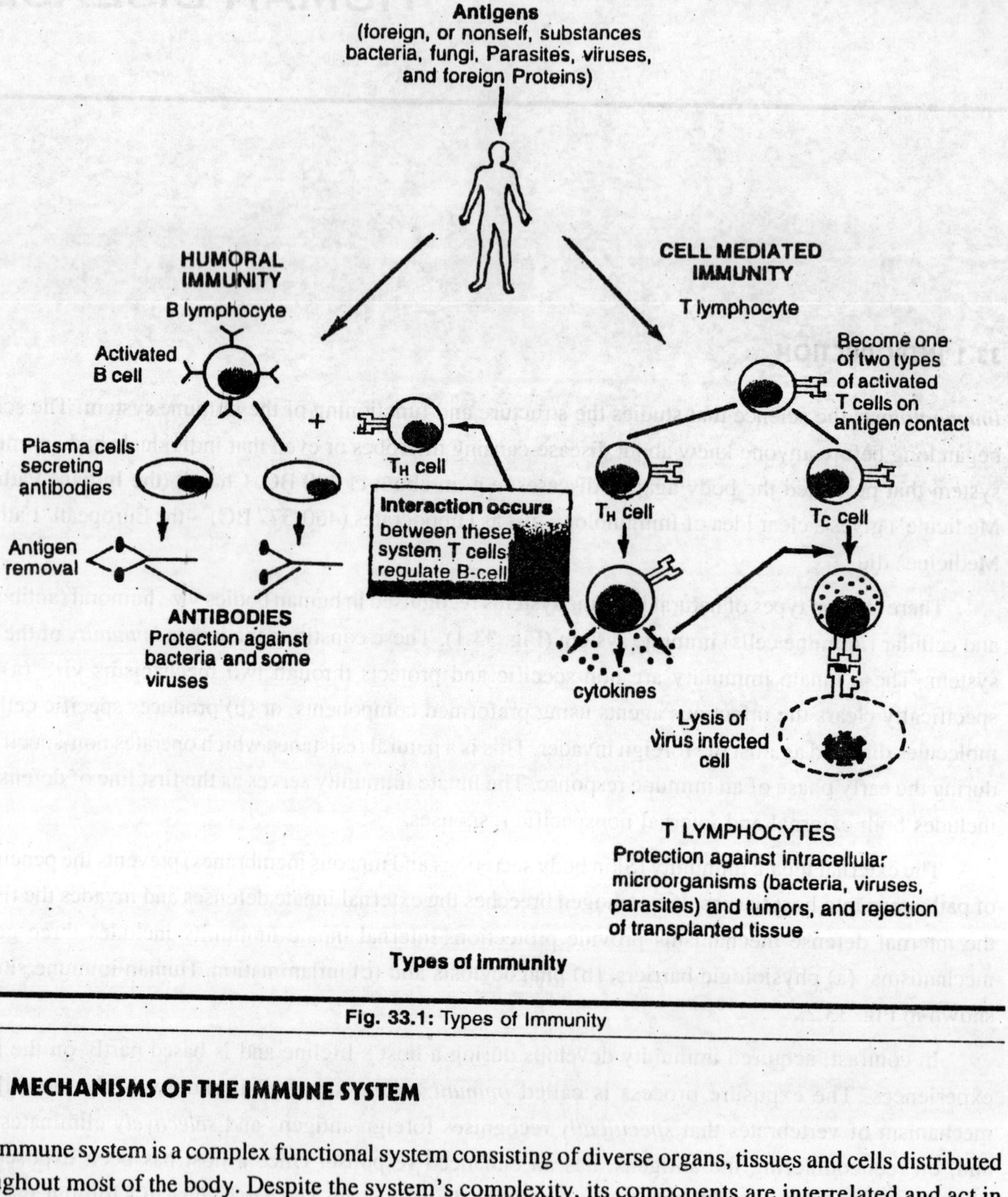

Fig. 33.1: Types of Immunity

33.2 MECHANISMS OF THE IMMUNE SYSTEM

The immune system is a complex functional system consisting of diverse organs, tissues and cells distributed throughout most of the body. Despite the system's complexity, its components are interrelated and act in a highly coordinated and specific manner when they recognise, eliminate and remember foreign macromolecules and cells.

Any foreign substance (living or non-living) that induces an immune response when introduced into a host is called an immunogen, or, more generally, an antigen. Most antigens are large, complex macromolecules not recognised as such. Only small parts of antigens, called *antigenic determinants* or epitopes, induce and, react with immune elements such as *antibodies* or lymphocytes. Antibodies recognize antigens through their surface characteristics, particularly by the antigen's pattern or shape and charge. The binding

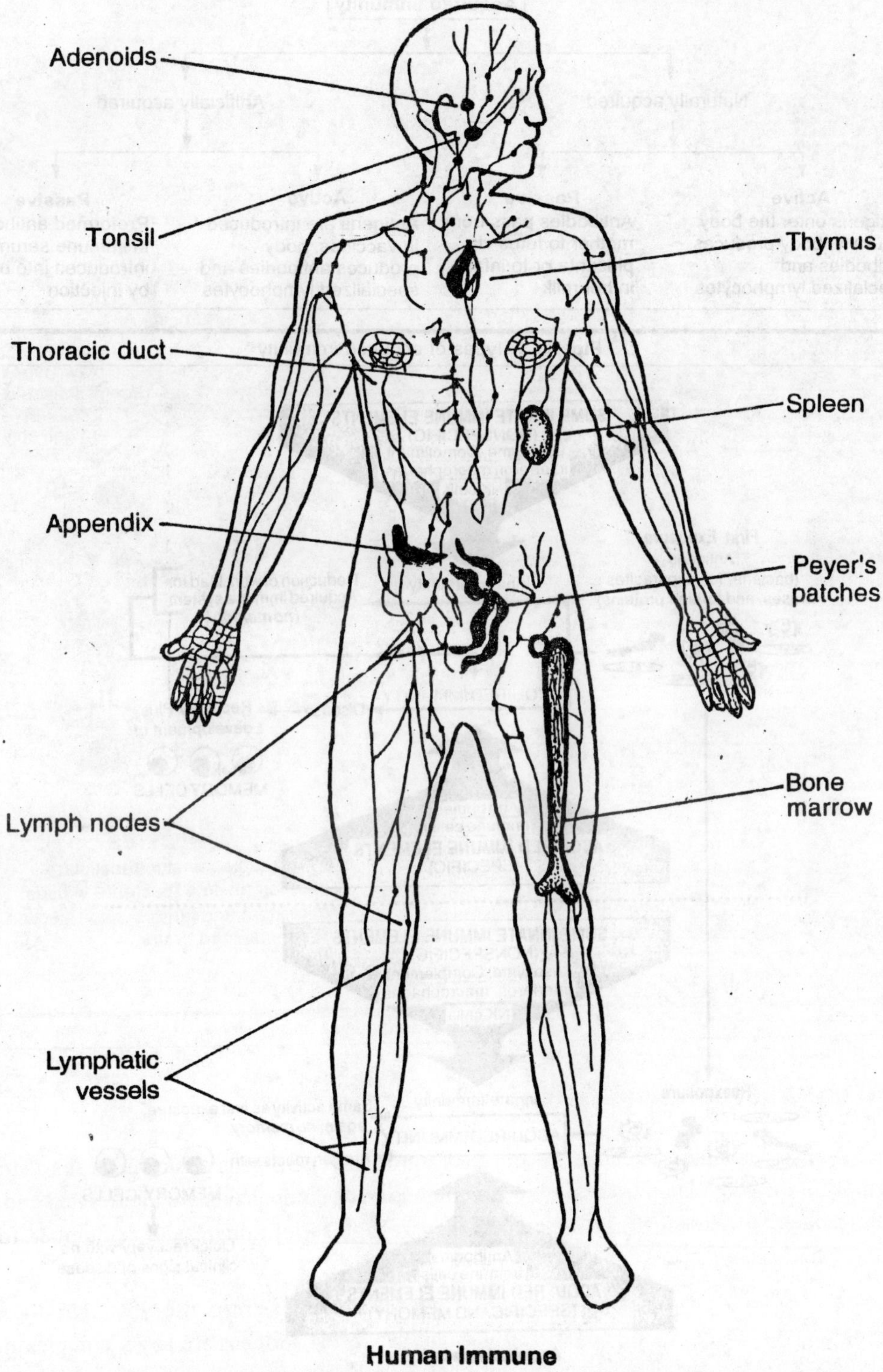

Fig. 33.2: Sites of Human Immunity Control

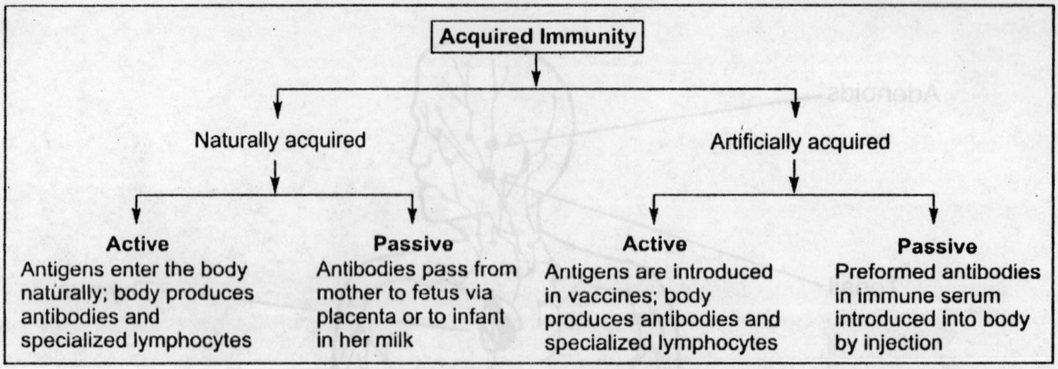

Fig. 33.3: Types of acquired immunity

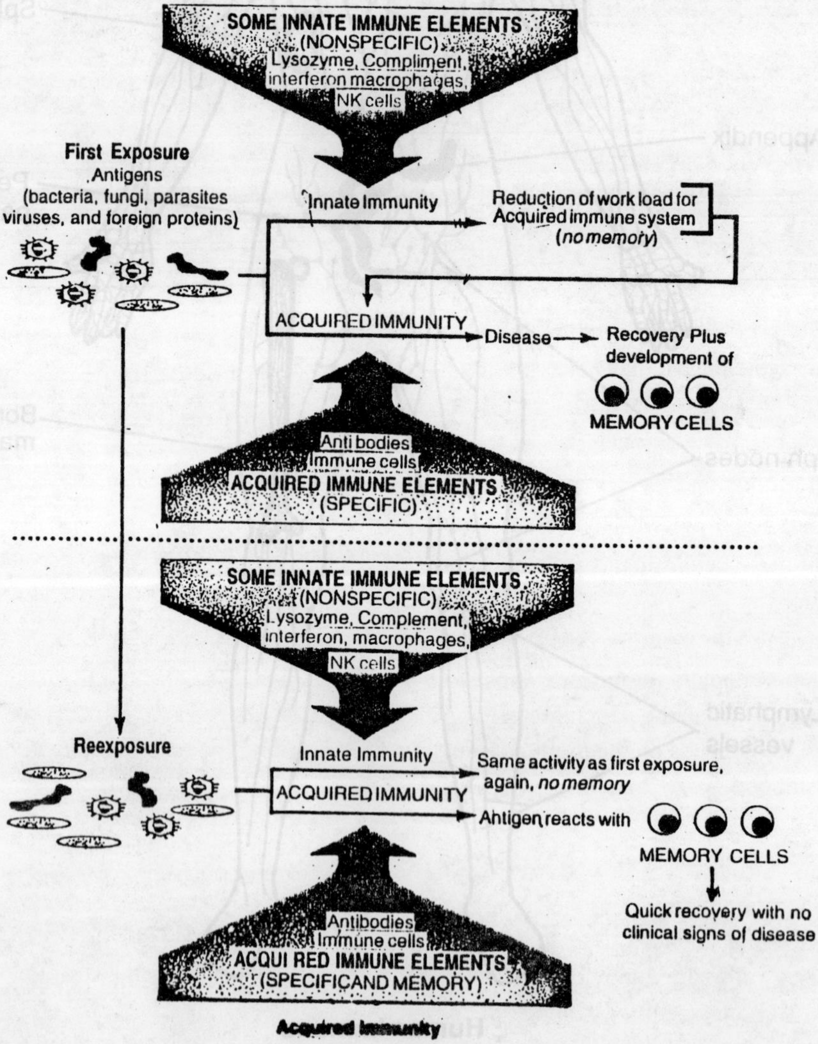

Fig. 33.4: Responses of Immunity

sites of the antibody are precisely complementary or specific for the right antigenic determinants. When antigens enters the body, it usually induces the production of antibodies that react only with that particular antigen. The immune system also has preexisting lymphocytes capable of reacting with the specific antigen. The first time we are exposed to an antigen, the result is a *primary immune response*. The second exposure to the same antigen leads to a *secondary immune response*, which is much faster and stronger. This phenomenon is moderated by immunologic memory and accounts for a person's long term immunity against infectious diseases.

To monitor against antigen intrusion anywhere within the body, the immune system stations the lymphoid system or immune system. The tissues are generally called lymphoid organs and include the bone marrow, thymus, lymph nodes, spleen and mucosa associated with lymphoid tissue. Two main groups of mononuclear leukocytes also participate in immune responses—lymphocytes and macrophages. Other cells crucial in the immune response are peripheral blood lymphocytes that do not express the classical characteristics of mature lymphocytes—they are called null cells.

33.3 ANTIGEN

An *antigen*, when introduced into a host, induces the formation of specific antibodies and T lymphocytes that are reactive against the antigen. Each antigen has four common characteristics—(a) High molecular weight, (b) Chemical complexity, (c) Solubility, and (d) Foreignness.

In each antigen there are usually two components, viz., *epitopes* (antigenic determinants), i.e., immunologically active portion, and *haptens*, i.e., non-immunogenic part of the antigen.

The chief characteristics of the antigenic determinants are:

(a) Antigenic determinants are small;

(b) All of the surface of a protein may be immunogenic and antigenic;

(c) Antigenic determinants must be accessible and are composed of assembled topographic determinants;

(d) Charge and polarity add to antigenic determinant immunogenicity;

(e) Antigenic determinants are conformation dependent and also have immuno-dominant building blocks;

(f) Antigen site mobility contributes to protein antigenicity;

(g) An individual's immune response to a protein antigen (a) is dictated by its genetic make-up; (b) depends on the structural differences between the antigen and the recipient's self-proteins, and (c) depends on the immunoregulatory mechanisms operating in that individual's immune systems.

33.4 ANTIBODY

The recognition proteins found in the serum and other body fluids of vertebrates that react specifically with the antigens that induced their formation are called *antibodies*. Antibodies belong to a family of globular proteins called immunoglobulins. The antibodies have three important features — (a) Binding versatility, (b) Binding specificity, and (c) Biological activity. A typical antibody molecule e.g. IgG is made up of four polypeptide chains with a molecular weight of about 150 kd. The four chains are divided into two identical *light chains* and two identical *heavy chains*. An antibody molecule is Y-shaped, with two identical antigen-binding sites at the ends of the arms of the Y (Fig. 33.5).

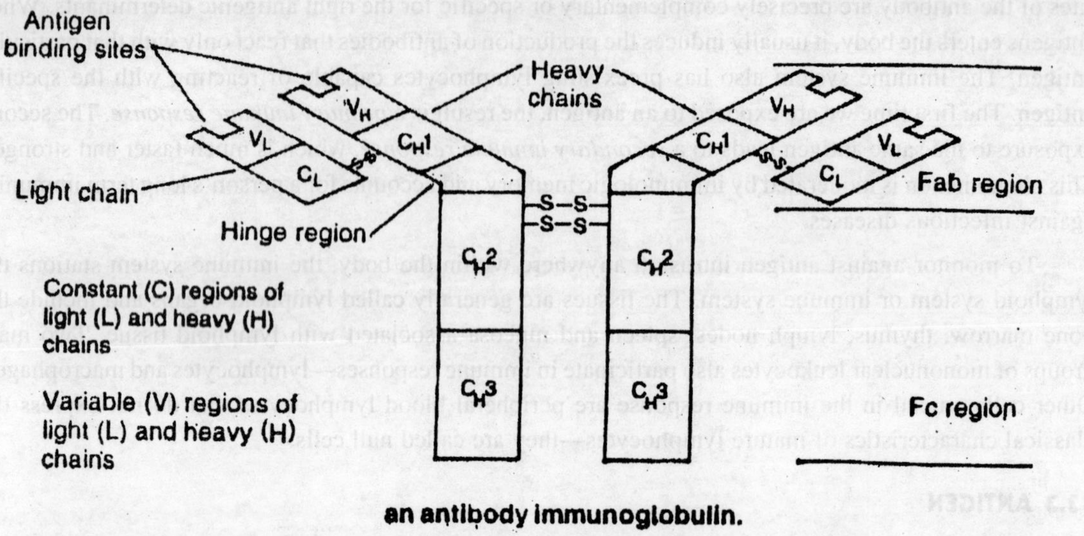

Fig. 33.5: Structure of immunoglobulin

The light and heavy chains contribute to the antigen-binding sites. Each antibody molecule can bind to two identical antigenic determinants.

There are five different classes of immunoglobulines—IgG, IgA, IgM, IgD and IgE—each with a distinctive heavy chain designated. IgG is found in blood, IgA in tissue fluid, IgM in glands, IgD in membrane and, finally, IgE also in blood.

When antibody forming cells are fused with myeloma cells, the resulting clone of cells is a *hybridoma.* Hybridoma cultures are the source of *monoclonal antibodies* of predefined antigen specificity.

33.5 ANTIGEN-ANTIBODY REACTIONS

Serology is the science dealing with the *in vitro* interactions of antibodies with antigens. The interaction of an antigenic determinant and an antibody molecule is called a primary antigen-antibody reaction. Cardinal characteristic of primary antigen-antibody reactions is that they are invisible. The conversion of invisible primary reactions to macroscopically visible ones leads to secondary antigen-antibody reactions such as *precipitation* and *agglutination*.

Primarily, antigen-antibody reactions can be measured using *fluorescence quenching, radioimmunasosay* (RIA), *enzyme linked immunosorbent assay* (ELISA) and *Immunofluorescence*. Precipitation test is done by ring test.

33.6 HYPERSENSITIVITY

Conventionally, there are four classes of hypersensitive reactions:

Type I : Immediate (anaphylactic) hypersensitivity.

Type II : Antibody-dependent cytotoxic hypersensitivity.

Type III : Immune complex mediated hypersensitivity.

Type IV : Cell mediated (delayed type) hypersensitivity.

The first three types of hypersensitivity are mediated by antibodies, and the last is mediated by T cells and activated macrophages. The details of hypersensitive reactions are shown in Table 33.1.

Table 33.1: Types of hypersensitivity

Types of reaction	Time before clinical signs	Characteristics	Example
Type I (anaphylactic)	< 30 min	IgE binds to mast cells or basophils; causes degranulation of mast cell or basophil and release of reactive substances such as histamine	Anaphylactic shock from drug injections and insect venom; common allergic condition, such as hay fever, asthma
Type II (cytotoxic)	5-12 hours	Antigen causes formation of IgM and IgG antibodies that bind to target cell; when combined with action of complement, destroys target cell	Transfusion reactions, Rh incompatibility
Type III (immune complex)	3-8 hours	Antibodies and antigens form complexes that cause damaging inflammation	Arthus reactions, serum sickness
Type IV (cell-mediated delayed-type	24-48 hours	Antigens cause formation of T_C that kill target cells	Rejection of transplanted tissues; contact dermatitis such as poison ivy; certain chronic diseases such as tuberculosis

33.7 AUTOIMMUNITY

It is the mirror image of tolerance; self-tolerance is the lack of reactivity to self, while autoimmunity reactivity to self, or the loss of tolerance to self. Self-reactivity (or auto-immunity) leads to immunity, but abnormal or excessive antiself responses may cause harm to the host. There are a number of diseases which are considered as autoimmune type. Examples of cell-mediated autoimmune diseases are Hashimoto's thyroiditis, and insulin-dependent diabetes mellitus, while antibody mediated autoimmune diseases are Grave's disease, rheumatoid arthritis and rheumatic fever etc.

33.8 IMMUNOLOGICAL TERMS

1. *Acquired immune deficiency syndrome (AIDS)*: A sexually-transmitted disease caused by a retrovirus called human immunodeficiency virus (HIV I). AIDS is characterised by a defective immune system and is generally fatal.
2. *Adjuvant*: A compound injected with an antigen which non-specifically enhances the immune responses to that antigen.
3. *Allergen*: Any substance capable of inducing an allergic response. Allergens are protein molecules, such as plant pollens and spores etc.
4. *Atopy*: A genetic tendency to develop immediate hypersensitivity states (allergies) such as hay fever and allergic asthma.
5. *Bacillus Calmette-Guerin (BCG)*: It is a strain of the tuberculosis organism found in cattle. It is now used as a living attenuated vaccine against the human form of tuberculosis.
6. *Hybridoma*: Cell line created by the *in vitro* fusing of lymphocytes and tumor cells—such as B cells and myeloma cells—to generate hybridomes producing monoclonal antibodies.

34 ALLERGY

34.1 INTRODUCTION

In the twentieth century, when allergies were first recognised, the public believed that they were caused by invisible emanations produced by new hay. A second generally accepted premise was that most cases of hay fever originated in the middle or upper classes of society. As Beard wrote in 1876, "Hay-fever is...a disease of the fashionable and thoughtful—the price of wealth and culture, a part of the penalty of a fine organisation and an in-door life".

Much more is known of the aetiology of allergy today, yet in 1990, 31 million sufferers in the United States from all walks of life can attest. Hay fever, largely caused by pollen and spores, afflicts no fewer than 13 million victims who endure physical and psychological misery at some time during the year, but mainly from spring until autumn. Among adults, 33 million disability days annually can be attributed to hay fever and asthma alone, while children lose no less than 36 million school days because of allergies. In North America, the greatest culprits are the ragweeds (Ambrosia) in the aster family (Asteraceae); in Europe and in more equatorial zones, however, grasses (Poaceae) are responsible for many cases of allergy. Since fungi are found everywhere, their spores are a constant cause of seasonal and perennial allergic rhinitis and asthma worldwide.

Probably one of the most common abnormalities of the immune system is the development of the allergic state. This atopic condition is often genetically predisposed and its evolution is related to factors such as allergen recognition through possession of specific receptor sites, changes in the proportion and types of immunoglobulins produced and differences in inherent tissue sensitivity to allergenically mediated substances. Depending on the overall mechanism involved and the speed and type of reactions observed, five types (I–V) have been distinguished (see Glossary under Hypersensitivity). Of these, examples of Types I to IV are discussed in the text, with Type II incorporated into the section on Drug Sensitivity Reactions.

34.2 ANTIBODY-MEDIATED (IMMEDIATE TYPE) HYPERSENSITIVITY (TYPE I)

Immediate hypersensitivity is related to the development of reaginic antibodies, usually Immunoglobulin E (IgE) and less often IgG, in sufficient quantities to evoke the symptoms of hay fever, asthma, hives, and anaphylactic shock. These antibodies are produced following allergen contact, locally within plasma cells embedded in the mucosa of target areas of the nasopharynx and the respiratory and gastrointestinal tracts, as well as elsewhere in the lymphatic system. After diffusing into the tissue fluid and serum the antibodies attach to mast cells, basophils and platelets. Allergen then binds to this cellbound reagin, triggering the release of histamine and other vasoactive substances (kinins, serotonin, slow reacting substances) that cause the allergic symptoms of bronchospasm, vasodilation, smooth muscle contraction and increased bronchial and nasal secretion. However, clinical manifestations are often not apparent until repeated exposures to the allergen have increased the number of cells involved.

An inciting allergen can be identified from among a host of possible candidates by several provocative *in vivo* tests. The most frequently employed are skin tests that involve the introduction of dilute concentrations of allergen into the skin by intradermal injection or through pricking. If a specific reaginic antibody is present in the skin, the characteristic wheal and erythema of a cutaneous anaphylaxis reaction appear within a few minutes. More rarely, tests using the conjunctiva or nasal mucosa as target organs have also been found effectual. Since the induction of systemic anaphylactic shock is always a possibility during such testing, however, it is recommended that the procedures always be carried out under careful medical supervision and that an antianaphylactic drug (i.e., epinephrine) be available.

Sensitive *in vitro* tests are also now employed. The radioallergosorbent test (RAST) is based on binding antibody to antigen in a system containing radiolabeled anti-antibody (anti-IgG, anti-IgE, anti-IgA). When necessary, the use of different radioactive iodine isotopes as labels allows for the simultaneous detection of reaginic IgE and blocking IgG in a patient's serum. Other tests have been devised that detect reaginic antibody through its ability to release histamine from mast cells or nonallergic donor leukocytes in the presence of allergen. Results of the passive leukocyte sensitizing test (PLS) have been used to link fluctuation in serum reagin content to the seasonal prevalence of allergen such as ragweed.

Allergic symptoms may be reduced by immunotherapy consisting of careful injection of small amounts of the allergen (often into site that is not the target area, e.g., upper arm instead of respiratory tract) to elicit specific blocking antibodies (IgA or IgG), or a white cell response. In some way, this response binds the allergens and prevents reagin-allergen complexes from forming (hyposensitisation), reduces the amount of IgE from being formed (tolerance) or, by neutralisation of reagin with excess antigen (desensitisation), reduces the amount of IgE antibody available in target areas.

In seasonal spore and pollen allergies, hyposensitisation may be attempted preseasonally, ensuring that blocking antibodies are at their highest titre during the provocative period, since treatment during the symptomatic period (coseasonal) is less satisfactory. For allergies that occur throughout the year, such as those elicited with fungal spores, it may be necessary to maintain treatment year-round.

34.3 ANAPHYLACTIC SHOCK

Systemic anaphylaxis, a rare event in man, is the most severe of all allergic reactions. Typically, it can occur after an inciting dose has been given intravenously to a hypersensitive individual, but it has also been noted after oral, subcutaneous, or intramuscular administration. It is characterised by sudden vasomotor collapse leading to shock, paroxysmal bronchoconstriction and if treatment is not undertaken immediately, death. Most cases have been associated with serum therapy (serum sickness), penicillin therapy and insect stings (especially the wasp, bee and hornet). However, the following case history underlines the danger of chamomile tea ingestion by an individual known to have ragweed and *Chrysanthemum* atopic disease and other allergies.

Within minutes of drinking a few sips of chamomile tea, a 35-year-old woman developed abdominal cramps, thickness of the tongue and a tight sensation in the throat. This was followed by angioedema of the lips and eyes, diffuse pruritus and a full sensation in the ears. There was no vomiting, diarrhea, sneezing or wheezing. Fortunately, diphenhydramine was administered immediately and a steroid shortly thereafter. The symptoms cleared gradually over the next few hours and disappeared overnight. A subsequent scratch test with chamomile tea produced a large wheal and flare reaction with pseudopod formation. Similar reactions were also elicited among other ragweed patients, although none gave a history of chamomile tea ingestion. This suggests that allergens are shared among members of the family of which ragweed and *Chrysanthemum* are a part.

Prevention is empirical, but atopic individuals should be aware of the increased risk and should avoid agents having known anaphylactic potential, such as penicillin or chamomile tea.

34.4 ALLERGIC RHINITIS (HAY FEVER) (TYPE II)

Characteristic symptoms of hay fever, induced following exposure of the nasal mucosa to the allergen through inhalation, include profuse watery nasal discharge with sneezing, frequently accompanied by redness, irritated and watery eyes and headache.

The inciting allergens are often found in windborne plant structures called aeroallergens. The spores from fungi and even certain algae may persist through the year, especially under warm humid conditions; but particularly in temperate regions, wind-pollinated plants elicit symptoms during certain flowering periods. In North America there are three peaks in the pattern of seasonal rhinitis: the first occurs in the spring when trees shed their pollen; the second, during the summer months, involves pollen from many grasses as well as late flowering trees and weeds; and the last peak, in the autumn, is typified by weed and secondarily by grass pollen grains. Ragweed pollen (*Ambrosia*) predominates during this time and is the most allergenic pollen found in North America.

In tropical areas both perennial and seasonal patterns can also be observed with this disorder. Fungal spores and grass pollen are common aeroallergens, whereas those from weeds and wind-pollinated trees are of secondary importance. However, the determination of aeroallergens in the more equatorial zones requires further study.

Although the majority of plants that induce allergic rhinitis are wind-pollinated, a number of plants that are typically pollinated by animals (insects, birds, bats) have also been implicated. For example, old-fashioned roses, which are infrequently found in gardens today, are often heavily scented and their anthers are exposed by the loose and open form of the floral bud. Thus their attractiveness frequently used to lead to sensitisation through inhalation of the pollen and the term rose-fever or rose-cold was used to describe plant-associated rhinitis.

34.5 POLLEN AND THEIR ALLERGENS

The morphological diversity of wind-borne pollen varies from smooth-walled grains having a single pore, as found among the grasses, to the very spiny grains of ragweed and the multiporate apertures typical of pigweed. Size and shape also vary. These features, as well as their wall structures, are related to dispersal mechanisms and their significance as aeroallergens.

The majority of allergens are found in the walls of pollen and spores, but their purpose is not to elicit allergy; rather, they act as recognition proteins to stimulate the growth of the sperm-containing pollen tubes on specific "female" parts of the flower. These proteins are concentrated below the apertures in the *intine* or inner wall as well as in the hard, rigid, outer wall or *exine* in minute micropores or even chambers. They are water soluble and when in contact with the "female" part of the flower, or the mucosa of man, the protein may be released within a few seconds.

Some of these proteins have been isolated only recently. For example, the fraction designated antigen E is the major allergenic protein of ragweed (*Ambrosia*). Others include antigens K, Ra, Ra3, and Ra5. Among the grasses, a number of allergens have also been identified, such as allergens A and B for timothy (*Pheleum pratense*) and alpha, beta and gamma allergens in rye (*Lolium perenne*). Also, a number of active fractions have been isolated from alder (*Alnus glutinosa*). However, the majority of specific allergens remain uncharacterised.

In India since 1980s, enormous studies were carried out to evaluate the potential of pollen induced allergic disorders. A good number of flowering plants now listed out, whose pollen grains are potentially allergenic (Table 34.1).

Table 34.1: Plants having allergenic pollen grains in India

Pollen allergen	Family	Vernacular name
Saccharum officinarum	Poaceae	Aankh
Azadirachta indica	Meliaceae	Neem
Phoenix sylvestris	Arecaceae	Khejur
Borassus flabellifer	Arecaceae	Taal
Cyperus rotundus	Cyperaceae	Mutha ghas
Areca catechu	Arecaceae	Supari
Catharanthus roseus	Apocynaceae	Nayantara
Cocos nucifera	Arecaceae	Narkel
Carica papaya	Caricaceae	Pepe
Parthenium hysterophorus	Asteraceae	Parthenium
Lantana camara	Verbenaceae	Putush
Datura metel	Solanaceae	Dhutura
Bombax ceiba	Bombacaceae	Shimul
Chenopodium album	Chenopodiaceae	Bethua saak
Langerstroemia speciosa	Lythraceae	Jarul
Delonix regia	Leguminosae	Golmohur
Moringa oleifera	Moringaceae	Sajina
Croton bonplandianum	Euphorbiaceae	Jaiphal
Mangifera indica	Anacardiaceae	Aam (Mango)
Amaranthus viridis	Amaranthaceae	Nota shaak
Trema orientalis	Urticaceae	Chikun
Tamarindus indicus	Leguminasea	Tentul
Cassia fistulosa	Leguminasea	Amaltas
Cassia sophera	Leguminasea	Kalkasunidi
Cynodon dactylon	Poaceae	Durbaghas
Ricinus communis	Euphorbiaceae	Bherenda (Castor oil)
Brassica compestris	Brassicaceae	Sorshe
Mimusops elengi	Sapotaceae	Bakul
Albizzia lebeck	Leguminosae	Sirish
Solanum xanthcarpum	Solanaceae	Kantikari
Xanthium strumarium	Astreraceae	Banokhra
Aegle marmelos	Rutaceae	Bel
Terminalia arjuna	Combretaceae	Arjun
Psidium guajava	Myrtaceae	Peyara
Putranjiva roxburghiana	Euphorbiaceae	Jio-putra
Alstonia scholaris	Apocynaceae	Chhatim

34.6 SPORE AND POLLEN INCITANTS IN MAN

Many kinds of fungi and flowering plants are responsible for allergic rhinitis. The most important fungal allergens are found in the Deuteromycetes, particularly the families Dematiaceae and Moniliaceae, which include such ubiquitous genera as *Alternaria, Cladosporium, Aspergillus* and *Penicillium*. Of these, *Alternaria* possesses the most allergenic substances and affects the greatest number of individuals, especially in the Midwestern United States. Even though they produce abundant amounts of windborne pollen, the Gymnosperms rarely elicit allergic rhinitis, whereas most windborne pollen grains from Angiosperms are common incitors. Overall, the most troublesome trees in North America are oaks (*Quercus*), hickories (*Carya*), and elms (*Ulmus*), but weedy urban box elder (*Acer*) pollen has recently been shown to have the highest level of allergenicity among tree pollen.

34.7 THERAPY

The control and treatment of allergic rhinitis falls into three major categories. The first is related to controlling contact with allergic material. Often referred to as environmental control, this function might include the elimination of the affecting plants, the filtering of aeroallergens by appropriate air-conditioning and, probably more important, moving away from the location of the offending allergens. It is surely relevant to consider the allergic person when plantings in urban developments are being designed.

Another prophylactic measure is hyposensitisation of the patient with the affecting allergen. This regimen, administered by an allergist, involves injection of increasing amounts of the allergen to develop a high titre of blocking antibodies prior to normal exposure. Today this is the most satisfactory form of prophylactic treatment, but its effect is directly related to the potency of the allergen in the antigens administered.

Finally, as an adjunct to hyposensitisation, acute symptoms are also treated by the administration of antihistamines and vasoconstrictors such as ephedrine (*Ephedra sinica*). Corticosteroid therapy, which may be employed to terminate a single severe attack, is often effective, though not curative. Prolonged use of steroids may be dangerous and possible benefits must always be weighed against deleterious effects.

34.8 ASTHMA

Attacks of bronchial asthma are usually precipitated by inhalation of the specific allergen and this form of allergy often has a more chronic course than that seen in allergic rhinitis even though the eliciting agents may be the same. Histamine and, perhaps, serotonin, are involved in symptoms that are characterised by bronchospasm and accompanied with excessive viscid bronchial secretion, which produces the asthmatic "wheezing". Extrinsic asthma occurs typically in children and young adults and is often aggravated by emotional factors. It is considered to be an IgE mediated disease caused by the inhalation of aeroallergens such as pollen, spores, feathers and animal danders. Although not a common aeroallergen, pollen from the lodgepole pine of Colorado (*Pinus contorta*) has also been known to cause bronchial asthma.

The mechanism for induction of intrinsic asthma is somewhat more obscure and is generally found in an older age group. The likely agents are allergic reactions to infectious materials, such as bacteria or viruses, or the inflammatory processes they elicit. Unlike extrinsic asthma antigens cannot be demonstrated and thus skin testing is of no value. The separation of purely extrinsic from intrinsic asthma can be diagnostically difficult whenever allergic phenomena are combined with infectious factors.

Possibly another IgE-mediated Type I disease is the coffeebean and castorbean workers disease that is characterised by rhinitis, asthma and dermatitis following inhalation of the hapten, chlorogenic acid.

As it is widespread in plants and is concentrated in coffeebeans and castorbeans, chlorogenic acid may act more as a universal allergen than was first suspected.

34.9 HYPERSENSITIVITY PNEUMONITIS (TYPE III)

Another type of allergic respiratory condition—known as hypersensitivity pneumonitis or extrinsic allergic alveolitis—is often associated with specific professions. In these instances, animal, vegetable or bacterial enzyme material may induce the disease. For example, inhalation of *Thermoactinomyces vulgaris* or fungal spores of *Microsporum faeni*, which can contaminate hay, moldy sugarcane, or mushroom compost, have been causally related to farmer's (thresher's) lung, bagassosis and mushroomworker's lung. In a similar way, *Cryptostroma corticale* has been associated with maple bark disease of woodworkers, *Penicillum caseii* to cheeseworker's disease, *Aspergillus clavatus* and *A. fumigatus* to brewer's lung disease and *Graphium* and *Aureobasidium pullulans* to sequoiosis. By inhalation of the enzyme of *Bacillus subtilis*, those who work with detergents may also develop an allergic pneumonitis. Diseases produced by inhalation of airborne algae such as *Gloecapsa* and *chlorella*, are of more general incidence, however. Wood and paper mill workers may also develop bronchial asthma through sawdust inhalation of the Gymnosperms, redwood (*Sequoia sempervirens*), western red cedar (*Thuja plicata*), cedar of Lebanon (*cedrus libani*) and the Angiosperms, iroko or African oak (*Chlorophora excelsa*), Nicaragua rosewood (*Dalbergia retusa*), and other exotic woods. The immunopathology suggests that a mixture of many types of immune or allergic reactions may be involved in extrinsic allergic alveolitis and thus is classified as Type III.

It is also possible that symptoms similar to those of allergic respiratory illness may be elicited by inhalation of airborne leaf hairs. Such a series of cases was recently reported among gardeners who had tended saplings of Oriental sycamores or the tree of Hippocrates (*Platanus orientalis*) at a medical school campus. It is interesting that, about 2000 years ago, Dioscorides (40-90 AD) had noted watery eyes, sneezing, an irritating sensation in the nasal passages, soreness of the throat, an irritating dry cough and other similar symptoms.

34.10 THERAPY

A regimen of environmental control and hyposensitisation is normally prescribed for pollinosis. Currently, three basic drugs have preempted ephedrine for use in the control of asthmatic attacks: epinephrine (adrenalin) and its congeners administered by aerosol, the methylxanthines administered intravenously for acute attacks and orally for chronic asthma, and the steroids (cortisone) for severe and intractable states, in combination with other drugs mentioned. In prophylaxis, particularly where hyposensitisation fails, cromolyn sodium is used; this new compound, obtained from the seeds of *Ammi visnaga* (Apiaceae), is believed to affect the release of vasoactive substances and, therefore, through a regimen of frequent inhalation, acts to prevent or modify the asthma. This plant, known as khella from its native Mediterranean region, has a long history of use as an antiasthmatic among the Arab peoples, who also believe it is useful in the treatment of angina pectoris.

Although other plants used in domestic medicine to treat asthma have rarely been studied, the recent Indian research using the leaves of *Tylophora indica* (Asclepiadaceae) furnishes an interesting exception. It is claimed that complete to moderate relief of nasobronchial allergic symptoms can be maintained up to one week after ingestion of but a few leaves of the plant. Typical of members of this family, *T. indica,* is, however, very toxic and also has blistering or vesicant properties.

34.11 INGESTANT ALLERGY

Symptoms from ingesting a potential allergen can vary from urticaria to vomiting, diarrhea and intestinal wall edema. It is sometimes difficult to differentiate between true atopy and toxicity, since clinical symptoms can be initiated by interaction of substances in foods with several different mediating systems. Therefore the appearance of hives (urticaria) after consumption of strawberries and citrus fruits can be traced to direct chemical mast cell degranulation; those of coeliac disease with intolerance to gluten (gliadin in wheat); and gastrointestinal upset associated with milk are traceable to inherited disaccharidase (lactase) deficiencies. Furthermore, several foodstuffs, particularly shellfish and mushrooms, contain notorious poisons. In susceptible persons, moreover, ricin (a phytotoxin from peanuts and castor beans), gossypol, aflatoxins, histamine or tyramine (in cheese or yeast products) may also produce symptoms. It is also possible that certain nonallergenic components in food can trigger built-in labilities of mediator systems or can activate the complement system (C1 and C3 : see Glossary) to generate anaphalatoxin-like agents.

Likewise, physiological age may also have a bearing on an individual's ability to absorb or reject certain allergens. Many food allergies of childhood are altered as the digestive system matures. Among cereals, for example, skin test reactions with rice indicate that there is a lower degree of reactivity among the few children affected; the opposite is true among adults and higher reactivity involving greater frequency is typical. Any ingestant may prove to have an allergic potential; skin testing, the usual method of determining such susceptibility, may not accurately reflect the true allergic state. Rather, susceptibility is better determined by demonstrating symptoms after deliberate feeding tests. Therapy related to the ingestant is used thereafter. Other techniques include using the rectal mucosa as a shock organ or feeding the test ingestant in dilute form and observing changes in the intestinal tract by X-ray.

Among the active allergens isolated from food, there is good evidence that tomato allergens fall into the same category of active glycoproteins described for inhalant allergens, as do the ovomucoids of egg white, whereas allergens of fish are simple amines.

The major symptom of ingestant hypersensitivity is urticaria, in which wheals and erythematous areas of the skin cause intense pruritus and discomfort. Local edema (angioedema) sometimes accompanies urticaria and this condition may be life-threatening if it affects the mucosa of the pharynx or larynx, since this may result in severe respiratory obstruction.

Atopic dermatitis, a common infant and childhood affliction, is often the first indicator of allergic predisposition. This infantile or atopic eczema may be clinically present as a persistent, pruritic dermatitis that may be papular, exudative or lichenified, involving the head, neck and flexor aspects of the trunk and extremities. Most provoking allergens are difficult to identify, varying from animal epidermal allergens (hair) to various foods in the diet.

THERAPY

Whenever it is possible to identify the allergen, avoidance is the best therapy. Otherwise, ephedrine is administered orally and topical steroids are applied if skin lesions are severe. Antihistamines are useful when pruritus complicates atopic dermatosis or if there is urticaria.

When angioedema is a complicating factor, epinephrine is most useful and can be used concurrently with intravenous antihistamine and steroids for severe cases that involve the larynx.

34.12 DERMATITIS

Irritant Dermatitis

Plants can mediate inflammatory reactions of the skin, which mimic in many ways the "wheal and erythematous flare" of immediate hypersensitivity, or the more severe reactions associated with the delayed response. In some instances this effect is attributable to the nature of the plant itself, that is, spines, thorns, bristles and hairs causing mechanical injury. Moreover, the needle-sharp calcium oxalate crystals found in the outer layers of many *Narcissus* species and hyacinth bulbs can elicit the formation of wheals (bulb fingers), a symptom suggestive of histamine release. However, most forms of irritation are related to specific substances produced by plants and the mechanism for the adverse reaction is usually unclear.

Of the many plants that transfer their toxins by means of stinging hairs, the nettles (*Urtica dioica* and *Laportea canadensis*) contain histamine-like substances found in bladders within the leaf. A more severe reaction is elicited by another species of Urticaceae, *Urera baccifera*, which is found in tropical America. The material in its spinelike hairs causes considerable pain followed by numbness lasting for several days. Similarly, the spurge nettle, *Cnidoscolus stimulosus* (Euphorbiaceae), can elicit painful irritation and itching following transfer of a caustic irritant. Perhaps the most painful reaction of all comes from *Mucuna pruriens* (Fabaceae), whose barbed spines covering the seed pods contain a highly irritating proteolytic enzyme called mucunain. Poisoning can take place long after the pods have dried, even from herbarium collections. However, the ultimate in adaptation is exemplified by *Gurania guaransenia* (Cucurbitaceae), which, in addition to its own stinging hairs, harbors a butterfly larva having similar devices.

The sap of other plants, itself caustic and corrosive, results in severe inflammatory reactions, including blistering of the skin on contact. Many of these plants are tropical, although a number are widely cultivated.

A characteristic of several of these irritant plant families is their milky or yellowish latex, usually found throughout the plant. Although some latex-possessing plants are harmless, a majority cause either irritant or contact dermatitis and as a general precaution they all should be avoided. Apart from dermatitis, a hazard to those who prune or tend horticultural varieties is the eye damage and blindness that can result from contact between the eye and the caustic sap.

Several of these irritant factors have been characterised. Members of the Brassicaceae contain sinigrin; this glucoside is harmless if dried, but it can be converted into an irritant mustard oil in the presence of water. A decomposition product of another glucoside, anemonin, has been isolated from the buttercup (*Ranunculus*) and produces blisters on the face and around the lips of children who may chew the leaves or stems of injured plants. Furthermore, the pineapple (*Ananas comosus,* Bromeliaceae) possesses a proteolytic enzyme, bromelain, which causes separation of the superficial layers of the skin and increases skin and capillary permeability in a manner not unlike that found in the allergic wheal and flare reaction.

34.13 CELLULAR-MEDIATED (DELAYED TYPE) HYPERSENSITIVITY (TYPE IV)

Delayed hypersensitivity may take days or weeks to develop; often prolonged contact with an antigen is necessary and the reaction depends on the formation of specifically modified lymphocytes from the thymus-dependent series (T cells). Through specific receptors or other mechanisms, these cells are capable of responding specifically to antigens deposited at a local site and also of mobilising non-sensitised phagocytic cells to localise there and participate with them in tissue destruction. Unlike immediate hypersensitivity, a reaction is not apparent for 12-24 hours, when inflammation and necrosis appear in the affected area.

To test for contact hypersensitivity, the candidate allergen is usually applied as a patch to the unbroken skin and observed for 24 hours for characteristic changes.

The tolerance by deliberate desensitisation through administration of repeated injection of antigen are often short-lived and frequently precipitate severe allergic reactions.

Substances of plant origin or chemicals including heavy metals can act as haptens to mediate a delayed hypersensitivity response after prolonged or repeated contact with the skin. Sensitisation is dependent on attachment of the chemical to the structural proteins of the skin, which results in a change of the skin proteins. Alone or through release of substances from epidermal cells, this type of antigenic alteration can stimulate a specific inflammatory response that produces, after 24-48 hours, symptoms of pruritis, burning or stinging, erythematous macules, papules, vesicles, exudation and crusting.

34.14 DRUG SENSITIVITY REACTIONS

Atopy that develops from the therapeutic administration of drugs may take many forms depending on the type of drug involved, the duration of contact and the method of administration. Moreover, from clinical symptoms, it may not always be possible to differentiate between drug toxicity or allergic phenomena.

Penicillin with its antigenic benzyl penicilloyl antigenic determinants and other haptenic moieties is perhaps the worst offender and some type of allergy appears in from 0.5% to 18% of patients using this drug therapeutically. The likelihood of these reactions occurring increases with age, the number of injections, the larger the doses given and the way it is administered (with a greater risk through injection rather than orally).

Type I reactions are the most common and usually appear shortly after administration as hives (2-48 hours) or generalised anahphylaxis (20 minutes). When large intravenous doses of benzylpenicillin have been given a hemolytic anemia suggestive of a Type II reaction has also been known to develop. The somewhat later appearing (3-5 weeks), maculopapular eruptions are, however, most likely Type III reactions, whereas the delayed cutaneous responses are of Type IV. The cephalosporins with their common 7-amino-cephalosporonic acid nucleus have cross-allergenicity with penicillin and have been known to cause thrombocytopenia, a Type II reaction. Similar Type II reactions have also been reported for quinidine, amidopyrine, chlorpromazine, chloramphenicol and the sulfonamides. Type IV reactions including maculopapular skin eruptions, fever and hepatic dysfunction have been associated with sulfonamide therapy as well as in patients receiving paraaminosalicylic acid (PAS). Because at least 5% of patients were so afflicted, ethambutol has now replaced PAS in tuberculosis therapy.

In patients with a history of antibiotic hypersensitivity it is not recommended to use the offending drug at all. Too often, allergic symptoms can only become aggravated or life-threatening. Nowadays, alternatives to penicillin are available that in most instances can be used as a satisfactory substitute without fear of eliciting an allergic response. For symptomatic treatment, antihistamines and corticosteroids are used and if systemic anaphylaxis occurs, epinephrine is administered. As over two-thirds of the cases of anaphylaxis develop obstructive edema of the respiratory tract, tracheostomy should be considered to prevent death.

35 ENVIRONMENTAL RISK ASSESSMENT

35.1 CONCEPT

Risk assessment and risk management give a framework for setting regulatory priorities and for making decisions that cut across different environmental areas. This kind of framework has become increasingly important in recent years for several reasons, one of which is the considerable progress made in pollution control in this country. In the 1980s, it wasn't hard to figure out where the first priorities should be. The worst pollution problems were all too obvious. Now that a number of cleanup areas are moving toward final control stages, the real priority problems and their solutions are not so obvious.

Environmental risk assessment may be broadly defined as a scientific enterprise in which facts and assumptions are used to estimate the potential for adverse effects on human health or the environment that may result from exposures to specific pollutants or other toxic agents. Risk management—as the term is used by the Environmental Protection Agency (EPA) and other regulatory agencies—refers to a decision-making process which involves such considerations as risk assessment, technological feasibility, economic information about cost and benefits, statutory requirements, public concerns and other factors. Risk communication is the exchange of information about risk.

35.2 RISK EVALUATION

How should this environmental risk be assessed and managed? In general, there are four steps in the evaluation of a risk:

- Hazard identification.
- Exposure assessment.
- Hazard assessment.
- Risk characterisation.

Hazard Identification

Hazard identification is an analysis to determine which of the discharges from a process or an accidental release will be a potential danger to public health. For instance, chemicals such as carbon monoxide, hydrogen chloride and unburned organics would be established as potential sources of risk for an incinerator.

The means of determining or identifying hazards is complex. Different methods are used to collect and evaluate toxic properties (those properties that indicate the potential to cause biological injury, disease or death under certain exposure conditions). One method is the use of epidemiological studies that deal with the incidence of disease among groups of people. Epidemiological studies attempt to correlate the incidence of cancer from an emission by an evaluation of people with a particular disease and people without the disease.

Long-term animal "bioassays" are the most common method of hazard determination. (A *bioassay* as referred to here is an evaluation of disease in a laboratory animal). Increased tumor incidence in laboratory animals is the primary health effect considered in animal bioassay. Exposure testing for a major portion of an animal's lifetime (2-3 years for rats and mice) provides information on disease susceptibility, primarily for carcinogenicity (the development of cancer).

Exposure Assessment

Exposure assessment determines how much of an emission actually reaches the receptor. Human exposure includes ingestion (eating or drinking), dermal contact (contact with the skin) and inhalation (breathing). These means of exposure are pathways to the human body.

There is no risk without exposure. The exposure assessment attempts to define the route, amount, frequency and duration of exposure of a given population to a toxic substance. The principal pathways of exposure generally considered in exposure assessment are atmospheric transport, surface and groundwater transport, ingestion of food materials etc.

Hazard Assessment

A hazard assessment is a determination of the relationship between the amount of exposure and the risk to public health. A person can drown in water, but that does not mean that a glass of water is a danger to health. Obviously, what defines a danger is not necessarily the presence of a particular chemical but rather the amount of that substance one is exposed to. This amount is known as a *dose*.

Hazard assessment is also referred to as *dose-response assessment*. Dose is usually expressed as milligrams of chemical received per kilogram of body weight per day. For toxic substances other than carcinogens, a threshold dose must be exceeded before a health effect will occur. For many substances, there is a dosage level below which there is no harm. However, at some given dosage, the threshold, a health effect will occur (or at least be detected).

It is generally assumed that for cancer-causing substances, there is no such threshold. Any substance that produces cancer is assumed to produce cancer at any concentration. It is, therefore, necessary to establish the risk to cancer and to determine whether that risk is acceptable. Analysis of a cancer risk is much more complicated than analysis of risk from noncarcinogenic substances. For noncarcinogenic substances, because there is a threshold level below which there is no danger, a system can be designed to generate an amount of that chemical below this danger level.

Risk Characterisation

Risk characterisation estimates the health risk associated with the process under investigation. The result of this characterisation is a number that represents the probability of adverse health effects from that process or from a substance released in that process. For instance, a risk characterisation for all effects from an incineration process might be expressed as one additional cancer case per 1 million people.

Once a risk characterisation is made, the meaning of that risk must be evaluated. Public health agencies generally only consider risks greater than 10 in 1 million to be significant risks warranting action.

There are two major types of risk:

- *Maximum Individual Risk (MIR)*: This is the maximum risk to an individual person. This individual is considered to have a 70-year lifetime of exposure to a process or a chemical. For an incinerator stack, for instance, the individual is considered to live downwind of the stack, never leaving this spot for every hour and every day of a 70-year life.

- *Population Risk (PR)*: This is a risk to a population, expressed as so many deaths per thousand or per million people.

The problem with risk estimates based on very conservative assumptions is that the substance or process may be calculated to present too high a risk. Unnecessary process control equipment may have to be installed on a facility to reduce the calculated risk.

In an effort to understand the significance of risk analyses, it is helpful to place the estimated risks in perspective with other everyday risks that have been determined by similar methodology. Table 35.1 lists a number of risks for comparison. These have been derived from actual statistics and reasonable estimates.

Table 35.1: Lifetime risks to life commonly faced by individuals

Cause of Risk	Lifetime (70-Year) Risk per Million
Cigarette smoking	252,000
All cancers	196,000
Construction	42,700
Agriculture	42,000
Police killed in line of duty	15,400
Air pollution (Eastern United States)	14,000
Motor vehicle accidents (traveling)	14,000
Home accidents	7,700
Frequent airplane traveller	3,500
Pedestrian hit by motor vehicle	2,940
Alcohol, light drinker	1,400
Background radiation at sea level	1,400
Peanut butter, four tablespoons per day	560
Electrocution	370
Tornado	42
Drinking water containing chloroform at maximum allowable EPA limit	42
Lightning	35
Living 70 years in zone of maximum impact from modern municipal incinerator	1
Smoking 1.4 cigarettes	1
Drinking 0.5 liters of wine	1
Traveling 10 miles by bicycle	1
Traveling 30 miles by car	1
Traveling 1,000 miles by jet plane (air crash)	1
Traveling 6,000 miles by jet plane (cosmic rays)	1
Drinking water containing trichloroethylene at maximum allowable EPA limit	0.1

Source: Charles T. Main. Inc. *Health Risk Assessment for Air Emissions of Metals and Organic Compounds from the PERC Municipal Waste to Energy Faculty,* Prepared for Penobscot Energy Recovery Company (PERC). Boston. MA. December 1985.

R. Wilson and E.A. Crouch, Risk Assessment and Comparisons: An Introduction, *Science*, April 1987.

Role of Economics in Risk Analysis

Environmental protection is designed to reduce risks to humans and the environment. The goal is to invest scarce resources to reduce risks to life and limb for people confronting threats from pollution or natural hazards and to reduce risk to ecosystem service and biological diversity affected by human actions. Paying for cleaner waterways can reduce the risk of cancer, paying for cleaner air can reduce the risk of asthma and other illness; precautionary investments in pollution control can reduce posed by transporting hazardous material and more insurance shifts resources to soften the shock of extreme weather events.

Traditional modelling approach was applied for evaluation of both exogenous and endogenous risks. The role of economics then is to understand how people react to these risks which means economists focus attention on the properties of risk preferences rather than on the technologies of risk control when explaining behaviour under risk.

35.3 PUBLIC PERCEPTION OF RISK

People often overestimate the frequency and seriousness of dramatic, sensational, dreaded, well-publicized causes of death and underestimate the risks from more familiar, accepted causes that claim lives one by one. Indeed, risk estimates by "experts" and lay people (or "the public") on many key environmental problems differ significantly. This problem and the reasons for it are extremely important, because in our society the public generally does not trust experts to make important risk decisions alone.

This situation was illustrated by an EPA study. In February 1987, the EPA released a report entitled *Unfinished Business: A Comparative Assessment of Environmental Problems.* EPA Administrator Lee M. Thomas had commissioned this study nearly a year earlier to "compare the risks currently associated with major environmental problems, given existing levels of control." Thomas' explicit premise was that "in a world of limited resources, it may be wise to give priority attention to those pollutants and problems that pose the greatest risks to our society."

To assess and compare these problems, the EPA created a special task force of about 75 career managers and experts from all EPA programmes. The task force compared four different types of risks existing now for each of 31 environmental problem areas: cancer risk, noncancer health risks, ecological effects and welfare effect (e.g., materials damage). While the task force did not try to "weight" or "add" the different types of risks for problem areas, they did develop rough rankings of problems within risk types (Tables 35.2 and 35.3). Beyond these rankings, the task force made no assertions about what EPA's priorities ought to be, noting that policy makers must consider many other factors besides risk when they set priorities, such as legislation, economics, technology and public mandate.

Table 35.2: How EPA experts rank environmental risks—highlights

Overall High/ Medium Risk	High Health; Low Ecological and Welfare Risk	Low Health; High Ecological and Welfare Risk	Overall Medium/Low Risk (Ground Water-Related Problems)	Mixed and/or Medium/Low Risk
"Criteria" air pollution from mobile and stationary sources (includes acid precipitation)	Hazardous/toxic air pollutants	Global warming	Hazardous waste sites, active (Resource Conservation and Recovery Act)	Contaminated sludge
	Indoor radon	Point and nonpoint sources of surface water pollution	Hazardous waste sites, inactive (Superfund)	Accidental releases of toxic chemicals

Contd.

Table 35.2: How EPA experts rank environmental risks—highlights

Overall High/ Medium Risk	High Health; Low Ecological and Welfare Risk	Low Health; High Ecological and Welfare Risk	Overall Medium/Low Risk (Ground Water-Related Problems)	Mixed and/or Medium/Low Risk
Stratospheric ozone depletion	Indoor air pollution other than radon	Physical alteration of aquatic habitat (including estuaries and wetlands) and mining waste	Other municipal and industrial waste sites	Accidental oil spills
Pesticide residues in or on foods	Drinking water as it arrives at the tap		Underground storage tanks	Biotechnology (environmental releases of genetically altered materials)
Runoff and air deposition of pesticides	Exposure to consumer products, Worker exposures to chemicals			

Source: U.S. EPA, Unfinished Business: A Comparative Assessment of Environmental Problems, 1987.

Table 35.3: How the public ranks selected environmental risks

High Risk	Medium Risk	Low Risk
Chemical waste disposal Water pollution	Oil tanker spills Exposure to pollutants on the job	Indoor air pollution Exposure to chemicals in consumer products
Chemical plant accidents Outdoor air pollution	Eating pesticide-treated foods Other pesticide risks Contaminated drinking water	Genetic engineering (biotechnology) Waste from strip mining Non-nuclear radiation "Greenhouse effect" (CO_2 and global warming)

Source: Original data drawn from 1984-1986 polls conducted by the Roper Organisation, Inc.

35.4 RISK COMMUNICATION

Frustrations abound when it comes to the way public opinion regards environmental risks and drives environmental protection. Some risks are large, sometimes frighteningly large, whereas others are small, sometimes vanishingly small. This is so whether they are placed on the measuring rod of total population life expectancy or on that of the probability of premature death for small numbers of exposed people. The same holds true of nonfatal disease and ecological harm. At least, this is what available scientific evidence suggests.

On the surface, it appears practical to remedy the most severe risks first, leaving the others until later—or may be, if small enough, never remedying the others at all. But the behaviour of individuals in everyday life often does not conform with this view. For example, there are toxic waste dumps where all evidence indicates that risks are minimal. Yet, the presence of such dumps can lead to numbing anxiety on the part of some, to loss of property values and to disruption of communities. Elsewhere, facilities to dispose safely of similar wastes may be resisted by all means possible, including threats of civil disobedience. And, at the same time, individuals may show little concern for hazardous products in ordinary commerce, resist efforts to protect wetlands vital to ecological integrity, not choose to test their homes for naturally occurring radon and ignore safe-use labels for pesticides in home use. Examples of hysteria in the face of apparently trivial risks—along with examples of apathy in the face of apparently serious ones—form an unsettling litany to risk managers.

There are no easy prescriptions for successful risk communication. However, those who have studied and participated in recent debates about risk generally agree on the following seven rules. These rules apply equally well to the public and private sectors.

Although many of the rules may seem obvious, they are continually and consistently violated in practice. Thus, a useful way to read these rules is to focus on why they are frequently not followed.

1. Accept and involve the public as a legitimate partner. A basic tenet of risk communication in a democracy is that people and communities have a right to participate in decisions that affect their lives, their property and the things they value.

Guidelines: Demonstrate your respect for the public and underscore the sincerity of your effort by involving the community early, before important decisions are made. Involve all parties that have an interest or a stake in the issue under consideration. If you are a government employee, remember that you work for the public. If you do not work for the government, the public still holds you accountable.

Point to Consider

- The goal of risk communication in a democracy should be to produce an informed public that is involved, interested, reasonable, thoughtful, solution-oriented and collaborative; it should not be to diffuse public concerns or replace action.

2. Plan carefully and evaluate your efforts. Risk communication will be successful only if carefully planned.

Guidelines: Begin with clear, explicit risk communication objectives, such as providing information to the public, motivating individuals to act, stimulating response to emergencies, or contributing to the resolution of conflict. Evaluate the information you have about the risks and know its strengths and weakness. Classify and segment the various groups in your audience. Aim your communications at specific subgroups in your audience. Recruit spokes people who are good at presentation and interaction. Train your staff, including technical staff, in communication skills; reward outstanding performance. Whenever possible, pretest your messages. Carefully evaluate your efforts and learn from your mistakes.

Points to consider

- There is no such entity as "the public"; instead, there are many publics, each with his or her own interests, needs, concerns, priorities, preferences and organisation.
- Different risk communication goals. audiences and media require different risk communication strategies.

3. Listen to the public's specific concerns. If you do not listen to people, you cannot expect them to listen to you. Communication is a two-way activity.

Guidelines: Do not make assumptions about what people know, think, or want done about risks. Take the time to find out what people are thinking: Use techniques such as interviews, focus groups and surveys. Let all parties that have an interest or a stake in the issue be heard. Identify with your audience and try to put yourself in their place. Recognise people's emotions. Let people know that you understand what they said, addressing their concerns as well as yours. Recognise the "hidden agendas," symbolic meanings and broader economic or political considerations that often underlie and complicate the task of risk communication.

Point to Consider

- People in the community are often more concerned about such issues as trust, credibility, competence, control, voluntariness, fairness, caring and compassion than about mortality statistics and the details of quantitative risk assessment.

4. Be honest, frank and open. In communicating risk information, trust and credibility are your most precious assets.

Guidelines: State your credentials, but do not ask or expect to be trusted by the public. If you do not know an answer or are uncertain, say so, get back to people with answers. Admit mistakes, disclose risk information as soon as possible (emphasising any reservations about reliability). Do not minimize or exaggerate the level of risk. Speculate only with great caution. If in doubt, lean toward sharing more information, not less, or people may think you are hiding something. Discuss data uncertainties, strengths and weaknesses, including the ones identified by other credible source. Identify worst-case estimates as such, and cite ranges of risk estimates when appropriate.

Point to Consider

- Trust and credibility are difficult to obtain. Once lost, they are almost impossible to regain completely.

5. Coordinate and collaborate with other credible sources. Allies can be effective in helping you communicate risk information.

Guidelines: Take time to coordinate all intra-organisational communications. Devote effort and resources to the slow, hard work of building bridges with other organisations. Use credible and authoritative intermediaries. Consult with others to determine who is best able to answer questions about risk. Try to issue communications jointly with other trustworthy sources (for example, credible university scientists, physicians or trusted local officials).

Point to Consider

- Few things make risk communication more difficult than do conflicts or public disagreements with other credible sources.

6. Meet the needs of the media. The media are a prime transmitter of information on risks; they play a critical role in setting agendas and in determining outcomes.

Guidelines: Be open with and accessible to reporters. Respect their deadlines. Provide risk information tailored to the needs of each type of media (for example, graphics and other visual aids for television). Prepare in advance and provide background material on complex risk issues. Do not hesitate to follow up on stories with praise or criticism, as warranted. Try to establish long-term relationships of trust with specific editors and reporters.

Point to Consider

- The media are frequently more interested in politics than in risk, more interested in simplicity than in complexity and more interested in danger than in safety.

7. Speak clearly and with compassion. Technical language and jargon are useful as professional shorthand, but they are barriers to successful communication with the public.

Guidelines: Use simple, nontechnical language. Be sensitive to local norms, such as speech and dress. Use vivid, concrete images that communicate on a personal level. Use examples and anecdotes that make technical risk data come alive. Avoid distant, abstract, unfeeling language about deaths, injuries and illnesses. Acknowledge and respond (both in words and with actions) to emotions that people express: anxiety, fear, anger, outrage and helplessness. Acknowledge and respond to the distinctions that the public views as important in evaluating risks: voluntariness, controllability, familiarity, dread, origin (natural or man-made), benefits, fairness and catastrophic potential. Use risk comparisons to help put risks in perspective, but avoid comparisons that ignore distinctions that people consider important. Always try to include a discussion of actions that are underway or that can be taken. Tell people what you cannot do. Promise only what you can do and be sure to do what you promise.

Points to Consider

- Regardless of how well you communicate risk information, some people will not be satisfied.
- Never let your efforts to inform people about risks prevent you from acknowledging and saying that any illness, injury, or death is a tragedy.
- If people are sufficiently motivated, they are quite capable of understanding complex risk information, even if they may not agree with you.

A schematic presentation of risk communication is shown in Fig. 35.1.

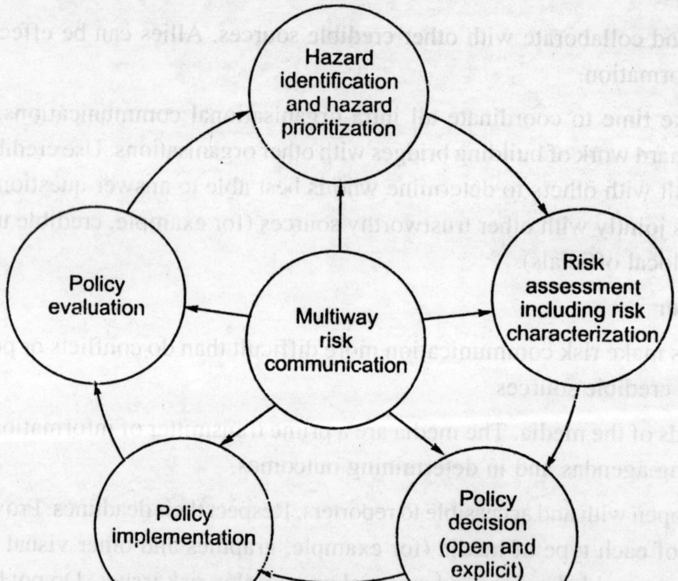

Fig. 35.1: The risk management cycle

SECTION VI

ENVIRONMENTAL QUALITY AND ENVIRONMENTAL IMPACT ASSESSMENT

36 STANDARDS FOR ENVIRONMENTAL QUALITY ASSESSMENT AND MONITORING

36.1 INTRODUCTION

The necessity for the development of realistic standards for environmental quality assessment and monitoring was felt for a long time. Thus various standards were adopted in different countries to establish the relationships between the intensity, duration and frequency of exposure to pollution and the risk or magnitude of an undesirable effect occurring in the environment.

Each kind of standard has a limit for three distinct aspects, viz.,:

(a) Limit set for a pollutant taken up by an organism or a population;

(b) Limit set for pollutant present in a specific environment medium or product (food or consumer products); and

(c) Limit set for discharges or emissions from polluting sources.

36.2 ENVIRONMENTAL PROTECTION STANDARDS IN INDIA

In India, the initiative for preparing standards for environmental protection was taken by the Bureau of Indian Standards (formerly Indian Standards Institution) long before legislation in this regard came into being. The work on water pollution started in the early seventies and on solid waste management in the late seventies.

On the basis of experience the world over, it is felt that two types of standards will best achieve the desired results. One of these falls in the category of environmental quality standards while the other can be described as discharge standards. The environmental quality standards lay down the goals of pollution control programmes. These are directly related to the maximum tolerance of human, plant, animal and aquatic life to a particular pollutant.

At any given point of time, these limits are determined by the available data and to safeguard the environmental quality, no relaxation can be allowed, whatever the technological or economic handicaps. In developed countries of Europe in particular, much work has been done in the field of environmental quality standards and data in this regard is readily available whereas, in India, little systematic efforts has been made so far in this area.

The discharge standards lay down limits for the discharge of gaseous, liquid or solid wastes into the environment. The limit may vary from industry to industry and from area to area. The discharge standards are linked with the nature of industry and pollutant, type of receiving media, area in which the industry is located, techniques and economics of treatment.

The first set of standards were published as Indian Standard 2490 in 1963. These standards set the tolerance limits for industrial effluent discharged into Inland surface waters. After that two more standards —viz., IS 3306 for discharge into public sewers, and IS 3307 for discharge on Land for irrigation purposes were made. Then in 1974 the IS 2490 was revised as industry specific standards, where Part I gave general limits and Part II onwards gave limits for specific industries. In 1976, IS 7968 was made for specifying the limits for discharge into marine coastal areas. Subsequently a number of standards published in the field of water, air, noise pollution and solid waste management.

Each standard has the following features:

(a) Methods for test, terminology;

(b) Guideline for waste treatment;

(c) Effluent/emission limits; and

(d) Receiving courses/ambient air quality limits.

The details of different environmental protection standards as promulgated in India is given in the Appendix I.

There are a number of regulations and legislation promulgated from time to time to protect the environmental changes by following the prescribed regulation of different standards. The standards for sewage and effluent discharges into streams, other water bodies and land are also given. The national drinking water quality, surface water quality, ambient air quality and noise quality are shown in Appendices II to VI.

36.3 INTERNATIONAL STANDARDS

In USA, Environmental Protection Agency (EPA) has prescribed tolerance limits for liquid effluents.

In UK, there is no national standard for liquid effluents. The effluents are controlled on a case-by-case system. Each discharge from a factory is considered on its merits such as capacity of receiving waters to absorb the water and the technology of abatement and finally a consent defining, the allowed volume and the concentration of the individual components is issued by the appropriate water authority.

In Japan, Environment Agency has prescribed separate standards for surface water for the protection of human health, the preservation of the living environment (rivers, lakes and waters), and liquid effluents discharged into inland surface water.

During 1990s in Finland, there are no consistent standards for the disposal of liquid effluents. According to the water law for Finland, the water courts set waste water standards for each plant separately and grant waste water permits for a fixed time. The criterion for water court permits are usually as maximum allowable concentration in waste water (mg/l) and/or maximum allowable amount per unit of time or of product (e.g. tonnes per day or per ton of production).

GLOBAL ENVIRONMENTAL MONITORING SYSTEM (GEMS)

GEMS is a worldwide, collective effort to monitor the global environment and make periodic assessments of the health of its constituents. Data are collected through monitoring and assessment activities covering most of the important environmental parameters. All told, 142 countries participate in at least one of these activities. The system involves the collaboration of hundreds of national and international organisations, of which the most important are the Food and Agriculture Organisation of the United Nations (FAO), the World Health Organisation (WHO), the World Meteorological Organisation (WMO), the United Nations Educational, Scientific and Cultural Organisation (UNESCO), and the International Union for the Conservation of Nature and Natural Resources (IUCN).

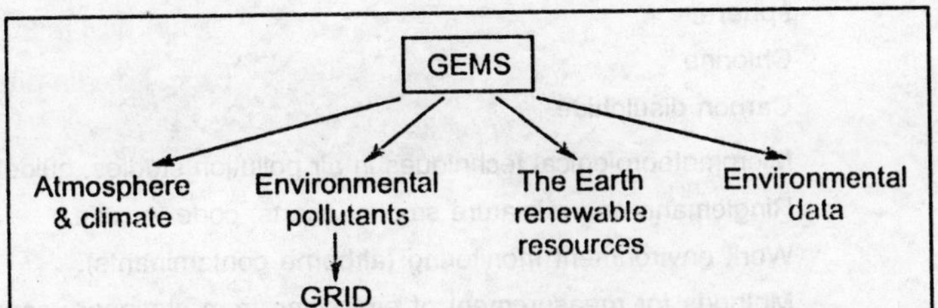

GEMS networks monitor changes in atmospheric composition and the climate system, freshwater and coastal pollution, air pollution, food contamination, deforestation, Ozone layer depletion, the build-up of greenhouse gases, acid rain, the extent of global ice cover and many issues related to biological diversity.

GEMS was first proposed in 1971 by the Scientific Committee on Problems of the Environment (SCOPE), later it was formally created in 1975, as a small secretariat of UNEP.

During the 1990s, GEMS took over a newer area of investigation, i.e., early warning system for environmental threats and disasters.

APPENDIX I

INDIAN STANDARDS ON ENVIRONMENTAL PROTECTION—AIR, WATER, NOISE AND SOLID WASTES MANAGEMENT

1. AIR

1.1 *Units and Terminology:*

IS : 4167—1980	Glossary of terms relating to air pollution
IS : 9260—1980	Guide for units used in air quality measurement

1.2 *Methods of Measurement of Pollution:*

IS : 5182	Methods of measurement of pollution
(Part 1)—1969	Dustfall
(Part 2)—1969	Sulphur dioxide
(Part 3)—1970	Radioactivity (particulate) in air
(Part 4)—1973	Suspended matter
(Part 5)—1975	Sampling of gaseous pollutants
(Part 6)—1975	Nitrogen oxide
(Part 7)—1973	Hydrogen sulphide
(Part 8)—1976	Sulphation rate
(Part 9)—1974	Oxidants
(Part 10)—1976	Carbon monoxide

Contd.

(Part 11)—1982	Benzene
(Part 12)—1974	Polynuclear aromatic hydrocarbons in air particulate matter.
(Part 13)—	Polynuclear aromatic hydrocarbons in water.
(Part 14)—1985	Guidelines for planning the sampling of atmosphere and location of monitoring station (first revision).
(Part 15)—1974	Mass concentration of particulate matter in the atmosphere.
(Part 16)—1980	Recommended practice for collection, filtration and determination of mass, number and optical sizing of atmospheric particulates.
(Part 17)—1979	C_1 to C_5 hydrocarbons in air by gas chromatography.
(Part 18)—1974	Continuous analysis and automatic recording of the oxidant content of the atmosphere.
(Part 19)—1982	Chlorine
(Part 20)—1982	Carbon disulphide
IS : 8829—1978	Micrometeorological techniques in air pollution studies, guidelines for monitoring.
IS : 9078—1979	Ringlemann and miniature smoke, charts, code for use.
IS : 9679—1980	Work environment monitoring (airborne contaminants).
IS : 11255	Methods for measurement of emissions from stationary sources.
(Part 1)—1985	Particulate matter
(Part 2)—1985	Sulphur dioxide
(Part 3)—1985	Flow rate
(Part 4)—1985	Hydrogen sulphide and carbon disulphide

1.3 EMISSION LIMITS

1.3.1 *Stationary Sources:*

IS : 8635—1977	Limits for gaseous emissions from sulphuric acid and phosphatic fertilizer industries.
IS : 8636—1983	Limits for gaseous emissions from petroleum refineries (first revision).
IS : 9005—1978	Limits for gaseous emissions from nitric acid and nitrogenous fertilizer industry.
IS : 9233	Limits for gaseous emissions from man-made fibre plants.
(Part 1)—1985	Cellulosic fibres (first revision)
IS : 10193	Limits for ambient air quality.
(Part 1)—1982	Sulphur dioxide and particulate matter.
IS : 10693	Limits for emissions from cement plants.
(Part 1)—1983	Kilns
IS : 11247—1985	Limits for emissions from iron and steel plants.
IS : 11450—1986	Method for determination of airborne asbestos fibre concentrations in work environment by light microscopy (membrane filter method).
IS : 11451—1986	Recommendation for safety and health requirements relating to occupational exposure to asbestos.

1.3.2 *Mobile Sources:*

IS : 8118—1976	Smoke emission levels for diesel vehicles.
IS : 9057—1979	Emission limits for carbon monoxide for vehicles powered by spark ignition engines.
IS : 10000	Methods of test for IC engines.
(Part 10)—1980	Tests for smoke levels, limits for smoke levels and corrections for variable speed engines.

1.4 *Code of Practice for Control:*

IS : 10179—1982	Code of practice for control of air pollution in petroleum refineries.
IS : 11304—1985	Code of practice for control of air pollution in iron and steel plant.

Contd.

2. WATER

2.1 *Terminology:*

IS : 7022 (Part 1)—1973	Glossary of terms relating to water, sewage and industrial effluents.
IS : 7022 (Part 2)	Glossary of terms relating to water, sewage and Industrial effluents.
IS : 10446—1983	Glossary of terms for water supply and sanitation.

2.2 *Methods of Sampling and Test of Water used in Industry:*

IS : 1622—1981	Microbiological examination of water.
IS : 3025—1964	Methods of sampling and test (physical and chemical) for water used in industry.
IS : 3025	Methods of sampling and test (physical and chemical) water and waste water.
(Part 4)—1983	Colour
(Part 5)—1983	Odour
(Part 6)—1983	Odour threshold
(Part 7)—1984	Taste threshold
(Part 8)—1984	Taste rating
(Part 9)—1984	Temperature
(Part 10)—1984	Turbidity
(Part 11)—1983	pH value
(Part 12)—1983	Density
(Part 13)—1983	Saturation index (with respect to calcium carbonate)
(Part 14)—1984	Specific conductance (Wheatstone bridges, conductance cell)
(Part 15)—1984	Total residue (total solids—disolved and suspended)
(Part 16)—1984	Filterable residue (total dissolved solids)
(Part 17)—1984	Non-filterable residue (total dissolved solids)
(Part 18)—1984	Volatile and fixed residue (total filterable and non-filterable)
(Part 19)—1984	Settleable matter
(Part 20)—1984	Dispersion characteristics (flow patterns)
(Part 21)—1983	Total hardness

2.3 **Tolerance Limits:**

2.3.1 *Tolerance Limits for industrial effluents:*

IS : 2490	Tolerance limits for industrial effluents discharged into inland surface waters.
(Part 1)—1981	General limits.
(Part 2)—1974	Distillary industry.
(Part 3)—1985	Tanning industry.
(Part 4)—1974	Strawboard industry.
(Part 5)—1974	Electroplating industry.
(Part 6)—1985	Dyestuff and dye intermediate manufacturing industry.
(Part 7)—1976	Coke ovens.
(Part 8)—1976	Phosphatic fertilizer industry.
(Part 9)—1977	Nitrogenous fertilizer industry.
(Part 10)—1985	Rubber industry.
IS : 4764—1973	Tolerance limits for sewage effluents discharged into inland surface waters.

2.3.2. *Composition of receiving waters:*

IS : 2296—1982	Tolerance limits for inland surface water subject to pollution.
IS : 7967—1976	Criteria for controlling pollution of marine coastal areas.

Contd.

2.4 *Treatment and disposal of Industrial Effluents:*

IS : 4903—1979	Guide for treatment of:
	Effluents of cane sugar industry
IS : 5061	Effluents of pulp. paper and board industries.
(Part 1) — 1978	For mills with chemical recovery systems.
	Guide for treatment and disposal of :
IS : 5183—1977	Effluents of tanning industry
IS : 7453—1974	Effluents of electroplating industry
IS : 8032—1976	Distillery effluents
IS : 8073—1976	Effluents of steel plant
IS : 8682—1977	Effluents of dairy industry
IS : 9508—1980	Effluents of cotton and synthetic textile industry
IS : 9509—1980	Effluents of viscose rayon industry
IS : 9841—1981	Effluents of fertilizer industry
IS : 10044—1981	Effluents of petroleum refining industry
IS : 10495—1983	Effluents of wool industry.

3. **Solid Wastes Management**

3.1 *Solid Wastes:*

IS : 9234—1979	Preparation of solid waste sample for chemical and microbiological analysis
IS : 9235—1979	Physical analysis and determination of moisture in solid waste (excluding industrial solid wastes)
IS : 9533—1980	Guidelines for selection of methods of urban solid waste disposal
IS : 9569—1980	Glossary of terms relating to solid wastes
IS : 9622—1980	Guidelines for management of solid wastes
IS : 10127—1982	Guidelines for utilisation and disposal of solid wastes from coal mining and washery industries
IS : 10152—1982	Guidelines for utilisation and disposal of solid wastes from paper industry
IS : 10158—1982	Methods of analysis of solid wastes (excluding industrial solid wastes)
IS : 10447—1983	Guidelines for utilisation and disposal of solid wastes from integrated steel plants

4. **Noise**

4.1 *Methods of Measurement:*

IS : 3028—1980	Noise emitted by moving road vehicles
IS : 4758—1968	Noise emitted by machine
IS : 6098—1971	Airborne noise emitted by rotating electrical machinery
IS : 6229—1980	Real-ear protection of hearing protectors and physical attenuation of earmuffs (first revision)
IS : 8146—1976	Reverberation time in auditoria, measurement
IS : 9167—1979	Ear protectors
IS : 9671—1980	Frequency weighing for the measurement of aircraft noise (D-weighing)
IS : 9779—1981	Sound level meters
IS : 9876—1981	Guide to the measurement of airborne acoustical noise and evaluation of its effects on man
IS : 9989—1981	Assessment of noise with respect to community responses
IS : 10399—1982	Measurement of noise emitted by stationary road vehicle
IS : 10423—1982	Personal sound exposure meter

Contd.

ANNEXURE II

RELEVANT INDIAN STANDARDS

Surface Water Quality Standards (as per IS : 2296)

Sl. No.	Parameters and Units	Cls-A	Cls-B	Cls-C	Cls-D	Cls-E
1.	Taste	Tless	—	—	—	—
2.	Odour	Unobj	—	—	—	—
3.	Colour (True) (Hazen Units)	10	300	300	—	—
4.	pH (max) (min : 6.5)	8.5	8.5	8.5	8.5	8.5
5.	Conductivity (25 °C) (umhos/cm)	—	—	—	1000	2250
6.	Dissolved oxygen (mg/L) (min)	6	5	4	4	—
7.	BOD (5 days at 20°C) (mg/L)	2	3	3	—	—
8.	Total Coliforms (MPN/100 mL)	50	500	5000	—	—
9.	TDS (mg/L)	500	—	1500	—	2100
10.	Oils and Grease (mg/L)	—	—	0.1	0.1	—
11.	Mineral Oil (mg/L)	0.01	—	—	—	—
12.	Total Hardness (mg/L as $CaCO_3$)	300	—	—	—	—
13.	Chlorides (mg/L as Cl)	250	—	600	—	600
14.	Sulphates (mg/L as SO_4)	400	—	400	—	1000
15.	Nitrates (mg/L as NO_3)	20	—	50	—	—
16.	Free Carbon Dioxide (mg/L CO_2)	—	—	—	6	—
17.	Free Ammonia (mg/L as N)	—	—	—	1.2	—
18.	Fluorides (mg/L as F)	1.5	1.5	1.5	—	--
19.	Calcium (mg/L as Ca)	80.10	—	—	—	—
20.	Magnesium (mg/L as Mg)	24.28	—	—	—	—
21.	Copper (mg/L as Cu)	1.5	—	1.5	—	—
22.	Iron (mg/L as Fe)	0.3	—	50	—	—
23.	Manganese (mg/L as Mn)	0.5	—	—	—	—
24.	Zinc (mg/L as Zn)	15	—	15	—	—
25.	Boron (mg/L as B)	—	—	—	—	2
26.	Barium (mg/L as Ba)	1	—	—	—	—
27.	Silver (mg/L as Ag)	0.05	—	—	—	—
28.	Arsenic (Total) (mg/L as As)	0.05	0.2	0.2	—	—
29.	Mercury (mg/L as Hg)	0.001	—	—	—	—
30.	Lead (mg/L as Pb)	0.1	—	0.1	—	—
31.	Cadmium (mg/L as Cd)	0.01	—	0.01	—	—
32.	Chromium (VI) (mg/L as Cr)	0.05	0.05	0.05	—	—
33.	Selenium (mg/L as Se)	0.01	—	0.05	—	—
34.	Cyanide (mg/L as CN)	0.05	0.05	0.05	—	—
35.	Phenolic Cmpds (mg/L C_6H_5OH)	0.002	0.005	0.005	—	—
36.	Anionic Detergents (mg/L MBAS)	0.2	1	1	—	—
37.	PAH (mg/L)	0.2	—	—	—	—
38.	Pesticides (mg/L)	0	—	—	—	—
39.	Insecticides (mg/L)	—	—	0	—	—
40.	Alpha Emitters (10^-6uc/mL)	0.001	0.001	0.001	0.001	0.001
41.	Beta Emitters (10^-6uc/mL)	0.01	0.01	0.01	0.01	0.01
42.	Percent Sodium (%)	—	—	—	—	60
43.	Sodium Absorption Ratio	—	—	—	—	26

Tless : Tasteless | Unobj : Unobjectionable

ANNEXURE III

Drinking Water Quality Standard (IS : 10500), 2012

Sl. No.	Parameters and Units	Standard
1.	Colour (True) (Hazen units)	5*
2.	Taste	Agrbl*
3.	Odour	Unobj*
4.	Turbidity (NTU)	5*
5.	pH	6.5-8.5*
6.	Total Coliforms (MFN/100 mL)	0
7.	TDS (mg/L)	500
8.	Mineral Oil (mg/L)	0.01
9.	Total Hardness (mg/L as $CaCO_3$)	300*
10.	Alkalinity (mg/L as $CaCO_3$)	200
11.	Chlorides (mg/L as Cl)	250*
12.	Sulphates (mg/L as SO_4)	200
13.	Nitrates (mg/L as NO_3)	45
14.	Free Residual Chlorine (mg/L)	0.2*
15.	Fluorides (mg/L as F)	1
16.	Calcium (mg/L as Ca)	75
17.	Copper (mg/L as Cu)	0.05
18.	Iron (mg/L as Fe)	0.3*
19.	Manganese (mg/L as Mn)	0.1
20.	Zinc (mg/L as Zn)	5
21.	Boron (mg/L as B)	1
22.	Aluminium (mg/L as AL)	0.03
23.	Arsenic (mg/L as As)	0.01
24.	Mercury (mg/L as Hg)	0.001
25.	Lead (mg/L as Pb)	0.05
26.	Cadmium (mg/L as Cd)	0.01
27.	Chromium (VI) (mg/L as Cr)	0.05
28.	Selenium (mg/L as Se)	0.01
29.	Cyanide (mg/L as CN)	0.05
30.	Phenolic Cmpds (mg/L C_6H_5OH)	0.001
31.	Anionic Detergents (mg/L MBAS)	0.2
32.	PAH (mg/L)	0
33.	Pesticides (ug/L)	0
34.	Alpha Emmitters (10^-6uc/mL)	0.0001
35.	Beta Emmitters (10^-6uc/mL)	0.001
36.	Pathogenic Organisms or Virus	0

Agrbl : Agreeable Unobj : Unobjectionable

* Essential characteristics

ANNEXURE IV

General Standards for Discharge of Effluents

Sl. No.	Parameter and Unit	Into Surface Water	Into Public Sewers	Into Irrigtn. to Land	Into Coastal Water
1.	Temperature (Degree C)	Note 1	—	—	Note 1
2.	Odour	Note 2	—	Note 2	Note 2
3.	Colour (True) (hazen units)	Note 2	—	Note 2	Note 2
4.	pH	5.5–9.0	5.5–9.0	5.5–9.0	5.5–9.0
5.	BOD (5 days at 20°C) (mg/L)	30	350	100	100
6.	COD (mg/L)	250			250
7.	TSS (mg/L)	100	600	200	* 100
8.	TDS (mg/L) (inorganic)	2100	—	2100	—
9.	Oils and Grease (mg/L)	10	20	10	20
10.	Chlorides (mg/L as Cl)	1000	1000	600	—
11.	Sulphates (mg/L as SO_4)	1000	1000	1000	—
12.	Nitrates (mg/L as NO_3)	10	—	—	20
13.	Total Residual Chlorine (mg/L)	1	—	—	1
14.	Free Ammonia (mg/L as N)	5	—	—	5
15.	Ammoniacal Nitrogen (mg/L N)	50	50	—	50
16.	TKN (mg/L as N)	100	—	—	100
17.	Fluorides (gm/L as F)	2	15	—	15
18.	Sulphide (mg/L as S)	2	—	—	5
19.	Dissolved Phosphates (mg/L P)	5	—	—	—
20.	Copper (mg/L as Cu)	3	3	—	3
21.	Iron (mg/L as Fe)	3	3	—	3
22.	Manganese (mg/L as Mn)	2	2	—	2
23.	Zinc (mg/L as Zn)	5	15	—	15
24.	Nickel (mg/L as Ni)	3	3	—	5
25.	Boron (mg/L as B)	2	2	2	—
26.	Arsenic (mg/L as As)	0.2	0.2	0.2	0.2
27.	Mercury (mg/L as Hg)	0.01	0.01	0.01	0.01
28.	Lead (mg/L as Fb)	0.1	1	—	1
29.	Cadmium (mg/L as Cd)	2	1	—	2
30.	Chromium (VI) (mg/L as Cr)	0.1	2	—	1
31.	Chromium (Total) (mg/L as Cr)	2	2	—	2
32.	Selenium (mg/L as Se)	0.05	0.05	—	0.05
33.	Vanadium (mg/L as V)	0.2	0.2	—	0.2
34.	Cyanide (mg/L as CN)	0.2	2	0.2	0.2
35.	Phenolic Cmpds (mg/L C_6H_5OH)	1	5	—	5
36.	Pesticides (ug/L) [Note4]	10	10	—	10
37.	Alpha Emmitters (uc/mL)	10^—7	10^—7	10^—8	10^—7
38.	Beta Emmitters (uc/mL)	10^—6	10^—6	10^—7	10^—6
39.	Percent Sodium (%)	—	60	60	—
40.	Resdl Sodium Carbonate (mg/L)	—	—	5	—
41.	Bio-assay (% 96-hrs Survival)	Note 3	Note 3	Note 3	Note 3
42.	SS Particle size (pass IS Sieve)	850	—	—	Note 5

Note 1 : Shall not exceed 5.0 above the receiving water temperature.

Note 2 : All efforts should be made to remove colour and unpleasant odour as far as practicable.

Note 3 : 90% survival of fish after 96 hours in 100% effluents.

Note 4 : Applicable to DDT. Endosulfan, Carbaryl, Malathion, Phenthoate, Methyl parathion, Phenitrothion, Phorate, Pyrethrum and Benzene hexachloride.

Note 5 : (a) Floatable solids : 3 mm. (b) Settleable solids : 850 micron.

* For cooling water efflunts 10% above TSS of influent.

ANNEXURE V

National Ambient Air Quality Standards

NOTIFICATION

Delhi, the 11th April 1994

S.O. 384 (E). — The Central Pollution Control Board in exercise of its powers conferred under Section 16(2) (h) of the Air (Prevention and Control of Pollution) Act, 1981 (14 of 1981) hereby notify the National Ambient Air Quality Standards with immediate effect.

SCHEDULE 1 (Prerevised)

Pollutant	Time Weighted average	Concentration in ambient air			Method of measurement
		Industrial Area	Residential Rural & other areas	Sensitive Area	
1	2	3	4	5	6
Sulphur Dioxide (SO_2)	Annual Average*	80 μg/m³	60 μg/m³	15 μg/m³	1. Improved West and Gaeke method
	24 hours**	120 μg/m³	80 μg/m³	30 μg/m³	2. Ultraviolet fluorescence
Oxides of Nitrogen as NO_2	Annual Average*	80 μg/m³	60 μg/m³	15 μg/m³	1. Jacob & Hochheiser modified (Na-Arsenite) Method
	24 hours**	120 μg/m³	80 μg/m³	30 μg/m³	2.Gas Phase Chemiluminescence
Suspended Particulate Matter (SPM)	Annual Average*	360 μg/m³	140 μg/m³	70 μg/m³	(Average flow rate not less than 1.1 m³/minute)
	24 hours**	500 μg/m³	200 μg/m³	100 μg/m³	
Respirable Particulate matter (size less than 10 μm) (RPM)	Annual Average*	120 μg/m³	60 μg/m³	50 μg/m³	
	24 hours**	150 μg/m³	100 μg/m³	75 μg/m³	
Lead (Pb)	Annual Average*	1.0 μg/m³	0.75 μg/m³	0.50 μg/m³	AAS Method after sampling using EPM 2000 or equivalent filter paper
	24 Hours**	1.5 μg/m³	1.00 μg/m³	0.75 μg/m³	
Carbon Monoxide (CO)	8 hours**	5.0 mg/m³	2.0 mg/m³	1.0 mg/m³	Non-dispersive infrared spectroscopy
	1 hour	10.0 mg/m³	4.0 mg/m³	2.0 mg/m³	

* Annual Arithmetic mean of minimum 104 measurements in a year taken twice a week 24 hourly at uniform interval.

** 24 hourly/8 hourly values should be met 98% of the time in a year. However, 2% of the time, it may exceed but not on two consecutive days.

Schedule VII
(See rule 3 (3B))

NATIONAL AMBIENT AIR QUALITY STANDARDS (Revised) NOTIFICATION
Delhi, the 16th November 2009

Sl. No.	Pollutant	Time Weighted Average	Concentration in Ambient Air		
			Industrial, Residential, Rural and Other Area	Ecologically Sensitive Area (notified by Central Government)	Methods of Measurement
1	2	3	4	5	6
1.	Sulphur Dioxide (SO_2), μg/m^3	Annual* 24 hours**	50 80	20 80	Improved West and Gaeke Ultraviolet fluorescence
2.	Nitrogen Dioxide (NO_2), μg/m^3	Annual* 24 hours**	40 80	30 80	Modified Jacob & Hoch-heiser (Na-Arsenite) Chemiluminescence
3.	Particulate Matter (size less than 10μm) or PM_{10} μg/m^3	Annual* 24 hours**	60 100	60 100	Gravimetric TOEM Beta attenuation
4.	Particulate Matter (size less than 2.5 μm) or $PM_{2.5}$ μg/m^3	Annual* 24 hours**	40 60	40 60	Gravimetric TOEM Beta attenuation
5.	Ozone (O_3) μg/m^3	8 hours** 1 hour**	100 180	100 180	UV photometric Chemiluminescence Chemical Method
6.	Lead (Pb) μg/m^3	Annual* 24 hours**	0.50 1.0	0.50 1.0	AAS/ICP method after sampling on EPM 2000 or equivalent filter paper ED-XRF using Teflon filter
7.	Carbon Monoxide (CO) mg/m^3	8 hours** 1 hour**	02 04	02 04	Non-Dispersive Infra Red (NDIR) spectroscopy
8.	Ammonia (NH_3) μg/m^3	Annual* 24 hours**	100 400	100 400	Chemiluminescence Indophenol blue method
9.	Benzene (C_6H_6) μg/m^3	Annual*	05	05	Gas chromatography based continuous analyzer Adsorption and Desorption followed by GC analysis
10.	Benzo(a) Pyrene (BaP) - particulate phase only, ng/m^3	Annual*	01	01	Solvent extraction followed by HPLC/GC analysis
11.	Arsenic (As), ng/m^3	Annual*	06	06	AAS/ICP method after sampling on EPM 2000 or equivalent filter paper
12.	Nickel (Ni), ng/m^3	Annual*	20	20	AAS/ICP method after sampling on EPM 2000 or equivalent filter paper

* Annual arithmetic mean of minimum 104 measurements in a year at a particular site taken twice a week 24 ho[illegible] at uniform intervals.

** 24 hourly or 08 hourly or 01 hourly monitored values, as applicable, shall be complied with 98% of th[illegible] year. 2% of the time, they may exceed the limits but not on two consecutive days of monitoring.

Note: Whenever and wherever monitoring results on two consecutive days of monitoring exce[illegible] fied above for the respective category, it shall be considered adequate reason to insti[illegible] ous monitoring and further investigation.

ANNEXURE VI

[1]SCHEDULE III
(See Rule 3)
AMBIENT STANDARDS IN RESPECT OF NOTICE

Area code	Category of Area	Limits in dB(A) Leq Day Time	Night Time
(A)	Industrial area	75	70
(B)	Commercial area	65	55
(C)	Residential area	55	45
(D)	Silence Zone	50	40

Note:

1. Day time is reckoned in between 6 a.m. and 9 p.m.
2. Night time is reckoned in between 9 p.m. and 6 a.m.
3. Silence zone is defined as areas up to 100 meters around such premises as hospitals, educational institutions and courts. The silence zones are to be declared by the Competent Authority.

   Use of vehicular horns, loudspeakers and bursting of crackers shall be banned in these zones.
4. Mixed categories of areas should be declared as one of the four above-mentioned categories by the Competent Authority and the corresponding standards shall apply.

[1] Schedule III inserted vide G.S.R. 1063(E) dt. 26-12-89. published in the Gazette No. 643 dt. 26-12-89.

36.4 ENVIRONMENTAL QUALITY MONITORING : ISO-14000

The International Standard Organisation (ISO) and International Electro-technical Commission (IEC) jointly set up a Strategic Advisory Group on the Environment (SAGE) in 1991. This Committee, at their meeting TC 207 in July 1993, agreed to set up a number of subcommittees (SCs) to deal with:

- SC1 — Environment management system;
- SC2 — Environmental auditing;
- SC3 — Environmental labelling;
- SC4 — Environmental performance evaluation;
- SC5 — Life cycle analysis; and
- SC6 — Term and definitions.

Each of the subcommittee further set up working groups (WGs) for appropriate functioning. The details are given in the Table 36.1. The ISO-14000 have five basic principles:

- Should focus on what needs to be done;
- Should formulate a plan to fulfil its environmental policy;
- Should develop the capabilities and support mechanisms necessary to achieve its environmental policy;
- Should measure, monitor and evaluate its environmental performance;
- Should review and continually improve its environmental management systems.

Table 36.1: Structure of ISO-14000 systems

ISO-14000 (TC-207)

SC1	SC2	SC3	SC4	SC5	SC6	Special
Environmental Management system	Environmental auditing	Environmental leveling	Environmental performance evaluation	Life Cycle assessment	Terms & definitions.	
WG1 Specification	WG1 Principles	WG1 Principles	WG1 Principles	WG1 Principle & Procedure		WG1 Environmental aspect of product standard
WG2 General guidelines	WG2 Procedures	WG2 Self-decleration claims	WG2 Operation procedure	WG2 Life cycle inventory analysis		
	WG3 Qualifications	WG3 Leveling Programme		WG3 Inventory analysis		
	WG4 Investigations			WG4 Impact assessment		
				WG5 Improvement assessment		

Key Elements of ISO 14001; Comparison With BS : 7750 (1994)

Element of ISO 14001	Brief Details	Correspondence with BS : 7750
4.1 Policy	* Documented * Communicated * Publicly available * Commitment to continual improvement	4.2 Requires PER
4.2.1 Evaluation of environmental aspects	* Identify significant aspects	4.4.2
4.2.2 Legal/Other requirements	* Identify and track applicable laws regulations and policies	4.4.3
4.2.3 Objectives and targets.	* Set at all relevant levels * Compliance with regulations * Consistent with policy	4.5
4.2.4 Environmental Management Programme	* Achieve objectives and targets * Designated responsibilities * Apply to new products/developments	4.6
4.3.1 Structure and responsibility	* Define and document roles responsibilities * Appoint management representative(s) * Provide resources	4.3.1, 4.3.2 & 4.3.3

Contd.

Key Elements of ISO 14001; Comparson With BS : 7750 (1994)

Element of ISO 14001	Brief Details	Correspondence with BS : 7750
4.3.2 Training awareness and competence	* Identify training needs * Provide training * Cover key specified elements	4.3.4
4.3.3 Communication	* Procedures for : receipt and documentation of and response to inquiries internal communication external communication	4.3.4 & 4.4.1
4.3.4, 4.3.5 and 4.4.3 Documentation and records	* Describe EMS * Procedures to maintain documents and records	4.7 and 4.9
4.3.6 Control Procedures	* Identify functions associated with significant impacts * Plan functions (including procedures, operating limit) * Address procurement/contracting.	4.3.2
4.3.7 Emergency Response	* Evaluate hazards * Prepare procedures	Require evaluation of accidents etc.
4.4.1 Monitoring & measurement	* Monitor key procedures * Record results	4.8.3
4.4.2 Corrective Action	* Define responsibility and authority to initiate corrective action. * Change procedures as needed	4.8.4
4.4.3 EMS Audits	* Verify conformance with standard * Consider system implementation and effectiveness	4.10.3
4.5 Management Review	* Periodic review of policy, objectives, targets, programmes	4.11

36.5 ISO 14000—IMPACT ON DEVELOPING COUNTRIES

During the Earth Summit in Rio de Janeiro in 1992, the Business Council for Sustainable Development (BCSD) emphasized that "business and industry needs tools to help measure environmental performance and to develop powerful environmental management techniques". ISO was specifically requested to step up its activities in the environmental field in response to such needs. The organisation set out to evaluate just what the market expected from an international environmental standardisation system, and to pinpoint the high priority issues.

Following the recommendations of its Strategic Advisory Group on Environment (SAGE)—which included representatives of 20 countries, 11 international organisations and more than 100 environmental experts—ISO created in 1993 its Technical Committee 207, Environmental Management, to develop standards in six areas for:

- Environmental management systems (EMS)
- Environmental auditing and related environmental investigations

- Environmental labelling
- Environmental performance evaluation
- Life cycle assessment
- Terms and definitions.

Progress in the development of environmental management and environmental auditing standards has been rapid, thanks to the enthusiasm of the committee members representing 70 countries and the support of the various stake-holders-industry, governments, inter-governmental and non-governmental organisations. The first two standards of the series-ISO 14001 Environmental Management Systems—specification with guidance for use, and ISO 14004 Environmental Management Systems—General Guidelines on principles, systems and supporting techniques were published in September 1996, three years after the establishment of TC 207 and in a record time for achieving international consensus.

Good progress was also seen in other areas of the ISO/TC 207 programme, notably environmental labelling, environmental performance evaluation, life cycle assessment and terms and definitions. The draft of a Guide for the inclusion of environmental aspects in product standards was approved for circulation to ISO member bodies.

The ISO 14001 Approach

While most governments have enacted legislation aimed at protecting the environment, it is generally recognised that "policing the compliance of industry to environmental legislation is no easy task in view of the large number of companies and the limited resources at the disposal of the enforcement agencies". Voluntary implementation by industrial firms of an environmental management system, according to ISO 14001, can go a long way towards ensuring compliance of firms with the legislation and continual improvement of their environmental performance.

Indeed, the ISO 14001 standard stipulates that the implementing organisation shall have an environmental policy defined by top management and communicated to all employees. This policy, which is available to the public, should be geared towards compliance with environmental legislation and continual improvement of environmental performance. The standard requires the organisation to establish a procedure for identifying aspects of its activities with significant impacts on the environment, as well as taking stock of legislation applicable to the environmental aspects of its activities, products or services as a basis for actions to improve the environmental performance. The standard further requires the organisation to establish environmental objectives consistent with its policy and carry out programmes to achieve these objectives.

As the case in other management systems standards (such as ISO 9000), ISO 14001 sets out procedures and measures that ensure that the environmental management system of the company is appropriate, effective and reliable. These measures include the clear definition of responsibility and authority, an enhanced level of awareness and competence of all employees achieved through training, effective internal and external communication, controlled documentation and procedures for initiating corrective action. The standard also requires the organisation procedures to identify potential for accidents and emergency situation and to institute emergency preparedness to prevent and mitigate their environmental impacts if they do occur.

Similarly, to quality management system standards, ISO 14001 institutes procedures for checking the effectiveness of the environmental management system, namely, record keeping, regular internal audits and a periodic management review of the continued effectiveness of the system.

It is expected that the widespread implementation of ISO 14001 by industrial firms will considerably boost their environmental performance and enhance compliance with environmental regulation, thanks to its comprehensive system approach that mobilises the employees of each implementing firm.

Impact on Developing Countries

Since the publication and widespread implementation of quality management system standards (ISO 9000) a debate has been raging in developing countries, if the latter cannot implement or cannot demonstrate that they are effectively implementing the standards. Sceptics in developing countries go to the extent of saying that ISO 9000 and ISO 14000 (as well as the whole of the GATT Agreements) are simply ploys invented by industrialised countries to stop the export of products from developing countries which compete with their own products. These claims are countered by voices from industrialised countries which say that, if developing countries do not follow the progress of technology in the areas of quality and preservation of the environment, they cannot expect their products to be welcome in other markets.

An Assessment

The real problem may be on another level—whether the awareness, the will and the necessary conditions exist for a widespread implementation of these standards in developing countries. To gain greater insight and to seek better ways to assist developing countries deal with the problem, the United Nations Industrial Development Organisation (UNIDO) conducted, in collaboration with ISO in 1995, a survey of the situation concerning ISO 9000 and ISO 14000 series of standards in developing countries and emerging economics in Latin America and the Caribbean, Africa, Asia and Eastern Europe.

The results of the survey show that, although a certain level of awareness of ISO 9000 exists in most regions and some efforts are being made to implement those standards with varying degrees of success, awareness of ISO 14000 standards clearly lags behind. For example, fewer respondents to the survey had seen a draft of ISO 14000 series while most respondents were familiar with ISO 9000. This situation was particularly difficult in Africa and catastrophic in some countries of Eastern Europe where the respondents had never read the standard. There appeared to have great difficulties in understanding ISO 14000 in most regions with the exception of Asia. Understanding was particularly poor in Africa. Even though almost all governments had supported the development of environmental awareness, very few had publicised ISO 14000. The lack of awareness of ISO 14000 standards and their potential benefits was considered as the most common factor deterring companies from implementing the standards. For SMEs the cost setting up an environmental management system as well as the lack of consultancy were considered as other weighty counter-arguments against ISO 14000 implementation.

36.6 OCCUPATIONAL HEALTH AND SAFETY MANAGEMENT SYSTEM (OHSAS 18001)

OHSAS 18001 has been established by the leading safety management systems organization from around the world. You'll need to check with your registrar to see if it's qualified to assist you with third-party registration to OHSAS 18001. Otherwise, your organization could make a self-determination and declaration of conformance with the OHSAS specification. Either way, the new standard represents a continuation of the ISO 9001 and ISO 14001 standards and can be integrated into existing management systems to help ensure the safety of employees and the organization.

When considering OHSAS 18001, we're dealing with a safety management standard. These standards don't usurp any local regulations concerning occupational safety and health issues but have been developed to ensure a system of managing the organization to reduce or eliminate risks to employees.

The basic framework of OHSAS 18001 is shown below:

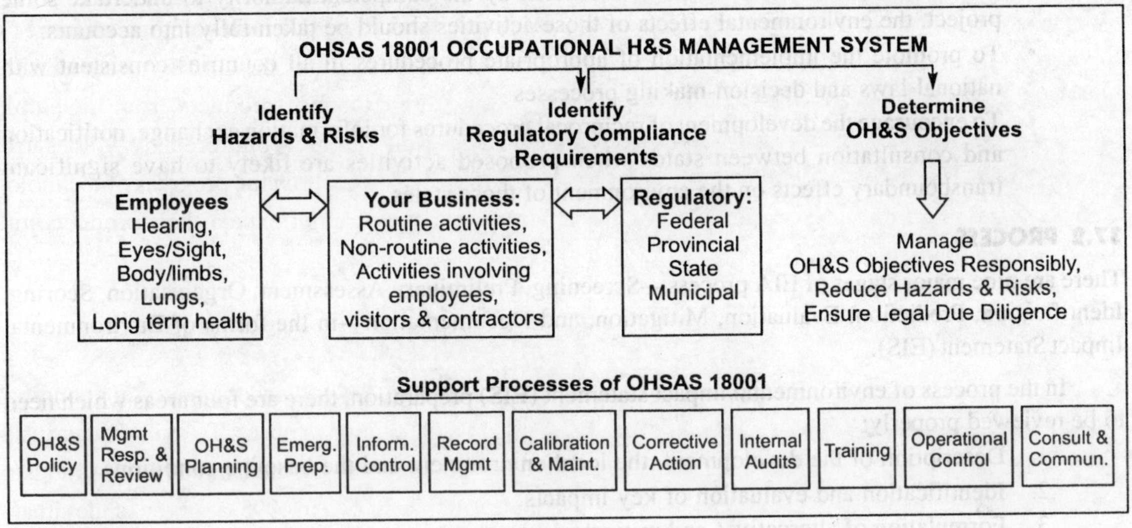

Conclusion

In view of these survey results, both ISO and UNIDO are now preparing plans to support the efforts of developing countries raise awareness and gain experience in implementing ISO 14000 standards. The ISO Programme for Developing Countries has implemented action to sponsor participation of developing country delegates at the meeting of ISO/TC 207 Environmental Management. Thanks to financial support from the Dutch and Finnish governments, some 50 delegates from countries of Africa, Asia, Latin America and Caribbean attended the meetings of TC 207 in Oslo, 1995 and Rio de Janeiro, 1992. A series of seminars of Environmental Management and ISO 14000 were planned to be held in different regions of the developing world.

37 ENVIRONMENTAL IMPACT ASSESSMENT

37.1 CONCEPT

Environmental Impact Assessment (EIA) means a formalised procedure for examination, analysis, and assessment of planned activities with a view to ensuring environmentally sound and sustainable development. The principal goals of EIA practices are:

- To establish that before decisions are taken by the competent authority to undertake some project, the environmental effects of those activities should be taken fully into accounts.
- To promote the implementation of appropriate procedures in all countries consistent with national laws and decision-making processes.
- To encourage the development of reciprocal procedures for information exchange, notification and consultation between states when proposed activities are likely to have significant transboundary effects on the environment of those states.

37.2 PROCESS

There are nine major stages of EIA process—Screening, Preliminary Assessment, Organisation, Scoring, Identification, Prediction, Evaluation, Mitigation, and Documentation—in the forms of Environmental Impact Statement (EIS).

In the process of environmental impact statement (EIS) preparation, there are four areas which need to be reviewed properly:

1. Description of the development, the local environment and the baseline conditions.
2. Identification and evaluation of key impacts.
3. Formulation of alternatives and mitigatory measures.
4. Impact interpretation and communication of results.

The purpose(s) of the development should be described as should the physical characteristics, scale and design. Quantities of materials needed during construction and operation should be included and where appropriate, a description of the production processes. In addition, the site land requirements of the development should be described and the duration of each land use. As and where practicable, specific mitigation measures should be put forward. Mitigation methods considered should include modification of the project, compensation and provision of alternative facilities as well as pollution control. Clear details of how the mitigation measures will be implemented and function over the time span for which they are necessary should be described in depth.

Each EIS should have an executive summary. The summary should be comprehensive, containing, at least, a brief description of the project and the environment, an account of the main mitigation measures to be undertaken by the developer and a description of any remaining or residual impacts. A brief explanation of the methods by which these data were obtained and an indication of the confidence which can be placed in them, should also be included.

In every project formulation, EIA preparation is one of the mandetory aspects as it forms an integral part of the project cycle (Fig. 37.1). On the whole the purpose of EIA is to give the environment its due

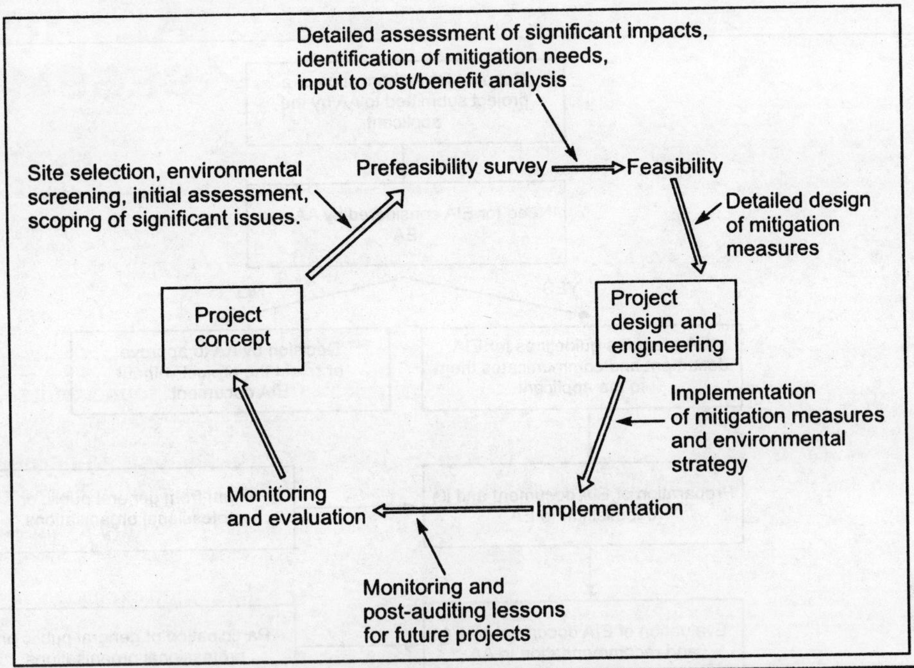

Fig. 37.1: A generalised scheme for project evaluation monitoring

weightege the decision making process by clearly evaluate the environmental consequances of a proposed activity. A simplified checklist for the procedure of EIA is given in Fig. 37.2.

In EIA process these couple of steps viz

1, Screening.
2. Scooping.
3. Baseline environmental data generation (physical, chemical, biological and social environment).
4. Impact prediction.
5. Impact evaluation.
6. Impact mitigation.
7. Environment management plan.

All these activities collective called EIA & EMP preparation. There are need for public hearing in proposed area and finally sent for environmental clearance of the project by environment appraisal committee set up by MOEFCC, New Delhi for 'A' category projects and environment appraisal committee set up by SDCB/SED for B category projects.

37.3 EVALUATION METHODOLOGY

Different methods currently used for EIA evaluation are summerised below with respect to their criteria suitability in Table 37.1.

Among these, four methods are often used for their various criteria suitability. A detailed account of criteria suitability is given in Table 37.2.

From the Table 37.2, it appears that for most of the project's impact evaluation matrix system is most suitable. Quite considerable number of matrix system application were known in recent years. Among these, Leopold and Lohani and Thanh's procedure is often applicable in most cases. A case study example of a model dam project impact evaluation is given in the Table 37.3.

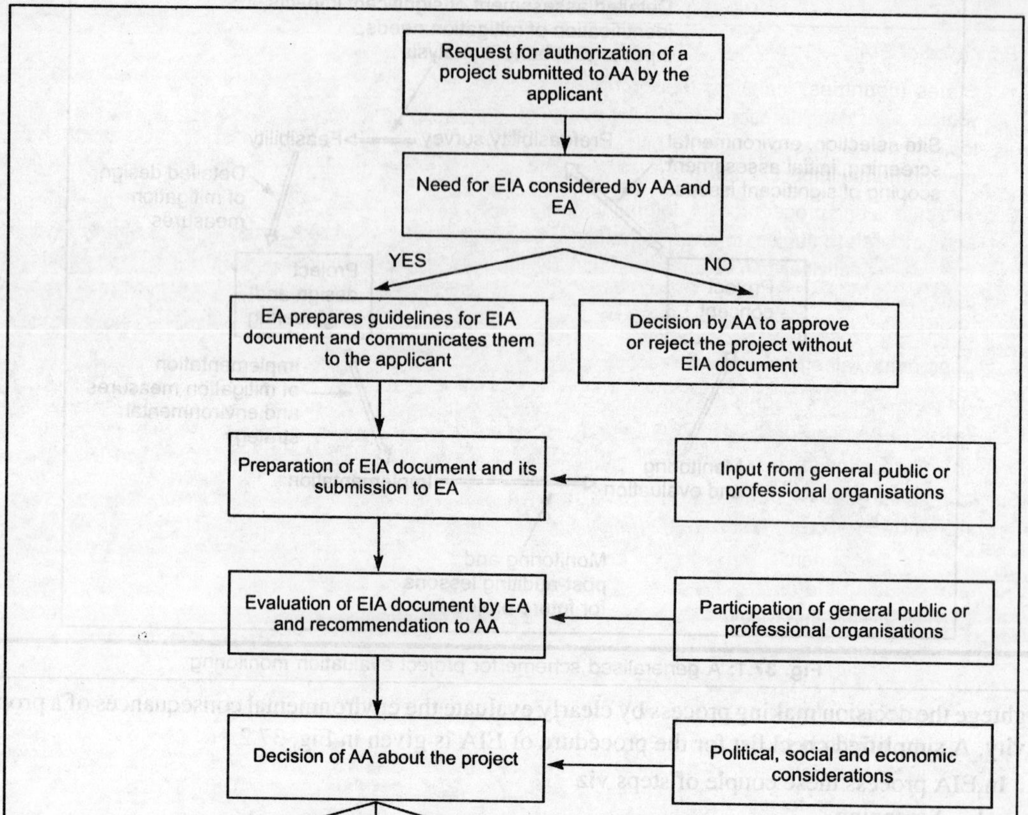

Fig. 37.2: Simplified flow chart of the EA procedure
(*Source: UNEP*. An approach the environmental impact assessment for projects affecting the coast and marine environment, Nairobi, Kenya, 1990, p. 10)

Box 37.1: UN Environment Programme (UNEP), 1975

Principles of EIA

1. States (countries, including their competent authorities) should not undertake or authorise activities without prior consideration, at an early stage, of their environmental effects. Where the extent, nature or location of a proposed activity is such that it is likely to significantly affect the environment, a comprehensive environmental impact assessment (EIA) should be undertaken in accordance with the following principles.
2. The criteria and procedures for determining whether an activity is likely to significantly affect the environment and is therefore subject to an EIA should be defined clearly by legislation, regulation, or other means, so that subject activities can be quickly and surely identified and EIA can be applied to the activity as it is being planned.
3. In the EIA process, the relevant significant environmental issues should be identified and studied. Where appropriate, all efforts should be made to identify these issues at an early stage in the process.
4. An EIA should include, at a minimum:
   (a) A description of the proposed activity;
   (b) A description of the potentially affected environment, including specific information necessary for identifying and assessing the environmental effects of the proposed activity;
   (c) A description of practical alternatives, as appropriate;
   (d) An assessment of the likely or potential environmental impacts of the proposed activity and alternatives, including the direct, indirect, cumulative, short-term and long-term effects;
   (e) An identification and description of measures available to mitigate adverse environmental impacts of the proposed activity and alternatives and an assessment of those measures;
   (f) An indication of the gaps in knowledge and uncertainties which may be encountered in compiling the required information;
   (g) An indication of whether the environment of any other State or areas beyond national jurisdiction is likely to be affected by the proposed activity or alternatives;
   (h) A brief, non-technical summary of the information provided under the above headings.
5. The environmental effects in an EIA should be assessed with a degree of detail commensurate with their likely environmental significance.
6. The information provided as part of an EIA should be examined impartially prior to the decision.
7. Before a decision is made on an activity, government agencies, members of the public, experts in relevant disciplines and interested groups should be allowed appropriate opportunity to comment on the EIA.
8. A decision as to whether a proposed activity should be authorised or undertaken should not be taken until an appropriate period has elapsed to consider comments pursuant to principles seven and twelve.
9. The decision on any proposed activity subject to an EIA should be in writing, state the reasons therefore and include the provisions, if any, to prevent, reduce or mitigate damage to the environment. This decision should be made available to interested persons and groups.
10. Where it is justified,, following a decision on an activity which has been subject to an EIA,, the activity and its effects on the environment or the provisions (pursuant to principle nine) of the decision on this activity should be subject to appropriate supervision.
11. States shouid endeavour to conclude bilateral, regional or multilateral arrangements,, as appropriate, so as to provide, on the basis of reciprocity, notification, exchange of information and agreed-upon consultation on the potential environmental effects of activities under their control or jurisdiction which are likely to significantly affect other States or areas beyond national jurisdiction.
12. When information provided as part of an EIA indicates that the environment within another State is likely to be significantly affected by a proposed activity, the State in which the activity is being planned should, to the extent possible:
    (a) Notify the potentially affected State of the proposed activity;
    (b) Transmit to the potentially affected State any relevant information from the EIA, the transmission of which is not prohibited by national laws or regulations.

Table 37.1: Currently used Impact evaluation methods

	Types	Detailed methodologies
I	Ad hoc	These methodologies provide minimal guidance for impact assessment beyond suggesting broad areas of possible impacts (e.g., impacts upon flora and fauna, lakes and forests), rather than defining the specific parameters within the impact area which should be investigated.
II	Simple check-lists	These methodologies present a specific lists of environmental parameters to be investigated for possible impacts, or a list of agency activities known to have caused environmental concern. They may have considerable value when many repetitive actions are carried out under similar circumstances. They do not, in themselves, establish a direct cause-effect link, but merely suggest lines of examination.
III	Overlays	These methodologies rely upon a set of maps of project area's environmental characteristics (physical, social, ecological, aesthetic). These maps are overlaid to produce a composite characterisation of the regional environment. Impacts are identified by noting the congruence of inherently antagonistic environmental characteristics within the project boundaries. The Geographic Information System (GIS), is modern development of this method.
IV	Matrices	The matrix methodologies incorporate both a list of project activities and a checklist of potentially impacted environmental characteristics. In a way, the matrix presents both alternatives from the check-list approach (i.e., both attributes and activities) to be considered simultaneously. The two lists are then related in a matrix which identifies cause and effect relationships between specific activities and impacts. Matrix methodologies may either specify which actions impact which environmental characteristics or may simply list the range of possible actions and characteristics in an open matrix to be completed by the analyst.
V	Networks	These methodologies work from a list of project activities to establish cause condition-effect relationships. They are an attempt to recognise that a series of impacts may be triggered by a project action. Their approaches generally define a set of possible networks and allow the user to identify impacts by selecting and tracing out the appropriate project actions.
VI	Combination computer-aided	These methodologies use a combination of matrices, networks, analytical models and a computer-aided systematic approach: — Identify activities associated with implementing major federal programmes; — Identify potential environmental impacts at different user levels; — Provide guidance for abatement and mitigation techniques; — Provide analytical models to establish cause-effect relationships to quantitatively determine potential environmental impacts; — Provide a methodology and a procedure to utilise this comprehensive information in responding to requirements of EIS preparation.

Table 37.2: EIA Methodology Evaluation

	Criteria	Simple check-list	Overlay	Network	Matrix
1.	Comprehensiveness	0.5	0.0	1.0	0.5
2.	Communicability	1.0	1.0	0.5	1.0
3.	Flexibility	1.0	0.0	1.0	1.0
4.	Objectivity	0.0	0.5	0.5	1.0
5.	Aggregation	0.0	0.5	0.0	0.0
6.	Multi-function	0.0	0.5	0.5	0.5
7.	Uncertainty	0.0	0.0	0.0	0.0
8.	Space-dimension	0.0	1.0	0.0	0.0
9.	Time dimension	0.5	0.0	0.0	0.0
10.	Economy	1.0	1.0	1.0	1.0
	Total score	4.0	4.5	4.5	5.0

[1.0 = Completely fulfilled; 0.5 = Partially fulfilled; 0.0 = negligibly fulfilled]

Table 37.3: Matrix for the impacts of the model dam project

Priority values	Proposed action / Impact	Immigration of labour	Dam construction	Transmission link	Reservoir filling	Heavy metal discharge	Growth of aquatic weed	Relocation of inhabitants	Total Leopold	Total Lohani & Thanh
10	Health	5 \ 8	4 \ 6		5 \ 8	4 \ 7	6 \ 6		24 \ 35	1680
8	Spawning of fish		3 \ 4		3 \ 6	3 \ 7	5 \ 5		14 \ 22	608
7	Archeological artifacts	4 \ 6			8 \ 8				12 \ 14	616
6	Tourism			7 \ 6	7 \ 6				14 \ 12	504
5	Downstream water pollution		7 \ 7		7 \ 8	2 \ 4			16 \ 19	565
4	Social and economic aspects							8 \ 7	8 \ 7	224
3	Forestry		4 \ 2						4 \ 2	24
9	Fishery		2 \ 5			2 \ 5			4 \ 10	180
1	Navigation				6 \ 5				6 \ 5	30
2	Aquatic plants				6 \ 6				6 \ 6	72
Total	Leopold	9 \ 14	20 \ 24	7 \ 6	42 \ 47	11 \ 23	11 \ 11	8 \ 7		
	Lohani & Thanh	64	103	42	286	67	61	56		

A semi-quantitative assessment of the impacts of the major activities of any project on the different environmental parameters has been attempted through a standard graded matrix system (Ran and Wooten 1980). The degree of anticipated impact of each activity on each of the environmental parameters has been graded as per index scale as:

Degree of Impact	Impact value
Low	0.5
Appreciable	1.0
Moderate	2.0
Significant	3.0
High/Major	4.0
Permanent	5.0

The score of each parameter has been converted to a probability value and then recalculated by multiplying each value by a factor of 1,000 to arrive at the parameter importance value (PIV). The impact score of each environmental parameter (SCi) has been calculated from the following formula:

$$SC_i = (PIV)_i \sum_{f-1}^{1} I_{ij}$$

where 'I' is the impact value (on the scale if 1 to 5, + or –) due to the project activity 'f' on an environmental parameter 'i' and 'n' is the total number of the project activities. The total impact score (TIS) is arrived at through the formula:

$$TIS = \sum_{i=1}^{m} SC_i$$

where m = total number of environmental parameters.

Since 1990s, however, a new type of 'Environmental Evaluation System' is often implemented, this is called "Battelle Environmental Evaluation System" (BEES). This process involves identification of environmental parameters likely to be affected by project implementation, estimation of resulting changes in the selected parameters and aggregating the changes in determining resultant environmental quality. Thus this process necessitates transformation of parameter estimates into Environmental Quality (EQ) on a scale ranging from zero to one. Such a transformation is achieved through value function graphs which provide functional relationships between the parameter estimates and environmental quality scale. In addition as each of the selected parameter represents only a part of the total environment, weights are assigned to the parameters to reflect their relative importance for ascertaining impact of the project on natural and socio-economic environment. These weights are derived by an interdisciplinary expert team using ranked pair-wise comparison technique and expressed as Parameter Importance Units (PIU). A total of 1,000 'PIU's are allocated amongst the selected parameters.

For evaluation of impact of project on environment, an index expressed as Environmental Impact Units (EIU) has been estimated for three alternative environmental conditions, viz., the baseline without project, without environmental management plan (EMP) and project with EMP. The index value is the summation of the product of EQ and PIU for individual parameters for above three alternatives as:

$$EIU_i = \sum_{i=1}^{i=n} (EQ)_i + PIU_i$$

where EIU_i = Environmental impact Unit for i-th environmental condition

n = Number of parameters considered for environmental impact evaluation

EQ_i = Environmental quality for i-th parameter

PIU_i = Parameter importance unit for i-th parameter.

The inference regarding impact of project is drawn on the basis of extent of positive or negative changes obtained in EIU from the difference in EIU of project with EMP (C) and the EIU for baseline Environmental conditions (A). EIU of project without EMP (B) is also assessed.

The exhaustive list of parameters selected for impact assessment by BEES methods is prepared alongwith PIUS indicated by numbers in parenthesis adjacent to parameters. The PIUs allocated amongst the selected environmental parameters for impact evaluation of the project, represent consensus in the opinion of experts of an interdisciplinary team. Depending on the type of projects the PIUs allocation and parameter selection differs widely. However a total of 1,000 units in most cases allocated in three to four major environmental categories viz., ecology, environmental pollution and human interests including socio-economics and aesthetics. The higher the number of PIUs, greater the relative importance.

A model Environmental impacts of LPG project with respective allocation of PIUs is given in Table 37.4.

The impact scaling in this process has beer, accomplished through the use of value functional relationships of identified 45 typical parameters (Dee et al, 1973). Functional relationships refer to

Table 37.4: Assigned weights for environmental parameters of LPG Plant impact assessment

Total parameter importance Unit : 1,000

- **Ecology PIU : 200**
  - Natural vegetation (40)
  - Crops (40)
  - Forests (30)
  - Species diversity (20)
  - Eutrophication (15)
  - Terrestrial Life (30)
  - Aquatic Life (25)
- **Environmental Pollutions : 450**
  - Water (50)
    - pH (10)
    - Turbidity (15)
    - BOD (20)
    - DO (20)
    - Oil & Grease (25)
    - TDS (10)
    - Phosphate (10)
    - Ammonia (10)
    - Faecal coliform (10)
    - Toxic substance (20)
  - Air (150)
    - SPM (40)
    - NO (60)
    - HC (50)
  - Land (90)
    - Landuse pattern (30)
    - Soil erosion (30)
    - Soil chemistry (15)
    - Soil fertility (15)
  - Noise (60)
    - Communication effect (20)
    - Occupational noise (20)
    - Physiological effect (20)
- **Aesthetics : 150**
  - Topographic character (30)
  - Visual quality:
    - Air (20)
    - Water (20)
  - Odour (20)
  - Native fauna (20)
  - Overall appearance (20)
  - Sound (20)
- **Socio-Economic : 200**
  - Economic benefit (50)
    - Employment (25)
    - Housing (15)
    - Education (10)
    - Sanitation (10)
    - Transportation and communication (10)
    - Health (10)
    - Rehabilitation (10)
    - Drinking water (10)
  - Fisheries (25)
  - Sericulture (25)

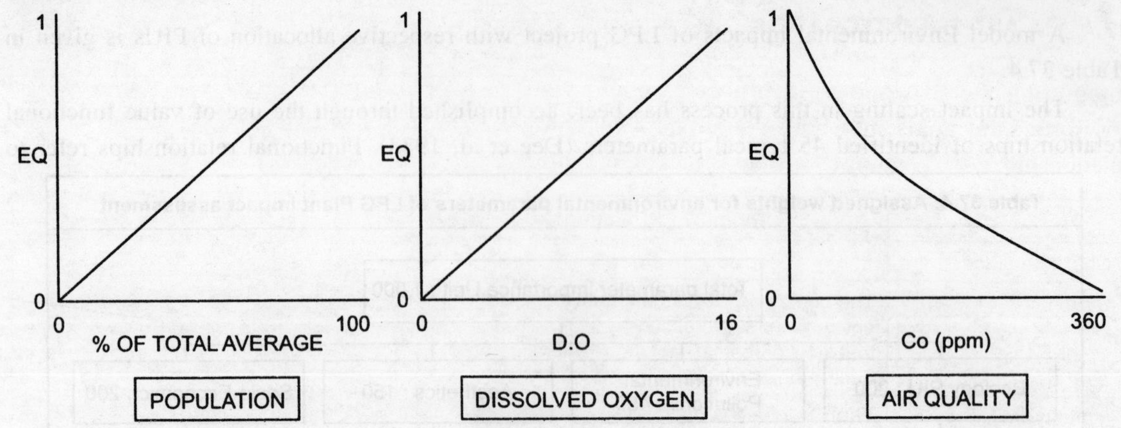

Fig. 37.3: Environmental quality index curve

transforming parameter measurements (baseline or predicted values) into subjective evaluations through graphical means. The typical functional relationships of some selected parameters are depicted in Fig. 37.3. Objective measurements are transformed into subjective interpretation of environmental quality (EQ) on a scale representing, a value of 1.0 for good quality and 0.0 for poor quality.

The summary of evaluation based on the above allocation is presented in Table 37.5. It is observed that the estimated change in EIU from baseline, due to the project without EMP is –111. The recommended EMP for the project charges the EIU value is +131. The overall change in EIU due to project over the baseline is +20.

Though Battelle Environmental Evaluation System (BEES) has been considered to be the best available evaluation technique in EIA evaluation, yet it suffers from inherent uncertainty in assigning relative weights, or PIU to different environmental parameters and subjectivity in formulation of Value Function Graphs (VFG).

Table 37.5: Summary of Environmental Impact Evaluation : LPG Plant

Parameter	Assigned parameter importance value (PIU)	Environmental Impact Unit			Change in EIU over baseline values		Net change in EIU due to EMP (C – B)
		Baseline without project (A)	Project without EMP(B)	Project with EMP (C)	Without EMP (B – A)	With EMP (C – A)	
Ecology	200	140	105	134	– 35	– 6	+ 29
Environmental pollution	450	324	242	309	– 82	– 15	+ 67
Aesthetics	150	103	77	98	– 26	– 5	+ 21
Socio-Economic	200	105	137	151	+ 32	+ 46	+ 14
Total	1,000	672	361	692	– 111	+ 20	+ 131

37.4 EIA AND EMS INTEGRATION

Integrating the EIA and EMS processes could yield new opportunities for infusing substantive environmental goals into both government planning and business decision making. The overall relationship of EIA and EMS system is shown in Table 37.6.

Table 37.6: Relationship of EIA & EMS

Characteristics	Typical EIA Processes	ISO 14000-Consistent EMS
Goal	The goal of an EIA is generally to provide environmental protection by ensuring that environmental factors are considered during the early planning process.	ISO's goal is to provide environmental protection by identifying impacts. Its continual improvement system is used to reduce these impacts.
Mandate	The EIA process often lacks a legally binding substantive mandate to protect the environment.	Substantive actions are expected to be taken, which lead to continual improvement in environmental protection.
Planning function	A comprehensive environmental planning process is followed, but it typically lacks an environmental quality system for ensuring that the decisions are properly implemented.	A planning function requires a system for ensuring that decisions are appropriately implemented. However, it does not prescribe a detailed process for performing the planning function.
External input	A detailed formal "public scoping" process for identifying significant impacts and eliminating nonsignificant issues is generally specified.	A procedure (not public) is required to record and respond to external parties, but it does not include specific steps for public involvement.
Other environmental requirements	Other environmental review processes and objectives such as pollution prevention measures, environmental justice, and biodiversity are either required to be integrated with EIA or may be done so voluntarily.	A top-level environmental policy is required, including an ongoing commitment to preventing pollution, which is very broadly defined. The policy does not specifically address integration of the EMS with other laws and policies.
Life cycle	An analysis of "reasonably foreseeable" impacts over the "life cycle" of the action is typically required.	Details of how to perform a life-cycle analysis are described in the ISO 14040 series.
Impact assessment requirements	EIA processes often include detailed directions for performing an analysis of direct, indirect, and cumulative impacts.	An in vestigation of "environmental aspects" is required. However, the requirements for performing this investigation generally lack specificity regarding scope or content.
Accumulated environmental experience	Many years of experience have been accumulated in the planning and analysis of significant environmental impacts through different EIA processes worldwide.	This is a relatively new process. Only limited experience has been accumulated in the planning and analysis of significant environmental issues.
Significance	Specific factors for determining the significance of environmental impacts are frequently specified in the EIA guidance documents or other relevant regulations.	No detailed direction for interpreting or determining the meaning of "significance" is provided.
Mitigation	Mitigation measures are generally required to be identified and analyzed as part of the EIA planning processes.	EMS provides a system for ensuring that mitigation measures are implemented during the operation of the project or policy.
Monitoring	EIA processes frequently encourage (but infrequently require) post-monitoring measures.	Monitoring is mandated as part of the continual improvement cycle.
Continual improvement	Most EIA processes lack a specific component for continually improving quality.	A continual improvement process is a basic concept inherent in an EMS.

A good EIA can serve to identify the key environmental issues and can recommend ways of avoiding or reducing adverse impacts, but if it lacks effective provisions for continuous monitoring and improvement, its objectives may not be met. An EMS may set forth a detailed process for monitoring and reducing adverse impacts, but if it is not supported by an objective assessment of potential impacts conducted via an EIA process, it may not effectively monitor the appropriate parameters or reduce the most significant adverse impacts. A conceptual framework for integrating an EIA with EMS is shown in Fig. 37.4.

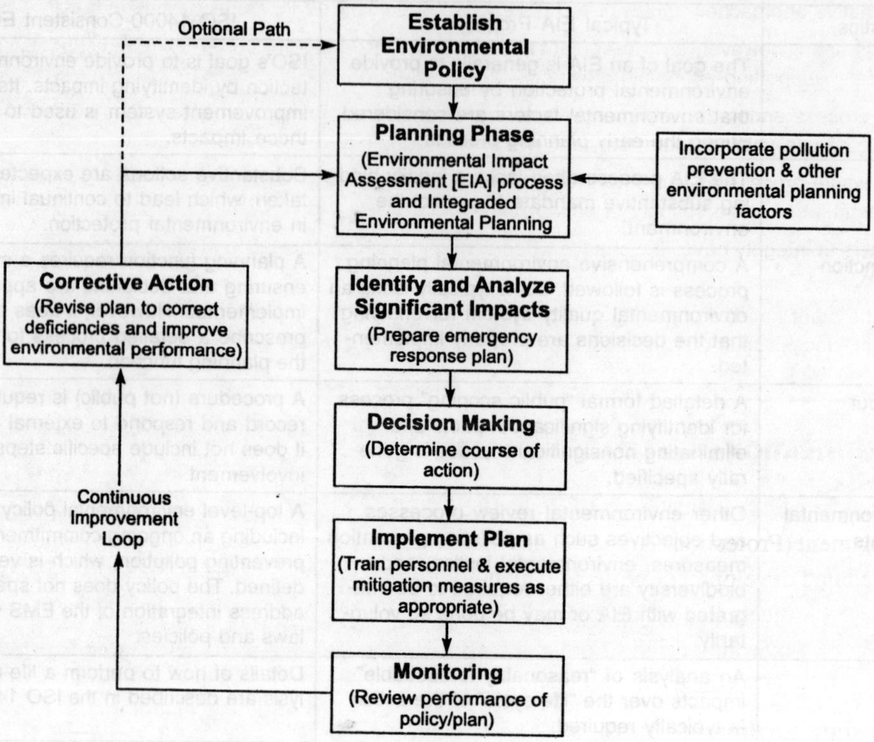

Fig. 37.4: Conceptual framework of EIA with EMS

Box 37.2: The Role of EIA in Society

1. Provides a procedure for the full consideration of the possible adverse environmental impacts of policies, programmes, activities, and projects before any decision to proceed; it precludes 'behind-closed doors' decision-making in the public and private sectors;
2. There is opportunity to present recommendations to the decision-maker on the suitability of the policy, programme (groups of projects, either sequential or concurrent), activity, or project, to proceed or not, on environmental grounds;
3. For proposals which proceed, there is the opportunity to present the incorporation of conditions of consent that should mitigate some of the adverse environmental effects;
4. It is an avenue for the public to contribute to the decision-making process, through written and oral contributions to the decision-maker(s)' appearances at public inquiries and hearings and possible participation in mediation processes;
5. The whole process of development is open to scrutiny for the benefit of all the key players: proponent (applicant), government and public, resulting in better projects more carefully thought-out;

Contd.

Box 37.2: The Role of EIA in Society

6. Basically unsatisfactory projects (including otherwise satisfactory projects on the wrong site) tend to weed themselves out before advancing far into the EIA process and certainly before reaching a public enquiry stage;
7. Conditions of approval may ensure monitoring, annual reporting by the proponent, post-project analyses (PPA) and independent auditing;
8. Alternative approaches, mixes of technology, and sites, can be thoroughly examined;
9. EIA is seen, however, as the servant of development: promoting better developments, at best, but basically supporting economic growth;
10. The process endorses waste discharges, the emission of greenhouse gases in many cases and the profligate use, mining, extraction, and processing of natural resources;
11. The whole process as a creature of government, is subject to political pressures; key players within government have no security of employment whatever;
12. Officers of integrity have little chance when confronted by a combination of hostile interests at a political level;
13. On the other hand, a vigilant public, skilled objectors and organisations with a range of legal rights to object, access to the courts, and a supportive media with some political sympathy, can exercise countervailing power and influence.

37.5 EIA NOTIFICATION 2006 (Amended in 2009)

Environmental Impact Notification S.O. 1533 (E), dated 14th September 2006 as amended 2009, issued under Environment (Protection) Act 1986, has made it mandatory to obtain prior environmental clearance (EC) for scheduled development projects. The notification has classified projects under two categories A & B. Category A projects (including expansion and modernization of existing projects) require clearance from Ministry of Environment and Forest (MoEF), Govt. of India (GoI) and for category B for State Environmental Impact Assessment Authority (SEIAA), constituted by Govt. of India.

The existing Environmental Impact Assessment manual of MoEF is common for all the sectors requiring prior environmental clearance. Considering the diversity in all the sectors related to infrastructure and industrial development. MoEF launched a programme for development of sector specific EIA guidance manual. The EIA guidance manual will help the project proponent and consultant in the preparation of EIA report. It also helps the regulatory authority while reviewing the report and also the public to be aware of the related environmental issues. This EIA guidance manual accordingly addresses their related environmental concern for the specific sector—ports & harbors. The sector specific manual consists of twelve chapters, which correspond to the generic structure as per the EIA Notification.

Chapter 1: Introduction

This chapter contains the general information on the port sector, major sources of environmental impact in respect of port projects and details of the environmental clearance process.

Chapter 2: Project Description

This chapter should cover the description of the project, such as the type of project, need for the project, project location, project layout, cargo handling methods, utilities and services, and the project implementation schedule, estimated cost of development etc.

Chapter 3: Analysis of Alternatives (Technology & Site)

This chapter should cover details of various alternatives both in respect of location of site and technologies to be deployed, in case the initial scoping exercise consider such a need.

Chapter 4: Description of Environment

This chapter designs the heart of the manual, since it is in this chapter that the proponent shall give the comprehensive data on the existing and additional data collected by him on the baseline environmental data in the study area as well as in the surrounding area that is likely to be affected by the proposed activity.

Chapter 5: Anticipated Environmental Impact and Mitigation Measures

This chapter should cover the anticipated impact on the environmental and mitigation measures. The method of assessment of impact including studies carried out, modeling techniques adopted to assess the impact where pertinent should be elaborated in this chapter. It should give the details of the impact on the baseline parameters, both during the construction and operational phases and the mitigation measures to be implemented by the proponent.

Chapter 7: Additional Studies

This chapter should cover the details of the additional studies, if any, required in addition to those specified in the TOR and which are necessary to cater to more specific issues applicable to the particular project. These studies may be suggested either by the proponent itself or the regulatory authority.

Chapter 8: Project Benefits

This chapter should cover the benefits accruing to the locality, neighbourhood, region and nation as whole. It should bring out details of benefits by way of improvements in the physical infrastructure, social infrastructure, employment potential and other tangible benefits.

Chapter 9: Environmental Cost Benefit Analysis

This chapter should cover Environmental Cost Benefit Analysis of the project, if recommended by the Expert Appraisal Committee at the scoping stage.

Chapter 10: Environmental Management Plan

This chapter should comprehensively present the Environmental Management Plan (EMP), which includes the administrative and technical setup, summary matrix of EMP, the cost involved to implement the EMP, both during the construction and operational phases.

Chapter 11: Summary and Conclusions

This chapter makes the summary of the full EIA report condensed to ten A4 size pages at the maximum. It should provide the overall justification for implementation of the project and should explain how the adverse effects are proposed to be mitigated.

Chapter 12: Disclosure of Consultants Engaged

This chapter should include the names of the consultants engaged with their brief resume and nature of consultancy rendered.

The contents of the manual are to be considered as version 1.0 (2010). An updating/revision of the manual will be taken up by the ministry as per the requirements. In case of interpretation of any question related to law, the provisions of the original laws and the Rules made thereunder with various Government directions/resolutions will have to be read and followed. In case of amendment to the original Act/Rules/Notifications made thereunder, the provisions as amended from time to time shall be applicable. Any obligations of international conventions, where GoI is a signatory and accepted for implementation are also to be followed.

Box 37.3: Term of Reference (TOR)

All 'A' category project when conceived by a developer, at first the developer submit a initial project proposal with preliminary environment status report of the concern project site with the help of accedated registrar consultant, for seeking the permission and advice of Environmental Appraisal Committee (EAC) of MOEFCC. Then as per term of reference given by EAC they can prepared the final EIA & EMP report for environmental clearance of the concerned project.

Box 37.4: Public Hearing

After complying of EIA & EMP report the EIA rule requires that the public must be informed and consulted on a proposal of newer development in their area. The State Pollution Control Board will conduct the public hearing meeting near the project site with prior notification. Local people can participate and developer present their proposal and seeking the views and aspiration of local. The developer has to address those issues and finally report was prepared with video graph by State Pollution Control Board and sent MOEFCC for environmental clearance of the project.

38 ENVIRONMENTAL AUDIT

38.1 CONCEPT

Environmental auditing is a management tool designed to provide information on environmental performance to the right people at the right time. This audit encompasses all kinds of activities related to environmental measures of an organisation.

An environmental audit is a systematic means of providing environmental management information:

(a) to all levels of management;

(b) for a variety of purposes.

The term 'environmental audit' is, therefore, used to refer to a number of different information and assessment activities. These can be categorised as:

(a) *A technical review*: Such an audit will involve the systematic collection of information about the existing and potential impact of the organisation's activities on the environment; it will normally cover compliance with pollution control and waste management legislation. It will not cover management practices.

(b) *A management review*: This will focus more on the management procedures and record- keeping and will also gather information on compliance with legislation. It may also review procedures in the context of company policies, programmes and other requirements. It will not examine the existing or likely impact of the operation on the surrounding environment from a technical standpoint.

(c) *Due diligence review*: This will examine the likely cost of implementing pollution control and site remediation actions and will take account of existing and future legislation. Such liability reviews are normally carried out in the context of mergers, acquisitions and long range company planning.

An environmental audit thus represent a management tool comprising a systematic, documented, periodic and objective evaluation of the performance of the organisation, management system and processes designed to protect the environment with the aim of:

(i) facilitating management control of practices which may have an impact on the environment;

(ii) assessing compliance with company environmental policies.

The Regulation specifies that an audit should include at least the following steps:

— Planning of the audit activities, including definition of responsibilities for the audit;

— Review of the environmental protection policy of the company;

— Assessment of the organisation, management and equipment;

— Gathering of data and of all relevant information;

— Evaluation of the overall performance;

— Identification of areas for improvement;

— Internal reporting to the top management;

An outline of Environmental management cycle is shown in Fig. 38.1.

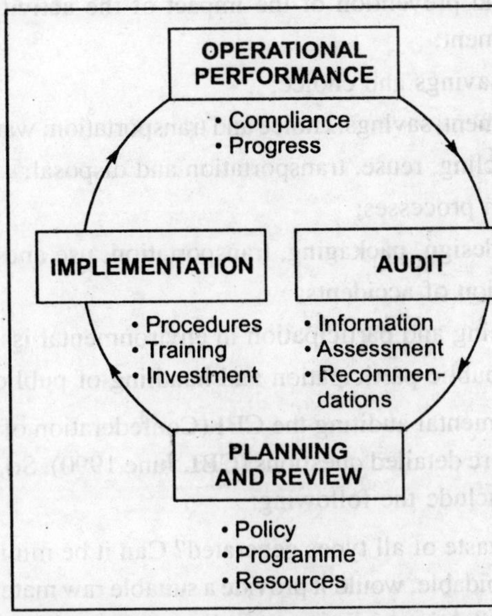

Fig. 38.1: Environmental management cycle

38.2 SETTING UP AN AUDIT PROGRAMME

Developing an audit programme requires decisions to be taken on:

— the scope of the audit: what information is to be collected;

— the frequency with which each site (or issue) is to be audited;

— who is to carry out the audit;

— what, if any, information is to be made available to the public.

It is also necessary to develop an audit protocol: that is, the detailed plan to be used by each auditor when carrying out the audit.

Scope and frequency of the audit

The scope of the audit will depend on the information that is required by management to monitor environmental performance. A full environmental audit will cover:

(i) compliance with environmental regulations,

(ii) implementation of company's environmental policies and procedures,

(iii) good environmental management practice, and

(iv) past activities.

In so far as it provides the information to determine an organisation's environmental performance it may be expected to cover not only waste streams from plants but also wider management and operational issues.

The Eco-Management and Audit Regulation suggests that the following should be covered within the framework of the wider environmental protection system:

— assessment, control and prevention of the impact of the activity concerned on the various sectors of the environment;
— energy management, savings and choice;
— raw materials, management, savings, choice and transportation; water management and savings;
— waste avoidance, recycling, reuse, transportation and disposal;
— selection of production processes;
— production planning (design, packaging, transportation, use and disposal);
— prevention and limitation of accidents;
— staff information, training and participation in environmental issues;
— external information, public participation and handling of public complaints.

In their guidance on environmental auditing the CBI (Confederation of British Industries) includes a similar listing, amplified into more detailed questions (CBI, June 1990). So, for example, on wastes and transport the detailed questions include the following:

Wastes: How and where is waste of all types generated? Can it be minimised/eliminated/recycled? Where production of waste is unavoidable, would it provide a suitable raw material for another organisation, either business or voluntary group? Are you losing recycling opportunities or increasing disposal costs by not segregating different types of waste? Who removes your waste—is it carried and disposed off responsibly?

Transport: Are you using the most efficient and environmentally sound systems for transporting your goods and materials? Would alternative systems make less demands on the environment? Do staff traveling on business use the most efficient system or merely the most 'convenient'? Would use of public transport bring about cost savings without significant increase in traveling times? Do you encourage staff to travel to work by public transport and could you increase availability of communal transport by providing additional company funded services?

Who should carry out the environmental audit?

To be of maximum value an environmental audit should be objective and the environmental auditor should be free from pressure from within the organisation; in other words it is important that the audit report should be objective and the auditor should feel able to make critical comments without fear of the effect of such comments on his future career within the organisation.

A number of major companies have established audit teams within their corporate environmental health and safety department. In some cases, the staffing of these teams relies on secondment from different departments within the company. In so far as individuals are then auditing different parts of the company this has a benefit that they understand the business but have no internal line of responsibility to the activity being audited.

Some organisations have used external consultants to carry out some or all of the audits; in this case it is often valuable for the external auditor to work with the company's team.

Environmental auditing requires an understanding of the activity to be audited (this may be a plant, an abandoned site or a corporate headquarters); an understanding of management systems, environmental regulations and permitting procedures; and also a broad understanding of environmental impacts. A qualified auditor will, in addition, need to be systematic, able to deal in a constructive way with a range of management and technical staff.

The audit protocol

The protocol represents the plan to be used by auditors in conducting an audit. It establishes what information is to be collected it provides a systematic basis to the audit; and it also gives a step-by-step guide to the environmental auditor on how evidence is to be collected. The protocol is an important tool in so far as it not only serves as the auditor's guide to conducting the audit but also acts as a record of the audit procedures and the notes completed by the team. In addition, a completed audit protocol provides a record for the rationale for any changes in audit procedures or deviations from plan if these prove necessary during the audit.

Specifically the protocol

(a) contains a basic questionnaire covering each environmental topic;

(b) provides the hard copy record to assist the auditors as the work progresses and provides a basis for referencing working papers and copies of documentation collected during the audit;

(c) the completed protocols with the supporting documents are the record of the audit and form the basis for the formal audit report.

The protocol is therefore the key step in establishing the audit programme in so far as it informs managers and others about the scope of the audit and provides the opportunity to request that the changes are made either to collect additional information or to delete information that will not be used.

Box 38.1: Elements of Environmental Auditing

Environmental audits have been defined by EPA as "a systematic, documented, periodic and objective review by regulated entities of facility operations and practices related to meeting environmental requirements" (Environmental, 1986). Audits differ widely among companies, but typically consist of review of permits and other environmental records, interviews with facility personnel, and a visual inspection of the facility. Key elements of an auditing programme generally include:

- defining audit objectives
- determining audit scope
- selecting or preparing audit protocol or checklist
- identifying auditors and key facility personnel
- developing report format and information management system
- managing audit scheduling and logistics
- tracking completion of audit action items.

38.3 TYPICAL AUDIT PROCESS

The following typical audit process reviews the work required prior to the audit; activities at a site; and post-audit activities. It is specifically applicable to a manufacturing site rather than, for example, a corporate headquarters. The issues that are likely to be addressed in the audit are shown in Table 38.1.

Table 38.1: Items to be addressed in audits

Environment	Safety	Occupational health	Product safety
Site history	Safety policy/procedures	Employee exposure to air contaminants	Product safety programme
Process materials used	Accident reporting Accident investigation	Exposures to physical agents e.g. noise, radiation, heat, etc.	Product quality control
Storage of materials above ground, below ground	Accident recording	Measurements of employee exposure	Product packaging, storage and shipping
Air emissions	Permit to work systems	Exposure records	Product recall-withdrawal procedures
Water discharges	Special procedures for:	Ventilation engineering controls	Customer information on product handling and quality
Solid wastes	confined space entry		
Liquid hazardous wastes	work on electrical equipment	Personal protective equipment	Regulatory compliance
Asbestos, PCBs	breaking into pipelines, etc.	Information and training on health hazards	Labelling
Waste disposal on site off site	Emergency response Fire fighting	Medical surveillance programme	Specifications for purchased materials products packaging
Oil/chemical spill prevention	Job safety analysis		
	Safety training	Hearing conversation	Material safety data
Permits, licences		First aid	Vendor qualification programme
Pollution control Contractors on site	Safety communications promotion	Regulatory requirements	QA testing and inspections
	Housekeeping		Record keeping
Past incidents	Regulations compliance		Product literature Process control

Pre-audit activities

(a) Select and schedule facility to audit based on:

— selection criteria

— priorities assigned.

(b) Select audit team members

— confirm their availability

— make travel and lodging arrangements

— assign audit responsibilities.

(c) Contact facility and plan audit

— discuss audit programme

— obtain background information

— administer questionnaire (if necessary)

— define scope

— determine applicable requirements

— note priority topics

— modify or adapt protocols

— determine resource needs

— identify facility staff to be interviewed, confirm availability.

Activities at site

Step 1: Identify and understand management control systems

— review background information

— opening meeting

— orientation tour of facility

— review audit plan

— confirm understanding of internal controls.

Step 2: Assess management control systems

— identify strengths and weaknesses of internal controls

— adapt audit plan and resource allocation

— define testing and verification strategies.

Step 3: Gather audit evidence

— apply testing and verification strategies

— collect data

— ensure protocol steps are completed

— review all findings and observations

— ensure that all findings are factual

— conduct further testing if required

— liaise with regulatory authorities.

Step 4: Evaluate audit findings

— develop complete list of findings

— assemble working papers and documents

— integrate and summarise findings

— prepare report for closing meeting.

Step 5: Verbal report of findings to facility

— present initial findings at closing meeting

— discuss findings with plant personnel.

Post-audit activities

(a) Issue draft report

— corrected closing report

— determine distribution list

— distribute draft report

— allow time for corrections.

(b) Issue final report

— corrected draft report

— distribute final report

— highlight requirement for action plan

— determine action plan preparation deadline.

(c) Action plan and preparation and implementation

— based on audit findings in final report.

(d) Follow-up in action plan.

38.4 CARRYING OUT THE AUDIT

Collecting Information

Environmental auditing requires the auditor to develop a full understanding of the controls, procedures and practices that are in place or are thought to be in place. These are likely to include formal procedures and practices; all forms of records (including monitoring); existing inspection and maintenance programmes; physical controls and other measures to contain spills and other incidents, etc. The auditor will need to understand how responsibilities are defined; how personnel are trained and how competent those on the site appear to be; how procedures are carried out and other standard management issues. In addition to the use of the detailed questionnaire (forming part of the protocol) the auditor will gain information through observation, interviewing individual staff, reviewing specific documentation and speaking to the appropriate regulatory authorities.

Audit Interviews

Since interviewing is one of the primary techniques used in gathering audit information, good interviewing skills are essential to the successful completion of the audit. Attention to the following elements of interviewing skills during the audit process will assist in establishing a good rapport with the facility personnel as well as obtaining the information required:

Take notes during the interview and record the name, title and job description of the person interviewed; be alert and responsive to the interviewee when asking questions and listening to responses; control the interview, but ensure that you appear relaxed and that your interviewee feels relaxed; go to the interviewee's workplace/office where possible; plan the interview so that you introduce yourself, explain the purpose of the interview and gather the appropriate information; know when to end the interview and thank the interviewee; end the interview on a positive note.

Documentation

It is a central principle that the audit should be fully documented; there should, therefore, be a clear, written, audit trail. The working papers from an audit will include completed protocols, associated questionnaires, auditor notes and copies of documents collected during the audit itself. It is important that each auditor ends the audit with a complete set of field notes and papers for each protocol step assigned to that auditor; that there is documentation (1) to ensure that an adequate audit was conducted and, (2) to substantiate compliance and non-compliance issues, both with respect to regulatory requirements and company procedures; that there is a record of reference copies of documents collected during the audit; and there is the data to support the audit report, which may be helpful in subsequent follow-up.

The working papers should, therefore, contain all the information that the auditor believes may be necessary to support the audit findings and should contain a description of environmental management and control systems—including flow charts, diagrams, copies of documents, etc.; a description of the action taken to complete each step of the audit protocol; details of answers to questions in the protocol and other questionnaires used; details of any 'tests' conducted during the audit (e.g., checking that consignments of hazardous waste had the relevant documentation for each disposal).

Public disclosure

Unlike financial audits, environmental audits are not carried out for the purposes of providing information to the regulatory authorities or the public at large. They are intended to assist management understand the impact their operations are having upon the environment and to provide a basis for taking action to remedy any problems. To this extent companies are likely to be unwilling to make the results of their environmental audits public; or, if the audit is carried out with this objective, it is likely that it will be a far less valuable management tool. The ICC (Indian Chamber of Commerce) has noted that if environmental audits are to be successfully used as part of company policy 'it is essential that the procedure should be seen as the responsibility of the company itself, should be voluntary and for company use only'.

The Eco-Audit proposal does however propose that certain information should be made available to the public on the environmental performance of the company. This would be an 'environmental statement' specific to each site audited and prepared with the same frequency as the environmental audits.

Such a statement would:

— describe the company's activities at the site;

— provide a detailed assessment of all the significant environmental issues of relevance;

— summarise the figures on pollutant emissions, waste generation, raw material, energy and water consumption and other significant environmental aspects;

— include the company's environmental policy programme as well as the specific objectives for the site considered;

— include an evaluation of the environmental performance of the environmental management system and note the date for the submission of the next statement;

— publish the name of the accredited environmental verifier for the Statement: this requirement for external, independent evaluation is a crucial component of the Regulation.

Simplified statements must be drawn up for the intervening years in the audit cycle. These must draw attention to significant changes from the previous statement and include details of pollutant emissions.

Where no significant changes have occurred, no further statement is required until the completion of the next audit.

It will be appreciated that for many plants and processes within the UK information on emissions, wastes and environmental effects is already on the public register (through the provisions of the *Environmental Protection Act,* 1990) and increasingly the larger companies are providing a clear statement of their environmental policies and programmes in order to make it clear to the public that they are aware of their environmental responsibilities.

So while the environmental audit remains the internal document (like management accounts) companies may wish to provide a general summary in order to demonstrate to the public that they are carrying out audits and are happy to make some of the information more widely available. It should perhaps be noted that such internal systematic reviews may also become public that, should there be a serious pollution incident at the plant or a civil action with the result that the documents are placed before an inquiry or any other form of legal hearing.

38.5 BENEFITS OF ENVIRONMENTAL AUDITING

A properly implemented environmental audit plan provides a range of benefits for an organisation. These are:

(i) It provides a framework for measuring (and therefore managing) environmental performance.

(ii) It reinforces accountability for the environmental dimension of the business: the audit process requires managers to be clear about their responsibilities and how these are being implemented.

(iii) It raises awareness of the importance of professional environment management throughout the organisation.

(iv) To the extent that external reporting will be required, it provides a sound basis that enables companies to feel sure that what is included in the reports is an accurate record.

(v) It provides valuable information for future planning. This includes the future design of processes and products as well as inputs into future financial plans where pollution control and other investments are to be made.

(vi) It allows senior managers to feel secure that all the environmental aspects of their business are being professionally managed: it will not remove all possibility of environmental incidents but will at least reduce their likelihood.

38.6 ENVIRONMENTAL AUDIT PROGRAMME IN INDIA

The concept of environmental auditing in industrial facilities in India, appears to have first got into meaningful discussions in the beginning of the nineties1990s. Efforts were initiated to see the practicability of this programme before it could be mandatory. This process finally resulted in the issuing of a gazette notification on 13 March 1992 (Appendix 12.1) through which submission of the environmental audit reports has been made mandatory. The industries are now required to submit their audit reports to the concerned state pollution control boards on or before 15th day of May every year beginning 1993.

A good number of polluting industries were identified for submission of regular environmental audit report. These are:

1. Cement factories (above 200 tonnes per day production capacity)

2. Thermal power plants
3. Fermentation/Distillery factories
4. Sugar factories
5. Fertiliser and sulphuric acid plants
6. Integrated iron and steel plants
7. Pulp & paper Industries (above 30 tonnes per day production)
8. Oil refineries
9. Caustic soda plants
10. Petrochemicals plants
11. Pesticide formulation and manufacturing plants
12. Leather processing industries including tanneries
13. Basic drugs and pharmaceuticals manufacturing plants
14. Dye and Dye intermediates
15. Zinc smelting industries
16. Copper smelting industries
17. Aluminium smelting industries
18. Lead smelting industries.

On the whole, the environmental audit studies were not intended for completing any regulatory requirements, but with the basic philosophy that at least those industries which need priority attention in the sense of pollution control, should know what it is, as well as why and how it is to be done, before the industries actually arrangê to set it done on their own as a mandatory requirement, as per the GOI's notification of 13 March 1992 (Annexure 12.1).

ANNEXURE 12.1

THE GAZETTE OF INDIA : EXTRAORDINARY

(PART II, SEC 3)

MINISTRY OF ENVIRONMENT AND FORESTS

NOTIFICATION

New Delhi, the 13th March 1992

G.S.R. 329(E)—In exercise of the powers conferred by Section 6 and 25 of the Environment (Protection) Act, 1986 (29 of 1986), the Central Government hereby makes the following rules further to amend the Environment (Protection) Rules, 1986, viz.:

1. (a) These rules may be called the Environment (Protection) Second Amendment Rules, 1992.

   (b) They shall come into force on the date of their publication in the Official Gazette.

2. In the Environment (Protection) Rules, 1986:

   (a) after Rule 13, the following rule shall be inserted, viz.:

   14. Submission of environmental audit report

Every person carrying on an industry operation or process requiring consent, under Section 25 of the Water (Prevention and Control of Pollution) Act, 1974 (6 of 1974) or under Section 21 of the Air (Prevention and Control of Pollution) Act, 1981 (14 of 1981) or both on authorisation under the Hazardous Wastes (Management and Handling) Rules, 1988 issued under the Environment (Protection) Act, 1986 (29 of 1986) shall submit an Environmental Audit report for the financial year ending the 31st March in Form V to the concerned State Pollution Control Board on or before the 15th day of May every year, beginning 1993.

(b) in Appendix 'A' after Form IV the following form shall be inserted, viz.:

FORM V

(See Rule 14)

Environmental Audit Report for the Financial Year ending the 31st March.

PART A

(i) Name and address of the owner/occupier of the industry operation or process

(ii) Date of the last environmental audit report submitted

PART B

Water and Raw Material Consumption

(i) Water consumption m^3/d

Process

Cooling

Domestic

Name of products	Water consumption per unit of products	
	During the previous financial year (1)	During the current financial year (2)
(1)		
(2)		
(3)		

(ii) Raw material consumption

Name of raw materials	Consumption of raw material per unit of output	
	During the previous financial year	During the current financial year

PART C

Pollution Generated

(Parameter as specified in the consent issued)

(i) Pollutants	Quantity of pollution Generated	Percentage of variation from Prescribed standards with reasons
(a) Water		
(b) Air		

PART D

Hazardous Wastes

(As specified under Hazardous Wastes/Management and Handling) Rules. 1989

Hazardous Wastes	Total Quantity (kg.)	
	During the previous financial year	During the current financial year
(a) From Process		
(b) From Pollution Control Facilities		

PART E

Solid Wastes

	Total Quantity	
	During the previous financial year	During the current fianancial year
(a) From Process		
(b) From Pollution Control Facilities		
(c) Quantity recycled or re-utilised		

PART F

Please specify the characteristics (in terms of concentration and quantum) of Hazardous as well as Solid Wastes and indicate disposal practice adopted for both these categories of wastes.

PART G

Impact of the pollution control measures on conservation of natural resources and consequently on the cost of production.

PART H

Additional investment proposal for environmental protection including abatement of pollution.

PART I

Miscellaneous

Any other particular in respect of environment protection and abatement of pollution.

(F. No.Q. 15015/1.90-CPA)

T. GEORGE JOSEPH. Jt. Secy.

39 LIFE CYCLE ASSESSMENT AND INDUSTRIAL ECOLOGY

39.1 CONCEPT

Life cycle assessment (LCA) was formalized during the early 1990s. It emerged from a conviction that it was important to conduct "cradle-to-grave" assessments of products, packages, processes and activities. The Ecobalance Study approach developed in Europe, emphasizing waste reduction to curtail water and air pollution and Resource and Environmental Profile Analysis work in the United States during the 1980s, were the forerunners of LCA. However, it was at a workshop organised by the Society of Environmental Toxicology and Chemistry (SETAC) in the United States during the early 1990s that the basic approach for LCA was developed. Thus life cycle assessment (LCA) is the process of evaluating the effects that a product has on the environment over the entire period of its life cycle.

Life cycle assessment is a decision support tool supplying information on the environmental effects of products. It furnishes information on the environmental effects of all the stages of a product's life cycle. This information can be used by governments and by companies as well as by NGOs and individual consumers when making decisions related to products. Ecolabelling, product and process improvements, and purchasing decisions, for example, can all be supported by life assessment.

39.2 STAGES IN LIFE CYCLE ANALYSIS

As indicated in Fig. 39.1, LCA can be divided into four stages or phases. Each is briefly considered:

- *Initiation*: This stage is very similar to scoping in impact assessment. The main purposes are to establish the objectives for and detail of an assessment to define the "system" to be assessed and to identify types of data needed.
- *Inventory*: Primary attention is allocated to collecting data about raw materials needed for inputs, including energy and water and about wastes produced as outputs during the process and at the end of the useful life of the product, process, package or related activity.
- *Impact*: This component is similar to the comparable phase in impact assessment. The focus here is upon identifying the effects and reaching a judgement about the significance of such effects. Effects on the environment, economy and health or well-being are usually included.
- *Improvement*: This can be characterised as the *cognitive* or *prescriptive* stage, in which attention is directed to possible actions to reduce or mitigate any negative impacts identified in the previous stage.

The four phases or components identified above and in Fig. 39.2 are intended to assist resource and environmental managers to encourage better product design, more effective processes regarding raw material inputs or waste outputs, improved transportation methods, more careful consumer use and better

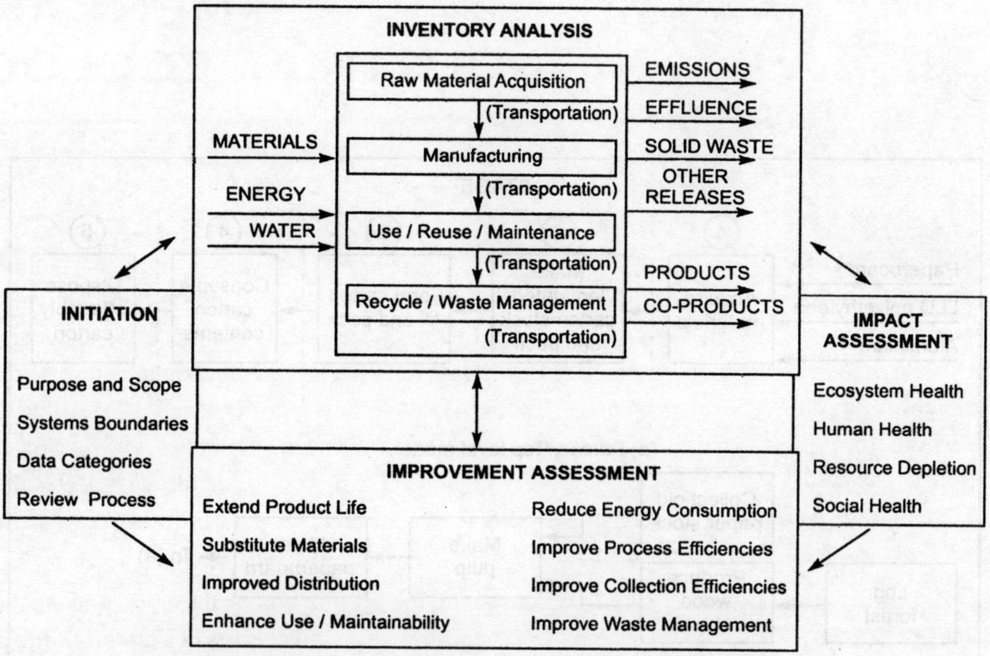

Fig. 39.1: The four phases of life-cycle assessment

waste disposal practices. More specifically, in the context of resource and environmental management, LCA is intended to lead to decisions which result in greater conservation of resources and the environment, increased energy conservation and decreased waste generation, improved industrial processes related to providing resource-based products and fewer problems in final disposal. Thus, regarding toxic wastes, for example, life-cycle analysis would consider whether a toxic chemical is needed to be used at all and, if it were to be used, what should be done to ensure its proper production, transportation, storage, use and disposal.

Some key aspects of each of the four components are considered in more detail below.

Initiation

Defining the *system boundaries* for the life cycle assessment is a critical aspect during initiation. For example, the scope of an LCA for a box-making plant in Germany should not only include the raw materials and energy required at the plant to manufacture each box and the wastes that must be disposed of. It should also examine the raw material and energy inputs for the other firms that provide component parts for the production of the boxes, even when those suppliers are located outside Germany. Furthermore, a decision must be taken as to whether the analysis focuses only on the primary manufacturers, or whether the assessment should include the forestry operations that provide the paperboard for the box, the natural gas industries that provide the resin and polyethylene, the aluminium industry that produces the aluminium foil and so on. Fig. 39.2 outlines in schematic form the various levels that could be involved in conducting a life cycle assessment of the process to produce laminated cartons. The choice of system boundaries is thus not always obvious or automatic. Furthermore, the definition of such boundaries can greatly complicate and extend the LCA and add to the time and cost to complete it.

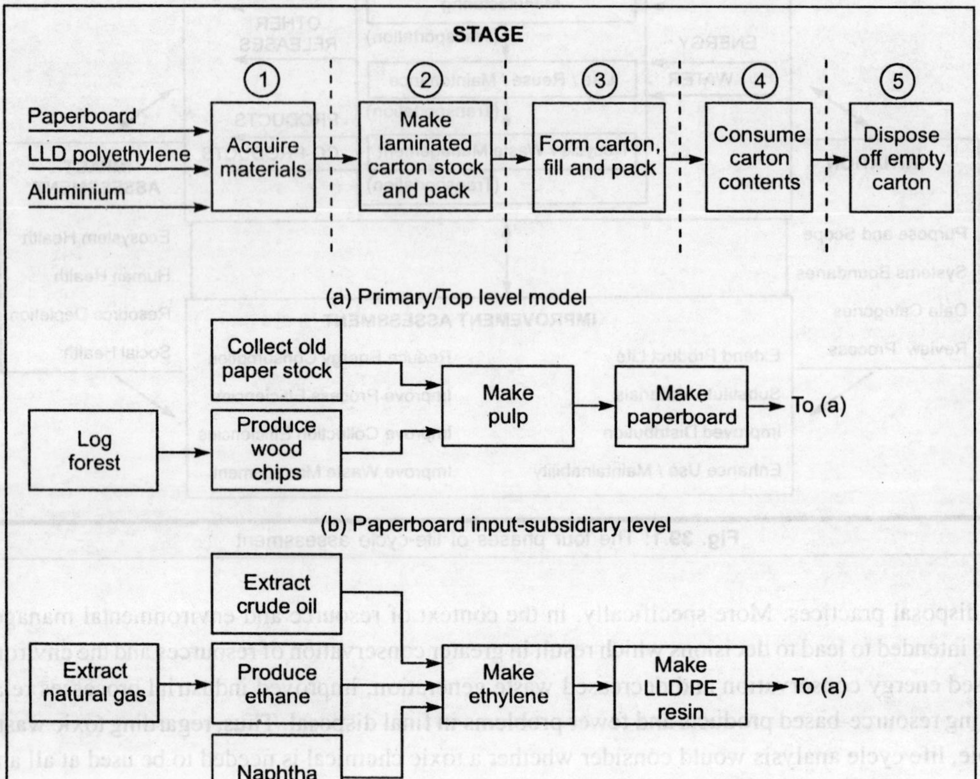

Fig. 39.2: Life cycle assessment for production of laminated cartons

Decisions also have to be taken regarding who will complete the life cycle assessment. There are at least three choices:

1. Employees of the firm could do the assessment. They probably are in the best position to be able to identify all of the inputs and outputs. However, the credibility of a self-evaluation can be low.
2. The alternative is to have outside experts conduct the assessment. If such an external team does the work with reference to some generally accepted standards, then the credibility of the assessment is usually greater.
3. The third option is to have a team involving both employees and outside experts conduct the task.

For practical reasons, it is often some form of the third alternative which is used, since a firm often does not have all of the expertise to conduct an LCA, yet the outside experts need the input and advice of people within the firm.

Inventory

As Fig. 39.2 indicates, inventory is required for raw material acquisition, manufacture, use/reuse/maintenance and recycling/waste management. Particular attention is given to the material, energy and water inputs and to the emission, effluent, solid waste and other release outputs, as well as to products and co-products resulting from recycling or waste management. A thread common to many of the activities is transportation, as it is necessary to determine which inputs and outputs are needed to move material from one stage to another. An inventory involves four major activities.

Impact assessment

The activities here are similar to what was described in the previous section dealing with environmental impact assessment. Particular concern usually focuses upon the implications of resource consumption associated with the creation of the product, material or packaging and the releases into the environment during each of the life cycle phases. Ideally, the information here is in the form of *impact indicators*, which indicate whether critical thresholds have been passed. The challenge, of course, is that for many outputs we do not yet know what the critical thresholds are and this aspect becomes even more problematic when the synergistic effects of two or more waste outputs are involved. One of the criticisms of LCA is that it is a very detailed and potentially very expensive procedure, but at the conclusion of the impact assessment stage there are still not enough environmental standards that are generally agreed upon to allow effective use of all the information which has been collected.

39.3 APPLICATION OF LIFE CYCLE ANALYSIS

The principal application of LCA are:

- Communications about the environmental aspects of products;
- Product and process improvement;
- Product and process design;
- Development of business strategies, including investment plans;
- Setting ecolabelling criteria;

- Developing product policies;
- Developing policy strategies;
- Purchasing decisions; and
- Development of life styles.

39.4 INDUSTRIAL ECOLOGY – CONCEPT

How industrial environment could be safely managed through understanding of industrial processes (including input and output of materials) towards better environmental protection. The entire phenomena come under industrial ecology. The major goal of industrial ecology must be to minimize or eliminate detrimental effects of anthropogenic activities in industrial processes in particular on other spheres of the environment. Beyond environmental protection and preservation, the practice of Industrial ecology should also improve and enhance environmental conditions.

39.5 COMPONENTS OF INDUSTRIAL ECOSYSTEM

Industrial ecosystems can be broadly defined to include all types of production, processing, and consumption. It is useful to define five major components of an industrial ecosystem, as shows in Fig. 39.3. These are: (1) a primary materials producer, (2) a source or sources of energy, (3) a materials processing and manufacturing sector, (4) a waste processing sector, and (5) a consumer sector. In such an idealized system, the flow of materials among the four major hubs is very high. Each constituent of the system evolves in a manner that maximizes the efficiency with which the system utilizes materials and energy.

It is convenient to consider the primary materials producers and the energy generators together because both materials and energy are required in order for the industrial ecosystem to operate. The primary materials producer or producers may consist of one or several enterprises devoted to providing the basic materials that sustain the industrial ecosystem. Most generally, in any realistic industrial ecosystem a significant fraction of the material processed by the system consists of virgin materials. In a number of cases, and increasingly so as pressures build to recycle materials, significant amounts of the materials come from recycling sources.

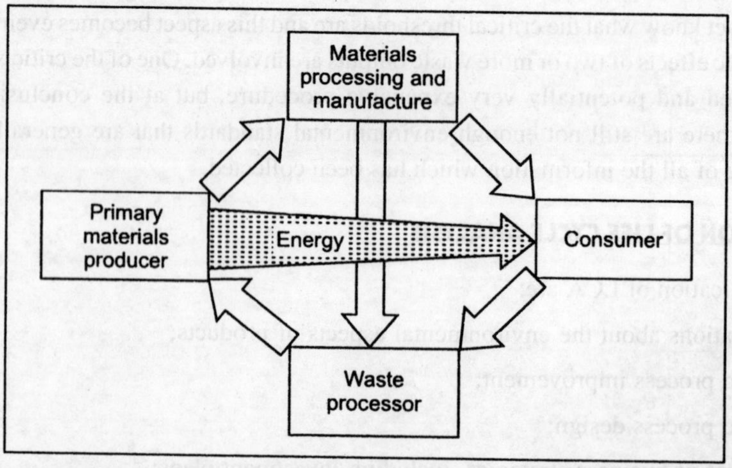

Fig. 39.3: The major constituents or hubs of an industrial ecosystem

Finished materials from primary materials producers are fabricated to make products in the materials processing and manufacturing sector, which is often a very complex system. For example, the manufacture of an automobile requires steel for the frame, plastic for various components, rubber in tires, lead in the battery, and copper in the wiring, along with a large number of other materials. Typically, the first step in materials manufacturing and processing is a forming operation. For example, sheet steel suitable for making automobile frames may be cut, pressed, and welded into the configuration needed to make a frame. At this step some wastes may be produced that require disposal. An example of such wastes consists of carbon fiber-epoxy composites left over form forming parts such as jet aircraft engine housings. Finished components from the forming step are fabricated into finished products that are ready for the consumer market.

In the consumer sector, products are sold or leased to the consumers who use them. The duration and intensity of use vary widely with the product; paper towels are used only once, whereas an automobile may be used thousands of times over many years. In all cases, however, the end of the useful lifetime of the product is reached and it is either (1) discarded, or (2) recycled. The success of a total industrial ecology system may be measured largely by the degree to which recycling predominates over disposal.

Recycling has become so widely practiced that an entirely separate waste processing sector of an economic system may now be defined. This sector consists of enterprises that deal specifically with the collection, separation, and processing of recyclable materials and their distribution to end users. Such operations may be entirely private or they may involve cooperative efforts with governmental sectors. They are often driven by laws and regulations that provide penalties against simply discarding used items and materials, as well as positive economic and regulator incentives for their recycle.

The strong analogy between natural ecosystems and efficient industrial systems was first clearly stated in 1989 by Frosch and Gallopoulos. A natural ecosystem, which is usually driven by solar energy and photosynthesis, consists of an assembly of mutually interacting organisms and their environment, in which materials are interchanged in a largely cyclical manner. It is possible to visualize an analogous industrial ecosystem in which materials are cycled, driven by an energy source (Table 39.1).

Table 39.1: Characteristics of Natural Ecosystem and Industrial Systems

Characteristic	Natural Ecosystems	Current Industrial Systems
Basic unit	Organism	Firm
Material pathways	Closed loops	Largely one way
Recycling	Essentially complete	Often very low
Material fate	Tend to concentrate, such as atmospheric CO_2 converted to biomass by photosynthesis	Dissipative to produce materials too dilute to use, but concentrated enough to pollute
Reproduction	A major function of organisms is reproduction	Production of goods and services is the prime objective, not reproduction *per se*

An idealized industrial ecosystem with no materials input and no wastes is illustrated in Fig. 39.4. The energy requirements of such a system are rather high and the materials flows within the system itself

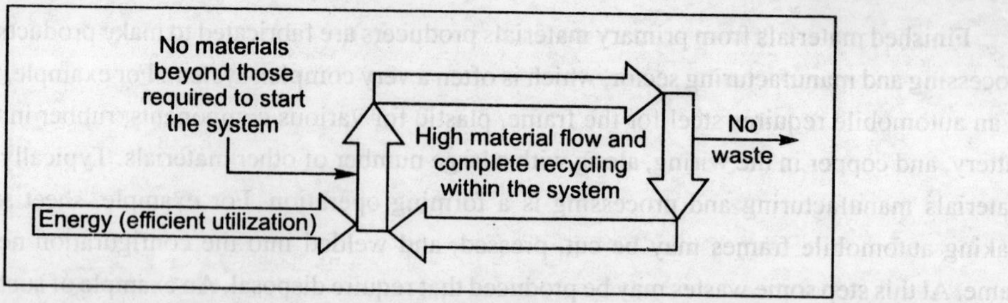

Fig. 39.4: Idealized industrial ecosystem in which no materials are required for input beyond those needed to start the system. Energy requirements are relatively high, and the material flow within the system is high and continuous

are quite high. Such a system is an idealized one that can never be realized in practice, but it serves as a useful goal around which more practical and achievable systems can be based.

SECTION VII

ENVIRONMENT, DEVELOPMENT TRADE AND ENVIRONMENTAL ECONOMICS

40 ENVIRONMENT AND DEVELOPMENT

40.1 INTRODUCTION

Disturbing visions of the human condition in the twentyfirst century are not difficult to imagine from the present situation. Population growth at the high end of the projected range, which corresponds to a continuation of current rates of decline of birth and death rates, would exacerbate already serious problems in many developing countries. The loss of agricultural land through erosion, salinisation and water-logging could undermine the livelihood of millions of cultivators and impede efforts to raise agricultural production. Thirty five per cent of the world's land area, including significant portions of the productive regions of the Northern Hemisphere, are at risk. In the industrialised world, disturbing environmental trends persist. From 1950 to 2000, worldwide emissions of CO_2 rose at an average rate of 6 per cent per year. There will be mean rise of surface temperatures of $8 \pm 1.5\,°C$ by the first quarter of this century. Such rise of global temperature may lead to a number of environmental changes.

40.2 CONCERN FOR DEVELOPMENT

With the rapid rise of global population, resources like forest wealth, water and minerals were rapidly depleted as well as polluted with times. Unequal mobility of resources in various section of people leads to poverty creation. The poverty and environmental connection is shown in Fig. 40.1.

Forest wealth is considered to be one of the major resources for human survival. The reasons for forest degradation and loss are manifold and thus needs appropriate analysis (Table 40.1).

Similarly, construction of dam for so-called efficient use of water resources for electricity generation and irrigation have multifacet short and long term environmental consequences. The details are given in Fig. 40.2.

Thus there is a great need for sustainable development. Such a developmental process has three bases, viz., scientific realities, consensus on ethical principles and considerations of long run self-interest.

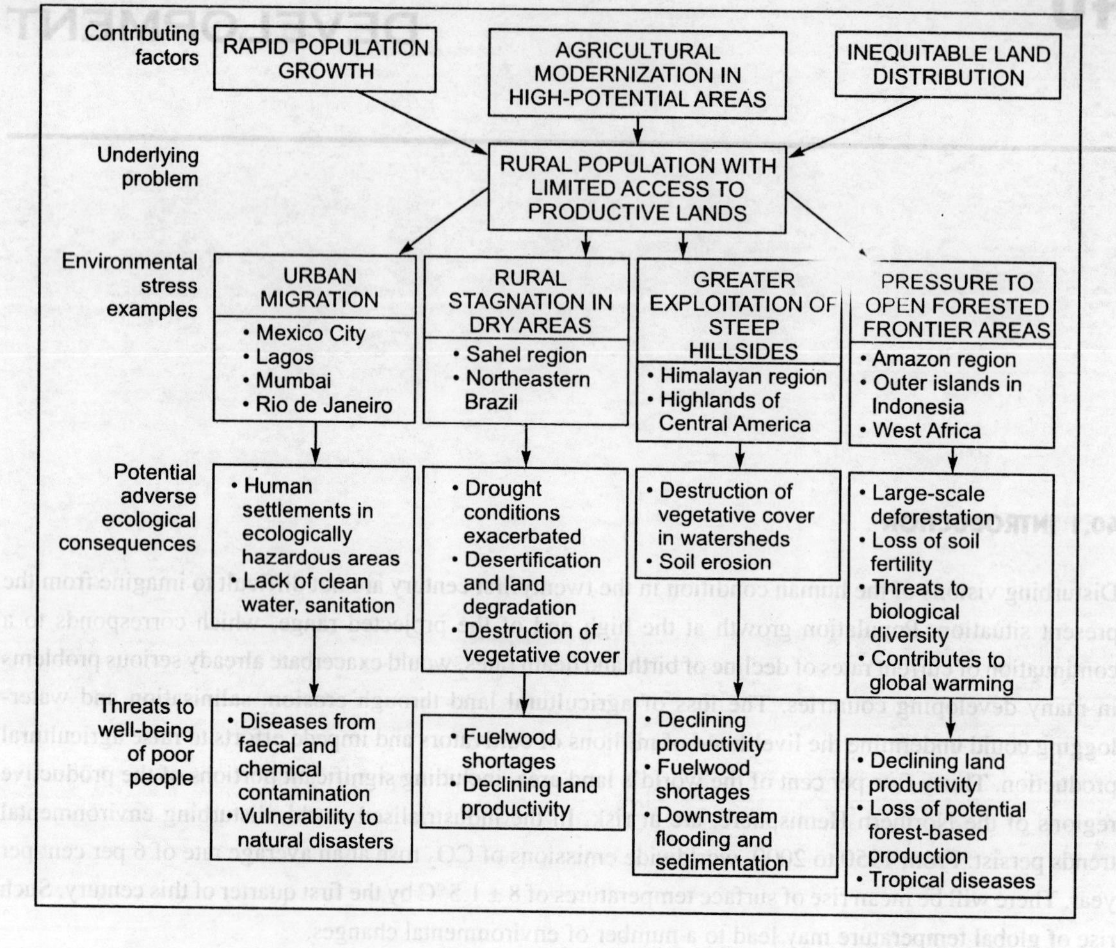

Fig. 40.1: The poverty and environment connection (*Source* : Leonard et al, 1989: 7(1))

Table 40.1: Reasons for Forest Degradation and Loss

Factor	Pressure exerted	Comment
Population increase	* demand for land * demand for fuelwood * demand for forest products * global or regional pollution	Not always a close link (Mather 1990 : 74), pressure exerted in a range of ways
Improved communications roads railways canals powerlines pipelines	* settlers get access * easier to remove products * pest organisms gain access * forest divided up; animal or plant movement hindered * poachers get access	Caufield (1982) suggested 25 per cent of Brazilian Amazon loss due to roads (in East Amazonia and Gabon, due to railways) Waves of exploitation most profitable species first, later the less valuable

Contd.

Table 40.1: Reasons for Forest Degradation and Loss

Factor	Pressure exerted	Comment
Large dams	* flooding of forest * influx of people * access roads * pollution/disturbance * downstream impacts on riverine forest/mangroves * industry may be attracted and cause pollution	Reservoir may flood several thousand km^2 Downstream impacts often missed in planning Richest forest in region may well be that flooded
Demand for pulp/ woodchips/veneer	* mangroves badly hit * plantations may be established	Likely to be mechanized, fast and ruthless Clearcutting likely. Indonesia is active
Smallholder agriculture govt.-assisted spontaneous shifting cultivation[b]	* clearance	Control may be a problem, settlers may relocate
	* gathering forest products	Can be a major cause of clearance (Steinlin, 1982)
	* hunting game * starting fires on land adjoining forest which 'nibbles away' remaining tree cover	Opened-up forest may be more prone to fires. Tenure problems lead landuser to neglect land or to be dislocated, and fail to sustain landuse
Large-scale ranching	* conversion to pasture	Major cause of clearing in Latin America and Botswana
Plantation crops	* conversion to crops	May be associated demand for fuelwood/ pollution associated with processing
	* agrochemical use may affect nearby forest/streams	Major cause in SE Asia and Caribbean
	* smallholders may get evicted for plantation land, move and clear forest, etc.	Pressures will increase if fuel/industrial crops expand
Fuelwood	* local use * commercial collector	More damaging in drier forests Access enables commercial gathering
	* smelting	Develops when forest already distrubed? Related to population increase?
Rattan and bamboo	* collection from wild damages trees and opens up forest	Little development of plantations yet Market for rattan expanding SE Asia problem
Failure to assist poor	* manifest through population increase; influx of settlers; poverty	May be 'driving force' but also forest is damaged where poverty is less of a problem
Warfare	* direct; defoliation/burning * demand for war materials * risk of nuclear war damage	Vietnam. In past 'scorched earth' in many countries Timber for warships denuded Mediterranean lands and Europe, later parts of USA and Caribbean

Contd.

Table 40.1: Reasons for Forest Degradation and Loss

Factor	Pressure exerted	Comment
Demand for timber	* clearcutting * selective logging * genetic erosion * logging roads open up forest	Major cause. From 1950 to 1982 a 16-fold increase in LDC tropical timber exports. Late 1980s logs and other forest products exports from LDCs was worth over US$7,500 million p/a (Wyatt-Smith, 1987)
Mining/Oil	* roads open up forests to settlers * may have associated air pollution * pipelines may open up access * mining may demand timber for props or fuel * mines may pollute and lead to railways	Problem on east slopes of Andes (Cleary, 1992) and SE Asia
Administrative causes	* tax laws encourage land clearing * desire to establish sovereignty * government encourages settlement/ clearing * poor conservation * poor logging control * failure to support alternatives to clearing	A cause in Nepal 1950s-60s (Little and Horrowitz, 1987: 373) Wide range of administrative causes, often important
Fires	* More a problem in drier forests * disturbed rainforest may be vulnerable during drought * natural/set by humans	Serious fires in disturbed rainforest in East Kalimantan in early 1980s
Global warming	* altered climate, sea-level, altered storm incidence, new pests/diseases?	Serious threat to conservation areas
Acid deposition	* dieback * soil changes	Europe, Central Europe, parts of USA, South America, China, India, spreading Some tropical soils vulnerable
Spread of pest organisms	* kill or compete * remove pollinator or seed disperser	Islands vulnerable, Florida has problems

Development strategies that attack poverty by emphasising employment growth; small scale, decentralised, quick-yielding investment programmes and wide access to land, credit, technology and other resources have resulted in faster overall growth in per capita income as well as more rapid elimination of poverty. A more equitable distribution of arable land and more attention to small holder agriculture typically result in higher agricultural productivity, as well as greater rural employment, wider availability of food crops and less pressure on marginal lands and forests.

Watershed protection programmes that introduce appropriate mixed farming systems to hilly areas—checking soil erosion by maintaining continuous vegetative cover and through better water control and improved animal husbandry—afford high economic returns in the form of improved productivity and income.

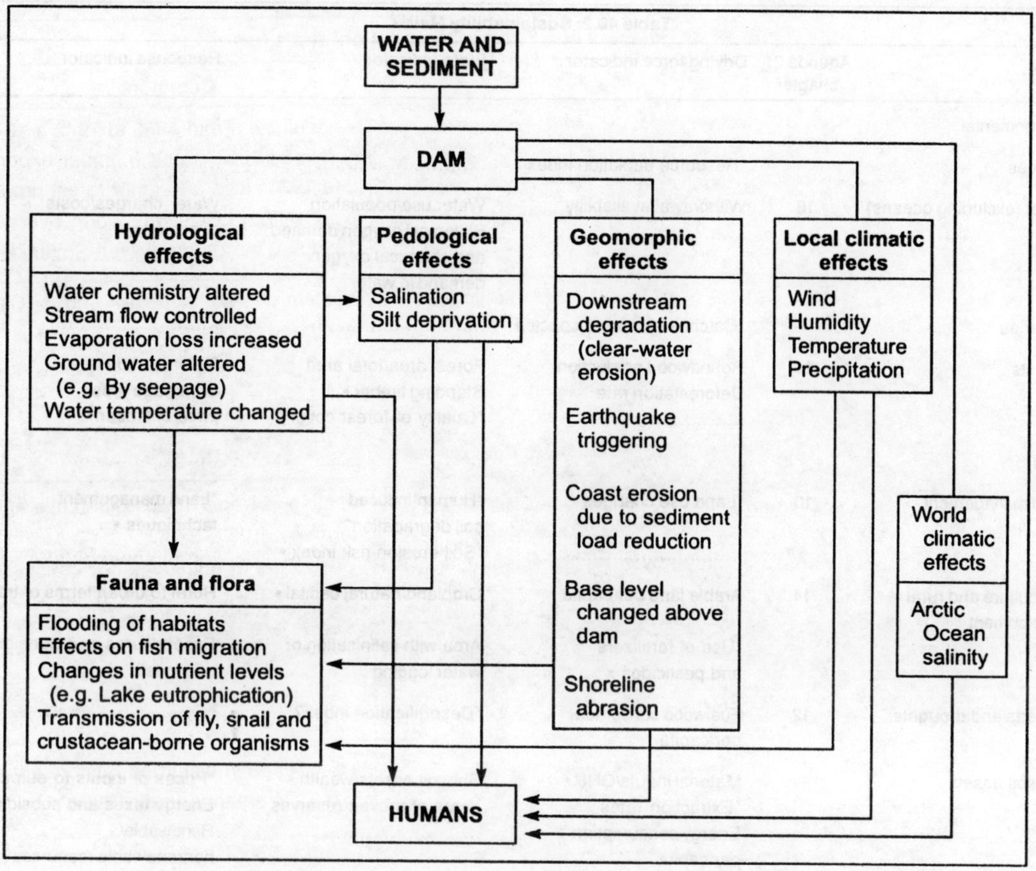

Fig. 40.2: The possible effects of dam construction on human life and various components of the environment. Redrawn after Goudie (1993b)

Energy conservation investments, despite the substantial gains that have already been made in many areas, remain the least-cost means of meeting demands for additional energy services, even before the environmental benefits are calculated. Identically, maintaining environmental quality through pollution abatement costs much less. In the industrial countries, estimates of the saving that can be achieved just by applying the severest strictures to sources that can reduce damage to the environment least expensively typically amount to 40-60% of the present costs of pollution control.

Thus economic factors do not block the attainment of a more desirable and sustainable future they favour it. The payoffs due to better management of resources are large. Many problems of resource deterioration, environmental decay, population growth and inadequate living standards persist or worsen not because they are impossible or difficult to solve but because they have not yet been seriously addressed. In the sustainability matrix, all these aspects has to be addressed in appropriate manner for better future human welfare (Table 40.2).

Table 40.2: Sustainability Matrix

Issue	Agenda 21 chapter	Driving force indicator	State indicator	Response indicator
Environmental				
Sources		*Resource depletion index x	....	....
Water (excluding oceans)	18	Withdrawal/availability	Water use/population Biological oxygen demand and chemical oxygen demand in water	Water charges/costs of provision ....
Fisheries	—	*Catches of marine species	....	
Forests	11	Roundwood production Deforestation rate	Forest area/total area Standing timber • *Quality of forest cover?	Reforestation rate Stumpage fees/ price of timber •
Land				....
Land management	10	*Land use changes	*Human-induced soil degradation *Soil erosion risk index •	*Land management techniques • ....
Agriculture and rural development	14	Arable land per capita *Use of fertilizers and pesticides x	Cropland/natural capital • Area with salinisation or water logging	Rural to urban terms of trade Expenditures on extension services?
Deserts and droughts	12	Fuelwood consumed per capita	*Desertification index?	
Subsoil assets	—	Material inputs/GNP • *Extraction rates Energy consumption per capita	Subsoil assets/wealth • Years of proven reserves	*Prices of inputs to outputs • Energy taxes and subsidies Renewable/ nonrenewable resources?
Sinks		*Pollution index x		
Solid waste	21	Industrial and municipal waste x	Waste disposal/ waste generation x	Expenditures on waste collection •
Toxics	19, 20, 22	*Generation of toxics •	*Area of contaminated land?	.... ?
Greenhouse gases	9	*Carbon dioxide and methane emissions	*Carbon dioxide and methane in atmosphere	Expenditures on abatement •
Stratospheric ozone	9	Production of CFCs	CFCs in atmosphere	Programmes to phase out ozone-depleting substances
Life support		*Ecosystem risk index x		
Biodiversity	15	*Rate of habital loss x *Rate of species extinction?	*Natural capital/wealth • *Number of threatened species	Protected area/total land area Protected areas/ sensitive areas •
Oceans and coastal zones	17			
Human health impact	6	*Index of environmental impact •		
Water quality and access	6	Household water use per capita	Access to safe water Fecal coliform *Lead in water?	Percentage of population with sanitary services

Contd.

Table 40.2: Sustainability Matrix

Issue	Agenda 21 chapter	Driving force indicator	State indicator	Response indicator
Air quality	6	*Pollution load •	*Ambient concentrations •	
Other	6		*Environment-related diseases?	
Social				
Demographies	5	Rate of population growth	Population density	Fertility rate
Health	6	Burden of disease (DALYs) •	Life expectancy	Health expenditures/GNP
		Calorie intake per capita	Infant mortality rate	....
Education	16	School enrolment	Adult literacy rate	Education expenditures per capita
		....	Educational Attainment •	....
Human settlements	7	Rate of urban population growth	Percentage of total population in urban areas	.... ?
		....	*Shelter index?	Housing expenditures/GNP
Housing	7	....	Marginal settlements?	
Infrastructure	7	Motor vehicles per capita	....	Infrastructure expenditures/ GNP
Economic				
Poverty	2, 3, 4	GNP/population growth rate	Headcount and poverty gap indices,	Labour-intensive growth?
		Distribution of wealth	Genuine saving/GNP •	
		*Production-consumption patterns?	Net primary school enrolment rate by poverty status and gender	
		Total fertility rate	Infant mortality rate	Expenditures for basic social servies/ total public expenditures
			Percentage of population using family planning	
Financial resources	33	Per capita wealth	Environmental protection expenditures per capita?	Investment/GNP
		....		Environmental taxes + Subsidies/GNP?
		....	....	New environmentally sustainable development funding?
Transfer of technology	34			
Productivity	—	NNP/wealth •	NNP/GNP •	Intermediate inputs/GNP •
		Unemployment rate	Manufacturing/GNP	Capital/Output ratio •
Institutional	8, 38-40	..................?	*Mandated environmental assessments?	*Ratification of international conventions?

Note: * = composite index; • = no internationally agreed compilation procedure exists but some work has been done; x = sources and methods suggested but no compilation done; ? = unsure how to do compilation; ..... = basic indicator research needed

40.3 SUSTAINABLE DEVELOPMENT—A SOCIAL MOVEMENT

Sustainable development can be viewed as a social movement—"a group of people with a common ideology who try together to achieve certain general goals". In an effort to encourage the creation of a broadly based social movement in support of sustainable development, UNCED was the first international, intergovernmental conference to provide full access to a wide range of nongovernmental organizations (NGOs) and to encourage an independent Earth Summit at a nearby venue. More than 1,400 NGOs and 8,000 journalists participated. One social movement launched from UNCED was the effort described above to create an Earth Charter, to ratify it, and to act upon its principles.

In 2002, 737 new NGOs and more than 8,046 representatives of major groups (business, farmers, indigenous peoples, local authorities, NGOs, the scientific and technological communities, trade unions, and women) attended the World Summit on Sustainable Development in Johannesburg. These groups organized themselves into approximately 40 geographical and issue-based caucuses.

But underlying this participation in the formal international sustainable development events are a host of social movements struggling to identify what sustainable development means in the context of specific places and peoples. One such movement is the effort of many communities, states, provinces, or regions to engage in community exercises to define a desirable sustainable future and the actions needed to attain it. Examples include Sustainable Seattle. Durban's Local Agenda 21 Programme, the Lancashire County Council Local Agenda 21 Strategy, and the Minnesota Sustainable Development Initiative.

Three related efforts are the sustainable livelihoods movement, the global solidarity movement, and the corporate responsibility movement. The movement for sustainable livelihoods consists of local initiative that seek to create opportunities for work and sustenance that offer sustainable and credible alternatives to current processes of development and modernization. Consisting primarily of initiative in developing countries, the movement has counterparts in the developed world, as seen, for example, in local efforts in the United States to mandate payment of a "living wage" rather than a minimum wage.

The global solidarity movement seeks to support poor people in developing countries in ways that go beyond the altruistic support for development funding. Their campaigns are expressed as antiglobalization or "globalization from below" in critical appraisals of major international institutions, in the movement for the cancellation of debt, and in critiques of developed-world policies—such as agricultural subsidies—that significantly impact developing countries and especially poor people.

The corporate responsibility movement has three dimensions: various campaigns by NGOs to change corporate environmental and social behavior; efforts by corporations to contribute to sustainable development goals and to reduce their negative environmental and social impacts; and international initiatives such as the UN Global Compact or the World Business Council for Sustainable Development that seek to harness the knowledge, energies, and activities of corporations to better serve nature and society. For instance, Global 100, the most sustainable corporations in the world were the top three corporations in the world—Toyota, selected for its leadership in introducing hybrid vehicles; Alcoa, for management of materials and energy efficiency; and British Petroleum for leadership in greenhouse gas emissions reduction, energy efficiency, renewables, and waste treatment and handling.

A related social movement focuses on excessive material consumption and its impacts on the environment and society and seeks to foster voluntary simplicity of one form or another. These advocates argue that beyond certain thresholds, ever-increasing consumption does not increase subjective levels of happiness, satisfaction, or health. Rather, it often has precisely the opposite effect. Thus, these efforts present a vision of "the good life" in which people work and consume less than is prevalent in today's consumer-driven affluent societies.

As with any social movement, sustainable development encounters opposition. The opponents of sustainable development attack from two very different perspectives: At one end of the spectrum are those that view sustainable development as a top-down attempt by the United Nations to dictate how the people of the world should live their lives—and thus as a threat to individual freedoms and property rights. At the other end are those who view sustainable development as capitulation that implies development as usual, driven by the interests of big business and multilateral institutions and that pays only lip service to social justice and the protection of nature.

Box 40.1: The Gaia Concept

The Gaia Thesis is the product of an unusual mind—the non-establishment scientist. Jim Lovelock is a geochemist and an inventor who can earn his keep by designing and patenting an imaginative piece or machinery. But his true love is 'real science'—the 'wondering about how the world works and the design of simple experiments to test the theories that thus come to mind' (Lovelock, 1992. p.15). Real science, like its companion artistic creativity, is best done quietly and inexpensively. He regrets the plethora of specialised environmental scientists, 'the brightly polished doctors of philosophy', who are most unlikely to achieve anything other than a secure and comfortable employment. Gaia is holist, not reductionist; integrative, not separatist; imaginative, not pedantic; open-ended, not closed.

Lovelock believes Gaia is a form of planetary medicine similar to the folk medicine of the pre-scientific era. Keeping a healthy planet with an equitable climate is as complicated as maintaining a healthy body free of disease. The Gaian idea is to design a mode of existence to let the Earth live in a healthy fashion—where possible, to let nature be the guide and the gyroscope of homeostasis. To ignore Gaia may mean that humans will have to live forever in a form of environmental conscription—looking after the planet as if it were on a kidney machine—functioning, but always circumscribed. Caring for Gaia means keeping a planetary healthy life—tending to the small things that accumulate across billions of thoughtful people into collective planetary health. This is Gaian anarchy. We behave appropriately because we just know others are doing so. Gaianism—the transpersonal lifestyle of a sustained world.

Schneider (1990) concludes that the Gaia notion is really five different ideas:

1. *Influential Gaia* simply states that biota have influence over the abiotic world, adjusting temperature and atmospheric chemistry.
2. *Co-evolutionary Gaia* believes that biotic and abiotic activities interact in a synergistic manner creating and sustaining the biosphere.
3. *Homeostatic Gaia* asserts that the co-evolutionary role is to maintain dynamic equilibrium of the biosphere, despite external or internal stresses.
4. *Teleological Gaia* holds that the atmosphere is kept in homeostasis not only by the biosphere but also for the biosphere.
5. *Optimising Gaia* concludes that the biosphere operates in such a manner as to create conditions that are optimal for biological survival.

These are listed in order of interconnectedness. The first two are relatively weak notions. The latter three suggest a purpose in life on Earth, namely to maintain itself in near optimal conditions. Take your pick.

Sources: Lovelock (1992); Schneider and Boston (Eds) (1991); Schneider (1990).

In UNCED global summit at Rio (1992), 22 principles were selected for sustainable development (Table 40.3), these criterion should be fulfilled for appropriate future development of mankinds.

Table 40.3: Selected Sustainable Development Principles from the Rio Declaration (UNCED, 1992)

Principle 1 :	Human beings are at the centre of concerns for sustainable development. They are entitled to a healthy and productive life in harmony with nature.
Principle 2 :	States have, in accordance with the Charter of the United Nations and the principles of international law, the sovereign right to exploit their own resources pursuant to their own environmental and development policies.
Principle 3 :	The right to development must be fulfilled so as to equitably meet development and environmental needs of present and future generations.
Principle 4 :	In order to achieve sustainable development,, environmental protection shall constitute an integral part of the development process and cannot be considered in isolation from it.
Principle 5 :	All States and all people shall cooperate in the essential task of eradicating poverty as an indispensable requirement for sustainable development.
Principle 8 :	To achieve sustainable development and a higher quality of life for all people States should reduce and eliminate unsustainable patterns of production and consumption and promote appropriate demographic policies.
Principle 9 :	States should cooperate to strengthen endogenous capacity-building for sustainable development by improving scientific understanding through exchanges of scientific and technological knowledge and by enhancing the development adaptions diffusion and transfer of technologies, including new and innovative technologies.
Principle 10 :	Environmental issues are best handled with the participation of all concerned citizens, at the relevant level. At the national level, each individual shall have appropriate access to information concerning the environment that is held by public authorities,, including information on hazardous materials and activities in their communities and the opportunity to participate in decision-making processes. States shall facilitate and encourage public awareness and participation by making information widely available.
Principle 15 :	In order to protect the environment, the precautionary approach shall be widely applied by States according to their capabilities. Where there are threats of serious or irreversible damage, lack of scientific certainty shall not be used as a reason for postponing cost-effective measures to prevent environmental degradation.
Principle 17 :	Environmental impact assessment, as a national instrument, shall be undertaken for proposed activities that are likely to have a significant adverse impact on the environment and are subject to a decision of a competent national authority.
Principle 20 :	Women have a vital role in environmental management and development. Their full participation is therefore essential to achieve sustainable development.
Principle 22 :	Indigenous people and their communities and other local communities have a vital role in environmental management and development because of their knowledge and traditional practices. States should recognise and duly support their identity, culture and interests and enable their effective participation in the achievement of sustainable development.

Source: The United Nations Conference on Environment and Development (1992). The Rio Declaration on Environment and Development, UNCED Secretariat, Geneva, Switzerland.

Box 40.2: Carbon Footprinting

Carbon footprinting is the technique of choice for substantiating claims about the impact of a product or an organisation on the climate. Simply a carbon footprint is the total emissions of carbon dioxide and its equivalents of other greenhouse gases for a defined system or activity. Every activities of our daily life can use carbon footprinting assessment to calculate the carbon impact of their operations (including goods and services) and as a basis for a future carbon reduction strategy.

Carbon footprint of a product

Product life cycle study is required to assess the potential of GHG emission

Raw material extraction ⟶ Processing ⟶ Manufacturing ⟶ Use ⟶ Disposal & Recycle

CARBON EMISSIONS FROM A CUP OF TEA (GRAMS)

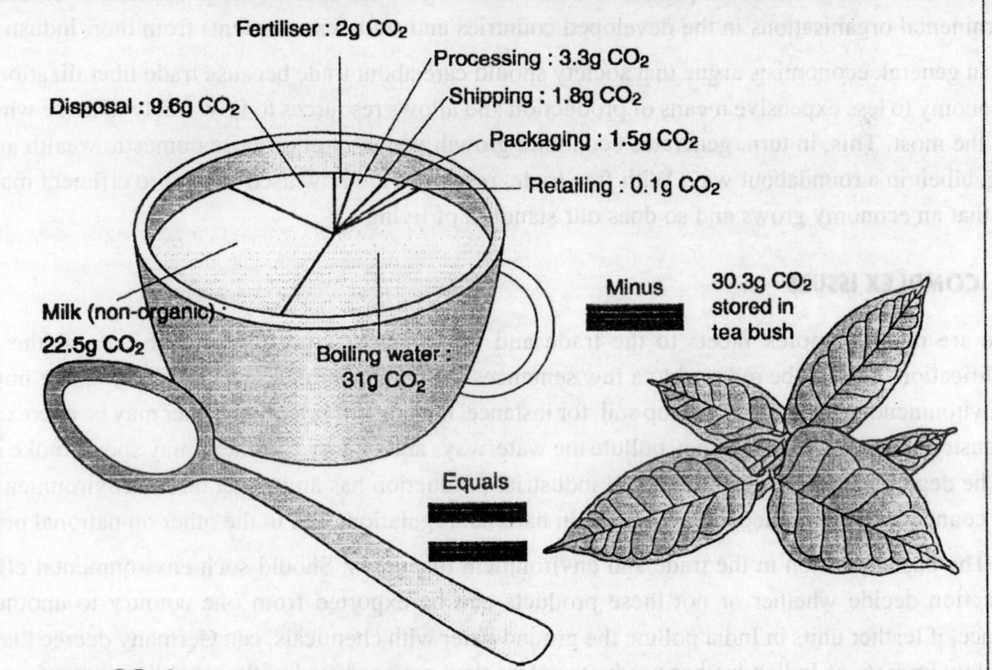

41.5g CO_2 in an average cuppa

Source: Details supplied by Nigel Melican, CEO Teacraft Ltd., www.teacraft.com

41 TRADE AND ENVIRONMENT

While the 1990s have seen a sharp divide between the rich and poor countries over what the world needs to do about climate change and biodiversity, in the years to come an even more acrimonious division will be in the arena of trade and environment.

The controversy is not a new one. But what is new is that pressures are now growing to frame substantive rules and regulations on this subject at the World Trade Organisation. And what sets this issue apart from the others is that the pressures are not coming so much from the governments as from non-governmental organisations in the developed countries and (to a lesser extent) from their industries.

In general, economists argue that society should care about trade because trade liberalization opens an economy to less expensive means of production and allows resources to flow freely to those who value them the most. This, in turn, generates economic growth which implies more domestic wealth and well being, albeit in a roundabout way. With free trade, resources are now used in a more efficient manner—such that an economy grows and so does our standard of living.

41.1 COMPLEX ISSUES

There are many complex facets to the trade and environment controversy, but even at the risk of simplification they can be reduced to a few sentences: When any product is made it has some impact on the environment. In agriculture, the top soil, for instance, may run off or ground water may be overexploited. In industry, untreated effluents may pollute the waterways and factory chimneys may spew smoke into the air. The degree to which agricultural and industrial production has an impact on the environment varies from country to country, depending in part on national regulations and in the other on national practices.

The basic question in the trade and environment dispute is: Should such environmental effects of production decide whether or not these products can be exported from one country to another? For instance, if leather units in India pollute the groundwater with chemicals, can Germany decree that it will not allow imports of Indian leather products unless they are produced without such environmental side-effects? Or, to take a real example, in 1996, citing its own laws, the US banned imports of shrimps from countries (including India) where turtles were killed during trawling for prawns.

Another aspect of the trade-environment dispute takes place amongst the rich countries itself. For example, the US argues that the European Union has been protecting its agriculture with a high level of subsidies, one effect of which is to encourage environmentally degrading forms of cultivation. The removal of subsidies, it is said, will not only facilitate free trade; it will also stop environmental degradation. Such issues are not discussed here, the focus being restricted to the North-South controversies.

The tussle is not only about some countries using their laws to influence environmental practices in others (as in the turtle-shrimp case). The concern of the developing countries is also that, under the guise of environmental considerations, "green protectionism" is being advanced and environment—which is a

Fig. 41.1: Banner over a pile of containers on an Australian freighter in Hong Kong in 1997 that was alleged to contain toxic waste being shipped to Asia

"non-trade" issue—is being brought into a trade body, the WTO. The argument is that the developed countries are talking about the environmental practices in the South merely to keep out imports from these countries. On its part, the North claims that unless environmental practices in individual countries are more integrated into world trade rules, there will be "a race to the bottom". That is, as countries try to out-do each other in the global market, each country will increasingly ignore the impact of production on the environment, ending up, ultimately, in a complete disregard for the environment.

41.2 DISPUTES AT WTO

The link between trade and the environment is not a new one. A number of global environmental agreements do cover this aspect. The Convention on International Trade in Endangered Species had, until recently, for example, banned trade in ivory on the ground that this was necessary to protect a dwindling population of elephants. But what is special now is that attempts are being made to bring it into the ambit of the WTO. NGOs in the North have taken it up in earnest because they feel that, unless the brakes are applied now, globalisation will end up destroying the environment the world over. They have focussed on the WTO because, in recent years, this body has been seen as an agent of globalisation. Partly in response to NGO pressure and partly because they lose no opportunity to bring in new trade issues, the governments of the North have readily articulated these concerns at the WTO.

To see how the trade-environment issue at the WTO has developed, a bit of recent history would help.

Towards the close of the far-reaching Uruguay Round of Multilateral Trade Negotiations (conducted under the WTO's predecessor—the GATT—General Agreement on Tariffs and Trade) some countries in the North did raise what are called the "environment-aspects of trade". Though the developing countries opposed any formal negotiations on the issue, a decision was taken in 1994 to establish the Committee on Trade and Environment (CTE) which would oversee a work (study) programme on the trade-environment links at the WTO.

In the seventeen years that the CTE has been in existence, the developing countries have been successful in pointing out that, while all countries are committed to sustainable development the best way, to achieve this in the South is to facilitate rapid economic growth, one requirement of which is opening markets to the developing countries. Second, the South has also managed to convincingly argue that it is the sovereign governments which have the right to establish national environment standards. And, third, member-countries have agreed at the CTE not to use environment protection as a guise to impose protectionist measures.

However, while the CTE continues its discussions (which have no implications for WTO rules on trade), the North has been getting impatient about the issue. So much so that a fair number of disputes since the WTO was set up in 1995 have involved environment aspects of trade, the most well-known of which is the shrimp-turtle case involving the US on one side and India, Malaysia, Pakistan, and Thailand on the other. In 1996 the US banned imports of shrimps from these countries, citing a 1989 US legislation which prohibited shrimps harvested in countries with technology that adversely affected sea turtles, five species of which were categorised in US law as endangered. (During trawling for shrimps, turtles can get caught in the nets. Fishing operators usually kill these turtles. A solution is believed to be the attachment of "turtles excluder devices" to the nets, which will allow the turtles to escape from the nets under water).

In 1998, a WTO Disputes Panel struck down the US ban as illegal. A WTO Appellate Body also held the US action as wrong. But what has opened a can of worms is that the Appellate Body ruled that the shrimp ban was illegal because the US did not follow the right procedures before taking action and that, under the WTO rules, countries "can and should" take action to protect the environment. In other words, if the US had followed the right procedures it would have been perfectly justified under WTO rules to ban shrimps imported from the four Asian countries—because they did not protect sea turtles in their waters.

At this point it would be useful to highlight the main issues involved in the trade-environment dispute and where it may be heading at the WTO.

The first issue is, of course, whether environment should figure at all at the WTO. The WTO, after all, is an organisation that is supposed to frame rules for global trade in goods and services and not on the environment. However, the North would like formal WTO rules to govern the "environment-aspects of trade". It would like, for instance, an amendment/new negotiations on this issue. The existing Articles of Agreement of the WTO do permit countries in exceptional circumstances to take action to protect the environment, but the rules are so tightly framed as to disallow such action by countries except in the most extreme of circumstances.

Article 20 of the WTO, which was cited by the Appellate Body in the shrimp-turtle case, is what the US, for instance, would like modified to permit WTO member-countries to take greater action on trade that has environmental implications. Quite naturally, the developing countries are strongly opposed to such a move.

The second and related issue is about what kinds of environment-related trade disputes should be covered by the WTO. All countries, including the developing ones, agree that there are certain environment issues over which the WTO has "jurisdiction". For example, if vegetables grown in New Zealand carry with them the residue of excessive pesticide application, an importing country—say the UK—is well within its rights to prohibit their entry. But such environmental/health issues are covered by a separate WTO agreement, called the Agreement on the Application of Sanitary and Phytosanitary Measures.

The third issue is: Where does the WTO stand vis-a-vis the provisions of the existing multilateral environment agreements (MEAs)? There are many kinds of global environment agreements that have already been negotiated under the aegis of other agencies and some of them do cover trade issues as well. To give a few examples, the Montreal Protocol on chlorofluorocarbons (CFCs) prohibits trade in banned CFCs; the Basel Convention on Toxic Wastes prohibits cross-country movement of wastes and, until recently, CITES (Convention on International Trade of Endangered Species) banned trade in ivory. (In theory, each of these bans contravene the provisions of WTO agreements.)

The view of the developing countries—India being a vocal country in this respect—is that since there are so many MEAs (Ministry of External Affairs) which also cover trade matters and, since none of them have been challenged at the WTO, it is best to leave the job of dealing with environment aspects of trade to these pacts and not bring the disputes to an organisation that is not equipped to deal with the issue. However, the support for this view does not run across all members of the WTO. The developed countries would like general rules to be administered by the WTO. For instance, there is no MEA—and there never will be one—on turtles caught by trawlers fishing for shrimps. The other issue is what if some WTO member-countries are not signatories to an MEA? The larger question here is that a few countries can negotiate an MEA on a subject amongst themselves and then threaten to take action against those non-signatories who are also members of the WTO.

These are just three of the very complex issues involved in the environment-trade debate. There are others like the use of 'eco-labeling' in some countries by voluntary agencies in order to certify imported products as having been produced in an environmentally benign fashion.

41.3 POLITICS OF DISPUTE

The issues are very much evolving since they are very new. But there is a lot of politics in them as well and the battlelines have already been drawn. The developing countries are very clear at the WTO that any move to make environment a formal part of the negotiating agenda will be opposed. (India is one of the leading proponents of the position of the developing countries.) The South sees this very much as 'a non-trade issue' that the North wants to bring to the negotiating table. The poor nations view the WTO moves on both environment and labour ('the social clause') in very similar terms. They see the initiative of the North on these issues as very much protectionist in nature. That, fearing the loss of markets, the governments of the North are raising these issues so as to keep certain products of the developing countries out of their markets.

There is little doubt that there are reasons other than a concern for the environment underlying the attempts of the developed countries to bring the issue to the WTO. And the developing countries are on a firmer ground when they argue that this non-trade issue has no place in the WTO. However, the scant regard that the governments of the developing countries have for the environment exposes them to the charge of hypocrisy when they say that they too are committed to environmental protection (Box 41.1, Box 41.2).

Box 41.1: ENVIRONMENTAL LABELLING

Increasingly important to many products, environmental labeling is of a somewhat different nature than depicted above. Environmental labeling is concerned with applying environmental standards to pre-determined product categories. While it sometimes only means that producers show the product's contents (and environmental impacts) on the pack, other ecolabels take into account the whole production process of the product in question, constantly researching whether or not the product can be produced in a more environmentally sound way. This includes ingredients as well as production processes and health and safety aspects (for both manufacturing employees and end-users). Environmental labeling is becoming increasingly important as a marketing tool in the European Union's (EU) member states.

Before starting the discussion concerning environmental labeling a clear distinction between the terms ecolabeling and hallmarking should be made. The Ecolabel is the name of the environmental label of the EU.

All national, branch or manufacturers' labels are called environmental hallmarks or environmental labels. In general, however, people refer to the term ecolabel for all environmental hallmarks.

The following sections deal with developments in the field of ecolabeling in the EU. As will become clear, ecolabels like environmental management systems can be used to distinguish between products and, therefore, they can be used as marketing instruments. The ability to use ecolabels as marketing instruments has led to the creation of a large number of such labels. Companies nowadays do not necessarily use the ecolabeling criteria in order to actually obtain the label (this is dependent on various factors, including costs, market share, public acceptance/recognition of the ecolabel etc.) but also to guide company-internal environmental management efforts, therefore using the ecolabeling guidelines to develop an environmental management system.

The ecolabels are grouped under a number of headlines:

1. **National ecolabels**: these are usually the national 'overall' ecolabels. The national ecolabeling institute has developed criteria for a larger number of different products. The focus is on purely environmental issues.
2. **Fair trade labels**: these labels mainly focus on the labour and living conditions of the workers/employees in developing countries who produce the product in question.

   Environmental issues play a (minor) role within the scope of these labels. Products for which fair trade labels exist are for example—fruits, vegetables and certain textiles.
3. **Labels for organic production**: these labels focus almost exclusively on agricultural produce. Fruits, vegetables, flowers and plants in the widest sense but also (in the near future) fishery products are eligible for such a label. Often, the ecolabeling institutes also require a link to be made with a more 'humane' philosophy towards treating our Earth.
4. **Product-specific labels**: These labels focus on environmental aspects of one product group only. An example is the labeling of forestry (products). In the scope of this chapter, these are labels which cannot easily be classified in groups 1 to 3 above.

Usually, only a small fraction of ecolabeled products originates in developing countries. The reason for this is that most ecolabels apply to products for which developing countries have only a small market share in world trade. Exceptions to this are organic agriculture, sustainable forestry and textiles.

Furthermore, it should be noted that raw materials originating in developing countries may be used by the (European) final manufacturer of a product. If he wants to obtain an ecolabel, he will require his suppliers to comply with the ecolabel criteria valid for their particular ingredient (not so much 'components' as this does not really play an important role for non-food products).

Besides the above-mentioned product ecolabels, there are also some (more or less) obligatory labels in several EU member states (e.g. Germany and Belgium) with regard to the labeling of product packaging. The labeling obligation in this case arises from the obligation for packers to partake in packaging waste collection systems under the EU's Directive on Packaging and Packaging Waste.

Box 41.2: DOHA DECLARATION, 1998

- The 4th Ministerial Conference in Doha agreed to launch negotiations on certain aspects of the trade and environment linkage.
- **Multilateral environmental agreement.** Ministers agreed to launch negotiations on the relationship between existing WTO rules and specific trade obligations set out in multilateral environmental agreements. The negotiations will address how WTO rules are to apply to WTO members that are parties to environmental agreements.
- **Information exchange.** Ministers agreed to negotiate procedures for regular information exchange between secretariats of multilateral environmental agreements and the WTO. The new information exchange procedures may expand the scope of existing cooperation.
- **Observer status.** Overall, the situation concerning the granting of observer status in the WTO to other international governmental organizations is currently blocked for political reasons. The negotiations will aim at developing criteria for observership in WTO.
- **Trade barriers on environmental goods and services.** Ministers also agreed to negotiations on the reductions or eliminations of tariff and non-tariff barriers to environmental goods and services.
- **Fisheries subsidies.** Ministers agreed to clarify and improve WTO rules that apply to fisheries subsidies. Ministers instructed the Trade and Environment Committee to pay particular attention to the following areas:
- **The effect of environmental measures on market access, especially for developing countries.**
- The purpose is not to get rid of the environmental requirements, but to strike an appropriate balance between trade and environmental objectives.
- **"Win-win-win" situations**
- WTO members recognize that removing trade restrictions and distortions can benefit both the multilateral trading system and the environment.
- **Intellectual property**
- WTO members acknowledge that the Trade-Related Aspects of Intellectual Property Rights (TRIPS) Agreement is essential for helping countries gain access to environmentally-sound technology and products and for the technology transfer in this area.
- Environmental labeling requirements. The Trade and Environment Committee is to look at the impact of ecolabeling on trade and examine whether existing WTO rules stand in the way of ecolabeling policies. Parallel discussions are to take place in the Technical Barriers to Trade (TBT) Committee.

41.4 CONCLUSIONS ON TRADE AND THE ENVIRONMENT

(a) Does free trade harm the environment ?

Because free trade has as its prime purpose the expansion of output and consumption, it will contribute to environmental degradation by expanding the demands that are made on natural resources and on land. However, restricting trade for this reason would be wholly wrong since environmental impacts can be controlled by vigorous and well-targeted environmental policy. There is a strong role for market-based incentives in this context. These aim to 'decouple' economic activity and environmental impact. In other words, the prime focus of policy must be on correcting the sources of environmental degradation in the country of export. Using trade restrictions is a last resort when all other means to correct the damage have failed. There is an exception for international agreements.

It is important to recognise, however, that free trade is not to be pursued at all costs. The environmental damage associated with expanded trade is a real cost that needs to be deducted from the (generally) large benefits of free trade. There is no justification in free trade arguments for ignoring environmental costs. Environmental costs change the conditions under which free trade maximises world human well-being.

(b) Does protection harm the environment ?

Many forms of protection can harm the environment, as exemplified by the Common Agricultural Policy. Less obviously, harm can be done to the sustainable development of countries supplying imports under special agreements, as with beef from Botswana to the EC. There is evidence to implicate the special agreement between the EC and Botswana in range degradation in Botswana due to the incentive to over-stock cattle.

(c) How does trade-related environmental degradation arise ?

Trade is linked to two forms of environmental damage. Damage can arise from the consumption of an imported product which gives rise to an environmental loss. Most usually such 'consumption externalities' are health-related, e.g. pesticides on imported fruit, treated products, etc.

The second form of damage arises because of the production of the goods in the exporting country. Tropical hardwood from unsustainable forestry, for example, imposes a cost on importing countries if the population of the importing country cares about deforestation (regardless of whether that population makes direct use of the forest). The same production process may also impose costs on the exporting country (e.g. loss of watershed functions due to soil erosion from deforestation). While these latter costs are the legitimate concern of the exporting country, importers may also feel concerned if, for example, minority groups suffer (e.g. forest-dwelling tribes).

The important point to note is that while consumption externalities have as their point of damage the importing country, many 'production externalities' will impose damage that is physically outside the importing countries. This raises the issue of the 'right' of importing countries to use trade restrictions to convey or impose their concern on countries who possess the resources in question.

A further category of damage arises from transboundary pollution, e.g. acid rain. Here the damage is physically located in the 'importing' country but caused by activity in the 'exporting' country. Trade restrictions as a weapon against the exporter are problematic since it may not be possible to target the product in question (e.g. electricity). GATT is generally consistent with environmentally motivated trade restrictions (under certain conditions—see below) on products, but is wholly inconsistent with generalised retaliation for an environmental harm done through some industrial process. Indeed, in terms of the distinction made above, GATT rules do not permit restrictions to be employed against processes or methods of production, only against products. In general, then, GATT does not acknowledge production externalities, only consumption externalities. This is confirmed by the recent ruling in the Mexican tuna fish case which found against the USA for prohibiting imports of Mexican tuna fish because of the loss of dolphins in the nets used. That is, GATT ruled against a production externality being relevant to trade restrictions. It also ruled against concern with extraterritorial physical impacts (the dolphin), although it appears that it is legitimate to employ an import tax against polluting goods if the same tax is applied domestically, regardless of the fact that the pollution from the imports may be physically outside the importing country.

If GATT was to be 'greened' to allow for production externalities it is clear that a massive potential would exist for countries to restrict trade on the grounds that they disapproved of the way other countries produce their products. From an economic perspective, however, such concerns, if legitimate, are no less real a cost than the damage done from consuming a polluted import.

(d) Are trade restrictions justified on environmental grounds?

GATT accepts that the trade weapon can be used where a product generates an environmental cost in the importing country. From an economic standpoint such a cost arises if the import contains 'more pollution' than the domestic product, i.e. fails to meet the domestic environmental standard. GATT, however, is wary of allowing domestic standards to be used as a weapon to raise the cost of imports. To date, there is considerable ambiguity over the use of standards as trade weapons. GATT recommends the international harmonisation of standards to ensure a 'level playing-field'. Economic analysis suggests that harmonised standards are not economically efficient since countries vary in their environmental preferences and in their resource endowments. But even if harmonisation is accepted as a general goal, it is not clear at what level harmonisation should occur. Left to a political process, any country failing to meet an agreed standard could be accused of providing an implicit subsidy to their exports. This might be challenged under the Subsidies Code, although no case of this has yet occurred. There is a need to clarify just what constitutes a subsidy. Moreover, subsidies are typically not consistent with the Polluter Pays Principle (PPP) of OECD.

Environmental taxes are, in principle, consistent with GATT. However, environmental taxes can be justified for processes as well as for products, whereas GATT is clear in opposing the extension to processes, although there are some debates about this. Moreover, environmental taxes ought, if they are to be efficient, to vary according to the pollution content of the import. But such a variable tax would be illicit under GATT since it must be non-discriminatory between 'like goods'. It can be argued that GATT needs clarification about the meaning of 'like goods' in the environmental context, i.e. that environmental characteristics should be used to distinguish goods. Once again, however, the prospect is one of virtually endless debate and dispute if this was to happen.

(e) GATT and international environmental agreements

There is an understandable concern that GATT's disallowance of trade restrictions—except in strict circumstances—could make compliance with international agreements more difficult. Such agreements might use trade restrictions as a weapon for compliance. Up to 1994, only 19 out of 170 international environment agreements make use of trade weapons to ensure compliance, but these include the Vienna Convention for the Protection of the Ozone Layer (1985); the Montreal Protocol on Substances that Deplete the Ozone Layer (1987); the Basel Convention on the Control of Transboundary Movements of Hazardous Wastes and Their Disposal (1989) and the Convention on International Trade in Endangered Species (CITES) (UNCTAD, 1991). It would seem worthwhile establishing clear guidelines in GATT which separate out this context of trade restriction from all others. The distinguishing features would seem to be:

1. That the restrictions are explicitly to secure compliance with , or the working of, an international agreement;
2. That they are necessary to the agreement in that other compliance incentives are not likely to be available;
3. That the trade restrictions would be collectively enforced by the signatory countries; and
4. That the 'same conditions prevail' clause of GATT differentiates non-signatory countries from parties to the agreement (i.e. non-signatories face different conditions).

In respect of (iv), however, a contracting party to GATT who is not a contracting party to an environmental agreement might challenge trade restrictions imposed through the environmental agreement on the grounds that GATT members may not vary the rights of third parties.

A possible mechanism for bringing international agreements into GATT is via Article XX(h) which provides an exception for measures undertaken:

In pursuance of obligations under any intergovernmental commodity agreement which conforms to criteria submitted to the Contracting Parties and not disapproved by them or which is itself so submitted and not disapproved.

It may, therefore, be the case that the extension of 'international commodity agreement' to 'international agreement' generally would provide the safeguard for the use of trade sanctions as weapons of compliance with such agreements. The option of using (XXh) in this way has been advocated by the German government and by the World Wide Fund for Nature. The Mexican tuna fish panel resolution also indicated that multilateral agreements were once of the options open for the USA to protect tuna outside its territory, suggesting that, if the Panel's rulings are adopted, that GATT is likely to approve of such agreement.

(f) Clarifying clause XX

Clause XX of GATT is the clause most in need of clarification under a new WTO from an environmental standpoint. The term 'environment' is not used, although it does appear in the Standards Code. Even a phrase like 'exhaustible resource' is unclear. Its original intent may have been to refer to non-renewable resources, but all resources are exhaustible in one sense or another. Clause XX(b) could be modified to include the phrase 'the environment', although the prior requirement (the 'chapeau') of the Article requires that all restrictions should be applied in a non-discriminatory way and must not be disguised protectionism.

(g) Overall features of GATT

The presumption in GATT is that the benefits of free trade exceed the environmental costs of free trade. But this is an empirical issue, and not an article of faith. GATT is on much stronger ground by insisting that trade restrictions are only justified when they are 'necessary' or 'related to' other objectives. Trade restrictions are generally unlikely to be the most efficient way of tackling environmental problems. But for compliance with international agreements they may well be the only way—given the absence of world government. As such, international agreements warrant special and explicit attention.

The remaining observations are that: (i) the lack of clarity in the wording of GATT and the associated Codes may work to the detriment of both free trade and the environment, and (ii) the strict focus on product damage omits most of the important environmental costs being generated by the production of goods for world trade. In general, however, reforming GATT to take account of production externalities would almost certainly do more harm than good.

41.5 THE WORLD TRADE ORGANISATION

In Jan. 1995 the GATT of old became the WTO. The WTO takes on the 'old' GATT provisions but a measure of the success of the environmental critique of the GATT is the preamble to the WTO which states that:

Their relations in the field of trade and economic endeavour should be conducted with a view to..... ensuring...... a large and steadily growing volume of real income and effective demand, while allowing for the optimal use of the world's resources in accordance with the objective of sustainable development; seeking both to protect and preserve the environment and to enhance the means for doing so in a manner consistent with their respective needs and concerns at different levels of economic development.

This suggests that the *presumption* of free trade is modified in the WTO: sustainable development and environmental protection issues matter more. What happens with the WTO and how it will interpret this revised general obligation remains to be seen.

42 ENVIRONMENTAL ECONOMICS

42.1 CONCEPT

An *economy* is a system of production, distribution and consumption of economic goods—any material items or services that satisfy people's needs. In an economy, individuals, businesses, and societies make *economic decisions* about what goods and services to produce, how to produce them, how much to produce, how to distribute them, and what to buy and sell.

"Environmental economics", a specialised branch of economics, has evolved in relatively recent years. It embraces the issue of pollution control, climate change, protection of natural environment, conservation of scarce resources, biodiversity and economic instruments—issues in the resolution of which market policy takes little or no part, but in which vast natural assets need to be allocated sensibly to the common good.

The kinds of capital used in an economy to produce material goods and services are called economic resources. They fall into three categories:

(a) *Earth Capital* or *Natural Resources*: (viz., plants, air, water and land, nutrients and minerals, biotic resources).

(b) *Manufactured Capital*: These include tools, machinery, equipments, etc.

(c) *Human Capital*: People's physical, intellectual and mental talents.

There are two major types of economic systems—centrally planned economy, and market-based or pure command economic system.

Virtually all economics seek *economic growth*, an increase in the capacity of the economy to provide goods and services for people's final use. Such growth is usually accomplished by maximising the flow of matter and energy resources (throughout) by means of population growth (more consumers), more consumption per person, or both. Economic growth is usually measured by the increase in a country's Gross National Product (GNP). Economists have never claimed that GNP indicators are good measures of environmental health and human welfare, but most governments and business leaders use them that way.

One of the chief tools cooperations and governments use in making economic decisions is *cost-benefit analysis*. This approach involves comparing the estimated short term and long term costs (losses) with estimated benefits (gains) for various alternative courses of action. Cost-benefit analysis can be useful guides for action and can indicate the cheapest way to go, but they can also be misused.

42.2 ENVIRONMENTAL GOODS

All environmental goods viz., climate protection, forest water, soil, nutrient cycle etc. are public goods, because its provision is both non-rival and non-excludable. These public goods may be local public good (i.e., abatement effort to clean up the river) or global public good (i.e., abatement effort to reduce carbon emissions feared to induce climate change. Public goods represent another form of market failure. Climate change is the classic example of a market failure: the flow of GHG emission accumulates into a global carbon stock that poses the risks to humanity. The GHGs remain in the atmosphere for hundreds of years. GHG concentration neglect long-term emissions; changes in any one years emissions have a trivial effect on current overall concentrations. Even significant reductions in emissions made today will not be evident in atmospheric concentrations for decades or more.

In case of a public good the individual takes the place of free rider. The individual may well use the good, but may not be willing to contribute to the cost of production/protection. Hence the problem of assessing the marginal valuation arises. To overcome such problems, the following approaches can be undertaken by the government agencies.

- The use of social welfare functions should be made and the environmental quality can be determined by maximising the function;
- Government agencies should work with CBOs to ensure that basic infrastructures and basic services are provided, maintained particularly in the area of health promotion and environmental protection.
- Waste reduction, reclamation or reuse should be encouraged by using participatory community based schemes.

42.3 ECONOMICS OF POLLUTION CONTROL

Most economists agree that controlling or preventing pollution and reducing resource waste require government intervention in the market place. Such governmental action can take the form of regulation, the use of market forces, or some combination of these approaches. *Regulation is a command and control approach* that involves passing and enforcing laws that set pollution standards, establish deadlines and penalties, regulate harmful activities, ban the release of toxic chemicals into the environment and require that certain irreplaceable or slowly replenished resources be protected from unsustainable use or from any use at all. *Market forces* can be used to improve environmental quality and reduce resource waste, mostly by using methods that encourage the internalisation of external costs. Another market approach is for the government to grant tradable pollution and resource use right. Identically, market-based approach is also to enact *green taxes* or *effluent fees* that would help internalise many of the harmful external costs of production and consumption. This method could include taxes on each unit of pollution discharged into the air or water, each unit of hazardous or nuclear waste produced, each unit of specified virgin resources used, each unit of pesticide use, each unit of fossil fuel used and each unit of solid waste produced; *charging user fees* is another market-based method.

Another market approach would require *business to post a pollution prevention bond when they open a mine, plant, incinerator or landfill*. After a set length of time, the deposit (with interest) would be returned minus actual or estimated environmental costs.

All these approaches described above have a number of advantages and disadvantages (Table 42.1):

Table 42.1: Economic solutions to pollution and resource waste

Solution	Internalises External costs	Innovation	International competitiveness	Administrative cost	Increase cost revised
1. Regulation	Partially	Can encourage	Decreased	High	No
2. Subsidies	No	Can encourage	Increased	Low	No
3. Withdrawing harmful subsidies	Yes	Can encourage	Decreased	Low	Yes
4. Tradable rights	Yes	Encourage	Decreased	Low	Yes
5. Green taxes	Yes	Encourage	Decreased	Low	Yes
6. User fees	Yes	Can encourage	Decreased	Low	Yes
7. Pollution prevention bonds	Yes	Encourage	Decreased	Low	No

Should not our goal be zero pollution? Ideally, Yes, in the real world, Not necessarily. *First*, natural processes can handle some of our wastes, so long as we don't destroy, degrade, or overload these processes. *Second*, as long as we continue to rely on pollution control, we can't offer zero pollution. After we have removed a certain proportion of the pollutants in air, water or soil, the clean-up cost per additional unit of pollutant rises sharply. Beyond a certain point, the clean up costs exceed the harmful costs of pollution.

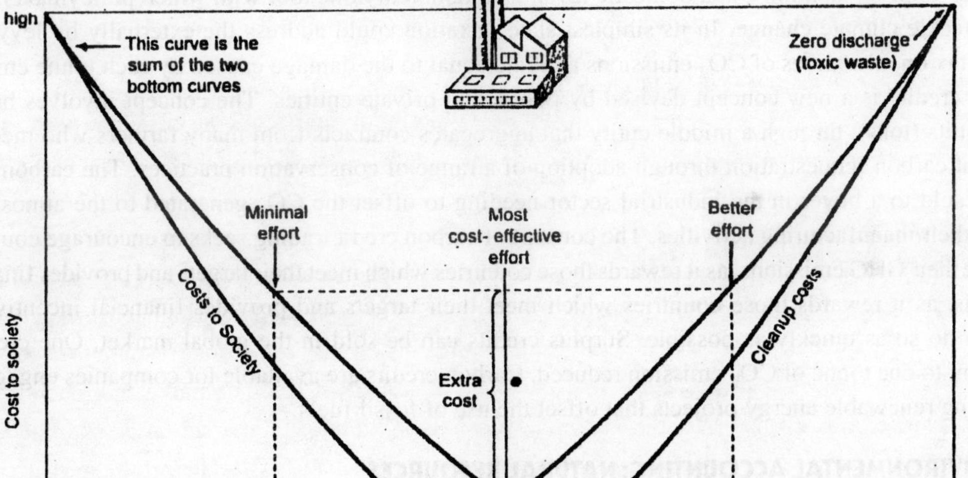

Fig. 42.1: Cost-effective analysis for finding the optimal level of pollution

To find the break-even point, economists plot two curves, a curve of the estimated economic costs of cleaning up pollution and a curve of the estimated social (external) costs of pollution. They then add the two curves together to get a third curve showing the total costs. The lowest point on this third curve is the break-even point or *optimal level of pollution* (Fig. 42.1). However, the environmentalists and business leaders often disagree in their estimates of the harmful costs of pollution.

Thus environmentalists believe that many environmental laws should be modified to place greater emphasis on pollution prevention, which avoids most of the regulatory problems and excessive costs of end of pipe pollution control.

Subsequently, in recent years, environmentalists justify their concerns by pointing out that the existing FTA, NAFTA and WTO trade agreements have lowered some environmental standards. Thus there is a need for making environmental concerns a key part of all trade agreements and of all loans made by international lending agreements.

One of the potent method of pollution management is the imposition of pollution takes. Many European countries imposes such kinds of direct pollution taxes under polluter pay principles. An alternative approach to pollution taxes as a way of achieving a target reduction in pollution is that of tradeable pollution permits (TPPs). The main idea behind TPPs is to allocate such rights, and make them tradable. This results in giving a market the right to pollute and consequently in the emergence of a market price for this right. Carbon dioxide trading is such type of TPP. Thus tradable pollution permits and pollution taxes are both capable, in theory, of achieving the least-cost solution to pollution control problems. "Carbon taxation", at a domestic level, is undoubtedly one tool with which policymakers will wish to tackle climate change. In its simplest state, taxation could address the externally by levying a uniform tax on all sources of CO_2 emissions at a rate equal to the damage caused by each tonne emitted. "Carbon credit" is a new concept devised by public and private entities. The concept involves buying carbon units (tons), through a middle entity that aggregates contracts from many farmers who meet the criteria of carbon sequestration through adoption of a range of conservation practices. The carbon units are then sold to a buyer in the industrial sector needing to offset the CO_2 generated to the atmosphere through their manufacturing activities. The concept of carbon credit trading seeks to encourage countries to reduce their GHG emissions, as it rewards those countries which meet their targets and provides financial emissions, as it rewards those countries which meet their targets and provides financial incentives to others to do so as quickly as possible. Surplus credits can be sold in the global market. One credit is equivalent to one tonne of CO_2 emission reduced. Carbon credits are available for companies engaged in developing renewable energy projects that offset the use of fossil fuels.

42.4 ENVIRONMENTAL ACCOUNTING: NATURAL RESOURCES

42.4.1 Introduction

Accounting of Natural Resource, i.e., incorporating environmental criteria in economic analysis, is gaining importance day by day in the entire world and the environmentalists are particularly concerned about the depletion and degradation of natural resources as also about the harmful impacts of economic activities. A number of groups have been constituted—both at national and international levels—to look into the problems arising out of unmindful handling of natural resources and a few countries are already preparing the Integrated Accounts taking into account the depletion, degradation and other harmful impacts of economic activities. Presently, the economic development of a country or a region is generally expressed in terms of the growth indicated by national income accounts which present systematic statistical statements reflecting the value of final goods and services produced in the economy during a particular year including change in stock. The System of National Accounts (SNA) takes into consideration only the economic activities within the specified production boundary and records only man-made assets as productive capital. The capital which is used up is written off as consumption of fixed capital against the value of production.

Present System of National Accounts does not keep proper account of natural assets—namely land, minerals, water and forest—in the sense that they are included in the SNA boundary insofar as they are under the effective control of any institutional unit. The cost of their use is not explicitly accounted for in production cost. It is wellknown that these natural resources are much more important than the man-made resources since, once used-up, these will not be available to future generations for further production. Day by day the need to have a set of accounts which provide information about the utilisation of these natural resources in a proper way is being increasingly felt and a number of countries have hence initiated to start with on an experimental basis, the exercises of preparing estimates which take into account the details of utilisation of the Natural Resources. The conventional system of national accounting suffers from two drawbacks: one, the neglect of new scarcities of natural resources which threaten the sustained productivity of the economy and two, the degradation of environmental quality, mainly from pollution and consequential effects on human health and welfare. In addition, some expenditures for maintaining environmental quality are accounted as increases in national income and product, despite the fact that such outlays could be considered as a maintenance cost to society, rather than social progress. In Natural Resource Accounting, attempt is made to convert these impacts in monetary terms. Thus the Natural Resource Accounting, when properly done, may present a completely different picture of the economy in the sense that an economy which might have been considered as a progressive economy on the basis of National Accounts may turn out to be an economy with zero or negative growth—if the depletion, degradation and other harmful impacts of economic activities are taken into account.

National Accounts Statistics (NAS) in India as on date are compiled following the guidelines set out in the United Nations System of National Accounts (SNA), 1968. Details of data sources used and methodology adopted are given in the Central Statisticai Organisation (CSO) publication—*National Accounts Statistics : Sources & Methods*, 1989. India is in the process of revising its national accounts series from the present base 1980-81 to 1990-91 and also simultaneously to adopt, to the extent feasible, the revised System of National Accounts, 1993 (1993 SNA), prepared under the auspices of the Inter-Secretariat Working Group on National Accounts comprising Commission of the European Communities-Eurostat, International Monetary Fund, Organisation for Economic Cooperation and Development, United Nations and World Bank. The revised System is a comprehensive, consistent and flexible set of macro-economic accounts intended to meet the needs of government and private sector analysts, policymakers and decision-takers. The 1993 SNA embodies the results of harmonising the SNA with other international statistical standards. It recommends preparation of sequence of accounts comprising Current and Accumulation Accounts.

42.4.2 Production Boundary of 1993 SNA

The production boundary of the System of National Accounts defines the range of economic activities recorded in the production accounts. Only the activities falling within the production boundary are considered to create output and value added. The production boundary not only defines what is production and what is not, but also determines indirectly what is income, consumption, investment, etc. For example, only the output generated within the production boundary could be consumed, invested, exported, etc. In general terms, production may be described as an activity in which an enterprise uses inputs to produce outputs. Thus purely natural process—for example the unmanaged growth of fish stocks in the international waters—is not production, whereas the activity of fish farming is production.

The production boundary of the 1993 SNA includes:

(i) The production of all individual or collective goods or services that are supplied to units or intended to be so supplied, including the production of goods and services used up in the process of producing such goods and services;

(ii) Own-account production of all goods that are retained by their producers for their own final consumption or gross capital formation;

(iii) Own-account production of housing services by owner-occupiers (ownership of dwellings) and of domestic and personal services produced by employing paid domestic staff.

The 1993 SNA includes the production of all goods within the production boundary. At the time the production takes place it may not even be known whether or not, in what proportions, the goods produced are destined for the market or for own use. With regards to own account production of goods by households, the 1993 SNA has removed the 1968 SNA limitations which excluded the production of goods not made from primary products, the processing of primary products by those who do not produce them and the production of other goods by households who do not sell any part of them on the market.

Towards an effort to compile the Environmental Economic Accounts, cost of use of all assets are to be accounted for which conventionally are not within the SNA asset boundary. In this context it is important to first describe the asset boundary of 1993 SNA.

42.4.3 Assets Boundary of 1993 SNA

Assets—as defined in the 1993 SNA—are entities that must be owned by some unit or unit(s) and from which economic benefits are derived by their owner(s) by holding or using them over a period of time. Every economic asset must function as a store of value that depends upon the amounts of the economic benefits that its owner can derive by holding or using it. With regard to the classification of assets, the 1993 SNA distinguishes at the first level of the classification between non-financial assets and financial assets/liabilities. Within non-financial assets, it distinguishes between produced and non-produced assets and within each of these between tangible and intangible assets. Produced assets are defined as non-financial assets that have come into existence as outputs from processes that fall within the production boundary of the system. This includes not only tangible fixed assets but also intangible fixed assets such as mineral exploration, computer software, entertainment, literary or artistic originals etc. Non-produced assets are defined as non-financial assets that have come into existence in ways other than the process of production. This includes tangible non-produced assets like land, sub-soil assets, etc. and intangible non-produced assets like patented entities, leases and other transferable contracts, purchased goodwill and other intangible non-produced assets.

Gross Fixed Capital Formation

Gross Fixed Capital Formation (GFCF) is measured by the total value of a producer's acquisition, less disposal, of fixed assets during the accounting period plus certain additions to the value of non-produced assets realised by the productive activity of institutional units. Fixed assets are those tangible or intangible assets produced as outputs from processes of production that are themselves used repeatedly or continuously in other processes of production for more than one year.

There is substantial diversity in the different types of gross fixed capital formation that may take place. The following main types may be distinguished:

(a) Acquisitions, less disposals, of new or existing tangible fixed assets, subdivided by types of asset into:

(i) Dwellings,
(ii) Other buildings and structures,
(iii) Machinery and equipment,
(iv) Cultivated assets—trees and livestock—that are used repeatedly or continuously to produce products such as fruit, rubber, milk etc.

(b) Acquisition, less disposal, of new and existing intangible fixed assets, sub-divided by type of assets into:

(i) Mineral exploration,
(ii) Computer software,
(iii) Entertainment, literary or artistic originals,
(iv) Other intangible fixed assets.

(c) Major improvements to tangible non-produced assets, including land,

(d) Costs associated with transfers of ownership of non-produced assets.

Various components of acquisitions and disposals of mixed assets, as referred to in categories (a) and (b) above, are:

(i) Value of fixed assets purchased,
(ii) Value of fixed assets acquired through barter,
(iii) Value of fixed assets received as capital transfers in kind,
(iv) Value of fixed assets retained by their producers for their own use, including the value of any fixed assets being produced on own account that are not yet completed or fully mature,

Less

(v) Value of existing fixed assets sold,
(vi) Value of existing fixed assets surrendered in barter,
(vii) Value of existing fixed assets surrendered as capital transfers in kind.

Acquisition of new assets covers not only complete assets but also any renovations, reconstruction or enlargements that significantly increase the productive capacity or extent the service life of an existing asset. In recognition of the newly increased capacity or newly extended service life, these improvements are treated as part of acquisitions of new assets even though physically they function as part of existing asset. Items (v), (vi) and (vii) above include disposals of assets that may cease to be used as fixed assets by their new owners: for example, vehicles sold by businesses to households for their personal use or assets that are scrapped or demolished by their new owners.

The 1993 SNA includes a third category of capital formation called "acquisitions less disposals of valuables". Valuables are defined as goods of considerable value that are not used primarily for purposes of production or consumption but are held as stores of value over time. The economic benefits that valuables bring are that their values are not expected to decline relatively to the general price level. They consist of precious metals and stones, jewellery, works of arts, etc.

Not all GFCF consists of acquisition less disposals of fixed assets. It is, therefore, convenient to describe the other components of gross fixed capital formation i.e., major improvements and costs of ownership transfers, as these may involve any type of assets. Classification of assets mentioned above may be summarised in Fig. 42.2.

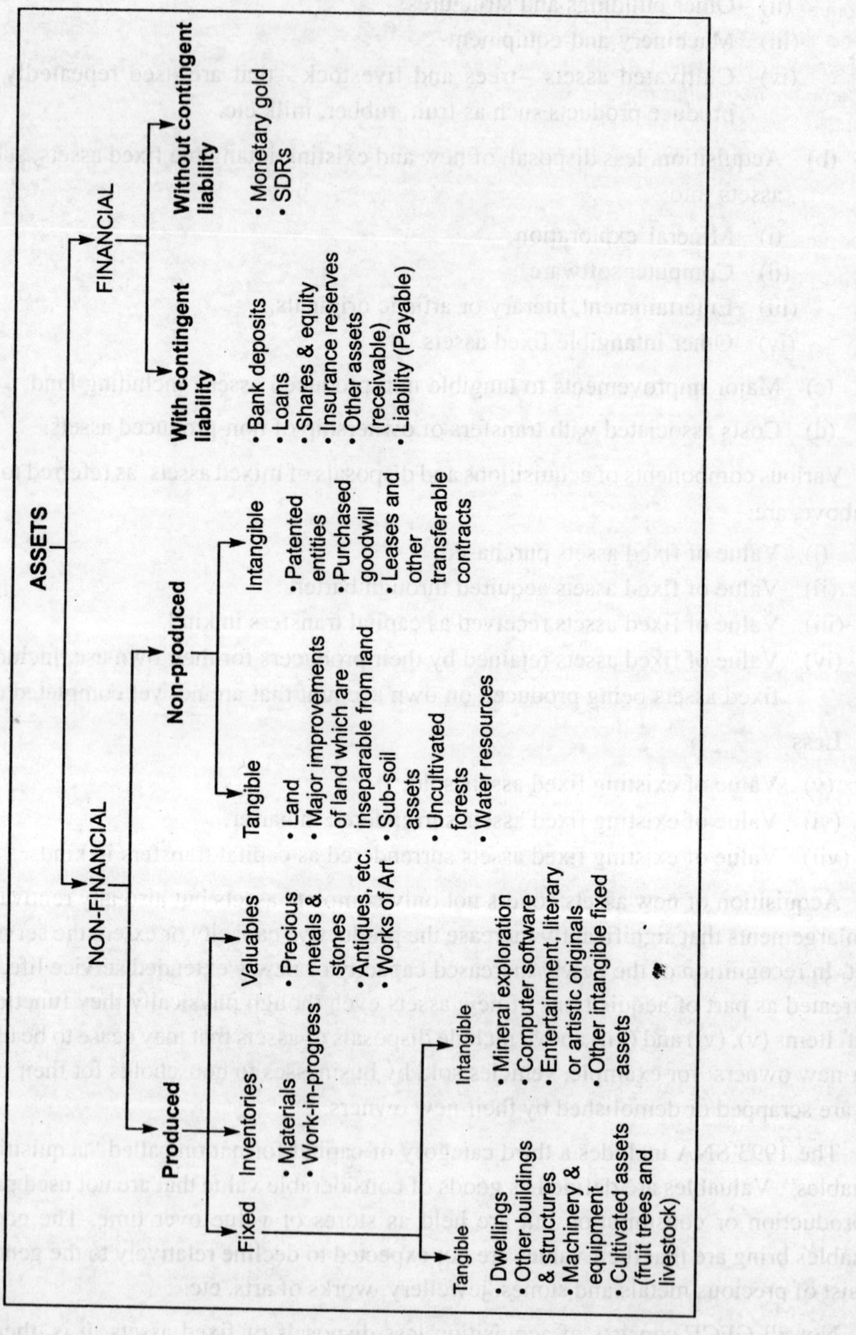

Fig. 42.2: Classification of assets

Box 42.1: Common Property Resource

Any resource or property, whether material or nonmaterial, used by every individual in the society for the benefit of one and all without having any characteristic feature of "private ownership" can be known as common property resource. Historically, air, water and other natural resources cannot be privately owned. This is because it is difficult and in most cases it is impossible to exclude people from the usage of these resources. No market exists for air where sellers with profit incentive can charge consumers for its use. Common ownership may be better characterized effectively as no ownership. All natural resources are publicly owned and one cannot claim ownership of these resources. At the same time, everyone intends to use or exploit these resources according to the whims and fancies of the person. The abuse of common property resource and the connected evils cannot be considered as a factor for converting the common property resource into a private property. This is neither possible nor feasible, due to the peculiar characteristic features of natural resources.

The primary features of common property resources are:

- Non-excludability
- Non-rivalry
- Scarcity of resources

Box 42.2: Tragedy of Commons

Right from olden days, the natural resources like air, water, etc., have been owned publicly and it is impossible to claim private ownership. We find that the resources are overexploited and misused as the individuals are provided with little cost or no cost. These individuals who exploit the resources reap full benefit but bear only the fraction of the damage cost of his pollution.

Thus, if there is absence of legal operations, every individual will ignore the drastic effects of his pollution on the environmental quality of the atmosphere. This will lead to a deterioration of the air quality and the society may end up worse off. Until and unless individuals maximize their self-interest, the cost of externality would not increase from the level of optimality. Unfortunately, the kind of perverse social situation is continued which has come to be known as the "tragedy of commons".

This, expression, "Tragedy of Commons" was propounded by professor Garret Hardin in describing environmental problems in 1968. Originally the commons referred to areas of pasture land in England that were provided free to all who wished to graze cattle. Despite the social virtue of the idea, the logic prevailed that the farmers who grazed most stood to benefit most out of the commons. And, if one refused to take benefit out of it, the other would.

Tangible non-produced assets are natural assets and these are available to the humankind as natural resources. Classification of natural resources has been dealt with in environmental literature (Judith Rees, 1985).

In 1993 SNA the environmental analysis has been dealt with as a functionally oriented satellite acc ount. The main central accounting framework has not been disturbed. A System of Environmental Economic Accounts (SEEA) has been developed by the UN and presented in the handbook *Integrated Environmental and Economic Accounting*.

There are three approaches to Environmental Accounting:

(i) Natural Resource Accounting in Physical Terms
(ii) Environmental Accounts in Monetary Terms
(iii) Welfare and Similar Approach as elaborated below.

Natural Resource Accounting in Physical Terms: This approach focuses on the physical asset balances i.e., Opening and Closing Stocks and changes therein of materials, energy and natural resources. It may also include changes in environmental quality of natural assets in terms of environmental (quality) indices. The SEEA, which shows the links between physical and monetary accounts, includes natural resource accounts as a module.

Environmental Accounts in Monetary Terms: This approach identifies the actual expenditure on environmental protection and deals with the treatment of environmental cost of natural and other assets caused by production activities in the calculation of net product. This approach is generally more limited in coverage than physical resource accounting. In monetary approach the GDP is adjusted for selected environmental costs like the cost of oil depletion, deforestation, depletion of fish stock and cost of soil erosion etc. The measurement of such costs is difficult, but considerable research work has been done in this direction, for example, Parikh (1991), Forsund and Strom (1989), Johansson (1989), Repetto (1989) and Tietenberg (1988).

Welfare and Similar Approach: This approach deals with the environmental effects borne by individuals and by producers other than the ones causing these effects. The effects on individuals and on others (that the producers) may be much larger than the cost caused and do not affect net product but rather net income through transfers of environmental services. The approach considers free environmental services provided by nature to producers and consumers and the subsequent damages borne by them. The environmental services provided free and the damages borne are considered as transfers by and to nature which increase or decrease environmentally adjusted net national income.

42.4.4 General Framework of Environmental Accounts and The SNA

Environmental analysis can be dealt with as a functionally oriented satellite account or in the context of a broadened framework by amending various concepts of the SNA to respond to the growing concern of incorporating environmental criteria in economic analysis. In the integrated economic and environmental satellite accounts, SNA aggregates are amended to treat natural resources as capital in the production of goods and services, to record the cost of using—i.e., depleting and degrading—those resources and to record the implicit transfers needed to account for the imputed cost and capital items.

Table 42.2: The basic structure of System of Environmental Economic Accounts (SEEA)

	Economic activities					Environment
	Production	Rest of World	Final consumption	Economic Assets		Other non-produced natural assets
				Produced assets	Non-produced assets	
	(1)	(2)	(3)	(4)	(5)	(6)
(i) Opening stock of assets				$KO_{p.ec}$ (Opening stock of produced assets)	$KO_{np.ec}$ (Opening stock of non-produced natural assets)	
(ii) Supply	P (Output)	M (Imports)				
(iii) Economic uses	C_I (Intermediate consumption)	X (Exports)	C (Final Consumption)	I_g (Gross capital formation)		
(iv) CFC	CFC			–CFC		
(v) NDP	NDP = $P - C_I - CFC$	X – M (Net exports)	C (Final Consumption)	$I = I_a - CFC$ (Net capital formation)		

Contd.

Table 42.2: The basic structure of System of Environmental Economic Accounts (SEEA)

	Economic activities					Environment
	Production	Rest of World	Final consumption	Economic Assets		Other non-produced natural assets
				Produced assets	Non-produced assets	
	(1)	(2)	(3)	(4)	(5)	(6)
(vi) Use of non-produced natural assets	Use_{np} (Use on non-produced natural assets in production)				$-Use_{np.ec}$ (Use on non-produced economic natural assets)	$-Use_{np.env}$ (Degradation of non-economic natural assets)
(vii) Other accumulation of non-produced natural assets					$I_{np.ec}$ (Change in stock of non-produced economic assets)	$-I_{np.env}$ (Reduction in natural assets other than economic assets)
(viii) Environmentally adjusted aggregates in monetary environmental accounting	EDP = C + ($A_{p.ec}$ + $A_{np.ec}$) − $A_{np.env}$ + (X−M)	X − M (Net exports)	C (Final Consumption)	$A_{p.ec}$ (Net accumulation of produced assets)	$A_{np.ec}$ (Net accumulation of non-produced economic assets)	$-A_{np.env}$ (Net accumulation of other non-produced natural assets)
(ix) Holding gains/losses				$Rev_{p.ec}$ (Holding gains/losses of produced assets)	$Rev_{np.ec}$ (Holding gains/losses of non-produced natural assets)	
(x) Other changes in volume of assets				$Vol_{p.ec}$ (Other changes in volume of produced assets)	$Vol_{np.ec}$ (Other changes in volume of non-produced natural assets)	
(xi) Closing stock of assets				$Kl_{p.ec}$ (Closing stock of produced assets)	$Kl_{np.ec}$ (Closing stock of non-produced natural assets)	

Table 42.2 presents the structure of SNA as amended to include the environmental accounts. The flow and stocks of the SNA are shown in the shaded portion and the non-shaded portion shows the additional elements that are needed to supplement the SNA concepts for the purpose of integrated environmental account. The use of the SEEA framework to present environmental accounting and the relationship to the SNA is convenient as the SEEA has been developed in immediate relationship to the SNA so that its concepts and classifications are linked to those of the SNA. The environmental cost and capital elements included in the SEEA can be interpreted in physical terms as well as in monetary terms. However, in view of the controversial issues surrounding valuation of the environmental cost and capital, much caution need to be exercised in the use of these elements in monetary terms and in the corresponding derivation of environmentally adjusted aggregates.

1993 *SNA Framework*: The columns of the Table 42.2 related to flows are a column (1) for production, covering output (P), intermediate consumption (C_i), consumption of fixed capital (CFC) and the net domestic product (NDP); a column (2) for the rest of the world, which includes exports (X) minus imports (M) and a column (3) for final consumption. The rows of the table referring to SNA flows are a row (ii) for supply, including outputs and imports; a row (iii) for economic uses, including elements of intermediate consumption, exports, final consumption and gross capital formation (I_g); a row (iv) for CFC and, finally, a row for (v) NDP which presents the elements that define the national accounts identity between NDP and the expenditure categories. The SNA column (4) for asset balances of produced assets includes the opening and closing stocks of produced assets ($KO_{p.ec}$ & $Kl_{p.ec}$) and the elements explaining the changes between the two i.e., net capital formation ($I - I_g - CFC$) holding gains/losses on produced assets ($Rev_{p.ec}$) and other changes in volume of produced assets ($Vol_{p.ec}$). The asset balances in the SNA area cover all economic assets and therefore, include the assets covered by column (5) for non-produced natural assets. The elements in this column, however, do not figure in the calculation of NDP as all changes in non-produced natural assets between opening and closing stocks ($KO_{np.ec}$ & $Kl_{np.ec}$) are explained in the SNA as holding gains/losses ($Rev_{np.ec}$) and other changes in volume of assets ($Vol_{np.ec}$).

SNA Framework extended to Environmental Accounts: The non-shaded areas of Table 42.2 include the additional elements that are needed to supplement the SNA concepts with data in physical terms on environmental cost and capital. There are two types of additional elements. The first group is included in an additional column (6) which records the effects of economic activities on non-produced natural assets such as air, water and virgin forests that are not included as economic assets in the SNA. The second group of elements is included in two additional rows (vi-vii) that include elements for the use of non-produced natural assets by depletion and degradation and for other accumulation of non-produced natural assets, which cover the transfer of natural assets to and between economic uses. Another row (viii) is included to derive an environmentally adjusted net domestic product (EDP) and other environmentally adjusted concepts. This row is only relevant in the case of monetary environmental accounting when additional SEEA elements are specified in value terms.

In row (vi) related to the use of non-produced natural assets, an additional element Use_{np} has been included in the column for production. This reflects the use of non-produced natural assets in production; it is sum of counterpart items in columns (5) and (6) representing respectively, the use of non-produced natural assets that are economic assets in the SNA sense, i.e. ($-Use_{np.ec}$) and the degradation of other natural assets that are not economic assets i.e., ($Use_{np.env}$). The use of non-produced economic assets ($-Use_{np.ec}$) includes the depletion of minerals, extraction of timber from forests that are economic assets and the effects on productivity of those forests and agricultural land of soil erosion, acid rain, etc. *The deteriorating effects of air pollution on buildings and structures and the effects of soil erosion on roads and other degrading effects on produced assets are not included as they are assumed to be reflected in the CFC.* The use of non-produced natural assets that are not economic assets ($Use_{np.env}$) covers the non-sustainable extraction of fishstock from oceans and rivers, extraction of firewood and lumber from tropical and other virgin forests of hunting animals living in the wild and also the effects of emission of residues on the quality of air, water, fishstock, wild forests and the effect of other economic activities (recreation, agriculture, transport, etc.) on ecosystems and species habitat.

Other accumulations in row (vii) records in physical or monetary terms the transfer of natural assets to economic uses as a change in the stock of non-produced economic assets ($I_{np.ec}$). The counterpart of this increase in economic asset is the reduction of natural assets other than economic assets ($-I_{np.env}$). Thus $I_{np.ec}$ would include the transfer of land to economic uses, the net additions to proven mineral reserves, the conservation of wild forests to timber tracts or agricultural land and the conversion of fishstocks to economic control. If the deterioration takes place at the same time that natural assets are incorporated as economic assets, the deterioration is not recorded in the other accumulation row, but is included as part of use of natural resources. If this deterioration takes place before the transfer, it is recorded as uses of natural resources that are economic assets ($-\text{Use}_{np.ec}$). As the elements of row (vi) for use (i.e. depletion or degradation) of non-produced natural assets, and row (vii) for other accumulation are included in the SNA in other volume changes, the content of other volume changes is reduced in the SEEA as compared with the SNA. *If the additional SEEA elements are valued in monetary terms, the incorporation of the use of non-produced natural assets (Use_{np}) as additional cost in the column for production results in an EDP, presented in row (viii), which is lower than NDP. The elements in row (vii) for other accumulations do not affect EDP.*

Corresponding to the monetary valuation of the additional SEEA elements, on the expenditure side a new concept called net accumulation is introduced in the SEEA to replace net capital formation in the SNA. It is presented in row (viii), separately for produced assets ($A_{p.ec}$), non-produced economic assets ($A_{np.ec}$), and other natural assets ($A_{np.env}$). For produced assets, it is same as net capital formation (i.e., $A_{p.ec} = I$), for non-produced economic assets, it reflects the net effects of negative depletion and degradation and positive additions of natural assets that are transferred to economic uses (i.e., $A_{np.ec} = -\text{Use}_{np.ec} + I_{np.ec}$). For natural assets other than economic assets it could be considered as the economic valuation of the impact of economic activities on the environment. It is, thus, sum of negative depletion and degradation effects ($-\text{Use}_{np.env}$) and negative effects of incorporating natural assets as economic assets (i.e., $-\text{Use}_{np.env} - I_{np.env}$).

If net accumulation replaces net capital formation when the additional SEEA elements are valued in monetary terms, the national accounts identity between NDP and final expenditures changes. In the SNA this identity, as reflected in row (v) of table 1 is:

$$\text{NDP} = C + I + (X - M) \tag{1}$$

and EDP is derived from column (1) as:

$$\text{EDP} = \text{NDP} - \text{Use}_{np} \tag{2}$$

If net capital accumulation in economic assets replaces net capital formation (I), the identity, as reflected in row (viii), becomes:

$$\text{EDP} = C + (A_{p\cdot ec} + A_{np\cdot ec}) - A_{np\cdot env} + (X - M) \tag{3}$$

Equation (3) can be proved to be equal to equation (1). We know from definition that:

$$\text{Use}_{np} = -\text{Use}_{np\cdot ec} - \text{Use}_{np\cdot env} \tag{4}$$

Also, $$A_{p\cdot ec} = I \tag{5}$$

and $$A_{np\cdot ec} = -\text{Use}_{np\cdot ec} + I_{np\cdot ec} \tag{6}$$

and $$-A_{np\cdot env} = -\text{Use}_{np\cdot env} - I_{np\cdot env} \tag{7}$$

Replacing LHS of equations (5), (6) and (7) in equation (3) we get:

$$\begin{aligned} \text{EDP} &= C + (I - \text{Use}_{np.ec} + I_{np.ec}) - \text{Use}_{np.env} - I_{np.env} + (X - M) \\ &= C + I + (X - M) - \text{Use}_{np.ec} - \text{Use}_{np.env} + I_{np.ec} - I_{np.env} \end{aligned}$$

Now as $I_{np.ec}$ is the counterpart of $-I_{np.env}$; therefore, last two terms cancel.

Thus,

$$\text{EDP} = C + I + (X - M) - \text{Use}_{np.ec} - \text{Use}_{np.env}$$

Now using (4), we get:

$$\begin{aligned} \text{EDP} &= C + I + (X - M) - \text{Use}_{np} \\ &= \text{NDP} - \text{Use}_{np} \end{aligned}$$

Box 42.3: Environmental Accounting

Environmental Accounting refers to accounting that includes environmental goods and services, which are valuable but not traded in markets. In other words, it is the process of recording and summarizing the value of environmental goods and services in monetary terms. The objective of the environmental accounting is to:

- Take the total stock of assets or reserves related to environmental issues and changes therein.
- Estimate of the total expenditure on protection or enhancement of environment.
- Identify that part of the gross domestic product, which reflects the cost necessary to compensate for the negative impact of economic growth, i.e., the defensive expenditure to protect environment.
- Assess the environmental costs and benefits: (1) the decrease (depletion) in natural resources due to their use in production and final demand, and (ii) the changes in environmental quality, resulting from pollution and other impacts of production and consumption and other natural events on one hand and the expenditure for environmental protection and enhancement of the environment, on the other.
- Elaborate and measure the indicators, relating to environmentally adjusted product and income, which are, disclosed EDP.

Practically, Environmental Accounting has two aspects:

(i) Accounting for the cost of impacts of organization's activities, products and services on natural environmental resources and society for which the organization is not financially responsible, known as Societal Costs.

(ii) Accounting for the costs, which the organization has incurred to prevent the degradation and deterioration in the quality and quantity of environmental resources from its activities, services and produces, known as Private Costs. It includes: Regulatory costs, Voluntary Costs, Upfront costs, Gray line Costs, Operational Costs, Break End Costs, Conventional/Usual Costs, Direct Costs, Hidden Costs, Overhead Costs, General and Administrative Cost, Research and Development, Exit Costs, Contingent Costs, Prospective Costs and Less Tangible Costs, and Environmental Liabilities.

Accountancy has traditionally measured financial reserves, but must now find ways to monitor the use and value of environmental resources both in terms of the raw materials consumed, the damage inflicted upon the environment by commercial activity and the waste disposal service which the environment provides.

The lack of accurate environmental measurement within the traditional accounting model means that environmental goods and services are often undervalued, which contributes to their misuse and explanation. With mounting natural calamities and wars, the need for incorporating environmental costs in the financial statements of both multinational corporations and domestic enterprises and to introduce the concept of environmental accountability and have environmental costs appear in Board of Director's reports, financial statement and the auditor's reports has been greatly felt.

Box 42.4: Green Marketing

Green marketing includes ecologically safer products, recyclable and biodegradable packaging, energy-efficient operations, and better pollution controls. Advances produced from green marketing include packaging made from recycled paper, phosphate-free detergents, refillable containers for cleaning products, and bottles using less plastic.

Marketing ecologically sound products to consumers at a premium price represents green marketing's ultimate challenge. Ottman (1993) uttered that consumers are demanding nontoxic products that are recyclable and environment friendly. Furthermore, she noted that there is a lucrative market for products that are responsive to the growing number of individuals with lifestyle concerns as environmentalism emerges as a social value. In view of this, companies are now in a new competence scene facing not only economic and social rules, but also ecological ones.

Unfortunately, a majority of people believe that green marketing refers solely to the promotion or advertising of products with environmental characteristics. Terms like 'Phosphate Free', 'Recyclable', 'Refillable', 'Ozone Friendly', 'Ammonia Free' and 'Environmentally Friendly' are some of the things consumers most often associate with green marketing. Thus green marketing incorporates a broad range of activities, including product modification, changes to the production process, packaging changes, as well as modifying advertising. Yet defining green marketing is not a simple task. Indeed the terminology used in this area has varied, it includes: Green Marketing, Environmental Marketing, and Ecological Marketing.

Marketing mainly deals with marketing-mix, i.e., product, price, place, promotion, packaging, etc. There are different concept which would give an idea to understand how to integrate environmental issues in marketing area of any business practice.

To integrate environmental issues in marketing, first we should put concentration on **'product design and materials selection'**. Often design criteria require large quantities of resources gathered through environmentally demanding material-extraction practices, such as strip-mining, clear-cutting of forests or energy dependence on fossil fuels.

42.4.5 Details of Environmental Amendments to SNA Framework, Concepts and Classifications

Assets Boundary and Classification: As per 1993 SNA the Natural Assets are included only if they provide economic benefits to the owner/controlling institution. This may mean explicit ownership and/or availability of a market price, such assets are referred to in SNA as economic assets. In SEEA, however, the asset boundary has been much widely defined and particularly include all the natural assets some of which may directly help in production while others may be affected by environmental impacts of economic activities. The asset boundary of SNA includes only economic assets as dealt with in columns (4) and (5) of Table 42.2. The SEEA asset boundary comprises all natural assets including those that are economic assets and are covered in columns (4) and (5) and other natural assets that are represented by column (6).

Table 42.3 shows that some assets categories in the SNA and SEEA are identical, some are closely related with different coverage, and some are not at all included in the SNA but are included only in the SEEA. The asset categories with an identical coverage are cultivated assets—such as orchards and plantations—and work in progress on cultivated assets including agriculture crops and livestock, which are treated as inventories in both the systems. Further, the sub-soil assets category is also identical in both the classifications. Air is one of the assets included in SEEA because it is affected by economic activities but is not included at all in the SNA. Land—although covered in both the systems—have widely different coverage in the systems in the sense that SNA land includes associated water surfaces such as lakes and rivers and, in some instances, also ground water. Since the SEEA deals with the effects of economic activities on the quality of water in rivers and lakes as also on the quality and quantity of available ground water it excludes these categories of water resource from land and includes them as sub-categories under the category of water. Thus the category of water in SEEA is much broader than the SNA category. The SNA category is mainly restricted to aquifers that are controlled by human activity while the SEEA land includes the ecosystem which are not explicitly included as economic assets in SNA. This classification of natural assets for environmental accounting is required in order to fully cover depletion, degradation and other accumulations, i.e., transfer of natural assets to economic activities. This classification will go whether physical units or monetary values are used for Environmental Accounting.

Table 42.3: Classification of Natural Assets in the SNA and the SEEA

SNA	SEEA
Produced Assets	
Fixed assets	
Cultivated assets	
Inventories	
Work-in-progress on cultivated assets	
Non-produced Assets	
Tangible Non-produced Assets	*Tangible Non-produced Assets*
Land including associated surface water	Land including ecosystems
Sub-soil assets	Sub-soil assets
Non-cultivated biological resources	Wild biota
Water resources	Water
	Air

Environmental Cost: SEEA identifies two types of environmental cost, viz., imputed cost for degradation and depletion and actual cost in the form of environmental protection expenses.

These costs are explained as:

The Cost of Use of Non-Produced Natural Assets: This cost has been introduced as an additional cost in SEEA and represents the cost of depletion and degradation in physical terms e.g., quantity of minerals extracted, quantity of timber cut, value of solid, liquid or gaseous wastes generated. Depletion is not restricted to economic assets only as in SNA and may cover depletion of uncontrolled forests for the purpose of firewood or the uncontrolled depletion of fish stocks. In these cases the assets are considered as non-economic in the sense that the exploitation of assets is uncontrolled. In such cases a depletion allowance is calculated in the SEEA except if the depletion in the case of wild biota is within the bounds of natural growth and the depleted assets would be naturally replaced. If expressed in monetary terms the depletion allowances could be considered as payments for the extraction of non-produced assets, such as minerals which are in stock either as economic assets or as Natural Assets. Depletion of non-produced natural assets is, therefore, considered in the SEEA as cost and, at the same time, treated as negative changes in inventories. Degradation effects in physical or monetary terms are included as additional cost elements in physical or monetary terms no matter whether these affect natural assets that are economic assets or other non-economic natural assets. The affects of degradation may include the deterioration of natural assets as a result of their use as sink for residues or as a consequence of other degrading activities. The deterioration of produced assets due to degradation is assumed to be reflected in CFC accounted for in the SNA. Environmental degradation by households is treated as a cost of internal household production activities and environmental degradation caused by fixed assets abandoned in the environment at the end of their useful economic life is allocated as cost of degradation to the producers owing those fixed assets. Both adjustments reduce EDP as compared to NDP in monetary environmental accounting. Another convention of SEEA relates to environmental degradation "imported from" or "exported to" neighbouring countries. Net of these is treated as a further environmental cost element.

Environmental Protection Expenses: Some actual expenses are incurred to avoid environmental degradation or to eliminate the effects after degradation takes place. Enterprises explicitly produce such services on a commercial basis. In many instances, however, the services are produced as ancillary activities. The ancillary activities are treated as separate establishment with an identifiable output and intermediate costs. The externalisation of ancillary environmental activities is done in order to measure and evaluate more comprehensively the efforts actually made to combat the environmental degradation and its effects. In general, the SEEA considers only those environmental protection expenses which are in immediate response to the effects caused by production. It excludes environmental protection expenditure responding to other environmental impacts. That is why household expenses for environmental protection are not considered—they respond to impacts borne by households and are dealt with in the SEEA in the same manner as in the SNA. An exception is made for environmental services produced as ancillary activities by government, which are included in the SNA as part of non-market output. They are identified in the SEEA as separate establishment.

Net Capital Accumulation: Capital formation in the SNA includes changes in the stock of produced assets used in production that are caused by economic decisions. Capital accumulation in the SEEA includes not only capital formation but also changes in non-produced natural assets that are explicitly used in production and are, thus, also related to economic decision. Such changes include the reductions in the capital stock as a result of depletion and degradation and also the incorporation of natural assets as economic assets and the transfer of natural assets between economic uses as a result of economic decisions taken in connection with production activities. Catastrophic losses due to events such as wars and natural disasters are not included in the concept of Net Capital Accumulation unless they are caused in connection with productive activities.

Valuation: Three alternative valuation methods are available, viz., market prices, valuation cost, and contingent valuation. In the SEEA the principles of market valuation are the same as in the SNA, they are applied to stocks of non-produced natural assets that are economic assets in the SNA, to changes therein as a result of uses of natural assets and also to the depletion of natural assets that are not economic assets such as the extraction of fish from the oceans and firewood from virgin forests. Market valuation, however, has been more extensively elaborated in the SEEA insofar as natural resources are concerned.

The only difference between the SNA and the SEEA is that the use of non-produced natural economic assets is recorded in the SNA as other volume changes, while in the SEEA cost of degradation and depletion is reflected in capital accumulation and deducted in the derivation of EDP. Opening and Closing stock of other non-produced natural assets which are non-economic assets in SNA are recorded in the SEEA in physical terms. Degradation of air and water is not valued in either system as these are uses of non-produced natural assets that are non-economic assets. However, while the SNA excludes, the SEEA includes in its market valuation all uses of natural assets that are not economic assets like the depletion of fish stock from the ocean, the extraction of firewood from virgin forests as the fish and firewood have a market value.

Two valuation methods are used for depletion—one is net rent approach which values the units extracted on the basis of the difference between output and all costs including labour cost and a normal profit margin incurred as a result of depletion. The other is user cost approach which values the units extracted on the basis of only a part of net rent, i.e., that part which, if reinvested, would generate a permanent income stream equal to the loss of income generation capacity through depletion. SEEA analyses the quality changes in assets, i.e., degradation, separately, deducting this value in principle from value added generated and capital accumulation.

The two other valuations used in the SEEA—viz., maintenance cost and contingency valuation—are only applied to uses and not to stocks of non-produced natural resources. Their use may, however, lead to a number of valuation inconsistencies within the SEEA.

However the effectiveness of economic valuation of environmental goods is, however, limited by the key methodological dilemmas:

- It is very hard to measure environmental benefits;
- Values are affected to pay (as well as inherent willingness to pay); and
- Estimates of willingness to pay depend on respondents knowledge about the good.

A variety of evaluative methods is available for the estimation of quasi-market values for non-market environmental goods (Fig. 42.3).

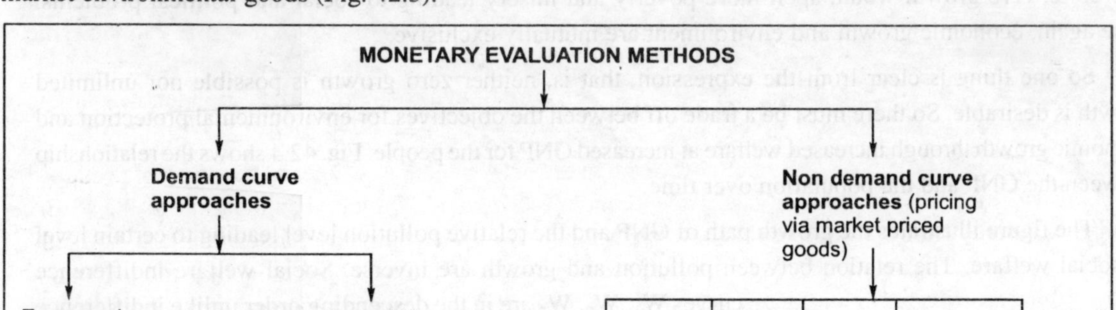

Fig. 42.3: Methods for estimating the value of environmental goods

42.4.6 Environmental Cost of Economic Growth

Environmental pollutants show a troublesome tendency in the world today. So it is time for the governance and the academics of the world to think in a systematic and scientific way the treatment and measure of nature in relation to man. After the Industrial Revolution begun in the !760s century the world is moving fast providing man better living condition which is termed as growth. But, at the same time, the world started degrading the level of purity in all forms. The system of growth hammers the natural resources through its innovative design technology to achieve higher rate of production in the quickest possible time. The process of growth and development disturbs the balance of the Ecosystem.

There is always a controversy between environmentalists and economists. The environmentalists argue that if all the eggs are broken and omeletts are made of it there will be no eggs for chicks, while economists argue that without breaking an egg an omelette the cannot be prepared.

Boulding argues that man must begin to see earth as a closed system, in contrast to the traditional conception that natural resources are boundless. Boulding's perspective is that the earth's resources should be viewed as essential and irreplaceable social capital, and the major purpose of economic activity should be to conceive this stock of capital of future generations. The overriding social goal would be the maintenance of the environmental resources and would entail acceptance of a lower rate of growth, even zero rate of current output towards the end.

The zero economic growth demanded by the ecologists has to be evaluated by two criteria. First, what kinds of opportunity costs are caused by zero growth? Then, can zero growth be a suitable measure to reduce pollution?

Though an increase in production may not seem to be an urgent need for rich nations, it is an absolute necessity for the less developed nations. The less developed nations have controlled birthrates and life expectancy is high. They have small but sick industrial bases with low per capita income. Therefore, zero growth would spell more poverty and misery leading to social and political problems. Here again, economic growth and environment are mutually exclusive.

So one thing is clear from the expression, that is, neither zero growth is possible nor unlimited growth is desirable. So there must be a trade off between the objectives for environmental protection and economic growth through increased welfare at increased GNP for the people. Fig. 42.4 shows the relationship between the GNP and the population over time.

The figure illustrates the growth path of GNP and the relative pollution level leading to certain level of social welfare. The relation between pollution and growth are inverse. Social welfare indifference curves W_1, W_2, W_3 are in the descending order unlike indifference curves and, hence, concave to the origin. Growth of GNP can be taken as consumption per capita of population and savings are held constant. The welfare curve indicates that it remains constant with different combinations of GNP and pollution. Both GNP and pollution move in the same direction.

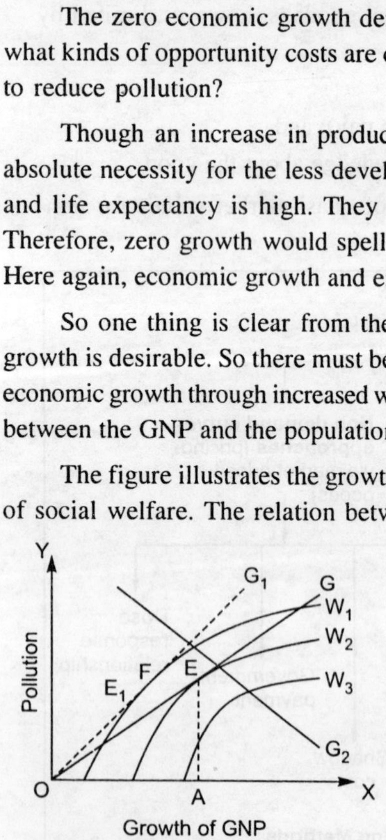

Fig. 42.4

In Fig. 42.4 the lower the level of the indifference curve, higher will be the level of welfare—while GNP level of pollution is lower. The growth path is shown by the OG line. The society is better off at point E at which the increase in GNP compensates the increase in pollution where OA level growth is reached. Now the economy has maximum level of welfare i.e., W_2 beyond OA if the economy grows the growth path shifts to OG, but the social welfare declines. This is shown by the movement from W_2 to W_1. W_1 tangents to OG at point E, where the pollution is greater than GNP growth causing reduction in welfare.

The problem of growth versus pollution finds solution by the attainment of sustainable development. The sustainable development implies that the development that meets the needs of the present generation without compromising the ability of future generations to meet their needs.

Thus sustainable development means that the growth of present times should not waste the resources, which are for the future generations. This is attained by reusing the by-products, finding substitutes for natural resources, and recycling the products.

Box 42.5: Certified Emission Reduction (CERs)

Certified Emissions Reductions (CERs) are a "certificate" just like a stock. A CER is given by the CDM Executive Board to projects in developing countries to certify they have reduced greenhouse gas emissions by one tonne of carbon dioxide per year. For example, if a project generates energy using wind power instead of burning coal, it can reduce 50 tonnes of carbon dioxide per year. There it can claim 50 CERs (as one CER is equivalent to one tonne of carbon dioxide reduced). The CDM is meant to work bottom up to proceed from individual proposals to approval by donor and recipient governments to the allocation of "Certified Emission Reduction" (CER) credits. In other words a CDM project has to follow a definite CDM Project Cycle. The CDM is supervised by an Executive Board which comprises 10 members, elected by the Conference of Parties (COP). All projects must result in a net **GHG reduction**. As defined by the CDM Executive Board, there are **15 sectorial scopes** under CDM.

15 Different Sectorial Scopes		
Energy Industries	Construction	Fugitive emission from Production & consumption of halocarbons & sulphur hexafluoride
Energy Distribution	Transport	Solvent use
Energy Demand	Mining/Mineral production	Waste handling & disposal
Manufacturing Industries	Metal Production	Afforestation & reforestation
Chemical Industries	Fugitive emission from fuels (Solid, Oil & Gas)	Agriculture

42.4.7 Comparison of Various Valuation Techniques and their Applicability

Thus a wide range of techniques available for attempting to value environmental effects—positive and negative—raises the questions as to whether some techniques are more advantageous than others. The details are given in Table 42.4.

Table 42.4: Valuation Methods of Environmental Components

1. **Productivity approach**—Here the product charges are imposed as inputs of economic activities, or on products and services themselves, as a means of indirectly controlling environmental impacts.
2. **Opportunity cost**—It is the cost of satisfying an objective, which is measured by the value of those resources that would have had another attractive alternative use.
3. **Preventive Expenditure/Replacement cost**—It is the value that people place on their environment. It can be estimated by observing how much is spent on preventing damage to it and on its enhancement.

Contd.

Valuation Methods of Environmental Components

4. **Surrogate market techniques**—Surrogate market techniques use actual market prices to value a quality of the environment that is not seperately marketed. These techniques aim to estimate the implicit price for environmental goods and services by observing the price paid for other related goods, which are marketed.

(a) **Hedonic price techniques**—This is a market approach technique, attempting to assess the value attributed by buyers to the environmental attribute of a dwelling.

(b) **Wage differential approach**—It is very similar to Hedonic techniques but it uses wage differentials in place of property prices as a guide to the values that people put on environmental qualities.

(c) **Travel cost approach**—It is applied for assessing the economic value of natural areas or recreational areas where no price is directly charged. In this case, the willingness to pay for an environmental amenity is assumed to be measured in the costs incurred by people when travelling to chosen locations.

5. **Survey based approaches**—Other valuation methods have been developed in an attempt to value environmental benefits in the absence of data on market prices. These methods rely on asking consumers directly about their willingness to pay for a benefit, or to be compensated for a loss.

6. **Contingent valuation**—It is a method of establishing a monetary value for a good or service by asking people what they are prepared to pay for it. The good or service in this context would be an environmental amenity, clearly of value but falling beyond the range of the market economy and, therefore, unpriced.

The method takes several forms:

(a) **Bidding games**—The simplest approach here is direct questioning, individuals being asked how much they are prepared to pay to enjoy a particular environmental amenity or bring about some environmental improvement.

(b) **Convergent direct quationing**—The questing may be in a number of persons about willing new to pay and then variations could be converge at an equilibrium.

(c) **Trade off games**—Each individual is asked to rank various combination of two objects—a sum of money and some environmental attribute.

(d) **Moneyless choice method**—Instread of using money as one of the objects, only specified commodities are used in combination with environmental attributes.

(e) **Priority valuation method**—In this method each individual is given a hypothetical sum of money to spend on conventional goods and environmental attributes at assumed prices.

7. **The Delphi Techniques**—This technique relies in the informed opinions of a group of experts, being commonly used in many complex planning and decision making situations. For economic evaluation the technique depends heavily on the knowledge and background of the experts and on the skill with which the technique is used.

The overall compatibilities of various techniques are:

	Techniques	Reliability	Data requirement	Application potential
1.	Productivity approach	High	Medium	High
2.	Opportunity cost	High	Medium	High
3.	Preventive expenditure	High	Medium	High
4.	Hedonic pricing	High	High	Medium
5.	Wage differential	Medium	High	Medium
6.	Travel cost	Medium	Medium	High
7.	Contingnet valuation	High	Medium	Low
8.	Delphi techniques	Medium	Low	Medium

Attempts are also made to categorize valuation techniques in terms of their suitability for particular application. The details are:

Methods	Health impact		Aesthetic impact	Ecosystem impact	Recreational impact	Production impact/ material damages
	Illness	Mortality				
1. Productivity approach	Yes	—	Yes	Yes	Yes	Yes
2. Opportunity cost	Yes	—	Yes	Yes		Yes
3. Preventive expenditure and replacement cost	Yes	Yes	Yes	Yes	—	Yes
4. Hedonic pricing		Yes	Yes	—	Yes	—
5. Wage differential	Yes	Yes	Yes	—	—	May be
6. Travel cost	—	—	Yes	—	Yes	—
7. Contingent valuation	Yes	Yes	Yes	Yes	Yes	Yes
8. Delphi techniques	May be	—	May be	Yes	May be	May be

However, an input impetus to valuation studies has been given by the rapid development of the regulating impact statement. So far, all the valuation methods we have outlined have focussed on eliciting preference for changes in environmental goods. The ecosystem service method takes a slightly different focus, and shares much with the production function methods. One ecological function of tropical forests is in the recharge of groundwater aquifers. Multitiered forests help this process through the collection of rainfall, the avoidance of soil erosion and by keeping soil permeable via the forest's root system. Kaiser and Raumasset (2002) study the contribution of Kóolam forest in Hawaii to the recharge of the Pearl Harbour aquifer. They reported estimated ground water recharge value of the forest as \$1.4 – \$2.6 billion.

42.4.8 Options for Achieving Global Climate Stabilization and Carbon Trading

In order to address the global externality of CO_2 emissions, carbon needs to be valued to represent the price (social cost) or penalty that would be paid by those who generate the CO_2 emissions. A range of approaches to value carbon is provided by economic theory. These can be categorised under two broad systems: a price could be imposed through a carbon tax, or could arise from a trading system (see Fig. 42.5).

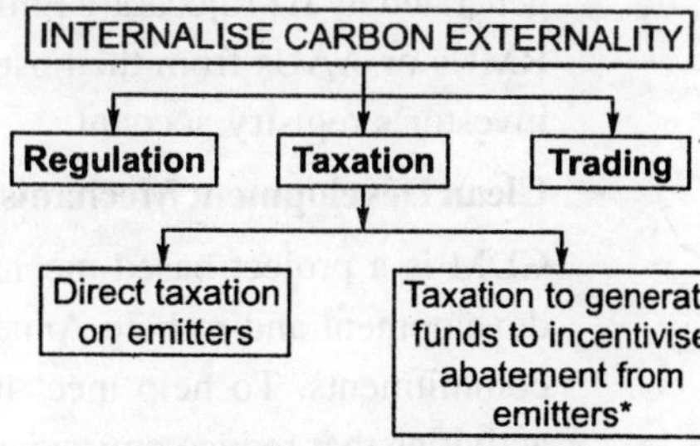

*This could include indirect taxation such as tariffs on the trade in timber and non-timber products

Fig. 42.5: Options for addressing the externality of CO_2 emissions from forests

Kyoto Protocol is a voluntary treaty signed (in Japan 1997) by 141 countries, including the European Union, Japan and Canada for reducing GHG emission by 5.2% below 1990 levels by 2012. However, the US, which accounts for one-third of the total GHG emission, is yet to sign this treaty. The preliminary phase of the Kyoto Protocol is to start in 2007 while the second phase starts from 2008. The penalty for noncompliance in the first phase is E40 per tonne of carbon dioxide (CO_2) equivalent. In the second phase, the penalty will be hiked to E100 per tonne of CO_2.

Credits traded under the three Kyoto mechanisms

The reduction targets set for each developed country are expressed in terms of a commitment. Participating countries are allocated emissions rights up to the quantity of emissions allowed by their commitment. These allowances can then be traded with other Annex I countries. Additionally, Annex I countries can obtain certified credits from emissions reductions via Joint Implementation or the CDM (Fig. 42.6).

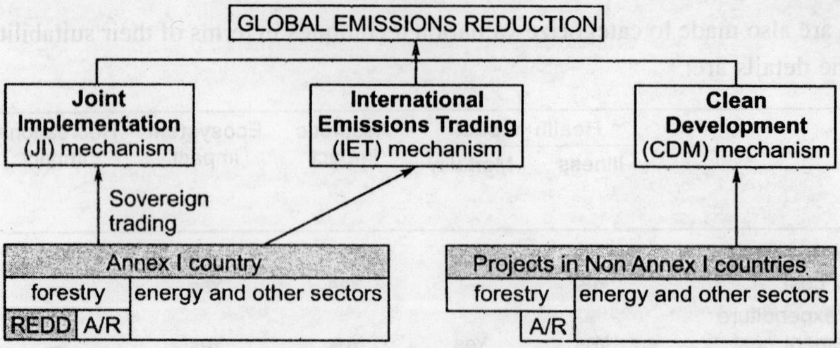

Fig. 42.6: The three Kyoto mechanisms

- **International Emissions Trading Mechanism (IET)**

  Annex I countries that exceed their emission targets are permitted to buy credits from other Annex I countries that have excess emissions rights to meet their commitment. There are two types of credits that can be traded in this way. Assigned Amount Units (AAUs) are tradable sovereign allowances to emit CO_2e. AAUs are issued by parties to the Kyoto Protocol to Annex I countries, calculated by reference to their base year emissions and their emission reduction commitment, or cap. Removal Units (RMUs) are tradable sovereign sequestration credits generated from land use, land use change and forestry (LULUCF), i.e., reforestation and afforestation activities, calculated against national baselines.

- **Joint Implementation Mechanism (JI)**

  JI is a project-based mechanism for Annex I countries. Under JI, an Annex I country may invest in an emission-reducing project or a project that enhances removals by sinks in another Annex I country. The credits earned from a JI project can be counted towards meeting the Kyoto target of the investing country. The private sector can be authorised to take part in JI projects. Credits generated by JI projects are Emissions Reduction Units (ERUs). ERUs are created by deducting RMUs or AAUs from the host country's registry account so that they can be credited to the investor's registry account.

- **Clean Development Mechanism (CDM)**

  CDM is a project-based mechanism to assist non-Annex I countries to achieve sustainable development and to help Annex I countries achieve their emission reduction and limitation commitments. To help meet its Kyoto target, an Annex I country may implement project activities that reduce emissions in non-Annex I countries. The tradable credits generated by non-forestry CDM are called Certified Emissions Reductions (CERs).

  CDM crediting can also take place for forestry activities. However, this only applies to afforestation and reforestation activities. Reduced emissions from deforestation and degradation (REDD) activities are excluded from crediting.

  Afforestation and reforestation projects under the CDM generate temporary and long-term CERs (tCERs and lCERs, respectively) that are equivalent to one metric tonne of CO_2e—but only for a finite period. A tCER expires at the end of the commitment period following the one in which it is issued, while an lCER expires at the end of the crediting period of the afforestation or reforestation project activity under the CDM for which it was issued.

SECTION VIII

ENVIRONMENTAL LAWS, RULES AND REGULATIONS

43 ENVIRONMENTAL LAWS

43.1 INTRODUCTION

The environment which surrounds us, consisting of living world (plants, animals and microbes) and also nonliving objects such as water, soil, air, light, temperature and so on. This environment is being damaged and disturbed (i.e., in peril). Entire human societal existence and also existence whole living world is now severely threatened. To avert such kind of disaster, a scientific approach not merely emotional and impulsive action is needed.

The continued degradation of environmental components is the result of the modern living technological advancement, industrialization and urbanisation. Contemporary scientific and technological revolution has significantly transformed the relationship between man-nature. That is why it is often stated that "Man is nature's best promise and worst enemy".

Nature has provide adequate resources to feed and beer the burden of the requirement of the mankind on planet earth. Once these resources are over exploited due to the undue pressure of human activities, it disturbs the equilibrium relationship between the man and the nature necessary for human existence. Consequently it gives rise to the problem of environmental pollution. It is really depressing to record that the water we drink, the air we breathe all are polluted. We are now facing the serious ecological crisis like acid rain, global warming, air and water pollution, extinction of species loss of biodiversity, ozone depletion, loss of crop productivity and induce changes in overall ecological balance. Thus the need of the day is therefore to bring greater awareness for harmonising population dynamics and socioeconomic development.

Country like India, where most of the population fall below the poverty line, in the absence of capital resources. The poor are dependant on the environment for their livelihood. Purity of air and water becomes irrelevant in the absence of good food and shelter. For them, preservation and protection of environment may not be a prime concern resulting in large quantities of waste being generated and released in air, dumping in soil or discharged in water bodies. This situation of air, water and soil pollution reached to a level of alarming conditions.

It is a serious concern that today for economic reason, man has forgotten the environmental and ecological significance of natural vegetation mainly forests and grasslands and has destroyed the forests so rapidly and alarmingly that the forest area at global, regional and local levels have so markedly decreased that several serious environmental problems have plagued the modern human society. From ecological point of view, at least 33% of land man of a country should be under rich forest cover but this general rule of environmental significance has been flouted in many of the countries.

43.2 ENVIRONMENTAL LEGISLATIONS IN INDIA

The basic principles of environmental protection are contained in the 130 year old Indian Penal Code (IPC) 1860, drafted by Lord Macaulay in Sections 268, 269, 277 and 278, incorporating penal provisions for offences relating to public nuisance, negligent acts likely to spread infection or diseases dangerous to life, fouling of water of public spring or reservoir and making atmosphere noxious to health. Sections 133 and 143 of code of Criminal Procedure (CrPC) also contained provisions to empower the District Magistrate or Sub-divisional Magistrate or any other executive Magistrate empowered by the state government or the District Magistrate in this behalf order any person restraining from committing a public nuisance as defined in the IPC or any special or local law or to remove the abovesaid public nuisance or to repeat or continue the same. The question of interpretation of these sections came before the Supreme Court of India in the Ratlam Municipality case.

Bengal Smoke Nuisance Act, 1905 (amended in 1973), is perhaps one of the oldest Environment Protection Act in India. This is followed by The Factories Act, 1948. Similarly, Mines and Minerals (Regulation and Development) Act, 1957, is another important environmental legislation formulated in post-Independent period. But most of the environmental legislations were instituted after 1970, when the Goverment of India committed for environmental protection and prevention of pollution as national policy. A series of Acts thereafter passed the Parliament. Subsequently, to facilitate the appropriate operations, rules and national policy on every aspect were also drafted by the high-power committees. Major environmental legislations (Acts & rules etc.) are summarised in Table 43.1:

Table 43.1: Environmental Legislations in India

Aspects	Acts	Rules
1. Water	1. The Water (Prevention & Control of Pollution) Act, 1974 (amended up to 1988)	1. The Water (Prevention and Control of Pollution) Rules, 1975
	2. The Water (Prevention & Control of Pollution) Cess Act, 1977 (as amended up to 1991)	2. The Water (Prevention and Control of Pollution) Cess Rules, 1978
2. Air	3. The Air (Prevention & Control of Pollution) Act, 1981 (amended up to 1987)	3. The Air (Prevention and Control of Pollution) Rules, (1982 & 1983)
3. Environment protection	4. The Environment (Protection) Act, 1986	4. Environment (Protection) Rules, 1992 & 1993: Environment Statement
		5. Environment (Protection) Rules, 1993: Environment Standard
		6. Environment (Protection) Rules, 1994: Environmental Clearance
4. Solid wastes (a) Hazardous wastes		7. The Hazardous Waste (Management and Handling) Rules, 1989
		8. Manufacture, Storage and Import of Hazardous Chemical Rules, 1989
(b) Municipal solid waste		9. Municipal Solid Waste (Handling and Management) Rules 2000
5. Microorganisms		10. Hazardous Microorganisms and Genetically Modified Organisms (Manufacture, Use, Import, Export & Storage) Rules, 1999 [Previously Manufacture, Use, Import, Export and Storage of Hazardous Microorganisms, Genetically Engineered Organisms or Cells Rules, 1989]
6. Biomedical wastes		11. Biomedical Waste Management and Handling Rules, 1998
7. Coastal zone protection		12. Coastal Zone Regulations, 1991
8. Factory	5. The Factories Act, 1948	

Cont.......

Table 43.1: Environmental Legislations in India

Aspects	Acts	Rules
9. Public liability	6. The Public Liability Insurance Act, 1991	13. The Public Liability Insurance Rules, 1991.
10. Environment Tribunal	7. National Environment Tribunal Act, 1995	
11. Forest	8. The Forest (Conservation) Act, 1980 (amended up to 1992).	14. Forest (Conservation) Rules, 1981, 1992
12. Wildlife	9. The Wildlife (Protection) Act, 1972 (amended 1991, 2002)	15. Wildlife (Transection and Taxidermy) Rules, 1973
		16. Wildlife (Stock Declearation) Central Rules, 1973
		17. Wildlife (Protection and Licensing) Rules, 1983
13. Biodiversity	10. The Biological Diversity Act, 2002	18. Biodiversity Rules, 1999
14. Mines & Minerals	11. Mines and Minerals (Regulation and Development) Act, 1957	19. Mineral Conservation and Development Rules, 1988
15. Monuments & Archaeological sites	12. Ancient Monuments and Archaeological Sites and Remains Act, 1958	
16. Energy	13. Energy Conservation Act, 2001	

In addition there are large number of laws in India, which are directly or indirectly dealing with subject of environment.

43.3. ENVIRONMENTAL ETHICS IN ANCIENT INDIA

India has a long tradition of protecting and worshipping the nature. The land, trees and even animals are placed on a high pedestal since vedic times. According to ancient Indian mythology the God prevails the whole universe. This point of view has helped Indian to maintain a constant link with nature. In ancient days the altitude towards all the objects of the universe recognised the divine being present in the **Panchabhuta** (the five elements) viz., air, water, fire, earth and space. Our ancestors have perceived divine existence in trees, medicines, river, lakes, mountains and living beings. Due to such extraordinary understanding about nature and ecological equilibrium, India was the most advanced country of the world in ancient times.

In ancient Indian way of life, before setting foot on the ground after waking in the morning, people addressed earth as mother and requested mother earths forgiveness for stepping on it. Similarly, rivers and forests are often worshiped as holy place. Religion is commonly pertains to the spiritual domain while the science and technology are viewed as systems pertaining to the material domain. By and large, religious traditions and beliefs in India played a vital role in protection of environment.

The Hindu religion and law enshrined respect for nature and environmental harmony and conservation. It instructed man to show reference for the presence of divinity in nature. Trees and plants had been regarded as indispensable in the life of human being and considered as the abode of various Gods and Goddesses (viz., Lotus symbolised for Laxmi (Goddess of wealth), Bawyan for Brahma (creator of universe) and Palasa for Brahma (maintenance of universe), etc. Similarly various animals are also associated with God and Goddesses viz., lion for Durga, elephant for Indra, eagle for Vishnu, owl for Laxmi, crocodile for Ganga, bull/snake for Shiva, rat for Ganesh and so on. Thus killing of plants and animals was against basic tenet of Hindu ways of life.

Bishnoie is a distinct section of Hindus who treated environment protection as a part of their religious duty. **Buddhism** came into existence as religion by the end of vedic period. The fundamental tenets of the Buddhism are truth, nonviolence, respect and love for living organism including trees and plants. **Jainism** or Jain religion maintaining harmony between man and nature, man and man but lays down even greater stress on minimum destruction of living and nonliving resources for man's satisfaction of material needs. According to religious belief of **Sikhs/Sikhism**, there is one way, one love, one beauty shines through air, water and fire.

In the **Islamic perspective**, the environmental and ecological activities are seen as a subject matter of a wide human effort to aster in a just and peaceful society. According to Islam, the God has created the earth and implanted therein mountains/hills and lakes, delta, valleys and slopes, rivers and springs, forests and river basins where diverse animals, birds, and diverse living creatures exists. All these are for the service of man, for his use and enjoyment, sustenance and development. But men should not forget that there are other living creatures that also have a right to exist and are entitled to use the resources and enjoy.

Thus anciently society in India, have deep regards for environmental protection and conservation.

43.4. PRE-CONSTITUTION LAWS RELATING TO ENVIRONMENTAL PROTECTION IN INDIA

It was already mentioned that British Raj, there is some landmark action taken by the ruler for control of air and water pollution as mentioned in various sections of IPC (1860), CrPC (1898) and Indian Easements Act (1882). Public nuisance leading to pollution of air, water and damage of environment was considered as offence. The important sections of IPC are as follows: Section 268, Section 269, Section 277, Section 278, Section 290, Section 425 and Section 511. The detailed provision are as follows:

Section 268: A person is guilty of a public nuisance, who does any act or is guilty of an illegal emission which causes any common injury, danger or annoyance to the public or to the people in general who dwell or occupy property in the vicinity, or which must necessarily cause injury, obstruction, danger or annoyance to persons who may have occasion to use any public right.

Section 269: Negligent act likely to spread infection of disease dangerous to life—whoever unlawfully or negligently does any act which is known and which he knows or has reason to believe to be, likely to spread the infection of any disease dangerous to life, shall be punished with imprisonment of either description for a term which may extend to six months, or with fine, or with both.

Section 277: Fouling water of public spring or reservoir—whoever voluntarily corrupts or fouls the water of any public spring or reservoir, so as to render it less fit for the purpose for which it is ordinarily used, shall be punished with imprisonment of either description for a term which may extend to three months, or with fine which may extend to five hundred rupees, or with both.

Section 278: Making atmosphere noxious to health—whoever voluntarily initiates the atmosphere in any place so as to make it noxious to the health of persons in general dwelling or carrying on business in the neighbourhood or passing along a public way, shall be punished with fine which may extend to five hundred rupees.

Section 290: It provides punishment for public nuisance (which includes pollution cases also) in case not otherwise provided for.

Section 425: The water pollution could also be punished if this activity causes wrongful loss or damage to public or to any person of if this act causes the destruction if any property or diminishes its value or utility.

Section 511: This section provide extension of provision nuisance of any person where an attempt to commit act of water pollution is made.

Some articles of CrPC (1898) viz., Section 133 and Section 143 are also considered as offence.

Section 133: Any unlawful obstruction/nuisance in public places, construction/disposal of toxics is a public nuisance certified by the executives viz. District Magistrate or Police Officials etc., cognigeable offence and thus punishable.

Section 143: A District Magistrate or sub-divisional Magistrate or any other Executive Magistrate empowered by the state government may order any person not to repeat or continue a public nuisance, as defined in the IPC or any special or local laws.

Identically two sections viz. **Section 15** and **Section 28(d)** of Indian Easement Act (1882), empowered limited right to individual for making some activity with limited pollution.

Before independence, there were some other Acts dealing with certain aspects of environmental protection. These are as follows:

1. The Bengal Smoke Nuisance Act, 1905.
2. The Bombay Smoke Nuisance Act, 1912.
3. The Poison Act, 1919.
4. The Indian Boiler's Act, 1923.
5. The Indian Forest Act, 1927.
6. The Factories Act, 1948.

43.5. CONSTITUTIONAL PROVISIONS FOR ENVIRONMENTAL PROTECTION IN INDIA

The Indian efforts for the protection of environment began only after the Stockholm Conference in 1972. In a way one can say that this Conference was the real milestone in awakening the concern of the Indian Government towards environmental pollution. Before this Conference, neither the Indian Constitution nor any of the Indian laws dealt with the problem of environment expressly. It was only after this Conference that both constitutional and legislative reforms have been effected in India to meet and combat the environmental problems in the country.

The Constitution of India, as adopted by the constitution makers in 1950, did not deal with 'Protection of environment' as a specific legislative subject. Nor did it contain any specific provision to embody India's commitment to the protection and preservation of environment. However, Indian Constitution was not totally silent in this regard as it contained (and still contains) several related subjects in its constitutional scheme of distribution of legislative powers. Thus, several legislative subjects such as forest, wildlife, population control, public health and sanitation, agriculture, land, water, atomic energy, oilfields, mines and interstate rivers, and fisheries, etc., were already provided for in the Constitution. It was only in 1976 that the Indian Government thought it necessary to get the Indian Constitution amended to expressly provide for the protection of environment by incorporating a few specific provisions in the Constitution. The result was the enactment of the Constitution (Forty-second Amendment) Act, 1976, providing for an express constitutional commitment to safeguard the environment in the country.

This amendment, coupled with the Indian higher Judiciary's innovative approach to the interpretation of the constitutional provisions, has been the chief source of the development of environmental jurisprudence in India. It may be appreciated that the Apex Court's liberal, innovative and dynamic interpretation of Articles 12, 14, 21, 32, 48A, 51A(g), 136, 142, 144 and 226 of the Indian Constitution has been cornerstone of the structure of the environmental jurisprudence in India.

It may be noted that Articles 48A and 51A(g) which embody India's constitutional concerns, find their place in Part IV of the Constitution, titled "Directive Principles of State Policy", which embody several unenforceable directives addressed to the State. While Article 48A imposes a constitutional obligation on the State to protect and improve the environment and to safeguard the forests and wildlife of the country, Article 51A(g) enjoins all the Indian citizens to protect and improve the natural environment, including forests, lakes, rivers and wildlife. Both these provisions, as will be seen in the later part of the chapter, not only highlight the national consensus on the importance of environmental protection but also lay the foundation for a 'judge induced' and 'judge led' and 'judicially-administered' jurisprudence of environmental protection in the country.

After focusing on the import and importance of Articles 48A and 51A(g) of the Indian Constitution Justice A.R. Lakshmanan, in *Intellectuals State of AP* observed:

These two Articles are not only fundamental in the governance of the Country but also it shall be the duty of the State to apply these Principles in making laws and further, these two articles are to be kept in mind in understanding the scope and purport of fundamental rights guaranteed by the Constitution of India and also the various laws enacted by the Parliament and the State Legislature.

Box 43.1: Constitutional Provisions

- Article 12 reads: "In this part, unless the context otherwise requires, 'the State' includes the Government and Parliament of the India and the Government and the legislature of each of the states and all local or other authorities within the territory of India or under the control of the Government of India".
- Article 14 declares: "The State shall not deny to any person equality before the law or the equal protection of the laws within the territory of India".
- Article 21 stipulates: "No person shall be deprived of his life or personal liberty except according to procedure established by law."
- Article 32 states: "(1) The right to move the Supreme Court by appropriate proceedings for the enforcement of the rights conferred by this part is guaranteed. (2) The Supreme Court shall have power to issue directions or orders or writs, including writs in the nature of habeas corpus, mandamus, prohibition, quowarranto and certiorari, which ever may be appropriate, for the enforcement of any of the rights conferred by this part.....".
- Article 48A declares: "The State shall endeavor to protect and improve the environment and to safeguard the forests and wildlife of the country".
- Articles 51A(g) says: "It shall be the duty of every citizen of India—to protect and improve the natural environment including forests, lakes, rivers and wildlife, and to have compassion for living creatures".
- Article 136 provides: "(1) Notwithstanding any thing in this chapter, the Supreme Court may, in its discretion, grant special leave to appeal from any judgement, decree, determination, sentence or order in any cause or matter passed or made by any court or tribunal in the territory of India......"
- Article 142 states: "The Supreme Court in the exercise of its jurisdiction may pass such decree or make such order as is necessary *for doing complete justice* in any cause or matter pending before it, and any decree so passed or order so made shall be enforceable throughout the territory of India in such manner as may be prescribed by or under any law made by parliament and, until provision in that behalf is so made, in such manner as the President may by order prescribe........" (emphasis added)
- Article 144 states: "All authorities, civil and judicial, in the territory of India shall act in aid of the Supreme Court."
- Article 226 declares: "(1) Not withstanding anything in Art. 32, every High Court shall have power, throughout the territories in relation to which it exercises jurisdiction, to issue to any person or authority, including in appropriate cases, any government, within those territories, directions, orders or writs, including writs in the nature of habeas corpus, mandamus, prohibition, quo warranto and certiorari, or any of them, for the enforcement of any of the rights conferred by part III and for any other purpose".

43.6. INTERNATIONAL LEGISLATIONS/TREATY FOR ENVIRONMENTAL PROTECTION

International concern for the protection, preservation and improvement of environment became pronounced only in the second-half of the last century. In the context of today's interdependent world, with the economic globalization as the emerging international norm, any act or omission on the part of one nation adversely affecting it national environment may have far reaching adverse consequences to the global environment. This fear has prompted the members of the family of nations to shed their attitude of indifference to the global environmental concerns and to become aware of the dangers of further environmental neglect. The result was the initiation of the international efforts to find international legislative and administrative strategies to address global environmental challenges.

Prior to 1972, there was no major international effort to tackle and face global environmental challenges, although there were several international legislative instruments dealing with environmental issues incidently which are still in operation. Some of these are:

- The nuclear weapons tests ban treaty of 1963
- The treaty for the prohibition of nuclear weapons in Latin America of 1967
- The treaty laying down the principles governing activities of states in the exploration and use of outer space including the moon and celestial bodies of 1967
- The treaty on the nonproliferation of nuclear weapons of 1968
- The treaty on the prohibition of the emplacement of nuclear weapons on the seabed, ocean floor and subsoil thereof 1971.

THE STOCKHOLM CONFERENCE, 1972

The major international effort for the protection of global environment began in 1972, when the international community convened the first historic United Nations Conference on Human Environment at Stockholm. This conference marked the culmination of the efforts that preceded the 'Stockholm Conference to place the issue of the protection of the environment on the official agenda of international policy and law as it represented the first major international effort to tackle the problem of the protection of environment and its improvement by international cooperation and agreement. The agenda of the conference was expanded to include the issues which recognized the relationship between development and environment. They were conceptualized as two sides of the same coin and it was realized that environmental protection was an essential element of the social and economic development. The conference adopted, among other things, three important nonbinding instruments which were:

1. A declaration containing seven home truths and twenty-six principles;
2. A resolution on institutional and financial arrangements;
3. An action plan containing 109 recommendations.

Nairobi Declaration, 1982

The tenth anniversary of the Stockholm Conference celebrated in Nairobi (Kenya) from 10-18 May 1982, and a declaration known as "Nairobi Declaration" was adopted on 18 May 1982. Expressing serious concern at the state of global environment at that time and recognizing the urgent necessity of intensifying the efforts at the global, regional and national levels to protect and improve environment, the Declaration required the state government and peoples to build on the progress achieved thus far. The Declaration also restated the principles of the Stockholm Conference.

The World Charter for Nature, 1982

In October 1982, the United Nations General Assembly adopted "The World Charter For Nature", which sought to lay down general principles of environmental protection, action plan and implementation scheme. It provided conservation principles, which were meant to guide all human conduct affecting nature. It declared that, "nature shall be respected and its essentials shall not be impaired". It required states to integrate conservation measures.

The Report Entitled "Our Common Future", 1987

The World Commission on Environment and Development was created, pursuant to the UN General Assembly, Resolution 36/161 in 1983, for suggesting and recommending legal principles based on Stockholm Conference and Nairobi Conference and many other existing international conventions and General Assembly Resolutions. Mrs. Gro Harlem Bruntland, Chairman of the World Commission on Environment and Development submitted a Report titled, "Our Common Future" in 1987 with a foreword by late Shri Rajiv Gandhi, the then Prime Minister of India. The Report showed that politicians, industrial leaders, and environmental groups around the world had endorsed sustainable development: "Meeting the needs of the present without compromising the ability of future generations to meet their own needs" ..... as a guiding principle for the future of their countries. The practical application of this vision would require profound changes in institutions and in the decision-making process of industrialized as well as developing and underdeveloped countries.

Vienna Convention of 1985 for the Protection of Ozone Layer

After Stockhold Conference of 1972, a Conference under the auspices of United Nations was held at Vienna on 22 March 1985 where a Convention was adopted for the protection of ozone layer. The Convention, known as Vienna Convention provided a foundation for global multilateral undertakings to protect the environment and public health from the potential adverse effects of depletion of Stratospheric Ozone. The Convention addresses this important environmental issue primarily by providing for international cooperation in research and exchange of information. The Convention was the product of more than three years of negotiations under the auspices of UNEP. The Convention sets out the general obligations of parties, including a requirement to take appropriate measures for the protection of ozone layer and the obligation to cooperate in research and information exchange. It specifies that parties will cooperate, as appropriate, in conducting the research and scientific assessments in a wide variety of areas, including chemical, biological, health and climatic effects. There is also a provision in the Convention for the exchange of socioeconomic commercial and legal information.

Montreal Protocol, 1987

The Conference of Plenipotentiaries met in Montreal from 14 to 16 September 1987 under the auspices of the UNEP and adopted a Protocol, which was signed by 47 participants including USA and EEC.

The Protocol established specific obligations to limit and reduce the use of chlorofluorocarbons and possibly other chemicals that deplete ozone. The substances that the Protocol currently addresses are listed in Annexture-A to the Protocol.

The ozone negotiations culminating in the Montreal Protocol seemed to show that the Vienna Convention created a framework of cooperation in which stages can reach agreement quickly when new scientific evidence became available, thus enabling them to respond to environmental problems before the damage becomes much worse or irreversible.

Although the Montreal Protocol sets reduction schedules for ozone-depleting chemical production, the Protocol allows member-states to overshoot the imposed limits by 10-15 per cent if they do so for purposes of "industrial rationalization", or "to meet the basic needs of the developing countries". Developing countries may defer compliance for up to ten years and may even increase production and consumption of ozone destructive Chlorofluorocarbons (CFCs) above 1986 levels. The Protocol provides for a multiple steps freeze and percentage reduction of CFC levels.

The Protocol also provides for payments by developed countries to underdeveloped countries to refrain from using CFCs and to purchase CFC substitutes. The Protocol exacerbates the inadequacy of the incentive structure and the side payments because it does not provide any monitoring of enforcement mechanism. Instead, it leaves these issues for resolution at the first meeting of the parties.

UNITED NATIONS CONFERENCE ON ENVIRONMENT AND DEVELOPMENT (UNCED), 1992 (RIO SUMMIT)

The second landmark event in the process of the evolution of the development of international concern for global environment was the United Nations Conference on Environment and Development held at Rio (Brazil) from 3 to 14 June 1992 to confront the twin problems of environment and development. The Conference, known as Rio Summit, was a spectacular event with 178 nations striving to evolve plans to save the planet earth from environmental degradation. In terms of numbers, it was one of the largest meeting ever held. The Conference was of great importance for its fivefold contribution to the international efforts to combat global environmental problems. The Conference resulted in the conclusion of a treaty on climate change with a general recognition of the importance of curbing emission of greenhouse gases. Another treaty on biodiversity aiming at the preservation of flora and fauna was also concluded. There was a "Statement of Principles on Forests", a nonbinding agreement for conservation and sustainable management of global forests. There was the most important "Rio Declaration", a nonbinding set of 27 principles, that deals with the rights and responsibilities of nations of the earth relating to environment and development. Besides these, a massive, 300 pages document titled as 'Agenda 21' known as 'blueprint for sustainable development', was adopted. It is supposed to be an action plan attempting to deal comprehensively with all aspects of environment. Although it is nonbinding, it may serve as the yardstick against which actions taken by each government in the following years are likely to be measured.

United Nations Framework Convention on Climate Change (UNFCCC), 1992

An international instrument showing international concern for the protection of global environment was the convention on the climate change, which was signed by 154 countries. The convention requires the states to prevent the global climate change, by taking appropriate steps to reduce their emissions of greenhouse gases believed to contribute to global warming. During the debate on the provisions on the convention, there was a wide divergence between the views of the developed countries and those of the developing countries. While the former wanted a 20 per cent cut in greenhouse gases emissions like carbon dioxide and methane by 2005 and a major shift from the use of coal and wood for energy, the latter blamed the former for excessive emissions over the past 150 years and wanted them to reduce them considerably. The developing countries refused to agree to any cut in their emissions, as it would hinder their development. Ultimately, consensus was that by the year 2000, the emission of carbon dioxide gases were to be reduced to the 1990 levels. Under the convention, the developed countries have agreed to provide financial help and technology to the third world nations, to help them deal with global warming. Commenting on the convention, Mr Boutros Ghali, the then UN Secretary General opined that the convention was a "major step forward and a platform with potential".

The Biodiversity Convention, 1992

This convention, signed by 153 countries, seeks to ensure that animals, plants and microorganisms as well as genetic variety and ecosystem, water, land and air in which they live, are properly protected. The treaty supports all guidelines to make economic use of natural resources more compatible with their planned conservation. The convention requires the signatory nations to regulate or manage biological resources important for the conservation or biological diversity whether within or outside protected areas, with a view to ensuring the conservation and sustainable use. It obligates the countries to promote the protection of ecosystems, natural habitat and the maintenance of viable populations of species in natural surroundings. The States have, under the convention, agreed to adopt measures relating to the use of biological resources to avoid or minimize adverse impacts on biological diversity.

While the convention is largely a 'framework' agreement, setting out principles for the protection and preservation of biodiversity, its wordings are somewhat weak as the obligations created under the convention are qualified. Besides, the convention does not prescribe specific methods or standards for their compliance nor does it prescribe enforcement mechanisms.

Post-Rio Summit Developments up to Johannesburg Summit

There have been several significant international efforts, since Rio Summit, for the protection of global environment. Some of the these are:

1. The Cairo Conference on Population and Development, 1994
2. SAARC Conference, 1995 1997 and 2002
3. The United Nations Conference on Human Settlements, 1996
4. Earths Summit + 5 Programme for Further Implementation of Agenda 21, 1997
5. The United Nations Millennium Declaration, 2000
6. The Kyoto Protocol, 1997 and
7. The Delhi Summit on Sustainable Development, 2002.

The Cairo Conference on Population and Development, 1994

Following close on the heals of the Earth Summit was another global conference held at Cairo. The International Conference on Population and Development (ICPD) was held in September 1994 in Cairo. Many environmentalists, especially those from the North, regarded overpopulation as the leading cause of environmental destruction around the world. They hoped that the lack of emphasis on population stabilization at the Earth Summit would be more than compensated by the issues, commanding preference at Cairo Conference. For one thing, the Cairo Conference changed the framework of debate from conventional issues of family planning to broader questions of reproductive health, empowerment of women and integration of population policy with environmental policy and development strategies. Representatives from 183 nations declared their consent to a programme of action that aimed at stabilizing human population by the year 2020, providing greater equality for women along with improved reproductive health care and the controlling of AIDS as part of population policy.

The Cairo Conference, like the Earth Summit, largely failed to integrate demographic, environmental and economic development issues. The World Human Rights Conference in Vienna in June 1993, the World Social Development Summit in Copenhagen in March 1995 have all failed to highlight the importance of the linkage between environment, population and development.

SAARC Conference, 1995, 1997 and 2002

Hand in hand with the global efforts, there have been regional efforts also to improve and protect environment at regional level. Thus a Conference of SAARC (South Asian Association for Regional Cooperation) was held at New Delhi from 2 to 4 May 1995. The SAARC countries, through Delhi Declaration, expressed their deep concern at the unabated degradation of the environment and recurrence of devastating natural disasters. They also underscored the risks and dangers involved in overlooking the challenges posed by these problems. They expressed their commitment to implement at all levels—national, bilateral, regional and global, programmes for the protection and preservation of environment and prevention of its degradation.

The United Nations Conference on Human Settlements, 1996

The United Nations Conference on Human Settlements was held in June 1996 in Turkey and Habitat Agenda, 1996 was adopted which lays down that the national governments at the appropriate levels should establish legal frameworks to facilitate the development and implementation of public plans and policies for sustainable urban development and rehabilitation, land utilization, housing and the improved management of urban growth. The Habitat Agenda, 1996 lays down, as its objectives, the following:

- Integrating urban planning and management in relation to housing, transport, employment opportunities, environmental conditions and community facilities.
- Providing adequate and integrated environmental infrastructure facilities in all settlements including continuous and safe freshwater supplies, sanitation and drainage and waste disposal services.
- Promoting the development of more balanced and sustainable human settlements by encouraging productive investments, social infrastructure development, harnessing and enhancing the efforts and potential of productive informal and private sectors.
- Giving priority attentions to human settlement programmes and policies to reduce urban pollution resulting especially from inadequate water supply, sanitation and drainage, poor industrial and domestic waste management, including solid waste management, and air pollution.
- Encouraging dialogue among public, private and non-governmental interested parties to develop an expanded concept of the "balance sheets". The concept would recognize that the economic, environmental, social and civic consequence for directly and indirectly affected parties including future generations should be taken into account in making decisions on the allocation of resources.
- Promoting optimal use of productive land in urban and rural areas and protecting fragile ecosystems and environmentally vulnerable areas from the negative impacts of human settlements.
- Promoting the redevelopment and reuse of already serviced but poorly utilized commercial and residential land in urban centres in order to revitalize them and reduce development pressures on productive agricultural land on the periphery.

The Kyoto Protocol, 1997

The Kyoto Protocol, 1997 to the United Nations Framework Convention on Climate Change was adopted in 1997. The Protocol, which consists of 28 provisions, requires the country parties to the protocol to endeavour to implement all the policies and measures (like enhancement of energy efficiency in relevant sectors of national economy; promotion and enhancement of sinks and reservoirs of greenhouse gases not controlled by the Montreal Protocol; promotion of sustainable forest management practices, afforestation and deforestation; promotion of sustainable forms of agriculture in the light of climate change considerations; encouragement of appropriate reforms in relevant sectors aimed at promoting policies and measures which limit or reduce emissions of greenhouse gases not controlled by the Montreal Protocol and measures to limit and/or reduce emissions of greenhouse gases in the transport sector, in such a way as to minimize adverse effects, including the effects of climate change, effects on international trade and social, environmental and economic impacts on other parties, especially developing country parties to the convention. The Protocol also called on the developed country parties to reduce, either individually or jointly, their emissions of the greenhouse gases by at least 5 per cent compared to 1990 levels by the period 2008-2012. They are also required to strive to implement their commitments mentioned above in such a way as to minimize adverse social, environmental and economic impacts on developing country parties, particularly those identified in paras 8 and 9 of Article 4 of the Climate Change Convention. The countries were given individual emission targets to be achieved by domestic action by a range of international cooperative mechanisms such as clean development mechanism, multilateral consultative process and emission trading which should be supplemental to domestic actions of developed country parties.

THE WORLD SUMMIT ON SUSTAINABLE DEVELOPMENT

The Johannesburg Summit on Sustainable Development was held at Johannesburg (South Africa) from 27 August to 4 September 2002, under the auspices of the Commission on Sustainable Development, which reviewed the progress made on the prospect of achieving the goal of sustainable development since the Rio Conference and recommended concrete steps for further implementation of the Rio Principles and Agenda 21. Reaffirming its commitment to the Rio Principles and Agenda 21, the international community expressed it commitment to undertake concrete actions and measures at all levels to enhancing international cooperation, taking into particular account, the principle of common but differentiated responsibilities as set out in Principle 7 of the Rio Declaration. The members of the international community were required to promote the integration of the three components of sustainable development, i.e., economic development, social development and environmental protection as interdependent and mutually reinforcing pillars. It was declared that poverty eradication, changing unsustainable patterns of production and consumption and protecting and managing the natural resource base of economic and social development are the overarching objectives of and essential requirements for sustainable development.

Earth Summit + 5 Programme for the Further Implementation of Agenda 21

A special session of the UN General Assembly, held from 23 to 28 June 1997 in New York to review the progress of the Rio Earth Summit called 'Earth Summit + 5', adopted a comprehensive document titled "Programme for the Further Implementation of Agenda 21". This was called to assess the progress in the five years since the 1992 Earth Summit at Rio and to chart out the future course for further action based upon Agenda 21 adopted at that Conference. It was acknowledged that, since Rio Conference, the global environment had continued to deteriorate with rising levels of polluting emissions, notably of greenhouse gases, toxic substances and waste volumes. It was also noted with concern that renewable resources, notably freshwater, forests, top soil and marine fish stock continued to be exploited at rates that were clearly unsustainable. To address and to take care of these problems, the governments resolved to take several steps at the special reaffirmed that Agenda 21 remained the fundamental programme of action for achieving sustainable development but also expressed their commitment to fully implement the Agenda Programme.

The United Nations Millennium Declaration, 2000

The United Nations Millennium Declaration was a landmark international 'soft' law instrument of the new century which was adopted at the Millennium Summit held in New York from 6 to 8 September 2000 and attended by 191 (approx.) nations, which was the largest ever gathering of the world leaders. The Declaration recognized that certain fundamental values including respect for nature are essential to international relations in the 21st century. Elaborating the content of the fundamental value of "respect of nature", the Declaration articulated that prudence must be shown in the management of all living species and natural resources, in accordance with precepts of sustainable development, and that only the, can be immeasurable riches provided to us by the nature be preserved for posterity and that the current unsustainable pattern of production and consumption must be changed in the interest of our future welfare and that of our descendants.

The Delhi Sustainable Development Summit (DSDS), 2002

A summit on Sustainable Development was held from 8 to 11 February 2002 at Delhi, India which was organized by Tata Energy Research Institute (TERI), a research-oriented nongovernmental organization with its headquarters in New Delhi, to discuss, examine and elaborate the dynamics of concept of Sustainable Development with a view to making recommendations for consideration at the World Summit on Sustainable Development (WSSD) to be held later in the year at Johannesburg (South Africa). The Delhi Summit was one of the regional conferences organized in the run-up to the main World Summit on Sustainable Development held at Johannesburg. The Delhi Summit sought to focus on poverty alleviation as the overriding concern to achieve sustainable development. It was realized that the path followed until now by governments and international institutions giving priority to economic growth does not necessarily bring about sustainable development, whose core-dimension was the alleviation of poverty by meeting the basic needs of the poorest of the poor which had remained the primary challenge in the developing countries.

The International Law Association (ILA) New Delhi Conference, 2002

The International Law Association held in 70th Biennial Conference from 2 to 7 April 2002 at Delhi (India), where the Reports of the various International Committees of the International law Association on a variety of topics where examined and discussed. The Committee on 'Legal Aspects of Sustainable Development' submitted its Fifth and Final Report for consideration of the Conference, which after a thorough discussion of the Report along with the earlier four reports of the Committee, came out with a Draft Declaration entitled "New Delhi ILA Declaration on Principles of International Law Relation to Sustainable Development". The Report and the Draft Declaration were submitted to the UN Secretary General with a request to forward the same to the UN Commission on Sustainable Development, UNCTAD, UNEP, the WTO, the World Bank and other relevant intergovernmental and nongovernmental organizations for their consideration, including organization at the regional level.

THE COPENHAGEN CONFERENCE AND COPENHAGEN ACCORD

The problem of emission of greenhouse gases resulting in global warming along with its inevitable consequence of climate change triggering in extreme whether conditions has been engaging the attention and concern of the international community. The problem has also assumed urgency and importance with the prospect of the Kyoto Protocol, 1997 getting expired by the end of December 2012. In view of this conspectus, the members of the international community have accelerated their international efforts to combat global warming by undertaking greenhouse gas emission cuts in their respective national jurisdictions. Towards this end, the first conference of the Parties to the Kyoto Protocol was held at Montreal (Canada) in 2005 where it was decided to establish a subsidiary body known as Ad Hoc Working Group of the Kyoto Protocol (AWQ KP) with a mandate to negotiate and decide on further commitments on the part of industrialized nations after the end of 2012 when the Kyoto Protocol ceases to operate. This was followed by another conference at Bali (Indonesia) in 2007 where Bali Action Plan was adopted.

United Nation Convention on Laws of Sea (UNCLOS), 1982

The 1982 UNCLOS was intended to be a comprehensive restatement of almost of all aspects of the Laws of the sea. The emergence of a more strongly expressed obligation to protect the marine environment is evidenced by articles 1925 of the 1982 UNCLOS, by regional treaties, and by other multilateral agreements negotiated progressively since 1954. These include the 1972 London dumping convention, the 1973/8 MARPOL convention which deals with pollution from ships and supersedes the earlier 1954 convention, and a variety of regional treaties requiring states to control land based sources of marine pollution, dumping, and seabed operations.

43.7. ENVIRONMENTAL PROTECTION THROUGH LEGISLATION IN INDIA

At the time of first earth summit (Stockholm Conference, 1972), the Prime Minister Mrs. Indira Gandhi said, "It is said that in country after country, progress should become synonymous with an assault on nature......Among the rest of mankind, we in India – inspite of Ashoka – have been guilty of wanton disregard for the courses of our sustenance". Soon after the Stockholm Conference, India set up a National Committee on Environment Planning and Coordination (NCEPC) in 1972. This committee was concerned with issues relating to appraisal of development projects, human settlements, planning, survey of ecosystems and spread of environment education. In addition 42nd Amendment of Indian Constitution is made in 1976. As per the guidelines of National conservation Strategy and Policy statements, a number of enactments/rules/notifications/regulations are framed in a sequence. There are three major categories of act thus framed and implemented thereafter through rules/notifications etc. These are primarily called as Conservation Laws/Acts, Pollution abatement Laws/Acts and other environmental management/regulatory laws. By and large most of the environmental laws are protective and preventive categories, planning legislation for environmental quality improvement is almost lacking in India.

There are some legislations where provision were given power to settle the environmental matter through setting up of environment related dispute disposal court. It may be stated that the area of environmental protection in India, through judicial process has come through Public Interest Litigation (PIL). The Supreme Court by relaxing the traditional rule of locus stands paved the way for NGO's and public spirited people to move the courts for the redressal of grievances, connected with environmental hazards under article 32 and article 226 of the Indian constitution. It is may be stated further that the apex court of India (i.e., Supreme Court) has sparked the imagination and developing the public law over past four decades a tide of public interest cases has been brought before the Supreme Court that has led the court to express its opinion on virtually every aspect of public life, including environmental protection viz., right to clean and pollution free environment, right to livelihood, right to maintaining ecological balance, protection against exposure to hazardous substances, protection of ancient monuments, road safety and traffic regulation, adherence to municipal and building laws by builders and so on. It may be further stated that the apex court of India, had widened the scope of Articles 21 of the constitution of India.

Thus Indian efforts to combat environmental problems of the country mitigated through two ways viz.,

- Constitutional and legislative measures
- Judicial initiatives

and subsequently for seedy trials of environmental cases the National Green tribunals were set up in distinct regions of the country as the provision of new Act, The National Green Tribunal Act 2010.

43.8. ACTS RELATING TO POLLUTION ABATEMENT AND OVER ALL ENVIRONMENT PROTECTION

Since 1974, several Bills passed by the parliament which then converted to Acts for the purpose of pollution abatement and over all environment protection. Subsequently rules and notifications were made against each acts for smooth implementation.

(a) The Water (Prevention and Control of Pollution) Act 1974, amended in 1988.

(b) The Water (Prevention and Control of Pollution) Cess Act, 1977, amended in 1991.

(c) The Air (Prevention and Control of Pollution) Act 1981, amended in 1988.

(d) The Environment (Protection) Act, 1986.

(e) The Public Liability Insurance Act, 1991.

In addition the Factory Act, 1948 was also operated in case of polluting industries. Each Acts and subsequent rules functions together through Regulatory Agency viz., Central Pollution Control Board of Centre & Union Territories and State Pollution Control Board in States. The detailed composition, power and functions are laid down in the Water Act 1974 and the Air Act 1981.

Each acts consists of preamble, short title, definition, and several sections. The details of some acts with amendment are given below:

43.9. SELECTED LAWS:

43.9.1 The Water (Prevention and Control of Pollution) Act, 1974

No. 6 of 1974

[23 March 1974]

An Act to provide for the prevention and control of water pollution and the maintaining or restoring of wholesomeness of water, for the establishment, with a view to carrying out the purposes aforesaid of Boards for the prevention, and control of water pollution, for conferring on and assigning to such Boards powers and functions relating thereto and for matters connected therewith.

Whereas it is expedient to provide for the prevention and control of water pollution and the maintaining or storing of wholesomeness of water, for the establishment, with a view to carrying out the purposes aforesaid, of Boards for the prevention and control of water pollution and for conferring on and assigning to such Boards powers and functions relating thereto;

And whereas Parliament has no power to make laws for the States with respect to any of the matters aforesaid except as provided in Articles 249 and 250 of the Constitution;

And whereas in pursuance of Clause (1) of Article 252 of the Constitution resolutions have been passed by all the Houses of the Legislatures of the States of Assam, Bihar, Gujarat, Haryana, Himachal Pradesh, Jammu and Kashmir, Karnataka, Kerala, Madhya Pradesh, Rajasthan, Tripura and West Bengal to the effect that the matters aforesaid should be regulated in those States by Parliament by law;

Be it enacted by Parliament in the Twenty-fifth Year of the Republic of India as follows:

Chapter I

PRELIMINARY

Short title, application and commencement

1. (1) This Act may be called the Water (Prevention and Control of Pollution) Act, 1974.

(2) It applies in the first instance to the whole of the States of Assam, Bihar, Gujarat, Haryana, Himachal Pradesh, Jammu and Kashmir, Karnataka, Kerala, Madhya Pradesh, Rajasthan, Tripura and West Bengal and the Union Territories; and it shall apply to such other State which adopts this Act by resolution passed in that behalf under Clause (1) of Article 252 of the Constitution.

(3) It shall come into force at once in the States of Assam, Bihar, Gujarat, Haryana, Himachal Pradesh, Jammu and Kashmir, Karnataka, Kerala, Madhya Pradesh, Rajasthan, Tripura and West Bengal and in the Union Territories, and in any other State which adopts this Act under Clause (1) of Article 252 of the Constitution on the date of such adoption and any reference in this Act to the commencement of this Act shall, in relation to any State or Union Territory, mean the date on which this Act comes into force in such State or Union Territory.

Definitions

2. In this Act, unless the context otherwise requires,

(a) "Board" means the Central Board or a State Board;

(b) "Central Board" means the Central Board for the Prevention and Control of Water Pollution constituted under Section 3;

(c) "member" means a member of a Board and includes the Chairman thereof;

(d) "occupier" in relation to any factory or premises means the person who has control over the affairs of the factory or the premises and where the said affairs are entrusted to a managing agent, such agent shall be deemed to be the occupier of the factory or the premises;

(e) "pollution" means such contamination of water or such alteration of the physical, chemical or biological properties of water or such discharge of any sewage or trade effluent or of any other liquid, gaseous or solid substance into water (whether directly or indirectly) as may, or is likely to, create a nuisance or render such water harmful or injurious to public health or safety, or to domestic, commercial, industrial, agricultural or other legitimate uses, or to the life and health of animals or plants or of aquatic organisms;

(f) "prescribed" means prescribed by rules made under this Act by the Central Government or, as the case may be, the State Government;

(g) "sewage effluent" means effluent from any sewerage system or sewage disposal works and includes sullage from open drains;

(h) "State Board" means a State Board for the Prevention and Control of Water Pollution constituted under Section 4;

(i) "State Government" in relation to a Union Territory means the Administrator thereof appointed under Article 239 of the Constitution;

(j) "Stream" includes:

  (i) river;

  (ii) water course (whether flowing or for the time being dry);

  (iii) inland water (whether natural or artificial);

  (iv) subterranean waters;

  (v) sea or tidal waters to such extent or, as the case may be, to such point as the State Government may, by notification in the Official Gazette, specify in this behalf;

(k) "trade effluent" includes any liquid, gaseous or solid substance which is discharged from any premises used for carrying on any trade or industry, other than domestic sewage.

Chapter II

THE CENTRAL AND STATE BOARDS FOR PREVENTION AND CONTROL OF WATER POLLUTION

Constitution of Central Board

3. (1) The Central Government shall, with effect from such date (being a date not later than six months of the commencement of this Act in the States of Assam, Bihar, Gujarat, Haryana, Himachal Pradesh, Jammu and Kashmir, Karnataka, Kerala, Madhya Pradesh, Rajasthan, Tripura and West Bengal and in the Union Territories) as it may, by notification in the Official Gazette, appoint, constitute a Central Board to be called the Central Board for the Prevention and Control of Water Pollution, to exercise the powers conferred on and perform the functions assigned to that Board under this Act.

(2) The Central Board shall consist of the following members, namely:

(a) a fulltime Chairman, being a person having special knowledge or practical experience in respect of matters relating to the use and conservation of water resources or the prevention and control of water pollution or a person having knowledge and experience in administering institutions dealing with the matters aforesaid, to be nominated by the Central Government.

(b) five officials to be nominated by the Central Government to represent that Government;

(c) such number of persons, not exceeding five, to be nominated by the Central Government, from amongst the members of the State Boards, of whom not exceeding two shall be from those referred to in clause (c) of Sub-section (2) of Section 4;

(d) three non-officials to be nominated by the Central Government, to represent the interests of agriculture, fishery or industry or trade or any other interest which, in the opinion of the Central Government, ought to be represented;

(e) two persons to represent the companies or corporations owned, controlled or managed by the Central Government, to be nominated by that Government;

(f) a fulltime Member-Secretary, qualified in public health engineering and having administrative experience, to be appointed by the Central Government.

(3) The Central Board shall be a body corporate with the name aforesaid having perpetual succession and a common seal with power, subject to the provisions of this Act, to acquire, hold and dispose of property and to contract, and may, by the aforesaid name, sue or be sued.

Constitution of State Boards

4. (1) The State Government shall with effect from such date (being a date not later than six months of the commencement of this Act in the State) as it may, by notification in the Official Gazette, appoint, constitute a State Board, under such name as may be specified in the notification, to exercise the powers confirmed on and perform the functions assigned to the Board under this Act.

(2) A State Board shall consist of the following members, namely:

(a) a fulltime Chairman, being a person having special knowledge or practical experience in respect of matters relating to the use and conservation of water resources or the prevention and control of water pollution or a person having knowledge and experience in administering institutions dealing with the matters aforesaid, to be nominated by the State Government;

(b) five officials to be nominated by the State Government to represent that Government;

(c) five persons to be nominated by the State Government from amongst the members of the local authorities functioning within the State;

(d) three non-officials to be nominated by the State Government to represent the interests of agriculture, fishery or industry or trade or any other interest which, in the opinion of the State Government, ought to be represented;

(e) two persons to represent the companies or corporations owned, controlled or managed by the State Government, to be nominated by that Government.

(f) a fulltime Member-Secretary, qualified in public health engineering and having administrative experience, to be appointed by the State Government.

(3) Every State Board shall be a body corporate with the name specified by the State Government in the notification under Sub-section (1), having perpetual succession and a common seal with power, subject to the provisions of this Act, to acquire, hold and dispose of property and to contract, and may, by the said name, sue or be sued.

(4) Notwithstanding anything contained in this Section, no State Board shall be constituted for a Union Territory and in relation to a Union Territory, the Central Board shall exercise the powers and perform the functions of a State Board for that Union Territory:

Provided that in relation to any Union Territory the Central Board may delegate all or any of its powers and functions under this Sub-section to such person or body of persons as the Central Government may specify.

Terms and conditions of service of members

5. (1) Save as otherwise provided by or under this Act, a member of a Board other than a Member-Secretary, shall hold office for a term of three years from the date of his nomination:

Provided that a member shall, notwithstanding the expiration of his term, continue to hold office until his successor enters upon his office.

(2) The term of office of a member of a Board nominated under Clause (b) of Sub-section (2) of Section 3 or Clause (b) of Sub-section (2) of Section 4 shall come to an end as soon as he ceases to hold the office under the Central Government or, as the case may be, the State Government, by virtue of which he was nominated.

(3) The Central Government or, as the case may be, the State Government, may, if it thinks fit, remove any member of a Board before the expiry of his term of office, after giving him a reasonable opportunity of showing cause against the same.

(4) A member of a Board, other than the Member-Secretary, may at any time resign his office by writing under his hand addressed —

(a) in the case of the Chairman, to the Central Government or, as the case may be, the State Government; and

(b) in any other case, to the Chairman of the Board; and the seat of the Chairman or such other member shall thereupon become vacant.

(5) A member of a Board, other than the Member-Secretary, shall be deemed to have vacated his seat if he is absent without reason, sufficient in the opinion of the Board, from three consecutive meetings of the Board, or where he is nominated under Clause (c) of Sub-section (2) of Section 3 or under Clause (c) of Sub-section (2) of Section 4, if he ceases to be a member of the State Board, or, as the case may be, of the local authority.

(6) A casual vacancy in a Board shall be filled by a fresh nomination and the person nominated to fill the vacancy shall hold office only for the remainder of the term for which the member in whose place he was nominated.

(7) A member of a Board shall not be eligible for renomination for more than two terms.

(8) The other terms and conditions of service of a member of a Board, other than the Chairman and Member-Secretary, shall be such as may be prescribed.

(9) The other terms and conditions of service of the Chairman shall be such as may be prescribed.

Disqualifications

6. (1) No person shall be a member of a Board, who —

(a) is, or at any time has been adjudged, insolvent or has suspended payment of his debts or has compounded with his creditors, or

(b) is of unsound mind and stands so declared by a competent court, or

(c) is, or has been, convicted of an offence which, in the opinion of the Central Government or, as the case may be, of the State Government, involves moral turpitude, or

(d) is, or at any time has been, convicted of an offence under this Act, or

(e) has directly or indirectly by himself or by any partner, any share or interest in any firm or company carrying on the business of manufacture, sale or hire of machinery, plant, equipment, apparatus or fittings for the treatment of a sewage or trade effluents, or

(f) is a Director or a Secretary, manager or other salaried officer or employee of any company or firm having any contract with the Board, or with the Government constituting the Board, or with a local authority in the State, or with a company or corporation owned, controlled or managed by the Government for the carrying out of sewerage schemes or for the installation of plants for the treatment of sewage or trade effluents, or

(g) has so abused, in the opinion of the Central Government or, as the case may be, of the State Government, his position as a member, as to render his continuance on the Board detrimental to the interest of the general public.

(2) No order of removal shall be made by the Central Government or the State Government, as the case may be, under this Section unless the member concerned has been given a reasonable opportunity of showing cause against the same.

(3) Notwithstanding anything contained in Sub-sections (1) and (7) of Section 5, a member who has been removed under this Section shall not be eligible for renomination as a member.

Vacation of seats by members

7. If a member of a Board becomes subject to any of the disqualifications specified in Section 6, his seat shall become vacant.

Meetings of Board

8. A Board shall meet at least once in every three months and shall observe such rules of procedure in regard to the transaction of business at its meetings as may be prescribed:

Provided that if, in the opinion of the Chairman, any business of an urgent nature is to be transacted, he may convene a meeting of the Board at such time as he thinks fit for the aforesaid purpose.

Constitution of committees

9. (1) A Board may constitute as many committees consisting wholly of members or wholly of other persons or partly of members and partly of other persons, and for such purpose or purposes as it may think fit.

(2) A committee constituted under this Section shall meet at such time and at such place, and shall observe such rules of procedure in regard to the transaction of business at its meetings, as may be prescribed.

(3) The members of a committee (other than the members of the Board) shall be paid such fees and allowances, for attending its meetings and for attending to any other work of the Board, as may be prescribed.

Temporary association of persons with Board for particular purposes

10. (1) A Board may associate with itself in such manner, and for such purposes, as may be prescribed, any person whose assistance or advice it may desire to obtain in performing any of its functions under this Act.

(2) A person associated with the Board under Sub-section (1) for any purpose shall have a right to take part in the discussions of the Board relevant to that purpose, but shall not have a right to vote at a meeting of the Board, and shall not be a member for any other purpose.

Vacancy in Board not to invalidate acts or proceedings

11. No act or proceeding of a Board or any committee thereof shall be called in question on the ground merely of the existence of any vacancy in, or any defect in the constitution of, the Board or such committee, as the case may be.

Member-Secretary and officers and other employees of Board

12. (1) The terms and conditions of service of the Member-Secretary shall be such as may be prescribed.

(2) The Member-Secretary shall exercise such powers and perform such duties as may be prescribed or as may, from time to time, be delegated to him by the Board or its Chairman.

(3) Subject to such rules as may be made by the Central Government or, as the case may be, the State Government in this behalf, a Board may appoint such officers and employees as it considers necessary for the efficient performance of its functions and the rules so made may provide for the salaries and allowances and other terms and conditions of service of such officers and employees.

(4) Subject to such conditions as may be prescribed, a Board may from time to time appoint any qualified person to be a consulting engineer to the Board and pay him such salaries and allowances and subject him to such other terms and conditions of service as it thinks fit.

Chapter III

JOINT BOARDS

Constitution of Joint Boards

13. (1) Notwithstanding anything contained in this Act, an agreement may be entered into—

(a) by two or more Governments of contiguous States, or

(b) by the Central Government (in respect of one or more Union Territories) and one or more Governments of States contiguous to such Union Territory or Union Territories,

to be in force for such period and to be subject to renewal for such further period, if any, as may be specified in the agreement to provide for the constitution of a Joint Board—

(i) in a case referred to in Clause (a), for all the participating States, and

(ii) in a case referred to in Clause (b), for the participating Union Territory or Union Territories and the State or States.

(2) An agreement under this Section may

(a) provide, in a case referred to in Clause (a) of Sub-section (1), for the apportionment between the participating States and in a case referred to in Clause (b) of that Sub-section, for the apportionments between the Central Government and the participating State Government or State Governments, of the expenditure in connection with the Joint Board;

(b) determine, in a case referred to in Clause (a) of Sub-section (1), which of the participating State Governments and in a case referred to in Clause (b) of that Sub-section, whether the

Central Government or the participating State Government (if there are more than one participating States, also which of the participating State Governments) shall exercise and perform the several powers and functions of the State Government under this Act and the references in this Act to the State Government shall be construed accordingly;

(c) provide for consultation, in a case referred to in Clause (a) of Sub-section (1), between the participating State Governments and in a case referred to in Clause (b) of that Sub-section, between the Central Government and the participating State Government or State Governments, either generally or with reference to particular matters arising under this Act;

(d) make such incidental and ancillary provisions, not inconsistent with this Act, as may be deemed necessary or expedient for giving effect to the agreement.

(3) An agreement under this Section shall be published, in a case referred to in Clause (a) of Sub-section (1), in the Official Gazette of the participating States and in a case referred to in Clause (b) of that Sub-section, in the Official Gazette of the participating Union Territory or Union Territories and the participating State or States.

Composition of Joint Boards

14. (1) A Joint Board, constituted in pursuance of an agreement entered into under Clause (a) of Sub-section (1) of Section 13, shall consist of the following members, namely:

(a) a full-time Chairman, being a person having special knowledge or practical experience in respect of matters relating to the use and conservation of water resources or the prevention and control of water pollution or a person having knowledge and experience in administering institutions dealing with the matters aforesaid, to be nominated by the Central Government;

(b) two officials from each of the participating States to be nominated by the concerned participating State Government to represent that Government;

(c) one person to be nominated by each of the participating State Governments from amongst the members of the local authorities functioning within the State concerned;

(d) one non-official to be nominated by each of the participating State Governments to represent the interests of agriculture, fishery or industry or trade in the State concerned or any other interest which, in the opinion of the participating State Government, is to be represented;

(e) two persons to be nominated by the Central Government to represent the companies or corporations owned, controlled or managed by the participating State Governments;

(f) a full-time Member-Secretary qualified in public health engineering and having administrative experience, to be appointed by the Central Government.

(2) A Joint Board constituted in pursuance of an agreement entered into under Clause (b) of Sub-section (1) of Section 13 shall consist of the following members, namely:

(a) a full-time Chairman, being a person having special knowledge or practical experience in respect of matters relating to the use and conservation of water resources or the prevention and control of water pollution or a person having knowledge and experience in administering institutions dealing with the matters aforesaid, to be nominated by the Central Government;

(b) two officials to be nominated by the Central Government from the participating Union Territory or each of the participating Union Territories, as the case may be, and two officials to be nominated, from the participating State or each of the participating States, as the case may be, by the concerned participating State Government;

(c) one person to be nominated by the Central Government from amongst the members of the local authorities functioning within the participating Union Territory or each of the participating Union Territories, as the case may be, and one person to be nominated, from amongst the members of the local authorities functioning within the participating State or each of the participating States, as the case may be, by the concerned participating State Government;

(d) one non-official to be nominated by the Central Government and one person to be nominated by the participating State Government or State Governments to represent the interests of agriculture, fishery or industry or trade in the Union Territory or in each of the Union Territories or the State or in each of the States, as the case may be, or any other interest which in the opinion of the Central Government or, as the case may be, of the State Government, is to be represented;

(e) two persons to be nominated by the Central Government to represent the companies or corporations owned, controlled or managed by the Central Government and situated in the participating Union Territory or Territories and two persons to be nominated by the Central Government to represent the companies or corporations owned, controlled or managed by the participating State Governments;

(f) a fulltime Member-Secretary qualified in public health engineering and having administrative experience, to be appointed by the Central Government.

(3) When a Joint Board is constituted in pursuance of an agreement under Clause (b) of Sub-section (1) of Section 13, the provisions of Sub-section (4) of Section 4 shall cease to apply in relation to the Union Territory for which the Joint Board is constituted.

(4) Subject to the provisions of Sub-section (3), the provisions of Sub-section (3) of Section 4 and Sections 5 to 12 (inclusive) shall apply in relation to the Joint Board and its Member-Secretary as they apply in relation to a State Board and its Member-Secretary.

(5) Any reference in this Act to the State Board shall, unless the context otherwise requires, be construed as including a Joint Board.

Special provision relating to giving of directions

15. Notwithstanding anything contained in this Act, where any Joint Board is constituted under Section 13—

(a) the Government of the State for which the Joint Board is constituted shall be competent to give any direction under this Act only in cases where such direction relates to a matter within the exclusive jurisdiction of the State;

(b) the Central Government alone shall be competent to give any direction under this Act where such direction relates to a matter within the territorial jurisdiction of two or more States or pertaining to a Union Territory.

Chapter IV

POWERS AND FUNCTIONS OF BOARDS

Functions of Central Board

16. (1) Subject to the provisions of this Act, the main function of the Central Board shall be to promote cleanliness of streams and wells in different areas of the States.

(2) In particular and without prejudice to the generality of the foregoing function, the Central Board may perform all or any of the following functions, namely:

(a) advise the Central Government on any matter concerning the prevention and control of water pollution;

(b) coordinate the activities of the State Boards and resolve disputes among them;

(c) provide technical assistance and guidance to the State Boards, carry out and sponsor investigations and research relating to problems of water pollution and prevention, control or abatement of water pollution;

(d) plan and organise the training of persons engaged or to be engaged in programmes for the prevention, control or abatement of water pollution on such terms and conditions as the Central Board may specify;

(e) organise through mass media a comprehensive programme regarding the prevention and control of water pollution;

(f) collect, compile and publish technical and statistical data relating to water pollution and the measures devised for its effective prevention and control and prepare manuals, codes or guides relating to treatment and disposal of sewage and trade effluents and disseminate information connected therewith;

(g) lay down, modify or annul, in consultation with the State Government concerned, the standards for a stream or well:

Provided that different standards may be laid down for the same stream or well or for different streams or wells, having regard to the quality of water, flow characteristics of the stream or well and the nature of the use of the water in such stream or well or streams or wells;

(h) plan and cause to be executed a nation-wide programme for the prevention, control or abatement of water pollution;

(i) perform such other functions as may be prescribed.

(3) The Board may establish or recognise a laboratory or laboratories to enable the Board to perform its functions under this Section efficiently, including the analysis of samples of water from any stream or well or of samples of any sewage or trade effluents.

Functions of State Board

17. (1) Subject to the provisions of this Act, the functions of a State Board shall be —

(a) to plan a comprehensive programme for the prevention, control or abatement of pollution of streams and wells in the State and to secure the execution thereof;

(b) to advise the State Government on any matter concerning the prevention, control or abatement of water pollution;

(c) to collect and disseminate information relating to water pollution and the prevention, control or abatement thereof;

(d) to encourage, conduct and participate in investigations and research relating to problems of water pollution and prevention, control or abatement of water pollution;

(e) to collaborate with the Central Board in organising the training of persons engaged or to be engaged in programmes relating to prevention, control or abatement of water pollution and to organise mass education programmes relating thereto;

(f) to inspect sewage or trade effluents, works and plants for the treatment of sewage and trade effluents and to review plans, specifications, or other data relating to plants set up for the treatment of water, works for the purification thereof and the system for the disposal of sewage or trade effluents or in connection with the grant of any consent as required by this Act;

(g) to lay down, modify or annul effluent standards for the sewage and trade effluents and for the quality of receiving waters (not being water in an inter-State stream) resulting from the discharge of effluents and to classify waters of the State;

(h) to evolve economical and reliable methods of treatment of sewage and trade effluents, having regard to the peculiar conditions of soils, climate and water resources of different regions, and more especially the prevailing flow characteristics of water in streams and wells which render it impossible to attain even the minimum degree of dilution;

(i) to evolve methods of utilisation of sewage and suitable trade effluents in agriculture;

(j) to evolve efficient methods of disposal of sewage and trade effluents on land, as are necessary on account of the predominant conditions of scanty stream flows that do not provide for major part of the year the minimum degree of dilution;

(k) to lay down standards of treatment of sewage and trade effluents to be discharged into any particular stream taking into account the minimum fair weather dilution available in that stream and the tolerance limits of pollution permissible in the water of the stream, after the discharge of such effluents;

(l) to make, vary or revoke any order —

 (i) for the prevention, control or abatement of discharges of waste into streams or wells;

 (ii) requiring any person concerned to construct new systems for the disposal of sewage and trade effluents or to modify, alter or extend any such existing system, or to adopt such remedial measures as are necessary to prevent, control or abate water pollution;

(m) to lay down effluent standards to be complied with by persons while causing discharge of sewage or sullage or both and to lay down, modify or annul effluent standards for the sewage and trade effluents;

(n) to advise the State Government with respect to the location of any industry the carrying on of which is likely to pollute a stream or well;

(o) to perform such other functions as may be prescribed or as may, from time to time, be entrusted to it by the Central Board or the State Government.

(2) The Board may establish or recognise a laboratory or laboratories to enable the Board to perform its functions under this Section efficiently, including the analysis of samples of water from any stream or well or of samples of any sewage or trade effluents.

Powers to give directions

18. In the performance of its functions under this Act—

(a) the Central Board shall be bound by such directions in writing as the Central Government may give to it; and

(b) every State Board shall be bound by such directions in writing as the Central Board or the State Government may give to it;

Provided that where a direction given by the State Government is inconsistent with the direction given by the Central Board, the matter shall be referred to the Central Government for its decision.

Chapter V

PREVENTION AND CONTROL OF WATER POLLUTION

Power of State Government to restrict the application of the Act to certain areas

19. (1) Notwithstanding anything contained in this Act, if the State Government, after consultation with, or on the recommendation of, the State Board, is of opinion that the provisions of this Act need not apply to the entire State, it may, by notification in the Official Gazette, restrict the application of this Act to such area or areas as may be declared therein as water pollution, prevention and control area or areas and thereupon the provisions of this Act shall apply only to such area or areas.

(2) Each water pollution, prevention and control area may be declared either by reference to a map or by reference to the line of any watershed or the boundary of any district or partly by one method and partly by another.

(3) The State Government may, by notification in the Official Gazette—

(a) alter any water pollution, prevention and control areas, whether by way of extension or reduction; or

(b) define a new water pollution, prevention and control area in which may be merged one or more water pollution, prevention and control area, or any part or parts thereof.

Power to obtain information

20. (1) For the purpose of enabling a State Board to perform the functions conferred on it by or under this Act, the State Board or any officer empowered by it in that behalf, may make surveys of any area and gauge and keep records of the flow or volume and other characteristics of any stream or well in such area, and may take steps for the measurement and recording of the rainfall in such area or any part thereof and for the installation and maintenance for those purposes of gauges or other apparatus and works connected therewith, and carry out steam surveys and may take such other steps as may be necessary in order to obtain any information required for the purposes aforesaid.

(2) A State Board may give directions requiring any person who in its opinion is abstracting water from any such stream or well in the area in quantities which are substantial in relation to the flow or volume of that stream or well or is discharging sewage or trade effluent into any such stream or well, to give such information as to the abstraction or the discharge at such times and in such form as may be specified in the directions.

(3) Without prejudice to the provisions of Sub-section (2), a State Board may, with a view to preventing or controlling pollution of water, give directions requiring any person in charge of any establishment where any industry or trade is carried on, to furnish to it information regarding the construction, installation or operation of such establishment or of any disposal system or of any extension or addition thereto in such establishment and such other particulars as may be prescribed.

Power to take samples of effluents and procedure to be followed in connection therewith

21. (1) A State Board or any officer empowered by it in this behalf shall have power to take for the purpose of analysis of samples of water from any stream or well or samples of any sewage or trade effluent which is passing from any plant or vessel or from or over any place into any such stream or well.

(2) The result of any analysis of a sample of any sewage or trade effluent taken under Sub-section (1) shall not be admissible in evidence in any legal proceeding unless the provisions of Sub-sections (3), (4) and (5), are complied with.

(3) Subject to the provisions of Sub-sections (4) and (5), when a sample (composite or otherwise as may be warranted by the process used) of any sewage or trade effluent is taken for analysis under Sub-section (1), the person taking the sample shall —

(a) serve on the person in charge of or having control over, the plant or vessel or in occupation of the place (which person is hereinafter referred to as the occupier) or any agent of such occupier, a notice, then and there in such form as may be prescribed, of his intention to have it so analysed;

(b) in the presence of the occupier or his agent, divide the sample into two parts;

(c) cause each part to be placed in a container which shall be marked and sealed and shall also be signed both by the person taking the sample and the occupier or his agent;

(d) send one container forthwith —

(i) in a case where such sample is taken from any area situated in a Union Territory, to the laboratory established or recognised by the Central Board under Section 16; and

(ii) in any other case, to the laboratory established or recognised by the State Board under Section 17;

(e) on the request of the occupier or his agent, send the second container—

(i) in a case where such sample is taken from any area situated in a Union Territory, to the laboratory established or specified under Sub-section (1) of Section 51; and

(ii) in any other case, to the laboratory established or specified under Sub-section (1) of Section 52.

(4) When a sample of any sewage or trade effluent is taken for analysis under Sub-section (1) and the person taking the sample serves on the occupier or his agent, a notice under Clause (a) of Sub-section (3) and the occupier or his agent wilfully absents himself, then, the sample so taken shall be placed in a container which shall be marked and sealed and shall also be signed by the person taking the sample and the same shall be sent forthwith by such person for analysis to the laboratory referred to in Sub-clause (i) or Sub-clause (ii), as the case may be, of Clause (e) of Sub-section (3) and such person shall inform the Government analyst appointed under Sub-section (1) or Sub-section (2) as the case may be, of Section 53, in writing about the wilful absence of the occupier or his agent.

(5) When a sample of any sewage or trade effluent is taken for analysis under Sub-section (1) and the person taking the sample serves on the occupier or his agent a notice under Clause (a) of Sub-section (3) and the occupier or his agent who is present at the time of taking the sample does not make a request for dividing the sample into two parts as provided in Clause (b) of Sub-section (3), then, the sample so taken shall be placed in a container which shall be marked and sealed and shall also be signal by the person taking the sample and the same shall be sent forthwith by such person for analysis to the laboratory referred to in Sub-clause (i) or Sub-clause (ii), as the case may be, of Clause (d) of Sub-section (3).

Reports of the result of analysis on samples taken under Section 21

22. (1) Where a sample of any sewage or trade effluent has been sent for analysis to the laboratory established or recognised by the Central Board or, as the case may be, the State Board, the concerned Board analyst appointed under Sub-section (3) of Section 53 shall analyse the sample and submit a report

in the prescribed form of the result of such analysis in triplicate to the Central Board or the State Board, as the case may be.

(2) On receipt of the report under Sub-section (1), one copy of the report shall be sent by the Central Board or the State Board, as the case may be, to the occupier or his agent referred to in Section 21, another copy shall be preserved for production before the court in case any legal proceedings are taken against him, and the other copy shall be kept by the concerned Board.

(3) Where a sample has been sent for analysis under Clause (e) of Sub-section (3) or Sub-section (4) of Section 21 to any laboratory mentioned therein, the Government analyst referred to in that Sub-section shall analyse the sample and submit a report in the prescribed form of the result of the analysis in triplicate to the Central Board or, as the case may be, the State Board which shall comply with the provisions of Sub-section (2).

(4) If there is any inconsistency or discrepancy between, or variation in the results of, the analysis carried out by the laboratory established or recognised by the Central Board or the State Board, as the case may be and that of the laboratory established or specified under Section 51 or Section 52, as the case may be, the report of the latter shall prevail.

(5) Any cost incurred in getting any sample analysed at the request of the occupier or his agent shall be payable by such occupier or his agent and, in case of default, the same shall be recoverable from him as arrears of land revenue or of public demand.

Power of entry and inspection

23. (1) Subject to the provisions of this Section, any person empowered by a State Board in this behalf shall have a right at any time to enter, with such assistance as he considers necessary, any place—

(a) for the purpose of performing any of the functions of the Board entrusted to him;
(b) for the purpose of determining whether and if so in what manner, any such functions are to be performed or whether any provisions of this Act or the rules made thereunder or any notice, order, direction or authorisation served, made, given, or granted under this Act, is being or has been complied with;
(c) for the purpose of examining any plant, record, register, document or any other material object or for conducting a search of any place in which he has reason to believe that an offence under this Act or the rules made thereunder has been or is being or is about to be committed and for seizing any such plant, record, register, document or other material object, if he has reason to believe that it may furnish evidence of the commission of an offence punishable under this Act or the rules made thereunder:

Provided that the right to enter under this Sub-section for the inspection of a well shall be exercised only at reasonable hours in a case where such well is situated in any premises used for residential purposes and the water thereof is used exclusively for domestic purposes.

(2) The provisions of the Code of Criminal Procedure, 1898 (5 of 1898) or, in relation to the State of Jammu and Kashmir, the provisions of any corresponding law in force in that State, shall, so far as may be, apply to any search or seizure under this Section as they apply to any search or seizure made under the authority of a warrant issued under Section 98 of the said Code, or, as the case may be, under the corresponding provisions of the said law.

Explanation—For the purposes of this Section, "place" includes vessel.

Prohibition on use of stream or well for disposal of polluting matter, etc.

24. (1) Subject to the provisions of this Section—

(a) no person shall knowingly cause or permit any poisonous, noxious or polluting matter determined in accordance with such standards as may be laid down by the State Board to enter (whether directly or indirectly) into any stream or well; or

(b) no person shall knowingly cause or permit to enter into any stream any other matter which may tend, either directly or in combination with similar matters, to impede the proper flow of the water of the stream in a manner leading or likely to lead to a substantial aggravation of pollution due to other causes or of its consequences.

(2) A person shall not be guilty of an offence under Sub-section (1), by reason only of having done or caused to be done any of the following acts, namely—

(a) constructing, improving or maintaining in or across or on the bank or bed of any stream any building, bridge, weir, dam, sluice, dock, pier, drain or sewer or other permanent works which he has a right to construct, improve or maintain;

(b) depositing any materials on the bank or in the bed of any stream for the purpose of reclaiming land or for supporting, repairing or protecting the bank or bed of such stream provided such materials are not capable of polluting such stream;

(c) putting into any stream, any sand or gravel or other natural deposit which has flowed from or been deposited by the current of such stream;

(d) causing or permitting, with the consent of the State Board, the deposit accumulated in a well, pond or reservoir to enter into any stream.

(3) The State Government may, after consultation with, or on the recommendation of the State Board, exempt, by notification in the Official Gazette, any person from the operation of Sub-section (1) subject to such conditions, if any, as may be specified in the notification and any condition so specified may, by a like notification, be altered, varied or amended.

Restrictions on new outlets and new discharges

25. (1) Subject to the provisions of this Section no person shall, without the previous consent of the State Board, bring into use any new or altered outlet for the discharge of sewage or trade effluent into a stream or well or begin to make any new discharge of sewage or trade effluent into a stream or well.

(2) An application for consent of the State Board under Sub-section (1) shall be made in the prescribed form and shall contain particulars regarding the proposed construction, installation or operation of the industrial or commercial establishment or of any treatment and disposal system or of any extension or addition thereto and such other particulars as may be prescribed.

(3) The State Board may make such inquiry as it may deem fit in respect of the application for consent referred to in Sub-section (1) and in making any such inquiry shall follow such procedure as may be prescribed.

(4) The State Board may grant its consent referred to in Sub-section (1), subject to such conditions as it may impose, being—

(a) in the case of a new or altered outlet, conditions as to the point of discharge into the stream or well or the construction of the outlet, or as to the use of that outlet or any other outlet for sewage or trade effluent from the same land or premises; and

(b) in the case of a new discharge, conditions as to the nature and composition, temperature, volume or rate of discharge of the effluent from the land or premises from which the new discharge is to be made,

and any such conditions imposed shall be binding on any person using the outlet, or discharging the effluent from the land or premises aforesaid.

(5) Where, without the consent of the State Board, a new or altered outlet is brought into use for the discharge of sewage or trade effluent into a stream or well, or a new discharge of sewage or trade effluent is made, the State Board may serve on the person making the outlet or making the discharge, as the case may be, a notice imposing any such conditions as it might have imposed on an application for its consent respect to such outlet or discharge.

(6) Every State Board shall maintain a register containing such particulars of the conditions imposed under this Section in relation to outlets or in relation to effluent from land or premises in its jurisdiction and as are for the time being in force (other than the conditions to be satisfied before an outlet is brought into use or a new discharge is made) and so much of the register as relates to any outlet, or to any effluent from such land or premises shall be open to inspection at all reasonable hours by any person interested in, or affected by, the outlet, or in the land or premises, as the case may be, or by any person authorised by him in this behalf and the conditions so contained in such register shall be conclusive proof that the consent was granted subject to such conditions.

(7) The consent referred to in Sub-section (1) shall, unless given or refused earlier, be deemed to have been given unconditionally on the expiry of a period of four months of the making of an application in this behalf complete in all respects to the State Board.

(8) For the purposes of this Section and Sections 27 and 30—

(a) the expression "new or altered outlet" means any outlet which is wholly or partly constructed on or after the commencement of this Act or which (whether so constructed or not) is substantially altered after such commencement;

(b) the expression "new discharge" means a discharge which is not, as respects the nature and composition, temperature, volume and rate of discharge of the effluent, substantially a continuation of a discharge made within the preceding twelve months (whether by the same or a different outlet), so however that a discharge which is in other respects a continuation of previous discharge made as aforesaid shall not be deemed to be a new discharge by reason of any reduction of the temperature or volume or rate of discharge of the effluent as compared with the previous discharge.

Provision regarding existing discharge of sewage or trade effluent

26. Where immediately before the commencement of this Act any person was discharging any sewage or trade effluent into a stream or well, the provisions of Section 25 shall, so far as may be, apply in relation to such person as they apply in relation to the person referred to in that Section subject to the modification that the application for consent to be made under Sub-section (2) of that Section shall be made within a period of three months of the constitution of the State Board.

Refusal or withdrawal of consent by State Board

27. (1) A State Board shall not grant its consent to the bringing into use of a new or altered outlet unless the outlet is so constructed as to comply with any conditions imposed by the Board to enable it to exercise its right to take samples of the effluent.

(2) A State Board may from time to time review any condition imposed under Section 25 (other than a condition to be satisfied before an outlet is brought into use or a new discharge is made), or under Section 26 and may serve on the person using the outlet or making the discharge, as the case may be, a notice making any reasonable variation of or revoking any such condition.

(3) Any condition imposed under Section 25 or Section 26 shall be subject to any variation made under Sub-section (2) and shall continue in force until revoked under that Sub-section.

Appeals

28. (1) Any person aggrieved by an order made by the State Board under Section 25, Section 26 or Section 27 may, within thirty days from the date on which the order is communicated to him, prefer an appeal to such authority (hereinafter referred to as the appellate authority) as the State Government may think fit to constitute:

Provided that the appellate authority may entertain the appeal after the expiry of the said expiry of thirty days if such authority is desired that the appellant was prevented by sufficient cause from filing the appeal in time.

(2) An appellate authority shall consist of three persons.

(3) The form and manner in which an appeal may be preferred under Sub-section (1), the fees payable for such appeal and the procedure to be followed by the appellate authority shall be such as may be prescribed.

(4) On receipt of an appeal preferred under Sub-section (1), the appellate authority shall, after giving the appellant and the State Board an opportunity of being heard, dispose of the appeal as expeditiously as possible.

(5) If the appellate authority determines that any condition imposed, or the variation of any condition, as the case may be, was unreasonable, then—

(a) where the appeal is in respect of the unreasonableness of any condition imposed, such authority may direct either that the condition shall be treated as annulled or that there shall be substituted for it such condition as appears to it to be reasonable;

(b) where the appeal is in respect of the unreasonableness of any variation of a condition, such authority may direct either that the condition shall be treated as continuing in force unvaried or that it shall be varied in such manner as appears to it to be reasonable.

Revision

29. (1) The State Government may at any time, either of its own motion or on an application made to it in this behalf, call for the records of any case where an order has been made by the State Board under Section 25, Section 26 or Section 27 for the purpose of satisfying itself as to the legality or propriety of any such order and may pass such order in relation thereto as it may think fit:

Provided that the State Government shall not pass any order under this Sub-section without affording the State Board and the person who may be affected by such order a reasonable opportunity of being heard in the matter.

(2) The State Governments shall not revise any order made under Section 25, Section 26 or Section 27 where an appeal against that order lies to the appellate authority, but has not been preferred or where an appeal has been preferred such appeal is pending before the appellate authority.

Power of State to carry out certain works

30. (1) Where under this Act any conditions have been imposed on any person for bringing into use any new or altered outlet for the discharge of sewage or trade effluent into a stream or well or for making any new discharge of sewage or trade effluent into a stream or well or on any person who, immediately before the commencement of this Act, was discharging any sewage or trade effluent in a stream or well and such conditions require such person to execute any work in connection therewith and such work has not been executed within such time as may be specified in this behalf, the State Board may serve on the person concerned a notice requiring him within such time (not being less than thirty days) as may be specified in the notice to execute the work specified therein.

(2) If the person concerned fails to execute the work as required in the notice referred to in Sub-section (1), then, after the expiration of the time specified in the said notice, the State Board may itself execute or cause to be executed such work.

(3) All expenses incurred by the State Board for the execution of the aforesaid work, together with interest, at such rate as the State Government may, by order, fix, from the date when a demand for the expenses is made until it is paid, may be recovered by that Board from the person concerned, as arrears of land revenue, or of public demand.

Furnishing of information to State Boards and other agencies in certain cases

31. (1) If at any place where any industry or trade is being carried on, due to accident or other unforeseen act or event, any poisonous, noxious or polluting matter is being discharged, or is likely to be discharged into a stream or well and, as a result of such discharge, the water in such stream or well is being polluted, or is likely to be polluted, then, the person in charge of such place shall forthwith intimate the occurrence of such accident, act or event to the State Board and to such other authorities or agencies as may be prescribed.

(2) Where any local authority operates any sewerage system or sewage works, the provisions of Sub-section (1) shall apply to such local authority as they apply in relation to the person in charge of the place where any industry or trade is being carried on.

Emergency measures in case of pollution of stream or well

32. (1) Where it appears to the State Board that any poisonous, noxious or polluting matter is present in any stream or well or has entered into that stream or well due to any accident or other unforeseen act or event, and if the Board is of opinion that it is necessary or expedient to take immediate action, it may for reasons to be recorded in writing, carry out such operations as it may consider necessary for all or any of the following purposes, that is to say—

(a) removing that matter from the stream or well and disposing it off in such manner as the Board considers appropriate;
(b) remedying or mitigating any pollution caused by its presence in the stream or well;
(c) issuing orders immediately restraining or prohibiting the person concerned from discharging any poisonous, noxious or polluting matter into the stream or well, or from making insanitary use of the stream or well.

(2) The power conferred by Sub-section (1) does not include the power to construct any works other than works of a temporary character which are removed on or before the completion of the operations.

Power of Board to make application to courts for restraining apprehended pollution of water in streams or wells

33. (1) Where it is apprehended by a Board that the water in any stream or well is likely to be polluted by reason of the disposal of any matter therein or of any likely disposal of any matter therein, or otherwise, the Board may make an application to a court, not inferior to that of a Presidency Magistrate or a Magistrate of the first class, for restraining the person who is likely to cause such pollution from so causing.

(2) On receipts of an application under Sub-section (1) the court may make such order as it deems fit.

(3) Where under Sub-section (2) the court makes an order restraining any person from polluting the water in any stream or well, it may in that order —

(i) direct the person who is likely to cause or has caused the pollution of the water in the stream or well, to desist from taking such action as is likely to cause pollution or, as the case may be, to remove from such stream or well, such matter, and

(ii) authorise the Board, if the direction under Clause (i) (being a direction for the removal of any matter from such stream or well) is not complied with by the person to whom such direction is issued, to undertake the removal and disposal of the matter in such manner as may be specified by the court.

(4) All expenses incurred by the Board in removing any matter in pursuance of the authorisation under Clause (ii) of Sub-section (3) or in the disposal of any such matter may be defrayed out of any money obtained by the Board from such disposal and any balance outstanding shall be recoverable from the person concerned as arrears of land revenue or of public demand.

Chapter VI

FUNDS, ACCOUNTS AND AUDIT

Contributions by Central Government

34. The Central Government may, after due appropriation made by Parliament by law in this behalf, make in each financial year such contributions to the Central Board as it may think necessary to enable the Board to perform its functions under this Act.

Contributions by State Government

35. The State Government may, after due appropriation made by the Legislature of the State by law in this behalf, make in each financial year such contributions to the State Board as it may think necessary to enable that Board to perform its functions under this Act.

Fund of Central Board

36. (1) The Central Board shall have its own fund, and all sums which may, from time to time, be paid to it by the Central Government and all other receipts (by way of gifts, grants, donations, benefactions or otherwise) of that Board shall be carried to the fund of the Board and all payments by the Board shall be made therefrom.

(2) The Central Board may expend such sums as it thinks fit for performing its functions under this Act, and such sums shall be treated as expenditure payable out of the fund of that Board.

Fund of State Board

37. (1) The State Board shall have its own fund, and the sums which may, from time to time, be paid to it by the State Government and all other receipts (by way of gifts, grants, donations, benefactions or otherwise) of that Board shall be carried to the fund of the Board and all payments by the Board shall be made therefrom.

(2) The State Board may expend such sums as it thinks fit for performing its functions under this Act, and such sum shall be treated as expenditure payable out of the fund of that Board.

Budget

38. The Central Board or, as the case may be, the State Board shall, during each financial year, prepare, in such form and at such time as may be prescribed, a budget in respect of the financial year next ensuing showing the estimated receipt and expenditure, and copies thereof shall be forwarded to the Central Government or, as the case may be, the State Government.

Annual report

39. (1) The Central Board shall, during each financial year, prepare, in such form and at such time as may be prescribed, an annual report giving a true and full account of its activities during the previous financial year and copies thereof shall be forwarded to the Central Government and that Government shall cause every such report to be laid before both Houses of Parliament within six months of the date on which it is received by that Government.

(2) The State Board shall, during each financial year, prepare, in such form and at such time as may be prescribed, an annual report giving a true and full account of its activities during the previous financial year and copies thereof shall be forwarded to the State Government and that Government shall cause every such report to be laid before the State Legislature within a period of six months of the date on which it is received by that Government.

Accounts and audit

40. (1) Every Board shall maintain proper accounts and other relevant records and prepare an annual statement of accounts in such form as may be prescribed by the Central Government or, as the case may be, the State Government.

(2) The accounts of the Board shall be audited by an auditor duly qualified to act as an auditor of companies under Section 226 of the Companies Act, 1956 (1 of 1956).

(3) The said auditor shall be appointed by the Central Government or, as the case may be, the State Government on the advice of the Comptroller and Auditor General of India.

(4) Every auditor appointed to audit the accounts of the Board under this Act shall have the right to demand the production of books, accounts, connected vouchers and other documents and papers and to inspect any of the offices of the Board.

(5) Every such auditor shall send a copy of his report together with an audited copy of the accounts to the Central Government or, as the case may be, the State Government.

(6) The Central Government shall, as soon as may be after the receipt of the audit report under Sub-section (5), cause the same to be laid before both Houses of Parliament.

(7) The State Government shall, as soon as may be after the receipt of the audit report under Sub-section (5), cause the same to be laid before the State Legislature.

Chapter VII

PENALTIES AND PROCEDURE

Failure to comply with directions under Sub-section (2) or Sub-section (3) of Section 20 or orders issued under Clause (c) of Sub-section (1) of Section 32

41. (1) Whoever fails to comply with any direction given under Sub-section (2) or Sub-section (3) of Section 20 within such time as may be specified in the direction or fails to comply with any orders issued under Clause (c) of Sub-section (1) of Section 32 shall, on conviction, be punishable with imprisonment for a term which may extend to three months or with fine which may extend to five thousand rupees or with both and in case the failure continues, with an additional fine which may extend to one thousand rupees for every day during which such failure continues after the conviction for the first such failure.

(2) Whoever fails to comply with any direction issued by a court under Sub-section (2) of Section 33 shall, on conviction, be punishable with imprisonment for a term which may extend to three months or with fine which may extend to five thousand rupees or with both and in case the failure continues, with an additional fine which may extend to one thousand rupees for every day during which such failure continues after the conviction for the first such failure.

Penalty for certain acts

42. (1) Whoever—

(a) destroys, pulls down, removes, injures or defaces any pillar, post or stake fixed in the ground or any notice or other matter put up, inscribed or placed, by or under the authority of the Board, or

(b) obstructs any person acting under the orders or directions of the Board from exercising his powers and performing his functions under this Act, or

(c) damages any works or property belonging to the Board, or

(d) fails to furnish to any officer or other employee of the Board any information required by him for the purpose of this Act, or

(e) fails to intimate the occurrence of any accident or other unforeseen act or event under Section 31 to the Board and other authorities or agencies as required by that Section, or

(f) in giving any information which he is required to give under this Act, knowingly or wilfully makes a statement which is false in any material particular,

(g) for the purpose of obtaining any consent under Section 25 or Section 26, knowingly or wilfully makes a statement which is false in any material particular,

shall be punishable with imprisonment for a term which may extend to three months or with fine which may extend to one thousand rupees or with both.

(2) Where for the grant of a consent in pursuance of the provisions of Section 25 or Section 26 the use of a meter or gauge or other measure or monitoring device is required and such device is used for the

purposes of those provisions, any person who knowingly or wilfully alters or interferes with that device so as to prevent it from monitoring or measuring correctly shall be punishable with imprisonment for a term which may extend to three months or with fine which may extend to one thousand rupees or with both.

Penalty for contravention of provisions of Section 24

43. Whoever contravenes the provisions of Section 24 shall be punishable with imprisonment for a term which shall not be less than six months but which may extend to six years and with fine.

Penalty for contravention of Section 25 or Section 26

44. Whoever contravenes the provisions of Section 25 or Section 26 shall be punishable with imprisonment for a term which shall not be less than six months but which may extend to six years and with fine.

Enhanced penalty after previous conviction

45. If any person who has been convicted of any offence under Section 24 or Section 25 or Section 26 is again found guilty of an offence involving a contravention of the same provision, he shall, on the second and on every subsequent conviction, be punishable with imprisonment for a term which shall not be less than one year but which may extend to seven years and with fine:

Provided that for the purpose of this Section no cognizance shall be taken of any conviction made more than two years before the commission of the offence which is being punished.

Publication of names of offenders

46. If any person convicted of an offence under this Act commits a like offence afterwards it shall be lawful for the court before which the second or subsequent conviction takes place to cause the offender's name and place of residence, the offence and the penalty imposed to be published at the offender's expense in such newspapers or in such other manner as the court may direct and the expenses of such publication shall be deemed to be part of the cost attending the conviction and shall be recoverable in the same manner as a fine.

Offences by companies

47. (1) Where an offence under this Act has been committed by a company, every person who at the time the offence was committed was in charge of, and was responsible to the company for the conduct of, the business of the company as well as the company, shall be deemed to be guilty of the offence and shall be liable to be proceeded against and punished accordingly:

Provided that nothing contained in this Sub-section shall render any such person liable to any punishment provided in this Act if he proves that the offence was committed without his knowledge or that he exercised all due diligence to prevent the commission of such offence.

(2) Notwithstanding anything contained in Sub-section (1), where an offence under this Act has been committed by a company and it is proved that the offence has been committed with the consent or

connivance of, or is attributable to any neglect on the part of, any director, manager, secretary or other officer shall also be deemed to be guilty of that offence and shall be liable to be proceeded against and punished accordingly.

Explanation — For the purposes of this Section—

(a) "company" means any body corporate, and includes a firm or other association of individuals; and

(b) "director" in relation to a firm means a partner in the firm.

Offences by Government Departments

48. Where an offence under this act has been committed by any Department of Government, the Head of the Department shall be deemed to be guilty of the offence and shall be liable to be proceeded against and punished accordingly:

Provided that nothing contained in this Section shall render such Head of the Department liable to any punishment if he proves that the offence was committed without his knowledge or that he exercised all due diligence to prevent the commission of such offence.

Cognizance of offences

49. (1) No court shall take cognizance of any offence under this Act except on a complaint made by, or with the previous sanction in writing of the State Board, and no court inferior to that of a Presidency Magistrate or a Magistrate of the first class shall try any offence punishable under this Act.

(2) Notwithstanding anything contained in Section 32 of the Code of Criminal Procedure, 1898 (5 of 1898), it shall be lawful for any Magistrate of the first class or for any Presidency Magistrate to pass a sentence of imprisonment for a term exceeding two years or of fine exceeding two thousand rupees on any person convicted of an offence punishable under this Act.

Members, officers and servants of Board to be public servants

50. All members, officers and servants of a Board when acting or purporting to act in pursuance of any of the provisions of this Act and the rules made thereunder shall be deemed to be public servants within the meaning of Section 21 of the Indian Penal Code (45 of 1860).

Chapter VIII

MISCELLANEOUS

Central Water Laboratory

51. (1) The Central Government may, by notification in the Official Gazette—

(a) establish a Central Water Laboratory; or

(b) specify any laboratory or institute as a Central Water Laboratory, to carry out the functions entrusted to the Central Water Laboratory under this Act.

(2) The Central Government may, after consultation with the Central Board, make rules prescribing—

(a) the functions of the Central Water Laboratory;

(b) the procedure for the submission to the said laboratory of samples of water or of sewage or trade effluent for analysis or tests, the form of the laboratory's report thereunder and the fees payable in respect of such report;

(c) such other matters as may be necessary or expedient to enable that laboratory to carry out its functions.

State Water Laboratory

52. (1) The State Government may, by notification in the Official Gazette—

(a) establish a State Water Laboratory; or

(b) specify any laboratory or institute as a State Water Laboratory, to carry out the functions entrusted to the State Water Laboratory under this Act.

(2) The State Government may, after consultation with the State Board, make rules prescribing—

(a) the functions of the State Water Laboratory;

(b) the procedure for the submission to the said laboratory of samples of water or of sewage or trade effluent for analysis or tests, the form of the laboratory's report thereon and the fees payable in respect of such report;

(c) such other matters as may be necessary or expedient to enable that laboratory to carry out its functions.

Analysts

53. (1) The Central Government may, by notification in the Official Gazette, appoint such persons as it thinks fit and having the prescribed qualifications to be Government analysts for the purpose of analysis of samples of water or of sewage or trade effluent sent for analysis to any laboratory established or specified under Sub-section (1) of Section 51.

(2) The State Government may, by notification in the Official Gazette, appoint such persons as it thinks fit and having the prescribed qualifications to be Government analysts for the purpose of analysis of samples of water or of sewage or trade effluent sent for analysis to any laboratory established or specified under Sub-section (1) of Section 52.

(3) Without prejudice to the provisions of Sub-section (5) of Section 12, the Central Board or, as the case may be the State Board may, by notification in the Official Gazette, and with the approval of the Central Government or the State Government, as the case may be, appoint such persons as it thinks fit and having the prescribed qualifications to be Board analysts for the purpose of analysis of samples of water or of sewage or trade effluent sent for analysis to any laboratory established or recognised under Section 16, or, as the case may be, under Section 17.

Reports of analysts

54. Any document purporting to be a report signed by a Government analyst or, as the case may be, a Board analyst may be used as evidence of the facts stated therein in any proceeding under this Act.

Local authorities to assist

55. All local authorities shall render such help and assistance and furnish such information to the Board as it may require for the discharge of its functions, and shall make available to the Board for inspection and examination such records, maps, plans and other documents as may be necessary for the discharge of its functions.

Compulsory acquisition of land for the State Board

56. Any land required by a State Board for the efficient performance of its functions under this Act shall be deemed to be needed for a public purpose and such land shall be acquired for the State Board under the provisions of the Land Acquisition Act, 1894 (1 of 1894), or under any other corresponding law for the time being in force.

Reports

57. The Central Board shall furnish to the Central Government, and a State Board shall furnish to the State Government and to the Central Board, such reports, returns, statistics, accounts and other information with respect to its fund or activities as that Government, or, as the case may be, the Central Board may, from time to time, require.

Bar of jurisdiction

58. No civil court shall have jurisdiction to entertain any suit or proceeding in respect of any matter which an appellate authority constituted under this Act is empowered by or under this Act to determine, and no injunction shall be granted by any court or other authority in respect of any action taken or to be taken in pursuance of any power conferred by or under this Act.

Protection of action taken in good faith

59. No suit or other legal proceedings shall lie against the Government or any officer of Government or any member or officer of a Board in respect of anything which is in good faith done or intended to be done in pursuance of this Act or the rules made thereunder.

Overriding effect

60. The provisions of this Act shall have effect notwithstanding anything inconsistent therewith contained in any enactment other than this Act.

Power of Central Government to supersede the Central Board and Joint Boards

61. (1) If at any time the Central Government is of opinion—

(a) that the Central Board or any Joint Board has persistently made default in the performance of the functions imposed on it by or under this Act; or

(b) that circumstances exist which render it necessary in the public interest so to do, the Central Government may, by notification in the Official Gazette, supersede the Central Board or such Joint Board, as the case may be, for such period, not exceeding one year, as may be specified in the notification:

Provided that before issuing a notification under this Sub-section for the reasons mentioned in Clause (a), the Central Government shall give a reasonable opportunity to the Central Board or such Joint Board, as the case may be, to show cause why it should not be superseded and shall consider the explanations and objections, if any, of the Central Board or such Joint Board, as the case may be.

(2) Upon the publication of a notification under Sub-section (1) superseding the Central Board or any Joint Board—

(a) all the members shall, as from the date of supersession, vacate their offices as such;

(b) all the powers, functions and duties which may, by or under this Act, be exercised, performed or discharged by the Central Board or such Joint Board shall, until the Central Board or the Joint Board, as the case may be, is reconstituted under Sub-section (3) be exercised, performed or discharged by such person or persons as the Central Government may direct;

(c) all property owned or controlled by the Central Board or such Joint Board shall until the Central Board or the Joint Board, as the case may be, is reconstituted under Sub-section (3) vest in the Central Government.

(3) On the expiration of the period of supersession specified in the notification issued under Sub-section (1), the Central Government may—

(a) extend the period of supersession for such further term, not exceeding six months, as it may consider necessary; or

(b) reconstitute the Central Board or the Joint Board, as the case may be, by fresh nomination or appointment, as the case may be, and in such case any person who vacated his office under Clause (a) of Sub-section (2) shall not be deemed disqualified for nomination or appointment:

Provided that the Central Government may, at any time before the expiration of the period of supersession, whether originally specified under Sub-section (1) or as extended under this Sub-section, take action under Clause (b) of this Sub-section.

Power of State Government to supersede State Board

62. (1) If at any time the State Government is of opinion—

(a) that the State Board has persistently made default in the performance of the functions imposed on it by or under this Act; or

(b) that circumstances exist which render it necessary in the public interest so to do,

the State Government may, by notification in the Official Gazette, supersede the State Board for such period, not exceeding one year, as may be specified in the notification:

Provided that before issuing a notification under this Sub-section for the reasons mentioned in Clause (a), the State Government shall give a reasonable opportunity to the State Board to show cause why it should not be superseded and shall consider the explanations and objections, if any, of the State Board.

(2) Upon the publication of a notification under Sub-section (1) superseding the State Board, the provisions of Sub-sections (2) and (3) of Section 61 shall apply in relation to the supersession of the State Board as they apply in relation to the supersession of the Central Board or a Joint Board by the Central Government.

Power of Central Government to make rules

63. (1) The Central Government may, simultaneously with the constitution of the Central Board, make rules in respect of the matters specified in Sub-section (2):

Provides that when the Central Board has been constituted, no such rule shall be made, varied, amended or repealed without consulting the Board.

(2) In particular, and without prejudice to the generality of the foregoing power, such rules may provide for all or any of the following matters, namely:

(a) the terms and conditions of service of the members (other than the Chairman and Member-Secretary) of the Central Board under Sub-section (8) of Section 5;

(b) the intervals and the time and place at which meetings of the Central Board or of any committee thereof constituted under this Act, shall be held and the procedure to be followed at such meetings, including the quorum necessary for the transaction of business under Section 8, and under Sub-section (2) of Section 9;

(c) the fees and allowances to be paid to such members of a committee of the Central Board as are not members of the Board under Sub-section (3) of Section 9;

(d) the manner in which and the purposes for which persons may be associated with a Board under Sub-section (1) of Section 10;

(e) the terms and conditions of service of the Chairman and the Member-Secretary of the Central Board under Sub-section (9) of Section 5 and under Sub-section (1) of Section 12;

(f) conditions subject to which a person may be appointed as a consulting engineer to the Central Board under Sub-section (4) of Section 12;

(g) the powers and duties to be exercised and performed by the Chairman and the Member-Secretary of the Central Board;

(h) the prohibition or regulation of bathing in any stream or well or the washing or cleaning therein of things of any class or description, or the putting of litter or other objectionable matter, whether poisonous, noxious or polluting or not, into any stream or well;

(i) the prohibition or regulation of the keeping or use, on any stream, of vessels provided with sanitary appliances from which polluting matter passes into the stream;

(j) the form of the report of the Central Board analyst under Sub-section (1) of Section 22;

(k) the form of the report of the Government analyst under Sub-section (3) of Section 22;

(l) the form in which, and the time within which, the budget and annual report of the Central Board may be prepared and forwarded to the Central Government under Sections 38 and 39;

(m) the form in which the accounts of the Central Board may be maintained under Section 40;

(n) any other matter relating to the Central Board, including the powers and functions of that Board in relation to Union Territories;

(o) any other matter which has to be, or may be, prescribed.

(3) Every rule made by the Central Government under this Act shall be laid, as soon as may be after it is made, before each House of Parliament while it is in session for a total period of thirty days whichever may be comprised in one session or in two or more successive sessions, and if, before the expiry of the session in which it is so laid or the successive sessions aforesaid, both Houses agree in making any modification in the rule or both Houses agree that the rule should not be made, the rule shall thereafter have effect only in such modified form or be of no effect, as the case may be; so, however, that any such modification or annulment shall be without prejudice to the validity of anything previously done under that rule.

Power of State Government to make rules

64. (1) The State Government may, simultaneously with the constitution of the State Board, make rules to carry out the purposes of this Act in respect of matters not falling within the purview of Section 63:

Provided that, when the State Board has been constituted, no such rule shall be made, varied, amended or repealed without consulting that Board.

(2) In particular, and without prejudice to the generality of the foregoing power, such rules may provide for all or any of the following matters, namely:

(a) the terms and conditions of service of the members (other than the Chairman and the Member-Secretary) of the State Board under Sub-section (8) of Section 5;

(b) the time and place of meetings of the State Board or of any committee of that Board constituted under this Act and the procedure to be followed at such meeting including the quorum necessary for the transaction of business under Section 8 and under Sub-section (2) of Section 9;

(c) the fees and allowances to be paid to such members of a committee of the State Board as are not members of the Board under Sub-section (2) of Section 9;

(d) the manner in which and the purposes for which persons may be associated with the State Board under Sub-section (1) of Section 10;

(e) the terms and conditions of service of the Chairman and the Member-Secretary of the State Board under Sub-section (9) of Section 5 and under Sub-section (1) of Section 12;

(f) the conditions subject to which a person may be appointed as a consulting engineer to the State Board under Sub-section (4) of Section 12;

(g) the powers and duties to be exercised and discharged by the Chairman and the Member-Secretary of the State Board;

(h) the form of the notice referred to in Section 21;

(i) the form of the report of the State Board analyst under Sub-section (1) of Section 22;

(j) the form of the report of the Government analyst under Sub-section (3) of Section 22;

(k) the form of application for the consent of the State Board under Sub-section (2) of Section 25, and the particulars it may contain;

(l) the manner in which inquiry under Sub-section (3) of Section 25 may be made in respect of an application for obtaining consent of the State Board and the matters to be taken into account in granting or refusing such consent;

(m) the form and manner in which appeals may be filed, the fees payable in respect of such appeals and the procedure to be followed by the appellate authority in disposing of the appeals under Sub-section (3) of Section 28;

(n) the form in which, and the time within which, the budget and annual report of the State Board may be prepared and forwarded to the State Government under Sections 38 and 39;

(o) the form in which the accounts of the State Board may be maintained under Sub-section (1) of Section 40;

(p) any other matter which has to be, or may be, prescribed.

THE WATER (Prevention and Control of Pollution) AMENDMENT ACT, 1978

No. 44 of 1978

(12th December, 1978)

An Act to amend the Water (Prevention and Control of Pollution) Act, 1978

Whereas, in pursuance of Clause (1) of Article 252 of the Constitution, the Water (Prevention and Control of Pollution) Act, 1974, had been passed by Parliament;

And whereas from the practical experience gained in the working of the aforesaid Act it is considered necessary to make certain amendments thereto;

And whereas, in pursuance of Clause (1) of Article 252 of the Constitution read with Clause (2) thereof, resolution have been passed by the Legislative Assemblies of the States of Assam, Haryana and West Bengal to the affect that the said Act should be amended by an Act of Parliament for the purposes hereinafter appearing;

Be it enacted by Parliament in the Twenty-ninth Year of the Republic of India as follows:

Short title, application and Commencement

1. (1) This Act may be called the Water (Prevention and Control of Pollution) Amendment Act, 1978.

(2) It applies in the first instance, to the whole of the States of Assam, Haryana and West Bengal and the Union Territories; and it shall apply to such other State which adopts this Act by resolution passed in that behalf under Clause (1) of Article 252 of the Constitution read with Clause (2) thereof.

(3) It shall come into force, at once in the States of Assam, Haryana and West Bengal and the Union Territories and in any other State which adopts this Act under Clause (1) of Article 252 of the constitution read with Clause (2) thereof on the date of such adoption and any reference in Section 22 of this Act of the commencement of this Act shall, in relation to any State or Union Territory, mean the date on which this Act comes into force in such State or Union Territory.

STATEMENT OF OBJECTS AND REASONS

In the process of implementation of the Water (Prevention and Control of Pollution) Act 1974, in various States, certain drawbacks have come to the notice of the Government and consequently it has become necessary to make certain amendments in the Act.

2. Section 44 of the Act requires that those State Governments which have adopted the Act should set up the State Boards within six months from the date of adoption. One of the fifteen States which have so far adopted the Act, has not set up the State Board within this time limit. It has been felt that the prescribed time limit of six months for constitution of the State Boards may be done away with. Further, it has come to the notice of the Government that some State Governments have set up the State Boards after the prescribed time limit of six months. It is, therefore, necessary to regularise the setting up of these Boards and also the action of the State Boards constituted after the prescribed time limit of six months.

3. Moreover, certain States are finding it difficult to provide a fulltime Chairman for the State Boards. It is, therefore, proposed to amend the Act to provide for the appointment of a Chairman of the State Board either on fulltime or on part-time basis, depending on the specific situation and as the State Government thinks fit.

4. It is felt that there should be an integrated approach for tackling the water and air pollution problem. It is, therefore, proposed that the existing Boards for the Prevention and Control of Water Pollution should also be authorised to perform functions relating to the prevention, control and abatement of air pollution. The Government has already introduced a Bill, namely the Air (Prevention and Control of Pollution) Bill, 1978, in the Lok Sabha on the 17th April 1978. It has thus become necessary to authorise the Central Board and the State Boards for the Prevention and Control of Water Pollution to spend from their funds for performing functions relating to prevention, control and abatement of air pollution also.

5. In addition, certain other minor amendments are also proposed in the Bill.

Amendment of Section 2

2. In Section 2 of the Water (Prevention and Control of Pollution) Act, 1974 (6 of 1974) (hereinafter referred to as a Principal Act) —

(a) after Clause (d), the following clause shall be inserted, namely:

(dd) "outlet" includes any conduit pipe or channel, open or closed, carrying sewage or trade effluent or any other holding arrangement which causes, or is likely to cause, "pollution";

(b) after Clause (g), the following clause shall be inserted, namely:

(gg) "sewer" means any conduit pipe or channel, open or closed, carrying sewage or trade effluent.

Amendment of Section 3

3. In Section 3 of the Principal Act, in Sub-section (2) —

(a) in Clause (a), for the words "matters relating to the use and conservation of water resources or the prevention and control of water pollution", the words "matters relating to environmental protection" shall be substituted;

(b) in Clause (b), for the words "five officials", the words "such number of officials, not exceeding five," shall be substituted;

(c) in Clause (d), for the words "three non-officials", the words "such number of non-officials, not exceeding three," shall be substituted;

(d) in Clause (f), for the words "and having administrative experience", the words "and having practical experience in respect of matters relating to environmental protection" shall be substituted.

Amendment of Section 4

4. In Section 4 of the Principal Act—

(a) in Sub-section (1), the brackets and words "Being a date not later than six months of the commencement of this Act in the State)" shall be omitted and shall be deemed always to have been omitted;

(c) in Sub-section (2)—

(i) in Clause (a):

(1) the word "full-time" shall be omitted;

(2) for the words "matters relating to the use and conservation of water resources or the prevention and control of water pollution" the words "matters relating to environmental protection" shall be substituted;

(3) the following proviso shall be inserted at the end namely:

"provided that the Chairman may be either wholetime or parttime as the State Government may think fit";

(ii) in Clause (b), for the words "five officials", the words "such number of officials, not exceeding five", shall be substituted;

(iii) in Clause (c), for the words "five persons", the words "such number of persons, not exceeding five", shall be substituted;

(iv) in Clause (d), for the words "three non-officials", the words "such number of non-officials, not exceeding three", shall be substituted;

(v) in Clause (f), for the words, "and having administrative experience", the words "and having practical experience in matters relating to environmental protection" shall be substituted.

Amendment of Section 5

5. In Section 5 of the Principal Act—

(a) for Sub-section (2), the following Sub-section shall be substituted, namely:

"(2) The term of office of a member of a Board nominated under Clause (b) or Clause (e) of Sub-section (2) of Section (3) or Clause (b) or Clause (e) of Sub-section (2) of Section 4 shall come to an end as soon as he ceases to hold the office under the Central Government or the State Government or, as the case may be, the company or corporation owned, controlled or managed by the Central Government or the State Government, by virtue of which he was nominated."

(b) in Sub-section (5), for the words, brackets, letters and figures "or where he is nominated under Clause (c) of Sub-section (2) of Section 3 or under Clause (c) of sub-section (2) of Section 4, if he ceases to be a member of the State Board, or as the case may be of the local authority", the following shall be substituted, namely:

"or where he is nominated under Clause (c) or Clause (e) or Sub-section (2) of Section 3 or under Clause (c) or Clause (e) of Sub-section (2) of Section 4, if he ceases to be a member of the State Board or of the local authority or, as the case may be, of the company or corporation owned, controlled or managed by the Central Government or the State Government and such vacation of seat shall, in either case, take effect from such date as the Central Government or, as the case may be, the State Government may, by notification in the Official Gazette, specify".

6. In Section 10 of the Principal Act, after Sub-section (2), the following Sub-section shall be inserted, namely:

(3) "A person associated with the Board under Sub-section (1) for any purpose shall be paid such fees and allowances, for attending its meetings and for attending to any other work of the Board, as may be prescribed".

Insertion of new Section 11A

7. After Section 11 of the Principal Act, the following Section shall be inserted, namely:

Delegation of powers to Chairman

"11A. The Chairman of a Board shall exercise such powers and perform such duties as may be prescribed or as may, from time to time, be delegated to him by the Board".

Amendment of Section 12

8. In Section 12 of the Principal Act —

(a) in Sub-section (3), the words "and the rules so made may provide for the salaries and allowances and other terms and conditions of service of such officers and employees" shall be omitted;

(b) after Sub-section (3), the following Sub-section shall be inserted, namely:

"(3A) the method of recruitment and the terms and conditions of service (including the scales of pay) of the officers (other than the Member-Secretary) and other employees of the Central Board or a State Board shall be such as may be determined by regulations made by the Central Board or, as the case may be, by the State Board:

Provided that no regulation made under this Sub-section shall take effect unless—

(a) in the case of a regulation made by the Central Board, it is approved by the Central Government; and

(b) in the case of a regulation made by the Central Board, it is approved by the State Government."

Amendment of Section 14

9. In Section 14 of the Principal Act—

(a) in Sub-section (1)—

(i) in Clause (a), for the words "matters relating to the use and conservation of water resources or the prevention and control of Water Pollution", the words "matters relating to environmental protection" shall be substituted;

(ii) in Clause (f), for the words "and having administrative experience", the words "and having practical experience in respect of matters relating to environmental protection" shall be substituted;

(b) in Sub-section (2)—

(i) in Clause (a), for the words "matters relating to the use and conservation of water resources or the prevention and control of water pollution", the words "matters relating to environmental protection" shall be substituted;

(ii) in Clause (f), for the words "having administrative experience", the words "and having practical experience in respect of matters relating to environmental protection" shall be substituted.

Amendment of Section 21

10. In Section 21 of the Principal Act, for Sub-section (4) the following Sub-section shall be substituted, namely:

(1) When a scheme of any sewage or trade effluent is taken for analysis under Sub-section (1) and the person taking the sample serves on the occupier or his agent, a notice under Clause (a) of Sub-section (3) and the occupier or his agent wilfully absents himself, then,

(a) the sample so taken shall be placed in a container which shall be marked and sealed and shall also be signed by the person taking the sample and the same shall be sent forthwith by such person for analysis to the laboratory referred to in Sub-clause (i) or Sub-clause (ii), as the case may be of Clause (e) of Sub-section (3) and such person shall inform the Government analyst appointed under Sub-section (1) or Sub-section (2), as the case may be, of Section 53, in writing about the wilful absence of the occupier or his agent; and

(b) the cost incurred in getting such sample analysed shall be payable by the occupier or his agent and in case of default of such payment, the same shall be recoverable from the occupier or his agent as the case may be, as an arrear of land revenue or of public demand:

Provided that no such recovery shall be made unless the occupier or, as the case may be, his agent has been given a reasonable opportunity of being heard in the matter".

Amendment of Section 23

11. In Section 23 of the Principal Act, in Sub-section (2)—

(a) for the words and figures "the Code of Criminal Procedure, 1898", (5 of 1898), the words and figures "the Code of Criminal Procedure, 1873" (2 of 1874) shall be substituted;

(b) for the word and figures "Section 98", the word and figures "Section 94," shall be substituted.

Amendment of Section 25

12. In Section 25 of the Principal Act—

(a) in Sub-section (1), for the words "stream or well" at both the places where they occur, the words "stream or well or sewer or on land" shall be substituted;

(b) in Sub-section (2), after the words, brackets and figure "under Sub-section (1)" the words "shall be accompanied with such fees as may be prescribed and" shall be inserted;

(c) in Sub-section (4), in Clause (a), for the words "stream or well" the words "stream or well or sewer or on land" shall be substituted;

(d) in Sub-section (5), for the words "stream or well", the words "stream or well or sewer or on land" shall be substituted.

Amendment of Section 26

13. In Section 26 of the Principal Act—

(a) for the words "stream or well", the words "stream or well or sewer or on land" shall be substituted;

(b) for the words "shall be made within a period of three months of the constitution of the State Board" the words "shall be made on or before such date as may be specified by the State Government by notification in this behalf in the Official Gazette" shall be substituted.

Amendment of Section 27

14. In Section 27 of the Principal Act, for Sub-section (2), the following Sub-section shall be substituted, namely:

"(2) A State Board may from time to time review—

(a) any condition imposed under Section 25 (other than a condition to be satisfied before an outlet is brought into use or a new discharge is made), or Section 26 and may serve on the person using the outlet or making the discharge, as the case may be, a notice, making any reasonable variation of or revoking any such condition;

(b) the refusal of any consent referred to in Sub-section (1) of Section 25 or Section 26 or the grant of such consent without any condition, and made such orders as it deemed fit".

Amendment of Section 28

15. In Section 28 of the Principal Act, for Sub-section (2), the following Sub-section shall be substituted, namely:

"(2) An appellate authority shall consist of a single person or three persons, as the State Government may think fit, to be appointed by that Government".

Amendment of Section 36

16. In Section 36 of the Principal Act—

(a) in Sub-section (1), after the word "benefactions", the word "fees" shall be inserted;

(b) in Sub-section (2) after the words "under this Act", the words "and, where any law for the time being in force relating to the prevention, control or abatement of air pollution provides for the performance of any function under such law by the Central Board, also for performing its functions under such law" shall be inserted.

Amendment of Section 37

17. In Section 37 of the Principal Act—

(a) in Sub-section (1), after the word "benefactions", the word "fees" shall be inserted;

(b) in Sub-section (2), after the word "under this Act", the words "and, where any law for the time being in force relating to the prevention, control or abatement of air pollution provides for the performance of any function under such law by the State Board, also for performing its functions under such law" shall be substituted.

Statutory obligation of industries as mentioned in section 25 and 26 should be carefully adhered.

Amendment of Section 39

18. In Section 39 of the Principal Act, in Sub-section (1), for the words "six months" the words "nine months" shall be substituted.

Amendment of Section 49

19. In Section 49 of the Principal Act—

(a) in Sub-section (1), for the words "Presidency Magistrate or a Magistrate", the words "Metropolitan Magistrate or a Judicial Magistrate" shall be substituted;

(b) in Sub-section (2)

(i) for the words and figures "Section 32 of the Code of Criminal Procedure, 1898," (5 of 1898) the words and figures "Section 29 of the Code of Criminal Procedure, 1898" (2 of 1898) shall be substituted.

(ii) for the words "Magistrate of the first class or for any Presidency Magistrate", the words, "Judicial Magistrate of the first class or for any Metropolitan Magistrate" shall be substituted.

Amendment of Section 63

20. In Section 63 of the Principal Act

(a) in Sub-section (2)

(i) for Clause (d), the following clause shall be substituted, namely:

"(d) the manner in which and the purposes for which persons may be associated with the Central Board under Sub-section (1) of Section 10 and the fees and allowances payable to such persons";

(ii) Clauses (h) and (i) shall be omitted;

(b) in Sub-section (3), for the words "before the expiry of the session in which it is so laid or the successive sessions aforesaid", the words "before the expiry of the session immediately following the session or the successive sessions aforesaid" shall be substituted.

Amendment of Section 64

21. In Section 64 of the Principal Act, in Clause (d) of Sub-section (2), the words "and the fees and allowances payable to such persons" shall be inserted at the end.

Validation

22. Notwithstanding anything contained in Section 4 of the Principal Act, as it stood immediately before the commencement of this Act, every State Board for the prevention and control of water pollution constituted under that Section after the expiry of a period of six months of the commencement of the Principal Act in the State concerned, shall be deemed to have been validly constituted and accordingly anything done or any action taken by such State Board before the commencement of this Act shall be deemed to have been validly done or taken and no such thing or action shall be called in question in any court merely on the ground that such State Board was constituted after the expiry of the period specified therefor under the said Section 4.

43.9.2 THE AIR (PREVENTION AND CONTROL OF POLLUTION) ACT, 1981

NO. 14 OF 1981

The Gazette of India (Extraordinary) Part II Section I (Page 55 to 80)
[29th March 1981]

An Act to provide for the prevention, control and abatement of air pollution, for the establishment, with a view to carrying out the aforesaid purposes, of Boards, for conferring on and assigning to such Boards powers and functions relating thereto and for matters connected therewith.

Whereas decisions were taken at the United Nations Conference on the Human Environment held in Stockholm in June 1972, in which India participated, to take appropriate steps for the preservation of the natural resources of the earth which, among other things, include the presevation of the quality of air and control of air pollution;

And whereas it is considered necessary to implement the decisions aforesaid in so far as they relate to the preservation of the quality of air and control of air pollution;

Be it enacted by Parliament in the Thirty-second Year of the Republic of India as follows:

Chapter I

Preliminary

1. (1) This Act may be called the Air (Prevention and Control of Pollution) Act, 1981.

(2) It extends to the whole of India.

(3) It shall come into force on such date as the Central Government may, by notification in the Official Gazette, appoint.

2. In this Act, unless the context otherwise requires—

(a) "air pollutant" means any solid, liquid or gaseous substance present in the atmosphere in such concentration as may be or tend to be injurious to human beings or other living creatures or plants or property of environment;

(b) "air pollution" means the presence in the atmosphere of any air pollutant;

(c) "approved appliance" means any equipment or gadget used for the burning of any combustible material or for generating or consuming any fume, gas or particulate matter and approved by the State Board for the purposes of this Act;

(d) "approved fuel" means any fuel approved by the State Board for the purposes of this Act;

(e) "automobile" means any vehicle powered either by internal combustion engine or by any method of generating power to drive such vehicle by burning fuel;

(f) "Board" means the Central Board or a State Board;

(g) "Central Board" means the Central Board for the Prevention and Control of Water Pollution constituted under Section 3 of the Water (Prevention and Control of Pollution) Act, 1974;

(h) "chimney" includes any structure with an opening or outlet from or through which any air pollutant may be emitted;

(i) "control equipment" means any apparatus, device, equipment or system to control the quality and manner of emission of any air pollutant and includes any device used for securing the efficient operation of any industrial plant;

(j) "emission" means any solid or liquid or gaseous substance coming out of any chimney, duct or flue or any other outlet;

(k) "industrial plant" means any plant used for any industrial or trade purposes and emitting any air pollutant into the atmosphere;

(l) "member" means a member of the Central Board or a State Board, as the case may be, and includes the Chairman thereof;

(m) "occupier", in relation to any factory or premises means the person who has control over the affairs of the factory or the premises and where the said affairs are entrusted to a managing agent, such agent shall be deemed to be the occupier of the factory or the premises;

(n) "prescribed" means prescribed by rules made under this Act by the Central Government or, as the case may be, the State Government;

(o) "State Board" means—

(i) in relation to a State in which the Water (Prevention and Control of Pollution) Act, 1974, is in force and the State Government has constituted for that State a State Board for the Prevention and Control of Water Pollution under Section 4 of that Act, the said State Board; and

(ii) in relation to any other State, the State Board for the Prevention and Control of Air Pollution constituted by the State Government under Section 5 of this Act.

Chapter II

Central and State Boards for the Prevention and Control of Air Pollution

3. The Central Board for the Prevention and Control of Water Pollution constituted under Section 3 of the Water (Prevention and Control of Pollution) Act, 1974, shall, without prejudice to the exercise and performance of its powers and functions under that Act, exercise the powers and perform the functions of the Central Board for the Prevention and Control of Air Pollution under this Act.

4. In any State in which the Water (Prevention and Control of Pollution) Act, 1974, is in force and the State Government has constituted for that State a State Board for the Prevention and Control of Water Pollution under Section 4 of that Act, such State Board shall be deemed to be the State Board for the Prevention and Control of Air Pollution constituted under Section 5 of this Act and accordingly that State Board for the Prevention and Control of Water Pollution shall, without prejudice to the exercise and performance of its powers and functions under that Act, exercise the powers and perform the functions of the State Board for the Prevention and Control of Air Pollution under this Act.

5. (1) In any State in which the Water (Prevention and Control of Pollution) Act, 1974, is not in force, or that Act is in force but the State Government has not constituted a State Board for the Prevention and Control of Water Pollution under that Act, the State Government shall, with effect from such date as it may, by notification in the Official Gazette, appoint, constitute a State Board for the Prevention and Control of Air Pollution under such name as may be specified in the notification, to exercise the powers conferred on, and perform the functions assigned to, that Board under this Act.

(2) A State Board constituted under the Act shall consist of the following members, namely:

(a) a Chairman, being a person having special knowledge or practical experience in respect of matters relating to environmental protection, to be nominated by the State Government: Provided that the Chairman may be either wholetime or parttime as the State Government may think fit;

(b) such number of officials, not exceeding five, as the State Government may think fit, to be nominated by the State Government to represent that Government;

(c) such number of persons, not exceeding five, as the State Government may think fit, to be nominated by the State Government from amongst the members of the local authorities functioning within the State;

(d) such number of non-officials, not exceeding three, as the State Government may think fit, to be nominated by the State Government to represent the interest of agriculture, fishery or industry or trade or labour or any other interest, which, in the opinion of that Government, ought to be represented;

(e) two persons to represent the companies or corporations owned, controlled or managed by the State Government, to be nominated by that Government;

(f) a fulltime Member-Secretary having practical experience in respect of matters relating to environmental protection and having administrative experience, to be appointed by the State Government;

Provided that the State Government shall ensure that not less than two of the members are persons having special knowledge or practical experience in respect of matters relating to the improvement of the quality of air or the prevention, control or abatement of air pollution.

(3) Every State Board constituted under this Act shall be a body corporate with the name specified by the State Government in the notification issued under Sub-section (1), having perpetual succession and a common seal with power, subject to the provisions of this Act, to acquire and dispose of property and to contract, and may by the said name sue or be sued.

6. No State Board shall be constituted for a Union Territory and in relation to a Union Territory, the Central Board shall exercise the powers and perform the functions of a State Board under this Act for that Union Territory:

Provided that in relation to any Union Territory the Central Board may delegate all or any of its powers and functions under this Section to such person or body of persons as the Central Government may specify.

7. (1) Save as otherwise provided by or under this Act, a member of a State Board constituted under this Act, other than the Member-Secretary, shall hold office for a term of three years from the date on which his nomination is notified in the Official Gazette:

Provided that a member shall, notwithstanding the expiration of his term, continue to hold office until his successor enters upon his office.

(2) The term of office of a member of a State Board constituted under this Act and nominated under Clause (b) or Clause (e) of Sub-section (2) of Section 5 shall come to an end as soon as he ceases to hold the office under the State Government or, as the case may be, the company or corporation owned, controlled or managed by the State Government, by virtue of which he was nominated.

(3) A member of a State Board constituted under this Act, other than the Member-Secretary, may at any time resign his office by writing under his hand addressed—

(a) in the case of the Chairman, to the State Government; and

(b) in any other case, to the Chairman of the State Board, and the seat of the Chairman or such other member shall thereupon become vacant.

(4) A member of a State Board constituted under this Act, other than the Member-Secretary, shall be deemed to have vacated his seat, if he is absent without reason, sufficient in the opinion of the State Board, from three consecutive meetings of the State Board or where he is nominated under Clause (c) of Sub-section (2) of Section 5, he ceases to be a member of the local authority and such vacation of seat shall, in either case, take effect from such date as the State Government may, by notification in the Official Gazette, specify.

(5) A casual vacancy in a State Board constituted under this Act shall be filled by a fresh nomination and the person nominated to fill the vacancy shall hold office only for the remainder of the term for which the member whose place he takes was nominated.

(6) A member of a State Board constituted under this Act shall be eligible for renomination but not for more than two terms.

(7) The other terms and conditions of service of the Chairman and other members (except the Member-Secretary) of a State Board constituted under this Act shall be such as may be prescribed.

8. (1) No person shall be a member of a State Board constituted under this Act, who—

(a) is, or at any time has been, adjudged insolvent, or
(b) is of unsound mind and has been so declared by a competent court, or
(c) is, or has been, convicted of an offence which, in the opinion of the State Government, involves moral turpitude, or
(d) is, or at any time has been, convicted of an offence under this Act, or
(e) has directly or indirectly by himself or by any partner, any share or interest in any firm or company carrying on the business of manufacture, sale, or hire of machinery, industrial plant, control equipment or any other apparatus for the improvement of the quality of air or for the prevention, control or abatement of air pollution, or
(f) is a director or a secretary, manager or other salaried officer or employee of any company or firm having any contract with the Board, or with the Government constituting the Board or with a local authority in the State, or with a company or corporation owned, controlled or managed by the Government, for the carrying out of programmes for the improvement of the quality of air or for the prevention, control or abatement of air pollution, or
(g) has so abused, in the opinion of the State Government, his position as a member, as to render his continuance on the State Board detrimental to the interests of the general public.

(2) The State Government shall, by order in writing, remove any member who is, or has become, subject to any disqualification mentioned in Sub-section (1):

Provided that no order of removal shall be made by the State Government under this Section unless the member concerned has been given a reasonable opportunity of showing cause against the same.

(3) Notwithstanding anything contained in Sub-section (1) or Sub-section (6) of Section 7, a member who has been removed under this Section shall not be eligible to continue to hold office until his successor enters upon his office, or, as the case may be, for renomination as a member.

9. If a member of a State Board constituted under this Act becomes subject to any of the disqualifications specified in Section 8, his seat shall become vacant.

10. (1) For the purpose of this Act, a Board shall meet at least once in every three months and shall observe such rules of procedure in regard to the transaction of business at its meetings as may be prescribed:

Provided that if, in the opinion of the Chairman, any business of an urgent nature is to be transacted, he may convene a meeting of the Board at such time as he thinks fit for the aforesaid purpose.

(2) Copies of the minutes of the meetings under Sub-section (1) shall be forwarded to the Central Board and to the State Government concerned.

11. (1) A Board may constitute as many committees consisting wholly of members or partly of members and partly of other persons and for such purpose or purposes as it may think fit.

(2) A committee constituted under this Section shall meet at such time and at such place, and shall observe such rules of procedure in regard to the transaction of business at its meetings, as may be prescribed.

(3) The members of a committee other than the members of the Board shall be paid such fees and allowances, for attending its meetings and for attending to any other work of the Board as may be prescribed.

12. (1) A Board may associate with itself in such manner, and for such purposes, as may be prescribed, any person whose assistance or advice it may desire to obtain in performing any of its functions under this Act.

(2) A person associated with the Board under Sub-section (1) for any purpose shall have a right to take part in the discussions of the Board relevant to that purpose, but shall not have a right to vote at a meeting of the Board and shall not be a member of the Board for any other purpose.

(3) A person associated with a Board under Sub-section (1) shall be entitled to receive such fees and allowances as may be prescribed.

13. No act or proceeding of a Board or any committee thereof shall be called in question on the ground merely of the existence of any vacancy in, or any defect in the constitution of, the Board or such committee, as the case may be.

14. (1) The terms and conditions of service of the Member-Secretary of a State Board constituted under this Act shall be such as may be prescribed.

(2) The Member-Secretary of a State Board, whether constituted under this Act or not, shall exercise such powers and perform such duties as may be prescribed:

(3) Subject to such rules as may be made by the State Government in this behalf, a State Board, whether constituted under this Act or not, may appoint such officers and other employees as it considers necessary for the efficient performance of its functions under this Act.

(4) The method of appointment, the conditions of service and the scales of pay of the officers (other than the Member-Secretary) and other employees of a State Board appointed under Sub-section (3) shall be such as may be determined by regulations made by the State Board under this Act.

(5) Subject to such conditions as may be prescribed, a State Board constituted under this Act may from time to time appoint any qualified person to be a consultant to the Board and pay him such salary and allowances or fees, as it thinks fit.

15. A State Board may, by general or special order, delegate to the Chairman or the Member-Secretary or any other officer of the Board subject to such conditions and limitations, if any, as may be specified in the order, such of its powers and functions under this Act as it may deem necessary.

Chapter III

Powers and Functions of Boards

16. (1) Subject to the provisions of this Act, and without prejudice to the performance of its functions under the Water (Prevention and Control of Pollution) Act, 1974, the main functions of the Central Board shall be to improve the quality of air and to prevent, control or abate air pollution in the country.

(2) In particular and without prejudice to the generality of the foregoing functions, the Central Board may—

(a) advice the Central Government on any matter concerning the improvement of the quality of air and the prevention, control or abatement of air pollution;

(b) plan and cause to be executed a nation-wide programme for the prevention, control or abatement of air pollution;

(c) coordinate the activities of the State Boards and resolve disputes among them;

(d) provide technical assistance and guidance to the State Boards, carry out and sponsor investigations and research relating to problems of air pollution and prevention, control or abatement of air pollution;

(e) plan and organise the training of persons engaged or to be engaged in programmes for the prevention, control or abatement of air pollution on such terms and conditions as the Central Board may specify;

(f) organise through mass media a comprehensive programme regarding the prevention, control or abatement of air pollution;

(g) collect, compile and publish technical and statistical data relating to air pollution and the measures devised for its effective prevention, control or abatement and prepare manuals, codes or guides relating to prevention, control or abatement of air pollution;

(h) lay down standards for the quality of air;

(i) collect and disseminate information in respect of matters relating to air pollution;

(j) perform such other functions as may be prescribed.

(3) The Central Board may establish or recognise a laboratory or laboratories to enable the Central Board to perform its functions under this Section efficiently.

(4) The Central Board may—

(a) delegate any of its functions under this Act generally or specially to any of the committees appointed by it;

(b) do such other things and perform such other acts as it may think necessary for the proper discharge of its functions and generally for the purpose of carrying into effect the purposes of this Act.

17. (1) Subject to the provisions of this Act, and without prejudice to the performance of its functions, if any, under the Water (Prevention and Control of Pollution) Act, 1974, the functions of a State Board shall be—

(a) to plan a comprehensive programme for the prevention, control or abatement of air pollution and to secure the execution thereof;

(b) to advise the State Government on any matter concerning the prevention, control or abatement of air pollution;

(c) to collect and disseminate information relating to air pollution;

(d) to collaborate with the Central Board in organising the training of persons engaged or to be engaged in programmes relating to prevention, control or abatement of air pollution and to organise mass-education programme relating thereto;

(e) to inspect, at all reasonable times, any control equipment, industrial plant or manufacturing process and to give, by order, such direction to such persons as it may consider necessary to take steps for the prevention, control or abatement of air pollution;

(f) to inspect air pollution control areas at such intervals as it may think necessary, assess the quality of air therein and take steps for the prevention, control or abatement of air pollution in such areas;

(g) to lay down, in consultation with the Central Board and having regard to the standards for the quality of air laid down by the Central Board, standards for emission of air pollutants into the atmosphere from industrial plants and automobiles or for the discharge of any air pollutant into the atmosphere from any other source whatsoever not being a ship or an aircraft:

Provided that different standards for emission may be laid down under this clause for different industrial plants having regard to the quantity and composition of emission of air pollutants into the atmosphere from such industrial plants;

(h) to advise the State Government with respect to the suitability of any premises or location for carrying on any industry which is likely to cause air pollution;

(i) to perform such other functions as may be prescribed or as may, from time to time, be entrusted to it by the Central Board or the State Government;

(j) to do such other things and to perform such other acts as it may think necessary for the proper discharge of its functions and generally for the purpose of carrying into effect the purposes of this Act.

(2) A State Board may establish or recognise a laboratory or laboratories to enable the State Board to perform its functions under this Section efficiently.

18. In the performance of its functions under this Act—

(a) the Central Board shall be bound by such directions in writing as the Central Government may give to it; and

(b) every State Board shall be bound by such directions in writing as the Central Board or the State Government may give to it:

Provided that where a direction given by the State Government is inconsistent with the direction given by the Central Board, the matter shall be referred to the Central Government for its decision.

Chapter IV

Prevention and Control of Air Pollution

19. (1) The State Government may, after consultation with the State Board, by notification in the Official Gazette, declare in such manner as may be prescribed, any area or areas within the State as air pollution control area or areas for the purposes of this Act.

(2) The State Government may, after consultation with the State Board, by notification in the Official Gazette—

(a) alter any air pollution control area whether by way of extension or reduction;

(b) declare a new air pollution control area in which may be merged one or more existing air pollution control areas or any part or parts thereof.

(3) If the State Government, after consultation with the State Board, is of opinion that the use of any fuel, other than an approved fuel, in any air pollution control area or part thereof, may cause or is likely to cause air pollution, it may, by notification in the Official Gazette, prohibit the use of such fuel in such area or part thereof with effect from such date (being not less than three months from the date of publication of the notification) as may be specified in the notification.

(4) The State Government may, after consultation with the State Board, by notification in the Official Gazette, direct that with effect from such date as may be specified therein, no appliance, other than an approved appliance, shall be used in the premises situated in an air pollution control area:

Provided that different dates may be specified for different parts of an air pollution control area or for the use of different appliances.

(5) If the State Government after consultation with the State Board, is of opinion that the burning of any material (not being fuel) in any air pollution control area or part thereof may cause or is likely to cause air pollution, it may, by notification in the Official Gazette, prohibit the burning of such material in such area or part thereof.

20. With a view to ensuring that the standards for emission of air pollutants from automobiles laid down by the State Board under Clause (g) of Sub-section (1) of Section 17 are complied with, the State Government shall, in consultation with the State Board, give such instructions as may be deemed necessary

to the concerned authority in charge of registration of motor vehicles under the Motor Vehicles Act, 1939, and such authority shall, notwithstanding anything contained in that Act or the rules made thereunder, be bound to comply with such instructions.

21. (1) Subject to the provisions of this Section, no person shall, without the previous consent of the State Board, operate any industrial plant for the purpose of any industry specified in the Schedule in an air pollution control area.

(2) An application for consent of the State Board under Sub-section (1) shall be accompanied by such fees as may be prescribed and shall be made in the prescribed form and shall contain the particulars of the industrial plant and such other particulars as may be prescribed:

Provided that where any person, immediately before the declaration of any area as an air pollution control area, operates in such area any industrial plant for the purpose of any industry specified in the Schedule, such person shall make the application under this Sub-section within such period (being not less than three months from the date of such declaration) as may be prescribed and where such person makes such application, he shall be deemed to be operating such industrial plant with the consent of the State Board until the consent applied for has been refused.

(3) The State Board may make such inquiry as it may deem fit in respect of the application for consent referred to in Sub-section (1) and in making any such inquiry, shall follow such procedure as may be prescribed.

(4) Within a period of four months after the receipt of the application for consent referred to in Sub-section (1), the State Board shall, by order in writing, either grant or refuse, for reasons to be recorded in the order, the consent applied for.

(5) Every person to whom consent has been granted by the State Board under Sub-section (4), shall comply with the following conditions, namely:

(i) the control equipment of such specifications as the State Board may approve in this behalf shall be installed and operated in the premises where the industry is carried on or proposed to be carried on;

(ii) the existing control equipment, if any, shall be altered or replaced in accordance with the directions of the State Board;

(iii) the control equipment referred to in Clause (i) or Clause (ii) shall be kept at all times in good running condition;

(iv) chimney, wherever necessary, of such specifications as the State Board may approve in this behalf shall be erected or re-erected in such premises;

(v) such other conditions as the State Board may specify in this behalf; and

(vi) the conditions referred to in Clauses (i), (ii), and (iv) shall be complied with within such period as the State Board may specify in this behalf:

Provided that in the case of a person operating any industrial plant for the purpose of any industry specified in the Schedule in an air pollution control area immediately before the date of declaration of such area as an air pollution control area, the period so specified shall not be less than six months:

Provided further that—

(a) after the installation of any control equipment in accordance with the specifications under Clause (i), or

(b) after the alteration or replacement of any control equipment in accordance with the directions of the State Board under Clause (ii), or

(c) after the erection or re-erection of any chimney under Clause (iv),

No control equipment or chimney shall be altered or replaced or, as the case may be, crected or re-erected except with the previous approval of the State Board;

(6) If due to any technological improvement or otherwise the State Board is of opinion that all or any of the conditions referred to in Sub-section (5) require or requires variation (including the change of any control equipment, either in whole or in part), the State Board shall, after giving the person to whom consent has been granted an opportunity of being heard, vary all or any of such conditions and thereupon such person shall be bound to comply with the conditions as so varied.

(7) Where a person to whom consent has been granted by the State Board under Sub-section (4) transfers his interest in the industry to any other person, such consent shall be deemed to have been granted to such other person and he shall be bound to comply with all the conditions subject to which it was granted as if the consent was granted to him originally.

22. No person carrying on any industry specified in the Schedule or operating any industrial plant, in any air pollution control area, shall discharge or cause or permit to be discharged the emission of any air pollutant in excess of the standards laid down by the State Board under Clause (g) of Sub-section (1) of Section 17.

23. (1) Where in any air pollution control area the emission of any air pollutant into the atmosphere in excess of the standards laid down by the State Board occurs, or is apprehended to occur due to accident or other unforeseen act or event, the person in charge of the premises from where such emission occurs or is apprehended to occur shall forthwith intimate the fact of such occurrence or the apprehension of such occurrence to the State Board and to such authorities or agencies as may be prescribed.

(2) On receipt of information with respect to the fact or the apprehension of any occurrence of the nature referred to in Sub-section (1), where through intimation under that Sub-section or otherwise, the State Board and the authorities or agencies shall, as early as practicable, cause such remedial measures to be taken as are necessary to mitigate the emission of such air pollutants.

(3) Expenses, if any, incurred by the State Board, authority or agency with respect to the remedial measures referred to in Sub-section (2) together with interest (at such reasonable rate, as the State Government may, by order, fix) from the date when a demand for the expenses is made until it is paid, may be recovered by that Board, authority or agency from the person concerned, as arrears of land revenue, or of public demand.

24. (1) Subject to the provisions of this Section, any person empowered by a State Board in this behalf shall have a right to enter, at all reasonable times with such assistance as he considers necessary, any place—

(a) for the purpose of performing any of the functions of the State Board entrusted to him;

(b) for the purpose of determining whether and if so in what manner, any such functions are to be performed or whether any provisions of this Act or the rules made thereunder or any notice, order, direction or authorisation served, made, given or granted under this Act is being or has been complied with;

(c) for the purpose of examining and testing any control equipment, industrial plant, record, register, document or any other material object or for conducting a search of any place in which he has reason to believe that an offence under this Act or the rules made thereunder has been or is being or is about to be committed and for seizing any such control equipment, industrial plant, record, register, document or other material object if he has reasons to believe that it may furnish evidence of the commission of an offence punishable under this Act or the rules made thereunder.

(2) Every person carrying on any industry specified in the Schedule and every person operating any control equipment or any industrial plant, in an air pollution control area shall be bound to render all assistance to the person empowered by the State Board under Sub-section (1) for carrying out the functions under that Sub-section and if he fails to do so without any reasonable cause or excuse, he shall be guilty of an offence under this Act.

(3) If any person wilfully delays or obstructs any person empowered by the State Board under Sub-section (1) in the discharge of his duties, he shall be guilty of an offence under this Act.

(4) The provisions of the Code of Criminal Procedure, 1973, or, in relation to the State of Jammu and Kashmir, or any area in which that Code is not in force, the provisions of any corresponding law in force in that State or area, shall, so far as may be, apply to any search or seizure under this Section as they apply to any search or seizure made under the authority of a warrant issued under Section 94 of the said Code or, as the case may be, under the corresponding provisions of the said law.

25. For the purposes of carrying out the functions entrusted to it, the State Board or any officer empowered by it in that behalf may call for any information (including information regarding the types of air pollutants emitted into the atmosphere and the level of the emission of such air pollutants) from the occupier or any other person carrying on any industry or operating any control equipment or industrial plant and for the purpose of verifying the correctness of such information, the State Board or such officer shall have the right to inspect the premises where such industry, control equipment or industrial plant is being carried on or operated.

26. (1) A State Board or any officer empowered by it in this behalf shall have power to take, for the purpose of analysis, samples of air or emission from any chimney, fluc or duct or any other outlet in such manner as may be prescribed.

(2) The result of any analysis of a sample of emission taken under Sub-section (1) shall not be admissible in evidence in any legal proceeding unless the provisions of Sub-sections (3) and (4) are complied with.

(3) Subject to the provisions of Sub-section (4), when a sample of emission is taken for analysis under Sub-section (1), the person taking the sample shall—

(a) serve on the occupier or his agent, a notice, then and there in such form as may be prescribed, of his intention to have it so analysed;

(b) in the presence of the occupier or his agent, collect a sample of emission for analysis;

(c) cause the sample to be placed in a container or containers which shall be marked and sealed and shall also be signed both by the person taking the sample and the occupier or his agent;

(d) send, without delay, the container or containers to the laboratory established or recognised by the State Board under Section 17 or, if a request in that behalf is made by the occupier or his agent when the notice is served on him under Clause (a), to the laboratory established or specified under Sub-section (1) of Section 28.

(4) When a sample of emission is taken for analysis under Sub-section (1) and the person taking the sample serves on the occupier or his agent, a notice under Clause (a) of Sub-section (3), then—

(a) in a case where the occupier or his agent wilfully absents himself, the person taking the sample shall collect the sample of emission for analysis to be placed in a container or containers which shall be marked and sealed and shall also be signed by the person taking the sample, and

(b) in a case where the occupier or his agent is present at the time of taking the sample but refuses to sign the marked and sealed container or containers of the sample of emission as required under Clause (c) of Sub-section (3), the marked and sealed container or containers shall be signed by the person taking the sample.

and the container or containers shall be sent without delay by the person taking the sample for analysis to the laboratory established or specified under Sub-section (1) of Section 28 and such person shall inform the Government analyst appointed under Sub-section (1) of Section 29, in writing, about the wilful absence of the occupier or his agent, or, as the case may be, his refusal to sign the container or containers.

27. (1) Where a sample of emission has been sent for analysis to the laboratory established or recognised by the State Board, the Board analyst appointed under Sub-section (2) of Section 29 shall analyse the sample and submit a report in the prescribed form of such analysis in triplicate to the State Board.

(2) On receipt of the report under Sub-section (1), one copy of the report shall be sent by the State Board to the occupier or his agent referred to in Section 26, another copy shall be preserved for production before the court in case any legal proceedings are taken against him and the other copy shall be kept by the State Board.

(3) Where a sample has been sent for analysis under Clause (d) of Sub-section (3) or Sub-section (4) of Section 26 to any laboratory mentioned therein, the Government analyst referred to in the said Sub-section (4) shall analyse the sample and submit a report in the prescribed form of the result of the analysis in triplicate to the State Board which shall comply with the provisions of Sub-section (2).

(4) Any cost incurred in getting any sample analysed at the request of the occupier of his agent as provided in Clause (d) of Sub-section (3) of Section 26 or when he wilfully absents himself or refuses to sign the marked and sealed container or containers of sample of emission under Sub-section (4) of that Section, shall be payable by such occupier or his agent and in case of default the same shall be recoverable from him as arrears of land revenue or of public demand.

28. (1) The State Government may, by notification in the Official Gazette—

(a) establish one or more State Air Laboratories; or

(b) specify one or more laboratories or institutes as State Air Laboratories to carry out the functions entrusted to the State Air Laboratory under this Act.

(2) The State Government may, after consultation with the State Board, make rules prescribing—

(a) the functions of the State Air Laboratory;

(b) the procedure for the submission to the said Laboratory of samples of air or emission for analysis or tests, the form of the Laboratory's report thereon and the fees payable in respect of such report;

(c) such other matters as may be necessary or expedient to enable that Laboratory to carry out its functions.

29. (1) The State Government may, by notification in the Official Gazette, appoint such persons as it thinks fit and having the prescribed qualifications to be Government analysts for the purpose of analysis of samples of air or emission sent for analysis to any laboratory established or specified under Sub-section (1) of Section 28.

(2) Without prejudice to the provisions of Section 14, the State Board may, by notification in the Official Gazette, and with the approval of the State Government, appoint such persons as it thinks fit and having the prescribed qualifications to be Board analysts for the purpose of analysis of samples of air or emission sent for analysis to any laboratory established or recognised under Section 17.

30. Any document purporting to be a report signed by a Government analyst or, as the case may be, a State Board analyst may be used as evidence of the facts stated therein in any proceeding under this Act.

31. (1) Any person aggrieved by an order made by the State Board under this Act may, within thirty days from the date on which the order is communicated to him, prefer an appeal to such authority (hereinafter referred to as the Appellate Authority) as the State Government may think fit to constitute:

Provided that the Appellate Authority may entertain the appeal after the expiry of the said period of thirty days if such authority is satisfied that the appellant was prevented by sufficient cause from filing the appeal in time.

(2) The Appellate Authority shall consist of a single person or three persons as the State Government may think fit to be appointed by the State Government.

(3) The form and the manner in which an appeal may be preferred under Sub-section (1), the fees payable for such appeal and the procedure to be followed by the Appellate Authority shall be such as may be prescribed.

(4) On receipt of an appeal preferred under Sub-section (1), the Appellate Authority shall, after giving the appellant and the State Board an opportunity of being heard, dispose of the appeal as expeditiously as possible.

Chapter V

Fund, Accounts and Audit

32. The Central Government may, after due appropriation made by Parliament by law in this behalf, make in each financial year such contributions to the State Boards as it may think necessary to enable the State Boards to perform their functions under this Act:

Provided that nothing in this Section shall apply to any State Board for the Prevention and Control of Water Pollution constituted under Section 4 of the Water (Prevention and Control of Pollution) Act, 1974, which is empowered by that Act, to expend money from its fund thereunder also for performing its functions, under any law for the time being in force relating to the prevention, control or abatement of air pollution.

33. (1) Every State Board shall have its own fund for the purposes of this Act and all sums which may, from time to time, be paid to it by the Central Government and all other receipts (by way of contributions, if any, from the State Government, fees, gifts, grants, donations, benefactions (or otherwise) of that Board shall be carried to the fund of the Board and all payments by the Board shall be made therefrom.

(2) Every State Board may expend such sums as it thinks fit for performing its functions under this Act and such sums shall be treated as expenditure payable out of the fund of that Board.

(3) Nothing in this Section shall apply to any State Board for the Prevention and Control of Water Pollution constituted under Section 4 of the Water (Prevention and Control of Pollution) Act, 1974, which is empowered by that Act to expend money from its fund thereunder also for performing its functions, under any law for the time being in force relating to the prevention, control or abatement of air pollution.

34. The Central Board or, as the case may be, the State Board shall, during each financial year, prepare, in such form and at such time as may be prescribed, a budget in respect of the financial year next ensuing showing the estimated receipt and expenditure under this Act, and copies thereof shall be forwarded to the Central Government or, as the case may be, the State Government.

35. (1) The Central Board shall, during each financial year, prepare, in such form and at such time as may be prescribed, an annual report giving full account of its activities under this Act during the previous financial year and copies thereof shall be forwarded to the Central Government and that Government shall cause every such report to be laid before both Houses of Parliament within six months of the date on which it is received by that Government.

(2) Every State Board shall, during each financial year, prepare, in such form and at such time as may be prescribed, an annual report giving full account of its activities under this Act during the previous financial year and copies thereof shall be forwarded to the State Government and that Government shall cause every such report to be laid before the State Legislature within a period of nine months of the date on which it is received by that Government.

36. (1) Every Board shall, in relation to its functions under this Act, maintain proper accounts and other relevant records and prepare an annual statement of accounts in such form as may be prescribed by the Central Government or, as the case may be, the State Government.

(2) The accounts of the Board shall be audited by an auditor duly qualified to act as an auditor of companies under Section 226 of the Companies Act, 1956.

(3) The said auditor shall be appointed by the Central Government or, as the case may be, the State Government on the advice of the Comptroller and Auditor-General of India.

(4) Every auditor appointed to audit the accounts of the Board under this Act shall have the right to demand the production of books, accounts, connected vouchers and other documents and papers and to inspect any of the offices of the Board.

(5) Every such auditor shall send a copy of his report together with an audited copy of the accounts to the Central Government or, as the case may be, the State Government.

(6) The Central Government shall, as soon as may be after the receipt of the audit report under Sub-section (5), cause the same to be laid before both Houses of Parliament.

(7) The State Government shall, as soon as may be after the receipt of the audit report under Sub-section (5), cause the same to be laid before State Legislature.

Chapter VI

Penalties and Procedure

37. (1) Whoever fails to comply with the provisions of Sub-section (5) of Section 21 or Section 22 or with any order or direction given under this Act shall, in respect of each such failure, be punishable with imprisonment for a term which may extend to three months or with fine which may extend to ten thousand rupees, or with both, and in case the failure continues, with an additional fine which may extend to one hundred rupees for every day during which such failure continues after the conviction for the first such failure.

(2) If the failure referred to in Sub-section (1) continues beyond a period of one year after the date of conviction, the offender shall be punishable with imprisonment for a term which may extend to six months.

38. Whoever—

(a) destroys, pulls down, removes, injures or defaces any pillar, post or stake fixed in the ground or any notice or other matter put up, inscribed or placed, by or under the authority of the Board, or

(b) obstructs any person acting under the orders or directions of the Board from exercising his powers and performing his functions under this Act, or

(c) damages any works or property belonging to the Board, or

(d) fails to furnish to the Board or any officer or other employee of the Board any information required by the Board or such officer or other employee for the purpose of this Act, or

(e) fails to intimate the occurrence of the emission of air pollutants into the atmosphere in excess of the standards laid down by the State Board or the apprehension of such occurrence, to the State Board and other prescribed authorities or agencies as required under Sub-section (1) of Section 23, or

(f) in giving any information which he is required to give under this Act, makes a statement which is false in any material particular, or

(g) for the purpose of obtaining any consent under Section 21, makes a statement which is false in any material particular,

shall be punishable with imprisonment for a term which may extend to three months or with fine which may extend to five hundred rupees or with both.

39. Whoever contravenes any of the provisions of this Act, for which no penalty has been elsewhere provided in this Act, shall be punishable with fine which may extend to five thousand rupees, and in the case of continuing contravention, with an additional fine which may extend to one hundred rupees for every day during which such contravention continues after conviction for the first such contravention.

40. (1) Where an offence under this Act has been committed by a company, every person who, at the time the offence was committed, was directly in charge of, and was responsible to the company for the conduct of the business of the company, as well as the company, shall be deemed to be guilty of the offence and shall be liable to be proceeded against and punished accordingly:

Provided that nothing contained in this Sub-section shall render any such person liable to any punishment provided in this Act, if he proves that the offence was committed without his knowledge or that he exercised all due diligence to prevent the commission of such offence.

(2) Notwithstanding anything contained in Sub-section (1), where an offence under this Act has been committed by a company and it is proved that the offence has been committed with the consent or connivance of, or is attributable to any neglect on the part of, any director, manager, secretary or other officer of the company, such director, manager, secretary or other officer shall also be deemed to be guilty of that offence and shall be liable to be proceeded against and punished accordingly.

Explanation—For the purposes of this Section—

(a) "company" means any body corporate, and includes a firm or other association of individuals; and

(b) "director", in relation to a firm, means a partner in the firm.

41. (1) Where an offence under this Act has been committed by any Department of Government, the Head of the Department shall be deemed to be guilty of the offence and shall be liable to be proceeded against and punished accordingly:

Provided that nothing contained in this Section shall render such Head of the Department liable to any punishment if he proves that the offence was committed without his knowledge or that he exercised all due diligence to prevent the commission of such offence.

(2) Notwithstanding anything contained in Sub-section (1), where an offence under this Act has been committed by a Department of Government and it is proved that the offence has been committed with the consent or connivance of, or is attributable to any neglect on the part of, any officer, other than the Head of the Department, such officer shall also be deemed to be guilty of that offence and shall be liable to be proceeded against and punished accordingly.

42. No suit, prosecution or other legal proceeding shall lie against the Government or any officer of the Government or any member or any officer or other employee of the Board in respect of anything which is done or intended to be done in good faith in pursuance of this Act or the rules made thereunder.

43. No court shall take cognizance of any offence under this Act except on a complaint made by, or with the previous sanction in writing of, the State Board, and no court inferior to that of a Metropolitan Magistrate or a Judicial Magistrate of the first class shall try any offence punishable under this Act.

44. All the members and all officers and other employees of a Board when acting or purporting to act in pursuance of any of the provisions of this Act or the rules made thereunder shall be deemed to be public servants within the meaning of Section 21 of the Indian Penal Code.

45. The Central Board shall, in relation to its functions under this Act, furnish to the Central Government, and a State Board shall, in relation to its functions under this Act, furnish to the State Government and to the Central Board such reports, returns, statistics, accounts and other information as that Government, or, as the case may be, the Central Board may, from time to time, require.

46. No civil court shall have jurisdiction to entertain any suit or proceeding in respect of any matter which an Appellate Authority constituted under this Act is empowered by or under this Act to determine, and no injunction shall be granted by any court or other authority in respect of any action taken or to be taken in pursuance of any power conferred by or under this Act.

Chapter VII

Miscellaneous

47. (1) If at any time the State Government is of opinion—

(a) that a State Board constituted under this Act has persistently made default in the performance of the functions imposed on it by or under this Act, or

(b) that circumstances exist which render it necessary in the public interest so to do,

the State Government may, by notification in the Official Gazette, supersede the State Board for such period, not exceeding six months, as may be specified in the notification;

Provided that before issuing a notification under this Sub-section for the reasons mentioned in Clause (a), the State Government shall give a reasonable opportunity to the State Board to show cause why it should not be superseded and shall consider the explanations and objections if any, of the State Board.

(2) Upon the publication of a notification under Sub-section (1) superseding the State Board—

(a) all the members shall, as from the date of supersession, vacate their offices as such;

(b) all the powers, functions and duties which may, by or under this Act, be exercised, performed or dishcarged by the State Board shall, until the State Board is reconstituted under Sub-section (3), be exercised, performed or discharged by such person or persons as the State Government may direct;

(c) all property owned or controlled by the State Board shall, until the Board is reconstituted under Sub-section (3), vest in the State Government.

(3) On the expiration of the period of supersession specified in the notification issued under Sub-section (1), the State Government may—

(a) extend the period of supersession for such further term, not exceeding six months, as it may consider necessary; or

(b) reconstitute the State Board by a fresh nomination or appointment, as the case may be, and in such case any person who vacated his office under Clause (a) of Sub-section (2) shall also be eligible for nomination or appointment;

Provided that the State Government may at any time before the expiration of the period of supersession, whether originally specified under Sub-section (1) or as extended under this Sub-section, take action under Clause (b) of this Sub-section.

48. Where the Central Board or any State Board constituted under the Water (Prevention and Control of Pollution) Act, 1974, is superseded by the Central Government or the State Government, as the case may be, under that Act, all the powers, functions and duties of the Central Board or such State Board under this Act shall be exercised, performed or discharged during the period of such supersession by the person or persons, exercising, performing or discharging the powers, functions and duties of the Central Board or such State Board under the Water (Prevention and Control of Pollution) Act, 1974, during such period.

49. (1) As and when the Water (Prevention and Control of Pollution) Act, 1974, comes into force in any State and the State Government constitutes a State Board for the Prevention and Control of Water Pollution under that Act, the State Board constituted by the State Government under this Act shall stand dissolved and the Board first-mentioned shall exercise the powers and perform the functions of the Board second-mentioned in that State.

(2) On the dissolution of the State Board constituted under this Act—

(a) all the members shall vacate their offices as such;

(b) all moneys and other property of whatever kind (including the fund of the State Board) owned by, or vested in, the State Board, immediately before such dissolution, shall stand transferred to and vest in the State Board for the Prevention and Control of Water Pollution;

(c) every officer and other employee serving under the State Board immediately before such dissolution shall be transferred to and become an officer or other employee of the State Board for the Prevention and Control of Water Pollution and hold office by the same tenure and at the same remuneration and on the same terms and conditions of service as he would have held the same if the State Board constituted under this Act had not been dissolved and shall continue to do so unless and until such tenure, remuneration and terms and conditions of service are duly altered by the State Board for the Prevention and Control of Water Pollution:

Provided that the tenure, remuneration and terms and conditions of service of any such officer or other employee shall not be altered to his disadvantage without the previous sanction of the State Government;

(d) all liabilities and obligations of the State Board of whatever kind, immediately before such dissolution, shall be deemed to be the liabilities or obligations, as the case may be, of the State Board for the Prevention and Control of Water Pollution and any proceeding or cause of action, pending or existing immediately before such dissolution by or against the State Board constituted under this Act in relation to such liability or obligation may be continued and enforced by or against the State Board for the Prevention and Control of Water Pollution.

50. (1) The Central Government may, of its own motion or on the recommendation of a Board, by notification in the Official Gazette, add to, or omit from, the Schedule any industry or alter the description of any industry and thereupon the Schedule shall be deemed to be amended accordingly.

(2) Every notification made under Sub-section (1) shall be laid, as soon as may be after it is made, before each house of Parliament.

51. (1) Every State Board shall maintain a register containing particulars of the persons to whom consent has been granted under Section 21, the standards for emission laid down by it in relation to each such consent and such other particulars as may be prescribed.

(2) The register maintained under Sub-section (1) shall be open to inspection at all reasonable hours by any person interested in or affected by such standards for emission or by any other person authorised by such person in this behalf.

52. Save as otherwise provided by or under the Atomic Energy Act, 1962, in relation to radioactive air pollution the provisions of this Act shall have effect notwithstanding anything inconsistent therewith contained in any enactment other than this Act.

53. (1) The Central Government may, in consultation with the Central Board, by notification in the Official Gazette, make rules in respect of the following matters, namely:

(a) the intervals and the time and place at which meetings of the Central Board or any committee thereof shall be held and the procedure to be followed at such meetings, including the quorum necessary for the transaction of business thereat, under Sub-section (1) of Section 10 and under Sub-section (2) of Section 11;

(b) the fees and allowances to be paid to the members of a committee of the Central Board, not being members of the Board, under Sub-section (3) of Section 11;

(c) the manner in which and the purposes for which persons may be associated with the Central Board under Sub-section (1) of Section 12;

(d) the fees and allowances to be paid under Sub-section (3) of Section 12 to persons associated with the Central Board under Sub-section (1) of Section 12;

(e) the functions to be performed by the Central Board under Clause (j) of Sub-section (2) of Section 16;

(f) the form in which and the time within which the budget and the annual report of the Central Board may be prepared and forwarded to the Central Government under Sections 34 and 35;

(g) the form in which the accounts of the Central Board may be maintained under Sub-section (1) of Section 36.

(2) Every rule made by the Central Government under this Act shall be laid, as soon as may be after it is made, before each House of Parliament, while it is in session, for a total period of thirty days which may be comprised in one session or in two or more successive sessions, and if, before the expiry of the session immediately following the session or the successive sessions aforesaid, both Houses agree in making any modification in the rule or both Houses agree that the rule should not be made, the rule shall thereafter have effect only in such modified form or be of no effect, as the case may be; so, however, that any such modification or annulment shall be without prejudice to the validity of anything previously done under that rule.

54. (1) Subject to the provisions of Sub-section (3), the State Government may, by notification in the Official Gazette, make rules to carry out the purposes of this Act in respect of matters not falling within the purview of Section 53.

(2) In particular, and without prejudice to the generality of the foregoing power, such rules may provide for all or any of the following matters, namely—

(a) the terms and conditions of service of the Chairman and other members (other than the Member-Secretary) of the State Board constituted under this Act under Sub-section (7) of Section 7;

(b) the intervals and the time and place at which meetings of the State Board or any committee thereof shall be held and the procedure to be followed at such meetings, including the quorum necessary for the transaction of business thereat, under Sub-section (1) of Section 10 and under Sub-section (2) of Section 11;

(c) the fees and allowances to be paid to the members of a committee of the State Board, not being members of the Board under Sub-section (3) of Section 11;

(d) the manner in which and the purposes for which persons may be associated with the State Board under Sub-section (1) of Section 12;

(e) the fees and allowances to be paid under Sub-section (3) of Section 12 to persons associated with the State Board under Sub-section (1) of Section 12;

(f) the terms and conditions of service of the Member-Secretary of a State Board constituted under this Act under Sub-section (1) of Section 14;

(g) the powers and duties to be exercised and discharged by the Member-Secretary of a State Board under Sub-section (2) of Section 14;

(h) the conditions subject to which a State Board may appoint such officers and other employees as it considers necessary for the efficient performance of its functions under Sub-section (3) of Section 14;

(i) the conditions subject to which a State Board may appoint a consultant under Sub-section (5) of Section 14;

(j) the functions to be performed by the State Board under Clause (i) of Sub-section (1) of Section 17;

(k) the manner in which any area or areas may be declared as air pollution control area or areas under Sub-section (1) of Section 19;

(l) the form of application for the consent of the State Board, the fees payable therefor, the period within which such application shall be made and the particulars it may contain, under Sub-section (2) of Section 21;

(m) the procedure to be followed in respect of an inquiry under Sub-section (3) of Section 21;

(n) the authorities or agencies to whom information under Sub-section (1) of Section 23 shall be furnished;

(o) the manner in which samples of air or emission may be taken under Sub-section (1) of Section 26;

(p) the form of the notice referred to in Sub-section (3) of Section 26;

(q) the form of the report of the State Board analyst under Sub-section (1) of Section 27;

(r) the form of the report of the Government analyst under Sub-section (3) of Section 27;

(s) the functions of the State Air Laboratory the procedure for the submission to the said

Laboratory of samples of air or emission for analysis on tests, the form of Laboratory's report thereon, the fees payable in respect of such report and other matters as may be necessary or expedient to enable that Laboratory to carry out its functions, under Sub-section (2) of Section 28;

(t) the qualifications required for Government analysts under Sub-section (2) of Section 29;

(u) the qualifications required for State Board analysts under Sub-section (2) of Section 29;

(v) the form and the manner in which appeals may be preferred, the fees payable in respect of such appeals and the procedure to be followed by the Appellate Authority in disposing of the appeals under Sub-section (3) of Section 31;

(w) the form in which and the time within which the budget and annual report of the State Board may be prepared and forwarded to the State Government under Sections 34 and 35;

(x) the form in which the accounts of the State Board may be maintained under Sub-section (1) of Section 36;

(y) the particulars which the register maintained under Section 51 may contain;

(z) any other matter which has to be, or may be, prescribed.

(3) After the first constitution of the State Board, no rule with respect to any of the matters referred to in Sub-section (2) [other than those referred to in Clause (a) thereof], shall be made, varied, amended or repealed without consulting that Board.

THE SCHEDULE

(See Sections 21, 22, 24 and 50)

1. Asbestos and asbestos products industries.
2. Cement and cement products industries.
3. Ceramic and ceramic products industries.
4. Chemical and allied industries.
5. Coal and lignite based chemical industries.
6. Engineering industries.
7. Ferrous metallurgical industries.
8. Fertilizer industries.
9. Foundries.
10. Food and agricultural products industries.
11. Mining industries.
12. Non-ferrous metallurgical industries.
13. Ores/mineral processing industries including beneficiation, pelletization, etc.
14. Power (coal, petroleum and their products) generating plants and boiler plants.
15. Paper and pulp (including paper products) industries.
16. Textile processing industries (made wholly or in part of cotton).
17. Petroleum refineries.
18. Petroleum products and petro-chemical industries.
19. Plants for recovery from and disposal of wastes.
20. Incinerators.

Statutory obligations under Air (Prevention and Control of Pollution) Act 1981, section 21 in particular should be followed. Section 8A of Bengal smoke nuisance Act 1903 was also followed.

43.9.3 THE ENVIRONMENT (PROTECTION) ACT, 1986

No. 29 of 1986

[23rd May 1986]

An Act to provide for the protection and improvement of environment and for matters connected therewith:

WHEREAS the decisions were taken at the United Nations Conference on the Human Environment held at Stockholm in June 1972, in which India participated, to take appropriate steps for the protection and improvement of human environment.

AND WHEREAS it is considered necessary further to implement the decisions aforesaid in so far as they relate to the protection and improvement of environment and the prevention of hazards to human beings, other living creatures, plants and property.

BE it enacted by Parliament in the Thirty-seventh Year of the Republic of India as follows:

Chapter I

PRELIMINARY

Short title, extent and commencement

1. (1) This Act may be called the Environment (Protection) Act, 1986.

(2) It extends to the whole of India.

(3) It shall come into force on such date as the Central Government may, by notification in the Official Gazette, appoint and different dates may be appointed for different provisions of this Act and for different areas.[1]

Definitions

2. In this Act, unless the context otherwise requires—

(a) "environment" includes water, air and land and the inter-relationship which exists among and between water, air and land, and human beings, other living creatures, plants, micro-organisms and property;

(b) "environmental pollutant" means any solid, liquid or gaseous substance present in such concentration as may be, or tend to be, injurious to environment;

(c) "environmental pollution" means the presence in the environment of any environmental pollutant;

(d) "handling", in relation to any substance, means the manufacture, processing, treatment, package, storage, transportation, use, collection, destruction, conversion, offering for sale, transfer or the like of such substance;

(e) "hazardous substance" means any substance or preparation which, by reason of its chemical or physico-chemical properties or handling, is liable to cause harm to human beings, other living creatures, plants, microorganisms, property or the environment;

(f) "occupier", in relation to any factory or premises, means a person who has control over the affairs of the factory or the premises and includes in relation to any substance, the person in possession of the substance;

(g) "prescribed" means prescribed by rules made under this Act.

[1]It came into force in the whole of India on 19th November 1986 vide Notification No. G.S.R. 1198(E) dated 12-11-86 published in the Gazette of India No. 525 dated 12-11-86.

Chapter II

GENERAL POWERS OF THE CENTRAL GOVERNMENT

Power of Central Government to take measures to protect and improve environment

3. (1) Subject to the provisions of this Act, the Central Government shall have the power to take all such measures as it deems necessary or expedient for the purpose of protecting and improving the quality of the environment and preventing, controlling and abating environmental pollution.

(2) In particular, and without prejudice to the generality of the provisions of Sub-section (1), such measures may include measures with respect to all or any of the following matters, namely:

- (i) coordination of actions by the State Governments, officers and other authorities—
  - (a) under this Act, or the rules made thereunder; or
  - (b) under any other law for the time being in force which is relatable to the objects of this Act;
- (ii) planning and execution of a nation-wide programme for the prevention, control and abatement of environmental pollution;
- (iii) laying down standards for the quality of environment in its various aspects;
- (iv) laying down standards for emission or discharge of environmental pollutants from various sources whatsoever;

Provided that different standards for emission or discharge may be laid down under this clause from different sources having regard to the quality or composition of the emission or discharge of environmental pollutants from such sources;

- (v) restriction of areas in which any industries, operations or processes or class of industries, operations or processes shall not be carried out or shall be carried out subject to certain safeguards;
- (vi) laying down procedures and safeguards for the prevention of accidents which may cause environmental pollution and remedial measures for such accidents;
- (vii) laying down procedures and safeguards for the handling of hazardous substances;
- (viii) examination of such manufacturing processes, materials and substances as are likely to cause environmental pollution;
- (ix) carrying out and sponsoring investigations and research relating to problems of environmental pollution;
- (x) inspection of any premises, plant, equipment, machinery, manufacturing or other processes, materials or substances and giving, by order, of such directions to such authorities, officers or persons as it may consider necessary to take steps for the prevention, control and abatement of environmental pollution;
- (xi) establishment or recognition of environmental laboratories and institutes to carry out the functions entrusted to such environmental laboratories and institutes under this Act;
- (xii) collection and dissemination of information in respect of matters relating to environmental pollution;
- (xiii) preparation of manuals, codes or guides relating to the prevention, control and abatement of environmental pollution;
- (xiv) such other matters as the Central Government deems necessary or expedient for the purpose of securing the effective implementation of the provisions of this Act.

(3) The Central Government may, if it considers it necessary or expedient so to do for the purpose of this Act, by order, published in the Official Gazette, constitute an authority or authorities by such name or

names as may be specified in the order for the purpose of exercising and performing such of the powers and functions (including the power to issue directions under Section 5) of the Central Government under this Act and for taking measures with respect to such of the matters referred to in Sub-section (2) as may be mentioned in the order and subject to the supervision and control of the Central Government and the provisions of such order, such authority or authorities may exercise the powers or perform the functions or take the measures so mentioned in the order as if such authority or authorities had been empowered by this Act to exercise those powers or perform those functions or take such measures.

Appointment of Officers and their powers and functions

4. (1) Without prejudice to the provisions of Sub-sections (3) of Section 3, the Central Government may appoint officers with such designation as it thinks fit for the purposes of this Act and may entrust to them such of the powers and functions under this Act as it may deem fit.

(2) The officers appointed under Sub-section (1) shall be subject to the general control and direction of the Central Government or, if so directed by that Government, also of the authority or authorities, if any, constituted under Sub-section (3) of Section 3 or of any other authority or officer.

Power to give directions

5. Notwithstanding anything contained in any other law but subject to the provisions of this Act, the Central Government[1] may, in the exercise of its powers and performance of its functions under this Act, issue directions in writing to any person, officer or any authority and such person, officer or authority shall be bound to comply with such directions.[2]

Explanation—For the avoidance of doubts, it is hereby declared that the power to issue directions under this Section includes the power to direct—

(a) the closure, prohibition or regulation of any industry, operation or process; or

(b) stoppage or regulation of the supply of electricity or water or any other service.

Rules to regulate environmental pollution

6. (1) The Central Government may, by notification in the Official Gazette, make rules in respect of all or any of the matters referred to in Section 3.

[1] The Central Government has delegated the powers vested in it under Section 5 of the Act to the State Governments of Andhra Pradesh, Assam, Bihar, Gujarat, Haryana, Himachal Pradesh, Karnataka, Kerala, Madhya Pradesh, Mizoram, Orissa, Rajasthan, Sikkim and Tamil Nadu subject to the condition that the Central Government may revoke such delegation of powers in respect of all or any one or more of the State Governments or may itself invoke the provisions of Section 5 of the Act, if in the opinion of the Central Government such a course of action is necessary in public interest, (Notification No. S.O. 152 (E) dated 10-2-88 published in Gazette No. 54 of the same date).

These Powers have been delegated to the following State Governments also on the same terms:

Meghalaya, Punjab and Uttar Pradesh vide Notification No. S.O. 389 (E) dated 14-4-88 published in the Gazette No. 205 dated 14-4-88;

Maharashtra vide Notification No. S.O. 488 (E) dated 17-5-88 published in the Gazette No. 255 dated 17-5-88;

Goa and Jammu & Kashmir vide Notification No. S.O. 881 (E) dated 22-9-88; published in the Gazette No. 749 dated 22.9.88;

West Bengal, Manipur vide Notification No. S.O. 408 (E) dated 6-6-89; published in the Gazette No. 319 dated 6-6-89;

Tripura vide Notification No. S.O. 479 (E) dated 25-7-91 published in the Gazette No. 414 dated 25-7-91.

[2] For issuing directions see r.4 of Environment (Protection) Rules, 1986.

(2) In particular, and without prejudice to the generality of the foregoing power, such rules may provide for all or any of the following matters, namely:

(a) the standards of quality of air, water or soil for various areas and purposes;[1]

(b) the maximum allowable limits of concentration of various environmental pollutants (including noise) for different areas;

(c) the procedures and safeguards for the handling of hazardous substances;[2]

(d) the prohibition and restrictions on the handling of hazardous substances in different areas;[3]

(e) the prohibition and restriction on the location of industries and the carrying on process and operations in different areas;[4]

(f) the procedures and safeguards for the prevention of accidents which may cause environmental pollution and for providing for remedial measures for such accidents.[5]

---

[1] See r. 3 of Environment (Protection) Rules, 1986, and Schedules thereto:

(i) Schedule I lists the standards for emission or discharge of environmental pollutants from the industries, processes or operations and their maximum allowable limits of concentration;

(ii) Schedule II lists general standards for discharge of effluents and their maximum limits of concentration allowable;

(iii) Schedule III lists ambient air quality standards in respect of noise and its maximum allowable limits; and

(iv) Schedule IV lists standards for emission of smoke, vapour etc. from motor vehicles and maximum allowable limits of their emission.

[2] See r.13 of Environment (Protection) Rules, 1986; and

(i) Hazardous Wastes (Management and Handling) Rules, 1989;

(ii) Manufacture, Storage and Import of Hazardous Chemicals Rules, 1989; and

(iii) Rules for the Manufacture, Use, Import, Export and Storage of Hazardous Microorganisms, Genetically-engineered organisms or Cells.

[3] Rule 13 SUPRA

[4] See r.5 of Environment (Protection) Rules, 1986.

[5] See r.12 of Environment (Protection) Rules and Schedule II, and relevant provisions of Hazardous Wastes (Management and Handling) Rules, Manufacture, Storage and Import of Hazardous Chemicals Rules and Rules for the Manufacture, Use, Import, Export and Storage of Hazardous Microorganisms, Genetically Engineered Organisms or Cells.

Chapter III

PREVENTION, CONTROL, AND ABATEMENT OF ENVIRONMENTAL POLLUTION

Persons carrying on industry operation, etc, not to allow emission or discharge of environmental pollutants in excess of the standards

7. No person carrying on any industry, operation or process shall discharge or emit or permit to be discharged or emitted any environmental pollutants in excess of such standards as may be prescribed.[1]

Persons handling hazardous substances to comply with procedural safeguards

8. No person shall handle or cause to be handled any hazardous substance except in accordance with such procedure and after complying with such safeguards as may be prescribed.[2]

Furnishing of information to authorities and agencies in certain cases

9. (1) Where the discharge of any environmental pollutant in excess of the prescribed standards occurs or is apprehended to occur due to any accident or other unforeseen act or event, the person responsible for such discharge and the person in charge of the place at which such discharge occurs or is apprehended to occur shall be bound to prevent or mitigate the environmental pollution caused as a result of such discharge and shall also forthwith—

(a) intimate the fact of such occurrence or apprehension of such occurrence; and

(b) be bound, if called upon, to render all assistance, to such authorities or agencies as may be prescribed.[3]

(2) On receipt of information with respect to the fact or apprehension of any occurrence of the nature referred to in Sub-section (1), whether through intimation under that Sub-section or otherwise, the authorities or agencies referred to in Sub-section (1) shall, as early as practicable, cause such remedial measures to be taken as are necessary to prevent or mitigate the environmental pollution.

(3) The expenses, if any, incurred by any authority or agency with respect to the remedial measures referred to in Sub-section (2), together with interest (at such reasonable rate as the Government may, by order, fix) from the date when a demand for the expenses is made until it is paid, may be recovered by such authority or agency from the person concerned as arrears of land revenue or of public demand.

Powers of entry and inspection

10. (1) Subject to the provisions of this Section, any person empowered by the Central Government in this behalf[4] shall have a right to enter, at all reasonable times with such assistance as he considers necessary, any place—

---

[1] See r.3 of Environment (Protection) Rules, 1986 and Schedule I.

[2] See r.13 of Environment (Protection) Rules, 1986 and

(i) Hazardous Wastes (Management and Handling) Rules, 1989;

(ii) Manufacture, Storage and Import of Hazardous Chemicals Rules, 1989; and

(iii) Rules for the Manufacture, Use, Import, Export and Storage of Hazardous Microorganisms, Genetically-Engineered organisms or Cells.

[3] For authorities or agencies see r.12 of Environment (Protection) Rules, 1986 and Schedule II.

[4] The Central Govt. has empowered 60 persons listed in the Table (p. 251) vide S.O.83 (E) published in the Gazette of India No. 66 dated 16-2-87 and S.O.63 (E) published in the Gazette of India No. 42 dated 18-1-88.

(a) for the purpose of performing any of the functions of the Central Government entrusted to him;

(b) for the purpose of determining whether and if so in what manner, any such functions are to be performed or whether any provisions of this Act or the rules made thereunder or any notice, order, direction or authorisation served, made, given or granted under this Act is being or has been complied with;

(c) for the purpose of examining and testing any equipment, industrial plant, record, register, document or any other material object or for conducting a search of any building in which he has reason to believe that an offence under this Act or the rules made thereunder has been or is being or is about to be committed and for seizing any such equipment, industrial plant, record, register, document or other material object if he has reason to believe that it may furnish evidence of the commission of an offence punishable under this Act or the rules made thereunder or that such seizure is necessary to prevent or mitigate environmental pollution.

(2) Every person carrying on any industry, operation or process of handling any hazardous substance shall be bound to render all assistance to the person empowered by the Central Government under Sub-section (1) for carrying out the functions under that Sub-section and if he fails to do so without any reasonable cause or excuse, he shall be guilty of an offence under this Act.

(3) If any person wilfully delays or obstructs any persons empowered by the Central Government under Sub-section (1) in the performance of his functions, he shall be guilty of an offence under this Act.

(4) The provisions of the Code of Criminal Procedure, 1973, or, in relation to the State of Jammu and Kashmir, or an area in which that Code is not in force, the provisions of any corresponding law in force in that State or area shall, so far as may be, apply to any search or seizures under this Section as they apply to any search or seizure made under the authority of a warrant issued under Section 94 of the said Code or as the case may be, under the corresponding provision of the said law.

Power to take sample and procedure to be followed in connection therewith

11. (1) The Central Government or any officer empowered by it in this behalf,[1] shall have power to take, for the purpose of analysis, samples of air, water, soil or other substance from any factory, premises or other place in such manner as may be prescribed.[2]

(2) The result of any analysis of a sample taken under Sub-section (1) shall not be admissible in evidence in any legal proceeding unless the provisions of Sub-sections (3) and (4) are complied with.

(3) Subject to the provisions of Sub-section (4), the person taking the sample under Sub-section (1) shall—

(a) serve on the occupier or his agent or person in charge of the place, a notice, then and there, in such form as may be prescribed, of his intention to have it so analysed;

(b) in the presence of the occupier or his agent or person, collect a sample for analysis;

[1] In exercise of powers conferred under Sub-section (i) of Section 11 the Central Government has empowered 60 officers listed in the Table (p. 254) vide S.O.84. (e) published in the Gazette No. 66 dated 16-2-87 and S.O.62 (E) published in the Gazette No. 42 dated 18-1-88.

[2] For procedure for taking samples see r. 6 of Environment (Protection) Rules, 1986, also.

(c) cause the sample to be placed in a container or containers which shall be marked and sealed and shall also be signed both by the person taking the sample and the occupier or his agent or person;

(d) send without delay, the container or the containers to the laboratory established or recognised by the Central Government under Section 12.

(4) When a sample is taken for analysis under Sub-section (1) and the person taking the sample serves on the occupier or his agent or person, a notice under Clause (a) of Sub-section (3), then—

(a) in a case where the occupier, his agent or person wilfully absents himself, the person taking the sample shall collect the sample for analysis to be placed in a container or containers which shall be marked and sealed and shall also be signed by the person taking the sample, and

(b) in a case where the occupier or his agent or person present at the time of taking the sample refuses to sign the marked and sealed container or containers of the sample as required under Clause (c) of Sub-section (3), the marked and sealed container or containers shall be signed by the person taking the samples,

and the container or containers shall be sent without delay by the person taking the sample for analysis to the laboratory established or recognised under Section 12 and such person shall inform the Government Analyst appointed or recognised under Section 13 in writing, about the wilful absence of the occupier or his agent or person, or, as the case may be, his refusal to sign the container or containers.

Environmental laboratories

12. (1) The Central Government[1] may, by notification in the Official Gazette—

(a) establish one or more environmental laboratories;

(b) recognise one or more laboratories or institutes as environmental laboratories to carry out the functions entrusted to an environmental laboratory under this Act.[2]

(2) The Central Government may, by notification in the Official Gazette, make rules specifying—

(a) the functions of the environmental laboratory;[3]

(b) the procedure for the submission to the said laboratory of samples of air, water, soil or other substance for analysis or tests, the form of the laboratory report thereon and the fees payable for such report;[4]

(c) such other matters as may be necessary or expedient to enable that laboratory to carry out its functions.

Government Analysts

13. The Central Government may, by notification in the Official Gazette, appoint or recognise such persons as it thinks fit and having the prescribed qualifications[5] to be Government Analysts for the purpose of analysis of samples of air, water, soil or other substance sent for analysis to any environmental laboratory established or recognised under Sub-section (1) of Section 12.

---

[1] The Central Government has delegated its powers under Clause (b) of Sub-section (i) of Section 12 and Section 13 of the Act to the Central Pollution Control Board vide Notification No. S.O.145 (E) dated 21-2-91; published in the Gazette No. 128 dated 27-2-91.

[2] The list of laboratories/institutes recognised as environmental laboratories and the persons recognised as Govt. Analysts is given in the Table (p. 223).

[3] See r. 9 of Environment (Protection) Rules, 1986.

[4] See r. 8 of Environment (Protection) Rules, 1986.

[5] For qualifications of Govt. Analyst see r. 10 of Environment (Protection) Rules, 1986.

Reports of Government Analysts

14. Any document purporting to be a report signed by a Government analyst may be used as evidence of the facts stated therein in any proceeding under this Act.

Penalty for contravention of the provisions of the Act and the rules, orders and directions

15. (1) Whoever fails to comply with or contravenes any of the provisions of this Act, or the rules made or orders or directions issued thereunder, shall, in respect of each such failure or contravention, be punishable with imprisonment for a term which may extend to five years with fine which may extend to one lakh rupees, or with both, and in case the failure or contravention continues, with additional fine which may extend to five thousand rupees for every day during which such failure or contravention continues after the conviction for the first such failure or contravention.

(2) If the failure or contravention referred to in Sub-section (1) continues beyond a period of one year after the date of conviction, the offender shall be punishable with imprisonment for a term which may extend to seven years.

Offences by companies

16. (1) Where any offence under this Act has been committed by a company, every person who, at the time the offence was committed, was directly in charge of, and was responsible to, the company for the conduct of the business of the company, as well as the company, shall be deemed to be guilty of the offence and shall be liable to be proceeded against and punished accordingly:

Provided that nothing contained in this Sub-section shall render any such person liable to any punishment provided in this Act, if he proves that the offence was committed without his knowledge or that he exercised all due diligence to prevent the commission of such offence.

(2) Notwithstanding anything contained in Sub-section (1), where an offence under this Act has been committed by a company and it is proved that the offence has been committed with the consent or connivance of, or is attributable to any neglect on the part of, any director, manager, secretary or other officer of the company, such director, manager, secretary or other officer shall also deemed to be guilty of that offence and shall be liable to be proceeded against and punished accordingly.

***Explanation*—For the purposes of this section—**

(a) "company" means any body corporate and includes a firm or other association of individuals;

(b) "director", in relation to a firm, means a partner in the firm.

(17). (1) Where an offence under this Act has been committed by any Department of Government, the Head of the Department shall be deemed to be guilty of the offence and shall be liable to be proceeded against and punished accordingly.

Provided that nothing contained in this section shall render such Head of the Department liable to any punishment if he proves that the offence was committed without his knowledge or that he exercise all due diligence to prevent the commission of such offence.

(2) Notwithstanding anything contained in sub-section (1), where an offence under this Act has been committed by a Department of Government and it is proved that the offence has been committed with the consent or connivance of, or is attributable to any neglect on the part of, any officer, other than the Head of the Department, such officer shall also be deemed to be guilty of that offence and shall be liable to be proceeded against and punished accordingly.

Chapter IV

MISCELLANEOUS

Protection of action taken in good faith

18. No suit, prosecution or other legal proceeding shall lie against the Government or any officer or other employee of the Government or any authority constituted under this Act or any member, officer or other employee of such authority in respect of anything which is done or intended to be done in good faith in pursuance of this Act or the rules made or orders or directions issued thereunder.

Cognizance of offences

19. No court shall take cognizance of any offence under this Act except of a complaint made by—

(a) the Central Government or any authority or officer authorised in this behalf by that Government,[1] or

(b) any person who has given notice of not less than sixty days, in the manner prescribed, of the alleged offence and of his intention to make a complaint, to the Central Government or the authority or officer authorised as aforesaid.

Information, reports or returns

20. The Central Government may, in relation to its function under this Act, from time to time, require any person, officer, State Government or other authority to furnish to it or any prescribed authority or officer any reports, returns, statistics, accounts and other information and such person, officer, State Government or other authority shall be bound to do so.

Members, officers and employees of the authority constituted under Section 3 to be public servants

21. All the members of the authority, constituted, if any, under Section 3 and all officers and other employees of such authority when acting or purporting to act in pursuance of any provisions of this Act or the rules made or orders or directions issued thereunder shall be deemed to be public servants within the meaning of Section 21 of the Indian Penal Code (45 of 1860).

Bar of Jurisdiction

22. No civil court shall have jurisdiction to entertain any suit or proceeding in respect of anything done, action taken or order or direction issued by the Central Government or any other authority or officer in pursuance of any power conferred by or in relation to its or his functions under this Act.

Powers to delegate

23. Without prejudice to the provisions of Sub-section (3) of Section 3, the Central Government may, by notification in the Official Gazette, delegate, subject to such conditions and limitations as may be specified in the notifications, such of its powers and functions under this Act [except the powers to constitute an authority under Sub-section (3) of Section 3 and to make rules under Section 25] as it may deem necessary or expedient, to any officer, State Government or other authority.

[1] In exercise of powers conferred under Clause (a) of Section 19, the Central Government has authorised the officers and authorities listed in the Table (p. 238) vide S.O. 394 (E) published in the Gazette No. 185 dated 16-4-87, S.O. 237 (E) published in the Gazette No. 171 dated 29-3-89 and S.O. 656 (E) dated 21-8-89 published in the Gazette No. 519 dated 21-8-89.

Effect of other laws

24. (1) Subject to the provisions of Sub-section (2), the provisions of this Act and the rules or orders made therein shall have effect notwithstanding anything inconsistent therewith contained in any enactment other than this Act.

(2) Where any act or omission constitutes an offence punishable under this Act and also under any other Act then the offender found guilty of such offence shall be liable to be punished under the other Act and not under this Act.

Power to make rules

25. (1) The Central government may, by notification in the Official Gazette, make rules for carrying out the purposes of this Act.

(2) In particular, and without prejudice to the generality of the foregoing power, such rules may provide for all or any of the following matters, namely—

(a) the standards in excess of which environmental pollutants shall not be discharged or emitted under Section 7;

(b) the procedure in accordance with and the safeguards in compliance with which hazardous substances shall be handled or caused to be handled under Section 8;

(c) the authorities or agencies to which intimation of the fact of occurrence or apprehension of occurrence of the discharge of any environmental pollutant in excess of the prescribed standards shall be given and to whom all assistance shall be bound to be rendered under Sub-section (1) of Section 9;

(d) the manner in which samples of air, water, soil or other substance for the purpose of analysis shall be taken under Sub-section (1) of Section 11;[1]

(e) the form in which notice of intention to have a sample analysed shall be served under Clause (a) of Sub-section (3) of Section 11;[2]

(f) the functions of the environmental laboratories,[3] the procedure for the submission to such laboratories of samples of air, water, soil and other substances for analysis or test;[4] the form of laboratory report; the fees payable for such report and other matters to enable such laboratories to carry out their functions under Sub-section (2) of Section 12;

(g) the qualifications of Government Analyst appointed or recognised for the purpose of analysis of samples of air, water, soil or other substances under Section 13;[5]

(h) the manner in which notice of the offence and of the intention to make a complaint to the Central Government shall be given under Clause (b) of Section 19;[6]

(i) the authority or officer to whom any reports, returns, statistics, accounts and other information shall be furnished under Section 20;

(j) any other matter which is required to be, or may be, prescribed.

---

[1] See r. 6 of Environment (Protection) Rules, 1986.

[2] See r. 7 of Environment (Protection) Rules, 1986.

[3] See r. 9 of Environment (Protection) Rules, 1986.

[4] For the procedure for submission of samples to laboratories and the form of laboratory report see r.8 of Environment (Protection) Rules, 1986.

[5] See r. 10 of Environment (Protection) Rules, 1986.

[6] See r. 11 of Environment (Protection) Rules, 1986.

43.9.4 RESTRICTIONS AND REGULATORY MEASURES THROUGH POLLUTION ABATEMENT ACTS: PRE-ESTABLISHMENT STAGE

Water Act, 1974

Any entrepreneur who wants to locate an industry has to check up whether the area in which he wants to locate an industry is in the area declared under Section 19 of the Water Act or not. In the absence of any notification excluding that area, the Water Act applies throughout the State and therefore he has to conform to the provisions of the Act. In so far as the Union Territory of Delhi is concerned, there is no such exclusion and therefore the Water Act can be enforced in the whole of the Union Territory of Delhi. Further under the amended Section 25 of the Water Act read with Section 26, unlike the unamended Sections, it requires *previous consent of the State Board* to establish or to take any steps to establish any industry. The amended Section 25 reads as follows:

Restrictions on new outlets and new discharges –

1. Subject to the provisions of this Section, no person shall, without the previous consent of the State Board –

(a) Establish or take any steps to establish any industry, operation or process, or any treatment and disposal system or any extension or addition thereto, which is likely to discharge sewage or trade effluent into a stream or well or sewage or land (such discharge being hereafter in this Section referred to as discharge of sewage), or

(b) bring into use any new or altered outlet for the discharge of sewage; or

(c) begin to make any new discharge of sewage;

Provided that a person in the process of taking any steps to establish any industry, operation or process immediately before the commencement of the Water (Prevention and Control of Pollution) Amendment Act, 1988, for which no consent was necessary prior to such commencement, may continue to do so for a period of three months from such commencement or, if he has made an application for such consent, within the said period of three months, till the disposal of such application.

2. An application for consent of the State Board under subsection (1) shall be made in such form, contain such particulars and shall be accompanied by such fees as may be prescribed.

3. The State Board may make such inquiry as it may deem fit in respect of the application for consent referred to in subsection (1) and in making any inquiry shall follow such procedure as may be prescribed.

4. The State Board may –

(a) grant its consent referred to in subsection (1) subject to such conditions as it may impose, being—

(i) in cases referred to in clause (a) and (b) of subsection (1) of Section 25, conditions as to the point of discharge of sewage or as to the use of that outlet of any other outlet for discharge of sewage;

(ii) in the case of a new discharge, conditions as to the nature and composition, temperature, volume or rate of discharge of the effluent from the land or premises from which the discharge or new discharge is to be made; and

(iii) that the consent will be valid only for such period as may be specified in the order, and any such conditions imposed shall be binding on any person establishing or taking any steps to establish any industry, operation or process, or treatment and disposal system or extension or addition thereto, or using the new of altered outlet, or discharging the effluent from the land or premises aforesaid; or

(b) refuse such consent for reasons to be recorded in writing.

5. Where, without the consent of the State Board, any Industry, operation or process, or any treatment and disposal system or any extension or addition thereto is established, or any steps for such establishment have been taken or a new or altered outlet is brought into use for the discharge of sewage or a new discharge of sewage is made the State Board may serve on the person who has established or taken steps to establish any industry operation or process, or any treatment and disposal system or any extension or addition thereto, or using the outlet, or making the discharge, as the case may be, a notice imposing any such conditions as it might have imposed on an application for its consent in respect of such establishment such outlet or discharge.

6. Every State Board shall maintain a register containing particulars of the conditions imposed under this Section and so much of the register as relates to any outlet or to any effluent, from any land or premises shall be open to inspection at all reasonable hours by any person interested in or affected by such outlet, land or premises, as the case may be, or by any person authorised by him in this behalf and the conditions as contained in such register shall be conclusive proof that the consent was granted subject to such conditions.

7. The consent referred to in subsection (1) shall, unless given or refused earlier, be deemed to have been given unconditionally on the expiry of a period of four months of the making of an application in this behalf complete in all respects to the State Board –

8. For the purposes of this section and section 27 and 30 –

(a) the expression "new or altered outlet" means any outlet which is wholly or partly constructed in or after the commencement of this Act or which (whether so constructed or not) is substantially altered after such commencement;

(b) the expression "new discharge" means a discharge which is not, as respects the nature and composition, temperature, volume and rate of discharge of the effluent substantially a continuation of a discharge made within the preceding twelve months (whether by the same or a different outlet), so however that a discharge which is in other respects a continuation of a discharge made within the preceding twelve months shall not be deemed to be a new discharge by reason of any reduction of the temperature or volume or rate of discharge of the effluent as compared with the previous discharge.

AIR ACT, 1981

Section 19 of the Act contemplates declaration of Air Pollution control areas. The State Government may after consultation with the State Board by notification in the Official Gazette declare any such area for the purposes of this Act. The entire Union Territory of Delhi, having been so declared, is an Air Pollution Control areas for the purposes of this Act.

Under Section 21 of the Air Act, only with the previous consent of the State Board, a person shall establish or operate any industrial plant in an air pollution control area. The amended Section makes a substantial change in omitting the specified industry in the Schedule. Therefore irrespective of the nature of the industry now, the pollution control measures are to be taken up. Such a consent under the Air Act is also a preestablishment measure. The same principle is applicable to the existing industries.

Under Section 22 of the Air Act, any person carrying on any industry or operating any industrial plant, in any air pollution control area is prohibited from discharging or causing or permitting to be discharged the emission of any air pollutants in excess of the standards laid down by the State Board.

ENVIRONMENT (PROTECTION) ACT, 1986

Section 3(2)(v) confers power on the Central Government to restrict the area in which any industry, operation or process or class of industries, operations or processes shall to be carried out or shall be carried subject to certain safeguards. Likewise under Section 6(e) of the Act, the Central Government is empowered to make rules which may provide for the prohibition and restrictions on the location of industries and carrying on the processes and operations in different areas.

Section 8 of the Act clearly provides that no person shall handle or cause to be handled any hazardous substance except in accordance with such procedure and after complying with such safeguards as may be prescribed.

The compliance of the abovesaid requirements are some of the preestablishment measures that are to be borne in mind by the entrepreneurs/industrialists.

POST-ESTABLISHMENT MEASURES

Under Section 25 and 26 of the Water Act and Section 21 of the Air Act, the State Boards are empowered to grant or reject consent orders, they may specify the period and also impose conditions in regard to the steps to establish industry, operations or processes and also lay down the standards of the trade effluents.

Under Section 3 and 6 of the Act, the Central Government is empowered to lay down standards for the quality of environment and also lay down standards for emission.

APPEALS

Any person aggrieved by the orders passed by the State Boards or Central Board in its capacity in respect of their consent application made under Section 25 and 26 of the Water Act or under Section 21 of the Air Act may prefer and Appeal before the Appellate Authority constituted under Section 28 of the Water Act and under Section 31 of the Air Act.

ENFORCEMENT

The Pollution Control Boards are given ample powers to enforce the provisions of these Acts. The enforcement power includes prosecution of a person who had failed and neglected to come under the provisions of the Act and also in respect of the persons though they have submitted themselves to the provisions of the Act but nevertheless do not conform to the standards stipulated under the consent conditions and fail to comply with the other conditions. The statutory responsibility of invoking the prosecution rests with the Boards. However, appreciating the modern trend of expanding the locus standi both under the Air Act and the Water Act, under Environment Protection Act, a private citizen can also invoke the jurisdiction of the Courts against the erring industry after giving 60 days notice to the Boards and Union of India. The Environment Protection Act also provides stringent measures to prosecute the erring industries who do not conform to the statutory legislations or who do not conform to the standards laid and notified under the provisions of the Environment Protection Act.

Section 24 of the Environment Protection Act provides for the effect of other laws while enforcing that Act. Under the subsection (1) subject to subsection (2), the provisions of Environment Protection Act and the Rules made therein shall have effect notwithstanding anything contained in any enactment other than this Act. Therefore it makes very clear that if the standards laid down under the Environment Protection Act are more stringent that would have effect even though under the Air Act or Water Act the standards are laid down. However, subsection 2 provides that where any act or emission subject to an offence punishable under this Act or also under any other Act that Entrepreneur shall be liable to be punished under the other Act and not under this Act.

PROCEDURAL SAFEGUARDS

For the proper implementations of the provisions of the Acts involving penal consequences on the persons for the failure to perform the functions under the Act, a very strict procedural safeguard is provided lest it would not by properly used by the Authorities. Under Section 21 and 22 of the Water Act and under Section 26 and 27 of the Air Act and under Section 11 of the Environment Protection Act, the procedure for taking samples and the procedure to be followed in connection therewith have been laid. Therefore to prosecute a person for the alleged violations of the standards, a strict compliance is necessary. The Pollution Control Board cannot also simply prosecute without any basis or for not complying with the procedure contemplated under the abovesaid Sections.

PROSECUTIONS AND PENALTY

For noncompliance of the provisions of the Act, Rules and Orders, the Act make out both summons cases and warrant cases depending on the nature. The provisions of the above Acts contemplate awarding of minimum sentences and the enhancement imprisonment and fine for continuous offences. The relevant Sections read as:

Water Act, 1974

"41. Failure to comply with directions under subsection (2) or subsection (3) of the Section 20, or orders issued under clause (c) of subsection (1) of Section 32 or directions issued under subsection (2) of Section 33 or Section 33A.

(1) Whoever fails to comply with any direction given under subsection (2) of subsection (3) of Section 20 within such time as may be specified in the direction shall, on conviction, be punishable with imprisonment for a term which may extend to three months or with fine which may extend to five thousand rupees or with both and in case the failure continues, with an addition fine which may extend to one thousand rupees for every day during which such failure continues after the conviction for the first such failure.

(2) Whoever fails to comply with any order issued under clause (c) of the subsection (1) of Section 32 or any direction issued by a Court under subsection (2) of Section 33 or any direction issued under Section 33A shall, in respect of each such failure and on conviction, be punishable with imprisonment for a term which shall not be less than one year and six months but which may extend to six years and with fine and in case the failure continues, with an additional fine which may extend to five thousand rupees for every day during which such failure continues after the conviction for the first such failure.

(3) If the failure referred to in subsection (2) continues beyond a period of one year after the date of conviction, the offender shall, on conviction, be punishable with imprisonment for a term which shall not be less than two years but which may extend to seven years and with fine.

42. Penalty for certain acts – (1) Whoever –

(a) destroys, pulls down, removes, injuries or defaces any pillar, post or stake fixed in the ground or any notice or other matter put up, inscribed or placed, by or under the authority of the Board, or

(b) obstructs any person acting under the orders or directions of the Board from exercising his powers and performing his function under this Act, or

(c) damages any works or property belonging to the Board or

(d) fails to furnish to any officer or other employee or the Board any information required by him of the purpose of this Act, or

(e) fails to intimate the occurrence of any accident or other unforeseen act or event under Section 31 to the Board and other authorities or agencies as required by that Section, or

(f) in giving any information which he is required to give under this Act, knowingly or wilfully makes a statement which if false in any material particular, or

(g) for the purpose of obtaining any consent under Section 25 or Section 26, knowingly or wilfully makes statement which is false in any material particular shall be punishable with imprisonment for a term where may extend to three months or with fine which may extend to ten thousand rupees or with both.

(2) Where for the grant of a consent in pursuance of the provisions of Section 25 or Section 26 the use of a meter of gauge or other measure or monitoring device is requires and such device used for the purposes of those provisions any persons who knowingly or wilfully alters or interface with that device so as to prevent it form monitoring to measuring correctly shall be punishable with imprisonment for a term which may extend to three months with fine which may extend to ten thousand rupees or with both.

43. Penalty for contravention of provisions of Section 24 – Whoever contravenes the provisions of Section 24 shall be punishable with imprisonment for a term which shall not be less than one year and six months but which may extend to six years and with fine.

44. Penalty for contravention of Section 25 or Section 26 – Whoever contravenes of Section 25 or Section 26 shall be punishable with imprisonment for a term which shall not be less than one year and six months but which may extend to six years and with fine.

45. Enhance penalty after previous conviction – If any person who has been convicted of any of offence under Section 24 or Section 25 or Section 26 is again found guilty of an offence involving a contravention of the same provision, he shall, no the second and on every subsequent conviction, be punishable with imprisonment for a term which shall be not be less than two years but which may extend to seven years and with fine;

Provided that for the purpose of this section no cognizance shall be taken of any conviction made more than two years before the commission of the offence which is being punished.

AIR ACT, 1981

37. Failure to comply with the provisions of Section 21(5) or Section 22 or with orders or directions issued under the Act –

(1) Whoever fails to comply with the provisions of subsection (5) of Section 21 or Section 22 or with any order or direction given under this Act shall, in respect of each such failure, be punishable with imprisonment for a term which may extend to ten thousand rupees, or with both, and in case the failure continues, with an additional fine which may extend to one hundred rupees for every day during which such failure continues after the conviction for the first such failure.

(2) If the failure referred to in subsection (1) continues beyond a period of one year after the date of conviction, the offender shall be punishable with imprisonment for a term which may extend to six months.

37A. Failure to comply with the provisions of Section 21 of Section 22 or with the directions issued under section 31A – (1) Whoever fails to comply with the provisions of Section 21 or Section 22 or directions issued under Section 31A shall, in respect of each such failure, be punishable with imprisonment for a term which shall not be less than one year and six months but which may extend to six years and with fine, and in case the failure continues, with an additional fine which may extend to five thousand rupees for every day during which such failure continues after the conviction for the first such failure.

(2) If the failure referred to in subsection (1) continues beyond a period of one year after the date of conviction, the offender shall be punishable with imprisonment for a term which shall not be less than two years but which may extend to seven years and with fine.

38. Penalties for certain acts – Whoever –

(a) destroys, pulls down, removes, injures or defaces any pillar, post or stake fixed in the ground or any notice or other matter put up, inscribed or placed, by or under the authority of the Board, or

(b) obstructs any person acting under the orders or directions of the Board from exercising his powers and performing his functions under this Act, or

(c) damages any works or property belonging to the Board, or

(d) fails to furnish to the Board or any officer or other employee of the Board any information required by the Board or such officer or other employee for the purposes of this Act, or

(e) fails to intimate the occurrence of the emission of air pollutants into the atmosphere in excess of the standards laid down by the State Board or the apprehension of such occurrence, to the State Board and other prescribed authorities or agencies as required under subsection (1) of Section 23, or

(f) in giving any information which he is required to give under this Act, makes a statement which is false in any material particular, or

(g) for the purpose of obtaining any consent under Section 21, makes a statement which is false in any material particular.

shall be punishable with imprisonment for a term which may extend to three months or with fine which may extend to ten thousand rupees or with both.

39. Penalty for contravention of certain provisions of the Act – Whoever contravenes any of the provisions of this Act or any order or direction issued thereunder, for which no penalty has been elsewhere provided in this Act, shall be punishable with imprisonment for a term which may extend to three months or with fine which may extend to ten thousand rupees or with both, and in the case of continuing contravention, with an additional fine which may extend to five thousand rupees for every day during which such contravention continues after conviction for the first such contravention.

ENVIRONMENT ACT, 1986

15. Penalty for contravention of the provisions of the Act and the Rules, Order and Directions – (1) Whoever fails to comply with or contravenes any of the provisions of this Act, or the rules made or orders or directions issued thereunder shall, in respect of each such failure or contravention, be punishable with imprisonment for a term which may extend to five years or with fine which may extend to one lakh rupees, or with both, and in case the failure or contravention continues, with additional fine which may extend to five thousand rupees for every day during which such failure or contravention continues after the conviction for the first such failure or contravention.

(2) If the failure or contravention referred to in subsection (1) continues beyond a period of one year after the date of conviction, the offender shall be punished with imprisonment for a term which may extend to seven years.

OFFENCES BY COMPANIES

To put it beyond any doubt the word 'Occupier' occurred in Section 2(d) of the Water Act is substituted by the following new definition:

"Occupier", in relation to any factory or premises, means the person who has control over the affairs of the factory or the premises and includes, in relation to any substance, the person in possession of the substance".

Under Section 47 of the Water Act, the offence committed by the Companies or a person who, at the time the offence was committed, was in case of, and was responsible to the Company for the conduct of the business of the Company, as well as the Company, shall be deemed to be guilty of the offence and shall be liable to be proceeded against and punished accordingly. Similar also is the case under the Air Act vide Section 40 red with definition of the Word 'Occupier' found in Section 2(m). The Environment Protection Act also contains the similar provisions in Section 16.

POWER TO ISSUE DIRECTIONS

Under Section 33-A[6] of the Water Act and under Section 31-A[7] of the Air Act and under Section 5[8] of the Environment Protection Act, the erring industries can be always issued directions and such directions would include directions to close down the units or directions to disconnect the electricity and stop the water supply. Several such directions have been issued against the erring industries all over the country both by the Government of India and various State Boards and the Central Board.

Apart from this, the Pollution Control Boards can also file applications before the Criminal Court for obtaining a restraint order against the erring industry.

PROSECUTION BY PUBLIC

Under the Amended Section 49(1)(b)[9] of the Water Act and under the Amended Section 43(1)(b)[10] of the Air Act and under Section 19(b)[11] of the Environment Act, any per who has given notice of not less than sixty days in the manner prescribed, of the alleged offence and of his intention to make a complaint, to the Central Government or the authority or officer authorised, can institute criminal proceedings for violations or the offences under the particular Statues. The Public participation in the prosecution has been in addition to the prosecution launched by the respective Statutory Authorities. The Public Interest Litigation originally restricted under Article 226 and 32 of The Constitution for enforcement of fundamental rights and other statutory rights has been broadened statutorily in respect of the prosecutions also.

WATER CESS ON CONSUMPTION OF WATER

The Parliament has enacted the Water (Prevention and Control of Pollution) Cess Act, 1977 for the levy and collection of a cess on water consumed by the persons carrying on certain industries and by local authorities. This is to augment the resources of the State and Central Pollution Control Board constituted under the Water Act. It may be borne in mind that the levy of cess is not related to the act of pollution caused by the industries liable to pay but it is based on the consumption of the water. The liability to pay cess is one the local bodies and every person carrying on any specified industry mentioned in the Schedule. However, under Section 7 of the Water Act, where any person or local authority who pays cess under this Act instals any plant for the treatment of sewage or trade effluent, such person or local authority is entitled to get a rebate of 75% of the cess payable. However, it must be borne in mind, it is not enough if the effluent treatment plant is installed but it must also successfully function. Under the provisions, there is no provision for giving rebate proportionate to the steps taken by them to instal treatment plant. Provision for Appeal is also provided against assessment orders passed by the Assessing Authority. Section 14 of the Water Cess Act also contains penal provisions against persons who fail to furnish return under the Act or fail to pay the cess. It is pertinent to note here that the burden cast upon the industries to put up the effluent treatment plant is substantially reduced by giving rebate to the tune of 75% if the effluent treatment plants is installed and it runs successfully. Apart from this, various concessions under the Income Tax Act and various State Acts also provide for concessions for the purposes to purchase and installation of machineries meant for effluent treatment.

NOISE POLLUTION CONTROL

No serious efforts have been far made to control the noise pollution. Though Section 6(2)(b) of the Environment (Protection) Act provides for making rules relating to the maximum allowable limits of concentration of environmental pollutants (including noise) for different areas, it is yet to be seriously implemented. The amended Air Act also provides for control of noise pollution.

43.9.5 THE FACTORIES ACT, 1948

Introduction

After the Bhopal tragedy, the Factories Act, 1948, was amended substantially to make it more comprehensive and effective. A list of hazardous processes was incorporated in it and permissible threshold limits of exposure of chemical and toxic substances were fixed. The existing schedule of notifiable (occupational) diseases was also revised.

It was made mandatory for the management to make full disclosures about the hazardous processes used in its factory and safety measures taken, maintain accurate and uptodate health medical records of workers exposed to such processes, provide for their regular medical check-up during the employment and even after that for a certain period. The factory management, certifying surgeons, and even any doctor treating a worker suffering from an occupational disease scheduled in the Act, is now required to inform the Chief Inspector of Factories about the incidence.

SCHEDULE

The First Schedule of the Act lists the industries involving hazardous processes which are:

1. Ferrous Metallurgical Industries
   - Integrated Iron and Steel
   - Ferrous-alloys
   - Special Steels
2. Non-ferrous Metallurgical Industries
   - Primary Metallurgical Industries, namely: zinc, lead, copper, manganese and aluminium
3. Foundries (ferrous and non-ferrous)
   - Castings and forgings including cleaning or smoothening/roughening by sand and shot blasting
4. Coal (including coke) industries
   - Coal, Lignite, Coke, etc.
   - Fuel Gases (including Coal Gas, Producer Gas, Water Gas)
5. Power Generating Industries
6. Pulp and paper (including paper products) industries
7. Fertilizer Industries
   - Nitrogenous
   - Phosphatic
   - Mixed
8. Cement Industries
   - Portland Cement (including slag cement, puzzolona cement and their products)
9. Petroleum Industries
   - Oil Refining
   - Lubricating Oils and Greases

10. Petro-chemical Industries
11. Drugs and Pharmaceutical Industries
    — Narcotics, Drugs and Pharmaceuticals
12. Fermentation Industries (Distilleries and Breweries)
13. Rubber (Synthetic) Industries
14. Paints and Pigment Industries
15. Leather Tanning Industries
16. Electro-plating Industries
17. Chemical Industries
    — Coke Oven by-products and Coal tar Distillation products
    — Industrial Gases (nitrogen, oxygen, acetylene, argon, carbon dioxide, hydrogen, sulphur dioxide, nitrous oxide, halogenated hydrocarbon, ozone, etc.)
    — Industrial Carbon
    — Alkalis and Acids
    — Chromates and dichromates
    — Leads and its compounds
    — Electrochemicals (metallic sodium, potassium and magnesium, chlorates, perchlorates and peroxides)
    — Electrothermal produces (artifical abrasive, calcium carbide)
    — Nitrogenous compounds (cyanides, cyanamides, and other nitrogenous compounds)
    — Phosphorus and its compounds
    — Halogens and Halogenated compounds (Chlorine, Fluorine, Bromine and Iodine)
    — Explosives (including industrial explosives and detonators and fuses)
18. Insecticides, Fungicides, Herbicides and other Pesticides Industries
19. Synthetic Resins and Plastics
20. Man-made fibre (Cellulosic and non-cellulosic) Industry
21. Manufacture and repair of electrical accumulators
22. Glass and Ceramics
23. Grinding or glazing of metals
24. Manufacture, handling and processing of asbestos and its products
25. Extraction of oils and fats from vegetable and animal sources
26. Manufacture, handling and use of benzene and substances containing benzene
27. Manufacturing process and operations involving carbon disulphide
28. Dyes and Dyestuff including their intermediates
29. Highly flammable liquids and gases.

Permissible levels of certain chemical substances in work environment

Substance	Permissible limits of exposure			
	Time-weighted average concentration (8 hrs)		Short-term exposure limit (15 min)*	
	ppm	mg/m³	ppm	mg/m³
Acetaldehyde	100	180	150	270
Acetic Acid	10	25	15	37
Acetone	750	1780	1000	2375
Acrelein	0.1	0.25	0.3	0.8
Acrylonitrile — skin	2	4.5	—	—
Aldrin — skin	—	0.25	—	0.75
Allyl Chloride	1	3	2	6
Ammonia	0.25	18	35	27
Aniline — skin	2	10	5	20
Anisidine (o, Pisons) skin	0.1	0.5	—	—
Arsenic & compounds (as As)	—	0.2	—	—
Benzene	10	20	25	75
Beryllium	—	0.002	—	—
Boron Trifluoride	0.1	0.3	—	—
Bromine	0.1	0.7	0.3	2
Butane	800	1900	—	—
2 Butanon (Methyl Ethyl Ketone—MEK)	200	590	300	885
n-Butyl acetate	150	710	200	950
n-Buytyl alcohol — skin	C50	C150	—	—
sec/tert. Butyl acetate	200	950	250	1190
Butyl Mercaptan	0.5	1.5	—	—
Cadmium-dust and salts (as Cd)	—	0.05	—	0.2
Calcium oxide		2	—	—
Carbaryl (ssevin)	—	5	—	10
Carbofuran (Furadan)	—	0.1	—	—
Carbon disulphide — skin	10	30	—	—
Carbon monoxide	50	40	400	440
Carbon tetrachloride — skin	m5	30	20	125
Carbonyl chloride (Phosgene)	0.1	0.4	—	—
Chlorobenxzene (monochlorobenzene)	75	350	—	—
Chloridance—skin	—	0.5	—	2

Contd.

Permissible levels of certain chemical substances in work environment				
Substance	Permissible limits of exposure			
	Time-weighted average concentration (8 hrs)		Short-term exposure limit (15 min)*	
	ppm	mg/m³	ppm	mg/m³
Chlorine	1	3	3	9
Chloroform	10	50	50	225
bis-Chloromethyl ether	0.001	0.005	—	
Chromic acid and chromates (as Cr)	—	0.05	—	—
Chromous Salts (as Cr)	—	0.05	—	—
Copper fume	—	0.2	—	—
Cotton dust, raw	—	0.2	—	0.6
Cresol, all isomers—skin	5	22	—	—
Cyanides (as CN)—skin	—	5	—	—
Cyanogen	10	20	—	—
DDT (Dichloro-diphenyl Trichloro-ethene)	—	1	—	3
Demeton—skin	0.01	0.1	0.03	0.3
Diazinon—skin	—	0.1	—	0.3
Dibutyl Phythalate	—	5	—	10
Dichlorvos (DDVP)—skin	0.1	1	0.3	3
Dieldrin—skin	—	0.25	—	0.75
Dinitrobenzene (all isomers)—skin	0.15	1	0.5	3
Dinitrotoluene—skin	—	1.5	—	5
Diophenyl	0.2	1.5	0.6	4
Endosulfan (Thiodan)—skin	—	0.1	—	0.4
Endrin—skin	—	0.1	—	0.3
Ethyl acetate	400	1400	—	—
Ethyl alcohol	1000	1900	—	—
Ethylamin	10	18	—	—
Flourides (as F)	—	2.5	—	—
Flourine	1	2	2	4
Formic Acid	5	9	—	—
Hydrazine—skin	0.1	0.1	—	—
Hydrogen Chloride	C5	C7	—	—
Hydrogen Cyanide—skin	C10	C10	—	—
Hydrogen Flouride (as F)	3	2.5	6	5
Hydrogen Peroxide	1	1.5	2	3

Contd.

Substance	Permissible limits of exposure			
	Time-weighted average concentration (8 hrs)		Short-term exposure limit (15 min)*	
	ppm	mg/m^3	ppm	mg/m^3
Hydrogen Sulphide	10	14	15	21
Gasoline	300	900	500	1500
Iodine	C0.1	C1	—	—
Iron Oxide fume (Fe_2O_3) (as Fe)	—	5	—	10
Isoamyl acetate	100	525	125	655
Isoamyl alcohol	100	300	125	450
Isobutyl alcohol	50	150	75	225
Lead, inorg., fumes and dust (as Pb)	—	0.15	—	0.45
Lindane—skin	—	0.5	—	1.5
Malathion—skin	—	10	—	—
Manganese (as Mn) dust and compounds	—	C05	—	—
Fume	—	1	—	0.3
Mercury (as Hg)—skin Alkyl compounds	—	0.01	—	0.03
All forms except alkyl vapour	—	0.05	—	—
Aryl and inorganic compounds	—	0.1	—	—
Methyl alcohol (methanol—skin)	200	260	250	310
Methyl cellosolve—skin (2 methoxy ethanol)	5	16	—	—
Methyl isobutyl Ketone—skin	50	205	75	300
Methyl Isocyanate	0.02	0.05	—	—
Naphthalene	10	50	15	75
Nickel carbonyl (as Ni)	0.05	0.35	—	—
Nitric acid	2	5	4	10
Nitric oxide	25	30	35	45
Nitrobenzene—skin	1	5	2	10
Nitrogen dioxide	3	6	5	10
Oil mist, minerals	—	5	—	10
Ozone	0.1	0.2	0.3	0.6
Parathion—skin	—	0.1	—	0.3
Phenol—skin	5	19	10	38
Phorate (Thimet)—skin	—	0.05	—	0.2
Phosgene (Carbonyl Chloride)	0.1	0.4	—	—
Phosphine	0.3	0.4	1	1
Phosphorus (yellow)	—	0.1	—	0.3
Phosphorus pentachloride	0.1	1	—	—

Permissible levels of certain chemical substances in work environment

Contd.

Permissible levels of certain chemical substances in work environment				
Substance	Permissible limits of exposure			
	Time-weighted average concentration (8 hrs)		Short-term exposure limit (15 min)*	
	ppm	mg/m^3	ppm	mg/m^3
Phosphorus trichloride	0.2	1.5	0.5	3
Picric acid—skin	—	0.1	—	0.3
Pyridine	5	15	10	30
Silane (silicon tetrahydride)	5	7	—	—
Sodium hydroxide	—	C2	—	—
Syrene monomer (phanylethylene)	50	215	100	425
Sulphur dioxide	2	5	5	10
Sulphur hexafluoride	1000	6000	1250	7500
Sulphuric acid	—	1	—	—
Toluene (Tuluol)	100	375	150	560
o-Tuluidine—skin	2	9		
Tributyl phosphate	0.2	2.5	0.4	5
Trichloroethylene	50	270	200	1080
Uranium, natural (as U)	—	0.2	—	0.5
Vinyl chloride	5	10	—	—
Welding fumes	$\frac{10}{\%\text{quartz}+3}$	5	—	—
Xylene (o –, m –, P– isomers)	100	435	150	655
Zirconium compounds (as Zr)	—	5	—	10

C denotes ceiling limit.

* Not more than 4 times a day with at least 60 min. interval between successive exposures.

Permissible levels of certain suspended particulates in work environment			
	Substance	Permissible time-weighted (8 hours)	Average concentration
(i)	Silica		
	(a) Crystalline		
	(b) Quartz		
	(1) In terms of count	$\frac{10,600}{\%\text{Quartz}+10}$	mg /m
	(2) In terms of respirable dust	$\frac{10}{\%\text{respirable quartz}+2}$	mg / mt
	(3) In terms of total dust		mg / mt
(ii)	Cristabalite	Half the limits given against quartz	
(iii)	Tridymite	Half the limits given against quartz	
(iv)	Silica fused	Same limit as for quartz	
(v)	(a) Tripoli	Same limit as in formula in item 2 given against quartz	
	(b) Amorphous	705 mpcm	

List of Notifiable Diseases

1. Lead poisoning, including poisoning by any preparation or compound of lead or their sequelae.
2. Lead tetra-ethyl poisoning.
3. Phosphorus poisoning or its sequelae.
4. Mercury poisoning or its sequelae.
5. Manganese poisoning or its sequelae.
6. Arsenic poisoning or its sequelae.
7. Poisoning by nitrous fumes.
8. Carbon bisulphide poisoning.
9. Benzene poisoning, including poisoning by any of its homologues, their nitro or amido derivatives or its sequelae.
10. Chrome ulceration or its sequelae.
11. Anthrax.
12. Silicosis.
13. Poisoning by halogens or halogen derivatives of the hydrocarbons of the aliphatic series.
14. Pathological manifestations due to—
    (a) Radium or other radioactive substances.
    (b) X-rays.
15. Primary epitheliomatous cancer of the skin.
16. Toxic anemia.
17. Toxic jaundice due to poisonous substances.
18. Oil acne or dermatitis due to mineral oils and compounds containing mineral oil base.
19. Byssionosis.
20. Asbestosis.
21. Occupational or contact dermatitis caused by direct contact with chemicals and paints. These are of two types—primary irritants and allergic sensitizers.
22. Noise induced hearing loss (exposure to high noise levels).
23. Beryllium poisoning.
24. Carbon monoxide.
25. Coal miners' pneumoconiosis.
26. Phosgene poisoning.
27. Occupational cancer.
28. Isocyanates poisoning.
29. Toxic nephritis.

43.10. CONSERVATION ACTS

There are three major related to conservation of wildlife, forest and biodiversity. These are stated below:

43.10.1 THE WILDLIFE (PROTECTION) ACT, 1972

(No. 53 of 1972) [9th September, 1972]

An Act to provide for the protection of wild animals and birds and for matters connected therewith or ancillary or incidental thereto

Whereas it is expedient to provide for the protection of wild animals and birds and for matters connected therewith or ancillary or incidental thereto;

And whereas Parliament has no power to make laws for the States with respect to any of the matters aforesaid except as provided in Arts. 249 and 250 of the Constitution;

And whereas in pursuance of Cl. (1) of Art. 252 of the Constitution resolutions have been passed by all the Houses of the Legislatures of the States of Andhra Pradesh, Bihar, Gujarat, Haryana, Himachal Pradesh, Madhya Pradesh, Manipur, Pubjab, Rajasthan, Uttar Pradesh and West Bengal to the effect that the matters aforesaid should be regulated in those States by Parliament by law;

Be it enacted by Parliament in the Twenty-third Year of Republic of India as follows:

Comments

Provision whether directory or mandatory—It is well established that the wordings of any provision are not determinative as to whether it is absolute or directory. Even the absence of penal provision for noncompliance does not lead to an inference that it is only directory. The Court, therefore, must carefully get into the underlying idea and ascertain the purpose to the achieved notwithstanding the text of the provision.[1]

Retrospective effect—It is well settled rule of construction that every statute or statutory rule is prospective unless it is expressly or by necessary implication made to have retrospective effect. Unless there are words in the statute or in the rules showing the intention to affect existing rights the rule must be held to be prospective. If a rule is expressed in language which is fairly capable of either interpretation it ought to be construed as prospective only. In the absence of any express provision or necessary intendment the rule cannot be given retrospective effect except in matter of procedure.[2]

CHAPTER 1

Preliminary

1. **Short title, extent and commencement**—(1) This Act may be called the Wild Life (Protection) Act, 1972.

(2) It extends, in the first instance, to the whole of the States of Andhra Pradesh, Bihar, Gujarat, Haryana, Himachal Pradesh, Madhya Pradesh, Manipur, Punjab, Rajasthan, Uttar Pradesh and West Bengal and to all the Union Territories; and it shall also extend to such other State as may adopt this Act by resolution passed in that behalf in pursuance of Cl. (1) of Art. 252 of the Constitution.

(3) It shall come into force[3] in a State of Union territory to which it extends, or may become

1. Kamal Leather Karmchari Senghatan (Regd) *v.* Liberty Footwear Company (Regd.), AIR 1990 S.C. 247 at p. 254
2. P. Mahendren *v.* State of Karnataka, AIR 1990 S.C. 405 at p. 408; see also Gurbachan Singh *v.* Satpal Singh, AIR 1990 S.C. 209 at p. 219; Y. Arul Nadar *v.* Authorised Officer, Land Reforms, Tanjavur, AIR 1990 Mad. 33 at p. 36.
3. The Act has been made applicable in various States and Union Territories on the following dates as under:
   1. Andaman and Nicobar, w.e.f. 1st July 1973, *vide* G.S.R. 332 (E), dated 1st July, 1973
   2. Andhra Pradesh, w.e.f. 1st August 1973, *vide* G.S.R. 371 (E), dated 1st August, 1973

extended in future, on such date as the Central Government may be notification, appoint, and different dates may be appointed for different provisions of this Act or for different States or Union territories.

2. **Definitions**—In this Act, unless the context otherwise requires—

(1) "animal" includes amphibians, birds, mammals and reptiles and their young, and also includes, in the cases of birds and reptiles, their eggs;

(2) "animal article" means an article made from any captive animal or wild animal, other than vermin, and includes an article or object in which the whole or any part of such animal has been used;

(3) "big game" means any animal specified in Sch. III;

(4) "Board" means the Wild Life Advisory Board constituted under subsection (1) of Sec. 6;

(5) "captive animal" means any animal, specified in Sch. I, Sch. II, Sch. III or Sch. IV, which is captured or kept or bred in captivity;

(6) "cattle" includes buffaloes, bulls, bullocks, camels, cows, domestic elephants, donkeys, goats, horses, mules, pigs and sheep, and also includes their young;

(7) "Chief Wild Life Warden" means the person appointed as such under C. (a) of subsection (1) of Sec. 4;

(8) "closed area" means the area which is declared under subsection (1) of Sec. 37 to be closed to hunting;

(9) "Collector" means the Chief officer in charge of the revenue administration of a district;

(10) "Commencement of this Act", in relation to—

(a) a State, means commencement of this Act in that State,

(b) any provision of this Act, means the commencement of that provision in the concerned State;

---

3. Arunachal Pradesh, w.e.f. 15th May, 1973
4. Assam w.e.f. 25th January, 1977, *vide* G.S.R. 33(E), dated 25th January, 1977
5. Bihar, w.e.f. 1st February, 1973, *vide* G.S.R. 40(E), dated 1st February, 1973
6. Dadra and Nagar Haveli, w.e.f. 1st September, 1973, *vide* G.S.R. 411(E), dated 1st September, 1973
7. Delhi, w.e.f. 1st June, 1973
8. Goa, Daman and Diu, w.e.f. 16th July, 1973, *vide* G.S.R. 349(E), dated 16th July 1978
9. Gujarat, w.e.f. 1st February, 1973, *vide* G.S.R. 62(E), dated 1st February, 1973
10. Haryana, w.e.f. 12th March, 1973, *vide* G.S.R. 63(E), dated 12th March, 1973
11. Himachal Pradesh, w.e.f. 2nd April, 1973, *vide* G.S.R. 190(E), dated 2nd April, 1973
12. Karala, w.e.f. 1st June, 1973, *vide* G.S.R. 293(F), dated 1st June, 1973
13. Lakshdweep, w.e.f. 1st November, 1973, *vide* G.S.R. 481(E), dated 1st November, 1973
14. Madhya Pradesh, w.e.f. 25th January, 1973, *vide* G.S.R. 28(E), dated 25th January, 1973
15. Maharashtra, w.e.f. 1st June, 1973, *vide* G.S.R. 296(E), dated 1st June, 1973
16. Manipur, w.e.f. 15th May, 1973, *vide* G.S.R. 269(E), dated 15th May, 1973
17. Meghalaya, w.e.f. 1st April, 1977
18. Mizoram, w.e.f. 1st October, 1974, *vide* G.S.R. 407(E), dated 1st October, 1974
19. Mysore, w.e.f. 1st July, 1973, *vide* G.S.R. 355(E), dated 1st July, 1973
20. Nagaland, w.e.f. 18th December, 1981, *vide* G.S.R. 668(E), dated 18th December, 1981
21. Orissa, w.e.f. 14th August, 1974, *vide* G.S.R. 364(E), dated 14th August, 1974
22. Punjab, w.e.f. 1st April, 1975
23. Rajasthan, w.e.f. 1st September, 1973, *vide* G.S.R. 410(E), dated 1st September, 1973
24. Sikkim, w.e.f. 1st May, 1976, *vide* G.S.R. 311(E), dated 1st May, 1976
25. Tamil Nadu, w.e.f. 1st January, 1974
26. Tripura, w.e.f. 2nd October, 1973, *vide* G.S.R. 465(E), dated 2nd October, 1973
27. Union Territory of Chadigarh, w.e.f. 2nd December, 1974, *vide* G.S.R. 674(E), dated 2nd December, 1974
28. Union Territory of Pondicherry, w.e.f. 1st March, 1975, *vide* G.S.R. 62(E), dated 1st March, 1975
29. Uttar Pradesh, w.e.f. 1st February, 1973
30. West Bengal, w.e.f. 1st May, 1973, *vide* G.S.R. 224(E), dated 1st May, 1973

(11) "dealer" means any person who carries on the business of buying and selling any captive animal, animal article, trophy, uncured trophy or meat;

(12) "Director" means the person appointed as Director of Wild Life Preservation under Cl. (a) of subsection (1) of Sec. 3;

(13) "game reserve" means an area declared under Sec. 36 to be a game reserve;

(14) "Government property" means any property means any property referred to in Sec. 39;

(15) "habitat" includes land, water or vegetation which is the natural home of any wild animal;

(16) "hunting", with its grammatical variations and cognate expressions, includes,

(a) capturing, killing, poisoning, snaring and trapping of any wild animal and every attempt to do so,

(b) driving any wild animal for any of the purposes specified in subclause (a),

(c) injuring or destroying or taking any part of the body of any such animal or, in the case of wild birds or reptiles, damaging the eggs or nests of such birds or reptiles;

(17) "land" includes canals, creeks and other water channels, reservoirs, rivers, streams and lakes whether artificial or natural, and also includes boulders ad rocks;

(18) "licence" means a licence granted under this Act;

(19) "manufacturer" of animal articles;

(20) "meat" includes blood, bones, sinew, eggs, fat and flesh, whether raw or cooked, or any wild animal, other than vermin;

(21) "National Park" means an area declared, whether under Sec. 35 or Sec. 38, or deemed, under subsection (3) of Sec. 66, to be declared, as a National Park;

(22) "notification" means a notification published in the Official Gazette;

(23) "permit" means a permit granted under this Act or any rule made thereunder;

(24) "person" includes a firm;

(25) "prescribed" means prescribed by rules made under this Act;

(26) "sanctuary" means an area declared, whether under Sec. 18 or Sec. 38, or deemed, under subsection (3) of Sec. 66, to be declared, as wild life sanctuary;

(27) "small game" means any animal specified in Sch. IV;

(28) "special game" means any animal specified in Sch. II;

(29) "State Government", in relation to a Union Territory, means the Administrator of the Union Territory appointed by the President under Art. 239 of the Constitution;

(30) "taxidermy", with its grammatical variations and cognate expressions, means the curing, preparation or preservation of trophies;

(31) "trophy" means the whole or any part of any captive animal or wild animal, other than vermin, which has been kept or preserved by any means, whether artificial or natural, and includes—

(a) rugs, skins and specimens of such animal mounted in whole or in part through a process of taxidermy, and

(b) antler, horn, rhinoceros horn, hair, feather, nail, tooth musk, eggs, and nests;

(32) "uncured trophy" means the whole or any part of any captive animal or wild animal, other than vermin, which has not undergone a process of taxidermy, and includes a freshly killed wild animal;

(33) "vehicle" means any conveyance used for movement on land, water or air and includes buffalo, bull, bullock, camel, donkey, elephant, horse and mule;

(34) "vermin" means any wild animal specified in Sch. V;

(35) "weapon" includes ammunition, bows and arrows, explosives, firearms, hooks, knives, nets, poison, snares and traps and any instrument or apparatus capable of anaesthetizing, decoying, destroying injuring or killing an animal;

(36) 'wild animal" means any animal found wild in nature and includes any animal specified in Sch. I, Sch. II, Sch. III, Sch IV or Sch. V, wherever found;

(37) "wild life" includes any animal, bees, butterflies, crustacea, fish and moths; and aquatic or land vegetation which forms part of any habitat;

(38) "Wild Life Warden" means the person appointed as such under Cl. (b) of subsection (1) of Sec. 4.

Comments

The word "animal article"—In the charge the learned Magistrate had referred to animal articles, that the word "animal article" was defined in Sec. 2 (2) of the said Act and a reading of this must make it clear that the word "animal article" covers article made only of those species of animals specified in the Schedules. That, therefore, there was no magic in contending that no particulars were afforded. that coupled with this was the admission of the accused himself which has gone on record, and in view of this, the question of setting aside the conviction cannot arise.[1]

Absence of notification and its effects—Section 18 requires that a declaration is to be made by notification. It is the case of the particulars that there is no publication of the notification in the Mizoram official Gazette. It was submitted that in view of the fact that the notification under Sec. 18 has not been published in the Official Gazette, action sought to be taken under the said Act by the impugned notices cannot be sustained. The notification is not found to be published in the Official Gazette. It is contended that as there is no publication of the notification under Sec. 18 of the Act, the purported declaration is honest. Therefore the impugned orders were set aside as they have been passed without jurisdiction.[2]

CHAPTER II

Authorities to be appointed or constituted under the Act

3. Appointment of Director and other officers—(1) The Central Government may, for the purposes of the Act, appoint—

(a) A Director of Wild Life Preservation;

(b) Assistant Directors of Wild Life Preservation, and

(c) such other officers and employees as may be necessary.

(2) In the performance of his duties and exercise of his powers by or under this Act, the Director shall be subject to such general or special directions, as the Central Government may, from time to time give.

(3) The Assistant Directors of Wild Life Preservation and other officers and employees appointed under this section shall be subordinate to the Director.

1. Rafique Ramzan Ali *v.* A.A. Jalgaonkar, 1984 Cr. L.J. 1460(2) at p. 1463 (bom); 1985 mah. L.R. 258; 1984 Cr. L.R. 262 (Mah.).
2. Jaladhar Chakma *v.* Deputy Commissioner, Aizawal, Mizoram, A.I.R. 1983 Gau. 18 at p. 20.

4. **Appointment of Chief Wild Life Warden and other officers**—(1) The State Government may, for the purposes of this Act, appoint—

(a) a Chief Wild Life Warden;

(b) Wild Life Wardens; and

(c) such other officers and employees as may be necessary.

(2) In the performance of his duties and exercise of his powers by or under this Act, the Chief Wild Life Warden shall be subject to such general or special directions, as the State Government may, from time to time, give.

(3) The Wild Life Warden and other officers and employees appointed under this section shall be subordinate to the Chief Wild Life Warden.

5. **Power to delegate**—(1) The Director may, with the previous approval of the Central Government, by order in writing, delegate all or any of his powers and duties under this Act to any officer subordinate to him subject to such conditions, if any, as may be specified in the order.

(2) The Chief Wild Life Warden may, with the previous approval of the State Government, by order in writing, delegate all or any of his powers and duties under this Act, except those under Cl. (a) of subsection (1) of Sec. 11, to any officer subordinate to him subject to such conditions, if any, as may be specified in the order.

(3) Subject to any general or special direction given or condition imposed by the Director or the Chief Wild Life Warden any person, authorised by the Director or the Chief Wild Life Warden to exercise any powers, may exercise those powers in the same manner and to the same effect as if they had been conferred on that person directly by this Act and not by way of delegation.

Comment

Complaint field after sanction is maintainable—In view of rule 31 of the Bihar Wild Life Protection Rules, 1973, even besides Chief Wild Life Warden the Divisional Forest Officer or the Deputy Conservator of forests are also entitled to file complaint. In the present case the complaint was filled after obtaining sanction from the Divisional Forest Officer who was authorised to file the complaint. It may be relevant to mention here the Sec. 5(2) of the Act also gives power to the authorities concerned to delegate his power to any of his subordinate officer. Therefore it could not be said that no other person except the Chief Wild Life Warden or such other officer who has been authorised by the State Government, can file a complaint upon which the cognizance can be taken.[1]

6. **Constitution of Wild Life Advisory Board**—(1) The State Government, or, in the case of a Union Territory, the Administration, shall, as soon as may be after the commencement of this Act, constitute a Wild Life Advisory Board consisting of the following members, namely:

(a) The Minister in charge of forests in the State of Union Territory, or, if there is no such Minister, the Chief Secretary to the State Government, or, as the case may be, the Chief Secretary to the Government or the Union Territory, who shall be the Chairman;

(b) two members of the State Legislature or, in the case of a Union Territory having a Legislature, two members of the Legislature of the Union Territory, as the case may be;

(c) Secretary to the State Government, or the Government of the Union Territory, in charge of forests;

1. Jagdish Singh *v.* State of Bihar, 1985 Cr. L.J. 1314 at p. 1315, 1316.

(d) Chief Conservator of Forests; *ex officio*;

(e) an officer to be nominated by the Director;

(f) Chief Wild Life Warden, *ex officio*;

(g) Such other officers and nonofficials, not exceeding fifteen, who, in the opinion of the State Government, are interested in the protection of wild life.

(2) The State Government shall appoint Chief Conservator of Forests or Chief Wild Life Warden as the Secretary of the Board.

(3) The term of office of the members of the Board referred to in Cl. (g) of subsection (1) and the manner of filling vacancies among them shall be such as may be prescribed.

(4) The members shall be entitled to receive such allowances in respect of expenses incurred in the performance of their duties as the State Government may prescribe.

7. **Procedure to be followed by the Board**—(1) The Board shall meet at least twice a year at such place as the State Government may direct.

(2) The Board shall regulate its own procedure (including the quorum).

(3) No act or proceeding of the Board shall be invalid merely by reason of the existence of any vacancy therein or any defect in the constitution thereof or any irregularity in the procedure of the Board affecting the merits of the case.

8. **Duties of Wild Life Advisory Board**—It shall be the duty of the Wild Life Advisory Board to advise the State Government—

(a) in the selection of areas to be declared as sanctuaries, National Parks, game reserves and closed areas and the administration thereof;

(b) in formulation of the policy in granting licences and permits under this Act;

(c) in any matter relating to the amendment of any Schedule; and

(d) in any other manner connected with the protection of wildlife which may be referred to it by the State Government.

CHAPTER III

Hunting of Wild Animals

9. **Hunting of wild animals**—(1) No person shall hunt any wild animal specified in Sch. 1.

(2) No person shall hunt any wild animal specified in Sch. II, Sch. III, or Sch. IV, except under, and in accordance with the conditions specified in a licence granted under subsection (5).

(3) Any person desiring to obtain a licence referred to in subsection (2) shall apply, in such form, and on payment of such fee, as may be prescribed, to the Chief Wild Life Warden or any other officer authorised by the State Government in this behalf (hereinafter referred to as the authorised officer).

(4) An application under subsection (3) may be made for all or any of the following kinds of licences, namely:

(a) Special game hunting licence.

(b) Big game hunting licence.

(c) Small game hunting licence.

(d) Wild animal trapping licence:

Provided that any such licence may be restricted to the hunting or trapping of such animal as may be specified therein.

(5) On receipt of an application under subsection (3), the Chief Wild Life Warden or the authorised officer may, after making such inquiry as he may think fit with regard to the fitness or otherwise of the applicant to receive the licence grant or refuse to grant such licence after recording in writing his reasons for such grant or refusal; and when the grant of a licence is refused, the fee paid therefore shall be refunded to the applicant.

(6) A licence granted under this section shall—

(a) be in such form as may be prescribed;

(b) be valid for such period as may be specified therein;

(c) be subject to such terms, conditions and restrictions as may be prescribed;

(d) not be transferable.

10. **Maintenance of records of wild animals killed or captured**—(1) The holder of every licence shall maintain a record, containing such particulars as may be prescribed, of the wild animals, other than vermin, killed or captured by him during the currency of the licence.

(2) When any animal specified in Sch. II or Sch. III is killed, wounded or captured by the holder of a licence, he shall, not latter than fifteen days from the date of such killing, wounding or capture, or before leaving the area specified in the licence, whichever is earlier, intimate in writing to the Chief Wild Life Warden or the authorised officer, the prescribed particulars of such animal killed, wounded or captured by him.

(3) The holder of every licence shall, not later than fifteen days from the date of expiry of such licence, surrender the records maintained by him under subsection (1) and the licence to the Chief Wild Life Warden or the authorised officer, and shall sign a declaration, in the prescribed form certifying the accuracy of the records maintained by him under subsection (1).

11. **Hunting of wild animals to be permitted in certain cases**—(2) Notwithstanding anything contained in any other law for the time being in force and subject to the provisions of Chapter IV—

(a) the Chief Wild Life Warden may, if he is satisfied that any wild animal specified in Sch. I has become dangerous to human life or is so disabled or diseased as to be beyond recovery, by order in writing and stating the reasons therefore, permit any person to hunt such animal or cause such animal to be hunted.

(b) the Chief wild Life Warden or the authorised officer may, if he is satisfied that any wild animal specified in Sch. II, Sch. III, or Sch. IV, has become dangerous to human life or to property (including standing crops on any land) or is so disabled or diseased as to be beyond recovery, by order in writing and stating the reasons therefor, permit any person to hunt such animal or cause such animal to be hunted.

(2) The killing or wounding in good faith of any wild animal in defence of oneself or of any other person shall not be an offence:

Provided that nothing in this subsection shall exonerate any person who, when such defence becomes necessary, was committing any act in contravention of any provision of this Act or any rule or order made thereunder.

(3) Any wild animal killed or wounded in defence of any person shall be Government property.

Comment

Conviction contrary to the provisions of Sec. 11 is liable to be set aside—To decide whether in shooting the accused acted in self-defense or not, the nature and ferocity of the animal will be relevant. A tiger is, what the Romans called, a *ferae naturae* by nature of dangerous ferocity, as distinguished from a *mansuatae naturae*, e.g. a dog or a horse, which have in individual cases given indication of a vicious or dangerous disposition. In the case of attack by a *ferae naturae* the victim cannot be expected to weigh the chances in golden scale and consequently, the inference that he was acting in defence of his own life, will be more easily drawn that in case of an attack by a *mansuatae naturae*, while no such inference may be drawn at all in cases of harmless wild life like birds. On the basis of facts and circumstances of the case there can be no doubt that the accused acted in defence of his life and his act did commensurate with defence. It is clearly a case of killing the tiger in good faith in defence of oneself and it cannot be said that the accused was committing any offence prior to shooting the tiger that charged at him. Therefore, he will be completely protected under subsection (2) of Sec. II. The impugned order of conviction and sentence is contrary to the provisions of Sec. 11 of the Act and as such it is liable to be set aside.[1]

12. **Grant of permit for special purposes**—Notwithstanding anything contained elsewhere in the Act, it shall be lawful for the Chief Wild Life Warden, to grant, [2][* * *] a permit, by an order in writing stating the reasons therefore, to any person, on payment of such fee as may be prescribed, which shall entitle the holder of such permit to hunt, subject to such conditions as may be specified therein, any wild animal specified in such permit, for the purpose of—

(a) education;

[3](b) scientific research;

(bb) scientific management;

Explanation—For the purpose of Cl. (bb), the expression "scientific management" means—

(i) translocation of any wild animals to an alternative suitable habitat; or

(ii) population management of wildlife, without killing or poisoning or destroying any wild animals; or

(c) collection of specimens for zoological gardens, museums an similar institutions:

[4][Provided that no such permit shall be granted—

(a) in respect of any wild animal specified in Sch. I, except with the previous permission of the Central Government, and

(b) in respect of any other wild animal, except with the previous permission of the State Government.]

13. **Suspension or cancellation of licence**—The Chief Wild Life Warden or the authorised officer may, subject to any general or special order of that State Government, for good and sufficient reasons, to be recorded in writing, suspend or cancel any licence granted under this Chapter:

Provided that no such suspension or cancellation shall be made except after giving the holder of the licence a reasonable opportunity of being heard.

1. Tilak Bahadur Rai *v.* State of Arunachal Pradesh, 1979 Cr. L.J. 1404 at p. 1405 (Gau.)
2. The words with the previous permission to the State Government, omitted by Act 23 of 1982, Sec. 2.
3. Subs. by *ibid*
4. Ins. by *ibid*

14. **Appeals**—(1) An appeal from an order refusing to grant a licence under Sec. 9, or an order suspending or cancelling a licence under Sec. 13, shall lie—

(a) if the order is made by the authorised officer, to the Chief Wild Life Warden; or

(b) if the order is made by the Chief Wild Life Warden, to the State Government.

(2) In the case of an order passed in appeal by the Chief Wild Life Warden under subsection (1), a second appeal shall lie to the State Government.

(3) Subject as aforesaid, every order passed in appeal under this section shall be final.

(4) No appeal shall be entertained unless it is preferred within fifteen days from the date of the communication to the applicant of the order appealed against:

Provided that the appellate authority may admit any appeal after the expiry of the period aforesaid, if it is satisfied that the appellant had sufficient cause for not preferring the appeal in time.

15. **Hunting of young and female of wild animal**—No person shall, unless specially authorised by a licence, hunt the young of any wild animal, other than vermin, or any female of such animal, or any deer with antlers in velvet.

16. **Declaration of closed time**—(1) The State Government may, by notification, declare the whole year or any part thereof, to be a closed time throughout the State, or any part thereof, for such wild animal as may be specified in the notification and no hunting of such animal shall be permitted, during the said period, in the area specified in the notification.

(2) The provisions of subsection (1) shall not apply to vermin unless otherwise specified by the State Government in this behalf.

Comment

Conviction—Setting aside of—On 5th February, 1954, at about 10 a.m. a rhinoceros was shot dead at village Hahpani. The petitioner along with a few others who have since been acquitted or discharged, were sent up for trial under Sec. 429 of the Indian Penal Code and Sec. 25 (g) of the Assam Forest Regulation. The charge under Sec. 25 (g) of the Forest Regulation failed because the occurrence was found to have taken place outside the game sanctuary. The petitioner was, however, convicted by the learned Magistrate under Sec. 429 of the Indian Penal Code for having shot and killed the rhinoceros with a gun, and that conviction has been upheld by the learned Sessions Judge on appeal. The only evidence against the petitioner was that of one prosecution witness who saw the petitioner firing at the rhinoceros with a single-barrelled muzzle loading gun, and later the animal was found dead. The evidence of this witness has been believed by the two Courts below who have, therefore, convicted the petitioner. Destruction of property is essential, where, therefore, no one has any property or right in an animal, the killing of that animal does not come within the meaning of Sec. 425 of the Indian Penal Code. And thus the provision of Sec. 429 of the Indian Penal Code also will not be attracted. The learned Sessions Judge appears to have overlooked the effect of Sec. 425 of the Indian Penal Code in the interpretation of Sec. 429. It as submitted that any rate an offence had been committed under Sec. 3 of the Wild Birds and Animals Protection Act, 1912 (8 of 1912), and therefore, the conviction of the petitioner should be altered to one under Sec. 4 of the Act and an appropriate sentence imposed on him. But this cannot be done for the obvious reason that there is nothing to show in this case that there was any such notification by the Government declaring that the offence, if any, was committed in a class season, as prohibited by the law. That being so, the conviction and sentence of the petitioner must be set aside.[1]

1. Nabin Chandra Gogoi v. State, A.I.R. 1961 Assam 18 at pp. 18, 19

17. Restrictions on hunting—(1) No person shall—

(a) hunt any wild animal, from or by means of, a wheeled or a mechanically propelled vehicle on water or land, or by aircraft;

(b) use an aircraft, motor vehicle or launch for the purpose of driving or stampeding any wild animal;

(c) hunt any wild animal with chemicals, explosives, nets, pitfalls, poison, poisoned-weapons, snares or traps, except in so far as they relate to the capture of wild animal under a Wild Animal Trapping Licence;

(d) hunt any special game or big game other than with a rifle unless specially authorised by the licence to hunt with a shotgun using single slug bullets;

(e) for the purpose of hunting, set fire to any vegetation;

(f) use any artificial light for the purpose of hunting, except when specially authorised to do so under licence in the case of carnivora over a kill;

(g) hunt any wild animal during the hours of night, that is to say between sunset and sunrise, except when specially authorised to do so under a licence in the case of carnivora over a kill;

(h) hunt any wild animal on a salt-lick or water hole or other drinking place or on path or approach to the same, except sand-grouse and water-birds;

(i) hunt any wild animal on any land not owned by Government, without the consent of the owner or his agent or the lawful occupier of such land;

(j) notwithstanding that he holds a licence of the purpose, hunt any wild animal during the closed time referred to in Sec. 16;

(k) hunt, with the help of dogs, any wild animal except water-bird, chakor, partridge or quail.

(2) The provision of subsection (1) shall not apply to vermin.

CHAPTER IV

Sanctuaries, National Parks, Game Reserves and Closed Areas

Sanctuaries

18. Declaration of sanctuary—(1) The State Government may, by notification, declare any area to be a sanctuary if it considers that such area is of adequate ecological, faunal, floral, geomorphological natural or zoological significance, for the purpose of protecting, propagating wild life or its environment.

(2) The notification referred to in subsection (1) shall specify, as nearly as possible, the situation and limits of such area.

Explanation—For the purposes of the section, it shall be sufficient to describe the area by roads, rivers, ridges or other well-known or readily intelligible boundaries.

Comments

Order without jurisdiction could be set aside—The petitioners have been staying within the area described in the notices before passing of the orders with their hearths and home and also having cultivable lands. The impugned orders were purporated to have been made under the provisions of the Wild Life

Sanctuary (Protection) Act, 1972, under which there may be declaration of sanctuary by virtue of the provision of Sec. 18 of the Act. It is found out from the records that under Sec. 18 of the Act a notification has been issued on 20th January, 1976, by the Development Commissioner, Ex-officio Secretary to the Government of Mizoram, declaring the area given in that notification as Dampa Wild Life Sanctuary. It appears that the respondents have not followed any of the provisions of Chapter IV of the Act. It is true that the respondents can take follow up actions for evicting persons from the area falling within the declared sanctuary in accordance with the provisions as contained in Chapter IV of the Act. None of these things have been found to be followed by the respondents in passing the orders prejudicially affecting the petitioners. Therefore, the impugned orders are not sustainable in law and accordingly they deserve to be quashed.[1]

19. **Collector to determine rights**—Whenever any area is declared to be a sanctuary, the Collector shall inquire into, and determine the existence nature and extent of the rights of any person in over the land comprised within the limits of the sanctuary.

20. **Bar of accrual of rights**—After the issue of a notification under Sec. 18, no right shall be acquired in, on or over the land comprised within the limits of the area specified in such notification, except by succession, testamentary or intestate.

21. **Proclamation of Collector**—When a notification has been issued under Sec. 18, the Collector shall publish in the regional language in every town and village in or in the neighbourhood of the area comprised therein, a proclamation—

(a) specifying, as nearly as possible, the situation and the limits of the sanctuary; and

(b) requiring any person, claiming any right mentioned in Sec. 19, to prefer before the Collector, within two months from the date of such proclamation, a written claim in the prescribed form specifying the nature and extent of such right with necessary details and the amount and particulars of the compensation, if any, claimed in respect thereof.

22. **Inquiry by Collector**—The Collector shall, after service of the prescribed notice upon the claimant, expeditiously inquire into—

(a) the claim preferred before him under Cl. (b) of Sec. 21, and

(b) the existence of any right mentioned in Sec. 19 and not claimed under Cl. (b) of Sec. 21,

so far as the same may be ascertainable from the records of the State Government, and the evidence of any person acquainted with the same.

23. **Powers of Collector**—For the purpose of such inquiry, the Collector may exercise the following powers, namely

(a) the power to enter in or upon any land and to survey, demarcate and make a map of the same or the authorise any other officer to do so;

(b) the same powers as are vested in a civil court for the trial of suits.

24. **Acquisition of rights**—(1) In the case of a claim to a right in or over any land referred to in Sec. 19, the Collector shall pass an order admitting or rejecting the same in whole or in part.

(2) If such claim is admitted in whole or in part, the Collector may either—

(a) exclude such land from the limits of the proposed sanctuary, or

1. Jaladhar Chakma *v* Deputy Commissioner, Aizawal, Mizoram, AIR 1983 Gau. 18 at pp. 20, 21.

(b) proceed to acquire such land or rights, except where by an agreement between the owner of such land or holder of rights and the Government the owner or holder of such rights has agreed to surrender his rights to the Government, in or over such land, and on payment of such compensation, as is provided in the Land Acquisition Act, 1894 (1 of 1894).

25. **Acquisition proceedings**—(1) For the purpose of acquiring such land, or rights in or over such land,—

(a) the Collector shall be deemed to be a Collector, proceeding under the Land Acquisition Act, 1894 (1 of 1894);

(b) the claimant shall be deemed to be a person interested and appearing before him in pursuance of a notice given under Sec. 9 of the Act;

(c) the provisions of the sections, proceeding Sec. 9 of the Act, shall be deemed to have been complied with;

(d) where the claimant does not accept the award made in his favour in the matter of compensation, he shall be deemed, within the meaning of Sec. 18 of that Act, to be a person interested who has not accepted the award, and shall be entitled to proceed to claim relief against the award under the provisions of Part III of the Act;

(e) the Collector, with the consent of the claimant, or the Court, with the consent of the both the parties, may award compensation in land or money or partly in land and partly in money; and

(f) in the case of the stoppage of a public way or a common pasture, the Collector may, with the previous sanction of the State Government, provide for an alternative public way or common pasture, as far as may be practicable or convenient.

(2) The acquisition under this Act of any land or interest therein shall be deemed to be acquisition for a public purpose.

26. **Delegation of Collector's powers**—The State Government may, by general or special order direct that the powers exercisable or the functions to be performed by the Collector under Secs. 19 to 25 (both inclusive) may be exercised and performed by such other officer as may be specified in the order.

27. **Restriction on entry in sanctuary**—(1) No person other than,—

(a) a public servant on duty,

(b) a person who has been permitted by the Chief Wild Life Warden or the authorised officer to reside within the limits of the sanctuary,

(c) a person who has any right over immoveable property within the limits of the sanctuary.

(d) a person passing through the sanctuary along a public highway, and

(e) the dependents of the person referred to in Cl. (a), Cl. (b) or Cl. (c), shall enter or reside in the sanctuary, except under and in accordance with the conditions of a permit granted under Sec. 28.

(2) Every person shall, so long as he resides in the sanctuary, be bound—

(a) to prevent the commission, in the sanctuary, of an offence against this Act;

(b) where there is reason to believe that any such offence against this Act has been, committed in such sanctuary, to help in discovering and arresting the offender;

(c) to report the death of any wild animal and to safeguard its remains until the Chief Wild Life Warden or the authorised officer takes charge thereof;

(d) to extinguish any fire in such sanctuary of which he has knowledge or information and to prevent from spreading, by any lawful means in his power, any fire within the vicinity of such sanctuary of which he has knowledge or information; and

(e) to assist any forest officer, Chief Wild Life Warden, Wild Life Warden or police officer demanding his aid for preventing the commission of any offence against this Act or in the investigation of any such offence.

28. **Grant of permit**—(1) The Chief Wild Life Warden may, on application, grant to any person a permit to enter or reside in a sanctuary for all or any of the following purposes, namely:

(a) investigation or study of wild life and purposes ancillary or incidental thereto;

(b) photography,

(c) scientific research;

(d) tourism;

(e) transaction of lawful business with any person residing in the sanctuary.

(2) A permit to enter or reside in a sanctuary shall be issued subject to such conditions and on payment of such fee as may be prescribed.

29. **Hunting in sanctuary without permit prohibited**—(1) Notwithstanding anything contained elsewhere in this Act, no person shall hunt any wild animal in a sanctuary or remove therefrom any wild animal, whether alive or dead, or any trophy, uncured trophy, or meat derived from such animal.

Provided that if the Chief Wild Life Warden is satisfied that it is necessary that any wild animal in a sanctuary should be hunted or removed,—

(a) for the better protection of wild life, or

(b) for any other good and sufficient reason, he may, with the previous approval of the State Government, grant a permit authorising any person to hunt or remove such wild animal under the direction of an officer authorised by him or cause it to be hunted or removed.

(2) A permit granted under subsection (1) shall specify the kind and number of wild animals that may be hunted or removed by the holder of such permit.

(3) The Chief Wild Life Warden may, for good and sufficient reason, to be recorded in writing cancel any permit granted under Sec. 28 or under this section:

Provided that no such cancellation shall be made except after giving the holder of the permit a reasonable opportunity of being heard.

(4) Any person aggrieved by the cancellation of a permit under subsection (3) may, within fifteen days from the date of such cancellation appeal to the State Government, whose decision shall be find:

Provided that the State Government may admit any appeal preferred after the expiry of the period aforesaid if it is satisfied that the appellant had sufficient cause for not preferring the appeal in time.

30. **Causing fire prohibited**—No person shall set fire to a sanctuary, or kindle any fire, or leave any fire burning, in a sanctuary, in such manner as to endanger such sanctuary.

31. **Prohibition of entry into sanctuary with weapon**—No person shall enter a sanctuary with any weapon except with the previous permission in writing of the Chief Wild Life Warden or the authorised officer.

32. **Ban on use of injurious substances**—No person shall use, in a sanctuary, chemicals, explosives or any other substances which may cause injury to, or endanger, any wild life in such sanctuary.

33. **Control of sanctuaries**—The Chief Wildlife Warden shall be the authority who shall control, manage and maintain all sanctuaries and for that purpose, within the limits of any sanctuary,—

(a) may construct such roads, bridges, buildings, fences or barrier gates, and carry our such other works as he may consider necessary for the purposes of such sanctuary;

(b) shall take such steps as will ensure the security of wild animals in the sanctuary and the preservation of the sanctuary and wild animals therein;

(c) may take such measures, in the interests of wild life, as he may consider necessary for the improvement of any habitat;

(d) may regulate, control or prohibit, in keeping with the interests of wild life, as he may consider necessary for the improvement of cattle;

(e) may regulate, control or prohibit, any fishing.

34. **Registration of certain persons in possession of arms**—(1) Within three months from the declaration of any area as a sanctuary, every person residing in or within ten kilometres of any such sanctuary and holding a licence granted under the Arms Act, 1959 (54 of 1959), for the possession of arms or exempted from the provisions of the Act and possessing arms, shall apply in such form, on payment of such fee and within such time as may be prescribed, to the Chief Wild Life Warden or the authorised officer, for the registration of his name.

(2) On receipt of an application under subsection (1) the Chief Wild Life Warden or the authorised officer shall register the name of the applicant in such manner as may be prescribed.

National Parks

35. **Declaration of National Parks**—(1) Whenever it appears to the State Government that an area, whether within a sanctuary or not, is by reason of its ecological, faunal, floral, geomorphological or zoological association or importance, needed to be constituted as a National Park for the purpose of protecting, propagating or developing wild life therein or its environment, it may, by notification, declare its intention to constitute such area as a National Park.

(2) The notification referred to in subsection (1) shall define the limits of the area which is intended to be declared as a National Park.

(3) Where any area is intended to be declared as a National Park, the provisions of Secs. 19 to 26 (both inclusive) shall, as far as may be, apply to the investigation and determination of claims, and extinguishment of rights, in relation to any land in such areas as they apply to the said matters in relation to any land in a sanctuary.

(4) When the following events have occurred, namely,—

(a) the period for preferring claims has elapsed, and all claims, if any, made in relation to any land in an area intended to be declared as a National Park, have been disposed of by the State Government, and

(b) all rights in respect of lands proposed to be included in the National Park have become vested in the State Government,

the State Government shall publish a notification specifying the limits of the area which shall be comprised within the National Park and declare that the said area shall be a National Park on and from such date as may be specified in the notification.

(5) No alteration of the boundaries of a National Park shall be made except on a resolution passed by the Legislature of the State.

(6) No person shall destroy, exploit or remove any wild life from a National Park or destroy or damage the habitat of any wild animal or deprive any wild animal of its habitat within such National Park except under and in accordance with a permit granted by the Chief Wild Life Warden and no such permit shall be granted unless the State Government, being satisfied that such destruction, exploitation or removal of wild life from the National Park is necessary for the improvement and better management of wild life therein, authorises the issue of such permit.

(7) No grazing of any cattle shall be permitted in a National Park and no cattle shall be allowed to enter therein except where such cattle is used as a vehicle by a person authorised to enter such National Park.

(8) The provisions of Secs. 27 and 28, Secs. 30 to 32 (both inclusive), and Cls. (a), (b) and (c) of Sec. 33 and Sec. 34 shall, as far as maybe apply, in relation to a National Park as they apply in relation to as sanctuary.

Game Reserve

36. **Declaration of game reserve**—(1) The State Government may, by notification, declare any area to be a game reserve.

(2) No hunting of any wild animal shall be permitted in such reserve except under and in accordance with a licence issued under this section by the Chief Wild Life Warden or the authorised officer.

Closed Area

37. **Declaration of closed area**—(1) The State Government may, be notification, declare any area closed to hunting for such period as may be specified in the notification.

(2) No hunting of any wild animal shall be permitted in a closed area during the period specified in the notification referred to in subsection (1).

Sanctuaries or National Parks declared by Central Government.

38. **Power of Central Government to declare areas as sanctuaries or National Park**—(1) Where the State Government leases or otherwise transfers any area under its control, not being an area within a sanctuary, to the Central Government, the Central Government may if it is satisfied that the conditions specified in Sec. 18 are fulfilled in relation to the area so transferred to it, declare such area, by notification, to be a sanctuary and the provisions of Secs. 19 to 35 (both inclusive), 54 and 55 shall apply in relation to such sanctuary as they apply in relation to a sanctuary declared by the State Government.

(2) The Central Government may, if it is satisfied that the conditions specified in Sec. 35 are fulfilled in relation to any area referred to in subsection (1), whether or to such area has been declared, to be a sanctuary by the Central Government or the State Government, declare such area, by notification to be a National Park and the provisions of Secs. 35, 54 and 55 shall apply in relation to a National Park declared by the State Government.

(3) In relation to a sanctuary or National Park declared by the Central Government, the powers and duties of the Chief Wild Life Warden under the section referred to in subsections (1) and (2), shall be exercised ad discharged by the Director or by such other officer as may be authorised by the Director in this behalf and references, in the sections aforesaid, to the State Government shall be construed as references to the Central Government and reference therein to the Legislature of the State shall be construed as a reference to Parliament.

CHAPTER V

Trade or Commerce in Wild Animals, Animal Articles and Trophies

39. Wild animals, etc. to be Government property—(1) Every

(a) wild animal, other than vermin, which is hunted under Sec. 11 or subsection (1) of Sec. 29 or subsection (6) of Sec. 35 or kept or bred in captivity in contravention of any provision of the Act or any rule or order made thereunder are found dead, or killed without a licence or by mistake; and

(b) animal article, trophy or uncured trophy or meat derived from any wild animal referred to in Cl. (a) in respect of which any offence against this Act or any rule or order made thereunder has been committed, shall be the property of the State Government, and, where such animal is hunted in a sanctuary of National Park declared by the Central Government, such animal or any article, trophy, uncured trophy or meat derived from such animal, shall be the property of the Central Government.

(2) Any person who obtaines, by any means, the possession of Government property, shall, within forty-eight hours of obtaining such possession to the nearest police station or the authorised officer and shall, if so required, hand over such property to the officer-in-charge of such police station or such authorised officer, as the case may be.

(3) No person shall, without the previous permission in writing of the Chief Wild Life Warden or the authorised officer—

(a) acquire or keep in his possession, custody or control, or

(b) transfer to any person, whether by way of gift, sale or otherwise, or

(c) destroy or damage, such Government property.

Comments

Defective charge—A reading of charge shows that all that had been referred to therein is that the petitioner had displayed for sale lizard and snake skin articles without declaring the same to the Chief Wild Life Warden. That no particulars whatsoever have been afforded as to what articles were seized, or which were the subject-matter of the charge, nor have any particulars been given as to the species of the lizards or snakes of which the articles are said to have been made, or for that matter no clue whatsoever has been afforded as to whether these articles referred to in the charge were made of any of those species specified in any of the Schedules of the Wild Life (Protection) Act, 1972. That the charges being defective, the plea of guilt of the accused would be immaterial and of no consequence whatsoever and the conviction must hence fail. Thus it could be said that in so far as the charge is concerned, that too admits of several infirmities.[1]

Ownership of the elephant who voluntarily quit the reserve forest and entered the private land where it dies—Determination of—It is clear whatever might be the position in regard to live animals, as regards, wild animals which are found dead in a private land the owner of the land is entitled to claim ownership of the dead animal, and that the bodies of wild animals which are found on a particular land become the absolute property of the owner of the soil even if killed by a trespasser, unless the trespasser chased the animal on the land of one person and killed it on the land of another. Admittedly, the animal was found dead in the *patta* land of the appellant. The dead elephant in question had not been chased or pursued by the forest authorities, but it had been shot at by some unknown persons and it voluntarily quit the reserve forest and entered the private land where it died. In this view, the order of the Lower Appellate Court cannot be sustained.[2]

1. Rafique Ramzan Ali *v.* A.A. Jalgaonkar, 1984 Cr. L.J. 1460(2) at p. 1462 (Bom.): 1985 Mah. L.R. 245: 1984 Cr. L.R. 626 (Mah.).
2. B. Nathmall Valid *v* State of Tamil Nadu, AIR 1979 Mad. 218 at p. 222.

40. **Declarations**—(1) Every person having at the commencement of this Act the control, custody or possession of any captive animal specified in Sch. I or Part II of Sch. II, or any uncured trophy from such animal or salted or dried skins of such animal or the musk of a musk deer or the horn of rhinoceros, shall, within thirty days from the commencement of this Act, declare to the Chief Wild Life Warden or the authorised officer the number and description of the animal, or article of the foregoing description under his control, custody or possession and the place where such animal of article is kept.

(2) No person shall, after the commencement of this Act, acquire, receive, keep in his control, custody or possession, sell, offer for sale or otherwise transfer or transport any animal specified in Sch. I or Part II of Sch. II or any uncured trophy or meat derived from such animal, or the salted or dried skins of such animal or the musk of a musk deer or the horn of a rhinoceros, except with the previous permission in writing of the Chief Wild Life Warden or the authorised officer.

(3) Nothing in subsection (1) or subsection (2) shall apply to a public museum or zoo.

(4) The State Government may, by notification, require any person to declare to the Chief Wild Life Warden or the authorised officer any animal, article or trophy (other than a musk of a musk deer or horn of a rhinoceros) or salted or dried skins derived from an animal specified in Sch. I or Part II of Sch. II in his control, custody or possession in such form, in such manner, and within such time as may be prescribed.

Comment

Offence—When could not be said to be made out—The admission by the petitioner accused can only extend to the factual part incorporated in the charge, viz, that certain articles were seized from him and that he did not have or posses any certificates or documents issued by the authorities or documents issued by the authorities under the Wild Life (Protection) Act, 1972. But this cannot mean that the petitioner accused admitted that the articles that were seized from him did require such documents as specified under Wild Life (Protection) Act, 1972. This could have only arisen provided the charge went further to state that the skins used in the said articles seized were of the species of lizards or snakes mentioned in the schedules. Looking to the nature of the complaint and the texture of the evidence led in this case, it is apparent that the petitioner never intended by his plea of guilty to admit more than the facts alleged against him were true. On the facts before the Court no offence has been made out, and the plea of guilty must be held to be immaterial in these circumstances.[1]

41. **Inquiry and preparation of inventories**—(1) On receipt of a declaration made under Sec. 40, the Chief Wild Life Warden or the authorised officer may, after such notice, in such manner and at such time, as may be prescribed,—

(a) enter upon the premises of a person referred to in Sec. 40;

(b) make inquiries and prepare inventories of animal articles, trophies, uncured trophies, salted and dried skins and captive animals specified in Sch. I and Part II of Sch. II and found thereon; and

(c) affix upon the animals, animal articles, trophies, uncured trophies identification marks in such manner as may be prescribed.

(2) No person shall obliterate or counterfeit any identification mark referred to in this Chapter.

40. **Certificate of ownership**—The Chief Wild Life Warden may, for the purpose of Sec. 40, issue a certificate of ownership in such form, as may be prescribed, to any person who, in his opinion, is in lawful possession of any wild animal or any animal article, trophy, uncured trophy and may, where possible, mark, in the prescribed manner, such animal article, trophy or uncured trophy for purposes of identification.

1. Rafique Ramzan Ali *v.* A.A. Jalgaonkar, 1984 Cr. L.J. 1460(2) at p. 1463 (Bom.): 1985 Mah. L.R. 258: 1984 Cr. L.R. 626 (Mah.).

43. Regulation of transfer of animal, etc.—(1) Subject to the provisions of subsections (2), (3) and (4), a person (other than a dealer) who does not possess a certificate of ownership shall not—

(a) sell or offer for sale or transfer whether by way of sale, gift or otherwise, any wild animal specified in Sch. I or Part II of Sch. II or any captive animal belonging to that category or any animal article, trophy uncured trophy or meat derived therefrom;

(b) make animal articles containing part or whole of such animal;

(c) put under a process of taxidermy an uncured trophy of such animal, except with the previous permission in writing of the Chief Wild Life Warden or the authorised officer.

(2) Where a person transfers or transports from the State in which he resides to another State or acquire by transfer from outside the State any such animal, animal article, trophy or uncured trophy as referred to in subsection (1) in respect of which he has a certificate of ownership, he shall within thirty days of the transfer or transport, report the transfer or transport to the Chief Wild Life Warden or the authorised officer within whose jurisdiction the transfer or transport if effected.

(3) No person who does not possess a certificate of ownership shall transfer or transport from one State to another State or acquire by transfer from outside the State any such animal, animal article, trophy or uncured trophy as is referred to in subsection (1) except with the previous permission in writing of the Chief Wild Life Warden or the authorised officer within whose jurisdiction the transfer is to be effected.

(4) Before granting any permission under subsection (1) or subsection (3), the Chief Wild Life Warden or the authorised officer shall satisfy himself that the animal or article referred to therein has been lawfully acquired.

(5) While permitting the transfer or transport of animal, animal article, trophy or uncured trophy, as is referred to in subsection (1), the Chief Wild Life Warden or the authorised officer—

(a) shall issue a certificate or ownership after such inquiry as he may deem fit;

(b) shall, where the certificate of ownership existed in the name of the previous owner, issue a fresh certificate of ownership in the name of the person to whom the transfer has been effected.

(c) ay affix an identification mark on any such animal, animal article, trophy or uncured trophy.

(6) Nothing in this section shall apply—

(a) to animal articles or trophies made out of feathers of peacocks;

(b) to any transaction entered into by a public museum or zoo with any other public museum or zoo.

44. Dealings in trophy and animal article without licence prohibited—(1) [1][Subject to the provisions of Chapter V-A, no person shall, except under, and in accordance with, a licence granted under subsection (4),—]

(a) commence or carry on the business as—

(i) a manufacture of, or dealer in, any animal article; or

[2][(i-a) a manufacture of, or dealer in, any article made of ivory imported into India;]

(ii) a taxidermist; or

(iii) a dealer in trophy or uncured trophy; or

(iv) a dealer in captive animal; or

(v) a dealer in meat; or

(b) cook or serve meat in any eating-house:

1. Subs. by Act 28 of 1986, Sec. 2, for the words "Except under, and in accordance with, a licence granted under subsection (4), no person shall" (w.e.f. 20th November, 1986)
2. Ins. by *ibid*, Sec. 2

Provided that nothing in this subsection shall prevent a person, who, immediately before the commencement of this Act was carrying on the business or occupation specified in this subsection, from carrying on such business or occupation for a period of thirty days from such commencement, or where he has made an application within that period for the grant of a licence to him, until the licence is granted to him or he is informed in writing that a licence cannot be granted to him:

[1][Provided further that nothing in this subsection shall apply to manufacturers of, and dealers in, article made of feathers of peacock.]

Explanation—For the purposes of this section, "eating-house" includes a hotel, restaurant or any other place where any eatable is served on payment, whether or not such payment is separately made for such eatable or is included in the amount charged for board and lodging.

(2) Every manufacturer of, or dealer in animal article, or every dealer in captive animals, trophies or uncured trophies, or every taxidermist shall, within fifteen days from the commencement of this Act, declare to the Chief Wild Life Warden his stocks of animals articles, captive animals, trophies and uncured trophies, as the case may be, as on the date of such declaration and the Chief Wild Life Warden or the authorised officer may place an identification mark on every animal article, captive animal, trophy or uncured trophy, as the case may be.

(3) Every person referred to in subsection (1) who intends to obtain a licence, shall, [2][* * *] make an application to the Chief Wild Life Warden or the authorised officer for the right grant of a licence.

(4) (a) Every application referred to in subsection (3) shall be made in such form and on payment of such fee as may be prescribed, to the Chief Wild Life Warden or the authorised officer.

[3][(b) No licence referred to the subsection (1) shall be granted unless the Chief Wild Life Warden or the authorised officer having regard to the antecedents and previous experience of the applicant, the implication which the grant of such licence would have on the status of wild life and to such other matters as may be prescribed in this behalf and after making such inquiry in respect of those matters as he may think fit, is satisfied that the licence should be granted.]

(5) Every licence granted under this section shall specify the premises in which and the conditions, if any, subject to which the licensee shall carry on his business.

(6) Every licence granted under this section shall—

(a) be valid for one year from the date of grant;

(b) not be transferable; and

(c) be renewable for a period not exceeding one year at a time.

(7) No application for the renewal of a licence shall be rejected unless the holder of such licence has been given a reasonable opportunity of presenting his case and unless the Chief Wild Life Warden or the authorised officer is satisfied that—

(i) the application for such renewal has been made after the expiry of the period specified thereof, or

(ii) any statement made by the application at the time of the grant or renewal of the licence was incorrect or false in material particulars, or

1. Subs. by *ibid.*
2. Omitted by Act 23 of 1982, Sec. 3, for the words "within fifteen days from the commencement of this Act".
3. Subs. by Act 23 of 1982, Sec. 3

(iii) the applicant has contravened any term or condition of the licence or any provision of this Act or any rule made thereunder, or

(iv) the applicant does not fulfill the prescribed conditions.

(8) Every order granting or rejecting and application for the grant or renewal of a licence shall be made in writing.

(9) Nothing in the foregoing subsection shall apply in relation to vermin.

Comment

Accused—When ought to have been discharged—The Wild Life (Protection) Act, 1972, was designed to afford protection to certain species of wild life as were from time to time included in the various schedules of the Act. That there were about 2,500 different types of lizards and over 3,000 types of snakes. That the Act did not extend to all types of lizards or snakes but only to those species of lizard or snakes specified in the schedule. That the Act provides that it was an offence to deal in skins or articles made out of skins of such protected species. That in this matter the prosecution could only proceed on the basis that the articles seized were made of skins of lizards and snakes of species protected by the Act and the petitioner had thereby committed an offence; yet neither the complainant nor the evidence led had spelt out any offence. That this being the position, no charge could have been framed. That the petitioner could never have been convicted on the material placed before the Court. That in the result, the order was liable to be set aside and the petitioner ought to have been discharged.[1]

Absence of licence for dealing in articles and its effect—It is the prosecution case that the petitioner carries on business in the firm name and style of "Jooti" at Hotel Sea Rock, Bandra. That on the 4th August, 1981, the petitioner was found exhibiting for sale articles made of lizard and snake skins. Hence on the 5th of August, 1981, respondent accompanied by one Assistant Conservator of Forests, raided the said shop. At this time they were accompanied by two *panchas*. During the course of this raid certain articles came to be seized and a *panchanama* was drawn up. The complaint shows that respondent asked the petitioner to produce a dealer's licence for dealing in articles or such documents as required to be kept under the Wild Life (Protection) Act, 1972, and the petitioner was unable to produce this. That it is in view of this that the petitioner was charged with having committed offences under Sec. 51 read with Sec. 39 (3), Sec. 51 read with Sec. 40(1), Sec. 51 read with Sec. 40(2), Sec. 51 read with Sec. 44, Sec. 51 read with Sec. 44(1), Sec. 51 read with Sec. 44(2) and Sec. 51 read with Sec. 49 of the said Act. That nowhere in the complaint has it been stated that the articles seized were made out of the skins of lizard or snakes such as were specified in any of the Sch. I, Sch. II, Sch. III, Sch. IV, or Sch. V or were otherwise covered by the Act and if the complaint did not make out the case, then *ex facie* no offence was disclosed and no further proceedings could lie nor could any subsequent proceedings continue, much less would the question of the petitioner putting in a plea of guilty or otherwise arise. Further the statement only shows that the petitioner admitted two facts: (1) that he was in possession of the articles that were seized from him, and (2) that he did not have any document issued under the provisions of the Wild Life (Protection) Act, 1972. But this statement can never mean that the petitioner accepted the position that the articles seized from him required any certificate or permission from the wild life protection authorities or that the articles were made up of skins of lizard and snakes covered under one or the other schedules. That this too cannot assist the prosecution in any manner whatsoever. Thus it could be said that on this basis no charge could have been framed and the accused ought to have been discharged.[2]

---

1. Rafique Ramzan Ali *v.* A.A. Jalgaonkar, 1984 Cr. L.J. 1460(2) at pp. 1462, 1463 (Bom.): 1985 Mah. L.R. 258: 1984 Cr. L.R 262 (Mah.).
2. *Ibid.*, at pp. 1461, 1462 (Bom.)

45. **Suspension or cancellation of licences**—Subject to any general or special order of the State Government, the Chief Wild Life Warden or the authorised officer may, for reasons to be recorded by him in writing, suspend or cancel any licence granted or renewed under Sec. 44:

Provided that no such suspension or cancellation shall be made except after giving the holder of the licence a reasonable opportunity of being head.

46. **Appeal**—(1) An appeal from an order refusing to grant or renew a licence under Sec. 44 or an order suspending or cancelling a licence under Sec. 45 shall lie—

(a) if the order is made by the authorised officer, to the Chief Wild Life Warden; or

(b) if the order is made by Chief Wild Life Warden, to the State Government.

(2) In the case of an order passed in appeal by the Chief Wild Life Warden under Cl. (a) of subsection (1), a second appeal shall lie to the State Government.

(3) Subject as aforesaid, every order passed in appeal under this section shall be final.

(4) An appeal under this section shall be preferred within thirty days from the date of the communication, to the applicant, of the order appealed against:

Provided that the appellate authority may admit any appeal preferred after the expiry of the period aforesaid if it is satisfied that the appellant had sufficient cause for not preferring the appeal in time.

47. **Maintenance of records**—A licensee under this Chapter shall—

(a) keep records, and submit such returns of his dealings, as may be prescribed,—

(i) to the Director or any other officer authorised by him in this behalf, and

(ii) to the Chief Wild Life Warden or the authorised officer; and

(b) make such records available on demand for inspection by such officers.

48. **Purchase of animal, etc. by licensee**—No licensee under this Chapter shall—

(a) keep in his control, custody or possession,—

(i) any animal, animal article, trophy or uncured trophy in respect of which a declaration under the provisions of subsection (2) of Sec. 44 has to be made but has not been made;

(ii) any animal or animal article, trophy, uncured trophy or meat which has not been lawfully acquired under the provisions of this Act or any rule or order made thereunder:

(b) (i) Capture any wild animal, or

(ii) acquire, receive, keep in his control, custody or possession, or sell, offer for sale or transport, any captive animal specified in Sch. I or Part II of Sch. II or any animal article, trophy, uncured trophy or meat derived therefrom or serve such meat, or put under a process; or taxidermy or make animal article containing part or whole of such animal,

except in accordance with such rules as may be made under this Act:

Provided that where the acquisition or, possession, control or custody of such animal or animal article, trophy or uncured trophy entails the transfer or transport from one State to another, no such transfer or transport shall be effected except with the previous permission in writing of the Director or any other officer authorised by him in this behalf:

Provided further that no such permission under the foregoing proviso shall be granted unless the Director or the officer authorised by him is satisfied that the animal or article aforesaid has been lawfully acquired.

49. **Purchase of captive animal, etc. by a person other than a licensee**—No person shall purchase, receive or acquire any captive animal, wild animal other than vermin, or any animal article, trophy, uncured trophy or meat derived therefrom otherwise than from a dealer or from a person authorised to sell or otherwise transfer the same under this Act:

Provided that nothing in this section shall apply to any transaction entered into by a public museum or zoo with any other public museum or zoo.

Comment

Absence of offence in the complaint and its effect—The complaint in the present case, *ex facie* does not disclose any offence whatsoever inasmuch as there is not a clue as to whether articles seized was made of skins of the species of lizards or snakes specified in any of the several schedules. The matter ought to have ended there. In so far as the evidence is concerned, here again one finds the same lacuna and unless the prosecution went further to establish that the articles seized were made of skins of those species specified in one or the other of the several schedules, the question of there being an offence cannot arise. So called confessional statement does not assist the prosecution. Therefore the charge could be quashed.[1]

[2]CHAPTER V-A

Prohibition of trade or commerce in trophies, animal articles, etc. derived from certain animals

49-A. **Definitions**—In this Chapter,—

(a) "scheduled animal" means an animal specified for the time being in Sch. I or Part II of Sch. II;

(b) "scheduled animal article" means an article made from any scheduled animal and includes an article or object in which the whole or any part of such animal has been used:

(c) "specified date" means—

(i) in relation to a scheduled animal on the commencement of the Wild Life (Protection) Amendment Act, 1986, the date of expiry of two months from such commencement; and

(ii) in relation to any animal added or transferred to Sch. I or Part II of Sch. II at any time after such addition or transfer.

49-B. **Prohibition of dealings in trophies, animal articles, etc. derived from scheduled animals**—

(1) Subject to the other provisions of this section, on and after the specified date, no person shall,—

(a) commence or carry on the business as—

(i) a manufacturer of, or dealer in, scheduled animal articles; or

(ii) a taxidermist with respect to any schedule animals or any parts of such animals; or

(iii) a dealer in trophy or uncured trophy derived from any scheduled animal; or

(iv) a dealer in any captive animals being scheduled animals; or

(v) a dealer in meat derived from any scheduled animal; or

(b) cook or serve meat derived from any scheduled animal in any eating house.

Explanation—For the purposes of this subsection, "eating-house" has the same meaning as in the Explanation below subsection (1) of Sec. 44.

1. Rafique Ramzan Ali *v.* A.A. Jalgaonkar, 1984 Cr. L.J. 1460(2) at p. 1463 (Bom.): 1985 Mah. L.R. 258: 1984 Cr. L.R. 262 (Mah.).
2. Ins. by Act 28 of 1986, Sec. 3 (w.e.f. 20th November, 1986)

(2) Subject to the other provisions of this section, no licence granted or renewed under Sec. 44 before the specified date shall entitle the holder thereof or any other person to commence or carry on the business referred to in Cl. (a) of subsection (1) of this section or the occupation referred to in Cl. (b) of that subsection after such date.

(3) Notwithstanding anything contained in subsection (1) or subsection (2), where the Central Government is satisfied that it is necessary or expedient so to do in the public interest, it may, by general or special order published in the official Gazette, exempt, for purposes of export, any corporation owned or controlled by the Central Government including a Government company within the meaning of Sec. 617 of the Companies Act, 1956 (1 of 1956), or any society registered under the Societies Registration Act, 1860 (21 of 1860), or any other law for the time being in force, wholly or substantially financed by the Central Government, from the provisions of subsections (1) and (2).

(4) Notwithstanding anything contained in subsection (1) or subsection (2), but subject to any rules which may be made in this behalf, a person holding a licence under Sec. 44 to carry on the business as a taxidermist may put under a process of taxidermy any scheduled animal or any part thereof,—

(a) for or on behalf of the Government or any corporation or society exempted under subsection (3), or

(b) with the previous authorisation in writing of the Chief Wild Life Warden, for and on behalf of any person for educational or scientific purposes.

49-C. **Declaration by dealers**—(1) Every person carrying on the business or occupation referred to in subsection (1) of Sec. 49-B shall, within thirty days from the specified date, declare to the Chief Wild Life Warden or the authorised officer,—

(a) his stocks, if any, as at the end of the specified date of—

(i) scheduled animal articles;

(ii) scheduled animals and parts thereof;

(iii) trophies and uncured trophies derived from scheduled animals;

(iv) captive animals, being scheduled animals;

(b) the place or places at which the stocks mentioned in the declaration are kept; and

(c) the description of such items, if any, of the stocks mentioned in the declaration which he desires to retain with himself for his *bona fide* personal use.

(2) On receipt of a declaration under subsection (1), the Chief Wild Life Warden or the authorised officer may take all or any of the measures specified in Sec. 41 and for this purpose, the provisions of Sec. 41 shall so far as may be, apply.

(3) Where, in a declaration made under subsection (1), the person making the declaration expresses his desire to retain with himself any of the items of the stocks specified in the declaration for his *bona fide* personal use, the Chief Wild Life Warden with the prior approval of the Director, may, if he is satisfied that the person is in lawful possession of such items, issue certificates of ownership in favour of such person with respect to all, or as the case may be, such of the items as in the opinion of the Chief Wild Life Warden, are required of the *bona fide* personal use of such person and affix upon such items identification marks in such manner as may be prescribed:

Provided that no such item shall be kept in any commercial premises.

(4) No person shall obliterate or counterfeit any identification mark referred to in subsection (3).

(5) An appeal shall lie against any refusal to grant certificate of ownership under subsection (3) and the provisions of subsection (2), (3) and (4) of Sec. 46 shall, so far as may be, apply in relation to appeals under this subsection.

(6) Where a person who has been issued a certificate of ownership under subsection (3) in respect of any item,—

(a) transfers such item to any person, whether by way of gift, sale or otherwise, or

(b) transfers or transport from the State in which he resides to another State any such item,

he shall, within thirty days of such transfer or transport, report the transfer or transport to the Chief Wild Life Warden or the authorised officer within whose jurisdiction the transfer or transport is effected.

(7) No person, other than a person who has been issued a certificate of ownership under subsection (3) shall, on and after the specified date, keep under his control, sell or offer for sale or transfer to any person any scheduled animal or a scheduled animal article.]

CHAPTER VI
Prevention and Detection of Offences

50. Power of entry, search, arrest and detention—(1) Notwithstanding anything contained in any other law for the time being in force, the Director or any other officer authorised by him in this behalf or the Chief Wild Life Warden or the authorised officer or any forest officer or any police officer not below the rank of a sub-inspector, may, if he has reasonable grounds for believing that any person has committed an offence against this Act,—

(a) require any such person to produce for inspection any captive animal, wild animal, animal article, meat, trophy or uncured trophy in his control, custody or possession, or any licence, permit or other document granted to him or required to be kept by him under the provisions of this Act;

(b) stop any vehicle or vessel in order to conduct search or inquiry or enter upon and search any premises, land, vehicle, or vessel, in the occupation of such person, and open and search any baggage or other things in his possession;

(c) seize any captive animal, wild animal, article, meat, trophy or uncured trophy in the possession of any person and appearing to him to be Government property, together with any trap, tool vehicle, vessel or weapon used for committing any such offence, and unless he is satisfied that such person will appear and answer any charge which may be preferred against him, arrest him without warrant, and detain him.

(2) Any officer of a rank not inferior to that of an Assistant Director of Wild Life Preservation or Wild Life Warden, who, or whose subordinate, has seized any trap, tool, vehicle, vessel or weapon under Cl. (c) of subsection (1), may release the same, on the execution by the owner thereof of a bond for the production of the property so released, if and when so required, before the Magistrate having jurisdiction to try the offence on account of which the seizure has been made.

(3) It shall be lawful for any of the officers referred to in subsection (1) to stop and detain any person, whom he sees doing any act for which a licence or permit is required under the provisions of this Act, for the purposes of requring such person to produce the licence or permit and if such person fails to produce the licence or permit, as the case may be, he may be arrested without warrant, unless he furnishes his name

and address, and otherwise satisfies the officer arresting him that he will duly answer any summons or other proceedings which may be taken against him.

(4) Any person detained, or things seized under the foregoing power, shall forthwith be taken before a Magistrate to be dealt with according to law.

(5) Any person who, without reasonable cause, fails to produce anything, which he is required to produce under this section, shall be guilty of an offence against this Act.

(6) (a) Where any meat or uncured trophy is seized under the provisions of this section, the Assistant Director of Wild Life Preservation or any other officer of a gazetted rank authorised by him in this behalf or the Chief Wild Life Warden or the authorised officer may arrange for the sale of the same and deal with the proceeds of such sale in such manner as may be prescribed.

(b) Where it is proved that the meat or uncured trophy seized under the provisions of this section is not Government property, the proceeds of the sale shall be returned to the owner.

(7) Whenever any person is approached by any of the officers referred to in subsection (1) for assistance in the prevention or detection of an offence against this Act, or in apprehending persons charged with the violation of his Act, or for seizure in accordance with Cl. (c) of subsection (1), it shall be the duty of such person or persons to render such assistance.

Comment

S.O. 1508, dated the 23rd February, 1988[1]**—**The undermentioned Regional Deputy Directors (Wild Life Preservation) Regional Wildlife Officers Ministry of Environment and Forests are hereby authorised to exercise powers under Sec. 50 of the Wild Life (Protection) Act, 1972.

Sl. No.	Name	Designation	Headquarters
1.	Shri S.S. Bist	Regional Deputy Director Wild Life Preservation (Northern Region)	New Delhi
2.	Shri Kalyan Chakraborty	do (Eastern Region)	Kolkata
3.	Shri P. Kannan	do (Western Region)	Mumbai
4.	Shri Promod Kant	do (Southern Region)	Chennai

51. Penalties—(1) Any person who[2] **[contravenes any provision of this Act (except Chapter V-A) or any rule order made thereunder or who commits a breach of any of the conditions of any licence or permit granted under this Act, shall be guilty of an offence against this Act, and shall, on conviction, be punished with imprisonment for a term which may extend to two years, or with fine which may extend to two thousand rupees, or with both:**

Provided that where the offence committed is in relation to any animal specified in Sch. I or Part II of Sch. II or meat of any such animal or animal article, trophy or uncured trophy derived from such animal or where the offence related to hunting in a sanctuary or a National Park, such offence shall be punishable with imprisonment for a term which shall not be less than six months but may extend to six years and also with fine which shall not be less than five hundred rupees:

1. Published in the *Gazette of India,* Pt II, Sec. 3(ii), dated 14th May, 1988
2. Subs. by Act 28 of 1986, Sec. 4, for certain words (w.e.f. 20th November, 1986)

Provided further that in the case of a second or subsequent offence of the nature mentioned in the foregoing proviso, the term of imprisonment shall not be less than one year and the amount of the fine shall not be less than one thousand rupees.

[1][(1-A) any person who contravenes any provisions of Chapter V-A shall be punishable with imprisonment for a term which shall not be less than one year but which may extend to seven years and also with fine which shall not be less than five thousand rupees.]

(2) When any person is convicted of an offence against this Act, the Court trying the offence may order that any captive animal, wild animal, animal article, trophy, uncured trophy or meat in respect of which the offence has been committed and any trap, tool, vehicle, vessel or weapon, used in the commission of the said offence be forfeited to the State Government and that any licence or permit, held by such person under the provisions of this Act, be cancelled.

(3) Such cancellation of licence or permit or such forfeiture shall be in addition to any other punishment that may be awarded for such offence.

(4) Where any person is convicted of an offence against this Act, the Court may direct that the licence, if any, granted to such person under the Arms Act, 1959 (54 of 1959), for possession of any arm with which an offence against this Act has been committed, shall be cancelled and that such person shall not be eligible for a licence under the Arms Act, 1959, for a period of five years from the date of conviction.

Comments

Question of sentence—In the undernoted case so far as the merit of the case is concerned, it is no doubt concluded by the concurrent findings of the Courts below and the High Court has got very limited power to interfere wit the concurrent findings of the Courts below in its revisional jurisdiction. So far as the sentence awarded to the petitioners is concerned, it is submitted that the occurrence took place in the year 1974, i.e. ten years ago and no fruitful purpose would be served in sending the petitioners, who are on bail, again to jail after a lapse of ten years. It was pleaded that the sentence may be converted into some fine. The ends of justice would be fully met if the sentence of regorous imprisonment of three months is altered into a fine of Rs. 50/-each. The fine must be paid by each of the petitioners within two months from today. If the petitioners fail to pay the fine within the time specified, the modification in the sentence would become inoperative.[2]

If it is an offence for a person holding a licence to go about with a loaded gun in his possession—The convict the accused under the Act, it is necessary to prove that he had either killed or attempted to kill one of the animals or birds mentioned in the schedule. But there is no evidence to show that he had actually fired at any bird or animal, much less one mentioned in the schedule of the Act. It is no offence for a person holding a licence to go about with a loaded gun in his possession. Therefore the conviction could be quashed.[3]

Conviction—When could be set aside—The petitioner was questioned and he is said to have made a statement admitting his guilt. This statement was reduced to writing. That a complaint was then filed in the Court of the Additional Chief Metropolitan Magistrate. The complaint was proceeded with and evidence of respondent and one of the *panchas* came to be recorded. On the basis of this evidence, the

1. Ins. by Act 28 of 1986, Sec. 4 (w.e.f. 20th November, 1986)
2. Jagdish Singh *v* State of Bihar, 1985 Cr. L.J. 1314 at p. 1316 (Pat.)
3. Batan Singh *v* Emperor, 29 Cr. L.J. 238 at p. 238 (Lah.).

learned Magistrate proceeded to frame a charge. On the charge being framed, the accused pleaded "guilty". By his order and judgement dated the 13th April, 1983, the learned Magistrate convicted the petitioner on various counts, and sentenced him. That in the evidence of prosecution witness there is no mention whatsoever that the articles seized were or are made of skin of such lizards or snakes, the species whereof have been enumerated in any one of the several schedules. That there were over 2,500 species of lizards and over 3,000 species of snakes, a position borne out by the *Encyclopaedia Britannica,* and if this be so, it was incumbent on the prosecution to lead evidence to show that the articles seized were made up of the skins of such lizards and snakes such as were specified in one or the other schedules of the said Act. That since the prosecution failed to establish his fact, the prosecution must fail. The *panchanama* has, of course, been put in, but this can only establish the fact of what was seized at the time of raid and no more and cannot advance the prosecution case any further. The provisions of the Wild Life (Protection) Act, 1972, have designed to prevent discrimination and commercial exploitation of rare species of animals and reptiles, and prosecutions need to be undertaken and conducted in all seriousness. In this matter, however, right from the complaint till the completion of the evidence the approach of the prosecution has been casual inasmuch as the prosecution has failed to adduce any evidence that the articles seized were made up of skins of such lizards or snakes as were enumerated in one of the other schedules. In the result, the conviction and sentence is set aside.[1]

In the evening of 8th May, 1975, a tiger was shot dead by accused petitioner at about 5.30 p.m. at Changlai Camp while on sentry duty. The accused, prior to shooting, informed the Guard Commander about the presence of the tiger and the Commander instructed that 2/3 rounds might be fired into the air to scare it away, but while this was being done the tiger, instead of fleeing, charged at the accused who fired two shots at it, one hitting the left shoulder blade and the other in the chest, and so hit, the tiger jumped and fell dead few yards away. In the instant case it is in record that the accused did not go out of his post while the tiger approached him. On the three shots being fired in the air, the animal, instead of fleeing away, charged at the accused seeing which the accused fired two shots to kill, which killed the tiger. The Deputy Commissioner while convicting the accused under Sec. 51(1) followed the spirit of the Act to protect the wild life but the provisions of Sec. 11 were not brought to his notice.[2]

Same offence under two different Acts—Prosecution under both Acts held possible—In the case of *State of Bihar* v. *Murad Ali Khan,*[3] the accusation against the three respondents is that on 8th June, 1986 at 2.00 p.m. they along with two others named in the complaint, shot and killed an elephant in compartment No. 13 of Kundurugutu Range Forest and removed the ivory tusks of the elephant. On 25th June, 1986 the Range Officer of Forest of that Range lodged a written complaint with the Judicial Magistrate, 1st class, Chaibasa, in this behalf alleging offences against respondents under Sec. 51 of the Wild Life (Protection) Act, 1972. The learned Magistrate took cognizance of the offence and ordered issue of process to the accused. It would appear that at the Police Station, Sonua, a case had been registered under Secs. 447, 429 and 379 (IPC) read with Secs. 54 and 39 of the Wild Life (Protection) Act, 1972, and that the matter was under investigation by the police. The respondents, who were amongst the accused, moved the High Court under Sec. 482, CrPC, for quashing of the order of the Magistrate taking congnizance of the alleged offence issuing summons. The High Court, accordingly, quashed the proceedings against the respondents.

1. Rafique Ramzan Ali *v.* A.A. Jalgaonkar, 1984 Cr. L.J. 1460(2) at pp. 1461, 1462, 1463 (Bom.): 1985 Mah. L.R. 258: 1984 Cr. L.R. 262 (Mah.).
2. Tilok Bahadur Rai *v.* State of Arunachal Pradesh, 1979 Cr. I.J. 1404 at pp. 1404, 1405 (Gau.).
3. AIR 1989 S.C. 1 at pp. 3, 4, 5, 6, 7, 9

From the orders under appeal it would appear that two grounds commended themselves for acceptance to the High Court. The first was that the learned Magistrate acted contrary to the provisions of Sec. 210. The High Court observed:

"The investigation is still continuing and pending in so far as the petitioner is concerned and the investigation shall continue. Obviously the judicial magistrate acted beyond jurisdiction in taking cognizance against the petitioner when for the same allegation the investigation was proceeding and pending. He acted contrary to the provisions of Sec. 210 of the Code of Criminal Procedure. The complaint was filed after long delay..."

The second ground was on the merits of the complaint. The High Court, *inter alia*, observed:

"On the face of the complaint petition of the first information report itself the facts alleged do not constitute the offence. The petitioner was never named in the first information report. There is no eye-witness in this case and there is no identification of the petitioner in any manner whatsoever to sustain the allegation even *prima facie* for the offence alleged."

Section 9(1) of the Act says that no person shall "hunt" any wild animal specified in Schedule I. Elephant is included in Schedule I. The expression "Wild animal" is defined in Sec. 2(36) to mean by "animal found wild in nature and includes any animal specified in Schedule I", etc. The expression "hunting" is defined in Sec. 2(16) in a comprehensive manner. Section 51 of the Act provides for penalties. Violation of Sec. 9(1) is an offence under Sec. 51(1). Section 55 deals with cognizance of offences. What emerges from a perusal of these provisions is that cognizance of an offence against the "Act" can be taken by a Court only on the complaint of the officer mentioned in Sec. 55. The person who lodged complaint dated 23rd June, 1986, claimed to be such an officer. In these circumstances even if the jurisdictional police purported to register a case for an alleged offence against the Act, Sec. 210(1) would not be attracted having regard to the position that cognizance of such an offence can only be taken on the complaint of the officer mentioned in the section. Even where a Magistrate takes cognizance of an offence instituted otherwise than on a police report and an investigation by the police is in progress in relation to same offence, the two cases do not lose their separate identity. The section seeks to obviate the anomalies that might arise from taking cognizance of the same offence more than one. But, where cognizance can be taken only in one way and that on the complaint of a particular statutory functionary, there is no scope or occasion for taking cognizance more than once and, accordingly, Sec. 210 has no role to play. The view taken by the High Court on the footing of Sec. 210 is unsupportable. The Supreme Court did not accept the contention of Shri R. F. Nariman that the specific allegation in the present case concerns the specific act of killing of an elephant, and that such an offence, at all events, falls within the overlapping areas between Sec. 429, IPC on the one hand and Sec. 9(1) read with Sec. 50(1) [*sic* should be Sec. 51(1)] of the Act on the other and therefore constitutes the same offence. Apart from the fact that this argument does not serve to support the order of the High Court in the present case this argument is, even on its theoretical possibilities, more attractive than sound. The expression "any act or omission which constitutes any offence under this Act" in Sec. 56 of the Act merely imports the idea that the same act or omission might constitute an offence under another law and could be tried under such other law or law also. It was held that the ingredients of an offence under Sec. 9(1) read with Sec. 50(1) [*sic* should be Sec. 51(1)] of the Act require for its establishment certain ingredients which are not part of the offence under Sec. 429 and *vice versa*.

No fault can be found with the acquittal where the evidence being insufficient—The respondents were charged for having illegally killed two tigers on 21/22nd March, 1978, by administering poison and deskinned them which was punishable under Sec. 51 of the Act. The prosecution case in brief was that the appellants had mixed aldrin into the water for drinking by animal which resulted in death of two tigers. Thereafter, they took away those tigers and deskinned them. On receiving this information the Ranger Officer Hanslal informed the police authorities. Later on, the tiger skin, nails and meat were seized from some of the respondents. During the trial first prosecution witness stated that the police did not interrogate the respondents in his presence. He also denied that anything was seized in his presence. According to him, he was called at the police station where he saw tiger skin, etc. He did not know who had brought the tiger skin, etc. in the police station. He also denied memorandums having been made in his presence though he admitted his signatures on the memorandums and seizure memos. The witness was, therefore, declared hostile and was permitted to be cross-examined. The second witness also denied that any of the respondents was interrogated in his presence. He further denied any seizure from the respondents. In spite of it, he accepted his signatures on memorandums and seizure memos. This witness was also declared hostile and cross-examined. In cross-examination, he admitted that the tiger skin was seized in his presence and he had signed the seizure memo. He also admitted that the respondents were his relations and belong to the village. He admitted that the memorandum and seizure memos were written in the Thana. Evidence of this witness is also not sufficient to hold that the respondents had given statements leading to discovery of seized articles. The third witness does not support the prosecution, as in cross-examination, he is not able to say how much of aldrin was sufficient to kill a tiger. He also did not know how much aldrin was found in the stomach of tigresses. According to him, aldrin is an agricultural insecticide easily available in the market. Except for these, there is no other witness to connect the respondents with the crime. The aforesaid evidence, however, does not indicate that the respondents had either mixed aldrin or had given any statement leading to recovery at their instance. The evidence being insufficient, no fault can be found with the acquittal of the respondents.[1]

52. **Attempts and abetment**—Whoever attempts to contravene, or abets the contravention of, any of the provisions of this Act or of any rule or order made thereunder shall be deemed to have contravened that provision or rule or order, as the case may be.

53. **Punishment for wrongful seizure**—If any person, exercising powers under this Act, vexatiously and unnecessarily seizes the property of any other person on the pretence of seizing it for the reasons mentioned in Sec. 50, he shall, on conviction, be punishable with imprisonment for a term which may extend to six months, or with fine which may extend to five hundred rupees, or with both.

54. **Power to compound offences**—(1) The State Government may, be notification, empower the Chief Wild Life Warden or any officer of a rank not inferior to that of a Deputy Conservator of Forests,—

(a) to accept, from any person against whom a reasonable suspicion exists that he has committed an offence against this Act, payment of a sum of money by way of composition of the offence which such person is suspected to have committed; and

(b) when any property has been seized is liable to be forfeited, to release the same on payment of the value thereof as estimated by such officer.

(2) On payment of such sum of money or such value, or both, as the case may be, to such officer, the suspected person, if in custody, shall be discharged, and the property, other than Government property, if any, seized, shall be released and no further proceedings in respect of the offence shall be taken against such person.

1. State of Madhya Pradesh *v.* Subke Baboo, 1986 (2) Crimes 232 at pp. 232, 233 (MP)

(3) The officer compounding any offence may order the cancellation of any licence or permit granted under this Act to the offender, or if not empowered to do so, may approach an officer so empowered, for the cancellation of such licence or permit.

(4) The sum of money accepted or agreed to be accepted as composition under Cl. (b) of subsection (1) shall, in no case exceed the sum of two thousand rupees:

Provided that no offence, for which a minimum period of imprisonment has been prescribed in subsection (1) of Sec. 51, shall be compound.

Comment

Jurisdiction to file the complaint—One Wild Life Warden in Palamau National Park submitted a forwarding report on 28th December, 1974 to the Sub-Divisional Judicial Magistrate, Latehar alleging, *inter alia*, that while he along with the staff was on patrolling duty in Betla Reserved Forest, he heard a sound of gunshot. When the patrolling party reached the junction of road Nos. 2 and 5 they heard some sound of talking of some persons and there they saw that a Bison was lying dead and petitioner was standing with a gun and other petitioners were skinning the dead animal. The patrolling party could catch only petitioners 1 to 3 and it was alleged that petitioner No. 4 could manage to escape. The complainant prepared a seizure list on the spot and arrested petitioners 1 to 3. Therefore a complaint was filed on the basis of which Sub-Divisional Judicial Magistrate, Latehar, took cognizance against the petitioners and they were put on trial. It was submitted that the entire conviction and sentence is bad in law and is to be set aside because the complaint has not been filed by an officer who was either the Chief Wild Life Warden or the officer authorised under the Act by the State Government. It may be relevant to mention here that Sec. 5(2) of the Act also gives power to the authorities concerned to delegate his power to any of his subordinate officer. Therefore it could not be said that the complainant had no jurisdiction or no authority to file the complaint.[1]

55. **Cognizance of offences**—No Court shall take cognizance of any offence against this Act except on the complaint of the Chief Wild Life Warden or such other officer as the State Government may authorise in this behalf.

56. **Operation of other laws not barred**—Nothing in this Act shall be deemed to prevent any person from being prosecuted under any other law for the time being in force, for any act or omission which constitutes an offence against this Act or from being liable under such other law to any higher punishment or penalty than that provided by this Act:

Provided that no person shall be punished twice for the same offence.

57. **Presumption to be made in certain cases**—Where, in any prosecution for one offence against this Act, it is established that a person is in possession, custody or control of any captive animal, article, meat, trophy or uncured trophy, it shall be presumed, until the contrary is proved, the burden of proving which shall lie on the accused, that such person is in unlawful possession, custody or control of such captive animal, animal article, meat, trophy or uncured trophy.

58. **Offences by companies**—(1)Where an offence against this Act has been committed by a company, every person who, at the time the offence was committed, was in charge of, and was responsible to, the company for the conduct of the business of the company as well as the company, shall be deemed to be guilty of the offence and shall be liable to be proceeded against and punished accordingly:

1. Jagdish Singh *v* State of Bihar, 1985 Cr. L.J. 1314 at pp. 1315-16 (Pat.)

Provided that nothing contained in this subsection shall render any such person liable to any punishment, if he proves that the offence was committed without his knowledge or that he exercised all due diligence to prevent the commission of such offence.

(2) Notwithstanding anything contained in subsection (1), where an offence against this Act has been committed by a company and it is proved that the offence has been committed with the consent or connivance of, or is attributable to any neglect on the part of, any director, manager, secretary or other officer of the company, such director, manager, secretary or other officer shall also be deemed to be guilty of that offence and shall be liable to be proceeded against and punished accordingly.

Explanation—For the purposes of this section,—

(a) "company" means any body corporate and includes a firm or other association of individuals; and

(b) "director", in relation to a firm, means a partner in the firm.

CHAPTER VII
Miscellaneous

59. **Officers to be public servants**—Every officer referred to in Chapter II and every other officer exercising any of the powers conferred by this Act shall be deemed to be a public servant within the meaning of Sec. 21 of the Indian Penal Code (45 of 1860).

60. **Protection of action taken in good faith**—(1) No suit, prosecution or other legal proceeding shall lie against any officer or other employee of the Central Government or the State Government for anything which is in good faith done or intended to be done under this Act.

(2) No suit or other legal proceeding shall lie against the Central Government or the State Government or any of its officers or other employees for any damage caused or likely to be caused by anything which is in good faith done or intended to be done under this Act.

61. **Power to alter entries in Schedules**—(1) The Central Government may, if it is of opinion that it is expedient so to do, by notification add any entry to any Schedule or transfer any entry from one Part of a Schedule to another Part of the same Schedule or from one Schedule to another.

(2) The State Government may, if it is of opinion that it is expedient so to do, by notification, transfer any entry from Sch. II, Sch. III, Sch. IV or Sch. V to Sch I and may also transfer any entry from Part I of Sch. II, or Sch. III, Sch. IV or Sch. V, to any other Schedule.

(3) On the issue of a notification under subsection (1) or subsection (2), the relevant Schedule shall be deemed to be altered accordingly, provided that every such alteration shall be without prejudice to anything done or omitted to be done before such alteration.

(4) If any alteration of any Schedule made by the State Government under subsection (2) is repugnant to any alteration made therein by the Central Government under subsection (1) then, the alteration made by the Central Government, whether made before or after the notification made by the State Government, shall prevail and the alteration made by the State Government shall, to the extend of the repugnancy, be void:

Provided that any such alteration made by the State Government, if it has been made with the previous consent of the Central Government, shall prevail in the State:

Provided further that nothing in the foregoing proviso shall prevent the Central Government from modifying or cancelling, at any time, the alteration made by the State Government.

62. Declaration of certain wild animals to be vermin—Subject to the provision of Sec. 61 the State Government may, by notification, declare any wild animal other than those specified in Sch. I and Part II of Sch. II to be vermin for any area and for such period as may be specified therein and so long as such notification is in force, such wild animal shall be deemed to have been included in Sch. V.

63. Power of Central Government to make rules—(1) The Central Government may, by notification, make rules for all or any of the following, namely;

(a) the form in which declaration shall be made under subsection (2) of Sec. 44;

[1](aa) the matters to be prescribed under Cl. (b) of subsection (4) of Sec. 44;]

(b) the terms and conditions which shall govern transaction referred to in Cl. (b) of Sec. 48;

(c) matters specified in subsection (2) of Sec. 64 in so far as they relate to sanctuaries and National Parks declared by the Central Government;

[2][(d) any other matter which is required to be, or may be prescribed.]

(2) Every rule made under this section shall be laid, as soon as may be after it is made, before each House of Parliament, while it is in session, for a total period of thirty days which may be comprised in one session or in two or more successive sessions, and if before the expiry of the session immediately following the session or the successive sessions aforesaid, both Houses agree in making any modification in the rule or both House agree that the rule should not be made, the rule shall thereafter have effect only in such modified form or be of no effect, as the case may be; so however, that any such modification or annulment shall be without prejudice to the validity of anything previously done under that rule.

64. Power of State Government to make rules—(1) The State Government may, be notification, makes rules for carrying out the provisions of this Act in respect of matters which do not fall within the purview of Sec. 63.

(2) In particular and without prejudice to the generality of the foregoing power, such rules may provide for all or any of the following matters, namely:

(a) the term of office of the members of the Board referred to in Cl. (g) of subsection (1) of Sec. 6 and the manner of filling vacancies among them;

(b) allowances referred to in subsection (4) of Sec. 6;

(c) the forms to be used for any application, certificate, claim, declaration licence, permit, registration, return or other document, made granted or submitted under the provisions of the Act and the fees, if any, therefore;

(d) the conditions subject to which any licence or permit may be granted under this Act;

(e) the particulars, of the record of wild animals (captured or killed) to be kept and submitted by the licensee;

(f) regulation of the possession, transfer and the sale of captive animals, meat, animal articles, trophies and uncured trophies;

(g) regulation of taxidermy;

(h) any other matter which has to be, or may be, prescribed under this Act.

---

1. Subs. by Act 23 of 1982, Sec. 4
2. Ins. by Act 28 of 1986, Sec. 5 (w.e.f. 20th November, 1986)

Comment

Delegation of power of authorities to file complaint—Cognizance can be taken of any offence against the Act either on the complaint of Chief Wild Life Warden or such other officer as the State Government may authorise in this behalf. Therefore, any officer who has been authorised by the State Government to file a complaint is quite competent under the Act to file such complaint. In this connection reference can be made to Bihar Wild Life (Protection) Rules, 1973, which have been framed under Sec. 64 of the Act. Section 64 of the Act gives power to the State Government to make rules. The State Government may under this power make any rule by notification for carrying out the provisions of the Act in respect of matters which do not fall within the purview of Sec. 63 of the Act. Section 63 of the Act gives power to the Central Government to frame rules by notification which is not concerned in this case. Therefore, it is now clear that the State Government has power to make rules under the rule making power of the Act and by virtue of this power, Rules, 1973, were framed and notified *vide* notification No. G.S.R. 127, dated 14th November, 1974. Therefore the complaint was maintainable.[1]

65. **Rights of Scheduled Tribes to be protected**—Nothing in this Act shall affect the hunting rights conferred on the Scheduled Tribes of the Nicobar Islands in the Union Territory of Andaman and Nicobar Islands by notification of Andaman and Nicobar Administration No. 40/97/F. No. G-635, Vol III, dated the 28th April, 1967, published at pages 1 to 5 of the *Extraordinary* issue of the Andaman and Nicobar Gazette, dated the 28th April, 1967.

66. **Repeal and savings**—(1) As from the commencement of this Act, every other Act relating to any matter contained in this Act and in force in a State shall, to the extent to which that Act or any provision contained therein corresponds, or is repugnant, to this Act or any provision contained in this Act, stand repealed:

Provided that such repeal shall not—

(i) affect the previous operation of the Act so repealed, or anything duly done or suffered thereunder;

(ii) affect any right, privilege, obligation or liability acquired, accrued or incurred under the Act so repealed;

(iii) affect any penalty, forfeiture or punishment incurred in respect of any offence committed against the Act so repealed; or

(iv) affect any investigation, legal proceeding or remedy in respect of any such right, privilege, obligation, liability, penalty, forfeiture, or punishment as aforesaid;

and any such investigation, legal proceeding or remedy may be instituted, continued or enforced, and any such penalty, forfeiture and punishment may be imposed, as if the aforesaid Act had not been repealed.

(2) Notwithstanding such repeal,—

(a) anything done or any action taken under the Act so repealed (including any notification, order, certificate, notice or receipt issued, application made, or permit granted) which is not inconsistent with the provision of this Act be deemed to have been done or taken under the corresponding provisions of this Act as if this Act were in force at the time such thing was done or action was taken, and shall continue to be in force, unless and until superseded by anything done or any action taken under this Act;

1. Jagadish Singh *v.* State of Bihar, 1985 Cr. I.J. 1314 at pp. 1315, 1316 (Pat.)

(b) every licence granted under any Act so repealed and in force immediately before the commencement of this Act shall be deemed to have been granted under the corresponding provisions of this Act and shall, subject to the provisions of this Act, continue to be in force for the unexpired portion of the period for which such licence had been granted.

(3) For the removal of doubts, it is hereby declared that any sanctuary or National Park declared by a State Government under any Act repealed under subsection (1) shall be deemed to be a sanctuary or National Park, as the case may be, declared by the State Government under this Act and where any right in or over any land in any such National Park which had not been extinguished under the said Act, at or before the commencement of this Act, the extinguishment of such rights shall be made in accordance with the provisions of this Act.

SCHEDULE I
(See Secs. 2, 8, 9, 11, 40, 41, 43, 48, 51, 61 and 62)
PART I
Mammals

[1][1. Andaman wild pig (*Susandamanensis*).]

[2][1-A. Bharal (*ovis nahura*).]

[2][1-B. Binturong (*Arctictis binturong*).]

2. Blank buck (*Antelope cervicapra*).

[2][2-A. * * * * * * * *]

3. Brow-antlered deer or thamin (*Cervus eldi*)

[3][3-A. Himalayan brown bear (*Ursus arcots*).]

[3][3-B. Capped langur (*Presbytis pileatus*).]

4. Caracal (*Felis caracal*).

[2][4-A. Catecean Spp.]

5. Cheetah (*Acinonyx jubatus*).

[4][5-A. Chinese pangoline (*Manispentadactyla*).]

[1][5-B. Chinkara or Indian Gazelle (*Gazalle gazetta bennetti*).]

6. Clouded leopard (*Neofelis nebulosa*).

[2][6-A. Crab-eating macaque (*Macacairus umbrosa*).

[2][6-B. Desert cat (*Felis libyca*).]

[3][6-C. Desert fox (*Vulpes bucopus*).

7. Dugong (*Dugong dugon*).

[2][7-A. Ermine (*Mustela erminea*).]

8. Fishing cat (*Felis viverrina*).

[1][8-A. Four-horned antelope (*Tetraceros quodricornis*).]

[2][8-B. * * *]

1. Vide Notification published in the *Gazette of India, Extraordinary,* Pt. II, Sec. 3(i), dated 5th October, 1977.
2. Vide Notification published in the *Gazette of India, Extraordinary,* Pt. II, Sec. 3(i), dated 2th October, 1980.
3. Vide S.O. 859(E), dated 24th November, 1986, published in the *Gazette of India, Extraordinary,* Pt. II, Sec. 3(ii), dated 24th November, 1986.
4. Vide Notification dated 29th August, 1977 published in the *Gazette of India, Extraordinary,* Pt. II, Sec. 3(i), dated 3rd September, 1977.

[1][8-C. * * *]
[1][8-D. Gangetic dolphin (*Platanista gangetia*).]
[1][8-E. Bison gaur (*Bas gaurus*).]
9. Golden cat (*Felis temmincki*).]
10. Golden Langur (*Presbytis geei*).
[1][10-A. Giant squirrels (*Ratufa macrouro*).]
[1][10-B. Himalayan ibex (*Capra ibex*).]
[1][10-C. Himalayan thar (*Hemitragus jemlahicus*).]
11. Hispid hare (*Caprolagus hispidus*).
[1][11-A. Hog bodger (*Arctonyx Gollaris*).]
12. Hoolock (*Hylobates hoolock*)
[2][12-A. * * * *]
[1][12-B. Indian elephant (*Elephas maximus*).]
13. Indian lion (*Panthera leo persica*).
14. Indian wild ass (*Equus hemionus khur*).
[3][15. Indian wolf (*Canis Iupus pallipes*).]
16. Kashmir stag (*Cervus elaphus hanglu*).
[2][16-A. Leaf monkey (*Presbytis phayrei*).]
[2][16-B. Leopard or Panther (*Panteera pardus*).]
17. Leopard cat (*Felis bengalensis*).
18. Lesser or Red panda (*Ailurus fulgens*).
19. Lion-tailed macaque (*Macaca silenus*).
20. Loris (*Loris tardigradus*).
[1][20-A. Little India Porpoise (*Necomeris phocaenoides*).]
21. Lynx (*Felis lynx isabellinus*).
22. Malbar civet (*Viverra megaspila*).
[4][22-A. Malay or sun bear (*Helarctos malayanus*).]
23. Marbled cat (*Felis marmorta*).
24. Markhor (*Capra falconeri*).
[4][24-A. Mouse deer (*Tragulus menrinual*).]
25. Musk deer (*Moschus moschiferus*).
[1][25-A. Nilgiri langur (*Presbytis johni*).
[1][25-B. Nilgiri thar (*Hemitragus hylocrius*).]
26. Ovis ammon or nyan (*Ovis ammon hodgosoni*).
27. Pallas's cat (*Felis manul*).
28. Pangolin (*Manis crassicaudata*).
29. Pygmy hog (*Sus salvanius*).
[2][29-A. Ratel (*Mellivora capensis*).]
30. Rhinoceros (*Rhinoceros unicornis*).
31. Rusty-spotted cat (*Felis rubiginosa*).
[4][31-A. Serow (*Capricornis sumatraensis*).]
[1][31-B. Otter (*Aonyx Cinerea*).]
[1][31-C. Sloth bear (*Melus susursinus*).]
32. Slow loris (*Nycticebus coucang*).
[2][32-A. Small Travancore flying squirrel (*Petinomys fuscocapillus*).

1. Vide S.O. 859(E), dated 24th November, 1986, published in the *Gazette of India, Extraordinary,* Pt. II, Sec. 3(ii), dated 24th November, 1986.
2. Vide Notification published in the *Gazette of India, Extraordinary,* Pt. II, Sec. 3(i), dated 2th October, 1980.
3. Vide Notification dated 29th August, 1977, published in the *Gazette of India, Extraordinary,* Pt. II, Sec. 3(ii), dated 3rd September, 1977.
4. Vide Notification published in the *Gazette of India, Extraordinary,* Pt. II, Sec. 3(i), dated 5th October, 1977

33. Snow leopard (*Panthera uncia*).
[1][33-A. Snubfin dolphin (*Orcaella brevirostris*).]
34. Spotted linsang (*Prionodon pardicolor*).
35. Swamp deer (All sub-species of *Cervus duvauceli*).
36. Takin or Mishmi takin (*Budorcas taxicolor*).
[3][36-A. Tibetan antelope or chiru (*Panthelops hodgsoni*).]
[1][36-B. Tibetan fox (*Valpes ferrilatus*).]
37. Tibetan gazelle (*Procapra picticaudata*).
38. Tibetan wild ass (*Equus hemionus klang*).
39. Tiger (*Panthera tigris*).
40. Urial or Shapu (*Ovis vignei*).
41. Wild buffalo (*Bubalus bubalis*).
[3][41-A. Wild yak (*Bos grunniens*).]
[1][41-B. Tibetan wolf (*Canis lupus chanco*).]

PART II
Amphibians and Reptiles

[3][1. Agra Monitor Lizard (*Varanus grieseus (Daudin)*).]
[1][1-A. * * * * * * *]
[1][1-B. Audithia turtle (*Pelochclys bibroni*).]
[1][1-C. Barred, oval or yellow monitor lizard (*Varanus flavescense*).]
[1][1-D. Crocodiles (including the estauarine or salt-water crocodile) [(*Crocodilus porosus and Crocodilus palustris*).]
[1][1-E. Batagur terrapin (*baska*).
1-F. Eastern hill terrapin (*Melanochelys triearinata*).]
2. Gharial (*Gavialis gangeticus*).
[4][3. Ganges soft-shelled turtle (*Trionyx gangeticus*).]
[1][3-A. Golden gecko (*Caloductyloides aureaus*).]
4. Green sea turtle (*Cheloniya mydas*).
5. Hansksbill turtle (*Eretmochelys imbrieata inlscata*).
[1][6. * * * * *]
7. Indian egg eating snake (*Elashistodon westermanni*).
8. Indian shoft-shelled turtle (*Lissemys punctata*).
9. Indian tent turtle (*Kchuga tecta tecta*).
10. Large Bengal monitor lizard (*Varnanus bengalensis*).
11. Leathery turtle (*Dermochelys coriacea*).
12. Logger head turtle (*Caretta caretta*).
13. Olive back logger head turtle (*Lepidochelys olivacea*).
14. Peacock marked soft-shelled turtle (*Trionyx hurum*).
[1][14-A. Pythons (*Gienus Python*).]
[1][14-B. Sail terrapin (*Kachuga kachuga*).
14-C Spotted black terrapin (*Geoclemys hemiltoni*).]
[2][15. * * * * *]
[2][16. * * *]
[2][17. * * *]
[4][17-A. Water lizard (*Varanus salvator*).]

---

1. Vide S.O. 859(E), dated 24th November, 1986, published in the *Gazette of India, Extraordinary*, Pt. II, Sec. 3(i), dated 24th November, 1986.
2. Vide Notification published in the *Gazette of India, Extraordinary*, Pt. II, Sec. 3(i), dated 2th October, 1980.
3. Vide Notification published in the *Gazette of India, Extraordinary*, Pt. II, Sec. 3(i), dated 5th October, 1977.
4. Vide Notification dated 29th August, 1977, published in the *Gazette of India, Extraordinary*, Pt. II, Sec. 3(i), dated 3rd September, 1977

PART III

Birds

[1][1. Andaman teal (*Anas gibberifrons albogularis*).

1-A. Assam bamboo partridge (*Bambusicola fytchii*).

[1][1-B. Bazas (*Aviceda jeordoni and Aviceda leuphotes*).]

[1][1-C. Bengal florican (*Eupodotis Bengalesis*).

1-D. Black-necked crane (*Grus Nigriocollis*).

1-E. Blood pheasants (*Ithaginis Cruentus tibetanus, Ithaginis Cruentus Kuseri*).]

[2][1-F. * * *]

2. Cheer pheasant (*Catreus wallichi*).

[2][2-A. Eastern white stork (*Ciconia ciconia boyciana*).]

[3][2-B. Forest-spotted owlet (*Athene blweitti*).]

[4][2-C. Frogmouth (*Genus Batrachostomus*).]

3. Great Indian Bustard (*Choriotis nigriceps*).

4. Great Indian Hornbill (*Buceros bicornis*).

[2][4-A. Hawks (*Accipitridae*).]

[3][4-B. Hooded crane (*Grus monacha*).]

[3][4-C. Hornbills (*Ptilolaemus tickelli, austeni, Aceros nipalensis, Rhyticeros undulatus ticehursti*).]

[3][4-D. Houbara bustard (*Chalamydotis undulata*).]

[3][4-E. humes bar backed pheasant (*Syrmaticus humiae*).]

[3][4-F. Indian pied hornbill (*Anthracoceros malabaricus*).]

5. Jerdon's courser (*Cursorius bitorguatus*).]

6. Lammergeier (*Gypaetus barbatus*).]

7. Large falcons (*Falco peregrimus, Falco biarmicus and Falco chicquera*).

[3][7-A. Large whistling teal (*Anatidae*).]

[4][7-B. Lesser florican (*Sypheotides indica*).]

[4][7-C. Monal pheasants (*Lophophorus impeyanus Lophophorus selateri*).]

8. Mountain quail (*Ophrysia superciliosa*).

9. Narcondom hornbill (*Rhyticeros (undulatus) naracondami*).

[4][9-A. Kerala forest terrapin (*Heosemys sylvatica*).]

10. Nicobar megapode (*Megapodius freycinet*).

[3][10-A. Nicobar pigeon (*Caloenas nicobarica pelewensis*).]

[2][10-B. Ospray or fish-eating eagle (*Pendion haliatetus*).]

[2][10-C. Peacock pheasants (*Polyplectron bicalcaratum*).]

11. Peafowl (*Pavo cristatus*).

12. Pink-headed duck (*Rhodonessa caryophyllacea*).

13. Sclater's monal (*Lophoophorus sclateri*).

14. Siberian white crane (*Grus leucogeranus*).

[2][14-A. * * *]

[4][14-B. Tibetan snow-cock (*Tetraogellus tibetanus*).]

15. Tragopan pheasants (*Tragopan melanocephalus, Tragopan blythii, Tragopan satyra, Tragopon temincki*).

16. White-bellied sea eagle (*Haliaetus leucogaster*).

---

1. Vide Notification dated 29th August, 1977, published in the *Gazette of India, Extraordinary,* Pt. II, Sec. 2(i), dated 3rd September, 1977.
2. Vide Notification published in the *Gazette of India, Extraordinary,* Pt. II, Sec. 3(i), dated 2th October, 1980.
3. Vide Notification dated 24th August, 1977, published in the *Gazette of India, Extraordinary,* Pt. II, Sec. 3(i), dated 3rd September, 1977.
4. S.O. 859(E), dated 24th November, 1986, published in the *Gazette of India, Extraordinary,* Pt. II, Sec. 3(i), dated 24th November, 1986.

17. White-eared phesant (*Crossoptilon crossoptilon*).

[1][17-A. White spoonbill (*Platalea leucorodia*).]

18. White-winged wood duck (*Cairina Scutulata*).

PART IV
Crustacea and Insects

[2][1. **Butterflies and Moths**:	**Common English name**
Family Amathusidae.	Duffer, banded.
Discophora deo deo.	Duffer, common.
Discophora sondaica muscina.	Pallid fauna.
Faunis faunula faunuloides.	
Family Danaidae	
Danaus gautama gautamoides.	Tigers.
Euploea crameri nicevillei.	Crow, spotted black.
Euploca midamus roepstortfi.	Cros, blue-spotted.
Family Lycaenidae.	
Allotinus drumila.	Darkie, crenulatc/great.
Allotinus fabious penormis.	Angled darkie.
Amblopala avidiena.	Hairstreak, Chinese.
Amblypodia ace arata.	Leaf Blue
Amblypodia alea constanceae.	Rosy Oakblue
Amblypodia amonariel.	Malayan Bush blue.
Amblypodia arvina ardea.	Purple brown tailless oakblue.
Amblypodia asopia.	Plain tailless oakblue.
Amblypodia comica.	Comic oakblue
Amblypodia oplima.	Opal oakblue
Amblupodia zeta.	Andaman tailless oakblue.
Biduanda melisa cyara.	Blue posy.
Collphyrs leechii.	Hairstreak Ferruginous.
Castalius rosimon alarbus.	Pierrot, common.
Charana cepheis.	Mandar in blue, cachar.
Chlioria othona.	tit, orchid.
Deudoryx epijarbas amatius	Cornelian, scarce.
Everes moorei.	Cupid, Moore's
Gerydus biggsii.	Bigg's Brownie
Gerydus symethus diopeithes.	Great Brownie
Heliophorus hybrida.	Sapphires.
Horaga albimacula.	Onyxes.
Jamides ferrari.	Caeruleans.
Liphyra brassolis.	Butterfly, Moth.
Listeria dudgenni.	Lister's hairstreak.
Logania watsoniana subfasciata.	Mottle, watsen's
Lycaenopsis binghami.	Hedge Blue
Lycaenopsis haraldus ananga.	Hedge Blue, Felder's
Lycaenopsis puspa prominens.	Common hedge blue.
Lycaenopsis quadriplaga dohertyl.	Naga hedge blue.
Nacaduba noreia hampsoni.	Lineblue, white-tipped.
Polymatus orbitulus leela.	Greenish mountain blue.

1. Vide Notification dated 29th August, 1977, published in the *Gazette of India, Extraordinary*, Pt. II, Sec. 3(i), dated 3rd September, 1977.
2. Vide S.O. 859(E), dated 24th November, 1986, published in the *Gazette of India, Extraordinary*, Pt. II, Sec. 3(ii), dated 24th November, 1986.

	Common English name
Pratapa icetas mishmia.	Royal, Dark blue.
Simiskina phalena harterti.	Brillian, Broadlanded.
Sinthusa virgo.	Spark, Pale.
Spindasis elwesi.	Silverline, Elwes's
Spindasis rukmini.	Silverline, Khaki.
Strymon mackwoodi.	Hairstreak, Mackwood's
Tajuria ister.	Royal, uncertain.
Tajuria luculentus nela.	Royal Chinese.
Tajuria yajna yajna.	Royal Chestnut and Black.
thecla ataxus zulla.	Wonderful hairstreak.
Thecla bieti mendera.	Indian Purple hairstreak.
Thecla letha.	Watsoon's hairstreak.
Thecla paona.	Paona hairstreak.
Thecla pavo.	Peacock hairstreak.
Virachola smilis.	Guava Blues.
Family Nymphalidae.	
Apatura ulupi ulupi.	Emperor, Tawny.
Argynnis hegenmone.	Silver-washed fritillary.
Calinage buddha.	Freak.
Charaxes durnfordi nicholi.	Rajah, chestnut.
Cirrochroa fasciata.	Yeomen.
Diagora nicevillei.	Siren, Scarce.
Dilipa morgiana.	Emperor, Golden.
Doleschallia bisaltide andamana	Autumn Leat.
Eriboea moorei sandakanus	Malayan Nawab.
Eriboea schreiberi.	Blue Nawab.
Eulaceura manipurensis.	Emperor, Tytler's
Euthalia durga splendens.	Barons/Counts/Duchesses.
Euthalia iva.	Duke, Grand.
Euthalia khama curvifascia.	Duke, Naga.
Euthalia telchinia.	Baron, Blue.
Helcyra hemina.	Emperor, White.
Hypolimnas missipus.	Eggfly, Danaid.
Limenitis austenia purpurascens.	Commodore, Grey.
Limenitis zulema.	Admirals.
Melitaea shandura.	Fritillaries/Silverstripes.
Neptis antilope.	Sailer, variegated.
Neptis aspasia.	Sailer, Great Hockystick.
Neptis columella kankena.	Sailer, Short banded.
Neptis cydippe kirbariiensis.	Sailer, Chinese yellow.
Neptis ebusa ebusa.	Sailer, Lascar.
Neptis Jaubah binghami.	Sailer, Chestnut-Streaked.
Neptis manasa.	Sailer, Pale Hockystick.
Neptis nycteus.	Sailer, Hockystick.
Neptis poona.	Lascar Tyter's
Neptis sankara nar.	Sailer, Broad-banded
Panthoporia jina jina	Bhutan sergeant.
Panthporia reta moorei.	Malay Staff sergeant.
Pothoe franckii regalis.	Begum, Blue
Sasakia funebris	Empress.
Sephisa chandra	Courtier, Eastern.
Symbrenthia silana.	Jester Scarce.
Vinessa antiopa yednnula.	Admirrables.

Family Papilionidae.	**Common English name**
Chilasa clytea clytea f. commixtus	Common mine.
Papilio elephenor.	Spangle, yellow-crested.
Papilio liomedon.	Swallowtail, malabar Banded.
Parnassiusaeco geminifer.	Apollo.
Parnassius delphius	Banded apollo.
Parnassius hannyngtoni.	Hannyngton's apollo.
Parnassius imperator augustus.	Imperio apollo.
Parnassius Stoliezanus.	Ladak Banded apollo.
Polydorus coonsambilanga.	Common clubtail.
Polydorus cerassipes	Black windmill.
Polydorus hector	Crimson rose.
Polydorus nevilli	Nevills windmill.
Polydorus plutonius pembertoni	Chinese windmill.
Polydorus polla	Deniceylle's windmill.
Family Pleridae	
Aporia harrietae harrietae.	Black veins.
Baltia butlerisikkima.	White butterfly.
Coliascolias thrasibulus.	Clouded yellows.
Colias dubia.	Dwarf-clouded yellow.
Deliassanaca.	Jexebel, Pale.
Pieris krueperidevta	Butterfly cabbaga/White-II
Family Satyridae.	
Coelitis mothis adamsoni.	Catseye, Scarce.
Cylloganesjaneae.	Evening Brown, Scarce.
Elymniaspeali.	Palmfly, Peal's.
Elymnias penanga chiiensis.	Palmfly, Painted.
Erebia annada annada.	Argus, Ringed.
Erebiana asinghana asingha	Argus, Mottled.
Lethe distans	Forester, Scarce Red.
Lethe dura qammiei.	Lilacfork, Scarce.
Lethe europa tamuna	Bamboo tree brown.
Lethe gemina gafuri.	Tytler's Treed brown.
Lethe gulunihal gulunihal.	Forester, Dull.
Lethe margarite.	Tree brown, Bhutan.
Lethe ocellata lyncus.	Mystic, Dismal.
Lethe remadeva.	Silverstripe Single.
Lethe satyabati	Forester, Pallid.
Mycalesis orseis nautilus.	Bushbrown, Purple.
Pa arge menava maeroides.	Wall Dark.
Ypthima dohertyi persimilis.	Five ring Gret.
Bigruslatro.	Coconut or Robber Carb.
Epiaplebialaidlawi	Dragon Fly.]

[1][1-A. Coconut or Robber crab (*Bigruslatro*).]

[1][2. Dragon fly (*Epioplebia laidlawi*).]

SCHEDULE II

(See Secs. 2, 8, 9, 10, 11, 40, 41, 43, 48, 51, 61 and 62)

Special Game

PART I

1. Giant squirrels (*Rafuta macroura, Ratufa indica and Ratufa bicolor*).

---

1. Vide Notification published in the *Gazette of India, Extraordinary,* Pt. II, Sec. 3(i), dated 2nd October, 1980.

[1][1-A. Assamese macaque (*Macaca assamensis*).]
[2][2. Himalayan brown bear (*Ursus arctos*).]
[2-A. Bengal Porcupine (*Atherurus macrourus assamensis*).]
3. Bison or Gaur (*Bos gaurus*).
[1][3-A. Bonnet macaque (*Macaca radiata*).]
[2][3-B. Himalayan black bear (*Selenarctos thibetanus*).]
[3][3-C. Cetacean species other than those listed in Schs. I and Sch. II, Part II.]
[2][4. * * * * * * * * *]
[1][4-A. Common langur (*Prescbytis entelus*).]
[4][5. Crab-eating macaque (*Macaca irus umbrosa*).]
[2][6. * * * * * * * * *]
7. Ferret badgers (*Melogale moschata and Melogale personata*).
[2][8. * * * * * * * * * * *]
[2][9. * * * * * * * * * * *]
[2][10. * * * * * * * * * * *]
11. Himalayan crestled Porcupine (*Hystrix hodgosoni*).]
[3][11-A. Himalayan newt or Salamandar (*Tyletotriton verrucosus*).]
[4][12. * * * *]
[4][13. * * * * * * * *]
[4][14. * * * * * *]
[2][15. * * * * * *]
16. Pig-tailed macaque (*Macaca remestrina*).
[2][17. * * * * * *]
[1][17-A. Rhesus macaque (*Macaca mulatta*).]
[4][18. * * * * * *]
19. Stump-tailed macaque (*Macaca speciosa*).
[4][20. * * * * * * * *]
21. Water lizard (*Varanus salvator*).
22. Wild dog or dhole (*Cuon alpinus*).
[4][23. * * * * * *]
[1][24. Chameleon (*Chameleon ealcaratus*).
25. Spingy Tailed lizard or Sanda (*Uromastrix hardwickii*).]

PART II

[4][1. Beetles

Family Caradiddae

Agonotrechus andrewesi
Amara brucei
Amara eleganfula
Brachinus atripennis.
Broscosoma gracile.
Brosous bipillifer.
Broter ovicolis
Caiathus amaroides.
Callistominus belli
Chlaenius championi.
Chlaenius kanarae.
Psylliodes shira.
Sebaethe cervina.
Sebaetha patkaia
Sphaeroderma brevicorne.

Family Cuaujidae

Carnophlocus raffrayi.
Cucujus bicolor.
Cacujus grouvelle.
Cacujus imperialis.
Heterojinus semilaetaneus.
Laemophloeus belli
Laemophloeus incertus.
Pediacus rufipes.

1. Vide Notification dated 29th August, 1977, published in the *Gazette of India, Extraordinary,* Pt. II, Sec. 3(i), dated 3rd September, 1977.
2. Vide Notification published in the *Gazette of India, Extraordinary,* Pt. II, Sec. 3(i), dated 5th October, 1977.
3. Vide S.O. 859(E), dated 24th November, 1986, published in the *Gazette of India, Extraordinary,* Pt. II, Sec. 3(i), dated 24th November, 1986.
4. Vide Notification published in the *Gazette of India, Extraordinary,* Pt. II, Sec. 3(i), dated 2nd October, 1980.

Chalenius masoni
Chlaenius masoni
Chlaenius nilgircus
Family Chrysomulla
Acrocrypta rotundata
Bimala indica
Clitea indica
Gopala pita.
Griva cyanipennis.
Nisotra cardoni.
Nisotra madurensis.
Nisotra nigripennis.
Nisotra semicoerullea.
Nisotra striatipennis.
Nonarthra patkaia.
Psylliodes plana.
Family Erycinidae
Abisara Kausambi.
Dodona adonira.
Dodona dipoea.
Dodona egeon.
Libythea lepita.
Family Hesperiidae
Baoris philippina.
Bebasa sena.
Halpe homolea
Family Lycaenidae
Allotinus subviolaceus manychus
Amblypodia abetrrans.
Amblypodia aenea.
Amblypodia agaba aurelia.
Amblypodia agrata.
Amblypodia alesia.
Amblypodia apidanus ahamus.
Amblypodia areste areste.
Amblypodia bazaloides.
Amblypodia camdeo.
Amblypodia ellisi.
Amblypodia fulla ignara.
Amblypodia genesa watsoni.
Amblypodia paraganesa zephpreeta
Amblypodia paralea.
Amblypodia silhetensis.
Amblypodia suffusa suffusa.
Amblypodia yendava
Apharitis lilacinus.
Araotes lapithis.
Artipe eryx.
Bidahara phocides.
Bothrinia chenellia
Castalius roxus manluena.
Catapoecilma delicatum
Catapoecilma elegans myositina.
Family Inopeplidau
Inopeplus albonotalus
Family Amathueiidae
Aemona amathusia amathusia
Amathusia phidippus andamanicus
Amathuxida amythaon
Discophara deo deodoides
Discophora lepida lepida.
Discophora timora andamanensi.
Enispe cycnus.
Faunis sumeus assama.
Sticopthalma nourmahal.
Thauria aliris amplifascia.
Family Danatdae.
Euploea melanaeluca.
Euploea midamus rogenhoferi.
Helipphorus androcles moorej.
Horage onyx.
Horaga viola.
Hypolycaena nilgirica.
Hypolycaena thecloides nicobarica.
Iraota rochana boswelliana.
Jamides slectokandulana.
Jamides exleodus para.
Jamides coerules.
Jamides kankerna.
Lampides boeticus.
Lilacea albocaerulea.
Lilacea atroguttata.
Lilacea lilacea
Lilacea melaena.
Lilacea minima
Logania massalia.
Lycaenesthes lycaeina.
Mahathala ameria.
Mahathala atkinsoni.
Magisba malaya presbyter.
Nacaduba aluta coelestis.
Nacaduba ancyra aberrans.
Nacaduba dubiosa fulva.
Nacaduba helicon.
Nacaduba hermus major.
Nacaduba pactolus.
Neucheritra febronia.
Niphanda cymbia
Orthomiella poutis.
Pithecops fulgens.
Polymmatus devanica devanica.
Polymmatus metallica metallica
Polymmatus orbitulus jaloka
Polymmatus yeonghusbandi
Poritia erycinoides elisel
Poritia hewitsoni

Charana jalindra.
Cheritrella truncipennis.
Chliaria kina.
Deudoryx hypargyria gaetulia.
Enchrysops enejus.
Everes kala.
Rapala chandrana chandrana.
Rapala nasaka.
Rapala regulgens.
Rapala rubida.
Rapala scintilla.
Rapala sphinx sphinx.
Rapala varuna.
Spindasis elima elima.
Spindasis lohita.
Spindasis nipalicus.
Suasa lisides
Surendra todara
Tajuria albiplaga.
Tajuria cippus cippus.
Tajuria culta.
Tajuria diaeus.
Tajuria illurigioides.
Tajuria illurgis.
Tajuria jangala anadamanica.
Tajuria melastigma.
Tajuria sebonga.
Tajuria thyia.
Tajuria yajna istroides.
Tajuria callinara.
Tarucus dharta.
Thaduka malticaudata kanara.
Thecla ataxus ataxus
Thecla bitel.
Thecla icana.
Thecla jakamensis.
Thecla kabreea.
Thecla khasia.
Thecla kirbariensis.
Thecla suroia.
Thecla syla assamica.
Thecla vittara.
Thecla ziha
Thecla zoa.
Una usta.
Ysoda tripunctala.
Family Nymphalidae
Adolias cyanipardus.
Adolias dirtea.
Kallima philarchus horsfieldii
Limenitis austenia austenia.
Limenitis damava.
Limenitis dudu.
Melitae robertsi lutko
Poritia plusrata geta
Pratapa bhetes.
Pratap blanka.
Pratapa deva.
Pratapa icetas.
Rapala buxaria.
Adolias khasiana.
Apatura chevana.
Apatura parvata.
Apatura soridida.
Apatura ullupi florencia.
Argynnis adippe pallida.
Argynnis altissima.
Argynnis clara clara.
Argynnis pales horla.
Atella alciope.
Calinaga buddha brahaman.
Charaxes aristogiton.
Charaxes fabius sulphureus.
Charaxes karruba
Charaxes marmax
Charaxes polyxena heman
Cheroonesia rahria rahrioides
Cyrestic cocles.
Diagora parsimilis.
Doleschallia bisaltide malabarica.
Eriboea athames andamanicus
Eriboea delphis.
Eriboea dolen.
Eriboea harcoea lissainei
Euripus consimilis.
Euripus halitherses.
Euthalia anosia.
Euthalia cocytus.
Euthalia duda.
Euthalia durga durga
Euthalia evalina landabilis.
Euthalia francia.
Euthalia garuda acontius.
Euthalia lepidea.
Euthalia merta eriphyle.
Euthalia nara nara
Euthalia patala taooana.
Euthalia teuta.
Horona marathus andamana.
Hyploimnas missipus.
Hyploimnas polynice birmana
Kallima albofasciata.
Kallima alompra.
papilio mayo.
Parnassius charltonius charltonius.
Parnassius epaphus hillensis.
Parnassius jacquemonti jacque-
monti

Neptis aurelia.
Neptis anjana nashona.
Neptis aurelia.
Neptis magadh khasiana.
Neptis nandina hamsoni.
Neptis narayana.
Neptis radha radha.
Neptis soma.
Neptis zaida.
Neurosigma doublodayi doubledayi
Partoporia kasura asura.
Pantoporia kanwa phorkys.
Pantoporia larymna siamensis.
Pantoporia pravara acutipennis.
Pantoporia ranga.
Parthenos sylvia.
Penthema lisarda.
Symbrenthia niphanda.
Vanesa egea agnicula.
Vanesa lalbum.
Vanesa polycholoros fervida.
Vanesa prarsoides dohertyi.
Vanesa urticoe rizama.
Family Papilionidae
Bhutanitis liderdalei liderdalei.
Chilasa ephycides ephycides.
Chilasa paradoxa telearchus.
Chilasa slateri slateri.
Graphium ariyteus anticrates.
Graphium arycles arycles.
Graphium eurpylus macronius
Graphium evemon albociliates.
Graphium gyas gyas.
Graphium megarus megarus.
Papilio bootes.
Papilio buddha.
Papilio fuscus andamanicus.
Papilio machaon verityi.
Lethe baladevr
Lethe brisanda.
Lethe goalpara goalpara.
Lethe insana insana.
Lethe jalaurida.
Lethe kabrua.
Lethe latiaris latiaris.
Lethe molleri moellri.
Lethe naga naga
Lethe nicetella.
Lethe pulaha.
Lethe scanda.
Lethe serbonis.
Lethe siderca.
Lethe sincrix
Lethe ristigmata.
Polydorus latreillei kabrua.
Polydorus plutonius tytleri.
Teinopalpus imperialis imprerialis.
Family Pieridae
Aporia nabellica.
Appias albina darada.
Appias indra shiva.
Appias lyncida latifasciata.
Appias wardi.
Baltia butleri butleri.
Cepora nadian remba.
Cepora nerissa dapha.
Colias eocandiea hinducucica.
Colias eogene.
Colias ladakensis.
Colias stoliczkana miranda.
Delias lativitta.
Dercas lycorias.
Euchloe charlonia lucilla.
Eurema andersoni ormistoni.
Metaporia agathon.
Pieris deota.
Pontia chloridice alipina.
Taletara panda chrysaca.
Valeria avatar avatar.
Family Satyridae
Anlocera brahminus.
Cyllogenes suradeva.
Elyminas malelas milamba.
Elymnias vasudeva.
Erebia annada suroia.
Erebia hygriva.
Erebia kalinda kalinda.
Erebia mani mani.
Erebia seanda opima
Erites falicipennis.
Hipparchis hoydenreichi shandura.
Lethe atkinsoni.
Mycalesis anaxias.
Mycalesis botama chamka.
Mycalesis heri.
Mycalesis lepcha bethami.
Mycalesis malarida.
Mycalesis misenus.
Mycalesis mystes.
Mycalesis suavolens.
Neorina hilda.
Neorina patria westwoodii.
Oeneis budha qurhwalica.
Parantirrhoes marshalia.
Parage eversmanni cashmirensis.
Pararge macrula maefula.
Ragadia crisilda crito.
Rhapicera sutriens kabrua.

Lethe violaceodicta, kanjupkula.
Lethe visrava.
Lethe yama.
Maniola davendra davendra
Metanitis zitenius.
Mycalesis admamsoni.
Ypthima bolanica.
Ypthima lycus.
Ypthima mathora mathoro.
Ypthima similis affectata.
Zipotis saitis.]

[1][1-A. Civets (all species of Viverridae except Malabar civet.]

[1][1-B. Common for (*Vulpes bengalensis*).]

[1][1-C. Flying squirrels—all species of the genus (*Bulopetes, petaurista, Belomys and Eupetaurus*).]

[1][1-D. Giant squirrels (*Ratufa macroura, Ratufa indica and Ratufa bicolor*).]

[1][2. * * * * *]

[2][2-A. Himalayan black bear (*Selenarctos thibetamus*).]

[1][2-B. Jackal.

[1][2-C. Jungle cat (*Felis Chaus*).]

[1][2-D. Ma mots (*Marmota bobak himalayana, Mamota caudata*).]

[1][2-E. Martens (*Martes foinaintermeda, Martes flavigula, Martes gwatkinsii*).

[1][3. * * * * *]

[1][4. * * *]

[1][4-A. Polo cats (*Vermela peregusna, Mustela putorius*).

4-B. Red fox (*Vulpes vulpes, Vulpes montana, Vulpes griffithi, Vulpes pusilla*).]

[2][5. Sloth bear (*Melursus ursinus*).]

[1][5-A. Sperm whale (*Physeter macrocephalus*).]

[1][6. * * * *]

[1][7. Weasels (*Mustela sibirica, Mustela kathian, Mustela altaica*).]

[1][8. Checked keelback (*Xanochrophispiscator*).

9. Dhamon or rat snake (*Ptyas mucousus*).

10. Dog-faced water snake (*Cerberus rhynchops*).

11. Indian cobras (all sub-species belonging to *genus Naja*).

12. King Cobra (*Ophiophagus hannah*).

13. Olivaceous keelback (*Artretiumschistosum*).

14. Russel's viper (*Vipera ruselli*).

15. Va anus species (*excluding yellow monitor lizard*).

SCHEDULE III

(See Secs. 2, 8, 9, 10, 11 and 61)

BIG GAME

[1][1. * * * *]

2. Barking deer or Muntjac (*Muntiacus muntijak*).

[2][3. * * * *]

[3][4. Otters (*Lutra lutra, Lutra perspicillata, Aonyx cinerea*).]

5. Chital (*Axis axis*).

[1][6. * * * *]

1. Vide S.O. 859(E), dated 24th November, 1986, published in the *Gazette of India, Extraordinary*, Pt. II, Sec. 3(ii), dated 24th November, 1986.
2. Vide Notification published in the *Gazette of India, Extraordinary*, Pt. II, Sec. 3(i), dated 2nd October, 1980.
3. Vide Notification published in the *Gazette of India, Extraordinary*, Pt. II, Sec. 3(i), dated 5th October, 1977.

7. Gorals (*Nemorhaedus goral, Nemorhaedus hodgsoni*).
[1][8. * * * * *]
[2][9. * * * * *]
[1][10. * * * * *]
11. Hog deer (*Axis porcinus*).
12. Hyaena (*Hyaena hyaena*).
[1][13. * * * * *]
14. Nilgai (*Boselaphus tragocamelus*).
[2][15. * * * * *]
16. Sambar (*Cervus unicolor*).
[1][17. * * * * * *]
[2][18. * * * * * *]
19. Wild pig (*Sus scrofa*).

SCHEDULE IV
(See Secs. 2, 8, 9, 11 and 61)
Small Game

[2][1. * * * * * *]
[1][1-A. * * *]
[3][2. * * *]
3. * * *]
[2][3-A. Five-striped palm squirrel (*Fanambulus Pennanti*).]
4. Hares (*Black naped, Common Indian, Desert, Himalayan Mousaehare*).
[3][4-A. Hedge-hog (*Hemiechinus auritus*).]
[3][4-B. * * * * * *]
[3][4-C. * * * *]
[3][4-D. * * *]
[3][4-E. Indian Porcupine (*Hystrix indica*).]
[3][5. * * * * * *]
[3][6. * * * * * *]
[4][6-A. Mongooses (all psecies of genus *Herpestes*).]
[1][6-B. Mongooses (all species of genus *Herpestes*).]
[1][7. * * * * * * * *]
[4][7-A. Pole cats (*Vormela peregusna, Mustela putorius*).]
[1][7-B. Pole cats (*Vormela peregusna, Mustela putorius*).]
8. Red fox (*Vulpes vulpes*).
[1][8-A. Red fox (*Vulpes monata, Vulpes griffithi and Vulpes pussila*).
[3][9. * * * * * * *]
[3][9-A. * * *]
10. Weasels (*Mustela sibirica, Mustela kathiah and Mustica altaica*).
[4][11. Birds [5][(other than those which appear in other Schedules)]:
1. Advavat (*Estrildinae*).
2. Avocet (*Ecurvirostridae*).

---

1. Vide Notification published in the *Gazette of India, Extraordinary*, Pt. II, Sec. 3(i), dated 5th October, 1980.
2. Vide Notification published in the *Gazette of India, Extraordinary*, Pt. II, Sec. 3(i), dated 2nd October, 1980.
3. Vide S.O. 859(E), dated 24th November, 1986, published in the *Gazette of India, Extraordinary*, Pt. II, Sec. 3(i), dated 24th November, 1986.
4. Vide Notification dated 29th August, 1977, published in the *Gazette of India, Extraordinary*, Pt. II, Sec. 3(i), dated 3rd September, 1977.
5. Vide S.O. 859(E), dated 24th November, 1986, published in the *Gazette of India, Extraordinary*, Pt. II, Sec. 3(i), dated 24th November, 1986.

3. Babblers (*Timaliinae*).
4. Barbets (*Capitonidae*).
5. Barbes Owls (*Tytoninae*).
6. Bittrns (*Ardeidae*).
7. Brown-headed gull (*Larus brunnicehalus*).
8. Bulbuls (*Eycononotidae*).
9. Buntings (*Emberizidae*).
10. Bustards (*Otidipae*).
11. Bustard-Quails (*Turnicidae*).
12. Chloropsis (*Irenidae*).
13. Comb duck (*Sarkidiornis melanotes*).
14. Coot (*Rallidae*).
15. Cormorants (*Phalacrocoracidae*).
16. Cranes (*Anatidae*).
17. Cuckoos (*Cuculidea*).
17-A. Curlews (*Scolopacinea*).
18. Darters (*Phalacr-oeracidae*).
19. Doves including the Emerald Dove (*Columbi-dae*).
20. Drongos (*Dicruridae*).
21. Ducks (*Anatidae*).
22. Egrets (*Ardeidae*).
23. Fairy Blue Bird (*Irenidae*).
24. Falcons (*Falconidae*), excepting the Shaheem and pergeren Falcons Falce peregrinus, the Shaker or Cherrug. Shanghar and Laggar Falcons, Falce biarmicus and the Red-headed Merlin, F. Chiquera.
25. Finches including the Chafinch (*Fringillidae*).
26. Flaminges (*Phoenicopteridae*).
27. Flowerpeekers (*Dicaedae*).
28. Flycatchers (*Muscicapidae*).
29. Geese (*Anatidae*).
30. Goldfinch and allies (*Carduelinane*).
31. Grebes (*Podicipedidae*).
32. Herons (*Ardeidae*).
33. Ibises (*Ph-reskiornithidae*).
34. Ioras-s (*Irenidae*).
35. Jays (*Corvidae*).
36. Jacanas (*Jacnidae*).
36-A. Jungle fowl (*Phasianidae*).
37. Kidgfishers (*Alcedinidae*).
38. Larks (*Alaudidae*).
39. Lorikeets (*Psittacidae*).
40. Magpies including the Hunting Magpie (*Corvidea*).
41. Mannikins (*Estrildinae*).
42. Magapodes (*Magapodidae*).
43. Minivets (*Campephagidae*).
44. Murias (*Eserildinae*).
45. Mynas (*Sturnidae*).
46. Nightjars (*Camprimalgiedae*).
47. Orioles (*Oriolidae*).
48. Owls (*Strigidae*).
49. Oystercatches (*Paematopodidae*).
50. Parakeet (*Psittacidae*).

51. Patiridges (*Phasianidae*).
52. Pelioans (*Pelecanidae*).
53. Pheasants (*Phasiandea*).
54. Pigeons (*Columbidae*) excepting the Blue Rock Pigeon (*Calumba Livia*).
55. Pipits (*Motacillidae*).
55-A. Pittas (*Pittidae*).
56. Plovers (*Charadriinae*).
57. Quails (*Phasianidae*).
58. Rails (*Rallidea*).
59. Rollers or Blue Jay (*Coracidae*).
60. Sandgrouses (*Petroclididae*).
61. Sandpipers (*Scolopacinae*).
62. Snipes (*Scolopacinae*).
63. Spurfowls (*Phasianidae*).
64. Starlings (*Sturnidae*).
65. Stone Curlew (*Burhinidae*).
66. Storks (*Ciconiidae*).
67. Stilt (*Recurvirostridae*).
68. Surbirds (*nectariniidae*).
69. Swans (*Anatidea*).
70. Teals (*Anatidae*).
71. Thrushes (*Turadinae*).
72. Tits (*Paridae*).
73. Tree-pies (*Corvidae*).
74. Trogons (*Trogonidae*).
75. Vultures (*Accipitridae*).
76. Waxbills (*Estrildinae*).
77. Weaver Birds or Bayas (*Ploceidae*).
78. White-eyes (*Zoster- pidae*).
79. Woodpeckers (*Picidae*).
80. Wrens (*Toglodytidae*).

[1][12. Snakes [2][other than those species listed in Sch. I, Pt. II, and Sch. II, Pt. II]:
(i) Amblycayhalidae.
(ii) Amilidae.
(iii) Boidae.
(iv) Colubridae.
(v) Dasypeptidae (Egg-eating snakes).
(vi) Elapidae (Cobras, Kraits and Coral snakes).
(vii) Glauconnidae.
(viii) Hydrobhilipae (Fesh water and sea snakes).
(ix) Ilysidae.
(x) Leptotyphlopidae.
(xi) Tyolopidae.
(xii) Uropeltidae.
(xiii) Viperidae.
(xiv) Xenapeltidae.]

[1][13. Fresh water frog (*Rana Spp.*).]

[1][14. Three-jelled turtle (*Geoemyda tricarinata*).]

1. Vide Notification published in the *Gazette of India, Extraordinary,* Pt. II, Sec. 3(i), dated 2nd October, 1980.
2. Vide S.O. 859(E), dated 24th November, 1986, published in the *Gazette of India, Extraordinary,* Pt. II, Sec. 3(i), dated 24th November, 1986.

[1][15. Tortoise (*Testudinidae, Trionyceidae*).]

[1][16. Viviparous toads (*Nectophyrynoides* Spp.)]

[1][17. Voles.]

[1][18. **Butterflies and moths:**

Family Danaidae

Euploca core simulatrix.

Duploca crassa.

Euploca dioclctianus ramsyai.

Euploca mulciber.

Family Hesperiidae

Baoris farri.

Hasora vitta.

Hyarotis Adrastus.

Oriens concinna.

Pelopidas assamensis.

Pelopidas sinensis.

Polytrema discreta.

Polytrema rubricans.

Thoressa horiorei.

Family Lycacenidae

Tarucus ananda.

Family Nymplialidae

Euthalia lubentina.

Family Pieridae

Appria agathon ariaca.

Appias libythea.

Appias nero galba.

Prioneris sita.]

SCHEDULE V

(See Secs. 2, 8, 61 and 62)

Vermin

1. Common crow.

[2][2. * * * *]

3. Fruit bats.

[4. Jackal.]

5. Mice.

6. Rats.

[1][7. * * *]

1. Vide Notification published in the *Gazette of India, Extraordinary,* Pt. II, Sec. 3(i), dated 2nd October, 1980.
2. Vide Notification, published in the *Gazette of India, Extraordinary,* Pt. II, Sec. 3(i), dated 5th October, 1977.

In the context of the ongoing denotification of sanctuaries, certain sections of The Wildlife (Protection) Act, 1972, are reproduced here.

Selected portion of amendment Chapter III of Act (1991) is discussed below:

A PROTECTION OF SPECIFIED PLANTS

17A Prohibition of picking, uprooting, etc. of specified plants

Save as otherwise provided in this Chapter, no person shall:

(a) Wilfully pick, uproot, damage, destroy, acquire or collect any specified plant from any forest land and area specified, by notification, by the Central Government;

(b) Possess, sell, offer for sale, or transfer by way of gift or otherwise, or transport any specified plant, whether alive or dead, or part or derivative thereof;

Provided that nothing in this section shall prevent a member of a Scheduled Tribe, subject to the provisions of Chapter IV, from picking, collecting or possessing in the district he resides any specified plant or part or derivative thereof for his bona fide personal use.

17B Grant of permit for special purposes

The Chief Wildlife Warden may, with the previous permission of the State Government, grant to any person a permit to pick, uproot, acquire or collect from a forest land or the area specified under Section 17A or transport, subject to such conditions as may be specified therein, any specified plant for the purpose of:

(a) education;

(b) scientific research;

(c) collection, preservation and display in a herbarium of any scientific institution; or

(d) propagation by a person or an institution approved by the Central Government in this regard.

17C Cultivation of specified plants without licence prohibited

1. No person shall cultivate a specified plant except under and in accordance with a licence granted by the Chief Wildlife Warden or any other officer authorised by the State Government in this behalf;

Provided that nothing in this section shall prevent a person, who, immediately before the commencement of the Wildlife (Protection) Amendment Act, 1991, was cultivating a specified plant, from carrying on such cultivation for a period of six months from such commencement, or where he has made an application within that period for the grant of a licence to him, until the licence is granted to him or he is informed in writing that a licence cannot be granted to him.

2. Every licence granted under this section shall specify the area in which and the conditions, if any, subject to which the licensee shall cultivate a specified plant.

17D Dealing in specified plants without licence prohibited

1. No person shall, except under and in accordance with a licence granted by the Chief Wildlife Warden or any other officer authorised by the State Government in this behalf, commence or carry on business or occupation as a dealer in a specified plant or part or derivative thereof;

Provided that nothing in this section shall prevent a person who, immediately before the commencement of the Wildlife (Protection) Amendment Act, 1991, was carrying on such business or occupation, from carrying on such business or occupation for a period of sixty days from such commencement, or where he has made an application within that period for the grant of a licence to him, until the licence is granted to him or he is informed in writing that a licence cannot be granted to him.

2. Every licence granted under this section shall specify the premises in which and the conditions, if any, subject to which the licensee shall carry on his business.

17E Declaration of stock

1. Every person cultivating, or dealing in, a specified plant or part or derivative thereof shall, within thirty days from the date of commencement of the Wildlife (Protection) Amendment Act, 1991 declare to the Chief Wildlife Warden or any other officer authorised by the State Government in this behalf, his stocks of such plants and part or derivative thereof, as the case may be, on the date of such commencement.

2. The provisions of sub-sections (3) to (8) (both inclusive) of Section 44, Section 45, Section 46 and Section 47 shall, as far as may be, apply in relation to an application and a licence referred to in Section 17C and Section 17D as they apply in relation to the licence or business in animals or animal articles.

17F Possession, etc., of plants by licensee

No licensee under this Chapter shall:

(a) keep in his control, custody or possession

(i) any specified plant, or part or derivative thereof in respect of which a declaration under the provisions of Section 17E has to be made but has not been made,

(ii) any specified plant, or part or derivative thereof which has not been lawfully acquired under the provisions of this Act or any rule or order made thereunder;

(b) (i) pick, uproot, collect or acquire any specified plant, or

(ii) acquire, receive, keep in his control, custody or possession, or sell, offer for sale or transport, any specified plant or part or derivative thereof,

except in accordance with the conditions subject to which the licence has been granted and such rules as may be made under this Act.

17G Purchase, etc. of specified plants

No person shall purchase, receive or acquire any specified plant or part or derivative thereof otherwise than from a licensed dealer;

Provided that nothing in this section shall apply to any person referred to in Section 17B.

17H Plants to be Government Property

1. Every specified plant or part or derivative thereof, in respect of which any offence against this Act or any rule or order made thereunder has been committed, shall be the property of the State Government and where such plant or part or derivative thereof has been collected or acquired from a sanctuary or National Park declared by the Central Government, such plant or part or derivative thereof shall be the property of the Central Government.

2. The provisions of sub-sections (2) and (3) of Section 39 shall, as far as may be, apply in relation to the specified plant or part or derivative thereof or they apply in relation to wild animals and articles referred to in sub-section (1) of that section.

CHAPTER IV

SANCTUARIES, NATIONAL PARKS AND CLOSED AREAS

18. Declaration of Sanctuary

1. The State Government may, by notification, declare its intention to constitute any area other than area comprised with any reserve forest or territorial waters as a sanctuary if it considers that such area is of adequate ecological, faunal, floral, geomorphological, natural or zoological significance, for the purpose of protecting, propagating or developing wildlife or its environment.

2. The notification referred to in sub-section (1) shall specify, as nearly as possible, the situation and limits of such area.

Explanation

For the purposes of this section, it shall be sufficient to describe the area by roads, rivers, ridges, or other wellknown or readily intelligible boundaries.

19. Collector to determine rights

When a notification has been issued under Sec. 18, the collector shall inquire into and determine the existence, nature and extent of the rights or any person in or over the land comprised within the limits of the sanctuary.

20. Bar of accrual of rights

After the issue of a notification under Sec. 18, no right shall be acquired in, or over the land comprised within the limits of the area specified in such notification, except by succession, testamentary or intestate.

21. Proclamation by Collector

When a notification has been issued under Sec. 18, the Collector shall publish in the regional language in every town and village in or in the neighbourhood of the area comprised therein, a proclamation :

(a) specifying, as nearly as possible, the situation and the limits of the sanctuary; and

(b) requiring any person, claiming any right mentioned in Sec. 19, to prepare before the Collector, within two months from the date of such proclamation, a written claim in the prescribed form specifying the nature and extent of such right, with necessary details and the amount and particulars of the compensation, if any, claimed in respect thereof.

22. Inquiry by Collector

The collector shall, after service of the prescribed notice upon the claimant, expeditiously inquire into:

(a) the claim preferred before him under Cl. (b) of Sec. 21, and

(b) the existence of any right mentioned in Sec. 19 and not claimed under Cl. (b) of Sec. 21.

So far as the same may be ascertainable from the records of the State Government and the evidence of any person acquainted with the same.

23. Powers of Collector

For the purpose of such inquiry, the Collector may exercise the following powers, namely:

(a) the power to enter in or upon any land and to survey, demarcate and make a map of the same or to authorise any other officer to do so;

(b) the same powers as are vested in a civil court for the trial of suits.

24. Acquisition of rights

1. In the case of a claim to a right in or over any land referred to in Sec. 19, the Collector shall pass an order admitting or rejecting the same in whole or in part.

2. If such claim is admitted in whole or in part, the Collector may either:

(a) exclude such land from the limits of the proposed sanctuary, or

(b) proceed to acquire such land or rights, except where by an agreement between the owner of such land or the holder of rights and the government the owner or holder of such rights has agreed to surrender his rights to the Government, in or over such land and on payment of such compensation, as is provided in the Land Acquisition Act, 1984 (1 of 1984);

(c) allow, in consultation with the Chief Wildlife Warden, the continuance of any right of any person in or over any land within the limits of the sanctuary.

25. Acquisition proceedings

1. For the purpose of acquiring such land, or rights in or over such land

(a) the Collector shall be deemed to be a Collector, proceeding under the Land Acquisition Act, 1894 (1 of 1894);

(b) the claimant shall be deemed to be a person interested and appearing before him in pursuance of a notice given under Sec. 9 of that Act;

(c) the provisions of the sections preceding Sec. 9 of that Act shall be deemed to have been complied with;

(d) where the claimant does not accept the award made in his favour in the matter or compensation, he shall be deemed, within the meaning of Sec. 18 of that Act, to be a person interested who has not accepted the award and shall be entitled to proceed to claim relief, against the award under the provision of Part III of that Act;

(e) the Collector, with the consent of the claimant, or the Court, with the consent of both the parties, may award compensation in land or money or partly in land and partly in money; and

(f) in the case of the stoppage of a public way or a common pasture, the Collector may, with the previous sanction of the State Government, provide for an alternative public way or common pasture, as far as may be practicable or convenient.

2. The acquisition under this Act of any land or interest herein shall be deemed to be acquisition for a public purpose.

26. Delegation of Collector's Powers

The State Government may, by general or special order, direct that the powers exercisable or the functions to be performed by the Collector under Secs. 19 to 25 (both inclusive) may be exercised and performed by such other officer as may be specified in the order.

26A Declaration of area as Sanctuary

1. When

(a) a notification has been issued under Sec. 18 and the period for preferring claim has elapsed and all claims, if any, made in relation to any land in an area intended to be declared as a sanctuary, have been disposed of by the State Government, or

(b) any area comprised within any reserve forest or any part of the territorial waters, which is considered by the State Government to be of adequate ecological, faunal, floral, geomorphological, natural or zoological significance for the purpose of protecting, propagating or developing wildlife or its environment, is to be included in a sanctuary, the State Government shall issue a notification specifying the limits of the area which shall be comprised within the sanctuary and declare that the said area shall be sanctuary on and from such date as may be specified in the notification.

Provided that where any part of the territorial waters is to be so included, prior concurrence of the Central Government shall be obtained by the State Government,

Provided further that the limits of the area of the territorial waters to be included in the sanctuary shall be determined in consultation with the Chief Naval Hydrographer of the Central Government and after taking adequate measures to protect the occupational interests of the local fishermen.

2. Notwithstanding anything contained in sub-section (1), the right of innocent passage of any vessel or boat through the territorial water shall not be affected by the notification issued under sub-section (1).

3. No alteration of the boundaries of a sanctuary shall be made except on a resolution passed by the Legislation of the State.

27. Restriction on entry in sanctuary

1. No person other than

(a) a public servant on duty

(b) a person who has been permitted by the Chief Wildlife Warden or the authorised officer to reside within the limits of the sanctuary,

(c) a person who has any right over immovable property within the limits of the sanctuary,

(d) a person passing through the sanctuary along a public highway, and

(e) the dependents of the person referred to in Cl. (a), (b) or (c), shall enter or reside in the sanctuary, except under and in accordance with the conditions of a permit granted under Section 28.

2. Every person shall, so long as he resides in the sanctuary, be bound

(a) to prevent the commission, in the sanctuary, of an offence against this Act;

(b) where there is reason to believe that any such offence against this Act has been committed in such sanctuary, to help in discovering and arresting the offender;

(c) to report the death of any wild animal and to safeguard its remains until the Chief Wildlife Warden or the authorised officer takes charge thereof;

(d) to extinguish any fire in such sanctuary of which he has knowledge or information and to prevent from spreading, by any lawful means in his power, any fire within the vicinity of such sanctuary of which he has knowledge or information; and

(e) to assist any forest officer, Chief Wildlife Warden, Wildlife Warden or police officer demanding his aid for preventing the commission of any offence against this Act or in the investigation of any such offence.

3. No person shall, with intent to cause damage to any boundary mark of a sanctuary or to cause any wrongful gain as defined in the Indian Penal alter, after destroy, move or deface such boundary-mark.

4. No person shall tease or molest any wild animal or litter the grounds of sanctuary.

28. Grant of permit

1. The Chief Wildlife Warden may, on application, grant to any person a permit to enter or reside in a sanctuary for all or any of the following purposes:

(a) investigation or study of wildlife and purposes ancillary or incidental thereto;

(b) photography;

(c) scientific research;

(d) tourism;

(e) transaction of lawful business with any person residing in the sanctuary.

2. A permit to enter or reside in a sanctuary shall be issued subject to such conditions and on payment of such fee as may be prescribed.

29. Destruction, etc. in a sanctuary prohibited without a permit

No person shall destroy exploit or remove any wildlife from a sanctuary or destroy or damage the habitat of any wild animal or deprive any wild animal of its habitat within such sanctuary except under and in accordance with permit granted by the Chief Wildlife Warden and no such permit shall be granted unless the State Government, being satisfied that such destruction, exploitation or removal of wildlife from the sanctuary is necessary for the improvement and better management of wildlife therein, authorises the issue of such permit.

30. Causing fire prohibited

No person shall set fire to a sanctuary, or kindle any fire, or leave any fire burning, in a sanctuary, in such manner as to endanger such sanctuary.

31. Prohibition of entry into sanctuary with weapon

No person shall enter a sanctuary with any weapon except with the previous permission in writing of the Chief Wildlife Warden or the authorised officer.

32. Ban on use of injurious substances

No person shall use in a sanctuary, chemicals, explosives or any other substances which may cause injury to, or endanger, any wildlife in such sanctuary.

Explanation

For the purpose of this section, grazing or movement of livestock permitted under Clause (d) of Sec. 33 shall not be deemed to be an act prohibited under this section.

33. Control of sanctuaries

The Chief Wildlife Warden shall be the authority who shall control, manage and maintain all sanctuaries and for that purpose, within the limits of any sanctuary

(a) may construct such roads, bridges, buildings, fences or barrier gates and carry out such other works as he may consider necessary for the purposes of such sanctuary;

(b) shall take such steps as will ensure the security of wild animals in the sanctuary and the preservation of the sanctuary and wild animals therein;

(c) may take such measures, in the interests of wildlife, as he may consider necessary for the improvement of any habitat;

(d) may regulate, control, or prohibit, in keeping with the interests of wildlife, the grazing or movement of livestock;

(e) (omitted 1991).

33-A Immunisation of Livestock

1. The Chief Wildlife Warden shall take such measures in such manner as may be prescribed, for immunisation against communicable diseases of the livestock kept in or within five kilometres of a sanctuary.

2. No person shall take, or cause, to be taken or grazed, any livestock in a sanctuary without getting it immunised.

34. Registration of certain persons in possession of arms

1. Within three months from the declaration of any area as a sanctuary, every person residing in or within ten kilometres of any such sanctuary and holding a licence granted under the Arms Act, 1959 (54 of 1959), for the possession of arms or exempted from the provisions of that Act and possessing arms, shall apply in such form, on payment of such fee and within such time as may be prescribed, to the Chief Wildlife Officer or the authorised officer, for the registration of his name.

2. On receipt of an application under Sub-clause (1), the Chief Wildlife Warden or the authorised officer shall register the name of the applicant in such manner as may be prescribed.

3. No new licences under the Arms Act, 1959, shall be granted within a radius of ten kilometres of sanctuary without the prior concurrence of the Chief Wildlife Warden.

35. Declaration of National Parks

1. Whenever it appears to the State Government that an area, whether within a sanctuary or not, is, by reason of its ecological, faunal, floral, geomorphological, or zoological association or importance, needed to be constituted as a National Park for the purpose of protecting, propagating, or developing wildlife therein or its environment, it may, by notification, declare its intention to constitute such area as a National Park.

Provided that where any part of the territorial waters is proposed to be included in such National Park, the provisions of Sec. 26-A shall, as far as may be, apply in relation to the declaration of a National Park as they apply in relation to the declaration of a sanctuary.

2. The notification referred to in sub-section (1) shall define the limits of the area which is intended to be declared as a National Park.

3. Where any area is intended to be declared as a National Park, the provisions of Sec. 19 to 26-a (both inclusive), except Clause (c) of sub-section (2) of Section 24.

4. When the following events have occurred, namely

(a) the period for preferring claims has elapsed and all claims, if any, made in relation to any land in an area intended to be declared as a National Park, have been disposed of by the State Government, and

(b) all rights in respect of lands proposed to be included in the National Park have become vested in the State Government.

The State Government shall publish a notification specifying the limits of the area which shall be comprised within the National Park and declare that the said area shall be a National Park on and from date as may be specified in the notification.

5. No alteration of the boundaries of a National Park shall be made except on a resolution passed by the Legislature of the State.

6. No person shall destroy, exploit, or remove any wildlife from a National Park, or destroy or damage the habitat of any wild animal, or deprive any wild animal of its habitat within such National Park except under and in accordance with a permit granted by the Chief Wildlife Warden and no such permit shall be granted unless the State Government, being satisfied that such destruction, exploitation, or removal of wildlife from the National Park is necessary for the improvement and better management of wildlife therein, authorises the issue of such permit.

7. No grazing of any livestock shall be permitted in a National Park and no livestock shall be allowed to enter therein except where such livestock is used as a vehicle by a person authorised to enter such National Park.

8. The provisions of Secs. 27 and 28, Secs. 30 to 32 (both inclusive) and Cls. (a), (b) and (c) of Sec. 33, 33-A and Sec. 34 shall, as far as may be, apply in relation to a National Park as they apply in relation to a Sanctuary.

36. Declaration of game reserve (omitted 1991)

37. Declaration of closed area

1. The State Government may, by notification, declare any area closed to hunting for such period as may be specified in the notification.

2. No hunting of any wild animal shall be permitted in a closed area during the period specified in the notification referred to in sub-section (1).

SANCTUARIES OR NATIONAL PARKS DECLARED BY CENTRAL GOVERNMENT

38. Power of Central Government to declare areas as Sanctuaries or National Parks

1. Where the State Government leases or otherwise transfers any area under its control, not being an area within a Sanctuary, to the Central Government, the Central Government may, if it is satisfied that the conditions specified in Sec. 18 are fulfilled in relation to the area so transferred to it, declare such are, by notification, to be Sanctuary and the provisions of Secs. 18 to 35 (both inclusive), 54 and 55 shall apply in relation to such Sanctuary as they apply in relation to a Sanctuary declared by the State Government.

2. The Central Government may, if it is satisfied that the conditions specified in Sec. 35 are fulfilled in relation to any area referred to in sub-section (1), whether or not such area has been declared, to be sanctuary by the Central Government or the State Government, declare such area, by notification, to be a National Park and the provisions of Secs. 18-35, 54 and 55 shall apply in relation to a National Park declared by the State Government.

Wildlife Conservation in India

Various activities relating to wild life conservation and implementation of the Wildlife (Protection) Act, 1972, were carried out by the Ministry during the year 2003-2004. The details are as follows:

Enforcement of Wildlife (Protection) Act, 1972, and Export-Import Policy

The Wildlife (Protection) Act, 1972, the provisions of the Convention of International Trade in Endangered Species (CITES) and Export and Import Policy of India were continued to be enforced through the offices of the Regional Deputy Directors of Wildlife Preservation located at Delhi, Mumbai, Kolkata and Chennai with the help of State Wildlife Departments, the State Police Departments, the Customs Departments, Border Security Force (BSF) and Coast-Guards. The Regional Deputy Directors detected several cases of poaching and illegal trade in wildlife products during the year.

During the year ban on export of 29 species of plant, plant portions and their derivatives obtained from wild were continued. Export of six species of exotic birds was continued subject to preshipment inspection and CITES permit wherever required.

Fig. 43.1 : 'Pitcher plant' *Nepenthes khasiana* Hook. f. — an endemic plant from Meghalaya

Wildlife (Protection) Act, 1972

The various amendments made to the Wildlife (Protection) Act, 1972, came into force from 1st April 2003 except for the provision relating to constitution of National Board for Wildlife.

National Board for Wildlife (NBWL)

As per the provisions of the Wildlife (Protection) Act, 1972, as amended in 2002, the National Board for Wildlife was constituted vide notification dated 21st September, 2003. The first meeting of the NBWL was held on 15th October, 2003 under the chairmanship of Hon'ble Prime Minister. The Standing Committee of the NBWL was also constituted and has met twice on 24th December 2003 and 18th March 2004, to discuss important aspects related to wildlife.

Animal Welfare

The Animal Welfare Division became a part of Ministry of Environment & Forests in the month of July 2002. Earlier the Division was under Ministry of Statistics & Programme Implementation. The Mandate of Animal Welfare Division is to prevent the infliction of unnecessary pain or suffering on animals. To accomplish this mission three-pronged approach is being adopted by the division:

Regulatory

The main task of the division is to implement effectively the various provision of Prevention of Cruelty to Animals Act, 1960. Under this Act, a number of Rules have been framed for various purposes. Some of the important Rules framed are:

- **Performing Animals Rules, 1973 & Performing Animals (Registration) Rules, 1972**
- **Transport of Animals Rules, 1978 & 2001**
- **Prevention of Cruelty (Slaughter Houses) Rules, 2000**

Fig. 43.2 : A herd of Cheetal at Bandipur Wildlife Sanctuary

- Prevention of Cruelty of Animals (Establishment and Regulation of Society for Prevention of Cruelty to Animals) Rules, 2001
- Animal Birth Control (Dogs) Rules, 2001; and
- Breeding of and Experiments on Animals (Control and Supervision) Rules, 1998 as amended in February, 2001

The following action has been taken for ensuring more effective regulation of these Rules:

- Rules under the Prevention of Cruelty to Animals Act, 1960, were notified by Animal Welfare Division as under:
    - Performing Animals (Registration) Rules, 2001 (Notified on 26th March, 2001)
    - Prevention of Cruelty to Animals (Transport of Animals on Foot) Rules, 2001 (Notified on 26th March, 2001)
    - Transport of Animals (Amendment) Rules, 2001 (Notified on 26th March 2001)
    - Prevention of Cruelty to Animals (Slaughter House) Rules, 2001 (Notified on 26th March, 2001)
    - Prevention of Cruelty to Animals (Establishment and Regulation of Societies for the Prevention of Cruelty to Animals) Rules, 2001 (Notified on 26th March, 2001)
    - Animal Birth Control (Dogs) Rules, 2001 (Notified on 24th December, 2001)
    - Performing Animals (Registration) Amendment Rules, 2001 (Notified on 8th January, 2002)
    - Experiments on Animals (Control and Supervision) (Amendment) Rules, 1998 (Notified on 26th August, 1998)
    - Breeding of and Experiments on Animals (Control and Supervision) Rules, 1998 (Notified on 15th December, 1998)
    - Breeding of and Experiments on Animals (Control and Supervision) Amendment Rules, 2001 (Notified on 15th February, 2001)

- Two hundred and twenty five number of animals from circuses viz. Ajanta Circus, Asian Circus, Kamala Circus etc. have been seized and transported to rescue centres at Chennai, Visakhapatnam, Bangalore and Tirupati
- Made timely intervention in likely cases of animal sacrifices in various states including Andhra Pradesh, Tamil Nadu, Orissa and Kerala and prevented the same
- A total of 307 Societies for Prevention of Cruelty to Animals (SPCAs) stand constituted. The number rose to this level from around 250
- Five proformae for monitoring compliance of Transport Rules, Slaughter House Rules, Performing Animals and Registration Rules, Registration of Cattle Premises Rules and Draught and Pack Rules have been sent to all State Department of Animal Husbandry, Home, Urban Development and Transport for furnishing quarterly performance report.

National Institute of Animal Welfare (NIAW)

An Institute is being constructed at Ballabhgarh, District, Faridabad, Haryana for providing various Graduate and Post Graduate courses to animal welfare. The institute would also offer intensive specialized diplomas in specific streams. Besides it would also provide guidance to research scholars.

The National Institute of Animal Welfare will work as a Teaching, Research and Extension Centre. It will impart teaching in disciplines like History of Animal Welfare Movement, Psychology of Animal Behavior, Animal diversity, Animal Laws, Animals Nursing, Animal Breeding, Zoo Management, Park Management, Aquatic Management, Animal Stress and Pain etc. It is proposed to start the academic session w.e.f. 1.10.2004. Before the institute in a full fledged manner, a number of issues are being to be firmed up. These are (a) assessment of market demand to ensure employability of candidates; (b) decision regarding degree or diploma and its due recognition; (c) evolving the course content and finalization of the curriculum; (d) deemed University status for the NIAW; (e) finalization of the Recruitment Rules for the faculty members & (f) appointment of Director & other core faculty members. As all this would take some more time, in the first year it is propsed to start with in service training. Simultaneously a road map is being developed for training the institute into one for degree courses. In service training course in the first year is proposed to assign to ED.Cil on turnkey basis.

STATUTORY BODIES

Animal Welfare Board of India (AWBI)

This is a statutory body under Section 4 of the Prevention of Cruelty to Animals Act, 1960, with headquarters at Chennai. Its basic mandate is to advise the Government on animal welfare issues, and create awareness in animal welfare.

Animal Welfare Board of India (AWBI) gives financial assistance to eligible Animal Welfare Organizations for Shelter Houses, Model Gaushalas for setting up Bio-Gas Plants, Famine/Drought Relief Earthquake Relief etc. in various states.

The number of Animal Welfare Organizations (AWOs) registered with AWBI went up to 2,100 during the year.

An amount of Rs. 4.95 crore under Plan scheme and Rs. 65 lakhs under non-Plan scheme have been released under the scheme during the year. A provision of Rs. 4.50 crores under Plan scheme and Rs. 68 lakhs under non-Plan scheme has been made in the budget estimate for Animal Welfare Board of India in 2008.

So far State Animal Welfare Boards (SAWBs) have been constituted in 24 States/UTs.

Committee for Purpose of Control & Supervision of Experiments on Animals (CPCSEA)

This is also a statutory body under Section 15 of Prevention of Cruelty to Animals Act, 1960, with headquarters at Chennai. The mandate of this committee is to register and monitor the animal breeders and institutions conducting experiments on animals. So far 820 units have been registered with this committee.

Project Tiger

"Project Tiger" was launched in 1973 with an objective "to ensure maintenance of a viable population of tigers in India for scientific, economic, aesthetic, cultural and ecological values, and to preserve for all times, areas of biological importance as a national heritage for the benefit, education and enjoyment of the people."

The Project has been successfully implemented, and in 2009 there are 28 Tiger Reserves in 17 states, covering an area of 37,761 sq. km. The selection of reserves was guided by the need to conserve unique ecosystem/habitat types across the geographic distribution of tigers in the country.

The network of tiger reserves include high mountainous terrains of Arunachal Pradesh, the heavy rainfall areas of Assam and West Bengal, the estuarine mangroves of Sunderbans, the dry forests of Rajasthan, the foothills of the Himalayas in Uttaranchal, Uttar Pradesh and Bihar, the Central Indian Highlands of Madhya Pradesh, Chattisgarh, and Maharashtra, and the plateau of Chota Nagpur (Jharkhand), the hilly tropical and evergreen forests of Orissa, the evergreen forests of Western Ghats in Kerala and Karnataka, the dry deciduous forests of Andhra Pradesh and the Southern moist deciduous forests of Tamil Nadu.

Project Elephant

Project Elephant was launched in February, 1992, to assist states having free ranging populations of wild elephants to ensure long term survival of identified viable populations of elephants in their natural habitats. The project is being implemented in 12 states, viz. Andhra Pradesh, Arunachal Pradesh, Assam, Jharkhand, Karnataka, Kerala, Meghalaya, Nagaland, Orissa, Tamil Nadu, Uttaranchal and West Bengal. States are being given financial as well as technical assistance in achieving the objectives of the project. Help is also provided to other states with small populations of elephants for the purpose of census, training of field staff and mitigation of human-elephant conflict.

Main Activities of Project Elephant

- Strengthening of measure for protection of wild elephants from poachers.
- Development of scientific and planned management for conservation of elephant habitats and viable population of wild Asiatic elephants in India.
- Ecological restoration of existing natural habitats and migratory routes of elephants.
- Promotion of measures for mitigation of man-elephant conflict in problem areas and moderating pressures of human and livestock on crucial elephant habitats.
- Eco-development.
- Veterinary care.
- Research on elephant issues relating to elephant conservation.
- Public education and awareness programme; and
- Capacity building of field staff, mahouts and veterinarians.

43.10.2 Forest (Conservation) Act, 1980 with Amendments Made in 1988

An Act to provide for the conservation of forests and for matters connected therewith or ancillary or incidental thereto. Be it enacted by Parliament in the Thirty-first Year of the Republic of India as follows:-

1. **Short title, extent and commencement.**
   (1) This Act may be called the Forest (Conservation) Act, 1980.
   (2) It extends to the whole of India except the State of Jammu and Kashmir.
   (3) It shall be deemed to have come into force on the 25th day of October, 1980.
2. **Restriction on the dereservation of forests or use of forest land for non-forest purpose.**

Notwithstanding anything contained in any other law for the time being in force in a State, no State Government or other authority shall make, except with the prior approval of the Central Government, any order directing-

(i) that any reserved forest (within the meaning of the expression "reserved forest" in any law for the time being in force in that State) or any portion thereof, shall cease to be reserved;

(ii) that any forest land or any portion thereof may be used for any non-forest purpose;

(iii) that any forest land or any portion thereof may be assigned by way of lease or otherwise to any private person or to any authority, corporation, agency or any other organisation not owned, managed or controlled by Government;

(iv) that any forest land or any portion thereof may be cleared of trees which have grown naturally in that land or portion, for the purpose of using it for reafforestation.

Explanation: For the porpose of this section, "non-forest purpose" means the breaking up or clearing of any forest land or portion thereof for-

(a) the cultivation of tea, coffee, spices, rubber, palms, oil-bearing plants, horticultural crops or medicinal plants;

(b) any purpose other than reafforestation;

but does not include any work relating or ancillary to conservation, development and management of forests and wildlife, namely, the establishment of check-posts, fire lines, wireless communications and construction of fencing, bridges and culverts, dams, waterholes, trench marks, boundary marks, pipelines or other like purposes.

3. **Constitution of Advisory Committee.**

The Central Government may constitute a Committee consisting of such number of persons as h may deem fit to advise that Government with regard to-

(i) the grant of approval. under Section 2; and

(ii) any other matter connected with the conservation of forests which may be referred to h by the Central Government.

3A. **Penalty for contravention of the provisions of the Act.**

Whoever contravenes or abets the contravention of any of the provisions of Section 2, shall be punishable with simple imprisonment for a period which may extend to fifteen days.

3B. **Offences by the Authorities and Government Departments.**

(1) Where any offence under this Act has been committed -

(a) by any department of Government, the head of the department; or

(b) by any authority, every person who, at the time the offence was committed, was directly in charge of, and was responsible to, the authority for the conduct of the business of the authority as well as the authority;

shall be deemed to be guilty of the offence and shall be liable to be proceeded against and punished accordingly:

Provided that nothing contained in this sub-section shall render the head of the department or any person referred to in clause (b), liable to any punishment if he proves that the offence was committed without his knowledge or that he exercised all due diligence to prevent the commission of such offence.

(2) Notwithstanding anything contained in sub-section (1), where an offence punishable under the Act has been committed by a department of Government or any authority referred to in clause (b) of sub-section (1) and it is proved that the offence has been committed with the consent or connivance of; or is attributable to any neglect on the part of any officer, other than the head of the department, or in the case of an authority, any person other than the persons referred to in clause (b) of sub-section (1), such officer or persons shall also be deemed to be guilty of that offence and shall be liable to be proceeded against and punished accordingly.

4. **Power to make rules.**

(1) The Central Government may, by notification in the Official Gazette, makes rules for carrying out the provisions of this Act.

(2) Every rule made under this Act shall be laid, as soon as may be after it is made, before each House of Parliament, while it is in session, for a total period of thirty days which may be comprised in one session or in two or more successive sessions, and if, before the expiry of the session immediately following the session or the successive sessions aforesaid, both Houses agree in making any modification in the rule or both Houses agree that the rule should not be made, the rule shall thereafter have effect only in such modified form or be of no effect, as the case may be; so, however, that any such modification or annulment shall be without prejudice to the validity of anything previously done under that rule.

5. **Repeal and saving.**

(1) The Forest (Conservation) Ordinance, 1980 is hereby replaced.

(2) Notwithstanding such repeal, anything done or any action taken under the provisions of the said Ordinance shall be deemed to have been done or taken under the corresponding provisions of this Act.

43.10.3 Forest (Conservation) Rules, 1981 (New Delhi, the 10th January, 2003)

Ministry of Environment and Forests
Notification

G.S.R.23(E): In exercise of the powers conferred by sub-section (1) of section 4 of the Forest (Conservation) Act, 1980 (69 of 1980), and in supersession of the Forest (Conservation) Rules, 1981, except as respects things done or omitted to be done before such supersession, the Central Government hereby makes the following rules, namely:

1. **Short title, extent and commencement**

(1) These rules may be called the Forest (Conservation) Rules, 2003.

(2) They shall extend to the whole of India except the State of Jammu and Kashmir.

(3) They shall come into force on the date of their publication in the Official Gazette.

2. **Definitions**: In these rules, unless the context otherwise requires:

(a) "Act" means the Forest (Conservation) Act, 1980 (69 of 1980);

(b) "Committee" means the Advisory Committee constituted under section 3 of the Act;

(c) "Chairperson" means the Chairperson of the Committee;

(d) "Member" means a member of the Committee;

(e) "Nodal Officer" means any officer not below the rank of Conservator of Forests, authorised by the State Government to deal with the forest conservation matters under the Act;

(f) "Regional Office" means a Regional Office of the Central Government in the Ministry of Environment and Forests established as part of the Ministry to deal with the forest conservation matters under the Act;

(g) "Section" means a section of the Act;

(h) "User Agency" means any person, organisation or Company or Department of the Central or State Government making a request for diversion or de-notification of forest land for non-forest purpose or using forest land for non-forest purpose in accordance with the permission granted by the Central Government under the Act or the rules.

3. **Composition of the Committee**

(1) The Committee shall be composed of the following members:

(i) Director General of Forests, Ministry of Environment and Forests - Chairperson.

(ii) Additional Director General of Forests, Ministry of Environment and Forests- Member.

(iii) Additional Commissioner (Soil Conservation), Ministry of AgricultureMember.

(iv) Three eminent experts in forestry and allied disciplines (non-officials)- Members.

(v) Inspector General of Forests (Forest Conservation), Ministry of Environment and Forests – Member Secretary

(2) Additional Director General of Forests shall act as the Chairperson in the absence of Director General of Forests.

4. **Terms of appointment of non-official members shall be as follows**

(i) a non-official member shall hold his office for a period of two years;

(ii) a non-official member shall cease to hold office if he becomes of unsound mind, becomes insolvent or is convicted by court of law on a criminal offence involving moral turpitude;

(iii) a non-official member may be removed from his office if he fails to attend three consecutive meetings of the Committee without any sufficient cause or reasons;

(iv) any vacancy in the membership caused by any reason mentioned in clauses (ii) and (iii) shall be filled by the Government for the unexpired portion of two years term.

(v) travelling and daily allowance shall be payable to the non-official members of the Committee at the highest rate admissible to the Government servants of Group 'A' under the rules and orders made by the Central Government and for the time being in force. Provided that the payment of travelling allowance and daily allowance to a member who is a Member of the Parliament or a Member of a State Legislature shall be regulated in accordance with the Salary, Allowances and Pension of Members of Parliament Act, 1954 (30 of 1954) or the respective provisions of law pertaining to the member of the concerned State Legislature.

5. **Conduct of business of the Committee**

(i) The Chairperson shall call the meeting of the Committee whenever considered necessary, but not less than once in a month.

(ii) The meeting of the Committee shall be held at New Delhi.

(iii) In a case where the Chairperson is satisfied that inspection of site or sites of forest land proposed to be used for non-forest purposes shall be necessary or expedient in connection with the consideration of the proposal or proposals received under sub-rule (3) of rule 6, he may direct that the meetings of the Committee to be held at a place other than New Delhi from where such inspection of site or sites is necessary.

(iv) The Chairperson shall preside over every meeting of the Committee at which he is present.

(v) Every question upon which the Central Government is required to be advised shall be considered in the meeting of the Committee provided that in urgent cases if the meeting cannot be convened within a month, the Chairperson may direct that papers may be circulated and sent to the members for their opinion within the stipulated time.

(vi) The quorum of the meeting of the Committee shall be three.

6. Submission of the proposals seeking approval of the Central Government under section 2 of the Act

(1) Every user agency, who wants to use any forest land for non-forest purposes shall make his proposal in the appropriate Form appended to these rules, i.e. Form 'A' for proposals seeking first time approval under the Act and Form 'B' for proposals seeking renewal of leases where approval of the Central Government under the Act had already been obtained earlier, to the concerned nodal officer authorized in this behalf by the State Government, alongwith requisite information and documents, complete in all respects, well in advance of taking up any non-forest activity on the forest land.

(2) Every State Government or other authority, after having received the proposal under sub-rule (1) and after being satisfied that the proposal requires prior approval under section 2 of the Act, shall send the proposal to the Central Government in the appropriate forms, within ninety days of the receipt of the proposal from the user agency for proposals seeking first time approval under the Act and within sixty days for proposals seeking renewal of leases where approval of the Central Government under the Act had already been obtained earlier: Provided that all proposals involving clearing naturally grown trees in forest land or portion thereof for the purpose of using it for reafforestation shall be sent in the form of Working Plan or Management Plan.

(3) The proposal referred to in sub-rule (2) above, involving forest land of more than forty hectare shall be sent by the State Government to the Secretary to the Government of India, Ministry of Environment and Forests, Paryavaran Bhavan, CGO Complex, Lodhi Road, New Delhi 110 003, with a copy of the proposal (with complete enclosures) to the concerned Regional Office.

(4) The proposal referred to in sub-rule (2) above, involving forest land up to forty hectare shall be sent to the Chief Conservator of Forests or Conservator of Forests of the concerned Regional Office of the Ministry of Environment and Forests.

(5) The proposal referred to in sub-rule (2) above, involving clearing of naturally grown trees in forest land or portion thereof for the purpose of using it for reafforestation shall be sent to the Chief Conservator of Forests or Conservator of Forests of the concerned Regional Office of the Ministry of Environment and Forests.

7. Committee to advise on proposals received by the Central Government

(1) The Central Government shall refer every proposal, complete in all respects, received by it under sub-rule (3) of rule 6 including site inspection report, wherever required, to the Committee for its advice thereon.

(2) The Committee shall have due regard to all or any of the following matters while tendering its advice on the proposals referred to it under sub-rule (1), namely:

(a) Whether the forests land proposed to be used for non-forest purpose forms part of a nature reserve, national park wildlife sanctuary, biosphere reserve or forms part of the habitat or any endangered or threatened species of flora and fauna or of an area lying in severely eroded catchment;

(b) Whether the use of any forest land is for agricultural purposes or for the rehabilitation of persons displaced from their residences by reason of any river valley or hydro-electric project;

(c) Whether the State Government or the other authority has certified that it has considered all other alternatives and that no other alternatives in the circumstances are feasible and that the required area is the minimum needed for the purpose; and

(d) Whether the State Government or the other authority undertakes to provide at its cost for the acquisition of land of an equivalent area and afforestation thereof.

(3) While tendering the advice, the Committee may also suggest any conditions or restrictions on the use of any forest land for any nonforest purpose, which in its opinion, would minimise adverse environmental impact.

8. **Action of the Central Government on the advice of the Committee**: The Central Government shall, after considering the advice of the Committee tendered under rule 7 and after such further enquiry as it may consider necessary, grant approval to the proposal with or without conditions or reject the same within sixty days of its receipt.

9. **Proceedings against persons guilty of offences under the Act**

(1) The Central Government may, by notification, authorize any officer not below the rank of Conservator of Forests or the concerned forest officer having territorial jurisdiction over the forest land in respect of which the said offence is said to have been committed, to file complaints against the person(s) prima-facie found guilty of offence under the Act or the violation of the rules made thereunder, in the court having jurisdiction in the matter. Provided that no complaint shall be filed in the court, without giving the person (s) or officer (s) or authority (s) against whom the allegations of offence exist, an opportunity to explain his or their conduct and to show cause, by issuing a notice in writing of not less than sixty days, as to why a complaint should not be filed in the court against him or them for alleged offences.

(2) The officer authorised by the Central Government in sub-rule (1) may require any State Government or its officer or any person or any other authority to furnish to it within a specified period any reports, documents, statistics and any other information related to contravention of the Act or the rules made thereunder, considered necessary for making a complaint in any court of jurisdiction and every such State Government or officer or person or authority shall be bound to do so.

APPENDIX

(See Rule 6)

FORM – 'A'

Form for seeking prior approval under section 2 of the proposals by the State Governments and other authorities

PART-I

(to be filled up by user agency)

1. Project details:

(i) Short narrative of the proposal and project/scheme for which the forest land is required.

(ii) Map showing the required forest land, boundary of adjoining forest on a 1:50,000 scale map.

(iii) Cost of the project:

(iv) Justification for locating the project in forest area.

(v) Cost-benefit analysis (to be enclosed).

(vi) Employment likely to be generated.

2. Purpose-wise break-up of the total land required:

3. Details of displacement of people due to the project, if any:

(i) Number of families.

(ii) Number of Scheduled Castes/Scheduled Tribe families

(iii) Rehabilitation plan. (to be enclosed)

4. Whether clearance under Environment (Protection) Act, 1986 required? (Yes/No).

5. Undertaking to bear the cost of raising and maintenance of compensatory afforestation and/or penal compensatory afforestation as well as cost for protection and regeneration of Safety Zone, etc. as per the scheme prepared by the State Government (undertaking to be enclosed).

6. Details of Certificates/documents enclosed as required under the instructions.

Signature
(Name in Block letters)
Designation
Address (of User Agency)
Date:
Place:
State serial No. of proposal
(To be filled up by the Nodal Officer with date of receipt)

PART-II

(To be filled by the concerned Deputy Conservator of Forests)

State serial No. of proposal

7. Location of the project/Scheme:

(i) State/Union Territory

(ii) District.

(iii) Forest Division

(iv) Area of forest land proposed for diversion (in ha.)

(v) Legal status of forest

(vi) Density of vegetation.

(vii) Species-wise (scientific names) and diameter class-wise enumeration of trees (to be enclosed. In case of irrigation / hydel projects enumeration at FRL, FRL-2 meter & FRL-4 meter also to be enclosed.)

(viii) Brief note on vulnerability of the forest area to erosion.

(ix) Approximate distance of proposed site for diversion from boundary of forest.

(x) Whether forms part of National Park, wildlife sanctuary, biosphere reserve, tiger reserve, elephant corridor, etc. (If so, the details of the area and comments of the Chief Wildlife Warden to be annexed).

(xi) Whether any rare/endangered/unique species of flora and fauna found in the area- if so details thereof.

(xii) Whether any protected archaeological/heritage site/defence establishment or any other important monument is located in the area. If so, the details thereof with NOC from competent authority, if required.

8. Whether the requirement of forest land as proposed by the user agency in col. 2 of Part-I is unavoidable and barest minimum for the project. If no, recommended area item-wise with details of alternatives examined.

9. Whether any work in violation of the Act has been carried out (Yes/No). If yes, details of the same including period of work done, action taken on erring officials. Whether work in violation is still in progress.

10. Details of compensatory afforestation scheme:

(i) Details of non forest area/degraded forest area identified for compensatory afforestation, its distance from adjoining forest, number of patches, size of each patch.

(ii) Map showing non-forest/degraded forest area identified for compensatory afforestation and adjoining forest boundaries.

(iii) Detailed compensatory afforestation scheme including species to be planted, implementing agency, time schedule, cost structure, etc.

(iv) Total financial outlay for compensatory afforestation scheme.

(v) Certificates from competent authority regarding suitability of area identified for compensatory afforestation and from management point of view. (To be signed by the concerned Deputy Conservator of Forests).

11. Site inspection report of the DCF (to be enclosed) especially highlighting facts asked in col. 7 (xi, xii), 8 and 9 above.

12. Division/District profile:

(i) Geographical area of the district.

(ii) Forest area of the district.

(iii) Total forest area diverted since 1980 with number of cases.

(iv) Total compensatory afforestation stipulated in the district/division since 1980 on (a) forest land including penal compensatory afforestation, (b) non-forest land.

(v) Progress of compensatory afforestation as on (date) ____ on (a) forest land; (b) non-forest land.

13. Specific recommendations of the DCF for acceptance or otherwise of the proposal with reasons.

Signature

Name

Official Seal

Date:

Place:

PART-III

(To be filled by the concerned Conservator of Forests)

14. Whether site, where the forest land involved is located has been inspected by concerned Conservator of Forests (Yes/No). If yes, the date of inspection & observations made in form of inspection note to be enclosed.

15. Whether the concerned Conservator of Forests agree with the information given in Part-B and the recommendations of Deputy Conservator of Forests.

16. Specific recommendation of concerned Conservator of Forests for acceptance or otherwise of the proposal with detailed reasons.

Signature

Name

Official Seal

Date:

Place:

PART-IV

(To be filled in by the Nodal Officer or Principal Chief Conservator of Forests or Head of Forest department)

17. Detailed opinion and specific recommendation of the State Forest Department for acceptance of otherwise of the proposal with remarks. (While giving opinion, the adverse comments made by concerned Conservator of Forests or Deputy Conservator of Forests should be categorically reviewed and critically commented upon).

Signature
Name & Designation
(Official Seal)
Date:
Place:

PART- V

(To be filled in by the Secretary in charge of Forest Department or by any other authorised officer of the State Government not below the rank of an Under Secretary)

18. Recommendation of the State Government:

(Adverse comments made by any officer or authority in Part-B or Part-C or Part-D above should be specifically commented upon)

Signature
Name & Designation
(Official Seal)
Date:
Place:

INSTRUCTIONS (for Part-I):

1. **The project authorities may annex a copy of the approved project/plan in addition to filling Col. 1 (i) e.g. IBM approved mining plan for major minerals/CMPDI plan with subsidence analysis reports, etc.**
2. **Map has to be in original duly authenticated jointly by project authorities and concerned DCF – Col. 1 (ii).**
3. **Complete details of alternative alignments examined especially in case of project like roads, transmission lines, railway lines, canals, etc. to be shown on map with details of area of forest land involved in each alternative to be given - Col. 1 (iii).**
4. **For proposals relating to mining, certificate from competent authority like District Mining Officer about non-availability of the same mineral in surrounding/nearby non-forest areas.**
5. **In case the same company/individual has taken forest land for similar project in the State, a brief detail of all such approvals/leases be given as an enclosure along with current status of the projects.**
6. **The latest clarifications issued by the Ministry under Forest (Conservation) Act, 1980 may be kept in mind. In case such information do not fit in the given columns, the same shall be annexed separately.**

GENERAL INSTRUCTIONS:

1. **On receipt of proposal, Nodal Officer shall issue a receipt to the user agency indicating therein the name of the proposal, user agency, area in hectare, serial number and date of receipt.**

2. **If the space provided above is not sufficient to specify any information, please attach separate details/documents.**

3. **While forwarding the proposal to the Central Government, complete details on all aspects of the case as per Form prescribed above read with the clarifications issued by the Ministry of Environment and Forests, Government of India, New Delhi should be given. Incomplete or deficient proposals shall not be considered and shall be returned to the State Government in original.**

4. **The State Government shall submit the proposal to the Central Government within stipulated time limits. In case of delay while forwarding, the reasons for the same to be given in the forwarding/covering letter.**

FORM – 'B'

(See Rule 6)

Form for seeking prior approval under section 2 of the proposals by the State Governments and other authorities in respect of renewal of leases, which have been earlier granted clearance under Forest (Conservation) Act, 1980

PART-I

(to be filled up by user agency)

1. Letter No. & date vide which clearance under Forest (Conservation) Act, 1980 accorded by the Central Government (copy to be enclosed):

2. Project details:

(i) Short narrative of the proposal and project/scheme for which the forest land is required.

(ii) Map showing the required forest land, boundary of adjoining forest on a 1:50,000 scale map.

(iii) Cost of the project:

3. Purpose-wise break-up of the total land required (already broken & to be broken):

4. Details of Certificates/documents enclosed as required under the instructions.

Signature
(Name in Block letters)
Designation
Address (of User Agency)
Date:
Place:
State serial No. of proposal
(To be filled up by the Nodal Officer with date of receipt)

PART-II

(To be filled by the concerned Deputy Conservator of Forests)

State serial No. of proposal

5. Location of the project/Scheme:

(i) State/Union Territory

(ii) District.

(iii) Forest Division

(iv) Area of forest land proposed for diversion (in ha.)

(v) Legal status of forest

(vi) Density of vegetation.

(vii) Species-wise (scientific names) and diameter class-wise enumeration of trees in unbroken area.

(viii) Whether forms part of National Park, wildlife sanctuary, biosphere reserve, tiger reserve, elephant corridor, etc. (If so, the details of the area and comments of the Chief Wildlife Warden to be annexed).

6. Whether any work in violation of the Act has been carried out (Yes/No). If yes, details of the same including period of work done, action taken on erring officials. Whether work in violation is still in progress.

7. Site inspection report of the DCF (to be enclosed) in respect to status of compliance of conditions stipulated during earlier approval.

8. Division/District profile:

(i) Geographical area of the district.

(ii) Forest area of the district.

(iii) Total forest area diverted since 1980 with number of cases.

(iv) Total compensatory afforestation stipulated in the district/division since 1980 on (a) forest land including penal compensatory afforestation, (b) non-forest land.

(v) Progress of compensatory afforestation as on (date) ____ on (a) forest land; (b) non-forest land.

9. Specific recommendations of the DCF for acceptance or otherwise of the proposal with reasons.

Signature
Name
Official Seal
Date:
Place:

PART-III

(To be filled by the concerned Conservator of Forests)

10. Whether site, where the forest land involved is located has been inspected by concerned Conservator of Forests (Yes/No). If yes, the date of inspection & observations made in form of inspection note to be enclosed.

11. Whether the concerned Conservator of Forests agree with the information given in Part-B and the recommendations of Deputy Conservator of Forests.

12. Specific recommendation of concerned Conservator of Forests for acceptance or otherwise of the proposal with detailed reasons.

Signature
Name
Official Seal
Date:
Place:

PART-IV

(To be filled in by the Nodal Officer or Principal Chief Conservator of Forests or Head of Forest department)

13. Detailed opinion and specific recommendation of the State Forest Department for acceptance of otherwise of the proposal with remarks.

(While giving opinion, the adverse comments made by concerned Conservator of Forests or Deputy Conservator of Forests should be categorically reviewed and critically commented upon).

Signature
Name & Designation
(Official Seal)
Date:
Place:

PART- V

(To be filled in by the Secretary in charge of Forest Department or by any other authorised officer of the State Government not below the rank of an Under Secretary)

14. Recommendation of the State Government:

(Adverse comments made by any officer or authority in Part-B or Part-C or Part-D above should be specifically commented upon)

Signature
Name & Designation
(Official Seal)
Date:
Place:

INSTRUCTIONS (for Part-I):

1. The project authorities may annex a copy of the approved project/plan in addition to filling Col. 2 (i) e.g. IBM approved mining plan for major minerals/CMPDI plan with subsidence analysis reports, etc.
2. Map has to be in original duly authenticated jointly by project authorities and concerned DCF – Col. 2 (ii).
3. In case the same company/individual has taken forest land for similar project in the State, a brief detail of all such approvals/leases be given as an enclosure along with current status of the projects.
4. Item-wise requirement (Col. 3) should be separately shown for broken up and fresh areas.
5. The latest clarifications issued by the Ministry under Forest (Conservation) Act, 1980 may be kept in mind. In case such information do not fit in the given columns, the same shall be annexed separately.

GENERAL INSTRUCTIONS:

1. On receipt of proposal, Nodal Officer shall issue a receipt to the user agency indicating therein the name of the proposal, user agency, area in hectare, serial number and date of receipt.
2. If the space provided above is not sufficient to specify any information, please attach separate details/documents.
3. While forwarding the proposal to the Central Government, complete details on all aspects of the case as per Form prescribed above read with the clarifications issued by the Ministry of Environment and Forests, Government of India, New Delhi should be given. Incomplete or deficient proposals shall not be considered and shall be returned to the State Government in original.
4. The State Government shall submit the proposal to the Central Government within stipulated time limits. In case of delay while forwarding, the reasons for the same to be given in the forwarding/covering letter."(File No. 5-5/98-FC)

Sd/-
(DR. V.K. BAHUGUNA)
Inspector General of Forests (Forest Conservation)

Note: The principal rules were published vide G.S.R. No. 719 dated the 1st August, 1981 in part II, Section 3, sub-section (i) of the Gazette of India and subsequently amended vide: (1) G.S.R. 14, dated the 28th December, 1987; (2) G.S.R. 640(E), dated the 26th June, 1989; (3) G.S.R. 563 (E), dated the 21st May, 1992.

43.10.4 THE BIOLOGICAL DIVERSITY ACT, 2002

NO. 18 OF 2003

[5th February 2003]

An Act to provide for conservation of biological diversity, sustainable use of its components and fair and equitable sharing of the benefits arising out of the use of biological resources, knowledge and for matters connected therewith or incidental thereto.

WHEREAS India is rich in biological diversity and associated traditional and contemporary knowledge system relating thereto.

AND WHEREAS India is a party to the United Nations Convention on Biological Diversity signed at Rio de Janeiro on the 5th day of June, 1992;

AND WHEREAS the said Convention came into force on the 29th December, 1993;

AND WHEREAS the said Convention reaffirms the sovereign rights of the States over their biological resources;

AND WHEREAS the said Convention has the main objective of conservation of biological diversity, sustainable use of its components and fair and equitable sharing of the benefits arising out of utilization of genetic resources;

AND WHEREAS it is considered necessary to provide for conservation, sustainable utilization and equitable sharing of the benefits arising out of utilization of genetic resources and also to give effect to the said Convention.

BE it enacted by Parliament in the Fifty-third Year of the Republic of India as follows:

Chapter I

PRELIMINARY

1. Short title, extent and commencement

1. This Act may be called the Biological Diversity Act. 2002.
2. It extends to the whole of India.
3. It shall come into force on such date as the Central Government may, by notification in the Official Gazette, appoint:

Provided that different dates may be appointed for different provisions of this Act and any reference in any such provision to the commencement of this Act shall be construed as a reference to the coming into force of that provision.

2. Definitions

In this Act, unless the context otherwise requires—

(a) "benefit claimers" means the conservers of biological resources, their byproducts, creators and holders of knowledge and information relating to the use of such biological resources, innovations and practices associated with such use and application;

(b) "bilogical diversity" means the variability among living organisms from all sources and the ecological complexes of which they are part, and includes diversity within species or between species and of eco-systems;

(c) "biological resources" means plants, animals and microorganisms or parts thereof, their genertic material and by-products (excluding value added products) with actual or potential use or value, but does not include human genetic material;

(d) "bio-survey and bio-utilization" means survey or collection of species, subspecies, genes, components and extracts of biological resource for any purpose and includes characterization, inventorisation and bioassay;

(e) "Chairperson" means the Chairperson of the National Biodiversity Authority or, as the case may be, of the State Biodiversity Board;

(f) "commercial utilization" means end uses of biological resources for commercial utilization such as drugs, industrial enzymes, food flavours, fragrance, cosmetics, emulsifiers, oleoresins, colours, extracts and genes used for improving crops and livestock through genetic intervention, but does not include conventional breeding or traditional practices in use in any agriculture, horticulture, poultry, dairy farming, animal husbandry or bee keeping;

(g) "fair and equitable benefit sharing" means sharing of benefits as determined by the National Biodiversity Authority under Section 21;

(h) "local bodies" means Panchayats and Municipalities, by whatever name called, within the meaning of Clause (1) of article 243B and Clause (1) or article 243Q of the Constitution and in the absence of any Panchayats or Municipalities, institutions of self-government constituted under any other provision of the Constitution of any Central Act or State Act;

(i) "member" means a member of the National Biodiversity Authority or a State Biodiversity Board and includes the Chairperson.

(j) "National Biodiversity Authority" means the National Biodiversity Authority established under Section 8;

(k) "prescribed" means prescribed by rules made 1 under this Act;

(l) "regulations" means regulations made under this Act;

(m) "research" means study or systematic investigation of any biological resource or technological application, that uses biological systems, living organisms or derivatives thereof to make or modify products or processes for any use:

(n) "State Biodiversity Board" means the State Biodiversity Board established under Section 22;

(o) "sustainable use" means the use of components of biological diversity in such manner and at such rate that does not lead to the long-term decline of the biological diversity thereby maintaining its potential to meet the needs and aspirations of present and future generations;

(p) "value added products" means products which may contain portions or extracts of plants and animals in unrecognizable and physically inseparable form.

Chapter II

REGULATION OF ACCESS TO BIOLOGICAL DIVERSITY

Certain persons not to undertake Biodiversity related activities without approval of National Biodiversity Authority

3. (1) No person referred to in Sub-section (2) shall, without previous approval of the National Biodiversity Authority, obtain any biological resource occurring in India or knowledge associated thereto for research or for commercial utilization or for bio-survey and bio-utilization.

(2) The persons who shall be required to take the approval of the National Biodiversity Authority under Sub-section (1) are the following, namely:

(a) a person who is not a citizen of India;

(b) a citizen of India, who is a non-resident as defined in Clause (3) of Section 2 of the Income-tax Act, 1961;

(c) a body corporate, association or organization—

(i) not incorporated or registered in India; or

(ii) incorporated or registered in India under any law for the time being in force which has any non-Indian participation in its share capital or management.

Results of research not to be transferred to certain persons without approval of National Biodiversity Authority

4. No person shall, without the previous approval of the National Biodiversity Authority, transfer the results of any research relating to any biological resources occurring in, or obtained from, India for monetary consideration or otherwise to any person who is not a citizen of India or citizen of India who is non-resident as defined in Clause (3) of Section 2 of the Income-tax Act, 1961 or a body corporate or organization which is not registered or incorporated in India or which has any non-Indian participation in its share capital or management.

Explanation—For the purposes of this Section, "transfer" does not include publication of research papers or dissemination of knowledge in any seminar or workshop, if such publication is as per the guidelines issued by the Central Government.

Sections 3 and 4 not to apply to certain collaborative research projects

5. (1) The provisions of Sections 3 and 4 shall not apply to collaborative research projects involving transfer or exchange of biological resources or information relating thereto between institutions, including Government sponsored institutions of India, and such institutions in other countries, if such collaborative research projects satisfy the conditions specified in Sub-section (3).

(2) All collaborative research projects, other than those referred to in Sub-section (1) which are based on agreements concluded before the commencement of this Act and in force shall, to the extent the provisions of agreement are inconsistent with the provisions of this Act or any guidelines issued under Clause (a) of Sub-section (3), be void.

(3) For the purposes of Sub-section (1), collaborative research projects shall—

(a) conform to the policy guidelines issued by the Central Government in this behalf;

(b) be approved by the Central Government.

Application for intellectual property rights not to be made without approval of National Biodiversity Authority

6. (1) No person shall apply for any intellectual property right, by whatever name called, in or outside India for any invention based on any research or information on a biological resource obtained from India without obtaining the previous approval of the National Biodiversity Authority before making such application.

Provided that if a person applies for a patent, permission of the National Biodiversity Authority may be obtained after the acceptance of the patent but before the seating of the patent by the patent authority concerned:

Provided further that the National Biodiversity Authority shall dispose of the application for permission made to it within a period of ninety days from the date of receipt thereof.

(2) The National Biodiversity Authority may, while granting the approval under this section, impose benefit sharing fee or royalty or both or impose conditions including the sharing of financial benefits arising out of the commercial utilization of such rights.

(3) The provisions of this Section shall not apply to any person making an application for any right under any law relating to protection of plant varieties enacted by Parliament.

(4) Where any right is granted under law referred to in Sub-section (3), the concerned authority granting such right shall endorse a copy of such document granting the right to the National Biodiversity Authority.

Prior intimation to State Biodiversity Board for obtaining biological resource for certain purposes

7. No person, who is a citizen of India or a body corporate, association or organization which is registered in India, shall obtain any biological resource for commercial utilization, or bio-survey and bio-utilization for commercial utilization except after giving prior intimation to the State Biodiversity Board concerned:

Provided that the provisions of this Section shall not apply to the local people and communities of the area, including growers and cultivators of biodiversity, and *vaids* and *hakims*, who have been practicing indigenous medicine.

Chapter III

NATIONAL BIODIVERSITY AUTHORITY

Establishment of National Biodiversity Authority

8. (1) With effect from such date as the Central Government may, by notification in the Official Gazette, appoint there shall be established by the Central Government for the purposes of this Act, a body to be called the National Biodiversity Authority.

(2) The National Biodiversity Authority shall be a body corporate by the name aforesaid, having perpetual succession and a common seal, with power to acquire, hold and dispose of property, both movable and unmovable, and to contract and shall by the said name sue and be sued.

(3) The head office of the National Biodiversity Authority shall be at Chennai and the National Biodiversity Authority may, with the previous approval of the Central Government, establish offices at other places in India.

(4) The National Biodiversity Authority shall consist of the following members:

(a) a Chairperson, who shall be an eminent person having adequate knowledge and experience in the conservation and sustainable use of biological diversity and in matters relating to equitable sharing of benefits, to be appointed by the Central Government;

(b) three ex-officio members to be appointed by the Central Government, one representing the Ministry dealing with Tribal Affairs and two representing the Ministry dealing with Environment and Forests or the Director General of Forests;

(c) seven ex-officio members to be appointed by the Central Government to represent, respectively, the Ministries of the Central Government dealing with—

(i) Agricultural Research and Education;
(ii) Biotechnology;
(iii) Ocean Development;
(iv) Agriculture and Cooperation;
(v) Indian Systems of Medicine and Homoeopathy;
(vi) Science and Technology;
(vii) Scientific and Industrial Research.

(d) five non-official members to be appointed from amongst specialists and scientists having special knowledge of, or experience in, matters relating to conservation of biological diversity, sustainable use of biological resources and equitable sharing of benefits arising out of the use of biological resources, representatives of industry, conservers, creators and know ledge-holders of biological resources.

Conditions of service of Chairperson and members

9. The term of office and conditions of service of the Chairperson and the other members other than *ex-officio* members of the National Biodiversity Authority shall be such as may be prescribed by the Central Government.

Chairperson to be Chief Executive of National Biodiversity Authority

10. The Chairperson shall be the Chief Executive of the National Biodiversity Authority and shall exercise such powers and perform such duties, as may be prescribed.

Removal of members

11. The Central Government may remove from the National Biodiversity Authority any member who, in its opinion, has—

(a) been adjudged as an insolvent; or
(b) been convicted of an offence which involves moral turpitude; or
(c) become physically or mentally incapable of acting as a member; or
(d) so abused his position as to render his continuance in office detrimental to the public interest; or
(e) acquired such financial or other interest as is likely to affect prejudically his functions as a member.

Meetings of National Biodiversity Authority

12. (1) The National Biodiversity Authority shall meet at such time and place and shall observe such rules of procedure in regard to the transaction of business at its meetings (including the quorum at its meetings) as may be prescribed.

(2) The Chairperson of the National Biodiversity Authority shall preside at the meetings of the National Biodiversity Authority.

(3) If for any reason the Chairperson is unable to attend any meeting of the National Biodiversity Authority, any member of the National Biodiversity Authority chosen by the members present at the meeting shall preside at the meeting.

(4) All questions which come before any meeting of the National Biodiversity Authority shall be decided by a majority of votes of the members present and voting and in the event of equality of votes, the Chairperson, or, in his absence, the person presiding, shall have and exercise a second or casting vote.

(5) Every member who is in any way, whether directly, indirectly or personally, concerned or interested in a matter to be decided at the meeting shall disclose the nature of his concern or interest and after such disclosure, the member concerned or interested shall not attend that meeting.

(6) No act or proceeding of the National Biodiversity Authority shall be invalidated merely by reason of—

(a) any vacancy in, or any defect in the constitution of the National Biodiversity Authority; or
(b) any defect in the appointment of a person acting as a member; or
(c) any irregularity in the procedure of the National Biodiversity Authority not affecting the merits of the case.

Committees of National Biodiversity Authority

13. (1) The National Biodiversity Authority may constitute a committee to deal with agro-biodiversity.

Explanation— For the purposes of this Sub-section, "agro-biodiversity" means biological diversity of agriculture related species and their wild relatives.

(2) Without prejudice to the provisions of Sub-section (1), the National Biodiversity Authority may constitute such number of committees as it deems fit for the efficient discharge of its duties and performance of its functions under this Act.

(3) A committee constituted under this Section shall co-opt such number of persons, who are not the members of the National Biodiversity Authority, as it may think fit and the persons so co-opted shall have the right to attend the meetings of the committee and take part in its proceedings but shall not have the right to vote.

(4) The persons appointed as members of th committee under Sub-section (2) shall be entitled to receive such allowances or fees for attending the meetings of the committee as may be fixed by the Central Government.

Officers and employees of National Biodiversity Authority

14. (1) The National Biodiversity Authority may appoint such officers and other employees as it considers necessary for the efficient discharge of its functions under this Act.

(2) The terms and conditions of service of such officers and other employees of the National Biodiversity Authority shall be such as may be specified by regulations.

Authentication of orders and decisions of National Biodiversity Authority

15. All orders and decisions of the National Biodiversity Authority shall be authenticated by the signature of the Chairperson or any other member authorized by the National Biodiversity Authority in this behalf and all other instruments executed by the National Biodiversity Authority shall be authenticated by the signature of an officer of the National Biodiversity Authority authorized by it in this behalf.

Delegation of powers

16. The National Biodiversity Authority may, by general or special order in writing, delegate to any member, officer of the National Biodiversity Authority or any other person subject to such conditions, if any as may be specified in the order, such of the powers and functions under this Act (except the power to prefer an appeal under Section 50 and the power to make regulations under Section 64) as it may deem necessary.

Expenses of National Biodiversity Authority to be defrayed out of the Consolidated Fund of India

17. The salaries and allowances payable to the members and the administrative expenses of the National Biodiversity Authority including salaries, allowances and pension payable to, or in respect of, the officers and other employees of the National Biodiversity Authority shall be defrayed out of the Consolidated Fund of India.

Chapter IV

FUNCTIONS AND POWERS OF THE NATIONAL BIODIVERSITY AUTHORITY

Functions and powers of National Biodiversity Authority

18. (1) It shall be the duty of the National Biodiversity Authority to regulate activities referred to in Sections 3, 4 and 6 and by regulations issue guidelines for access to biological resources and for fair and equitable benefit sharing.

(2) The National Biodiversity Authority may grant approval for undertaking any activity referred to in Sections 3, 4 and 6.

(3) The National Biodiversity Authority may—

(a) advise the Central Government on matters relating to the conservation of biodiversity, sustainable use of its components and equitable sharing of benefits arising out of the utilization of biological resources;

(b) advise the State Governments in the selection of areas of biodiversity importance to be notified under Sub-section (1) of Section 37 as heritage sites and measures for the management of such heritage sites;

(c) perform such other functions as may be necessary to carry out the provisions of this Act.

(4) The National Biodiversity Authority may, on behalf of the Central Government, take any measures necessary to oppose the grant of intellectual property rights in any country outside India on any biological resource obtained from India or knowledge associated with such biological resource which is derived from India.

Chapter V

APPROVAL BY THE NATIONAL BIODIVERSITY AUTHORITY

Approval by National Biodiversity Authority for undertaking certain activities

10. (1) Any person referred, to in Sub -section (2) of Section 3 who intends to obtain any biological resource occurring in India or knowledge associated thereto for research or for commercial utilization or

for bio-survey and bio-utilization or transfer the results of any research relating to biological resources occurring in, or obtained from, India, shall make application in such form and payment of such fees as may be prescribed, to the National Biodiversity Authority.

(2) Any person who intends to apply for a patent or any other form of intellectual property protection wherether in India or outside India referred to in Sub-section (1) of Section 6, may make an application in such form and in such manner as may be prescribed to the National Biodiversity Authority.

(3) On receipt of an application under Sub-section (1) or Sub-section (2) the National Biodiversity Authority may, after making such enquiries as it may deem fit and if necessary after consulting an expert committee constituted for this purpose, by order, grant approval subject to any regulations made in this behalf and subject to such terms and conditions as it may deem fit, including the imposition of charges by way of royalty or for reasons to be recorded in writing, reject the application:

Provided that no such order for rejection shall be made without giving an opportunity of being heard to the person affected.

(4) The National Biodiversity Authority shall give public notice of every approval granted by it under this Section.

Transfer of biological resource or knowledge

20. (1) No person who has been granted approval under Section 19 shall transfer any biological resource or knowledge associated thereto which is the subject matter of the said approval except with the permission of the National Biodiversity Authority.

(2) Any person who intends to transfer any biological resource or knowledge associated thereto referred to in Sub-section (1) shall make an application in such form and in such manner as may be prescribed to the National Biodiversity Authority.

(3) On receipt of an application under Sub-section (2), the National Biodiversity Authority may, after making such enquiries as it may deem fit and if necessary after consulting an expert committee constituted for this purpose, by order, grant approval subject to such terms and conditions as it may deem fit, including the imposition of charges by way of royalty or for reasons to be recorded in writing, reject the application:

Provided that no such order for rejection shall be made without giving an opportunity of being heard to the person affected.

(4) The National Biodiversity Authority shall give public notice of every approval granted by it under this Section.

Determination of equitable benefit sharing by National Biodiversity Authority

21. (1) The National Biodiversity Authority shall, while granting approvals under Section 19 or Section 20, ensure that the terms and conditions subject to which approval is granted secures equitable sharing of benefits arising out of the use of accessed biological resources, their by-products, innovations and practices associated with their use and applications and knowledge relating thereto in accordance with mutually agreed terms and conditions between the person applying for such approval, local bodies concerned and the benefit claimers.

(2) The National Biodiversity Authority shall, subject to any regulations made in this behalf, determine the benefit sharing which shall be given effect in all or any of the following manner, namely:

(a) grant of joint ownership of intellectual property rights to the National Biodiversity Authority, or where benefit claimers are identified, to such benefit claimers;

(b) transfer of technology;

(c) location of production, research and development units in such areas which will facilitate better living standards to the benefit claimers;

(d) association of Indian scientists, benefit claimers and the local people with research and development in biological resources and bio-survey and bio-utilization;

(e) setting up of venture capital fund for aiding the cause of benefit claimers;

(f) payment of monetary compensation and non-monetary benefits to the benefit claimers as the National Biodiversity Authority may deem fit.

(3) Where any amount of money is ordered by way of benefit sharing, the National Biodiversity Authority may direct the ammout to be deposited in the National Biodiversity Fund:

Provided that where biological resource or knowledge was a result of access from specific individual or group of individuals or organizations, the National Biodiversity Authority may direct that the amount shall be paid directly to such individual or group of individuals or organizations in accordance with the terms of any agreement and in such manner as it deems fit.

(4) For the purposes of this Section, the Nation Biodiversity Authority shall, in consultation with the Central Government, by regulations, frame guidelines.

Chapter VI

STATE BIODIVERSITY BOARD

Establishment of State Biodiversity Board

22. (1) With effect from such date as the State Government may, by notification in the Official Gazette, appoint in this behalf, there shall be established by that Government for the purposes of this Act, a Board for the State to be known as the (name of the State) Biodiversity Board.

(2) Notwithstanding anything contained in this Section, no State Biodiversity Board shall be constituted for a Union Territory and in relation to a Union Territory, the National Biodiversity Authority shall exercise the powers and perform the functions of a State Biodiversity Board for that Union Territory:

Provided that in relation to any Union Territory, the National Biodiversity Authority may delegate all or any of its powers or functions under this Sub-section to such person or group of persons as the Central Government may specify.

(3) The Board shall be a body corporate by the name aforesaid, having perpetual succession and a common seal, with power to acquire, hold and dispose of property, both movable and immovable, and to contract, and shall by the said name sue and be sued.

(4) The Board shall consist of the following members, namely:

(a) a Chairperson who shall be an eminent person having adequate knowledge and experience in the conservation and sustainable use of biological diversity and in matters relating to equitable sharing of benefits, to be appointed by the State Government;

(b) not more than five ex-officio members to be appointed by the State Government to represent the concerned Departments of the State Government;

(c) not more than five members to be appointed from amongst experts in matters relating to conservation of biological diversity, sustainable use of biological resources and equitable sharing of benefits arising out of the use of biological resources.

(5) The head office of the State Biodiversity Board shall be at such place as the State Government may, by notification in the Official Gazette, specify.

Functions of State Biodiversity Board

23. The functions of the State Biodiversity Board shall be to—

(a) advise the State Government, subject to any guidelines issued by the Central Government, on matters relating to the conservation of biodiversity, sustainable use of its components and equitable sharing of the benefits arising out of the utilization of biological resources;

(b) regulate by granting of approvals or otherwise requests for commercial utilization or bio-survey and bio-utilization of any biological resource by Indians;

(c) perform such other functions as may be necessary to carry out the provisions of this Act or as may be prescribed by the State Government.

Power of State Biodiversity Board to restrict certain activities

24. (1) Any citizen of India or a body corporate, organization or association registered in India intending to undertake any activity referred to in Section 7 shall give prior intimation in such form as may be prescribed by the State Government to the State Biodiversity Board.

(2) On receipt of an intimation under Sub-section (1) the State Biodiversity Board may, in consultation with the local bodies concerned and after making such enquires as in conservation, may deem fit by order, prohibit or restrict any such activity if it is of opinion that such activity is detrimental or contrary to the objectives of conservation and sustainable use of biodiversity or equitable sharing of benefits arising out of such activity.

Provided that no such order shall be made without giving an opportunity of being heard to the person affected.

(3) Any information given in the form referred to in Sub-section (1) for prior intimation shall be kept confidential and shall not be disclosed, either intentionally or unintentionally, to any person not concerned thereto.

Provisions of Sections 9 to 17 to apply with modifications to State Biodiversity Board

25. The provisions of Sections 9 to 17 shall apply to a State Biodiversity Board and shall have effect subject to the following modifications, namely:

(a) references to the Central Government shall be construed as references to the State Government;

(b) references to the National Biodiversity Authority shall be construed as references to the State Biodiversity Board;
provisions of Sections 9 to 17 to apply with modifications to State Biodiversity Board.

(c) reference to the Consolidated Fund of India shall be construed as reference to the Consolidated Fund of the State.

Chapter VII

FINANCE, ACCOUNTS AND AUDIT OF NATIONAL BIODIVERSITY AUTHORITY

Grants or loans by the Central Government

26. The Central Government may, after due appropriation made by Parliament by law in this behalf, pay to the National Biodiversity Authority by way of grants or loans such sums of money as the Central Government may think fit for being utilized for the purposes of this Act.

Constitution of National Biodiversity Fund

27. (1) There shall be constituted a Fund to be called the National Biodiversity Fund and there shall be credited thereto—

(a) any grants and loans made to the National Biodiversity Authority under Section 26.
(b) all charges and royalties received by the National Biodiversity Authority under this Act; and
(c) all sums received by the National Biodiversity Authority from such other sources as may be decided upon by the Central Government.

(2) The Fund shall be applied for—
(a) channeling benefits to the benefit claimers;
(b) conservation and promotion of biological resources and development of areas from where such biological resources or knowledge associated thereto has been accessed;
(c) socio-economic development of areas referred to in clause (b) in consultation with the local bodies concerned.

Annual Report of National Biodiversity Authority

28. The National Biodiversity Authority shall prepare, in such form and at such time each financial year as may be prescribed, its annual report, giving a full account of its activities during the previous financial year and furnish, to the Central Government, before such date as may be prescribed, its audited copy of accounts together with auditors' report thereon.

Budget accounts and Audit

29. (1) The National Biodiversity Authority shall prepare a budget, maintain proper accounts and other relevant records (including the accounts and other relevant records of the National Biodiversity Fund) and prepare an annual statement of account in such form as may be prescribed by the Central Government in consultation with the Comptroller and Auditor-General of India.

(2) The accounts of the National Biodiversity Authority shall be audited by the Comptroller and Auditor-General of India at such intervals as may be specified by him and any expenditure incurred in connection with such audit shall be payable by the National Biodiversity Authority to the Comptroller and Auditor-General of India.

(3) The Comptroller and Auditor-General of India and any other person appointed by him in connection with the audit of the accounts of the National Biodiversity Authority shall have the same rights and privileges and authority in connection with such audit as the Comptroller and Auditor-General

generally has in connection with the audit of the Government accounts and, in particular, shall have the right to demand the production of books, accounts, connected vouchers and other documents and papers and to inspect any of the offices of the National Biodiversity Authority.

(4) The accounts of the National Biodiversity Authority as certified by the Comptroller and Auditor-General of India or any other person appointed by him in this behalf together with the audit report thereon shall be forwarded annually to the Central Government.

Annual Report to be laid before Parliament

30. The Central Government shall cause the annual report and auditor's report to be laid, as soon as may be after they are received, before each House of Parliament.

Chapter VIII

FINANCE, ACCOUNTS AND AUDIT OF STATE BIODIVERSITY BOARD

Grants of money by State Government to State Biodiversity Board

31. The State Government may, after due appropriation made by the State Legislature by law in this behalf, pay to the State Biodiversity Board by way of grants or loans such sums of money as the State Government may think fit for being utilized for the purposes of this Act.

Constitution of State Biodiversity Fund

32. (1) There shall be constituted a Fund to be called the State Biodiversity Fund and there shall be credited thereto—

(a) any grants and loans made to the State Biodiversity Board under Section 31;
(b) any grants or loans made by the National Biodiversity Authority;
(c) all sums received by the State Biodiversity Board from such other sources as may be decided upon by the State Government.

(2) The State Biodiversity Fund shall be applied for—

(a) the management and conservation of heritage sites;
(b) compensating or rehabilitating any section of the people economically affected by notification under Sub-section (1) of Section 37;
(c) conservation and promotion of biological resources;
(d) socio-economic development of areas from where such biological resources or knowledge associated thereto has been accessed subject to any order made under Section 24, in consultation with the local bodies concerned;
(e) meeting the expenses incurred for the purposes authorized by this Act.

Annual Report of State Biodiversity Board

33. The State Biodiversity Board shall prepare, in such form and at such time in each financial year as may be prescribed, its annual report, giving a full account of its activities during the previous financial year, and submit a copy thereof to the State Government.

Audit of accounts of State Biodiversity Board

34. The accounts of the State Biodiversity Board shall be maintained and audited in such manner as may, in consultation with the Accountant-General of the State, be prescribed and the State Biodiversity Board shall furnish, to the State Government, before such date as may be prescribed, its audited copy of accounts together with auditor's report thereon.

Annual Report of State Biodiversity Board to be laid before State Legislature

35. The State Government shall cause the annual report and auditor's report to be laid, as soon as may be after they are received, before the House of State Legislature.

Chapter IX

DUTIES OF THE CENTRAL AND THE STATE GOVERNMENTS

Central Government to develop National strategies, plans, etc., for conservation, etc., of biological diversity

36. (1) The Central Government shall develop national strategies, plans, programmes for the conservation and promotion and sustainable use of biological diversity including measures for identification and monitoring of areas rich in biological resources, promotion of *in situ*, and *ex situ*, conservation of biological resources, incentives for research, training and public education to increase awareness with respect to biodiversity.

(2) Where the Central Government has reason to believe that any area rich in biological diversity, biological resources and their habitats is being threatened by overuse, abuse or neglect, it shall issue directives to the concerned State Government to take immediate ameliorative measures, offering such State Government any technical and other assistance that is possible to be provided or needed.

(3) The Central Government shall, as far as practicable wherever it deems appropriate, integrate the conservation, promotion and sustainable use of biological diversity into relevant sectoral or cross-sectoral plans, programmes and policies.

(4) The Central Government shall undertake measures—

(i) wherever necessary, for assessment of environmental impact of that project which is likely to have adverse effect on biological diversity, with a view to avoid or minimize such effects and where appropriate provide for public participation in such assessment:

(ii) to regulate, manage or control the risks associated with the use and release of living modified organisms resulting from biotechnology likely to have adverse impact on the conservation and sustainable use of biological diversity and human health.

(5) The Central Government shall endeavour to respect and protect the knowledge of local people relating to biological diversity, as recommended by the National Biodiversity Authority through such measures, which may include registration of such knowledge at the local, State or national levels, and other measures for protection, including *sui generis* system.

Explanation— For the purposes of this Section—

(a) "*ex situ* conservation" means the conservation of components of biological diversity outside their natural habitats;

(b) to regulate, manage or control the risks associated with the use and release of living modified organisms resulting from biotechnology likely to have adverse impact on the conservation and sustainable use of biological diversity and human health.

(5) The Central Government shall endeavour to respect and protect the knowledge of local people relating to biological diversity, and recommended by the National Biodiversity Authority through such measures, which may include registration of such knowledge at the local, State or national levels, and other measures for protection, including *sui generis* system.

Explanation— For the purposes of this Section—

(a) "*ex situ* conservation" means the conservation of components of biological diversity outside their natural habitats;

(b) "*in situ* conservation" means the conservation of ecosystems and natural habitats and the maintenance and recovery of viable populations of species in their natural surroundings and, in the case of domesticated or cultivated species, in the surroundings where they have developed their distinctive properties.

Biodiversity heritage sites

37. (1) Without prejudice to any other law for the time being in force, the State Government may, from time to time in consultation with the local bodies, notify in the Official Gazette, areas of biodiversity importance as biodiversity heritage sites under this Act.

(2) The State Government, in consultation with the Central Government, may frame rules for the management and conservation of all the herritage sites.

(3) The State Government shall frame schemes for compensating or rehabilitating any person or section of people economically affected by such notification.

Power of Central Government to notify threatened species

38. Without prejudice to the provisions of any other law for the time being in force, the Central Government, in consultation with the concerned State Government, may from time to time notify any species which is on the verge of extinction or likely to become extinct in the near future as a threatened species and prohibit or regulate collection thereof for any purpose and take appropriate steps to rehabilitate and preserve those species.

Power of Central Government to designate repositories

39. (1) The Central Government may, in consultation with the National Biodiversity Authority, designate institutions as repositories under this Act for different categories of biological resources.

(2) The repositories shall keep in safe custody the biological material including voucher specimens deposited with them.

(3) Any new taxon discovered by any person shall be notified to the repositories or any institution designated for this purpose and he shall deposit the voucher specimens with such repository or institution.

Power of Central Government to exempt certain biological resources

40. Notwithstanding anything contained in this Act, the Central Government may, in consultation with the Nationl Biodiversity Authority, by notification in the Official Gazette, declare that the provisions of this Act shall not apply to any items, including biological resources, normally traded as commodities.

Chapter X

BIODIVERSITY MANAGEMENT COMMITTEES

Constitution of Biodiversity Management Committee

41. (1) Every local body shall constitute a Biodiversity Management Committee within its area for the purpose of promoting conservation, sustainable use and documentation of biological diversity including preservation of habitats, conservation of land races, folk varieties and cultivars, domesticated stocks and breeds of animals and microorganisms and chronicling of knowledge relating to biological diversity.

Explanation— For the purposes of this Sub-section—

(a) "cultivar" means a variety of plant that has originated and persisted under cultivation or was specifically bred for the purpose of cultivation;

(b) "folk variety" means a cultivated variety of plant that was developed, grown and exchanged informally among farmers;

(c) "landrace" means primitive cultivar that was grown by ancient farmers and their successors.

(2) The National Biodiversity Authority and the State Biodiversity Boards shall consult the Biodiversity Management Committees while taking any decision relating to the use of biological resources and knowledge associated with such resources occurring within the territorial jurisdiction of the Biodiversity Management Committee.

(3) The Biodiversity Management Committees may levy charges by way of collection fees from any person for accessing or collecting any biological resource for commercial purposes from areas falling within its territorial jurisdiction.

Chapter XI

LOCAL BIODIVERSITY FUND

Grants to Local Biodiversity Fund

42. The State Government may, after due appropriation made by State legislature by law in this behalf, pay to the Local Biodiversity Funds by way of grants or loans such sums of money as the State Government may think fit for being utilized for the purposes of this Act.

Constitution of Local Biodiversity Fund

43. (1) There shall be constituted a Fund to be called the Local Biodiversity Fund at every area notified by the State Government where any institution of self-government is functioning and there shall be credited thereto—

(a) any grants and loans made under Section 42;

(b) any grants or loans made by the National Biodiversity Authority;

(c). any grants or loans made by the State Biodiversity Boards;

(d) fees referred to in Sub-section (3) of Section 41 received by the Biodiversity Management Committees;

(e) all sums received by the Local Biodiversity Fund from such other sources as may be decided upon by the State Government.

Application of Local Biodiversity Fund

44. (1) Subject to the provisions of Sub-section (2), the management and the custody of the Local Biodiversity Fund and the purposes for which such Fund shall be applied, be in the manner as may be prescribed by the State Government.

(2) The Fund shall be used for conservation and promotion of biodiversity in the areas falling within the jurisdiction of the concerned local body and for the benefit of the community in so far such use is consistent with conservation of biodiversity.

Annual Report of Biodiversity Management Committees

45. The person holding the custody of the Local Biodiversity Fund shall prepare, in such form and during each financial year at such time as may be prescribed, its annual report, giving a full account of its activities during the previous financial year, and submit a copy thereof to the concerned local body.

Audit of accounts of Biodiversity Management Committees

46. The accounts of the Local Biodiversity Fund shall be maintained and audited in such manner as may, in consultation with the Accountant-General of the State be prescribed and the person holding the custody of the Local Biodiversity Fund shall furnish, to the concerned local body, before such date as may be prescribed, its audited copy of accounts together with auditor's report thereon.

Annual report etc. of the Biodiversity Management Committee to be submitted to District Magistrate

47. Every local body constituting a Biodiversity Management Committee under Sub-section (1) of Section 41 shall cause the annual report and audited copy of accounts together with auditor's report thereon referred to in Sections 45 and 46, respectively, and relating to such Committee to be submitted to the District Magistrate having jurisdiction over the area of the local body.

Chapter XII

MISCELLANEOUS

National Biodiversity Authority to be bound by the directions given by Central Government

48. (1) Without prejudice to the foregoing provisions of this Act, the National Biodiversity Authority shall, in the discharge of its functions and duties under this Act, be bound by such directions on questions of policy as the Central Government may give in writing to it from time to time.

Provided that the National Biodiversity Authority shall, as far as practicable, be given opportunity to express its views before any direction is given under this Sub-section.

(2) The decision of the Central Government whether a question is one of policy or not shall be final.

Power of State Government to give Directions

49. (1) Without prejudice to the foregoing provisions of this Act, the State Biodiversity Board shall, in the discharge of its functions and duties under this Act, be bound by such directions on questions of policy as the State Government may give in writing to it from time to time:

Provided that the State Biodiversity Board shall, as far as practicable, be given an opportunity to express its views before any direction is given under this Sub-section.

(2) The decision of the State Government whether a question is one of policy or not shall be final.

Settlement of disputes between State Biodiversity Boards

50. (1) If a dispute arises between the National Biodiversity Authority and a State Biodiversity Board the said Authority or the Board, as the case may be, may prefer an appeal to the Central Government within such time as may be prescribed.

(2) Every appeal made under Sub-section (1) shall be in such form as may be prescribed by the Central Government.

(3) The procedure for disposing of an appeal shall be such as may be prescribed by the Central Government:

Provided that before disposing of an appeal, the parties shall be given a reasonable opportunity of being heard.

(4) If a dispute arises between the State Biodiversity Boards, the Central Government shall refer the same to the National Biodiversity Authority.

(5) While adjudicating any dispute under Sub-section (4), the National Biodiversity Authority shall be guided by the principles of natural justice and shall follow such procedure as may be prescribed by the Central Government.

(6) The National Biodiversity Authority shall have, for the purposes of discharging its functions under this Section, the same powers as are vested in a civil court under the Code of Civil Procedure, 1908, in respect of the following matters, namely—

(a) summoning and enforcing the attendance of any person and examining him on oath;
(b) requiring the discovery and production of documents;
(c) receiving evidence on affidavits;
(d) issuing commissions for the examination of witnesses or documents;
(e) reviewing its decisions, dismissing an application for default or deciding it *ex parte*;
(f) setting aside any order of dismissal of any application for default or any order passed by it *ex parte*;
(g) any other matter which may be prescribed.

(7) Every proceeding before the National Biodiversity Authority shall be deemed to be a judicial proceeding within the meaning of Sections 193 and 228, and for the purpose of Section 196, of the Indian Penal Code, and the National Biodiversity Authority shall be deemed to be a civil court for all the purposes of Section 195 and Chapter XXVI of the Code of Criminal Procedure, 1973.

Members, officers, etc. of National Biodiversity Authority and State Biodiversity Board deemed to be public servants

51. All members, officers and other employees of the National Biodiversity Authority or the State Biodiversity Board shall be deemed, when acting or purporting to act in pursuance of any of the provisions of this Act, to be public servants within the meaning of Section 21 of the Indian Penal Code.

Appeal

52. Any person, aggrieved by any determination of benefit sharing or order of the National Biodiversity Authority or a State Biodiversity Board under this Act, may file an appeal to the High Court within thirty days from the date of communication to him, of the determination or order of the National Biodiversity Authority or the State Biodiversity Board, as the case may be:

Provided that the High Court may, if it is satisfied that the appellant was prevented by sufficient cause from filing the appeal within the said period, allow it to be filed within a further period not exceeding sixty days.

Execution of determination or order

53. Every determination of benefit sharing or order made by the National Biodiversity Authority or a State Biodiversity Board under this Act or the order made by the High Court in any appeal against any determination or order of the National Biodiversity Authority or a State Biodiversity Board shall, on a certificate issued by any officer of the National Biodiversity Authority or a State Biodiversity Board or the Registrar of the High Court, as the case may be, deemed to be decree of the civil court and shall be executable in the same manner as a decree of that court.

Explanation— For the purposes of this Section and Section 52, the expression "State Biodiversity Board" includes the person or group of persons to whom the powers or functions under Sub-section (2) of Section 22 have been delegated under the proviso to that Sub-section and the certificate relating to such person or group of persons under this Section shall be issued by such person or group of persons, as the case may be.

Protection of action taken in good faith

54. No suit, prosecution or other legal proceedings shall lie against the Central Government or the State Government or any officer of the Central Government or the State Government or any member, officer or employee of the National Biodiversity Authority or the State Biodiversity Board for anything which is in good faith done or intended to he done under this Act or the rules or regulations made thereunder.

Penalties

55. (1) Whoever contravenes or abets the contravention of the provisions of Section 3 or Section 4 or Section 6 shall be punishable with imprisonment for a term which may extend to five years, or with fine which may extend to ten lakh rupees and where the damage caused exceeds ten lakh rupees such fine may commensurate with the damage caused, or with both.

(2) Whoever contravenes or attempts to contravene or abets the contravention of the provisions of Section 7 or any order made under Sub-section (2) of Section 24 shall be punishable with imprisonment for a term which may extend to three years, or with fine which may extend to five lakh rupees, or with both.

Penalty for contravention of directions or orders of Central Government, State Government, National Biodiversity Authority and State Biodiversity Boards

56. If any person contravenes any direction given or order made by the Central Government, the State Government, the National Biodiversity Authority or the State Biodiversity Board for which no punishment has been separately provided under this Act, he shall be punished with a fine which may extend to one lakh rupees, and in case of a second or subsequent offence, with fine which may extend to two lakh rupees and in the case of continuous contravention with additional fine which may extend to two lakh rupees everyday during which the default continues.

Offences by companies

57. (1) Where an offence or contravention under this Act has been committed by a company, every person who at the time the offence or contravention was committed was in charge of, and was responsible to the company for the conduct of the business of the company, as well as the company, shall be deemed to be guilty of the offence or contravention and shall be liable to be proceeded against and punished accordingly.

Provided that nothing contained in this Sub-section shall render any such person liable to any punishment provided in this Act, if he proves that the offence or contravention was committed without his knowledge or that he had exercised all due diligence to prevent the commission of such offence or contravention.

(2) Notwithstanding anything contained in Sub-section (1), where an offence or contravention under this Act has been committed by a company and it is proved that the offence or contravention has been committed with the consent or connivance of, or is attributable to, any neglect on the part of any director, manager, secretary or other officer of the company, such director, manager, secretary or other officer shall also be deemed to be guilty of the offence or contravention and shall be liable to be proceeded against and punished accordingly.

Explanation— For the purposes of this Section—

(a) "company" means any body corporate and includes a firm or other association of individuals; and

(b) "director" in relation to a firm, means a partner in the firm.

Offences to be cognizable and non-bailable

58. The offences under this Act shall be cognizable and non-bailable.

Act to have effect in addition to other Acts

59. The provisions of this Act shall be in addition to, and not in derogation of, the provisions in any other law, for the time being in force, relating to forests or wildlife.

Power of Central Government to give directions to State Government

60. The Central Government may give directions to any State Government as to the carrying into execution in the State of any of the provisions of this Act or of any rule or regulation or order made thereunder.

Cognizance of offences

61. No Court shall take cognizance of any offence under this Act except on a complaint made by—

(a) the Central Government or any authority or officer authorized in this behalf by that Government; or

(b) any benefit claimer who has given notice of not less than thirty days in the prescribed manner, of such offence and of his intention to make a complaint, to the Central Government or the authority or officer authorized as aforesaid.

Power of Central Government to make rules

62. (1) The Central Government may, by notification in the Official Gazette, make rules for carrying out the purposes of this Act.

(2) In particular, and without prejudice to the generality of the foregoing power, such rules may provide for A or any of the following matters, namely—

(a) terms and conditions of service of the Chairperson and members under Section.

(b) powers and duties of the Chairperson under Section 10.

(c) procedure under Sub-section (1) of Section 12 in regard to transaction of business at meetings.

(d) form of application and payment of fees for undertaking certain activities under Sub-section (1) of Section 19;

(e) the form and manner of making an application under Sub-section (2) of Section 19;

(f) form of application and the manner for transfer of biological resource or knowledge under Sub-section (2) of Section 20;

(g) form in which, and the time of each financial year at which the annual report of the National Biodiversity Authority shall be prepared and the date before which its audited copy of accounts together with auditor's report thereon shall be furnished under Section 28;

(h) form in which the annual statement of account shall be prepared under Sub-section (1) of Section 29;

(i) the time within which and the form in which, an appeal may be preferred, the procedure for disposing of an appeal and the procedure for adjudication, under Section 50;

(j) the additional matter in which the National Biodiversity Authority may exercise powers of the civil court under Clause (h) of Sub-section (6) of Section 50;

(k) the manner of giving notice under Clause (b) of Section 61;

(l) any other matter which is to be, or may be, prescribed or in respect of which provision is to be made, by rules.

(3) Every rule made under this Section and every regulation made under this Act shall be laid, as soon as may be after it is made, before each House of Parliament, while it is in session or a total period of thirty days which may be comprised in one session or in two or more successive sessions, and if, before the expiry of the session immediately following the session or the successive sessions aforesaid, both Houses agree in making any modification in the rule or regulation or both Houses agree that the rule or regulation should not be made, the rule or regulation shall thereafter have effect only in such modified form or be of

no effect, as the case may be; so, however, that any such modification or annulment shall be without prejudice to the validity of anything previously done under that rule or regulation.

Power of State Government to make rules

63. (1) The State Government may, by notification in the Official Gazette, make rules for carrying out the purposes of this Act.

(2) In particular, and without prejudice to the generality of the foregoing power, such rules may provide for all or any of the following matters, namely—

(a) the other functions to be performed by the State Biodiversity Board under Clause (c) of Section 23;

(b) the form in which the prior intimation shall be given under Sub-section (1) of Section 24;

(c) the form in which, and the time of each financial year at which, the annual report shall be prepared under Section 33;

(d) the manner of maintaining and auditing the accounts of the State Biodiversity Board and the date before which its audited copy of the accounts together with auditor's report thereon shall be furnished under Section 34;

(e) management and conservation of national heritage sites under Section 37;

(f) the manner of management and custody of the Local Biodiversity Fund and the purposes for which such Fund shall be applied under Sub-section (1) of Section 44;

(g) the form of annual report and the time at which such report shall be prepared during each financial year under Section 45;

(h) the manner of maintaining and auditing the accounts of the Local Biodiversity Fund and the date before which its audited copy of the accounts together with auditor's report thereon shall be furnished under Section 46;

(i) any other matter which is to be, or may be, specified.

(3) Every rule made by the State Government under this Section shall be laid, as soon as may be after it is made, before each House of the State Legislature where it consists of two Houses, or where such Legislature consists of one House, before that House.

Power to remove difficulties

64. The National Biodiversity Authority shall, with the previous approval of the Central Government, by notification in the Official Gazette, make regulations for carrying out the purposes of this Act.

65. (1) If any difficulty arises in giving effect to the provisions of this Act, the Central Government may, by order, not inconsistent with the provisions of this Act, remove the difficulty:

Provided that no such order shall be made after the expiry of a period of two years from the commencement of this Act.

(2) Every order made under this section shall be laid, as soon as may be, after it is made, before each House of Parliament.

43.11. OTHER ALLIED ACTS

43.11.1 THE NATIONAL ENVIRONMENT TRIBUNAL ACT, 1992

Introduction

A Bill to provide for strict liability for damages arising out of any accident occurring while handling any hazardous substance and for the establishment of a National Environment Tribunal for effective and expeditious disposal of cases arising from such accident, with a view to giving relief and compensation for damages to persons, property and the environment and for matters connected therewith or incidental thereto.

Whereas decisions were taken at the United Nations Conference on Environment and Development held at Rio de Janeiro in June 1992, in which India participated, calling upon the States to develop national laws regarding liability and compensation for the victims of pollution and other environmental damages;

And whereas it is considered expedient to implement the decisions of the aforesaid Conference so far as they relate to the protection of environment and payment of compensation for damage to persons, property and the environment while handling hazardous substances.

Be it enacted by Parliament in the Fortythird Year of the Republic of India as follows:

CHAPTER I

PRELIMINARY

1. (a) This Act may be called the *National Environment Tribunal Act, 1992.*

Short Title and Commencement

(b) It shall come into force on such date or dates as the Central Government may, by notification, appoint and different dates may be appointed for different States and any reference in any provision of this Act to the commencement of this Act shall be construed in relation to any State or part thereof as a reference to the coming into force of that provision in that State or part thereof.

Definitions

2. In this Act, unless the context otherwise requires —

(a) "accident" means an accident involving a fortuitous or sudden or unintended occurrence while handling any hazardous substance resulting in continuous or intermittent or repeated exposure to death of or injury to any person or damage to any property or environment but does not include an accident by reason only of war or radioactivity;

(b) "Bench" means a Bench of the Tribunal;

(c) "Chairman" means the Chairman of the Tribunal;

(d) "environment" includes water, air and land and the inter-relationship which exists among and between water, air and land, and human beings, other living creatures, plants, microorganisms and property;

(e) "handling", in relation to any hazardous substance, means the manufacture, processing, treatment, package, storage, transportation by vehicle, use, collection, destruction, conversion, offering for sale, transfer or the like of such hazardous substance;

29 of 1986, 6 of 1991

(f) "hazardous substance" means any substance or preparation which is defined as hazardous substance in the Environment (Protection) Act, 1986, and exceeding such quantity as specified by the Central Government under the Public Liability Insurance Act, 1991;

(g) "Judicial Member" means a Member of the Tribunal appointed as such under this Act and includes the Chairman or a Vice-Chairman who possesses any of the qualifications specified in sub-section (3) of Section 11;

(h) "Member" means a Member (whether Judicial or Technical) of the Tribunal and includes the Chairman and a Vice-Chairman;

(i) "notification" means a notification published in the Official Gazette;

(j) "prescribed" means prescribed by rules made under this Act;

(k) "rules" means the rules made under this Act,

(l) "Technical Member" means a Member of the Tribunal who is not a Judicial Member within the meaning of clause (g);

(m) "Tribunal" means the National Environment Tribunal established under Section 9;

(n) "Vice-Chairman" means the Vice-Chairman of the Tribunal;

Explanation — In the case of the Tribunal having two or more Vice-Chairmen, references to the Vice-Chairman in this Act shall be construed as a reference to each of those Vice-Chairmen; and

(o) "owner" means a person who owns, or has control over handling, any hazardous substance at the time of accident and includes:

(i) in the case of a firm, any of its partners;

(ii) in the case of an association, any of its members; and

(iii) in the case of a company, any of its directors, managers, secretaries or other officers who are directly in charge of and is responsible to the company for the conduct of the business of the company.

CHAPTER II

COMPENSATION FOR DEATH OF, OR INJURY TO A PERSON AND DAMAGE TO PROPERTY AND ENVIRONMENT

Liability to pay compensation in certain cases on principle of no fault

3. (a) Where death of or injury to any person (other than a workman) or damage to any property or environment has resulted from an accident, the owner shall be liable to pay compensation for such death, injury or damage under all or any of the heads specified in the Schedule.

(b) In any claim for compensation under sub-section (1) the claimant shall not be required to plead and establish that the death, injury or damage in respect of which the claim has been made was due to any wrongful act, neglect or default of any person.

Explanation—For the purposes of this section—

(i) "workman" has the meaning assigned to it in the Workmen's Compensation Act, 1923; *8 of 1923*

(ii) "injury" includes permanent total or permanent partial disability or sickness resulting out of an accident.

(c) If the death, injury or damage caused by an accident cannot be attributed to any individual activity but is the combined or resultant effect of several such activities, apportions and processes, the Tribunal may apportion the liability for compensation amongst those responsible for such activities, apportions and processes on an equitable basis.

4. The Central Government may, by notification, exempt from the operation of this Act any owner, namely :

(a) by the person who has sustained the injury;

(b) any State Government;

(c) any corporation owned or controlled by the Central Government or a State Government; or

(d) any local authority.

Application for claim for compensation

5. (a) An application for claim for compensation may be made —

(i) by the person who has sustained the injury;

(ii) by the owner of the property to which the damage has been caused;

(iii) where death has resulted from the accident, by all or any of the legal representatives of the deceased;

(iv) by any agent duly authorised by such person or owner of such property or all or any of the legal representatives of the deceased, as the case may be;

(v) by any representative body or organisation, functioning in the field of environment and recognised in this behalf by the Central Government, under all or any of the heads specified in the Schedule; or

(vi) by the Central Government or a State Government or a local authority under all or any of the heads specified in the Schedule;

Provided that where all the legal representatives of the deceased have not joined in any such application for compensation, the application shall be made on behalf of or for the benefit of all the legal representatives of the deceased and the legal representatives who have not so joined shall be impleaded as respondents to the application.

(b) Any claimant making an application under sub-section (1) may also make an application before the Tribunal for such relief as is provided in the Public Liability Insurance Act, 1991;

6 of 1991

Provided that no such application shall be made if the relief has been received by the claimant earlier or an application made by the claimant to the Collector under the said Act is pending and has not been withdrawn.

6 of 1991

(c) The Tribunal shall have and exercise the same jurisdiction, powers and authority in respect of the matters specified in the Public Liability Insurance Act, 1991, as the Collector has and may exercise and for this purpose the provisions of that Act shall have effect subject to the modification that the references therein to the Collector shall be construed as including a reference to the Tribunal.

(d) Every application under sub-section (1) shall be made to the Tribunal and shall contain such particulars and shall be accompanied by such documents and such fee, not exceeding one thousand rupees, as may be prescribed;

Provided that no fee shall be payable by a person whose annual income is below the prescribed limit or by a representative body or organisation referred to in clause (e) of subsection (1) or by the Central Government, a State Government or a local authority.

(e) No application for compensation shall be entertained unless it is made within five years of the occurrence of the accident.

Procedure and powers of Tribunal

6. (a) On receipt of an application under sub-section (1) of Section 5, the Tribunal may, after such inquiry as it may deem fit, reject the application summarily.

(b) Where the Tribunal does not reject the application under sub-section (1) the Tribunal may, after giving notice of the application to the owner and after giving the parties an opportunity of being heard, hold an inquiry into the claim or each of the claims and may make an award determining the amount of compensation which appears to be just and specifying the person or persons to whom such amount of compensation shall be paid.

5 of 1908

(c) The Tribunal shall not be bound by the procedure laid down by the Code of Civil Procedure, 1908, but shall be guided by the principles of natural justice and subject to the other provisions of this Act and of any rules, the Tribunal shall have power to regulate its own procedure including the fixing of places and times of its inquiry.

5 of 1908

(d) The Tribunal shall have, for the purpose of discharging its functions under this Act, the same powers as are vested in a civil court under the Code of Civil Procedure, 1908, while trying a suit, in respect of the following matters, namely:

(i) summoning and enforcing the attendance of any person and examining him on oath;

(ii) requiring the discovery and production of documents;

(iii) receiving evidence on affidavits;

1 of 1872

(iv) subject to the provisions of Sections 123 and 124 of the Indian Evidence Act, 1872, requisitioning any public record or document or copy of such record or document from any office;

(v) issuing commissions for the examination of witnesses or documents;

(vi) reviewing its decisions;

(vii) dismissing an application for default or deciding it *ex parte*;

(viii) setting aside any order of dismissal of any application for default or any order passed by its *ex parte*; and

(ix) any other matter which is to be or may be prescribed.

Conditions as to making of interim orders

7. Notwithstanding anything contained in any other provisions of this Act or in any other law for the time being in force, no interim order (whether by way of injunction or stay or in any other manner) shall be made on, or in any proceedings relating to, an application unless

(a) copies of such application and of all documents in support of the plea for such interim order are furnished to the party against whom such application is made or proposed to be made; and

(b) opportunity is given to such party to be heard in the matter;

Provided that the Tribunal may dispense with the requirements of clauses (a) and (b) and make an interim order as an exceptional measure if it is satisfied for reasons to be recorded in writing, that it is necessary so to do for preventing any loss or damage being caused to the applicant which cannot be adequately compensated in money but any such interim order shall, if it is not sooner vacated, cease to have effect on the expiry of a period of fourteen days from the date on which it is made unless the said requirements have been complied with before the expiry of that period and the Tribunal has continued the operation of the interim order.

Reduction of amount of relief paid under any other law

8. Where in respect of death of or injury to, any person or damage to any property, the owner, liable to pay compensation under this Act, is also liable to pay any amount as relief under the Public Liability Insurance Act, 1991, or any other compensation under any other law, the amount of compensation payable under this Act shall be reduced by the amount of relief and other compensation paid under any other law.

CHAPTER III

ESTABLISHMENT OF NATIONAL ENVIRONMENT TRIBUNAL AND BENCHES THEREOF

Establishment of National Environment Tribunal

9. The Central Government shall, by notification, establish a Tribunal to be known as the National Environment Tribunal to exercise the jurisdiction, powers and authority conferred on it by or under this Act.

Composition of Tribunal and Benches thereof

10. (1) The Tribunal shall consist of a Chairman and such number of Vice-Chairmen, Judicial Members and Technical Members as the Central Government may deem fit and subject to the other provisions of this Act, the jurisdiction, powers and authority of the Tribunal may be exercised by Benches thereof.

(2) Subject to the other provisions of this Act, a Bench shall consist of one Judicial Member and one Technical Member.

(3) Notwithstanding anything contained in sub-section (1) the Chairman —

(a) may, in addition to discharging the functions of the Judicial Member or the Technical Member of the Bench to which he is appointed, discharge the functions of the Judicial Member or, as the case may be, the Technical Member of any other Bench;

(b) may transfer the Vice-Chairman or other Member from one Bench to another Bench;

(c) may authorise the Vice-Chairman or the Judicial Member or the Technical Member appointed to one Bench to discharge also the functions of the Vice-Chairman or as the case may be, the Judicial Member or the Technical Member of another Bench; and

(d) may, for the purpose of securing that any case or cases which, having regard to the nature of the questions involved, requires or require, in his opinion or under the rules made by the Central Government in this behalf, to be decided by a Bench composed of more than two members, issue such general or special orders, as he may deem fit;

Provided that every Bench constituted in pursuance of this clause shall include at least one Judicial Member and one Technical Member.

(4) Notwithstanding anything contained in the foregoing provisions of this section, it shall be competent for the Chairman or any other Member authorised by the Chairman in this behalf to function as a Bench consisting of a single Member and exercise the jurisdiction, powers and authority of the Tribunal in respect of such classes of cases or such matters pertaining to such classes of cases as the Chairman may, by general or special order, specify;

Provided that if at any stage of the hearing of any such case or matter, it appears to the Chairman or such Member that the case or matter is of such a nature that it ought to be heard by a Bench consisting of two Members, the case or matter may be transferred by the Chairman or as the case may be, referred to him for transfer to such Bench as the Chairman may deem fit.

(5) Subject to the other provisions of this Act, the Benches of the Tribunal shall ordinarily sit at New Delhi (which shall be known as the principal Bench) and at such other places as the Central Government may, by notification, specify.

Qualifications for appointment as Chairman, Vice-Chairman or other Member

11. (1) A person shall not be qualified for appointment as Chairman unless he —

(a) is, or has been, or is qualified to be a Judge of a High Court; or

(b) has, for at least two years, held the office of Vice-Chairman.

(2) a person shall not be qualified for appointment as the Vice-Chairman unless he —

(a) is, or has been, a Judge of a High Court; or

(b) has, for at least two years, held the post of a Secretary to the Government of India or any other post under the Central or a State Government carrying a scale of pay which is not less than that of a Secretary to the Government of India; or

(c) (i) has, for at least five years, held the post of an Additional Secretary to the Government of India or any other post under the Central or a State Government carrying a scale of pay which is not less than that of an Additional Secretary to the Government of India; and

(ii) has adequate knowledge of, or experience in, legal, administrative, scientific or technical aspects of the problems relating to environment; or

(d) has, for at least three years, held office as a Judicial Member or a Technical Member.

(3) A person shall not be qualified for appointment as a Judicial Member unless he —

(a) is, or has been, or is qualified to be a Judge of a High Court; or

(b) has been a member of the Indian Legal Service and has held a post in Grade I of that Service for at least three years.

(4) a person shall not be qualified for appointment as a Technical Member unless he has adequate knowledge of, or experience in, or capacity to deal with, administrative, scientific or technical aspects of the problems relating to environment.

(5) Subject to the provisions of the sub-sections (6) and (7); the Chairman, Vice-Chairman and every other Member of the Tribunal shall be appointed by the President.

(6) No appointment of a person possessing the qualifications specified in this section as the Chairman or the Vice-Chairman shall be made except after consultation with the Chief Justice of India.

(7) No appointment of a person as a Judicial Member or a Technical Member shall be made except on the recommendation of a Selection Committee appointed by the Central Government consisting of the following, namely:

(a) Chairman of the Tribunal—Chairman of the Committee *ex-officio*;

(b) Secretary to the Government of India in the Ministry of Environment and Forests; — member *ex-officio*;

(c) Secretary to the Government of India in the Ministry of Law, Justice and Company Affairs (Department of Legal Affairs); — member *ex-officio*;

(d) Director-General, Council of Scientific and Industrial Research; — member *ex-officio*;

(e) an Environmentalist to be nominated by the Central Government — member.

Vice-Chairman to act as Chairman or to discharge his functions in certain circumstances

12. (1) In the event of the occurrence of any vacancy in the office of the Chairman by reason of his death, resignation or otherwise, the Vice-Chairmen or, as the case may be, such one of the Vice-Chairmen, as the Central Government may by notification, authorise in this behalf, shall act as the Chairman until the date on which a new Chairman, appointed in accordance with the provisions of this Act to fill such vacancy, enters upon his office.

(2) When the Chairman is unable to discharge his functions owing to absence, illness or any other cause, the Vice-Chairman or, as the case may be, such one of the Vice-Chairmen, as the Central

Government may by notification, authorise in this behalf, shall discharge the functions of the Chairman until the date on which the Chairman resumes his duties.

Term of office

13. The Chairman, Vice-Chairman and other Members shall hold office as such for a term of five years from the date on which he enters upon his office, but shall be eligible for re-appointment for another term of five years;

Provided that no Chairman, Vice-Chairman, or other Member shall hold office as such after he has attained—

(a) in the case of the Chairman, the age of seventy years;

(b) in the case of the Vice-Chairman, the age of sixty-five years; and

(c) in the case of any other Member, the age of sixty-two years.

Resignation and removal

14. The Chairman, Vice-Chairman or other Member, may, by notice in writing under his hand addressed to the President, resign his office:

Provided that the Chairman, Vice-Chairman or other Member shall, unless he is permitted by the President to relinquish his office sooner, continue to hold office until the expiry of three months from the date of receipt of such notice or until a person duly appointed as his successor enters upon his office or until the expiry of his term of office, whichever is the earliest.

(2) The Chairman, Vice-Chairman or any other Member shall not be removed from his office except by an order made by the President on the ground of proved misbehaviour or incapacity after an inquiry made by a Judge of the Supreme Court in which such Chairman, Vice-Chairman or other Member had been informed of the charges against him and given a reasonable opportunity of being heard in respect of those charges.

(3) The Central Government may, by rules, regulate the procedure for the investigation of misbehaviour or incapacity of the Chairman, Vice-Chairman or other Members referred to in sub-section (2).

Salaries and allowances and other terms and conditions of service of Chairman, Vice-Chairman and other Members

15. The salaries and allowances payable to and the other terms and conditions of service (including pension, gratuity and other retirement benefits) of the Chairman, Vice-Chairman and other Members shall be such as may be prescribed;

Provided that neither the salary and allowances nor the other terms and conditions of service of the Chairman, Vice-Chairman or other Members shall be varied to his disadvantage after his appointment.

Provision as to the holding of offices by Chairman, etc. on ceasing to be such Chairman, etc.

16. On ceasing to hold office —

(a) the Chairman of the Tribunal shall be ineligible for further employment either under the Government of India, or under the Government of a State;

(b) the Vice-Chairman of the Tribunal shall, subject to the other provisions of this Act, be eligible for appointment as the Chairman of the Tribunal, but not for any other employment either under the Government of India or under the Government of a State;

(c) a Member (other than the Chairman or Vice-Chairman) of the Tribunal shall, subject to the other provisions of this Act, be eligible for appointment as the Chairman or Vice-Chairman of the Tribunal or as the Chairman, Vice-Chairman or Member of any other Tribunal but not for any other employment either under the Government of India or under the Government of a State;

(d) the Chairman, Vice-Chairman or other Members shall not appear, act or plead before the Tribunal.

Explanation: For the purposes of this section, employment under the Government of India or under the Government of a State includes employment under any local or other authority within the territory of India or under the control of the Government of India or under any corporation or society owned or controlled by the Government.

Financial and administrative powers of Chairman

17. The Chairman shall exercise such financial and administrative powers over the Benches as may be vested in him under the rules;

Provided that the Chairman shall have authority to delegate such of his financial and administrative powers as he may think fit to the Vice-Chairman or any other officer of the Tribunal, subject to the condition that the Vice-Chairman or such officer shall, while exercising such delegated powers, continue to act under the direction, control and supervision of the Chairman.

Staff of the Tribunal

18. (1) The Central Government shall determine the nature and categories of the officers and other employees required to assist the Tribunal in the discharge of its functions and provide the Tribunal with such officers and other employees as it may think fit;

(2) The officers and other employees of the Tribunal shall discharge their functions under the general superintendence of the Chairman.

(3) The salaries and allowances and conditions of service of the officers and other employees of the Tribunal shall be such as may be prescribed.

Distribution of business amongst the Benches

19. (1) Where any Benches of the Tribunal are constituted, the Central Government may, from time to time, by notification, make provisions as to the distribution of the business of the Tribunal amongst the Benches and specify the matters which may be dealt with by each Bench;

(2) If any question arises as to whether any matter falls within the purview of the business allocated to a Bench, the decision of the Chairman shall be final.

Explanation: For the removal of doubts, it is hereby declared that the expression "matters" includes applications for interim relief.

CHAPTER IV

JURISDICTION AND PROCEEDINGS OF THE TRIBUNAL

Bar of jurisdiction

20. On and from the commencement of this Act, no court or other authority except the Tribunal shall have, or be entitled to exercise, any jurisdiction, powers or authority to entertain any application or action for any claim for compensation which may be entertained or dealt with by the Tribunal.

Power of Chairman to transfer cases from one Bench to another

21. On the application of any of the parties and after notice to the parties, and after hearing such of them as he may desire to be heard, or on his own motion without such notice, the Chairman may transfer any case pending before one Bench for disposal to any other Bench.

Decision to be taken by majority

22. If the Members of a Bench differ in opinion on any point, the point shall be decided according to the opinion of the majority, if there is a majority, but if the Members are equally divided, they shall state the point or points on which they differ and make a reference to the Chairman who shall either hear the point or points himself or refer the case for hearing on such point or points by one or more of the other Members and such point or points shall be decided according to the opinion of the majority of the Members who have heard the case including those who first heard it.

Deposit amount payable for damage to environment

23. (1) Where any amount of compensation is ordered to be paid under any award by the Tribunal on the ground of any damage to environment, that amount shall be remitted to the authority specified under sub-section (3) of Section 7A of the Public Liability Insurance Act, 1991, for being credited to the Environmental Relief Fund established under that section;

6 of 1991

(2) The amount of compensation credited to the Environmental Relief Fund under sub-section (1) may be utilised by such person or authority, in such manner and for such purposes of environment as may be prescribed.

Execution of award or order of Tribunal

24. (1) An award made by the Tribunal under this Act shall be executable by the Tribunal as a decree of Civil Court and for this purpose, the Tribunal shall have all the powers of a Civil Court;

(2) Notwithstanding anything contained in sub-section (1), the Tribunal may transmit to the Collector having jurisdiction over the area in which the accident has occurred the copy of the order made by it for payment of relief as provided in the Public Liability Insurance Act, 1991, and the Collector shall execute the order in the same manner as if it were an order made by him under that Act.

6 of 1991

(3) Where the owner against whom the award or order is made by the Tribunal fails to make the payment or deposit the amount as directed by the Tribunal within the period specified in the award or order, such amount shall be recoverable from the owner as arrears of land revenue or of public demand.

Appeals

25. (1) Save as provided in sub-section (2) and notwithstanding anything contained in the Code of Civil Procedure, 1908, or in any other law, an appeal shall lie against any award or other order, not being an interlocutory order of the Tribunal to the Supreme Court on one or more of the grounds specified in Section 100 of that Code.

5 of 1908

(2) No appeal shall lie against an award or other order made by the Tribunal with the consent of the parties;

(3) Every appeal under this section shall be preferred within a period of ninety days from the date of the award or other order appealed against;

Provided that no appeal by the person who is required to pay any amount in terms of such award shall be entertained by the Supreme Court unless he has deposited with it the amount so awarded in the manner directed by the Supreme Court;

Provided further that the Supreme Court may entertain the appeal after the expiry of the said period of ninety days, if it is satisfied that the appellant was prevented by sufficient cause from preferring the appeal in time.

CHAPTER V

MISCELLANEOUS

Penalty for failure to comply with orders of Tribunal

26. Whoever fails to comply with any order made by the Tribunal, he shall be punishable with imprisonment for a term which may extend to three years, or with fine which may extend to ten lakh rupees, or with both.

Offences by companies

27. (1) Where any offence under this Act has been committed by a company, every person who, at the time the offence was committed, was directly in charge of and was responsible to, the company for the conduct of the business of the company, as well as the company, shall be deemed to be guilty of the offence and shall be liable to be proceeded against and punished accordingly;

Provided that nothing contained in this sub-section shall render any such person liable to any punishment provided in this Act, if he proves that the offence was committed without his knowledge or that he exercised all due diligence to prevent the commission of such offence;

(2) Notwithstanding anything contained in such-section (1) where an offence under this Act has been committed by a company and it is proved that the offence has been committed with the consent or connivance of, or is attributable to any neglect on the part of, any director, manager, secretary or other officer of the company, such director, manager, secretary or other officer shall also be deemed to be guilty of that offence and shall be liable to be proceeded against and punished accordingly.

Explanation—For the purposes of this section —

(a) "company" means any body corporate and includes a firm or other association of individuals; and

(b) "director", in relation to a firm, means a partner in the firm.

Proceedings before the Tribunal to be judicial proceedings

28. All proceedings before the Tribunal shall be deemed to be judicial proceedings within the meaning of Sections 193, 219 and 228 of the Indian Penal Code.

45 of 1860

Members and staff of Tribunal to be public servants

29. The Chairman, Vice-Chairman and other Members and the officers and other employees of the Tribunal shall be deemed to be public servants within the meaning of Section 21 of the Indian Penal Code.

45 of 1860

Protection of action taken in good faith

30. No suit, prosecution or other legal proceeding shall lie against the Central Government or against the Chairman, Vice-Chairman or other Member of the Tribunal or any other person authorised by the Chairman, Vice-Chairman or other Member for anything which is in good faith done or intended to be done in pursuance of this Act or any rule or order made thereunder.

6 of 1991

Act to have overriding effect

31. Save as provided in the Public Liability Insurance Act, 1991, the provisions of this Act shall have effect notwithstanding anything inconsistent therewith contained in any other law for the time being in force or in any instrument having effect by virtue of any law other than this Act.

Power to make rules

32. (1) The Central Government may, by notification, make rules for carrying out the purpose of this Act.

(2) In particular and without prejudice to the generality of the foregoing power, such rules may provide for all or any of the following matters, namely:

(a) the particulars which an application shall contain, the documents and the fee which shall be accompanied with it and the limit of annual income of a person so as to enable him to make application without paying any fee, under sub-section (4) of Section 5;

(b) any such matter in respect of which the Tribunal shall have powers of a civil court, under clause (i) of sub-section (4) of Section 6;

(c) the case or cases which, having regard to the nature of the questions involved, requires or require to be decided by a bench of more than two members, under clause (d) of sub-section 3 of Section 10;

(d) procedure for the investigation of misbehaviour or incapacity of the Chairman, Vice-Chairman or other Members of the Tribunal under sub-section (3) of Section 14;

(e) the salaries and allowances payable to, and the other terms and conditions of service of, the Chairman, Vice-Chairman and other Members under Section 15;

(f) financial and administrative powers of the Chairman over the Benches under Section 17;

(g) the salaries and allowances and conditions of service of the officers and other employees of the Tribunal under sub-section (3) of Section 18;

(h) the person or the authority by whom, the manner in which and the purposes of environment for which the amount of compensation credited to the Environmental Relief Fund shall be utilised under sub-section (2) of Section 23; and

(i) any other matter which is required to be, or may be prescribed.

(3) Every notification issued under Section 4 and every rule made under other provisions of this Act shall be laid, as soon as may be after it is made, before each House of Parliament, while it is in session for a total period of thirty days which may be comprised in one session or in two or more successive sessions and if, before the expiry of the session immediately following the session or the successive sessions aforesaid, both Houses agree in making any modification in the notification or the rule or both Houses agree that the notification or the rule should not be made, the notification.

THE SCHEDULE
[See Section 3(1)]

Heads under which compensation for damages may be claimed

(a) Death;

(b) Permanent, temporary, total or partial disability or other injury or sickness;

(c) Loss of wages due to total or partial disability or permanent or temporary disability;

(d) Medical expenses incurred for treatment of injuries or sickness;

(e) Damages to private property;

(f) Expenses incurred by the Government or any local authority in providing relief, aid and rehabilitation to the affected person(s) or the rule shall thereafter have effect only in such modified form or be of no effect, as the case may be; so, however, that any such modification or annulment shall be without prejudice to the validity of anything previously done under that notification or the rule.

(g) Expenses incurred by Government for any administrative or legal action or to cope with any harm or damage, including compensation for environmental degradation and restoration of the quality of environment;

(h) Loss to Government or local authority arising out of, or connected with, the activity causing any damage;

(i) Claims on account of any harm, damage or destruction to flora including milch and draught animals and aquatic fauna;

(j) Claims on account of any harm, damage or destruction to flora including aquatic flora, crops, vegetables, trees and orchards;

(k) Claims including cost of restoration on account of any harm or damage to environment including pollution of soil, air, water, land and eco-systems;

(l) Loss and destruction of any property other than private property;

(m) Loss of business or employment or both;

(n) Any other claim arising out of, or connected with, any activity of handling of hazardous substance.

Statement of Objects and Reasons

1. The principle of strict civil liability in accident cases arising from the activities involving hazardous substances has been highlighted in a case by the Supreme Court of India. An enterprise engaged in activities with potential threat to the health and safety of the persons residing in the surrounding areas of the factory owes an absolute duty to the community to ensure that no harm is caused to anyone on account of hazardous and inherently dangerous nature of such activities. Cases seeking compensation for damages to human health, property and the environment, particularly contamination of sub-surface water, are increasing. There is also an increasing trend in the number of industrial disasters.

2. The United Nations Conference on Environment and Development held at Rio de Janeiro in June 1992, in which India participated, has also called upon the States to develop National laws regarding liability and compensation for the victims of pollution and other environmental damage.

3. It is deemed expedient to develop and codify the principle of strict civil liability in respect of all such cases where damage is caused while handling hazardous substances. It is proposed to establish a National Environmental Tribunal for effective and expeditious disposal of cases arising from certain industrial accidents and disasters with a view of providing effective and expeditious relief and compensation for damages to human health, property and the environment.

4. The Bill seeks to achieve the aforesaid objects.

New Delhi — KAMAL NATH

The 5th August, 1992.

President's Recommendation Under Article 117 of the Constitution of India

[Copy of letter No. 8/6/91PL (CPW), dated 6th August, 1992, from Shri Kamal Nath, Minister of State of the Ministry of Environment and Forests to the Secretary-General, Lok Sabha]

The President, having been informed of the proposed Bill to provide for effective and expeditious relief, restitution and compensation for damages to human health, property and the environment to the persons affected by the accidents occurring while handling hazardous substance and for matters connected therewith or incidental thereto, recommends under Article 117(1) and (3) of the Constitution of India the introduction and consideration of the National Environment Tribunal Bill, 1992 by Lok Sabha.

43.11.2 THE SCHEDULED TRIBES AND OTHER TRADITIONAL FOREST DWELLERS (RECOGNITION OF FOREST RIGHTS) ACT, 2006

ACT NO. 2 OF 2007

[29th December, 2006.]

An Act to recognise and vest the forest rights and occupation in forest land in forest dwelling Scheduled Tribes and other traditional forest dwellers who have been residing in such forests for generations but whose rights could not be recorded; to provide for a framework for recording the forest rights so vested and the nature of evidence required for such recognition and vesting in respect of forest land.

WHEREAS the recognised rights of the forest dwelling Scheduled Tribes and other traditional forest dwellers include the responsibilities and authority for sustainable use, conservation of biodiversity and maintenance of ecological balance and thereby strengthening the conservation regime of the forests while ensuring livelihood and food security of the forest dwelling Scheduled Tribes and other traditional forest dwellers;

AND WHEREAS the forest rights on ancestral lands and their habitat were not adequately recognised in the consolidation of State forests during the colonial period as well as in independent India resulting in historical injustice to the forest dwelling Scheduled Tribes and other traditional forest dwellers who are integral to the very survival and sustainability of the forest ecosystem;

AND WHEREAS it has become necessary to address the long standing insecurity of tenurial and access rights of forest dwelling Scheduled Tribes and other traditional forest dwellers including those who were forced to relocate their dwelling due to State development interventions.

BE it enacted by Parliament in the Fifty-seventh Year of the Republic of India as follows:–

CHAPTER 1

PRELIMINARY

1. **Short title, extent and commencement:**

(1) This Act may be called the Scheduled Tribes and Other Traditional Forest Dwellers (Recognition of Forest Rights) Act, 2006.

(2) It extends to the whole of India except the State of Jammu and Kashmir.

(3) It shall come into force on such date[1] as the Central Government may, by notification in the Official Gazette, appoint.

2. **Definitions:** In this Act, unless the context otherwise requires –

(a) "community forest resource" means customary common forest land within the traditional or customary boundaries of the village or seasonal use of landscape in the case of pastoral communities, including reserved forests, protected forests and protected areas such as Sanctuaries and National Parks to which the community had traditional access;

(b) "critical wildlife habitat" means such areas of National Parks and Sanctuaries where it has been specifically and clearly established, case by case, on the basis of scientific and objective criteria, that such areas are required to be kept as inviolate for the purposes of wildlife conservation as may be determined and notified by the Central Government in the Ministry of Environment and Forests after open process of consultation by an Expert Committee, which includes experts from the locality appointed by that Government wherein a representative of the Ministry of Tribal Affairs shall also be included, in determining such areas according to the procedural requirements arising from subsections (1) and (2) of section 4;

(c) "forest dwelling Scheduled Tribes" means the members or community of the Scheduled Tribes who primarily reside in and who depend on the forests or forest lands for bona fide livelihood needs and includes the Scheduled Tribe pastoralist communities;

(d) "forest land" means land of any description falling within any forest area and includes unclassified forests, undemarcated forests, existing or deemed forests, protected forests, reserved forests, Sanctuaries and National Parks;

(e) "forest rights" means the forest rights referred to in section 3;

(f) "forest villages" means the settlements which have been established inside the forests by the forest department of any State Government for forestry operations or which were converted into forest villages through the forest reservation process and includes forest settlement villages, fixed demand holdings, all types of *taungya* settlements, by whatever name called, for such villages and includes lands for cultivation and other uses permitted by the Government;

1. 31st December, 2007, vide notification No. S. O. 2224(E), dated 31st December, 2007, see Gazette of India Extraordinary, Part II, sec 3(ii)

(g) "Gram Sabha" means a village assembly which shall consist of all adult members of a village and in case of States having no Panchayats, Padas, Tolas and other traditional village institutions and elected village committees, with full and unrestricted participation of women;

(h) "habitat" includes the area comprising the customary habitat and such other habitats in reserved forests and protected forests of primitive tribal groups and pre-agricultural communities and other forest dwelling Scheduled Tribes;

(i) "minor forest produce" includes all non-timber forest produce of plant origin including bamboo, brush wood, stumps, cane, tussar, cocoons, honey, wax, lac, tendu or kendu leaves, medicinal plants and herbs, roots, tubers and the like;

(j) "nodal agency" means the nodal agency specified in section 11;

(k) "notification" means a notification published in the Official Gazette;

(l) "prescribed" means prescribed by rules made under this Act;

(m) "Scheduled Areas" means the Scheduled Areas referred to in clause (1) of article 244 of the Constitution;

(n) "sustainable use" shall have the same meaning as assigned to it in clause (o) of section 2 of the Biological Diversity Act, 2002 (18 of 2003);

(o) "other traditional forest dweller" means any member or community who has for at least three generations prior to the 13th day of December, 2005 primarily resided in and who depend on the forest or forests land for bona fide livelihood needs.

Explanation: For the purpose of this clause, "generation" means a period comprising of twentyfive years;

(p) "village" means–

(i) a village referred to in clause (b) of section 4 of the Provisions of the Panchayats (Extension to the Scheduled Areas) Act, 1996 (40 of 1996); or

(ii) any area referred to as a village in any State law relating to Panchayats other than the Scheduled Areas; or

(iii) forest villages, old habitation or settlements and unsurveyed villages, whether notified as village or not; or

(iv) in the case of States where there are no Panchayats, the traditional village, by whatever name called;

(q) "wild animal" means any species of animal specified in Schedules I to IV of the Wild Life (Protection) Act, 1972 (53 of 1972) and found wild in nature.

CHAPTER II

FOREST RIGHTS

3. Forest rights of Forest dwelling Scheduled Tribes and other tradional forest dwellers.

(1) For the purposes of this Act, the following rights, which secure individual or community tenure or both, shall be the forest rights of forest dwelling Scheduled Tribes and other traditional forest dwellers on all forest lands, namely:

(a) right to hold and live in the forest land under the individual or common occupation for habitation or for self-cultivation for livelihood by a member or members of a forest dwelling Scheduled Tribe or other traditional forest dwellers;

(b) community rights such as nistar, by whatever name called, including those used in erstwhile Princely States, Zamindari or such intermediary regimes;

(c) right of ownership, access to collect, use, and dispose of minor forest produce which has been traditionally collected within or outside village boundaries;

(d) other community rights of uses or entitlements such as fish and other products of water bodies, grazing (both settled or transhumant) and traditional seasonal resource access of nomadic or pastoralist communities;

(e) rights including community tenures of habitat and habitation for primitive tribal groups and pre-agricultural communities;

(f) rights in or over disputed lands under any nomenclature in any State where claims are disputed;

(g) rights for conversion of *Pattas* or leases or grants issued by any local authority or any State Government on forest lands to titles;

(h) rights of settlement and conversion of all forest villages, old habitation, unsurveyed villages and other villages in forests, whether recorded, notified or not into revenue villages;

(i) right to protect, regenerate or conserve or manage any community forest resource which they have been traditionally protecting and conserving for sustainable use;

(j) rights which are recognised under any State law or laws of any Autonomous District Council or Autonomous Regional Council or which are accepted as rights of tribals under any traditional or customary law of the concerned tribes of any State;

(k) right of access to biodiversity and community right to intellectual property and traditional knowledge related to biodiversity and cultural diversity;

(l) any other traditional right customarily enjoyed by the forest dwelling Scheduled Tribes or other traditional forest dwellers, as the case may be, which are not mentioned in clauses (a) to (k) but excluding the traditional right of hunting or trapping or extracting a part of the body of any species of wild animal;

(m) right to in situ rehabilitation including alternative land in cases where the Scheduled Tribes and other traditional forest dwellers have been illegally evicted or displaced from forest land of any description without receiving their legal entitlement to rehabilitation prior to the 13th day of December, 2005.

(2) Notwithstanding anything contained in the Forest (Conservation) Act, 1980 (69 of 1980), the Central Government shall provide for diversion of forest land for the following facilities managed by the Government which involve felling of trees not exceeding seventy-five trees per hectare, namely:

(a) schools;

(b) dispensary or hospital;

(c) *anganwadis*;

(d) fair price shops;

(e) electric and telecommunication lines;

(f) tanks and other minor water bodies;

(g) drinking water supply and water pipelines;

(h) water or rain water harvesting structures;

(i) minor irrigation canals;

(j) non-conventional source of energy;

(k) skill upgradation or vocational training centres;

(l) roads; and

(m) community centres:

Provided that such diversion of forest land shall be allowed only if,

(i) the forest land to be diverted for the purposes mentioned in this sub-section is less than one hectare in each case; and

(ii) the clearance of such developmental projects shall be subject to the condition that the same is recommended by the Gram Sabha.

CHAPTER III

RECOGNITION, RESTORATION AND VESTING OF FOREST RIGHTS AND RELATED MATTERS

4. Recognition of, and vesting of, forest rights in forest dwelling Scheduled Tribes and other traditional forest dwellers

(1) Notwithstanding anything contained in any other law for the time being in force, and subject to the provisions of this Act, the Central Government hereby recognises and vests forest rights in–

(a) the forest dwelling Scheduled Tribes in States or areas in States where they are declared as Scheduled Tribes in respect of all forest rights mentioned in section 3;

(b) the other traditional forest dwellers in respect of all forest rights mentioned in section 3.

(2) The forest rights recognised under this Act in critical wildlife habitats of National Parks and Sanctuaries may subsequently be modified or resettled, provided that no forest rights holders shall be resettled or have their rights in any manner affected for the purposes of creating inviolate areas for wildlife conservation except in case all the following conditions are satisfied, namely:

(a) the process of recognition and vesting of rights as specified in section 6 is complete in all the areas under consideration;

(b) it has been established by the concerned agencies of the State Government, in exercise of their powers under the Wild Life (Protection) Act, 1972 (53 of 1972) that the activities or impact of the presence of holders of rights upon wild animals is sufficient to cause irreversible damage and threaten the existence of said species and their habitat;

(c) the State Government has concluded that other reasonable options, such as, co-existence are not available;

(d) a resettlement or alternatives package has been prepared and communicated that provides a secure livelihood for the affected individuals and communities and fulfils the requirements of such affected individuals and communities given in the relevant laws and the policy of the Central Government;

(e) the free informed consent of the Gram Sabhas in the areas concerned to the proposed resettlement and to the package has been obtained in writing;

(f) no resettlement shall take place until facilities and land allocation at the resettlement location are complete as per the promised package:

Provided that the critical wildlife habitats from which rights holders are thus relocated for purposes of wildlife conservation shall not be subsequently diverted by the State Government or the Central Government or any other entity for other uses.

(3) The recognition and vesting of forest rights under this Act to the forest dwelling Scheduled Tribes and to other traditional forest dwellers in relation to any State or Union territory in respect of forest land and their habitat shall be subject to the condition that such Scheduled Tribes or tribal communities or other traditional forest dwellers had occupied forest land before the 13th day of December, 2005.

(4) A right conferred by sub-section (1) shall be heritable but not alienable or transferable and shall be registered jointly in the name of both the spouses in case of married persons and in the name of the single head in the case of a household headed by a single person and in the absence of a direct heir, the heritable right shall pass on to the next-of-kin.

(5) Save as otherwise provided, no member of a forest dwelling Scheduled Tribe or other traditional forest dweller shall be evicted or removed from forest land under his occupation till the recognition and verification procedure is complete.

(6) Where the forest rights recognised and vested by sub-section (1) are in respect of land mentioned in clause (a) of sub-section (1) of section 3 such land shall be under the occupation of an

individual or family or community on the date of commencement of this Act and shall be restricted to the area under actual occupation and shall in no case exceed an area of four hectares.

(7) The forest rights shall be conferred free of all encumbrances and procedural requirements, including clearance under the Forest (Conservation) Act, 1980 (69 of 1980), requirement of paying the 'net present value' and 'compensatory afforestation' for diversion of forest land, except those specified in this Act.

(8) The forest rights recognised and vested under this Act shall include the right of land to forest dwelling Scheduled Tribes and other traditional forest dwellers who can establish that they were displaced from their dwelling and cultivation without land compensation due to State development interventions, and where the land has not been used for the purpose for which it was acquired within five years of the said acquisition.

5. **Duties of holders of forest rights**: The holders of any forest right, Gram Sabha and village level institutions in areas where there are holders of any forest right under this Act are empowered to–

(a) protect the wild life, forest and biodiversity;

(b) ensure that adjoining catchments area, water sources and other ecological sensitive areas are adequately protected;

(c) ensure that the habitat of forest dwelling Scheduled Tribes and other traditional forest dwellers is preserved from any form of destructive practices affecting their cultural and natural heritage;

(d) ensure that the decisions taken in the Gram Sabha to regulate access to community forest resources and stop any activity which adversely affects the wild animals, forest and the biodiversity are complied with.

CHAPTER IV

AUTHORITIES AND PROCEDURE FOR VESTING OF FOREST RIGHTS

6. **Authorities to vest forest rights in forest dwelling Scheduled Tribes and other traditional forest dwellers and procedure thereof**

(1) The Gram Sabha shall be the authority to initiate the process for determining the nature and extent of individual or community forest rights or both that may be given to the forest dwelling Scheduled Tribes and other traditional forest dwellers within the local limits of its jurisdiction under this Act by receiving claims, consolidating and verifying them and preparing a map delineating the area of each recommended claim in such manner as may be prescribed for exercise of such rights and the Gram Sabha shall, then, pass a resolution to that effect and thereafter forward a copy of the same to the Sub-Divisional Level Committee.

(2) Any person aggrieved by the resolution of the Gram Sabha may prefer a petition to the Sub-Divisional Level Committee constituted under sub-section (3) and the Sub-Divisional Level Committee shall consider and dispose of such petition:

Provided that every such petition shall be preferred within sixty days from the date of passing of the resolution by the Gram Sabha:

Provided further that no such petition shall be disposed of against the aggrieved person, unless he has been given a reasonable opportunity to present his case.

(3) The State Government shall constitute a Sub-Divisional Level Committee to examine the resolution passed by the Gram Sabha and prepare the record of forest rights and forward it through the Sub-Divisional Officer to the District Level Committee for a final decision.

(4) Any person aggrieved by the decision of the Sub-Divisional Level Committee may prefer a petition to the District Level Committee within sixty days from the date of decision of the Sub-Divisional Level Committee and the District Level Committee shall consider and dispose of such petition:

Provided that no petition shall be preferred directly before the District Level Committee against the resolution of the Gram Sabha unless the same has been preferred before and considered by the SubDivisional Level Committee:

Provided further that no such petition shall be disposed of against the aggrieved person, unless he has been given a reasonable opportunity to present his case.

(5) The State Government shall constitute a District Level Committee to consider and finally approve the record of forest rights prepared by the Sub-Divisional Level Committee.

(6) The decision of the District Level Committee on the record of forest rights shall be final and binding.

(7) The State Government shall constitute a State Level Monitoring Committee to monitor the process of recognition and vesting of forest rights and to submit to the nodal agency such returns and reports as may be called for by that agency.

(8) The Sub-Divisional Level Committee, the District Level Committee and the State Level Monitoring Committee shall consist of officers of the departments of Revenue, Forest and Tribal Affairs of the State Government and three members of the Panchayati Ráj Institutions at the appropriate level, appointed by the respective Panchayati Raj Institutions, of whom two shall be the Scheduled Tribe members and at least one shall be a woman, as may be prescribed.

(9) The composition and functions of the Sub-Divisional Level Committee, the District Level Committee and the State Level Monitoring Committee and the procedure to be followed by them in the discharge of their functions shall be such as may be prescribed.

CHAPTER V

OFFENCES AND PENALTIES

7. **Offences by members or officers of authorities and Committees under this Act**: Where any authority or Committee or officer or member of such authority or Committee contravenes any provision of this Act or any rule made thereunder concerning recognition of forest rights, it, or they, shall be deemed to be guilty of an offence under this Act and shall be liable to be proceeded against and punished with fine which may extend to one thousand rupees:

Provided that nothing contained in this sub-section shall render any member of the authority or Committee or head of the department or any person referred to in this section liable to any punishment if he proves that the offence was committed without his knowledge or that he had exercised all due diligence to prevent the commission of such offence.

8. **Cognizance of offences**: No court shall take cognizance of any offence under section 7 unless any forest dwelling Scheduled Tribe in case of a dispute relating to a resolution of a Gram Sabha or the Gram Sabha through a resolution against any higher authority gives a notice of not less than sixty days to the State Level Monitoring Committee and the State Level Monitoring Committee has not proceeded against such authority.

CHAPTER VI

MISCELLANEOUS

9. **Members of authorities, etc., to be public servants**: Every member of the authorities referred to in Chapter IV and every other officer exercising any of the powers conferred by or under this Act shall be deemed to be a public servant within the meaning of section 21 of the Indian Penal Code (45 of 1860).

10. **Protection of action taken in good faith**

(1) No suit, prosecution or other legal proceeding shall lie against any officer or other employee of the Central Government or the State Government for anything which is in good faith done or intended to be done by or under this Act.

(2) No suit or other legal proceeding shall lie against the Central Government or the State Government or any of its officers or other employees for any damage caused or likely to be caused by anything which is in good faith done or intended to be done under this Act.

(3) No suit or other legal proceeding shall lie against any authority as referred to in Chapter IV including its Chairperson, members, member-secretary, officers and other employees for anything which is in good faith done or intended to be done under this Act.

11. **Nodal agency**:The Ministry of the Central Government dealing with Tribal Affairs or any officer or authority authorised by the Central Government in this behalf shall be the nodal agency for the implementation of the provisions of this Act.

12. **Power of Central Government to issue directions**: In the performance of its duties and exercise of its powers by or under this Act, every authority referred to in Chapter IV shall be subject to such general or special directions, as the Central Government may, from time to time, give in writing.

13. **Act not in derogation of any other law**: Save as otherwise provided in this Act and the Provisions of the Panchayats (Extension to the Scheduled Areas) Act, 1996 (40 of 1996), the provisions of this Act shall be in addition to and not in derogation of the provisions of any other law for the time being in force.

14. **Power to make rules**

(1) The Central Government may, by notification, and subject to the condition of previous publication, make rules for carrying out the provisions of this Act.

(2) In particular, and without prejudice to the generality of the foregoing powers, such rules may provide for all or any of the following matters, namely:

(a) procedural details for implementation of the procedure specified in section 6;

(b) the procedure for receiving claims, consolidating and verifying them and preparing a map delineating the area of each recommended claim for exercise of forest rights under sub-section (1) of section 6 and the manner of preferring a petition to the Sub-Divisional Committee under sub-section (2) of that section;

(c) the level of officers of the departments of Revenue, Forest and Tribal Affairs of the State Government to be appointed as members of the Sub-Divisional Level Committee, the District Level Committee and the State Level Monitoring Committee under sub-section (8) of section 6;

(d) the composition and functions of the Sub-Divisional Level Committee, the District Level Committee and the State Level Monitoring Committee and the procedure to be followed by them in the discharge of their functions under sub-section (9) of section 6;

(e) any other matter which is required to be, or may be, prescribed.

(3) Every rule made by the Central Government under this Act shall be laid, as soon as may be after it is made, before each House of Parliament, while it is in session, for a total period of thirty days which may be comprised in one session or in two or more successive sessions, and if, before the expiry of the session immediately following the session or the successive sessions aforesaid, both Houses agree in making any modification in the rule or both Houses agree that the rule should not be made, the rule shall thereafter have effect only in such modified form or be of no effect, as the case may be; so, however, that any such modification or annulment shall be without prejudice to the validity of anything previously done under that rule.

43.12 MINISTRY OF LAW, JUSTICE AND COMPANY AFFAIRS

(Legislative Department)

New Delhi, the 1st October, 2001/Asvina 9, 1923 (Saka)

The following Act of Parliament received the assent of the President on the 29th September, and is hereby published for general information:

43.12.1 THE ENERGY CONSERVATION ACT, 2001

NO. 52 OF 2001

[29th September 2001]

An Act to provide for efficient use of energy and its conservation and for matters connected therewith or incidental thereto.

BE it enacted by Parliament in the Fifty-second Year of the Republic of India as follows:

CHAPTER I

PRELIMINARY

Short title, extent and commencement

1. (1) This Act may be called the Energy Conservation Act, 2001.

(2) It extends to the whole of India except the state of Jammu and Kashmir.

(3) It shall come into force on such dates as the Central Government may, by notification in the Official Gazette, appoint; and different dates may be appointed for different provisions of this Act and any reference in any such provision to the commencement of this Act shall be construed as a reference to the coming into force of that provision.

Definitions

2. In this Act, unless the context otherwise requires:

(a) "accredited energy auditor" means an auditor possessing qualifications specified under clause (p) of subsection (2) of Section 13;

(b) "Appellate Tribunal" means Appellate Tribunal for Energy Conservation established under Section 30;

(c) "building" means any structure or erection or part of a structure or erection, after the rules relating to energy conservation building codes have been notified under clause (a) of Section 15 of clause (1) of subsection (2) of Section 56, which is having a connected load of 500 kW or contract demand of 600 kVA and above and is intended to be used for commercial purposes;

(d) "Bureau" means the Bureau of Energy Efficiency established under subsection (1) of Section 3;

(e) "Chairperson" means the Chairperson of the Governing Council;

(f) "designated agency" means any agency designated under clause (d) of Section 15;

(g) "designated consumer" means any consumer specified under clause (e) of Section 14;

(h) "energy" means any form of energy derived from fossil fuels, nuclear substances or materials, hydro-electricity and includes electrical energy or electricity generated from renewable sources of energy or biomass connected to the grid;

(i) "energy audit" means the verification, monitoring and analysis of use of energy including submission of technical report containing recommendations for improving energy efficiency with cost benefit analysis and an action plan to reduce energy consumption;

(j) "energy conservation building codes" means the norms and standards of energy consumption expressed in terms of per square meter of the area wherein energy is used and includes the location of the building;

(k) "energy consumption standards" means the norms for process and energy consumption standards specified under Clause (a) of Section 14;

(l) "Energy Management Centre" means the Energy Management Centre set up under the Resolution of the Government of India in the erstwhile Ministry of Energy, Department of Power No. 7(2)/87-EP (Vol. IV), dated the 5th July, 1989 and registered under the Societies Registration Act, 1860; 21 of 1860

(m) "energy manager" means any individual possessing the qualifications prescribed under Clause (m) of Section 14;

(n) "Governing Council" means the Governing Council referred to in Section 4;

(o) "member" means the member of the Governing Council and includes the Chairperson;

(p) "notification" means a notification in the Gazette of India or, as the case may be, the Official Gazette of a State;

(q) "prescribed" means prescribed by rules made under this Act;

(r) "regulations" means regulations made by the Bureau under this Act;

(s) "schedule" means the Schedule of this Act;

(t) "State Commission" means the State Electricity Regulatory Commission established under subsection (1) of Section 17 of the Electricity Regulatory Commissions Act, 1998; 14 of 1998

9 of 1940
54 of 1948
14 of 1998

(u) words and expressions used and not defined in this Act but defined in the Indian Electricity Act, 1910, or the Electricity (Supply) Act, 1948, or the Electricity Regulatory Commissions Act, 1998, shall have meanings respectively assigned to them in those Acts.

CHAPTER II
BUREAU OF ENERGY EFFICIENCY

3. (1) With effect from such date as the Central Government may, by notification, appoint, there shall be established, for the purposes of this Act, a Bureau to be called the **Bureau of Energy Efficiency**. **Establishment and incorporation of Bureau of Energy Efficiency**

(2) The Bureau shall be a body corporate by the name aforesaid having perpetual succession and a common seal, with power subject to the provisions of this Act, to acquire, hold and dispose of property, both movable and immovable, and to contract, and shall, by the said name, sue or be sued.

(3) The head office of the Bureau shall be at Delhi.

(4) The Bureau may establish offices at other places in India.

4. (1) The general superintendence, direction and management of the affairs of the Bureau shall vest in the Governing Council which shall consists of not less than twenty, but not exceeding twenty-six members to be appointed by the Central Government. **Management of Bureau**

(2) The Governing Council shall consist of the following members, namely:

(a) the Minister in charge of the Ministry or Department of the Central Government dealing with the Power	*ex officio* Chairperson;
(b) the Secretary to the Government of India, in charge of the Ministry or Department of the Central Government dealing with the Power	*ex officio* member;
(c) the Secretary to the Government of India, in charge of the Ministry or Department of the Central Government dealing with the Petroleum and Natural Gas	*ex officio* member;
(d) the Secretary to the Government of India, in charge of the Ministry or Department of the Central Government dealing with the Coal	*ex officio* member;
(e) the Secretary to the Government of India, in charge of the Ministry or Department of the Central Government dealing with the Non-conventional Energy Sources	*ex officio* member;
(f) the Secretary to the Government of India, in charge of the Ministry or Department of the Central Government dealing with the Atomic Energy	*ex officio* member;
(g) the Secretary to the Government of India, in charge of the Ministry or Department of the Central Government dealing with the Consumer Affairs	*ex officio* member;
54 of 1948 (h) Chairman of the Central Electricity	*ex officio* member;

		Authority established under the Electricity (Supply) Act, 1948	
Karnataka Act 17 of 1960	(i)	Director-General of the Central Power Research Institute registered under the Karnataka Societies Act, 1960	*ex officio* member;
XXI of 1860	(j)	Executive Director of the Petroleum Conservation Research Association, a society registered under the Societies Registration Act, 1860	*ex officio* member;
1 of 1956	(k)	Chairman-cum-Managing Director of the Central Mine Planning and Design Institute Limited, a company incorporated under the Companies Act, 1956	*ex officio* member;
63 of 1986	(l)	Director-General of the Bureau of Indian Standards established under the Bureau of Indian Standards Act, 1986	*ex officio* member;
	(m)	Director-General of the National Test House, Department of Supply, Ministry of Commerce and Industry, Kolkata	*ex officio* member;
1 of 1956	(n)	Managing Director of the Indian Renewable Energy Development Agency Limited, a company incorporated under the Companies Act, 1956	*ex officio* member;
	(o)	one member each from five power regions representing the States of the region to be appointed by the Central Government	member;
	(p)	such number of persons, not exceeding four as may be prescribed, to be appointed by the Central Government as members from amongst persons who are in the opinion of the Central Government capable of representing industry, equipment and appliance manufacturers, architects and consumers	member;
	(q)	such number of persons, not exceeding two as may be nominated by the Governing Council as members	member;
	(r)	Director-General of Bureau	*ex officio* member; – secretary

(3) The Governing Council may exercise all powers and do all acts and things which may be exercised or done by the Bureau.

(4) Every member referred to in clauses (o), (p) *and* (q) of subsection (2) shall hold office for a term of three years from the date on which he enters upon his office.

(5) The fee and allowances to be paid to the members referred to in clauses (o), (p) *and* (q) of subsection (2) and the manner of filling up of vacancies and the procedure to be followed in the discharge of their functions shall be such as may be prescribed.

Meetings of Governing Council

5. (1) The Governing Council shall meet at such times and places, and shall observe such rules of procedure in regard to the transaction of business as its meetings (including quorum of such meetings) as may be provided by regulations.

Meetings of Governing Council

(2) The Chairperson or, if for any reason, he is unable to attend a meeting of the Governing Council, any other member chosen by the members present from amongst themselves at the meeting shall preside at the meeting.

(3) All questions which come up before any meeting of the Governing Council shall be decided by a majority vote of the members present and voting, and in the event of an equality of votes, the Chairperson or his absence, the person presiding, shall have second or casting vote.

Vacancies etc., not to invalidate proceedings of Bureau, Governing Council or Committee

6. No act or proceeding of the Bureau or the Governing Council or any Committee shall be invalid merely by reason of

(a) any vacancy in, or any defect in the constitution of, the Bureau or the Governing Council or the Committee; or

(b) any defect in the appointment of a person acting as a Director-General or Secretary of the Bureau or a member of the Governing Council or the Committee; or

(c) any irregularity in the procedure of the Bureau or the Governing Council or the Committee not affecting the merits of the case.

Removal of member from office

7. The Central Government shall remove a member referred to in clauses (o), (p) *and* (q) *of* subsection (2) of Section 4 from office if he

(a) is, or at any time has been, adjudicated as insolvent;

(b) is of unsound mind and stands so declared by a competent court;

(c) has been convicted of an offence which, in the opinion of the Central Government, involves a moral turpitude;

(d) has, in the opinion of the Central Government, so abused his position as to render his continuation in office detrimental to the public interest: Provided that no member shall be removed under this clause unless he has been given a reasonable opportunity of being heard in the matter.

Constitution of Advisory Committees and other committees

8. (1) Subject to any regulations made in this behalf, the Bureau shall, within six months from the date of commencement of this Act, constitute Advisory Committees for the efficient discharge of its functions.

(2) Each Advisory Committee shall consist of a Chairperson and such other members as may be determined by regulations.

(3) Without prejudice to the powers contained in subsection (1), the Bureau may constitute, such number of technical committees of experts for the formulation of energy consumption standards or norms in respect of equipment or processes, as it considers necessary.

Director General of Bureau

9. (1) The Central Government shall, by notification, appoint a Director-General from amongst persons of ability and standing, having adequate

knowledge and experience in dealing with the matters relating to energy production, supply and energy management standardisation and efficient use of energy and its conservation

(2) The Central Government shall, by notification appoint any person not below the rank of Deputy Secretary to the Government of India as Secretary of the Bureau

(3) The Director-General shall hold office for a term of three years from the date on which he enters upon his office or until he attains the age of sixty years, whichever is earlier

(4) The salary and allowances payable to the Director-General and other terms and conditions of his service and other terms and conditions of service of the Secretary of the Bureau shall be such as may be prescribed

(5) Subject to general superintendence, direction and management of the affairs by the Governing Council, the Director-General of the Bureau shall be the Chief Executive Authority of the Bureau

(6) The Director-General of the Bureau shall exercise and discharge such powers and duties of the Bureau as may be determined by regulations.

Officers and employees of Bureau

10. (1) The Central Government may appoint such other officers and employees in the Bureau as it considers necessary for the efficient discharge of its functions under this Act.

(2) The terms and conditions of service of officers and other employees of the Bureau appointed under subsection *(1)* shall be such as may be prescribed.

Authentication of orders and decisions of Bureau

11. All orders and decisions of the Bureau shall be authenticated by the signature of the Director-General or any other officer of the Bureau authorised by the Director-General in this behalf.

CHAPTER III

TRANSFER OF ASSETS, LIABILITIES ETC. OF ENERGY MANAGEMENT CENTRE TO BUREAU

Transfer of assets, liabilities and employees of Energy Management Centre

12. (1) On and from the date of establishment of the Bureau

(a) any reference to the Energy Management Centre in any law other than this Act or in any contract or other instrument shall be deemed as a reference to the Bureau;

(b) all properties and assets, movable and immovable of, or belonging to, the Energy Management Centre shall vest in the Bureau;

(c) all the rights and liabilities of the Energy Management Centre shall be transferred to, and be the right and liabilities of, the Bureau;

(d) without prejudice to the provisions of clause (c), all debts, obligations and liabilities incurred, all contracts entered into and all matters and things engaged to be done by, with or for the Energy Management Centre immediately before that date for or in connection with the purposes of the said Centre shall be deemed to have been incurred, entered into, or engaged to be done by, with or for, the Bureau;

(e) all sums of money due to the Energy Management Centre immediately before that date shall be deemed to be due to the Bureau;

(f) all suits and other legal proceedings instituted or which could have been instituted by or against the Energy Management Centre immediately before that date may be continued or may be instituted by or against the Bureau; and

(g) every employee holding any office under the Energy Management Centre immediately before that date shall hold his office in the Bureau by the same tenure and upon the same terms and conditions of service as respects remuneration, leave, provident fund, retirement or other terminal benefits as he would have held such office if the Bureau had not been established and shall continue to do so as an employee of the Bureau or until the expiry of six months from the date if such employee opts not to be the employee of the Bureau within such period.

(2) Not withstanding anything contained in the Industrial Disputes Act, 1947, or in any other law for the time being in force, the absorption of any employees by the Bureau in its regular service under this section shall not entitle such employees to any compensation under that Act or other law and no such claim shall be entertained by any court, tribunal or other authority. 14 of 1947

CHAPTER IV
POWERS AND FUNCTIONS OF BUREAU

Powers and functions of Bureau

13. (1) The Bureau shall, effectively coordinate with designated consumers, designated agencies and other agencies, recognise and utilise the existing resources and infrastructure, in performing the functions assigned to it by or under this Act

(2) The Bureau may perform such functions and exercise such powers as may be assigned to it by or under this Act and in particular, such functions and powers include the function and power to

(a) recommend to the Central Government the norms for processes and energy consumption standards required to be notified under clause (a) of Section 14;

(b) recommend to the Central Government the particulars required to be displayed on label on equipment or on appliances and manner of their display under clause (d) of Section 14;

(c) recommend to the Central Government for notifying any user or class of users of energy as a designated consumer under clause (e) of Section 14;

(d) take suitable steps to prescribe guidelines for energy conservation building codes under clause (p) of Section 14;

(e) take all measures necessary to create awareness and disseminate information for efficient use of energy and its conservation;

(f) arrange and organize training of personnel and specialists in the techniques for efficient use of energy and its conservation;

(g) strengthen consultancy services in the field of energy conservation;

(h) promote research and development in the field of energy conservation;

(i) develop testing and certification procedure and promote testing facilities for certification and testing for energy consumption of equipment and appliances;

(j) formulate and facilitate implementation of pilot projects and demonstration projects for promotion of efficient use of energy and its conservation;

(k) promote use of energy efficient processes, equipment, devices and systems;

(l) promote innovative financing of energy efficiency projects;

(m) give financial assistance to institutions for promoting efficient use of energy and its conservation;

(n) levy fee, as may be determined by regulations, for services provided for promoting efficient use of energy and its conservation;

(o) maintain a list of accredited energy auditors as may be specified by regulations;

(p) specify, by regulations, qualifications for the accredited energy auditors;

(q) specify, by regulations, the manner and intervals of time in which the energy audit shall be conducted;

(r) specify, by regulations, certification procedures for energy managers to be designated or appointed by designated consumers;

(s) prepare educational curriculum on efficient use of energy and its conservation for educational institutions, boards, universities or autonomous bodies and coordinate with them for inclusion of such curriculum in their syllabus;

(t) implement international cooperation programmes relating to efficient use of energy and its conservation as may be assigned to it by the Central Government;

(u) perform such other functions as may be prescribed.

CHAPTER V
POWER OF CENTRAL GOVERNMENT TO FACILITATE AND ENFORCE EFFICIENT USE OF ENERGY AND ITS CONSERVATION

Power of Central Government to enforce efficient use of energy and its conservation

14. The Central Government may, by notification, in consultation with the Bureau—

(a) specify the norms for processes and energy consumption standards for any equipment, appliances which consumes, generates, transmits or supplies energy;

(b) specify equipment or appliances or class of equipments or appliances, as the case may be, for the purposes of this Act;

(c) prohibit manufacture or sale or purchase or import of equipment or appliance specified under clause (b) unless such equipment or appliances conforms to energy consumption standards;

Provided that no notification prohibiting manufacture or sale or purchase or import of equipment or appliance shall be issued within two years from the date of notification issued under clause (a) of this section;

(d) direct display of such particulars on label on equipment or on appliance specified under clause (b) and in such manner as may be specified by regulations;

(e) specify, having regarding to the intensity or quantity of energy consumed and the amount of investment required for switching over to energy efficient equipments and capacity of industry to invest in it and availability of the energy efficient machinery and equipment required by the industry, any user or class of users of energy as a designated consumer for the purposes of this Act;

(f) alter the list of Energy Intensive Industries specified in the Schedule;

(g) establish and prescribe such energy consumption norms and standards for designated consumers as it may consider necessary:

Provided that the Central Government may prescribe different norms and standards for different designated consumers having regard to such factors as may be prescribed;

(h) direct, having regard to quantity of energy consumed or the norms and standards of energy consumption specified under clause (a) the energy intensive industries specified in the Schedule to get energy audit conducted by an accredited energy auditor in such manner and intervals of time as may be specified by regulations;

(i) direct, if considered necessary for efficient use of energy and its conservation, any designated consumer to get energy audit conducted by an accredited energy auditor;

(j) specify the matters to be included for the purposes of inspection under subsection (2) of Section 17;

(k) direct any designated consumer to furnish to the designated agency, in such form and manner and within such period, as may be prescribed, the information with regard to the energy consumed and action taken on the recommendation of the accredited energy auditor;

(l) direct any designated consumer to designate or appoint energy manager in charge of activities for efficient use of energy and its conservation and submit a report, in the form and manner as may be prescribed, on the status of energy consumption at the end of every financial year to designated agency;

(m) prescribe minimum qualification for energy managers to be designated or appointed under clause (1);

(n) direct every designated consumer to comply with energy consumption norms and standards;

(o) direct any designated consumer, who does not fulfil the energy consumption norms and standards prescribed under clause (g), to prepare a scheme for efficient use of energy and its conservation and implement such scheme keeping in view of the economic viability of the investment in such form and manner as may be prescribed;

(p) prescribe energy conservation building codes for efficient use of energy and its conservation in the building or building complex;

(q) amend the energy conservation building codes to suit the regional and local climatic conditions;

(r) direct every owner or occupier of the building or building complex, being a designated consumer to comply with the provisions of energy conservation building codes for efficient use of energy and its conservation;

(s) direct, any designated consumer referred to in clause *(r)*, if considered necessary, for efficient use of energy and its conservation in his building to get energy audit conducted in respect of such building by an accredited energy auditor in such manner and intervals of time as may be specified by regulations;

(t) take all measures necessary to create awareness and disseminate information for efficient use of energy and its conservation;

(u) arrange and organise training of personnel and specialists in the techniques for efficient use of energy and its conservation;

(v) take steps to encourage preferential treatment for use of energy efficient equipment or appliances: Provided that the powers under clauses (p) and (s) shall be exercised in consultation with the concerned State.

CHAPTER VI

POWER OF STATE GOVERNMENT TO FACILITATE AND ENFORCE EFFICIENT USE OF ENERGY AND ITS CONSERVATION

Power of State Government to enforce certain provisions for efficient use of energy and its conservation

15. The State Government may, by notification, in consultation with the Bureau

(a) amend the energy conservation building codes to suit the regional and local climatic conditions and may, by rules made by it, specify and notify energy conservation building codes with respect to use of energy in the buildings;

(b) direct every owner or occupier of a building or building complex being a designated consumer to comply with the provisions of the energy conservation building codes;

(c) direct, if considered necessary for efficient use of energy and its conservation, any designated consumer referred to in clause (b) to get energy audit conducted by an accredited energy auditor in such manner and at such intervals of time as may be specified by regulations;

(d) designate any agency as designated agency to coordinate, regulate and enforce provisions of this Act within the State;

(e) take all measures necessary to create awareness and disseminate information for efficient use of energy and its conservation;

(f) arrange and organise training of personnel and specialists in the techniques for efficient use of energy and its conservation;

(g) take steps to encourage preferential treatment for use of energy efficient equipment or appliances;

(h) direct, any designated consumer to furnish to the designated agency, in such form and manner and within such period as may be specified by rules made by it, information with regard to the energy consumed by such consumer;

(i) specify the matters to be included for the purposes of inspection under subsection (2) of Section 17.

Establishment of Fund by State Government

16. (1) The State Government shall constitute a Fund to be called the State Energy Conservation Fund for the purposes of promotion of efficient use of energy and its conservation within the State.

(2) The Fund shall be credited all grants and loans that may be made by the State Government or Central Government or any other organization or individual for the purposes of this Act.

(3) The Fund shall be applied for meeting the expenses incurred for implementing the provisions of this Act.

(4) The Fund created under subsection (1) shall be administered by such persons or any authority and in such manner as may be specified in the rules made by the State Government.

Power of inspection

17. (1) The designated agency may appoint, after the expiry of five years from the date of commencement of this Act, as many inspecting officers as may be necessary for the purpose of ensuring compliance with energy consumption standard specified under clause (a) of Section 14 or ensure display of particulars on label on equipment or appliances specified under clause (b) of Section 14 or for the purpose of performing such other functions as may be assigned to them.

(2) Subject to any rules made under this Act, an inspecting officer shall have power to

(a) inspect any operation carried on or in connection with the equipment or appliance specified under clause (b) of Section 14 or in respect of which energy standards under clause (a) of Section 14 have been specified;

(b) enter any place of designated consumer at which the energy is used for any activity and may require any proprietor, employee, director, manager or secretary or any other person who may be attending in any manner to or helping in, carrying on any activity with the help of energy

(i) to afford him necessary facility to inspect

(A) any equipment or appliance as he may require and which may be available at such place;

(B) any production process to ascertain the energy consumption norms and standards;

(ii) to make an inventory of stock of any equipment or appliance checked or verified by him;

(iii) to record the statement of any person which may be useful for, or relevant to, for efficient use of energy and its conservation under this Act.

(3) An inspecting officer may enter any place of designated consumer

(a) where any activity with the help of energy is carried on; and

(b) where any equipment or appliance notified under clause (b) of Section 14 has been kept, during the hours at which such places is open for production or conduct of business connected therewith.

(4) An inspecting officer acting under this section shall, on no account, remove or cause to be removed from the place wherein he has entered, any equipment or appliance or books of accounts or other documents.

Power of Central Government or State Government to issue directions

18. The Central Government or the State Government may, in the exercise of its powers and performance of its functions under this Act and for efficient use of energy and its conservation, issue such directions in writing as it deems fit for the purposes of this Act to any person, officer, authority or any designated consumer and such person, officer or authority or any designated consumer shall be bound to comply with such directions.

Explanation – For the avoidance of doubts, it is hereby declared that the power to issue directions under this section includes the power to direct

(a) regulation of norms for process and energy consumption standards in any industry or building or building complex; or

(b) regulation of the energy consumption standards for equipment and appliances.

CHAPTER VII

FINANCE, ACCOUNTS AND AUDIT OF BUREAU

Grants and loans by Central Government

19. The Central Government may, after due appropriation made by Parliament by law in this behalf, make to the Bureau or to the State Government grants and loans of such sums or money as the Central Government may consider necessary.

Establishment of Fund by Central Government

20. (1) There shall be constituted a Fund to be called as the Central Energy Conservation Fund and there shall be credited thereto

(a) any grants and loans made to the Bureau by the Central Government under Section 19;

(b) all fees received by the Bureau under this Act;

(c) all sums received by the Bureau from such other sources as may be decided upon by the Central Government.

(2) The Fund shall be applied for meeting

(a) the salary, allowances and other remuneration of Director-General, Secretary officers and other employees of the Bureau,

(b) expenses of the Bureau in the discharge of its functions under Section 13;

(c) fee and allowances to be paid to the members of the Governing Council under subsection (5) or Section 4;

(d) expenses on objects and for purposes authorised by this Act.

Borrowing powers of Bureau

21. (1) The Bureau may, with the consent of the Central Government, or in accordance with the terms of any general or special authority given to it by the Central Government, borrow money from any source as it may deem fit for discharging all or any of its functions under this Act.

(2) The Central Government may guarantee, in such manner as it thinks fit, the repayment of the principal and the payment of interest thereon with respect to the loans borrowed by the Bureau under subsection (l).

Budget

22. The Bureau shall prepare, in such form and at such time in each financial year as may be prescribed, its budget for the next financial year, showing the estimated receipts and expenditure of the Bureau and forward the same to the Central Government.

Annual Report

23. The Bureau shall prepare, in such form and at such time in each financial year as may be prescribed, its annual report, giving full account of its activities during the previous financial year, and submit a copy thereof to the Central Government.

Annual Report to be laid before Parliament

24. The Central Government shall cause the annual report referred to in Section 23 to be laid, as soon as may be after it is received, before each House of Parliament.

Accounts & Audit

25. (1) The Bureau shall maintain proper accounts and other relevant records and prepare an annual statement of accounts in such form as may be prescribed by the Central Government in consultation with the Comptroller and Auditor-General of India.

Accounts and Audit

(2) The accounts of the Bureau shall be audited by the Comptroller and Auditor-General of India at such intervals as may be specified by him and any expenditure incurred in connection with such audit shall be payable by the Bureau to the Comptroller and Auditor-General.

(3) The Comptroller and Auditor-General of India and any other person appointed by him in connection with the audit of the accounts of the Bureau shall have the same rights and privileges and authority in connection with such audit as the Comptroller and Auditor-General generally has in connection with the audit of the Government accounts and, in particular, shall have the right to demand the production of books, accounts, connected vouchers and other documents and papers and to inspect any of the offices of the Bureau.

(4) The accounts of the Bureau as certified by the Comptroller and Auditor-General of India or any other person appointed by him in this behalf together with the audit report thereon shall forward annually to the Central Government and that Government shall cause the same to be laid before each House of Parliament.

CHAPTER VIII
PENALTIES AND ADJUDICATION

Penalty

26. (1) If any person fails to comply with the provision of clause (c) or the clause (d) or clause (h) or clause (i) or clause (k) or clause (l) or clause (n) or clause (r) or clause (s) of Section 14 or clause (b) or clause (c) or clause (h) of Section 15, he shall be liable to a penalty which shall not exceed ten thousand rupees for each such failures and, in the case of continuing failures, with an additional penalty which may extend to one thousand rupees for every day during which such failures continues:

Provided that no person shall be liable to pay penalty within five years from the date of commencement of this Act.

(2) Any amount payable under this section, if not paid, may be recovered as if it were an arrear of land revenue.

Power to adjudicate

27. (1) For the purpose of adjudging Section 26, the State Commission shall appoint any of its members to be an adjudicating officer for holding an inquiry in such manner as may be prescribed by the Central Government, after giving any person concerned a reasonable opportunity of being heard for the purpose of imposing any penalty.

(2) While holding an inquiry the adjudicating officer shall have power to summon and enforce the attendance of any person acquainted with the facts and circumstances of the case to give evidence or produce any document which, in the opinion of the adjudicating officer, may be useful for or relevant to the subject-matter of the inquiry, and if, on such inquiry, he is satisfied that the person has failed to comply with the provisions of any of the clauses of the sections specified in Section 26, he may impose such penalty as he thinks fit in accordance with the provisions of any of those clauses of that section:

Provided that where a State Commission has not been established in a State, the Government of that State shall appoint any of its officer not below the rank equivalent to a Secretary dealing with legal affairs in that State to be an adjudicating officer for the purposes of this section and such officer shall cease to be an adjudicating officer immediately on the appointment of an adjudicating officer by the State Commission on its establishment in that State:

Provided further that where an adjudicating officer appointed by a State Government ceased to be an adjudicating officer, he shall transfer to the adjudicating officer appointed by the State Commission all matters being adjudicated by him and thereafter the adjudicating officer appointed by the State Commission shall adjudicate the penalties on such matters.

Factors to be taken into account by adjudicating officer

28. While adjudicating the quantum of penalty under Section 26, the adjudicating officer shall have due regard to the following factors, namely:

(a) the amount of disproportionate gain or unfair advantage, wherever quantifiable, made as a result of the default;

(b) the repetitive nature of the default.

Civil court not to have jurisdiction

29. No civil court shall have jurisdiction to entertain any suit or proceeding in respect of any matter which an adjudicating officer appointed under this Act or the Appellate Tribunal is empowered by or under this Act to determine and no injunction shall be granted by any court or other authority in respect of any action taken or to be taken in pursuance of any power conferred by or under this Act.

CHAPTER IX
APPELLATE TRIBUNAL FOR ENERGY CONSERVATION

Establishment of Appellate Tribunal

30. The Central Government shall, by notification, establish an Appellate Tribunal to be known as the Appellate Tribunal for Energy Conservation to hear appeals against the orders of the adjudicating officer or the Central Government or the State Government or any other authority under this Act.

Appeal to Appellate Tribunal

31. (1) Any person aggrieved, by an order made by an adjudicating officer or the Central Government or the State Government or any other authority under this Act, may prefer an appeal to the Appellate Tribunal for Energy Conservation:

Provided that any person appealing against the order of the adjudicating officer levying any penalty, shall while filing the appeal, deposit the amount of such penalty:

Provided further that where in any particular case, the Appellate Tribunal is of the opinion that the deposit of such penalty would cause undue hardship to such person, the Appellate Tribunal may dispense with such deposit subject to such conditions as it may deem fit to impose so as to safeguard the realisation of penalty.

(2) Every appeal under subsection (1) shall be filed within a period of forty-five days from the date on which a copy of the order made by the adjudicating officer or the Central Government or the State Government or any other authority is received by the aggrieved person and it shall be in such form, verified in such manner and be accompanied by such fee as may be prescribed:

Provided that the Appellate Tribunal may entertain an appeal after the expiry of the said period of forty-five days if it is satisfied that there was sufficient cause for not filing it within that period.

(3) On receipt of an appeal under subsection (1), the Appellate Tribunal may, after giving the parties to the appeal an opportunity of being heard, pass such orders thereon as it thinks fit, confirming, modifying or setting aside the order appealed against.

(4) The Appellate Tribunal shall send a copy of every order made by it to the parties to the appeal and to the concerned adjudicating officer or the Central Government or the State Government or any other authority.

(5) The appeal filed before the Appellate Tribunal under subsection (1) shall be dealt with by it as expeditiously as possible and endeavour shall be made by it to dispose of the appeal finally within one hundred and eighty days from the date of receipt of the appeal:

Provided that where an appeal could not be disposed of within the said period of one hundred and eighty days, the Appellate Tribunal shall record its reasons in writing for not disposing of the appeal within the said period.

(6) The Appellate Tribunal may, for the purpose of examining the legality, propriety or correctness of any order made by the adjudicating officer or the Central Government or the State Government or any other authority under this Act, as the case may be in relation to any proceeding, on its own motion or otherwise, call for the records of such proceedings and make such order in the case as it thinks fit.

Composition of Appellate Tribunal

32. (1) The Appellate Tribunal shall consist of a Chairperson and such number of Members not exceeding four, as the Central Government may deem fit.

(2) Subject to the provisions of this Act,

(a) the jurisdiction of the Appellate Tribunal may be exercised by Benches thereof;

(b) a Bench may be constituted by the Chairperson of the Appellate Tribunal with two or more Members of the Appellate Tribunal as the Chairperson of the Appellate Tribunal may deem fit;

Provided that every Bench constituted under this clause shall include at least one Judicial Member and one Technical Member;

(c) The Benches of the Appellate Tribunal shall ordinarily sit at Delhi and such other places as the Central Government may, in consultation with the Chairperson of the Appellate Tribunal, notify;

(d) The Central Government shall notify the areas in relation to which each Bench of the Appellate Tribunal may exercise jurisdiction.

(3) Notwithstanding anything contained in subsection (2), the Chairperson of the Appellate Tribunal may transfer a Member of the Appellate Tribunal from one Bench to another Bench

Explanation – For the purposes of this Chapter,

(i) "Judicial Member" means a Member of the Appellate Tribunal appointed as such under item (i) or item (ii) or clause (b) of subsection (1) of Section 33, and includes the Chairperson of the Appellate Tribunal;

(ii) "Technical Member" means a Member of the Appellate Tribunal appointed as such under item (iii) or item (iv) or item (v) or item (vi) of clause (b) of subsection (1) of Section 33.

Qualifications for appointment of Chairperson and Members of Appellate Tribunal

33. (1) A person shall not be qualified for appointment as the Chairperson of the Appellate Tribunal or a Member of the Appellate Tribunal unless he

(a) in the case of Chairperson of the Appellate Tribunal, is or has been, a judge of the Supreme Court or the Chief Justice of a High Court; and

(b) in the case of a Member of the Appellate Tribunal,

(i) is, or has been, or is qualified to be, a Judge of a High Court; or

(ii) is, or has been, a Member of the Indian Legal Service and has held a post in Grade I in that service for at least three years; or

(iii) is, or has been, a Secretary for at least one year in Ministry or Department or the Central Government dealing with the Power, or Coal, or Petroleum and Natural Gas, or Atomic Energy; or

(iv) is, or has been Chairman of the Central Electricity Authority for at least one year; or

(v) is, or has been, Director-General of Bureau or Director-General of the Central Power Research Institute or Bureau of Indian Standards for at least three years or has held any equivalent post for at least three years; or

(vi) is, or has been, a qualified technical person of ability and standing having adequate knowledge and experience in dealing with the matters relating to energy production and supply, energy management, standardisation and efficient use of energy and its conservation, and has shown capacity in dealing with problems relating to engineering, finance, commerce, economics, law or management

Term of Office

34. The Chairperson of the Appellate Tribunal and every Member of the Appellate Tribunal shall hold office as such for a term of five years from the date on which he enters upon his office:

Provided that no Chairperson of the Appellate Tribunal or Member of the Appellate Tribunal shall hold office as such after he has attained, —

(a) in the case of the Chairperson of the Appellate Tribunal, the age of seventy years;

(b) in the case of any Member of the Appellate Tribunal, the age of sixty-five years.

Terms and conditions of service

35. The salary and allowances payable to and the other terms and conditions of service of the Chairperson of the Appellate Tribunal, Members of the Appellate Tribunal shall be such as may be prescribed:

Provided that neither the salary and allowances nor the other terms and conditions of service of the Chairperson of the Appellate Tribunal or a Member of the Appellate Tribunal shall be varied to his disadvantage after appointment.

Vacancies

36. If for reason other than temporary absence any vacancy occurs in the office of the Chairperson of the Appellate Tribunal or the Member of the Appellate Tribunal, the Central Government shall appoint another person in accordance with the provisions of this Act to fill the vacancy and the proceedings may be continued before the Appellate Tribunal from the stage at which the vacancy is filled.

Registration and removal

37. (1) The Chairperson or a Member of the Appellate Tribunal may, by notice in writing under his hand addressed to the Central Government, resign his office: Provided that the Chairperson of the Appellate Tribunal or a Member of the Appellate Tribunal shall, unless he is permitted by the Central Government to relinquish his office sooner, continue to hold office until the expiry of three months from the date of receipt of such notice or until a person duly appointed as his successor enters upon his office or until the expiry of term of office, whichever is the earliest.

(2) The Chairperson of the Appellate Tribunal or a Member of the Appellate Tribunal shall not be removed from his office except by an order by the Central Government on the ground of proved misbehaviour or incapacity after an inquiry made by such persons as the President may appoint for this purpose in which the Chairperson or a Member of the Appellate Tribunal concerned has been informed of the charges against him and given a reasonable opportunity of being heard in respect of such charges.

Member to act as Chairperson in certain circumstances

38. (1) In the event of the occurrence of vacancy in the office of the Chairperson of the Appellate Tribunal by reason of his death, resignation or otherwise, the senior-most member of the Appellate Tribunal shall act as the Chairperson of the Appellate Tribunal until the date on which a new Chairperson appointed in accordance with the provisions of this Act to fill such vacancy enters upon his office.

(2) When the Chairperson of the Appellate Tribunal is unable to discharge his functions owing to his absence, illness or any other cause, the senior-most Member of the Appellate Tribunal shall discharge the functions of the Chairperson of the Appellate Tribunal until the date on which the Chairperson of the Appellate Tribunal resumes his duties.

Staff of Appellate Tribunal

39. (1) The Central Government shall provide the Appellate Tribunal with such officers and employees as it may deem fit.

(2) The officers and employees of the Appellate Tribunal shall discharge their functions under the general superintendence of the Chairperson of the Appellate Tribunal as the case may be.

(3) The salaries and allowances and other conditions of service of the officers and employees of the Appellate Tribunal shall be such as may be prescribed.

Procedure and powers of Appellate Tribunal

5 of 1908 40. (1) The Appellate Tribunal shall not be bound by the procedure laid down by the Code of Civil Procedure, 1908, but shall be guided by the principles of natural justice and subject to the other provisions of this Act, the Appellate Tribunal shall have powers to regulate its own procedure.

5 of 1908 (2) The Appellate Tribunal shall have, for the purposes of discharging its functions under this Act, the same powers as are vested in the civil court under the Code of Civil Procedure 1908, while trying to suit in respect of the following matters, namely:

(a) summoning and enforcing the attendance of any person and examining him on oath;

(b) requiring the discovery and production of documents;

(c) receiving evidence of affidavits;

1 of 1872 (d) subject to the provisions of Sections 123 and 124 of the Indian Evidence Act, 1872, requisitioning any public record or document or copy of such record or document from any office;

(e) issuing commissions for the examination of witnesses or documents;

(f) reviewing its decisions;

(g) dismissing a representation of default or deciding it, *ex parte;*

(h) setting aside any order of dismissal or any representation for default or any order passed by it, *ex parte;*

(i) any other matter which may be prescribed by the Central Government.

(3) An order made by the Appellate Tribunal under this Act shall be executable by the Appellate Tribunal as a decree of civil court and, *for* this purpose, the Appellate Tribunal shall have all the powers of a civil court.

(4) Notwithstanding anything contained in subsection (3), the Appellate Tribunal may transmit any order made by it to a civil court having local jurisdiction and such civil court shall execute the order as if it were a decree made by the that court.

45 of 1860 (5) All proceedings before the Appellate Tribunal shall be deemed to be judicial proceedings within the meaning of Sections 193 and 228 of the Indian Penal Code and the Appellate Tribunal shall be deemed to be civil court for the purposes of Sections 345 and 346 of the Code of Criminal

2 of 1974 Procedure, 1973.

Distribution of business amongst Benches

41. Where Benches are constituted, the Chairperson of the Appellate Tribunal may, from time to time, by notification, make provisions as to the distribution of the business of the Appellate Tribunal amongst the Benches and also provide for the matters which may be dealt with by each Bench.

Power of Chairperson to transfer cases

42. On the application of any of the parties and after notice to the parties, and after hearing such of them as he may desire to be heard, or on his own motion without such notice, the Chairperson of the Appellate Tribunal may transfer any case pending before one Bench for disposal, to any other Bench.

Decision to be by majority

43. If the Members of the Appellate Tribunal of a Bench consisting of two Members differ in opinion on any point, they shall state the point or points on which they differ, and make a reference to the Chairperson of the Appellate Tribunal who shall either hear the point or points himself or refer the case for hearing on such point or points by one or more of the other Members of the Appellate Tribunal and such point or points shall be decided according to the opinion of the majority of the Members of the Appellate Tribunal who have heard the case, including those who first heard it.

Right to appellant to take assistance of legal practitioner or accredited auditor and of Government to appoint presenting officers

44. (1) A person preferring an appeal to the Appellate Tribunal under this Act may either appear in person or take assistance of a legal practitioner or an accredited energy auditor of his choice to present his case before the Appellate Tribunal, as the case may be.

(2) The Central Government or the State Government may authorise one or more legal practitioners or any of its officers to act as presenting officers and every person so authorised may present the case with respect to any appeal before the Appellate Tribunal as the case may be.

Appeal to Supreme Court

45. Any person aggrieved by any decision or order of the Appellate Tribunal may, file an appeal to the Supreme court within sixty days from the date of communication of the decision or order of the Appellate Tribunal to him, on any one or more of the ground specified in Section 100 of the Code of Civil Procedure, 1908: 5 of 1908

Provided that the Supreme Court may, if it is satisfied that the appellant was prevented by the sufficient cause from the filing the appeal within the said period, allow it to be filed within a further period of not exceeding sixty days.

CHAPTER X
MISCELLANEOUS

Power of Central Government to issue directions to Bureau

46. (1) Without prejudice to the foregoing provisions of this Act, the Bureau shall, in exercise of its powers or the performance of its functions under this Act, be bound by such directions on questions of policy as the Central Government may give in writing to it from time to time: Provided that the Bureau shall, as far as practicable, be given an opportunity to express his views before any direction is given under this subsection,

(2) The decision of the Central Government, whether a question is one of policy or not, shall be final.

Power of Central Government to supersede Bureau

47. (1) If at any time the Central Government is of opinion —

(a) that on account of grave emergency, the Bureau is unable to discharge the functions and duties imposed on it by or under the provisions of this Act; or

(b) that the Bureau has persistently made default in complying with any direction issued by the Central Government under this Act or in discharge of the functions and duties imposed on it by or under the provisions of this Act and as a result of such default, the financial position of the Bureau had deteriorated or the administration of the Bureau had deteriorated; or

(c) that circumstances exist which render it necessary in the public interest so to do, the Central Government may, by notification, supersede the Bureau for such period, not exceeding six months, as may be specified in the notification.

(2) Upon the publication of a notification under subsection (1) superseding the Bureau

(a) all the members referred to in clauses (o), (p) and (q) of subsection (2) of Section 4 shall, as from the date of supersession, vacate their offices as such;

(b) all the powers, functions and duties which may, by or under the provisions of this Act, be exercised or discharged by or on behalf of the Bureau, shall until the Bureau is reconstituted under subsection (3), be exercised and discharged by such person or persons as the Central Government may direct; and

(c) all property owned or controlled by the Bureau shall, until the Bureau is reconstituted under subsection (3), vest in the Central Government.

(3) On the expiration of the period of supersession specified in the notification issued under subsection (1), the Central Government may reconstitute the Bureau by a fresh appointment and in such case any person or persons who vacated their offices under clause (a) of subsection (2), shall not be deemed disqualified for appointment:

Provided that the Central Government may, at any time, before the expiration of the period of supersession, take action under this subsection.

(a) the Central Government shall cause a notification issued under subsection (1) and full report of any action taken under this section and the circumstances leading to such action to be laid before each House of Parliament at the earliest.

Default by companies

48. (1) Where a company makes a default in complying with the provisions of clause (c) or clause (d) or clause (h) or clause (i) or clause (k) or clause (l) or clause (n) or clause (r) or clause (s) of section 14 or clause (b) or clause (c) or clause (h) of Section 15, every person who at the time of such contravention was incharge of, and was responsible to the company for the conduct of the business of the company, as well as the company, shall be deemed to have acted in contravention of the said provisions and shall be liable to be proceeded against and imposed penalty under Section 26 accordingly:

Provided that nothing contained in this subsection shall render any such person liable for penalty provided in this Act if he proves that the contravention of the aforesaid provisions was committed without his knowledge or that he exercised all due diligence to prevent the contravention of the aforesaid provision.

(2) Notwithstanding anything contained in subsection (1), where any contravention of the provisions of clause (c) or clause (d) or clause (h) or clause (i) or clause (k) or clause (l) or clause (n) or clause (r) or clause (s) of Section 14 or clause (b) or clause (c) or clause (h) of Section 15 has been committed with the consent or connivance of, or in attributable to, any neglect on the part of, any director, manager, secretary or other officer of the company, such director, manager, secretary or other officer shall also be deemed to have contravened the said provisions and shall be liable to be proceeded for imposition of penalty accordingly.

Explanation – For the purposes of this section, "company" means a body corporate and includes a firm or other association of individuals.

43 of 1961

Exemption from tax on income

49. Notwithstanding anything contained in the Income Tax Act, 1961, or any other enactment for the time being in force relating to tax on income, profits or gains –

(a) the Bureau;

(b) the existing Energy Management Centre from the date of its constitution to the date of establishment of the Bureau, shall not be liable to pay any income tax or any tax in respect of their income, profits or gains derived.

Protection of action taken in good faith

50. No suit, prosecution or other legal proceedings shall lie against the Central Government or Director-General or Secretary or State Government or any officer of those Governments or State Commission or its members or any member or officer or other employee of the Bureau for anything which is in good faith done or intended to be done under this Act or the rules or regulations made thereunder.

Delegation

51. The Bureau may, by general or special order in writing, delegate to any member, member of the committee, officer of the Bureau or any other person subject to such conditions, if any, as may be specified in the order, such of its powers and functions under this Act (except the powers under Section 58) as it may deem necessary.

Power to obtain information

52. Every designated consumer or manufacturer of equipment or appliances specified under clause (b) of Section 14 shall supply the Bureau with such information, and with such samples of any material or substance used in relation to any equipment or appliance, as the Bureau may require.

Power to exempt

53. If the Central Government or the State Government is of the opinion that it is necessary or expedient so to do in the public interest, it may, by

notification and subject to such conditions as may be specified in the notification, exempt any designated consumer or class of designated consumers from application of all or any of the provisions of this Act:

Provided that the Central Government or the State Government, as the case may be, shall not grant exemption to any designated consumer or class of designated consumers for the period exceeding five years:

Provided further that the Central Government or State Government, as the case may be, shall consult the Bureau of Energy Efficiency before granting such exemption.

Chairperson, Members, officers and employees of the Appellate Tribunal, Members of State Comission, Director-General, Secretary, members, officers and employees to be public servants

54. The Chairperson of the Appellate Tribunal or the Members of the Appellate Tribunal or officers or employees of the Appellate Tribunal or the members of the State Commission or the members, Director-General, Secretary, officers and other employees of the Bureau shall be deemed, when acting or purporting to act in pursuance of any of the provisions of the Act, to be public servants within the meaning of Section 21 of the Indian Penal Code. 45 of 1860

Power of Central Government to issue directions

55. The Central Government may give directions to a State Government or the Bureau as to carrying out into execution of this Act in the State.

Power of Central Government to make rules

56. (1) The Central Government may, by notification, make rules for carrying out the provisions of this Act.

(2) In particular, and without prejudice to the generality of the foregoing power, such rules may provide for all or any of the following matters, namely:

(a) such number of persons to be appointed as members by the Central Government under clauses (o), (p) and (q) of subsection (2) of Section 4;

(b) the fee and allowances to be paid to the members under subsection (5) of Section 4;

(c) the salary and allowances payable to the Director-General and other terms and conditions of his service and other terms and conditions of service of the Secretary of the Bureau under subsection (4) of Section 9;

(d) the terms and conditions of service of officer and other employees of the Bureau under subsection (2) of Section 10;

(e) performing such other functions by the Bureau, as may be prescribed, under clause (u) of subsection (2) or Section 13;

(f) the energy consumption norms and standards for designated consumers under clause (g) of Section 14;

(g) prescribing the different norms and standards for different designated consumers under the proviso to clause (g) of Section 14;

(h) the form and manner and the time within which information with regard to energy consumed and the action taken on the recommendations of the accredited energy auditor be furnished under clause (k) of Section 14;

(i) the form and manner in which the status of energy consumption be submitted under clause (l) of Section 14;

(j) the minimum qualification for energy managers under clause (m) of Section 14;

(k) the form and manner for preparation of scheme and its implementation under clause (o) of Section 14;

(l) the energy conservation building codes under clause (p) of Section 14;

(m) the matters relating to inspection under subsection (2) of Section 17;

(n) the form in which, and the time at which, the Bureau shall prepare its budget under Section 22;

(o) the form in which, and the time at which, the Bureau shall prepare its annual report under Section 23;

(p) the form in which the accounts of the Bureau shall be maintained under Section 25;

(q) the manner of holding inquiry under subsection (l) of Section 27;

(r) the form of and fee for filing such appeal under subsection (2) of Section 31;

(s) the salary and allowances payable to and other terms and conditions of service of the Chairperson of the Appellate Tribunal and Member of the Appellate Tribunal under Section 35;

(t) the salary and allowances and other conditions of service of the officers and other employees of the Appellate Tribunal under subsection (3) of Section 39;

(u) the additional matters in respect of which the Appellate Tribunal may exercise the powers of a civil court under clause (i) of subsection (2) of Section 40;

(v) any other matters which is to be, or may be, prescribed, or in respect of which provision is to be made, or may be made by rules.

Power of State Governments to make rules

57. (1) The State Government may, by notification, makes rules for carrying out the provisions of this Act and not inconsistent with the rules, if any, made by the Central Government.

(2) In particular, and without prejudice to the generality of the foregoing power, such rules may provide for all or any of the following matters, namely:

(a) energy conservation building codes under clause (a) of Section 15;

(b) the form, the manner and the period within which information with regard to energy consumption shall be furnished under clause (h) of Section 15;

(c) the person or any authority who shall administer the Fund and the manner in which the Fund shall be administered under subsection (4) of Section 16;

(d) the matters to be included for the purposes of inspection under subsection (2) of Section 17;

(e) any other matter which is to be, or may be, prescribed, or in respect of which provision is to be made, or may be made, by rules.

Power of Bureau to make regulations

58. (1) The Bureau may, with the previous approval of the Central Government and subject to the condition of previous publication, by notification, make regulations not inconsistent with the provisions of this Act and the rules made thereunder to carry out the purposes of this Act.

(2) In particular, and without prejudice to the generality of the foregoing power, such regulations may provide for all or any of the following matters, namely:

(a) the times and places of the meetings of the Governing Council and the procedure to be followed at such meetings under subsection (1) of Section 5;

(b) the members of advisory committees constituted under subsection (2) of Section 8;

(c) the powers and duties that may be exercised and discharged by the Director-General of the Bureau under subsection (6) of Section 9;

(d) the levy of fee for services provided for promoting efficient use of energy and its conservation under clause (n) of subsection (2) of Section 13;

(e) the list of accredited energy auditors under clause (o) of subsection (2) of Section 13;

(f) the qualifications for accredited energy auditors under clause (p) of subsection (2) of Section 13;

(g) the manner and the intervals or time in which the energy audit shall be conducted under clause (q) of subsection (2) of Section 13;

(h) certification procedure for energy managers under clause (r) of subsection (2) of Section (13);

(i) particulars required to be displayed on label and the manner of their display under clause (d) of Section 14;

(j) the manner and the intervals of time for conduct of energy audit under clause (h) or clause (s) of Section 14;

(k) the manner and the intervals of time for conducting energy audit by an accredited energy auditor under clause (c) of Section 15;

(l) any other matter which is required to be, or may be, specified.

Rules and regulations to be laid before Parliament and State Legislature

59. (1) Every rule made by the Central Government and every regulation made under this Act shall be laid, as soon as may be after it is made, before each House of Parliament while it is in session, for a total period of thirty days which may be comprised in one session or in two or more successive session, and if, before the expiry of the session immediately following the session or the successive sessions aforesaid, both Houses agree in making any modification in the rule or regulation, or both Houses agree that the rule or regulation should not be made, the rule or regulation shall thereafter have effect only in such modified form or be of no effect, as the case may be; so however that any such modification or annulment shall be without prejudice to the validity of anything previously done under that rule or regulation.

(2) Every rule made by the State Government shall be laid, as soon as may be after it is made, before each House of the State Legislature where it consists of two Houses, or where such Legislature consists of one House, before that House.

Application of other laws not barred

60. The provisions of this Act shall be in addition to, and not in derogation of, the provisions of any other law for the time being in force.

Provisions of Act not to apply in certain cases

61. The provisions of this Act shall not apply to the Ministry or Department of the Central Government dealing with Defence, Atomic Energy or such other similar Ministries or Departments, undertakings or Boards or institutions under the control of such Ministries or Departments as may be notified by the Central Government.

Power to remove difficulty

62. (1) If any difficulty arises in giving effect to the provisions of this Act, the Central Government may, by order, published in the Official Gazette, make such provisions not inconsistent with the provisions of this Act as may appear to be necessary for removing the difficulty:

Provided that no such order shall be made under this section after the expiry of two years from the date of commencement of this Act.

(2) Every order made under this section shall be laid, as soon as may be after it is made, before each House of Parliament.

43.12.2 ENVIRONMENTAL LEGISLATIONS IN USA

There are four categories of US Environmental legislations : viz., Administrative laws, Information laws, Natural resource laws, and Pollution control laws. There is only one *Administrative law* viz., Administrative Procedures Act (APA). There are four *information laws viz.* :

- The National Environmental Policy Act, 1969, 1992;
- The Freedom of Information Act;
- The Occupational Safety and Health Act;
- The Emergency Planning & Community Right to Know Act, 1986.

The *natural resource* laws are

- The Endangered Species Act, 1973.
- The Coastal Zone Management Acts, 1972, 1990.

The *Pollution Control laws* are :

- The Clean Air Act, 1970, 1977, 1990.
- The Resource Conservation and Recovery Act, 1976, 1992.
- The Comprehensive Environmental Response, Compensation & Liability Act, 1992.
- The Noise Control Act, 1972, 1992.
- The Safe Drinking Water Act, 1974.
- The Federal Water Pollution Control Act, 1972, 1990.
- The Toxic Substance Control Act, 1992.
- The Federal Insecticide, Fungicide and Rodenticide Act, 1972.
- The Pollution Prevention Act, 1990.

Federal Environmental Protection Agencies (FEPA) are acting as monitoring or enactment agencies in all the matters relating to environmental protection and conservations.

Green Bench

The Supreme Court of India felt the need of establishing the Green Bench in different high courts in India with the objectives of dealing with the environmental issues in the states. There are two Green Benches—one at Kolkata and the other at Chennai was established for dealing environmental cases. In the eastern region of Supreme Court transfer the cases of tannery effluent issues and of 68 categories of industries to Calcutta High Court for passing necessary order. Accordingly the Green Bench was established in the Calcutta High Court and the newly formed Green Bench started functioning since June 1996. The Green Bench dealt with about cases up to 2010. These cases related to various kinds of environmental issues such as:

(a) Industries without pollution control system and allegedly violating norms
(b) Illegal filling of water bodies/tanks
(c) Cutting and felling of trees
(d) Auto emission
(e) Biomedical waste and health hazards
(f) Dumping of garbage on streets and non-clearance
(g) Pollution generated from morgues in the state
(h) Regeneration of lakes and parks

Green Justice : Selected Environmental Cases

The Taj Mahal Case

Besides the obvious health and environmental impacts, the almost intolerable levels of air pollution in India's cities are wreaking havoc on the country's renowned architectural heritage. International concern has been generated in light of the noticeable deterioration of the famous Taj Mahal in Agra, Uttar Pradesh. Legal history was made in 1993 when the Supreme Court of India ordered the closure of 212 industrial units in and around Agra that were found to be in willful default of air emission standards and thus thought to be contributing directly to the deterioration of this historic seventeenth-century monument.

By virtue of the fact that public interest litigation is allowed under Section 19 of the Environmental Protection Act (1986), the environmental lawyer M. C. Mehta filed a writ with the Supreme Court of India demanding that 'suitable order/directions be issued to restrain air pollution being caused by the polluting industries located in the area' of the

Contd.

Taj Mahal. Subsequently the Supreme Court ordered the Uttar Pradesh Pollution Control Board (UPPCB) to undertake a survey:

In order to identify the actual industries and foundries operating in the region and further directed that 'the Board, after completing the survey, should issue notices to all the foundries and industries in that region to ensure that the necessary anti-pollution measures were undertaken by said industries and foundries.

In its report to the Supreme Court the UPPCB stated that notices had been issued to 511 industries in and around Agra. The Court then directed the UPPCB to issue notice to these industries to instal antipollution mechanisms. Some months later the UPPCB filed an affidavit stating that 212 industries had ignored the notice and the court ordered the immediate closure of these units. Industries that had filed replies were given time to instal pollution-control devices. As noted by Mehta:

It was shocking to know that none of the 511 factories surveyed by the Board had any pollution control devices. This showed clearly that the (UPPCB) and the Government of India did not take any action in the past against these industries and they were allowed to pollute an ecologically fragile area recklessly.

A further 29 units that had made no progress towards the installation of air pollution-control systems were directed to close, meaning that a total of 241 industries were required to shut down. One hundred and thirtysix of these have since installed pollution-control systems and eight have removed the source of air pollution. These 144 industries have been allowed to reopen, but the remaining 97 are still closed. Out of the remaining 270 industries (511 minus 241), 257 have installed pollution-control systems and three have removed the source of pollution. Installation of such systems is under way in ten industries. Thus, out of a total of 511 industries, 393 have installed air-pollution-control systems, 10 are constructing such systems, eleven have removed the source of pollution and 97 remain closed. In 1994 the Supreme Court also ordered the Central Ministry of Environment and Forests to establish a special division to develop a green belt around the Taj Mahal.

The Ganga Pollution Cases

In 1985, as a reflection of growing public concern regarding the highly polluted state of one of India's most important rivers, the environmental lawyer M. C. Mehta filed a writ under Article 32 of the Indian Constitution stating that neither the relevant government agencies nor industry were taking adequate steps to mitigate the flow of pollution into the Ganga. The writ requested the Court to order government authorities and tanneries in Jajmau near Kanpur to halt the discharge of effluent into the river until appropriate treatment facilities had been installed. The relevant authorities and industries were ordered to show cause as to why the petition should not be adhered to. As a result of the process, 29 tanneries were ordered to close for failing to provide even primary treatment of their waste. This action may also have motivated greater diligence on the part of the environmental authorities.

In the Ganga basin area of Uttar Pradesh (excluding Kanpur) between 1987 and 1993, 860 industries were identified by the Uttar Pradesh Pollution Control Board (UPPCB) as guilty of water pollution. Of these, 191 failed to respond to notices issued to them by the UPPCB in September 1993 to submit a report on their pollution-control status. These 'willful defaulters' were subsequently ordered to close by the Supreme Court of India. Another 43 that had made no progress towards installing effluent treatment plants were directed to close. Therefore, a total of 234 industries were ordered to close. However, 142 subsequently installed treatment facilities and were permitted to reopen. Of the remaining 626 water polluting industries, 493 have installed effluent treatment plants, 261 of which meet the required standards but 232 remain in default. These defaulters are required to meet the standards either by upgrading their technology or by proper operation and maintenance of the existing treatment facilities. Out of the remaining 133 industries, 66 are in the process of installing waste treatment equipment and 67 have closed.

The Delhi Oleum Gas Leak Case

In 1985, the environmental lawyer M. C. Mehta filed a writ with the Supreme Court of India alleging that pollution being generated by the Shri Ram Foods and Fertiliser Company in a densely populated area of Delhi was an infringement of the Constitutional right to life (Article 21 of the Constitution). In this instance the right to life was seen as including the right to live in a healthy environment. The petition sought to have the manufacturer's caustic chlorine and sulphuric acid plants relocated to a more suitable industrial area.

On 4 December 1985, one month after the writ was filed and one day after the first anniversary of the Bhopal Tragedy, the fumes from an oleum leak from Sri Ram's sulphuric acid plant struck down a number of nearby residents and caused widespread panic. In a precedent-setting decision by the Supreme Court, the Shri Ram Foods and Fertilisers Company was ordered to pay compensation to those affected by the leak. The Court's action is considered revolutionary in legal terms because, as Mehta explains :

The power of the Supreme Court to grant remedial relief for a proved infringement of a Fundamental Right (in this case Article 21) includes the power to award compensation. Thus the Court not only widened the scope of Article 21 (by including in it protection of the environment) but also included a liability in tort for those harmed by pollution.

Though initially recalcitrant to the Court's directions, the Shri Ram Foods and Fertilisers Company is now going beyond simple compliance with environmental standards and is employing a waste minimisation approach in order to lean up its operations, increase production efficiency, and quell public anxiety.

44 MINISTRY OF ENVIRONMENT AND FORESTS—RULES AND NOTIFICATIONS RELATING TO HAZARDOUS MICROORGANISMS AND GENETICALLY MODIFIED ORGANISMS OR CELL

New Delhi, the 9th February, 1999

G. S. R. 98(E)—In exercise of the powers conferred by Sections 6, 8, 10 and 25 of the Environment (Protection) Act, 1986 (29 of 1986), the Central Government hereby publishes the following draft of certain rules in supersession of the *Rules for Manufacture, Use, Import, Export and Storage of Hazardous Microorganisms/Genetically Modified Organisms or Cells, 1993,* except as respects things done or omitted to be done before such supersession, for the information of all persons likely to be affected thereby and notice is hereby given that the said draft rules will be taken into consideration after the expiry of a period of 60 days from the date on which the Gazette copies containing this notification are made available to the public.

Any objection or suggestion which may be received from any person in respect of the said draft rules before expiry of the period specified above, will be taken into consideration by the Central Government.

The objections or suggestions may be addressed to the Secretary, Ministry of Environment and Forests, Paryavaran Bhawan, C. G. O. Complex, Lodhi Road, New Delhi.

Draft rules

G. S. R. No.—In exercise of the powers conferred by Sections 6, 8, 10 and 25 of the Environment (Protection) Act, 1986 (29 of 1986); the Central Government hereby makes the following rules, namely:

1. **Short title and commencement**

   (1) These Rules may be called the Hazardous Microorganisms and Genetically Modified Organisms (Manufacture, Use, Import, Export and Storage) Rules, 1999.

   (2) They shall come into force on the date of their publication in the Official Gazette.

2. **Application**

   (1) These rules shall be applicable to:

   (a) hazardous microorganisms and genetically modified organisms including their products for the purpose of

   (i) import and export.

   (ii) manufacture, use and storage, if the capacity is of 20 litres and above culture volume.

   (b) new gene technologies apart from those referred to in clause (h) of rule 3 and to micro-organisms generated by the utilisation of such other gene technologies and to substances and products of which such organisms form part.

(2) These rules shall not apply to the pesticides governed by the Insecticides Act, 1968 (46 of 1968).

3. **Definitions**: In these rules, unless the context otherwise requires:

(a) "accident" has the meaning assigned to it by clause (a) of Section 2 of the Public Liability Insurance Act, 1991 (6 of 1991);

(b) "Act" means the Environment (Protection) Act, 1986 (29 of 1986);

(c) "biotechnology" means the exploitation of biological processes for industrial and other purposes, especially genetic manipulation of microorganisms (for the production of antibiotics, hormones and the like);

(d) "contained use" means any operation in which microorganisms are genetically modified or in which such genetically modified microorganisms are cultured, stored, used, transported, destroyed or disposed of and for which physical barriers or a combination of physical barriers together with chemical and/or biological barriers, are used to limit contact with the environment;

(e) "deliberate release" means any intentional introduction into the environment of hazardous microorganisms or genetically modified organisms or their combination without provisions for containment such as physical barriers or a combination of physical barriers together with chemical and/or biological barriers used to limit their contact with the general population and the environment;

(f) "export" with its grammatical variations and cognate expressions, means taking out of India to a place outside India;

(g) "Form" means form annexed to these rules;

(h) "genetically modified microorganisms (GMOs)" means a microorganism in which the genetic material has been altered, in a way that does not occur naturally by mating and/or recombination, by genetic modification through the use of the techniques listed in Part I of Schedule I but does not include a modification through a technique listed in Part II or Part III of the said Schedule;

(i) "hazardous microorganism" means any microscopic living entity—cellular or non-cellular—capable of replication or of transferring genetic material hazardous to health and environment as listed in Schedule II.

(j) "import" with its grammatical variations and cognate expression, means bringing into India from a place outside India;

(k) "product" means a preparation consisting of, or containing a hazardous microorganism or GMO or a combination of these and the resulting material placed on the market;

(l) "Schedule" means a schedule annexed to these rules.

4. **Expert Committees**

(1) Recombinant DNA Advisory Committee (RDAC)

(i) The Central Government may, by order, constitute a committee to be called the Recombinant DNA Advisory Committee (RDAC), in the Department dealing with biotechnology.

(ii) The RDAC shall review developments in biotechnology at national and international levels and recommend appropriate safety guidelines for India in recombinant research, use and applications from time to time.

(2) Review Committee on Genetic Manipulation (RCGM)

(i) The Review Committee on Genetic Manipulation (RCGM) shall consist of

(a) representatives of

(1) Department dealing with Biotechnology.

(2) Indian Council of Medical Research.

(3) Indian Council of Agricultural Research.

(4) Council of Scientific and Industrial Research; and

(b) other expert or experts nominated by the Department dealing with biotechnology in his or their individual capacity.

(ii) The RCGM shall function in the Department dealing with Biotechnology to monitor the safety related aspect in respect of ongoing research projects and activities involving hazardous microorganisms or GMOs.

(iii) The RCGM shall bring out Manuals or Guidelines specifying the procedure for regulatory process with respect to activities involving genetically modified organisms in research and their use and application in industry and other areas with a view to ensure environmental safety.

(iv) The RCGM shall also review all ongoing projects involving high risk category and controlled field experiments to ensure that adequate precautions and containment conditions are followed as per its guidelines.

(v) The RCGM shall lay down procedures restricting or prohibiting production, sale, importation and use of such hazardous microorganisms or GMOs as mentioned in Schedule II.

(vi) On the recommendation of the RCGM, the Central Government may, by notification in the official Gazette:

(a) recognise one or more laboratories or institutes for conducting tests on hazardous microorganisms or GMOs; and

(b) prescribe special tests to be conducted for hazardous microorganisms or GMOs.

(3) Institutional Bio-Safety Committee (IBSC)

(i) The Institutional Bio-safety Committee (IBSC) shall be constituted by an occupier including research institutions handling hazardous microorganisms or GMOs.

(ii) The IBSC shall be constituted as required under para (i) of this sub-rule:

(a) in case of a new activity before that activity is commenced; and

(b) in case of an existing activity, within ninety days of coming into operation of these rules.

(iii) The IBSC in respect of an institution shall consist of head of that institution, scientists engaged in the work on hazardous microorganisms or GMOs, a medical expert and a nominee of the Department dealing with biotechnology.

(iv) The occupier including research institutions handling hazardous microorganisms or GMOs, shall prepare, with the assistance of IBSC, an up-to-date on-site emergency plan according to the guidelines issued from time-to-time.

(v) The occupier including research institutions shall make available copies of its plan to the concerned District Level Committee and State Biotechnology Coordination Committee.

(vi) The occupier including research institutions handling hazardous microorganisms or GMOs shall prepare on-site plan required under para (v).

(a) in case of a new industrial or research activity before that activity is commenced; and

(b) in case of an existing industrial or research activity, within ninety days of coming into operation of these rules.

(4) Genetic Engineering Approval Committee (GEAC):

(i) The Central Government may constitute a Committee to be called the Genetic Engineering Approval Committee (GEAC) in the Ministry of Environment and Forests consisting of the following, namely:

1. Additional Secretary or Joint Secretary, Ministry of Environment and Forests—Chairman.
2. Nominee of the Department of Biotechnology—Co-Chairman.
3. A representative each of the Ministry of Industrial Development and the Department dealing with Biotechnology and the Department dealing with Atomic Energy—Members.
4. (a) A representative of: (1) Director-General, Indian Council of Medical Research; (2) Director-General, Indian Council of Agricultural Research; (3) Director-General, Council of Scientific and Industrial research; (4) Director-General, Health Services; (5) Plant Protection Adviser, Directorate of Plant Protection, Quarantine and Storage, (6) Chairman, Central Pollution Control Board; and

   (b) three outside experts in biotechnology nominated by the Central Government in individual capacity—Members.
5. An officer of the Ministry of Environment and Forest—Member-Secretary.

(ii) The GEAC may, with the approval of its Chairman, co-opt members/experts.

(iii) The GEAC shall be responsible for approval of activities involving large-scale (20 litres and above culture volume) manufacture, use and storage of hazardous microorganisms or GMOs in research and industry from environmental angle.

(iv) The GEAC shall also be responsible for environmental approval of proposals relating to large-scale release (20 litres and above culture volume) of hazardous microorganisms or GMOs and products including deliberate release in the environment including experimental or field trials.

(v) The GEAC shall also be responsible for environmental approval of proposals relating to import and export of hazardous microorganisms or GMOs.

(5) State Biotechnology Coordination Committee (SBCC):

(i) The State Government may constitute a State Biotechnology Coordination Committee (SBCC) consisting of

(a) Chief Secretary	Chairman
(b) Secretary Department of Environment	Member
(c) Secretary Department of Health	Member

(d) Secretary Department of Agriculture	Member
(e) Secretary Department of Industries	Member
(f) Secretary Department of Labour	Member
(g) Secretary Department of Forests	Member
(h) Secretary Department of Public Works	Member
(i) State Microbiologist and Pathologist	Member
(j) Member-Secretary State Pollution Control Board or an officer of SPCB authorized by it.	Member Secretary

(ii) The SBCC shall coordinate and monitor the inspection and investigation of violations of the provisions of the Act and these rules and advise the concerned State Pollution Control Board to take necessary corrective steps.

(iii) The SBCC shall review periodically the safety and control measures in various industries and institutions handling hazardous microorganisms or GMOs.

(iv) The SBCC may, with the approval of its Chairman, co-opt members/experts.

(6) District Level Committee (DLC):

(i) The District Level Committee shall consist of such members as the State Government may deem fit.

(ii) In such districts where installations engaged in the activity involving hazardous micro-organisms or GMOs are located, the State Government may constitute a District Level Committee to monitor the safety regulations in installations engaged in the use of hazardous microorganisms or GMOs and its applications in the environment.

(iii) The DLC may, with the permission of the occupier, visit the installations engaged in the activity involving hazardous microorganisms or GMOs, collect information, find out hazards and risks associated with each of these installations and coordinate activities with a view to meet any emergency.

5. **Classification of hazardous microorganisms or GMOs**

(i) The hazardous microorganisms or GMOs shall, for the purpose of these rules, be classified to the risk group(s) as specified in Schedule II.

(ii) If any of the hazardous microorganisms or GMOs falls within the limits of more than one risk group as specified in Schedule II, it shall be deemed to be of the higher order of such group.

6. Containment measures

(1) In order to ensure safety in handling hazardous microorganisms or GMOs, principles of good microbiological practices, occupational safety and hygiene and other containment measures as provided in the guidelines issued by the Government, from time to time, shall be applied.

(2) The safety practices and containment measures applied shall be periodically reviewed by the occupier/Biosafety Committee or the person handling hazardous microorganisms or GMOs to take into account new scientific and technical knowledge related to risk assessment, treatment and disposal of wastes.

7. Hazardous microorganisms or GMOs not to be handled without approval

(1) No person shall manufacture, use, import, export, store or undertake deliberate release of any hazardous microorganisms or GMOs without obtaining the approval of GEAC.

(2) A person seeking approval of GEAC, as required under sub-rule (1), shall submit his application in Form I for clinical, veterinary and food products, in Form II for transgenic plants, and in Form III for other products including products used in pollution control.

(3) An application for approval shall be disposed of by the GEAC after considering all material facts: Provided that no such application shall be rejected without giving the applicant a reasonable opportunity of being heard.

(4) The data to be generated by a person before undertaking a deliberate release of hazardous microorganisms or GMOs or any combination of them shall follow step-by-step principle, that is where the scale of release increases evaluation of the earlier step becomes necessary.

(5) The experiments for generation of data referred to in sub-rule (4) shall be done sequentially from controlled laboratory conditions to the growth chamber and the greenhouse testing to small-scale field trials and finally to large-scale field trials. The GEAC/RCGM shall, if necessary, carry out tests in the approved laboratories or inspections for control purposes.

(6) In the event of any modification of the deliberate release of hazardous microorganisms or GMOs or a combination of any two or more of them which could have consequences with regard to the risks of human health or the environment or if new information has become available on such risks, either while the data is being examined by the GEAC or after the GEAC has given the environmental clearance, the person seeking approval shall immediately

(a) revise the measures specified in the report;

(b) inform the GEAC of any modification as soon as the new information is available; and

(c) take the measures necessary to protect human health and the environment.

(7) (a) An application for approval, complete in all respects, shall be disposed of by the GEAC within 90 days.

(b) Where the GEAC requires any additional information in case of an application or the application is incomplete in any respect, it shall require the applicant to supply such additional information or complete the application form in all respects within a period specified by it, before such application is disposed of.

Explanation—The approval under this sub-rule shall not be construed as a clearance, approval or licence as may be required under any other law.

(8) The GEAC may give directions to the occupier to determine and take measures concerning the discharge of hazardous microorganisms or GMOs mentioned in Schedule II including prohibition of such discharges and laying down measures to be taken to prevent such discharges.

8. Terms and conditions of Approval

(1) In connection with the granting of approval under rule 7, the terms and conditions shall be stipulated, including terms and conditions relating to the control to be exercised by the applicant, supervision, restriction on use, layout of the enterprise and the information to be furnished to the RCGM, SBCC and to the DLC.

(2) The environmental approval of GEAC shall be for a specified period not exceeding four years at the first instance, renewable for two years at a time. The GEAC shall have powers to revoke such approval in the following cases, namely:

(a) if there is any new information as to the harmful effects of the hazardous microorganisms or GMOs; or

(b) if the hazardous microorganisms or GMOs cause such damage to the environment, nature or health as could not be envisaged when the approval was given; or

(c) non-compliance of any of the terms and conditions stipulated by the GEAC.

9. Supervision

(1) The GEAC shall supervise the implementation of the terms and conditions laid down in connection with the approval accorded by it.

(2) The GEAC may carry out the supervision with the assistance of SBCC or the Central or State Pollution Control Boards, RCGM or DLC.

10. Responsibility to notify interruptions or accidents

(1) Any person who under rules 7 and 8 is responsible for complying with the terms and conditions of the approval shall immediately notify the GEAC, DLC and SBCC and the State Pollution Control Board of any interruptions or accidents that may lead to discharge of hazardous microorganisms or GMOs which may be harmful to the environment.

(2) Any notice given under sub-rule (1) shall not absolve the person who is responsible to minimise or prevent the effects of interruptions of operations or accidents from such responsibilities.

11. Preparation of off-site Emergency Plan by the DLC

(1) It shall be the duty of the DLC to prepare an off-site emergency plan detailing how emergencies relating to a possible major accident at a site will be dealt with and in preparing the plan, the DLC shall see that there are adequate containment facilities with the occupier.

(2) The DLC shall prepare the off-site emergency plan required under sub-rule (1), in the case of:

(a) a new activity before that activity has commenced; and

(b) an existing activity within nine months of coming into operation of these rules.

(3) For the purpose of enabling the DLC to prepare the off-site emergency plan required under sub-rule (1) the occupier shall provide the DLC with such information relating to the steps taken for containment and handling of hazardous microorganisms or GMOs under his control as the DLC may require including the nature, extent and likely off-site effects of a possible

major accident and the preventive steps and the DLC shall provide the occupier with any information from the off-site emergency plan which relates to his duties under rule 10.

12. **Inspections**

(1) Any member of the SBCC, GEAC or DLC duly empowered under Section 10 of the Act may, at all reasonable times, enter and inspect any premises for carrying out the functions assigned to the committee to which he is a member.

(2) The SBCC or GEAC or DLC may call from the occupier information, data or report about its activities relating to manufacture, use, import, export and storage of any hazardous micro-organisms or GMOs.

13. **Maintenance of Records**

The occupier or any person handling hazardous microorganisms or GMOs shall be required to keep records of the work carried out by him and be made available to the concerned authority on request.

14. **Appeal**

(1) Any person aggrieved by a decision made by the GEAC or SBCC or DLC in pursuance of these rules may within thirty days from the date on which the decision is communicated to him, prefer an appeal to the Central Government:

Provided that an appeal may be admitted after expiry of the said period if the appellant has sufficient reasons for not preferring the appeal within stipulated time but no such appeal shall be admitted after 45 days.

(2) An appeal under sub-rule (1) shall be addressed and sent to the Secretary to the Government of India, Ministry of Environment and Forests, New Delhi, and shall clearly and precisely specify the grounds on which the appeal is preferred. The appeal shall, after giving an opportunity of being heard to the appellant, be disposed of as expeditiously as possible.

SCHEDULE I
[See rule 3(h)]

PART I

Techniques of genetic modification are:

(1) recombinant DNA techniques involving in vitro introduction of different segments of DNA (one being the vector and others normally unrelated DNA sequences) that are capable of replication in a host cell either autonomously or as an integral part of host's genome and maintenance of their continued propagation;

(2) techniques involving the direct introduction into a microorganism of heritable material prepared outside the microorganism including microinjection, macroinjection and microencapsulation;

(3) cell fusion or hybridization techniques where live cells with new combinations of heritable genetic material are formed through the fusion of two or more cells by means of methods that do not occur naturally; and

(4) genetic engineering including self-cloning and deletion, transformation and other types of virus or pathogen introduction into unnatural hosts.

PART II

Techniques which are not considered to result in genetic modification, on condition that they do not involve the use of recombinant-DNA molecules or genetically modified organisms:

(1) in vitro fertilization;

(2) conjugation, transduction, transformation or any other natural process; and

(3) polyploidy induction.

PART III

Techniques of genetic modification to be excluded from the Rules, on condition that they do not involve the use of genetically modified microorganisms as recipient or parental organisms:

(1) mutagenesis;

(2) the construction and use of somatic animal hybridoma cells (e.g. for the production of monoclonal antibodies);

(3) cell fusion (including protoplast fusion) of cells from plants which can be produced by traditional breeding methods; and

(4) self-cloning of non-pathogenic naturally occurring microorganisms.

SCHEDULE II
[See rules 3 (i) & 5]

PART I : ANIMAL AND HUMAN PATHOGENS

A. BACTERIAL AGENTS

(a) Risk Group II

1. Acinetobacter calcoaceticus
2. Actinobacillus—all species except A. mallei, which is in Risk Group III
3. Aeromonas hydrophila
4. Arizona hinshawil—all serotypes
5. Bacillus anthracis
6. Bordetella—all species
7. Borrella recurrentis B. Vincenti
8. Campylobacter fetus
9. Campylobacter jejuni
10. Chlamydia psittaci
11. Chlamydia trachomatis

12. Clostridum chauvoel, C. difficile, C. fallax, C. haemolyticum, C. histolyticum. C. novyi (=C. oedematiens), C. perfringen, C. septicum, C. sordellii

13. Corynebacterium diphtheriae, C. equi, C. haemolyticum. C. pseudotuberculosis, C. pyogenes, C. renale

14. Diplococcus (Streptococcus) pneumoniae

15. Edwardsiela tarda

16. Erysipelothrix insidiosa

17. Escherichia Coli—all enteropathogenic serotypes, enterotoxigenic

18. Haemophilus ducreyi, H. influenzae, H. pneumoniae

19. Herellea vaginicola

20. Klebsiella—all species and all serotypes

21. Legionella pneumophila

22. Lettonella

23. Leptospira interrogans—all serotypes reported in India

24. Listeria—all species

25. Mima polymorpha

26. Moraxella—all species

27. Mycobacteria—all species including Mycobacterium avium

28. M. bovis, M. tuberculosis, M. leprae

29. Mycoplasma—all species except M. mycoides and M. agalactiae

30. Neisseria gonorrhoeae, N. meningitis.

31. Pasteurella—all species except those listed in Risk Group III

32. Salmonella—all species and all serotypes

33. Shigella—all species and all serotypes

34. Sphaerophorus necrophorus

35. Staphylococcus aureus

36. Streptobacillus moniliformis

37. Streptococcus pneumoniae

38. Streptococcus pyogenes, S. equai

39. Streptomyces madurae, S. Pelletierii, S. somaliensis

40. Treponema carateum, T. pallidum and T. pertenue

41. Vibrio foetus

42. V. parahaemolyticus

43. Vibrio cholerae [or V. comma, including two biotypes-eltor (= El Tor) and cholerae (= classical biotype)]

(b) Risk Group III

44. Actinobacillus mallei
45. Bartonella—all species
46. Brucella—all species
47. Clostridium botulinum, Cl. tetani
48. Francisella tularensis
49. Mycobacterium avium, M. bovis, M. tuberculosis. M. leprae
50. Pasteurella multocida type B ("buffalo" and other foreign virulent strains)
51. Pseudomonas pseudomallei
52. Yersinta pestis
53. Actinomycetes (including Nocardia SP, Actinomyces species and Arachnia propionica)

B. FUNGAL AGENTS

(a) Risk Group II

54. Aspergillus fumigatus
55. Blastomyces dermatitidis
56. Cryptococcus neoformans, C. fersiminosos
57. Epidermophyton madurella, Microsporum
58. Paracoccidioides brasilliensis
59. Sporothrix
60. Trichoderma
61. Trichophyton

(b) Risk Group III

62. Coccidiodes immitis
63. Histoplasma capsulatum
64. Histoplasma capsulatum var duboisii

C. PARASITIC AGENTS

(a) Risk Group II

65. Entamoeba histolytica
66. Lieshmania species
67. Naegleria gruberi

68. Plasmodium theilera, P. fabesia, P. falciparum

69. Plasmodium fabesia

70. Schistosoma

71. Toxoplasma gendii

72. Toxocara canis

73. Trichinella spiralis

74. Trichomonas

75. Trypanosoma cruzi

(b) Risk Group III

76. Schistosoma mansoni

D. VIRAL, RICKETTSIAL AND CHLAMYDIAL AGENTS

(a) Risk Group II

77. Adenoviruses—Human, all types

78. Avian leukosis

79. Cache Valley virus

80. CELO (avian adenovirus)

81. Coxsackie A and B viruses

82. Corona viruses

83. Cytomegalo viruses (CMV)

84. Dengue virus, when used for transmission experiments

85. Echo viruses—all types

86. Encephalomyocarditis virus (EMCV)

87. Flanders virus

88. Hart Park virus

89. Hepatitis associated antigen material—hepatitis A and B viruses, non-A and non-B hepatitis viruses, HDV (Delta virus)

90. Herpes viruses—except herpes virus simiae (monkey B virus) which is in Risk Group IV.

91. Infectious Bovine Rhinotracheitis virus (IBR)

92. Infectious Bursal diseases of poultry and infectious Bronchitis

93. Infectious Laryngotracheitis (ILT)

94. Influenza virus—all types, except A/PRS/34 which is in Risk Group I

95. Langur virus Leucosis Complex

96. Lymphogranuloma venereum agent

97. Marck's Disease virus

98. Measles virus

99. Mumps virus

100. Newcastle disease virus (other than licensed strain for vaccine use)

101. Parainfluenza viruses—all types except parainfluenza virus 3, SF4 strain, which is in Risk Group I

102. Polio viruses—all types, wild and attenuated

103. Pox viruses—all types except Alastrim, monkey pox, sheep pox and white pox, which, depending on experiments, are in Risk Group III or IV

104. Rabies virus—all strains except rabies street virus, which should be classified in Risk Group III when inoculated into carnivores

105. Reoviruses—all types

106. Respiratory syncytial virus

107. Rhinoviruses—all types

108. Rinderpest (other than vaccine strain in use)

109. Rubella virus

110. Simian viruses—all types except herpes virus simiae (Monkey Virus) which is in Risk Group IV

111. Simian virus 40 Ad7 SV 40 (Defective)

112. Sindbis virus (SIN Virus)

113. Tensaw virus

114. Turlock virus

115. Vaccinia virus

116. Varicella virus

117. Vole rickettsia

118. Yellow fever virus, 17 D vaccine strain

(b) Risk Group III

119. African Horse Sickness (attenuated strain except animal passage)

120. Alastrim, monkey pox and white pox, when used in vitro

121. Arboviruses—all strains except those in Risk Group II and IV

122. Blue-tongue virus (only serotypes reported in India)

123. Ebola fever virus

124. Epstein-Barr Virus (EBV)

125. Feline Leukaemia virus (Fel V)

126. Feline sarcoma viruses

127. Foot and Mouth Disease virus (all serotypes and subtypes)

128. Gibbon Ape Lymphosarcoma

129. Herpesvirus ateles

130. Herpes virus saimiri

131. Herpes simplex 2

132. HIV-I & HIV-2 and strains of SIV

133. Infectious Equine Anaemia

134. Lymphocytic choriomeningitis virus (LCM)

135. Monkey pox, when used in vitro

136. Non-defective Adeno-2 SV-40 hybrids

137. Psittacosis-ornithosis-trachoma group of agents

138. Pseudorabies virus

139. Rabies street virus, when used for inoculations of carnivores

140. Rickettsia—all species except Vole rickettsia and Coxiella burnetii, when used for vector transmission or animal inoculation experiments

141. Sheep pox (Field strain)

142. Swine Fever virus

143. Vesicular stomatitis virus

144. Woolly monkey Fibrosarcoma

145. Yaba pox virus.

(C) RISK GROUP IV

146. Alastrim monkey pox, white pox, when used for transmission or animal inoculation experiments

147. Haemorrhagic fever agents including Crimean haemorrhagic fever (Congo)

148. Korean haemorrhagic fever and others as yet undefined

149. Herpes virus simiae (monkey B virus)

150. Tick-borne encephalitis virus complex, including—Russian Spring-Summer Encephalitis, Kyasanur Forest Disease, Omsk haemorrhagic fever and Central European encephalitis viruses.

E. SPECIAL CATEGORY

(a) Bacterial

151. Contagious Equine Metritis (H. equigenitalis)

152. Peste des-petits ruminants.

(b) Viral, Rickettsial and Chlamydial

153. African Horse Sickness virus (Serotypes not reported in India and challenge strains)

154. African swine fever

155. Bat rabies virus

156. Blue-tongue virus (serotypes not reported in India)

157. Exotic FMD virus types and sub-types

158. Junin and Machupo viruses

159. Lassa virus

160. Marburg virus

161. Murray Valley encephalitis virus

162. Rift Valley fever virus

163. Smallpox virus—Archieval storage and propagation

164. Swine vesicular disease

165. Venezuelan equine encephalitis virus—epidemic strains

166. Western Equine encephalitis virus

167. Yellow fever virus—Wild strain

168. Other Arboviruses causing epizootics and so far not recorded in India.

PART II PLANT PESTS

Any living stage (including active and dormant forms) of insects, mites, nematodes, slugs, snails, bacteria, fungi, protozoa, other parasitic plants or reproductive parts thereof; viruses, or any organisms similar to or allied with any of the foregoing; or any infectious agents or substances, which can directly or indirectly injure or cause disease or damage in or to any plants or parts thereof, or any processed, manufactured or other products of plants are considered plant pests. Organisms belonging to all lower-taxa contained within the group listed are also included.

A. VIRUSES

All Viroids

All bacterial, fungal, algal, plant, insect and nematode viruses; special care should be taken for

(i) Geminiviruses

(ii) Caulimoviruses

(iii) Nuclear polyhedrosis viruses

(iv) Granulosis viruses

(v) Cytoplasmic polyhedrosis viruses

B. BACTERIA

1. Family Pseudomonadaceae

(i) Genus Pseudomonas

(ii) Genus Xanthomonas

2. Family Azotobacteriaceae
   (i) Genus Azotobacter
3. Family rhizobiaceae
   (i) Genus Rhizobium/Azorhizobium
   (ii) Genus Bradyrhizobium
   (iii) Genus Agrobacterium
   (iv) Genus Phyllobacterium
4. Family Enterobacteriaceae
   (i) Genus Erwinia
   (ii) Genus Enterobacter
   (iii) Genus Klebsiellea
5. Family Spirillaceae
   (i) Genus Azospirillum
   (ii) Genus Aquaspirillum
   (iii) Genus Oceanospirillum
6. Family Streptomycetaceae
   (i) Genus Streptomyces
   (ii) Genus Nocardia
7. Family Actinomycetaceae
   (i) Genus Actinomyces
8. Coryneform Group
   (i) Genus Clavibacter
   (ii) Genus Arthrobacter
   (iii) Genus Curtobacterium
   (iv) Genus Bdellovibrio
9. Family Rickettsiaceae
   (i) Rickettsial-like organisms associated with insect diseases
   (ii) Gram-negative phloem—limited bacteria associated with plant diseases
   (iii) Gram-negative xylem—limited bacteria associated with plant diseases
   (iv) Cyanobacteria—all members of blue-green algae
   (v) Mollicutes
10. Family Sptroplasmataceae
   (i) Mycoplasma-like organisms associated with plant diseases
   (ii) Mycoplasma-like organisms associated with insect diseases

C. ALGAE

11. Family Chlorophyceae
12. Family Euglenophyceae
13. Family Pyrrophyceae
14. Faimly Chrysophyceae
15. Family Phacophyceae
16. Family Rhodophyceae

D. FUNGI

17. Family Plasmodiophoraceae
18. Family Chytridiaceae
19. Family Olpidiopsidaceae
20. Family Synchytriaceae
21. Family Catenariaceae
22. Family Coelomycetaceae
23. Family Saprolegniaceae
24. Family Zoopagaceae
25. Family Albuginaceae
26. Family Peronosporaceae
27. Family Pythiaceae
28. Family Mucoraceae
29. Family Choaneophoraceae
30. Family Mortierellaceae
31. Faimily Endogoniaceae
32. Family Syncephalastraceae
33. Family Dimargaritaceae
34. Family Kickxellaceae
35. Family Saksenaeaceae
36. Family Entomophthoraceae
37. Family Ecerinaceae
38. Family Taphrinaceae
39. Family Endomycetaceae
40. Family Saccharomycetaceae
41. Family Eurotiaceae

42. Family Gymnoascaceae
43. Family Ascosphaeriaceae
44. Family Onygenaceae
45. Family Micgoascaceae
46. Family Protomycetaceae
47. Family Elsinoaceae
48. Family Myriangiaceae
49. Family Dothidiaceae
50. Family Chaetothyrlaceae
51. Family Parmulariaceae
52. Family Phillipslellaceae
53. Family Hysterlaceae
54. Family Pleosporaceae
55. Family Melanomataceae
56. Family Ophiostomataceae
57. Family Ascosphaeriaceae
58. Family Erysiphaceae
59. Family Meliolaceae
60. Family Xylariaceae
61. Family Diaporthaceae
62. Family Hypocreaceae
63. Family Clavicipitaceae
64. Family Phacidiaceae
65. Family Ascocorticiaceae
66. Family Hemiphacidiaceae
67. Family Dermataceae
68. Family Sclerotiniaceae
69. Family Cyttariaceae
70. Family Heiotiaceae
71. Family Sarcosomataceae
72. Family Sarcoscyphaceae
73. Family Auriculariaceae
74. Family Ceratobasidiaceae

75. Family Corticiaceae
76. Family Hymenochaetaceae
77. Family Echinodontiaceae
78. Family Fistulinaceae
79. Family Clavariaceae
80. Family Polyporaceae
81. Family Tricholomataceae
82. Family Ustilaginaceae
83. Family Sporobolomycetaceae
84. Family Uredirraceae
85. Family Agaricaceae
86. Family Graphiolaceae
87. Family Pucciniaceae
88. Family Melampsoraceae
89. Family Gandodeniatiaceae
90. Family Laboulbeniaceae
91. Family Sphaeropsidaceae
92. Family Melaneonlaceae
93. Family Tuberculariaceae
94. Family Domattaceae
95. Family Moniliaceae
96. Family Aganomycetaceae

E. PARASITIC WEEDS

97. Family Balanosporaceae—parasitic species
98. Family Cuscutaceae—parasitic species
99. Family Ttydonoraceae—parasitic species
100. Family Lauraceae—parasitic species Genus Cassytha
101. Family Lennoaceae—parasitic species
102. Family Loranthaceae—parasitic species
103. Family Myzodendraceae—parasitic species
104. Family Olacaceae—parasitic species
105. Family Orobanchaceae—parasitic species
106. Family Rafflesiaceae—parasitic species

107. Family Santalaceae—parasitic species
108. Family Scrophulariaceae—parasitic species

F. PROTOZOA

109. Genus Phytomonas (Family Trypanasomatidae)
110. All protozoa associated with insect diseases

G. NEMATODES

111. Family Anguinidae
112. Family Belonolaimidae
113. Family Caloosiidae
114. Family Criconematidae
115. Family Dolichodoridae
116. Family Fergusobiidae
117. Family Hemicycliophoridae
118. Family Heteroderidae
119. Family Haplolaimidae
120. Family Meloidogynidae
121. Family Neotylenchidae
122. Family Nothotylenchidae
123. Family Paidlylenchidae
124. Family Pratylonchidae
125. Family Tylenchidae
126. Family Tylenchulidae
127. Family Aphelenchoidae
128. Family Longidoridae
129. Family Trichodoridae

H. MOLLUSCA

130. Superfamily Planorbacea
131. Superfamily Achatinacea
132. Superfamily Arionaceae
133. Superfamily Limacacea
134. Superfamily Helicacea
135. Superfamily Veromcellacea

I. ARTHROPODA

136. Superfamily Ascoidea
137. Superfamily Dermanyssoidea
138. Superfamily Eriophyoidea
139. Superfamily Tetranychoidea
140. Superfamily Eupodoidea
141. Superfamily Tydeoidea
142. Superfamily Erythraenoidea
143. Superfamily Trombidioidea
144. Superfamily Hydryphantoidea
145. Superfamily Tarsonemoidea
146. Superfamily Pyemotoidea
147. Superfamily Hemisarcoptoidea
148. Superfamily Acaroidea
149. Order Polydesmida
150. Family Sminthuridae
151. Family Forficulidae
152. Order Isoptera
153. Family Thysanoptera
154. Family Acrididyea
155. Family Gryllidae
156. Family Gryllacrididae
157. Family Gryllotalpidae
158. Family Phymatidae
159. Family Ronalcidae
160. Family Tettigoniidae
161. Family Tetrigidae
162. Family Thaumastocoridae
163. Superfamily Piesmatoidea
164. Superfamily Lygacoidea
165. Superfamily Idiostoloidea
166. Superfamily Coreoidea
167. Superfamily Pentatomoidea

168. Superfamily Pyrrhocoroidea
169. Superfamily Tingoidea
170. Superfamily Miroidea
171. Order Homoptera
172. Family Anobiidae
173. Family Apionidae
174. Family Anthribidae
175. Family Bostrichidae
176. Family Brenthidae
177. Family Bruchidae
178. Family Buprestidae
179. Family Byturidae
180. Family Cantharidae
181. Family Carabidae
182. Family Cerambycidae
183. Family Chrysomelidae
184. Family Coccinellidae
185. Family Curculionidae
186. Family Dermestidae
187. Family Elateridae
188. Family Hydrophilidae
189. Family Lyctidae
190. Family Meloidae
191. Family Mordellidae
192. Family Platypodidae
193. Family Scarabacidae
194. Family Scolytidae
195. Family Seblytidae
196. Order Lepidoptera
197. Family Agromyzidae
198. Family Anthomyiidae
199. Family Cecidomyidae
200. Family Chloropidae

201. Family Ephydridae

202. Family Lonchaeidae

203. Family Muscidae

204. Family Otitidae

205. Family Syrphidae

206. Family Tephritidae

207. Family Tipulidae

208. Family Apidae

209. Family Cephidae

210. Family Chaloididae

211. Family Cynipidae

212. Family Eurytomidae

213. Family Formicidae

214. Family Psilidae

215. Family Siricidae

216. Family Tenthredinidae

217. Family Torymidae

218. Family Xylocopidae and also unclassified organisms and/or organisms whose classification is unknown and all other organisms associated with plant and insect diseases.

Note: The risk groups given in this schedule conform to the World Health Organisation classification.

FORM I

[See rule 7(2)]

APPLICATION FOR ENVIRONMENTAL APPROVAL OF CLINICAL, VETERINARY AND FOOD PRODUCTS BASED ON HAZARDOUS MICROORGANISMS/GMO

PART A

(a) Not all the points included will apply to every case. It is to be expected, therefore, that individual applicant will address only the particular parameters that are appropriate to individual situations. In each case where it is not technically possible or it does not appear necessary to give the information, the reasons shall be stated.

(b) The details required in response to each parameters is also likely to vary according to the nature and scale of the proposed release.

(c) The description of the methods used or the reference to the standardised or internationally recognised methods shall also be mentioned in the form together with the name of the body or bodies responsible for carrying out the studies.

(d) A 1-2 page summary of the proposal shall be appended with application.

PART B

1. Name of the Applicant
2. Name of the organisation/firm
3. Approval Required
   (i) Manufacture
   (ii) Import
   (iii) Marketing
4. Quantity per year of the product to be manufactured imported/marketed

PART C

1. Name of the product
   (i) Human Medicine
   (ii) Veterinary Medicine
   (iii) Food Product
   (iv) Others (please specify)
2. Intended Use (Pathological, Metabolic and Immune Response)
3. SAFETY CONCERN
   (i) Donor (heterologous Nucleic acid segments from sources)
   (ii) Vector (DNA molecules to which heterologous nucleic acid segments are joined for transfer to hosts)
   (iii) Hosts (living cells or organisms into which r-DNA molecules are introduced).

4. Production method (the production method should give the details of the cell lines used, the information on the production of the recombinant proteins within or outside the cells, the concentration of the products in units per litre of the fermented broth as well as the concentration in physical weights. The description should also include in brief the methods applied for reducing the genomic particles, proteins and living contaminants including virus, bacteria, fungi, parasite, etc).

(a) Characterisation of the system of production used

(i) Details of the expression system

* Description of the host cell line.
* Identification of the genera and the species.
* The risks involved in handling the cell lines.
* The classification of the cell line according to the Government of India's Biosafety guidelines or any other accepted Recombinant cell line usage guideline.
* The method(s) of maintenance and growth of the cell line.
* The nature and hazards of using substrates, inducing agents, etc.

(ii) Characteristics of the target gene and vector

* **The full description of the source of the target gene.**
* **The composition of the vector used indicating the promoter sequence as well as the regulatory mechanisms utilised in the expression cassette.**
* **Schematic diagrams of the expression cassette to describe fully the marker genes used.**
* **The restriction sites related to specific endonucleases and the cell lines used for shuttling and amplification of the expression cassette.**
* **The method of constructing the target gene alongwith all the sequences added or deleted.**
* **The extent of target gene amplification into the host genome.**
* **The target gene should also contain, alongwith the nucleotide sequence, the description of the amino acids below the codons.**

(iii) Approaches adopted for expression of the gene

* **The description of the transcripted messenger RNA with its analysis of sequence and identification procedure.**
* **The translation information indicating whether the protein product is chimaeric, whether the expression is found as inclusion bodies or as intracellular protein or whether the protein is secreted out.**
* **The extent of the target protein produced as the percentage of the total cell dry mass as well as its percentage compared to the total proteins of the cell.**
* **The quantity of the target protein produced after the cell growth per litre of the fermented broth.**
* **The full sequence of the recombinant gene alongwith the promoters the marker genes and the terminater sequence.**

(b) Description of the production process

(i) Production set up

* **The handling of the stock cell lines.**
* **The evaluation and uses of the cell lines in pre-fermentor processes.**
* **The preparation of the seed vessel.**
* **The main production fermentation conditions need to be described in brief.**

(ii) Growth Kinetics

* **Graphical plots of the versus substrate usage, optical density change, biomass formation, protein products formation, inducing agents used and their effects on the target protein formation.**
* **The effects of change of standard parameters like pH, dissolved oxygen, temperature, etc. in the main production fermenter.**

(iii) Fermentation parameters

* **The pH, temperature, aeration and rpm of the shaft.**

* Volume of seed to the volume of main fermenter.
* Main substrates used, inducing agents used, if any.
* The fermentation time and the concentration of the target protein in the fermented broth.

(c) In-process control methods

* All the analytical methods used for this in-process control must be described to ensure regulatory authorities that the chances of entry of unwanted products have been minimised.

(d) Description of the raw and processing materials used

* Description of all the raw materials and processing materials used starting from the stock cells handling to the finished dosage form must be in a tabular form.
* Grades of materials used and their usual sourcing.

(e) Description of the Plant and Machinery used

* List of all the equipment alongwith the indicative capacity.
* Names of the manufacturers
* List of main production equipment and the quality control equipment.

(f) Description of the building and facilities created for the manufacture

* The manufacturing area.
* The flow of the process/operation starting from the stock culture area to the finishing area.
* The plot plan of various floors used in the manufacture and processing.
* The building diagram in the plan and/or elevation in the outline indicating the cleanliness maintained in different sections such as class 10,000, class 1,000 and class 100 areas etc.
* Description of the air handling system which is very important.
* A separate write-up indicating the movement of operating personnel in the area.

5. Approaches for the extraction and purification of the product. The purification methods should particularly indicate the modes of elimination of viruses, bacteria, fungi, parasites etc., if the cell lines are mammalian origin. It should also include methods for the removal of genomic particles and proteins irrespective of cell lines/hosts used for production, removal of adjuvants, reagents, chemicals including vectors, donor and recipient organisms, foreign DNA and adventitious materials associated with production methods.

* The methods of handling the cells.
* The methods of isolating the cell soup containing the target protein, the methods of enzyme treatment, if any.
* The methods of concentration, precipitation (if applied), reconstitution, salt separation (if applicable), absorption and desorption methods, chromatographic methods used, if any, ultracentrifugation methods if used.
* Sterilization and final dosage formulation.
* Processes to obtain the product in the finished dosage form.

6. Quality Control and Quality Assurance

(a) Bulk material

* Test relating to the identification of the stock cells.
* The plasmid construct retention in the cells during cell growth.
* The media used.
* The procedures adopted for testing and the percentage retention of the target plasmid.
* Description of the contamination test of the stock culture to assess and monitor the microbial contamination including the media used and the procedures adopted.
* The criteria of acceptance of contamination free culture characterisation of the bioactive molecule produced by monitoring physical and chemical properties of the molecules by the following or more authentic methods should be provided.

(a) Microscopic examination including light, phase contract and electron microscopic studies.

(b) UV Spectroscopic analysis to show the absorption spectra of the product and comparison of this with authentic material.

(c) Density gradient centrifugation to show single peak.

(d) Pattern in high performance liquid chromatography to indicate how many peaks or if only one peak is noticed in the produced and purified bulk.

(e) Isoelectric focussing analysis to show the pl-values.

(f) Western Blot and SDS-PAGE to map the peptide/target protein.

(g) Existence of special bonds like disulfide bonds by breaking the protein and subjecting to SDS-PAGE mapping.

(h) Immunodiffusion Test to find out the absence of contaminants or the extent of the presence of contaminants.

(i) Amino acid composition of the purified protein and its comparison with the authentic material.

(j) N-Terminal Amino Acid Sequence analysis and its comparison with the gene construct used.

(k) Determination of the biological activity in the animal model.

(l) Determination of contaminants per milligram or any convenient unit of the manufacturing bulk purified protein is also to be carried out to indicate the extent of contaminant nucleic acid stretches, proteins, carbohydrates, lipids, detergents, salts and other processing chemicals used in the purification. The impacts of the presence of these contaminants are also to be indicated with authentic reference, if any, to ensure that the risks associated with their presence are minimal. The limits of contaminants and the acceptance criteria need to be quantified for each contaminant.

(b) Formulated Material

* The ingredients incorporated subsequently in the manufacture. The specification of the final product.
* The Quality Control Department would have to certify that the final product has the results within the specifications prescribed and accepted by the regulatory Authorities.

* The documentation should therefore indicate the following specifically:
  - (i) Product presentation
  - (ii) Physical appearance
  - (iii) Product inserts, literature and label claim
  - (iv) Volume/Quantity per pack/dose
  - (v) Potency of the product
  - (vi) Particulate matter limits for liquid
  - (vii) Preservative usage percentage
  - (viii) pH
  - (ix) Other extraneous materials used, their extent, such as contents of DNA, RNA, carbohydrates, lipids, processing materials like detergents, salts, etc.
  - (x) Protein content in the product
  - (xi) Sterility status
  - (xii) Toxicity information
  - (xiii) Pyrogen status

7. Possible Hazards to environment from release of GMO/nucleic acid/microorganisms or products.

8. Clinical field trials done in India/abroad (whether permission of Institutional Ethics Committee obtained).

9. Stability and Shelf Life of product.

10. Method of disposal of vials of syringes/waste matter.

11. Regulatory status in India/abroad (enclose certificates like free sale certificates/GMP certificate/ Certificates granted by Health and Environment Authorities of the country of origins signed by applicant. In case the certificate is issued by the concerned authority of country of origin, the certificate should be endorsed/authenticated by Indian Embassy/High Commission/Consulate in that country).

* Every certificate shall be accompanied by other statutory information like manufacturing batch no., date of manufacture, date of analysis, date of release of the certificate, signatory to the certificate etc.

* The final formulated product needs to be certified for acceptance by the manufacturer.

12. Quantity of the product to be imported and the foreign exchange involved.

PART D

Additional Information Required in the case of Microorganism or GMO-based product is to be placed in market.

1. The following information shall be provided for placing on the market of products:
   - (i) Name of the product and names of GMOs contained therein.
   - (ii) Name of the manufacturer or distributor and his address.
   - (iii) Specificity of the product, exact conditions of use including, when appropriate, the type of environment and/or the geographical area(s).

2. The following information shall also be provided

   (i) Measures to take in case of unintended release or misuse.

   (ii) Specific instructions or recommendations for storage and handling.

   (iii) Estimated production and/or imports.

   (iv) Proposed packaging and labelling requirement.

Applicant's Signature with Seal

FORM II

[See rule 7(2)]

APPLICATION FOR ENVIRONMENTAL APPROVAL OF TRANSGENIC PLANTS (GENETICALLY MODIFIED PLANTS)

PART A

(a) An applicant will address only the particular parameters that are appropriate in his case. In each case where it is not technically possible or it does not appear necessary to give the information, the reasons shall be stated or the column may be filled up by stating "Not Applicable".

(b) The details required in response to each parameter is also likely to vary according to the nature and scale of the proposed release.

(c) The description of the methods used or the reference to the standardised or internationally recognised methods shall also be mentioned in the proforma together with the name of the body or bodies responsible for carrying out the studies.

(d) A 1-2 page summary of the proposal shall be appended with application.

PART B

1. Name of the Applicant
2. Name of Organisation/Firm
3. Type of Release:
   (i) Movement
      (a) Import
      (b) Export
   (ii) Release of
      (a) Imported Material
      (b) Indigenously Developed Material
4. Identification of the Transgenic Plant
   (i) Donor species
   (ii) Recipient Plant Species

5. Risk Assessment
   (i) Are they harmful species or closely related to any such species?
   (ii) Are they harmful to man, animal, plant or environment?
6. Anticipated or actual expression of the incorporated genetic material in the plant
   (i) Modified plant
   (ii) Non-modified plant
      (a) Morphology
      (b) Physiology
      (c) Number of copies of the genes incorporated
      (d) Products
7. Purpose of the experimental design:
   (i) Resistance to pesticides
   (ii) Resistance to biotic stress
   (iii) Resistance to abiotic stress
   (iv) Metabolites
   (v) Any other
8. Details of the experimental system
   (i) Nature of gene
   (ii) Vector/Vector agent used for gene transfer
   (iii) Testing procedure for trans-gene
9. Safeguards required for field release
   (i) Biological
   (ii) Physical
10. Site for Field Trials
11. Quantity, schedule and number of releases
12. Mode of Propagation/Pollination
13. Regulatory status in India/abroad (enclose certificates duly signed by applicant)

Applicant's signature with Seal

FORM III

[See rule 7(2)]

APPLICATION FOR ENVIRONMENTAL APPROVAL OF HAZARDOUS MICROORGANISMS OR GMOS USED IN POLLUTION CONTROL

PART A

(a) An applicant will address only the particular parameters that are appropriate in his case. In each case where it is not technically possible or it does not appear necessary to give the information, the reasons shall be stated or the column may be filled up by stating "Not applicable".

(b) The details required in response to each parameter is also likely to vary according to the nature and scale of the proposed release.

(c) The description of the methods used or the reference to the standardised or internationally recognised methods shall also be mentioned in the proforma together with the name of the body or bodies responsible for carrying out the studies.

(d) A 1-2 page summary of the proposal shall be appended with application.

PART B: CONTAINED USE

1. Characteristics of 'the donor, recipient or (where appropriate) parent' organism(s)

(a) names and designation;

(b) degree of relatedness;

(c) sources of the organism(s);

(d) information on reproductive cycles (sexual/asexual) or the parental organism(s) or, where applicable, of the recipient microorganism;

(e) history of prior genetic modification,

(f) stability of parental or of recipient organism in terms of relevant genetic traits;

(g) nature of parthogenicity and virulence, infectivity, toxicity and vectors of disease transmission;

(h) nature of indigenous vector:

— sequence,

— frequency of mobilisation,

— specificity,

— presence of genes which confer resistance;

(i) host range;

(j) other potentially significant physiological traits;

(k) stability of these traits;

(l) natural habitat and geographic distribution, climatic characteristics of original habitats;

(m) significant involvement in environmental processes (such as nitrogen fixation or pH regulation);

(n) interaction with and effects on, other organisms in the environment (including likely competitive or symbiotic properties);

(o) ability to form survival structures (such as spores or sclerotia).

2. Characteristics of the modified microorganism

(a) The description of the modification including the method for introducing the vector-insert into the recipient organism or the method used for achieving the genetic modification involved;

(b) the function of the genetic modification and/or of the new nucleic acid;

(c) nature and source of the vector;

(d) structure and amount of any vector and/or donor nucleic acid remaining in the final construction of the modified microorganism;

(e) stability of the microorganism in terms of genetic traits;

(f) frequency of mobilisation of inserted vector and/or genetic transfer capability;

(g) rate and level of expression of the new genetic material, method and sensitivity of measurement;

(h) activity of the expressed protein.

3. Health considerations

(a) toxic or allergenic effects of non-viable organisms and/or their metabolic products;

(b) product hazards;

(c) comparison of the modified microorganism to the donor, recipient or (where appropriate) parental organism regarding pathogenicity;

(d) capacity for colonisation;

(e) if the microorganism is pathogenic to humans who are immunocompetent;

(i) diseases caused and mechanism of pathogenicity including invasiveness and virulence;

(ii) communicability;

(iii) infective dose;

(iv) host range, possibility of alteration;

(v) possibility of survival outside of human host;

(vi) presence of vectors or means of dissemination;

(vii) biological stability;

(viii) antibiotic-resistance patterns;

(ix) allergenicity;

(x) availability of appropriate therapies.

4. Environmental considerations

(a) factors affecting survival, multiplication and dissemination of microorganism or GMO in the environment (self application);

(b) available techniques for detection, identification and monitoring of microorganism or GMO;

(c) available techniques for detecting transfer of the new genetic material to other organisms;

(d) known and predicted habitats of the microorganism or GMO;

(e) description of ecosystems to which the microorganism could be accidentally disseminated;

(f) anticipated mechanism and result of interaction between the modified microorganism or microorganisms which might be exposed in case of release into the environment;

(g) known or predicted effects on plants and animals such as pathogenicity, infectivity, toxicity, virulence, vector of pathogen, allergenicity, colonisation;

(h) known or predicted involvement in biogeochemical processes;

(i) availability of methods for decontamination of the area in case of release to the environment.

PART C: Additional Information required for Deliberate Release

1. Potential Environmental Impact

(a) potential for excessive population increase in the environment;

(b) competitive advantage of the GMOs in relation to the unmodified recipient of potential organism(s);

(c) identification and description of the target organism;

(d) anticipated mechanism and result of interaction between the released microorganisms or GMOs and the target organism;

(e) identification and description of non-target organisms which may be affected unwittingly;

(f) likelihood or post-release shifts in biological interactions or in host range;

(g) known or predicted effects on non-target organisms in the environment, impact on population levels of competitors; preys, hosts, symbionts, predators, parasites and pathogens;

(h) known or predicted involvement in biogeochemical processes;

(i) other potentially significant interactions with the environment;

(j) potential of vertical transmission of microorganism or GMOs and effect on the offspring;

(k) extent of transfusion of the test material into the animal products like milk, meat, skin (hides), eggs etc. and likely effect of their consumption/contact with by man/animals including other members of the biosphere.

The impact on human/animal health from the continued consumption over a specified period/ periods.

2. Information on Monitoring, Control, Waste Treatment and Emergency Response Plans

(a) Monitoring techniques

(i) methods for tracing the microorganisms or GMOs and for monitoring their effects;

(ii) specificity (to identify the GMOs, and to distinguish them from the donor, recipient or, where appropriate, the parental organisms), sensitivity and reliability of the monitoring techniques;

(iii) techniques for detecting transfer of the donated genetic material to other organisms;

(iv) duration and frequency of the monitoring.

(b) Control of the Release:

(i) methods and procedures to avoid and/or minimise the spread of the microorganism or GMOs beyond the site of release or the designated area for use;

(ii) methods and procedures to protect the site from intrusion by unauthorised individuals;

(iii) methods and procedures to prevent other organisms from entering the site.

(c) Waste Treatment:

(i) type of waste generated;

(ii) expected amount of waste;

(iii) possible risks;

(iv) description of treatment envisaged;

(v) techniques of testing the efficacy of the waste treatment;

(vi) mode of final disposal of the treated waste and impact on environment.

(d) Emergency Response Plans:

(i) methods and procedures for controlling the microorganisms or GMOs in case of unexpected spread;

(ii) methods for decontamination of the areas affected, e.g., eradication of the microorganisms or GMOs;

(iii) methods for disposal or sanitation of plants, soils, etc. that were exposed during or after the spread;

(iv) methods for isolation of the area affected by the spread;

(v) plans for protecting human health, all life forms and the environment in case of the occurrence of an undesirable effect.

PART D: Additional Information required in the case of Microorganisms or GMOs based product is to be placed in Market

1. The following information shall be provided for placing on the market of products:

(i) name of the product and names of microorganisms or GMOs contained therein;

(ii) name of the manufacturer or distributor and his address;

(iii) specificity of the product, exact conditions of use including, when appropriate, the type of environment and/or the geographical area(s);

(iv) type of expected use: industry, agriculture and skilled trades, consumer use by public at large.

2. The following information shall also be provided:

(i) measures to take in case of unintended release or misuse;

(ii) specific instructions or recommendations for storage and handling;

(iii) estimated production and/or imports;

(iv) proposed packaging requirement. This must be appropriate so as to avoid unintended release of microorganisms or GMOs during storage, or at a later stage;

(v) proposed labelling requirements.

3. Regulatory status in India/abroad (enclose certificates duly signed by the applicant).

Applicant's Signature with Seal

[No. 1 (20)/86-PL/HSMD]

V. RAJAGOPALAN, Jt. Secy.

45 BIO-MEDICAL WASTE (MANAGEMENT AND HANDLING) RULES, 1998

MINISTRY OF ENVIRONMENT AND FORESTS

NOTIFICATION

New Delhi, 20th July, 1998

S.O. 630 (E)—Whereas a notification in exercise of the powers conferred by Sections 6, 8 and 25 of the *Environment (Protection) Act, 1986 (29 of 1986) was published in the Gazette vide S.O. 746 (E) dated 16 October, 1997 inviting objections from the public within 60 days from the date of the publication of the said notification on the Bio-Medical Waste (Management and Handling) Rules, 1998,* and whereas all objections received were duly considered.

Now, therefore, in exercise of the powers conferred by Section 6, 8 and 25 of the Environment (Protection) Act, 1986 the Central Government hereby notifies the rules for the management and handling of bio-medical waste.

1. Short Title And Commencement

(1) These rules may be called the *Bio-Medical Waste (Management and Handling) Rules*, 1998

(2) They shall come into force on the date of their publication in the official Gazette.

2. Application: These rules apply to all persons who generate, collect, receive, store, transport, treat, dispose, or handle bio-medical waste in any form.

3. Definitions: In these rules unless the context otherwise requires:

(1) "Act" means the Environment (Protection) Act, 1986 (29 of 1986);

(2) "Animal House" means a place where animals are reared/kept for experiments or testing purposes;

(3) "Authorisation" means permission granted by the prescribed authority for the generation, collection, reception, storage, transportation, treatment, disposal and/or any other form of handling of bio-medical waste in accordance with these rules and any guidelines issued by the Central Government;

(4) "Authorised person" means an occupier or operator authorised by the prescribed authority to generate, collect, receive, store, transport, treat, dispose and/or handle bio-medical waste in accordance with these rules and any guidelines issued by the Central Government;

(5) "Bio-medical waste" means any waste, which is generated during the diagnosis, treatment or immunisation of human beings or animals or in research activities pertaining thereto or in the production or testing of biologicals and including categories mentioned in Schedule I;

(6) "Biologicals" means any preparation made from organisms or microorganisms or product of metabolism and biochemical reactions intended for use in the diagnosis, immunisation or the treatment of human beings or animals or in research activities pertaining thereto;

(7) "Bio-medical waste treatment facility" means any facility wherein treatment, disposal of bio-medical waste or processes incidental to such treatment or disposal is carried out;

(8) "Occupier" in relation to any institution generating bio-medical waste, which includes a hospital, nursing home, clinic, dispensary, veterinary institution, animal house, pathological laboratory, blood bank by whatever name called, means a person who has control over that institution and/or its premises;

(9) "Operator of a bio-medical waste facility" means a person who owns or controls or operates a facility for the collection, reception, storage, transport, treatment, disposal or any other form of handling of bio-medical waste;

(10) "Schedule" means schedule appended to these rules.

4. Duty of Occupier: It shall be the duty of every occupier of an institution *generating bio-medical waste* which includes a hospital, nursing home, clinic, dispensary, veterinary institution, animal house, pathological laboratory, blood bank by whatever name called, to take all steps to ensure that such waste is handled without any adverse effect to human health and the environment.

5. Treatment and Disposal

(1) Bio-medical waste shall be treated and disposed of in accordance with Schedule I, and in compliance with the standards prescribed in Schedule V.

(2) Every occupier, where required, shall set up in accordance with the time-schedule in *Schedule VI,* requisite bio-medical waste treatment facilities like incinerator, autoclave, microwave system for the treatment of waste, or, ensure requisite treatment of waste at a common waste treatment facility or any other waste treatment facility.

6. Segregation, Packaging, Transportation and Storage

(1) Bio-medical waste shall not be mixed with other wastes.

(2) Bio-medical waste shall be segregated into containers/bags at the point of generation in accordance with Schedule II prior to its storage, transportation, treatment and disposal. The containers shall be labelled according to Schedule III.

(3) If a container is transported from the premises where bio-medical waste is generated to any waste treatment facility outside the premises, the container shall, apart from the label prescribed in Schedule III, also carry information prescribed in Schedule IV.

(4) Notwithstanding anything contained in the Motor Vehicles Act, 1988, or rules thereunder, untreated bio-medical waste shall be transported only in such vehicle as may be authorised for the purpose by the competent authority as specified by the government.

(5) No untreated bio-medical waste shall be *kept stored beyond a period of 48 hours*: Provided that if for any reason it becomes necessary to store the waste beyond such period, the authorised person must take permission of the prescribed authority and take measures to ensure that the waste does not adversely affect human health and the environment.

7. Prescribed Authority

(1) The Government of every State and Union Territory shall establish a prescribed authority with such members as may be specified for granting authorisation and implementing these rules. If the prescribed authority comprises of more than one member, a chairperson for the authority shall be designated.

(2) The prescribed authority for the State or Union Territory shall be appointed within one month of the coming into force of these rules.

(3) The prescribed authority shall function under the supervision and control of the respective government of the State or Union Territory.

(4) The prescribed authority shall *on receipt of Form I make such enquiry as it deems fit and if it is satisfied that the applicant possesses the necessary capacity to handle bio-medical waste in accordance with these rules, grant or renew an authorisation as the case may be.*

(5) An authorisation shall be granted for a period of three years, including an initial trial period of one year from the date of issue. Thereafter, an application shall be made by the occupier/operator for renewal. All such subsequent authorisation shall be for a period of three years. A provisional authorisation will be granted for the trial period, to enable the occupier/operator to demonstrate the capacity of the facility.

(6) The prescribed authority may, after giving reasonable opportunity of being heard to the applicant and for reasons thereof to be recorded in writing, refuse to grant or renew authorisation.

(7) Every application for authorisation shall be disposed of by the prescribed authority within ninety days from the date of receipt of the application

(8) The prescribed authority may cancel or suspend an authorisation, if for reasons, to be recorded in writing, the occupier/operator has failed to comply with any provision of the Act or these rules:

Provided that no authorisation shall be cancelled or suspended without giving a reasonable opportunity to the occupier/operator of being heard.

8. Authorisation

(1) Every occupier of an institution generating, collecting, receiving, storing, transporting, treating, disposing and/or handling bio-medical waste in any other manner, except such occupier of clinics, dispensaries, pathological laboratories, blood banks providing treatment/service to less than 1,000 (one thousand) patients per month, shall make an application in Form I to the prescribed authority for grant of authorisation.

(2) Every operator of a bio-medical waste facility shall make an application in Form I to the prescribed authority for grant of authorisation.

(3) Every application in Form I for grant of authorisation shall be accompanied by a fee as may be prescribed by the Government of the State or Union Territory.

9. Advisory Committee: The Government of every State/Union Territory shall constitute an advisory committee. The committee will include experts in the field of medical and health, animal husbandry and veterinary sciences, environmental management, municipal administration and any other related department or organisation including non-governmental organisation. The State Pollution Control Board/Pollution Control Committee shall be represented. As and when required, the committee shall advise the government of the State/Union Territory and the prescribed authority about matters related to the implementation of these rules.

10. Annual Report: Every occupier/operator shall submit an annual report to the prescribed authority in *Form II by* 31 January every year, to include information about the categories and quantities of bio-medical wastes handled during the preceding year. The prescribed authority shall send this information in a compiled form to the Central Pollution Control Board by 31 March every year.

11. Maintenance of Records

(1) *Every authorised person shall* maintain records related to the generation, collection, reception, storage, transportation, treatment, disposal and/or any form of handling of bio-medical waste in accordance with these rules and any guidelines issued.

(2) All records shall be subject to *inspection* and *verification* by the *prescribed* authority at any time.

12. Accident Reporting: When any accident occurs at any institution or facility or any other site where bio-medical waste is handled or during transportation of such waste, the authorised person shall report the accident in Form III to the prescribed authority forthwith.

13. Appeal: Any person aggrieved by an order made by the prescribed authority under these rules may, within thirty days from the date on which the order is communicated to him, prefer an appeal to such authority as the Government of State/Union Territory may think fit to constitute:

Provided that the authority may entertain the appeal after the expiry of the said period of thirty days if it is satisfied that the appellant was prevented by sufficient cause from filling the appeal in time.

SCHEDULE I

(See Rule 5)

CATEGORIES OF BIO-MEDICAL WASTE

Option	Waste Category	Treatment & Disposal
Category No.1	Human Anatomical Waste (human tissues, organs, body parts)	incineration[1]/deep burial**
Category No. 2	Animal Waste (animal tissues, organs, body parts, carcasses, bleeding parts, fluid, blood and experimental animals used in research, waste generated by veterinary hospitals, colleges, discharge from hospitals, animal houses)	incineration[1]/deep burial**
Category No. 3	Microbiology & Biotechnology Waste (wastes from laboratory cultures, stocks or specimens of microorganisms, live or attenuated vaccines, human and animal cell culture used in research and infectious agents from research and industrial laboratories, wastes from production of biologicals, toxins, dishes and devices used for transfer of cultures)	local autoclaving/ microwaving/incineration[1]
Category No. 4	Waste sharps (needles, syringes, scalpels, blades, glass, etc. that may cause puncture and cuts. This includes both used and unused sharps)	disinfection (chemical treatment/autoclaving/ microwaving[11] and mutilation/shredding##
Category No. 5	Discarded Medicines and Cytotoxic drugs (wastes comprising of outdated, contaminated and discarded medicines)	incineration[1]/destruction and drugs disposal in secured landfills*

Contd.

Option	Waste Category	Treatment & Disposal
Category No. 6	Solid Waste (Items contaminated with blood, and body fluids including cotton, dressings, soiled plaster casts, lines, beddings, other material contaminated with blood)	incineration[1] autoclaving/microwaving[11]
Category No. 7	Solid Waste (wastes generated from disposable items other than the waste sharps such as tubings, catheters, intravenous sets etc.)	disinfection by chemical treatment@ autoclaving/microwaving and mutilation/shredding##
Category No. 8	Liquid Waste (waste generated from laboratory and washing, cleaning, housekeeping and disinfecting activities)	disinfection by chemical treatment@ and discharge into drains
Category No. 9	Incineration Ash (ash from incineration of any bio-medical waste)	disposal in municipal landfill
Category No. 10	Chemical Waste (chemicals used in production of biologicals, chemicals used in disinfection, as insecticides, etc.)	chemical treatment@ and discharge into drains for liquids and secured landfill for solids

@ Chemicals treatment using at least 1% hypochlorite solution or any other equivalent chemical reagent. It must be ensured that chemical treatment ensures disinfection.

11 Autoclaving/microwaving

Mutilation shredding must be such so as to prevent unauthorised reuse.

1 There will be no chemical pretreatment before incineration. Chlorinated plastics shall not be incinerated.

* Secured landfill

** Deep burial shall be an option available only in towns with population less than five lakhs and in rural areas.

SCHEDULE II

(See Rule 6)

COLOUR CODING AND TYPE OF CONTAINER FOR DISPOSAL OF BIO-MEDICAL WASTES

Colour Coding	Type of Container	Waste Category	Treatment options as per Schedule I
Yellow	Plastic bag	Cat. 1, Cat. 2, and Cat. 3, Cat. 6	Incineration/deep burial
Red	Disinfected containe./ plastic bag	Cat. 3, Cat. 6, Cat. 7	Autoclaving/Microwaving/ Chemical Treatment
Blue/White translucent	Plastic bag/puncture proof container	Cat. 4, Cat. 7	Autoclaving/Microwaving/ Chemical Treatment and Destruction/Shredding
Black	Plastic bag	Cat. 5, and Cat. 9 and Cat. 10. (solid)	Disposal in secured landfill

Notes:

1. Colour coding of waste categories with multiple treatment options as defined in Schedule I, shall be selected depending on treatment option chosen which shall be as specified in Schedule I.
2. Waste collection bags for waste types needing incineration shall not be made of chlorinated plastics.
3. Categories 8 and 10 (liquid) do not require containers/bags.
4. Category 3 if disinfected locally need not be put in containers/bags.

SCHEDULE III

(See Rule 6)

LABEL FOR BIO-MEDICAL WASTE CONTAINERS/BAGS

BIOHAZARD SYMBOL

BIOHAZARD

CYTOTOXIC HAZARD SYMBOL

CYTOTOXIC

HANDLE WITH CARE

Note : Label shall be non-washable and prominently visible

SCHEDULE IV

(See Rule 6)

LABEL FOR TRANSPORT OF BIO-MEDICAL WASTE CONTAINERS/BAGS

Day .... Month ....

Year ....

Waste category No. .... Date of generation ....

Waste class

Waste description

Sender's Name & Address | **Receiver's Name & Address**

Phone No. .... | Phone No. ....

Telex No. .... | Telex No. ....

Fax No. .... | Fax No. ....

Contact Person .... | Contact Person ....

In case of emergency please contact:

Name & address

Phone No.

Note: Label shall be non-washable and prominently visible.

SCHEDULE V

(See Rule 5 and Schedule I)

STANDARDS FOR TREATMENT AND DISPOSAL OF BIO-MEDICAL WASTES

STANDARDS FOR INCINERATORS:

All incinerators shall meet the following operating and emission standards:

A. Operating Standards

1. **Combustion efficiency (CE) shall be at least 99.00%**
2. **The Combustion efficiency is computed as follows:**

$$C.E. = \frac{\% CO_2}{\% CO_2 + \% CO} \times 100$$

3. **The temperature of the primary chamber shall be 800 ± 50 deg. C.**
4. **The secondary chamber gas residence time shall be at least 1 (one) second at 1,050 ± 50°C with minimum 3% C in the stack gas.**

Emission Standards

Parameters	Concentration mg/Nm^3 at (12% CO_2 correction)
(i) Particulate matter	150
(ii) Nitrogen Oxides	450
(iii) HCL	50

(iv) **Minimum stack height shall be 30 metres above ground**

(v) **Volatile organic compounds in ash shall not be more than 0.01%**

Note: Suitably designed pollution control devices should be installed/retrofitted with the incinerator to achieve the above emission limits, if necessary.

Wastes to be incinerated shall not be chemically treated with any chlorinated disinfectants.

Chlorinated plastics shall not be incinerated.

Toxic metals in incineration ash shall be limited within the regulatory quantities as defined under the Hazardous Waste (Management and Handling Rules,) 1989.

Only low sulphur fuel like L. D. O./L. S. H. S./Diesel shall be used as fuel in the incinerator.

Standards for Waste Autoclaving:

The autoclave should be dedicated for the purposes of disinfecting and treating bio-medical waste,

(I) **When operating a gravity flow autoclave, medical waste shall be subjected to:**

(i) **a temperature of not less than 121°C of 15 pounds per square inch (psi) for an autoclave residence time of not less than 60 minutes; or**

(ii) **a temperature of not less than 135°C and a pressure of 31 psi for an autoclave residence time of not less than 45 minutes; or**

(iii) a temperature of not less than 149°C and a pressure of 52 psi for an autoclave residence time of not less than 30 minutes.

(II) When operating a vacuum autoclave, medical waste shall be subjected to a minimum of one pre-vacuum pulse to purge the autoclave of all air. The waste shall be subjected to:

(i) a temperature of not less than 121°C and a pressure of 15 psi per an autoclave residence time of not less than 45 minutes; or

(ii) a temperature of not less than 135°C and a pressure of 31 psi per an autoclave residence time of not less than 30 minutes;

(III) Medical waste shall not be considered properly treated unless the time, temperature and pressure indicators indicate that the required time, temperature and pressure were reached during the autoclave process. If, for any reasons, time, temperature or pressure indicator indicates that the required temperature, pressure or residence time was not reached, the entire load of medical waste must be autoclaved again until the proper temperature, pressure and residence time were achieved.

(IV) Recording of operational parameters

Each autoclave shall have graphic or computer recording devices which will automatically and continuously monitor and record dates, time of day, load identification number and operating parameters throughout the entire length of the autoclave cycle.

(V) Validation test

Spore testing:

The autoclave should completely and consistently kill the approved biological indicator at the maximum design capacity of each autoclave unit. Biological indicator for autoclave shall be Bacillus stearothermophilus spores using vials or spore strips, with at least 1×10^4 spores per millilitre. Under no circumstances will an autoclave have minimum operating parameters less than a residence time of 30 minutes, regardless of temperature and pressure, a temperature less than 121°C or a pressure less than 15 psi.

(VI) Routine Test

A chemical indicator strip/tape that changes colour when a certain temperature is reached can be used to verify that a specific temperature has been achieved. It may be necessary to use more than one strip over the waste package at different location to ensure that the inner content of the package has been adequately autoclaved.

Standards for Liquid Waste

The effluent generated from the hospital should conform to the following limits:

PARAMETERS	PERMISSIBLE LIMITS
pH	6.5-9.0
Suspended solids	100 mg/1
Oil and grease	10 mg/1
BOD	30 mg/1
COD	250 mg/1
Bio-assay test	90% survival of fish after 96 hours in 100% effluent.

These limits are applicable for those hospitals which are either connected with sewers without terminal sewage treatment plant or not connected to public sewers. For discharge into public sewers with terminal facilities, the general standards as notified under the Environment (Protection) Act, 1986, shall be applicable.

Standards of Microwaving

1. Microwave treatment shall not be used for cytotoxic, hazardous or radioactive wastes, contaminated animal carcasses, body parts and large metal items.
2. The microwave system shall comply with efficacy test/routine tests and a performance guarantee may be provided by the supplier before operation of the unit.
3. The microwave should completely and consistently kill the bacteria and other pathogenic organisms that is ensured by approved biological indicator at the maximum design capacity of each microwave unit. Biological indicators for microwave shall be Bacillus Subtilis spores using vials or spore strips with at least 1×10^4 spores per milliliter.

Standards for Deep Burial

1. A pit or trench should be dug about 2 meters deep. It should be half filled with waste, then covered with lime within 50 cm of the surface, before filling the rest of the pit with soil.
2. It must be ensured that animals do not have any access to burial sites. Covers of galvanised iron/wire meshes may be used.
3. On each occasion, when wastes are added to the pit, a layer of 10 cm of soil shall be added to cover the wastes.
4. Burial must be performed under close and dedicated supervision.
5. The deep burial site should be relatively impermeable and no shallow well should be close to the site.
6. The pits should be distant from habitation and sited so as to ensure that no contamination occurs of any surface water or ground water. The area should not be prone to flooding or erosion.
7. The location of the deep burial site will be authorised by the prescribed authority.
8. The installation shall maintain a record of all pits for deep burial.

SCHEDULE VI
(See Rule 5)

SCHEDULE FOR WASTE TREATMENT FACILITIES LIKE INCINERATOR/AUTOCLAVE/ MICROWAVE SYSTEM

A. Hospitals and nursing homes in towns with population of 30 lakhs and above	by 31st December, 1999 or earlier
B. Hospitals and nursing homes in towns with population of below 30 lakhs	
(a) with 500 beds and above	by 31st December, 1999 or earlier
(b) with 200 beds and above but less than 500 beds	by 31st December, 2000 or earlier
(c) with 50 beds and above but less than 200 beds	by 31st December, 2001 or earlier
(d) with less than 50 beds	by 31st December, 2002 or earlier
C. All other institutions generating bio-medical waste not included in A and B above	by 31st December, 2002 or earlier

FORM I

(See Rule 8)

APPLICATION FOR AUTHORISATION

(To be submitted in duplicate)

To,

The Prescribed Authority
(Name of State Govt/UT Administration)

Address.

1. **Particulars of Applicant**
   (i) **Name of the Applicant**
   (In block letters & in full)
   (ii) **Name of the Institution:**
   Address:
   Tele No.
   Fax No.
   Telex No.
2. **Activity for which authorisation is sought:**
   (i) **Generation**
   (ii) **Collection**
   (iii) **Reception**
   (iv) **Storage**
   (v) **Transportation**
   (vi) **Treatment**
   (vii) **Disposal**
   (viii) **Any other form of handling**
3. **Please state whether applying for fresh authorisation or for renewal:**
   (In case of renewal, previous authorisation—number and date)
4. (i) **Address of the institution handling bio-medical wastes;**
   (ii) **Address of the place of the treatment facility;**
   (iii) **Address of the place of disposal of the waste;**
5. (i) **Mode of transportation (if any) of bio-medical waste;**
   (ii) **Mode(s) of treatment;**
6. **Brief description of method of treatment and disposal (attach details):**
7. (i) **Category (See Schedule I) of waste to be handled**
   (ii) **Quantity of waste (category-wise) to be handled per month**

8. Declaration

I do hereby declare that the statements made and information given above are true to the best of my knowledge and belief and that I have not concealed any information.

I do also hereby undertake to provide any further information sought by the prescribed authority in relation to these rules and to fulfil any conditions stipulated by the prescribed authority.

Date: Signature of applicant

Place: Designation of the applicant

FORM II

(See Rule 10)

ANNUAL REPORT

(To be submitted to the prescribed authority by 31 January every year)

1. Particulars of the applicant:

   (i) Name of the authorised person (occupier/operator):

   (ii) Name of the institution:

   Address

   Tel. No.

   Telex No.

   Fax No.

2. Categories of waste generated and quantity on a monthly average basis:
3. Brief details of the treatment facility:

   In case of off-site facility:

   (i) Name of the operator;

   (ii) Name and address of the facility:

   Tel. No.

   Telex No.

   Fax No.

4. Category-wise quantity of waste treated:
5. Mode of treatment with details:
6. Any other information:
7. Certified that the above report is for the period from..

   ....

Date: .... Signature............................

Place: .... Designation ....

FORM III

(See Rule 12)

ACCIDENT REPORTING

1. **Date and time of accident:**
2. **Sequence of events leading to accident:**
3. **The waste involved in accident:**
4. **Assessment of the effects of the accidents on human health and the environment:**
5. **Emergency measures taken:**
6. **Steps taken to alleviate the effects of accidents:**
7. **Steps taken to prevent the recurrence of such an accident:**

Date: .... **Signature:** ....

Place: .... **Designation:** ....

(F. No. 23-2/96-HSMD)

VIJAI SHARMA. Jt. Secy.

The Biomedical Waste (Management and Handling) Rules was revised and amended in 2015. With more emphasis was given on duties of the operator of a common biomedical waste treatment facility.

46 MUNICIPAL SOLID WASTES (MANAGEMENT AND HANDLING) RULES, 2000

MINISTRY OF ENVIRONMENT AND FORESTS

NOTIFICATION

New Delhi, the 27th September, 1999

S.O. 783(E) — The following draft of the Municipal Wastes (Management & Handling) Rules, 1999, which the Central Government proposes to make in exercise of the powers conferred by Sections 3, 6 and 25 of the Environment (Protection) Act, 1986 (29 of 1986), are hereby published for information of all persons likely to be affected hereby; and notice is hereby given that the said draft rules shall be taken into consideration after the expiry of period of sixty days from the date on which copies of the Gazette of India containing this notification are made available to the public;

Any person desirous of making any suggestion or objection in respect to the said draft rules, may forward the same for consideration of the Central Government within the period so specified to the Secretary, Ministry of Environment & Forests, Paryavaran Bhawan, CGO Complex, Lodi Road, New Delhi 110 003.

The final version of rule read as Municipal Solid Wastes (Management and Handling) Rule, 2000.

DRAFT RULES

1. **Short title and commencement**

   (1) These rules may be called the **Municipal Solid Wastes (Management and Handling) Rules,** 2000.

   (2) They shall come into force on the date of their publication in the official Gazette.

2. **Application**: These rules shall apply to every Municipal Authority responsible for collection, segregation, storage, transportation, processing and disposal of municipal solid wastes.

3. **Definitions**: In these rules, unless the context otherwise requires:

   (i) "**anaerobic digestion**" means digestion of organic matter in absence of molecular oxygen;

   (ii) "**biodegradable substance**" means a substance that can be degraded by microorganisms;

   (iii) "**collection**" means collection and removal of solid wastes from different collection points;

   (iv) "**composting**" means a controlled process involving microbial degradation of organic matter;

   (v) "**demolition and construction waste**" means wastes from building materials debris and rubble resulting from construction, remodelling, repair and demolition operation;

   (vi) "**disposal**" means final disposal of municipal solid waste;

   (vii) "**Form**" means Form appended to these rules.

   (viii) "**landfilling**" means disposal of solid wastes by spreading it in layers over a lined surface or land, compacting it to the smallest practical volume and covering it by impervious soil layer at the end of the day or more frequently. A landfill is operated to prevent leachate from contaminating ground water and maintaining ambient air quality;

(ix) **"leachate"** means liquid that seeped through solid waste or other medium and has extracted dissolved or suspended material from it;

(x) **"lysimeter"** A device used to measure rate of movement of water through or from a soil layer or used to collect percolated water for quality analysis;

(xi) **"Municipal authority"** means Municipal Corporation, Municipality, Nagar Palika, Nagar Nigam, Nagar Panchayat/Municipal Council or any other local body and, where the management of municipal solid waste is entrusted to an agency; such agency;

(xii) **"municipal solid waste"** includes commercial and residential wastes generated from a community;

(xiii) **"operator of the facility"** Municipal authority or any other agency appointed by the municipal authority for management and handling of municipal solid waste in the area;

(xiv) **"processing"** means processed for making use by adopting process like recycling or composting, or both;

(xv) **"recycling"** means the process by which waste materials are transformed into new products in such a manner that the original products lose their identity;

(xvi) **"schedule"** means a schedule appended to these rules;

(xvii) **"vadose water"** Water which occurs between the ground, surface and the water table i.e. the unsaturated zone;

(xviii) **"vermicomposting"** is a process of using earthworms as a natural bioreactor to convert biodegradable waste into biofertilizer (vermicasting).

4. **Responsibility of Municipal authority**

(1) Every municipal authority shall, within the territorial area of the municipality, be responsible for the implementation of the provisions of these rules, and for any infrastructure development for collection, storage, segregation, transportation, processing and disposal of municipal solid wastes.

(2) The Municipal authorities shall comply with these rules as per the implementation schedule laid down in **Schedule I**.

(3) The Municipal authority shall furnish its annual report in **Form I** to the District Magistrate or the Deputy Commissioner, who shall forward the report to the State Pollution Control Board or as the case may be to the Pollution Control Committee in Union Territories.

5. **Power of the District Magistrate or the Deputy Commissioner**: The District Magistrate or the Deputy Commissioner of the concerned district shall have the overall responsibility for the enforcement of the provisions of these rules.

6. **State Boards to monitor enforcement of rules**: The State Pollution Control Board or, as the case may be, the Pollution Control Committee in the Union Territory, shall monitor the collection of samples, their analysis and compliance with standards regarding ground water, ambient air and leachate quality, and from time to time require the District Magistrate or the Deputy Commissioner of the concerned District or the Municipal Authority to take steps as may be necessary for proper implementation of these rules.

7. **Central Pollution Control Board to coordinate with State Boards and Committees:** The Central Pollution Control Board shall coordinate with the State Pollution Control Boards and the Pollution Control Committees in the matters of municipal solid waste disposal and its management and handling.

8. **Management of Municipal Solid Waste:**

(1) Any municipal solid waste generated in a city or a town, shall be managed in accordance with the compliance criteria and the procedure laid down in Schedule II.

(2) The disposal of municipal solid wastes of the specified categories shall be through landfill as per specifications and standards laid down in Schedule III.

(3) The standards for compost and disposal of treated leachate to be followed by Municipal authorities shall be as laid down in **Schedule IV**.

9. **Annual Reports:**

(1) The State Pollution Control Boards and Pollution Control Committees of Union Territories shall prepare and submit to the Central Pollution Control Board and Annual Review Report with regard to the implementation of these rules every year on or before 30th June in **Form II**.

(2) The Central Pollution Control Board shall prepare the consolidated annual review report on management of municipal solid wastes and forward it to the Central Government alongwith its recommendations before 15th September every year.

10. **Accident Reporting**: When any accident occurs at any municipal solid wastes collection, segregation, storage, processing treatment and disposal facility or landfill site or during the transportation of such wastes, the Municipal authority shall forthwith report the accident in **Form III** to the District Magistrate or the Deputy Commissioner of the concerned District.

SCHEDULE I
[see rule 4(2)]
Implementation Schedule

Compliance Criteria	Cities/towns with population			
	more than 10 lakhs	1 to 10 lakhs	0.5 to 1.0 lakhs	less than 0.5 lakhs
Setting up of suitable composting facilities to make use of waste	by 31.12.2001 or earlier	by 31.12.2001 or earlier	by 31.12.2001 or earlier	by 31.12.2001 or earlier
Monitoring of disposal facilities set up to meet laid down standards	Once in four months on yearly basis	Once in six months on yearly basis	Once in a year on annual basis	Once in a year on annual basis
Existing landfill sites to be improved as per existing provisions of the rules	by 31.12.2001 or earlier	by 31.12.2001 or earlier	by 31.12.2001 or earlier	by 31.12.2001 or earlier
Identification of landfill sites for future use	by 31.12.2000 or earlier	by 31.12.2000 or earlier	by 31.12.2000 or earlier	by 31.12.2000 or earlier

SCHEDULE II
[see rule 8(1)]
Management of Municipal Solid Wastes

<table>
<tr><th>S. No.</th><th>Parameters</th><th>Compliance Criteria</th></tr>
<tr><td>1.</td><td>Collection of municipal solid wastes</td><td>1. Littering of municipal solid waste shall be prohibited in cities, towns and in urban areas notified by the Government. To prohibit littering, following steps shall be taken, namely:
(i) Organising house to house collection of garbage through any of the methods, like containerised collection, community bin collection (central bin), house to house collection, collection on regular pre-informed timings and scheduling by using bell ringing/musical vehicle (without exceeding permissible noise levels).
(ii) Collection of waste from slums and squatter areas/localities including hotels/restaurants/office complexes and commercial areas shall be devised in consultation with municipal authority.
(iii) Wastes from slaughter houses, fruits and vegetable markets, which are biodegradable in nature, shall be managed to make use of such wastes.
(iv) Bio-medical wastes and industrial wastes shall not be mixed with municipal solid wastes and such wastes shall follow rules specified separately for the purpose.
(v) Collected waste from residential and other areas shall be transferred to community bin by hand-driven containerised Carts.
(vi) Horticultural and construction/demolition wastes/debris shall be separately collected and disposed off following proper norms. Similarly, activities relating to dairies (milking of cows/buffaloes) shall be regulated in accordance with State laws.
(vii) Waste (garbage, dry leaves) shall not be burnt.
(viii) Stray animals shall not be allowed to move around waste storage facilities or at any other place in city/town and shall be managed as per State laws.
2. Municipal authority shall notify waste collection schedule and the likely method to be adopted for public benefit in a city/town.</td></tr>
<tr><td>2.</td><td>Segregation of Municipal solid wastes</td><td>In order to encourage public, municipal authority shall organise awareness programmes for segregation of wastes and shall encourage recycling/reuse of segregated materials. Municipal authority shall undertake phased programme to ensure that community is fully involved in waste segregation.</td></tr>
<tr><td>3.</td><td>Storage of municipal solid wastes</td><td>Municipal authorities shall establish and maintain storage facilities in such a manner as they do not create unhygienic/insanitary conditions around it. Following criteria shall be taken into account while establishing and maintaining storage facilities:
(i) Storage facilities shall be created/established by taking into account quantities of waste generation in a given area and the population densities. A storage facility shall be so placed that user finds it easy to approach within the walking range.
(ii) Storage facilities to be set up by Municipal authorities or any other agency shall be so designed that waste stored shall not be exposed to open atmosphere and shall be aesthetically acceptable and user-friendly.
(iii) Storage facilities or 'bins' shall have 'easy to operate' design for handling, transfer and transportation of waste.
(iv) Manual handling of waste shall be prohibited. If unavoidable due to constraints, manual handling shall be carried out under proper precaution with due care for safety of workers.</td></tr>
</table>

Contd.

S. No.	Parameters	Compliance Criteria
4.	Transportation of municipal solid wastes	Vehicles used for transportation of wastes shall be covered. Waste should not be visible to public, nor exposed to open environment preventing their scattering. The following criteria shall be made: (i) The storage facilities set up by Municipal authorities shall be daily attended for clearing of wastes. (ii) Collection and transportation vehicles shall be so designed that multiple handling of wastes, prior to final disposal, is avoided.
5.	Processing of municipal solid wastes	Municipal authorities shall adopt suitable technology or combination of such technologies to make use of wastes so as to minimise burden on landfill. Following criteria shall be adopted: (i) The biodegradable wastes, invariably segregated waste and waste not containing any toxic contaminants, shall be processed by composting, vermicomposting, anaerobic digestion or any other appropriate biological processing for stabilisation of waste. It shall be ensured that compost shall be free from contamination due to heavy metals, pesticides or any other contaminants. Standards in respect of compost are laid down in **Schedule IV** for compliance. (ii) Waste containing recoverable material shall follow the route of recycling.
6.	Disposal of municipal solid wastes	Landfilling shall be restricted to non-biodegradable, inert waste and other waste that are not suitable either for recycling or for biological processing. Landfilling shall also be carried out for residues of waste processing facilities as well as preprocessing rejects from waste processing facilities. Landfilling of mixed waste shall be avoided unless same is found unsuitable for waste processing. Under unavoidable circumstances or till installation of alternate facilities, landfilling shall be done following proper norms. Landfilling shall meet the following criteria: (i) Landfill siting and construction shall be done after proper care. However, in respect of cities having population over five lakh, proper environmental impact assessment shall be conducted by Municipal authority before selecting a site. (ii) Provision for future landfilling cities shall be included in the land use plan of city/ town. (iii) The future and existing landfill site shall comply with the norms for control of air and water (ground and surface water) pollution and other environment norms as laid down specifications/standards (**Schedule II**). (iv) Waste at disposal site shall not be burnt. Sites where waste is to be burnt as an interim measure, shall not cause air pollution. Ambient air quality shall be monitored for compliance.

SCHEDULE III

[see rule 8(2)]

SPECIFICATIONS FOR LANDFILL SITES

1. The Municipal authority of the concerned district and town shall identify, develop and maintain one or more landfill sites as the case may be for the disposal of municipal solid wastes. Prior to commencement of landfilling, the landfill site shall be planned and designed with proper documentation of a phased construction plan as well as a closure plan.

2. The Municipal authority shall identify the disposal site after environment impact assessment in a city/town having population more than five lakh and taking into consideration the views of the competent town planning authority. The identified sites shall be brought into public notice for their views.

3. The landfill sites shall be selected to make use of nearby wastes processing facility. Otherwise, wastes processing facility shall be planned as an integral part of the landfill site.

4. The existing landfill sites that continue to be used for more than five years, shall be improved in accordance of these specifications.

5. The landfill sites in municipal bodies having more than 5 lakh population shall be used only for those municipal solid wastes that are not suitable for bio-composting or recycling.

6. Bio-medical waste, slaughter house waste (including fish, meat, poultry processing), industrial wastes, sludge and other hazardous wastes shall not be dumped at landfill site meant for disposal of non-hazardous and non-biodegradable wastes. Bio-medical wastes shall be disposed off as per the Bio-medical Wastes (Management and Handling) Rules, 1998. Hazardous wastes shall be managed as per the Hazardous Wastes (Management and Handling) Rules, 1989.

Site Selection

7. The landfill site shall be large enough to last for 20-25 years and preferably within 5 km from present city limits.

8. The site shall be at least 0.5 km away from habitation clusters, forest areas, monuments, National Parks, Wetlands and places of important cultural, historical or religious interest.

9. Landfill site shall be atleast 20 km away from airport including airbase. Municipal authorities shall obtain approval of airport/airbase authorities prior to the setting up of the landfill site.

10. A 500 meter wide buffer zone of no-development be maintained around landfill site and shall be incorporated in the Town Planning Department's land-use plans.

Facilities at the Site

11. Landfill site shall be fenced/hedged and provided with proper gate to monitor incoming vehicles/ or other modes of transportation.

12. The landfill site shall be well protected to prevent entry of stray animals.

13. Approach and other internal roads for free flow of vehicles and other machinery shall exist at the landfill site.

14. The landfill site shall have wastes inspection facility to monitor wastes brought in for landfill, office facility for record keeping and shelter for keeping equipment and machinery including pollution monitoring equipment.

15. Provisions like weigh bridge to measure quantity of waste brought at landfill site, fire protection equipments and other facilities as required, shall be provided for all cities with more than 5 lakh population. For setting up of the weigh bridge, following schedule shall apply:

(a) cities having population of more than ten lakh shall provide weigh bridge by 31st December, 2002 or earlier;

(b) cities having population between one and ten lakh shall install weigh bridge by 31st December, 2003 or earlier; and

(c) cities having population between fifty thousand and one lakh shall make / arrangements for weigh-ment of wastes collected and brought at landfill site by 2004 or earlier.

16. Utilities, such as drinking water (preferably bathing facilities to workers) and lighting arrangements for easy landfill operations when carried out in night hours, shall be provided.

17. Safety provisions including health inspections of workers at landfill site shall be periodically made.

Specifications for Landfilling

18. Wastes subjected to landfilling shall be compacted in thin layers using landfill compactors to achieve high density of the wastes.

19. Wastes shall be covered immediately or at the end of each working day with 7.5-10 cm of soil/ sweet earth. In case of non-availability of such material, demolition and construction material shall be used.

20. Prior to the commencement of monsoon season, an intermediate cover of 40-65 cm thickness of soil shall be placed on the landfill with proper compaction and grading to prevent infiltration during monsoon. Proper drainage berms shall be constructed to divert.

21. After completion of landfill, a final cover shall be designed to minimise infiltration and erosion. The final cover shall meet the following specifications:

(a) The final cover shall have a barrier soil layer comprising of 60 cm of clay/amended soil with permeability less that 1×10^{-7} cm/sec.

(b) On top of the barrier soil layer, there shall be a drainage layer of 15 cm.

(c) On top of the drainage layer, there shall be a vegetative layer of 45 cm to support natural plant growth and to minimise erosion.

Pollution prevention

22. In order to prevent pollution problems from landfill operations, the following provisions shall be made:

(a) Diversion of storm water drains to minimise leachate generation and prevent pollution of surface water and also for avoiding flooding and creation of marshy conditions;

(b) Construction of a non-permeable lining system at the base and wall of waste disposal area. The low permeability lining system must have barrier soil layer (clay/amended soil) of minimum 60 cm thickness with permeability not greater than 1×10^{-7} cm per second if, waste reaching the landfill is non-biodegradable and inert. For landfill receiving residues of waste processing facilities or mixed waste of waste having contamination of hazardous materials (such as aerosols, bleaches, polishes, batteries, waste oils, paint products and pesticides) minimum liner specifications shall be a composite barrier having 1.5 mm high density polyethylene (HDPE) geomembrane (or equivalent) overlying 90 cm of soil (clay/amended soil) having permeability not greater than 1×10^{-7} cm/sec. The highest level of water table shall be at least 2.0 meter below the base of clay/amended soil barrier layer.

(c) Provisions for management of leachates collection and treatment. The treated leachates shall meet the standards laid down in Schedule IV.

(d) Prevention of runoff from landfill area entering any stream, river, lake or pond.

Water Quality Monitoring

23. Monitoring of ground water quality shall be done in cities having population of more than ten lakh during pre-landfill operations, filling and also during operational and post-operational period at such locations preferably, to cover one upgradient and two downgradient monitoring wells.

24. The monitoring schedule for cities having population less than ten lakhs, shall be worked out by Pollution Control Boards in States and Pollution Control Committees in Union Territories in consultation with municipal bodies.

25. Ground water in and around landfill site shall not be used for drinking purposes or otherwise, unless it meets specified standards for that particular use. The ground water quality shall conform to following maximum acceptable limits:

S. No.	Parameters	Maximum Acceptable limit
1.	Arsenic	0.02 mg/l
2.	Cadmium	0.01 mg/l
3.	Chromium (hexavalent)	0.02 mg/l
4.	Fluoride	1.5 mg/l
5.	Lead	0.05 mg/l
6.	Mercury	0.001 mg/l
7.	Nitrate	10.0 mg/l
8.	pH	6.5-7.5
9.	Conductivity	100-200 μ mhos/cm
10.	Total dissolved solids	500 mg/l
11.	Chlorides	250 mg/l
12.	Sulphates	1000 mg/l
13.	Colour	5 Hazon units
14.	Bio-chemical oxygen demand	30 mg/l or lower
15.	Chemical oxygen demand	250 mg/l or lower

26. Where ground water is deep below the landfill, vadose zone (Vadose water) monitoring shall be undertaken using pore-liquid samplers (lysimeters) and pore-gas samplers.

Ambient Air Quality Monitoring

27. Installation of landfill gas control system including gas collection system shall be made at landfill site to minimise odour generation, prevent off-site migration of gases and to protect vegetation planted on the rehabilitated landfill surface.

28. The concentration of methane gas generated at landfill site shall not exceed 25 per cent of the lower explosive limit (LEL).

29. The landfill gas from the collection facility at a landfill site shall be utilised for either direct thermal applications or power generation, as per viability. Otherwise, landfill gas shall be burnt (flared), and shall not be allowed to directly escape to the atmosphere or for illegal tapping.

30. Ambient air quality at the landfill site and at the vicinity shall be monitored to meet the following prescribed standards:

S. No.	Parameters	Acceptable Levels
(i)	Sulphur dioxide	120 μg/m³ (24hours)
(ii)	Suspended Particulate Matter	500 μg/m³ (24 hours)
(iii)	Methane	Not to exceed 25 per cent of the lower explosive limit (equivalent to 650 μg/m³)
(iv)	Ammonia Daily average (Sample duration 24 hrs)	0.4 mg/m³ (400 μg/m³)

31. The ambient air quality monitoring shall be carried out by the concerned authority as per the following schedule:

(a) Six times in a year for cities having population of more than fifty lakh.

(b) Four times in a year for cities having population between ten and fifty lakh.

(c) Two times in a year for cities having population between one and ten lakh.

Plantation at Landfill Site

32. A vegetative cover shall be provided over the completed site and following guidelines shall be followed:

(a) Selection of locally adopted perennial plants that are resistant to drought and extreme temperatures;

(b) Root of the plant shall not disrupt the low-permeability layer;

(c) Selected plants shall have ability to thrive on low-nutrient soil with minimum nutrient addition;

(d) Plantation to be made in sufficient density to minimise soil erosion; and

Closure of Landfill Site and Post-care

33. The post-closure care of landfill site shall be conducted for at least fifteen years and long-term monitoring/care plan shall consist of the following:

(a) Maintaining the integrity and effectiveness of final cover, making repairs and preventing runon and runoff from eroding or otherwise damaging the final cover;

(b) Monitoring leachate collection system in accordance with the requirement;

(c) Monitoring of ground water in accordance with requirements and maintaining ground water quality;

(d) Maintaining and operating the landfill gas collection system to meet the standards.

34. Use of closed landfill sites for human settlement or otherwise shall be considered after ensuring that gaseous and leachate analysis complied with the laid down standards.

Special provisions for hilly areas and smaller towns

35. Cities and towns located on hills, shall have location-specific methods evolved for final disposal of solid wastes by the Municipal Authority and the concerned State Pollution Control Board. As land filling in hilly areas is not an environmentally safe option, the Municipal Authority shall set up processing facilities for utilisation of biodegradable wastes. The inert and non-biodegradable waste shall be used for building roads or filling-up of appropriate areas on hills.

SCHEDULE IV

[see rule 8(3)]

Standards for Compost

1. In order to ensure safe application of compost, the following standards for production of compost may be ensured:

Parameters	Maximum acceptable Concentration (in parts per million)
Arsenic	20
Cadmium	20
Chromium	300
Copper	500
Lead	500
Mercury	10
Nickel	100
Zinc	2,500

2. Standards for disposal of treated leachates

Sl. No.	Parameter	Standards (Mode of Disposal)		
		Inland surface water	Public sewers	Land disposal
1.	Suspended solids, mg/l, Max	100	600	200
2.	Dissolved solids (inorganic) mg/l, Max.	2100	2100	2100
3.	pH value	5.5 to 9.0	5.5 to 9.0	5.5 to 9.0
4.	Ammonical nitrogen (as N), mg/l, Max.	50	50	–
5.	Total Kjeldahl nitrogen (as N), mg/l, Max	100	–	–
6.	Biochemical oxygen demand (3 days at 27°C) Max. (mg/l)	30	350	100
7.	Chemical Oxygen demand, mg/l, Max	250	–	–
8.	Arsenic (as As), mg/l, Max	0.2	0.2	0.2
9.	Mercury (as Hg), mg/l, Max	0.01	0.01	–
10.	Lead (as Pb), mg/l, Max	0.1	1.0	–
11.	Cadmium (as Cd), mg/l, Max	2.0	1.0	–
12.	Total Chromium (as Cr), mg/l, Max.	2.0	2.0	–
13.	Copper (as Cu), mg/l, Max.	3.0	3.0	–
14.	Zinc (as Zn), mg/l, Max.	5.0	15	–
15.	Nickel (as Ni), mg/l, Max.	3.0	3.0	–
16.	Cyanide (as CN), mg/l, Max.	0.2	2.0	0.2
17.	Chloride (as Cl), mg/l, Max.	1000	1000	600
18.	Fluoride (as F), mg/l, Max.	2.0	1.5	–
19.	Phenolic compounds (as C_6H_5OH), mg/l, Max.	1.0	5.0	–

This rule is further revised and amended and designated as Solid Waste Management Rules, 2015. With more emphasis waste source segregation, waste recycle & reuse to create wealth. In addition Plastic Waste (Management and Handling) Rules 2011, and e-waste (Management and Handling) Rules 2011 were also reused and amended in 2018 and 2016 respectively.

47 ENVIRONMENTAL CLEARANCES AND PREVENTION AND CONTROL OF POLLUTION

PART I
ENVIRONMENTAL CLEARANCES

Environmental clearance of polluting or degrading development activities is done by the Central and/or the State Governments, with the following objectives:

— Optimal utilisation of finite natural resources through use of better technologies and management packages.

— Incorporating suitable remedial measures at the project formulation stage.

Till January 1994, obtaining environmental clearance from the Central Ministry was only an administrative requirement intended for mega projects undertaken by the Government or Public Sector Undertakings. However, the new Notification (which shall be referred to as the EIA Notification from now on) issued by the Ministry in January 1994 (as amended in May 1994) makes Environment Impact Assessment statutory for 29 different identified activities. This EIA Notification also includes details of procedures for obtaining environmental clearance and for public involvement besides setting time schedules for decision-taking (Annexure I).

47.1 COVERAGE OF PROJECTS WHICH REQUIRE ENVIRONMENTAL CLEARANCE

All the projects listed under Schedule I of the EIA Notification at Annexure I are required to obtain environmental clearance from the Central Government. It is worth pointing out here that even if any of these projects fall under the delicenced category of the New Industrial Policy, they are still required to get environmental clearance from the Central Government to be able to get established and function.

Besides this, site specific projects such as mining, pit-head thermal power stations, hydro-power, major irrigation projects, ports and harbours will also have to obtain separate site clearance from the Central Government (or the State Government as the case may be) as specified in the EIA Notification.

Over and above this, the Government of India (Ministry of Environment and Forests) notifies certain areas as ecologically sensitive/fragile areas from time to time and all developmental projects which are to be located in these notified areas need to obtain environmental clearance from the Central Government irrespective of whether they are listed under Schedule I of the EIA Notification or not. Some of the identified ecologically sensitive/fragile areas notified under the Environment (Protection) Act, 1986, so far include:

— Doon Valley, Murud-Janjira, Dahanu Taluka, Aravaliranges in Gurgaon district of Haryana and Alwar district of Rajasthan, etc.

— Identified coastal areas as per the Coastal Regulation Zone Notification (Part III).

— Forests, Wildlife Sanctuaries, National Parks, Wetlands, Mangroves, Biosphere Reserves, Hill and Mountain areas, etc.

The 29 projects listed in Schedule I of the EIA Notification can be broadly categorised under the following sectors for the sake of convenience:

— Industries
— Mining
— Thermal Power Plants
— River Valley
— Ports, Harbours and Airports
— Communication
— Atomic Energy
— Transport (Rail, Road, Highway)
— Tourism (including hotels, beach resorts).

47.1.1 Industrial Projects

Details of industrial projects that are required to obtain environmental clearance from the Union Ministry of Environment and Forests are:

A. The following industrial activities where the investment is Rs. 50 crores or above:

— Petroleum Refineries including Crude and Product Pipelines;
— Chemical Fertilisers (Nitrogenous and Phosphatic) other than single superphosphate;
— Petrochemical complexes and petrochemical intermediates such as DMT, Caprolactum, LAB etc. and production of basic plastics such as LLPDE, HPDE, PP PVC;
— Exploration for oil and gas and their production, transportation and storage;
— Synthetic Rubber;
— Hydrocyanic Acid and its derivatives;
— Primary metallurgical industries; Electric Arc Furnaces;
— Chlor Alkali Industry;
— Viscose staple fibre and filament yarn;
— Storage batteries integrated with manufacture of oxides of lead and lead antimony alloy;
— Pulp, paper and newsprint;
— Cement.

B. The following industrial activities where the investment is Rs. 1 crore and above:

— Pesticides (technical);
— Bulk drugs and pharmaceuticals;
— Asbestos and asbestos products;

— Integrated paint complex including manufacture of resins and basic raw materials required in the manufacture of paints;

— Distilleries;

— Raw skins and hides;

— Dyes;

— Foundries (individual);

— Electroplating.

C. All industrial projects to be located in any of the notified ecologically fragile/sensitive areas mentioned in para 5 at page 4 (Schedule I of Annexure I).

The industrial licensing mechanism of the Government of India provides that:

— The investor obtains a Letter of Intent (LOI) for certain specific products to be manufactured (Annexure 2).

— The LOI is to be converted into an Industrial Licence subject to certain conditions.

A list of industries that still required licensing is given at Annexure 2.

The investor, therefore, first approaches the Ministry of Industry and obtains LOI. With the LOI, the investor then has to approach the concerned State Pollution Control Board (SPCB) (and the State Forest Department, if the location involves use of forest land). The SPCB evaluates and assesses the quantity and quality of effluents likely to be generated by the proposed unit as well as the efficacy of the control measures proposed by the investor to meet the prescribed standards. If the SPCB is satisfied that the proposed unit will meet all the prescribed effluent and emission standards, it issues a NOC or a permission as the case may be, which is valid for 15 years. It is worth pointing out here that the effluent and emission standards applicable in India are either comparable to or a little less stringent than those prevailing in most of the industrialised countries. Once a NOC is obtained, the LOI is converted into an Industrial Licence by the State authorities themselves, if the activity does not fall under the 29 activities mentioned under Schedule I of the EIA Notification. If, however, the project is listed under Schedule I, then the investor is required to approach the Union Ministry of Environment and Forests for obtaining environmental clearance.

In case of delicenced industries, all the procedural formalities are completed at the State level itself except for those activities which are listed in Schedule I of the EIA Notification or if the proposed activity is to be located in any of the notified ecologically sensitive/fragile areas mentioned.

Obtaining the necessary approvals is facilitated by following a set of "Siting Guidelines" which have been prepared by the Ministry (Chapter 52).

Separate guidelines for projects under various sectors such as River Valley, Thermal Power Plants, Mining, etc. have been prepared by the Ministry and are available on request.

Flow Chart Nos. 1 to 5 (pages 1079-1083) describe the various steps that an investor has to go through for obtaining the necessary permits and licenses for establishing and operating an industrial unit in India.

47.1.2 Mining Projects

Only those mining projects involving more than 5 hectares of land are required to obtain clearance from the Central Government. Other projects involving leases less than 5 ha of land are required to approach the concerned State Government departments for obtaining necessary clearances.

In the case of mining projects, a separate "Site Clearance" is required which is also issued either by the Central Government or the State Government, depending upon the area of land leased. Sites involving less than 5 ha can be cleared at the State level itself and only those above 5 ha of land are required to come to the Central Government.

47.1.3 Thermal Power Plants

All those Thermal Power Plants where the investment is Rs. 50 crores or above, are required to obtain environmental clearance from the Central Government. In case of those TPPs where the investment is less than Rs. 50 crores, environmental clearance from the Central Government is not required and they can approach the concerned State Government departments/agencies. The investor has to obtain a separate 'Site Clearance' from the Central Government, in case of pit-head thermal power stations, which is also applicable only to those projects involving an investment of Rs. 50 crores or above.

47.1.4 River Valley Projects

All River Valley Projects including hydel power, major irrigation and their combination including flood control, where the investment is Rs. 50 crores or above, only need to obtain environmental clearance from the Central Government. All other projects need to approach the concerned State Government departments/agencies only for the necessary clearances and permits.

47.1.5 Tourism Projects including Hotels and Beach Resorts

All tourism projects between 200-500 meters of High Tide Line and at locations with an elevation of more than 1,000 metres and involving an investment of Rs. 5 crores and above are required to obtain environmental clearance from the Central Government.

Hotels, beach resorts and other tourism projects have to comply with certain conditions detailed in the Coastal Regulation Zone (CRZ) Notification of 19th February 1991. The CRZ Notification stipulates that the proposed construction of Hotels and Beach Resorts has to be 200-500 meters away from the High Tide Line depending on the ecological sensitivity of the area. The total covered areas should not exceed 33% of the plot area and the maximum height should not be more than 9 meters and only two floors are permitted.

47.2 GUIDELINES FOR INVESTORS

To enable the investors to prepare the requisite documents such as Environmental Impact Assessment Reports/Environmental Management Plans, etc. as well as to choose the most appropriate sites for their projects, the Ministry has prepared comprehensive guidelines for each of the sectors. Detailed questionnaires have also been prepared for submission of project proposals in different sectors. The siting guidelines and the questionnaire for Industrial Projects have been enclosed in Chapter 48, as examples. The other guidelines and questionnaires can be had from the Ministry on request.

47.3 NEEDED DOCUMENTATION

While this is specified in the EIA Notification reproduced at Annexure I, the documents required are listed below once again to facilitate the investor:

— Feasibility/Project Report;

— Site clearance (only for site-specific projects mentioned in the EIA Notification);

— No Objection Certificate from the SPCBs and other local authorities;

— Environment Appraisal Questionnaire (available at the Ministry of Environment and Forests and/or Application Forms as prescribed in Schedule II of the EIA Notification);

— Environment Impact Assessment Report/Environmental Management Plan;

— Risk Analysis/Emergency Preparedness Plan (only in the case of projects involving hazardous substances);

— Rehabilitation plans where large-scale displacement of people is anticipated.

47.4 ENVIRONMENTAL APPRAISAL PROCEDURE

The documents submitted by an investor are first scrutinised by a multi-disciplinary staff complement functioning in the Ministry of Environment and Forests who may also undertake site-visits wherever required, interact with the investors and hold consultations with experts on specific issues as and when necessary.

After this preliminary scrutiny, the proposals are placed before specially constituted Committees of Experts whose composition is specified in Schedule III of the EIA Notification. Such committees, known as Environmental Appraisal Committees, have been constituted for each sector such as River Valley, Industries, Mining etc. and these committees meet regularly to appraise the proposals received in the Ministry. Besides evaluating and assessing the documents submitted by the investors, the committee may also undertake site-visits on specific cases and interact with the affected people and environmental groups directly if it is felt necessary. In case of certain very special/controversial projects which have aroused considerable public interest, the committee may also decide to arrange for public hearings on those projects to ensure public participation in developmental decisions. Announcements for such public hearing shall be made at least 30 days before through newspapers.

On the basis of the exercise described in the foregoing paragraphs, the Appraisal Committees make their recommendations for approval or rejection of particular projects. The recommendations of the Committees are then processed in the Ministry of Environment and Forests for approval or rejection.

47.5 ISSUE OF ENVIRONMENTAL CLEARANCE/REJECTION LETTER

47.5.1 Single Window Clearance

When a project requires both environmental clearance as well as approval under the Forest (Conservation) Act, 1980, proposals for both are required to be given simultaneously to the concerned divisions of the Ministry. The processing is done simultaneously for clearance/rejection, although separate letters may be issued. If the project does not involve diversion of forest land, the case is processed only for environmental clearance. (For more details on forest clearance, please see Part III as well as Flow Chart No. 5).

47.5.2 Time-Frame

Once all the requisite documents and data from the project authorities are received and public hearings (where required) have been held, assessment and evaluation of the project from the environment angle is completed within 90 days and the decision of the Ministry shall be conveyed within 30 days thereafter.

47.6 POST-PROJECT MONITORING

It is to be noted that whenever a project is given environmental clearance, a set of recommendations and conditions are stipulated by the Appraisal Committee on a case-to-case basis, which have to be complied with by the investor once the project is commissioned. The investors are, therefore, required to submit a half-yearly compliance report to the Ministry after the project is commissioned, to enable the Ministry to monitor the implementation of the recommendations and conditions stipulated by the Appraisal Committee, subject to which the environmental clearance has been given. The six Regional Offices located at Shillong, Bhubaneshwar, Chandigarh, Bangalore, Lucknow and Bhopal help the Ministry in post-project monitoring of the cleared projects.

Cases of non-compliance of the recommendations and conditions by cleared projects/units are brought to the notice of the concerned SPCB, which may then initiate action against the project authorities as described under Part II as well as in Flow Chart No. 3.

PART II
PREVENTION AND CONTROL OF POLLUTION

The Policy Statement for Abatement of Pollution, issued by the Ministry of Environment and Forests in 1992, provides instruments in the form of legislation and regulations, fiscal incentives, voluntary agreements, educational programmes and information campaigns in order to prevent, control and reduce environmental pollution. The establishment and functioning of any industry will be governed by the following Acts of the Ministry of Environment and Forests besides the local zoning and land use laws of the States and UTs:

1. The Water (Prevention and Control of Pollution) Act, 1974, as amended from time to time (Water Act);
2. The Water (Prevention and Control of Pollution) Cess Act, 1977, as amended (Water Cess Act);
3. The Air (Prevention and Control of Pollution) Act, 1981, as Amended (Air Act);
4. The Environment (Protection) Act, 1986 (EPA);
5. The Public Liability Insurance Act, 1991, as amended (PLI Act).

Once an industry has been set up, during the process of manufacture and operation, the industry is also required to meet the standards of emissions, effluents and noise levels prescribed under the Environment (Protection) Rules framed under the Environment (Protection) Act of 1986 which is an umbrella legislation for the protection of environment in the country.

1.0 EXPLANATORY NOTES FOR THE ACTS

1.1 The Water (Prevention and Control of Pollution) Act, 1974

The main provisions of this Act aim at prevention and control of water pollution as well as restoration of water quality, through the establishment of State Pollution Control Boards.

Some salient features of this Act are:

Subject to the provisions in the Act

(a) No person shall knowingly cause or permit any poisonous, noxious or polluting matter determined in accordance with such standards as may be laid down by the State Board to enter (whether directly or indirectly) into any stream or well or sewer or on land;

(b) No person shall knowingly cause or permit to enter into any stream any other matter which may tend, either directly or in combination with similar matters, to impede the proper flow of the water of the stream in a matter leading or likely to lead to a substantial aggravation of pollution due to other causes or of its consequences.

(c) No person shall, without the previous consent of the State Board:

(i) establish or take any steps to establish any industry, operation or process, or any treatment and disposal system or an extension or addition thereto which is likely to discharge sewage or trade effluent into a stream or well or sewer or on land (such discharge being hereafter in this section referred to as discharge of sewage); or

(ii) bring into use any new or altered outlets for the discharge of sewages; or

(iii) begin to make any new discharge of sewage.

An investor intending to set up an industry is required to apply in the prescribed form to the SPCB concerned to obtain the consent to establish as well as the consent to operate the industry after establishment. While granting the consent, the SPCB also stipulates specific conditions relating to the temperature, volume, composition, rate and point of discharge of emissions, effluents etc. The consent to operate an industry is granted for a specific period after which the conditions stipulated at the time of granting consent are reviewed by the SPCB. Even before the expiry of the consent period, the SPCB is authorised to carry out random checks on any industry to check if the standards prescribed are being complied with by the industry; in case the standards are not being met, the SPCB is authorised to serve a notice to the concerned person. (Please see Flow Chart Nos. 2 and 3).

The owner of a defaulting industry may be required to construct a sewage/effluent treatment system. In the event of non-compliance of the standards, the SPCB may issue directions for disconnecting electricity and water supply or any other services to the industry, in extreme cases even to close down the unit.

Stringent penalties are prescribed in this Act for those who operate their industry without the valid consent or in violation of consent conditions.

Any person aggrieved by an order of the SPCB in the above context may appeal to the Appellate Authority constituted by the concerned State Government. As per the provisions under the Water Act and the Air Act, all State Governments are required to constitute Appellate Authorities for addressing the appeals received against SPCB. While some State Governments have separate Appellate Authorities under the two Acts, a single Appellate Authority may also exist in other states.

1.2 The Water (Prevention and Control of Pollution) Cess Act, 1977

The purpose of this Act is to levy and collect cess on water consumed by certain categories of industry specified in the Schedule appended to the Act. Local authorities may also specify the categories of such industries. The money thus collected is used by the Central and State Pollution Control Boards to prevent and control water pollution. Some salient features of this Act are:

— The assessing authority under the Act levies and collects cess based on the amount of water consumed by these industries; the rate is also determined by the purpose for which the water is used.

— The concerned industries are required to install standard water meters at such places as may be required by the concerned authority for measuring and recording the quantity of water consumed by the industry.

— Based on the cess returns to be furnished by the industry every month, the amount of cess is assessed by the assessing authorities.

— Aggrieved persons may appeal against the assessment to the Appellate Authority.

— The Act also provides for a 25% rebate on the cess payable to those industries who consume water within the quantity prescribed for that category of industries and also comply with the effluents standards prescribed under the Water Act or the EPA.

1.3 The Air (Prevention and Control of Pollution) Act, 1981

The objective of the Air Act is to prevent, control and reduce air pollution including noise pollution and to establish Boards at the States/UTs for this. Under the provisions of this Act, no person shall establish or operate any industrial plant without the consent of the SPCB/PCC

— For obtaining consent to establish an industry, the investor has to apply to the SPCB/PCC in the prescribed form accompanied by the prescribed fees. The Board is required to grant consent within 4 months of receipt of the application. The consent would contain conditions relating to specifications of pollution control equipment to be installed. (Please see Flow Chart No. 2).

Other provisions of the Act are similar to those of the Water Act, 1974.

1.4 The Environment (Protection) Act, 1986

The Act is an umbrella legislation providing a single focus in the country for the protection of environment and seeks to plug the loopholes of earlier legislations relating to environment. Several sets of Rules relating to various aspects of management of hazardous chemicals, wastes, microorganisms etc. have been notified under this Act.

Some important points of this Act are:

— The Central Government may also put restrictions on an area in which any industry, operation or process or class of industries or operations shall not be carried out. If they are to be carried out, they may be permitted with certain safeguards.

— Emissions and effluent standards in respect of 61 categories (list can be made available if required) of industries have been evolved and notified so far.

— The standards in respect of pollutants are to be achieved within a period of one year from the date of their notification, specially by those industries identified as highly polluting.

— However, if a particular SPCB may so desire, it may reduce the time limit and also specify more stringent standards in respect of a specified category of industries within their jurisdiction. The SPCB, however, cannot relax either the time limit or the standards.

— Those industries who require consent under the Water Act, Air Act, or both, or authorisation under the Hazardous Waste (Management and Handling) Rules, 1989, are required to submit an environmental audit report to the concerned SPCB/PCC on or before 30th September every year.

2.0 HAZARDOUS WASTES

Persons handling specific categories of hazardous wastes (18 categories of hazardous wastes have been identified so far) are required to obtain authorisation for handling such wastes from the concerned SPCB/PCC. (Please see Flow Chart No. 4).

As per the provisions of the Environment (Protection) Act, 1986, Rules regarding Management and Handling of Hazardous Wastes have been notified. Main points are:

— Every occupier generating hazardous waste and intending to collect, treat, transport, store and dispose of such waste, is required to make an application in Form I (provided in the above quoted Rules) to the SPCB/PCC. The SPCB/PCC issues the authorisation alongwith stipulations for treatment and disposal of such wastes.

— Rules for the Management, Storage and Import of Hazardous Chemicals have also been notified in 1989 under the EPA. As per these Rules, the occupier shall not undertake any industrial activity involving any of the 17 (lists will be provided if necessary) identified hazardous chemicals unless he/she has prepared a Safety Report containing the requisite information specified in the Rules. New industries are required to prepare the Safety Report within 5 years of coming into operation. The occupier of industries handling any of the chemicals specified in Schedules 2 and 3 of the Rules is also required to prepare and maintain an uptodate on-site emergency plan for dealing with major accidents.

2.1 Other important features of these Rules

1. Owners of industries utilising any of the identified hazardous chemicals are required to identify likely hazards and their danger potentials and are expected to take adequate steps to prevent such occurrences.
2. Preparation of Materials Safety Data Sheets (MSDS) is essential for all chemicals handled.
3. Workers on-site are required to be provided with information, training and necessary equipment to ensure their safety.
4. On-site Emergency Plan is to be prepared before commencement of activity.
5. Information regarding off-site spill, if any, of an accident envisaged on-site, needs to be provided to the District Collector for preparation of an off-site emergency plan before commencement of the activity.
6. Before commencement of activity, the public in the vicinity of the plant needs to be informed of the nature of a major accident that might occur on-site and the Do-s and Don't-s to be followed in case of such an occurrence.

3.0 HAZARDOUS MICROORGANISMS AND GENETICALLY ENGINEERED ORGANISMS

The Department of Bio-Technology under the Ministry of Science and Technology is the nodal agency for granting licences for manufacture, import and export of microorganisms and genetically engineered organisms. However, such licences are issued only after the proposal is cleared from the environmental angle by the Ministry of Environment and Forests.

For, according to such clearances, the Ministry has notified the "Rules for the Manufacture, Use, Import and Storage of Hazardous Microorganisms and Genetically Engineered Organisms or Cells" in 1989 under the EPA, 1986.

The investor is required to submit an application in the prescribed proforma to the Ministry of Environment and Forests (Flow Chart No. 4).

The Genetic Engineering Approval Committee (GEAC) constituted by the Ministry of Environment and Forests examines the application form and makes recommendations.

All industries involving hazardous microorganisms or genetically engineered organisms are required to constitute an Institutional Bio-Safety Committee (IBSC) to oversee the planning and execution of various operations in a safe manner and to prepare an on-site emergency plan to deal with any accident which may occur due to a hazardous microorganism.

4.0 THE PUBLIC LIABILITY INSURANCE ACT, 1991, AS AMENDED

This Act, unique to India, imposes on the owner the liability to provide immediate relief in respect of death or injury to any person or damage to any property resulting from an accident while handling any of the notified hazardous chemicals. This relief has to be provided on "no fault" basis. To be able to meet this liability, the owner handling hazardous chemicals has to take an insurance policy of an amount equal to its "paid-up capital" or up to Rs. 500 million, whichever is less. The policy has to be renewed every year. New undertakings will have to take this policy before starting their activity. For existing plants the policy was to be taken within one year of the Act coming into force, i.e., up to 3 March 1992. The owner also has to pay an amount equal to its annual premium to the Central Government's Environment Relief Fund (ERF). The reimbursement of relief to the extent of Rs. 25,000 per person is admissible in case of fatal accidents in addition to the reimbursement of medical expenses up to Rs. 12,500. The liability of the insurer is limited to Rs. 50 million per accident up to Rs. 150 million per year or up to the tenure of the policy. Any claims in excess to this liability will be paid from the ERF. In case the award still exceeds, the remaining amount shall have to be met by the owner. The payment under the Act is only for the immediate relief, owners shall have to provide the final compensation, if any, arising out of legal proceedings.

5.0 FISCAL INCENTIVES TO ENCOURAGE CONTROL AND PREVENTION OF POLLUTION

The Government of India offers several incentives to ensure that industries are motivated to comply with the various environmental standards prescribed under different Acts and Rules to control and prevent pollution. Some major fiscal incentives are:

5.1 Exemption from Income Tax

In order to ensure enthusiastic public involvement—particularly of the corporate sector and private individuals—donation given by a tax payer to any association or institution for programmes on conservation of nature and natural resources will be exempt from income tax. The Secretary, Ministry of Environment and Forests, approves such institutions and associations.

5.2 Depreciation Allowance

Depreciation allowance at 30% is allowed on devices and systems installed by industrial units for minimising pollution or for conservation of natural resource.

5.3 Investment Allowance

An investment allowance at the rate of 35% of the actual costs of new machinery or plant expected to assist in the control of pollution and protection of environment is granted. (The Central Government has notified a list of such machinery and plants on which investment allowance is granted).

5.4 Exemption from Tax on Capital Gains

The purpose of this incentive is to encourage industries to shift from congested urban areas. Capital gains arising from transfer of buildings or lands used for the purpose of business are exempt from tax if these are used for acquiring land or for constructing buildings for the purpose of business at a new place.

5.5 Excise and Duty Exemption on utilisation of fly ash, phosphogypsum etc.

- — Excise duty is exempted on the production of low-cost building materials and components.
- — Excise duty is exempted on the production of building materials using fly ash or phosphogypsum in 25% or more quantities as raw materials.
- — Custom duty is exempted on the import of equipment, machinery and capital goods required for the production of building materials such as bricks, light-weight aggregates, light-weight concrete elements etc. using fly ash and phosphogypsum.
- — Reduction in excise duty up to 15% is allowed on pre-fab components required for housing.

For more details on this subject, the relevant portions of the Income Tax and related Acts and Rules may be consulted.

PART III
CONSERVATION OF NATURAL RESOURCES

1.0 FOREST CONSERVATION

The Forest (Conservation) Act, 1980, as amended from time to time, checks the indiscriminate diversion of forest land for non-forest purposes. The State Governments/Union Territories are required to submit formal proposals to the Central Government (Ministry of Environment and Forests) for diversion of forest land for non-forest purposes in the prescribed proforma alongwith details such as flora, fauna, map of the area, compensatory afforestation proposed etc. As the Government of India does not favour the diversion of forest land for non-forest purposes, clearances under the Forest Act are very difficult to obtain. The rare exceptions carry stipulations for compensatory afforestation and the like as laid down in the Act and the National Forest Policy, 1988. Therefore, as far as possible, non-forest land should be selected for projects.

As per the amended Forest (Conservation) Rules in force now, the Regional Chief Conservators of Forests have the powers to decide proposals involving forest land up to 5 hectares. Proposals involving forest land between 5-20 ha shall be processed by the Regional Chief Conservator in consultation with a State Advisory Group consisting of representatives of the concerned State Government. Proposals involving more than 20 ha of forest land are required to be placed before the Advisory Committee constituted by the Ministry of Environment and Forests (Flow Chart No. 5).

1.1 Some important guidelines laid down in Forest (Conservation) Act and Rules for diversion of forest lands for non-forest purposes

These guidelines ensure that ecological considerations are in no way undermined while examining investment proposals. The State Government or the Union Territory administration concerned is required to envisage in the initial stage itself, the different likely impacts of the proposed project, starting with the proposed site. The guidelines specify certain criteria for the siting of an industry, which state that no projects should be in the vicinity of the following:

— National Parks, Wildlife Sanctuaries and Core areas of the Biosphere Reserves.

— Scenic landscapes, areas of geomorphological significance, unique and representative biomes and ecosystems, heritage sites/structures and areas of cultural heritage and importance.

— Fragile ecosystems such as mountains, areas rich in coral formations as well as marine, coastal, desert, wetland, riverine and island ecosystems.

— Areas rich in biological diversity, genepool and other natural resources.

Other important points are:

— Investors in industrial projects involving forest land are required to undertake detailed Environmental Impact Assessment Studies of their projects with respect to physical resources, hydrology and water quality, socio-economic aspects, human use values, etc.

— Introduction of exotic species of plants and animals without adequate investigations are to be avoided.

— Monoculture and plantation of dominating and exotic species shall be discouraged without sufficient experimentation.

— Wild varieties of animals and plant species are not to be disturbed.

— Besides forest lands, diversion of prime agricultural lands for non-forest purposes are also restricted.

— Investigations and surveys carried out in connection with hydro-electric projects, transmission lines, seismic surveys, exploration for oil drilling or mining will not attract the provisions of the Forest Act provided these surveys are restricted to only clearing of bushes and lopping of tree branches and do not involve clearing of forests or cutting of trees.

— Such surveys, investigations and explorations are prohibited in Sanctuaries, National Parks, Biosphere Reserves and in the preservation and sample plots under the Wildlife (Protection) Act, 1972.

— The permission to survey, explore or prospect does not imply any commitment on the part of the Central Government for any subsequent use of the forest land.

2.0 SUBMISSION OF PROPOSALS INVOLVING DIVERSION OF FOREST LAND

There are prescribed procedures for submission of proposals involving dereservation of reserved forests or use of forest land for non-forest purposes. (Flow Chart No. 5 shows the various steps involved in this exercise). The investor is required to furnish a brief note giving essential details of the project relating to:

(a) Cost and outlay

(b) Justification for locating the project in the forest area, indicating alternative sites that were examined and reasons for their rejection.

(c) Financial and social benefits

(d) Total population benefited

(e) Employment generated etc.

A comprehensive land-use plan of the area required should be attached if land is required for more than one purpose. The building plan should also be attached if the area is required for construction purposes.

3.0 CONSERVATION OF COASTAL ZONES

Under the EPA, 1986, the Government of India has declared the coastal stretches of seas, bays, estuaries, creeks, rivers and backwaters which are influenced by tidal action in the landward side up to 500 metres from the High Tide Line (HTL) and the land between the Low Tide Line (LTL) and HTL as the Coastal Regulation Zone (CRZ).

Certain activities are prohibited in the CRZ, such as:

— Setting up of new industries or expansion of existing industries except those directly related to water-front or directly needing foreshore facilities.

— Manufacturing, handling, storage or disposal of hazardous substances.

— Setting up and expansion of units/mechanisms for disposal of wastes and effluents etc.

FLOW CHART No. 1

Industrial Licensing Procedure
(Applicable only to those Industries listed at Annexure 2)

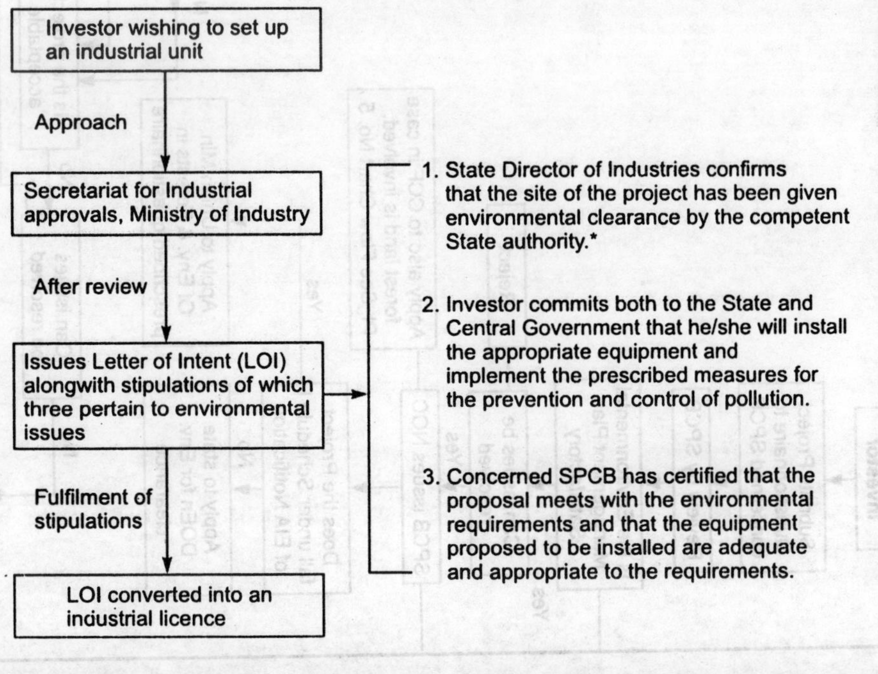

For fulfilling the above three stipulations, the procedure is described in Flow Chart Nos. 2-5.

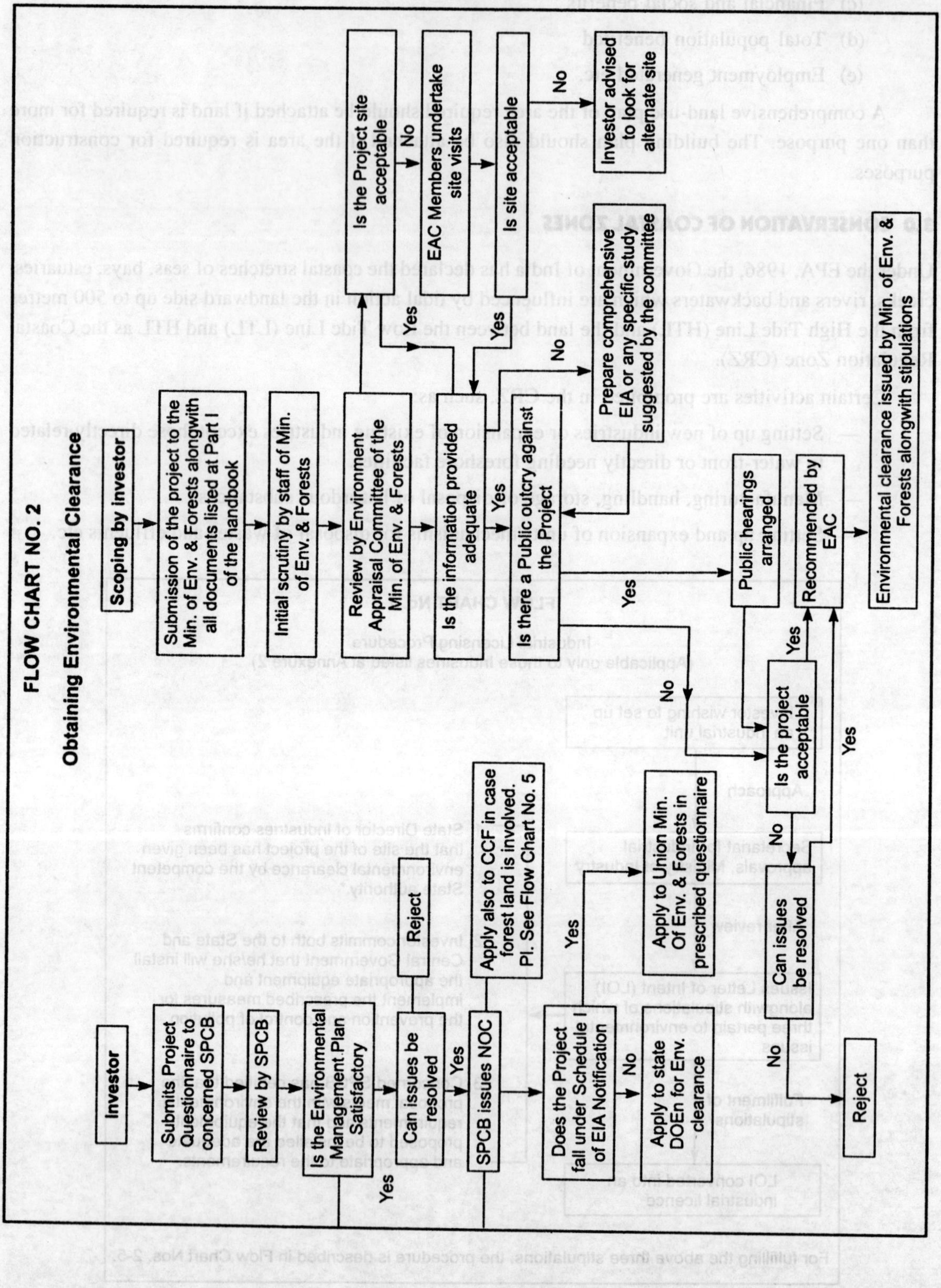
FLOW CHART NO. 2
Obtaining Environmental Clearance
Investor
Submits Project Questionnaire to concerned SPCB
Review by SPCB
Is the Environmental Management Plan Satisfactory
Yes
Can issues be resolved
Yes
Reject
SPCB issues NOC
Apply also to CCF in case forest land is involved. Pl. See Flow Chart No. 5
Does the Project fall under Schedule I of EIA Notification
No
Apply to state DOEn for Env. clearance
Yes
Apply to Union Min. Of Env. & Forests in prescribed questionnaire
Scoping by investor
Submission of the project to the Min. of Env. & Forests alongwith all documents listed at Part I of the handbook
Initial scrutiny by staff of Min. of Env. & Forests
Review by Environment Appraisal Committee of the Min. of Env. & Forests
Is the information provided adequate
Yes
No
Prepare comprehensive EIA or any specific study suggested by the committee
Is there a Public outcry against the Project
Yes
Public hearings arranged
No
Is the project acceptable
Yes
Recommended by EAC
Environmental clearance issued by Min. of Env. & Forests alongwith stipulations
No
Can issues be resolved
Yes
No
Reject
Is the Project site acceptable
Yes
No
EAC Members undertake site visits
Is site accetable
Yes
No
Investor advised to look for alternate site

FLOW CHART NO. 3

Consent to operate an Industrial Unit

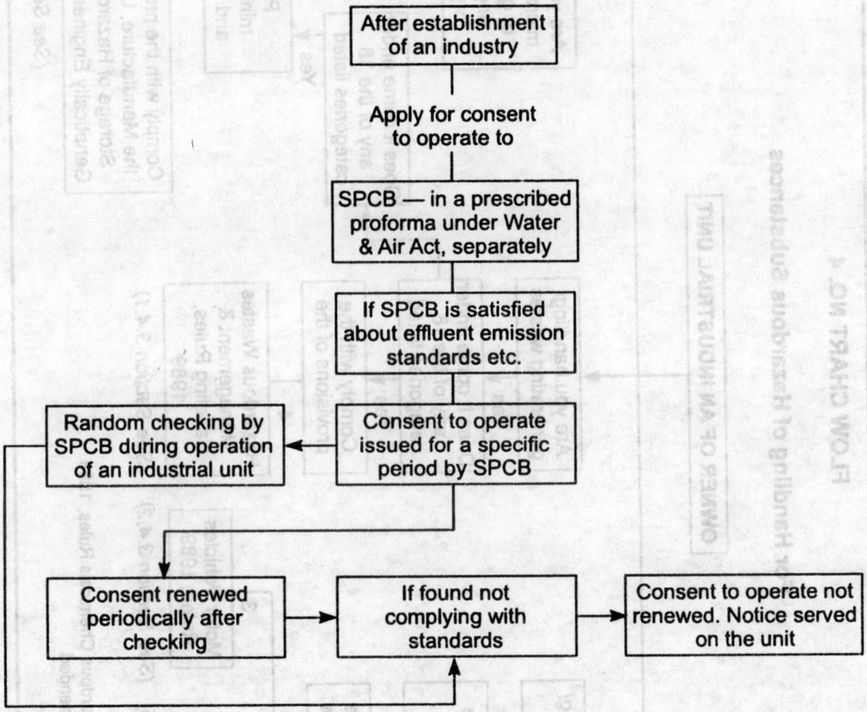

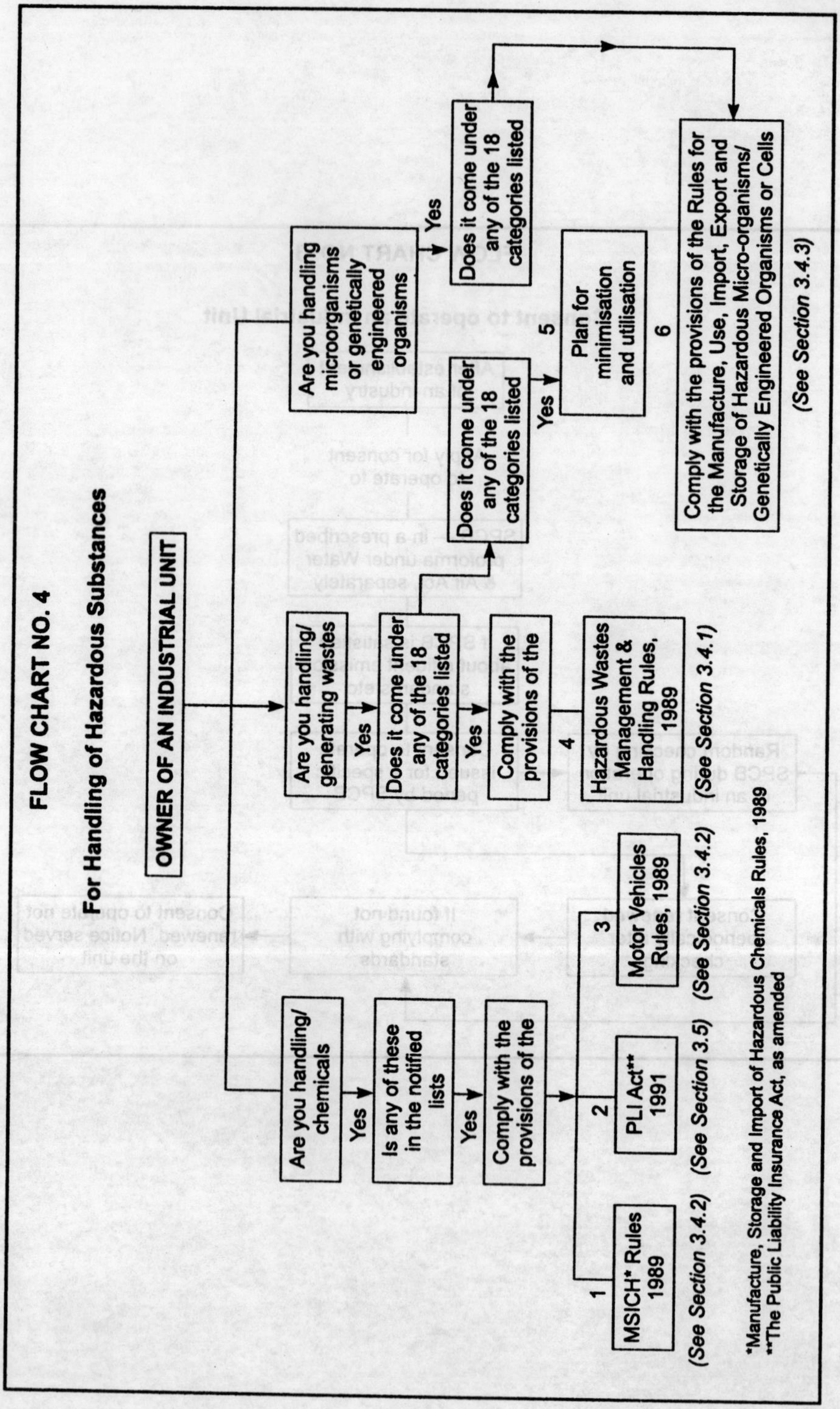

FLOW CHART NO. 4

For Handling of Hazardous Substances

*Manufacture, Storage and Import of Hazardous Chemicals Rules, 1989

**The Public Liability Insurance Act, as amended

FLOW CHART NO. 5

Obtaining Forest Clearance

Flow chart showing various steps involved in examination of cases received under FC Act

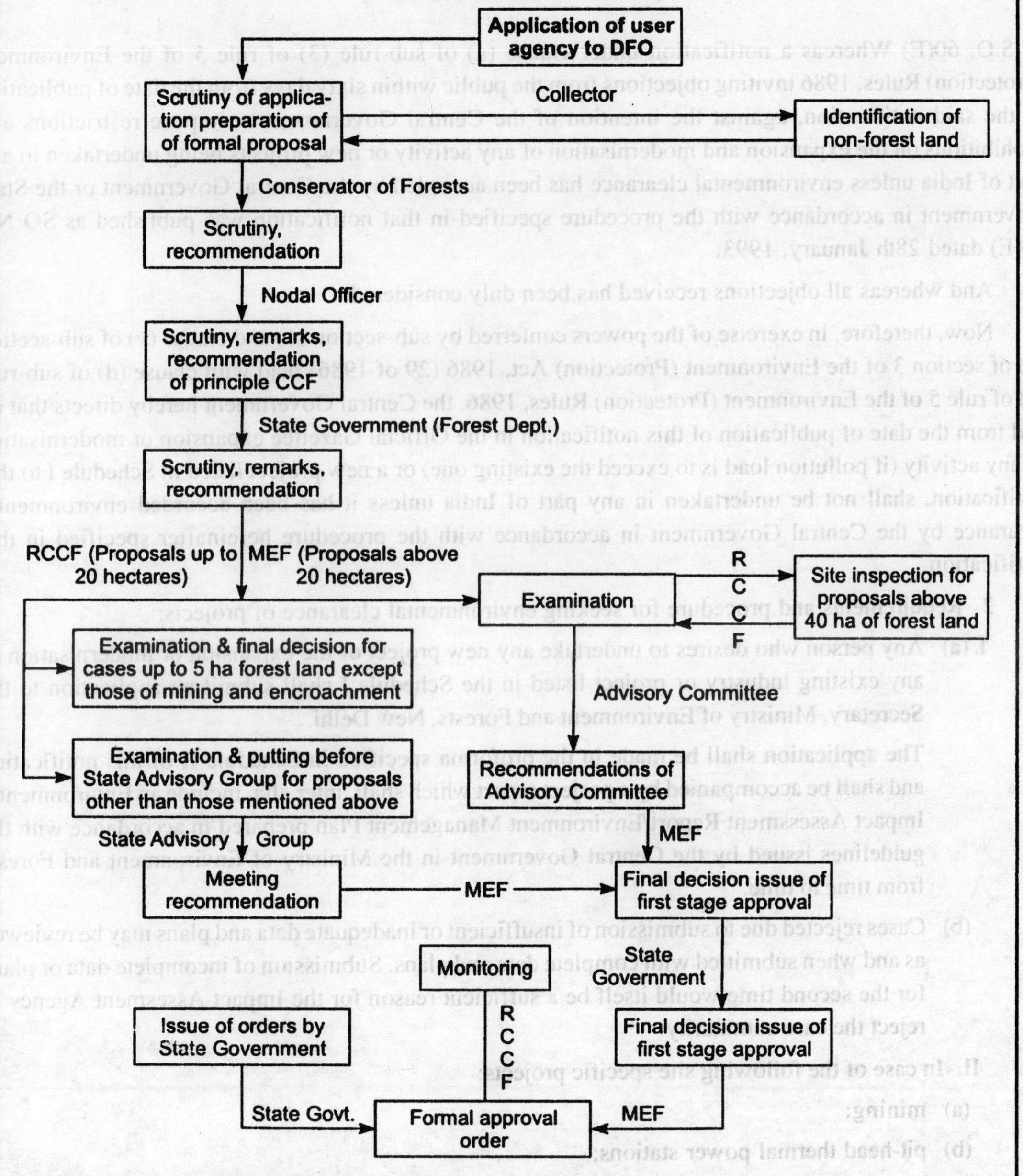

Annexure I

MINISTRY OF ENVIRONMENT AND FORESTS

Notification on

Environmental Impact Assessment of Development Projects

New Delhi, the 27th January, 1994

(As amended on 04/05/1994)

1. S.O. 60(E) Whereas a notification under clause (a) of sub-rule (3) of rule 5 of the Environment (Protection) Rules, 1986 inviting objections from the public within sixty days from the date of publication of the said notification, against the intention of the Central Government to impose restrictions and prohibitions on the expansion and modernisation of any activity or new projects being undertaken in any part of India unless environmental clearance has been accorded by the Central Government or the State Government in accordance with the procedure specified in that notification was published as SO No. 80(E) dated 28th January, 1993;

And whereas all objections received has been duly considered,

Now, therefore, in exercise of the powers conferred by sub-section (1) and clause (v) of sub-section (2) of section 3 of the Environment (Protection) Act, 1986 (29 of 1986) read with clause (d) of sub-rule (3) of rule 5 of the Environment (Protection) Rules, 1986, the Central Government hereby directs that on and from the date of publication of this notification in the Official Gazettee expansion or modernisation of any activity (if pollution load is to exceed the existing one) or a new project listed in Schedule I to this notification, shall not be undertaken in any part of India unless it has been accorded environmental clearance by the Central Government in accordance with the procedure hereinafter specified in this notification.

2. Requirements and procedure for seeking environmental clearance of projects:

I. (a) Any person who desires to undertake any new project or the expansion or modernisation of any existing industry or project listed in the Schedule I shall submit an application to the Secretary, Ministry of Environment and Forests, New Delhi.

The application shall be made in the proforma specified in Schedule II of this notification and shall be accompanied by a project report which shall, inter alia, include an Environmental Impact Assessment Report/Environment Management Plan prepared in accordance with the guidelines issued by the Central Government in the Ministry of Environment and Forests from time to time.

(b) Cases rejected due to submission of insufficient or inadequate data and plans may be reviewed as and when submitted with complete data and plans. Submission of incomplete data or plans for the second time would itself be a sufficient reason for the Impact Assesment Agency to reject the case summarily.

II. In case of the following site specific projects:

(a) mining;

(b) pit-head thermal power stations;

(c) hydropower, major irrigation projects and/or their combination including flood control;

(d) ports and harbours (excluding minor ports);

(e) prospecting and exploration of major minerals in areas above 500 ha.

The project authorities will intimate the location of the project site to the Central Government in the Ministry of Environment and Forests while initiating any investigation and surveys. The Central Government in the Ministry of Environment & Forests will convey a decision regarding suitability or otherwise of the proposed site within a maximum period of thirty days. The said site clearance shall be granted for a sanctioned capacity and shall be valid for a period of five years for commencing the construction, operation or mining.

III. (a) The reports submitted with the application shall be evaluated and assessed by the Impact Assessment Agency, and if deemed necessary it may consult a committee of Experts, having a composition as specified in Schedule III of this Notification. The Impact Assessment Agency (IAA) would be the Union Ministry of Environment and Forests. The committee of experts mentioned above shall be constituted by the IAA or such other body under the Central Government authorised by the IAA in this regard.

(b) The said Committee of experts shall have full right of entry and inspection of the site or, as the case may be factory premises at any time prior to, during or after the commencement of the operations relating to the project.

(c) The Impact Assessment Agency shall prepare a set of recommendations based on technical assessment of documents and data, furnished by the project authorities, supplemented by data collected during visits to sites or factories if undertaken, and interaction with affected population and environmental groups, if necessary. Summary of the reports, the recommendation and the conditions, subject to which environmental clearance is given, shall be made available subject to the public interest to the concerned parties or environmental groups on request. Comments of the public may be solicited, if so decided by Impact Assessment Agency, within thirty days of receipt of proposal, in public hearing arranged for the purpose after giving thrity days' notice of such hearings in at least two newspapers. Public shall be provided access, subject to the public interest, to the summary of the reports/ Environmental Management Plans at the Headquarters of the Impact Assessment Agency.

The assessment shall be completed within a period of ninety days from receipt of the requisite documents and data from the project authorities and completion of the public hearing, where required, and decision conveyed within thirty days thereafter.

The clearance granted shall be valid for a period of five years for commencement of the construction or operation.

No construction work, preliminary or otherwise, relating to the setting up of the project may not be undertaken till the environmental and/or site clearance is obtained.

IV. In order to enable the Impact Assessment Agency to monitor effectively the implementation of the recommendations and conditions subject to which the environmental clearance has been given, the project authorities concerned shall submit a half-yearly report to the Impact Assessment Agency. Subject to the public interest, the Impact Assessment Agency shall make compliance reports publicly available.

V. If no comments from the Impact Assessment Agency are received within the time limit, the project would be deemed to have been approved as proposed by project authorities.

3. Nothing contained in this Notification shall apply to:

(a) any item falling under entry Nos. 3, 18 and 20 of the Schedule I to be located or proposed to be located in the areas covered by the Notifications S.O. No. 102(E) dated 1st February, 1989; S.O. 114(E) dated 20th February, 1991; S.O. No. 416(E) dated 20th June, 1991 and S.O. No. 319(E) dated 7th May, 1992.

(b) any item falling under entry Nos. 1, 2, 3, 4, 5, 7, 9, 10, 12, 13, 14, 16, 17, 19, 21, 25 and 27 of Schedule I if the investment is less than Rs. 50 crores.

(c) any item reserved for Small Scale Industrial Sector with investments less than Rs. 1 crore.

4. Concealing factual data or submission of false, misleading data/reports, decisions or recommendations would lead to the project being rejected. Approval, if granted earlier on the basis of false data, would also be revoked. Misleading and wrong information will cover the following:

— False information

— False data

— Engineered reports

— Concealing of factual data

— False recommendations or decisions.

[No. Z-12013/4/89-IA-I]

R. Rajamani, Secy. (E & F)

SCHEDULE I

(See para 1 and 2)

List of Projects Requiring Environmental Clearance from the Central Government

1. Nuclear Power and related projects such as Heavy Water Plants, Nuclear Fuel Complex, Rare Earths.

2. River Valley projects including hydel power, major irrigation and their combination including flood control.

3. Ports, Harbours, Airports (except minor ports and harbours).

4. Petroleum Refineries including crude and product pipelines.

5. Chemical Fertilisers (Nitrogenous and Phosphatic other than single superphosphate).

6. Pesticides (Technical).

7. Petrochemical complexes (Both Olefinic and Aromatic) and Petrochemical intermediates such as DMT, Caprolactam LAB etc. and production of basic plastics such as LLDPE, HDPE, PP, PVC.

8. Bulk drugs and pharmaceuticals.

9. Exploration for oil and gas and their production, transportation and storage.

10. Synthetic Rubber.

11. Asbestos and Asbestos products.

12. Hydrocyanic acid and its derivatives.

13. (a) Primary metallurgical industries (such as production of Iron and Steel, Aluminium, Copper, Zinc, Lead and Ferro Alloys).

(b) Electric arc furnaces (Mini Steel Plants).

14. Chlor alkali industry.

15. Integrated paint complex including manufacture of resins and basic raw materials required in the manufacture of paints.

16. Viscose Staple fibre and filament yarn.

17. Storage batteries integrated with manufacture of oxides of lead and lead antimoney alloys.

18. All tourism projects between 200m-500 meters of High Water Line and at locations with an elevation of more than 1000 meters with investment of more than Rs. 5 crores.

19. Thermal Power Plants.

20. Mining projects (major minerals) with leases more than 5 hectares.

21. Highway projects.

22. Tarred Roads in Himalayan and or Forest areas.

23. Distilleries.

24. Raw Skins and Hides.

25. Pulp, paper and newsprint.

26. Dyes.

27. Cement.

28. Foundries (individual).

29. Electroplating.

SCHEDULE II

(Sub Sub-para I (a) of Para 2)

Application Form

1. (a) Name and Address of the Project proposed:

(b) Location of the project:

Name of the place:

District, Tehsil:

Latitude / Longitude:

Nearest Airport/Railway Station:

(c) Alternate sites examined and the reasons for selecting the proposed site:

(d) Does the site conform to stipulated land use as per local land use plan:

2. Objectives of the project:

3. (a) Land Requirement:

Agriculture Land:

Forest land and Density of vegetation:

other (specify):

(b) (i) Land use in the Catchment/within 10 km radius of the proposed site:

(ii) Topography of the area indicating gradient, aspects and altitude:

(iii) Erodibility classification of the proposed land:

(c) Pollution sources existing in 10 km radius and their impact on quality of air, water and land:

(d) Distance of the nearest National Park/Sanctuary/Biosphere Reserve/Monuments/Heritage site/Reserve Forest:

(e) Rehabilitation plan for quarries/burrow areas:

(f) Green belt plan:

(g) Compensatory afforestation plan:

4. Climate and Air Quality:

(a) Windrose at site:

(b) Max/Min/Mean annual temperature:

(c) Frequency of inversion:

(d) Frequency of cyclones/tornadoes/cloud burst:

(e) Ambient air quality data:

(f) Nature & concentration of emission of SPM, Gas (CO, CO_2, NO_x, CH_4 etc.) from the project:

5. Water balance:

(a) Water balance at site:

(b) Lean season water availability:

Water Requirement:

(c) Source to be tapped with competing users (River, Lake, Ground, Public Supply):

(d) Water quality:

(e) Changes observed in quality and quantity of ground water in the last 15 years and present charging & extraction details:

(f) (i) Quantum of waste water to be released with treatment details:

(ii) Quantum of quality of water in the receiving body before and after disposal of solid wastes:

(iii) Quantum of waste water to be released on land and type of land:

(g) (i) Details of reservoir water quality with necessary Catchment Treatment Plan:

(ii) Command Area Development Plan:

6. Solid wastes:

(a) Nature and quantity of solid wastes generated:

(b) Solid waste disposal method:

7. Noise and Vibrations:

(a) Sources of Noise and Vibrations:

(b) Ambient noise level:

(c) Noise and Vibration control measures proposed:

(d) Subsidence problem if any with control measures:

8. Power requirement indicating source of supply: Complete environmental details to be furnished separately, if captive power unit proposed:

9. Peak labour force to be deployed giving details of:

— Endemic health problems in the area due to waste water/air/soil borne diseases:

— Health care system existing and proposed:

10. (a) Number of villages and population to be displaced:

(b) Rehabilitation Master Plan:

11. Risk Assessment Report and Disastar Management Plan:

12. (a) Environmental Impact Assessment:

(b) Environment Management Plan:

(c) Detailed Feasibility Report:

(d) Duly filled in questionnaire.

Report prepared as per guidelines issued by the Central Government in the MOEF from time to time.

13. Details of Environmental Management Cell:

I hereby give an undertaking that the data and information given above are true to the best of my knowledge and belief and I am aware that if any part of the data/information submitted is found to be false or misleading at any stage, the project be rejected and the clearance given, if any, to the project is likely to be revoked at our risk and cost.

Signature of the applicant
with name and full address

Given under the seal of
Organisation on behalf of
whom the applicant is signing

Date:

Place:

In respect to item for which data are not required or is not available as per the declaration of project proponent, the project would be considered on that basis.

SCHEDULE III

[See sub para III(a) of Para 2]

Composition of the Expert Committees for Environmental Impact Assessment

1. The committees will consist of experts in the following disciplines:

(i) Eco-System Management

(ii) Air/Water Pollution Control

(iii) Water Resource Management

(iv) Flora/Fauna conservation and management

(v) Land Use Planning

(vi) Social Sciences/Rehabilitation

(vii) Project Appraisal

(viii) Ecology

(ix) Environmental Health

(x) Subject Area Specialists

(xi) Representatives of NGOs/persons concerned with environmental issues.

2. The Chairman will be an outstanding and experienced ecologist or environmentalist or technical professional with wide managerial experience in the relevant development sector.

3. The representative of Impact Assessment Agency will act as a Member-Secretary.

4. Chairman and Members will serve in their individual capacities except those specifically nominated as representatives.

5. The Membership of a Committee shall not exceed 15.

EXPLANATORY NOTE

1. **Expansion and modernisation of existing projects**: A project proponent is required to seek environmental clearance for a proposed expansion/modernisation activity if the resultant pollution load is to exceed the existing levels. The words "Pollution Load" will in this context cover emissions—liquid effluents and solid or semi-solid wastes generated. A project proponent may approach the concerned State Pollution Control Board (SPCB) for certifying whether the proposed modernisation/expansion activity as listed in Schedule I to the Notification is likely to exceed the existing pollution load or not. If it is certified that no increase is likely to occur in the existing pollution load due to the proposed expansion or modernisation, the project proponent will not be required to seek environmental clearance, but a copy of such certificate issued by the SPCB will have to be submitted to the Impact Assessment Agency (IAA) for information. The IAA will, however, reserve the right to review such cases in the public interest if material facts justifying the need for such review come to light.

2. **Availability of Summary Feasibility Report, EIA/EMP Report etc. to concerned parties or groups**: The project proponent will have to submit an executive summary incorporating in brief the essence of project details and findings of environmental impact assessment study which could be made available to concerned parties or environmental groups on request.

3. **Clarification about concerned parties or environmental groups** : The concerned parties or environmental groups will be the bona fide residents located at or around the project site or site of displacement or site of alleged adverse environmental impact.

4. **Public Hearing** : Public hearings could be called for in case of projects involving large displacement or having severe environmental ramifications.

5. **Requisite information required for site clearance/project clearance**

(a) **Site clearance** : Site clearance will be given for site specific projects as mentioned in para-2(ii) of the Notification. Project proponents will be required to furnish information according to the environmental appraisal questionnaires for site clearance, as may be prescribed by the IAA from time to time. Additional information whenever required by the IAA will be communicated immediately to the project proponents who will then be required to furnish the same within the time frame specified.

(b) **Project clearance**: In addition to the application form as mentioned in Schedule II to the notification, project proponents are required to furnish the following information for environmental appraisal :

(i) EIA/EMP report (20 copies)

(ii) Risk Analysis Report (20 copies); however, such reports if normally required for a particular category of project, project proponents can state so accordingly, but the IAA's decision in this regard will be final;

(iii) NOC from the State Pollution Control Board;

(iv) Commitment regarding availability of water and electricity from the competent authority;

(v) Summary of Project Report/Feasibility Report (one copy);

(vi) Filled in questionnaire (as prescribed by the IAA from time to time) for environmental appraisal of the project;

(vii) Comprehensive rehabilitation plan, if more than 1,000 people are likely to be displaced, otherwise a summary plan would be adequate.

As a Comprehensive EIA report will normally take at least one year for its preparation, project proponents may furnish Rapid EIA report to the IAA based on one season data (other than monsoon), for examination of the project. Comprehensive EIA report may be submitted later, if so asked for by the IAA.

The requirement of EIA can be dispensed with by the IAA, in case of projects which are unlikely to cause significant impacts on the environment. In such cases, project proponent will have to furnish full justification for such exemption, for submission of EIA. Where such exemption is granted, project proponents may be asked to furnish such additional information as may be required.

6. **Submission of insufficient or inadequate data**: Regarding cases liable to be rejected due to inadequacy of data, it is clarified that the IAA will make such rejection within 30 days from the date of submission of the proposal. While rejecting a proposal due to insufficient or inadequate data after the first evaluation, the IAA may also stipulate additional requirement of information/clarification for impact assessment purposes. If deemed essential due to the specific nature of location of the proposed project whose data as prescribed is not available, the IAA can examine the project on the basis of available data.

7. **Application form**

(i) In order to remove any hardship to the project proponent in providing any information, the project proponent may, where some information is not available or would cause inordinate delay, mention this in their application form. The IAA may consider the project proposal based on the information available.

(ii) Quality and quantity of ground water

If 15 years' data on the quantity and quality variation of ground water is not available with the concerned Department or Authorities, the project proponent may mention this accordingly in the application form prescribed in Schedule II to the notification. Further, in case of projects where ground water is not to be used, and effluents are not to be discharged on the land, the requirement of ground water variation data for the previous 15 years will be dispensed with.

(iii) A project proponent may write the words "Not Applicable" while filling the application form as mentioned in Schedule II to the Notification in respect of items which are not relevant for the purposes of the proposed project.

8. **Exemption for projects already initiated**: For projects listed in Schedule I to the Notification in respect of which the requirement land has been acquired and all relevant clearances of the State Government including NOC from the respective State Pollution Control Boards have been obtained before 27th January, 1994, a project proponent will not be required to seek environmental clearance from the IAA. However, those units who have not as yet commenced production will inform the IAA.

Annexure II

INDUSTRIES WHICH REQUIRE INDUSTRIAL LICENSING

1. Coal and Lignite

Coal.

Lignite.

2. Petroleum (other than crude) and its distillation products

Petroleum oils, other than crude.

Petroleum gases and other gaseous hydrocarbons.

Petroleum waxes and other similar products obtained through distillation of petroleums.

Petroleum coke and other residues of petroleum oils.

3. Distillation and brewing of alcoholic drinks

Beer made from malt.

Wine of fresh grapes, including fortified wines.

Vermouth and other wine of fresh grapes flavoured with plant or aromatic substances.

Other fermented beverages (for example, cider, perry, mead).

Undenatured ethyl alcohol of an alcoholic strength by volume of less than 80% vol; spirits, liquors and other spirituous beverages, compound alcoholic perparation of a kind used for the manufacture of beverages.

4. Sugar

Cane Sugar, refined.

Other sugar, including centrifugal sugar.

5. Animal fats and oils

Animal fats and oils, partly or wholly hydrogenated.

Edible mixtures of preparations of animal fats and oils.

Inedible mixtures or preparations of animals fats and oils.

6. Cigars and cigarettes of tobacco and manufactured tobacco substitutes

Cigars, cheroots, cigarillos and cigarettes, of tobacco or of tobacco substitutes.

7. Asbestos and asbestos-based products

Articles of asbestos-cement of cellulose fibre cement or the like.

Fabricated asbestos fibres; mixtures with a basis of asbestos or with a basis of asbestos and magnesium carbonate; articles thereof.

Asbestos friction materials.

8. Plywood, decorative veneers and other wood based products such as particle board, medium density fibre board, and block board

Veneer sheets, plywood and other wood sawn lengthwise of a thickness not exceeding 6 mm.

Particle board and other similar board of wood or other ligneous materials.

Fibre board of wood or other ligneous materials.

Plywood, veneered panels and similar laminated wood.

Densified wood, in blocks, plates strips and other profile shapes.

9. Leather

Chamois (including combination chamois) leather.

10. Toned or dressed furskins.

11. Paper and Newsprint except bagasse based unit (i.e. except units based on minimum 75% pulp from agricultural residues, bagasse and other non-conventional raw materials).

Mechanical wood pulp.

Chemical wood pulp, dissolving grades.

Chemical wood pulp, soda or sulphate, other than dissolving grades.

Chemical wood pulp, sulphite, other than dissolving grades.

Semi-chemical wood pulp.

Newsprint, in rolls or sheets.

Uncoated paper of a kind used for writing, printing or other graphic purposes, in rolls or sheets.

Paper of a kind used for household or sanitary purposes, in rolls or sheets.

Uncoated kraft paper, in rolls or sheets.

Other uncoated paper, in rolls or sheets.

Vegetable parchment, greaseproof papers, tracing papers and the like in rolls or sheets.

Composite paper in rolls or sheets.

Paper, corrugated, creped, crinkled, embossed or perforated, in rolls or sheets.

Carbon paper, self copy paper and other copying or transfer papers, in rolls or sheets.

Paper, coated with Kaolin or other inorganic substances, in rolls or sheets.

Other coated or impregnated paper, in rolls or sheets.

Filter blocks, slabs and plates, of paper pulp.

Cigarette paper.

12. Electronic aerospace and defence equipment : all types

Tanks and other armoured fighting vehicles.

Defence aircraft, spacecraft, and parts thereof.

Warships—all kinds.

Arms and ammunition; parts and accessories thereof.

13. Industrial explosive including detonating fuses, safety fuses, gun powder, nitrocellulose and explosives; pyrotechnic products; pyrophoric alloys; certain combustible preparations

Undenatured ethyl alcohol of an alcoholic strength by volume of 80% vol or higher ethyl alcohol and other spirits, denatured, or any strength (Industrial alcohol).

Hydrocyanic acid and its derivatives.

Phosgene and its derivatives.

Sodium Hydroxide (Caustic soda) : solid.

Sodium Hydroxide (Caustic soda) : in aqueous solution.

Ethylene.

Propane (propylene).

Butadienes.

Benzene.

Toluene.

O-xylene.

M-xylene.

P-xylene.

Mixed xylene isomers.

Ethylene glycol (ethanediol)/ethylene oxide.

Industrial alcohol.

Meta amino phenol.

Isocyanates and diisocyanates of hydrocarbon, not elsewhere specified (example : Methyl isocyanate).

Aluminium Phosphide.

Dimethoate.

Quinalphos.

Carbaryl, Phorate and Fenitrophion.

Polyethylene having a specific gravity of less than 0.94.

14. Drugs and Pharmaceuticals (according to Drug Policy)

Provitamins and vitamins, natural or reproduced by synthesis (including natural concentrates), derivatives thereof used primarily as vitamins, and inter-mixtures of the forgoing. (Subject to the Drug Policy).

Hormones; natural or reproduced by synthesis; derivatives thereof, used primarily as hormones; other steroids used primarily as hormones. (Subject to the Drug Policy).

Glycosides, natural or reproduced and synthesis, and their salts, ethers, esters and other derivatives. (Subject to the Drug Policy).

Antibiotics. (Subject to the Drug Policy).

Other synthetic drugs, not elsewhere specified or included. (Subject to the Drug Policy).

Pharmaceutical products. (Subject to the Drug Policy).

15. Entertainment electronics (VCRs, colour TVs, CD players, tape recorders)

Compact disc players.

Tape recorders, casette-type.

Tape recorders, other than cassette-type.

Video recording or reproducing apparatus.

Colour television receivers.

MINISTRY OF ENVIRONMENT AND FORESTS
NOTIFICATION

New Delhi, the 14th September, 2006

1. **S.0.1533(E)**—Whereas, a draft notification under Sub-rule (3) of Rule 5 of the Environment (Protection) Rules, 1986 for imposing certain restrictions and prohibitions on new projects or activities, or on the expansion or modernisation of existing projects or activities based on their potential environmental impacts as indicated in the Schedule to the notification, being undertaken in any part of India[1], unless prior environmental clearance has been accorded in accordance with the objectives of National Environment Policy as approved by the Union Cabinet on 18th May, 2006 and the procedure specified in the notification, by the Central Government or the State or Union Territory Level Environment Impact Assessment Authority (SEIAA), to be constituted by the Central Government in consultation with the State Government or the Union Territory Administration concerned under Sub-section (3) of Section 3 of the Environment (Protection) Act, 1986 for the purpose of this notification, was published in the Gazette of India, Extraordinary, Part II, Section 3, Sub-section (ii) *vide* number S.O. 1324(E), dated the 15th September, 2005 inviting objections and suggestions from all persons likely to be affected thereby within a period of sixty days from the date on which copies of Gazette containing the said notification were made available to the public;

And whereas, copies of the said notification were made available to the public on 15th September, 2005;

And whereas, all objections and suggestions received in response to the above mentioned draft notification have been duly considered by the Central Government;

Now, therefore, in exercise of the powers conferred by sub-section (1) and clause (v) of sub-section (2) of section 3 of the Environment (Protection) Act, 1986, read with clause (d) of sub-rule (3) of rule 5 of the Environment (Protection) Rules, 1986 and in supersession of the notification number S.O. 60 (E) dated the 27th January, 1994, except in respect of things done or omitted to be done before such supersession, the Central Government hereby directs that on and from the date of its publication the required construction of new projects or activities or the expansion or modernization of existing projects or activities listed in the Schedule to this notification entailing capacity addition with change in process and or technology shall be undertaken in any part of India only after the prior environmental clearance from the Central Government or as the case may be, by the State Level Environment Impact Assessment Authority, duly constituted by the Central Government under sub-section (3) of section 3 of the said Act, in accordance with the procedure specified hereinafter in this notification.

2. **Requirements of prior Environmental Clearance (EC)**: The following projects or activities shall require prior environmental clearance from the concerned regulatory authority, which shall hereinafter referred to be as the Central Government in the Ministry of Environment and Forests for matters falling under Category 'A' in the Schedule and at State level the State Environment Impact Assessment Authority (SEIAA) for matters falling under Category 'B' in the said Schedule, before any construction work, or preparation of land by the project management except for securing the land, is started on the project or activity:

(i) All new projects or activities listed in the Schedule to this notification;

(ii) Expansion and modernisation of existing projects or activities listed in the Schedule to this notification with addition of capacity beyond the limits specified for the concerned sector, that is, projects or activities which cross the threshold limits given in the Schedule, after expansion or modernisation;

(iii) Any change in product–mix in an existing manufacturing unit included in Schedule beyond the specified range.

3. State Level Environment Impact Assessment Authority:

(1) A State Level Environment Impact Assessment Authority hereinafter referred to as the SEIAA shall be constituted by the Central Government under sub-section (3) of section 3 of the Environment (Protection) Act, 1986 comprising of three Members including a Chairman and a Member-Secretary to be nominated by the State Government or the Union territory Administration concerned.

(2) The Member-Secretary shall be a serving officer of the concerned State Government or Union territory administration familiar with environmental laws.

(3) The other two Members shall be either a professional or expert fulfilling the eligibility criteria given in Appendix VI to this notification.

(4) One of the specified Members in sub-paragraph (3) above who is an expert in the Environmental Impact Assessment process shall be the Chairman of the SEIAA.

(5) The State Government or Union territory Administration shall forward the names of the Members and the Chairman referred in sub-paragraph 3 to 4 above to the Central Government and the Central Government shall constitute the SEIAA as an authority for the purposes of this notification within thirty days of the date of receipt of the names.

(6) The non-official Member and the Chairman shall have a fixed term of three years (from the date of the publication of the notification by the Central Government constituting the authority).

(7) All decisions of the SEIAA shall be unanimous and taken in a meeting.

4. Categorisation of projects and activities:

(i) All projects and activities are broadly categorized in to two categories–Category A and Category B, based on the spatial extent of potential impacts and potential impacts on human health and natural and man made resources.

(ii) All projects or activities included as Category 'A' in the Schedule, including expansion and modernisation of existing projects or activities and change in product mix, shall require prior environmental clearance from the Central Government in the Ministry of Environment and Forests (MoEF) on the recommendations of an Expert Appraisal Committee (EAC) to be constituted by the Central Government for the purposes of this notification;

(iii) All projects or activities included as Category 'B' in the Schedule, including expansion and modernization of existing projects or activities as specified in sub paragraph (ii) of paragraph 2, or change in product mix as specified in sub paragraph (iii) of paragraph 2, but excluding those which fulfill the General Conditions (GC) stipulated in the Schedule, *will* require prior environmental clearance from the State/Union territory Environment Impact Assessment

Authority (SEIAA). The SEIAA shall base its decision on the recommendations of a State or Union Territory level Expert Appraisal Committee (SEAC) as to be constituted for in this notification. In the absence of a duly constituted SEIAA or SEAC, a Category 'B' project shall be treated as a Category 'A' project.

5. **Screening, Scoping and Appraisal Committees:**

The same Expert Appraisal Committees (EACs) at the Central Government and SEACs (hereinafter referred to as the (EAC) and (SEAC) at the State or the Union Territory level shall screen, scope and appraise projects or activities in Category 'A' and Category 'B' respectively. EAC and SEAC's shall meet at least once every month.

(a) The composition of the EAC shall be as given in Appendix VI. The SEAC at the State or the Union Territory level shall be constituted by the Central Government in consultation with the concerned State Government or the Union Territory Administration with identical composition;

(b) The Central Government may, with the prior concurrence of the concerned State Governments or the Union Territory Administrations, constitutes one SEAC for more than one State or Union Territory for reasons of administrative convenience and cost;

(c) The EAC and SEAC shall be reconstituted after every three years;

(d) The authorised members of the EAC and SEAC, concerned, may inspect any site(s) connected with the project or activity in respect of which the prior environmental clearance is sought, for the purposes of screening or scoping or appraisal with prior notice of at least seven days to the applicant, who shall provide necessary facilities for the inspection;

(e) The EAC and SEACs shall function on the principle of collective responsibility. The Chairperson shall endeavour to reach a consensus in each case, and if consensus cannot be reached, the view of the majority shall prevail.

6. **Application for Prior Environmental Clearance** (EC):

An application seeking prior environmental clearance in all cases shall be made in the prescribed Form 1 annexed herewith and Supplementary Form 1A, if applicable, as given in Appendix II, after the identification of prospective site(s) for the project and/or activities to which the application relates, before commencing any construction activity, or preparation of land, at the site by the applicant. The applicant shall furnish, along with the application, a copy of the pre-feasibility project report except that, in case of construction projects or activities (item 8 of the Schedule) in addition to Form 1 and the Supplementary Form 1A, a copy of the conceptual plan shall be provided, instead of the pre-feasibility report.

7. **Stages in the Prior Environmental Clearance (EC) Process for New Projects:**

7 (i) The environmental clearance process for new projects will comprise of a maximum of four stages, all of which may not apply to particular cases as set forth below in this notification. These four stages in sequential order are:

- Stage (1) Screening (Only for Category 'B' projects and activities)
- Stage (2) Scoping
- Stage (3) Public Consultation
- Stage (4) Appraisal

I. Stage (1) – Screening

In case of Category 'B' projects or activities, this stage will entail the scrutiny of an application seeking prior environmental clearance made in Form 1 by the concerned State level Expert Appraisal Committee (SEAC) for determining whether or not the project or activity requires further environmental studies for preparation of an Environmental Impact Assessment (EIA) for its appraisal prior to the grant of environmental clearance depending up on the nature and location specificity of the project. The projects requiring an Environmental Impact Assessment Report shall be termed Category 'Bl' and remaining projects shall be termed Category 'B2' and will not require an Environment Impact Assessment Report. For categorisation of projects into Bl or B2 except item 8 (b), the Ministry of Environment and Forests shall issue appropriate guidelines from time to time.

II. Stage (2) – Scoping

(i) "Scoping" refers to the process by which the Expert Appraisal Committee in the case of Category 'A' projects or activities, and State level Expert Appraisal Committee in the case of Category 'Bl' projects or activities, including applications for expansion and/or modernisation and/or change in product mix of existing projects or activities, determine detailed and comprehensive Terms Of Reference (TOR) addressing all relevant environmental concerns for the preparation of an Environment Impact Assessment (EIA) Report in respect of the project or activity for which prior environmental clearance is sought. The Expert Appraisal Committee or State level Expert Appraisal Committee concerned shall determine the Terms of Reference on the basis of the information furnished in the prescribed application Form 1/ Form 1A including Terms of Reference proposed by the applicant, a site visit by a sub-group of Expert Appraisal Committee or State level Expert Appraisal Committee concerned only if considered necessary by the Expert Appraisal Committee or State Level Expert Appraisal Committee concerned, Terms of Reference suggested by the applicant if furnished and other information that may be available with the Expert Appraisal Committee or State Level Expert Appraisal Committee concerned. All projects and activities listed as Category 'B' in Item 8 of the Schedule (Construction/Township/Commercial Complexes /Housing) shall not require Scoping and will be appraised on the basis of Form I/ Form 1A and the conceptual plan.

(ii) The Terms of Reference (TOR) shall be conveyed to the applicant by the Expert Appraisal Committee or State Level Expert Appraisal Committee as concerned within sixty days of the receipt of Form 1. In the case of Category A Hydroelectric projects Item l(c) (i) of the Schedule the Terms of Reference shall be conveyed along with the clearance for pre-construction activities. If the Terms of Reference are not finalized and conveyed to the applicant within sixty days of the receipt of Form 1, the Terms of Reference suggested by the applicant shall be deemed as the final Terms of Reference approved for the EIA studies. The approved Terms of Reference shall be displayed on the website of the Ministry of Environment and Forests and the concerned State Level Environment Impact Assessment Authority.

(iii) Applications for prior environmental clearance may be rejected by the regulatory authority concerned on the recommendation of the EAC or SEAC concerned at this stage itself. In case of such rejection, the decision together with reasons for the same shall be communicated to the applicant in writing within sixty days of the receipt of the application.

III. Stage (3) - Public Consultation

(i) "Public Consultation" refers to the process by which the concerns of local affected persons and others who have plausible stake in the environmental impacts of the project or activity are ascertained with a view to taking into account all the material concerns in the project or activity design as appropriate. All Category 'A' and Category Bl projects or activities shall undertake Public Consultation, except the following:

(a) modernisation of irrigation projects (item l(c) (ii) of the Schedule).

(b) all projects or activities located within industrial estates or parks (item 7(c) of the Schedule) approved by the concerned authorities, and which are not disallowed in such approvals.

(c) expansion of Roads and Highways (item 7 (f) of the Schedule) which do not involve any further acquisition of land.

(d) all Building/Construction projects/Area Development projects and Townships (item 8);

(e) all Category 'B2' projects and activities.

(f) all projects or activities concerning national defence and security or involving other strategic considerations as determined by the Central Government.

(ii) The Public Consultation shall ordinarily have two components comprising of:

(a) a public hearing at the site or in its close proximity—district-wise, to be carried out in the manner prescribed in Appendix IV, for ascertaining concerns of local affected persons;

(b) obtain responses in writing from other concerned persons having a plausible stake in the environmental aspects of the project or activity.

(iii) the public hearing at, or in close proximity to, the site(s) in all cases shall be conducted by the State Pollution Control Board (SPCB) or the Union Territory Pollution Control Committee (UTPCC) concerned in the specified manner and forward the proceedings to the regulatory authority concerned within 45 (forty five) of a request to the effect from the applicant.

(iv) in case the State Pollution Control Board or the Union Territory Pollution Control Committee concerned does not undertake and complete the public hearing within the specified period, and/or does not convey the proceedings of the public hearing within the prescribed period directly to the regulatory authority concerned as above, the regulatory authority shall engage another public agency or authority which is not subordinate to the regulatory authority, to complete the process within a further period of forty five days.

(v) If the public agency or authority nominated under the sub-paragraph (iii) above reports to the regulatory authority concerned that owing to the local situation, it is not possible to conduct the public hearing in a manner which will enable the views of the concerned local persons to be freely expressed, it shall report the facts in detail to the concerned regulatory authority, which may, after due consideration of the report and other reliable information that it may have, decide that the public consultation in the case need not include the public hearing.

(vi) For obtaining responses in writing from other concerned persons having a plausible stake in the environmental aspects of the project or activity, the concerned regulatory authority and the State Pollution Control Board (SPCB) or the Union Territory Pollution Control Committee (UTPCC) shall invite responses from such concerned persons by placing on their website the

Summary EIA report prepared in the format given in Appendix IIIA by the applicant along with a copy of the application in the prescribed form, within seven days of the receipt of a written request for arranging the public hearing. Confidential information including non-disclosable or legally privileged information involving Intellectual Property Right, source specified in the application shall not be placed on the web site. The regulatory authority concerned may also use other appropriate media for ensuring wide publicity about the project or activity. The regulatory authority shall, however, make available on a written request from any concerned person the Draft EIA report for inspection at a notified place during normal office hours till the date of the public hearing. All the responses received as part of this public consultation process shall be forwarded to the applicant through the quickest available means.

(vii) After completion of the public consultation, the applicant shall address all the material environmental concerns expressed during this process, and make appropriate changes in the draft EIA and EMP. The final EIA report, so prepared, shall be submitted by the applicant to the concerned regulatory authority for appraisal. The applicant may alternatively submit a supplementary report to draft EIA and EMP addressing all the concerns expressed during the public consultation.

IV. Stage (4) - Appraisal

(i) Appraisal means the detailed scrutiny by the Expert Appraisal Committee or State Level Expert Appraisal Committee of the application and other documents like the Final EIA report, outcome of the public consultations including public hearing proceedings, submitted by the applicant to the regulatory authority concerned for grant of environmental clearance. This appraisal shall be made by Expert Appraisal Committee or State Level Expert Appraisal Committee concerned in a transparent manner in a proceeding to which the applicant shall be invited for furnishing necessary clarifications in person or through an authorized representative. On conclusion of this proceeding, the Expert Appraisal Committee or State Level Expert Appraisal Committee concerned shall make categorical recommendations to the regulatory authority concerned either for grant of prior environmental clearance on stipulated terms and conditions, or rejection of the application for prior environmental clearance, together with reasons for the same.

(ii) The appraisal of all projects or activities which are not required to undergo public consultation, or submit an Environment Impact Assessment report, shall be carried out on the basis of the prescribed application Form 1 and Form 1A as applicable, any other relevant validated information available and the site visit wherever the same is considered as necessary by the Expert Appraisal Committee or State Level Expert Appraisal Committee concerned.

(iii) The appraisal of an application shall be completed by the Expert Appraisal Committee or State Level Expert Appraisal Committee concerned within sixty days of the receipt of the final Environment Impact Assessment report and other documents or the receipt of Form 1 and Form 1 A, where public consultation is not necessary and the recommendations of the Expert Appraisal Committee or State Level Expert Appraisal Committee shall be placed before the competent authority for a final decision within the next fifteen days. The prescribed procedure for appraisal is given in Appendix V.

7 (ii). Prior Environmental Clearance (EC) process for Expansion or Modernisation or Change of product mix in existing projects:

All applications seeking prior environmental clearance for expansion with increase in the production capacity beyond the capacity for which prior environmental clearance has been granted under this notification or with increase in either lease area or production capacity in the case of mining projects or for the modernisation of an existing unit with increase in the total production capacity beyond the threshold limit prescribed in the Schedule to this notification through change in process and/or technology or involving a change in the product-mix shall be made in Form I and they shall be considered by the concerned Expert Appraisal Committee or State Level Expert Appraisal Committee within sixty days, who will decide on the due diligence necessary including preparation of EIA and public consultations and the application shall be appraised accordingly for grant of environmental clearance.

8. Grant or Rejection of Prior Environmental Clearance (EC):

(i) The regulatory authority shall consider the recommendations of the EAC or SEAC concerned and convey its decision to the applicant within forty five days of the receipt of the recommendations of the Expert Appraisal Committee or State Level Expert Appraisal Committee concerned or in other words within one hundred and five days of the receipt of the final Environment Impact Assessment Report, and where Environment Impact Assessment is not required, within one hundred and five days of the receipt of the complete application with requisite documents, except as provided below.

(ii) The regulatory authority shall normally accept the recommendations of the Expert Appraisal Committee or State Level Expert Appraisal Committee concerned. In cases where it disagrees with the recommendations of the Expert Appraisal Committee or State Level Expert Appraisal Committee concerned, the regulatory authority shall request reconsideration by the Expert Appraisal Committee or State Level Expert Appraisal Committee concerned within forty five days of the receipt of the recommendations of the Expert Appraisal Committee or State Level Expert Appraisal Committee concerned while stating the reasons for the disagreement. An intimation of this decision shall be simultaneously conveyed to the applicant. The Expert Appraisal Committee or State Level Expert Appraisal Committee concerned, in turn, shall consider the observations of the regulatory authority and furnish its views on the same within a further period of sixty days. The decision of the regulatory authority after considering the views of the Expert Appraisal Committee or State Level Expert Appraisal Committee concerned shall be final and conveyed to the applicant by the regulatory authority concerned within the next thirty days.

(iii) In the event that the decision of the regulatory authority is not communicated to the applicant within the period specified in sub-paragraphs (i) or (ii) above, as applicable, the applicant may proceed as if the environment clearance sought for has been granted or denied by the regulatory authority in terms of the final recommendations of the Expert Appraisal Committee or State Level Expert Appraisal Committee concerned.

(iv) On expiry of the period specified for decision by the regulatory authority under paragraph (i) and (ii) above, as applicable, the decision of the regulatory authority, and the final recommendations of the Expert Appraisal Committee or State Level Expert Appraisal Committee concerned shall be public documents.

(v) Clearances from other regulatory bodies or authorities shall not be required prior to receipt of applications for prior environmental clearance of projects or activities, or screening, or scoping, or appraisal, or decision by the regulatory authority concerned, unless any of these is sequentially dependent on such clearance either due to a requirement of law, or for necessary technical reasons.

(vi) Deliberate concealment and/or submission of false or misleading information or data which is material to screening or scoping or appraisal or decision on the application shall make the application liable for rejection, and cancellation of prior environmental clearance granted on that basis. Rejection of an application or cancellation of a prior environmental clearance already granted, on such ground, shall be decided by the regulatory authority, after giving a personal hearing to the applicant, and following the principles of natural justice.

9. **Validity of Environmental Clearance (EC)**

The "Validity of Environmental Clearance" is meant the period from which a prior environmental clearance is granted by the regulatory authority, or may be presumed by the applicant to have been granted under sub-paragraph (iv) of paragraph 7 above, to the start of production operations by the project or activity, or completion of all construction operations in case of construction projects (item 8 of the Schedule), to which the application for prior environmental clearance refers. The prior environmental clearance granted for a project or activity shall be valid for a period of ten years in the case of River Valley projects (item l(c) of the Schedule), project life as estimated by Expert Appraisal Committee or State Level Expert Appraisal Committee subject to a maximum of thirty years for mining projects and five years in the case of all other projects and activities. However, in the case of Area Development projects and Townships [item 8(b)], the validity period shall be limited only to such activities as may be the responsibility of the applicant as a developer. This period of validity may be extended by the regulatory authority concerned by a maximum period of five years provided an application is made to the regulatory authority by the applicant within the validity period, together with an updated Form 1, and Supplementary Form 1 A, for Construction projects or activities (item 8 of the Schedule). In this regard the regulatory authority may also consult the Expert Appraisal Committee or State Level Expert Appraisal Committee as the case may be.

10. **Post Environmental Clearance Monitoring**

(i) It shall be mandatory for the project management to submit half-yearly compliance reports in respect of the stipulated prior environmental clearance terms and conditions in hard and soft copies to the regulatory authority concerned, on 1st June and 1st December of each calendar year.

(ii) All such compliance reports submitted by the project management shall be public documents. Copies of the same shall be given to any person on application to the concerned regulatory authority. The latest such compliance report shall also be displayed on the web site of the concerned regulatory authority.

11. **Transferability of Environmental Clearance (EC)**

A prior environmental clearance granted for a specific project or activity to an applicant may be transferred during its validity to another legal person entitled to undertake the project or activity on application by the transferor, or by the transferee with a written "no objection" by the transferor, to, and by the regulatory authority concerned, on the same terms and conditions under which the prior

environmental clearance was initially granted, and for the same validity period. No reference to the Expert Appraisal Committee or State Level Expert Appraisal Committee concerned is necessary in such cases.

12. Operation of EIA Notification, 1994, till disposal of pending cases

From the date of final publication of this notification the Environment Impact Assessment (EIA) notification number S.O.60 (E) dated 27th January, 1994 is hereby superseded, except in suppression of the things done or omitted to be done before such suppression to the extent that in case of all or some types of applications made for prior environmental clearance and pending on the date of final publication of this notification, the Central Government may relax any one or all provisions of this notification except the list of the projects or activities requiring prior environmental clearance in Schedule I, or continue operation of some or all provisions of the said notification, for a period not exceeding one year from the date of issue of this notification.

SCHEDULE

(See paragraph 2 and 7)

List of Projects or Activities Requiring Prior Environmental Clearance

Project or Activity		Category with threshold limit		Conditions, if any
		A	B	
1		**Mining, extraction of natural resources and power generation (for a specified production capacity)**		
(1)	(2)	(3)	(4)	(5)
1 (a)	Mining of minerals	≥ 50 ha of mining lease area Asbestos mining irrespective of mining area	< 50 ha ≥ 5 ha of mining lease area	General condition shall apply **Note**: Mineral prospecting (not involving drilling) are exempted provided the concession areas has got previous clearance for physical survey
1(b)	Offshore and onshore oil and gas exploration, development & production	All projects		**Note**: Exploration surveys (not involving drilling) are exempted provided the concession areas have got previous clearance for physical survey
1(c)	River valley projects	(i) ≥ 50 MW hydroelectric power generation; (ii) ≥ 10,000 ha of culturable command area	(i) <50 MW ≥ 25 MW hydroelectric power generation; (ii) < 10,000 ha of culturable command area	General condition shall apply
1(d)	Thermal Power Plants	≥ 500 MW (coal/lignite/ naphtha & gas based); ≥ 50 MW (Pet coke, diesel and all other fuels)	< 500 MW (coal/lignite/ naphtha & gas based); < 50 MW ≥ 5 MW (Pet coke, diesel and all other fuels)	General condition shall apply
1(e)	Nuclear power projects and processing of nuclear fuel	All projects		

Contd.

Project or Activity		Category with threshold limit A	Category with threshold limit B	Conditions if any
2			**Primary Processing**	
(1)	(2)	(3)	(4)	(5)
2(a)	Coal washeries	≥ 1 million ton/annum throughput of coal	< 1 million ton/annum throughput of coal	General condition shall apply (If located within mining area the proposal shall be appraised together with the mining proposal)
2(b)	Mineral beneficiation	≥ 0.1 million ton/annum mineral throughput	< 0.1 million ton/annum mineral throughput	General condition shall apply (Mining proposal with Mineral beneficiation shall be appraised together for grant of clearance)
3			**Materials Production**	
3(a)	Metallurgical industries (ferrous & non-ferrous)	(a) Primary metallurgical industry All projects (b) Sponge iron manufacturing ≥ 200 TPD (c) Secondary metallurgical processing industry All toxic and heavy metal producing units ≥ 20,000 tonnes/annum	Sponge iron manufacturing < 200 TPD Secondary metallurgical processing industry (i) All toxic and heavy metal producing units < 20,000 tonnes/annum (ii) All other non-toxic secondary metallurgical processing industries > 5,000 tonnes/annum	General condition shall apply for sponge iron manufacturing
3(b)	Cement plants	≥ 1.0 million tonnes/annum production capacity	< 1.0 million tonnes/annum production capacity. All stand alone grinding units	General condition shall apply
4			**Materials Processing**	
4(a)	Petroleum refining industry	All projects	—	—
4(b)	Coke oven plants	≥ 250,000 tonnes/annum	< 250,000 & ≥ 25,000 tonnes/annum	—
4(c)	Asbestos milling and asbestos based products	All projects	—	—
4(d)	Chlor-alkali industry	≥ 300 TPD production capacity or a unit located outside the notified industrial area/estate	< 300 TPD production capacity and located within a notified industrial area/estate	Specific condition shall apply No new Mercury Cell based plants will be permitted and existing units converting to membrane cell technology are exempted from this notification

Contd.

Project or Activity		Category with threshold limit A	B	Conditions if any
(1)	(2)	(3)	(4)	(5)
4(e)	Soda ash industry	All projects	—	—
4(f)	Leather/skin/hide processing industry	New projects outside the industrial area or expansion of existing units outside the industrial area	All new or expansion of projects located within a notified industrial area/estate	Specific condition shall apply
5		Manufacturing / Fabrication		
5(a)	Chemical fertilizers	All projects	—	—
5(b)	Pesticides industry and pesticide specific intermediates (excluding formulations)	All units producing technical grade pesticides	—	—
5(c)	Petrochemical complexes (industries based on processing of petroleum fractions & natural gas and/or reforming to aromatics)	All projects	—	—
5(d)	Manmade fibres manufacturing	Rayon	Others	General condition shall apply
5(e)	Petrochemical based processing (processes other than cracking & reformation and not covered under the complexes)	Located outside the notified industrial area/estate	Located in a notified industrial area/estate	Specific condition shall apply
5(f)	Synthetic organic chemicals industry (dyes & dye intermediates; bulk drugs and intermediates excluding drug formulations; synthetic rubbers; basic organic chemicals, other synthetic organic chemical and chemical intermediates)	Located outside the notified industrial area/estate	Located in a notified industrial area/estate	Specific condition shall apply
5(g)	Distilleries	(i) All molasses based distilleries (ii) All cane juice/non-molasses based distilleries ≥ 30 KLD	All cane juice/non-molasses based distilleries < 30 KLD	General condition shall apply

Contd.

Project or Activity		Category with threshold limit		Conditions if any
		A	B	
(1)	(2)	(3)	(4)	(5)
5(h)	Integrated paint industry	—	All projects	General condition shall apply
5(i)	Pulp & paper industry excluding manufacturing of paper from waste paper and manufacture of paper from ready pulp without bleaching	Pulp manufacturing and Pulp & Paper manufacturing industry	Paper manufacturing industry without pulp manufacturing	General condition shall apply
5(j)	Sugar industry	—	≥ 5000 TCD cane crushing capacity	General condition shall apply
5(k)	Induction/arc furnaces/cupola furnaces 5TPH or more	— —	All projects	General condition shall apply
6			Service Sectors	
6(a)	Oil & gas transportation pipe line (crude and refinery/petrochemical products), passing through national parks/ sanctuaries/coral reefs/ecologically sensitive areas including LNG Terminal	All projects		
6(b)	Isolated storage & handling of hazardous chemicals (As per threshold planning quantity indicated in column 3 of schedule 2 & 3 of MSIHC Rules 1989 amended 2000)		All projects	General condition shall apply
7		Physical infrastructure including environemntal services		
7(a)	Air ports	All projects	—	—
7(b)	All ship breaking yards including ship breaking units	All projects	—	—

 Contd.

Project or Activity		Category with threshold limit A	B	Conditions if any
(1)	(2)	(3)	(4)	(5)
7(c)	Industrial estates/ parks/complexes/ areas, export processing zones (EPZs), Special Economic Zones (SEZs), Biotech Parks, Leather Complexes	If at least one industry in the proposed industrial estate falls under the Category A, entire industrial area shall be treated as Category A, irrespective of the area Industrial estates with area greater than 500 ha and housing at least one Category B industry	Industrial estates housing at least one Category B industry and area < 500 ha Industrial estates of area > 500 ha and not housing any industry belonging to Category A or B.	Special conditions shall apply Note: Industrial Estate of area bleow 500 ha and not housing any industry of Category A and B does not require clearnace
7(d)	Common hazardous waste treatment, storage and disposal facilities (TSDFs)	All integrated facilities having incineration & landfill or incineration alone	All facilities having landfill only	General condition shall apply
7(e)	Ports, Harbours	≥ 5 million TPA of cargo handling capacity (excluding fishing harbours)	< 5 million TPA of cargo handling capacity and/or ports/harbours ≥ 10,000 TPA of fish handling capacity	General condition shall apply
7(f)	Highways	(i) New National Highways; and (ii) Expansion of National Highways greater than 30 KM, involving additional right of way greater than 20 m involving land acquisition and passing through more than one State	(i) New State Highways; and (ii) Expansion of National / State Highways greater than 30 km involving additional right of way greater than 20 m involving land acquisition	General condition shall apply
7(g)	Aerial ropeways		All projects	General condition shall apply
7(h)	Common Effluent Treatment Plants (CETPs)		All projects	General condition shall apply
7(i)	Common Municipal Solid Waste Management Facility (CMSWMF)		All projects	General condition shall apply
8		**Building/Construction projects/Area Development projects and Townships**		
8(a)	Building and construction projects		≥ 20,000 sq mtrs and < 150,000 sq mtrs of built up area#	#(built up area for covered construction; in the case of facilities open to the sky, it will be the activity area)
8(b)	Townships and Area Development projects		Covering an area ≥ 50 ha and or built up area ≥ 150,000 sq mtrs[++]	[++] All projects under Item 8(b) shall be appraised as Category B1

Contd.

Note:

General Condition (GC): Any project or activity specified in Category 'B' will be treated as Category A, if located in whole or in part within 10 km from the boundary of: (i) Protected Areas notified under the Wildlife (Protection) Act, 1972, (ii) Critically Polluted areas as notified by the Central Pollution Control Board from time to time, (iii) Notified Eco-sensitive areas, (iv) inter-State boundaries and international boundaries.

Specific Condition (SC): If any Industrial Estate/Complex/Export Processing Zones/Special Economic Zones/Biotech Parks / Leather Complex with homogeneous type of industries such as Items 4(d), 4(f), 5(e), 5(f), or those Industrial estates with pre-defined set of activities (not necessarily homogeneous, obtains prior environmental clearance, individual industries including proposed industrial housing within such estates/complexes will not be required to take prior environmental clearance, so long as the Terms and Conditions for the industrial estate/complex are complied with (Such estates/complexes must have a clearly identified management with the legal responsibility of ensuring adherence to the Terms and Conditions of prior environmental clearance, who may be held responsible for violation of the same throughout the life of the complex/estate).

[No. J-11013/56/2004-IA-II(I)]

R. CHANDRAMOHAN, Jt. Secy.

APPENDIX I
(See paragraph 6) FORM 1

(I) Basic Information

Name of the Project:

Location/site alternatives under consideration:

Size of the Project:

Expected cost of the Project:

Contact Information: Screening Category:

- Capacity corresponding to sectoral activity (such as production capacity for manufacturing, mining lease area and production capacity for mineral production, area for mineral exploration, length for linear transport infrastructure, generation capacity for power generation etc.)

(II) Activity

1. Construction, operation or decommissioning of the Project involving actions, which will cause physical changes in the locality (topography, land use, changes in water bodies, etc.):

Sl. No.	Information/Checklist confirmation	Yes/No	Details thereof (with approximate quantities/rates, wherever possible) with source of information data
1.1	Permanent or temporary change in land use, land cover or topography including increase in intensity of land use (with respect to local land use plan)		
1.2	Clearance of existing land, vegetation and buildings?		

Contd.

Sl. No.	Information/Checklist confirmation	Yes/No	Details thereof (with approximate quantities /rates, wherever possible) with source of information data
1.3	Creation of new land uses?		
1.4	Pre-construction investigations e.g. bore houses, soil testing?		
1.5	Construction works?		
1.6	Demolition works?		
1.7	Temporary sites used for construction works or housing of construction workers?		
1.8	Above ground buildings, structures or earthworks including linear structures, cut and fill or excavations		
1.9	Underground works including mining or tunneling?		
1.10	Reclamation works?		
1.11	Dredging?		
1.12	Offshore structures?		
1.13	Production and manufacturing processes?		
1.14	Facilities for storage of goods or materials?		
1.15	Facilities for treatment or disposal of solid waste or liquid effluents?		
1.16	Facilities for long-term housing of operational workers?		
1.17	New road, rail or sea traffic during construction or operation?		
1.18	New road, rail, air waterborne or other transport infrastructure including new or altered routes and stations, ports, airports etc.?		
1.19	Closure or diversion of existing transport routes or infrastructure leading to changes in traffic movements?		
1.20	New or diverted transmission lines or pipelines?		
1.21	Impoundment, damming, culverting, realignment or other changes to the hydrology of water courses or aquifers?		
1.22	Stream crossings?		
1.23	Abstraction or transfers of water from ground or surface waters?		
1.24	Changes in water bodies or the land surface affecting drainage or runoff?		
1.25	Transport of personnel or materials for construction, operation or decommissioning?		
1.26	Long-term dismantling or decommissioning or restoration work?		
1.27	Ongoing activity during decommissioning which could have an impact on the environment?		
1.28	Influx of people to an area in either temporarily or permanently?		
1.29	Introduction of alien species?		
1.30	Loss of native species or genetic diversity?		
1.31	Any other actions?		

2. **Use of Natural resources for construction or operation of the Project (such as land, water, materials or energy, especially any resources which are nonrenewable or in short supply):**

Sl. No.	Information/Checklist confirmation	Yes/No	Details thereof (with approximate quantities /rates, wherever possible) with source of information data
2.1	Land, especially undeveloped or agricultural land (ha)		
2.2	Water (expected source & competing users) unit: KLD		
2.3	Minerals (MT)		
2.4	Construction material – stone, aggregates, and/soil (expected source – MT)		
2.5	Forests and timber (source – MT)		
2.6	Energy including electricity and fuels (source, competing users) Unit: fuel (MT), energy (MW)		
2.7	Any other natural resources (use appropriate standard units)		

3. **Use, storage, transport, handling or production of substances or materials, which could be harmful to human health or the environment or raise concerns about actual or perceived risks to human health:**

Sl. No.	Information/Checklist confirmation	Yes/No	Details thereof (with approximate quantities /rates, wherever possible) with source of information data
3.1	Use of substances or materials, which are hazardous (as per MSIHC rules) to human health or the environment (flora, fauna, and water supplies)		
3.2	Changes in occurrence of disease or affect disease vectors (e.g. insect or water borne diseases)		
3.3	Affect the welfare of people e.g. by changing living conditions?		
3.4	Vulnerable groups of people who could be affected by the project e.g. hospital patients, children, the elderly etc.		
3.5	Any other causes		

4. **Production of solid wastes during construction or operation or decommissioning (MT/month):**

Sl. No.	Information/Checklist confirmation	Yes/No	Details thereof (with approximate quantities /rates, wherever possible) with source of information data
4.1	Spoil, overburden or mine wastes		
4.2	Municipal waste (domestic and or commercial wastes)		
4.3	Hazardous wastes (as per Hazardous Waste Management Rules)		
4.4	Other industrial process wastes		
4.5	Surplus product		
4.6	Sewage sludge or other sludge from effluent treatment		

Contd.

Sl. No.	Information/Checklist confirmation	Yes/No	Details thereof (with approximate quantities /rates, wherever possible) with source of information data
4.7	Construction or demolition wastes		
4.8	Redundant machinery or equipment		
4.9	Contaminated soils or other materials		
4.10	Agricultural wastes		
4.11	Other solid wastes		

5. Release of pollutants or any hazardous, toxic or noxious substances to air (kg/hr):

Sl. No.	Information/Checklist confirmation	Yes/No	Details thereof (with approximate quantities /rates, wherever possible) with source of information data
5.1	Emissions from combustion of fossil fuels from stationary or mobile sources		
5.2	Emissions from production processes		
5.3	Emissions from materials handling including storage or transport		
5.4	Emissions from construction activities including plant and equipment		
5.5	Dust or odours from handling of materials including construction materials, sewage and waste		
5.6	Emissions from incineration of waste		
5.7	Emissions from burning of waste in open air (e.g. slash materials, construction debris)		
5.8	Emissions from any other sources		

6. Generation of noise and vibration, and emissions of light and heat:

Sl. No.	Information/Checklist confirmation	Yes/No	Details thereof (with approximate quantities/rates, wherever possible) with source of information data
6.1	From operation of equipment e.g. engines, ventilation plant, crushers		
6.2	From industrial or similar processes		
6.3	From construction or demolition		
6.4	From blasting or piling		
6.5	From construction or operational traffic		
6.6	From lighting or cooling systems		
6.7	From any other sources		

7. Risks of contamination of land or water from releases of pollutants into the ground or into sewers, surface waters, groundwater, coastal waters or the sea:

Sl. No.	Information/Checklist confirmation	Yes/No	Details thereof (with approximate quantities/rates, wherever possible) with source of information data
7.1	From handling, storage, use or spillage of hazardous materials		
7.2	From discharge of sewage or other effluents to water or the land (expected mode and place of discharge)		
7.3	By deposition of pollutants emitted to air into the land or into water		
7.4	From any other sources		
7.5	Is there is risk of long term build up of pollutants in the environment from these sources?		

8. Risk of accidents during construction or operation of the Project, which could affect human health or the environment:

Sl. No.	Information/Checklist confirmation	Yes/No	Details thereof (with approximate quantities/rates, wherever possible) with source of information data
8.1	From explosions, spillages, fires etc. from storage, handling, use or production of hazardous substances		
8.2	From any other causes		
8.3	Could the project be affected by natural disasters causing environmental damage (e.g. floods, earthquakes, landslides, cloudburst etc.)?		

9. Factors which should be considered (such as consequential development) which could lead to environmental effect or the potential for cumulative impacts with other existing or planned activities in the locality:

Sl. No.	Information/Checklist confirmation	Yes/No	Details thereof (with approximate quantities/rates, wherever possible) with source of information data
9.1	Lead to development of supporting, cities, ancillary development or development stimulated by the project which could have impact on the environment e.g.: • Supporting infrastructure (roads, power supply, waste or waste water treatment, etc.) • housing development • extractive industries • supply industries • other		
9.2	Lead to after-use of the site, which could have an impact on the environment		
9.3	Set a precedent for later developments		
9.4	Have cumulative effects due to proximity to other existing or planned projects with similar effects		

(III) Environmental Sensitivity

Sl. No.	Area	Name/ Identity	Aerial distance (within 15 km) Proposed project location boundary
1.	Areas protected under international conventions, national or local legislation for their ecological, landscape, cultural or other related value		
2.	Areas which are important or sensitive for ecological reasons – Wetlands, watercourses or other water bodies, coastal zone, biospheres, mountains, forests		
3.	Areas used by protected, important or sensitive species of flora or fauna for breeding, nesting, foraging, resting, over wintering, migration		
4.	Inland, coastal marine or underground waters		
5.	State, National boundaries		
6.	Routes or facilities used by the public for access to recreation or other tourist, pilgrim areas		
7.	Defence installations		
8.	Densely populated or built-up area		
9.	Area occupied by sensitive man-made land uses *(hospitals, schools, places of worship, community facilities)*		
10.	Areas containing important, high quality or scarce resources *(ground water resources, surface resources, forestry, agriculture, fisheries, tourism, minerals)*		
11.	Areas already subjected to pollution or environmental damage *(those where existing legal environmental standards are exceeded)*		
12.	Areas susceptible to natural hazard which could cause the project to present environmental problems *(earthquakes, subsidence, landslides, erosion, flooding or extreme or adverse climatic conditions)*		

(IV) Proposed Terms of Reference for EIA studies.

APPENDIX II

(See paragraph 6)

FORM-1 A (only for construction projects listed under item 8 of the Schedule)

CHECK LIST OF ENVIRONMENTAL IMPACTS

(Project proponents are required to provide full information and wherever necessary attach explanatory notes with the Form and submit along with proposed environmental management plan & monitoring programme)

1. LAND ENVIRONMENT

(Attach panoramic view of the project site and the vicinity)

1.1. Will the existing land use get significantly altered from the project that is not consistent with the surroundings? (Proposed land use must conform to the approved Master Plan/Development

Plan of the area. Change of land use if any and the statutory approval from the competent authority be submitted). Attach Maps of (i) site location, (ii) surrounding features of the proposed site (within 500 meters) and (iii) the site (indicating levels & contours) to appropriate scales. If not available attach only conceptual plans.

1.2. List out all the major project requirements in terms of the land area, built up area, water consumption, power requirement, connectivity, community facilities, parking needs etc.

1.3. What are the likely impacts of the proposed activity on the existing facilities adjacent to the proposed site? (Such as open spaces, community facilities, details of the existing land use, disturbance to the local ecology).

1.4. Will there be any significant land disturbance resulting in erosion, subsidence & instability? (Details of soil type, slope analysis, vulnerability to subsidence, seismicity etc. may be given).

1.5. Will the proposal involve alteration of natural drainage systems? (Give details on a contour map showing the natural drainage near the proposed project site)

1.6. What are the quantities of earthwork involved in the construction activity-cutting, filling, reclamation etc.? (Give details of the quantities of earthwork involved, transport of fill materials from outside the site etc.)

1.7. Give details regarding water supply, waste handling etc. during the construction period.

1.8. Will the low lying areas & wetlands get altered? (Provide details of how low lying and wetlands are getting modified from the proposed activity)

1.9. Whether construction debris & waste during construction cause health hazard? (Give quantities of various types of wastes generated during construction including the construction labour and the means of disposal)

2. WATER ENVIRONMENT

2.1. Give the total quantity of water requirement for the proposed project with the breakup of requirements for various uses. How will the water requirement be met? State the sources & quantities and furnish a water balance statement.

2.2. What is the capacity (dependable flow or yield) of the proposed source of water?

2.3. What is the quality of water required, in case the supply is not from a municipal source? (Provide physical, chemical, biological characteristics with class of water quality)

2.4. How much of the water requirement can be met from the recycling of treated wastewater? (Give the details of quantities, sources and usage)

2.5. Will there be diversion of water from other users? (Please assess the impacts of the project on other existing uses and quantities of consumption)

2.6. What is the incremental pollution load from wastewater generated from the proposed activity? (Give details of the quantities and composition of wastewater generated from the proposed activity)

2.7. Give details of the water requirements met from water harvesting. Furnish details of the facilities created.

2.8. What would be the impact of the land use changes occurring due to the proposed project on the runoff characteristics (quantitative as well as qualitative) of the area in the post construction phase on a long term basis? Would it aggravate the problems of flooding or water logging in any way?

2.9. What are the impacts of the proposal on the ground water? (Will there be tapping of ground water; give the details of ground water table, recharging capacity, and approvals obtained from competent authority, if any)

2.10. What precautions/measures are taken to prevent the runoff from construction activities polluting land & aquifers? (Give details of quantities and the measures taken to avoid the adverse impacts)

2.11. How is the storm water from within the site managed? (State the provisions made to avoid flooding of the area, details of the drainage facilities provided along with a site layout indication contour levels)

2.12. Will the deployment of construction labourers particularly in the peak period lead to unsanitary conditions around the project site? (Justify with proper explanation)

2.13. What on-site facilities are provided for the collection, treatment & safe disposal of sewage? (Give details of the quantities of wastewater generation, treatment capacities with technology & facilities for recycling and disposal)

2.14. Give details of dual plumbing system if treated wastewater used is used for flushing of toilets or any other use.

3. VEGETATION

3.1. Is there any threat of the project to the biodiversity? (Give a description of the local ecosystem with its unique features, if any)

3.2. Will the construction involve extensive clearing or modification of vegetation? (Provide a detailed account of the trees & vegetation affected by the project)

3.3. What are the measures proposed to be taken to minimise the likely impacts on important site features? (Give details of proposal for tree plantation, landscaping, creation of water bodies etc. along with a layout plan to an appropriate scale)

4. FAUNA

4.1. Is there likely to be any displacement of fauna—both terrestrial and aquatic or creation of barriers for their movement? Provide the details.

4.2. Any direct or indirect impacts on the avifauna of the area? Provide details.

4.3. Prescribe measures such as corridors, fish ladders etc. to mitigate adverse impacts on fauna.

5. AIR ENVIRONMENT

5.1. Will the project increase atmospheric concentration of gases & result in heat islands? (Give details of background air quality levels with predicted values based on dispersion models taking into account the increased traffic generation as a result of the proposed constructions)

5.2. What are the impacts on generation of dust, smoke, odorous fumes or other hazardous gases? Give details in relation to all the meteorological parameters.

5.3. Will the proposal create shortage of parking space for vehicles? Furnish details of the present level of transport infrastructure and measures proposed for improvement including the traffic management at the entry & exit to the project site.

5.4. Provide details of the movement patterns with internal roads, bicycle tracks, pedestrian pathways, footpaths etc., with areas under each category.

5.5. Will there be significant increase in traffic noise & vibrations? Give details of the sources and the measures proposed for mitigation of the above.

5.6. What will be the impact of DG sets & other equipment on noise levels & vibration in & ambient air quality around the project site? Provide details.

6. AESTHETICS

6.1. Will the proposed constructions in any way result in the obstruction of a view, scenic amenity or landscapes? Are these considerations taken into account by the proponents?

6.2. Will there be any adverse impacts from new constructions on the existing structures? What are the considerations taken into account?

6.3. Whether there are any local considerations of urban form & urban design influencing the design criteria? They may be explicitly spelt out.

6.4. Are there any anthropological or archaeological sites or artefacts nearby? State if any other significant features in the vicinity of the proposed site have been considered.

7. SOCIO-ECONOMIC ASPECTS

7.1. Will the proposal result in any changes to the demographic structure of local population? Provide the details.

7.2. Give details of the existing social infrastructure around the proposed project.

7.3. Will the project cause adverse effects on local communities, disturbance to sacred sites or other cultural values? What are the safeguards proposed?

8. BUILDING MATERIALS

8.1. May involve the use of building materials with high-embodied energy. Are the construction materials produced with energy efficient processes? (Give details of energy conservation measures in the selection of building materials and their energy efficiency)

8.2. Transport and handling of materials during construction may result in pollution, noise & public nuisance. What measures are taken to minimise the impacts?

8.3. Are recycled materials used in roads and structures? State the extent of savings achieved.

8.4. Give details of the methods of collection, segregation & disposal of the garbage generated during the operation phases of the project.

9. ENERGY CONSERVATION

9.1. Give details of the power requirements, source of supply, backup source etc. What is the energy consumption assumed per square foot of built-up area? How have you tried to minimize energy consumption?

9.2. What type of, and capacity of, power backup do you plan to provide?

9.3. What are the characteristics of the glass you plan to use? Provide specifications of its characteristics related to both short wave and long wave radiation?

9.4. What passive solar architectural features are being used in the building? Illustrate the applications made in the proposed project.

9.5. Does the layout of streets & buildings maximise the potential for solar energy devices? Have you considered the use of street lighting, emergency lighting and solar hot water systems for use in the building complex? Substantiate with details.

9.6. Is shading effectively used to reduce cooling/heating loads? What principles have been used to maximise the shading of Walls on the East and the West and the Roof? How much energy saving has been effected?

9.7. Do the structures use energy-efficient space conditioning, lighting and mechanical systems? Provide technical details. Provide details of the transformers and motor efficiencies, lighting intensity and air-conditioning load assumptions. Are you using CFC and HCFC free chillers? Provide specifications.

9.8. What are the likely effects of the building activity in altering the microclimates? Provide a self assessment on the likely impacts of the proposed construction on creation of heat island & inversion effects.

9.9. What are the thermal characteristics of the building envelope? (a) roof; (b) external walls; and (c) fenestration? Give details of the material used and the U-values or the R-values of the individual components.

9.10. What precautions & safety measures are proposed against fire hazards? Furnish details of emergency plans.

9.11. If you are using glass as wall material provide details and specifications including emissivity and thermal characteristics.

9.12. What is the rate of air infiltration into the building? Provide details of how you are mitigating the effects of infiltration.

9.13. To what extent the non-conventional energy technologies are utilised in the overall energy consumption? Provide details of the renewable energy technologies used.

10. **Environment Management Plan**: The Environment Management Plan would consist of all mitigation measures for each item-wise activity to be undertaken during the construction, operation and the entire life cycle to minimise adverse environmental impacts as a result of the activities of the project. It would also delineate the environmental monitoring plan for compliance of various environmental regulations. It will state the steps to be taken in case of emergency such as accidents at the site including fire.

APPENDIX III
(See paragraph 7)

GENERIC STRUCTURE OF ENVIRONMENTAL IMPACT ASSESSMENT DOCUMENT

Sl. No	EIA Structure	Contents
1.	Introduction	• Purpose of the report • Identification of project & project proponent • Brief description of nature, size, location of the project and its importance to the country, region • Scope of the study—details of regulatory scoping carried out (As per Terms of Reference)
2.	Project Description	• Condensed description of those aspects of the project (based on project feasibility study), likely to cause environmental effects. Details should be provided to give clear picture of the following: • Type of project • Need for the project • Location (maps showing general location, specific location, project boundary & project site layout) • Size or magnitude of operation (incl. Associated activities required by or for the project) • Proposed schedule for approval and implementation • Technology and process description • Project description. Including drawings showing project layout, components of project etc. Schematic representations of the feasibility drawings which give information important for EIA purpose • Description of mitigation measures incorporated into the project to meet environmental standards, environmental operating conditions, or other EIA requirements (as required by the scope) • Assessment of New & untested technology for the risk of technological failure
3.	Description of the Environment	• Study area, period, components & methodology • Establishment of baseline for valued environmental components, as identified in the scope • Base maps of all environmental components
4.	Anticipated Environmental Impacts & Mitigation Measures	• Details of Investigated Environmental impacts due to project location, possible accidents, project design, project construction, regular operations, final decommissioning or rehabilitation of a completed project • Measures for minimising and/or offsetting adverse impacts identified • Irreversible and Irretrievable commitments of environmental components • Assessment of significance of impacts (Criteria for determining significance, Assigning significance) • Mitigation measures
5.	Analysis of Alternatives (Technology & Site)	• In case the scoping exercise results in need for alternatives: • Description of each alternative • Summary of adverse impacts of each alternative • Mitigation measures proposed for each alternative, and • Selection of alternative

Contd.

Sl. No	EIA Structure	Contents
6.	Environmental Monitoring Programme	• Technical aspects of monitoring the effectiveness of mitigation measures (incl. Measurement methodologies, frequency, location, data analysis, reporting schedules, emergency procedures, detailed budget & procurement schedules)
7.	Additional Studies	• Public Consultation • Risk assessment • Social Impact Assessment. R & R Action Plans
8.	Project Benefits	• Improvements in the physical infrastructure • Improvements in the social infrastructure • Employment potential—skilled; semiskilled and unskilled • Other tangible benefits
9.	Environmental Cost Benefit Analysis	If recommended at the Scoping stage
10.	EMP	• Description of the administrative aspects of ensuring that mitigative measures are implemented and their effectiveness monitored, after approval of the EIA
11.	Summary & Conclusion (This will constitute the summary of the EIA Report)	• Overall justification for implementation of the project • Explanation of how adverse effects have been mitigated
12.	Disclosure of Consultants engaged	• The names of the Consultants engaged with their brief resumee and nature of Consultancy rendered

APPENDIX III A
(See paragraph 7)

CONTENTS OF SUMMARY ENVIRONMENTAL IMPACT ASSESSMENT

The Summary EIA shall be a summary of the full EIA Report condensed to ten A4 size pages at the maximum. It should necessarily cover in brief the following Chapters of the full EIA Report:

1. Project Description
2. Description of the Environment
3. Anticipated Environmental Impacts and Mitigation Measures
4. Environmental Monitoring Programme
5. Additional Studies
6. Project Benefits
7. Environment Management Plan

APPENDIX IV
(See paragraph 7)

PROCEDURE FOR CONDUCT OF PUBLIC HEARING

1.0 The Public Hearing shall be arranged in a systematic, time bound and transparent manner ensuring widest possible public participation at the project site(s) or in its close proximity District -wise, by the concerned State Pollution Control Board (SPCB) or the Union Territory Pollution Control Committee (UTPCC).

2. 0 The Process:

2.1 The Applicant shall make a request through a simple letter to the Member Secretary of the SPCB or Union Territory Pollution Control Committee, in whose jurisdiction the project is located, to arrange the public hearing within the prescribed statutory period. In case the project site is extending beyond a State or Union Territory, the public hearing is mandated in each State or Union Territory in which the project is sited and the Applicant shall make separate requests to each concerned SPCB or UTPCC for holding the public hearing as per this procedure.

2.2 The Applicant shall enclose with the letter of request, at least 10 hard copies and an equivalent number of soft (electronic) copies of the draft EIA Report with the generic structure given in Appendix III including the Summary Environment Impact Assessment report in English and in the local language, prepared strictly in accordance with the Terms of Reference communicated after Scoping (Stage 2). Simultaneously the applicant shall arrange to forward copies, one hard and one soft, of the above draft EIA Report along with the Summary EIA report to the Ministry of Environment and Forests and to the following authorities or offices, within whose jurisdiction the project will be located:

(a) District Magistrate/(s)
(b) Zila Parishad or Municipal Corporation
(c) District Industries Office
(d) Concerned Regional Office of the Ministry of Environment and Forests.

2.3 On receiving the draft Environmental Impact Assessment report, the abovementioned authorities except the MoEF, shall arrange to widely publicize it within their respective jurisdictions requesting the interested persons to send their comments to the concerned regulatory authorities. They shall also make available the draft EIA Report for inspection electronically or otherwise to the public during normal office hours till the Public Hearing is over. The Ministry of Environment and Forests shall promptly display the Summary of the draft Environmental Impact Assessment report on its website, and also make the full draft EIA available for reference at a notified place during normal office hours in the Ministry at Delhi.

2.4 The SPCB or UTPCC concerned shall also make similar arrangements for giving publicity about the project within the State/Union Territory and make available the Summary of the draft Environmental Impact Assessment report (Appendix III A) for inspection in select offices or public libraries or panchayats etc. They shall also additionally make available a copy of the draft Environmental Impact Assessment report to the above five authorities/offices viz, Ministry of Environment and Forests, District Magistrate etc.

3.0 Notice of Public Hearing:

3.1 The Member-Secretary of the concerned SPCB or UTPCC shall finalise the date, time and exact venue for the conduct of public hearing within 7 (seven) days of the date of receipt of the draft Environmental Impact Assessment report from the project proponent, and advertise the same in one major National Daily and one Regional vernacular Daily. A minimum notice period of 30 (thirty) days shall be provided to the public for furnishing their responses;

3.2 The advertisement shall also inform the public about the places or offices where the public could access the draft Environmental Impact Assessment report and the Summary Environmental Impact Assessment report before the public hearing;

3.3 No postponement of the date, time, venue of the public hearing shall be undertaken, unless some untoward emergency situation occurs and only on the recommendation of the concerned District Magistrate the postponement shall be notified to the public through the same National and Regional vernacular dailies and also prominently displayed at all the identified offices by the concerned SPGB or Union Territory Pollution Control Committee;

3.4 In the above exceptional circumstances fresh date, time and venue for the public consultation shall be decided by the Member-Secretary of the concerned SPCB or UTPCC only in consultation with the District Magistrate and notified afresh as per procedure under 3.1 above.

4.0 The Panel

4.1 The District Magistrate or his or her representative not below the rank of an Additional District Magistrate assisted by a representative of SPCB or UTPCC, shall supervise and preside over the entire public hearing process.

5.0 Videography

5.1 The SPCB or UTPCC shall arrange to video film the entire proceedings. A copy of the videotape or a CD shall be enclosed with the public hearing proceedings while forwarding it to the Regulatory Authority concerned.

6.0 Proceedings

6.1 The attendance of all those who are present at the venue shall be noted and annexed with the final proceedings.

6.2 There shall be no quorum required for attendance for starting the proceedings.

6.3 A representative of the applicant shall initiate the proceedings with a presentation on the project and the Summary EIA report.

6.4 Every person present at the venue shall be granted the opportunity to seek information or clarifications on the project from the Applicant. The summary of the public hearing proceedings accurately reflecting all the views and concerns expressed shall be recorded by the representative of the SPCB or UTPCC and read over to the audience at the end of the proceedings explaining the contents in the vernacular language and the agreed minutes shall be signed by the District Magistrate or his or her representative on the same day and forwarded to the SPCB/UTPCC concerned.

6.5 A Statement of the issues raised by the public and the comments of the Applicant shall also be prepared in the local language and in English and annexed to the proceedings.

6.6 The proceedings of the public hearing shall be conspicuously displayed at the office of the Panchyats within whose jurisdiction the project is located, office of the concerned Zila Parishad, District Magistrate, and the SPCB or UTPCC. The SPCB or UTPCC shall also display the proceedings on its website for general information. Comments, if any, on the proceedings may be sent directly to the concerned regulatory authorities and the Applicant concerned.

7.0 Time period for completion of public hearing

7.1 The public hearing shall be completed within a period of 45 (forty five) days from date of receipt of the request letter from the Applicant. Therefore the SPCB or UTPCC concerned shall send the public hearing proceedings to the concerned regulatory authority within 8 (eight) days of the completion of the public hearing. The applicant may also directly forward a copy of the approved public hearing proceedings to the regulatory authority concerned along with the final Environmental Impact Assessment report or supplementary report to the draft EIA report prepared after the public hearing and public consultations.

7.2 If the SPCB or UTPCC fails to hold the public hearing within the stipulated 45 (forty five) days, the Central Government in Ministry of Environment and Forests for Category 'A' project or activity and the State Government or Union Territory Administration for Category 'B' project or activity at the request of the SEIAA, shall engage any other agency or authority to complete the process, as per procedure laid down in this notification.

APPENDIX V
(See paragraph 7)

PROCEDURE PRESCRIBED FOR APPRAISAL

1. The applicant shall apply to the concerned regulatory authority through a simple communication enclosing the following documents where public consultations are mandatory:

- Final Environment Impact Assessment Report [20 (twenty) hard copies and 1 (one) soft copy]
- A copy of the video tape or CD of the public hearing proceedings
- A copy of final layout plan (20 copies)
- A copy of the project feasibility report (1 copy).

2. The Final EIA Report and the other relevant documents submitted by the applicant shall be scrutinized in office within 30 (Thirty) days from the date of its receipt by the concerned Regulatory Authority strictly with reference to the TOR and the inadequacies noted shall be communicated electronically or otherwise in a single set to the Members of the EAC/SEAC enclosing a copy each of the Final EIA Report including the public hearing proceedings and other public responses received along with a copy of Form 1 or Form 1A and scheduled date of the EAC/SEAC meeting for considering the proposal.

3. Where a public consultation is not mandatory and therefore a formal EIA study is not required, the appraisal shall be made on the basis of the prescribed application Form 1 and a pre-feasibility report in the case of all projects and activities other than Item 8 of the Schedule. In the case of Item 8 of the Schedule, considering its unique project cycle, the EAC or SEAC concerned shall appraise all Category B projects or activities on the basis of Form 1, Form 1A and the conceptual plan and stipulate the conditions for environmental clearance. As and when the applicant submits the approved scheme/building plans complying with the stipulated environmental clearance conditions with all other necessary statutory approvals, the EAC /SEAC shall recommend the grant of environmental clearance to the competent authority.

4. Every application shall be placed before the EAC /SEAC and its appraisal completed within 60 days of its receipt with requisite documents/details in the prescribed manner.

5. The applicant shall be informed at least 15 (fifteen) days prior to the scheduled date of the EAC/ SEAC meeting for considering the project proposal.

6. The minutes of the EAC /SEAC meeting shall be finalised within 5 working days of the meeting and displayed on the website of the concerned regulatory authority. In case the project or activity is recommended for grant of EC, then the minutes shall clearly list out the specific environmental safeguards and conditions. In case the recommendations are for rejection, the reasons for the same shall also be explicitly stated.

APPENDIX VI
(See paragraph 5)

COMPOSITION OF THE SECTOR/ PROJECT SPECIFIC EXPERT APPRAISAL COMMITTEE (EAC) FOR CATEGORY A PROJECTS AND THE STATE/UT LEVEL EXPERT APPRAISAL COMMITTEES (SEACs) FOR CATEGORY B PROJECTS TO BE CONSTITUTED BY THE CENTRAL GOVERNMENT

1. The Expert Appraisal Committees (EAC(s)) and the State/UT Level Expert Appraisal Committees (SEACs) shall consist of only professionals and experts fulfilling the following eligibility criteria:

Professional: The person should have at least (i) 5 years of formal University training in the concerned discipline leading to a MA/MSc Degree, or (ii) in case of Engineering /Technology/Architecture disciplines, 4 years' formal training in a professional training course together with prescribed practical training in the field leading to a B.Tech/B.E./B.Arch. Degree, or (iii) Other professional degree (e.g. Law) involving a total of 5 years of formal University training and prescribed practical training, or (iv) Prescribed apprenticeship/articleship and pass examinations conducted by the concerned professional association (e.g. Chartered Accountancy),or (v) a University degree, followed by 2 years of formal training in a University or Service Academy (e.g. MBA/IAS/IFS). In selecting the individual professionals, experience gained by them in their respective fields will be taken note of.

Expert: A professional fulfilling the above eligibility criteria with at least 15 years of relevant experience in the field, or with an advanced degree (e.g. Ph.D.) in a concerned field and at least 10 years of relevant experience.

Age: Below 70 years. However, in the event of the non-availability of/paucity of experts in a given field, the maximum age of a member of the Expert Appraisal Committee may be allowed up to 75 years.

2. The Members of the EAC shall be Experts with the requisite expertise and experience in the following fields /disciplines. In the event that persons fulfilling the criteria of "Experts" are not available, Professionals in the same field with sufficient experience may be considered:

- **Environment Quality Experts**: Experts in measurement/monitoring, analysis and interpretation of data in relation to environmental quality.
- **Sectoral Experts in Project Management**: Experts in Project Management or Management of Process/Operations/Facilities in the relevant sectors.
- **Environmental Impact Assessment Process Experts**: Experts in conducting and carrying out Environmental Impact Assessments (EIAs) and preparation of Environmental Management Plans (EMPs) and other Management plans and who have wide expertise and knowledge of predictive techniques and tools used in the EIA process.
- **Risk Assessment Experts**
- **Life Science Experts in floral and faunal management**
- **Forestry and Wildlife Experts**
- **Environmental Economics Expert with experience in project appraisal**

3. The Membership of the EAC shall not exceed 15 (fifteen) regular Members. However, the Chairperson may co-opt an expert as a Member in a relevant field for a particular meeting of the Committee.

4. The Chairperson shall be an outstanding and experienced environmental policy expert or expert in management or public administration with wide experience in the relevant development sector.

5. The Chairperson shall nominate one of the Members as the Vice Chairperson who shall preside over the EAC in the absence of the Chairman /Chairperson.

6. A representative of the Ministry of Environment and Forests shall assist the Committee as its Secretary.

7. The maximum tenure of a Member, including Chairperson, shall be for 2 (two) terms of 3 (three) years each.

8. The Chairman/Members may not be removed prior to expiry of the tenure without cause and proper enquiry.

It is noted that this notification is once again amended in 2009.

48 SITTING GUIDELINES FOR INDUSTRIES

1. INTRODUCTION

1.1 Industrial development significantly contributes towards economic growth. However, industrial progress brings along with it a host of environmental problems. Many of these problems could be avoided if industries are located on the basis of environmental considerations. Injudicious siting of industry can seriously effect the environmental features such as air, water, land, flora, fauna, human settlements and health of people. The entrepreneur should be fully aware of these implications and he should take necessary steps while setting up the industry so as to minimise the possible adverse effects on the environmental resources and quality of life. Often, an entrepreneur finds it very costly to install pollution control equipment and other mitigative measures after the industry is already set up. As such, preventive steps are needed at the time of siting rather than going in for curative measures at a later stage.

1.2 The Industrial Policy Statement of July 1980 recognised the need for preserving ecological balance and improving living conditions in the urban centres of the country. On the basis of this Policy, indiscriminate expansion of the existing industries and setting up of new industrial undertakings within the limits of metropolitan cities and the larger towns should not be permitted. However, the Policy has not touched upon the implications of setting up an industry in sensitive areas—both ecological or otherwise—which would have an effect on the overall development process.

1.3 At present, industries are being located on the basis of raw material availability, access to the market, transport facilities and such other techno-economic considerations without adequate attention to environmental considerations are recognised as an important criterion for setting of industry.

1.4 To prevent air, water and soil pollution arising out of industrial projects, the industrial licensing procedure requires that the entrepreneurs before setting up the industry should obtain clearance from Central/State Air and Water Pollution Control Boards. The Central/State Pollution Control Boards stipulate that air (gases) and water (effluents) emanating from the industry should adhere to certain quality standards. However, these stipulations do not prevent the industry from affecting the total environment by wrong siting. Also, the cumulative effect of a number of industries at a particular place is not being studied upon, with the result that an industry or an industrial area over a period of time could cause significant damage to the surrounding environment and ecological features.

1.5 In respect of certain industrial development projects it is not only necessary to install suitable pollution control equipment but also to identify appropriate sites for their location. To give a concrete shape to this requirement, a select group of 20 industries has been notified by the Department of Industrial Development. A formalised procedure has been stipulated for site selection from environmental angle with regard to these projects.

1.6 According to this procedure for the select group of industries, the Letters of Intent should be converted to Industrial Licenses only after the following conditions have been fulfilled:

(i) The State Director of Industries confirms that the site of the project has been approved from environmental angle by the competent State Authority.

(ii) The entrepreneur commits both to the State Government and Central Government that he will install the appropriate equipment and implement the prescribed measures for the prevention and control of pollution.

(iii) The concerned State Pollution Control Board has certified that the proposal meets with the environmental requirements and that the equipment installed or proposed to be installed are adequate and appropriate to the requirement.

1.7 The State Department of Environment will be the competent authority for approval of project sites from environmental angle. In those states where such Departments have not yet been set up, approval should be obtained from the nodal agency designated for looking after environmental matters. With regard to projects where support from the Central Government/ International Agencies is envisaged and which come under the purview of Industrial Licensing, approval of the project site from environmental angle should be obtained from the Ministry of Environment & Forests, Government of India. The entrepreneur should provide the details of proposed project site, pollution abatement measures and such other relevant information as required for review from environmental angle.

1.8 The entrepreneur will be required to submit half-yearly progress report on installation of pollution control devices to the respective State Pollution Control Boards.

1.9 Depending on the nature and location of the project, the entrepreneur will be required to submit comprehensive Environmental Impact Assessment Report and Environmental Management Plans.

2. ENVIRONMENTAL GUIDELINES FOR INDUSTRIES

2.1 In order to help the concerned authorities and the entrepreneurs, it is necessary to frame certain broad guidelines for siting an industry. It is also necessary to identify the parameters that should be taken into account while setting up an industry. With this in view, the following environmental guidelines are recommended for siting of industries to ensure optimum use of natural and man-made resources in sustainable manner with minimal depletion, degradation and/or destruction of environment. Those are in addition to those directives that are already in existence under the Industries (Development and Regulation) Act.

2.2 Areas to be avoided

In siting industries, care should be taken to minimise the adverse impact of the industries on the immediate neighbourhood as well as distant places. It is well-known that some of the natural life-sustaining systems and some specific land use are sensitive to industrial impacts because of the nature and extent of fragility. With a view to protecting such an area industrial sites shall maintain the following distances from the areas listed:

(a) Ecologically and/or otherwise sensitive areas: at least 25 km; depending on the geo-climatic conditions the requisite distance shall have to be increased by the appropriate agency.

(b) Coastal areas: at least 1/2 km from High Tide Line.

(c) Flood Plain of the Riverine Systems: at least 1/2 km from flood plain or modified flood plain affected by dam in the upstream or by flood control systems.

(d) Transport/Communication System: at least 1/2 km from highway and railway.

(e) Major Settlements (300,000 population): distance from settlements is difficult to maintain because of urban sprawl. At the time of siting of the industry if any major settlement's notified limit is within 50 km the spatial direction of growth of the settlement for at least a decade must be assessed and the industry shall be sited at least 25 km from the projected growth boundary of the settlement.

Note: Ecological and/or otherwise sensitive areas include: (i) Religious and Historic Places; (ii) Archaeological Monuments (e.g. identified zone around Taj Mahal); (iii) Scenic Areas; (iv) Hill Resorts; (v) Beach Resorts; (vi) Health Resorts; (vii) Coastal areas rich in Coral, Mangroves, Breeding Grounds of Specific Species; (viii) Estuaries rich in Mangroves, Breeding Ground of Specific Species; (ix) Gulf Areas; (x) Biosphere Reserves; (xi) National Parks and Sanctuaries; (xii) Natural Lakes, Swamps; (xiii) Seismic Zones; (xiv) Tribal Settlements; (xv) Areas of Scientific and Geological interest; (xvi) Defence Installations, specially those of security importance and sensitive to pollution; (xvii) Border Areas (International); and (xviii) Airports.

Pre-requisite: State and Central Governments are required to identify such areas on priority basis:

2.3 **Siting Criteria**

Economic and social factors are recognised and assessed while siting an industry. Environmental factors must be taken into consideration in industrial siting. Proximity of water sources, highway, major settlements, markets for products and raw material resources is desired for economy of production, but all the abovelisted systems must be away for environmental protection. Industries are, therefore, required to be sited, striking a balance between economic and environmental considerations. In such a selected site, the following factors must be recognised:

— No forest land shall be converted into non-forest activity for the sustenance of the industry (Ref.: Forest Conservation Act, 1980).

— No prime agricultural land shall be converted into industrial site.

— Within the acquired site the industry must locate itself at the lowest location to remain obscured from general sight.

— Land acquired shall be sufficiently large to provide space for appropriate treatment of waste water still left for treatment after maximum possible reuse and recycle. Reclaimed (treated) wastewater shall be used to raise green belt and to create water body for aesthetics, recreation and, if possible, for aquaculture. The green belt shall be 1/2 km wide around the battery limit of the industry. For industry having odour problem it shall be 1 km wide.

— The green belt between two adjoining large-scale industries shall be 1 km.

— Enough space should be provided for storage of solid wastes so that these could be available for possible reuse.

— Layout and form of the industry that may come up in the area must conform to the landscape of the area without affecting the scenic features of that place.

— Associated township of the industry must be created at a space having physiographic barrier between the industry and the township.

— Each industry is required to maintain three ambient air quality measuring stations within 120 degree angle between stations.

3. ENVIRONMENTAL IMPACT ASSESSMENT (EIA)

3.1 The purpose of Environmental Impact Assessment (EIA) is to identify and evaluate the potential impacts (beneficial and adverse) of development projects on the environmental system. It is an useful aid for decision-making based on understanding of the environmental implications including social, cultural and aesthetic concerns which could be integrated with the analysis of the project costs and benefits. This exercise should be undertaken early enough in the planning stage of projects for selection of environmentally compatible sites, process technologies and such other environmental safeguards.

3.2 While all industrial projects may have some environmental impacts all of them may not be significant enough to warrant elaborate assessment procedures. The need for such exercises will have to be decided after initial evaluation of the possible implications of a particular project and its location.

The projects which could be the candidates for detailed Environmental Impact Assessment includes:

(i) Those which can significantly alter the landscape, land use pattern and lead to concentration of working and service population;

(ii) Those which need upstream development activity like assured mineral and forest products supply or downstream industrial process development;

(iii) Those involving manufacture, handling and use of hazardous materials;

(iv) Those which are sited near ecologically sensitive areas, urban centres, hill resorts, places of scientific and religious importance;

(v) Industrial Estates with constituent units of various types which could cumulatively cause significant environmental damage.

3.3 The Environmental Impact Assessment (EIA) should be prepared on the basis of the existing background pollution levels vis-a-vis contributions of pollutants from the proposed plant. The EIA should address some of the basic factors listed below:

(a) Meteorology and air quality

Ambient levels of pollutants such as sulphur dioxide, oxides of nitrogen, carbon monoxide, suspended particulate matters, should be determined at the centre and at 3 other locations on a radius of 10 km with 120 degrees angle between stations. Additional contribution of pollutants at the locations are required to be predicted after taking into account the emission rates of the pollutants from the stacks of the proposed plant, under different meteorological conditions prevailing in the area.

(b) Hydrology and water quality

(c) Site and its surroundings

(d) Occupational safety and health

(e) Details of the treatment and disposal of effluents (liquid, air and solid) and the methods of alternative uses

(f) Transportation of raw material and details of material handling

(g) Impact on sensitive targets

(h) Control equipment and measures proposed to be adopted.

3.4 Preparation of environmental management plans required for formulation, implementation and monitoring of environmental protection measures during and after commissioning of projects. The plans should indicate the details as to how various measures have been or are proposed to be taken including cost components as may be required. Cost of measures for environmental safeguards should be treated as an integral component of the project cost and environmental aspects should be taken into account at various stages of the projects:

(a) Conceptualisation: preliminary environmental assessment.

(b) Planning: detailed studies of environmental impacts and design of safeguards.

(c) Execution: implementation of environmental safety measures.

(d) Operation: Monitoring of effectiveness of built-in safeguards.

The management plans should be necessarily based on consideration of resource conservation and pollution abatement, some of which are enumerated as under:

3.5 **Liquid Effluents**

(i) Effluents from the industrial plants should be treated well to the standards as prescribed by the Central/State Water Pollution Control Boards.

(ii) Soil permeability studies should be made prior to effluents being discharged into holding tanks or impoundments and steps taken to prevent percolation and groundwater contamination.

(iii) Special precautions should be taken regarding flight patterns of birds in the area. Effluents containing toxic compounds, oil and grease have been known to cause extensive death of migratory birds. Location of plants should be prohibited in such type of sensitive areas.

(iv) Deep well burial of toxic effluents should not be resorted to as it can result in resurfacing and groundwater contamination. Resurfacing has been known to cause extensive damage to crops and livestocks.

(v) In all cases, efforts should be made for reuse of water and its conservation.

3.6 **Air Pollution**

(i) The emission levels of pollutants from the different stacks should conform to the pollution control standards prescribed by Central or State Boards.

(ii) Adequate control equipment should be installed for minimising the emission of pollutants from the various stacks.

(iii) In-plant control measures should be taken to contain the fugitive emission.

(iv) Infrastructural facilities should be provided for monitoring the stack emissions and measuring the ambient air quality including micro-meteorological data (wherever required) in the area.

(v) Proper stack height as prescribed by the Central/State Pollution Control Boards should be provided for better dispersion of pollutants over a wider area to minimise the effect of pollution.

(vi) Community buildings and townships should be built up-wind of plant with $\frac{1}{2}$ km greenbelt in addition to physiographical barrier.

3.7 Solid Wastes

(i) The site for waste disposal should be checked to verify permeability so that no contaminants percolate into the groundwater or river/lake.

(ii) Waste disposal areas should be planned down-wind of villages and townships.

(iii) Reactive materials should be disposed of by immobilising the reactive materials with suitable additives.

(iv) The pattern of filling disposal site should be planned to create better landscape and be approved by appropriate agency and the appropriately pretreated solid wastes should be disposed according to the approved plan.

(v) Intensive programmes of tree plantation on disposal sites should be undertaken.

3.8 Noise and Vibration

Adequate measures should be taken for control of noise and vibrations in the industries.

3.9 Occupational Safety and Health

Proper precautionary measures for adopting occupational safety and health standards should be taken.

3.10 Prevention, Maintenance and Operation of Environmental Control Systems

(i) Adequate safety precautions should be taken during preventive maintenance and shut-down of the control systems.

(ii) A system of inter-locking with the production equipment should be implemented where highly toxic compounds are involved.

3.11 House-keeping

Proper house-keeping and cleanliness should be maintained both inside and outside the industry.

3.12 Human Settlements

(i) Residential colonies should be located away from the solid and liquid waste dumping areas. Meteorological and environmental conditions should be studied properly before selecting the site for residential areas in order to avoid air pollution problems.

(ii) Persons who are displaced or have lost agricultural lands as a result of locating the industries in the area should be properly rehabilitated.

3.13 Transport Systems

(i) Proper parking places should be provided for the trucks and other vehicles by the industries to avoid any congestion or blocking of roads.

(ii) Siting of industries on the highways should be avoided as it may add to more road accidents because of substantial increase in the movements of heavy vehicles and unauthorised shops and settlements coming up around the industrial complex.

(iii) Spillage of chemicals/substances on roads inside the plant may lead to accidents. Proper road safety signs—both inside and outside the plant—should be displayed for avoiding road accidents.

3.14 Recovery-Reuse of waste products

Efforts should be made to recycle or recover the waste materials to the extent possible. The treated liquid effluents can be conveniently and safely used for irrigation of lands, plants and fields for growing non-edible crops.

3.15 Vegetal Cover

Industries should plant trees and ensure vegetal cover in their premises. This is particularly advisable for those industries having more than 10 acres of land.

3.16 Disaster Planning

Proper disaster planning should be done to meet any emergency situation arising due to fire, explosion, sudden leakage of gas etc. Fire fighting equipment and other safety appliances should be kept ready for use during disaster/emergency situation including natural calamities like earthquake/flood.

3.17 Environmental Management Cell

Each industry should identify within its set up a Department/Section/Cell with trained personnel to take up the model responsibility of environmental management as required for planning and implementation of the projects.

49 COASTAL REGULATIONS: NOTIFICATION

Notification under Section 3(1) and Section 3(2)(v) of the Environment (Protection) Act, 1986, and Rule 5(3)(d) of Environment (Protection) Rules, 1986, declaring Coastal Stretches as Coastal Regulation Zone (CRZ) and Regulating Activities in the CRZ.

New Delhi, the 19th February, 1991

49.1. S.O. 114(E)—Whereas a Notification under Section 3(1) and Section 3(2) (v) of the Environment (Protection) Act, 1986, inviting objections against the declaration of Coastal Stretches as Coastal Regulation Zone (CRZ) and imposing restrictions on industries, operations and processes in the CRZ was published vide S.O. No. 944 (E) dated 15th December, 1990.

And whereas all objections received have been duly considered by the Central Government;

Now, therefore, in exercise of the powers conferred by Clause (d) of sub-rule (3) of Rule 5 of the Environment (Protection) Rules, 1986, and all other powers vesting in its behalf, the Central Government hereby declares the coastal stretches of seas, bays, estuaries, creeks, rivers and backwaters which are influenced by tidal action (in the landward side) up to 500 metres from the High Tide Line (HTL) and the land between the Low Tide Line (LTL) and the HTL as Coastal Regulation Zone and imposes with effect from the date of this Notification, the following restrictions on the setting up and expansion of industries, operations or processes etc. in the said Coastal Regulation Zone (CRZ). For purposes of this Notification, the High Tide Line (HTL) will be defined as the line up to which the highest high tide reaches at spring tides.

Note. The distance from the High Tide Line (HTL) to which the proposed regulations will apply in the case of rivers, creeks and backwaters may be modified on a case-by-case basis for reasons to be recorded while preparing the Coastal Zone Management Plans (referred to below); however, this distance shall not be less than 100 metres or the width of the creek, river or backwater, whichever is less.

49.2. **Prohibited Activities**: The following activities are declared as prohibited within the Coastal Regulation Zone, namely:

(i) setting up of new industries and expansion of existing industries, except those directly related to water front or directly needing foreshore facilities;

(ii) manufacture or handling or storage or disposal of hazardous substances as specified in the Notification of the Government of India in the Ministry of Environment & Forests No. S.O. 594(E) dated 28th July, 1989, S.O. 966(E) dated 27th November, 1989 and GSR 1037(E) dated 5th December, 1989;

(iii) setting up and expansion of fish processing units including warehousing (excluding hatchery and natural fish drying in permitted areas);

(iv) setting up and expansion of units mechanisms for disposal of waste and effluents, except facilities required for discharging treated effluents, into the water-course with approval under the Water (Prevention and Control of Pollution) Act, 1974, and except for storm water drains;

(v) discharge of untreated wastes and effluents from industries, cities or towns and other human settlements. Schemes shall be implemented by the concerned authorities for phasing out the existing practices, if any, within a reasonable time period not exceeding three years from the date of this notification;

(vi) dumping of city or town waste for the purposes of landfilling or otherwise; the existing practice, if any, shall be phased out within a reasonable time not exceeding three years from the date of this Notification;

(vii) dumping of ash or any wastes from thermal power stations;

(viii) land reclamation, bunding or disturbing the natural course of sea water with similar obstructions, except those required for control of coastal erosion and maintenance or clearing of waterways, channels and ports and for prevention of sandbars and also except for tidal regulators, storm water drains and structures for prevention of salinity ingress and for sweet water recharge;

(ix) mining of lands, rocks and other substrata materials, except those rare minerals not available outside the CRZ areas;

(x) harvesting or drawal of ground water and construction of mechanisms therefor within 200 m of HTL; in the 200 m to 500 m zone it shall be permitted only when done manually through ordinary wells for drinking, horticulture, agriculture and fisheries;

(xi) construction activities in ecologically sensitive areas as specified in Annexure I of this Notification;

(xii) any construction activity between the Low Tide Line and High Tide Line except facilities for carrying treated effluents and waste water discharges into the sea, facilities for carrying sea water for cooling purposes, oil, gas and similar pipelines and facilities essential for activities permitted under this Notification; and

(xiii) dressing or altering of sand dunes, hills, natural features including landscape changes for beautification, recreational and other such purposes; except as permissible under this Notification.

49.3. **Regulation of Permissible Activities**: All other activities, except those prohibited in para 2 above, will be regulated as under:

(1) Clearance shall be given for any activity within the Coastal Regulation Zone only if it requires waterfront and foreshore facilities.

(2) The following activities will require environmental clearance from the Ministry of Environment & Forests, Government of India, namely:

(i) Construction activities related to Defence requirements for which foreshore facilities are essential (e.g. slipways, jetties etc.); except for classified operational component of defence projects for which a separate procedure shall be followed. (Residential buildings, office buildings, hospital complexes, workshops shall not come within the definition of operational requirements except in very special cases and, hence, shall not normally be permitted in the CRZ);

(ii) Operational constructions for ports and harbours and lighthouses requiring water frontage; jetties, wharves, quays, slipways etc. (Residential buildings & office buildings shall not come within the definition of operational activities except in very special cases and, hence, shall not normally be permitted in the CRZ);

(iii) Thermal power plants (only foreshore facilities for transport of raw materials facilities for intake of cooling water and outfall for discharge of treated waste water cooling water); and

(vi) All other activities with investment exceeding rupees five crores.

(3) (i) The coastal States and Union Territory Administrations shall prepare, within a period of one year from the date of this Notification, Coastal Zone Management Plans identifying and classifying the CRZ areas within their respective territories in accordance with the guidelines given in Annexures I and II of the Notification and obtain approval (with or without modifications) of the Central Government in the Ministry of Environment & Forests;

(ii) Within the framework of such approved plans, all development and activities within the CRZ other than those covered in para 2 and para 3(2) above shall be regulated by the State Government, Union Territory Administration or the local authority as the case may be in accordance with the guidelines given in Annexures I and II or the Notification; and

(iii) In the interim period till the Coastal Zone Management Plans mentioned in para 3(3) above are prepared and approved, all developments and activities within the CRZ shall not violate the provisions of this Notification. State Governments and Union Territory Administrations shall ensure adherence to these regulations and violations, if any, shall be subject to the provisions of the Environment (Protection) Act, 1986.

49.4. **Procedure for monitoring and enforcement**: The Ministry of Environment & Forests and the Government of State or Union Territory and such other authorities at the State or Union Territory levels, as may be designated for this purpose, shall be responsible for monitoring and enforcement of the provisions of this notification within their respective jurisdictions.

[N. K-15019/1/84-IA-III (Vol. II)]

R. RAJAMANI, Secy.

ANNEXURE I

COASTAL AREA CLASSIFICATION AND DEVELOPMENT REGULATIONS

Classification of Coastal Regulation Zone

6(1) For regulating development activities the coastal stretches within 500 metres of High Tide Line of the landward side are classified into four categories, namely:

Category I (CRZ I)

(i) Areas that are ecologically sensitive and important, such as National Parks/marine parks, sanctuaries, reserve forests, wildlife habitats, mangroves, corals/coral reefs, areas close to breeding and spawning grounds of fish and other marine life, areas of outstanding natural beauty/historical/heritage areas rich in genetic diversity, areas likely to be inundated due to rise in sea level consequent upon global warming and such other areas as may be declared by the Central Government or the concerned authorities at the State/Union Territory level from time to time.

(ii) Area between the Low Tide Line and the High Tide Line.

Category II (CRZ II)

The areas that have already been developed up to or close to the shoreline. For this purpose, "developed area" is referred to as that area within the municipal limits or in other legally designated urban areas which is already substantially built up and which has been provided with drainage and approach roads and other infrastructural facilities, such as water supply and sewerage mains.

Category III (CRZ III)

Areas that are relatively undisturbed and those which do not belong to either Category I or II. These will include coastal zone in the rural areas (developed and undeveloped) and also areas within Municipal limits or in other legally designated urban areas which are not substantially built up.

Category IV (CRZ IV)

Coastal stretches in the Andaman & Nicobar, Lakshadweep and small islands except those designated as CRZ I, CRZ II or CRZ III.

Norms for Regulation of Activities

6(2) The development or construction activities in different categories of CRZ areas shall be regulated by the concerned authorities at the State/Union Territory level, in accordance with the following norms:

CRZ I

No new construction shall be permitted within 500 metres of the High Tide Line. No construction activity, except as listed under 2(xii), will be permitted between the Low Tide Line and the High Tide Line.

CRZ II

(i) Buildings shall be permitted neither on the seaward side of the existing road (or roads proposed in the approved Coastal Zone Management Plan of the area) nor on seaward side of existing authorised structures. Buildings permitted on the landward side of the existing and proposed roads, existing authorised structures shall be subject to the existing local Town and Country Planning Regulations including the existing norms of FSI/FAR.

(ii) Reconstruction of the authorised buildings to be permitted subject with the existing FSI/FAR norms and without change in the existing use.

(iii) The design and construction of buildings shall be consistent with the surrounding landscape and local architectural style.

CRZ III

(i) The area up to 200 metres from the High Tide Line is to be earmarked as 'No Development Zone'. No construction shall be permitted within this zone except for repairs of existing authorised structures not exceeding existing FSI, existing plinth area and existing density. However, the following uses may be permissible in this zone—agriculture, horticulture, gardens, pastures, parks, playfields, forestry and salt manufacture from sea water.

(ii) Development of vacant plots between 200 and 500 metres of High Tide Line in designated areas of CRZ III with prior approval of MEF permitted for construction of hotels/beach resorts for temporary occupation of tourists/visitors subject to the conditions as stipulated in the guidelines at Annexure II.

(iii) Construction/reconstruction of dwelling units between 200 and 500 metres of the High Tide Line permitted so long it is within the ambit of traditional rights and customary uses such as existing fishing villages and goathans. Building permission for such construction/reconstruction will be subject to the conditions that the total number of dwelling units shall not be more than twice the number of existing units; total covered area on all floors shall not exceed 33 per cent of the plot size; the overall height of construction shall not exceed 9 metres and construction shall not be more than 2 floors (ground floor plus one floor).

(iv) Reconstruction/alterations of an existing authorised building permitted subject to (i) to (iii) above.

CRZ IV

Andaman & Nicobar Islands:

(i) No new construction of buildings shall be permitted within 200 metres of the HTL;

(ii) The buildings between 200 and 500 metres from the High Tide Line shall not have more than 2 floors (ground floor and 1st floor), the total covered area on all floors shall not be more than 50 per cent of the plot size and the total height of construction shall not exceed 9 metres;

(iii) The design and construction of buildings shall be consistent with the surrounding landscape and local architectural style.

(iv) Corals and sand from the beaches and coastal waters shall not be used for construction and other purposes;

(v) Dredging and underwater blasting in and around coral formations shall not be permitted; and

(vi) However, in some of the islands, coastal stretches may also be classified into categories CRZ I or II or III with the prior approval of Ministry of Environment and Forests and in such designated stretches, the appropriate regulations given for respective categories shall apply.

Lakshadweep and small islands:

(i) For permitting construction of buildings, the distance from the High Tide Line shall be decided depending on the size of the islands. This shall be laid down for each island, in consultation with the experts and with approval of the Ministry of Environment & Forests, keeping in view the land use requirements for specific purposes vis-a-vis local conditions including hydrological aspects, erosion and ecological sensitivity;

(ii) The buildings within 500 metres from the HTL shall not have more than 2 floors (ground floor and 1st floor), the total covered area on all floors shall not be more than 50 per cent of the plot size and the total height of construction shall not exceed 9 metres;

(iii) The design and construction of buildings shall be consistent with the surrounding landscape and local architectural style.

(iv) Corals and sand from the beaches and coastal waters shall not be used for construction and other purposes;

(v) Dredging and underwater blasting in and around coral formations shall not be permitted; and

(vi) However, in some of the islands, coastal stretches may also be classified into categories CRZ I or II or III, with the prior approval of Ministry of Environment & Forests and in such designated stretches, the appropriate regulations given for respective categories shall apply.

ANNEXURE II

Guidelines for Development of Beach Resorts/Hotels in the Designated Areas of CRZ-III for Temporary Occupation of Tourists/Visitors, with Prior Approval of the Ministry of Environment & Forests.

7(1) Construction of beach resorts/hotels with prior approval of MEF in the designated areas of CRZ III for temporary occupation of tourists/visitors shall be subject to the following conditions:

(i) The project proponents shall not undertake any construction (including temporary constructions and fencing or such other barriers) within 200 metres (in the landward side) from the High Tide Line and within the area between the Low Tide and High Tide Line;

(ii) The total plot size shall not be less than 0.4 hectares and the total covered area on all floors shall not exceed 33 per cent of the plot size, i.e., the FSI shall not exceed 0.33. The open area shall be suitably landscaped with appropriate vegetal cover;

(iii) The construction shall be consistent with the surrounding landscape and local architectural style;

(iv) The overall height of construction, up to the highest ridge of the roof, shall not exceed 9 metres and the construction shall not be more than 2 floors (ground floor plus one upper floor);

(v) Ground water shall not be tapped within 200 m of the HTL; within the 200 metres–500 metres zone it can be tapped only with the concurrence of the Central/State Ground Water Board;

(vi) Extraction of sand, levelling or digging of sandy stretches except for structural foundation of building, swimming pool shall not be permitted within 500 metres of the High Tide Line;

(vii) The quality of treated effluents, solid wastes, emissions and noise levels etc. from the project area must conform to the standards laid down by the competent authorities including the Central/State Pollution Control Board and under the Environment (Protection) Act, 1986;

(viii) Necessary arrangements for the treatment of the effluents and solid wastes must be made. It must be ensured that the untreated effluents and solid wastes are not discharged into the water or on the beach; and no effluent/solid waste shall be discharged on the beach;

(ix) To allow public access to the beach, at least a gap of 20 metres width shall be provided between any two hotels/beach resorts; and in no case shall gaps be less than 500 metres apart; and

(x) If the project involves diversion of forest land for non-forest purposes, clearance as required under the Forest (Conservation), Act, 1980 shall be obtained. The requirements of other Central and State laws as applicable to the project shall be met with.

(xi) Approval of the State/Union Territory Tourism Department shall be obtained.

7(2) In ecologically sensitive areas (such as marine parks, mangroves, coral reefs, breeding and spawning grounds of fish, wildlife habitats and such other areas as may be notified by the Central/State Government/Union Territories) construction of beach resorts/hotels shall not be permitted.

MINISTRY OF ENVIRONMENT AND FORESTS
CORRIGENDUM

New Delhi, the 18th March, 1991

S.O. 190(E). In partial modification of the Gazette Notification S.O. No. 114(E) dated 20th February, 1991, the following amendments may be made:

Sl. No.	REFERENCE	FOR	READ
1	2	3	4
1.	Page 5, Para 2(viii) line-5	.... cleansing of waterways	clearing of waterways ....
2.	Page 6, Ann. 1 Para 6(1) line-3	.... line of the	.... line on the ....
3.	Page 6, Ann. 1 Para 6(1) line-5	Category 1 (CRZ-T)	Category 1 (CRZ-I)
4.	Page 7, Para 6(2) CRZ-II(ii) line-2	.... subject with the	.... subject to the..........
5.	Page 7, Para 6(2) CRZ-III(ii) line-4	.... prior approval of MEF	.... prior approval of Ministry of Environment and Forests (MEF) ....
6.	Page 8, Lakshadweep and small islands, Paras (i) to (vi)	May be omitted as the same has been given earlier in the text on pages 7-8	

[File No. K-15019/1/84-IA. III]

S. MAUDGAL, Adviser–IA

This notification is amended several times, the latest one in January 2011.

COASTAL REGULATION ZONE NOTIFICATION
MINISTRY OF ENVIRONMENT AND FORESTS
(Department of Environment, Forests and Wildlife)

S.O.19(E): WHEREAS a draft notification under sub-section (1) of section and clause (V) of sub-section (2) of section 3 of the Environment (Protection) Act, 1986 was issued inviting objections and suggestions for the declaration of coastal stretches as Coastal Regulation Zone and imposing restrictions on industries, operations and processes in the CRZ was published vide S.O.No.2291 (E), dated 15th September, 2010.;

AND WHEREAS, copies of the said Gazette were made available to the public on 15th September, 2010.;

AND WHEREAS, the suggestions and objections received from the public have been considered by the Central Government.;

Now, therefore, in exercise of the powers conferred by sub-section (1) and clause (v) of sub-section (2) of section 3 of the Environment (Protection) Act, 1986 (29 of 1986), the Central Government, with a view to ensure livelihood security to the fisher communities and other local communities, living in the coastal areas, to conserve and protect coastal stretches, its unique environment and its marine area and to promote development through sustainable manner based on scientific principles taking into account the dangers of natural hazards in the coastal areas, sea level rise due to global warming, does hereby, declare the coastal stretches of the country and the water area upto its territorial water limit, excluding the islands of Andaman and Nicobar and Lakshadweep and the marine areas surrounding these islands upto its territorial limit, as Coastal Regulation Zone (hereinafter referred to as the CRZ) and restricts the setting up and expansion of any industry, operations or processes and manufacture or handling or storage or disposal of hazardous substances as specified in the Hazardous Substances (Handling, Management and Transboundary Movement) Rules, 2009 in the aforesaid CRZ.; and

In exercise of powers also conferred by clause (d) and sub rule (3) of rule 5 of Environment (Protection) Act, 1986 and in supersession of the notification of the Government of India in the Ministry of Environment and Forests, number S.O.114(E), dated the 19th February, 1991 except as respects things done or omitted to be done before such supercession, the Central Government hereby declares the following areas as CRZ and imposes with effect from the date of the notification the following restrictions on the setting up and expansion of industries, operations or processes and the like in the CRZ,

(i) the land area from High Tide Line (hereinafter referred to as the HTL) to 500 metres on the landward side along the sea front.

(ii) CRZ shall apply to the land area between HTL to 100 metres or width of the creek whichever is less on the landward side along the tidal influenced water bodies that are connected to the sea and the distance upto which development along such tidal influenced water bodies is to be regulated shall be governed by the distance upto which the tidal effects are experienced which shall be determined based on salinity concentration of 5 parts per thousand (ppt) measured during the driest period of the year and distance upto which tidal effects are experienced shall be clearly identified and demarcated accordingly in the Coastal Zone Management Plans (hereinafter referred to as the CZMPs).

Explanation: For the purposes of this sub-paragraph the expression tidal influenced water bodies means the water bodies influenced by tidal effects from sea, in the bays, estuaries, rivers, creeks, backwaters, lagoons, ponds connected to the sea or creeks and the like.

(iii) the land area falling between the hazard line and 500 metres from HTL on the landward side, in case of seafront and between the hazard line and 100 metres line in case of tidal influenced water body the word 'hazard line' denotes the line demarcated by Ministry of Environment and Forests (hereinafter referred to as the MoEF) through the Survey of India (hereinafter referred to as the SoI) taking into account tides, waves, sea level rise and shoreline changes.

(iv) land area between HTL and Low Tide Line (hereinafter referred to as the LTL) which will be termed as the intertidal zone.

(v) the water and the bed area between the LTL to the territorial water limit (12 Nm) in case of sea and the water and the bed area between LTL at the bank to the LTL on the opposite side of the bank, of tidal influenced water bodies.

2. For the purposes of this notification, the HTL means the line on the land upto which the highest water line reaches during the spring tide and shall be demarcated uniformly in all parts of the country by the demarcating authority(s) so authorized by the MoEF in accordance with the general guidelines issued at Annexure-I. HTL shall be demarcated within one year from the date of issue of this notification.

3. Prohibited activities within CRZ: The following are declared as prohibited activities within the CRZ

(i) Setting up of new industries and expansion of existing industries except,

(a) those directly related to waterfront or directly needing foreshore facilities;

Explanation: The expression "foreshore facilities" means those activities permissible under this notification and they require waterfront for their operations such as ports and harbours, jetties, quays, wharves, erosion control measures, breakwaters, pipelines, lighthouses, navigational safety facilities, coastal police stations and the like;

(b) projects of Department of Atomic Energy;

(c) facilities for generating power by non-conventional energy sources and setting up of desalination plants in the areas not classified as CRZ-I(i) based on an impact assessment study including social impacts;

(d) development of green field Airport already permitted only at Navi Mumbai;

(e) reconstruction, repair works of dwelling units of local communities including fishers in accordance with local town and country planning regulations.

(ii) manufacture or handling oil storage or disposal of hazardous substance as specified in the notification of Ministry of Environment and Forests, No. S.O.594 (E), dated the 28th July 1989, S.O.No.966(E), dated the 27th November, 1989 and GSR 1037 (E), dated the 5th December, 1989 except,

(a) transfer of hazardous substances from ships to ports, terminals and refineries and vice versa;

(b) facilities for receipt and storage of petroleum products and liquefied natural gas as specified in Annexure-II appended to this notification and facilities for regasification of Liquefied Natural Gas (hereinafter referred to as the LNG) in the areas not classified as CRZ- I(i) subject to implementation of safety regulations including guidelines issued by the Oil Industry Safety Directorate in the Ministry of Petroleum and Natural Gas and guidelines issued by MoEF and subject to further terms and conditions for implementation of ameliorative and restorative measures in relation to environment as may be stipulated by in MoEF.

Provided that facilities for receipt and storage of fertilizers and raw materials required for manufacture of fertilizers like ammonia, phosphoric acid, sulphur, sulphuric acid, nitric acid and the like, shall be permitted within the said zone in the areas not classified as CRZ-I(i).

(iii) Setting up and expansion of fish processing units including warehousing except hatchery and natural fish drying in permitted areas.

(iv) Land reclamation, bunding or disturbing the natural course of seawater except those,

(a) required for setting up, construction or modernisation or expansion of foreshore facilities like ports, harbours, jetties, wharves, quays, slipways, bridges, sealink, road on stilts, and such as meant for defence and security purpose and for other facilities that are essential for activities permissible under the notification;

(b) measures for control of erosion, based on scientific including Environmental Impact Assessment (hereinafter referred to as the EIA) studies;

(c) maintenance or clearing of waterways, channels and ports, based on EIA studies;

(d) measures to prevent sand bars, installation of tidal regulators, laying of storm water drains or for structures for prevention of salinity ingress and freshwater recharge based on carried out by any agency to be specified by MoEF.

(v) Setting up and expansion of units or mechanism for disposal of wastes and effluents except facilities required for,

(a) discharging treated effluents into the water course with approval under the Water (Prevention and Control of Pollution) Act, 1974 (6 of 1974);

(b) storm water drains and ancillary structures for pumping;

(c) treatment of waste and effluents arising from hotels, beach resorts and human settlements located in CRZ areas other than CRZ-I and disposal of treated wastes and effluents;

(vi) Discharge of untreated waste and effluents from industries, cities or towns and other human settlements. The concerned authorities shall implement schemes for phasing out existing discharge of this nature, if any, within a time period not exceeding two years from the date of issue of this notification.

(vii) Dumping of city or town wastes including construction debris, industrial solid wastes, fly "4 ash for the purpose of land filling and the like and the concerned authority shall implement schemes for phasing out any existing practice, if any, shall be phased out within a period of one year from date of commencement of this notification.

Note: The MoEF will issue a separate instruction to the State Governments and Union territory Administration in respect of preparation of Action Plans and their implementation as also monitoring including the time schedule thereof, in respect of paras (v), (vi) and (vii).

(viii) Port and harbour projects in high eroding stretches of the coast, except those projects classified as strategic and defence related in terms of EIA notification, 2006 identified by MoEF based on scientific studies and in consultation with the State Government or the Union territory Administration.

(ix) Reclamation for commercial purposes such as shopping and housing complexes, hotels and entertainment activities.

(x) Mining of sand, rocks and other sub-strata materials except,

(a) those rare minerals not available outside the CRZ area,

(b) exploration and exploitation of Oil and Natural Gas.

(xi) Drawl of groundwater and construction related thereto, within 200 metres of HTL; except the following:

(a) in the areas which are inhabited by the local communities and only for their use.

(b) In the area between 200 metres-500 metres zone the drawl of groundwater shall be permitted only when done manually through ordinary wells for drinking, horticulture, agriculture and fisheries and where no other source of water is available.

Note: Restrictions for such drawl may be imposed by the Authority designated by the State Government and Union territory Administration in the areas affected by sea water intrusion.

(xi) Construction activities in CRZ-I except those specified in para 8 of this notification.

(xiii) Dressing or altering the sand dunes, hills, natural features including landscape changes for beautification, recreation and other such purpose.

(xiv) Facilities required for patrolling and vigilance activities of marine/coastal police stations.

4. Regulation of permissible activities in CRZ area: The following activities shall be regulated except those prohibited in para 3 above:

(i) (a) clearance shall be given for any activity within the CRZ only if it requires waterfront and foreshore facilities;

(b) for those projects which are listed under this notification and also attract EIA notification, 2006 (S.O.1533 (E), dated the 14th September, 2006), for such projects clearance under EIA notification only shall be required subject to being recommended by the concerned State or Union territory Coastal Zone Management Authority (hereinafter referred to as the CZMA).

(c) Housing schemes in CRZ as specified in paragraph 8 of this notification;

(d) Construction involving more than 20,000sq mts built up area in CRZ-II shall be considered in accordance with EIA notification, 2006 and in case of projects less than 20,000sq mts built up area shall be approved by the concerned State or Union territory Planning authorities in accordance with this notification after obtaining recommendations from the concerned CZMA and prior recommendations of the concern CZMA shall be essential for considering the grant of environmental clearance under EIA notification, 2006 or grant of approval by the relevant planning authority;

(e) MoEF may under a specific or general order specify projects which require prior public hearing of project affected people;

(f) construction and operation for ports and harbours, jetties, wharves, quays, slipways, ship construction yards, breakwaters, groynes, erosion control measures.

(ii) The following activities shall require clearance from MoEF, namely:

(a) those activities not listed in the EIA notification, 2006;

(b) construction activities relating to projects of Department of Atomic Energy or Defence requirements for which foreshore facilities are essential such as, slipways, jetties, wharves, quays; except for classified operational component of defence projects. Residential buildings, office buildings, hospital complexes, workshops of strategic and defence projects in terms of EIA notification, 2006;

(c) construction, operation of lighthouses;

(d) laying of pipelines, conveying systems, transmission line;

(e) exploration and extraction of oil and natural gas and all associated activities and facilities thereto;

(f) Foreshore requiring facilities for transport of raw materials, facilities for intake of cooling water and outfall for discharge of treated wastewater or cooling water from thermal power plants. MoEF may specify for category of projects such as at (f), (g) and (h) of para 4;

(g) Mining of rare minerals as listed by the Department of Atomic Energy;

(h) Facilities for generating power by non-conventional energy resources, desalination plants and weather radars;

(i) Demolition and reconstruction of (a) buildings of archaeological and historical importance, (b) heritage buildings; and buildings under public use which means buildings such as for the purposes of worship, education, medical care and cultural activities.

4.2. Procedure for clearance of permissible activities: All projects attracting this notification shall be considered for CRZ clearance as per the following procedure, namely:

(i) The project proponents shall apply with the following documents seeking prior clearance under CRZ notification to the concerned State or the Union territory Coastal Zone Management Authority:

(a) Form-1 (Annexure-IV of the notification);

(b) Rapid EIA Report including marine and terrestrial component except for construction projects listed under 4(c) and (d);

(c) Comprehensive EIA with cumulative studies for projects in the stretches classified as low and medium eroding by MoEF based on scientific studies and in consultation with the State Governments and Union territory Administration;

(d) Disaster Management Report, Risk Assessment Report and Management Plan;

(e) CRZ map indicating HTL and LTL demarcated by one of the authorized agency (as indicated in para 2) in 1:4000 scale;

(f) Project layout superimposed on the above map indicated at (e) above;

(g) The CRZ map normally covering 7km radius around the project site.

(h) The CRZ map indicating the CRZ-I, II, III and IV areas including other notified ecologically sensitive areas;

(i) No Objection Certificate from the concerned State Pollution Control Boards or Union territory Pollution Control Committees for the projects involving discharge of effluents, solid wastes, sewage and the like.

(ii) The concerned CZMA shall examine the above documents in accordance with the approved CZMP and in compliance with CRZ notification and make recommendations within a period of sixty days from date of receipt of complete application-

(a) MoEF or State Environmental Impact Assessment Authority (hereinafter referred to as the SEIAA) as the case may be for the project attracting EIA notification, 2006;

(b) MoEF for the projects not covered in the EIA notification, 2006 but attracting para 4(ii) of the CRZ notification;

(iii) MoEF or S E I A A shall consider such projects for clearance based on the recommendations of the concerned CZMA within a period of sixty days.

(vi) The clearance accorded to the projects under the CRZ notification shall be valid for the period of five years from the date of issue of the clearance for commencement of construction and operation.

(v) For Post clearance monitoring –

(a) it shall be mandatory for the project proponent to submit half-yearly compliance reports in respect of the stipulated terms and conditions of the environmental clearance in hard and soft copies to the regulatory authority(s) concerned, on 1st June and 31st December of each calendar year and all such compliance reports submitted by the project proponent shall be published in public domain and its copies shall be given to any person on application to the concerned CZMA;

(b) the compliance report shall also be displayed on the website of the concerned regulatory authority.

(vi) To maintain transparency in the working of the CZMAs it shall be the responsibility of the CZMA to create a dedicated website and post the agenda, minutes, decisions taken, clearance letters, violations, action taken on the violations and court matters including the Orders of the Hon'ble Court as also the approved CZMPs of the respective State Government or Union territory.

5. Preparation of Coastal Zone Management Plans.

(i) The MoEF may obtain the CZMPs prepared through the respective State Government or Union territory;

(ii) The CZMPs may be prepared by the coastal State Government or Union territory by engaging reputed and experienced scientific institution(s) or the agencies including the National Centre for Sustainable Coastal Management (hereinafter referred to as the NCSCM) of MoEF and in consultation with the concerned stakeholders;

(iii) The hazard line shall be mapped by MoEF through SoI all along the coastline of the country and the hazard line shall be demarcated taking into account, tide, waves, sea level rise and shoreline changes;

(iv) For the purpose of depicting the flooding due to tides, waves and sea level rise in the next fifty and hundred years, the contour mapping of the coastline shall be carried out at 0.5m interval normally upto 7km from HTL on the landward side, and the shoreline changes shall be demarcated based on historical data by comparing the previous satellite imageries with the recent satellite imageries;

(v) Mapping of the hazard line shall be carried out in 1:25,000 scale for macro level planning and 1:10,000 scale or cadastral scale for micro level mapping and the hazard line shall be taken into consideration while preparing the land use plan of the coastal areas;

(vi) The coastal States and Union Territory will prepare within a period of twenty four months from the date of issue this notification, draft CZMPs in 1:25,000 scale map identifying and classifying the CRZ areas within the respective territories in accordance with the guidelines given in Annexure-I of the notification, which involve public consultation;

(vii) The draft CZMPs shall be submitted by the State Government or Union territory to the concerned CZMA for appraisal, including appropriate consultations, and recommendations in accordance with the procedure(s) laid down in the Environment (Protection) Act, 1986;

(viii) The State Government or Union territory CZMA shall submit the draft CZMPs to MoEF alongwith its recommendations on the CZMP within a period of six months after incorporating the suggestions and objections received from the stakeholders;

(ix) MoEF shall thereafter consider and approve the CZMPs within a period of four months from the date of receipt of the CZMPs complete in all respects;

(x) All developmental activities listed in this notification shall be regulated by the State Government, Union Territory Administration, the local authority or the concerned CZMA within the framework of such approved CZMPs as the case may be in accordance with provisions of this notification;

(xi) The CZMPs shall not normally be revised before a period of five years after which, the concerned State Government or the Union territory may consider undertaking revision of the maps following the above procedures;

(xii) The CZMPs already approved under CRZ notification, 1991 shall be valid for a period of twenty four months unless the aforesaid period is extended by MoEF by a specific notification subject to such terms and conditions as may be specified therein.

6. Enforcement of the CRZ, notification, 2011 –

(a) For the purpose of implementation and enforcement of the provisions this notification and compliance with conditions stipulated thereunder, the powers either original or delegated are available under Environment (Protection) Act, 1986 with the MoEF, State Government or the Union territory Administration NCZMA and SCZMAs;

(b) The composition, tenure and mandate of NCZMA and State Government or the Union territory CZMAs have already been notified by MoEF in terms of Orders of Hon'ble "8 Supreme Court in Writ Petition 664 of 1993;

(c) the State Government or the Union territory CZMAs shall primarily be responsible for enforcing and monitoring of this notification and to assist in this task, the State Government and the Union territory shall constitute district level Committees under the Chairmanship of the District Magistrate concerned containing atleast three representatives of local traditional coastal communities including from fisherfolk;

(d) The dwelling units of the traditional coastal communities including fisherfolk, tribals as were permissible under the provisions of the CRZ notification, 1991, but which have not obtained formal approval from concerned authorities under the aforesaid notification shall be considered by the respective Union territory CZMAs and the dwelling units shall be regularized subject to the following condition, namely-

(i) these are not used for any commercial activity;

(ii) these are not sold or transferred to non-traditional coastal community.

7. Classification of the CRZ – For the purpose of conserving and protecting the coastal areas and marine waters, the CRZ area shall be classified as follows, namely:-

(i) CRZ-I–

A. The areas that are ecologically sensitive and the geomorphological features which play a role in the maintaining the integrity of the coast-

(a) Mangroves, in case mangrove area is more than 1000 sq mts, a buffer of 50meters along the mangroves shall be provided;

(b) Corals and coral reefs and associated biodiversity;

(c) Sand Dunes;

(d) Mudflats which are biologically active;

(e) National parks, marine parks, sanctuaries, reserve forests, wildlife habitats and other protected areas under the provisions of Wild Life (Protection) Act, 1972 (53 of 1972), the Forest (Conservation) Act, 1980 (69 of 1980) or Environment (Protection) Act, 1986 (29 of 1986); including Biosphere Reserves;

(f) Salt Marshes;

(g) Turtle nesting grounds;

(h) Horse shoe crabs habitats;

(i) Sea grass beds;

(j) Nesting grounds of birds;

(k) Areas or structures of archaeological importance and heritage sites.

B. The area between Low Tide Line and High Tide Line;

(ii) CRZ-II: The areas that have been developed upto or close to the shoreline.

Explanation: For the purposes of the expression "developed area" is referred to as that area within the existing municipal limits or in other existing legally designated urban areas which are substantially built up and has been provided with drainage and approach roads and other infrastructural facilities, such as water supply and sewerage mains;

(iii) CRZ-III: Areas that are relatively undisturbed and those do not belong to either CRZ-I or II which include coastal zone in the rural areas (developed and undeveloped) and also areas within municipal limits or in other legally designated urban areas, which are not substantially built up.

(iv) CRZ-IV:

A. the water area from the Low Tide Line to twelve nautical miles on the seaward side;

B. shall include the water area of the tidal influenced water body from the mouth of the water body at the sea upto the influence of tide which is measured as five parts per thousand during the driest season of the year.

(v) Areas requiring special consideration for the purpose of protecting the critical coastal environment and difficulties faced by local communities,-

A. (i) CRZ area falling within municipal limits of Greater Mumbai;

(ii) the CRZ areas of Kerala including the backwaters and backwater islands;

(iii) CRZ areas of Goa.

B. Critically Vulnerable Coastal Areas (CVCA) such as Sunderbans region of West Bengal and other ecologically sensitive areas identified as under Environment (Protection) Act, 1986 and managed with the involvement of coastal communities including fisherfolk.

8. Norms for regulation of activities permissible under this notification,

(i) The development or construction activities in different categories of CRZ shall be regulated by the concerned CZMA in accordance with the following norms, namely:

Note: The word existing use hereinafter in relation to existence of various features or existence of regularisation or norms shall mean existence of these features or regularisation or norms as on 19.2.1991 wherein CRZ notification, was notified.

I. CRZ-I:

(i) no new construction shall be permitted in CRZ-I except:

(a) projects relating to Department of Atomic Energy;

(b) pipelines, conveying systems including transmission lines;

(c) facilities that are essential for activities permissible under CRZ-I;

(d) installation of weather radar for monitoring of cyclones movement and prediction by Indian Meteorological Department;

(e) construction of trans harbour sea link and without affecting the tidal flow of water, between LTL and HTL.

(f) development of green field airport already approved at only Navi Mumbai;

(ii) Areas between LTL and HTL which are not ecologically sensitive, necessary safety measures will be incorporated while permitting the following, namely:

(a) exploration and extraction of natural gas;

(b) construction of dispensaries, schools, public rainshelter, community toilets, bridges, roads, jetties, water supply, drainage, sewerage which are required for traditional inhabitants living within the biosphere reserves after obtaining approval from concerned CZMA;

(c) necessary safety measure shall be incorporated while permitting such developmental activities in the area falling in the hazard zone;

(d) salt harvesting by solar evaporation of seawater;

(e) desalination plants;

(f) storage of non-hazardous cargo such as edible oil, fertilizers and food grain within notified ports;

(g) construction of trans harbour sea links, roads on stilts or pillars without affecting the tidal flow of water.

II. CRZ-II:

(i) buildings shall be permitted only on the landward side of the existing road, or on the landward side of existing authorized structures;

(ii) buildings permitted on the landward side of the existing and proposed roads or existing authorized structures shall be subject to the existing local town and country planning regulations including the 'existing' norms of Floor Space Index or Floor Area Ratio:

Provided that no permission for construction of buildings shall be given on landward side of any new roads which are constructed on the seaward side of an existing road:

(iii) reconstruction of authorized building to be permitted subject with the existing Floor Space Index or Floor Area Ratio Norms and without change in present use;

(iv) facilities for receipt and storage of petroleum products and liquefied natural gas as specified in Annexure-II appended to this notification and facilities for regasification of Liquefied Natural Gas subject to the conditions as mentioned in sub-paragraph (ii) of paragraph 3;

(v) desalination plants and associated facilities;

(vi) storage of non-hazardous cargo, such as edible oil, fertilizers and food grain in notified ports;

(vii) facilities for generating power by non-conventional power sources and associated facilities;

III. CRZ-III:

A. Area upto 200 metres from HTL on the landward side in case of seafront and 100 metres along tidal influenced water bodies or width of the creek whichever is less is to be earmarked as "No Development Zone (NDZ)",

(i) the NDZ shall not be applicable in such area falling within any notified port limits;

(ii) No construction shall be permitted within NDZ except for repairs or reconstruction of existing authorized structure not exceeding existing Floor Space Index, existing plinth area and existing density and for permissible activities under the notification including facilities essential for activities; Construction/reconstruction of dwelling units of traditional coastal communities including fisherfolk may be permitted between 100 and 200 metres from the HTL along the seafront in accordance with a comprehensive plan prepared by the State Government or the Union territory in consultation with the traditional coastal "11 communities including fisherfolk and incorporating the necessary disaster management provision, sanitation and recommended by the concerned State or the Union territory CZMA to NCZMA for approval by MoEF;

(iii) however, the following activities may be permitted in NDZ –

(a) agriculture, horticulture, gardens, pasture, parks, play field, and forestry;

(b) projects relating to Department of Atomic Energy;

(c) mining of rare minerals;

(d) salt manufacture from seawater;

(e) facilities for receipt and storage of petroleum products and liquefied natural gas as specified in Annexure-II;

(f) facilities for regasification of liquefied natural gas subject to conditions as mentioned in sub-paragraph (ii) of paragraph 3;

(g) facilities for generating power by non conventional energy sources;

(h) Foreshore facilities for desalination plants and associated facilities;

(i) weather radars;

(j) construction of dispensaries, schools, public rain shelters, community toilets, bridges, roads, provision of facilities for water supply, drainage, sewerage, crematoria, cemeteries and electric sub-station which are required for the local inhabitants may be permitted on a case to case basis by CZMA;

(k) construction of units or auxiliary thereto for domestic sewage, treatment and disposal with the prior approval of the concerned Pollution Control Board or Committee;

(l) facilities required for local fishing communities such as fish drying yards, auction halls, net mending yards, traditional boat building yards, ice plant, ice crushing units, fish curing facilities and the like;

(m) development of green field airport already permitted only at Navi Mumbai.

B. Area between 200 metres to 500 metres-

The following activities shall be permissible in the above areas:

(i) development of vacant plot in designated areas for construction of hotels or beach resorts for tourists or visitors subject to the conditions as specified in the guidelines at Annexure-III ;

(ii) facilities for receipt and storage of petroleum products and liquefied natural gas as specified in Annexure-II;

(iii) facilities for regasification of liquefied natural gas subject to conditions as mentioned in sub-paragraph (ii) of paragraph 3;

(iv) storage of non-hazardous cargo such as, edible oil, fertilisers, food grain in notified ports;

(v) foreshore facilities for desalination plants and associated facilities;

(vi) facilities for generating power by non-conventional energy sources;

(vii) construction or reconstruction of dwelling units so long it is within the ambit of traditional rights and customary uses such as existing fishing villages and goathans. Building permission for such construction or reconstruction will be subject to local town and country planning rules with overall height of construction not exceeding 9 metres with two floors (ground + one floor);

(viii) Construction of public rain shelters, community toilets, water supply drainage, sewerage, roads and bridges by CZMA who may also permit construction of "12 schools and dispensaries for local inhabitants of the area for those panchayats, the major part of which falls within CRZ if no other area is available for construction of such facilities;

(ix) reconstruction or alteration of existing authorised building subject to sub-paragraph (vii). (viii);

(x) development of green field airport already permitted only at Navi Mumbai.

(IV) In CRZ-IV areas,

The activities impugning on the sea and tidal influenced water bodies will be regulated except for traditional fishing and related activities undertaken by local communities as follows:

(a) No untreated sewage, effluents, ballast water, ship washes, fly ash or solid waste from all activities including from aquaculture operations shall be let off or dumped. A comprehensive plan for treatment of sewage generating from the coastal towns and cities shall be formulated within a period of one year in consultation with stakeholders including traditional coastal communities, traditional fisherfolk and implemented;

(b) Pollution from oil and gas exploration and drilling, mining, boat house and shipping;

(c) There shall be no restriction on the traditional fishing and allied activities undertaken by local communities.

V. Areas requiring special consideration,

1. CRZ areas falling within municipal limits of the Greater Mumbai.

(i) Developmental activities in the CRZ area of the Greater Mumbai because of the environmental issues, relating to degradation of mangroves, pollution of creeks and coastal waters, due to discharge of untreated effluents and disposal of solid waste, the need to provide decent housing to the poor section of society and lack of suitable alternatives in the inter connected islands of Greater Mumbai shall be regulated as follows, namely:

A. Construction of roads - In CRZ-I areas indicated at sub-paragraph (i) of paragraph 7 of the notification the following activities only can be taken up:-

(a) Construction of roads, approach roads and missing link roads approved in the Developmental Plan of Greater Mumbai on stilts ensuring that the free flow of tidal water is not affected, without any benefit of CRZ-II accruing on the landward side of such constructed roads or approach roads subject to the following conditions:

(i) All mangrove areas shall be mapped and notified as protected forest and necessary protection and conservation measures for the identified mangrove areas shall be initiated.

(ii) Five times the number of mangroves destroyed/cut during the construction process shall be replanted.

B. Solid waste disposal sites shall be identified outside the CRZ area and thereafter within two years the existing conventional solid waste sites shall be relocated outside the CRZ area.

(iii) In CRZ-II areas-

(a) The development or redevelopment shall continue to be undertaken in accordance with the norms laid down in the Town and Country Planning Regulations as they existed on the date of issue of the notification dated the 19th February, 1991, unless specified otherwise in this notification.

(b) SLUM REHABILITATION SCHEMES,

1. In the Greater Mumbai area there are large slum clusters with lakhs of families residing therein and the living conditions in these slums are deplorable and the civic agencies are not able to provide basic infrastructure such as drinking water, electricity, roads, drainage and the like because the slums come up in an unplanned and congested manner and the slums in the coastal area are at great risk in the event of cyclones, storm surges or tsunamis, in view of the difficulties in providing rescue, relief and evacuation.

2. To provide a safe and decent dwelling to the slum dwellers, the State Government may implement slum redevelopment schemes as identified as on the date of issue of this notification directly or through its parastatal agencies like Maharashtra Housing and Area Development Authority (MHADA), Shivshahi Punarvasan Prakalp Limited (SPPL), Mumbai Metropolitan Region Development Authority (MMRDA) and the like.:

Provided that–

(i) such redevelopment schemes shall be undertaken directly or through joint ventures or through public private partnerships or other similar models ensuring that the stake of the State Government or its parastatal entities shall be not less than 51%;

(ii) the Floor Space Index or Floor Area Ratio for such redevelopment schemes shall be in accordance with the Town and Country Planning Regulations prevailing as on the date on which the project is granted approval by the competent authority;

(iii) it shall be the duty of the project proponent undertaking the redevelopment through conditions (i) (2) above along with the State Government to ensure that all legally regularized tenants are provided houses in situ or as per norms laid down by the State Government in this regard.

(c) REDEVELOPMENT OF DILAPIDATED, CESSED AND UNSAFE BUILDINGS:

1. In the Greater Mumbai, there are, also a large number of old and dilapidated, cessed and unsafe buildings in the CRZ areas and due to their age these structures are extremely

vulnerable and disaster prone and therefore there is an urgent need for the redevelopment or reconstruction of these identified buildings.

2. These projects shall be taken up subject to the following conditions and safeguards:

(i) such redevelopment or reconstruction projects as identified on the date of issue of this notification shall be allowed to be taken up involving the owners of these buildings either above or with private developers in accordance with the prevailing Regulation, directly or through joint ventures or through other similar models;

(ii) the Floor Space Index or Floor Area Ratio for such redevelopment schemes shall be in accordance with the Town and Country Planning Regulations prevailing as on the date on which the project is granted approval by the competent authority;

(iii) suitable accommodation to the original tenants of the specified buildings shall be ensured during the course of redevelopment or reconstruction of the buildings by the project proponents, undertaking the redevelopment through condition 2(i) above.

(d) Notwithstanding anything contained in this notification, the developmental activities for slums and for dilapidated, cessed and unsafe buildings as specified at paras (b) and (c) above shall be carried out in an accountable and transparent manner by the project proponents mentioned therein which shall include the following pre-condition measures, wherever applicable;

1. (i) applicability of the Right to Information Act, 2005 to all redevelopment or reconstruction projects granted clearance by the Competent Authorities;

(ii) MoEF shall issue an order constituting the CPIO and the first Appellate Authority of appropriate ranks in consultation with Government of Maharashtra;

(iii) details of the Slum Rehabilitation Scheme, including the complete proposal and the names of the eligible slum dwellers will be declared suo-moto as a requirement of Section 4 of compliance of the Right to Information Act, 2005 by the appropriate authority in the Government of Maharashtra in one month before approving it;

(iv) the implementing or executing agency at the State Government with regard to projects indicated at sub-item (b) and (c) of item (iii) of sub-paragraph V shall display on a large notice boards at the site and at the office of the implementing or executing agency the names of the eligible builders, total number of tenements being made, names of eligible slum dwellers who are to be provided the dwelling units and the extra area available for free sale.

(v) Projects being developed under sub-items (b) and (c) of item (iii) of sub-paragraph V shall be given permission only if the project proponent agree to be covered under the Right to Information Act, 2005.

2. MoEF may appoint statutory auditors, who are empanelled by the Comptroller and auditor General (hereinafter referred to as the C&AG) to undertake performance and fiscal audit in respect of the projects relating to redevelopment of dilapidated, cessed and unsafe buildings and the projects relating to Slum Rehabilitation Scheme shall be audited by C&AG.

3. A High Level Oversight Committee may be set up by the Government of Maharashtra for periodic review of implementation of V(iii)(b) and (c) which shall include eminent representatives of various Stakeholders, like Architects, Urban Planner, Engineers, and Civil Society, besides the local urban bodies, the State Government and the Central Government.

4. The individual projects under V(iii)(b) and (c) shall be undertaken only after public consultation in which views of only the legally entitled slum dweller or the legally entitled tenent of the dilapidated or cessed buildings shall be obtained in accordance with the procedures laid down in EIA notification, 2006.

(e) In order to protect and preserve the 'green lung' of the Greater Mumbai area, all open spaces, parks, gardens, playgrounds indicated in development plans within CRZ-II shall be categorized as CRZ-III, that is, 'no development zone'.

(f) the Floor Space Index upto 15% shall be allowed only for construction of civic amenities, stadium and gymnasium meant for recreational or sports related activities and the residential or commercial use of such open spaces shall not be permissible.

(g) Koliwada namely, fishing settlement areas as identified in the Development Plan of 1981 or relevant records of the Government of Maharashtra, shall be mapped and declared as CRZ-III so that any development, including construction and reconstruction of dwelling units within these settlements shall be undertaken in accordance with applicable as per local Town and Country Planning Regulations.

(h) Reconstruction and repair works of the dwelling units, belonging to fisher communities and other local communities identified by the State Government, shall be considered and granted permission by the Competent Authorities on a priority basis, in accordance with the applicable Town and Country Planning Regulations.

2. CRZ for Kerala

In view of the unique coastal systems of backwater and backwater islands alongwith space limitation present in the coastal stretches of the State of Kerala, the following activities in CRZ shall be regulated as follows, namely:

(i) all the islands in the backwaters of Kerala shall be covered under the CRZ notification;

(ii) the islands within the backwaters shall have 50 metres width from the High Tide Line on the landward side as the CRZ area;

(iii) within 50 metres from the HTL of these backwater islands existing dwelling units of local communities may be repaired or reconstructed however no new construction shall be permitted;

(iv) beyond 50 metres from the HTL on the landward side of backwater islands, dwelling units of local communities may be constructed with the prior permission of the Gram panchayat;

(v) foreshore facilities such as fishing jetty, fish drying yards, net mending yard, fishing processing by traditional methods, boat building yards, ice plant, boat repairs and the like, may be taken up within 50 metres width from HTL of these backwater islands.

3. CRZ of Goa.

In view of the peculiar circumstances of the State Goa including past history and other developments, the specific activities shall be regulated and various measures shall be undertaken as follows:

(i) the Government of Goa shall notify the fishing villages wherein all foreshore facilities required for fishing and fishery allied activities such as traditional fish processing yards, boat building or repair yards, net mending yards, ice plants, ice storage, auction hall, jetties may be permitted by Grama Panchayat in the CRZ area;

(ii) reconstruction, repair works of the structures of local communities including fishermen community shall be permissible in CRZ;

(iii) purely temporary and seasonal structures customarily put up between the months of September to May;

(iv) the eco sensitive low lying areas which are influenced by tidal action known as khazan lands shall be mapped;

(v) the mangroves along such as khazan land shall be protected and a management plan for the khazan land prepared and no developmental activities shall be permitted in the khazan land;

(vi) sand dunes, beach stretches along the bays and creeks shall be surveyed and mapped. No activity shall be permitted on such sand dune areas;

(vii) the beaches such as Mandrem, Morjim, Galgiba and Agonda has been designated as turtle nesting sites and protected under the Wildlife Protection Act, 1972 and these areas "16 shall be surveyed and management plan prepared for protection of these turtle nesting sites;

(viii) no developmental activities shall be permitted in the turtle breeding areas referred to in sub-paragraph (vii).

4. (a) Critical Vulnerable Coastal Areas (CVCA) which includes Sunderbans and other identified ecological sensitive areas which shall be managed with the involvement of the local coastal communities including the fisher folk;-

(b) the entire Sunderbans mangrove area and other identified ecologically important areas such as Gulf of Khambat and Gulf of Kutchchh in Gujarat, Malvan, Achra-Ratnagiri in Maharashtra, Karwar and Coondapur in Karnataka, Vembanad in Kerala, Gulf of Mannar in Tamil Nadu, Bhaitarkanika in Orissa, Coringa, East Godavari and Krishna in Andhra Pradesh shall be declared as Critical Vulnerable Coastal Areas (CVCA) through a process of consultation with local fisher and other communities inhabiting the area and depend on its resources for their livelihood with the objective of promoting conservation and sustainable use of coastal resources and habitats;

(c) the process of identifying planning, notifying and implementing CVCA shall be detailed in the guideline which will be developed and notified by MoEF in consultations with the stakeholders like the State Government, local coastal communities and fisherfolk and the like inhabiting the area;

(d) the Integrated Management Plans (IMPs) prepared for such CVCA shall interalia keep in view the conservation and management of mangroves, needs of local communities such as, dispensaries, schools, public rain shelter, community toilets, bridges, roads, jetties, water supply, drainage, sewerage and the impact of sea level rise and other natural disasters and the IMPs will be prepared in line with the para 5 above for preparation of Coastal Zone Management Plans;

(e) till such time the IMPs are approved and notified, construction of dispensaries, schools, public rain shelters, community toilets, bridges, roads, jetties, water supply, drainage, sewerage which are required for traditional inhabitants shall be permitted on a case to case basis, by the CZMA with due regards to the views of coastal communities including fisherfolk.

[F.No.11-83/2005-IA-III]

J. M. MAUSKAR, Addl. Secy

ANNEXURE I

GUIDELINES FOR PREPARATION OF COASTAL ZONE MANAGEMENT PLANS

I. A. Demarcation of High Tide Line

1. Demarcation of High Tide Line (HTL) and Low Tide Line (LTL) shall be carried out by one of the agencies authorised by MoEF based on the recommendations of the National Centre for Sustainable Coastal Management (NCSCM).
2. Demarcation of the High Tide Line or LTL shall be made on the Coastal Zone Management (CZM) Maps of scale 1:25,000 prepared by the agencies identified by the MoEF.
3. Local level CZM Maps shall be prepared for use of officials of local bodies for determination of the CRZ.
4. The local level CZM Maps shall be prepared on a Cadastral scale in accordance with the CZM Maps approved by the Central Government.

B. Preparation of CZM Maps

5. Base Maps of 1:25,000 scale shall be acquired from the Survey of India (SOI) and wherever 1:25,000 maps are not available, 1:50,000 maps shall be enlarged to 1:25,000 for the purpose of base map preparation and these maps will be of the standard specification given below:

Unit	:	7.5 minutes X 7.5minutes
Numbering	:	Survey of India Sheet Numbering System
Horizontal Datum	:	Everest or WGS 84
Vertical Datum	:	Mean Sea Level (MSL)
Topography	:	Topography in the SOI maps will be updated using latest satellite imageries or aerial photographs

6. The High Water Level (HWL) and Low Water Level (LWL) marked on the Base maps will be transferred to the CZM maps.
7. Coastal geomorphological signatures in the field or satellite imageries or aerial photographs will be used for appropriate adjustment, in the HWL or LWL for demarcating HTL or LTL in accordance with the CRZ notification.
8. The following geomorphological features shall be considered while demarcating in HTL or LTL:
   Landward (monsoonal) berm crest in the case of sandy beaches
   Rocks, Headlands, Cliffs
   Seawalls or revetments or embankments
9. 500 meter and 200 metre lines will be demarcated with respect of HTL.
10. HTL (as defined in the CRZ notification) and LTL shall also be demarcated in the CZM maps along the banks of tidal influenced inland water bodies with the help of the geomorphological signatures or features.
11. Classification of different coastal zones shall be done as per the CRZ notification
12. Standard national or international colour codes shall be used to highlight sub-classification of data.

C. Local level CZM Maps

Local level CZM Maps are for the use of local bodies and other agencies to facilitate implementation of the Coastal Zone Management Plans

13. Cadastral (village) maps in 1:3960 or the nearest scale, shall be used as the base maps.

14. These maps are available with revenue Authorities and are prepared as per standard norms.
15. HTL (as defined in the CRZ notification) and LTL will be demarcated in the cadastral map based on detailed physical verification using coastal geomorphological signatures or features in accordance with the CZM Maps approved by the Central Government.
16. 500 metres and 200 metres lines shall be demarcated with respect to the HTL thus marked.
17. HTL (as defined in the CRZ notification, 1991) and LTL will also be demarcated along the banks of tidal influenced inland water bodies with the help of geomorphological signatures or features.
18. Classifications shall be transferred into local level CZM maps from the CZM Plans.
19. Symbols will be adopted from CZM Maps.
20. Colour codes as given in CZM Maps shall be used.
21. Demarcation of cadastral maps will be done by local agencies approved by the Central Government. The local agencies shall work under the guidance of the concerned State Government or Union Territory Coastal Zone Management Authorities.

D. Hazard mapping:

II. Classification of CRZ areas

1. The CZM Maps shall be prepared in accordance with para 5 of the CRZ notification demarcating CRZ I, II, III, IV and V.
2. The CZM Maps shall clearly demarcate the land use plan of the area and lists out the CRZ-I areas. All the CRZ-I areas listed under para 7(I)A and B shall be clearly demarcated and colour codes given so that each of the CRZ-I areas can be clearly identified.
3. Buffer zone along mangrove areas of more than 1000sq mts shall be stipulated with a different colour distinguishing from the mangrove area.
4. The buffer zone shall also be classified as CRZ-I area.
5. The hazard line to be drawn up by MoEF shall be superimposed on the CZM maps in 1:25,000 scale and also on the cadastral scale maps.
6. The CRZ-II areas shall be those areas which have been substantially built up with a ratio of builtup plots to that of total plots is more than 50%.
7. In the CRZ areas, the fishing villages, common properties of the fishermen communities, fishing jetties, ice plants, fish drying platforms or areas infrastructure facilities of fishing and local communities such as dispensaries, roads, schools, and the like, shall be indicated on the cadastral scale maps. States shall prepare detailed plans for long-term housing needs of coastal fisher communities in view of expansion and other needs, provisions of basic services including sanitation, safety, and disaster preparedness.
8. No developmental activities other than those listed above shall be permitted in the areas between the hazard line and 500 metres or 100 metres or width of the creek on the landward side. The dwelling unit of the local communities including that of the fishers will not be relocated if the dwelling units are located on the seaward side of the hazard line. The State Government will provide necessary safeguards from natural disaster to such dwelling units of local communities.
9. The water areas of CRZ IV shall be demarcated and clearly demarcated if the water body is sea, lagoon, backwater, creek, bay, estuary and for such classification of the water bodies the terminology used by Naval Hydrographic Office shall be relied upon.
10. The fishing Zones in the water bodies and the fish breeding areas shall be clearly marked.

11. The water area shall be demarcated indicating the pollution levels as per Central Pollution Control Board standards on water quality.
12. In the CRZ V areas the land use maps shall be superimposed on the Coastal Zone Management Plan and clearly demarcating the CRZ I, II, III, IV.
13. The existing authorized developments on the sea ward side shall be clearly demarcated.
14. The features like cyclone shelters, rain shelters, helipads and other infrastructure including road network may be clearly indicated on the CZM Maps for the purpose of rescue and relief operations during cyclones, storms, tsunami and the like.

III. CZMPs approved by MoEF in accordance with CRZ notification, 1991

1. While preparing the CZMPs under CRZ notification, 2011, the CZMPs that have been approved under the CRZ Notification, 1991 shall be compared. A justification shall be provided by the concerned CZMA in case the CZMPs prepared under CRZ notification, 2011 varies with respect to the approved CZMP prepared under CRZ notification, 1991.

IV. Public Views on the CZMP.

(a) The draft CZMPs prepared shall be given wide publicity and suggestions and objections received in accordance with the Environment (Protection) Act, 1986. Public hearing on the draft CZMPs shall be held at district level by the concerned CZMAs.

(b) Based on the suggestions and objections received the CZMPs shall be revised and approval of MoEF shall be obtained.

(c) The approved CZMP shall be put up on the website of MoEF, concerned website of the State, Union Territory CZMA and hard copy made available in the panchayat office, District collector office and the like.

V. Revision of Coastal Zone Management Plans

1. Whenever there is a doubt the concerned State or Union territory Coastal Zone Management Authority shall refer the matter to the National Centre for Sustainable Coastal Management who shall verify the CZMP based on latest satellite imagery and ground truthing.
2. The rectified map would be submitted to MoEF for its record.

ANNEXURE II

List of petroleum and chemical products permitted for storage in [CRZ except CRZ-I(A)]

(i) Crude oil;
(ii) Liquefied Petroleum Gas;
(iii) Motor spirit;
(iv) Kerosene;
(v) Aviation fuel;
(vi) High speed diesel;
(vii) Lubricating oil;
(viii) Butane;
(ix) Propane;
(x) Compressed Natural Gas;
(xi) Naphtha:
(xii) Furnace oil;
(xiii) Low Sulphur Heavy Stock;
(xiv) Liquefied Natural Gas;
(xv) Fertilizers and raw materials for manufacture of fertilizers.

ANNEXURE III

Guidelines for development of beach resorts or hotels in the designated areas of CRZ-III and CRZ-II for occupation of tourist or visitors with prior approval of the Ministry of Environment and Forests

I. Construction of beach resorts or hotels with prior approval of MoEF in designated areas of CRZ-II and III for occupation of tourist or visitors shall be subject to the following conditions, namely:

(a) the project proponent shall not undertake any construction within 200 metres in the landward side of High Tide Line and within the area between Low Tide Line and High Tide Line;

(b) the proposed constructions shall be beyond the hazard line or 200 metres from the High Tide Line whichever is more;

(c) live fencing and barbed wire fencing with vegetative cover may be allowed around private properties subject to the condition that such fencing shall in no way hamper public access to the beach;

(d) no flattening of sand dunes shall be carried out;

(e) no permanent structures for sports facilities shall be permitted except construction of goal posts, net posts and lamp posts;

(f) construction of basement may be allowed subject to the condition that no objection certification is obtained from the State Ground Water Authority to the effect that such construction will not adversely affect fee flow of groundwater in that area;

(g) the State Ground Water Authority shall take into consideration the guidelines issued by Central Government before granting such no objection certificate;

(h) though no construction is allowed in the no development zone for the purposes of calculation of Floor Space Index, the area of entire plot including the portion which falls within the no development zone shall be taken into account;

(i) the total plot size shall not be less than 0.4 hectares and the total covered area on all floors shall not exceed 33 percent of the plot size i.e., the Floor Space Index shall not exceed 0.33 and the open area shall be suitably landscaped with appropriate vegetal cover;

(j) the construction shall be consistent with the surrounding landscape and local architectural style;

(k) the overall height of construction upto the highest ridge of the roof, shall not exceed 9metres and the construction shall not be more than two floors (ground floor plus one upper floor);

(l) groundwater shall not be tapped within 200metre of the High Tide Line; within the 200metre-500metre zone it can be tapped only with the concurrence of the Central or State Ground Water Board;

(m) extraction of sand, leveling or digging of sandy stretches except for structural foundation of building, swimming pool shall not be permitted within 500metres of the High Tide Line;

(n) the quality of treated effluents, solid wastes, emissions and noise levels and the like, from the project area must conform to the standards laid down by the competent authorities including the Central or State Pollution Control Board and under the Environment (Protection) Act, 1986;

(o) necessary arrangements for the treatment of the effluents and solid wastes must be made and it must be ensured that the untreated effluents and solid wastes are not discharged into the water or on the beach; and no effluent or solid waste shall be discharged on the beach;

(p) to allow public access to the beach, atleast a gap of 20metres width shall be provided between any two hotels or beach resorts; and in no case shall gaps be less than 500metres apart; and

(q) if the project involves diversion of forestland for non-forest purposes, clearance as required under the Forest (Conservation) Act, 1980 shall be obtained and the requirements of other Central and State laws as applicable to the project shall be met with; and

(r) approval of the State or Union territory Tourism Department shall be obtained.

II. In ecologically sensitive areas (such as marine parks, mangroves, coral reefs, breeding and spawning grounds of fish, wildlife habitats and such other area as may be notified by the Central or State Government Union territories) construction of beach resorts or hotels shall not be permitted

ANNEXURE IV

Form – I for seeking clearance for project attracting CRZ notification

Basic information:

Name of the Project:

Location or site alternatives under consideration:

Size of the project (in terms of total area):

CRZ classification of the area:

Expected cost of the project:

Contact Information:

(II) Activity

1. Construction, operation or decommissioning of the Project involving actions, which will cause physical changes in the locality (topography, land use, changes in water bodies, and the like)

S. No.	Information/Checklist confirmation	Yes/No	Details thereof (with approximate quantities /rates, wherever possible) with source of information data
1.1.	Permanent or temporary change in land use, land cover or topography including increase in intensity of land use (with respect to local land use plan)		
1.2.	Details of CRZ classification as per the approved Coastal Zone Management Plan?		
1.3.	Whether located in CRZ-I area?		
1.4.	The distance from the CRZ-I areas.		
1.5.	Whether located within the hazard zone as mapped by Ministry of Environment and Forests/National Disaster Management Authority?		
1.6.	Whether the area is prone to cyclone, tsunami, tidal surge, subduction, earthquake etc.?		
1.7.	Whether the area is prone for saltwater ingress?		
1.8.	Clearance of existing land, vegetation and buildings?		

S. No.	Information/Checklist confirmation	Yes/No	Details thereof (with approximate quantities /rates, wherever possible) with source of information data
1.9.	Creation of new land uses?		
1.10.	Pre-construction investigations e.g. bore hole, soil testing?		
1.11.	Construction works?		
1.12.	Demolition works?		
1.13.	Temporary sites used for construction works or housing of construction workers?		
1.14.	Above ground buildings, structures or earthworks including linear structures, cut and fill or excavations		
1.15.	Underground works including mining or tunneling?		
1.16.	Reclamation works?		
1.17.	Dredging/reclamation/land filling/disposal of dredged material etc.?		
1.18.	Offshore structures?		
1.19.	Production and manufacturing processes?		
1.20.	Facilities for storage of goods or materials?		
1.21.	Facilities for treatment or disposal of solid waste or liquid effluents?		
1.22.	Facilities for long-term housing of operational workers?		
1.23.	New road, rail or sea traffic during construction or operation?		
1.24.	New road, rail, air waterborne or other transport infrastructure including new or altered routes and stations, ports, airports etc?		
1.25.	Closure or diversion of existing transport routes or infrastructure leading to changes in traffic movements?		
1.26.	New or diverted transmission lines or pipelines?		
1.27.	Impoundment, damming, culverting, realignment or other changes to the hydrology of watercourses or aquifers?		
1.28.	Stream and river crossings?		
1.29.	Abstraction or transfers of water form ground or surface waters?		
1.30.	Changes in water bodies or the land surface affecting drainage or runoff?		
1.31.	Transport of personnel or materials for construction, operation or decommissioning?		
1.32.	Long-term dismantling or decommissioning or restoration works?		
1.33.	Ongoing activity during decommissioning which could have an impact on the environment?		
1.34.	Influx of people to an area in either temporarily or permanently?		
1.35.	Introduction of alien species?		
1.36.	Loss of native species or genetic diversity?		
1.37.	Any other actions?		

2. Use of Natural resources for construction or operation of the Project (such as land, water, materials or energy, especially any resources which are non-renewable or in short supply)

S. No.	Information/Checklist confirmation	Yes/No	Details thereof (with approximate quantities /rates, wherever possible) with source of information data
2.1.	Land especially undeveloped or agricultural land (ha)		
2.2.	Water (expected source & competing users) unit: KLD		
2.3.	Minerals (MT)		
2.4.	Construction material – stone, aggregates, sand/soil (expected source – MT)		
2.5.	Forests and timber (source – MT)		
2.6.	Energy including electricity and fuels (source, competing users) Unit: fuel (MT), energy (MW)		
2.7.	Any other natural resources (use appropriate standard units)		

3. Use, storage, transport, handling or production of substances or materials, which could be harmful to human health or the environment or raise concerns about actual or perceived risks to human health.

S. No.	Information/Checklist confirmation	Yes/No	Details thereof (with approximate quantities /rates, wherever possible) with source of information data
3.1.	Use of substances or materials, which are hazardous (as per MSIHC rules) to human health or the environment (flora, fauna, and water supplies)		
3.2.	Changes in occurrence of disease or affect disease vectors (e.g. insect or water borne diseases)		
3.3.	Affect the welfare of people e.g. by changing living conditions?		
3.4.	Vulnerable groups of people who could be affected by the project e.g. hospital patients, children, the elderly etc.,		
3.5.	Any other causes, that would affect local communities, fisherfolk, their livelihood, dwelling units of traditional local communities etc.		

4. Production of solid wastes during construction or operation or decommissioning (MT/month)

S. No.	Information/Checklist confirmation	Yes/No	Details thereof (with approximate quantities /rates, wherever possible) with source of information data
4.1.	Spoil, overburden or mine wastes		
4.2.	Municipal waste (domestic and or commercial wastes)		
4.3.	Hazardous wastes (as per Hazardous Waste Management Rules)		
4.4.	Other industrial process wastes		
4.5.	Surplus product		
4.6.	Sewage sludge or other sludge from effluent treatment		
4.7.	Construction or demolition wastes		
4.8.	Redundant machinery or equipment		
4.9.	Contaminated soils or other materials		
4.10.	Agricultural wastes		
4.11.	Other solid wastes		

5. Release of pollutants or any hazardous, toxic or noxious substances to air (Kg/hr)

S. No.	Information/Checklist confirmation	Yes/No	Details thereof (with approximate quantities /rates, wherever possible) with source of information data
5.1.	Emissions from combustion of fossil fuels from stationary or mobile sources		
5.2.	Emissions from production processes		
5.3.	Emissions from materials handling including storage or transport		
5.4.	Emissions from construction activities including plant and equipment		
5.5.	Dust or odours from handling of materials including construction materials, sewage and waste		
5.6.	Emissions from incineration of waste		
5.7.	Emissions from burning of waste in open air (e.g. slash materials, construction debris)		
5.8.	Emissions from any other sources		

6. Generation of Noise and Vibration, and Emissions of Light and Heat:

S. No.	Information/Checklist confirmation	Yes/No	Details thereof (with approximate quantities /rates, wherever possible) with source of information data
6.1.	From operation of equipment e.g. engines, ventilation plant, crushers		
6.2.	From industrial or similar processes		
6.3.	From construction or demolition		
6.4.	From blasting or piling		
6.5.	From construction or operational traffic		
6.6.	From lighting or cooling systems		
6.7.	From any other sources		

7. Risks of contamination of land or water from releases of pollutants into the ground or into sewers, surface waters, groundwater, coastal waters or the sea:

S. No.	Information/Checklist confirmation	Yes/No	Details thereof (with approximate quantities /rates, wherever possible) with source of information data
7.1.	From handling, storage, use or spillage of hazardous materials		
7.2.	From discharge of sewage or other effluents to water or the land (expected mode and place of discharge)		
7.3.	By deposition of pollutants emitted to air into the land or into water		
7.4.	From any other sources		
7.5.	Is there a risk of long-term build up of pollutants in the environment from these sources?		

8. **Risk of accidents during construction or operation of the Project, which could affect human health or the environment**

S. No.	Information/Checklist confirmation	Yes/No	Details thereof (with approximate quantities /rates, wherever possible) with source of information data
8.1.	From explosions, spillages, fires etc from storage, handling, use or production of hazardous substances		
8.2.	From any other causes		
8.3.	Could the project be affected by natural disasters causing environmental damage (e.g., floods, earthquakes, landslides, cloudburst etc)?		

9. **Factors which should be considered (such as consequential development) which could lead to environmental effects or the potential for cumulative impacts with other existing or planned activities in the locality**

S. No.	Information/Checklist confirmation	Yes/No	Details thereof (with approximate quantities /rates, wherever possible) with source of information data
9.1.	Lead to development of supporting lities, ancillary development or development stimulated by the project which could have impact on the environment e.g.: Supporting infrastructure (roads, power supply, waste or waste water treatment, etc.) housing development extractive industries supply industries other		
9.2.	Lead to after-use of the site, which could have an impact on the environment		
9.3.	Set a precedent for later developments		
9.4.	Have cumulative effects due to proximity to other existing or planned projects with similar effects		

III. Environmental Sensitivity

S. No.	Areas	Name/ Identity	Aerial distance (within 15 km.) Proposed project location boundary
1.	Areas protected under international conventions, national or local legislation for their ecological, landscape, cultural or other related value		
2.	Areas which are important or sensitive for ecological reasons - Wetlands, watercourses or other water bodies, coastal zone, biospheres, mountains, forests		
3.	Areas used by protected, important or sensitive species of flora or fauna for breeding, nesting, foraging, resting, over wintering, migration		
4.	Inland, coastal, marine or underground waters		
5.	State, National boundaries		
6.	Routes or facilities used by the public for access to recreation or other tourist, pilgrim areas		
7.	Defence installations		

S. No.	Areas	Name/ Identity	Aerial distance (within 15 km.) Proposed project location boundary
8.	Densely populated or built up area		
9.	Areas occupied by sensitive man-made land uses (hospitals, schools, places of worship, community facilities)		
10.	Areas containing important, high quality or scarce resources (ground water resources, surface resources, forestry, agriculture, fisheries, tourism, minerals)		
11.	Areas already subjected to pollution or environmental damage. (those where existing legal environmental standards are exceeded)		
12.	Areas susceptible to natural hazard which could cause the project to present environmental problems (earthquakes, subsidence, landslides, erosion, flooding or extreme or adverse climatic conditions)		

The revised and amended guideline of coastal regulation was introduced in 2011. With revised classification of CRZ and their prohibition activities.

50 BRIEF NOTE ON STATUTORY OBLIGATIONS OF INDUSTRIES IN RESPECT OF ENVIRONMENTAL LAWS IN WEST BENGAL

Since 1974, a number of Acts were enacted to protect environment from pollution, environmental disasters and accidents. Naturally, industries have become liable to follow a number of statutory obligations in respect of their manufacturing process, discharge, storage, transportation, exporting or importing their products, by-products and raw materials including the safety of the industrial workers and people at large.

After the enactment of Water (Prevention & Control of Pollution) Act, 1974, in most of the States, State Pollution Control Boards were constituted and Departments of Environment were also designed by the Central Government and some State Governments. Presently the State Board and the Central Board are executing a number of Acts which are:

(i) Water (Prevention & Control of Pollution) Act, 1974.

(ii) Air (Prevention & Control of Pollution) Act, 1981.

(iii) Water (Prevention & Control of Pollution) Cess Act, 1977.

(iv) Environment (Protection) Act, 1986.

(v) Public Liability Insurance Act, 1991.

In addition, in the State of West Bengal, Bengal Smoke Nuisance Act, 1905, as amended in the year 1978 is also operative.

Statutory obligations under Water (Prevention & Control of Pollution) Act, 1974

As per the Section 25(i)(a) no person shall without the previous consent of the State Board establish or to take any steps to establish any industry, or process, or any treatment and disposal system or any extension or addition thereto, which is likely to discharge sewage or trade effluent into stream or well or sewer on land.

It is also provided under Section 25(i)(b)(c) that no person shall without the previous consent of the State Board begin to use any new or alternative outlet for the discharge of sewage.

Statutory obligations under Air (Prevention & Control of Pollution) Act, 1981

It is mentioned in Section 21 that no person shall without the previous consent of the State Board establish or operate any industrial plant in an air pollution control area.

It is also provided in the statute that no industrial unit can discharge any general substance in the air beyond permissible limits as laid down by the State Board.

Statutory obligations under Bengal Smoke Nuisance Act, 1905, as amended in the year 1973

It is provided in Section 8A that no furnace, flue, chimney or smoke abating unit shall be created or recreated or altered or added without the previous sanction of the plants by the Chief Inspector of Smoke Nuisances. Every grant of permission shall be valid for a period of one year and may be renewed from time to time within such period as may be prescribed by the statute.

Statutory obligations under Environment (Protection) Act, 1986

No industrial units can discharge any solid, liquid or gaseous substances beyond the permissible limit as laid down by the Central Government or any other authority authorised by the Central Government on its behalf.

It is also provided that Central Government or its authorised person can give any direction as provided under the statute and such directions are binding upon the industries.

As per the provision of Section 3 all specified industries under Water (Prevention & Control of Pollution) Cess Act, 1977 are liable to pay cess in the prescribed rate made under the statute. It is provided under Section 5 that every specified industry or local authority is liable to furnish cess return in such manner to such authority as may be prescribed under the statute.

For the purpose of measuring and recording the quantity of water consumption all specified industries and local authority shall affix meters of such standards at such places as may be prescribed.

According to the latest policy statement the distribution of cess is made on the following principles:

25% of the total amount is retained by the Central Government to encourage R & D on clean technology and assist local bodies in Class II & III towns to set up sewage treatment systems according to need.

The remaining 75% is released to the concerned State Boards according to the following guidelines:

(i) 25% to be utilised according to the decision of the State Pollution Control Boards for pollution control measures.

(ii) 25% to be utilised according to the decisions of the State Pollution Control Boards for assisting industrial units in the adoption of clean process technologies and sewage treatment systems in class II and III towns.

(iii) The remaining 25% will be reimbursed to the State Pollution Control Boards on the basis of collection of the target for that year/quarter i.e., 10% for collecting 85% of the target for collection of cess, at the minimum and additional 1% for each additional percent of the cess collected. The amount will be utilised according to the decisions of the State Pollution Control Boards.

The West Bengal Pollution Control Board is extending assistance for development of technology through R & D cell for helping small scale sector. The PCB has initiated a scheme for assistance for sanitation works in Class II and Class III towns.

Annex 1: Schedule of Specified Industries for Water Cess
1. Ferrous metallurgical industry
2. Non-ferrous metallurgical industry
3. Mining industry
4. Ore processing industry
5. Petroleum industry

Contd.

Annex 1: Schedule of Specified Industries for Water Cess
6. Petrochemical industry
7. Chemical industry
8. Ceramic industry
9. Cement industry
10. Textile industry (including cotton, synthetic and semi-synthetic fibres manufactured from these fibres)
11. Paper industry
12. Fertiliser industry
13. Coal (including coke) industry
14. Power (thermal, diesel and hydel) generating industry
15. Processing of animal or vegetable products industry (including processing of milk, meat, hides and skins, all agricultural products and their wastes)
16. Engineering industry.

Annex 2: Prescribed Rates of Water Cess

Purpose for which water is consumed	Rate of Cess Under Sub-section (2) of Section 3	*Rate of Cess Under Sub-section (2A) of Section 3
1. Industrial cooling, spraying in mine pits or boiler feeds	One and half paise per kilolitre	Two and one fourth paise per kilolitre
2. Domestic Purpose	Two paise per kilolitre	Three paise per kilolitre
3. Processing whereby water gets polluted and the pollutants are easily bio-degradable	Four paise per kilolitre	Seven and a half paise per kilolitre
4. Processing whereby water gets polluted and the pollutants are not easily bio-degradable and are toxic	Five paise per kilolitre	Nine and a half paise per kilolitre.

*These rates shall be applicable only in case of non-compliance with any of the prescribed consent conditions or any of the standards for discharge of effluent prescribed by the Central Government under Environment (Protection) Act, 1986, otherwise, as specified under column (2).

Statutory obligations under Manufacture, Storage and Import of Hazardous Chemicals Rules, 1989 Read with Environment (Protection) Act, 1986

Chemicals occupy an important part in our economy and are the source of several benefits to the country. Their number, variety and complexity are increasing day by day. Many of these chemicals which are used in the industries are toxic, highly reactive, explosive and inflammable. Because of these properties they have the potential threat to the overall environment. It is necessary to handle these chemicals with great care at all stages of manufacturing, processing treatment, storage, transportation and use. Rules have been prescribed under Environment (Protection) Act, 1986, and responsibility have been fixed on several agencies for safe handling of those hazardous chemicals.

In exercise of powers conferred by Sections 6, 8 and 25 of The Environment (Protection) Act, 1986, the Central Government has notified Rules called "Manufacture, Storage and Import of Hazardous Chemicals Rules 1989". As per Section 13 and 14 of this Rule it is mandatory to prepare on-site emergency plans by the industries.

The main elements of on-site plan are:

(i) Preparation of site plan area, place and type of likely hazardous materials are to be indicated in on-site plan. Surrounding areas outside the boundary of the factory premises indicating roads, residential colonies etc. are to be shown.

(ii) Appointment of key-personnel: Site Controller, Incident Controller, Shift Executive-in-Charge, Communication Officer, Safety Officers, Fire and Security Officer, Utilities and Services-in-Charge, and Traffic Controller are the key-persons. Specific duties of each of them are to be laid down.

(iii) Emergency Control Centre—Emergency Control Centre is to be set up and marked on the site plan. Control Centre will be the focal point in case of an emergency from where the operations to handle the emergency are directed and coordinated. It will control the site activities and should be furnished with external and internal telephone connections, list of essential telephone numbers, list of key-persons and their addresses and a copy of Chemical Fact Sheet for the factory. This will also act as an operating point for Public Address System.

Under Rule 10, an occupier shall not undertake any industrial activity to which this rule applies unless he has prepared a safety report on that industrial activity containing the information specified in the Schedule 8 of this Rules and such information should be submitted before the State Pollution Control Boards or any other authorities prescribed in the Rules.

As per Rule 11, the occupier shall not make any modification to the industrial activity furnished in the safety report unless he has made a further report to the concerned authority at least 90 days before making those modifications.

Statutory obligations under Hazardous Waste (Management and Handling) Rules, 1989 Read with Environment (Protection) Act, 1986

The growth of chemical industries are resulting in extensive use of chemicals and release of huge quantities of wastes into the environment in the form of solids, liquids and gases. A substantial amount of these waste are potentially hazardous to the living organisms including human beings. A potential risk to human health is the contamination of surface and ground-water supplies due to leaching at hazardous waste dumping sites. The situation in the developing countries is worse because apart from non-availability of appropriate technology, adequate facilities to treat waste generated by such industries are also lacking. An effective control of hazardous waste is of paramount importance for proper health and Environmental Protection and Natural Resource Management.

Large-scale production of a variety of chemicals during the past four decades in India has led to the release of hazardous wastes into the environment. Toxic compounds, which were hitherto unknown, are being detected in increasing numbers in sewage and water supplies due to the indiscriminate discharge of waste materials on land and into water bodies.

The Environment (Protection) Act, 1986, was enacted to provide a single focus for all environmental issues and to plug loopholes in the earlier enactments. Section 3(2)(vi) and (vii) gives the responsibility to the Central Government to lay down procedures and safeguards for handling hazardous substances, which include hazardous chemicals, hazardous wastes and hazardous microorganisms.

The Hazardous Wastes (Management & Handling) Rules were notified in 1989 under the Environment (Protection) Act, 1986. These rules aim at providing control for the generation, collection, treatment, transport, import, storage and disposal of hazardous wastes. The principal objective of these regulations is to establish a control mechanism for the management of hazardous wastes.

The Hazardous Wastes (Management & Handling) Rules, 1989, provided for an effective inventorisation and controlled handling and disposal of hazardous wastes. The applicability of these rules have been stated in Rules 2, 18 categories of wastes have been identified along with regulatory quantities and the Rules shall apply specifically to these categories of wastes.

RESPONSIBILITY OF GENERATOR OF HAZARDOUS WASTES

(i) To report to the State agencies the hazardous wastes being handled (Rule 4).

(ii) To apply for authorisation for handling hazardous wastes in Form U provided in the Rules (Rule 5).

(iii) Proper packaging, labelling and transportation of hazardous wastes in accordance with the provisions in the Motor Vehicles Act, 1988 (Rule 7).

(iv) To maintain records of the operator involved in the generation and disposal of hazardous wastes in Form 3.

(v) To file annual return (Form 4).

(vi) To report accidents to State Pollution Control Boards (Rule 10).

Non-compliance of the statutory provisions under Environmental Laws referred to above shall make the industries liable for prosecution leading even to closure to be ordered by the West Bengal Pollution Control Board or by Judicial Authorities.

Statutory obligations under the Public Liability Insurance Act, 1991

The Public Liability Insurance Act, 1991, was enacted to provide for public liability insurance for the purpose of providing immediate relief to the persons affected by accidents occurring while handling any hazardous substances. The Act provides that every owner as defined in the Act should take an insurance policy so that in case of accidents, the victims can be given immediate relief without their having to go to any court of law.

Facing some difficulties in implementing the Act, the Government of India had promulgated an Ordinance namely Public Liability Insurance (Amendment) Ordinance 1992 and on 3rd of March, 1992 the Parliament replaced the Ordinance by introducing Public Liability Insurance Bill, 1992.

The Government of India vide Notification dated 1st May, 1991 brought out Rules under the Act. The salient features of the Rules are:

(a) An application for claim can be made in a prescribed Form (Form No. 1) alongwith documents to the Collector.

(b) The Collector will have all the powers of a civil court for the purpose defined in the Rules.

(c) The owner shall create and establish a public liability Insurance Fund with the Bank for setting the liability arising out of any claim which will be operated by the administrator nominated by the owner.

Later, on 24th March, 1992, the Government of India specified that owners handling hazardous substances defined in the Rules shall take insurance policy as per the provision of the Act.

In this notification about 179 chemicals and corresponding qualifying quantities have been listed.

For environmental clearance of projects under Category B, all guideline will be followed MoEF notification of 2006.

51 POLLUTION REGULATION STATUS IN WEST BENGAL

In West Bengal, there are about 10,000 registered factories and over one lakh small scale industries. Only 10% of them regularly take permission of West Bengal Pollution Control Board (WBPCB) for operation of their activities. This is primarily due to lack of infrastructural facilities of the WBPCB.

At present, there are three categories of industries from the standpoint of pollution generation. Most polluting (red categories), Moderately polluting (orange categories) and Non-polluting (green categories). From January 2000, WBPCB planned to operate a new system of monitoring. They delegated power to Executive Officer, District industry centre for giving permission for operation of green category industries for a period of 5 years at a time with provision of periodic checkup if there is public complaint. Permission for operation to orange category industries would be done by WBPCB for a period of 3 years with provision for periodic checkup. Lastly, permission for operation to red category industries will be done by WBPCB for year to year renewal basis.

CATEGORIES OF INDUSTRIES

LIST OF INDUSTRIES UNDR "RED" CATEGORY

1. All mining activities including quarries
2. Aluminium smelter
3. Asbestos and asbestos-based industries
4. Basic Drugs
5. Bleaching and Dyeing
6. Bone dust
7. Cement
8. Chemical, petrochemical and electrochemicals including manufacture of acids such as Sulphuric Acid, Phosphoric Acid, etc.
9. Chlor-alkali industry
10. Chlorates, perchlorates and peroxides
11. Chlorine, fluorine, bromine, iodine and their compounds
12. Coke making, coal liquefaction, coaltar distillation or fuel gas making
13. Composite woollen mill including dewaxing of raw wool
14. Copper smelter
15. Distillation of wood

16. Distillery including Fermentation industry
17. Dry cell battery manufacturing
18. Dry coal processing/Mineral processing industries like ore sintering palletization, etc.
19. Dyes and Dye-intermediates
20. Explosives including detonators, fuses, etc.
21. Fermentation industry including manufacture of yeast, beer, etc.
22. Fertiliser (Basic) (excluding formulation)
23. Fire crackers
24. Fish feed and poultry feed
25. Food and food processing (excluding Home Scale—of Rs. 2 lakh capital investment, or 5 tpm production)
26. Glass and fibre glass production and processing
27. Glue and gelatine
28. Gold and Silver smithy (above 5 workers)
29. Hydrocyanic acid and its derivatives
30. Incineration plants
31. Industrial carbon including electrodes and graphite blocks, activated carbon, carbon black, etc.
32. Industrial or inorganic gases
33. Industry or process involving electroplating operations
34. Industry or process involving foundry operations
35. Industry or process involving metal treatment or process such as pickling, surface coating (excluding manual brush painting), paint banking, paint stripping, heat treatment, phosphating or finishing etc.
36. Iron and Steel Industries
37. Lead processing and battery reconditioning & manufacturing (including lead smelting)
38. Lime manufacturing
39. Lubricating oils, greases or petroleum-based products
40. Milk processing and dairy products (Integrated Project)
41. Oil refinery (Mineral oil or Petro-refineries)
42. Paints and varnishes (excluding blending/mixing)
43. Pesticides
44. Petrochemicals (Manufacture of and not merely use of as raw material)
45. Petroleum products involving storage, transfer or processing
46. Pharmaceuticals (excluding formulation)

47. Phenolic products
48. Phosphate rock processing plants
49. Phosphorous and its compounds
50. Photographic films and chemicals
51. Pigments and intermediates
52. Potable alcohol (MFL) by blending or distillation of alcohol
53. Power generating plants (excluding D. G. Sets)
54. Processes involving chlorinated hydrocarbons
55. Pulp & Paper (Paper manufacturing with or without pulping)
56. Radioactive elements
57. Refractories
58. Rolling mill
59. Rubber chemicals
60. Shellac process
61. Slaughter houses and meat processing units
62. Spice grinding (above 20 HP motor)
63. Steel and steel products including coke plants involving use of any of the equipment such as blast furnaces, open hearth furnace, induction furnace or arc furnace, etc. or any of the operations or processes such as heat treatment, acid picking, rolling or galvanising etc.
64. Stone crushing (this has been considered under the Orange Category units)
65. Sugar (excluding khandsari)
66. Surgical and medical products involving prophylatics and latex
67. Synthetic detergent (excluding formulation) and soap (with steam boiling)
68. Synthetic fibre including rayon, tyre cord, polyester filament yarn
69. Synthetic resins
70. Synthetic rubber
71. Tanneries
72. Thermal power plants
73. Tobacco products including cigarettes and tobacco processing
74. Tyres and tubes vulcanisation/hot retreading (using boiler)/moulding
75. Vegetable oils including solvent extracted oils, hydrogenated oils
76. Yarn and textile processing involving scouring, bleaching, dyeing, printing or any effluent/ emission generating process
77. Zinc smelter

Industries mentioned in the list above can't be permitted in Municipal areas falling under Kolkata Metropolitan Area (KMA). These can, however, be set up beyond KMA with adequate pollution abatement system subject to site clearance by local bodies.

LIST OF INDUSTRIES UNDER "ORANGE" CATEGORY

1. Automobile servicing and preparing stations
2. Coke briquette (sun-dried)
3. Cotton spinning and weaving
4. Drilling, grinding etc.
5. Engineering and fabrication units
6. Fish processing
7. Flour mills (excluding Domestic Aatta Chakki)
8. Food additives, nutrients and flavours
9. Food including fruits and vegetable processing
10. Fragrances and industrial perfumes
11. Gold and silver smithy (5 workers)
12. Hatchery and piggery products
13. Hotels and Restaurants
14. Instant tea/coffee, coffee processing
15. Laboratory chemicals involving distillation, purification process
16. Laboratory-wares
17. Leather cutting and stitching
18. Malted food
19. Malted food other than brewery
20. Manufacture of mirror from sheet glass
21. Non-alcoholic beverages (soft drinks)
22. NPK Fertilisers/Granulation
23. Organic nutrients
24. Paint (Bail mill)
25. Pesticides/Insecticides/Fungicides/Herbicides/Agrochemical formulation
26. Pharmaceuticals formulation—suggested (plant and machinery cost Rs. 2 lakhs)
27. Pulping and fermenting of coffee beans
28. Saw mill
29. Spice grinding (less than 20 HP motor)

30. Stone crushers—(this may be considered under the Red Category)
31. Surgical and medical products not involving effluent/emission generating processes
32. Synthetic detergents (non-phosphatic)
33. Tyres and tubes hot retreading (without boiler)
34. Tyres and tubes vulcanisation, retreading, moulding
35. Wire drawing (cold process) and bailing straps.

\# Industries mentioned in the above list can be permitted in all Municipal areas other than KMC and HMC or within Industrial Estates in case of KMC & HMC area with adequate pollution control measures subject to the conditions that the site clearance should be obtained from the Municipal authorities for setting up of new units.

LIST OF INDUSTRIES UNDER "GREEN" CATEGORY

1. Aatta chakkis
2. Apparel making
3. Assembly of air coolers, conditioners
4. Assembly of bicycles, baby carriage and other small non-motorised vehicles
5. Assembly of domestic electrical appliances
6. Bakery products, biscuits and confectionery
7. Bamboo and cane products (only dry operations)
8. Block making for printing
9. Candles
10. Cardboard or corrugated box and paper products (excluding paper or pulp manufacturing)
11. Carpentry (excluding saw mill)
12. Carpet weaving
13. Chilling paints and cold storage
14. Cotton and woollen hosiery
15. Dal mills
16. Electronic and electrical goods
17. Electronic equipment (Assembly)
18. Footwear (rubber)—(excluding the use of boiler)
19. Fountain pens
20. Furniture (wooden and steel)—(without spray painting)
21. Garments stitching, tailoring
22. Gold & Silver smithy (5 workers)
23. Gold & Silver thread zari work

24. Groundnut decorticating (dry)
25. Handloom weaving
26. Ice-cream or ice-making
27. Insulation and other coated papers (paper or pulp manufacturing excluding)
28. Jobbing and machining
29. Leather footwear and leather products excluding tanning and hide processing
30. Manufacture of steel trunks and suitcases
31. Medical and surgical instruments
32. Mineralised water
33. Musical instruments manufacturing
34. Oil ginning/expelling (no hydrogenation/refining)
35. Optical frames
36. Paint (by mixing process only)
37. Paper pins and U-clips
38. Plastic processed goods
39. Ploythene, plastic and P.V.C. goods through extrusion moulding
40. Powerlooms/handlooms (without dyeing & bleaching)
41. Printing press
42. Radio assembling
43. Rice mills
44. Rope (cotton and plastic)
45. Scientific and mathematical instruments
46. Shoelace manufacturing
47. Steeping and processing of grains
48. Sports goods
49. Supari (Betelnut) and masala grinding
50. Tailoring and garment making
51. Thermometer making
52. Toys
53. Washing of used sand by hydraulic discharge
54. Water softening and demineralised plants

List of industries that can be permitted in any area with adequate pollution control measures subject to site clearance by local body.

The following declearation made by WBPCB on June 2002:

You have enabled us to perform

- Entire state under Air Act.
- Consent administration for small scale industries considerably decentralised through General Managers of District Industries Centres.
- New private non-commercial vehicles in CMA made to conform to EURO I (India 2000) norms with effect from 01.11.1999.
- All new vehicles in the state made to conform to EURO I (India 2000) norms with effect from 1.04.2000.
- Unleaded petrol (containing 0.013% lead) made available from 01.02.2000 all over the state.
- Sulphur content in diesel reduced to 0.25% throughout the state by 01.02.2000.

You have given us the strength to propose

- Clear siting policy for entry level restriction in congested and critically polluted areas.
- Decentralisation of large and medium green industries consent through Department of Commerce and Industries.
- Stricter emission standards for boilers in Kolkata and for significant polluting clusters in Kolkata and other critically polluted areas.
- Agreements with some gross polluting units and cluster of small industries for going beyond compliance.
- Introduction of EURO II (Bharat Stage II).
- Reduction of sulphur content of automobile fuel to 0.05% in KMA.
- Improved Inspection & Maintenance for in-use vehicles.

website: www.wbpcb.gov.in

52 NATIONAL CONSERVATION STRATEGY AND POLICY STATEMENT ON ENVIRONMENT AND DEVELOPMENT

52.1 PREAMBLE

1. The survival and well-being of a nation depend on sustainable development. It is a process of social and economic betterment that satisfies the needs and values of all interest groups without foreclosing future options. To this end, we must ensure that the demand on the environment from which we derive our sustenance, does not exceed its carrying capacity for the present as well as future generations.

2. In the past, we had a great tradition of environmental conservation which taught us to respect nature and to take cognisance of the fact that all forms of life—human, animal and plant—are closely interlinked and that disturbance in one give rise to an imbalance in other's. Even in modern times, as is evident in our constitutional provisions and environmental legislation and planning objectives, conscious efforts have been made for maintaining environmental security along with developmental advances. The Indian Constitution has laid a new important trail in the Section on Directive Principles of State Policy by assigning the duties for the State and all citizens through *Article 48 A* and *Article 51 A(g)* which state that the "State shall endeavour to protect and improve the environment and to safeguard the forests and wildlife in the country" and "to protect and improve the natural environment including forests, lakes and rivers and wildlife and to have compassion for the living creatures."

3. Nevertheless, over the years, there has been progressive pressure on the environment and the natural resources, the alarming consequences of which are becoming evident in increasing proportions. These consequences detract from the gains of development and worsen the standard of living of the poor who are directly dependent on natural resources. It is in this context that we need to give a new thrust towards conservation and sustainable development.

4. The National Conservation Strategy and the Policy Statement on Environment and Development are in response to the need for laying down the guidelines that will help to weave environmental considerations into the fabric of our national life and of our development process. It is an expression of our commitment for reorienting policies and action in unison with the environmental perspective.

52.2 ENVIRONMENTAL PROBLEMS: NATURE AND DIMENSIONS

1. Environmental problems in India can be classified into two broad categories:

(a) those arising as negative effects of the very process of development; and

(b) those arising from conditions of poverty and underdevelopment.

The first category has to do with the impact of efforts to achieve rapid economic growth and development and continuing pressures of demand generated by those sections of society who are economically more advanced and impose great strains on the supply of natural resources. Poorly planned developmental projects are also often environmentally destructive. The second category has to do with the impact on the health and integrity of our natural resources (land, soil, water, forests, wildlife, etc.) as a result of poverty and the inadequate availability, for a large section of our population, of the means to fulfil basic human needs (food, fuel, shelter, employment, etc.). Needless to say, the two problems are interrelated.

2. Population is an important resource for development, yet it is a major source of environmental degradation when it exceeds the threshold limits of the support systems. Unless the relationship between the multiplying population and life support systems can be stabilised, development programmes, however innovative, are not likely to yield the desired results. It is possible to expand the 'carrying capacity' through technological advances and spatial distribution. But neither of these can support unlimited population growth. Although technological progress will add to the capabilities for sustaining a large number of population, the need for a vigorous drive for population control can hardly be overemphasised in view of the linkage between poverty, population growth and the environment.

3. Even today, over 250 million children, women and men suffer from undernutrition. The scenario for the coming years is alarming and we are likely to face a food crisis unless we are in a position to increase crop and animal productivity on a continuing basis, since the only option open to us for increasing production is productivity improvement. Also, access to food will have to be ensured through opportunities for productive employment.

4. A growth in domesticated animal population has been accompanied by a loss of area under grasslands and pastures. Hardly 3.5 per cent of our geographical area is under grasslands, while our domesticated animal population numbers nearly 500 million. The livelihood security of majority of our people depends on land and water based occupations such as crop and animal husbandry, forestry and fisheries.

5. Out of the total area of India of about 329 million hectares, 175 million hectares of land require special treatment to restore such land to productive and profitable use. The degradation is caused by water and wind erosion (150 million ha), salinity and alkalinity (8 million ha) and river action and other factors (7 million ha).

6. Our forest wealth is dwindling due to overgrazing, overexploitation—both for commercial and household needs, encroachments, unsustainable practices including certain practices of shifting cultivation and developmental activities such as roads, buildings, irrigation and power projects. The recorded forest cover in the country is 75.01 million ha which works out to 19.5% of the total geographical area against the broad national goal of 33% for the plain areas and 66% for hilly regions. Even within this area, only 11% constitute forests with 40% or more of crown cover. According to the State of Forest Report, 2001, the actual forest cover in the country was 76.52 million hectares during 1998. The loss of habitat is leading to the extinction of plant, animal and microbial species. According to the Botanical and Zoological Surveys of India, over 1,500 plant and animal species are in the endangered category. The biological impoverishment of the country is a serious threat to sustainable advances in biological productivity. Gene erosion also erodes the prospects for deriving full economic and ecological benefits from recent advances in molecular biology and genetic engineering.

7. Our unique wetlands, rich in aquatic and bird life, providing food and shelter as also the breeding and spawning ground for the marine and fresh water fishes, are facing problems of pollution and over-exploitation. The major rivers of the country are also facing problems of pollution and siltation. Our long

coastline (7,500 km) is under similar stress. Our coastal areas have been severely damaged due to indiscriminate construction near the water-line. Coastal vegetation including mangroves and sea grasses is getting denuded. Our mountain ecosystems are under threat of serious degradation. Extensive deforestation leading to the erosion of valuable topsoil is threatening the livelihood security of millions of hill people. Equally serious is the downstream effects of the damage done upstream. Indo-Gangetic agriculture, often described as a 'potential bread basket' in the world is being damaged beyond repair as a result of soil degradation. Some areas are facing problems of water-logging and rising water tables because of poorly planned and ill-executed irrigation. In other areas, the water table is receding because of overexploitation of ground water. Furthermore, the quality of groundwater is being affected due to chemical pollution and in coastal areas, due to the ingress of sea water. The excessive use of fertilisers and pesticides impose threat to human health, to the genetic stocks and reduces the natural soil fertility in the long run. The absence of an integrated land and water use policy for the country is taking a heavy toll on these basic natural assets.

8. Coral reefs are the most productive marine ecosystems and provide habitat for diverse flora and fauna. These ecosystems are adversely affected by indiscriminate exploitation of coral for production of lime, recreational use and for ornamental trade. Similarly, the fragile environs of island ecosystems have been subjected to pressures of various forms including migration of people from the mainland.

9. Global atmospheric changes resulting in altered temperature and precipitation and rising ocean levels are no longer within the realm of mere theoretical possibilities. Combination of local subsidence, greenhouse induced sea-level rise and coastal environmental degradation may lead to periodic floods, incursion of salt water, melting of glaciers and river flooding. Local changes of average rainfall will severely affect agriculture and water supply, especially in semi-arid areas.

10. Compounding these human-inflicted wounds on natural ecosystems and life-support mechanisms, we are facing serious problems of pollution and unsanitary conditions especially in urban areas. Pollution arising from toxic wastes and non-biodegradable consumer articles is tending to increase.

11. Lack of opportunities for gainful employment in villages and the ecological stresses is leading to an ever increasing movement of resource poor families to towns. Megacities are emerging and urban slums are expanding. Illiteracy and child labour are persisting. There has been a substantial urban growth in the last four decades. This has resulted in congestion and squatter settlements with millions of people having no access to the basic needs of civic amenities. The green cover in our urban centres has been largely destroyed and once beautiful garden cities have become concrete jungles. The man-made heritage in India has been often gravely and even irrevocably damaged.

12. A large number of industries and other development projects have been incorrectly sited, leading, on the one hand, to over-congestion and over-pollution in our urban centres and, on the other hand, to diversion of population and economic resources from the rural areas. This has also resulted in the pollution of most of our water bodies which are major constituents of our life support systems. Pollution of water bodies, in turn, has adversely affected the growth of aquatic fauna and flora which in an environmentally undesirable phenomenon for any ecosystem. The problems of women in villages is compounded in this whole scenario of energy, environmental and developmental imbalance. The incidence of malaria is high in many parts of the country. Safe drinking water is still a luxury in many villages. Liver ailments and gastro-intestinal diseases are common due to unclean drinking water.

13. It is difficult to clearly delineate the causes and consequences of environmental degradation in terms of simple one-to-one relationships. The causes and effects are often interwoven in complex webs of social, technological and environmental factors. For instance, from a purely scientific and technological

standpoint, soil erosion would result from the cultivation of marginal lands. However, from the point of view of a comprehensive environmental impact analysis, it is important to go further back and analyse the circumstances that force people to cultivate marginal lands. Viewed in this light, it becomes clear that a concern for the environment is essentially a desire to see that national development proceeds along rational, sustainable lines. Environmental conservation is, in fact, the very basis of all development.

14. The overriding impact of adverse demographic pressures on our resources and ecosystems due to poverty and overpopulation of man and livestock has to be highlighted. Unless there is a curb on population growth and even a reduction of such populations and a corresponding improvement in land use policies, the current trend of overexploitation and ecological degradation is not likely to improve.

15. Thus, we are faced with the need for accelerating the pace of development for alleviation of poverty which is, to a great extent, responsible for many of our environmental problems. On the other hand, we have to avoid proceeding along paths with environmental costs so high that these activities cannot be sustained. Development has to be sustainable and all round, whether for the poor or the not so-poor or for the village folk or for the town people. The development models followed so far need to be reviewed.

52.3 ACTIONS TAKEN

In recognition of the felt need for environmental protection, various regulatory and promotional measures have been taken in our country over the past twenty years. These include:

52.4 LEGAL

— The Wildlife (Protection) Act, 1972, amended in 1983, 1986 and 1991.

— The Water (Prevention and Control of Pollution) Act, 1974, amended in 1988.

— The Water (Prevention and Control of Pollution) Cess Act, 1977, amended in 1991.

— The Forest (Conservation) Act, 1980, amended in 1988.

— The Air (Prevention and Control of Pollution) Act, 1981, amended in 1988.

— The Environment (Protection) Act, 1986.

— The Motor Vehicles Act, 1938, amended in 1988.

— The Public Liability Insurance Act, 1991.

— A Notification on Coastal Regulation Zone, 1991.

52.5 INSTITUTIONS

Department of Environment in 1980 and the integrated Ministry of Environment & Forests in 1985, Department of Science and Technology, Department of Agriculture and Cooperation, Department of Biotechnology, Department of Ocean Development, Department of Space, Department of Non-Conventional Energy Sources, Energy Management Centre, Council of Scientific and Industrial Research etc. at the Centre, Departments of Environment at the State and Union Territory level:

— Central Pollution Control Board and State Pollution Control Boards.

— Central Forestry Board.

— Indian Council of Forestry Research and Education with specialised institutions for research in arid zone, forestry, moist and deciduous forests, wood technology, genetics and tree breeding and deciduous forests.
— Forest Survey of India (FSI) and the Wildlife Institute of India (WII) in addition to the existing organisations like Botanical Survey of India (BSI) and Zoological Survey of India (ZSI).
— National Landuse and Wasteland Development Council.
— National Wastelands Development Board.
— Indian Board of Wildlife.
— National Museum of Natural History, Centre for Environmental Education, Institute for Himalayan Environment and Development and Centres of Excellence in specialised subject areas are among the various institutions set up.

52.6 PREVENTION AND CONTROL OF POLLUTION

— Water and air quality monitoring stations in selected areas.
— Use-based zoning and classification of major rivers.
— Notification and enforcement of standards for polluting industries through the Central and State Pollution Control Boards.
— Rules for manufacture, storage, transportation and disposal of hazardous substances.
— On-site and off-site emergency plans for preparedness against chemical accidents.
— Fiscal incentives for installation of pollution control devices.
— Ganga Action Plan to prevent pollution of the river and restore its water quality which could be expanded to cover other major river systems subject to availability of resources.

Identification of critically polluted areas and of highly polluting industries.

52.7 CONSERVATION OF FORESTS AND WILDLIFE

Adoption of a new Forest Policy (1988) with the principal aim of ensuring ecological balance through conservation of biological diversity, soil and water management, increase of tree cover, meeting the requirements of the rural and tribal population, increase in the productivity, efficient utilisation of forest produce, substitution of wood and people's involvement for achieving these objectives.

— Under the Forest (Conservation) Act, 1980, stringent provisions for preventing diversion of forest land for any other purpose.
— Setting up of the National Wastelands Board to guide and oversee the wastelands development programme by adopting a mission approach for enlisting people's participation, harnessing the inputs of science and technology and achieving interdisciplinary coordination in programme planning and implementation.
— Formulation of a National Wildlife Action Plan.
— An exercise for preparation of a National Forestry Action Programme.
— Establishment of National Parks and Sanctuaries covering about 4% of the country's area.
— Eco-development plans for sanctuaries and National Parks.

— Identification of biogeographical zones in the country for establishing a network of protected areas including seven Biosphere Reserves set up so far.

— Management Plans for identified wetlands, mangrove areas and coral reefs.

— Formulation of a National River Action Plan.

52.8 LAND AND SOIL

— Surveys by the All India Soil and Land-Use Survey Organisation. Treatment of catchment in selected river valley projects and integrated watershed management projects in catchment of flood prone rivers.

— Assistance to States to control shifting cultivation.

— Assistance for reclamation and development of ravine areas.

— Drought prone areas programme.

— Desert development programme.

52.9 ENVIRONMENTAL IMPACT ASSESSMENT

— Establishment of procedures for environmental impact assessment and clearance with regard to selected types of projects requiring approval of the Government of India.

— Prior clearance of projects requiring diversion of forests for non-forest purposes under the Forest (Conservation) Act, 1980.

— Formulation of Environmental guidelines for projects in various sectors.

52.10 OTHER ACTIVITIES

— Eco-Task Forces of ex-servicemen for ecological restoration through afforestation and soil conservation.

— National Environmental Awareness Campaigns for creating environmental awareness through non-governmental organisations.

— Surveys and research studies.

— Training programmes, workshops and seminars for building up professional competence and for creation of awareness.

52.11 CONSTRAINTS AND AGENDA FOR ACTION

1. The modest gains made by the steps taken during the past few years leave no room for complacency when viewed in the context of enormous challenges. We can meet the challenges only by redirecting the thrust of our developmental process so that the basic needs of our people are fulfilled by making judicious and sustainable use of our natural resources. Conservation, which covers a wide range of concerns and activities, is the key element of the policy for sustainable development. Framing a conservation strategy is, therefore, an imperative first step. Development requires the use and modification of natural resources; conservation ensures the sustainability of development for the present and in the future. The conservation strategy is to serve as a management guide for integrating environmental concerns with developmental imperatives.

2. The primary purpose of the strategy and the policy statement is to reinforce our traditional ethos and to build up a conservation society living in harmony with Nature and making frugal and efficient use of resources guided by the best available scientific knowledge.

3. The agenda for action in this regard will include:

— to ensure sustainable and equitable use of resources for meeting the basic needs of the present and future generations without causing damage to the environment;

— to prevent and control future deterioration in land, water and air which constitute our life-support systems;

— to take steps for restoration of ecologically degraded areas and for environmental improvement in our rural and urban settlements;

— to prevent further damage to and conserve natural and man-made heritage;

— to ensure that development projects are correctly sited so as to minimise their adverse environmental consequences;

— to ensure that the environment and productivity of coastal areas and marine ecosystems are protected;

— to conserve and nurture the biological diversity, genepool and other resources through environmentally sustainable development and management of ecosystems, with special emphasis on our maintain, marine and coastal, desert, wetlands, riverine and island ecosystems; and

— to protect the scenic landscapes, areas of geomorphological significance, unique and representative biomes and ecosystems and wildlife habitats, heritage sites/structures and areas of cultural heritage importance.

4. To address the above stated agenda, the instruments for action will include:

— to carry out environmental impact assessment of all development projects right from the planning stage and integrate it with their cost-benefit considerations. Appropriate costs of environmental safeguards and regeneration would continue to form an integral part of the projects;

— to ensure that all projects above a certain size and in certain ecologically sensitive areas should require compulsory prior environmental clearance;

— to incorporate environmental safeguards and protection measures, in policies, planning, site selection, choice of technology and implementation of development projects like agriculture, water resource development, industry, mineral extraction and processing, energy, forestry, transport and human settlements;

— to encourage research, development and adoption of environmentally compatible technologies; and to promote application of the modern tools of science and technology for conservation, bridging of large gaps in supply and demand as well as control and monitoring of natural resources;

— to elicit and ensure participation of people in programmes for environmental improvement and for integrating the environmental concerns in planning and implementation of development programmes;

— to create environmental cons iousness through education and mass awareness programmes;

— to aim at moderation of process of demand unleashed by the developmental process itself by taking measures to recycle waste materials and natural resources, conserve energy, conserve use of natural resources in industrial products by measures like wood substitution and generally try to reach moderations in life styles consistent with sustainability and human dignity;

— to develop appropriate organisational structures and a pool of professional manpower to serve as the cadre for environmental management service; and

— to effectively implement the various environmental laws and regulations for environmental protection through creation or strengthening of the requisite enforcement machinery.

52.12 PRIORITIES AND STRATEGIES FOR ACTION

1. Population Control

Unabated population growth (1.93 per cent in 2001), as at present, not only adds to the economic burden for all developmental activities, but also reduces the impact of economic growth on our society. Therefore, for the success of our planning, population control becomes the most urgent necessity. A comprehensive programme, with strong political backing and appropriate socio-economic measures, fully utilising the available scientific know-how, simultaneously making efforts for developing new methodologies and supported by modern communication technology and managerial and organisational skills, is essential for success in this most difficult area. Population control should be a national mission for the next decade. Despite efforts of several years, population control projects have not met with success. More stern measures such as legislative and better incentives are needed.

Alongwith the development programmes to improve the living conditions, action must be directed towards stabilisation of population including the following measures:

— Launching a time-bound national campaign for population stabilisation with the small family as a socially responsible objective;

— Increased support for female education, female employment and of social security programmes;

— Easier access to the means of family planning and health care facilities;

— Added incentives in terms of taxation and other benefits for family planning;

— Environmental sanitation, prevention and control of communicable diseases through integrated vector control and health education; and

— Adoption of decentralised renewable energy devices that enhance quality of life in remote pockets while taking special care of the health needs of women.

2. Conservation of Natural Resources

2.1 Land and Water

2.1.1 An integrated land and water management approach is extremely important to sustain the food production, animal husbandry and other activities.

2.1.2 Amelioration of water-logged and salt affected lands, command area development, protection of good agricultural land against diversion to urban and other uses, prevention of land fragmentation, maintenance of sustained productivity of soil and conservation of lands with forests and vegetal cover are the integral components of sustainable management.

2.1.3 The importance of water as a finite, though a renewable resource, must be clearly recognised. Land and water use are to be considered together, particularly in the context of recurring droughts and floods. Water conservation measures; discipline on use of water: economising the consumption of water in households, agriculture and industry; and appropriate recycling would be essential.

2.1.4 The steps to be taken for sustainable use of land and water should include the following:

— Classification, zoning and apportionment of land for designated uses such as agriculture, forestry, grassland, green areas, industrial activities, catchment areas and watersheds and human settlements based on assessment of their capabilities and environmental considerations;

— Enactment of laws for appropriate land uses to protect the soil from erosion, pollution and degradation;

— Protection of land near water bodies and prevention of construction thereupon;

— Measures to ensure equitable access to and responsibility for sustainable use of land and water resources;

— Micro-level planning to develop appropriate methodology and implementation of action plan by involving the people at the village level in social forestry programmes, landuse planning, afforestation etc.;

— Countrywide campaign to minimise soil and run-off losses by carrying out extensive works like contour trenching, contour bunding, terracing, construction of small storages, catchment treatment and protection of the vegetal cover in the catchments and watersheds. This is to be a specific charge of project authorities in all irrigation, power, road and agricultural projects;

— Restoration and reclamation of degraded areas including weed-infested areas, mined areas, grazing lands and salt affected soils;

— Measures for preventing wind erosion by undertaking special programmes of conservation and afforestation in desert areas;

— Development of suitable agro-silvipastoral techniques with special emphasis on hilly areas and in arid and semiarid zones;

— Building up a network for assessment and monitoring of soil and water (surface and ground water) quality throughout the country which should be on a permanent basis as in the case of meteorological stations;

— Measures for water conservation, recycling and optimal conjunctive use of surface and ground water for specific uses;

— Legislative measures to check overexploitation of surface and ground water for various uses;

— Conservation of wetlands for ensuring sustainable ecological and economic benefits;

— Encouragement to and improvement in traditional methods of rain water harvesting and storage;

— Stringent measures for prevention and control of pollution due to indiscriminate disposal of solid wastes, effluents and hazardous substances in land and water courses;

— Control and abatement of pollution of water bodies from municipal and industrial wastes generated from urban habitats by intercepting and diverting such wastes away from water bodies;

— Classification, zoning and regulations for maintaining the quality of the water bodies to protect and enhance their capabilities to support the various designated uses; and

— Adoption of low cost sanitation technology for prevention and control of pollution in water courses.

2.2 Atmosphere

For prevention and control of atmospheric pollution including noise pollution, the thrust will be on:

— Use of clean fuels and clean technologies, energy efficient devices and air and noise pollution control systems;

— Setting up of source specific and area-wise air quality standards and time bound plans to prevent and control pollution;

— Proper location of projects to minimise the adverse impact on people and environment;

— Incentives for environmentally benign substitutes, technologies and energy conservation;

— Raising of green belts with pollution tolerant species;

— Developing coping mechanisms for future climatic changes as a result of increased emission of carbon dioxide and greenhouse gases; and

— Appropriate action to control adverse impact on Indian continent due to ozone depletion and other gaseous effects in the atmosphere at global level.

2.3 Biodiversity

About 90% of the world-food comes from 20 plant species. The plant breeders find that they have to turn more and more to the wild species to introduce into the cultivated forms desired qualities of resistance to pests and diseases and the ability to withstand adverse soil and weather conditions. India's biological diversity is very rich but unfortunately its wealth is being eroded due to various reasons. This diversity needs to be preserved and the immediate task will be to devise and enforce time bound plans for saving the endangered plant and animal species as well as habitats of biological resources.

Action for conservation must be directed to:

— Intensification of surveys and inventorisation of biological resources in different parts of the country including the island ecosystems. The survey should include information on distribution pattern of particular species/population communities and the status of ethnobiologically important groups;

— Conservation of biodiversity through a network of protected areas including Biosphere Reserves, Marine Reserves, National Parks, Sanctuaries, Gene Conservation Centres, Wetlands, Coral Reefs and such other natural habitats of biodiversity. This should include taxonomic and ecological studies on the flora and fauna with adequate emphasis placed on the lower vertebrate, invertebrate and microflora which are important in contributing to the healthy maintenance of ecosystems;

— Full and correct rehabilitation of rural poor tribals displaced due to creation of national parks/biosphere reserves/tiger reserves;

— Conservation of microfauna and microflora which help in reclamation of wastelands and revival of biological potential of the land;

— Protection and sustainable use of plant and animal genetic resources through appropriate laws and practices;

— Protection of domesticated species/varieties of plants and animals in order to conserve indigenous genetic diversity;

— Maintenance of corridors between national parks, sanctuaries, forests and other protected areas;

— Emulation and support for protecting traditional skills and knowledge for conservation;

— Development of methodologies to multiply, breed and conserve the threatened and endangered species through modern techniques of tissue culture and biotechnology;

— Discouragement of monoculture and plantation of dominating and exotic species, in areas unsuited for them and without sufficient experimentation; and

— Restriction on introduction of exotic species of animals without adequate investigations.

2.4 BIOMASS

For the vast majority of our rural people, the foremost need is for fuelwood, timber, fodder, fibre, etc. The issue of sustainable resource utilisation, therefore, has to be specially addressed first from the point of view of the biomass requirements of the rural poor. Action must be directed to:

— Devising ways and means by which local people can conserve and use thereafter the resources of the common lands and degraded forests, so that they have a stake in the continuing productivity of the resources;

— Encouraging private individuals and institutions to regenerate and develop their wastelands;

— Raising of fuelwood species and provision of alternatives to reduce dependence on fuelwood;

— Taking measures to increase the production of fodder and grasses to bridge the wide gap between supply and demand;

— Raising of bamboo and species providing small timber for local house-construction and agricultural implements;

— Increasing biomass to meet essential requirement of biomass based industry;

— Promoting direct relationship between forest-based industry and farmers to raise needed raw materials, provided this does not result in diversion of prime agricultural lands and displacement of small and marginal farmers;

— Extensive research and development in forestry for better regeneration and improved productivity;

— Development of technologies for enhancing the productivity and efficiency of use of all biomass resources (both terrestrial and marine);

— Institutional and technological systems to enable rural artisans to sustain biomass based crafts; and,

— Curtailment of the supply of subsidised biomass resources to industrial consumers.

52.13 DEVELOPMENT POLICIES FROM ENVIRONMENTAL PERSPECTIVES

Implementation of the aims and objectives of conservation and sustainable development will require integration and internalisation of environmental considerations in the policies and programmes of development in various sectors.

Curtailment of consumerism and shift towards use of environment-friendly products and processes and low waste generating technologies through conscious efforts and appropriate economic policies including pricing of natural resources as well as fiscal incentives and disincentives will be the guiding factors for ensuring conservation and sustainable development.

For environmental conservation and sustainable development, the steps which need to be taken in some of the key sectors of development activities are outlined in the following sections.

1. Agriculture and Irrigation

For sustainable management of agriculture and irrigation, the action points should include:

1.1 **Agriculture**

— Development of pesticides and insecticides policy for the country;

— Development of integrated pest management and nutrient supply system;

— Development and promotion of methods of sustainable farming, especially organic and natural farming;

— Efficient use of inputs including agro-chemicals with minimal degradation of environment;

— Phasing out and stoppage of persistent and toxic pesticides and their substitution by environmentally safe and appropriate pesticides;

— Promotion of environmentally compatible cropping practices, biofertilisers and biopesticides,

— Restriction on diversion of prime agricultural land for other purposes;

— Ensuring land for different uses based upon land capability and land productivity;

— Evolving cost effective and efficient methods of water conservation and use;

— Incentives for cultivation of crops with high nutritive value and those with lesser demands on water and energy inputs;

— Encouraging crop rotation patterns;

— Strengthening of local bodies like Zilla Parishads, Panchayats and Samitis to ensure effective decentralisation and optimal resource managements; and

— Anticipatory programmes and contingency plans for disasters such as drought, flood and climate change.

1.2 **Irrigation**

— Priority to small projects to meet the requirements of irrigation without causing significant alteration in the environmental conditions;

— Revival of traditional water management systems and development of alternate irrigation systems such as harvesting and conservation of run-off rain water;

— Measures for increasing the efficiency of wateruse, water conservation and recycling;

— Measures for provision of drainage as an integral component of irrigation projects and to prevent water logging and leaching;

— Watershed management through catchment treatment of the drainage areas, protection of vegetal cover and measures to prevent siltation in an integrated manner with the irrigation authorities being fully accountable; catchment treatment would be so designed as to have a direct impact on the life of the reservoir, hydrological regime and life support systems. It would depend on the location specific conditions in each case;

— Adoption of command area development approach for all irrigation projects to ensure optimal utilisation;

— Critical assessment of irrigation projects and delivery systems to ensure optimal utilisation of water resources along with measures to mitigate environmental and social damage;

— Focus on decentralised network of small irrigation and water projects with minimum environmental disruption which will be of great value to local communities and yet capable of generating surplus for other areas at low cost;

— Design and implementation of irrigation projects which are environmentally sustainable, based on lessons learnt from a critical analysis of all past projects; and

— Continuous and ongoing evaluation and monitoring of all projects.

2. Animal Husbandry

The activities relating to animal husbandry should concentrate on:

— Development of an animal husbandry policy for the country;

— Intensification of sterilisation programme for containing unsustainable growth in livestock population;

— Improvement in genetic variability of indigenous population;

— Distribution of animals like goats under the Integrated Rural Development Programme strictly consistent with the availability of pasture lands to reduce pressure on the lands;

— Propagation of wildlife and wildlife resources management on sustainable basis;

— Selective breeding of animals used for draught power to conserve fuel;

— Promotion of stall feeding and rotational grazing;

— Restoration and protection of grazing lands;

— Involvement of local people in the policy planning on pasture lands and stall feeding to avoid fodder scarcity; and

— Incentive for growing fodder crops and establishment of fodder banks.

3. Forestry

Concerted efforts should be made for raising the forest cover and for conservation of existing forests which constitute an essential life support system and an important source of food, fibre, fodder, fuel and medicines etc. For attaining the goal of having at least one third of our land area under forest cover, intensified measures on a mission mode are required to be taken along with commensurate mobilisation of resources for this purpose. As outlined in the National Forest Policy (1988), the action points should include the following:

— Maintenance of environmental stability through preservation and, where necessary, restoration of the ecological balance that has been adversely disturbed by serious depletion of the forests of the country;
— Conserving the natural heritage of the country by preserving the remaining natural forests with the vast majority of flora and fauna, which represent the biological diversity and genetic resources of the country;
— Increasing substantially the forest/tree cover in the country through massive afforestation and social forestry programmes, especially on all denuded, degraded and unproductive lands involving the local people in this endeavour by giving them tangible economic motives and employment opportunities;
— Meeting the rights and concessions for requirements of fuelwood, fodder, minor forest produce and small timber of the rural and tribal population with due cognisance of the carrying capacity of forests;
— Increasing the productivity of forests to meet the essential national needs;
— Encouraging efficient utilisation of forest produce;
— Restriction on diversion of forest lands for non-forest uses and compensatory afforestation in case where diversion is unavoidable;
— Afforestation on common lands by the local communities through usufruct-sharing schemes;
— Motivation of farmers/land owners to resort to tree farming in similar manner of crop based farming;
— Substitution of wood by other materials, alternative sources of energy and fuel efficient stoves;
— Permission to forest-based enterprises after a thorough scrutiny regarding the availability of raw materials;
— Supply of forest produce to the industrial consumers only at its true market value and not at concessional prices;
— Involvement of local people and dedicated grassroot non-governmental organisations in the afforestation programme and for protection of existing forests; and
— Creation of land banks for compensatory afforestation.

52.14 ENERGY GENERATION AND USE

For prevention and control of pollution and environmental hazards in energy generation and use as also for encouraging popularisation of environmentally benign energy systems, the following measures should be taken:

— Environmental impact assessment prior to investment decisions and site selection; choice of practicable clean technologies for energy production and processes including waste utilisation, treatment and disposal of solid wastes, effluents and emissions;
— Location of energy generation projects based on environmental considerations including pollution, displacement of people and loss of biodiversity;

— Decentralised small projects for meeting the rural energy needs and incentives for use of non-conventional energy sources;

— Incentives and punitive measures (including proper pricing) to prevent abuse and to promote the use of energy efficient devices in the production and distribution systems and for energy conservation in all sectors including households, agriculture, industry, power and transportation;

— Concerted efforts for development and propagation of non-conventional renewable energy generation systems; and

— Setting up of biogas plants based on cow-dung, human excreta and vegetable wastes.

52.15 INDUSTRIAL DEVELOPMENT

Environmental considerations should be integrated while encouraging industrial growth. The action points in this regard should include a mix of promotional and regulatory steps:

— Incentives for environmentally clean technologies, recycling and reuse of wastes and conservation of natural resources;

— Operationalisation of "polluter pays principle" by introducing effluent tax, resource cess for industry and implementation of standards based on resource consumption and production capacity.

— Fiscal incentives to small-scale industries for pollution control and for reduction of wastes;

— While deciding upon site, priority to compatible industries so that, to the extent possible, wastes from one could be used as raw material for the other and thus the net pollution load is minimised;

— Location of industries as per environmental guidelines for siting of industry;

— Enforcement of pollution control norms in various types of industrial units depending on their production processes/technologies and pollution potential; particular attention to be paid to highly polluting industries;

— Encouragement for use of environmentally benign automobiles/motor vehicles and reduction of auto-emissions;

— Collective efforts for installation and operation of common effluent treatment facilities in industrial estates and in areas with a cluster of industries;

— Introduction of "Environmental Audit" and reports thereof to focus on environment related policies, operations and activities in industrial concern with specific reference to pollution control and waste management;

— Dissemination of information for public awareness on environmental safety aspects and stringent measures to ensure safety of workers and general population against hazardous substances and processes;

— Preparation of on-site emergency plans for hazardous industries and off-site emergency plans for districts in which hazardous units are located;

— Public liability insurance against loss or injury to life or property;

— Setting up of environment cells in industries for implementing environmental management plans and for compliance of the requisites of environmental laws;

— Internalising the environmental safeguards as integral component of the total project cost;

— Environmental impact assessment from the planning stage and selection of sites for location of industries; and

— Clearance by Ministry of Environment & Forests of all projects above a certain size and in certain fragile areas.

52.16 MINING AND QUARRYING

To prevent and to mitigate environmental repercussions in mining and quarrying operations, action must be directed to:

— Mined area rehabilitation and implementation of the environmental management plans concurrently with the ongoing mining operations to ensure adequate ecological restoration of the affected areas;

— Rehabilitation of the abandoned mined areas in a phased manner so that scarce land resources can be brought back under productive use;

— Laying down of requisite stipulations for mining leases regarding tenure, size, shape and disposition with reference to geological boundaries and other mining conditions to ensure systematic extraction of minerals along with environmental conservation;

— Emphasis on production of value added finished products from mining so as to reduce indiscriminate extraction;

— Upgradation and beneficiation of minerals at the source, to the extent possible in order to ensure utilisation of low-grade mineral resources and to reduce the cost of transportation, processing and utilisation;

— Environmentally safe disposal of the by-products of mining;

— Restriction of mining and quarrying activities in sensitive areas such as hill slopes, areas of natural springs and areas rich in biological diversity;

— Discouraging selective mining of high grade ores and recovery of associated lower-grade ores during mining; and

— Environmental impact assessment prior to selection of sites for mining and quarrying activities.

52.17 TOURISM

To ensure sustainable growth of tourism without causing irreversible damage to the natural environment, activities relating to tourism should take care of:

— Promotion of tourism based on careful assessment of the carrying capacity and support facilities such as transport, fuel, water and sanitation;

— Development of tourism in harmony with the environmental conditions and without affecting the lifestyles of local people; and

— Restriction on indiscriminate growth of tourism and strict regulation of the tourist activities in sensitive areas such as hill slopes, islands, coastal stretches, National Parks and Sanctuaries.

52.18 TRANSPORTATION

For prevention of pollution and for development of environmentally compatible transportation systems, the following steps should be taken:

— Improvement in mass transport system to reduce increasing consumption of fuel, traffic congestion and pollution;

— Improved transport system based on bio-energy and other non-polluting energy sources;

— Rail transport and pipeline transport instead of road transport, wherever possible, by appropriate freight pricing so as to reduce congestion, fuel consumption and environmental hazards;

— Transportation of hazardous substances through pipelines;

— Improvement in traffic flow through proper maintenance of roads, updated traffic regulation and strict enforcement of prescribed standards;

— Enforcement of smoke emission standards for containing vehicular exhausts, at the manufacturer and user level;

— Phasing out the use of lead in motor spirit; and

— Regulations for environmental safety in transportation of hazardous substances.

52.19 HUMAN SETTLEMENTS

To check unplanned growth of human settlements and to ensure a better quality of life for the rural and urban population, the action points should include:

— Creation of gainful employment opportunities and provision for meeting the basic needs through better communications, entertainment, medical and educational facilities in rural areas to check rural-urban migration;

— Decentralisation of urbanisation through establishment of secondary cities and towns with requisite infrastructural services and employment opportunities by developing human settlement perspective plan at National and State level;

— Disincentives for industrial and job location in existing urban centres which have exceeded their carrying capacity;

— Improvement of infrastructural facilities such as water supply, sewerage, solid waste disposal, energy recovery systems and transportation in an integrated manner;

— Promoting the use of indigenous building materials and appropriate construction technologies by revising building and planning codes supporting small scale production, skill upgradation of artisans and people-oriented delivery systems;

— Conservation of heritage sites and buildings, through regulation to ensure that these are not demolished, encroached upon and affected by indiscriminate construction and pollution;

— Stock-taking of buildings, areas, monuments of heritage value in the country;

— Recycling of existing building stock to save green open compounds and save building material;

— Planting of shade giving and fruit bearing and ornamental trees along the roadside, in the compounds of schools, hospitals, Government as well as private office buildings, places of worship, places meant for public fairs, assemblies and markets and the periphery of playgrounds and water bodies;

— Botanical gardens representing the local flora;

— Raising of gardens, parks and open spaces in the towns and cities for public use and for promotion of environmental consciousness;

— Laying down a system for the propagation and protection of urban forestry by assigning responsibility amongst the various authorities;

— Deterrent measures to discourage indiscriminate growth of human settlement and polluting industries in vulnerable areas such as hilly regions and coastal stretches;

— Environmental appraisal of projects related to urban development and regional planning, preparation of environmental/ecodevelopment plans for sensitive regions and subregions for evolving desirable norms and space standards;

— Prevention of environmental health problems and associated communicable and non-communicable diseases by educating people on personal hygiene, sanitation and use of potable water;

— Creation/strengthening of health care facilities for all sections of society both in rural and urban areas; and

— Establishment of monitoring systems and epidemiological data to ensure adequate early warning system for prevention and control of diseases.

52.20 INTERNATIONAL COOPERATION

A major threat to sustainable development has been visualised, in recent years, from environmental problems of a global nature—ozone layer depletion, global warming and climate change, destruction of biological diversity, transboundary air pollution, marine pollution and land-based marine pollution, transboundary movement of hazardous substances. On a philosophical plain, the scientific proof of such problems provides an opportunity to reconsider the development path ushered in by the Industrial Revolution (from about 1760) and the blinkered pursuit of lifestyles which place extreme pressures on the natural resource base. But, at a practical level, it means pressures on developing countries to take measures which they can ill-afford.

It has been India's firm conviction that it is the process of industrialisation and the continued profligacy of industrialised economies that have created the problems which threaten our planet and its lifeforms. Not only do they use up non-renewable natural resources in disproportionate quantities, but create discharge and emissions which disturb delicate balances in ecosystems and atmospheric equilibrium. It is true, of course, that this has not been done consciously or intentionally (except in matters such as dumping of hazardous wastes or the use of nuclear and chemical weapons). Nevertheless, the responsibility is clearly established, as also the need for urgent and effective action, by the developed world, to prevent global disaster. This includes not only direct action, but also indirect measures such as creation of an economic order which helps developing countries to exert less pressure on their own natural resources.

The Indian approach to global environmental problems is generally in keeping with other developing countries and has the following basic elements:

— Our economic development cannot be hampered in the name of the global environment, which we have done nothing to damage and can do little to save. Our resources are required to meet our developmental needs such as education, nutrition, health services, drinking water, housing, sanitation, agriculture, industry, infrastructure, even all of which we find it difficult to provide having been behind in the race for development. Without this development, threats to the environment will, in any case, grow. In the short run, this developmental effort could even add to the discharges and emissions which cause global problems—but these are minuscule compared to the quantities which industrialised countries have already contributed. In any case, such emissions etc. can easily be compensated for a marginal reduction of the same in the industrialised world;

— For environmental protection and improvement, we will do our best with the resources available in the country. With new and additional funding support and transfer of environmentally sound technologies from the developed countries, we will be in a position to augment our capacity to deal with the environmental problems; and,

— Regulatory international regimes can be useful in some areas—such as ozone depletion or even climate change—provided the special situation of developing countries is fully addressed. But in other sectors—such a forestry—such as regime is neither workable nor acceptable. In such sectors, what is required is a reduction of international, economic and commercial pressures which generate unsustainable exploitation and additional financial resources to tackle the damage already done.

India's traditional lifestyle—still followed by a vast majority of its population—has always emphasised conservation of plant and animal life, waste minimisation, recycling, simplicity in food habits and other such environment-friendly attitudes. There is no doubt that with economic development will also come lifestyles which require more intensive resource use. On the one hand, we have to minimise the adverse environmental impacts of development (e.g., through legislation and control, impact assessment and monitoring, education and awareness). On the otherhand, we have to continuously see how far the traditional Indian ethos can be reflected in modern lifestyles. If these efforts are supported by adequate financial resources from the international community, as well as the transfer (and development) of environmentally sound technology. India can contribute significantly to the international action to deal with global environmental problems.

52.21 SUPPORT POLICIES AND SYSTEMS

Implementation of the aims and objectives of environmental policy will need support policies and systems for filling up of the gaps in the existing institutional set up, legislative instruments and enforcement mechanisms, research and development, mobilisation of financial resources, creation of public awareness and training of professionals.

1. Strengthening of Institutions and Legislation

1.1 It will require strengthening of existing institutions at different levels. It will need a close linkage among the compartmentalised sectors which have been historically dealt with by separate organisations. It will call for a change in the institutional mechanism for enlisting public participation. It will necessitate quick decision making on development projects based on assessment of their potential of rendering long term sustainable benefits to the society at large, particularly vulnerable sections. It will also require effective implementation of laws and regulations for environmental protection through strengthening of and closer interaction among the regulatory bodies and administrative machinery.

1.2 Existing laws and enforcement mechanisms should be subjected to periodic review to evaluate their adequacy and efficacy in the light of changed circumstances and experience.

2. Natural Resource Accounting

2.1 As economic policies form the framework for a range of sectoral development, it will be necessary to consider how these policies affect the quality and productivity of environmental resources. This will require a system of resource accounting along with the other exercises of cost-benefit analyses.

2.2 In essence, indicators of growth such as GNP and GDP should include a measure of depletion cost and value judgements in terms of environmental resources. It will require instruments and expertise for evaluation and conscious trade offs, where unavoidable, to meet the legitimate development needs.

2.3 The Government will prepare, each year, a natural resources budget which will reflect the state and availability of resources like land, forests, water etc. and which will rationally allocate these resources in keeping with the principles of conservation and sustainable development.

3. Training and Orientation Programmes

3.1 Available management resources in the enterprises/projects would be oriented towards environmental considerations and expertise to be developed through appropriate training programmes.

3.2 Formal education and training programme in specialised areas of pollution control and environmental management will be a continuing need. For this purpose, intensive programmes for education and training will need to be introduced in the universities, IITs and other professional institutions. Environmental education at the school level including training of teachers shall be an important component of educational programmes.

4. Promoting Environmental Awareness

To raise public awareness and involvement in environmental activities, the mass media ranging from local folklores to electronic media should serve a vital role. To raise public awareness on environmental issues and to promote people's participation in environmental activities and conservation of natural resources, development of environmental education resource material and use of traditional and modern media of communication need to be strengthened. Scope and functions of the existing environmental education centres should be further strengthened and enlarged to develop a network of infrastructure for environmental education including development of orientation centres and provision of education material for visitors at the national parks, sanctuaries and tiger reserves.

5. Promoting Appropriate Environmental Technologies

Existing research and development efforts need to be strengthened to develop the appropriate low cost technologies considering the possibilities, opened up by biotechnology, genetic engineering, information and material technologies and remote sensing, tailored to the local environmental and socio-economic conditions.

6. Rehabilitation of Project Oustees

6.1 While implementing the projects in various sectors, conscious efforts should be made to avoid displacement of local people. Where it is unavoidable, necessary measures should be taken to ensure their rehabilitation by providing suitable facilities.

6.2 The Government will formulate a comprehensive national rehabilitation policy which, apart from other things, ensures that the oustees are economically better off than before and above the poverty line as a result of their rehabilitation.

7. Role of Non-Governmental Organisations

7.1 Implementation of the conservation strategy would be impossible without active participation of the people. Non-Governmental Organisations (NGOs) can play an important role in mobilising the people at grassroots. This will need a network among NGOs and interface between the people and the Government to work on community involvement, providing information on environmental surveillance and monitoring, transmitting development in science and appropriate technology to the people at large.

7.2 Environmental Information Centres should be set up at the district level to generate knowledge regarding traditional and endogenous system management practices. NGOs at the district level should be involved in the management and dissemination of the environmental information.

7.3 Non-Governmental Organisations, citizen groups and village level institutions like forests, panchayats, and Gram Sabhas should be empowered with locus standi and support for mobilisation of public opinion and participation in development activities.

7.4 Managerial capacity of the NGOs should be strengthened. Training programmes for NGOs on regional basis should be organised. An advisory cell for rural NGOs should be made available at all district headquarters.

8. Women and Environment

Women at the grassroot level should be actively involved in the conservation programmes which should be income generating and self-financing and sustainable on a long-term basis and the Government Ministries/Departments should have an NGs cell or at least Liaison Officer for interaction with the NGOs.

9. Partnership Role of Centre and State Governments

Effective implementation of necessary measures, as outlined in the statement, will be facilitated by a partnership role of the Central and State Governments. Many environmental problems assume national significance. Hence, the policies and programmes at the State and Central level should be drawn up keeping in view overall national policy considerations. A monitoring mechanism involving Central and State Government representatives will be set up for interaction as required for implementation of the policy initiatives.

52.22 CONCLUSION

1. It is only through such initiatives—the contours of which have been highlighted in preceding paragraphs—we will be in a position to resolve the conflicts which often arise between the environmental concerns and developmental pursuits that have a direct bearing on the very fabric of our society and lifestyles.
2. The task before us would be daunting if it were not for the many positive factors that are emerging: people's movements to conserve their own environment, greater public and media concern for environmental issues and spread of environmental awareness among the children and youth.
3. It is up to us, as State and citizens, to undertake development processes in keeping with our heritage and the traditional conservation ethos and in harmony with the environmental imperatives of this land.

53 POLICY STATEMENT FOR ABATEMENT OF POLLUTION

53.1 PREAMBLE

The commitment of the Government on abatement of pollution for preventing deterioration of the environment is stated here. The policy elements seek to shift emphasis from defining objectives for each problem area towards actual implementation, but the focus is on the long-term, because pollution particularly affects the poor. The complexities are considerable given the number of industries, organisations and government bodies involved. To achieve the objectives, maximum use will be made of a mix of instruments in the form of legislation and regulation, fiscal incentives, voluntary agreements, educational programmes and information campaigns. The emphasis will be on increased use of regulations and an increase in the development and application of financial incentives.

53.2 THE PROBLEM

1. There is an increasing trend in environmental pollution. Water is polluted by four kinds of substances: traditional organic waste, waste generated from industrial processes, chemical agents for fertilizers and pesticides for crop protection, and silt from degraded catchments. While it is estimated that three-fourths by volume of the waste water generated is from municipal sources, industrial waste—though small in volume—contributes over one-half of the total pollutant load, and the major portion of this is coming from large and medium industries. For Class 1 cities of the country, less than five per cent of the total waste water generated is collected and less than one-fourth of this is treated.

2. Ambient air quality trends in the major cities indicate that levels of suspended particulate matter (SPM) are higher than the prescribed standards or limits, especially in summer months. Levels of nitrogen dioxide are increasing in urban centres with growing vehicle emissions.

3. Environmental problems are becoming larger in scale. The chemical industry generates an increasing quantity of substances every year; adversely affecting essential aspects of the composition of the atmosphere, soil and water. In the industrial high density areas, in addition to the effects on local health and impact on nature, we are confronted with damage to the social and economic functions of the environment.

4. With restrictions on releases to air and waste water, hazardous chemical wastes are getting diverted to land for their disposal. Earlier concerns with pollution that was visible and degradable are giving way to new types of pollution with very small quantities of synthetic chemicals that are not so visible and are injurious to health and damage the environment because of widespread use, persistence and toxicity. Reducing the hazards from toxic chemicals is now a primary public concern.

5. Human activities are also influencing the composition of the atmosphere. Despite uncertainties and insufficient knowledge, political and scientific decisions concerning environmental change will increasingly be necessary.

6. The state of the environment continues to deteriorate. The growth in scientific and technical knowledge has made it possible to use an ever-increasing quantum of natural resources. The increase in population is further enhancing the pressure on the environment. The depletion of forests (In India, forest cover is meagre 20% of land-area, whereas it should be about one-third—33.3%) has been accompanied by increasing amount of pollution affecting atmosphere, soil and water. Some of the damage is irreversible. In seeking a higher quality of life, while developed countries need to focus on changing the composition of their processes and products, developing countries will need to obtain the benefits of economic growth. The policy statement on Abatement of Pollution thus complements the Forest Policy Statement. The Government seeks to ensure that its policies in every sector are based on a set of principles that harmonise economic development and environmental imperatives.

53.3 FUTURE DIRECTIONS AND OBJECTIVES

1. It is not enough for the Government to notify laws which are to be complied with. A positive attitude on the part of everyone in society is essential for the prevention of pollution and wide consultation has been held with those who will ultimately implement the policy.

2. A comprehensive approach is taken to integrate the environmental and economic aspects in development planning; stress is laid on preventive aspects for pollution abatement and promotion of technological inputs to reduce industrial pollutants; and through reliance upon public cooperation in securing a clean environment to respond to the coming challenges.

3. The objective is to integrate environmental considerations into decision-making at all levels. To achieve this, steps have to be taken to:

- prevent pollution at source;
- encourage, develop and apply the best available practicable technical solutions;
- ensure that the polluter pays for the pollution and control arrangements;
- focus protection on heavily polluted areas and river stretches; and
- involve the public in decision-making.

53.4 CRITICALLY POLLUTED AREAS

1. Mechanisms should be evolved to reduce local concentration of pollutants in complex industrial sites. Strategies should be developed for areas with high pollution loads where the accumulative effect of the various types of pollutants would be taken into account including pollution of groundwater. Existing units in these areas will be targeted for effective action. New units in these areas will be required to comply with location-specific standards for stringent environmental quality objectives. These will include matching waste generators with waste buyers, with the objective of solving waste disposal.

2. Setting up of industrial estates and clusters of small industrial units in rural areas will include pollution abatement measures as an essential component of infrastructure. In the past, the absence of adequate provisions of space for installing treatment facilities and arrangements for disposal of wastes led to severe pollution of agricultural land and rivers.

3. There has been a steady increase in the amount of waste water produced from urban communities and industries. In the coming years, due to rapid growth in population, urbanisation, industrial development and better water supply, the amount of waste water may increase manifold. Generally, these waters are discharged into lagoons or dumped on low-lying areas without any pre-treatment, thereby creating sewage pools, contaminating ground waters, salinising good quality lands around cities, acting as a source of foul smell and breeding grounds for mosquitoes and other pathogens. At many places this

waste water is discharged into drains and rivers causing serious water pollution. However, awareness has now grown and more attention is being paid to develop systems to treat sewage waters. For a country like India, conventional treatment plants are costly. In fact, these are beyond the financial means of many small towns. Biological waste water treatment, on land disposal using suitable vegetative cover and resource recovery technologies cannot only be an attractive alternative, but also economical, safe and socially acceptable.

4. Mining operations should not ordinarily be taken up in ecologically fragile areas. Every mining project shall be accompanied by a mining plan, including an environmental management plan and time-bound reclamation programme for controlling the environmental damage and for restoration of mined areas.

53.5 ASSISTANCE FOR ADOPTION OF CLEAN TECHNOLOGIES BY SMALL-SCALE INDUSTRIES

1. Small-scale industries are a special feature of our economy. The Governments are implementing a scheme for providing assistance for promoting combined facilities for treatment of effluents and solid wastes generated in clusters of small-scale units. This scheme should be extended to provide necessary technical support as well.

2. While the large and medium industrial units should remain totally responsible for control of their pollution, assistance should be provided to small-scale industrial units, particularly those located in rural areas, to aid the implementation of pollution control measures. This will be achieved by promoting development and adoption of cleaner technologies including environmentally friendly biotechnology.

53.6 STANDARDS

1. The present standards are based on the concentration of pollutants in effluents and in emissions. The norms will be revised to lay down mass-based standards, which will set specific limits to encourage the minimisation of waste, promote recycling and reuse of materials, as well as conservation of natural resources, particularly water. Since the standards will be source-related, they will require for the most polluting industrial processes, particularly those using toxic substances, application of the best available technological solutions and also be an instrument for technological upgradation.

2. To act against potential problems in the future, new units will have to conform to stricter standards. They will need to select technologies that produce no or low quantities of wastes and recycle or reuse waste products. Progressively, more strict vehicle emission standards will also be evolved to deal with environmental hazards caused by vehicular traffic.

3. Standards will not merely be a regulatory tool but will be a mechanism to promote technological upgradation to prevent pollution, conserve resources and regulate waste. For this purpose, codes of practice and guidelines will be evolved for specific processes.

4. The environmental effects—from production to disposal of products that are hazardous and toxic—will be taken into account in the regulations. Chemicals will be reviewed according to the level of risk, and, where safer alternatives have become available, restrictions will be imposed. Regulations for liability and compensation for damages will supplement standards, to promote greater care and caution, particularly in the management of hazardous wastes and remedial action in case of contamination of soil and ground water.

53.7 FISCAL MEASURES

1. While regulatory measures remain essential for the effectiveness of the policy, new approaches for considering market choices will be introduced. The aim is to give industries and consumers clear signals about the cost of using environmental and natural resources. The expectation is that market-oriented price mechanisms will influence behaviour to avoid excessive use of natural resources.

2. There are at present several fiscal incentives for installation of pollution control equipment and for shifting polluting industries from congested areas. The items for which excise and customs rebate are allowed will be reviewed. This will stimulate the advancement of abatement technologies and create increased demands for the products.

3. Economic instruments will be investigated to encourage the shift from curative to preventive measures, internalise the costs of pollution and conserve resources, particularly water. A direct economic signal is offered by an effluent charge based on the nature and volume of releases to the environment. The level will be based on the cost of treatment and the flow discharged, in order to provide an incentive to set-up treatment plants. The scope of the charges will also be extended to emissions and solid waste. Charges provide a continuing incentive towards optimal releases.

4. These instruments will also have a distributive effect as the revenues will be used for enforcement, collective treatment facilities, research and promoting new investment.

5. The precise choice of economic instruments adopted will be determined by the ease with which releases can be measured, as well as prospective changes in technology and market structures. To deal with the range of pollution problems, a mix of regulatory and economic measures will be adopted.

53.8 INTEGRATION

1. Critical policy areas for control of pollution come under different departments and levels of Government. Sectoral Ministries, State Governments, local bodies and agencies responsible for planning and implementation of development projects will be required to integrate environmental concerns more effectively in all policy areas. Local authorities play a key role in abatement of pollution and environmental concerns need to be built into the way they operate. Steps will have to be taken to strengthen governmental and institutional structures dealing with environmental management, especially within the ministries dealing with the sectors of energy, industry, water resources, transport and agriculture and who would develop specific programmes in regard to pollution prevention.

2. Policy-making, legislation and law enforcement influence each other. The increase in the number of regulations increases difficulties in enforcement. Legislation regulating particular activities will be amended to incorporate and eliminate clashes with environmental criteria. Traditional instruments for monitoring of compliance and investigation of offences are becoming over-burdened. An integrated overview and organisational structure for decentralised environment impact assessments and environmental law enforcement based on cooperation with local authorities will be sought.

3. While pollution from specific sources including towns and industries have been addressed, non-point pollution from run-off of agricultural inputs such as pesticides, insecticides, fertilisers, etc. has not been dealt with. This is gaining increasing proportions, which is polluting not only our water bodies but even sub-soil water resources and would affect the health of human beings. A long-term policy for pesticides use, including the introduction of environmentally acceptable pesticides—particularly

biopesticides and non-persistent biodegradable ones and integrated pest management together with the phasing out of the proven harmful toxic and persistent ones—would be formulated in collaboration with the concerned Ministries and infrastructure involved for its effective implementation. A similar policy for fertiliser use will also need to be formulated.

4. Plant and vegetation in general play a vital role in controlling pollution by regulating the climate and atmospheric equilibrium, protecting the soil and maintaining the hydrological regime. Hence, existing forests and natural vegetation should be fully protected. The forest and vegetal cover should be restored and increased wherever possible, specially on hill slopes, in catchment areas of rivers, lakes and reservoirs, ocean shores, semi-arid and arid tracts, in and around urban centres and industrial establishments. It is necessary to encourage the planting of trees alongside roads, rail lines, canals and on other unutilised lands under State/corporate, institutional or private ownership. Green belts should be raised in urban and industrial areas as well as in arid tracts. Such a programme will also check erosion, desertification as well as improve the micro-climate.

5. The Annual Administration Reports of the Ministries will include a chapter on the action taken to follow up the policy statement and other environmental initiatives they have taken or are proposing.

53.9 ENVIRONMENTAL AUDIT

Industrial concerns and local bodies should feel that they have a responsibility for abatement of pollution. The procedure of an environmental statement will be introduced in local bodies, statutory authorities and public limited companies to evaluate the effect of their policies, operations and activities on the environment, particularly compliance with standards and the generation and recycling of waste. An annual statement will help in identifying and focusing attention on areas of concern, practices that need to be changed and plans to deal with adverse effects. This will be extended to an environmental audit. The measures will provide better information to the public.

53.10 ENVIRONMENTAL STATISTICS

Authoritative statistical data on the environment is vital for Developmental decision-making. Resource accounting will be used to give an idea as to how economic policies are affecting the environment. Current economic accounts are concerned mainly with the volume of economic activity; they ignore expenditures to protect the environment and encourage inefficient use of resources.

The collection and integration of environmental, economic and health data will be done to determine the status and to develop a concise set of environmental indicators for monitoring the effects of pollution. Information and access to the public are essential so that everyone knows what is happening to the environment.

53.11 PUBLIC PARTNERSHIP

1. The public must be made aware in order to be able to make informed choices. A high governmental priority will be to educate citizens about environmental risks, the economic and health dangers of resource degradation and the real cost of natural resources. Information about the environment will be published periodically. Affected citizens and non-governmental organisations play a role in environmental monitoring and, therefore, allowing them to supplement the regulatory system and recognising their expertise where such exists and their commitments and vigilance, will also be cost effective. Access to information to enable public monitoring of environmental concerns will be provided for.

2. Public interest litigations has successfully demonstrated that responsible non-governmental organisations and public spirited individuals can bring about significant pressure on pollution units for adopting abatement measures. This commitment and expertise will be encouraged and their practical work supported.

3. Householders, as consumers, make large number of relatively small individual contributions, whose cumulative effect is considerable. A system of certification of goods that are "environmentally friendly" will be set up to make available information to encourage environmental consciousness amongst consumers. This advice will also encourage manufacturers to produce goods that are environmentally more friendly as well as encourage recycling and adequate waste management. Consumer awareness would also be encouraged by involvement of consumer organisations in cooperative testing, and dissemination of information relating to environmental friendliness of these products.

4. As the present system of jurisprudence does not provide for compensation to individuals for environmental damage, including effects on health and environmental damage caused by pollution, it is proposed to set up special legal institutions to redress this deficiency and also make adequate arrangements for interim relief.

5. Greater emphasis will be placed on promoting awareness, undertaking and competence in schools, colleges and training institutions. Professional and non-governmental bodies will be encouraged to be more active in environmental training and building awareness.

6. Society has accepted many practices which cause pollution. Reckless use of loudspeakers, dumping in water bodies, and scattering of wastes are common. Noise nuisance requires specific devices as well as greater consideration for neighbours and there is growing concern that litter has increased in recent years. Social action in these matters by voluntary organisations and individuals will be promoted through knowledge, education, training camps and public information campaigns.

This statement is based on considerations of effectiveness, efficiency and availability of financial resources. The responsibility for abatement of pollution is not a duty of the Government alone, it is an obligation on all of us.

The approach mentioned above should indicate how everyone can help in achieving a safe and environmentally appropriate environment in our country.

54 NATIONAL ENVIRONMENTAL POLICY, 2006

INTRODUCTION

A diverse developing society such as ours provides numerous challenges in the economic, social, political, cultural, and environmental arenas. All of these coalesce in the dominant imperative of alleviation of mass poverty, reckoned in the multiple dimensions of livelihood security, health care, education, empowerment of the disadvantaged and elimination of gender disparities. The present national policies for environmental management are contained all aspect of environmental management. All of these policies have recognized the need for sustainable development in their specific contexts and formulated necessary strategies to give effect to such recognition. The National Environment Policy seeks to extend the coverage, and fill in gaps that still exist, in light of present knowledge and accumulated experience. It does not displace, but builds on the earlier policies.

Across the political spectrum of the country there has been recognition of the vital role natural resources play in providing livelihoods, and securing life support ecological services. In this perspective a need for a comprehensive policy statement has been evident for some time in order to infuse a common approach to the various sectoral and cross-sectoral, including fiscal, approaches to environmental management. As our development challenges have evolved and our understanding of the centrality of environmental concerns in development has sharpened, there is also a need to review the earlier objectives, policy instruments, and strategies. This dynamic requires an evolving and flexible policy framework, with a built in system for monitoring and review, and where necessary, revision. Sustainable development concerns in the sense of enhancement of human well-being, broadly conceived, are a recurring theme in India's development philosophy. The present day consensus reflects three foundational aspirations.

For this to occur there is a need for balance and harmony between economic, social and environmental needs of the country. India also plays an important role in several significant international initiatives concerned with the environment. It is a party to the key multilateral agreements, and recognizes the interdependencies among, and transboundary character of, several environmental problems. The National Environment Policy (NEP) is also intended to be a statement of India's commitment to making a positive contribution to international efforts.

This is a response to our national commitment to a clean environment, mandated in the Constitution in Articles 48A and 51A (g), strengthened by judicial interpretation of Article 21. It is recognised that maintaining a healthy environment is not the state's responsibility alone, but also that of every citizen. A spirit of partnership should thus be realised throughout the spectrum of environmental management in the country. While the state must galvanize its efforts, there should also be recognition by each individual —natural or institutional, of its responsibility towards maintaining and enhancing the quality of the environment. The National Environment Policy has been motivated by the above considerations and is intended to mainstream environmental concerns in all development activities, it briefly describes the key environmental challenges currently and prospectively facing the country, the objectives of environment policy, normative principles underlying policy action, strategic themes for intervention, broad indications of the legislative and institutional development, needed to accomplish the strategic themes, and mechanisms for implementation and review. It has been prepared through a process of extensive consultation with experts, as well as diverse stakeholders, and this process is also documented. The National Environment Policy is intended to be a guide to action: in regulatory reform, programmes and projects for environmental conservation; and review and enactment of legislation, by agencies of the Central, State, and Local Governments. The policy also seeks to stimulate partnerships of different stakeholders, i.e. public agencies, local communities, academic and scientific institutions, the investment community, and international development partners, in harnessing their respective resources and strengths for environmental management.

Key Environmental Challenges: Causes and Impacts

The key environmental challenges that the country faces relate to the nexus of environmental degradation with poverty in its many dimensions, and economic growth. These challenges are intrinsically connected with the state of environmental resources, such as land, water, air, and their flora and fauna. The proximate drivers of environmental degradation are population growth, inappropriate technology and consumption choices, and poverty, leading to changes in relations between people and ecosystems, and development activities such as intensive agriculture, polluting industry and unplanned urbanization. However, these factors give rise to environmental degradation only through deeper causal linkages, in particular, institutional failures, resulting in lack of clarity or enforcement of rights of access and use of environmental resources, policies which provide disincentives for environmental conservation (and which may have origins in the fiscal regime), market failures (which may be linked to shortcomings in the regulatory regimes), and governance constraints.

Environmental degradation is a major causal factor in enhancing and perpetuating poverty, particularly among the rural poor, when such degradation impacts soil fertility, quantity and quality of water, air quality, forests, wildlife and fisheries. The dependence of the rural poor, in particular, tribal societies, on their natural resources, especially biodiversity is self-evident. Women in particular face greater adverse impacts of degradation of natural resources, being directly responsible for their collection and use.

The commitment of time and effort in collection of these resources has a direct impact on the capacity of rural women to devote time to raising and educating children, enhancing their earning skills, or participating in gainful livelihoods. The poor are also more vulnerable to loss of resilience in ecosystems. Large reductions in resilience may mean that the ecosystems, on which livelihoods are based, break down, causing distress. The loss of the environmental resource base can result in certain groups of people being made destitute, even if overall, the economy shows strong growth. Further, urban environmental degradation, through lack of (or inappropriate) waste treatment and sanitation, industry and transport related pollution, adversely impacts air, water, and soil quality, and differentially impacts the health of the urban poor. This, in turn, affects their capability to seek and retain employment, attend school, and enhances gender inequalities, all of which perpetuate poverty. Poverty itself can accentuate environmental degradation, given that institutional failures persist.

For the poor, several environmental resources are complementary in production and consumption to other commodities (e.g. water in relation to agricultural production, fuel wood in relation to consumption of food), while a number of environmental resources are a source of income or food (e.g. fisheries, non-timber forest produce). This is frequently a source of cumulative causation, where poverty, gender inequalities, and environmental degradation mutually reinforce each other. Poverty and environmental degradation are also reinforced by and linked to population growth, which in turn, depends on a complex interaction of diverse causal factors and stages of development. The social and economic context of population growth has been detailed in the National Population Policy 2000, which recognises stabilization of population as a necessary condition for sustainable development. Economic growth, in its turn, bears a dichotomous relationship to environmental degradation. On the one hand, growth may result in "excessive" environmental degradation through use of natural resources and generation of pollution aggravated by institutional failures. If impacts on the environmental resource base are neglected, an incorrect picture is obtained from conventional monetary estimates of national income. On the other hand, economic growth permits improvement in environmental quality by making available the necessary resources for environmental investments, and generating societal pressures for improved environmental behaviour, and institutional and policy change. Unsustainable consumption patterns, particularly in industrialized countries also have serious adverse impacts on the environment, both local, and global. The global impacts are largely manifest in developing countries, and further accentuate poverty. It is increasingly evident that poor environmental quality has adversely affected human health. Environmental factors are estimated as being responsible in some cases, for nearly 20 percent of the burden of disease in India, and a number of environment-health factors are closely linked with dimensions of poverty (e.g. malnutrition, lack of access to clean energy and water). It has been shown that interventions such as reducing indoor air pollution, protecting sources of safe drinking water, protecting soil from contamination, improved sanitation measures, and better public health governance, offer tremendous opportunities in reducing the incidence of a number of critical health problems. It is also evident that these environmental protection measures would be difficult to accomplish without extensive awareness rising, and education, on good practices with respect to public and private behaviour. Institutional failures, referring to unclear or insufficiently enforced rights of access to, and use of, environmental resources, result in environmental degradation because third parties primarily experience impacts of such degradation, without cost to the persons responsible for the damage. Such rights, both community based and individual, are critical institutions mediating the relationships between humans and the use of the environment. Traditionally, village commons water sources, grazing grounds, local forests, fisheries, etc., have been protected by local communities from overexploitation through various norms, which may include penalties for disallowed behaviour. These norms, may, however, be degraded through the very process of development, including urbanization, and population growth resulting from sharp reduction in mortality, and also through state actions which may create conditions for the strengthening of individual over communitarian rights, and in doing so allow market forces to press for change that has adverse environmental implications. If such access to the community resources under weakened norms continues, the resources would be degraded, and the livelihoods of the community would suffer. Policy failures can emerge from various sources, including the use of fiscal instruments, such as explicit and implicit subsidies for the use of various resources, which provide incentives for excessive use of natural resources. Inappropriate policy can also lead to changes in commonly managed systems, with adverse environmental outcomes. Another major set of challenges arises from emerging global environmental concerns such as climate change, stratospheric ozone depletion, and biodiversity loss. The key is to operationalize the principle of common but differentiated responsibility of countries in relation to these problems. Multilateral regimes and programmes responding to these global environmental issues must not adversely impact the development opportunities of developing countries. Further, the sharing of global natural resources must proceed only on the basis of equal sharing per capita across all countries.

Objectives of the National Environment Policy

The principal Objectives of this policy are enumerated below. These Objectives relate to current perceptions of key environmental challenges. They may, accordingly, evolve over time: To protect and conserve critical ecological systems and resources, and invaluable natural and man-made heritage, which are essential for life support, livelihoods, economic growth, and a broad conception of human well-being. To ensure equitable access to environmental resources and quality for all sections of society, and in particular to ensure that poor communities, which are most dependent on environmental resources for their livelihoods, are assured secure access to these resources. To ensure judicious use of environmental resources to meet the needs and aspirations of the present and future generations. To integrate environmental concerns into policies, plans, programmes, and projects for economic and social development.

(i) Conservation of Critical Environmental Resources
(ii) Intra-generational Equity: Livelihood Security for the Poor
(iii) Inter-generational Equity
(iv) Integration of Environmental Concerns in Economic and Social Development
(v) Efficiency in Environmental Resource Use
(vi) Environmental Governance
(vii) Enhancement of Resources for Environmental Conservation

To ensure efficient use of environmental resources in the sense of reduction in their use per unit of economic output, to minimise adverse environmental impacts. To apply the principles of good governance (transparency, rationality, accountability, reduction in time and costs, participation, and regulator independence) to the management and regulation of use of environmental resources. To ensure higher resource flows, comprising finance, technology, management skills, traditional knowledge, and social capital, for environmental conservation through mutually beneficial multistake holder partnerships between local communities, public agencies, the academic and research community, investors, and multilateral and bilateral development partners.

Principles

This policy has evolved from the recognition that only such development is sustainable, which respects ecological constraints, and the imperatives of justice. The Objectives stated above are to be realized through various strategic interventions by different public authorities at Central, State, and Local Government levels. They would also be the basis of diverse partnerships. These strategic interventions, besides legislation and the evolution of legal doctrines for realization of the Objectives, may be premised on a set of unambiguously stated Principles depending upon their relevance, feasibility in relation to costs, and technical and administrative aspects of their application. The following Principles, may accordingly, guide the activities of different actors in relation to this policy. Each of these Principles has an established genealogy in policy pronouncements, jurisprudence, international environmental law, or international State practice:

Human beings are at the centre of concerns for sustainable development. They are entitled to a healthy and productive life in harmony with nature. The right to development must be fulfilled so as to equitably meet developmental and environmental needs of present and future generations.

(i) Human Beings are at the Centre of Sustainable Development Concerns

(ii) The Right to Development

(iii) Environmental Protection is an Integral part of the Development Process

(iv) The Precautionary Approach

(v) Economic Efficiency

(vi) Entities with "Incomparable" Values

In order to achieve sustainable development, environmental protection shall constitute an integral part of the development process and cannot be considered in isolation from it. Where there are credible threats of serious or irreversible damage to key environmental resources, lack of full scientific certainty shall not be used as a reason for postponing cost-effective measures to prevent environmental degradation. In various public actions for environmental conservation, economic efficiency would be sought to be realised. This Principle requires that the services of environmental resources be given economic value, and such value to count equally with the economic values of other goods and services, in analysis of alternative courses of action. Further implications of this Principle are as follows.

Impacts of acts of production and consumption of one party may be visited on third parties who do not have a direct economic nexus with the original act. Such impacts are termed externalities. If the costs (or benefits) of the externalities are not revisited on the party responsible for the original act, the resulting level of the entire sequence of production or consumption, and externality, is inefficient. In such a situation, economic efficiency may be restored by making the perpetrator of the externality bear the cost (or benefit) of the same. The policy will, accordingly, promote the internalisation of environmental costs. including through the use of incentives based policy instruments, taking into account the approach that the polluter should, in principle, bear the cost of pollution, with due regard to the public interest, and without distorting international trade and investment. Where the environmental benefits of a course of action cannot, for methodological or conceptual reasons, be imputed economic value (as in the case of "Incomparable Entities"), in any event the economic costs of realising the benefits should be minimised. Efficiency of resource use may also be accomplished by the use of policy instruments that create incentives to minimize wasteful use and consumption of natural resources. The principle of efficiency also applies to issues of environmental governance by streamlining processes and procedures in order to minimise costs and delays.

Significant risks to human health, life, and environmental life-support systems, besides certain other unique natural and man-made entities, which may impact the well-being, broadly of large numbers of persons, may be considered as "Incomparable" in that individuals or societies would not accept these risks for compensation in money or conventional goods and services. A conventional economic cost-benefit calculus would not, accordingly, apply in their case, and such entities would have priority in allocation of societal resources for their conservation without consideration of direct or immediate economic benefit. The cardinal principle of equity or justice requires that human beings cannot be treated differently based on irrelevant differences between them.

Equity norms must be distinguished according to context, i.e. "procedural equity", relating to fair rules for allocation of entitlements and obligations, and "end-result equity", relating to fair outcomes in terms of distribution of entitlements and obligations. Each context, in addition, must be distinguished in terms of "intra-generational equity", relating to justice within societies, and in particular, providing space for the participation of the underprivileged, and "inter-generational equity", relating to justice between generations. Equity, in the context of this policy refers to both equity in entitlements to, and participation of the relevant publics, in processes of decision-making over use of environmental resources.

(vii) Equity

(viii) Legal Liability:

(a) Fault Based Liability

(b) Strict Liability

(ix) Public Trust Doctrine

(x) Decentralisation

(xi) Integration

(xii) Environmental Standard Setting

The present environmental redressal mechanism is predominantly based on doctrines of criminal liability, which have not proved sufficiently effective, and need to be supplemented. Civil liability for environmental damage would deter environmentally harmful actions, and compensate the victims of environmental damage. Conceptually, the principle of legal liability may be viewed as an embodiment in legal doctrine of the "polluter pays" approach, itself deriving from the principle of economic efficiency. The following alternative approaches to civil liability may apply: In a fault based liability regime a party is held liable if it breaches a preexisting legal duty, for example, an environmental standard. Strict liability imposes an obligation to compensate the victim for harm resulting from actions or failure to lake action, which may not necessarily constitute a breach of any law or duty of care. The State is not an absolute owner, but a trustee of all natural resources, which are by nature meant for public use and enjoyment, subject to reasonable conditions, necessary to protect the legitimate interest of a large number of people, or for matters of strategic national interest. Decentralisation involves ceding or transfer of power from a Central Authority to State and Local Authorities, in order to empower public authorities having jurisdiction at the spatial level at which particular environmental issues are salient, to address these issues. Integration refers to the inclusion of environmental considerations in sectoral policy-making, the integration of the social and natural sciences in environment related policy research, and the strengthening of relevant linkages among various agencies at the Central, State, and Local Self-Government levels, charged with the implementation of environmental policies. Environmental standards must reflect the economic and social development situation in which they apply. Standards adopted in one society or context may have unacceptable economic and social costs if applied without discrimination in another society or context. Setting environmental standards would involve several considerations, i.e. risks to human health, risks to other environmental entities, technical feasibility, costs of compliance, and strategic considerations. It is preferable to prevent environmental damage from occurring in the first place, rather than attempting to restore degraded environmental resources after the fact. There is a general obligation to protect threatened or endangered species and natural systems that are of special importance to sustaining life, providing livelihoods, or general well-being. If for exceptional reasons of overriding public interest such protection cannot be provided in particular cases, cost-effective offsetting measures must be undertaken by the proponents of the activity, to restore as nearly as may be feasible, the lost environmental services to the same publics.

(xiii) Preventive Action

(xiv) Environmental Offsetting

STRATEGIES AND ACTIONS

Regulatory Reforms

The foregoing statement of policy Objectives and Principles are to be realised by concrete actions in different areas relating to key environmental challenges. A large number of such actions are currently under way and have been for several years, in some cases, for many decades. In some aspects new themes would need to be pursued to realise the Principles and Objectives. Action plans would need to be prepared on identified themes by the concerned agencies at all levels of Government Central, State/UT, and Local. In particular, the State and Local Governments would be encouraged to formulate their own strategies or action plans consistent with the National Environment Policy.

Regulatory Reforms

Empowerment of Panchayats and the Urban Local Bodies, particularly, in terms of functions, functionaries, funds, and corresponding capacities, will require greater attention for operationalising some of the major provisions of this policy. Integration of environmental concerns in all relevant development processes is among the Objectives of this policy. Further, inclusion of environmental considerations in sectoral policy making has also been recognised as among the Principles underpinning the policy. In order to operationalize these, a mechanism for ensuring necessary due diligence at all levels of government, will be institutionalised. The following Strategic Themes, and outlines of actions to be taken in each, focus on ongoing activities, functions, and roles, as well as new initiatives that are necessary. However, they are not necessarily a complete enumeration in each case.

Revisiting the Policy and Legislative Framework

The regulatory regimes for environmental conservation comprise a legislative framework, and a set of regulatory institutions. Inadequacies in each have resulted in accelerated environmental degradation on the one hand, and long delays and high transaction costs in development projects on the other. Apart from legislation which is categorically premised on environmental conservation, a host of sectoral and cross-sectoral laws and policies, including fiscal regimes, also impact environmental quality (some of these are discussed in the succeeding sections). The present legislative framework is broadly contained in the umbrella: which contained management of forests and biodiversity. There are several other enactments, which complement the provisions of these basic enactments.

The following specific actions would be taken:

(a) Institutionalise a holistic and integrated approach to the management of environmental and natural resources, explicitly identifying and integrating environmental concerns in relevant sectoral and cross-sectoral policies through review and consultation, in line with the National Environment Policy.

(b) Identify emerging areas for new legislation, due to better scientific understanding, economic and social development, and development of multilateral environmental regimes, in line with the National Environment Policy.

(c) Review the body of existing legislation in order to develop synergies among relevant statutes and regulations, eliminate obsolescence, and amalgamate provisions with similar objectives, in line with the National Environment Policy. Further, encourage and facilitate review of legislation at the level of State and Local Governments with a view to ensuring their consistency with this policy.

(d) Take steps to adopt and institutionalise techniques for environmental assessment of sector policies and programmes to address any potential adverse impacts, and enhance potential favorable impacts.

(e) Ensure accountability of the concerned levels of Government (Centre, State, Local) in undertaking the necessary legislative changes in a defined time-frame, with due regard to the Objectives and Principles of National Environment Policy, in particular, ensuring the livelihoods and well-being of the poor by ensuring improved access to the necessary environmental resources.

PROCESS AND FORMULATION OF THIS POLICY

(i) Approach

(ii) Framework for Legal Action

The recommendations of The Govindarajan Committee, which identified delays in environment and forest clearances as the largest source of delays in development projects, will be followed for reviewing the existing procedures for granting clearances and other approvals under various statutes and rules. These include the objective is to reduce delays and levels of decision-making, realise decentralisation of environmental functions, and ensure greater transparency and accountability. In addition, the following actions will be taken:

(a) In order to ensure faster decision making with greater transparency, and access to information, use of information technology based tools will be promoted, together with necessary capacity-building, under all action plans.

(b) In order to realise greater decentralisation, State level agencies may be given greater responsibility for environmental regulation and management. Such empowerment must, however, be premised on increased transparency, accountability, scientific and managerial capacity, and independence in regulatory decision making and enforcement action. Accordingly, States would be encouraged to set up Environment Protection Authorities on this basis.

(c) Mechanisms and processes would be set up to identify entities of "Incomparable Value" in different regions. It would be ensured that all regulatory mechanisms are legally empowered to follow the principles of good governance. The present approach to dealing with environmentally unacceptable behaviour in India has been largely based on criminal processes and sanctions. Although criminal sanctions, if successful, may create a deterrent impact, in reality they are rarely fruitful for a number of reasons. On the other hand, giving unfettered powers to enforcement authorities may lead to rent-seeking. Civil law, on the other hand, offers flexibility, and its sanctions can be more effectively tailored to particular situations. The evidentiary burdens of civil proceedings are less daunting than those of criminal law. It also allows for preventive policing through orders and injunctions. Accordingly, a judicious mix of civil and criminal processes and sanctions will be employed in the legal regime for enforcement, through a review of the existing legislation. Civil liability law, civil sanctions, and processes, would govern most situations of noncompliance. Criminal processes and sanctions would be available for serious, and potentially provable, infringements of environmental law, and their initiation would be vested in responsible authorities. Recourse may also be had to the relevant provisions in the Indian Penal Code, and the Criminal Procedure Code. Both civil and criminal penalties would be graded according to the severity of the infraction. Environmental Impact Assessment (EIA) will continue to be the principal methodology for appraising and reviewing new projects. The assessment processes are under major revision in line with the Govindarajan Committee recommendations. Under the new arrangements, there would be significant devolution of powers to the State/UT level. However, such devolution, to be effective, needs to be accompanied by adequate development of human and institutional capacities. Further, in order to make the clearance processes more effective, the following actions will be taken:

(a) Encourage regulatory authorities, Central and State, to institutionalize regional and cumulative environmental impact assessments (R/CEIAs) to ensure that environmental concerns are identified and addressed at the planning stage itself.

(b) Specifically assess the potential for chemical accidents of relevant projects as part of the environmental appraisal process.

(c) Give due consideration, to the quality and productivity of lands which are proposed to be converted for development activities, as part of the environmental clearance process. Projects involving large-scale diversion of prime agricultural land would require environmental appraisal.

(d) Encourage clustering of industries and other development activities to facilitate setting up of environmental management infrastructure, as well as monitoring and enforcing environmental compliance. Emphasise post project monitoring and implementation of environmental management plans through participatory processes, involving adequately empowered relevant levels of government, industry, and the potentially impacted community.

(e) Restrict the diversion of dense natural forests and areas of high endemism of genetic resources, to non-forest purposes, only to site-specific cases of vital national interest. No further regularisation of encroachment on forests should be permitted.

(f) Ensure that in all cases of diversion of forest, the essential minimum needed for the project or activity is diverted. The diverted area must not be cleared until the actual construction starts.

(g) Ensure provision for environmental restoration after decommissioning of industries, in particular mine closure in all approvals of mining plans, and institutionalise a system of post-monitoring of such projects.

(h) Formulate, and periodically update, codes of "good practices" for environmental management for different categories of regulated activities.

(i) Environment and Forests Clearances

(ii) Coastal Areas

(iii) Living Modified Organisms (LMOs).

Development activities in the coastal areas are regulated by means of the Coastal Regulation Zone notifications and Integrated Coastal Zone Management (ICZM) plans made under them.

However, there is need to ensure that the regulations are firmly founded on scientific principles, including the physical, natural, and social sciences. This is necessary to ensure effective protection to valuable coastal environmental resources, without unnecessarily impeding livelihoods, or legitimate coastal economic activity, or settlements, or infrastructure development. Islands offer unique ecosystems and coastal planning and regulation in their case needs to take into account features such as their geological nature, settlement patterns, volcanic or coral nature of nature of the island, size of the habitations, unique cultures, livelihood patterns, etc. Adequate environmental safeguards should be built into development projects in the islands, in particular those relating to tourism, high value agriculture, deep sea fishing, prospecting for oil and natural gas, etc. It is recognized that states will require both technical and financial resources for preparation of ICZM plans. The following actions will be taken:

(a) Revisit the Coastal Regulation Zone (CRZ) notifications to make the approach to coastal environmental regulation more holistic, and thereby ensure protection to coastal ecological systems, coastal waters, and the vulnerability of some coastal areas to extreme natural events and potential sea level rise. The Integrated Coastal Zone Management (ICZM) plans need to be comprehensive, and prepared on strong scientific basis by experts with the participation of the local communities both in formulation and implementation. The ICZM, plans should be reviewed at predetermined intervals to take account of changes in geomorphology, economic activities, settlement patterns, and coastal and marine environmental conditions.

(b) Decentralize, to the extent feasible, the clearance of specific projects to State level environmental authorities, exempting activities, which do not cause significant environmental impacts, and are consistent with approved ICZM plans. Biotechnology has immense potential to enhance livelihoods and contribute to the economic development of the country. On the other hand, LMOs may pose significant risks to ecological resources, and perhaps, human and animal health.

In order to ensure that development of biotechnology does not lead to unforeseen adverse impacts, the following actions will be taken:

(a) Review the regulatory processes for LMOs so that all relevant scientific knowledge is taken into account, and ecological, health, and economic concerns are adequately addressed.

(b) Periodically review the National Bio-safety Guidelines, and Bio-safety Operations Manual to ensure that these are based on current scientific knowledge.

(c) Ensure the conservation of bio-diversity and human health when dealing with LMOs in transboundary movement in a manner consistent with the multilateral Bio-safety Protocol. Environmentally Sensitive Zones may be defined as areas with identified environmental resources having "Incomparable Values" which require special attention for their conservation. In order to conserve and enhance these resources, without impeding legitimate socioeconomic development of these areas, the following actions will be taken:

1. Identify and give legal status to Environmentally Sensitive Zones in the country having environmental entities with "Incomparable values" requiring special conservation efforts.

2. Formulate area development plans for these zones on a scientific basis, with adequate participation by the local communities.

3. Create local institutions with adequate participation for the environmental management of such areas, to ensure adherence to the approved area development plans, which should be prepared in consultation with the local communities.

(iv) Environmentally Sensitive Zones

Weak enforcement of environmental compliance is attributed to inadequate technical capacities, monitoring infrastructure, and trained staff in enforcement institutions. In addition, there is insufficient involvement of the potentially impacted local communities in the monitoring of compliance, and absence of institutionalized public-private partnerships in enhancement of monitoring infrastructure.

The following actions will be taken:

(a) Take measures, including capacity development initiatives to enable Panchayati Raj Institutions and urban local bodies to undertake monitoring of compliance with environmental management plans. Measures will also be taken to encourage municipalities to annually report their environmental performance to their governing bodies.

(b) Develop feasible models of public-private partnerships to leverage financial, technical, and management resources of the private sector in setting up and operating infrastructure for monitoring of environmental compliance, with ironclad safeguards against possible conflict of interest or collusion with the monitored entities. It is necessary that the costs associated with the degradation and depletion of natural resources be incorporated into the excisions of economic actors at various levels, to reverse the tendency to treat these resources as "free goods" and to pass the costs of degradation to other sections of society, or to future generations of the country.

(v) Monitoring of Compliance

(vi) Use of Economic Principles in Environmental Decision-making

At the macro-level, a system of natural resource accounting is required to assess whether in the course of economic growth we are drawing down, or enhancing, the natural resource base of production, including relevant depletable assets. In addition, the environmental costs and benefits associated with various activities, including sectoral policies, should be evaluated to ensure that these factors are duly taken into account in decision-making. The current near exclusive reliance on fiats based instruments for environmental regulation do not permit individual actors to minimise their own costs of compliance. This leads, on the one hand, to noncompliance in many cases, and, on the other, unnecessary diversion of societal resources from other pressing needs. Economic instruments, of which a large, feasible suite, has emerged through practical experience in several developed and developing countries, work by aligning the interests of economic actors with environmental compliance, primarily through application of "polluter pays". This may ensure that for any given level of environmental quality desired, the society-wide costs of meeting the standard are minimised. However, in some cases, use of economic instruments may require intensive monitoring, which too may entail significant societal costs. On the other hand, use of existing policy instruments, such as the fiscal regime, may significantly reduce or eliminate the need for enhanced institutional capacities to administer the incentive based instruments. In future, accordingly, a judicious mix of incentives and fiats based regulatory instruments would be considered for each specific regulatory situation.

The following actions will be taken:

(a) Strengthen, including through capacity building, the initiatives, taken by the Central Statistical Organization in the area of natural resource accounting, with a view to its adoption in the system of national income accounts. Further strengthen in all respects, the system of collection, collation and analysis of all significant and relevant environmental monitoring data.

(b) Develop and promote the use of standardised environmental accounting practices and norms in preparation of statutory financial statements for large industrial enterprises, in order to encourage greater environmental responsibility in investment decision-making, management practices, and public scrutiny.

(c) Encourage financial institutions to adopt appropriate appraisal practices, so that environmental risks are adequately considered in the financing of projects.

(d) Facilitate the integration of environmental values into cost-benefit analysis, to encourage more efficient allocation of resources while making public investment decisions.

(e) Prepare and implement an action plan on the use of economic instruments for environmental regulation in specified contexts, including those relating to unsustainable production and consumption.

(f) Consider creation of a National Environment Restoration Fund from the net proceeds of economic instruments, user fees for access to specified natural resources, and voluntary contributions. The Fund may be used for restoration of environmental resources, including clean-up of toxic and hazardous waste legacies.

Enhancing and Conserving Environmental Resources

Land Degradation

Perverse production and consumption practices are the immediate causes of environmental degradation, but an exclusive focus on these aspects alone is insufficient to prevent environmental harm. The causes of degradation of environmental resources lie ultimately in a broad range of policy, and institutional, including regulatory shortcomings, leading to the direct causes. However, the range of policies, and legal and institutional regimes, which impact the proximate factors, is extremely wide, comprising fiscal and pricing regimes, and sectoral and cross-sectoral policies, laws, and institutions. Accordingly, apart from programmatic approaches, review and reform of these regimes to account for their environmental consequences is essential. In addition, there is lack of awareness of the causes and effects of environmental degradation, and how they may be prevented, among both specialised practitioners of the relevant professions, including policymakers, as well as the general public, which needs to be redressed. In this subsection, in respect of major categories of environmental resources, the proximate and deeper causes of their degradation and specific initiatives for addressing them are outlined. The degradation of land, through soil erosion, alkali-salinisation, water logging, pollution, and reduction in organic matter content has several proximate and underlying causes. The proximate causes include loss of forest and tree cover (leading to erosion by surface water runoff and winds), unsustainable grazing, excessive use of irrigation (in many cases without proper drainage, leading to leaching of sodium and potassium salts), improper use of agricultural chemicals (leading to accumulation of toxic chemicals in the soil), diversion of animal wastes for domestic fuel (leading to reduction in soil nitrogen and organic matter), and disposal of industrial and domestic wastes on productive land.

These proximate causes of land degradation in turn, are driven by implicit and explicit subsidies for water, power, fertiliser and pesticides. Grazing lands are usually common property resources, and insufficient empowerment of local institutions for their management leads to overexploitation of the biomass base. The absence of conducive policies and persistence of certain regulatory practices reduces people's incentives for afforestation, and leads to reduced levels of green cover. It is essential that the relevant fiscal, tariffs, and sectoral policies take explicit account of their unintentional impacts on land degradation, if the fundamental basis of livelihoods for the vast majority of our people is not to be irreparably damaged.

In addition, to such policy review, the following specific initiatives would be taken:

(a) Encourage adoption of science based, and traditional sustainable land use practices, through research and development, extension of knowledge, pilot scale demonstrations, and large scale dissemination, including farmer's training, and, where necessary, access to institutional finance.

(b) Promote reclamation of wasteland and degraded forestland through formulation and adoption of multi-stake holder partnerships, involving the land owning agency, local communities, and investors.

(c) Prepare and implement thematic action plans incorporating watershed management strategies, for arresting and reversing desertification, and expanding green cover.

(d) Promote sustainable alternatives to shifting cultivation where it is no longer ecologically viable, ensuring that the culture and social organisation of the local people are not disrupted.

(e) Encourage agro-forestry, organic farming, environmentally sustainable cropping patterns, and adoption of efficient irrigation techniques. The arid and semiarid region of India covers 127.3 mha (38.8%) of India's geographical area and spreads over 10 states. The Indian desert fauna is extremely rich in species diversity of mammals and winter migratory birds. However the pressures of a rapidly increasing population on the natural resource base necessitate adoption of innovative and integrated measures for conservation of desert ecosystems. These pressures are enhanced by practices which lead to land degradation, as described above.

The needed measures include:

(a) Intensive water and moisture conservation through practices based on traditional and science based knowledge, and relying on traditional infrastructure.

(b) Enhancing and expanding green cover based on local species.

(c) Reviewing the agronomic practices in these areas, and promoting agricultural practices and varieties, which are well adapted to the desert ecosystem.

Desert Ecosystems

Forests and Wildlife

(i) Forests

Forests provide a multiplicity of environmental services. Foremost among these is the recharging of mountain aquifers, which sustain our rivers. They also conserve the soil, and prevent floods and drought. They provide habitat for wildlife and the ecological conditions for maintenance and natural evolution of genetic diversity of flora and fauna. They are the homes of traditional forest dependent communities. They yield timber, fuel wood, and other forest produce, and possess immense potential for economic benefits, in particular for local communities, from sustainable eco-tourism. On the other hand, in recent decades, there has been significant loss of forest cover, although there are now clear signs of reversal of this trend. The principal direct cause of forest loss has been the conversion of forests to agriculture, settlements, infrastructure, and industry. In addition, commercial extraction of fuel wood, illegal felling, and grazing of cattle, has degraded forests. These causes, however, have their origins in the fact that the environmental values provided by forests are not realised as direct financial benefits by various parties, at least to the extent of exceeding the monetary incomes from alternative uses, including those arising from illegal use. Moreover, while since antiquity forest dwelling communities had generally recognized traditional community entitlements over the forests, on account of which they had strong incentives to use the forests sustainably and to protect them from encroachers, following the commencement of formal forest laws and institutions in 1865, these entitlements were effectively extinguished in many parts of the country. Such disempowerment has led to the forests becoming open access in nature, leading to their gradual degradation in a classic manifestation of the. It has also led to perennial conflict between the forest dependent communities and the Forest Department, constituting a major denial of justice. The relevant provisions of may provide a framework for restoration of the key traditional entitlements.

It is essential that women play a greater role in the management of natural resources. While they have to bear the burden of natural resource degradation, they have little control over the management of these resources. Relevant provisions of the National Policy for the Empowerment of Women provide a framework for incorporating elements of proposed actions.

It is possible that some site-specific non-forest activities may yield overall societal benefits significantly exceeding that from the environmental services provided by the particular tract of forest. However, large scale forest loss would lead to catastrophic, permanent change in the country's ecology, leading to major stress on water resources and soil erosion, with consequent loss of agricultural productivity, industrial potential, living conditions, and the onset of natural disasters, including drought and floods. In any event, the environmental values of converted forests must be restored, as nearly as may be feasible, to the same publics, as well as the regulations under it, provide a comprehensive basis for forest conservation. The set up in 2003 is reviewing the policy, legislative and institutional basis of forest management. Nevertheless, it is necessary, looking to some of the underlying causes of forest loss, to take some further steps. These include:

(a) Give legal recognition of the traditional entitlements of forest dependent communities taking into consideration the provisions of the (PESA). This would remedy a serious historical injustice, secure their livelihoods, reduce possibilities of conflict with the Forest Departments, and provide longterm incentives to these communities to conserve the forests.

(b) Formulate an innovative strategy for increase of forest and tree cover from the 2003 level of 23.69 per cent of the country's land area, to 33 per cent in 2012, through afforestation of degraded forest land, wastelands, and tree cover on private or revenue lands. Key elements of the strategy would include: (i) the implementation of multi-stakeholder partnerships involving the Forest Department, land owning agencies, local communities, and investors, with clearly defined obligations and entitlements for each partner, following good governance principles, to derive environmental, livelihood, and financial benefits, (ii) rationalization of restrictions on cultivation of forest species outside notified forests, to enable farmers to undertake social and farm forestry where their risk-return-term profiles are more favourable than cropping, (iii) universal adoption of community based practices such as Joint Forest Management, Van Panchayats and their variants, in forest management, with assured participation of women, throughout the country, (iv) focus public investments on enhancing the density of natural forests, mangroves conservation, and universalisation of community based practices.

(c) Formulate an appropriate methodology for reckoning and restoring the environmental values of forests, which are unavoidably diverted to other uses.

(d) Formulate and implement a "Code of Best Management Practices" for dense natural forests, to realise the Objectives and Principles of National Environment Policy. Forests of high indigenous genetic diversity should be treated as entities with Incomparable Value.

(e) Denotify Bamboo and similar other species as 'Forest Species' under the, to facilitate their cultivation outside notified forests, and encourage their productive utilisation in economic activities.

(f) Promote plantation of only such species as are conducive to the conservation and sustainability of given ecosystems.

(ii) Wildlife

The status of wildlife in a region is an accurate index of the state of ecological resources, and thus of the natural resource base of human wellbeing. This is because of the interdependent nature of ecological entities, in which wildlife is a vital link. Moreover, several charismatic species of wildlife embody "Incomparable Values", and, at the same time, comprise a major resource base for sustainable eco-tourism. Conservation of wildlife, accordingly, involves the protection of entire ecosystems. However, in several cases, delineation of and restricting access to such Protected Areas (PAs), as well as disturbance by humans on these areas has led to man-animal conflicts. While physical barriers and better policing may temporarily reduce such conflict, it is also necessary to address their underlying causes. These may largely arise from the non-involvement of relevant stakeholders in identification and delineation of PAs, as well as the loss of traditional entitlements of local people, especially tribals, over the PAs. There is also a strong need for creation of corridors to ensure proper genetic flows across habitats. Since wildlife does not remain confined to particular areas, there is also need to ensure greater protection, and habitat enhancement outside the PAs.

In respect of Wildlife Conservation, the following actions will be pursued:

(a) Expand the Protected Area (PA) network of the country, including Conservation and Community Reserves, to give fair representation to all biogeographic zones of the country. In doing so, develop norms for delineation of PAs in terms of the Objectives and Principles of the National Environment Policy, in particular, participation of local communities, concerned public agencies, and other stakeholders, who have a direct and tangible stake in protection and conservation of wildlife, to harmonise ecological and physical features with needs of socioeconomic development.

(b) Revisit the norms, criteria and needs of data for placing particular species in different schedules of the Wildlife Protection Act.

(c) Formulate and implement programmes for conservation of endangered species outside protected areas, while reducing the scope for man-animal conflict.

(d) Empower, build capacities, and facilitate access to finance and technology, for local people, in particular tribals, who are relocated from PAs, or live in the fringe areas, for provision of ecotourism services in the PAs.

(e) Paralleling multi stakeholder partnerships for afforestation, formulate and implement similar partnerships for enhancement of wildlife habitat in Conservation Reserves and Community Reserves, to derive both environmental and eco-tourism benefits.

(f) Promote site-specific eco-development programmes in fringe areas of PAs, to restore livelihoods and access to forest produce by local communities, owing to access restrictions in PAs.

(g) Strengthen capacities and implement measures for captive breeding and release into the wild, identified endangered species.

(h) Review and tighten the provisions of relevant legislation to enhance their deterrence. Further, strengthen institutional measures and capacities of enforcement authorities, with respect to intelligence collection, investigation, and prosecution, to deal with wild life crime.

(i) Ensure that human activities on the fringe areas of PAs do not degrade the habitat or otherwise significantly disturb wildlife. Conservation of genetic diversity is crucial for development of improved crop varieties resistant to particular stresses, new pharma products, etc., apart from ensuring the resilience of ecosystems. Traditional Knowledge (TK), referring to ethno-biology knowledge possessed by local communities, is the basis of their livelihoods, and also a potent means of unlocking the value of genetic diversity through reduction in search costs. Natural heritage sites, including endemic "biodiversity hotspots", sacred groves and landscapes, are repositories of significant genetic and ecosystem diversity, and the latter are also an important basis for eco-tourism. They are nature's laboratories for evolution of wild species in response to change in environmental conditions. India is fortunate in having, through the efforts of dedicated scientists over many decades, developed vast inventories of floral and faunal resources, as well as ethno-biology knowledge. India is, thus well-placed to tap this enormous resource base for benefits for the country as a whole, and local communities in particular, provided that the genetic resources are conserved, and appropriate Intellectual Property Rights (IPRs) conferred on local communities in respect of their ethno-biology knowledge. A largescale exercise has been completed for providing inputs towards a National Biodiversity Strategy and Action Plan. These inputs would be reviewed in terms of the Objectives and Principles of National Environment Policy, scientific validity, financial and administrative feasibility, and legal aspects.

In any event, the following measures would be taken:

Biodiversity, Traditional Knowledge, and Natural Heritage

(a) Strengthen the protection of areas of high endemism of genetic resources ("biodiversity hot spots"), while providing alternative livelihoods and access to resources to local communities who may be affected thereby.

(b) Pay explicit attention to the potential impacts of development projects on biodiversity resources and natural heritage. In appraisal of such projects by cost-benefit analysis, assign values to biodiversity resources at or near the upper end of the range of uncertainty. In particular, ancient sacred groves and "biodiversity hotspots" should be treated as possessing "Incomparable Values".

(c) Enhance conservation of genetic resources in designated gene banks across the country. Genetic material of threatened species of flora and fauna must be conserved on priority.

(d) The Patents Act, 1970, addresses several aspects of the issue of disclosure. The Act mandatorily requires disclosure of source and geographical origin of biological material used in an invention while applying for patents. Failure to disclose and wrongful disclosure are allowed as grounds for opposition to the grain of patent and patent may be revoked on these grounds. The Patent Act also requires the applicant to furnish a declaration with regard to having obtained necessary permission of the competent authority to use the biological material from

India. There is need to harmonise these provisions with the Biodiversity Conservation Act, in particular to enable local communities holding traditional knowledge of use of such biological material to benefit from providing access to such knowledge.

(e) There is need to formulate an appropriate system for Prior Informed Consent and Fair and Equitable Benefit sharing in respect of biological material and traditional knowledge of use of such biological material to enable the country and local communities respectively to derive economic benefits from providing access. These issues are complex and, therefore, modalities for their implementation need to be carefully worked out. Towards this end, attempt would be made to attain greater congruence between these Trade Related Aspects of Intellectual Property Rights.

India's freshwater resources comprise the single most important class of natural endowments enabling its economy and its human settlement patterns. The freshwater resources comprise the river systems, groundwater and wetlands. Each of these has a unique role, and characteristic linkages to other environmental entities. India's river systems typically originate in its mountain ecosystems, and deliver the major part of their water resources to the populations in the plains. According to the Third Assessment Report of the Intergovernmental Panel on Climate Change (IPCC, 2001) almost 67% of the glaciers in the Himalayan mountain range have retreated in the past decade.

(i) **River Systems**

Freshwater Resources

Gangotri glacier is retreating by about 30 meters per year. Elevated global mean temperature may increase net melting rates resulting in glacial retreat and consequent adverse impact on flows in major rivers. Rivers are also subject to siltation from sediment loads due to soil loss, itself linked to loss of forest and tree cover. They are also subject to significant net water withdrawals along their course, due to agricultural, industrial, and municipal use; as well as pollution from human and animal waste, agricultural runoffs, and industrial effluents. Although the rivers possess significant natural capacity to assimilate and render harmless many pollutants, the existing pollution inflows in many cases substantially exceed such natural capacities. This fact, together with progressive reductions in stream flows, ensures that the river water quality in the vast majority of cases declines as one goes downstream. The results include loss of aquatic flora and fauna, leading to loss of livelihoods for river fisher folk, significant impacts on human health from polluted water, increased drudgery for poor rural women in collecting drinking water from distant water bodies, loss of habitat for many bird species, and loss of inland navigation potential. Apart from these, India's rivers are inextricably linked with the history and religious beliefs of its peoples, and the degradation of important river systems accordingly offends their spiritual, aesthetic, and cultural sensibilities. The direct causes of river degradation are, in turn, linked to several policies and regulatory regimes. These include tariff policies for irrigation systems and industrial use, which, through inadequate cost recovery, provide incentives for overuse near the head works of irrigation systems, and drying up of irrigation systems at the tail-ends. This results in excessive cultivation of water intensive crops near the head works, which may lead to inefficient water use, water logging and soil salinity and alkalinity. The irrigation tariffs also do not yield resources for proper maintenance of irrigation systems, leading to loss in their potential. In particular, resources are generally not available for lining irrigation canals to prevent seepage loss. These factors result in reduced flows in the rivers. Pollution loads are similarly linked to pricing policies leading to inefficient use of agricultural chemicals, arid municipal and industrial water use. In particular, revenue yields for the latter two are insufficient to install and maintain sewage and effluent treatment plants, respectively. Pollution regulation for industries is typically not based on formal spatial planning to facilitate clustering of industries to realise scale economies in effluent treatment, resulting in relatively high costs of effluent treatment, and consequent increased incentives for non-compliance. There is accordingly need to review the relevant pricing policy regimes and regulatory mechanisms, in terms of their likely adverse environmental impacts.

The following comprise elements of an action plan for river systems:

(a) Promote research in glaciology to evaluate the impacts of climate change on glaciers and river flows.

(b) Promote integrated approaches to management of river basins by the concerned river authorities, considering upstream and downstream inflows and withdrawals by season, interface between land and water, pollution loads and natural regeneration capacities, to ensure maintenance of adequate flows, in particular for maintenance of in-stream ecological values, and adherence to water quality standards throughout their course in all seasons.

(c) Consider and mitigate the impacts on river and estuarine flora and fauna, and the resulting change in the resource base for livelihoods, of multipurpose river valley projects, power plants, and industries.

(d) Consider mandating the installation of water saving closets and taps in the building byelaws of urban centres, and other available regulatory mechanisms.

(e) Integrate conservation and wise use of wetlands into river basin management involving all relevant stakeholders, in particular local communities, to ensure maintenance of hydrological regimes and conservation of biodiversity.

(f) Incorporate a special component in afforestation programmes for afforestation on the banks and catchments of rivers and reservoirs to prevent soil erosion and improve green cover.

Groundwater is present in underground aquifers in many parts of the country. Aquifers near the surface are subject to annual recharge from precipitation, but the rate of recharge is impacted by human interference. Deep aquifers, on the other hand, occur below a substratum of hard rock. The deep aquifers generally contain very pure water, but since they are recharged only over many millennia, must be conserved for use only in periods of calamitous drought, such as may happen only once in several hundred years. The boundaries of groundwater aquifers do not generally correspond to the spatial jurisdiction of any local public authorities or private holdings, nor are they easily discernible, nor can withdrawals be easily monitored, leading to the unavoidable situation of groundwater being an open access resource. The water table has been falling rapidly in many areas of the country in recent decades. This is largely due to withdrawal for agricultural, industrial, and urban use, in excess of annual recharge. In urban areas, apart from withdrawals for domestic and industrial use, housing and infrastructure such as roads prevent sufficient recharge. In addition, some pollution of groundwater occurs due to leaching of stored hazardous waste and use of agricultural chemicals, in particular, pesticides. Contamination of groundwater is also due to geogenic causes, such as leaching of arsenic and fluoride from natural deposits. Since groundwater is frequently a source of drinking water, its pollution and contamination leads to serious health impacts. The direct causes of groundwater depletion have their origin in the pricing policies for electricity and diesel. In the case of electricity, wherever individual metering is not practiced, a flat charge for electricity connections makes the marginal cost of electricity effectively zero. Subsidies for diesel also reduce the marginal cost of extraction to well below the efficient level. Given the fact that groundwater is an open access resource, the user then "rationally" (i.e. in terms of his individual perspective), extracts groundwater until the marginal value to him equals his now very low marginal cost of extraction. The result is inefficient withdrawals of groundwater by all users, leading to the situation of falling water tables. Support prices for several water intensive crops with implicit price subsidies aggravate this outcome by strengthening incentives to take up these crops rather than less water intensive ones. In coastal areas this overexploitation and inadequate recharge of ground water may also cause serious problem of saline ingress, leading to adverse health impacts and loss of land productivity. Falling water tables have several perverse social impacts, apart from the likelihood of mining of deep aquifers, the capital costs of pump sets and bore wells for groundwater extraction when water tables are very deep may be relatively high, with no assurance that water would actually be found. In such a situation, a user who may be a marginal farmer able to borrow the money only at usurious rates of interest, may, in case water is not found, find it impossible to repay his debts. This may lead to destitution, or worse. Even if the impacts were not so dire, there would be excessive use of electricity and diesel. The efficient use of groundwater would, accordingly, require that the practice of non-metering of electric supply to farmers be discontinued in their own enlightened self-interest. It would also be essential to progressively ensure that the environmental impacts are taken into account in setting electricity tariffs, and diesel pricing. Increased runoff of precipitation in urban areas due to impermeable structures and infrastructure prevents natural groundwater recharge. This is an additional cause of falling water tables in urban areas as large quantities of ground water are pumped out for urban use. In rural areas several cost-effective rainwater, groundwater harvesting and artificial recharge techniques have been proven to enhance groundwater recharge. A number of effective traditional water management techniques to recharge groundwater have been discontinued by the local communities due to the onset of pump sets extraction, and need to be revived. Finally, increase in tree cover, is also effective in enhancing soil moisture retention and preventing soil erosion. Pollution of groundwater from agricultural chemicals is also linked to their improper use, once again due to pricing policies, especially for chemical pesticides, as well as agronomic practices, which do not take the potential environmental impacts into account. While transiting through soil layers may considerably eliminate organic pollution loads in groundwater, this is not true of several chemical pesticides. The pesticides themselves may become a source of pollution when it leaches into the ground water.

The following actions will be taken:

(a) Take explicit account of impacts on groundwater tables of electricity tariffs and pricing of diesel.

(b) Promote efficient water use techniques, such as sprinkler or drip irrigation, among farmers. Provide necessary pricing, inputs, and extension support to feasible and remunerative alternative crops which may be raised by efficient water use.

(c) Ensure availability of ground water potential maps through a designated institution.

(d) Support practices of rain water harvesting and artificial recharge and revival of traditional methods for enhancing groundwater recharge.

(e) Mandate water harvesting and artificial recharge in all new constructions in relevant urban areas, as well as design techniques for road surfaces and infrastructure to enhance groundwater recharge. Promote capacity development of relevant stakeholders and provide web based information on water harvesting techniques.

(f) Prepare and implement a comprehensive strategy for regulating use of ground water by large industrial and commercial establishments on the basis of a careful evaluation of aquifer capacity and annual recharge.

(g) Support R&D in cost effective techniques suitable for rural drinking water projects for remedial measures and removal of arsenic fluoride, and other toxic substances and mainstream their adoption in rural drinking water schemes in relevant areas.

(h) Improve productivity per unit of water consumed in industrial processes, by making water assessments and water audits mandatory in identified industries and utilities.

(i) Suitable sites for dumping the toxic waste material may be identified and remedial measures may be taken to prevent the movement of the toxic waste in the ground water.

(j) Excessive use of fertilisers, pesticides and insecticides are the main non point source of the pollution. These pollutants contribute to the pollution of the ground water as well as surface water. The optional utilisation of fertilisers, pesticides and insecticides should be encouraged for improving the water quality.

(ii) Wetlands

Wetlands, natural and manmade, freshwater or brackish, provide numerous ecological services. They provide habitat to aquatic flora and fauna, as well as numerous species of birds, including migratory species. The density of birds, in particular, is an accurate indication of the ecological health of a particular wetland. Several wetlands have sufficiently unique ecological character as to merit international recognition as Ramsar Sites. Wetlands also provide freshwater for agriculture, animal husbandry, and domestic use, drainage services, and provide livelihoods to fisher folk. Larger wetlands may also comprise an important resource for sustainable tourism and recreation. They may be employed as an alternative to power, technology, and capital intensive municipal sewage plants; however, if used for this purpose without proper reckoning of their assimilative capacity, or for dumping of solid and hazardous waste, they may become severely polluted, leading to adverse health impacts. The inadvertent introduction of some alien species of flora in wetlands has also degraded their ecology. Wetlands are under threat from drainage and conversion for agriculture and human settlements, besides pollution. This happens because public authorities or individuals having jurisdiction over wetlands derive little revenues from them, while the alternative use may result in windfall financial gains to them. However, in many cases, the economic values of wetlands environmental services may significantly exceed the value from alternative use. On the other hand, the reduction in economic value of their environmental services due to pollution, as well as the health costs of the pollution itself, are not taken into account while using them as a waste dump. There also does not yet exist a formal system of wetland regulation outside the international commitments made in respect of Ramsar sites. A holistic view of wetlands is necessary, which looks at each identified wetland in terms of its causal linkages with other natural entities, human needs, and its own attributes.

The following actions will be taken:

(a) Set up a legally enforceable regulatory mechanism for identified valuable wetlands, to prevent their degradation and enhance their conservation. Develop a national inventory of such wetlands.

(b) Formulate conservation and prudent use strategies for each significant catalogued wetland, with participation of local communities, and other relevant stakeholders.

(c) Formulate and implement eco-tourism strategies for identified wetlands through multi-stakeholder partnerships involving public agencies, local communities, and investors.

(d) Take explicit account of impacts on wetlands of significant development projects during the environmental appraisal of such projects; in particular, the reduction in economic value of wetland environmental services should be explicitly factored into cost-benefit analyses.

(e) Consider particular unique wetlands as entities with "Incomparable Values", in developing strategies for their protection.

(f) Integrate wetland conservation, including conservation of village ponds and tanks, into sectoral development plans for poverty alleviation and livelihood improvement, and link efforts for conservation and sustainable use of wetlands with the ongoing rural infrastructure development and employment generation programmes.

Promote traditional techniques and practices for conserving village ponds. Mountain ecosystems play a key role in providing forest cover, feeding perennial river systems, conserving genetic diversity, and providing an immense resource base for livelihoods through sustainable tourism. At the same time, they are among the most fragile of ecosystems in terms of susceptibility to anthropogenic shocks. There has been significant adverse impact on mountain ecosystems by way of deforestation, submergence of river valleys, pollution of freshwater sources, despoliation of landscapes, degradation of human habitat, loss of genetic diversity, plantation of species not conducive to conservation of the mountain environment, retreat of glaciers, and pollution. The most significant proximate causes of these are illegal logging and commercial fuel wood collection, besides faulty construction of infrastructure, unplanned urbanisation and lack of enforcement of building bye-laws, absence or disrepair of sanitation systems, setting up of polluting industries, large scale mining of low unit value minerals, climate change, and excessive use of agricultural chemicals. The underlying causes relate to absence of conducive policies to enable local communities to derive adequate financial returns from afforestation and non-consumptive use of forest resources, pricing policies for agricultural chemicals.

Mountain Ecosystems

Enforcement of pollution standards, poor institutional capacities for urban and regional planning and municipal regulatory functions and preparation of environmental impact assessments of infrastructure; besides absence of consensus on means of financing municipal infrastructure. Clearly, there is need to address these shortcomings through review of the relevant sectoral and cross sectoral policies, and institutional capacity building. Additionally, the following elements of an Action Plan will be taken up:

(a) Adopt appropriate land use planning and watershed management practices for sustainable development of mountain ecosystems.

(b) Adopt "best practice" norms for infrastructure construction in mountain regions to avoid or minimise damage to sensitive ecosystems and despoiling of landscapes.

(c) Encourage cultivation of traditional varieties of crops and horticulture by promotion of organic farming, enabling farmers to realise a price premium.

(d) Promote sustainable tourism through adoption of "best practice" norms for tourism facilities and access to ecological resources, and multi stakeholder partnerships to enable local communities to gain better livelihoods, while leveraging financial, technical, and managerial capacities of investors.

(e) Take measures to regulate tourist inflows into mountain regions to ensure dial these remain within the carrying capacity of the mountain ecology.

(f) Consider particular unique mountain scapes as entities with "Incomparable Values", in developing strategies for their protection. Coastal environmental resources comprise a diverse set of natural and manmade assets, including mangroves, coral reefs, estuaries, coastal forests, genetic diversity, sand dunes, geomorphologies and beaches, land for agriculture and human settlements, coastal infrastructure, and heritage sites. These provide habitats for marine species, which, in turn comprise the resource base for large numbers of fisher folk, protection from extreme weather events, a resource base for sustainable tourism, and agricultural and urban livelihoods. In recent years there has been significant degradation of coastal resources, for which the proximate causes include poorly planned human settlements, improper location of industries and infrastructure, pollution from industries and settlements, and overexploitation of living natural resources. In the future, sea level rise due to climate change may have major adverse impacts on the coastal environment. The deeper causes of these proximate factors lie in inadequate institutional capacities for and participation of local communities in, formulation and implementation of coastal management plans, the open access nature of many coastal resources, and lack of consensus on means of provision of sanitation and waste treatment.

The following further actions will be taken up:

(a) Mainstream the sustainable management of mangroves into the forestry sector regulatory regime, ensuring that they continue to provide livelihoods to local communities.

(b) Disseminate available techniques for regeneration of coral reefs, and support activities based on application of such techniques.

(c) Explicitly consider sea-level rise and vulnerability of coastal areas to climate change and geological events, in coastal management plans, as well as infrastructure planning and construction norms.

(d) Adopt a comprehensive approach to Integrated Coastal Management by addressing linkages between coastal areas, wetlands, and river systems, in relevant policies, regulations, and programs.

(e) Develop a strategy for strengthening regulations, and addressing impacts, of ship-breaking activities on human health, and coastal and near marine resources. Pollution is the inevitable generation of waste streams from the production and consumption of anything. Pollution directly impacts the quality of the receiving medium, i.e. air, water, soil, or electromagnetic spectrum, and when this impaired medium acts upon a receptor, say, a living being, also impacts the receptor. In general, the impacts on the receptor are adverse, but not always. Typically, ecosystems have some natural capacities to assimilate pollution: however, these vary considerably with the nature of the pollutant and the ecosystem. In general, it is cheaper to reduce the emissions of pollution than to mitigate it after generation, or to treat the receiving medium or receptor. The impacts of pollution may differentially impact the poor, or women, or children, or developing regions, who may also have relatively low contributions to its generation, and, accordingly, the costs and benefits of abatement may have important implications for equity. Air pollution may have adverse impacts on human health, as well the health of other living entities, manmade heritage, and life-support systems, such as global climate. Depending upon the lifetime of the pollutants, the location of the source, and the prevailing air currents, the receptors may be located at homestead, local, regional, or global levels, at time intervals from near instantaneous, to several decades.

The direct causes of air pollution are emissions from the use of fossil energy, and other industrial processes, and some consumption activities. The deeper causes arise in a multiplicity of policy, and institutional, including regulatory shortcomings, in particular, inefficient pricing of fossil fuel based energy. Indoor air pollution, a special case, arises from the low societal status of women, leading to continued use of polluting, inefficient biomass stoves, besides pricing policies for agricultural chemicals which lead to substitution of biomass based fertiliser by chemicals, the biomass then being used inefficiently as fuel. These deeper causes need to be addressed through policies and programmes for redressing women's status, and dialogue aimed at consideration of the environmental impacts of pricing policies for agricultural chemicals.

In addition, the following specific actions will be taken:

(a) Take an integrated approach to energy conservation and adoption of renewable energy technologies, including hydropower, by appropriately linking efforts to improve conversion, transmission, distribution, and end-use efficiency, and R&D in and dissemination of renewable energy technologies. Remove policy, legal, and regulatory barriers to setting up decentralised generation and distribution systems for power and other secondary energy forms, based on local primary energy resources.

(b) Accelerate the national programmes of dissemination of improved fuel wood stoves, and solar cookers, suited to local cooking practices and biomass resources.

(c) Strengthen the monitoring and enforcement of emission standards for both point and non-point sources.

(d) Prepare and implement action plans for major cities for addressing air pollution for both point and non-point sources, relying on a judicious combination of fiats and incentive based instruments.

(e) Formulate a national strategy for urban transport to ensure adequate investment, public and private, in low pollution mass transport systems.

(f) Promote reclamation of wastelands by energy plantations for rural energy through multi stakeholder partnerships involving the land owning agencies, local communities, and investors.

(g) Strengthen efforts for partial substitution of fossil fuels by biofuels, through promotion of biofuels plantations, promoting relevant research and development, and streamlining regulatory certification of the new technologies.

The direct and indirect causes of pollution of surface (river, wetlands) water sources, groundwater and coastal areas have been discussed above. The following comprise further elements of an action plan:

(a) Develop and implement, initially on a pilot scale, public-private partnership models for setting up and operating effluent and sewage treatment plants. Once the models are validated, progressively use public resources, including external assistance, to catalyze such partnerships. Enhance the capacities of municipalities for recovery of user charges for water and sewage systems.

(b) Prepare and implement action plans for major cities for addressing water pollution, comprising regulatory systems relying on a appropriate combination of fiats and incentive based instruments, projects implemented through public agencies as well as public-private partnerships for treatment, reuse, and recycle where applicable, of sewage and wastewater from municipal and industrial sources, before final discharge to water bodies.

(c) Take measures to prevent pollution of water bodies from other sources, especially waste disposal on lands.

(d) Enhance capacities for spatial planning among the State and Local Governments, with adequate participation by local communities, to ensure clustering of polluting industries to facilitate setting up of common effluent treatment plants, to be operated on cost recovery basis. Ensure that legal entity status is available for common effluent treatment plants to facilitate investments, and enable enforcement of standards.

(e) Promote R&D in development of low cost technologies for sewage treatment at different scales, in particular, replication of the East Kolkata wetlands and other bio-processing based models for sewage treatment, to yield multiple benefits.

(f) Take explicit account of groundwater pollution in pricing policies of agricultural inputs, especially pesticides, and dissemination of agronomy practices. Encourage Integrated Pest Management (IPM) and use of biodegradable pesticides. Similarly, the immediate and deeper causes of soil pollution have been considered above. Management of industrial and municipal waste is the major cause of soil pollution and is a serious challenge in terms of magnitude and required resources.

An Action Plan will comprise:

(a) Develop and implement viable models of public-private partnerships for setting up and operating secure landfills, incinerators, and other appropriate techniques for the treatment and disposal of toxic and hazardous waste, both industrial and biomedical, on payment by users, taking the concerns of local communities into account. The concerned local communities and State Governments must have clear entitlements to specified benefits from hosting such sites, if access is given to non-local users. Develop and implement strategies for clean up of toxic and hazardous waste dump legacies, in particular in industrial areas, and abandoned mines, and reclamation of such lands for future, sustainable use.

(b) Survey and develop a national inventory of toxic and hazardous waste dumps, and an online monitoring system for movement of hazardous wastes. Strengthen capacities of institutions responsible for monitoring and enforcement in respect of toxic and hazardous wastes.

(c) Strengthen the legal arrangements and response measures for addressing emergencies arising out of transportation, handling, and disposal of hazardous wastes, as part of the chemical accidents regime.

(d) Strengthen the capacities of local bodies for segregation, recycling, and reuse of municipal solid wastes-recognizing inter-alia the positive impacts it may have on the welfare of *safai-karamcharis,* and setting up and operating sanitary landfills, in particular through competitive outsourcing of solid waste management services.

(e) Give legal recognition to and strengthen the informal sector systems of collection and recycling of various materials. In particular, enhance their access to institutional finance and relevant technologies.

(f) Promote organic farming of traditional crop varieties through research in and dissemination of techniques for reclamation of land with prior exposure to agricultural chemicals, facilitating marketing of organic produce in India and abroad, including by development of transparent, voluntary, science-based labelling schemes.

(g) Promote biodegradable and recyclable substitutes for non-biodegradable materials, and develop and implement strategies for their recycle, reuse, and final environmentally benign disposal, including thorough promotion of relevant technologies, and use of incentive based instruments.

(h) Develop and enforce regulations and guidelines for management of e-waste, as part of the hazardous waste regime.

(i) Promote, through incentives, removal of barriers, and regulation, the beneficial utilisation of generally non-hazardous waste streams such as fly ash, bottom ash, red mud, and slag, including in cement and brick making, and building railway and highway embankments.

Persistent exposure to elevated noise levels has been established to result in significant adverse health impacts. While, in many instances, the identification of a particular sound as "noise" is unambiguous, in the case of others, such as "music" or "chanting" or "fireworks displays", it is inherently subjective. In all such cases, societal well-being would require that exposure levels to third parties be maintained below levels at which significant adverse health impacts may occur. At the same time, it needs to be understood that certain environments in which people choose to live and work necessarily involve a certain level of noise.

The following will comprise elements of an Action Plan on abatement of noise pollution:

(a) Make appropriate distinctions between different environments in terms of setting ambient noise standards, e.g. rural versus urban; educational and medical establishments versus other areas, nighttime versus daytime in residential areas, areas in the vicinity of road, rail, airport infrastructure, and protected areas, etc.

(b) Distinguish between noise standards in the context of occupational exposure (with protection measures), and environmental exposure to third parties.

(c) Formulate noise emissions norms (e.g. loudspeaker, automobile horns, fire works ratings) appropriate to various activities to ensure that exposure levels to third parties who are not participants in the activity do not exceed prescribed ambient standards.

(d) Include ambient noise as among the environmental quality parameters to be routinely monitored in specified urban areas.

(e) Encourage dialogue between State/Local Authorities and religious/community representatives on the adoption of enforceable durations, timings, and use of loudspeakers/fireworks, etc. in case temporary exceedance of prescribed ambient noise standards for observance of traditional religious/cultural/social events cannot be avoided. Man-made heritage reflects the prehistory, history, ways of living, and culture, of people. In the case of India, such heritage is at the core of our national identity.

At the same time, considerable economic value, and livelihoods may be derived from conservation of man-made heritage and their sustainable use, through realisation of their tourism potential.

Conservation of Man-made Heritage

The criteria for, and processes of identification of heritage sites, besides legislation and fiscal measures to ensure that they are not damaged or converted by direct human interference, are outside the scope of the National Environment Policy. However, the impact of environmental quality on their conservation is an environmental policy concern. Heritage sites may be impacted by pollution, or they may face threats of inundation or conversion by development projects. Several prominent heritage sites may be held to possess "Incomparable Values".

The following action points will be pursued:

(a) In setting ambient environmental standards, especially for air quality, the potential impacts on designated heritage sites must be taken into account.

(b) Heritage sites considered to have "Incomparable Values" would merit stricter standards than otherwise comparable situations, and particular attention should be paid to monitoring and enforcement environmental standards in their case. Integrated regional development plans should be drawn up, with participation of the local community, to shift polluting activities or render them much less polluting, to treat waste streams, to review transportation options, and adopt building norms which maintain the overall heritage ambience of the area.

(c) Impacts on designated heritage sites must be considered at the stage of developing the terms of reference for environmental impact assessments of projects, and consideration given to the potential impacts during appraisal, in terms of the Objectives and Principles of the National Climate change, resulting from anthropogenic emissions of a suite of gases (called "greenhouse gases" or GHGs) due to fossil fuel use, certain agricultural land industrial activities, and deforestation, leading to their increasing concentrations in the atmosphere, has the potential, over the next few generations, to significantly alter global climate. This would result in large changes in ecosystems, leading to possibly catastrophic disruptions of livelihoods, economic activity, living conditions, and human health. On the other hand, abatement of GHGs, would involve significant economic costs.

Climate Change

While climate change is a global environmental issue, different countries bear different levels of responsibility for increase in atmospheric GHGs concentrations. Further, the adverse impacts of climate change will fall disproportionately on those who have the least responsibility for causing the problem, in particular, developing countries, including India. India's GHG emissions in 1994 were 1228 million ton (Mt) CO equivalent, which is below 3% of global GHG emissions. In per capita terms, it is 23 per cent of the global average, and 4 per cent of USA, 8 per cent of Germany, 9 per cent of UK, and 10 per cent of Japan, per capita emissions in 1994. In terms of the GHG intensity of the economy, in Purchasing Power Parity terms. India emitted a little above 0.4 ton CO equivalent per 1000 US dollars in 2002, which is lower than the global average. In terms of primary energy use, India's share of renewable energy (being an on GHG emitting energy form) at 36 per cent is far higher than industrialised countries can hope to reach in many decades. Since GHG emissions are directly linked to economic activity, India's economic growth will necessarily involve increase in GHG emissions from the current extremely low levels. Any constraints on the emissions of GHG by India, whether direct, by way of emissions targets, or indirect, will reduce growth rates. On the other hand, India's policies for sustainablc development, by way of promotion of energy efficiency, appropriate mix of fuels and primary energy sources including nuclear, hydro and renewable sources, energy pricing, pollution abatement, afforestation, mass transport, besides differentially higher growth rates of less energy intensive services sectors as compared to manufacturing, results in a relatively GHGs benign growth path. Anthropogenic climate change, significant responsibility for which clearly does not lie with India or other developing countries, may, on the other hand, have likely adverse impacts on India's precipitation patterns, ecosystems, agricultural potential, forests, water resources, coastal and marine resources, besides increase in range of several disease vectors. Largescale resources would clearly be required for adaptation measures for climate change impacts, if catastrophic human misery is to be avoided. Accordingly, the following will comprise essential elements of India's response to climate change:

(a) Adherence to the principle of common but differentiated responsibilities and respective capabilities of different countries in respect of both mitigation of GHGs and adaptation measures.

(b) Reliance on multilateral approaches, as opposed to bilateral or plurilateral or unilateral measures.

(c) Equal per capita entitlements of global environmental resources to all countries.

(d) Over-riding priority of the right to development.

(e) Identify key vulnerabilities of India to climate change, in particular impacts on water resources, forests, coastal areas, agriculture, and health.

(f) Assess the need for adaptation to future climate change, and the scope for incorporating these in relevant programmes, including watershed management, coastal zone planning and regulation, forestry management, agricultural technologies and practices, and health programmes.

(g) Encourage Indian Industry to participate in the Clean Development Mechanism (CDM) through capacity building for identifying and preparing CDM projects, including in the financial sector.

(h) Participate in voluntary partnerships with other countries—both developed and developing—to address the challenges of sustainable development and climate change, consistent with the provisions of the Framework Convention on Climate Change. Environmental Standards of air, water, noise, soil etc. both to the acceptable levels of specified environmental quality parameters at different categories of pollutants ("ambient standards"), as well as permissible levels of discharges of specified waste streams by different types of activities ("emission standards"). It is now well understood that environmental standards cannot be universal, and each country should set standards in terms of its national priorities, policy objectives, and resources. These standards may, of course, vary (in general, become more stringent) as a country develops, and has greater access to technologies and financial resources for environmental management. While within the country different States, UTs and local bodies may adopt stricter standards, based on local considerations, they would require concurrence of the Central Government to ensure adherence to the provisions of this policy. Environmental standards also need to relate to other measures for risk mitigation in the country, so that a given societal commitment of resources for achieving overall risk reduction yields the maximum aggregate reduction in risk.

Environmental Standards, Management Systems, Certification, and Indicators:

Environmental Standards

Specific considerations for setting ambient standards in each category of location (residential, industrial, environmentally sensitive zones, etc.) include the reductions in potential aggregate health risks (morbidity and mortality combined in a single measure) to the exposed population; the risk to sensitive, valuable ecosystems and man-made assets; and the likely societal costs, of achieving the proposed ambient standard. Similarly, emissions standards for each class of activity need to be set on the basis of general availability of the required technologies, the feasibility of achieving the applicable environmental quality standards at the location (specific or category) concerned with the proposed emissions standards, and the likely unit costs of meeting the proposed standard. It is also important that the standard is specified in terms of quantities of pollutants that may be emitted, and not only by concentration levels, since the latter can often be easily met through dilution, with no actual improvement in ambient quality. The tendency to prescribe specific abatement technologies should also be eschewed, since these may unnecessarily increase the unit and societal costs of achieving the ambient environmental quality, and, in any case, because a technology that is considered ideal for meeting a given emission standard may not be acceptable on other relevant parameters—including possibly other sources of societal risk.

The following specific actions will be taken:

(a) Set up a permanent machinery comprising experts in all relevant disciplines to review notified ambient and emissions standards in the light of new scientific and technological information as they become available, and changing national circumstances, ensuring adequate participation by potentially impacted communities, and industry associations.

(b) Strengthen the testing infrastructure and network for monitoring ambient environmental quality, including thorough participation by local communities, and public-private partnerships. Progressively ensure realtime, and online availability of the monitoring data. Environmental Management Systems (EMS), such as ISO 14000, by requiring the adoption of standardised environmental management practices, documenting their actual use, and credible third party verification of the fact, may significantly ease the public burden of monitoring and enforcement of prescribed emissions standards. On the other hand, their adoption may involve transaction costs, which, for small and medium enterprises may be significant in relation to their total investment. Global harmonisation of EMS, however, is a safeguard against adoption of arbitrary national EMS regimes to serve as non-tariff barriers.

Environmental Management Systems, Ecolabelling and Certification

Ecolabelling (and other voluntary certification mechanisms) differ from the EMS in that they address the preferences of environmentally conscious consumers, rather than ensuring adherence to national environmental standards. They may involve review of the entire product cycle, from sourcing raw materials, to final disposal of the product after use, and since they are concerned primarily with consumer preferences, may relate to external or adhoc, rather than national environmental standards. Further, at present, non-public bodies have established several labelling schemes in India's export destinations, with no satisfactory evidence of being based on scientific knowledge, or participation by the potentially affected producers. Moreover, they may be based on prescriptions of production processes, and not only of the product characteristics, and for this reason, their mandatory application is inconsistent with provisions of the

WTO regime. Ecolabels, etc., clearly have the potential to be employed as trade barriers, atleast by competing firms in the export destinations, if not directly by their Governments. The obtaining of an ecolabel—especially one granted by an agency located in a developed country—may involve large transactions costs. However, ecolabelled products may command significant price premia—as well as ease of entry to markets. On the other hand, the fact of a large, rapidly growing, environmentally conscious consumer base, in India itself, may provide the required leverage to realise significant advantages from mutual recognition of ecolabelling schemes.

The following actions will be taken:

(a) Encourage industry associations to promote the adoption of ISO 14000 among their members, through provision of technical and training support. Mainstream promotion of ISO 14000 in the small-scale sector in the various promotion schemes for the sector.

(b) Encourage adoption of EMS through purchase preference for ISO 14000 goods and services for Government procurement, except for items reserved for the small-scale sector at any given time, Mandate ISO 14000 when a sufficient number of domestic suppliers for each good or service have ISO 14000 certification.

(c) Formulate "Good Practice Guidelines" for ecolabels to enhance their scientific basis, transparency, and requirements of participation. Promote the mutual recognition of Indian and foreign ecolabels, which adhere to the Good Practice Guidelines, to ensure that Indian exporters enhance their market access at lower costs.

(d) Promote "good practice" norms in all relevant sectors to conserve natural resources and reduce adverse environmental impacts, covering sitting, choice of materials, use of appropriate energy efficiency and renewable energy options, and addressing solid wastes generation, effluents and sewage handling, gaseous emissions, and noise. Clean technologies, as distinct from "end-of-pipe" abatement technologies minimise the generation of waste streams in the production processes and utilise waste from other consumption goods and production processes, rather than treating the waste after generation. In general, clean technologies are less intensive in use of raw materials and energy, than conventional technologies, which rely on pollution abatement after generation. For this reason, they may also offer significant cost advantages to the producer.

Clean Technologies and Innovation

Barriers to the adoption of clean technologies are, first, the fact that many of them are proprietary, and protected by strong patent regimes held abroad. The vendors, accordingly, would be able to extract large premiums in the absence of competitive substitutes. In such cases a project-specific approach will be adopted in respect of enhancing market access. Second, lack of capacity in development financial institutions for appraisal of proposals for switching existing production facilities to clean technologies. Third, the lack of coordination in R&D efforts in India aimed at developing a shelf of commercially viable clean technologies. The last should also be viewed against the fact that in future, almost all commercial transfers of production technology worldwide may be for clean technologies.

The following will comprise elements of an Action Plan:

(a) Encourage capacity building in the financial sector for appraising clean technology switchover project proposals.

(b) Set up a mechanism to network technology research institutions in the country, public and private, for cooperation in technology research and development and adaptation, information, and evaluation of clean technologies. Create a database of such technologies, and promote dissemination of new technologies developed both, in India and abroad.

(c) Consider use of revenue enhancing fiscal instruments to promote shifts to clean technologies in both existing and new units.

(d) Promote adoption of clean technologies by industry, in particular in the small and medium sector, through regulatory and fiscal measures, and standards setting.

Environmental Awareness, Education, and Information

Enhancing environmental awareness is essential to harmonise patterns of individual behaviour with the requirements of environmental conservation. This would minimise the demands placed on the monitoring and enforcement regimes; in fact, large scale non-compliance would simply overwhelm any feasible regulatory machinery. Awareness relates to the general public, as well as specific sections, e.g. the youth, adolescents, urban dwellers, industrial and construction workers, municipal and other public employees. Awareness involves not only internalisation of environmentally responsible behaviour, but also enhanced understanding of the impacts of irresponsible actions, including to public health, living conditions, sanitation, and livelihood. Prospects Environmental education is the principal means of enhancing such awareness, both among the public at large, and among focused groups. Such education may be formal, or informal, or a combination of both. It may rely on educational institutions at different levels; the print, electronic, or live media; and various other formal and informal settings. Several steps to expand and enrich the content of the environment awareness and education programmes have been taken. The Supreme Court has also mandated that environmental education must be imparted at all levels, including higher education in the formal system. However, there is need for further strengthening the existing programmes and making them more inclusive and participatory. Access to environmental information is the principal means by which environmentally conscious stakeholders may evaluate compliance by the concerned parties with environmental standards, legal requirements, and covenants. They would thereby be enabled to stimulate necessary enforcement actions, and through censure, motivate compliance. Access to information is also necessary to ensure effective, informed participation by potentially impacted publics in various consultation processes, such as for preparation of environmental impact assessments, and environment management plans of development projects. The National Natural Resources Management System (NNRMS) was set up in 1983 for optimally managing the natural resources and environment of the country using an optimal mix of remote sensing and conventional techniques. Remote Sensing and data—both satellite and aerial—is being used extensively in the country for mapping and managing the natural resources and environment, over the past three decades.

The following actions will be taken:

(a) Develop and operate an online, real line, publicly accessible environmental information system to provide all relevant information on key environmental resources and parameters, including ambient quality, as well as major point sources of pollution, and make archival data available in convenient format.

(b) Further promote the use of Remote Sensing data to provide valuable inputs on the extent and quality of forests, wildlife habitats, biodiversity, wastelands, wetlands, groundwater, deserts, rivers, etc. and monitor pollution and its impacts.

(c) Mainstream scientifically valid environment content in the curricula of formal education—at primary, secondary, tertiary, and professional levels—focusing on the content appropriate at each stage, and without increasing the course load overall, besides non-formal programmes, such as adult education. Special mid-career training programmes may be conducted for groups with special responsibilities, e.g. the judiciary, policymakers, legislators, industrial managers, city and regional planners, and voluntary and community based organisations.

(d) Prepare and implement a strategy for enhancing environmental awareness among the general public, and special groups, by professional production and airing of information products through diverse media catering to the different target groups. The media products should, as far as possible, eschew focusing on the achievements of public agencies, but instead document real world events of human interest. The production, as well as dissemination may involve public, private, and voluntary agencies. Ensure that adequate financial resources are available for the purpose. Conservation of the environment requires the participation of multiple stakeholders, who may bring to bear their respective resources, competencies, and perspectives, so that the outcomes of partnerships are superior to those of each acting alone. Implementing and policy making agencies of the Government, at Central, State, Municipal, and Panchayat levels; the legislatures and judiciary; the public and private corporate sectors; financial institutions; industry associations; academic and research institutions; independent professionals and experts; the media; youth clubs; community based organisations; voluntary organisations; and multilateral and bilateral development partners may each play important roles in partnerships for the formulation, implementation, and promotion of measures for environmental conservation. Institutions of local self-government have an important role in management of the environment and natural resources. The 73 and 74 Constitutional amendments provide the framework for their empowerment. Further policy and legislative changes are necessary to enable them to actually realise such a role, and participate in various partnerships in this context.

Partnerships and Stakeholder Involvement

A number of specific themes for partnerships have been identified above. A generic classification of some, not exhaustive, possible partnerships is as follows:

(a) Public-Community Partnerships, by which public agencies and local communities cooperate in the management of a given environmental resource, each partner bringing agreed resources, assuming specified responsibilities, and with defined entitlements, e.g. Joint Forestry Management.

(b) Public-Private Partnerships, by which specified public functions with respect to environmental management are contracted out competitively to private providers, e.g. monitoring of environmental quality.

(c) Public Community-Private Partnerships, in terms of which the partners assume joint responsibility for a particular environmental function, with defined obligations and entitlements for each, with competitive selection of the private sector partner, e.g. afforestation of degraded forests.

(d) Public-Voluntary Organisation Partnerships, similar to public-private partnerships, in respect of functions in which voluntary organisations may have a comparative advantage over others, the voluntary organisations, in turn, being selected competitively, e.g. environmental awareness raising.

(e) Public-Private-Voluntary Organisation Partnerships, in which the provision of specified public responsibilities is accomplished on competitive basis by the private sector, and the provision is monitored by competitively selected voluntary organisations, e.g. "Build, Own, Operate" sewage and effluent treatment plants.

(f) It is also essential that all partnerships arc realised through and are carried out in terms of the principles of good governance, in particular transparency, accountability, cost effectiveness, and efficiency.

(g) Youth constitute the most vibrant segment of the country's population. Their strength needs to be harnessed and channelised in protection and conservation of environment. They also need to be involved in relevant stakeholder participation. The multi-stakeholder character of environmental issues and continuous developments in the field of environment make it necessary to have a continuing focus on capacity building in all concerned institutions: public, private, voluntary, academic, research, and the media.

Capacity Building

The following actions are necessary:

(a) Review the present institutional capacities at the Central and State levels, in respect of enforcement of environmental laws and regulations. Prepare and implement suitable programmes for enhancement of the capacities, as required.

(b) Incorporate in all environmental programmes a capacity development component, with sufficient earmarked funds.

(c) Ensure continuous upgradation of knowledge and skills of the scientific and technical personnel involved in environmental management in public institutions at all levels—Central, State, and Local—through dedicated capacity building programmes. In order to rapidly advance scientific understanding of environmental issues, it is necessary to promote properly focussed research by competent institutions. A continuous engagement with the scientific community, in government, academic, and private institutions, will provide important insights for policy making and regulation, including in the field of multilateral negotiations, and help realise deeper and broader skills in the scientific community. Key areas of research are as follows (not in order of priority, which is changeable over time): Taxonomies of living natural resources. Research leading to better understanding of ecological processes and pathways. Research which provides direct inputs to policy making, R&D in technologies for environmental management and clean production.

The following actions would be taken:

1. Periodically identify and prioritise areas for research.
2. Establish a research programme in priority areas within the Government, with expected outputs clearly specified.
3. Encourage research in priority areas outside the Government, with necessary financial and institutional support India has participated in major international events on the

environment, since 1972. The country has contributed to, and ratified several key multilateral agreements on environmental issues in recognition of the transboundary nature of several environmental problems, and has complied with its commitments. It has also participated in numerous regional and bilateral programmes for environmental cooperation. Given the need to enhance our own capacities to comply with our commitments, and ensure sustained flows of resources for environmental management, the following steps would be taken:

Research and Development

(a) Implementation of multilateral and bilateral cooperation programmes, for capacity building for environmental management, and activities in relation to commitments under multilateral instruments.

(b) Planning to participate in mechanisms and arrangements under multilateral agreements for enhancing flows of resources for sustainable development.

(c) Provide assistance to other developing countries, in particular for scientific and technical capacity building for environmental management. We live in a rapidly changing global community, in a rapidly developing, highly diverse country. The environmental issues that are salient as of now may evolve over time, and new ones may take their place. Scientific understanding of environmental matters would advance rapidly. Changes in economic structure, technologies, resource availability—in each case nationally as well as globally—are likely, as are evolution of global environmental regimes, and norms arising from jurisprudence. To set forth an immutable National Environment Policy in this dynamic situation would be unwise.

International Cooperation

A prudent course would be to provide for updating every few years in light of new knowledge and developments and a comprehensive review in about a decade. The following provisions are, accordingly made for review, updating, and renewal of the National Environment Policy:

(a) Undertake consultations every three years with groups of diverse stakeholders, i.e. researchers and experts, community based organisations, industry associations, and voluntary organisations, and update the National Environment Policy.

(b) In the third of the three-year reviews, undertake a more comprehensive examination of the scientific and policy understanding of environmental issues, redefine the Objectives and Principles, and recast the Strategic Themes for Action. A new National Environment Policy should be the outcome. Any policy is only as good as its implementation. The National Environment Policy outlines a significant number of new and continuing initiatives for enhancing environmental conservation. These require the coordinated actions of diverse actors, for the major part organised and stimulated by one or more public agencies. While coordination and review mechanisms are necessary in respect of the individual action plans under each of the strategic themes at relevant operational levels, a formal, periodic high level review of implementation of the different elements of the National Environment Policy is essential. This would enhance accountability of the different public agencies responsible for implementation. It would also reveal practical issues in implementation, including absence of political will at concerned levels, or official indifference. Accordingly, the Cabinet or a nominated Committee of the Cabinet may be requested to review the implementation of the National Environment Policy, once a year, within three months from the close of the previous fiscal year. The findings of the review should be publicly disclosed, so that stakeholders are assured of the seriousness of the Government in ensuring implementation of the Policy.

Conclusion

The preparation of this Policy has involved inputs and consultations with diverse experts, and stakeholders. A draft of the National Environment Policy was prepared through an intensive process of consultation within the Government and inputs from experts. The draft, in English and Hindi, was posted on the website of the Ministry of Environment and Forests and responses were invited from individuals and organisations, through advertisements in national and regional newspapers. The draft was open for public consultation from 21 August 2004 to 31 December 2004. Consultations were held with concerned Ministries of the Central Government, and all State/UT governments at meetings of the State Environment Ministers and senior officials. The latter were encouraged to undertake local level public consultations. The draft was also provided to the Members of Parliament and their views and suggestions were invited. The Ministry of Environment and Forests also held consultations with representatives of major academic and research institutions, and key Industry Associations, Voluntary Organisations, and individuals who are well known in the field. Detailed summaries of responses were prepared and the various concerns expressed by the respondents were addressed. Many cf the suggestions received have been incorporated in the Policy.

Summary

On the whole, citizens must learn the national environmental policy and combination provisions of the various environmental laws. They are the major protectant of their own environment from the consequences of unsustainable projects around them. Well-informed citizens not only have rights but also have a duty to perform in this regard. They can join action groups to develop a lobby to strengthen the environmental movements in the country. Individuals can also take one or several actions when they observe offenders who for their own interest damage the environment for others living in the area.

SECTION IX

ENVIRONMENTAL MODELLING, INDUSTRIAL POLLUTION MANAGEMENT AND ENVIRONMENTAL PLANNING

55 MODELLING IN ENVIRONMENTAL SCIENCE

55.1 MATHEMATICAL MODELS

Mathematical models are constructed by translating *conceptual models* into the formal logic of mathematics, i.e., the process of transforming physical, chemical, and/or biological *concepts* about a system into a set of mathematical relationships. In this process, mathematical symbols provide a useful shorthand to describe complex systems, while equations permit formal statements of how system components are likely to interact.

These sets of mathematical relationships are commonly based on four fundamental mathematical elements: the state of the system is described by sets of state variables representing the condition of the system at any particular time; transfer functions are used to represent the functional relationships in terms of the *flows* (transfers) or *interactions* between the components (or compartments) of the system; *inputs* (of matter and/or energy) to the system, or *factors* which can be thought of as affecting the system by influencing the conditions under which it operates, but which are not themselves affected by the components of the system, are modeled by equations known as forcing functions; finally, *constants* used in the formulation of the mathematical equations are called parameters. The values of these parameters may be determined by the theory on which the model is based or they may be derived empirically from observation and experience. Models where the values of parameters are derived from underlying theory are termed fully specified models (non-empirical models). Those which require calibration, i.e. the estimation of parameter values from empirical observation or experiment are termed partially specified models (empirical models). The set of mathematical relationships comprising a model is itself often referred to as a system (i.e. a system of equations) and the manipulation of this mathematical system can be referred to as systems analysis. The mathematical system is, of course, a model and, as such, is an imperfect and abstract representation of reality. The degree to which the mathematical statements correspond to the physical, chemical or biological concepts of the system they are intended to represent is a measure of the realism of the model.

However, it is the ability to *predict* which makes mathematical models preferable to conceptual models; indeed, mathematical models are more often developed for prediction of dynamic change over time rather than for mere description. The ability of the model to predict numerical change, and to mimic the data on which it is based, are measures of the precision of the model. The extent to which the model can be applied to different situations, or the number of situations where it holds good, is a measure of the generality of the model. Conceptual models represent a number of concepts or ideas about a system, but are not able to test these ideas in any formal way. On the other hand, the validity of a mathematical model can be assessed by comparing its predictions against the empirical observations. The failure of the model to predict change is in itself useful, in that it may point out flaws in the conceptual framework from which the model was developed, thereby contributing to the development of theory.

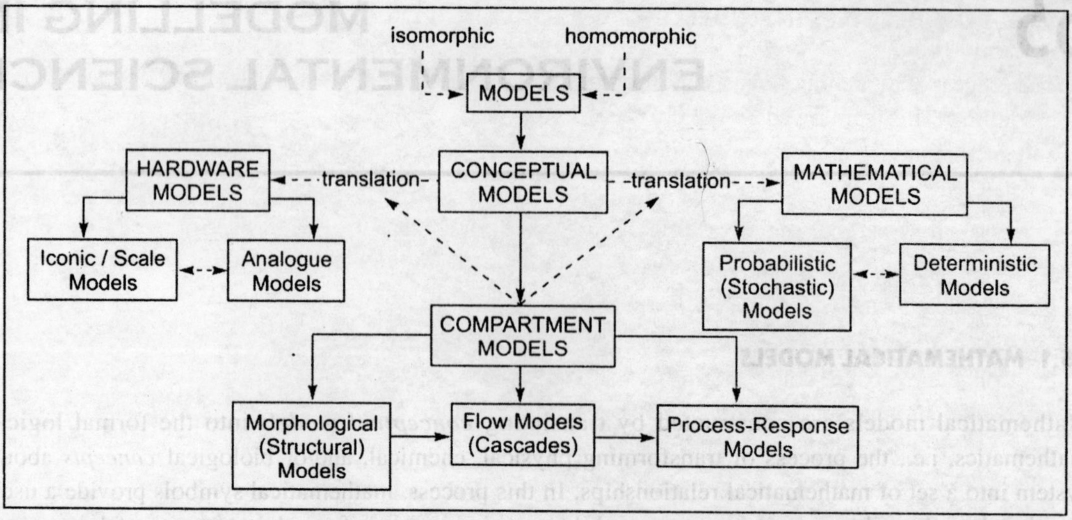

Fig. 55.1: Categories of mathematical models

The equations and the mathematical relationships which define the model can take several forms and the kinds of mathematical manipulation employed to develop the model can also vary. For example, the symbolism of *matrix algebra* is often used to display system relationships and to manipulate those relationships, while *difference* and *differential equations* are the basis of many models which aim to describe quantitatively the way systems change over time (see the surface-water balance equation in Chapter 12 and the treatment of primary production and population growth in Chapters 19 and 25, respectively). In more advanced models, relationships may be modelled *statistically* and the basis of the logical operation of the model may lie in *probability theory*. Here we may recognise an important—but to some degree arbitrary—distinction which is usually made when considering mathematical models.

Where the model is expressed in such a way that the outcome of the operation of the model (the experiment) is fully determined by the initial values of the state variables, input variables, parameters and factors, it is called a deterministic model. In such models the outcome of the operation of the system is fully predicted by the model, so that the precision of the model is only constrained by its realism. Where the model is formulated in such a way that it acknowledges an element of uncertainty and attempts to include the effects of random variation in forcing functions and parameters, it is called probabilistic or stochastic model. Such models, therefore, do not predict the outcome of the operation of the system (the experiment), but instead predict the probability of the outcome with some degree of confidence defined by the model.

There are two reasons for using stochastic models. *First*, there may be good *a priori* theoretical reasons to believe that the real world system possesses inherent elements of uncertainty, or most of its behaviour is, at least in part, random. *Secondly*, although in theory a system may be capable of being fully specified and therefore of being described by a deterministic model, in practice that cannot be achieved. This is usually because there is a mismatch between the complexity of the system and the degree of realism of the model we can formulate, i.e., our deterministic model, is very coarse. In these situations the uncertainty introduced by lack of realism can be partially overcome by incorporating a random element into the model, i.e., using a stochastic model. The major categories of mathematical models are shown in Fig. 55.1.

55.2 POSSIBLE STEPS IN MODELLING APPROACHES

The process of modelling consists of different steps. The first step is always that of descriptive or conceptual modelling. This is followed by field observation and data generation. Subsequently, both physical models and mathematical models can be developed in various ways through interaction analysis at various levels.

Usually potential model users have to make a choice out of a number of models available for description/production of the behaviour of a chemical, ranging from simplified screening models to highly sophisticated dynamic 3-dimensional water quality models etc.

As a first step in the process of model building, a descriptive model is shown of the process that controls the exposure concentrations of chemicals in water. It lists the process to be accounted for in a mathematical simulation model. The next modeling step is to express the rates of these processes in mathematical formulations. Ideally, these formulations should express the mass rate of change of the chemical resulting from each of the individual processes in terms of characteristics of the environment and basic properties of the chemical. The basic form of these expressions may be:

$$v\ dc/dt = kc$$

where, v = volume,

c = concentration of the chemical,

k = first order rate constant of the process.

Finally, the mathematically expressed process rates are to be integrated into equations describing the spatial and temporal distribution of the chemical. This is generally done by using the principle of conservation of mass. The general expression for the mass balance equation is:

$$V_i\ dC_i/dt = E_i + A_i + D_{ij} + R_i$$

where, V_i = volume in compartment i,

C_i = Concentration of the chemical in compartment i,

E_i = Emission of the chemical into compartment i,

A_i = Net sum of advective import and export for compartment i,

D_{ij} = Net sum of dispursive exchange between compartment i and the other compartment j,

R_i = Sum of transformation reactions.

This equation describes the mass rate of change of the chemical because of the next effect of the various fluxes and transformations. The term compartment is used here in physical sense and refers to atmosphere, water, suspended particulate matter, sediment etc.

A useful way to gain understanding with a model involves systematic variation of parameters, in order to find the parameters to which the model output is most sensitive. The procedure of sensitivity analysis helps to identify the key processes and pathways for the chemical.

55.3 LIMITATIONS OF MODEL APPLICATION

Both scientists and decision-makers often raise doubts on the confidence limit of a model, because the predictions of the model differ widely from the real world. This is primarily due to proper inadequacy of visualisation of descriptive model and lack of its subsequent validation through field observations.

Majority of the models now used—either with respect of the fate of chemicals in the environment or risk assessments etc.—are primarily based on descriptive features translated into quantitative forms using

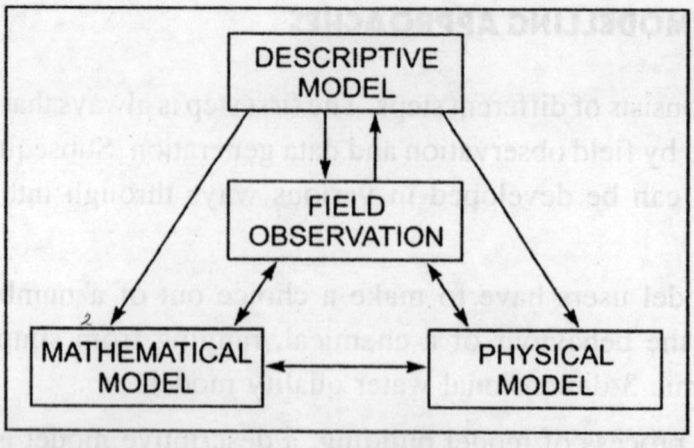

Fig. 55.2: Possible modelling approaches and their relationships

data of field observations/physical experimentations. Most of these models are developed in temperate countries. Many simulation models which were predicted for the tropics appear to be futile because of inadequacy of descriptive model and proper assessment of field observations. For instance, the prediction of methane emission rates—as suggested through simulation model in USA—is far behind the real-world situation.

Besides considerable materials and financial resources, running sophisticated models requires detailed, high quality input. Without adequate supporting (site specific) data, through testing and not at least without a high level of skill of the user, sophisticated models yield nothing but a false sense of security.

55.3.1 Modelling the Fate of Chemicals in Aquatic Environment

As a first step in the process of model building, a descriptive model of the processes that control the exposure concentrations of chemicals in water is given in Fig. 55.3. It lists the processes to be accounted for in a mathematical simulation model.

The next modelling step is to express the rates of these processes in mathematical formulations. Ideally, these formulations should express the mass rate of changes of the chemical resulting from each of the individual processes in terms of characteristics of the environment and basic properties of the chemical. The basic form of these expressions may be:

$$V\,dc/dt = k\,c$$

1. where, v = volume,
   c = concentration of the chemical,
   k = (pseudo-) first order rate constant of the process.

Finally, the mathematically expressed process rates are to be integrated into equations describing the spatial and temporal distribution of the chemical. This is generally done by using the principle of conservation of mass. The general expression for the mass-balance equation is:

$$V_i\,dC_i/dt = E_i + A_i + D_{ij} + R_i$$

2. where, V_i = volume of compartment i,
   C_i = concentration of the chemical into compartment i,
   E_i = emission of the chemical into compartment i,
   A_i = net sum of advective import and export for compartment i,
   D_{ij} = net sum of dispersive exchange between compartment i and the other compartment j,
   R_i = sum of transformation reactions.

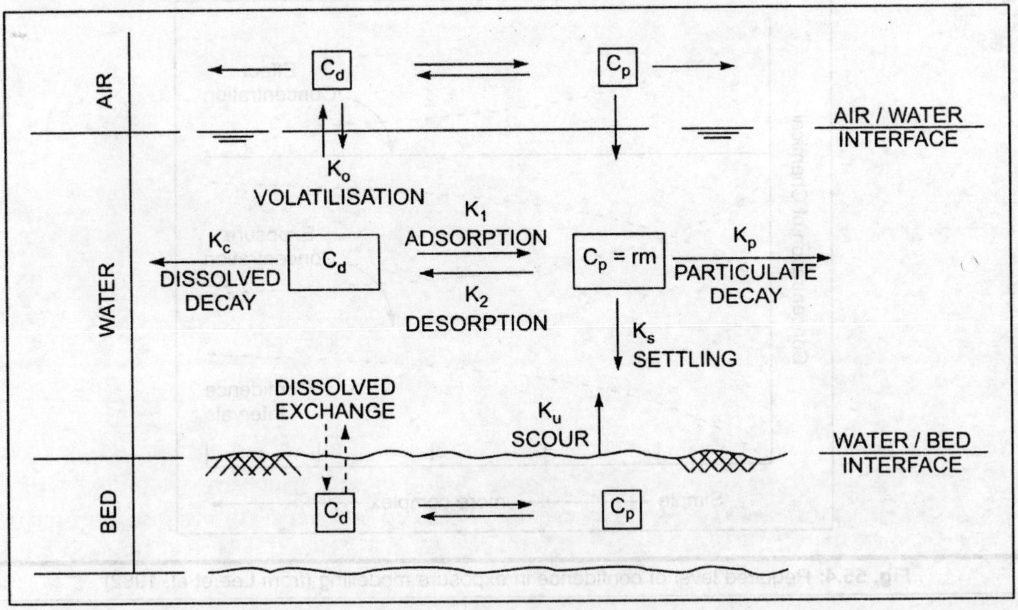

Fig. 55.3: Descriptive model of the behaviour of chemicals in water (from O'Connor and St. John, 1982)

This equation describes the mass rate of change of the chemical because of the net effect of the various fluxes and transformation. The term "compartment" is used here in physical sense and refers to atmosphere, water (dissolved phase), suspended particulate matter, sediment, etc. Mass-balance equations can be written for all the compartments. This body of differential equations is called "the model". Solution of these equations is usually done by numerical techniques, by means of computers. The result of this computation is time series of concentrations in all compartments.

However, in many cases, one is not interested in the temporal variations of the concentrations. For instance, if the time scale of the variations is short in comparison with the time scale of the study, it leads to apparently steady concentrations. If the system is at the steady state, all mass-balances become equal to zero. If also the process expressions take a simple linear form, like equation (1), the model becomes a very simple set of linear expressions, which can be solved algebraically. If the number of compartments is small, the computation can be done manually, using a pocket calculator. An excellent example of this type of modelling is given by DiToro et al (1982).

55.3.2 Required Level of Sophistication

Usually potential model users have to make a choice out of a number of models available for description/ prediction of the behaviour of a chemical, ranging from (over-) simplified screening models to highly sophisticated dynamic, 3-dimensional water quality models. Selection of the minimum level of sophistication necessary for a specific modelling task should be based on some objective criteria:

1. Purpose of the modelling activity: This should always be the key criterium. Possible modelling objectives are:

   * Prediction of the problems that may occur as a result of the introduction and the inevitable release of a new chemical into the environment.
   * Prediction of the concentrations that may result of a given release (or a proposed reduction of a release) at a specific place and time.

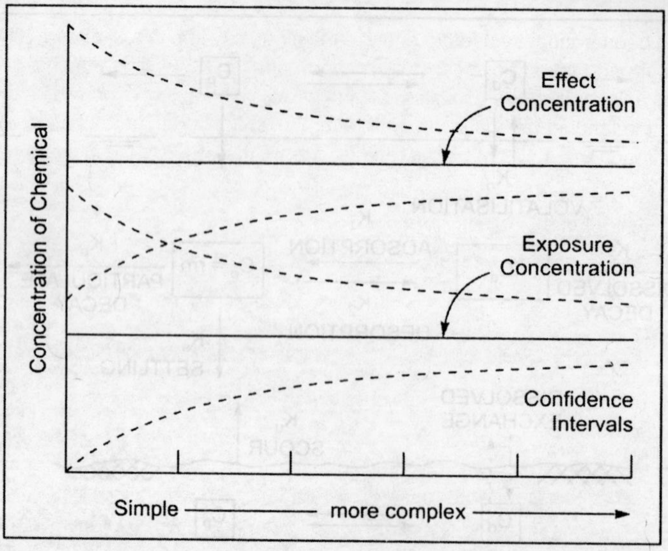

Fig. 55.4: Required level of confidence in exposure modelling (from Lee et al, 1982)

For the first category, very simple screening models like the "unit world models"—as introduced by Mackay and others—may be adequate, whereas the second category may require a higher level of sophistication.

2. Level of uncertainty to be tolerated: Simple models give only rough estimates of the concentrations that can be expected. Use of sophisticated models may result in more accurate predictions.

The required level of certainty should follow from the way the results are to be used. In hazard assessment, a predicted exposure concentration is to be compared with some sort of effect concentration. As shown in Fig. 55.4, the required level of certainty and, thus, the required level of sophistication of the model, is determined mainly by the size of the margin between these concentrations.

3. Availability of resources: Besides considerable material and financial resources, running sophisticated models requires detailed, high quality input. Without adequate supporting (site specific) data, thorough testing and not—at least—without a high level of skill of the user, sophisticated models yield nothing but a false sense of security.

Box 55.1: BOX-MODEL APPROACHES

A simple atmospheric dispersion model, called a "box-model", can be used to calculate groundlevel concentrations of specific air pollutants of concern emitted from the project-activity. A box-model is based on the assumption that pollutants emitted to the atmosphere are uniformly mixed in a volume, or "box", of air (Canter, 1985). The most critical aspect of the usage of the box-model is to establish, with rationale, the downwind, crosswind, and vertical dimensions of the box.

In addition, the time period over which pollutant emissions will be considered must be established; a typical time period is 1 hr. The time and physical dimension considerations are based on the assumption of steady state conditions; that is, it is supposed that the emissions, wind speed, and characteristics of air available for dilution will not vary over time (Ortolano, 1985). A box-model is also based on the assumption that discharges mix completely and instantaneously with the air available for dilution and the released material is chemically stable and remains in the air.

Contd.

Box-models have been frequently used to analyze the air quality impacts of airports (line sources of emissions from aircraft-landing and takeoff cycles) (Nelson and LaBelle, 1975). The dimensions of a box-model for an airport runway, as an example, are established as follows: (1) the height is determined to be 1,100 m based on emissions during the landing-takeoff cycle of an aircraft, (2) the length is a function of the type of aircraft and its flight angles through the 1,100 m height, and (3) the width is 1,600 m (representing calm wind conditions).

The box-model can be used for single-point, multiple-point, area, or line, or "hybrid-type" sources of air pollutants. It can also be used in valley settings.

The basic box-model is depicted in Fig. 55.5 and mathematically expressed as (Ortolano, 1985; Canter, 1985):

$$C = \frac{Qt}{xyz}$$

where, C = average concentration of gas or particulate < 20 μm in size, throughout box, including at ground level, $\mu g/m^3$,

Q = release rate of gas, or particulates < 20 μm in size, from source type(s), μg/sec (related to emission-factor information described in step 1),

t = time period over which assumption of uniform mixing in box holds valid, sec (typical period, 1 hr),

x = downwind dimension of box, m (chosen based on average wind speed and physical aspects of terrain),

Y = crosswind dimension of box, m (chosen based on average wind speed, source configuration, and physical aspects of terrain),

z = vertical dimension of box, m (chosen based on limiting inversion heights in area and physical aspects of terrain).

Contd.

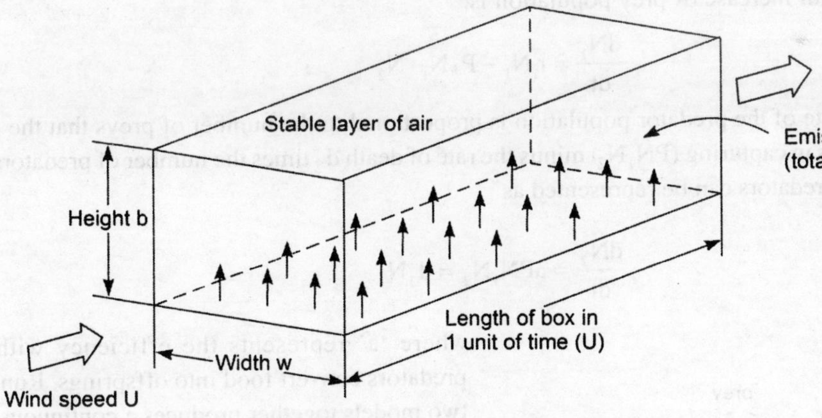

$$\frac{\text{Volume of air}}{\text{Time}} = b(L) \bullet w(L) \bullet U(L/T)$$

$$\frac{\text{Mass of residuals}}{\text{Time}} = P(M/T)$$

where, L = length units, M = mass units, T = time units.

Fig. 55.5: Air available for dilution in a Simple Box Model (Ortolano, 1985)

The box-model approach could be focused on the key pollutants of concern identified using mass-balance approaches. One method of establishing the dimensions of the box and the time period for the emissions is to use the data on limiting meteorological conditions as developed in step 2. In this approach, the box-model calculations are focused on worst-case conditions.

The box-model results can be interpreted on a pollutant-by-pollutant basis, in relation to existing ambient air quality and the relevant standard. It is important to compare the sum of the existing pollutant concentration and the concentration from the proposed project or activity, as calculated from the box-model, to the value given in the applicable standard. Greater significance should be attached to, and greater attention should be given to, those pollutants which do not meet or are near the allowable ambient air concentrations.

55.4 APPLICATION OF SELECTED MODELS

(a) Prey-Predator Model

Suppose the number of individuals of prey population is N_1 and their intrinsic rate of population increase r_1.

where, $r_1 = (b - m)$,

b = birth rate per individual,

m = mortality rate per individual.

Then the rate of change of the prey population size (without considering immigration and emigration) will be:

$$\frac{dN_1}{dt} = r_1 N_1 \quad (1)$$

However, if a predator population (number of individuals = N_2) is present, it will kill and consume a proportion of the prey it encounters. Lotka and Volterra assumed the encounter to be the product of the prey and predator population ($N_1 \cdot N_2$) times a coefficient of perdition, P.

Thus the overall increase of prey population is:

$$\frac{dN_1}{dt} = r_1 N_1 - P \cdot N_1 \cdot N_2 \quad (2)$$

The growth rate of the predator population is proportional to the number of preys that the predator population succeeds in capturing (PN_1N_2) minus the rate of death d_2 times the number of predators. So the growth rate of the predators can be represented as

$$\frac{dN_2}{dt} = aPN_1N_2 = d_2N_2 \quad (3)$$

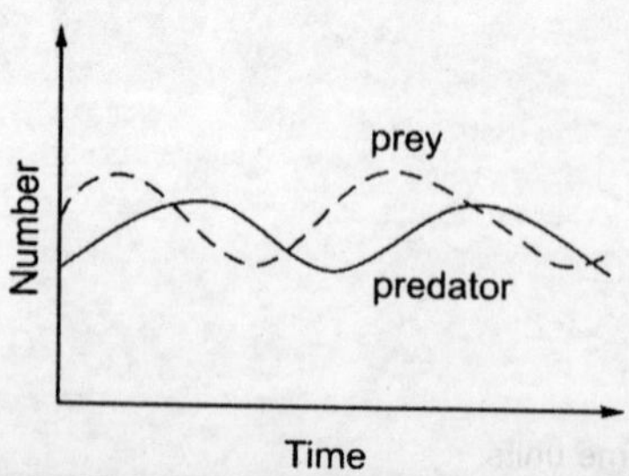

Fig. 55.6: The population dynamics of a predator & its prey, obtained from the Lotka-Volterra model

where 'a' represents the efficiency with which predators convert food into offsprings. Running the two models together produces a continuous cycling of both populations with a lag period. A large prey population allows the predator numbers to grow, but this leads to a decline in the prey abundance. Predator numbers follow this decline since food becomes increasingly scarce. Eventually the predator numbers are so low that the prey population is able to recover and start the cycle once again (Fig. 55.6). The cycling will continue with a period and amplitude set by the terms of the equation. This is maintained unless the system is disturbed, when a new amplitude is set by the size of the disturbance. The stability of the system can be increased by introducing the carrying capacity of the environment for the prey population which limits the population growth in the absence of the predators, so:

$$\frac{dN_1}{dt} = rN\left(1 - \frac{N}{K}\right) - P \cdot N_1 N_2 \tag{4}$$

Further increase in stability will result from limiting the predator population. The simplest way is to place a limit on the number of prey consumed, by assigning to it one of the three types of functional response. Functional response is the type of response of a predator to an increase in pest numbers. For biological control a predator with a sigmoid response is generally used. In this type there is a brief lag in the rise of consumption (Fig. 55.7), but then the rate of consumption increases rapidly. The initial lag is due to a predator learning prey image and then adjusting to the behavioral mechanism to find, catch, kill and consume an individual prey. This time decreases with the age of the predator and thus age distribution is an important aspect.

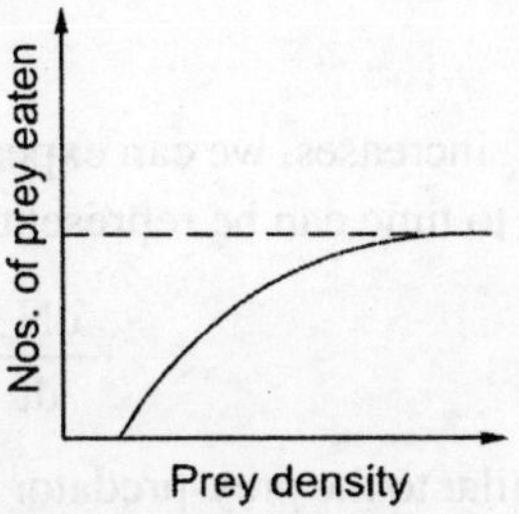

Fig. 55.7: The sigmoidal response of a predator population. The lag time represents learning a prey image and subsequent adjustments to find, catch and kill prey

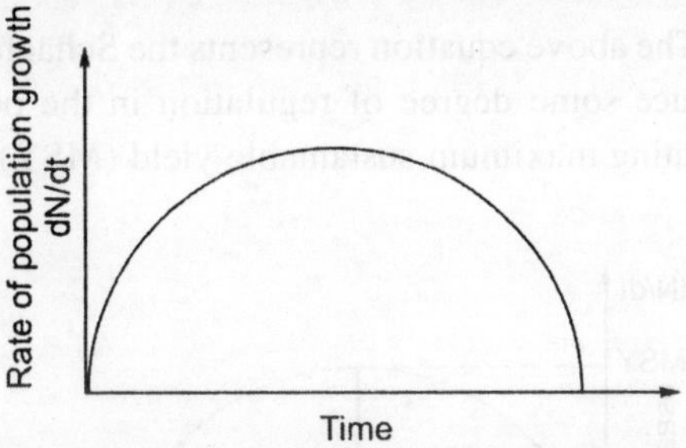

Fig. 55.8: The rate of increase in population size with time. This rises as N increased but begins to decline as density-dependent factors slow population growth when N approaches K

(b) Models in Fisheries

Proper management of fisheries aims at achieving a sustainable yield which has proved difficult due to economics leading to overfishing (Fig. 55.8).

In this model the death rate consists of two components—natural mortality (M), and the fish taken in the catch (F). If the population is not to decline, these losses have to be met by recruitment of new fish (R).

If the biomass of the stock is to be maintained, tissue growth (G) and recruitment have to match these losses:

$$F + M = R + G \tag{1}$$

F, the instantaneous fishing mortality rate, can be computed as

$$F = qE \tag{2}$$

where, q = the catchability coefficient depending on the type of gear used for fishing,
and E = fishing effort (boat-hours per unit area).

In effect, q is the proportion of the catchable stock caught per unit effort, or the efficiency of the fishing operation. So qE is the fishing mortality rate (F) for a given level of effort (E).

The yield Y is, therefore, equal to F multiplied by the size of the catchable stock (N_c) or

$$Y = FN_c \tag{3}$$

The yield may be considered to be at equilibrium when the population can maintain itself by replacing the harvest through G and R.

The catch per unit effort (U) is given by

$$U = \frac{Y}{E} \tag{4}$$

$$\text{or,} \quad U = \frac{FN_c}{E} = \frac{qEN_c}{E} \quad \text{or,} \quad U = qN_c$$

$$\text{or,} \quad N_c = \frac{U}{q} \tag{5}$$

Thus, if N_c increases, we can expect more fish to be caught for each unit effort. The catchable stock N_c with respect to time can be represented as

$$\frac{dN_c}{dt} = rN_c\left(1 - \frac{N_c}{K}\right) - qEN_c \tag{6}$$

This is similar to the prey-predator model, and an equilibrium is achieved when the rate of harvesting is balanced by the rate of replacement, or, when

$$rN_c\left(1 - \frac{N_c}{K}\right) - qEN_c = 0$$

The above equation represents the Schaefer's model and has played an important part in attempts to introduce some degree of regulation in the otherwise open access fisheries, and is currently used in calculating maximum sustainable yield (MSY).

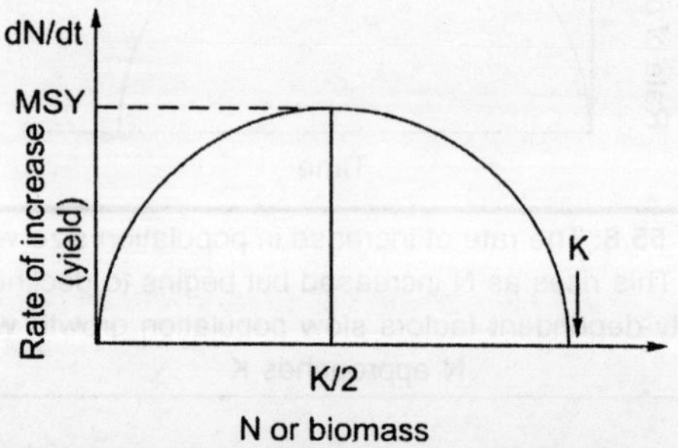

Fig. 55.9: The rate of increase in numbers or biomass also rises and falls as N approaches the carrying capacity. The maximum rate of increase represents the maximum sustainable yield (MSY) & this is equivalent to K/2

The logistic equation gives a sigmoid population growth curve (Fig. 55.9). However, the rate of change of a population dN/dt plotted against time produces a parabola because the rate of growth of the population decreases as it approaches the carrying capacity of the environment. A similar graph is obtained if dN/dt is plotted against N (Fig. 55.8). The graph shows that the maximum population growth (or the rate at which new individuals are added to the population) is at the point K/2, and also represents the limit of harvest without the population declining. Thus the maximum rate of increase represents the maximum sustainable yield (MSY).

Although the K and MSY can be calculated for closely managed aquatic systems but is impossible in the case of sea fisheries. Thus managers have to decide on MSY using the evidence of their catch sizes. Using past records for seasons in which the same fishing gear has been used, the relation between catch per unit effort U and fishing effort E is established using statistical analysis (Fig. 55.10(a)). From this the yield can be predicated as a function of the fishing effort and then the MSY can be derived (Fig. 55.10(b)).

However, there are certain weak points of the model. It does not take into assumption the age distribution of the fish population, and secondly it is unrealistic to assume that K remains constant for any length of time. Pitcher and Hart (1982) gives a clear description of the collapse of the Peruvian anchovy

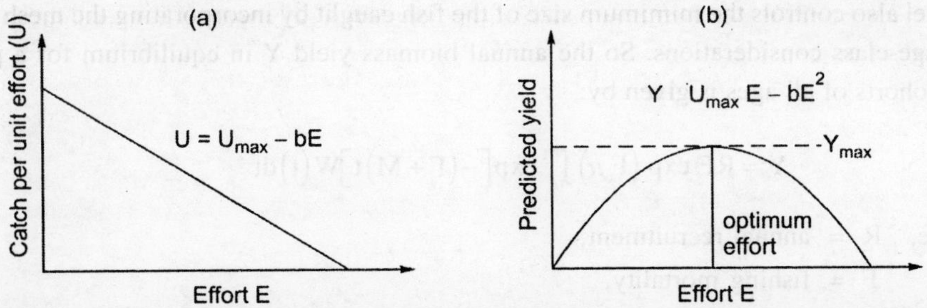

Fig. 55.10: (a) A line of best fit for catch per unit effort against effort E. This is calculated using linear regression equation to derive the equation for the line using data from fishing records. (Using the same fishing gear). U_{max} is the maximum catch per unit effort, the intercept on the y-axis and b is the slope of the line (b) The prediction of maximum yield (Y_{max}) from fishing effort E

fishery because of a simple reliance on MSY's (Fig. 55.11) with due allowance for the changing environmental systems.

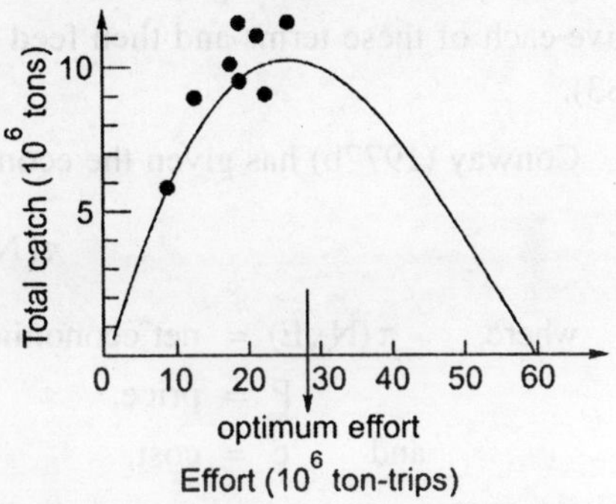

Fig. 55.11: The prediction of the yield from the Peruvian achovy fishery before its collapse in 1972 (from Boerema and Gulland, 1973)

The second approach is the dynamic pool model of Beverton and Holt (1957). In this model the total yield is derived as a series of simple steps: the number caught in each age-class is the fishing mortality rate for that age class mulitiplied by the number of fish in it (F_tN_t). Multiplying this catch size by the average weight of a fish in that cohort gives the biomass it yields to the fishery. The total yield is the sum of these values for all cohorts in the catchable stock:

$$Y = \sum_{t=t_c}^{t_m} F_t N_t W_t \tag{7}$$

where, Y = the total yield of the fishery,
N = number of fishes in age-class t,
W = average weight in age-class t,
F_t = the instantaneous fishing mortality mortality rate for age-class t,
t_c = the age at which an individual is recruited to the fishery,
and t_m = maximum age of the fish in the stock.

Thus, to derive the total yield, estimates of N, F and W are needed for each age-class. The number in a particular cohort (N_t) will depend upon the number of recruits entering the catchable stock and their subsequent survival. The exponential model is used to derive this as:

$$N_t = Re^{-(F+M)t} \tag{8}$$

where, N_t = number of fish still alive after t years in the catchable stock,
and R = the number of recruits that originally entered the pool in this cohort.

This model also controls the mimimum size of the fish caught by incorporating the mesh size of the net based on age-class considerations. So the annual biomass yield Y in equilibrium for a population consisting of cohorts of all ages is given by:

$$Y = RF\exp\left(F_t\mu\right)\int_{t\mu}^{N_t}\exp\left[-\left(F+M\right)t\right]W\left(t\right)dt \qquad (9)$$

where, R = annual recruitment,

F = fishing mortality,

M = natural mortality,

$W_{(t)}$ = Weight computed by the von Bertalanffy function (Chapter 4),

μ = mesh size.

So, t_μ is the age when the fish are first caught. The model can thus be used to prescribe management strategies in terms of fishing mortality and mesh size. Generally commercial fisheries run submodels to derive each of these terms and then feed these predictions into the Beverton and Holt model (Gulland, 1983).

Conway (1977b) has given the economic formulation:

$$\pi\left(N_c, E\right) = PqEN_c - cE \qquad (10)$$

where, $\pi(N_c, E)$ = net economic revenue,

P = price,

and c = cost.

For the open access fishery the supply curve for sustained yield is of the form

$$Y = rc\left(1 - c / pK\right) / P \qquad (11)$$

High prices thus leads to biological overfishing (Fig. 55.12).

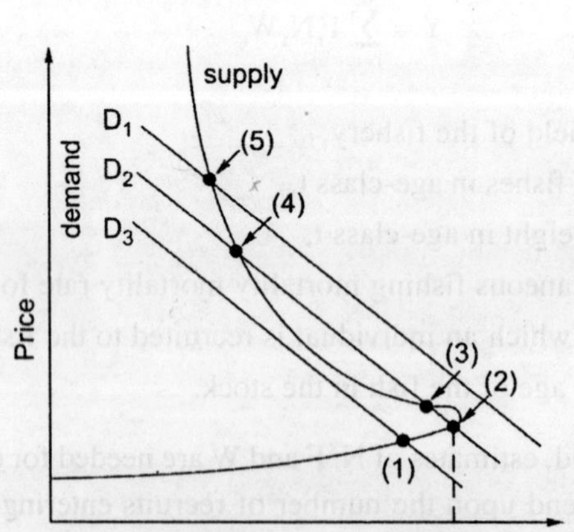

Fig. 55.12: Instability of supply and demand in the open access fishery as revealed by the Schaefer model. For demand D_2 the equilibrium is at 2 or 4—depending on the starting population. However, a slight demand shift to D_3 results in inevitable overfishing (from Conway 1977)

There are economic reasons for not fishing too close to the MSY. If the unit price of the yield is fixed and the price of the fish remains the same whatever the supply, the revenue from the catch will simply follow the yield curve. Additional fishing effort will incur additional costs: in this case, costs are assumed to rise in a simple manner with fishing effort (Fig. 55.12). The profits from the catch are the differences between the revenue received (price) and the costs. The greatest profits are made by fishing at some level before the MSY is reached. Thus the maximum sustainable yield does not necessarily produce the maximum economic yield.

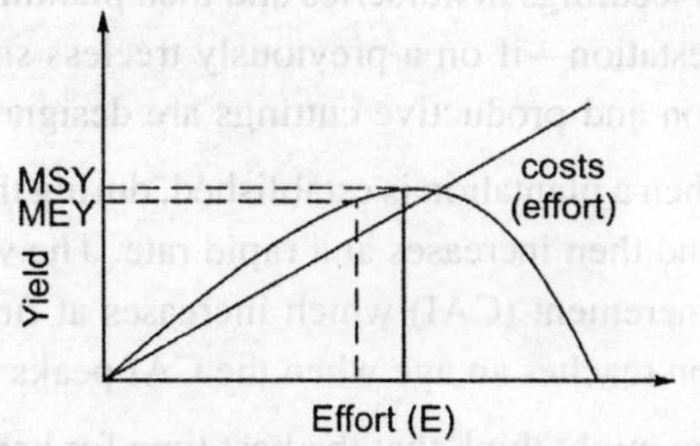

Fig. 55.13: As the yield rises and falls so does the revenue which it generates. Cost increases with fishing effort (E). The difference between revenue and cost is the profit. The greafest difference occurs somewhat below MSY. This is the maximum economic yield (MEY) (from beeby 1993)

Proper management programmes involves the regulation of such aspects as:

(a) limiting the total permitted annual catch
(b) limiting the size of the fishing fleet
(c) limiting the length of the fishing season
(d) limiting the type and size of the fishing gear
(e) limiting the areas in which the fish may be caught
(f) prohibiting the landing of undersized fishers
(g) closing nursery areas to fishing
(h) prescribing minimum mesh or hook sizes.

The suitability of the different methods of control will vary with the nature of the fishery. In the oceans the highest productivity is near the continental shelf because these are the zones of nutrient upwelling. Thus concentration of fish stocks are near the shore. Since the late 1970s the ocean extending 200 miles (320 km) off the shore of each country has been the 'Exclusive Economic Zone' of the country, meaning that it legally controls the fishing within that zone. This covers most of the world's fishing, since most fishing is within 200 miles of the shore, leaving a small proportion in international waters where agreements on control of fishing require negotiations between governments.

(c) Forest Model

This is a model used for assessment of sustainable yield. A forest, in a physical sense, does not move from place to place, but moves in the sense of growth. Having once established, it can grow if it has room. Periodic harvest of increment furnishes this room: Commonly plantations are created by

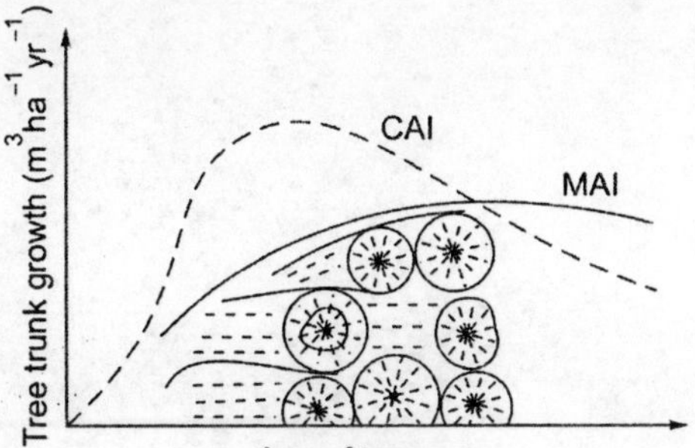

Fig. 55.14: Relation of tree trunk growth to plantation age. CAI—current annual increment, MAI—mean annual increment (from Newman, 1993)

growing seedlings in nurseries and then planting them in the harvested places. Technically, this is known as afforestation—if on a previously treeless site, and reforestation if it is to reclothe a logged-off area. Protection and productive cuttings are designed to maintain and increase productivity.

When a plantation is established, during the early years, growth is slow (following a sigmoid growth curve) and then increases at a rapid rate. The wood production per hectare per year is called the current annual increment (CAI) which increases at first and then decreases. However, between this period the plantation reaches an age when the CAI peaks (Fig. 55.14) and then starts to fall.

One might think that the best time for harvesting is when the CAI is at its maximum, but this is not so. Harvesting should be carried out when the mean annual increment (MAI) is at a maximum. This is represented as:

$$\text{MAI} = \frac{\text{Volume of timber in stand}}{\text{age of stand}}$$

This is what determines the timber yield per 1,000 years if there is a continuous series of rotations on a site. Like the CAI, the MAI also rises to a maximum as the stand ages and then starts to decline, but the maximum occurs later than the CAI.

56 INDUSTRIAL POLLUTION MANAGEMENT: THERMAL POWER PLANT

56.1 INTRODUCTION

As per notification of Ministry of Environment and Forest (January, 1994), 29 different activities are needed to undertake EIA studies before the planning for new setting up or expansion. All these activities are broadly categorised as:

— Industries

— Mining

— Thermal power plants

— River valley projects

— Ports, Harbours and Airports

— Communication

— Atomic energy

— Transport (Rail, Road, Highway)

— Tourism (including hotels, beach resorts)

To enable the investors to prepare the requisite documents—such as Environmental Impact Assessment Reports/Environmental Management Plans etc.—as well as to choose the most appropriate sites for their projects, the Ministry has prepared comprehensive guidelines for each of the sectors.

For getting clearance from the Ministry, the following documents are to be submitted :

— Feasibility/Project Report

— Site clearance

— No objection certificate from SPCBs and other local authorities

— Environmental appraisal questionnaire

— EIA report/EMP layout

— Risk analysis/Energy preparation plan

— Rehabilitation plans if large scale displacement of people is anticipated.

The documents submitted by a developer are first scrutinised by a Multi-disciplinary staff competent for functioning in the Ministry of Environment and Forests who may also undertake site visits wherever required, interact with the developers and hold consultations with experts on specific issues as and when necessary. Public hearings could be called for in case of projects involving large displacement or having severe environmental ramifications.

56.2 ENVIRONMENTAL GUIDELINE FOR POWER PLANTS IN INDIA

The power sector in India has registered remarkable growth since India gained independence in 1947. Power generation increased from 1,362 MW in 1947 to about 98,000 MW in 2000. Power generation from thermal, hydro, and nuclear sources in 2000 was 71,341 MW (73 per cent), 23,816 MW (24 per cent), and 2,680 MW (3 per cent), respectively. Coal, gas, and diesel are contributing 84 per cent, 13 per cent, and 3 per cent of thermal power generation, respectively. Coal is the major fossil fuel for power generation in India, and all utility boilers use a pulverized firing system for combustion of coal.

As demand for power generation increases, there is a growing need to ensure acceptable environmental quality. Environmental issues need to be integrated into the planning, development, and operational processes of power plants so that these plants maximize positive socioeconomic impacts and minimize negative environmental impacts. Appropriate environmental guidelines play an important role in establishing the base for environmentally sound development by facilitating developer adoption of environmentally sound technology and better management practices.

The Ministry of Environment and Forests (MoEF) of the Government of India has promulgated environmental guidelines for thermal power plants, which include siting criteria, as well as requirements regarding environmental impact statements (EIS) and environmental management.

The Central Pollution Control Board (CPCB) has also prescribed a set of guidelines for permissible flue gas emissions and liquid effluents for thermal power plants using coal, oil, and natural gas as fuel. In 1994, CPCB issued revised National Ambient Air Quality Standards (under the Air Act of 1981) for industrial, residential, rural, and other areas and sensitive areas.

MoEF has issued guidelines for the siting of thermal power plants in India. The siting guidelines are intended to protect ecologically sensitive areas, archaeological sites, national parks and sanctuaries, settlements, and other areas. Proper siting of thermal power plants not only reduces the cost of required pollution control measures, it can also limit the damage that these stations could cause to the natural and human environment.

Environmental impact assessment (EIA) is the process by which environmental factors are integrated into project planning and decision making to achieve ecologically sustainable development. The best-practice EIA identifies environmental risks, lessens conflict by promoting community participation, minimizes adverse environmental effects, informs decision makers, and helps lay the foundation for environmentally sound projects.

In India, the Central Pollution Control Board (CPCB) has prescribed emission standards for particulates and gaseous pollutants from power plants using coal or natural gas fuel. Emission standard (1986) is given below:

Particulate Matter (mg/Nm^3)	SO_2 (mg/Nm^3)	NO_x (mg/Nm^3)
150 (210 MW and above)	220 m stack height (200 to 500 MW)	50 ppm (Natural gas plants of 400 MW and above)
350 (<210 MW)	275 m stack height (500 MW and above)	75 ppm (Natural gas plants of 100-400 MW)
		100 ppm (Natural gas plants of less than 100 MW)
		100 ppm (Naphtha plants)

The thermal power plant also produces waste water and fly ash. The fly ash must be used or disposed safely. The details are discussed latter. Regarding waste water discharge standards the CPCB guideline is given below:

Table 56.1: Standards for discharge of liquid effluents

Source	Pollutants	Concentration
1. Condenser cooling water (once through cooling system)	pH	6.5-8.5
	Temperature	more than 5°C than the intake water temperature
	Free available Chlorine	0.5 mg/l
2. Boiler blow down	Suspended solids	100 mg/l
	Oil and grease	20 mg/l
	Copper (total)	1.0 mg/l
	Iron (total)	1.0 mg/l
3. Cooling tower blow down	Free available chlorine	0.5 mg/l
	Zinc	1.0 mg/l
	Chromium	0.2 mg/l
	Phosphate	5.0 mg/l
	Other corrosion of materials	Limit to be established on case by case basis
4. Ash pond effluent	pH	6.5-8.5
	SS	100 mg/l
	Oil & grease	20 mg/l
No limits for heavy metals are given at present		

56.3 THERMAL POWER PLANT, KOLAGHAT (W. B.) — A CASE STUDY

Kolaghat Thermal Power Station (KTPS) is located on the bank of Rupnarayan near Kolaghat and Mechada Railway Station (Dist. Midnapore). This is a Power Plant of West Bengal Power Development Corporation Ltd. (WBPDCL), having six 210 MW Power generating units. This is a coal-based power plant, which occupies a total area of 1159.738 acres. About 45 million tonnes of coal is consumed annually in six pulverised coal-fire boilers and thus generates about 15 million tonnes of bottom ash and fly ash. Pollution potential of a typical thermal power plant is shown in Fig. 56.1.

As a part of typical EIA data generation the following activities were made:

— Ambient air quality monitoring

— Ambient noise monitoring

— Micro-metrological data generation

— Water quality monitoring

— Soil quality monitoring

— Land use and land cover studies

— Terrestrial and aquatic ecological studies; and

— Socioeconomic studies.

Subsequently, impact prediction on target system, mitigatory plan in the form EMP was developed in the report.

Some of the salient environmental data are presented in the Tables 56.1, 56.2, 56.3 and 56.4.

Table 56.2: Hazardous Chemical and Gases Consumed at KTPS

Item	Annual consumption (approx)	Stored in
(a) Acid — HCl	1,850 t	Acid tank
(b) Acid H_2SO_4	890 t	Do
(c) Hydrazine hydrate	4.2 t	200 L drums
(d) Chlorine gas	110 t	Cylindrical container of 1 t cap
(e) H_2 gas	4,200 t	Stored in 1 t cylinder
(f) Cyclohexylamine	3,600 kg	200 t drum

Table 56.3: Gaseous Pollutants Discharged to Atmosphere through KTPS stalks (1995-96)

Unit	Stalk discharge M^3/day	Emission conc. $\mu g/M^3$			Total emission t/day		
		PM	SO_2	NO_X	PM	SO_2	NO_X
I	22.86 Million	207.1	532.5	284.5	4.73	12.17	6.50
II	24.45 Million	391.5	527.9	265.8	9.57	12.91	6.50
III	25.33 Million	273.7	570.0	259.0	6.93	14.44	6.56
IV	23.53 Million	89.5	600.0	303.0	2.11	14.12	7.11
V	30.46 Million	95.5	620.0	273.0	2.91	18.89	8.31
VI	28.52 Million	90.0	590.0	290.0	2.57	16.83	8.27

PM — Particulate matters

Table 56.4: Results of Chemical Analysis of Fly Ash, Solid Waste and Soil

Parameters	Fly Ash	Bottom Ash (solid waste)	Soil (crop field)
pH (1 : 2.5)	7.7	8.2	7.8
EC (1 : 5) ($d_s\ m^{-1}$)	0.231	0.264	0.155
Organic matter (%)	2.06	1.17	1.55
Nitrogen (%)	0.059	0.060	0.096
Phosphorus (%)	0.015	0.020	0.046
Sulphur (%)	0.070	0.076	0.084
Iron (%)	1.42	3.61	1.81
Copper (ppm)	21.1	23.5	19.50
Manganese (ppm)	209.3	736.0	328.9
Zinc (ppm)	63.6	50.0	49.70
Lead (ppm)	NA	NA	NA
Cadmium (ppm)	NA	NA	NA
Chromium (ppm)	23.9	23.9	12.8
Nickel (ppm)	40.2	37.0	50.0
Mercury (ppm)	NA	NA	NA
Arsenic (ppm)	NA	NA	NA

On the basis of salient data of power plant sites, the impact prediction can be made by the conventional checklist method.

Further environmental management plan for control/disposal/usage of solid wastes, liquid effluents and air pollutants to improve the environment, ecology and aesthetic quality, a number of measures were suggested. KTPS has promised to undertake the said EMP as an ecofriendly measure. However, one of the major constraints was the problems relating to safe disposal of enormous quality of fly ash.

The flow process of a typical coal fired plant is shown in Fig. 56.2.

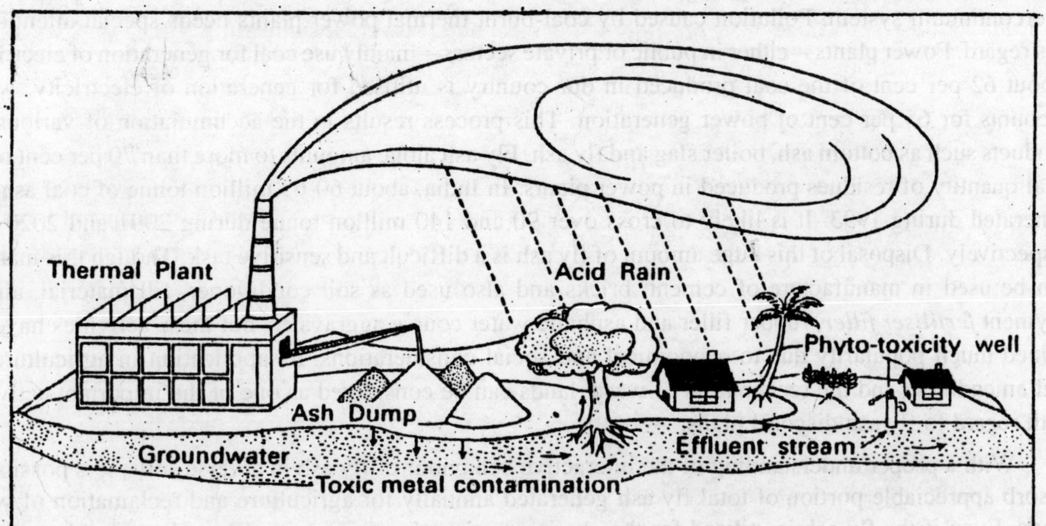

Fig. 56.1: Pollution potential of Thermal Power Plant

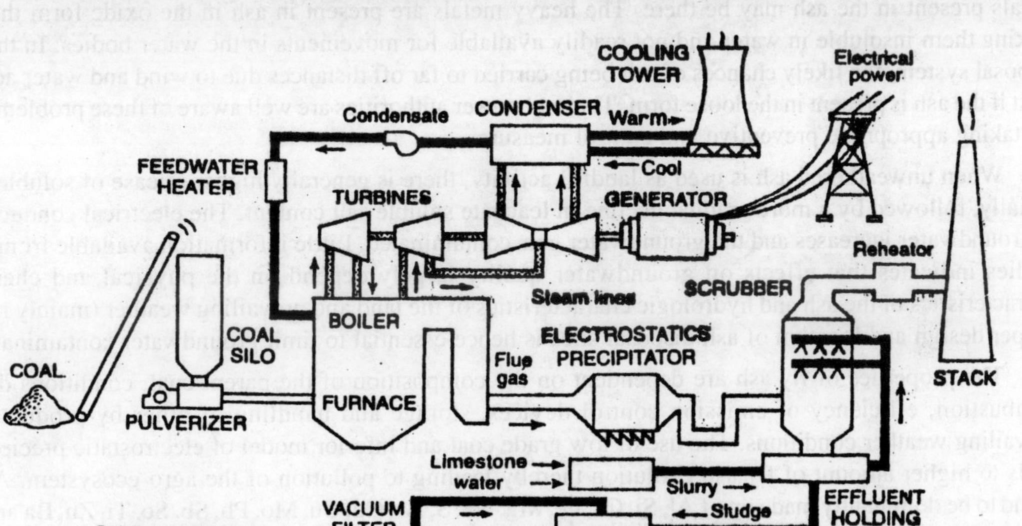

Fig. 56.2 : Flow process of a typical coal fired plant

56.4 ENVIRONMENTAL ISSUES ASSOCIATED WITH COAL ASH

The environment degradation is a major concern due to increased population and industrialised activities and is thus attracting attention of ecologists, scientists, engineers, town planners and politicians. The level of environmental awareness has also increased and needs collective thinking for evolving management options for checking environmental pollution. In this regard, emphasis needs to be laid on working environmental impact/assessment for any existing or coming project and defining disposal/recycling management options of effluent or sewage wastes for minimising the pollution of air-water-soil

intercontinuum system. Pollution caused by coal-burnt thermal power plants needs special attention in this regard. Power plants—either in public or private sectors—mainly use coal for generation of electricity. About 62 per cent of the coal produced in our country is utilised for generation of electricity, which accounts for 65 per cent of power generation. This process results in the accumulation of various by-products such as bottom ash, boiler slag and fly ash. Fly ash alone amounts to more than 70 per cent of the total quantity of residues produced in power plants. In India, about 60-65 million tonne of coal ash was generated during 1993. It is likely to cross over 90 and 140 million tonne during 2001 and 2020 AD, respectively. Disposal of this huge amount of fly ash is a difficult and sensitive task. Though this material can be used in manufacture of cement, bricks and also used as soil conditioner, fill material, airport payment *fertiliser filler*, rubber filler and asphatic water course aggravates, but these activities have not gained much popularity due to economical and social considerations. Its application in agriculture for soil amendment and the reclamation of waste lands can be considered as one of the important activities with regard to fly ash disposal plans.

With a proper understanding of the interactive behaviour of fly ash in diverse soils, it is possible to absorb appreciable portion of total fly ash generated annually for agriculture and reclamation of waste lands. Even if the fly ash is utilised for the above-mentioned activities, it will not be possible to utilise even 30-40 per cent of the ash produced. Thus there is a need to store the ash produced in such a way as to have minimum damage to air, water and soil bodies. The ash disposal is done either in *wet slurry* or *dry form*. In the wet slurry form, the possibility of ground and surface waters pollution primarily due to heavy metals present in the ash may be there. The heavy metals are present in ash in the oxide form thereby making them insoluble in water and not readily available for movements in the water bodies. In the dry disposal system, the likely chances of ash being carried to far off distances due to wind and water actions exist if the ash is present in the loose form. Thermal power authorities are well aware of these problems and are taking appropriate preventive and control measures.

When unweathered ash is used as landfill activity, there is generally higher release of soluble salts initially, followed by a more gradual decline in leachate soluble salt content. The electrical conductivity of groundwater increases and the groundwater gets contaminated. Little information available from field studies indicates that effects on groundwater quality largely depend on the physical and chemical characteristics of the ash and hydrologic characteristics of the land and prevailing weather (mainly rains). Proper design and location of ash disposal sites is hence essential to limit groundwater contamination.

The properties of fly ash are dependent on the composition of the parent coal, conditions during combustion, efficiency of emission control devices, storage and handling of other by-products and prevailing weather conditions. The use of low grade coal and inferior model of electrostatic precipitator leads to higher amount of fly ash evolution thereby leading to pollution of the agro-ecosystem. Ash is found to be dominantly made up of Al, Si, Ca, Fe, Mg, Na, S, As, Cd, Cu, Mo, Pb, Sb, Sc, Ti, Zn, Ba and Se. Concentrations of B, Mo and Se biologically toxic elements in fly ash greatly exceed their concentration in soil. Carbon and nitrogen are usually present in small amounts in fly ash, whereas the available P and K levels are generally higher as compared to the parent material i.e. coal. The fly ash particles have size distribution with median value around 25 μm while the bottom ash has relatively higher value around 70 μm.

A super thermal power plant built on about 400 hect of land normally requires 500 hect of land for ash disposal. On the basis of the ash production trends, the area requirement for dumping of the ash is around 40,000 hectares. Power plants are preferably placed away from the human settlements and moreover on wastelands, but with course of time some of the cultivable area is also covered for ash mound site. Fly ash is dumped in huge quantities in the ash mound area around plant site. The fly ash constitutes very fine and light particles, dominant in silt and clay size fractions and prone to erosion primarily due to wind and water. The fly ash particles, on account of their low particle density, can be carried to far off distances.

There is a definite settling gradient of these particles with maximum deposition being closest to the site and a gradual and continuous decrease in settle density with increasing distance from the source. Fraction of ash particles gets deposited over tree and crop canopies and may result in a physical barrier in canopy-atmosphere system impeding plant processes like transpiration, photosynthesis and respiration and can thus have negative affect on crop growth and subsequent yield. The portion of ash getting incorporated in soil can alter physical, chemical and biological characteristics of the soil. Although research studies have shown that addition of ash in the soil below a certain threshold level can be beneficial for crop growth and yield but higher amounts added lead to texture and structural changes in the upper soil layers, surface crust formation impeding the water intake capacity of the soil, addition of toxic elements and alteration of physico-chemical properties as pH, cation exchange capacity and electrical conductivity. These changes can affect the crop growth with regard to soil moisture and nutrient availabilities, seedling emergence and crop stand establishment, root growth in terms of depth and distribution, shoot growth and subsequent yields.

Revegetation of fly ash filled lands/sites often proceeds slowly due to modifications in the physical and/or chemical characteristics which is deleterious to plant survival. Existing tree species around a thermal power plant continuously get dust deposition on the canopies, thereby resulting in biophysical and physiological changes. Bio-modifications coupled with the ash presence in the soil-plant-atmospheric system needs to be investigated in detail for understanding the fly ash effects on agriculture. Several research studies conducted in different agro-climatic zones for understanding the fly ash incorporation effects on soil and crop productivity showed increased yields for most of the agricultural crops up to a given level of addition in the soil and thereafter decrease in trend with further increased application rate. But for drawing inference, there is a need to conduct experiments to study the longterm effects of fly ash on soil health and agricultural productivity, keeping in view the possible consequences of heavy metals toxicity and poor soil characteristics. With the thermal power authorities taking full precautions in checking the pollution caused in the agro-ecosystems, the dust/ash fall around the complex is reduced to a great extent and effects on the agricultural productivity are not foreseen. The increased crop productivity in various crops could be attributed to the increased availability of Boron. Decreased transpiration and photosynthesis in case of heavily dusted plants could have been due to a physical barrier created by a thicker layer of fly ash forming a boundary layer.

Presence of ash particulates in the atmosphere is of major concern to the people living close to the plant site. This is particularly severe in summers due to prevailing high wind speeds. The finer fractions (around 1μm) of fly ash are potentially harmful as they get deposited in lungs/pulmonary tissues of respiratory tract when inhaled. These fly ash fractions are, in fact, found to be most mutagenic and carcinogenic. A 210 MW capacity of power plant emits through the stack about 50 to 80 tonne of fly ash, 350 to 430 kg of oxides of sulfur and 150 to 200 kg of oxides of nitrogen everyday. Research is going on to increase the efficiency of Electro Static Precipitator (ESP) which would also help in reducing the dust emission and ensuring a cleaner environment. Continuous monitoring of the atmosphere at several locations around the plant site is done by the authorities to ensure atmospheric condition well below the pollution standards as laid down by the Central/State Pollution Control Boards. To ensure minimal escape of the ash from the ash mounds, management practices such as bitumen spray and sprinkler water application and subsequent compaction to form hard surface are being practised. This activity has to be in tune with the fly ash production rate to minimise the availability of ash in the loose form for being easily lost to the surroundings due to climatic actions. Growing vegetation on ash mound structures for utilising the space as a picnic resort or grazing lands will improve the stability of the structure further in checking the losses. Grass growth on the surface layer protects the whole composite material against climatic erosion and makes this layer acceptable from the aesthetic point of view.

Water pollution from coal-based thermal power stations has acquired significance over the years due to huge requirement of water in different processes and subsequent discharge of effluents in the neighbouring environment. The present trend in most of the thermal power plants is to dispose of fly-and bottom-ash in wet slurry form to a nearby ash pond site where ash settles and clear water is allowed to overflow. By virtue of the leaching characteristics of the ash, the heavy metals contained in the ash present in the oxide form after slow and gradual degradation may get leached down and percolate to the ground water-bodies and thus affecting the human and aquatic life. A detailed study in this regard is desired as most of the existing literature speaks about leaching under different dilution levels in the columns. Soil hydrology, ash characteristics, degradation—as related to physical, chemical and biological characteristics of soil and ground water aquifer characteristics—are to be considered for understanding of the heavy metals addition into the ground waters.

Large area requirement for disposing the huge amount of ash sometimes produced need acquiring of the agricultural lands, which may pose problems with regard to giving compensation in cash or employment to the dependants of the affected persons. The carriage of the ash to the surrounding areas may cause poor visibility, respiratory problems and reduced silted fodder diets by the animals. Authorities are taking full precautions in ensuring the carriage of the ash to the surroundings to the minimum possible extent. Several rehabilitation programmes which include road construction, agro-forestry, education and medical aids have been launched from time to time. NTPC authorities have sponsored surveys and research studies with organisations like IITs, JNU, DST, IARI to evaluate the problems associated with fly ash and design management options to overcome the problems/constraints associated with fly ash. However, the details of fly ash conversion products and other related issues are shown in Table 56.5, 56.6 and 56.7.

56.5 CONCLUSION

Disposal of huge amount of fly ash produced is a major concern. Its use in manufacture of cement, concrete and bricks needs to be encouraged and requires research and development efforts from cost consideration. This material has to be used for landfill activity at full swing. Fly ash as an amendment for reclaiming problem soils needs to be investigated in depth and subsequently ways for its implementation, after convincing the farms, have to be formulated. Ensuring stable structure to the ash mounds, using mechanical procedures and encouraging agro-forestry over the mounds for minimising particulate to the mounds areas has to be in time with the rate of ash produced, changes in the soil characteristics and the cropping systems response to addition of fly ash particulates need to be studied in details. Identification of crops and their cultivators is needed in relation to the changed scenario arising due to presence of fly ash particulates and evolving suitable agroeconomic management option. In recent notification of Ministry of Environment and Forests (2003) all coal based thermal power plants are directed to make arrangement of fly ash disposal by making agreement with companies for making bricks, tiles and pozzolana cement.

Table 56.5: Fly Ash Conversion Products

Table 56.5: Fly Ash Conversion Products
A. Bricks
1. Clay Fly Ash Bricks
2. Lime Fly Ash Bricks
3. Fly Ash Sand-Lime Bricks
4. Cement sand fly ash Bricks
5. Lime-Fly Ash-Gypsum Bricks
6. Calcium silicate Bricks
7. Insulation Bricks
8. Acid resistant Bricks
B. Cement & Cement products
1. Portland Pozzolana Cement
2. Raw material for OPC clincker making

Contd.

3. Ready Mixed Fly Ash Concrete
4. Lime Fly Ash Cellular Concrete
5. Precast Fly Ash Concrete Building Unit
6. Sintered Fly Ash Lightweight Aggregate

C. Land reclamation & Soil stabilisation

1. Sub-base of Roads & Runways
2. Filling of Mines
3. Canal embankment
4. Reclamation of Low/Marshy Land

D. Mineral/Metal extraction (not commercialised)

1. Alumina
2. Magnesium
3. Germanium (fly ash is only source)
4. Zinc
5. Selenium

E. Others

1. Mineral Wool
2. Distemper (not commercialised)
3. Emulsion paints

Table 56.6: Constrains In Large Scale Utilisation

A. Technical

1. Varying nature of fly ash depending upon the coal and process combustion, hence the same research result is not applicable to other.
2. Non-availability of adequate data regarding the fly ash.
3. Supply of fly ash conforming to proper grade is not stable.

B. Psychological

The product of the fly ash conversion plant has a difference in look with the conventional building materials.

C. Economical

1. High initial investment in research laboratory and the conversion plant.
2. Absence of the marketing agencies and the market survey.

D. Administrative

Absence of clean govt. policy to promote the fly ash conversion plant.

Table 56.7: Expectation from Government

A. Restriction or ban on selling of conventional building material within a specified radius of the fly ash conversion plant.

B. Exemption or reduction in taxes and duties for the building material converted from fly ash, to make it attractive to users.

C. Formulate and publish market survey and research report depending upon the various ash analyses.

D. Use of fly ash to back-fill the abandoned coal mine and replacement of conventional construction material for making the road.

E. Design suitable means to transport the fly ash by rail.

56.6 BACKGROUND OF INDIAN STANDARD FOR PULVERIZED FUEL ASH (IS 3812 PART I, 2003 & PART 2, 2003)

Pulverised fuel ash is a residue resulting from the combustion of ground or powdered or crushed bituminous coal or subbituminous coal (lignite). About 80 per cent of the total ash is finely divided and get out of boiler along with flue gases and is collected by suitable technologies. This ash, generally, and in this standard, is termed fly ash. It is sometimes referred as chimney ash and hopper ash. The balance, about 20 per cent of ash, gets collected at the bottom of the boiler and is taken out by suitable technologies and is referred as bottom ash. Fly ash is collected and stored in dry condition. When fly ash alone or alongwith bottom ash is carried to storage or deposition lagoon or pond in the form of water slurry and deposited, it is termed pond ash. Whereas if fly ash, alone or alongwith bottom ash is carried to a storage or deposition site in dry form and deposited, it is termed mound ash.

Pulverised fuel ash is available in large quantities in the country as a waste product from a number of thermal power stations and industrial plants using pulverized or crushed or ground coal or lignite as fuel for boilers. The effective use of pulverised fuel ash as a pozzolana in the manufacture of and for part replacement of cement, as an admixture in cement, cement mortar and concrete, lime pozzolana mixture and products such as fly ash lime bricks, autoclaved aerated concrete blocks, etc. have been further established in the country in recent years. Recent investigations of Indian pulverised fuel ashes have indicated greater scope for their utilisation as a construction material. Greater utilisation of pulverised fuel ash will lead to not only saving of scarce construction materials but also assist in solving the problem of disposal of this waste product. The recent investigations have also indicated the necessity to provide proper collection methods for fly ash so as to yield fly ash of quality and uniformity which are prime requirements of fly ash for use as a construction material.

This standard was first published in 1966 in three parts to cater to the requirements of fly ash for three specific uses: Part I covering use of fly ash as a pozzolana, Part II covering use of fly ash as an admixture for concrete, and Part III covering use of fly ash as fine aggregate for mortar and concrete. The Sectional Committee responsible for the formulation of this standard subsequently felt that the performance of fly ash as a pozzolana or an admixture or a fine aggregate, was complementary and not separable and hence requirements of fly ash for these uses should be covered by a single standard. This standard was, therefore, revised in 1981 by combining the three parts into a single standard, also incorporating the modifications found necessary based on the experience gained with the use of earlier standards. This revision classified fly ash in two grades—Grade I for incorporation in cement, mortar and concrete and in lime pozzolana mixture, and for manufacture of Portland pozzolana cement; and Grade 2 for incorporation in cement, mortar and concrete and in lime pozzolana mixture.

Improvements have taken place over time in combustion technologies and ash collection technologies. These technological developments have resulted in improvement in ash qualities; specially the fineness and loss on ignition. Technologies have also been developed for a large number of utilisations of wide range of pulverised fuel ash. Application of technologies in the collection, transportation and deposition of ash have also resulted in availability of pulverised fuel ash in four forms—fly ash, bottom ash, pond ash and mound ash. Though the last revision of the standard pertained only to fly ash, number of important applications of fly ash were less uncovered. The Sectional Committee, therefore, felt necessary that a comprehensive specification should be brought out. Separate Indian Standards for fly ash for use in different end applications such as lime pozzolana mixture applications, sintered applications, geotechnical applications and agricultural applications are also being developed.

Standard details:

Part I: For use as Pozzolana in cement, cement mortar and concrete

Part II: For use as admixture in cement mortar and concrete.

For coal thermal power plant, revised notification of emission standard and efficient water use was made in 2015. The details given below:

MINISTRY OF ENVIRONMENT, FOREST AND CLIMATE CHANGE

NOTIFICATION

New Delhi, the 7th December, 2015

S.O. 3305(E): In exercise of the powers conferred by sections 6 and 25 of the environment (Protection) Act, 1986 (29 of 1986), the Central Government hereby makes the following rules further to amend the Environment (Protection) Rules, 1986, namely:

1. (1) These rules may be called the Environment (Protection) Amendment Rules, 2015.

   (2) They shall come into force on the date of their publication in the Official Gazette.

2. In the Environment (Protection) Rules, 1986, in Schedule – I

   (a) after serial number 5 and entries relating thereto, the following serial number and entries shall be inserted, namely:

Sr. No.	Industry	Parameter	Standards
1	2	3	4
"5A.	**Thermal Power Plant (Water consumption limit)**	**Water consumption**	I. All plants with Once Through Cooling (OTC) shall install Cooling Tower (CT) and achieve specific water consumption upto maximum of 3.5m^3/MWh within a period of two years from the date of publication of this notification. II. All existing CT-based plants reduce specific water consumption upto maximum of 3.5m^3/MWh within a period of two years from the date of publication of this notification. III. New plants to be installed after 1st January, 2017 shall have to meet specific water consumption upto maximum of 2.5m^3/MWh and achieve zero waste water discharged".

   (b) for serial number 25, and the entries related thereto, the following serial number and entries shall be substituted, namely:

Sr. No.	Industry	Parameter	Standards
1	2	3	4
"25.	**Thermal Power Plant**	**TPPs (units) installed before 31st December, 2003***	
		Particulate Matter	100 mg/Nm^3
		Sulphur Dioxide (SO_2)	600 mg/Nm^3 (Units Smaller than 500MW capacity units) 200 mg/Nm^3 (for units having capacity of 500MW and above)

Contd.

		Oxides of Nitrogen (NO_x)	600 mg/Nm^3
		Mercury (Hg)	0.03 mg/Nm^3 (for units having capacity of 500MW and above)
		TPPS (units) installed after 1st January, 2003, upto 31st December, 2016*	
		Particulate Matter	50 mg/Nm^3
		Sulphur Dioxide (SO_2)	600 mg/Nm^3 (Units Smaller than 500MW capacity units) 200 mg/Nm^3 (for units having capacity of 500MW and above)
		Oxides of Nitrogen (NO_x)	300 mg/Nm^3
		Mercury (Hg)	0.03 mg/Nm^3
		TPPS (units) to be installed from 1st January, 2017**	
		Particulate Matter	30 mg/Nm^3
		Sulphur Dioxide (SO_2)	100 mg/Nm^3
		Oxides of Nitrogen (NO_x)	100 mg/Nm^3
		Mercury (Hg)	0.03 mg/Nm^3

***TPPs (units) shall meet the limits within two years from date of publication of this notification.**

****Includes all the TPPs (units) which have been accorded environmental clearance and are under construction".**

[F. No. Q-15017/40/2007-CPW]

Dr. RASHID HASAN, Advisor

57 ENVIRONMENTAL PLANNING

57.1 INTRODUCTION

Environmental planning is concerned with society's collective stewardship of the earth's resources. The 'environment' refers to the physical and biological systems which provide our basic life support and which contribute to our psychological well-being. 'Planning' is a generic activity of purposeful anticipation of, and provision for the future. The ultimate aim of environmental planning is to help us achieve the sustainable growth of the society with appropriate life support system availability.

Environmental planners are always concerned with a number of issues (Table 57.1) that needs to be addressed all the time for better quality of life. Environmental planners are also constantly working within the context of policy agenda framed at various tiers in the public and private sector hierarchies.

The importance of the environment in land-use planning decisions has been recognised for a long time, and the early-twentieth-century pioneers of town planning were alert to the significance of ecological pattern and process. However, our understanding of the centrality of environmental dynamics to planning decisions has greatly increased in recent years. In particular, the new focus on sustainable development rather than on purely 'green' issues had important consequencs. A key role of planning in the twenty-first century appears to be that of helping society cross the 'sustainability transition'. Sustainable development is now an overarching agenda for planning providing a meta-narrative within which the main principles of decision making are couched.

In pursuit of these environmental goals, planners have a range of powers. As important as their statutory powers, however, is their potential for engaging society in debates and collective decisions about sustainable futures. In practice, rational and deliberative processes are starting to merge, as decisions are increasingly expected to reflect a broader social consensus. Environmental planning can only flourish and retain its importance within a wider context of concern for and participation in issues which affect the quality of life of present and future generations.

Table 57.1: The Principal "Environmental Agenda"

• **Population and health:**	Accelerating population growth, wide disparities in health standard, deterioration of life quality.
• **Human settlements:**	Rapid urbanisation, land use changes, pollution problems, and various anthropogenic pressures.
• **Food & agriculture:**	Growing level of hunger & malnutrient, sustainability in yield, restoration of soil fertility.
• **Forest and farmlands:**	Decreased tropical forest, soil erosion, wasteland creation, monocultural practices in afforestation.
• **Wildlife and habitats:**	Species losses, illegal trade of endangered and rare species.
• **Energy:**	Economic dependence on forest fuels and nuclear power, shortage of fuelwoods, non-conventional energy resources.
• **Freshwater:**	Water scarcity, salinization, pollution of river, lake, reservoir, destruction of wetland, ground water pollution.
• **Oceans and coasts:**	Overexploitation of fishes, marine pollution, coastal habitat destruction, sea level rises & its impact.
• **Atmosphere and climate:**	Air pollution, global warming, Ozone depletion, El Nino, tropical cyclone, etc.

57.2 PERSPECTIVE OF ENVIRONMENTAL PLANNING

There are a couple of issues to be considered by the environmental planner:

1. Natural resources management planning;
2. Landscape ecological planning;
3. Urban ecological planning;
4. Environmental quality improvement planning;
5. Planning of implementation of legal and administrative network.

On the whole, all planners need to be aware of the complex systems which sustain the earth's 'free services'. Traditionally, the environment has been treated as a distinctive and specialised domain of planning, requiring an occasional input from environmental scientists. It is now clear that scientific expression alone is insufficient as a basis for solving environmental dilemmas. Our rationalisation of the use of environmental resources requires an awareness of economic principles and means of analysing policy performance. Thus, social science and natural science methods merge imperceptibly in the search for sustainable ways of living. Equally, the purely rational approach has its limitations and planners must be alert to the more subtle human believes and reactions which condition out perception of natural and industrial hazards.

The economic perspective is nowadays central to understanding our use of the environment. In essence, planning is necessary because of market failures, due especially to the existence of "externalities", i.e., external effects which distort investment decisions and prevent the "unseen hand" of the market producing outcomes which are of the greatest benefit to society. Planning is based on the notion of internalising externalities in the land market, by including the social interest within decisions surrounding the allocation and use of land. The broader idea of actual or imaginary (shadow) markets being associated with any environmental good has become integral to the farming and implementation of environmental policy.

All planners need to be aware of the complex systems which sustain the earth's 'free services'. In terms of environmental dynamics, these include fundamental life support systems associated with the cycling of rocks, minerals and water, as well as the flow of energy through ecosystem. These processes are relatively robust, and the earth has remarkable compensatory mechanisms to dampen the impacts of human activity. However, these systems clearly can be destabilised and even overwhelmed by insensitive use. Whilst planners in isolation cannot prevent nuisance, they can perform less a central role in promoting environmentally friendly development.

The complexity of environmental interests means that intervention by public agencies in the use of the environment can only be achieved through a variety of areas of legislation and tiers of administration. This means that planners must be aware of:

- the policy context in which environmental actions are framed;
- the legislation and administration understanding environmental action, especially the legal instruments and executive agencies which enable objectives to be implemented; and
- opportunities for integrating environmental actions.

These are clearly very wide-ranging issues, and the account here deliberately focuses upon those aspects of general environmental policy and legislation which are most pertinent to the spheres in which planners operate.

A major area covered by environmental legislation is that of pollution control, especially the permitted levels of contaminants emitted to air and water. These affect both in-pipe concentrations and end-of-pipe emissions. A strategy of dispersal is normally adopted for airprone pollutants by the use of high chimney stacks, above the temperature inversion layer, a dilution approach is taken to aqueous discharges involving the outfall being placed at a point where water volume is sufficient to reduce concentrations and far enough from other outfalls to avoid mixing of pollutants. Of course, when contaminants display a high degree of persistence, dilution and dispersal strategies do not remove pollution, they merely delay or transfer it. Though planners have few statutory responsibilities related to pollution control, their role in contributing to pollution management strategies cannot be ignored.

57.3 LAND RESOURCE DEVELOPMENT PLANNING

Planners have been centrally concerned with the stewardship of the land resource. From the point of view of environmental resources, three particular qualities of land are especially important : the ecological (i.e., biodiversity and nature conservation values), the visual (i.e., scenic and aesthetic quality), and the positional (i.e., the location of land relative to a source of demand for specific uses such as recreation or house building). These qualities can be related closely to the sustainable development notions of critical and constant natural capital, where assets which have taken millennia and longer to develop or which have unique positional attributes can be treated as essentially irreplaceable and thus absolutely worthy of safeguard (critical capital), whilst those which are more recent or common-place may still be important, but could perhaps be subject to tradeoffs and compensation (constant capital). Sustainable development of countryside resources has been related to the adoption of long-term strategies, vision, giving greater recognition to the character of individual areas; and using new methods to measure and protect what people see as valuable.

Yet the land-use planning system can exert relatively little control over these major land-use changes. Indeed, some key transformations in the use and management of rural land have effectively fallen outside any form of statutory control. But between them, planning, agriculture, forestry and conservation agencies can exercise a range of direct controls and indirect influence over many environmental changes. Planners are widely concerned with these, even if only because they inherit many of the consequences of change.

57.4 PLANNING AND MANAGING THE NATURAL RESOURCES

A major focus of environmental planning is the management of natural resources in the countryside. These resources can broadly be summarised as 'productive' and 'protective' : productive uses are principally those of agriculture, forestry and minerals; protective uses include nature and landscape conservation.

One area of natural resource management in which planners are centrally involved is in the development of minerals. Significantly, this is a principal category of nonrenewable resources and, thus, raises distinctive and intractable issues of sustainability. Indeed, the tenets of sustainable development are forcing us to reappraise more generally our use of natural resources. Our understanding of what is optional in terms of resource use is being transformed from one of maximum sustainable yield and supply management, to one of multi-objective planning within a policy context of demand management. This transition is as yet nowhere near complete, and policy-makers have been taking only fairly faltering first steps. However, it is likely to be a dominant policy theme of the early part of the twenty-first century.

57.5 LANDSCAPE ECOLOGICAL PLANNING

The landscape is a complex entity, which is both naturally and socially constructed, and which has multiple scientific and cultural meanings. The landscape must be understood—not simply as a resource of fine scenery and recreational opportunity—but also as a dynamic living system driven by fundamental environmental processes, and an archive of history and cultural treasure. Our traditional approach to landscape planning has been to fragment the countryside for administrative purposes, and to plan or manage different elements separately. Although this approach now seems less than satisfactory, it has nevertheless provided significant levels of protection to areas of critical capital which could otherwise have been debased for ever.

Urban areas — in the developed world especially — are notorious for the scale of their ecological footprint. Whilst planning controls have helped to contain the land-use impact of urbanisation, the often unseen impact of cities spread far beyond their physical boundaries. Air and water pollution, solid waste generation, derelict land, congested roads, waste energy and disruption of hydrological and biological processes — all reflect the dysfunction of a current state of urban ecology. The urban environmental management is, however, an area of intense activity both at the research and policy levels. After decades of indecision, societies are now starting to accept the need for much more vigorous demand management and are turning to a variety of regulatory and market-based instruments to shape our use of natural resources more efficiently.

57.6 INFORMATION AND DECISION OF ENVIRONMENTAL PLANNING

During the last quarter of the twentieth century, there was a veritable explosion of environmental information, owing much to the advent of satellite imagery, geographic information systems and the Internet. This complemented the continued mapping and evaluation of rural resources by more traditional methods such as field survey, and the growth of statistical techniques to assist in the sampling and summary of data. Much of this information is highly technical, but considerable efforts have been made to communicate selected 'headline' indicators to wider audiences. Equally, many attempts have been made to incorporate lay knowledge into datasets and decisions.

Whilst this is allegedly something of a movement away from rationalistic styles of planning, there is nevertheless an evident need to weigh the arguments for and against developmental options very carefully. To assist with this, a range of environmental decision-making techniques have been refined over a substantial period. Some of these, such as environmental impact assessment, have become adopted as standard practice, whereas others, such as capacity planning, are still illdefined and experimental. Ideally, environmental planning methods should have formal ways of incorporating public viewpoints, and not merely be treated as expert tools.

It is clearly naive to believe that improved information and decision-making techniques alone can deliver sustainability. Much more profound debates must take place at a social level in order to create the changes necessary forn truly environmentally led planning to take place.

57.7 LAND-USE POLICY IN INDIA

Since Independence very little emphasis was given on land-use planning for any area on long term basis. Today there is some short term populatic measures. However, for the land, holding or ownership aspects was considerably well-planned for distribution of review land to the landless people. In many land-use planning was not based on environmental consideration.

However, for long term sustained land-use policy there is need for more closed observation on merit and demerit of current land-use pattern.

57.7.1 Urban Land-use Planning

Land management for the urban sector thus requires planning not only for residential, institutional and public utility buildings, but also for purification of water, disposal of wastes, treatment of sewer water facilities for recreation, etc. There are then problems relating to the stability of the ground — foundation conditions for construction of building, availability of construction material, condition of soils, rocks suitable for laying down of roads, sewerlines, pipelines, etc.

Another very important consideration is the freedom or safety from natural hazards like flood, ground subsidence, landslides, etc. Third vital problem relates to the unending tide of migration from the rural to urban centres.

Over 30% of the Indian population resides in the urban areas, which are ever increasing. All the urban areas gradually become producers of pollution and heave themselves deteriorated into slums as a result of failure of proper land-use planning.

Urban land-use planning in India was attempted through setting up a committee on Urban Land-use Policy in 1965. The Town and Country Planning Act was then promulgamated with a view of urban development and improvement of quality of life. The major urban areas have their own development authority, viz., Delhi Development Authority (DDA), Kolkata Metropolitan Development Authority (KMDA), Madras Development Authority (MDA), etc., were framed for proper planning of urban development.

The uncontrolled growth and expansion of the city all along the fringes and development of slums in vital locations are the two important problems of any urban area. The appropriate land-use planning is essential for future sustained growth of city.

57.7.2 Rural Land-use Planning

It is well-known that over 60% Indian population live in rural areas of the country, where major activity is agriculture. About 44% of the geographic area is under the plough, 12% under forest and 4% in grassland. With the passage of time the land-use pattern is very much altered. There are land ownership selling in urban and rural areas too, but there is strict legislation of land-use conversion from one category to others. As such, orchards, wetlands, crop field are very much altered with the passage of time.

In recent decades, with use of GIS application, resource-based land-use planning was developed in each district for optimization of yield of the land categories. Microwater-shed based development is one of the major activities in many rural areas of the country.

58 PHYSICAL ENVIRONMENT—WEATHER AND CLIMATE

Earth is one of the smaller of the planets (Mercury, Venus, Earth, Mars, Jupiter, Saturn, Uranus, Neptune, and dwarf planets—Pluto, Charon, Eris, Ceres) forming the solar system. Among all the planets, two solid terrestrial planets are Earth and Mars. Earth is of almost unimaginable (4.5 billion years) age and yet its modern character and geological processes can be traced directly to its astronomic origins. Many geological processes are imperceptibly slow—measured against our own life spans—but earthquakes and volcanic eruptions are sharp reminders of Earth's relentless evolution. The accretion of a cocktail of planetary matter cover 4.6 billion years ago set in train enduring processes of chemical fractionation, heat generation and cooling which progressively refined our planet's constituent parts. Some processes are necessarily familiar to us— the weathering of simple nutrients from more complex rocks, the evaporation of H_2O from ocean, solutions, the abstraction and reformation of carbon, hydrogen, oxygen from a variety of sources to from living cells, and even the segregation of different isotopes of these fractionates (^{14}C ,^{16}O, ^{18}O, etc.). All require energy, and the nature and rate of their reactions are also determined by temperature and pressure.

By regarding the familiar atmosphere, hydrosphere and biosphere as Earth's outer fractionates we can more easily understand the mantle, asthenosphere and lithosphere inner fractionates with which they are linked. This lays the foundations for appreciating the morphotectonic and rock cycles which shape our earth, its principal architecture of oceans, continents, mountains and sedimentary basins, etc., and surface geomorphic processes and land systems. The sun's energy represents the prime source of the earth's climatic system.

The atmosphere around the earth is highly complex system, and the effects of changes in any single property tend to be transmitted to many other properties. Thus, the heating and cooling of the air cause adjustments in relative humidity and buoyancy; they may cause condensation and evaporation, cloud formation and the development of storms.

58.1 ATMOSPHERIC CIRCULATION

Earth's atmosphere is in perpetual motion: movement which is striving to eradicate the constant differences by temperature and pressure between different parts of the globe. It is this motion which produces the winds and storms with which we are all familiar. It is this circulation which plays a basic part in maintaining a steady state in the atmosphere and generating the climatic zones which characterize the earth. So far we have considered the upward movements which transfer energy from the surface to the atmosphere.

The basic force causing atmospheric motion is the pressure gradient; this gradient arises from the unequal heating of the atmosphere by solar radiation. At the equator—the "firebox" of the circulation, as it has been called—solar radiation is converted into heat. The air expands and rises and flows out towards the poles. Cool, demure air from the poles returns to replace it. At depth the flow is reversed. So long as this unequal heating is continued, the cellular flow is maintained.

The air circulation within the tropics consists of two cells. Air blows in towards the low-pressure belts of the equator (the equatorial trough) across the subtropical seas. The equatorial air then diverges and flows polewards, so the potential energy is exported to higher latitude.

There are four main surface wind belts which can be distinguished, viz.,

1. The *equatorial trough*, which is a shallow trough of low pressure generally situated near the equator.
2. The *trade winds*, which lies between the equatorial trough and the subtropical highs. This zone occupies nearly half the globe, much of it ocean, and within that area the steady trade wind provides a stable and relatively constant climate.
3. The *Westerlies* which lies in mid-latitudes, and produces cyclones and monsoon.
4. *Polar Easterlies*, which are linked with shallow polar anticyclones.

The surface wind of any location can be depicted by *"windrose"* which represent wind speed, direction, % of calm period over the season in any location. There are a variety of weather events that occur every year. These are floods, tornadoes, hurricanes, blizzards or even heatwaves. It is to be further noted that weather is a description of physical conditions in the atmosphere (humidity, temperature, pressure, wind and precipitation), then climate is the pattern of weather in a region over long time periods. The interactions of atmospheric systems are so complex that climatic conditions are never exactly the same at any given location from one time to the next. While it is possible to discern patterns of average conditions over a season, year, decade or century, complex fluctuations and cycles within cycles make generalizations difficult and forecasting hazardous. We always wonder whether anomalies in local weather patterns represent normal variations, a unique abnormality, or the beginnings of a shift to a new region. When climatic change is gradual, species may have time to adapt or migrate to more suitable locations. Where climatic change is relatively abrupt, many organisms are unable to respond before conditions exceed their tolerance limits, whole community may be destroyed and if the climatic change is widespread, many species may become extinct.

58.2 WEATHER PREDICTIONS

To forecast values of pressure and wind, the globe is subdivided into a grid consisting of about 60,000 squares, each with 217 points from one pole to the other and 288 around most latitude circles. For each point, the upper atmosphere is subdivided into 20 levels and the values of critical atmospheric properties are determined—mainly from satellite information. The physical equations of motion, continuity and thermodynamics are applied to each grid point at each level to predict the new value a short time period ahead. The new dataset then provides the starting point for the next set of predictions and so on every 15 minutes until the 24-hour or 48-hour forecast is produced.

58.3 CLOUD

When air becomes saturated water vapour condenses and *clouds* are formed. The type of cloud produced depends greatly on the process by which the air is brought to saturation (Table 58.1):

Table 58.1: Cloud Types and Formation Processes

Processes		Common Cloud Types
A. Without cooling : air mass mixing		Stratocumulus (Sc)
B. Cooling without vertical motions		
radiative cooling		Radiative fog
advective cooling		Advective fog
C. Cooling with vertical motions :	orographic uplift	Stratus (St), Altostratus (As)
	frontal uplift	All types
	airflow confluence	Cirrus (Cr), Stratus (St), Altostratus (As)
	convection	Cumulus (Cu), Cumulonimbus (Cb)

On the whole, it implies that as soon as the air reached saturation, the condensation and cloud formation occurred. However, rarely does condensation occurs immediately.

58.4 TROPICAL WEATHER AND CLIMATE

Tropical climates are usually characterised as climates where there are no true temperature distinction between summer and winter. They normally occur between 30°N and 30°S. This is the area of the Hadley cell, so that much of the weather is controlled by the prevailing barotropic conditions.

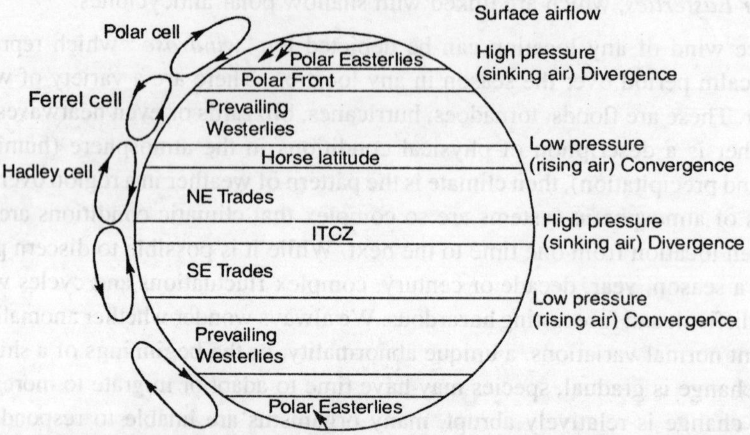

Fig. 58.1: The three cell circulation of the atmosphere

A Hadley cell, part of the general circulation (Fig. 58.1), occur in each hemisphere. The weather is very closely related to the airflow patterns. Air flows equator-ward at low levels in each cell. These airflows converge at the Intertropical Convergence Zone (ITCZ) and rise to create cloud and precipitation. The latent heat released during this ascent provides much of the energy needed to continue the whole cellular circulation. The air ascends towards the tropopause where it diverges and begins to flow polewards. As it flows, it cools by long-wave radiation loss, increase in density and descends. Although the descent covers a circular area it is particularly concentrated around 30° latitude. In the upper troposphere at these latitudes the interaction between Hadley cell air and that converging from mid-latitudes, combined with a strong Coriolis force, leads to the creation of a subtropical jet stream. This has a major influence on mid-latitude weather. Within the Hadley cell the descending air is adiabatically warmed, so that it arrives back at the surface as a dry, cloudless airstream. Once at the surface the air diverges, some flowing as the trade winds towards the ITCZ to complete the cell. The trade winds are, of course, influenced by the Earth's rotation, although there is no complete deflection, so that in the Northern hemisphere the near-surface flow is the northeast trades. The airflow aloft, known as the counter trades, is from the southwest.

In the Southern hemisphere the corresponding winds are the southeast trades and northwest counter trades. The major area of cloud formation in the tropics is around the ITCZ, which is a region of instability. Virtually all of the rest of the tropics is subject to more or less persistent subsidence invasions in which warm descending air traps somewhat cooler air right at the surface. This feature is usually best developed and closest to the surface, at the outer, poleward limits of the Hadley cell. However, it persists throughout much of the trade-wind area although rising to a higher level as the equator is approached. This feature is known as the *trade wind inversion*. Any tendency towards instability caused by surface heating can rarely overcome this dynamically created feature. Although the surface trade winds are able to pick up a great deal of moisture when they flow over the ocean, they cannot release it, or the associated latent energy, until they reach the ITCZ. Hence high humidities but low precipitation amounts are a feature of several tropical areas (Fig. 58.2).

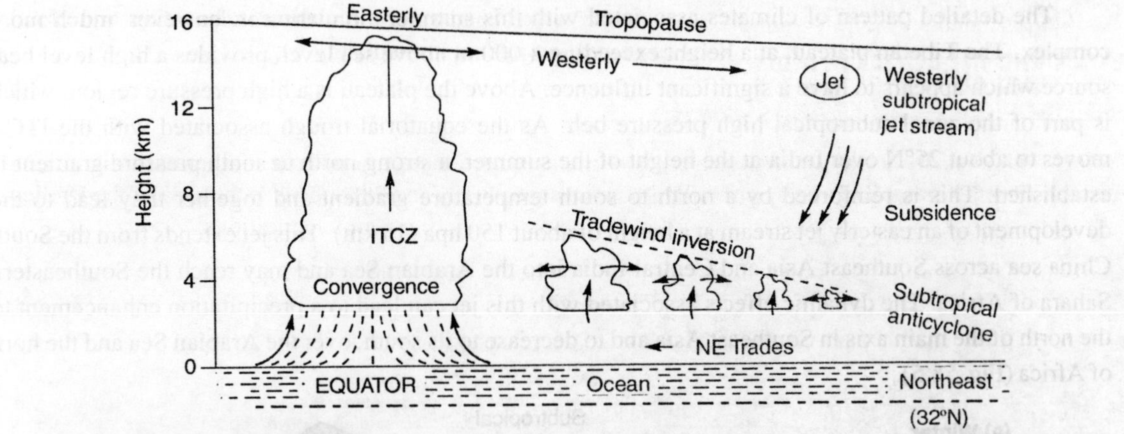

Fig. 58.2: Schematic diagram of the air flow and associated weather in the Hadley cell regime

58.5 THE ASIATIC MONSOON

The Asiatic monsoon is itself the result of a complex interaction between the distribution of land and water, topography and tropical and midlatitude circulation. The simple model (Fig. 58.3) reflecting the seasonal pressure changes over Asia, provides a good first approximation. This is particularly true for summer, where the circulation pattern is driven by low pressure centres over the northern part of the Indian subcontinent and northern Southeast Asia (Fig. 58.4). Warm moist air is drawn in to the thermally created low pressure areas of the continental interior, where it rises, releasing both precipitation to create the wet season and latent heat to provide the energy necessary to continue the system.

Over the oceans is a compensating descent of cold dry air. Once this pattern is established the onshore winds bring the monsoon rains. The moisture-laden winds are highly susceptible to orographic influence, so that the coastlands of India, for example, which are backed by mountain ranges, receive the most rainfall. The interior, of course, also receives rain since the whole area is one of rising air. The Himalayas effectively provide a barrier to the north, confining the circulation to the area south of them.

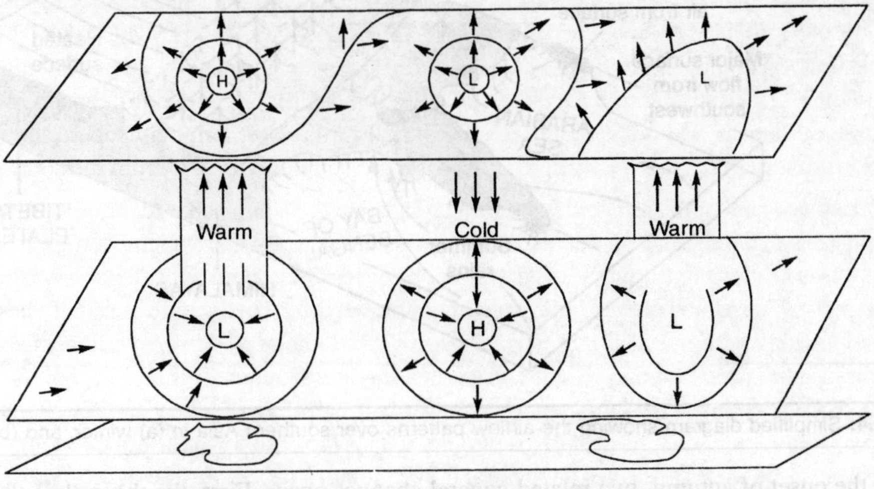

Fig. 58.3: Idealised representation of the monsoon circulation

The detailed pattern of climates associated with this summer circulation is, however, much more complex. The Tibetan plateau, at a height exceeding 4,000 m above sea level, provides a high level heat source which appears to have a significant influence. Above the plateau is a high pressure region, which is part of the zonal subtropical high pressure belt. As the equatorial trough associated with the ITCZ moves to about 25°N over India at the height of the summer, a strong north to south pressure gradient is established. This is reinforced by a north to south temperature gradient and together they lead to the development of an easterly jet stream at a height of about 150 hpa (15 km). This jet extends from the South China sea across Southeast Asia and Central India into the Arabian Sea and may reach the Southeastern Sahara of Africa. The dynamic effects associated with this jet can lead to a precipitation enhancement to the north of the main axis in Southeast Asia and to decrease to its south, over the Arabian Sea and the horn of Africa (Fig. 58.5).

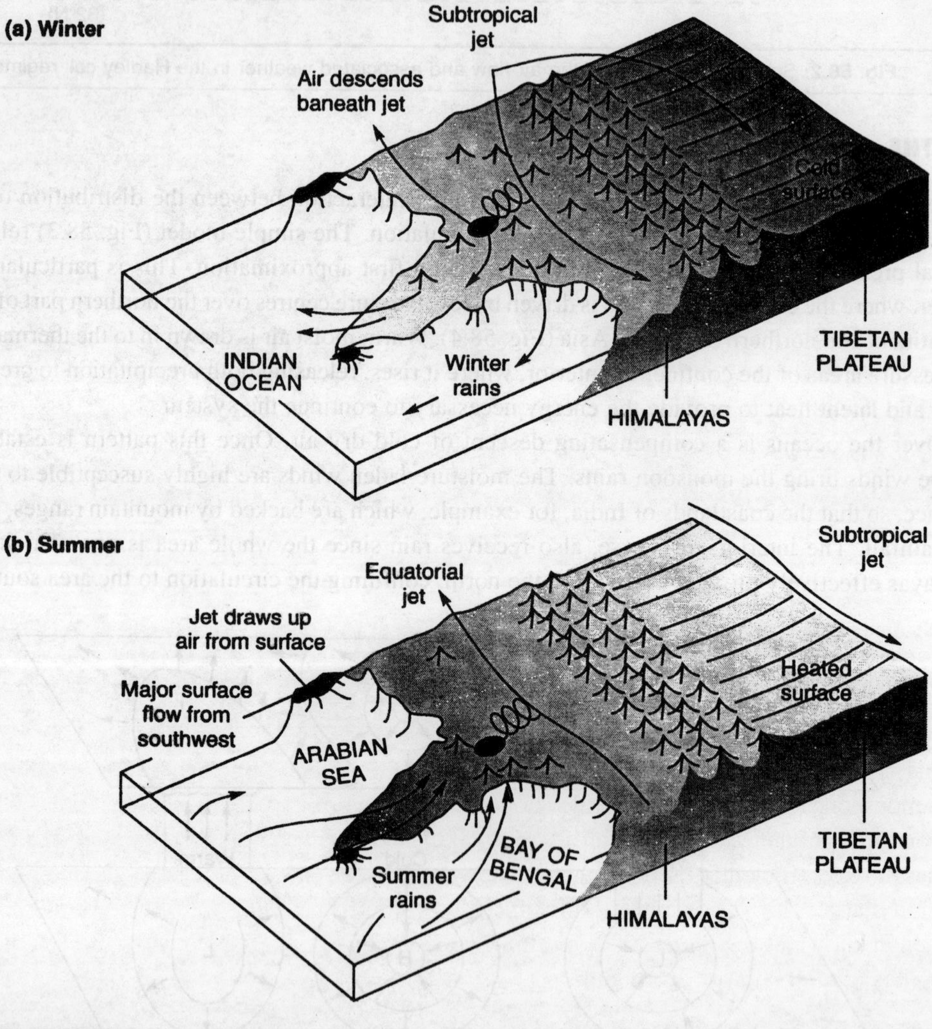

Fig. 58.4: Simplified diagram showing the airflow patterns over southern Asia in (a) winter, and (b) summer

With the onset of autumn, two related general changes occur. First, the thermal contrasts between land and water decrease, weakening the circulation. Second, the Westerlies of mid-latitudes begin to migrate southwards. Eventually, part of this Westerly airstream blows to the south of the Himalayas and completely disrupts the tropical circulation. The easterly jet is replaced by strong Westerly winds aloft. At the surface a north to south pressure gradient extending from Siberia almost unbroken to the equator is established, with a high pressure entered over Siberia. The resulting airflow near the surface is from the north or northeast. This air combines with that resulting from subsidence below the Westerlies to ensure that dry air covers the region, leading to the relatively cold conditions as noted for Hue. Over much of India, air has blown downslope of the Himalayas. Although this air has been warmed adiabatically it gives relatively cold conditions.

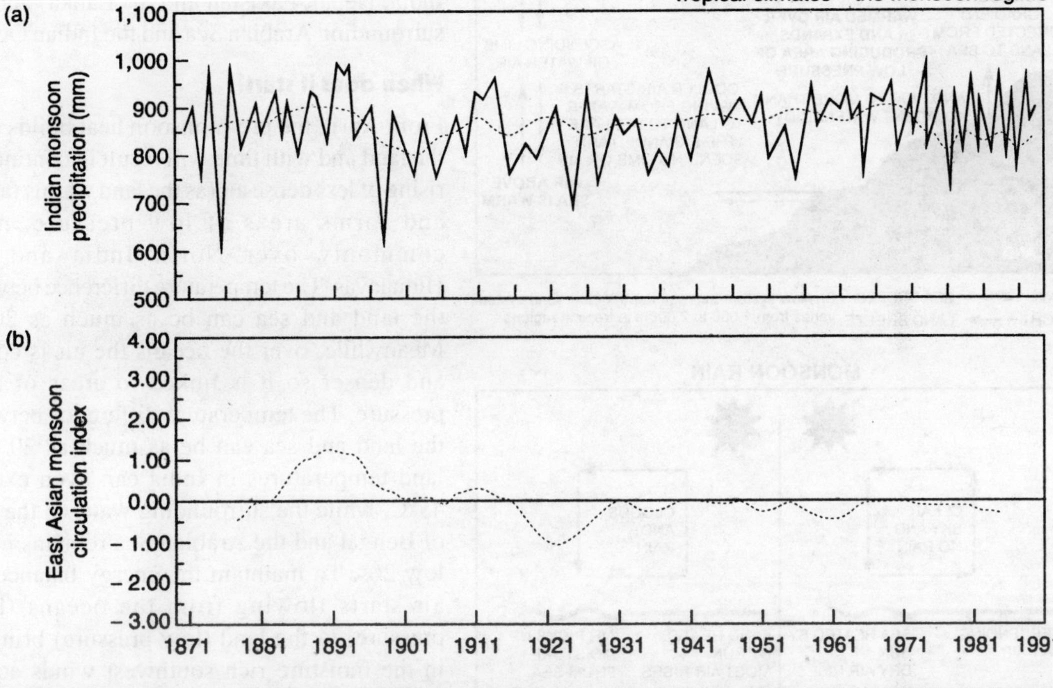

Fig. 58.5: Annual summer values of (a) average monsoon precipitation over India, based on the average of 366 observing stations, and (b) East Asian monsoon circulation index, based on pressure differences between 110º E and 160º E. In both cases the dotted lines indicate annual values, the dashed ones decadal running means. ((a) From Parthasarathy *et al*, 1991 : (b) From Shi and Zhu, 1996)

This type of pattern persists through the spring. As summer approaches, the conditions for the summer monsoon circulation are slowly developed. The onset of the actual circulation is delayed, however, by the continued influence of the Westerlies south of the mountain barrier. As soon as the main line of the westerlies moves sufficiently far north that the flow south of the Himalayas ceases, the summer monsoon circulation starts. The very rapid establishment of warm moist airflow from the south creates the phenomenon known as the 'burst' of monsoon. Cool, dry conditions are replaced by warm, humid ones, with copious rainfall, almost overnight.

MONSOON

Monsoon is used to describe seasonal reversals of wind direction, caused by temperature differences between the land and sea. They occur in a number of countries around the world. However, the most well known of these is the Asian Monsoon.

The cause

In some respects it is a large version of the 'land-sea breeze', where on a sunny day at the beach, the land warms more quickly than the ocean. As the hot air rises over the land, it is replaced by the cooler air from over the water. At night, however, the land cools at a quicker rate than the water, so the wind shifts, blowing from the land to the warmer water. So our two key ingredients for the Monsoon are a large land mass and a large ocean—namely southern Asia (Bangladesh, India, Nepal, Pakistan and Sri Lanka) and the surrounding Arabian Sea and the Indian Ocean.

LAND AND SEA BREEZE

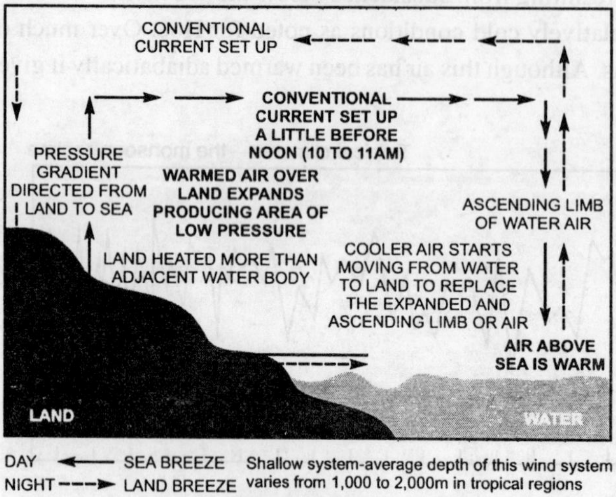

MONSOON RAIN

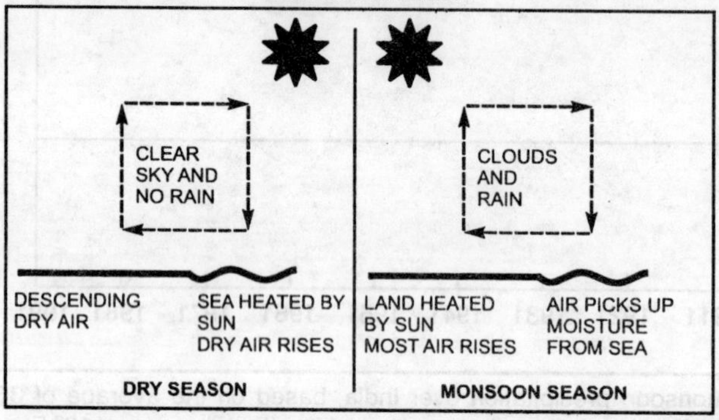

When does it start?

From April, the pre-monsoon heat builds over the land and with time will result in continuous rising of less dense air (as the land warms faster) and forms areas of low pressure, most commonly over North India and the Himalayas. The temperature difference between the land and sea can be as much as 20°C. Meanwhile, over the oceans the air is cooler and denser so it is linked to areas of high pressure. The temperature difference between the land and sea can be as much as 20°C—land temperatures in India can even exceed 45°C, while the surrounding water in the Bay of Bengal and the Arabian Sea remains in the low 20s. To maintain the energy balance the air starts flowing from the oceans (high pressure) to the land (low pressure) bringing in the moisture rich southwest winds across southern Asia the—'wet' phase. When the 'wet' phase starts, around 25 May, it has two arms. One coming into Sir Lanka and the other one moving up from the Bay of Bengal into parts of NE India and Bangladesh.

The retreat

As the land and ocean begin to cool in late summer and into autumn, the land loses heat quicker than the ocean. The wind reverses during this 'dry' phase, becoming north easterly. The withdrawal is usually completed by late December.

Forecasting the arrival of the monsoon

Every year the distribution and the pattern of precipitation in monsoon is different. The monsoon is very important to India and rest of south Asia, as its economy is based on agriculture. Accurate forecasting of the timing of the onset is therefore very important, as it then allows farmers to pick the best time—plant crops in order to take advantage of the rains. Too much or too little rain can have disastrous effects on the people and the economy. The importance of monsoon forecasting can therefore not be underestimated. Computer models of the monsoon are becoming more complex and increasingly accurate. However, the impact that global warming may have on the monsoon is not yet fully understood. Scientific research will have to be increased in this area to determine these effects, which may have important implications for the economy, health and agriculture, not only in southern Asia, but across the whole planet.

MUMBAI VORTEX

An unusual weather system called "mesoscale offshore vortex" over north Mumbai and low pressure area in eastern Madhya Pradesh explains the catastrophic rain over Mumbai and Konkan. The weather office at Santa Cruz in north Mumbai recorded 94.4 cm of rain over 24 hours on 26 July 2005–2.5 times higher than the last record of 37.5 cm in 1974.

The postmortem of this unprecedented rain reveals a whirling mass of air that sucks enormous moisture especially from the adjoining Arabian Ocean, towards its centre–leading to heavy rains. The typical result is an extremely localised heavy downpour, spread over as little as 30 sq. km.

After the deluge, the National Centre for Medium Range Weather Forecasting (NCMRWF) sent out mails with data to weather stations to see if they were able to forecast what Mumbai had in store. The only station that came close was the UK Meteorological Department. When they ran the model again, it came up with 80 cms of rain for the city.

Experts say this phenomenon was unusual for Mumbai but not for India. It was the intensity that took everyone by surprise. The cloud column that's normally supposed to be 6-7 kms during the monsoon was as high as 15 km that day. Essentially, it was this column that drowned the city.

This dangerous offshore vortex was formed when moisture-laden strong winds came in from the Arabian Sea but ran into an obstacle—the Western Ghats. Unable to cross over directly, it began flowing in a channel from the side, slightly northwards of the Western Ghats. The moisture-laden winds took a turn forming a sort of a wedge. This wind trough became a vortex when the winds took a full 360 degree turn, lifting the air upwards in an anti-clockwise direction. What made it deadlier was the low pressure area over central Madhya Pradesh. Though the amount of rain is similar to a cloudburst, it is different as the cloud column is constantly being fed with moisture from the Arabian Sea making it more lethal.

This vortex is not an unknown phenomenon: it is just that the intensity and the location created havoc. It was so localised that even in Mumbai, it was North Mumbai and its suburbs that bore the brunt.

Other examples: On May 4, 2004 a station in Minicoy had 116.5 cm of rain. On the same day, Virar lake had 104.5 cm of rain.

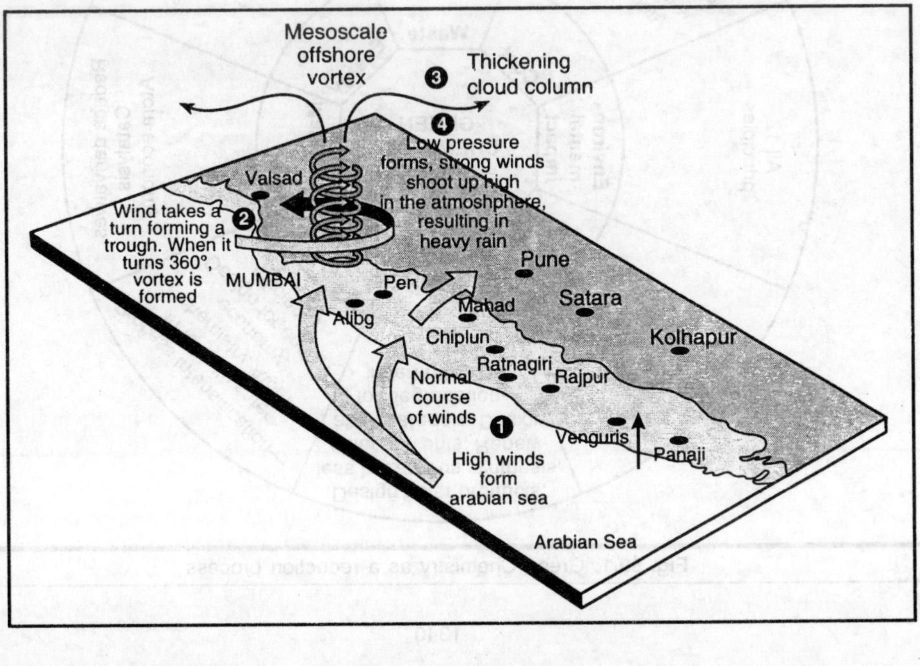

59 GREEN CHEMISTRY

59.1 INTRODUCTION

Green chemistry or environmentally benign chemistry is focused on processes and products that reduce or eliminate the use and generation of hazardous sources. During the early 1990s the US Environmental Protection Agency (EPA) coined the phrase Green Chemistry "To promote innovative chemical technologies that reduce or eliminate the use or generation of hazardous substances is the design, manufacture and use of chemical products. The 12 principles of Green Chemistry (Box 59.1) help show how this can be achieved. Green chemistry is a reduction process (Fig. 59.1).

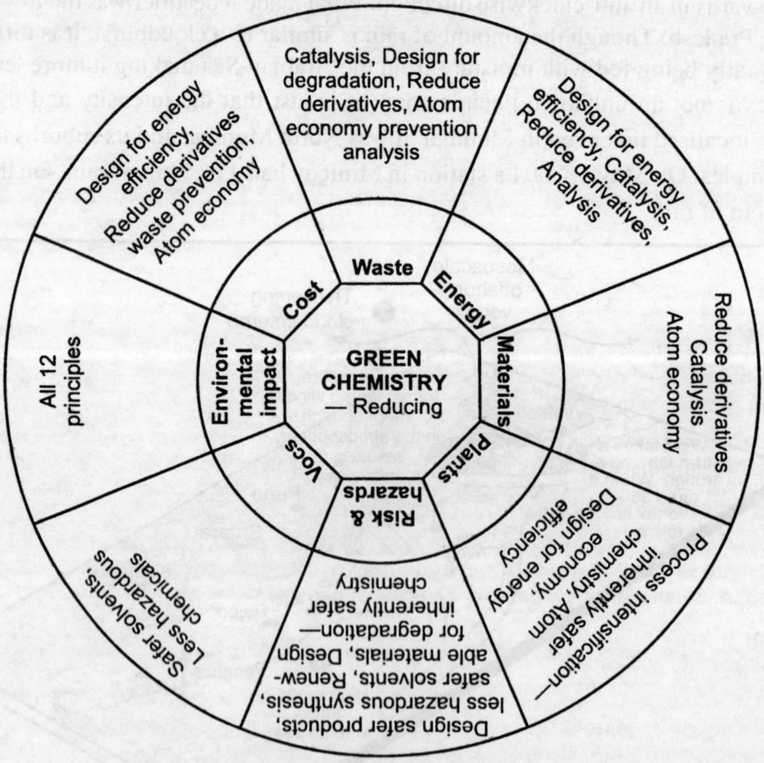

Fig. 59.1: Green Chemistry as a reduction process

One of the fundamental and most important principles of Green Chemistry is that of **atom economy**. This essentially is a measure of how many atoms of reactants end up in the final product and how many end up in by-products or waste. The percentage atom economy can be calculated as 100 times the relative molecular mass (RMM) of all atoms used to make wanted product divided by the RMM of all reactants—Box 59.2. The real benefit of atom economy is that it can be calculated at the reaction planning stage from a balanced reaction equation. Taking the following theoretical reaction:

$$X + Y = P + U$$

the reaction between X and Y to give product P may proceed in 100% yield with 100% selectivity but because the reaction also produces unwanted materials U its atom economy will be less than 100%.

Chemical industries that deal with large volumes of chemicals (such as commodity chemical and petrochemical industries) tend to incorporate a high percentage of the atoms in their starting materials into the desired final product (good atom economy), thus minimising waste by-products, However, chemists who work in the areas of fine chemicals and pharmaceuticals, as well as academic chemists, have traditionally opted for reactions that have high selectivity and high yields.

Manufacturing and service industries currently use organic solvents for precision cleaning. These industries use more than 30 billion lb of halogenated and other organic solvent per year. The CO_2 surfactant technology, thereby, eliminates the need for halogenated solvents (ODS).

Pharmaceutical drugs like analgesics, antiinflammatory drugs are widely used. Recent addition of greener syntheses of Ibuprofen produces good quality drugs with less waste and fewer by-products.

Worldwide, more than 1.9 billion kg of adipic acid are produced each year for manufacture of nylon, polyurethane, lubricants and plasticizers. Now the use of genetically altered *Eschesichia Coli* (GMOs) can synthesise adequate quantity of adipic acid, catechol or DHS from renewable feedstocks.

Production and use of biopestides and biofertilisers are said to be the use of green chemistry in sustainable agriculture. By and large the green chemistry and its application potential is very much compatable with the ecofriendly management of environment.

Box 59.1: The Twelve Principles of Green Chemistry

1. **Prevention**: It is better to prevent waste than to treat or clean up waste after it has been created.
2. **Atom Economy**: Synthetic methods should be designed to maximise the incorporation of all materials used in the process into the final product.
3. **Less hazardous chemical synthesis**: Wherever practicable, synthetic methods should be designed to use and generate substances that possess little or no toxicity to people or the environment.
4. **Designing safer chemicals**: Chemical product should be designed to effect their desired function while minimising their toxicity.
5. **Safer solvents and auxiliaries**: The use of auxiliary substances (e.g., solvents or separation agents) should be made unnecessary whenever possible and innocuous when used.
6. **Design for energy efficiency**: Energy requirements of chemical processes should be recognised for their environmental and economic impacts and should be minimised. If possible, synthetic methods should be conducted at ambient temperature and pressure.
7. **Use of renewable feedstock**: A raw material or feedstock should be renewable rather than depleting whenever technically and economically practicable.
8. **Reduce derivatives**: Unnecessary derivatisation (use of blocking groups, protection/de-protection and temporary modification of physical/chemical processes) should be minimised or avoided if possible, because such steps require additional reagents and can generate waste.

Contd.

Box 59.1: The Twelve Principles of Green Chemistry

9. **Catalysis**: Catalytic reagents (as selective as possible) are superior to stoichiometric reagents.
10. **Design for degradation**: Chemical products should be designed so that at the end of their function they breakdown into innocuous degradation products and do not persist in the environment.
11. **Real-time Analysis for Pollution prevention**: Analytical methodologies need to be further developed to allow for real-time, in-process monitoring and control prior to the formation of hazardous substances.
12. **Inherently safer chemistry for accident prevention**: Substances and the form of a substance used in a chemical process should be chosen to minimise the potential for chemical accidents, including releases, explosions, and fires.

(Source: Adopted after Anastas and Warners, 1998)

59.2 GREEN ENGINEERING

Green engineering is an approach to engineering, design, and production based upon sustainability, minimum environmental impact, and favourable economic and social aspects. Green engineering may be conducted within a framework of 12 principles, each of which is discussed here:

- **Inherent sustainability and reduction of hazard**. The sustainability of a process can be enhanced and its hazard reduced by **circumstantial control**, in which the conditions under which the process is carried out are carefully regulated. For example, the environmental and safety problems posed by large SUVs can be reduced by equipping them with emissions control devices and requiring drivers to obey speed limits and other traffic laws. Such circumstantial controls are prone to failure; emissions control devices wear out and fail, and drivers routinely break speed limits and other regulations designed to reduce their hazard to themselves and others. **Inherency**, in contrast, involves the design of systems that, by their nature, are safer and more sustainable and environmentally friendly. In the case of transportation, for example, this may involve construction of rail transportation systems run electrically. Such systems are demonstrably safer and less polluting than private vehicles such as SUVs.

- **Prevention vs treatment and remediation**. Until recently it was often cheaper in strictly monetary terms to use processes that generated wastes and simply disposed of them to the environment. Now, as the costs of treating and remediating wastes are taken into full account, it has become much more cost-effective to prevent the generation of wastes rather than making or treating or remediating them. An important principle of waste prevention is to use processes in which maximum fractions of material put into the process end up in the final product.

- **Facilitate separation**. The recovery, recycle, and reuse of materials and components are often made much more complicated and expensive—sometimes impossible—by difficulty in separating them. At the component level, careful consideration should be given to facilitating separation of reusable components. For example, from the standpoint of later recovery, it is much more desirable to fasten components together with screws or other mechanical fasteners than with adhesive.

- **Maximise efficiency in the utilisation of materials, energy, space, and time**. This aspect of green engineering means that operations are performed with minimum consumption of materials and energy as rapidly as possible and in as little space as possible. These goals may often be accomplished by process intensification, for example, in chemical synthesis, using a small reactor operated under intense conditions of pressure, temperature, and agitation, thus maximising through put while minimising space and time required. These conditions must be balanced with the need to avoid hazards, such as increased danger of explosion, that come with process intensification.

- **Driving processes by pulling output**. Well known as a basic rule of chemistry, Le Châtelier's principle states that a system placed under stress will respond in a manner that relieves the stress. In the case of a chemical synthesis, the time to produce a product can be lowered and the quantity of product produced may be increased by increasing quantities of one of the reagents required. This "input-pushing" of a reaction can result in large quantities of wastes from the reagent used in excess. The same end can be achieved in a manner more consistent with green engineering by an "output-pulled" approach. For example, a chemical reaction can be driven more strongly to completion by removing a product, such as the one desired, or a by-product such as water. Just-in-time manufacturing, in which a product is made as needed by the consumer, is an example of output-pulled processing.
- **Conservation of complexity.** Much of the cost, including environmental cost, of making a product lies in fabricating relatively complex components or assemblies. Relatively higher efficiencies can be attained by retaining as much as is practical of the complexity of the product. This normally means favouring reuse of a product over recycling the materials in it. For example, an automobile cylinder head is a moderately complex engine component requiring careful casting, honing, and assembly. If a worn cylinder head is simply melted down and recycled for its iron, all of the complexity in it is lost. However, if it is replaned and refurbished, the complexity is retained for much greater efficiency.
- **Durability, but not immortality**. Durability of a product and the materials in it means that it will last through its useful lifetime. However, a product should be degradable and not environmentally immortal. An example is that of biodegradable polymers, such as polylactic acid, which are durable in the applications for which they are used but do eventually biodegrade when discarded.
- **Meet the need without excess**. Overdesign can be costly in terms of resources and even environmental damage. A prime example is the large vehicles favoured by some people, often traveling almost empty while consuming far more fuel than necessary to haul their passengers to their destinations. Chlorofluorocarbons (Freons) and PCBs were much more degradation resistant than their uses required. The former ended up depleting stratospheric ozone, whereas the latter ended up as very persistent environmental pollutants. Excess can be avoided by high specificity. Finely targeted insecticides that kill specific insects while not harming other organisms are examples of high specificity.
- **Use of fewer kinds of materials**. Recycling is greatly complicated by the use of a wide variety of materials that make it difficult to isolate pure products for recycling. For example, the diverse nature of plastics in automobiles can make it very difficult to recover plastic materials in a form suitable for recycling. This difficulty can be alleviated by using only a few kinds of readily identified plastics.
- **Utilisation of available materials and energy**. Efficiencies can be obtained by utilising locally available sources of materials and energy, which are cheaper and require less transportation. Utilisation of locally available construction materials, such as wood in lumber-

producing regions, can enable less costly and more environmentally friendly construction. In chemical production, heat generated by exothermic processes can be captured to provide energy for other processes that require heat. Rather than dissipating energy from braking as useless heat, hybrid electric vehicles use the electric motor for braking and produce electricity that is stored in a battery for later use.

- **Design for later use**. Some efficiencies can be obtained by designing components for commercial afterlife. A commendable trend in urban redevelopment has been to convert old factories and warehouses into restaurants, microbreweries, and even apartment housing. Such a conversion can be facilitated by designing the buildings in the beginning with an afterlife in mind. Computers often have short replacement cycles, but are perfectly suitable for many more years of use by people who find the capabilities of computers that are several years old perfectly adequate for their needs.
- **Maximise use of renewable materials and energy rather than depleting resources**. The most common examples of renewable resources are those made by biological processes. However, recyclable materials that originally came from depleting resources—aluminium, for example—can be considered at least partially renewable. The practice of green engineering requires maximum use of renewable materials and minimisation of depleting resources.

59.3 GREEN POLYMERS: AN APPLICATION OF GREEN CHEMISTRY

An area of high potential for the application of green chemical and engineering approaches is the area of polymers, such as those used to make plastics, textiles, and rubber products. Polymers are made in enormous quantities for a huge variety of applications, and even small advances in making polymers in a more environment friendly and sustainable manner can have significant beneficial consequences. Most green synthetic processes rely upon renewable raw materials in contrast to the petroleum-based substances commonly used to make polymers. Whereas the traditional syntheses of polymers and the monomers from which they are made commonly involve severe, energy-consuming, potentially hazardous conditions, green syntheses are usually biological processes that only function under moderate conditions. Biologically synthesised polymers tend to be biodegradable, an important attribute of green polymers.

An example of a green polymer made by green processes is DuPont's Sorona polymer of polypropylene terephthalate, used in large quantities to make textile fibers and other applications. The DuPont Company was the recipient of one of the 2003 Presidential Green Chemistry Challenge Awards for development of a biologically based process for making the 1, 3-propanediol monomer used to synthesise the polymer. Starting with petroleum based, flammable, toxic ethylene oxide and toxic formaldehyde, 1, 3-propanediol is conventionally made by several steps involving hydrogenation over catalysts at pressures around 100 times atmospheric pressure. DuPont biochemists developed an alternate one step biosynthesis of 1, 3-propanediol from glucose feedstock with *Escherichia Coli* K12 bacteria, incorporating genes from *Klebsiella penumoniae* bacteria and *Saccharomyces cerevisiae* yeast. Not only are the synthesis conditions much more benign than the traditional chemical route, but the feedstock also can be obtained from a renewable source.

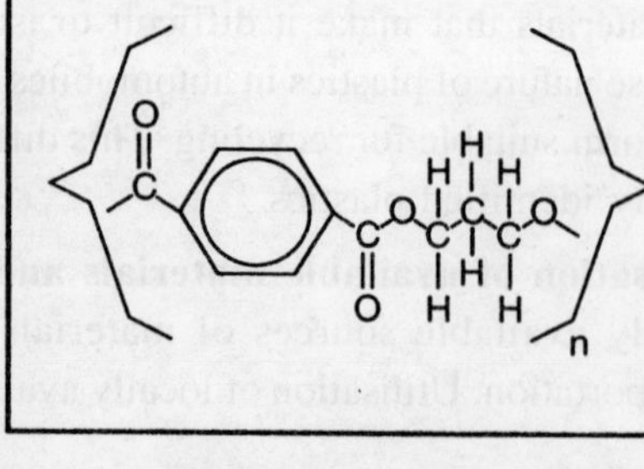

Polypropylene terephthalate

SECTION X

ANALYTICAL TECHNIQUES, STATISTICAL APPLICATIONS IN ENVIRONMENTAL SCIENCE, INSTRUMENTATION TECHNIQUES

60 ANALYTICAL TECHNIQUES IN ENVIRONMENTAL QUALITY ASSESSMENT

60.1 AIR QUALITY MONITORING TECHNIQUES

(A) Sulphur dioxide Determination

Background

Sulfur dioxide is one of the important gaseous air pollutants. Major source of this is from combustion of fossil fuels and specific industrial process such as smelting, sulphuric acid manufacture and petroleum refineries.

Sulphur dioxide is non-inflammable, colourless and can exist in air either as a gas (SO_2 and SO_3) or dissolved in humidity (small water droplets). SO_2 oxidises in air to SO_3 which—if burnt speedily—combines with H_2O vapour available in the atmosphere and form H_2SO_4. Thus this gas has detrimental effects on all living and non-living things, including human beings, livestock vegetation and building material—both metallic and non-metallic.

Quantitative Analysis

West and Gaeke method for quantitative analysis of SO_2 *has been recommended by both EPA as well as ISI (IS :* 5182) *(Part II)* 1969. In this, ambient air absorbed in Sodium Tetrachloromercurate and estimated by colour produced when p-rosaniline hydrochloride is added to the solution in presence of formaldehyde. The colour is estimated by reading from spectrophotometer for which a calibration curve has already been prepared.

Lower limit of detection of this method as recommended in the IS Code is 0.005 ppm, *equivalent to* 13 μg/m^3. EPA also recommend this method (EPM 1975) where collection effeciency is over 98% up to concentration of 25 μg/m^3 (CFR-1975) has been reported. Below this concentration the absorption efficiency of TCM falls off and thus this becomes the lower limit of SO_2 by this method. Thus any calculated value obtained less than lower detectable limit has to be reported as below 25 μg/m^3 as per EPA (that is, if strength of TCM used for sampling was 0.04M).

Similarly, for ISI, recommended methods where strength of sampling reagent is 0.1 M, all results need to be reported as below 13 μg/m^3.

Principle

When air containing SO_2 is bubbled through Sodium Tetrachloromercurate solution (absorbent) taken in the impinger, SO_2 forms a stable dichlorosulphito-mercurate complex (DCSM). This complex is not oxidised by the oxygen of air or that remains dissolved in the absorbent. Once formed, DCSM also resists oxidation by ozone. The absorbent containing DCSM is then treated with pararosaniline and formaldehyde to form an intense red-violet colour. The intensity of this colour is directly related to the amount of SO_2 absorbed and is measured colourimetrically by spectrophotometer. The quantity of SO_2 is then obtained from a calibration curve prepared earlier. The absorbed samples are relatively stable. Losses of SO_2 from the sample may occur at a rate of one per cent per day at 22°C. No measurable loss is found to occur when stored at 5°C for 30 days.

Chemistry of SO_2 Analysis

The p-rosaniline formaldehyde of West and Gaeke

The SO_3 is allowed to react with p-rosaniline in presence of formaldehyde. The reaction, which is similar to the Schiff reaction, is highly sensitive and specific with minimum detection limit of 1 μg of SO_2 in 12 ml of a solution.

Mechanism

The mechanism of the reaction between SO_2 p-rosaniline and formaldehyde is very much similar to the Schiff reaction. Main postulates can be given:

(i) Formation of Schiff base:

$$R-NH_2 + HCHO \xrightarrow{SO_2} R-N=CH_2 + H_2O$$

(ii) Formation of an amino sulphuric acid followed by addition of the aldehyde:

$$R.NH_2 + SO_2 \longrightarrow R-NH_2-SO_2H$$

$$R-NH_2-SO_2H + HCHO \longrightarrow R-NH_2-SO_2CH_2OH$$

(iii) Formation of Sulphuric Acid:

$$R-NH_2 + SO_2 + HCHO \xrightarrow{H_2O} R-NH-CH_2SO_3H$$

Ultraviolet and visible absorption spectra indicate that when HCl is added to p-rosaniline the decolourisation occurs because of the formation of a trianilinum chloronulthane:

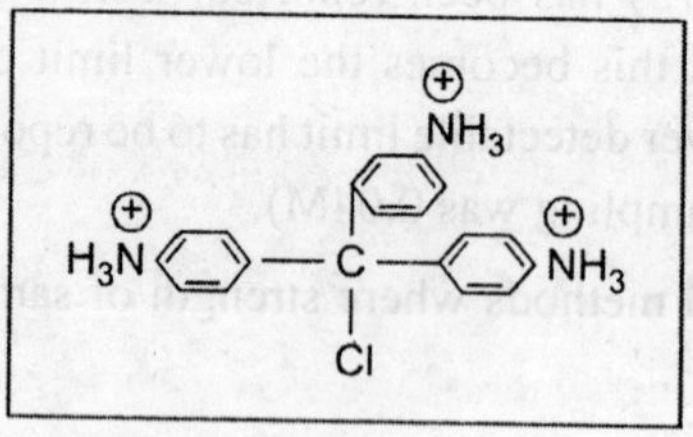

This has an absorption spectrum very similar to that of either toluene or triphenyl methane, both of which are colourless. The addition of SO_2 in available form in the presence of formaldehyde leads to the synthesis of p-rosaniline methyl sulfonic acid which gives a adsorption spectrum similar to that of p-rosaniline but which requires a higher acid concentration than does unsubstituted p-rosaniline for formation of tranilinium form.

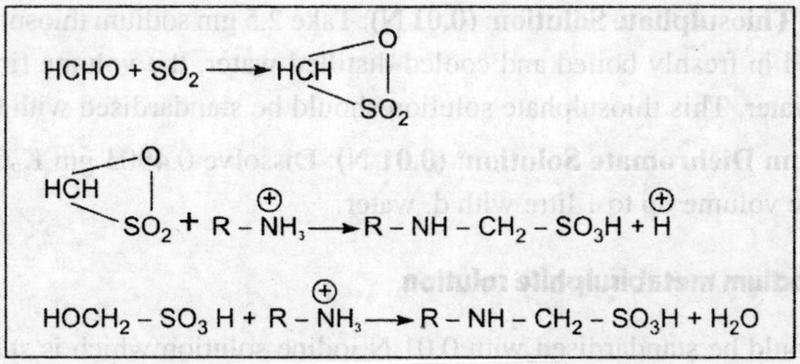

As a result of the above factors, the formation of the colour of the pararosaniline methyl sulfonic acid is considered to be the basis for the quantitative estimation of sulphur dioxide.

When, as in the quantitative procedure for SO_2, the SO_2 is collected as the dichlorosulfitomercurate (II) complex, $[HgCl_2SO_3]^{2-}$ is only the first step of the reaction and thus only the reaction's speed, is changed. Under these conditions the first reaction step is probably:

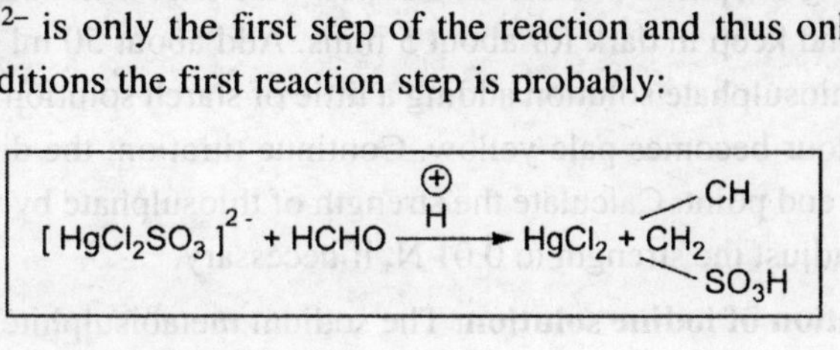

and the remainder of the reactions are as indicated above.

REAGENTS

(i) **Absorbing Reagent**: Take 27.2 gms of mercuric chloride ($HgCl_2$) in one litre flask, add 11.7 gms sodium chloride dissolve in distilled water and make up to one liter.

(ii) **P-rosaniline Hydrochloride**: Take 200 mg of p-rosaniline hydrochloride, make up to 100 ml by distilled water; leave the solution for 48 hours. Filter the solution and take 20 ml of this, add 6 ml of concentrated HCl, wait for 10 minutes and then make up to 100 ml by distilled water.

(iii) **Sodium Sulphite Solution**: 640 mg of sodium metabisulphite $Na_2S_2O_5$ (assay 65.5% as SO_2) dissolved in 1 litre of d.water 1 ml of this solution yield 0.04 mg/SO_2. The sulphite solution should be standardised with standard 0.01 N iodine solution and the strength adjusted to 0.0123 N. 1 ml of this 0.0123 N sulphite solution then is equivalent to 150 microliters of SO_2 at 25°C and 760 mm Hg.

(iv) **Starch**: A thin paste of 1.25 gm of soluble starch is prepared with cold water and then slowly added into 500 ml of boiling water. Boiling continues till solution is clear, cool the solution and store in glass-stoppered bottle.

(v) **Stock Iodine Solution**: **(0.1 N)**: Take 12.7 gm iodine in a 250 ml beaker, add 40 gm potassium iodide (KI) and 25 ml of d. water. Stir until all is dissolved, then dilute to one liter with d. water.

(vi) **Formaldehyde Solution**: Take 5 ml of 40% Formaldehyde and dilute it up to one litre.

(vii) **Iodine Solution**: **(0.01 N)**: Prepare approximately 0.01 N iodine by diluting 50 ml of stock solution to 500 ml with d. water.

(viii) **Sodium Thiosulphate Solution: (0.01 N)**: Take 2.5 gm sodium thiosulphate ($Na_2S_2O_3.5H_20$) dissolved in freshly boiled and cooled distilled water, the volume finally made 1 litre with this d. water. This thiosulphate solution should be standardised with 0.01 N.

(ix) **Potassium Dichromate Solution: (0.01 N)**: Dissolve 0.4903 gm $K_2Cr_2O_7$ (over-dried) and make the volume up to 1 litre with d. water.

Standardisation of sodium metabisulphite solution

Sulphite solution should be standardised with 0.01 N iodine solution which is standardised earlier with 0.01 N sodium thiosulphate solution. Sodium thiosulphate should be standardised beforehand with 0.01 N $K_2Cr_2O_7$ solution as primary standard:

(i) **Standardisation of thiosulphate solution**: Take 10 ml of 0.01 N $K_2Cr_2O_7$ in an Orlenmeyer flask, add 0.5 g KI, pour about 50 ml distilled water and then add (1 + 9) conc. H_2SO_4, shake the content and keep in dark for about 5 mins. Add about 50 ml more of distilled water and titrate with thiosulphate solution adding a little of starch solution (indicator) towards the end when the colour becomes pale yellow. Continue titration; the disappearance of blue colour indicates the end point. Calculate the strength of thiosulphate by using the relationship $N_1V_1 = N_2V_2$ and adjust the strength to 0.01 N, if necessary.

(ii) **Standardisation of iodine solution**: The sodium metabisulphite solution is standardised by using excess iodine and then back-titrating the unreacted iodine with standardised thiosulphate solution.

Take 50 ml of the standardised 0.01 N iodine solution into each of two 500 ml beakers A and B. To the beaker A (blank) add 25 ml distilled water, while add 25 ml of sulphite solution into beaker B. Allow 5 mins for the reaction to become complete. Titrate each beaker with thiosulphate solution using starch as indicator towards the end.

Disappearance of blue colour indicates end-point. Calculate the strength of the sulphate solution by the equation:

$$\text{Strength of sulphite solution} = \frac{(a-b)N}{25}$$

a = ml of thiosulphate needed for the blank;

b = ml of thiosulphate for the sulphite sample;

N = normality of thiosulphate solution;

25 = volume of sulphite sample taken for titration, ml

The strength of this sulphite solution should preferably be adjusted to 0.0123.

Procedure

The sample is collected by bubbling through a dilute aqueous solution of sodium tetrachloromercurate (II), which can trap SO_2 quantitatively at concentrations as low as 0.002 ppm:

$$SO_2 + HgCl_4^{2-} + H_2O \rightarrow [HgCl_2SO_3]^{2-} + 2H^+ + 2Cl^-$$

After collection of 1-hour or 24 hour average sample in the field, the solution is returned to the laboratory. Sulfamic acid is added to reduce nitrogen dioxide, which otherwise would destroy some of the colour reagent. Then the dichlorosulfitomercurate (II) is reacted with formaldehyde and pararosaniline in a solution whose pH has been controlled using H_3PO_4. The pararosaniline methylsulfonic acid formed has an absorption maximum at 575 nm at pH 1.

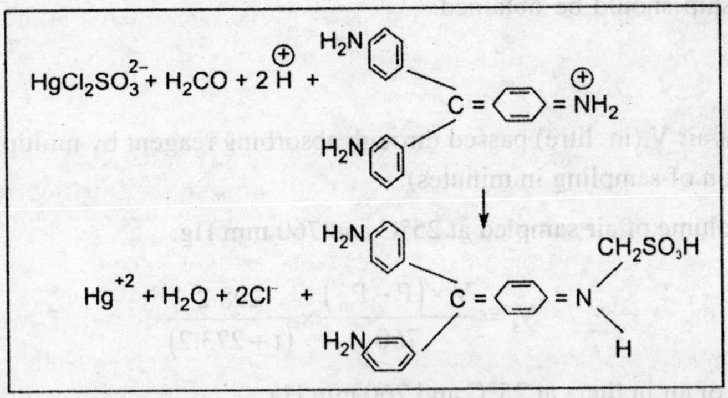

If the concentration of $[HgCl_2SO_3]^{2-}$ is too high, more than one on the pararosaniline dye may react and Beer's law will no longer be obeyed. This limits the method to solution concentrations of sulphur dioxide between 0.05 and 5 mg/m^3. The range of atmospheric concentrations may be extended by appropriate dilutions of concentrated samples from 0.002 to 5 ppm.

Sample Preparation

After collection, if a preceptitate is observed in the sample, filter it, or centrifuge it, and take the filtrate for analysis.

Determination

For each set of determination prepare a reagent blank by taking 10 ml of unexposed *TCM solution* into 25 ml flask. Similarly, a control should also be prepared using 10 ml standard known concentration of sulphite solution.

Pipett out 10 ml sampling reagent in 25 ml flasks as done for blank & control. Add 1.0 ml each of *p-rosaniline* and *formaldehyde solution* and shake well. Leave above solution for 30 minutes for development of red-violet colour.

Measure absorbance at 560 nm with the reagent blank and find out SO_2 concentration μl/ml from calibration curve.

Preparation of calibration curve

(a) Transfer 2 ml of 0.0123 N Sulphite solution (solution preparation and standardisation—please see details in Part II) into 100 ml flask and fill up to the mark using absorbing reagent (TCM). 1 ml of this diluted solution will be equivalent to 3 ml of SO_2.

(b) Take accurately 0.5, 1.0, 1.5, 2.0 ml of above diluted solution in a series of 25 ml volumetric flask and add 9.5, 9.0, 8.5 and 8.0 ml exactly of absorbing reagents in the flask.

(c) Add 1.0 ml each of p-rosaniline and formaldehyde solutions to each of these 25 ml flasks and shake well. Take 10 ml reagent blank similarly in 25 ml flask without adding sulphite solution.

(d) After 30 mins measure the absorbance (optical density) of the standard samples with a spectrophotometer at 560 nm with reagent blank as reference.

(e) Plot the calibration curve with absorbance alongwith ordinate against µl ml of SO_2. A linear relationship should be obtained.

Calculation

Obtain the volume of air V (in litre) passed through absorbing reagent by multiplying average flow rate into the time (duration of sampling in minutes).

Compute the volume of air sampled at 25°C and 760 mm Hg:

$$V_s = \frac{V \times (P - P_m)}{760} \times \frac{298.2}{(t + 273.2)}$$

where, V_s = Volume of air in liters at 25°C and 760 mm Hg.

V = Volume of air sampled as measured by flow meter in litre.

P = Barometeric Pressure mm Hg.

P_m = Suction Pressure mm Hg.

t = Temperature of sampled air °C.

Concentration of SO_2 can be calculated as:

$$SO_2 (\mu g / m^3) = \frac{X \times A \times 64 \times 1.000}{V_s \times 24.470}$$

where, X = Calibration curve value in µl/ml,

A = Total volume of absorbing reagent in ml,

V_s = Corrected volume of air passed through absorbing reagent in liters.

Interferences

Ozone and NO_2 interfere when present in air at concentrations greater than SO_2. The interference from NO_2 may be eliminated by using 1 ml of 0.06 per cent sulphamic acid (H_2NSO_3H) in the absorbing reagent. In this case absorbed samples should be analysed—preferably within 48 hours—because of probable SO_2 loss at greater rate. The interfering effect of absorbed O_3 is eliminated by delaying the addition of colourimetric reagents by 20 mins O_3 decomposes during this period.

Oxides of heavy metals, especially of Mn, Fe and Cr, may cause interference by oxidising the DCSM complex. The use of Na EDTA and phosphoric acid is suggested to eliminate this interference, if expected.

(B) OXIDES OF NITROGEN

Background

Of the various oxides of nitrogen only nitric oxide (NO) and nitrogen dioxide (NO_2) are the most important air pollutants. These gaseous pollutants are formed mostly during high temperature combustion of all kinds of fossil fuels. The relative concentration of NO in the combustion emissions is much higher than NO_2; but, upon being discharged to the atmosphere, NO oxidises to NO_2. This NO_2 is very reactive, highly oxidising and an absorber of sunlight. NO_2 is reddish-orange-brown in colour and has a characteristic pungent odour.

NO and NO_2 can be sampled and analysed either separately or collectively by either manual or automated technique. In a manual method, air is bubbled through an absorbing solution and then analysed separately; but sampling and analysis are done automatically in automated techniques.

Manual procedure for determination of NO_2 is presented here. Many methods are available for this purpose. Of these, Sodium Arsenate Modification of Jacobs-Hocchieser Method is the recommended manual method for determination of NO_2.

Principle — (Jacobs-Hocchieser, Arsenite Modification)

NO_2 is absorbed in an alkaline solution (NaOH-sodium arsenite absorbent) wherein it forms sodium nitrite which is quite stable. The solution is then freed of possible SO_2 interference by treatment with H_2O_2 and acidified. The nitrite ions reacts with sulphanilamide phosphatic acid solution to form a diazonium salt which couples with NEDA to form a deeply coloured azodye. Absorbance due to this colour is measured in spectrophotometer against a blank.

Chemistry of NO_2 Determination

NO_2 from air is absorbed in a solution of NaOH, thereby converting into a mixture of nitrite and nitrate.

$$2NO_2 + 2NaOH \rightarrow NaNO_2 + NaNO_3 + H_2O$$

Absorption is also done in NaOH containing arsenite to increase the efficiency of absorption. The nitrite formed is determined by classical diazotisation method. Nitrite reacts with primary aromatic amine to form diazonium salts. The salt is then coupled with an aromatic compound containing amine or phenol groups. The coupling furnishes an azodye which is intensely coloured and suitable for spectrophotometric measurement. Diazotisation should be carried out in presence of HCl or acitic acid at room temperature or with cooling. The colour is stable and sensitive. Sulphanilamide is primary amine which is diazotised and the diazotised salt is coupled with NEDA hydrochloride (NEDA).

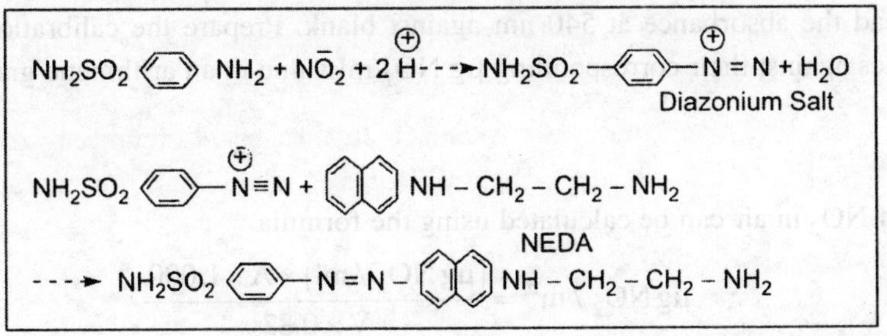

REAGENTS

(i) **Absorbing Reagent**: Dissolve 4.0 gm sodium hydroxide in distilled water, add 1 gm of sodium arsenite and dilute to 1,000 ml with distilled water.

(ii) **Sulphanilamide (Melting Point of 165° to 167°C)**: Dissolve 10 gms of Sulphanilamide in 350 cc of of distilled water, add 25 cc of phosphoric acid (85%) and make up to 500 cc.

(iii) **H_2O_2 Solution**: 0.2 ml of 30% H_2O_2 mix with 250 cc of distilled water.

(iv) **NEDA**: Dissolve 0.5 gms of NEDA in 500 ml of distilled water. This solution is stable for a month if refrigerated and protected from light. On storing, if colour of the solution is changed to brown, do not use for analysis.

(v) **Standard Nitrite Solution**: Dissolve sufficient desiccated sodium nitrite and dilute with distilled water to 1,000 ml so that this solution contains 1,000 μg NO_2 ml. The amount of $NaNO_2$ to use is calculated as:

$$G = \frac{1.500 \times 100}{A}$$

G = Amount of $NaNO_2$ (gm.)

A = Assay per cent

1.500 = Gravimetric factor in converting NO_2 into $NaNO_2$.

Analysis Procedure

Take 10 ml of exposed absorbent into conical flask. Add 1.0 ml of H_2O_2 solution, 10.0 ml of sulphanilamide solution and 1.4 ml of NEDA with thorough mixing after the addition of each reagent. A 10 ml unexposed absorbent should also be treated similarly, it serves as reagent blank for colorimetry. After 20 minutes' colour development period, the absorbance of the exposed sample is measured with the spectrophotometer at 540 nm against the reagent blank. Microgram of NO_2 present per ml of the solution is read from the calibration curve.

Calibration Curve Preparation

5.0 ml of the standard sodium nitrite solution containing 1,000 μg NO_2/ml is diluted to 200 ml with the absorbing reagent. This solution contains 25 μg NO_2/ml Transfer by pipetting 1, 2, 5 and 15 ml of this dilute nitrite solution into 50, 100 and 250 ml volumetric flasks and dilute to the mark with *absorbing reagent*. These solutions contain 0.50, 1.00, 1.25 and 1.50 μg NO_2/ml, respectively. Take 10 ml from each of these solutions and add 1.0 ml H_2O_2, 10 ml sulphanilamide and 1.4 ml NEDA; prepare one blank also similarly and read the absorbance at 540 nm against blank. Prepare the calibration curve by plotting absorbance values against their corresponding μg NO_2/ml value in an arithmetic graph paper.

Calculation

Concentration of NO_2 in air can be calculated using the formula

$$\mu g\, NO_2 / m^3 = \frac{(\mu g\, NO_2 / ml) \times A \times 1{,}000}{V \times 0.82}$$

A = Volume of absorbing reagent.

V = Volume of air passed through absorbing reagents in litre.

0.82 = Factor for correction efficiency.

μg NO_2/ml = Read from the calibration curve corresponding to the absorbance value of the exposed samples.

Range and Sensitivity

The nominal range of the method is 9-750 µg NO_2/m^3 (0.005 to 0.4 ppm). The range of the analysis is 0.04 to 2.0 µg NO_2/ml following Beer's law throughout this range (0 to 1.0 absorbance units) under the specified conditions of 50 ml of absorbing reagent, a sampling rate of 200 cc/min for 24 hours and sampling efficiency of 0.82, the range of the method is, therefore, 9 to 420 µg NO_2/m^3 (0.005 to 0.22 ppm). Nitrogen dioxide concentration in the range of 420-750 µg/m^3 (0.22 to 0.4 ppm) are accurately measured by 1 : 1 dilution of the collected sample.

Interference

By way of interference NO can raise the NO_2 response by 5 to 15% of NO_2 actually present in the air sampled. The possibility of SO_2 interference is eliminated by adding H_2O_2 to the exposed absorbent prior to addition of azo dye forming reagents H_2O_2 converts SO_2 to sulphate ions.

60.2 WATER QUALITY MONITORING

(A) BACTERIOLOGICAL EXAMINATION OF WATER (IS : 1622 - 1981)

3.1 General Equipment

3.1.1 Equipment

3.1.1.1 *General*: It is essential for accurate and satisfactory laboratory work that good equipment in proper working order be provided. Thus, the minimum laboratory equipment listed must be available in an approved laboratory and all items should meet the minimum requirements given. Additional items of equipment not listed will be required in an approved laboratory and they should meet similar standards of quality and operation.

3.1.1.2 *Incubators*: Incubators should maintain a uniform and constant temperature (35-37°C or 44-45°C) at all times in all parts. This can be accomplished by the use of a water-jacketed or anhydric type of incubator, with thermostatically controlled low-temperature electric heating units properly insulated and located in or adjacent to walls or floor of chamber and preferably equipped with mechanical means of circulating air.

Incubators should also be provided with shelves spaced to ensure uniformity of temperature throughout the chamber. The inside dimensions of the chamber should be at least 50 × 50 cm at the base and 60 cm high, to accommodate a maximum of 200 Petri dishes; 2.5 cm space should be provided between adjacent stacks of plates and between walls and stacks.

Accurate thermometers, with bulb continuously immersed in liquid (glycerine, water or mineral oil), should be maintained within the incubator and daily readings of the temperatures should be recorded. In addition, it is desirable to maintain a maximum and minimum registering thermometer within the incubator on the middle shelf to record temperature variations over a 24-hour period. Temperature variations within the incubator filled to maximum capacity should be determined at intervals. It is recommended that a recording thermometer be installed in every incubator whenever possible, so that a permanent record of temperature variations within the incubating chamber may be kept.

Incubators equipped with high-temperature heating units are unsatisfactory, since such sources of heat frequently cause localised overheating. Incubators so heated may be made to operate satisfactorily by replacing the high temperature units by suitable wiring, arranged to operate at a lower temperature and by installing mechanical air circulation. It is desirable, where ordinary room temperature varies excessively, that laboratory incubators be kept in special rooms which may be maintained at a few degrees below the recommended incubator temperature.

3.1.1.3 *Water-baths*: Water-baths are useful for carrying out the 44°C fermentation test. They should be capable of maintaining a temperature of 44°C-45°C. They should be equipped with mercury-toluol or other reliable thermometer sensitive regulation of the temperature and should be adequately insulated against heat loss. An accurate thermometer should be provided, with its bulb placed at the level of the medium in the fermentation tubes. A continuous-recording thermometer is advisable.

3.1.1.4 ***Sterilisers***

(a) *Ovens*: Hot-air sterilising ovens should be of sufficient size to prevent crowding of the interior and constructed to give uniform and adequate sterilising termperatures and equipped with suitable thermometers capable of registering accurately in the range 160°-180°C. The use of temperature-recording instrument is optional.

(b) *Autoclaves*: Autoclaves should be of sufficient size to prevent crowding of the interior and constructed to provide uniform temperatures within chambers up to and including the sterilising conditions of 102 ± 0.03 kg/cm^3 gauge pressure (15 ± 0.5 psi gauge pressure, 120°C temperature approximately). They should be equipped with pressure gauges, properly adjusted, safety valves and accurate thermometers with bulb properly located on exhaust line, so as to register minimum temperature within sterilising chambers (temperature recording instrument optional). In emergencies, a pressure-cooker may be substituted for an autoclave, if results have previously been demonstrated to be satisfactory with this method.

3.1.1.5 *Glassware*: Clean glassware is critical to ensure valid results. Previously used or new glassware should be thoroughly cleaned with phosphorus-free laboratory detergent and hot water, followed by at least three rinses with laboratory pure water. Generally, pipettes, dilution bottles and petri dishes are required.

(a) *Pipettes*: Pipettes may be of any convenient size (generally 1 ml or 10 ml) provided it is found by actual test that they deliver accurately the required amount in the manner in which they are used. The error of calibration should not exceed 2.5 per cent. Pipettes with unbroken tips and with graduations distinctively marked should be used. Pipettes with damaged tips should be repaired or discarded.

(b) *Dilution bottles*: Bottles or tubes of resistant glass, preferably Pyrex, closed with glass stoppers, rubber stoppers, or screw caps equipped with liners that do not produce toxic or bacteriostatic compounds on sterilisation, should be used. Cotton plugs shall not be used as closures. Graduation levels should be indelibly marked on the side.

(c) *Petri dishes*: Petri dishes 100 mm in diameter, with the side wall at least 15 mm high, should be used with glass or porous tops, as preferred. The bottoms of the dishes should be free from bubbles and scratches and should be flat so that the medium will be of uniform thickness throughout the plate. When available, sterilised disposable plastic petri dishes may be used as an alternative.

3.1.1.6 *Inoculating needle and loop*: It shall be 4 mm in diameter formed at one end of a length of 0.375 mm thick wire of nichrome, platinum or platinum-iridium alloy. It shall be 38 mm long from the loop to the holder. The holder consists of a thin metal rod or tube. It shall be sterilised by flaming.

3.1.1.7 *Refrigerators*: An approved laboratory should have a refrigerator of sufficient capacity for the required work load and capable of maintaining a continuous temperature between 0°C and 5°C. An electrically operated refrigeration unit provides the most efficient service.

3.1.1.8 *Colony counter*: An effective device for examining colonies, providing a magnification of 3 X, should be available. In general, a Quebec or similar colony counter will be suitable for this purpose.

3.2 Standard Plate Count

3.2.1 *General* : Standard plate count (which is an empirical method) serves to indicate the efficiency of certain processes in water treatment, particularly coagulation, filtration and disinfection and the cleanliness of the mains, reservoirs, etc. It provided an estimate of the general hygienic quality of water, which is important where large scale preparation of food and drink is concerned. Low counts are of importance for avoiding food spoilage, while rising plate counts give the earliest sign of pollution.

3.2.1.1 The standard plate count method is a direct measurement of the viable aerobic and facultative aerobic bacteria in a water environment capable of growth on the selected plating medium. The procedure does not allow the more fastidious aerobes or obligate anaerobes to develop. Also the bacteria of possible importance in water such as Crino-thrix, sphaerotilus and actinomycetes will not develop within the incubation period specified for potable water. Clumps of organisms in the water sample which are not broken up by shaking result in underestimate of bacterial density. Since, an aggregate of cells will appear as one colony on the growth medium. The number of types of bacteria that develop are influenced by the time and comparative temperature of incubation, the pH of the medium, the level of oxygen, the presence of specific nutrients on the growth medium competition among cells for nutrients, antibioisis, medation, etc.

3.2.2 In solid medium, counting of organisms depends on the fact that living cells will proceed to multiply and in time will produce sufficient progeny to form a colony visible to naked eye. Since bacteria occur in water as single cells, pair groups, chains or even dense clumps, not every individual living cell will develop into a separate colony on incubation. Therefore, number of colonies appearing on a plate does not necessarily represent the total number of organisms present in test volume. The results are expressed as number of colonies per ml.

3.2.3 Medium and Reagent

3.2.3.1 *Nutrient agar*: Dissolve 1g glucose, 5.0 g of peptone and 3.0 g of beef extract in 1,000 ml of distilled water. Adjust the pH to 7.2, distribute in required quantity and add 1.5 per cent agar powder. Sterilise at 1.02 ± 0.03 kg/cm^3 gauge pressure (15 ± 0.5 psi gauge pressure, 120°C temperature approximately) for 15 minutes in the autoclave.

3.2.3.2 *Dilution water*

(a) *Buffered dilution water*: To prepare stock phosphate buffer solution, dissolve 34 g of potassium dihydrogen phosphate (KH_2PO_4) in 500 ml of distilled water, adjust pH to 7.2 with sodium hydroxide solution (1N) and dilute to 1 litre with distilled water.

Add 1.25 ml of stock phosphate buffer solution to 1 litre of distilled water. Dispense in amounts that will provide 18 ± 0.4 ml or 9 ± 0.2 ml in 150 × 25 mm or 150 × 18 mm test tubes, respectively. Sterilise in auto-clave at 1.02 ± 0.03 kg/cm^3 gauge pressure (15 ± 0.5 psi gauge pressure, 120°C temperature) approximately for 15 minutes.

(b) *Quarter strength singer's solution*: Dissolve 9.00 g of sodium chloride, 0.42 g of potassium chloride, 0.48 g of calcium chloride and 0.20 g of sodium bicarbonate in 1 litre of water. This solution is known as Singer's solution. Dilute 500 ml of this solution to 2 litres to obtain quarter strength Ringer's solution. Dispense in amounts that will provide 1.8 ± 0.4 ml or 9 ± 0.2 ml in 150 × 25 mm or 150 × 18 mm test tubes, respectively. Sterilise in auto-clave at 1.02 ± 0.03 kg/cm^3 gauge pressure (15 ± 0.5 psi gauge pressure, 120°C temperature approximately) for 15 minutes.

3.2.4 Procedure

3.2.4.1 *Preparation and dilution*: Shake the samples about 25 times. Withdraw required portion with a sterile pipette and introduce into the petri dish or dilution tube.

3.2.4.2 *Plating*: Place 1.0 ml or 0.1 ml of other suitable dilution to be used for plating in the petri dish first. Then add to the petri dish 10 to 15 ml of melted nutrient Gagar medium at a temperature of 40-45°C (tolerable to the skin). The nutrient agar and the sample shall be thoroughly mixed over the bottom of the petri dish by tilting and rotating the dish several times. Allow the plate to solidify and place immediately in the incubator in an inverted position.

3.2.4.3 *Incubation*: Incubate the plates at 37°C for 24 hours.

3.2.4.4 *Counting*: In preparing plates, plant such amounts of water for dilution which will give from 30 to 300 colonies on a plate. Always have two or more plates for each dilution. Report the result as the average of all plates falling within limits. It is not desirable to plant more than 1.0 ml in a plate. If the colonies are more than 300 or less than 30 from 1 ml sample, disregard it. In practice, counts less than 30 occur when chlorinated water samples are plated. When the number of colonies is more than 300 in a plate, report the count at 'TNC' (too numerous to count). Counting shall be done with an approved counting aid, such as colony counter. Record the number of colonies to the nearest 5 units per ml and report the temperature of incubations.

3.3 Test for Coliforms

3.3.0 The coliform group includes all of the aerobic and facultative anaerobic gram negative, non-spore forming rod-shaped bacteria which ferment lactose with gas formation within 48 hours at 37°C. The standard test for the estimation of number of the coliform groups may be carried out either by the multiple tube dilution test (presumptive test, confirmed test, or completed test) or by the membrane filter technique.

3.3.1 *Multiple Tube Dilution Test (MTD)*: The presumptive, confirmed and completed tests are presented as total independent procedures. In using these procedures, the worker must know what is to be the stage at which the test is to be ended and details of the procedure throughout. Thus, if the worker knows that the test will be ended at the confirmed test, he will stop at the confirmed test stage only. All the necessary information regarding the sample should be recorded. It is convenient to express the results of the examination of replicate tubes and dilutions in terms of most probable number. This term is actually an estimate based on certain probability formulae. The most satisfactory information is obtained when the largest portion examined shows no gas in all or a majority of the tubes. The MPN value for a given sample is obtained by the use of MPN tables. Standard practice in water analysis is to plant five tubes for each dilution and a minimum three differents dilutions are employed. The results are to be recorded in the proper form. Table of MPN are given in Appendix I.

Appendix I: MPN/100 ml for various combinations of positive results 5 tubes each of 10, 1 and 0.1 ml sample fractions are used

Combinations	MPN/100 ml	Combinations	MPN/10
0-0-0	> 2	4-3-0	27
0-0-1	2	4-3-1	33
0-1-0	2	4-4-0	34
0-2-0	4	5-0-0	23
1-0-0	2	5-0-1	31
1-0-1	4	5-0-2	43
1-1-0	4	5-1-0	33
1-1-1	6	5-1-1	46
1-2-0	6	5-1-2	63
2-0-0	5	5-2-0	49
2-0-1	7	5-2-1	70
2-1-0	7	5-2-2	94
2-1-1	9	5-3-0	79
2-2-0	9	5-3-1	110
2-3-0	12	5-3-2	140
3-0-0	8	5-3-3	180
3-0-1	11	5-4-0	130
3-1-0	11	5-4-1	170
3-1-1	14	5-4-2	220
3-2-0	14	5-4-3	280
3-2-1	17	5-4-4	350
4-0-0	13	5-5-0	240
4-0-1	17	5-5-1	350
4-1-0	17	5-5-2	540
4-1-1	21	5-5-3	920
4-1-2	26	5-5-4	1,600
4-2-0	22	5-5-5	> 2,400
4-2-1	26		

3.3.1.1 *Media and reagents*

(a) *Dilution water: See 3.2.3.2.*

(b) *MacConkey broth*: This is used as a presumptive medium for the enumeration of coliform bacteria in water samples. Its composition is:

Peptone	20 g
Lactose	10 g
Sodium chloride	5 g
Bile salt	5 g
Distilled water	1,000 ml

In place of bile salt, which is a commercial product, sodium taurocholate or sodium tauroglycocholate may be used.

Dissolve all the ingredients and adjust the pH to 7.4. After adjusting the pH, add 1 ml of 1 per cent alcoholic solution of bromocresol purple or 5 ml of 1 per cent aqueous solution of neutral red. This will be the single strength medium. Distribute 10 ml of the medium into 150 × 15 mm test tubes and add a Durham's tube (25 × 5 mm) in an inverted position. Plug the tubes with non-absorbent cotton and sterilise at 115°C for 10 minutes in the autoclave. This medium is used for 1 ml and the decimal dilutions of the water sample. For 10 ml and larger aliquots a double strength medium is used. For the double strength medium add the above ingredients in double the quantities in 1,000 ml of distilled water. This medium is dispensed into 10 ml quantities in 150 × 18 mm test tubes added with Durham's tube and sterilised.

(c) *Brilliant Green bile lactose broth (RGB)*: This medium is used as a confirmatory test for coliforms as well as for fecal coliforms. Its composition is:

Peptone	20 g
Lactose	10 g
Bile salt	5 g
Distilled water	1,000 ml

Dissolve all the ingredients and adjust the pH to 7.4. Add 1.33 ml of 1 per cent aqueous solution of brilliant green indicator. Distribute 4 ml quantities into 150 × 12 mm test tubes and add a Durham's tube to each. After plugging with non-absorbent cotton, sterilise at 1.02 ± 0.03 kg/cm^3 gauge pressure (15 ± 0.5 psi gauge pressure, 120°C temperature approximately) for 15 minutes in the autoclave.

(d) *Peptone water*: This is used for indole test or for preparing a liquid culture of an organism. Its composition is:

Peptone	20 g
Sodium chloride	5 g
Distilled water	1,000 ml

Dissolve all the ingredients. Adjust the pH to 7.4. Dispense 4 ml medium into 100 × 12 mm tubes and plug with non-absorbent cotton. Sterilise in the autoclave at 1.02 ± 0.03 kg/cm^2 gauge pressure (15 ± 0.5 psi gauge pressure, 120°C temperature approximately) for 15 minutes.

(e) *MacConkey agar*: The medium is used for the completed test or for IMVIC classification of coliforms. Its composition is:

Peptone	20 g
Lactose	10 g
Sodium chloride	5 g
Bile salt	5 g
Distilled water	1,000 ml

Dissolve all the ingredients and adjust the pH to 7.4. Add 10 ml of 1 per cent aqueous solution of neutral red indicator and 15 g of agar. Steam the medium for 15 to 30 minutes so that agar is dissolved properly and sterilise in autoclave at 1.02 ± 0.03 kg/cm^2 gauge pressure (15 ± 0.5 psi gauge pressure, 120°C temperature approximately) for 15 minutes. After sterilisation, cool to 45°C and prepare the plates by pouring 15 ml of melted agar per plate. Allow to solidify, invert and incubate at 37°C for drying as well as for sterility test.

(f) *Nutrient agar slants*: Prepare the nutrient agar as prescribed in 3.2.3.1. Dispense while in the marked condition about 10 ml quantity into each tube (150 mm × 15 mm). Sterilise in the autoclave at 1.02 ± 0.03 kg/cm^2 gauge pressure (15 ± 0.5 psi gauge pressure, 120°C temperature approximately) for 15 minutes. After sterilisation the slants are prepared by keeping the tubes in a slanting position and allow them to solidify. Unless they are to be used, they should be stored in a refrigerator.

(g) *Kosac's reagent*: It is used for indole test. Its composition is:

Paradimethyl aminobenzaldehyde	5 g
Amyl alcohol or n-butanol	75 ml
Concentrated hydrochloric acid	25 ml

Dissolve paradimethyl aminobenzaldehyde in amyl alcohol and then add 25 ml hydrochloric acid. The reagent shall be yellowish in colour. Store in amber coloured glass stoppered bottle.

(h) *Gram staining reagents*

(i) Crystal violet is used as a primary stain:

Solution A —	Crystal violet (85 per cent dye content)	2 g
	Ethyl alcohol (95 per cent)	20 ml
Solution B —	Ammonium oxalate	0.8 g
	Water	80 ml

Mix solutions A and B in equal parts. It is sometimes found, however, that this gives so concentrated a stain that gram-negative organisms do not properly decolorise. To avoid this, dilute solution A as much as ten times. Use 20 ml of this diluted solution and mix with solution B.

(ii) *Lugol's iodine*: Dissolve 1 g of iodine crystals and 2 g of potassium iodide in 300 ml of distilled water.

(iii) Safranin is used as a counter-stain. Dissolve 25 g of safranin dye in 100 ml of 95 per cent ethyl alcohol. Add 10 ml of the solution to 100 ml of distilled water.

(iv) Ethyl alcohol—95 per cent.

3.3.1.2 *Procedure*: Shake the water samples thoroughly before making dilutions or before inoculation.

(a) *Presumptive test*

(i) Use MacConkey broth. Inoculate a series of fermentation tubes with appropriate measured quantities of the water to be tested. The concentration of nutritive ingredients in the mixture should be sufficient and according to requirements. 10 ml and above aliquots should be inoculated in double strength and 1 ml and its dilution should be inoculated into single strength medium.

(ii) Incubate all tubes at 37°C for 24 to 48 hours. Examine each tube at the end of 24 ± hours for gas production and, if no gas has been formed, reincubate for another 24 hours and, at the end of 48 hours, examine again. Record the presence of or absence of gas at each examination of the tubes regardless of the amount.

(iii) Formation of the gas within 48 ± 3 hours in any amount, in the inner fermentation tubes, constitutes a possible presumptive test. The absence of gas formation at the end of 48 ± 3 hours of incubation constitutes a negative test.

(b) *Confirmed test*: The medium used for confirmed test is brilliant green bile lactose broth (BGB).

(i) Submit all primary fermentation tubes showing any amount of gas at the end of 24 hours incubation to the confirmed test. If additional primary fermentation tubes show gas at the end of 48 hours incubation, these too shall be submitted to the confirmed test. Use a sterile metal loop, 3 to 4 mm in diameter, to transfer one or two loopful of medium from the presumptive positive tubes to a tube of BGB broth. When making such transfers, gently shake the tube first or mix by rotating. Incubate the inoculated tubes at 37°C for 48 ± 3 hours.

(ii) The formation of gas in any one of the Durham's tubes of BGB tube at any time within 48 ± 3 hours constitutes a positive confirmed test.

(c) *Completed test*

(i) It may be applied to positive BGB tubes. Shake the tube and streak with the help of a loop on the MacConkey agar plates as soon as possible in such a way so as to get discrete colonies. Incubate the plates at 37°C for 24 ± 2 hours.

(ii) From each plate pick up typical or atypical colonies and inoculate lactose broth and nutrient agar slants. Incubate at 37°C for 24 to 48 hours.

(iii) Nutrient agar slants can be used for gram stain. If organisms are gram-negative, non-spore forming bacilli and if gas is produced in lactose broth the test is considered completed, and the presence of coliform organisms is demonstrated.

(iv) *Gram-stain technique*: Prepare a thin smear of the growth on the agar slant on clean glass slide. Air dry, fix by passing the slide through a flame and stain for 1 minute with ammonium oxalate-crystal violet solution. Wash the slide in water, immerse in Lugol's iodine solution for 1 minute. Wash the slide in water, blot dry decolorise with ethyl alcohol for 30 seconds, using gentle agitation. Blot and cover with counter stain for 10 seconds with safranin, then wash, dry and examine under oil immersion.

Cells which decolorise and accept the safranin stain are pink in colour and defined as gram-negative in reaction. Cells which do not decolorise but retain the crystal violet stain are deep blue in colour and are defined as gram-positive.

3.3.1.3 *Computing and reading of MPN*: The number of positive findings of coliform group organisms (either presumptive, confirmed, or completed) resulting from the multiple portion decimal dilution planting should be computed as combination of the positives and recorded in terms of Most Probable Number. The Most Probable Number for a variety of planting series and results is given in Appendix I.

3.3.2 Membrane Filter (MF) Technique

3.3.2.1 *Outline of the method*: The membrane filter technique in water analysis is becoming more and more popular due to its advantages over the multi-tube dilution technique. Results are confined within 24 or 48 hours as compared to 48 to 96 hours by multi-tube dilution technique. Much larger volume and hence more representative sample can be tested. Results are obtained with much greater precision and require less laboratory space, equipment is not bulky and involves less labour. The limitations of this technique are few. Samples with high turbidity and less indicator bacterial count will be difficult to examine. Samples having high number of non-indicator organisms will give less count.

3.3.2.2 *Description of MF assembly*

(a) There are many varities of MF assemblies. The common one is millipore standard hydrosol filter holder. Most components are made of stainless steel. These are locking ring, fine mesh stainless steel screen for supporting the filter membrane and the funnel assembly.

(b) Only those filter membranes may be used which have been found, through complete laboratory tests certified by manufacturer, to provide full bacterial retention, stability in use, freedom from chemicals inimical to the growth and development of bacteria and satisfactory speed of filtration. They should preferably be gridmarked in such a way that material grown is neither inhibited nor stimulated along the grid lines. The membrane filters of 47 mm in diameter and 0.45 micron pore size is used.

(c) Absorbent pads for nutrients should consist of discs or filter paper or other materials known to be of high quality and free of sulphites or other substances that could inhibit bacterial growth. These should be approximately 45 mm in diameter and of thickness sufficient to absorb 1.8 to 2.2 ml of nutrient.

(d) Sterilisation of filter assembly, filters and absorbent pads is carried out at 1.02 ± 0.03 kg/cm^2 gauge pressure (15 ± 0.5 psi gauge pressure, 120°C temperature approximately) for 15 minutes. After sterilisation, immediately release steam in the autoclave by opening the outlet.

3.3.2.3 *Selection of sample sizes*: The size of the sample is governed by the expected bacterial density. An ideal quantity should result in the growth of 20 colonies and not more than 200 colonies of all types. Always filter sample in duplicate. If water is heavily contaminated use less quantity of water. When less than 20 ml is to be filtered, dilute the portion to a minimum of 30 ml before filtration.

3.3.2.4 *Filtrations of sample*: Using sterile forceps, place a sterile filter over the porous plate or stainless steel mesh of the apparatus, grid side up. Place the funnel unit carefully over the receptacle and lock it in place. Then pass the sample through the filter under vacuum. Rinse the filter by filtration—two to three times with 20 to 30 ml of sterile buffer water. Unlock the assembly, remove the filter by sterile forceps and place it on the sterile pad or agar with a rolling motion to avoid the entrapment of air.

3.3.2.5 *Medium*: Medium used for enumerating coliforms by membrane filter technique is known as M. Endo broth. The composition of the medium is:

Tryptone or polypeptone	10.0 g
Thiopeptone or thiotone	5.0 g
Casitone or Trypticase	5.0 g
Yeast extract	1.5 g
Lactose	12.5 g
Sodium chloride	5.0 g
Dipotassium hydrogen phosphate	4.375 g
Potassium dihydrogen phosphate	1.375 g
Sodium lauryl sulphate	0.05 g
Sodium desoxycholate	0.1 g
Sodium sulphite	2.1 g
Basic fuchsin	1.05 g

Dissolve the above ingredients in 1,000 ml of distilled water containing 20 ml of ethyl alcohol (95 per cent). Heat the medium to boiling point. Do not heat for a long time or do not submit to steam under pressure. The final pH should be between 7.1 to 7.3.

3.3.2.6 *Procedure*: Saturate the pad in a small petridish with M. Endo broth and place the filter on it. Remove excessive medium by tilting. Invert the plates and incubate at 37°C for 24 hours under humid chamber. All colonies which produce a dark red colony with a metallic shine within 24 hours incubation are considered members of coliform group and are counted. The count is made by a low power stereo microscope.

$$\text{Coliform density (coliforms / 100 ml)} = \frac{\text{Coliform colonies counted} \times 100}{\text{Volume in ml of the sample filtered}}$$

3.3.3 Test for Fecal Coliforms

3.3.3.1 *General*: This procedure is used to differentiate coliforms of fecal origin from those on non-fecal origin. Fecal coliforms are those coliforms which can ferment lactose at 44.5°C within 24 ± 2 hours with the production of gas. Use brilliant green bile lactose broth medium for this test.

3.3.3.2 Subculture all presumptive positive tubes of the coliform test, at the end of 24 and 48 hours into BGB medium (see 3.3.1.1) and incubate at 44.5°C for 24 hours in a water-bath. Gas formation within 24 hours is considered a positive reaction for fecal coliforms.

3.3.4 Test for E. Coli

3.3.4.1 E. Coli is one of the members of fecal coliforms which ferments lactose with the production of gas at 44.5°C within 24 hours, as well as produce indole from tryptophone at 44.5°C within 24 hours. Subculture from all the positive tubes of BGB broth at 44.5°C (fecal coliforms) into tubes of peptone water. Incubate at 44.5°C for 24 ± 2 hours. At the end of the incubation period, test for indole production by adding a few drops of Kovac's reagent. Positive test will give pink colour while negative test will give yellow colour.

3.3.5 Delayed Incubation Method (Total coliforms)

3.3.5.1 *General*: The delayed incubation MF method is useful in survey, monitoring or emergency situations when the single step coliform test cannot be performed at the sampling site or when time and temperature limits for sample storage cannot be met. The advantages are: (a) The method permits confirmation and biochemical identification of organisms as necessary; and (b) The method eliminates field processing and equipment needs.

The coliform bacteria can be kept for up to 72 hours with little effect on final counts. However, it is desirable that the holding period should be kept to the minimum. The applicability of the delayed incubation procedure for specific water source should be determined by comparative test procedures with conventional methods. The delayed incubation procedure is not a substitute for the immediate incubation test and should be used only when other alternatives are not applicable.

3.3.5.2 *Principle*: A specific volume of water sample is filtered through membrane filter. The filter retaining the microorganisms is placed on an absorbent pad saturated with M. Endo preservative medium or M. Coliform Holding Broth in a tight-lidded petridish and transported from field site to the laboratory. The holding medium maintains the viability of the coliform organisms and generally does not permit visible growth during transport. In the laboratory the filter is transferred to M. Endo growth medium and incubated at 35°C for 18-24 h. Sheen colonies are counted in total coliforms/100 ml.

3.3.5.3 *Media*

(a) M. Endo holding medium is prepared by adding 3.2 ml of solution benzoate solution (12 per cent) per 100 ml of M. Cofliform broth, prepared as given below 0.4 ml of the cycloheximide solution may be added, if required.

(i) *M. Coliform broth*

Tryptone or polypeptone	10.0 g
Casitone or Trypticase	5.0 g
Lactose	12.5 g
Dipotassium hydrogen phosphate	4.375 g
Sodium lauryl sulphate	0.075 g
Basic fuchsin	1.05 g
Thiopeptone or thiotone	10.0 g
Yeast extract	1.5 g
Sodium chloride	5.0 g
Potassium dihydrogen phosphate	1.375 g
Sodium desoxycholate	0.10 g

Add 48 g of this medium to 1 liter of laboratory pure water containing 20 ml of 95 per cent ethanol. Denatured alcohol should not be used. Heat in boiling water bath for solution. Store prepared medium in the dark at 4°C. Discard the unused medium after 96 hours.

(ii) *Sodium benzoate solution*: Dissolve 12 g of sodium benzoate in about 85 ml of laboratory pure water, then bring to 100 ml final volume. Sterilise by autoclaving or filtration. Discard the solution after 6 months.

(iii) *Cycloheximide solution (optional)*: Prepare an aqueous solution containing 1.25 grams of cycloheximine/100 ml laboratory pure water. Store solution in refrigerator and discard after 6 months.

Note: Cycloheximide is used for samples that have shown problems of overgrowth with fungi. It is a powerful skin irritant and should be handled with care.

(b) M. Coliform holding medium (LES holding medium) may be used as an alternative holding medium. To prepare the medium, add the following reagents in 1 litre of laboratory pure water and mix (do no heat) to dissolve:

Composition

Tryptone or Trypticase Peptone	3.0 g
Dipotassium hydrogen phosphate	3.0 g
Sulphanilamide	1.0 g
Cycloheximide	0.5 g
M. Endo Broth MF	3.0 g
Sodium benzoate	1.0 g
Paraminobenzoic acid	1.2 g

3.3.5.4 *Procedure*: Saturate the sterile absorbent pads with about 2.0 ml of M-Endo Holding Medium or LES Holding Medium, prepared as outlined above. Pour off excess broth. Using a sterile forceps, place a membrane filter on the filter base, grid side up. Attach the funnel to the base of the filter unit; the membrane filter is now held between the funnel and base. Shake the sample vigorously about 25 times and measure into the funnel with the vacuum off. If the sample is less than 10 ml, add 10 ml of sterile dilution water to the membrane filter before adding the sample. The sample volume should be such as to produce counts of 20-80 coliform colonies. Filter the sample through the membrane and rinse the sides of the funnel walls at least twice with 20-30 ml sterile dilution water. Turn off the vacuum and remove the funnel from the base of the filter unit. With the flame-sterilised forceps remove the filter from the filterbase and place grid side up on an absorbent pad saturated with M. Endo Holding Medium of LES Holding Medium, using a rolling action at one edge. Exercise care to avoid trapping air bubbles under the membrane. Place top on petri dish and proceed with filtration of next volume. Clearly mark the lid of each petri dish, indicating location, time of collection, time of incubation, sample number and sample volume. Use a waterproof felt tip marker or grease pencil.

Inspect each membrane in the petri dish for uniform contact with the saturated pad. If air bubbles are present under the filter (indicated by bulges) remove filter with sterile forceps and roll onto the absorbent pad again. Seal the petri dish by firmly pressing down the top. Place the culture dish in shipping container and send it to the examining laboratory. At the examining laboratory replace the membrane from the holding medium. Place it in another dish containing M. Endo Broth or agar medium and complete testing as already described.

3.3.6 Delayed Incubation Test for Fecal Coliforms

3.3.6.1 The delayed incubation procedure for fecal coliforms is same as that for total coliforms, as given in 3.3.5, except that M-VFC holding medium is used. The composition of the medium is:

Casitone, Vitamin-free	0.2 g
Sodium benzoate	4.0 g
Sulphanilamide	0.5 g
Final pH	6.7 ± 0.2

Preparation—Add 4.7 g of medium (as given above) per litre of laboratory pure water containing 10 ml of 95 per cent ethanol. Denatured alcohol should not be used. Heat slightly to dissolve the ingredients, then sterilise by membrane filtration (0.22 pm). Store prepared medium at 4°C. Discard after 1 month.

DISSOLVED OXYGEN: (Winkler's Iodometric Method)

Principle

The manganous sulphate reacts with the alkali (KOH or NaOH) to form a white precipitate of manganous hydroxide which, in the presence of oxygen, gets oxidised to a brown colour compound. In the strong acid medium, manganic ions are reduced by iodide ions which get converted to iodine equivalent, to the original concentration of oxygen in the sample. The iodine can be titrated against thiosulphate using starch as an indicator.

Reagents

A. **Sodium thiosulphate, 0.025N**: Dissolve 24.82 g of $Na_2S_2O_3.5H_2O$ in boiled distilled water and make up the volume to 1 liter. Add 0.4 g of borax or a pallet of NaOH as stabiliser. This is 0.1 N stock solution. Dilute it to 4 times with boiled distilled water to prepare 0.025 N solution (250 - 1,000 ml). Keep in a brown glass-stoppered bottle.

B. **Alkaline potassium iodide solution**: Dissolve 100 g of KOH and 50 g of KI in 200 ml of boiled distilled water.

C. **Manganous sulphate solution**: Dissolve 100 g of $MnSO_4.4H_2O$ in 200 ml of boiled distilled water and filter.

D. **Starch solution**: Dissolve 1 g of starch in 100 ml of warm (80°C - 90°C) distilled water and add a few drops of formaldehyde solution.

E. **Sulphuric acid**: H_2SO_4, conc. (sp. gr. 1.84).

Procedure

1. Fill the sample in a glass stoppered bottle (BOD Bottle) of known volume (100-300 ml), carefully, avoiding any kind of bubbling and trapping of the air bubbles in the bottle after placing the stopper.
2. Pour 1 ml of each $MnSO_4$ and alkaline KI solutions (in case, the volume of the sample is about 300 ml instead of 1 ml of reagents, add 2 ml solutions of each), well below the surface from the walls. The reagents can also be poured at the bottom of the bottle with the help of special pipette syringes to ensure better mixing of the reagents with the sample. Always use separate pipettes for these two reagents. A precipitate will appear.
3. Place the stopper and shake the contents well by inverting the bottle repeatedly. Keep the bottle for some time to settle down the precipitate. If the titration is to be prolonged for few days, keep the sample at this stage with the precipitate.
4. Add 1-2 ml of concentrated H_2SO_4 and shake well to dissolve the precipitate.
5. Remove either the whole contents, or a part of them (50-100 ml) in conical flask for titration. Prevent any bubbling to avoid further mixing of oxygen.
6. Titrate the contents within one hour of dissolution of the precipitate against sodium thiosulphate solution using starch as an indicator. At the end-point, initial dark blue colour changes to colourless.

Calculation

When whole contents have been titrated:

$$\text{Diss. Oxygen mg/l} = \frac{(\text{ml} \times \text{N}) \text{ of titrant} \times 8 \times 1{,}000}{V_1 - V}$$

When only a part of the contents has been titrated:

$$\text{Diss. Oxygen mg/l} = \frac{(\text{ml} \times \text{N}) \text{ of titrant} \times 8 \times 1{,}000}{V_2\left(\dfrac{V_1 - V}{V_1}\right)}$$

where, V_1 = volume of sample bottle after placing the stopper.

V_2 = volume of the part of the contents titrated.

V = volume of $MnSO_4$ and KI added.

In oceanography, the unit ml/l is preferred over mg/l. It can be obtained by dividing the value in mg/l by 1.43.

(B) BIOCHEMICAL OXYGEN DEMAND (BOD) DETERMINATION

Principle

The principle of the method involves measuring the difference of the oxygen concentration between the sample and after incubating it for 5 days at 20°C.

Apparatus and Reagents

A. **BOD bottles**

B. **BOD incubator**: Having temperature control at 20°C.

C. **Phosphate buffer**: Dissolve each 8.5 g KH_2PO_4, 21.75 g K_2HPO_4, 33.4 g $Na_2HPO_4.7H_2O$ and 1.7 g NH_4Cl in distilled water to prepare 1 litre of solution. Adjust pH to 7.2.

D. **Magnesium sulphate**: Dissolve 82.5 g $MgSO_4.7H_2O$ in distilled water to prepare 1 litre of solution.

E. **Calcium chloride**: Dissolve 27.5 g of anhydrous $CaCl_2$ in distilled water to prepare 1 litre of solution.

F. **Ferric chloride**: Dissolve 0.25 g $FeCl_3.6H_2O$ in distilled water to prepare 1 litre of solution.

G. **Sodium sulphite solution, 0.025 N**: Dissolve 1.575 g Na_2SO_3 and dilute to 1,000 ml Solution should be prepared freshly.

Procedure

1. Prepare dilution water in a glass container by bubbling compressed air in distilled water for about 30 minutes.
2. Add 1 ml each of phosphate buffer, magnesium sulphate, calcium chloride and ferric chloride solutions for each litre of dilution water and mix thoroughly.
3. Neutralise the sample to pH around 7.0 by using 1 N NaOH or H_2SO_4.
4. Since the DO in the sample is likely to be exhausted; it is usually necessary to prepare a suitable dilution of the sample according to the expected B. O. D. range. See Table 60.1 for dilution of the sample.
5. Prepare dilutions in a bucket or a large glass trough, mix the contents thoroughly. Fill 2 sets of the BOD bottles.
6. Keep one set of the bottles in BOD incubator at 20°C for 5 days and determine the DO content in another set immediately.
7. Determine DO in the sample bottles immediately after the completion of 5 days' incubation.

Table 60.1: Preparation of dilutions for various ranges of BOD in the samples

Range of BOD mg/l O_2	Dilution (%)	Sample volume in 1 litre of mixture
0-6	No dilution	1,000
4-12	50	500
10-30	20	200
20-60	10	100
40-120	5	50
100-300	2	20
200-600	1	10
400-1,200	0.5	5
1,000-3,000	0.2	2
2,000-6,000	0.1	1
Above 6,000	0.05	0.5

Calculation

$$\text{BOD, mg/l} = (D_0 - D_5) \times \text{dilution factor}$$

where, D_0 = Initial D_0 in the sample

D_5 = D_0 after 5 days

Note: If the sample contains free residual chlorine, it is necessary to remove it due to its toxic effects on microorganisms of the sample. For removal of residual chlorine take 50 ml of the sample and acidify with 10 ml acetic acid (1 + 1). Add about 1 g KI. Titrate with $Na_2S_2O_3$ (0.025 N) using starch as an indicator. Calculate the volume of $Na_2S_2O_3$ required per ml of the sample and add accordingly to the sample to be tested for BOD.

It is advisable to rely on COD of the samples, suspected of having toxic materials.

(C) CHEMICAL OXYGEN DEMAND (COD)

Principle

Chemical Oxygen Demand (COD) is the measure of oxygen consumed during the oxidation of the oxidisable organic matter by a strong oxidising agent. Potassium Dichromate in the presence of sulphuric acid is generally used as an oxidising agent in determination of COD.

The sample is refluxed with $K_2Cr_2O_7$ and H_2SO_4 in presence of mercuric sulphate to neutralise the effect of chlorides and silver sulphate (catalyst). The excess of potassium dichromate is titrated against ferrous ammonium sulphate using ferroin as an indicator. The amount of $K_2Cr_2O_7$ used is proportional to the oxidisable organic matter present in the sample.

Reagents

A. **Potassium dichromate solution, 0.25 N**: Dissolve 12.259 g of dried A. R. grade $K_2Cr_2O_7$ in distilled water to make 1 litre of solution.

B. **Potassium dichromate solution, 0.025 N**: Dilute 0.25 N $K_2Cr_2O_7$ 10 times (100 → 1,000 ml)

C. **Ferrous ammonium sulphate, 0.1 N**: Dissolve 39.2 g of $Fe(NH_4)_2(SO_4)_2.6H_2O$ in water adding 20 ml conc. H_2SO_4 to make 1 litre of solution, standardise this solution with $K_2Cr_2O_7$. For standardisation, dilute 10.0 ml of $K_2Cr_2O_7$ to about 100 ml, add 30 ml of conc. H_2SO_4, and titrate with ferrous ammonium sulphate using ferroin as an indicator.

D. **Ferrous ammonium sulphate, 0.01 N**: Dilute 0.1 N ferrous ammonium sulphate to 10 times (100 → 1,000 ml).

E. **Ferroin indicator**: Dissolve 1.485 g of 1,10 - phenonthroline and 0.695 g of ferrous sulphate ($FeSO_4.7H_2O$) in distilled water to make 100 ml of solution.

F. **Sulphuric acid**: H_2SO_4, conc. (sp. gr. 1.84)

G. **Mercuric sulphate**: $HgSO_4$, solid

H. **Silver sulphate**: Ag_2SO_4, solid

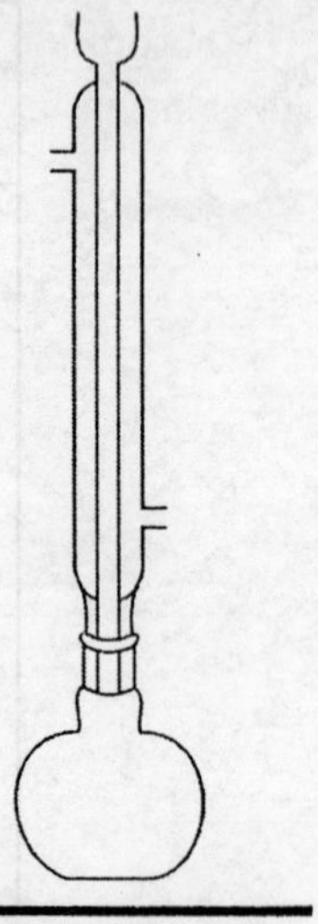

Fig. 60.1: COD reflux assembly

Procedure

1. Take 20 ml of sample in a 250-500 ml COD flask (Round-bottom or Erlenmeyer flask with a ground joint for Liebig reflux condenser (Fig. 60.1).
2. If the sample is expected to have COD more than 50 mg/l, add 10 ml of 0.025 N potassium dichromate solution. Extreme care should be taken in case of low COD samples. A small trace of organic matter in glassware may contribute a significant error.
3. Add a pinch of Ag_2SO_4 and $HgSO_4$. If the sample contains chlorides in higher amount, $HgSO_4$ is added in the ratio of 10 : 1, to the chlorides. COD cannot be determined accurately if the sample contains more than 2,000 mg/l of chlorides.
4. Add 30 ml of sulphuric acid.
5. Reflux at least for 2 hours on a water bath or a hot plate. The reflux water to make the final volume to about 140 ml.
6. Add 2-3 drops of ferroin indicator, mix thoroughly and titrate with 0.1 N ferrous ammonium sulphate (with 0.01 N ferrous ammonium sulphate if 0.025 N $K_2Cr_2O_7$ has been used).
7. Run a blank with distilled water using same quantity of the chemicals.

Calculation

$$\text{COD, mg/l} = \frac{(b-a) \times N \text{ of } K_2Cr_2O_7 \times 1{,}000 \times 8}{\text{ml sample}}$$

where, a = ml of titrant with sample,
b = ml of titrant with blank.

(D) INORGANIC PHOSPHORUS

Principle

The phosphates in water react with ammonium molybdate and form complex heteropoly acid (molybydophosphoric acid), which gets reduced to a complex of blue colour in the presence of $SnCl_2$.

The absorption of light by this blue colour can be measured at 690 nm to calculate the concentration of phosphates.

Reagents

A. **Ammonium molybdate solution**
   (a) Dissolve 25.0 g of ammonium molybdate in 175 ml of distilled water.
   (b) Add 280 ml of concentrated H_2SO_4 to 400 ml of distilled water and cool. Mix the two solutions (a) and (b) and dilute to 1 litre.

B. **Stannous chloride solution**: Dissolve 2.5 g of stannous chloride in 100 ml glycerol by heating on a water bath for rapid dissolution.

C. **Standard phosphate solution**: Dissolve 4.388 g of dried anhydrous potassium hydrogen phosphate K_2HPO_4 in distilled water and make up the volume to 1 litre. Dilute this solution to 100 times (10 → 1,000 ml). This is standard phosphate solution containing 10 mg P/1 (1 ml = 0.01 mg P).

Procedure

1. Take 50 ml of filtered clear sample in a clean Erlenmeyer flask. If the sample contains colour and colloidal impurities, they can be removed by adding a spoonful of activated charcoal and then filtering the sample.
2. Add 2 ml of ammonium molybdate followed by 5 drops of $SnCl_2$ solution.
3. A blue colour will appear. Take reading at 690 nm on a spectrophotometer using a distilled water blank with the same amount of the chemicals. Take the reading after 5 minutes but before 12 minutes of the addition of the last reagent.
4. Find out the concentration with the help of the standard curve.

Preparation of Standard Curve

1. Make various dilutions at the interval of 0.1 mg P/1 from the standard phosphate solution according to Table 10.
2. Take 50 ml of each dilution and add 2 ml ammonium molybdate and 5 drops of $SnCl_2$ solution.
3. Take the reading at 690 nm and plot a graph between 'absorbance' or 'transmittance' and concentration.

Table 60.2: Preparation of various dilutions of phosphorus for the standard curve

ml of standard solution	Diluted to (ml)	Concentration of $PO_4 - P$, mg/l
5.0	50	1.0
4.5	50	0.9
4.0	50	0.8
3.5	50	0.7
3.0	50	0.6
2.5	50	0.5
2.0	50	0.4
1.5	50	0.3
1.0	50	0.2
0.5	50	0.1

(E) NITRATE: (Phenol disulfonic acid method)

Principle

Nitrate reacts with phenol disulfonic acid to form a nitro-derivative which in alkaline medium develops a yellow colour. The concentration of NO_3 can be determined colorimetrically, since the colour so formed obeys the Bee's law.

Reagents

A. **Phenol disulfonic acid**: Dissolve 25 g of white pure phenol in 150 ml of Conc. H_2SO_4 and add 75 ml of farming H_2SO_4. Heat for two hours on a water bath and keep in dark bottle. Commercially phenol disulfonic acid is also available.

B. **Silver sulfate solution**: Dissolve 4.4 g of Ag_2SO_4 in distilled water to prepare the solution.

C. **Liquior ammonia** (30%).

D. **Standard nitrate solution** (1 mg N/lit): Dissolve 0.7222 g of KNO_3 in distilled water and make up the volume to 1 lit. This solution contains 100 mg N/lit. Dilute it to 100 times to prepare a solution having 1 mg N/lit (10 → 1000 ml).

Procedure

1. Take 50 ml of filtered sample (water/soil aliquot) containing not more than 1 mg/lit NO_3–N in an Erlenmeyer flask.
2. Add an equivalent amount of silver sulfate solution to remove chlorides (1 mg/litre).
3. Heat slightly and filter the precipitate.
4. Evaporate the solution in a bath to make dryness.
5. Cool and dissolve the residue in 2 ml disul phenol disulfonic acid and dilute the contents to 50 ml.
6. Add 6 ml of liquid ammonia to develop a yellow colour.
7. Take the reading at 410 nm.
8. Prepare a standard curve with known concentration of NO_3–N solution and compare the unknown concentration from standard curve (standard curve may be prepared with 0.1 mg/N lit to 1 mg/N per lit).

Calculation

Finally calculate the NO_3–N lebel from the sample aliquot analysed and expressed as mg/lit.

(F) SULPHATE (Gravimetric method)

Principle

Sulphate is precipitated as barium sulphate in the hydrochloric acid medium by addition of barium chloride at boiling temperature. The precipitate is filtered and dried and expressed as $BaSO_4$.

Reagents

A. Methyl red indicator: Dissolve 100 mg methyl red sodium salt in 100 ml distilled water.

B. Hydrochloric acid (1:1 dit).

C. Barium chloride solution: Dissolve 100 gm $BaCl_2\ 2H_2O$ in distilled water to prepare 1 lit solution.

D. Silver nitrate-Nitric acid reagent: Dissolves 85 g of $AgNO_3$ and 0.5 ml concentrated HNO_3 in distilled water to prepare 500 ml reagent.

Procedure

1. 50 ml of sample (water or extracted aliquot) is taken in a conical flask to which a few drops of methyl red indicator is added.
2. The pH of the solution is adjusted at 4.5 to 5.0 by addition of HCl solution.
3. Boil the solution and add $BaCl_2$ solution slowly in excess, till precipitation completed.
4. Filter the precipitate, dry at 70-80°C for 2 hours.
5. Dry the filter paper containing precipitate containing and ignite it in a crucible at 800°C for about 1 hrs. Cool and weigh the precipitate of $BaSO_4$.

Calculation

$$SO_4(\text{mg/lit}) = \frac{\text{mg } BaSO_4 \times 411.5}{\text{ml sample}}$$

(G) CHLORIDE

Principle

Silver nitrate reacts with chloride to form very slightly soluble white precipitate of AgCl. At the end point, when all the chlorides get precipitated, free silver ions react with chromate to form silver chromate of reddish brown colour.

Reagents

A. **Silver nitrate** (0.02N): Dissolve 3.4 gm of dried $AgNO_3$ (AR) in distilled water to make 1 lit soln. and keep in a dark bottle.

B. **Potassium chromate (5%)**: Dissolve 5 g of K_2CrO_4 in 100 ml of distilled water.

Procedure

1. Take 50 ml of sample (water/aliquot) in an Erlenmeyer flask and add 2 ml of $K_2Cr_2O_7$ solution.
2. Titrate the content against 0.02N $AgNO_3$ until a persistent red tinge appear.

Calculation

$$\text{Chloride}(\text{mg/l}) = \frac{(\text{ml} \times \text{N of } AgNO_3 \text{ Soln.}) \times 1{,}000 \times 35.5}{\text{ml sample}}$$

(H) HARDNESS

Principle

It indicates the presence of calcium and magnesium in sample. It can be measured by complexometric analysis using EDTA soln. at higher pH (>10) using Eriochrome Black T indicator.

Reagents

A. **EDTA solution (0.01M)**: Dissolve 3.723 g of disodium salt of EDTA in distilled water to make 1 lit soln.

B. **Buffer solution**: Dissolve 16.9 g NH_4Cl in 143 ml of concentrated NH_4OH soln.

C. **Eriochrome Black T indicator**: Mix 0.40 gm of Eriochrome Black T with 100 mg NaCl (AR) and grind the powder.

D. **Sodium sulphide solution**: Dissolve 5.0 g $Na_2S, 9H_2O$ dissolved in 100 ml distilled water and kept in tight bottle to avoid oxidation.

Procedure

1. Take 50 ml of sample (water or extracted aliquot) in a conical flask.
2. Add 1 ml buffer solution.
3. Add 1 ml of Na_2S solution.
4. Add 100 mg of Eriochrome Black T powder in the mix and shake, solution turn wine red.
5. Titrate the content against EDTA solution. The end point colour changed to blue.

Calculation

$$\text{Hardness}\left(\text{mg / l CaCO}_3\right) = \frac{\text{ml EDTA used} \times 1{,}000}{\text{ml sample}}$$

(I) IRON

Principle

All the iron is converted into ferrous state by boiling with hydrochloric acid and hydroxylamine. The reduced iron chelates with 1, 10-phenonthroline at pH 3.2 to 3.3 to form a complex of orange red colour.

Reagents

A. **Hydrochloric acid** (concentrated).

B. **Hydroxylamine hydrochloride solution**: Dissolve 10 gm of hydroxylamine hydrochloride in 100 ml distilled water;

C. **Ammonium acetate buffer solution**: Dissolve 250 gm of NH_4CH_2COOH in 15 ml distilled water and add 700 ml glacial acetic acid;

D. **Phenonthroline solution**: Dissolve 100 ml of 1.10 phenonthroline monohydrate in 100 ml distilled water through stirring and heating at 80°C.

E. **Iron stock solution** (200 mg Fe/lit): 1.404 g ferrous ammonium sulphate is dissolved in 20 ml Conc. H_2SO_4 and add 50 ml distilled water. Add 0.1 N $KMnO_4$ soln. till a persistent faint pink colour appears. Dilute to 1 lit. Standard iron solution (10 mg Fe/lit) is prepared from stock solution (50-1,000 ml).

Procedure

1. Take 50 ml of sample (water or extracted aliquot) in a flask and add 2 ml conc. HCl and 1 ml of hydroxylamine soln;
2. Boil the contents till volume reduced to half;
3. Cool and add 100 ml ammonium acetate buffer and 2 ml phenonthroline solution;
4. Finally make the volume up to 100 ml. After 10 min. take reading at 510 nm on a spectrophotometer;
5. Prepare a standard curve with known volume of iron soln. (1 to 4 mg/lit) and read the unknown Fe content directly from standard curve by comparison.
6. Calculate the iron content of the sample directly.

(J) PRIMARY PRODUCTIVITY MEASUREMENT

General

Primary productivity is the rate at which the sun's radiant energy is stored by photosynthetic activity of producer organisms (various forms of algae and higher plants in water) in the form of organic substances which can be used as food material. The primary productivity is thus the basis of the whole metabolic cycle in natural aquatic ecosystems, the remainder is consumption and decay. The organic matter synthesised by primary producers is utilised by the consumers inhabiting the system.

All the material synthesised by the producers, however, is not available to *consumers* (all other life forms except producers). Part of it is utilised by the producers themselves in their maintenance (respiration). Some part of it is wasted and is used by the decomposers (non-photosynthetic bacteria, fungi etc). The remainder is consumed by the *herbivores* (organisms using plant material as food). Herbivores transfer some amount to the *carnivores* (organisms using living animal material as food). The accumulation of biomass in all other organisms except in producers is referred to as *secondary production.*

There are four successive steps of production process:

A. **Gross Primary Productivity (GPP)**: It is the rate of photosynthesis and includes the organic matter used up in the respiration during the measurement period.

B. **Net Primary Productivity (NPP)**: It is the rate of storage of organic matter in plant tissues in excess of the respiratory utilisation by the producers during the period of measurement.

C. **Net Community Productivity (NCP)**: It is the rate of organic matter not used by heterotrophs (that is NPP-heterotrophic consumption) during the period under consideration.

D. **Secondary Productivity**: It is the rate of energy storage at consumer level.

The respiration rate of primary producers is referred to as R and that of heterotrophs as RH. The whole situation can be understood by these equations:

$$GPP = NPP + R$$

$$NPP = GPP - R$$

$$NCP = NPP - RH$$

Primary production studies are of paramount interest in understanding the effect of pollution on the system's efficiency. High rates of production—both in natural and cultural ecosystems—occur when physico-chemical factors are favourable. Pollution of water in the long run leads to a reduction in primary productivity. Pollution also affects the production (P) respiration (R) ratio, a proper level of which is very essential for the sustenance of the system. In a non-polluted water, the P usually exceeds R but, in organically polluted systems, R exceeds P, and no organic material is left available for the bioactivity of the system leading to the system's impairment.

Measurement of primary productivity

The basic equation of photosynthesis is:

$$6CO_2 + 6H_2O \xrightarrow[\text{chlorophyl}]{\text{light}} C_6H_{12}O_6 + 6O_2$$

Hence, to estimate primary productivity, one can measure the carbon uptake as well as the oxygen production or the formation of organic compounds or the gain of chemical energy of the system. In the aquatic ecosystems, the primary production is mainly due to phytoplankton and aquatic macrophytes. The methods of measuring primary production of phytoplankton are given here.

LIGHT AND DARK BOTTLE METHOD (GAARDER AND GRAN):

Principle and General Considerations

It is well known that the amount of carbon dioxide consumed in carbon assimilation is proportional to the oxygen liberated by plants so that the amount of carbon assimilated can be calculated from the amount of oxygen provided.

The Gaarder and Gran's method of 'light and dark bottle' involves collecting water sample from the water of which productivity is to be estimated, filling light and dark bottles with it and resuspending them at the same place for some duration. The initial concentration of oxygen in the water is estimated separately by filling one more bottle (initial bottle). After the chosen exposure period, the bottles are removed and their oxygen contents are measured. It is expected that in the light bottle (where sunlight is reaching) the oxygen content will increase due to photosynthesis, but it will fall in the dark bottle where only respiration is taking place. If other processes of oxygen consumption (e.g., photorespiration etc.) are absent, then the difference in oxygen content of the light and initial bottle is net photosynthesis; the difference between initial and dark bottle is respiration. Since some oxygen is also consumed by the heterotroph-bacteria, zooplanktons etc., this respiration is actually *community respiration*, CR (Their contribution, however, is not more than 10% of the total respiration). The difference between light and dark bottles is total photosynthesis or gross primary production.

Important considerations in the method are:

Sampling: Sampling of water for primary productivity is same as for water analysis. Additional consideration is drawing sufficient volumes to fill at least 5 bottles of 300 ml and still leaving some water for chemical analysis (if required).

Enclosures: BOD bottles (made up of good transparent glass) serve the purpose best. Though 100 ml capacity ground stoppered 'corning' flasks can also be used. In any case, containers of less than 100 ml capacity should not be used.

The bottles can be darkened by using either black paint or a detachable black cloth.

Suspension: The bottles can be suspended in water by attachment to a line or wire which is held in position by an anchored buoy or float. It is better to suspend the bottles in a horizontal position.

For calculation of per m^2 productivity, bottles are suspended at various depths until the limit of euphotic zone and the obtained productivity values are integrated. For this purpose, the water is drawn from the same depths. If the water is well mixed, then the water from only surface layer may be collected and then suspended at various depths.

Time of exposure: Since the algae inside these enclosures may sediment at the bottom and the bacteria on the walls of the bottle may multiply and thus influence oxygen concentration, it is better to expose the bottles for short duration (2 to 3 hrs). If the water is highly productive, the bottles may get supersaturated with oxygen in a short time; in such cases 1/2 hr exposure is adequate. It is better to estimate productivity for 3-4 times for short durations between dawn and sunset and estimate an average hourly value (e.g., 8-9 a.m., 11-1 p.m., 2-4 p.m., 5-6 p.m.).

Apparatus and Reagents

A. BOD bottles, 300 ml (Dark few bottles by using black paint).

B. Water sampler.

C. Reagents and accessories for measuring dissolved oxygen (see page 50).

D. Secchi disc.

Procedure

1. Determine depth of the euphotic zone by using Secchi disc. Fix depth intervals for suspension of bottles. For extremely shallow waters, select only surface (0-20 cm) and bottom.
2. Collect water by using a sampler, transfer it gently to a bucket or trough. Fill avoiding agitation, 3 sets of bottles—initial, light and dark. Suspend light and dark bottles at desired depths and determine oxygen in the initial bottle immediately (see dissolve oxygen determination).
3. Incubate for selected number of hours. After the expiry of incubation period, carefully remove the bottles and determine oxygen in them immediately.

Calculation

Net primary productivity:

$$O_2 \, mg/l/hr = \frac{Dl - Di}{h}$$

Gross primary productivity:

$$O_2 \, mg/l/hr = \frac{Dl - Dd}{h}$$

Community respiration:

$$O_2\,mg/l/hr = \frac{Di - Dd}{h}$$

where, Di = Dissolved oxygen in the initial bottle in mg/l,

Dl = Dissolved oxygen in the light bottle in mg/l,

Dd = Dissolved oxygen in the dark bottle in mg/l,

h = Duration of exposure in hrs.

The values can be converted to carbon by using the formula

$$GPP, NPP \text{ or } CR, gC/m^3/hr = GPP, NPP \text{ or } CR, \text{ in mg } O_2/l/hr \times 0.375.$$

Calculation of Daily Rates

Daily rates can be obtained by multiplying the average hourly rate with the sunshine values (obtained for that day from the nearest meteorological station). Some authors prefer to multiply the maximum hourly rate obtained during the day by 9 to get the daily rates.

Calculation of Per Metre Square Productivity and Community Respiration

Two methods can be employed for this purpose. In *Graphic method*, the depthwise production values are plotted on a graph paper and then graphic integration is made. In another method, the average m^3 values are multiplied by the mean depth of the waterbody (volume of the waterbody/surface area) to get m^2 values.

Calculation of Annual Production of Whole Waterbody

In large waterbodies, different areas may have different productivity rates. In such cases, several sites must be selected and the productivity must be estimated in each area at a 15 days interval for the whole year. The waterbody is then divided into several parts having comparable productivity value and at least one sampling site should fall in this area. The area of each part is determined by using a planimeter. The annual production m^2 in each part is calculated by multiplying the average daily rate with the number of days in a year (for this the productivity has to be estimated regularly at a fortnightly interval which will also include the cloudy and rainy days). The annual production /m^2 is multiplied by the area of one part in the lake. The productivity values of various parts are totalled and annual productivity (expressed as tonnes of carbon) for the whole waterbody is obtained. Dividing this value by the area of the waterbody will give average per m^2 productivity of that waterbody.

Note 1. If the rate of respiratory consumption of oxygen is more than the production of oxygen (as may happen in very highly polluted waters or during very low sunlight); the values of oxygen in light bottle may be less than that in the initial bottles. In such cases, the values of GPP and NPP can be represented as negative.

2. When presenting primary productivity and community respiration data, always report degree of cloudiness on the measurement day (in % on the basis of visual observation), wind intensity and land direction, water and air temperature etc.

60.3 DETERMINATION OF SOIL QUALITY

(A) pH

Principle

pH of soil is the measure of the hydrogen ion activity and depends largely on relative amounts of the adsorbed hydrogen and metallic ions.

Equipment

pH meter

Procedure

1. 50 g of soil is suspended in 250 ml distilled water for 2 hour;
2. Filter the suspension and filtrate and in pH determination using calibrated pH meter directly.

(B) CONDUCTIVITY

Principle

It is a measure of the current carrying capacity indicating the ionic distribution.

Equipment

Conductivity meter.

Procedure

Filter soil suspension prepared in the same manner as in pH measurement case, and conductivity directly read from equipment.

(C) HARDNESS

Soil hardness is measured by extracting calcium and magnesium from 100 gms of soil sample with the use of ammonium acetate solution (pH 7.0) as extract out (dilute 57 ml of glacial acetic acid to 800 ml with distilled water and then neutralize to pH 7.0 with conc. NH_4OH. Make the volume 1 lit). Extraction is made by shaking for 2-3 hours. All other procedure is same as in water.

(D) AVAILABLE PHOSPHORUS

With the help of 0.1N H_2SO_4 solution soil is digested (1 gm soil in 200 ml 0.002 N H_2SO_4) for 1 to 2 hours in a shaker. Then filter the solution and measure phosphorus by same method as in water.

(E) NITRATE (Nitrogen)

Nitrate of soil extracted with 1N copper sulphate solution (dissolve 125 g $CuSO_4$, $5H_2O$ in 1 lit distilled water and mix 0.6% silver sulphate solution in a ratio of 1:4). Filter the extract after shaking forl to 2 hours and filtrate in same manner as in water.

(F) ORGANIC MATTER (Walkley and Black Method)

Principle

The organic matter present in the soil is digested with excess of potassium dichromate and sulphuric acid, and the residual unutilised dichromate is then titrated with ferrous ammonium sulphate.

The elementry carbon present as graphite, charcoal etc. is not attacked in this method and only organic carbon is determined. The recovery of the carbon in this method is not 100 per cent. Only about 60 - 90% of the total organic matter is recovered depending upon the kind. For example, in most cases, the proteins remain unaffected by this method.

Reagents

A. **Potassium dichromate solution, 1 N**: Dissolve 49.04 g of $K_2Cr_2O_7$ in distilled water to prepare 1 liter of solution.

B. **Sulphuric acid**: H_2SO_4 (sp. gr. 1.84).

C. **Phosphoric acid**: H_3PO_4, conc. (sp. gr. 1.71).

D. **Ferrous ammonium sulphate 0.4 N**: Dissolve 156.86 g $Fe(NH_4)_2(SO_4)_2.6H_2O$ in distilled water, adding 14 ml conc. H_2SO_4 to prepare 1 liter of solution.

E. **Diphenylamine indicator**: Dissolve with care 0.5 g diphenylamine in a mixture of 20 ml distilled water and 100 ml conc. sulphuric acid.

Procedure

1. Take oven-dried or freeze-dried soil sample and pass through a 0.5 mm non-ferrous screen.
2. Weigh a suitable quantity of soil not exceeding 10 g (containing about 10-25 mg carbon) and transfer to a dried 500 ml conical flask.
3. Add 10 ml 1N $K_2Cr_2O_7$ solution and 20 ml conc. H_2SO_4 and mix by gentle swirling.
4. Keep the flask to react the mixture for about 30 minutes.
5. After the reaction is over, dilute the contents with 200 ml of distilled water and add 10 ml phosphoric acid followed by 1 ml of diphenylamine indicator.
6. Titrate the sample with 0.4 N ferrous ammonium sulphate; at the end-point colour changes to brilliant green.
7. If more than 8 ml of the 10 ml added $K_2Cr_2O_7$ is consumed (ml titrant less than 5 ml), repeat with less quantity of the sample.
8. Run a blank with same quantity of the chemicals but without soil.

Calculation

(a) $$*\%C = \frac{3.951}{g}\left(1 - \frac{T}{S}\right)$$

(b) % Organic matter = %C × 1.724

where, g = weight of sample in g.

S = ml ferrous solution with blank titration.

T = ml ferrous solution with sample titration.

The factor 1.724 is based on the assumption that carbon is only 58% of the organic matter.

* A factor for average recovery of about 75% organic matter by this method has been taken into consideration in the above formula. Some laboratories do not take into consideration the recovery factor. However, in such cases, the results should be accompanied with a mention of the Walkley and Black method.

(G) EXCHANGEABLE CALCIUM AND MAGNESIUM

Principle

Cations present in the exchange complex of the soils can be removed by leaching the soil with ammonium acetate solution. Different exchangeable cations are then estimated separately in this ammonium acetate leachate.

For determination of only exchangeable cations, the soil is first washed with ethyl alcohol to remove the soluble fraction. However, when exchangeable plus soluble fraction is required, there is no necessity of such alcohol washing. If only soluble fraction is to be estimated a 1 : 5 soil solution is made to estimate the cations.

Reagents

A. **Ethyl alcohol, 40%**: Mix 600 ml of distilled water with 400 ml of absolute or 95% alcohol.

B. **Absolute alcohol.**

C. **Ammonium acetate solution**: Dilute 57 ml of glacial acetic acid to 800 ml with distilled water and then neutralise to pH 7.0 with concentrated NH_4OH. Make up final volume to 1 litre.

D. All the reagents used in determination of calcium and magnesium are:

- EDTA solution
- Buffer solution
- Erichrome indicator.

Procedure

1. Take 50 g of air-dried soil in 500 ml beaker and add about 100 ml of 40% alcohol. Shake well and keep for 15 min.
2. Filter the suspension through Whatman No. 50 filter paper using Buchner funnel and vacuum pump.
3. Wash the soil 4-5 times with 50 ml portions of 40% alcohol. Perform the final washing with 50 ml of absolute alcohol to dry the soil.
4. Remove the filter paper and scrap the soil in a 250 ml beaker. Wash the Buchner funnel and the filter paper with 100 ml ammonium acetate solution for removing any adhered portion of the soil.
5. Stir the suspension and keep overnight.

6. Filter the supernatant, and finally the soil, with additional ammonium acetate through Whatman No. 42 filter paper using Buchner funnel and vacuum pump. Leach the soil 4-5 times more with portions of ammonium acetate and make up the final volume of the filtrate to 500 ml in a volumetric flask.
7. Find out the concentration of calcium and magnesium following the EDTA method.
8. In case of total calcium and magnesium (exchangeable + soluble) or when soluble salts are in very little quantity, do not apply the alcohol washings and add 100 ml of ammonium acetate directly to 50 g of soil and follow from the step 5 onwards.

Calculation

$$\text{Calcium}(\%) = \frac{A \times 400.8 \times V}{v \times 10{,}000 \times S}$$

$$\text{Calcium, mg}/100\text{g} = \frac{A \times 400.8 \times V}{v \times 10 \times S}$$

$$\text{Calcium, meq}/100\text{g} = \frac{A \times 400.8 \times V}{v \times 20.04 \times 10 \times S}$$

$$\text{Magnesium}(\%) = \frac{(B - A) \times 400.8 \times V}{v \times 1.645 \times 10{,}000 \times S}$$

$$\text{Magnesium mg}/100\text{g} = \frac{(B - A) \times 400.8 \times V}{v \times 1.645 \times 10 \times S}$$

$$\text{Magnesium, meq}/100\text{g} = \frac{(B - A) \times 400.8 \times V}{v \times 1.645 \times 12.16 \times 10 \times S}$$

where, A = Volume of EDTA used for calcium determination (ml)

B = Volume of EDTA used for calcium + magnesium determinaion (ml)

V = Total volume of soil extract prepared (in this case 500 ml)

S = Weight of soil taken (in this case 50 gms)

v = volume of soil extract titrated (ml).

60.4 PLANKTON ANALYSIS

Microscopic free-floating planktons (algae i.e. phytoplanktons, protozoa, rotifer, copepods etc. as zooplanktons) are collected by plankton net (40 μm) through filtering of water (50-100 lits). Planktons are preserved after centrifugation in small vial with addition of FAA preservatives for microscopic identification.

Plankton load is measured microscopically with the help of Sedwick-Raffer counter or Haemocytometer. Species diversity index was computed by Shannon & Weiner index formulae.

Plankton analysis is a good indicator of water quality and primary productivity.

61 APPLICATION OF STATISTICS IN ENVIRONMENTAL SCIENCE

Once the biological/environmental data have been collected, we will have a mass of raw data. Such raw data have to be condensed or summarised statistically and then presented in the form of tables and graphs or in various measures of dispersion. All these summarisation methods are commonly known as **descriptive statistics**.

Summarisation of data starts with classification of the data. Arranging of data according to some common features is known as classification. Broadly, classification of data is done according to the type of data—either attribute or measurable data. With attribute data, we count items with specified characters and put them in appropriate categories. But if the data are measureable, they are classified according to classes.

61.1 DATA SUMMARISATION

The data could be summarised in various ways. A brief description is given:

(a) Frequency distribution

Sorting of data into categories or classes will lead to formation of frequency distributions. Frequency distribution gives the number of times a category or class occurs. The frequency is denoted by 'f'. The class is denoted by 'x'.

The following tables are examples of frequency distribution:

Table 61.1: Frequency distribution of seed yield of 100 sesamum plants

Seed yield (x) (g)	Number of plants (f)
2.5 — 3.5	4
3.5 — 4.5	6
4.5 — 5.5	10
5.5 — 6.5	26
6.5 — 7.5	24
7.5 — 8.5	15
8.5 — 9.5	10
9.5 — 10.5	5
Total	100

Table 61.2: Frequency distribution of number of seeds germinated out of five in each of 50 pots

No. of seeds germinated (x)	No. of pots (f)
0	4
1	13
2	16
3	9
4	5
5	3
Total	50

Table 61.3: Frequency distribution of petal colour in linseed flowers

Colour	No. of plants
Blue	174
White	58
Pink	18
Total	250

In Table 61.1, the boundary figures of the classes 2.5 – 3.5, 3.5 – 4.5 etc. are called class limits. The lower and upper bound values are lower limit and upper limit of the class, respectively. The distribution between lower and upper limits of a class is known as class interval. These are all oneway tables.

If the table contains two way variables one can obtain a twoway *correlation table*. A twoway table with two attributes is known as *contingency* table. Tables 61.4 and 61.5 are examples for *correlation* and *contingency* table, respectively.

Table 61.4: The distribution of length of ear and yield of grain of a wheat variety

Yield per	Ear length (cm)			
plant (g)	5-8	8-11	11-14	Total
10 —15	4	32	10	46
15 —20	28	160	12	200
20 —25	5	59	20	84
Total	37	251	42	330

Table 61.5: The distribution of bacterial population (10^5/g of soil) in the rhizosphere of certain ornamental plants in different physiological stages

Rhizosphere	Physiological stages			
	Vegetative stage	Flowering stage	Seed setting stage	Total
Cock's comb	167	130	33	330
Sabia	144	60	53	257
Zinnia	97	33	34	164
Aster	56	78	10	144
Total	464	301	130	895

In the same way, one can construct multiway tables depending upon the number of variables involved.

In constructing a frequency distribution table the following general guidelines may be followed:

First, the number of classes are to be decided. The appropriate number of classes may be decided by Yule's formula $2.5 \times n^{1/4}$, where n is the total number of observations. For example, if $n = 256$, then $(256)^{1/4} = (16)^{1/2} = 4$. Then the number of classes will be $2.5 \times 4 = 10$.

The class-interval is then found out by using the relationship

$$C = \frac{(\text{Maximum value in the data set} - \text{Minimum value in the data set})}{\text{Number of classes}}$$

It is convenient to have the class intervals in multiples of round figures like 0.05, 0.1, 5, etc. Also, to avoid difficulties later in analysis, we should have equal class-interval for all the classes. Further, open-end classes like "more than 100" or "less than 5" should be avoided.

In practice, first we form discontinuous classes like 2.5 - 3.4, 3.5 - 4.4, etc. This will avoid difficulties in assigning an item to an appropriate class. Once the counting is over they are made continuous classes like 2.5 - 3.5, 3.5 - 4.5, etc. In certain cases, continuous classes are formed by taking the mean of the upper bound of the previous class and the lower bound of the succeeding class. For example, consider the tentative classes 2.5- 3.4 and 3.5 - 4.4. The mean of 3.4 and 3.5 is $(3.4 + 3.5) \div 2 = 3.45$. The upper bound for the classes will then be 3.45, 4.45 etc. We have to make an adjustment in the lower limit of the first class and upper limit of the final class so that the class intervals are not changed for these classes.

The final table should have a self-explanatory title in order to make the content of data comprehensive.

The classes should be given in increasing or decreasing order. In case of qualitative data, the items may be arranged alphabetically, chronologically or by some conventional order.

The classes and frequencies should be clearly indicated. For example, the class may be "yield in gms" and the frequencies may be "number of plants".

Example 61.1

The following data relate to the grain yield (in g per plot) of a sorghum variety from 100 experimental plots of equal area:

196	169	126	181	174	164	209	143	65	165
194	129	166	164	154	139	128	120	80	168
150	186	156	179	153	157	155	115	176	171
118	143	191	148	152	187	129	119	139	177
191	214	167	165	186	111	155	164	125	99
86	170	111	169	141	164	89	180	225	139
127	136	144	165	154	74	156	142	162	160
171	139	177	178	168	165	188	131	154	107
139	156	176	150	142	144	153	190	183	180
161	170	195	136	91	187	152	145	98	166

The number of classes for this data is

$(2.5)\ (100)^{1/4} = (2.5)\ (10)^{1/2} = (2.5)\ (3.162) = 7.9 = 8$

$$\text{The class-interval} = \frac{\text{Maximum value} - \text{Minimum value}}{\text{Number of classes}}$$

$$= \frac{225-65}{8} = \frac{160}{8} = 20$$

In practice we may not get such exact values. Some readjustments might be done.

In the next step the frequencies are counted by using tally of marks :

Class	Tally mark	Total
65 — 84	III	3
85 — 104	~~IIII~~	5
105 — 124	~~IIII~~ II	7
125 — 144	~~IIII~~ ~~IIII~~ ~~IIII~~ ~~IIII~~	20
145 — 164	~~IIII~~ ~~IIII~~ ~~IIII~~ ~~IIII~~ IIII	24
165 — 184	~~IIII~~ ~~IIII~~ ~~IIII~~ ~~IIII~~ ~~IIII~~ I	26
185 — 204	~~IIII~~ ~~IIII~~ II	12
205 — 224	II	2
225 — 244	I	1
Total		100

The final table is:

Table 61.6: The frequency distribution of grain yield of sorghum from 100 experimental plots

Grain yield (g)	No. of plots
65 - 85	3
85 - 105	5
105 - 125	7
125 - 145	20
145 - 165	24
165 - 185	26
185 - 205	12
205 - 225	2
225 - 245	1
Total	100

Sometimes, it is useful to have cumulative frequency distribution. The *Cumulative frequency* of a class is the total frequency up to and including that class. The table of cumulative frequencies is called a cumulative frequency distribution table. Again, there are two types of cumulative frequency distribution. The cumulative frequency distribution of all values greater than or equal to the lower limit of each class is called *more than cumulative frequency distribution*. The cumulative frequency distribution of all values less than or equal to the upper limit of each class is called *less than cumulative frequency distribution*. Whenever we refer to cumulative frequency distributions without qualification, the "less than" type is implied. The results in Table 61.6 are presented as cumulative distributions in Tables 61.7 and 61.8.

Table 61.7: Less than cumulative frequency distribution for the data in Table 61.6

Less than (<) class	Cumulative frequency
< 85	3
< 105	8 (= 3 + 5)
< 125	15 (= 8 + 7)
< 145	35 (= 15 + 20)
< 165	59 (= 35 + 24)
< 185	85 (= 59 + 26)
< 205	97 (= 85 + 12)
< 225	99 (= 97 + 2)
< 245	100 (= 99 + 1)

Table 61.8: More than cumulative frequency distribution for the data in Table 61.6

More than (>) class	Cumulative frequency
> 65	100
> 85	97 (= 100 – 3)
> 105	92 (= 97 5)
> 125	85 (= 92 – 7)
> 145	65 (= 85 – 20)
> 165	41 (= 65 – 24)
> 185	15 (= 41 – 26)
> 205	3 (= 15 – 12)
> 225	1 (= 3 – 2)

(b) Diagrams and Graphs

The important diagrams are simple bar diagram, multiple bar diagram, component bar diagram, percentage bar diagram, pie diagram and histogram. Frequency polygon, line graph and frequency curve are the important graphs. Selection of a diagram or graph depends upon the nature of classification, number of characters involved and the type of comparison required to be made.

In all the diagrams and graphs, the groups of classes are represented on the X-axis and the volumes or frequencies are represented in the Y-axis. The titles of these visual aids should be self-explanatory:

(i) **Simple bar diagram**: If the classification is based on attributes and if the attributes are to be compared with respect to a single character we use simple bar diagram. For example, the area under different crops in a state, the foodgrain production in different years, etc., can be represented by simple bar diagram. In case of experimental data, the yield performance of different varieties of a crop, the effects of different treatments, etc., can be compared using simple bar diagrams.

Simple bar diagrams consist of vertical bars of equal width. The heights of these bars are proportional to the volume or magnitude of the attribute. All bars stand on the same baseline. The bars are separated from each other by equal intervals. The bars may be coloured or marked.

(ii) **Multiple bar diagram**: If the data is classified by attributes and if two or more characters or groups are to be compared within each attribute we use multiple bar diagrams. If only two characters are to be compared within each attribute, then the resultant bar diagram used is known as double bar diagram.

The multiple bar diagram is simply the extension of simple bar diagram. For each attribute two or more bars representing separate characters or groups are to be placed side by side. Each bar within an attribute will be marked or coloured differently in order to distinguish them. Same type of marking or colouring should be done under each attribute. A footnote has to be given explaining the markings or colourings.

Example 61.2

The average removal of nitrogen by weeds (kg/ha) under different weedicides in three different days after sowing (DAS) were:

Weedicide	Removal of nitrogen by weeds (kg/ha)		
	25 DAs	55 DAS	At harvest
Oxadiazon	2.4	1.3	1.8
Fluchloralin	6.1	2.5	4.4
Isoprofuron	10.2	3.0	7.8
Unweeded control	27.2	51.9	64.0

The multiple bar diagram for the data in Example 61.2 is given in Fig. 61.1:

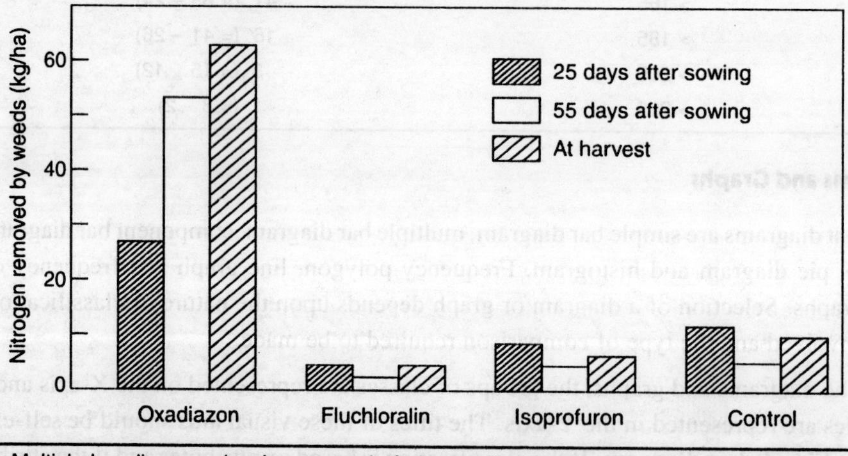

Fig. 61.1: Multiple bar diagram showing removal of nitrogen by weeds under different weedicide treatments

(iii) **Component bar diagram**: Sometimes the multi-character data may possess additive feature. For example, the production of foodgrains may be classified as rice, millets, etc. Each group is a component of foodgrain. When yields of different paddy varieties are compared, within each variety the yield under different nitrogen levels may be compared. From weed control trials with different treatments, we may get the weight of weeds classified by type of weeds. Here the type of weeds are components of weed under a treatment. Thus, each attribute may possess a number of components with additive feature.

For the type of data stated above we may use component bar diagrams. Instead of placing the bars for each component side by side we may place these one on top of the other. This will result in a component bar diagram.

Example 61.3

The component bar diagram for the data in Table 61.5 is given in Fig. 61.2:

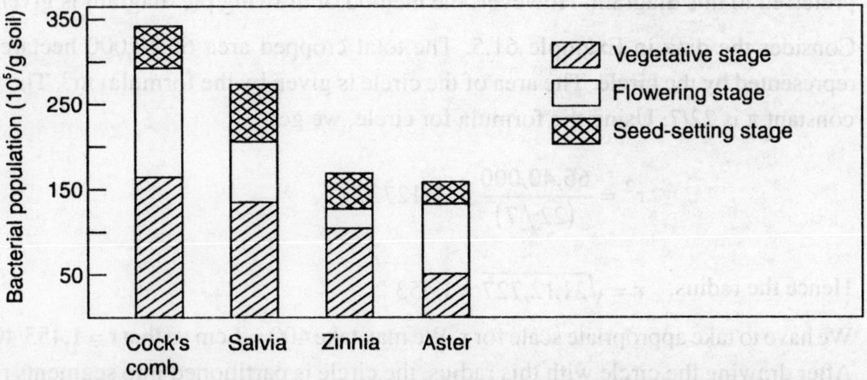

Fig. 61.2: Component bar diagram showing the distribution of bacterial population in the rhizosphere of certain ornamental plants in different physiological stages

(iv) **Percentage bar Diagram**: Sometimes the volumes of different attributes may be greatly different. For making meaningful comparisons, the components of the attributes are reduced to percentages. In that case each attribute will have 100 as its maximum volume. This sort of component bar chart is known as percentage bar diagram.

Example 61.4

The percentage for the data in Table 61.5 are:

	Vegetative stage	Flowering stage	Seed-setting stage	Total
Cock's comb	50.6	39.4	10.0	100
Salvia	56.0	23.3	20.7	100
Zinnia	59.1	20.1	20.8	100
Aster	38.9	54.2	6.9	100

The above data are shown in Fig. 61.3.

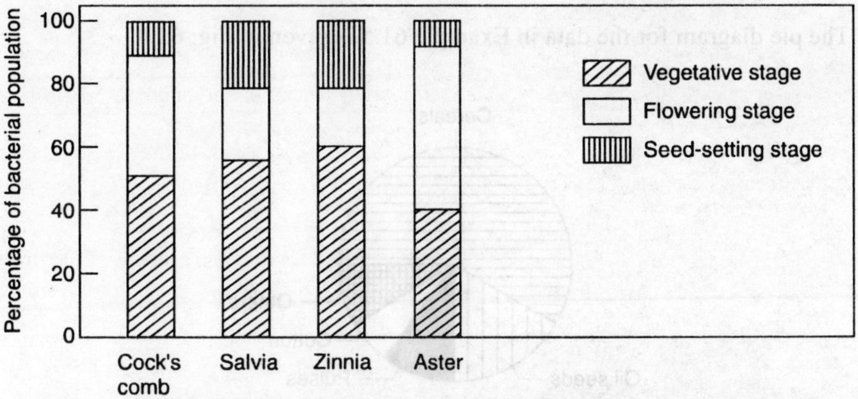

Fig. 61.3: Percentage bar diagram showing the distribution of bacterial population in the rhizosphere of certain ornamental plants in different physiological stages

(v) **Pie Diagram**: Pie diagram is a circular diagram. It may be used in place of bar diagrams. Drawing a pie diagram is a tedious process. Hence, usually the appropriate bar diagrams are preferred to this diagram. However, the method of drawing pie diagram is given:

Consider the data in Example 61.5. The total cropped area 66,40,000 hectares has to be represented by the circle. The area of the circle is given by the formula πr^2. The value of the constant π is 22/7. Using the formula for circle, we get

$$r^2 = \frac{66,40,000}{(22/7)} = 2112727$$

Hence the radius, $r = \sqrt{21,12,727} = 1,453$

We have to take appropriate scale for r. We may take 400 = 1 cm so that r = 1,453/400 = 3.6 cm. After drawing the circle with this radius, the circle is partitioned into segments representing each crop. The segments are derived based on the angles. The circle contains 360°. Hence each 10,000 hectares will be represented by 360/664 = 0.54°. The appropriate angles for each crop are then worked out. The angles will be 214° for cereals, 64° for oil seeds, 25° for pulses, 14° for cotton and 44° for others. The larger segment will be represented first, and the others in their order. However, the 'miscellaneous' or 'other' segment is drawn last. Usually the horizontal radius to the right is taken as the starting point.

Example 61.5

The cropping pattern in Odisha in the year 1978-79 was:

Crops	Area (in 1,000 hectares)
Cereals	3,940
Oilseeds	1,165
Pulses	464
Cotton	249
Other	822
Total	6,640

The pie diagram for the data in Example 61.5 is given in Fig. 61.4.

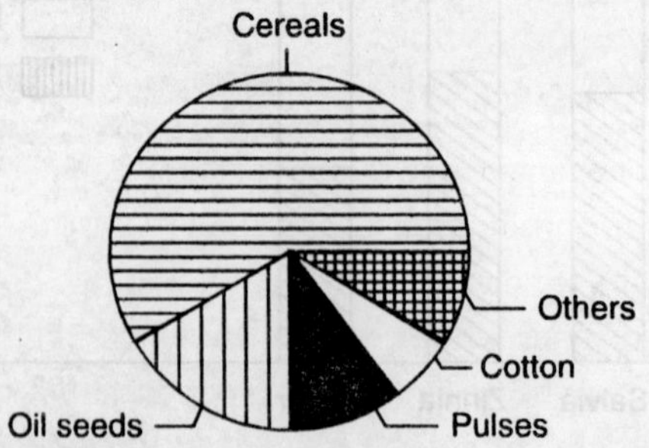

Fig. 61.4: Pie diagram showing the cropping pattern in Tamil Nadu in the year 1974-75

(vi) **Histogram**: When the data are classified based on the class-intervals, it can be represented by a histogram. Histogram is just like a simple bar diagram with minor differences. There is no gap between the bars since the classes are continuous. The bars are drawn only in outline without colouring or marking as in the case of simple bar diagrams.

Histograms can be obtained for frequency distribution with equal as well as unequal class-intervals. If the class-intervals are not equal, the frequency corresponding to the classes are obtained by dividing each class frequency by the corresponding class-interval. The resultant values are then represented against the corresponding class-intervals. This is done on the basis that the total area represented by the bar is equal to the frequency.

Example 61.6

For the data in Table 61.1, the histogram is given in Fig. 61.5:

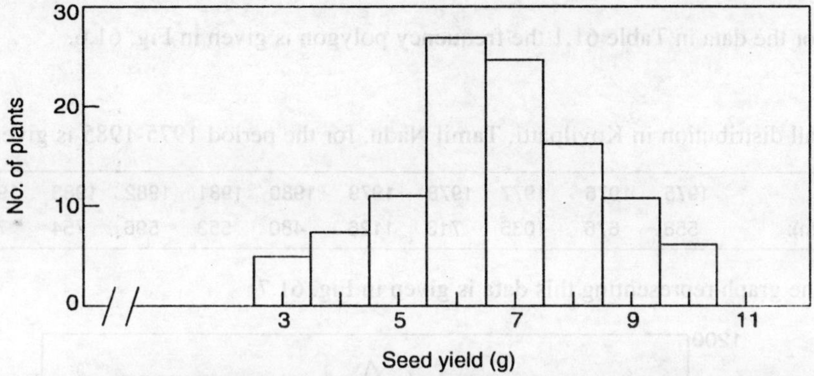

Fig. 61.5: Histogram showing the frequency distribution of seed yield of sesamum plants

(vii) **Frequency Polygon and Line Graph**: The type of data required for frequency polygon is the same as that for histogram.

The frequencies of the classes are plotted by dots against the mid-points of each class. The adjacent dots are then joined by straight lines. The resulting graph is known as frequency polygon.

Drawing of frequency polygon is simpler and quicker than drawing histograms. When two or more frequency distributions are to be compared, frequency polygon is more useful. Histogram cannot be used for such purposes. In case of two or more frequency polygons, the lines are marked differently.

Sometimes the data may be taken repeatedly over-time. Such chronological data are known as time series data. For example, the rainfall data for standard weeks of a year or over a period of time, the foodgrain production over a period of time, etc., are time series data. In experimental situations we may observe the growth rate of insects over a period of time or we may observe the responses of a crop at different growth stages. In some experiments the responses of a crop may be observed corresponding to a quantitative factor like nitrogen levels. Though such data are not the frequency distributions based on class-intervals, the frequency polygon type of graphs are used to represent them. This kind of graphs are known as line graphs.

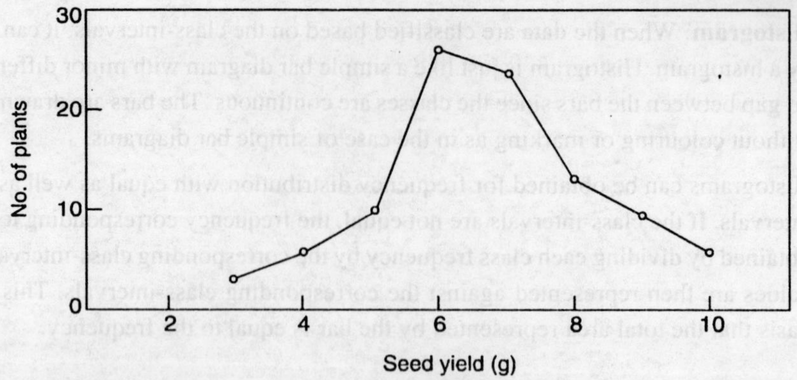

Fig. 61.6: Frequency polygon showing the frequency distribution of seed yield of sesamum plants

Example 61.6

For the data in Table 61.1 the frequency polygon is given in Fig. 61.6.

Example 61.7

The rainfall distribution in Kovilpatti, Tamil Nadu, for the period 1975-1985 is given:

Year:	1975	1976	1977	1978	1979	1980	1981	1982	1983	1984	1985
Rainfall (mm):	558	676	1035	713	1126	480	553	596	754	780	614

The graph representing this data is given in Fig. 61.7:

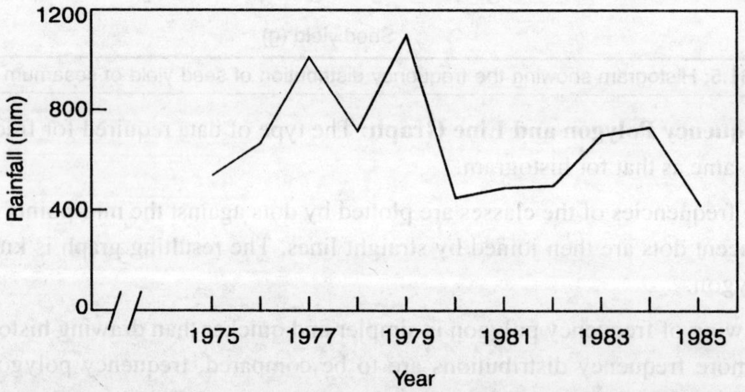

Fig. 61.7: Graph showing rainfall distribution at Kovilpatti during 1975-1985

(viii) **Frequency Curve**: The procedure for drawing a frequency curve is same as for frequency polygon. But the points are joined by smooth or free-hand curve. The curve may not pass through all points as in the case of frequency polygon.

Example 61.8

The frequency curve for the data in Table 64.1 is given in Fig. 61.8.

The frequency curve need not be symmetrical as is shown in Fig. 3.10. They may be as shown in Fig. 61.10.

Sometimes the change in the variable may be in geometric progression as in cell division, bacterial growth, population growth, etc. In such cases our interest may be in relative change rather than in absolute change. It may be presented in semi-log graph sheets, if they are available. Otherwise, we can use ordinary graph sheet by marking log values of the frequencies on Y-axis.

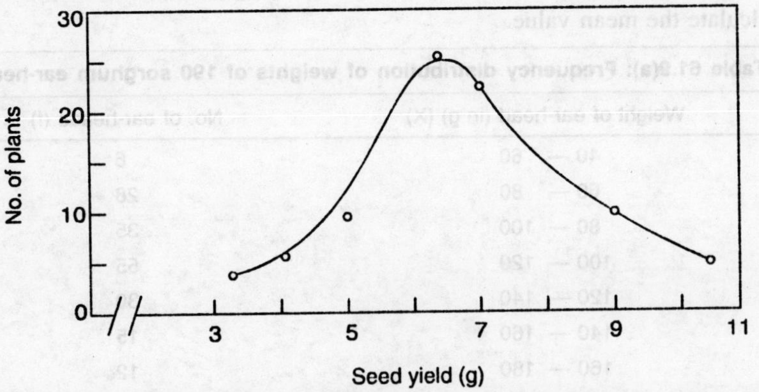

Fig. 61.8: Frequency curve showing the frequency distribution of seed yield of sesamum plants

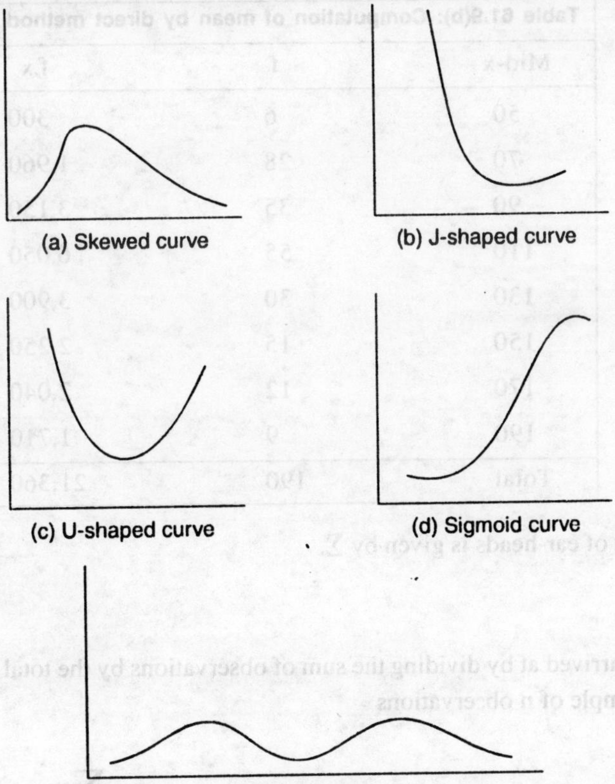

(a) Skewed curve

(b) J-shaped curve

(c) U-shaped curve

(d) Sigmoid curve

(e) Bimodal curve

Fig. 61.9: Certain characteristic shapes of frequency curves

$$\overline{X} = A + \left[\frac{\sum fd}{n} \times C \right]$$

Example 61.9

For the frequency distribution of weights of sorghum ear-heads given in Table 61.9(a), calculate the mean value.

Table 61.9(a): Frequency distribution of weights of 190 sorghum ear-heads

Weight of ear-head (in g) (X)	No. of ear-heads (f)
40 — 60	6
60 — 80	28
80 — 100	35
100 — 120	55
120 — 140	30
140 — 160	15
160 — 180	12
180 — 200	9
Total	190

Table 61.9(b): Computation of mean by direct method

Mid-x	f	f.x
50	6	300
70	28	1,960
90	35	3,150
110	55	6,050
130	30	3,900
150	15	2,250
170	12	2,040
190	9	1,710
Total	190	21,360

The mean weight of ear-heads is given by Σ.

(c) Mean

The mean is the value arrived at by dividing the sum of observations by the total number of observations. Symbolically, for a sample of n observations

$$\overline{x} \text{ (readas'x bai')} = \frac{x_1 + x_2 + \cdots + x_n}{n} = \frac{\sum_{i=1}^{n} x_1}{n}$$

The symbol Σ (read as 'sigma') means sum the individual values $x_1, x_2 \ldots$ of the variable, x. Usually, the limits of the summations are not written, since it is always understood that the summation is over all n values. Hence, we can write

$$\bar{x} = \frac{\sum x}{n}$$

For example, if the weights of 5 ear-heads of sorghum are 100, 102, 118, 124 and 126 g, then the mean weight is given by

$$\bar{x} = \frac{100+102+118+124+126}{5} = \frac{570}{5} = 114\,g.$$

If we have grouped data (frequency distribution) instead of raw data, then the mean is calculated by the formula

$$\bar{x} = \frac{\sum fx}{n}$$

where, f = the frequency.
x = mid-class value.

In short-cut method the following formula is used:

$$\bar{x} = A + \left[\frac{\sum fd}{n} \times C\right]$$

where,

A = assumed mean,
f = frequency,
d = (x − A)/C,
C = class-interval,
n = total number of observations.

In calculating the mean value by direct method the following steps are followed:

1. Verify whether the classes are continuous or not. If they are not continuous we have to make them continuous.
2. Find the mid-values of the classes. For this purpose add the lower and upper limits of the first class and divide by 2. For the subsequent classes go on adding the class interval.
3. Multiply the mid-values of the classes by the corresponding frequencies and add them up to get **Σfx**.
4. Use the formula.

For short-cut method the following steps are followed: Steps 1 and 2 are as in the preceding case.

Step 3. Fix the assumed mean, A. The assumed mean can be the mid-value for which the frequency is maximum or it may be the middle-most mid-value.

Step 4. Write down the deviations which can be found as

$$d = \frac{x - A}{C}$$

Step 5. Multiply the d values by the corresponding frequencies and find out their sum to get **Σfd.**

Step 6. Substitute the values in the formula.

$$\bar{x} = \frac{\sum fx}{n} = 21{,}360/190 = 112.4\,\text{g}.$$

Table 61.9(c): Computation of mean by short-cut method

x	f	d	fd
50	6	– 3	– 18
70	28	– 2	– 56
90	35	– 1	– 35
110	55	0	0
130	30	1	30
150	15	2	30
170	12	3	36
190	9	4	36
Total	190		(–109 + 132 = 23)

In this example, the maximum frequency is 55 and the mid-value against this value is 110. Hence, A = 110. The other steps are shown in the Table 61.9(c).

The mean weight of sorghum ear-heads is then,

$$\bar{x} = A + \left[\frac{\sum fd}{n} \times C\right]$$

$$= 110 + \left(\frac{23}{190} \times 20\right) = 110 + \frac{46}{19} = 112.4\,\text{g}$$

The result is same as in the direct method.

(d) Mode

The mode is defined as that item which occurs most frequently. For a frequency distribution the mode may or may not exist. Even if it exists it may be unimodal, bimodal or multimodal.

The frequency curves for no mode, unimodal and bimodal distributions are given in Fig. 61.10.

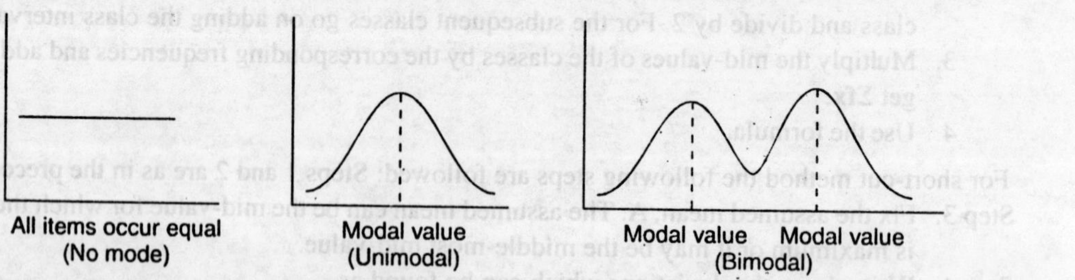

Fig. 61.10: Frequency curves

Example 61.10

If the yield of paddy from different fields are 6.0, 4.9, 6.0, 5.8, 6.2, 6.0, 6.3, 4.8, 6.0, 5.7 and 6.0 tonnes per hectare, the modal value is 6.0 tonnes per hectare.

Example 61.11

For the frequency distribution of weights of sorghum ear-heads given in Table 61.10 calculate the mode.

Table 61.10: The frequency distribution of weights of 190 sorghum ear-heads

Weight of ear-head (in g)	No. of ear-heads
40 — 60	6
60 — 80	28
80 — 100	35 → f_p
100 — 120	55
120 — 140	30 → f_s
140 — 160	15
160 — 180	12
180 — 200	9
Total	190

For calculating the mode for grouped data the following steps are followed:

Step 1 As in the calculation of mean.

Step 2 Find out the modal class. The modal class is the class against the maximum frequency. In our example, the maximum frequency is 55 and hence the modal class is 100 - 120.

Step 3 Use the formula,

$$\text{Mode} = l + \left(\frac{f_s}{f_p + f_s} \times C\right)$$

where,

l = lower limit of the modal class,
f_p = the frequency of the class preceding the modal class,
f_s = the frequency of the class succeeding the modal class, and
C = class-interval.

In our example, $l = 100$, $f_p = 35$ and $f_s = 30$. Hence,

$$\text{Modal} = 100 + \left(\frac{30}{35+30} \times 20\right)$$

$$= 100 + \left(\frac{600}{65}\right) = 109.23 \text{ g.}$$

(e) Median

The median is the middle-most item that divides the distribution into two equal parts when the items are arranged in ascending order.

In case of raw data, the median is the $\left(\frac{n+1}{2}\right)^{th}$ item where, n is the total number of observations. In case of grouped data, it is given by the formula,

$$\text{Median} = l + \left[\frac{(n/2) - cf}{f} \times C\right]$$

where,

l = lower limit of the medianal class, in which $(n/2)^{th}$ item falls,
n = total number of observations,
cf = cumulative frequency up to l,
f = frequency of the medianal class,
C = class-interval.

Example 61.12

If the weights of sorghum ear-heads are 45, 60, 48, 100, 65 g, then the data arrangement will be 45, 48, 60, 65, 100. Since, there are 5 items, the median is $\frac{(5+1)}{2}^{th}$ item, that is, 3rd item. It is 60 g.

Example 61.13

If the weights of sorghum ear-heads are 45, 48, 60, 65, 65, 100 g, then the median is $\frac{(6+1)}{2} = (3.5)^{th}$ item. That is,

Median = 3rd item + (4th item − 3rd item) × (0.5)
= 60 + (65 − 60) (0.5)
= 60 + 2.5 or 62.5 g.

Example 61.14

For the data given in Example 61.11, find out the median.

Table 61.11: Frequency distribution of weights of 190 sorghum ear-heads

Weight of ear-heads (in g)	No. of ear-heads	Less than class	Cumulative frequency
40 — 60	6	< 60	6
60 — 80	28	< 80	34
80 — 100	35	< 100	69 → $\frac{n}{2}$ = 95
100 — 120	55	< 120	124
120 — 140	30	< 140	154
140 — 160	15	< 160	169
160 — 180	12	< 180	181
180 — 200	9	< 200	190
Total	190		

The steps in the computation of median for grouped data are:

Step 1 As in the calculation of mean
Step 2 Write down the less than classes
Step 3 Write down the cumulative frequency
Step 4 Find out the medianal class by locating $(n/2)^{th}$ item in the cumulative frequency column
Step 5 Use the fomula.

In our example, (n/2) = (190/2) = 95, this value lies in between 69 and 124, and less than classes corresponding to these values are 100 and 120, respectively. Hence the medianal class is 100 - 120 and the lower limit of this class is 100. The cumulative frequency up to 100 is 69 and the frequency of the medianal class, 100 - 120 is 55. Therefore

$$\text{Median} = 100 + \left[\frac{(95-69)}{55} \times 20\right]$$

$$= 100 + \left[\frac{26}{55} \times 20\right]$$

$$= 100 + 9.45 \text{ or } 109.45\text{g.}$$

(f) Geometric Mean

If there are n items $x_1, x_2, \ldots x_n$, then the geometric mean (G.M.) is defined as the nth root of the product of the items. Symbolically,

$$\text{G.M.} = (x_1, x_2, \ldots x_n)^{1/n}$$

In practice, we take the logarithms of the values and then proceed as in the case of mean. The antilog of the mean of log values will give us geometric mean. This method holds good for frequency distributions also. It should be noted that G.M. cannot be calculated if any item is either zero or negative.

Example 61.15

Let 2, 4, 8, 16 be the 4 items. Their geometric mean is calculated as:

x	Log x
2	0.3010
4	0.6021
8	0.9031
16	1.2041
Total	3.0103

$$\text{Mean (of log values)} = \frac{\sum \log x}{n}$$

$$= \frac{3.0103}{4} = 0.7526$$

Antilog (0.7526) = 5.66

Therefore, G.M. = 5.66

In case of frequency distribution,

$$\text{G.M.} = \text{Antilog}\left[\frac{\sum f(\log x)}{n}\right]$$

(g) Harmonic Mean

The harmonic mean (H.M.) is the reciprocal of the mean of the reciprocal of the values. Symbolically:

$$H.M. = \frac{1}{\left(\frac{1}{x_1} + \frac{1}{x_2} + ... + \frac{1}{x_n}\right)\left(\frac{1}{n}\right)}$$

$$= \frac{n}{\sum (1/x)}$$

Example 61.16

There are 5 agricultural labourers. They can complete weeding operations on a 100 square metre land in 4.5, 5, 6 and 7 hours, respectively. If these 5 labourers are employed for weeding in 500 square metre area, in how many hours will they complete the work?

For such problems we compute harmonic mean:

$$HM = \frac{5}{(1/4) + (1/5) + (1/5) + (1/6) + (1/7)}$$

$$= \frac{5}{(0.25 + 0.20 + 0.20 + 0.167 + 0.143)}$$

$$= \frac{5}{0.960} = 5.21 \text{ hours}$$

The harmonic mean is rarely computed for a frequency distribution.

(h) Weighted Mean

If there are n observations, $x_1, x_2, x_3, ... x_n$ with corresponding weights $w_1, w_2, w_3, ... w_n$, then the weighted mean is given by

$$\bar{x}_w = \frac{\sum wx}{\sum w}$$

Example 61.17

The average yield of IR 20 paddy from two localities are 55 quintals and 65 quintals per hectare, respectively. The averages were based on 20 hectares and 10 hectares, respectively. What is the combined average yield per hectare?

The simple average of 55 are 65 quintals will give us (55 + 65)/2 = 60 quintals per hectare. This is not correct. We have to compute the weighted mean:

$$\bar{x}_w = \frac{(55 \times 20) + (65 \times 10)}{20 + 10}$$

$$= \frac{1,000 + 650}{30} = \frac{1,750}{30}$$

$$= 58.3 \text{ quintals per hectare}$$

In computing the mean, we take the frequency of a class as its weight. That is $\bar{x} = \frac{\sum fx}{\sum f}$. Hence, it is a special case of weighted mean.

PERCENTILES AND QUARTILES

Just like the median there are other partitional values called percentiles and quartiles, which divide the set of observations into desired partitions. Such values help us in knowing the relative standing of an individual observation within the total distribution.

The percentiles are the values which divide the whole distribution into hundred equal parts. The xth percentile is that value below which x per cent of values in the distribution fall.

For raw data, first arrange the n observations in increasing order. Then the xth percentile is given by

$$P_x = \frac{x(n+1)}{100} \text{ th item}$$

For a frequency distribution the xth percentile is given by

$$P_x = l + \left[\frac{(x.n/100) - cf}{f} \times C\right]$$

where,

l = lower limit of the percentile class which contains the xth percentile value (x.n/100)
cf = cumulative frequency up to l
f = frequency of the percentile class
C= class interval
n = total number of observations

The quartiles are the values which divide the whole distribution into four equal parts. The lower quartile, denoted as Q_1 is the value which separates bottom 25 per cent from the remainder of the distribution. The upper quartile, denoted as Q_3, is the value which separates top 25 per cent from the remainder of the distribution. In effect Q_1 is the 25th percentile value and Q_3 is the 75th percentile value.

It can be verified that the median is the 50th percentile value. For the sake of uniformity we may designate median as Q_2.

Example 64.18

The following are the paddy yields (kg/plot) from 14 plots: 30, 32, 35, 38, 40, 42, 48, 49, 52, 55, 58, 60, 62 and 65 (after arranging in ascending order). The computation of 25th percentile (Q_1) and 75th percentile (Q_3) are:

$$P_{25}\,(\text{or } Q_1) = \frac{25(14+1)}{100} \text{ th item.}$$

$$= (3¾) \text{ th item}$$

$$= \text{3rd item} + (\text{4th item} - \text{3rd item})\left(\frac{3}{4}\right)$$

$$= 35 + (38 - 35)\left(\frac{3}{4}\right)$$

$$= 35 + 3\left(\frac{3}{4}\right) = 37.25\,\text{kg}$$

$$P_{75}\,(\text{or } Q_3) = \frac{75(14+1)}{100}\text{ th item}$$

$$= \left(11\frac{1}{4}\right)^{th}\text{ item}$$

$$= 11^{th}\text{ item} + (12^{th}\text{ item} - 11^{th}\text{ item})\left(\frac{1}{4}\right)$$

$$= 55 + (58 - 55)\left(\frac{1}{4}\right)$$

$$= 55 + 3\left(\frac{1}{4}\right) = 55.75\,\text{kg}$$

(i) Measures of Dispersion

The simplest measure of dispersion is the range. The range is the difference between the minimum and maximum values in a group of observations. For example, suppose that the kapas yields (kg per plot) of a cotton variety from five plots are 8, 9, 8, 10 and 11. The range is 11 – 8 = 3 kg. In practice the range is indicated as 8 – 11 kg.

Range takes only the maximum and minimum values into account and not all the values. Hence it is a very unstable or unreliable indicator of the amount of deviation. It is affected by extreme values. In the above example, if we have 15 instead of 1, the range will be 8 – 15 (= 7 kg).

A good number of measures were often done for the analysis of dispersion of data.

(i) Mean deviation: Mean deviation is the mean of the deviations of individual values from their average. The average may be either mean or median. For raw data the mean deviation from the median is the least. Therefore, median is considered to be most suitable for raw data. But usually the mean is used to find out the mean deviation.

The mean deviation is given by

$$\text{M.D.} = \frac{\Sigma|x - \bar{x}|}{n}\ \text{for raw data, and}$$

$$\text{M.D.} = \frac{\Sigma f|x - \bar{x}|}{n}\ \text{for grouped data.}$$

All positive and negative differences are treated as positive values. Hence we use the modulus symbol ||. We have to read $|x - \bar{x}|$ as "modulus $x - \bar{x}$". If we take $(x - \bar{x})$ as such, the sum of the deviations, $\Sigma(x - \bar{x})$ will be 0. Hence, if the signs are not eliminated, the mean deviation will always be 0, which is not correct.

Example 61.19

The kapas yields (in kg. per plot) of a cotton variety from seven plots are 5, 6, 7, 7, 7, 8 and 9. The mean deviation for this data is computed as follows:

$$\text{Mean} = \frac{5+6+7+7+7+8+9}{7} = \frac{49}{7} = 7 \text{ kg}$$

$$\text{M.D.} = \frac{1}{7}(|5-7| + |6-7| + |7-7| + |7-7| + |7-7| + |8-7| + |9-7|)$$

$$= \frac{1}{7}(2+1+0+0+0+1+2) = \frac{6}{7} \text{ kg}$$

Example 61.20

From a field with large number of sesamum plants of a variety, 100 plants were selected at random. The seed yields (in g) per plant were recorded. The results are presented in Table 61.12. Find the mean deviation for the data.

Table 61.12: Frequency distribution of seed yield of 100 sesamum plants

Seed yield (X) in g	No. of plants (f)	Mid class (x)	$\lvert x-\bar{x}\rvert = \lvert x-6.7\rvert$	$f\lvert x-\bar{x}\rvert$
2.5 — 3.5	4	3	3.7	14.8
3.5 — 4.5	6	4	2.7	16.2
4.5 — 5.5	10	5	1.7	17.0
5.5 — 6.5	26	6	0.7	18.2
6.5 — 7.5	24	7	0.3	7.2
7.5 — 8.5	15	8	1.3	19.5
8.5 — 9.5	10	9	2.3	23.0
9.5 —10.5	5	10	3.3	16.5
Total	100			132.4

The steps of computation are:

Step 1 **If the classes are not continuous we have to make them continuous. In this case they are continuous.**

Step 2 **Find out the mid-values of the classes (mid-X = x).**

Step 3 **Compute the mean.**

Step 4 **Find out $|x-\bar{x}|$ for all values of x.**

Step 5 **Multiply each $|x-\bar{x}|$ by the corresponding frequencies.**

Step 6 **Use the formula.**

Up to Step 5, the results are presented in Table 64.12 itself.

Now,

$$\text{M.D.} = \frac{\sum f|x-\bar{x}|}{n} = \frac{132.4}{100} = 1.324 \text{ g}$$

The mean deviation takes all the values into consideration. It is fairly stable compared to range or quartile deviation. But it is not stable as standard deviation which is defined in the next section. Since the mean deviation ignores signs of deviations it is not possible to use it for further statistical analysis.

(ii) **Standard Deviation**: We have seen that the signs of the deviations should be eliminated before averaging them. But ignoring the signs of the deviations is mathematically not correct. We may square the deviations to make a negative value as positive. After calculating the average squared deviations, it can be expressed in original units by taking its square root. This type of the measure of variation is known as Standard Deviation.

The standard deviation is defined as the square root of the mean of the squared deviations of individual values from their mean. Symbolically,

$$S.D. = \sqrt{\frac{\Sigma(x-\bar{x})^2}{n}}$$

This is called standard deviation because of the fact that it indicates a sort of group standard spread of values around their mean.

In later analysis, we are going to use the sample standard deviation to estimate the population standard deviation. The sample standard deviation should be an unbiased estimate of the population standard deviation. For this we substitute n − 1 for n in the formula. Thus, the standard deviation is written as

$$s = \sqrt{\frac{\Sigma(x-\bar{x})^2}{n-1}}$$

For practical purposes, the formula can be rewritten as

$$s = \sqrt{\frac{\Sigma x^2 - \frac{(\Sigma x)^2}{n}}{n-1}}$$

For grouped data it is given by

$$s = \sqrt{\frac{\Sigma fx^2 - \frac{(\Sigma fx)^2}{n}}{n-1}}$$

Alternatively,

$$s = \sqrt{\left[\frac{\Sigma fd^2 - \frac{(\Sigma fd)^2}{n}}{n-1}\right]} \times C$$

where,

C = class interval

$d = \frac{x-A}{C}$ as given under mean.

The square of the standard deviation is known as the *variance*. In the analysis of variance technique, the term $\Sigma x^2 - \frac{(\Sigma x)^2}{n}$ is called the *sum of squares*, and the variance is called the *mean square*. The standard deviation is denoted by s in case of sample, and by σ (read 'sigma') in case of population.

Example 61.21

For the data in Example 64.19., the standard deviation is computed as:

$$\bar{x} = \frac{5+6+7+7+7+8+9}{7} = 7\,\text{kg}$$

The deviations ($x - \bar{x}$) are (5 – 7), (6 – 7), (7 – 7), (7 – 7), (7 –7), (8 – 7) and (9 –7). Therefore,

$$s = \sqrt{\frac{(-2)^2 + (-1)^2 + (0)^2 + (0)^2 + (0)^2 + (1)^2 + (2)^2}{7-1}}$$

$$= \sqrt{\frac{4+1+1+4}{6}} = \sqrt{\frac{10}{6}} = \sqrt{1.67} = 1.29\,\text{kg}$$

The second method is,

$$s = \sqrt{\frac{\sum x^2 - \frac{(\sum x)^2}{n}}{n-1}}$$

$$= \sqrt{\frac{(5^2+6^2+7^2+7^2+7^2+8^2+9^2) - \frac{(49)^2}{7}}{7-1}}$$

$$= \sqrt{\frac{353 - \frac{2401}{7}}{6}} = \sqrt{\frac{353-343}{6}} = \sqrt{\frac{10}{6}} = \sqrt{1.67} = 1.29\,\text{kg}$$

The variance is given by, $s^2 = 1.67\ \text{kg}^2$

Example 61.22

For the data given in Table 61.13 compute the standard deviation and variance.

Table 61.13: Computation of standard deviation

Seed yield (X) g	No. of plants (f)	Mid. class x	$\frac{x-A}{C} = d$	fd	$fd.d = fd^2$
2.5 — 3.5	4	3	–3	–12	36
3.5 — 4.5	6	4	–2	–12	24
4.5 — 5.5	10	5	–1	–10	10
5.5 — 6.5	26	6	0	0	0
6.5 — 7.5	24	7	1	24	24
7.5 — 8.5	15	8	2	30	60
8.5 — 9.5	10	9	3	30	90
9.5 — 10.5	5	10	4	20	80
Total	100			70	324

The steps up to the computation of fd are same as those given for the computation of mean by shortcut method. Here, A = 6 and C = 10. The last column fd^2 is the only addition. Using the undermentioned formula we get the standard deviation. Thus,

$$s = \sqrt{\frac{\sum fd^2 - \frac{(\sum fd)^2}{n}}{n-1}} \times C$$

$$= \sqrt{\frac{324 - \frac{(70)^2}{100}}{100-1}} \times 1$$

$$= \sqrt{\frac{324 - \frac{4,900}{100}}{99}} \times 1$$

$$= \sqrt{\frac{324-49}{99}} \times 1 = \sqrt{\frac{275}{99}} \times 1$$

$$= \sqrt{2.7676} \times 1 = (1.66) \times 1 = 1.66\,g$$

Therefore Variance is $s^2 = 2.77\ g^2$

The standard deviation is the most widely used measure of dispersion. It takes all the items into consideration. It is more stable compared to other measures. However, it will be inflated by extreme items as is the mean.

The standard deviation has some additional special characteristics. It is not affected by adding or subtracting a constant value to each observed value. It is affected by multiplying or dividing each observation by a constant. When the observations are multiplied by a constant, the resulting standard deviation will be equivalent to the product of the actual standard deviation and the constant. (Note that division of all observations by a constant, C, is equivalent to multiplication by its reciprocal, 1/C. Subtracting a constant C is equivalent of adding a constant, − C.)

(iii) **Relative Dispersion**: Suppose that the two distributions to be compared are expressed in the same units and their means are equal or nearly equal. Then their variability can be compared directly by using their standard deviations. However, if their means are widely different or if they are expressed in different units of measurement, we cannot use the standard deviationss as such for comparing their variability. We have to use the relative measures of dispersion in such situations.

There are relative dispersions in relation to range, the quartile deviation, the mean deviation, and the standard deviation. Of these, the coefficient of variation which is related to the standard deviation is important. The coefficient of variation is given by

$$C.V. = \frac{S.D.}{Mean} \times 100$$

The C.V. is a unit-free measure. It is always expressed as percentage. The C.V. will be small if the variation is small.

Of the two groups, the one with less C.V. is said to be more consistent.

The coefficient of variation is unrealiable if the mean is near zero. Also it is unstable if the measurement scale used is not ratio scale. The C.V. is informative if it is given along with the mean and standard deviation. Otherwise, it may be misleading.

Example 61.23

Consider the distribution of the yields (per plot) of two paddy varieties. For the first variety, the mean and standard deviation are 60 kg and 10 kg, respectively. For the second variety, the mean and standard deviation are 50 kg and 9 kg, respectively.
Then we have, for the first variety,

$$C.V. = \frac{10}{60} \times 100 = 16.7\%$$

For the second variety,

$$C.V. = \frac{9}{50} \times 100 = 18.0\%$$

It is apparent that the variability in the first variety is less as compared to that in the second variety. But in terms of standard deviation, the interpretation could be reverse.

Example 61.24

Consider the measurements on yield and plant height of a paddy variety. The mean and standard deviation for yield are 50 kg and 10 kg, respectively. The mean and standard deviation for plant height are 55 cm and 5 cm, respectively.
Here the measurements for yield and plant height are in different units. Hence, the variabilities can be compared only by using coefficient of variation. For yield,

$$C.V. = \frac{10}{50} \times 100 = 20\%$$

For plant height,

$$C.V. = \frac{5}{55} \times 100 = 9.1\%$$

The yield is subject to more variation than the plant height.

(j) Skewness and Kurtosis

We have seen that the averages and measures of dispersion can help in describing the frequency distribution. However, they are not sufficient to describe the nature of the distribution. For this purpose we use other concepts known as Skewness and Kurtosis.

(i) **Skewness**: Skewness means lack of symmetry. A distribution is said to be *symmetrical* when the values are uniformly distributed around the mean. For example, the following distribution is symmetrical about its mean 3.

x	: 1	2	3	4	5
frequency, f	: 5	9	12	9	5

In a symmetrical distribution the mean, median and mode coincide, that is, mean = median = mode. The symmetrical and skewed distributions are shown by curves in Fig. 61.12.

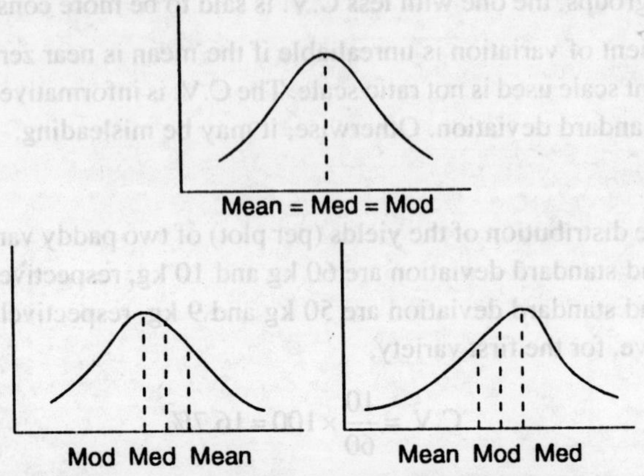

Fig. 61.11: The curves for different distributions

Several measures are used to express the direction and extent of skewness of a dispersion. The important measures are that given by Pearson. The first one is the Coefficient of Skewness:

$$S_k = \frac{3(\text{mean} - \text{median})}{\text{Standard deviation}}$$

For a symmetrical distribution, $S_k = 0$. If the distribution is negatively skewed then S_k is negative, and if it is positively skewed then S_k is positive. The range for S_k is from − 3 to + 3. The other measure uses the β (read 'beta') coefficient which is given by

$$\beta_1 = \frac{\mu_3^2}{\mu_2^3}$$

where, μ_2 and μ_3 are the second and third *central moments*. The second central moment μ_2 is nothing but the variance. The sample estimate of this coefficient is

$$b_1 = \frac{m_3^2}{m_2^3}$$

where, m_2 and m_3 are the sample central moments given by

$$m_2 = \text{variance} = \frac{\Sigma(x - \bar{x})^2}{n - 1} \text{ or } \frac{\Sigma f(x - \bar{x})^2}{n - 1}$$

and,

$$m_3 = \frac{\Sigma(x - \bar{x})^3}{n - 1} \text{ or } \frac{\Sigma f(x - \bar{x})^3}{n - 1}$$

For a symmetrical distribution, $b_1 = 0$. Skewness is positive or negative depending upon whether m_3 is positive or negative.

(ii) **Kurtosis**: A measure of the peakedness or convexity of a curve is known as *Kurtosis*. Consider the curves in Fig. 61.12.

It will be clear from the Fig. 61.12 that all the three curves (1), (2) and (3) are symmetrical about the mean. Still they are not of the same type. One has different peak as compared to that of others. Curve (1) is known as *mesokurtic* (normal curve); curve (2) is known as *leplokurtic* (leaping curve), and curve (3) is known as *platykurtic* (flat curve).

Kurtosis is measured by Pearson's coefficient, β_2 (read 'beta-two').

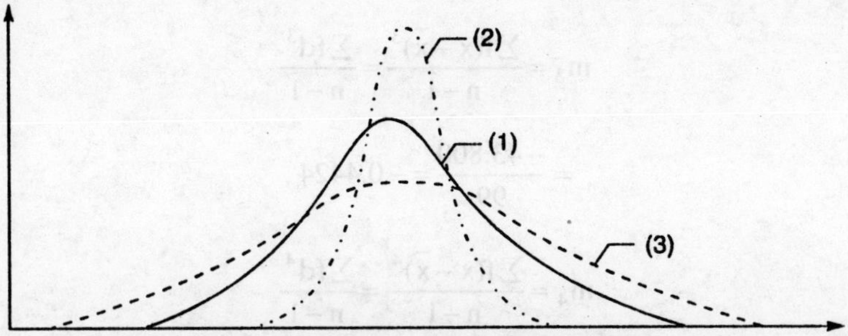

Fig. 61.12: The nature of distributions

It is given by

$$\beta_2 = \frac{\mu_4}{\mu_2^2}$$

The sample estimate of this coefficient is

$$b_2 = \frac{m_4}{m_2^2}$$

where, m_4 is the fourth central moment given by

$$m_4 = \frac{\Sigma(x-\bar{x})^4}{n-1} \text{ or } \frac{\Sigma f(x-\bar{x})^4}{n-1}$$

The distribution is called normal if $b_2 = 3$. When b_2 is more than 3 the distribution is said to be leptokurtic. If b_2 is less than 3 the distribution is said to be platykurtic.

Example 61.25

The measures of skewness and kurtosis are given below for the data in Table 61.14.

Table 61.14: Computation of moments

x	f	$(x-\bar{x}) = d$	f.d	$f.d^2$	$f.d^3$	$f.d^4$
3	4	– 3.7	– 14.8	54.76	– 202.612	749.6644
4	6	– 2.7	– 16.2	43.74	– 118.098	318.8646
5	10	– 1.7	– 17.0	28.90	– 49.130	83.5210
6	26	– 0.7	– 18.2	12.74	– 8.918	6.2426
7	24	0.3	7.2	2.16	0.648	0.1944
8	15	1.3	19.5	25.35	32.955	42.8415
9	10	2.3	23.0	52.90	121.670	279.8410
10	5	3.3	16.5	54.45	179.685	592.9605
Total	100		0	275.00	– 43.800	2,074.1300

$$m_2 \text{ or Variance } = \frac{\Sigma f(x-\bar{x})^2}{n-1} = \frac{\Sigma fd^2}{n-1}$$

$$= \frac{275.000}{99} = 2.7777$$

$$m_3 = \frac{\Sigma f(x-\bar{x})^3}{n-1} = \frac{\Sigma fd^3}{n-1}$$

$$= \frac{-43.800}{99} = -0.4424$$

$$m_4 = \frac{\Sigma f(x-\bar{x})^4}{n-1} = \frac{\Sigma fd^4}{n-1}$$

$$= \frac{2074.1300}{99} = 20.9508$$

$$b_1 = \frac{(-0.4424)^2}{(2.7777)^3} = \frac{0.1957}{21.4376} = 0.0091$$

$$b_2 = \frac{20.9508}{(2.7777)^2} = \frac{20.9508}{7.7156} = 2.7153$$

Since, b_1 is 0.0091, it is only slightly skewed. It is negatively skewed since m_3 is negative.

The value of b_2 is 2.7153 which is less than 3. Hence the distribution is platykurtic.

(k) Probability Distribution

The probability is a measure of chance that an event occurs. It is defined in several ways. It suffices to remember that it is a relative frequency of an event to the total number of events. For example, suppose there are 20 plants in a plot and 5 of them are diseased. The probability that a randomly selected plant from the plot will be a diseased plant is 5/20. The probability lies in the range 0 to 1.

The reader may deepen his knowledge of probability by reading of more technical accounts of it.

Simple probabilities can be computed from elementary consideration. However, in dealing with probabilities of whole classes of events, we have to consider more efficient ways of analysis of probability. For this purpose we should know the concept of a probability distribution.

Probability distribution is defined for random variables. A variable X which takes values $x_1, x_2, ..., x_n$ with associated probabilities, $p_1, p_2, ..., p_n$ respectively, is called a random variable. The qualification 'random' is made because the values which X takes depend on chance and these are not known beforehand. If the variable is continuous, it is called continuous random variable. If the variable is discrete, it is called discrete random variable.

Suppose that there are 5 coconut seedlings. Also suppose that the seedlings are classified as 'good' or 'not good'. Let the selection of good seedling be the random variable X. Now X can take any value between 0 and 5 (x_0, x_1, x_2, x_3, x_4 and x_5. The value is 0 when none of the seedlings is good, 1, if one is good, ... and

5 if all are good. The number of possible events is 2, i.e., good or not good. Since there are 5 seedlings, the sample space will be $2^5 = 32$. The possibilities of selecting no good seedling, 1 good seedling, etc. are given by $\binom{5}{0}\binom{5}{1}$ etc. The probabilities are worked out at the relative frequencies viz., $\binom{5}{0}/32$, $\binom{5}{1}/32$, and so on. The probabilities are presented in Table 61.15.

Table 61.15: The random variable and the probabilities

No. of good seedlings, x	Probability, p (x)
0	$\binom{5}{0}/32 = 1/32$
1	$\binom{5}{1}/32 = 5/32$
2	$\binom{5}{2}/32 = 10/32$
3	$\binom{5}{3}/32 = 10/32$
4	$\binom{5}{4}/32 = 5/32$
5	$\binom{5}{5}/32 = 1/32$
Total	$\frac{32}{32} = 1$

For continuous random variable, we can consider the example of sesamum yield given in Table 61.16:

Table 61.16: Frequency distribution and probabilities of sesamum yield

Seed yield (g per plot) X	No. of plots f	Probability f(x)
65 — 105	8	$p(65 \le X < 105) = 8/100 = 0.08$
105 — 145	27	$p(105 \le X < 145) = 27/100 = 0.27$
145 — 185	50	$p(145 \le X < 185) = 50/100 = 0.50$
185 — 225	15	$p(185 \le X < 225) = 15/100 = 0.15$
Total	100	1.00

The function p(x) is called *Probability Mass function*, and f(x) is called *Probability Density function*. The summary of the distinct values x_i of a random variable X together with their probabilities p(x) or f(x) is known as the *Probability Distribution* of the random variable.

In practice a detailed list as given in the examples will not be presented. Instead, a model or formula will be used. These models are named normal distribution, binomial distribution, Poisson distribution, etc. If the random variable is discrete, the model is known as *discrete probability distribution*. It is known as *continuous probability distribution* if the random variable is continuous.

NORMAL DISTRIBUTION

The most important and widely used probability distribution is normal distribution. It is also known as Gaussian distribution. Most of the distributions occurring in practice—for instance binomial, Poisson—can be approximated by normal distribution. Further, many of the sampling distributions like student's t distribution, F distribution and X^2 distribution lend to normality for large samples. Therefore, the normal distribution finds an important place in statistical inference.

Normal distribution means commonly occurring distribution. Earlier, it was believed that every real-life data must conform to the bell-shaped curve. If the curve did not have that shape the data was suspected for its quality. It is in this context the term 'normal' was used.

The normal distribution is used to represent the probability distribution of a continuous random variable. Its probability density function is expressed by the relation

$$f(x) = \frac{1}{\sigma\sqrt{2\pi}} . e^{½}\left(\frac{x - \mu}{\sigma}\right)^2$$

In the above formula,

π = a constant equaling 22/7

e = Napierian base equaling 2.7183

μ = population mean

σ = population standard deviation

x = a given value of the random variable in the range $-\infty \leq x \leq \infty$

For a normal distribution the frequency curve will be symmetrical or bell-shaped. However, not all symmetrical curves are normal. The shape of the normal curve is completely determined by two parameters μ and σ. For any given σ, there can be a number of normal curves each with a different μ. Likewise, for any given μ, there can be a number of normal curves each with a different σ. In order to make all such distributions readily comparable with each other, their individuality as expressed by their mean and standard deviation has to be suppressed. This is done by transforming the normal variable into a standard normal variable. The standard normal variable is denoted by Z and is given by

$$Z = \frac{x - \mu}{\sigma}$$

The distribution of the standard normal variable is known as standard normal distribution. It is given by

$$f(x) = \frac{1}{\sqrt{2\pi}} e - \frac{s^2}{2}, \text{ in the range } -\infty \leq z \leq \infty$$

$$= \varphi\,(Z) \quad (\varphi \text{ is read as 'phi'}).$$

For a standard normal distribution, $\mu = 0$ and $\sigma = 1$.

Tables are readily available for different values of z. Because of the symmetrical nature of the normal distribution the tables are presented only for the positive values of z.

From the table of normal curves it can be seen that 68.26 per cent of the area lies within the range of $\mu \pm \sigma$ (read as 'μ plus or minus σ'), 95.46 per cent within the range of $\mu \pm 2\sigma$, and 99.74 per cent within the range of $\mu \pm 3\sigma$. This is an important property of normal distribution which is frequently used in statistical inference.

The normality of a distribution can be tested using several methods. Three important methods can be mentioned. They are graphic method, Pearson's test and *goodness of fit*. The graphic method requires normal probability papers which may not be available easily. The third method involves testing of hypothesis. Hence, it is discussed under the chapter on testing of hypothesis.

On the basis of Pearson's coefficients, β_1 and β_2, we can say whether a distribution is normal or not. If $\beta_1 = 0$ and $\beta_2 = 3$, the distribution is said to be normal.

For the data in Table 61.13, the β_1 and β_2 have been computed in Example 61.1. The value of β_1 is 0.0091 which is very close to 0. The value of β_2 is 2.7153 which is only slightly less than 3. Hence we can say that the distribution is nearly normal.

BINOMIAL DISTRIBUTION

The simplest probability distribution is binomial distribution. It is used to represent the probability distribution of discrete random variables. Binomial means two categories. For example, a plant may be categorised as healthy or pest-infested, a flower may be classified as pigmented or non-pigmented, an insect may be classified by sex as male or female, and a gene may be classified as dominant or recessive. In general, the successive repetition of an observation (trial) may result in an outcome which possesses or which does not possess a specified character. Our primary interest will be either of these possibilities. Conventionally, the outcome of primary interest is termed as *success*. The alternative outcome is termed as *failure*. These terminologies are used irrespective of the nature of the outcome. For example, non-germination of a seed may be termed as success.

The experiments consisting of a series of trials resulting in only two outcomes are called *Bernoulli trials*, if the trials are repeated a finite number of times (n), if the trials are independent and if at each trial the probability of success (p) remains same.

The variable X which represents the count of the number of successes in Bernoulli trials will be a *discrete random variable*. The probability distribution of such discrete random variable X is called the *binomial distribution*.

The binomial distribution is given by the probability mass function

$$p(x) = \binom{n}{x} p^x . q^{n-x}$$

for all possible values of x.

In the formula,

$n =$ number of trials

$x =$ number of successes in a trial

$n - x =$ number of failures in a trial

$p =$ probability of success $= x/n$

$q = I - p =$ probability of failure

$\binom{n}{x} =$ the possible number of ways in which x successes can occur.

The binomial distribution is determined by the two parameters n and p. The mean and variance of the binomial distribution are, respectively, np and npq.

Suppose that an experiment consists of n trials and that the experiment is repeated N times. We can then get the frequency function of the binomial distribution by multiplying the probability function p(x) by N.

The binomial distribution approaches normal distribution as the number of trials n tends to be large ($n \to \infty$) for any fixed value of p. A rule of thumb is that for $p < 0.5$, the normal approximation is adequate if $np > 15$. Departures from the given conditions result in less accurate approximations.

When n is very large and p is very small ($n \to \infty$ and $p \to 0$) the binomial distribution approaches Poisson distribution.

Example 61.26

Using binomial distribution we can get the probability value easily. Consider the example of coconut seedlings. The probability that a coconut seedling is good is ½. For n = 5, p = ½, q = 1 − p = ½, and x = 0, 1, 2, 3, 4 and 5, the probabilities can be worked out by using the binomial distribution:

$$p(0) = \binom{5}{0} (1/2)^0 (1/2)^5 = 1(1)((1/2)\times(1/2)\times(1/2)\times(1/2)\times(1/2)) = 1/32$$

$$p(1) = \binom{5}{1} (1/2)^1 (1/2)^4 = 5(1/2)^5 = 5(1/32) = 5/32$$

$$p(2) = \binom{5}{2} (1/2)^2 (1/2)^3 = 10(1/2)^5\ 10(1/32) = 10/32$$

Similarly, p(3) = 10/32, p(4) = 5/32 and p(5) = 1/32. These results can be compared with those given in Table 61.16.

The mean, np = 5(1/2) = 2.5

The variance, npq = 5(1/2) (1/2) = 1.25

$$\text{(i)}\quad p(2) = \binom{10}{2}\left(\frac{2}{5}\right)^2\left(\frac{3}{5}\right)^8$$

$$= 45\ (0.16)\ (0.01680) = 0.12096 = 0.1210$$

(ii) $p(\geq 2) = 1 - p(0) - p(1)$

$$p(0) = \binom{10}{0}\left(\frac{2}{5}\right)^0\left(\frac{3}{5}\right)^{10} = 0.006047$$

$$p(1) = \binom{10}{1}\left(\frac{2}{5}\right)^1\left(\frac{3}{5}\right)^9 = 0.040320$$

Hence,

$$p(\geq 2) = 1 - 0.00647 - 0.040320 = 0.953633 = 0.9536$$

POISSON DISTRIBUTION

The Poisson distribution is also used to represent the probability distribution of a discrete random variable. It is employed in describing random events that occur rarely over a continuum of time or space. The number of insects caught in a trap in one hour is an example for random events in time. The incidence of stem fly in blackgram in a field, and the number of wilt-affected coconut trees in a locality are examples of random distribution of objects in space.

The Poisson distribution bears a close similarity to the binomial distribution. Suppose that we are interested in the number of occurrences of an event A in a time period of length t. This time period can be split into n equal intervals, each of length t/n. These n intervals can be treated as n independent trials by Bernoulli process. But there is a difficulty. Since the event occurs at various points of time, it can occur twice or more in one of the trials of length t/n. In case of binomial distribution the event is dichotomous, and, hence, there is no possibility of such multiple occurrence within a single trial. In order to overcome this difficulty we make n larger and larger. When n is large, the trials are shorter in terms of length of time. As a result, the probability of occurrence of an event in a single trial would be smaller. It is equivalent of saying that it is a rare event. The binomial distribution can still be used to represent the distribution of such random events. However, the computations become tedious since n is very large. This can be explained by an example.

Suppose that the number of insects caught in a trap is being studied and that the data are collected on the number of insects caught per hour. Assume that the probability that an insect will be caught in any single minute is 0.06. Assume further that the events of insects being trapped are mutually independent and the probability p = 0.06 remains same for all the minutes. We may use the binomial distribution to calculate the number of insects caught per hour by considering each minute as a separate Bernoulli trial. If x is the number of insects caught in a minute then we have

$$p(x) = \binom{60}{x} (0.06)^x \, (0.94)^{60-x}$$

Instead of dividing the hour into minutes the seconds may be used as basic units. Then the value of p would be reduced to, $p = \frac{0.06}{60} = 0.001$. Considering each second as a Bernoulli trial, we would have a sample size 60 × 60 = 3,600 for the period of one hour. The binomial distribution would now be

$$p(x) = \binom{3{,}600}{x} (0.001)^x \, (0.999)^{3600-x}$$

Thus when n becomes larger and larger the computations using binomial become tedious. Fortunately, it has been shown by Poisson that the value of $\binom{n}{x} p^x q^{n-x}$ approaches the value of $\frac{(np)^x . e^{-np}}{x!}$, when n becomes large and p becomes small in such a way that the equality np = m is maintained.

The Poisson distribution is given by the probability mass function

$$p(x) = \frac{e^{-m} . m^x}{x!}$$

In the formula,
m = np = mean number of times an event occurs,
x = the number of times the event occurs,
The value of e^{-m} can be obtained directly from mathematical tables.

In case of Poisson distribution the counts of alternative events, i.e., failures, are not of interest. This is a contrast between binomial and Poisson distributions. For Poisson distribution all that we need is np, the mean number of successes. We need not know about n and p individually. Thus, the Poisson distribution is determined by the parameter m.

The special property of Poisson distribution is that its mean and variance are equal to m.

Example 61.27

In previous experiments it was observed that the incidence of stem fly in blackgram was 6 per cent. Suppose we examine 50 blackgram plants in a field at random. What is the probability that at the most 3 plants will be found to be affected by stem fly?

The probability that a plant is affected by stem fly is given as 0.06. The number of plants observed = n = 50. Hence mean = m = np = 50 (0.06) = 3. The required probability is

$$p(\leq 3) = p(0) + p(1) + p(2) + p(3)$$

$$p(x) = \frac{e^{-m}.m^x}{x!}$$

Hence,

$$p(0) = \frac{e^{-3}.3^0}{0!} = e^{-3}(1)$$

$$p(1) = \frac{e^{-3}.3^1}{1!} = e^{-3}(3)$$

$$p(2) = \frac{e^{-3}.3^2}{2!} = e^{-3}\left(\frac{9}{(1)(2)}\right) = e^{-3}.(9/2)$$

$$p(3) = \frac{e^{-3}.3^3}{3!} = e^{-3}\left(\frac{27}{(1)(2)(3)}\right) = e^{-3}.(9/2)$$

Hence,

$$p(\leq 3) = e^{-3} + 3.e^{-3} + (9/2)\,e^{-3} + (9/2)\,e^{-3}$$
$$= e^{-3}(1 + 3 + (9/2) + (9/2))$$
$$= e^{-3}\left[\frac{(2+6+9+9)}{2}\right]$$
$$= e^{-3}(26/2) = e^{-3}(13)$$

Then from the mathematical tables it can be found that $e^{-3} = 0.0498$

Therefore, $p(\leq 3) = (0.0498)\ (13)$
$= 0.6474$

(l) Sampling Distribution

We have seen that from a population of size N, we can draw $\binom{N}{n}$ number of random samples of same size n. Consider all such possible samples. For each of the sample drawn we can compute a statistic. These $\binom{N}{n}$ values may be grouped into a frequency distribution. The probability distribution is then obtained from the frequency distribution. The probability distribution of a statistic so obtained is known as the sampling distribution of the statistic. Thus we can have the sampling distribution of $\overline{X}$, s, p, etc.

In practice we need not construct the empirical sampling distribution of a statistic by drawing all possible samples of the same size. Prior to drawing any samples at all, the sampling distributions can be obtained theoretically from the properties of random samples.

Once the sampling distribution of a statistic is obtained, explicit methods can be developed for estimating population parameters, and for testing hypotheses about these parameters. Thus, the sampling distribution provides a conceptual bridge between descriptive statistics and inferential statistics.

The form of the sampling distribution depends upon the shape of the distribution of the population from which the sample is drawn and the size of the sample. Generally, it is required that the underlying population should be normal. However, the *central limit theorem* comes to the rescue.

The central limit theorem states that, irrespective of the form of the distribution of the population from which the sample is drawn, the sampling distribution of a statistic will be approximately normal provided that the sample size is sufficiently large. If the population distribution departs more and more from normality, larger and larger samples are required for the theorem to hold. The statement of this theorem is not precise with respect to what is meant by 'sufficiently large'. Conventionally, a rule of thumb being used is that the sample size should be at least 30.

STANDARD ERROR

As in the case of population distributions the characteristics of the sampling distributions are also described by some constants like mean and standard deviation. Since a statistic is a random variable, the mean of the sampling distribution of a statistic is called the expected value of the statistic. The standard deviation of the sampling distribution of a statistic is called the *standard error* of the statistic. The square of the standard error is known as the *variance* of the statistic.

The standard error plays an important role in the inferential statistics. The reciprocal of the standard error is taken as the measure of precision for statistics whose distribution tends to normality. Hence the magnitude of the standard error is an index of the precision of the estimate of the parameter. The standard error is used to determine the limits within which the population parameter may be expected to lie. Further, the standard error forms the basis of the testing of hypotheses.

The standard error depends upon the sample size, method of sampling and the nature of population from which the samples are drawn. For our discussions we consider only simple random samples from infinite population.

(m) Testing of Hypothesis

The testing of statistical hypothesis involves the choice of a region on the sampling distribution of statistic. If the statistic falls within this region, the null hypothesis is rejected; otherwise it is accepted. This region is called the *critical region*. The concept of critical region is explained below.

Let our null hypothesis be $H_0 : \mu_1 = \mu_2$ and its alternative be $H_1 : \mu_1 \neq \mu_2$. Suppose H_0 is true. Based on the sample data, it may be observed that the statistic $(\bar{x}_1 - \bar{x}_2)$ follows a normal distribution given by

$$Z = \frac{(\bar{x}_1 - \bar{x}_2) - (\mu_1 - \mu_2)}{SE(\bar{x}_1 - \bar{x}_2)}$$

We know that about 95 per cent values of the statistic from repeated samples will fall in the range $(\bar{x}_1 - \bar{x}_2) \pm 1.96$ times SE $(\bar{x}_1 - \bar{x}_2)$. In other words, this range is the 95 per cent confidence limits for the parameter $(\mu_1 - \mu_2)$. This range can be represented by a diagram as in Fig. 61.13.

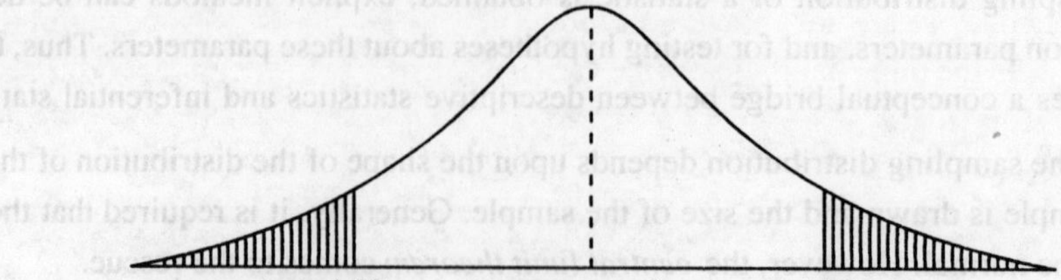

Fig. 61.13 : Diagram showing 95 per cent confidence limits for $\mu_1 - \mu_2$

The value ± 1.96 is the critical value of Z. The area beyond the critical value (shaded area) is known as the *critical region* or *region of rejection*. The remaining area is known as the *region of acceptance*. If the statistic falls in the critical region, we reject the null hypothesis; if it falls in the region of acceptance we accept the null hypothesis. In other words, if the calculated value of a test statistic (Z, t, X^2, etc.) is more than the critical value in magnitude we reject H_0 and otherwise we accept H_0. The critical values for different distributions (Z, t, X^2, etc.) are available from statistical tables.

LEVEL OF SIGNIFICANCE

The probability that the statistic will fall in the critical region is $\frac{\alpha}{2} + \frac{\alpha}{2} = \alpha$. This α is nothing but the probability of committing type I error. Technically, the probability of committing type I error is known as *level of significance*.

Conventionally, the level of significance is taken as 5 per cent (0.05) or 1 per cent (0.01). However, no one is bound to adhere to these standards if he has a good reason for adopting some other levels, say, 0.001 or 0.10. In choosing the level of significance no precise guidelines can be given. The researcher has the freedom to choose his own criterion in fixing the level of significance. The criterion may depend on consideration of factors such as availability of material and money to pursue an experiment, time, seriousness of type I error, and the nature of the problem. For example, suppose that a new variety of paddy appears to give higher yield than a ruling variety. Further, the new variety may possess some other advantages such as resistance to pests and other diseases, has low cost of cultivation etc. Such factors cannot be expressed in terms of numerical probabilities, but the researcher can make allowance for them by taking the level of significance as 10 per cent or even 20 per cent. On the other hand, if the new variety appears to be better than the ruling variety, but the ruling variety possesses some advantages such as we have just mentioned, then the researcher may decide in favour of the new variety by taking 1 per cent or 0.1 per cent level of significance.

Whatever be the level of significance, it should be decided upon at the time of designing the investigation.

ONE AND TWO-TAILED TESTS

The nature of the original hypothesis determines the nature of the alternative hypothesis H_1. If the original hypothesis states that two groups will differ with respect to means, then H_1 is that $\mu_1 \neq \mu_2$.

It implies that either $\mu_1 > \mu_2$ or $\mu_1 < \mu_2$. Here the direction of the difference is not known. But if the hypothesis predicts, the direction of the difference as one group will have larger mean than the other, then H_1 may be either $\mu_1 < \mu_1 > \mu_2$. For example, suppose we want to compare the mean yield of a new paddy variety (μ_1) with that of an old variety (μ_1) with that of an old variety (μ_2). Unless the new variety is more promising than the old one, we are not going to accept the new variety based on yield performance. In this case our H_1 is $\mu_1 > \mu_2$. On the other hand, if both the varieties are new our interest will be to choose the best of the two. Here our H_1 is $\mu_1 \neq \mu_2$.

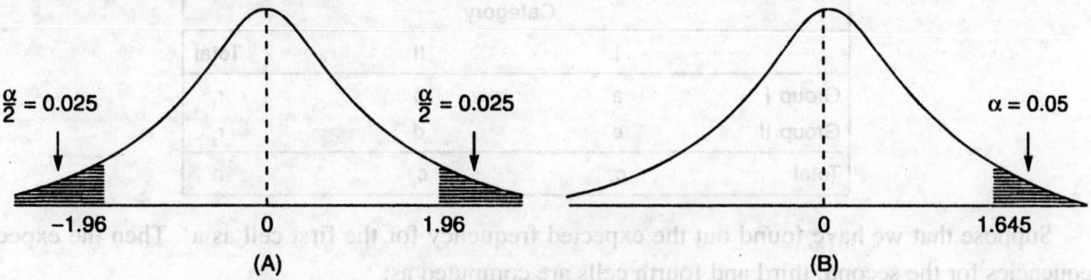

Fig. 61.14 : (A) Critical region for two-tailed test, (B) Critical region for one-tailed test

In case the direction of the difference is known, the critical region will be on one end of the sampling distribution. Hence the test is known as *one-tailed test*. If the direction of the difference is not known, the critical region will be on either end of the sampling distribution. Therefore, the test is known as two-tailed test.

One-tailed and two-tailed critical regions for 5 per cent level of significance ($\alpha = 0.05$) in case of standard normal distribution are illustrated in Fig. 61.14.

The critical value at 5 percent level of significance for one-tailed test is ± 1.645, whereas it is ± 1.96 for two-tailed test.

From Fig. 61.14, it is clear that for a specified level of significance the critical value for one-tailed test will be different from that for two-tailed test.

The tables of critical values may be presented for two-tailed tests. In such tables, we have to refer the column against 10 per cent level of significance in order to achieve a true 5 per cent level critical value for one-tailed test. Similarly, to achieve a true 1 per cent value for one-tailed test, we have to refer the tables under 2 per cent column. If the tables are given for one-tailed test, it is necessary to enter the table with half the value of the intended level of significance (that is, if the intended $\alpha = 0.05$, it is referred under $\alpha = 0.025$).

DEGREES OF FREEDOM

The number of degrees of freedom is the number of observation that are free to vary after certain restriction have been placed on the data. If there are n observations in the sample, for each restriction imposed upon the original observations the number of degrees of freedom is reduced by one. The concept of degrees of freedom can be explained by some examples.

In computing the sample variance s^2, first we find out the deviations of individual observations from the mean, that is, we find out $(x - \bar{x})$. We know that $\Sigma(x - \bar{x}) = 0$. If the observations are 1, 3, 9, 5, 7 we get $\bar{x} = 5$. The deviations of the individual observations from $\bar{x}$ are −4, −2, 4, 0 and 2. One of these deviations

must be such that, when added to the others, the sum is 0. If the first four deviations are known we have– 4 – 2 + 4 + 0 + (5th deviation value) = 0. Therefore, the value of fifth deviation is 2. The number of deviations that are free to vary is 5 – 1 = 4. Here the degrees of freedom is 4. In general, in computing the sample variance we impose the restriction that the deviations are to be taken from the mean value. Hence, if there are n observations the degrees of freedom will be n – 1.

In case of x^2 test the number of expected frequencies that are computed independently will be the degrees of freedom. For example, consider a 2 × 2 contingency table with four cells:

	Category		
	I	II	Total
Group I	a	b	r_1
Group II	c	d	r_2
Total	c_1	c_2	n

Suppose that we have found out the expected frequency for the first cell as a′. Then the expected frequencies for the second, third and fourth cells are computed as:

$b' = r_1 - a'$

$c' = c_1 - a'$

$d' = c_2 - b'$ or $r_2 - c'$

These three values are dependent on the value of the first cell. Thus, there is only one expected frequency which is computed independently. Hence the degrees of freedom is 1. In general, there will be (r – 1) (c – 1) degrees of freedom in case of x^2 test for a contingency table with r rows and c columns.

Tests based on x^2 Distribution

The x^2 distribution has a number of applications in testing of hypotheses. It is used to test (1) the independence of attributes, (2) the goodness of fit, (3) the homogeneity of variances, (4) the homogeneity of correlation coefficients, and (5) the linkage in genetic problems. For applying x^2 test procedure we have to use only the actual observed frequencies and not the percentages or ratios. Further, the samples should be independent and the observations within a sample should be non-overlapping. The expected frequency in each category should be more than 5. The number of observations should be large, say, more than 50.

x^2 TEST FOR GOODNESS OF FIT

In testing of hypothesis our objective may be to test whether a sample has come from a population which has a specified theoretical distribution like normal, binomial and Poisson. In other words, it may be necessary to test whether an obtained frequency distribution resembles a theoretical distribution. In plant genetics, our interest may be to test whether the observed segregation ratios deviate significantly from the Mendelian ratios. In such situations we want to test the agreement between the observed and theoretical frequencies. Such test is called test of *goodness of fit*.

The null hypothesis in goodness of fit tests is that there is no disagreement between the observed and theoretical distributions, or the observed distribution fits to the theoretical distribution.

The test statistic to test this null hypothesis is

$$x^2 = \sum \frac{(0-E)^2}{E}$$

In general, the degrees of freedom associated with this X^2 is k – 1, where k is the number of classes or groups. In case of normal fit, binomial fit, etc., the degrees of freedom is ((k – 1)—the number of parameters estimated). For example, in testing the normality of a distribution μ and σ^2 would be estimated by $\bar{x}$ and s^2. The degrees of freedom would, therefore, be (k – 1) – 2. In case of Poisson distribution, m = μ would be estimated by $\bar{x}$. Hence the degrees of freedom would be (k – 1) – 1.

The expected frequencies are computed based on the appropriate theoretical distributions like normal, binomial, Poisson, etc. The expected frequencies may be derived based on the scientific principles like Mendel's law of inheritance of certain characteristics, sex-ratio of insects, etc. Sometimes we may assume that all the classes are equally frequent. For example, the number of insects caught in a trap in different times of a day may be expected to be equal. In such cases the expected frequency is computed as

$$E = \frac{\text{Total of the observed frequencies}}{\text{Number of groups}} = \frac{n}{k}$$

It may be noted that the expected frequencies computed on the basis of prediction formulas like regression equation cannot be used in testing the goodness of fit.

Example 61.28

In a cross between parents of the genetic constitution AAbb and aaBB, the phenotypes in F_2 sample is classified as:

AB	Ab	aB	ab	Total
87	29	32	12	160

They are expected to occur in a 9 : 3 : 3 : 1 ratio. Does the segregation agree with the theoretical ratio?

H_0 : the segregation agrees with the theoretical ratio.
H_1 : the segregation does not agree with the theoretical ratio.
$\alpha = 0.05$

The test statistic is

$$X^2 = \Sigma\frac{(0-E)^2}{E}$$

The expected frequencies are computed on the basis of the theoretical segregation ratio 9 : 3 : 3 :1. The total is 9 + 3 + 3 + 1 = 16. We expect 9 out of 16 to belong to AB group, that is, the probability of AB is $\frac{9}{16}$. The expected frequency of AB is, therefore, $\frac{9}{16}\times 160 = 90$. Similarly, the expected frequency of Ab is $\frac{3}{16}\times 160 = 30$; the expetected frequency of aB is $\frac{3}{16}\times 160 = 30$; and the expected frequency of ab is $\frac{1}{16}\times 160 = 10$.

The expected frequencies and the computation of X^2 are given in Table 61.17.

Table 61.17: Computation of X²

Phenotype	0	E	(0 s E)²/E
AB	87	90	0.300
Ab	29	30	0.033
aB	32	30	0.133
ab	12	10	0.400
Total	160	160	0.866

The calculated X^2 value is 0.866. Entering the X^2 table for $(4-1)=3$ d.f. and $\alpha = 0.05$, we get the critical value of X^2 as 7.815. The X^2 is, therefore, not significant. We accept the null hypothesis and conclude that the observed phenotypic ratio conforms to the theoretical segregation ratio 9 : 3 : 3 : 1.

Example 61.29

From a cotton field 50 plots were selected at random. The number of plants affected by root rot in each plot was observed. The results are given in Table 61.18. Test whether the distribution of disease-affected plants follows a Poisson distribution.

Table 61.18: Distribution of root rot affected cotton plants

Number of diseased plants per plot:	0	1	2	3	4	5	6
Number of plots:	21	13	10	4	1	0	1

$$\text{The mean} = m = \frac{(0\times 20)+(1\times 13)+......+(6\times 1)}{21+13+......+0+1} = \frac{55}{50} = 1.1$$

The expected frequencies are computed using the formula $\frac{e^{-m}m^x}{x!}(n)$. The other computations are shown in Table 61.19.

Table 61.19: Computation of X² for data in Table 61.18

x	0	E	(0 – E)²/E
0	21	$50.e^{-1.1}.(1.1)^0/01 = 16.6$	1.1663
1	13	$50.e^{-1.1}(1.1)^1/11 = 18.3$	1.5350
2	10	$50.e^{-1.1}(1.1)^2/21 = 10.1$	0.0010
3	4	$50.e^{-1.1}(1.1)^3/31 = 3.7$	0.2000
4	1	$50.e^{-1.1}(1.1)^4/41 = 1.0$	
5	0	$50.e^{-1.1}(1.1)^5/51 = 0.2$	
6	1	$50.e^{-1.1}(1.1)^6/61 = 0.1$	
			$X^2 = 2.9023$

The last four expected frequencies are combined since they are less than 5. Hence 3 degrees of freedom is lost. The degrees of freedom is, therefore, $(7-1-1)-3=2$. For 2d.f. and $\alpha = 0.05$, the critical value of X^2 is 5.991. The X^2 is found to be not significant.

We conclude that the observed frequency distribution fits the Poisson distribution.

(n) Correlation

So far we have confined to the distributions involving only one variable, that is, univariate distributions. Often we obtain information not on one variable alone but on some related variables. Such information is usually referred to as concomitant variation. For example, the yield and yield attributes of a crop, incidence of pests and weather factors, crop yield and pest incidence, size of farm and income, etc., are concomitant variables.

In case of two attributes, we have used x^2 test to study the association between them. Such type of study is not ideal in situations where both variables are continuous. For example, we may be interested in finding out the relationship between the incidence of pests and temperature. For studying the relationship between such continuous variables, we use correlation.

When there are two continuous variables which are concomitant their joint distribution is known as bivariate normal distribution. If there are more than two such variables their joint distribution is known as multivariate normal distribution. In case of bivariate or multivariate normal distributions, we may be interested in discovering and measuring the magnitude and direction of the relationship between two or more variables. For this purpose we use the statistical tool known as *correlation.*

Suppose we have two continuous variables X and Y. If the change in one variable affects the change in the other variable, the variables are said to be correlated. In other words, the systematic interrelationship between the variables is termed as correlation. When only two variables are involved the correlation is termed as simple correlation. If more than two variables are involved the correlation is known as multiple correlation. In this chapter, we confine our discussions to simple correlation.

An increase in one variable may cause an increase in the other variable, or a decrease in one variable may cause a decrease in the other variable. For example, the soluble protein nitrogen and total chlorophyll in the leaves of paddy have such relationship. When the variables move in the same direction like this they are said to be positively correlated. The positive correlation may be termed as direct correlation. If a decrease in one variable causes an increase in the other, variable or vice versa, the variables are said to be negatively correlated. For example, the pest incidence and crop yield have such relationship. The negative correlation may be termed as inverse correlation. In case the two variables are not at all related they are said to be independent.

We have to take extreme care in solving correlation problems. The correlation is not merely a mutual association between twó variables. It implies a cause and effect relationship between the variables. For example, in case of relative humidity and pest incidence, the correlation indicates a cause and effect relationship. On the other hand the relationship between increased use of tractors and increased pest incidence does not indicate cause and effect relationship. However, in correlation studies we need not know which is the cause and which is the effect. The causal relations may be revealed by intuitive inferences or repeated experimentation.

61.2 SCATTER DIAGRAM

In correlation problems, first we have to investigate whether there is any relation between the variables, say, X and Y. For this purpose we use scatter diagram.

Let $(x_1, y_1), (x_2, y_2), ..., (x_n, y_n)$ be n pairs of observations. If the values of the variables X and Y are plotted along the X-axis and Y-axis, respectively, in the XY-plane of a graph sheet, the resultant diagram of dots is known as scatter diagram,. The scatter diagram may be any one of the forms given in Fig. 61.15.

From the scatter diagram we can say whether there is any correlation between X and Y, whether the correlation is positive or negative and where the correlation is linear or curvilinear. The diagrams a, b, and c in the Fig. 61.15 show positive correlation, negative correlation and no correlation, respectively. When the trend is linear as in a and b of Fig. 61.15, the relationship between X and Y is called *linear correlation*. When the trend is curvilinear as in d of Fig. 61.15, the relationship between X and Y is called *curvilinear correlation* or *non-linear correlation*. The non-linear relationship may be quadratic, cubic, etc. We will confine to the study of linear correlation in this chapter.

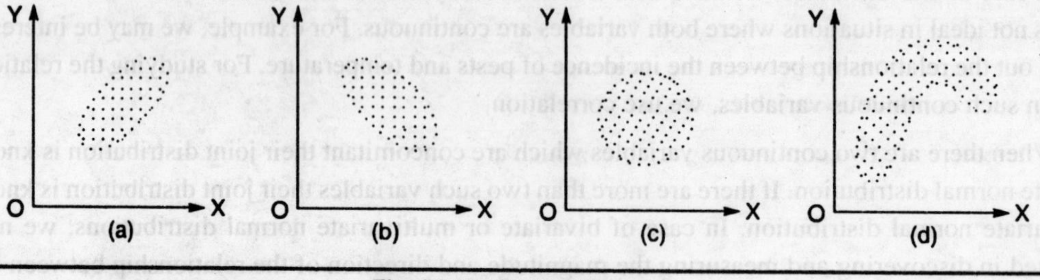

Fig. 61.15: Scatter diagram

61.3 CORRELATION COEFFICIENT

The scatter diagram will give only a vague idea about the presence or absence of correlation and the nature (positive or negative) of correlation. It will not indicate about the strength or degree of relationship between two variables. The index of the degree of relationship between two continuous variables is known as *correlation coefficient*. The correlation coefficient is symbolised as r in case of a sample, and as ρ (read 'rho') in case of population. The correlation coefficient, r, is known as Pearson's Correlation coefficient, since it was developed by Karl Pearson. It is often referred to as *Product-moment correlation* in order to distinguish it from other measures of inter-relationship.

The correlation coefficient, r, is given as the ratio of covariance of the variables X and Y to the product of the standard deviations of X and Y. Symbolically

$$r = \frac{\frac{1}{n-1}(\Sigma(x-\bar{x})(y-\bar{y}))}{\sqrt{\frac{1}{n-1}\Sigma(x-\bar{x})^2}\sqrt{\frac{1}{n-1}\Sigma(y-\bar{y})^2}}$$

This can be simplified as

$$r = \frac{\sum xy - \frac{\sum x \sum y}{n}}{\sqrt{\sum x^2 - \frac{(\sum x)^2}{n}}\sqrt{\sum y^2 - \frac{(\sum y)^2}{n}}}$$

The numerator is termed as the *sum of products of X and Y* and abbreviated as SP (XY). In the denominator, the first term is called the sum of squares of X or SS(X) and the second term is called the sum of squares of Y or SS(Y). The above simplified formula is used for computational purposes.

The denominator in the above formula is always positive. The numerator may be positive or negative, making r to be either positive or negative.

The correlation coefficient r is used under certain assumptions. They are, (1) the variables under study are continuous random variables and they are normally distributed; (2) the relationship between the variables is linear; and (3) each pair of observation is unconnected with other pairs. When the assumptions for the applicability of r are not met, the other measures of association have to be employed. These measures are described in later sections of this chapter.

Example 61.30

Find out the correlation coefficient between seed yield per plant and plant height of sesamum using data in the following table:

Table 61.20: Seed yield per plant and plant height of 24 sesamum plants

Seed yield per plant (in g) (y)	Plant height (in cms) (x)	(y)	(x)
5.22	94.2	5.17	95.7
8.13	69.3	7.22	83.1
6.52	115.3	10.08	85.5
4.16	83.3	6.20	66.3
8.98	85.4	8.39	74.4
3.05	68.1	4.08	60.4
3.49	50.7	6.90	81.2
5.40	96.2	6.05	83.5
2.39	76.1	4.76	94.5
2.71	52.0	2.87	80.1
3.97	82.1	7.21	76.9
7.56	81.3	7.06	75.1

$\Sigma y = 138.02$; $\Sigma x = 1910.7$; $\Sigma y^2 = 897.2384$; $\Sigma x^2 = 156929.25$; $\Sigma xy = 11193.4020$

$$SS(Y) = \sum y^2 - \frac{(\sum y)^2}{n} = 897.2384 - \frac{(138.02)^2}{24}$$

$$= 897.2384 - \frac{19049.5204}{24}$$

$$= 897.2384 - 793.7300 = 103.5084$$

$$SS(X) = \sum x^2 - \frac{(\sum x^2)}{n} = 156929.25 - \frac{(1910.7)^2}{24} = 4813.6462$$

$$SP(XY) = \sum xy - \frac{\sum x \sum y}{n} = 11193.4020 - \frac{(1910.7)(138.02)}{24} = 205.2847$$

$$r = \frac{SP(XY)}{\sqrt{SS(X)}\sqrt{SS(Y)}} = \frac{205.2847}{\sqrt{4813.6462}\sqrt{103.5084}}$$

$$= \frac{205.2847}{(69.3804)(103.5084)} = \frac{205.2847}{705.8693} = 0.2908$$

61.4 REGRESSION ANALYSIS

Of two variables under study one may represent the cause and the other may represent the effect. The variable representing the cause is known as independent variable and it is denoted by X. The variable X is sometimes called predictor variable or regressor. The variable representing the effect is known as dependent variable and is denoted by Y. The variable Y is sometimes called the *predicted variable*. The relationship between the independent and dependent variables may be expressed as a function. Such functional relationship between two variables is termed as *regression*.

When only two variables are involved the functional relationship is known as simple regression. If the relationship between the two variables is a straight line, it is known as *simple linear regression*, otherwise it is called simple non-linear regression. When there are more than two variables and one of them is assumed to be dependent upon the others, the functional relationship between the variables is known as *multiple regression*. In this chapter we discuss the simple linear regression.

CURVE FITTING

In general, when two variables are studied simultaneously, our interest will be to know the functional relationship between them. The scatter diagram is used to identify the type of relationship between the two variables. For example, when n pairs of observations $(x_1, y_1), (x_2, y_2), ... (x_n, y_n)$ are plotted on a graph sheet, it may result in the form shown in Fig. 61.16.

The scatter diagram for a given data may result in non-linear forms as given in Fig. 61.16. The functional relationships may be given by algebraic expressions like

$$y = a + bx$$

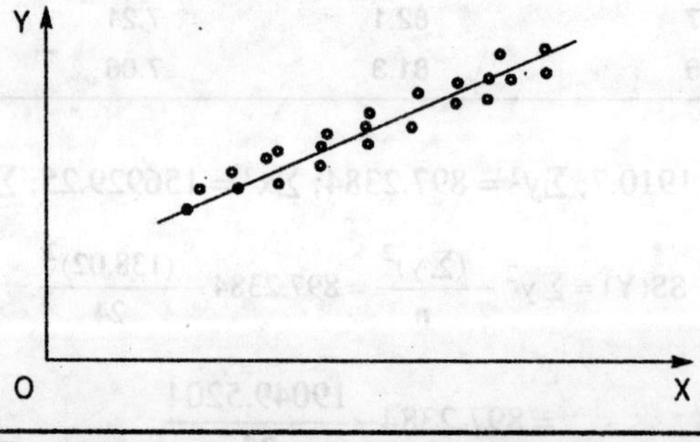

Fig. 61.16: Scatter diagram showing linear relationship

The above algebraic expression is known as curve. The method of finding such relationship is known as *curve fitting*. Fitting of curves to a set of numerical data is useful in the study of regression.

Suppose the line is represented by the equation $y = a + bx$. For each observed value of the variable X we can find out the value of Y. The computed values of Y are knows as the *expected values* of Y. If the observed value of Y is denoted by y and the expected value by $\hat{y}$, the difference between the, $y - \hat{y}$, is knows as *error* or *residual*. The residual is denoted by e. The residuals may be positive or negative or zero. While fitting a curve, our aim is to fit a best line. A *best fitting* line is one for which the sum of the squares of the residuals, Σe^2, is minimum. For this purpose the principle known as *method of least squares* is used.

According to the principle of least squares, one should select a and b such that Σe^2 will be as small as possible, that is, we minimise

$$\Sigma e^2 = S = \Sigma[y - (a + bx)]^2$$

To minimise this function, first we take the partial derivatives of S with respect to a and b. Then the partial derivatives are equated to zero separately. These will result in the equations known as *normal equations*.

For the straight line y = a + bx the normal equations are

$$\Sigma y = na + bx$$
$$\Sigma xy = a\Sigma x + b\Sigma x^2$$

On solving these normal equations, we can get the values of a and b.

For the second degree polynomial, $y = a + bx + cx^2$ the normal equations are

$$\Sigma y = na + b\Sigma x + c\Sigma x^2$$
$$\Sigma xy = a\Sigma x + b\Sigma x^2 + c\Sigma x^3$$
$$\Sigma x^2 y = a\Sigma x^2 + b\Sigma x^3 + c\Sigma x^4$$

The normal equations for other types of equations can easily be written in the same way.

SIMPLE LINEAR REGRESSION

The simple linear regression of Y and X in the population is given by

$$y = \alpha + \beta x + e$$

where,

α = y intercept

β = slope of the line or regression coefficient

e = Error $(y - \hat{y})$.

For any given value of X, the sum of the error term will be 0. Hence, the mean value of e is 0. The simple linear regression is, therefore, rewritten as

$$y = \alpha + \beta x$$

The slope of a linear regression line may be positive, negative or zero as shown in Fig. 61.17.

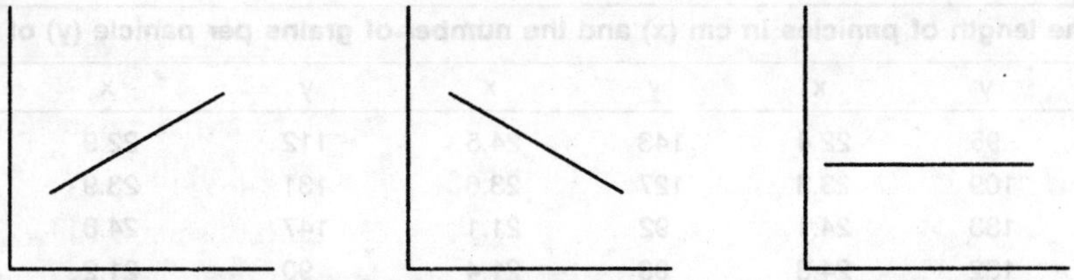

(a) Positive slope (b) Negative slope (c) Zero slope

Fig. 61.17: The slopes of linear regression lines

For a given slope there may be a large number of lines all of them parallel as shown in Fig. 61.18.

Hence, to define a regression line uniquely we have to specify any one point on the line along with the slope. Usually, this point is chosen on the line where x = 0. The value of Y at this point is called Y-intercept which is denoted by α. For example, the points 15, 5 and –5 in Fig. 61.18 are y-intercepts.

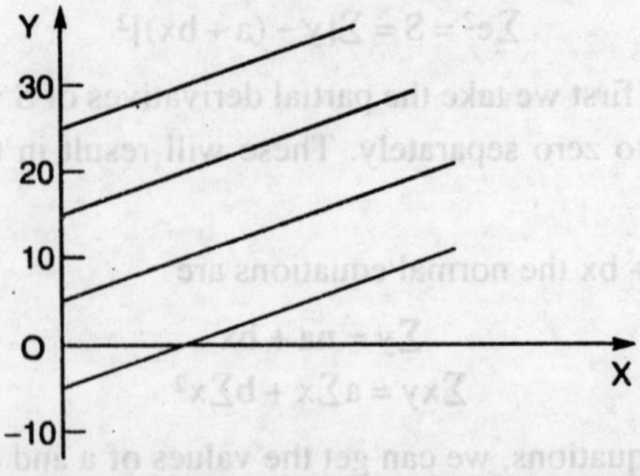

Fig. 61.18: The regression lines with a given slope

The Y-intercept α and the slope ß are the population parameters. We generally obtain the estimates of α and ß from the sample. The estimators of α and ß are denoted by a and b, respectively. The fitted regression line is thus

$$y = a + bx$$

The values of a and b are obtained using the method of least squares. It can be verified that the best estimate of ß is given by

$$b = \frac{\sum xy - \frac{\sum x \sum y}{n}}{\sum x^2 - \frac{(\sum x)^2}{n}} = \frac{SP(XY)}{SS(X)}$$

The best estimate of α is given by

$$a = \bar{y} - b\,\bar{x}$$

The regression line indicates the average value of the dependent variable, Y associated with a particular value of the independent variable, X. The slope b, hereafter referred to as regression coefficient, indicates the change in Y with a one-unit change in X.

Example 61.31

From a paddy field, 36 plants were selected at random. The length of panicles (x) and the number of grains per panicle (y) of the selected plants were recorded. The results are given in Table 61.21. Fit a regression line of Y and X.

Table 61.21: The length of panicles in cm (x) and the number of grains per panicle (y) of paddy plants

y	x	y	x	y	x
95	22.4	143	24.5	112	22.9
109	23.3	127	23.6	131	23.9
133	24.1	92	21.1	147	24.8
132	24.3	88	21.4	90	21.2
136	23.5	99	23.4	110	22.2
116	22.3	129	23.4	106	22.7
126	23.9	91	21.6	127	23.0
124	24.0	103	21.4	145	24.0
137	24.9	114	23.3	85	20.6
90	20.0	124	24.4	94	21.0
107	19.8	143	24.4	142	24.0
108	22.0	108	22.5	111	23.1

$$\Sigma y = 4174;\ \Sigma y^2 = 496258;\ \bar{y} = 115.94$$
$$\Sigma x = 822.9;\ \Sigma x^2 = 18876.83;\ \bar{x} = 22.86$$
$$\Sigma xy = 96183.4$$

$$SS(Y) = \Sigma y^2 - \frac{(\Sigma y)^2}{n} = 496258 - \frac{(4174)^2}{36} = 12305.8889$$

$$SS(X) = \Sigma x^2 - \frac{(\Sigma x)^2}{n} = 18876.83 - \frac{(822.9)^2}{36} = 66.7075$$

$$SP(XY) = \Sigma xy - \frac{\Sigma x \Sigma y}{n} = 96183.4 - \frac{(822.9)(4174)}{36} = 772.7167$$

$$b = \frac{SP(XY)}{SS(X)} = \frac{772.7167}{66.7075} = 11.5837$$

$$a = \bar{y} - b\bar{x} = 115.94 - (11.5837)(22.86) = -148.8396$$

The fitted regression line is

$$y = -148.8396 + 11.5837x$$

Multiple regression analysis

In simple regression analysis we considered the regression of a dependent variable (Y) on a single independent variable (X). In agricultural research, we come across the dependent variables which are influenced by a number of causal variables than by a single variable. For example, the crop yield depends on the plant characters, the climatic factors, soil factors, etc. In such cases the simple linear regression is inadequate. The inadequacy of a simple linear regression line is indicated by smaller value of the coefficient of determination (r^2) even though the model is found to be appropriate by lack of fit test.

The problem of predicting the value of a dependent variable by incorporating a number of independent variables simultaneously is known as multiple regression analysis. The multiple regression is the extension of simple regression. A multiple regression model may be linear or non-linear.

We will consider the multiple linear regression first.

Suppose we have p independent variables $X_1, X_2, ..., X_p$.

$$y = \alpha + \beta_1 X_1 + \beta_2 X_2 + \beta_3 X_3 + ..., + \beta_p X_p + E$$

where,

α = the Y intercept

$\beta_1, \beta_2, ... \beta_p$ = the partial regression coefficients

E = the error term

As in the case of simple linear regression, we will omit E from the equation. The value β_1 measures the change in the mean value of Y per unit change in X_1 holding $X_2, X_3, ..., X_p$ constant. The values β_2, β_3, ... have similar meaning. For this reason βs are called partial regression coefficients of Y on X's.

The quantities $\alpha, \beta_1, \beta_2, ..., \beta_p$ are the parameters and they are unknown. They are estimated by a, b_1, b_2, ..., b_p, respectively, from the given sample. Hence the sample regression model is given by

$$y = a + b_1 x_1 + b_2 x_2 + ... + b_p x_p.$$

In addition to the assumptions made in the simple linear regression one more assumption is made in the multiple regression. The assumption is that the independent variables, Xs, are independent from each other and their effects are additive.

The steps involved in fitting a multiple linear regression line can be explained with an example.

Example 61.32

The results from the study on the relationship between seed yield and plant characters of sesamum are taken for illustration. In this example,

Y = Seed yield (g)
X_1 = Plant height (cms)
X_2 = Length of primaries (cms)
X_3 = Length of secondaries (cms)
X_4 = Days to first flowering

The steps involved in multiple linear regression analysis are:

Step 1 : Compute the mean and SS values of the variables along with their inter-correlations.

$\bar{y}$ = 5.9477; SS(Y) = 1980.9750

$\bar{x}_1$ = 84.7109; $SS(X_1)$ = 113978.6901

$\bar{x}_2$ = 143.3922; $SS(X_2)$ = 1426443.3395

$\bar{x}_3$ = 13.6470; $SS(X_3)$ = 142021.5139

$\bar{x}_4$ = 38.6572; $SS(X_4)$ = 5655.2955

r_y^1 = 0.4632; $r_{12} = r_{21}$ = 0.5683

r_y^2 = 0.6590; $r_{13} = r_{31}$ = 0.2856

r_y^3 = 0.5535; $r_{14} = r_{41}$ = 0.4673

r_y^4 = 1778

$r_{23} = r_{32}$ = 0.6185 n = 423

$r_{24} = r_{42}$ = 0.3735

$r_{34} = r_{43}$ = 0.2795

Step 2 : Write the correlation matrix (without involving y) and invert it using Doolittle solution as shown in Table 61.22. The Doolittle solution is:

(i) Enter the given correlation matrix in appropriate places and write the unit matrix by its side (lines 1 to 4).

(ii) Copy line 1 and denote it as 1′ (line 5).

(iii) Divide each entry in line 5 by the first number in line 5 (that is, by 1) Denote it as 1″ (line 6).

(iv) Copy line 2 (line 7)

(v) Multiply each entry of 1′ line by the element of the second column in 1″ line with sign changed (that is, by – .5683) to get line 8.

(vi) Add lines 7 and 8 to get 2′ line (line 9).

(vii) Divide each entry in 2′ line by the first non-zero number in that line (that is, by .6770). Denote it by 2″ (line 10).

(viii) Similar steps are followed by copying next lines in the correlation matrix. In general, copy line j. When line j is copied j^{th} column values against the previous i″ lines (i = 1, 2, 3, ... but less than j) are taken as the multiplier for i′ lines with the sign changed. For example, when 3rd line is copied the values in column 3 of 1″ and 2″ lines (that is, .2856 and .6739) with changed sign (that is, -.2856 and -.6739) are taken as multipliers for 1′ and 2′ lines.

Similarly, complete all the lines of correlation matrix (in the present case there are 4 lines).

Table 61.22: Doolittle Solution

Line Step	X_1	X_2	X_3	X_4	Check (x_1)	l_1	l_2	l_3	l_4	Check (l_i)
1. X_1	1	.5683	.2856	.4673	2.3212	1	0	0	0	1
2. X_2	.5683	1	.6185	.3735	2.5603	0	1	0	0	1
3. X_3	.2856	.6185	1	.2795	2.1836	0	0	1	0	1
4. X_4	.4673	.3735	.2795	1	2.1203	0	0	0	1	1
5. 1′ = 1	1	.5683	.2856	.4673	2.3212	1	0	0	0	1
6. 1″ = 1′ + 1	1	.5683	.2856	.4673	2.3212	1	0	0	0	1
7. 2	.5683	1	.6185	.3735	2.5603	0	1	0	0	1
8. 1′ (–.5683)	–.5683	–.3230	–.1623	–.2656	–1.3192	–.5683	0	0	0	–.5683
9. 2′ = 7 + 8	0	.6770	.4562	.1079	1.2411	–.5683	1	0	0	.4317
10. 2″ = 2′ + (.6770)	0	1	.6739	.1594	1	.8394	1.4771	0	0	.6377
11. 3	.2856	.6185	1	.2795	2.1836	0	0	1	0	1
12. 1′ (–.2856)	–.2856	–.1623	–.0816	–.1335	–.6630	–.2856	0	0	0	–.2856
13. 2′ (–.6739)	0	–.4562	–.3072	–.0727	–.8363	–.3830	–.6739	0	0	–.2909
14. 3′ = (11 + 12 +13)	0	0	.6110	.0733	.6843	.0974	–.6739	1	0	.4235
15. 3″ = 3′ + (.6110)	0	0	1	.1200	1.1199	.1594	–1.1029	1.6367	0	.6931
16. 4	.4673	.3735	.2795	1	2.1203	0	0	0	1	1
17. 1′ (–.4673)	–.4673	–.2656	–.1335	–.2184	–1.0848	–.4673	0	0	0	–.4673
18. 2′ (–.1594)	0	–.1079	–.0727	–.0172	–.1978	.0906	–.1594	0	0	–.0688
19. 3′ (–.1200)	0	0	–.0733	–.0083	–.0821	–.0117	.0809	–.1200	0	–.0508
20. 4′ = 16 + 17 + 18 + 19	0	0	0	.7556	.7556	–.3884	–.0785	–.1200	1	.4131
21. 4″ = 4′ (.7556)	0	0	0	1	1	–.5140	–.1039	–.1588	1.3235	.5467

When some operation is done on the left hand side matrix, the same operation is performed on the Unit, that is, l matrix also.

Under the check column, the total of each line is written. As a check, the operations performed on the entries of a line are done on check column entries. For example, the total of line 7 is 2.5603. The value against line 8 under the check column is (2.5603) (–.5683) = –1.3192. It should tally with the total of line 8. The value against line 9 under check column is the sum of lines 7 and 8, that is, 2.5603 + (–1.3192) = 1.2411. The value against line 10 under check column is 1.2411 + 6770 = 1.8333. This is tallied with the total of the line 10.

Same procedure is followed for l_i check column.

When the total of a line does not tally with the check column value, it is an indication that our computations in that line or previous lines are wrong.

(ix) Write down 1′ and 1″ matrices from the Doolittle solution, that is, copy the l′ and i″ lines under l matrix. In our case,

$$[1'] = \begin{bmatrix} 1 & 0 & 0 & 0 \\ -.5683 & 1 & 0 & 0 \\ .0974 & -.6739 & 1 & 0 \\ -.3884 & -.0785 & -.1200 & 1 \end{bmatrix}$$

$$[1''] = \begin{bmatrix} 1 & 0 & 0 & 0 \\ -.8394 & 1.4771 & 0 & 0 \\ .1594 & -1.1029 & 1.6367 & 0 \\ -.5140 & -.1039 & -.1588 & 1.3235 \end{bmatrix}$$

(x) Obtain the inverse of the correlation matrix, and denote it as $[c_{ij}]$ matrix.

$$[c_{ij}] = [1']^T [1'']$$

where, $[1']^T$ = transpose of 1′ matrix.

In our case,

$$[1']^T = \begin{bmatrix} 1 & -.5683 & .0974 & -.3884 \\ 0 & 1 & -.6739 & -.0785 \\ 0 & 0 & 1 & -.1200 \\ 0 & 0 & 0 & 1 \end{bmatrix}$$

$$[c_{ij}] = [1']^T [1''] = \begin{bmatrix} 1.6922 & -.9065 & .2211 & -.5140 \\ -.9065 & 2.2285 & -1.0905 & -.1039 \\ .2211 & -1.0904 & 1.6558 & -.1588 \\ -.5140 & -0.1039 & -0.1588 & 1.3235 \end{bmatrix}$$

Step 3 : Compute standardised partial regression coefficients (ß's) using matrix algebra:

$$\begin{bmatrix} \beta_1 \\ \beta_2 \\ \beta_3 \\ \cdots \\ \beta_p \end{bmatrix} = \begin{bmatrix} c_{11} & c_{12} & c_1 \cdots & c_{1p} \\ c_{21} & c_{22} & c_{23} \quad \cdots & c_{2p} \\ c_{31} & c_{32} & c_{33} \quad \cdots & c_{3p} \\ \cdots & \cdots & \cdots \quad \cdots & \cdots \\ c_{p1} & c_{p2} & c_{p3} \quad \cdots & c_{pp} \end{bmatrix} \begin{bmatrix} r_{y1} \\ r_{y2} \\ r_{y3} \\ \cdots \\ r_{yp} \end{bmatrix}$$

For our example,

$$\begin{bmatrix} \beta_1 \\ \beta_2 \\ \beta_3 \\ \beta_4 \end{bmatrix} = \begin{bmatrix} 1.6922 & -.9065 & .2211 & -.5140 \\ -.9065 & 2.2285 & -1.0905 & -.1039 \\ .2211 & -1.0904 & 1.6558 & -.1588 \\ -.5140 & -.1039 & -.1588 & 1.3235 \end{bmatrix} \begin{bmatrix} .4632 \\ .6590 \\ .5535 \\ .1778 \end{bmatrix}$$

$$= \begin{bmatrix} .2174 \\ .4266 \\ .2721 \\ -.1591 \end{bmatrix}$$

That is, $\beta_i = 0.2174$; $\beta_2 = 0.4266$, $\beta_3 = 0.2721$ and $\beta_4 = -0.1591$

Step 4 : Compute partial regression coefficients (b's) from ß's

$$b_i = \beta_i \times \sqrt{\frac{SS(Y)}{SS(X_i)}}$$

For example,

$$b_1 = \beta_i \times \sqrt{\frac{SS(Y)}{SS(X_1)}}$$

$$= .2174 \times \sqrt{\frac{1980.9750}{113978.6901}} = 0.0287$$

Similarly, $b_2 = 0.0159$, $b_3 = 0.0321$, and $b_4 = -0.0942$.

Step 5 : Compute the intercept, a.

$a = \bar{y} - b_1 \bar{x}_1 - b_2 \bar{x}_2 - \cdots - b_p x_p$

$= (5.9477) - (.0287 \times 84.7109) - (.0159 \times 143.3922) - (.0321 \times 13.6470) - (-.0942 \times 38.6572)$

$= 5.9477 - 2.4312 - 2.2799 - .4381 + 3.6415 = 4.4400$

The regression equation is then written as

$y = a + b_1 x_1 + b_2 x_2 + \cdots$

$y = 4.4400 + 0.0287x_1 + 0.0159x_2 + 0.0321x_3 - 0.0942x_4$

Step 6 : Compute R^2.

$$R^2 = \beta_1 r_{y1} + \beta_2 r_{y2} + \beta_3 r_{y3} + \cdots$$

$$= (.2174 + .4632) + (.4266 + .6590)$$

$$+ (.2721 + .5535) + (-.1591 + .1778) = 0.5041$$

Step 7 : Test the significance of R using analysis of variance technique. The analysis of variance for multiple correlation with p independent variables will be given as

Source	d.f.	SS	MS	F
Regression	p	R^2	s^2_r	s^2_r/s^2_e
Error	n – p – 1	1 – R^2	s^2_e	
Total	n – 1	1		

For our example, the analysis of variance is

Source	d.f.	SS	MS	F
Regression	4	0.5041	0.12602	106.30
Error	418	0.4959	0.001186	
Total	422	1.0000		

If R^2 is significant, it may be concluded that the fitted model is adequate one. In other words, it may be concluded that at least some of the ßs are significant. Then we can proceed to test the significance of individual β's.

Step 8 : Test the significance of individual β's. The test criteria is

$$t_i = \frac{\beta_i}{SE(\beta_i)} \text{ with } n-p-1 \text{ d.f.}$$

$$SE(\beta_i) = \sqrt{s_e^2 \times C_{ii}}$$

s_e^2 = error mean square in Step 7

C_{ii} = diagonal values in the $[C_{ij}]$ matrix.

Thus,

$$SE(\beta_1) = \sqrt{S_e^2 . C_{11}} = \sqrt{(0.001186)(1.6922)} = 0.0448$$

$$SE(\beta_2) = \sqrt{S_e^2 . C_{22}} = \sqrt{(0.001186)(2.2285)} = 0.0514$$

$$SE(\beta_3) = \sqrt{S_e^2 . C_{33}} = \sqrt{(0.001186)(1.6558)} = 0.0443$$

$$SE(\beta_4) = \sqrt{S_e^2 . C_{44}} = \sqrt{(0.001186)(1.3235)} = 0.0396$$

$$t_1 = \frac{\beta_1}{SE(\beta_1)} = \frac{0.2174}{0.0448} = 4.8527$$

$$t_2 = \frac{\beta_2}{SE(\beta_2)} = \frac{0.4266}{0.0514} = 8.3016$$

$$t_3 = \frac{\beta_3}{SE(\beta_3)} = \frac{0.2721}{0.0443} = 6.1400$$

$$t_4 = \frac{\beta_4}{SE(\beta_4)} = \frac{-0.1591}{0.0396} = -4.0202$$

If β_i is significant, the corresponding b_i is also significant. However, the SE of b's should be computed to give confidential interval statements for bs.

Step 9 : Compute the SE of bs

$$SE(b_i) = \frac{b_i}{t_i}$$

For our example,

$$SE(b_1) = \frac{b_1}{t_1} = \frac{0.0287}{4.8527} = 0.005914$$

$$SE(b_2) = \frac{b_2}{t_2} = \frac{0.0159}{8.3016} = 0.001915$$

$$SE(b_3) = \frac{b_3}{t_3} = \frac{0.0321}{6.1400} = 0.005228$$

$$SE(b_4) = \frac{b_4}{t_4} = \frac{-0.0942}{-4.0202} = 0.023432$$

Step 10 : Give the consolidated results

b_i	$SE(b_i)$	t_i	
0.0287	0.005914	4.8527	(Significant at
0.0159	0.001915	8.3016	1% level)
0.0321	0.005228	6.1400	
- 0.0942	0.023432	4.0202	
a = 4.400	R = .71	R^2 = 0.5041	

61.5 ANALYSIS OF VARIANCE

We know that variance of a series of observations is the mean (A. M.) of the squares of the deviations of the observations from their mean, and *variance* is used as a measure of variation. Student's t-test is used to test whether there is any significant difference (or variation) in the *means of two populations*. This test fails when we consider the *means of three or more populations*. The equality of several population means can be tested by comparing the *sample variances using F-distribution*, i.e., *by analysis of variance*. The *analysis of variance* is one of the most powerful tools of statistical analysis and this technique is used to test whether the difference between the means of three or more populations is significant or not. The systematic procedure of this statistical technique was first developed by R. A. Fisher and the F-distribution was named in his honour. Earlier, this technique was used in agricultural experiments. Now *analysis of variance* is widely used in natural social and physical sciences. For example, (1) we can explain by using the technique of the analysis of variance whether different varieties of seeds or fertilisers or soils differ significantly or not as regards average yields of land; (2) a manager of a firm may use this technique to know whether there is a significant difference in the average sale figures of different salesmen employed by the firm; (3) the differences in various types of drugs manufactured by a company to cure a particular disease may be studied through this technique.

TECHNIQUE OF ANALYSIS OF VARIANCE

The observations may be classified according to *one factor* (criterion) or *two factors*. The classifications according to *one factor* and *two factors are respectively called one-way classification* and *two-way classification*.

We shall discuss the technique of analysis of variance separately for: (1) One-way classification and (2) Two-way classification.

Variance and its different components may be obtained in each of the two types of classification by (a) Direct Method, (b) Shortcut Method, (c) Coding Method.

One-way Classification

In *one-way classification*, the observations (or data) are classified according to *one factor* (or *criterion*). For example, the yields of several plots of land may be classified according to one or more types of fertilisers.

Direct Method

Let the *Null Hypothesis* be $H_0 : \mu_1 = \mu_2 = ... = \mu_k$ where $\mu_1, \mu_2, ..., \mu_k$ are the arithmetic means of the populations from which k samples are randomly drawn.

The *Alternative Hypothesis* is H_1 : all μ_i (i = 1,2, ..., k) are not equal. The steps in carrying out the analysis are:

I. Calculation of Variance between the Samples or Groups (Calculation of the Sum of the Squares of Variations between the Samples)

It is the sum of the squares of the deviations of the means of various samples (or groups) from the grand mean. To calculate *Variance between the Samples*, we proceed as:

(i) Calculate the *sample means* $\overline{X}_1, \overline{X}_2, ..., \overline{X}_k$ of all the k samples.

(ii) Calculate the *grand mean* $\overline{X}$ by using the formula: $\overline{X} = \frac{\overline{x}_1 + \overline{x}_2 ......\overline{x}_k}{k}$ or $\frac{T}{N}$, where T = Grand total of all observations and N = total no. of observations in all k samples = $\sum_{i=1}^{k} n_i$ where n_i is the size of the ith sample.

(iii) Find the deviations of the means of the different samples from the grand mean $\overline{X}$, i.e., find $\overline{X}_1 - \overline{X}, \overline{X}_2 - \overline{X}, ..., \overline{X}_k - \overline{X}$.

(iv) Square these deviations and multiply by the number of items in the corresponding sample and then obtain the total. This total gives *the sum of the squares of the deviations for variance between the samples* (or *between the columns*) and it is denoted by SSB or SSC.

Thus, $SSB = \sum_{i=1}^{k} n_i (\overline{X}_i - \overline{X})^2$.

Variance or Mean Square Between Samples (MSB or MSC)

If we divide the total obtained in (iv) by the degrees of freedom $v_1 = (k - 1)$, we get the *Variance* or *the Mean Square Between the Sample* which is denoted by MSB or MSC. In this case, the degrees of freedom v_1 is the number of samples minus one. Thus, for 4 samples, the degree of freedom will be 4 – 1, i.e., 3.

MSB indicates the degree of explained variance due to sampling fluctuations.

II. Calculation of Variance within the Samples
(Calculation of the Sum of the Squares of Variations within the Samples)

Variance within the samples is the total of the sum of the squares of the deviations of the various items from the mean values of the respective samples, and its calculation is.

(i) Calculate the sample means $\overline{X}_1$, $\overline{X}_2$, ..., $\overline{X}_k$ of all the k samples.

(ii) Find the deviations of the various items of k samples from the mean values of the respective samples.

(iii) Square all the deviations obtained in (ii) and find the total of these squared deviations. This total is the *Sum of the Square of Deviations within the Samples* or *Sum of the Squares due to Error*. It is denoted by SSW or SSE.

Thus, $SSW = (X_1 - \overline{X}_1)^2 + \Sigma(X_2 - \overline{X}_2)^2 + ... + \Sigma(X_k - \overline{X}_k)^2$.

Variance or Mean Square within the Sample (MSW or MSE)

If we divide the total obtained in (iii) by the degrees of freedom $v_2 = (N - k)$, we get the *Variance* or the *Mean Square within the Samples* which is denoted by MSW or MSE. In this case, the degrees of freedom is the total number of items in all the samples minus the number of samples. MSW indicates chance variation and it cannot be explained in terms of variation in the population.

III. Calculation of the Test Statistic F

When H_0 is true, both the Mean Squares (or Variances) MSB and MSW are the independent unbiased estimates of the same population variance σ^2. Hence, the *test statistic* is

$$F = \frac{MSB}{MSW}, \text{ or } \frac{MSC}{MSE}$$

which follows *F-distribution with degrees of freedom* $v_1 = k - 1$ and $v_2 = N - k$. In general, MSB > MSW. But if MSB < MSW, we shall take the inverse ratio for F, i.e., F = MSW/MSB.

The technique of *analysis of variance* is called ANOVA and a table showing the Source of Variation, Sum of Squares, Degrees of Freedoms, Mean Squares and Mean Square (or Variance) Ratio is called ANOVA Table.

IV. Decision

Compare the calculated value of F with the tabulated (critical) value of F for (k – 1. N – k) degrees of freedom at $\propto$% level (may be 5% or 1% level). *If the calculated value > the tabulated value, reject H_0 and conclude that all the population means are not equal. Otherwise, accept H_0 and conclude that the population means are all equal.*

Let us now summarise calculations for Sums of Squares together with the corresponding degrees of freedom. Mean Squares and Variance Ratio in an ANOVA Table 61.23.

Table 61.23: Analysis of Variance (or ANOVA) Table

Source of Variation	Sum of Squares (SS)	Degrees of Freedom (d.f.)	Mean Squares (MS)	Test Statistic (Variance Ratio)
Between (Col.)	SSB	k − 1	MSB = SSB/(k − 1)	
				F = (MSB/MSW)
Within (Errors)	SSW	N − k	MSW = SSW/(N − k)	
Total	SST	N − 1		

SST = Total sum of Squares of Variations = SSB + SSW. In the above table, SSB and SSW may be replaced by SSC and SSE, and MSB and MSW may be replaced by MSC and MSE.

Check

To check up the calculations, the sum of the squares of the deviations for total variation should be calculated. The total variation is calculated by taking the squares of the deviations of all the individual items from the grand average. This total should be equal to SSB + SSW. This explains the additive property of the analysis of variation. The degrees of freedom (d.f.) for total variance = N − 1 = (No. of items in all the samples) − 1.

Examples 61.33

Fill in the blanks of the following ANOVA table:

Source	d.f.	SS	MSS	F
Between	2	*	5	
Within	*	14	*	*
Total	9	*		

Solution : d.f. within = 0 − 2 = 7, i.e., k − 1 = 2 and N − k = 7.

Now, $MSB = \dfrac{SSB}{k-1}$, or, $5 = \dfrac{SSB}{2}$, or, SSB = 10 and $MSW = \dfrac{SSW}{N-k} = \dfrac{14}{7} = 2$.

Total SS = SSB + SSW = 10 + 14 = 24; $F = \dfrac{MSB}{MSW} = \dfrac{5}{2} = 2.5$.

The required complete ANOVA table is:

Table 61.24: ANOVA Table

Source	d.f.	SS	MSS	F
Between	2	10	5	
Within	7	14	2	2.5
Total	9	24		

Example 61.34

The following data give the yields on 12 plots of land in three samples, each of a 4 plots under three varieties of fertilisers A, B and C.

A	B	C
25	20	24
22	17	26
24	16	30
21	19	20

Is there any significant difference in the average yields of land under the three varieties of fertilizers?

Solution. Let the Null Hypothesis be H_0 : there is no significant difference in the average yields under the three varieties.

The Alternative Hypothesis is H_1 : the difference is significant.

The calculations for sample means, the Variance between and within the samples are:

Table 61.25: Calculations for Sample Means

Sample I X_1	Sample II X_2	Sample III X_3
25	20	24
22	17	26
24	16	30
21	19	20
$92 = \Sigma X_1$	$72 = \Sigma X_2$	$100 = \Sigma X_3$

$$\overline{X}_1 = \frac{\Sigma X_1}{n_1} = \frac{92}{4} = 23,\ \overline{X}_2 = \frac{\Sigma X_2}{n_2} = \frac{72}{4} = 18,\ \overline{X}_3 = \frac{\Sigma X_3}{n_3} = \frac{100}{4} = 25.$$

$$\text{Grand mean } (\overline{X}) = \frac{\overline{X}_1 + \overline{X}_2 + \overline{X}_3}{3} = \frac{66}{3} = 22.$$

SSB = Sum of the Squares between the samples

$$= \Sigma n_i (\overline{X}_i - \overline{X})^2 = 4(23-22)^2 + 4(18-22)^2 + 4(25-22)^2 \qquad [\because n_1 = n_2 = n_3 = 4.]$$

$$= 4 + 64 + 36 = 104,$$

and v_1 = degrees of freedom = k – 1 = 3 – 1 = 2.

$$\therefore \text{MSB} = \text{Mean Square between the samples} = \frac{\text{SSB}}{v_1} = \frac{104}{2} = 52.$$

Table 61.26: Calculations for SSW

Sample I		Sample II		Sample III	
X_1	$(X_1 - \overline{X}_1)^2$	X_2	$(X_2 - \overline{X}_2)^2$	X_3	$(X_3 - \overline{X}_3)^2$
25	4	20	4	24	1
22	1	17	1	26	1
24	1	16	4	30	25
21	4	19	1	20	25
Total	$10 = \Sigma(X_1 - \overline{X}_1)^2$	—	$10 = \Sigma(X_2 - \overline{X}_2)^2$	—	$52 = \Sigma(X_3 - \overline{X}_3)^2$

SSW = Sum of the Squares within the samples

$$= \Sigma(X_1 - \bar{X}_1)^2 + \Sigma(X_2 - \bar{X}_2)^2 + \Sigma(X_3 - \bar{X}_3)^2 = 10 + 10 + 52 = 72,$$

and v_2 = degrees of freedom = N − k = 12 − 3 = 9.

$$\therefore \text{MSW} = \text{Mean Square within the samples} = \frac{\text{SSW}}{v_2} = \frac{72}{9} = 8.$$

Table 61.27: Analysis or Variance (ANOVA) Table

Source of Variation	Sum of Squares (SS)	Degrees of freedom (d.f.)	Mean Square (MS)	Test Statistic
Between Samples	104	2	52	
				F = 52/8 = 6.5
Within Samples	72	9	8	
Total	176 = SST	11 = N − 1		

From the table given at the end of the book, the value of F for $v_1 = 2$ and $v_2 = 9$ at 5% level is 4.26. We see that *the calculated value 6.5 of F is greater than the tabulated value* 4.26. Hence, *we reject the Null Hypothesis at 5% level and conclude that the difference in the average yields under the three varieties is* significant.

Shortcut Method

The procedure shown above in *Direct Method* is lengthy and time-consuming and, hence, it is not generally used in practice. A *Shortcut Method* based on the sum of the squares of the individual values is usually used. In this method the computational work is considerably reduced and the method is more convenient when the sample means and the grand mean are fractions.

The various steps of *Shortcut Method* are:

(i) Calculate the *total T of all the observations in all k samples*. Thus $T = \Sigma X_1 + \Sigma X_2 + ... + \Sigma X_k$.

(ii) Find the *Correction factor* by using the formula: Correction factor $\frac{T^2}{N}$.

(iii) Calculate the sum of the squares of all the values (or observations) in k samples and subtract the correction factor $\frac{T^2}{N}$ from this sum to obtain the *Total sum of the Squares of deviations* SST. Thus $\text{SST} = \Sigma X_1^2 + \Sigma X_3^2 + ... + \Sigma X_k^2 - \frac{T^2}{N}$.

(iv) Find the square of the total of each sample and divide each such squared value by the number of values of the corresponding sample and then calculate the total of all the results thus obtained and subtract the correction factor from this total. This final result gives the *Sum of the Squares of Deviations between the Samples*. Thus

$$\text{SSB} = \left[\frac{(\Sigma X_1)^2}{n_1} + \frac{(\Sigma X_2)^2}{n_2} + ... + \frac{(\Sigma X_k)^2}{n_k}\right] - \frac{T^2}{N}.$$

(v) Calculate SSW, i.e., the *Sum of the Squares within the samples by subtracting SSB from SST. Thus SSW = SST – SSB.*

Now proceed as in Direct Method to obtain MSB, MSW and F and to arrive at the final decision. The procedure of Shortcut Method will be well-understood from the following illustration:

Example 61.35

Let us solve the previous problem (Example 61.34) by Shortcut Method.

Table 61.28: Calculations for Shortcut Method

Sample I		Sample II		Sample III	
X_1	X_1^2	X_2	X_2^2	X_3	X_3^2
25	625	20	400	24	576
22	484	17	289	26	676
24	576	16	256	30	900
21	441	19	361	20	400
$92 = \Sigma X_1$	$2126 = \Sigma X_1^2$	$72 = \Sigma X_2$	$1306 = \Sigma X_2^2$	$100 = \Sigma X_3$	$2{,}552 = \Sigma X_3^2$

T = Sum of all the values of the three samples $= \Sigma X_1 + \Sigma X_2 + \Sigma X_3 = 92 + 72 + 100 = 264.$

Correction factor $= \frac{T^2}{N} = \frac{(264)^2}{12} = 5{,}808.$ [Here N = 12]

SST = Total Sum of Squares

$$= \Sigma X_1^2 + \Sigma X_2^2 + \Sigma X_3^2 - \frac{T^2}{N} = 2{,}126 + 1{,}306 + 2{,}552 - 5{,}808 = 176.$$

SSB = Sum of the Squares Between the Samples

$$= \frac{(\Sigma X_1)^2}{n_1} + \frac{(\Sigma X_2)^2}{n_2} + \frac{(\Sigma X_3)^2}{n_3} - \frac{T^2}{N} = \frac{(92)^2}{4} + \frac{(72)^2}{4} + \frac{(100)^2}{4} - 5{,}808$$

$= 2{,}116 + 1{,}296 + 2{,}500 - 5{,}808 = 104;$

$v_1 = \text{d.f.} = k - 1 = 3 - 1 = 2.$

SSW = Sum of the Squares within the samples

$= \text{SST} - \text{SSB} = 176 - 104 = 72$

$v_2 = \text{d.f.} = N - k = 12 - 3 = 9.$

Now proceed as in Example 61.34 to arrive at the final decision.

62 PRINCIPLES OF ANALYTICAL METHODS AND INSTRUMENTATION TECHNIQUES

62.1 SEPARATION METHODS IN BIOCHEMISTRY

It is well-known that in biochemical analysis, a wide variety of compounds needs to be separated and characterised. The organic molecules vary widely in their chemical nature, solublity, and molecular weight.

A good number of routine analysis is often made to understand the nature and properties as well as quantity of the substances. Some major separation methods are listed in Table 62.1 and described later:

Table 62.1: Separation Techniques and Phase Systems

	Technique	Phase system
1.	Solvent system	Liquid - Solid Liquid - Liquid
2.	Liquid (column) Chromatography	Liquid - Solid Liquid - Liquid
3.	Gas Chromatography	Gas - Solid Gas - Liquid
4.	Supercritical Fluid Extraction and Chromatography	Liquid - Solid
5.	Gel Permeation Chromatography	Liquid - Solid
6.	Electrophoresis	Liquid - Solid
7.	Thin Layer Chromatography	Liquid - Solid Liquid - Liquid
8.	Paper Chromatography	Liquid - Liquid
9.	Ion-exchange Chromatography	Liquid - Solid
10.	Precipitation	Liquid - Solid

1. CHROMATOGRAPHIC ANALYSIS

Principle

The term 'chromatography' embraces a family of closely related separation methods based on experiments described by Day and Tswett in 1897-1906. The feature distinguishing chromatography from most other physical and chemical methods of separation is that two mutually immiscible phases are brought into contact wherein one phase is stationary and the other mobile. The sample mixture, introduced into the mobile phase, undergoes a series of interactions (partitions) many times between the stationary and mobile phases as it is being carried through the system by the mobile phase. Interactions exploit differences in the physical or chemical properties of the components in the sample. These differences govern the rate of migration of the individual components under the influence of a mobile phase moving through a column containing the stationary phase. Separated components emerge in the order of increasing interaction with the stationary phase. The least retarded component elutes first, the most strongly retained material elutes last. Separation is obtained when one component is retarded sufficiently to prevent overlap with the zone of an adjacent solute as sample components elute from the column.

The column is the heart of a chromatograph and provides versatility in the types of analyses that can be obtained with a single instrument. This versatility, due to the wide choice of materials for the stationary and mobile phases, makes it possible to separate molecules that differ only slightly in their physical and chemical properties. It can also be a problem when it comes time to select an appropriate chromatographic method.

The mobile phase can be a gas or a liquid, whereas the stationary phase can only be a liquid or a solid. The field of liquid column chromatography (LCC) embraces several distinct types of interaction. When the separation involves predominantly a simple partitioning between two immiscible liquid phases—one stationary and the other mobile—the process is called Liquid-Liquid (or partition) Chromatography (LIC). When physical surface forces are mainly involved in the retentive ability of the stationary phase, the process is denoted Liquid-Solid (or adsorption) chromatography (LSC). Two other chromatographic methods differ somewhat in their mode of action. In Ion-Exchange Chromatography (IEC), ionic components of the sample are separated by selective exchange with counterions of the stationary phase. The use of exclusion packings as the stationary phase brings about a classification of molecules based largely on molecular geometry and size. Exclusion chromatography (EC) is referred to as gel-permeation chromatography by polymer chemists and as gel filtration by biochemists. If the mobile phase is a gas, the methods are called Gas-Liquid Chromatography (GLC) and Gas-Solid Chromatography (GSC). Gas and liquid mobile phase methods will be treated individually.

1. PARTITION CHROMATOGRAPHY

Two immiscible liquid phases are passed through the supporting phase-containing materials. This is also called liquid-liquid chromatography. Another type of partition chromatography is gas-liquid partition chromatography, where carrier gas such as N_2, helium, argon, H_2, and CO_2 are not carrier gases used as mobile phase.

In partition chromatography one solvent acts as stationary and the other as mobile phase:

Stationary Phase	Mobile Phase
1. Water	1. Butanol
2. Aqueous acids	2. Pentanol
3. Aqueous alkalis	3. Benzyl alcohol
4. Buffer sol.	4. Phenol
5. Salt sol.	5. Benzene
6. Aq. methanol etc.	6. Toluene etc.

The materials of Chromatographic phase are:

(1) Silica gel, (2) Kieselguhr, (3) Cellulose, (4) Starch, (5) Dextrin and (6) Rubber and Chlrorinated rubber.

Application—This method is used for the following analyses:

1. Amino acids and peptides separation;
2. Protein separation;
3. Separation of carbohydrates;
4. Organic acids;
5. Steroids.

2. ION EXCHANGE CHROMATOGRAPHY

In this method ion exchange materials are used as the stationary phase in chromatography. A variety of materials have been used as ion exchanger. The most important categories of ion exchangers are:

(a) Natural and artificial siliceous minerals;
(b) Sulphonated carbonaceous materials;
(c) Synthetic polymers;
(d) Derivatives of cellulose;
(e) Miscellaneous.

3. ADSORPTION CHROMATOGRAPHY

In this method certain adsorbents are used for separation technique. Many substances of widely different chemical nature have been used as adsorbents.

The essential apparatus for a chromatographic separation is very simple. It consists of a tube of suitable dimensions, with means for supporting the packed adsorbent and openings for the admission and collection of mobile phase. Recently, multiple coupled filter column is used for this purpose.

The major types of adsorbents are:

(a) Polar: Alumina (Al_2O_3), $CaCO_3$, Cellulose, Kaolin, MgO, Magnesium Silicate, Silicic acid,
(b) Nonpolar: Carbon.

Application: It is used for analysis of porphyrins separation.

4. GEL PERMEATION CHROMATOGRAPHY

Principle

Gel permeation chromatography is a liquid chromatographic method known variously as gel filtration and molecular exclusion chromatography. The gel structure contains pores of varying diameters up to a maximum size. The test molecules are washed through a column of the gel and molecules larger than the largest pores in the gel are excluded from the gel structure. Smaller molecules, however, penetrate the gel to a varying extent depending upon their size and this retards their progress through the column. Elution, therefore, is in order of decreasing size (Fig. 62.1).

There are a variety of products available for gel permeation chromatography (Table 62.2) and they are usually classified according to their exclusion limit, i.e., the relative molecular mass above which all molecules are excluded from the gel structure. Sephadex, the original medium, is based on dextran (a linear glucose polymer) that is modified to give varying degrees of cross-linking which determines the pore size of the material. It is a strongly hydrophilic polymer and swells considerably in water; before a column is prepared, the gel must be fully hydrated.

Polyacrylamide beads are also available with a wide range of pore sizes and are prepared commercially in a manner similar to that described for polyacrylamide electrophoresis. They are also hydrophilic and swell significantly in aqueous solutions but are chemically more stable than the dextran gels. Agarose gels are particularly useful when gels with a very large pore size are required. They are also hydrophilic but are usually sold in the swollen form. A major problem is the fact that they soften at temperatures above 30°C. Mixed gels of polyacrylamide and agarose are available (Ultragel, LKB) in which the polyacrylamide provides a three-dimensional structure which supports the interstitial agarose gel.

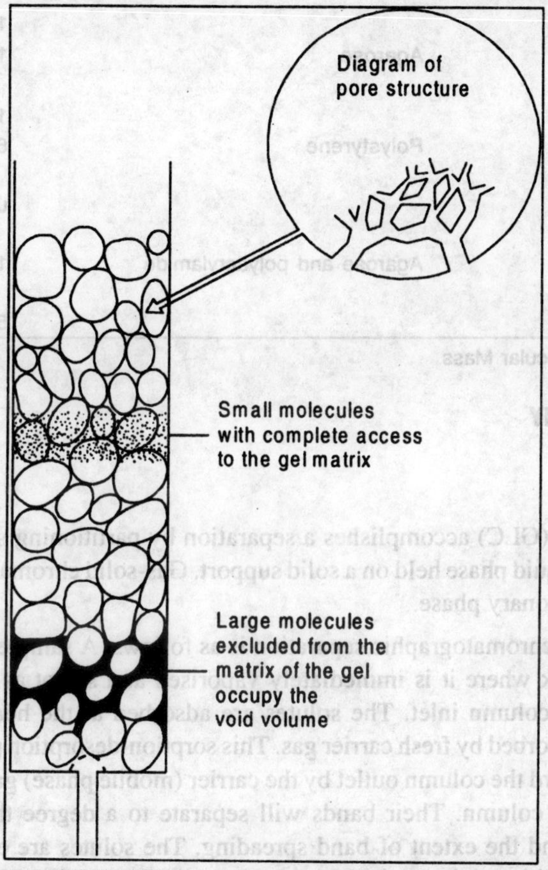

Fig. 62.1: Separation by Gel Permeation Chromatography

Polystyrene gels are hydrophobic and as a result are used primarily with non-aqueous solvents and for organic chemical applications rather than with biological samples.

Table 62.2: Lists of Gel Permeation Media

Trade name and manufacturer	Chemical nature	Fractionation range (RMM)
Pharmacia		
Sephadex G 10	Cross-linked dextrans	up to 700
to		
G 200		5,000-800,000

Contd.

Sepharose 6B	Agarose	up to 4 × 10^6
to		
2B		
Sepharose C1-6B		up to 4 × 10^6
to		
C1-2B		up to 40 × 10^6
Biorad		
Biogel P2	Cross-linked polyacrylamide	200-2,500
to		
P300		100,000–400,000
Biogel A 0.5	Agarose	10,000–500,000
to		
A 150		1 × 10^6 – 150 × 10^6
Biobeads S × 1	Polystyrene	600–14,000
to		
S × 12		up to 400
LKB		
Ultragel AcA22	Agarose and polyacrylamide	100,000–1.2 × 10^6
to		
AcA54		5,000–70,000

RMM = Relative Molecular Mass

5. GAS CHROMATOGRAPHY

Principle

Gas-liquid chromatography (GLC) accomplishes a separation by partitioning solutes between a mobile gas phase and a stationary liquid phase held on a solid support. Gas-solid chromatography (GSC) employs a solid adsorbent as the stationary phase.

The sequence of a gas chromatographic separation is as follows. A sample containing the solutes is injected into a heating block where it is immediately vaporised and swept as a plug of vapour by the carrier gas stream into the column inlet. The solutes are adsorbed at the head of the column by the stationary phase and then desorbed by fresh carrier gas. This sorption-desorption process occurs repeatedly as the sample is moved toward the column outlet by the carrier (mobile phase) gas. Each solute will travel at its own rate through the column. Their bands will separate to a degree that is determined by the individual partition ratios and the extent of band spreading. The solutes are eluted sequentially in the increasing order of their partition ratios and enter a detector attached to the column exit. If a recorder is used, the signals appear on the chart as a plot of time versus the composition of the carrier gas stream. The time of emergence of a peak is characteristic for each component; the peak area is proportional to the concentration of the component in the mixture. Although gas chromatography is limited to volatile materials—about 15% of all organic compounds—the availability of column temperatures up to 450°C, pyrolytic techniques, and the possibility of converting nonvolatile materials into a volatile derivative extend somewhat the applicability of the method.

Basically, a gas chromatograph consists of six parts: (1) a supply of carrier gas in a high pressure cylinder with attendant pressure regulators and flow meters, and a valve to introduce extra make-up gas to some detectors, (2) a sample injection system, (3) the separation column, (4) the detector, (5) an electrometer and strip-chart recorder (and integrator perhaps), and (6) separate thermostated compartments for housing the column and the detector so as to regulate their temperature, or to programme the column temperature. The components are shown schematically in Fig. 62.2.

The most exacting problem in gas chromatography is presented by the sample injection system. The sample must be introduced as a vapour in the smallest possible volume and in a minimum of time without decomposition or fraction occurring. Generally, two basic type of columns viz., packed or capillary columns, are used for analytical purpose.

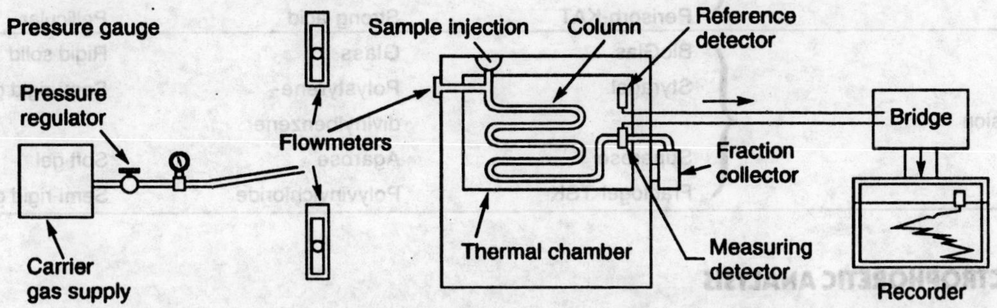

Fig. 62.2: Schematic View of a Gas Chromatograph

A variety of detector systems is used for analytical purpose viz., Photoionization (PID), Flame ionization (FID), Electron capture (ECD), Thermal conductivity (TCD) and Flame photometric (FPD). Among them thermal conductivity detectors were first and are still widely used; their simplicity is an advantage and they are nondestructive. In contrast, for high-sensitivity analyses of organic compounds, the hydrogen flame ionisation detector is used.

6. HIGH PERFORMANCE LIQUID CHROMATOGRAPHY (HPLC)

Principle

This a specialised version of column chromatography, where much more precision analysis of organics will be available. The wide applicability, speed and sensitivity of HPLC have resulted in it becoming the most popular form of chromatography and virtually all types of biological molecules have been analysed or purified using the technique. There are four distinct modes of HPLC separation methods viz., Adsorption, Partition, Ion Exchange and Exclusion. All of them vary with the nature of stationary phase.

Some examples of HPLC stationary phases are given in Table 62.3.

Table 62.3: Some examples of HPLC stationary phases

Chromatographic separation principle	Commercial name	Nature of stationary phase	Type of support
Adsorption	Partisil C_8	Octylsilane	Porous
	Corasil	Silica	Pellicular
	Pellumina	Alumina	Pellicular
	Partisil	Silica	Microporous
	MicroPak A1	Alumina	Microporous
Partition	Bondapak-C_{18}/Corasil	Octadecylsilane (ODS)	Pellicular
	μBondapak-C_{18}	Octadecylsilane	Porous
	ULTRApak TSK ODS	Octadecylsilane	Porous
	μBondapak-NH_2	Alkylamine	Porous
	ULTRApak TSK-NH_2	Alkylamine	Porous

Contd.

Ion-exchange	Partisil SAX	Strong base	Porous
	MicroPak-NH_2	Weak base	Porous
	Partisil-SCX	Strong acid	Porous
	AS Pellionex-SAX	Strong base	Pellicular
	Zipak-WAX	Weak base	Pellicular
	Perisorb-KAT	Strong acid	Pellicular
Exclusion	BioGlas	Glass	Rigid solid
	Styragel	Polystyrene-divinylbenzene	Semi-rigid gel
	Superose	Agarose	Soft gel
	Fractogel TSK	Polyvinylchloride	Semi-rigid gel

7. ELECTROPHORETIC ANALYSIS

Principle

The term 'electrophoresis' describes the migration of a charged particle under the influence of an electric field. Many important biological molecules—such as amino acids, peptides, proteins, nucleotides and nucleic acids—possess ionisable groups and, therefore, at any given pH, exist in solution as electrically charged species either as cations (+) or anions (–). Under the influence of an electric field these charged particles will migrate either to the cathode or to the anode, depending on the nature of their net charge.

The equipment required for electrophoresis consists basically of two items—a power pack and an electrophoresis unit. Electrophoresis units are available for running either vertical or horizontal gel systems. The power pack supplies a direct current between the electrodes in the electrophoresis unit. All electrophoresis is carried out in an appropriate buffer, which is essential to maintain a constant state of ionisation of the molecules being separated. Any variation in pH would alter the overall charge and hence the mobilities of the molecules being separated.

There are a variety of support media—viz., "Agarose Gels", Polyacrylamide gels, Sodium dedecyl sulphate polyacrylamide gel, and cellulose acetate—used in different types of electrophoresis.

Applications

Electrophoresis is commonly used for separation of proteins, nucleic acids and DNA sequencing etc.

8. DIALYSIS AND ULTRAFILTRATION

Principle

Dialysis is a technique which permits separation of micromolecules from smaller solute molecules. The word refers to the diffusion of the solute molecules through a membrane which, owing to its pore size, restricts the movement of large molecules. The passage of the small molecules is due solely to a concentration gradient across the membrane. In ultrafiltration the solvent and solutes are forced through the membrane under pressure with the movement of the large molecules again being restricted by the pore size.

Cellophane is frequently used for dialysis and it has a pore size of approximately 4 – 8 μm and is impermeable to molecules with a relative molecular mass in excess of about 10,000. The development of a variety of membrane materials—in which the pore size is much more rigorously controlled—has led to wider applications of ultrafiltration (Table 62.4). Various cellulose and polycarbonate membranes are available with pore sizes down to 5 nm which are capable of excluding molecules with a relative molecular mass of about 50. The internal structure of such membranes, as well as the pore size, determine their exclusion range and as a result precise specifications of membranes vary from one manufacturer to another.

Table 62.4: Applications of Ultrafiltration

Particle size (nm)	Particles	Application
0.1 – 1.0	Salts, monomers	Water purification
1 – 100	Proteins, viruses	Fractionation, desalting
100 – 1000	Bacteria	Sterilisation

9. ULTRACENTRIFUGATION

Principle

Although the use of high rotor speeds is not always necessary, most ultracentrifuges are designed to operate safely at rotor speeds of up to 60,000 or even 100,000 rpm. Despite the use of high speed, the sedimenting forces involved are only small and the technique is extremely gentle, providing a valuable method for the separation of large, labile molecules or particles. Preparative ultracentrifugation uses rotors with relatively large capacities to separate various components of a mixture while analytical ultracentrifugation involves the measurement of sedimentation velocities in small volumes under carefully controlled conditions.

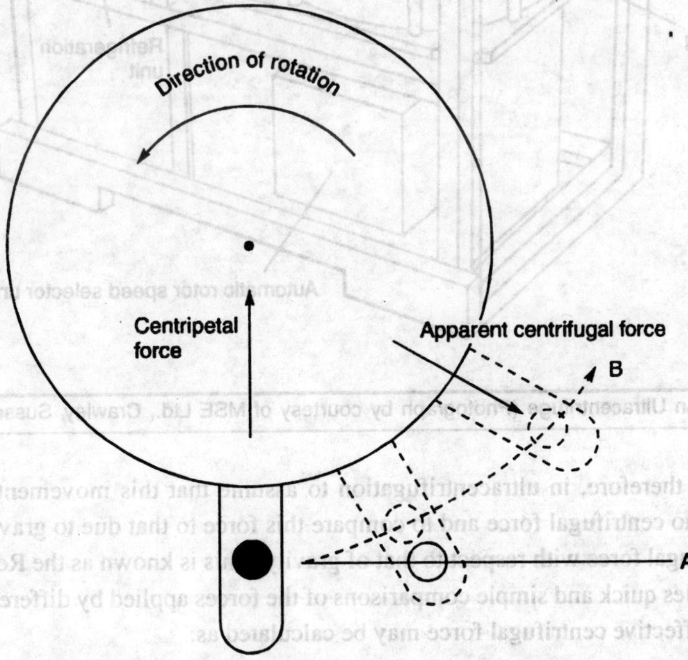

Fig. 62.3: Centrifugal Effects. The path of a particle unimpeded by frictional effects is at a tangent (A) to the circle described by the rotation of the centrifuge tube. When impeded by friction, its path (B) more closely resembles circular movement with a reduced rate of sedimentation in the tube

Because particles sediment in a tube during centrifugation, it may appear that a force is acting radially outwards whereas, in fact, there is no such force. Any particle which is rotating in a circle will always tend to move away at a tangent to the circle and to prevent this a force has to be applied towards the centre of rotation. This is known as the centripetal force and is provided in a centrifuge by the rotor arms and spindle. Particles suspended in solution, when rotated in a centrifuge tube, are not connected to the axis of rotation and, therefore, tend to move at a tangent to the circle. Their movement, however, relative to the centrifuge tube is radially down the tube and appears to be determined by a force equal and opposite to the centripetal force (Fig. 62.3).

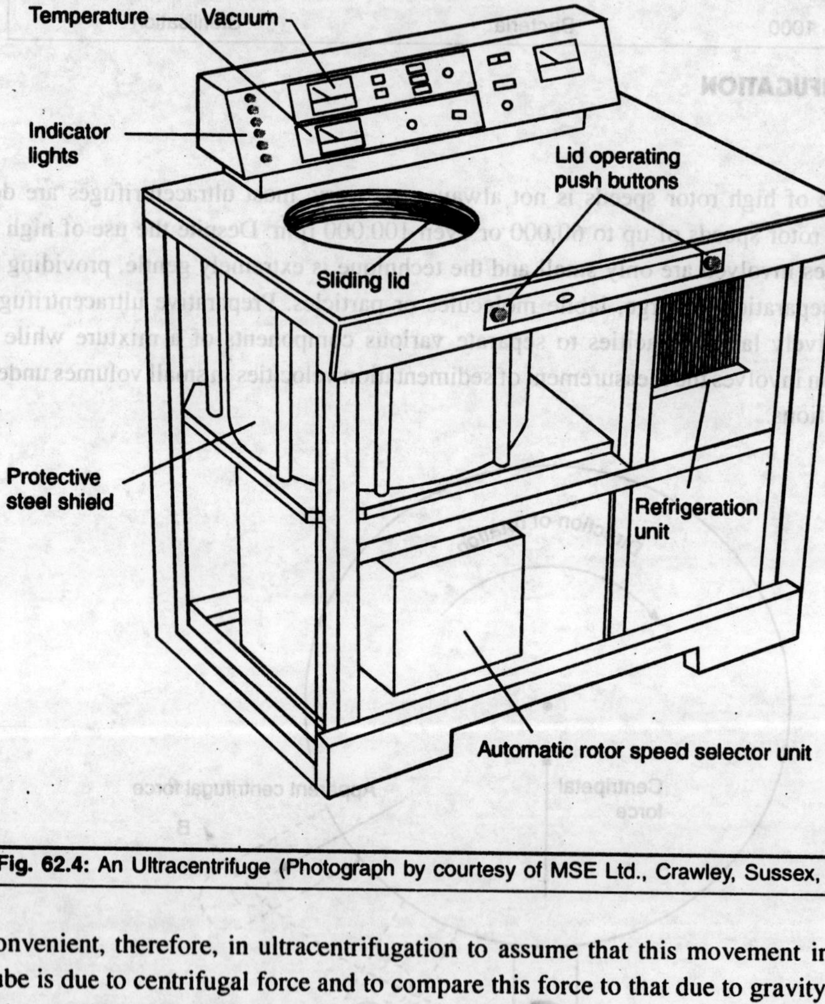

Fig. 62.4: An Ultracentrifuge (Photograph by courtesy of MSE Ltd., Crawley, Sussex, UK)

It is convenient, therefore, in ultracentrifugation to assume that this movement in an apparently stationary tube is due to centrifugal force and to compare this force to that due to gravity by quoting the intensity of the centrifugal force with respect to that of gravity. This is known as the Relative Centrifugal Force (RCF) and enables quick and simple comparisons of the forces applied by different centrifuges and different rotors. The effective centrifugal force may be calculated as:

$$F = M\omega^2 r$$

where M is the mass of the particle, ω (omega) is the angular velocity in radians per second, and r is the radius of rotation in centimeters.

The effect of centrifugation is, therefore, dependent upon the mass of the particle and the convenience of using the relative centrifugal force is that the value for the mass of the particle is eliminated from the equation, permitting more general comparisons of centriguge performance:

$$RCF = \frac{fc}{fg} = \frac{M\omega^2 r}{Mg} = \omega^2 r \times g^{-1}$$

It is more usual to monitor the speed of a centrifuge in terms of revolutions per minute (rpm) rather than in radians per second. Hence:

$$RCF = \left(\frac{2\pi\, r.p.m}{60}\right)^2 r \times g^{-1}$$

In comparing the relative centrifugal forces developed by different centrifuges it is important to appreciate that the values vary considerably from the top to the bottom of a centrifuge tube.

For a particle to sediment it must displace an equal volume of the solvent from beneath it. This can only be achieved by centrifugation if the mass of the particle is greater than the mass of the solvent displaced. Hence, density as well as mass is an important consideration and explains the phenomenon of low density particles floating rather than sedimenting during centrifugation (Fig. 62.4).

62.2 QUANTITATIVE ANALYTICAL METHODS

Diverse analytical methods are often used for quantitative determination/estimations of different organic and inorganic substances. In this section, an attempt is made to describe their principles and application potentials.

1. TITRIMETRIC ANALYSIS

Principle

This is a widely used analytical method, where quantitative analysis of one substance is measured by another substance which is capable of reacting with the previous substance (whose concentration will be determined). In this method, the equivalent point of reaction between two substance will be determined in presence/absence of indicator. This is a fairly good method of analysis of large number of organic and inorganic parameters. The sensitivity of measurement ranges between 1 mg to 0.1 mg. Many ions, or molecules form complexes in stepwise fashion with amino polycarboxylic acid viz., EDTA, NTA, and DCTA. Each complex is characterised by its own stability constant. By applying this principle in titrimetry, complexometic titration is applied in a variety of ion determination. The end point of titration is detected by indicators.

E.D.T.A. (1, 2 diamino ethane tetra-acetic acid) is one of a class of aminopolycarboxylic acids which form soluble strong complexes with many metal ions. The metal ion may be divalent, (Ca^{+}, Mg^{++}) tri, tetra valent, even monovalent metals such as silver (I) will form complex with EDTA. Besides EDTA, other amino polycarboxylic acids, which are used for this purpose—*NTA* (nitrilotriaceatic acid), *DCTA* (transdiaminocyclohexanetetra acetic acid), DTPA (diethylenetrinitrilopenta acetic acid); EGTA [brs (aminoethyl) glocolether - N, N, N', N"— tetra acetic acid, & TTHA (triethylenetetratranitrilohexa acetic acid).

The end points of titrations with EDTA are usually detected by use of *metallochromic indicators* (organic dyestuffs). Most metallochromic indicators also possess acid-base indicator properties — hence a fairly strict control of pH is essential.

Example of some indicators —

1. *Murexide :* (ammonium purpurate)

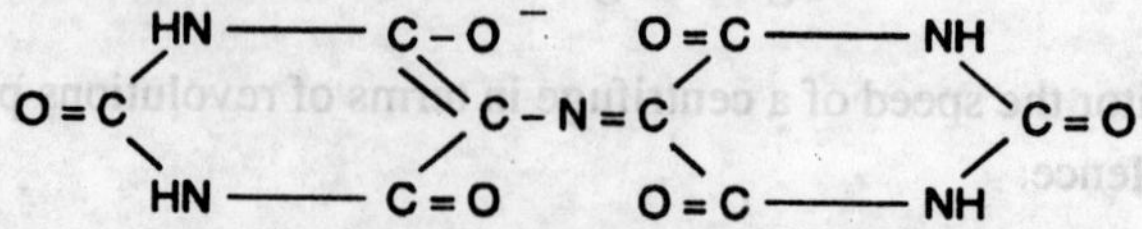

2. *Eriochrome Black T :* [Solochrome Black WDFA or T]

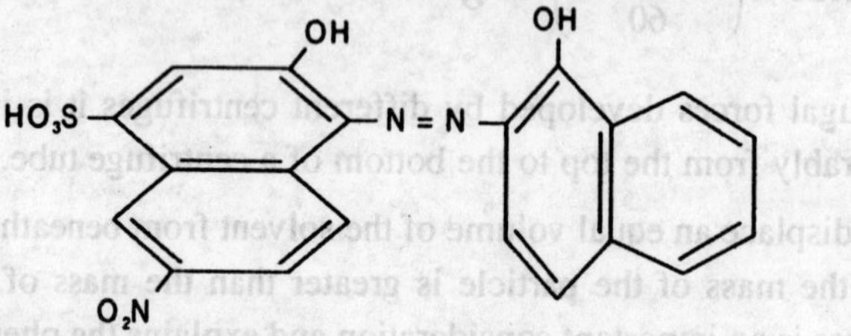

Wine red to blue (pH 6.3)
blue to orange (pH 11.3)

3. *Pyrocatechol violet :*

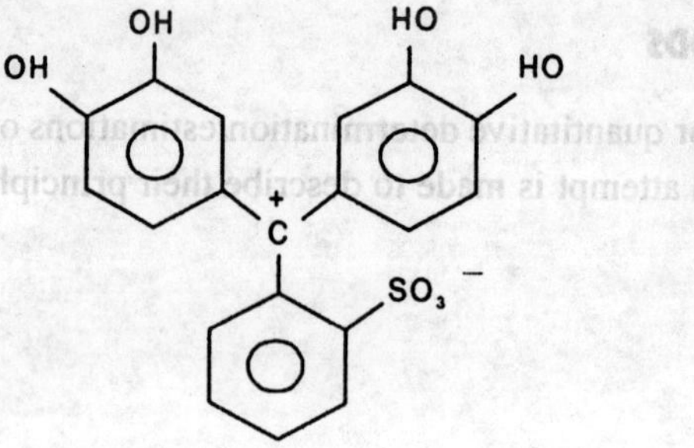

Red (Below pH 1.5)
Yellow (pH_2 – 6)
Violet = (above – pH6)

4. *Zincon :*

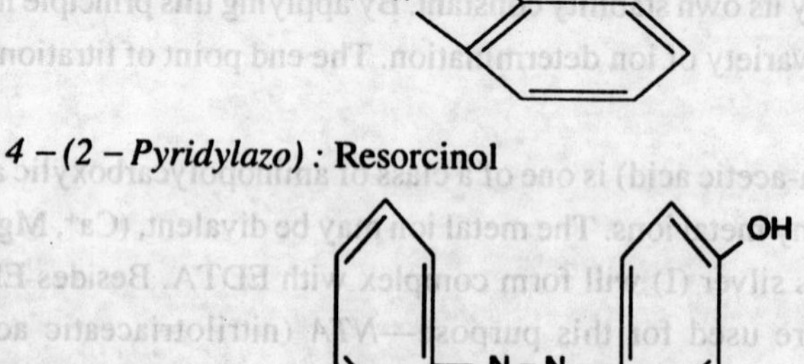

Brilliant blue with
Zn^{++} at pH 9 – 10
Indicator colour – Yellow

5. *4 – (2 – Pyridylazo) :* Resorcinol

6. *Xylenol Orange :*

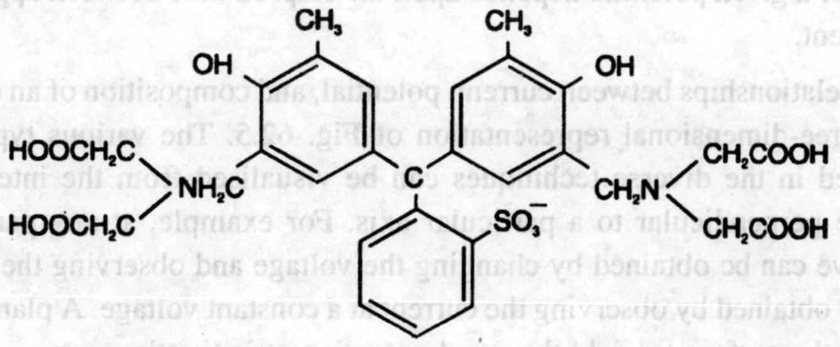

2. POTENTIOMETRIC ANALYSIS

Principle

The measurement of the potential of an appropriate indicator electrode has been used for many years as a method of detecting the equivalence point in a variety of titrations. When a potentiometric titration is being performed the major interest is focussed on changes in the emf of the electrochemical cell as a titrant of precisely known concentration which is added to a solution of the text element. The method can be applied to any titrimetric reaction for which an indicator electrode is available to follow the activity of at least one of the substances involved. The chief advantages of potentiometric titrations are applicability to turbid fluorescent, opaque or coloured solutions, or when suitable visual indicators are unavailable or inapplicable.

The principles governing the major types of potentiometric titrations are:

(i) Oxidation-reduction reactions;

(ii) In combination reactions;

(iii) Acid-Base reactions.

3. ELECTROMETRIC ANALYSIS

Principle

Electroanalytical chemistry encompasses a wide variety of techniques, each based upon a particular phenomenon occurring within an electrochemical cell. Each basic electrical measurement—current, resistance, and voltage—has been used alone or in combination for analytical purposes. If these electrical properties are measured as a function of time, many other techniques are possible. Table 62.5 contains a brief classification of electrochemical methods arranged according to the quantity measured.

Electrometric methods can also be grouped into two categories: *steady-state* or equilibrium method (nonpolarised), *transient* or *dynamic* (polarised or diffusion-controlled) methods. In the first category, equilibrium is effectively assured by vigorously stirring the solution or rotating the electrode or both so that concentration gradients at the electrode are completely or nearly completely eliminated. In such cases, the electrode potential is related to concentration by the Nernst equation. In transient methods, both the electrode and solution are static, in which case, after electrolysis begins, the concentration gradients at the electrodes are time-dependent or diffusion-controlled. This category includes such methods as chronopotentiometry, chronoamperometry, and polarography (voltametry). When either impressed potential or current are controlled, the dependent variable is a function of both concentration and time. If data are recorded as current versus potential, time still enters the experiment because the value of the current observed for a given potential depends upon the elapsed time between application of the potential and the measurement.

The general relationships between current, potential, and composition of an electroactive system are depicted in the three-dimensional representation of Fig. 62.5. The various types of two-dimensional variations observed in the diverse techniques can be visualised from the intersection with the solid surface of a plane perpendicular to a particular axis. For example, at any particular concentration a polarographic curve can be obtained by changing the voltage and observing the current. Amperometric titration curves are obtained by observing the current at a constant voltage. A plane corresponding to zero current interesects the surface to yield the usual potentiometric titration curve.

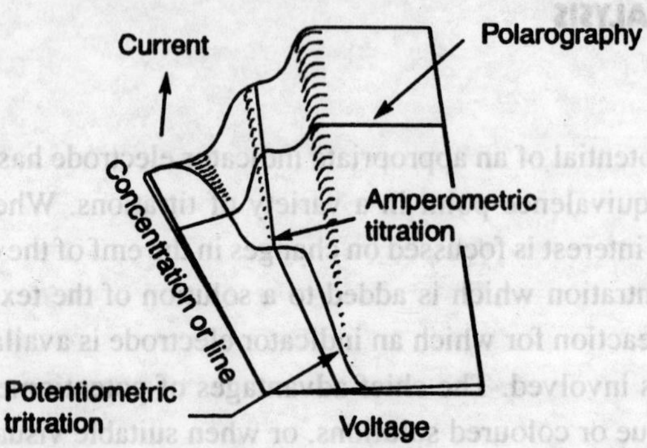

Fig. 62.5: Three-dimensional representation of the relationship between current, potential, and composition of a system

Electrometric methods are characterised by a high degree of sensitivity, selectivity, and accuracy. Highly refined ways of making electrical measurements permit reliable determination.

Table 62.5: A Summary of Electrometric Methods of Analysis

Quantity Measured	Variable Controlled	Name or Description of Method
E versus volume of reactant	i = 0	Potentiometric titrations
	i	Potentiometric titration at constant current
E	i = 0	1. Measurement of ionic activities or concentrations 2. Null-point potentiometry
E versus time	i	Chronopotentiometry
Weight of separated phase	i or E	Electrogravimetry
1/R (conductance)	E	Concentration measured after calibration with known mixtures
1/R (conductance) versus volume of reagent	E	Conductometric titrations
i versus E	Concentration	1. Polarography or volametry 2. Stripping volametry
i versus volume of reagent	E	Amperometric titrations
Coulombs (current × time)	E	Coulometry at controlled electrode potential
	i	1. Coulometry at constant current 2. Stripping analysis

4. GRAVIMETRIC ANALYSIS

Principle

This is a method of quantitative determination of substance whereby suitable reaction substance is precipitated and the content is measured by weight. The elements like iron and sulphur are usually determined by this method. This method is sensitive up to mg level.

Conductance Analysis

Principle

Conductance measurements were among the first to be used for determining solubility products, dissociation constants, and other properties of electrolyte solutions. Conductance is an additive property of a solution depending on all the ions present. Solution conductance measurements, therefore, are nonspecific. This nonspecificity restricts the quantitative analytical use of this technique to situations where only a single electrolyte is present or where the total ionic species needs to be ascertained. In these special situations, however, conductance measurements are capable of extreme sensitivity.

The possibility of using conductance to locate end points in titrations was also recognised early in the development of instrumental methods. Changes in the slope of conductance versus titrant volume occur because ionic mobilities vary and also because of the formation of insoluble or nonionised materials. Accordingly, conductometric indication of end points is possible. A related technique, high-frequency conductometric titration, was developed in recent years. High-frequency measurements permit the determination of changes in conductance, or dielectric constant, without the introduction of electrodes into direct contact with the solution.

Electrolytic conductivity is a measure of the ability of a solution to carry an electric current. Solutions of electrolytes conduct an electric current by the migration of ions under the influence of an electric field. Like a metallic conductor, they obey Ohm's law. Exceptions to this law occur only under abnormal conditions; for example, very high voltages or high-frequency currents. Thus, for an applied electromotive force E, maintained constant but at a value that exceeds the decomposition voltage of the electrolyte, the current is flowing between the electrodes immersed in the electrolyte will vary inversely with the resistance of the electrolytic solution R. The reciprocal of the resistance 1/R is called the *conductance*, and is expressed in reciprocal ohms, or mhos.

The standard unit of conductance is specific conductance k, which is defined as the reciprocal of the resistance in ohms of a 1-cm cube of liquid at a specified temperature. The units of specific conductance are reciprocal ohm cm (or mho/cm). The observed conductance of a solution depends inversely on the distance d between the electrodes and directly upon their area A:

$$\frac{1}{R} = k\frac{A}{d}$$

The electrical conductance of a solution is a summation of contributions from all the ions present. It depends upon the number of ions per unit volume of the solution and upon the velocities with which these ions move under the influence of the applied electromotive force. As a solution of an electrolyte is diluted, the specific conductance in will decrease. Fewer ions to carry the electric current are present in each cubic centimeter of solution. However, in order to express the ability of individual ions to conduct a function called the *equivalent conductance* is employed. It may be derived that A is equal to the area of two large parallel electrodes set 1 cm apart and holding between them a solution containing one equivalent of solute. If C_s is the concentration of the solution in gram equivalents per litre, then the volume of solution in cubic centimetres per equivalent is equal to 1,000/C_s, so that the equation becomes

$$A = 1{,}000\frac{x}{C_s}$$

At infinite dilution the ions theoretically are independent of each other and each ion contributes its part to the total conductance, thus

$$A_\infty = \Sigma(\lambda_+) + (\lambda_-)$$

where λ_+ and λ_- are the ionic conductance of cations and anions, respectively, at infinite dilution. Values for the limiting ionic conductances for selected ions in water at 25°C . The ionic conductance is a definite constant for each ion in a given solvent, its value depending only on the temperature. Since these are actually equivalent conductances, symbols such as ½ Ba^{2+} are sometimes employed. At finite concentrations interionic forces generally lower the ionic mobilities.

Electrolytic conductance measurements usually involve determination of the resistance of a segment of solution between two parallel electrodes by means of Ohm's Law. These electrodes are of platinum metal that has been coated with a deposit of platinum black to increase the surface area and reduce the polarisation resistance. A typical commercial conductivity bridge is designed to measure electrolytical conductance in micro mhos and resistance in ohms. A built-in-generator provides bridges current at frequencies of 100 and 1,000 Hz.

Generally, the lower frequency should be used when the measured resistance is high, and the higher frequency when the measured resistance is low. The instrument is balanced with the aid of a phase-sensitive detector and a null meter. Usually oscilloscopes are used for balance indication. The null condition is indicated as a straight horizontal line on the oscilloscope, whereas resistive and reactive imbalance are shown, respectively, by tilting of the line and by widening to an ellipse.

Although the electrical conductance of a solution is a general property and is not specific for any particular ion, a number of analyses can be made by means of a measurement of conductivity. In general, the success of a measurement depends upon relating the property of the sample that is to be estimated to the conductance of some highly conducting ion.

5. SPECTROPHOTOMETRY: VISIBLE AND ULTRAVIOLET (UV)

Principle

The absorption spectrum of a compound may be shown as a plot of the light absorbed (extinction) by that compound against wavelength. The absorption spectra in the ultraviolet (200 to 400 nm) and visible regions (400 to 700 nm) are due to energy transitions of both bonding and non-bonding outer electrons of the molecule. To obtain an absorption spectrum, the extinction of a substance must be measured as a series of wave-lengths. Absorption in the visible and ultraviolet regions may be detected by eye or by a photographic emulsion. However, the technique of spectrophotometry utilises the fact that the potential developed in a photoelectric cell is proportional to the intensity of radiation impinging upon that photocell. The main components of a simple spectrophotometer are illustrated as:

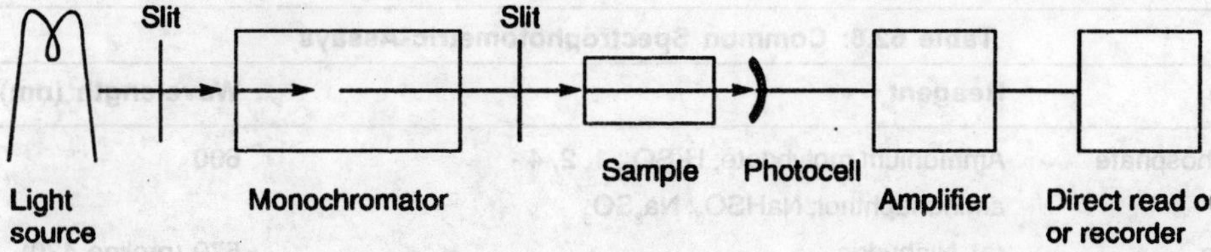

Fig. 62.6: The main components of a Simple Spectrophotometer

Applications

Many substances which do not possess significant extinction coefficients in the visible region will react quantitatively with some other reagent to give a coloured product. This property is used as a basis for assaying such substances. The colour (*chromophore*) is produced under standard conditions from known quantities of the substance, and the extinctions of these samples are measured, using a reagent blank in the reference cuvette which contains all of the reagents except the substance being estimated. The extinction reading is set to zero using this blank by amplification of the photocell current or potential. The measured extinctions are plotted against the quantity or concentration of the test substance producing the colour. This graph is known as the *calibration curve*. Unknown quantities of the substance may then be assayed by producing the colour under the same standard conditions, measuring the extinctions and using the calibration curve to determine the quantity of the substance which produces that extinction. Since it is often possible to produce comparatively high extinctions with relatively small amounts of material, spectrophotometry is widely used in biochemistry to assay a wide range of biologically important molecules. Table 62.6 summarises some of the most commonly used assays.

Some important points to bear in mind when carrying out colorimetry are:

(i) unlike straight-forward spectrophotometry, colorimetry is a destructive technique, i.e., that portion of the sample analysed cannot be recovered;

(ii) a chromophore exhibits the complementary colour to that which it absorbs, i.e., a yellow compound appears yellow because it absorbs blue light and, therefore, it must be estimated in the blue region of the spectrum;

(iii) colorimetric assays are usually most sensitive at the extinction peak of the chromophore produced. If this is not known the spectrum should be determined before commencing the assay;

(iv) the reference cunvette should contain all the reagents other than the substance being assayed in the same concentrations as in the test cunvettes;

(v) the reference cunvette and its contents require, if anything, more care because any error in these will be reflected in all of the values obtained;

(vi) assays should normally be performed in duplicate and individual values, not means, should be plotted. This procedure allows the experimenter to omit erroneous extinction values justifiably from the calibration curve if the duplicate value lies on the curve produced from the other extinction values;

(vii) the best line through the points should be drawn, not necessarily the best line through the origin and the other points. This is because it is just as feasible to obtain an erroneous value for reagent blanks as for any other sample;

Table 62.6: Common Spectrophotometric Assays

Substance	Reagent	Wavelength (nm)
Inorganic phosphate	Ammonium molybdate; H_2SO_4; 1, 2, 4 - aminonaphthol; $NaHSO_3$; Na_2SO_3	600
Amino acids	(a) Ninhydrin	570 (proline 420)
	(b) Cupric salts	620
Peptide bonds	Biuret (alkaline tartrate buffer, cupric salt)	540
Phenols, tyrosine	Folin (phosphomolybdate, phosphotungstate, cupric salt)	660 or 750 (750, more sensitive)
Protein	(a) Folin	660
	(b) Biuret	540
	(c) BCA reagent (Bicinchoninic Acid)	562
	(d) Coomassie Blue	595
Carbohydrate	(a) Phenol, H_2SO_4	Varies e.g., glucose 490, xylose 480
	(b) Anthrone (anthrone, H_2SO_4)	620 or 625
Reducing sugars	Dinitrosalicylate, alkaline tartrate buffer	540
Pentoses	(a) Bial (orcinol, ethanol, $FeCl_3$, HCl)	665
	(b) Cysteine, H_2SO_4	380-415
Hexoses	(a) Carbazole, ethanol, H_2SO_4	540 or 440
	(b) Cysteine, H_2SO_4	380-415
	(c) Arsenomolybdate	Usually 500-570
Glucose	Glucose oxidase, peroxidase, o-dianisidine, phosphate buffer	420
Ketohexose	(a) Resorcinol, thiourea, ethanoic acid, HCl	520
		560
	(b) Carbazole, ethanol, cysteine, H_2SO_4	
	(c) Diphenylamine, ethanol, ethanoic acid, HCl	635
Hexosamines	Ehrlich (dimethylaminobenzaldehyde, ethanol, HCl)	530
DNA	Diphenylamine	
RNA	Bial (orcinol, ethanol, $FeCl_3$, HCl)	665
α-oxo acids	Dinitrophenylhydrazine, Na_2CO_3, ethyl acetate	435
Sterols	Liebermann-Burchardt reagent (acetic anhydride, H_2SO_4, chloroform)	625
Steroid hormones	Liebermann-Burchardt reagent	425
Cholesterol	Cholesterol oxidase, peroxidase, 4-amino-antipyrine, phenol	500

(viii) the calibration curve may vary between batches of both reagent and standard and hence a new calibration curve should be constructed each time the assay is used, unless there is evidence to suggest that the previously obtained curve is still valid. Hence all calibration curves cannot be used as standard curves;

(ix) calibration curves should never be extrapolated beyond the highest extinction value measured. Thus, it may sometimes be expedient to prepare a curve with extinction values up to 1.0. However, it is always more accurate to repeat the assay at a concentration which falls within the most accurate region of the calibration curve. This is usually from 0.0 to 0.6 because of the logarithmic nature of the extinction scale;

(x) if the extinction value of a sample lies outside the range of the calibration curve, the sample should be reassayed at a concentration that falls within this range. Alternatively, if the system obeys the Beer-Lambert Law, it may be possible to use curvettes of shorter optical light path;

(xi) in colorimetric assays, the reproducibility of the methods is far more important than the Beer-Lambert relationship. As with all analytical procedures, it is the reproducibility of the assay in the experimentalist's hands that determines the accuracy of the final data, not what the schedule or reference states.

Qualitative Analysis: Visible and ultraviolet spectra may be used to identify classes of compounds in both the pure state and in biological preparations, e.g., proteins, nucleic acids, cytochromes, chlorophylls. This technique may also be used to indicate chemical structures and intermediates occurring in a system.

Quantitative Spectrophotometric Analysis: A number of important classes of biological compounds may be measured semi-quantitatively using ultraviolet and visible spectrophotometers, e.g., proteins at 280 nm and nucleic acids at 260 nm. Corrections may be made for the presence of impurities by also measuring the extinction at wavelengths where the impurities absorb more than the compound under investigation and applying the appropriate algebraic formula, e.g., the Morton and Stubbs correction for the amount of vitamin A in saponified extracts of natural oils, and the $E_{280/260}$ ratio for protein estimation in the presence of nucleic acids. The amount of two components with overlapping spectra—such as chlorophylls a and b in ether—may be estimated if their extinction coefficients at two wavelengths are known. For n components, extinction data at n wavelengths are required. When measuring the concentration of large DNA molecules at 260 nm the interference of *Rayleigh light scattering* may be corrected by extrapolation from the extinction values obtained at non-absorbent regions of the DNA spectrum (e.g. 330 to 430 nm).

Enzyme Assays and Kinetic Analyses: Ultraviolet and visible spectrophotometry provide the most commonly used assays of enzymes and their substrates.

Turbidometry and Nephelometry: Very dilute suspensions may be assayed by *turbidometry*, that is by measuring their apparent extinction at a wavelength which is not absorbed. Fortunately, scattering of radiation varies much less with wavelength than does absorption. Relationships are, however, non-linear and standardisation of this technique is difficult since the particle size is critical, but it may be used to obtain an estimate of the concentration of bacterial cells at 600 nm and of proteins in solution.

In contrast, *nephelometry* measures the intensity of radiation *scattered* by a suspension and is commonly used for estimating the concentration of microorganisms (Note: strictly speaking, nephelometry is *not*, therefore, a spectrophotometric technique).

6. SPECTROFLUORIMETRY

Principle

Fluroresence is the phenomenon whereby a molecule, after absorbing radiation, emits radiation of a longer wavelength. Thus, a compound may absorb radiation in the ultraviolet region and emit visible light. This increase in wavelength is known as the *Stokes' shift*.

At room temperature most organic molecules are in the ground state (S_0V_0). Absorption of photons elevates an electron in these molecules to a higher energy state (S_1, S_2, etc.) in less than 10^{-15} seconds. After absorption, energy is lost very rapidly by collision degradation (as heat), resulting in the energy of the excited molecules falling rapidly to that of minimal vibrational energy in the lowest excited state (S_1V_0). The energy emitted from these molecules in regaining the ground state within a period of less than 10^{-8} seconds gives rise to a fluorescent peak, showing the Stokes' shift. Although many organic molecules absorb in the ultraviolet and visible regions, only a few fluoresce. Whereas knowledge of the structure of an organic molecule permits predictions about its absorption spectrum, structural information cannot be used to predict that a compound will fluoresce. However, whereas flexible molecules such as aliphatic compounds tend to photodissociate rather than fluoresce, aromatic molecules containing delocalised π-electrons sometimes flurorescе. Fortunately for the biochemist a relatively high proportion of biological molecules do fluoresce.

Since the emitted radiation can have a number of closely related values depending on the final vibrational and rotational energy levels attained, fluorescence spectra are band spectra. They are usually independent of the wavelength of the radiation absorbed and have a mirror image relationship to the absorption peak with the greatest wavelength.

Obviously, fluorescence spectra can supply information only about events which take less than 10^{-8} seconds to occur. The *quantum efficiency, Q*, which is equal to the number of quanta fluoresced divided by the number of quanta absorbed, is usually independent of the wavelength of the activating light. At low concentrations, the intensity of fluroescence (I_f) is related to the incident radiation (I_0) by the following simple relationship:

$$I_f = I_0 2.3 \epsilon_\lambda \text{ cdQ}, \quad \text{i.e., } I_f \propto c$$

where, c = concentration of the fluorescing solution (molar)

d = light path in the fluorescing solution (cm)

ϵ_λ = molar extinction coefficient for the absorbing material at wavelength λ (in units of $dm^3\ mol^{-1}\ cm^{-1}$)

Simple electronics are adequate in fluorimeters due to the direct relationship between concentration and fluorescence intensity. Spectrofluorimetry is most accurate at very low concentrations when absorption spectrophotometry is least accurate, for example 100 pg of catecholamines, or NADH, may be detected by spectrofluorimetry, as compared to the 100 µg of the catecholamines, serotonin and adrenaline detectable by absorption spectrophotometry. The sensitivity of fluorimeters is usually easily adjusted over a large range by amplification of the current produced in the photocell circuit. Spectrofluorimeters enable the utilisation of great spectral selectively since, due to the Stokes' shift, two monchromators may be used, one selecting the activating wavelength and the other the fluorescent wavelength. No reference curvette is required, but a calibration curve must be constructed.

Disadvantages of spectrofluorimetry include its susceptibility to pH, temperature and solvent polarity, and the virtual impossibility of predicting whether a compound will fluoresce. The major problem with fluorimetry is *quenching*, whereby the energy, which could be emitted as fluorescence, is transferred to adjacent molecules. It should also be realised that detergents, stopcock grease, filter paper and some laboratory tissues may cause interference in fluorimetric assays by the release of strongly fluorescing materials. The main components of a complete spectrofluorimeter are indicated in Fig. 62.7, and include: a continuous spectrum source, for example a mercury lamp or Xenon arc; a monochromator (M_1) for irradiating the specimen with any chosen wavelength; a second monochromator (M_2) which, under conditions of constant irradiating wavelengths, enables the determination of the fluorescent spectrum of the specimen; and a detector which is usually a sensitive photocell, for example a red sensitive photomultiplier for wavelengths greater than 500 nm. Since the intensity of fluorescence may decrease by 10 to 50% if the sample temperature falls from 30°C to 20°C, adequate temperature control is essential for accurate fluorimetric estimations.

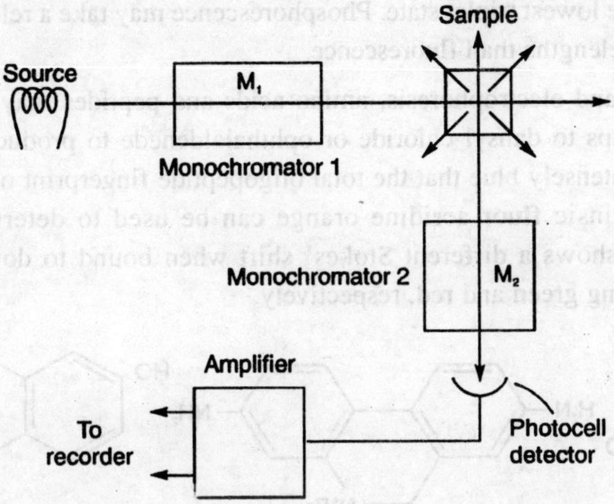

Fig. 62.7: The basic component of a spectrofluorimeter, set up for 90º illumination

Pre-filter absorption reduces the amount of radiation reaching fluorescent molecules farthest from the light source, and post-filter effects reduce the amount of fluoresced radiation escaping from the curvette (Fig. 62.8). Use of microcells reduces both pre-filter and post-filter effects which occur with concentrated solutions.

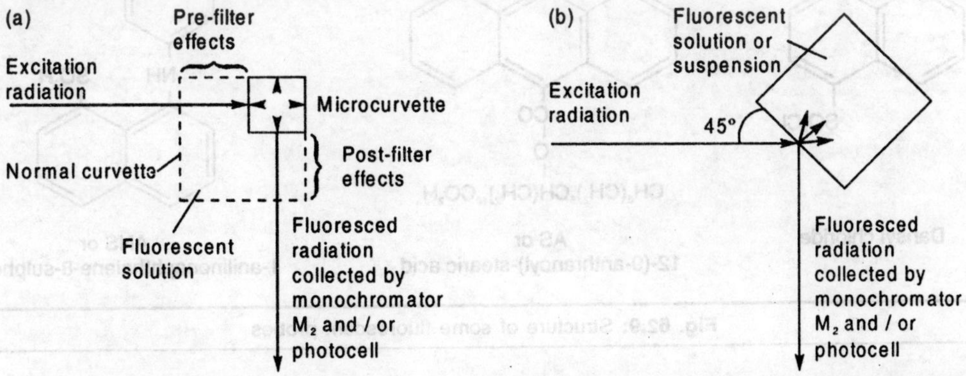

Fig. 62.8: Reduction of filter effects: **(a)** using microcurvette; and **(b)** using front face illumination

Instead of 90° illumination as illustrated in Fig. 62.8, Front Face Illumination (FFI), which is essential for suspensions, may be used to prevent filter effects. FFI only requires curvettes with one optical surface, so normal spectrophotometer curvettes may be used. The geometry of the system is arranged so that fluorescence may be measured through the curvette face that allowed entry of the excitation radiation (Fig. 62.8). FFI is usually less sensitive than 90° illumination.

Applications

Qualitative Analysis: The determination and comparison of both excitation and fluorescence spectra of a compound may help to identify it. The effect of pH and solvent composition on the fluorescence of a compound, and the polarisation of its fluorescence, may contribute information about its structure. The measurement of the *phosphorescence* and *phosphorescence lifetime* of a compound may contribute information to its identification. Phosphorescence is a light emission phenomenon which involves intersystem crossing to the lowest triplet state. Phosphorescence may take a relatively long time to decay and occurs at longer wavelengths than fluorescence.

In chromatography and electrophoresis, amino acids and peptides may be detected by coupling their primary amino groups to dansyl chloride or ophthalaldehede to produce fluorescene. The latter conjugates fluoresce so intensely blue that the total oligopeptide fingerprint of only 10^{-5} g protein may be determined. The extrinsic fluor acridine orange can be used to determine the strandeness of polynucleotides since it shows a different Stokes' shift when bound to double and single stranded polynucleotides, fluorescing green and red, respectively.

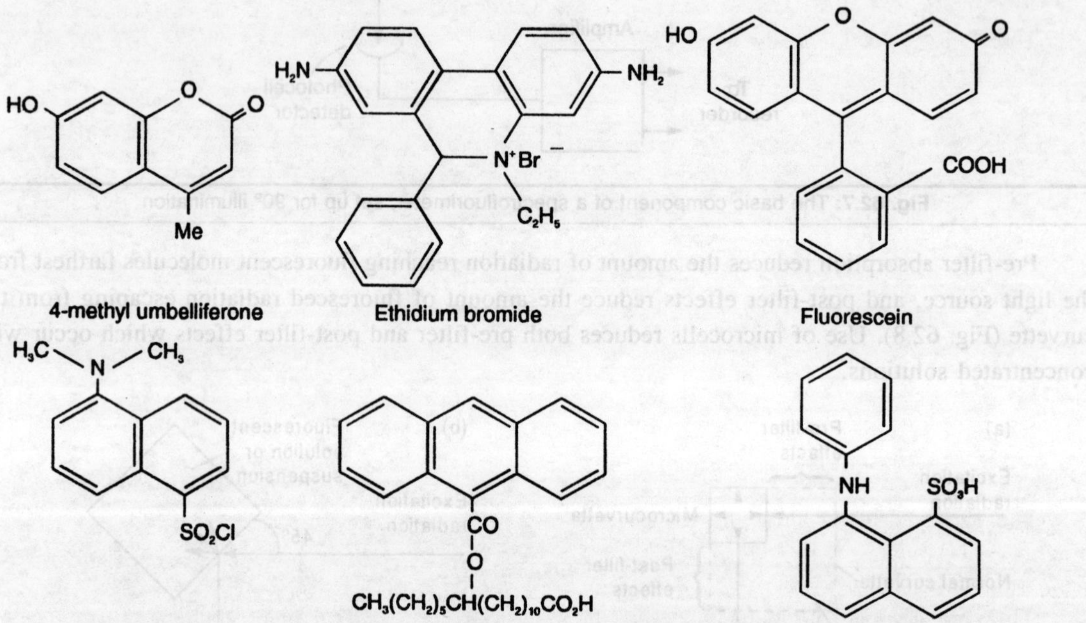

Fig. 62.9: Structure of some fluorescent probes

Quantitative Analysis: This is really the major use of fluorimetry in biochemistry, because of the accuracy, reproducibility and sensitivity of the technique, at concentrations too low for absorption spectral analysis. Typical applications include the assay of vitamin B_1 in foodstuff, NADH, hormones such as cortisol and oestradiol, drugs such as lysergic acid and barbiturates in blood, organophosphorous pesticides and carcinogens. Chlorophyll, cholesterol, porphyrins and some metal ions may also be assayed spectrofluorimetrically. The use of internal standards, i.e., measuring the fluorescence of an unknown quantity of pure compound before and after addition of a known quantity of standard, permits both self and contaminant quenching to be allowed for. Ethidium bromide binds tightly to DNA resulting in a large increase in fluorescence and allows the amount of DNA to be estimated. This is particularly useful in conjunction with gel electrophoresis.

Instead of studying a molecule by measuring its intrinsic fluorescence it is often possible to couple or bind the molecule to a fluorescent probe or fluor, and to measure the so-called *extrinsic fluorescence* of the conjugate produced (similar to the use of reporter groups in absorption spectrophotometry). Ideally, the fluor should be tightly bound at a specific site in the molecule and its fluorescence should be sensitive to environment conditions. Obviously, the fluor should not affect the system being studied. The structures of some fluorescent probes are illustrated in Fig. 62.9.

Quin-2, which preferentially binds Ca^{2+}, may be used to measure unbound cytoplasmic calcium because the fluorescence of quin-2 increases about fivefold when it binds Ca^{2+}. Cells are loaded with the membrane-permeable quin-2 acetoxymethyl ester and cytoplasmic esterates hydrolyse the ester and leave the membrane-impermeant quin-2 anion trapped in the cytoplasm. Calibration curves for this fluorescence assay are linear over the range 10^{-4} to 10^{-2} M Ca^{2+}. Apparently the recently developed probes fura-2 and indo-1 are thirty times more sensitive than quin-2 in the fluorescent assay of Ca^{2+}.

The fluorescent probe quin-1 can be similarly used to monitor intracellular pH changes since it shows a thirtyfold increase in fluorescence as the pH changes from 5 to 9.

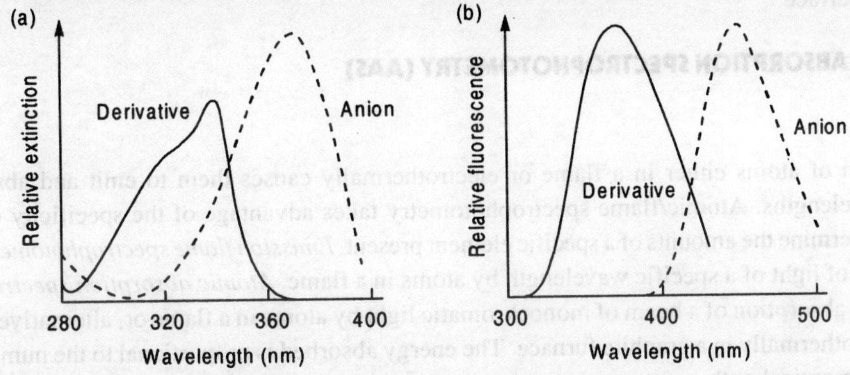

Fig. 62.10: Spectra of the methylumbelliferone anion and derivatives of 4-methylumbelliferone at pH 10. **(a)** Absorption spectra; **(b)** Fluroescence spectra

Enzyme Assays and Kinetic Analysis: Group specific hydrolases may be readily assayed by measuring the rate of appearance of fluorescence at 450 nm of the anion of 4-methylumbelliferone when the enzyme acts upon an ether or ester derivative of 4-methylumbelliferone. The absorption and fluorescence spectra of these compounds are indicated in Figs. 62.9 and 62.10, respectively, which indicates that if the assay system is irradiated with 350 to 400 nm light virtually all of the fluorescence measured around 450 to 500 nm is due to the anion product of the enzyme. The fluorimetric assay of ß-galactosidase using fluorescein di-(ß-D-galactopyranoside) as substrate is so sensitive that the presence of a single enzyme molecule can be detected. This allows the number of molecules of this enzyme in a single bacterial cell to be determined and, hence, the induction of synthesis of the enzyme in individual cells of a bacterial population.

Since NADH and NADPH fluoresce, spectrofluorimetry can be applied to numerous metabolic reactions that are or can be coupled to the oxidation of NADH or NADPH, or the reduction of NAD^+ or $NADP^+$.

Owing to the sensitivity of the technique, an enzyme can often be studied under conditions similar to those found *in vivo*, for example low concentrations of substrate, enzyme and cofactors. Sometimes exogenous cofactors need not be added. Instead, the kinetics of oxidation and reduction of the endogenous material in intact organelles or cells may be measured, for example mitochondrial NADH.

Studies of Protein Structure: Proteins that contain fluorescent chromophores such as tryptophan and FAD fluoresce. The binding of substances such as inhibitors, coenzymes, and allosteric effectors close to such intrinsic fluors can be measured by the changes in their fluorescence spectra. This, in turn, may give information about the conformation, denaturation and aggregation of the proteins. These studies may be extended to proteins lacking suitable intrinsic fluors by coupling them to extrinsic fluors such as anilinonaphthalene-8-sulphonate (ANS), dansyl chloride and derivatives of fluorescein or rhodamine (Fig. 65.9). Examples include the study of conformational changes in enzymes on binding their substrates and determination of bound fatty acids in so-called purified bovine serum albumin.

Fluorescent Probes and Studies on Membrane Structure: Since the fluorescent properties of a molecule vary with its mobility and the polarity of its environment, such properties may be monitored in the vicinity of a fluorescent probe by measuring changes in its fluorescence. Various fluorescent probes have been designed to sample specific locations in both artificial and natural membranes. ANS and N-methyl-2-anilino-6-naphthalene sulphonate (MNS) contain both charged and hydrophobic areas and locate at the water-lipid interface. Studies using such compounds with membranes yield information about the interface.

7. ATOMIC ABSORPTION SPECTROPHOTOMETRY (AAS)

Principle

Volatilisation of atoms either in a flame or electrothermally causes them to emit and absorb light of specific wavelengths. Atomic/flame spectrophotometry takes advantage of the specificity of these line spectra to determine the amounts of a specific element present. *Emission flame spectrophotometry* measures the emission of light of a specific wavelength by atoms in a flame. *Atomic absorption spectrophotometry* measures the absorption of a beam of monochromatic light by atoms in a flame or, alternatively, by atoms heated electrothermally in a graphite furnace. The energy absorbed is proportional to the number of atoms present in the optical path.

The most readily emitted and absorbed wavelengths in the spectrum of an element are those in which the energy change is minimal, such as the orange radiation of the D line transition of the sodium spectrum, involving the small 3s-3p transition. Since the transitions available to the electrons in any atom are specified by the available energy levels and the occupied energy levels, atomic spectra are absolutely characteristic of the element involved. Theoretically, for simple atoms, one could deduce their electronic structure from their line spectra. The wavelengths of radiation emitted on volatilisation of elements in a flame may be readily resolved into simple line spectra in a spectroscope, spectrograph or direct reading spectrophotometer (which use as detectors the human eye, photographic emulsion and photoelectric cells, respectively).

The amount of radiation emitted is proportional to the number of excited atoms present, which depends on the temperature and composition of the flame. Standard solutions must always be used to calibrate the system. Flame composition is very important since sodium gives a very high background emission. Sodium is usually measured first, in order that a similar amount may be added to all the standard solutions. Chloride usually gives rise to the most volatile salts and, hence, the solutions studied should contain excess hydrochloric acid. Alkali metals cause an enhancement of the emission of other atoms whereas phosphate, silicate, and aluminate give rise to nondissociable salts and, hence, suppress the emission of calcium and magnesium. Fortunately, this may be relieved by the addition of lanthanum or strontium salts.

Cyclic analysis may be used to allow for interference by other ions in a complex mixture. First each (interfering) component in the sample is estimated. Then the standards for all the components to be assayed are prepared such that they contain the estimated concentration of interfering components. By repeating all of the estimations a number of times and modifying the overall composition of the standards each time—such that composition of all other components in the standard solution approaches that in the unknown—constant and accurate values for each component may be obtained. Usually only one or two such cycles of determining and making up of standards are required. Because of flame instability, assays are usually made in triplicate. Calibration curves should be checked or reconstructed when the assays are carried out.

A more accurate procedure is that of *bracketing* in which a standard solution, containing almost the same concentration of the element as the sample solution, is assayed before and after the sample solution. Whenever possible, internal standards should be used. Lithium is often used as a general internal standard. Since even well-cleaned, best quality, glassware absorbs and releases metal ions, samples and standards are best stored in polythene bottles whenever possible.

When assaying metal in biological samples it is usual to remove organic molecules by ashing. Care must be taken to ensure that the more volatile elements are not inadvertently sublimed off. This may be achieved by ashing at relatively low temperatures in a steam of oxygen or by liquid ashing, i.e., by oxidative digestion of the sample in hydrogen peroxide/concentrated sulphuric acid solution. Selenium sulphate is added as a catalyst and lithium sulphate to raise the boiling point.

(a) Atomic emission Spectrophotometry

The basic components of an atomic (flame) emission spectrophotometer are indicated in Fig. 62.11. *Nebulisers Catomisers* are usually of the scent spray type, in which a forced stream of air passes over a capillary tube dipping into the test solution. Large and small drops of solution are produced which, in direct injection systems, pass with the air stream into the burner. Unfortunately, large drops may not remain in the hottest part of the flame long enough for their constituents to be volatilised and excited.

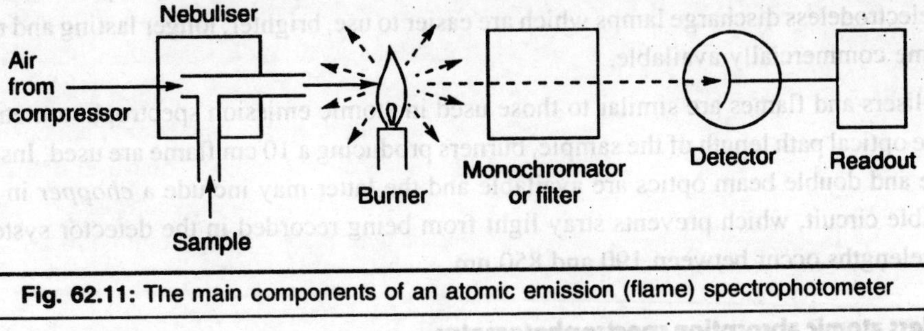

Fig. 62.11: The main components of an atomic emission (flame) spectrophotometer

This may be avoided by passing the drops through a cloud chamber which removes large drops from the air stream by gravitation. Sodium and potassium can be assayed at 1,500°C which can be attained by combustion of a mixture of air and natural gas. Calcium can be assayed at this temperature but 2,000-2,500°C is preferable. Magnesium and iron require a mixture of air and acetylene to reach the necessary

Table 62.7: The detection limits for various elements in emission and absorption flame spectrophotometry, flameless absorption spectrophotometry, and ion-selective electrodes

	Emission		Absorption			Ion-selective electrode
Element	Detection limit (parts per million)	Wavelength (nm)	Detection limit (parts per million) (flame)	(flameless)	Wavelength (nm)	Detection limit (parts per million)
Calcium	0.005	442.7	0.1	0.00007	442.7	0.02
Copper	0.1	324.8	0.1	0.0001	324.8	0.0006
Iron	0.5	372.0	0.2	0.0001	248.3	
Lead			0.5	0.0002	283.3	0.21
Lithium	0.001	670.7	0.03	0.0001	670.7	
Magnesium	0.1	285.2	0.01	0.00001	285.2	
Manganese	0.02	403.3	0.05	0.00004	279.5	
Mercury			10.0	0.018	253.8	
Potassium	0.001	766.5	0.03	0.00003	766.5	0.04
Sodium	0.0001	589.0	0.03	0.00001	589.0	0.02
Strontium	0.01	460.7	0.06	0.0001	460.9	

2,500°C. For simple routine analysis of sodium, potassium and calcium, the monochromator may be replaced by a simple filter, but a monochromator is required for more precise work. The most accurate instruments have a resolution of 0.1 to 0.2 nm over the range 200 to 1,000 nm. The wavelengths used in the analysis of a number of metals are listed in Table 62.7 along with their detection limits. Photocells are usually used, but, unfortunately, the instability of the emission of most flames decreases the potential accuracy of the photocell. Multi-channel polychromators are used in some routine procedures to measure the emission from up to six elements at one time.

(b) Atomic absorption spectrophotometry

In order to produce a beam of radiation with a very narrow band width, either a source of white light plus a double monochromator, or *a hollow cathode discharge lamp,* is used. The discharge lamps are specific to the element being assayed. A sample of the element to be excited is held in a metal cup, a tungsten anode carrier gas at a low pressure and a high voltage are used to produce an arc spectrum of the element. Recently, electrodeless discharge lamps which are easier to use, brighter, longer lasting and more sensitive have become commercially available.

Nebulisers and flames are similar to those used in atomic emission spectrophotometry. In order to increase the optical path length of the sample, burners producing a 10 cm flame are used. Instruments with both single and double beam optics are available and the latter may include a *chopper* in the light path and a suitable circuit, which prevents stray light from being recorded in the detector system. The most useful wavelengths occur between 190 and 850 nm.

(c) Flameless atomic absorption spectrophotometry

1 to 100 mm^3 of sample or standard analyte is deposited in a graphite tube in the presence of an inert gas and the temperature raised electrothermally to 3,000°C—either continuously or in discrete steps. Monochromatic light specific to the element being determined is produced by either a hollow cathode

discharge lamp or an electrodeless discharge lamp and shone through the graphite tube. The absorption of this light is measured continuously as the temperature is raised. If the system is controlled by a programmable microcomputer the results may be obtained on a visual display unit as the superimposition of the absorption and temperature profiles of the graphite tube contents as a function of time. This allows the optimum conditions for the analysis to be found and used in future analytical programmes.

The flameless technique is often more than 100 times as sensitive as flame atomic absorption spectrophotometry. Because the flameless system does not use combustible gases it may be safely left to run unattended which allows the relative slowness of such a system compared to flame spectrophotometry to be offset by automation.

(d) Atomic fluorescence spectrophotometry

This technique is based on an emission phenomenon which relies upon excitation of atoms by radiation instead of by thermal excitation. It is analogous to molecular fluorescence, but occurs with atoms in the vaporised state, not molecules in solution. An intense beam of light is required, but it needs to be less spectrally pure than that required for atomic absorption spectrophotometry, since only the resonant wavelengths will be absorbed and, hence, may lead to fluorescence. Modulation of the detector amplifier to the same frequency as the primary source prevents the direct emission from the flame being recorded.

Although currently limited to a few metals, this technique offers great refinement in the limits of their detection; for example, zinc and cadmium may be detected at levels as low as 1 and 2 parts 10^{-10}, respectively.

Applications

Emission flame *filter photometers* are used routinely to assay sodium, potassium and calcium. Fullscale deflection can be readily obtained for sodium and potassium, using standard solutions containing 1 and 3 parts per million, respectively. Emission flame spectrophotometers are used to assay more than twenty elements in biological samples, particularly calcium, magnesium, and manganese. Absorption flame spectrophotometers are usually more sensitive than atomic emission spectrophotometers, although not for the alkali metals, and can detect less than 1 part per million of more than twenty elements. The reproducibility of the technique is about 1% and the best working range is 20 to 200 times the detection limit (Table 62.7). Flame photometry is used most extensively in clinical biochemistry laboratories to assay the composition of body fluids such as blood, urine, saliva, milk and cerebrospinal fluid. Departure of the composition of such fluids from the range of normal values can often be used in the diagnosis of particular clinical states, and the monitoring of such information can indicate the success or otherwise of a course of treatment. Similar monitoring of the body fluids of animals is used in physiological and pharmacological research. Sodium, potassium, calcium, magnesium, cadmium and zinc levels are measured directly, but copper, iron, lead and mercury must first be extracted from biological fluids. These techniques are also used extensively in soil and plant analyses after suitable extraction of metal salts from the soil or plant samples. After suitable ashing procedures have been carried out, these techniques may also be used to assay the metal composition of macromolecules, organelles, cells and tissues.

8. NUCLEAR MAGNETIC RESONANCE (NMR) SPECTROMETRY

Principle

This is a technique for detecting atoms which have nuclei that possess a magnetic moment. These are usually atoms containing an odd number of protons in their nuclei. In the same way that pairs of electrons in the same atomic orbital have opposite spin and no resultant magnetic moment, pairs of protons in a nucleus do not have a magnetic moment. However, an odd proton in a nucleus imparts a magnetic moment to the molecule which can interact with an applied magnetic field. This interaction is the basis of nuclear magnetic resonance spectrometry. An unpaired proton in such a magnetic field can exist in one of two states: either in a low energy state with the nuclear spin aligned parallel with the field, or in a higher energy state antiparallel to the field. In order to change from the low energy state to the high energy state, such nuclei must absorb the appropriate quantum of energy. In a magnetic field of several hundred millitesla (several thousand gauss) such nuclei absorb radiation in the radiowave region of the electromagnetic spectrum, giving rise to the phenomenon known as *Nuclear Magnetic Resonance* (NMR). Most studies are conducted using the lightest isotope of hydrogen, ^{1}H (hence the term *proton magnetic resonance*, PMR), but ^{13}C, ^{15}N, ^{19}F and ^{31}P isotopes are also used in biochemical studies. These nuclei possess a nuclear spin value of 1/2. Other nuclei common in biological molecules—such as ^{12}C, ^{16}O and ^{32}S—possess zero spin and, hence, lack a magnetic moment and do not give NMR signals.

The frequency of the radiowaves absorbed during NMR depends on both the isotopes being studied and the magnitude of the applied magnetic field. Usually the absorption of a monochromatic radiowave frequency by the sample is measured as the applied magnetic field is varied, rather than using a constant magnetic field and variable radiation frequency. Thus, NMR spectra are usually plotted as energy absorbed against the magnetic field strength applied (not wavelength or frequency). The range of magnetic field scanned is very small compared to the total magnetic field applied. Hence, the total field applied and the frequency of the radiowaves absorbed are stated on such spectra. A radiowave frequency of 40 MHz is commonly used to achieve resonance of the ^{1}H nucleus.

The actual field experienced by a proton in a particular compound depends upon its molecular environment since the applied field induces secondary fields (15 to 20×10^{-4} tesla) by interacting with the bonding electrons adjacent to the proton. If this induced field opposes the applied field, a slightly higher applied field must be used to make the nucleus resonate. Such nuclei are said to be *shielded*. The magnitude of the shielding decreases with increasing electron-withdrawing power of nearby substituents. If the induced field augments the applied field, the nucleus is said to be *deshielded* and a smaller applied field is required to achieve resonance. Such displacements in the spectral lines are known as *chemical shifts*.

The scale usually used for measuring such shifts records them in parts per million of the applied magnetic field with reference to the absorption peak of a reference compound. Usually tetramethylsilane, $Si(CH_3)_4$, is used as internal reference in organic solvents, and sodium trimethylsilylpropanesulphonate, $(CH_3)_3SiCH_2CH_2SO_3^{-}Na^{+}$, in aqueous system.

Most modern NMR spectrometer scales are calibrated in tau (τ) units. On this scale the $Si(CH_3)_4$ peak occurs at τ 10, since most proton shifts occur between τ 0 and τ 10. As a result of chemical shifts the presence of protons in functional groups such as ^{1}H-O-, ^{1}H-CH, ^{1}H-CH_2 may be detected by the occurrence of particular peaks in the NMR spectrum. In addition, the relative number of ^{1}H atoms in each type of chemical group within the sample is indicated by the intensity of the appropriate absorption peaks, i.e., by the areas under those peaks.

Nuclear magnetic resonance spectra are further complicated by interaction between like or different spins through the bonding electrons. This *spin-spin coupling* may extend to nuclei which are 4 to 5 bonds apart. It is manifested as a splitting of the absorption peaks already separated by the chemical shift (*hyperfine splitting*).

The basic components of an NMR spectrometer are shown in Fig. 62.12. Electromagnets weighing 10^3 to 10^4 kg, and stable to better than ± 0.1%, are used to produce fields of 1 to 10 tesla. An auxiliary sweep generator is used to vary the magnetic field over about 10^{-2} tesla. A radio frequency transmitter in place of a Klystron source produces the monochromatic radiation used to irradiate the sample. For PMR the sample must be dissolved to a relatively high concentration in a solvent which lacks protons, such as D_2O or $CDCl_3$. To minimise variations in the magnetic field the sample is contained in a tube of high precision diameter and is usually rotated at high speed by an air turbine. The absorption signal, detected by a radio receiver, is amplified and recorded.

Computer enhancement (Computer Averaging Techniques, CAT), in which the read-out from 10^1 to 10^3 spectra of the same sample are superimposed, minimises background noise and is very useful for studying weakly absorbing biological samples.

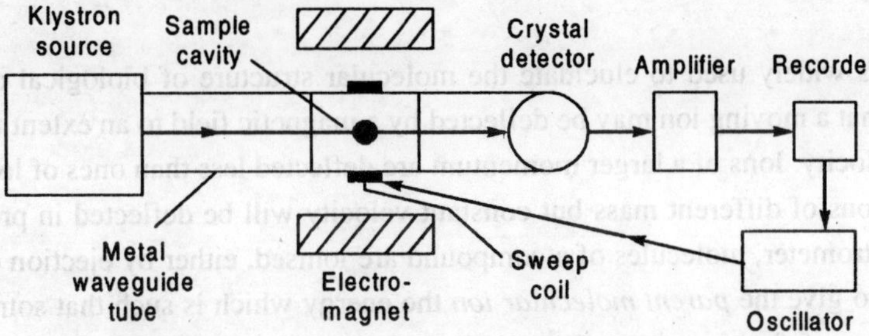

Fig. 62.12: The main components of NMR-Spectrometer

Applications

The main use of NMR is for the study of the molecular structure of relatively simple organic molecules. Thus, structural information relevant to the biological action of the antibiotics, gramicidin and valinomycin has been obtained from NMR studies. In addition, the effects of alamethicin and cholesterol on the mobility of lecithin, in both artificial and erythrocyte membranes, has been studied using NMR as has the relative motion of different parts of the fatty acid side chains of lecithin in lipid bilayers using lecithins substituted in different positions with ^{19}F and ^{13}C.

The use of NMR to study biological macromolecules is basically very difficult. Molecules such as proteins may contain several thousand protons, all of which resonate in a similar region. Only with the introduction of high resolution instruments and the development of computer techniques has the potential of NMR begun to be realised. Even so, there appears to be an upper molecular weight limit of 20,000 daltons for obtaining complete hyperfine structural information. Thus, in a study of the catalytic mechanism of ribonuclease by NMR, it was possible to identify proton resonance from the four histidine residues present in the molecule since their NMR signals were shifted from those of the main bulk of protons. It was possible to identify which two of the four were involved in the catalytic site. In fact, this information was already known from chemical and X-ray diffraction studies, but it demonstrates the possibilities. The main advantage of NMR over X-ray diffraction is that it is capable of monitoring conformational changes of 100 pm. Thus, it is invaluable in the study of small conformational changes accompanying membrane dynamics and the binding of enzymes to substrates, drugs to receptors, and antigens to antibodies.

NMR has also been particularly useful in the study of phosphate metabolism. The sensitivity of ^{31}P NMR is such that the concentration of inorganic phosphate, AMP, ADP, ATP, phosphocreatine and metabolic sugar phosphates can be measured and, hence, their metabolism studied in living cells and tissues. Intracellular and extracellular inorganic phosphate concentrations may be determined because the chemical shift of inorganic phosphate varies with pH. Development of NMR focusing techniques and larger, scanning NMR machines allow the study of whole animals including man. Studies on the effects of exercise on normal, athletic and unhealthy individuals have shown that some healthy people can function even when their ATP level has been depleted by 60% and that others can sustain a muscle pH decrease from the normal range of 7.0 - 7.4 to 6.0. In particular, detailed studies are revealing patients with previously unknown disorders such as mitochondrial myopathies.

9. MASS SPECTROMETRY

Principle

Mass Spectrometry is widely used to elucidate the molecular structure of biological compounds. It is based upon the fact that a moving ion may be deflected by a magnetic field to an extent that is dependent upon its mass and velocity. Ions of a larger momentum are deflected less than ones of lower momentum, whilst a mixture of ions of different mass but constant velocity will be deflected in proportion to their mass. In a mass spectrometer, molecules of a compound are ionised, either by ejection of an electron or capture of a proton, to give the *parent molecular ion* the energy which is such that some fragmentation occurs to give a series of *fragment ions*. Knowledge of the mass of the molecular ion and its major fragment ions is frequently sufficient to enable the structure of the parent compound to be uniquely deduced. The method is very sensitive and uses as little as 10^{-6} to 10^{-9} g of material. A mass spectrum is a plot of the abundance of the fragment and molecular ions against mass and, hence, is not at all like the various types of electromagnetic spectra.

The majority of ions produced during the initial ionisation procedure have a single positive charge, i.e., one electron is removed from the molecule or fragment so that the mass to charge ratio (*m/e*) is numerically equal to the mass. Thus, the ions produced differ only in their mass. Occasionally, however, molecules lose more than one electron and multicharged ions are produced. When producing ions by electron bombardment the degree of fragmentation of the molecule depends on the energy of the bombarding electrons. At low energies (1 to 20×10^{-18} J), only one electron is removed from the molecule. The resulting positively charged molecular ion has an *m/e* value corresponding to the molecular weight of the parent compound. At the usual operating electron beam energy of 10^{-17} J, the molecules are cleaved into positively charged fragments of different masses. The way in which the compound fragments to give a mass spectrum is characteristic of the compound being analysed and is called the *fragmentation pattern*. From this pattern the structure of the molecules can be deduced. Tables have been compiled for the fragmentation patterns of a wide range of compounds and reference to these is particularly helpful in elucidating the structure of an unknown compound. Most modern instruments have built-in data systems which enable the experimental fragmentation data to be compared with stored reference data.

Mass spectra show a series of peaks or lines corresponding to the *m/e* values of the positive ions produced from the compound. The height of the peaks corresponds to the relative abundance of the ions. A reference ion of similar *m/e* value to that of the parent ion is used to calibrate the mass axis (abscissa) of the spectrum. The *parent ion* is the peak with the greatest mass, although it is not necessarily the most abundant (base peak). Ion intensities in a mass spectrum are usually recorded as percentages of the intensity of the base peak.

Carbon dioxide is ionised and fragmented to yield CO_2^+, CO^+, O_2^+, O^+ and C^+, resulting in a mass spectrum with major peaks at *m/e* values of 44, 28, 32, 16 and 12, respectively. Other minor peaks are also present corresponding to the other natural isotopes of the above fragments, for example $^{13}C^+$ (*m/e*, 13) and $^{13}CO_2^+$ (*m/e*, 45).

The basic components of a double focusing mass spectrometer, essential for high resolution studies, are illustrated in Fig. 62.13. It is essential that the analysis is carried out under a very high vacuum in order to minimise the loss or transformation of ions by collision with any unionised molecules present. The major factor which governs the amenability of a sample to mass spectrometry is its vapour pressure at the temperature of the ion source. If the compound has a vapour pressure of approximately 1.3 Pa (1.9×10^{-4} p.s.i.) at 100 to 200°C, it can be injected into a reservoir attached to the ion source. As the pressure in the reservoir is higher than that in the ion source—which is under a vacuum of about 10^{-5} Pa (1.5×10^{-9} p.s.i.) —a small stream of sample vapour is drawn through a minute aperture into the ionisation chamber. Samples of lower volatility or small amounts of other samples may be introduced directly into the ionisation chamber by means of a quartz probe, so that the small amount of vapour produced enters the ionisation chamber directly.

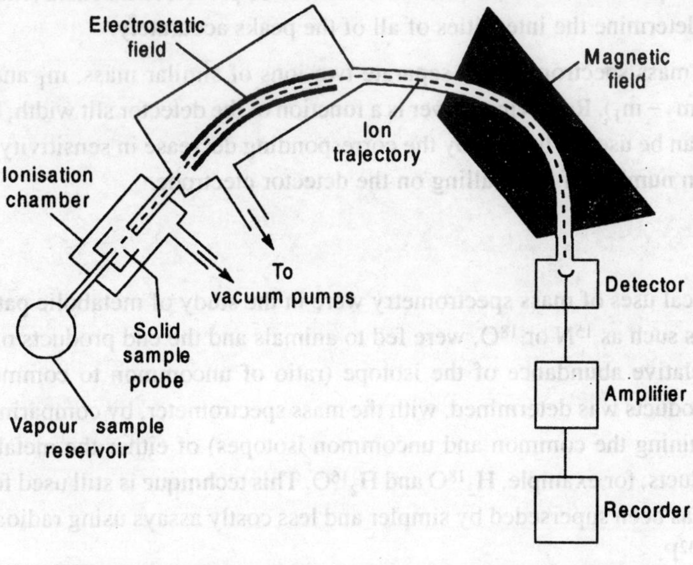

Fig. 62.13: The basic components of a Mass Spectrometer

The most common means of ionising organic compounds is by *electron bombardment*. The electrons are obtained by thermionic emisson from a heated tungsten filament and are accelerated by application of a 50 to 100 V potential. If the mass spectrometer is coupled to a gas-liquid chromatograph, a lower accelerating voltage is used to avoid the ionisation of the helium carrier gas. In all cases, the energy associated with the electrons is sufficient to cause both ionisation and fragmentation of the organic compound so that fragment ions predominate and the molecular ion is weak. An alternative form of ionisation is *chemical ionisation* in which ionisation of the organic compound is induced by ions of simple hydrocarbon molecules such as methane and butane. The hydrocarbon is introduced into an ionisation chamber at a pressure of up to 200 Pa (3×10^{-2} p.s.i.) and the ions so produced are immediately mixed with the molecules of the organic compound. Proton transfer from the hydrocarbon ions (e.g. CH_5^+ and $C_4H_{11}^+$) takes place with high efficiency to give the M + 1 ion of the organic compound. Ion fragmentation is less common than in electron bombardment, so chemical ionisation spectra are simpler than, and complementary to, electron ionisation spectra—since they give more detail of the molecular ion. *Field ionisation*, which uses potentials of up to 10 kV, resembles chemical ionisation, in that it too gives rise predominantly to the molecular ion of the organic compound.

The ions are separated in an analyser according to their *m/e* values by being accelerated to constant velocity in a vacuum by a series of negatively charged electrostatic plates and are then deflected from their initial trajectory by a magnetic field. For ions carrying the same charge the amount of deflection is dependent on the mass of the ion, so that the ions of smallest mass undergo the greatest degree of deflection. By varying either the magnetic field or the accelerating voltage the ion trajectories may be varied so that fragment ions of a given *m/e* value impinge on the collector slit of the detector. The number of ions of a given mass impinging on the detector is a measure of the abundance of that particular ion.

The detector is usually either a *simple electrode* (Faraday cup) or an *electron multiplier*. The ion current is amplified and recorded. Due to the vast range of ion intensities occurring in a single mass spectrum, it is common practice to record them at a number of preselected sensitivities, usually varying 300-fold, in order to determine the intensities of all of the peaks accurately.

The ability of a mass spectrometer to separate two ions of similar mass, m_1 and m_2, is called the *resolving power*, $m_1/(m_2 - m_1)$. Resolving power is a function of the detector slit width, but the narrowness of the slit width that can be used is limited by the corresponding decrease in sensitivity of the instrument, due to the reduction in number of ions falling on the detector electrode.

Applications

The earliest biochemical uses of mass spectrometry were in the study of metabolic pathways. Substrates enriched with isotopes such as ^{15}N or ^{18}O, were fed to animals and the end products of their metabolism were isolated. The relative abundance of the isotope (ratio of uncommon to common isotope) in the different metabolic products was determined, with the mass spectrometer, by comparing the intensities of the parent ions (containing the common and uncommon isotopes) of either the metabolites directly, or their degradation products; for example, $H_2{}^{18}O$ and $H_2{}^{16}O$. This technique is still used for ^{15}N and ^{18}O, but wherever possible it has been superseded by simpler and less costly assays using radioactive isotopes, for example ^{14}C, ^{3}H and ^{32}P.

The major use of mass spectrometry in biochemistry remains the determination of chemical structure and, hence, the identification of a compound, i.e., qualitative analysis of small amounts of relatively complex organic molecules. Because all elements have a non-whole number atomic weight, the accurate determination of the molecular ion (to 4 places of decimals) gives a unique molecular formula. By studying the accompanying fragment ions and their relative abundance, functional groups, and frequently the whole molecular structure, can be deduced e.g., steroids, ubiquinones and triglycerides.

The interpretation of the structure of any compound by mass spectrometry necessitates having the compound in a relatively pure state. It is, therefore, particularly advantageous to couple a mass spectrometer to a gas-liquid chromatograph. A mixture of compounds may be applied to the GLC and separated into the individual components. As each compound emerges after a different time interval from the GLC, it is fed into the mass spectrometer, analysed, and its fragmentation pattern recorded in the usual manner. This requires rapid scanning mass spectrometers, for example the *m/e* range 10^1 to 10^3 should be scanned in 2 to 5 seconds. Such rapidly scanning instruments allow the identification of samples inadequately resolved by GLC, by scanning both the front and rear of overlapping peaks. Excellent mass spectra can often be obtained from less than 10^7 g of compound present in the sample originally introduced into the GLC. The maximum gas flow into the ion source of the mass spectrometer is 1 mm^3 s^{-1} if a high vacuum is to be maintained. Thus, it is usual to remove the bulk of the carrier gas from the outlet of the GLC, before passing the sample into the spectrometer, by taking advantage of the rapid diffusion of carrier gas through very fine sinters.

Mass spectrometry has been used to determine the amino acid sequence of oligopeptides derived from protein hydrolysates and other sources. Originally, the peptide was made more volatile by acetylation and permethylation. The peptide bonds were then most susceptible to cleavage on electron bombardment. Moreover, fragmentation occurred from the C-terminal end of the peptide. Since no two commonly occurring amino acids have the same molecular weight (other than leucine and isoleucine), the difference in mass between adjacent major peaks in the spectrum unambiguously identifies the C-terminal amino acid residue of the heavier fragment ion. More recently *fast atom bombardment* (FAB) has been used to sequence oligopeptides without derivatisation. A paste of the peptide and sodium chloride is introduced into the ionisation chamber and bombarded with excited argon atoms, resulting in ionisation and fragmentation of the peptide. Computers have been programmed to determine the amino acid sequence of the parent ion by determining the mass difference between each of the major peaks of such a mass spectrum. The limits of this approach in determining the primary structure of proteins are governed only by the rate at which suitable oligopeptides may be isolated from the proteins and by the total mass of the oligopeptide.

10. LUMINOMETRY

Principles

Luminometry is a photometric technique in which light is emitted as a result of a chemical reaction (*luminescence*) in contrast to the result of a physical reaction (fluorescence or atomic emission). It is measured in a luminometer. It is not really a spectroscopic technique since it does not usually involve a monochromator, but it is discussed here as a technique involving light emission that is becoming increasingly important to biologists.

When a chemical reaction yields a fluroescent product in an excited state, light is emitted as the excited electrons regain the ground state (chemiluminescence). The chemiluminescent spectrum of a reaction such as that of luminol with oxygen to produce 3-aminophthalate is the same as the fluorescent spectrum of the product. Bioluminescence is characterised by light emission produced in enzyme catalysed reactions. The colour of the light emitted is characteristic of the origin of the luciferase, and varies between greenish yellow (560 nm) and red (620 nm). Under favourable conditions the quantum yield is nearly 100%.

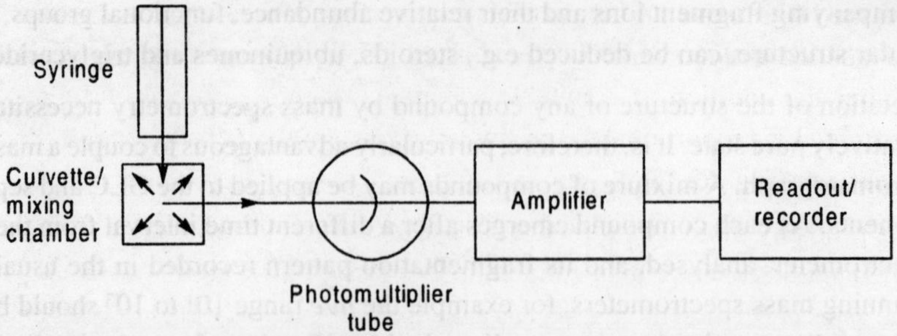

Fig. 62.14: A diagram of the main components of a Simple Luminometer

Luminometers are relatively simple photometers, complicated only slightly by the need to amplify and record the signal from the photocell. Obviously temperature control is important because of the temperature sensitivity of enzyme catalysed reactions. The main components of a luminometer (Fig. 62.14) include a photomultiplier tube with a well-stabilised high-voltage power supply to ensure sensitive, reproducible measurement of light emission; a direct current amplifier with a wide range of sensitivity and linear response; and a reaction chamber which allows temperature control, adequate mixing of reactants and protection from extraneous light.

Applications

The Firefly Luciferase System (for Determining ATP Concentration): This ATP assay is rapid to perform and can be as accurate as spectrophotometric and fluorimetric assays but it is much more sensitive (10^{-15} M). The linear range of the assay is 10^{-12} to 10^{-6} MATP. This assay system may be extended to the determination of ADP, AMP and c-AMP in a single assay by successive additions of the appropriate enzymes. These include pyruvate kinase, to convert ADP to ATP, adenylate kinase to convert AMP to ADP and phosphodiesterase to convert c-AMP to AMP. In principle, all enzymes and metabolites participating in ATP conversion reactions may be assayed by this method, including the enzymes creatine kinase, hexokinase and ATP-sulphurylase, and reactions involving creatine phosphate, glucose, GTP, PEP and 1, 3 diphosphoglycerate.

The Bacterial Luciferase System: The major uses of this system are in the determination of NADH, NADPH and $FMNH_2$ in the 10^{-9} to 10^{-12} M range, which is much more sensitive than the corresponding spectrophotometric and spectrofluorimetric assays. However, the NADPH assay is 20 times less sensitive than that for NADH.

62.3 IMMUNOASSAY TECHNIQUES

1. Enzyme Linked Immunosorbent Assay (ELISA)

Principle

The name ELISA is an acronym coined by the developers of the method and stands for Enzyme Linked ImmunoSorbent Assay. It was developed primarily to ease the process by which the bound antigen is separated from the free and also to use a label other than a radio-isotope.

The method involves the binding of an antibody to the walls of a plastic tube and subsequently adding the test and labelled antigens which compete for the available antibody. After a suitable reaction time, the tube is washed to remove all the reagents except those which are bound to the antibody. The amount of enzyme-labelled antigen which is bound to the tube wall can be measured by monitoring the enzymic reaction after the addition of a suitable substrate. Under carefully controlled conditions the amount of product formed is proportional to the amount of enzyme present and inversely proportional to the amount of test antigen (Fig. 62.15).

The method is probably more useful in what is known as double antibody or sandwich form which uses enzyme-labelled antibody rather than antigen. In this form, the method is not competitive in nature and the quantitative aspects become more direct as a result. The specific antibody is bound to the walls of the vessel but, instead of using a labelled antigen, the test antigen is added and allowed to react. After washing the tube well to remove all the other components, an enzyme conjugate of the same specific antibody is added which binds to the antigen molecules already held by the solid-phase antibody producing the so-called sandwich (Fig. 62.16). The excess enzyme-antibody conjugate is washed away and a suitable substrate added, and the amount of product formed in a given time is directly proportional to the amount of antigen originally present.

The enzymes commonly used as labels include alkaline phosphatase, (EC 3.1.3.1.), peroxidase (EC 1.11.1.7), ß-galactosidase (EC 3.2.1.23) and glucose oxidase (EC 1.1.3.4) and it is essential that the enzyme is capable of being covalently linked to the antigen or antibody without loss of its catalytic activity. Glutaraldehyde is most frequently used as a linking agent while enzymes which are glycoproteins, such as peroxidase, may be linked via the carbohydrate group using periodic acid to form a reactive aldehyde group.

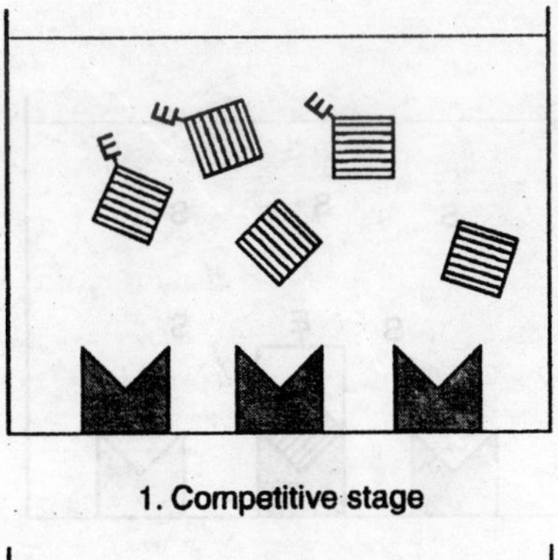

1. Competitive stage

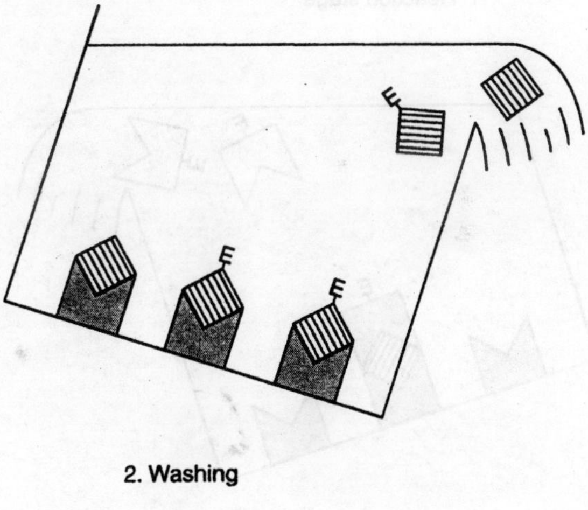

2. Washing

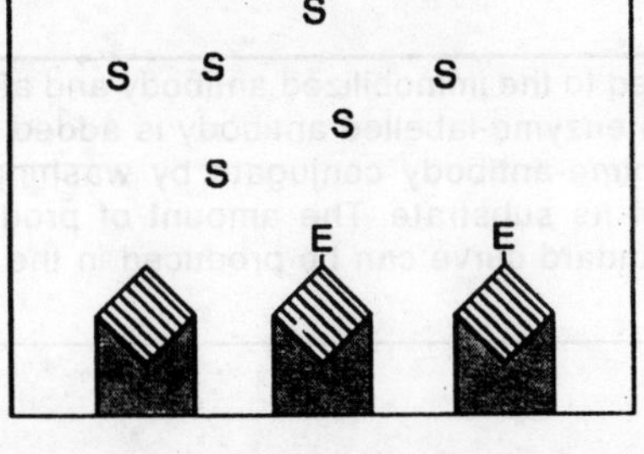

3. Assay stage

Fig. 62.15 : Enzyme linked immunosorbent assay (ELISA). The enzyme-labelled antigen and the test antigen are introduced into a tube containing the immobilised antibody. After several hours, the unbound components are removed by washing and the substrate for the enzyme label is added. The amount of product formed (P_t) is proportional to the amount of labelled antigen which was bound. A control assay without test antigen gives the maximum amount of product (P_0) and the difference between the two values ($P_0 - P_t$) is proportional to the amount of unlabelled (test) antigen present

A major factor in the choice of an enzyme is the availability of a substrate which can be measured easily. The use of p-nitrophenyl phosphate and 0-nitrophenyl-ß-galactoside as substrates for alkaline phosphatase and ß-galactosidase, respectively, is eminently suitable for such techniques because the absorption maxima of both products are in the visible region of the spectrum. Enzymes involving the use of two-stage reactions to a coloured product are sometimes employed, e.g., glucose oxidase which requires the presence of peroxidase and a chromogen in order to detect the hydrogen peroxide initially formed.

The ELISA technique can be used in a variety of ways. In the manner described it is suitable for the detection and quantitation of an antigen. However, by coating the tube with the antigen it is possible to detect the presence of specific antibodies by the use of an enzyme-labelled anti-gamma globulin. In this mode it is extremely useful for screening for antibodies against such antigens as rubella virus (German measles) and hepatitis virus.

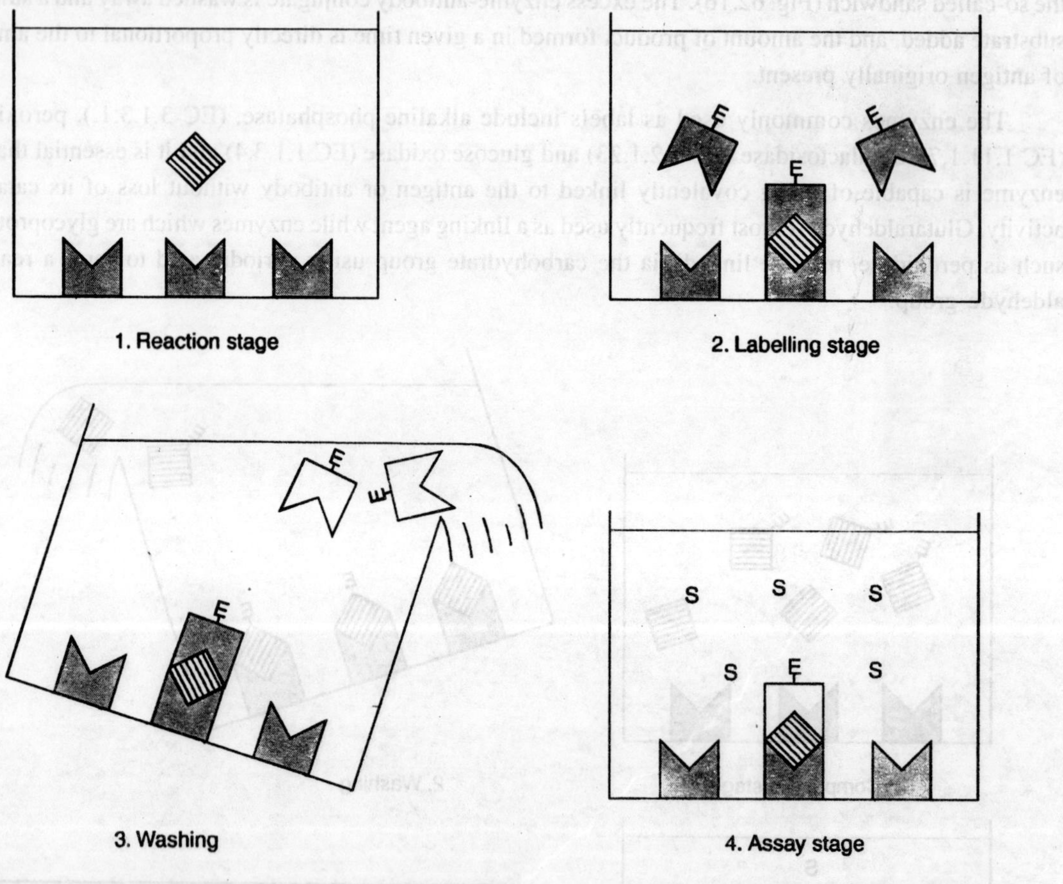

Fig. 62.16 : ELISA — sandwich method. Unlabelled antigen is added to the immobilized antibody and allowed to react. The excess unbound antigen is removed by washing and enzyme-labelled antibody is added which binds to the remaining antigen. After removing the excess enzyme-antibody conjugate by washing, the enzyme label can be detected and measured by the addition of its substrate. The amount of product is proportional to the amount of antigen originally present, and a standard curve can be produced in the usual way, providing the basis for a direct quantitative assay

SECTION XI

ENVIRONMENTAL EDUCATION, MOVEMENT, CONVENTIONS AND TREATY

63 ENVIRONMENTAL EDUCATION

63.1 BACKGROUND

Our life-styles are among the determining factors and precursors for the pattern of development in vogue in contemporary civilisation. The younger generation—especially students of higher education and the future leaders and decision-makers—may opt for changes in their life-styles. These changes will certainly be some determining factors in the renewal of development, especially with respect to the products as well as demands of all types of industry and agriculture. Higher educational institutions and their teaching staff members have major responsibilities in shaping the above-mentioned changes directly and/or indirectly. Higher education students should be served wisely in this mission. This would mean the provision of a sound environmental education at university level during which students form and reform their philosophy which will serve them as a context for their decisions and actions during active life and work.

In EE the transfer of knowledge and skills about the environment and its allied problems is essential but not enough. Knowledge has to lead to positive changes in attitudes, commitments and civic actions and, above all, to ethical values and responsibilities with respect to the protection and improvement of the environment with due consideration given to future generations. This is a major responsibility for education in general and for higher education as the producers of future decision-makers in particular.

Furthermore, the future leaders, politicians, planners, economists, scientists, technologists, industrialists, agriculturalists and other decision-makers as the product of current higher education system must be oriented to see the holistic needs of humanity and the holistic nature of the environment with its finite resources as well as the dependence of humanity upon the environment. Similarly, it is essential to consider development as a sustainable process for the rational use and management of the environment and its resources for meeting basic human needs for the present and future generations. Otherwise careless intervention in the environment through the use of available technology and excessive power coupled with human greed for lavish consumption will lead to environmental degradation and resources depletion which could serve as seeds for future local, national and international unrests and conflicts.

Specially the following are the justifications for the development of environmental education:

63.1.1 Knowledge about the Environment

Knowledge about the environment—especially its holistic nature and the laws that govern the ongoing processes in the environment particularly in an ecosystem and the interdependence of different elements in the environment—is a priority need for a better understanding, protection and improvement of the environment. This knowledge needs to be provided through formal and nonformal environmental education to decision-makers, politicians, students, teachers and the general public. For example, the very concept of the environment has meant different things to different people at different times. To some it is the home, to the others it is the village, city, country or the world.

Environment has been considered as the sum total of all the factors that effect an organism. Environment is also defined as "the biosphere and its implications for mankind" (UNESCO and International Association of Universities, 1986). The most comprehensive definition of the environment came out in the work of the Intergovernmental Conference of Environmental Education, Tbilisi, in the then USSR, October 1977. In this context, the environment is considered to consist of a "whole set of natural and social systems in which man and other organisms live and from which they draw their sustenance. This concept embraces the natural and man-made resources and products whereby human needs are satisfied. The natural environment consisting of the four interlocking systems of the atmosphere, the hydrosphere, the lithosphere and the biosphere are in constant change but the nature and system of such changes are affected by human activities. The social environment includes human groups, the material infrastructures built by man and the production relationships and institutional systems that he has devised. The social environment shows the way in which human societies have organised themselves and how they function in order to satisfy, first of all, needs relating to food, shelter, health, education and work." (UNESCO, 1977).

In brief, environment has a holistic nature consisting of its physical, biological, social, cultural economic, political, etc. dimensions which are interrelated with and interdependent on each other. The holistic nature of the environment has gradually developed through its within-ongoing-forces and changes since its inception. This holistic nature is dynamic which continually undergoes changes and develops towards a more stable holistic state.

Knowledge of the environment and its definition becomes essential at national level for establishing environmental acts and laws. For example, the Environment (Protection) Act, 1986, of India defines that "Environment includes water, air and land, the interrelationship which exists among and between water, air and land and human beings, other living creatures, plants, microorganisms and property" (Ministry of Law and Justice, 1986).

Similarly, a definition of the environment should be adopted to facilitate understanding and handling environmental issues and problems and formulating needed environmental legislation at national and international levels as well. Knowledge in the above context produced by scientific studies and research, needs to be translated into appropriate messages through environmental education for various target groups. Thus, the need for environmental education.

63.1.2 Knowledge about Humanity—Environment Relationship

Humans needs to become aware, conscious and knowledgeable about their place in the environment. There is almost a balanced relationship between humanity and the environment and it has to be maintained. For a long time humanity cherished the concept that they were the masters of the environment. One can trace the teaching of such concepts in the science textbooks taught in primary and secondary schools even in the 1950s. Humanity not only wanted to be the master of the environment but cherished the idea and took action to conquer it. This concept and its associated actions have complications that humanity and environment are against each other. This has been the most erroneous and damaging concept developed with respect to humanity-environment relationship.

Environment was also considered to be limitless with its resources. This is another erroneous concept about the environment that, perhaps, led to planning and development processes of the present civilisation into the wrong direction.

There is an urgent need for humanity to understand the environment and its governing principles as well as humanity's close relationship with and dependence upon it. In simple terms, humanity has to come to terms with the environment as the life support system—specially through sustainable and sound development of its finite resources.

In other words, humanity is an integral part of the environment and there is a two-way reaction between them. That is to say, humanity and environment shape each other to reach a dynamic and stable humanity-environment relationship. The present state of this relationship is the result of thousands of years of interaction between humanity and the environment. It could easily be disturbed by forceful natural catastrophes and/or by humanity's continued unsound interventions in the environment. It is to the advantage of humanity if this dynamic and stable humanity-environment relationship is maintained. Otherwise rapid disequilibria will put stress on one or both sides which could lead to different consequences resulting to humanity's disadvantage. Putting it differently, humanity depends on the various living and non-living elements of the environment, and not vice versa. This must be considered as an important discovery and very essential knowledge for humanity to possess. The emergence of this domain of knowledge and the need for its transfer to the various school and out-of-school target groups is another strong justification for the development of formal and nonformal environmental education.

63.1.3 Knowledge about Environment and Population Growth

It is time now that humanity becomes aware of population growth and its implications for present and future generations. For example, world population was 3 billion in 1960; 3.7 billion in 1970; 4 billion in 1975; 4.8 billion in 1985 (Bulletin of the UNESCO Regional Office for Education in Asia and the Pacific, 1982). It passed 5 billion in 1987 and 6 billion in 2000. Now it is 7 billion and may reach 8 billion in 2025. The continent of Asia had a population of 3 billion in 1986 and 3.7 billion in 2000 which means a net increase of 50 million per year in Asia and the Pacific (Roy Singh, 1986). This could be interpreted as implanting an invisible new country of 50 million people among the existing countries of Asia and the Pacific year after year! Humanity ought to become knowledgeable about this trend and its implication—especially for the future.

Furthermore, humanity's basic and luxury needs are on an enormous increase which consequently lead to corresponding fast environmental resources consumption. For instance, for the ever-increasing population of the world, one needs to provide more food and clothes, make more homes and furnitures, schools, villages and cities, buses, cars, tractors, airplanes as well as roads and highways, airports and many more similar items, each one of which reduces the fertile lands, forests and foster consumption of resources of the environment on an increasing rate every single day. Furthermore, a good many of human needs such as having cars using petroleum are becoming more and more stereotype leading to imbalanced utilisation of certain resources, thus, enhancing their depletion faster. Humanity must become aware and knowledgeable as to how to meet their basic and luxury needs because world's resources are finite, some of which are subject to depletion—such as natural gas and petroleum.

The question to pose here is how far and for how long the present trend of population growth and humanity's growing needs and fast resource consumption can continue. In the first decade of the 21st century, we find out that despite the marvelous scientific and technological achievements, humanity's miseries of poverty, hunger, sickness, homelessness, illiteracy are on the increase for a good portion of the global population—especially those in the developing countries and scientist have not been able to fulfil those promises made in the 1950s. In other words, scientific and technological development have not been responsive to the ever-increasing needs and demands of humanity whose numbers are on the increase every minute around the planet.

Another factor which cannot be ignored in the study of population and environment is the trend of movement of people from the rural to urban areas. It is estimated that more than half of the population of the world, especially in the developing countries, are already crowded in the big cities by 2000. They are crowded in the big cities in search of imaginary jobs, education, health facilities and food which will not be found easily by everybody.

Another observation to be made is the fact that despite the global need for population control, population policies in different countries do not recognise universal trend of decrease in population growth. In certain countries such as India, China, Indonesia and many other developing countries, attempts are made to reduce population growth while in certain industrialised countries—such as France and Canada—attempts are made to augment population growth by providing families with subsidies to have more children in order to avoid population decline and the shift of population pyramid towards the aged group. One can easily predict that, as the above, national population policies have future reversibility possibilities in order to meet population or manpower needs of a country.

The fundamental global issue here is to hit the optimum balance between global population increase to meet national and global manpower needs and world resources availability and/or use for the present and future generations in the context of social, cultural, religious and political values of the different societies of the human family for a sustainable population and development growth. In brief, population growth and demands and environmental resources should be made compatible and means and methods should be looked for achieving it by concerned national and international institutions and organisations, with perspectives for human survival and meeting manpower needs nationally and globally. This is a fundamental justification for the development of environmental education in the context of all types and levels of education especially at the tertiary level. In other words, environmental education is a fundamental measure for enhancing compatibility among population growth, environment and development with the perspectives of sustainability.

63.1.4 Knowledge about the Solution and Prevention of Environmental Problems

Environmental education is a necessary tool for making appropriate decisions concerning the solution and prevention of environmental problems which may be considered a disequilibrium caused by certain factors in the established relationship between humanity, animals, plants and other organisms and the environment. It has been in the last few decades that impact of human activities on the environment has become more accelerated and pronounced. Pressures from rapid population growth, uncontrolled and lavish consumption, urbanisation, industrial expansion and advances in science and technology and their application coupled with huge energy utilisation have caused these accelerated changes leading to serious environmental problems. Humanity is the cause and maker of the current environmental problems. Humanity is hurt on individual and community basis by these problems. So, humanity needs to be educated for the understanding, solution and prevention of these problems.

Environmental problems range from local to global. Global environmental problems call for worldwide action and management based on international conventions. Examples include threats to the ozone layer, the impact of weather modification, exhaustion of petroleum resources and the extinction of certain wildlife species. Environmental problems at the regional level require the agreement of the affected countries on a plan of action in solving problems of common concern such as the destruction of tropical rainforests, advancing desertification, acid rain, etc.

Developing countries throughout the world are confronted with problems due to underdevelopment and problems which arise in the course of development. Environmental problems resulting from underdevelopment are poor conditions of human settlements, loss of productivity through disease and malnutrition, vulnerability to natural disasters such as an earthquake and loss of natural resources, for instance, forest destruction and soil erosion.

Industrialised or developed countries are faced with more complex environmental problems. These problems may arise from the intensive application of science and technology resulting in industrial pollution and high-rate resources use. For example, large-scale mining may bring in big profits fast but may speed up resource depletion, pollution and social disruption.

In brief, whether we like it or not, environmental problems exist at local, national, regional and global levels. What is needed is to understand the causes and effects of these problems and devise ways and means for their solution and prevention in the context of the totality of the environment where humanity is an integral part of it. A dynamic measure that could bring appropriate responses to these problems would be to educate the people of all ages on the environment for making better decisions. In this context, environmental education becomes a cornerstone of public awareness about the environment and the solution and prevention of its problems. Thus, the need and justification for environmental education.

Environmental problems as referred to above are numerous and too many to be treated here but a glance is made at those which have global impact. UNEP reported during the International Congress on Environmental Education and Training, August 1987, Moscow (in the then USSR), that major pressing environmental problems include the loss of productivity of land; present and future climatic changes due to carbon dioxide build-up in the atmosphere; deletion of tropical forests; mismanagement, pollution and shortage of water responsible for 80% of diseases in the world; mismanagement of marine resources; threat to human health and the environment by the hazards of 70,000 chemicals in use with knowledge about the long-term effects of only 10% of them; inadequate energy resources leading to the use of wood, for fuel for domestic uses furthering deforestation, erosion, floods and famine especially in developing countries; environmental degradation due to fast urbanisation with huge population such as Mexico City with 20 million; atmospheric pollution and its impact on ecosystems, etc.

Pollution of the air, water and land is considered one of the most widespread and threatening environmental problems of our time. Pollution is the state of availability of a substance that is beyond the need and/or tolerance of an organism. Pollution persisting for a certain duration may lead to changes in an organism which could be detrimental to its survival. Numerous plants as well as insects located on the sides of roads with heavy traffic producing exhaust pollution can show unhealthy signs. Certain trees may even stop to blossom. In other words, "pollution of the environment can be defined as the interference with, or the interruption of, biological cycles" (UNESCO and International Association of Universities, 1986) with disastrous impacts. It is again the biological functions, for example, when toxic "substances get into the food-chain, passing into herbivores and carnivores, the accumulation may lead to large-scale mortality in sometimes distant animal communities. The carnivores at the end of the food-chain suffer most and it should be remembered that, biologically speaking, man is, though an omnivore to some extent, one of those large carnivores at the end of the food-chain." (UNESCO and International Association of Universities, 1986).

A concrete example of atmospheric pollution with possibly lasting consequences on all living organisms on earth is the wide use of chlorofflurocarbons (CFCs) as aerosol, in electric industry and in cosmetics which has been identified as a possible responsible chemical interacting with the ozone layer of the earth leading to its depletion. It has been asserted that the earth's ozone layer has been reduced by 5% and a hole the size of Europe has been observed, through satellites, over the Antarctica. The reduction of ozone layer around the planet may result in the receipt of heavy doses of ultraviolet light on earth which could lead to skin cancers and possible genetic changes. The impact of ozone over photosynthesis as the widest and most important chemical reaction in plants on earth is attracting the attention of researchers around the world. Furthermore, international and governmental organisations as well as industries are formulating strategies and actions to combat or control CFC atmospheric pollution for safeguarding the ozone layer. In this context an international meeting was organised in Montreal, Canada, in 1987 (by UNEP) as well as a meeting in Paris, France, in January 1988 by chemical industries which announced their intention to gradually reduce the use of CFC and find a substitute for it. In 1987, global CFC industry involves about eight billion dollars. Thus, redirecting such a global undertaking—in which many have vested interest—takes courage for effective decisions. A similar case is the current impact of acid rain, especially in industrialised countries where the damage has been severe on the forests and lakes as well as on historical monuments.

Pollution control is essential for environmental protection and it is developing as a lucrative business in numerous countries. In 1986, an estimate has put the value of the world market in pollution control and other types of environmental goods and services at some \$100 billion. New firms have been set up to meet the demand and many of the older established companies such as Exxon, Dow, Krupp, ICI, Shell and Philips have now created new divisions to exploit this new market.

63.1.5 Rational Use of Resources

There is intense need for the better management and rational use of the environment and its resources—especially with the perspectives of sustainability. Consumers, industries and agriculture play an important role in this context. Environmental education is needed to bring the message of rational use of resources particularly to the consumers.

63.2 ENVIRONMENTAL EDUCATION (EE)

63.2.1 Background

It was during the United Nations Conference on Human Environment, Stockholm, Sweden, 1972, that for the first time governments at the highest level came together to take stock of what mankind had done to the environment of which he was an integral part. Industrialised countries expressed their shock of what had happened to them in the process of industrialisation and fast development while developing countries expressed that poverty and underdevelopment were the major factors of their environmental degradation (United Nations, 1972). Thus, a global concern was expressed and needed actions were formulated.

Recommendation 96 of the Stockholm Conference stated that "the Secretary-General, the Organisations of the United Nations System, especially the United Nations Educational, Scientific and Cultural Organisation and the other international agencies concerned, should, after consultation and agreement, take the necessary steps to establish an international programme in environmental education, interdisciplinary in approach, in-school and out-of-school, encompassing all levels of education and directed towards the general public, in particular the ordinary citizen living in rural and urban areas, youth and adults alike with a view to educating him as to the simple steps he might take within his means, to manage and control his environment" (United Nations, 1972). Thus, environmental education was recognised and its development was recommended by the world community as a measure for the understanding, protection and improvement of the environment and its quality.

In response to the above recommendation, UNESCO in cooperation with UNEP launched the UNESCO-UNEP International Environmental Education Programme (IEEP) in 1975 which is housed in UNESCO Headquarters in Paris. IEEP's objectives are to assist governments, national regional and international institutions to incorporate Environmental Education (EE) into formal and nonformal education systems and programmes in order to:

— make people aware of the nature of the relationship between humanity and the environment on which he depends;

— impart knowledge and skills to understand and solve environment and development related problems; and

— enable people to acquire the attitudes and motivations leading to sound decisions and civic actions for the protection and improvement of the environment and its quality.

For the clarification of the concept, goals, objectives and guiding principles of environmental education and formulation of strategies for its development at national, regional and international levels, UNESCO in cooperation with UNEP organised, in the context of IEEP activities, the first Ministerial Intergovernmental Conference on Environmental Education in 1977 in Tbilisi (in the then USSR).

63.2.2 Goals

The goals of EE formulated and adopted at the Tbilisi Conference (UNESCO-UNEP, 1977) are:

— to foster clear awareness of and concern about economic, social, political and ecological interdependence in urban and rural areas;

— to provide every person with opportunities to acquire the knowledge, values, attitudes, commitment and skills needed to protect and improve the environment; and

— to create new patterns of behaviour of individuals, groups and society as a whole towards the environment.

63.2.3 Objectives

The objectives of EE endorsed at the Tbilisi Conference (UNESCO-UNEP, 1977) are:

— *Awareness*: to help social groups and individuals acquire an awareness of and sensitivity to the total environment and its allied problems.

— *Knowledge*: to help social groups and individuals gain a variety of experience in, and acquire a basic understanding of, the environment and its associated problems.

— *Attitudes*: to help social groups and individuals acquire a set of values and feelings of concern for the environment and the motivation for actively participating in environmental improvement and protection.

— *Participation*: to provide social groups and individuals with an opportunity to be actively involved at all levels in working toward resolutions of environmental problems.

63.2.4 Guiding Principles

The guiding principles of EE developed and endorsed at Tbilisi Conference (UNESCO-UNEP, 1977) are:

— To consider the environment in its totality—natural and built, technological and social (economic, political, technological, cultural, historical, moral, aesthetic);

— To be a continuous lifelong process, beginning at the pre-school level and continuing through all formal and non-formal states;

— To be interdisciplinary in its approach, drawing on the specific content of each discipline in making possible a holistic and balanced perspective;

— To examine major environmental issues from local, national, regional and international point of view so that students receive insights into environmental conditions in other geographical areas;

— To focus on current and potential environmental situations while taking into account the historical perspective;

— To promote the value and necessity of local, national and international cooperation in the prevention and solution of environmental problems;

— To explicitly consider environmental aspects in plans for development and growth;

— To enable learners to have a role in planning their learning experiences and provide an opportunity for making decisions and accepting their consequences;

— To relate environmental sensitivity, knowledge, problem-solving skills and values clarification to every age, but with special emphasis on environmental sensitivity to the learner's own community in early years;

— To help learners discover the symptoms and real causes of environmental problems;

— To emphasise the complexity of environmental problems and thus the need to develop critical thinking and problem-solving skills;

— To utilise diverse learning environments and a broad array of educational approaches to teaching/learning about and from the environment with due stress on practical activities and firsthand experience.

In the light of the above, environmental education is the educational process through which is imparted to its target groups in schools, homes, factories, farms, etc. the environmental sensitivity, awareness, knowledge, understanding, attitudes, skills, commitments and civic actions needed for understanding protection and improvement of the environment and the prevention of solutions of its allied problems. In other words, the mission of environmental education is to foster environmental literacy among all citizens who should make sound judgement and decisions and have ethical responsibilities about the environment and its related issues.

In the above context, environmental education serves as a common denominator in educational renewal at all levels of education—by enhancing interdisciplinarity among the different subjects and by relating education to the real life of its target groups. Environmental education advocates problem-solving approach as a skill which could help and guide its target groups in the solution of current and future environment and related problems.

63.3 STRATEGIES FOR EE DEVELOPMENT

63.3.1 Authorisation

For the incorporation of EE into education process it is essential to establish the needed mandate or authorisation and responsibility for educational institutions to abide with. Without authorisation and responsibility, EE will stay as a peripheral subject or activity and it may never be established as a needed part of the core of education curriculum for students and teachers. The formulation of the mandate and guidelines for environmentalisation of education may be achieved by a high level education committee. This committee has to be assigned by the decision makers at the highest level.

63.3.2 Curriculum Renewal

The next stage to focus is the renewal of the curriculum. For renewal of curriculum, it is, again, essential to designate a committee consisting of curriculum experts, experienced teachers, environmental science experts and community representatives. Their task will be

— to formulate EE objectives for education;

— to examine the curriculum of the primary and middle schools and identify the environmental education concepts already covered so that it could serve as one of the basis upon which the environmentalisation of school curriculum could be based;

— to identify local environmental problems, additional EE contents required, appropriate activities, time needed for different subjects and topics, instructional methodologies, appropriate instructional techniques.

The amalgamation of the above elements and their organisation at grade or age levels would form the environmental education dimension of the curriculum. The nature of identified EE objectives and pertinent environmental contents and problems will give the curriculum its essential local dimension or character.

The above EE dimension of curriculum may be delivered through different mechanics depending on the priorities of education institutions for environmentalisation of their school curriculum and teacher education and the needed personnel and facilities at their disposal. The first mechanism is to improve the teaching of the identified environmental topics in the ongoing curriculum. This requires the development of new EE teaching objectives, teaching methods and their inclusion in the daily lesson plans. Modules may also be prepared on certain concepts such as air, water, land, plants, energy, etc. and infused into the teaching/learning process in the context of the implementation of existing curriculum.

Introduction of a compulsory EE subject or topic is considered a high priority. This may be achieved by introducing a new compulsory course on EE. If the curriculum is too crowded it can be studied for substitution of an obsolete course by the EE course or by the merging of two or three old courses to make room for an EE course or courses. If the substitution or merging of courses cannot be achieved, the other alternative is to infuse the compulsory environmental topics through units or modules into all teaching subjects.

63.3.3 Teachers Training Renewal

There is urgent need to renovate teacher training with the incorporation of environmental education. It is the teacher who makes a curriculum function. This can be done if the teacher is trained through pre-service and in-service programmes. If teachers are trained in environmental education they can also help in the formulation and development of suitable curriculum and teaching materials referred to earlier. The involvement of teachers in curriculum and materials renewal will lead to teachers' commitment and responsibility for the effective implementation of the curriculum and teaching materials. Teachers should be trained through in-service and pre-service environmental education programmes.

It may be reasonable to give priority to the in-service training of teachers in order to have an immediate environmental impact on students especially those who are in the last year of their schooling. In-service teacher training may be achieved through well prepared one- or two-week workshops or summer institutes.

It is very essential that when curriculum renewal with respect to environmental education is undertaken, it should include teacher education renewal as well despite the usual separate administrative setups responsible for school curriculum development and teacher education.

Pre-service teacher education should be based on the courses and elements of the environmental education dimension of the curriculum for schools. This core may further develop to suit the level of the teacher with due consideration to their needs in the cognitive, affective and psychomotor domains.

63.3.4 Teaching Methods

Teaching methods that may be used in environmental education are lecture, demonstration, discussion, field trips, projects, games, simulations, debates, case studies, competitions, exercises and laboratory work. One or more methods may be opted for that may match and facilitate the teaching/learning of a curriculum element.

However, EE advocates the use of the environment as a living laboratory in the teaching/learning process at all levels of education. This approach has its unique place in education if learning by doing is opted for as a teaching method and strategy.

Learning by doing can be achieved in the environment by undertaking experiments at micro and/or macro levels with air, water, soil, plants, animals and their interactions with and interdependence on each other. Environment is always the field for the unknown to be discovered by the learners in accordance with their age level. If activities are well planned in this respect, learners will not only understand the environment better but they will be initiated in research thinking, techniques and activities and will also become acquainted with problem-solving approach.

Furthermore, learning can be facilitated and made more appealing and possibly become more permanent if it is achieved through research and experimentation in the indoor and outdoor environment requiring problem identification, observation, exploration, data collection and interpretation, formulating solutions and reaching conclusions; for example, on the holistic nature of the environment, the interdependencies of the different elements in an ecosystem, the simultaneous physical, chemical and biological changes that are in action in an ecosystem or in the environment in general. This orientation ought to be accommodated in the curriculum development (objectives, contents, methods, activities and evaluation) and teachers should be well-trained through pre-service and in-service programmes on the understanding and use of the environment, as mentioned above, in the teaching/learning processes.

63.3.5 Evaluation

Evaluation in this context is the process of finding out if EE objectives in the context of teaching/ learning have been fulfilled. Results of an evaluation are feedback indicating the achievement and weakness of the student and the productivity of the efforts of the teachers.

Feedbacks are essential to guide the students to improve their learning efforts, to help the teacher to adapt his teaching for better learning and to improve the school curriculum if certain difficulties are encountered. Evaluation instruments in education may be revised and renewed to accommodate environmental questions based on the objectives and contents of environmental education that have been incorporated into education. Particular emphasis is put on the conclusion of environmental dimension in the examinations that entitle the students for passing to higher grades and/or receive achievement certificates. This is an important factor in obliging the teachers and students to give enough time and attention to the teaching and learning of environmental dimension in education. If the environmental dimension is not tested or evaluated in the development of environmental education, it will always remain as a peripheral activity with the possibility of being dropped someday.

Evaluation of knowledge and skills may be done satisfactorily through essays, objective tests, true and false questions, short answer questions, project assessment, laboratory work and report, survey, case studies etc.

The fundamental task in the teaching of environmental education is to inculcate sound environmental attitudes and ethical responsibilities conducive to sound civic actions. This task is achieved through a long period of time and there are no standardised evaluation instruments that could measure this type of affective learning. The immediate need is to develop effective instruments for evaluating this type of achievement. This is of high priority in the development of environmental education.

63.4 MODELS FOR FUTURE ENVIRONMENTAL EDUCATION SYSTEMS

It has been established that environmental education has the task of addressing an extremely wide-ranging and bewildering array of content. This is a dynamic, ever-changing content, characterised by highly complex inter-relationships, priority problem causes, impacts and solutions. The knowledge-base of environmental education is made all the more difficult to comprehend because the human race often simply cannot understand environmental issues or their potential resolutions in any definitive or permanent sense. It is a highly value-laden content, and one person's solution may be another's catastrophe. It is a content that incorporates aesthetic, spiritual, social, political and economic dimensions alongside (not separate from) the purely scientific. Furthermore, it is a content that does not and should not focus solely on environmental disasters and negative issues. Environmental education is not simple about 'saving the whale' or indeed 'saving the world'. It is equally about the development of an appreciation of the wonders and beauty of the world and of a sense of *wanting* to save it—in short, the development of ecological thinking or of an environmental ethics.

It has also been established that there are a number of different ideologies or perspectives on the root causes of environmental problems, which give rise to various implications for education. In particular, students should be exposed to a plurality of ideologies in order that they might develop their own set of environmental beliefs and values and can then go on to pursue actions that are considered necessary and justifiable (according to their own reasoned thinking) to achieving environmental reform in accordance with the personal ideological position that has been adopted.

This position (i.e. relating to both content and ideologies) is in conflict with current educational practice. A series of inconsistencies and limitations has been established that suggest that the gap between the rhetoric and the reality of environmental education is wide. Various statements of internationally agreed guidelines and goals have been well-publicised, yet, at best, attempts are made to implement them as unchallengeable 'givens' and, more often than not, educators possess neither the time nor the motivation and training to implement (or appraise) them. A number of exciting and forward-thinking case studies have illuminated this text, yet regrettably they remain the shining examples of good practice rather than the global norm. Research shows that the influence of formal 'quality' environmental education is not as successful as it could or ought to be for two reasons: firstly, the logistical problems of its widespread implementation ensure that there is not a sufficient amount of it and, secondly, even when well-designed and seemingly appropriate programmes of environmental education do exist, it would appear that other influences on people's lives are far more significant than education programmes in the development of their environmental understanding, awareness, concern and action—and formal programmes take little or no account of these. Indeed it is extremely difficult to match up the complexities of environmental education with the intricacies and demands of formal education curricula and with the principles and processes of learning—at any stage of education, be it the early years of schooling or within teacher education programmes.

So what are the priorities for attempting to solve these problems? For some, as we have seen, the answers lie in major paradigm shifts: substantial changes in thinking and practice that may apply not solely to environmental education, but perhaps to education itself. Alternatively, solutions may be approached by the adoption of less radical changes in practice, for example, by taking account of and building upon prior knowledge and formative experiences; by setting locally appropriate goals; and, most importantly, by planning coverage of all of the core components of environmental teaching and learning.

For this latter task to be accomplished, a framework for guidance is surely necessary. So we return to an overall model for planning teaching and learning, which provides an elaboration of the model presented in Part IV and takes account of other critical factors that have since been introduced.

In the first instance, the curriculum's essential 'elements' (empirical, ethical, aesthetic) deserve emphasis; hence stage 1 of the revised model (Fig. 63.1) includes them. As explained in Part IV, these components of the model are useful in that they help make the critical links between the three-fold framework of environmental education and the dimensions of learning. It should be noted that a fourth element (previously introduced as the 'synoptic' element) is omitted since, surely, this is an all-embracing concept. Indeed, the whole of the model is concerned with helping learners to realise the complex nature of the environment and making sure that they are introduced to the inseparable nature of the various components of an environment and to the inter-relations of these. The overarching complexity should be central to all planning.

Any curriculum review needs to consider whether the essential elements are adequately covered by asking such questions as:

- do the pupils/students have as many opportunities as possible for empirical investigations *in* the environment; for observation, measuring, recording, interpreting and discussing what has been observed? Research has shown conclusively how important '*in* the environment' experiences are;
- are pupils made aware of the complex nature of the environment? Do studies help pupils realise the interrelationships and complexities within and among issues; the inseparable nature of the various components of the environment, and the unique position of human life in relation to other things?

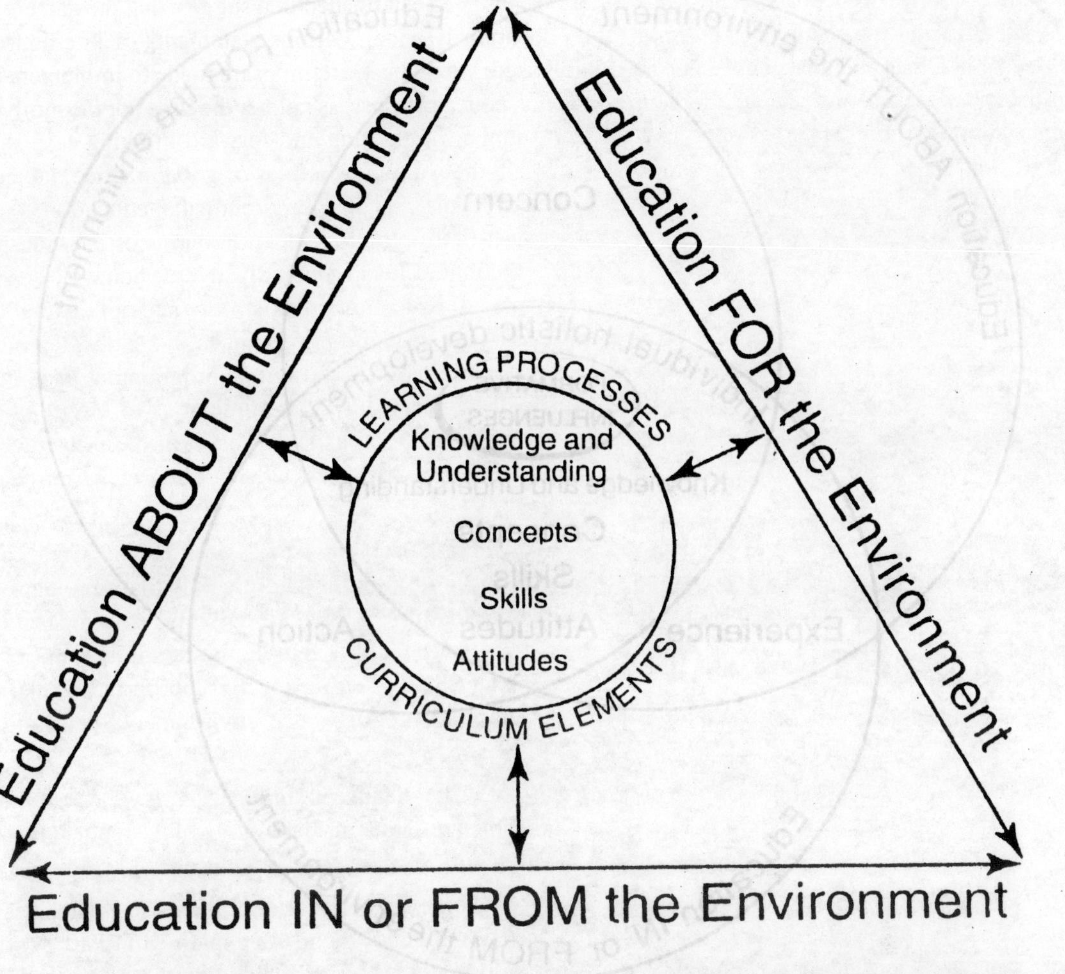

Fig. 63.1: Model for teaching & learning in environmental education: components of the planning task

- is there a focus on aesthetic dimensions and qualitative study? '*In* the environment' experiences do not necessarily have to focus on quantitative investigations. Research shows the critical importance of aesthetic and spiritual experiences, of realising that there is no right or wrong answer in absolute terms to aesthetic questions;
- are the pupils/students introduced to the ideas of personal responsibility for the environment and the concept of stewardship? In other words, does the programme contain ethical dimensions; encouraging learners to ask if the criteria of proposed actions are based on morally justifiable values?

Whilst this static model (Fig. 63.1) shows the three elements as primarily located within one of the three approaches to teaching and learning (about, in and for the environment), it should be emphasised that, inevitably, they merge and overlap. Such are the limitations of two-dimensional representation. How useful it would be if the diagram's three circles were revolving spheres, whose surface elements similarly rotated and blended with each other.

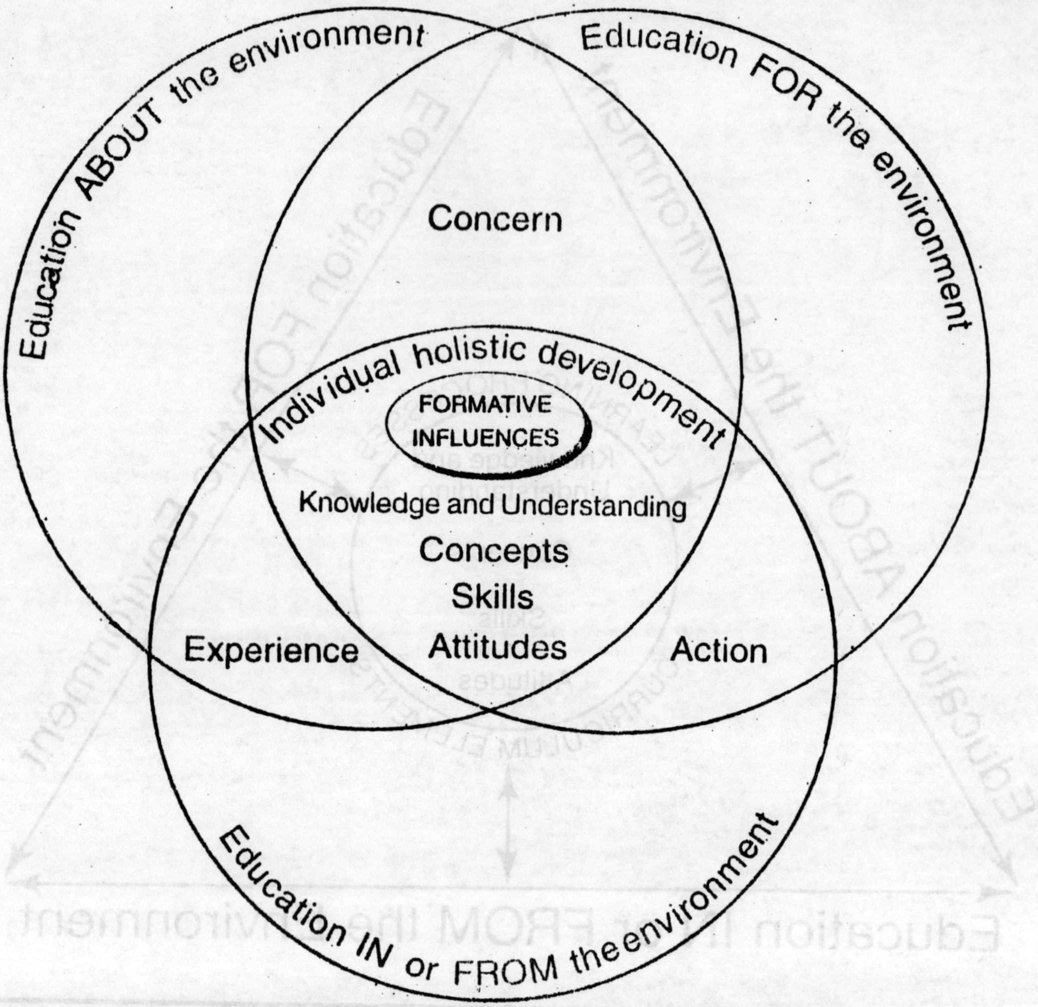

Fig. 63.2: Model for teaching and learning in environmental education: complete planning framework (1)

A second and extremely critical elaboration of the original model is also required. Fig. 63.1 and Fig. 63.2 represent components of the planning task, yet there is a major omission on these diagrams, namely the wealth of formative influences or significant life experiences that individuals bring *to* their further learning. Research has demonstrated that these may indeed be more significant than planned formal educational programmes in the development of environmental understanding and concern and, wherever possible, educators need to be aware of prior knowledge and build upon it in a meaningful way. Thus the centre, or the 'core', of the diagram ought to contain this powerful base of formative influences upon which all further learning depends and which most educational programmes and practice seem to ignore (Fig. 63.2). From such formative influences, through a combination of 'life' (i.e. further significant experiences) and formal education programmes, hopefully, individuals may acquire the necessary range of knowledge and understanding, skills, attitudes and values that foster personal concern and enable the ability to act in pro-environmental ways.

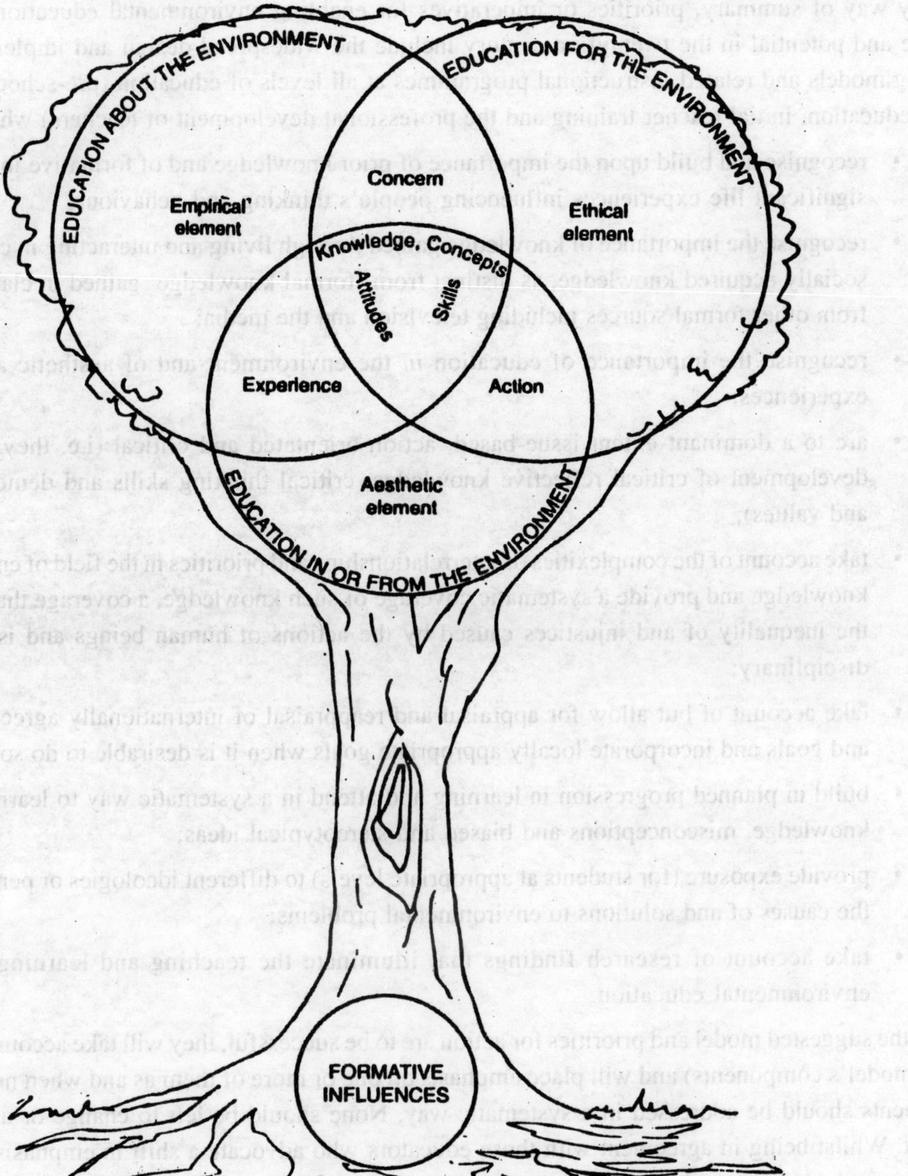

Fig. 63.3: Model for teaching & learning in environmental education: complete planning framework (2)

Perhaps one further—and indeed 'final'—revision of the model should be made, which returns us to the issue of the limitations of two-dimensional representation. So far, the model's components have been presented solely in two dimensions, suggesting that relationships between them are comparable. In reality this is not the case, since the 'core' of formative influences is *pre-supposed* and cannot be planned for in the same sense as other components. The core is independent of—yet interrelates with—the experiences of planned programmes, and so is better regarded as a 'foundation' that continues to feed into and nurture other experiences (Fig. 63.3).

By way of summary, priorities or imperatives for enabling environmental education to meet its promise and potential in the twentyfirst century include the widespread design and implementation of planning models and related instructional programmes at all levels of education (pre-school through to higher education, initial teacher training and the professional development of teachers) which:

- recognise and build upon the importance of prior knowledge and of formative influences and significant life experiences influencing people's thinking and behaviour;
- recognise the importance of knowledge gained through living and interacting in communities, socially acquired knowledge, as distinct from 'formal knowledge' gained in classrooms and from other formal sources including television and the media;
- recognise the importance of education *in* the environment and of aesthetic and spiritual experiences;
- are to a dominant extent issue-based, action-orientated and critical (i.e. they involve the development of critical reflective knowledge, critical thinking skills and democratic skills and values);
- take account of the complexities of interrelationships and priorities in the field of environmental knowledge and provide a systematic coverage of such knowledge, a coverage that recognises the inequality of and injustices caused by the actions of human beings and is truly interdisciplinary;
- take account of but allow for appraisal and reappraisal of internationally agreed guidelines and goals and incorporate locally appropriate goals when it is desirable to do so;
- build in planned progression in learning and attend in a systematic way to learners' gaps in knowledge, misconceptions and biased and stereotypical ideas;
- provide exposure (for students at appropriate levels) to different ideologies or perspectives on the causes of and solutions to environmental problems;
- take account of research findings that illuminate the teaching and learning process in environmental education.

If the suggested model and priorities for action are to be successful, they will take account of balance (of the model's components) and will place emphasis on one or more of them as and when necessary. *All* components should be addressed in a systematic way. None should be left to change or an opportune moment. Whilst being in agreement with those educators who advocate a shift in emphasis away from environmental education as being *about* the environment (grounded in the scientific domain) towards an all-encompassing education *for* the environment, it is argued here that whilst a shift of emphasis is to be welcomed and indeed promoted, a total rejection of other approaches is not. What is needed is an appropriate balance. Perhaps the pendulum has swung too far in favour of calls to educate *for* the environment—an education that focuses exclusively upon a particular issue as an organising principle and that draws upon knowledge opportunistically rather than systematically. I agree that education *about* the environment should go *within* socially critical education, but surely, if any degree of true and comprehensive understanding of the world is to be achieved (even in small measures), then education *about* the environment should go *alongside*, inter-linked with, issue-based, action-orientated, socially critical education. So too can education *in* the environment—that essential foundation for formative influences of the personal, aesthetic kind. Research findings and case studies of good practice cited within this text show that all of these aspects of the environmental education framework have a significant role to play alongside and inter-linking with each other.

The immediate practical consequences of these recommendations are:

- the need for every educational institution to have a well-designed policy document for environmental education that includes details of aims, objectives, components of the planning task, methods, style and timing of teaching, content (details of issues, knowledge, skills, concepts to be addressed), resources and organisation of resources, assessment, record-keeping and evaluation (see also Palmer and Neal, 1994);
- the need for every institution to have a coordinator for environmental education;
- the need for institutions to establish adequate mechanisms for review and revisions of policy and practice;
- the need for on-going programmes of in-service professional development for teachers;
- the need for widespread attention to be paid to environmental education in the initial training of teachers;
- the need for on-going research in the field and for the outcomes of research studies to be accessible to all concerned with environmental education—policy-makers, curriculum developers and educators alike.

Finally—and in some ways by way of postscript—an imperative for the twenty-first century is the internalisation of the realistic view that formal programmes of environmental education alone cannot educate people to 'save the planet'. Strange as it may sound, this view is rarely, if at all, articulated in the relevant educational literature which focuses almost entirely on how to plan, implement and evaluate environmental education programmes so that they succeed in their task of educating people 'for sustainable living'.

There is no intention whatsoever here of delivering the totally unexpected message that formal programmes of environmental education are, in colloquial terms, a 'waste of time'. It is, however, being suggested that until such times as a major paradigm shift in education (in the direction of ecological thinking) takes place—and even the most optimistic amongst us would surely say this was not imminent. Then formal environmental education programmes can and will form only a small part of an individual's education relating to the environment. The sooner this is recognised—by governments, NGOs, educators and community groups alike—the better.

Environmental education, in its broadest sense, is about 'empowerment' and developing a sense of 'ownership', improving the capacity for people to address environment and development issues in their own communities. It is about touching people's beliefs and attitudes so that they *want* to live sustainably, providing sufficient information to support these beliefs and to translate attitudes and values into action. Of course this process includes formal teaching, but it also encompasses another whole dimension—that which has been placed at the 'core' of the teaching and learning model—the dimension of 'informal' education sources. As we have seen, these include communication and information that results from living and interacting in a particular locality and community, newspapers, television, radio, other media forms, 'events' in one's life and in the wider world and interacting with relatives and close friends. Environmental education depends upon the totality of an individual's experiences relating to the world —of which formal education is but a small part. *Caring for the Earth* (IUCN/UNEP/WWF, 1991) emphasises the need for all sections of the community to engage in education and communication relating to sustainable living:

The effective implementation of the world ethics will only be possible if it is taken up by all sections of the community. The range of actions that could be taken is very large, but should include:

- parents teaching their children to act with respect for other people and other species;
- educators incorporating the world ethic into their teaching;
- children helping their parents to become aware and change their behaviour by explaining at home the new ideas they learned at school;
- artists in all media using their creative skills to inspire people with a new understanding and respect for nature and a wish to conserve it;
- scientists improving understanding of ecosystems, their sensitivity to human impact and their capacity to meet human needs and at the same time ensuring that findings are communicated accurately and applied reasonably;
- lawyers evaluating the legal implication of the world ethic and drafting the laws required to support it;
- technologists, economists and industrialists entering into dialogue with the coalitions and groups (concerned with respecting and caring for the community of life and with the consequent personal and social obligations) and establishing new technologies and business approaches to implement the world ethic;
- politicians, other policy-makers and public administrators working similarly to evaluate the changes needed in public policy and then to put them into effect.

Clearly, 'education' and 'communication' are inseparable processes that impact upon people's thinking and actions. As I have argued elsewhere, since education and communication both deal with the transfer or exchange of ideas, information and skills in a two-way process, they inevitably shade into one another. Environmental education (formal as well as informal) builds the motivation, skills and understanding on which environmental citizenship may be based. Environmental communication is aimed at changing practices and behaviour, and inviting participation or action in relation to environmental issues. It has a shortterm, action-specific goal (Palmer et al, 1995). A published, edited collection of papers (Palmer et al, 1995), delivered at a workshop on 'Changing Personal Attitudes and Practices' at the IUCN General Assembly of 1994, provides a series of case study examples of good practice that go some way towards answering the three critical questions:

- how can education and communication be used to change the attitudes and behaviour of people?
- how can education and communication be focused effectively on sustainability?
- how can we promote, plan and carry out education more effectively?

A number of key messages portrayed by the case studies (from both developed and developing nations) go some way towards answering these questions. Firstly, both education and communication have a very different starting point from science and other disciplines and from research. They begin with the perceptions of their target audiences and try to improve these perceptions by providing better and more reliable information. In seeking to change knowledge, attributes or behaviour, scientific facts and data are in themselves insufficient. They may even be counter-productive. They have to be translated into concepts and messages that make sense to a target audience and to be relevant to them. In many cases, communication experts have proved that knowledge of facts alone is of no consequence when it comes to changing behaviour. Linking information to personal benefit is more important (Palmer, 1994).

> Secondly, action for environmental education and communication has to be integrated into wider policies and into plans for environmental care and for sustainable resource use.
>
> It has to be combined with legal, economic and technical instruments for change of different kinds of communication and support for education are needed at different stages in the action plan to start with, people need to be enthused to join in discussion of the problems and how they may be solved. Then they need to be motivated to join in the action.
>
> (Holdgate, 1996, commenting on Palmer *et al*, 1995)

Thirdly, communication strategies must be flexible and adaptive. New insights must be used to improve policy. Education and communication programmes need a four-phase, cyclical approach leading from research through planning to implementation and on-going evaluation (Palmer, 1994; Holdgate, 1996).

Fourthly, people need to feel 'ownership' of problems and solutions and to derive benefits from the new policies they adopt as a result of the programme of education and communication (Palmer, 1994; Palmer *et al*, 1995).

For information alone does not necessarily lead to changed behaviour: the latter is far more likely if the change will bring personal benefit. This can take many forms, including social acceptance, enjoyment, greater environmental or financial security and a healthier life for children.

To be effective, education and communication have to be seen as relevant and they have to offer the prospect of an improvement in people's lives.

As we have seen through the accounts and case studies within this text (see also Palmer *et al*, 1995, and Holdgate, 1996), progress in establishing environmental education around the world is uneven in many respects. In developed nations, it tends to be far more institutionalised, with much stronger support from governments. Education and communication tend to be separated, with 'formal' environmental education being seen as the task of schools, colleges and universities and 'informal' environmental education being left to change in the hands of the media and NGOs. In the developing world, on the other hand, the existence of structured, comprehensive programmes of formal environmental teaching and learning is still quite rare; where such programmes exist, they are relatively new. Here, more attention has been given to informal education, via the media and through local community projects, often supported by a government. As Holdgate points out, where formal programmes do exist in developing countries, they tend to be weakly structured, weakly developed and poorly supported socially. As a result, a 'negative feedback loop' is established.

Environmental education does not pervade the whole educational system, because it lacks high level institutional support; it is not informed by efficient educational research because the system is not attuned to using the results of such studies; and it becomes discredited further because what is done does not produce useful results.

Where examples of success are found (in both developed and developing countries) they tend to fall into two categories, admirably illustrated by the case studies in Palmer et al, (1995). These are: successful action to establish organisational structures for environmental education; and successful actions for conservation, based on education and communication at community levels, which provide public information for action.

The world's *most* successful programmes in the twentyfirst century will surely be those in which the formal and informal elements of education are supported alongside each other. Many countries have made and continue to make excellent progress in the establishment of organisational structures for environmental education—obviously a fact to be welcomed. Yet formal programmes cannot 'do it all' and neither should they. Formative influences and informal educative experiences do not simply precede and 'feed into' formal programmes that can then 'take over the task' in a linear progression. Formative influences occur before, within and after formal programmes and interact with them. It is a combination of these 'life' experiences, general maturation and exposure to 'environmental education' at a variety of levels and degrees of formality that will lead to individuals gaining a sense of personal concern for the environment, a desire to live in sustainable ways and the ability to act appropriately.

Our task as environmental educators in the twentyfirst century is to recognise this complexity of experience and to assist the overall process—by striving to implement programmes of education that *inform* our students about the complexities of the environment in which they are growing up; *empower* them to address environment and development issues in their own lives and provide them with opportunities to be *inspired* by the joys, wonders and mysteries of the natural world and human achievement.

64 ENVIRONMENTAL MOVEMENT

64.1 GENESIS OF GLOBAL ENVIRONMENTAL MOVEMENT

The global environmental movement had no clear beginning. There were no landmark events which sparked mass outrage, no specific leaders who inspired a mass movement and no sudden changes in human thinking. Rather, the global movement grew out of a series of independent responses to local issues, in different places at different times. Individuals came together to form local groups since the mid-19th century, responding to different issues for often different reasons. Local groups slowly became national movements, combined after the Second World War (1939-45) to become a multinational movement and have today become a large complex and disparate global movement. However, the roots of the environmental movement lay in the technological, economic and social changes brought by the industrial revolution in Europe and North America; these included improved scientific understanding, increased personal mobility and the spread of human settlement. In early 20th century, organised movement was initiated for sharing of knowledge across national boundaries for preservation of nature and natural resources and also for preservation movement.

By the close of the 19th century, the first attempts were being made to promote greater international cooperation, particularly in the field of 'Wildlife Protection'. The protection of birds and mammals provided the first focus of attention. In 1909, European nature protectionists met in Paris at the International Congress for the Protection of Nature. Reviewing the record of nature protection in Europe, they argued the need for a multinational response and proposed the creation of an international nature protection body.

The work of various conservationist groups viz., Society for the Preservation of the Wild Fauna of the Empire (SPWFE), International Office for the Protection of Nature (IOPN) and International Committee for Bird Protection (ICBP) and the signing of the 1900 and 1933 African conventions and Western Hemisphere convention represented the seeds of an interest in international cooperation.

Later, the Second World War transformed attitudes towards internationalism. But since the UN-intervention, the conservation movement received the proper focus of movement. The creation of International Union for Conservation of Nature (IUCN), the World Wildlife Fund (WWF) and the International Union for the Protection of Nature (IUPN) are the steps towards such activities. The overall structure of IUCN is shown in Fig. 64.1.

64.2 THE ENVIRONMENTAL REVOLUTION—PUBLIC PARTICIPATION

In the late summer of 1962, Rachel Carson's book entitled "Silent Spring" struck a chord with its readers and subsequently organised mass resistance on pesticide use in agriculture. This event prompted the then American President to creation of an advisory panel on Pesticides. In April 1970, over 300,000 Americans took part in Earth Day Celebration, this is perhaps the largest environmental demonstration in history.

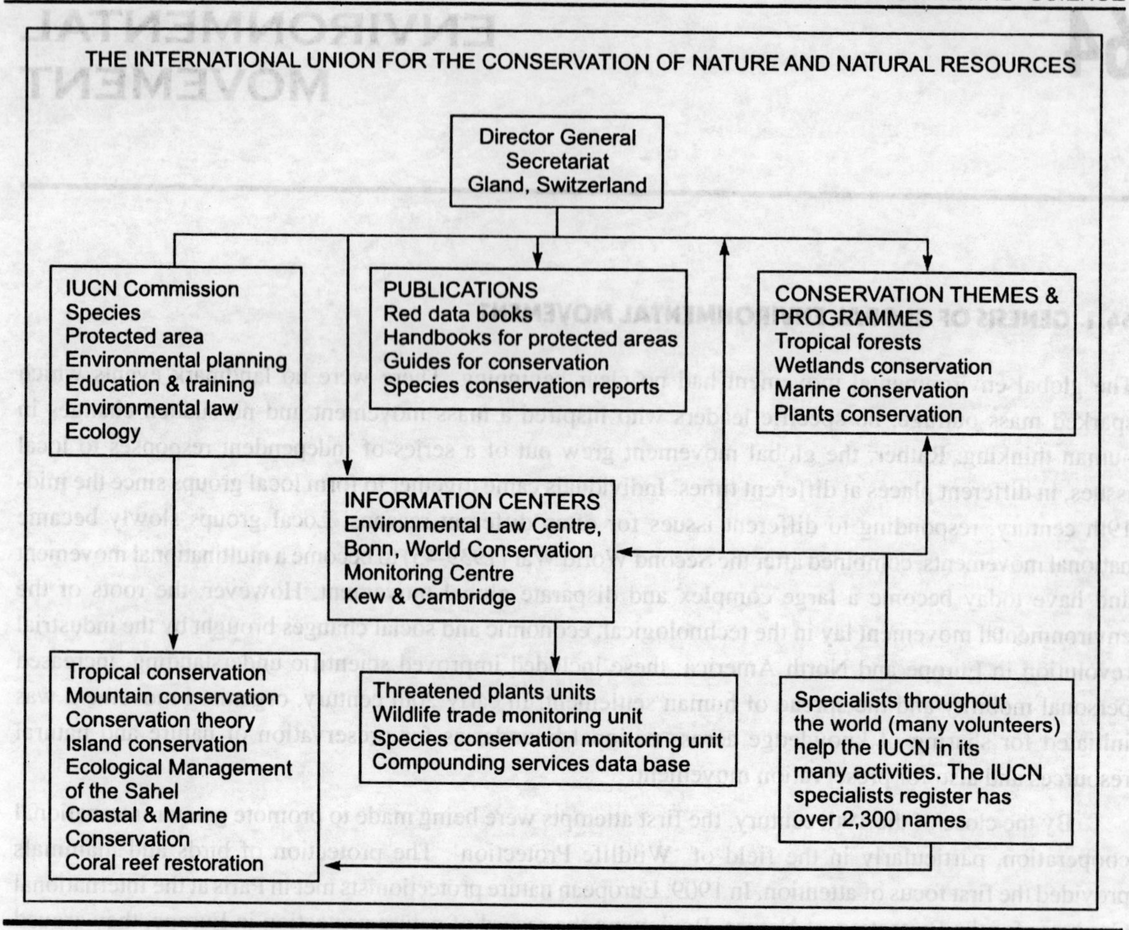

Fig. 64.1: Administrative network of IUCN

But since 1970s, new Environmentalism grew out of industrial society and they are the combined groups of activities, politicians, social reformers etc. whose aim is to get immediate objectives. Subsequently, nuclear testing, star war and the threat of global destruction was realised over the years, thus global public opinion was focussed on banning of such destructive testing systems. But, in reality, such an arm race is growing day-by-day in a very alarming rate.

The scientific understanding of Environmental Policy at National and International level emerged as an arena of sound growth and thus green viable alternative technology came up for abatement of several environmental problems.

The new environmentalism later linked with roots of several social and cultural revolution relating to human right issues. Between 1968-1972, two international conferences met to assess the problems of the global environment, and more importantly, to suggest corrective action. Later The Stockholm Conference on global environment in June 1972 was the pivotal event in the growth of the global environment movement. It was the first occasion on which the political, social and economic problems of the global environment were discussed at an inter-governmental forum with a view to actually taking corrective action. It aimed to "create a basis for comprehensive consideration within the United Nations of the problems of the human environment" and to focus the attention of governments and public opinion to various countries on the importance of the problem. It resulted in the creation of the United Nations

Environment Programme (UNEP) and it marked an important transition from the emotional and occasionally native New Environmentalism of the 1960s to the more rational, political and global perspective of the 1970s. Above all, it brought the debate between poor and rich countries—with their different perceptions of environmental priorities—into the open and caused a fundamental shift in the direction of environmentalism.

Deep Ecology

The deep ecology movement began in the early 1970s. It was named by the Norwegian, Arne Naess, in 1973. Naess's original article has spawned an ecological cult. It set the deep ecologists against the shallow ecologists. The central objective of the shallow ecologists—in fighting against pollution and resource depletion—is the health and affluence of people in developed countries. The deep ecology manifesto, as originally penned by Naess, was embodied in a set of slogans (Table 64.1). The differing attitudes of deep ecologists and shallow ecologists are summarized in Table 64.2. However, some commentrators believe that deep and shallow ecological perspectives are a somewhat crude division of an envrionmental-perspectives continuum that has grown from a long-standing critique of Western development (e.g. Jacob, 1994).

Table 64.1: Slogans for deep and shallow ecology

Shallow ecology	*Deep ecology*
1. Natural diversity is valuable as a resource	1. Natural diversity has its own intrinsic value
2. Nature is there for humans to exploit as they will, even if that should cause some damage	2. Nature does not belong to humans and humans have no right to destroy natural features of the planet
3. It is nonsense to talk about value except as value for humankind	3. Equating value with value for humans reveals a radical prejudice
4. Animals are valuable only as resources for people	4. Animals have their own intrinsic value and have a right to live, even if they should be of no use to humans
5. Plant species should be saved because they are valuable genetic reserves for agriculture and medicine	5. Plant species should be saved because of their intrinsic value
6. Pollution should be decreased if it should threaten economic growth	6. Decrease of pollution has priority over economic growth
7. Third World population growth threatens ecological equilibrium	7. World population at the present level threatens ecosystems, the major threat being posed by the people and attitude of industrial states more than by those of many others. Human population is today excessive
8. 'Resource' means resource for humans	8. 'Resource' means resource for living beings
9. People will not tolerate a broad decrease in their standard of living	9. People should not tolerate a broad decrease in quality of life, but they should be ready to accept a reduction in the living standard in overdeveloped countries
10. Nature is cruel and necessarily so	10. Humans are cruel but not of necessity

Source: After Naess (1973)

Table 64.2: Different attitudes of deep and shallow ecologists

Environmental problem	*Shallow ecology*	*Deep ecology*
Pollution	Advocate a technology that seeks to purify the air and water and to spread pollution more evenly; laws limit possible pollution; polluting industries are preferably exported to developing countries	Evaluate pollution from an ecospheric point of view, not focusing on its effects on human health, but on the effects on life as a whole, including life conditions of every species and ecosystem
Resources	Emphasise resources for human beings, especially for the present generation in affluent societies. Resources belongs to those who have the technology to exploit them. Resources will not be depleted because, as they get rarer, higher market prices conserve them, and substitutes will be found through technological progress. Plants, animals, and natural objects are valuable only as resources for humans; if they should have no use to humans, they can be destroyed	Deeply concerned with resources and habitats for all life-forms for their own sake; no natural object is regarded solely as a resource; this view leads to a critical examination of human modes of production and consumption
Population	See overpopulation as a problem of developing countries. The question of the optimum human population is discussed without reference to the optimum population of other life forms. The destruction of wild habitats is accepted as a necessary evil	Recognise that excessive pressures on planetary life stem from the human population explosion. The pressure stemming from industrial societies is a major factor, and population reduction in these societies, as well as in the developing countries, is imperative
Cultural diversity and appropriate technology	Regard industrialisation of the kind manifested in the West as an example for developing nations	Concerned with maintaining cultural diversity, and limiting the impact of Western society upon presently existing non-industrial societies, defending the fourth world against foreign domination, and applying local, soft technologies where appropriate
Source: After Naess (1986)		

The main focus of the Stockholm Declaration lies on the following facts:

1. Natural resources should be conserved, the earth's capacity to produce renewable resources should be maintained and non-renewable resources should be shared.
2. Development and environmental concern should go together and less developed countries should be given every assistance and incentive to promote rational environmental management.
3. Each state should establish its own standards of environmental movement as it wished, but should not endanger other states. There should be international cooperation aimed at improving the state of the environment.
4. Pollution should not exceed the capacity of the environment to clean itself and marine pollution should be prevented.
5. Science, technology, education and research should all be used to promote environmental protection.

The Stockholm Conference is eventfully marked by the presence of large number of NGO groups, this indicates the era of public participation in the work of government network.

UNEP experienced mixed fortunes since inception. Its prospects were made no easier by the fact that it had one of the most difficult jobs in the entire UN system. The performance of UNEP during the first decade was assessed by environmental groups (Table 64.3), that indicates it has good progress in certain areas but limited progress or very slow progress in other areas too.

Table 64.3: Performance of UNEP, 1972 - 1982

	Sierra Club	Augesta Group
Good Progress	Regional Seas conservation, Conventions/Protocols, INFOTERRA, Global monitoring	Regional Seas conservation/ Protocols; Periodic Reports
Fair Progress	IRPTC Microbiology Rural Technology Industries Inter-state relations Education Periodic Reports	IRPTC Microbiology Rural Technology Industries Inter-state relations Global monitoring INFOPERRM
Limited Progress	Advice to governments Development Planning Communications Desertification	Advice to governments Development Planning Communications Education
Slow Progress	Health Disaster warning system Technical assistance Management	Health Disaster warning management Technical assistance Managements Desertification

A decade after Stockholm, attitudes of both arenas—i.e. environment vs development—were more conciliatory. The interests of economic development were no longer seen as incorruptible and it was widely agreed that compromise between the two was both possible and essential. The new environmental slogan was 'sustainable development'. Since Stockholm Conference, in India, there are thousands of citizen groups, from local grassroots movement to broader national organisations. The power of grassroots movements is epitomised by the record of the Dasohli Gram Swarajya Mandal (DGSM). DGSM in the subsequent period organised a people's movement named Chipko Andolan or Movement to hug trees—attracted worldwide attention and admiration for its non-violent Gandhian style.

64.3 POLITICS OF GREEN MOVEMENTS

In March 1983, a new dimension was added to environmental politics in Western Europe with the arrival in the West German Bundestag of 27 members representing Die Griinen (the Greens). Although they represent the environmentist block, they also considered themselves the political voice of citizens movements opposed to nuclear power and supporting peace, feminism and LDC issues. Subsequently, by 1988, 15 West European countries had national Green parties, 118 Green representatives sat in nine national assemblies (Austria, Belgium, Finland, Italy, the then West Germany, Luxembourg, Portugal, Sweden and Switzerland). The spread of Green politics in Europe and USA further extended in subsequent period.

By the mid 1970s, few countries, if any, could claim to be unaffected either by environmental problems or by the rise of the environmental movement. Whether rich or poor, industrial or agrarian, authoritarian or democratic, socialist or capitalist—almost every society felt compelled to reassess its attitudes towards resource management and the condition of human environment. There is no denial of the fact that Stockholm has provided a compelling boost to internationalism of environmental issues.

The concern of World Environment Report described as ' Global 2000' indicate the following major issues of great concern:

- World population would increase by half—the greatest growth being in by developed countries;
- The gap between the richest and the poorest as per GNP would widen;
- There would be fewer national resources available;
- Important life supporting ecosystems—such as forest, wildlife, air, water etc. would be reduced;
- Prices of many vital resources would increase;
- The world would be more vulnerable to natural disaster and to disruptions from human causes.

One of the major efforts for combating the growing environmental problems at international level were several conventions, treaties, agreements made during the post-Stockholm decade in particular (Table 64.4).

The debate over achieving a balance between environmental management and economic development continued to build during the 1980s and it became increasingly obvious that much more thought needed to be given to the means of achieving such a balance. Rich and poor nations continued to have very different perceptions about the relative place of environment on the policy agenda and about the relative priorities of short term growth and longterm planning.

The report of World Commission made some categorical recommendation on the issue of implementation of environmental policy as International level. These are:

- National environmental protection agencies needed urgent strengthening, particularly in LDCs;
- UNEP's work needed to be reinforced and extended;
- Monitoring and assessment needed better focus and coordination;
- Policy-makers needed to work more closely with NGOs and industry;
- Law and International conventions needed strengthening and better implementation;
- The UN should work towards a universal declaration and later a convention on environmental protection and sustainable development.

As a follow up of the Commission's recommendation Rio Summit was arranged in 1992. Both government agencies and WGOs were invited in two separate platforms for the policy adoption or perception evaluation on the above issues.

The outcome of 'Rio' lies on five key issues:

- *The framework Convention climate change* i.e. debate relating to global warming issues;
- *The convention on Biological diversity* i.e. guideline for global biological conservation issues;
- *Agenda 21*, i.e. action plan for sustainable development through integration of goals of environmental protection and economic development;
- *The forest principle*, i.e. protection against exploitation of forest resource in rampant manner;
- *Rio declaration on Environment and Development*—general 27 principles.

Rio perplexed many inside and outside the environmental movement. It drew unparalleled levels of public attention to the problems of the environment and represented another major step along the road to a workable resolution of the tensions between environmental management and economic development.

Table 64.4: Some important International Conventions

Subject Area Name of convention	Subject Matter
Water	
The Rhine Convention	
Convention on the protection of the Rhine against chemical pollution.	Drawn up by Switzerland, France, West Germany, Luxembourg and the Netherlands to protect the Rhine against pollution; provides for strict control of discharge. Agreed in Bonn, 1976. The European Community has also become a party.
The Strasbourg Convention	
Draft European Convention for the protection of international water-courses against pollution.	Initially drawn up by the Council of Europe in May 1969, but complicated by overlap between its proposals and those in the European Community's own action programme on the environment. Still not adopted.
The Paris Convention	
Convention for the prevention of marine pollution from land-based sources.	Drawn up in 1974 to prevent pollution of the northeast Atlantic, including the North Sea. Initial signatories were those whose coasts bordered these waters. The EC now participates. The scope has extended to cover atmospheric inputs to the sea.
Bonn Agreement	
Agreement for cooperation in dealing with pollution of the North Sea by oil.	Drawn up in 1969 by countries with coasts on the North Sea. Provides for manpower, supplies, equipment and scientifiec advice in the event of discharges of oil or other noxious or hazardous substances in the North Sea. The EC is now a party.
Oslo Convention	
Convention for the prevention of marine pollution by dumping from ships and aircraft	Drawn up in 1972, came into force in 1974. Covers north-east Atlantic. Countries with coasts bordering this area have undertaken to stop dumping certain materials, and only to allow dumping of others with a specific permit.
Barcelona Convention	
Convention for the Mediterranean Sea against pollution.	Programme of action drawn up by countries bordering the Mediterranean in 1975 under the United Nations Environment Programme.
Cartagena Convention	
Convention for the protection and development of the marine environment of the wider-Caribbean region.	Similar to Barcelona Convention in respect of wider Caribbean region.
Helsinki Convention	
Convention on the protection of the marine environment of the Baltic Sea area.	Adopted by several countries bordering the Baltic Sea in 1980. Seeks to control and restrict discharges of harmful substances into the area.
Geneva Convention	
Convention on the long range transboundary air pollution.	Drawn up under the auspices of the United Nations Economic Commission for Europe. Adopted in 1979, came into force in 1983. Seeks protection from and gradual reduction of, air pollution. Provides for exchange of information, research, monitoring and development of policies. A Protocol signed at Helsinki required parties to reduce emissions of sulphur dioxide by 30% by 1993 (the 30% Club). The UK had not signed this protocol.

Contd.

Vienna Convention	
Convention for the protection of the ozone layer.	The first global convention concerned with the atmosphere. The Montreal Protocol, which came into force in 1989, requires countries to cut consumption of chlorofluorocarbons (CFCs).
Wildlife	
Ramsar Convention	
Convention on wetlands of international importance, especially as water foul habitat.	Adopted in 1971, came into force in 1975. Aims to stem encroachment on and loss of wetlands.
Cites	
Convention on International Trade in Endangered species (Washington)	Drawn up in 1973. Institutes a system of licensing for trade in endangered species, and prevents trade in most endangered species.
Berne Convention	
Convention on the conservation of European wildlife and natural habitats.	Drawn up by Council of Europe in 1979. Seeks to conserve wild flora and fauna in their natural habitats when conservation requires the cooperation of several states.
Bonn Convention	
Convention on the conservation of migratory species of wild animals.	Drawn up in 1979. States within whose borders there are threatened populations of migratory species should take concerted action to ensure conservation and management.

Source: Haigh, N. (1987) EEC Environmental Policy and Britain, 2nd edition, Longman, London.

64.4 MAJOR ENVIRONMENTAL MOVEMENTS IN INDIA

64.4.1 The Chipko Movement

The 'Chipko Movement'—which focussed world attention on the environmental problems of the Alakananda catchment area in the mid-Western Himalayas—was a movement of the local people to resist the deforestation attempt on the hill slopes in the name of developmental programme of the area. In 1960, from the point of view of border security, a vast network of roads was constructed in this area besides taking up projects of various other types. All this was disastrous for the forests and also total environment of the area. Felling of trees and rolling them down hills loosened the upper soil which eroded further during rain. All this resulted in the devastating flood in the Alakananda in July 1970, which caused havoc in the upper catchment area. It was here the Dasholi Gram Sarajya Mandal (DGSM), Gopeswar, which was engaged in social work in the Uttarakhand, stepped in the relief works during the 1970s flood. The volunteers of the Mandal realised them that forest and land and forest and man were intricately linked. Then they started educating the people about the ill-effects of deforestation in the hill slopes and gradually they geared themselves into a movement. Then the "Chipko Movement" was started. Initially there were severe conflicts with the local Forest Department. In 1974, the Forest Department began marking trees for felling in the Perg Murnenda Jungle, 430 hectares were auctioned for Rs. 471,000. However, volunteers of Chipko Movement were able to fight back for such tree-felling plan. Then this movement attracted the attention of the Government, scientists and experts of national and international bodies.

Though the movement initially started to educate the local people about afforestation, yet later it was possible to involve people in greater issues of ecology and environmental protection. The Chipko Movement also took the lead in introducing smokeless chulah and gobar gas plant in the area. The men behind the organisation of this kind of successful movement were Chandi Prasad Bhatt and Sundarlal Bahuguna.

Now the movement organisers not only look after the environmental problem of entire Garwal Himalayas, but also liaison with other such movements in the country and outside.

64.4.2 Silent Valley Movement

The Kerala Sastra Sahitya Parishad (KSSP), an NGO, was working for three decades among masses of Kerala for growing environmental awareness. In fact they were the pioneer in raising voice on Silent Valley Hydral Project in 1978. The campaign to save Silent Valley turned out to be a public education programme in many respects. The movement in many ways saved the ecosystem of Silent Valley area.

64.4.3 Appiko Movement

Karnataka's Uttar Kannada, which forms part of the Western Ghats, is known as the "forest district". The area has rich forest wealth with a typical micro-climate for cash crops such as black pepper and cardamon. During the colonial rule, the rich forest resources were exploited; the teak trees were felled to build ships and timber and fuelwood were sent to Mumbai. After independence, the government also began felling trees for revenue and the Forest Department, which continued the colonial forest policy, converted the primeval tropical forests into monoculture teak and eucalyptus plantations. This kind of development activities led to a drastic drop in the area under forest cover from 82% (in 1947) to 20% (in 1983).

A group of youth in Balegadde village, protesting against moves to establish teak plantations, wrote to forest officials asking them to stop clearing the natural forest. But this appeal was ignored. Then the villagers decided to launch a movement. They invited Sundarlal Bahuguna, the architect of Chipko Movement, and gathered local people to take up oath to protect trees by embracing them. In September 1983, when the axe-men came for felling the Kalase forests, people embraced the trees and thus the "Appiko Movement" was launched. This simple, non-violent action then became popular statewide and people joined this movement spontaneously in many forest areas of the Western Ghats. Appiko activists trekked along the Western Ghats from Coorg to Goa, spreading the movements message. Padayatras, slide shows and street plays created mass awareness. 'Yakshagana'—a form of folk dance—was used effectively to convey the message of conservation. The Appiko Movement had three main objectives:

- Ulisu (to conserve);
- Belsu (to grow), and
- Balasu (to make rational use).

It can hardly be denied that the Appiko Movement has created a mass awakening about conservation in South India and particularly in Karnataka, emphasising the need to involve the local people in saving the Western Ghats.

64.4.4 Narmada Bachao Andolan

During 1970s there was a plan to build 30 massive dams on the river Narmada. At present, the construction of two major dams: the Sardar Sarovar Project in Bharuch district of Gujarat and Indira Sagar Project in Khandwa district of Madhya Pradesh, raised tremendous conflict among common masses of the Narmada Valley. Thus the Narmada Bachao Andolan was developed gradually. Baba Amte, leader of Narmada Bachao Andolan, makes a case against the Sardar Sarovar Project and Indira Sagar Project in Gujarat and Madhya Pradesh.

According to the report prepared by Ministry of Environment and Forests, the environmental cost of loss of forest due to the Indira Sagar Project will be Rs. 30,923 crores and that due to the Sardar Sarobar Project Rs. 8,190 crores. The Forest Conservation Advisory Committee of the Ministry demanded, in January 1990, that the work on the project be halted immediately as it violated specific guidelines. The Committee said that the ecological balance of the area would be seriously affected due to the cutting down of 80 million trees at one go. The climate in the area will change, the rain cycle will be upset and the vital oxygen contents in the air in and around the area will decrease.

According to official figures, this project will submerge about 130,482 hectares of which 55,681 hectares is prime agricultural land and 56,066 hectares are forests. Apart from this direct submergence, there is also the land which is required for cultivation, fuel and timber for displaced human population and for grazing the massive numbers of livestock.

The Narmada Bachao Andolan has, with the help of experts of various fields, worked out the broad outlines of an alternative strategy both for water and power. For power, the Andolan is working out an alternative energy scenario for Madhya Pradesh, Maharashtra and Gujarat. The preliminary results indicate that a combination of energy conservation measures, decentralised power generation and centralised natural gas plants will provide much cheaper power, at a much lower cost of human displacement and environmental damage than the power provided by the Narmada dams. The thrust of the new strategy for water is the conservation and replenishment of the fast-dwindling resources through a large number of small watershed development projects. These projects help preserve and revive groundwater as also the water of innumerable small and big rivers. They do so in a manner which is non-destructive, beneficial to the environment and less costly. These are to be implemented with the full involvement of the locals who will derive benefits in a shorter time frame. Several such projects are already being implemented in the country and are yielding excellent results.

In fact, till date, this conflict is yet to be solved properly. The project is neither stopped or modified; as such there is every chance of fresh mass agitation on the same issue.

THE CHIPKO ANDOLAN

The Chipko Andolan—the movement to hug trees—is probably the world's most well-known grassroots ecodevelopment movement. Chipko activists have been subjects to newspaper articles and television programmes from India to the United States. They have attracted the attention of environmentalists, Gandhians and various people interested in the process of social change. Though pitted against the forest policies of the State of Uttar Pradesh, the movement has been praised by successive Prime Ministers. But despite its fame, the movement's objective of changing the State Government's forest policy and the current relationship of the hill people with forest resources, remains unfulfilled. Observers continue to be confused about what the movement actually stands for.

The Chipko Movement was born one morning in March 1973 in the remote hill town of Gopeshwar in Chamoli District. On that fateful day representatives from a sports goods factory situated in Allahabad reached Gopeshwar to cut 10 ash trees near village Mandal. The villagers courteously told them not to do so but when the contractors persisted, they hit upon the idea of hugging the earmarked trees. The next day the sports goods manufacturers had to return empty-handed.

Some weeks later the same contractor surfaced at Rampur Phata, another village some 80 km away from Gopeshwar, with a fresh allotment from the forest department. As soon as the villagers of Gopeshwar learned of this, they marched to Rampur Phata with drums and songs, gathering more people on the way. A confrontation ensued and the agitators hugged the earmarked trees to foil the contractors once again.

The Chipko Movement reached its climax in 1974 when the women of village Reni, some 65 km from Joshimath, got involved in a dramatic way. One day when their men were away in Joshimath protesting against the auction of a forest neighbouring Reni, the contractor arrived at the village to begin felling, taking this as an opportune moment. Undaunted by the number of men or their axes, the women of Reni, led by Gaura Devi, an illiterate woman of 50, barred the path to the forest which went through the village. As the women stood there, they sang: "This forest is our mother's home, we will protect it with all our might".

The genesis of the Chipko Movement has both an ecological and an economic background. The Alkananda Valley in which the movement originated was the scene of an unprecedented flood in 1970. The tragic aftermath of this flood left a deep impression on the hillfolk and, with it, soon followed the appreciation of the vital ecological role that forests play in their lives.

Contd.

The villagers here have also seen and resent the manner in which successive governments—beginning with the British—have taken away their forest wealth and turned it into a resource bank for faraway urban markets. Even for minor forest produce and articles of daily necessity like firewood, the local people have been forced to become thieves in their own homeland. Slowly, the entire ecology of the region has changed. The local people prefer the broad-leaved oak. But with the demands of the industrial culture increasing, oak forests have been destroyed and replaced extensively by the chir pine. Today, when the local villagers want to get oak wood to make ploughs, they are allotted pine trees,, whose wood is useless for this purpose.

When the Dasohli Gram Swarajya Mandal—the Gopeshwar-based organisation that has been at the forefront of the Chipko Movement since its early days—decided to set up a small factory to use the pine resin, it was persistently discriminated against in the supply of the raw material. While the allotted quantity was small, the price was much higher than the prices at which the resin would be allotted to the larger factories down in the plains. No amount of persuasion or running to Lucknow seemed to work. The factory still limps along.

The non-violent, action-oriented Chipko Movement has greatly helped to unite the people and focus attention on the mismanagement of forest resources. Its Gandhian character has brought it considerable sympathy. The expert committee set up by the State Government to enquire into whether the Reni forest should be felled found that the Reni women were more right from a scientific point of view than the Forest Department. This gave the movement considerable respectability. The committee concluded that because of the highly sensitive nature of the watersheds situated deep in the Himalayas, all felling should be banned to allow regeneration. These developments have not made the Forest Department change its forest policy but at least in the Chamoli district it is no longer in a position to implement its policy of selling forests to private contractors.

Meanwhile, as the years have gone by, the movement itself has acquired two distinct streams of thought marked by its two leaders, Chandi Prasad Bhatt from Gopeshwar who pioneered the movement, and Sundarlal Bahuguna from Silyara in the Tehri region. The operational style of both these leaders is totally different. While Bhatt is a grassroots worker and believes mainly in organising the people, Bahuguna, a journalist, is a publicist par excellence. Though Bahuguna has also organised some protest activities in his region—for instance, Chipko activists in Henwal Ghati once went to the forest to bandage the wounded trees with mud and sacking to protest against the indiscriminate tapping of pine trees—his main focus has been on spreading the message of Chipko far and wide. In 1981, Bahuguna started on a foot march from Kashmir to Kohima to campaign against deforestation.

Bhatt, on the other hand, has dug deep roots in the Chamoli region. He is, as a result, far less known than Bahuguna. Bhatt has realised that if the local village communities have the right to control their surrounding resources, they must also undertake to conserve and develop those resources. So he has organised the country's largest voluntary afforestation programme through ecodevelopment camps sponsored by the Dasohli Gram Swarajya Mandal. These bring together local villagers, students and social workers who have planted over a million trees. The survival rate of these Chipko plantations has been an astonishing 85-90 per cent in most cases. Bahuguna, however, tends to dismiss this activity as irrelevant at this stage of the movement against deforestation and concentrates all his writing and speaking power aginst the Forest Departments.

The two leaders differ not just in their operational styles but also in their philosophy with respect to the use of forests. Bahuguna is fiercely ecological in approach. Regreening of forests is top priority—a matter of national defence—for him. For instance, he argues that the main objective of forest management in Himachal Pradesh should be soil and water conservation; forests do not bear timber, resin and foreign exchange but soil, water and pure air. The self-sufficiency of the hill people in food, clothing and shelter is important to Bahuguna but secondary to the major ecological objectives. He is suspicious of all types of forest-based industries, including those set up by the local people. He wants all commercial green felling to be stopped and no new contracts to be entered into with industrialists to supply raw material.

For Bhatt, however, the search for a new ecodevelopment process for the region and the involvement of the local people are primary issues. "Saving the trees is only the first step in the Chipko Movement", says Bhatt, "Saving ourselves is the real goal. Our future is tied up with them". The economic conditions in the hills are today worse than even the average conditions of poverty in India. No conservation is possible without the cooperation of the people. Ecodevelopment strategies will inspire the poor when they are based not just on ecology but also on the needs of people.

Bhatt, therefore, wants forest resources to be used in a manner that is both environmentally and developmentally sound—in other words, while the environment is preserved, the benefits of the controlled exploitation accrue to the local people, a process in which decentralised economic growth and ecological conservation go hand in hand.

Contd.

It is a fact that whereas tree felling has continued at the instance of the Forest Department in Tehri Garhwal, it is not possible to do so any longer in the Chamoli region because of the higher level of mobilisation of the local people there.

Notwithstanding the divergent opinions of these leaders, the real strength of the movement are the women of the region. Except for a few "organised" events as at Rampur Phata or Henwal Ghati, the Chipko Movement essentially consists of a string of spontaneous confrontations in which none of the so-called leaders were present. Women, acting entirely on their own, rose up on the spur of the moment. While in Reni this protest was against a timber contractor, in all other cases the protest was against their own cash-hungry men, who could not care less if the forest was detroyed while their women had to walk for many more miles to collect their daily load of fuel and fodder.

Two dramatic incidents have taken place in the Chamoli region: the first in 1978 in the Bhyundar Valley, the area that leads on to the famous Valley of Flowers, and the second in 1980 in the village Dungari-paitoli situated in the Pindar watershed, which is so remote that the women there had not even heard of the Chipko Movement.

In Bhyundar, the women were confronted by their own relatives from a neighbouring village who were going to cut trees to meet the fuel needs of the 250,000 odd pilgrims that visit the famous Badrinath Temple every year. As the men of the Bhyundar village had refused to cut the trees earlier, the Forest Department had stealthily given the contract to a cooperative society in a neighbouring village. The Bhyundar women stole the axes of the men and refused to return them until they agreed to go back.

In Dungari-Paitoli, the battle was much more bitter and set wife against husband and mother against son. In 1978, the Government's horticulture department negotiated with the male-dominated panchayat for the acquisition of a nearby community forest to set up a potato seed farm. The men were led to believe that they would get employment in the farm and with its construction, the village would get a motor road, electricity, health centre and its primary school would be upgraded to a high school. When the women learnt that the forest had been sold, they protested strongly. The destruction of the forest would have meant walking at least another five km everyday to fetch fuel and fodder. Emboldened by the support they received from the Chipko activists like Bhatt, the women refused to let the forest be destroyed. After a bitter struggle in which the women were strongly opposed by their men and threatened with arrests by the district administration, the women finally won. None of the men's tactics, including threats to Chipko workers and spreading of rumours that bowing to the women would mean the village youth would not get employment in the army, worked in the end.

At its heart, thus, the Chipko Movement is very much a feminist movement. It has not only brought forth in a dramatic manner a greatly increased understanding of the divergent interests of local communities and State bureaucracies in the management of local resources, it is now finding that the interests of men and women within the same community can differ greatly. As long as the leadership of the Chipko Movement remains sensitive to this learning process, the movement is bound to grow in strength. The latest demand to emerge from the women of Chamoli is that it is they who should be elected to the Forest Panchayats and not their men.

NARMADA DAM — AT A GLANCE
(Two major Dam projects viz., Sardar Sarovar and Narmada Sagar is considered here)

	Information	Sardar Sarovar Project	Narmada Sagar Project
1.	Location	Bharuch district (Gujarat)	Indore district (Madhaya Pradesh)
2.	Height	138.68 meter	
3.	Water storage capacity	0.95 million hectare	NA
4.	Electricity generation capacity	1,200 megawatt (from dam) 250 megawatt (from canals)	225 megawatt
5.	Canal details	Main canal—458 km long (with water capacity 40,000 cusec) and 42 branch canals of 66,000 km long.	NA
6.	Expected sinking area in catchment	Agricultural land : 11,279 hectares Forest land—13,582 hectare 245 villages (193 in MP, 33 in Maharashtra and 19 in Gujarat)	44,363 hectare, 40,332 hectares, 149 villages
7.	Irrigation area	17.93 lac hectare	23,000 hectares
8.	Drinking water availability	0.86 million cff	NA

* NA — Data Not Available

India's Dam Dilemma

India is a land of rivers. From the Ganges in the north to the Cauvery in the south, most devout Hindus still consider its waterways sacred. The 'Puranas' (Hindu religious texts) hold that the Ganges River purifies the bather from sins; merely catching the sight of central India's Narmada River was said to do the same.

About 14 large, 44 medium, and 55 minor rivers run through India, and the river basins constitute about 90 per cent of the total drainage area. India's all major river basins are interstate in character, and 77.7 per cent of the nation's geographical area falls within these basins. India's rivers carry a total of 395 cubic miles of water annually; the Ganges alone carries 30 per cent of it, benefiting 400 million farmers in the Gangetic plains. But the survival of many rivers is at stake now due to aggressive development activities involving major dams.

India is the most prolific dam builder in the world today, with 4,525 large and medium sized dams. It ranks third in the world in the number of completed megadams, following China and the United States. India's megadams have also contributed to the destruction of natural ecosystems and the displacement of millions of people. Big dams built up-river in Karnataka virtually stopped the flow, especially at the tail end of one of the Cauvery River's tributaries in Tamil Nadu, where the water normally enters the Bay of Bengal near Poombuhar.

However, environmentalists and human rights activists argue that megadams displace millions of people and have debilitating effects on riverine ecosystems, affecting floodplains, woodlands, wildlife, fisheries, and forests that local communities rely on to secure their livelihoods. India's Sardar Sarovar dam on the Narmada became one of the world's most contentious dam projects mainly because of the displacement of people. According to the nongovernmental group International Rivers, after the government increased the height of the Sardar Sarovar dam to 122 m in December 2006, the total number of displaced people reached 320,000!! Most of the affected appeared to be poor, indigenous tribal communities. Several environmental and antidam groups, especially the Narmada Bachao Andolan (Save the Narmada Movement), continue to advocate for the affected, and hundreds of legal cases were launched against the government to fight for the rights of those displaced.

Statewide distribution of India's large dams (as of April 2002)

State or Union Territory	Completed dams	Dams under construction	Total
Andaman and Nicobar	0	1	1
Andhra Pradesh	161	24	185
Arunachal Pradesh	0	1	1
Assam	2	1	3
Bihar	24	5	29
Chhattishgarh	247	7	254
Goa	5	2	7
Gujarat	470	97	567
Haryana	0	0	0
Himachal Pradesh	5	1	6
Jammu and Kashmir	8	2	10
Jharkhand	48	28	76
Karnataka	203	28	231
Kerala	50	4	54
Madhya Pradesh	793	10	803
Maharashtra	1,453	198	1,651
Manipur	2	3	5
Meghalaya	6	0	6
Mizoram	0	0	0
Nagaland	0	0	0
Odisa	143	16	159
Punjab	11	1	12
Rajasthan	180	8	188
Sikkim	0	1	1
Tamil Nadu	92	8	100
Tripura	1	0	1
Uttar Pradesh	113	17	130
Uttaranchal	11	6	17
West Bengal	22	6	28
Total	**4,050**	**475**	**4,525**

Source: Central Water Commission, Dam Safety Dirctorate, Government of India, 2002

THE NARMADA VALLEY PROJECT, INDIA

The Narmada Valley Project in central-west India is one of the largest development projects ever begun in the world and it may take a hundred years to complete. Approved for commencement by the Indian Government in 1987, the project involves the construction of about 3,200 dams, thirty of them are major projects. Of these dams, five will produce a total of 2,750 megawatts of electricity, and nineteen will provide vast areas with vital irrigation water.

Unfortunately, the first two large dams to commence, the Narmada Sagar and Sardar Sarovar, are demonstrating some likely future negative consequences. The Narmada Sagar Dam will submerge more land than it will irrigate and eight communities based around tribal villages have been relocated. These people include farmers and tribal people who live on fish and forest produce. Because the tribal people own no land they receive no compensation from the government. Moving the expected one million or more people off their land for the larger dams is not easy and where can they go? All fertile land in India is already being occupied.

The Indian Government talks of the benefits to 11.5 million local villagers, who will gain greater productivity from irrigation, and be connected to electricity supply. People protesting against the dams have been ignored, or in some cases gaoled. The Environmental Impact Statement, prepared for the government, warns of salinity and water logging problems resulting from irrigation of unsuitable soils (known as "black cotton soil") in an area of high evaporation rate.

64.4.5 Tehri Dam Conflict

The 260.5 metre high Tehri Dam on the Bhagirathi in the Garhwal-Himalayas is being publiscised as one of the most promising projects — a symbol of the country's prosperity and prestige. It will bring prosperity by generating 2,400 MW peaking power, which, according to the dam builders, will help in establishing 140 industrial cities such as Modinagar (near Delhi), provide 300 cusec of water to meet the water supply needs of Delhi and irrigate 2.7 lakh hectares of land in Western Uttar Pradesh. The project has generated controversy since its inception. In spite of objections of several scientists of national and international repute, the project is yet to be modified or stopped.

TEHRI DAM — AT A GLANCE

1. Location	Tehri town of U.P. at the confluence of river Bhagirathi and Vilganga.
2. Height of Dam	260.5 meter
3. Electricity generation capacity	1,000 megawatt
4. Irrigation facility available	910,000 hectare
5. Drinking water facilities available	500 cusec
6. Water cover area of reservoir	4,250 hectare
7. Downstream electricity generation capacity (at Koteshwar)	400 megawatt
8. Expected area under inundation	1,600 hectare cultivated land, 2,000 hectare forest & 23 villages with over 50,000 people
9. Tehri dispute onset (By Tehri Dam opposition struggle committee)	1977

64.4.6 Conflict over Sharing of Water Resources

(A) Inter-State Conflicts

Conflicts occur when states which share river basins disagree on the usage of water. In most cases the upstream states have an upper hand. A number of such conflicts have been going on in the country for years, some for over a century. For bringing an amicable settlement to the interstate disputes, various treaties and tribunals have been established. While some of the settlements have been successful, many are still waiting an award.

TAMIL NADU-KARNATAKA

The most well-known of the Indian water conflicts, the Cauvery water dispute, every year promises a new drama. In India, the term water conflict has become synonymous with the Cauvery, where the states of Tamil Nadu and Karnataka are the conflicting parties. Violence, political battles, and strikes characterise the Cauvery dispute.

The Cauvery River dispute emerged in 1974 alter a 50-year-old agreement between the then Madras Presidency and the princely state of Mysore collapsed. Karnataka avowed that the agreement signed in 1924 prescribed discontinuation of water supply to Tamil Nadu after 50 years. The conflict is actually more than a century old, involving farmers in Tamil Nadu and Karnataka. Tamil Nadu does not control any of the Cauvery headwaters, but maximum irrigation development has taken place in the delta region within the state. The Krishna Raja Sagar in Karnataka and Mettur in Tamil Nadu are the two important storage dams. Before these dams were constructed, tanks and check bunds were common means of storing water for irrigation. About 95 per cent of the water of Cauvery, the highest for any river in India is utilised for irrigation. During a good monsoon year, there is peace between the neighbours. The conflict resurfaces every time the monsoon fails.

Until now, no long term solution for the problem has been found. In 1990, Tamil Nadu asked the Central Government for a Tribunal, and, on the direction of the Supreme Court, the Cauvery Water Disputes Tribunal was constituted. Karnataka argued that the pre-independence agreements were void and no longer applicable. The state claimed that Tamil Nadu has a good monsoon, while drought-prone areas in Karnataka were increasing, and were almost double the drought-prone areas of Tamil Nadu. Tamil Nadu countered this by stating that the old agreements could not be termed void, and stated that Tamil Nadu is completely dependent on the Cauvery for its agriculture. After hearing arguments from both states, the Tribunal directed Karnataka to release 205 thousand million cubic feet (TMC) of water so that it is available in Mettur reservoir from June to May every year. Karnataka was asked not to increase its irrigation area beyond the existing 11.2 lakh acres. Karnataka felt aggrieved by the orders of the Tribunal and refused to accept its decision, particularly since it felt that the Tribunal was not completely objective.

There was regular flow of water to Mettur until 1994. In 1995, there was a severe drought in parts of Karnataka. Water in the reservoirs of the Cauvery Basin was at 23.2 thousand million cubic feet (TMC) while the total requirement for Karnataka in 1994 was 24 TMC of water. Southern Karnataka and Bangalore faced a drinking water crisis. Karnataka refused to release water to Tamil Nadu. The state appealed to the Supreme Court, and, after hearing Tamil Nadu's plea, the Tribunal asked Karnataka to release 11 TMC of water. Karnataka refused to follow the orders of the Tribunal. The apex court asked the Prime Minister to intervene. In early 1996, the Prime Minister asked Karnataka to release 6 TMC of water to Tamil Nadu to save the standing crops. The Prime Minister also announced the formation of an expert council to assess the status of the standing crop in both states and estimate the amount of water needed to sustain the crops. This conflict keeps surfacing almost every year but there is no longterm solution to the problem in sight yet. Recently, Tamil Nadu has agreed to distress sharing during times of drought.

MAJOR INTER-STATE DISPUTES

Rivers and Projects	States involved
Ravi-Beas	Punjab and Haryana
Yamuna	Delhi, Uttar Pradesh, Uttaranchal and Haryana
Sutlej-Yamuna Link Canal	Punjab, Rajasthan and Haryana
Sone	Bihar and Uttar Pradesh
Damodar	West Bengal and Jharkhand
Krishna	Karnataka, Maharashtra and Andhra Pradesh
Telegu-Ganga	Karnataka and Andhra Pradesh
Cauvery	Karnataka and Tamil Nadu

Complied by: WWI Research Team

PUNJAB-HARYANA-RAJASTHAN

The Punjab Government passed the Punjab Termination Act 2004. By this Act, Punjab annulled all water sharing pacts it had with its neighbouring state in spite of Supreme Court orders to the contrary. This termination can have far reaching implications for the whole country as other states involved in inter-state disputes might follow suit and terminate their water sharing treaties. By passing the Bill in the Assembly, the Punjab Government has set a dangerous precedence for disregarding the directives of the Supreme Court.

The cause of the extreme step taken by the Punjab government was a Supreme Court order directing the state to hand over construction of the Sutlej Yamuna Link Canal (SYL) to a central agency as it refused to go ahead with the construction. The SYL is to feed water to other states, especially Haryana, from the Ravi and Beas rivers. Sharing of water between Punjab, Haryana and Rajasthan has been a cause for dispute right from 1970s when the foundation stone for the SYL was laid. When the state of Haryana was formed, it demanded its share of the water from the Ravi, Sutlej and Beas, which was reserved for India as allocated by the Indus Treaty with Pakistan. There were much bickering and several litigations were filed between the states. The Central Government intervened and, in 1981, a deal was brokered whereby water was to be shared between the various states. This allowed the sharing of surplus water of 15.85 million acre feet (maf), with Punjab getting 4.22 maf, Haryana 3.5 maf, Rajasthan 8 maf, Jammu and Kashmir 0.65 maf and Delhi 0.20 maf. The construction of SYL was started but has been halted many times as Punjab claimed that there was no surplus water and, if water was transferred through SYL, 9 lakh acres of land would go dry in Punjab. In 1987, Punjab rejected a commission report for supplying water to Haryana through SYL. The matter reached the Supreme Court, which in 2002 asked Punjab to construct the SYL canal within a year. Punjab filed a petition against the order, which the apex court rejected and asked the state to hand over construction of the canal to the Centre. When the deadline was approaching, Punjab passed the Termination Act to annul all deals on water with other states.

This created ripples and the Centre, along with the President of India, referred the matter to the Supreme Court. The validity of the Act is also being questioned. Meanwhile, the Himachal Pradesh Government has moved the Supreme Court to protect its interest that gets affected due to the Termination Act. There is no solution in sight and the matter is being passed back and forth between the court and the State and Central governments.

INTERNATIONAL DISPUTES

Due to its geographical position, India shares a number of rivers with neighbouring countries such as Indus with Pakistan, Ganga with Bangladesh and Kosi with Nepal. The allocation and use of these river waters has been agreed upon in various treaties signed with these nations. Some of the important treaties are the Indus Waters Treaty of 1960 between India and Pakistan, Mahakali Treaty of 1996 between India and Nepal, Kosi Bilateral Agreement of 1954, revised in 1966, between India and Nepal, and Farakka Treaty of 1996 between India and Bangladesh. The treaties are normally followed but there has always been space for disagreement among the signatories. Every now and then differences arise. India is the upstream country in most cases. By constructing barrages across the common rivers, water flow is regulated, which is not always accepted by the lower riparian neighbours.

INDIA-PAKISTAN

One of the most successful resolutions to an impending dispute between India and Pakistan has been the signing of the Indus Water Treaty in 1960. After the partition of the country, the boundary was drawn across the Indus basin with Pakistan being the lower riparian. This cut off two irrigation headworks—Madhopur on Ravi and Ferozepur on Sutlej. These headworks were in the Indian Territory and Pakistan Punjab was dependent on them. A dispute arose regarding the utilisation of water from the rivers that start in India and flow into Pakistan. Pakistan appealed internationally and negotiations held through the World Bank resulted in the signing of the Indus Water Treaty. According to the Treaty, the eastern flowing rivers Ravi, Beas and Sutlej are allocated to India and the western flowing rivers Indus, Jhelum and Chenab are mainly for the use of Pakistan. India can use water from the western rivers to irrigate about 3 lakh hectares. The country has very little consumptive rights but has full rights for non-consumptive use. India has to release water first for Pakistan's use and the country's water storage cannot exceed more than that prescribed under the treaty. The treaty provides for the formation of a permanent Indus Commission constituted with two Commissioners, one from each country. The Commission maintains cooperative arrangements, answers questions from both countries, settles issues and disagreements arising between

Fig. 64.2: Water of the Indus River is shared by India and Pakistan for irrigation to Leh, Jammu and Kashmir

the countries. The Commission also periodically inspects the rivers for ascertaining facts connected with various development works. It meets regularly at least once a year, alternately in India and Pakistan, and submits a status report to both before the first of June of every year. Both countries exchange data on flows and utilisation of the waters of the rivers of the Indus Basin. India supplies data of 27 hydel projects to Pakistan and also data of irrigated cropped area from the western rivers. A control room regularly informs the Pakistan control room about flood flows from the rivers. This treaty is a clear demarcation of water resources and until recently has never caused any serious disagreement between the two countries despite fighting a series of wars.

The Bagligarh Hydropower Project on the Chenab in the Doda district of Jammu and Kashmir was conceived in 1992, and the construction was started in 1999. The project will generate 450 MW power in the first phase and later will be doubled to 900 MW. This project is expected to benefit Jammu and Kashmir and help in its economic development. However, Pakistan has opposed the project stating that it violates the Indus Water Treaty, since the treaty accords the waters of the Chenab for the use of Pakistan. Pakistan has raised doubts against the purpose of the project stating that the 450 feet high dam is much higher than that required for power generation. The high dam will result in excess water being stored, more than the amount stated in the treaty. Also, the gates of the dam prevent water from flowing into Pakistan, affecting irrigation. It has suggested that the maximum capacity for the Bagligarh project should be 130 megawatt (MW). India, on its part, denies these allegations and states that the project is a run of the river project and will not involve water storage. Moreover, power projects are permitted under the Indus Water Treaty. Pakistan is not satisfied with India's response and has decided to take the matter to the World Bank and demand the establishment of a neutral party for resolving the dispute. Another conflict between the two countries that is yet to be settled is over the construction of Tulbul Navigation Project on the Jhelum.

INDUS WATER TREATY

River	Flood Storage Capacity (MAF)	General Storage Capacity (MAF)	Power Storage Capacity (MAF)
Indus	0.25	0.15	Nil
The Jhelum (excluding the Jhelum Main)	0.50	0.25	0.75
The Jhelum Main	Nil	Nil	As per the treaty
The Chenab (excluding the Chenab Main)	0.50	0.60	Nil
The Chenab Main	Nil	0.60	Nil

INDIA-BANGLADESH

The Farakka Barrage is one of the biggest and oldest discords between upstream India and downstream Bangladesh. The barrage has been constructed on the Ganga at Farakka in Malda district of West Bengal to increase water flow to the river Bhagirathi by diverting 40,000 cusec of water and to simultaneously flush the silt off the important Kolkata port, plus increase navigation in the Hoogly-Bhagirathi stretch. The barrage was commissioned in 1963 and was ready by 1975. Bangladesh was against the construction of the barrage and experts in India too opposed it—arguing that it would bring in ecological degradation and increase water logging upstream, and dry places downstream. Bangladesh perceived this as unilateral diversion of waters of the Ganga by India which was detrimental to Bangladesh and its ecology. Soon after the completion of the barrage, it was observed that during the dry seasons, the barrage blocked

Fig. 64.3: Upstream of Farakka Barrage on the Ganga

natural flow of water into Bangladesh, causing water shortages. Many places began to dry up and it is alleged that desertification in Natore and the drying of Gorai—a distributory of the Ganges—was due to the barrage. Navigation, agriculture and fisheries too were affected. The Sundarbans and its mangroves were affected by salinity. In the rainy season, water releases from Farakka caused floods in Bangladesh.

In 1976, Bangladesh took the matter to the United Nations, which asked both the countries to arrive at a fair and expeditious settlement. In 1996, both the countries signed a treaty on the sharing of waters from the Ganges. Under this agreement, India was expected to release water for Bangladesh from Farakka. This 30-year treaty emphasises water-sharing arrangements primarily for the dry season. These arrangements are to be reviewed every five years. Though the agreement is in force and water is released every year, Bangladesh is not happy with the situation and on and off accuses India of not releasing the amount of water agreed upon. Bangladesh is now seeking a review of the Farakka treaty.

Besides causing a trans-boundary conflict, the Farakka Barrage has destroyed the ecology in the region. Observations in recent years uphold the predictions made by Bangladesh when plans were being made to construct the barrage. One of the observations made by experts on the Indian side, both upstream in Jharkhand and downstream in West Bengal, is the steady decline of the depth of the river Ganga. In 1975, the depth of the river at the barrage was 25 m while in 1997 it was only 4 m. Experts say that the bed of the river Ganga has risen by 20 m due to silt accumulation caused by the barrage. The Central Water Commission in 1995 stated that in West Bengal, Bihar, Jharkand and Uttar Pradesh, two and a half acres of land in the proximity of the Ganga remains submerged under water for most of the year. The soil is turning alkaline and acres of fertile land in Jharkand have turned barren.

It has been observed that the water tends to cut into the land more sharply now resulting in soil erosion in villages and towns on the banks of the Bhagirathi. In Murshidabad district, more than 600 square kilometres of fertile land has been swept away and more than 5 lakh people have been displaced in the past 22 years. Recently, there have been signs of the Ganga trying to change its course due to the barrage and merging with its tributary Pagla in Malda district. There are fears that the town of Malda will be inundated in a few years as the river continues to cut into the land. Embankments constructed to stop erosion have not helped.

The Farakka barrage has obstructed the migration of marine fishes leading to the near absence of several popular varieties in entire northern India. There are many fishes that live in fresh water but breed in marine water, which is no longer possible due to the barrage.

The Farraka barrage constructed in order to save the Kolkata port has not served its purpose as it did not help much in desilting the port area, and consequently the Government of West Bengal has made Haldia the main port. The barrage has created two types of conflicts—a trans-boundary dispute with Bangladesh, and the more acute ecological conflict with nature.

ALLOCATION FROM FARAKKA BARRAGE

Availability at Farakka	Share of India	Share of Bangladesh
70,000 cusecs or less	50 per cent	50 per cent
70,000-75,000 cusecs	Balance of flow	35,000 cusecs
5,000 cusecs or more	40,000 cusecs	Balance of flow

Source: http//www.thewaterpage.com, Water Web Management Ltd. 1 Dome Hill, Caterham, Surrey CR3 6EE, UK

INDIA-NEPAL

This is one trans-boundary dispute where India is the lower riparian. The Mahakali treaty was signed in 1996 between India and Nepal for the development of the Tanakpur Barrage and Pancheshwar Dam. Misgivings from both sides preceded the signing of the treaty, and disagreements and conflicts continue to this day. India has regularly accused Nepal of flooding the Indian states of Uttar Pradesh and Bihar by releasing water from Kosi and Gandak. Nepal, on its part, has felt deprived of using the river waters. India constructed barrages over Ghagra and Mahakali. Nepal too built barrages over the Kamala and Bagmati. In 1991, the two countries signed a treaty for the Tanakpura Barrage Project. Under the Memorandum of Understanding, the barrage built in India will cover parts of the embankment in Nepal. But the Supreme Court of Nepal stated that the agreement needed ratification in the Parliament. In 1996, the treaty for an integrated development of Mahakali river was ultimately prepared, which included Sarda and Tanakpur Barrages along with the Pancheshwar Dam. The 315 m proposed dam with a gross storage capacity of 12.3 bcm water and 6,480 MW power generation capacity is the highlight of the project. The treaty quantifies Nepal's portion of water, but does not state India's share. Nepal is not happy with this arrangement and they argue that since the quantity to be used by India is not mentioned, the country can use all the water of Mahakali, especially in the lean season. Nepal's stand is that the Mahakali is a border river and this fact should guide the decisions and allocations, a Mahakali Commission should be set up, the sale of electricity should be based on avoided cost principle, and there should be equal usage of water for both countries. Currently, the status of the project is dormant with the detailed project report that was supposed to be completed in six months, lying incomplete even after eight years, while the disputes over sharing of water continue!

INDIA-CHINA

The recent controversy of Dam creation by China in Tibet by diverting water of upper part Brahmaputra, raised great controversy.

Fig. 64.4: Numerous such rivulets flow from Nepal into India

CONCLUSIONS

Disputes over water resources arise mainly due to lack of regulatory mechanisms for allocation and rational use of these resources. Although institutional mechanisms are in place such as tribunals, pollution control boards, and other regulatory bodies, these institutions are either not equipped to handle conflicts or have been ineffective in enforcing the rules and regulations. As a new crisis emerges every year, short-term makeshift solutions are found, but efforts are not made to resolve the disputes on a permanent basis. This is also largely due to vested interests that not only turn a blind eye to the need for regulation, but also issue the permits required to exploit the water resources for commercial gain.

A well-defined, objective, legal mechanism is required to enforce permanent solutions. Apart from this, it is important to focus on reducing these conflicts, which can happen only by increasing water efficiency and regulating the use of water. The wastage in all sectors, be it agriculture, industry or domestic consumption needs to be controlled for conserving this already scarce resource and minimising disputes. Comprehensive policy reforms are required, including revision of subsidies that encourage the exploitation of water for irrigation, regulation of water use by industrial and commercial sectors which obtain water freely from the environment and recognising and safeguarding the rights of local communities to water resources.

64.4.7 Ideological Trends in Indian Environmentalism

Social action in the three generic modes outlined above (i.e. struggle, publicity, and restoration) constitutes the bedrock of the Indian environmental movement. While such activism has characteristically been localised—with most groups working within one district—the links between the micro and macro spheres have been made most explicit (recent initiatives like the JVA excepted) through the environmentalists' critique of the ruling ideology of Indian democracy, that of imitative industrialisation. For environmentalists have insistently claimed that the intensification of natural resources conflict is a direct consequence of the resource—and capital-intensive pattern of economic development, modeled on the Western experience, followed since Independence. The resource, illiteracy of development planning, the claim, are directly responsible for the impoverishment of the resource base and of the millions of rural people who depend on it (JVA, 1990).

While there is widespread agreement within the environmental movement as regards the failures of the present development model, there is little consensus on plausible alternatives, all responding to the range of conflicts we have analysed above, but advocating widely varying proposals for mitigating these conflicts. It is, however, possible to identify three distinct ideological perspectives within the movement. It is, of course, entirely possible that none of the ideologies so identified is present in a particular struggle, or indeed that adherents of all three viewpoints might participate in unison in a specific initiative. However, close study and discussions with groups spread all over India does suggest that the three strands analysed below are the dominant ideologies of Indian environmentalism.

The first, which we may call *Crusading Gandhian*, relies heavily on a moral/religious idiom in its rejection of the modern way of life. Here, environmental degradation and social conflict are viewed above all as a *moral* problem, their origins lying in the wider acceptance of the ideology of materialism and consumerism, which draw humans away from nature even as it encourages wasteful lifestyles. Crusading Gandhians argue that the essence of 'Eastern' cultures is their indifference, even hostility to economic gain: thus, if India were to abandon its pursuit of Western models of economic development, it would only be returning to its cultural roots. These environmentalists call, therefore, for a return to precolonial (and precapitalist) village society, which they uphold as the example of social and ecological harmony. Gandhi's own invocation of Ram Rajya (the mythical but benign rule of King Rama) is here being taken literally, rather than metaphorically. In this regard Crusading Gandhians frequently cite Hindu scriptures as exemplifying a 'traditional' reverence for nature and lifeforms.

Crusading Gandhians have worked hard in carrying their message of moral regeneration across the country and indeed across the globe. They have sharply attacked the stranglehold of modernist philosophies—particularly those upholding rationalism and economic growth—on the Indian intelligentsia; through the written and spoken word, they propagate an alternative, non-modern philosophy whose roots lie in Indian tradition (Bahuguna, 1983; Nandy, 1987, 1989; Shiva, 1988).

The second trend, in many ways the polar opposite of the first, is Marxist in inspiration. Marxists see the problem in political and economic terms, arguing that it is unequal access to resources, rather than the question of values, which better explains the patterns and processes of environmental degradation and social conflict. In this sharply stratified society, the rich destroy nature in the pursuit of profit, while the poor do so simply to survive (the Crusading Gandhians would tend to deny altogether that the poor also contribute to environmental degradation). For Ecological Marxists, therefore, the creation of an economically just society is a logical precondition of social and ecological harmony. In their practical emphasis, socialist activists concentrate on organising the poor for collective action, working towards their larger goal of the redistribution of economic and political power. While including various Naxalite and radical Christian groupings, Ecological Marxists in the Indian context are perhaps most closely identified with People's Science Movements (PSMs)—the best known of which is the Kerala Sastra Sahitya Parishad—whose initial concern with taking 'science to the people' has been widened to include environmental protection. Ecological Marxists can be distinguished from Gandhians in two significant respects: their unremitting hostility to tradition (and corresponding faith in modernity and modern science) and in their relatively greater emphasis on confrontational movements (KSSP, 1984).

Crusading Gandhians and Ecological Marxists can be seen as representing the 'ideological' and 'political' extremes of the Indian environmental movement, respectively. Because of their ideological purity and consistency, their arguments are often compelling, albeit to different sets of people. In between these two extremes and occupying the vast middle ground, lies a third tendency, which may be termed (less controversialy) Appropriate Technology. Less strident than the Gandhian in its opposition to industrial society, this strand of the environmental movement strives for a working synthesis of agriculture and industry, big and small units and Western and Eastern (or modern and traditional) technological traditions. Both in its ambivalence about religion and in its criticism of traditional social hierarchies it is markedly influenced by Western socialism. Yet, in its practical emphasis on constructive work, it taps another vein in the Gandhian tradition. Thus Appropriate Technologists have done pioneering work in the generation and diffusion of resource conserving, labour intensive and socially liberating technologies. Their emphasis is not so much, *pace* the Marxists, on challenging the 'system'—or, *pace* the Gandhians, the system's ideological underpinnings—as in demonstrating in practice a set of socio-technical alternatives to the centralising and degrading technologies presently in operation (Bhatt, 1992; Reddy, 1982).

All three tendencies are represented in that most celebrated of environmental initiatives, the Chipko Movement (Guha, 1989a). The Gandhian trend, associated above all with the figure of Sunderlal Bahuguna, is best known outside the Himalaya. The Marxist trend within Chipko has been represented by the Uttarakhand Sangarsh Vahini, a youth organisation that has organised popular movements against commercial forestry, unregulated mining and the illegal liquor trade. Finally, the Appropriate Technologists are represented by the organisation under whose auspices the movement began, the Dashauli Gram Swarajya Mandal, whose fine work in ecological restoration has already been alluded to.

These contrasting perspectives may be further clarified by examining each strand's attitudes towards equity and science, as well as their style and scale of activism. Most Crusading Gandhians reject socialism as a Western concept: this leads some among them to gloss over inequalities in traditional Indian society and others even to justify them. Clearly the Marxists have been most forthright in their denunciations of inequality across the triple axes of class, caste and gender. The Appropriate Technologists have been sufficiently influenced by Marxism to acknowledge the presence and pervasiveness of inequality, but have rarely shown the will to challenge social hierarchies in practice. Attitudes towards modern science and technology also vary widely. The Gandhians consider science to be a brick in the edifice of industrial society and responsible for some of its worst excesses. Marxists yield to no one in their admiration of modern science and technology, viewing science and the 'scientific temper' as an indispensable ally in the construction of a new social order. Here, the Appropriate Technologists are the most judicious, calling for a pragmatic reconciliation between modern and traditional knowledge and technique, to fulfil the needs of social equity, local self-reliance and environmental sustainability.

On the scale of activism, Appropriate Technologists prefer to work on a micro scale—a group of contiguous villages at best—in demonstrating the viability of an alternative model of economic development. The Gandhians have the largest attempted reach, carrying their crusade on worldwide lecture tours: they have often tended to think globally and act globally, even as the Appropriate Technologists have acted locally and, occasionally, thought locally too. The Marxist groupings work in the intermediate range, at the level of a district perhaps, or (as in the case of the KSSP) the level of a state. Finally, the three strands also differ in their preferred sectors of activism. Their rural romanticism has led the Gandhians to exclusively emphasise agrarian environmental problems, a preference reinforced by their well-known hostility to modern industry. While Appropriate Technologists do recognise that some degree of industrialisation (though not of the present resource-intensive kind) is inevitable, in practice they too have worked largely on technologies aimed at relieving the drudgery of work in the village. Here it is the Ecological Marxists, with their natural constituency among miners and workers, who have been most alert to questions of industrial pollution and workplace safety.

Crusading Gandhians, Appropriate Technologists, and Ecological Marxists represent the three most forceful strands in the Indian environmental debate; but we should also take account, however briefly, of two other points of view. First, we have the Indian variant of that vibrant strand in global environmentalism, the Wilderness Movement. Indian naturalists have provided abundant documentation of the decline of natural forests and their plant and animal species, urging the government to take remedial action (Krishnan, 1975). Although their earlier efforts were directed almost exclusively towards the protection of large mammals, more recently wildlife preservationists have used the scientific rhetoric of biological diversity and the moral arguments in favour of 'species equality' in pursuit of a more extensive system of parks and sanctuaries and a total ban on human activity in protected areas (Guha, 1989b).

So we come, finally, to an influential strand of thinking within the state and state agencies, which might be termed Scientific Conservation. Preeminent here is the work of B.B. Vohra, a senior bureaucrat who was one of the first to draw public attention to land and water degradation. In a pioneering and impressively thorough paper (Vohra, 1973), he documented the extent of erosion, waterlogging and other forms of land degradation. There was, he noted, no countrywide organisation or policy to deal with these problems; nor was there coordination between concerned government departments. For Vohra, as for the early scientific conservationists (Hays, 1957), the solution lies in the creation of new ministries and departments to deal with problems of environmental degradation. The Central Government, he has written, 'has no option but to obtain a commanding position for itself in the field of land and soil management through financial and administrative measures' (Vohra, 1973; see also Vohra, 1980, 1982).

Neither wilderness protection nor scientific conservation command a popular following, yet each has had a considerable influence on government policy. Both tendencies look upon the state as the ultimate guarantor of environmental protection and their energetic lobbying has informed stringent legislation in pursuit of this ideal, such as the Wildlife Protection Act of 1972 (modified in 1991), the Forest Conservation Act of 1980 and the Environment Protection Act of 1986. However, in so far as neither group is cognisant of the social roots of environmental use and abuse, they tend to be dismissed as 'elite' conservationists by environmentalists owing allegiance to Gandhian or Marxist traditions.

64.5 ENVIRONMENTAL AWARENESS

In recent years environmental concern have been at the forefront of public attention. Public awareness of environmental issues is a major driving force leading to calls for the development of national and international policies to protect the environment. Growing public participation in community organisations, lobbying groups and local action groups also reflect an increasing environmental awareness.

64.5.1 Public Opinion

Between February 1988 and June 1989 a survey of public and leadership attitudes towards environmental issues was conducted in 16 countries by Louis Harris and Associates on behalf of UNEP. The results of the survey pointed to deep public concern about the quality of the environment, coupled with the belief that environmental protection should be a major governmental priority. In all countries but one, most members of the public and leaders surveyed believed that their environments had become worse over the past 10 years.

64.5.2 Members of Non-governmental Organisations

Since 1970s non-government organisations (NGOs) have played an increasingly important role in promoting environmental awareness. It is difficult to estimate how many people worldwide are members of environmental organisations. Non-government organisations are as varied as the issues they confront, ranging from small groups organised in response to a particular issue of local concern to large organisations sapanning several countries and working on global issues. This latter category includes the World Conservation Union (IUCN), the World Resources Institute (WRI) and the International Institute for Environment and Development (IIED). Many NGOs have similar interests and coordination of activities is seen to be increasingly important. Major Environmental NGOs (ENGOs) are listed in Table 64.5 below:

Table 64.5: Major ENGOs of the World

Name	Notes
Friends of Earths (FOE)	Established a tropical forest campaign in 1984, linking UK and European timber companies and tropical deforestation. Organised a boycott against tropical hardwoods. Lobbies on trade and aid issues. Carried out work on illegal use of aid funding, the potential for sustainable management of tropical forests and the misuse of plantations as carbon stores.
Greenpeace	Set up a forest campaign in the 1990s, although has worked on pulp and paper issues for years. Blockaded some illegal logging in Siberia and involved in campaign against logging of Clayoquot Sound in Vancouver Island, Canada.
International Union for the Conservation of Nature (IUCN)	Research organisation although also runs many field projects and has a policy brief. Involved in tropical forestry and has drawn up guidelines for management of reserves and for timber operations; now interested in all forests.
Native Forests Network	Recently established international NGO, working on native forests and with a leaning towards direct action.
Rainforest Foundation	To support indigenous people and traditional populations of the World's Rainforests for securing natural resources and their culture.
Survival International	Operates in forest areas. Area of origin is the UK.
World Rainforest Movement (WRM)	Strong emphasis on local and indigenous peoples' rights and on sustainable use of the rainforest. Occasional publications and lobbying role; closely connected with The Ecologist magazine. Active critic of the Tropical Forestry Action Plan.
World Wildlife Fund for Nature (WWF)	Has run field projects in forest conservation for 30 years; policy work began on tropical forest issues in the 1980s and extended to all forest in the early 1990s. Influential in promoting independent timber certification and establishing the Forest Stewardship Council. Many national organisations also involved in forest conservation.
Source: Equilibrium 1994	

The more developed Western nations are not alone in experiencing a risk in the number of—and participation in—NGOs. Developing regions have a history of action by non-government groups in environmental and development issues. Often based around local communities, these groups are having an increasingly important impact on environmental outcomes.

In Indonesia, for example, there are an estimated 600 independent development groups working in environmental protection. In Colombia, farmers have organised themselves into local "green councils". In Eastern Europe there has also been growing environmental awareness; an estimated 62 independent environmental groups have been organised in Poland since 1980, focusing mainly on air and water pollution and forest destruction (During, 1989).

The public opinion poll, conducted by Louis Harris and Associates (1990) referred to above, found that 74 per cent of people surveyed felt that voluntary and community organisations which are not a part of government could play a major part in solving environmental problems. The poll found that 81 per cent of leaders surveyed agreed that NGOs had an important role to play. From these responses it can safely be assumed that NGOs will continue to be a major driving force in solving both local and global environmental problems.

In India, thousands of Environmental Non-Governmental Organisations (ENGOs) have raised awareness and mobilized people to take action on environment-development issues. The importance given to NGOs is denoted by the fact that the National Environmental Council has five representatives from NGOs. An effective means of advocacy that NGOs in India have adopted is peoples' tribunals. For instance the Permanent Peoples Tribunals (PPTs) hears cases filed by individuals or communities affected by environmental degradation whose judgments are widely publicised. Programmes related to community forestry have been initiated under capital Eco-Development Programme. Panchyati Raj cooperatives and self-help groups facilitate community-based management of the environment.

In India, ENGOs have made important initiatives in rural resource related problems, but the complexities of urban issues have blocked their progress in cities. They lack technical capacity to tackle environmental threats like pollution and fuel quality. Development and environmental NGOs, for example, are learning that they can be more effective, and their work can have more positive effects, if they work with the actual communities and help them to empower themselves.

When urban managers, environmental regulators, and voluntary environmental programmes for industry fail to address critical environmental problems, the citizenry must step in to take action to resolve environmental issues. For this reason United States-Asia Environmental Partnership (US-AEP) has identified a number of critical areas in which to demonstrate the importance and value of civil society participation in India. The critical issues include drought mitigation and the management of water resources, river pollution, urban environmental devastation of fragile Himalayan cities, the connection between environmental management and poor labour practices, the development of standards for the management of toxic wastes, and the necessity to involve women in environmental decision making. These six problems or issues are widespread and occur in many varying forms in every corner of India. Through its support for targeted opportunities to demonstrate successful citizen participation in environental decision making US-AEP would help develop and disseminate model citizen participatory approaches.

Activities of few selected ENGOs are given:

- **The Centre for Science and Environment** does research, investigative and educational work in the field of pollution, forest, wildlife, land and water use. The activities are carried out through lectures, field trips, publications, exhibitions on the various issues they take up, meetings and workshops. Their publications *Down to Earth*—a fortnightly environment magazine—and the annual *Citizens' State of Environment Report*—to spread messages across which focus on specific issues such as urban pollution and flood management in nontechnical language—are quite popular. CSE has also introduced Green Rating Project for industrial sector in India.

- **Dasholi Gram Swarajya Mandal** was set up at Gopeshwar, Chamoli District, UP, in 1964. Its aims are to encourage forest conservation and the use of forest products for self-employment; to cut down soil erosion by encouraging volunteers to build embankments in the catchment areas and to plant trees. It gives a great deal of importance to forest conservation in the Himalayas. It has promoted environmental awareness in these area and carried out major eco-regeneration programmes. The world famous **Chipko Movement** was spearheaded by this organisation. The Chipko Movement grew out of the fact that women are the most potent campaigners through the Gandhian method of nonviolent resistance in the Himalayan forests of Uttar Pradesh against the forest contractors.

- **Development Alternatives**, based at Delhi, work in all parts of the country. It was established in 1983 to design options and promote sustainable development through programmes of economic efficiency, equity and social justice, resource conservation and self-reliance. They are working in the field of pollution monitoring and control; waste recycling management, wasteland development; and appropriate use of clean technology in industrial processes.
- **Gandhi Peace Foundation's Environment Cell** began functioning at Delhi from June 1979. It was set up mainly to promote the environmental activities of rural development agencies; to disseminate environmental information through the publication of up to date reports on environmental issues; to organise workshops and seminars for environmental experts, policy makers, individuals and organisations working for environmental issues. Their activities include researching the role of women in community forestry and rural development; conducting studies in soil erosion, water-logging, drainage and seepage around select dams; planting fast growing varieties of trees.
- **The Energy and Resources Institute (TERI)**, established in 1974, is an independent non-profit research institute. Its mission is to build up and promote technologies, policies and institutions for efficient and sustainable use of natural resources. It has been imparting environmental education through projects, workshops, audio visual aids and quiz competitions. It deals with policy related work in the energy sector, research on environmental subjects development on renewable energy technologies and promotion of energy-efficiency techniques in the industry and transport segment. TERI also has a major programme in biotechnology, the applications of which are oriented towards increased biomass production, conversion of waste into useful products and mitigating the harmful environmental impacts of several economic activities.

64.6 ENVIRONMENTAL CONFERENCES

Unlike the early days of man where he lived in harmony with nature, development of human civilisation led to population explosion, industrialisation, depletion of natural resources, nuclear fallouts, border conflicts leading to war, and wide economic disparity. It was in this background that the United Nations Environmental Agency organised the Internation Conferences.

1. **The Stockholm Conference (1972)**: It is known as International Conference on Human Environment, held at Stockholm from June 5 to 14, 1972. It was attended by representatives of 114 Nations. 150 act plans and 20 principles were approved to protect the delicate balance of the ecosystem and also to preserve them for the coming generation also. The conference adopted the motto "Only One Earth" for the entire humanity. The conference declared June 5 as the World Environment Day.

2. **Nairobi Conference (1982)**: This is the UN conference on "World Environment" held at Nairobi to make an assessment of stock-taking exercise of the world environmental scenario.

3. **The Rio Summit (1992)**: The UN convened "Earth Summit" at Rio de Janeiro to forecast "Our common future" from June 3 to 14, 1992. The conference was attended by 115 heads of states, 10,000 government missions and 20,000 NGOs. The six basic conspicuous issues were discussed viz., greenhouse gas emission, forests, population, technology transfer, finance sharing for environment protection (Global environmental facility) and degradation. The Earth Summit ended with the adoption of the 'Rio Declaration' and Agenda 21—a blueprint for sustainable development.

4. **Kyoto Summit (1997)**: 180 countries gathered in Kyoto (Japan) in search of coordinated international response to global warming.

5. **The Johannesburg Earth Summit (2002)**: This UN conference on "Sustainable Development", focussed on the following issues:

(a) Addressing key development—environmental issues;

(b) Facing longterm challenges;

(c) Strengthening partnerships and financing.

SUMMIT 2012

64.7 INTERNATIONAL AGREEMENTS

During the past several decades growing environmental awareness has encouraged national governments to legislate for better protection and management of the environment. In addition, recognition of the global nature of problems such as ozone layer depletion and climate change has resulted in cooperation between governments in negotiating international treaties and agreements.

The UNEP Environmental Law and Institutions Unit in Other Agreements a 'Register of International Treaties' (UNEP 1989). This listing is submitted annually to the General Assembly of the United Nations, which has expressed concern that many existing international agreements have not yet received wide acceptance and application they deserve. UNEP continues to urge all states entitled to become parties to existing conventions and protocols to do so as soon as possible. A comprehensive listing of existing international treaties and agreements on the environment was provided in the 1989 edition of this report (UNEP, 1989).

The late 1980s saw the signing of several important conventions and protocols concerned with the environment. Agreements covering air pollution, endangered wildlife, Antarctica, depletion of the ozone layer and hazardous wastes have set the stage for further international negotiation and cooperation on environmental issues. The following sections provide an overview of the status of three major conventions and protocols: the Convention on Long-range Transboundary Air pollution and its related protocols, the Montreal Protocol on the Substances that Deplete the Ozone Layer and the Basle Convention on the Control of Transboundary Movements of Hazardous Wastes and their Disposal.

In the Earth Summit held at Rio, Brazil (1992), two important treaties were signed by a number of countries. They are namely

- Convention of Biodiversity, and
- UN Convention on Climate Change.

64.8 RIO EARTH SUMMIT: CONVENTION ON BIODIVERSITY

Preamble

The Contracting Parties,

Conscious of the intrinsic value of biological diversity and of the ecological, genetic, social, economic, scientific, educational, cultural recreational and aesthetic values of biological diversity and its components.

Conscious also of the importance of biological diversity for evolution and for maintaining life sustaining systems of the biosphere.

Affirming that the conservation of biological diversity is a common concern of humankind.

Reaffirming that States have sovereign rights over their own biological resources.

Reaffirming also that States are responsible for conserving their biological diversity and for using their biological resources in a sustainable manner.

Concerned that biological diversity is being significantly reduced by certain human activities.

Aware of the general lack of information and knowledge regarding biological diversity and of the urgent need to develop scientific, technical and institutional capacities to provide the basic understanding upon which to plan and implement appropriate measures.

Noting that it is vital to anticipate, prevent and attack the causes of significant reduction or loss of biological diversity at source.

Noting also that where there is a threat of significant reduction or loss of biological diversity, lack of full scientific certainty should not be used as a reason for postponing measures to avoid or minimise such a threat.

Noting further that the fundamental requirement for the conservation of biological diversity is the *in-situ* conservation of ecosystems and natural habitats and the maintenance and recovery of viable populations of species in their natural surroundings.

Noting further that *ex-situ* measures, preferable in the country of origin, also have an important role to play.

Recognising the close and traditional dependence of many indigenous and local communities embodying traditional lifestyles on biological resources and the desirability of sharing equitably benefits arising from the use of traditional knowledge, innovations and practices relevant to the conservation of biological diversity and the sustainable use of its components.

Recognising also the vital role that women play in the conservation and sustainable use of biological diversity and affirming the need for the full participation of women at all levels of policy-making and implementation for biological diversity conservation.

Stressing the importance of and the need to promote, international, regional and global cooperation among States and intergovernmental organisations and the non-governmental sector for the conservation of biological diversity and the sustainable use of its components.

Acknowledging that the provision of new and additional financial resources and appropriate access to relevant technologies can be expected to make a substantial difference in the world's ability to address the loss of biological diversity.

Acknowledging further that special provision is required to meet the needs of developing countries, including the provision of new and additional financial resources and appropriate access to relevant technologies.

Noting in this regard the special conditions of the least developed countries and small island States.

Acknowledging that substantial investments are required to conserve biological diversity and that there is the expectation of a broad range of environmental, economic and social benefits from those investments.

Recognising that economic and social development and poverty eradication are the first and overriding priorities of developing countries.

Aware that conservation and sustainable use of biological diversity is of critical importance for meeting the food, health and other needs of the growing world population, for which purpose access to and sharing of both genetic resources and technologies are essential.

Noting that, ultimately, the conservation and sustainable use of biological diversity will strengthen friendly relations among States and contribute to peace for humankind.

Desiring tô enhance and complement existing international arrangements for the conservation of biological diversity and sustainable use of its components, and

Determined to conserve and sustainably use biological diversity for the benefit of present and future generations,

Have agreed as follows:

Article 1. Objectives

The objectives of this Convention, to be pursued in accordance with its relevant provisions, are the conservation of biological diversity, the sustainable use of its components and the fair and equitable sharing of the benefits arising out of the utilisation of genetic resources, including by appropriate access to genetic resources and by appropriate transfer of relevant technologies, taking into account all rights over those resources and to technologies and by appropriate funding.

Article 2. Use of Terms

For the purposes of this Convention:

"Biological diversity" means the variability among living organisms from all sources including, inter alia, terrestrial, marine and other aquatic ecosystems and the ecological complexes of which they are part; this includes diversity within species, between species and of ecosystems.

" Biological resources" includes genetic resources organisms or parts thereof, populations, or any other biotic component of ecosystems with actual or potential use or value for humanity.

"Biotechnology" means any technological application that uses biological systems, living organisms, or derivatives thereof, to make or modify products or processes for specific use.

"Country of origin of genetic resources" means the country which possesses those genetic resources in *in-situ* conditions.

"Country providing genetic resources" means the country supplying genetic resources collected from *in-situ* sources, including populations of both wild and domesticated species, or taken from *ex-situ* sources, which may or may not have originated in that country.

"Domesticated or cultivated species" means species in which the evolutionary process has been influenced by humans to meet their needs.

"Ecosystem" means a dynamic complex of plant, animal and microorganism communities and their non-living environment interacting as a functional unit.

"*Ex-situ* conservation" means the conservation of components of biological diversity outside their natural habitats.

"Genetic material" means any material of plant, animal, microbial or other origin containing functional units of heredity.

"Genetic resources" means genetic material of actual or potential value.

"Habitat" means the place or type of site where an organism or population naturally occurs.

"*In-situ* conditions" means conditions where genetic resources exist within ecosystems and natural habitats and in the case of domesticated or cultivated species, in the surroundings where they have developed their distinctive properties.

"*In-situ* conservation" means the conservation of ecosystems and natural habitats and the maintenance and recovery of viable populations of species in their natural surroundings and in the case of domesticated or cultivated species, in the surroundings where they have developed their distinctive properties.

"Protected area" means a geographically defined area which in designated or regulated and managed to achieve specific conservation objectives.

"Regional economic integration organisation" means an organisation constituted by sovereign States of a given region, to which its member States have transferred competence in respect of matters governed by this Convention and which has been duly authorised, in accordance with its internal procedures, to sign, ratify, accept, approve or accede to it.

"Sustainable use" means the use of components of biological diversity in a way and at a rate that does not lead to the longterm decline of biological diversity, thereby maintaining its potential to meet the needs and aspirations of present and future generations.

"Technology" includes biotechnology.

Article 3. Principle

States have, in accordance with the Charter of the United Nations and the principles of international law, the sovereign right to exploit their own resources pursuant to their own environmental policies and the responsibility to ensure that activities within their jurisdiction or control do not cause damage to the environment of other States or of areas beyond the limits of national jurisdiction.

Article 4. Jurisdictional Scope

Subject to the rights of other States and except as otherwise expressly provided in this Convention, the provisions of this Convention apply, in relation to each Contracting Party:

(a) In the case of components of biological diversity, in areas within the limits of its national jurisdiction; and

(b) In the case of processes and activities, regardless of where their effects occur, carried out under its jurisdiction or control, within the areas of its national jurisdiction or beyond the limits of national jurisdiction.

Article 5. Cooperation

Each contracting party shall, as far as possible and as appropriate, cooperate with other Contracting Parties, directly or, where appropriate, through competent international organisations, in respect of areas beyond national jurisdiction and on other matters of mutual interest, for the conservation and sustainable use of biological diversity.

Article 6. General Measures for Conservation and Sustainable Use

Each Contracting Party shall, in accordance with its particular conditions and capabilities:

1. Develop national strategies, plans or programmes for the conservation and sustainable use of biological diversity or adapt for this purpose existing strategies, plans or programmes which shall reflect, *inter alia*, the measures set out in this convention relevant to the Contracting Party concerned; and
2. Integrate, as far as possible and as appropriate, the conservation and sustainable use of biological diversity into relevant sectoral or cross-sectoral plans, programmes and policies.

Article 7. Identification and Monitoring

Each Contracting Party shall, as far as possible and as appropriate, in particular for the purposes of Articles 8 to 10:

1. Identify components of biological diversity important for its conservation and sustainable use having regard to the indicative list of categories set down in Annex I;
2. Monitor, through sampling and other techniques, the components of biological diversity identified pursuant to subparagraph (a) above, paying particular attention to those requiring urgent conservation measures and those which offer the greatest potential for sustainable use;
3. Identify processes and categories of activities which have or are likely to have significant adverse impacts on the conservation and sustainable use of biological diversity and monitor their effects through sampling and other techniques; and
4. Maintain and organise, by any mechanism, data derived from identification and monitoring activities pursuant to subparagraphs (a), (b) and (c) above.

Article 8. In-situ Conservation

Each Contracting Party shall, as far as possible and as appropriate:

1. Establish a system of protected areas or areas where special measures need to be taken to conserve biological diversity;
2. Develop, where necessary, guidelines for the selection, establishment and management of protected areas or areas where special measures need to be taken to conserve biological diversity;
3. Regulate or manage biological resources important for the conservation of biological diversity whether within or outside protected areas, with a view to ensuring their conservation and sustainable use;
4. Promote the protection of ecosystems, natural habitats and the maintenance of viable populations of species in natural surroundings;
5. Promote environmentally sound and sustainable development in areas adjacent to protected areas with a view to furthering protection of these areas;
6. Rehabilitate and restore degraded ecosystems and promote the recovery of threatened species, *inter alia*, through the development and implementation of plans or other management strategies;

7. Establish or maintain means to regulate, manage or control the risks associated with the use and release of living modified organisms resulting from biotechnology which are likely to have adverse environmental impacts that could affect the conservation and sustainable use of biological diversity, taking also into account the risks to human health;
8. Prevent the introduction of control or eradicate those alien species which threaten ecosystems, habitats or species;
9. Endeavour to provide the conditions needed for compatibility between present uses and the conservation of biological diversity and the sustainable use of its components;
10. Subject to its national legislation, respect, preserve and maintain knowledge, innovations and practices of indigenous and local communities embodying traditional lifestyles relevant for the conservation and sustainable use of biological diversity and promote their wider application with the approval and involvement of the holders of such knowledge, innovations and practices and encourage the equitable sharing of the benefits arising from the utilisation of such knowledge, innovations and practices;
11. Develop or maintain necessary legislation and/or other regulatory provisions for the protection of threatened species and populations;
12. Where a significant adverse effect on biological diversity has been determined pursuant to Article 7, regulate or manage the relevant processes and categories of activities; and
13. Cooperate in providing financial and other support for *in-situ* conservation outlined in subparagraphs (a) to (i) above, particularly to developing countries.

Article 9. Ex-situ Conservation

Each Contracting Party shall, as far as possible and as appropriate and predominantly for the purpose of complementing *in-situ* measures:

1. Adopt measures for the ex-situ conservation of components of biological diversity, preferably in the country of origin of such components;
2. Establish and maintain facilities for ex-situ conservation of and reserarch on plants, animals and microorganisms, preferably in the country of origin of genetic resources;
3. Adopt measures for the recovery and rehabilitation of threatened species and for their reintroduction into their natural habitats under appropriate conditions;
4. Regulate and manage collection of biological resources from natural habitats for *ex-situ* conservation purposes so as not to threaten ecosystems and *in-situ* populations of species, except where special temporary *ex-situ* measures are required under subparagraph (c) above; and
5. Cooperate in providing financial and other support for *ex-situ* conservation outlined in subparagraphs (a) to (d) above and in the establishment and maintenance of *ex-situ* conservation facilities in developing countries.

Article 10. Sustainable Use of Components of Biological Diversity

Each Contracting Party shall, as far as possible and as appropriate:

1. Integrate consideration of the conservation and sustainable use of biological resources into national decision-making;

2. Adopt measures relating to the use of biological resources to avoid or minimise adverse impacts on biological diversity;
3. Protect and encourage customary use of biological resources in accordance with traditional cultural practices that are compatible with conservation or sustainable use requirements;
4. Support local populations to develop and implement remedial action in degraded areas where biological diversity has been reduced; and
5. Encourage cooperation between its governmental authorities and its private sector in developing methods for sustainable user of biological resources.

Article 11. Incentive Measures

Each Contracting Party shall, as far as possible and as appropriate, adopt economically and socially sound measures that act as incentives for the conservation and sustainable use of components of biological diversity.

Article 12. Research and Training

The Contracting Parties, taking into account the special needs of developing countries, shall:

1. Establish and maintain programmes for scientific and technical education and training in measures for the identification, conservation and sustainable use of biological diversity and its components and provide support for such education and training for the specific needs of developing countries;
2. Promote and encourage research which contributes to the conservation and sustainable use of biological diversity, particularly in developing countries, *inter alia*, in accordance with decisions of the Conference of the Parties taken in consequence of recommendations of the Subsidiary Body on Scientific, Technical and Technological Advice; and
3. In keeping with the provisions of Articles 16, 18 and 20, promote and cooperate in the use of scientific advances in biological diversity research in developing methods for conservation and sustainable use of biological resources.

Article 13. Public Education and Awareness

The Contracting Parties shall:

1. Promote and encourage understanding of the importance of and the measures required for, the conservation of biological diversity, as well as its propagation through media and the inclusion of these topics in educational programmes; and
2. Cooperate, as appropriate, with other States and international organisations in developing educational and public awareness programmes, with respect to conservation and sustainable use of biological diversity.

Article 14. Impact Assessment and Minimising Adverse Impacts

1. Each Contracting Party, as far as possible and as appropriate, shall:

   (a) Introduce appropriate procedures requiring environmental impact assessment of its proposed projects that are likely to have significant adverse effects on biological diversity with a view to avoiding or minimising such effects and, where appropriate, allow for public participation in such procedures;

(b) Introduce appropriate arrangements to ensure that the environmental consequences of its programmes and policies that are likely to have significant adverse impacts on biological diversity are duly taken into account;

(c) Promote, on the basis of reciprocity, notification, exchange of information and consultation on activities under their jurisdiction or control which are likely to significantly affect adversely the biological diversity of other States or areas beyond the limits of national jurisdiction, by encouraging the conclusion of bilateral, regional or multilateral arrangements, as appropriate;

(d) In the case of imminent or grave danger or damage, originating under its jurisdiction or control, to biological diversity within the area under jurisdiction of other States or in areas beyond the limits of national jurisdiction, notify immediately the potentially affected States of such danger or damage, as well as initiate action to prevent or minimise such danger or damage; and

(e) Promote national arrangements for emergency responses to activities or events, whether caused naturally or otherwise, which present a grave and imminent danger to biological diversity and encourage international cooperation to supplement such national efforts and where appropriate and agreed by the States or regional economic integration organisations concerned, to establish joint contingency plans.

2. The Conference of the Parties shall examine, on the basis of studies to be carried out, the issue of liability and redress, including restoration and compensation, for damage to biological diversity except where such liability is a purely internal matter.

Article 15. Access to Genetic Resources

1. Recognising the sovereign rights of States over their natural resources, the authority to determine access to genetic resources rests with the national Governments and is subject to national legislation.

2. Each Contracting Party shall endeavour to create conditions to facilitate access to genetic resources for environmentally sound uses by other Contracting Parties and not to impose restrictions that run counter to the objectives of this Convention.

3. For the purpose of this Convention, the genetic resources being provided by a Contracting Party, as referred to in this Article and Articles 16 and 19, are only those that are provided by Contracting Parties that are countries of origin of such resources or by the Parties that have acquired the genetic resources in accordance with this Convention.

4. Access, where granted, shall be on mutually agreed terms and subject to the provisions of this Article.

5. Access to genetic resources shall be subject to prior informed consent of the Contracting Party providing such resources, unless otherwise determined by that Party.

6. Each Contracting Party shall endeavour to develop and carry out scientific research based on genetic resources provided by other Contracting Parties with the full participation of and where possible in, such Contracting Parties.

7. Each Contracting Party shall take legislative, administrative or policy measures, as appropriate and in accordance with Articles 16 and 19 and, where necessary, through the financial mechanism established by Articles 70 and 21 with the aim of sharing in a fair and equitable way the results of research and development and the benefits arising from the commercial and other utilisation of genetic resources with the Contracting Party providing such resources. Such sharing shall be upon mutually agreed terms.

Article 16. Access to and Transfer of Technology

1. Each Contracting Party, recognising that technology includes biotechnology and that both access to and transfer of technology among Contracting Parties are essential elements for the attainment of the objectives of this Convention undertakes, subject to the provisions of this Article, to provide and/or facilitate access for and transfer to other Contracting Parties of technologies that are relevant to the conservation and sustainable use of biological diversity or make use of genetic resources and do not cause significant damage to the environment.
2. Access to and transfer of technology referred to in paragraph 1 above to developing countries shall be provided and/or facilitated under fair and most favourable terms, including on concessional and preferential terms where mutually agreed and where necessary, in accordance with the financial mechanism established by Articles 20 and 21. In the case of technology subject to patents and other intellectual property rights, such access and transfer shall be provided on terms which recognise and are consistent with the adequate and effective protection of intellectual property rights. The application of this paragraph shall be consistent with paragraphs 3, 4 and 5 below.
3. Each Contracting Party shall take legislative, administrative or policy measures, as appropriate, with the aim that Contracting Parties, particularly those that are of developing countries, which provide genetic resources, are provided access to and transfer of technology which makes use of those resources, on mutually agreed terms, including technology protected by patents and other intellectual property rights, where necessary, through the provisions of Articles 20 and 21 and in accordance with international law and consistent with paragraphs 4 and 5 below.
4. Each Contracting Party shall take legislative, administrative or policy measures as appropriate, with the aim that the private sector facilitates access to, joint development and transfer of technology referred to in paragraph 1 above for the benefit of both governmental institutions and the private sector of developing countries and in this regard shall abide by the obligations included in paragraphs 1, 2 and 3 above.
5. The Contracting Parties, recognising that patents and other intellectual property rights may have an influence on the implementation of this Convention, shall cooperate in this regard, subject to national legislation and international law, in order to ensure that such rights are supportive of and do not run counter to its objectives.

Article 17. Exchange of Information

1. The Contracting Parties shall facilitate the exchange of information from all publicly available sources, relevant to the conservation and sustainable use of biological diversity, taking into account the special needs of developing countries.

2. Such exchange of information shall include exchange of results of technical, scientific and socio-economic research, as well as information on training and surveying programmes, specialised knowledge, indigenous and traditional knowledge as such and in combination with the technologies referred to in Article 16, paragraph 1. It shall also, where feasible, include repatriation of information.

Article 18. Technical and Scientific Cooperation

1. The Contracting Parties shall promote international, technical and scientific cooperation in the field of conservation and sustainable use of biological diversity, where necessary, through the appropriate international and national institutions.
2. Each Contracting Party shall promote technical and scientific cooperation with other Contracting Parties, in particular developing countries, in implementing this Convention, *inter alia*, through the development and implementation of national policies. In promoting such cooperation, special attention should be given to the development and strengthening of national capabilities, by means of human resources development and institution building.
3. The Conference of the Parties, at its first meeting, shall determine how to establish a clearing-house mechanism to promote and facilitate technical and scientific cooperation.
4. The Contracting Parties shall, in accordance with national legislation and policies, encourage and develop methods of cooperation for the development and use of technologies, including indigenous and traditional technologies, in pursuance of the objectives of this Convention. For this purpose, the Contracting Parties shall also promote cooperation in the training of personnel and exchange of experts.
5. The Contracting Parties shall, subject to mutual agreement, promote the establishment of joint research programmes and joint ventures for the development of technologies relevant to the objectives of this Convention.

Article 19. Handling of Biotechnology and Distribution of its Benefits

1. Each Contracting Party shall take legislative, administrative or policy measures, as appropriate, to provide for the effective participation in biotechnological research activities by those Contracting Parties, especially developing countries, which provide the genetic resources for such research, and where feasible in such Contracting Parties.
2. Each Contracting Party shall take all practicable measures to promote and advance priority access on a fair and equitable basis by Contracting Parties, especially developing countries, to the results and benefits arising from biotechnologies based upon genetic resources provided by those Contracting Parties. Such access shall be on mutually agreed terms.
3. The Parties shall consider the need for and modalities of a protocol setting out appropriate procedures, including, in particular, advance informed agreement, in the field of the safe transfer, handling and use of any living modified organism resulting from biotechnology that may have adverse effect on the conservation and sustainable use of biological diversity.
4. Each Contracting Party shall, directly or by requiring any natural or legal person under its jurisdiction providing the organisms referred to in paragraph 3 above, provide any available information about the use and safety regulations required by that Contracting Party in handling

such organisms, as well as any available information on the potential adverse impact of the specific organisms concerned to the Contracting Party into which those organisms are to be introduced.

Article 20. Financial Resources

1. Each Contracting Party undertakes to provide, in accordance with its capabilities, financial support and incentives in respect of those national activities which are intended to achieve the objectives of this Convention, in accordance with its national plans, priorities and programmes.
2. The developed country Parties shall provide new and additional financial resources to enable developing country Parties to meet the agreed full incremental costs to them of implementing measures which fulfil the obligations of this Convention and to benefit from its provisions and which costs are agreed between a developing country Party and the institutional structure referred to in Article 21, in accordance with policy, strategy, programme priorities and eligibility criteria and an indicative list of incremental costs established by the Conference of the Parties. Other Parties, including countries undergoing the process of transition to a market economy, may voluntarily assume the obligations of the developed country Parties. For the purpose of this Article, the Conference of the Parties shall, at its first meeting, establish a list of developed country Parties and other Parties which voluntarily assume the obligations of the developed country Parties. The Conference of the Parties shall periodically review and if necessary amend the list Contributions from other countries and sources on a voluntary basis would also be encouraged. The implementation of these commitments shall take into account the need for adequacy, predictability and timely flow of funds and the importance of burden sharing among the contributing Parties included in the list.
3. The developed country Parties may also provide and developing country Parties avail themselves of financial resources related to the implementation of this Convention through bilateral, regional and other multilateral channels.
4. The extent to which developing country Parties will effectively implement their commitments under this Convention will depend on the effective implementation by developed country Parties of their commitments under this Convention related to financial resources and transfer of technology and will take fully into account the fact that economic and social development and eradication of poverty are the first and overriding priorities of the developing country Parties.
5. The Parties shall take full account of the specific needs and special situation of least developed countries in their actions with regard to funding and transfer of technology.
6. The Contracting Parties shall also take into consideration the special conditions resulting from the dependence on distribution, and location of, biological diversity within developing country Parties, in particular small island States.
7. Consideration shall also be given to the special situation of developing countries, including those that are most environmentally vulnerable, such as those with arid and semi-arid zones, coastal and mountainous areas.

Article 21. Financial Mechanism

1. There shall be a mechanism for the provision of financial resources to developing country Parties for purposes of this Convention on a grant or concessional basis the essential elements

of which are described in this Article. The mechanism shall function under the authority and guidance of, and be accountable to, the Conference of the Parties for purposes of this Convention. The operations of the mechanism shall be carried out by such institutional structure as may be decided upon by the Conference of the Parties at its first meeting. For purposes of this Convention, the Conference of the Parties shall determine the policy, strategy, programme priorities and eligibility criteria relating to the access to and utilisation of such resources. The contributions shall be such as to take into account the need for predictability, adequacy and timely flow of funds referred to in Article 20 in accordance with the amount of resources needed to be decided periodically by the Conference of the Parties and the importance of burden-sharing among the contributing Parties included in the list referred to in Article 20, paragraph 2. Voluntary contributions may also be made by the developed country Parties and by other countries and sources. The mechanism shall operate within a democratic and transparent system of governance.

2. Pursuant to the objective of this Convention, the Conference of the Parties shall, at its first meeting, determine the policy, strategy and programme priorities, as well as detailed criteria and guidelines for eligibility for access to and utilisation of the financial resources including monitoring and evaluation on a regular basis of such utilisation. The Conference of the Parties shall decide on the arrangements to give effect to paragraph 1 above after consultation with the institutional structure entrusted with the operation of the financial mechanism.

3. The Conference of the Parties shall review the effectiveness of the mechanism established under this Article, including the criteria and guidelines referred to in paragraph 2 above, not less than two years after the entry into force of this Convention and thereafter on a regular basis. Based on such review, it shall take appropriate action to improve the effectiveness of the mechanism if necessary.

4. The Contracting Parties shall consider strengthening existing financial institutions to provide financial resources for the conservation and sustainable use of biological diversity.

Article 22. Relationship with Other International Conventions

1. The provisions of this Convention shall not affect the rights and obligations of any Contracting Party deriving from any existing international agreement, except where the exercise of those rights and obligations would cause a serious damage or threat to biological diversity.

2. Contracting Parties shall implement this Convention with respect to the marine environment consistently with the rights and obligations of States under the Law of the Sea.

Article 23. Conference of the Parties

1. A Conference of the Parties is hereby established. The first meeting of the Conference of the Parties shall be convened by the Executive Director of the United Nations Environment Programme not later than one year after the entry into force of this Convention. Thereafter, ordinary meetings of the Conference of the Parties shall be held at regular intervals to be determined by the Conference at its first meeting.

2. Extraordinary meetings of the Conference of the Parties shall be held at such other times as may be deemed necessary by the Conference, or at the written request of any Party, provided that, within six months of the request being communicated to them by the Secretariat, it is supported by at least one-third of the Parties.

3. The Conference of the Parties shall by consensus agree upon and adopt rules of procedure for itself and for any subsidiary body it may establish, as well as financial rules governing the funding of the Secretariat. At each ordinary meeting, it shall adopt a budget for the financial period until the next ordinary meeting.
4. The Conference of the Parties shall keep under review the implementation of this Convention and for this purpose, shall:
   (a) Establish the form and the intervals for transmitting the information to be submitted in accordance with Article 26 and consider such information as well as reports submitted by any subsidiary body;
   (b) Review scientific, technical and technological advice on biological diversity provided in accordance with Article 25;
   (c) Consider and adopt, as required, protocols in accordance with Article 28;
   (d) Consider and adopt, as required, in accordance with Article 29 and 30, amendments to this Convention and its annexes;
   (e) Consider amendments to any protocol, as well as to any annexes thereto and if so decided, recommend their adoption to the parties of the protocol concerned;
   (f) Consider and adopt, as required, in accordance with Article 30, additional annexes to this Convention;
   (g) Establish such subsidiary bodies, particularly to provide scientific and technical advice, as are deemed necessary for the implementation of this Convention;
   (h) Contact, through the Secretariat, the executive bodies of conventions dealing with matters covered by this Convention with a view to establishing appropriate forms of cooperation with them; and
   (i) Consider and undertake any additional action that may be required for the achievement of the purposes of this Convention in the light of experience gained in its operation.
5. The United Nations, its specialised agencies and the International Atomic Agency, as well as any State not Party to this Convention, may be represented as observers at meetings of the Conference of the Parties. Any other body or agency, whether governmental or non-governmental, qualified in fields relating to conservation and sustainable use of biological diversity, which has informed the Secretariat of its wish to be represented as an observer at a meeting of the Conference of the Parties, may be admitted unless at least one-third of the Parties present object. The admission and participation of observers shall be subject to the rules of procedure adopted by the Conference of the Parties.

Article 24. Secretariat

1. A Secretariat is hereby established. Its functions shall be:
   (a) To arrange for and service meetings of the Conference of the Parties provided for in Article 23;
   (b) To perform the functions assigned to it by any protocol;
   (c) To prepare reports on the execution of its functions under this Convention and present them to the Conference of the Parties;

(d) To coordinate with other relevant international bodies and in particular to enter into such administrative and contractual arrangements as may be required for the effective discharge of its functions; and

(e) To perform such other functions as may be determined by the Conference of the Parties.

2. At its first ordinary meeting, the Conference of the Parties shall designate the secretariat from amongst those existing competent international organisations which have signified their willingness to carry out the secretariat functions under this Convention.

Article 25. Subsidiary Body on Scientific, Technical and Technological Advice

1. A subsidiary body for the provision of scientific, technical and technological advice is hereby established to provide the Conference of the Parties and, as appropriate, its other subsidiary bodies with timely advice relating to the implementation of this Convention. This body shall be open to participation by all Parties and shall be multidisciplinary. It shall be comprised of government representatives competent in the relevant field of expertise. It shall report regularly to the Conference of the Parties on all aspects of its work.

2. Under the authority of and in accordance with guidelines laid down by the Conference of the Parties and upon its request, this body shall:

(a) Provide scientific and technical assessments of the status of biological diversity;

(b) Prepare scientific and technical assessment of the effects of types of measures taken in accordance with the provisions of this Convention;

(c) Identify innovative, efficient and state-of-the-art technologies and knowhow relating to the conservation and sustainable use of biological diversity and advise on the ways and means of promoting development and/or transferring such technologies;

(d) Provide advice on scientific programmes and international cooperation in research and development related to conservation and sustainable use of biological diversity; and

(e) Respond to scientific, technical, technological and methodological questions that the Conference of the Parties and its subsidiary bodies may put to the body.

3. The functions, terms of reference, organisation and operation of this body may be further elaborated by the Conference of the Parties.

Article 26. Reports

Each Contracting Party shall, at intervals to be determined by the Conference of the Parties, present to the Conference of the Parties, reports on measures which it has taken for the implementation of the provisions of this Convention and their effectiveness in meeting the objectives of this Convention.

Article 27. Settlement of Disputes

1. In the event of a dispute between Contracting Parties concerning the interpretation or application of this Convention, the parties concerned shall seek solution by negotiation.

2. If the parties concerned cannot reach agreement by negotiation, they may jointly seek the good offices of, or request mediation by, a third party.

3. When ratifying, accepting, approving or acceding to this Convention, or at any time thereafter, a State or regional economic integration organisation may declare in writing to the Depositary

that for a dispute not resolved in accordance with paragraph 1 or paragraph 2 above, it accepts one or both of the following means of dispute settlement as compulsory;

(a) Arbitration in accordance with the procedure laid down in Part I of Annex II;

(b) Submission of the dispute to the International Court of Justice.

4. If the parties to the dispute have not, in accordance with paragraph 3 above, accepted the same or any procedure, the dispute shall be submitted to conciliation in accordance with Part II of Annex II unless the parties otherwise agree.

5. The provisions of this Article shall apply with respect to any protocol except as otherwise provided in the protocol concerned.

Article 28. Adoption of Protocols

1. The Contracting Parties shall cooperate in the formulation and adoption of protocols in this Convention.
2. Protocols shall be adopted at a meeting of the Conference of the Parties.
3. The text of any proposed protocol shall be communicated to the Contracting Parties by the Secretariat at least six months before such a meeting.

Article 29. Amendment of the Convention of Protocols

1. Amendments to this Convention may be proposed by any Contracting Party. Amendments to any protocol may be proposed by any Party to that protocol.
2. Amendments to this Convention shall be adopted at a meeting of the Conference of the Parties. Amendments to any protocol shall be adopted at a meeting of the Parties to the Protocol in question. The text of any proposed amendment to this Convention or to any protocol, except as may otherwise be provided in such protocol, shall be communicated to the Parties to the instrument in question by the secretariat at least six months before the meeting at which it is proposed for adoption. The secretariat shall also communicate proposed amendments to the signatories to this Convention for information.
3. The Parties shall make every effort to reach agreement on any proposed amendment to this Convention or to any protocol by consensus. If all efforts at consensus have been exhausted and no agreement reached, the amendment shall, as a last resort, be adopted by a two-third majority vote of the parties to the instrument in question present and voting at the meeting shall be submitted by the Depositary to all Parties for ratification, acceptance or approval.
4. Ratification, acceptance or approval of amendments shall be notified to the Depositary in writing. Amendments adopted in accordance with paragraph 3 above shall enter into force among Parties having accepted them on the nineteenth day after the deposit of instruments of ratification, acceptance or approval by at least two-thirds of the Contracting Parties to this Convention or of the Parties to the protocol concerned, except as may otherwise be provided in such protocol. Thereafter the amendments shall enter into force for any other Party on the ninetieth day after that Party deposits its instrument of ratification, acceptance or approval of the amendments.
5. For the purposes of this Article, "Parties present and voting" means Parties present and casting an affirmative or negative vote.

Article 30. Adoption and Amendment of Annexes

1. The annexes to this Convention or to any protocol shall form an integral part of the Convention or of such protocol, as the case may be, and, unless expressly provided otherwise, a reference to this Convention or its protocols constitutes at the same time a reference to any annexes thereto. Such annexes shall be restricted to procedural, scientific, technical and administrative matters.
2. Except as may be otherwise provided in any protocol with respect to its annexes, the following procedure shall apply to the proposal, adoption and entry into force of additional annexes to this Convention or of annexes to any protocol:
   (a) Annexes to this Convention or to any protocol shall be proposed and adopted according to the procedure laid down in Article 29;
   (b) Any party that is unable to approve an additional annex to this Convention or an annex to any protocol to which it is party shall so notify the Depositary, in writing, within one year from the date of the communication of the adoption by the Depositary. The Depositary shall, without delay, notify all parties of any such notification received. A party may at any time withdraw a previous declaration of objection and the annexes shall thereupon enter into force for that party subject to subparagraph (c) below;
   (c) On the expiry of one year from the date of communication of adoption by the Depositary, the annex shall enter into force for all parties to this convention or to any protocol concerned which have not submitted a notification in accordance with the provisions of subparagraph (b) above.
3. The proposal, adoption and entry into force of amendments to annexes to this convention or to any protocol shall be subject to the same procedure as for the proposal, adoption and entry into force of annexes to the convention or annexes to any protocol.
4. If an additional annex or an amendment to an annex is related to an amendment to this convention or to any protocol, the additional annex or amendment shall not enter into force until such time as the amendment to the convention or to the protocol concerned enters into force.

Article 31. Right to Vote

1. Except as provided for in paragraph 2 below, each Contracting Party to this Convention or to any protocol shall have one vote.
2. Regional economic integration organisations, in matters within their competence, shall exercise their right to vote with a number of votes equal to the number of their member States which are Contracting Parties to this convention or the relevant protocol. Such organisations shall not exercise their right to vote if their member States exercise theirs and *vice versa*.

Article 32. Relationship Between This Convention and its Protocols

1. A State or a regional economic integration organisation may not become a party to a protocol unless it is, or becomes at the same time, a Contracting Party to this Convention.
2. Decisions under any protocol shall be taken only by the parties to the protocol concerned. Any Contracting Party that has not ratified, accepted or approved a protocol may participate as an observer in any meeting of the parties to that protocol.

Article 33. Signature

This Convention shall be open for signature at Rio De Janeiro by all States and any regional economic integration organisation from 5 June, 1992 until 14 June, 1992 and at the United Nations Headquarters in New York from 15 June, 1992 to 4 June, 1993.

Article 34. Ratification, Acceptance or Approval

1. This Convention and any protocol shall be subject to ratification, acceptance or approval by States and by regional economic integration organisations. Instruments of ratification, acceptance or approval shall be deposited with the Depositary.
2. Any organisation referred to in paragraph 1 above which becomes a Contracting Party to this convention or any protocol without any of its member States being a contracting party shall be bound by all the obligations under the convention or the protocol, as the case may be. In the case of such organisations one or more of whose member States is a contracting party to this convention or relevant protocol, the organisation and its member States shall decide on their respective responsibilities for the performance of their obligations under the convention or protocol, as the case may be. In such cases, the organisation and the member States shall not be entitled to exercise rights under the convention or relevant protocol concurrently.
3. In their instruments of ratification, acceptance or approval, the organisations referred to in paragraph 1 above shall declare the extent of their competence with respect to the matters governed by the convention or the relevant protocol. These organisations shall also inform the Depositary of any relevant modification in the extent of their competence.

Article 35. Accession

1. This Convention and any protocol shall be open for accession by States and by regional economic integration organisations from the date on which the convention or the protocol concerned is closed for signature. The instruments of accession shall be deposited with the Depositary.
2. In their instruments of accession, the organisations referred to in paragraph 1 above shall declare the extent of their competence with respect to the matters governed by the Convention or the relevant protocol. These organisations shall also inform the Depositary of any relevant modification in the extent of their competence.
3. The provisions of Article 34, paragraph 2, shall apply to regional economic integration organisations which accede to this Convention or any protocol.

Article 36. Entry Into Force

1. This Convention shall enter into force on the ninetieth day after the date of deposit of the thirtieth instrument of ratification, acceptance, approval or accession.
2. Any protocol shall enter into force on the ninetieth day after the date of deposit of the number of instruments of ratification, acceptance, approval or accession specified in that protocol, has been deposited.
3. For each Contracting Party which ratifies, accepts or approves this convention or accedes thereto after the deposit of the thirtieth instrument of ratification, acceptance, approval or accession, it shall enter into force on the ninetieth day after the date of deposit be such contracting party of its instrument of ratification, acceptance, approval or accession.

4. Any protocol, except as otherwise provided in such protocol, shall enter into force for a contracting party that ratifies, accepts or approves that protocol or accedes thereto after its entry into force pursuant to paragraph 2 above, on the ninetieth day after the date on which that contracting party deposits its instrument of ratification, acceptance, approval or accession, or on the date on which this convention enters into force for that Contracting Party, whichever shall be the later.
5. For the purposes of paragraphs 1 and 2 above, any instrument deposited by a regional economic integration organisation shall not be counted as additional to those deposited by member States of such organisation.

Article 37. Reservations

No reservations may be made to this Convention.

Article 38. Withdrawals

1. At any time after two years from the date on which this Convention has entered into force for a Contracting Party, that contracting party may withdraw from the Convention by giving written notification to the Depositary.
2. Any such withdrawal shall take place upon expiry of one year after the date of its receipt by the Depositary, or on such later date as may be specified in the notification of the withdrawal.
3. Any Contracting Party which withdraws from this convention shall be considered as also having withdrawn from any protocol to which it is party.

Article 39. Financial Interim Arrangements

Provided that it has been fully restructured in accordance with the requirements of Article 21, the Global Environment Facility of the United Nations Development Programme, the United Nations Environment Programme and the International Bank for the Reconstruction and Development shall be the institutional structure referred to in Article 21 on an interim basis, for the period between the entry into force of this convention and the first meeting of the conference of the parties or until the conference of the parties decides which institutional structure will be designated in accordance with Article 21.

Article 40. Secretariat Interim Arrangements

The secretariat to be provided by the Executive Director of the United Nations Environment Programme shall be the secretariat referred to in Article 24, paragraph 2, on an interim basis for the period between the entry into force of this convention and the first meeting of the Conference of the Parties.

Article 41. Depositary

The Secretary-General of the United Nations shall assume the functions of Depositary of this Convention and any protocols.

Article 42. Authentic Texts

The original of this Convention of which the Arabic, Chinese, English, French, Russian and Spanish texts are equally authentic, shall be deposited with the Secretary-General of the United Nations.

In witness whereof the undersigned, being duly authorised to that effect, have signed this Convention. Done at Rio de Janeiro on this fifth day of June, one thousand nine hundred and ninety two.

ANNEX I : Identification and Monitoring

1. Ecosystems and habitats: containing high diversity, large numbers of endemic or threatened species, or wilderness; required by migratory species; of social, economic, cultural or scientific importance or which are representative unique or associated with key evolutionary or other biological processes;
2. Species and communities which are: threatened; wild relatives of domesticated or cultivated species; of medicinal, agricultural or other economic value; or social, scientific or cultural importance; or importance for research into the conservation and sustainable use of biological diversity, such as indicator species; and
3. Described genomes and genes of social, scientific or economic importance.

ANNEX II — PART I

Article 1 — Arbitration

The claimant party shall notify the secretariat that the parties are referring to a dispute to arbitration pursuant to Article 27. The notification shall state the subject-matter of arbitration and include, in particular, the articles of the Convention or the protocol, the interpretation or application of which are at issue. If the parties do not agree on the subject matter of the dispute before the President of the tribunal is designated, the arbitral tribunal shall determine the subject matter. The secretariat shall forward the information thus received to all Contracting Parties to this Convention or to the protocol concerned.

Article 2

1. In disputes between two parties, the arbitral tribunal shall consist of three members. Each of the parties to the dispute shall appoint an arbitrator and the two arbitrators so appointed shall designate by common agreement the third arbitrator who shall be the President of the tribunal. The latter shall not be a national of one of the parties to the dispute, nor have his or her usual place of residence in the territory of one of these parties, not be employed by any of them, nor have dealt with the case in any other capacity.
2. In disputes between more than two parties, parties in the same interest shall appoint one arbitrator jointly by agreement.
3. Any vacancy shall be filled in the manner prescribed for the initial appointment.

Article 3

1. If the President of the arbitral tribunal has not been designated within two months of the appointment of the second arbitrator, the Secretary-General of the United Nations shall, at the request of a party, designate the President within a further two-month period.
2. If one of the parties to the dispute does not appoint an arbitrator within two months of receipt of the request, the other party may inform the Secretary-General who shall make the designation within a further two-month period.

Article 4

The arbitral tribunal shall render its decisions in accordance with the provisions of this Convention, any protocols concerned and international law.

Article 5

Unless the parties to the dispute otherwise agree, the arbitral tribunal shall determine its own rules of procedure.

Article 6

The arbitral tribunal may, at the request of one of the parties, recommend essential interim measures of protection.

Article 7

The parties to the dispute shall facilitate the work of the arbitral tribunal and in particular, using all means at their disposal, shall:

(a) Provide it with all relevant documents, information and facilities, and
(b) Enable it, when necessary, to call witnesses or experts and receive their evidence.

Article 8

The parties and the arbitrators are under an obligation to protect the confidentiality of any information they receive in confidence during the proceedings of the arbitral tribunal.

Article 9

Unless the arbitral tribunal determines otherwise because of the particular circumstances of the case, the costs of the tribunal shall be borne by the parties to the dispute in equal shares. The tribunal shall keep a record of all its costs and shall furnish a final statement thereof to the parties.

Article 10

Any Contracting Party that has an interest of a legal nature in the subject-matter of the dispute which may be affected by the decision in the case, may intervene in the proceedings with the consent of the tribunal.

Article 11

The tribunal may hear and determine counterclaims arising directly out of the subject-matter of the dispute.

Article 12

Decisions both on procedure and substance of the arbitral tribunal shall be taken by a majority vote of its members.

Article 13

If one of the parties to the dispute does not appear before the arbitral tribunal or fails to defend its case, the other party may request the tribunal to continue the proceedings and to make its award. Absence of a party or a failure of a party to defend its case shall not constitute a bar to the proceedings. Before rendering its final decision, the arbitral tribunal must satisfy itself that the claim is well founded in fact and law.

Article 14

The tribunal shall render its final decision within five months of the date on which it is fully constituted unless it finds it necessary to extend the time limit for a period which should not exceed five more months.

Article 15

The final decision of the arbitral tribunal shall be confined to the subject-matter of the dispute and shall state the reasons on which it is based. It shall contain the names of the members who have participated and the date of the final decision. Any member of the tribunal may attach a separate or dissenting opinion to the final decision.

Article 16

The award shall be binding on the parties to the dispute. It shall be without appeal unless the parties to the dispute have agreed in advance to an appellate procedure.

Article 17

Any controversy which may arise between the parties in the dispute as regards the interpretation or manner of implementation of the final decision may be submitted by either party for decision to the arbitral tribunal which rendered it.

PART II — CONCILIATION

Article 1

A conciliation commission shall be created upon the request of one of the parties to the dispute. The commission shall, unless the parties otherwise agree, be composed of five members, two appointed by each Party concerned and a President chosen jointly by those members.

Article 2

In disputes between more than two parties, parties in the same interest shall appoint their members of the commission jointly by agreement. Where two or more parties have separate interests or there is a disagreement as to whether they are of the same interest, they shall appoint their members separately.

Article 3

If any appointments by the parties are not made within two months of the date of the request to create a conciliation commission, the Secretary-General of the United Nations shall, if asked to do so by the party that made the request, make those appointments within a further two-month period.

Article 4

If a President of the conciliation commission has not been chosen within two months of the last of the members of the commission being appointed, the Secretary of the United Nations shall, if asked to do so by a party, designate a President within a further two-month period.

Article 5

The conciliation commission shall take its decisions by majority vote of its members. It shall, unless the parties to the dispute otherwise agree, determine its own procedure. It shall render a proposal for resolution of the dispute, which the parties shall consider in good faith.

Article 6

A disagreement as to whether the conciliation commission has competence shall be decided by the commission.

64.9 UNITED NATIONS CONVENTION ON CLIMATE CHANGE

The Parties to this Convention

Acknowledging that change in the Earth's climate and its adverse effects are a common concern of humankind.

Concerned that human activities have been substantially increasing the atmospheric concentrations of greenhouse gases, that these increases enhance the natural greenhouse effect and that this will result on average in an additional warming of the Earth's surface and atmosphere and may adversely affect natural ecosystems and humankind.

Noting that the largest share of historical and current global emissions of greenhouse gases has originated in developed countries, that per capita emissions in developing countries are still relatively low and that the share of global emissions originating in developing countries will grow to meet their social and development needs.

Aware of the role and importance in terrestrial and marine *ecosystems* of sinks and reservoirs of greenhouse gases,

Noting that there are many uncertainties in predictions of climatic change, particularly with regard to the timing, magnitude and regional patterns thereof.

Acknowledging that the global nature of climate change calls for the widest possible cooperation by all countries and their participation in an effective and appropriate international response, in accordance with their common but differentiated responsibilities and respective capabilities and their social and economic conditions.

Recalling the pertinent provisions of the Declaration of the United Nations Conference on the Human Environment, adopted at Stockholm on 16 June, 1972.

Recalling that States have, in accordance with the Charter of the United Nations and the principles of international law, the sovereign right to exploit their own resources pursuant to their own environmental and developmental policies and the responsibility to ensure that activities within their jurisdiction or control do not cause damage to the environment of other States or of areas beyond the limits of national jurisdiction.

Reaffirming the principle of sovereignty of States in international cooperation to address climate change.

Recognising that States should enact effective environmental legislation, that environmental standards, management objectives and priorities should reflect the environmental and developmental context to which they apply and that standards applied by some countries may be inappropriate and of unwarranted economic and social cost to other countries, in particular developing countries.

Recalling the provisions of the General Assembly resolution 44/228 of 22 December, 1989 on the United Nations Conference on Environment and Development; and resolutions 43/53 of 6 December, 1988, 44/207 of 22 December, 1989, 45/212 of 21 December, 1990 and 46/169 of 19 December, 1991 on protection of global climate for present and future generations of mankind.

Recalling also the provisions of the General Assembly resolution 44/206 of 22 December, 1989 on the possible adverse effects of sea level rise on islands and coastal areas particularly low lying coastal areas and the pertinent provisions of the General Assembly resolution 44/172 of 19 December, 1989 on the implementation of the Plan of Action to Combat Desertification,

Recalling further the Vienna Convention for the Protection of the Ozone Layer, 1985 and the Montreal Protocol on Substances that Deplete the Ozone Layer, 1987 as adjusted and amended on 29 June, 1990.

Noting the Ministerial Declaration of the Second World Climate Conference adopted on 7 November, 1990.

Conscious of the valuable analytical work being conducted by many States on climate change and of the important contributions of the World Meteorological Organisation, the United Nations Environment Programme and other organs, organisations and bodies of the United Nations system, as well as other international and inter governmental bodies, in exchange of results of scientific research and the coordination of research,

Recognising that steps required to understand and address climate change will be environmentally, socially and economically most effective if they are based on relevant scientific, technical and economic considerations and continually re-evaluated in the light of new findings in these areas,

Recognising that various actions to address climatic change can be justified economically in their own right and can also help in solving other environmental problems.

Recognising also the need for developed countries to take immediate action in a flexible manner on the basis of clear priorities, as a first step towards comprehensive response strategies at the global, national and, where agreed, regional levels that take into account all greenhouse gases, with due consideration of their relative contributions to the enhancement of the greenhouse effect.

Recognising further that low lying and other small island countries and low-lying coastal, arid and semi-arid areas or areas liable to floods, drought and desertification and developing countries with fragile mountainous ecosystems are particularly vulnerable to the adverse effects of climatic change.

Recognising the special difficulties of those countries, especially developing countries, whose economies are particularly dependent on fossil fuel production, use and exploitation, as a consequence of action taken on limiting greenhouse gas emissions.

Affirming that responses to climate change should be coordinated with social and economic development in an integrated manner with a view to avoiding adverse impacts on the latter, taking into full account the legitimate priority needs of developing countries for the achievement of sustained economic growth and the eradication of poverty.

Recognising that all countries, especially developing countries, need access to resources required to achieve sustainable social and economic development and that, in order for developing countries to progress towards that goal, their energy consumption will need to grow taking into account the possibilities for achieving greater energy efficiency and for controlling greenhouse gas emissions in general, including through the application of new technologies on terms which make such an application economically and socially beneficial.

Determined to protect the climate system for present and future generations.

Have agreed as follows:

Article 1

Definitions

For the purposes of this Convention:

1. "Adverse effects of climate change" means changes in the physical environment or biota resulting from climatic change which have significant deleterious effects on the composition, resilience or productivity of natural and managed ecosystems or on the operation of socio-economic systems or on human health and welfare.
2. "Climate change" means a change of climate which is attributed directly or indirectly to human activity that alters the composition of the global atmosphere and which is in addition to natural climate variability observed over comparable time periods.
3. "Climate system" means the totality of the atmosphere, hydrosphere, biosphere and geosphere and their interactions.
4. "Emissions" means the release of greenhouse gases and/or their precursors into the atmosphere over a specified area and period of time.
5. "Greenhouse gases" mean those gaseous constituents of the atmosphere, both natural and anthropogenic, that absorb and re-emit infra-red radiation.
6. "Regional economic integration organisation" means an organisation constituted by sovereign States of a given region which has competence in respect of matters governed by this Convention or its protocols and has been duly authorised, in accordance with its internal procedures, to sign, ratify, accept, approve or accede to the instruments concerned.
7. "Reservoir" means a component or components of the climate system where a greenhouse gas or a precursor of a greenhouse gas is stored.
8. "Sink" means any process, activity or mechanism which removes a greenhouse gas, an aerosol or a precursor of a greenhouse gas from the atmosphere.
9. "Source" means any process or activity which releases a greenhouse gas, an aerosol or a precursor of a greenhouse gas into the atmosphere.

Article 2. Objective

The ultimate objective of this Convention and any related legal instruments that the Conference of the Parties may adopt is to achieve, in accordance with the relevant provisions of the Convention, stabilisation of greenhouse gas concentrations in the atmosphere at a level that would prevent dangerous anthropogenic interference with the climate system. Such a level should be achieved within a time frame sufficient to allow ecosystems to adapt naturally to climate change, to ensure that food production is not threatened and to enable economic development to proceed in a sustainable manner.

Article 3. Principles

In their actions to achieve the objective of the Convention and to implement its provisions, the Parties shall be guided, *inter alia*, by the following:

1. The Parties should protect the climate system for the benefit of present and future generations of humankind, on the basis of equity and in accordance with their common but differentiated responsibilities and respective capabilities. Accordingly, the developed country Parties should take the lead in combating climatic change and the adverse effects thereof.
2. The specific needs and special circumstances of developing country Parties, especially those that are particularly vulnerable to the adverse effects of climate change and of those Parties, especially developing country Parties, that would have to bear a disproportionate or abnormal burden under this Convention, should be given full consideration.
3. The Parties should take precautionary measures to anticipate, prevent or minimise the causes of climatic change and mitigate its adverse effects. Where there are threats of serious or irreversible damage, lack of full scientific certainty should not be used as a reason for postponing such measures, taking into account that policies and measures to deal with climatic change should be cost-effective so as to ensure global benefits at the lowest possible cost. To achieve this, such policies and measures should take into account different socio-economic contexts, be comprehensive, cover all relevant sources, sinks and reservoirs of greenhouse gases and adaptation and comprise all economic sectors. Efforts to address climatic change may be carried out cooperatively by interested Parties.
4. The Parties have a right to and should, promote sustainable development. Policies and measures to protect the climate system against human-induced change should be appropriate for the specific conditions of each Party and should be integrated with national development programmes, taking into account that economic growth is essential for adopting measures to address climatic change.
5. The Parties should cooperate to promote a supportive and open international economic system that would lead to sustainable economic growth and development in all Parties, particularly developing country Parties, thus enabling them better to address the problems of climatic change. Measures taken to combat climatic change, including unilateral ones, should not constitute a means of arbitrary or unjustifiable discrimination or a disguised restriction on international trade.

Article 4. Commitments

1. All Parties, taking into account their common but differentiated responsibilities and their specific national and regional development priorities, objectives and circumstances, shall :
   (a) Develop, periodically update, publish and make available to the Conference of the Parties in accordance with Article 12, national inventories of emissions from sources and removals by sinks of all greenhouse gases not controlled by the Montreal Protocol, using comparable methodologies to be agreed upon by the Conference of the Parties;
   (b) Formulate, implement, publish and regularly update national and where appropriate, regional programmes containing measures to mitigate climatic change by addressing anthropogenic emissions, by sources and removals by sinks of all greenhouse gases not controlled by the Montreal Protocol and measures to facilitate adequate adaptation to climatic change;
   (c) Promote and cooperate in the development, application and diffusion, including transfer of technologies, practices and processes that control, reduce or prevent anthropogenic emissions of greenhouse gases in all relevant sectors, including the energy, transport, industry, agriculture, forestry and waste management sectors;

(d) Promote sustainable management and promote and cooperate in the conservation and enhancement, as appropriate, of all sinks and reservoirs of all greenhouse gases not controlled by the Montreal Protocol, including biomass, forests and oceans as well as other terrestrial, coastal and marine ecosystems;

(e) Cooperate in preparing for adaptation to the impacts of climate change: develop and elaborate appropriate and integrated plans for coastal zone management, water resources and agriculture and for the protection and rehabilitation of areas, particularly in Africa, affected by drought and desertification, as well as floods;

(f) Take climate change considerations into account, to the extent feasible, in their relevant social, economic and environmental policies and actions and employ appropriate methods, for example impact assessments, formulated and determined nationally, with a view to minimising adverse effects on the economy, on public health and on the quality of the environment of projects or measures undertaken by them to mitigate or adapt to climate change;

(g) Promote and cooperate in scientific, technological, technical, socio-economic and other research, systematic observation and development of data archives related to the climate system and intended to further the understanding and to reduce or eliminate the remaining uncertainties regarding causes, effects, magnitude and timing of climate change and the economic and social consequences of various response strategies;

(h) Promote and cooperate in the full, open and prompt exchange of relevant scientific, technological, technical, socio-economic and legal information related to the climate system and climate change system and climate change and to the economic and social consequences of various response strategies;

(i) Promote and cooperate in education, training and public awareness related to climate change and encourage the widest participation in this process, including that of nongovernmental organisations; and

(j) Communicate to the Conference of the Parties, information related to implementation, in accordance with Article 12.

2. The developed country Parties and other Parties included in Annex I commit themselves specifically as provided for in the following:

(a) Each of these Parties shall adopt national policies and take corresponding measures on the mitigation of climate change, by limiting its anthropogenic emissions of greenhouse gases and protecting and enhancing its greenhouse gas, sinks and reservoirs. These policies and measures will demonstrate that developed countries are taking the lead in modifying longerterm trends in anthropogenic emissions consistent with the objective of this Convention, recognising that the return by the end of the present decade to earlier levels of anthropogenic emissions of carbon dioxide and other greenhouse gases not controlled by the Montreal Protocol would contribute to such modification and taking into account the differences in these Parties' starting points and approaches, economic structures and resource bases, the need to maintain strong and sustainable economic growth, available technologies and other individual circumstances, as well as the need for equitable and appropriate contributions by each of these Parties to the global effort regarding that objective. These Parties may implement such policies and measures jointly with other Parties and may assist other Parties in contributing to the achievement of the objective of the Convention and in particular that of this subparagraph;

(b) In order to promote progress to this end, each of these Parties shall communicate, within six months of the entry into force of the Convention for it and periodically thereafter and in accordance with Article 12, detailed information on its policies and measure referred to in subparagraph (a) above, as well as on its resulting projected emissions from sources and removals by sinks of greenhouse gases for the period referred to in that subparagraph, with the aim of returning individually or jointly to their 1990 levels these anthropogenic emissions of carbon dioxide and other greenhouse gases. This information will be reviewed by the Conference of the Parties, at its first session and periodically thereafter, in accordance with Article 7;

(c) Calculations of emissions by sources and removals by sinks of greenhouse gases for the purposes of subparagraph (b) above should take into account the best available scientific knowledge, including the effective capacity of sinks and the respective contributions to global warming of such gases. The Conference of the Parties shall consider and agree on methodologies for these calculations at its first session and review them regularly thereafter;

(d) The Conference of the Parties shall, at its first session, review the adequacy of subparagraphs (a) and (b) above. Such review shall be carried out in the light of the best available scientific information and assessment on climate change and its impacts, as well as relevant technical, social and economic information. Based on this review, the Conference of the Parties shall take appropriate action, which may include the adoption of amendments to the commitments in subparagraphs (a) and (b) above. The Conference of the Parties, at its first session, shall also take decisions regarding criteria for joint implementation as indicated in subparagraph (a) above. A second review of subparagraphs (a) and (b) shall take place not later than 31 December, 1998, and thereafter at regular intervals determined by the Conference of the Parties, until the objective of the Convention is met;

(e) Each of these Parties shall:

(i) coordinate as appropriate with other such Parties, relevant economic and administrative instruments developed to achieve the objective of the Convention; and

(ii) identify and periodically review its own policies and practices which encourage activities that lead to greater levels of anthropogenic emissions of greenhouse gases not controlled by the Montreal Protocol than would otherwise occur;

(f) The Conference of the Parties shall review, not later than 31 December, 1998, available information with a view to taking decisions regarding such amendments to the lists in Annexes I and II as may be appropriate, with the approval of the Party concerned;

(g) Any Party not included in Annex I may, in its instrument of ratification, acceptance, approval or accession, or at any time thereafter, notify the Depositary that it intends to be bound by subparagraphs (a) and (b) above. The Depositary shall inform the other signatories and Parties of any such notification.

3. The developed country Parties and other developed Parties included in Annex II shall provide new and additional financial resources to meet the agreed full costs incurred by developing country Parties in complying with their obligations under Article 12, paragraph 1. They shall also provide such financial resources, including for the transfer of technology, needed by the developing country Parties to meet the agreed full incremental costs of implementing measures that are covered by paragraph 1 of this Article and that are agreed between a developing country party and the international entity or entities referred to in Article 11, in accordance

with that Article. The implementation of these commitments shall take into account the need for adequacy and predictability in the flow of funds and the importance of appropriate burden sharing among the developed country Parties.

4. The developed country Parties and other developed Parties included in Annex II shall also assist the developing country parties that are particularly vulnerable to the adverse effects of climate change in meeting costs of adaptation to these adverse effects.

5. The developed country Parties and other developed Parties included in Annex II shall take all practicable steps to promote, facilitate and finance, as appropriate, the transfer of, or access to, environmentally sound technologies and know-how to other Parties, particularly developing country Parties, to enable them to implement the provisions of the Convention. In this process, the developed country Parties shall support the development and enhancement of endogenous capacities and technologies of developing country Parties. Other Parties and organisations in a position to do so may also assist in facilitating the transfer of such technologies.

6. In the implementation of their commitments under paragraph 2 above, a certain degree of flexibility shall be allowed by the Conference of the Parties to the Parties included in Annex I undergoing the process of transition to a market economy, in order to enhance the ability of these Parties to address climate change, including with regard to the historical level of anthropogenic emissions of greenhouse gases not controlled by the Montreal Protocol chosen as a reference.

7. The extent to which developing country Parties will effectively implement the commitments under the Convention will depend on the effective implementation by developed country Parties of their commitments under the Convention related to financial resources and transfer of technology and will take fully into account that economic and social development and poverty eradication are the first and overriding priorities of the developing country Parties.

8. In the implementation of the commitments in this Article, the Parties shall give full consideration to what actions are necessary under the Convention, including actions related to funding, insurance and the transfer of technology, to meet the specific needs and concerns of developing country Parties arising from the adverse effects of climate change, and/or the impact of the implementation of response measures, especially on:

   (a) Small island countries;

   (b) Countries with low-lying coastal areas;

   (c) Countries with arid and semi-arid areas, forested areas and areas liable to forest decay;

   (d) Countries with areas prone to natural disasters;

   (e) Countries with areas liable to drought and desertification;

   (f) Countries with areas with fragile ecosystems including mountainous ecosystems;

   (g) Countries whose economies are highly dependent on income generated from the production, processing and export, and/or on consumption of fossil fuels and associated energy-intensive products; and

   (h) land-locked and transit countries.

   Further, the Conference of the Parties may take actions, as appropriate, with respect to this paragraph.

9. The Parties shall take full account of the specific needs and special situations of the least developed countries in their actions with regard to funding and transfer of technology.
10. The Parties shall, in accordance with Article 10, take into consideration in the implementation of the commitments of the Convention the situation of Parties, particularly developing country Parties, with economies that are vulnerable to the adverse effects of the implementation of measures to respond to climate change. This applies notably to Parties with economies that are highly dependent on income generated from the production, processing and export, and/or consumption of fossil fuels and associated energy-intensive products and/or the use of fossil fuels for which such Parties have serious difficulties in switching to alternatives.

Article 5. Research and Systematic Observation

In carrying out their commitment under Article 4, paragraph 1 (g), the Parties shall:

1. Support and further develop, as appropriate, international and intergovernmental programmes and networks or organisations aimed at defining, conducting, assessing and financing research, data collection and systematic observation, taking into account the need to minimise duplication of effort;
2. Support international and intergovernmental efforts to strengthen systematic observation and national scientific and technical research capacities and capabilities, particularly in developing countries and to promote access to and the exchange of data and analyses thereof obtained from areas beyond national jurisdiction; and
3. Take into account the particular concerns and needs of developing countries and cooperate in improving their endogenous capacities and capabilities to participate in these efforts referred to in subparagraphs (a) and (b) above.

Article 6. Education, Training and Public Awareness

In carrying out their commitments under Article 4, paragraph 1(i), the Parties shall:

1. Promote and facilitate at the national and, as appropriate, subregional and regional levels and in accordance with national laws and regulations and within their respective capacities:
   (a) The development and implementation of education and public awareness programmes on climate change and its effects;
   (b) Public access to information on climate change and its effects;
   (c) Public participation in addressing climate change and its effects and developing adequate responses; and
   (d) Training of scientific, technical and managerial personnel.
2. Cooperate in and promote, at the international level and where appropriate, using existing bodies:
   (a) The development and exchange of educational and public awareness material on climate change and its effects; and
   (b) The development and implementation of education and training programmes, including the strengthening of national institutions and the exchange or secondment of personnel to train experts in this field, in particular for developing countries.

Article 7. Conference of the Parties

1. A Conference of the Parties is hereby established.
2. The Conference of the Parties as the supreme body of this Convention shall keep under regular review the implementation of the Convention and any related legal instruments that the Conference of the Parties may adopt, and shall make, within its mandate, the decisions necessary to promote the effective implementation of the Convention. To this end, it shall:

(a) Periodically examine the obligations of the Parties and the institutional arrangements under the Convention, in light of the objective of the Convention, the experience gained in its implementation and the evolution of scientific and technological knowledge;

(b) Promote and facilitate the exchange of information on measures adopted by the Parties to address climate change and its effects, taking into account the differing circumstances, responsibilities and capabilities of the Parties and their respective commitments under the Convention;

(c) Facilitate, at the request of two or more Parties, the coordination of measures adopted by them to address change and its effects, taking into account the differing circumstances, responsibilities and capabilities of the Parties and their respective commitments under the Convention;

(d) Promote and guide, in accordance with the objective and provisions of the Convention, the development and periodic refinement of comparable methodologies, to be agreed on by the Conference of the Parties, *inter alia,* for preparing inventories of greenhouse gas emissions by sources and removal by sinks and for evaluating the effectiveness of measures to limit the emissions and enhance the removal of these gases;

(e) Assess, on the basis of all information made available to it in accordance with the provisions of the Convention, the implementation of the Convention by the Parties, the overall effects of the measures taken pursuant to the Convention, in particular environmental, economic and social effects as well as their cumulative impacts and the extent to which progress towards the objective of the Convention is being achieved;

(f) Consider and adopt regular reports on the implementation of the Convention and ensure their publication;

(g) Make recommendations on any matters necessary for the implementation of the Convention;

(h) Seek to mobilise financial resources in accordance with Article 4, paragraphs 3, 4 and 5 and Article 11;

(i) Establish such subsidiary bodies as are deemed necessary for the implementation of the Convention;

(j) Review reports submitted by its subsidiary bodies and provide guidance to them;

(k) Agree upon and adopt by consensus, rules of procedure and financial rules for itself and for any subsidiary bodies;

(l) Seek and utilise, where appropriate, the services and cooperation of and information provided by, competent international organisations and inter-governmental and non-governmental bodies; and

(m) Exercise such other functions as are required for the achievement of the objective of the Convention as well as all functions assigned to it under the other Convention.

3. The Conference of the Parties shall, at its first session, adopt its own rules of procedures as well as those of the subsidiary bodies established by the Convention, which shall include decision-making procedures for matters not already covered by decision-making procedures stipulated in the Convention. Such procedures may include specified majorities required for the adoption of particular decisions.

4. The first session of the Conference of the Parties shall be convened by the interim secretariat referred to in Article 21 and shall take place not later than one year after the date of entry into force of the Convention. Thereafter, ordinary sessions of the Conference of the Parties shall be held at least every year unless otherwise decided by the Conference of the Parties.

5. Extraordinary sessions of the Conference of the Parties shall be held at such other times as may be deemed necessary by the Conference, or at the written request of any Party, provided that, within six months of the request being communicated to the Parties by the secretariat, it is supported by at least one-third of the Parties.

6. The United Nations, its specialised agencies and the International Atomic Energy Agency, as well as any State member thereof or observer thereto not Party to the Convention, may be represented at sessions of the Conference of the Parties as observers. Any body or agency, whether national or international, governmental or non-governmental, which is qualified in matters covered by the Convention and which has informed the secretariat of its wish to be represented at a session of the Conference of the Parties as an observer, may be so admitted unless at least one-third of the Parties present object. The admission and participation of observers shall be subject to the rules of procedure adopted by the Conference of the Parties.

Article 8. Secretariat

1. A secretariat is hereby established.

2. The functions of the secretariat shall be:

   (a) To make arrangements for sessions of the Conference of the Parties and its subsidiary bodies established under the Convention and to provide them with services as required;

   (b) To compile and transmit reports submitted to it;

   (c) To facilitate assistance to the Parties, particularly developing country Parties, upon request. In the compilation and communication of information required in accordance with provisions of the Convention;

   (d) To prepare reports on its activities and present them to the Conference of the Parties;

   (e) To ensure the necessary coordination with the secretariats of other relevant international bodies;

   (f) To enter, under the overall guidance of the Conference of the Parties, into such administrative and contractual arrangements as may be required for the effective discharge of its functions; and

   (g) To perform the other secretariat functions specified in this Convention and in any of its Protocols and such other functions as may be determined by the Conference of the Parties.

3. The Conference of the Parties, at its first session, shall designate a permanent secretariat and make arrangements for its functioning.

Article 9. Subsidiary Body for Scientific and Technological Advice

1. A subsidiary body for scientific and technological advice is hereby established to provide the Conference of the Parties and as appropriate, its other subsidiary bodies with timely information and advice on scientific and technological manners relating to the Convention. This body shall be open to participation of all the Parties and shall be multidisciplinary. It shall comprise government representatives competent in the relevant field of expertise. It shall report regularly to the Conference of the Parties on all aspects of its work.

2. Under the guidance of the Conference of the Parties, and drawing upon existing competent international bodies, this body shall:

   (a) Provide assessments of the state of scientific knowledge relating to climate change and its effects;

   (b) Prepare scientific assessments on the effects of measures taken in the implementation of the Convention;

   (c) Identify innovative, efficient and state-of-the-art technologies and know-how and advise on the ways and means of promoting development and/or transferring such technologies;

   (d) Provide advice on scientific programmes, international cooperation in research and development related to climate change, as well as on ways and means of supporting endogenous capacity building in developing countries; and

   (e) Respond to scientific, technological and methodological questions that the Conference of the Parties and its subsidiary bodies may put to the body.

3. The functions and terms of reference of this body may be further elaborated by the Conference of the Parties.

Article 10. Subsidiary Body for Implementation

1. A subsidiary body for implementation is hereby established to assist the Conference of the Parties in promoting and reviewing the effective implementation of the Convention. This body shall be open to participation by all Parties and comprise government representatives who are experts on matters related to climate change. It shall report regularly to the Conference of the Parties on all aspects of its work.

2. Under the guidance of the Conference of the Parties, this body shall:

   (a) Consider the information communicated in accordance with Article 12, paragraph 1, to assess the overall aggregated effect of the steps taken by the Parties in the light of the latest scientific assessments concerning climate change;

   (b) Consider the information communicated in accordance with Article 12, paragraph 2, in order to assist the Conference of the Parties in carrying out the reviews required by Article 4, paragraph 2(d); and

   (c) Assist the Conference of the Parties, as appropriate, in the preparation and implementation of its decisions.

Article 11. Financial Mechanism

1. A mechanism for the provision of financial resources on a grant or concessional basis, including for the transfer of technology, is hereby defined. It shall function under the guidance of and be

accountable to the Conference of the Parties, which shall decide on its policies, programme priorities and eligibility criteria related to this Convention. Its operation shall be entrusted to one or more existing international entities.

2. The financial mechanism shall have an equitable and balanced representation of all Parties within a transparent system of governance.

3. The Conference of the Parties and the entity or entities entrusted with the operation of the financial mechanism shall agree upon arrangements to give effect to the above paragraphs, that shall include the following:

   (a) Modalities to ensure that the funded projects to address climate change are in conformity with the policies, programme priorities and eligibility criteria established by the Conference of the Parties;

   (b) Modalities by which a particular funding decision may be reconsidered in light of these policies, programme priorities and eligibility criteria;

   (c) Provision by the entity or entities of regular reports to the Conference of the Parties on its funding operations, which is consistent with the requirement for accountability set out in para 1 above; and

   (d) Determination in a predictable and identifiable manner of the amount of funding necessary and available for the implementation of this Convention and the conditions under which that amount shall be periodically reviewed.

4. The Conference of the Parties shall make arrangements to implement the above-mentioned provisions at its first session, reviewing and taking into account the interim arrangements referred to in Article 21, Paragraph 3 and shall decide whether these interim arrangements shall be maintained. Within four years thereafter, the Conference of the Parties shall review the financial mechanism and take appropriate measures.

5. The developed country Parties may also provide and developing country Parties avail themselves of financial resources related to the implementation of this Convention through bilateral, regional and other multilateral channels.

Article 12. Communication of Information Related to Implementation

1. In accordance with Paragraph 1 of Article 4, each Party shall communicate to the Conference of the Parties, through the secretariat, the following elements of information:

   (a) A national inventory of anthropogenic emissions from sources and removals by sinks of all greenhouse gases not controlled by the Montreal Protocol, to the extent of its capacities permit, using comparable methodologies to be promoted and agreed upon by the Conference of the Parties;

   (b) A general description of steps taken or envisaged by the Party to implement the Convention; and

   (c) Any other information that the party considers relevant to the achievement of the objective of the Convention and suitable for inclusion in its communication, including, if feasible, material relevant for calculations of global emission trends.

2. Each developed country Party and each other party included in Annex I shall incorporate in its communication the following elements of information:
   (a) A detailed description of the policies and measures that it has adopted to implement its commitment under subparagraphs 2(a) and 2(b) of Article 4;
   (b) A specific estimate of the effects that the policies and measures referred to in subparagraph (a) immediately above will have an anthropogenic emissions from its sources and removals by its sinks of greenhouse gases during the period referred to in subparagraph 2(a) of Article 4.
3. In addition, each developed country Party and each other developed Party included in Annex II shall incorporate details of measures taken in accordance with Article 4, paragraphs 3, 4 and 5.
4. Developing country Parties may, on a voluntary basis, propose projects for financing, including specific technologies, materials, equipment, techniques or practices that would be needed to implement such projects, along with, if possible, an estimate of all incremental costs, of the reductions of emissions and increments of removals of the greenhouse gases, as well as an estimate of the consequent benefits.
5. Each developed country Party included in Annex I shall make its initial communication within six months of the entry into force of the Convention for that Party. Each Party not so listed shall make its initial communication within three years of the entry into force of the Convention for that Party, or of the availability of financial resources in accordance with Paragraph 3 of Article 4. Parties that are least developed countries may make their initial communication at their discretion. The frequency of subsequent communications by all Parties shall be determined by the Conference of the Parties, taking into account the differentiated time table set by this paragraph.
6. Information communicated by Parties under this Article shall be transmitted by the secretaries as soon as possible to the Conference of the Parties and to any subsidiary bodies concerned. If necessary the procedures for the communication of information may be further considered by the Conference of the Parties.
7. From its first session, the Conference of the Parties shall arrange for the provision to developing country Parties of technical and financial support, on request, in compiling and communicating information under this Article; as well as in identifying the technical and financial needs associated with proposed projects and response measures under Article 4. Such assistance may be provided by other Parties, by competent international organisations and by the secretariat, as appropriate.
8. Any group of Parties may, subject to guidelines adopted by the Conference of the Parties and to prior notification to the Conference of the Parties, make a joint communication in fulfilment of their obligations under this Article, provided that such a communication includes information on the fulfilment by each of these parties of its individual obligations under this Convention.
9. Information received by the secretariat that is designated by a Party as confidential. In accordance with criteria to be established by Conference of the Parties, shall be aggregated by the secretariat to protect its confidentiality before being made available to any of the bodies involved in the communication and review of information.

10. Subject to Paragraph 9 above and without prejudice to the ability of any Party to make public its communication at any time, the secretariat shall make communications by Parties under this Article publicly available at the time they are submitted to the Conference of the Parties.

Article 13. Resolution of Questions Regarding Implementation

The Conference of the Parties shall, at its first session, consider the establishment of a multilateral consultative process for the resolution of questions regarding the implementation of the Convention.

Article 14. Settlement of Disputes

1. In the event of a dispute between any two or more Parties concerning the interpretation or application of the Convention, the Parties concerned shall seek a settlement of the dispute through negotiation or any other peaceful means of their own choice.
2. When ratifying, accepting, approving or acceding to the Convention, or at any time thereafter, a Party which is not a regional economic integration organisation may declare in a written instrument submitted to the Depositary, that, in respect of any dispute concerning the interpretation or application of the Convention, it recognises as compulsory *ipso facto* and without special agreement, in relation to any Party accepting the same obligation:
   (a) Submission of the dispute to the International Court of Justice, and/or
   (b) Arbitration in accordance with procedures to be adopted by the Conference of the Parties as soon as practicable in an annex on arbitration.

   A Party which is a regional economic integration organisation may make a declaration with like effect in relation to arbitration in accordance with the procedures referred to in subparagraph (b) above.
3. A declaration made under Paragraph 2 above shall remain in force until it expires in accordance with its terms or until three months after written notice of its revocation has been deposited with the Depositary.
4. A new declaration, a notice of revocation or the expiry of a declaration shall not in any way affect proceedings pending before the International Court of Justice or the Arbitral Tribunal, unless the Parties to the dispute otherwise agree.
5. Subject to the operation of Paragraph 2 above, if after 12 months following notification by one Party to another that a dispute exists between them, the Parties concerned have not been able to settle their dispute through the means mentioned in paragraph 1 above, the dispute shall be submitted, at the request of any of the Parties to the dispute, to conciliation.
6. A conciliation commission shall be created upon the request of one of the Parties to the dispute. The commission shall be composed of an equal number of members appointed by each Party concerned and a chairman chosen jointly by the members appointed by each Party. The commission shall render a recommendatory award, which the Parties shall consider in good faith.
7. Additional procedures relating to conciliation shall be set out by the Conference of the Parties, as soon as practicable, in an annex on conciliation.
8. The provisions of this Article shall apply to any related legal instrument which the Conference of the Parties may adopt, unless the instrument provides otherwise.

Article 15. Amendments to the Convention

1. Any Party may propose amendments to the Convention.
2. Amendments to the Convention shall be adopted at an ordinary session of the Conference of the Parties. The text of any proposed amendment to the Convention shall be communicated to the Parties by the secretariat at least six months before the meeting at which it is proposed for adoption. The secretariat shall also communicate proposed amendments to the Signatories to the Convention and for information, to the Depositary.
3. The Parties shall make every effort to reach agreement on any proposed amendment to the Convention by consensus. If all efforts at consensus have been exhausted and no agreement reached, the amendments shall be adopted by a three-fourths majority vote of the Parties present and voting at the meeting. The adopted amendment shall be communicated by the secretariat to the Depositary, who will circulate it to all Parties for their acceptance.
4. Instruments of acceptance in respect of an amendment shall be deposited with the Depositary. An amendment adopted in accordance with Paragraph 3 above shall enter into force for those Parties having accepted it on the ninetieth day after the date of receipt by the Depositary of an instrument of acceptance by at least three-fourths of the Parties to the Convention.
5. The amendment shall enter into force for any other Party on the ninetieth day after the date on which that Party deposits with the Depositary its instrument of acceptance of the said amendment.
6. For the purposes of this Article, 'Parties present and voting' means Parties present and casting an affirmative or negative vote.

Article 16. Adoption and Amendment of Annexes to the Convention

1. The Annexes to the Convention shall form an integral part thereof and, unless otherwise expressly provided, a reference to the Convention constitutes at the same time a reference to any Annexes thereto. Without prejudice to the provisions of Article 14, Paragraphs 2(b) and 7, such Annexes shall be restricted to lists, forms and any other material of a descriptive nature that is of a scientific, technical, procedural or administrative character.
2. Annexes to the Convention shall be proposed and adopted in accordance with the procedure set forth in Article 15, paragraphs 2, 3 and 4.
3. Annexes that have been adopted in accordance with Paragraph 2 above, shall enter into force for all Parties to the convention six months after the date of the communication by the Depositary to such Parties of the adoption of the Annex, except for those Parties that have notified the Depositary, in writing, within that period of their non-acceptance of the Annex. The Annex shall enter into force for Parties which withdraw their notification of non-acceptance on the ninetieth day after the date on which withdrawal of such notification has been received by the Depositary.
4. The proposal, adoption and entry into force of amendments to Annexes to the Convention shall be subject to the same procedure as that for the proposal, adoption and entry into force of Annexes to the Convention in accordance with Paragraphs 2 and 3 above.
5. If the adoption of an annex or an amendment to an annex involves an amendment to the Convention, that annex or amendment to an annex shall not enter into force until such time as the amendment to the Convention enters into force.

Article 17. Protocols

1. The Conference of the Parties may, at any ordinary session, adopt protocols to the Convention.
2. The text of any proposed protocol shall be communicated to the Parties by the secretariat at least six months before such a session.
3. The requirements for the entry into force of any protocol shall be established by that instrument.
4. Only Parties to the Convention may be Parties to a protocol.
5. Decisions under any protocol shall be taken only by the Parties to the protocol concerned.

Article 18. Right to Vote

1. Each Party to the Convention shall have one vote, except as provided for in Paragraph 2 below.
2. Regional economic integration organisations, in matters within their competence, shall exercise their right to vote with a number of votes equal to the number of their member States that are Parties to the Convention. Such an organisation shall not exercise its right to vote if any of its member States exercises its right and vice versa.

Article 19. Depositary

The Secretary-General of the United Nations shall be the Depositary of the Convention and of Protocols adopted in accordance with Article 17.

Article 20. Signature

The Convention shall be open for signature by States that are Members of the United Nations or of any of its Specialised Agencies or that are Parties to the Statute of the International Court of Justice and by regional economic integration organisations at Rio de Janeiro, during the United Nations Conference on Environment and Development and thereafter at United Nations Headquarters in New York from 20 June, 1992 to 19 June, 1993.

Article 21. Interim Arrangements

1. The secretariat functions referred to in Article 8 will be carried out on an interim basis by the secretariat established by the General Assembly of the United Nations in its resolution 45/212 of 21 December, 1990, until the completion of the first session of the Conference of the Parties.
2. The head of the interim secretariat referred to in paragraph 1 above will cooperate closely with the Intergovernmental Panel on Climate Change to ensure that the Panel can respond to the need for objective, scientific and technical advice. Other relevant scientific bodies could also be consulted.
3. The Global Environment Facility of the United Nations Development Programme and the International Bank for Reconstruction and Development shall be the international entity entrusted with the operation of the financial mechanism referred to in Article 11 on an interim basis. In this connection, the Global Environment Facility should be appropriately restructured and its membership made universal to enable it to fulfil the requirements of Article 11.

Article 22. Ratification, Acceptance, Approval or Accession

1. The Convention shall be subject to ratification, acceptance, approval or accession by States and by regional economic integration organisations. It shall be open for accession from the day after the date on which the Convention is closed for signature. Instruments of ratification, acceptance, approval or accession shall be deposited with the Depositary.
2. Any regional economic integration organisation which becomes a Party to the Convention without any of its member States being a Party shall be bound by all the obligations under the Convention. In the case of such organisations, one or more of whose member States is a Party to the Convention, the organisation and its member States shall decide on their respective responsibilities for the performance of their obligations under the Convention. In such cases, the organisation and the member States shall not be entitled to exercise rights under the Convention concurrently.
3. In their instruments of ratification, acceptance, approval or accession, regional economic integration organisations shall declare the extent of their competence with respect to the matters governed by the Convention. These organisations shall also inform the Depositary, who shall in turn inform the Parties, of any substantial modification in the extent of their competence.

Article 23. Entry into Force

1. The Convention shall enter into force on the ninetieth day after the date of deposit of the fiftieth instrument of ratification, acceptance, approval or accession.
2. For each State or regional economic integration organisation that ratifies, accepts or approves the Convention or accedes therein after the deposit of the fiftieth instrument of ratification, acceptance, approval or accession, the Convention shall enter into force on the ninetieth day after the date of deposit by such State or regional economic integration organisation of its instrument of ratification, acceptance, approval or accession.
3. For the purposes of Paragraphs 1 and 2 above, any instrument deposited by a regional economic integration organisation shall not be counted as additional to those deposited by member States of the organisation.

Article 24. Reservations

No reservations may be made to the Convention.

Article 25. Withdrawal

1. At any time after three years from the date on which the Convention has entered into force for a Party, that Party may withdraw from the Convention by giving written notification to the Depositary.
2. Any such withdrawal shall take effect upon expiry of one year from the date of receipt by the Depositary of the notification of withdrawal, or on such later date as may be specified in the notification of the withdrawal.
3. Any Party that withdraws from the Convention shall be considered as also having withdrawn from any protocol to which it is a Party.

Article 26. Authentic Texts

The original of the Convention, of which the Arabic, Chinese, English, French, Russian and Spanish texts are equally authentic, shall be deposited with the Secretary-General of the United Nations.

In Witness whereof the undersigned, being duly authorised to that effect, have signed this Convention.

Done at New York this 9th day of May, 1992.

64.10 ENVIRONMENTAL COOPERATION IN SOUTH ASIA

One-fifth of the world's population and nearly half of the world's poor live in South Asia. The region also shows high rates of population growth. Despite rapid economic growth during the 1990s, the region has among the lowest per capita incomes in the world. Growing subsistence needs impose larger demands on water supplies, arable land, forests, and coastal habitats. Almost all of the seven countries in the region are already suffering from a series of environmental problems in the form of deforestation, soil erosion, and a scarcity of fresh water.

South Asia has long been a region of instability and unrest. It is home to two of the world's most intractable ethnic conflicts: in the Indian state of Kashmir and in Sri Lanka.

With massive population growth and rampant poverty in both economic and human terms (see Table 64.6), the countries of South Asia—primarily India, Pakistan, Bangladesh, Nepal, and Bhutan-already suffer from large-scale deforestation, soil erosion, desertification, and depletion of water resources.

South Asia is highly vulnerable to this environmental conflict scenario due to its high population density and one of the world's highest levels of poverty. As stated earlier, one-fifth of the world's population and nearly half of the world's poor live in South Asia.

Table 64.6: Human Development Statistics for South Asia

Country	HDI rank	Population, 1998 (millions)	Annual population growth rate, 1998-2015 (estimated, %)	GDP per capita, 1998 (PPP, US$)
Sri Lanka	84	18.5	1.0	2,979
Maldives	89	0.3	2.6	4,083
India	128	982.2	1.2	2,077
Pakistan	135	148.2	2.4	1,715
Bhutan	142	2.0	2.6	1,536
Nepal	144	22.8	2.1	1,157
Bangladesh	146	124.8	1.5	1,361

Source: United Nations Development Programme (UNDP), *Human Development Report* 2000

Some discussions are already made earlier under DAM conflict about the water sharing intercountry conflict too, yet regional cooperation issues are discussed here.

India, Pakistan, and the Indus River

The potential for real military conflict exists between India and Pakistan; the issue of nuclear capabilities has added a further chilling dimension to these countries' tense fifty-year relationship. Surprisingly, however, the Indo-Pakistani dispute over the sharing of the Indus River system has not been as contentious as one might expect. The Indus Waters Treaty of 1960 between India and Pakistan is one of the few examples of successful resolution of a major dispute over an international river basin.

Although it took nine years for the World Bank to bring the two countries into agreement, the Indus Waters Treaty was finally signed on September 19, 1960, in the Pakistani port city of Karachi.

India, Bangladesh, and the Ganges

India and Bangladesh are the two major riparian countries in the eastern Himalayan region. These two neighbours share two large Himalayan river systems, the Ganges and the Brahmaputra. The major focus of dispute between the two countries has been the Ganages, as the Brahmaputra has not yet been substantially tapped.

The dispute over sharing the dry-season flow of the Ganges originated between India and Pakistan in 1951, when India started planning to construct a barrage at Farakka, eighteen kilometers upstream from the border with East Pakistan (later Bangladesh). The plan include a 38-kilometer canal that could divert up to 40,000 cusecs from the dam to supplement the waters of the Bhagirathi-Hooghly. The scheme to divert water from the Ganges to the distributary Bhagirathi-Hooghly was intended to flush out silt and keep the port of Kolkata navigable. The domestic and industrial demands of the city of Kolkata and irrigation schemes in the state of West Bengal also induced the Indian government to develop the project. The Pakistani objection to this plan was very vocal but failed to block India's unilateral decision to begin dam construction in 1962.

In December 1996, the prime ministers of India and Bangladesh signed a new Ganges water-sharing agreement, after a gap of eight years. Instead of their usual short-term approach of sharing the dry-season flow at the Farakka barrage, they agreed to a thirty-year arrangement. The treaty stipulates that when flow rates fall below 70,000 cusecs, India and Bangladesh will each receive half of the water. At flow rates above 75,000 cusecs, India is guaranteed a share of 40,000 cusecs, with the balance of the flow going to Bangladesh.

India, Nepal, and the Mahakali River

Indo-Nepalese relations have been smooth and friendly in comparison to Indo-Pakistani or Indo-Bangladeshi relations. However, sharing the waters of the major rivers originating in Nepal and flowing into India has strained the bilateral relationship. Negotiations regarding projects on the shared river systems have been dominated by controversies because of a lack of mutual trust. Nepalese rivers have tremendous potential for hydropower generation, but Nepal lacks the capital and technology required for such large projects and also needs a buyer for the surplus hydropower. For various reasons, India is the only country that can provide the needed assistance. Thus, India's direct involvement in the use of the river water has been crucial. In general, however, the Nepalese feel that they have not been treated equitably under the various water-source development agreements with India, including those governing the waters of the Sarada (1920), the Kosi (1954), and Gandak (1959) rivers.

The Mahakali treaty also paved the way for the construction of the Pancheshwar multipurpose project. The agreement set the basis for developing the 6,480 megawatt Pancheshwar hydropower project on a stretch of the Mahakali that crosses the border between the two countries. When completed, the 315-meter-high, rock-fill structure will be second highest dam in the world (after the Rogun Dam in Ukraine). The 12-billion-cubic-meter reservoir created will store the flow of the Mahakali River to yield a discharge of 885 cubic meters per second below the powerhouse, which will produce up to 3,240 megawatts of power on each side of the river. The reservoir will inundate 134 square kilometers of land (54 km^2 in Nepal and 80 km^2 in India) and is likely to directly displace 65,000 people (15,000 in Nepal and 50,000 in India). The Mahakali treaty is undoubtedly a very positive step toward further bilateral cooperation on water-resource development between India and Nepal (although controversies around the agreement endure, as discussed below). Both sides have gradually come to realise that their own interests are better served through mutual cooperation. Nepal's vast hydropower potential and regulated releases of water therefrom to Indian agricultural fields could immensely benefit both countries. The treaty has also enabled the establishment of the Mahakali River Commission, which has a relatively broad mandate. The comprehensive nature of the treaty shows the determination of India and Nepal to move beyond the unilateral water-development strategy to which they had previously adhered and instead promote and strengthen their bilateral cooperation in order to receive the maximum benefits of shared river waters.

In 1985, seven South Asian countries officially formed a regional organisation, the South Asian Association for Regional Cooperation (SAARC), to promote cooperation in science, education, technology, cultural exchanges, and other matters of mutual concern and interest. At the third meeting of SAARC heads of state, held in 1987, it was decided to study and then issue recommendations on how regional cooperation could respond to rapid and continuing environmental degradation.

Agenda 21

In World Summit (1992) held at Rio de Jenario (Brazil), the following resolutions have been taken up:

1. To adopt measures for International cooperation to accelerate sustainable development in developing countries and related domestic policies;
2. To adopt policies for combating poverty;
3. To take up measures for changing consumption pattern;
4. To adopt measures for demographic dynamics and sustainability;
5. To adopt measures for protecting and promoting human health;
6. To adopt measures for promoting sustainable human settlement and development;
7. To integrate environment and development in decision making;
8. To take up measures for protecting the atmosphere;
9. To undertake integrated approach to the planning and management;
10. To adopt measures for combating deforestation;
11. To manage fragile ecosystems: combating desertification and drought;
12. To manage fragile ecosystem: sustainable mountain development;
13. To promote sustainable agriculture and rural development;
14. To conserve biological diversity;
15. To adopt environmentally sound management of biotechnology;
16. To adopt measures for protection of the oceans, all kinds of seas including enclosed and semi-enclosed seas and coastal areas and the protection, rational use and development of their living resources;
17. To adopt measures for protection of the quality and supply of freshwater resources: application of integrated approaches to the development, management and use of water resources;
18. To take up measures for environmentally sound management of toxic chemicals including prevention of illegal international traffic in toxic and dangerous products;
19. To adopt measures for environmentally sound management for hazardous waste including precaution of illegal international traffic in hazardous waste;
20. To adopt measures for environmentally sound management of solid waste and sewage related issues.

Box 64.1: Appendix : International Environmental Agreements

Dates represent date of adoption.

Global climate change

- Kyoto Protocol on Climate Change, November 1997
- United Nations Framework Convention on Climate Change (UNFCCC), May 9, 1992
- UN Conference on Environment and Development
  — Agenda 21, June 3-14, 1992
  — Rio Declaration, June 3-14, 1992

Stratospheric ozone depletion

- Adjustments and Amendment to the Montreal Protocol on Substances that Deplete the Ozone Layer, November 23–December 5, 1992
- Adjustments and Amendment to the Montreal Protocol on Substances that Deplete the Ozone Layer, June 29, 1990
- Montreal Protocol on Substances that Deplete the Ozone Layer, September 16, 1987
- Vienna Convention for the Protection of the Ozone Layer, March 22, 1985

Desertification and land cover change

- United Nations Convention to Combat Desertification in those Countries Experiencing Serious Drought and/or Desertification, Particularly in Africa, September 12, 1994
- United Nations Conference on Desertification (UNCOD) Plan of Action to Combat Desertification and General Assembly Resolutions, August 29, 1977

Conservation of biological diversity

- Agreement Establishing the Inter-American Institute for Global Change Research, May 13, 1992
- Convention on Biological Diversity, June 5, 1992
- Protocol on Environmental Protection to the Antarctic Treaty, October 4, 1991
- European Convention for the Protection of Pet Animals, November 13, 1987
- ASEAN Agreement on the Conservation of Nature and Natural Resources, July 9, 1985
- International Tropical Timber Agreement, November 18, 1983
- Protocol to Amend the Convention on Wetlands of International Importance Especially as Waterfowl Habitat, December 3, 1982
- Benelux Convention on Nature Conservation and Landscape Protection, June 8, 1982
- World Charter for Nature, 1982
- Convention on the Conservation of Antarctic Marine Living Resources, May 20, 1980
- Convention on the Conservation European Wildlife and Natural Habitats, September 19, 1979
- Convention on the Conservation of Migratory Species of Wild Animals, June 23, 1979
- Agreement on Conservation of Polar Bears, November 15, 1973
- Convention on International Trade in Endangered Species of Wild Fauna and Flora, March 3, 1973
- Convention for the Protection of the World Cultural and Natural Heritage, November 23, 1972
- Convention for the Conservation of Antarctic Seals, June 1, 1972
- Convention on Wetlands of International Importance Especially as Waterfowl Habitat, February 2, 1971
- Benelux Convention on the Hunting and Protection of Birds (as amended), June 10, 1970
- International Convention for the Protection of Birds, October 18, 1950

Transboundary air pollution

- Convention on the Transboundary Effects of Industrial Accidents, March 17, 1992
- Convention on Environmental Impact Assessment in a Transboundary Context, February 25, 1991
- Basel Convention on the Control of Transboundary Movements of Hazardous Wastes and their Disposal, March 22, 1989

Contd.

Box 64.1: Appendix: International Environmental Agreements

- Protocol to the 1979 Convention on Long-range Transboundary Air Pollution on Longterm Financing of Cooperative Program for Monitoring and Evaluation of the Long-range Transmissions of Air Pollutants in Europe (EMEP), September 28, 1984
- Protocol Concerning Cooperation in Combating Pollution in Cases of Emergency, March 21, 1981
- Convention on Long-range Transboundary Air Pollution, November 13, 1979

Hazardous waste management

- Bamako Convention on the Ban of the Import into Africa and the Control of Transboundary Movements and Management of Hazardous Wastes Within Africa, January 30, 1991
- Basel Convention on the Control of Transboundary Movements of Hazardous Wastes and their Disposal, March 22, 1989

Ocean resources

- Agreement for the Implementation of the Provisions of the United Nations Convention on the Law of the Sea of December 10, 1982 Relating to the Conservation and Management of Straddling Fish Stocks and Highly Migratory Fish Stocks, August 4, 1995
- Agreement Relating to the Implementation of Part XI of the 1982 United Nations Convention on the Law of the Sea, July 28, 1994
- Agreement to Promote Compliance with International Conservation and Management Measures by Fishing Vessels on the High Seas, November 29, 1993
- Agreement Establishing the South Pacific Regional Environment Program (SPREP), June 16, 1993
- Convention for the Conservation of Southern Bluefin Tuna, May 10, 1993
- Annex III to the Protocol of February 17, 1978 Relating to the International Convention for the Prevention of Pollution from Ships of November 2, 1973 (MARPOL 73/78), as amended on October 30, 1992
- Protocol on Protection of the Black Sea Marine Environment Against Pollution from Land-based Sources, April 21, 1992
- Convention for a North Pacific Marine Science Organization (PICES), December 12, 1990
- International Convention on Oil Pollution Preparedness, Response and Cooperation, November 29, 1990
- Amendment to the Annex to the Convention for the Prevention of Marine Pollution by Dumping of Wastes and Other Matter, November 3, 1989
- Protocol Concerning Marine Pollution Resulting from Exploration and Exploitation of the Continental Shelf, March 29, 1989
- Protocol for the Prevention of Pollution of the South Pacific Region by Dumping, November 25, 1986
- Protocol Concerning Cooperation in Combating Pollution Emergencies in the South Pacific Region, November 25, 1986
- Convention for the Protection of the Natural Resources and Environment of the South Pacific Region, November 24, 1986
- Protocol to Amend the International Convention on Civil Liability for Oil Pollution Damage, May 25, 1984
- Protocol to Amend the International Convention on the Establishment of an International Fund for Compensation for Oil Pollution Damage, May 25, 1984
- Agreement for Cooperation in Dealing with Pollution of the North Sea by Oil and Other Harmful Substances, September 13, 1983
- United Nations Convention on the Law of the Sea, December 10, 1982
- Convention for the Conservation of Salmon in the North Atlantic Ocean, March 2, 1982
- Amendments to the Convention on the Prevention of Marine Pollution by Dumping of Wastes and Other Matter concerning Settlement of Disputes, October 12, 1978
- Protocol of 1978 Relating to the International Convention for the Prevention of Pollution from Ships, February 17, 1978
- Protocol to the International Convention on Civil Liability for Oil Pollution Damage, November 19, 1976

Contd.

Box 64.1: Appendix: International Environmental Agreements

- Protocol to the International Convention on the Establishment of an International Fund for Compensation for Oil Pollution Damage, November 19, 1976
- Interim Convention on Conservation of North Pacific Fur Seals, February 9, 1957 (as amended on May 7, 1976)
- Protocol for the Prevention of Pollution of the Mediterranean Sea by Dumping from Ships and Aircraft, February 16, 1976
- Convention on the Prevention of Marine Pollution from Land based Sources, June 4, 1974
- Nordic Environmental Protection Convention, February 19, 1974
- Protocol Relating to Intervention on the High Seas in Cases of Pollution by Substances Other than Oil, November 2, 1973
- International Convention for the Prevention of Pollution from Ships, November 2, 1973
- Convention on Fishing and Conservation of Living Resources in the Baltic Sea and Belts, September 13, 1973
- Convention on the Prevention of Marine Pollution by Dumping of Wastes and Other Matter, December 29, 1972
- Convention for the Prevention of Marine Pollution by Dumping from Ships and Aircraft (as amended), 15 February 1972
- International Convention on the Establishment of an International Fund for Compensation for Oil Pollution Damage, December 18, 1971
- Agreement Concerning Cooperation in Taking Measures Against Pollution of the Sea by Oil, September 16, 1971
- International Convention on Civil Liability for Oil Pollution Damage, November 29, 1969
- International Convention for the Conservation of Atlantic Tunas, May 14, 1966
- Convention for the International Council for the Exploration of the Sea (as amended), September 12, 1964
- Agreed Measures for the Conservation of Antarctic Fauna and Flora, June 2, 1964
- Fisheries Convention, March 9, 1964
- Treaty Banning Nuclear Weapon Tests in the Atmosphere, in Outer Space, and Under Water, October 10, 1963
- Convention on Fishing and Conservation of the Living Resources of the High Seas, April 29, 1958
- Convention on the Continental Shelf, April 29, 1958
- Interim Convention on Conservation of North Pacific Fur Seals, February 9, 1957
- International Convention for the Prevention of Pollution of the Sea by Oil (as amended on April 11, 1962 and October 21, 1969), May 12, 1954
- International Convention for the Regulation of Whaling, December 2, 1946
- MARPOL Optional Annex—Annex IV: Regulations for the Prevention of Pollution by Sewage from Ships

International trade and the environment

- Rio Declaration and Agenda 21, June 3-14, 1992
- Bamako Convention on the Ban of the Import into Africa and the Control of Transboundary Movements and Management of Hazardous Wastes Within Africa, January 30, 1991
- Basel Convention on the Control of Transboundary Movements of Hazardous Wastes and their Disposal, March 22, 1989
- Convention on the Regulation of Antarctic Mineral Resource Activities, June 2, 1988
- International Tropical Timber Agreement, November 18, 1983
- European Convention for the Protection of Animals Kept for Farming Purposes, March 10, 1976
- Convention on International Trade in Endangered Species of Wild Fauna and Flora, March 3, 1973
- Convention for the Protection of the World Culture and Natural Heritage, November 2, 1972

Sources: Fridtjof Nansen Institute (1996) : UN (1997);
http ://sedac.ciesin.org/pidb/texts-subject.html (April 20, 2000)

drought: Condition in which an area does not get enough water because of lower than normal precipitation, higher than normal temperatures that increase evaporation, or both.

dust dome: Dome of heated air that surrounds an urban area and traps pollutants, especially suspended particulate matter. See also *urban heat island*.

dust plume: Elongation of a dust dome by winds that can spread a city's pollutants hundreds of kilometers downwind.

earth capital: The earth's natural resources and processes that sustain us and other species. Compare *human capital, manufactured capital, solar capital*.

earthquake: Shaking of the ground resulting from the fracturing and displacement of rock, producing a fault, or from subsequent movement along the fault. See *plc'es*.

earth-sustaining economy: Economic system in which the number of people and the quantity of goods are maintained at some constant level. This level is ecologically sustainable over time and meets at least the basic needs of all members of the population.

earth-wisdom revolution: Cultural change involving halting population growth and altering lifestyles, political and economic systems, and the way we treat the environment so that we can help sustain the earth for ourselves and other species. This involves working with the rest of nature by learning more about how nature sustains itself. See *earth-wisdom worldview*.

earth-wisdom society: Society based on working with nature by recycling and reusing discarded matter; by pollution prevention; by conserving matter and energy resources through reducing unnecessary waste and use; by not degrading renewable resources, by building things that are easy to recycle, reuse, and repair; by not allowing population size to exceed the carrying capacity of the environment; and by preserving biodiversity. See *earth-wisdom worldview*. Compare *matter-recycling society, planetary management worldview, spaceship-earth worldview*.

earth-wisdom worldview: Beliefs that nature exists for all the earth's species, not just for us; we are not in charge of the rest of nature; there is not always more, and it's not all for us; some forms of economic growth are beneficial, and some are harmful; our goals should be to design economic and political systems that encourage earth-sustaining forms of growth and discourage or prohibit earth-degrading forms; our success depends on learning to cooperate with one another and with the rest of nature instead of trying to dominate and manage the earth's life-support systems primarily for our own use. Compare *spaceship-earth worldview, planetary management worldview*.

ecological diversity: The variety of forests, deserts, grasslands, oceans, streams, lakes, and other biological communities interacting with one another and with their nonliving environment. See *biological diversity*. Compare *genetic diversity, species diversity*.

ecological integrity: The conditions and natural processes (such as the flow of materials and energy through ecosystems) that generate and maintain biodiversity and allow evolutionary change as a key mechanism for adapting to changes in environmental conditions.

ecological land-use planning: Method for deciding how land should be used; development of an integrated model that considers geological, ecological, health, and social variables.

ecologically sustainable development: Development in which the total human population size and resource use in the world (or in a region) are limited to a level that does not exceed the carrying capacity of the existing natural capital and is therefore sustainable. Compare *economic development, economic growth, sustainable development*.

ecological niche: Total way of life or role of a species in an ecosystem. It includes all physical, chemical, and biological conditions a species needs to live and reproduce in an ecosystem. See *fundamental niche, realised niche*.

ecological population density: Number of individuals of a population per unit area of habitat. Compare *population density*.

ecological succession: Process in which communities of plant and animal species in a particular area are replaced over time by a series of different and often more complex communities. See *primary succession, secondary succession*.

ecology: Study of the interactions of living organisms with one another and with their nonliving environment of matter and energy; study of the structure and functions of nature.

economic decision: Deciding what goods and services to produce, how to produce them, how much to produce, and how to distribute them to people.

economic depletion: Exhaustion of 80% of the estimated supply of a nonrenewable resource. Finding, extracting, and processing the remaining 20% usually costs more than it is worth; may also apply to the deple-

APPENDICES

APPENDIX I

Relative Atomic Masses

Element	Symbol	Atomic No.	Atomic Weight
Actinium	Ac	89	(227)
Aluminium	Al	13	26.981539
Americium	Am	95	(243)
Antimony	Sb	51	121.75
Argon	Ar	18	39.948
Arsenic	As	33	74.92159
Astatine	At	85	(210)
Barium	Ba	56	137.327
Berkelium	Bk	97	(247)
Beryllium	Be	4	9.012182
Bismuth	Bi	83	208.98037
Boron	B	5	10.811
Bromine	Br	35	79.904
Cadmium	Cd	48	112.411
Caesium	Cs	55	132.90543
Calcium	Ca	20	40.078
Californium	Cf	98	(251)
Carbon	C	6	12.011
Cerium	Cc	58	140.115
Chlorine	Cl	17	35.4527
Chromium	Cr	24	51.9961
Cobalt	Co	27	58.93320
Copper	Cu	29	63.546
Curium	Cm	96	(247)
Dysprosium	Dy	66	162.50
Einsteinium	Es	99	(254)
Erbium	Er	68	167.26
Europium	Eu	63	151.965
Fermium	Fm	100	(257)

Element	Symbol	Atomic No.	Atomic Weight
Fluorine	F	9	18.9984032
Francium	Fr	87	(223)
Gadolinium	Gd	64	157.25
Gallium	Ga	31	69.723
Germanium	Ge	32	72.61
Gold	Au	79	196.96654
Mercury	Hg	80	200.59
Molybdenum	Mo	42	95.94
Neodymium	Nd	60	144.24
Neon	Ne	10	20.1797
Neptunium	Np	93	(237)
Nickel	Ni	28	58.69
Niobium	Nb	41	92.90638
Nitrogen	N	7	14.00674
Nobelium	No	102	(255)
Osmium	Os	76	190.2
Oxygen	O	8	15.9994
Palladium	Pd	46	106.42
Phosphorus	P	15	30.973762
Platinum	Pt	78	195.08
Plutonium	Pu	94	(244)
Polonium	Po	84	(209)
Potassium	K	19	39.0983
Prascodymium	Pr	59	140.90765
Promethium	Pm	61	(145)
Protactinium	Pa	91	231.035
Radium	Ra	88	226.0254
Radon	Rn	86	(222)
Rhenium	Rc	75	186.207

Element	Symbol	Atomic No.	Atomic Weight	Element	Symbol	Atomic No.	Atomic Weight
Rhodium	Rh	45	102.90550	Lead	Pb	82	207.2
Rubidium	Rb	37	85.4678	Lithium	Li	3	6.941
Ruthenium	Ru	44	101.07	Lutetium	Lu	71	174.967
Samarium	Sm	62	150.36	Magnesium	Mg	12	24.3050
Scandium	Sc	21	44.955910	Manganese	Mn	25	54.93805
Selenium	Se	34	78.96	Mendelevium	Md	101	(258)
Silicon	Si	14	28.0855	Technetium	Te	43	(97)
Silver	Ag	47	107.8682	Tellurium	Tc	52	127.60
Sodium	Na	11	22.989768	Terbium	Tb	65	158.92534
Strontium	Sr	38	87.62	Thallium	Tl	81	204.3833
Sulphur	S	16	32.066	Thulium	Tm	69	168.93421
Tantalum	Ta	73	180.9479	Thorium	Th	90	232.0381
Hafnium	Hf	72	178.49	Tin	Sn	50	118.710
Helium	He	2	4.002602	Titanium	Ti	22	47.88
Holmium	Ho	67	164.93032	Tungsten	W	74	183.85
Hydrogen	H	1	1.00794	Uranium	U	92	238.0289
Iodine	I	53	126.90447	Vanadium	V	23	50.9415
Indium	In	49	114.82	Xenon	Xe	54	131.29
Iridium	Ir	77	192.22	Ytterbium	Yb	70	173.04
Iron	Fe	26	55.847	Yttrium	Y	39	88.90585
Krypton	Kr	36	83.80	Zinc	Zn	30	65.38
Lanthanum	La	57	138.9055	Zirconium	Zr	40	91.224
Lawrencium	Lr	103	(260)				

Notes:

1. This table is scaled to the relative atomic mass $A_r(^{12}C) = 12$.
2. Values in parentheses refer to the isotope of longest known half-life for radioactive elements.
3. Information provided here is based mainly upon the Report of the Commission on Relative Atomic Masses.

Pure and Applied Chemistry, 1986, 58 1678

APPENDIX II

Standard Solution of Metals for Atomic Absorption Spectrophotometry

1. Antimony (Sb) : Dissolve 2.7426 g of antimony potassium tartarate (A.R. grade) $K(SbO)C_4H_4O_6.1/2H_2O$ in 1000 ml water; 100 ml = 1.00 mg Sb

2. Bismuth (Bi) : Dissolve 1.000 g bismuth metal in a minimum volume of 1+1 HNO_3. Dilute to 1000 ml with 2% (V/V) HNO_3; 1.00 ml = 1.00 mg Bi.

3. Cadmium (Cd) : Dissolve 1.000 g cadmium metal in a minimum volume of 1+1 HCl. Dilute to 1000 ml with water; 1.00 ml = 1.00 mg Cd.

4. Calcium (Ca) : To 2.4972 g $CaCO_3$ add 50 ml water and add dropwise a minimum volume of conc. HCl (about 10 ml) to complete solution. Dilute to 1000 ml with water, 1.00 ml = 1.00 mg Ca.

5. Cesium (Cs) : Dissolve 1.267 g cesium chloride CsCl in 1000 ml water; 1.00 ml = 1.00 mg Cs.

6. Chromium (Cr) : Dissolve 2.828 g anhydrous potassium dichromate, $K_2Cr_2O_7$ in 200 ml water, add 1.5 ml conc. HNO_3 and dilute to 1000 ml with water; 1.00 ml = 1.00 mg Cr.

7. Cobalt (Co) : Dissolve 1.407 g cobaltic oxide, Co_2O_3 in 20 ml hot conc. HCl. Cool and dilute to 1000 ml with water; 1.00 ml = 1.00 mg Co.

8. Copper (Cu) : Dissolve 1.000 g copper metal in 15 ml of 1+1 HNO_3 and dilute to 1000 ml with water; 1.00 ml = 100 mg Cu.

9. Gold (Au) : Dissolve 0.1000 g gold metal in a minimum volume of aqua regia. Evaporate to dryness, dissolve residue in 5 ml conc. HCl, cool and dilute to 100 ml with water; 1.00 ml = 100 mg Au.

10. Iridium (Ir) : Dissolve 1.147 g ammonium chloroiridate, $(NH_4)_2IrCl_6$ in a minimum volume of 1% (v/v) HCl and dilute to 100 ml with 1% (v/v) HCl; 1.00 ml = 5.00 mg Ir.

11. Iron (Fe) : Dissolve 1.00 g iron wire in 50 ml of 1+1 HNO_3 and dilute to 1000 ml with water; 1.00ml 1 = 1.00 mg Fe.

12. Lead (Pb) : Dissolve 1.598 g lead nitrate $Pb(NO_3)_2$, in about 200 ml water, add 1.5 ml conc. nitric acid, and dilute to 1000 ml with water; 1.00 ml = 1.00 mg Pb.

13. Lithium (Li) : Dissolve 5.324 g lithium carbonate Li_2CO_3 in a minimum volume of 1+1 HCl and dilute to 1000 ml with water. 1.00 ml = 1.00 mg Li.

14. Manganese (Mn) : Dissolve 3.076 g manganous sulfate, $MnSO_4.H_2O$ in about 200 ml water, add 1.5 ml conc. HNO_3 and dilute to 1000 ml with water; 1.00 ml = 1.00 mg Mn.

15. Magnesium (Mg) : Dissolve 4.952 g magnesium sulfate in 200 ml water, add 1.5 ml conc. HNO_3 and dilute to 1000 ml with water; 1.00 ml = 1.00 mg Mg.

16. Nickel (Ni) : Dissolve 1.273 g nickel oxide, NiO, in a minimum volume of 10% (v/v) HCl and dilute to 1000 ml with water; 1.00 ml = 1.00 mg Ni.

17. Platinum (Pt) : Dissolve 0.1000 g platinum metal in a minimum volume of aqua regia and evaporate to dryness. Add 5 ml conc. HCl and 0.1 g NaCl and again evaporate to dryness. Dissolve residue in 20 ml of 1+1 HCl and dilute to 100 ml with water, 1.00 ml = 1.00 mg Pt.

18. Potassium (K) : Dissolve 1.907 g potassium chloride (KCl) in water and make up to 1000 ml; 1.00 ml = 1.00 mg K.

19. Rhodium (Rh) : Dissolve 0.412 ammonium hexachlororhodate $(NH_4)_3$ Rh. Cl_6 1.5 H_2O, in a minimum volume of 10% (v/v) HCl and dilute to 100 ml with 10% (v/v) HCl; 1.00 ml = 1.00 mg Rh.

20. Ruthenium (Ru) : Dissolve 0.2052 g ruthenium chloride ($RuCl_3$) in a minimum volume of 20% (v/v) HCl and dilute to 100 ml with 20% (v/v) HCl; 1.00 ml = 1.00 mg Ru.

21. Silver (Ag) : Dissolve 1.575 silver nitrate ($AgNO_3$) in water, add 1.5 ml conc. HNO_3 and make up to 1000 ml; 1.00 ml = 1.00 mg Ag.

22. Strontium (Sr) : Dissolve 2.415 g strontium nitrate Sr $(NO_3)_2$ in 1000 ml of 1% HNO_3 (v/v); 1.00 ml = 1.00 mg Sr.

23. Thallium (TI) : Dissolve 1.303 g thallium nitrate ($TINO_3$) in water. Add 10 ml conc. HNO_3 and dilute to 1000 ml with water; 1.00 ml = 1.00 mg TI.

24. Tin (Sn) : Dissolve 1.000 g tin metal in 100 ml conc. HCl and dilute to 1000 ml with water. 1.00 ml = 1.00 mg Sn.

25. Zinc (Zn) : Dissolve 1.000 g zinc metal in 20 ml 1+1 HCl and dilute to 1000 ml with water; 1.00 ml = 1.00 Zn.

water : deionized water

When determining calcium or magnesium, dilute and mix 100 ml sample with 10 ml Lanthanum solution before aspirating.

When determining iron or manganese, mix 100 ml with 25 ml of calcium solution before aspirating.

When determining chromium, mix 1 ml of 30% H_2O_2 with 100 ml sample before aspirating.

Calcium solution : Dissolve 630 mg calcium carbonate in 50 ml of 1+5 HCl. If necessary, heat and boil gently to obtain complete solution. Cool and dilute to 1000 ml with water.

Lanthanum solution : Dissolve 58.65 g lanthanum oxide (La_2O_3) in 250 ml conc. HCl. Add acid slowly until the material is dissolved and dilute to 1000 ml with water.

Aqua regia : Add 3 volumes conc. HCl to 1 volume conc. HNO_3.

APPENDIX III

Air Pollutant Units

Pollutant	1 ppm = x ng/m³ 1 ppb = x μg/m³	1 ng/m³ = x ppm 1 μg/m³ = x ppb
SO_2	2.87	0.38
NO_x	1.91	0.52
O_3	2.00	0.50
HF	0.83	1.20
NH_3	0.71	1.41

high-waste society: The situation in most advanced industrialised countries, in which ever-increasing economic growth is sustained by maximising the rate at which matter and energy resources are used, with relatively little emphasis on pollution prevention, recycling, reuse, reduction of unnecessary waste, and other forms of resource conservation. Compare *earth-wisdom society, matter-recycling society.*

homeostasis: Maintenance of favourable internal conditions in a system despite changes in external conditions. See *constancy, inertia, resilience.*

host: Plant or animal on which a parasite feeds.

human capital: Physical and mental talents of people used to produce, distribute, and sell an economic good. Compare *earth capital, manufactured capital, solar capital.*

humus: Slightly soluble residue of undigested or partially decomposed organic material in topsoil. This material helps retain water and water-soluble nutrients, which can be taken up by plant roots.

hunter-gatherers: People who get their food by gathering edible wild plants and other materials and by hunting wild animals and fish.

hydrocarbon: Organic compound of hydrogen and carbon atoms. The simplest hydrocarbon is methane (CH_4), the major component of natural gas.

hydroelectric power plant: Structure in which the energy of falling or flowing water spins a turbine generator to produce electricity.

hydrologic cycle: Biogeochemical cycle that collects, purifies, and distributes the earth's fixed supply of water from the environment to living organisms and then back to the environment.

hydropower: Electrical energy produced by falling or flowing water. See *hydroelectric power.*

hydrosphere: The earth's liquid water (oceans, lakes and other bodies of surface water, and underground water), frozen water (polar ice caps, floating ice caps, and ice in soil known as permafrost), and small amounts of water vapour in the atmosphere.

identified resources: Deposits of a particular mineral-bearing material of which the location, quantity, and quality are known or have been estimated from direct geological evidence and measurements. Compare *undiscovered resources.*

IFF (Intergovernmental Forum on Forests): The nineteenth special session of the UN General Assembly held in June 1997 decided to continue the IPFs (see IPF) intergovernmental policy dialogue on forests through the establishment of an ad hoc open-ended IFF under the UN Commission of Sustainable Development.

igneous rock: Rock formed when molten rock material (magma) wells up from the earth's interior, cools, and solidifies into rock masses. Compare *metamorphic rock, sedimentary rock.* See *rock cycle.*

immature community: Community at an early stage of ecological succession. It usually has a low number of species and ecological niches and cannot capture and use energy and cycle critical nutrients as efficiently as more complex, mature ecosystems. Compare *mature community.*

immigrant species: Species that migrate into an ecosystem or that are deliberately or accidentally introduced into an ecosystem by humans. Some of these are beneficial, whereas others can take over and eliminate many native species. Compare *indicator species, keystone species, native species.*

immigration: Migration of people into a country or area to take up permanent residence.

indicator species: Species that serve as early warnings that a community or an ecosystem is being degraded. Compare *immigrant species, keystone species, native species.*

indigenous peoples: As defined by the United Nations and International Labour Organisation (Convention 169 concerning indigenous and tribal peoples in independent countries, 1989) are "(a) tribal peoples in independent countries whose social, cultural and economic conditions distinguish them from other sections of the national community and whose status is regulated wholly or partially by their own customs or traditions or by special laws or regulation; (b) people in independent countries who are regarded as indigenous on account of their descent from the populations which inhabited the country, or a geographical region to which the country belongs, at the time of conquest or colonisation or the establishment of present state boundaries and who, irrespective of their legal status, retain some or all of their own social, economic, cultural and political institution."

industrialised agriculture: Using large inputs of energy from fossil fuels (especially oil and natural gas), water, fertiliser, and pesticides to produce large quantities of crops and livestock for domestic and foreign sale. Compare *subsistence farming.*

industrial revolution: Use of new sources of energy from fossil fuels and later from nuclear fuels, and use of new technologies, to grow food and manufacture products.

GLOSSARY

abiotic: Nonliving. Compare *biotic*.

absolute humidity: Amount of water vapour found in a certain mass of air (usually expressed as grams of water per kilogram of air). Compare *relative humidity*.

acclimation: Adjustment to slowly changing new conditions. Compare *threshold effect*.

acid deposition: The falling of acids and acid-forming compounds from the atmosphere to the earth's surface. Acid deposition is commonly known as *acid rain*—a term that refers only to wet deposition of droplets of acids and acid-forming compounds.

acid rain: See *acid deposition*.

acid solution: Any water solution that has more hydrogen ions (H^+) than hydroxide ions (OH^-); any water solution with a pH less than 7. Compare *basic solution, neutral solution*.

active solar heating system: System that uses solar collectors to capture energy from the sun and store it as heat for space heating and heating water. A liquid or air pumped through the collectors transfers the captured heat to a storage system such as an insulated water tank or rock bed. Pumps or fans then distribute the stored heat or hot water throughout a dwelling as needed. Compare *passive solar heating system*.

adaptation: Any genetically controlled structural, physiological, or behavioural characteristic that enhances members of a population chances of surviving and reproducing in its environment. It usually results from a beneficial adaptation. See *biological evolution, differential reproduction, mutation, natural selection*.

adaptive radiation: Period of time (usually millions of years) during which numerous new species evolve to fill vacant and new ecological niches in changed environments, usually after a mass extinction.

adaptive trait: See *adaptation*.

advanced sewage treatment: Specialised chemical and physical processes that reduce the amount of specific pollutants left in wastewater after primary and secondary sewage treatment. This type of treatment is usually expensive. See also *primary sewage treatment, secondary sewage treatment*.

aerobic respiration: Complex process that occurs in the cells of most living organisms, in which nutrient organic molecules such as glucose ($C_6H_{12}O_6$) combine with oxygen (O_2) and produce carbon dioxide (CO_2), water (H_2O), and energy. Compare *photosynthesis*.

afforestation: The establishment of a tree crop on an area from which it has always been or for very long been absent. Compare *reforestation*.

age structure: Percentage of the population (or the number of people of each sex) at each age level in a population.

Agenda 21: The internationally agreed plan of action on implementing sustainable development which emerged from the 1992 Earth Summit.

agricultural revolution: Gradual shift from small, mobile hunting and gathering bands to set-tied agricultural communities, in which people survived by learning how to breed and raise wild animals and to cultivate wild plants near where they lived. It began 10,000-12,000 years ago. Compare *industrial revolution*.

agroforestry: Land-use system in which tree crops are planted on the same land used for raising agricultural crops or animals.

air pollution: One or more chemicals in high enough concentrations in the air to harm humans, other animals, vegetation, or materials. Excess heat or noise can also be considered forms of air pollution. Such chemicals or physical conditions are called air pollutants. See *primary pollutant, secondary pollutant*.

albedo effect: Reflectivity for solar energy: high albedo means high reflectivity.

alien species: See *immigrant species*.

alleles: Slightly different molecular forms found in a particular gene.

alley cropping: Planting of crops in strips with rows of trees or shrubs on each side.

alpha particle: Positively charged matter, consisting of two neutrons and two protons, that is emitted as a form of radioactivity from the nuclei of some radioisotopes. See also *beta particle, gamma rays*.

altitude: Height above sea level. Compare *latitude*.

anaerobic respiration: Form of cellular respiration in which some decomposers get the energy they need through the breakdown of glucose (or other nutrients) in the absence of oxygen. Compare *aerobic respiration*.

ancient forest: See *old-growth forest*.

animal manure: Dung and urine of animals that can be used as a form of organic fertilizer. Compare *green manure*.

animals: Eukaryotic, multicelled organisms such as sponges, jellyfishes, arthropods (insects, shrimp, lobsters), mollusks (snails, clams, oysters, octopuses), fish, amphibians (frogs, toads, salamanders), reptiles (turtles, lizards, alligators, crocodiles, snakes), birds, mammals (kangaroos, bats, cats, rabbits, elephants, whales, porpoises, monkeys, apes, humans). See *carnivores, herbivores, omnivores*.

Annual Allowable Cut or AAC: The amount of timber that is permitted to be cut annually from a particular area. It is used to ensure a sustainable supply of timber. It averages roughly the amount of new growth each year minus deductions for losses due to fire, insects and diseases.

aquaculture: Growing and harvesting of fish and shellfish for human use in freshwater ponds, irrigation ditches, and lakes, or in cages or fenced-in areas of coastal lagoons and estuaries. See *fish farming, fish ranching*.

aquatic: Pertaining to water. Compare *terrestrial*.

aquifer: Porous, water-saturated layers of sand, gravel, or bed rock that can yield an economically significant amount of water. See *confined aquifer, unconfined aquifer*.

arable land: Land that can be cultivated to grow crops.

arid: Dry. A desert or other area with an arid climate has little precipitation.

asthenosphere: Portion of the earth's mantle that is capable of solid flow. See *crust, lithosphere, mantle*.

atmosphere: The whole mass of air surrounding the earth. See *stratosphere, troposphere*.

atomic number: Number of protons in the nucleus of an atom. Compare *mass number*.

atoms: Minute units made of subatomic particles that are the basic building blocks of all chemical elements and, thus, all matter; the smallest unit of an element that can exist and still have the unique characteristics of that element. Compare *ion, molecule*.

autotroph: See *producer*.

background extinction: Normal extinction of various species as a result of changes in local environmental conditions. Compare *mass extinction*.

bacteria: Prokaryotic, one-celled organisms. Some transmit diseases. Most act as decomposers and get the nutrients they need by breaking down complex organic compounds in the tissues of living or dead organisms into simpler inorganic nutrient compounds.

barrier islands: Long, thin, low, offshore islands of sediment that generally run parallel to the shore along some coasts.

basic solution: Water solution with more hydroxide ions (OH^-) than hydrogen ions (H^+); water solution with a pH greater than 7. Compare *acid solution, neutral solution*.

beneficiation: Separation of an ore mineral from the waste mineral material (gangue). See *tailings*.

beta particle: Swiftly moving electron emitted by the nucleus of a radioactive isotope. See also *alpha particle, gamma rays*.

biodegradable pollutant: Material that can be broken down into simpler substances (elements and compounds) by bacteria or other decomposers. Paper and most organic wastes such as animal manure are biodegradable but can take decades to biodegrade in modern landfills. Compare *degradable pollutant, nondegradable pollutant, slowly degradable pollutant*.

biodiversity: See *biological diversity*.

biofuel: Gas or liquid fuel (such as ethyl alcohol) made from plant material (biomass).

biogeochemical cycle: Natural processes that recycle nutrients in various chemical forms from the nonliving environment to living organisms and then back to the nonliving environment. Examples are the carbon, oxygen, nitrogen, phosphorus, sulfur, and hydrologic cycles.

biological amplification: Increase in concentration of DDT, PCBs, and other slowly degradable, fat-soluble chemicals in organisms at successively higher trophic levels of a food chain or web.

biological community: See *community*.

biological diversity: Variety of different species (*species diversity*), genetic variability among individuals within each species (*genetic diversity*), and variety of ecosystems (*ecological diversity*).

biological evolution: Change in the genetic makeup of a population of a species in successive generations. If continued long enough, it can lead to the formation of a new species. Note that populations not individuals evolve. See also *adaptation, differential reproduction, natural selection, theory of evolution*.

Biochemical Oxygen Demand (BOD): Amount of dissolved oxygen needed by aerobic decomposers to break down the organic materials in a given volume of water at a certain temperature over a specified time period.

biological pest control: Control of pest populations by natural predators, parasites, or diseasescausing bacteria and viruses (pathogens).

biomass: Organic matter produced by plants and other photosynthetic producers; total dry weight of all living organisms that can be supported at each trophic level in a food chain or web; dry weight of all organic matter in plants and animals in an ecosystem; plant materials and animal wastes used as fuel.

biome: Terrestrial regions inhabited by certain types of life, especially vegetation. Examples are various types of deserts, grasslands, and forests.

bioregion: A unique life-place with its own soils, landforms, watersheds, climates, native plants and animals, and many other distinct natural characteristics.

biosphere: Zone of the earth where life is found. It consists of parts of the atmosphere (the troposphere), hydrosphere (mostly surface water and groundwater), and lithosphere (mostly soil and surface rocks and sediments on the bottoms of oceans and other bodies of water) where life is found. It is also called the *ecosphere*.

biotic: Living organisms in an ecosystem. Compare *abiotic*.

biotic potential: Maximum rate at which the population of a given species can increase when there are no limits of any sort on its rate of growth. See *environmental resistance*.

birth rate: See *crude birth rate*.

bitumen: Gooey, black, high-sulfur, heavy oil extracted from tar sand and then upgraded to synthetic fuel oil. See *tar sand*.

boreal forest (or taiga): The predominantly coniferous forests growing in the northern climatic zone characterised by short summers and long freezing winters.

breeder nuclear fission reactor: Nuclear fission reactor that produces more nuclear fuel than it consumes by converting nonfissionable uranium-238 into fissionable plutonium-239.

broad-leaved forest: A forest composed predominantly of broad-leaved species, also known as hardwood, non-coniferous or angiosperm species. The species classification is botanical, based on the structure of seed and wood—not all broad-leaved species have broad leaves or hard wood. Examples of broad-leaved species include acacia, eucalyptus and oak.

calorie: Unit of energy; amount of energy needed to raise the temperature of 1 gram of water 1°C (unit on Celsius temperature scale). See also *kilocalorie*.

cancer: Group of more than 120 different diseases—one for each type of cell in the human body. Each type of cancer produces a tumour in which cells multiply uncontrollably and invade surrounding tissue. See *metastasis*.

canopy: Tree crown cover.

capitalism: See *pure market economic system*.

carbon cycle: Cyclic movement of carbon in different chemical forms from the environment to organisms and then back to the environment.

carbon sequestration: The removal and storage of carbon from the atmosphere by green plants through the process of photosynthesis, in which carbon dioxide is combined with water to form carbohydrates that can be stored in plant tissues.

carbon sink: An area where the rate of carbon uptake by living organisms exceeds the rate of carbon release, so that carbon is sequestered in organic or inorganic forms.

carcinogen: Chemicals, ionising radiation, and viruses that cause or promote the development of cancer. See *cancer, mutagen, teratogen*.

carnivore: Animal that feeds on other animals. Compare *herbivore, omnivore*.

carrying capacity (K): Maximum population of a particular species that a given habitat can support over a given period of time.

cell: Smallest living unit of an organism. Each cell is encased in an outer membrane or wall and contains genetic material (DNA) and other parts to perform its life function. Organisms such as bacteria consist of only one cell, but most of the organisms we are familiar with contain many cells. See *eukaryotic cell, prokaryotic cell*.

centrally planned economy: See *pure command economic system.*

certification : The process of evaluation or audits by independent third parties of forest practices carried out by forest owners and managers primarily engaged in wood production. The evaluation is usually carried out in relation to widely accepted standards of sustainable forest management. Some processes focus on certifying products, while others focus on forest management systems.

clear-cutting or clear-felling: Logging technique that involves the complete felling and removal of a stand of trees. Compare with *selective logging.*

CFCs: See *chlorofluorocarbons.*

chain reaction: Multiple nuclear fissions, taking place within a certain mass of a fissionable isotope, that release an enormous amount of energy in a short time.

chemical: One of the millions of different elements and compounds found naturally and synthesised by humans. See *compound, element.*

chemical change: Interaction between chemicals in which there is a change in the chemical composition of the elements or compounds involved. Compare *physical change.*

chemical evolution: Formation of the earth and its early crust and atmosphere, evolution of the biological molecules necessary for life, and evolution of systems of chemical reactions needed to produce the first living cells. These processes are believed to have occurred about 1 billion years before biological evolution. Compare *biological evolution.*

chemical formula: Shorthand way to show the number of atoms (or ions) in the basic structural unit of a compound. Examples are H_2O, NaCl, $C_6H_{12}O_6$ etc.

chemical reaction: See *chemical change.*

chemosynthesis: Process in which certain organisms (mostly specialised bacteria) extract inorganic compounds from their environment and convert them into organic nutrient compounds without the presence of sunlight. Compare *photosynthesis.*

chlorofluorocarbons (CFCs): Organic compounds made up of atoms of carbon, chlorine, and fluorine. An example is Freon-12 (CCl_2F_2), used as a refrigerant in refrigerators and air-conditioners and in making plastics such as Styrofoam. Gaseous CFCs can deplete the ozone layer when they slowly rise into the stratosphere and their chlorine atoms react with ozone molecules. Use of CFCs is being phased out.

chromosome: A grouping of various genes and associated proteins in plant and animal cells that carry certain types of genetic information. See *genes.*

clear-cutting: Method of timber harvesting in which all trees in a forested area are removed in a single cutting. Compare *seed-tree cutting, shelterwood cutting, selective cutting, strip cutting.*

climate: General pattern of atmospheric or weather conditions, seasonal variations, and weather extremes in a region over a long period—at least 30 years; average weather of an area. Compare *weather.*

climate change: An alteration to measured quantities (e.g., precipitation, temperature, radiation, wind and cloudiness) within the climate system that departs significantly from previous average conditions and is seen to endure, bringing about corresponding changes to ecosystems and socioeconomic activity. Associated with greenhouse gas emissions, of which fossil fuel combustion and forest decline are contributors.

climax community: See *mature community.*

closed forest: Forest with a minimum of 30% tree canopy cover.

coal: Solid, combustible mixture of organic compounds with 30% to 98% carbon by weight, mixed with various amounts of water and small amounts of sulfur and nitrogen compounds. It is formed in several stages as the remains of plants are subjected to heat and pressure over millions of years.

coal gasification: Conversion of solid coal to synthetic natural gas (SNG).

coal liquefaction: Conversion of solid coal to a liquid hydrocarbon fuel such as synthetic gasoline or methanol.

coastal wetland: Land along a coastline, extending inland from an estuary that is covered with salt water all or part of the year. Examples are marshes, bays, lagoons, tidal flats, and mangrove swamps. Compare *inland wetland.*

coastal zone: Relatively warm, nutrient-rich, shallow part of the ocean that extends from the high-tide mark on land to the edge of a shelflike extension of continental land masses known as the continental shelf. Compare *open sea.*

coevolution: Evolution when two or more species interact and exert selective pressures on each other that can lead each species to undergo various adaptations. See *biological evolution, natural selection.*

cogeneration: Production of two useful forms of energy, such as high-temperature heat or steam and electricity, from the same fuel source.

commensalism: An interaction between organisms of different species in which one type of organism benefits while the other type is neither helped nor harmed to any great degree. Compare *mutualism*.

commercial extinction: Depletion of the population of a wild species used as a resource to a level at which it is no longer profitable to harvest the species.

commercial inorganic fertiliser: Commercially prepared the structures of plant nutrients such as nitrates, phosphates, and potassium applied to the soil to restore fertility and increase crop yields. Compare *organic fertilizer*.

common-property resource: Resource that people are normally free to use; each user can deplete or degrade the available supply. Most are potentially renewable and are owned by no one. Examples are clean air, fish in parts of the ocean not under the control of a coastal country, migratory birds, gases of the lower atmosphere, and the ozone content of the upper atmosphere. See *tragedy of the commons*.

community: Populations of all species living and interacting in an area at a particular time.

community development: See *ecological succession*.

community forest: A forest area primarily designated to serve community needs.

community forestry (or community forest management): Used as an all encompassing operational term. The common denominator in community forestry is the focus on the role of local communities (usually forest-dependent) in managing neighbouring forests and sharing the benefits from those forests. An important role is to foster productive relationships between communities, government agencies and NGOs.

competition: Two or more individual organisms of a single species (*intraspecific competition*) or two or more individuals of different species (*interspecific competition*) attempting to use the same scarce resources in the same ecosystem.

competitive exclusion: Situation in which no two species can occupy exactly the same fundamental niche indefinitely in a habitat where there is not enough of a particular resource to meet the needs of both species. See *ecological niche, fundamental niche, realized niche*.

compost: Partially decomposed organic plant and animal matter that can be used as a soil conditioner or fertilizer.

compound: Combination of atoms, or oppositely charged ions, of two or more different elements held together by attractive forces called chemical bonds. Compare *element*.

concentration: Amount of a chemical in a particular volume or weight or air, water, soil, or other medium.

condensation nuclei: Tiny particles on which droplets of water vapour can collect.

confined aquifer: Aquifer between two layers of relatively impermeable earth materials, such as clay or shale. Compare *unconfined aquifer*.

coniferous trees: Cone-bearing trees, mostly evergreens, that have needle-shaped or scalelike leaves. They produce wood known commercially as softwood. Compare *deciduous plants*.

concession: Lease or contract to use a forest for the production of forest products. Concessions often take the form of commercial agreements granting loggers the right to harvest a given area of public forest land for industrial wood.

coniferous forest: Forest composed predominantly of needle-leaved species (softwoods or gymnosperms) such as pine, spruce and larch. Mostly evergreen with a few exceptions such as larch.

consensus science: Scientific data, models, theories, and laws that are widely accepted. This aspect of science is very reliable. Compare *frontier science*.

conservation biology: Multidisciplinary science created to deal with the crisis of maintaining the genes, species, communities, and ecosystems that make up earth's biological diversity. Its goals are to investigate human effects on biodiversity and to develop practical approaches to preserving biodiversity and ecological integrity.

conservation-tillage farming: Crop cultivation in which the soil is disturbed little (minimum-tillage farming) or not at all (no-till farming) to reduce soil erosion, lower labour costs, and save energy. Compare *conventional-tillage farming*.

constancy: Ability of a living system, such as a population, to maintain a certain size. Compare *inertia, resilience*. See *homeostasis*.

consumer: Organism that cannot synthesize the organic nutrients, it needs and gets its organic nutrients by feeding on the tissues of producers or of other consumers; generally divided into *primary consumers* (herbivores), *secondary consumers* (carnivores), *tertiary* (*higher-level*) *consumers*, *omnivores*, and *detritivores* (decomposers and detritus feeders). In economics, one who uses economic goods.

contour farming: Ploughing and planting across the changing slope of land, rather than in straight lines, to help retain water and reduce soil erosion.

controlled burning: Deliberately set, carefully controlled surface fires to reduce flammable litter and decrease the changes of damaging crown fires. See *crown fire, ground fire, surface fire*.

conventional-tillage farming: Crop cultivation involving making a planting surface by ploughing land, breaking up the exposed soil and then smoothing the surface. Compare *conservation-tillage farming*.

convergent plate boundary: Area where the earth's lithospheric plates are pushed together. See *subduction zone*. Compare *divergent plate boundary, transform fault*.

coral reef: Formation produced by massive colonies containing billions of tiny coral animals, called polyps, which secrete a stony substance (calcium carbonate) around themselves for protection. When the corals die, their empty outer skeletons form layers that cause the reef to grow. They are found in the coastal zones of warm tropical and subtropical oceans.

core: Inner zone of the earth. It consists of a solid inner core and a liquid outer core. Compare *crust, mantle*.

cost-benefit analysis: Estimates and comparison of short-term and long-term costs (losses) and benefits (gains) from an economic decision. If the estimated benefits exceed the estimated costs, the decision to buy an economic good or provide a public good is considered worthwhile.

critical mass: Amount of fissionable nuclei needed to sustain a nuclear fission chain reaction.

criteria and indicators: Criteria are rules, standards, norms, conditions, or behaviour which are considered to be 'good' or 'ideal'. Indicators suggest what kind of measures are needed to judge the success or failure of a program.

crop rotation: Planting a field, or an area of a field, with different crops from year to year to reduce depletion of soil nutrients. A plant such as corn, tobacco, or cotton, which removes large amounts of nitrogen from the soil, is planted one year. The next year a legume—such as soyabeans, which adds nitrogen to the soil—is planted.

crown fire: Extremely hot forest fire that burns ground vegetation and tree tops. Compare *controlled burning, ground fire, surface fire*.

crude birth rate: Annual number of live births per 1,000 persons in the population of a geographical area at the midpoint of a given year. Compare *crude death rate*.

crude death rate: Annual number of deaths per 1,000 persons in the population of a geographical area at the midpoint of a given year. Compare *crude birth rate*.

crude oil: Gooey liquid consisting mostly of hydrocarbon compounds and small amounts of compounds containing oxygen, sulfur, and nitrogen. Extracted from underground accumulations, it is sent to oil refineries, where it is converted to heating oil, diesel fuel, gasoline, tar and other materials.

crust: Solid outer zone of the earth. It consists of oceanic crust and continental crust. Compare *core, mantle*.

cultural eutrophication: Overnourishment of aquatic ecosystems with plant nutrients (mostly nitrates and phosphates) because of human activities such as agriculture, urbanisation, and discharges from industrial plants and sewage treatment plants. See *eutrophication*.

cyanobacteria: Single-celled, prokaryotic, microscopic organisms. Before being reclassified as monera they were called blue-green algae.

DDT: Dichlorodiphenyltrichloroethane, a chlorinated hydrocarbon that has been widely used as a pesticide but is now banned in some countries.

death rate: See *crude death rate*.

debt-for-nature swap: Agreement in which a certain amount of foreign debt is cancelled in exchange for local currency investments that will improve natural resource management or protect certain areas in the debtor country from harmful development.

deciduous plants: Trees, such as oaks and maples, and other plants that survive during dry seasons or cold seasons by shedding their leaves. Compare *coniferous trees, succulent plants*.

deciduous: Exhibiting discontinuous, often seasonal tree leaf cover. Opposite of evergreen.

decomposer: Organism that digests parts of dead organisms and cast-off fragments and wastes of living organisms by breaking down the complex organic molecules in those materials into simpler inorganic compounds and then absorbing the soluble nutrients. Most of these chemicals are returned to the soil and water for reuse by producers. Decomposers consist of various bacteria and fungi. Compare *consumer, detritivore, producer*.

defoliation: Loss of tree leaf cover.

deforestation: Removal of trees from a forested area without adequate replanting.

degradable pollutant: Potentially polluting chemical that is broken down completely or reduced to acceptable levels by natural physical, chemical, and biological processes. Compare *biodegradable pollutant, nondegradable pollutant, slowly degradable pollutant*.

degree of urbanisation: Percentage of the population in the world, or a country, living in areas with a population of more than 2,500 people (higher in some countries). Compare *urban growth*.

democracy: Government 'of the people, for the people, by the people' through their elected officials and appointed representatives. In a *constitutional democracy*, a constitution provides the basis of governmental authority and puts restraints on governmental power through free elections and freely expressed public opinion.

demographic transition: Hypothesis that countries, as they become industrialised, have declines in death rates followed by declines in birth rates.

depletion time: How long it takes to use a certain fraction—usually 80%—of the known or estimated supply of a nonrenewable resource at an assumed rate of use. Finding and extracting the remaining 20% usually costs more than it is worth.

desalination: Purification of salt water or brackish (slightly salty) water by removing dissolved salts.

desert: Biome where evaporation exceeds precipitation and the average amount of precipitation is less than 25 centimeters (10 inches) a year. Such areas have little vegetation or have widely spaced, mostly low vegetation. Compare *forest, grassland.*

desertification: Conversion of rangeland, rainfed cropland, or irrigated cropland to desertlike land, with a drop in agricultural productivity of 10% or more. It is usually caused by a combination of overgrazing, soil erosion, prolonged drought, and climate change.

detritivore: Consumer organism that feeds on detritus, parts of dead organisms and cast-off fragments and wastes of living organisms. The two principal types are *detritus feeders* and *decomposers.*

detritus: Parts of dead organisms and cast-off fragments and wastes of living organisms.

detritus feeder: Organism that extracts nutrients from fragments of dead organisms and their cast-off parts and organic wastes. Examples are earth-worms, termites, and crabs. Compare *decomposer.*

deuterium (D; hydrogen-2): Isotope of the element hydrogen, with a nucleus containing one proton and one neutron, and a mass number of 2. Compare *tritium.*

developed countries: See *more developed countries.*

developing countries: See *less developed countries.*

dieback: Sharp reduction in the population of a species when its numbers exceed the carrying capacity of its habitat. See *carrying capacity.*

differential reproduction: Phenomenon in which individuals with adaptive genetic traits produce more living offspring than do individuals without such traits. See *natural selection.*

dioxins: Family of 75 different toxic chlorinated hydrocarbon compounds formed as by-products in chemical reactions involving chlorine and hydrocarbons, usually at high temperatures.

discount rate: The economic value a resource will have in the future compared with its present value.

dissolved oxygen (DO) content: Amount of oxygen gas (O_2) dissolved in a given volume of water at a particular temperature and pressure, often expressed as a concentration in parts of oxygen per million parts (ppm) of water.

designated forest: A forest area legally set aside for protection, production, or some other specified use.

divergent plate boundary: Area where the earth's lithospheric plates move apart in opposite directions. Compare *convergent plate boundary, transform fault.*

DNA: (deoxyribonucleic acid) Large molecules in the cells of organisms that carry genetic information in living organisms.

dose: The amount of a potentially harmful substance an individual ingests, inhales, or absorbs through the skin. Compare *response.* See *dose-response curve, lethal dose, median lethal dose.*

dose-response curve: Plot of data showing effects of various doses of a toxic agent on a group of test organisms. See *dose, response.*

doubling time: The time it takes (usually in years) for the quantity of something growing exponentially to double. It can be calculated by dividing the annual percentage growth rate into 70.

drainage basin: See watershed.

dredge spoils: Materials scraped from the bottoms of harbours and streams to maintain shipping channels. They are often contaminated with high levels of toxic substances that have settled out of the water. See *dredging.*

dredging: Type of surface mining in which chain buckets and draglines scrape up sand, gravel, and other surface deposits covered with water. It is also used to remove sediment from streams and harbours to maintain shipping channels. See *dredge spoils.*

drift-net fishing: Catching fish in huge nets that drift in the water.

drought: Condition in which an area does not get enough water because of lower than normal precipitation, higher than normal temperatures that increase evaporation, or both.

dust dome: Dome of heated air that surrounds an urban area and traps pollutants, especially suspended particulate matter. See also *urban heat island.*

dust plume: Elongation of a dust dome by winds that can spread a city's pollutants hundreds of kilometers downwind.

earth capital: The earth's natural resources and processes that sustain us and other species. Compare *human capital, manufactured capital, solar capital.*

earthquake: Shaking of the ground resulting from the fracturing and displacement of rock, producing a fault, or from subsequent movement along the fault. See *plates.*

earth-sustaining economy: Economic system in which the number of people and the quantity of goods are maintained at some constant level. This level is ecologically sustainable over time and meets at least the basic needs of all members of the population.

earth-wisdom revolution: Cultural change involving halting population growth and altering lifestyles, political and economic systems, and the way we treat the environment so that we can help sustain the earth for ourselves and other species. This involves working with the rest of nature by learning more about how nature sustains itself. See *earth-wisdom worldview.*

earth-wisdom society: Society based on working with nature by recycling and reusing discarded matter; by pollution prevention; by conserving matter and energy resources through reducing unnecessary waste and use; by not degrading renewable resources, by building things that are easy to recycle, reuse, and repair; by not allowing population size to exceed the carrying capacity of the environment; and by preserving biodiversity. See *earth-wisdom worldview.* Compare *matter-recycling society, planetary management worldview, spaceship-earth worldview.*

earth-wisdom worldview: Beliefs that nature exists for all the earth's species, not just for us; we are not in charge of the rest of nature; there is not always more, and it's not all for us; some forms of economic growth are beneficial, and some are harmful; our goals should be to design economic and political systems that encourage earth-sustaining forms of growth and discourage or prohibit earth-degrading forms; our success depends on learning to cooperate with one another and with the rest of nature instead of trying to dominate and manage the earth's life-support systems primarily for our own use. Compare *spaceship-earth worldview, planetary management worldview.*

ecological diversity: The variety of forests, deserts, grasslands, oceans, streams, lakes, and other biological communities interacting with one another and with their nonliving environment. See *biological diversity.* Compare *genetic diversity, species diversity.*

ecological integrity: The conditions and natural processes (such as the flow of materials and energy through ecosystems) that generate and maintain biodiversity and allow evolutionary change as a key mechanism for adapting to changes in environmental conditions.

ecological land-use planning: Method for deciding how land should be used; development of an integrated model that considers geological, ecological, health, and social variables.

ecologically sustainable development: Development in which the total human population size and resource use in the world (or in a region) are limited to a level that does not exceed the carrying capacity of the existing natural capital and is therefore sustainable. Compare *economic development, economic growth, sustainable development.*

ecological niche: Total way of life or role of a species in an ecosystem. It includes all physical, chemical, and biological conditions a species needs to live and reproduce in an ecosystem. See *fundamental niche, realised niche.*

ecological population density: Number of individuals of a population per unit area of habitat. Compare *population density.*

ecological succession: Process in which communities of plant and animal species in a particular area are replaced over time by a series of different and often more complex communities. See *primary succession, secondary succession.*

ecology: Study of the interactions of living organisms with one another and with their nonliving environment of matter and energy; study of the structure and functions of nature.

economic decision: Deciding what goods and services to produce, how to produce them, how much to produce, and how to distribute them to people.

economic depletion: Exhaustion of 80% of the estimated supply of a nonrenewable resource. Finding, extracting, and processing the remaining 20% usually costs more than it is worth; may also apply to the depletion of a potentially renewable resource, such as a species of fish or tree.

economic development: Improvements in the efficiency of resource use so that the same or greater output of goods and services is produced with smaller throughputs of natural, manufactured, and human capital. Compare *ecologically sustainable development, economic growth, sustainable development*.

economic growth: Increase in the real value of all final goods and services produced by an economy; an increase in real GNP. Compare *ecologically sustainable development, economic development, sustainable development*.

economic resources: Natural resources, capital goods, and labour used in an economy to produce material goods and services. See *earth capital, human capital, manufactured capital*.

economic system: Method that a group of people uses to choose what goods and services to produce, how to produce them, how much to produce, and how to distribute them to people. See *mixed economic system, pure command economic system, pure market economic system*.

economy: System of production, distribution, and consumption of economic goods.

ecosphere: The earth's collection of living organisms interacting with one another and their non-living environment (energy and matter) throughout the world; all the earth's ecosystems. Also called the *biosphere*.

ecosystem: Community of different species interacting with one another and with the chemical and physical factors making up its nonliving environment.

ecosystem services: See *earth capital*.

ecotone: Transitional zone in which one type of ecosystem tends to merge with another ecosystem.

efficiency: Measure of how much output of energy or of a product is produced by a certain input of energy, materials, or labour. See *energy efficiency*.

electromagnetic radiation: Forms of kinetic energy travelling as electromagnetic waves. Examples are radio waves, TV waves, microwaves, infrared radiation, visible light, ultraviolet radiation, X-rays, and gamma rays. Compare *ionizing radiation, non-ionizing radiation*.

electron (e): Tiny particle moving around outside the nucleus of an atom. Each electron has one unit of negative charge (–) and almost no mass. Compare *neutron, proton*.

element: Chemicals, such as hydrogen (H), iron (Fe), sodium (Na), carbon (C), nitrogen (N), or oxygen (O), whose distinctly different atoms serve as the basic building blocks of all matter. There are 92 naturally occurring elements. Another 18 have been made in laboratories. Two or more elements combine to form compounds that make up most of the world's matter. Compare *compound*.

El Niño: Which is also called the Southern Oscillation, or ENSO (El Niño Southern Oscillation) is the largest climatic fluctuation observed on earth. It occurs sporadically with a period between 2 and 10 years and is accompanied by unusually high sea surface temperatures, air pressures, and rainfall in the Eastern Pacific near the equator. ENSO influences both the atmospheric climate and the ocean conditions.

endangered species: Wild species with so few individual survivors that the species could soon become extinct in all or most of its natural range. Compare *threatened species*.

endemism: Endemic species are those whose distribution is restricted to a particular area or region.

energy: Capacity to do work by performing mechanical, physical, chemical, or electrical tasks or to cause a heat transfer between two objects at different temperatures.

energy budget: The character of the energy flows which drive all natural processes from forest growth and photosynthesis to wind and rain.

energy efficiency: Percentage of the total energy input that does useful work and is not converted into low-quality, usually useless, heat in an energy conversion system or process. See *energy quality, net energy*.

energy quality: Ability of a form of energy to do useful work. High-temperature heat and the chemical energy in fossil fuels and nuclear fuels are concentrated high-quality energy. Low-quality energy such as low-temperature heat is dispersed or diluted and cannot do much useful work. See *high-quality energy, low-quality energy*.

enhanced oil recovery: Removal of some of the heavy oil left in an oil well after primary and secondary recovery. Compare *primary oil recovery, secondary oil recovery*.

environment: All external conditions and factors, living and nonliving (chemicals and energy), that affect an organism or other specified system during its lifetime; the earth's life-support systems for us and for all other forms of life—in effect another term for describing solar capital and earth capital.

environmental degradation: Depletion or destruction of a potentially renewable resource such as soil, grassland, forest, or wildlife by using it at a faster rate than it is naturally replenished. If such use continues, the resource can become non-renewable on a human time scale or nonexistent (extinct). See also *sustainable yield*.

environmental resistance: All the limiting factors jointly acting to limit the growth of a population. See *biotic potential, limiting factor*.

environmental revolution: See *earth-wisdom revolution*.

environmental science: Study of how we and other species interact with one another and with the nonliving environment (matter and energy). It is a physical and social science that integrates knowledge from a wide range of disciplines, including physics, chemistry, biology (especially ecology), geology, geography, resource technology and engineering, resource conservation and management, demography (the study of population dynamics), economics, politics, sociology, psychology, and ethics. In other words, it is a study of how the parts of nature and human societies operate and interact—a study of connections and interactions.

environmental worldview: How individuals think the world works, what they think their role in the world should be, and what they believe is right and wrong environmental behavior (ethics). See *earth-wisdom worldview, planetary management worldview, spaceship-earth worldview*.

EPA: Environmental Protection Agency, responsible for managing federal efforts in the United States to control air and water pollution, radiation and pesticide hazards, environmental research, hazardous waste, and solid waste disposal.

epidemiology: Study of the patterns of disease or other harmful effects from toxic exposure within defined groups of people to find out why some people get sick and some do not.

epiphytes: Plants that use their roots to attach themselves to branches high in trees, especially in tropical forests.

erosion: Process or group of processes by which loose or consolidated earth materials are dissolved, loosened, or worn away and removed from one place and deposited in another. See *weathering*.

estuary: Partially enclosed coastal area at the mouth of a river where its fresh water, carrying fertile silt and runoff from the land, mixes with salty seawater.

ethics: Our beliefs about what is right or wrong behaviour.

eukaryotic cell: Cell containing a *nucleus*, a region of genetic material surrounded by a membrane. Membranes also enclose several of the other internal parts found in a eukaryotic cell. Compare *prokaryotic cell*.

eutrophication: Physical, chemical, and biological changes that take place after a lake, an estuary, or a slow-flowing stream receives inputs of plant nutrients—mostly nitrates and phosphates—from natural erosion and runoff from the surrounding land basin. See *cultural eutrophication*.

eutrophic lake: Lake with a large or excessive supply of plant nutrients, mostly nitrates and phosphates. Compare *mesotrophic lake, oligotrophic lake*.

evaporation: Physical change in which a liquid changes into a vapour or gas.

even-aged management: Method of forest management in which trees, sometimes of a single species in a given stand, are maintained at about the same age and size and are harvested all at once. Compare *uneven-aged management*.

evergreen plants: Plants that keep some of their leaves or needles throughout the year. Examples are ferns and cone-bearing trees (conifers) such as firs, spruces, pines, redwoods, and sequoias. Compare *deciduous plants, succulent plants*.

exhaustible resource: See *nonrenewable resource*.

exotic species: Foreign or non-active species introduced from another ecozone. Opposite of indigenous species.

exploitation competition: Situation in which two competing species have equal access to a specific resource but differ in how quickly or efficiently they exploit it. See *interference competition, interspecific competition*. Compare *intraspecific competition*.

exponential growth: Growth in which some quantity, such as population size or economic output, increases by a fixed percentage of the whole in a given time period; when the increase in quantity over time is plotted, this type of growth yields a curve shaped like the letter J. Compare *linear growth*.

external benefit: Beneficial social effect of producing and using an economic good that is not included in the market price of the good. Compare *external cost, full cost, true cost*.

external cost: Harmful social effect of producing and using an economic good that is not included in the market price of the good. Compare *external benefit, full cost, true cost*.

externalities: Social benefits ("goods") and social costs ("bads") not included in the market price of an economic good. See *external benefit, external cost*. Compare *full cost, internal cost*.

extinction: Complete disappearance of a species from the earth. This happens when a species cannot adapt and successfully reproduce under new environmental conditions or when it evolves into one or more new species. Compare *speciation*. See also *endangered species, threatened species*.

family planning: Providing information, clinical services, and contraceptives to help individuals or couples choose the number and spacing of children they want to have.

famine: Widespread malnutrition and starvation in a particular area because of a shortage of food, usually caused by drought, war, flood, earthquake, or other catastrophic event that disrupts food production and distribution.

feedback loop: Circuit of sensing, evaluating, and reacting to changes in environmental conditions as a result of information fed back into a system; it occurs when one change leads to some other change, which then eventually either reinforces or slows the original change. See *negative feedback loop, positive feedback loop*.

feedlot: Confined outdoor or indoor space used to raise hundreds to thousands of domesticated livestock. Compare *rangeland*.

fertilizer: Substance that adds inorganic or organic plant nutrients to soil and improves its ability to grow crops, trees, or other vegetation. See *commercial inorganic fertilizer, organic fertilizer*.

first law of energy: See *first law of thermodynamics*.

first law of human ecology: We can never do merely one thing. Any intrusion into nature has numerous effects, many of which are unpredictable.

first law of thermodynamics: In any physical or chemical change, no detectable amount of energy is created or destroyed, but in these processes energy can be changed from one form to another; you can't get more energy out of something than you put in; in terms of energy quantity, you can't get something for nothing (there is no free lunch). This law does not apply to nuclear changes, in which energy can be produced from small amounts of matter. See also *second law of thermodynamics*.

fishery: Concentrations of particular aquatic species suitable for commercial harvesting in a given ocean area or inland body of water.

fish farming: Form of aquaculture in which fishes are cultivated in a controlled pond or other environment and harvested when they reach the desired size. See also *fish ranching*.

fish ranching: Form of aquaculture in which members of a fish species such as salmon are held in captivity for the first few years of their lives, released, and then harvested as adults when they return from the ocean to their fresh-water birthplace to spawn. See also *fish farming*.

fissionable isotope: Isotope that can split apart when hit by a neutron at the right speed and thus undergo nuclear fission. Examples are uranium-235 and plutonium-239.

floodplain: Flat valley floor next to a stream channel. For legal purposes, the term is often applied to any low area that has the potential for flooding, including certain coastal areas.

flyway: Generally fixed route along which water-fowl migrate from one area to another at certain seasons of the year.

food additive: A natural or synthetic chemical deliberately added to processed foods to retard spoilage, to provide missing amino acids and vitamins, or to enhance flavour, colour, and texture.

food chain: Series of organisms, each eating or decomposing the preceding one. Compare *food web*.

food web: Complex network of many interconnected food chains and feeding relationships. Compare *food chain*.

forest: Biome with enough average annual precipitation (at least 76 centimeters, or 30 inches) to support growth of various species of trees and smaller forms of vegetation. Compare *desert, grassland*.

forest capital: A weighted measure of the extent and integrity of forest resources in a given region (say a country). Primary forests may be deemed to have greater integrity than other types of forests, and, therefore, be given more weight than modified, fragmented or planted forests.

forest capital credits: The notion that, since forest capital is of global value due to the global ecological services provided by forests, the forest capital concept could form the basis of a global market in tradeable credits for the global services provided by forests. The forest capital credits system could build on or complement experiences with the emerging international market in carbon emissions trading.

forest capital index: A comparative measure of the forest capital of different regions or countries.

forest capital trend: A comparative measure of the degree of forest decline experienced within a given region over a period of time, expressed in percentage terms.

forest decline: A blanket term encompassing deforestation (conversion of forests to other uses) and forest degradation (deterioration in health and quality).

forest degradation: A reduction in the biomass, productivity or biodiversity of a forest, and alterations in ecological functions such as habitat, climate regulation and soil and water protection. Sometimes also referred to as a lowering in the quality or integrity of a forest ecosystem. One example is a change in forest class, say from closed to open forest, which negatively affects the stand or site or lowers the production capacity.

forest fallow: The woody vegetation which follows after shifting cultivation.

forest goods and services: All benefits provided by forests, whether material (such as timber or non-timber forest products) or non-material (such as shade, biodiversity, climate regulation or aesthetic enjoyment).

forest health: A characteristic of forests that are naturally resilient to damage, based on factors such as biological diversity and ecosystem stability.

forest practices: Any forest management activity applied directly to forest land, including silvicultural and harvesting treatments.

Forest Principles: The non-legally binding set of principles for managing forests which emerged from the 1992 Earth Summit. Despite calls to make these principles legally binding, governments only reached agreement on adopting them as voluntary measures. The Principles were therefore published separately from the text contained in Agenda 21.

forest products: All timber and non-timber goods derived from forests.

fossil fuel: Products of partial or complete decomposition of plants and animals that occur as crude oil, coal, natural gas, or heavy oils as a result of exposure to heat and pressure in the earth's crust over millions of years. See *coal, crude oil, natural gas.*

fossils: Skeletons, bones, shells, body parts, leaves, seeds, or impressions of such items that provide recognizable evidence of organisms that lived long ago.

Freons: See *chlorofluorocarbons.*

frontier science: Preliminary scientific data, hypotheses, and models that have not been widely tested and accepted. Compare *consensus science.*

full cost: Cost of a good when its internal costs and its estimated short and long-term external costs are included in its market price. Compare *external cost, internal cost.*

fundamental niche: The full potential range of the physical, chemical and biological factors a species can use if there is no competition from other species. See *ecological niche.* Compare *realised niche.*

fungi: Eukaryotic, mostly multicelled organisms such as mushrooms, molds, and yeasts. As decomposers, they get the nutrients they need by secreting enzymes that break down the organic matter in the tissue of other living or dead organisms. Then they absorb the resulting nutrients.

fungicide: Chemical that kills fungi.

Gaia hypothesis: Proposal that the earth is alive and can be considered a system that operates and changes by feedback of information between its living and nonliving components.

game species: Type of wild animal that people hunt or fish for, for sport and recreation and sometimes for food.

gamma rays: A form of ionizing, electromagnetic radiation with a high energy content emitted by some radioisotopes. They readily penetrate body tissues.

gangue: Undesired or waste material in an ore. See *ore.*

gasohol: Vehicle fuel consisting of a mixture of gasoline and ethyl or methyl alcohol—typically 10-23% ethanol or methanol by volume.

gene pool: The sum total of all genes found in the individuals of the population of a particular species.

generalist species: Species with a broad ecological niche. They can live in many different places, eat a variety of foods, and tolerate a wide range of environmental conditions. Examples are flies, cockroaches, mice, rats, and human beings. Compare *specialist species.*

genes: Coded units of information about specific traits that are passed on from parents to offspring during reproduction. They consist of segments of DNA molecules found in chromosomes.

genetic adaptation: Changes in the genetic makeup of organisms of a species that allow the species to reproduce and gain a competitive advantage under changed environmental conditions. See *biological evolution, differential reproduction, mutation, natural selection.*

genetic diversity: Variability in the genetic makeup among individuals within a single species. See *biodiversity.* Compare *ecological diversity, species diversity.*

geographic isolation: Separation of populations of a species for fairly long times into areas with different environmental conditions.

geothermal energy: Heat transferred from the earth's underground concentrations of dry steam (steam with no water droplets), wet steam (a mixture of steam and water droplets), or hot water trapped in fractured or porous rock.

GNP: See *gross national product, per capita GNP.*

global commons: Natural resources which belong primarily to all living beings rather than to individual nations. These are generally considered to include the oceans and outer space. The inclusion of forests in this category is a contentious issue.

global ecological services: The role forests play in regulating global climate, energy and hydrological flows, and other key biophysical functions.

grassland: Biome found in regions where moderate annual average precipitation (25 to 76 centimeters, or 10 to 30 inches) is enough to support the growth of grass and small plants, but not enough to support large stands of trees. Compare *desert, forest*.

greenhouse effect: A natural effect that traps heat in the atmosphere (troposphere) near the earth's surface. Some of the heat flowing back toward space from the earth's surface is absorbed by water vapour, carbon dioxide, ozone, and several other gases in the lower atmosphere (troposphere) and then radiated back toward the earth's surface. If the atmospheric concentrations of these greenhouse gases rise and are not removed by other natural processes, the average temperature of the lower atmosphere will gradually increase.

greenhouse gases: Gases in the earth's lower atmosphere (troposphere) that cause the greenhouse effect. Examples are carbon dioxide, chlorofluorocarbons, ozone, methane, water vapour, and nitrous oxide.

green manure: Freshly cut or still-growing green vegetation that is plowed into the soil to increase the organic matter and humus available to support crop growth. Compare *animal manure*.

green revolution: Popular term for introduction of scientifically bred or selected varieties of grain (rice, wheat, maize) that, with high enough inputs of fertilizer and water, can greatly increase crop yields.

Gross National Product (GNP): Total market value in current dollars of all goods and services produced by an economy for final use—usually during a year. Compare *per capita GNP, real per capita GNP, real per capita NEW, real GNP*.

gross primary productivity: The rate at which an ecosystem's producers capture and store a given amount of chemical energy as biomass in a given length of time. Compare *net primary productivity*.

ground fire: Fire that burns decayed leaves or peat deep below the ground surface. Compare *crown fire, surface fire*.

groundwater: Water that sinks into the soil and is stored in slowly flowing and slowly renewed underground reservoirs called aquifers; underground water in the zone of saturation, below the water table. See *confined aquifer, unconfined aquifer*. Compare *runoff, surface water*.

growing stock: All live trees of commercial species, except rough or rotten trees.

gully reclamation: Restoring land suffering from gully erosion by seeding gullies with quick-growing plants, building small dams to collect silt and gradually fill in the channels, and building channels to divert water away from the gully.

habitat: Place or type of place where an organism or a population of organisms lives. Compare *ecological niche*.

half-life: Time needed for one half of the nuclei in a radioisotope to emit its radiation. Each radioisotope has a characteristic half-life, which may range from a few millionths of a second to several billion years.

hardwood: A broad-leaved or angiosperm tree species, or its wood (See 'broadleaved forest').

harvesting pattern: The spatial distribution of felled areas and reserve (not felled) areas across the forest landscape.

hazard: Something that can cause injury, disease, economic loss, or environmental damage.

hazardous chemical: Chemical that can cause harm because it is flammable or explosive, or that can irritate or damage the skin or lungs (such as strong acidic or alkaline substances) or cause allergic reactions of the immune system (allergens). See *toxic chemical*.

hazardous waste: Any solid, liquid, or containerized gas that can catch fire easily, is corrosive to skin tissue or metals, is unstable and can explode or release toxic fumes, or has harmful concentrations of one or more toxic materials that can leach out. See also *toxic waste*.

herbicide: Chemical that kills a plant or inhibits its growth.

herbivore: Plant-eating organism. Examples are deer, sheep, grasshoppers, and zooplankton. Compare *carnivore, omnivore*.

heterotroph: See *consumer*.

high-quality energy: Energy that is organised or concentrated and has great ability to perform useful work. Examples are high-temperature heat and the energy in electricity, coal, oil, gasoline, sunlight, and nuclei or uranium-235. Compare *low-quality energy*.

high-quality matter: Matter that is organised, concentrated, and contains a high concentration of a useful resource. Compare *low-quality matter*.

high-throughput society: See *high-waste society*.

high-waste society: The situation in most advanced industrialised countries, in which ever-increasing economic growth is sustained by maximising the rate at which matter and energy resources are used, with relatively little emphasis on pollution prevention, recycling, reuse, reduction of unnecessary waste, and other forms of resource conservation. Compare *earth-wisdom society, matter-recycling society*.

homeostasis: Maintenance of favourable internal conditions in a system despite changes in external conditions. See *constancy, inertia, resilience*.

host: Plant or animal on which a parasite feeds.

human capital: Physical and mental talents of people used to produce, distribute, and sell an economic good. Compare *earth capital, manufactured capital, solar capital*.

humus: Slightly soluble residue of undigested or partially decomposed organic material in topsoil. This material helps retain water and water-soluble nutrients, which can be taken up by plant roots.

hunter-gatherers: People who get their food by gathering edible wild plants and other materials and by hunting wild animals and fish.

hydrocarbon: Organic compound of hydrogen and carbon atoms. The simplest hydrocarbon is methane (CH_4), the major component of natural gas.

hydroelectric power plant: Structure in which the energy of falling or flowing water spins a turbine generator to produce electricity.

hydrologic cycle: Biogeochemical cycle that collects, purifies, and distributes the earth's fixed supply of water from the environment to living organisms and then back to the environment.

hydropower: Electrical energy produced by falling or flowing water. See *hydroelectric power*.

hydrosphere: The earth's liquid water (oceans, lakes and other bodies of surface water, and underground water), frozen water (polar ice caps, floating ice caps, and ice in soil known as permafrost), and small amounts of water vapour in the atmosphere.

identified resources: Deposits of a particular mineral-bearing material of which the location, quantity, and quality are known or have been estimated from direct geological evidence and measurements. Compare *undiscovered resources*.

IFF (Intergovernmental Forum on Forests): The nineteenth special session of the UN General Assembly held in June 1997 decided to continue the IPFs (see IPF) intergovernmental policy dialogue on forests through the establishment of an ad hoc open-ended IFF under the UN Commission of Sustainable Development.

igneous rock: Rock formed when molten rock material (magma) wells up from the earth's interior, cools, and solidifies into rock masses. Compare *metamorphic rock, sedimentary rock*. See *rock cycle*.

immature community: Community at an early stage of ecological succession. It usually has a low number of species and ecological niches and cannot capture and use energy and cycle critical nutrients as efficiently as more complex, mature ecosystems. Compare *mature community*.

immigrant species: Species that migrate into an ecosystem or that are deliberately or accidentally introduced into an ecosystem by humans. Some of these are beneficial, whereas others can take over and eliminate many native species. Compare *indicator species, keystone species, native species*.

immigration: Migration of people into a country or area to take up permanent residence.

indicator species: Species that serve as early warnings that a community or an ecosystem is being degraded. Compare *immigrant species, keystone species, native species*.

indigenous peoples: As defined by the United Nations and International Labour Organisation (Convention 169 concerning indigenous and tribal peoples in independent countries, 1989) are "(a) tribal peoples in independent countries whose social, cultural and economic conditions distinguish them from other sections of the national community and whose status is regulated wholly or partially by their own customs or traditions or by special laws or regulation; (b) people in independent countries who are regarded as indigenous on account of their descent from the populations which inhabited the country, or a geographical region to which the country belongs, at the time of conquest or colonisation or the establishment of present state boundaries and who, irrespective of their legal status, retain some or all of their own social, economic, cultural and political institution."

industrialised agriculture: Using large inputs of energy from fossil fuels (especially oil and natural gas), water, fertiliser, and pesticides to produce large quantities of crops and livestock for domestic and foreign sale. Compare *subsistence farming*.

industrial revolution: Use of new sources of energy from fossil fuels and later from nuclear fuels, and use of new technologies, to grow food and manufacture products.

industrial smog: (Smoke + Fog) Type of air pollution consisting mostly of a mixture of sulfur dioxide, suspended droplets of sulfuric acid formed from some of the sulfur dioxide, and a variety of suspended solid particles. Compare *photochemical smog.*

inertia: Ability of a living system to resist being disturbed or altered. Compare *constancy, resilience.*

infant mortality rate: Number of babies out of every 1,000 born each year that die before their first birthday.

infiltration: Downward movement of water through soil.

information feedback: Process by which information is fed back into a system, causing it to change. See *negative feedback loop, positive feedback loop.*

inland wetland: Land away from the coast, such as a swamp, marsh, or bog, that is covered all or part of the time with fresh water, excluding streams and lakes. Compare *coastal wetland.*

inorganic fertiliser: See *commercial inorganic fertiliser.*

input: Matter, energy, or information entering a system. Compare *output, throughput.* See *information feedback.*

input pollution control: See *pollution prevention.*

insecticide: Chemical that kills insects.

Integrated Pest Management (IPM): Combined use of biological, chemical, and cultivation methods in proper sequence and timing to keep the size of a pest population below the size that causes economically unacceptable loss of a crop or livestock animal.

intercropping: Growing two or more different crops at the same time on a plot. For example, a carbohydrate-rich grain that depletes soil nitrogen and a protein-rich legume that adds nitrogen to the soil may be intercropped. Compare *monoculture, polyculture, polyvarietal cultivation.*

industrial roundwood: All unprocessed wood fit for commercial use other than fuelwood.

integrated resource management: The identification and consideration of all values, including the social, economic and environmental, in natural resource management.

intensive forest management (or intensification): The management of forests to maximise a particular outcome within a given area—usually timber or pulp fibre production.

IPF (Intergovernmental Panel on Forests): An open-ended adhoc body established in 1995 by the UN Commission on Sustainable Development (CSD) to pursue consensus on the management, conservation and sustainable development of all types of forests. The fourth and final session of the IPF was held in February 1997.

interference competition: Situation in which one species limits access of another species to a resource, regardless of whether the resource is abundant or scarce. See *exploitation competition, interspecific competition.* Compare *intraspecific competition.*

intermediate goods: See *manufactured capital.*

internal cost: Direct cost paid by the producer and the buyer of an economic good. Compare *external cost.*

Interplanting: Simultaneously growing a variety of crops on the same plot. See *agroforestry, intercropping, polyculture, polyvarietal cultivation.*

interspecific competition: Members of two or more species trying to use the same limited resources in an ecosystem. See *competition, competitive exclusion, exploitation competition, interference competition, intraspecific competition.*

intraspecific competition: Two or more individual organisms of a single species trying to use the same limited resources in an ecosystem. See *competition, interspecific competition.*

inversion: See *thermal inversion.*

invertebrates: Animals that have no backbones. Compare *vertebrates.*

ion: Atom or group of atoms with one or more positive (+) or negative (–) electrical charges. Compare *atoms, molecule.*

ionising radiation: Fast-moving alpha or beta particles or high-energy radiation (gamma rays) emitted by radioisotopes. They have enough energy to dislodge one or more electrons from atoms they hit, forming charged ions in tissue that can react with and damage living tissue.

isotopes: Two or more forms of a chemical element that have the same number of protons but different mass numbers due to different numbers of neutrons in their nuclei.

J-shaped curve: Curve with a shape similar to that of the letter J; can represent prolonged exponential growth.

Joint Forest Management (JFM): A variant of community forestry (see 'community forestry') widely adopted in India for managing government-owned forests in which responsibility and benefits are shared between local user groups and government forestry departments.

kerogen: Solid, waxy mixture of hydrocarbons found in oil shale rock. When the rock is heated to high temperatures, the kerogen is vapourised. The vapour is condensed, purified, and then sent to a refinery to produce gasoline, heating oil, and other products. See also *soil shale, shale oil*.

keystone species: Species that play roles affecting many other organisms in an ecosystem. Compare *immigrant species, indicator species, native species*.

kilocalorie (kcal): Unit of energy equal to 1,000 calories. See *calorie*.

kilowatt (kW): Unit of electrical power equal to 1,000 watts. See *watt*.

kinetic energy: Energy that matter has because of its motion and mass. Compare *potential energy*.

k-strategists: Species that produce a few, often fairly large offspring but invest a great deal of time and energy to ensure that most of those offspring will reach reproductive age. Compare *r-strategists*.

kwashiorkor: Type of malnutrition that occurs in infants and very young children when they are weaned from mother's milk to a starchy diet low in protein. See *marasmus*.

Kyoto Protocol: Is a treaty negotiated in Kyoto, Japan, in 1997 between 159 countries setting out legally binding reduction targets for six greenhouse gases increasing 5% emissions below 1990 levels for industrialised countries between 2008 and 2012.

As of April 2010, 191 states have signed & ratified the protocol.

The Protocol was adopted on 11.12.1997 but entered into force on 16.02.2005.

labour: See *human capital*.

lake: Large natural body of standing fresh water formed when water from precipitation, land runoff, or groundwater flow fills a depression in the earth movement, volcanic activity, or a giant meteorite. See *eutrophic lake, mesotrophic lake, oligotrophic lake*.

landfill: See *sanitary landfill*.

landscape: The geographical and ecological integrity and resilience of a particular land area, not merely its aesthetic qualities. Landscape is not just a geographical concept, but includes human, cultural, and traditional values that are associated with the land.

landscape management: Involves managing an area or region by integrating a variety of values associated with the land that may be ecological, social, cultural or economic.

land-use planning: Process for deciding the best present and future use of each parcel of land in an area.

latitude: Distance from the equator. Compare *altitude*.

law of conservation of energy: See *first law of thermodynamics*.

law of conservation of matter: In any physical or chemical change, matter is neither created nor destroyed, but merely changed from one form to another; in physical and chemical changes, existing atoms are rearranged into either different spatial patterns (physical changes) or different combinations (chemical changes).

law of tolerance: The existence, abundance, and distribution of a species in an ecosystem are determined by whether the levels of one or more physical or chemical factors fall within the range tolerated by the species. See *threshold effect*.

LD_{50}: Amount of a toxic material per unit of body weight of test animals that kills half the test population in a certain time.

LDC: See *less developed country*.

leaching: Process in which various chemicals in upper layers of soil are dissolved and carried to lower layers and, in some cases, to groundwater.

Less Developed Country (LDC): Country that has low-to-moderate industrialisation and low-to-moderate per capita GNP. Most are located in Africa, Asia, and Latin America. Compare *more developed country*.

life-cycle cost: Initial cost plus lifetime operating costs of an economic good.

life expectancy: Average number of years a newborn infant can be expected to live.

limiting factor: Single factor that limits the growth, abundance, or distribution of the population of a species in an ecosystem. See *limiting factor principle*.

limiting factor principle: Too much or too little of any abiotic factor can limit or prevent growth of a population of a species in an ecosystem, even if all other factors are at or near the optimum range of tolerance for the species.

linear growth: Growth in which a quantity increases by some fixed amount during each unit of time. Compare *exponential growth*.

Liquefied Natural Gas (LNG): Natural gas converted to liquid form by cooling to a very low temperature.

Liquefied Petroleum Gas (LPG): Mixture of liquefied propane (C_3H_8) and butane (C_4H_{10}) gas removed from natural gas.

lithosphere: Outer shell of the earth, composed of the crust and the rigid, outermost part of the mantle outside the asthenosphere; material found in the earth's plates. See *crust, mantle*.

loams: Soils containing a mixture of clay, sand, silt, and humus. Good for growing most crops.

logging: Felling and extraction of wood, especially as logs (excludes fuelwood harvested non-commercially).

low-quality energy: Energy that is disorganized or dispersed and has little ability to do useful work. An example is low-temperature heat. Compare *high-quality energy*.

low-quality matter: Energy that is disorganised, dilute, or dispersed, or contains a low concentration of a useful resource. Compare *high-quality matter*.

low-waste society: See *earth-wisdom society*.

LPG: See *liquefied petroleum gas*.

magma: Molten rock below the earth's surface.

malnutrition: Faulty nutrition. Caused by a diet that does not supply an individual with enough protein, essential fats, vitamins, minerals, and other nutrients needed for good health. Compare *overnutrition, undernutrition*.

mangrove swamps: Swamps found on the coastlines in warm tropical climates. They are dominated by mangrove trees, any of about 55 species of trees and shrubs that can live partly submerged in the relatively salty environment of coastal swamps.

mantle: Zone of the earth's interior between its core and its crust. Compare *core, crust*. See *lithosphere*.

manufactured capital: Manufactured items made from earth capital and used to produce and distribute economic goods and services bought by consumers. These include tools, machinery, equipment, factory buildings, and transportation and distribution facilities. Compare *earth capital, human capital, solar capital*.

manure: See *animal manure, green manure*.

marasmus: Nutritional-deficiency disease caused by a diet that does not have enough calories and protein to maintain good health. See *kwashiorkor, malnutrition*.

market equilibrium: State in which sellers and buyers of an economic good agree on the quantity to be produced and the price to be paid.

mass: The amount of material in an object.

mass extinction: A catastrophic, widespread—often global—event in which major groups of species are wiped out over a relatively short time compared to normal (background) extinctions. Compare *background extinction*.

mass number: Sum of the number of neutrons and the number of protons in the nucleus of an atom. It gives the approximate mass of that atom. Compare *atomic number*.

mass transit: Buses, trains, trolleys, and other forms of transportation that carry large numbers of people.

matter: Anything that has mass (the amount of material in an object) and takes up space. On the earth, where gravity is present, we weigh an object to determine its mass.

matter quality: Measure of how useful a matter resource is based on its availability and concentration. See *high-quality matter, low-quality matter*.

matter-recycling society: Society that emphasises recycling the maximum amount of all resources that can be recycled. The goal is to allow economic growth to continue without depleting matter resources and without producing excessive pollution and environmental degradation. Compare *earth-wisdom society, high-waste society*.

mature community: Fairly stable, self-sustaining community in an advanced stage of ecological succession; usually has a diverse array of species and ecological niches; captures and uses energy and cycles critical chemicals more efficiently than simpler, immature communities. Compare *immature community*.

maximum sustainable yield: See *sustainable yield*.

MDC: See *more developed country*.

meltdown: The melting of the core of a nuclear reactor.

mesosphere: Third layer of the atmosphere; found above the stratosphere. Compare *stratosphere, thermosphere, troposphere*.

mesotrophic lake: Lake with a moderate supply of plant nutrients. Compare *eutrophic lake, oligotrophic lake*.

metabolism: Ability of a living cell or organism to capture and transform matter and energy from its environment to supply its needs for survival, growth, and reproduction.

metamorphic rock: rock produced when a preexisting rock is subjected to high temperatures (which may cause it to melt partially), high pressures, chemically active fluids, or a combination of these agents. Compare *igneous rock, sedimentary rock*. See *rock cycle*.

metastasis: Spread of malignant (cancerous) cells from a cancer to other parts of the body.

microorganisms: Organisms that are so small that they can be seen only by using a microscope.

mineral: Any naturally occurring inorganic substance found in the earth's crust as a crystalline solid. See *mineral resource*.

mineral resource: Concentration of naturally occurring solid, liquid, or gaseous material, in or on the earth's crust, in such form and amount that its extraction and conversion into useful materials or items is currently or potentially profitable. Mineral resources are classified as metallic (such as iron and aluminum ores) or nonmetallic (such as fossil fuels, sand, and salt).

minimum-tillage farming: See *conservation-tillage farming*.

mixed economic system: Economic system that falls somewhere between pure market and pure command economic systems. Virtually all the world's economic systems fall into this category, with some closer to a pure market system and some closer to a pure command system. Compare *pure command economic system, pure market economic system*.

mixture: Combination of one or more elements and compounds.

model: See *scientific model*.

modified forest: Non-primary, ecologically-altered forest.

molecule: Combination of two or more atoms of the same chemical element (such as O_2) or different chemical elements (such as H_2O) held together by chemical bonds.

monera: See *bacteria, cyanobacteria*.

monoculture: Cultivation of a single crop, usually on a large area of land. Compare *polyculture, polyvarietal cultivation*.

monopoly or oligopoly: An economic arrangement in which one or a few interests capture a disproportionate amount of control over the supply of a resource.

more developed country (MDC): Country that is highly industrialised and has a high per capita GNP. Compare *less developed country*.

multiple use: Principle of managing public land, such as a national forest, so that it is used for a variety of purposes, such as logging, mining, recreation, grazing, wildlife preservation, and soil and water conservation. See also *sustainable yield*.

municipal solid waste: Solid materials discarded by homes and businesses in or near urban areas. See *solid waste*.

mutagen: Chemical, or form of ionising radiation, that causes inheritable changes in the DNA molecules in the genes found in chromosomes (mutations). See *carcinogen, mutation, teratogen*.

mutation: A random change in DNA molecules making up genes that can yield changes in anatomy, physiology, or behaviour in offspring. See *mutagen*.

mutualism: Type of species interaction in which both participating species generally benefit. Compare *commensalism*.

native species: Species that normally live and thrive in a particular ecosystem. Compare *immigrant species, indicator species, keystone species*.

natural forest: A forest composed of indigenous tree species: non-exotic.

natural gas: Underground deposits of gases consisting of 50-90% by weight methane gas (CH_4) and small amounts of heavier gaseous hydrocarbon compounds such as propane (C_3H_8) and butane (C_4H_{10}).

natural ionising radiation: Ionising radiation in the environment from natural sources.

natural radioactive decay: Nuclear change in which unstable nuclei of atoms spontaneously shoot out particles (usually alpha or beta particles), energy (gamma rays), or both at a fixed rate.

natural recharge: Natural replenishment of an aquifer by precipitation, which percolates downward through soil and rock. See *recharge area*.

natural resource capital: See *earth capital*.

natural resources: Area of the earth's solid surface; nutrients and minerals in the soil and deeper layers of the earth's crust; water; wild and domesticated plants and animals; air; and other resources produced by the earth's natural processes. Compare *human capital, manufactured capital*. See *earth capital*.

natural selection: Process by which a particular beneficial gene (or set of genes) is reproduced in succeeding generations more than other genes. The result of natural selection is a population that contains a greater proportion of organisms better adapted to certain environmental conditions. See *adaptation, biological evolution, differential reproduction, mutation*.

negative feedback loop: Flow of information into a system in which one change leads to a lessening of that change. Compare *positive feedback loop*.

negawatt: A watt of electrical power saved by improving energy efficiency.

nematocide: Chemical that kills nematodes (roundworms).

Net Economic Welfare (NEW): Measure of annual change in quality of life in a country. It is obtained by subtracting the value of all final products and services that decrease the quality of life from a country's GNP. See *per capita NEW*.

net energy: Total amount of useful energy available from an energy resource or energy system over its lifetime minus the amount of energy used (the first energy law), automatically wasted (the second energy law), and unnecessarily wasted in finding, processing, concentrating, and transporting it to users.

net primary productivity: Rate at which all the plants in an ecosystem produce net useful chemical energy; equal to the difference between the rate at which the plants in an ecosystem produce useful chemical energy (primary productivity) and the rate at which they use some of that energy through cellular respiration.

neutral solution: Water solution containing an equal number of hydrogen ions (H^+) and hydroxide ions (OH^-); water solution with a pH of 7. Compare *acid solution, basic solution*.

neutron (n): Elementary particle in the nuclei of all atoms (except hydrogen-1). It has a relative mass of 1 and no electric charge. Compare *electron, proton*.

NEW: See *net economic welfare*.

niche: See *ecological niche*.

nitrogen cycle: Cyclic movement of nitrogen in different chemical forms from the environment to organisms and then back to the environment.

nitrogen fixation: Conversion of atmospheric nitrogen gas into forms useful to plants, by lightning, bacteria, and cyanobacteria; it is part of the nitrogen cycle.

noise pollution: Any unwanted, disturbing, or harmful sound that impairs or interferes with hearing, causes stress, hampers concentration and work efficiency, or causes accidents.

nondegradable pollutant: Material that is not broken down by natural processes. Examples are the toxic elements lead and mercury. Compare *biodegradable pollutant, degradable pollutant, slowly degradable pollutant*.

nonionising radiation: Forms of radiant energy such as radio wave, microwaves, infrared light, and ordinary light that do not have enough energy to cause ionisation of atoms in living tissue. Compare *ionising radiation*.

nonpersistent pollutant: See *degradable pollutant*.

nonpoint source: Large or dispersed land areas such as crop fields, streets, and lawns that discharge pollutants into the environment over a large area. Compare *point source*.

nonrenewable resource: Resource that exists in a fixed amount (stock) in various places in the earth's crust and has the potential for renewal only by geological, physical, and chemical processes taking place over hundreds of millions to billions of years. Examples are copper, aluminium, coal, and oil. We classify these resources as exhaustible because we are extracting and using them at a much faster rate than they were formed. Compare *potentially renewable resource*.

non-timber forest products: Medicinal plants, resins, mushrooms, rattans, wildlife and other non-wood goods obtained from forests.

nontransmissible disease: A disease that is not caused by living organisms and that does not spread from one person to another. Examples are most cancers, diabetes, cardiovascular disease, and malnutrition. Compare *transmissible disease*.

no-till farming: See *conservation-tillage farming*.

nuclear change: Process in which nuclei of certain isotopes spontaneously change, or are forced to change, into one or more different isotopes. The three principal types of nuclear change are natural radioactivity, nuclear fission, and nuclear fusion. Compare *chemical change*.

nuclear energy: Energy released when atomic nuclei undergo a nuclear reaction such as the spontaneous emission of radioactivity, nuclear fission, or nuclear fusion.

nuclear fission: Nuclear change in which the nuclei of certain isotopes with large mass numbers (such as uranium-235 and plutonium-239) are split apart into lighter nuclei when struck by a neutron. This process releases more neutrons and a large amount of energy. Compare *nuclear fusion*.

nuclear fusion: Nuclear change in which two nuclei of isotopes of elements with a low mass number (such as hydrogen-2 and hydrogen-3) are forced together at extremely high temperatures until they fuse to form a heavier nucleus (such as helium-4). This process releases a large amount of energy. Compare *nuclear fission*.

nucleus: Extremely tiny centre of an atom, making up most of the atom's mass. It contains one or more positively charged protons and one or more neutrons with no electrical charge (except for a hydrogen-1 atom, which has one proton and no neutron in its nucleus).

nutrient: Any food or element an organism must take in to live, grow, or reproduce.

nutrient cycle : See *biochemical cycle*.

oil: See *crude oil*.

oil shale: Fine-grained rock containing various amounts of kerogen, a solid, waxy mixture of hydrocarbon compounds. Heating the rock to high temperatures converts the kerogen into a vapour that can be condensed to form a slow-flowing heavy oil called shale oil. See *kerogen, shale oil*.

oil-growth forest: Virgin and old, second-growth forest containing trees that are often hundreds, sometimes thousands, of years old. Examples include forests of Douglas fir, western hemlock, giant sequoia, and coastal redwoods in the western United States. Compare *second-growth forest, tree* *farm.*

open forest: Forest with between 10 and 30% tree canopy cover. Includes many savannah, tundra and dry forests.

oligotrophic lake: Lake with a low supply of plant nutrients. Compare *eutrophic lake, mesotrophic lake.*

omnivore: Animal that can use both plants and other animals as food sources. Examples are pigs, rats, cockroaches, and humans. Compare *carnivore, herbivore.*

open-pit mining: Removal of minerals such as gravel, sand, and ores of metals (such as iron and copper) by digging them out of the earth's surface and leaving an open pit.

open sea: The part of an ocean that is beyond the continental shelf. Compare *coastal zone.*

ore: Part of a metal-yielding material that can be economically and legally extracted at a given time. An ore typically contains two parts: the ore mineral, which contains the desired metal, and waste mineral material (gangue).

organic farming: Producing crops and livestock naturally by using organic fertiliser (manure, legumes, compost) and natural pest control (bugs that eat harmful bugs, plants that repel bugs, and environmental controls such as crop rotation) instead of using commercial inorganic fertilisers and synthetic pesticides and herbicides.

organic fertiliser: Organic material such as animal manure, green manure, and compost, applied to cropland as a source of plant nutrients. Compare *commercial inorganic fertiliser.*

organism: Any form of life.

other resources: Identified and undiscovered resources not classified as reserves. See *identified resources, resources, undiscovered resources.*

output: Matter, energy, or information leaving a system. Compare *input, throughput.* See *information feedback.*

output pollution control: See *pollution cleanup.*

overburden: Layer of soil and rock overlying a mineral deposit; removed during surface mining.

overconsumption: Situation in which some people consume much more than they need at the expense of those who cannot meet their basic needs and at the expense of the earth's present and future life-support systems.

overfishing: Harvesting so many fish of a species, especially immature fish, that there is not enough breeding stock left to replenish the species, such that it is not profitable to harvest them.

overgrazing: Destruction of vegetation when too many grazing animals feed too long and exceed the carrying capacity of a rangeland area.

overnutrition: Diet so high in calories, saturated (animal) fats, salt, sugar, and processed foods, and so low in vegetables and fruits that the consumer runs high risks of diabetes, hypertension, heat disease, and other health hazards. Compare *malnutrition, undernutrition.*

overpopulation: State in which there are more people than can live on the earth or in a geographic region in comfort, happiness, and health and still leave the planet or region a fit place for future generations. It is a result of growing numbers of people, growing affluence (resource consumption), or both. See *carrying capacity, overshoot.*

overshoot: Condition in which population size of a species temporarily exceeds the carrying capacity of its habitat. This leads to a sharp reduction in its population. See *carrying capacity, dieback.*

oxygen-demanding wastes: Organic materials that are usually biodegraded by aerobic (oxygen-consuming) bacteria if there is enough dissolved oxygen in the water. See also *biological oxygen demand.*

ozone layer: Layer of gaseous ozone (O_3) in the stratosphere that protects life on the earth by filtering out harmful ultraviolet radiation from the sun.

PANs: Peroxyacyl nitrates. Group of chemicals found in photochemical smog.

parasite: Consumer organism that lives on or in and feeds on a living plant or animal, known as the host, over an extended period of time. The parasite draws nourishment from and gradually weakens its host; it may or may not kill the host. See *parasitism.*

parasitism: Interaction between species in which one organism, called the parasite, preys on another organism, called the host, by living on or in the host. See *host, parasite.*

parts pre billion (ppb): Number of parts of a chemical found in one billion parts of a particular gas, liquid, or solid.

parts per million (ppm): Number of parts of a chemical found in one million parts of a particular gas, liquid, or solid.

participatory forest management: Forest management in which the decision-making follows from a process of broad public involvement.

passive solar heating system: System that captures sunlight directly within a structure and converts it into low-temperature heat for space heating or for heating water for domestic use without the use of mechanical devices. Compare *active solar heating system.*

pathogen: Organism that produces disease.

PCBs: See *polychlorinated biphenyls.*

per capita GNP: Annual gross national product (GNP) of a country divided by its total population. See *gross national product, real per capita GNP.*

per capita NEW: Annual Net Economic Welfare (NEW) of a country divided by its total population. See *net economic welfare, real per capita NEW.*

permafrost: Perennially frozen layer of the soil that forms when the water there freezes. It is found in arctic tundra soils.

permeability: The degree to which underground rock and soil pores are interconnected with each other, and thus a measure of the degree to which water can flow freely from one pore to another. Compare *porosity.*

persistence: See *inertia.*

persistent pollutant: See *slowly degradable pollutant.*

pest: Unwanted organism that directly or indirectly interferes with human activities.

pesticide: Any chemical designed to kill or inhibit the growth of an organism that people consider to be undesirable. See *fungicide, herbicide, insecticide.*

pesticide treadmill: Situation in which the cost of using pesticides increases while their effectiveness decreases, mostly because the pest species develop genetic resistance to the pesticides.

petrochemicals: Chemicals obtained by refining (distilling) crude oil. They are used as raw materials in the manufacture of most industrial chemicals, fertilisers, pesticides, plastics, synthetic fibres, paints, medicines, and many other products.

petroleum: See *crude oil.*

pH: Numeric value that indicates the relative acidity or alkalinity of a substance on a scale of 0 to 14, with the neutral point at 7. Acid solutions have pH values lower than 7, and basic or alkaline solutions have pH values greater than 7.

phosphorus cycle: Cyclic movement of phosphorus in different chemical forms from the environment to organisms and then back to the environment.

photochemical smog: Complex mixture of air pollutants produced in the lower atmosphere by the reaction of hydrocarbons and nitrogen oxides under the influence of sunlight. Especially harmful components include ozone, peroxyacyl nitrates (PANs), and various aldehydes. Compare *industrial smog.*

photosynthesis: Complex process that takes place in cells of green plants. Radiant energy from the sun is used to combine carbon dioxide (CO_2) and water (H_2O) to produce oxygen (O_2) and carbohydrates (such as glucose, $C_6H_{12}O_6$) and other nutrient molecules. Compare *aerobic respiration, chemosynthesis.*

photovoltaic cell: Device in which radiant (solar) energy is converted directly into electrical energy. Also called a *solar cell.*

physical change: Process that alters one or more physical properties of an element or a compound without altering its chemical composition. Examples are changing the size and shape of a sample of matter (crushing ice and cutting aluminium foil) and changing a sample of matter from one physical state to another (boiling and freezing water). Compare *chemical change.*

phytoplankton: Small, drifting plants, mostly algae and bacteria, found in aquatic ecosystems. Compare *plankton, zooplankton.*

pioneer community: First integrated set of plants, animals, and decomposers found in an area undergoing primary ecological succession. See *immature community, mature community.*

pioneer species: First hardy species, often microbes, mosses, and lichens, that begin colonizing a site as the first stage of ecological succession. See *ecological succession, pioneer community.*

planetary management worldview: Beliefs that we are the planet's most important species; there is always more, and it's all for us; all economic growth is good, more economic growth is better, and the potential for economic growth is limitless; our success depends on how well we can understand, control, and manage the earth's life-support systems for our own benefits. See *spaceship-earth worldview.* Compare *earth-wisdom worldview.*

plankton: Small plant organisms (phytoplankton) and animal organisms (zooplankton) that float in aquatic ecosystems.

plantation: Tree crop of one or a few species, usually planted and managed intensively for industrial wood production, whether timber or fibre. Sometimes plantations are also managed to produce fuelwood.

plantation agriculture: Growing specialized crops such as bananas, coffee, and cocoa in tropical LDCs, primarily for sale to MDCs.

planted forest: Any forest planted by humans.

plants (plantae): Eukaryotic, mostly multicelled organisms such as algae (red, blue, and green), mosses, ferns, flowers, cacti, grasses, beans, wheat, rice, and trees. These organisms use photosynthesis to produce organic nutrients for themselves and for other organisms feeding on them. Water and other inorganic nutrients are obtained from the soil for terrestrial plants and from the water for aquatic plants.

plates: Various-sized areas of the earth's lithosphere that move slowly around with the mantle's flowing asthenosphere. Most earthquakes and volcanoes occur around the boundaries of these plates. See *asthenosphere, lithosphere, plate tectonics*.

plate tectonics: Theory of geophysical processes that explains the movements of lithospheric plates and the processes that occur at their boundaries. See *lithosphere, plates*.

plywood: A composite material made of thin sheets of wood or veneer, glued or bonded together.

point source: A single identifiable source that discharges pollutants into the environment. Examples are the smokestack of a power plant or an industrial plant, the drainpipe of a meat-packing plant, the chimney of a house, or the exhaust pipe of an automobile. Compare *nonpoint source*.

poison: A chemical that in one dose kills exactly 50% of the animals (usually rats and mice) in a test population (usually 60 to 200 animals) within a 14-day period. See LD_{50}.

policy: a rule or norm usually prescribed by governments to help direct [be]haviour or decision-making.

politics: Process through which individuals and groups try to influence or control the policies and actions of governments that affect the local, state, national, and international communities.

pollutant: A particular chemical or form of energy that can adversely affect the health, survival, or activities of humans or other living organisms. See *pollution*.

pollution: An undesirable change in the physical, chemical, or biological characteristics of air, water, soil, or food that can adversely affect the health, survival, or activities of humans or other living organisms.

pollution cleanup: Device or process that removes or reduces the level of a pollutant after it has been produced or has entered the environment. Examples are automobile emission-control devices and sewage treatment plants. Compare *pollution prevention*.

pollution prevention: Device or process that prevents a potential pollutant from forming or from entering the environment or that sharply reduces the amounts entering the environment. Compare *pollution cleanup*.

polychlorinated biphenyls (PCBs): Group of 209 different toxic, oily, synthetic chlorinated hydrocarbon compounds that can be biologically amplified in food chains and webs.

polyculture: Complex form of intercropping in which a large number of different plants maturing at different times are planted together. See also *intercropping*. Compare *monoculture, polyvarietal cultivation*.

polyvarietal cultivation: Planting a plot of land with several varieties of the same crop. Compare *intercropping, monoculture, polyculture*.

population: Group of individual organisms of the same species living within a particular area.

population change: An increase or decrease in the size of a population. It is equal to (births + immigration) − (deaths + emigration).

population crash: Large number of deaths over a fairly short time, brought about when the number of individuals in a population is too large to be supported by available environmental resources.

population density: Number of organisms in a particular population found in a specified area.

population dispersion: General pattern in which the members of a population are arranged throughout its habitat.

population distribution: Variation of population density over a particular geographical area. For example, a country has a high population density in its urban areas and a much lower population density in rural areas.

population dynamics: Major abiotic and biotic factors that tend to increase or decrease the population size and age and sex composition of a species.

population size: Number of individuals making up a population's gene pool.

porosity: The pores (cracks and spaces) in rocks or soil, or the percentage of the rock's or soil's volume not occupied by the rock or soil itself. Compare *permeability*.

positive feedback loop: Situation in which a change in a certain direction provides information that causes a system to change further in the same direction. This can lead to a runaway or vicious cycle. Compare *negative feedback loop*.

potential energy: Energy stored in an object because of its position or the position of its parts. Compare *kinetic energy*.

potentially renewable resource: Resource that theoretically can last indefinitely without reducing the available supply because it is replaced more rapidly through natural processes than are nonrenewable resources or because it is essentially inexhaustible (solar energy). Examples are trees in forest, grasses in grasslands, wild animals, fresh surface water in lakes and streams, most groundwater, fresh air, and fertile soil. If such a resource is used faster than it is replenished, it can be depleted and converted into a nonrenewable resource. Compare *nonrenewable resource*. See also *environmental degradation*.

poverty: Inability to meet basic needs for food, clothing, and shelter.

ppb: See *parts per billion*.

ppm: See *parts per million*.

precautionary principle: The maxim that prevention is better than cure—where there are threats of serious or irreversible damage, lack of full scientific certainty shall not be used as a reason to postpone action.

precipitation: Water in the form of rain, sleet, hail, and snow that falls from the atmosphere onto the land and bodies of water.

predation: Situation in which an organism of one species (the predator) captures and feeds on parts or all of an organism of another species (the prey).

predator: Organism that captures and feeds on parts or all of an organism of another species (the prey).

predator-prey relationship: Interaction between two organisms of different species in which one organism, called the predator, captures and feeds on parts or all of another organism, called the prey.

prey: Organism that is captured and serves as a source of food for an organism of another species (the predator).

primary consumer: Organism that feeds directly on all or part of plants (herbivore) or other producers. Compare *detritivore, omnivore, secondary consumer*.

private interests: Interests of individuals, landowners, communities, corporations, and other groups and stakeholders.

primary oil recovery: Pumping out the crude oil that flows by gravity or under gas pressure into the bottom of an oil well. Compare *enhanced oil recovery, secondary oil recovery*.

primary pollutant: Chemical that has been added directly to the air by natural events or human activities and occurs in a harmful concentration. Compare *secondary pollutant*.

primary sewage treatment: Mechanical treatment of sewage in which large solids are filtered out by screens and suspended solids settle out as sludge in a sedimentation tank. Compare *advanced sewage treatment, secondary sewage treatment*.

primary succession: Sequential development of communities in a bare area that has no soil or bottom sediment. Compare *secondary succession*.

principle of connectedness: Everything is connected to and intermingled with everything else; we are all in it together.

principle of multiple use: See *multiple use*.

prior appropriation: Legal principle by which the first user of water from a stream establishes a legal right to continued use of the amount originally withdrawn.

primary forest: A forest in a mature succession phase, whose structure and composition have resulted from unrestrained ecological processes rather than from human activity.

probability: A mathematical statement about how likely it is that something will happen.

producer: Organism that uses solar energy (green plant) or chemical energy (some bacteria) to manufacture the organic compounds it needs as nutrients from simple inorganic compounds obtained from its environment. Compare *consumer, decomposer*.

production forest: A forest area primarily designated to achieve economic objectives such as the sustained production of forest products. Forest structure is usually less diverse than a protected forest.

prokaryotic cell: Cell that doesn't have a distinct nucleus. Other internal parts are also not enclosed by membranes. Compare *eukaryotic cell*.

protected forest: A forest area primarily designated to serve ecological, social or cultural objectives. A protected forest is usually more diverse in species and structure than a production forest.

protists (protista): Eukaryotic, mostly single-celled organisms such as diatoms, amoebas, some algae (golden brown and yellow-green), protozoans, and slime molds. Some protists produce their own organic nutrients through photosynthesis. Others are decomposers, and some feed on bacteria, other protists, or cells of multicellular organisms.

proton (p): Positively charged particle in the nuclei of all atoms. Each proton has a relative mass of 1 and a single positive charge. Compare *electron, neutron*.

public interest: The broader interest transcending individual or group interest for the good of the whole.

pulp: The liquefied material used for making paper, prepared by chemical or mechanical means from woodchips.

pure capitalism: See *pure market economic system.*

pure command economic system: System in which all economic decisions are made by the government or some other central authority. Compare *mixed economic system, pure market economic system.*

pure market economic system: System in which all economic decisions are made in the market, where buyers and sellers of economic goods freely interact, with no government or other interference. Compare *mixed economic system, pure command economic system.*

pyramid of biomass: Diagram representing the biomass, or total dry weight of all living organisms, that can be supported at each trophic level in a food chain or food web. See *pyramid of energy flow, pyramid of numbers.*

pyramid of energy flow: Diagram representing the flow of energy through each trophic level in a food chain or food web. With each energy transfer, only a small part (typically 10%) of the usable energy entering one trophic level is transferred to the organisms at the next trophic level. Compare *pyramid of biomass, pyramid of numbers.*

pyramid of numbers: Diagram representing the number of organisms of a particular type that can be supported at each trophic level from a given input of solar energy at the producer trophic level in a food chain or food web. Compare *pyramid of biomass, pyramid of energy flow.*

radiation: Fast-moving particles (particulate radiation) or waves of energy (electromagnetic radiation).

radioactive decay: Change of a radioisotope to a different isotope by the emission of radioactivity.

radioactive isotope: See *radioisotope.*

radioactive waste: Radioactive waste products of nuclear power plants, research, medicine, weapons production, or other processes involving nuclear reactions. See *radioactivity.*

radioactivity: Nuclear change in which unstable nuclei of atoms spontaneously shoot out "chunks" of mass, energy, or both, at a fixed rate. The three principal types of radioactivity are gamma rays and fast-moving alpha particles and beta particles.

radioisotope: Isotope of an atom that spontaneously emits one or more types of radioactivity (alpha particles, beta particles, gamma rays).

rain shadow effect: Low precipitation on the downwind side (leeward side) of a mountain when prevailing winds flow up and over a high mountain or range of high mountains. This creates semiarid and arid conditions on the leeward side of a high mountain range.

range condition: Estimate of how close a particular area of rangeland is to its potential for producing vegetation that can be consumed by grazing or browsing animals.

rainforest: Tropical or temperate forest growing in regions of high rainfall where the dry season is short or absent, characterised by a high number of plant and animal species and an abundance of climbers and epiphytes.

rangeland: Land that supplies forage or vegetation (grasses, grasslike plants, and shrubs) for grazing and browsing animals and that is not intensively managed. Compare *feedlot.*

range of tolerance: Range of chemical and physical conditions that must be maintained for populations of a particular species to stay alive and grow, develop, and function normally. See *law of tolerance.*

real GNP: Gross national product adjusted for inflation. Compare *gross national product, per capita GNP, real per capita GNP.*

real per capita GNP: Per capita GNP adjusted for inflation. See *per capita GNP.*

real per capita NEW: Per capita NEW adjusted for inflation. See *net economic welfare, per capita NEW.*

realised niche: Parts of the fundamental niche of a species that are actually used by that species. See *ecological niche, fundamental niche.*

recharge area: Any area of land allowing water to pass through it and into an aquifer. See *aquifer, natural recharge.*

recycling: Collecting and reprocessing a resource so that it can be made into new products. An example is collecting aluminium cans, melting them down, and using the aluminium to make new cans or other aluminium products. Compare *reuse.*

reforestation: Renewal of trees and other types of vegetation on land where trees have been removed; can be done naturally by seeds from nearby trees or artificially by planting seeds or seedlings.

relative humidity: Measure (as a percentage) of the amount of water vapour in a certain mass of air compared with the maximum amount it could hold at that temperature.

renewable resource: See *potentially renewable resource.*

replacement-level fertility: Number of children a couple must have to replace themselves. The average for a country or the world is usually slightly higher than 2 children per couple (2.1 in the United States and 2.5 in some LDCs) because some children die before reaching their reproductive years. See also *total fertility rate*.

reproduction: Production of offspring by one or more parents.

reproductive isolation: Long-term geographic separation of members of a particular sexually reproducing species.

reserve-to-production ratio: Number of years reserves of a particular nonrenewable mineral will last at current annual production rates. See *reserves*.

reserves: Resources that have been identified and from which a usable mineral can be extracted profitably at present prices with current mining technology. See *identified resources, other resources, undiscovered resources*.

resilience: Ability of a living system to restore itself to original condition after being exposed to an outside disturbance that is not too drastic. See *constancy, inertia*.

resource: Anything obtained from the living and nonliving environment to meet human needs and wants. It can also be applied to other species.

resource partitioning: Process of dividing up resources in an ecosystem so that species with similar requirements (overlapping ecological niches) use the same scarce resources at different times, in different ways, or in different places. See *ecological niche, fundamental niche, realised niche*.

respiration: See *aerobic respiration*.

response: The amount of health damage caused by exposure to a certain dose of a harmful substance or form of radiation. See *dose, dose-response curve*.

reuse: To use a product over and over again in the same form. An example is collecting, washing, and refilling glass beverage bottles. Compare *recycling*.

riparian zones: Thin strips and patches of vegetation that surround streams. They are very important habitats and resources for wildlife.

risk: The probability that something undesirable will happen from deliberate or accidental exposure to a hazard. See *risk analysis, risk assessment, risk-benefit analysis, risk management*.

risk analysis: Identifying hazards, evaluating the nature and severity of risks (*risk assessment*), using this and other information to determine options and make decisions about reducing or eliminating risks (*risk management*), and communicating information about risks to decision-makers and the public (*risk communication*).

risk assessment: Process of gathering data and making assumptions to estimate short and long-term harmful effects on human health or the environment from exposure to hazards associated with the use of a particular product or technology. See *risk, risk-benefit analysis*.

risk-benefit analysis: Estimate of the short and long-term risks and benefits of using a particular product or technology. See *risk*.

risk communication: Communicating information about risks to decision makers and the public. See *risk, risk-analysis, risk-benefit analysis*.

risk management: Using risk assessment and other information to determine options and make decisions about reducing or eliminating risks. See *risk, risk analysis, risk-benefit analysis, risk communication*.

rock: Any material that makes up a large, natural, continuous part of the earth's crust. See *mineral*.

rock cycle: Largest and slowest of the earth's cycles, consisting of geologic, physical, and chemical processes that form and modify rocks and soil in the earth's crust over millions of years.

rodenticide: Chemical that kills rodents.

rotation: Planned number of years between establishment of a forest stand and its felling.

r-strategists: Species that reproduce early in their lifespan and that produce large numbers of usually small and short-lived offspring in a short period of time. Compare *K-strategists*.

rule of 70: Doubling time in years = 70/ percentage growth rate. See *doubling time, exponential growth*.

runoff: Fresh water from precipitation and melting ice that flows on the earth's surface into nearby streams, lakes, wetlands, and reservoirs. See *surface runoff, surface water*. Compare *groundwater*.

rural area: Geographical area with a population of less than 2,500 people per sq. km. area. The number of people used in this definition may vary in different countries. Compare *urban area*.

salinity: Amount of various salts dissolved in a given volume of water.

salinisation: Accumulation of salts in soil that can eventually make the soil unable to support plant growth.

saltwater intrusion: Movement of salt water into freshwater aquifers in coastal and inland areas as groundwater is withdrawn faster than it is recharged by precipitation.

sanitary landfill: Waste disposal site on land in which waste is spread in thin layers, compacted, and covered with a fresh layer of clay or plastic foam each day.

scavenger: Organism that feeds on dead organisms that either were killed by other organisms or died naturally. Examples are vultures, flies, and crows. Compare *detritivore*.

science: Attempts to discover order in nature and then use that knowledge to make predictions about what will happen in nature. See *scientific data, scientific hypothesis, scientific law, scientific methods, scientific model, scientific theory*.

scientific data: Facts obtained by making observations and measurements. Compare *scientific hypothesis, scientific law, scientific methods, scientific model, scientific theory*.

scientific hypothesis: An educated guess that attempts to explain a scientific law or certain scientific observations. Compare *scientific data, scientific law, scientific methods, scientific model, scientific theory*.

scientific law: Description of what scientists find happening in nature over and over in the same way, without known exception. See *first law of thermodynamics, second law of thermodynamics, law of conservation of matter*. Compare *scientific data, scientific hypothesis, scientific law, scientific model*.

scientific methods: The ways scientists gather data and formulate and test scientific hypotheses, models, theories, and laws. See *scientific data, scientific hypothesis, scientific law, scientific model, scientific theory*.

scientific model: A simulation of complex processes and systems. Many are mathematical models that are run and tested using computers.

scientific theory: A well-tested and widely accepted scientific hypothesis. Compare *scientific data, scientific hypothesis, scientific law, scientific model*.

secondary consumer: Organism that feeds only on primary consumers. Most secondary consumers are animals, but some are plants. Compare *detritivore, omnivore, primary consumer*.

secondary oil recovery: Injection of water into an oil well after primary oil recovery to force out some of the remaining, usually thicker, crude oil. Compare *enhanced oil recovery, primary oil recovery*.

secondary pollutant: Harmful chemical formed in the atmosphere when a primary air pollutant reacts with normal air components or with other air pollutants. Compare *primary pollutant*.

secondary sewage treatment: Second step in most waste treatment systems, in which aerobic bacteria break down up to 90% of degradable, oxygen-demanding organic wastes in wastewater. This is usually done by bringing sewage and bacteria together in trickling filters or in the activated sludge process. Compare *advanced sewage treatment, primary sewage treatment*.

secondary succession: Sequential development of communities in an area in which natural vegetation has been removed or destroyed but the soil is not destroyed. Compare *primary succession*.

second-growth forest: Stands of trees resulting from secondary ecological succession. Compare *ancient forest, old-growth forest, tree farm*.

second law of energy: See *second law of thermodynamics*.

second law of thermodynamics: In any conversion of heat energy to useful work, some of the initial energy input is always degraded to a lower-quality, more dispersed, less useful energy—usually low-temperature heat that flows into the environment; you can't break-even in terms of energy quality. See *first law of thermodynamics*.

secondary forest: Forest ecosystems that regenerate after substantial natural or human caused disturbances (e.g. flood, fire, extensive logging or slash-and-burn farming), with relatively few mature trees.

sedimentary rock: Rock that forms from the accumulated products of erosion and in some cases from the compacted shells, skeletons, and other remains of dead organisms. Compare *igneous rock, metamorphic rock*. See *rock cycle*.

seed-tree cutting: Removal of nearly all trees on a site in one cutting, with a few seed-producing trees left uniformly distributed to regenerate the forest. Compare *clear-cutting, selective cutting, shelterwood cutting, strip cutting*.

selective cutting: Cutting of intermediate-aged, mature, or diseased trees in an uneven-aged forest stand, either singly or in small groups. This encourages the growth of younger trees and maintains an uneven-aged stand. Compare *clear-cutting, seed-tree cutting, shelterwood cutting, strip cutting*.

selective logging: Method of logging that only removes some trees from a forest stand, for instance, commercially important species or certain age classes. Compare *clear-cutting*.

septic tank: Underground tank for treatment of wastewater from a home in rural and suburban areas. Bacteria in the tank decompose organic wastes and the sludge settles to the bottom of the tank. The effluent flows out of the tank into the ground through a field of drainpipes.

sewage sludge: See *sludge.*

shale oil: Slow-flowing, dark brown, heavy oil obtained when kerogen in oil shale is vapourised at high temperatures and then condensed. Shale oil can be refined to yield gasoline, heating oil, and other petroleum products. See *kerogen, oil shale.*

shelterbelt: Strip of trees providing shelter, generally from wind.

shelterwood cutting: Removal of mature, marketable trees in an area in a series of partial cuttings to allow regeneration of a new stand under the partial shade of older trees, which are later removed. Typically, this is done by making two or three cuts over a decade. Compare *clear-cutting, seed-tree cutting, selective cutting, strip cutting.*

shifting cultivation: Clearing a plot of ground in a forest, especially in tropical areas, and planting crops on it for a few years (typically two to five years) until the soil is depleted of nutrients or until the plot has been invaded by a dense growth of vegetation from the surrounding forest. Then a new plot is cleared and the process is repeated. The abandoned plot cannot successfully grow crops for 10-30 years. See also *slash-and-burn cultivation.*

silviculture: The science and practice of cultivating forest crops.

slash-and-burn cultivation: Cutting down trees and other vegetation in a patch of forest, leaving the cut vegetation on the ground to dry, and then burning it. The ashes that are left add nutrients to the nutrient-poor soils found in most tropical forest areas. Crops are planted between tree stumps. Plots must be abandoned after a few years (typically two to five years) because of loss of soil fertility or invasion of vegetation from the surrounding forest. See also *shifting cultivation.*

slowly degradable pollutant: Material that is slowly broken down into simpler chemicals or reduced to acceptable levels by natural physical, chemical, and biological processes. Compare *biodegradable pollutant, degradable pollutant, non-degradable pollutant.*

sludge: Gooey mixture of toxic chemicals, infectious agents, and settled solids removed from wastewater at a sewage treatment plant.

smelting: Process in which a desired metal is separated from the other elements in an ore mineral.

smog: Originally a combination of smoke and fog but now used to describe other mixtures of pollutants in the atmosphere. See *industrial smog, photochemical smog.*

soil: Complex mixture of inorganic minerals (clay, silt, pebbles, and sand), decaying organic matter, water, air, and living organisms.

soil conservation: Methods used to reduce soil erosion, to prevent depletion of soil nutrients, and to restore nutrients already lost by erosion, leaching, and excessive crop harvesting.

soil erosion: Movement of soil components, especially topsoil, from one place to another, usually by exposure to wind, flowing water, or both. This natural process can be greatly accelerated by human activities that remove vegetation from soil.

soil horizons: Horizontal zones that make up a particular mature soil. Each horizon has a distinct texture and composition that varies with different types of soils.

soil permeability: Rate at which water and air move from upper to lower soil layers. Compare *porosity.*

soil porosity: See *porosity.*

soil profile: Cross-sectional view of the horizons in a soil.

soil structure: How the particles that make up a soil are organised and clumped together. See also *soil permeability, soil texture.*

soil texture: Relative amounts of the different types and sizes of mineral particles in a sample of soil.

soil water: Underground water that partially fills pores between soil particles and rocks within the upper soil and rock layers of the earth's crust, above the water table. Compare *groundwater.*

solar capital: Solar energy from the sun reaching the earth. Compare *earth capital.*

solar cell: See *photovoltaic cell.*

solar collector: Device for collecting radiant energy from the sun and converting it into heat. See *active solar heating system, passive solar heating system.*

solar energy: Direct radiant energy from the sun and a number of *indirect* forms of energy produced by the direct input. Principal indirect forms of solar energy include wind, falling and flowing water (hydropower), and biomass (solar energy converted into chemical energy stored in the chemical bonds of organic compounds in trees and other plants).

solar pond: Fairly small body of freshwater or salt water from which stored solar energy can be extracted because of the temperature difference between the hot surface layer exposed to the sun during daylight and the cooler layer beneath it.

solid waste: Any unwanted or discarded material that is not a liquid or a gas. See *municipal solid waste*.

spaceship-earth worldview: The earth is viewed as a spaceship—a machine that we can understand, control, and change at will by using advanced technology. See *planetary management worldview*. Compare *earth-wisdom worldview*.

specialist species: Species with a narrow ecological niche. They may be able to live in only one type of habitat, tolerate only a narrow range of climatic and other environmental conditions, or use only one or a few types of food, e.g., pandas of China. Compare *generalist species*.

speciation: Formation of two species from one species as a result of divergent natural selection in response to changes in environmental conditions; usually takes thousands of years. Compare *extinction*.

species: Group of organisms that resemble one another in appearance, behaviour, chemical makeup and processes, and genetic structure. Organisms that reproduce sexually are classified as members of the same species only if they can actually or potentially interbreed with one another and produce fertile offspring.

species diversity: Number of different species and their relative abundances in a given area. See *biological diversity*. Compare *ecological diversity, genetic diversity*.

spoils: Unwanted rock and other waste materials produced when a material is removed from the earth's surface or subsurface by mining, dredging, quarrying, and excavation.

S-shaped curve: Leveling off of an exponential, J-shaped curve when a rapidly growing population exceeds the carrying capacity of its environment and ceases to grow in numbers. See also *overshoot, population crash*.

stability: Ability of a living system to withstand or recover from externally imposed changes or stresses. See *constancy, interia, resilience*.

stewardship: View that because of our superior intellect and power or because of our religious beliefs we have an ethical responsibility to manage and care for domesticated plants and animals as well as for the rest of nature. Compare *earth-wisdom worldview, planetary management worldview*.

stratosphere: Second layer of the atmosphere, extending from about 17 to 48 kilometers (11-30 miles) above the earth's surface. It contains small amounts of gaseous ozone (O_3), which filters out about 99% of the incoming harmful ultraviolet (UV) radiation emitted by the sun. Compare *troposphere*.

steam: Flowing body of surface water. Examples are creeks and rivers.

strip cropping: Planting regular crops and close-growing plants, such as hay or nitrogen-fixing legumes, in alternating rows or bands to help reduce depletion of soil nutrients.

strip cutting: A variation of clear-cutting in which a strip of trees is clear-cut along the contour of the land, with the corridor narrow enough to allow natural regeneration within a few years. After regeneration, another strip is cut above the first, and so on. Compare *clear-cutting, seed-tree cutting, selective cutting, shelterwood cutting*.

strip mining: Form of surface mining in which bulldozers, power shovels, or stripping wheels remove large chunks of the earth's surface in strips. See *surface mining*. Compare *subsurface mining*.

subatomic particles: Extremely small particle—electrons, protons, and neutrons—that make up the internal structure of atoms.

subduction zone: Area in which oceanic lithosphere is carried downward (subducted) under the island area or continent at a convergent plate boundary. A trench ordinarily forms at the boundary between the two converging plates. See *convergent plate boundary*.

subsidence: Slow or rapid sinking down of part of the earth's crust that is not slope-related.

subsistence farming: Supplementing solar energy with energy from human labour and draft animals to produce enough food to feed oneself and family members; in good years there may be enough food left over to sell or put aside for hard times. Compare *industrialised agriculture*.

subsurface mining: Extraction of a metal ore or fuel resource such as coal from a deep underground deposit. Compare *surface mining*.

succulent plants: Plants, such as desert cacti, that survive in dry climates by having no leaves, thus reducing the loss of scarce water. They store water and use sunlight to produce the food they need in the thick fleshy tissue of their green stems and branches. Compare *deciduous plants, evergreen plants*.

sulfur cycle: Cyclic movement of sulfur in different chemical forms from the environment to organisms and then back to the environment.

superinsulated house: House that is heavily insulated and extremely airtight. Typically, active or passive solar collectors are used to heat water, and an air-to-air heat exchanger is used to prevent buildup of excessive moisture and indoor air pollutants.

surface fire: Forest fire that burns only undergrowth and leaf litter on the forest floor. Compare *crown fire, ground fire*.

surface mining: Removing soil, subsoil, and other strata, and then extracting a mineral deposit found fairly close to the earth's surface. Compare *subsurface mining*.

surface runoff: Water flowing off the land into bodies of surface water.

surface water: Precipitation that does not infiltrate the ground or return to the atmosphere by evaporation or transpiration. See *runoff*. Compare *groundwater*.

survivorship curve: Graph showing the number of survivors in different age groups for a particular species.

sustainable agriculture: Method of growing crops and raising livestock based on organic fertilizers, soil conservation, water conservation, biological control of pests, and minimal use of nonrenewable fossil-fuel energy.

sustainable development: Forms of economic growth and activities that do not deplete or degrade natural resources on which present and future economic growth and life depend. Compare *economic development, economic growth, ecologically sustainable development*.

sustainable forest management (sfm): A process, rather than a fixed state. As such, universal prescriptions are oversimplistic. Elements may include participatory and equitable approaches to decision-making geared toward maintaining ecosystem and landscape functions while also meeting broad economic, social and cultural needs. The management context is all-important.

sustainable living: Taking no more potentially renewable resources from the natural world than can be replenished naturally and not overloading the capacity of the environment to cleanse and renew itself by natural processes.

sustainable society: A society that manages its economy and population size without doing irreparable environmental harm by overloading the planet's ability to absorb environmental insults, replenish its resources, and sustain human and other forms of life over a specified period—usually hundreds to thousands of years. During this period, it satisfies the needs of its people without depleting earth capital and thereby jeopardizing the prospects of current and future generations of humans and other species.

sustainable yield (sustained yield): Highest rate at which a potentially renewable resource can be used without reducing its available supply throughout the world or in a particular area. See also *environmental degradation*.

symbiosis: Any intimate relationship or association between members of two or more species. See *symbiotic relationship*.

symbiotic relationship: Species interaction in which two kinds of organisms live together in an intimate association. Members of the participating species may be harmed by, benefit from, or be unaffected by the interaction. See *commensalism, interspecific competition, mutualism, parasitism, predation*.

synergistic interaction: Interaction of two or more factors or processes so that the combined effect is greater than the sum of their separate effects.

synfuels: Synthetic gaseous and liquid fuels produced from solid coal or sources other than natural gas or crude oil.

Synthetic Natural Gas (SNG): Gaseous fuel containing mostly methane produced from solid coal.

taiga (or boreal forest): The predominantly coniferous forests growing in the northern climatic zone characterised by short summers and long freezing winters.

tailings: Rock and other waste materials removed as impurities when waste mineral material is separated from the metal in an ore.

tar sand: Deposit of a mixture of clay, sand, water, any varying amounts of a tar-like heavy oil known as bitumen. Bitumen can be extracted from tar sand by heating. It is then purified and upgraded to synthetic crude oil. See *bitumen*.

technology: Creation of new products and processes intended to improve our efficiency, our chances for survival, our comfort level, and our quality of life. Compare *science*.

temperate forests: Forests found in the mid-latitudes, between the tropical and boreal forests.

temperature inversion: See *thermal inversion*.

tenure: Ownership or use rights.

teratogen: Chemical, ionising agent, or virus that causes birth defects. See *carcinogen, mutagen*.

terracing: Planting crops on a long, steep slope that has been converted into a series of broad, nearly level terraces with short vertical drops from one to another that run along the contour of the land to retain water and reduce soil erosion.

terrestrial: Pertaining to land. Compare *aquatic*.

tertiary (higher-level) consumers: Animals that feed on animal-eating animals. They feed at high tropic levels in food chains and webs. Examples are hawks, lions, bass, and sharks. Compare *detritivore, primary consumer, secondary consumer*.

tertiary oil recovery: See *enhanced oil recovery.*

tertiary sewage treatment: See *advanced sewage treatment.*

theory of evolution: Widely accepted idea the all life-forms developed from earlier life-forms. Although this theory conflicts with the creation stories of most religions, it is the way biologists explain how life has changed over the past 3.8-3.6 billion years and why it is so diverse today.

thermal inversion: Layer of dense, cool air trapped under a layer of less dense, warm air. This prevents upward-flowing air currents from developing. In a prolonged inversion, air pollution in the trapped layer may build up to harmful levels.

thermocline: Zone of gradual temperature decrease between warm surface water and colder deep water in a lake, reservoir, or ocean.

thermosphere: Fourth layer of the atmosphere; found above the mesosphere. Compare *mesosphere, stratosphere, troposphere.*

threatened species: Wild species that is still abundant in its natural range but is likely to become endangered because of a decline in numbers. Compare *endangered species.*

threshold effect: The harmful or fatal effect of a small change in environmental conditions that exceeds the limit of tolerance of an organism or population of a species. See *law of tolerance.*

throughput: Rate of flow of matter, energy, or information through a system. Compare *information feedback, input, output.*

throwaway society: See *high-waste society.*

time delay: Lapse between the input of a stimulus into a system and the response to the stimulus.

timber licence: An agreement that provides for the establishment of timber harvesting rights and, in some cases, management and forest regeneration responsibilities in a prescribed forest area.

timber supply area (TSA): An area specified by the land owning authority (usually governments) from which timber could be extracted. TSAs generally have a prescribed annual allowable cut.

timber utilisation: The dimensions and quality of wood that are actually cut and removed from an area. Low utilisation generally implies high wastage, and vice versa.

tolerance limits: Minimum and maximum limits for physical conditions (such as temperature) and concentrations of chemical substances beyond which no members of a particular species can survive. See *law of tolerance.*

total fertility rate (TFR): Estimate of the average number of children that will be born alive to a woman during her lifetime if she passes through all her childbearing years (ages 15-44) conforming to age-specific fertility rates of a given year. In simpler terms, it is an estimate of the average number of children a woman will have during her childbearing years.

totally planned economy: See *pure command economic system.*

toxic chemical: Chemical that is fatal to humans in low doses, or fatal to over 50% of test animals at stated concentrations. Most are neurotoxins, which attack nerve cells. See *carcinogen, hazardous chemical, mutagen, teratogen.*

toxicity: Measure of how harmful a substance is.

toxicology: Study of the adverse effects of chemicals on health.

toxic waste: Form of hazardous waste that causes death or serious injury (such as burns, respiratory diseases, cancers, or genetic mutations). See *hazardous waste.*

traditional commercial forest management: An approach to forest management focused on the production of industrial wood.

traditional intensive agriculture: Producing enough food for a farm family's survival and perhaps a surplus that can be sold. This type of agriculture requires higher inputs of labour, fertiliser, and water than traditional subsistence agriculture. See *traditional subsistence agriculture.*

traditional subsistence agriculture: Production of enough crops or livestock for a farm family's survival and, in good years, a surplus to sell or put aside for hard times. Compare *traditional intensive agriculture.*

tragedy of the commons: Depletion or degradation of a resource to which people have free and unmanaged access. An example is the depletion of commercially desirable species of fish in the open ocean beyond areas controlled by coastal countries. See *common-property resource.*

transfer pricing: The practice of placing artificially low export prices on timber sold to foreign subsidiaries or associates, usually for the purpose of avoiding taxes.

transform fault: Area where the earth's lithospheric plates move in opposite but parallel directions along a fracture (fault) in the lithosphere. Compare *convergent plate boundary, divergent plate boundary.*

transmissible disease: A disease that is caused by living organisms such as bacteria, viruses, and parasitic worms and that can spread from one person to another by air, water, food, body fluids, or, in some cases, by insects or other organisms. Compare *nontransmissible disease.*

transpiration: Process in which water is absorbed by the root systems of plants, moves up through the plants, passes through pores (stomata) in their leaves or other parts, and then evaporates into the atmosphere as water vapour.

tree farm: Site planted with one or only a few tree species in an even-aged stand. When the stand matures, it is usually harvested by clear-cutting and replanted. Normally used to grow rapidly growing tree species for fuelwood, timber, or pulpwood. See *even-aged management.* Compare *old-growth forest, second-growth forest, uneven-aged management.*

tritium (T; hydrogen-3): Isotope of hydrogen with a nucleus containing one proton and two neutrons, thus having a mass number of 3. Compare *deuterium.*

trophic level: All organisms that are the same number of energy transfers away from the original source of energy (for example, sunlight) that enters an ecosystem. For example, all producers belong to the first trophic level, and all herbivores belong to the second trophic level in a food chain or a food web.

tropical dry forest: Forest found in tropical regions with low rainfall.

tropical moist forest: Forest found in tropical regions with moderate to high rainfall. Includes tropical rainforest and tropical moist deciduous forest. Mostly low-lying, generally closed.

tropical rainforest: Evergreen forest situated in the more humid and low-lying parts of the tropical-most forest zone.

troposphere: Innermost layer of the atmosphere. It contains about 75% of the mass of the earth's air and extends about 17 kilometers (11 miles) above sea level. Compare *stratosphere.*

true cost: See *full cost.*

unconfined aquifer: Collection of groundwater above a layer of material (usually rock or clay) through which water flows very slowly (low permeability). Compare *confined aquifer.*

undernutrition: Consuming insufficient food to meet one's minimum daily energy requirement, for a long enough time to cause harmful effects. Compare *malnutrition, overnutrition.*

undiscovered resources: Potential supplies of a particular mineral resource, believed to exist because of geologic knowledge and theory, though specific locations, quality, and amounts are unknown. Compare *identified resources, other resources, reserves.*

uneven-aged management: Method of forest management in which trees of different species in a given stand are maintained at many ages and sizes to permit continuous natural regeneration. Compare *even-aged management.*

upwelling: Movement of nutrient-rich bottom water to the ocean's surface. This can occur far from shore but usually occurs along certain steep coastal areas where the surface layer of ocean water is pushed away from shore and replaced by cold, nutrient-rich bottom water.

urban area: Geographic area with a population of 2,500 or more people per sq. km. area. The number of people used in this definition may vary, with some countries setting the minimum number of people at 10,000 - 50,000.

urban growth: Rate of growth of an urban population. Compare *degree of urbanization.*

urban heat island: Buildup of heat in the atmosphere above an urban area. This heat is produced by the large concentration of cars, buildings, factories, and other heat-producing activities. See also *dust domer.*

urbanization: See *degree of urbanization.*

vertebrates: Animals with backbones. Compare *invertebrates.*

volcano: Vent or fissure in the earth's surface through which magma, liquid lava, and gases are released into the environment.

water cycle: See *hydrologic cycle.*

waterlogging: Saturation of soil with irrigation water or excessive precipitation so that the water table rises close to the surface.

water pollution: Any physical or chemical change in surface water or groundwater that can harm living organisms or make water unfit for certain uses.

watershed: Land area that delivers the water, sediment, and dissolved substances via small streams to a major stream (river).

water table: Upper surface of the zone of saturation, in which all available pores in the soil and rock in the earth's crust are filled with water.

watt: Unit of power, or rate at which electrical work is done. See *kilowatt.*

weather: Short-term changes in the temperature, barometric pressure, humidity, precipitation, sunshine (solar radiation), cloud cover, wind direction and speed, and other conditions in the troposphere at a given place and time. Compare *climate.*

weathering: Physical and chemical processes in which solid rock exposed at the earth's surface is changed to separate solid particles and dissolved material, which can then be moved to another place as sediment. See *erosion*.

wetland: Land that is covered all or part of the time with salt water or fresh water, excluding streams, lakes, and the open ocean. See *coastal wetland, inland wetland*.

wilderness: Area where the earth and its community of life have not been seriously disturbed by humans and where humans are only temporary visitors.

wildlife: All free, undomesticated species. Sometimes the term is used to describe only free, undomesticated species of animals.

wildlife management: Manipulation of populations of wild species (especially game species) and their habitats for human benefit, the welfare of other species, and the preservation of threatened and endangered wildlife species.

wildlife resources: Species of wildlife that have actual or potential economic value to people.

windbreak: Row of trees or hedges planted to partially block wind flow and reduce soil erosion on cultivated land.

wind farm: Cluster of small to medium-sized wind turbines in a windy area to capture wind energy and convert it into electrical energy.

work: What happens when a force is used to move a sample of matter over some distance or to raise its temperature. Energy is defined as the capacity to do such work.

worldview: How individuals think the world works and what they think their role in the world should be. See *earth-wisdom worldview, planetary management worldview, spaceship-earth worldview*.

zero population growth (ZPC): State in which the birth rate (plus immigration), equals the death rate (plus emigration) so that the population of a geographical area is no longer increasing and decreasing.

zone of saturation: Area where all available pores in soil and rock in the earth's crust are filled by water. See *water table*.

zoning: Regulating how various parcels of land can be used.

zooplankton: Animal plankton. Small floating herbivores that feed on plant plankton (phytoplankton). Compare *phytoplankton*.

BIBLIOGRAPHY

Adams, M.S. & F. McManus, 1994. *Noise & Noise Law—A Practical Approach,* Wiley Chancery Law, London.

Ahluwalia, V.K. 2009, *Green Chemistry—Benign Reactions.* Ane Books Ptv. Ltd. P-210

Anonymous 1992. *Pollution Control Acts, Rules and Notifications* issued thereunder, Central Pollution Control Board, New Delhi, P-501.

Barry, R.G. and Chorley, Richard, J. 2010. *Atmosphere, Weather, and Climate.* 8th Ed. Oxford. P-421

Benjamin, A. Pierce 2012. *Genetics: A Conceptual Approach,* 4th Edition. W.H. Freeman and Company, N.Y., P-721.

Bradbury, Ian, 1991. *The Biosphere,* New York, Belhaven Press.

Brady, Nyle C 1990. *The Nature and Properties of Soils,* 10th Edition, New York, Macmillan.

Bridgeman, Howard, 1991. *Global Air Pollution: Problems for the 1990s,* New York, Belhaven Press.

Bridggs, Shirley and the Rachel Carson Council, 1992. *Basic Guide to Pesticides: Their Characteristics & Hazards,* Washington, DC, Taylor & Francis.

Brudtland, G.H. *et al.* 1987. *Our Common Future: World Commission on Environment and Development,* New York, Oxford University Press.

Canter, L.W. 1994. *Environmental Impact Assessment,* (2nd Edition). New York, McGraw-Hill.

Carless, Jennifer, 1993. *Renewable Energy: A Concise Guide to Green Alternative,* New York, Walker.

Carson, Rachel, 1962. *Silent Spring, Boston,* Houghton Mifflin.

Clayton, AG Spinardi and R. Williams, 1999. *Policies for Cleaner Technology—A new agenda for Government & Industry.* Earth Scan, London.

Colinvaux, Paul A. 1986. *Ecology,* New York, John Wiley.

Desjardins, J. R. 1993. *Environmental Ethics,* Belmont, Calif : Wadsworth.

Dugan, P. J. (ed), 1990, *Wetland Conservation: A review of current issues & required action,* IUCN, P-96.

Flowler, John M. 1984, *Energy & the Environment,* (2nd Ed.), New York, McGraw-Hill.

Gates, David. M. 1993. *Climate Change and its Biological Consequences,* Sunderland, Mass, Sinauer Asso.

Gore, Al. 1992. *Earth in the Balance: Ecology & the Human Spirit, Boston:* Houghton Mifflin.

Goudie, A. 1993. *The Human Impact on the Natural Environment* (3rd Ed.), Cambridge, Mass, MIT Press. Grime 1979.

Grilbert, M. & R. Gould, 1998. *Achieving Environmental Standards* (2nd Ed.), Pitman Publishing, London.

Gray, N.E. 1992. *Biology of Wastewater Treatment,* New York, Oxford University Press.

Holloway, A.M. and Richard P. Wayne 2010. *Atmospheric Chemistry.* RSC Publisher, P-271.

Hunter, Malcolm L. Jr. 1990. *Wildlife, Forests and Forestry : Principles of Managing Forests for Biodiversity,* Englewood Cliffs. N. J., Prentice Hall.

INDP 1992. *Global Biodiversity Strategy,* Washington, DC, World Resource Institute.

IUCN, UNEP, WWE 1991. *Caring for the Earth: A Strategies for Sustainable Living,* London, Earthscan.

Jones J. AA. 1997. *Global Hydrology: Processes, Resources and Environmental Management,* Longman, Essex, England.

Karpagam, M; Geetha Jaikumar, 2010. *Green Management: Theory & Application.* Ane Books Pvt. Ltd. P-410.

Keller, Edward A. 1988. *Environmental Geology.* (5th Edition), Columbus, Ohio: Charles, E Merrill.

Levinton, Jeffrey 2011. *Marine Biology—Function, Biodiversity and Ecology.* Oxford, P-588.

Lindsey, K. and M. G. K. Jones. 1989. *Plant Biotechnology in Agriculture,* John Wiley & Sons, New York.

Mandal, F.B. 2010. *Text Book of Animal Behaviour.* PHI Learning Pvt. Ltd. P-302.

Manahan, S. E. 1983. *Environmental Chemistry,* Willaw Grant Press, Boston, USA.

Manahan Stanley, E. 2011. *Environmental Science & Technology—A Sustainable Approach to Green Science and Technology* (Special Indian Edition). Taylor & Francis & CRC Press, P-646.

McEldownery, S; D. J. Hardman and Stephen Waite. 1993. *Pollution: Ecology & Biotreatment.* Longman Scientific & Technology, England.

Moeller, Dave, W. 1992. *Environmental Health.* Cambridge, Mass : Harvard University Press.

Newman, Michael C. 2010. *Fundamentals of Ecotoxicology* (3rd Ed.) CRC press, P-539.

Odum, E. P. 1980. *Fundamental Principles of Ecology.*

Odum, Eugene, P. 1993. *Ecology & Our Endangered Life Support Systems.* (2nd Ed.), Sunderland Mass : Sinauer.

Pacione, M. (ed) 1999. *Applied Geography: Principles & Practice—An Introduction to Useful Research in Physical, Environmental and Human Geography,* Routledge, London.

Paul Hatcher and Nick Battey, 2011. *Biological Diversity: Exploiters and Exploited.* Wiley–Blackwell, P-427.

Pelczar, Jr. Michael, J.; ECS Chan. and Noel R. Krieg 2010. *Microbiology: Application Based Approach.* Tata McGraw-Hill, Aug. 2009.

Perry, G. 1980. *Introduction of Environmental Toxicology,* Elsevier, Netherland.

Santra, S. C. 1994. *Ecology: Basic & Applied,* M. D. Publication, New Delhi. P-250.

Santra, S. C. 2018. *Fundamental of Ecology and Environmental Biology.* 2nd Edition, New Central Book Agency (P) Ltd., Kolkata (In Press).

Sharma, A and A. Roychowdhury 1996. *Slow Murder—The Deadly Story of Vehicular Pollution in India,* Centre for Science & Environment, New Delhi, P-148.

Smith, Robert L. 1992. *Elements of Ecology,* (3rd Ed.), New York, Harper & Row.

Smith, B. J. 1982. *Acoustics & Noise Control,* Addison Wesley Longmans Limited, Essex, England.

Trivedy, R. K. (eds) 1995. Advances in Environmental Science and Technology, Ashish Publishing House, New Delhi, P-342.

Tudge, Colin, 1991. *Global Ecology,* New York, OUP.

Turner, R.K. *et al.,* 1994. *Environmental Economics: An Elementary Introduction.* Baltiaeare, Johns Hopkins University.

Valdiya, G.S. 1987. *Environmental Geology—Indian Context.* TMH Pub. Comp. Ltd., New Delhi.

WCMC 1992. *Global Biodiversity: Status of the Earth's Living Resources,* Chapman & Hall.

Wathern, Peter (ed) 1988. *Environmental Impact Assessment: Theory & Practice,* Routledge, London, World Research Institute, 1986.

Wilson, E. O. (ed) 1988. *Biodiversity,* Washington, DC, National Academy Press.

INDEX

A

B

C

D

E

F

G

H

T

U

V

W

Z